Comprehensive Human Physiology: From Cellular
Mechanisms to Integration, Vol. 2

Springer
Berlin
Heidelberg
New York
Barcelona
Budapest
Hong Kong
London
Milan
Paris
Santa Clara
Singapore
Tokyo

R. Greger   U. Windhorst (Eds.)

Comprehensive Human Physiology

From Cellular Mechanisms to Integration

Volume 2

With 722 Figures and 189 Tables

 Springer

Prof. Dr. Rainer Greger
Universität Freiburg
Physiologisches Institut
Hermann-Herder-Str. 7
79104 Freiburg
Germany

Prof. Dr. Uwe Windhorst
University of Calgary
Faculty of Medicine
Department of Clinical Neurosciences
3330 Hospital Drive N.W.
Calgary, Alberta
Canada T2N 4N1

ISBN 3-540-58109-X Springer-Verlag Berlin Heidelberg New York

Library of Congress Cataloging-in-Publication Data. Comprehensive human physiology: from cellular mechanisms to integration/R. Greger, U. Windhorst (eds.). p.   cm.   Includes bibliographical references and index. ISBN 3-540-58109-X (hardcover: alk. paper) 1. Human physiology.   I. Greger, Rainer. II. Windhorst, Uwe, 1946–   . QP34.5.C656 1996 612—dc20 96-34793

© Springer-Verlag Berlin Heidelberg 1996
Printed in Germany

Cover design: Springer-Verlag, Design & Production

Typesetting: Best-set Typesetter Ltd., Hong Kong

SPIN: 10038380   27/3130/SPS – 5 4 3 2 1 0 – Printed on acid-free paper

Preface

Adding yet another physiology text to the many good textbooks that are available around the world seems like carrying coal to Newcastle. In the physiological sciences, there is no lack of good texts written at different levels and targeting different groups of readers. Our motivation to prepare this new book arose from another, yet very traditional desire, namely to provide a textbook that includes recent developments in the field. These developments confront us not only with an enormous growth of knowledge in quantitative terms, but also with the necessity of revising classical concepts and classifications.

Two diverging trends characterize recent developments in our field. One originates from advances in cellular physiology that have provided us deep and fascinating insights into the function of the cell and its organelles. The success of the reductionist approach of identifying ever more refined details is conspicuous and undeniable as evidenced by the flood of data on individual membrane receptors, ion channels, and other membrane proteins which, with ever more sophisticated tools, can now be studied in isolation. Regulatory processes in the individual cell are now understood at several molecular levels, spanning acute modifications of existing matter to selective gene expression. In neurophysiology, methodological progress has made it possible to identify, purify, and localize an ever-increasing number of ion channels, transmitters, cotransmitters, and neuromodulators. Perhaps surprisingly, this success of reductionism that might have been expected to lead to an ever more diversified parcellation of knowledge, has also provided an unanticipated basis for unification. The reductionist approach continues to reveal that diverse organ functions are based on much less diverse basic cellular mechanisms, i.e., many special mechanisms are variations of far fewer basic themes. This makes it much easier to understand the bewildering array of new findings. These developments thus promise to reverse trends towards segregation of specialties, such as cell biology, biochemistry, molecular biology, and biophysics, which may in future be reunited under the wider umbrella of physiology. In addition, these attractive and fascinating new disciplines are of importance not only from a theoretical, basic sciences viewpoint, but increasingly for the understanding and treatment of diseases. Each student of physiology will thus have to understand such basic mechanisms and their control. And so will physicians. These relations are therefore reflected in sections on pathophysiology throughout this book.

While the results of extremely sophisticated and specialized studies continue to accumulate, the trend towards integration is also gaining strength. Indeed, this is a necessary reaction and a countermovement to dissection and division. The ultimate challenge to physiologists is to cope with the new results of reductionist research and at the same time to deal with the fundamental question of how the parts and pieces interact. Our goal remains to understand how cells are coordinated to function as an organ, how organs cooperate in systems, and finally, how system functions are integrated in the somatomotor and neurovegetative behavior of the whole organism when adapting to internal and external needs. Physiology has never been and will not be a mere set of facts, but rather a highly dynamic science based on functional thinking and embedded in the fascinating, ever-continuing process of learning more about life. Our intention not only to include, but also to combine the molecular and integrative aspects of physiology comes at a time when our science is in a state of transition between classical, partly still valid concepts, and entirely new approaches.

As a consequence the book reflects this fruitful conflict. A new horizon is emerging in the field of integrative research in that we are learning that different control systems are in part governed by common substrates. For example, the recognition of the wealth of newly identified transmitters and related substances opens unexpected outlooks as to how common neuronal substrates can play a role in differentiated control functions. A hierarchy of biological rhythms superimposed on homeostatic autoregulatory feedback circuits play an important role in the coordination of various body functions. General rules can be disclosed, applicable to diverse phenomena in various systems and frequency domains. Dynamic approaches begin to replace static classical concepts. Again, in this discipline, physiology starts to cross borders to other sciences, such as the new concepts and mathematics of rhythmic self-organizing systems.

In this situation we felt it would be useful to compile our current knowledge on peripheral and molecular mechanisms and on their integrative regulation. This book is intended to provide the reader with an update on both aspects. For this reason, we asked our authors to touch upon and to combine, wherever possible, both issues in each single chapter. In addition, there is frequent overlap between the different chapters, as indicated by numerous common references and cross-references. This serves to emphasize the interrelations between the macroscopic and microscopic levels of the organism, between its different parts and to thus underscore the general concept of the book.

This book is not an introductory text, but aims at a more advanced readership. It therefore does not oversimplify just for the merit of didactic presentation and easy readability. On the contrary, we have encouraged our authors to point out open and controversial issues. In fact, for this very same purpose we recruited authors all around the world, respected experts in their fields. On the other hand, to warrant some uniformity of the entire text, the number of authors has been limited. Original data and references are provided to permit easy access to relevant original literature. As might have been expected in these times of rapid, in part turbulent, changes in science and elsewhere, this book has experienced a history with many twists and turns. Many of these were related to the problems which must be encountered in putting together such an extensive textbook and dealing with an often capricious body of authors. As if that would not have been enough, however, the emergence of this book was badly struck by personal losses. The book was initiated and pushed ahead by Wilfried Mommaerts' never-failing enthusiasm. He was then joined by Hans-Peter Koepchen, whose intense interest in the application of the new integrative concepts of synergetics provided much of the impetus to this book's outline. When it became clear that age and disease would take their toll, the present two editors joined the team and took over the brunt of the editing work after Koepchen and Mommaerts' deaths, which occurred within a year of each other. We owe them deep thanks, appreciation, and awe. This book is dedicated to their memory and spirit.

There are more thanks to be distributed. The first goes to our authors who, after some measure of arm twisting provided chapters at a time when they are being forced to overcommit themselves to ensure scientific survival in an adverse environment. We are grateful to all the publishers who granted permission to reproduce previously published material. We express our gratitude to Springer-Verlag for venturing into this publication, on the verge of an era without printed textbooks. We thank Dr. R. Lange, Ms. A. Clauss, Ms. A. Hepper, Ms. B. Schmidt-Löffler and their colleagues at Springer for their continued help, support, and hard work. We are deeply indebted to our wives, Beate and Sigrid, for their indulgence and for enduring their husbands' long working hours, frustation, anger, and occasional fits of enthusiasm. We would like to thank countless colleagues for encouragement. U.W. is grateful to the Alberta Heritage Foundation for Medical Research, the Medical Research Council of Canada, and the Faculty of Medicine at the University of Calgary that have supported him over the past few years.

R. GREGER
U. WINDHORST

Contents

Volume 1

Volume 2

IV Acute Vital Functions; Heart; Circulation and Respiration

K Blood and Immune Defense

L Heart and Circulation

V   The Life Cycle

O   Reproduction

List of Contributors

ANTONI, H.
Physiologisches Institut der Universität Freiburg, Hermann-Herder-Str. 7,
79104 Freiburg, Germany

BARTH, C.A.
Deutsches Institut für Ernährungsforschung, Arthur-Scheunert-Allee 114–116,
14558 Bergholz-Rehbrücke, Germany

BAUER, C.
Physiologisches Institut der Universität Zürich, Winterthurer Str. 190, 8057 Zürich,
Switzerland

BLEICH, M.
Physiologisches Institut der Universität Freiburg, Hermann-Herder-Str. 7,
79104 Freiburg, Germany

BRADSHAW, R.A.
University of California at Irvine, Department of Biological Chemistry,
College of Medicine, Irvine, CA 92717, USA

BRANDIS, M.
Universitäts-Kinderklinik, Abteilung Allgemeine Kinderheilkunde, Mathildenstr. 1,
79106 Freiburg, Germany

BRECKWOLDT, M.
Universitäts-Frauenklinik, Abteilung Frauenheilkunde und Geburtshilfe II,
Hugstetter Str. 55, 79106 Freiburg, Germany

BÜHLER, H.
Freie Universität Berlin, Universitätsklinikum Benjamin Franklin,
Institut für Klinische Physiologie, Hindenburgdamm 30, 12200 Berlin, Germany

CARPENTER, R.H.S.
University of Cambridge, The Physiological Laboratory, Downing Street,
Cambridge CB2 3EG, UK

CASTEELS, R.
Laboratorium voor Fysiologie, Campus Gasthuisberg K.U.L., Herestraat,
3000 Leuven, Belgium

CHIAPPELLI, F.
University of California at Los Angeles, Department of Anatomy and Cell Biology,
UCLA Medical Center, 10833 Le Conte Avenue, Los Angeles, CA 90024-1763, USA

CONIGRAVE, A.D.
University of Sydney, Faculty of Medicine, Edward Ford Building (A27), Sydney,
NSW 2006, Australia

Cook, D.I.
University of Sydney, Department of Physiology (F13), Sydney, NSW 2006, Australia

Cooper, K.E.
The University of Calgary, Faculty of Medicine, Health Sciences Centre, Department of Medical Physiology, 3330 Hospital Drive N.W., Calgary, Alberta, Canada T2N 4N1

Crone, C. (†)
University of California at Davis, School of Medicine, Department of Human Physiology, Davis, CA 95616, USA

Dale, B.
Stazione Zoologica "Anton Dohrn", Laboratory of Cell and Developmental Biology, Villa Communale 1, 80121, Naples, Italy

Davison, J.S.
University of Calgary, Faculty of Medicine, Department of Medical Physiology, 3330 Hospital Drive N.W., Calgary, Alberta, Canada T2N 4N1

Dawson, C.A.
Medical College of Wisconsin, Department of Clinical Physiology, VA Medical Center, Research Service 151, 5000 West National Avenue, Milwaukee, WI 53295-1000, USA

Dowling, J.E.
Harvard University, The Biological Laboratories, Department of Cellular and Developmental Biology, 16 Divinity Avenue, Cambridge, MA 02138, USA

Drews, G.
Universität Tübingen, Pharmazeutisches Institut, Auf der Morgenstelle 8, 72076 Tübingen, Germany

Ehrenstein, W.H.
Institut für Arbeitsphysiologie an der Universität Dortmund, Abteilung Sinnes- und Neurophysiologie, Ardeystr. 67, 44139 Dortmund, Germany

Faissner, M.
Universität Heidelberg, Institut für Neurobiologie, Im Neuenheimer Feld 364, 69120 Heidelberg, Germany

Finke, R.
Freie Universität Berlin, Universitätsklinikum Benjamin Franklin, Institut für Klinische Physiologie, Hindenburgdamm 30, 12200 Berlin, Germany

Fisher, D.A.
Coining Nichols Institute, 33608 Ortega Highway, San Juan, Capistrano, CA 92690, USA

Florez, J.C.
Northwestern University, Department of Neurobiology and Physiology, 2153 N Campus Drive, Evanston, IL 60208-3520, USA

Forte, J.G.
University of California at Berkeley, Department of Molecular and Cell Biology, Division of Cell and Developmental Biology, Berkeley, CA 94720, USA

Fox, S.E.
SUNY Health Sciences Center, Department of Physiology, Box 31, 450 Clarkson Avenue, Brooklyn, NY 11203, USA

FRANCESCHI, C.
University of Modena, Faculty of Pathology, Department of Immunology,
41100 Modena, Italy

FROTSCHER, M.
Anatomisches Institut der Universität Freiburg, Abteilung Anatomie I,
Albertstr. 17, 79104 Freiburg, Germany

GREGER, R.
Physiologisches Institut der Universität Freiburg, Hermann-Herder-Str. 7,
79104 Freiburg, Germany

GUMMER, A.W.
Universitäts-Hals-Nasen-Ohren-Klinik, Hörforschungslaboratorien, Silcherstr. 5,
72076 Tübingen, Germany

HARPER, R.M.
UCLA School of Medicine, Department of Anatomy, Los Angeles, CA 90024, USA

HÄUSSINGER, D.
Klinik für Gastroenterologie und Infektiologie, Heinrich-Heine-Universität,
Moorenstr. 5, 40225 Düsseldorf, Germany

HENATSCH, H.D.
Georg-August-Universität Göttingen, Zentrum Physiologie und Pathophysiologie,
Abteilung Neuro- und Sinnesphysiologie, Humboldtallee 23, 37073 Göttingen,
Germany

HERLITZE, S.
HNO-Universitätsklinik Tübingen, Sektion Sensorische Biophysik, Röntgenweg 11,
72076 Tübingen, Germany

HIERHOLZER, K.
Freie Universität Berlin, Universitätsklinikum Benjamin Franklin,
Institut für Klinische Physiologie, Hindenburgdamm 30, 12200 Berlin, Germany

HINSSEN, H.
Universität Bielefeld, Fakultät Biologie, Biochemische Zellbiologie, 33501 Bielefeld,
Germany

HOLTZ, J.
Martin-Luther-Universität Halle, Institut für Pathophysiologie, Magdeburger Str. 6,
06112 Halle, Germany

JACKSON, M.B.
University of Wisconsin School of Medicine, Department of Physiology,
1300 University Avenue, Madison, WI 53706, USA

JÄNIG, W.
Universität Kiel, Physiologisches Institut, Olshausenstr. 40, 24118 Kiel, Germany

JENSEN, A.
Universitäts-Frauenklinik, Knappschaftskrankenhaus, In der Schornau 23–25,
44892 Bochum, Germany

JERRY, M.
Tom Baker Cancer Center, 1331, 29th Street N.W., Calgary, Alberta,
Canada T2N 4N2

JOCKUSCH, B.M.
Technische Universität Braunschweig Carolo Wilhelmina, Zoologisches Institut,
Zellbiologie, Spielmannstr. 7, 38106 Braunschweig, Germany

JOCKUSCH, H.
Universität Bielefeld, Entwicklungsbiologie, W7, 33501 Bielefeld, Germany

JUNGE, D.
UCLA School of Medicine, Department of Physiology, 10833 Le Conte Avenue,
Los Angeles, CA 90024-1751, USA

JUNGERMANN, K.
Universität Göttingen, Biochemisches Institut, Humboldtallee 23, 37073 Göttingen,
Germany

KARCK, U.
Klinikum der Universität Freiburg, Universitäts-Frauenklinik, Abteilung
Frauenheilkunde und Geburtshilfe II, Hugstetter Str. 55, 79106 Freiburg, Germany

KECK, C.
Klinikum der Universität Freiburg, Universitäts-Frauenklinik, Abteilung
Frauenheilkunde und Geburtshilfe II, Hugstetter Str. 55, 79106 Freiburg, Germany

KELLOW, J.E.
University of Sydney, Department of Medicine, The Royal North Shore Hospital,
Pacific Highway, St. Leonards, NSW 2065, Australia

KETTENMANN, H.
Max-Delbrück-Zentrum für Molekulare Medizin, Zelluläre Neurowissenschaften,
Robert-Rössle-Str. 10, 13122 Berlin, Germany

KOEPCHEN, H.P. (†)
Freie Universität Berlin, Institut für Physiologie, Arnimallee 22, 14195 Berlin,
Germany

KOIZUMI, K.
SUNY Downstate Medical Center, Department of Physiology,
450 Clarkson Avenue, Box 31, Brooklyn, NY 11203-2098, USA

KOOPMANS, H.S.
The University of Calgary, Department of Medical Physiology, Gastrointestinal
Research Group, 3330 Hospital Drive N.W., Calgary, Alberta, Canada T2N 4N1

KOURY, M.J.
Vanderbilt University, Department of Medicine, Hematology Division,
Nashville, TN 37232-2287, USA

KUSCHINSKY, W.
I. Physiologisches Institut der Universität Heidelberg, Im Neuenheimer Feld 326,
69120 Heidelberg, Germany

LANG, F.
Physiologisches Institut der Universität Tübingen, Gmelinstr. 5, 72076 Tübingen,
Germany

LANGER, G.A.
UCLA School of Medicine, Cardiovascular Research Laboratories,
Department of Physiology, Macdonald Research Laboratory Building,
675 Circle Drive South, Los Angeles, CA 90095, USA

LINDENMANN, J.
Institut für Immunologie und Virologie, Gloriastr. 30, 8028 Zürich, Switzerland

Logan, C.G.
The University of Iowa College of Medicine, Department of Neurology,
200 Hawkins Dr., Iowa City, IA 52242-1053, USA

Lopes da Silva, F.H.
University of Amsterdam, Faculty of Biology, Graduate School
for the Neurosciences, Institute of Neurobiology, Kruislaan 320,
1098 SM Amsterdam, The Netherlands

McEwen, B.S.
Rockefeller University, Laboratory of Neuroendocrinology, 1230 York Avenue,
New York, NY 10021, USA

Mommaerts, W.F.H.M. (†)
UCLA School of Medicine, Center for the Health Sciences, Department of
Physiology, 10833 LeConte Avenue, Los Angeles, CA 90024-1751, USA

Neulen, J.
Universitäts-Frauenklinik, Abteilung Frauenheilkunde und Geburtshilfe II,
Hugstetter Str. 55, 79106 Freiburg, Germany

Nieschlag, E.
Westfälische Wilhelms-Universität, Klinische Forschergruppe
für Reproduktionsmedizin der Max-Planck-Gesellschaft, Dogmagkstr. 11,
48149 Münster, Germany

Nilius, B.
Laboratorium voor Fysiologie, Campus Gasthuisberg K.U.L., Herestraat,
3000 Leuven, Belgium

Ottaviani, E.
University of Modena, Department of Animal Biology, 41100 Modena,
Italy

Piiper, J.
Max-Planck-Institut für experimentelle Medizin, Hermann-Rein-Str. 3,
37075 Göttingen, Germany

Pittman, Q.J.
University of Calgary, Health Sciences Center, Department of Medical Physiology,
3330 Hospital Drive N.W., Calgary, Alberta, Canada T2N 4N1

Plinkert, P.
Universitäts-Hals-Nasen-Ohren-Klinik, Silcherstr. 5, 72076 Tübingen,
Germany

Renkin, E.M.
University of California at Davis, School of Medicine,
Department of Human Physiology, Davis, CA 95616, USA

Richter, D.W.
Zentrum Physiologie und Pathophysiologie, Abteilung Neuro- und
Sinnesphysiologie, Humboldtallee 23, 37073 Göttingen, Germany

Rüegg, J.C.
II. Physiologisches Institut der Universität Heidelberg, Im Neuenheimer Feld 326,
69120 Heidelberg, Germany

RUPPERSBERG, J.P.
HNO-Universitätsklinik Tübingen, Sektion Sensorische Biophysik, Röntgenweg 11,
72076 Tübingen, Germany

SCHILD, D.
Universität Göttingen, Zentrum Physiologie und Pathophysiologie, Abteilung
Neuro- und Sinnesphysiologie, Humboldtallee 23, 37073 Göttingen, Germany

SCHLÄFKE, M.E.
Ruhr-Universität Bochum, Abteilung für Angewandte Physiologie,
Universitätsstr. 150, 44780 Bochum, Germany

SCHLEUSENER, H.
Freie Universität Berlin, Universitätsklinikum Benjamin Franklin,
Institut für Klinische Physiologie, Hindenburgdamm 30, 12200 Berlin, Germany

SCHMID-SCHÖNBEIN, H.
Rheinisch-Westfälische Technische Hochschule Aachen, Medizinische Fakultät,
Institut für Physiologie, Pauwelsstraβe, 52074 Aachen, Germany

SHARKEY, K.A.
University of Calgary, Health Sciences Center, Department of Medical Physiology,
3330 Hospital Drive N.W., Calgary, Alberta, Canada T2N 4N1

SHEPHERD, G.M.
Yale University School of Medicine, Section of Neurobiology, 333 Cedar Street,
PMB-236, New Haven, CT 06510-8001, USA

SIEGEL, G.
Institut für Physiologie der Freien Universität Berlin, Fachbereich Natur und
sozialwissenschaftliche Grundlagenmedizin und medizinische Ökologie,
Arnimallee 22, 14195 Berlin, Germany

SIEGEL, J.M.
University of California at Los Angeles, School of Medicine,
Department of Psychiatry, Neurobiology Research 151A3, Sepulveda VAMC,
North Hills, CA 91343, USA

SOKOLOFF, L.
National Institute of Mental Health, Laboratory of Cerebral Metabolism,
Building 36, Bethesda, MD 20892, USA

SOLOMON, G.F.
University of California at Los Angeles, Department of Psychiatry
and Biobehavioral Sciences, Los Angeles, CA 90024-1759, USA

SPILLMANN, L.
Universität Freiburg, Institut für Biophysik und Strahlenbiologie, Hansastr. 9,
79104 Freiburg, Germany

SPYER, K.M.
Royal Free Hospital, School of Medicine, Department of Physiology,
Rowland Hill Street, London NW3 2PF, UK

STAUB, N.C.
University of California, Cardiovascular Research Institute,
Department of Physiology, Box 0130, San Francisco, CA 94143, USA

SUNDBERG, J.
Kung Tekniska Högskolan, Department of Speech Communication and Music
Acoustic, Drotning Kristinas väg 31, Box 700 14, 10044 Stockholm, Sweden

TAKAHASHI, J.S.
Northwestern University, Department of Neurobiology and Physiology,
2153 N Campus Drive, Evanston, IL 60208-3520, USA

TAYLOR, A.N.
University of California at Los Angeles, Department of Neurobiology, Los Angeles,
CA 90024-1763, USA

TER KEURS, H.E.D.J.
The University of Calgary, Department of Medicine, 3330 Hospital Drive N.W.,
Calgary, Alberta, Canada T2N 4N1

THOMPSON, R.F.
University of Southern California, Program for Neural, Informational and
Behavioral Sciences, HDCO Neurosciences Building, HNB 122, Los Angeles,
CA 90089-2520, USA

TIMIRAS, P.S.
University of California at Berkeley, Department of Molecular and Cell Biology,
Division of Cell and Developmental Biology, 221 Life Sciences Addition,
Berkeley, CA 94720, USA

TROTTER, J.
Universität Heidelberg, Institute für Neurobiologie, Im Neuenheimer Feld 364,
69120 Heidelberg, Germany

TYBERG, J.V.
The University of Calgary, Department of Medicine, 3330 Hospital Drive N.W.,
Calgary, Alberta, Canada T2N 4N1

WARD, S.A.
South Bank University, Division of Human and Exercise Science,
School of Applied Science, 103 Borough Rd., London SE1 0AA, UK

WEINBAUER, G.F.
Westfälische Wilhelms-Universität, Klinische Forschergruppe für
Reproduktionsmedizin der Max-Planck-Gesellschaft, Dogmagkstr. 11,
48149 Münster, Germany

WHIPP, B.J.
St. George's Hospital Medical School, Department of Physiology, Cranmer Terrace
(Tooting), London SW17 ORE, UK

WILLIS, JR., W.D.
The University of Texas Medical Branch at Galveston, School of Medicine,
Department of Anatomy and Neurosciences, Marine Biomedical Institute,
200 University Boulevard, Galveston, TX 77550-2772, USA

WINDHORST, U.
University of Calgary, Faculty of Medicine, Departments of Clinical Neurosciences
and Medical Physiology, 3330 Hospital Drive N.W., Calgary, Alberta,
Canada T2N 4N1

WUILLEMIN, W.
Hämatologisches Zentrallabor, Universitatsspital Bern, Inselspital,
3010 Bern, Switzerland

YAMAMOTO, W.S.
The George Washington University School of Medicine,
Department of Computer Medicine, 2300 K St. N.W., Washington, DC 20037, USA

YOUNG, J.A.
University of Sydney, Faculty of Medicine, Edward Ford Building (A27), Sydney,
NSW 2006, Australia

ZAHRADNIK, H.P.
Universitäts-Frauenklinik II, Hugstetter Str. 55, 79106 Freiburg, Germany

ZENNER, H.-P.
Universitäts-Hals-Nasen-Ohren-Klinik, Hörforschungslaboratorien,
Silcherstr. 5, 72076 Tübingen, Germany

III   Intake and Excretion

I Epithelial Transport and Gastrointestinal Physiology

59 Epithelial Transport

R. GREGER

Contents

59.1 Introduction: Why Polarity?

Epithelia are the borders between the "milieu intérieur" and the "milieu extérieur" [6]. Such borders have two functions:

- They separate the two compartments.
- They control communication.

The skin is one such border (epithelium): it is designed to prevent losses of fluid, electrolytes, heat, etc., but at the same time it can excrete water in the form of sweat if needed and thus discharge heat. The many other functions of this barrier are beyond the scope of this chapter. The renal tubules are another example of an epithelium. They are the border between the urinary space and the blood. The tubules are highly specialized epithelia, which perform absorption and excretion. Chapters 74 and 75 will deal with these functions. The epithelia of the gastrointestinal tract serve to digest food and absorb its constituents. They con-

trol water and electrolyte intake. These functions will be discussed explicitly in coming chapters. The exocrine glands produce secretions that contain water, electrolytes, enzymes, etc. The primary process consists in the secretion of NaCl and enzymes, and a second step in the modification of these secretions. This will also be discussed in Chaps. 64 and 65. The airways are covered by still other epithelia, which are largely heterogeneous as we progress through the nose, trachea, bronchi and bronchioli to the alveoli. In the former epithelia, cleaning, warming and moistening of the inspired air is one task, while the latter have the tasks of actual gas exchange and the production and secretion of surfactant. Two specific epithelial barriers between blood and brain, the blood – brain barrier, formed by specialized endothelial cells, and the choroid plexus, will be discussed in a separate chapter (Chap. 27). This additional barrier within the body segregates the brain from the rest of the body, guaranteeing more stability and fewer disturbances by extraneous influences for this organ. At the same time, this barrier raises the problem of adequate import of fuel and export of waste.

The above tasks and problems require highly specific solutions. However, one thing common to all these epithelia is their general organization as polar cells. They all have an apical (luminal or lumenal) membrane which is different in its properties from the basolateral (basal and lateral) membrane. The individual epithelial cells are held together by tight junctions. The different properties of the two poles of the epithelial cell ensure that the transport is controlled in two steps. If needed the transport can be energized by pumps in direct or indirect fashion (cf. Chap. 8). Hence, epithelia are not passive, i.e., isolating or finitely permeable (permissive) barriers, but active borders through which controlled transport occurs in a regulated fashion [98]. to make this point more clearly let us start with an example. Consider that the body may be threatened by Na$^+$ losses. By mechanisms discussed explicitly in Chaps. 77 and 82, this would lead to the secretion of aldosterone by the glomerulosa cells of the adrenal cortex. Aldosterone acts on the kidney by maximizing Na$^+$ absorption (and enhanc-

R. Greger/U. Windhorst (Eds.)
Comprehensive Human Physiology, Vol. 2
© Springer-Verlag Berlin Heidelberg 1996

ing K+ secretion) [80]. Most importantly, the same altered transcription program is initiated by aldosterone in sweat gland ducts, colon, and respiratory epithelium. Therefore, Na+ losses are minimized wherever they might occur at the borders with the outside world. The circulating hormone distributes the message to all relevant borders, and the response is an increase in the number of active Na+ channels in the apical membrane of these epithelia, an increase in mitochondrial enzymes to produce more ATP, and an increase in the number of basolateral (Na++K+) pumps [80].

Initially, the knowledge of epithelial function developed slowly over the past 40 years. The pioneering work of Ussing was pivotal in this respect [61,97]. He designed the two serial barriers model for the frog skin [98]. The frog skin has since served as a paradigm for other epithelia. Rapid progress was made 20–30 years ago in the understanding of the function of renal tubule epithelia. The epithelia of the gastrointestinal tract were studied at approximately the same time, but detailed knowledge has generally lagged behind that on the kidney. Other epithelia, such as exocrine glands, have been studied with ever increasing intensity, while those of the respiratory tract had not been studied well until the past 10–15 years. Initially, the level of investigation was examination of the barrier and the transport function of the intact epithelium. This was followed by examination of the properties of the membranes and of the individual components of the two membranes, and by examination and cloning of the individual transport proteins. Parallel to this development, cell cultures of many epithelia became available. Primary cultures usually preserve at least some of the properties of their source tissue, although fairly rapid alterations of properties have also been noted. Permanent cultures of epithelial tissues obtained by transformation or obtained from tumors usually differ in many respects from their source tissue, but they may be extremely valuable for study of one specific property.

At this stage it appears appropriate to briefly discuss the methods with which epithelia can be examined.

59.2 Methods of Examining Epithelial Function

A first step is the examination of the overall transport properties of a sheet of epithelium. The skin of a frog or a toad may appear especially suitable for this [61].

Ussing Chamber (Fig. 59.1). After some rough dissection, a piece of frog or toad skin can be mounted in an Ussing chamber, which is made like a sandwich: two halves with the skin in the middle. The solutions in the chamber will then be separated by the frog skin. One side might be called the blood side (basolateral, inside) solution and the other corresponds to the outside world (mucosa, apical, outside). If we were to examine ion transport of this epithelium we might add radioactive ^{22}Na+ to the outside [apical

(a) side] and measure the tracer flux to the other [basolateral (b)] side. We can express this flux (Φ_{a-b}) in distintegrations per minute appearing at the basolateral side per unit of time or, after division by the specific activity, in moles per unit of time and area. Now we can add the tracer to the blood side and repeat the experiment, and hence obtain Φ_{b-a}. In our example Φ_{a-b} exceeds Φ_{b-a}, and we can write:

$$\Phi_{net} = \Phi_{a-b} - \Phi_{b-a} \qquad (59.1)$$

Na+ is absorbed at a rate of Φ_{net}. Obviously our next concern is the question of whether this ion transport can be accounted for by passive mechanisms or whether it is caused by an active transport event. The answer is quite simple: the solutions on both sides have been chosen to be identical. Therefore, no chemical driving force can be responsible for this net flux. Now we examine whether there is any electrical driving force. We measure the transepithelial voltage with two identical electrodes mounted on either side and with a high-resistance electrometer. We find that the apical side is negative with respect to the basolateral side by 90 mV. This active transport voltage [33] therefore has a polarity that cannot cause Na+ absorption, but could be caused by active Na+ absorption. Therefore, we conclude at this stage that Na+ is probably absorbed actively by this epithelium. The movement of the counter-ion, Cl−, could be passive, because the electrical driving force would favor absorption of this ion.

To examine this further we decide to remove the electrical driving force by clamping it at zero. For this we need an adjustable battery, two current electrodes and a more or less sophisticated feedback device that provides just sufficient charges to keep the voltage exactly at 0 mV. We include an ampèremeter in the current loop (cf. Fig. 59.1). Now active absorption will move the actively transported ion species across the epithelium, whilst the passive (counter-ion) transport will have to occur through the external circuit (and ampèremeter). The current loop is closed: the current through the instrument must be identical, and of opposite charge, to the current moved actively through the epithelium (I_{sc}). In our example the current could have been determined at 30 μA/cm². Now we measure the net transport of Na+ under the same circumstances and we may determine 18.6 nmol/cm² · min. If we now simply try to equate both fluxes:

$$J_{Na} = I_{sc}/zF \qquad (59.2)$$

we recognize that, in fact, J_{Na} can fully account for I_{sc}. How does the Cl− move? Under open circuit conditions active Na+-transport builds up a voltage of, e.g., −90 mV (apical side negative), which drives Cl− passively across the epithelium. Under voltage clamp conditions, the Cl− ions are delivered to the current electrode on the outside and set free on the current electrode on the inside. The beauty of Ussing's method is the direct comparison of the fluxes and the current. However, with this method the epithelium is

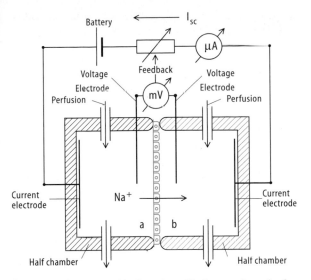

Fig. 59.1. The Ussing chamber. An epithelium such as the frog skin is mounted between two halves of a lucid chamber. Two electrodes are used to measure the voltage (mV), and two more to measure current. The epithelium is bathed by the perfusion solutions on the apical (*a*) and basolateral (*b*) sides. With this device the voltage can be clamped to 0 mV by means of an adjustable battery. This adjustment is done by a feedback control to a resistor. Under voltage-clamped conditions the Na^+ ion is still transported actively through the epithelium. A corresponding current I_{sc} flows through the current meter (μA)

altered in its function, because the voltages of the cells and the ionic compositions will be altered.

One can also examine the tissue electrically under normal, zero current, conditions. This is done by determining the transepithelial resistance (R_{te}) by short current pulses and by recording the corresponding changes in V_{te} (ΔV_{te}) as well as the basal V_{te} continuously. The equivalent short circuit current (I_{sc}') can now be written according to Ohm's law:

$$I_{sc}' = V_{te}/R_{te} \qquad (59.3)$$

I_{sc}' is not the short circuit current (I_{sc}), but a virtual current, corresponding to but not identical to that obtained with the above clamp protocol. Consider Fig. 59.2, where I_{cc} is a circular current that moves as a Na^+ current through the cell and as a passive, dissipative current through the paracellular shunt or through shunting cells with different properties. Therefore, as in real life, the current is a circular current. To translate I_{cc} into the short circuit current, it is necessary to know the resistance of the paracellular or shunt path, because V_{te} drives a current equal to the active component through the paracellular or shunt pathway:

$$I_{sc} = V_{te}/R_s \qquad (59.4)$$

According to Kirchhoff's law we can write:

$$R_{te} = \frac{R_s(R_a + R_b)}{R_a + R_b + R_s} \qquad (59.5)$$

where R_s, R_a and R_b denote the resistances of the paracellular shunt, the apical and the basolateral membrane, respectively. Inserting Eq. 59.5. into Eq. 59.4, we can now calculate the short circuit current as:

$$I_{sc} = V_{te} \frac{R_a + R_b + R_{te}}{R_s(R_a + R_b)} \qquad (59.6)$$

This simple consideration, of course, is only valid to the extent that the individual resistance values are the same in both measuring conditions (open circuit and short circuit conditions). This need not be the case, because the driving forces are grossly changed by voltage clamping in the short circuit measurements. In our example we may have found that R_{te} is 3600 $\Omega \cdot cm^2$, and we may have determined from analyses to be discussed below that R_s is 4000 $\Omega \cdot cm^2$. Then I_{sc}' is 1.1 times I_{sc}.

Analyses of this kind can of course be much more detailed and define

- The current–voltage relation
- The flux ratio [97]
- The ions carrying the current by replacement studies
- Specific inhibitors
- Regulatory mechanisms.

The determination of R_{te} is rather straightforward in planar tissues: a current pulse is injected and the corresponding voltage deflection is measured. In tubule structures, such as that of renal tubules or glandular ducts, it is much more difficult to determine transepithelial resistance: cable equations have to be formulated to calculate the R_{te} from the input resistance, the tubule radius, and/or the length constant [41,96].

In various epithelia, including the small tubule-like structures of exocrine glands and kidney, net flux studies can also be performed by simply comparing the amount of one substance at one site to that present at a measured distance away [96]. Such studies can be carried out under in vivo conditions by removing small samples of tubule fluid and analyzing their composition [43] or performed as perfusion studies [66] with a given perfusate of which the composition is examined at another site downstream. The analyses are usually concentration measurements. Therefore, to obtain the mass balance:

$$M_1 = M_2 + M_3 \qquad (59.7)$$

where M_1, M_2, and M_3 are delivered amount, the recovered amount, and the amount removed by absorption, it is necessary to add some sort of a volume (perfusion rate) marker, i.e., a substance that is not transported. In the kidney inulin (in) (cf. Chap. 73) is used for this purpose. We can write:

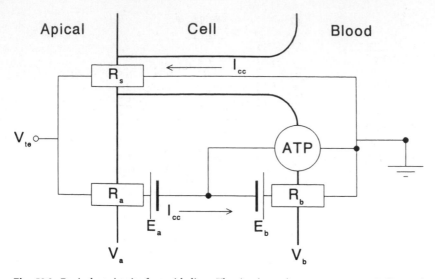

Fig. 59.2. Equivalent circuit of an epithelium. The circuit consists of two electrochemical driving forces of the two membranes (E_a and E_b). In addition, the ATP pump generates a current (I_{act}). Owing to its coupling ratio of 3 Na^+ per 2 K^+. The circuit consists of three resistors: R_s, R_a, and R_b for the paracellular shunt pathway, apical membrane, and basolateral membrane, respectively. E_a, E_b, and I_{act} generate a circular current (I_{cc}), which produces the voltages V_a (apical membrane voltage), V_b (basolateral membrane voltage) and V_{te} (transepithelial voltage) across the respective resistors. A circuit of this type can be analyzed by Ohm's and Kirchoff's laws:

$$V_a = \frac{E_a\,(R_b + R_s) - E_b \cdot R_a + R_b \cdot R_a \cdot I_{act}}{R_s + R_a + R_b}$$

$$V_b = \frac{E_b\,(R_a + R_s) - E_a \cdot R_b - R_b \cdot I_{act}(R_a + R_s)}{R_s + R_a + R_b}$$

These equations can be resolved for E_a and E_b [46]

$$V_1 \cdot c_1 = V_2 \cdot c_2 + M_3 \tag{59.7a}$$

where V and c denote the respective flow rates (volumes/time) and concentrations of the substance under study. For V_2 we can write:

$$V_2 = V_1 \cdot in_1 / in_2 \tag{59.7b}$$

where in_1 and in_2 are the respective inulin (volume marker) concentrations. Insertion reveals:

$$M_3 = V_1(c_1 - c_2 \cdot in_1 / in_2) \tag{59.7c}$$

M_3 is the amount reabsorbed per examined tubule length (area) and time. This rate can of course be compared to (for example) the equivalent I_{sc} obtained in the same preparation. This is only possible if the composition of the luminal fluid is not altered substantially by reabsorption, and if the I_{cc} is known from a resistance analysis (cf. above). Then we can write:

$$I_{cc}/z \cdot F = k \cdot J_x \tag{59.8}$$

where I_{cc} is the circular current, corresponding to I_{sc} (A/cm²); z and F have their usual meaning; J_x is the ion flux in mol/s · cm², and k is the unknown coupling ratio of the two fluxes.

Measurement of Cytosolic Composition. At the next level of investigation, information on the cytosolic composition (measured with ion-selective electrodes, electron probe or chemically) and on the properties of the two cell membranes is necessary. To obtain this in an intact preparation, the cells can be impaled by microelectrodes [69]. These can be simple KCl-filled microelectrodes with fine tips or ion-selective microelectrodes [26,44,54]. With the first type, the voltage across the individual cell membranes can be measured. It is only necessary to measure across two barriers, e.g., across the basolateral membrane (V_b) and across the epithelium (V_{te}). The voltage across the third can then be calculated:

$$V_{te} = V_b - V_a \tag{59.9}$$

where V_a is the voltage across the apical (luminal) membrane.

Figure 59.3 gives an example from the thick ascending limb of the loop of Henle (TAL) of rabbit kidney [39,46,47]. The tubule segment is perfused in vitro with identical Ringer-type solutions on both sides, and V_{te} and V_b are measured. In addition, short current pulses are injected to measure the voltage deflections caused by the individual resistances. V_{te} is lumen positive (+6mV)and V_b is −72mV. Therefore, V_a (luminal membrane) is −78mV (with reference to the lumen). The voltage deflection across the basolateral membrane is approximately 35% of that across

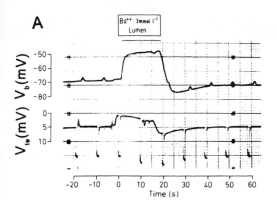

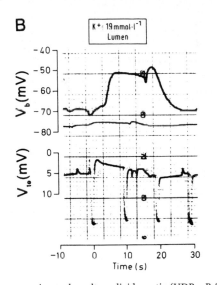

Fig. 59.3A,B. Transepithelial (V_{te}) and intracellular voltage (V_b) measurements in isolated in vitro perfused rabbit cortical thick ascending limb segments of the loop of Henle (*TAL*). (Note that the traces of the pen recorder are shifted to each other by three small boxes.) The lumen and bath perfusate contain identical NaCl-Ringer's solution. Current pulses are injected through the luminal perfusion pipette. This leads to deflections of V_{te}. Ohm's law and cable equations [38] are used to determine the transepithelial resistance (R_{te}) and the transepithelial voltage deflection at the site of cellular impalement. The tubule is impaled with KCl-filled (2.7 mol/l) fine-tipped ($\approx$70 nm, 200 MΩ) Ling-Gerard microelectrodes across the basolateral membrane. A Under control conditions V_{te} is lumen positive ($\approx$ +5 mV). This voltage cannot be accounted for by ion diffusion, because the perfusates on both sides are identical. Therefore this voltage is called active transport voltage. R_{te} in this example is 30 $\Omega \cdot$cm^2. Hence the equivalent short circuit current, I_{sc}', is 167 μA/cm^2. V_b across the basolateral membrane is -70 mV (cytosol negative with respect to bath). This suggests that the voltage of the impaled cell is largely determined by the E_K. The voltage across the luminal membrane (V_l) can now be calculated from $V_b - V_{te} = -75$ mV. The fact that the voltage changes of both cell membranes are similar is caused by the coupling (partial short-circuiting) through the paracellular pathway. Note the small pulses in the V_b trace. These are caused by the intratubular current injections. They can be used

to estimate the voltage divider ratio (VDR = R_l/R_b). In this example VDR is 3. The *horizontal bar* indicates a rapid change in the luminal perfusate to a solution containing Ba^{2+}. This leads to a partial collapse of V_{te} to +2 mV and a marked increase in the transepithelial voltage deflections, indicating an increase in R_{te} to 36 $\Omega \cdot$ cm^2. Hence I_{sc}' falls to 56 μA/cm^2. In addition, Ba^{2+} depolarizes V_b and V_l (not shown) by 23 and 2 mV, respectively. This is due to the inhibition of luminal K$^+$ channels. At the same time the small pulses disappear, which indicates that VDR has become infinite. This experiment proves that the luminal membrane of the TAL segment has a K$^+$ conductance, which apparently dominates the conductance properties of this membrane, and which is required for active transport. B In a similar experiment to that shown in A the luminal perfusate is rapidly exchanged for one containing 19 mmol/l K$^+$ (instead of 4 mmol/l, as in the control solution). This leads to a fall in V_{te}, which now cannot be interpreted in terms of an active transport voltage, because the solutions on the two sides are now different. In fact, this fall in V_{te} by 3 mV is caused by a K$^+$ diffusion voltage across the shunt and across the luminal membrane. The current-induced voltage deflections become smaller, which is caused by the fall in luminal membrane resistance. V_b depolarizes by 18, and V_l by 20 mV. This depolarization of V_l (ΔV_l) is caused by the change in E_l (ΔE_l). The quantitative analysis of these data is discussed in Sect. 59.2

the entire epithelium. From the input resistance and the internal tubule diameter and with the use of cable equations R_{te} can be determined as 34 $\Omega \cdot$ cm^2. To obtain absolute values for R_b and R_l (luminal membrane), the ratio of the two (the voltage divider ratio, VDR, in our example $R_l/R_b = 65\%/35\% = 1.8$) and one absolute value have to be known. This can be obtained by two subsequent measurements of this kind with one resistance changed experimentally and the two others kept constant. One such experiment in our example is the luminal application of Ba^{2+}, which only increases R_l to very high values, because this membrane, as we shall see, contains only K$^+$ channels. Therefore we can write in a slightly simplified analysis that R_{te}' (with Ba^{2+} present) is equal to R_s. Let us assume that we have measured R_{te}' as 47 $\Omega \cdot$ cm^2, then we can calculate from VDR and Eq. 59.6 that R_l is 88 and R_b 47 $\Omega \cdot$ cm^2 [46,47].

Determination of Ion Selectivity of Cell Membranes. The next question regards the ion selectivity properties of the two cell membranes and of the paracellular shunt. To obtain these, ion replacement studies can be performed, and the voltages have to be measured. For example, an increase in the luminal K$^+$ concentration, as shown in Fig. 59.3B, leads to a marked depolarization of V_l but has little effect on V_{te}. The analysis of the voltage change in terms of the fractional conductance (cf. Chap. 8) is not straightforward, because the change in the electromotive force (E) across the luminal membrane induces an eddy (circular) current (CC) which leads to a voltage drop across each barrier. We can write for the current loop induced voltage deflection across the shunt:

$$\Delta V_{te} = CC \cdot R_s \qquad (59.10)$$

for CC we can write:

$$CC = \Delta E_l / (R_l + R_s + R_b) \qquad (59.11)$$

combined to:

$$\Delta V_{te} = \Delta E_l \cdot R_s / (R_l + R_s + R_b) \qquad (59.10a)$$

For ΔV_b we can write by analogy:

$$\Delta V_b = \Delta E_l \cdot R_b / (R_l + R_s + R_b) \qquad (59.10b)$$

and for ΔV_l, according to Eq. 59.8:

$$\Delta V_l = \Delta E_l (R_b + R_s) / (R_l + R_s + R_b) \qquad (59.10c)$$

In our example, ΔV_l is +20 mV. Therefore, with the use of the above values for the individual resistances, we determine ΔE_l as 40 mV. Now we can go to the Nernst equation to obtain the fractional conductance for K^+ of the luminal membrane(f^K_l):

$$f^K_l = \Delta E_l / \Delta E^K \qquad (59.11)$$

where ΔE^K is the change in the chemical potential for K^+. In this case it was 42 mV. Therefore, we conclude that f^K_l is 0.95. In other words, the vast majority of the conductance of this membrane is caused by K^+. Similar analyses can be performed for the other barriers. In the example of the TAL these analyses reveal that the paracellular shunt is cation selective with a relative permeability for Na^+ 3 times that for Cl^- [38]. The basolateral membrane is almost exclusively Cl^- conductive [39,46]. If all resistances have been determined and if the voltages are known one can directly go one step further and use Kirchhoff laws (cf. equations in legend to Fig. 59.2) to determine the E-values of the two membranes (E_l and E_b) [46].

Measurement of Cytosolic Ion Activities. The next step is the measurement of ion activities in the cell by ion-selective microeletrodes. These kinds of electrodes are designed to be selective for one ion. Unfortunately selectivities are only finite for most ion exchangers, and the interferences by other ions have to be controlled for very carefully [44,49,54]. Furthermore, with single-barrelled ion-selective microelectrodes the electrochemical potential difference for the ion under study rather than the chemical potential is measured. Therefore, wherever possible, ion activities are measured with double-barrelled microelectrodes [48]. One barrel is filled with KCl and used to measure the membrane voltage, and the other barrel is used to monitor the chemical potential. A summary of such measurements in TAL segments is given below:

f^K_l	E_K (mV)	V_l (mV)	f^{Cl}_b	E_{Cl} (mV)	V_b (mV)
1.0	−90	−78	1.0?	−35	−72

These data indicate that K^+ ions are secreted into the lumen

with a driving force of 12 mV, and that the driving force for Cl^- exit across the basolateral membrane is around 37 mV. Such analyses can of course be complemented by studies using inhibitors [22,45,86,100], and they can be performed in various regulatory states [53,85]. The final goal at this level is a model containing all the relevant transporters responsible for the transepithelial transport of the substance under study [39,51].

Vesicle Technique. At the next level of investigation one wants to examine the respective transporters in isolated membranes. One such approach is the vesicle technique [72]. By a detergent and Ca^{2+} or Mg^{2+} precipitation technique, membrane vesicles can be produced from the luminal and from the basolateral membrane [59]. The preparation is, of course, accompanied by a validation procedure in which the origin of the vesicular membrane is monitored. This method has been available for almost 20 years now [59], and it has been extremely helpful in the description, especially, of the various carrier systems. Two examples are given in Figs. 11.8 and 75.20.

Patch Clamp Analysis. Another very valuable method is the patch clamp analysis of ion channels [81]. This can be performed on the cell with the epithelium intact, or in excised membrane patches. Many examples are given in several chapters (e.g., Fig. 8.3; Figs. 11.3, 75.18) In excised inside-out patches the cytosolic surface becomes available for direct experimentation: ion replacement, addition of inhibitors, regulatory factors, etc.

New Techniques. At the level of isolated membrane preparations and membrane fractions, transport proteins can also be purified and reconstituted in liposomes or lipid bilayers [5,64]. Even for purified preparations the analysis of the amino acid sequence is hardly feasible at the protein level. Here, recent advances in cloning techniques have paved the way for entirely new approaches, such as expression cloning [56]. Once the cDNA is available [79], further steps include the expression in other cells (oocytes, Chinese hamster ovary cells, fibroblasts, insect cells such as SF9 cells, and many more [58,93]). Now the transporter can be studied in a system that usually does not express this protein. Antibody studies can be extremely valuable in the labelling of respective transporters in various tissues, inasmuch as they can disclose the cell distribution and organ distribution of a given transport protein. The transporter itself can be engineered at the genomic level, and a correlation between amino acid sequence and function can be drawn [3]. Most recently, transgenic animals (usually mice) have become available, in which one transporter is altered stably and the consequences for organ and body function can be examined. One such example is the knock out mouse carrying the cystic fibrosis defect, which is naturally only found in man as a mutation at the respective gene locus on chromosome 7 [15].

The repertoire of methods has increased very rapidly over the past 10 years, and as a consequence we can now replace some of the arrows in the various cell models by precise information on the respective protein. However, the interpretation of the function of a given and cloned transporter is still very difficult.

Punctual mutations, produced by the experimenter to study the function of a transport protein, are usually based on guesses rather than detailed knowledge. The basis of such guesses is the secondary/tertiary structure obtained from various techniques, such as Doolittle plots (hydropathy plots), glycosylation sites, antibody binding, partial hydrolysis by enzymes, and comparison with known proteins. For a very few transporters, crystallization has been successful and steric information has been yielded by X-ray diffraction studies [19,101].

There is a danger that we are getting ahead of ourselves or may lose solid ground. It appears appropriate, and probably even necessary, to try to interpret all findings at all levels on the basis of the intact cell and even at the organ level. Like all other techniques, genetic engineering in itself is a tool of physiology, but does not tell us the "logic of life" [9].

To conclude this section, the spectrum of methods available for the study of epithelial function has expanded very rapidly. It is predictable that we shall be able to understand many epithelia in terms of cell models and individual transporters in the foreseeable future. It is also likely that we shall possess the genomic information relating to most of these transporters within a very few years. It will take much longer, however, to understand the regulatory mechanisms at the various levels of coordination.

59.3 The Tight Junctions, a Back-leak or a Route of Transport?

Tight junctions are a specific feature of epithelia. Figure 59.4 shows the schematic structure of a tight junction and also gives an example of freeze-fracture study of one such tight junction. It has been known for some time that these junctions present a specific border in epithelia [28]. They can be rather complex, with a larger number of strands, or rather primitive, with only a few strands. In these strands the membranes come very close to each other (Fig. 59.4). The tight junctions can be looked at as a belt or a series of belts holding the individual cells together. Tight junctions are internally stabilized by the cytoskeleton (cf. Figs. 4.2, 4.3). It has been suggested that preformed junctional material is inserted by exocytosis [36]. Besides the tight junctions, cells are of course held together by desmosomes and other junctions (cf. Chap. 4).

The tight junctions, depending on their complexity [28], are permeable for small charged and uncharged molecules. The importance and magnitude of this pathway has been determined in epithelia with simple and multi-stranded tight junctions. Figure 59.5 gives an example of the *Necturus* gallbladder, which has rather "primitive" tight junctions. In this epithelium the tight junctions are the main route of current path [32]. This has been shown very elegantly by scanning with an electrode over the epithelium. The highest current densities are found over the tight junctions. These studies were a turning point, inasmuch as they proved for the first time that passive ion transport in this and similar epithelia (proximal tubule of the kidney, other intestinal epithelia) proceeds preferentially through the tight junctions.

With the advances in the techniques for measurement of transepithelial resistance, it has become popular to classify the epithelia into those called leaky and others called tight [32,71,89,97]. Obviously such a distinction only makes sense if we do not simply base it on absolute numbers for permeability or conductance (or the inverse: resistance). This terminology is much more appropriate if we base it on the ratio of the conductance (resistance) of the paracellular (shunt) pathway (R_s) versus that across the cell ($R_a + R_b$). Table 59.1 gives examples for various epithelia and correlates R_s with the tightness of the tight junctions. It is apparent, for example, that the proximal tubule of the kidney has to be classified as a leaky epithelium with primitive strands, that the thick ascending limb of the loop of Henle is an example of the intermediate case, and that the collecting tubule, like the colon has to be classified as a rather tight epithelium.

Although the tight junctions in leaky and intermediately leaky epithelia are rather permeable, they usually expose some ion selectivity which deviates from free solution mobility. For instance, the tight junction of the proximal tubule with a resistance of some $5\,\Omega\cdot cm^2$ has a permselectivity of around ≈ 1.5 for Na^+/Cl^- and that of the TAL, one of ≈ 3, whilst the corresponding free solution mobility ratio is ≈ 0.6. It is believed that this selectivity is caused by fixed negative charges, which provide a hindrance for the permeation of anions.

Obviously the tight junctions are only the entrance into the lateral space, and it might be assumed that this space also provides some resistance to the movement of ions, substances and water. From a theoretical point of view this must in fact be true [24], and there have been many attempts to quantify the resistance of this space and the standing gradients that build up as a consequence [16]. There is no consensus beyond the generally accepted view that the specific morphology of epithelia, like the proximal tubule, guarantees that both such resistances and the corresponding standing gradients must be small. Whether such gradients are in the millimolar range or lower is obviously very difficult to determine.

Tight junctions are probably not static. Their permselectivity is variable: high concentrations of Ca^{2+}, Ba^{2+}, and protons reduce the cation selectivity [21,109,110]. In addition, some studies have led to claims that hormones can vary the conductance of the paracellular shunt [65]. A very interesting case is that of TAL, where it has been shown that PTH enhances the conductive movement of Ca^{2+} probably via a cAMP-dependent regulation of the tight junctions [106]. As we shall see below, the paracellular pathway is an important route of ion transport in leaky and intermediately leaky epithelia, and probably also to some extent a route of water transport.

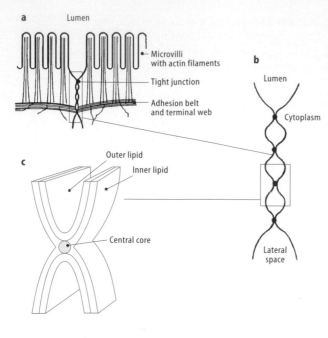

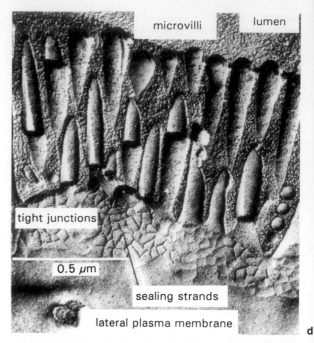

Fig. 59.4a–d. The structure of tight junctions. **a** Apical pole of two epithelial cells from small intestine. The tight junction is rather leaky, with ≤6 strands (for simplification only 4 are shown). The strand is easily seen in transmission electromicrographs or freeze-fracture micrographs (**d**). For comparison, there are only 1 or 2 strands in the proximal tubule and 4 between the acini of the exocrine pancreas. **b** Tight junctions shown schematically at higher magnification. **c** Hypothetical arrangement of a single strand (after [17]). Note that the outer lipids do not communicate. The strand contains the central core, which consists of an inverted cylindrical phospholipid micelle. **d** Freeze-fracture micrograph of tight junctions of rat small intestine. Microvilli and the tight junctions are easily apparent. (Modified from [2])

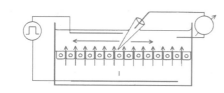

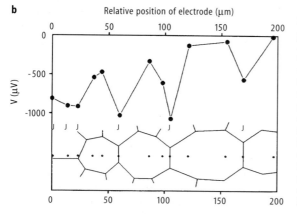

Fig. 5a,b. Voltage scanning of the current path in *Necturus* gallbladder [32]. **a** Schematic of the experimental design. The epithelium is mounted in an Ussing-type chamber. An upper and a lower current electrode are used for current pulse injections. The voltage scanning electrode with a very fine tip is moved longitudinally over the surface of the gallbladder. The current (I) moves preferentially through the funnels of the paracellular pathway. Whenever the scanning electrodes reaches an area of high current density this leads to a corresponding voltage deflection. **b** The voltage profile obtained from scanning over a distance of 200 μm. A sketch of the microscopic picture is shown in the *lower part*, The points of measurement are indicated by the *small open circles*. The *upper half* shows the voltage deflections. Note that negative voltage deflections are obtained over each tight junction (*J*). The voltage deflections are small over the cell soma. These data prove that the current flows mostly through the lateral spaces and tight junctions. (Modified from [32])

Table 59.1. The transepithelial resistance (R_{te}), transepithelial voltage (V_{te}), and complexity of tight junctions in various epithelia. (Modified from [97])

Epithelium	R_{te} (Ω cm^2)	ΔV_{te} (mV)	Tight junction
Gallbladder	30	0	Leaky
Ileum	100	4	Leaky
Colon	500	20	Tight to intermediate
Proximal tubule	5	1–2	Leaky
Thick ascending limb of loop of Henle	34	6	Intermediate
Collecting duct	800	25	Tight
Urinary bladder	1500	60	Tight
Frog skin	3600	90	Tight

Tight junctions are necessary to preserve full epithelial polarity [12–14]. When tight junctions are interrupted by cell separation, e.g., in Ca^{2+}-free media, polarity is disturbed for some, but not all, proteins. The role of the tight junctions in the generation of polarity is less clear. The occurrence of tight junctions seems to parallel the polarization of cultured renal epithelial (MDCK) cells, but they are not responsible for polarization [14]. The issue of polarization has already been discussed in Chap. 4.

59.4 Transcellular Transport Usually Requires ATP-driven Pumps

As has already been stated in Sect. 59.1, transepithelial transport requires a driving force to proceed in one direction. Furthermore, transepithelial gradients have to be built up by transport in many (mostly tight) epithelia. These processes require some fuelling. This usually occurs by way of one of the ion pumps or ATPases. These pumps have been discussed in Chap. 8. The (Na^++K^+)-ATPase exports Na^+ and imports K^+. Therefore, this pump provides the driving force for Na^+ uptake or any substrate, which is coupled to Na^+. To obtain vectorial transport this pump is usually present in the basolateral membrane. However, there are exceptions to this rule: namely the choroid plexus (cf. Chap. 27) and in the eye, pigment epithelium, where the pump is present in the luminal membrane. Other well-known pumps that can energize epithelial transport are the two types of proton pumps: the vacuolar H^+-ATPase and H^+/K^+-ATPase [29,99], and the plasma membrane Ca^{2+}-ATPase [11]. The different systems of energy coupling have been addressed in Chap. 8 and will be memtioned only briefly below.

59.4.1 NaCl Reabsorption to Ascertain Salt Intake and Avoid NaCl Losses

As an example of reabsorptive mechanisms, the absorption of NaCl, which occurs in many epithelia, will be briefly discussed here. Figure 59.6 summarizes two cell models for NaCl absorption. Figure 59.6A refers to the proximal tubule of the kidney, and Fig. 59.6B is prototypic for the surface cells in colon, principal cells of the collecting duct (CD), and frog skin. The important common feature is the presence of the (Na^++K^+)-ATPase in the basolateral membrane. The uptake of Na^+ occurs via Na^+/H^+ exchange (probably NHE3, cf. Chap. 8 [72,94]). in the proximal tubule and proximal intestine, and it occurs via Na^+-selective channels [10,70] in the CD (cf. Chaps. 74, 75). Epithelia such as the gallbladder, upper intestine, and proximal tubule of the kidney (PT) are characterized by a very permeable tight junction (cf. Table 59.4). Therefore, these epithelia build up very small transepithelial gradients for Na^+ and for Cl^-. In the case of the PT these gradients amount to only a few millivolts. Still, such small gradients can drive very substantial amounts of ions through the paracellular pathway, which has good conductance. An estimation of the contributions of transcellular versus paracellular transport fractions has revealed that almost the entire amount of Cl^- absorbed in the PT crosses the paracellular pathway, and that as much as two-thirds of the Na^+ absorbed takes the same route [33]. This explains why the average ratio for the transport of NaCl over the consumption of ATP of the kidney is as high as 5, whilst the stoichiometry of the pump itself is 3 (cf. Chap. 74). In the intestine the situation is probably different, because there NaCl is absorbed transcellularly via a double exchange mechanism: Na^+/H^+ exchange and Cl^-/HCO_3^- exchange [68] (Chap. 62). From a more general perspective, the mechanisms in the PT are therefore tuned to achieve large bulk transport at small cost.

Figure 59.6B illustrates the other extreme. Like the frog skin, the colon and the CD, especially in the presence of aldosterone (cf. above), build up very large ionic gradients and voltages. A frog must preserve its body composition against fresh water, which contains an extremely low ion concentration. Consider Fig. 59.6B and accept the approximations that the pump usually loses its ability to bind Na^+ at cytosolic concentrations of a few millimoles per liter. How high an Na^+ gradient can be built up by one of these cells? In the frog skin the voltage across the luminal membrane may be close to -20 mV, and the transepithelial voltage may be -75 mV. Then, a positive driving force for Na^+ uptake would persist until the luminal Na^+ concen-

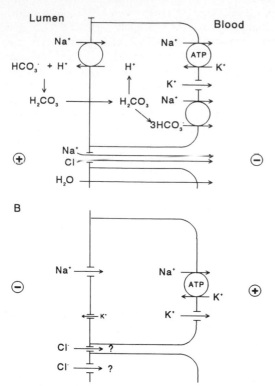

A

Lumen   Blood

B

Fig. 59.6A,B. NaCl absorption in **A** the proximal tubule and **B** the principal cell of the collecting duct. *Circle with ATP*, ATP-driven pump; *circle*, carrier mechanism; *arrow*, ion channel. **A** Note that the uptake of Na⁺ occurs via the Na⁺/H⁺ exchanger. The process is energized by the basolateral (Na⁺+K⁺)-ATPase. K⁺ recycles through basolateral K⁺ channels. The transcellular ion movement serves mostly the absorption of Na⁺ and HCO₃⁻. This leads to electrochemical gradients favoring the absorption of Na⁺ and Cl⁻ via the paracellular shunt. (See Chap. 74 for further details.) Note that this mechanism enables the proximal tubule to absorb more Na⁺ than is moved through the cell. In fact , the ratio is 2 equivalents moving through the shunt to 1 equivalent moving through the cell [33]. This mechanism is thus very economical but cannot build up any substantial gradient. **B** In tight epithelia Na⁺ absorption occurs via amiloride-sensitive Na⁺ channels. Na⁺ is pumped out by the basolateral (Na⁺+K⁺)-ATPase. K⁺ recycles through basolateral K⁺ channels. In addition, some K⁺ is secreted into the lumen. The active transcellular Na⁺ transport generates Na⁺ gradients and a lumen-negative voltage. This voltage drives Cl⁻ through a shunt pathway, which is either the paracellular route or another cell type. This mode of Na⁺ absorption is prototypic for all epithelia that have to build up and sustain large Na⁺ gradients. Compared with the scheme in **A**, this mode is obviously less economical, since all Na⁺ is transported across the cell

tration falls below 1 mmol/l. Owing to the serial arrangement of very many cells in the skin, the gradient can in reality be even much bigger [97]. This consideration tells us that this system is constructed to enable the body to avoid Na⁺ losses. It goes without saying that a permeable paracellular pathway (tight junction) would be disastrous for this system, because it would lead to back-leak of Na⁺ and ATP would be consumed in a futile cycle.

There are several other basic mechanisms for NaCl reabsorption. They are covered in other chapters and are therefore simply listed below:

- Na⁺/H⁺-exchange, Na⁺HCO₃⁻ absorption like in PT [7,72]
- Na⁺-channels in parallel with K⁺ channels and Cl⁻ conductances like in colon, sweat duct, airway epithelium, CD [8,10,62,70,74,77]
- Na⁺Cl⁻ cotransport, e.g., in distal tubule [92]
- Na⁺2Cl⁻K⁺ cotransport, e.g., in TAL [39]
- Na⁺/H⁺-exchange and Cl⁻/HCO₃⁻-exchange, e.g., in gallbladder and intestine [68,78].

The options of these various systems for generating large gradients (static heads) or carrying out what one might call bulk or mass transport (level flow) are different for each individual system. The two extremes have been discussed above.

59.4.2 NaCl Secretion, a Key Component of Exocrine Secretion

There are several exocrine glands or glandular tissues in which large quantities of secretions are produced every day. In the entire gastrointestinal tract the volume is remarkable, with approximately 8 l/day. These secretions come from salivary glands, stomach, liver, pancreas, small Brunner and Lieberkühn glands, and the crypts in jejunum, ileum, and colon. The secretions, except for those of the stomach, are rich in Na⁺ and usually also in HCO₃⁻. Similarly, some portions of the respiratory duct, the subepithelial glands, possess the ability to secrete NaCl, and the primary sweat produced in the sweat gland coils also consists of NaCl (cf. Chaps. 61, 62, 64, 65, 67, 112).

One rather general mechanism of production of such secretions is depicted in Fig. 59.7. This scheme was originally elaborated for a non-mammalian gland, namely the rectal gland of the small shark *Squalus acanthias* [48,51,53], and it was found shortly thereafter that this concept may also be applicable to many exocrine glands [40,76]. The basic feature of the model is again the basolateral export of Na⁺ by the (Na⁺+K⁺)-ATPase. This may appear surprising, because this tissue is meant to secrete, and not to reabsorb, Na⁺. Furthermore, for effective secretion one might have thought that the (Na⁺+K⁺)-ATPase should be localized in the luminal (apical) but not in the basolateral membrane. Closer inspection reveals that the model achieves secretion, and that with an unexpectedly high degree of economy. Na⁺ is taken up across the basolateral membrane by the Na⁺2Cl⁻K⁺ cotransporter [51]. Na⁺ is continuously recycled across the basolateral membrane. K⁺ taken up by the carrier and by the pump is also recycled by way of a basolateral K⁺ conductance [34,42]. Cl⁻ leaves the cell via a Cl⁻ conductance in the luminal membrane [35,50,52]. The conductive K⁺ exit across the basolateral and that of Cl⁻ across the luminal membrane generate a circular current loop, which polarizes the lumen negatively and drives Na⁺

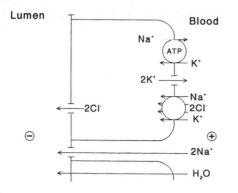

Fig. 59.7. NaCl secretion in exocrine glands. *Circle with ATP,* ATP-driven pump; *circle,* carrier mechanism; *arrow,* ion channel. The process is energized by the basolateral (Na^++K^+)-ATPase. Cl^- is taken up across the basolateral membrane by a cotransporter that couples the flux of 2 Cl^-, 1 Na^+, and 1 K^+. K^+ recycles through basolateral K^+ channels. Na^+ is pumped out by the basolateral (Na^++K^+)-ATPase. Cl^- leaves the cell through luminal Cl^- channels. The exit of K^+ across the basolateral membrane and that of Cl^- across the luminal membrane produces a transepithelial lumen-negative voltage, which drives paracellular Na^+ secretion. Note that this process is highly economical, since more Na^+ is secreted than is pumped by the (Na^++K^+)-ATPase [51]

passively across the paracellular shunt. Therefore, the economy of this system is 6 NaCl/ATP, which is twice as high as would have been achieved by transcellular Na^+ secretion [51]. Again, a close functional coupling of the properties of the paracellular shunt pathway and the transcellular transport is necessary to obtain such a surprisingly simple and economical system. It is also worth noting that the essential transport components are very similar to those found in reabsorptive tissues, such as the TAL [40]. Obviously the repertoire of individual transporters (with all their variants) is finite, and the various components are put together in different ways to achieve the appropriate solution. It is not surprising that this archaic secretory mechanism has been well preserved in evolution [51]. There are, however, also other mechanisms of NaCl secretion in several glands. These will be discussed in the appropriate chapters (cf. Chaps. 64, 65).

59.4.3 HCO_3^-/H^+ Transport for Acid Secretion and for Control of Acid/Base Status

Some extrusion of acid and HCO_3^- absorption can be achieved by the epithelial type of the Na^+/H^+-exchanger. This has been discussed above for the proximal renal tubule. The maximal pH gradient produced by this exchanger is in the order of a little more than 1 pH unit. At steeper pH gradients the system runs backward and secretes Na^+. In several epithelia much larger pH gradients are required. One such example is the kidney [27], where about 20–30% of PT HCO_3^- and most of the distal tubular acidification are driven by a vacuolar type H^+ pump, which, when required, is exocytosed into the luminal membrane.

Another example is shown in Fig. 59.8 for the stomach. Here, a H^+/K^+-ATPase serves to secrete H^+ and generates a pH gradient of 6 units or slightly higher [20]. For this pump to operate, several additional transporters have to be present. As discussed in Chap. 61, it is likely that the luminal membrane, in addition to the proton pump, possesses K^+, Cl^- channels and a KCl cotransporter. In the basolateral membrane a band-3-type protein serves for Cl^- uptake. The other three components probably fulfil housekeeping functions, inasmuch as the pH of these cells must be closely watched due to the enormous proton extrusion capacity. Similar types of H^+/K^+-ATPases are present in the colon and in the collecting duct [57,103].

59.4.4 Fuel Reabsorption

Intestinal epithelia and renal epithelia absorb metabolic fuels such as D-glucose, other sugars, and amino acids [18,91,108]. The detailed mechanisms of these reabsorptive processes are discussed in the respective chapters (Chaps. 8, 62, 75). It is important to note that this specialized function is quantitatively very important. In the renal tubules, some 150 g D-glucose is absorbed per day, and in the intestine the amount absorbed is even larger. The specific task, namely to transport the substrates across the luminal membrane in an Na^+-dependent fashion has been discussed in Chap. 8 (Fig. 8.7). The coupling to Na^+ makes it possible to generate large transepithelial gradients. The coupling ratio (Na^+/substrate) is variable and exceeds unity in some cases: D-glucose absorption in intestine and in the S2, S3 segments of the proximal tubule [95,108]. It is

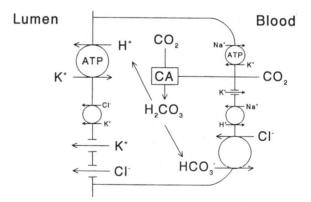

Fig. 59.8. Proton secretion in the oxyntic cell of the stomach. *Circle with ATP,* ATP-driven pump; *circle,* carrier mechanism; *arrow,* ion channel. CO_2 produced and taken up by diffusion is hydrated to carbonic acid by carbonic anhydrase (*CA*). Carbonic acid dissociates into bicarbonate and protons. The protons are secreted by a H^+/K^+-ATPase. K^+ recycles across the luminal membrane via a KCl cotransporter and K^+ channels. Bicarbonate leaves the cell at the basolateral pole via a Cl^-/HCO_3^- exchanger. Cl is secreted by the KCl cotransporter and by Cl^- channels. The pH homeostasis of these cells also requires the presence of a Na^+/H^+ exchanger in the basolateral membrane (housekeeper). Na^+ is removed from the cell via the basolateral (Na^++K^+)-ATPase. The K^+ taken up by the pump recycles through K^+ channels in the basolateral membrane

unclear at this stage how the transport pathways are separated from the metabolic pathways.

To make this point, consider the early proximal tubule of the kidney. This segment is flooded by D-glucose across the luminal membrane. Still, D-glucose is not utilized as a metabolic substrate [104]. D-Glucose bypasses the metabolism and is absorbed across the epithelium, whilst at the same time metabolism is fuelled by short fatty acids taken up across the basolateral membrane. In fact, if the fuel availability exceeds the acute needs, this extra fuel is utilized in gluconeogenesis. In other words, the proximal tubule cells behave like honest bank clerks, who trade enormous amounts of money but do not reroute (steal) any of it to increase their salary. Fuel uptake for the metabolic cost of basal turnover and transport occurs exclusively from the basolateral side [107]. The metabolism of epithelia such as renal tubules is very high and the majority of O_2 consumption is concerned with the transepithelial transport of Na^+, Cl^-, HCO_3^-, and water (cf. Chap. 74).

59.4.5 Water Transport by Osmosis

Like any other cell in the body, epithelia must preserve their volume by water exchange with the extracellular fluid. This exchange occurs via the water permeability of the basolateral membrane. The luminal membrane water permeability of epithelia is variable and determines the degree to which water is absorbed or secreted by epithelia. This is easily evident for the collecting duct, where the luminal water permeability is controlled by antidiuretic hormone [ADH, used synonymously with arginine vasopressin (AVP)] by way of water channels [1,55,60]. If ADH (AVP) is present, water is reabsorbed following the osmotic gradient directed from the interstitium to the lumen [37]. The water channels inserted into the luminal membrane have been identified recently as a specific channel protein [1,25,73]. Their insertion into the luminal membrane is probably controlled by cAMP-dependent exocytosis [55].

In leaky epithelia it has long been the subject of controversy whether water moves across or between cells. A specifically interesting case is the proximal tubule of the kidney, where some 100 l is absorbed every day. Originally it was proposed that most of the water was transported via the tight junctions and the paracellular shunt pathway [33]. Later it was shown that this is unlikely, since the cross-sectional area of the tight junctions is so much smaller than that of the luminal membrane of the proximal tubule. Now it is assumed that a large fraction of the water reabsorption occurs via the water permeation pathways of the luminal and the basolateral membrane [82,83,88].

The osmotic gradients driving water absorption in leaky epithelia are difficult to determine. In fact, analysis of the absorbed fluid and that on the blood side has always revealed that the differences are very small [16,23]; in most, though not all studies, they have not been detectable. The analysis has still not been convincingly resolved. The reason for this dilemma is that nature has set water permeability so high that trivial osmotic gradients of even a few milliosmoles per liter are sufficient to drive the entire water transport. An osmotic gradient may even be absent, and water transport can still continue if the "effective" osmolality of the *trans*-side is higher than that on the *cis*-side. In the proximal tubule this situation may be realized, since the absorbate contains more HCO_3^- etc. than the luminal fluid [84]. From the experimenter's perspective this is a very unfortunate constellation, because a quantitative description of water permeation still appears impossible. From a teleological perspective, however, this appears entirely plausible, because it makes water transport energetically inexpensive.

59.5 Organization and Maintenance of Epithelial Polarity

Epithelial polarity exists because the membrane transport proteins are sorted in such a way that some are present only in the luminal membrane and others only in the basolateral membrane. One example is the constellation shown in Fig. 59.6. Na^+ uptake mechanisms, such as the Na^+ channel, and various Na^+ contransport and antiport systems are localized in the luminal membrane, and the Na^+ extrusion mechanism, the (Na^++K^+)-ATPase, in the basolateral membrane. The mechanisms of sorting have been discussed in Chap. 4. It has been shown there that sorting can be dependent on the intactness of the tight junction, and that it can be determined by a signal sequence addressing the peptide to the luminal membrane. In a very general way, the luminal membrane can be regarded as the specific achievement of epithelia, whilst the basolateral membrane corresponds to the plasma membrane of apolar cells. Epithelial polarization is lost if epithelial cells are cultured, and it returns with the beginning of cell–cell interaction. The process of polarization is generally aided if cells grow on a permeable (filter membrane), as opposed to a glass, support. Some properties, such as the epithelial Na^+ channel, localized in the luminal membrane are only apparent if the epithelium forms tight junctions (author's unpublished observations). Observations of this kind prompt some caution with respect to the interpretation of data obtained in cultured epithelial cells. Specific properties of epithelial cells are rather labile and are easily lost in cultures. This also implies that the understanding of epithelial function will always require the examination of intact epithelia.

59.6 Conclusion: Flush Through and Homeostasis

One specific problem present in epithelia is the perturbation caused by transepithelial transport. Consider a TAL cell with a height of only $1-2\,\mu m$. This cell absorbs at a rate of $300\,\mu A/cm^2$. Translated into a turnover rate, this is equivalent to a complete exchange within 1–2 s [45]. Such high turnover requires "homocellular regulation" [87,90], i.e., an adjustment of the export of Na$^+$ by the (Na$^+$+K$^+$)-ATPase to the import mechanisms. These import mechanisms in epithelia are:

- Na$^+$/H$^+$ exchange, occurring (e.g.) in the luminal membrane of the proximal tubule and small intestine
- Na$^+$ coupled uptake of Cl$^-$, as it occurs in the thick ascending limb of the loop of Henle
- Uptake of Na$^+$ via Na$^+$-channels
- Na$^+$-coupled uptake of substrates such as D-glucose.

For instance, the uptake can increase very rapidly if a large Na$^+$ load floods the thick ascending limb cell, as it does in volume expansion, or if more Na$^+$ is delivered to the collecting duct principal cell, for example, as with volume expansion and pharmacological inhibition of Na$^+$ reabsorption in the thick ascending limb of the loop of Henle. Most importantly, it can also occur if a large load of D-glucose floods the small intestine. It is not surprising that in all these cases an increase in the cytosolic Na$^+$ activity/concentration has been measured with ion-selective microelectrodes or other types of intracellular measurements. However, the increases were unexpectedly small compared with the change in transport rate [90]. Furthermore, although the transport continued as needed at increased rates as long as the Na$^+$ and/or substrate load was increased, the cytosolic Na$^+$ activity (concentration) fell to values very close to those obtained under control conditions. This precludes the simple explanation that the rate of the pump is mainly geared by the cytosolic Na$^+$ activity. The pump is readjusted after only a short delay and works at higher rates at only slightly increased or normal cytosolic Na$^+$ activity. It is unclear how this communication is achieved, but it is likely that it involves regulatory loops with second messengers.

One might go on and ask why this sort of adjustment is present. Part of the answer has been given above. The absolute rates of transport are so large compared with cell size that the absence of such fine regulatory mechanisms would immediately flood the cell with Na$^+$. This, in addition, would cause Cl$^-$ import in the case of the Na$^+$2Cl$^-$K$^+$ cotransporter and the other two electrogenic import pathways, because in all instances the cell is depolarized with increased Na$^+$ uptake [31,39,63]. The increase in cellular content of NaCl then induces cell swelling. If there were not the activation of Na$^+$ extrusion mechanisms, the cell would be endangered by this increase in volume and the possible concomitant physical damage. It has been argued convincingly that the presence of very sensitive volume-regulatory mechanisms in epithelial cells is designed to prevent and balance volume changes occurring as a consequence of incomplete "homocellular" Na$^+$ regulation [87,88].

Another aspect of homocellular regulation is the adjustment of the K$^+$ conductance, which usually resides in the basolateral membrane, to the same rates as the (Na$^+$+K$^+$)-ATPase. Such an adjustment appears desirable, because the K$^+$ uptake by the pump requires a corresponding recy-

cling. At some point, it has even been speculated that pump and channel maybe directly coupled units [87]. This is not the case: pump and channels are entirely different membrane proteins, and it has been explained in Chap. 8 that the density of the pump is higher by orders of magnitude than the density of the K$^+$ channels. Nonetheless, close coupling has been demonstrated [67]. Several transduction pathways contribute to this readjustment of the K$^+$ conductance. It is worth emphasizing here that the increase in K$^+$ conductance usually observed with increased Na$^+$ transport helps to

- Repolarize the cell
- Extrude Cl$^-$
- Reduce cell volume
- Recycle extra K$^+$ taken up by the (Na$^+$+K$^+$)-ATPase.

Besides the adjustment of pump and K$^+$ conductance, negative feedback loops prevent excessive Na$^+$ overload. One such mechanism is the down-regulation of the luminal Na$^+$ conductance by cytosolic Ca^{2+} and protons [75,102]. The mechanism of this regulation of the Na$^+$ is probably complex, inasmuch as inhibition by Ca^{2+}, for example, cannot be demonstrated in excised Na$^+$ channels.

Finally, another specific aspect of epithelia should also be addressed. Epithelia have to absorb or secrete and become aware of more of less disturbed cellular contents, yet they must still continue to ignore such alterations. One example has been discussed above for early proximal renal tubule cells. They reabsorb large quantities of D-glucose, but do not use it for metabolism. They export this valuable substrate on the basolateral pole and make their "living" on short fatty acids taken up from the blood. Another very interesting case is that of the Ca^{2+}-absorbing epithelia, such as intestine and thick ascending limb of the loop of Henle. In these epithelia the transported Ca^{2+} must be "shielded" in the cytosol to avoid a serious misunderstanding, which would occur if increased cytosolic Ca^{2+} activity were understood as a tertiary messenger signal. This problem becomes even more serious if transport rates are increased dramatically by hormones such as PTH [105]. One solution to this problem is the binding of Ca^{2+} to specific transport proteins within the cell; another solution is the paracellular transport of this divalent cation [106]. Nonetheless, some changes in cytosolic Ca^{2+} activity have been claimed for distal tubule cells in the presence of hormones [30].

To conclude this chapter, it will presumably have become apparent that epithelial transport faces several specific problems: one is the polarity and its maintenance; another concerns the design of tight junctions, which have to be tight in some epithelia but must be "speedways" in others; a third problem concerns the complex regulatory network, which is necessary to cope with the large transport rates observed in several epithelia. The processes of exocytosis, endocytosis, and transcytosis have not been addressed in this chapter; they are discussed in Chaps. 4 and 8.

References

1. Agre P, Preston GM, Smith BL, Jung JS, Raina S, Moon C, Guggino WB, Nielsen S (1993) Aquaporin CHIP: the archetypal molecular water channel. Am J Physiol 265:F463–F476
2. Alberts B, Bray D, Lewis J, Raff M, Roberts K, Watson JD (1983) Molecular biology of the cell. Garland, New York, p 687
3. Anderson MP, Gregory RJ, Thompson S, Souza DW, Paul S, Mulligan RC, Smith AE, Welsh MJ (1991) Demonstration that CFTR is a chloride channel by alteration of its anion selectivity. Science 253:202–205
4. Baumgarten CM (1981) An improved liquid ion exchanger for chloride ionselective microelectrodes. Am J Physiol 241:C258–C263
5. Bear CE, Li C, Kartner N, Bridges RJ, Jensen TJ, Ramjeesingh M, Riordan JR (1992) Purification and functional reconstitution of the cystic fibrosis transmembrane conductance regulator (CFTR). Cell 68:809–818
6. Bernard C (1878) Leçons sur les phénomènes de la vie, communs aux animaux et aux végétaux, vol 1. Ballière, Paris
7. Boron WF, Boulpaep EL (1989) The electrogenic Na/HCO3 cotransporter. Kidney Int 36:392–402
8. Boucher RC, Stutts MJ, Knowles MR, Cantley L, Gatzy JT (1986) Na+ transport in cystic fibrosis respiratory epithelia. J Clin Invest 78:1245–1252
9. Boyd CAR, Noble D (1993) The logic of life. Oxford University Press, Oxford
10. Canessa CM, Horisberger JD, Rossier BC (1993) Epithelial sodium channel related to proteins involved in neurodegeneration. Nature 361:467–470
11. Carafoli E (1991) The calcium pumping ATPase of the plasma membrane. Annu Rev Physiol 53:531–547
12. Cereijido M, Contreras RG, Gonzalez-Mariscal L (1989) Development and alteration of polarity. Annu Rev Physiol 51:785–795
13. Cereijido M, Gonzalez-Mariscal L, Contreras G (1989) Tight junctions: barrier between higher organisms and environment. NIPS 4:72–75
14. Cereijido M, Contreras RG, Gonzales-Mariscal L (1989) Development and alteration of polarity. Annu Rev Physiol 51:785–795
15. Clarke LL, Grubb BR, Gabriel SE, Smithies O, Koller BH, Boucher RC (1992) Defective epithelial chloride transport in a gene-targeted mouse model of cystic fibrosis. Science 257:1125–1128
16. Curci S, Frömter E (1979) Micropuncture of lateral intercellular spaces of Necturus gallbladder to determine space fluid K+ concentration. Nature 278:355–357
17. Da Silva PP, Kachar B (1982) On tight junction structure. Cell 28:441–450
18. Deetjen P, v. Baeyer H, Drexel H (1992) Renal glucose transport. In: Seldin DW, Giebisch G (eds) The kidney: physiology and pathophysiology. Raven, New York, pp 2873–2888
19. Deisenhofer J, Michel H, Huber R (1985) The structural basis of photosynthetic light reactions in bacteria. Trends Biochem Sci 10:243–248
20. Demarest JR, Loo DD (1990) Electrophysiology of the parietal cell. Annu Rev Physiol 52:307–319
21. Di Stefano A, Wittner M, Gebler B, Greger R (1988) Increased Ca^{2+} or Mg^{2+} concentration reduces relative tight-junction permeability to Na+ in the cortical thick ascending limb of Henle's loop of rabbit kidney. Renal Physiol Biochem 11:70–79
22. Di Stefano A, Wittner M, Schlatter E, Lang HJ, Englert H, Greger R (1985) Diphenylamine-2-carboxylate, a blocker of the Cl−-conductive pathway in Cl−-transporting epithelia. Pflugers Arch 405 [Suppl 1]:S95–S100
23. Diamond JM (1979) Osmotic water flow in leaky epithelia. J Membr Biol 51:195–216
24. Diamond JM, Bossert WH (1967) Standing-gradient osmotic flow. A mechanism for coupling of water and solute transport in epithelia. J Gen Physiol 50:2061–2083
25. Echevarria M, Frindt G, Preston GM, Milovanovic S, Agre P, Fischbarg J, Windhager EE (1993) Expression of multiple water channel activities in xenopus oocytes injected with mRNA from rat kidney. J Gen Physiol 101:827–841
26. Edelman A, Curci S, Samarzija I, Frömter E (1978) Determination of intracellular K+ activity in rat kidney proximal tubular cells. Pflugers Arch 378:37–45
27. Emmett M, Alpern RJ, Seldin DW (1992) Metabolic acidosis. In: Seldin DW, Giebisch G (eds) The kidney: physiology and pathophysiology. Raven, New York, pp 2759–2836
28. Farquar MG, Palade GE (1963) Junctional complexes in various epithelia. J Cell Biol 17:375–412
29. Forgac M (1989) Structure and function of vacuolar class of ATP-driven proton pumps. Physiol Rev 69:765–796
30. Friedman PA, Gesek FA (1993) Calcium transport in renal epithelial cells. Am J Physiol 264:F181–F198
31. Frömter E (1981) Electrical aspects of tubular transport of organic substances. In: Greger R, Lang F, Silbernagl S (eds) Renal transport of organic substances. Springer, Berlin Heidelberg New York, pp 30–44
32. Frömter E, Diamond J (1972) Route of passive ion permeation in epithelia. Nature 235:9–13
33. Frömter E, Rumrich G, Ullrich KJ (1973) Phenomenologic description of Na+, Cl− and HCO3− absorption from proximal tubules of the rat kidney. Pflugers Arch 343:189–220
34. Gögelein H, Greger R, Schlatter E (1987) Potassium channels in the basolateral membrane of the rectal gland of Squalus acanthias. Regulation and inhibitors. Pflugers Arch 409:107–113
35. Gögelein H, Schlatter E, Greger R (1987) The "small" conductance chloride channel in the luminal membrane of the rectal gland of the dogfish (Squalus acanthias). Pflugers Arch 409:122–125
36. Gonzalez-Mariscal L, Contreras RG, Bolivar JJ, Ponce A, De Ramirez BC, Cereijido M (1990) Role of calcium in tight junction formation between epithelial cells. Am J Physiol 259:C978–C986
37. Grantham JJ, Burg MB (1966) Effect of vasopressin and cyclic AMP on permeability of isolated collecting tubules. Am J Physiol 211:255–259
38. Greger R (1981) Cation selectivity of the isolated perfused cortical thick ascending limb of Henle's loop of rabbit kidney. Pflugers Arch 390:30–37
39. Greger R (1985) Ion transport mechanisms in thick ascending limb of Henle's loop of mammalian nephron. Physiol Rev 65:760–797
40. Greger R (1986) Chlorid-transportierende Epithelien. In: Physiologie aktuell, vol 2. Fischer, Stuttgart, pp 47–58
41. Greger R (1990) An electrophysiological approach to the study of isolated perfused tubules. Methods Enzymol 191:289–302
42. Greger R, Gögelein H, Schlatter E (1987) Potassium channels in the basolateral membrane of the rectal gland of the dogfish (Squalus acanthias). Pflugers Arch 409:100–106
43. Greger R, Lang F, Knox FG, Lechene C (1978) Analysis of tubule fluid. In: Martinez-Maldonado M (ed) Methods in pharmacology. Plenum, New York, pp 105–140
44. Greger R, Oberleithner H, Schlatter E, Cassola AC, Weidtke C (1983) Chloride activity in cells of isolated perfused cortical thick ascending limbs of rabbit kidney. Pflugers Arch 399:29–34
45. Greger R, Schlatter E (1983) Cellular mechanism of the action of loop diuretics on the thick ascending limb of Henle's loop. Klin Wochenschr 61:1019–1027
46. Greger R, Schlatter E (1983) Properties of the basolateral membrane of the cortical thick ascending limb of Henle's loop of rabbit kidney. A model for secondary active chloride transport. Pflugers Arch 396:325–334

47. Greger R, Schlatter E (1983) Properties of the lumen membrane of the cortical thick ascending limb of Henle's loop of rabbit kidney. Pflugers Arch 396:315–324

48. Greger R, Schlatter E (1984) Mechanism of NaCl secretion in the rectal gland of spiny dogfish (Squalus acanthias). I. Experiments in isolated in vitro perfused rectal gland tubules. Pflugers Arch 402:63 75

49. Greger R, Schlatter E (1986) Electrolyte activities in Cl⁻-transporting epithelia: Cortical thick ascending limb of rabbit nephron and rectal gland tubules of the spiny dogfish, Squalus acanthias. In: Kessler M, Harrison DK, Höper J (eds) Ion measurements in physiology and medicine. Springer, Berlin Heidelberg New York, pp 301–308

50. Greger R, Schlatter E, Gögelein H (1985) Cl⁻-channels in the apical cell membrane of the rectal gland "induced" by cAMP. Pflugers Arch 403:446–448

51. Greger R, Schlatter E, Gögelein H (1986) Sodium chloride secretion in rectal gland of dogfish Squalus acanthias. NIPS 1:134–136

52. Greger R, Schlatter E, Gögelein H (1987) Chloride channels in the luminal membrane of the rectal gland of the dogfish (Squalus acanthias). Properties of the "larger" conductance channel. Pflugers Arch 409:114–121

53. Greger R, Schlatter E, Wang F, Forrest JN Jr (1984) Mechanism of NaCl secretion in rectal gland tubules of spiny dogfish (Squalus acanthias). III. Effects of stimulation of secretion by cyclic AMP. Pflugers Arch 402:376–384

54. Greger R, Weidtke C, Schlatter E, Wittner M, Gebler B (1984) Potassium activity in cells of isolated perfused cortical thick ascending limbs of rabbit kidney. Pflugers Arch 401:52–57

55. Hays RM (1991) Cell biology of vasopressin. In: Brenner BM, Rector FC (eds) The kidney. Saunders, Philadelphia, pp 424–444

56. Hediger MA, Coady MJ, Ikeda TS, Wright EM (1987) Expression cloning and cDNA sequencing of the Na⁺/glucose cotransporter. Nature 333:379–381

57. Jaisser F, Coutry N, Farman N, Binder HJ, Rossier BC (1993) A putative H⁺-K⁺-ATPase is selectively expressed in surface epithelial cells of rat distal colon. Am J Physiol 265:C1080–C1090

58. Kartner N, Hanrahan JW, Jensen TJ, Naismith AL, Sun S, Ackerley CA, Reyes EF, Tsui LC, Rommens JM, Bear CE, Riordan JR (1991) Expression of the cystic fibrosis gene in non-epithelial invertebrate cells produces a regulated anion conductance. Cell 64:681–691

59. Kinne-Saffran EM, Kinne R (1990) Isolation of lumenal and contralumenal plasma membrane vesicles from kidney. Methods Enzymol 191:450–479

60. Kirk KL, Schafer JA (1992) Water transport and osmoregulation by antidiuretic hormone in terminal nephron segments. In: Seldin DW, Giebisch G (eds) The kidney: physiology and pathophysiology. Raven, New York, pp 1693–1725

61. Koefoed-Johnsen V, Ussing HH (1958) The nature of the frog skin potential. A P S 42:298–308

62. Koeppen BM, Giebisch G (1985) Cellular electrophysiology of potassium transport in the mammalian cortical collecting tubule. Pflugers Arch 405 [Suppl 1]:S143–S146

63. Koeppen BM, Stanton BA (1992) Sodium chloride transport: distal nephron. In: Seldin DW, Giebisch G (eds) The kidney: physiology and pathophysiology. Raven, New York, pp 2003–2040

64. Koepsell H, Seibicke S (1991) Reconstitution and fractionation of renal brush border transport proteins. In: Fleischer S, Fleischer B (eds) Methods in enzymology. Academic, San Diego, pp 583–605

65. Krasny EJ, DiBona AI, Frizzell RA (1982) Regulation of paracellular permselectivity in flounder intestine. Bull Mt Desert Isl Biol Lab 22:82–84

66. Lang F, Greger R, Lechene C, Knox FG (1978) Micropuncture techniques. In: Martinez-Maldonado M (ed) Methods in pharmacology. Plenum Press, New York, pp 75–103

67. Lang F, Rehwald W (1992) Potassium channels in renal epithelial transport regulation. Physiol Rev 72:1–32

68. Liedtke CM, Hopfer U (1982) Mechanism of Cl⁻ translocation across small intestinal brush border membrane. I. Absence of Na⁺-Cl⁻ cotransport. Am J Physiol 242:G263–G271

69. Ling GN, Gerard RW (1949) The normal membrane potential of frog sartorius fibers. J Cell Comp Physiol 34:383–396

70. Lingueglia E, Voilley N, Waldmann R, Lazdunski M, Barbry P (1993) Expression cloning of an epithelial amiloride-sensitive Na⁺ channel. Fed Eur Biochem Soc 318:95–99

71. Macknight ADC, DiBona DR, Leaf A (1980) Sodium transport across toad urinary bladder: a model "tight" epithelium. Physiol Rev 60:615–715

72. Murer H, Hopfer U, Kinne R (1976) Sodium/proton antiport in brush border membrane vesicles isolated from rat small intestine and kidney. Biochem J 154:597–604

73. Nielsen S, Smith B, Christensen EI, Agre P (1993) Distribution of the aquaporin CHIP in secretory and resorptive epithelia and capillary endothelia. Proc Natl Acad Sci USA 90:7275–7279

74. Palmer LG (1990) Epithelial Na channels: the nature of the conducting pore. Renal Physiol Biochem 13:51–58

75. Palmer LG, Frindt G (1987) Effects of cell Ca and pH on Na channels from rat cortical collecting tubule. Am J Physiol 253:F333–F339

76. Petersen OH, Findlay I (1987) Electrophysiology of the pancreas. Physiol Rev 67:1054–1107

77. Quinton PM (1990) Cystic fibrosis: a disease in electrolyte transport. FASEB J 4:2709–2717

78. Reuss L (1989) Ion transport across gallbladder epithelium. Physiol Rev 69:503–545

79. Riordan JR, Rommens JM, Kerem B-S, Alon N, Rozmahel R, Grzelczak Z, Zielenski J, Lok S, Plavsic N, Chou J-L, Drumm ML, Iannuzzi MC, Collins FS, Tsui L-C (1989) Identification of the cystic fibrosis gene: cloning and characterization of complementary DNA. Science 245:1066–1073

80. Rossier BC, Palmer LG (1992) Mechanisms of aldosterone action on sodium and potassium transport. In: Seldin DW, Giebisch G (eds) The kidney, physiology and pathophysiology. Raven Press, New York, pp 1373–1409

81. Sakmann B, Neher E (1984) Patch clamp techniques for studying ionic channels in excitable membranes. Annu Rev Physiol 46:455–472

82. Schafer JA (1990) Transepithelial osmolality differences, hydraulic conductivities, and volume absorption in the proximal tubule. Annu Rev Physiol 52:709–726

83. Schafer JA, Andreoli TE (1986) Principles of water and nonelectrolyte transport across membranes. In: Andreoli TE, Hoffman JF, Fanestil DD, Schultz SG (eds) Membrane physiology. Plenum, New York, pp 177–190

84. Schafer JA, Troutman SL, Andreoli TE (1974) Volume reabsorption, transepithelial potential differences, and ionic permeability properties in mammalian superficial proximal straight tubules. J Gen Physiol 64:582–607

85. Schlatter E, Greger R (1985) cAMP increases the basolateral Cl⁻-conductance in the isolated perfused medullary thick ascending limb of Henle's loop of the mouse. Pflugers Arch 405:367–376

86. Schlatter E, Greger R, Weidtke C (1983) Effect of "high ceiling" diuretics on active salt transport in the cortical thick ascending limb of Henle's loop of rabbit kidney. Correlation of chemical structure and inhibitory potency. Pflugers Arch 396:210–217

87. Schultz SG (1981) Homocellular regulatory mechanisms in sodium-transporting epithelia: avoidance of extinction by "flush-through". Am J Physiol 241:F579–F590

88. Schultz SG (1989) Volume preservation: then and now. NIPS 4:169–172

89. Schultz SG, Frizzell RA, Nellans HN (1974) Ion transport by mammalian small intestine. Annu Rev Physiol 36:51–91

90. Schultz SG, Hudson RL (1986) How do sodium-absorbing cells do their job and survive. NIPS 1:185–188

91. Silbernagl S (1992) Amino acids and oligopeptides. In: Seldin DW, Giebisch G (eds) The kidney: physiology and pathophysiology. Raven, New York, pp 2889–2920

92. Stokes JB (1989) Electroneutral NaCl transport in the distal tubule. Kidney Int 36:427–433

93. Tabcharani JA, Chang XB, Riordan JR, Hanrahan JW (1991) Phosphorylation-regulated Cl^- channel in CHO cells stably expressing the cystic fibrosis gene. Nature 352:628–631

94. Tse CM, Brant SR, Walker MS, Pouysségur J, Donowitz M (1992) Cloning and sequencing of a rabbit cDNA encoding an intestinal and kidney-specific Na^+/H^+ exchanger isoform (NHE-3). J Biol Chem 267:9340–9346

95. Turner RJ, Moran A (1982) Stoichiometric studies of the renal outer cortical brush border membrane D-glucose transporter. J Membr Biol 67:73–80

96. Ullrich KJ, Greger R (1985) Approaches to the study of tubule transport functions. In: Seldin DW, Giebisch G (eds) The kidney, physiology and sathophysiology. Raven, New York, pp 427–469

97. Ussing HH, Erlij D (1974) Transport pathways in biological membranes. Annu Rev Physiol 36:17–49

98. Ussing HH (1994) Does active transport exist? J Membr Biol 137:91–98

99. van Uem TJF, de Pont JJHHM (1992) Structure and function of gastric H, K-ATPase. In: de Pont JJHHM (ed) Molecular aspects of transport proteins. Elsevier, Amsterdam, pp 27–55

100. Wangemann P, Wittner M, Di Stefano A, Englert HC, Lang HJ, Schlatter E, Greger R (1986) Cl^--channel blockers in the thick ascending limb of the loop of Henle. Structure activity relationship. Pflugers Arch 407 [Suppl 2]:S128–S141

101. Weiss MS, Abele U, Weckesser J, Welte W, Schiltz E, Schulz GE (1991) Molecular architecture and electrostatic properties of a bacterial porin. Science 254:1627–1630

102. Windhager EE, Frindt G, Yang JM, Lee CO (1986) Intracellular calcium ions as regulators of renal tubular sodium transport. Klin Wochenschr 64:847–852

103. Wingo CS, Cain BD (1993) The renal H-K-ATPase: physiological significance and role in potassium homeostasis. Annu Rev Physiol 55:323–347

104. Wirthenson G, Guder WG (1990) Metabolism of isolated kidney tubule segments. Methods Enzymol 191:325–340

105. Wittner M, Di Stefano A (1990) Effects of antidiuretic hormone, parathyroid hormone and glucagon on transepithelial voltage and resistance of the cortical and medullary thick ascending limb of Henle's loop of the mouse nephron. Pflugers Arch 415:707–712

106. Wittner M, Mandon B, Roinel N, De Rouffignac C, Di Stefano A (1993) Hormonal stimulation of Ca^{2+} and Mg^{2+} transport in the cortical thick ascending limb of Henle's loop of the mouse: evidence for a change in the paracellular pathway permeability. Pflugers Arch 423:387–396

107. Wittner M, Weidtke C, Schlatter E, Di Stefano A, Greger R (1984) Substrate utilization in the isolated perfused cortical thick ascending limb of rabbit nephron. Pflugers Arch 402:52–62

108. Wright EM (1993) The intestinal Na^+/glucose cotransporter. Annu Rev Physiol 55:575–589

109. Wright EM, Barry PH, Diamond JM (1971) The mechanism of cation permeation in rabbit gallbladder: conductances, the current-voltage relation, the concentration dependence of anion cation discrimination, and the calcium competition effect. J Membr Biol 4:331–357

110. Wright EM, Diamond JM (1968) Effects of pH and polyvalent cations on the selective permeability of gallbladder epithelium to monovalent ions. Biochim Biophys Acta 163:57

60 Mastication and Swallowing

J.E. Kellow

Contents

60.1 Introduction

Mastication (chewing) requires the co-ordinated action of the muscles of the jaw, tongue, cheeks and palate. The end result is that a food bolus is periodically forced backwards by the tongue against the palate into the oropharynx, thereby initiating swallowing. The act of swallowing (deglutition) is a programmed motor response; it originates from the medullary swallowing center and encompasses the subsequent motor events in the oropharynx and esophagus, at the same time affording protection to the airway.

Three main functional regions of the esophagus are recognized – the *upper esophageal sphincter* (UES), the *esophageal body*, and the *lower esophageal sphincter* (LES). Upper and lower sphincteric function is intimately coordinated with pharyngeal and gastric motor activity, respectively, and motor activity in the striated and smooth muscle regions of the esophagus is closely integrated.

Several methods of investigation have led to a vast increase in our understanding of normal human *swallowing and esophageal motility*; in general, motility in these regions is more clearly defined than that in other parts of the gastrointestinal tract. The most well-established technique is esophageal manometry. This can now be performed on an ambulatory basis, combined with distal esophageal pH monitoring, allowing definition of physiological esophageal activity over 24 h [12,15]. Newer techniques related to swallowing and oropharyngeal motor function include specialized manometric techniques (e.g., perfused sleeve sensor, strain gauge catheter), videoradiography, ultrasonography, and scintigraphy. Videofluoroscopy provides qualitative and quantitative information on the tim-ing of swallow events and measures of laryngeal motion and UES opening [5]. The perfused sleeve sensor records UES pressure profiles faithfully [9], while solid state transducers record pharyngeal contractions more reliably [7]. Ultrasonography of the oral cavity during swallowing can assess oral and lingual function and the overall duration of the swallow [16]. Scintigraphy, using liquid boluses labeled with ^{99}Tc sulfur colloid, is particularly useful for objective measurement of percentage bolus clearance [8].

60.2 Functional Anatomy

60.2.1 Oral Cavity

The striated *muscles of mastication* – chiefly the masseter, temporalis and lateral and medial pterygoids – are innervated by the mandibular division of the trigeminal nerve. The extrinsic musculature of the tongue, i.e., the hypoglossus, genioglossus, styloglossus and palatoglossus muscles, and also the intrinsic tongue muscles, are innervated by the hypoglossal nerve. The muscles of the face, such as the buccinator and orbicularis oris, are also involved in chewing and are innervated by the facial nerve. Four basic types of teeth accomplish mastication: incisors are used for cutting; canines for piercing and slashing; premolars for crushing and shearing; and molars for crushing and grinding. Chewing is usually voluntary, initiated by input from higher centers. A type of reflex chewing, however, can probably originate from the pontine reticular formation and trigeminal nucleus; the reflex is modulated by orofacial afferent input and descending projections from higher centers.

60.2.2 Pharynx and Upper Esophageal Sphincter

In the oropharynx and hypopharynx the superior, middle and inferior constrictor muscles produce the major propulsive forces in the pharynx, and the last-named muscle abuts the upper border of the esophagus at the UES. The UES has an oval configuration with a transverse long axis when closed; it is 2–4.5 cm in length and is formed mainly by the cricopharyngeus muscle, with a smaller contribution from the inferior constrictor. Between swallows, there

R. Greger/U. Windhorst (Eds.)
Comprehensive Human Physiology, Vol. 2
© Springer-Verlag Berlin Heidelberg 1996

is a constant cholinergic nicotinic excitation to the UES, resulting in a degree of basal tone; the pressure is highest in the sagittal orientation (radial asymmetry) owing to the slit-like nature of the *sphincter*. A reflex increase in basal tone can be provoked by slow distension of the upper esophagus or the presence of acid in this region; a decrease in tone occurs during a belch or in response to abrupt esophageal distension. The striated musculature of the sphincter generates continuous electrical spike activity, which fluctuates with respiration; the number of spike potentials is in direct proportion to the resting pressure [2].

60.2.3 Esophageal Body

This is a physiologically defined region of the esophagus, 20–22 cm in length. The striated muscle portion constitutes the upper 10% and commences at the inferior border of the cricopharyngeus muscle. The distal half of the esophagus consists of smooth muscle alone; the thicker inner circular layer extends more proximally than the outer longitudinal layer. All pressures within the esophageal body are transmitted passively by respiratory variations or cardiovascular pulsations; the intraesophageal pressure closely reflects the intrapleural pressure, changing from −5 to −15 mmHg at inspiration to −2 to +5 mmHg at expiration. Electromyographically, the esophageal body is quiescent at rest.

60.2.4 Lower Esophageal Sphincter

The LES is a high-pressure zone 2–4 cm in length, and a thicker ring of the circular smooth muscle can be demonstrated. At rest, the LES is tonically closed (average pressure approximately 20 mmHg, but with wide interindividual variability). The tone fluctuates according to raised intraabdominal pressure, and during fasting in sequence with the phases of the migrating motor complex (highest during phase 3, lowest during phase 1; – see Sect. 63.2.2). Many gut peptides, substances such as histamine, prostaglandins, dopamine, and serotonin, and various drugs can affect LES tone. The physiological significance of the effects of most of these agents remains unclear. The LES does relax transiently at times independently of swallowing. Normally, some gastroesophageal reflux occurs during these relaxations. (This so-called physiological acid reflux can be quantitated by ambulatory intraesophageal pH monitoring.) A number of reflux episodes occur, especially postprandially, with the total time at which the pH is below 4 being less than 4% of any 24-h period. An excessive number of *transient inappropriate relaxations*, which are probably neurally mediated, appears to be the major pathogenetic mechanism in gastroesophageal reflux disease [6]. A major stimulus to these events appears to be gastric distension.

Myenteric plexuses are present in both the striated and the smooth muscle portions of the esophagus. In the striated muscle, they appear to have mainly a sensory role. In the smooth muscle, two important *effector neurons* are present – excitatory cholinergic neurons and non-adrenergic, noncholinergic (*NANC*) *inhibitory* neurons. Sympathetic nerves can be identified within the myenteric plexuses and appear to function only as modulatory fibers. The extrinsic esophageal innervation coordinates the swallowing sequence in the oropharyngeal region and the striated muscle portion of the esophagus and also modulates peristalsis within the smooth muscle esophagus. Information from mucosal receptors in the oropharyngeal area enters the medullary swallowing center through extravagal cranial nerves and vagal nerve pathways; the most potent stimuli are light touch or pressure in the region of the tonsils and posterior pharynx. Sensory information from the esophagus travels via the vagus, with the cell bodies in the nodose ganglion, or via sympathetic pathways to spinal cord segments T3–12. Each level of the esophagus, at least in the cat, has a distinct but overlapping sensory projection to the spinal cord, and afferents from all parts of the esophagus overlap the distribution of cardiac afferents [3]. Initiation and organization of the swallowing sequence, and coordination of the interneuronal connections to the appropriate effector neurons, probably take place in the nucleus tractus solitarius. The motor neurons for the striated muscle esophagus are present mainly in the trigeminal, facial, glossopharyngeal, and hypoglossal nuclei and the nucleus ambiguus; those for the smooth muscle are present in the dorsal motor nucleus of the vagus, and the efferent output from these travels via the vagus. Sympathetic efferents arise predominantly in the cervical and paravertebral ganglia, reaching the esophagus via the vascular supply and the connections to the vagus nerves. There also appears to be a cortical swallowing center situated in the prefrontal cortex [13], which integrates orofacial movements with swallowing. These cortical (and subcortical) pathways, however, are not essential for deglutition.

60.3 Mastication

The main functions of mastication are:

- To disrupt food mechanically to facilitate the action of digestive enzymes
- To mix food with saliva to initiate carbohydrate digestion by salivary α-amylase
- To stimulate afferent receptors that trigger the cephalic phase of digestion
- To form the food into a bolus in preparation for the onset of swallowing.

The primary *muscles of mastication* produce rhythmic separation and apposition of the maxilla and the mandible. They also enable limited forward, backward and lateral movements of the mandible to occur during chewing, ensuring appropriate contact of the food with each of the types of teeth. For jaw closure, the medial pterygoids are the first muscles to be activated, followed by contraction of the posterior temporals and the masseters. The mylohyoid, digastric and lateral pterygoids bring about the jaw-opening phase, and the lateral pterygoids also produce rotational grinding movements. The muscles of the tongue and cheeks also play an important part in maintaining the food between the chewing surfaces.

The rate of *chewing* depends on the type of food, but is about 1–2 strokes/s on average, with wide interindividual variations. During one complete cycle of the mandible, the following sequence can be recognized [14]: a preparatory phase, contact with the food bolus, a crushing phase, tooth contact, a grinding phase, and centric occlusion. The precise path followed by the mandible depends on the amount of lateral slide occurring at the onset of opening. The biting force between the upper and lower dentition is considerable, reportedly up to 90 kg, although values in the literature vary considerably. The maximal biting force can, of course, greatly exceed the forces that develop during normal chewing. Overall occlusal contact area appears to be one of the most important parameters in *masticatory efficiency*, and the presence of molars is critical, as they constitute a large proportion of this area. Other factors, such as the number and condition of the teeth, the function of the tongue and facial musculature, and the chewing time, also influence the masticatory efficiency. The duration and extent to which a mouthful of food is masticated varies greatly among individuals and in general does not appear to influence the process of digestion greatly even if chewing time is very short, despite obvious limitation of the exposure of the bolus to salivary amylase, and of cephalic phase duration. The end result of mastication is the *formation of a bolus* in preparation for swallowing: the chewed food is mixed with air and saliva and fashioned into a soft lubricated mass with a volume of 5–15 ml.

60.4 Swallowing

The main functions of swallowing are:

- To transport the prepared food bolus (solid or liquid) from the pharynx into the stomach
- To prevent esophagopharyngeal reflux and gastroesophageal reflux.

During eating, both conscious and unconscious sensory cues are responsible for eliciting rapid *sequences of swallows*. At other times, swallows are usually triggered involuntarily by salivary stimulation of oropharyngeal sensory receptors. Most individuals swallow up to 1000 or more times per day, varying from about 70 times per hour between meals to a virtual absence during deep sleep. There are four main phases of swallowing: preparatory, oral, pharyngeal and esophageal [4]. The preparatory phase is voluntary and involves bolus formation and lubrication during mastication.

60.4.1 Oral and Pharyngeal Phases

In the *oral phase of swallowing*, the bolus is propelled into the pharynx by a finely coordinated progressive contact of the tongue against the palate in a posterior direction. Respiration is inhibited at this stage, and this inhibition persists throughout the pharyngeal phase. Graded tongue actions are directly responsible for the ability to accommodate boluses of varied volume and for generating the increased intrabolus pressure associated with vigorous bolus expulsion during larger volume swallows [11].

The *pharyngeal phase* commences as the tail of the bolus passes the fauces. Manometrically, this event is marked by a single contraction peak, usually greater than 100 mmHg, coinciding with the beginning of the peristaltic wave. The soft palate elevates and seals off the nasopharynx, preventing postnasal regurgitation. The larynx ascends and the epiglottis is tilted downwards. Laryngeal elevation is vital for protection of the airway, as it facilitates closure of the laryngeal vestibule and removes the laryngeal inlet from the direct path of the oncoming bolus. UES relaxation commences with the initiation of the pharyngeal phase. Both cessation of tonic neural excitation to the sphincter and elevation and forward movement of the cricoid cartilage act to diminish sphincter resting pressure for 0.5–1 s and open the UES during swallowing [10]. Electromyographically, UES relaxation is accompanied by the cessation of all spike activity. The final degree to which the UES opens is related both to the intrabolus pressure and to the intrinsic compliance of the UES. Even though the mechanics of pharyngeal peristalsis are relatively constant, some factors can influence both receptive accommodation of the pharynx and UES opening. Thus, increasing the bolus volume results in earlier anterior movement of the hyoid and larynx, earlier sphincter relaxation and opening, and prolonged sphincter opening. Increasing the viscosity of the bolus prolongs the oral and pharyngeal phases, delays sphincter relaxation and opening, and prolongs the duration of UES opening [4].

60.4.2 Esophageal Phase

As the UES closes, *primary peristalsis* occurs – a progressive circular contraction beginning in the upper esophagus and proceeding distally (Fig. 60.1). In the absence of a bolus, this peristaltic wave may not progress beyond the oropharyngeal phase. *Secondary peristalsis* is a progressive contraction wave in the esophageal body, which is not induced by a swallow but is provoked by stimulation of sensory receptors in the esophageal body, e.g., through distension by a bolus. The peristaltic wave progresses (Fig. 60.1) along the length of the esophagus within 6–8 s, travelling at a velocity of 3–4 cm/s [2]. The velocity is slower in the proximal portion than in the distal portion. The amplitude of contraction is up to about 150 mmHg, but interindividual variation is great; the amplitude is greatest

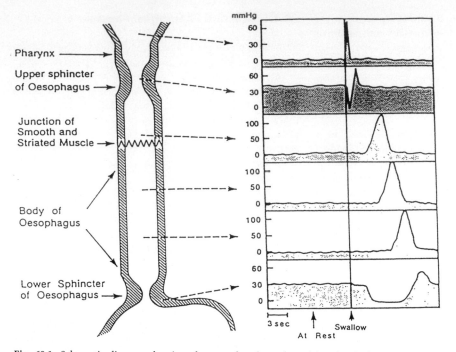

Fig. 60.1. Schematic diagram showing pharyngeal and esophageal intraluminal pressures at rest and after a swallow. Contractions propagate from pharynx to lower esophageal sphincter, which relaxes after the onset of the swallow. *Pressure scales* show only approximate normal values. (From [1], with permission)

in the distal 5 cm of the esophagus, and lowest below the UES, where striated and smooth muscle proportions are most nearly equivalent. The duration of an individual phasic contraction is usually 2–7 s and tends to increase distally. Occasionally, double-peaked contractions and non-propagated contractions occur, apparently increasing with advancing age. A *wet* swallow (liquid bolus) increases the amplitude and duration of contractions and decreases the propagation velocity. A contraction wave that closely follows a preceding contraction wave may have a diminished amplitude and propagation. In humans, *swallow-induced peristalsis* is cholinergic and results from sequencing and activation of the intramural excitatory cholinergic neurons. It is probable that the central control mechanisms exert the major influence on these neurons for initiation and coordination of the peristaltic wave.

The LES relaxes during swallowing, minimizing resistance to flow across the segment. This relaxation may be delayed for 2–3 s after initiation of swallowing, is complete or almost complete, lasts 5–10 s, and is followed by an after-contraction lasting 7–10 s [2]. LES relaxation occurs with virtually every swallow, even where definite esophageal body motor activity is not demonstrable. The mechanism of relaxation appears to involve both active inhibition by NANC inhibitory neurons, and cessation of tonic neural excitation to the sphincter. Truncal and selective vagotomy do not appreciably affect LES tone or relaxation.

References

1. Christensen J (1983) The oesophagus. In: Christensen J, Wingate DL (eds) A guide to gastrointestinal motility. Wright-PSG, Bristol, pp 198–214
2. Christensen J (1987) Motor functions of the pharynx and esophagus. In: Johnson LR (ed) Physiology of the gastrointestinal tract. Raven, New York, pp 595–612
3. Collman PI, Tremblay L, Diamant NE (1992) The distribution of spinal and vagal sensory neurons that innervate the esophagus of the cat. Gastroenterology 103:817–822
4. Cook IJ (1991) Normal and disordered swallowing. In: Dent J (ed) Practical issues in gastrointestinal motor disorders. Baillieres Clin Gastroenterol 5:245–267
5. Cook IJ, Dodds WJ, Dantas RO, Kern MK, Massey BT, Shaker R, Hogan WJ (1989) Timing of videofluoroscopic manometric events and bolus transit during the oral and pharyngeal phases of swallowing. Dysphagia 4:8–15
6. Dent J, Dodds WJ, Friedman RH (1980) Mechanism of gastroesophageal reflux in recumbent asymptomatic human subjects. J Clin Invest 65:256–67
7. Dodds, WJ, Hogan WJ, Lydon SB, Stewart ET, Stef JJ, Arndorfer RC (1975) Quantitation of pharyngeal motor function in normal human subjects. J Appl Physiol 39:692–696
8. Hamlet SL, Muz J, Patterson R, Jones L (1989) Pharyngeal transit time: assessment with videofluoroscopic and scintigraphic techniques. Dysphagia 4:4–7
9. Kahrilas PJ, Dent J, Dodds WJ, Hogan WJ, Arndorfer RC (1987) A method for continuous monitoring of upper esophageal sphincter pressure. Dig Dis Sci 32:121–128
10. Kahrilas PJ, Dodds WJ, Dent J, Logemann JA, Shaker R (1988) Upper esophageal sphincter function during deglutition. Gastroenterology 95:52–62

11. Kahrilas PJ, Lin S, Logemann JA, Ergun GA, Facchini F (1993) Deglutitive tongue action: volume accommodation and bolus propulsion. Gastroenterology 104:152–162
12. Kruse-Anderson S, Wallin L, Madsen T (1991) Ambulatory 23 hour recording of intraesophageal pressures in normal volunteers: a propagation analysis from one proximal and two distal recording sites. Gut 32:1270–1274
13. Miller AJ, Bowman JP (1977) Precentral cortical modulation of mastication and swallowing. J Dent Res 56:1154
14. Murphy JR (1965) Timing and mechanism of human masticatory stroke. Arch Oral Biol 10:981–992
15. Smout AJPM, Breedijk M, Van der Zouw C, Akkermans LMA (1989) Physiological gastroesophageal reflux and esophageal motor activity studied with a new system for 24-hr recording and automated analysis. Dig Dis Sci 34:372–378
16. Sonies BC, Weiffenbach J, Atkinson JC et al (1987) Clinical examination of motor and sensory functions of the adult oral cavity. Dysphagia 1:178–186

61 Gastric Function

J.G. FORTE

Contents

61.1 Introduction

The vertebrate stomach is a sac-like organ, in the upper portion of the gastrointestinal tract between the esophagus and intestine, serving a number of important functions for the initiation of the digestive process. The motor activities of the stomach relate to its functions as a storage organ for food, in the milling and mixing of food with gastric secretions, and in regulating the amount of food reaching the intestine. The secretions of the stomach initiate digestion by acid denaturation of the ingested food and by promoting enzymatic hydrolysis of proteins. In addition to its direct role in digestion, the strongly acidic nature of gastric juice serves to diminish the number of microorganisms that invade the body through the mouth, and one component of the juice, called intrinsic factor, promotes the absorption of vitamin B_{12}, which is essential for normal maturation of red blood cells.

61.2 Structure of the Stomach and Glandular Histology

61.2.1 General Organization of the Stomach

Like most of the gastrointestinal tract, the stomach is organized in several concentric tissue layers (Fig. 61.1) consisting of

- The mucosa lining the lumen of the stomach
- The submucosa
- The muscle coat, composed of several layers of smooth muscle
- The serosa facing the peritoneal cavity.

The stomach is innervated by parasympathetic neurons from the vagus nerve and sympathetic neurons originating from the celiac ganglion. Generally parasympathetic innervation serves to enhance gastric motor and secretory activities, while sympathetic innervation reduces gastric functions. The stomach, like the rest of the gastrointestinal tract, has its own intrinsic system of neuronal organization that sends signals to gastric smooth muscle, secretory cells and endocrine cells. This so-called enteric nervous system is composed of two distinct, but interconnecting, networks of neurons organized between tissue layers: the submucosal plexus within the submucosal layer, and the myenteric plexus between the longitudinal and circular layers of the smooth muscle. The enteric nervous system provides a means of longitudinal communication between different regions of the stomach and along the entire gut wall, as well as of relaying information to and from the stomach via the parasympathetic and sympathetic neural paths.

61.2.2 Gastric Mucosa and Gastric Glands

Gastric secretory activity occurs within the mucosa, which actually consists of three components

R. Greger/U. Windhorst (Eds.)
Comprehensive Human Physiology, Vol. 2
© Springer-Verlag Berlin Heidelberg 1996

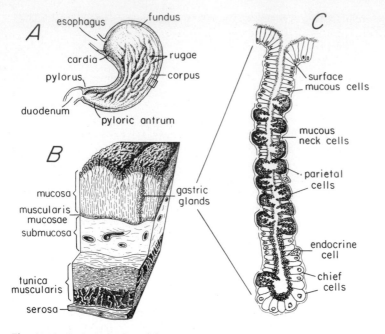

Fig. 61.1A–C. Organization of the stomach and gastric tissue. **A** Macroscopic view indicating major anatomic regions of the stomach. Rugal folds (rugae) of the mucosal lining are also indicated. **B** Schematic representation of the major gastric tissues from section through the wall of the corpus, indicating mucosa, submucosa, tunica muscularis (longitudinal and circular muscle coats), and serosa. **C** Epithelial organization of gastric glands from the fundus or corpus. Columnar surface mucous cells predominate at the luminal surface of the stomach and line the invaginating pits. Parietal cells and cuboidal mucous neck cells abound in the upper regions of the glands. Chief cells, and somewhat sparser parietal cells, are found in lower regions. A number of endocrine cells, including enterochromaffin-like cells, which secrete histamine, and D cells, which secrete somatostatin, can also be found at the most basal aspect of the glandular epithelial cells. (Adapted from [30])

- The epithelium
- The lamina propria
- The muscularis mucosae.

The epithelium is a single layer of cells, but is quite complex in form and is made up of several different types of exocrine secretory cells. The lamina propria is a connective tissue matrix supporting the gastric epithelium, including the mucosal blood supply and containing local endocrine (paracrine) cells that help regulate secretory activity. The specific function of the thin layer of muscularis mucosae is uncertain.

More extensive discussion of the gastric epithelial histology and detailed description of the cell types can be found in several excellent reviews [26,30,31]; here we will provide a brief overview of the histological organization (Fig. 61.1). The luminal surface of the mucosa is covered by a layer of columnar epithelial cells, called surface mucous cells, that secrete mucus and an alkaline fluid. The surface is studded with numerous invaginations, or pits, that serve as conduits for secretions from the subadjacent tubular gastric glands. Three types of glands are found in gastric mucosa, designated cardiac glands, oxyntic glands, and pyloric glands, according to their general anatomical location.

Cardiac glands occur in the delimited region surrounding the gastro-esophageal junction, and are relatively short, mucus-secreting, glands.

The much longer oxyntic glands are present throughout the fundus and main body of the stomach and are responsible for secreting most of the digestive juice. As schematically depicted in Fig. 61.1, oxyntic glands contain three principal epithelial cell types:

- Parietal cells
- Chief cells
- Mucous neck cells.

The large acid-secreting cells are called parietal cells because of their histological orientation with their basal aspects projecting out to the wall of the gland; they have also been called oxyntic cells, based on their functional acid secretory activity, derived from the Greek *oxyntos* (to form acid). Parietal cells are found throughout the length of the gland, but tend to be more abundant in the neck region of the gland. Parietal cells account for more than one third of the total cells present in the oxyntic mucosa, and because of their large size, they represent about 50%–60% of the mass of the oxyntic tissue (Table 61.1). Chief cells are the principal source of pepsinogen, a precursor form of the gastric proteolytic enzyme pepsin, and are confined to the base of the glands. Mucous neck cells are small cells, located in the region of transition at the base of the pits extending throughout the neck of the glands. Mucous neck cells secrete a mucus glycoprotein that is distinct from that of surface epithelial cells, and they also secrete some pepsinogen. There are also a number of endocrine and paracrine cells within the epithelium that play important regulatory roles in gastric secretory function.

Pyloric glands occur in the terminal portion of the stomach referred to as the pyloric antrum, or gateway to the intestine. The pyloric glands are primarily composed of surface epithelial cells and mucous neck cells, but they also contain a very important cell, the G cell, which is responsible for the endocrine secretion of the gastric regulatory hormone gastrin.

Gastric epithelial cells, like cells in other regions of the gastrointestinal tract, are in a constant state of turnover. A population of small undifferentiated stem cells located in the upper neck region of the glands represent the progenitor cells for all gastric epithelial cell types. Stem cells divide and differentiate into surface epithelial cells, which migrate up the pit and over the luminal surface. These same stem cells are the proliferative source of glandular cells, which migrate down the length of the gland. Parietal cells and mucous neck cells differentiate directly from the stem cells; further differentiation of mucous neck cells gives rise to the chief cells toward the base of the gland. The life time, or turnover, of gastric epithelial cells is different for each cell type. Careful studies monitoring the incorporation of ^{3}H-thymidine in the mouse stomach provide an indication of the average cell turnover times [32–34]. Surface epithelial cells have a relatively short life of about 3 days. Parietal cells are longer lived, with an average turnover time of 71 days. Mucous neck cells spend about 40 days in the neck region of the gland, and as they migrate to the base they transform into chief cells, which have an average life of about 250 days.

61.3 Composition of Gastric Juice

The fluid that is secreted into the stomach is called gastric juice. Secretion of gastric juice is regulated, through specific neural and hormonal pathways, by the act of eating and by the presence of food in the stomach and intestine. Gastric juice is the admixture of secretions from the various specialized epithelial cells, both on the surface and within the glands. Three major constituents of gastric juice are the mucus component, the enzyme component and the aqueous component. For the most part, these components are secreted by separate cells, and differential rates of secretion account for the composite variation.

61.3.1 Mucus and HCO$_3^-$

Mucus is a viscous, slippery, gel that covers most of the mucosal surfaces throughout the gastrointestinal tract [1,41]. The consistency and specialized properties of mucus are primarily due to its constituent gel-forming glycoprotein molecules, which are referred to as mucins. In the stomach, mucins are the major organic secretory product of the surface epithelial cells, forming a gelatinous coating over the mucosal surface (Fig. 61.2). Mucous neck cells within the gastric glands also secrete a mucin, but one that is chemically distinct from that of the surface cells.

The output, or secretion, of gastric mucins is under both local and neural control [41]. Irritation of the mucosal surface, either by direct mechanical means, such as rubbing the surface, or by exposure to chemical agents that damage the mucosa, effectively activate mucin secretion. Stimulation of the vagus nerve, or splanchnic nerves, or administration of parasympathomimetic drugs induces the release of copious amounts of viscous mucus into the stomach. Gastrin and histamine, which are effective secretagogues for other gastric epithelial cells, do not stimulate secretion of mucins.

Mucin monomers are glycoproteins with a molecular mass of approximately 50 kDa; they are very heavily glycosylated, with only about 15%–20% of mass represented by the protein core [1]. Some 100–200 oligosaccharides are linked along the length of the protein core via hydroxyl groups (O-linked glycosylation) of serine and threonine residues. Each of the monomers has the molecular appearance of a bottle brush, with the protein as the central core and the oligosaccharides as the projecting bristles. Intact gastric mucin molecules consist of four mucin monomers linked as a tetrad by disulfide bridges, and as secreted by surface cells this tetrameric mucin forms a highly viscous gel layer through which protein molecules, such as pepsin, cannot penetrate. The heavy glycosylation provides protection from peptic hydrolysis along the length of the core protein, but the oligosaccharide-free tetrad center is sensitive to pepsinolysis at the luminal surface of the gel. Thus, steady-state maintenance of the protective gel layer requires continued synthesis and secretion of mucus commensurate with luminal surface degradation by pepsin.

Table 61.1. Quantitative analysis of cells in the body mucosa of five human stomachs. (Data from [29])

Cellular composition as % of mucosal cells (range)			
Parietal cells 32 (24–38)[a]	Chief cells 26 (20–30)	Surface mucous cells 17 (12–20)	Mucous neck cells 6 (4–9)

Total weight of stomach: 143 ± 18 g; wt. of body of stomach as % of stomach wt.: 8 ± 2%; wt. of body mucosa as % of somach wt.: 38 ± 2%

[a] Parietal cells constitute about one third of the mucosa cell population, but because of their relatively larger size they represent about 50–60% of the mucosa mass

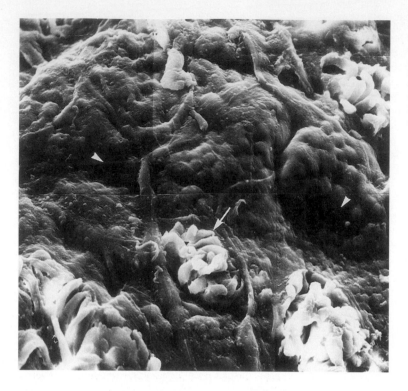

Fig. 61.2. Scanning electron micrograph of the surface of a bullfrog gastric mucosa. Surface mucous cells can be identified by their cobblestone appearance beneath a layer of mucus that covers the entire luminal surface. Strands of a different type of mucus can be seen emanating from several gastric pits (arrow), while some pits are free of mucus (*arrow heads*). ×1000

In addition to the visible cloudy mucus, the surface mucous cells secrete a fluid that is rich in $NaHCO_3$. The $NaHCO_3$ secretory activity of surface mucous cells appears to be similar to HCO_3^- secretion by cells of the duodenal epithelium, being dependent upon carbonic anhydrase activity and stimulated by low mucosal pH, cholinergic agonists, and elevated Ca^{2+} levels. It has been proposed that this gastric $NaHCO_3$ secretion, acting in concert with the layer of surface adherent mucus, provides a protective environment against the low pH and peptic conditions of the gastric lumen. The interstices of the loosely meshed mucus gel layer are impermeable to proteins, but not to acid; although the measured H^+ diffusion rate is somewhat slower through mucus than diffusion in free solution (2- to 3-fold), H^+ (and HCO_3^-) can diffuse through the gel. As a defense against high acidity, the function of the adherent mucus is to provide a stable unstirred layer that prevents convective mixing and rapid dissipation of the small amounts of HCO_3^- secretion at the mucosal surface [19,44]. Studies with microelectrodes to probe the pH within the gel matrix have demonstrated that the relatively low steady secretion of HCO_3^- serves as a neutralizing force and thus sets up a pH gradient within the unstirred layer of mucus. Even with a bulk luminal pH of 1–3, a pH of 5–7 is maintained directly at the surface mucous cells [44].

The viscoelastic properties of mucin molecules make them an ideal substratum for lubricating biological surfaces [1,41]. Moreover, non-digested food, inert particles, bacteria and sloughed mucosal cells become coated with mucin, which facilitates the slippage of these materials through the gastrointestinal tract. It has been shown experimentally that a thick continuous layer of mucus is also important to protect the gastric mucosa from the damaging effects of acid, pepsin, alcohol, and other noxious luminal agents. However, the loosely meshed mucus gel layer is not an absolute permeability barrier to acid within the stomach, as H^+ can readily diffuse through the gel. Rather, it has been proposed that the primary function of the adherent layer of mucus in protecting against acid is to provide a stable unstirred layer at the mucosal surface, supporting the neutralization of H^+ by a low level of HCO_3^--rich secretion from the surface epithelial cells [19,44]. The layer of mucus thus acts as a barrier against convective mixing, preventing the rapid dissipation of the small amounts of HCO_3^- secretion with the large amounts of acid in the luminal juice.

61.3.2 Pepsins

The principal enzyme of gastric juice is pepsin, although several other enzymes are present in smaller amounts. The less abundant enzymes include a gastric lipase, which is maximally effective against triglycerides with short-chain fatty acids, and the proteolytic enzymes cathepsin and gelatinase, both of which, like pepsin, are derived from chief cells. Gastric pepsin is actually a heterogeneous group of proteins responsible for the proteolytic activity of gastric juice [51]. These peptic proteins are secreted in the form of inactive precursor zymogens called pepsinogens. The pepsinogens can be broadly divided into two immunochemically distinct types, pepsinogen I (PG I) and

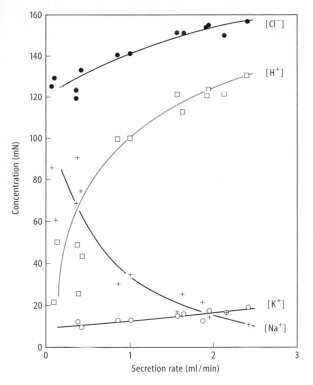

Fig. 61.3. Relation between electrolyte content and secretion rate of gastric juice. At low secretory flow the juice is low in acidity, and high in Na⁺ and Cl⁻. Increased secretory rate induced by infusion of histamine causes a marked rise in H⁺ and an inverse fall in Na⁺ concentration. The secretion-dependent rise in Cl⁻ concentration is matched by a fall in HCO_3^- (not shown). There is a modest, but significant, increase in K⁺ concentration with increased secretory rate. (Adapted from [41])

ther digestive function. The catalytic activity of pepsin is that of an *endoprotease*, preferentially cleaving peptide linkages involving aromatic amino acids (e.g., phenylalanine, tryptophan or tyrosine) and an adjacent amino acid, thus producing polypeptide digestion products of very diverse size.

61.3.3 Hydrochloric Acid

In the adult human, the stomach will typically secrete about 2–3 l of gastric juice per day, and in some individuals up to 3–4 l/day is secreted. The acidity and ionic composition of the gastric secretory product is not constant, but varies with the rate of volume flow, or secretory rate, as demonstrated in Fig. 61.3. The concentrations of H⁺ and Na⁺ show the closest dependence on secretory rate: H⁺ concentration asymptotically approaches a maximum of about 150–160 mmol/l (pH ∼ 0.8) at high secretory rates, with a concomitant decline in Na⁺ concentration of the juice. In addition, there are smaller increases in the concentrations of Cl⁻ and K⁺ as the flow of juice increases [42]. These interrelations between ionic composition and secretory rate are conveniently explained in terms of a two-component theory of gastric juice formation: a small, relatively constant, flow of a Na⁺-rich alkaline juice from surface epithelial cells and other nonoxyntic cells; and a variable flow of isotonic HCl from parietal cells [35]. The volume of the oxyntic HCl secretion is dependent upon the nature and intensity of gastric secretory stimuli. At high rates of glandular stimulation, and thus high secretory flow, the gastric juice approaches the composition of isotonic HCl.

61.3.4 Intrinsic Factor

Based on experimental observation and tests, Castle proposed that absorption of *vitamin B₁₂* from the diet required some factor that is secreted into gastric juice, which he called intrinsic factor [7]. In the absence of intrinsic factor a severe, and potentially fatal, anemia develops because of failure of red blood cells to mature (*pernicious anemia*). We now know that intrinsic factor is a glycoprotein of 55 000 D, which in the human is secreted by parietal cells along with HCl. The same secretagogues and intracellular messengers that stimulate HCl secretion also promote the secretion of intrinsic factor [17]. Intrinsic factor binds to vitamin B₁₂, forming a complex that is resistant to digestion and binds to surface receptors in the ileum to promote the absorption of B₁₂. Patients with pernicious anemia are characterized by achlorhydria, the inability to secrete HCl, and by chronic atrophic gastritis with progressive degeneration of gastric glandular cells and extreme thinning of the gastric epithelium. The condition is one of a group of autoimmune diseases in which antibodies to parietal cell proteins, including intrinsic factor, are found in the serum. For such individuals, parenteral injection of vitamin B₁₂ is required to circumvent the anemia.

pepsinogen II (PG II). In the human PG I is synthesized, stored and secreted by chief cells and mucous neck cells only in the oxyntic region of the mucosa, whereas PG II is also secreted by mucous neck cells in the antral and pyloric mucosa. Both PG I and PG II consist of molecular variants (isozymes) that differ in net charge and/or molecular weight, as can be readily demonstrated by a number of molecular separating techniques. The heterogeneity of the pepsinogens (and pepsins) appears to arise from several factors, including variable numbers of pepsinogen genes, allelic variation within the genes, and post-translational modifications that may occur during processing [55].

Independently of the heterogeneity, all the pepsinogens share the feature of conversion to an enzymatically active form catalyzed by the acidity of gastric juice: the lower the pH the more rapid the conversion, which is almost instantaneous below pH 2. As the pH falls, pepsinogen with a molecular weight of about 42 000 D begins to unfold, and the N-terminal portion splits off, yielding the active enzyme pepsin of about 35 000 D (precise sizes differ with the isozyme). Pepsin is stable at low pH, and has optimal proteolytic activity in the same pH range (i.e., pH 1–3). When gastric juice is neutralized as it passes into the duodenum, pepsin is denatured and thus eliminated from fur-

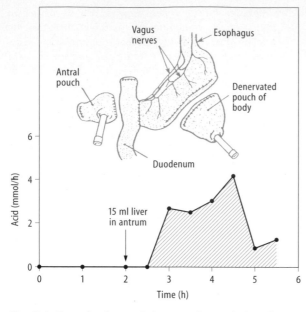

Fig. 61.4. Example of a surgical preparation made in a dog to study gastric secretory activity and regulation. In this experiment, adapted from [14], denervated pouches were prepared from both the body and the antrum of the stomach and exteriorized through canulae so that fluids could be introduced and secretions conveniently collected. When 15 ml of homogenized beef liver was placed into the antral pouch (*arrow*), acid secretory activity from the body pouch commenced within 30 min (*hatched area*)

61.4 Physiological Regulation of Gastric Secretion

61.4.1 Methods for Study of Gastric Secretory Control

Gastric secretory function is under neural and humoral control that is regulated by the alimentary intake of food and the location and nature of the ingested materials within the gastrointestinal tract. Our current understanding of the mechanisms of gastric secretory control comes from a whole spectrum of experimental preparations, extending from the intact stomach to isolated cells and membranes. A great deal of information can be gleaned from the least invasive preparation, in which a subject simply swallows a tube that is used to aspirate the gastric contents. A variation frequently used in whole animal studies involves the surgical creation of a fistula, or hole, through the gastric wall and exteriorization to the abdominal surface by a cannula that can be closed, or open to collect gastric juice. In fact, several cases of accidental gastric fistulae in humans have been used for study. An abdominal gunshot wound resulted in a permanent gastric fistula in Alexis St.-Martin, a French-Canadian trapper whose gastric secretory and digestive functions were studied by William Beaumont in the early nineteenth century, providing an historic base of information [2].

Surgically prepared pouches from the oxyntic gland area with an exteriorized cannula have been used to study gas-

tric secretion without contamination by salivary juice, food, or intestinal regurgitation. The pouches are prepared with and without parasympathetic or sympathetic innervation to allow experimental discrimination among neural and hormonal pathways regulating secretion (e.g., see Fig. 61.4). The innervated pouch (also called the Pavlov pouch) retains all autonomic connections, whereas parasympathetic innervation is interrupted in the vagally denervated pouch (called the Heidenhain pouch), and all neural connections are severed in the totally denervated pouch. In addition to the relatively intact gastric preparations mentioned above, the use of isolated gastric tissue, cells and organelle preparations has provided the basis for identifying and localizing receptors and secretory processes to specific cellular and subcellular loci.

61.4.2 Cephalic Phase of Secretion

Functional activity within the stomach is carefully coordinated with alimentation and digestive function throughout the entire gastrointestinal tract. For convenience of discussion, gastric secretory output can be divided into three phases of control [14,15]:

- The cephalic phase
- The gastric phase
- The intestinal phase.

The cephalic phase initiates and accounts for about 30% of the response to a meal. As its name implies, the cephalic phase is directly controlled by the brain and it is mediated through efferent fibers of the vagus nerve. Afferent stimuli from taste and smell receptors converge on the vagal nucleus in the medulla, which regulates parasympathetic output to the stomach, as depicted in Fig. 61.5A. Vagal nerve impulses excite postganglionic fibers in the myenteric and submucous plexi, which in turn liberate acetylcholine in the region of the secretory cells in the main body of the stomach. Acetylcholine has both direct and indirect effects in promoting secretion by gastric exocrine cells. One of the indirect effects of acetylcholine is to promote the local release of histamine from a group of enterochromaffin-like cells (ECL cells) in the oxyntic mucosa. Histamine acts as a powerful paracrine stimulant of HCl output by parietal cells.

Vagal efferent impulses also travel to the gastric antrum, where postganglionic neurons release a transmitter peptide called gastrin-releasing peptide (GRP). GRP acts on a population of G cells in the antral mucosa to effect the endocrine release of the peptide hormone gastrin (Fig. 61.6), which enters the circulation and stimulates receptors on parietal cells and chief cells [18,56]. All elements of the cephalic phase are eliminated by vagotomy, since the efferent nerve traffic to the stomach is carried via the vagus. Atropine is an antagonist of acetylcholine, and thus will abolish the vagal response in the main body of the stomach; however, atropine does not alter the vagal response in the antrum, which uses GRP as the transmitter. The

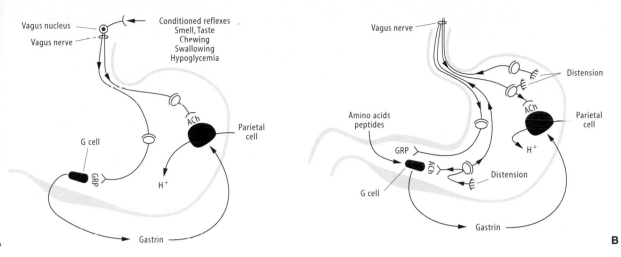

A   B

Fig. 61.5A,B. Pathways involved in the stimulation of gastric acid secretion. **A** Cephalic phase of secretion. Activation of taste and smell chemoreceptors, or even even the thought of a meal, relays impulses to the hypothalamus, then to vagal nucleus in the medulla. Efferent impulses from the vagal nucleus travel to the stomach via the vagus nerve. *1* some fibers directly innervate postganglionic neurons to stimulate parietal cells (*P*) via acetylcholine release and cholinergic receptors. *2* Other vagal efferent fibers to the antrum cause postganglionic neurons to release gastrin releasing peptide and the secretion of gastrin into the circulation where it acts on gastric glands by an endocrine pathway. **B** Gastric phase of secretion initiated by the presence of food

in the stomach. The physical presence of food provides a distending force for activating receptors in the mucosa. Distension receptors in the fundus and corpus (*3*) and in the antrum (*4*) activate afferent neuronal traffic to the vagal nucleus, which in turn relays efferent impulses back to the stomach (*1,2*). Since both the afferent and efferent limbs use the vagus nerve, this is called a vago-vagal reflex. Distension also has small direct effect on postganglionic intramural neurons (*5,6*). *7* Peptides and amino acids, from partial digestion of protein in the food, have a profound effect in directly stimulating the release of gastrin from antral G cells, and activating the endocrine pathway

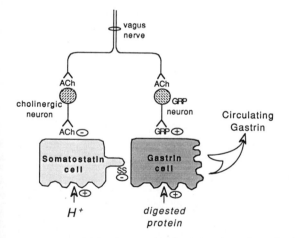

Fig. 61.6. Model for the control of gastrin release from G cells. Vagal stimulation to the antrum causes presynaptic release of acetylcholine (*ACh*). ACh causes the release of gastrin releasing peptide (*GRP*) from GRP post-synaptic neurons, which directly stimulates the endocrine release of gastrin from G cells. G cells are also stimulated to release gastrin in response to protein digestion products on the luminal surface. The release of somatostatin (*SS*) from nearby D cells ordinarily plays a negative influence on gastrin secretion. Release of SS is enhanced by high luminal acidity (*H+*), which provides a negative feedback for the endocrine pathway. Release of SS is inhibited by vagal cholinergic neurons. (Adapted from [18])

cephalic phase is responsible for initiating the response to a meal and will usually begin within a few minutes after the appropriate afferent stimuli. It will also occur in response to stimuli that have been conditioned to be associated with feeding, e.g., the sight of a well-prepared meal or the sound of bacon sizzling in the pan.

61.4.3 Gastric Phase of Secretion

The gastric phase of secretion is regulated by events within the stomach and usually provides the largest part, about 60%, of the response to a meal [14]. The stimulus can be due to the simple physical presence of the food and/or to the chemical nature of the food, and involves both neural and hormonal components (Fig. 61.5B). Distension of the stomach activates intrinsic neurons of the plexuses, leading to both an intramural local reflex and an extramural vagal reflex response. The local reflex involves direct stimulation of postganglionic neurons and supports very little secretory response unless potentiated by other secretagogues; it is insensitive to vagotomy, but can be abolished by atropine. The extramural response to distension is called a vago-vagal reflex because it uses the vagus nerve to transmit afferent impulses to the medulla and return via vagal efferent neurons to stimulate secretion. All aspects of the vagal efferent response for the gastric phase are similar to those described for the cephalic phase.

The nature of the food within the antrum has a profound effect on secretion through the hormone gastrin, and this aspect of the gastric secretory phase is not abolished by vagotomy or atropine. *Gastrin* is a peptide hormone, 17 amino acids in length, that is synthesized and stored in a group of cells within the antral and pyloric mucosa called G cells [16,25]. Studies with synthetic peptides have shown that only the last four C-terminal acids, sometimes called tetragastrin, are required for the gastric secretory activity of the hormone. Peptides from partially digested protein directly stimulate G cells within the antral and pyloric mucosa to release gastrin, which enters the circulation and stimulates acid and pepsinogen secretion. Carbohydrate and fat, like undigested protein, do not stimulate gastrin release except by distension. Lowering of pH (pH 2 or less) at the surface of the antral mucosa greatly inhibits the gastric phase of secretion, and this can be attributed to the prevention of gastrin release by another peptide called *somatostatin*. Somatostatin is contained within and secreted by endocrine-like cells, called D cells, located throughout most of the gastric mucosa [58]. D Cells in the antral and pyloric mucosa respond to low luminal pH by releasing somatostatin, which acts in a local paracrine manner to turn off gastrin secretion by G cells. This mechanism of inhibiting the release of gastrin by low antral mucosal pH, as schematically outlined in Fig. 61.6, is an important aspect of negative feedback regulation of gastric HCl output.

61.4.4 Intestinal Phase of Secretion

There is very little stimulation associated with the intestinal phase of gastric secretion, less than 10% of the secretory response to a meal. Immunocytochemical data show that a few G cells spread through the pyloric area into the duodenum, and these may be responsible for some weak stimulatory effects in a manner similar to that described for the gastric phase. There are also reports that another hormone, tentatively called *enterooxyntin*, may respond to duodenal distension and stimulate the oxyntic mucosa. However, the most important regulatory aspects of the intestinal phase are those associated with inhibition of both gastric secretion and gastric emptying. The duodenum provides a means of feedback inhibition and termination of gastric secretion after a meal. The nature of the partially digested food materials that enter the duodenum, particularly solutions of high acidity, high fat content, or high osmotic pressure, represent the major stimuli for feedback inhibition of gastric secretion. On the other hand, if the duodenum is surgically bypassed and gastric contents are emptied further down in the intestine, the secretory output of the oxyntic mucosa is greatly enhanced.
The principal means of feedback regulation is through the action of several hormones that are released from the duodenal mucosa. For example, acid in the duodenum causes the release of the peptide hormone *secretin*, which, in addition to its major role to stimulate the output of HCO_3^- by the pancreas and liver (see Chap. 20), has an inhibitory effect on both gastric acid secretion and motility. Secretin inhibits acid secretion by causing the release of

somatostatin, which in turn attenuates the output of gastrin by G cells [11,50,58]. Additionally, acid in the duodenum feeds back to inhibit gastric acid secretion via an intramural nervous reflex that might also operate via somatostatin release. Fats in the duodenum, especially triglyceride digestion products such as fatty acids and monoglycerides, cause the release of two additional hormones, known as *cholecystokinin* (CCK) and *gastric inhibitory polypeptide* (GIP). CCK is the major hormone regulating pancreatic enzyme secretion and biliary output (see Chap. 20), but it also has both stimulatory and inhibitory actions on gastric secretion. CCK is a 33-amino-acid polypeptide containing the same five C-terminal amino acids as does gastrin, and CCK can be shown to act on parietal cells both as a poor agonist for gastrin and, in higher doses, as an antagonist to gastrin [47]. Because of the dose range there may be little physiological significance for these actions of CCK on gastric acid secretion. However, CCK does stimulate chief cells to secrete pepsinogen, and CCK has a possible role in inhibition of gastric emptying by enhancing pyloric constriction. GIP, which has some sequence homology with secretin, came to be accepted as a distinct peptide hormone only after adequate procedures had been developed for separating it from other hormones present in extracts of duodenal mucosa, such as secretin and CCK [6]. GIP inhibits both parietal cell secretion and the output of gastrin by its action in promoting the paracrine release of somatostatin in the antrum and body of the stomach. In the antrum, somatostatin prevents the release of gastrin; in the fundic and body mucosa, somatostatin has a direct inhibitory effect on parietal cells, as discussed below.

61.4.5 Integrated Secretory Response to a Meal

In the interdigestive period the stomach has a basal secretory activity in the absence of extrinsic digestive stimulatory input [15]. The amount of basal secretion varies among species, and even shows a circadian rhythm, but in a healthy human basal secretion does not represent more than 10% of the maximal secretory response to a meal. Basal secretion can be reduced by atropine and histamine-H2-receptor antagonists, showing that there is some background level of stimulation. In the absence of food, and the buffering capacity that the food represents, the pH of the gastric contents can be quite low, often less than 2.0. The acidic luminal pH operates to minimize any additional secretion. High activity of H^+ on the mucosal surface stimulates D-cells to release somatostatin locally in both the antrum and the main body of the stomach. Somatostatin inhibits release of gastrin by G-cells, and also has some direct inhibitory effect on parietal cells, so that the normal stomach is maintained in a minimal basal secretory state.
The ingestion of a meal has a series of primary, secondary and tertiary effects in turning on secretory activity and eventually restoring the resting state. The immediate effects are those afferent stimuli that activate vagally medi-

ated cephalic phase, including direct cholinergic stimulation, and the action of vagal efferents to inhibit somatostatin cells and thus release G-cells and parietal cells from somatostatin inhibition. The food that enters the stomach provides a buffering action, raising gastric luminal pH and releasing the inhibitory effects of somatostatin. Physical distension and the consequent local responses initiate the gastric phase of secretion. Peptide digestion products directly stimulate gastrin release from G-cells, providing the large hormonal stimulation of gastric secretory output.

As gastric digestion proceeds, several gastric and intestinal processes act to reduce HCl secretory output [15]. Alimentary satiety, operating through a hypothalamic center, abolishes the cephalic phase of stimulation. The buffer capacity of the meal is exceeded and gastric pH falls, stimulating the local release of somatostatin, which inhibits gastrin output. The partially digested food moves into the duodenum where the acidity, fats and osmolarity produce the neural and hormonal responses of the intestinal phase that reduce H^+ secretion and return the stomach to the resting state.

61.5 The Cellular Basis of Secretory Mechanisms

61.5.1 HCl Secretion by Parietal Cells

61.5.1.1 Secretagogues and Receptors for Acid Secretion

There are three primary physiological stimulants, or secretagogues, of acid secretion:

- Acetylcholine, a neurocrine stimulant released by postganglionic neurons of the vagus nerve
- Gastrin, a classic endocrine stimulant released by G-cells in the antrum of the stomach
- Histamine, a paracrine stimulant released by enterochromaffin-like cells that are in close proximity to the basal aspect of parietal cells.

All three secretagogues most certainly operate to stimulate acid secretion, but there is considerable variation among different animal species as to the relative abundance of the receptor types specifically on parietal cells, and this has been the source of a long-standing controversy on the physiological role of histamine. Views on the specific role of histamine have ranged from one extreme, with uncertainty as to whether it has any physioligical function at all, to the extreme hypothesis that histamine is the sole final common mediator for parietal cell activation, with acetylcholine and gastrin operating by the local release of histamine from the histamine-containing cells in the lamina propria [12].

The discovery that HCl secretion was inhibited by specific histamine receptor antagonists of the H2 type (H2 antagonists) was a milestone in gastric secretory physiology, providing an effective treatment for conditions of hyperacidity and peptic ulcers without surgical intervention [4]. The use of H2 antagonists to inhibit acid secretion in vivo stimulated by acetylcholine and gastrin, as well as by histamine, definitively demonstrated that histamine was a physiological stimulant and offered some support for the final common mediator hypothesis. The histamine hypothesis was further supported by data from several in vitro gastric preparations, in which histamine was shown to be the most effective stimulant (i.e., maximal rates); the in vitro data also show that acetylcholine and/or gastrin can cause the local release of histamine [3]. However, potentiating effects of multiple secretagogues are well known. Moreover, it has been clearly shown in some species that parietal cells have receptors for all three secretagogues, acetylcholine, gastrin and histamine, and that an acid secretory response can occur without the involvement of histamine [52]. In fact, these varied observations are consistent with a more moderate hypothesis proposing a convergence of biochemical information from different secretagogue receptors at the basal surface to a common intracellular step for parietal cell activation [8,23,53]. Different animal species or types of experimental preparations may have widely differing densities of the three receptor types, but if the means for parietal cell activation converged to the same final events of cellular transformation, a scheme such as that shown in Fig. 61.7 would explain preparation-dependent variation in response to secretagogues, as well as providing the basis for the observed potentiating responses when more than one secretagogue is given.

61.5.1.2 Cellular Activation of Secretory Function

Stimulation of gastric HCl secretion by the primary secretagogues is mediated by at least two distinct sets of biochemical pathways, or cascades, each operating through its own intracellular, or second, messenger (see also Chap. 6). Although there are differences in receptors and second-messenger signalling systems, evidence strongly suggests that there is convergence within the terminal steps to effect the final secretory processes. The primary parietal cell secretagogues and their second messenger pathways are shown schematically in Fig. 61.7.

H2 receptors on the basal surface of parietal cells are linked to the activation of adenylate cyclase through stimulatory G proteins, G_s [8]. Thus, stimulation of gastric secretion by histamine, but not by acetylcholine or gastrin, leads to enhanced formation of cyclic AMP (cAMP) as a principal second messenger in parietal cells. Additional and unequivocal support for the cAMP hypothesis comes from observations that gastric acid production can be stimulated by:

- Membrane-permeable derivatives of cAMP
- Blockers of cAMP catabolism, such as caffeine
- Agents that directly activate adenylate cyclase, such as forskolin

The stimulatory action of cAMP is mediated by activation of cAMP-dependent protein kinases (PKA) whose function in parietal cells is to promote a cascade of protein phosphorylations and cell activation. The so-called histamine/cAMP/PKA pathway has been shown to increase phosphorylation of several parietal cell proteins, including cytoskeletal proteins and possible K^+ conductance proteins [9,21]; however, details as to how these and other phosphoproteins initiate the HCl secretory process is a subject of intense current investigation.

Both direct and indirect studies to monitor intracellular Ca^{2+} concentration have demonstrated that acetylcholine and its agonists cause an increase in cytosolic Ca^{2+} activity [10,40]. Thus, the parietal cell conforms to the model established in other tissues in which cholinergic receptor stimulation works via a Ca^{2+}-dependent pathway. The action of gastrin on parietal cells appears to be similar to cholinergic stimulation, involving an increase in cytosolic Ca^{2+}, and no change in cAMP concentration. This is distinct from the effect of histamine, which, in addition to stimulating the cAMP/PKA pathway described above, has been shown to cause a spike in intracellular Ca^{2+} concentration, although a much more modest one than appears with cholinergic agonists [10].

When Ca^{2+} ionophores are used to elevate cytosolic Ca^{2+} no parietal cell secretion occurs if other secretagogues are not present, whereas the secretory response to histamine is slightly potentiated [59]. The mechanism by which elevated cytosolic Ca^{2+} levels facilitate HCl secretion is uncertain, but it is likely to require concomitant changes in the phosphorylation of specific proteins. Increased levels of cytosolic Ca^{2+} are themselves the result of another major intracellular signalling system involving receptor activation of phospholipid hydrolysis, producing inositol (1,4,5)-trisphosphate and diacylglycerol from phosphatidyl inositol (cf. Chap 5; Fig. 61.7). Inositol trisphosphate is responsible for liberating Ca^{2+} from intracellular bound stores and for increasing entry of Ca^{2+} from the external milieu. Diacylglcerol acts as a second messenger to activate protein kinase C (PKC), which like PKA promotes protein phosphorylation. Whether, and to what extent, specific gastric phosphoproteins produced by PKC and PKA are similar, and how they converge on a final common pathway, remains to be determined.

Figure 61.7 also indicates that there are several inhibitory receptor pathways on parietal cells. Somatostatin acts as a paracrine agent to inhibit parietal cell secretion, being liberated by local D cells. Somatostatin receptors activate an

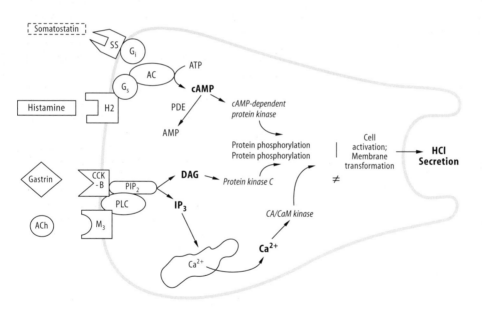

Fig. 61.7. Secretagogue–receptor relationships and signal transduction pathways for activation of HCl secretion by parietal cells. The most prominent stimulation is via histamine and the so-called cyclic AMP/protein kinase A (*cAMP/PKA*) pathway. Histamine acts on H2 receptors to activate stimulatory G proteins (G_s) and adenylylcyclase (*AC*). The resulting production of cAMP activates PKA, which catalyzes the phosphorylation of several phosphoproteins. The specific role of these phosphoproteins in parietal cell activation is uncertain, but one of them, ezrin, is a membrane cytoskeletal linker possibly involved in membrane recycling of the H,K-ATPase. Breakdown of cAMP occurs via phosphodiesterase (*PDE*); inhibition of PDE, e.g., by caffeine, has the same effect as activation of AC. Somatostatin, and also prostaglandins and epidermal growth factor (not shown), interact with parietal cell receptors to effect an inhibition of cell function,

by activating an inhibitory G protein (G_i) which turns off AC. Acetylcholine (*ACh*) and gastrin have a major stimulatory effect by causing the local liberation of histamine from enterochromaffin-like cells (paracrine stimulation), and thus promote the cAMP/PKA pathway. The parietal cell also has receptors for ACh (M_3 receptors) and gastrin (CCK-B receptors), which activate phospholipase C (PLC) through a different class of G proteins. PLC hydrolyzes phosphatidylinostitol to its products, inositol (1,4,5)-trisphosphate (IP_3) and diacylglycerol (*DAG*). IP_3 mobilizes intracellular Ca^{2+}, and together with DAG, activates protein kinase C to catalyze phosphorylation of other proteins. Ca^{2+}/calmodulin-mediated protein kinases may also be involved via Ca^{2+} mobilization. Ultimately, the protein phosphorylation pathways converge on the critical, but currently uncertain, cell-activation processes

inhibitory G protein, G_i, that lowers adenylylate cyclase activity and cAMP concentration. Output of somatostatin ordinarily occurs at a sustained level, and is inhibited by cholinergic stimulation. This suggest that parietal cells are under a tonic state of inhibition that is relaxed by suppressing D-cell secretion. Epidermal growth factor (EGF) has an negative effect on histamine-mediated HCl secretion. This most probably ocurs via an inhibitory phosphorylation pathway, possibly through a tyrosine kinase. Prostaglandins can be shown to inhibit HCl output, but the required concentrations are high and it is difficult to sort out specific parietal cell effects from other mucosal protective actions of prostaglandins (see Sect. 61.6).

61.5.1.3 Cellular and Molecular Basis of HCl Secretion

The H^+/K^+-ATPase, the Primary Gastric Proton Pump. The transport of gastric H^+ is powered by a membrane-bound pump that uses ATP to drive an electroneutral one-for-one H^+/K^+ exchange, with a stoichiometry for $H^+:K^+:ATP$ of 1:1:1 [24,49], (see also Chap. 8). This gastric proton pump enzyme is called the H^+/K^+-ATPase; a schematic representation of the operation of the gastric H^+/K^+-ATPase is shown in Fig. 61.8. Although the H^+/K^+-ATPase is highly enriched in, and unique to, parietal cells, it is closely related to the Na^+/K^+-ATPase and the Ca^{2+}-ATPase, sharing a number of structural and functional features. All three of these cation-transporting systems belong to a larger family of ATPases, called P-type ATPases since the

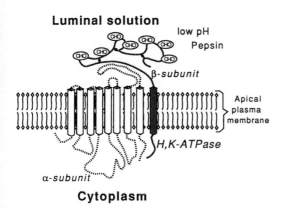

Luminal solution

low pH
Pepsin

β-subunit

Apical plasma membrane

H,K-ATPase

α-subunit

Cytoplasm

Fig. 61.8. Schematic representation of the gastric proton pump. The proton pump is called the H^+/K^+-ATPase because it uses energy from ATP to drive the electroneutral exchange of K^+ for H^+. The functional enzyme consists of two integral membrane peptide subunits. The α-subunit has eight transmembrane spanning segments and a large cytoplasmic domain that includes the ATP-binding site and energy transducing machinery. The β-subunit has a single membrane-spanning segment (*shaded loop*), and most of its mass resides in the extracytoplasmic domain, including seven potential glycosylation sites (*CHO*) and three structurally important disulfide bonds. The extensive glycosylation may play a part in protecting the enzyme from the harsh luminal conditions of low pH and pepsin

enzymes undergo phosphorylation/dephosphorylation and ligand-dependent conformational rearrangements during their transport/catalytic cycles. These is an especially high degree of homology between the H^+/K^+-ATPase and the (Na^++K^+)-ATPase: the primary amino acid sequence shows about 60% identity; both pumps work as exchangers, transporting H^+ or Na^+ in exchange for K^+; and they operate in the membrane as heterodimers, being composed of α- and β-subunits.

The Membrane Recycling Hypothesis for HCl Secretion. While the H^+/K^+-ATPase provides the molecular mechanism for the transduction of metabolic energy into secreted gastric H^+, the physiological activity of the pump is intimately related to the changes in parietal cell morphology that are associated with the turning on and off of HCl secretion at the time of the meal. A schematic representation of the ultrastructure of parietal cells in the resting and secreting states is shown in Fig. 61.9. Parietal cells are among the most metabolically active cells in the body, with numerous, large mitochondria to support the enormous energy demand required to secrete H^+ against a gradient of more than a million-fold, i.e., from a cytosolic pH ~7.0 to luminal pH of less than 1.0. However, the most distinctive structural features are the highly specialized membranes, at the apical cell surface and within the cytoplasm, whose stimulation-dependent structural changes form the basis for the membrane recyling hypothesis of HCl secretion [22]. The apical surface of the parietal cell has a unique arrangement of invaginations, or secretory canaliculi, that extend to and radiate through the cell. The secretory canaliculi of the resting, or nonsecreting, cell are relatively narrow, and the entire apical surface is covered with short, stubby microvilli containing an organized array of actin microfilaments. Parietal cells also have an extensive system of cytoplasmic membranes, called tubulovesicles. In resting cells, during the interdigestive period, the H^+/K^+-ATPase is localized to the tubulovesicles.

When parietal cells are stimulated there is a profound morphological change whereby the tubulovesicles fuse with the apical membrane, leading to a 5- to 10-fold expansion of the apical surface area within the secretory canaliculi. According to the membrane recycling hypothesis these dynamic membrane transformations provide access for the intrinsic proton pumps to the gland lumen. Parietal cell stimulation, primarily under the direction of the cAMP/PKA system, not only incorporates the H^+/K^+-ATPase to the apical plasma membrane, but also results in the activation of two ionic pathways at the apical membrane, K^+ channels and and Cl^- channels [20,46,57]. Membrane fractionation studies strongly support the membrane recycling hypothesis. In resting parietal cells H^+/K^+-ATPase-rich membranes are isolated in the form of tubulovesicles with low intrinsic ionic (K^+ and Cl^-) permeability, and thus poor proton transport (H^+/K^++ exchange) capability. For stimulated parietal cells H^+/K^+-ATPase is associated with large vesicles that (1) are derived from the apical canalicular plasma membrane, (2) have high intrinsic conductive pathways for K^+ and Cl^-, and (3) effectively

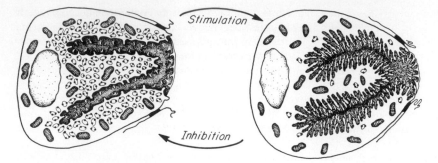

Fig. 61.9. The morphological basis for the membrane recycling hypothesis of HCl secretion. The apical surface of the parietal cell is specialized in a unique system of invaginated canals, called secretory canaliculi. In the resting state (*left*) the canaliculi are relatively narrow with short microvillar projections. The resting cell is also characterized by an enormous pool of intracellular membranes, called tubulovesicles, that are highly enriched in H$^+$/

K$^+$-ATPase. In the stimulated parietal cell (*right*), tubulovesicles fuse with the canalicular membrane, translocating the H$^+$/K$^+$-ATPase to the apical surface and resulting in a greatly expanded surface area of elongated microvilli. Removal of the stimulus results in recycling of the H$^+$/K$^+$-ATPase back into the tubulovesicle compartment

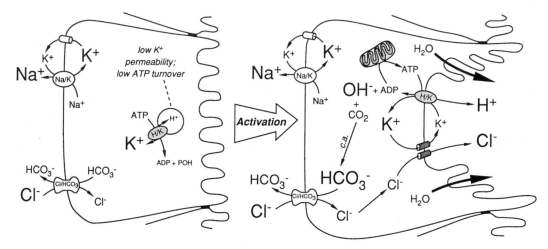

Fig. 61.10. Schematic representation of parietal cell transport processes in the transformation from rest to active HCl secretion. In the resting cell (*left*) the tubulovesicles are rich in H$^+$/K$^+$-ATPase (*H/K*), but because of the low intrinsic permeability to K$^+$ (and Cl$^-$) there is very little pumping of H$^+$ and very low ATP turnover. Parietal cell activation (*right*) leads to transfer of H$^+$/K$^+$-ATPase to the apical surface by fusion of tubulovesicles with the plasma membrane, and to activation of passive conductive pathways for K$^+$ and Cl$^-$ at the apical membrane. The parallel arrangement of these apical transport systems provides the means for passive movement of KCl from cytoplasm to canalicular lumen,

recycling of K$^+$ back into the cytoplasm in exchange for H$^+$ by the H$^+$/K$^+$-ATPase, and net HCl secretion driven by ATP turnover. Water flux into the expanded canaliculus is driven by net solute flux. Important transport components are also present at the basolateral surface of the cell. The (Na$^+$+K$^+$) pump (*Na/K*) maintains high intracellular K$^+$ levels through basolateral K$^+$ recycling, and the Cl$^-$/HCO$_3^-$ exchanger provides the pathway for the loss of excess HCO$_3^-$ and uptake of Cl$^-$ for the secretory product. The production of HCO$_3^-$ is catalyzed by carbonic anhydrase (*c.a.*) from CO$_2$ and the cytoplasmic base that is produced by the operation of the H$^+$/K$^+$-ATPase

accumulate large proton gradients in the presence of K$^+$, Cl$^-$ and ATP.

A functional representation of the resting/stimulated transition for parietal cells is given in Fig. 61.10. In the resting cell, low ATP turnover in maintained by relative impemeability of tubulovesicle membranes. In the stimulated cell, where tubulvesicles have fused with the apical membrane, passive flux of K$^+$ and Cl$^-$ from cell to lumen occurs in parallel to K$^+$-for-H$^+$ exchange by the pump, resulting in the recycling of K$^+$ and a net flow of HCl into the lumen accompanied by an osmotic equivalent of water. For each mole of H$^+$ secreted by the pump, an equivalent

amount of base must also be produced, and it must be eliminated from the cell. The parietal cell uses CO$_2$ and carbonic annhydrase to convert the base into HCO$_3^-$, which is then transported out of the cell by an efficient basolateral membrane Cl$^-$/HCO$_3^-$ exchanger [39,43]. This mechanism provides the source of Cl$^-$ that will accompany H$^+$ in gastric juice, as well as the source of HCO$_3^-$ that is released into the plasma as part of the alkaline tide of the meal. Withdrawal of the stimulus from the parietal cell reverses the activation process; the apical surface area is diminished by mass endocytosis and tubulovesicles reform within the apical pole of the cytoplasm [20].

61.5.2 Stimulation of Chief Cells

Receptors and Activation Pathways. Chief cells synthesize pepsinogen through a very active protein synthetic machinery. Newly formed pepsinogen granules are stored in the apical pole of the cell until the appropriate stimuli come to bear, when the granules fuse with the apical membrane and release their contents into the lumen of the gastric gland, much like zymogen release from pancreatic acini.

A scheme of receptor/activation pathways in chief cells is shown in Fig. 61.11. Release of pepsinogen is strongly regulated by cholinergic receptors, which operate as they do in parietal cells by stimulating phospholipid turnover and generating the second messengers, inositol trisphosphate and diacylglycerol, elevating cytosolic Ca^{2+} and activating PKC [45]. The specific proteins (e.g., phosphoproteins) regulating the pepsinogen release process are not known. Gastrin is not a potent stimulant of pepsinogen release, but chief cells are stimulated by two other peptide hormones secreted by the duodenum, cholecystokinin and secretin. Interestingly, each of the duodenal hormones operates through a different intracellular activation pathway (Fig. 61.11). Cholecystokinin uses the PKC and Ca^{2+} pathway, while secretin receptors activate the cAMP/PKA pathway. Thus, as in parietal cells there appear to be separate, and possibly convergent, pathways for activating chief cells. Studies with isolated chief cells have also demonstrated the presence of adrenergic receptors that elevate cAMP, but their physiological relevance in vivo is uncertain. Chief cells do not have histamine receptors, and the small increase in output of pepsinogen that is seen in response to injection of histamine has been attributed to a washout of pepsinogen with the volume flow of acid.

61.6 Pathophysiology of Gastric Secretion

One of the most interesting questions regarding the overall functional activity of the stomach concerns its resistance to autodigestion; that is, why does the stomach not digest itself? Gastric digestive secretions can destroy almost any cell and hydrolyze almost any biological protein. In fact, the gastric mucosa itself is rapidly destroyed and hydrolyzed if the acid and pepsin normally present in gastric juice are placed on the serosal side, but the tissue is endowed with special mechanisms, collectively referred to as the gastric mucosal barrier, that protect the luminal surface from autodigestion. However, there are conditions in which the gastric mucosal barrier is challenged or functionally defective, and this leads to painful conditions of gastritis and erosion and the potentially fatal degenerative conditions of ulceration.

Gastric and duodenal ulcers form when endogenous mucosal protective devices are overwhelmed by the damaging properties of acid and pepsin. Both gastric and duodenal ulcers are generically called peptic ulcer disease, because acid and pepsin is required for the ulcer forma-

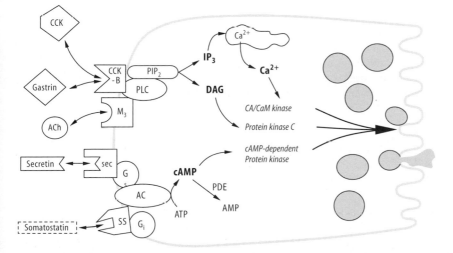

Fig. 61.11. Secretagogue–receptor activation of the chief cell. Activation, fusion and release of pepsinogen granules is a function of phosphorylation of certain critical proteins, which in turn are regulated by independent pathways via specific protein kinases. The major stimulatory pathway involves cholinergic M_3-type receptors and CCK-B type receptors, which activate phospholipase C (*PLC*) to hydrolyze phosphatidylinositol (*PIP$_2$*), yielding the intracellular messengers inositol (1,4,5)-trisphosphate (*IP$_3$*) and diacylglycerol (*DAG*). IP$_3$ mobilizes intracellular Ca^{2+} and, together with DAG, activates protein kinase C. Ca^{2+} mobilization may also be involved in activating Ca^{2+}/calmodulin-mediated protein kinase. A separate pathway occurs via secretin receptors (or β-adrenergic receptors, not shown) to activate stimulatory G protein (*G$_s$*) and adenylyl cyclase (*AC*). The resulting production of cAMP activates cAMP-dependent protein kinase (*PKA*). Stimulation by multiple pathways leads to convergence and potentiation of the activation process. Somatostatin has an inhibitory influence on pepsinogen output by activating an inhibitory G protein (*G$_i$*) which turns off AC

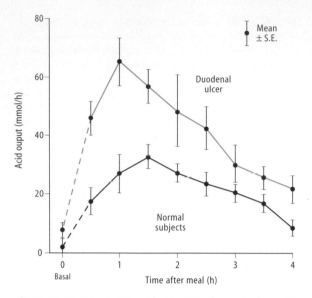

Fig. 61.12. Human gastric acid secretion after a steak meal in normal subjects and those with a diagnosed duodenal ulcer. (Reproduced from [5])

tion, but the two conditions differ in etiology and locus within the mucosal lining. In general, gastric ulcers in the body or antrum of the stomach are more typically the result of a weakening of the defense mechanisms, while duodenal ulcers in the prepyloric and duodenal mucosa are more often related to abnormally high secretory levels.

Although duodenal and prepyloric ulcers are generally associated with high endogenous rates of acid secretion (Fig. 61.12) there is a lot of variance in the population. High secretory rates are due to any one or any combination of the following factors: increased circulating levels of gastrin (hypergastrinemia), an especially high sensitivity to gastrin, and increased parietal cell mass. Individuals with duodenal ulcers usually do not have hypergastrinemia in the non-feeding state, but their responses to a meal, in both gastrin output and sensitivity of parietal cell secretion, are greater than normal. There is also an interesting correlation between the measured levels of pepsinogen I circulating in the serum and the occurrence of duodenal ulcers in patients. Although there is no explanation for the elevated serum pepsinogen, and at the pH of blood the enzyme is totally inactive, it has been proposed that the serum pepsinogen level might be used as a predictor of susceptibility [51].

Acid hypersecretion and the presence of duodenal ulcers occur in individuals with Zollinger-Ellison syndrome. This syndrome is the result of tumors that secrete gastrin (gastrinomas), usually occurring in the pancreas, but also found in duodenum and stomach. The sustained hypergastrinemia produces high acid output and severe ulcers. High levels of gastrin also have a trophic effect on parietal cell growth, which can further exacerbate the problem.

Excess gastric acid can be countered to some extent by ingestion of antacids or buffering mixtures, but advances in receptor biology and some fundamental understanding of the gastric proton pump have led to the development of drugs that are extraordinarily effective in controlling the output of HCl by parietal cells. H2-receptor antagonists, such as cimetidine and ranitidine, block the action of histamine on parietal cells with very little effect on non-gastric histaminergic sites, and they have therefore been extensively used to treat peptic ulcer disease. More recently, drugs have been developed that directly inhibit the H^+/K^+-ATPase [49]. One of these, omeprazole, obtains its specificity by being concentrated and activated in the acidic environment of the parietal cell canaliculus, where it covalently modifies the pump enzyme. Additional H^+/K^+-ATPase inhibitors that compete reversibly with pump ligands, such as K^+, are currently being developed with a view to increasing their specificity of action.

Physiological defense within the stomach is the result of a complicated and cooperative group of factors, collectively known as the gastric mucosal barrier, and better understood on the basis of causal experimentation than from fundamental mechanisms. Experimental tests have shown that competency of barrier resistance depends upon physiological integrity of the tight junctions between epithelial cells, mucus and bicarbonate secretion by surface mucous cells, adequacy of mucosal blood flow, and processes of cell renewal, as well as being negatively affected by damaging drugs and bacterial infection [28]. It is also reasonable to conclude that the exterior surface of apical plasma membranes of gastric epithelial cells must have highly specialized structural and biochemical adaptations that resist the caustic acidic and peptic conditions, yet the nature of these specializations remains elusive.

As noted above, mucus forms a viscoelastic meshwork covering the surface of the mucosa, serving to protect the surface epithelium from abrasion by food particles, and acting in concert with HCO_3^- secretion to minimize the concentration of H^+ at the surface epithelial cells. Conditions that destroy the layer of mucus or inhibit HCO_3^- secretion lead to greater H^+-induced surface damage. A number of drugs, or even endogenous components like bile salts moving retrogradely from duodenum to stomach, cause concentration-dependent surface erosion by damage to surface cells and tight junctions. Aspirin, and the related group of non-steroidal anti-inflammatory drugs (NSAIDs), are a frequent cause of gastric ulceration because of their wide use. The precise mechanism of injury is unknown, but it almost certainly occurs by virtue of the reduced prostanoid synthesis caused by the NSAIDs. Gastric mucosal injury induced by NSAIDs and a variety of noxious agents, including alcohol and boiling water, can be attenuated by the luminal application of prostaglandins [48]. Such observations have led to the suggestion of treatment with prostanoid drugs for certain degenerative conditions.

In the past several years gastroenterologists have become increasingly aware of the possibility that peptic ulcers may have an infectious origin. In 1984 Marshall and Warren [37] proposed that an unusual bacterium was the cause of peptic ulcer disease. This microorganism, now known as *Helicobacter pylori*, thrives only in the unusual environment of the surface of the human gastric epithelium, beneath the mucous coat, and has an active urease that hydrolyzes urea to produce abundant levels of ammonia. Known infections with *H. pylori* are associated with chronic gastritis and thus the organism may be another "barrier breaker" leading to erosion and ulceration. Recent studies have shown that, although antisecretory drugs are effective in healing peptic ulcers over the short term, long-term prevention of recurrence was more closely correlated with eradicating the infection as well [27].

61.7 Gastric Motility

61.7.1 Muscular and Neural Components

The stomach has three major motile functions associated with digestion:

* To serve as a reservoir as the meal is ingested (the stomach can accommodate up to 1.5l)
* To mix the ingested food with gastric secretions
* To empty gastric contents into the duodenum.

These motile functions are accomplished by coordinated activity of three layers of smooth muscle: an outermost longitudinal layer, a middle circular layer, and an inner oblique layer [38]. The longitudinal and oblique layers are incomplete, and not represented along the full length of the stomach. The longitudinal layer is present only in the distal two-thirds of the stomach, while the oblique layer of muscle can be distinguished only in the proximal half of the stomach. The circular layer is prominent throughout, although its thickness is obviously increased in the antrum of the stomach, where the force of contraction is greatest. The fundamental characteristics of excitability and contractility for gastric smooth muscle are similar to those described in Chap. 63 for intestinal smooth muscle. Although innervation is intrinsic to the smooth muscle cells themselves, coordination is highly dependent upon the enteric neural plexuses, especially the myenteric plexus, and the intensity of contractile activity is under the influence of parasymathetic and sympathetic efferent neural activity.

The motile functions of the stomach can be divided into two general classes of activity:

* Adjustments to the volume of the meal
* Peristaltic mixing and propulsion.

Furthermore, these activities can be categorized with respect to the anatomical region [36,38]. The proximal stomach behaves as a reservoir and accomodates to its volume by modulating tonic contractile activity. In contrast, the distal stomach generates phasic peristaltic waves of contraction for mixing, grinding and propelling the contents (Fig. 61.13).

61.7.2 Adjustments to the Meal –
Isotonic Relaxation and Contraction

It is important for normal digestion that the stomach be able to store food, initiating the digestive process and allowing the passage of food only when the intestine is prepared to receive it. Fundamental to serving as a reservoir is the ability of the muscle to undergo isotonic relaxation and contraction as the contents of the stomach radically change. During feeding, the stomach can greatly expand in

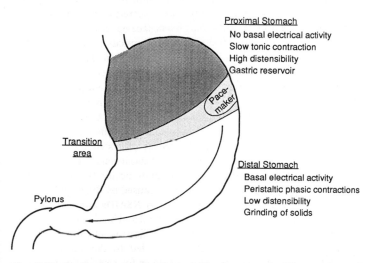

Fig. 61.13. Regional motor functions within the stomach. The proximal stomach shows no basal electrical activity, but it serves as a highly "distensible" gastric reservoir via tonic contraction and relaxation that accommodates the filling and emptying associated with a meal. The distal stomach has basal electrical slow waves that originate from a pacemaker region and lead to phasic peristaltic contractions, as well as the more vigorous grinding patterns that occur in the distal antrum

volume with virtually no change in the steady-state intragastric pressure. The accommodation of large volume changes comes about by active relaxation of smooth muscle in the more proximal half of the stomach, and has been called receptive relaxation, or isotonic relaxation. Receptive relaxation is mediated by a vago-vagal reflex, involving vagal afferent signals that travel from gastric stretch receptors to the central nervous system, and return via efferent signals to the muscle also through the vagus. The complementary half of this total reflex activity involves the gradual decrease in volume via isotonic contraction of the gastric smooth muscle and restoration of tone as the stomach empties. Because of the role of the vagus, isotonic relaxation and contraction are predictably abolished by vagotomy; however, since it is not very much altered by atropine, the reflex response is probably not cholinergic, but the specific neurotransmitter is unknown [38].

61.7.3 Peristalsis and Mixing

As food is ingested it tends to remain in relatively unmixed layers within the proximal half of the stomach because of the absence of mixing waves. Integrated and regular waves of contractile activity are a fundamental property of the smooth muscle in the distal half of the stomach. These contractions are peristaltic in nature and serve as mixing waves, but they also tend to move the contents toward the pylorus. The contractile waves originate in mid-stomach with very little force, and gradually build up in strength and velocity as they migrate through the antrum to the pylorus. High pressures are developed in the region of the pylorus, forcing a small amount of the contents through the sphincter into the duodenum and propelling the remainder back into the stomach, thus serving to pulverize as well as thoroughly mix the contents (Fig. 61.14). The mixing waves of peristaltic contraction are initiated by the intrinsic electrical activity of the smooth muscle cells and occur with regular frequency, e.g., about three per minute in humans. The frequency of the peristaltic pressure wave is based on the pacesetter activity in smooth muscle cells of the corpus [54]. Comparisons of the electrical activity of progressive regions from proximal to distal stomach are shown in Fig. 61.15. For contraction to occur, the depolarizing slow wave must reach the threshold to produce a spike potential [54]. The amplitude or strength of contraction is a function of the magnitude of the plateau of the slow wave of depolarization and the frequency of spike potentials during the plateau. In the fasted stomach, contractile activity is almost imperceptible, and although slow waves of depolarization occur with about the same frequency as in the fed state, they are not sufficient to trigger spikes and the contractile event. Thus, rhythmicity of the slow waves is an intrinsic property of the smooth muscle cells, but the intensity of activity and translation into peristaltic waves of contraction depend on the digestive conditions, such as food within the stomach. The neural and hormonal activities associated with the early stages of digestion markedly alter the amplitude of the slow waves,

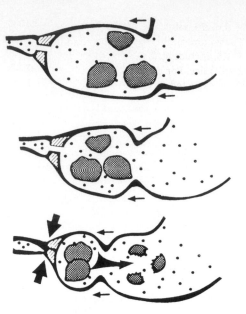

Fig. 61.14. Antropyloric propulsion/retropulsion mechanism. At the initiation of a peristaltic contraction in mid-stomach the pylorus is open, but its luminal diameter is much smaller than that of the antrum. The ring of contraction propagating through the antrum develops pressure to propel the antral contents toward the pylorus, where liquid and small particles pass through the limited opening. As the wave of contraction approaches the pylorus, the pylorus closes and the terminal antrum produces a mass contraction that retropulses the antral contents against the constricted antral ring. This forceful ejection of the antral contents back into the body of the stomach produces a shearing effect that helps to fragment food particles. (From [36])

the generation of spike potentials, and the consequent force of peristaltic contraction.

Stimulation of the vagus nerve increases the force and frequency of gastric contraction. The neurotransmitter for this vagal effect is acetylcholine, and it is totally blocked by atropine. The local effect of acetylcholine on gastric smooth muscle cells is to increase the amplitude and duration of the slow waves of depolarization, promoting spikes and enhancing the peristaltic force of contraction, as exemplified in Fig. 61.16a [54]. Stimulation of the sympathetic nerves inhibits gastric motility, most probably with norepinephrine as transmitter (e.g., Fig. 61.16c). Norepinephrine decreases the amplitude and duration of the rhythmic slow waves in the smooth muscle cells. Gastric smooth muscle is also influenced by certain digestive hormones. For example, gastrin and CCK have been shown experimentally to stimulate contractile activity (e.g., Fig. 61.16b), while secretin and GIP decrease it. However, it is uncertain whether, and to what extent, these modulating effects operate at the normal circulating levels of the hormones [38].

61.7.4 Gastric Emptying

The emptying of the stomach following a meal is regulated by a number factors, generally related to

1254

optimizing the ultimate digestion and absorption of the material. The volume of the stomach after a meal may be as much as 1–1.5, including the ingested solids and liquids and the gastric juice. For a typical balanced meal this volume might be expected to empty into the duodenum in

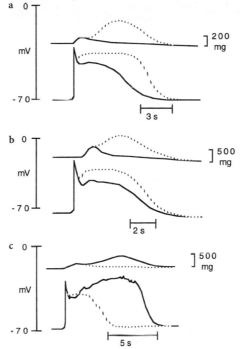

Fig. 61.16a–c. Effects of **a** acetylcholine, **b** gastrin, and **c** norepinephrine on electromechanical coupling of dog gastric smooth muscle. For each experiment the *solid line* represents the control trace and the *dotted line* represents treatment with the indicated drug. Contractile force shown in *top traces*; intracellular action potential in *bottom traces*. (Adapted from [54])

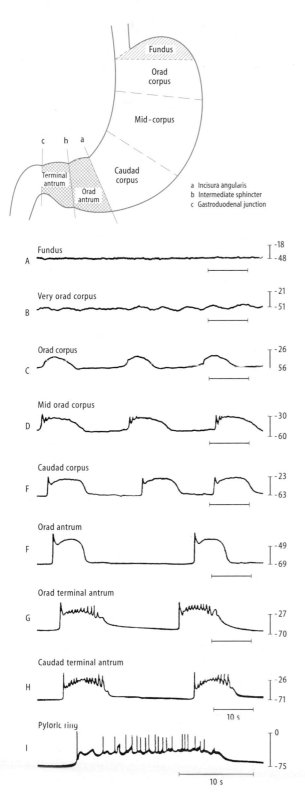

about 3 h. The emptying is a function of the force of the peristaltic wave moving through the antrum and squeezing particles of food through the resistance of the pyloric constriction. Code et al. [13] have provided a classic description of this process in what has been called the antropyloric propulsion/retropulsion mechanism, as depicted in Fig. 61.14. In fact, the rate of emptying depends on the volume of material in the stomach, the physical state of the contents, and the chemical nature of the food itself. Other factors being equal, the greater the volume within the stomach, the more rapid the emptying. This response is an intrinsic property of gastric smooth muscle and the coordination provided by the myenteric plexus. As would be expected, the physical consistency is an important

◄─────────────────────────

Fig. 61.15. Intracellular recordings of electrical activity of smooth muscle strips from various regions of a dog stomach. Note that rhythmic slow waves are absent in fundus, weak in the orad corpus, and increase in strength toward the antrum. Spikes of action potentials are seen in mid-corpus, and their frequency, on the plateau of slow waves, increases in the terminal portions of the antrum, where contractile force is greatest. In these records from isolated muscle strips there are slight regional differences in the intrinsic slow wave frequency. In the intact stomach, slow waves have the same frequency in all parts of the stomach, because they are all driven by the same pacemaker. (From [54])

1255

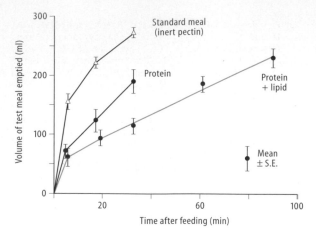

Fig. 61.17. Effect of protein and fat on the rate of emptying of the human stomach. Subjects were fed 300-ml liquid meals. (From [5])

determinant. Liquids empty faster than solids, and efficient exit occurs when the particle diameter is 1 mm or less. The rate of gastric emptying is also greatly influenced by the chemical nature of the material reaching the duodenum, especially the lipid content, pH and osmolarity of the duodenal contents [36]. Food rich in carbohydrate clears the stomach relatively quickly, a protein-rich diet more slowly, and emptying is slowest for a meal rich in fat (Fig. 61.17).

The precise neural and hormonally mediated pathways controlling the rate of gastric emptying are poorly defined, but the process clearly involves duodenal and jejunal receptors sensing the chemical nature of the material reaching the duodenum. As fats reach the duodenum CCK is released, and apart from its hepatopancreatic effects CCK induces contraction of the pyloric musculature, reducing the exit probability for gastric particles. Other duodenal influences, such as low pH and altered osmolarity of the contents, are known to slow gastric emptying possibly via inhibitory effects of secretin and GIP on motility.

References

1. Allen A (1989) Gastrointestinal mucus. In: Schultz SG, Forte JG (eds) Handbook of physiology, section 6: the gastrointestinal system, vol III. Salivary, gastric, pancreatic and hepatobiliary secretion, American Physiological Society, Bethesda, pp 359–382
2. Beaumont W (1847) The physiology of digestion. Chauncey Goodrich, Burlington, VT
3. Berglindh T (1984) The mammalian gastric parietal cell in vitro. Annu Rev Physiol 46:377–392
4. Black JW, Duncan WAM, Durant, Ganellin CR, Parsons ME (1972) Definition and antagonism of histamine H2-receptors. Nature 236:385–390
5. Brooks FP (1974) Integrative lecture: response of the GI tract to a meal. Under graduate teaching project. American Gastroenterological Association
6. Brown JC, Buchan AMJ, McIntosh CHS, Pederson RA (1989) Gastric inhibitory polypeptide. In: Schultz SG, Makhlouf GM (eds) Handbook of physiology, section 6: the gastrointestinal system, vol II, neural and endocrine biology. American Physiological Society, Bethesda, pp 403–430
7. Castle WB (1975) Observations on the etiologic relationship of achylia gastrica to pernicious anemia. Am J Med Sci 178:748–764
8. Chew CS (1989) Intracellular activation events for parietal cell hydrochloric acid secretion. In: Schultz SG, Forte JG (eds) Handbook of physiology, section 6: the gastrointestinal system, vol III. Salivary, gastric, pancreatic and hepatobiliary secretion. American Physiological Society, Bethesda, pp 255–266
9. Chew CS, Brown MR (1987) Histamine increases phosphorylation of 27- and 40-kDa parietal cell proteins. Am J Physiol 253:G823–G829
10. Chew CS (1986) Cholecystokinin, carbachol, gastrin, histamine, and forskolin increase $[Ca^{2+}]_i$ in gastric glands. Am J Physiol 250:G814–G823
11. Chey WY, Chang T (1989) Secretin. In: Schultz SG, Makhlouf GM (eds) Handbook of physiology, section 6: the gastrointestinal system, vol II, neural and endocrine biology. American Physiological Society, Bethesda, pp 359–402
12. Code CF (1965) Histamine and gastric secretion: a late look. Fed Proc 24:1311–1321
13. Code CF, Carlson HC (1967) Motor activity of the stomach. In: Code CF (ed) Handbook of physiology, section 6: the alimentary canal, vol IV. American Physiological Society, Washington DC, pp 1903–1916
14. Davenport HW (1971) Physiology of the digestive tract. Year Book Medical Publisher, Chicago
15. Debas HT (1989) Peripheral regulation of gastric acid secretion. In: Johnson LR (ed) Physiology of the gastrointestinal tract, vol 2, 2nd edn. Raven, New York, pp 931–945
16. Dockray GJ, Gregory RA (1989) Gastrin. In: Schultz SG, Makhlouf GM (eds) Handbook of physiology, section 6: the gastrointestinal system, vol II, neural and endocrine biology. American Physiological Society, Bethesda, pp 311–336
17. Donaldson RM (1989) Intrinsic factor and the transport of cobalamin. In: Johnson LR (ed) Physiology of the gastrointestinal tract, vol 2, 2nd edn. Raven, New York, pp 959–973
18. Du Val JW, Saffouri B, Weir GC, Walsh JH, Arimura A, Makhlouf GM (1981) Stimulation of gastrin and somatostatin secretion from the isolated rat stomach by bombesin. Am J Physiol 241:G242–G247
19. Flemstrom G, Garner A (1989) Secretion of bicarbonate by gastric and duodenal mucosa. In: Schultz SG, Forte JG (eds) Handbook of physiology, section 6: the gastrointestinal system, vol III. Salivary, gastric, pancreatic and hepatobiliary secretion. American Physiological Society, Bethesda, pp 309–326
20. Forte JG, Black JA, Forte TM, Machen TE, Wolosin JM (1981) Ultrastructural changes related to functional activity in gastric oxyntic cells. Am J Physiol 241:G349–G358
21. Forte JG, Hanzel DK, Urushidani T, Wolosin JM (1989) Pumps and pathways for gastric HCl secretion. Ann NY Acad Sci 574:145–158
22. Forte TM, Machen TE, Forte JG (1977) Ultrastructural changes in oxyntic cells associated with secretory function: a membrane recycling hypothesis. Gastroenterology 73:941–955
23. Forte JG, Soll AH (1989) Cell biology of hydrochloric acid secretion. In: Schultz SG, Forte JG (eds) Handbook of physiology, section 6: the gastrointestinal system, vol III. Salivary, gastric, pancreatic and hepatobiliary secretion. American Physiological Society, Bethesda, pp 207–228
24. Forte JG, Wolosin JM (1987) HCl secretion by the gastric oxyntic cell. In: Johnson LR (ed) Physiology of the gastrointestinal tract, vol 2, 2nd edn. Raven, New York, pp 853–863
25. Gregory RA, Tracy HJ (1961) The preparation and properties of gastrin. J Physiol (Lond) 156:523–543
26. Helander HF (1981) The cells of the gastric mucosa. Int Rev Cytol 70: 217–289

27. Hentschel E, Brandstätter G, Dragosics B, Hirschl AM, Nemec H, Schütze K, Taufer M, Wurzer H (1993) Effect of ranitidine and amoxicillin plus metronidazole on the eradication of helicobacter pylori and the recurrence of duodenal ulcer. N Engl J Med 328:308–312

28. Hirst B (1989) Cell biology of hydrochloric acid secretion. In: Schultz SG, Forte JG (eds) Handbook of physiology, section 6: the gastrointestinal system, vol III. Salivary, gastric, pancreatic and hepatobiliary secretion. American Physiological Society, Bethesda, pp 279–308

29. Hogben CAM, Kent TH, Woodward PA, Sill AJ (1974) Quantitative histology of the gastric mucosa: man, dog, cat, guinea pig and frog. Gastroenterology 67:1143–1154

30. Ito S (1967) Anatomic structure of the gastric mucosa. In: Code CF (ed) Handbook of physiology, section 6: alimentary canal, vol II. Secretion. American Physiological Society, Washington DC, pp 705–779

31. Ito S (1987) Functional gastric morphology. In: Johnson LR (ed) Physiology of the gastrointestinal tract, vol 2, 2nd edn. Raven, New York, pp 817–851

32. Karam SM (1993) Dynamics of epithelial cells in the corpus of the mouse stomach. IV. Bidirectional migration of parietal cells ending in their gradual degeneration and loss. Anat Rec 236:314–332

33. Karam SM, Leblond CP (1993) Dynamics of epithelial cells in the corpus of the mouse stomach. I. Identification of proliferative cell types and pinpointing of the stem cell. Anat Rec 236:259–279

34. Karam SM, Leblond CP (1993) Dynamics of epithelial cells in the corpus of the mouse stomach. III. Inward migration of neck cells followed by progressive transformation into zymogenic cells. Anat Rec 236:297–313

35. Makhlouf GM (1981) Electrolyte composition of gastric secretion. In: Johnson LR (ed) Physiology of the gastrointestinal tract. Raven, New York, pp 551–556

36. Malagelada JR, Azpiroz F (1989) Determinants of gastric emptying ans transit in the small intestine. In: Schultz SG, Wood JD (eds) Handbook of physiology, section 6: the gastrointestinal system, vol I. Motility and circulation. American Physiological Society, Bethesda, pp 909–937

37. Marshall BJ, Warren JR (1984) Unidentified curved bacilli in the stomach of patients with gastritis and peptic ulceration. Lancet 1:1311–1315

38. Meyer JH (1987) Motility of the stomach and gastroduodenal junction. In: Johnson LR (ed) Physiology of the gastrointestinal tract, vol 2, 2nd edn. Raven, New York, pp 613–629

39. Muallem S, Burnham C, Blissard D, Berglindh T, Sachs G (1985) Electrolyte transport across basolateral membrane of parietal cells. J Biol Chem 260:6644–6653

40. Muallem S, Fimmel CJ, Pandol SJ, Sachs G (1986) Regulation of free cytosolic Ca^{2+} in the peptic and parietal cells of the rabbit gastric gland. J Biol Chem 261:2660–2667

41. Neutra MR, Forstner JF (1987) Gastrointestinal mucus: synthesis, secretion and function. In: Johnson LR (ed) Physiology of the gastrointestinal tract vol 2, 2nd edn. Raven, New York, pp 975–1009

42. Nordgren B (1963) The rate of secretion and electrolyte content of normal gastric juice. Acta Physiol Scand Suppl 202, pp 1–83

43. Paradiso AM, Tsien RY, Demarest JR, Machen TE (1987) Na-H and $Cl-HCO_3$ exchange in rabbit oxyntic cells using fluorescence microscopy. Am J Physiol 253:C30–C36

44. Quigley EMM, Turnberg LA (1987) pH of the microclimate lining human gastric and duodenal mucosa in vivo: studies in control subjects and duodenal ulcer patients. Gastroenterology 92:1876–1884

45. Raufman JP (1992) Gastric chief cells: receptors and signal transduction mechanisms. Gastroenterology 102:699–710

46. Reenstra WW, Forte JG (1990) Characterization of K^+ and Cl^- conductances in apical membrane vesicles from stimulated rabbit oxyntic cells. Am J Physiol 259:G850–G858

47. Rehfeld JF (1989) Cholecystokinin. In: Schultz SG, Makhlouf GM (eds) Handbook of physiology, section 6: the gastrointestinal system, vol II, neural and endocrine biology. American Physiological Society Bethesda, pp 337–358

48. Robert A (1977) Cytoprotection by prostaglandins. Gastroenterology 77:761–767

49. Sachs G, Kaunitz J, Mendlein J, Wallmark B (1989) Biochemistry of gastric acid secretion: H^+-K^+-ATPase. In: Schultz SG, Forte JG (eds) Handbook of physiology, section 6: the gastrointestinal system, vol III. Salivary, gastric, pancreatic and hepatobiliary secretion. American Physiological Society Bethesda, pp 229–253

50. Saffouri B, DuVal JW, Arimura A, Makhlouf GM (1984) Effects of vasoactive intestinal peptide and secretin on gastrin and somatostatin secretion in the perfused rat stomach. Gastroenterology 86:839–842

51. Samloff IM (1989) Peptic ulcer: the many proteinases of agression. Gastroenterology 96:586–595

52. Soll AH (1978) The interaction of histamine with gastrin and carbamylcholine on oxygen uptake by isolated ammmalian parietal cells. J Clin Invest 61:381–389

53. Soll AH, Berglindh T (1987) Physiology of isolated gastric glands and parietal cells: receptors and effectors regulating function. In: Johnson LR (ed) Physiology of the gastrointestinal tract, vol 2, 2nd edn. Raven, New York, pp 883–909

54. Szurszewski JH (1987) Electrical basis for gastrointestinal motility. In: Johnson LR (ed) Physiology of the gastrointestinal tract, vol 2, 2nd edn. Raven, New York, pp 383–422

55. Taggart RT, Samloff IM (1987) Immunochemical, electrophoretic, and genetic heterogeneity of pepsinogen I. Gastroenterology 92:143–150

56. Walsh JH (1989) Bombesin-like peptides. In: Schultz SG, Makhlouf GM (eds) Handbook of physiology, section 6: the gastrointestinal system, vol II, neural and endocrine biology. American Physiological Society, Bethesda, pp 587–610

57. Wolosin JM, Forte JG (1984) Stimulation of oxyntic cell triggers K^+ and Cl^- conductances in apical (H^++K^+)-ATPase membrane. Am J Physiol 246:C537–C545

58. Yamada T, Chiba T (1989) Somatostatin. In: Schultz SG, Makhlouf GM (eds) Handbook of physiology, section 6: the gastrointestinal system, vol II, neural and endocrine biology. American Physiological Society, Bethesda, pp 431–453

59. Yao X, Thibodeau A, Forte (1993) Ezrin-calpain I interactions in gastric parietal cells. Am J Physiol 265 (Cell Physiol 34):C36–C46

62 Function of the Intestine

A.D. Conigrave and J.A. Young

Contents

62.1 Primary Functions of the Small and Large Intestines

In addition to its motor functions (see Chap. 60), the small intestine has the following primary functions:

- Enzymatic digestion of ingested nutrients
- Absorption of water, electrolytes and the products of nutrient digestion
- Secretion of mucus and digestive juices
- Exclusion of indigestible material and its delivery into the colon.

These functions require the controlled delivery of chyme and liquids from the distal stomach into the duodenum. They also depend on the controlled addition of pH-neutralising fluid and enzymes secreted by the exocrine pancreas and bile secreted by the liver and bile ducts. Finally, they depend on the controlled progression of luminal contents through the small intestine, as determined by intrinsic intestinal motor activity, modulated by the activity of the intrinsic and extrinsic nerves of the intestine and the various hormones secreted by it.

The primary (non-motor) function of the large intestine is the absorption of salt and water from the faeces, which occurs in the proximal colon. This reduces daily faecal water content from 1000 to 100 ml, containing less than 5 mmol Na^+/l, 9 mmol K^+/l and 2 mmol Cl^-/l; in other words, approximately 80% of the water and 90% of the electrolytes which enter the colon each day are absorbed. In addition, colonic crypts have a secretory function, forming an isotonic juice rich in K^+ and HCO_3^-.

62.1.1 Optimising the Primary Functions of the Intestine

Enteroendocrine Cells. Intestinal function is optimised in a number of important ways. These include:

- Coupling of motility with digestion and absorption
- Neuroendocrine control of gastric emptying and small and large bowel motility
- Neuroendocrine control of the release of bile and the secretion of pancreatic juice
- Endocrine control of the hepatic storage of absorbed nutrients.

The non-motor primary functions of the intestine are performed in the intestinal lumen (the early stages of digestion) and by its lining epithelium (absorption of water, electrolytes and nutrients in the later stages of digestion, and the exclusion of indigestible material). The optimising

R. Greger/U. Windhorst (Eds.)
Comprehensive Human Physiology, Vol. 2
© Springer-Verlag Berlin Heidelberg 1996

functions, however, are due to a range of muscular, neural and endocrine activities. Neural control involves reflexes dependent on sensory nerves arising in the subepithelial space and the motor ganglion cells of the enteric nervous system, as well as central co-ordination involving the autonomic nervous system, which is largely responsible for the co-ordination of digestion and absorption with gastrointestinal motility. Hormones secreted by enteroendocrine cells in the stomach and small intestine also act to control gastric emptying and the coupling of digestive and absorptive functions in the small intestine with motility, and to modulate the release of insulin from the β-cells in the islets of Langerhans so as to facilitate the hepatic storage of absorbed glucose as it passes from the intestine to the liver in the hepatic portal veins. Finally, endocrine cells in the small intestine may release satiety hormones that suppress appetite.

62.2 Structure–Function Associations in the Small Intestine

Villi and Brush-Border Membrane. The function of the small intestine is greatly enhanced by its structural organisation. The absorptive and membrane-dependent digestive surface is amplified about 500 times when compared to a cylinder of the same length and internal diameter because of extensive folding at three different levels (see Figs. 62.1, 62.2):

- The epithelium is folded at a macroscopic level to create valve-like flaps known as the valvulae conniventes.
- The epithelium and the submucosa that supports it are organised into myriads of flattened, leaf-like projections known as the intestinal villi.
- The luminal plasma membrane of each enterocyte consists of a densely packed system of microvilli known as the brush-border membrane.

62.3 The Crypt–Villus Axis

Enterocytes, Enteroendocrine Cells and Goblet Cells. As indicated above, the adult small intestinal epithelium is a cellular monolayer, much of it organised into villi. The epithelial lining of each villus encloses a core of connective tissue (Fig. 62.2) that supports blood vessels, a major lymph vessel (the central lacteal) and sensory and secretomotor nerves, which terminate near the villus epithelium. The major cell type forming the epithelial covering of the villi, the enterocyte, provides the absorptive surface for nutrients, electrolytes and water. The enterocytes also express digestive enzymes in their luminal plasma (brush-border) membranes which operate on the breakdown products of luminal digestion. In addition, there are two other cell types present in the epithelium covering the villi, enteroendocrine cells, most of which sense the presence of nutrients in the lumen and respond by the release of hormones or parahormones, and goblet cells, which secrete mucus to protect the epithelial surface.

Crypts of Lieberkühn. Interspersed amongst the villi are the crypts of Lieberkühn (Fig. 62.2), simple tubular glands lying in the subepithelial space which secrete water and electrolytes to form the intestinal juice [64]. In the duodenum there are about 14 crypts surrounding the base of each villus, whereas in the ileum there are only about six. The stem cell precursors for the villus enterocytes are found in the epithelium lining the base of the crypts, so that each villus is supplied with cells migrating up from more than one crypt (see Fig. 62.2). There are no villi in the colon, but crypts remain conspicuous.

Paneth Cells. The three cell types seen in the epithelium covering the villi are also found in the crypts but, in addition, a fourth cell type, the Paneth cell, is found at the crypt base in humans and most other mammals [65]. Paneth cells are believed to play an important role in antimicrobial defence. They are characterised by the presence of large

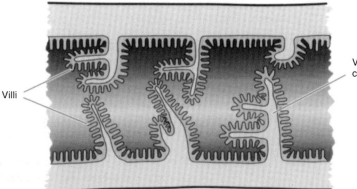

Fig. 62.1. Longitudinal section through the small intestine showing the gross organisation of the epithelium into valve-like flaps, the valvulae conniventes, and leaf-like projections, the villi

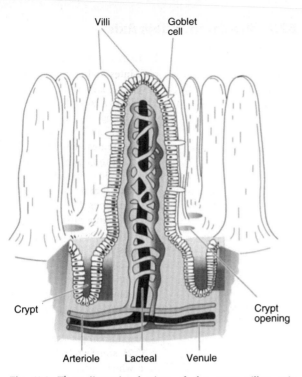

Villi   Goblet cell

Crypt

Crypt opening

Arteriole   Lacteal   Venule

Fig. 62.2. Three-dimensional view of the crypt–villus axis. Epithelial cells arise from stem cells in the crypts and migrate upwards at a rate of about 1–2 cell positions per h. After leaving the mouth of a crypt, they reach the base of a neighbouring villus and then migrate upwards over the its surface. In the duodenum there are about 14 crypts, which are really just simple tubular glands, adjacent to any given villus, but in the ileum there are only about six. The central lacteal of a villus, surrounded by a capillary bed, is shown draining into an intestinal lymph duct

eosinophilic secretion granules and are known to release lysozyme, immunoglobulin A (IgA) and potent antimicrobial peptides that are structurally related to the defensins secreted by neutrophils and macrophages [47]. Two closely related defensin-like peptides, cryptdins 1 and 2 (both consisting of 35 amino acid residues), have recently been isolated from Paneth cells [20]. Apical secretion granules in Paneth cells are also known to contain EGF (epidermal growth factor), phospholipase A$_2$, dipeptidyl-peptidase, and TNF (tumour necrosis factor). The presence of some of these constituents suggests that Paneth cells also have a digestive function (perhaps digesting bacterial membranes) and the presence of others suggests a local role in the control of epithelial cell division. Exocytosis of apical secretory vesicles from Paneth cells is controlled by the bacterial flora in the intestine: the cells are quiescent in the germ-free state and degranulation occurs when the intestine is inoculated with bacteria. It appears that cholinergic nerves acting on muscarinic receptors are involved in the transduction of this process [47].

Migration. All the cell types found in crypts and villi arise from a small pool of common stem cells (located within the crypts), which themselves are the offspring of a common progenitor cell [42]. Enterocytes, some enteroendocrine

cells and goblet cells arise as descendants of crypt daughter cells and migrate in vertical bands up the villi to reach an exclusion zone at the tip of the villus from which they are shed into the lumen at a rate of approximately 10^{11} cells per day (in humans) for the entire small intestine [43]. The details of this migration process are not well understood. The cells adhere to a thin basement lamina, but whether they migrate over it or are drawn upwards by its movement is not clear. One clue suggesting that movement of the basement lamina may be involved is the observation that subepithelial myofibroblasts co-migrate with the epithelial cells. The whole process takes about 5 days in humans. Although most enteroendocrine cells migrate onto the villi, some migrate to the lowermost portions of the crypts. Paneth cells tend to occupy only the very bases of the crypts, where they turn over at a much slower rate (about 4 weeks) than the other progeny of the crypt stem cells.

Oligosaccharides and Oligopeptidases. The functional role played by the enterocytes changes dramatically as the process of migration occurs. In the crypts, the enterocytes are secretory, whereas on the villi they are absorptive. In the crypts, the enterocytes express very low levels of oligosaccharidases and oligopeptidases in their brush-border membranes, but after migration onto the villi they express high levels of these enzymes. The oligosaccharidases include maltase, isomaltase-sucrase (which is a single polypeptide gene product that is further processed by limited proteolysis within the brush-border membrane to yield separate isomaltase and sucrase proteins) and lactase. Lactase (150 kDa), the enzyme that hydrolyses the milk disaccharide lactose to its constituent monosaccharides glucose and galactose is an integral glycoprotein of the brush-border membrane which is normally not expressed in human enterocytes after weaning. Atypically, it persists in adult Northern Europeans. Lactase deficiency in human infants can arise either from a failure of mRNA transcription or from a partial block in the transport of the nascent polypeptide from the Golgi to the plasma membrane [37].

Enteroendocrine cells represent less than 1% of terminally differentiated cells in the intestinal mucosa and are found in both the small and large intestine. They and the hormones they secrete are discussed in Sect. 62.6.

Biology of Crypt–Villus Axis. Studies on the biology of the crypt–villus axis have drawn on techniques originally developed in a variety of other disciplines. These include the following:

- The development of suitable transgenic and chimeric mice strains, which has permitted the precise identification of cells of known lineage [29]
- Advanced cellular separation techniques, which have allowed the isolation of subsets of crypt and villus cells after their separation from the epithelium [27]
- Immunohistochemistry, which has permitted precise identification of the various cells that synthesise peptide

1261

hormones and amines so that their distribution within the gastrointestinal tract can be mapped

- Labelled cDNA probes, which have been used successfully to test for the expression of mRNA encoding specific enzymes or transporters
- ^{3}H-labelled thymidine incorporation studies, which have been used to track the movement of cells from crypts to villi
- Advanced electrophysiological techniques, which have been used to identify the ion channels that support epithelial transport of ions and water in secretory and absorptive enterocytes
- Biochemical studies on the digestive enzymes, which have been used to elucidate the mechanisms involved in the small intestinal digestion of proteins, lipids and carbohydrates.

Na$^+$-Dependent Glucose Transporters and Pentose Carriers. Enterocytes are critical to the passage of electrolytes and water across the intestinal epithelium. In the crypts, they are equipped with ion transport machinery (pumps, transporters and channels) that promotes the secretion of an isotonic fluid. This machinery can be activated by secretomotor nerves arising in the submucous ganglia or by endocrine and paracrine agents. As the enterocytes pass along the crypt–villus axis, they lose the ability to secrete electrolytes and water and acquire, instead, the ability to absorb them. For this to happen, the cell must selectively express transporters and/or receptors appropriate for absorption and suppress the expression of those specific to secretion. Simultaneously, there is an increase in the expression of (Na$^+$ + K$^+$)-ATPase in the basolateral plasma membranes [66] and, in the luminal membranes, of carriers for glucose and galactose (the sodium-dependent glucose transporter SGLT-1) and for fructose (the pentose carrier) [45]. The expression of at least some of the transporters and brush-border enzymes found in the enterocytes is modulated by nutrient levels. For example, the uptake of glucose by the intestinal brush-border is activated by increased dietary carbohydrate intake resulting in increased numbers of transporters on the enterocytes, as assessed by phlorrhizin-binding site density [22].

Neural Activity. Absorption of fluid and electrolytes by villus enterocytes is also controlled by neural activity as well as by endocrine and paracrine agents. The submucous plexus is important in the control of net electrolyte and fluid movements. Neural activity can promote net absorption by stimulating the rate of absorption by the villus. Alternatively, it can promote secretion by inhibiting absorption by the villus and by evoking crypt secretion. Normally, there is net absorption in the small intestine. This implies that the quantity of fluid and electrolytes secreted by the crypt enterocytes (plus the products of the compound glands that drain into the small intestinal lumen) is exceeded by the quantity absorbed by the villus enterocytes (which includes the bulk of ingested water and electrolytes).

62.4 Digestion, Absorption and Secretion

Most of the nutrients actually absorbed by the enterocytes (i.e., monosaccharides, amino acids and fatty acids), are derived by digestion of dietary polysaccharides, proteins and triacylglycerols, since these complex substances cannot normally be absorbed intact across the intestinal epithelium. Digestion in the small intestine occurs simultaneously in three separate chemical phases:

- The aqueous phase, in which the initial hydrolysis of proteins and polysaccharides in the luminal contents occurs
- The oil–water interface formed by emulsification of dietary fat in the stomach and duodenum, at which the hydrolysis of dietary triacylglycerols, phospholipids and cholesterol esters occurs
- The brush-border–aqueous interface, at which the hydrolysis of oligosaccharides and oligopeptides on the luminal surface of the villus enterocytes occurs.

These digestive processes take place optimally at neutral pH. Because the chyme entering the duodenum is strongly acidic (pH $\leq$ 3), it is necessary for the pH of the intestinal contents to be raised (and buffered) by mixture with an alkaline fluid. In the small intestine, this alkaline fluid is derived from two main sources. First, in the duodenum, HCO$_3^-$-rich fluids are secreted by the exocrine pancreas and the biliary tree, as well as by duodenal submucosal glands (Brunner's glands), in response to the release of secretin and other peptide hormones from specialised enteroendocrine cells. Second, a HCO$_3^-$-containing fluid is secreted by the intestinal crypts of the small and large intestine.

62.4.1 Secretion and Absorption of Salt and Water

Each day, about 8–9 l water (and 50–100 g electrolytes) undergo net absorption by the small intestine and about 1.0 l by the large intestine. Of this fluid, only 1.5 l comes from the diet. The remainder consists of 1.0 l saliva, 1.5 l gastric juice, 3.0 l intestinal secretion, 2.0 l pancreatic juice and 0.6 l bile. Water diffuses rapidly across the lining epithelium, the rate and direction of flow being determined by the osmolality difference between the intestinal lumen and the plasma. Thus, the basis of water transport by the small intestine is the ability of the intestinal mucosa to control the osmolality of the intestinal contents. This is determined by the balance of two opposing processes in the small bowel: one is the absorption of NaCl and other osmotically active compounds by the epithelium covering the villi, and the other is the secretion of K$^+$ and/or anions (Cl$^-$ and HCO$_3^-$, accompanied passively by Na$^+$) by the crypt epithelium. Under normal circumstances, the absorptive processes are quite active and only about 1.0 l water enters the colon each day from the small intestine. In patients with cholera, where the secretory process is greatly stimulated, up to 20 l fluid enter the colon each day from the small intestine, thereby causing a massive, life-threatening dehydration due to the fluid loss through diarrhoea.

62.4.1.1 Secretion Mechanisms

Transport Elements. As indicated above, secretion of salt and water in the intestine is a function of enterocytes lining the crypts, and the bulk of the fluid secreted in this way is reabsorbed by the enterocytes lining the villi (small intestine) and the cells of the surface epithelium (colon). For technical reasons, secretion by the crypts has been difficult to study and important insights into the underlying mechanisms first came from studies on colonic carcinoma cell lines that have retained their secretory capacity. The principal secretory mechanism operating in crypt cells in both the small and large intestine depends on a furosemide-sensitive Na^+-K^+-$2Cl^-$ co-transporter in the basolateral membrane, a K^+ channel in the basolateral membrane, and several Cl^- channel types in the luminal membrane, one of which is powerfully activated by cyclic adenosine monophosphate (cyclic AMP). These transport elements drive Cl^- secretion transcellularly, and Na^+ accompanies the Cl^- current paracellularly. The mechanism is thus similar to the common secretory mechanism seen in some salivary glands (see Chap. 64) and in the rectal salt gland of the shark.

Cystic Fibrosis and Secretory Diarrhoeas. It appears that functional Cl^- channels are lost when crypt secretory cells are transformed into enterocytes on the villi or on the surface colonic epithelium, so that the cells become absorptive rather than secretory. It is of interest to note that the mRNA for the CFTR Cl^- channel (cystic fibrosis transmembrane conductance regulator), the transcript of the "healthy" or "wild-type" form of the cystic fibrosis gene, is expressed in colonic and small intestinal crypt cells but not in colonic surface epithelial cells or villus enterocytes [59] and that in cystic fibrosis, when the gene product is defective, crypt cell secretion does not occur [6]. It is also of interest to note that in many secretory diarrhoeas associated with microbial toxins (e.g., cholera), Cl^- secretion is strongly stimulated as a result of the sustained over-production of cyclic AMP within the enterocytes [23,24,62].

62.4.1.2 Absorption Mechanisms

Absorption in the intestine depends on the operation of one or other of five mechanisms, four involving active Na^+ absorption and one involving active K^+ absorption. The mechanisms whereby salt, water and nutrients are absorbed in the intestine are similar to those operating in the kidney.

Active Cellular Uptake of Glucose. Mechanism I (Fig. 62.3) is conspicuous in the jejunum. Na^+ ions cross the luminal plasma membrane of the epithelial cells from lumen to cytosol by means of one or other of an array of cotransport proteins that couple the energy of the Na^+ gradient to the active cellular uptake of glucose, amino acids or peptides.

Na^+/H^+ Exchangers. Mechanism II (Fig. 62.4) is also conspicuous in the jejunum. The luminal plasma membranes of the absorptive cells contain Na^+/H^+ exchangers providing a pathway for Na^+ (and HCO_3^-) to enter the cell across the luminal plasma membrane.

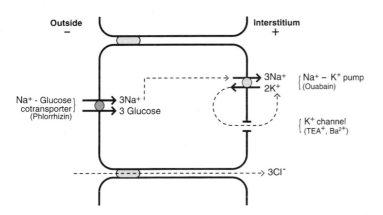

Fig. 62.3. Model for NaCl and glucose absorption dependent on the presence of a Na^+-coupled glucose carrier in the luminal brush-border membrane of an enterocyte (mechanism I). Na^+ ions enter the cytosol down an electrochemical gradient via a Na^+-coupled glucose carrier (blockable by phlorrhizin) and are pumped out to the interstitium by the basolateral (Na^++K^+)-ATPase. K^+ ions carried by the ATPase are recycled across the basolateral membrane via basolateral K^+ channels (blockable by tetraethylammonium [*TEA*], and Ba^{2+} ions). The transepithelial ion current is balanced by a paracellular flow of Cl^- ions from lumen to interstitium, driven by the transepithelial potential difference (lumen negative). The pathway for glucose movement across the basolateral membrane has not been shown. Three moles of NaCl and of glucose are absorbed for every mole of ATP hydrolysed. This mechanism, which is also seen in kidney proximal tubules, is prominent in the jejunum

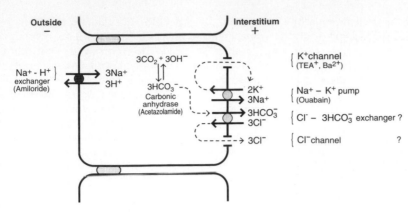

Fig. 62.4. Model for NaHCO₃ absorption dependent on the presence of a Na⁺/H⁺ exchanger in the luminal brush-border membrane of an enterocyte (mechanism II). Na⁺ ions enter the cytosol from the lumen via an amiloride-sensitive Na⁺/H⁺ exchanger and are pumped out to the interstitium via the basolateral (Na⁺+K⁺)-ATPase. The expulsion of H⁺ ions by the Na⁺/H⁺ exchanger leaves an excess of OH⁻ ions in the cytosol which react with CO₂ (cata-

lysed by carbonic anhydrase) to form HCO₃⁻ ions. K⁺ ions carried by the ATPase are recycled across the basolateral membrane via basolateral K⁺ channels (blockable by tetraethylammonium [*TEA*] and Ba²⁺ ions). The mechanism shown for the exit of HCO₃⁻ ions to the interstitium is hypothetical; it is at least as likely that an electrogenic Na⁺-2HCO₃⁻ carrier in the basolateral membrane is involved. This mechanism is prominent in the jejunum

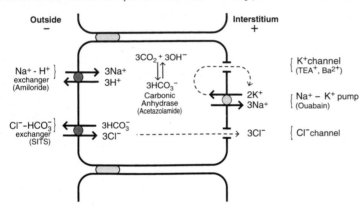

Fig. 62.5. Model for NaCl absorption dependent on the presence of paired Na⁺/H⁺ and Cl⁻/HCO₃⁻ exchangers in the luminal brush-border membrane of an enterocyte or a colonic epithelial cell (mechanism III). Na⁺ ions enter the cytosol from the lumen via an amiloride-sensitive Na⁺/H⁺ exchanger and are pumped out to the interstitium via the basolateral (Na⁺+K⁺)-ATPase. The expulsion of H⁺ ions by the Na⁺/H⁺ exchanger leaves an excess of OH⁻ ions

in the cytosol which react with CO₂ (catalysed by carbonic anhydrase) to form HCO₃⁻ ions. These, in turn, exit from the cytosol to the lumen via a Cl⁻/HCO₃⁻ exchanger (blockable by stilbene disulfonic acids). The Cl⁻ ions entering the cytosol then enter the interstitium via basolateral Cl⁻ channels. Three moles of NaCl are absorbed per mole of ATP hydrolysed. This mechanism is prominent in the ileum and distal colon

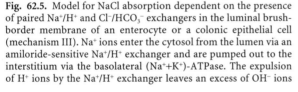
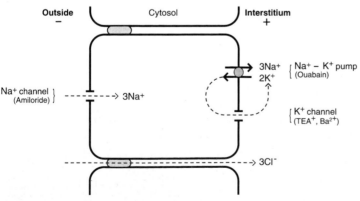

Fig. 62.6. Model for NaCl absorption dependent on the presence of an amiloride-sensitive Na⁺ selective channel in the luminal brush-border membrane of a colonic epithelial cell (mechanism IV). Na⁺ ions enter the cytosol from the lumen via the channel and are pumped out to the interstitium by the basolateral (Na⁺+K⁺)-ATPase. K⁺ ions carried by the ATPase are recycled across the basolateral membrane via basolateral K⁺ channels (blockable by

tetraethylammonium [*TEA*] and Ba²⁺ ions). The transepithelial ion current is balanced by a paracellular flow of Cl⁻ ions from lumen to interstitium, driven by the transepithelial potential difference (lumen negative). This mechanism, which is also seen in kidney collecting tubules, is prominent in the distal colon. Three moles of NaCl are absorbed per mole of ATP hydrolysed

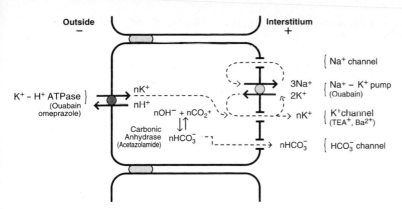

Fig. 62.7. Model for KHCO₃ absorption dependent on the presence of a (K⁺+H⁺)-ATPase in the luminal membrane of a colonic epithelial cell. K⁺ ions in the lumen are actively taken up into the cytosol by means of a (K⁺+H⁺)-ATPase in the luminal membrane (blockable in the colon by ouabain) and leave the cytosol for the interstitium via basolateral K⁺ channels, through which K⁺ ions taken up by the basolateral (Na⁺+K⁺)-ATPase also pass. The expulsion of H⁺ ions into the lumen by the (K⁺+H⁺)-ATPase leads to the accumulation of OH⁻ ions in the cytosol which react with CO₂ (catalysed by carbonic anhydrase) to form HCO₃⁻ ions, which, in turn, exit to the interstitium. The mechanism shown, i.e., a basolateral HCO₃⁻ channel, is purely hypothetical. One mole of KHCO₃ is absorbed for each mole of ATP hydrolysed by the (K⁺+H⁺)-ATPase. It should be noted that the basolateral (Na⁺+K⁺)-ATPase has no essential role to play in the process and there is no necessary stoichiometric relation between the rates of operation of the two ATPases. This has been indicated by the use of *n* rather than an integer to denote the activity of the (K⁺+H⁺)-ATPase relative to the (Na⁺+K⁺)-ATPase. *TEA,* tetraethylammonium

Paired Na⁺/H⁺ and Cl⁻/HCO₃-Exchangers. Mechanism III (Fig. 62.5) is conspicuous in the ileum and proximal colon. The luminal plasma membranes of the absorptive cells contain paired Na⁺/H⁺ and Cl⁻/HCO₃⁻ exchangers providing pathways for Na⁺ and Cl⁻ to enter the cell across the luminal plasma membrane.

Amiloride-Sensitive Na⁺ Channels. Mechanism IV (Fig. 62.6) is conspicuous in the distal colon. Na⁺ ions enter the cells via amiloride-sensitive Na⁺ channels in the luminal plasma membrane.

(K⁺ + H⁺)-ATPase. Mechanism V (Fig. 62.7) is conspicuous throughout the colon. K⁺ ions are actively absorbed from lumen to cytosol by means of a (K⁺+H⁺)ATPase in the luminal plasma membrane of the surface epithelial cells of the colon.

Ca²⁺ Absorption. Ca²⁺ is mainly absorbed in the duodenum. It diffuses across the luminal plasma membrane by an illdefined mechanism and is then pumped out of the cell across the basolateral plasma membrane by Ca²⁺-ATPases or Na⁺/Ca²⁺ exchangers. Vitamin D acts to increase Ca²⁺ absorption by increasing the permeability of the luminal plasma membrane to Ca²⁺ and the activity of the Ca²⁺-ATPases. It also increases the synthesis of a cytosolic Ca²⁺-binding protein, the function of which is to increase the rate of Ca²⁺ diffusion across the cytosol from one side of the cell to the other.

Water Absorption. As in the case of transepithelial secretion, in absorption, water follows net transepithelial solute flow by osmosis whenever the epithelium is water permeable, the flow being predominantly transcellular rather than paracellular.

62.4.1.3 Control of Absorption

Noradrenaline, Somatostatin, Mineralocorticoids, Glucocorticoids and Angiotensin. In contrast to secretion, absorption of salt and water takes place at a relatively steady rate. Consequently, neither intestinal peptide hormones nor autonomic nerve activity produces the dramatic changes in transport rate that are seen with secretion. Nevertheless, noradrenaline, acting on α-receptors, can increase the rate of fluid absorption in the intestine by a direct action on enterocytes, as can somatostatin and several peptide neurotransmitters. Long-term regulation of salt and water transport is brought about by mineralocorticoids (such as aldosterone) and glucocorticoids, as well as peptides such as angiotensin.

62.4.1.3.1 Nerual Control

Innervation of the Intestinal Mucosa. The mucosa is innervated by non-myelinated axons derived primarily from ganglion cells in the submucous plexus. Approximately half of the neurones supplying the mucosa are cholinergic (acting on muscarinic receptors) and half contain the peptide transmitter VIP (formerly known as vasoactive intestinal peptide); both are excitatory of secretion. The cholinergic neurones may be further subdivided on the basis of which peptide co-transmitters, if any, they also contain: some contain substance P, others CCK-8, and yet others, somatostatin, but in some no peptides have been identified thus far. The VIP-containing neurones may also contain other peptides. Some idea of which cells a particular neurone can activate can be gained from knowing what types of neuroreceptor they possess. The secretory enterocytes of the crypts have receptors for acetylcholine, VIP, substance P, and CGRP (calcitonin gene-related peptide), and the absorptive enterocytes of the intestinal villi have receptors for somatostatin, NPY (neuropeptide Y), dynorphin, β-endorphin and noradrenaline.

Intramural Reflexes Controlling Intestinal Secretory Activity. Although they cannot be distinguished on the basis of their morphology, there are three types of sensory receptors in the submucosa: chemoreceptors, mechanoreceptors and thermoreceptors. There are several chemoreceptors that are activated specifically by stimuli such as the presence of alkaline or acidic solutions, or glucose or amino acids, in the intestinal lumen, and these are presumably involved in controlling luminal composition. The mechanoreceptors in the submucosa respond to tactile stimulation, whereas those in the smooth muscle layers respond to stretch and so presumably monitor luminal volume and/or wall tension. The thermoreceptors respond to cold (10°–36°C) or to heat (38°–50°C), but what physiological role they could possibly play is difficult to see.

Role of Extrinsic Nerves in the Control of Secretion. In humans, the vagus (parasympathetic preganglionic) nerve does not influence the secretory activity of the small or large intestine, but the sympathetic nerves exert a tonic inhibitory effect on secretion. Since these nerves do not project to epithelial cells, but only to ganglion cells in the submucous plexus, the effect cannot be a direct one and must involve inhibition of submucous excitatory ganglion cells.

62.4.1.3.2 Endocrine Control

Of the substances found in enteroendocrine cells, secretin, *gastrin*, GIP (formerly known as gastric inhibitory peptide), motilin, enteroglucagon/glicentin, serotonin, CCK (formerly known as cholecystokinin), and neurotensin increase the rate of secretion of crypt cells, whilst somatostatin and noradrenaline, acting on α-receptors, enhance absorption by the villus enterocytes and the cells of the colonic surface epithelium. In general, agents that increase cytosolic cyclic AMP concentrations enhance secretion in crypt cells and simultaneously inhibit absorption in villus and surface epithelial cells. Cyclic AMP acts on crypt cells by opening luminal plasma membrane Cl⁻ channels and on villus cells by reducing the rate of entry of NaCl into the cytosol from the lumen.

Prostaglandins and Volatile Fatty Acids. Increased levels of aldosterone, as well as some glucocorticoids, increase Na^+ and water absorption in the small intestine, whereas arachidonic acid and E prostaglandins (PG_E) inhibit active NaCl absorption and stimulate active Cl⁻ secretion. Prostaglandins also stimulate duodenal HCO_3^- secretion. The kinins bradykinin and kallidin are potent intestinal secretomotor agonists, the actions of which may be mediated at least in part by prostaglandins. Angiotensin II stimulates colonic NaCl absorption. Adrenergic agonists,

acting via both α- and β-receptors, stimulate colonic NaCl absorption, possibly by increasing the rate of NaCl cotransport or the rate of paired Na^+/H^+ and Cl^-/HCO_3^- exchange. Volatile fatty acids produced by bacteria in the colon also stimulate colonic Na^+ absorption, and the presence of bacteria in the colon is important in maintaining normal colonic function because of their role in determining the chemical composition of the luminal contents.

62.4.1.4 Control of Colonic Secretion

Enterotoxins and Laxatives. Colonic secretion of KCl and water is evoked by acetylcholine, VIP, PG_{E2}, and some leukotrienes. Several species of micro-organism including *Escherichia coli*, *Shigella flexneri*, *Salmonella typhimurium* and *Vibrio cholerae* produce enterotoxins which probably induce secretion by activating G proteins regulating membrane-bound adenylate cyclase: this causes secretory diarrhoea, which, particularly in cases caused by cholera toxin, can be so severe as to be life threatening. The effects are seen in both small and large intestine. Many commercial laxatives act by reducing net colonic NaCl and water absorption, even to the point of reversing it to net secretion. Some of these, such as phenolphthalein, bisacodyl and ricinoleic acid (the active ingredient in castor oil), inhibit fluid absorption in both the small and the large intestine. Others, such as the anthraquinones (e.g., sennacot), are ingested as inactive glycosides and only become active after the glycoside group has been cleaved off by the action of colonic bacteria.

62.4.1.5 Intestinal Defence Against Micro-Organisms and Enterotoxins

Gut-Associated Lymphoid Tissue and Peyer's Patches. The small and large intestine are in constant contact with micro-organisms and their products. Many micro-organisms belong to the normal flora (in the colon), but others are acquired transiently from the environment and must be eliminated. Micro-organisms are usually acquired by ingestion with food and water but also with swallowed air and, from the nasopharynx and respiratory tract, with swallowed mucus. The acidic environment of the stomach contents provides the first line of defence; in addition, the small intestine presents a barrier to pathogenic organisms and their products. This consists of the epithelium itself, its secreted mucus and the secretory products of the Paneth cells in the crypts of Lieberkühn, supported by gut-associated lymphoid tissue (GALT), neutrophils and macrophages. GALT consists of discrete nodules (follicles) of lymphoid tissue scattered through the submucosal space, as well as isolated lymphocytes lying both superficial (intra-epithelial lymphocytes) and deep (lamina propria lymphocytes) to the epithelial basement lamina. Intestinal lymphocytes circulate from the submucosal follicles via the mesenteric lymph nodes, the thoracic duct and the systemic blood to return to the intestine as isolated intra-epithelial and lamina propria lymphocytes. Lymphocyte sensitisation is thought to occur in the follicles after antigen transfer across modified epithelial M (for "membrane") cells that lie in the epithelium, interposed between the lumen and the follicles. In the ileum, which is exposed to a higher load of micro-organisms than more proximal regions of the small bowel, follicles are frequently massed together in collections known as Peyer's patches.

T and B Lymphocytes. Both T and B lymphocytes have been identified in the small intestine. B lymphocytes are specialised for the secretion of secretory IgA, IgG and IgD, and T lymphocyte subclasses are distributed between intra-epithelial (predominantly cytotoxic suppressor cells) and lamina propria (predominantly helper-inducer cells) sites, but the full significance of these findings is far from clear.

62.4.2 Protein Digestion and Absorption

Oligopeptidases. In humans, more than 100 g protein are ingested each day as part of a normal diet. These ingested proteins are partially hyrolysed in the stomach by the action of pepsin, and the process is continued and completed in the small intestine by the action of pancreatic proteases (Table 62.1) and the brush-border and cytoplasmic oligopeptidases expressed by villus enterocytes. The major pancreatic proteases, their modes of activation and their preferred sites of peptide bond attack are given in Table 62.1. The function of the pancreas is described in Chap. 65.

Pancreatic Proteases. These are released by the action on pancreatic acinar cells of CCK (from intestinal enteroendocrine cells) and of acetylcholine (from intrapancreatic postganglionic nerves). CCK is synthesised and released from duodenal enteroendocrine cells in response to the delivery of amino acids into the duodenum. Pancreatic proteases are released initially as inactive pro-enzymes or zymogens. This has the effect of protecting the membranes of the pancreatic acini and ducts from protease attack. Critical to the co-ordinated activation of the pancreatic proteases is the cleavage from trypsinogen at its N terminus, between Lys_8 and Ile_9, of a peptide, which includes a terminal $(Asp)_4$-Lys sequence, by a specific brush-border peptidase, enteropeptidase (previously known as enterokinase), to yield trypsin. This enzyme is not only an important protease in its own right but is the common activator of other important pancreatic zymogens, including the endoprotease precursors chymotrypsinogen (which yields several different forms of chymotrypsin as a result of sequential limited proteolysis by trypsin and then chymotrypsin) and pro-elastases 1 and 2, and the exoprotease precursors pro-carboxypeptidases A and B.

Trypsin, Chymotrypsin, Elastases (1 and 2) and Protease E. These are related serine proteases that contain the same three critical amino acid residues at the catalytic site: serine, histidine and aspartate (see Sect. 62.4.4.2.2). Differ-

ences in the substrate specificities of the different serine proteases arise from small differences in the amino acids that project into the catalytic pocket, e.g., two glycines in the case of chymotrypsin, threonine and valine residues (which prevent amino acids with bulky side-chains from binding) in the case of elastase 1 (which is not expressed in adult humans), and a negatively charged aspartate residue in the case of trypsin [11,52].

Carboxypeptidases A and B. These are structurally related zinc metalloproteases with quite different substrate specificities (see Table 62.1). Carboxypeptidase A, but not carboxypeptidase B, forms complexes with other proteases. In the human, pro-carboxypeptidase A can form binary complexes with either a glycosylated form of protease E or, alternatively, pro-elastase 2 [39]. These complexes appear to persist after activation by trypsin and may result in sequential proteolytic attack on substrates. The activation of monomeric carboxypeptidase A is slow because, after cleavage, the activation peptide binds to the peptidase with high affinity and inhibits it. The protease is then only fully activated after further cleavage of the activation peptide by trypsin [61].

Pro-phospholipase A_2 and Pro-colipase. In addition to its role in the controlled activation of protein digestion, trypsin also has an important role in the control of lipid digestion because it activates pro-phospholipase A_2 and pro-colipase (see below). Given the central importance of trypsin to protein and lipid hydrolysis, it is interesting to note that the pancreas also secretes an endogenous trypsin antagonist, the 6-kDa protein, pancreatic secretory trypsin inhibitor (PSTI), which delays trypsin activation until the delivery of the pancreatic juice into the lumen of the small intestine.

Brush-Border Oligopeptidases. The oligopeptides released by the action of the pancreatic peptidases are further broken down by the action of brush-border oligopeptidases. As a result, free amnio acids and some short-chain peptides (di-, tri- and tetrapeptides) can be bound to specific carrier proteins in the luminal membranes of the villus enterocytes and then absorbed. The brush-border oligopeptidases include enteropeptidase (which cleaves a unique Lys_8-Ile_9 peptide bond near the N terminus of trypsinogen to release trypsin, the active endopeptidase), aminopeptidase A (which hydrolyses bonds involving acidic amino acids at the N terminus), aminopeptidase N (which acts on bonds involving neutral amino acids at the N terminus), dipeptidyl aminopeptidase IV (which acts on bonds involving proline or alanine at the N terminus), γ-glutamyl transpeptidase, folate conjugase (which attacks polyglutamylfolates to facilitate folate absorption), glutathione dipeptidase, glycyl-leucine peptidase (which attacks dipeptides, especially glycine-leucine), and zinc-stable aspartyl-lysine peptidase (which attacks dipeptides, especially aspartate-lysine) [3]. There are also several cytoplasmic peptidases in villus enterocytes which may facilitate the absorption of di- and

Table 62.1. Activation and actions of the major pancreatic enzymes

Enzyme (EC codes, 1992)	Mode of zymogen activation	Preferred cleavage
Proteases (3.4)		
Trypsin (3.4.21.4)	Brush-border enteropeptidase cleaves off an N-terminal peptide of trypsinogen between Lys_8 and Ile_9	On C side of Arg or Lys residues
Chymotrypsin (3.4.21.1)	Trypsin cleaves off an N-terminal tetrapeptide of chymotrypsinogen; further cleavage by chymotrypsin yields additional active forms	On C side of aromatic amino acids or residues with bulky non-polar side chains
Elastase-1[a] (3.4.21.36)	Trypsin cleaves off an N-terminal decapeptide [50]	On C side of small aliphatic amino acids
Elastase-2 (3.4.21.71)	Trypsin cleaves off an N-terminal duodecapeptide [51]	On C side of residues with bulky non-polar side chains
Protease E (3.4.21.70)	Trypsin cleaves off an N-terminal undecapeptide from pro-protease [49]	On C side of small aliphatic amino acids; protease E does not attack elastin
Carboxypeptidase A (3.4.17.1)	Trypsin liberates a 94-residue activation peptide [15]	Aromatic or aliphatic amino acids at the C terminus
Carboxypetidase B (3.4.17.2)	Trypsin liberates a 95-residue activation peptide [12]	Basic amino acids at the C terminus
Lipases (3.1)		
Pancreatic lipase (3.1.1.3)	Secreted in active form	Hydrolysis of C-1 and C-3 glycerol ester bonds
Phospholipase A_2 (3.1.1.4)	Trypsin cleaves off an N-terminal heptapetide	Hydrolysis of **sn**-2 (or C-2) ester bonds of 1,2-diacylglycero-phospholipids
Non-specific carboxylesterase (3.1.1.1)	Secreted in active form	Hydrolysis of all esters
Other enzymes (3.1 and 3.2)		
Pancreatic α-amylase (3.2.1.1)	Secreted in active form	Hydrolysis of α_{1-4} glucosidic bonds in starch, amylopectin and glycogen
RNase (3.1.27.5)	Secreted in active form	Hydrolysis of phosphate ester bond in RNA
DNase (I) (3.1.21.1)	Secreted in active form	Hydrolysis of DNA at 3′ end of phosphate ester bond
DNase (II) (3.1.21.4)	Secreted in active form	Hydrolysis of DNA at 5′ end of phosphate ester bond

[a] Elastase-1 is not expressed in humans

tripeptides. They include an aminotripeptidase, an aminodipeptidase and a proline dipeptidase.

Amino Acid Transporters. The details of amino acid absorption across the small intestine vary according to species and according to which small intestinal segment is being investigated. Distinct carrier systems have been described for at least four classes of amino acid: neutral, basic, acidic and secondary amino acids. Separate Na^+-dependent and Na^+-independent carrier systems have also been identified. On the basis of experiments in rabbit jejunum, the brush-border amino acid transporters are known to number at least five, three of which are Na^+ dependent [55]. The Na^+-dependent transporters include enterocyte-specific transporters for neutral amino acids (NBB), for the secondary amino acid proline (IMINO), and for phenylalanine and methionine (PHE). The Na^+-independent transporters include a transporter (y^+) for the basic amino acids lysine and arginine and a transporter for leucine and branched-chain neutral amino acids (L). Nevertheless, the distribution of amino acid transporters in the small intestine appears to vary considerably among species. In the rat, for example, cysteine and valine transport is greatest in the jejunum, whereas leucine transport is greatest in the ileum [60].

Hartnup's Disease. Similarities between the recognised intestinal brush-border transporters and renal brush-border transporters have been identified. Consistent with this, some patients with selective amino-acidurias have been shown also to exhibit defects in the absorption of amino acids across the small intestine. Nevertheless, differences between the transporters expressed in the small intestine and the kidney have been demonstrated. For example, in patients with Hartnup's disease, who have an inherited defect affecting the intestinal transport of neutral amino acids, methionine is poorly absorbed by the intestine, even though urinary excretion of the amino acid is not affected [3].

Oligopeptide Transporters. Peptides absorbed across the luminal membranes of villus enterocytes may be further degraded by the action of cytoplasmic di- and tripeptidases or delivered intact into the portal venous blood. Transporters for oligopeptides represent an alternative route by which amino acids may be absorbed. A transporter for the dipeptide carnosine (β-Ala-His), for example, a compound found in high concentration in skeletal muscle, represents an alternative route by which histidine may be absorbed [40]. One tripeptide that can be absorbed intact across the rat small intestine is glutathione (γ-Glu-Cys-Gly). Interestingly, this transport process can be activated by α-adrenergic agonists [30].

62.4.3 Carbohydrate Digestion and Absorption

Starch and Amylopectin. In humans, more than 300 g carbohydrate are ingested each day as part of a normal diet. This is composed mainly of complex polysaccharides (64% starch, 0.5% glycogen) and disaccharides (26% sucrose, 6.5% lactose). Monosaccharides (3%, largely fructose) make up the remainder. After complete digestive hydrolysis, the diet would yield three monosaccharides: glucose (80%), fructose (14%) and galactose (5%). The complex dietary carbohydrates take the form of polymeric chains of covalently linked monosaccharides, particularly glucose. Two of these polymeric forms of glucose are susceptible to the action of human small intestinal enzymes: the straight-chain polymer starch (otherwise known as amylose) and a polymer consisting of short straight-chain segments linked at intermittent branch points, amylopectin. The covalent bonds linking successive glucose monomers in starch (α_{1-4}) and amylopectin (α_{1-4} and α_{1-6}) are susceptible to the action of pancreatic (and parotid) α-amylase, which attacks α_{1-4} bonds in straight-chain segments of glucose polymers, liberating maltose and maltotriose (both α_{1-4}), and α-limit dextrins (α_{1-4} and α_{1-6}). Subsequently, the brush-border enzymes, maltase (acting on α_{1-4} bonds) and isomaltase (acting on α_{1-6} bonds), attack the oligosaccharide products of the luminal digestion of starch and amylopectin to liberate glucose monomers.

Cellulose. Although starch and amylopectin are polysaccharides of plant origin, the major storage form of glucose in plants is cellulose. Like starch, it is a straight-chain polymer of glucose, but unlike starch it is composed of repeating glucose units covalently linked in the β_{1-4} configuration, bonds that are not susceptible to cleavage by α-amylase. Glycogen, on the other hand, is an important dietary polysaccharide of animal origin. It resembles amylopectin in that it consists of straight-chain segments of glucose monomers connected by α_{1-4} bonds and linked up at intermittent branch points by α_{1-6} bonds.

Sucrose and Lactose. In addition to these dietary polysaccharides, two important dietary disaccharides are also readily digestible in the small intestine: sucrose (glucose α_{1-2} fructose) and lactose (galactose β_{1-4} glucose). These disaccharides are broken down to their respective monosaccharides by the action of two brush-border enzymes, sucrase and lactase. As a result, three major monosaccharide species are absorbed by the small intestine: glucose, galactose and fructose. Prior to weaning, glucose and galactose are of equal importance, but after weaning glucose and fructose are of considerably greater significance.

Na^+-Dependent Hexose Transporter. Glucose and galactose compete for the same Na^+-dependent hexose transporter (SGLT-1) in the brush-border membranes of villus enterocytes. This transporter is a member of a family of proteins which includes Na^+-dependent transporters for nucleosides and *myo*-inositol [41] and is sensitive to blockade by phlorrhizin. Fructose, on the other hand, is absorbed across the brush-border membrane by a distinct electro-neutral pentose transporter.

Glucose Transporter. After they have entered the villus enterocyte, glucose, galactose and fructose diffuse through the cytoplasm, cross the basolateral membrane utilising a variety of carrier mechanisms and enter the portal blood. A stereo-selective D-glucose transporter that is sensitive to phloretin and cytochalasin B has been identified in basolateral membrane vesicles from rat duodenum and jejunum. Despite its stereo-selectivity, this carrier has been found to transport a broad spectrum of hexoses including mannose and galactose. It has not been found in basolateral membrane vesicles from rat ileum [67], but recent developments on the molecular biology of human glucose transporters suggest that the GLUT-2 carrier, which can transport glucose, galactose and fructose, corresponds to this transporter [5]. The mRNA for GLUT-2 has been detected in enterocytes using specific cDNA clones and has been shown to be most conspicuous in the mid-villus region and in the proximal as opposed to the distal small intestine [45]. In keeping with this observation, the absorption of monosaccharides is known under normal circumstances to be complete by the time the intestinal contents leave the jejunum.

62.4.4 Lipid Digestion and Absorption

The dietary intake of lipids in humans can vary widely according to dietary preference and food availability. It averages 60–100 g per day, but can be as low as 10 g and as high as 250 g per day. The major mammalian lipids are:

- Fatty acids, which are either esterified, in phospholipids, sphingolipids, triacylglycerols and cholesterol esters, or unesterified ("free")
- Triacylglycerols (also known as triglycerides), which are fatty acid esters of glycerol, the fatty acids being attached via ester bonds to the 3-carbon backbone of glycerol
- Cholesterol and fatty acid esters of cholesterol, in which the fatty acids are esterified to the hydroxyl group on the C-3 position of cholesterol
- Membrane lipids, which include the sphingolipids, formed from serine backbones, and the glycerol phospholipids, formed from glycerol backbones.

The structures of some of these lipids are given in Fig. 62.8.

Fat-Soluble Vitamins. For the sake of completeness, we should also make reference to the fat-soluble vitamins (vitamins A, D, E and K), which are absorbed by enterocytes along with fatty acids from the cores of bile salt-stabilised micelles in the intestinal lumen and are transported around the body by various lipoproteins.

62.4.4.1 Mammalian Lipids

62.4.4.1.1 Fatty Acids

Fatty acid molecules have a carbon-chain backbone of variable length terminating in a negatively charged carboxylate group with an associated proton. In general, the backbones of naturally occurring fatty acids consist of an even number of carbon atoms, since they are usually synthesized by the successive condensation of two-carbon units derived from acetyl coenzyme A (acetyl-CoA). Nevertheless, fatty acids having an odd number of carbon atoms do occur naturally, and mitochondria are equipped with enzymes which permit the utilisation, via the citric acid cycle, of all the carbon atoms in the chain.

Saturated and Unsaturated Fatty Acids. Fatty acids are described as saturated when the carbon chain contains no double bonds and unsaturated when double bonds are present. Typically, naturally occurring unsaturated fatty acids have one or more double bonds in the *cis* configuration. In contrast to *trans* double bonds, *cis* bonds introduce a kink in the carbon chain, which lowers the melting point of the free fatty acid and, in biological membranes, interferes with the packing of the phospholipids within the hydrophobic zone.

Palmitate and Arachidonate. Fatty acid synthesis leads initially to the formation of the C-16 saturated fatty acid, palmitate, which may then be modified by chain lengthening and by the insertion of one or more double bonds into the carbon chain, although mammals are incapable of inserting double bonds beyond the C-9/C-10 bond in the carbon chain. Consequently, biologically important fatty acids that contain double bonds beyond C-9/C-10, such as arachidonate (C-20), have to be synthesised from pre-formed unsaturated fatty acids ingested as part of the diet.

Oleate. Fatty acids are incorporated into triacylglycerols or cholesterol esters for storage inside cells or for transport via lipoproteins. They can also be transported in the plasma in a "free" form, in which they are actually bound non-covalently to albumin. Palmitate (C-16) forms a major fatty acid ester of cholesterol in low-density lipoproteins (LDL), but oleate (C-18) esters predominate in intracellular fat stores.

62.4.4.1.2 Triacylglycerols

Triacylglycerols are the major (90%) storage form of fat. They are, in consequence, the major form of fat derived from the ingestion of animal products. Adipose tissue, for example, is composed of cells, the cytoplasm of which is laden with triacylglycerol-rich lipid droplets. Nevertheless, triacylglycerols are not found exclusively in adipocytes: hepatocytes and enterocytes are also important sites of synthesis. For their energy potential to be utilised, stored triacylglycerols must be broken down to free fatty acids and glycerol by the action of hormone-sensitive lipase. Both fatty acids (bound to albumin) and glycerol are then released into the blood-stream and used as a source of energy by many tissues. Alternatively, triacylglycerols syn-

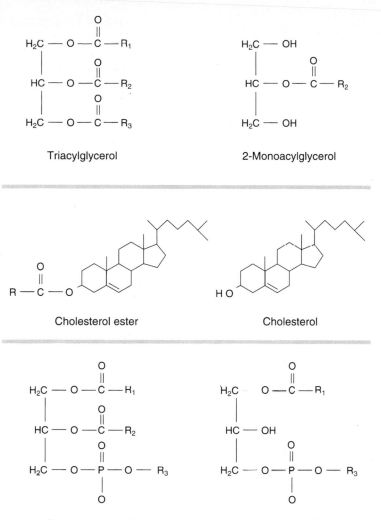

Triacylglycerol

2-Monoacylglycerol

Cholesterol ester

Cholesterol

Glycerol phospholipid

Lysophospholipid

Fig. 62.8. Structures of some major dietary lipid species. Cleavage of the C-1 and C-3 ester bonds of triacylglycerols by pancreatic (and gastric) lipases yields fatty acids and 2-monoacylglycerols. Cleavage of the ester bond at C-3 of cholesterol esters by bile salt-activated lipase yields fatty acids and unesterified (free) cholesterol. Cleavage of the C-2 ester bond of glycerol phospholipids by pancreatic phospholipase A_2 yields fatty acids and lysophospholipids

thesised in hepatocytes and enterocytes may be packaged into lipoproteins (very low density lipoproteins, VLDL, and chylomicrons, respectively) and transported intact (see Table 62.2). Triacylglycerols carried in the cores of lipoproteins cannot, however, be taken up by cells in the intact, esterified form; this is true also for ingested triacylglycerols (in the cores of the emulsion droplets formed in the duodenal lumen) and their immediate target cells, the enterocytes of the duodenal and jejunal villi. Rather, they must be broken down to free fatty acids, 2-monoacylglycerols and glycerol by the action of one of the major lipases: lipoprotein lipase, which acts on chylomicrons and VLDL, hepatic lipase, which acts on high-density lipoproteins (HDL) and VLDL remnants (formed by the action of lipoprotein lipase on VLDL), and pancreatic lipase, which acts on emulsion droplets in the intestinal chyme.

62.4.4.1.3 Cholesterol

Synthesis of Bile Salts and Steroid Hormones. Cholesterol is an important component of most biological membranes and is an essential substrate in the synthesis of bile salts in hepatocytes. Bile salt synthesis in hepatocytes is essential for the emulsification of dietary triacylglycerols and cholesterol esters and for the micellisation of their hydrophobic breakdown products, cholesterol and fatty acids. Cholesterol is also required for the synthesis of steroid hormones in skin cells (vitamin D), adrenal cortex (mineralocorticoids, glucocorticoids and adrenal androgens), gonads (estradiol, testosterone), corpus luteum and placenta (progesterone).

Table 62.2. The properties of the principal lipoprotein classes

Lipoprotein class	Major lipids	Apoproteins	Density (g/ml)	Diameter (nm)	Origin	Destination
Chylomicrons	TAG and CE from diet	A_I, A_{II}, $\mathbf{B_{48}}$ C_{II}, C_{III}, E	<0.94	80–500	Enterocytes	Capillary beds in adipose tissue
Chylomicron remnants	CE from diet	A_I, A_{II}, $\mathbf{B_{48}}$ C_{II}, C_{III}, $\mathbf{E}$	<1.006	40–100	Capillary beds	Hepatocytes
VLDL	TAG from liver	$\mathbf{B_{100}}$, C_{II} C_{III}, E	<1.006	30–80	Hepatocytes	Peripheral capillary beds
VLDL remnants	TAG and CE	$\mathbf{B_{100}}$, C_{III}, E	<1.019	25–35	VLDL	LDL
LDL	CE	$\mathbf{B_{100}}$	1.019–1.063	15–25	VLDL remnants	Peripheral tissues and hepatocytes
HDL	Cholesterol, CE and TAG	A_I, A_{II}, E	1.063–1.21	5–12	Enterocytes	Hepatocytes

Where an apoprotein has a demonstrated function with respect to a particular lipoprotein class, the code is shown in boldface. Only apoproteins discussed in the text have been included; others have been identified, but their functions have not yet been established unequivocally
CE, cholesterol ester; TAG, triacylglycerol; VLDL, very low density lipoproteins; LDL, low density lipoproteins; HDL, high density lipoproteins

62.4.4.1.4 Membrane Lipids

Glycerol Phospholipids and Sphingolipids. Glycerol and serine form the backbones of the two major groups of membrane lipids, the glycerol phospholipids and the sphingolipids. These are synthesised in the endoplasmic reticulum and spontaneously form bimolecular sheets (lipid bilayers). Although the glycerol phospholipids and sphingolipids are distinct biochemical species, they assume similar structural configurations in biological membranes. Each can be viewed as having a three-carbon backbone supporting two hydrocarbon tails derived from fatty acids and a polar head group projecting outwards to face the extracellular medium or inwards to face the cytosol. Both membrane lipid species are heterogeneous. This heterogeneity arises primarily from differences in the fatty acids linked at two of the carbon positions in the backbone, but heterogeneity also arises from differences in the composition of the head groups. Head groups found in glycerol phospholipids include choline, ethanolamine, serine, inositol and glycerol itself (as well as other alcohols), whereas those found in sphingolipids include choline (in the case of sphingomyelin) and various sugars or sugar chains. The head groups form hydrogen bonds with water molecules and also undergo ionic interactions with electrolytes dissolved in the surrounding aqueous phases. The two fatty acid tails, on the other hand, are aligned side by side at right angles to the surface to form the substance of the membrane. Because biological membranes form bilayers, the thickness of each membrane is approximately twice the length of the fatty acid tail. Membrane specialisation is determined in part by differences in the composition and disposition of membrane lipids. For example, sugar-containing sphingolipids (glycolipids) are concentrated exclusively in the outer leaflet of plasma membranes, whereas the glycerol phospholipid phosphatidylserine, is concentrated in the inner leaflet.

62.4.4.2 Lipases

62.4.4.2.1 Gastric Lipase

Triacylglycerols are hydrolysed to free fatty acids and 2-monoacylglycerols by the action of triacylglycerol lipases, first in the stomach and subsequently in the small intestine. There has been considerable uncertainty about the origin of gastric lipase: some authors have been prepared to accept the existence only of a lingual lipase (of importance in early post-natal life; see Chap. 64) which is active in the acidic environment of the gastric lumen. Recently, however, several research groups have identified a triacylglycerol lipase in gastric mucosa as well as in gastric aspirates [4,7,13,58]. Its activity is highest in the fundic mucosa and it appears to be released from mucus-secreting cells in the fundic glands. It has a pH optimum in the range 4–6, but the optimum varies within this range according to the substrate. The pH optimum for the hydrolysis of triacylglycerols containing long-chain fatty acids is about 4, but for medium- and short-chain triacylglycerols the pH optima are about 5 and 6, respectively [13]. As a consequence, the enzyme is active against medium- and short-chain triacylglycerols in the duodenum, where it complements the action of pancreatic lipase, as well as in the stomach, where it preferentially attacks long-chain triacylglycerols. Unlike pancreatic lipase, it is stable under the acidic conditions of the stomach. Estimates of its relative importance suggest that gastric lipase may normally hydrolyse as much as 10%–30% of dietary triacylglycerols in adult humans [14], but this figure varies widely among mammalian species [18]. Gastric and pancreatic lipase are structurally very different. Human gastric lipases has an amino acid sequence consisting of 379 residues (43 kDa), whereas human pancreatic lipase has 449 residues, but the two lipases show a short region of structural homology (-Gly-Xaa-Ser-Xaa-Gly-) which includes an essential serine residue.

62.4.4.2.2 Pancreatic Lipase

Clearly, despite the recently recognised importance of gastric lipase, pancreatic lipase remains the most important enzyme in lipid digestion. The manner of its release and action is also better understood. It is released from pancreatic acinar cells (along with the other pancreatic enzymes) by the action of CCK (from enteroendocrine cells) and acetylcholine (from nerve terminals). Its activity is promoted by agents that increase the surface area of the oil–water interface, such as bile salts and lysophosphatidyl-choline. Lysophosphatidyl-choline is itself a product of lipid digestion in the small intestine, being derived by the action of the pancreatic enzyme phospholipase A$_2$ on the phosphoglyceride, phosphatidyl-choline.

Catalytic Amino Acids. With the aid of site-directed mutagenesis of the DNA for human pancreatic lipase in a COS-1 cell transfection system, three amino acids in the peptide have been shown to be essential for human pancreatic lipase activity: Ser$_{153}$, His$_{264}$ and Asp$_{177}$ [38]. This catalytic triad is identical to that found in the serine proteases, suggesting that pancreatic lipase and the serine proteases (which include trypsin) have a similar ontogeny.

62.4.4.2.3 Pancreatic Colipase

Pancreatic lipase has a native pH optimum of about 9, but is unstable in the presence of bile salts. Its pH optimum is shifted toward the neutral pH conditions of the duodenum and jejunum, and its stability in the presence of bile salts is greatly improved by the presence of colipase (96 amino acids), a co-enzyme secreted by the pancreas. The activation of pancreatic lipase involves a complex sequence of events which includes its interaction with colipase, interfacial binding of the lipase–colipase complex to the surface of the lipid emulsion droplet and then catalysis, whereby molecules of water are brought into close apposition with the ester bonds of the triacylglycerol substrate.

Pro-colipase. Like the zymogen precursors of the pancreatic proteases, colipase is also secreted in an inactive form. This inactive precursor, pro-colipase (101 amino acids), is a substrate for trypsin. Indeed, in experiments in which human pancreatic juice was incubated with enteropeptidase in vitro, pro-colipase breakdown was detected before that of the other trypsin-sensitive substrates, pro-phospholipase A$_2$ and chymotrypsinogen [8].

Enterostatin. When pro-colipase in split by the action of trypsin, a second biologically active peptide in addition to colipase is also released. This pentapeptide, called enterostatin (Ala-Pro-Gly-Pro-Arg in humans), which is derived from the N terminus of pro-colipase, is apparently absorbed during the meal [21] and has been reported to cause a marked reduction in subsequent food intake; oral administration of pro-colipase leads to a similar effect. Enterostatin, like CCK (see Sect. 62.6.2), is thus of considerable interest as a potential satiety hormone.

62.4.4.2.4 Bile Salt-Activated (Non-Specific) Lipase (Cholesterol Esterase)

Cholesterol esters are also important dietary lipid components, but cholesterol can only be absorbed readily by the small intestine after the fatty acid esterified at the C-3 position has been liberated. The pancreatic enzyme involved in this process is commonly referred to as bile salt-activated lipase because of its requirement for bile salts when hydrolysing insoluble fatty acid esters, but it is also known as cholesterol esterase and non-specific lipase. In addition to hydrolysing cholesterol esters, it liberates fatty acids from phospholipids, triacylglycerols and diacylglycerols, as well as retinyl fatty acid esters. Thus, it supplements the actions of phospholipase A$_2$ and pancreatic lipase, as well as further digesting the breakdown products they liberate [28,35,36]. Furthermore, the same enzyme appears to have lipoamidase removes as well as esterase activity. Similar enzymes have been described in human milk, serum and liver [32,63].

62.4.4.2.5 Phospholipase A$_2$

Phospholipids are the major lipid constituents of biological membranes and are, therefore, an additional source of fatty acids in the diet. The pancreas secretes two enzymes active in the breakdown of phospholipids: phospholipase A$_2$, which removes the fatty acid at the C-2 position to yield lysophospholipids, and bile salt-activated lipase (see above), which then removes the fatty acid at the C-1 position.

62.4.4.2.6 Absorption of the Products of Fat Digestion

Micelles. The digestion of triacylglycerols, cholesterol esters and phospholipids results in the liberation, inter alia, of free fatty acids, 2-monoacylglycerols and cholesterol. They are incorporated into mixed micelles, mainly with the glycine- and taurine-conjugated forms of bile salts, and transferred by diffusion to the brush-border membranes of the villus enterocytes, where they are taken up by the enterocytes and re-esterified prior to export. Monoacylglycerols are believed to diffuse passively across the brush-border plasma membrane.

Sphingomyelinase. On the basis of experiments on CaCo-2 cells [16], cholesterol absorption appears to be regulated by the sphingomyelin content of the brush-border membrane. Thus, after the depletion of plasma membrane sphingomyelin by the action of sphingomyelinase, cholesterol absorption is suppressed. Pancreatic juice contains sphingomyelinase, which thus may be involved in the control of cholesterol uptake.

Fatty Acid-Binding Proteins. Fatty acids, on the other hand, are absorbed by a high-affinity (K_a approximately 10^8 M^{-1}), saturable uptake mechanism in both jejunal brush-border and hepatocyte plasma membranes. The fatty acid-binding protein (FABP) responsible for this transport (40 kDa) has been isolated from both tissues and is referred to as $FABP_{PM}$ to distinguish it from other fatty acid-binding proteins [44]. After the fatty acids have been transferred to the cytosol, they are sequestered by further fatty acid-binding proteins (two separate 14- to 15-kDa forms are expressed in enterocytes [56]). It is likely that this is important for the transfer of fatty acids to specific intracellular compartments, in particular to the endoplasmic reticulum for re-esterification by the enzymes, monoacylglycerol acyltransferase and diacylglycerol acyltransferase, to yield intracellular triacylglycerol. Resynthesised triacylglycerols are incorporated into lipoprotein particles known as chylomicrons (see Sect. 62.5) and released into the lymph via the central lacteals of the villi. The intestine also uses diacylglycerol-phosphate (phosphatidic acid) as a substrate for the synthesis of triacylglycerols [68].

Enterohepatic Circulation. Intestinal bile salts are not absorbed along with the products of lipid digestion in the jejunum and proximal ileum. Intsead, they remain in the intestinal lumen until they reach the terminal segment of the ileum. Enterocytes in this segment of the small intestine, which lacks villi, express a high-affinity Na^+-coupled carrier protein in their luminal membranes which facilitates reabsorption of the bile salts and their return, via the portal blood, to the liver where their re-uptake by the hepatocytes completes an enterohepatic circulation (Fig. 62.9). The recirculation of bile salts is highly efficient: each bile salt molecule is recycled on average about 18 times before being lost in the faeces, and the whole pool recycles several times during the intestinal processing of a single meal. As a result, the amount of hepatic cholesterol required for new bile salt synthesis in humans is limited to about 0.5 g per day.

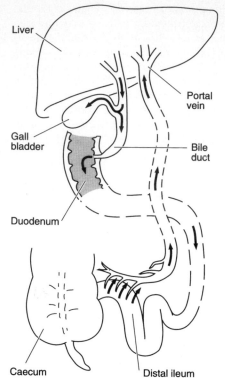

Fig. 62.9. The enterohepatic circulation of bile salts. Bile salts synthesised in hepatocytes along with recycled bile salts from the portal blood are secreted into the biliary canaliculi and thence into the major ducts of the biliary tree. After entering the small intestine, bile salts facilitate the digestion of ingested lipid and the absorption of lipid breakdown products. Bile salts traverse the small intestine, enter the distal ileum and are reabsorbed there by the enterocytes so that only about 0.5% of bile salts delivered to the distal ileum are lost to the large bowel. Reabsorbed bile salts are returned to the liver in the portal blood and extracted once more by the hepatocytes. On average, each bile salt molecule recycles 18 times before being lost into the colon

62.5 The Lipoproteins

The digestion and absorption of lipids represents just one aspect of lipid transport occurring in the body, and the chylomicron is but one of the lipoprotein particles participating in this process. Lipoproteins are a class of heterogeneous lipid-protein particles (stabilised emulsion droplets) that serve to transport lipids in a highly selective fashion from one tissue to another via the blood or the lymph. They are classified on the basis of their size, density, electrophoretic mobility, predominant lipid species (whether triacylglycerol or cholesterol ester) and associated proteins (apolipoproteins or apoproteins). Since chylomicrons and chylomicron remnants are two of the six major lipoprotein types, it is convenient to discuss them all at this point.

Structural Design of Lipoproteins. The major subclasses of lipoproteins are chylomicrons, chylomicron remnants, very low density lipoprotein (VLDL), VLDL remnants (formerly called intermediate-density lipoproteins or IDL), low-density lipoproteins (LDL) and high density lipoproteins (HDL). The lipoproteins of all classes conform to the same basic structural design, an amphiphilic surface coat and a hydrophobic core (Figs. 62.10, 62.11). The surface coat is a phospholipid-cholesterol monolayer, and the class-associated apoproteins are either inserted into or at-

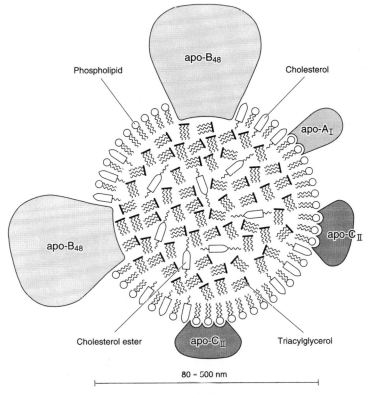

Phospholipid

Cholesterol

apo-B$_{48}$

apo-A$_I$

apo-C$_{II}$

apo-B$_{48}$

Cholesterol ester

apo-C$_{II}$

Triacylglycerol

80 – 500 nm

Fig. 62.10. Basic structural design of a chylomicron. Note the outer coat composed of a monolayer of phospholipids, cholesterol and proteins and the core, which, in the case of a chylomicron, is rich in both triacylglycerols and cholesterol esters. Only apo-B$_{48}$, apo-C$_{II}$ and apo-A$_I$ are depicted, but apo-A$_{II}$, apo-C$_{III}$ and apo-E may also be present in chylomicrons

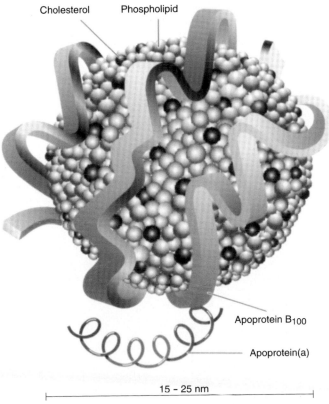

Cholesterol

Phospholipid

Apoprotein B$_{100}$

Apoprotein(a)

15 – 25 nm

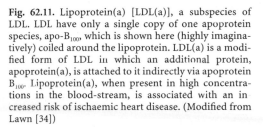

Fig. 62.11. Lipoprotein(a) [LDL(a)], a subspecies of LDL. LDL have only a single copy of one apoprotein species, apo-B$_{100}$, which is shown here (highly imaginatively) coiled around the lipoprotein. LDL(a) is a modified form of LDL in which an additional protein, apoprotein(a), is attached to it indirectly via apoprotein B$_{100}$. Lipoprotein(a), when present in high concentrations in the blood-stream, is associated with an increased risk of ischaemic heart disease. (Modified from Lawn [34])

tached to it. In structure, it is directly analogous to the outer leaflet of the plasma membrane. Whereas most lipoprotein species include several different apoproteins, usually in multiple copies, in the case of LDL and a subfraction of HDL (sometimes designated HDL$_3$) the lipoprotein particle contains only one copy of a single apoprotein species, apo-B$_{100}$ and apo-A$_I$, respectively.

Hydrophobic Lipids. Beneath the surface of the enveloping phospholipid-cholesterol monolayer is a generally spherical core consisting of hydrophobic lipids (mainly triacylglycerols and cholesterol esters). The core lipids represent the "cargo" of the lipoprotein particle. Data on the six major classes of lipoprotein, including their principal apoproteins and core lipid species, are summarised in Table 62.2.

62.5.1 Apoproteins

The apoproteins support the structural integrity of the different lipoprotein species, but, more significantly, they participate in molecular interactions between lipoprotein particles and specific protein targets. These targets include:

- Enzymes, particularly in the vascular space (whether free in plasma or immobilised on the surface of capillary endothelial cells)
- Specific receptors for the internalisation of lipoprotein particles by various cell types.

Nomenclature. Numerous apoproteins have been identified (Tables 62.2, 62.3). The original system of nomenclature supposed that lipoprotein families could be identified which would be characterised by an exclusive association with a particular lipoprotein species [1]. If this system had held, HDL apoproteins would all have been assigned to class A, LDL apoproteins to class B, and VLDL apoproteins to class C, but it is now clear that the same apoproteins are often associated with more than one lipoprotein species. The outcome of this initial misapprehension, unfortunately, is a complex and arbitrary nomenclature that cries out for rationalisation.

Exchangeable Apoproteins. In fact, for two reasons, very few apoproteins are subclass specific. First, particles belonging to one lipoprotein class may evolve into particles of another class (e.g., VLDL particles evolve by the action of lipoprotein lipase to VLDL remnants, and VLDL remnants evolve to LDL) and particular apoproteins may or may not remain following the transformation. Second, most apoproteins are exchangeable, i.e., they swap more or less readily, not only among particles of the same species, but also among particles of unrelated species. For example, apo-A$_I$ is found associated both with HDL and with chylomicrons. Similarly, apo-E is associated with chylomicrons, chylomicron remnants, VLDL, VLDL remnants and some HDL, and apo-C$_{II}$ is associated with chylomicrons and VLDL. The ability of apoproteins to exchange between different lipoprotein species (and perhaps between lipoproteins and the plasma) suggests that there is a dynamic turnover of apoproteins which, at least in part, is independent of the particles themselves. As a result, the apoprotein composition of some lipoproteins may vary according to the metabolic circumstances.

Table 62.3. Functions of some apoproteins

Apoprotein	Molecular mass (kDa)	Functions
Apo-A$_I$	28.0	Structural component of HDL; co-enzyme for LCAT; promotes receptor-mediated cholesterol uptake by HDL
Apo-A$_{II}$ (monomer)	8.5	Blocks cholesterol esterification in HDL by LCAT
Apo-B$_{48}$	250.0	Facilitates chylomicron entry into lacteals
Apo-B$_{100}$	514.0	LDL receptor ligand; promotes VLDL secretion by hepatocytes (?)
Apo-C$_{II}$	9.0	Activates lipoprotein lipase
Apo-C$_{III}$	9.0	Inhibits uptake of chylomicron remnants; inhibits lipoprotein lipase
Apo-E	35.0	Promotes hepatic recognition and uptake of chylomicron remnants, VLDL remnants and HDL

HDL, high density lipoproteins; LCAT, lecithin: cholesterol acetyltransferase; LDL, low density lipoproteins; VLDL, very low density lipoproteins

Non-exchangeable Apoproteins. Apo-B_{100} (which is found on VLDL, VLDL remnants and LDL) and apo-B_{48} (which is found on chylomicrons and chylomicron remnants) are examples of non-exchangeable apoproteins [48] (Fig. 62.11). They bind tightly to the lipids of their respective lipoproteins, in this respect behaving like the intrinsic proteins of the plasma membrane. Each appears to be essential for the initial secretion of the parent lipoprotein particle (apo-B_{100} for VLDL and apo-B_{48} for chylomicrons) and both are important ligands for the receptor-mediated uptake of the respective daughter particles, LDL and chylomicron remnants. Apo-B_{48} and the N terminus of apo-B_{100} exhibit immune cross-reactivity and are now known to be products of the same gene. Apo-B_{48} is synthesised by enterocytes, in which a tissue-specific stop codon is present in the mRNA transcript, whereas apo-B_{100} is synthesised by hepatocytes, in which the stop codon is not expressed in the mRNA. Until recently, there had been considerable doubt about the molecular weights of the two species so that a scheme of relative molecular weights had to be adopted [26]. According to this system, the larger B apoprotein was arbitrarily assigned a relative molecular weight of 100 and the other B apoprotein was found to have a molecular weight about 48% of that of the larger particle. The names apo-B_{100} and apo-B_{48} have persisted, although accurate estimates of the molecular weights of the two species are now available [17,33] (see Table 62.3). The fact that these proteins are non-exchangeable provides evidence for the successive evolution of VLDL to VLDL remnants and from VLDL remnants to LDL. Apoprotein(a) is a separate protein that binds selectively to apo-B_{100} in a small subfraction of LDL known as lipoprotein(a) (Fig. 62.11). It is important because of its association with an increased risk of myocardial infarction.

The functions of many apoproteins have been determined (Table 62.3; Sect. 62.5.3), but the roles of others await clarification.

62.5.2 Enzymes

Numerous enzymes, both intracellular and extracellular, participate in lipoprotein metabolism.

62.5.2.1 Extracellular Enzymes

The extracellular enzymes fall into two main categories:

- Triacylglycerol esterases (or lipases), including the endothelial enzymes, lipoprotein lipase and hepatic lipase, as well as pancreatic lipase (see Sect. 62.4.4)
- Lecithin:cholesterol acyltransferase (LCAT), which acts to esterify cholesterol in the surface coat of HDL particles.

Lipoprotein Lipase. This 50.4-kDa enzyme plays a primary role in triacylglycerol metabolism [9]. Its physiological site of action is the luminal membrane of capillary endothelial cells, where it hydrolyses triacylglycerols to free fatty acids and 2-monoacylglycerols. It is present in high density in capillary endothelium in adipose tissue (both white and brown), heart, lactating mammary gland, red but not white skeletal muscle, adrenal cortex and ovary. It is not present in the liver, except in neonates, and is not present in the brain except for the hippocampus. Lipoprotein lipase exhibits a high degree of selectivity for chylomicrons and VLDL, presumably because it is activated by apo-C_{II}, which is unique to these two lipoprotein species.

Although lipoprotein lipase activity is expressed on the luminal membranes of capillary endothelial cells, it is not synthesised by them. Rather, it is synthesised in the parenchymal cells of the tissue surrounding the capillaries [9,10]. For example, the site of lipoprotein lipase synthesis in adipose tissue is in the adipocyte. After synthesis, lipoprotein lipase is exported from the parenchymal cells into the interstitium and from there across the capillary endothelium to take up a position on the luminal plasma membranes of the endothelial cells. There, it is held in place by ionic interactions with a proteoglycan species enriched in heparan sulphate which, in turn, is covalently linked to a membrane phospholipid.

Hepatic Lipase. Hepatic lipase closely resembles lipoprotein lipase in the sense that it is also synthesised in parenchymal cells (the hepatocytes) and is subsequently transported to the luminal plasma membranes of capillary endothelial cells. Although hepatic lipase also hydrolyses triacylglycerols to form free fatty acids and 2-monoacylglycerols and shares structural homology with lipoprotein lipase (and pancreatic lipase), it is a separate gene product, is not activated by apoprotein C_{II}, and is selective for HDL and VLDL remnants rather than for chylomicrons and VLDL. It is also tissue specific, being found only in the liver. Hepatic lipase may also be involved in the conversion of VLDL remnants to LDL. Hepatic lipase, lipoprotein lipase and pancreatic lipase are believed to have a common evolutionary origin [31,69].

Lecithin:Cholesterol Acyltransferase. LCAT is a plasma enzyme that transfers fatty acids from phosphatidylcholine (lecithin), a component of the surface coat of HDL, to cholesterol, another component of the surface coat. It is stimulated by apo-A_I in HDL particles. Since the cholesterol esters so formed are more hydrophobic than either of the substrates, they move into the core of the particles, which increase steadily in size (and are transformed in shape from discoidal "primitive" HDL to spherical HDL particles). Alternatively, the cholesterol esters may be transferred to VLDL and VLDL remnants by means of a specific transfer protein, lipid transfer protein-1 (LTP-1; see Fig. 62.13).

62.5.2.2 Intracellular Enzymes

The intracellular enzymes also fall broadly into two groups:

- Those involved in the synthesis (mono- and diacylglycerol acyltransferases in hepatocytes, adipocytes and enterocytes) and breakdown (hormone-sensitive lipase, mainly in adipocytes) of triacylglycerols
- Those involved in the synthesis (acyl-CoA:cholesterol acyltransferase, ACAT) and breakdown (lysosomal cholesterol esterase) of cholesterol esters

Only hormone-sensitive lipase will be considered further here.

Hormone-Sensitive Lipase. The activity of this enzyme in adipocytes controls the rate at which unesterified fatty acids are released from intracellular triacylglycerol stores into the blood-stream. It is activated by glucagon and adrenaline (acting on β_1-adrenergic receptors), acting via a cyclic AMP-dependent protein kinase, and is inhibited by insulin. The transfer of unesterified fatty acids across the adipocyte plasma membrane is, therefore, dependent on the activity of lipoprotein lipase, which makes fatty acids and monoacylglycerol available for uptake on the extracellular side of the membrane, and hormone-sensitive lipase, which makes these compounds available on the intracellular side.

62.5.3 Lipoprotein Pathways

Cholesterol, cholesterol esters, triacylglycerols and fatty acids are all hydrophobic. Consequently, they must be transported around the body from tissue to tissue and organ to organ in the blood and lymph in a solubilised form. As indicated earlier, fatty acids circulate bound to plasma albumin, but the other lipids are carried in lipoproteins, which permit their transport from sites of synthesis to sites of storage or metabolic need. In general, the classes of lipoproteins fall into two broad groups: those of lower density (chylomicrons and VLDL) and those of higher density (chylomicron remnants, VLDL remnants, LDL and HDL). The lower-density lipoprotein species are enriched in the lower-density lipids, the triacylglycerols, whereas the higher-density lipoprotein species are enriched in the higher-density lipids, the cholesterol esters. Two of the three parent lipoprotein particles, chylomicrons and VLDL, have their origins in enterocytes and hepatocytes, respectively, and their progeny, chylomicron remnants, VLDL remnants and LDL, are formed from the parent types in the blood-stream. The site of origin of primitive HDL is uncertain, but it may be in the enterocytes; the progeny of primitive HDL, the various HDL subfractions, are formed in the blood-stream. Lipoprotein movements may be viewed as following one of three distinct pathways (Fig. 62.12):

- From the small intestine to adipose tissue to the liver (chylomicrons and chylomicron remnants)
- From the liver to the peripheral tissues and sometimes back to the liver (VLDL, VLDL remnants and LDL)
- From the enterocytes to the blood, thence to the peripheral tissues and finally to the liver (HDL).

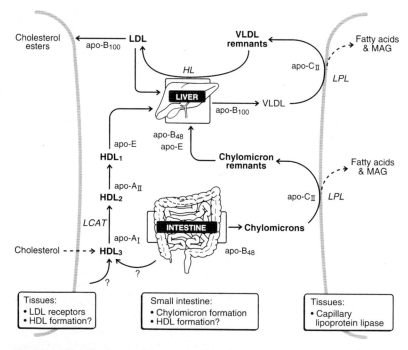

Fig. 62.12. The three normal lipoprotein pathways. High-density lipoproteins (*HDL*) are depicted here (and in Fig. 62.13) as evolving through three stages: HDL$_3$ have only a single copy of one apoprotein, apo-A$_I$, whereas HDL$_2$ have two apoprotein types, apo-A$_I$ and apo-A$_{II}$, each in multiple copies, and HDL$_1$ also have a third apoprotein type, apo-E. *HL*, hepatic lipase; *LPL*, lipoprotein lipase; *LCAT*, lecithin:cholesterol acyltransferase; *MAG*, monoacylglycerols; *LDL*, low density lipoproteins; *VLDL*, very low density lipoproteins

62.5.3.1 Chylomicron Pathway (Fig. 62.12)

Chylomicrons are formed in intestinal epithelial cells from triacylglycerols (resynthesised from absorbed 2-monoacylglycerols) and cholesterol esters (resynthesised from absorbed cholesterol). They are extruded into the interstitium, but instead of entering the blood capillaries, they enter the lacteals, the terminal ductules of the lymph system, directed there by an apo-B_{48}-dependent process, and pass thence, via the thoracic duct and the left subclavian vein to the systemic blood. In this way, triacylglycerols, carried in the cores of the chylomicrons, are made directly available to tissues expressing high levels of lipoprotein lipase activity, particularly adipose tissue. In these tissues, the triacylglycerols present in chylomicrons are broken down in the presence of apo-C_{II} to form 2-monoacylglycerols and free fatty acids, which are then released from the lipoprotein and taken up by the endothelial cells. Another apoprotein, apo-C_{III}, exerts fine control over the process of lipoprotein lipase activation and chylomicron remnant uptake by the liver. It inhibits remnant uptake, applying a size limit to the particles that can be taken up. Particles below a critical size cannot support apo-C_{III} and are therefore subject to hepatic uptake. Apo-C_{III} also inhibits lipoprotein lipase activity and so acts to increase the chylomicron half-life in the plasma. From the endothelial cells, these products are transferred to the interstitium and thence to the target cells (e.g., adipocytes), although the carriers involved in facilitating these transfer steps have not yet been identified.

Chylomicron Remnants. The lipoprotein particles remaining after this process of enzymatic liberation and depletion of the triacylglycerol breakdown products, now carrying mainly cholesterol esters in their cores, are known as chylomicron remnants. They are released from their docking sites on the endothelial cell surfaces by an unknown mechanism and remain, apparently inert, in the blood-stream until they reach the liver. Hepatocytes recognise and internalise the remnant particles at receptors that recognise and bind apo-E, the uptake process also being facilitated by apo-B_{48}. In this way, chylomicron remnants deliver dietary cholesterol to the liver.

Exogenous Pathway. The transport of chylomicrons to the tissues and of their remnants to the liver may be viewed as a single lipoprotein cycle, a cycle governed by apo-B_{48} (for extrusion of chylomicrons and uptake of chylomicron remnants), apo-C_{II} (for activation of lipoprotein lipase) and apo-E (for hepatic recognition and uptake of remnants). It is sometimes referred to as the "exogenous pathway" because it distributes triacylglycerols of dietary origin to the adipose tissue and cholesterol esters of dietary origin to the liver and the peripheral tissues.

62.5.3.2 The VLDL-LDL Pathway (Fig. 62.12)

Low-Density Lipoproteins. The release of VLDL by the liver, a process that may be assisted by apo-B_{100}, provides a continuous supply of triacylglycerols to the peripheral tissues. There, as a result of the action of capillary lipoprotein lipase (activated by apo-C_{II}), VLDL are converted to VLDL remnants and the breakdown products of the triacylglycerols are taken up by the parenchymal cells. The VLDL remnants thus formed are converted, in turn, to LDL, perhaps by the action of hepatic lipase. LDL are the major carriers of cholesterol to the peripheral tissues, where they are recognised by receptors for apo-B_{100} and taken up by endocytosis, although they also convey cholesterol back to the liver, since they can also be recognised by similar receptors on hepatocytes.

Coated Pits. LDL particles (Fig. 62.11) are enriched in fatty acid (palmitate) esters of cholesterol, and each particle contains a single copy of only one apoprotein species, apo-B_{100} [25]. The cellular uptake of the cholesterol esters depends on the receptor-mediated internalisation of LDL particles, for which apo-B_{100} is required. Two alternative pathways for the endocytosis of LDL are currently recognised, one for uptake of native LDL and another, the scavenger pathway, for the uptake of a modified form of LDL. In the first (physiological) pathway, native LDL particles bind at apo-B_{100} receptor sites on the parenchymal cell surface. A vesicle-forming coat protein, clathrin, then binds to the cytoplasmic domain of the occupied LDL receptor [53] and the receptor–clathrin complexes are transferred to discrete invaginations of the cell surface known as coated pits, where they cluster prior to internalisation. Upon activation of the internalisation process, the LDL particles associated with each pit enter the cell in the core of a clathrin-coated vesicle. The coat is removed from the vesicle, which is then transferred to the endosome/lysosome system. Within the lysosomes, cholesterol esters are liberated from LDL particles and de-esterified by the action of an acidic cholesterol esterase. In most cells, cholesterol is then released from the lysosomes and either transferred to the plasma membrane, perhaps by vesicular transport, or re-esterified by the action of ACAT, and stored. The major fatty acid involved in the intracellular re-esterification of cholesterol is oleate.

Familial Hypercholesterolemia. Reduced uptake of native LDL particles by receptor-mediated endocytosis in hepatocytes and other cells is responsible for an inherited form of hypercholesterolemia known as familial hypercholesterolemia, although this autosomal dominant condition is now known to be heterogeneous in origin. The majority of patients have a 50% decrease in functional receptors, but in a small subgroup, LDL receptor binding is normal. In these individuals there is defective internalisation as a result of a mutation in the cytoplasmic domain of the LDL receptor. As a result, transfer of the occupied LDL–receptor complex to coated pits does not occur.

The uptake of native LDL particles as described above completes the second lipoprotein pathway from VLDL to VLDL remnants to LDL, a pathway governed by apo-B_{100} (for uptake of LDL) and apo-C_{II} (for activation of lipoprotein lipase in the peripheral tissues).

Scavenger Pathway. A variant pathway by which LDL particles may be taken up from the plasma, the so-called scavenger pathway, is implicated in the development of atherosclerosis. This pathway terminates in monocytes and macrophages, which recognise and internalise a subfraction of LDL in which apo-B_{100} has been pathologically modified at its lysine residues.

62.5.3.3 The HDL Pathway (Fig. 62.13)

The principal function of HDL seems to be to take up free cholesterol from peripheral tissues and, after esterifying it, to transfer the cholesterol esters to the liver or to transfer them to VLDL, and so indirectly to the liver or to the peripheral tissues via LDL. At the same time as HDL transfer cholesterol esters to VLDL, they acquire triacylglycerols from them, and these are conveyed to the liver in the HDL, along with some cholesterol esters.

It is not clear how or where HDL arise, perhaps because they arise from several different sites. Since all HDL subfractions contain apo-A_I [10], which is known to be synthesised in enterocytes, it has been argued that HDL, like chylomicrons, must have their origins in the enterocytes. It is possible, however, that primitive (discoidal) HDL acquire apo-A_I in the blood-stream by direct transfer from chylomicrons. If so, then the origin of primitive HDL is unknown: they might possibly be recycled from HLD particles after they have been processed by the liver.

Lipid Transfer Protein-1. HDL particles are converted, probably in sequence, from primitive (discoidal) HDL to HDL_3 particles containing apo-A_I, which take up cholesterol from peripheral tissues and allow it to be esterified by LCAT, to HDL_2 particles containing both apo-A_I and apo-A_{II}, in which further esterification of cholesterol seems to be prevented by the acquisition of apo-A_{II}, and, finally, to HDL_1 particles containing apo-A_I, apo-A_{II} and apo-E, which are processed by hepatocytes, which have receptors for recognition of apo-E. Primitive HDL and HDL_3 containing only apo-A_I acquire free cholesterol as they circulate through the peripheral tissues, although it is not clear whether the particles do this directly by docking onto cell membranes or whether they acquire cholesterol after it has

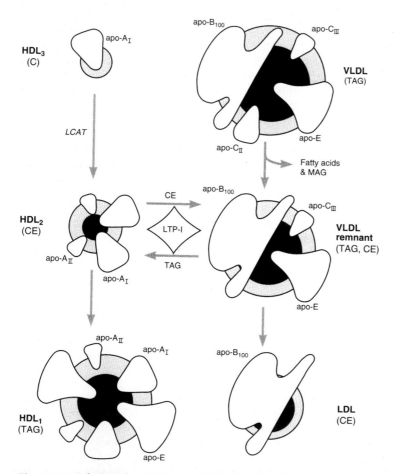

Fig. 62.13. Evolution of very high density lipoproteins (*VLDL*) via VLDL remnants to low-density lipoproteins (*LDL*) and the evolution of high-density lipoproteins (*HDL*) through three stages. Also shown is the interaction that takes place between HDL and VLDL remnants, mediated via lipid transfer protein-1 (*LTP-1*). *C*, cholesterol; *CE*, cholesterol esters; *TAG*, triacylglycerols; *MAG*, monoacylglycerols; *LCAT*, lecithin:cholesterol acyltransferase

been released into the interstitium. The cholesterol taken up into the outer envelope of these particles is then fixed within the lipoprotein pool by esterification through the action of LCAT in the plasma, which is activated by apo-A_I. The acquisition of free cholesterol followed by its esterification ceases when the particles acquire apo-A_{II}. Cholesterol esters are then transferred to VLDL by the action of a specific protein known as cholesterol ester transfer protein (CETP) or, better, lipid transfer protein-1 (LTP-1), since it also transfers triacylglycerols from VLDL to HDL. After the acquisition of apo-E, HDL particles can provide the liver with an additional supply of triacylglycerols along with cholesterol esters [2,57]. It seems likely that apo-E facilitates the hepatic uptake of HDL_1, just as it is known to do for chylomicron and VLDL remnants. The transfer of cholesterol from peripheral tissues to HDL and, after esterification, from HDL to VLDL is a process known as reverse cholesterol transport.

The movement of HDL from the peripheral tissues, where they acquire cholesterol, to the blood-stream, where they exchange cholesterol esters for triacylglycerols and phospholipids, and from the blood-stream to the liver, where triacylglycerols are hydrolysed by hepatic lipase to form free fatty acids and 2-monoacylglycerols, constitutes the third lipoprotein pathway. It is governed by apo-A_I (for uptake of free cholesterol), apo-A_{II} (for halting cholesterol esterification), and apo-E (to enable hepatic recognition). It is also dependent on LTP-1 and hepatic lipase. Whether HDL particles that have been attacked by hepatic lipase are subsequently endocytosed by the liver (perhaps at apo-E receptors) or are returned to the blood-stream as recycled, primitive HDL particles, composed largely of apo-A_I, is unclear.

62.6 Gastrointestinal Hormones and Related Substances

Enteroendocrine Cells. Throughout the intestine, the lining epithelium and its associated glands contain specialised cells characterised by the presence of secretion granules in the basal region destined for discharge into the interstitium. These are the enteroendocrine cells. On the basis of immunohistochemical identification, they appear to be organised into about 15 regionally distributed subsets. Cells containing serotonin (5-hydroxytryptamine) constitute the largest subgroup (amine precursor uptake and decarboxylation or APUD cells), and, in consequence, many enteroendocrine tumours produce large amounts of serotonin (giving rise to a condition called the carcinoid syndrome). Many of these serotonin-containing cells also contain substance P and related tachykinins [46]. Among the other enteroendocrine cells are cells containing gastrin (in G cells), CCK (in I cells), secretin (in S cells), GIP (in K cells), enteroglucagon (in L cells), somatostatin (in D cells), motilin (in M cells), neurotensin (in N cells), PP (formerly known as pancreatic polypeptide; in F cells) or histamine (in ECl cells). Most, if not all of these cells have an endocrine function and release their hormones in response to neural and hormonal stimuli as well as certain chemicals (called secretagogues) in the lumen of the gastrointestinal tract. The hormones act on secretory and/or smooth muscle cells to evoke secretion and contraction (or relaxation), respectively. Some of these compounds act also as trophic hormones, promoting epithelial cell growth.

62.6.1 Gastrin

C-Terminal Tetrapeptide. Gastrin is synthesised and secreted by G cells, two thirds of which are found in the pyloric glands and one third in the duodenal mucosa. It is synthesised as pre-pro-gastrin, consisting of a 21-residue signal sequence, a 58-residue intermediate peptide and the 34-residue unit that constitutes gastrin proper (G-34). The hormone exists in two major active forms, G-34 (or big gastrin) and G-17 (or little gastrin), but the biological activity of both forms resides exclusively in an amidated C-terminal tetrapeptide (G-4): -Trp-Met-Asp-Phe-NH_2. The two forms differ in their distribution within the intestine (G-17 predominates in the antrum, G-34 in the duodenum) and in their biological half-lives (G-17 is cleared from the plasma in less than 3 min, whereas G-34 remains for 9–15 min). The mechanism of gastrin clearance from the body is not fully understood, but the C-terminal tetrapeptide is known to be cleared in the liver.

Release of Gastrin. Gastrin is released into the blood-stream in response to the presence of peptides and some amino acids (mainly the aromatic amino acids) in the gastric lumen, provided that free Ca^{2+} is also present. It is also released by vagal nerve activity (involving gastric ganglion cells containing a peptide transmitter called bombesin or GRP, gastrin-releasing peptide) and by high levels of circulating catecholamines. Gastrin release is inhibited whenever the pH of the contents of the gastric antrum falls below about 3 (an example of negative feedback control) and is also inhibited by circulating somatostatin. It is thought that the presence of proteins in the duodenal lumen may cause release of gastrin from the duodenal G cells.

Actions of Gastrin. Gastrin activates receptors in the plasma membranes of its target cells, which, in turn, activate the phospholipase C signal transduction pathway via a G protein, leading to an increase in the formation of inositol trisphosphate (IP_3), which causes Ca^{2+} to be released from intracellular stores, and diacylglycerol, which activates protein kinase C. Its actions are as follows:

* It acts directly on parietal cells to cause secretion of HCl; it is more potent than histamine, although its potency is greatly diminished if the histamine (H_2) receptor is blocked.

1281

- It can also cause discharge of pepsinogen from the chief cells of the oxyntic glands, although it is not a very powerful agonist in this respect.
- It acts on gastrin receptors on the smooth muscle of the distal stomach, where it increases the force and frequency of gastric peristaltic contractions.
- It acts as a trophic (i.e., growth-promoting) hormone on the gastric and duodenal epithelia. Consequently, after surgical resection of the antrum, mucosal atrophy occurs. Conversely, G cell tumours, which secrete excess quantities of gastrin, are associated with mucosal hypertrophy.
- It has been reported to stimulate pancreatic acinar and biliary secretion and to cause gall-bladder contraction, probably because it mimics many of the actions of CCK.

Achlorhydria and the Zollinger-Ellison Syndrome. The clinical condition known as achlorhydria (literally, a condition of no HCl production) is commonly associated with increased circulating levels of gastrin. It appears that the primary defect in this condition is in the parietal cells, which undergo atrophy and are unresponsive to gastrin, so that the normal negative feed-back regulation of gastrin release, mediated by the luminal pH in the antrum, fails. A somewhat rarer condition, known as the Zollinger-Ellison syndrome, is seen when G cell tumours develop. It is characterised by a high rate of gastrin secretion and an increased production of HCl by the parietal cells. This leads to gastric mucosal ulceration, malabsorption and diarrhoea (due to acid inactivation of pancreatic lipase and precipitation of bile salts).

62.6.2 CCK (see also Sect. 65.4.1)

CCK (formerly known as cholecystokinin) is synthesised in I cells in the epithelium of the small intestine, particularly in the duodenum and jejunum. It is synthesised as pre-pro-CCK, consisting of a 20-residue signal sequence, a 25-residue linking sequence, a 58-residue active sequence (CCK-58) and a 12-residue C-terminal extension. The normal circulating forms consist of CCK-58 and CCK-33 (formed by cleaving off a 22-residue segment from the N terminus of CCK-58), but the amidated C-terminal octapeptide (CCK-8) is found as a transmitter in nerve tissue. The ratio of CCK-58 to CCK-33 is about 50:50 in the proximal duodenum, increasing to about 90:10 in the distal jejunum. The maximum biological activity of all forms resides in the amidated C-terminal octapeptide, i.e. -Asp-Tyr(SO_3)-Met-Gly-Trp-Met-Asp-Phe-NH_2, but the full spectrum of activity can be elicited with the C-terminal heptapeptide CCK-7. It is noteworthy that the C terminal pentapeptide of this octapeptide is identical to the corresponding C-terminal pentapeptide of gastrin. Once released into the blood-stream, all forms of CCK are inactivated within a few minutes by plasma and tissue aminopeptidases.

Release of CCK. CCK is released in response to the presence of fatty acids, peptides and aromatic amino acids in the duodenal lumen. Release is also evoked by the presence of glucose (plus Ca^{2+} and Mg^{2+}) in the duodenal lumen. A negative feed-back control of CCK release, based on the level of free trypsin in the intestinal lumen, also operates. Thus, whenever the level of free trypsin (i.e. trypsin not bound to ingested proteins or other substrates) in the duodenum increases, CCK release is inhibited, and whenever trypsin levels are low, so that dietary protein is present in excess, CCK release is increased.

Actions of CCK. CCK, like gastrin, works via the phospholipase C signal transduction pathway. Its actions are as follows:

- It acts on pancreatic acinar cells to evoke secretion of a neutral, Cl^--rich juice containing pancreatic enzymes.
- It acts on pancreatic duct cells, where it potentiates the action of secretin in causing secretion of an alkaline, HCO_3^--rich juice.
- It acts on the chief cells in gastric glands to cause pepsinogen release.
- It is a weak stimulant of HCO_3^- secretion by the gastric and duodenal surface epithelia and by the bile ducts.
- It inhibits gastric HCl secretion, possibly by evoking somatostatin secretion or perhaps by competing with gastrin for it receptor, for which CCK is only a partial agonist.
- It is a potent stimulant of gall-bladder contraction whilst simultaneously relaxing the sphincter of Oddi.
- It evokes release of many pancreatic hormones including insulin, glucagon, somatostatin and PP.
- It is trophic to its target tissues, particularly the pancreas. In populations in which soya flour, which contain a powerful trypsin inhibitor, forms part of the dietary staple, the ingestion of the trypsin inhibitor, which reduces the concentration of free trypsin in the duodenal lumen, leads to hypersecretion of CCK with development of pancreatic hypertrophy. This, in turn, is associated with an increased frequency of pancreatic carcinoma.
- It is thought to act on the medullary feeding centres as a satiety hormone, signalling that feeding should cease.

Recently, an intestinal cell line (STC-1) that secretes CCK has been described. These cells are activated by depolarisation in response to the closure of Ba^{2+}- and tetraethylammonium (TEA)-sensitive K^+ channels. Ca^{2+} ions then enter the cell as a result of the opening of voltage-sensitive, L-type Ca^{2+} channels [54]. The link between exposure of the epithelium to undigested fats and peptides, on the one hand, and the release of CCK, on the other, has not yet been identified.

62.6.3 Secretin

Secretin, which was the first hormone ever to be described, is synthesised in S cells in the epithelium of the duodenum and jejunum. It is synthesised as a single-chain, 27-residue peptide which shows close structural homology with glucagon, enteroglucagon, GIP and VIP. The entire polypeptide is necessary for maximum biological activity, but the full spectrum of activity can be elicited with a 14-residue N-terminal peptide, S_{1-14}. It is cleared from the plasma by the kidneys with a half-time of about 3 min. Secretin is released in response mainly to acidification of the duodenal contents, the amount released being correlated with the length of the acidified duodenal segment. Release of secretin is enhanced by the presence of bile salts in the duodenal lumen and inhibited by the presence of pancreatic juice. Its release is not influenced by vagal activity.

Actions of Secretin. In contrast to gastrin and CCK, secretin works by activation of the adenylate cyclase signal transduction pathway. Its principal actions are as follows:

- Its major action is to evoke formation of an alkaline juice by the pancreatic ducts, an action that is potentiated by the presence of CCK.
- It also evokes formation of an alkaline juice by the bile ducts and Brunner's glands.
- It inhibits salt and water absorption by the gall-bladder.
- It slows down the rate of gastric emptying by reducing the activity of the gastric musculature.
- It is antitrophic to the gastric epithelium.

62.6.4 Enteroglucagon, Glucagon-Like Peptides and the "Incretin" Effect

Structurally, enteroglucagon resembles pancreatic glucagon, although it is a somewhat larger molecule and its actions are quite different. Pre-pro-glucagon (180 residues) is produced both in α-cells of pancreatic islets and in L enteroendocrine cells in the epithelium of the distal ileum and the colon. In both cases, it is processed to pro-glucagon by the cleavage of a 20-residue N-terminal signal peptide (Fig. 62.14).

Glucagon and GLP-I. In α-cells, pro-glucagon is further processed by limited proteolysis to form glucagon (29 residues), a 37-residue peptide called glucagon-like peptide-I (GLP-I) and two other peptides of unknown function (Fig. 62.14).

Enteroglucagon (Glicentin and Oxyntomodulin). In L-enteroendocrine cells, pro-glucagon is converted instead to a 69-residue peptide called glicentin and several shorter peptides including GLP-I. Glicentin is further processed by cleavage of the N-terminal peptide to yield oxyntomodulin (37 residues), which consists of pancreatic glucagon extended by eight residues at the C terminus (Fig. 62.14). Glicentin and oxyntomodulin cross-react immunologically and, together, are referred to as enteroglucagon. They are released by the presence of glucose and fats in the lumen of

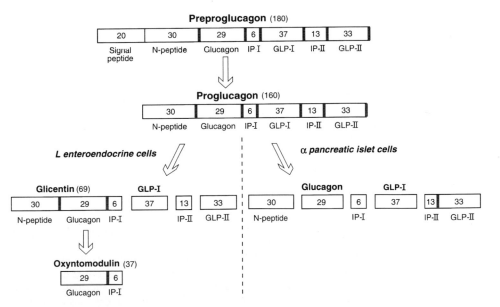

Fig. 62.14. The peptides liberated by pancreatic islet α-cells and L enteroendocrine cells during the processing of pre-pro-glucagon to yield glucagon (α-pancreatic islet cells) and glicentin and oxyntomodulin (L enteroendocrine cells), together with glucagon-like peptide (GLP)-I from both cell types. The numbers on each bar or in parentheses indicate the number of amino acid residues in each peptide: the solid vertical bars denote excision dipeptides. It should be noted that glicentin and oxyntomodulin show immunological cross-reactivity and are referred to, collectively, as enteroglucagon. IP, intermediate peptide

the lower ileum so that, normally, their plasma concentrations are only increased appreciably in cases of upper intestinal malabsorption. Their only known functions are to inhibit gastric and intestinal motility and gastric acid secretion and to regulate the growth of cells in the intestinal crypts. They do not mimic any of the major actions of pancreatic glucagon.

"Incretin". GLP-I is biologically inactive, but can be processed into one of two forms, either of which can cause insulin release in response to the ingestion of carbohydrates. This is known as the "incretin effect", because is was formerly thought to be due to an impurity in preparations of secretin which was provisionally given the name "incretin". In fact, when released into the blood-stream in picomolar concentrations, several peptides, including modified GLP-I and GIP (see Sect. 62.6.5), evoke insulin release from pancreatic β-cells in this way, as well as potentiating the response of the β-cells to glucose.

62.6.5 GIP

GIP (formerly known as gastric inhibitory peptide and also known as glucose-dependent insulin-releasing peptide) is synthesised and secreted by K cells found throughout the small intestine, particularly in the jejunum. It is a single-chain, 42-residue peptide showing homology to secretin and glucagon. It is released in response to the presence of glucose, fat and amino acids in the upper small intestine, as well as by a fall in luminal pH. GIP has one major function: to enhance insulin release from β-cells in the pancreatic islets of Langerhans (hence the name "glucose-dependent insulin-releasing peptide"), a phenomenon known as the "incretin effect" (see above, Sect. 62.6.4). Although there is some doubt about the physiological importance of this action of GIP, since it has only been induced by administration of rather high concentrations of the peptide, antibodies to GIP have been shown to inhibit the incretin effect in rats. In addition, as its original name, "gastric inhibitory peptide", implies, it inhibits gastric HCl secretion and reduces gastric motor activity. It also stimulates intestinal secretion.

62.6.6 VIP

VIP (formerly known as vasoactive intestinal peptide) is a 28-residue peptide showing very close homology with secretin. Although it is widely distributed in the intestine (particularly the colon), it appears to be exclusively a neurotransmitter peptide and is found in excitatory enteric ganglion cells, either alone or co-localised with acetylcholine. When administered to humans parenterally, it mimics all the major actions of secretin and is, in fact, far more potent than secretin itself. In addition, VIP is a powerful relaxant of smooth muscle in many organs (hence its original name). VIP-secreting tumours are occasionally encountered in humans: they give rise to a condition known as pancreatic cholera or the WDHA (watery diarrhoea, hypokalaemia, acidosis) syndrome.

62.6.7 Somatostatin

Somatostatin (SS) is synthesised as a pre-pro-hormone of 116 amino acids which has an N-terminal signal sequence of 24 residues followed by an intermediate peptide of 64 residues, with somatostatin making up the last 28 residues at the C terminus. The pro-hormone can be cleaved to release either SS-28 or SS-14. The two forms differ in their biological potencies, SS-14 being ten times more potent as an inhibitor of gastric acid secretion. Somatostatin is found in nerve terminals and ganglion cells in the enteric nervous system and in D cells in the pancreatic islets and scattered throughout the intestinal epithelium. The presence of protein, fat, glucose and bile salts in the small intestine causes plasma somatostatin levels to increase. Plasma somatostatin levels are also increased after commencement of a migrating motor complex. Circulating somatostatin is cleared by the kidneys and liver with a half-time of 3 min for SS-14 and 30 min for SS-28.

Somatostatin has four major functions:

- It is a powerful inhibitor of gastric acid and pepsin secretion and of gastrin release by G cells in the glands of the gastric antrum. It perhaps provides the mechanism by which fat in the duodenum inhibits gastric secretion of acid.
- It also inhibits pancreatic enzyme secretion, the intestinal absorption of amino acids and the secreto-motor effects of CCK.
- It inhibits the trophic effects of gastrin on the gastric mucosa.
- It reduces acetylcholine release from enteric neurones and thus inhibits the normal intestinal responses to vagal stimulation, thereby altering intestinal motility. (For example, it counteracts the effects of motilin and delays the normal occurrence of the migrating motor complex.)

62.6.8 Motilin

Motilin is a linear 22-amino acid peptide in which the entire molecule is necessary for biological activity. It is found in M cells in the upper small intestine. It is released during the occurrence of the migrating motor complex and in response to acid, alkali, fat or pancreato-biliary juice in the duodenum. Its release is inhibited by somatostatin. It appears to control the onset of the migrating motor complex. It also appears to increase gastric motor activity, whilst relaxing the pylorus at the end of the gastric phase of digestion, thus being the only hormone described as directly promoting gastric emptying.

62.6.9 Neurotensin

Neurotensin is a 13-residue peptide found in the N cells of the ileal epithelium. It is released in response to the ingestion of fat and, like somatostatin, is a candidate mediator for the inhibition of gastric acid secretion by the presence of fat in the intestine.

62.6.10 PP and NPY

PP (formerly known as pancreatic polypeptide) is a 36-residue peptide. Its major source is the F cell in pancreatic islets, particularly in islets in the uncinate process and head of the pancreas. It is released in response to a protein meal and by cholinergic reflexes such as those induced by sham feeding. It is cleared mainly by the kidneys. Its major biological action is to inhibit pancreatic secretion, both of enzymes and HCO_3^-.

NPY (formerly known as neuropeptide Y) is another 36-residue peptide showing close structural homology with PP. Unlike PP, however, it functions only as a neurotransmitter, usually in sympathetic neurones, where it is co-localised with noradrenaline, the actions of which it potentiates.

62.6.11 Substance P

Tachykinins. Substance P is an 11-amino acid peptide belonging to a class of peptides known as the tachykinins. Although it functions as a hormone in some lower animals, in humans it appears to function only as a neurotransmitter. Its biological activity resides in the amidated C-terminal pentapeptide, i.e., -Phe-Phe-Gly-Leu-Met-NH$_2$. In some species, e.g., rat, when substance P is injected into the blood-stream it acts as an exceptionally potent secreto-motor agonist and causes marked secretion by virtually every exocrine gland in the body and contraction of smooth muscle in most sites. In other species, e.g., mouse, substance P has little or no action. Cells containing substance P seem to be found scattered in the epithelium throughout the small intestine, particularly in the villi. It seems to co-localise with serotonin. In salivary glands, it has been co-localised with acetylcholine in postganglionic parasympathetic secreto-motor nerve terminals, but elsewhere in the enteric nervous system it is found in afferent neurones. It is found also in the central nervous system.

62.6.12 GRP (Bombesin)

GRP (formerly known as gastrin-releasing peptide), which is also called bombesin, is a 27-residue peptide found in the terminals of some fibres of the intrinsic nervous system of the stomach. Its activity resides in the C-terminal nonapeptide. Its principal action is to evoke gastrin release from G cells in the gastric antrum and duodenum. The GRP-containing nerves that evoke gastrin release are activated via the vagus.

62.6.13 Histamine

Histamine is formed by the decarboxylation of the amino acid histidine. It is synthesised and stored in granules throughout the body in mast cells (and is involved in allergic reactions in which it acts on H_1 receptors). In addition, it is synthesised in ECl cells in the oxyntic glands of the stomach. It is released from these cells in response to vagal stimulation, and it acts on H_2 receptors on parietal cells to cause secretion of HCl and on chief cells to cause release of pepsinogen.

62.6.14 5-Hydroxytryptamine (5-OHT) or Serotonin

Serotonin is formed by the decarboxylation of tryptophan. It is synthesised in myenteric neurones and in enteroendocrine cells located in the gastric glands and throughout the small intestine. In the enteroendocrine cells, it is found alone or in association with one or more of a range of intestinal peptides, particularly substance P. When injected intravenously, serotonin increases secretion of water and electrolytes in the small intestine by stimulating cholinergic secreto-motor neurones in the submucous plexus but the physiological significance of this is unknown.

References

1. Alaupovic P, Lee DM, McConathy WJ (1972) Studies on the composition and structure of plasma lipoproteins. Distribution of lipoprotein families in major density classes of normal human plasma lipoproteins. Biochim Biophys Acta 260:689–707
2. Albers JJ (1990) Lipid transfer proteins. In: Carlson LA (ed) Disorders of HDL. Smith-Gordon, London, pp 25–30
3. Alpers DH (1985) Uptake and fate of absorbed amino acids and peptides in the mammalian intestine. FASEB J 45:2261–2267
4. Aoubala M, Douchet I, Laugier R, Hirn M, Verger R, Caro AD (1993) Purification of human gastric lipase by immuno-affinity and quantification of this enzyme in the duodenal contents using a new ELISA procedure. Biochim Biophys Acta 1169:183–188
5. Bell GI, Kayano T, Buse JB, Burant CF, Takeda J, Lin D, Fukumoto H, Seino S (1990) Molecular biology of mammalian glucose transporters. Diabetes Care 13:198–208
6. Bijman J, Veeze H, Kansen M, Tilly B, Scholte B (1991) Chloride transport in the cystic fibrosis enterocyte. Adv Exp Med Biol 290:287–294
7. Bodmer MW, Angal S, Yarranton GT, Harris TJR, Lyons A, King DJ, Pieroni G, Riviere C, Verger R, Lowe PA (1987) Molecular cloning of a human gastric lipase and expression of the enzyme in yeast. Biochim Biophys Acta 909:237–244
8. Borgström A, Erlanson-Albertsson C, Borgström B (1993) Human pancreatic proenzymes are activated at different rates in vitro. Scand J Gastroenterol 28:455–459

9. Braun JEA, Severson DL (1992) Regulation of the synthesis, processing and translocation of lipoprotein lipase. Biochem J 287:337–347

10. Brewer HB, Rader D, Fojo S, Hoeg JM (1990) Frontiers in the analysis of HDL structure, function, and metabolism. In: Carlson LA (ed) Disorders of HDL. Smith-Gordon, London, pp 51–58

11. Brown WE, Wold F (1973) Alkyl isocyanates as active-site-specific reagents for serine proteases. Reaction properties. Biochemistry 12:828–834

12. Burgos FJ, Salva M, Villegas V, Soriano F, Mendez E, Aviles FX (1991) Analysis of the activation process of porcine procarboxypeptidase B and determination of the sequence of its activation segment. Biochemistry 30:4082–4089

13. Carriere F, Moreau H, Raphel V, Laugier R, Benicourt C, Junien J, Verger R (1991) Purification and biochemical characterization of dog gastric lipase. Eur J Biochem 202:75–83

14. Carriere F, Barrowman JA, Verger R, Laugier R (1993) Secretion and contribution to lipolysis of gastric and pancreatic lipases during a test meal in humans. Gastroenterology 105:876–888

15. Chapus C, Kerfelec B, Foglizzo E, Bonicel J (1987) Further studies on the activation of bovine pancreatic procarboxy-peptidase A by trypsin. Eur J Biochem 166:379–385

16. Chen H, Born E, Mathur SN, Johlin FC, Field FJ (1992) Sphingomyelin content of intestinal cell membranes regulates cholesterol absorption. Evidence for pancreatic and intestinal cell sphingomyelinase activity. Biochem J 286:771–777

17. Chen S, Habib G, Yang C, Gu Z, Lee BR, Weng S, Silberman SR, Cai S, Deslypere JP, Rosseneu M, Grotto AM, Li W, Chan L (1987) Apolipoprotein B-48 is the product of a messenger RNA with an organ-specific in-frame stop codon. Science 238:363–366

18. DeNigris SJ, Hamosh M, Kasbekar DK, Lee TC, Hamosh P (1988) Lingual and gastric lipases: species differences in the origin of prepancreatic digestive lipases and the localization of gastric lipase. Biochim Biophys Acta 959:38–45

19. Eckel RH (1989) Lipoprotein lipase. A multifunctional enzyme relevant to common metabolic diseases. N Engl J Med 320:1060–1068

20. Eisenhauer PB, Harwig SSSL, Lehrer RI (1992) Cryptdins: antimicrobial defensins of the murine small intestine. Infect Immun 60:3556–3565

21. Erlanson-Albertsson C (1992) Enterostatin: the pancreatic colipase activation peptide. A signal for regulation of fat intake. Nutr Rev 50:307–310

22. Ferraris RP, Diamond J (1992) Crypt-villus site of glucose transporter induction by dietary carbohydrate in mouse intestine. Am J Physiol 262:G1069–G1073

23. Field M, Frizzell RA (eds) (1991) Handbook of physiology. The gastrointestinal system. Intestinal absorption and secretion, sect 6, vol IV. American Physiological Society, Bethesda

24. Field M, Semrad CE (1993) Toxigenic diarrheas, congenital diarrheas, and cystic fibrosis. Disorders of intestinal ion transport. Annu Rev Physiol 55:631–655

25. Fielding CJ (1992) Lipoprotein receptors, plasma cholesterol metabolism, and the regulation of cellular free cholesterol concentration. FASEB J 6:3162–3168

26. Fisher WR, Schumaker VN (1986) Isolation and characterization of apolipoprotein B-100. Methods Enzymol 128:247–262

27. Filint N, Cove FL, Evans GS (1991) A low-temperature method for the isolation of small-intestinal epithelium along the crypt-villus axis. Biochem J 280:331–334

28. Gjellesvik DR (1991) Fatty acid specificity of bile salt-dependent lipase: enzyme recognition and super-substrate effects. Biochim Biophys Acta 1086:167–172

29. Gordon JI, Schmidt GH, Roth KA (1992) Studies of intestinal stem cells using normal, chimeric, and transgenic mice. FASEB J 6:3039–3050

30. Hagen TM, Bai C, Jones DP (1991) Stimulation of glutathione absorption in rat small intestine by α-adrenergic agonists. FASEB J 5:2721–2727

31. Hide WA, Chan L, Li W (1992) Structure and evolution of the lipase superfamily. J Lipid Res 33:167–178

32. Hui DY, Hayakawa K, Oizumi J (1993) Lipoamidase activity in normal and mutagenized pancreatic cholesterol esterase (bile salt-stimulated lipase). Biochem J 291:65–69

33. Knott TJ, Pease RJ, Powell LM, Wallis SC, Rall SC, Innerarity TL, Blackhart B, Taylor WH, Marcel Y, Milne R, Johnson D, Fuller M, Lusis AJ, McCarthy BJ, Mahley RW, Levy-Wilson B, Scott J (1986) Complete protein sequence and identification of structural domains of human apolipoprotein B. Nature 323:734–738

34. Lawn RM (1992) Lipoprotein(a) in heart disease. Sci Am 1992:54–60

35. Lombardo D, Guy O (1980) Studies on the substrate specificity of a carboxyl ester hydrolase from human pancreatic juice. II. Action on cholesterol esters and lipid-soluble vitamin esters. Biochim Biophys Acta 611:147–155

36. Lombardo D, Fauvel J, Guy O (1980) Studies on the substrate specificity of a carboxyl ester hydrolase from human pancreatic juice. I. Action on carboxyl esters, glycerides and phospholipids. Biochim Biophys Acta 611:136–146

37. Lorenzsonn V, Lloyd M, Olsen WA (1993) Immunocytochemical heterogeneity of lactase-phlorizin hydrolase in adult lactase deficiency. Gastroenterology 105:51–59

38. Lowe ME (1992) The catalytic site residues and interfacial binding of human pancreatic lipase. J Biol Chem 267:17069–17073

39. Moulard M, Michon T, Kerfelec B, Chapus C (1990) Further studies on the human pancreatic binary complexes involving procarboxypeptidase A. FEBS Lett 261:179–183

40. Navab F, Beland SS, Cannon DJ, Texter EC (1984) Mechanisms of transport of L-histidine and β-alanine in hamster small intestine. Am J Physiol 247:G43–G51

41. Pajor AM, Wright EM (1992) Cloning and expression of a mammalian Na$^+$/nucleoside cotransporter: a member of the SGLT family. J Biol Chem 267:3557–3560

42. Ponder BAJ, Schmidt GH, Wilkinson MM, Wood MJ, Monk M, Reid A (1985) Derivation of mouse intestinal crypts from single progenitor cells. Nature 313:689–691

43. Potten CS, Loeffler M (1990) Stem cells: attributes cycles, spirals, pitfalls and uncertainties. Lessons for and from the crypt. Development 110:1001–1020

44. Potter BJ, Sorrentino D, Berk PD (1989) Mechanisms of cellular uptake of free fatty acids. Annu Rev Nutr 9:253–270

45. Rand EB, Depaoli AM, Davidson NO, Bell GI, Burant CF (1993) Sequence, tissue distribution, and functional characterization of the fructose transporter GLUT5. Am J Physiol 264:G1169–G1176

46. Roth KA, Gordon JI (1990) Spatial differentiation of the intestinal epithelium: analysis of enteroendocrine cells containing immunoreactive serotonin, secretin, and substance P in normal and transgenic mice. Proc Natl Acad Sci U S A 87:6408–6412

47. Satoh Y, Ishikawa K, Oomori Y, Takeda S, Ono K (1992) Bethanechol and a G-protein activator, NaF/AlCl$_3$, induce secretory response in Paneth cells of mouse intestine. Cell Tissue Res 269:213–220

48. Segrest JP, Jones MK, Loof HD, Brouillette CG, Venkatachalapathi YV, Anantharamaiah GM (1992) The amphipathic helix in the exchangeable apolipoproteins. J Lipid Res 33:141–166

49. Shen W, Fletcher TS, Largman C (1987) Primary structure of human pancreatic protease E determined by sequence analysis of the cloned mRNA. Biochemistry 26:3447–3452

50. Shirasu Y, Yoshida H, Mikayama T, Matsuki S, Tanaka J, Ikenaga H (1986) Isolation and expression in *Escherichia coli* of a cDNA clone encoding porcine pancreatic elastase. J Biochem (Tokyo) 99:1707–1712

51. Shirasu Y, Yoshida H, Matsuki S, Takemura K, Ikeda N, Shimada Y, Ozawa T, Mikayama T, Iijima H, Ishida A, Sato Y, Tamai Y, Tanaka J, Ikenaga H (1987) Molecular cloning and expression in *Escherichia coli* of a cDNA encoding human pancreatic elastase 2. J Biochem (Tokyo) 102:1555–1563

52. Shotton DM, Watson HC (1970) Three dimensional structure of tosylelastase. Nature 225:811–816

53. Smythe E, Warren G (1991) The mechanism of receptor-mediated endocytosis. Eur J Biochem 202:689–699

54. Snow ND, Mangel AW, Sharara AI, Liddle RA (1993) Potassium channels regulate cholecystokinin secretion in STC-1 cells. Biochem Biophys Res Commun 195:1379–1385

55. Stevens BR, Kaunitz JD, Wright EM (1984) Intestinal transport of amino acids and sugars: advances using membrane vesicles. Annu Rev Physiol 46:417–433

56. Sweetser DA, Heuckeroth RO, Gordon JI (1987) The metabolic significance of mammalian fatty-acid binding proteins. Annu Rev Nutr 7:337–359

57. Swenson TL (1991) The role of cholesteryl ester transfer protein in lipoprotein metabolism. Diabetes Metab Rev 7:139–153

58. Tiruppathi C, Balasubramanian KA (1982) Purification and properties of an acid lipase from human gastric juice. Biochim Biophys Acta 712:692–697

59. Tresize AEO, Buchwald M (1991) In vivo cell-specific expression of the cystic fibrosis transmembrane conductance regulator. Nature 353:434–437

60. Vadgama JV, Evered DF (1992) Characteristics of L-citrulline transport across rat small intestine in vitro. Pediatr Res 32:472–478

61. Vendrell J, Cuchillo C, Aviles FX (1990) The tryptic activation pathway of monomeric procarboxypeptidase A. J Biol Chem 265:6949–6953

62. Walters RJ, Obrien JA, Valverde MA, Sepulveda FV (1992) Membrane conductance and cell volume changes evoked by vasoactive intestinal polypeptide and carbachol in small intestinal crypts. Pflugers Arch 421:598–605

63. Wang C, Hartsuck JA, Downs D (1988) Kinetics of acylglycerol sequential hydrolysis by human milk bile salt activated lipase and effect of taurocholate as fatty acid acceptor. Biochemistry 27:4834–4840

64. Welsh MJ, Smith PL, Fromm M, Frizzell RA (1982) Crypts are the site of intestinal fluid and electrolyte secretion. Science 218:1219–1221

65. Wheeler EJ, Wheeler JK (1964) Comparative study of Paneth cells in vertebrates. Anat Rec 148:350

66. Wild GE, Murray D (1992) Alterations in quantitative distribution of Na, K-ATPase activity along the crypt-villus axis in animal model of malabsorption characterized by hyperproliferative crypt cytokinetics. Dig Dis Sci 37:417–425

67. Wright EM, Van Os CH, Mircheff AK (1980) Sugar uptake by intestinal basolateral membrane vesicles. Biochim Biophys Acta 597:112–124

68. Yang LY, Kuksis A (1991) Apparent convergence (at 2-monoglycerol level) of phosphatidic acid and 2-monoacylglycerol pathways of synthesis of chylomicron triacylglycerols. J Lipid Res 32:1173–1186

69. Yatani A, Codina J, Imoto Y, Reeves JP, Birnbaumer L, Brown AM (1987) A G protein directly regulates mammalian cardiac calcium channels. Science 238:1288–1292

63 Gastrointestinal Motility and Defecation

J.E. KELLOW

Contents

63.1 Introduction

During the last decade, renewed interest in motility of the gastrointestinal tract has been generated by a greater understanding of enteric neuroanatomy and neuropharmacology. Improvements in recording systems and techniques, particularly those applicable to humans, have led to more definitive descriptions of motility in all regions of the gut.

The term *motility* encompasses both motor activity and transit. *Motor activity* refers to the intestinal contractions and relaxations of the gut and to intestinal tone. These result in the propulsion of digesta along the gastrointestinal tract, i.e., *transit*, or in mixing of the digesta. A number of special features of the motor system of the gastrointestinal tract (smooth muscle and enteric nervous system) enable these events to occur, namely: the specific anatomical arrangement of smooth muscle layers and neural plexuses, the spontaneous electrical rhythmicity of the gastrointestinal smooth muscle (cf. Chap. 44), the "pacemaker" sites within the stomach and intestines, and the functional integration of the enteric nervous system with the central nervous system. These features provide for:

- Alteration of motility according to local conditions (e.g., presence or absence of luminal contents)
- Co-ordination of motility within and between gut regions (e.g., stomach and duodenum)
- Integration of motility with other physiological states (e.g., wakefulness/sleep)
- Appropriate interaction between motility and other gut functions (e.g., absorption/secretion)

In this chapter, the more applied aspects of motility in each specific organ are emphasized, with particular reference to the above areas. Where information in humans is not yet complete, extrapolations from other species are discussed.

63.2 Regional Gastrointestinal Motility

63.2.1 Motility of the Stomach

The *stomach* has several important roles as a *motor* organ:

- Postprandially, to act as a short-term storage site for ingested food and as a receptacle that can grind, triturate and sieve food
- To deliver chyme into the small bowel at rates appropriate for optimum digestion, at the same time preventing retrograde passage of small bowel contents into the antral region
- During fasting, to ensure the continual delivery of secretions and swallowed saliva into the duodenum and to empty the indigestible solids remaining in the stomach after a meal.

Regional motor specialization of the stomach has been recognized since Cannon's early fluoroscopic observations of gastric motility in the cat [15]. Thus, the proximal one-third of the stomach (fundus and proximal corpus) acts as a reservoir, maintaining a constant low pressure sufficient to transport the gastric contents towards the antrum (Fig. 63.1); this region plays an important part in the emptying of liquids. The distal stomach (the greatest part of the corpus, the antrum and the pylorus) displays phasic contractions which sweep from the mid-stomach aborally to the pylorus, breaking down and homogenizing solid food. Techniques that have led to the elucidation of human gastric motor function can be divided into three main categories (Table 63.1). The most widely used approach is *radioisotopic scintigraphy*; it is non-invasive, can assess emptying of solid and liquid phases simultaneously, and can even delineate the spatial sequence of gastric contractions [116]. In this technique, a meal is ingested in which one or several components have been labelled with a different radioisotopic γ- emitting marker, e.g., ^{99m}Tc for the solid component and ^{113m}In (chelated with a water-soluble carrier) for the aqueous (liquid) phase. An external γ-camera simultaneously quantifies the disappearance of each radiolabel from the gastric region, and this corresponds to the rate of emptying of the labeled component. The selection of meal, tracer, scintigraphic technique and method of analysis is extremely variable. The most detailed information on gastric motor physiology, however, is obtained by using several techniques simultaneously, e.g.,

R. Greger/U. Windhorst (Eds.)
Comprehensive Human Physiology, Vol. 2
© Springer-Verlag Berlin Heidelberg 1996

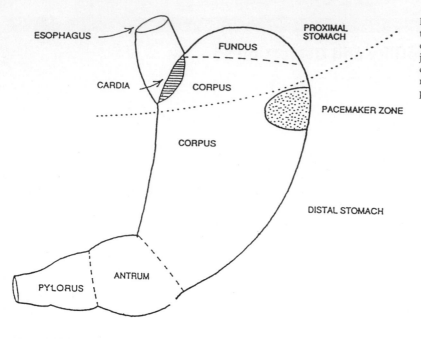

Fig. 63.1. Schematic showing the anatomical and motor subdivisions of the empty stomach. *Dotted line* shows the junction between the proximal (accommodative) and the distal (peristaltic) regions of the stomach. (From [121], with permission)

Table 63.1. Techniques used to assess motor functions of the human stomach

Motor function	Technique	References
Gastric motor activity	Gastropyloroduodenal manometry – perfused catheter/sleeve sensor	[42,46,47,68]
	Gastric barostat	[6]
Gastric emptying	Radioisotopic scintigraphy	[14,17,18,22,84]
	Ultrasonography	[11,41,62]
	Epigastric impedance	[73]
	Applied potential tomography	[2]
	Magnetic resonance imaging	[99]
	Paracetamol absorption	[80]
	Alcohol absorption	[45]
	Radio-opaque markers	[32]
	Gastric aspiration	[37]
Gastric myoelectrical activity	Electrogastrography (surface, mucosal)	[1]

antropyloroduodenal manometry combined with barostat recording and sophisticated scintigraphic or ultrasonic techniques.

Gastric Myoelectrical Activity. As in the remainder of the digestive tract, gastric smooth muscle is composed of two main layers. The outer longitudinal layer is in continuity with the pylorus, while the inner circular muscle is present only as far as the distal antrum. There is also a third (inner) layer of muscle, immediately beneath the mucosa of the proximal stomach, along the aspect of the lesser curve and in continuity with the muscle of the lower esophageal sphincter. At the pylorus, the circular muscle bundles thicken to form proximal and distal muscle bundles. Depolarization of the resting transmembrane electrical potential maintained by each individual smooth muscle cell triggers contraction of gastric smooth muscle. Smooth muscle cells of the proximal gastric region demonstrate no spontaneous depolarization, while cells of the distal gastric region exhibit spontaneous action potentials [110]. The frequency, duration and character of these fluctuations in membrane potential vary according to the particular location along the stomach, the frequency being greatest in the most *orad* extremity of the distal gastric region [75].

When extracellular recordings are made from the serosal surface of the fundus and orad one third of the corpus, no electrical activity is discernible. In the antrum and distal two thirds of the corpus, however, *electrical activity* is characterized by the presence of *slow waves*, also called *electrical control activity* (ECA), *basal electrical rhythm*, or *pace-setter potential* (Fig. 63.2). The frequency of gastric slow waves is approximately 3/min in man. This frequency

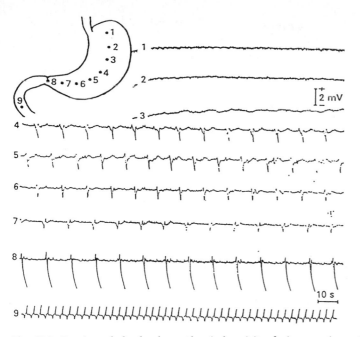

Fig. 63.2. Gastric and duodenal myoelectrical activity during motor quiescence in the dog. The fundus and orad corpus are electrically silent, whereas a regular, distally propagating depolarization – the slow wave – is present in the distal stomach. The slow wave frequency is approximately 5/min. (From [61], with permission)

is uniform throughout the stomach, since although there is a decreasing cellular frequency gradient from the orad to the aboral extremity of the distal region, an area of greater frequency of spontaneous depolarization will entrain an adjacent region of slower frequency to the higher rate. The gastric pacemaker appears to be localized to the area around the junction of the orad and middle thirds of the corpus, along the greater curvature [60]. The *pacemaking activity* appears to be due to the network of the highly specialized interstitial cells of Cajal. The velocity of propagation of slow waves increases distally, from 0.1 cm/s in the corpus to 4 cm/s in the terminal antrum. Slow waves propagate faster in the transverse than in the longitudinal direction, resulting in a wave of gastric depolorization, which propagates circumferentially and repetitively from the corpus to the pylorus.

Gastric contractions occur *only* during the period of depolarization represented by the slow wave. In this instance, fast action potentials termed *spike bursts*, or *electrical response activity* (ERA), are superimposed on the slow wave [110]. Acetylcholine and adrenergic agonists, respectively, stimulate and inhibit spike bursts. Gastric ECA, but not ERA, can be recorded from surface electrodes on the skin (surface electrogastrography). Alterations in the slow wave rhythm have been detected, including *tachygastria* (frequency higher than 5 cycles/min), *bradygastria* (frequency lower than 2 cycles/min) or *arrhythmia* (irregular, disorganized slow wave rhythm) [1]. Gastric dysrhythmias can occur spontaneously, or can be provoked pharmacologically by a variety of substances, including epinephrine, glucagon, metenkephalin, prostaglandin E_2 and others. They may also be induced in some healthy subjects by motion sickness.

Postprandial Gastric Motor Activity. During ingestion of a meal, a reflex relaxation of the proximal stomach occurs in response to swallowing; this is termed *receptive relaxation* and prepares the proximal stomach to receive the swallowed bolus from the esophagus (Fig. 63.3). The afferent neurons of this reflex arise in the wall of the pharynx and upper esophagus, and in addition to esophageal distension, liquid boluses (wet swallows) and sham feeding elicit the reflex. It is followed by a further gastric relaxatory response, termed *gastric accommodation*, whereby the proximal stomach progressively relaxes to accommodate increasing volumes, with very little associated increase in intragastric pressure. The distension-sensing afferent neurons of this reflex arise in the stomach wall. In both reflexes, the efferent neurons are vagal fibers projecting to inhibitory non-adrenergic, non-cholinergic (NANC) myenteric ganglion cells. Enterogastric reflexes also operate; thus, nutrients perfused into the proximal small bowel (especially liquid) and into the distal small bowel (especially carbohydrate) relax the proximal stomach [3,5].

The second phase of postprandial proximal gastric motor activity is represented by a period of continuous *tonic contractions*; these maintain a gastroduodenal pressure gradient, ensuring that solids progress into the distal stomach for grinding and sieving. This phase appears to be largely dependent on vagal efferent projections to excitatory cholinergic myenteric ganglion cells.

As filling of the stomach progresses, powerful antral *peristaltic waves* are initiated from the pacemaker region (Fig. 63.4). These ring contractions follow the pattern of propagation of the slow waves; their amplitude depends on the particular neurohumoral milieu at the time, as well as on the changing physical consistency of the gastric chyme.

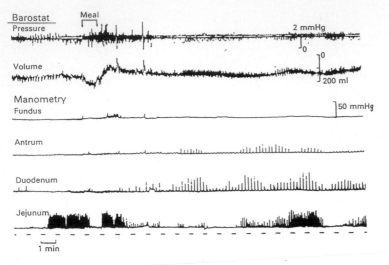

Fig. 63.3. Gastric motor response to feeding, measured by barostat and intraluminal manometry in the dog. The barostat consists of a large-capacity thin-walled flaccid bag introduced into the stomach and inflated with air to fill the proximal gastric region. The bag is connected via a double-lumen tube to an injection withdrawal pump system that maintains a constant intra-bag pressure of 2 mmHg. Gastric tonic contraction causes the barostat to withdraw air to maintain the pressure, while relaxation triggers air injection to maintain pressure. Thus, the intra-bag volume of barostat (*second tracing*) inversely measures proximal gastric tone. After ingestion of the meal, an increase in volume (receptive relaxation) of the proximal stomach, followed by an increase in tone, is observed. Changes in fundic tone are not detected by manometry. Meal stimulates phasic antral and small bowel contractions. (From [4], with permission)

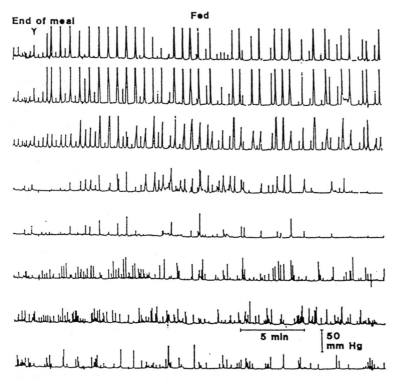

Fig. 63.4. Manometric perfused catheter recording of normal postprandial gastrointestinal motor activity in the upper gut of a healthy subject. Each of *channels* 1–4 is separated from the next by 1 cm, and all show antral phasic contractions. *Channels 5–8* demonstrate phasic contractions in the pyloric region and small bowel. (From [68], with permission)

Some contractions begin proximally and become less intense in the mid-antrum, while others progressively increase in intensity as they approach the pylorus. They are non-occlusive and propel gastric contents adjacent to the mucosa towards the duodenum, continuing as long as food remains in the stomach. Larger particles accumulate away from the gastric walls, where the direction of flow is reversed; these tend to be pressed between the contraction

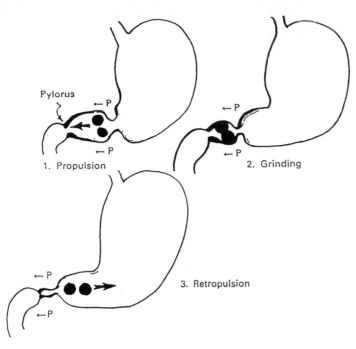

Fig. 63.5. Schematic illustration of effect of antral peristalsis on solids. An advancing contraction carries solids distally (*P* indicates direction of propulsion) and they are retained by the pylorus. The terminal antral contraction grinds the solids and finally retropels them into the proximal antrum. (From [59], with permission)

ring and retropelled into the proximal stomach (Fig. 63.5). The contraction wave travels with increasing velocity until the 3–4 cm of antrum is reached, when the terminal antrum and pylorus appear to contract simultaneously, generating pressures of up to 100 mmHg or more (*terminal antral contraction*). This contraction, combined with the shear forces generated by retropulsion, grinds and disrupts digestible solids and mixes them with gastric juice. Transpyloric fluid flow appears to cease approximately 2 s before the terminal antral contraction [62]. The timing of pyloric closure appears to be a constant phenomenon irrespective of the type of meal or rate of gastric emptying, and can be explained by the unique electromechanical properties of pyloric muscle [110]. Relaxation of the terminal antrum is followed by pyloric opening and emptying of liquid and suspended particles. Digestible solids empty from the stomach only after they have been reduced to small particles, usually smaller than 1 mm [75]. Neither pyloroplasty nor pylorectomy, however, significantly affects the size distribution of solid particle emptying, while, at least in dogs, pylorus-sparing antrectomy preserves sieving [43,76]. Thus, control mechanisms are such that the antrum alone is capable of selectively retaining solids, and even the corpus can triturate solids providing the pylorus is intact.

Just as the gastric barostat has enabled a more direct assessment of proximal gastric motility, so has use of the pyloric sleeve sensor enhanced the direct recording of pyloric motor activity. It is now clear that the *pylorus* is a *distinct functional component of the gastroduodenal region*. Pyloric contractions occur either in temporal association with the antrum and/or duodenum, or independently during motor suppression of antral and duodenal contractions. These latter pyloric contractions are usually phasic and are present at the gastric frequency of 3/min. They can be detected manometrically as *isolated pyloric pressure waves*, often in association with a small tonic elevation of basal pyloric pressure (Fig. 63.6). Both isolated pyloric pressure waves and the basal pyloric pressure are confined to a narrow band at the pylorus, usually less than 6 mm in length. Isolated pyloric pressure waves appear to be stimulated via duodenal and small intestinal luminal nutrient receptors [42], and this pyloric resistance to emptying is probably one factor limiting the rate of nutrient emptying from the stomach, as isolated pyloric pressure waves have been demonstrated to cause sustained localized pyloric luminal closure and cessation of transpyloric flow [115]. Moreover, in contrast to the antrum, specific and rapidly adapting alterations in the phasic and tonic motor responses of the pylorus to the changing nutrient composition in the duodenum can be demonstrated [30].

Interdigestive Gastric Motor Activity. During fasting, a cyclical pattern of motor activity occurs in the stomach every 90–120 min. This pattern – the gastric component of the *migrating motor complex* (MMC) – features a band of intense gastric phasic contractions, which then migrates from the distal stomach along the small bowel (see sect. 63.2.2). Both the proximal and the distal gastric regions are involved, and the cycle continues until interrupted by feeding. Phase 1, as it is termed, is the longest part of the cycle, during which slow waves continue to sweep the antrum unassociated with muscle contractions. At this time, although there are no phasic contractions, the proximal

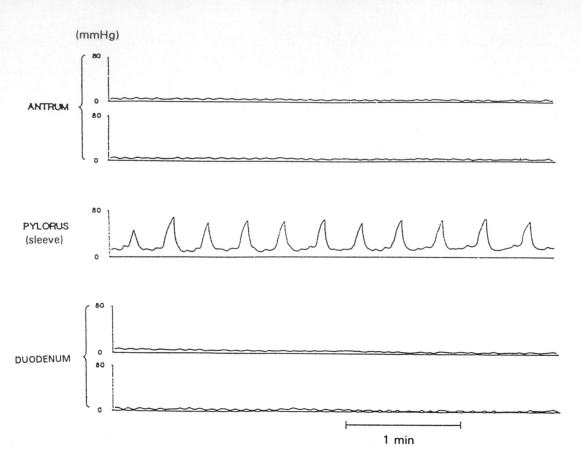

(mmHg)

ANTRUM

PYLORUS
(sleeve)

DUODENUM

1 min

Fig. 63.6. Manometric recording of isolated pyloric pressure waves (frequency approximately 3/min) stimulated by intraduodenal lipid infusion in a healthy subject. Recording made with a catheter incorporating a pyloric sleeve sensor (length 45 mm) with antral side-holes 5 and 35 mm orad of the proximal end of the sleeve and duodenal side-holes at the aborad sleeve margin and 30 mm aborad from this. Contractions in antrum and duodenum are completely suppressed, and pyloric tone is demonstrable. (From [42], with permission)

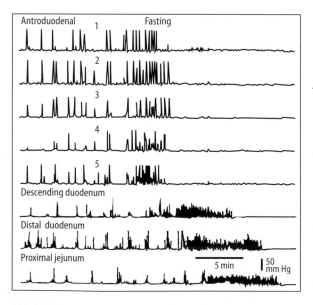

Fig. 63.7. Manometric recording of normal gastrointestinal motility in the upper gut during fasting, showing phase 3 of the migrating motor complex present in the gastric antrum (channels *1–4*) and small intestine. (From [68], with permission)

stomach maintains an intermediate tone, rather than undergoing complete relaxation [3,4]. During phase 2 of the MMC, phasic contractions begin to appear with increasing frequency after each spike burst. During phase 3 (*activity front*), which lasts for approximately 10–20 min, nearly every spike burst is associated with powerful co-ordinating contractions in the proximal and distal stomach (Fig. 63.7). In the intervals between fundic contractions, the distal stomach is also quiescent. The gastric phase 3 coincides with late phase 2 and phase 3 in the duodenum and, during gastric contractions, duodenal contractions are inhibited. The pylorus contracts in coordination with fasting antral contractions; antropyloroduodenal contractions are the most frequent type during fasting and occur in association with the phases of the gastric MMC cycle. Pyloric tone appears to be absent during phase 1 and also between phase 2 contractions; resistance to flow is likely to be minimal at these times. Indigestible solid residue that has been retained in the stomach is emptied during the subsequent activity front of the interdigestive cycle. The maximum diameter of particles that can be emptied from the stomach is about 25 mm, and the antral contractions that accomplish the emptying produce a sequential lumen occlusion of the proximal antrum, associated with late closure of the terminal antropyloric segment. Gastric bezoar formation in diabetic or post-vagotomy gastroparesis is associated with lack of a normal gastric component of the MMC.

The control of the gastric component of the *MMC* is not completely understood. There appears to be at least some *hormonal* regulation, as cyclical activity in autotransplanted stomach pouches remains coordinated with the small bowel MMC. Indeed, it has been proposed that gastric activity fronts are initiated by cyclic variations of plasma *motilin* concentration, although the possibility that the intense motor activity is itself responsible for the increased plasma motilin concentrations has not been excluded. The switch from the fasting to the postprandial or "fed" pattern appears to be mediated, at least in part, by the *vagus nerves* – acute vagal cooling produces cyclical bursts of activity similar to phase 3 activity fronts, while chronic vagotomy can inhibit conversion to the postprandial pattern or at least shorten its duration [75].

Gastric Emptying. Postprandially, gastric emptying commences while mixing and grinding is taking place, but the rate at which nutrients enter the small intestine is highly controlled (cf. Chap. 61). A range of scintigraphic techniques has confirmed that there are important differences in the rates at which various food components (e.g., digestible solids, indigestible solids, liquids and fats), empty from the stomach (Fig. 63.8). A clear distinction between solid and liquid phases cannot always be inferred in these studies, however, as for some meals it is only a liquid containing fine particulate solids in suspension that needs to pass through the pylorus. On the other hand, certain viscous liquid meals may behave more like meals containing larger solid particles.

The sustained, albeit low, tension exerted by the proximal stomach appears to play a major part in the emptying of liquids from the stomach (*hydrodynamic concept of liquid emptying*) [75]. Determinants of the emptying rate appear to be both the force exerted upon the fluid mass by the stomach and the variations in flow resistance determined by motor activity in the distal antrum, pylorus and proximal duodenum. Liquids are emptied essentially exponentially, at a rate proportional to the intragastric volume. Models that have been applied for analysis of gastric emptying include a power exponential function and a two-component linear analysis (i.e., where rapid linear emptying occurs initially, followed by a slower rate of linear emptying). The concentration of various nutrients or acidity in liquid meals affects the rate of gastric emptying; thus, fatty acids (particularly medium and long chain) inhibit gastric emptying, as do monosaccharides and amino acids such as tryptophan, phenylalanine, glutamate, arginine and cysteine. Surgical alterations to the gastric anatomy influence the time-course of emptying and can give a clue to mechanisms of normal emptying. Thus, truncal, or proximal, gastric vagotomy, like antral or pyloric resection, accelerate the initial rapid emptying phase.

The emptying of the *solid phase* is slower than that of the liquid phase, and its rate is inversely proportional to the diameter of the particles present in the stomach. Thus, a *lag period*, due to the triturative process in the antrum as well as to such factors as a changing viscosity of the contents, follows ingestion of many meals and precedes the onset of emptying of the solid phase. The subsequent emptying follows a slow, nearly linear course. After ingestion of a mixed meal the solid, but not the liquid, phase may be retained in the proximal stomach; there follows gradual transfer of the solid phase to the antrum and the establishment of a steady state where the quantity entering the antrum and leaving the pylorus is identical. Thus, the proximal stomach does regulate the transfer of solid food to the distal stomach, but little information is available on the relative importance of fundic tone on antral filling. Studies of the specific emptying pattern of the lipid phase of a meal are limited. It appears, however, that the majority of extracellular fat empties as an oil phase, more slowly than the aqueous phase of a meal, while the majority of intracellular fat empties with the solid phase [22]. In patients who have undergone truncal vagotomy and pyloroplasty or gastroenterostomy, solids empty more slowly than usual.

Gastric emptying of both solids and liquids is regulated by *small-bowel receptors* for luminal nutrients. In the case of liquids, nutrient sensors for glucose, and probably fat, are equally potent in the proximal and distal small bowel [66], but those for acid appear to be confined to the proximal bowel. For solids, gastric emptying is potently inhibited by the presence of glucose in the ileum rather than in the jejunum, suggesting that carbohydrate sensors in the ileum are of major importance. Modulations in the sensitivity of small-bowel feedback control may be influenced by the preceding nutrient intake; for example, gastric emptying of glucose is faster after prior dietary supplementation with glucose, while prior consumption of a high-fat diet for several days is associated with more rapid emptying of a fatty test meal [23,24]. Thus, it is possible that the large inter and intraindividual variability of normal gastric emptying can be accounted for, at least in part, by differences in diet.

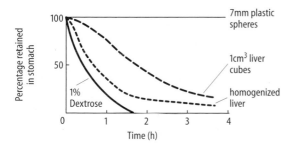

Fig. 63.8. Schematic of the emptying of particles of various sizes and of liquids from the stomach: 1 cm³ liver cubes empty after a lag phase, while the nutrient liquid (1% dextrose) empties in an exponential function. The indigestible solids (7-mm plastic spheres) do not empty within the 4-h period. (From [29], with permission)

Duodenogastric Reflux. Reflux of some duodenal content occurs normally, in short intermittent pulses, during both the fasting and the postprandial gastric motor states, and does not appear to depend on the phases of the interdigestive cycle. Bile reflux, however, is greatest during phase 2 and postprandially. The mechanisms associated with this duodenogastric reflux are not well defined, but retrograde peristalsis and isolated proximal contractions in the duodenum are probably important. Postprandial reflux is clearly not harmful, whereas fasting reflux may be injurious to the gastric mucosa in some, as yet unknown, circumstances.

Vomiting Reflex. This reflex depends on the coordination of autonomic and somatic motor activity and can be initiated by a variety of sensory inputs such as:

- Noxious intestinal stimuli acting via visceral afferent fibers running in the vagus
- Blood-borne stimuli acting on the chemoreceptor trigger zone in the medulla
- Psychophysiological stimuli (e.g., emotions) acting on the vestibular apparatus
- Higher centers

There are four main stages of the *vomiting reflex*:

- Initially, the diaphragm descends while the glottis remains closed, leading to the development of negative intrathoracic and intraesophageal pressure
- After about 0.5 s the stomach and the lower esophageal sphincter relax and the abdominal wall muscles contract, propelling the gastric contents through the lower esophageal sphincter
- Contraction of the esophageal longitudinal muscle then shortens the esophagus, and the thoracic cage expands, causing a further lowering of the intrathoracic pressure;
- Finally, the gastric antrum contracts and the upper esophageal sphincter relaxes so that the vomitus is expelled

Retrograde propulsion in the proximal small intestine can be documented before emesis: giant contractions travel rapidly toward an open pylorus and are preceded by relaxation of the circular muscle over an extended length of bowel [65]. This pattern of reverse peristalsis appears to be consistent, irrespective of how the vomiting is induced, and is likely to represent reverse wiring of the same basic circuits as normally generate aboral propulsion of contents.

Regulation of Gastric Motility. Control of motility in the gastropyloroduodenal region is achieved by a hierarchy of extrinsic neural (vagus and splanchnic nerves), intrinsic enteric, and humoral mechanisms. The complexity of these interactions makes it difficult to evaluate the precise contribution of one or more of these elements in vivo.

The efferent *parasympathetic nerve supply* to the stomach is via the vagus nerves, with either preganglionic cho-

linergic, postganglionic noradrenergic, or *NANC fibers*. Stimulation of preganglionic parasympathetics causes increased gastric contractility, pyloric relaxation and enhanced gastric emptying. The efferent *sympathetic supply* originates in the thoracic spinal cord (T6–T9) and passes from these to the celiac ganglion and stomach. Sympathetic efferents inhibit gastric contraction, presumably via modulation of the myenteric plexus. The vagus nerves, however, contain approximately 80% afferent fibers – in the stomach these originate in either stretch receptors or chemoreceptors. The antropyloric region contains a high density of *stretch receptors*, and their threshold for firing is lower in the pylorus than in the adjacent antrum. It is also possible that antral and pyloric stretch receptors respond to contractile activity of the wall but not to physiological levels of distension, whereas receptors in the proximal stomach respond to distension only [39]. The afferent fibers appear able to modulate efferent vagal firing, thus influencing vagal tone in the antroduodenal region.

The *myenteric* and, to a lesser extent, the *submucosal plexuses*, appear to be the major originators of the various types of *programmed motor activity* discussed earlier. The cholinergic preganglionic axons apparently synapse with a number of intramural nerves, which contain a range of neurotransmitters. Postganglionic noradrenergic nerves in the myenteric plexuses innervate the smooth muscle both indirectly by synapsing with myenteric ganglia, and directly by entering the inner circular, but not the outer longitudinal, muscle. Distension of the proximal stomach increases antral peristalsis (antral reflex), and this effect is reduced by vagotomy [106]; intrinsic relexes, however, also appear to play a role. Distension of the duodenum inhibits gastric tone and distal peristalsis (*enterogastric reflex*); this effect can be partially blocked by vagotomy or splanchnectomy, and completely abolished by both, and the reflex appears to consist of afferents in the splanchnic nerves and efferents in the vagus [40].

A large number of *gut-brain peptides* can be shown to modify gastric motor activity, but in most cases their physiological role remains uncertain. This is because of the difficulty in elucidating the complex interactions between the many peptides released postprandially and the associated neural reflexes. In particular, whether the peptides are tested in vivo or in vitro appears to modify the response observed. Thus, for example, in the case of gastrin in the canine stomach, a high-affinity action on nerves is predominant in vivo, whereas in vitro the effect is absent and a direct smooth muscle effect is observed [34].

Cholecystokinin (CCK) delays gastric emptying at low doses and appears to have a physiological role in the regulation of gastric emptying and secretion. Its action is exerted partly by a direct interaction with receptors on smooth muscle cells and partly by an effect on cholinergic excitatory myenteric ganglion cells; it may inhibit spontaneous contraction of gastric smooth muscle and yet increases the contractility of antral circular muscle [120].

Secretin, released by infusion of physiological amounts of acid into the duodenum, may play a part in the normal regulation of gastric emptying. It delays gastric emptying by relaxing fundal and antral smooth muscle, and also contracts the pylorus, perhaps by diminishing the responsiveness of gastric smooth muscle to other neural and humoral stimuli.

Gastrin can delay gastric emptying via relaxation of the fundus and antral contractions [19], but these effects occur with pharmacological doses.

Motilin is the only peptide to date that has been shown to stimulate gastric emptying, apparently via a direct effect on both proximal and distal gastric smooth muscle; its role in initiation of the gastric component of the MMC is controversial (see Sect. 63.2.2).

Neurotensin, somatostatin, gastric inhibitory peptide, pancreatic polypeptide and glucagon can all affect gastric smooth muscle activity, but a physiological role for these peptides has not been established.

63.2.2 Motility of the Small Intestine

Postprandially, motor activity of the small intestine has several vital functions:

- To mix food with digestive secretions
- To circulate chyme so that mucosal contact is maximal
- To propel contents in a net aboral direction
- To clear residua left over from the digestive process
- To transport continuing secretions during fasting from the upper gut

The *patterns of motility* required to accomplish these functions are complex and have received a great deal of attention since the precise descriptions of interdigestive motility, including the MMC, 25 years ago [109]. In contrast to the stomach, however, the demonstration of regional motor specialization of the small bowel is a relatively recent phenomenon, largely because suitable recording techniques were not available earlier. Thus, the jejunum (approximately the proximal two-fifths of small bowel) acts primarily as a mixing and conduit segment, while the ileum (distal three-fifths of small bowel), perhaps because of its specialized absorptive properties, retains chyme until digestion and absorption are largely complete. The terminal ileum and the ileocolonic junction control emptying of ileal contents into the colon in such a way that the absorptive capacity of the colon is not compromised. This junctional region also minimizes coloileal reflux.

Techniques enabling assessment of small bowel motor function in man can be divided into those that measure motor or myoelectrical activity and those that measure transit of contents (Table 63.2). As in the stomach, the most detailed information can be gained from the simultaneous assessment of both of these aspects. Except for scintigraphy, the techniques used to measure small-bowel transit are limited in that they only enable assessment of the movement of the head of the bolus; thus, only the movement of the marker, and not necessarily of the foodstuff, can be determined.

Small-bowel Myoelectrical Activity. The muscularis externa of the small intestine consists of two layers, the thicker inner layer of circular smooth muscle and the thinner outer longitudinal layer. When spike bursts are superimposed on the *slow wave potentials, rhythmic muscular contractions* occur with the same frequency as the slow wave. The slow wave frequency of intestinal smooth muscle declines progressively from duodenum to terminal ileum (Fig. 63.9). Since adjacent muscle cells within a given bundle are electrotonically coupled, the bundle as a whole exhibits the slow wave fluctuations, which, in the case of circularly arranged bundles, results in the formation of a contraction ring. In the case of longitudinally arranged bundles, those with the fastest slow wave frequencies form a pacemaker zone driving more remote (slower) bundles, thereby leading to propagation of slow waves with the frequency of the pacemaker zone. There comes a point along the bowel, however, at which the pacemaker frequencies are completely out of phase with the intrinsic slow wave frequency of the muscle bundles, so that the entrained slow waves are cancelled out and contractions become irregular. Just beyond this point, the smooth muscle bundles are free to exhibit their own intrinsic slow wave rhythm and thus function as a new pacemaker for more distal regions [16,27].

Thus, slow waves, and the resultant contractions, normally propagate only in the direction of the slow wave frequency gradient, that is from duodenum to ileum.

Table 63.2. Techniques used to assess motor functions of the human small intestine

Motor function	Technique	References
Small-bowel motor and myoelectrical activity	Small-bowel manometry	
	– Perfused catheter	[55,56,68,117]
	– Strain gauge catheter (stationary, ambulatory)	[38,57]
	Small-bowel barostat	[89,90]
	Small-bowel electromyography	[33]
	Small-bowel radiotelemetry	[112,113]
Small-bowel transit	Radioisotopic scintigraphy	[69]
	Breath hydrogen testing	[20]
	Sulfapyridine absorption (plasma, saliva)	[10,54]

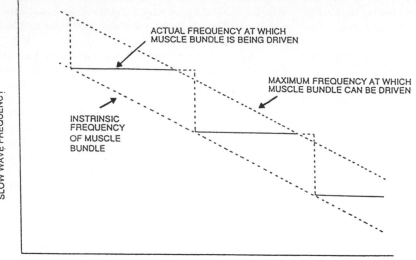

SLOW WAVE FREQUENCY

ACTUAL FREQUENCY AT WHICH
MUSCLE BUNDLE IS BEING DRIVEN

MAXIMUM FREQUENCY AT WHICH
MUSCLE BUNDLE CAN BE DRIVEN

INSTRINSIC
FREQUENCY
OF MUSCLE
BUNDLE

DISTANCE ALONG BOWEL

Fig. 63.9. Diagram showing how the intrinsic frequency of the upper intestinal pacemaker drives the slow wave rhythm of intestinal smooth muscle in the aborad direction. The interruptions (indicated by the *vertical broken lines*) occur whenever the intrinsic slow wave frequency of the muscle is too slow for it to be driven by the pacemaker region, so that this region contracts at its own intrinsic frequency and functions as a pacemaker for regions distal to it. (Redrawn from [27], with permission)

Postprandial Small-bowel Motor Activity. Three main types of contractions can be recognized in the small bowel. *Segmentation contractions* mix gut contents and propel them caudally. These contraction rings appear and reappear regularly, repeatedly subdividing luminal contents; their site of appearance fluctuates, and there is no obvious coordination with contraction rings occurring at successive points along the viscus. Although segmentation contractions have an obvious mixing function, they are also propulsive, leading to a steady aboral progression of gut contents. Rhythmic contractions of the longitudinal muscle bundles also occur, giving rise to the so-called *pendular contractions*. The effect of these contractions is to move the bowel over its contents, rather than to subdivide the contents, so inevitably their action is propulsive. When successive contractions are coordinated along a substantial length of the gut, propulsion is conspicuous, and the phenomenon is termed *peristalsis*. In the gut, each propagated contraction is preceded by a propagated relaxation, and it is this coordinated precession of relaxation followed by a contraction, termed the peristaltic reflex, that ensures propulsion if effective. *Peristaltic reflexes* are initiated by the contents of the lumen; distension is a sufficient stimulus, but the amplitude of the contractions appears also to be influenced by the chemical nature of the contents.

Postprandially, contractions of variable amplitude occur irregularly, but consistently, in the small bowel (Fig. 63.4) replacing the cyclical interdigestive motor activity. This "fed" pattern has received little detailed analysis. Studies in dogs have demonstrated that various nutrients induce different *contractile patterns in the jejunum* [97]. Some contractile episodes migrate for short distances, others for intermediate or long distances; rates of transit are variable and in general appear to be related to the distance of contractile propagation. In man, an increase in caloric load can also affect some parameters, such as the individual contractile duration [31]. The physicochemical composition of food appears to be more important in determining the fed pattern duration than the volume or caloric load of the meal; fat has a more potent effect (longer fed pattern duration) than does carbohydrate, which is more potent than protein. Oral intake of nutrients is not essential for establishment of the fed pattern, since jejunal or ileal infusions of nutrients inhibit the MMC. Also, the duration of the fed pattern is not influenced by the method of administration, since oral and intraduodenal meals produce fed patterns of similar lengths. After a meal, interdigestive cycles in humans almost always reappear first in the proximal jejunum, the site at which cycles occur most frequently during prolonged periods of fasting (see below).

It is not surprising that motor patterns differ along the length of the small bowel, since the fluid and nutrient loads presented differ considerably according to site, and different absorptive processes occur. In the ileum, for example, unique patterns of contraction occur in response to various stimuli. Few studies have attempted to define the *motor response of the terminal ileum* to feeding – the "gastro-ileal reflex" – in humans. This response, an increase in the incidence of irregular phasic contractions that can be documented in man, does not appear to require the stomach; it is seen in animals after the administration of food through a duodenal fistula. The ileal response occurs before food arrives in the distal ileum, is preserved following extrinsic and intrinsic denervation, but is abolished by atropine [100].

Ambulatory recording techniques, which enable meals to be eaten at normal times and in the home environment, have further characterized normal patterns of small intestinal motor activity. In healthy subjects eating a conventional diet (three meals a day) only isolated interdigestive cycles occur during the day, often a single activity front occurring shortly before the next meal in each case. Further, when subjects pursue their usual activities and select their own meals, free from the environment of a hospital or laboratory, the overall *motor pattern depends on the frequency of eating*. Thus, a greater proportion of fed motor activity is seen in subjects who eat snacks between meals, even to the point of there being no diurnal interdigestive cycles [57].

Interdigestive Small-bowel Motor Activity. The *interdigestive myoelectrical complex* and its most dramatic mechanical component, phase 3 of the MMC, were first described most clearly in the dog [109]. The MMC occurs cyclically in the small intestine of humans [117], as well as many other species. The motor equivalents of the interdigestive myoelectrical complex are phase 1 (motor quiescence), phase 2 (a period of irregular and intermittent

contractions), and phase 3 (a brief sequence of uninterrupted, rhythmic contractions) (Fig. 63.7). In addition to the periodic appearance of this sequence at any single locus, the entire cycle migrates caudad along the fasting bowel. Thus, the cycle passes from antroduodenum towards ileum every 60–120 min in the fasting state.

Quantitative mapping of MMC cycles along the length of the human small intestine has established several important differences from other species (Fig. 63.10). The relative infrequency of ileal MMC propagation is in contrast to the situation in species such as the dog, where MMCs traverse the entire small bowel; however, in both the pig and the rat, up to 50% of MMCs do not reach the terminal ileum. Further, whereas in the dog ileal motility is highly organized, the predominant motor pattern in the human ileum during fasting is that of irregular, and apparently random, contractions with some degree of clustering. MMC velocity also slows in the ileum, the duration of phase 3 is greater, and as expected, the frequency of the maximum contractile rate slows from duodenum to ileum (Table 63.3).

Specific motor patterns other than the MMC can be recognized in the small bowel. Regular bursts or clusters of con-

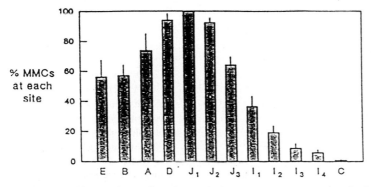

Fig. 63.10. Incidence of interdigestive migrating motor complexes along the human bowel [incidence in proximal jejunum (J_1 arbitrarily assigned 100%]. Note that approximately half are seen in the esophagus (*E*) and gastric body (*B*) and that most are present in the proximal jejunum (jejunal sensors J_1–J_3); few are seen in the distal ileum (ileal sensors, I_1–I_4); *C* cecum; *D*, duodenum; *A*, gastric antrum. (From [55], with permission)

Table 63.3. General characteristics of phase 3 of the migrating motor complex in the human upper gastrointestinal tract. (Modified from [55], with permission)

Site	Velocity of Propagation (cm min^{-1})	Maximum frequency of contractions (per min)	Duration (min)
Duodenum	–	11.7 ± 0.1	8.7 ± 1.0
Jejunum 1	4.3 ± 0.6	11.3 ± 0.1	9.1 ± 0.6
2	2.8 ± 0.4	10.7 ± 0.1	11.7 ± 0.6
3	2.0 ± 0.5	10.4 ± 0.2	14.7 ± 1.2
Ileum 1	1.3 ± 0.2	9.8 ± 0.2	15.6 ± 1.6
2	0.9 ± 0.1	9.3 ± 0.2	15.3 ± 1.0
3	0.7 ± 0.2	8.9 ± 0.2	13.9 ± 1.1
4	0.6 ± 0.2	8.5 + 0.2	13.8 ± 1.5
Cecum	–	6.1 ± 0.2	

Data from 16/healthy subjects, each with manometric recording sensors spanning duodenum to proximal cecum, with three positioned in the jejunum and four in the ileum. Data expressed as mean ± SEM

tractions, at times propagated rapidly in an aboral direction, can occur in the proximal small intestine of apparently healthy subjects. Like the MMC, the overall incidence of these, their duration, and their propagation are seen to be variable when recordings are made over prolonged periods. This pattern appears to represent the mechanical equivalent of the myoelectrical *minute rhythm* [33]; its functional significance remains to be established, even though a propulsive potential seems likely. In the ileum, striking contractile sequences of relatively long duration (greater than 12 s), large amplitude, and rapid propagation traverse the ileo-cecal region infrequently [55]. Comparable events can be recorded in the canine ileum with serosal strain gauges. In the dog they are clearly propulsive, emptying the ileum of its contents, and can be provoked by chemical and mechanical (luminal) stimulation [52]. Thus, short-chain fatty acids present in refluxed colonic material appear to stimulate ileal chemoreceptors, restricting coloileal reflux.

There are prominent *circadian variations* in interdigestive small bowel motility [55,57]. The periodicity of the MMC cycle is significantly shorter at night than during the day, and becomes more regular, with the most frequent cycle lengths in the range of 30–90 min. During the day the distribution is biphasic, with the most common cycle lengths in the ranges of 60–120 min and 120–150 min. The mean cycle length at night is about 65 min, compared with about 100 min during the daytime; there is wide inter- and intraindividual variability. Phase 2 motor activity lessens or even disappears in the jejunum and ileum of most healthy subjects during sleep. Phase 3 is longer, and its propagation velocity markedly reduced, during sleep. Although a relationship between the phases of cortical sleep and the ultradian rhythm of the MMC has been suggested, no good correlation between the two phenomena has been documented.

Only recently, since the advent of appropriate technology ("sleeve" manometry, barostat), has the physiology and regulation of intestinal tone been investigated in man. In the proximal jejunum, changes in tone can be induced reflexly by pulse distensions of the intestine in either antegrade or retrograde fashion [89]. However, the responsiveness of the jejunum to these intestinal reflexes fades distally, whereas intestinal sensitivity (perception) to distension remains uniform [90]. Tone at the ileocecal sphincter is difficult to demonstrate. Basal tone in this region, however, is augmented by phasic pressure waves propagated from the distal ileum into the ileocolonic sphincter [83].

Transit Through the Small Intestine and Relationship to Motor Activity. The actual movement of material within the lumen of the small bowel is complex. During fasting, flow varies according to the specific phase of the interdigestive cycle. It is most pronounced at the time of phase 3 of the MMC, but is not insignificant even during the motor quiescence of phase 1. Flow rates increase postprandially, fluctuating rapidly – by the time the fasting pattern returns, several hours after a liquid meal, almost all of the chyme has reached the terminal ileum. Solids and liquids travel along the small intestine at similar rates. In humans, ileocolonic flow of liquids and solid residue occurs predominantly in the form of bolus transfers, and these are often grouped around the time of the subsequent meal. Ileocolonic transit increases again approximately 2 h postprandially, when meal residue has reached the distal small bowel and ileal flow is greatest. Functionally, the distal ileum has unique propulsive properties in vitro. Thus, feline ileal segments spontaneously propel fluid loads in an aborad direction, whereas comparable net fluid transit cannot be demonstrated in duodenal and jejunal segments [118]. As with the stomach, nutrients present in the human ileum exert negative feedback on jejunal motility and transit – this *ileal brake* operates for fat and partially digested carbohydrate, and probably for other nutrient groups [104,105]. More detailed analysis of the relationships among motor patterns and flow has required animal models and the use of computer analysis of the contractile events constituting the fed pattern. The results suggest that transit is fastest when localized propulsive events are closely coordinated and propagate distally [97]. Overall, there is a correlation between the small-bowel half-emptying time of a solid meal and the contraction frequency in the proximal jejunum, but no correlation between transit and the contraction frequency in the ileum.

Regulation of Small-bowel Motility. The control mechanisms of small-bowel motility remain poorly understood, although the central nervous system (CNS), the extrinsic (autonomic) nerves, the *enteric nervous system* (ENS) and humoral factors are all involved. The periodicity and cyclical nature of the MMC are probably generated within the ENS itself, with the two main programs reflecting the absence or presence of nutrient in the lumen. However, the effects of sleep and extrinsic denervation (for example vagotomy) also indicate that cyclical activity is modulated by *higher centers*. Thus, vagal blockade significantly shortens or abolishes phase 2 activity in the duodenum and jejunum and slows phase 3 propagation velocity [112]. Psychological stress can also interrupt fasting motor complexes. The function of the ENS as a "minibrain" that coordinates intestinal motility is most obvious when peristalsis, segmentation, and adynamic ileus – the local or regional systems of transit – are examined neurophysiologically. Thus, each of these distinct motor states can be accounted for by intrinsic neural mechanisms.

The *inhibitory effects of food on the MMC* appear to be mediated through both humoral and neural pathways. When infused exogenously, the foregut peptides *gastrin*, *secretin and CCK* disrupt fasting patterns; however, it is not known whether the irregular patterns induced by these peptides are the same as those found after feeding. Conversion to the fed state in the distal small intestine could also involve peptides located primarily in this region, such as neurotensin or enteroglucagon, which are released when chyme reaches the ileum. Despite uncertainties over the role of gastrointestinal peptides in establishing or maintaining the fed pattern, CCK remains a possible candidate. When the motor response of the gallbladder is used as a "simultaneous bioassay," motor activity in the jejunum is increased at "physiological" doses of CCK, but that in the ileum is reduced [58]. CCK also has inhibitory effects in the duodenum of some species, while regional differences can be demonstrated in the canine small intestine. The simultaneous onset of the fed pattern along the small bowel suggests *neural* as well as humoral *influences*. After vagotomy, initiation of the fed pattern requires larger caloric loads, its onset is delayed, it is of shorter duration, and the amplitude and frequency of the contractions are decreased [112]. Initiation and maintenance of postprandial motor patterns, with concurrent inhibition of the fasting cycle, requires an intact vagus; during vagal blockade in the fed state, the entire small bowel in the dog exhibits migrating electrical spike bursts, which have been termed postprandial *vagally independent complexes*, although the ileum appears to be less affected than the jejunum. These bursts occur with the same periodicity as would be expected of spontaneous MMCs, as if nutrients had not been administered. With regard to the ileocolonic sphincter, although vagal and sympathetic inputs appear to play a modulating role, integrity of extrinsic nerves is not essential for the maintenance of tone. Whereas evidence for the role of vagal innervation is contradictory, there is stronger support for an excitatory alpha adrenergic mechanism, contrasting with the inhibitory effects of this transmitter on the adjacent bowel. The most important inhibitory responses of the sphincter, however, appear to be from the input of *NANC nerves*.

63.2.3 Motility of the Large Intestine

The motor functions of the colon are varied and complex, including:

- Mixing the contents to promote absorption of water and electrolytes, short-chain fatty acids and bacterial metabolites
- Maintaining an appropriate intraluminal bacterial mass
- Transporting contents in a net aboral direction

- Storing residual fecal material until defecation
- Emptying appropriate colonic contents rapidly during defecation

Because of its relative inaccessibility, the colon is still the least well understood region of the gastrointestinal tract. Nevertheless, it is clear that the different regions of the colon subserve these different functions. Thus, the cecum, ascending colon and rectum usually act as *reservoirs for storage of feces*, while the remainder of the colon (transverse, descending and sigmoid colons) act predominantly to *propel feces intermittently* from the first reservoir into the second reservoir (rectum) as a prelude to defecation. Techniques used to assess *colonic motility*, as in the small intestine, can be divided into those enabling assessment of motor or myoelectrical activity and those that measure transit (Table 63.4). Radio-opaque markers are the traditional transit markers, but only allow measurement of total mouth-to-anus transit time. Precise non-invasive measurements of colonic transit can be obtained using isotopic scintigraphy: the subject ingests radiolabeled pellets, which disintegrate in the terminal ileal region. The *geometric center* calculation is then used to indicate the mean fecal progression at a given time.

Large-bowel Myoelectrical Activity. Bundles of the outer longitudinal muscle layer of the colon are grouped into three thick bands, the *taeniae of the colon*. The thin longitudinal muscle layer between the three taeniae, together with elongation of the muscle bundles of the *inner circular muscle* coat, results in the saccular appearance of the colon. At regular intervals, bands of contraction of the circular muscle layer occur, producing the typical haustral appearance. Human colonic myoelectrical activity consists of:

- Slow waves (electrical control activity, ECA)
- Superimposed spike bursts, and
- Oscillatory activity.

The last two are the electrical equivalent of *contractile activity*. In man, serosal electrodes record slow waves to be omnipresent, although variable and irregular, and almost

Table 63.4. Techniques used to assess human large intestinal and anorectal motor function

Motor function	Technique	References
Colonic/anorectal motor and myoelectrical activity	Colonic/anorectal manometry	
	– Perfused catheter	[8,71,79,86]
	– Strain gauge catheter (stationary, ambulatory)	[103]
	Colonic barostat	[107]
	Colonic/anorectal electromyography	[7,101,102]
Colonic transit	Radio-opaque markers	[44,74]
	Radioisotopic scintigraphy	[63,82,104]
	Defecating proctography	[67]
Defecation	Combined dynamic assessment of anorectal function	[119]

completely phase-unlocked, unlike those in the upper gut [93]. This variability and irregularity of the colonic slow wave may explain the differences in reported estimates of ECA frequency. There appear to be two main frequency ranges, a low-frequency component with 2–11 cycles/min and a high-frequency component with 9–13 cycles/min (Table 63.5). Spike bursts usually occur at the time of maximal depolarization of the ECA; thus, the colonic slow wave acts as a pacesetter, and, at any given time phasic contractions may occur at the rate of the dominant frequency [49,50].

The *mid-transverse colon region* appears to function as the colonic *pacemaker*, and microelectrode physiological studies have recently demonstrated that it is represented by the interconnecting network of the interstitial cells of Cajal,

situated between the submucosa and the circular muscle layer [48,91].

Two types of *colonic spike activity* can be differentiated (Fig. 63.11). Short spike bursts (about 3 s duration) are stationary and appear to originate in the circular muscle layer, superimposed on individual slow wave cycles. These spike bursts may be the myoelectrical equivalent of higher frequency phasic contractions [9]. Long spike bursts (about 10–30 s duration) are either stationary or propagating (migrating), originate in both muscle layers, and may last for one or more slow wave cycles. They appear to be the electrical counterpart of the lower frequency phasic contractions and are not controlled by the ECA [35]. *Oscillatory activity* (also called contractile electrical complex) is electrical activity, which occurs, with a frequency of 25–45 cycles/min, superimposed on the ECA and can be migrating in nature; this can also be the electrical equivalent of low-frequency (and longer duration) contractions.

Large-bowel Motor Activity. In contrast to the upper gastrointestinal tract, the colon is in a continuous "digestive" state. Contractions in the colon occur at irregular intervals and there are two main types: individual *phasic contractions* and high-amplitude propagated contractions or *giant migrating contractions*.

The phasic contractions can be of either short (less than 15 s) or long duration (40–60 s), occurring independently or in combination. As they are poorly coordinated spatially, they accomplish mixing and propulsion but can also propel feces in a retrograde fashion from the transverse into the ascending colon.

Giant migrating contractions occur either singly or in groups, infrequently (once or twice daily), and are believed to be the major propulsive motor events in the colon. They

Table 63.5. Frequency range and percentage presence of human colonic electrical control (slow wave) activity. (Modified from [92], with permission)

Lower frequency[a] range	Higher frequency[a] range	References
	8–11 (5)	[21]
3 (10)	6 (20)	[101]
2–4 (3–20)	6–11 (17–51)	[111]
2–4 (46–60)	6–12 (20–50)	[108]
2–4 (61)	6–9 (69)	[51]
2–9 (97)	9–13 (33)	[95]
3 (18)	6 (5)	[36]
2–7 (11)	10–12 (20)	[12]

[a] Numbers in parentheses represent the percentage presence of the frequency component. The total percentage presence of frequency components in both frequency ranges exceeds 100 in some cases, indicating that frequency components were present in both ranges concurrently

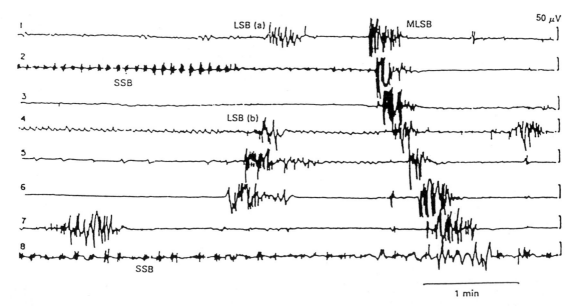

Fig. 63.11. Electromyogram of the large intestine recorded by intraluminal electrodes from the right colon [1] down to the rectosigmoid junction [8]. Different types of spike activity are seen: short spike bursts (SSB) are stationary, long spike bursts (LSB) are either stationary or propagating, migrating in an orad [*LSB(b)*] or aborad [*LSB(a)*] direction. A migrating long spike burst (*MLSB*) migrates from the right colon to the rectosigmoid. (From [35], with permission)

1302

often occur in the morning after waking, or in the late postprandial period [79,103], and are of large amplitude (greater than 100 mmHg); those that reach the distal colon are sometimes accompanied by the urge to defecate. They can be provoked by stimulant laxatives, and different agents provoke these contractions preferentially in different regions of the colon. Their myoelectrical counterpart has not been conclusively demonstrated.

Organized groups of contractions, termed *colonic migrating and non-migrating motor complexes*, have also been recognized in the colon [94]. This cyclical motor activity consists of alternating periods of quiescence and bursts of phasic contractions, and has been demonstrated in humans and in some other species [103]. Very few details are currently available on the specific features of this pattern in man or its functional significance.

Prolonged recordings of colonic motor (and myoelectrical) activity indicate that, as in the small bowel, motility follows a *circadian pattern*: it increases after waking and is markedly inhibited during sleep, in all regions of the colon (Fig. 63.12) [8,79,103]. After eating, colonic motor activity increases (*gastrocolic reflex*) in both the proximal and the distal colon [102,103], although the distal colon responds to a greater extent and with a slow and more sustained increase [26]. In the canine colon three components can be recognized: an immediate response occurring when food comes into contact with the gastric mucosa; an early response, following the immediate response but before the chyme enters the colon; and a late response occurring after the chyme enters the colon. These components have not been fully characterized in man, but motor activity is stimulated in the human colon within 10 min of the ingestion of a meal and continues for at least 60 min [79,102]. There appears to be no cephalic phase of the colonic motor response to eating; sham feeding does not stimulate colonic motor activity. Fat is the most potent stimulus, while protein and amino acids also modulate or even inhibit it. Caloric load is also important: a 1000 kcal meal significantly increases colonic motility, whereas a 350 kcal meal has minimal effects [102]. Gastric or duodenal mucosal contact is necessary for the immediate and early responses, and both are likely to be neurally mediated by cholinergic and opioid pathways. A physiological role for humoral mediators in initiating or maintaining the colonic motor response to feeding has not been established. Changes in tone in the human colon have recently been studied by means of a barostat system [107]. Alterations in tone are not always associated with phasic contractions, and a clear decrease in colonic tone occurs during sleep. Immediately after nutrient ingestion, colonic tone increases.

Transit Through the Large Intestine and Relationship to Motor Activity. The relationship between the spatial and temporal organization of contractile activity and transit is poorly understood in humans. The major colonic motor events – giant migrating contractions – clearly represent the motor correlate of mass movements, but information on the propulsive nature or otherwise of individual phasic contractions is extremely limited. In the dog, parameters of contraction, such as the migration index (based on the distance and direction of migration of contractile events) and the total duration of contractile activity, only correlate

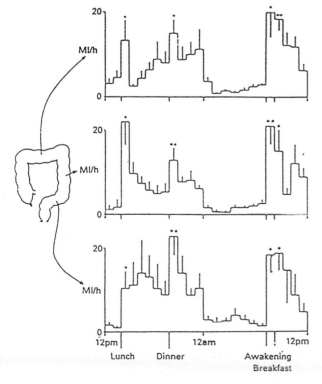

Fig. **63.12.** Total motor activity ("motility index", *MI*) per hour, recorded by intraluminal manometry in human transverse (*top panel*), descending (*middle panel*) and sigmoid (*bottom panel*) colons. A significant increase of motor activity (•) occurs after lunch, after dinner, after waking, and after breakfast. During the night, motor activity is reduced. (From [79], with permission)

moderately with colonic transit. This supports the concept that most of the action of these contractions is of a mixing, rather than a propulsive nature.

Propulsion through the colon results from the net effect of slow and fast, and antegrade and retrograde, movements. Under resting conditions, in the absence of mass movements, colonic contents probably move at an average rate of about 1 cm/h in good health [88]. Quantitative measurements of overall colonic transit are now available from radio-opaque marker studies and scintigraphy. The mean transit time measured with the aid of radio-opaque markers is approximately 33 h in men and 47 h in women; scintigraphic data indicate that three-quarters of the ileal effluent entering the colon is emptied within 48 h [63]. There is great inter- and intraindividual variability, however, and it is almost impossible to define a precise normal upper limit. Moreover, there is no correlation between the number of bowel movements per day and transit time, as fecal material is stored until defecation can occur at a convenient time.

In radio-opaque marker studies, no significant differences in transit between the right, left and rectosigmoid colons are detectable [74], each region contributing about one-third of the overall colonic transit time. By scintigraphy, however, the composition of ileal effluent can be shown to influence the rate of transit through the right colon [63], although the total colonic transit time remains relatively unaffected. In response to eating, the right colon transfers contents in a distal direction only, whereas the splenic flexure region transfers contents both distally and proximally [81]; both antegrade and retrograde isotopic movements increase after eating, but the proportion of antegrade movements is greater. Thus it appears that during fasting the right colon acts as a storage area, and that eating induces right colonic emptying. In contrast, in the transverse colon, postprandial antegrade and retrograde movements originate from the right colon, and the splenic flexure appears to be responsible for the mixing of intraluminal contents. Movements in the descending and rectosigmoid colon are more limited: no obvious propulsive response to the meal is seen in these regions, although some retrograde transit of contents may occur in response to eating [78].

Regulation of Colonic Motility. The colonic smooth muscle is innervated essentially by excitatory and inhibitory neurons, with the cell bodies lying mainly in the myenteric plexus and the other minor plexuses. Postsynaptic excitatory enteric neurons receive input from both the *vagus and the pelvic nerves* at nicotinic receptors. Preganglionic vagal fibers synapse on pure cholinergic post-synaptic enteric neurons; this is known because the stimulatory motor response to vagal stimulation is completely blocked by atropine. By contrast, the motor response to pelvic nerve stimulation is only partially blocked by atropine, suggesting that the fibers may synapse on two types of excitory enteric neurons: *cholinergic and non-cholinergic*. The latter do not seem to be involved in spontaneous colonic motor activity, because all contractions in

vivo are blocked by atropine. Postsynaptic enteric inhibitory neurons in the colon do not appear to receive input from the vagus or the pelvic nerves, except in the rectum. The *role of the vagus* in the control of colonic motility is not well defined. Anatomical evidence suggests it innervates at least half the proximal colon, and possibly the entire colon [96]. Indeed, studies in monkeys have demonstrated that the vagi exert a major direct or indirect influence on fasting and postprandial colonic motor activity along the entire colon [25]: efferent vagal stimulation in the presence of cholinergic and adrenergic blockade revealed the existence of an *inhibitory NANC pathway* under vagal control.

The *prevertebral ganglia* provide a reflex pathway for modulation of colonic motor activity. An afferent and/or efferent pathway from the colon to the superior and inferior mesenteric ganglia has been established in several species; the afferent limb of this reflex is composed of peptidergic and cholinergic neurons, while the efferent limb consists of noradrenergic and somatostatinergic neurons. The likely function of these prevertebral reflexes is to maintain a *continuous inhibition of colonic motor activity*. The *peristaltic reflex* can also be shown to occur in the colon, although its physiological function is not clear as, unlike the small bowel, the colon is usually evenly filled with fecal material. Descending inhibition may, however, precede a giant migrating contraction [53], thus facilitating rapid propulsion.

Other reflex pathways exist to initiate ascending and descending excitatory reflexes in response to chemical or mechanical stimulation of the colonic wall. Thus, colorectal distension inhibits proximal small-bowel motor activity [56], perhaps providing a mechanism for a delay in the small-bowel transit of nutrients until defecation occurs. Much recent interest has centered on the role of 5 HT as a neurotransmitter in the colon and the rest of the gastrointestinal tract: at least four subtypes of 5 HT receptors are present in the CNS and peripheral neurons [72]. Reports of its effects in vivo are inconsistent: some groups report that it stimulates the proximal but inhibits the distal colon, while others report that it inhibits motor activity of the entire colon. Recently, $5 HT_3$ receptor antagonists have been shown to slow colonic transit.

Central mechanisms modulating clonic motility are most obvious during *defecation*. Emotional factors and stress, however, can also modulate colonic motor function, as in other regions of the gut. Thus, motor activity of the sigmoid colon is increased during short-term physical or mental stress, and in rats, restraint stress increases colonic transit and fecal pellet output, an effect apparently mediated by corticotropin-releasing factor.

As in the other regions of the gastrointestinal tract, the functions and sites of action of many of the *gut peptides* in the control of colonic motor activity remain an enigma. Intravenous CCK-octapeptide stimulates, while secretin inhibits, distal colonic motor activity [28,70]. In contrast, CCK appears to inhibit motor activity of the proximal colon in man. Pentagastrin has not been shown to affect human rectosigmoid motor activity [77], but neurotensin, which is also released after feeding, increases motor activity in the proximal and to a lesser extent the distal colon [114]. Little, if any, information is available on the other gut peptides present in the colon.

63.2.4 Anorectal Motility and Defecation

The *internal anal sphincter* is a thickening of the inner circular smooth muscle layer of the rectum at the level of the dentate line. Nerve fibers innervating the sphincter include cholinergic excitatory nerves from the pelvic plexus, adrenergic excitatory nerves from the pelvic plexuses and pudendal nerves, and intramural NANC inhibitory nerves, which also arise from the pelvic plexuses [13]. The relative importance of each of these fibers in normal functioning of the anorectum remains unclear.

The *external anal sphincter* is a striated muscle surrounding the internal sphincter in several bundles and merging with the outer longitudinal smooth muscle coat. Motor innervation of this sphincter and the sensory supply of the lower anal canal are via the pudendal nerve [2–4]; no autonomic innervation has been described, and reflexes involving the external anal sphincter are mediated by external nerves and the higher centers.

Delivery of colonic contents into the rectum is intermittent, and the rectum is likely to remain empty at times. Even at rest, however, there is a basal anal canal pressure, which undergoes regular fluctuations at a rate of 10–20/min. The major contributor to this basal pressure appears to be the internal anal sphincter. The tone of this sphincter is generated by the slow wave activity (frequency 15–35 cycles/min) of its smooth muscle cells; unlike the activity in the colon, each slow wave cycle is associated with a contraction in the absence of superimposed spike burst activity. In the external anal sphincter, however, there is continuous striated muscle spike activity demonstrable by electromyography, and this activity increases with an increase in intra-abdominal pressure, raising tone to help preserve continence. Recently, the rectum, but not the anal canal, has been shown to exhibit periodic contractile activity [64].

The detection of material delivered to the rectum results in a series of both voluntary and involuntary reflexes that ultimately result in *defecation*. Defecation cannot occur unless the normal resistance factors that maintain continence are decreased, and important events accomplishing this are summarized in Table 63.6. When the rectum is distended by a balloon, rectal contractions are initiated; as the maximum tolerable volume is reached there is failure of the accommodative process and intrarectal pressures rise, with associated discomfort. A rapid and short-lived relaxation of the internal anal sphincter occurs with a low distending volume (*recto-anal inhibitory reflex*), as shown in Fig. 63.13 [98]. This reflex arc includes the stretch receptors in the muscularis, an afferent spinal reflex arc, the NANC inhibitory nerves to the internal anal sphincter smooth muscle, and the excitatory nerves to the external anal sphincter. Relaxation of the internal anal sphincter with reduction of pressure in the proximal anal canal in

Table 63.6. Major phases of the normal defecation process. (Modified from [87], with permission)

Delivey	Transit of colonic contents
	– Giant migrating contractions
	– Propagating contractions
	Sigmoid emptying into rectum
Detection	Accommodation
	Sensory innervation
Decreased resistance	Rectoanal inhibitory reflex
	Opening of anorectal angle
	– Hip flexure
	– Relaxation of puborectalis and
	– pubococcygeus
	Perineal descent
	Relaxation of the external anal sphincter
Discharge	Valsalva maneuver
	Rectal contraction
	Sigmoid peristalsis

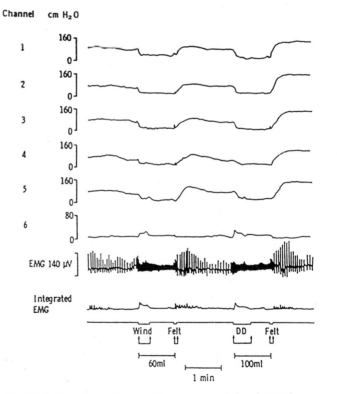

Fig. 63.13. Recordings of anorectal pressure and the electrical activity of the EAS and IAS in a healthy subject before, during, and after inflation of a rectal balloon with 60 and 100 ml of air. *Channels 1–6* represent recording sensors situated 0.5, 1.0, 1.5, 2.0, 2.5 and 4.5 cm from the anal verge. Note that rectal distension induces a relaxation in sphincter pressure. This is associated with abolition of the electrical oscillations produced by IAS activity, and an increase in the electrical activity of the EAS. Deflation produces a rebound increase in pressure, which is associated with a marked increase in the slow-wave oscillations. Rectal sensations experienced by the subject are shown. *DD*, desire to defecate. (From [85], with permission)

response to rectal filling allows the rectal contents to come in contact with the mucosa in the anal canal. This enables discrimination between solid stool, liquid feces, and flatus. It is also feasible, however, that gas or liquid rectal contents stimulate rapidly adapting rectal stretch receptors while solid feces are discriminated by stretch receptors in the pelvic floor. Fecal continence during micturition and the passage of flatus is maintained by contractile activity both in the anal canal and external anal sphincter.

If defecation is to be deferred, a *voluntary contraction of the pelvic floor and external anal sphincter* reinforces the sphincteric mechanisms and returns rectal contents to the rectal ampulla from the upper anal canal, and the "call to stool" subsides. If defecation is to proceed, relaxation of the muscles of the pelvic floor – primarily the puborectalis and pubococcygeus – results in perineal descent. Further opening of the anorectal angle is accomplished by hip flexure, in the sitting or squatting position. External anal sphincter relaxation occurs voluntarily, and the final discharge of rectal contents is produced by rectal contraction and increased intra-abdominal pressure in response to a *Valsalva maneuver*. Giant migrating contractions provide the major force for evacuation of feces from the sigmoid colon (and beyond), stimulated by chemoreceptors and mechanoreceptors in the colonic wall with or without input from higher centers. At the end of defecation a contraction of the sphincteric and pelvic floor musculature – the *closing reflex* – returns the pelvic floor to the resting position and reconstitutes the resting anorectal angle.

References

1. Abell TL, Malagelada J-R (1988) Electrogastrography: current assessment and future perspectives. Dig Dis Sci 33:982–002
2. Avill R, Mangnall YF, Bird NC (1987) Applied potential tomography. A new non-invasive technique for measuring gastric emptying. Gastroenterology 92:1019–1026
3. Azpiroz F, Malagelada J-R (1985) Intestinal control of gastric tone. Am J Physiol 249:G501–509
4. Azpiroz F, Malagelada J-R (1985) Physiological variations in canine gastric tone measured by an electronic barostat. Am J Physiol 248:G229–G237
5. Azpiroz F, Malagelada J-R (1986) Vagally mediated gastric relaxation induced by intestinal nutrients in the dog. Am J Physiol 251:G727–G735
6. Azpiroz F, Malagelada JR (1987) Gastric tone measured by an electronic barostat in health and postsurgical gastroparesis. Gastroenterology 92:934–943
7. Bartolo DCC, Roe AM, Locke Edmonds JC, Virgee J, Mortensen NJMcC (1986) Flap-valve theory of anal continence. Br J Surg 73:1012–1014
8. Bassotti G, Betti G, Imbimbo BP, Peli MA, Morelli A (1989) Colonic motor response to eating: a manometric investigation in proximal and distal portions of the viscus in man. Am J Gastroenterol 84:118–122
9. Bassotti G, Gaburri M (1988) Manometric investigation of high-amplitude propagated contractile activity of the human colon. Am J Physiol 255:G660–G664
10. Bergmann JF, Dhote D, Leglise P, Chassany O, Conort O, Caulin C (1993) Oro-cecal transit time assessed by sulphapyridine appearance in saliva. Gastroenterology 104: 477 (abstract)
11. Bolondi L, Bortolotti M, Santi V (1985) Measurement of gastric emptying time by real-time ultrasonography. Gastroenterology 89:752–759
12. Bueno L, Fioramonti J, Ruckebusch Y, Frexinos J, Coulom P (1980) Evaluation of colonic myoelectrical activity in health and functional disorders. Gut 21:480–485
13. Burleigh DE, D'Mello A (1983) Neural and pharmacologic factors affecting motility of the internal anal sphincter. Gastroenterology 84:409–417
14. Camilleri M, Colemont LJ, Phillips SF (1989) Human gastric emptying and colonic filling of solids characterized by a new method. Am J Physiol 257:G284–G290
15. Cannon WB (1898) The movements of the stomach studied by means of Roetngen rays. Am J Physiol 1:359–382
16. Christensen J, Schedl HP, Clifton JA (1966) The small intestinal basic electrical rhythm (slow wave) frequency gradient in normal man and in patients with a variety of diseases. Gastroenterology 50:309–315
17. Collins PJ, Horowitz M, Chatterton BE (1988) Proximal, distal and total stomach emptying of a digestible solid meal in normal subjects. Br J Radiol 61:12–18
18. Collins PJ, Horowitz M, Cook DJ (1983) Gastric emptying in normal subjects – a reproducible technique using a single scintillation camera and computer system. Gut 24:1117–1125
19. Cooke AR, Chvasta TE, Weisbrodt NW (1972) Effect of pentagastrin on emptying of liquids in dogs. Am J Physiol 223:934–938
20. Corbett CL, Thomas S, Read NW, Hobson N, Bergman I, Holdsworth CD (1981) Electrochemical detector for breath hydrogen determination: measurement of small bowel transit time in normal subjects and patients with the irritable bowel syndrome. Gut 22:836–840
21. Courturier D, Roze C, Turpin MH, Debray C (1969) Electromyography of the colon in situ. An experimental study in man and in the rabbit. Gastroenterology 56:317–322
22. Cunningham KM, Baker R, Horowitz M (1991) Use of technetium 99 m(v) thiocyanate to measure gastric emptying of fat. J Nucl Med 32:878–881
23. Cunningham KM, Daly J, Horowitz M, Read NW (1991) Gastrointestinal adaptation to diets of differing fat composition in human volunteers. Gut 32:483–486
24. Cunninham KM, Horowitz M, Read NW (1991) The effect of short term dietary supplementation with glucose on gastric emptying in humans. Br J Nutr 65:15–19
25. Dapoigny M, Cowles VE, Zhu YR, Condon RE (1992) Vagal influence on colonic motor activity in conscious nonhuman primates. Am J Physiol 262:G231–G236
26. Dapoigny M, Trolese JF, Bommelaer G, Tournut R (1988) Myoelectric spiking activity of right colon, left colon, and rectosigmoid of healthy humans. Dig Dis Sci 33:1007–1012
27. Diamant NE, Bortoff A (1969) Nature of the intestinal slow wave frequency gradient. Am J Physiol 216:301–307
28. Dinoso VP Jr, Meshknpour H, Lorber SH, Gutierrez JG, Chey WY (1973) Motor responses of the sigmoid colon and rectum to exogenous cholecystokinin and secretin. Gastroenterology 65:438–444
29. Dubois A (1983) The stomach. In: Christensen J, Wingate D (eds) A guide to gastrointestinal motility. Wright, Boston, pp 215–232
30. Edelbroek M, Horowitz M, Fraser R, Wishart J, Morris H, Dent J, Akkermans L (1992) Adaptive changes in the pyloric motor response to intraduodenal dextrose in normal subjects. Gastroenterology 103:1754–1761
31. Evans PR, Bak Y-T, Jones MP, Kellow JE (1992) Postprandial human small bowel motility: variability in response to differing coloric loads. Gastroenterology 102:446 (abstract)
32. Feldman M, Smith HJ (1987) Effect of cisapride on gastic emptying of indigestible solids in patients with gastroparesis diabeticorum: a comparison with metoclopramide and placebo. Gastroenterology 92:171–174

33. Fleckenstein P (1978) Migrating electrical spike activity in the fasting human small intestine. Am J Dig Dis 23:769–775

34. Fox JET, Daniel EE, Jury J, Fox AE, Collins SM (1983) Sites and mechanisms of action of neuropeptides on canine gastric motility differ in vivo and in vitro. Life Sci 33:817–825

35. Frexinos J, Bueno L, Fioramonti J (1985) Diurnal changes in myoelectric spiking activity of the human colon. Gastroenterology 88:1104–1110

36. Frieri G, Parisi F, Corazziari E, Caprilli R (1983) Colonic electromyography in chronic constipation. Gastroenterology 84:737–740

37. George JD (1968) New clinical method for measuring the rate of gastric emptying: the double sampling test meal. Gut 9:237–242

38. Gill RC, Kellow JE, Browning C, Wingate DL (1990) The use of intraluminal strain-gauges for recording small bowel motility. Am J Physiol 258:G61–G615

39. Grundy D (1988) Vagal control of gastrointestinal function. Baillieres Clin Gastroenterol 2:23–43

40. Grundy D, Scratcherd T (1982) A splanchno-vagal component of the inhibition of gastric motility. In: Wienbeck M (ed) Motility of the digestive tract. Raven, New York, pp 39–43

41. Hausken T, Odegaard S, Berstad A (1991) Antroduodenal motility studied by real-time ultrasonography. Effect of enprostil. Gastroenterology 100:59–63

42. Heddle R, Dent J, Toouli J, Read NW (1988) Antropyloroduodenal motor responses to intraduodenal lipid infusion in healthy volunteers. Am J Physiol 254: 6671–6679

43. Hinder RA, San-Garde BA (1982) Individual and combined roles of the pylorus and the antrum in the canine gastric emptying of a liquid and a digestible solid. Gastroenterology 84:281–286

44. Hinton JM, Lennard-Jones JE, Young AC (1969) A new method for studying gut transit times using radioopaque markers. Gut 10:842–847

45. Horowitz M, Maddox A, Bochner M (1989) Relationships between gastric emptying of solid and caloric liquid meals and alcohol absorption. Am J Physiol 257:G291–G298

46. Houghton LA, Read NW, Heddle R (1988) The relationship of the motor activity of the antrum, pylorus and duodenum to gastric emptying of a solid-liquid mixed meal. Gastroenterology 94:1285–1291

47. Dent J (1976) A new technique for continuous sphincter pressure measurement. Gastroenterology 71:263–267

48. Huizinga JD, Berezin I, Daniel EE, Chow E (1990) Inhibitory innervation of colonic smooth muscle cells and intestinal cells of Cajal. Can J Pharmacol 69:447–454

49. Huizinga JD, Stern HS, Chow E, Diamant NE, El-Sharkawy TY (1985) Electrophysiologic control of motility in the human colon. Gastroenterology 88:500–511

50. Huizinga JD, Stern HS, Chow E, Diamant NE, El-Sharkawy TY (1986) Electrical basis of excitation and inhibition of human colonic smooth muscle. Gastroenterology 90:1197–1204

51. Hyland JMP, Darby CF, Hammond P, Taylor I (1980) Myoelectrical activity of the sigmoid colon in patients with diverticular disease and the irritable colon syndrome suffering from diarrhoea. Digestion 20:293–299

52. Kamath PS, Hoepfner MT, Phillips SF (1987) Short-chain fatty acids stimulate motility of the canine ileum. Am J Physiol 253:G427–G433

53. Karaus M, Sarna SK (1987) Giant migrating contractions during defecation in the dog colon. Gastroenterology 92: 925–933

54. Kellow JE, Borody TJ, Brown ML, Haddad AC, Phillips SF (1986) Sulfapyridine appearance in plasma after salicylazosulfapyridine: another simple measure of intestinal transit. Gastroenterology 91:396–400

55. Kellow JE, Borody TJ, Phillips SF, (1986) Human interdigestive motility: variations in patterns from esophagus to colon. Gastroenterology 91:386–395

56. Kellow JE, Gill RC, Wingate DL (1987) Modulation of human upper gastrointestinal motility by rectal distension. Gut 28:864–868

57. Kellow JE, Gill RC, Wingate DL (1990) Prolonged ambulant recordings of small bowel motility demonstrate abnormalities in the irritable bowel syndrome. Gastroenterology 98:1208–1218

58. Kellow JE, Miller LJ, Phillips SF, Haddad AC, Zinsmeister AR, Charboneau JW (1987) Sensitivities of human jejunum, ileum, proximal colon and gallbladder to cholecystokinin octapeptide. Am J Physiol 252:G345–G356

59. Kelly KA (1981) Motility of the stomach and gastroduodenal junction. In: Johnson LR (ed) Physiology of the gastrointestinal tract. Raven, New York, pp 393–410

60. Kelly KA, Code CF (1971) Canine gastric pacemaker. Am J Physiol 220:112–118

61. Kelly KA, Code CF, Elveback LR (1969) Patterns of canine gastric electrical activity. Am J Physiol 217:461–470

62. King PM, Adam RD, Pryde A, McDicken WN, Heading RC (1984) Relationships of human antroduodenal motility and transpyloric fluid movement: non-invasive observations with real-time ultrasound. Gut 25:1384–1391

63. Krevsky B, Malmud LS, D'Ercole F (1986) Colonic transit scintigraphy. Gastroenterology 91:1102–1112

64. Kumar D, Williams NS, Waldron D, Wingate DL (1989) Prolonged manometric recording of anorectal motor activity in ambulant human subjects: evidence of periodic activity. Gut 30:1007–1011

65. Lang IM, Sarna SK, Condon RE (1986) Gastrointestinal motor correlates of vomiting in the dog: quantification and characterization as an independent phenomenon. Gastroenterology 90:40–47

66. Lin HC, Kim BH, Elashoff JD, Doty JE, Gu Y-G, Meyer JH (1992) Gastric emptying of solid food is most potently inhibited by carbohydrate in the canine distal ileum. Gastroenterology 102:793–801

67. Mahieu P, Pringot J, Bodart P (1984) Defecography. I. Description of a new procedure and results in normal patients. Gastrointes Radiol 9:247–251

68. Malagelada J-R, Camilleri M, Stanghellini V (1986) Manometric diagnosis of gastrointestinal motility disorders. Thieme, New York

69. Malagelada J-R, Robertson JS, Brown ML, Remington M, Duenes JA, Thomforde GM, Carryer PW (1984) Intestinal transit of solid and liquid components of a meal in health. Gastroenterology 87:1255–1263

70. Mangel AW (1984) Potentiation of colonic contractility to cholecystokinin and other peptides. Eur J Pharmacol 100:285–290

71. Matheson DM, Keighley MRB (1981) Manometric evaluation of rectal prolapse and faecal incontinence. Gut 22:126–129

72. Mawe GM, Branchek TA, Gershon MD (1986) Peripheral neural serotonin receptors: identification and characterization with specific antagonists and agonists. Proc Natl Acad Sci USA 83:9799–9803

73. McLelland GR, Sutton JA (1985) Epigastric impedance: a non-invasive method for the assessment of gastric emptying and motility. Gut 6:607–614

74. Metcalf AM, Phillips SF, Zinsmeister AR (1987) Simplified assessment of segmental colonic transit. Gastroenterology 92:40–47

75. Meyer J (1987) Motility of the stomach and gastroduodenal junction. In: Johnson LR (ed) Physiology of the gastrointestinal tract. Raven, New York, pp 613–629

76. Meyer JH, Thomson JB, Cohen MB, Shadchehr A, Mandiolla S (1979) Sieving of food by the canine stomach and sieving after gastric surgery. Gastroenterology 76:804–813

77. Misiewica JJ, Holdstock DJ, Waller SL (1967) Motor responses of the human alimentary tract to near-maximal infusions of pentagastrin. Gut 8:463–469

78. Moreno-Osset E, Bazzocchi G, Lo S (1989) Association between postprandial changes in colonic intraluminal pressure and transit. Gastroenterology 96:1265–1273

79. Narducci F, Bassotti G, Gaburri M, Morelli A (1987) Twenty four hour manometric recording of colonic activity in healthy man. Gut 28:17–25

80. Nimmo WS (1976) Drugs, diseases and altered gastric emptying. Clin Pharmacokinet 1:189–203

81. Picon L, Lemann M, Flourie B, Rambaud J-C, Rain J-D, Jian R (1992) Right and left colonic transit after eating assessed by a dual isotopic technique in healthy humans. Gastroenterology 103:80–85

82. Proano M, Camilleri M, Phillips SF, Brown ML, Thomforde GM (1990) Transit of solids through the human colon: regional quantitation in the unprepared bowel. Am J Physiol 258:G856–G862

83. Quigley EMM, Borody TJ, Phillips SF (1984) Motility of the terminal ileum and ileocecal sphincter in healthy man. Gastroenterology 87:857–866

84. Read NW, Miles CA, Fisher D (1980) Transit of a meal through the stomach, small intestine, and colon in normal subjects and its role in the pathogenesis of diarrhea. Gastroenterology 79:1276–1282

85. Read NW, Sun W-M. Disorders of the anal sphincters. In: Snape WJ (ed) Pathogenesis of functional bowel disease. Plenum, New York, pp 289–310

86. Read NW, Harford WV, Schmulen AC, Read MG, Santa Ana C, Fordtran JS (1979) A clinical study of patients with faecal incontinence and diarrhoea. Gastroenterology 76:747–756

87. Reynolds JC (1989) Mechanisms and management of chronic constipation. In: Snape WJ (ed) Pathogenesis of functional bowel disease. Plenum, New York, pp 199–225

88. Ritchie JA (1968) Colonic motor activity and bowel function. Gut 9:442–456

89. Rouillon J-M, Azpiroz F, Malagelada J-R (1991) Sensorial and intestinointestinal reflex pathways in the human jejunum. Gastroenterology 101:1606–1612

90. Rouillon J-M, Azpiroz F, Malagelada J-R (1991) Reflex changes in intestinal tone: relationship to perception. Am J Physiol 261:G280–G286

91. Sanders KM, Stevens R, Burke E, Ward SM (1990) Slow waves actively propagate at submucosal surface of circular layer in canine colon. Am J Physiol 260:G258–G263

92. Sarna SK (1991) Physiology and pathophysiology of colonic motor activity. Dig Dis Sci 36:998–1018

93. Sarna SK, Bardakjian BL, Waterfall WE, Lind JF (1980) Human colonic electrical control activity (ECA). Gastroenterology 78:1526–1536

94. Sarna SK, Condon R, Cowles V (1984) Colonic migrating and non migrating motor complexes in dogs. Am J Physiol 246:G355–G360

95. Sarna SK, Latimer P, Campbell D, Waterfall W (1982) The effects of stress, meal and neostigmine on rectosigmoid motility in normals. In: Wienbeck M (ed) Motility of the digestive tract. Raven, New York, pp 499–511

96. Satomi H, Yamamoto F, Ise H, Takatama H (1978) Origins of the parasympathetic preganglionic fibres to cat intestine as demonstrated by the horseradish peroxidase method. Brain Res 151:571–578

97. Schemann M, Ehrlein H-J (1986) Postprandial patterns of canine jejunal motility and transit of luminal content. Gastroenterology 90:991–1000

98. Schuster MM, Hendrix TR, Mendeloff AI (1963) The internal anal sphincter response: manometric studies on its normal physiology, neural pathways, and alteration in bowel disorders. J Clin Invest 42:196–207

99. Schwizer W, Maecke H, Fried M (1992) Measurement of gastric emptying by magnetic resonance imaging in humans. Gastroenterology 103:369–376

100. Sillin LF, Schulte WJ, Woods JH et al (1979) Electromotor feeding responses to primate ileum and colon. Am J Surg 137:99–105

101. Snape WJ Jr, Carlson GM, Cohen S (1976) Colonic myoelectric activity in the irritable bowel syndrome. Gastroenterology 70:326–330

102. Snape WJ Jr, Matrazzo SA, Cohen S (1978) Effect of eating and gastrointestinal hormones on human colonic myoelectrical motor activity. Gastroenterology 75:373–378

103. Soffer EE, Scalabrini P, Wingate DL (1989) Prolonged ambulant monitoring of human colonic motility. Am J Physiol 256:G601–G806

104. Spiller RC, Brown ML, Phillips SF (1986) Decreased fluid tolerance, accelerated transit, and abnormal motility of the human colon induced by oleic acid. Gastroenterology 91:100–107

105. Spiller RC, Trotman IF, Higgins BE (1984) The ileal brake – inhibition of jejunal motility after ileal fat perfusion in man. Gut 25:365–374

106. Staadas J, Aune S (1970) Intragastric pressure/volume relationship before and after vagotomy. Acta Chir Scand 136:611–615

107. Steadman CJ, Phillips SF, Camilleri M, Haddad AC, Hanson RB (1991) Variation of muscle tone in the human colon. Gastroenterology 101:373–381

108. Stoddard CJ, Duthie HL, Smallwood RH, Linkens DH (1979) Colonic myoelectric activity in man: comparison of recording techniques and methods of analysis. Gut 29:476–483

109. Szurszewski JH (1969) A migrating electric complex of the canine small intestine. Am J Physiol 217:1757–1763

110. Szurszewski JH (1987) Electrical basis for gastrointestinal motility. In: Johnson LR (ed) Physiology of the gastrointestinal tract. Raven, New York, pp 383–422

111. Taylor I, Darby C, Hammond P, Basu P (1978) Is there a myoelectrical abnormality in the irritable colon syndrome? Gut 19:391–395

112. Thompson DG, Ritchie HD, Wingate DL (1982) Patterns of small intestinal motility in duodenal ulcer patients before and after vagotomy. Gut 21:500–506

113. Thompson DG, Wingate DL, Archer L (1980) Normal patterns of human upper small bowel motor activity recorded by prolonged radiotelemetry. Gut 21:500–506

114. Thor K, Rosell S (1986) Neurotensin increases colonic motility. Gastroenterology 90:27–31

115. Tougas G, Anvari M, Dent J, Somers S, Richards D, Stevenson GW (1992) Relationship of pyloric motility to pyloric opening and closure in healthy subjects. Gut 33:466–471

116. Urbain J-LC, Van Cutsem E, Siegel JA, Mayeur S, Vandecruys A, Janssens J, De Roo M, Vantrappen G (1990) Visualization and characterization of gastric contractions using a radionuclide technique. Am J Physiol 259:G1062–G1067

117. Vantrappen G, Janssens J, Hellemans J (1977) The interdigestive motor complex in normal subjects and patients with bacterial overgrowth of the small intestine. J Clin Invest 59:1158–1166

118. Weems WA, Seygel GE (1981) Fluid propulsion by cat intestinal segments under conditions requiring hydrostatic work. Am J Physiol 240:G147–G156

119. Womack NR, Williams NS, Holmfield JHM, Morrison JFB, Simpkins KC (1985) New method for the dynamic assessment of anorectal function in constipation. Br J Surg 72:994–998

120. Yamagishi T, Debas HT (1978) Cholecystokinin inhibits gastric emptying by acting on both proximal stomach and pylorus. Am J Physiol 234:E375–E378

121. Young JA, Cook DI (1991) Gastrointestinal motility. In: Young JA, Cook DI, Conigrave AD, Murphy CR (eds) Gastrointestinal physiology. Globe, Melbourne, pp 45–80

64 The Major Salivary Glands

J.A. YOUNG and D.I. COOK

Contents

64.1 Introduction

In humans there are three pairs of major salivary glands, the *parotid* glands, the *submandibular* glands (called the mandibular glands in all other mammals) and the *sublingual* glands. The parotid is wholly *serous* secreting, i.e., it secretes a juice consisting of water, electrolytes and an enzyme (amylase), whereas the sublingual gland is wholly *mucous* secreting, i.e., it secretes water and electrolytes together with mucins. The submandibular glands are usually mixed serous and mucous, with some cells secreting a watery solution of enzymes and others secreting mucins.

In addition to the major salivary glands, there are numerous other compound glands beneath the epithelium of the oral mucosa which secrete mucins. Some of these glands are relatively constant in location and large enough to receive names in some mammals, e.g., the zygomatic glands of dogs. In humans, the minor salivary glands associated with the circumvallate papillae at the back of the tongue – von Ebener's glands – are important because they secrete a non-specific lipase (lingual lipase) that assists in fat digestion (in the stomach), particularly in the newborn.

We have recently published an extensive review of the literature on salivary gland function, on which the present chapter is based. Students seeking more detailed information are referred to this review [10] and to our earlier monographs on salivary gland structure [48] and function [8,49].

64.2 Functions of Saliva

The principal functions of the salivary glands are as follows:

- They provide a fluid medium in which to dissolve some of the ingested food and provide a lubricant to aid in chewing and swallowing what cannot readily be dissolved. The salivary water itself acts as a *lubricant*, but in addition, saliva contains secreted *mucus*.
- They keep the buccal cavity moist and clean and *prevent growth of infectious agents* in the mouth. This is achieved primarily by irrigation of the mouth and teeth but , in addition, saliva contains lysozyme, peroxidase and immunoglobulin (IgA), which have antibacterial and/or antiviral properties. A moist buccal cavity is essential for clear speech.
- They secrete *digestive enzymes*, which help dislodge food residues impacted around the teeth.
- They secrete various *growth-promoting factors*, in particular NGF (nerve growth factor) and EGF (epidermal growth factor).
- In fur-bearing mammals, they aid in *thermal regulation*: saliva can be evaporated from the upper respiratory tract and the tongue by panting (e.g., in dogs) or from the surface of saliva-wetted fur (e.g., in cats).
- In some non-mammalian species, e.g., snakes, the salivary glands *secrete venom*, and in others, e.g., crocodiles, they act as salt-excreting glands.
- In mammalian infants, saliva provides a *fluid seal* between lips and nipple, thereby *permitting suckling*, the very activity that gives the order Mammalia its name.

R. Greger/U. Windhorst (Eds.)
Comprehensive Human Physiology, Vol. 2
© Springer-Verlag Berlin Heidelberg 1996

64.3 Morphology

In contrast to the exocrine pancreas, which exhibits a certain morphological homogeneity among all the vertebrates, the salivary glands are characterized by extraordinary diversity in location, development, microscopic structure, and function [10,48].

64.3.1 Secretomotor Innervation

In most mammalian species, the major salivary glands are well supplied with both parasympathetic and sympathetic *secretomotor nerve fibers* (cf. Sect. 64.7.5). Preganglionic parasympathetic fibres arise in the ipsilateral nucleus reticularis parvocellularis. Nerve fibers for the parotid may accompany the glossopharyngeal nerve to the otic ganglion or they may reach the parotid via a buccal branch of the trigeminal nerve. Fibers projecting onto ganglion cells in or near the mandibular and the sublingual glands mostly follow the facial nerve and the chorda tympani, but some travel via the lingual branch of the trigeminal nerve. Postganglionic sympathetic fibers originate in the bilateral superior cervical ganglia and follow the blood vessels supplying the gland.

64.3.2 Microscopic Anatomy

As is the case for all compound exocrine glands, the salivary gland parenchyma consists of *secretory endpieces* and a *branching duct system*, arranged in lobules separated by septa of interlobular connective tissue, which contain the larger excretory ducts, blood vessels, nerve bundles, and small ganglia. The secretory endpieces may vary in size and shape (i.e., acini, tubules or tubulo-acini), even within a single lobule. Typical mucous endpieces are tubular, whereas serous endpieces are mostly acinar (i.e., berry shaped). Whereas the endpieces in the parotid glands of most mammals appear very similar histologically and histochemically, the same cannot be said for the submandibular glands, in which, in many species, two different types of secretory cell occur, either within the same endpiece, e.g., as mucous tubules with serous demilunes, or in different endpieces. The terms "mucous," "serous," and "seromucous" are rather loosely applied according to what proportion of the secretory proteins in the storage granules are made up of carbohydrate-rich mucins [48].

The duct system of the major salivary glands consists of *intercalated*, *striated*, and *excretory* ducts (Fig. 64.1). Intercalated and striated ducts are located within the lobules, whereas the excretory ducts lie embedded in the

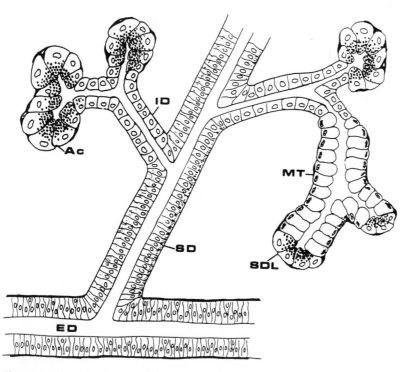

Fig. 64.1. Schematic diagram of the microscopic anatomy of a mammalian mandibular gland. The secretory endpieces, represented here by three acini (*Ac*) and a mucous tubule (*MT*) capped by a seromucous demilune (*SDL*), drain via intercalated ducts (*ID*) into intralobular (striated) ducts (*SD*). These are called striated because of the prominent striations in the basolateral regions of the epithelial cells comprising the ducts. They drain into a converging system of extralobular ducts (*ED*), which drain into a single main excretory duct opening into the floor of the mouth. (From [10])

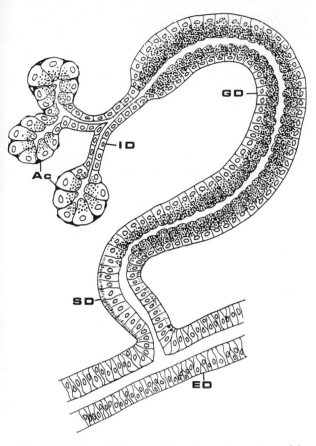

Fig. 64.2. Schematic diagram of the microscopic anatomy of the mandibular gland of the male mouse. The secretory endpieces, represented here by three acini (*Ac*), drain via two generations of intercalated ducts (*ID*) into an intralobular duct, known in this gland as a granular duct (*GD*) on account of the presence of numerous secretion granules packed into the luminal region of the epithelial cells comprising the ducts. The most proximal region of the intralobular duct (*SD*) is less markedly granulated than elsewhere and exhibits basal striations characteristic of the intralobular ducts of other mammalian mandibular glands (including the mandibular glands of female mice). The intralobular ducts drain into a converging system of extralobular ducts as they do in other glands. (From [10])

interlobular stroma. The duct system is usually reduced in the minor salivary glands. In many species, the striated ducts are modified by the presence of secretion granules in the cells (Fig. 64.2). Sexual dimorphism is conspicuous in the mandibular glands of a number of species, particularly in some rodents (e.g., mouse), the pig, and various insectivores.

Myoepithelial cells are present in most salivary glands, usually associated with the secretory endpieces and, to a lesser extent, with intercalated ducts (Fig. 64.3). They are particularly well developed in glands that produce a highly viscous mucus [47]. Their principal function seems to be the prevention of endpiece distension, which tends to occur during secretion as a consequence of the accompanying increase in intraluminal pressure. This is achieved in part by provision of physical support for the endpiece epi-

thelium, and in part by causing a widening and shortening of the intralobular ducts, which has the effect of lowering the outflow resistance.

64.3.3 Ultrastructure

The fine structure of the cells of salivary secretory endpieces conforms to the characteristic pattern for protein-secreting exocrine cells; in other words, they have a well-developed *rough endoplasmic reticulum* (RER) in the basal cytoplasm and numerous stored *secretion granules* in the apical region of the cell. The diversity in types of secretory protein produced in different glands, or even by different cells within the same gland, results in considerable variation in the ultrastructural appearances of such cells, particularly with respect to the secretion granules and the relative abundance of RER.

The basolateral plasma membranes usually show some infolding, whereas the luminal membrane may be relatively smooth or studded with short *microvilli*. *Secretion canaliculi*, i.e., intercellular extensions of the endpiece lumen, are commonly found between serous (or seromucous) secretory cells, but are rare between mucous cells.

The epithelium of the intercalated ducts shows few distinguishing features, although the presence of secretion granules has been reported in some glands. In contrast to the relatively simple infolding of basolateral plasma membranes in the intercalated duct epithelium, the striated duct epithelium shows characteristic deep infoldings as-

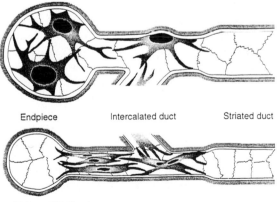

Fig. 64.3. Schematic diagram showing how myoepithelial cells are arranged around the parenchymal structures in salivary glands. Typically, stellate myoepithelial cells are clustered around the endpieces, looking like a coarse-meshed basket, and aligned along the intercalated ducts. In some glands, notably the rat parotid gland, myoepithelial cells are associated only with the intercalated ducts, and take on an elongated spindle-like shape. By contracting during evoked secretion, the myoepithelial cells prevent the endpieces from distending and they lower outflow resistance in the intercalated ducts by shortening them and widening their lumens. (From [48])

sociated with numerous mitochondria [39]. In the submandibular glands of many species, the apical cytoplasm of the striated duct cells contains small secretion granules. In many rodents, including mouse and rat, the duct cells are packed with fairly large, electron-dense granules; hence the name "*granular ducts*" for such modified striated ducts (Fig. 64.2). Since the granular ducts are also longer than the original striated ducts and therefore more convoluted, these ducts are also referred to as "granular convoluted tubules." The epithelium of the extralobular excretory ducts also shows plasma membrane infolding and a relative abundance of mitochondria, although these are less pronounced than in the striated ducts.

Of physiological interest is the extent to which *tight junctions* (zonulae occludentes) are developed in salivary epithelia. Studies in which tracers, such as lanthanum hydroxide and horseradish peroxidase, or freeze-fracture techniques have been used suggest that the tight junctions between *endpiece cells* are usually rather *leaky* [48]. In contrast, freeze-fracture studies on the *striated duct* epithelium have revealed very well-developed zonulae occludentes [35], suggesting that the junctions in the ducts are probably not leaky.

64.4 Stimulus–Secretion Coupling in Salivary Glands

Salivary secretion is evoked when neurohumoral agonists bind to surface receptors on the cells of the secretory endpieces and generate intracellular second messengers which, in turn, activate the cellular mechanisms responsible for secretion (cf. Chaps. 5, 6). Receptors for several neurohumoral agonists have been demonstrated in salivary glands by means of binding studies with labelled receptor ligands or by immunohistochemistry. In this way, receptors for *acetylcholine* (M_1 and M_3 muscarinic receptors), *noradrenaline* (α_1, α_2, β_1 and β_2 receptors), various *purines* (P_{2z} receptors), *VIP* (vasoactive intestinal peptide), *substance P* (NK1 receptors), *neurotensin*, various nucleotides, *EGF* (epidermal growth factor), *parathyroid hormone*, and *prostaglandin E_2* have been demonstrated [10]. Not all these receptors are present in every gland type, however: for example, substance P receptors are present in the mandibular glands of the rat but not in those of the mouse [11]. Salivary glands within a species may also have differing receptor populations: for example, substance P receptors are expressed at very much higher levels in the sublingual and mandibular glands of the rat than they are in the parotid glands. Furthermore, ducts and endpieces within a gland may have differing receptor populations [11].

Although non-adrenergic, non-cholinergic agonists have a role in controlling salivary secretion, the neurotransmitters of greatest physiological importance are clearly *acetylcholine*, released from parasympathetic nerves and acting on M_3 (and M_1) muscarinic receptors, and *noradrenaline*, released from sympathetic nerves and

acting on α_1, α_2, β_1 and β_2 receptors. It is not currently possible to say precisely what roles the other agonists have, although it is clear that substance P and VIP are true neurotransmitters. Acetylcholine, acting on M_3 muscarinic receptors, activates phospholipase C, which catalyzes the formation of inositol-1,4,5-trisphosphate (IP_3) which, in turn, leads to an increase in the free Ca^{2+} concentration in the cytosol of the secretory cells (Fig. 64.4; cf. Chap. 5). Noradrenaline (acting on adrenergic receptors) and VIP (acting on VIP receptors) activate adenylate cyclase and in this way increase intracellular levels of cyclic AMP (Fig. 64.5).

Among the agonists for which a role as a neurotransmitter has not yet been demonstrated, *ATP* is of particular interest. In salivary tissues it activates at least two distinct receptor types, one of which is a receptor of the P_{2z} type [31]; when present, this receptor may activate a Ca^{2+}-permeable non-selective cation channel directly, and thereby increase intracellular free Ca^{2+} [37].

It appears that receptor activation leads to secretion by one of two pathways, one resulting in the mobilization of Ca^{2+} and the other resulting in increased production of cyclic AMP. Unfortunately, there is no tidy division of these pathways with respect to the control of fluid and protein secretion. Although fluid secretion in most salivary glands is evoked by activation of the Ca^{2+}-dependent pathway, in some glands, e.g., the rat mandibular gland, fluid secretion is also evoked by activation of the cyclic AMP pathway. Similarly, protein secretion can result from activation of either the cAMP-dependent or the Ca^{2+}-dependent pathway. The details of both these signal transduction systems are very well known and are summarized in Figs. 64.4 and 64.5.

64.4.1 Cyclic AMP-Dependent Secretion

Cyclic AMP appears to be the second messenger mediating the β-adrenergic response in salivary glands, as in many other tissues. The processes involved have been defined in a variety of tissues (Fig. 64.5), and the available evidence overwhelmingly indicates that the same steps are involved in β-adrenergic stimulation in salivary glands as in other tissues [10]. A change in the level of protein phosphorylation is one of the most common means available for control of intracellular reactions. Control may be exerted through *protein kinases*, which phosphorylate proteins, or through *phosphatases* which dephosphorylate them. Cyclic AMP-dependent protein kinases are present in salivary glands, and it seems clear that they are involved in stimulus-secretion coupling.

64.4.2 Calcium-Dependent Secretion

There is a substantial body of evidence to support the belief that salivary endpiece cells, when they have appropriate receptors, respond to muscarinic, adrenergic and substance P stimulation by activation of phosphatidylinositide

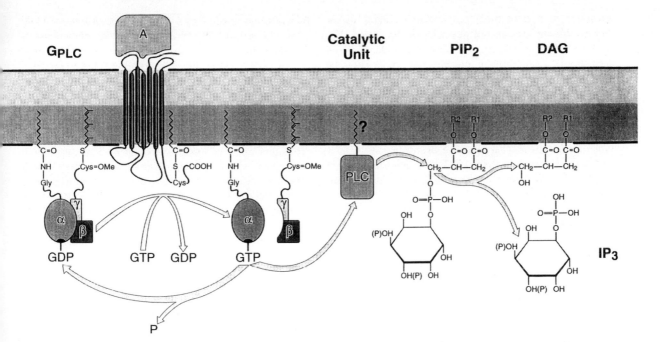

Fig. 64.4. Schematic diagram showing how stimulus–response coupling operates via the phospholipase C pathway. A cell surface receptor, shown here with seven membrane-spanning polypeptide α-helices, has been engaged (and thereby activated) by an extracellular agonist (*A*). The activated receptor, in turn, activates a G protein, a guanine nucleotide (*GDP*)-binding protein (G$_{PLC}$), causing it to disaggregate into two fragments, one called βγ, whose function is not understood, and another called α$_{PLC}$, which activates the catalytic unit, phospholipase C (*PLC*), a membrane-associated enzyme. PLC then hydrolyzes a membrane phospholipid, phosphatidylinositol-4,5-bisphosphate (*PIP$_2$*), to liberate two products, inositol trisphosphate (*IP$_3$*), which enters the cytosol, and a diacylglycerol (*DAG*), which remains in the plasma membrane. Both products function as second messengers. (From [10])

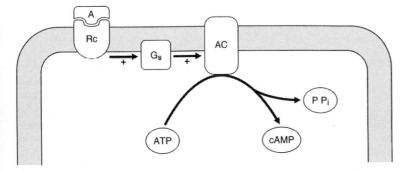

Fig. 64.5. Schematic diagram showing how stimulus–response coupling operates via the adenylate cyclase pathway. As with the PLC pathway shown in Fig. 64.4, a membrane-bound receptor protein (*Rc*) is shown engaged (and activated) by an extracellular agonist (*A*). The activated receptor, in turn, activates a G protein (*Gs*), a guanine nucleotide (GDP)-binding protein, causing it to disaggregate into βγ and α$_s$ fragments (see Fig. 64.4 for details). The α$_s$ fragment, to which GTP is bound, activates the catalytic unit, in this case adenylate cyclase, thought to be a membrane-bound enzyme. This enzyme catalyzes the conversion of cytosolic ATP to cyclic adenosine monophosphate (*cAMP*), which functions in the cytosol as a second messenger. (From [10])

metabolism and the *mobilization of Ca²⁺*. The increase in intracellular free Ca²⁺ following muscarinic stimulation (and stimulation with these other agonists) originates from three sources:

- Release of Ca²⁺ from intracellular stores that are sensitive to IP₃

- Release of Ca²⁺ from intracellular stores triggered by Ca²⁺-induced Ca²⁺ release
- Influx from the extracellular medium

The pathway appears to be activated by depletion of Ca²⁺ from the IP₃-sensitive store rather than by the elevation of inositol phosphates per se, although the mechanism by which depletion of these Ca^{2+} stores regulates plasma membrane Ca^{2+} permeability remains obscure.

The mechanisms reducing the intracellular free Ca^{2+} concentration include uptake into the intracellular IP_3-sensitive store and extrusion across the plasma membrane (Chap. 5). Uptake into the IP_3-sensitive store is driven by a Ca^{2+}-ATPase, and the site of this uptake is spatially separate from the site of IP_3-induced Ca^{2+} release. Efflux across the plasma membrane is driven principally by an electrogenic Ca^{2+}-ATPase that is activated by cAMP and by calmodulin, although the physiological importance of this efflux is doubtful since there is no evidence that muscarinic stimulation activates Ca^{2+} extrusion. Na^+/Ca^{2+} exchange across the plasma membrane also occurs but is of only minor significance.

Unlike cyclic AMP, which usually works via protein kinases, there appear to be several different ways in which Ca^{2+} can act as a modulator of cellular events:

- It can interact directly with effector proteins.
- Together with diacylglycerol, it can activate protein kinase C, which can phosphorylate effector proteins.
- It can bind to calmodulin, which then acts, either by phosphorylation or by direct interaction with the effector mechanism.
- It can evoke the release of arachidonic acid from membrane phospholipids by activation of phospholipases.
- It may act by increasing the rate of production of cyclic GMP.

Of these five mechanisms, direct interactions with membrane proteins and activation of protein kinase C appear to be of particular importance in controlling salivary secretion.

64.5 Secretion of Organic Substances

64.5.1 Secretion of Exportable Proteins

In contrast to the process of fluid secretion, which is essentially one of selective transport of salts from the interstitium to the lumen with water following salt movement by osmosis, the process of *secretion of exportable protein* is more complex, since it involves uptake of amino acids by the secretory cells followed by protein synthesis and eventual discharge of the proteins from the cells, usually by exocytosis. The mechanisms whereby these processes occur are essentially the same in all protein-secreting cells, so they need only a brief treatment here.

The first step in the synthesis of exportable proteins is the uptake of the required amino acids out of the tissue fluid across the basolateral membrane of the glandular cell. In salivary glands at least four *amino acid uptake systems* are present [52]: one for basic amino acids, one for acidic amino acids, and two for amino neutral acids. The neutral systems consist of an Na^+-dependent short-chain (ASC) system and an Na^+-independent long-chain (L) system. In contrast to most other tissues, there appears to be no spe-

cific alanine-preferring (A) system, at least in the cat submandibular gland [3]. With the aid of specific aminoacyl tRNA-synthetases, the amino acids are then attached to tRNA molecules specific for each amino acid and carried to the site of synthesis on a ribosome.

The two proteins destined to make up a ribosome become attached to mRNA, and the ribosome translates the mRNA code and accepts the appropriate amino acids from tRNA, adding them to the growing polypeptide chain that emerges from the ribosome (cf. Chap. 4). When the protein being synthesized is destined for export, the first 20 or so amino acids in the forming polypeptide chain, which constitute the so-called *signal sequence*, bind to a cytoplasmic nucleopeptide called the signal recognition particle (or SRP). Translation is then halted until the ribosome binds to a receptor protein on the RER called the SRP receptor (or docking protein). Once this occurs, the SRP is discarded and translation resumes in such a way that the growing polypeptide chain passes through the RER membrane and enters the cisternal space.

By the time translation is complete, the signal peptide sequence is cleaved off the polypeptide chain by a peptidase in the RER membrane and the newly synthesized protein is released into the cisternal space whilst the ribosome and the mRNA strand separate from one another and move into the cytosol once more. After their release into the cisternal space, the newly formed polypeptides move to the edges of the RER (the transitional sheets), where they are released into the cytosol packaged in small membrane-enclosed vesicles that move towards and fuse with the *cis* membranes of the Golgi apparatus aided by the presence of a specific cytoplasmic protein.

The *Golgi apparatus* is the site at which post-translational changes, such as *protein glycosylation*, take place. After such modifications have occurred, the synthesized proteins in their final forms are released from the *trans* membranes of the last layers of the Golgi stacks in membrane-enclosed vesicles called condensing vacuoles, structures in which further transformation and concentration of the secretory proteins occur. Eventually the condensing vacuoles become transformed into mature secretion granules, which accumulate in the apical region of the secretory cell. *Exocytosis* involves fusion of the granule membrane with the apical plasma membrane of the secretory cell followed by rupture of the fused membranes, which allows release of the granule contents into the lumen. The process is continuous in most cells, but it can be greatly accelerated by neural or hormonal stimulation.

The processes that lead directly to *exocytosis*, either via protein phosphorylation or through the effect of locally raised Ca^{2+} levels, are still poorly understood. There is little doubt that exocytosis must involve an interaction with the *actin filaments* of the *terminal web*, and both actin and Ca^{2+} have been shown to be involved in the reverse process, membrane retrieval after exocytosis. Nevertheless, notwithstanding the histochemical demonstration of actin around zymogen granules in exocrine secretory cells, there is as yet no conclusive evidence that actin is directly involved in exocytosis.

The rate of synthesis of secretory protein by the rat and mouse parotid endpiece cells is largely independent of the rate of secretion, although β-adrenoceptor agonists will evoke both synthesis and secretion. The control for such β-adrenergic stimulation is at the translation level [19]. In contrast, cholinergic and α-adrenergic agonists, which also evoke protein secretion, inhibit incorporation of ³[H] leucine into secretory proteins [33]. Even in the rat sublingual gland, which lacks a parenchymal sympathetic innervation, and in which only cholinergic agonists evoke secretion, synthesis of the secretory mucins is still inhibited by parasympathomimetic drugs [32]. Insulin stimulates synthesis of secretory proteins in the rat mandibular gland but it does not evoke exocytosis [1].

64.5.2 Exportable Proteins in Saliva

The two largest groups of exportable proteins secreted by the salivary glands (and by glands elsewhere in the alimentary canal) are *mucins* and *digestive enzymes*. Digestive enzymes will be dealt with in the appropriate chapters on each individual digestive organ, but it is convenient to comment here on mucins in general, since they are secreted in all regions of the alimentary canal including, of course, the mandibular and sublingual glands as well as all of the minor salivary glands.

Mucins. Mucins are glycoproteins, polypeptides with carbohydrate side-chains attached, in which carbohydrates make up 60–80% of the total molecular weight. They serve many functions:

- Mechanical protection for the epithelium
- Prevention of its dehydration
- Lubrication for solid food
- Protection of the epithelial lining of the stomach and small intestine from the actions of digestive enzymes and gastric acid
- Trapping of micro-organisms

A mucin monomer consists of a single polypeptide with numerous oligosaccharide side-arms attached; the oligosaccharides usually consist of 2–15 carbohydrate units with either *fucose* or *sialic acid* (neuraminic acid) as a terminal unit. When the final units consist mainly of sialic acids, and/or the oligosaccharides contain sulfate residues, the resultant presence of numerous negative charges prevents close coiling of the molecules, so that they adopt an extended conformation. This conformation, together with the fact that the carbohydrate side-arms are strongly hydrated, means that mucin solutions are highly viscous. In addition, mucin molecules usually possess a carbohydrate-free section at one end which can form cross-links with other mucin molecules via disulfide bridges so as to create the molecular network of a soft gel, which can cover the surface epithelium (in the stomach and duodenum) to form so-called mucosal protection barriers.

Digestive Enzymes. In man, the principal digestive enzyme in saliva is the starch-digesting enzyme, α-*amylase*, which is secreted mainly by the parotid glands but also, in some other species, by the intralobular ducts of the mandibular glands. There is sufficient amylase in saliva to digest all the starch present in a normal diet, but swallowing is usually so rapid that most salivary amylase is inactivated in the stomach long before starch digestion is complete, the result being that starch digestion depends largely on pancreatic amylase secreted into the duodenum. The main role of salivary amylase and proteases seems to be to promote oral hygiene by facilitating the dislodgement of food particles impacted around the teeth. *Non-specific lipase*, secreted by Von Ebner's lingual glands, is important in the newborn since it promotes milk fat digestion in the stomach.

64.5.3 Small Non-electrolytes

Small organic molecules, such as *glucose* and *urea*, usually enter the saliva only to a limited extent, and their capacity to do so is inversely related to their molecular size, and directly related to their liposolubility [2]. This may appear somewhat of a paradox, since an inverse dependence of excretion on molecular size would imply that the non-electrolytes move through aqueous pores in the epithelium, whereas a direct dependence on liposolubility implies that they diffuse through the lipid phase of the epithelium, i.e., the plasma membranes. The most likely resolution of this paradox is that the compounds diffuse by at least two separate pathways, one pathway across the basolateral and luminal membranes of the cells (i.e., transcellular) involving diffusion through the lipid plasma membranes, and the other pathway via aqueous pores, which could be the intercellular junctions (i.e., an intercellular pathway), channels in the plasma membranes (i.e., a second transcellular route), or both.

An analysis of how nine different non-electrolytes of widely varying size and liposolubility are excreted in saliva is illustrated in Fig. 64.6 [4], in which the concentration of each non-electrolyte in saliva relative to its concentration in the interstitial fluid is plotted as a function of the rate at which the fluid component of the saliva is secreted. Some compounds (ethanol and antipyrine), which are highly soluble in both lipid and water, enter the saliva so readily that their salivary and interstitial concentrations are almost identical no matter how fast the salivary fluid is secreted. Such compounds can evidently cross both the aqueous and lipid phases of the secretory epithelium so readily that equilibrium between lumen and interstitium is always maintained. More strongly hydrophilic compounds, such as urea and sucrose, tend to come to equilibrium only at very low salivary secretory rates, however, and their concentrations show a hyperbolic dependency on salivary secretory rate (cf. Fig. 64.6). Substances of this type are thought to enter the saliva by a combination of molecular diffusion and solvent drag through aqueous pores.

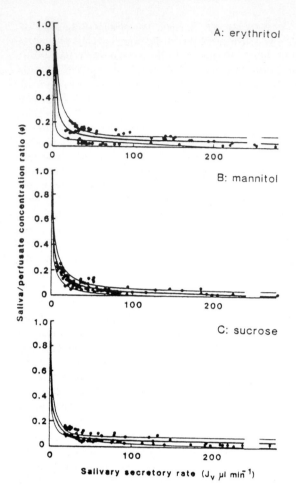

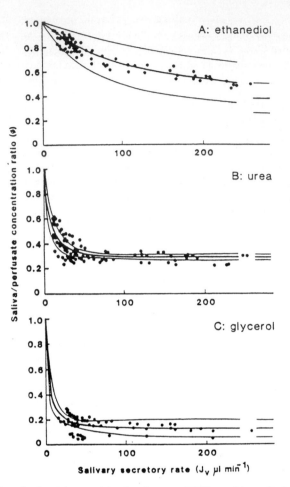

Fig. 64.6. Dependence on salivary fluid secretory rate of the salivary concentrations, expressed as ratios of the interstitial concentrations, of six organic non-electrolytes. The data have been fitted by an equation derived from non-equilibrium thermodynamics based on the assumption that the compounds enter the saliva by diffusion and solvent drag (solute–water interaction). The *horizontal asymptote* of each *curve* (±95% confidence limits) is shown to the the *right* of each *curve*. By application of hydrodynamic theory to the data from these and similar experiments, it has been possible to deduce that polar non-electrolytes enter the saliva through aqueous pores having an effective radius of 0.4 nm. (From [4])

By applying hydrodynamic theory to the analysis of these experiments, one can calculate that the pores (or slits) through which the polar molecules enter the saliva must have an effective radius of about 0.4 nm. In experiments of this kind, qualitatively similar behavior is observed in the excretion of small polar molecules (such as urea), which are able to cross the plasma membranes of secretory cells and enter the cytoplasm, and larger molecules (such as sucrose and mannitol), which cannot. Since the larger molecules cannot enter the cytoplasm of the secretory cells, it is necessary to postulate that they enter the saliva paracellularly, probably by "solvent drag" due to the "frictional" interaction of the solute with that component of the secreted water that passes paracellularly. Since the smaller molecules, which can enter the cytoplasm, are excreted in the saliva in a qualitatively similar way, it seems likely that solvent drag is also responsible for their entry into the saliva and that their much greater excretion rates arise because secreted water passes both transcellularly and

paracellularly, and that these molecules can be dragged via both routes.

A second argument supporting the view that the aqueous pores through which the larger non-electrolytes enter the saliva are located in the junctional membranes rather than in the plasma membranes is that their presence in the luminal membrane would confer an extraordinarily high conductance on it. We estimate [10] that a luminal membrane with an area of 59 cm², containing aqueous pores of 0.4 nm radius in a density sufficient to account for the observed excretion of non-electrolytes, would have a conductance to monovalent electrolytes of about 80 mS/cm². Such a high value for the plasma membrane of an epithelial cell would be most unusual since, in general, the conductance of such membranes is about 100th of this. On the other hand, if the aqueous pores were located in the junctional complexes, the implied conductance would be quite in keeping with the transepithelial conductances recorded for many leaky epithelia, including salivary secretory endpieces [25]. Wherever the pores are located, it appears that their size can be increased by autonomic agonists [22], but how this comes about is unclear.

64.6 Secretion of Water and Electrolytes

With few exceptions, salivary glands do not secrete in the absence of a secretomotor stimulus and, when stimulated, normally secrete at high rates for only brief periods. Since salivary glands, like other exocrine glands, can secrete in the complete absence of any blood pressure and of vascular perfusion, it is evident that *active transport* must underlie the *secretory process* (cf. Chap. 59), although filtration, driven by the blood pressure, may play a supplementary role, at least in some glands [51].

64.6.1 Two-stage Model of Salivary Secretion

Exocrine glands are frequently classified according to the osmolality of their final secretions. We speak of isotonic-secreting glands, such as the pancreas, in which the end-secretion is always isotonic regardless of the rate of gland secretion, of hypertonic-secreting glands, such as the shark rectal salt gland, and hypotonic-secreting glands, such as

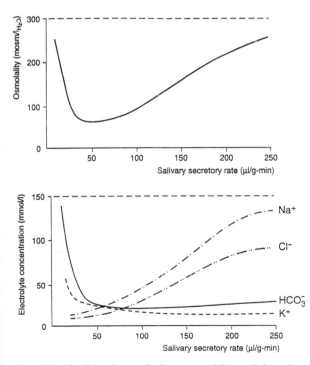

Fig. 64.7. The dependence of salivary osmolality and electrolyte composition on the rate of salivary fluid secretion. These patterns, which are idealized abstractions from experimentally derived data, can be accounted for in terms of a two-stage secretion model [40] as explained in Fig. 64.9

the human sweat glands and most major salivary glands in mammals. Unlike isotonic-secreting glands, the secretions of the latter two groups do not have constant osmolality; rather, the degree of departure of the osmolality from isotonicity varies as a function of the rate of glandular secretion. Heidenhain first drew attention to this flow dependence in the 1870s (see [49] for details; also Chap. 1), but it was not until 1954 that a satisfactory explanation was advanced to account for the phenomenon. In that year Thaysen and his colleagues [41] studied electrolyte concentrations in human parotid saliva following para-sympathomimetic stimulation. In saliva collected at low secretory rates, they observed that Na^+ and Cl^- were present only in very low concentrations and that their concentrations increased to approach plateau values as salivary flow rate increased (see Fig. 64.7). They also noticed that K^+ concentrations were high at low secretory rates and that they fell towards plateau values as secretory rate increased. To explain these results, Thaysen [40] put forward a two-stage model. In the first stage, he postulated that a *primary juice* having approximately plasma-like concentrations of Na^+, K^+, Cl^- and HCO_3^- was formed in the *secretory endpieces*, and that autonomic stimulation increased the rate at which this juice was secreted without altering its composition. In the second stage, which occurred as the primary juice passed along the gland *duct system*, he postulated that the primary juice was modified by processes of Na^+ and Cl^- absorption and of K^+ and HCO_3^- secretion across a water-impermeable epithelium. Since the rate of Na^+ and Cl^- absorption was greater than the rate of K^+ and HCO_3^- secretion, the result was the formation of a hypotonic final saliva, poor in Na^+ and Cl^- and rich in K^+ and HCO_3^- (see Fig. 64.7).

This model, modified to take account of factors controlling ductal transport activity, still serves to explain how the composition of saliva is determined, but it says nothing about the mechanism of formation of the primary secretion, or about the way the ducts produce their modifications. In the mid-1950s, however, Lundberg published a series of fine papers on secretion by perfused cat salivary glands, which afforded us our first real insight into the mechanism of the secretory process [25]. These studies demonstrated that secretion depended on the anionic composition of the extracellular fluid and led to the conclusion that the process was driven by an active anion uptake mechanism located in the basolateral plasma membranes of the secretory cells.

64.6.2 Spontaneous and Evoked Secretion

Although most salivary glands appear not to secrete unless stimulated, some quite clearly do [5,16]. Babkin was the first to draw attention to the occurrence of *spontaneous secretion* and to distinguish it from *evoked secretion* (which he called continuous secretion); he reserved the term spontaneous for secretory processes, whether continuous or not, occurring in the absence of any neural or hormonal secretomotor stimuli. Although this distinction seems quite clear-cut, in practice it is difficult to make. On the one hand, it is now appreciated that secretomotor nerves can contain non-adrenergic, non-cholinergic fibers, so that one cannot easily be confident that all nervous stimuli have been eliminated and that the residual secretion seen after administration of a cocktail of all known neurotransmitter blockers is not due to persisting neural activity. On the other hand, it is now realized that salivary ducts absorb some fluid, so that small volumes of primary secretion may be completely reabsorbed by the salivary ducts, thereby creating the false impression that no spontaneous salivary secretion is taking place.

Regardless of the presence of a true spontaneous secretion, it is clear that, with few exceptions, significant volumes of saliva are produced only in response to *autonomic nerve stimulation*, principally parasympathetic but sometimes also sympathetic. While it is possible that autonomically evoked secretion is formed merely by an acceleration of the process responsible for formation of the spontaneous secretion, this is unproven and, on the basis of the evidence available, can be said to be unlikely.

64.6.3 Electrolyte Composition of the Primary Saliva

Thaysen [40] postulated intuitively that the *primary secretion* would have a plasma-like electrolyte composition, basing his argument on what he saw as analogous absorptive behavior in the kidney (with respect to glucose) and salivary glands (with respect to Na^+ and Cl^-). He reasoned that the primary fluid ought to resemble the glomerular filtrate in composition, although he nevertheless appreciated that, unlike the glomerular fluid, the primary fluid was not formed by filtration. It was not until 1966, however, that direct evidence in support of this hypothesis was forthcoming. In that year, Martinez et al. [29] published the results of the first micropuncture studies to be performed on an exocrine gland. They showed that the primary fluid of the rat mandibular gland was isotonic, with an Na^+ concentration of around 145 mmol/l. These studies were subsequently extended by Young's group [44,46], who showed that the K^+ concentration in the primary fluid was about 7–8 mmol/l and the Cl^- concentration about 105 mmol/l. Since that time, similar micropuncture studies have been conducted on the salivary glands of numerous mammals [49].

In the interpretation of gland micropuncture studies it should be stressed that it now appears that salivary glands may be classified into two groups on the basis of the anion composition of their primary fluids. By far the best studied by micropuncture methods are those glands in which the primary fluid is *Cl^--rich*. In glands of this type, the primary fluid osmolality normally ranges between 290 and 310 msmol/kg_{H_2O}, i.e., it is either isotonic or slightly hypertonic, and the Na^+ concentration ranges between 125 and 160 mmol/l. The primary K^+ concentration ranges from 4 to 15 mmol/l and is always greater than in plasma, whereas the Cl^- concentration is close to that in the inter-

stitial fluid (100–120 mmol/l). Only rarely have primary HCO_3^- concentrations been measured, but from the known Na^+, K^+, and Cl^- concentrations, it can be inferred that HCO_3^- content must be similar to that of plasma or lower. Parasympathomimetic stimulation tends to cause a slight fall in Cl^- (and perhaps also K^+) concentration, but no change in Na^+ concentration or total osmolality [49]. β-Adrenergic stimulation causes small falls in Na^+ and Cl^- concentrations and a rise in K^+ concentration, but no definite change in osmolality. We can thus conclude that the primary secretion in this group of glands, although similar in composition to a plasma ultrafiltrate, is not identical to one, being richer in K^+ in particular. Furthermore, although autonomic stimulation alters the primary fluid composition, the changes are small.

A second group of glands produces a final secretion that becomes HCO_3^--rich and Cl^--poor as the secretion rate is increased; the group includes the parotid glands of sheep, cattle, and kangaroos, and the mandibular and parotid glands of many primates, including *Macaca mulatta* and *Papio anubis*, as well as the human parotid gland [9,10]. These glands were once thought to produce a Cl^--rich secretion that was then made HCO_3^--rich solely as a result of secondary ductal modification, but studies on their secretory mechanisms [9] suggest that they form a HCO_3^--rich, rather than a Cl^--rich, primary fluid, even if their ducts also secrete HCO_3^-. Only one micropuncture study has been performed on a gland of this type, viz, the sheep parotid [6]. It showed, at least in Na^+-replete sheep, that the primary fluid Na^+ concentration was 120 mmol/l at rest and increased slightly during stimulation, the K^+ concentration was 11.5 mmol/l at rest and decreased slightly during stimulation, and the osmolality was 311.5 msmol/kg_{H_2O} and decreased slightly during stimulation. Although these findings are in line with observations made in glands in which the primary fluid is Cl^--rich, the anion composition of the primary fluid in the sheep parotid gland was quite different: the Cl^- concentration was only 50 mmol/l and the phosphate concentration, 14 mmol/l. The workers who published this study [6] did not measure the HCO_3^- concentration in the primary fluid, however, so it is not known with certainty that HCO_3^- was the residual anion.

64.6.4 Mechanism of Secretion

Currently accepted models of epithelial secretion all invoke mechanisms for the active transport of solute across the epithelium (cf. Chap. 59). Since the principal solutes of saliva and of plasma are Na^+, K^+, Cl^- and HCO_3^-, it is to be expected that secretion will depend on the active transport of one or more of these, and that water will enter the saliva by osmosis, and most other solutes by diffusion and solvent drag. It was believed during the 1970s that that secretion depended primarily on the active transport of Na^+ across the luminal membrane, although it was always appreciated that this would require the presence of a luminal membrane Na^+ pump working to drive Na^+ out of the cell in antagonism to the *basolateral (Na^++K^+)-ATPase*, which drives Na$^+$ into the interstitium. In theory, it would be far more economical for secretion to be dependent on K$^+$ transport, since that would allow the basolateral (Na$^+$+K$^+$)-ATPase to energize the ion flow directly, without the intervention of any other active transport step. This mechanism has been proposed as the major driving force for secretion by the rat lacrimal gland [30,38], and it may well support a component of secretion in many salivary glands [7]. With few exceptions, however, glandular secretions are rich in Na$^+$ and not in K$^+$, so if secretion were to be largely driven by K$^+$ transport, a subsidiary transport step to exchange secreted K$^+$ for Na$^+$ from the interstitium would be required. Such systems do exist, one example being the secretion of HCl by gastric oxyntic cells (see Chap. 61) in which K$^+$ transport is an essential step, but they have not so far been shown to be important in salivary secretion.

Since the publication by Silva and his colleagues of a mechanism for secretion by the shark rectal gland which invoked the *secondary active transport of Cl$^-$* into the cytosol across the basolateral membrane as the underlying secretory mechanism [17,36], it has become widely accepted that exocrine secretion is usually driven by the secondary active transport of anions. In the model as originally described, it was postulated that a Na$^+$-Cl$^-$ cotransporter in the basolateral plasma membrane used the energy of the Na$^+$ gradient generated by the (Na$^+$+K$^+$)-ATPase in the basolateral membrane to maintain cytosolic Cl$^-$ above its electrochemical equilibrium. Subsequently, however, it was realized that the basolateral cotransporter had *Na$^+$-K$^+$-Cl$^-$* rather than Na$^+$-Cl$^-$ stoichiometry [20]. Regardless of the stoichiometry of the cotransporter, however, Cl$^-$ is thought to flow down its electrochemical gradient into the luminal compartment through *Cl$^-$ channels* in the luminal membrane, and the current thus flowing out of the lumen is thought to be balanced by the electrogenic movement of Na$^+$ into the lumen through cation-selective intercellular junctions. In this modified model, the current

flowing into the cell due to Cl$^-$ efflux across the luminal membrane is balanced partially (83%) by K$^+$ efflux through *K$^+$ channels* in the basolateral membrane and partially (17%) by the activity of the electrogenic (Na$^+$+K$^+$)-ATPase in the basolateral membrane. Six molecules of NaCl would thus be secreted into the lumen for each molecule of ATP hydrolyzed by the basolateral *(Na$^+$+K$^+$)-ATPase* (Fig. 64.8). In salivary glands, it is necessary to modify the mechanism postulated for the shark rectal gland by: (1) introducing a greater variety of basolateral secondary active transport systems, including Na$^+$-H$^+$ and Cl$^-$-HCO$_3^-$ exchangers; (2) permitting electrogenic HCO$_3^-$ flux across the luminal membrane in addition to electrogenic Cl$^-$ flux; and (3) including a significant luminal membrane conductance to K$^+$. The effects of these modifications are that salivary glands can:

- Secrete NaCl as a result of active Cl$^-$ uptake, either through the action of the Na$^+$-K$^+$-2Cl$^-$ cotransporter or through the paired operation of Na$^+$/H$^+$ and Cl$^-$/HCO$_3^-$ exchangers
- Secrete NaHCO$_3$ as a result of the accumulation HCO$_3^-$ in the cytosol due to the operation of the basolateral Na$^+$/H$^+$ exchanger
- Secrete K$^+$ as well as Cl$^-$

64.7 Function of Salivary Ducts

64.7.1 Ductal Protein Secretion

In addition to the organic components of saliva synthesized and secreted by endpiece cells, some proteins are synthesized, stored, and secreted by the cell of the salivary duct system. These include *growth factors*, such as nerve

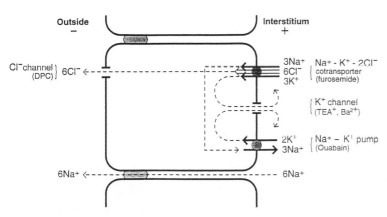

Fig. 64.8. Schematic diagram of a salivary secretory cell showing how NaCl is secreted by secondary active Cl$^-$ transport, involving a furosemide-sensitive basolateral Na$^+$-K$^+$-2Cl$^-$ cotransporter. Na$^+$, K$^+$ and Cl$^-$ ions enter the cytosol from the interstitium via a cotransporter that couples ion movement in a fixed stoichiometry of 1Na$^+$, 1K$^+$ and 2Cl$^-$. The driving force for the ion fluxes comes ultimately from the Na$^+$ gradient created by the membrane-bound (Na$^+$+K$^+$)-ATPase. Cl$^-$ is accumulated in the cytosol against its electrochemical gradient and enters the lumen via anion channels sensitive to blockade with diphenylamine-2-carboxylic acid (DPC). To preserve electroneutrality, Na$^+$ enters the lumen from the interstitium via cation-selective intercellular junctions

growth factor (NGF), epidermal growth factor (EGF) and erythropoietin, and *enzymes*, such as ribonucleases, various proteases, including renin and the homeostatic esteroproteases, kallikrein and tonin, and some α-amylase. It is also possible that a number of regulatory peptides, such as glucagon and somatostatin, are produced by the ducts, although local synthesis has not been proved for this group. Most of the secretory IgA in saliva passes through the duct epithelium, IgA itself being synthesized by interstitial immunocytes, and secretory component (sc), by the duct cells.

The growth factors and enzymes are stored in *secretion granules*, which can be seen in the striated duct cells of almost all salivary glands, although the intralobular ducts in the mandibular glands of many rodents exhibit such an abundance of secretory granules that they are called granular rather than *striated ducts*, because the ducts have lost the basolateral striations that normally characterize intralobular ducts and the cytoplasmic granules have become their most striking morphological feature. Interestingly, in mouse mandibular glands, these granular ducts are much more extensively developed in the male than in the female, which is reflected in the amounts of active peptides present in the glands of the two sexes. As might be expected, the content of these peptides in the mouse glands is hormone-dependent (androgens, thyroxine). The contents of the ductal granules are discharged into the saliva following stimulation with α-adrenergic and, to a lesser extent, cholinergic (muscarinic) agonists [11].

As mentioned above, several *regulatory peptides* have also been reported to be present in salivary glands, such reports being based either on radioimmunoassays of gland extracts or on immunohistochemical evidence. These peptides include glucagon immunologically related to intestinal glucagon, and somatostatin. Immunohistochemical studies have shown that the peptides are located mainly in the striated and excretory duct cells of the mandibular and sublingual glands. In contrast to the growth factors and homeostatic proteases, the regulatory peptides appear to be stored diffusely in the cytoplasm rather than in the secretion granules of granular duct cells. Growth factors and homeostatic proteases (such as kallikrein, renin, and tonin) are thought to act, respectively, on tissue cells and circulating plasma proteins. To do so they must first enter the circulation. The fact that salivary glands contain large amounts of these biologically active peptides and secrete them into the saliva does not necessarily imply that they are also released into the circulation. To establish such a role for the peptides, three questions must be answered: (1) Does sialoadenectomy affect plasma levels? (2) Do plasma levels reflect sex differences in the salivary gland content of the peptides? (3) Is evoked secretion of the peptides into the saliva accompanied also by a rise in the plasma levels?

Removal of the mandibular gland does not appear to affect plasma levels of NGF, EGF, or renin, nor are there any significant differences in plasma levels between male and female mice. Stimulation of salivary secretion of EGF and renin by administration of α-adrenergic agonists (or by

provoking aggressive behavior in the experimental animals) results in a marked rise in the plasma levels of these polypeptides, although, at least in the case of EGF, the rise in plasma level is quite slow, not peaking until after 60 min. Stimulation causes plasma levels of these peptides to rise much higher in male than in female mice. Curiously, although stimulation with α-adrenergic agonists causes marked secretion of NGF in the saliva, it does not result in any significant rise in the NGF plasma level. These studies suggest that under "normal" conditions, i.e., in the absence of strong pharmacological or neural stimulation, the mouse mandibular gland does not contribute to the plasma levels of NGF, EGF or renin. This is in keeping with results showing only a slow turnover of secretory proteins in mouse and rat granular duct proteins.

There are three possible routs by which EGF and the homeostatic proteases could enter the circulation: (1) by direct endocrine release across the basolateral plasma membrane of the duct cells; (2) by reabsorption from the duct lumen across the epithelium; and (3) following swallowing, by reabsorption in the stomach or intestine. Experiments in which kallikrein and tonin levels were determined in the venous outflow from salivary glands seem to indicate that swallowing followed by reabsorption in stomach or intestine cannot represent a major route for entry of kallikrein and tonin into the circulation [10]. One study suggests that exocytosis takes place across the basolateral membranes of granular duct cells, but the delayed rise in EGF plasma levels seen after stimulation (see above) argues against a very significant role for such a direct type of endocrine secretion [10]. Ductal reabsorption, on the other hand, remains a distinct possibility, since retrograde injection of ^{125}[I]-kallikrein has been shown to lead to an increase of labelled kallikrein in the plasma.

With the possible exception of secreted IgA, which may contribute to oral hygiene, it is difficult to say what physiological function these various ductal peptides might have, and no study yet published has demonstrated an essential function for any of them. Why, for instance, the ducts of mandibular glands should synthesize and secrete small quantities of α-amylase when the endpieces of parotid glands secrete so much, remains a mystery.

64.7.2 Ductal Transport of Water and Electrolytes

For many years is was simply assumed that the salivary ducts fulfilled only the passive function of conveying saliva from its site of formation in the endpieces to its final destination in the mouth. As indicated earlier, it was Thaysen [40,41], in the 1950s, who first made it plain how the ducts might act to modify the composition of the saliva prior to its discharge into the oral cavity. His insight was to realize that the ducts must reabsorb Na^+ and Cl^- if he was correct in his deduction that the primary saliva had an electrolyte composition similar to that of plasma. As pointed out above, micropuncture studies [29,46] soon confirmed

Thaysen's supposition about the composition of the primary fluid and thereby strengthened his inference about the function of the ducts. Direct evidence that the ducts *reabsorb Na⁺ and Cl⁻* came from duct micropuncture experiments and from experiments in which the main excretory duct of the rat submandibular gland was perfused in vivo.

It is now clear that saliva is formed in at least two distinct stages and that the ducts are active in the final stage (Fig. 64.9). When Thaysen and subsequent authors wrote about ductal transport activity, they thought in terms of the intralobular (striated) ducts, which are the most conspicuous ductal structures in salivary glands. For technical reasons, however, the perfusion studies carried out by all investigators have been confined to the main excretory (extralobular) duct, and, until recently, it was only by inference that we could ascribe similar transport properties to the intralobular ducts, although the indirect evidence that justified this inference was strong. Recently, however, techniques have been developed for isolating intralobular ducts, and these have enabled research workers to study the ion channels in the plasma membranes of the duct cells using patch-clamp techniques and the changes in cytosolic electrolyte concentrations that occur in response to autonomic stimulation using ion-sensitive fluorescent dyes. Thus far, the evidence suggests that the model developed from perfusion studies on the *extralobular ducts* [24] is also applicable to the *intralobular ducts* [11,13–15,42].

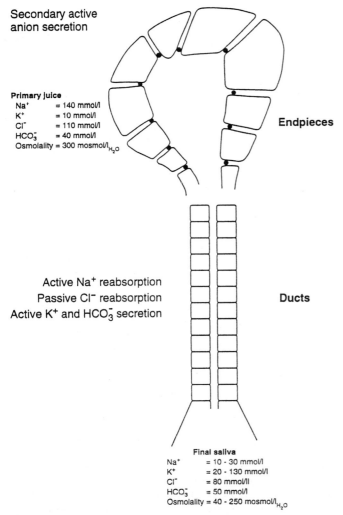

Fig. 64.9. Schematic diagram to explain Thaysen's [40] two-stage model of salivary secretion. Saliva is postulated to be formed first as an isotonic primary juice by the secretory endpieces. This juice has approximately the same electrolyte composition as plasma, and does not change appreciably when secretory rate is increased: it is a true secretion, however, not a filtrate. As the primary juice passes down the gland duct system, Na⁺ is reabsorbed actively (and Cl⁻ passively) whilst K⁺ and HCO₃⁻ are secreted actively at somewhat lower rates. Since the ducts are impermeable to water, the result is the formation of a hypotonic final saliva, poor in NaCl and rich in KHCO₃

1321

64.7.3 Salivary Excretion Patterns for the Common Electrolytes

If we assume that *primary saliva* (more or less like plasma in composition) is formed only in the *endpieces*, then it follows that a comparison of the composition of the primary secretion with that of the *final saliva*, and a study of how salivary composition changes when the secretory rate is increased, will provide information on the role of the ducts in determining the composition to the final saliva (Fig. 64.7). Studies of this kind have been performed on the major salivary glands of at least ten mammalian species, including man. Since there are several reviews dealing with the significance of these excretion patterns in considerable detail [43,45,49], only a summary of what has been concluded in these reviews is presented here:

Sodium. Salivary Na^+ concentrations are usually low when salivary secretory rate is low, but they rise to approach an asymptote equal to or somewhat lower than the concentration in the primary fluid as secretory rates increase (Fig. 64.7). This is thought to be because the salivary ducts absorb Na^+ by a mechanism having a limited maximum transport capacity that saturates as the rate of formation of primary saliva increases. The extent of this ductal absorption varies greatly among species and among glands within a given species, so that saturation is scarcely noticeable in some glands, whereas in others ductal transport is minimal and salivary Na^+ concentrations remain high even at low fluid secretory rates.

Potassium. Salivary excretion patterns for K^+ suggest that the duct systems of all glands can secrete this ion against an electrochemical gradient. As in the case of Na^+ transport, however, the K^+ secretory rate does not increase in parallel with the increase in the rate of fluid secretion evoked by stimulation of the endpieces, so that salivary K^+ concentrations tend to be very high only at low flow rates and to fall hyperbolically to a minimum value close to the primary level at high flow rates (Fig. 64.7). The asymptote to which the concentration falls varies between about 6 and 40 mmol/l depending on the species studied.

Bicarbonate. The excretion patterns for HCO_3^- are extremely variable among different glands and can only be explained by assuming that the ducts in some glands absorb HCO_3^-, that in others they secrete it, and that in yet others the direction of transport is dependent on autonomic nerve activity. In addition, of course, account must be taken of the possibility that the primary secretion in some glands is HCO_3^- rich.

Chloride. In general, salivary Cl^- concentrations move in parallel with Na^+, implying that Cl^-, like Na^+, is reabsorbed in the ducts. There is no evidence to indicate that Cl^- is reabsorbed actively.

Other Ions. Although numerous studies have been performed on the salivary excretion of Ca^{2+}, Mg^{2+}, and phosphate (for review see [45]), very few micropuncture studies have been performed, so that an accurate assessment of the ductal transport of these ions is not possible. It appears that the ducts do not absorb or secrete Mg^{2+} or phosphate and that they absorb Ca^{2+} only to a slight extent. The ducts do influence the excretion patterns of these ions, however, but only secondarily, as a consequence of water absorption.

Water. Salivary ducts have water permeabilities as low as those seen in the collecting ducts of the mammalian kidney during water diuresis, but unlike the collecting ducts, salivary ducts are not affected by vasopressin. Although ductal water permeabilities are low, a significant amount of water is absorbed by the ducts in some glands at low secretory rates when fluid-duct contact time is long. This can be inferred from the observation that the salivary urea concentration in rat and rabbit mandibular saliva can rise to more than three times the primary fluid concentration when the secretory rate is very low. In the rabbit mandibular gland, ductal water reabsorption becomes appreciable when the secretory rate falls below about 10% of the maximum [4], but in the parotid glands of the rat, mouse, and rabbit, water reabsorption seems to be negligible, even at the lowest fluid secretory rates.

64.7.4 Transport Studies on Perfused Salivary Ducts

Although it is clear that the luminal surface area of the *main excretory ducts* of mandibular glands is too small for the ducts to be important in determining the composition of the saliva, their morphology and transport properties suggest that they should function similarly to the rest of the excretory duct system. Since they are easy to perfuse, they have been studied extensively [10,49]. Most studies have been carried out in the rat and rabbit, but it is not easy to generalize since the species differences are quite substantial. The rat duct is a vigorous transporter of K^+ and has been preferred for in vivo studies, whereas the rabbit duct secretes little K^+: both ducts absorb Na^+ strongly and are highly impermeable to water.

Steady-State Conditions. Both rat and rabbit ducts are capable of absorbing Na^+ and of secreting K^+ and HCO_3^- actively. This is best demonstrated by examining the transepithelial ion and potential gradients under steady-state conditions [49]. For instance, in the rat duct (in vivo), the composition (mmol/l) of the luminal fluid at steady state is $Na^+ = 2$, $K^+ = 135$, $HCO_3^- = 55$, $Cl^- = 78$, and the transepithelial voltage is 11 mV, lumen negative [49]. It is clear that only Cl^- is at electrochemical equilibrium and that active absorption of Na^+ and secretion of K^+ and HCO_3^- must have taken place.

Na+ and Cl- Absorption. Na^+ ions can enter duct cells from the lumen by two alternative pathways, an *Na^+-selective channel* and an *Na^+/H^+ exchanger*, and they leave the cell across the basolateral membrane via the $(Na^+ + K^+)$-ATPase. On the luminal plasma membrane, the presence of a conspicuous Na^+ conductance can be demonstrated by eliminating the shunting effect of an almost equally conspicuous anion conductance. Thus, replacing Cl^- in the luminal fluid with SO_4^{2-} unmasks a classical Nernstian relation between voltage and luminal Na^+ concentration. The underlying conductance can discriminate between Na^+ and K^+ (and choline) but it is quite permeable to Li^+; it can be blocked by luminal application of *amiloride* (cf. Chaps. 59,74,112), which has been used to demonstrate its presence in intralobular ducts [12,13] as well as the more accessible extralobular ducts. [49].

There is good evidence to indicate that there is an additional pathway available for Na^+ to cross the luminal plasma membrane involving an electroneutral mechanism exchanging Na^+ for H^+ [24]. Since this mechanism permits Na^+ entry to the cell in electroneutral exchange for H^+, it leads to HCO_3^- absorption, although this is normally obscured by the simultaneous activity of a K^+-coupled H^+ absorptive mechanism (K^+/H^+ exchanger) in the luminal plasma membrane (see below). Like the conductive Na^+ channel, the H^+-coupled *Na$^+$ exchanger* can be blocked by higher concentrations of amiloride. As a result of the operation of the exchanger, ductal Na^+ absorption can be stimulated by including HCO_3^- in the luminal fluid.

As indicated above, Na^+, having gained access to the cell from the lumen, appears to exit across the basolateral membrane via a $(Na^+ + K^+)$-ATPase. *Ouabain* (5×10^{-6} mol/l) abolishes the active transport potential rapidly and reversibly when applied from the interstitial but not from the luminal surface of the epithelium, and the effect of the drug can be antagonized competitively by K^+ in the interstitial bathing solution. In addition, removal of K^+ from the interstitial fluid reduces the transport potential.

As Na^+ ions are reabsorbed from the lumen, other ion movements must occur to balance the charge flow. As indicated above, some Na^+ leaves the lumen in exchange for H^+ via an Na^+/H^+ exchanger, but the bulk of the Na^+ is absorbed electrogenically, accompanied by a parallel flow of Cl^- ions. There is good evidence to show that the electrogenic Cl^- flux is not paracellular but transcellular, passing via a large *Cl$^-$ conductance* in the luminal plasma membrane [18]. The properties of the matching Cl^- conductance in the basolateral plasma membrane are less well defined, but it has been demonstrated in intralobular ducts by use of patch-clamp techniques [14].

In the course of our own studies on the *intralobular ducts* of the mouse submandibular gland [11–15], we have found that the size of the Na^+ and Cl^- conductances in the plasma membrane is influenced by the level of the cytosolic Cl^- concentration [15]. As the cytosolic Cl^- concentration is increased from 5 to 150 mmol/l, the size of the *Na$^+$ conductance* declines steadily, the effect being half maximal at a Cl^- concentration of about 50 mmol/l. In contrast, as the cytosolic Cl^- concentration is increased, the size of the *Cl$^-$ conductance* remains very low until the Cl^- concentration exceeds 80 mmol/l, when the size of the Cl^- conductance begins to increase steeply. Since cytosolic Cl^-

concentration is closely correlated with cell volume in many epithelia, it is tempting to speculate that the relation between Cl^- concentration and Cl^- and Na^+ conductances provides a mechanism whereby ductal Na^+ and Cl^- transport can be adjusted to maintain a constant cell volume. Since these ducts can secrete exportable protein by *exocytosis* when stimulated with autonomic agonists (see above), it is also possible that an increase in cytosolic Cl^- may accompany secretomotor stimulation of exocytosis, and that this switches the duct from a resting state, in which it reabsorbs NaCl, to a secreting state, in which it secretes Cl^- and exportable protein.

K^+ Secretion. When Young and Schögel [46] first studied rat mandibular secretions collected by micropuncture from various regions of the excretory duct tree, they were struck by the extremely high K^+ concentrations that they encountered. Their subsequent perfusion studies showed that the rat submandibular duct could secrete this ion actively at a rate about half that at which Na^+ can be absorbed. When the secretory process was allowed to establish a steady state, the limiting electrochemical gradient for K^+ was approximately 7.5 kJ/mol (cf. 10 kJ/mol for Na^+). The development of such a steep gradient depends not only on an active transport mechanism, but also on a low ductal permeability to passive back-flux of K^+. The site of this permeability barrier is the luminal plasma membrane, not the basolateral membrane, which, in fact, is quite permeable to the cation [49]. The source of active transport work appears to be the basolateral $(Na^+ + K^+)$-ATPase [23,49].

HCO_3^- Secretion. An examination of the steady-state HCO_3^- gradient developed by the mandibular duct makes it clear that the ion can be secreted actively [49]. Nevertheless, when perfused with HCO_3^--saline, the rat duct secretes HCO_3^- very slowly [24,27,50] and the rabbit duct actually absorbs it [28].

An important insight into the mechanism of ductal secretion of HCO_3^- in the rat has come from Knauf's observation that whereas the rate of ductal HCO_3^- secretion is normally lower than the rate of K^+ secretion, the two ions are secreted at equal rates whenever Na^+ absorption is completely blocked [24]. From this he reasoned that K^+ and HCO_3^- are secreted together in equimolar amounts by a luminal membrane K^+/H^+ exchanger, but that most of the secreted HCO_3^- is normally absorbed simultaneously owing to the operation of the luminal Na^+/H^+ exchanger, so that net HCO_3^- secretion is considerably less than net K^+ secretion.

Summary Overview of Ductal Transport. The available information on transport by the rat duct can best be accounted for in terms of the transport model depicted in Fig. 64.10. In a general way the same model can be used for the rabbit duct, although several features of the rabbit duct will have to be explored further before it becomes possible to conclude that the two ducts have essentially the same transport mechanisms. The model depicted in Fig. 64.10 incorporates many of the features of a model first proposed

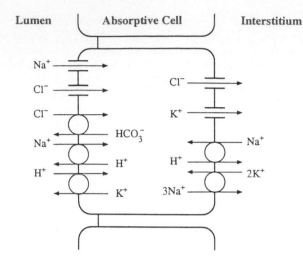

Fig. 64.10. Schematic diagram of a salivary duct cell, illustrating some of the transport elements involved in the absorption of NaCl and the secretion of KHCO$_3$. The driving force for absorption comes from the basolateral (Na$^+$+K$^+$)-ATPase, which maintains a low cytosolic Na$^+$ concentration and a high cytosolic K$^+$ concentration. Na$^+$ enters the cytosol from the lumen either via a Na$^+$ channel or via a Na$^+$/H$^+$ exchanger, and Cl$^-$ enters via a Cl$^-$ channel or a Cl$^-$/HCO$_3^-$ exchanger. H$^+$, entering the lumen via the Na$^+$/H$^+$ exchanger, may leave again via a K$^+$/H$^+$ exchanger, thereby leading to apparent electroneutral exchange of K$^+$ for Na$^+$. Na$^+$ leaves the cell for the interstitium via the (Na$^+$+K$^+$)-ATPase and Cl$^-$ leaves via an anion channel. K$^+$ enters the cell from the interstitium via the (Na$^+$+K$^+$)-ATPase and may leave via the luminal K$^+$/H$^+$ exchanger of a basolateral K$^+$ channel. To be complete, this model also requires a basolateral Na$^+$/H$^+$ exchanger, for which there is some experimental evidence, and a K$^+$/Cl$^-$ cotransporter, for which, as yet, there is no evidence. (Modified from [24])

by Knauf et al. [24], including the K$^+$/H$^+$ and Na$^+$/H$^+$ exchangers, but it locates the anion conductances in the luminal and basolateral membranes rather than in the junctional membrane.

64.7.5 Autonomic Control of Ductal Electrolyte Transport

An examination of salivary electrolyte excretion patterns makes it clear that ductal electrolyte transport maxima change during *nerve stimulation* [49], and there is an abundance of morphological data to show that salivary ducts are innervated by both *sympathetic* and *parasympathetic* fibers [48]. Microperfusion studies have provided ample direct evidence that salivary duct electrolyte transport is under autonomic control [49].

Cholinergic agonists are able to *inhibit Na$^+$ reabsorption* and depolarize the duct potential when applied from the interstitial, but not the luminal surface of the mandibular main duct in both rat and rabbit [10]. The drugs are effective in extremely low concentrations; for instance, in the rabbit duct perfused in vitro, Na$^+$ absorption is reduced when acetylcholine is given in concentrations as low as 10^{-11} mol/l. Similar effects on ductal electrolyte transport are

seen when the parasympathetic secretomotor nerves are stimulated [34].

The actions of *adrenergic agonists* and sympathetic nerve stimulation have also been investigated in some detail [10]. It appears that *α-adrenergic agonists* and strong sympathetic nerve stimulation mimic all the actions of cholinergic agonists on ductal electrolyte transport, but *β-adrenergic agonists*, provided they are administered in low concentrations, actually stimulate Na$^+$ absorption.

64.7.6 Endocrine Control of Ductal Electrolyte Transport

Although numerous papers have been published on the effects of endocrine disturbances on salivary electrolyte output, few of these have shed any light on the possible roles of hormones in the control of salivary composition. Firm conclusions can be drawn only in the case of the *mineralocorticoids*, and a few tentative hypotheses might be advanced in the case of some polypeptides, including *angiotensin* and the gastrointestinal hormones. Strangely, no good study has been published on the effects of vasopressin on ductal function, probably because it has none.

Mineralocorticoids. It is well established that salivary electrolyte composition reflects body mineralocorticoid status [49]. Administration of adrenocorticotropin (ACTH) or a mineralocorticoid reduces salivary Na$^+$ and increases salivary K$^+$ concentrations, and similar changes are seen in *salt depletion*, in *hyper-aldosteronism*, and in pregnancy. Changes in the reverse direction are seen in *Addison's disease*. Comparison of the composition of primary saliva (from micropuncture studies) with that of final saliva makes it plain that the salivary ducts, and not the endpieces, are the principal targets for aldosterone.

Angiotensin. Although many salivary glands contain large amounts of *renin*, there is no unequivocal evidence to indicate that salivary gland renin has any physiological role in determining systemic levels of angiotensin, although a local, intraglandular role for it cannot be excluded. Angiotensin, however it is generated, seems to have no direct effect on electrolyte transport by salivary endpieces, but it does alter Na$^+$ transport by the excretory ducts [21]. What physiological role this ductal action of angiotensin may have is obscure.

Gastrointestinal and Related Peptides. Naturally occurring peptides, such as VIP, eledoisin, physalaemin and substance P, are able to evoke salivary secretory responses, at least in some species [10]. The effects of a few of these on salivary duct function have been investigated. Physalaemin, applied from the interstitial but not the luminal surface of the main ducts of both rat and rabbit mandibular gland, can stimulate Na$^+$ transport, but the related peptide, *caerulein*, has no effect. *VIP* and *GIP* inhibit Na$^+$ transport, but *secretin* has no effect except at very high concentrations. Functional receptors for VIP and *neurotensin* have been identified in a human submandibular duct cell line. As in the case of angiotensin, it is difficult to say what physiological importance these effects might have, but since VIP is demonstrable in secretomotor nerve endings in at least some salivary glands [26] it seems likely that this peptide will prove to have a physiological role in secretion.

References

1. Anderson LC, Shapiro BL (1980) Calcium and the response of the rat submandibular gland to insulin stimulation in vitro. J Dent Res 59:1989–1990
2. Burgen ASV, Emmelin NG (1961) Physiology of the salivary glands. Arnold, London
3. Bustamante JC, Mann GE, Yudilevich DL (1981) Specificity of neutral amino acid uptake at the basolateral side of the epithelium in the cat salivary gland in situ. J Physiol (Lond) 313:65–79
4. Case RM, Cook DI, Hunter M, Steward MC, Young JA (1985) Transepithelial transport of non-electrolytes in the rabbit mandibular salivary gland. J Membr Biol 84:239–248
5. Coats DA, Wright RD (1957) Secretion by the parotid gland of the sheep: the relationship between salivary flow and composition. J Physiol (Lond) 135:611–622
6. Compton JS, Nelson J, Wright RD, Young JA (1980) A micropuncture investigation of electrolyte transport in the parotid glands of sodium-replete and sodium-depleted sheep. J Physiol (Lond) 309:429–446
7. Cook DI, Young JA (1989) Effect of K^+ channels in the apical plasma membrane on epithelial secretion based on secondary active Cl^- transport. J Membr Biol 110:139–146
8. Cook DI, Young JA (1989) Fluid and electrolyte secretion by salivary glands. In: Forte JG (ed) Handbook of physiology. The gastrointestinal system. Salivary, pancreatic, gastric and hepatobiliary secretion, section 6, vol III. American Physiological Society, Bethesda, pp 1–23
9. Cook DI, Young JA (1993) Epithelial secretion driven by anions other than chloride. News Physiol Sci 8:91–93
10. Cook DI, Van Lennep EW, Roberts ML, Young JA (1994) Secretion by the major salivary glands. In: Johnson L, Christensen J, Jackson M, Jacobson E, Walsh J (eds) Physiology of the gastrointestinal tract, vol 2, 3rd edn. Raven, New York, pp 1061–1117
11. Dinudom A, Poronnik P, Allen DG, Young JA, Cook DI (1993) Control of intracellular Ca^{2+} by adrenergic and muscarinic agonists in mouse mandibular ducts and endpieces. Cell Calcium 14:631–638
12. Dinudom A, Poronnik P, Young JA, Cook DI (1993) Patch-clamp and fura-2 studies on granular intralobular ducts in mouse mandibular glands. In: Ussing HH, Fischbarg J, Sten-Knudsen O, Hviid-Larsen E, Willumsen NW, Thaysen JH (eds) Isotonic transport in leaky epithelia. Munksgaard, Copenhagen, pp 85–102 (Alfred Benzon Symposium no 34)
13. Dinudom A, Young JA, Cook DI (1993) Amiloride-sensitive Na^+ current in the granular duct cells of mouse mandibular glands. Pflugers Arch 423:164–166
14. Dinudom A, Young JA, Cook DI (1994) Ion channels in the basolateral membrane of intralobular duct cells of mouse mandibular glands. Pflugers Archiv 428:202–208
15. Dinudom A, Young JA, Cook DI (1993) Na^+ and Cl^- conductances are controlled by cytosolic Cl^- concentration in the intralobular duct cells of mouse mandibular glands. J Membr Biol 135:289–295
16. Emmelin N (1953) On spontaneous secretion of saliva. Acta Physiol Scand 30:34–58
17. Eveloff J, Kinne R, Kinne-Saffran E, Murer H, Silva P, Epstein FH, Stoff J, Kinter WB (1978) Coupled sodium and chloride transport into plasma membrane vesicles prepared from dogfish rectal gland. Pflugers Arch 378:87–92
18. Frömter E, Diamond JM (1972) Route of passive ion permeation in epithelia. Nature 235:9–13
19. Grand RJ, Gross PR (1970) Translation-level control of amylase and protein synthesis by epinephrine. Proc Natl Acad Sci U S A 65:1081–1088
20. Hannafin J, Kinne-Saffran E, Friedman D, Kinne R (1983) Presence of a sodium-potassium chloride cotransport system in the rectal gland of *Squalus acanthias*. J Membr Biol 75:73–83
21. Healy JK, Fraser PA, Young JA (1976) Inhibition of sodium transport by angiotensin II in the main duct of the rabbit mandibular gland isolated and perfused in vitro. Pflugers Arch 363:69–73
22. Howorth AJ, Case RM, Steward MC (1987) Effects of acetylcholine and forskolin on the non-electrolyte permeability of the perfused rabbit mandibular gland. Pflugers Arch 408:209–214
23. Knauf H (1972) The minimum requirements for the maintenance of active sodium transport across the isolated salivary duct epithelium of the rabbit. Pflugers Arch 333:326–336
24. Knauf H, Lübcke R, Kreutz W, Sachs G (1982) Interrelationships of ion transport in rat submaxillary duct epithelium. Am J Physiol 242:F132–F139
25. Lundberg A (1958) Electrophysiology of salivary glands. Physiol Rev 38:21–40
26. Lundberg JM, Änggard A, Fahrenkrug J (1982) Complimentary role of vasoactive intestinal polypeptide (VIP) and acetylcholine for cat submandibular gland blood flow and secretion. Acta Physiol Scand 114:329–337
27. Martin CJ, Young JA (1971) A microperfusion investigation of the effects of a sympathomimetic and a parasympathomimetic drug on water and electrolyte fluxes in the main duct of the rat submaxillary gland. Pflugers Arch 327:303–323
28. Martin CJ, Frömter E, Gebler E, Knauf B, Young JA (1973) The effects of carbachol on water and electrolyte fluxes and transepithelial electrical potential differences of the rabbit submaxillary main duct perfused in vitro. Pflugers Arch 341:131–142
29. Martinez JR, Holzgreve H, Frick A (1966) Micropuncture study of submaxillary glands of adult rats. Pflugers Arch 290:124–133
30. Marty A, Tan YP, Trautmann A (1984) Three types of calcium-dependent channels in rat lacrimal glands. J Physiol (Lond) 357:293–325
31. McMillian MK, Soltoff SP, Cantley LC, Rudel RA, Talamo BR (1993) Two distinct cytosolic calcium responses to extracellular ATP in rat parotid acinar cells. Br J Pharmacol 108:453–461
32. Nieuw Amerongen AV, Aarsman MEG, Vreugdenhil AP, Roukema PA (1981) Comparison in vitro of the incorporation of [^{3}H]-leucine and N-acetyl-[^{14}C]-mannosamine into proteins and glycoproteins of the parotid, submandibular and sublingual glands of the mouse. Arch Oral Biol 26:651–656
33. Nieuw Amerongen AV, Aarsman MEG, Bos-Vreugdenhil AP, Roukema PA (1982) Influence of isoproterenol on the incorporation in vitro of [^{3}H]-leucine and N-acetyl-[^{14}C]-mannosamine into proteins and glycoproteins of the parotid gland of the mouse. Arch Oral Biol 27:659–665
34. Schneyer LH (1977) Parasympathetic control of Na, K transport in perfused submaxillary duct of the rat. Am J Physiol 233:F22–F28
35. Shimono M, Yamamura T, Fumagalli G (1980) Intercellular junctions in salivary glands: freeze-fracture and tracer studies of normal rat sublingual gland. J Ultrastruct Res 72:286–299

36. Silva P, Stoff J, Field M, Fine L, Forrest JN, Epstein FH (1977) Mechanism of active chloride secretion by shark rectal gland: role of Na-K-ATPase in chloride transport. Am Physiol 233:F298–F306

37. Soltoff SP, McMillian MK, Cragoe EJ, Cantley LC, Talamo BR (1990) Effects of extracellular ATP on ion transport systems and $[Ca^{2+}]_i$ in rat parotid acinar cells: comparison with the muscarinic agonist carbachol. J Gen Physiol 95:319–346

38. Tan YP, Marty A, Trautmann A (1992) High density of Ca^{2+}-dependent K^+ and Cl^- channels on the luminal membrane of lacrimal acinar cells. Proc Natl Acad Sci U S A 89:11229–11233

39. Tandler B, Phillips CJ, Toyoshima K, Nagato T (1989) Comparative studies of the striated ducts of mammalian salivary glands. Prog Clin Biol Res 295:243–248

40. Thaysen JH (1960) Handling of alkali metals by exocrine glands other than the kidney. In: Ussing HH, Kruhöffer P, Thaysen JH, Thorn NA (eds) The alkali metal ions in biology. Springer, Berlin Heidelberg New York, pp 424–463 (Handbuch der experimentellen Pharmakologie, vol 13)

41. Thaysen JH, Thorn NA, Schwartz IL (1954) Excretion of sodium, potassium, chloride and carbon dioxide in human parotid saliva. Am J Physiol 178:155–159

42. Valdez IH, Turner RJ (1991) Effects of secretagogues on cytosolic Ca^{2+} levels in rat submandibular granular ducts and acini. Am J Physiol 261:G359–G363

43. Young JA (1979) Salivary secretion of inorganic electrolytes. In: Crane RK (ed) International review of physiology, gastrointestinal physiology III, vol 19. University Park Press, Baltimore, pp 1–58

44. Young JA, Martin CJ (1971) The effect of a sympatho- and a parasympathomimetic drug on the electrolyte concentrations of primary and final saliva of the rat submaxillary gland. Pflugers Arch 327:285–302

45. Young JA, Schneyer CA (1981) Composition of saliva in mammalia. Aust J Exp Biol Med Sci 59:1–53

46. Young JA, Schögel E (1966) Micropuncture investigation of sodium and potassium excretion in rat submaxillary saliva. Pflugers Arch 291:85–98

47. Young JA, Van Lennep EW (1977) Morphology and physiology of salivary myoepithelial cells. In: Crane RK (ed) International review of physiology, gastrointestinal physiology II, vol 12. University Park Press, Baltimore, pp 105–125

48. Young JA, Van Lennep EW (1978) The morphology of salivary glands. Academic, London

49. Young JA, Van Lennep EW (1979) Transport in salivary and salt glands. Part I: Salivary glands. In: Giebisch G, Tosteson DC, Ussing HH (eds) Membrane transport in biology, vol 4B: transport organs. Springer, Berlin Heidelberg New York, pp 563–674

50. Young JA, Martin CJ, Asz M, Weber FD (1970) A microperfusion investigation of bicarbonate secretion by the rat submaxillary gland. The action of a parasympathomimetic drug on electrolyte transport. Pflugers Arch 319:185–199

51. Young JA, Chapman BE, Cook DI, Healey AP, Kuchel PW, Lingard JM, Nicol M, Novak I, Seow F (1982) Salivary secretory mechanism: recent advances and concepts. In: Quinton PM, Martinez JR, Hopfer U (eds) Fluid and electrolyte transport in exocrine glands in cystic fibrosis. San Francisco Press, San Francisco, pp 102–124

52. Yudilevich DL, Sepulveda FV, Bustamante JC, Mann GE (1979) A comparison of amino acid transport and ouabain binding in brain endothelium and salivary epithelium studied in vivo by rapid paired tracer dilution. J Neural Transm 15 [Suppl]:15–27

65 Function of the Exocrine Pancreas

D.I. Cook and J.A. Young

Contents

65.1 Anatomical Structure

Acini and Centro-acinar Cells. The exocrine pancreas is a compound gland consisting of secretory endpieces (acini) draining into a converging duct system [5,17,30]. The acini, which are composed of cells that synthesise digestive enzymes and store them as zymogen granules, are found at the terminations of the intercalated ducts, but also, in some species, at intermediate points along the ducts, so that an acinus may surround an intercalated duct part way along its course. In most species, the individual cells form truncated pyramids so that, when they are aggregated to form a secretory endpiece, the endpiece is shaped like a berry (Latin: *acinus*) rather than being tubular [64]. Each acinus envelops a layer of intercalated duct cells which, in consequence, are often called centro-acinar cells, although there is no evidence to suggest that they differ morphologically or functionally from cells elsewhere in the interca-

lated ducts. Acinar cells make up 77%–90% of the total cell volume in the pancreas. As would be expected of cells adapted for protein secretion, the apical poles of the acinar cells are packed with mature secretion granules, there are prominent Golgi complexes on the apical side of the nucleus, and there are conspicuous stacks of rough endoplasmic reticulum in the cytosol lying basal and lateral to the nucleus. The apical membrane has numerous microvilli which project into the acinar lumen. The cells are joined by elaborate junctional complexes composed of occluding junctions (zonulae occludentes) close to the lumen, belt desmosomes (zonulae adherentes) at the level of the terminal web and spot desmosomes along the lower parts of the lateral plasma membranes [30].

Ducts. The duct system (Fig. 65.1), which makes up between 4% (guinea pig) and 11% (rat) of the total cell volume, can be subdivided into a number of generations. Intercalated ducts, into which the acini discharge their contents, drain into intralobular ducts, which in turn drain into interlobular (extralobular) ducts running in the connective tissue between the pancreatic lobules. Finally, the interlobular ducts drain into the main pancreatic duct, which, in some species (including man and the rat), opens into the second part of the duodenum, but in others (e.g., the guinea pig) opens into the third part of the duodenum. In man, there may be one or more accessory pancreatic ducts, bypassing the main duct and opening into the duodenum separately from the common pancreato-biliary duct. Although this description of a hierarchically organised duct tree draining peripherally located acini is widely accepted, it should be pointed out that wax-cast studies on the human pancreas show that the smallest duct structures and the endpieces exhibit an anastomosing tubular arrangement rather than a traditional acinar ("grapes on a bunch") arrangement (Fig. 65.1) [4,5]. The traditional picture is more typical of the rat pancreas.

Intercalated Ducts. The cells of intercalated ducts form a flat or low-cuboidal epithelium with only a few microvilli on the apical surface, often with only a singly cilium. They contain little rough endoplasmic reticulum, but large numbers of mitochondria, and the basolateral membranes are elaborately folded and interdigitated with those of the neighbouring cells. The morphology of the cells in the larger ducts is similar to that of the intercalated ducts except that they have conspicuous secretion vesicles contain-

R. Greger/U. Windhorst (Eds.)
Comprehensive Human Physiology, Vol. 2
© Springer-Verlag Berlin Heidelberg 1996

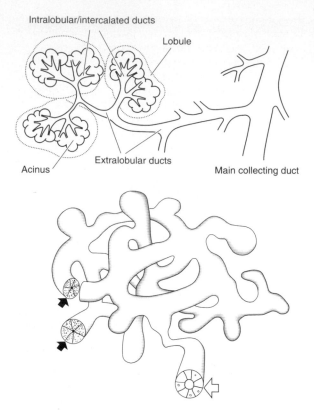

Intralobular/intercalated ducts

Lobule

Extralobular ducts

Acinus

Main collecting duct

Fig. 65.1. *Top*, anatomical arrangement of secretory end-pieces and ducts in the rat pancreas. In the rat, the berry-shaped end-pieces (acini) are clustered at the ends of numerous intercalated ducts, which drain into a converging duct system of interlobular (or extralobular) ducts that terminates in a single main excretory duct opening into the duodenum. *Bottom*, in the human pancreas the distinction between acini and intercalated ducts is not so clear. The secretory endpieces, which are more tubular than acinar in shape, are found along the course of the intercalated ducts, and the ducts themselves form an anastomosing network. The *solid arrows* indicate the tubular secretory endpieces and the *open arrow* indicates an intercalated duct. (Modified from [4])

ing mucoproteins packed into their apical cytoplasm. In the interlobular and main ducts there are also occasional goblet cells and scattered endocrine cells which stain for hormones such as insulin, glucagon and pancreatic polypeptide (PP).

Myoepithelial Cells. An interesting feature of pancreatic morphology is the complete absence of myoepithelial cells [63,64]. In other exocrine glands, myoepithelial cells provide support for the acini undergoing secretomotor stimulation (see Sect. 64.3.2 and Fig. 64.3) when the intraluminal pressure inevitably rises to overcome the outflow resistance; their absence in the pancreas means that the acini are more likely to rupture during secretion, with release of enzymes into the interstitial fluid and, consequently, the development of pancreatitis. This is most likely to occur when the outflow resistance of the pancreatic duct system increases above normal. For instance, in cystic fibrosis (see Chap. 112), when the pancreatic juice is abnormally viscous, chronic pancreatitis is thought to develop because of

the slow leakage of pancreatic enzymes into the tissues from ruptured acini and/or intercalated ducts. Similarly, acute pancreatitis is most commonly caused by impaction of gallstones in the common pancreato-biliary duct.

65.2 Secretory Proteins (see Table 62.1)

Digestive Enzymes and Zymogens. The human pancreas delivers between 6 and 20 g protein per day to the digestive tract. Whilst digestive enzymes constitute a major fraction of the secreted proteins, some non-enzymic proteins, such as immunoglobulins and mucoproteins, are also secreted. The acinar cells are the source of the digestive enzymes and zymogens of pancreatic juice, as well as non-enzymic proteins, such as pancreatic secretory trypsin inhibitor (PSTI), and some glycoproteins, such as lithostathine [8]. The duct cells are a major source of mucin in the pancreatic juice, particularly the *Muc*-1 gene product [41].

Acute Pancreatitis. Damage to pancreatic tissue by its own secreted enzymes (autodigestion) is normally avoided because most of the digestive enzymes are secreted as inactive zymogens (e.g., trypsinogen as the precursor of trypsin) and activation normally occurs extra-pancreatically. Thus, since trypsinogen activation normally begins only when trypsinogen is acted on by enteropeptidase, an enzyme secreted by the cells of the small intestinal epithelium, trypsin is usually first formed only after pancreatic juice enters the duodenal lumen. Since the other zymogens in the pancreatic juice rely on trypsin for their activation, it follows that they too will not normally be activated within the pancreas and the entire cascade of zymogen activation should take place only within the lumen of the small intestine. Unfortunately, trypsinogen can sometimes be activated prematurely, for instance by the action of the lysosomal enzyme, cathepsin-B, and, once formed, trypsin-1 and trypsin-3 can catalyse the further formation of trypsin from trypsinogen autocatalytically, albeit at a much slower rate than occurs with enteropeptidase. This can lead to an explosive cascade of zymogen activation intrapancreatically, precipitating the potentially life-threatening syndrome of acute pancreatitis. Should such premature activation of trypsinogen occur, a very important protective mechanism relies on the secretion by the acinar cells of the trypsin inhibitor, PSTI, which is capable of blocking up to one fifth of all the potentially available tryptic activity in pancreatic juice. Another protective mechanism arises from the pH and Ca^{2+} sensitivity of one of the three isoforms of trypsin (trypsin-2), since, under the conditions that usually prevail in the lumen of the pancreatic duct system (pH 8, $Ca^{2+} <$ 1 mmol/l), trypsin-2, if formed, initiates rapid degradation rather than activation of trypsinogen, thereby removing the potential source of further trypsin formation before the supply of PSTI is exhausted. Because trypsin is necessary for the activation of the other pancreatic zymogens, this will halt the enzyme activation cascade [51]. In addition, the pancreatic juice appears to contain trace quantities of another, unidentified enzyme, which, when activated by trypsin, degrades a wide range of zymogens [51].

The following paragraphs on the pancreatic enzymes are based on two recent review articles [39,51], but reference should also be made to Sect. 62.4.4, in which lipid digestion is discussed in more detail, and to Table 62.1, in which the properties of the principal pancreatic enzymes are summarised.

65.2.1 Serine Proteases

Endopeptidases. Serine proteases are endopeptidases characterised by the presence of a serine residue projecting into the reaction pocket of the enzyme which is essential for the catalytic cleavage of peptide chains. This serine residue reacts irreversibly with diisopropylfluorophosphate, a general inhibitor of this enzyme class. The serine proteases found in pancreatic juice include the following:

- Trypsin
- Chymotrypsin
- Protease E
- Elastase
- Kallikrein.

All are synthesised as inactive zymogens. In many cases, different isozymes can be distinguished on the basis of their net charge, and when this is the case they are numbered in accordance with their electrophoretic mobility, consecutively from the anode to the cathode.

Trypsin(ogen). This enzyme (EC 3.4.21.4) is produced in three distinct forms: trypsinogen-1 (cationic trypsinogen), trypsinogen-2 (neutral trypsinogen) and trypsinogen-3 (anionic trypsinogen), which account for approximately 65%, 5% and 30%, respectively, of the total tryptic activity in pancreatic juice. The three forms are closely homologous and all have molecular masses of approximately 25 kDa. The pH optima of the active enzymes lie in the range 7.5–8.5, and they attack peptide bonds on the carboxyl sides of arginine and lysine residues. The zymogens are activated by enteropeptidase, an intestinal enzyme that cleaves the N terminus of the zymogen between Lys_8 and Ile_9 at a rate a thousand times faster than trypsin (see Sect. 62.4.2). The exact amino acid sequence of trypsinogen varies among species and among the different forms synthesised by a given species, but the presence of four aspartate residues at positions 7, 6, 5 and 4, adjacent to Lys_8, particularly favours enteropeptidase activity [39]. The anionic and cationic forms of trypsin are inhibited by PSTI and by Kunitz-type trypsin inhibitor (Trasylol, aprotinin), but the neutral from (trypsin-2) is not blocked by PSTI or by Trasylol.

Chymotrypsin(ogen). This enzyme (EC 3.4.21.1) is produced in at least two forms, chymotrypsinogen A and chymotrypsinogen B. Chymotrypsinogen A has a molecular mass of 24 kDa, and chymotrypsinogen B a molecular mass of 27 kDa. Both forms are activated by trypsin to form chymotrypsin A and B, although no activation peptides are released because the cleaved peptides remain bound to the chymotrypsin molecule via a disulphide bridge. The functional difference between the A and B forms is unclear, and chymotrypsin A may merely be a degradation peptide of chymotrypsin B. The pH optimum of both active forms is about 8, and they catalyse the hydrolysis of peptide bonds on the carboxyl sides of phenylalanine, tyrosine and tryptophan residues, but will also attack bonds involving leucine and methionine.

(Pro)protease E. In human pancreatic juice this enzyme (EC 3.4.21.70) was previously called proelastase-1. Its substrate-binding site is homologous to that of elastase-1 in other species, but it can be distinguished from the true elastases in that it is anionic rather than cationic and does not hydrolyse elastin. It has a molecular mass of 29 kDa. It is activated by trypsin and attacks peptide bonds on the carboxyl side of alanine, isoleucine, valine and hydroxyamino acids.

(Pro)elastase. Two types of proelastase occur in pancreatic juice. Although found in infants, proelastase-1 (EC 3.4.21.36) is not found in adult humans. Proelastase-2 (EC 3.4.21.71) exists in two forms, 2A and 2B, which yield differing activation peptides, but are otherwise closely homologous, with molecular masses of about 30 kDa. Proelastases are activated by trypsin, and, as with chymotrypsin, activation does not cause release of activation peptides because the cleaved peptides remain bound to the activated enzymes. The enzymes attack peptide bonds on the carboxyl side of alanine, isoleucine, valine and hydroxyamino acids. Unlike protease E, they will also attack elastin.

Kallikrein(ogen). This enzyme (EC 3.4.21.35) makes up less than 0.1% of protein in pancreatic juice and its role in pancreatic function is unclear. Preferentially, it hydrolyses bonds in small peptides, acting on the carboxyl side of arginine residues.

65.2.2 Exopeptidases

(Pro)carboxypeptidase. Pancreatic juice contains zymogens for two exopeptidases, carboxypeptidase A and B, which are formed by the action of trypsin on the precursor zymogens. Carboxypeptidase A (EC 3.4.17.1) is a zinc-containing enzyme that excises the C-terminal amino acids of substrate proteins and peptides. It acts most rapidly when the C-terminal residue of its substrate is an aromatic amino acid. The zymogen has a molecular mass of 45–47 kDa, and the active enzyme a molecular mass of 35 kDa. At least three forms have been described (A_1, A_2 and A_3): A_1 and A_2 are monomers, whereas A_3 is a binary complex with proprotease E. Carboxypeptidase B (EC 3.4.17.2) also has two distinct variants (B_1 and B_2). The molecular mass of the zymogen is 47 kDa, and that of the active enzyme 35 kDa. Like carboxypeptidase A, carboxypeptidase B contains one atom of zinc and is activated by trypsin. It differs, however, in being directed against C-terminal arginine and lysine residues.

65.2.3 Lipolytic Enzymes

The lipolytic pancreatic enzymes include phospholipase A_2, pancreatic (triacylglycerol) lipase and non-specific carboxylesterase.

Pancreatic Phospholipase A_2. This enzyme (EC 3.1.1.4) is secreted as a zymogen (prophospholipase A_2) with a molecular mass of 14 kDa. It is activated by trypsin which cleaves off an N-terminal heptapeptide. It hydrolyses the *sn*-2 (or C-2) fatty acyl ester bond of phosphoglycerides, to produce, for example, lysolecithin from lecithin. It acts at lipid–water interfaces, but, unlike the phospholipases found in snake venoms, does not attack cell membranes.

Pancreatic Lipase. This enzyme (EC 3.1.1.3) is secreted in the active form and has a molecular mass of 48 kDa. It hydrolyses C-1 and C-3 glycerol ester bonds particularly rapidly, but is active against all fatty acid esters. Its activity is inhibited by bile salts and phospholipids except in the presence of colipase (see below). It is structurally related to the serine proteases and is thought to have evolved with these from a common ancestral enzyme (see Sect. 62.4.4.2.2).

Non-specific Carboxylesterase. This enzyme (EC 3.1.1.1) is also known as carboxyl ester hydrolase, lysophospholipase, cholesterol esterase, non-specific lipase, lysophospholipase (EC 3.1.1.5) and cholesterol ester hydrolase (3.1.27.5); it is a glycoprotein with a molecular mass of 105 kDa. It hydrolyses a wide variety of lipid esters including cholesteryl esters, retinyl esters and lysophosphatidylglycerols as well as mono-, di- and triacylglycerols. For the hydrolysis of long-chain triacylglycerols it requires the presence of bile salts. It has wide substrate specificity, as reflected in its alternative names.

65.2.4 Other Enzymes Secreted by the Pancreas

Pancreatic α-Amylase, Pancreatic Ribonuclease and Desoxyribonuclease. Pancreatic α-amylase is a glycoprotein with a molecular mass of 50 kDa. It hydrolyses $α_{1,4}$-glucosidic linkages. At least two isozymes of pancreatic α-amylase have been described, although these probably arise from post-translational modification of a single peptide (see also Sect. 62.4.3). Pancreatic ribonuclease (EC 3.1.27.5) has a molecular mass of 15 kDa and hydrolyses the phosphate bonds in RNA. Desoxyribonuclease I and II

(EC 3.1.21.1 and 3.1.21.4) have molecular masses of about 33–38 kDa and hydrolyse phosphate bonds in DNA.

Lysosomal Enzymes. One surprising feature of pancreatic secretion is the presence of a number of lysosomal enzymes, such as arylsulfatase and N-acetyl-β-glucosaminidase, in pancreatic juice [51]. The levels of these enzymes increase following secretomotor stimulation in much the same way as do levels of the normally recognised pancreatic secretory enzymes. This is said to occur because there is incomplete segregation of proteins into the secretory and lysosomal pathways during synthesis. However, not all lysosomal enzymes show this behaviour; cathepsin B and β-glucuronidase, for example, do not. As lysosomal enzymes are only active under acid conditions, it is unlikely that they have an important phsiological role to play in pancreatic juice.

65.2.5 Protein Products Other than Enzymes

Pancreatic juice contains a number of proteins that are not enzymes. These include PSTI, colipase and lithostathine, which are products of the acinar cells, and mucin, which is a product of duct cells.

Pancreatic Secretory Trypsin Inhibitor. PSTI, also called the "Kazal" pancreatic inhibitor, is a single-chain peptide with 56 amino acid residues and three disulphide bonds. PSTI and trypsin (or trypsinogen) form stable complexes involving a bond between a residue within the reaction pocket of trypsin and two residues ($-Lys_{18}-Ile_{19}-$) at the reaction site of the inhibitor. PSTI does not inhibit chymotrypsin, kallikrein or trypsin-2.

Co-lipase. This protein has a molecular mass of 10 kDa and is secreted in a precursor form, procolipase, which is activated by trypsin (see Sect. 62.4.4.2.3). It has no intrinsic enzymatic activity, but binds at lipid–water interfaces on bile salt micelles and emulsion droplets, where it complexes with and activates pancreatic lipase.

Lithostathine. This protein is also called the G4 protein or pancreatic thread protein; it is a glycoprotein with 144 amino acid residues and a molecular mass of 16 kDa which makes up 5%–10% of the total amount of secreted pancreatic protein [8]. It is synthesised in acinar cells and can be cleaved by trypsin between Arg_{11} and Ile_{12} to form lithostathine S_1 and an N-terminal undecapeptide, lithostathine H_2. Although lithostathine was originally isolated from pancreatic stones, its normal action is to prevent the precipitation of $CaCO_3$, which is present in supersaturated concentrations in pancreatic juice. It does this both by preventing nucleation and by retarding crystal growth. Its activity appears to reside in the N-terminal undecapeptide (lithostathine H_2), whereas lithostathine S_1 polymerises to form fibrils and may seed pancreatic stones. Lithostathine secretion is increased in subjects on a high-protein diet, possibly reflecting a common ontogenetic origin with the serine proteases.

Mucins [41] (see also Sect. 64.5.2). The small intralobular and larger interlobular ducts secrete the *Muc*-1 gene product. This protein has 1272 amino acid residues and contains numerous identical 20-amino acid tandem repeats in the central two thirds of the molecule. It contains large numbers of serine and threonine groups, the majority of which have *O*-glycosidic linkages to *N*-acetylgalactosaminyl residues. The *N*-acetylgalactosaminyl groups are in turn linked to chains of two to 15 saccharides, including *N*-acetylglucosamine, galactose and fucose, and may be sulphated and branched. The *Muc*-1 gene is also expressed by the acinar cells, but the product seems to be differently glycosylated to the product formed in duct cells, and its particular role may be in packaging secretory enzymes rather than serving as an exportable secretion product for some other purpose. The duct goblet cells express only the *Muc*-2 gene product.

65.2.6 Dietary Modulation of Pancreatic Protein Production and Secretion

There is considerable evidence to suggest that, although the total quantity of pancreatic enzyme secreted changes little with diet, the proportions of the different enzyme types present in the pancreatic juice can vary in the longer term according to dietary patterns [10]. For example, increased levels of saturated or unsaturated fat in the diet lead to increased synthesis and secretion of pancreatic lipase and co-lipase. Similarly, an increased protein content leads to an increased synthesis and secretion of chymotrypsin and other proteases (except trypsin-1 and proelastase-2), and increased levels of dietary carbohydrate lead to increased amylase secretion. These changes are mediated by CCK (cholecystokinin) in the case of protein, secretin and GIP (gastric inhibitory peptide) in the case of lipid, and insulin in the case of carbohydrate. In man, the changes occur only in response to prolonged changes in dietary composition, and although the total amount of pancreatic enzymes secreted in response to an individual meal depends on the composition of the food ingested in that meal, the proportions of the different enzymes present in the pancreatic juice are not altered unless the change in dietary composition is sustained over days or weeks.

65.3 Fluid and Electrolyte Secretion

Humans. The pancreas secretes about 2 l fluid each day. In humans, there is only a low flow rate (0.2–0.3 ml/min) in the interdigestive period (resting secretion), which increases by a factor of about 10 during digestion. Secretin evokes a marked increase in the rate of formation of an isotonic juice, rich in Na^+, and with K^+ in concentrations

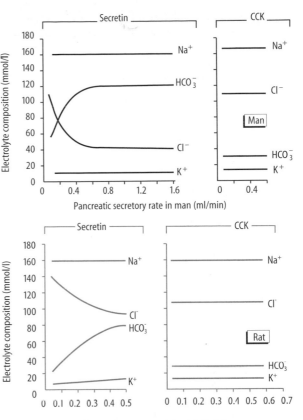

Fig. 65.2. The relation between the concentrations of Na^+, K^+, Cl^- and HCO_3^- in the juice and the rate of fluid secretion by the pancreas during stimulation with secretin (*left*) or CCK (*right*). The *upper panels* show typical patterns for the pancreas in humans, cats and dogs, in which there is a low resting secretory rate and secretin evokes a more vigorous secretory response than CCK. The *lower panels* show typical patterns in the rat, in which there is a more substantial resting secretion and CCK evokes a more vigorous response than secretin. In all species, the concentrations of Na^+ and K^+ are approximately plasma-like and do not change greatly with increasing secretory rate. In resting juice and in juice evoked by CCK, the Cl^- and HCO_3^- concentrations are also plasma-like and independent of the secretory rate, but the secretin-evoked juice has a high HCO_3^- concentration (>100 mmol/l in man, cat and dog; >70 mmol/l in the rat) and a correspondingly low Cl^- concentration. In consequence, since the HCO_3^--rich juice evoked by secretin must necessarily be mixed with the Cl^--rich resting juice, the anion concentrations seen in the final juice show some flow dependency. The concentrations seen at low levels of secretin stimulation will reflect the fact that two juices of different anion composition are being mixed, but as the intensity of secretin stimulation is increased, the anion concentrations will come more and more nearly to reflect what would be seen in a "pure" secretin-evoked juice (i.e., uncontaminated by a Cl^--rich resting secretion). A similar explanation probably accounts for the flow-dependent increase in K^+ concentration seen in rat pancreatic juice evoked in response to secretin (*lower left panel*). The K^+ concentration in resting and in CCK-evoked juice is 4 mmol/l, whereas in secretin-evoked juice it appears to be more than twice this value [54]; consequently, the K^+ concentration in the mixture increases as the intensity of secretin stimulation is increased, whereas it does not change with increasing intensity of CCK stimulation

slightly above those in plasma (Fig. 65.2); during maximal secretin stimulation, the juice has a HCO_3^- concentration almost six times higher (up to 135 mmol/l) than in plasma and a correspondingly low Cl^- concentration. CCK has only a small effect on the fluid secretory rate (Fig. 65.2) and no effect on electrolyte concentrations. The effects of vagal stimulation on the secretion rate are intermediate between those of secretin and CCK [10,34].

Cat, Dog, Rat and Pig. The pancreatic secretory responses of the cat and dog are similar to those of humans, but the responses of some other species show markedly different patterns. In the rat, which feeds almost continuously when not asleep, there is a high rate of formation of the resting secretion, and secretin stimulation evokes only a small increase in fluid secretion rate and a modest increase in HCO_3^- concentration (to about 70 mmol/l) with a correspondingly modest fall in Cl^- concentration (Fig. 65.2) [54]. These changes in anion concentration are not as marked as in humans, and the flow rates in the rat, when corrected for differences in gland mass, are only 20% of those observed in humans. In the rat, however, CCK produces a vigorous fluid secretory response (Fig. 65.2), the effect being greater than that evoked by secretin, and the evoked juice is Cl^--rich rather than HCO_3^- rich. The pig pancreas, like the human pancreas, secretes copious amounts of a HCO_3^--rich juice in response to secretin, but its secretory response to CCK is much greater than in humans. Even so, the Cl^--rich juice evoked in the pig by CCK is secreted only at about one tenth the rate evoked by secretin [23,24].

Pancreatic Juice Composition. The sites of production of pancreatic juice have been studied using micropuncture techniques [52] and in experiments in which the effects of the differential destruction of duct or acinar tissue were examined, for example, using a copper-deficient diet to destroy acinar tissue [2]. On the basis of these studies, it is generally accepted that the Cl^--rich fluid associated with CCK or vagal stimulation is derived from the acini [1,6]. It is also generally agreed that the HCO_3^--rich juice evoked by secretin comes from the small interlobular (extralobular) ducts, but some doubt remains about the HCO_3^- secretory capacity of the intercalated ducts and the centro-acinar cells [1,6].
When the pancreas secretes at a high rate in response to secretin stimulation, the HCO_3^- concentration of the emergent juice is very high, but when secretory rates are lower, the HCO_3^- concentration is also lower (Fig. 65.2). The concentrations of Cl^- vary in inverse relation to those of HCO_3^-, the sum of their concentrations being constant at all flow rates and equal to the sum of the concentrations of Na^+ and K^+, neither of which varies much with flow rate. The juice is approximately isotonic at all flow rates. This pattern of behaviour arises because the pancreas secretes two isotonic juices, a NaCl-rich juice containing enzymes from the acini and a $NaHCO_3$-rich juice from the ducts [1,6,54]. In the unstimulated gland, both resting secretions are produced at a low flow rate, but the NaCl-rich juice from the acini predominates, so that the Cl^- concentration

in the mixture is relatively high, whereas during stimulation with secretin, the volume of the $NaHCO_3$-rich juice increases so that the concentration of HCO_3^- in the mixture rises progressively as the rate of formation of duct secretion increases. Simultaneously, of course, the concentration of Cl^- must fall since the total cation concentration does not change (Fig. 65.2). In contrast, if secretion is evoked with CCK (or acetylcholine), the HCO_3^- concentration in the juice does not change because the CCK-evoked juice has a similar electrolyte composition to that of the resting secretion (Fig. 65.2).

65.3.1 Ductal Fluid and Electrolyte Secretion

The mechanism (Figs. 65.3, 65.4) by which the interlobular ducts secrete a HCO_3^--rich fluid has been the subject of intense investigation in the rat [3,44,45], and important insights into ductal secretory mechanisms have been obtained, although, as pointed out above, it needs to be stressed that the rat pancreas is not typical of the mammalian pancreas in general. The simplest possible model for $NaHCO_3$ secretion (Fig. 65.3) does not operate in the rat pancreas, but might prove to be applicable in other species. The cell model developed for secretion by the interlobular ducts of the rat pancreas (Fig. 65.4) postulates a basolateral Na^+/H^+ exchanger that uses the energy of the Na^+ gradient, maintained by the basolateral $(Na^+ + K^+)$-ATPase, to pump H^+ out of the cytosol. The H^+ ions are replenished by the action of cytosolic carbonic anhydrase, which produces HCO_3^- from CO_2 and water. Thus, in the cytosol of the duct cells, the pH and the HCO_3^- concentration are maintained above electrochemical equilibrium. The HCO_3^- then leaves the cell across the luminal membrane in exchange for Cl^- through a Cl^-/HCO_3^- exchanger and the Cl^- recycles across the apical membrane through an apical membrane Cl^- channel. There is thus net movement of HCO_3^- into the lumen, giving it a negative charge and causing Na^+ and K^+ to enter the lumen across cation-selective tight junctions to preserve electroneutrality. Recent microfluorescence studies on microdissected segments of rat pancreatic ducts have confirmed the presence of a basolateral Na^+/H^+ exchanger and an apical Cl^-/HCO_3^- exchanger [65]. (A Cl^-/HCO_3^- exchanger has also been demonstrated in the basolateral membrane.) Interestingly, these studies suggest that HCO_3^- accumulation across the basolateral membrane is not solely due to Na^+/H^+ exchange, but is partly due to a bafilomycin-sensitive H^+-ATPase and possibly also to a Na^+-dependent HCO_3^- co-transporter [65].
In this model, the secretory rate is dependent on the activity of the apical Cl^- channels as well as the Cl^-/HCO_3^- exchangers, the basolateral K^+ channels and the Na^+/H^+ exchangers. The apical membrane Cl^- channel has been shown to have properties similar to those of the cystic fibrosis transmembrane conductance regulator (CFTR) channel, which is expressed in intralobular and small interlobular pancreatic duct cells [22]. Thus, the pancreatic duct Cl^- channel (in the rat and in humans) has a linear current–voltage relation with a single-channel conduct-

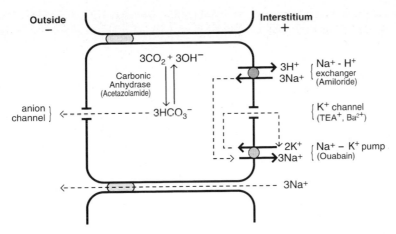

Fig. 65.3. A simple model to account for transepithelial secretion of NaHCO₃, dependent on a Na⁺/H⁺ exchanger in the basolateral membrane of the secretory cell. The Na⁺ gradient created by the basolateral (Na⁺+K⁺)-ATPase energises the exchanger, causing the intracellular pH to rise. HCO₃⁻ is then formed by the hydration of CO₂ and catalysed by carbonic anhydrase; it diffuses down its electrochemical gradient into the lumen via an anion-selective channel in the apical membrane. Na⁺ enters the lumen paracellularly via cation-selective tight junctions, thereby preserving electroneutrality. This model may be simpler than is actually encountered in the pancreatic ducts of any species, but variations on it probably account for HCO₃⁻ secretion across most mammalian epithelia. *TEA*, tetraethylammonium

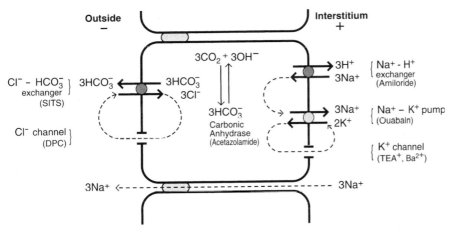

Fig. 65.4. Model describing the mechanism of HCO₃⁻ secretion by the rat pancreatic duct, in which HCO₃⁻ secretion is known to be dependent on the luminal concentration of Cl⁻. As in the model depicted in Fig. 65.3, HCO₃⁻ is generated in the cytosol by a process involving basolateral Na⁺/H⁺ exchangers, basolateral (Na⁺+K⁺)-ATPases and cytosolic carbonic anhydrase. In this model, however, instead of entering the lumen via an anion-selective channel in the apical membrane, HCO₃⁻ enters via a Cl⁻/HCO₃⁻ exchanger (sensitive to disulfonic stilbenes), the activity of which depends on a supply of luminal Cl⁻ delivered to it from the cytosol by a recycling process utilising an apical membrane Cl⁻ channel (sensitive to diphenylamine-2-carboxylic acid). As in the model depicted in Fig. 65.3, Na⁺ enters the lumen across cation-selective intercellular junctions (*TEA*, tetraethylammonium; *DPC*, diphenylalanine-2-carboxylate; *SITS*, 4-acetamido-4'-isothiocyanato-2,2'-disulfonic acid stilbene). (Based on [44,45])

ance of 4 pS and is not activated by changes in membrane potential; it has a selectivity sequence of Cl⁻ = Br⁻ > I⁻, is blocked by typical CFTR channel blockers such as the aryl-alkyl-amino-benzoates (but not by disulfonic stilbenes) and is phosphorylated (and activated) by protein kinase A [18]. It thus seems likely that the apical Cl⁻ channel in the duct cells and the CFTR channel are one and the same. If so, this provides an explanation for the reduced pancreatic HCO₃⁻ secretion observed in patients with cystic fibrosis, a disease caused by mutations in the gene coding for the CFTR protein (see also Chap. 112). Duct cells (in the

mouse) also contain a Ca²⁺-activated Cl⁻ conductance, which may be important in the duct response to agonists such as acetylcholine, which act by increasing intracellular free Ca²⁺ [19]. This Ca²⁺-activated Cl⁻ conductance, if also present in humans, might be important in ameliorating the effects of an altered function of the CFTR Cl⁻ channel in cystic fibrosis.

As mentioned above, the electrolyte composition of the pancreatic secretion evoked by secretin in rats is unusual in having only a moderately high HCO₃⁻ concentration and a correspondingly higher concentration of Cl⁻ than is seen in human pancreatic juice. These observations are clearly concordant with the above model, but it has not yet been established whether the model, with its apical Cl⁻/HCO₃⁻ exchanger and Cl⁻-recycling mechanism, can also be applied to humans and carnivores such as the cat and the dog, in which the pancreatic juice evoked in response to secretin stimulation is very low in Cl⁻. The simpler model depicted in Fig. 65.3, or another variant of it, might well be more appropriate in these species.

65.3.2 Fluid and Electrolyte Secretion by the Pancreatic Acini

The mechanism by which pancreatic acini secrete fluid in response to CCK (and acetylcholine) is disputed, although some form of secondary active Cl⁻ transport seems likely to form a major part of any acceptable secretory model. As mentioned above, in humans there is little or no fluid secretion in response to CCK [10,34], but in pigs there is a marked fluid response to CCK and to vagal stimulation [23]. This may be driven by an Na⁺-2Cl⁻-K⁺ co-transporter (see Chap. 59) similar to that in salivary endpiece cells, although this has not yet been fully investigated and it may be that, as in some salivary glands [49], other basolateral transporters, such as paired Na⁺/H⁺ and Cl⁻/HCO₃⁻ exchangers, may have a role.

In the rat, the species in which most studies have been carried out, the fluid secretory response of the pancreas to CCK is greater than the response to secretin (Fig. 65.2) [54]. From the time of the earliest organ perfusion studies on the mechanism of CCK-induced fluid secretion in the rat pancreas, it has been known that, although secretion is blocked by loop diuretics such as bumetanide, piretanide

and furosemide, the rank order of potency of these agents does not correspond with their potency as blockers of Na⁺-2Cl⁻-K⁺ co-transport in other glands [12,53]. Furthermore, pancreatic secretion is blocked at least partially by amiloride, by disulfonic stilbenes and by methazolamide, suggesting the involvement of Na⁺/H⁺ and Cl⁻/HCO₃⁻ exchangers as well as carbonic anhydrase (Fig. 65.5). Interestingly, the effects of these blockers were greater in the absence of HCO₃⁻, suggesting that HCO₃⁻ can activate an additional transport system [53]. These conclusions have been strengthened by recent studies employing a pH-sensitive fluorescent dye to monitor cell pH, which have confirmed the presence of Na⁺/H⁺ and Cl⁻/HCO₃⁻ exchangers in pancreatic acinar cells, together with an Na⁺/HCO₃⁻ cotransporter, which can account for the HCO₃⁻-dependent component of secretion [42,43].

A major difficulty in arriving at a model for secretion by rat pancreatic acini, however, is caused by the fact that electrophysiological studies in the rat (and mouse) have shown that, although the acinar cells have a resting K⁺ conductance, CCK stimulation does not appear to activate it; rather, stimulation activates a Cl⁻ current and a non-selective cation current, carried by Ca²⁺-activated Cl⁻ channels and Ca²⁺-activated non-selective cation channels (Ca-NS), respectively [48,50]. The activation of a non-selective cation conductance is incompatible with an efficient model of pancreatic acinar secretion based on Na⁺-coupled secondary active Cl⁻ transport as first postulated by Seow et al. [53] (Fig. 65.5) or, for that matter, with any other similar transport model. This is because the driving force for Cl⁻ exit across the apical membrane is provided by the electrogenic efflux of K⁺ across the basolateral membrane, which maintains the cell potential more negative than the Nernst potential for Cl⁻. The presence of non-selective

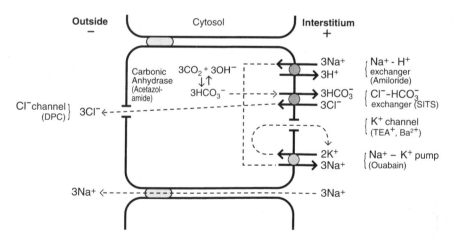

Fig. 65.5. Possible model to account for secretion by rat pancreatic acini, based on Na⁺-coupled secondary active Cl⁻ transport. The Na⁺ gradient generated by the basolateral (Na⁺+K⁺)-ATPase energises a basolateral Na⁺/H⁺ exchange carrier. This raises the cytosolic pH and promotes formation of HCO₃⁻ by hydration of CO₂, catalysed by carbonic anhydrase (see Fig. 65.3). The HCO₃⁻ so formed exits to the interstitium in exchange for Cl⁻ via a Cl⁻/HCO₃⁻ exchanger in the basolateral membrane and the Cl⁻ exits to the lumen via Cl⁻-selective channels in the apical membrane. (An

alternative to the operation of paired Na⁺/H⁺ and Cl⁻/HCO₃⁻ exchangers would be an Na⁺-2Cl⁻-K⁺ cotransporter.) Na⁺ enters the cell paracellularly, thereby preserving electroneutrality. How the working of this model would be disrupted by incorporation of a non-selective cation channel in the basolateral membrane is shown in Fig. 65.6. (TEA, tetraethylammonium; DPC, diphenylalanine-2-carboxylate; SITS, 4-acetamido-4'-isothiocyanato-2,2'-disulfonic acid stilbene). (Based on [53])

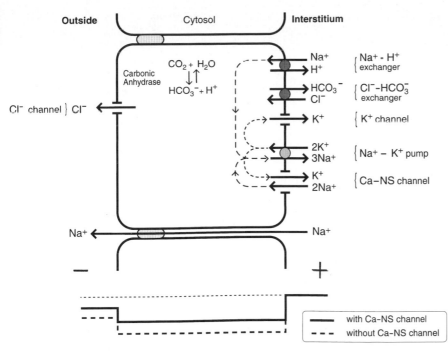

Fig. 65.6. Deleterious impact of introducing a non-selective cation channel (Ca-NS) into the basolateral membrane on the secondary active Cl⁻ transport model depicted in Fig. 65.5. By providing an alternative pathway for Na⁺ entry to the cytosol, the Ca-NS channel reduces the effectiveness of the Na⁺/H⁺ exchanger, and, by depolarising the cell potential, it reduces the drive for Cl⁻ efflux across the apical membrane. In this entirely arbitrary example, the Ca-NS channel accounts for half of the K⁺ exit from the cytosol and two thirds of the Na⁺ entry into the cytosol, thereby reducing NaCl secretion to one third of its potential maximum rate. The electrical potential profiles are shown below the cell diagram

cation channels in the basolateral membrane impairs the efficiency of a secretion model based on Na⁺-coupled secondary active Cl⁻ transport because, by depolarising the cell potential, it removes the driving force for electrogenic Cl⁻ exit and, by permitting Na⁺ entry, it dissipates the Na⁺ gradient, which energises the basolateral anion uptake system. The deleterious impact of introducing a non-selective cation conductance in the basolateral membrane on secondary active Cl⁻ transport is illustrated in Fig. 65.6.

Push–Pull Mechanism. These difficulties have led to the suggestion that rat (and mouse) pancreatic acini may secrete by a different mechanism (Fig. 65.7), a so-called "push–pull" mechanism, in which intracellular free Ca²⁺ concentrations oscillate during sustained secretomotor stimulation, rising and falling alternately at the apical and basolateral poles of the acinar cells [29]. When intracellular free Ca²⁺ rises at the apical pole (and is low at the basolateral pole), apical membrane Cl⁻ channels are activated, but the basolateral Cl⁻ channels and non-selective cation channels remain inactive, so that the cell potential hyperpolarises due to the activity of a resting basolateral K⁺ conductance. Consequently, Cl⁻ moves down its electrochemical gradient into the lumen. When the intracellular free Ca²⁺ rises at the basolateral pole of the cell (and falls at the apical pole), the apical membrane Cl⁻ channels close, and the basolateral Cl⁻ channels and non-selective

cation channels become active. The non-selective cation channels then depolarise the cell potential and move the basolateral membrane potential to a value more positive than the Nernst potential for Cl⁻; Cl⁻ then flows into the cell from the interstitium, down its electrochemical gradient.

This model provides a role for the non-selective cation channels in driving secretion, although the role it postulates (driving electrogenic Cl⁻ influx across the basolateral membrane of the acinar cells) could equally well be performed by the Na⁺-coupled transport systems known to be present in the basolateral membrane of rat pancreatic acinar cells [12,42,43,53]. Since the push–pull model still requires the presence of a K⁺ conductance on the basolateral membrane, the most important advantage it offers over the secondary active transport model is that it removes the deleterious interaction between Cl⁻ efflux across the apical membrane and the non-selective cation conductance in the basolateral membrane by postulating that these are active at different times. The presence of a K⁺ channel in the basolateral membranes of the acinar cells of the rat pancreas has proved difficult to demonstrate, although a recent whole-cell study [60] has shown a transient voltage-sensitive K⁺ current in rat pancreatic acinar cells: it remains to be shown whether this is located in the basolateral membrane, as would be necessary for either secretion model.

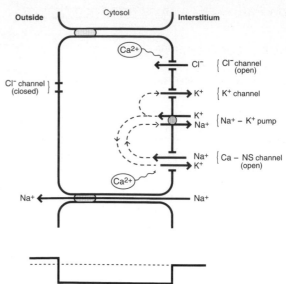

Push phase   **Pull phase**

Fig. 65.7. The "push–pull" model for secretion by rat pancreatic acini [29]. In the push phase (*left*), extracellular stimulation with CCK is supposed to have activated the phospholipase C pathway and increased the concentration of inositol-1,4,5-triphosphate (IP_3) in the cytosol. This leads to intracellular release of Ca^{2+} from IP_3-sensitive Ca^{2+} stores postulated to be located near the apical membrane. The Ca^{2+} then activates Cl^- channels in the apical membrane, permitting Cl^- to exit to the lumen down its electrochemical gradient (with a corresponding exit of K^+ ions across the basolateral membrane so as to preserve cytosolic electroneutrality). Some of the Ca^{2+} released from the IP_3-sensitive pool diffuses towards the basolateral pole of the cell (*right*), where it initiates the pull phase. In this phase, just as the IP_3-activated pool near the apical region becomes exhausted, Ca^{2+} activates a Ca^{2+}-sensitive Ca^{2+} pool close to the basolateral membrane, releasing more Ca^{2+}. This leads to activation of Ca^{2+}-sensitive Cl^- and Ca-NS channels in the basolateral membrane, thus depolarising the cell potential and allowing Cl^- to enter the cytosol down its electrochemical gradient. Note that a basolateral K^+ channel is active in both the push and the pull phases. The electrical potential profiles are shown below the cell diagrams

Ca2+ Pools. The question of how the Ca^{2+} oscillations are caused needs addressing. Petersen [61] postulates that there is an inositol-1, 4, 5-triphosphate (IP_3)-sensitive Ca^{2+} pool near the apical membrane which is activated immediately after the acini are stimulated and that this leads to activation of the apical membrane Cl^- channels (push phase). The free Ca^{2+} in the apical region then spreads towards the basolateral pole and activates a Ca^{2+}-sensitive Ca^{2+} pool near the basolateral membrane, which, in turn, activates the basolateral Cl^- and non-selective cation channels (pull phase) just as the IP_3-sensitive Ca^{2+} pool is emptied and the apical Ca^{2+} concentration begins to fall. For this cycle to repeat itself, one must also postulate that each burst of IP_3-induced Ca^{2+} release is self limiting and can only recur after a latent period.

In summary, it can be said that the availability of a substantial body of evidence based on blocker studies and pH measurements [12,42,43,53] supports the idea that Na^+-coupled secondary active anion transport plays a role in acinar secretion. An argument in favour of the push–pull model, however, is the fact that it accommodates the presence of a basolateral non-selective cation channel, and the appropriate cytosolic oscillations in Ca^{2+} concentration

in rat pancreatic acini have been observed [46] (although the physiological significance of these oscillations has been questioned [20,21]).

65.3.3 Physiological Role of Alkaline Pancreatic Juice

Steatorrhoea. The alkaline fluid raises the pH of the duodenal contents from the low values set by HCl secreted in the stomach to higher values compatible with the activity of the digestive enzymes present in the duodenum. If insufficient alkaline fluid is secreted for maximum enzyme activity, malabsorption and malnutrition will result. This occurs in cystic fibrosis, in which the pancreas undergoes fibrotic degeneration as a result of chronic pancreatitis, consequent on duct obstruction by thick viscous secretions. Lipases and bile salts are particularly sensitive to the environmental pH, with the result that steatorrhoea is a common problem both in cystic fibrosis patients, who secrete insufficient alkali in the pancreatic juice, and in patients with gastrinomas (Zollinger–Ellison syndrome), who secrete excess acid in the stomach [11].

65.4 Membrane Receptors on Pancreatic Acinar and Duct Cells

65.4.1 CCK and Analogues

Trophic Actions. Pancreatic acinar cells contain at least two classes of CCK receptor in their basolateral membranes (see also Sect. 62.6.2). The predominant receptor belongs to the CCK_A (or peripheral) receptor class, but CCK_B receptors (also activated by gastrin and often called CCK_B/gastrin receptors), which are more conspicuous in the central nervous system, are seen in the pancreatic acinar cells of some mammals, including man [16,28]. They will not be discussed further here since activation of CCK_B receptors does not evoke a secretory response and their role in pancreatic function is unclear. Although the CCK_B receptor seems to be linked to a G protein, it does not appear to activate phospholipase C. It might perhaps have a role in mediating the trophic actions of CCK.

CCK_A. These receptors also exert their actions via a G protein, which activates phospholipase C, thereby generating IP_3, which increases intracellular free Ca^{2+} and activates protein kinase C. Functionally, pancreatic acini behave as if they had two distinct classes of CCK_A receptor, one having a high affinity for CCK-8 (the active C-terminal octapeptide of all naturally occurring forms of CCK) and the other having an affinity almost a thousand times lower. Activation of the high-affinity (CCK_{Ah}) receptor is correlated with activation of acinar enzyme secretion, whereas activation of the low-affinity receptor (CCK_{Al}) is correlated with a partial inhibition of the enzyme secretion seen during maximal CCK stimulation. Since CCK_A receptors studied in vitro in isolated plasma membranes no longer exhibit these two distinct classes of receptor, it seems likely that CCK_{Ah} and CCK_{Al} are not two different proteins but, rather, the same molecule, reacting differently because of differences in the normal local environments within the intact cell. For example, a single receptor type might be associated with two different signal transduction pathways involving different G proteins or different catalytic units. CCK receptors are competitively inhibited by several different classes of blocker [16,28], but L-364,718, a benzodiazepine derivative, is specific for CCK_A receptors (both the CCK_{Ah} and CCK_{Al} classes, as might be expected if they are the same receptor protein). It is effective in nanomolar concentrations and is now the most widely used CCK_A antagonist.

Although the most biologically active form of CCK is CCK-8, i.e., Asp-Tyr (SO_3)-Met-Gly-Trp-Met-Asp-Phe(NH_2), the entire spectrum of CCK activity can be reproduced by the C-terminal heptapeptide (CCK-7). Three shorter C-terminal peptides, CCK-6, CCK-5 and CCK-4, retain the ability to evoke maximum enzyme secretion (i.e., they activate CCK_{Ah} receptors maximally) but fail to show the partial inhibition of enzyme secretion seen with high CCK-8 concentrations (presumably because they do not activate CCK_{Al} receptors). CCK-3 and CCK-2 act as CCK antagonists, being able to bind to the CCK_{Ah} receptor but not to evoke enzyme secretion. Modifying CCK-7 by sulfating the Tyr_1 residue, or adding an Asp residue to the N-terminal, increases the affinity for the CCK_{Ah} receptor by factors of 1000 and 10, respectively. Nevertheless, CCK-8 is not an important secretomotor agonist in the pancreas; the major agonists are longer-chain peptides incorporating CCK-8 at the C terminus, such as CCK-33 (CCK itself), CCK-39 and CCK-58 (see Sect. 62.6.2). Of particular importance is a synthetic analogue of CCK-7 called CCK-JMV-180, i.e., BOC-Tyr (SO_3)-Nle-Gly-Trp-Nle-Asp-phenylethyl ester, which acts as a high-affinity agonist on CCK_{Ah} receptors and as a competitive antagonist on CCK_{Al} receptors [14,59]. Increasingly, this compound is being used to elucidate the mechanisms controlling pancreatic acinar secretion.

65.4.2 Acetylcholine

M4 Muscarinic Receptor Gene. Pancreatic acinar cells have muscarinic receptors of the glandular type, corresponding to the M4 muscarinic receptor gene. (The nomenclature for the five recognised muscarinic receptor types has become very confused: we will follow Jensen [28] who refers to the glandular receptor type as M_4, the nomenclature used by molecular biologists working in this field.) Like CCK_A receptors, the M_4 muscarinic receptors exist in high- and low-affinity forms (M_{4h} and M_{4l}), occupation of the high-affinity form being associated with activation of enzyme secretion and occupation of the low-affinity form being associated with partial inhibition of maximum enzyme secretion [9,38]. As with CCK_A receptors, the M_4 receptors in pancreatic acinar cells work via a G protein to activate phospholipase C, leading to increased production of IP_3, increased intracellular free Ca^{2+} and activation of protein kinase C. Recently, pancreatic duct cells have also been shown to have functional muscarinic receptors, which, when activated, increase intracellular free Ca^{2+} and thereby lead to activation of the Ca^{2+}-sensitive Cl^- channels in the luminal membrane [37,47].

65.4.3 Substance P

Substance P (see also Sect. 62.6.11) interacts with a single class of receptors on pancreatic acinar cells. In contrast to CCK and muscarinic agonists, when the concentration of substance P is increased, the rate of evoked enzyme secretion increases to a maximum, but then does not decline at higher agonist concentrations [62]. Like CCK and muscarinic receptors, however, substance P receptors, when activated, work via a G protein linked to phospholipase C, with a consequent increase in IP_3, leading to an increase in intracellular free Ca^{2+} and activation of protein kinase C. The order of potency of the various tachykinins acting on pancreatic substance P receptors is: physalaemin > substance P > eledoisin > kassinin. Substance P receptors are blocked by micromolar concentrations of D-amino acid-substituted analogues of substance P, such as spantide [16].

65.4.4 Gastrin-Releasing Peptide (GRP)
(see also Sect. 62.6.12)

Bombesin. GRP is the mammalian analogue of bombesin, a peptide isolated from the skin of the frog *Bombina bombina*, and GRP receptors, even in humans, are commonly referred to as bombesin receptors. GRP binds to a single class of receptor on pancreatic acinar cells and evokes enzyme secretion by activating phospholipase C. Analogues of GRP are available which block the receptor competitively in nanomolar concentrations. In contrast to CCK and muscarinic agonists, but like substance P, GRP does not induce a partial inhibition of enzyme secretion when administered in high concentrations [16].

65.4.5 Secretin

Secretin receptors are found on both pancreatic acinar cells and on pancreatic duct cells. The intrinsic biological activity of secretin (see also Sect. 62.6.3) resides at the N-terminal end of the molecule and secretin$_{1-14}$ (S_{1-14}) has the same efficacy (albeit with a 1000-fold lower affinity) as the native molecule (S_{1-27}). When secretin binds to its receptor, a G_s protein is activated, which in turn activates adenylate cyclase and increases cyclic adenosine monophosphate (cyclic AMP) production.

65.4.6 Vasoactive Intestinal Peptide (VIP)

Vasoactive intestinal peptide (VIP) (see also Sect. 62.6.6) is structurally related to secretin and binds to secretin receptors as well as to VIP receptors (which are defined as having a low affinity for secretin and a high affinity for VIP). VIP receptors, like secretin receptors, activate adenylate cyclase and increase intracellular cyclic AMP. PHI (peptide histidine-isoleucine-amide), PHM (peptide histidine-methionine-amide) and helodermin (a peptide extracted from Gila monster venom), which are all structurally related to VIP, also bind to and activate VIP receptors. C-terminal analogues of secretin, such as S_{5-27}, act as blockers of both secretin and VIP receptors.

65.4.7 Other Receptor Types

Pancreatic acinar cells have been shown to have receptors with a high affinity for gastrin (and also for CCK-8) which, like CCK_A receptors, are linked to the phospholipase C pathway via a G protein. They are distinguished from CCK_A receptors by their much higher affinity for gastrin [16]. They also have receptors for CGRP (calcitonin gene-related peptide), which increases acinar cyclic AMP and evokes protein secretion, insulin and epidermal growth factor (EGF), which potentiate enzyme secretion, and somatostatin, which inhibits enzyme secretion by activating a G_i protein and inhibiting adenylate cyclase.

Adrenergic Receptors. There has been considerable doubt whether adrenergic receptors are present on the parenchymal cells of the exocrine pancreas [16,28]; it is usually held that the acini do not receive a direct adrenergic innervation and that the inhibitory effects of adrenergic stimulation on pancreatic secretion arise from modulation of the activity of cholinergic ganglion cells (see Chap. 70). One study on the isolated perfused rat pancreas has presented unequivocal evidence that β-adrenergic stimulation with isoproterenol can evoke secretion of an HCO_3^--rich fluid (presumably from duct cells) and that the effect can be blocked with propranolol [36]. While it is possible that this effect could have been indirect, with isoproterenol activating VIPergic nerves supplying the ducts, it seems much more likely that the effect was direct and that the duct cells have β-adrenergic receptors in their basolateral membranes.

65.5 Control of Pancreatic Secretion

65.5.1 Extrinsic and Intrinsic Innervation

Parasympathetic and Sympathetic Fibres. The preganglionic parasympathetic fibres supplying the pancreas, which run in the vagus nerve, are cholinergic, and the effects of vagal nerve stimulation can be blocked completely by nicotinic ganglionic blocking agents such as hexamethonium [26]; the vagal fibres may run to the pancreas directly or via the coeliac ganglion, through which they pass without interruption. The sympathetic fibres, which run in the splanchnic nerves, are postganglionic noradrenergic fibres which arise in the coeliac and mesenteric ganglia. In addition to these morphologically defined nerve trunks, in a number of animals, including man, nerve fibres enter the pancreas through the upper and lower borders of the organ from adjacent tissues. It has recently been shown that these include nerve fibres projecting directly from ganglion cells in the enteric nervous system onto ganglion cells within the pancreas [32].

Intrinsic Nervous System. The pancreas has an extensive intrinsic nervous system, which comprises the perivascular, periacinar and peri-insular plexuses [40]. Embryologically, the intrinsic pancreatic nerves arise from vagal neural crest cells that form a subset of the cell group from which the enteric nervous system also develops [13]. The pancreatic intrinsic nerves are best thought of as being analogous to the enteric nervous system. In other words, we should consider the (extrinsic) vagal preganglionic fibres as projecting onto pancreatic ganglion cells, rather than describing the pancreatic ganglion cells as postganglionic vagal fibres. The pancreatic ganglia lie scattered along nerve bundles within the interlobular connective tissue: each contains only relatively few ganglion cells. Studies using neuronal tracing compounds [32] show that only a few ganglion cells project directly to the epithelial cells of the acini and ducts; the great majority project to ganglion cells in other pancreatic ganglia, a finding that accords with earlier electrophysiological studies [31]. It is thus clear that the pancreatic ganglia and the intra-pancreatic nerve bundles constitute a true intrinsic nervous system and that the ganglion cells are not to be regarded merely as postganglionic fibres relaying impulses from the preganglionic parasympathetic fibres to the parenchymal cells. The function of this intrinsic nervous system remains largely a matter for speculation: clearly, such a neuronal network could facilitate the co-ordinated delivery of secretomotor impulses to all the secretory units; equally, the network could facilitate interactions between the endocrine and the exocrine pancreatic tissue or a variety of other interactions.

Neurotransmitters. As in the enteric nervous system, neurones in the intrinsic nervous system of the pancreas are characterised by the presence of a variety of neurotransmitters. The most important of these are acetylcholine, VIP (which, like acetylcholine, is found in a great many intrapancreatic ganglion cells), PHI (or PHM in humans), neuropeptide Y (NPY), GRP and 5-hydroxytryptamine (serotonin) [40]: in some species, substance P, galanin and CCK also appear to be important neurotransmitters. These neurotransmitters commonly act as co-transmitters; for example, VIP and PHI (or, in humans, PHM) are expressed in the same neurones.

Vagal Fibres. As indicated above, vagal (preganglionic) fibres terminate on the intrapancreatic ganglion cells, and stimulation of the vagus (whether by direct electrical stimulation of the nerve trunks, by the production of central neuroglucopenia with insulin or 2-deoxyglucose or by sham feeding) produces marked increases in pancreatic enzyme and fluid secretion which can be up to 50% of the maximum responses evoked by infusion of CCK-8 or secretin, respectively. Many of these vagal effects, including increased enzyme secretion, are mediated by cholinergic ganglion cells in the intrinsic nervous system and can be blocked by atropine; others, however, are not blocked by atropine and can be presumed to be mediated by non-cholinergic transmitters. For example, the increased HCO_3^- and fluid secretion evoked in some species by vagal stimulation is at least partially attributable to the release of neurotransmitters such as VIP and PHI (or PHM). It should also be noted that the cholinergic intrinsic innervation of the pancreas may mediate some of the pancreatic effects of agents that are normally regarded as hormones. For example, the effect of low doses of GRP on pancreatic enzyme secretion can be blocked by atropine, suggesting that it is mediated by the intrinsic nervous system rather than being due to the direct action of GRP on the acinar cells. The acinar cells appear to be supplied mainly by cholinergic ganglion cells, and the role of the peptidergic (VIPergic) ganglion cells is unclear. In some species at least, VIPergic ganglion cells project to ducts and are responsible for a substantial component of vagally induced fluid and HCO_3^- secretion. They may also exert a modulatory effect on cholinergic ganglion cells.

Adrenergic Innervation. Pancreas adrenergic innervation is entirely extrinsic, consisting of post ganglionic sympathetic fibres arising from the prevertebral ganglia; as in the enteric nervous system, adrenergic fibres project onto ganglion cells rather than directly onto parenchymal cells. The available evidence suggests that the role of the sympathetic innervation is to modulate the responses of the exocrine pancreas to vagal and hormonal stimuli; its effect is usually to inhibit pancreatic secretion, although, as mentioned above, there is good evidence to suggest that pancreatic duct cells in the rat have β-adrenergic secretomotor receptors [36]. As in the enteric nervous system [15], NPY seems to be co-expressed in those adrenergic neurones that are concerned with the control of blood flow [25,27,55].

Vago-humoural and Vago-vagal Reflexes. In addition to the direct effects of the parasympathetic nervous system on pancreatic secretion, there are at least two other ways in which vagal activity might influence pancreatic secretion. First, because many endocrine cells, both in the islets and in the gut wall, are innervated, vagal activity may evoke the release of hormones into the bloodstream which can then modulate pancreatic secretomotor activity (vago-humoural reflexes). For example, vagal activity is closely correlated with release from islet cells of PP which inhibits pancreatic exocrine secretion [56]. Since administration of neutralising antibodies against PP concurrently with vagal stimulation produces marked increases in the pancreatic secretion rate, it seems likely that PP functions as a physiologically important modulator of pancreatic activity. Second, there appear to be numerous vago-vagal reflexes linking pancreatic activity with events in the proximal intestine. For example, the presence of amino acids and fatty acids in the lumen of the upper small intestine produces a reflex increase in pancreatic enzyme secretion, which, although primarily attributable to the release of CCK from I cells in the wall of the small intestine, seems also to be due to a vago-vagal reflex, which can be inhibited by atropine or by vagal section [58]. There is also evidence to suggest that the increase in pancreatic HCO_3^- secretion evoked by a low pH in the duodenal lumen is partly attributable to a vago-vagal reflex in addition to the well-known effect of the release of secretin from S cells in the gut wall [58]. The evidence for these vago-vagal reflexes is not conclusive, however, and experiments are needed in which the effects of vagal section are studied while blood levels of CCK and secretin are monitored.

65.5.2 Hormones

Pancreatic exocrine secretion is evoked by a variety of hormones (see also Sect. 62.6) including:

- CCK
- Secretin
- Gastrin
- Neurotensin

- Motilin
- Insulin.

CCK is released from I cells in the upper small intestine in response to the presence of digestive products such as long-chain fatty acids (e.g., oleate), amino acids (e.g. tryptophan and valine) and a low pH in the lumen. It can also be released by vagal stimulation and by GRP. Studies using the CCK antagonist loxiglumide suggest that CCK release accounts for approximately 50% of the pancreatic response to the entry of chyme into the duodenum after a meal. Secretin is released from S cells in the upper small intestine in response to a low pH in the lumen, and, to a lesser extent, in response to the presence of long-chain fatty acids. Secretin has been claimed to mediate up to 80% of the fluid and HCO_3^- secretory response to a meal, although this depends on the relative importance of vagal stimulatory pathways, which varies from species to species. As noted above, gastrin, although pharmacologically able to evoke pancreatic enzyme secretion, does not usually play a major role under physiological conditions. Neurotensin secretion is evoked by feeding, particularly by ingested fat, and may play a role in evoking pancreatic enzyme secretion. Insulin release from β-cells is evoked postprandially, and, at least in the rat and rabbit, it potentiates pancreatic fluid and enzyme secretion evoked primarily by secretin and CCK. Motilin also evokes pancreatic fluid and enzyme secretion and may be responsible for the increased pancreatic secretion that accompanies the migrating motor complex (MMC) in the interdigestive period.

Several peptides inhibit pancreatic exocrine secretion:

- Somatostatin
- PP
- Enkephalins
- Glucagon.

Somatostatin inhibits acinar and ductal secretion, both directly, via an action on pancreatic epithelial cells, and indirectly, by inhibiting release of other hormones, such as secretin. PP is a physiological inhibitor of enzyme secretion. Its release is evoked by CCK, gastrin and secretin and by vagal stimulation. Enkephalins have been shown to inhibit both enzyme and HCO_3^- secretion by the pancreas, but a physiological role for them has not definitely been established. Glucagon inhibits pancreatic ductal secretion.

65.5.3 Pancreatic Secretion in Response to Feeding

The pancreatic response to a meal, like the gastric response, falls into three overlapping phases: cephalic, gastric and intestinal. The cephalic phase is the phase preceding entry of food into the stomach, and the gastric phase begins when food enters the stomach; the intestinal phase begins when chyme first enters the duodenum.

Cephalic Phase. In the cephalic phase, the sight, smell and taste of food and, particularly, chewing cause the pancreas to secrete enzymes at up to 55% of its maximum rate, and the effect can last for 60 min or longer, even if food is not actually ingested. This phase is mediated by the vagus nerve. It is often overlooked that, in the cephalic phase, acid delivery from the stomach into the duodenum also takes place and that this produces a further increase in pancreatic secretion, so that secretory rates often approach maximum values towards the end of the cephalic phase. As mentioned earlier, the effects of acid entering the duodenal lumen on fluid secretion may be mediated both by the release of secretin and by a vago-vagal reflex.

Gastric Phase. In the gastric phase, gastric distension evokes pancreatic enzyme secretion by means of a vago-vagal reflex, and the presence of peptides and amino acids in the lumen of the gastric antrum evokes pancreatic secretion by a gastrin-mediated mechanism, although the physiological importance of this in relation to pancreatic secretion (as distinct from secretion by the gastric glands) has been questioned [33,57].

Intestinal Phase. In the intestinal phase, pancreatic ductal HCO_3^- secretion is strongly stimulated by the presence of acid, fatty acids and monoacylglycerols in the duodenal lumen. These effects are mediated by the release of secretin and probably also by a vago-vagal reflex. Simultaneously, pancreatic acinar secretion (of enzymes) is strongly stimulated by fatty acids in the intestinal lumen (long-chain fatty acids such as oleate have a greater effect than medium- or short-chain acids), monoacylglycerols, peptides and amino acids (particularly phenylalanine, valine, methionine and tryptophan), high osmolality and high levels of free Ca^{2+} or Mg^{2+} in the luminal contents, whereas luminal acid and luminal glucose have only slight stimulatory effects. These effects are mediated by the release of CCK and probably by a vago-vagal reflex. In the intestinal phase of the pancreatic response to food ingestion, there is also release of hormones such as PP and somatostatin, which inhibit pancreatic secretion.

The effects of chyme in the upper small intestine on pancreatic secretion also depend on the following:

- The length of the gut segment exposed to the secretagogues (the greater the surface area exposed to stimulation, the greater the effect)
- The presence of bile salts in the lumen (these reduce the stimulatory effect of fatty acids by increasing the rate of fatty acid absorption and, conversely, reduce the stimulatory effects of amino acids by inhibiting amino acid absorption [10])
- The physical properties of the ingested food (a meal containing unhomogenised solids and liquids has a more prolonged effect than one that is homogenised, presumably because the homogenised meal leaves the stomach more rapidly)

- The action of pancreatic enzymes (which are needed for the release of secretagogues such as amino acids and fatty acids from ingested food).

The distal small intestine and the large intestine also appear able to regulate pancreatic secretion. For example, the presence of carbohydrates or fats in the lumen of the distal small intestine and the proximal colon inhibits pancreatic secretion, although the mechanism of these effects is unclear. It may be hormonal, for example, mediated by peptide YY or neurotensin, or it may involve nervous reflexes.

65.5.4 Pancreatic Secretion in the Interdigestive Period

Secretory Cycle. In humans and many other mammals, the pancreas shows cyclical changes in secretory activity during the interdigestive period [6,7,10,57] which are closely correlated with the changes in upper gastrointestinal motility, referred to as the MMC, in which there is regular alternation of motor quiescence (phase I), irregular motor activity of increasing frequency and magnitude (phase II) and bursts of peristaltic contractions (phase III) which last for approximately 10 min before the upper gastrointestinal tract returns to quiescence. The entire cycle repeats itself about once every 2 h until the next feeding period begins. Associated with each period of increased motility (phase II) is a phase of increased pancreatic secretion of fluid and enzymes, which peaks just before the onset of phase III of the MMC [35]. The causes of these changes in pancreatic secretion are not known with certainty and may vary in different species, but it is likely that they arise from cyclical alterations in blood levels of motilin (which activates pancreatic secretion and upper gastrointestinal motility), and PP (which inhibits pancreatic secretion). Vago-vagal reflexes and activity in the pancreatic intrinsic nervous system, as well as extrinsic sympathetic activity, also appear to play a role [6,7,10,57]. In humans eating three meals a day, the only time during which interdigestive gastrointestinal activity is likely to be observed is late at night (during sleep), and even then it may not be observed if the evening meal is sufficiently large. In humans, during waking hours, the pancreas secretes almost continuously at close to maximum rates, particularly in affluent societies, where eating between meals is increasingly common.

65.5.5 Negative Feedback Control of Pancreatic Enzyme Secretion

Pancreatic Monitor Peptide and CCK-Releasing Peptide. It has been repeatedly observed in rats that diversion of the pancreatic juice from the duodenal lumen results in an increase in the rate of pancreatic enzyme secretion and that pancreatic secretion rates are inversely related to the levels of free trypsin in the duodenal lumen. This is believed to have two separate causes. One is the presence in pancreatic juice of pancreatic monitor peptide (PMP), which has an amino acid sequence similar to that of PSTI [7,10]. PMP is able to evoke the release of CCK from the upper small intestinal mucosa. Since it can be broken down by trysin, the amount of PMP will increase if the level of free trypsin in the duodenal lumen falls (for example, when large amounts of ingested protein are present). This leads to increased CCK release and increased pancreatic enzyme secretion. The second mechanism, which has been best studied in the rat, involves another trypsin-degradable peptide, CCK-releasing peptide, which, in contrast to PMP, is secreted by cells of the duodenal epithelium, not of the pancreas [7,10]. There has been considerable argument whether similar mechanisms linking duodenal free trypsin activity with pancreatic enzyme secretion exist in animals other than the rat, but it now seems clear that feedback regulation of this type does occur, at least in humans and dogs, although it may only be of importance under extreme conditions. In humans, the effect appears to be mediated by vago-vagal pathways rather than by CCK [10].

Secretin-Releasing Peptide. In the rat, a trypsin-degradable peptide called secretin-releasing peptide is also secreted by the duodenal mucosa [7].

References

1. Argent BE, Case RM (1994) Cellular mechanism and control of bicarbonate secretion. In: Johnson LR (ed) Physiology of the gastrointestinal tract, vol 2, 3rd edn. Raven, New York, pp 1473–1497
2. Arkle S, Lee CM, Cullen MJ, Argent BE (1986) Isolation of ducts from the pancreas of copper-deficient rats. Q J Exp Physiol 71:249–265
3. Ashton N, Argent BE, Green R (1991) Characteristics of fluid secretion from isolated rat pancreatic ducts stimulated with secretin and bombesin. J Physiol (Lond) 435:533–546
4. Brockman DE (1978) Anastomosing tubular arrangement of dog exocrine pancreas. Cell Tissue Res 189:497–500
5. Brockman DE (1993) Anatomy of the pancreas. In: Go VLW, DiMagno EP, Gardner JD, Lebenthal E, Reber HA, Scheele GA (eds) The pancreas. Biology, pathobiology, and disease, 2nd edn. Raven, New York, pp 1–8
6. Case RM, Argent BE (1993) Pancreatic duct secretion: control and mechanism of transport. In: Go VLW, DiMagno EP, Gardner JD, Lebenthal E, Reber HA, Scheele GA (eds) The pancreas. Biology, pathobiology, and disease, 2nd edn. Raven, New York, pp 301–350
7. Chey WY (1993) Hormonal control of pancreatic exocrine secretion. In: Go VLW, DiMagno EP, Gardner JD, Lebenthal E, Reber HA, Scheele GA (eds) The pancreas. Biology, pathobiology, and disease, 2nd edn. Raven, New York, pp 403–424
8. Dagorn JC (1993) Lithostathine. In: Go VLW, DiMagno EP, Gardner JD, Lebenthal E, Reber HA, Scheele GA (eds) The pancreas. Biology, pathobiology, and disease, 2nd edn. Raven, New York, pp 253–263
9. Dehaye JP, Winand J, Poloczek P, Christophe J (1984) Characterization of muscarinic cholinergic receptors on rat pancreatic acini by ^{3}[H]N-methylscopolamine binding. Their

relationship with $^{45}[Ca^{2+}]$ efflux and amylase secretion. J Biol Chem 259:294–300

10. DiMagno EP, Layer P (1993) Human exocrine pancreatic enzyme secretion. In: Go VLW, DiMagno EP, Gardner JD, Lebenthal E, Reber HA, Scheele GA (eds) The pancreas. Biology, pathobiology, and disease, 2nd edn. Raven, New York, pp 275–300

11. DiMagno EP, Layer P, Clain JE (1993) Chronic pancreatitis. In: Go VLW, DiMagno EP, Gardner JD, Lebenthal E, Reber HA, Scheele GA (eds) The pancreas. Biology, pathobiology, and disease, 2nd edn. Raven, New York, pp 665–706

12. Evans LAR, Pirani DP, Cook DI, Young JA (1986) Intra-epithelial current flow in rat pancreatic secretory epithelia. Pflugers Arch 407 [Suppl 2]:S107–S111

13. Fontaine J, Le Lievre C, Le Douarin NM (1977) What is the developmental fate of the neural crest cells which migrate into the pancreas in the avian embryo? Gen Comp Endocrinol 33:394–404

14. Fulcrand P, Rodriguez M, Galas MC, Lignon MF, Laur J, Aumelas A, Martinez J (1988) 2-Phenylethyl ester and 2-phenylethyl amide derivative analogues of the C-terminal hepta- and octapeptide of cholecystokinin. Int J Pept Protein Res 32:384–395

15. Furness JB, Costa M (1987) The enteric nervous system. Churchill-Livingstone, Edinburgh

16. Gardner JD, Jensen RT (1993) Receptors for secretagogues on pancreatic acinar cells. In: Go VLW, DiMagno EP, Gardner JD, Lebenthal E, Reber HA, Scheele GA (eds) The pancreas. Biology, pathobiology, and disease, 2nd edn. Raven, New York, pp 151–166

17. Gorelick FS, Jamieson JD (1994) The pancreatic acinar cell. Structure-function relationships. In: Johnson LR (ed) Physiology of the gastrointestinal tract, vol 2, 3rd edn. Raven, New York, pp 1353–1376

18. Gray MA, Pollard CE, Harris A, Coleman L, Greenwell JR, Argent BE (1990) Anion selectivity and block of the small-conductance chloride channel on pancreatic duct cells. Am J Physiol 259:C752–C761

19. Gray MA, Winpenny JP, Porteous DJ, Dorin JR, Argent BE (1994) CFTR and calcium-activated chloride currents in pancreatic duct cells of a transgenic CF mouse. Am J Physiol 266:C213–C221

20. Habara Y, Kanno T (1991) Dose-dependency in spatial dynamics of $[Ca^{2+}]$ in pancreatic acinar cells. Cell Calcium 12:533–542

21. Habara Y, Satoh Y, Kanno T (1993) The direction of Ca^{2+} wave propagation is dependent on the dose of carbachol and acinar topography in rat pancreas. Biomed Res 14:377–384

22. Harris A, Chalkley G, Goodman S, Coleman L (1991) Expression of the cystic fibrosis gene in human development. Development 113:305–310

23. Hickson JCD (1970) The secretion of pancreatic juice in response to stimulation of the vagus nerves in the pig. J Physiol (Lond) 206:275–297

24. Hickson JCD (1970) The secretory and vascular response to nervous and hormonal stimulation in the pancreas of the pig. J Physiol (Lond) 206:299–322

25. Holst JJ (1993) Neural regulation of pancreatic exocrine function. In: Go VLW, DiMagno EP, Gardner JD, Lebenthal E, Reber HA, Scheele GA (eds) The pancreas. Biology, pathobiology, and disease, 2nd edn. Raven, New York, pp 381–402

26. Holst JJ, Schaffalitzky de Muckadell OB, Fahrenkrug J (1979) Nervous control of pancreatic exocrine secretion in pigs. Acta Physiol Scand 105:33–51

27. Holst JJ, Ørskov C, Knuhtsen S, Sheikh S, Nielsen OV (1989) On the regulatory functions of neuropeptide Y (NPY) with respect to vascular resistance and exocrine and endocrine secretion in the pig pancreas. Acta Physiol Scand 136:519–526

28. Jensen RT (1994) Receptors on pancreatic acinar cells. In: Johnson LR (ed) Physiology of the gastrointestinal tract, vol 2, 3rd edn. Raven, New York, pp 1377–1446

29. Kasai H, Augustine GJ (1990) Cytosolic Ca^{2+} gradients triggering unidirectional fluid secretion from exocrine pancreas. Nature 348:735–738

30. Kern HF (1993) Fine structure of the human exocrine pancreas. In: Go VLW, DiMagno EP, Gardner JD, Lebenthal E, Reber HA, Scheele GA (eds) The pancreas. Biology, pathobiology, and disease, 2nd edn. Raven, New York, pp 9–19

31. King BF, Love JA, Szurszewski JH (1989) Intracellular recordings from pancreatic ganglia of the cat. J Physiol (Lond) 419:379–403

32. Kirchgessner AL, Gershon MD (1990) Innervation of the pancreas by neurons in the gut. J Neurosci 10:1626–1642

33. Köhler E, Beglinger C, Eysselein V, Grötzinger U, Gyr K (1987) Gastrin is not a physiological regulator of pancreatic exocrine secretion in the dog. Am J Physiol 252:G40–G44

34. Kopelman H, Durie P, Gaskin K, Weizman Z, Forstner G (1985) Pancreatic fluid secretion and protein hyperconcentration in cystic fibrosis. N Engl J Med 312:329–334

35. Lee KY, Shiratori K, Chen YF, Chang TM, Chey WY (1986) A hormonal mechanism for the interdigestive pancreatic secretion in dogs. Am J Physiol 251:G759–G764

36. Lingard JM, Young JA (1983) β-Adrenergic control of exocrine secretion by perfused rat pancreas in vitro. Am J Physiol 245:G690–G696

37. Lingard JM, Alnakkash L, Argent BE (1994) Acetylcholine, ATP, bombesin, and cholecystokinin stimulate I^{125} efflux from a human pancreatic adenocarcinoma cell line (BxPC-3). Pancreas 9:599–605

38. Louie DS, Owyang C (1986) Muscarinic receptor subtypes on rat pancreatic acini: secretion and binding studies. Am J Physiol 251:G275–G279

39. Lowe ME (1994) The structure and function of pancreatic enzymes. In: Johnson LR (ed) Physiology of the gastrointestinal tract, vol 2, 3rd edn. Raven, New York, pp 1531–1542

40. Mawe GM (1995) Prevertebral, pancreatic and gall bladder ganglia: non-enteric ganglia that are involved in gastrointestinal function. In: Burnstock G (ed) The autonomic nervous system, vol 6. Hanwood Academic, Chur, Switzerland, pp 397–444

41. Metzgar RS, Hollingsworth MA, Kaufman B (1993) Pancreatic mucins. In: Go VLW, DiMagno EP, Gardner JD, Lebenthal E, Reber HA, Scheele GA (eds) The pancreas. Biology, pathobiology, and disease, 2nd edn. Raven, New York, pp 351–367

42. Muallem S, Loessberg PA (1990) Intracellular pH-regulatory mechanisms in pancreatic acinar cells. I. Characterization of H^+ and HCO_3^- transporters. J Biol Chem 265:12806–12812

43. Muallem S, Loessberg PA (1990) Intracellular pH-regulatory mechanisms in pancreatic acinar cells. II. Regulation of H^+ and HCO_3^- transporters by Ca^{2+}-mobilizing agonists. J Biol Chem 265:12813–12819

44. Novak I, Greger R (1988) Electrophysiological study of transport systems in isolated perfused pancreatic ducts: properties of the basolateral membrane. Pflugers Arch 411:58–68

45. Novak I, Greger R (1988) Properties of the luminal membrane of isolated perfused rat pancreatic ducts. Effect of cyclic AMP and blockers of chloride transport. Pflugers Arch 411:546–553

46. Osipchuk YV, Wakui M, Yule DI, Gallacher DV, Petersen OH (1990) Cytoplasmic Ca^{2+} oscillations evoked by receptor stimulation, G-protein activation, internal application of inositol trisphosphate or Ca^{2+} simultaneous microfluorimetry and Ca^{2+} dependent Cl^- current recording in single pancreatic acinar cells. EMBO J 9:697–704

47. Pahl C, Novak I (1993) Effect of vasoactive intestinal peptide, carbachol and other agonists on the membrane voltage of pancreatic duct cells. Pflugers Arch 424:315–320

48. Petersen OH (1993) Electrophysiology of acinar cells. In: Go VLW, DiMagno EP, Gardner JD, Lebenthal E, Reber HA, Scheele GA (eds) The pancreas. Biology, pathobiology, and disease, 2nd edn. Raven, New York, pp 191–218

49. Pirani D, Evans LAR, Cook DI, Young JA (1987) Intracellular pH in the rat mandibular salivary gland: and role of Na-H and Cl-HCO$_3$ antiports in secretion. Pflugers Arch 408:178–184

50. Randriamampita C, Chanson M, Trautmann A (1988) Calcium and secretagogues-induced conductances in rat exocrine pancreas. Pflugers Arch 411:53–57

51. Rinderknecht H (1993) Pancreatic secretory enzymes. In: Go VLW, DiMagno EP, Gardner JD, Lebenthal E, Reber HA, Scheele GA (eds) The pancreas. Biology, pathobiology, and disease, 2nd edn. Raven, New York, pp 219–251

52. Schulz I, Yamagata A, Weske M (1969) Micropuncture studies on the pancreas of the rabbit. Pflugers Arch 308:277–290

53. Seow FKT, Lingard JM, Young JA (1986) The anionic basis of fluid secretion by rat pancreatic acini in vitro. Am J Physiol 250:G140–G148

54. Sewell WA, Young JA (1975) Secretion of electrolytes by the pancreas of the anaesthetized rat. J Physiol (Lond) 252:379–396

55. Sheikh SP, Roach E, Fuhlendorff J, Williams JA (1991) Localization of Y1 receptors for NPY and PYY on vascular smooth muscle cells in rat pancreas. Am J Physiol 260:G250–G257

56. Shiratori K, Lee KY, Chang TM, Jo YH, Coy DH, Chey WY (1988) Role of pancreatic polypeptide in the regulation of pancreatic exocrine secretion in dogs. Am J Physiol 255:G535–G541

57. Singer MV (1993) Neurohormonal control of pancreatic enzyme secretion in animals. In: Go VLW, DiMagno EP, Gardner JD, Lebenthal E, Reber HA, Scheele GA (eds) The pancreas. Biology, pathobiology, and disease, 2nd edn. Raven, New York, pp 425–448

58. Solomon TE (1994) Control of exocrine pancreatic secretion. In: Johnson LR (ed) Physiology of the gastrointestinal tract, vol 2, 3rd edn. Raven, New York, pp 1499–1529

59. Stark HA, Sharp CM, Sutliff VE, Martinez J, Jensen RT, Gardner JD (1989) CCK-JMV-180: a peptide that distinguishes high-affinity cholecystokinin receptors from low-affinity cholecystokinin receptors. Biochim Biophys Acta 1010:145–150

60. Thorn P, Petersen OH (1994) A voltage-sensitive transient potassium current in mouse pancreatic acinar cells. Pflugers Arch 428:288–295

61. Thorn P, Lawrie AM, Smith PM, Gallacher DV, Petersen OH (1993) Ca^{2+} oscillations in pancreatic acinar cells. Spatio-temporal relationships and functional implications. Cell Calcium 14:746–757

62. Uhlemann ER, Rottman AJ, Gardner JD (1979) Actions of peptides isolated from amphibian skin on amylase release from dispersed pancreatic acini. Am J Physiol 236:E571–E576

63. Young JA, Van Lennep EW (1977) Morphology and physiology of salivary myoepithelial cells. In: Crane RK (ed) International review of physiology, gastrointestinal physiology II, vol 12. University Park Press, Baltimore, pp 105–125

64. Young JA, Van Lennep EW (1978) The morphology of salivary glands. Academic, London

65. Zhao H, Star RA, Muallem S (1994) Membrane localization of H$^+$ and HCO$_3^-$ transporters in the rat pancreatic duct. J Gen Physiol 104:57–85

66 Endocrine Pancreas

G. DREWS

Contents

66.1 Introduction

Islets of Langerhans. In 1869, Paul Langerhans described small areas of histologically distinct cells in the pancreas of the rabbit [85]. At this time the physiological function of these "islets" was unknown. In 1890, Mering and Minkowski [98] were the first to demonstrate a relation between the removal of the pancreas and the occurrence of diabetes mellitus. Today, it is well established that the *islets of Langerhans* represent an endocrine organ which comprises different hormone-producing cells:

- Glucagon-secreting A cells
- Insulin-secreting B cells
- Somatostatin-secreting D cells
- Pancreatic polypeptide-secreting PP cells

Structural Aspects. In the human pancreas, about one million islets are scattered throughout the exocrine pancreas (Fig. 66.1). They are principally located along the main arteries. The islets contribute to about 1%–2% of the total volume of the pancreas, and their size varies in diameter between 75 and 225 µm [119]. B cells (see Fig. 66.2) are the most abundant cells within an islet (around 80% of the islet volume) and form a core which is surrounded by a mantle of A and D cells [109]. A cells contribute to about 10%–20% of the islet cell mass, while D and PP cells are less numerous. According to the intraislet organization, most B cells are surrounded by other B cells. In the periphery, however, the B cells are in contact with other endocrine cells. Depending on the topography, each cell may be influenced by different signals from neighboring cells via gap junctions or via paracrine effects [119]. Arterial blood reaches the islets through arterioles branching into fenestrated capillaries, which pass into the B cell-rich core. At the secretory poles of the B cells, insulin is released and taken up into the capillaries. The vessels then turn back to the surface of the islets, passing the A and D cells [151]. The endocrine pancreas is innervated by sympathetic–adrenergic and parasympathetic–cholinergic nerves, and some fibers terminate close to the endocrine cells. In addition to the fibers containing the classical neurotransmitters, peptidergic nerve terminals have also been found in islets.

The rate of insulin, glucagon, and somatostatin release depends on the nutrient concentration, mainly glucose, in the blood:

- Glucose stimulates insulin and somatostatin release.
- Glucose inhibits glucagon secretion.

Functional Considerations. The two most prominent hormones produced by the endocrine pancreas, insulin and glucagon, are involved in the control of glucose homeostasis. The major physiological function of islet cells is to recognize and to regulate the nutrient concentration in the blood. In most respects both hormones act antagonistically, whereby insulin is the only hormone that effectively counteracts hyperglycemia by reducing the blood glucose concentration.

Insulin Receptor. In the target cells, insulin binds to its receptor. The receptor consists of two identical units connected by a disulfide bond. Each unit is composed of an α- and a β-subunit. The α-subunits are exposed to the extracellular side and contain the binding sites for insulin.

R. Greger/U. Windhorst (Eds.)
Comprehensive Human Physiology, Vol. 2
© Springer-Verlag Berlin Heidelberg 1996

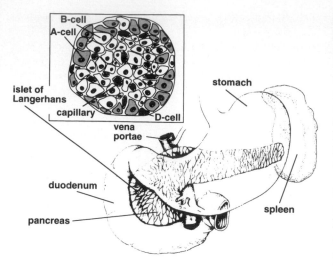

Fig. 66.1. Pancreas and its position in the body. The *inset* shows an islet with the distribution of the different endocrine cells

Insulin binding to the receptor results in an autophosphorylation of tyrosine sites located on the β-subunits which protrude into the cytoplasmic side of the molecule. The activated tyrosine kinase catalyzes the phosphorylation of tyrosine residues in target proteins, which in turn phosphorylate or dephosphorylate other enzymes influencing metabolic pathways. Insulin promotes the storage and prevents the breakdown of nutrients by its anabolic action on liver, muscle, and adipose tissue in the following ways:

- It increases the rate of glucose transport across the cell membrane of adipose tissue and muscle.
- In liver cells it enhances glycogen synthesis and storage, protein and triglyceride synthesis, and very low density lipoprotein (VLDL) formation and glycolysis, and it inhibits glycogen breakdown and gluconeogenesis (also see Chap. 76).
- In muscle cells it promotes protein and glycogen synthesis.
- In adipose tissue it augments triglyceride storage and inhibits lipolysis.

Glucagon. In contrast, the main function of glucagon is to provide energy for the organism by the breakdown of fuel stores in a situation where no nutrient intake occurs. Its main target is the liver. Receptor binding of glucagon on hepatocytes results in the activation of *adenylate cyclase* and the formation of cyclic adenosine monophosphate (cAMP) (see Chap. 5). It stimulates glycogenolysis, gluconeogenesis from amino acids and proteolysis. Glucagon also increases lipolysis in adipose tissue, which subsequently causes enhanced production of ketone bodies from triglycerides in the liver.

Catecholamines, Glucocorticoids, Growth Hormone, and Somatomedins. Besides glucagon, several other hormones such as catecholamines, glucocorticoides, and growth hormone are known to counterregulate the hypoglycemic action of insulin. Therefore, overproduction of these hormones can induce glucose intolerance or even overt diabetes. Insulin-like growth factors (IGF), which are mainly produced in liver, have been shown to possess insulin-like activity. These proteins, also known as somatomedins, are structurally related to proinsulin. However, their insulin-like activity is too low to substitute for insulin, the only hormone that effectively reduces the blood glucose concentration.

Somatostatin and Pancreatic Polypeptide. Somatostatin owes its name to its ability to inhibit the release of growth hormone from the pituitary. However, it is unclear to what extent somatostatin released from the pancreas contributes to the effects of circulating somatostatin. Likewise, the physiological significance of pancreatic polypeptide produced in the human pancreas is not well understood. Somatostatin and pancreatic polypeptide released from the endocrine cells appear to play a role in the inhibition of secretion from the exocrine pancreas.

Tumor Cell Lines. Much of the current knowledge about the physiology and pathophysiology of the endocrine pancreas has been derived from studies with animals and tumor cell lines. The use of cell lines is helpful if large amounts of cell material are needed. However, some of their properties differ from those observed in normal cells. Thus, this chapter preferentially refers to studies with normal human and animal cells.

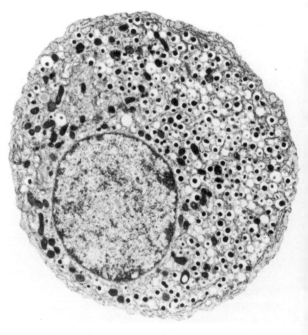

Fig. 66.2. Electron microscopic photograph showing a single B cell isolated from an islet of Langerhans. Numerous secretory vesicles are visible in the cytoplasm around the nucleus. Magnification, ×4400. (From [109], with permission)

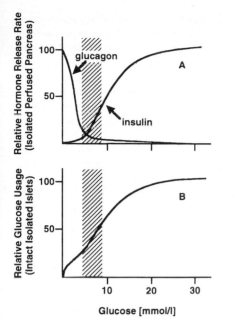

Fig. 66.3A,B. Secretory capacities and glucose usage of pancreatic islet cells. **A** Relative secretion rates of insulin and glucagon from the isolated perfused pancreas as a function of the extracellular glucose concentration. **B** Glucose usage measured in intact isolated islets as a function of the glucose concentration. The *cross-hatched areas* represent the physiological blood glucose concentration. (Reproduced with minor modifications from [95], with the permission of the American Society for Clinical Investigation)

66.2 B Cells

In contrast to the other endocrine cells of the pancreas, the mechanisms of hormone secretion and its regulation have been extensively studied in B cells. It is not surprising that interest has focused on B cells, because of the close relationship between the function of these cells and diabetes mellitus. Moreover, B cells are much more abundant in the islets than other endocrine cells.

66.2.1 Insulin Secretagogues

Primary Stimuli. Substances which are able to initiate insulin secretion in the absence of any other secretagogues are called primary stimuli [62]. Glucose is certainly the most important stimulus of insulin release under physiological conditions. However, besides D-glucose, D-glyceraldehyde, D-mannose, and L-leucine and its derivative α-ketoisocaproic acid, inosine and pharmacological agents such as sulfonylureas are able to initiate hormone secretion.

Secondary Stimuli. These are substances which are without effect in the absence of an initiator, but which increase insulin secretion in the presence of a primary stimulus [62]. Potentiators of insulin secretion are D-fructose, fatty

acids, ketone bodies, arginine, alanine, and certain hormones, for example acetylcholine and glucagon.

Hormones and amino acids are important modulators of insulin secretion. These secondary stimuli do not simply potentiate the effect of glucose. Their effectiveness depends on the glucose concentration, i.e., their potency is low at a low glucose concentration and vice versa. This is of physiological importance for the regulation of glucose homeostasis in vivo. A small decline in the plasma glucose concentration (during fasting) could decrease the effectiveness of a secondary stimulus and thus avoid inappropriate insulin secretion. On the other hand, a slight increase in the glucose concentration (after a meal) can enhance the effectiveness of a secondary stimulus, thus providing an additional safeguard against hyperglycemia.

66.2.2 Dependence of Insulin Secretion on Glucose Concentration

Concentration–Response Curve. Figure 66.3A shows the relationship between the extracellular glucose concentration and the relative rate of insulin and glucagon secretion. The curve representing insulin release is sigmoidal, with a steep increase in the physiological range of glucose concentration, which fluctuates between 4 and 8 mmol/l in humans. Glucose concentrations below 3 mmol/l do not significantly influence the secretion rate. At concentrations above 20 mmol glucose/l, insulin secretion has reached its maximum.

Biphasic Insulin Secretion. A sudden increase in the extracellular glucose concentration is followed by a biphasic rise in insulin secretion. During the first 5–10 min, insulin release rapidly increases and falls again to secretion rates somewhat higher than basal ones. This first phase is followed by a prolonged second phase that characteristically shows a slow, sustained increase in insulin secretion (Fig. 66.4) [62]. This biphasic response is not

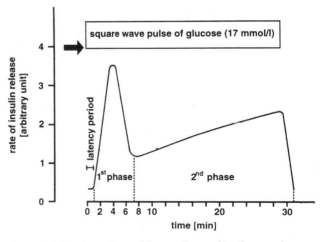

Fig. 66.4. Time dependency of the two phases of insulin secretion after a rapid rise in the extracellular glucose concentration. (From [62] with minor changes, with permission)

observed after a meal, when blood glucose concentration slowly rises, but it can be reproduced by a rapid increase of the plasma glucose concentration as it occurs after i.v. glucose infusion [53,120]. In a glucose infusion test, the first phase is found to be impaired in patients with non-insulin-dependent diabetes mellitus (NIDDM) [120].

The two phases of insulin secretion could be due to a biphasic rise in second messenger molecules produced by glucose, such as Ca^{2+}, and/or the existence of distinct pools of secretory granules which are more or less easily released [106].

After 2–3 h of exposure to a stimulatory glucose concentration, a third phase of insulin secretion begins, characterized by a spontaneous decline of release to about 20% of peak secretion. In experiments with animal cells, this phase of reduced secretion is sustained for up to 48 h. The important conclusion from these experiments is that longer-lasting hyperglycemia itself may impair the response to glucose. The molecular mechanisms underlying this desensitization are unknown, but it has been shown that it is not restricted to glucose and that it is not due to insufficient insulin production [53].

66.2.3 Glucose Sensor

Insulin secretion from B cells is not an "all or nothing" response to glucose. Secretion is tightly adjusted to the actual fuel concentration. This implies that a glucose concentration-sensing device must exist within the B cells.

Glucose Uptake. Glucose uptake into the B cells occurs via a high-capacity, high K_m hexose transporter, GLUT-2 [79] (also see Chap. 8). The formation of GLUT-2 mRNA is

stimulated by glucose [24,154]. The intracellular glucose concentration quickly follows changes in the extracellular concentration of the sugar. Consequently, glucose is equally distributed between the extracellular space and the cell interior [94]. Thus, glucose entry is not a rate-limiting step for glucose utilization. The glucose transport rate has to be reduced by more than 90% to markedly influence glucose metabolism. This points to an intracellular localization of the glucose-sensing system.

Glucokinase as a Glucose Sensor. Glycolysis is the major pathway of glucose metabolism in B cells. The first step of glucose phosphorylation in B cells is controlled by at least two enzymes, a hexokinase (probably hexokinase I) and glucokinase (Fig. 66.5) [87,95,96]. The islet hexokinase has a high affinity to glucose (K_m, 0.05–0.15 mmol/l). The low K_m value excludes the possibility that hexokinase might be the coupling factor between changes in the glucose concentration in a millimolar range and changes in metabolic flux initiating insulin secretion. Moreover, in intact islets hexokinase is markedly inhibited by phosphorylated intermediates of glycolysis such as glucose-6-phosphate, glucose- 1,6-diphosphate, and 6-phosphogluconate [95]. In contrast, glucokinase has a high K_m for glucose. In addition to hepatocytes, the B cell is the only cell with a glucose-phosphorylating enzyme with a low affinity for the sugar. With a K_m of about 10 mmol/l, the glucokinase is able to control glycolysis in the physiological range of the glucose concentration. Glucokinase plays a pacemaker role for glycolysis in pancreatic B cells, and this enzyme is now called the glucose sensor [87,95,96]. Other characteristics are in accordance with this concept [96]: glucokinase has an affinity for glucose similar to that of islet glucose utilization (see Fig. 66.3B), and the activity of glucokinase under

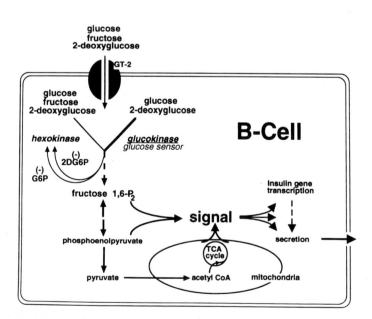

Fig. 66.5. Metabolic signal production in the B cell. A proposed schematic outline of B cell glucose metabolism as it relates to production of signals for insulin synthesis and secretion *GT-2*, GLUT-2; *G6P*, glucose 6-phosphate; *2DG6P*, 2-deoxyglucose 6-phosphate; *TCA*, tricarboxylic acid. (From [45] with minor changes, with permission)

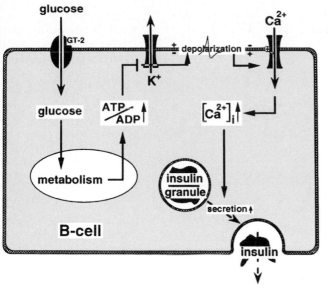

Fig. 66.6. Metabolic and electrical events between the rise in glucose concentration and insulin secretion. *GT-2*, GLUT-2; *ATP*, adenosine triphosphate; *ADP*, adenosine diphosphate

physiological conditions is consistent with the rate of glucose metabolism in islets. It distinguishes between anomers of aldohexoses in the order of their metabolic utilization by islets and their insulin-releasing capacity. Inhibition of glucokinase, for example by manno-heptulose, blocks islet glucose metabolism and insulin secretion.

In both forms of diabetes, insulin-dependent (IDMM) and NIDDM, genetic components are involved (see Sect. 66.7.2). Interestingly, there is accumulating evidence that mutations in the glucokinase gene are related to a special form of NIDDM, maturity-onset diabetes of the young (MODY) [95,114,138].

66.2.4 Link Between Metabolism and Electrical Activity

It is a peculiarity of the B cell that glucose induces electrical activity, i.e., rhythmic changes of the membrane potential (see Fig. 66.8 and Sect. 66.2.5). These oscillations are closely related to insulin secretion. Both parameters show a similar dependence on extracellular glucose concentration, and inhibition of electrical activity abolishes insulin secretion. Changes in membrane potential are not directly caused by glucose, but depend on glucose metabolism [4,99]. One link between glucose metabolism and electrical activity is adenosine triphosphate (ATP) or rather the ratio of the ATP to adenosine diphosphate (ADP) concentrations in the cytoplasm. Changes in the ATP to ADP ratio occur in the range of physiological glucose concentrations [91]. Thus, the demonstration that a distinct K^+ channel regulating electrical activity in B cells can be blocked by ATP [4,28,124] was a milestone in the understanding of stimulus–secretion coupling in B cells.

The important steps in coupling extracellular glucose concentration to insulin secretion are summarized in Fig. 66.6:

- Glucose influx and metabolism
- An increase of the intracellular ATP to ADP ratio
- Depolarization of the cell membrane due to closure of K^+_{ATP} channels
- Opening of voltage-dependent Ca^{2+} channels
- Ca^{2+} influx followed by an increase of the cytoplasmic Ca^{2+} concentration
- Exocytosis of insulin-containing granules.

66.2.5 Ion Channels and Electrical Activity

Oscillations of the Membrane Potential. The oscillations of the membrane potential induced by glucose are caused by rhythmic openings and closures of different ion channels. A variety of ion channels have been characterized in B cells. In particular, B cells possess voltage-dependent ion channels like those known to be present in "classical" excitable tissues, such as neurons, heart, and muscle cells (see Chap. 12). The Ca^{2+} channels in B cells belong to this group of voltage-dependent channels. Their opening leads to Ca^{2+} action potentials, which in some animal species and in humans are preceded by Na^+ action potentials evoked by the opening of voltage-dependent Na^+ channels [101]. In addition to the ATP-sensitive K^+ channel (K^+_{ATP} channel), at least one voltage-dependent K^+ channel and a K^+ channel activated by intracellular Ca^{2+}, $K^+_{(Ca)}$ are present in B cells. This chapter will focus on the changes in membrane potential and their relation to insulin secretion. Details about the properties of ion channels in B cells, which are beyond the scope of this chapter, are presented in two excellent reviews by Ashcroft and Rorsman [5] and Ashcroft and coworkers [7].

K^+_{ATP} Channels. At a low glucose concentration, the membrane potential of B cells is stable and around −70 mV. The resting potential is determined by a high K^+ conductance

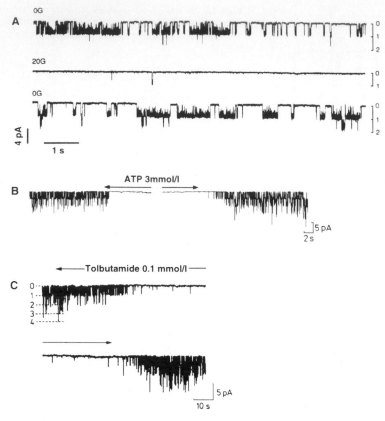

Fig. 66.7A–C. Influence of glucose, adenosine triphosphate (*ATP*), and the sulfonylurea tolbutamide on the K$^+_{ATP}$ channel. **A** Patch-clamp recording from a single B cell in the cell-attached mode without glucose (*upper trace*), with 20 mmol/l glucose in the bath solution (*middle trace*), and after washout of glucose (*lower trace*). (From [4], with permission). **B** Influence of 3 mmol ATP/l on the channel activity in an inside-out patch. (From [124], with permission). **C** Effect of 0.1 mmol tolbutamide/l is demonstrated in an inside-out patch. (From [144], with permission). In all three types of experiments, bath and pipette solution contained a high concentration of K$^+$, i.e., the K$^+_{ATP}$ current was inwardly directed

[65], which is attributed to a high open probability of K$^+_{ATP}$ channels under these conditions. With increasing glucose concentration, the membrane potential slowly depolarizes due to the closure of ATP-sensitive K$^+$ channels (K$^+_{ATP}$ channels) until the threshold potential for the opening of voltage-dependent Ca^{2+} channels is reached. Figure 66.7A,B shows patch-clamp experiments demonstrating the influence of glucose and ATP on the activity of single K$^+_{ATP}$ channels. In the cell-attached patch configuration (where cell metabolism is not disturbed), the addition of glucose to the bath solution inhibited the channel activity (Fig. 66.7A). Along these lines, addition of ATP to the bath solution in an inside-out patch (in which the cytosolic side of the channel faces the bath solution) blocked the current through the K$^+_{ATP}$ channel (Fig. 66.7B).

The K$^+_{ATP}$ channel is also the target for oral antidiabetic drugs of the sulfonylurea group, which are used in the treatment of NIDDM. As shown in Fig. 66.7C, with tolbutamide in an inside-out patch these drugs directly block K$^+_{ATP}$ channels. In an intact cell, the subsequent membrane depolarization and influx of Ca^{2+} would stimulate insulin secretion. Likewise, the specific K$^+_{ATP}$ channel opener diazoxide, which is used in the treatment of insulinomas, directly acts on the channel. It specifically opens K$^+_{ATP}$ channels, hyperpolarizes the membrane, and thus decreases insulin secretion.

Ca^{2+} Action Potentials. Figure 66.8 shows the membrane potential of a B cell at different glucose concentrations. In human B cells, the threshold concentration of glucose at which electrical activity and insulin secretion occur is around 5 mmol/l. At this glucose concentration, the cytoplasmic ATP concentration is sufficiently enhanced so that enough K$^+_{ATP}$ channels are closed to depolarize the cells to the potential at which the first Ca^{2+} action potentials appear. Above the threshold concentration, the membrane potential oscillates. Burst phases with action potentials alternate with interburst phases, during which the membrane potential is repolarized. The phases with bursts are prolonged and the interburst phases are shortened with increasing glucose concentration until continuous activity is reached at glucose concentrations above 20 mmol/l. During the spike phases with Ca^{2+} action potentials, Ca^{2+} from the extracellular space enters the cells. There is a good correlation between the time spent at the plateau potential and insulin secretion at different glucose concentrations

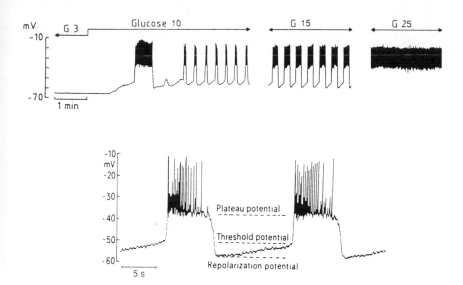

Fig. 66.8. The electrical activity of B cells. *Top*, the membrane potential of a B cell measured with an intracellular microelectrode at 3, 10, 15, and 25 mmol/l glucose in the extracellular solution. (From [71], with permission.) *Bottom*, two burst phases and an interval recorded in 10 mmol glucose/l with an extended time scale. (From [65], with permission)

[97]. The close relationship between electrical activity and insulin secretion is also confirmed by the observation that both processes strongly depend on extracellular Ca^{2+}. Ca^{2+} channel blockers or removal of extracellular Ca^{2+} suppresses both electrical activity and insulin secretion [67].

Burst. Each individual burst is initiated by the first Ca^{2+} action potential, starting when the threshold potential is reached (see Fig. 66.8). Each action potential is terminated by the repolarization to the plateau potential. This repolarization is mainly caused by the opening of voltage-activated K^+ channels [134]. The mechanism that ends a burst phase is less clear, although it can be ruled out that an opening of Ca^{2+}-sensitive K^+ channels terminates the burst phase, since the burst duration is not influenced by blocking the $K^+_{(Ca)}$ channels [70,82]. Most likely, two counteracting current components are involved: the hyperpolarizing current through K^+_{ATP} channels that are still open at the actual ATP concentration and the depolarizing current through slowly inactivating Ca^{2+} channels. Due to inactivation, the Ca^{2+} current slowly decreases during a burst phase until it can no longer balance the hyperpolarizing influence of the K^+_{ATP} current. According to this concept the termination of the interburst phase is due to recovery of the Ca^{2+} current from inactivation [5,7,29].
All our basic knowledge about the physiological role and regulation of ion channels in B cells was obtained from animal studies. However, recent data confirm that the principle mechanisms are very similar in human B cells [6,100,101].

Insulinotropic Effect of Glucose Independent of K^+_{ATP} Channels. For many years it was unequivocally accepted that the closure of K^+_{ATP} channels and the subsequent membrane depolarization is a prerequisite for nutrient-induced insulin secretion. However, several studies indicate that glucose can trigger insulin secretion independent of K^+_{ATP} channels [21,43,129]. Under appropriate conditions, glucose stimulates insulin secretion either when K^+_{ATP} channels are kept in the open state or when they are completely blocked. This K^+_{ATP} channel-independent insulinotropic effect of glucose is not mimicked by nonmetabolizable sugars, suggesting that the metabolism of glucose is involved. The effect may be correlated to changes in the energy state of B cells [44].

66.2.6 Intracellular Ca^{2+} and Insulin Secretion

$[Ca^{2+}]_i$ Increase. The influx of Ca^{2+} into the B cells is essential for the initiation of exocytosis. However, for a long time it was controversial whether the first phase of insulin secretion occurs due to the glucose-induced release of Ca^{2+} from intracellular stores. It has now been established that the first phase depends solely on Ca^{2+} influx and that glucose does not mobilize intracellular Ca^{2+} [56,121]. Although the pivotal role of $[Ca^{2+}]_i$ for insulin secretion has long been accepted [63,121,153], the exact mechanism by which $[Ca^{2+}]_i$ triggers the extrusion of the secretory vesicles is still unclear.
$[Ca^{2+}]_i$ in an unstimulated B cell is around 60–100 nmol/l. In mouse B cells, the relationship between the increase of $[Ca^{2+}]_i$ and the glucose concentration is represented by a sigmoidal curve, very similar to that obtained for the relationship between insulin secretion and glucose concentration (see Fig. 66.3A). Even at a very high glucose concentration of 20 mmol/l, $[Ca^{2+}]_i$ activity is only double that in unstimulated cells [57]. Thus, the amplitude of changes in $[Ca^{2+}]_i$ in the range of physiological glucose concentrations is very low.

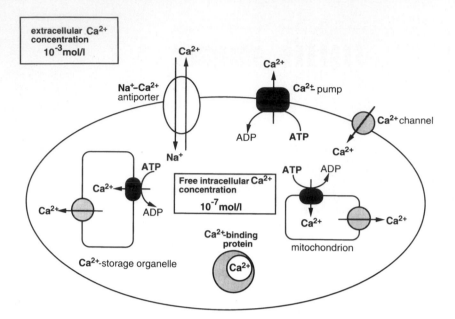

Fig. 66.9. B cell Ca^{2+} homeostasis, the major mechanisms involved in the regulation of [Ca^{2+}]$_i$ in pancreatic B cells. *ATP*, adenosine triphosphate; *ADP*, adenosine diphosphate. (From [8] with minor modifications, with the permission of Oxford University Press)

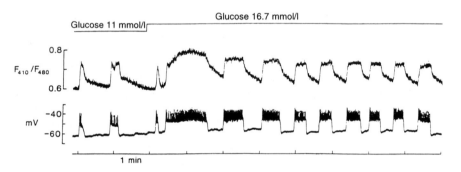

Fig. 66.10. Simultaneous measurement of [Ca^{2+}]$_i$ with the fluorescent dye indo-1 (*upper trace*) from a single pancreatic islet and the membrane potential with an intracellular microelectrode (*lower trace*) from a single B cell within the islet in the presence of 11 and 16.7 mmol glucose/l. (From [128], with permission)

Ca^{2+} Homeostasis. Glucose not only stimulates Ca^{2+} influx, but also facilitates Ca^{2+} removal from the cytoplasm [64,121]. Figure 66.9 shows the different systems contributing to Ca^{2+} homeostasis. Ca^{2+} influx occurs through voltage-dependent Ca^{2+} channels (see Sect. 66.2.5). Ca^{2+} removal is achieved by Ca^{2+} outward transport via a Ca^{2+} ATPase and a Ca^{2+}/Na$^+$ exchange system and by uptake of Ca^{2+} into the intracellular stores. Although glucose does not directly act on intracellular stores, the mobilization of intracellular Ca^{2+} is a mechanism by which some hormones enhance insulin secretion (see Sect. 66.2.8).

Ca^{2+} Oscillations. With improved fluorescence techniques for the measurement of the cytoplasmic Ca^{2+} activity, it was possible to demonstrate that glucose induces oscillations of [Ca^{2+}]$_i$ in single B cells [50,51]. These oscillations have a low frequency of several minutes. In intact islets, oscillations of [Ca^{2+}]$_i$ have a frequency identical to the frequency

of the burst phases of electrical activity [128]. This is shown in Fig. 66.10, with electrical activity and [Ca^{2+}]$_i$ recorded simultaneously from the same islet. The physiological significance of [Ca^{2+}]$_i$ oscillations in B cells has not yet been definitively established. However, the observation that [Ca^{2+}]$_i$ and electrical activity change in parallel in response to an increase in the glucose concentration points to a frequency modulation of the Ca^{2+} signal rather than an amplitude modulation. It has been shown that the oscillations of [Ca^{2+}]$_i$ in B cells from different areas of an islet are synchronous [47]. Sustained depolarization of the B cells by tolbutamide, arginine, or an elevation of the K$^+$ concentration as well as hyperpolarization by diazoxide causes sustained increases or decreases of [Ca^{2+}]$_i$ [47].

Oscillations of Insulin Secretion. Since the discovery that the plasma insulin concentration in humans and other species oscillates, attention was focused on the question of

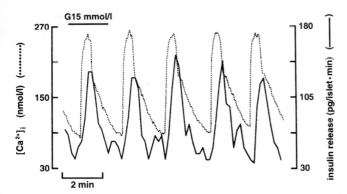

Fig. 66.11. Simultaneous monitoring of $[Ca^{2+}]_i$ oscillations measured with the fluorescent dye fura-2 (*dashed line*) and oscillations of insulin release (*continuous line*) from a single pancreatic islet in the presence of 15 mmol glucose/l. (From [48], with permission)

whether pulsatile hormone release is coupled to the oscillations of electrical activity in B cells [86]. In 1977, Goodner and coworkers [49] clearly established that insulin and glucagon secretion in monkeys is pulsatile. Plasma insulin concentrations in humans oscillate with a mean period of about 13 min and a mean amplitude of 1.6 mU/l [84]. Accordingly, plasma glucose concentration cycles with a very similar frequency and with a mean amplitude of 0.05 mmol/l. Blood glucose oscillations precedes insulin oscillations by 2 min. The physiological role of an oscillatory pattern of hormone release is to prevent downregulation of peripheral receptors and thus to augment hormone action. The nature of a probable pacemaker controlling pulsatile insulin secretion is not yet known; however, it is known that adrenergic, muscarinic, or endorphin receptors are not involved. Interestingly, insulin secretion measured in groups of collected intact islets in response to a sudden increase in extracellular glucose concentration oscillates with a similar periodicity as found in vivo [17,107]. Evidently, the oscillatory pattern of insulin secretion does not require the neuronal network. These oscillations with a very low frequency are not consistent with the more rapid bursting of electrical activity. In isolated islets, much shorter oscillations of insulin secretion have been demonstrated [16,48]. As shown in Fig. 66.11, the oscillations of $[Ca^{2+}]_i$ are paralleled by fluctuations of insulin release [48]. Taken together, there is a close relationship between the oscillations of electrical activity, $[Ca^{2+}]_i$, and insulin secretion. Further knowledge about how oscillations of insulin secretion are triggered is required, in particular because abnormal oscillatory secretion of the hormone is thought to be an early phenomenon in the development of NIDDM [108].

66.2.7 Role of Other Second Messengers in the Regulation of Insulin Secretion

In this section two systems will be discussed which amplify the response to glucose. Hormones whose action is mediated by these systems are presented in the next section.

Cyclic Adenosine Monophosphate–Protein Kinase A System. Although glucose causes a slight increase in the cytoplasmic cyclic adenosine monophosphate (cAMP) concentration, the role of cAMP in stimulus–secretion coupling in B cells is not a primary one. cAMP does not initiate, but modulate glucose-induced insulin secretion [75,90,121]. Studies with purified B cells have shown that a minimum concentration of cAMP is required for the maintenance of a normal secretory response to glucose [118,130]. In purified B cells, the cytoplasmic cAMP concentration is reduced and glucose fails to increase it. The poor secretory response of purified B cells to glucose can be markedly improved by elevation of the cAMP concentration, either by the addition of glucagon or by the addition of A cells, which reaggregate with B cells.

The main physiological significance of cAMP in the stimulus–secretion coupling in B cells is its amplification of the response to various secretagogues other than glucose [75,90,121]. A receptor-mediated increase of the cAMP concentration is caused by several hormones, e.g., glucagon, glucagon-like peptide-1, and gastric-inhibitory polypeptide. It is important to emphasize that these hormones only enhance insulin secretion in the presence of a stimulatory glucose concentration. They are not able to initiate the exocytotic process.

Adenylate cyclase is coupled to several membrane receptors. Receptor activation stimulates guanosine triphosphate (GTP)-binding proteins, which in turn activate or inhibit the adenylate cyclase according to the G protein subtype (see Chap. 5). Activation of adenylate cyclase leads to the formation of cAMP from ATP, and cAMP binds to its target, protein kinase A. The insulinotropic action of hormones which increase the cAMP concentration presumably results from the phosphorylation by protein kinase A of proteins involved in the secretory process. However, although several substrates which are phosphorylated by protein kinase A have been described, the exact nature and the role of these proteins in the secretory process remain to be identified [25,58,76].

Cyclic Adenosine Monophosphate-Induced Potentiation of Insulin Release. Several mechanisms may contribute to the cAMP-induced potentiation of insulin release. Agents which increase the concentration of cAMP such as forskolin, which stimulates adenylate cyclase, or methylxanthines, which inhibit the cAMP-degrading enzyme phosphodiesterase, augment $[Ca^{2+}]_i$. The increase in insulin secretion caused by agents that elevate the cAMP concentration is preceded by an increase in electrical activity and consequently enhanced Ca^{2+} influx [38,66,68]. $^{45}Ca^{2+}$ flux measurements suggested an activation of the voltage-dependent Ca^{2+} channel as the mechanism underlying the increase in electrical activity [66]. Using the patch-clamp technique, a direct effect of an increase in the cAMP concentration on the Ca^{2+} ion channel has been reported [3]. Forskolin increased $[Ca^{2+}]_i$ and the cell capacitance, which can be taken as a measure of exocytotic activity. The rise in $[Ca^{2+}]_i$ is mainly due to a slowing of Ca^{2+} channel inactivation. However, the rise in $[Ca^{2+}]_i$ is not a pivotal step for the action of cAMP. Studies in which measurements of $[Ca^{2+}]_i$ and membrane capacitance have been combined demonstrate that a rise in the cAMP concentration can increase membrane capacitance independent of $[Ca^{2+}]_i$ [3,46]. Even in permeabilized cells, in which $[Ca^{2+}]_i$ can be fixed, cAMP, IBMX (3-iso-butyl-1-methylxanthine), or forskolin augment insulin secretion independent of changes in $[Ca^{2+}]_i$ and ion currents [81,139]. Most likely, activation of protein kinase A stimulates insulin secretion by enhancing the sensitivity of the secretory machinery to a given amount of Ca^{2+}. However, nothing is known so far about the underlying mechanism. Taken together, at least two distinct mechanisms may contribute to the enhancement of glucose-induced insulin release by protein kinase A activation:

- Phosphorylation of the Ca^{2+} channel, leading to increased Ca^{2+} influx
- Phosphorylation of a still unidentified substrate which renders the secretory machinery more sensitive to Ca^{2+}.

Phospholipase C, protein Kinase C, and Diacylglycerol. Phospholipase C (PLC) activation results in the formation of inositol 1,4,5-trisphosphate (IP_3) and diacylglycerol (DAG), which both act as second messengers in a variety of signaling systems (see Chap. 5). IP_3 is known to release Ca^{2+} from the endoplasmic reticulum [18]. DAG activates a Ca^{2+}/phospholipid-dependent protein kinase known as protein kinase C (PKC) [105], and glucose and other nutrients activate PLC and increase the concentration of IP_3 and DAG [20,102,115,121]. Nutrients which are not metabolized do not mimic this effect. However, these second messengers, like cAMP, do not initiate insulin secretion, but provide a second system of signal amplification besides the cAMP–protein kinase A system.

Particularly the role of PKC in glucose-stimulated insulin secretion has long been a matter of dispute, because many studies have shown that agents which stimulate PKC, such as phorbol esters and DAG analogues, also stimulate insu-

lin secretion. On the other hand, several reports have demonstrated that inhibitors of the PKC block glucose-induced insulin secretion [115]. These experiments have been taken as arguments in support of the hypothesis that PKC plays an essential role in the secretory response of islet cells to glucose. However, inhibitors of PKC are not truly specific agents, and thus the results must be interpreted with caution. In another series of studies, PKC was downregulated by prolonged exposure of the islets to phorbol esters. Again, the results obtained were controversial, and it is certainly a disadvantage of this approach that downregulation is not always complete. However, for normal B cells there is prevailing evidence that downregulation of PKC does not alter the response of the cells to glucose, suggesting that the activation of PKC is not crucial for glucose-stimulated hormone release [115].

The physiological relevance of PLC activation lies in the amplification of the insulin response to secretagogues other than glucose, in particular to cholinergic agonists. A rapid increase of IP_3 formation has been shown for the cholinergic agonist carbamylcholine [19] and the peptide hormones cholecystokinin (CCK) [155], oxytocin [41], and vasopressin [40]. It is technically more difficult to estimate the DAG concentration, and thus little information is available about the effects of insulin secretagogues on the production of DAG. Changes in the formation of DAG have been reported after stimulation of islets with carbachol [116].

Interestingly, the DAG produced in response to carbachol and to glucose are of different composition. The DAG formed after carbachol stimulation contains more arachidonate and less palmitate than that formed in response to glucose [116]. This points to a de novo synthesis of DAG after glucose stimulation, while DAG produced after exposure to carbachol is a product of inositol phospholipid hydrolysis. This observation again underlines the different action of nutrient and non-nutrient secretagogues on the second messenger systems in B cells. The activation of PKC leads to the phosphorylation of specific proteins which probably influence exocytosis. A variety of substrates for PKC have been found in islet cells; however, the exact nature and function of these protein substrates remain to be identified [115]. Since PKC activation enhances insulin secretion from permeabilized islets in which the Ca^{2+} concentration has been fixed [80,139], it is assumed that PKC, like protein kinase A, sensitizes the secretory machinery to Ca^{2+}.

In summary, two distinct systems exist in B cells by which non-nutrient secretagogues, in particular hormones that bind to special receptors, amplify glucose-induced insulin secretion:

- Activation of adenylate cyclase, which stimulates protein kinase A via cAMP
- Activation of PLC, resulting in the formation of IP_3, which mobilizes $[Ca^{2+}]_i$ and DAG, which in turn activates PKC .

66.2.8 Influence of Hormones, Neurotransmitters, and Neuropeptides on Insulin Secretion

Although nutrients, in particular glucose, are the main stimuli, insulin secretion is influenced by a variety of hormones, neurotransmitters, and neuropeptides. These agents can be subdivided into stimulators and inhibitors of glucose-induced insulin secretion. Best known are the actions of the classical neurotransmitters acetylcholine and norepinephrine, which are released after parasympathetic or sympathetic nerve stimulation and of which the former stimulates and the latter inhibits insulin release. In recent years, many neuropeptides have been discovered which are released from the endocrine cells within the pancreas or from the nerve endings surrounding the islet cells. Only those hormones, transmitters, and peptides are mentioned which are likely to be of physiological importance.

66.2.8.1 Stimulators of Insulin Secretion

Acetylcholine. This ester is a potent stimulator of insulin secretion in the presence of glucose. It is released from the parasympathetic nerve endings during the cephalic phase of insulin release, i.e., shortly after food intake and prior to glucose absorption. Its stimulatory action is mediated by the activation of muscarinic receptors [2]. Acetylcholine activates the PLC pathway, resulting in the stimulation of PKC and mobilization of intracellular Ca^{2+} via IP_3 [121]. However, in the absence of extracellular Ca^{2+}, acetylcholine triggers only a short transient peak of insulin secretion, indicating that for a sustained effect of the neurotransmitter additional influx of Ca^{2+} is required [69]. The extracellular removal of Na^+ suppresses the stimulatory effect of acetylcholine. Acetylcholine increases the Na^+ conductance of B cells, which might explain the depolarization caused by acetylcholine. Subsequently, this depolarization might at least contribute to enhanced Ca^{2+} influx through voltage-dependent Ca^{2+} channels [69].

Glucagon. The insulinotropic effect of glucagon is commonly attributed to its ability to activate adenylate cyclase and to increase the cAMP concentration [90] (see Sect. 66.2.7). However, it is not clear whether this is the sole mechanism by which it amplifies nutrient-induced insulin secretion. Glucagon released from A cells is likely to influence insulin secretion, although direct proof of a paracrine action of glucagon is still lacking.

Epinephrine and Norepinephrine. The catecholamines epinephrine and norepinephrine are released from the pancreatic nerve endings or reach the islets through the circulation. They bind to both α- and β-adrenoreceptors, and their effect thus depends on the distribution of receptors and sensitivity to the agonists. The most important effect of catecholamines in this context is a suppression of glucose-induced insulin secretion (see below).

However, β-adrenoreceptor activation stimulates insulin secretion in vivo, although it hardly influences release in vitro. A study with purified islet cells demonstrated that A cells respond with changes in cAMP concentration and hormone release to stimulation by selective β-adrenergic agonists, but not by α-adrenergic ones, and B cells to activation by α-receptor agonists, but not by β-receptor agonists [131]. This suggests that B cells are only equipped with functional α-receptors, and A cells with β-receptors. Thus, stimulation of β-adrenoreceptors might increase the cAMP concentration in A cells, which would increase glucagon release; this, in turn, might augment insulin secretion by a paracrine effect on B cells.

Glucagon-Like Peptide-1, Gastric Inhibitory Peptide, and Cholecystokinin. After a normal, mixed meal, the blood glucose concentration only slightly increases, representing only a weak stimulus for secretion. The insulin response to a given amount of glucose is much higher after oral glucose intake than after intravenous glucose ingestion. This so-called incretin effect is attributed to the release of several hormones after oral nutrient intake which markedly enhance insulin secretion. The most potent of these "incretins" are glucagon-like peptide-1 (GLP-1), which is cleaved from proglucagon, gastric inhibitory peptide (GIP), and probably CCK [37,112]. These hormones are secreted from endocrine cells located in the mucosa of the gastrointestinal tract. GLP-1 and GIP increase the cAMP concentration in B cells, and CCK activates the PLC; other mechanisms which are less clear or even unidentified also participate in the stimulatory action of these hormones.

Oxytocin and Vasopressin. These hormones enhance insulin secretion in the presence of a stimulatory concentration of glucose. Both hormones activate the phosphoinositide metabolism in islet cells [40,41]. Interestingly, oxytocin, which exerts its main physiological effect on the uterus and the mammary gland, is present in pancreatic nerve endings.

Gastrin-Releasing Peptide and Vasoactive Intestinal Polypeptide. Two other neuropeptides which have been localized in the intrapancreatic nerves are gastrin-releasing peptide (GRP) and vasoactive intestinal polypeptide (VIP). GRP is structurally related to bombesin and is also present in the gastrointestinal tract. Both peptides stimulate insulin secretion [37,147,148] and electrical activity and Ca^{2+} uptake in B cells [147,148]. However, the exact mode of action and the physiological significance of these peptides are still unclear.

66.2.8.2 Inhibitors of Insulin Secretion

Epinephrine, Norepinephrine, Galanin, and Somatostatin. The catecholamines epinephrine and norepinephrine and the peptides galanin and somatostatin are known to be potent inhibitors of nutrient-induced insulin secretion.

Norepinephrine is the main neurotransmitter released after sympathetic nerve stimulation. Epinephrine reaches the islet cells via the circulation. One important physiological role of catecholamines released during a stress situation is to increase the availability of glucose. Apart from the inhibition of insulin secretion, this is achieved by activation of glucagon release and glycogen mobilization in the liver as well as the secretion of glucocorticoids. Galanin is present in nerve fibers innervating the islets and is co-released with norepinephrine after sympathetic nerve stimulation [36]. Interestingly, the actions of catecholamines, galanin, and somatostatin on B cells are very similar. One important mechanism by which these agents suppress insulin secretion is membrane hyperpolarization and inhibition of glucose-induced electrical activity [35,113]. This hyperpolarization is due to the opening of K^+ channels, which are distinct from the K^+_{ATP} channels [127]. Exposure of B cells to these agents decreases $[Ca^{2+}]_i$ [15]. This effect may be secondary to hyperpolarization, since membrane hyperpolarization decreases the activity of voltage-dependent Ca^{2+} channels and subsequently Ca^{2+} influx. Inhibition of insulin secretion by catecholamines follows activation of α_2-receptors in B cells. Stimulation of these receptors also exerts effects which are not related to changes in membrane potential. All effects are mediated by pertussis toxin-sensitive G proteins. Catecholamines decrease the concentration of cAMP by inhibition of adenylate cyclase. Moreover, an interference with the exocytotic process at a stage after the generation of second messengers has been described [15]. However, the relative contribution of all these effects to the inhibition of insulin secretion is still disputed. The same holds true for galanin and somatostatin. Despite the striking similarities in the effects of catecholamines and galanin on B cells, they act via different receptors [35].

Pancreastatin, Substance P, Calcitonin Gene-Related Peptide, Insulin-Like Growth Factor, and Peptide YY. These and several other peptides (including opioids), which are in part located directly in different endocrine cells of the pancreas, inhibit insulin secretion. So far, neither the underlying mechanisms nor the physiological relevance of these observations has been established.

66.2.9 Secretory Granules

Within the secretory granules, insulin is cleaved from its precursor, stored, and transported to the plasma membrane, which fuses with the vesicle membrane to release the hormone. Insulin diffuses through the capillary wall into the blood capillaries. These granules, which have a diameter of between 200 and 300 nm, are complicated microorganelles in which more than 100 polypeptides have been detected [77].

Insulin Biosynthesis. This process depends on the extracellular glucose concentration in a sigmoidal relationship similar to that of insulin release. Gene transcription and mRNA translation are regulated by nutrients. The effect of glucose is not mimicked by sugars which are not metabolized. However, with 2–4 mmol glucose/l, the threshold for the onset of biosynthesis is lower than that for secretion. This may represent a safeguard to ensure the adequate supply of insulin [73]. In humans, the insulin gene is located on the short arm of chromosome 11. After transcription to mRNA and adequate splicing, preproinsulin is synthesized. Insulin gene transcription only occurs in the B cells of the pancreas. In the endoplasmic reticulum, preproinsulin is cleaved to proinsulin, which is composed of an A chain with 21 amino acids, a B chain with 30 amino acids, and the connecting peptide sequence (C peptide) between the two chains, which is secreted in equal amounts to insulin (Fig. 66.12) [55]. The C peptide probably ensures the correct folding of the molecule.

Maturation of the Granules. Proinsulin is transported in microvesicles to the Golgi apparatus. Conversion of proinsulin into insulin starts in clathrin-coated vesicles, which bud off from specialized regions in the trans-Golgi apparatus. During the process of maturation, the vesicles convert into noncoated secretory granules with an electron-dense core [78,110]. Several enzymes, including two endopeptidases and carboxypeptidase H, are involved in the cleavage of C peptide from proinsulin and the subsequent formation of insulin. These enzymes are dependent on Ca^{2+} and pH. The free Ca^{2+} concentration in the granules is in the millimolar range, and the pH optimum of the peptidases is around 5.5, which corresponds to the pH of the mature granules [33,111]. Newly formed coated vesicles are relatively neutral, and pH decreases with maturation [78]. Accumulation of H^+ within the vesicles is achieved by an ATP-dependent proton pump. Conversion of proinsulin to insulin is prevented when the pH gradient across the vesicle membrane is destroyed, for example by ionophores [111]. The proton pump generates a potential across the granule membrane of around 45 mV, inside positive. This may promote the accumulation within the vesicles of biogenic amines, such as 5-hydroxytryptamine and dopamine, which are cosecreted with insulin [55,78]. At low pH, insulin is insoluble and is stored in crystals. Six insulin molecules are bound to two central Zn^{2+} ions stabilizing the hexamer complex. The pH and Ca^{2+} dependency of the peptidases ensures that the conversion of proinsulin to insulin only occurs within the granules.

Granule Translocation. The elements of the cytoskeleton, the microfilaments and microtubules, not only stabilize the structure of cells, but also contribute to transport processes. Microtubules and actin and myosin filaments are present in B cells in relatively high amounts [74]. The hypothesis that the cytoskeleton is involved in the transport of the secretory granules in B cells is based on the observation that colchicine inhibits insulin secretion [83]. Interestingly, microtubule polymerization and depolymerization suppress hormone release [74]. Although the pivotal role of $[Ca^{2+}]_i$ in glucose-induced insulin secretion is unques-

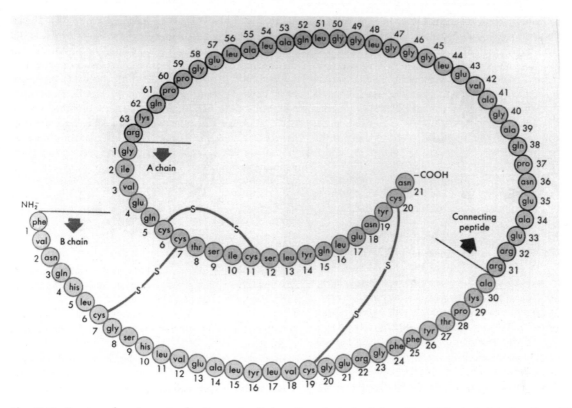

Fig. 66.12. Structure of porcine proinsulin. Human insulin has a threonine molecule in position B30 instead of alanine. (From [23], with permission)

tionable, the precise molecular mechanisms by which $[Ca^{2+}]_i$ induces exocytosis are poorly understood. The interaction of Ca^{2+} with elements of the cytoskeleton may trigger the movement of the granules towards the plasma membrane. There is some evidence that this involves phosphorylation of microtubule and microfilament subunits by a Ca^{2+}-calmodulin-dependent protein kinase [27,74]. Furthermore, the interaction between the granule membrane and the cytoplasmic side of the plasma membrane is modulated by Ca^{2+} and calmodulin [150]. This suggests that phosphorylation by a Ca^{2+}-calmodulin-dependent protein kinase also directly affects the fusion process.

66.3 A Cells

For an optimal regulation of blood glucose concentration, it is important that the ratio between released insulin and glucagon is tightly regulated according to the metabolic state of the organism. After a mixed meal, the ratio of insulin to glucagon secretion normally rises to counteract hyperglycemia. In contrast, during starvation or stress the secretion of glucagon in relation to that of insulin is enhanced, so that the organism's glucose requirements could be provided by increased glycogenolysis and gluconeogenesis.

In comparison to B cells, our knowledge about the mechanisms involved in the regulation of hormone secretion in A cells is limited. Since A cells represent only a small percentage of total islet cells, it is necessary for measurements of intracellular signals or electrophysiological measurements to identify them and to separate them from other cells. This is possible, but technically difficult.

As in B cells, the most prominent regulator of glucagon secretion is glucose, which suppresses hormone release from A cells in a concentration-dependent manner (see Fig. 66.3A). This effect is mimicked by other metabolizable sugars. Catecholamines increase glucagon secretion and thus contribute to a rise in blood glucose concentration. This ensures that glucose supply during a stress situation is sufficient in those organs which can hardly use any fuels other than the sugar, e.g., the brain. However, not all substances influence insulin and glucagon secretion in opposite directions. Amino acids, such as glutamine, alanine, and arginine, and some neuropeptides, such as VIP and GRP, stimulate insulin as well as glucagon secretion [2,117]. Somatostatin suppresses secretion from A and B cells.

Electrophysiological Considerations. Only very few electrophysiological data exist about A cells. Patch-clamp experiments on isolated A cells have shown that these cells produce spontaneous action potentials [125]. They start at membrane potentials of around −65 mV and peak at

+20 mV. Like B cells, A cells are equipped with voltage-dependent Na⁺, Ca²⁺, and K⁺ channels. Glucagon secretion is dependent on extracellular Ca²⁺ [117]. Inwardly directed Na⁺ and Ca²⁺ currents contribute to depolarization, and activation of an outward K⁺ current to repolarization. However, the physiological significance of these observations for stimulus–secretion coupling remains unclear, since the action potential frequency is influenced by the secretagogue alanine, but is unaffected by changes in the glucose concentration.

γ-Aminobutyric Acid and Glucose-Induced Inhibition of Glucagon Secretion. γ-Aminobutyric acid (GABA) is known as an inhibitory neurotransmitter in the nervous system. Activation of GABA receptors induces a Cl⁻ current, which hyperpolarizes the nerve cells and suppresses electrical excitability. GABA was also found to be localized in the B cells of the pancreas. Rorsman and coworkers [126] showed that GABA inhibited arginine-induced glucagon secretion and suppressed action potentials in A cells by inducing a Cl⁻ current. From these observations they proposed the following model: GABA is released with insulin after glucose stimulation of B cells. GABA reduces electrical activity in A cells by increasing the Cl⁻ permeability via activation of GABA receptor Cl⁻ channels. Consequently, Ca²⁺ influx would be decreased and this might inhibit glucagon secretion. However, it remains questionable whether this elegant concept is sufficient to explain glucose-induced inhibition of glucagon release, since glucagon secretion is inhibited at glucose concentrations that do not stimulate insulin secretion (see Fig. 66.3A). Moreover, glucose inhibits glucagon release from purified A cells [117].

Taken together, the mechanisms by which glucose and other nutrients suppress glucagon release are still poorly understood.

66.4 D Cells and PP Cells

Cells mainly located in the periphery of the islets produce somatostatin (D cells) and pancreatic polypeptide (PP cells).

Somatostatin Release. Almost all known stimulators of insulin secretion, e.g., glucose, arginine, gastrointestinal hormones, and tolbutamide, enhance somatostatin release from the D cells. Sympathetic nerve stimulation inhibits somatostatin release. Two forms of somatostatin with 14 and 28 amino acids, respectively, are cleaved from the prohormone. Their physiological action is similar. Somatostatin is known to inhibit the release of the growth hormone somatotropin from the pituitary gland, to inhibit the release of gastrointestinal hormones, to decrease gastric motility, and to suppress secretion from the exocrine pancreas. However, it is not clear to what extent somatostatin secreted by the pancreas contributes to these actions of somatostatin. The possible paracrine actions of somatostatin are discussed in Sect. 66.5.

Pancreatic Polypeptide. Secretion of pancreatic polypeptide, a hormone composed of 36 amino acids, is enhanced after a meal and is stimulated by protein rather than carbohydrate intake. Secretion is also augmented by sympathetic nerve stimulation. Most likely, the main physiological action is inhibition of secretion from ducts and acinus cells of the exocrine pancreas. Whether it influences the release of other hormones from the endocrine pancreas is not clear. The molecular mechanisms involved in the secretion of somatostatin and pancreatic polypeptide are unknown.

66.5 Interactions Between the Islet Hormones

It is well known that the hormones released from the endocrine pancreas influence the secretory activity of the other endocrine cells (Fig. 66.13) [93]:

- Somatostatin inhibits the secretion of insulin and glucagon.
- Glucagon enhances the release of insulin and somatostatin.
- Insulin blocks glucagon secretion.

Blood Supply of Islet Cells. Islet hormones can influence the secretion rate of the other endocrine cells by a paracrine pathway, i.e., through the interstitium without having passed the microcirculation of the islet or through the vascular system. Islets of Langerhans have a direct arterial blood supply, and most islets are located near the main stem of the pancreatic arteries. The blood first passes the islets before reaching the exocrine pancreas [151]. There is now convincing evidence that in human islets [135], as in other species [93], vascular perfusion occurs from the core to the mantle. B cells in the core are perfused first, followed by perfusion of the A cells in the mantle; finally, blood reaches the D cells. Thus, the direction of the blood flow

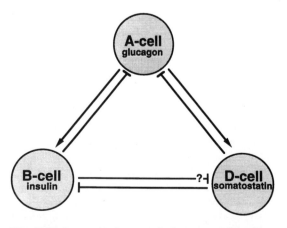

Fig. 66.13. Interaction between the hormones released from the different endocrine cells of the pancreas

through the islet vessels can easily explain the fact that insulin influences glucagon and somatostatin secretion and that glucagon affects somatostatin release. On the other hand, it seems unlikely that somatostatin influences glucagon or insulin secretion via the vascular system. Within an islet, a few D cells may be located in the direct vicinity of A and B cells. Although somatostatin is a very potent inhibitor of glucagon and insulin secretion, it is unclear whether it affects glucagon or insulin release under physiological conditions via a paracrine effect [93]. The physiological role of somatostatin released from the endocrine pancreas may rather be a suppressive modulation of exocrine pancreatic secretion.

Paracrine Action of Glucagon. There is evidence that glucagon augments insulin secretion via a paracrine mechanism, although direct proof is still lacking [93]. Pipeleers and coworkers have demonstrated that the response of purified B cells to glucose is impaired and that the response can be improved by the addition of external glucagon, membrane-permeant cAMP, or purified A cells [118]. Islets from the dorsal pancreas contain a significantly higher amount of A cells than those from the ventral part. It has been shown that under identical perfusion conditions and with similar insulin content the response of the dorsal islets to glucose was markedly higher than that of the ventral ones [143]. These two elegant studies with rat islets indicate that glucose-induced insulin secretion critically depends on glucagon secreted by the A cells in the vicinity.

66.6 Coordinated Effects of Insulin and Glucagon

In healthy mammals, the blood insulin concentration rises in the absorptive phase after food intake. The major function of insulin is to promote the storage of ingested nutrients by influencing carbohydrate, lipid, and protein metabolism. Accordingly, the main target organs for insulin are the liver, muscle cells, and adipose tissue. In the postabsorptive phase (especially during overnight fasting), the insulin concentration decreases and the glucagon concentration increases. In most respects glucagon acts antagonistically to insulin. Glucagon stimulates the breakdown of stored nutrients to make energy available to the tissues during the phases when no nutrient intake occurs. The most important target organ for the action of glucagon is the liver.

Insulin Receptor. The effects of insulin and glucagon are mediated by binding of the hormones to membrane receptors. Insulin receptor [39,133] belongs to a family of growth factor-binding receptors with internal tyrosine kinase activity; it is a glycosylated integral membrane protein which is widely distributed in the plasma membrane of mammalian cells. The receptor contains two subunits, an α- and a β-subunit with an apparent molecular weight of 135 000 and 95 000. The functional receptor is a heterotetramer consisting of two α- and two β-subunits. The α- and β-subunits as well as both α-subunits are connected by disulphide bridges. Insulin binds from the extracellular side to one of the α-subunits. Only one molecule of insulin binds with high affinity. Although the α-subunits are apparently equivalent, binding to one subunit dramatically decreases the affinity of the unoccupied subunit. The exact nature of the binding site has not yet been established. The β-subunit is a transmembrane protein, and the part facing the cell interior contains the tyrosine-specific protein kinase which is activated by insulin. Receptor stimulation results in autophosphorylation of specific sites at the β-subunit. Two sites with several tyrosine residues which become phosphorylated have now been identified on the intracellular part of the receptor, one within the kinase domain and one at the C-terminal region. Phosphorylation of the tyrosine cluster within the kinase domain seems to be critical for kinase activation. Insulin-dependent autophosphorylation only occurs in the heterotetrameric assembly of the subunits. Tyrosine-specific protein phosphatases are assumed to be involved in the termination of the kinase activity. Receptor activation by insulin induced the endocytotic internalization of the insulin–receptor complex. In the endosomal compartment, insulin dissociates from its receptor because of the acid pH; the internalized insulin is largely degraded within lysosomes, and the receptors are recycled to a large extent from the endosome system to the plasma membrane. Receptor internalization and probably increased receptor degradation is thought to be involved in downregulation of hormone receptors when the ambient insulin concentration is persistently high.

Glucagon. The hormone glucagon also binds extracellularly to a transmembrane-spanning receptor which is associated to a G protein. G proteins have a heterotrimeric structure consisting of an α-, a β-, and a γ-subunit. Activation of the receptor by the hormone results in the binding of GTP in exchange for GDP. The β- and γ-subunit dissociate from the α-subunit, and the γ-subunit activates membrane-associated adenylate cyclase, which catalyzes the formation of cAMP from ATP. cAMP is the second messenger molecule which triggers the hormone effects in the cell. The steps by which cAMP influences glycogen metabolism have been intensively studied. In brief, cAMP activates protein kinase A. Protein kinase A phosphorylates the glycogen synthase and thus inactivates it, but it also stimulates another protein kinase by phosphorylation, phosphorylase kinase, which in turn phosphorylates glycogen phosphorylase and converts it from the inactive into the active form. Insulin is supposed to counteract the effects of glucagon by the inhibition of adenylate cyclase and by activation of phosphodiesterase, which hydrolyzes cAMP [72].

Action of Insulin on Glucose Uptake. In contrast to liver cells, the rate of glucose uptake in muscle cells and adipose tissue is stimulated by insulin [32,88]. In muscle and fat cells, glucose uptake is mediated by the GLUT-4 glucose transporter (see Chap. 8). It is assumed that, in the absence of insulin, glucose transporter molecules are stored in membrane vesicles beneath the plasma membrane. By a mechanism that is as yet unknown, insulin binding to its receptor stimulates the fusion of these vesicles with the plasma membrane, thus enhancing the number of transport molecules and the rate of glucose uptake into the cells. Lowering the insulin concentration reverses this exocytotic process by endocytosis of the vesicles [32,88]. Thus, when the insulin concentration decreases in the postabsorptive phase, glucose uptake in muscle and adipose cells is reduced and other substrates such as fatty acids are used as fuels. This glucose-sparing effect contributes to the maintenance of the blood glucose concentration in the postabsorptive phase.

Actions of Insulin and Glucagon in the Target Organs. In the liver, insulin enhances glycogen formation and storage and inhibits glycogen breakdown. As in the pancreatic B cell (see Sect. 66.2.3), glucose is oxidized in the liver by a hexokinase with low glucose affinity, glucokinase. Consequently, in the physiological glucose concentration range, the activity of the enzyme and thus the rate of glucose-6-phosphate formation is proportional to the blood glucose concentration. Moreover, in contrast to glucokinase in B cells, liver glucokinase is positively regulated by insulin [14]. Insulin also inhibits gluconeogenesis and promotes glycolysis. Glycolysis provides precursor products for protein and triglyceride synthesis, which is also stimulated by insulin. In contrast, glucagon stimulates glycogenolysis and enhances hepatic glucose output by the stimulation of gluconeogenesis from amino acids. The main glycogen stores in mammalian organisms are the liver and the muscles. However, the muscle glycogen stores cannot be used as a glucose source to counterregulate a decrease in blood glucose concentration, because in muscle cells the enzyme glucose-6-phosphatase, which hydrolyzes glucose-6-phosphate, is lacking. Thus one important mechanism which contributes to the maintenance of blood glucose concentration is "glucose buffering" by the liver, i.e., uptake and storage of the sugar in the absorptive phase (high insulin to glucagon ratio) and glucose output in the postabsorptive phase (low insulin to glucagon ratio). Glucagon also enhances the formation of ketone bodies in the liver from fatty acids. Ketone bodies are used to ensure the energy supply in a variety of tissues such as the brain and the heart when the blood glucose concentration is low.

In muscle cells, insulin stimulates the formation of proteins by activating amino acid transport and ribosomal protein synthesis. It promotes glycogen storage and inhibits glycogen breakdown. Glucagon stimulates protein breakdown.

In adipose tissue, insulin promotes the storage of fat by activating the lipoprotein lipase bound to endothelial cells and by inhibiting intracellular lipase. Lipolysis of circulating lipoproteins and subsequent uptake of free fatty acids, together with the enhanced uptake of glucose, provides substrates for triglyceride formation, while the breakdown of intracellular triglyceride stores is inhibited. In the postabsorptive phase, the rising glucagon concentration activates intracellular lipase and thus stimulates the formation of free fatty acids from the adipose tissue.

The intracellular action of insulin is far from understood. A variety of intracellular molecules which change their functional state in response to insulin have been identified, the functions of which are only partly known [32]. One mechanism contributing to the actions of insulin and glucagon on nutrient storage and breakdown is a change in the kinetics of some key enzymes of the intermediary metabolism depending on the ratio of insulin to glucagon. Glucagon promotes phosphorylation, and insulin predominantly dephosphorylation, of these interconvertible enzymes. However, the acetyl-CoA (coenzyme A) carboxylase in liver, muscle, and adipose tissue is phosphorylated by insulin, resulting in activation and thus triglyceride synthesis [32]. Hormone-induced changes in the phosphorylation–dephosphorylation state of enzymes provide a relatively fast (minutes to hours) control of metabolism. A slower control mechanisms (hours to days) is the induction (or repression) of the synthesis of key enzymes by these hormones.

Glycogen synthase and glycogen phosphorylase are reversely regulated by insulin and glucagon. Glycogen phosphorylase is active in the phosphorylated form (low insulin to glucagon ratio) and inactive in the dephosphorylated form (high insulin to glucagon ratio). In contrast, glycogen synthase is active in the dephosphorylated form and inactive in the phosphorylated one. Glucose enhances the effect of the hormones by directly acting on the enzymes regulating the glycogen metabolism. Binding of glucose to the active phosphorylase *a* makes the enzyme sensitive for a phosphatase which converts it into the inactive form, phosphorylase *b*. At the same time phosphatase activates glycogen synthase.

Insulin stimulates glycolysis by the dephosphorylation (activation) of phosphofructokinase, pyruvate kinase, and pyruvate dehydrogenase as well as the induction of glucokinase, phosphofructokinase, and pyruvate kinase. Likewise, lipogenic enzymes are stimulated and the activity of gluconeogenic enzymes is reduced when the insulin to glucagon ratio is high. The intracellular hormone-sensitive lipoprotein lipase is inactive in adipose tissue in the dephosphorylated state (high insulin to glucagon ratio), thus preventing hydrolysis of triglyceride stores. All these effects are reversed with the lowering of the insulin to glucagon ratio in the postabsorptive phase.

Cell Volume. Most cells have developed mechanisms which enable them to counteract changes in cell volume by the activation of ion transport systems. In recent years evidence has accumulated that changes in cell volume influence various metabolic pathways, i.e., that cell volume acts as a signal-transducing factor [1,60,61]. This relationship has been extensively studied in liver cells, but it may also hold true for other cells. In hepatocytes insulin enhances the cell volume, while glucagon and cAMP cause the cell to shrink. Insulin activates the Na^+/H^+ exchanger, the loop diuretic-sensitive $Na^+2Cl^-K^+$ cotransport system, and the (Na^++K^+)-ATPase. This results in the intracellular accumulation of K^+, Na^+, and Cl^- and thus cell swelling. Conversely, glucagon and cAMP cause a decrease in the intracellular ion concentration and cell shrinkage by activation of K^+ and Cl^- channels. It has been shown that liver cell swelling inhibits glycogenolysis and proteolysis and stimulates glycogen synthesis, amino acid uptake, and protein synthesis. In contrast, cell shrinkage enhances lactate, pyruvate, and glucose output from liver cells and thus increases glycogenolysis. Likewise, glycogen synthesis is inhibited and protein breakdown is augmented. Thus, cell swelling seems to activate anabolic pathways, while cell shrinkage seems to act as a catabolic signal. From these observations it has been concluded that alterations in cell volume at least partly mediate the well-known effects of insulin and glucagon on metabolism [60,61]. However, further steps in the cascade between changes in cell volume and regulation of metabolism are largely unknown. Insulin receptor activation leads to the stimulation of several serine/threonine kinases, including mitogen-activated protein kinases (MAP kinases), which are assumed to be involved in the stimulation of glycogen synthesis and inhibition of glycogenolysis. Direct involvement of cell swelling in the activation of MAP kinases has been suggested [1]. As mentioned above, it has been shown that insulin inhibits proteolysis in liver cells (via cell swelling), and it is thought that cell swelling alkalizes the lysosomal pH in hepatocytes [146]. Since proteolysis is controlled by the lysosomal pH, these observations indicate that the antiproteolytic effect of insulin and cell swelling might be mediated by alkalization of lysosomes.

66.7 Diabetes Mellitus

Diabetes mellitus is characterized by the clinical syndrome of an elevated blood glucose concentration. According to the WHO criteria, fasting blood glucose concentrations above 6.7 mmol/l constitute diabetes. This disorder represents a severe public health problem, as up to 5% of the population in Western countries develops a form of this disease. Two types of the disorder can be distinguished which differ in the cause and course of disease, genetic susceptibility, incidence, and pathology.

Several common classifications of the two forms of diabetes mellitus exist which are often used synonymously:

- Juvenile-onset diabetes and maturity-onset diabetes, according to the age of onset
- Insulin-dependent diabetes (IDDM) and non-insulin-dependent diabetes (NIDDM), according to the therapy
- Type I diabetes and type II diabetes, according to the pathology, i.e., the occurrence of antibodies.

Insulin-Dependent Diabetes Mellitus. In IDDM antibodies against islet cells are commonly found at the onset of the disease, and B cells are destroyed by immune cells.

Insulin secretion is subsequently reduced. In most cases the onset of disease occurs below 30 years, and patients are not obese. The clinical symptoms develop within a few weeks; the blood glucose concentration is high and often ketosis occurs. Therapy with insulin is required.

Non-Insulin-Dependent Diabetes Mellitus. In NIDDM no islet cell antibodies are observed. The onset is normally at an age over 30 years. Patients are frequently obese. Insulin secretion is often reduced, but it can also be enhanced, and resistance to insulin by the target organs commonly occurs. The symptoms develop slowly and the blood glucose concentration is usually moderately raised. Ketosis is rare. Treatment with special diets or oral antidiabetic drugs is possible.

Such a strict classification may be useful for practical reasons but it normally oversimplifies the real situation. Characteristics attributed to one of the two types often overlap, and each type of diabetes comprises several subtypes with heterogeneous clinical manifestation.

Secondary Diabetes. Several disorders which are associated with the excess production of insulin-counter-regulating hormones can cause diabetes symptoms. For example, overproduction of growth hormone (acromegaly), glucocorticoids (Cushing disease), and catecholamines (pheochromocytoma) can lead to impaired glucose tolerance or even overt diabetes with a decreased insulin responsiveness of peripheral target tissues. Catecholamines directly inhibit insulin secretion (see Sect. 66.2.8). In some women the metabolic and hormonal changes associated with gestation lead to the occurrence of glucose intolerance or diabetes during pregnancy. Impaired glucose tolerance during pregnancy may be attributed to the enhanced production of human placental lactogen, cortisol, and progesterone (see also Chap. 116).

66.7.1 Consequences of Hyperglycemia

Insulin deficiency is often paralleled by an excess of glucagon. This phenomenon is probably due to the reduced inhibitory action of insulin on A cells. The imbalance of hormone secretion causes hyperglycemia because it results in reduced glucose uptake into the cells and enhanced glycogenolysis and gluconeogenesis. Sustained hyperglycemia leads to the classical symptoms of diabetes mellitus, polyuria and polydipsia. Both are the consequence of osmotic diuresis due to glucose in the urine. Diuresis occurs when the blood glucose concentration is above the renal threshold, which is around 10–14 mmol/l. This causes loss of water and electrolytes in the urine. Hyperglycemia and water loss contribute to an increase in extracellular osmolarity, which in turn causes thirst. Insulin deficiency induces the breakdown of proteins, glycogen, and triglyceride stores. The formation of ketone bodies due to the enhanced breakdown of triglyceride stores is usually linked to IDDM. The mobilization of triglycerides leads to the formation of large amounts of acetyl-CoA which cannot enter the tricarboxylic acid cycle because the concentration of oxaloacetate is not sufficient. In mammals, acetoacetic acid, hydroxybutyric acid, and acetone are formed from acetyl-CoA under these conditions. Accumulation of ketone bodies can lead to severe metabolic acidosis (see Chap. 78). The elimination of ketone bodies in salt form in the urine leads to an additional loss of Na$^+$ and K$^+$.

In the course of the disease, most diabetic patients develop secondary pathologic changes. The long-term complications associated with diabetes mellitus that are most often observed are cardiovascular disease, nephro- and neuropathy, and ophthalmologic problems. Two features related to hyperglycemia particularly contribute to the development of these secondary complications: alterations in sorbitol, myoinositol, and phosphoinositide metabolism and nonenzymatic protein glycosylation [22].

Nonenzymatic Glycosylation. Glucose is known to attach to amino groups in proteins, resulting in the formation of unstable Schiff base adducts (aldimines). These adducts undergo slow chemical rearrangement, a process lasting several weeks. First, more stable ketoamines or "Amadori products" are formed, which are finally converted into "advanced glycosylation end products" (AGE). The AGE are irreversibly attached to proteins. The rate of AGE formation is proportional to the glucose concentration. The products accumulate with time, and their concentration remains elevated after the blood glucose concentration is reduced to normoglycemia. AGE formation has a variety of consequences on the vascular system which explain, at least in part, the main vascular complications associated with diabetes: abnormal leakage of proteins from the circulation and the progressive narrowing of the lumen of small and large vessels. Particularly in the kidney and eye, the occlusion of small vessels has severe consequences. The capillaries in the renal glomerulus and in the retina are "end" capillaries which scarcely communicate with each other. If a capillary is destroyed, the surrounding tissue will become ischemic and die.

The attachment of AGE to proteins in the basement membrane of vessels leads to covalent cross-links between matrix proteins, especially collagen, and between matrix proteins and plasma proteins. These cross-links cause a thickening of the basement membrane, whereby the elasticity of the vessel wall is reduced and the resistance is increased. Cross-linked proteins are more resistant to normal enzymatic degradation, which contributes to wall thickening. AGE induce the secretion of tumor necrosis factor and interleukin-1 from macrophages (see Chap. 6). These growth-stimulating monokines induce matrix and vessel cell proliferation, which also promotes vessel wall thickening. Moreover, AGE-induced cross-links lower the binding affinity of the basement membrane to growth-inhibiting proteoglycans.

Anionic proteoglycans normally help to prevent negatively charged molecules from leaving the vessel lumen, and therefore their loss increases vascular permeability. Abnormal self-assembly of basement membrane components

due to the presence of AGE and the effect of the released monokines on the tightness of the endothelium contribute to the increase in vascular permeability.

Glycosylation also occurs in other proteins, e.g., hemoglobin. Glycohemoglobin is produced by a ketoamine reaction between glucose and the N-terminal amino acids of the β-chains of the hemoglobin molecule. Normally 4%–6% of the major glycohemoglobin HbA$_{1c}$ is formed. Glycohemoglobin is elevated in chronic hyperglycemia. Most importantly, however, it reflects the state of glycemia and thus of effectiveness of metabolic control in diabetics over a period of 8–12 weeks.

Sorbitol and Myoinositol. In tissues in which the enzyme aldose reductase is present, excess glucose leads to the virtually irreversible formation of sorbitol, which cannot be removed from the cells and which thus causes osmotic problems. In the lens cells, swelling due to sorbitol accumulation can promote the development of a cataract.

Neuropathy. Insulin deficiency leads to a reduction in (Na$^+$+K$^+$)-ATPase activity [52,152]. It appears that the uptake of myoinositol into cells is competitively inhibited by glucose because of the structural similarity of both molecules. As a consequence, the synthesis of phosphatidylinositol in the cells and the formation of second messengers such as IP$_3$ and DAG are reduced. Finally, the activity of PKC, which is stimulated by DAG, is diminished. This in turn influences (Na$^+$+K$^+$)-ATPase activity, which is normally stimulated by the activation of PKC. In myelinated nerve fibers, (Na$^+$+K$^+$)-ATPase is aggregated at the Ranvier nodes. Inhibition of (Na$^+$+K$^+$)-ATPase causes the accumulation of Na$^+$ in the area of the nodes and may thus depolarize the nerve fiber and reduce excitability [52]. Simultaneous swelling of the Ranvier node due to Na$^+$ accumulation probably causes the "axoglial disjunction" observed in diabetic patients and animals. Due to this disruption of myelin, Na$^+$ channels migrate to the internodal region, a process that would also contribute to the delay of signal transmission along the nerve fiber.

66.7.2 Genetic Aspects

Although there is overwhelming evidence that IDDM and NIDDM have a genetic component, the genes involved have not yet been identified. Most likely in both disorders the susceptibility to the disease is associated with more than one gene. In predisposed subjects, environmental factors such as stress, exposure to toxic chemical reagents, and infections may contribute to the development of diabetes [145,149].

Studies with identical twins have demonstrated that the probability that both siblings develop the disease is higher for NIDDM than for IDDM. One study with twins in England revealed that 89% of twins with NIDDM were concordant for the disease compared to only 55% of twins with IDDM [122]. It has been confirmed by other studies that NIDDM has a stronger inherited component than IDDM.

Insulin-Dependent Diabetes Mellitus. IDDM is an autoimmune disease which results in a T cell-mediated autoimmune destruction of pancreatic B cells. The HLA gene cluster on the short arm of chromosome 6 constitutes the major histocompatibility complex (MHC). Proteins encoded in this region (class I, II, and III antigens) are involved in immunologic recognition and defense. The MHC class I molecules comprise HLA-A, HLA-B, and HLA-C antigens. Their main function is to present foreign antigens to cytotoxic T cells which destroy them. The MHC class II molecules include the HLA-DP, HLA-DQ, and HLA-DR antigens, each containing genes for at least one α- and one β-chain. Because of the large number of MHC genes and the high degree of polymorphism, there is a very large number of possible combinations within the population and great variability among individuals. However, due to "linkage disequilibrium," certain combinations of MHC antigens occur more often than others. MHC class II molecules present foreign antigens to T-helper cells, which interfere with B lymphocytes to produce antibodies. In autoimmune disease the complex processes involved in recognition and discrimination between self and foreign substances are defective. Thus the immune system treats its own proteins as foreign antigens. Almost all auto-immune diseases are associated with particular HLA types. Susceptibility to IDDM was first found to be related to MHC class I molecules B8 and B15 [104]. Later an association with the MHC class II molecules DR3 and DR4 was discovered. DR3/DR4 heterozygous individuals have been found to have a particular high risk of developing IDDM. In contrast, other DR regions, e.g., DR2, seem to be linked to resistance to IDDM. An even stronger susceptibility and resistance to IDDM has been found to be related to certain genes of the DQ region. Thus, it appears that a distinct combination of HLA genes confers an increased or decreased risk of developing diabetes [12,123,132,141,142]. It has been suggested that HLA region-encoded genes are required, but not sufficient to induce the disease and that other genetic factors are involved. The predominant HLA antigens associated with IDDM vary in different racial groups.

Non-Insulin-Dependent Diabetes Mellitus. Although it is commonly accepted that NIDDM is a genetic disease, the genes involved or genes associated with the susceptibility to the disease are unknown. NIDDM is not linked to specific HLA genes. There is one rare form of NIDDM known as maturity-onset diabetes of the young (MODY) which is inherited in an autosomal dominant mode. There is evidence that MODY is associated with mutations in the glucokinase gene [95,114,138]. However, for the majority of NIDDM patients there is no evidence of dominant inheritance of the disease. Mutations in the insulin gene can lead to reduced binding of the hormone to the receptor or to impaired processing of the prohormone. However, these mutations are rare and account for only a very few cases of NIDDM [136]. Several very rare syndromes of severe insulin resistance are known which are related to different mutations in the insulin receptor [140].

66.7.3 Immunological Aspects of Insulin-Dependent Diabetes Mellitus

Islet Cell Antibodies. In IDDM, cells of the immune system reaching the islet cells have lost the ability to discriminate between self and foreign substances. As a consequence, they induce cytotoxic reactions against islet antigens, finally leading to the destruction of B cells. After acute onset of IDDM, inflammatory infiltrates in or around the pancreatic islets are observed in the majority of patients. Lymphocytic infiltration, mainly by T lymphocytes, is detected shortly after onset, but disappears again with the concomitant destruction of B cells [42]. This suggests that lymphocytic infiltration is specifically directed against B cells. In addition to cellular immune abnormalities, antibodies to islet cell components are commonly observed in IDDM [9,26,59]. Islet cell antibodies (ICA) have been demonstrated to occur with a high prevalence in newly diagnosed patients. About 80% of patients with recent onset of IDDM possess antibodies with cytoplasmic immunoreactivity which specifically bind to islets, while less than 5% of healthy controls are positive. ICA is the original immune marker of recent-onset and preclinical IDDM. ICA titers normally decline with the onset of diabetes and concomitant loss of B cells. Furthermore, ICA have been detected in individuals several years before the development of overt diabetes. This suggests that autoimmunity to the islet cells may precede the acute onset of the disease by several years [26]. However, the target antigens for ICA have not been identified. One target might be a glycolipid containing sialic acid [103]. Other antibodies associated with IDDM other than ICA have been detected [9,26,59].

Autoantibodies Against Insulin. In about 50% of diabetic (IDDM) patients, circulating autoantibodies against insulin (IAA) are observed at or before the onset of diabetes. IAA production is positively correlated to HLA-DR3 and/or HLA-DR4, the same haplotypes of MHC class II antigens that confer an increased risk of developing IDDM [10]. However, because of the relatively low frequency of IAA in IDDM patients, IAA alone is not a reliable serological marker for the disease, but the combined presence of IAA and ICA might enhance the predictive value of ICA.

Glutamic Acid Decarboxylase. Recently, a protein with a molecular mass of 64 kDa was discovered that specifically binds to antibodies from sera of diabetic patients. This protein was not found in tissues other than islets and seems to be B cell specific. Antibodies can be detected in about 80% of patients with newly diagnosed IDDM, and often they can be detected several years before onset of the disease [26]. The human diabetes-associated, 64-kDa antigen has been identified as glutamic acid decarboxylase (GAD) [13], an enzyme involved in the synthesis of the neurotransmitter GABA. Apart from its expression in GABAergic neurons of the central nervous system, the enzyme is located in pancreatic B cells. It is thought to play a role in the glucose-mediated regulation of hormone secretion from A cells [126]. However, further characterization of the 64-kDa B cell antigen and knowledge about its interaction with immune cells are needed to elucidate the role of GAD immunity in the pathology of IDDM.

Antibodies Against Other Tissues. In addition to the islet-specific antibodies, a variety of antibodies directed against antigens in other, nonpancreatic tissue have been detected in diabetic patients [34]. This could at least partly be associated with the fact that diabetic patients have a higher risk of developing a further autoimmune disease than control individuals.

Interleukin-1 and Nitric Oxide. During B cell injury the secretion of cytokines with cytotoxic and antiviral activity appears to be elevated. It has been shown that cytokines such as interleukin-1 exert cytotoxic effects on B cells [92]. In animals, oxygen free radical scavengers were found to protect islets against damage by cytokines [137]. Moreover, cytokines have been shown to stimulate free radical production in macrophages. Nitric oxide (NO) is one possible candidate involved in the destructive action of cytokines, since the deleterious effects of interleukin-1β on B cells are suppressed by inhibitors of NO synthase. Pancreatic B cells possess an extremely low capacity to react against oxidative stress due to a low activity of endogenous free radical scavengers [89]. Therefore, oxidative stress has been proposed as an alternative mechanism leading to the destruction of B cells in diabetes [11,30].

66.7.4 Pathophysiological Considerations in Non-Insulin-Dependent Diabetes Mellitus

NIDDM is characterized by impaired insulin secretion on the one hand and resistance to insulin by the target organs on the other. However, the sequence of events is still a matter of debate: low responsiveness of the B cells to glucose could be the primary lesion which precedes glucose intolerance, or insulin resistance could be the first event which causes insulin hypersecretion and subsequently progressive fatigue of the B cells with reduced hormone secretion. Either effect can probably be primary, but neither is sufficient to induce the disease.

Disturbance of Insulin Release. In NIDDM, in contrast to IDDM, no severe morphological changes of the islets cells are observed which can account for decreased insulin release. Patients with NIDDM retain the ability to synthesize, store, and secrete insulin. Interestingly, the response to secretagogues other than glucose, e.g., arginine, is normal. However, there are several striking changes in the glucose-induced pattern of hormone secretion. In NIDDM patients, the first phase of insulin secretion is reduced [120] and the regular oscillatory pattern of release is disturbed [108]. These observations seem to be a consequence of the impaired ability of the B cells to sense glucose, but it is not clear how they are linked to the development of diabetes. Chronic hyperglycemia itself may reduce the ability of B cells to respond to an acute hyperglycemic challenge. In vivo and in vitro experiments have proved that longer periods of hyperglycemia reduce the response to glucose, a phenomenon known as "glucose toxicity." This desensitization also occurs when B cells are exposed to nonglucose stimulators which cause hypersecretion. This suggests that a possible intracellular potentiating mechanism of insulin secretion is impaired. It appears that product(s) of the phosphoinositol pathway might be involved in B cell desensitization [54].

Insulin Resistance. The second classical symptom in NIDDM is insulin resistance by the insulin-dependent target organs, muscle and adipose tissue. It is assumed that mainly effects secondary to insulin receptor binding are responsible for insulin resistance in NIDDM [31]. As a consequence, most likely steps in glucose uptake as well as in glucose metabolism are disturbed and contribute to enhanced blood glucose concentration. Diminished insulin receptor tyrosine kinase activity, decreased glucose transport, impaired glycogen synthase activity, and reduced pyruvate dehydrogenase activity are thought to be possible reasons.

Obesity. The most common situation in which insulin resistance and insulin hypersecretion occur is obesity. However, two thirds of these individuals remain normoglycemic despite increased hormone secretion. On the other hand, up to 85% of NIDDM patients are obese. Thus, in individuals with impaired B cell function (perhaps due to genetic factors), the development of insulin resistance may have much more dramatic consequences on the blood glucose concentration than in subjects with normal rates of hormone secretion. In these predisposed individuals, the increased insulin release that results from insulin resistance leads to a progressive exhaustion of the B cells with reduced hormone release.

References

1. Agius L, Peak M, Beresford G, Al-Habori M, Thomas TH (1994) The role of ion content and cell volume in insulin action. Biochem Soc Trans 22:516–532
2. Ahrén B, Taborsky GJ, Porte D (1986) Neuropeptidergic versus cholinergic and adrenergic regulation of islet hormone secretion. Diabetologia 29:827–836
3. Ämmälä C, Ashcroft FM, Rorsman P (1993) Calcium-independent potentiation of insulin release by cyclic AMP in single β-cells. Nature 363:356–358
4. Ashcroft FM, Harrison DE, Ashcroft SJH (1984) Glucose induces closure of single potassium channels in isolated rat pancreatic B-cells. Nature 312:446–448
5. Ashcroft FM, Rorsman P (1989) Electrophysiology of the pancreatic β-cell. Prog Biophys Mol Biol 54:87–143
6. Ashcroft FM, Kakei M, Gibson J, Gray D, Sutton R (1989) The ATP- and tolbutamide-sensitivity of the ATP-sensitive K-channel from human pancreatic islet cells. Diabetologia 31:591–598
7. Ashcroft FM, Williams B, Smith PA, Fewtrell CMS (1992) Ion channels involved in the regulation of nutrient-stimulated insulin secretion. In: Flatt PR (ed) Nutrient regulation of insulin secretion. Portland, London, pp 193–212

8. Ashcroft FM, Ashcroft SJH (1992) Mechanism of insulin secretion. In: Ashcroft FM, Ashcroft SJH (eds) Insulin molecular biology to pathology. Oxford University Press, Oxford, pp 97–150

9. Atkinson MA, Maclaren NK (1993) Islet cell autoantigens in insulin-dependent diabetes. J Clin Invest 92:1608–1616

10. Atkinson MA, Maclaren NK, Riley WG, Winter WE, Fish DD, Spillar RP (1986) Are insulin autoantibodies markers for insulin-dependent diabetes mellitus? Diabetes 35:894–898

11. Baynes JW (1991) Role of oxidative stress in development of complications in diabetes. Diabetes 40:405–412

12. Baisch JM, Weeks T, Giles R, Hoover M, Stastny P, Capra JD (1990) Analysis of HLA-DQ genotypes and susceptibility in insulin-dependent diabetes mellitus. N Engl J Med 322:1836–1882

13. Beakkeskov S, Aanstoot HK, Christgau S, Reetz A, Solimena M, Cascalho M, Folli F, Richter-Olesen H, De Camilli P (1990) Identification of the 64k autoantigen in insulin-dependent diabetes as the GABA-synthesizing enzyme glutamic acid decarboxylase. Nature 347:151–156

14. Bedoya FJ, Matschinsky FM, Shimizu T, O'Neil JJ, Appel MC (1986) Differential regulation of glucokinase activity in pancreatic islets and liver of the rat. J Biol Chem 261:10760–10764

15. Berggren P-O, Rorsman P, Efendic S, Östenson C-G, Flatt PR, Nilsson T, Arkhammar P, Juntti-Berggren L (1992) Mechanisms of action of entero-insular hormones, islet peptides and neural input on the insulin secretory process. In: Flatt PR (ed) Nutrient regulation of insulin secretion. Portland, London, pp 289–318

16. Bergsten P, Hellman B (1993) Glucose-induced cycles of insulin release can be resolved into distinct periods of secretory activity. Biochem Biophys Res Commun 192:1182–1188

17. Bergstrom RW, Fujimoto WY, Teller DC, de Haën C (1989) Oscillatory insulin secretion in perifused isolated rat islets. Am J Physiol 257:E479–E485

18. Berridge MJ, Irvine RF (1989) Inositol phosphates and cell signalling. Nature 341:197–205

19. Best L, Malaisse WJ (1983) Stimulation of phosphoinositide breakdown in rat pancreatic islets by glucose and carbamylcholine. Biochem Biophys Res Commun 116:9–16

20. Best L, Dunlop M, Malaisse WJ (1984) Phospholipid metabolism in pancreatic islets. Experientia 40:1085–1091

21. Best L, Yates AP, Tomlinson S (1992) Stimulation of insulin secretion by glucose in the absence of diminished potassium ($^{86}Rb^+$) permeability. Biochem Pharmacol 43:2483–2485

22. Brownlee M, Cerami A, Vlassara H (1988) Advanced glycosylation end products and the biochemical basis of diabetic complications. N Engl J Med 318:1315–1321

23. Chance RE, Ellis RM, Bromer WW (1968) Porcine proinsulin: characterization and amino acid sequence. Science 161:165–167

24. Chen L, Alam T, Johnson JH, Hughes S, Newgard CB, Unger RH (1990) Regulation of B-cell glucose transporter gene expression. Proc Natl Acad Sci U S A 87:4088–4092

25. Christie MC, Ashcroft SJH (1985) Substrates for cyclic AMP-dependent protein kinase in islets of Langerhans. Biochem J 227:727–736

26. Christie MR (1992) Aetiology of type I diabetes: immunological aspects. In: Ashcroft FM, Ashcroft SJH (eds) Insulin molecular biology to pathology. Oxford University Press, Oxford, pp 306–346

27. Colca JR, Brooks CL, Landt M, McDaniel ML (1983) Correlation of Ca^{2+} and calmodulin-dependent protein kinase activity with secretion of insulin from islets of Langerhans. Biochem J 212:819–827

28. Cook DL, Hales N (1984) Intracellular ATP directly blocks K^+ channels in pancreatic B-cells. Nature 311:271–273

29. Cook DL, Satin LS, Hopkins WF (1991) Pancreatic B-cells are bursting, but how? Trends Neurosci 14:411–414

30. Corbett JA, McDaniel ML (1992) Does nitric oxide mediate autoimmune destruction of β-cells? Diabetes 41:897–903

31. DeFronzo RA, Bonadonna RC, Ferrannini E (1992) Pathogenesis of NIDDM. Diabetes Care 15:318–368

32. Denton RM, Tavaré JM (1992) Mechanisms whereby insulin may regulate intracellular events. In: Ashcroft FM, Ashcroft SJH (eds) Insulin molecular biology to pathology. Oxford University Press, Oxford, pp 235–262

33. Docherty K, Steiner DF (1982) Posttranslational proteolysis in polypeptide hormone biosynthesis. Annu Rev Physiol 44:625–638

34. Drell DW, Notkins AL (1987) Multiple immunological abnormalities in patients with type I (insulin-dependent) diabetes mellitus. Diabetologia 30:132–143

35. Drews G, Debuyser A, Nenquin M, Henquin JC (1990) Galanin and epinephrine act on distinct receptors to inhibit insulin release by the same mechanisms including an increase in K^+ permeability of the B-cell membrane. Endocrinology 126:1646–1653

36. Dunning BE, Taborsky GJ (1988) Galanin-sympathetic neurotransmitter in endocrine pancreas? Diabetes 37:1157–1162

37. Ebert R, Creutzfeldt W (1987) Gastrointestinal peptides and insulin secretion. Diabetes Metab Rev 3:1–26

38. Eddlestone GT, Oldham SB, Lipson LG, Premdas FH, Beigelman PM (1985) Electrical activity, cAMP concentration and insulin release in mouse islets of Langerhans. Am J Physiol 248:C145–C153

39. Gammeltoft S (1984) Insulin receptors: binding kinetics and structure-function relationship of insulin. Physiol Rev 64:1321–1378

40. Gao Z-Y, Drews G, Nenquin M, Plant TD, Henquin JC (1990) Mechanisms of the stimulation of insulin release by arginine-vasopressin in normal mouse islets. J Biol Chem 265:15724–15730

41. Gao Z-Y, Drews G, Henquin JC (1991) Mechanisms of the stimulation of insulin release by oxytocin in normal mouse islets. Biochem J 276:169–174

42. Gebts W, De Mey J (1978) Islet cell survival determined by morphology: an immunocytochemical study of the islets of Langerhans in juvenile diabetes mellitus. Diabetes 27 [Suppl 1]:251–261

43. Gembal M, Gilon P, Henquin JC (1992) Evidence that glucose can control insulin release independently from its action on ATP-sensitive K^+ channels in mouse B-cells. J Clin Invest 89:1288–1295

44. Gembal M, Detimary P, Gilon P, Gao Z-Y, Henquin JC (1993) Mechanisms by which glucose can control insulin release independently from its action on adenosine triphosphate-sensitive K^+ channels in mouse B cells. J Clin Invest 91:871–880

45. German MS (1993) Glucose sensing in pancreatic islet beta cells: the key role of glucokinase and the glycolytic intermediates. Proc Natl Acad Sci U S A 90:1781–1785

46. Gillis KD, Misler S (1993) Enhancers of cytosolic cAMP augment depolarization-induced exocytosis from pancreatic B-cells: evidence for effects distal to Ca^{2+} entry. Pflugers Arch 424:195–197

47. Gilon P, Henquin JC (1992) Influence of membrane potential changes on cytoplasmic Ca^{2+} concentration in an electrically excitable cell, the insulin-secreting pancreatic B-cell. J Biol Chem 267:20713–20720

48. Gilon P, Shepherd RM, Henquin JC (1993) Oscillations of secretion driven by oscillations of cytoplasmic Ca^{2+} as evidenced in single pancreatic islets. J Biol Chem 268:22265–22268

49. Goodner CJ, Walike BC, Koerker DJ, Ensinck JE, Brown AC, Chideckel EW, Palmer J, Kalnasy L (1977) Insulin, glucagon, and glucose exhibit synchronous sustained oscillations in fasting monkeys. Science 195:177–179

50. Grapengiesser E, Gylfe E, Hellman B (1988) Glucose-induced oscillations of cytoplasmic Ca^{2+} in the pancreatic β-cells. Biochem Biophys Res Commun 151:1299–1304

51. Grapengiesser E, Gylfe E, Hellman B (1989) Three types of cytoplasmic Ca^{2+} oscillations in stimulated pancreatic β-cells. Arch Biochem Biophys 268:404–407

52. Greene DA, Lattimer SA, Sima AAF (1987) Sorbitol, phosphoinositides, and sodium-potassium-ATPase in the pathogenesis of diabetic complications. N Engl J Med 316:599–606

53. Grodsky GM (1989) A new phase of insulin secretion. How will it contribute to our understanding of β-cell function? Diabetes 38:673–678

54. Grodsky GM, Bolaffi JL (1992) Desensitization of the insulin-secreting beta cell. J Cell Biochem 48:3–11

55. Guest PC, Hutton JC (1992) Biosynthesis of insulin secretory granule proteins. In: Flatt PR (ed) Nutrient regulation of insulin secretion. Portland, London, pp 59–82

56. Gylfe E (1988) Glucose-induced early changes in cytoplasmic calcium of pancreatic β-cells studied with time-sharing dual-wavelength fluorometry. J Biol Chem 263:5044–5048

57. Gylfe E (1988) Nutrient secretagogues induce bimodal early changes in cytoplasmic calcium of insulin-releasing ob/ob mouse β-cells. J Biol Chem 263:13750–13754

58. Harrison DE, Ashcroft SJH, Christie MC, Lord JM (1984) Protein phosphorylation in the pancreatic B-cell. Experientia 40:1075–1084

59. Harrison LC (1992) Islet cell antigens in insulin-dependent diabetes: Pandora's box revisited. Immunol Today 13:348–352

60. Häussinger D, Lang F (1991) Cell volume in the regulation of hepatic function: a mechanism for metabolic control. Biochem Biophys Acta 1071:331–350

61. Häussinger D, Lang F (1991) Cell volume – a "second messenger" in the regulation of metabolism by amino acids and hormones. Cell Physiol Biochem 1:121–130

62. Hedeskov CJ (1980) Mechanism of glucose-induced insulin secretion. Physiol Rev 60:442–509

63. Hellman B (1986) Calcium transport in pancreatic β-cells: implications for glucose regulation of insulin release. Diabetes Metab Rev 2:215–241

64. Hellman B, Gylfe E, Grapengiesser E, Lund P-E, Marcström A (1992) Cytoplasmic calcium and insulin secretion. In: Flatt PR (ed) Nutrient regulation of insulin secretion. Portland, London, pp 213–246

65. Henquin JC, Meissner HP (1984) Significance of ionic fluxes and changes in membrane potential for stimulus-secretion coupling in pancreatic B-cells. Experientia 40:1043–1052

66. Henquin JC, Meissner HP (1984) The ionic, electrical, and secretory effects of endogenous cyclic adenosine monophosphate in mouse pancreatic B cells: studies with forskolin. Endocrinology 115:1125–1134

67. Henquin JC (1987) Regulation of insulin release by ionic and electrical events in B cells. Horm Res 27:168–178

68. Henquin JC, Bozem M, Schmeer W, Nenquin M (1987) Distinct mechanisms for two amplification systems of insulin release. Biochem J 246:393–399

69. Henquin JC, Garcia M-C, Bozem M, Hermans MP, Nenquin M (1988) Muscarinic control of pancreatic B cell function involves sodium-dependent depolarization and calcium influx. Endocrinology 122:2134–2142

70. Henquin JC (1990) Role of voltage- and Ca^{2+}-dependent K^+ channels in the control of glucose-induced electrical activity in pancreatic B-cells. Pflugers Arch 416:568–572

71. Henquin JC (1990) Les mécanismes de contrôle de la sécrétion d'insuline. Arch Int Physiol Biochim 98:A61-A80

72. Houslay MD (1986) Insulin, glucagon and the receptor-mediated control of cyclic AMP concentrations in liver. Biochem Soc Trans 14:183–193

73. Howell SL, Bird GSJ (1989) Biosynthesis and secretion of insulin. Br Med Bull 45:19–36

74. Howell SL (1984) The mechanism of insulin secretion. Diabetologia 26:319–327

75. Hughes SJ, Christie MR, Ashcroft SJH (1987) Potentiators of insulin secretion modulate Ca^{2+} sensitivity in rat pancreatic islets. Mol Cell Endocrinol 50:231–236

76. Hughes SJ, Ashcroft SJH (1992) Cyclic AMP, protein phosphorylation and insulin secretion. In: Flatt PR (ed) Nutrient regulation of insulin secretion. Portland, London, pp 271–288

77. Hutton JC (1984) Secretory granules. Experientia 40:1091–1097

78. Hutton JC (1989) The insulin secretory granule. Diabetologia 32:271–281

79. Johnson JH, Newgard CB, Milburn JL, Lodish HF, Thorens B (1990) The high K_m glucose transporter of islet of Langerhans is functionally similar to the low affinity transporter of liver and has an identical primary sequence. J Biol Chem 265:6548–6551

80. Jones PM, Stutchfield J, Howell SL (1985) Effects of Ca^{2+} and a phorbol ester on insulin secretion from islets of Langerhans permeabilized by high-voltage discharge. FEBS Lett 191:102–106

81. Jones PM, Fyles JM, Howell SL (1986) Regulation of insulin secretion by cAMP in rat islets of Langerhans permeabilized by high-voltage discharge. FEBS Lett 205:205–209

82. Kukuljan M, Goncalves A, Atwater I (1991) Charybdotoxin-sensitive $K_{(Ca)}$ channel is not involved in glucose-induced electrical activity in pancreatic β-cells. J Membr Biol 119:187–195

83. Lacy PE, Howell SL, Young DA, Fink CJ (1968) A new hypothesis of insulin secretion. Nature 219:1177–1179

84. Lang DA, Matthews DR, Peto PJ, Turner RC (1979) Cyclic oscillations of basal plasma glucose and insulin concentrations in human beings. N Engl J Med 301:1023–1027

85. Langerhans P (1869) Beiträge zur mikroskopischen Anatomie der Bauchspeicheldrüse. Thesis, Wilhelm-Friedrich University, Berlin

86. Lefèbvre PJ, Paolisso G, Scheen AJ, Henquin J-C (1987) Pulsatility of insulin and glucagon release: physiological significance and pharmacological implications. Diabetologia 30:443–452

87. Lenzen S, Panten U (1988) Signal recognition by pancreatic B-cells. Biochem Pharmacol 37:371–378

88. Lienhard GE (1983) Regulation of cellular membrane transport by the exocytotic insertion and endocytotic retrieval of transporters. TIBS 8:125–127

89. Malaisse WJ, Malaisse-Lagae F, Sener A, Pipeleers DG (1982) Determinants of the selective cytotoxicity of alloxan to the pancreatic B cell. Proc Natl Acad Sci U S A 79:927–930

90. Malaisse WJ, Malaisse-Lagae F (1984) The role of cyclic AMP in insulin release. Experientia 40:1068–1075

91. Malaisse WJ, Sener A (1987) Glucose-induced changes in cytosolic ATP content in pancreatic islets. Biochim Biophys Acta 927:190–195

92. Mandrup-Poulsen T, Spinas GA, Prowse SJ, Hansen BS, Jørgensen DW, Bendtzen K, Nielsen JH, Nerup J (1987) Islet cytotoxity of interleukin 1. Influence of culture conditions and islet donor characteristics. Diabetes 36:641–647

93. Marks V, Samols E, Stagner G (1992) Intra-islet interactions. In: Flatt PR (ed) Nutrient regulation of insulin secretion. Portland, London, pp 41–57

94. Matschinsky FM, Ellerman JE (1968) Metabolism of glucose in the islets of Langerhans. J Biol Chem 243:2730–2736

95. Matschinsky F, Liang Y, Kesavan P, Wang L, Froguel P, Velho G, Cohen D, Permutt MA, Tanizawa Y, Jetton TL, Niswender K, Magnuson MA (1993) Glucokinase as pancreatic B cell glucose sensor and Diabetes gene. J Clin Invest 92:2092–2098

96. Meglasson MD, Matschinsky FM (1984) New perspectives on pancreatic islet glucokinase. Am J Physiol 246:E1–E13

97. Meissner HP, Preissler M (1979) Glucose-induced changes of the membrane potential of pancreatic B-cells: their significance for the regulation of insulin release. In: Rafael A,

Camerini-Davalos A, Hanover B (eds) Treatment of early diabetes. Plenum, New York, pp 97–107

98. Mering J, Minkowski O (1890) Diabetes mellitus nach Pankreasextirpation. Arch Exp Pathol Pharmakol 26:371–387

99. Misler S, Falke LC, Gillis K, McDaniel ML (1986) A metabolite-regulated potassium channel in rat pancreatic B cells. Proc Natl Acad Sci U S A 83:7119–7123

100. Misler S, Barnett DW, Pressel DM, Gillis KD, Scharp DW, Falke LC (1992) Stimulus-secretion coupling in β-cells of transplantable human islets of Langerhans. Evidence for a critical role for Ca²⁺ entry. Diabetes 41:662–670

101. Misler S, Barnett DW, Gillis KD, Pressel DM (1992) Electrophysiology of stimulus-secretion coupling in human β-cells. Diabetes 41:1221–1228

102. Morgan NG, Montague W (1992) Phospholipids and insulin secretion. In: Flatt PR (ed) Nutrient regulation of insulin secretion. Portland, London, pp 125–155

103. Nayak RC, Omar MAK, Rabizadeh A, Srikanta S, Eisenbarth GS (1985) Cytoplasmic islet cell antibodies. Evidence that the target antigen is a sialoglycoconjugate. Diabetes 34:617–619

104. Nerup J, Platz P, Anderssen OO (1974) HL-A antigens and diabetes mellitus. Lancet ii:864–866

105. Nishizuka Y (1984) The role of protein kinase C in cell surface signal transduction and tumour promotion. Nature 308:693–698

106. O'Connor MDL, Landahl H, Grodsky GM (1980) Comparison of storage- and signal-limited models of pancreatic insulin secretion. Am J Physiol 238:R378–R389

107. Opara EC, Atwater I, Go VLW (1988) Characterization and control of pulsatile secretion of insulin and glucagon. Pancreas 3:484–487

108. O'Rahilly S, Turner RC, Matthews DR (1988) Impaired pulsatile secretion of insulin in relatives of patients with non-insulin-dependent diabetes. N Engl J Med 318:1225–1230

109. Orci L (1982) Banting lecture 1981. Macro- and micro-domains in the endocrine pancreas. Diabetes 31:538–565

110. Orci L, Ravazzola M, Amherdt M, Madsen O, Vassalli J-D, Perrelet A (1985) Direct identification of prohormone conversion site in insulin-secreting cells. Cell 42:671–681

111. Orci L, Ravazzola M, Amherdt M, Madsen O, Perrelet A, Vassalli J-D, Anderson RGW (1986) Conversion of proinsulin to insulin occurs coordinately with acidification of maturing secretory vesicles. J Cell Biol 103:2273–2281

112. Ørskov C (1992) Glucagon-like peptide-1, a new hormone of the entero-insular axis. Diabetologia 35:701–711

113. Pace CS, Murphy M, Conant S, Lacy PE (1977) Somatostatin inhibition of glucose-induced electrical activity in cultured rat islet cells. Am J Physiol 233:C164–C171

114. Permutt MA, Chiu KC, Tanizawa Y (1992) Glucokinase and NIDDM. A candidate gene that paid off. Diabetes 41:1367–1372

115. Persaud SJ, Jones PM, Howell SL (1992) The role of protein kinase C in insulin secretion. In: Flatt PR (ed) Nutrient regulation of insulin secretion. Portland, London, pp 247–269

116. Peter-Riesch B, Fathi M, Schlegel W, Wollheim CB (1988) Glucose and carbachol generate 1,2-diacylglycerols by different mechanisms in pancreatic islets. J Clin Invest 81:1154–1161

117. Pipeleers DG, Schuit FC, Van Schravendijk CFH, Van De Winkel M (1985) Interplay of nutrients and hormones in the regulation of glucagon release. Endocrinology 117:817–823

118. Pipeleers DG, Schuit FC, In't Veld PA, Meas E, Hooghe-Peters EL, Van de Winkel M, Gebts W (1985) Interplay of nutrients and hormones in the regulation of insulin release. Endocrinology 117:824–833

119. Pipeleers D, Kiekens R, In't Veld P (1992) Morphology of the pancreatic B-cell. In: Ashcroft FM, Ashcroft SJH (eds) Insulin molecular biology to pathology. Oxford University Press, Oxford, pp 5–31

120. Porte D (1991) Banting Lecture 1990. β-Cells in type II diabetes mellitus. Diabetes 40:166–180

121. Prentki M, Matschinsky FM (1987) Ca²⁺, cAMP, and phopholipid-derived messengers in coupling mechanisms of insulin secretion. Physiol Rev 67:1185–1248

122. Pyke DA (1979) Diabetes: the genetic connections. Diabetologia 17:333–343

123. Risch N (1989) Genetics of IDDM: evidence for complex inheritance with HLA. Genet Epidemiol 6:143–148

124. Rorsman P, Trube G (1985) Glucose dependent K⁺-channels in pancreatic B-cells are regulated by intracellular ATP. Pflugers Arch 405:305–309

125. Rorsman P, Hellman B (1988) Voltage-activated currents in guinea pig pancreatic α2 cells. Evidence for Ca²⁺-dependent action potentials. J Gen Physiol 91:223–242

126. Rorsman P, Berggren P-O, Bokvist K, Ericson H, Möhler H, Östenson C-G, Smith PA (1989) Glucose-inhibition of glucagon secretion involves activation of GABAₐ-receptor chloride channels. Nature 341:233–236

127. Rorsman P, Bokvist K, Ämmälä C, Arkhammar P, Berggren P-O, Larsson O, Wåhlander K (1991) Activation by adrenaline of a low-conductance G protein-dependent K⁺ channel in mouse pancreatic B cells. Nature 349:77–79

128. Santos RM, Rosario LM, Nadel A, Garcia-Sancho J, Soria B, Valdeolmillos M (1989) Widespread synchronous [Ca²⁺]ᵢ oscillations due to bursting electrical activity in single pancreatic islets. Pflugers Arch 418:417–422

129. Sato Y, Aizawa T, Komatsu M, Okada N, Yamada T (1992) Dual functional role of membrane depolarization/Ca²⁺ influx in rat pancreatic B-cell. Diabetes 41:438–443

130. Schuit FC, Pipeleers DG (1985) Regulation of adenosine 3',5'-monophosphate levels in the pancreatic B-cell. Endocrinology 117:834–840

131. Schuit FC, Pipeleers DG (1986) Differences in adrenergic recognition by pancreatic A and B cells. Science 232:875–877

132. Sheehy MJ, Scharf SJ, Rowe JR, Neme de Gimenez MH, Meske LM, Erlich HA, Nepom BS (1989) A diabetes-susceptible HLA haplotype is best defined by a combination of HLA-DR and -DQ alleles. J Clin Invest 83:830–835

133. Siddle K (1992) The insulin receptor. In: Ashcroft FM, Ashcroft SJH (eds) Insulin molecular biology to pathology. Oxford University Press, Oxford, pp 191–234

134. Smith PA, Bokvist K, Arkhammar P, Berggren P-O, Rorsman P (1990) Delayed rectifying and calcium-activated K⁺ channels and their significance for action potential repolarization in mouse pancreatic β-cells. J Gen Physiol 95:1041–1059

135. Stagner JI, Samols E (1992) The vascular order of islet cellular perfusion in the human pancreas. Diabetes 41:93–97

136. Steiner DF, Tager HS, Chan SJ, Nanjo K, Sanke T, Rubenstein AH (1990) Lessons learned from molecular biology of insulin-gene mutations. Diabetes Care 13:600–609

137. Sumoski W, Baquerizo H, Rabinovitch A (1989) Oxygen free radical scavengers protect rat islet cells from damage by cytokines. Diabetologia 32:792–796

138. Sun F, Knebelmann B, Pueyo ME, Zouali H, Lesage S, Vaxillaire M, Passa P, Cohen D, Velho G, Antignac C, Froguel P (1993) Deletion of the donor splice site of intron 4 in the glucokinase gene causes maturity-onset diabetes of the young. J Clin Invest 92:1174–1180

139. Tamagawa T, Niki H, Niki A (1985) Insulin release independent of a rise in cytosolic free Ca²⁺ by forskolin and phorbol ester. FEBS Lett 183:430–432

140. Taylor SI (1992) Lilly lecture: Molecular mechanism of insulin resistance. Lessons from patients with mutations in the insulin-receptor gene. Diabetes 41:1473–1490

141. Thorsby E, Rønningen KS (1993) Particular HLA-DQ molecules play a dominant role in determing susceptibility or resistance to type 1 (insulin-dependent) diabetes mellitus. Diabetologia 36:371–377

142. Todd JA, Bell JI, McDevitt HO (1987) HLA DQ β gene contributes to susceptibility and resistance to insulin-dependent diabetes mellitus. Nature 329:599–604

143. Trimble ER, Hallban PA, Wollheim CB, Renold AE (1982) Functional differences between rat islets of dorsal and ventral pancreatic origin. J Clin Invest 69:405–413

144. Trube G, Rorsman P, Ohno-Shosaku T (1986) Opposite effects of tolbutamide and diazoxide on the ATP-dependent K$^+$ channel in mouse pancreatic β-cells. Pflugers Arch 407:493–499

145. Turner R, Neil A (1992) Introduction to diabetes. In: Ashcroft FM, Ashcroft SJH (eds) Insulin molecular biology to pathology. Oxford University Press, Oxford, pp 268–284

146. Völkl H, Busch G, Häussinger D, Lang F (1994) Alkalinization of acidic cellular compartments following cell swelling. FEBS Lett 338:27–30

147. Wahl MA, Plehn RJ, Landsbeck EA, Verspohl EJ, Ammon HPT (1991) Gastrin releasing peptide augments glucose mediated $^{45}Ca^{2+}$ uptake, electrical activity, and insulin secretion of mouse pancreatic islets. Endocrinology 128:3247–3252

148. Wahl MA, Straub SG, Ammon HPT (1993) Vasoactive intestinal polypeptide-augmented insulin release: actions on ionic fluxes and electrical activity of mouse islets. Diabetologia 36:920–925

149. Wassmuth R, Kockum I, Karlsen A, Hagopian W, Bärmeier H, Dube S, Lernmark Å (1992) Aetiology of type I diabetes: genetic aspects. In: Ashcroft FM, Ashcroft SJH (eds) Insulin molecular biology to pathology. Oxford University Press, Oxford, pp 285–305

150. Watkins DT, Cooperstein SJ (1983) Role of calcium and calmodulin in the interaction between islet cell secretion granules and plasma membranes. Endocrinology 112:766–768

151. Weir GC, Bonner-Weir S (1990) Islets of Langerhans: the puzzle of intraislet interactions and their relevance to diabetes. J Clin Invest 85:983–987

152. Winegrad AI (1987) Banting Lecture 1986. Does a common mechanism induce the diverse complications of diabetes? Diabetes 36:396–406

153. Wollheim CB, Sharp GWG (1981) Regulation of insulin release by calcium. Physiol Rev 61:914–973

154. Yasuda K, Yamada Y, Inagaki N, Yano H, Okamoto Y, Tsuji K, Fukumoto H, Imura H, Seino S, Seino Y (1992) Expression of GLUT1 and GLUT2 glucose transporter isoforms in rat islets of Langerhans and their regulation by glucose. Diabetes 41:76–81

155. Zawalich WS, Takuwa N, Takuwa Y, Diaz VA, Rasmussen H (1987) Interactions of cholecystokinin and glucose in rat pancreatic islets. Diabetes 36:426–433

67 Physiological Functions of the Liver

D. Häussinger

Contents

67.1 General Considerations and Hepatic Architecture

The liver has a strategically important position in the circulation. It is the first organ that comes in contact with the blood after its exposure to the intestine. This implies not only that the liver faces absorbed nutrients, xenobiotics, toxins and gut-derived microorganisms, which need to be processed before entering the systemic circulation, first, but also suggests a hepatic role in the excretion of compounds into the intestinal lumen, in addition to the sensing and processing of gut-derived signals that then may trigger the function of other organs. Thus, in a wide sense the liver acts as a preparation plant for both hardware and software delivery to other organs.

67.1.1 Parenchymal and Non-parenchymal Cells

In a wide sense, the term "hepatocyte" means all resident liver cells, but frequently the term "hepatocyte" is used to mean liver parenchymal cells only. This latter convention will also be used in the present chapter. In functional terms, the liver is composed of microcirculatory units, the so-called acini or metabolic lobuli [63]. Although the acinar and lobular concepts predict different geometric shapes [43] of these units, this difference is not really relevant for functional purposes. Different cell types are arranged in a sophisticated way in these functional units, which extend from an afferent terminal portal venule along the sinusoids to an efferent hepatic venule. These units can be seen as tubes formed by parenchymal cells (about 20–30 cells long), whose inner wall is coated with endothelial and Kupffer cells separating the sinusoidal space from the perisinusoidal space of Dissé (Fig. 67.1). This perisinusoidal space is freely accessible to blood plasma but not to erythrocytes, and harbors fat-storing cells (ITO cells) and Pit cells. Depending on their anatomical position parenchymal cells differ in their enzyme equipment and metabolic function, despite their similar light microscopic appearance (so-called functional hepatocyte heterogeneity or metabolic zonation; see Chap. 68). Parenchmyal liver cells constitute about 60%, Kupffer cells 25–30%, endothelial cells about 10%, Pit and ITO cells less than 5% of all liver cells.

67.1.2 Liver Hemodynamics

The liver receives blood by way of the portal vein (75–80%) and the hepatic artery (20–25%). In man, hepatic blood flow is about 1500 ml/min; i.e., one quarter of cardiac output. Portal venous blood is derived from the intestine (75%) and the spleen (25%). Portal blood flow is primarily determined by the vascular resistances in the intestine and the spleen, whereas hepatic arterial blood flow is determined by the intrahepatic vascular resistance. The portal vein pressure (normal range 7–12 mmHg) is modified by the inflow resistance of the portal vein tract within the liver. It increases following sympathetic nerve stimulation and adrenaline administration, because the vascular smooth muscle cells contain only α-adrenoceptors. Arteriolar sphincters are present before the arterial blood drains into the sinusoids, in order to throttle the arterial pressure. In addition, sphincters in the sinusoids have been postulated, which allow the blood flow through an individual sinusoid to be changed, Hepatic arterial resistance increases following sympathetic nerve and α-

R. Greger/U. Windhorst (Eds.)
Comprehensive Human Physiology, Vol. 2
© Springer-Verlag Berlin Heidelberg 1996

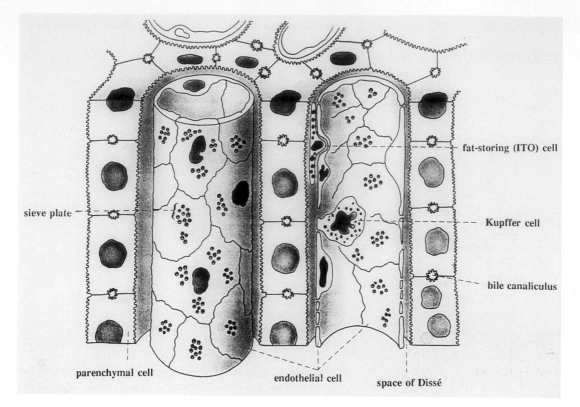

Fig. 67.1. Organization of parenchymal and non-parenchymal liver cells in an acinar section. The acinus extends from the terminal portal venule along the sinusoids to the draining terminal hepatic venule and is formed by tubes (20–30 cells long) of parenchymal cells. Endothelial and Kupffer cells form an inner coat of these tubes, thereby separating the sinusoidal space from the perisinusoidal space of Dissé. The two spaces communicate via endothelial fenestrae and allow access of blood plasma (but not erythrocytes) to parenchymal cells. (Modified from [63])

adrenoceptor-stimulation, but decreases under the influence of glucagon, vasopressin, and β_2-adrenoceptor stimulation (at physiologically low adrenaline concentrations). Hepatic arterial blood flow changes reciprocally with portal blood flow in an attempt to maintain a constant total liver perfusion and oxygen supply. Nitric oxide may play a role in this reciprocal regulation of liver perfusion, because endothelial cells in both the portal and the arterial vessels contain constitutively nitric oxide synthase. At pharmacological doses many compounds, such as biogenic amines and eicosanoids, can affect the vascular resistance in both the portal and the arterial system [4,78]. They frequently exhibit "transhepatic responses"; that is to say, application of an agent to the portal vein will also modulate hepatic arterial blood flow and vice versa.

67.1.3 Innervation

The mammalian liver is richly supplied with afferent and efferent nerves of the autonomous (sympathetic and parasympathetic) nervous system from two communicating plexuses around the portal vein and the hepatic artery. Hepatic afferent nerves appear to be connected to putative chemo-, osmo- and baroreceptors; they are also involved in the regulation of short-term feeding behavior. Efferent nerves participate in the regulation of hepatic blood flow. Other efferent nerves terminate on the hepatocytes and exert control on hepatocellular metabolic function. For example, electrical stimulation of the ventromedial hypothalamic nucleus stimulates glycogenolysis in the liver, whereas stimulation of the vagus promotes glycogenesis. In contrast to rat liver, the human liver shows a high density of parenchymal adrenergic fibers. Because there is an inverse relationship between the density of intraparenchymal nerve fibers and the frequency of gap junctions in different species, it is thought likely that the lack of innervation such as is seen in rat liver is compensated for by direct intercellular communicative gap junctions propagating the nerve signals from an innervated hepatocyte to others [17].

67.2 Non-parenchymal Cell Function in Liver

The functions of hepatic nonparenchymal cells [77] are summarized in Table 67.1. Endothelial cells are self-proliferating cells with numerous fenestrae perforating their cytoplasm, which are organized in sieve plates. The fenestral diameter (normal 150–175 nm) changes dynami-

Table 67.1. Functions of non-parenchymal liver cells

Cell type	Functions
Kupffer cells	Phagocytosis (particles > 0.1 µm) of microorganisms, cell debris, tumor cells, aged erythrocytes Endocytosis (endotoxins, immune complexes) Antigen processing (antigen sequestration and presentation) Cytotoxicity (production of superoxide, antitumor effects) Production of signal molecules for intercellular communication (eicosanoids, interleukins 1 and 6, tumor necrosis factor α, interferon α, β)
Endothelial cells	Barrier function between blood and hepatocytes Receptor-mediated uptake of modified HDL, LDL via scavenger receptor Receptor-mediated uptake of mannose- and galactose-terminated glycoproteins, N-acetyl-glucosamine, N-acetylneuraminic acid (transferrin, ceruloplasmin, transcobalamine II, lysosomal hydrolases) Pinocytosis (polyvinylpyrrolidone colloidal carbon) Presentation of ectoenzymes (lipase) Synthesis of effector molecules (prostacyclin, prostaglandin E_2, cytokines) Endocytosis (particles < 0.1 µm)
Fat-storing (ITO) cells	Vitamin A storage Synthesis of extracellular matrix proteins Contractility, regulation of sinusoidal blood flow Expression and secretion of growth factors
Pit cells	Activated natural killer cells protecting against viral infections, metastatic tumor cells

cally in response to ethanol, pressure, and hormones such as serotonin. The fenestrae allow communication between the sinusoidal and the perisinusoidal space, i.e. free access of solutes, but not of particles larger than 0.2 µm, such as erythrocytes or unprocessed large chylomicrons. This filtering effect regulates fat uptake by parenchymal cells. Endothelial cells possess specialized endocytotic mechanisms and receptors, which allow the specific uptake of transferrin, ceruloplasmin, modified lipoproteins, apoprotein E-containing remnants, lysosomal hydrolases and other particles (<0.1 µm).

Kupffer cells are resident liver macrophages, which are self-proliferating but can also be recruited from extrahepatic sources. They are located predominantly periportally, i.e., at the sinusoidal inflow, and their main function is to phagocytose particulate material (cellular debris, parasites, viruses, bacteria) and to take up macromolecules (immune complexes, bacterial endotoxins) by receptor-mediated endocytosis. There are receptors for immune complexes (Fc and C3b receptors), fibronectin, galactose, mannose, N-acetylglucosamine, and positively charged proteins. Thus, Kupffer cells provide a powerful and important phagocytotic barrier for gut-derived toxins and microorganisms. Indeed, when portal blood bypasses the liver via portocaval anastomoses, as it does in patients with liver cirrhosis, systemic endotoxinemia develops, with such sequelae as induction of nitric oxide synthase in peripheral vascular endothelia and smooth muscle cells leading to vasodilation and hyperdynamic circulation [73]. Kupffer cell activation by endotoxin results in an increased production of cytokines and eicosanoids, i.e., signals acting on other cell types in the liver [5,12]. Kupffer cells play an important role in antigen processing. They digest potentially immunogenic material, thereby preventing immune responses to dietary proteins. On the other hand, during inflammation and infection they can act as antigen-presenting cells initiating T- and B-lymphocyte-mediated-immunity.

Pit cells are the perisinusoidal equivalent of large granular lymphocytes and natural killer cells and exhibit similar functions to these cell types.

ITO cells contain a high percentage of the body's vitamin A content. These cells can produce components of the extracellular matrix, such as collagen (types I, III, IV) and laminin; they can transform to myofibroblast-like cells and proliferate under the influence of CCl_4. ITO cells, which are located in the space of Dissé, have long branching processes underlying the endothelium. They contract under the influence of thromboxanes and endothelin and may play a decisive part in the regulation of sinusoidal blood flow [39].

67.3 Metabolic Parenchymal Cell Function

Liver parenchymal cells ("hepatocytes") have important roles in:

- Processing of absorbed nutrients and xenobiotics
- Maintenance of glucose, amino acid, ammonia and bicarbonate homeostasis in the body
- Synthesis of most plasma proteins
- Bile acid synthesis and bile formation
- Storage and processing of signal molecules.

Sensitive control is exerted on these processes at different levels, allowing the rapid adaptation of hepatic metabolism to altered needs of the organism.

67.3.1 Regulatory Principles

Only the most important regulatory principles will be discussed in this chapter; for more in-depth treatment of the subject the reader is referred to textbooks of biochemistry.

Transport and Compartmentation. In the liver acinus metabolic pathways are compartmentalized at the intercellular and intracellular level. Depending on their location in the acinus, hepatocytes exhibit a specific enzyme pattern, resulting in a predominant location of metabolic pathways in either:

• the periportal area (at the inflow of the sinusoidal bed) or
• the perivenous area (at the venous outflow of the sinusoidal bed; so-called "metabolic zonation" or "functional hepatocyte heterogeneity" [24,36,72]). For further details the reader is referred to Chap. 68.

Within an individual hepatocyte, enzymes are frequently restricted to specific subcellular compartments, such as the cytosol, mitochondria or endoplasmatic reticulum. Such compartmentation introduces further sites of metabolic control resulting from the necessity for substrate transport across membranes or the regulation of an organelle-specific "milieu intérieur" by ion and substrate transport systems in their bordering membranes. For example, the pH in lysosomes, the cytosol and mitochondria is adjusted to values of 5–6, 6.9–7.3 and 7.4–7.6, respectively, due to the action of H^+-translocating systems. Extracellular substrates must be transported across the plasma membrane and the membrane of organelles by specific transport systems to become accessible for metabolizing enzymes [42]. Whereas some substrates, such as NH_3, can enter the cell by simple or nonionic diffusion, others, such as glucose, are taken up into hepatocytes by facilitated diffusion via specific transport systems. Other substrates, however, enter the cell in an energy-driven way. Here,

primary, secondary and tertiary active transport systems can be distinguished, which are directly coupled to ATP hydrolysis or indirectly driven by the electrochemical Na^+ gradient being built up by the (Na^++K^+)-ATPase (Fig. 67.2; see also Chap. 8). These transporters can built up an intra/extracellular substrate gradient. Examples of Na^+-gradient driven transporters in the sinusoidal plasma membrane are the amino acid transport systems A (small neutral amino acids, such as alanine, glycine, proline, serine) and N (glutamine, asparagine), the sinusoidal transporters for conjugated bile acids and fatty acids, and dicarboxylates.

There is unequivocal evidence that transport across biological membranes is an important site of metabolic flux control [42]. Here, three different modes can be envisaged.

• When the flux through a metabolizing enzyme is much faster than the rate of substrate transport across the plasma membrane, flux control of metabolism by the transporter activity is obvious; this holds for the metabolism of physiological alanine concentrations delivered via the portal vein.
• However, also when the rate of substrate transport apparently far exceeds the rate of flux through an enzyme-regulated metabolic pathway, transport can become the controlling factor. An example of this is glutamine hydrolysis by glutaminase, which takes place inside the mitochondria. This enzyme has a K_m (glutamine) of about 28 mmol/l. Despite an extracellular concentration of this amino acid of only about 0.6 mmol/l, the enzyme operates at glutamine concentrations close to its K_m value. This is achieved by the concentrating activities of the Na^+-dependent glutamine transporting system N in the plasma membrane and the H^+-gradient-driven glutamine transporter in the mitochondrial membrane, resulting in steady state glutamine concentrations of about 8 mmol/l in the cytosol and of about 20 mmol/l inside the mitochondria [27]. Thus, although the rate of unidirectional glutamine transport across the plasma and mitochondrial membranes by far exceeds flux

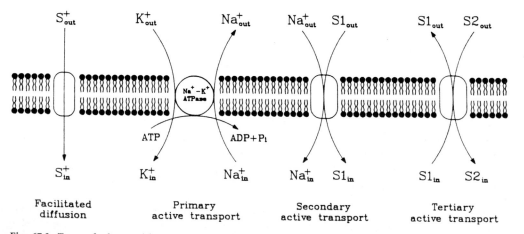

Fig. 67.2. Types of substrate (S) transport across biological membranes. (From [42])

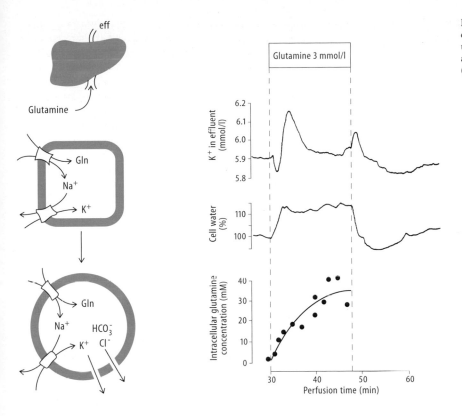

Fig. 67.3. Liver cell swelling during cumulative glutamine uptake via Na⁺-dependent amino acid transport system N. (From [27])

through glutaminase, these transporters control flux through this enzyme by determining the steady state concentration of the amino acid inside the mitochondrial matrix.

- The concentrative, mainly Na⁺-driven uptake of substrates into the hepatocyte leads to an osmotic water shift into the cell. This increase in cellular hydratation in turn affects a variety of metabolic processes, which need not necessarily be related to the metabolism of the transported substrate (see below). Thus, the role of concentrative plasma membrane transporters is twofold: they not only serve for substrate translocation across a membrane, but also modify hepatocellular function by altering the cellular hydratation state.

The activity of transport systems in the plasma membrane is regulated by the nutritional state and hormones. For example, starvation increases the activity of the alanine-transporting system A. This supports an increased utilization of muscle-borne alanine for hepatic gluconeogenesis during starvation.

Hormones. Most hormones exert their regulatory action on hepatic metabolism at different levels simultaneously. They do not only have short-term effects on the activity of various enzymes or transport systems by generation of second-messenger molecules such as cAMP or inositol-1,4,5-trisphosphate, elevation of cytosolic free Ca^{2+}, initiation of complex cascades of protein phosphorylation/dephosphorylation or alterations of the cellular hydrata-

tion state; they also act on a long-term basis at the level of gene expression. For example, glucagon increases the activity of the amino acid transport system A in the liver within minutes by hyperpolarizing the membrane (which increases the driving force for Na⁺-coupled transporters) and in the long term by stimulating the synthesis of new transporter molecules. It also acutely activates glutaminase and induces urea cycle enzymes, serine dehydratase, tyrosine aminotransferase, aspartate aminotransferase and gluconeogenic enzymes in addition to a cAMP-triggered cell shrinkage, which stimulates hepatic proteolysis. Glucagon also decreases protein synthesis in extrahepatic tissues and stimulates amino acid release from the muscle. Thus, the concerted actions of the hormone augment an increased uptake of amino acids by the liver and utilization for gluconeogenesis and ureogenesis.

Hepatocellular Hydratation. The hepatocellular hydratation state can change within minutes under the influence of substrates and hormones; i.e., upon alterations of the composition of portal blood [29]. This is because many substrates are concentrated inside the hepatocytes by Na⁺-coupled transport systems. Thus, intra/extracellular substrate concentration gradients of up to 30 are generated, triggering osmotic water flow into the cells. Indeed, hepatocytes swell up to 12% when exposed to amino acids such as glutamine, despite activation of volume-regulatory mechanisms which only prevent cell swelling to become excessive (Fig. 67.3). Insulin also increases the hepatocellular hydratation state within minutes by accumulating Na+, K+ and Cl− inside the cells due to a hormone-induced stimulation of (Na+ + K+)-ATPase, Na+/H+ antiport and Na+2Cl−K+ cotransport (Fig. 67.4). Conversely, glucagon decreases hepatocellular hydratation by depleting cellular K+, Na+ and Cl−. It was recently recognized that hormone and substrate-induced alterations of the hepatocellular hydratation state act like a second-messenger triggering metabolic liver cell function in concert with other hormone-activated intracellular signalling pathways, such as the formation of second-messenger molecules or initiation of complex phosphorylation cascades [26,45]. The metabolic alterations in response to cell swelling are summarized in Table 67.2. Precisely the opposite metabolic responses are triggered by cell shrinkage. Apparently, an increase of the hepatocellular hydratation state acts like an anabolic, proliferative signal, whereas cell shrinkage is catabolic and antiproliferative. The intracellular signalling paths that link the cellular hydratation state to metabolic cell function have not yet been elucidated, but they involve effects on the cytoskeleton, intracellular membrane flow, intracellular ion concentrations, stretch-induced activation of membrane-associated signalling systems, and, most importantly, alterations in protein phosphorylation. In rat hepatocytes, hypoosmotic swelling results in a rapid activation of mitogen-activated protein (MAP) kinases. This activation is sensitive to G-protein inhibitors and blockers of tyrosine kinases, suggesting that cell swelling results in a G-protein- and tyrosine-kinase-dependent activation of MAP kinases [64a]. Interruption of this signal transduction cascade by inhibitors, such as pertussis toxin, genistein or erbstatin blocks the swelling-induced stimulation of

transcellular taurocholate transport. Thus, the intracellular signalling events turned on in response to increases of cell hydration mediate at least some of the metabolic responses and resemble those activated by growth factors.

It should be emphasized that alterations of the hepatocellular hydratation state, which occur in response to physiological fluctuations of the portal hormone and nutrient load, provide an elegant mechanism by which the liver adapts different metabolic functions to an altered environment.

Table 67.2. Adaptation of metabolic liver function in response to increases in the hepatocellular hydratation state ("cell swelling") such as occur during cumulative substrate uptake or under the influence of hormones such as insulin

Stimulation of	Inhibition of
Protein synthesis	Proteolysis
Glycogen synthesis	Glycogenolysis
	Glycolysis
Amino acid uptake	
Amino acid catabolism	
Pentose phosphate shunt	
Release of reduced glutathione into the sinusoidal space	Release of oxidized glutathione into bile
Bile acid secretion	
	Acidification of endocytotic vesicles
	Viral replication
mRNA induction for β-actin, tubulin, c-jun	mRNA induction for PEPCK, tyrosine amino-transferase

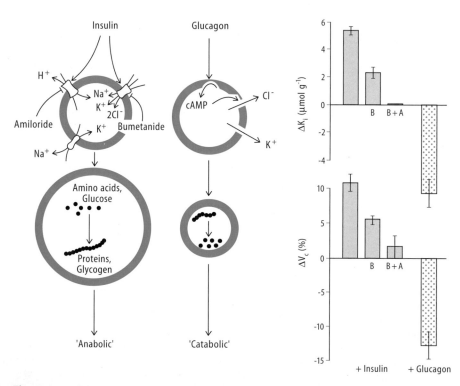

Fig. 67.4. Modulation of the hepatocellular hydratation state by insulin and glucagon. (From [29])

Enzyme Activity and Substrates. Substrate availability for a certain enzyme is not only controlled by transport across hepatocellular membranes, but also by the extent of delivery from extrahepatic tissues, e.g., intestine, muscle, adipose tissue or kidney. Accordingly, the rate of hepatic gluconeogenesis ultimately depends on the amino acid and lactate supply derived from proteolysis or glycolysis in skeletal muscle or from intestinal absorption.

Enzyme activity is regulated on a short-term basis by activator or inhibitor molecules, phosphorylation/dephosphorylation cycles or association/dissociation of enzymes from intracellular membranes. Here complex regulatory cascades can evolve, as shown by the example given in Fig. 67.5. In the long term enzyme activities are regulated at the level of gene expression by hormones, substrates and the cellular hydratation state. For example, glucagon, corticosteroids and cell shrinkage lead to an induction of phosphoenolpyruvate carboxykinase, whereas insulin and cell swelling have opposing effects.

Sometimes, enzymes catalyzing sequential reactions of a metabolic pathway, such as pyrimidine biosynthesis [20], are organized as multienzyme complexes. In such "metabolons" the product formed by one enzyme is channeled into the next reaction. Here, metabolites are handled from one enzyme to the next without mixing with the general cellular metabolite pool. Such phenomena explain why kinetic properties of isolated enzymes may sometimes differ markedly from their in situ kinetics. For example (Fig. 67.5) ammonia derived from mitochondrial glutaminase is channeled into the carbamoylphosphate synthetase reaction [48], which increases the efficiency of urea synthesis from glutamine-derived ammonia [26] (see also Chap. 68).

Substrate cycles contribute to a rapid and effective regulation of metabolic fluxes, especially in the control of pathways in which the direction of net flux is reversible. They have a crucial regulatory role in the switch from net gluconeogenesis to net glycolysis and vice versa (see Sect. 67.3.2).

Zonal Heterogeneity and Cell–Cell Interactions. Liver parenchymal cell function is also controlled by nonparenchymal cells. Endotoxin stimulates Kupffer cells to produce eicosanoids such as prostaglandins, which can bind to specific receptors on the parenchymal cell surface and trigger inositol-1,4,5-trisphosphate formation, elevation of intracellular Ca^{2+}, and consequently a glycogenolytic response. Interleukin-6 produced by Kupffer cells triggers the synthesis of acute phase proteins in parenchymal cells [3]. Further, hormonal modulation of the diameter of endothelial cell fenestrae determines the access of lipoproteins to the parenchymal cell and controls hepatic cholesterol uptake from chylomicron remnants. Moderate doses of ethanol open the fenestrae; this explains in part the plasma cholesterol-lowering effect of ethanol, but also supports the development of fatty liver. Severe chronic alcohol abuse, on the other hand, defenestrates the endothelial cell lining, contributing to the development of

hyperlipidemia. In addition, there are also interactions between different nonparenchymal cell populations. For example, endothelin released from endothelial cells causes contraction of ITO cells. Parenchymal–parenchymal cell interactions have also been demonstrated, in that metabolic products from upstream periportal hepatocytes, such as ammonia, can be delivered to downstream perivenous hepatocytes for further processing [23]. The role of efferent liver nerves in regulating metabolic liver function [18] has already been mentioned in Sect. 67.1.3.

67.3.2 Carbohydrate Metabolism

The Liver as a Glucostat. The liver plays an important part as a "glucostat" in the organism [34]: it removes glucose, if it is present in excess, via glycogen synthesis, glycolysis and lipogenesis and liberates glucose if needed via glycogenolysis and gluconeogenesis. This is important, because the function of some organs, such as the brain, depends largely on a balanced glucose supply. Bidirectional glucose transport across the hepatocellular plasma membrane occurs by way of the Glut-2 transporter mediating facilitated diffusion of the sugar [55] (cf. Chap. 8). However, plasma membrane transport of glucose is not a major site of regulation of hepatic glucose metabolism. During a carbohydrate-rich meal in man, glucose is absorbed in the intestine at a rate of approximately 40 g/h, and more than 50% of the absorbed glucose (25 g/h) is taken up by the liver and converted to glycogen or fatty acids or oxidized. During a short-term fast, the liver releases glucose at a rate sufficient to cover the glucose requirements of the brain, i.e., 7.5 g/h, two thirds of which is derived from glycogenolysis and the remainder from gluconeogenesis. Following long-term starvation the hepatic glycogen stores are depleted; this occurs within 24 h of fasting. In this case cerebral metabolism adapts: glucose requirements decrease to about 2 g/h, whereas ketone bodies are increasingly used as energy fuel. Both the glucose and the ketone body supply are maintained by the liver through hepatic gluconeogenesis and ketogenesis. Physical exercise in the early postprandial phase markedly increases glucose utilization in skeletal muscle, which is provided predominantly by hepatic glycogenolysis. Among the factors controlling the reversible switch between glycogenolysis/gluconeogenesis in the postabsorptive state to glycogen synthesis/glycolysis during absorption, substrate concentrations, hormone levels, hepatic nerves, the hepatocellular hydratation state (Table 67.2) and zonal hepatocyte heterogeneity play decisive roles (see also Chap. 68). Glycogen synthesis (Fig. 67.6) is stimulated predominantly by increases in the portal glucose concentration with insulin and parasympathetic nerves being auxiliary factors. Glycogen breakdown is activated by glucagon and sympathetic nerve activity, but is inhibited by high glucose concentrations. Glycolysis is activated by high portal glucose concentrations, by insulin, whereas gluconeogenesis is activated primarily by glucagon.

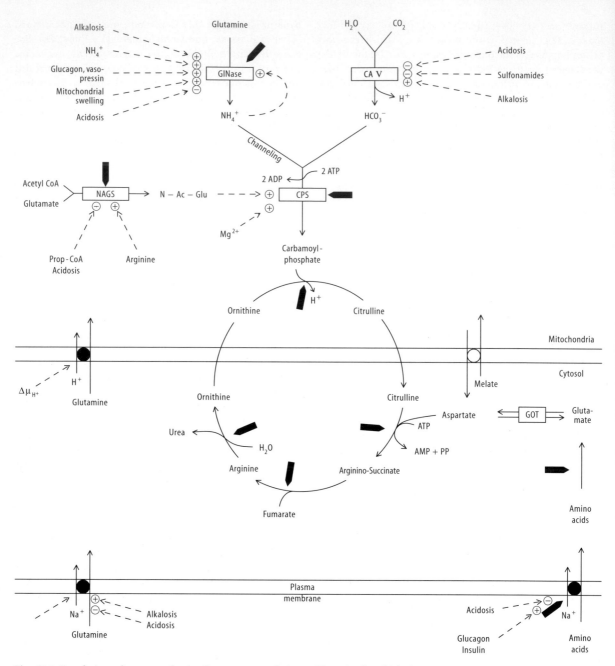

Fig. 67.5. Regulation of urea synthesis. Short-term regulation (*dotted arrows*): in the presence of physiological substrate concentrations, the rate-controlling step of the urea cycle is mitochondrial carbamoylphosphate synthase (*CPS*). Owing to its high K_M values for its substrates NH_4^+ and HCO_3^- (1–2 mmol/l and 5 mmol/l, respectively) flux through CPS is largely determined by the mitochondrial substrate concentration. Provision of these substrates is largely controlled by the activity of mitochondrial glutaminase and carbonic anhydrase V, which underlie sensitive regulation by the acid–base status and hormones. Most importantly, mitochondrial glutaminase is activated by ammonia in physiological concentrations. Thus, the enzyme has the unique distinction of being product-activated and it functions as a mitochondrial ammonia-amplifying system. The activity of this ammonia amplifier is a crucial determinant for urea cycle flux.

The mitochondrial glutamine concentration is controlled by its concentrative transport across the plasma and mitochondrial membrane. Plasma membrane glutamine transport is controlled by the acid–base status, while transport across the mitochondrial membrane depends upon the energization of the mitochondria ("proton-motive force", $\Delta\mu H^+$). CPS is allosterically activated by N-acetylglutamate (*N-Ac-Glu*; half-life about 20 min); its synthesis by N-acetylglutamate synthase (*NAGS*) is stimulated by arginine. Further sites of control of urea synthesis at the level of metabolite transport across the mitochondrial membrane are omitted for clarity. Long-term regulation (*red arrows*): all urea cycle enzymes, amino acid transporters, several amino acid metabolizing enzymes and NAGS are induced by glucagon and ingestion of a high-protein diet. Degradation of glutaminase protein is inhibited by ammonia and glutaminase is induced by glucagon and ammonia.

Regulation of Hepatic Gluconeogenesis/Glycolysis. Figure 67.6 depicts the pathways of glycolysis and gluconeogenesis, which involve three substrate cycles, i.e., the glucose/glucose-6-P cycle, the fructose-6-P/fructose-1, 6-bisphosphate cycle and the pyruvate/phosphoenolpyruvate (PEP) cycle. The enzymes of these substrate cycles exhibit coordinated short- and long-term regulation by hormones, allosteric activators, phosphorylation and substrates [56], which allows a rapid switch of flux through the pathway determining whether the liver becomes a net producer or net consumer of glucose (Table 67.3). Starvation increases the plasma concentrations of glucagon, glucocorticoids and catecholamines, which in turn increase the activity of phosphoenolpyruvate carboxykinase (PEPCK), fructose-1,6-bisphosphatase and glucose-6-phosphatase with a concomitant decrease in pyruvate kinase, 6-phosphofructo-1-kinase and glucokinase. These alterations drive the pathway depicted in Fig. 67.6 in the direction of gluconeogenesis. Reciprocal activity changes occur when high amounts of carbohydrates are ingested and plasma insulin concentrations are high but those of glucagon, glucocorticoids and catecholamines are not. Glucagon effects on the pathway are mediated by cAMP-stimulated phosphorylation of pyruvate kinase, 6-phosphofructo-1-kinase and fructose-1,6-bisphosphatase, decreased formation of fructose-2,6-bisphosphate, and cAMP regulation of gene expression (Table 67.3). Fructose-2, 6-bisphosphate (Fru-2, 6-P_2) is an allosteric activator of 6-phosphofructokinase and an inhibitory of fructose-1,6-bisphosphatase. Thus, at low levels of Fru-2,6-P_2 (glucagon, starvation), gluconeogenesis is high. Fru-2,6-P_2 levels are controlled by a bifunctional enzyme 6-phosphofructo-2-kinase/fructose-2,6-phosphatase, which is itself a substrate of cAMP-dependent protein kinase: phosphorylation inhibits the kinase activity but stimulates the phosphatase activity.

67.3.3 Amino Acid and Ammonia Metabolism

The liver acts as a metabolic buffer in the control of plasma amino acid concentrations; however, other organs also contribute to the maintenance of amino acid homeostasis [10,25,42]. The importance of the liver in maintaining the balance of free amino acids rapidly becomes evident after hepatectomy or in fulminant liver failure: in these situations the concentrations of almost all amino acids increase excessively. When the isolated rat liver is perfused with a medium devoid of amino acids the liver releases amino acids, and after 90 min almost constant concentrations are established for all individual amino acids, a pattern resembling that found in the plasma in vivo [64]. In this latter case amino acids are derived from hepatic proteolysis, a process stimulated by glucagon and amino acid deprivation and inhibited by amino acids and insulin [49]. In part, these hormonal and amino acid effects on hepatic protein breakdown are mediated by alterations of the hepatocellular hydratation state [29]. Conversely, in the postprandial absorptive state amino acids are delivered to the liver in high concentrations (up to 30-fold basal) and are efficiently extracted and metabolized there. Effective hepatic amino acid extraction explains why the fluctuations of most arterial plasma amino acids during absorption from the intestine are only slight [25]. Highest rates of amino acid extraction are observed for the gluconeogenic amino acids alanine, serine, and threonine. In addition, most essential amino acids are degraded by the liver. Exceptions are the branched chain amino acids leucine, valine, and isoleucine, which are only used for protein synthesis in the liver, and are not catabolized. The major site of their breakdown is skeletal muscle, but the liver efficiently takes up and oxidizes the α-keto-analogues of branched amino acids, which can be released from muscle tissue. Citrulline, which is formed and released by the intestinal mucosa, is also not taken up by the liver; it is delivered to the kidney as a precursor of arginine biosynthesis [76]. An increased portal amino acid load to the liver induces hepatocellular swelling, which in turn is known to increase amino acid flux across the plasma membrane, to stimulate amino acid breakdown and utilization for protein synthesis and glycogen synthesis, and simultaneously inhibits amino acid generation from proteolysis (Table 67.2). This swelling-induced basic metabolic pattern is further modified by hormones, which are released during intestinal absorption. In starvation, the muscle becomes a major site of amino acid release. A glucose/alanine cycle between liver and muscle exists in this situation: alanine released from the muscle is taken up by the liver and used for gluconeogenesis; glucose is delivered to the muscle and enters the glycolytic pathway, and the pyruvate generated there is transaminated to alanine, which flows back again to the liver. For the hepatic metabolism of individual amino acids, the reader is referred to biochemistry textbooks.

The portal venous blood contains high concentrations of ammonia (0.2–0.5 mmol/l) owing to ammonia generation in the intestinal mucosa from glutamine and due to the action of intestinal microorganisms [76]. Ammonia is also produced inside the hepatocytes during the breakdown of amino acids. Ammonia is detoxified in the liver by both the liver-specific pathway of urea synthesis (Fig. 67.5) and by glutamine synthesis. The structural and functional organization of these pathways in the liver acinus and the role of urea synthesis, not only in detoxifying ammonia but also in the maintenance of systemic acid base homeostasis, are discussed in detail in Chap. 68. Failure of the liver to eliminate ammonia, a potent neurotoxin, leads to hepatic encephalopathy, a neuropsychiatric disorder.

67.3.4 Protein Synthesis and Secretion

Except for immunoglobulins, most circulating plasma proteins are synthesized in the liver (Table 67.4). They are produced by parenchymal cells, except for the von Willebrand factor, which is produced by endothelial cells only. However, non-parenchymal cells can contribute to the synthesis of plasma proteins; synthesis of retinol-bind-

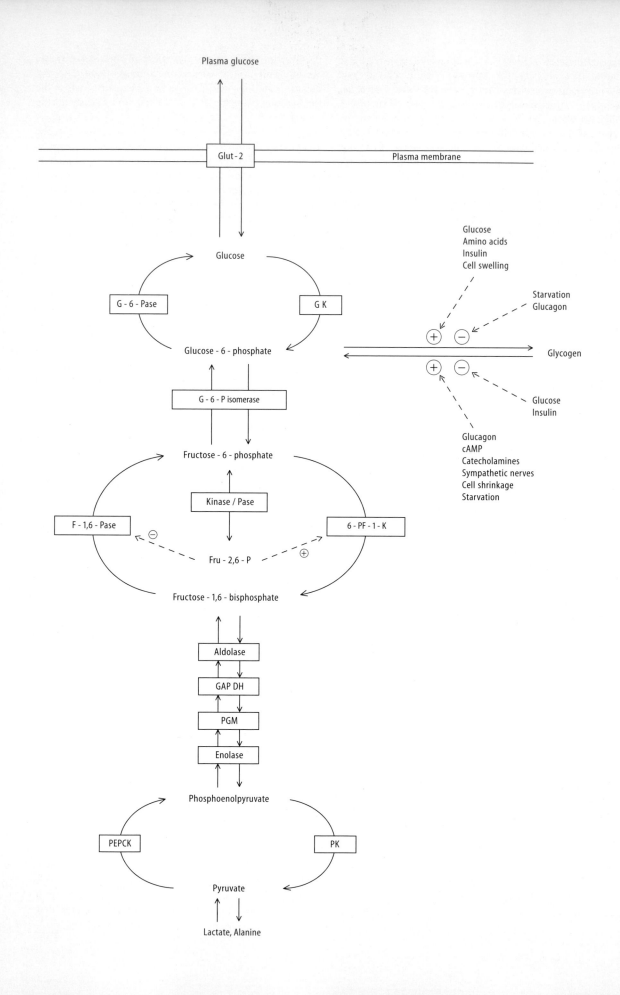

ing protein and α_1-antitrypsin can also occur in ITO and Kupffer cells, respectively. The half-lives of plasma proteins are highly variable and range from hours (clotting factors) to 10–14 days (albumin, cholinesterase). Accordingly, an acute hepatic failure to synthesize plasma proteins first becomes manifest as a blood coagulation defect. The capacity of the liver to synthesize albumin is high: the amount produced daily in humans is about 12 g, but synthesis can increase up to four-fold following albumin losses.

The individual parenchymal cell is not specialized on the synthesis of one specific plasma protein; rather, each hepatocyte is able to synthesize the whole spectrum of plasma proteins [16], as evidenced by in situ hybridization of specific mRNAs. However, parenchymal cell heterogeneity may be seen to exist for the synthesis of some plasma proteins when assessed by immunohistochemistry. For example, immunhistochemically albumin is randomly found in some but not all hepatocytes, whereas albumin mRNA is detectable in each parenchymal liver cell.

Table 67.3. Hormonal regulation of hepatic glucose metabolism

Enzyme[a]	Regulation at the level of			
	Gene expression (long term)[b]		Enzyme activity (short term)[c]	
	Insulin	cAMP	Phosphorylation/ dephosphorylation	Allosteric
Glucokinase	+	−	No	No
6-PF-1-K	+	−	No	Yes
Pyruvate kinase	+	−	Yes	Yes
6-PF-2-K/Fru-2,6-Pase	+	−	Yes	Yes
PEPCK	−	+	No	No
Fru-1,6-Pase	−	+	No	Yes
G-6-Pase	?	?	No	?

[a] Fru-1,6-Pase, fructose-1,6-phosphatase; Fru-2,6-Pase, fructose-2,6-bisphosphatase; G-6-Pase, glucose-6-phosphatase; PEPCK, phosphoenol pyruvate carboxy-kinase; 6-PF-1-K; 6-phosphofructo-1-kinase; 6-PF-2-K, bifunctional enzyme 6-phosphofructokinase fructose 2,6-bisphosphatase.
[b] + Activation, − inhibition of gene expression
[c] Effects of insulin on gene expression of pyruvate kinase and 6-PF-2-K/Fru-2,6-Pase require the presence of glucose

Table 67.4. Functions of plasma proteins synthesized and secreted by the liver

Function	Secreted plasma protein
Oncotic pressure in plasma	albumin
Binding and plasma transport of	
Lipids	Apoproteins (VLDL, HDL), albumin (fatty acids, lysolecithin)
Bilirubin	Albumin
Divalent cations	Transferrin (Fe^{2-}), ceruloplasmin (Cu^{2+}), albumin (Zn^{2+})
Vitamins	Retinol-binding protein (vitamin A), transcalciferrin (vitamin D)
Hormones	
Thyroid hormones	Thyroxine-binding globulin, thyroxine-binding prealbumin, albumin
Steroid hormones	Corticoid-binding globulin (transcortin), sex hormone-binding globulin (testosterone, estradiol), orosomucoid = α_1-acid glycoprotein (synthetic estrogens), albumin (aldosterone, conjugates of steroid hormones)
Blood coagulation	
Clotting factors	All coagulation factors (except factor VIII, which is also produced by extrahepatic tissues), prekallikrein, fibrinogen
Anticoagulatory activity	Plasminogen, α_2-antiplasmin, antiproteases blocking clotting factors (antithrombin III, protein C and S)
Host defense	Complement factors, acute phase proteins
Secretion of plasma enzymes	Pseudocholinesterase, liver lipase, α_1-Antitrypsin, α_1-antichymotrypsin,
Inactivation of proteases	inter-α-trypsin inhibitor α_2-antiplasmin, anti-thrombin III, activated protein C inhibitor, C1 inhibitor

◀

Fig. 67.6. Regulation of hepatic gluconeogenesis/glycolysis by substrate cycles *Glut-2*, glucose transporter; *G-6-Pase*, glucose-6-phosphatase; *GK*, glucokinase; *F-1,6-P₂ase*, fructose-1,6-bisphosphatase; bifunctional enzyme: *6-PF-2-K/F-2,6 -Pase,* phosphofructokinase/fructose-2,6-bisphosphatase ("kinase Pase"); *6-PF-1-K*, 6-phosphofructo-1-kinase; *PEPCK*, phosphoenolpyruvate carboxykinase; *PK*, pyruvate kinase; *GAPDH*, glyceraldehyde-3-phosphate dehydrogenase; *PGM*, phosphoglycerate mutase

Except for albumin and C-reactive protein, all proteins secreted by the liver are glycoproteins, in which an oligosacharide chain is linked to the amide side chain of asparagine (*N*-glycosidic link) or to the OH groups of threonine or serine (*O*-glycosidic link). For the various steps of protein synthesis, such as transcription, translation, co-translational and post-translational modification and protein targeting the reader is referred to textbooks of biochemistry. Specific receptors for galactose- and mannose-terminated glycoproteins also allow the liver to play an important part in the plasma clearance of glycoproteins [22].

Hepatocytes are constitutively protein-secreting cells. They continuously synthesize and secrete proteins and there are no stores of newly synthesized proteins that could be released from the cell upon external stimuli. Thus, the hepatocellular protein secretory rate is primarily influenced by alterations in the rate of protein synthesis, which in turn is controlled at the translational and transcriptional levels. However, the transport rate of different newly synthesized proteins from the endoplasmic reticulum to the Golgi apparatus and the retention time within this latter compartment are not uniform for all proteins. Here, rapidly secreted proteins (albumin, fibronectin, α_1-protease inhibitor) can be distinguished from slowly secreted proteins (fibrinogen, transferrin).

Protein synthesis in the liver decreases upon amino acid deprivation (starvation) or lowering of the hepatocellular hydratation state, and under the influence of glucagon and vasopressin. These conditions simultaneously stimulate proteolysis and RNA breakdown, whereas RNA synthesis is inhibited. Conversely, amino acids, cell swelling, and insulin stimulate protein synthesis and inhibit proteolysis. Hepatic synthesis of plasma proteins is also stimulated by thyroid and growth hormone.

The liver also is the major site of synthesis of acute phase proteins; i.e. a heterogeneous group of plasma proteins whose concentrations rapidly change in response to injury and infection even when set in extrahepatic tissues. Cytokines secreted by inflammatory cells, such as interleukin-1 (IL-1), interleukin-6 (IL-6) and tumor necrosis factor (TNF), mediate the induction of acute phase proteins in the liver [31]. Cytokine-induced induction depends on the presence of glucocorticoids, whereas glucocorticoids themselves are ineffective (so-called permissive role). IL-6 appears to be the most important mediator; its secretion from monocytes, macrophages and endothelial cells is stimulated by IL-1, TNF, endotoxin and viruses. IL-6, which has a half-life in plasma of only minutes, binds to specific receptors on the hepatocyte surface and increases transcription of genes coding for acute phase proteins. Albumin mRNA levels decrease owing to diminished transcription under the influence of IL-6. Thus, albumin is seen as a "negative" acute phase protein. The function of acute phase proteins, whose plasma concentrations can increase only 1.5-fold (ceruloplasmin, complement C3) or up to several hundred-fold (C-reactive protein, serum amyloid A) is variable. Some of the acute phase proteins are protease inhibitors (α_1-antitrypsin. α_1-antichymotrypsin) and act to restrict the proteolytic activity of enzymes secreted by inflammatory cells to the site of inflammation. Others have immunomodulatory action (α_1-acid glycoprotein), facilitate the clearance of foreign material and microorganisms (C-reactive protein, serum amyloid A) or play a part in host defense (complement system).

67.3.5 Lipid and Lipoprotein Metabolism

The liver has a central role in lipoprotein metabolism. Plasma lipoproteins are complex particles with a surface made up of cholesterol, phospholipids, and specific apoproteins and a hydrophobic core of cholesteryl esters and triglycerides. The different lipoprotein classes differ in the relative amounts of these constituents and are routinely distinguished on the basis of their electrophoretic mobility or their flotation behavior during ultracentrifugation. According to their densities, lipoproteins are classified as:

- HDL = high-density lipoproteins
- IDL = intermediate-density lipoproteins
- LDL = low-density lipoproteins
- VLDL = very-low-density lipoproteins
- CM = chylomicrons

The apoproteins [46] are not only essential structural components of lipoprotein particles, but also act as recognition sites for receptor-mediated endocytosis of lipoproteins and as activators of enzymes involved in lipid metabolism [30] (Table 67.5). In addition to the synthesis and secretion

Table 67.5. Features of some apoproteins (apo)

Apoprotein	Synthesis	Lipoprotein	Function
apoA-I	Liver, intestine	HDL, CM, VLDL	LCAT activation
apoA-II	Intestine	HDL, CM, VLDL	Activation of hepatic triglyceride lipase
apoB-48	Intestine	CM, VLDL	LCAT activation
apoB-100	Liver	VLDL, IDL, LDL	LDL receptor binding
apoC-II	Liver	VLDL, CM	Lipoprotein lipase activation
apoE	Liver, spleen, kidney macrophages	CM, VLDL, HDL	Binding to apoE receptor (remnants)

CM, Chylomicrons; HDL, high-density lipoproteins; IDL, intermediate-density lipoproteins; LDL, low-density lipoproteins; VLDL, very-low-density lipoproteins; LCAT, lecithin; cholesterol acyltransferase

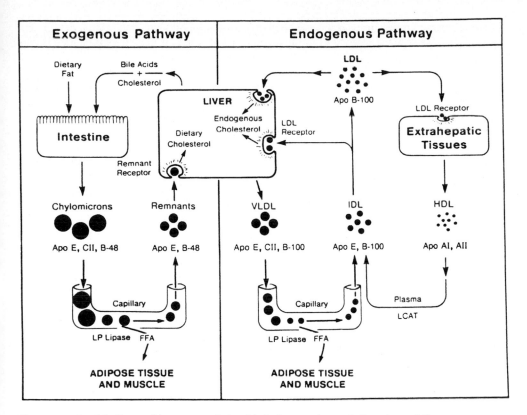

Fig. 67.7. Role of the liver and interorgan relationship in lipoprotein metabolism. (From [8])

of apoproteins, the liver also secretes lipid-metabolizing enzymes, such as lecithin:cholesterol acyltransferase (LCAT) and hepatic triglyceride lipase. This latter enzyme binds to the surface of hepatic endothelial cells and is activated in the same way as lipoprotein lipase in other vascular endothelia by heparin. The role of the liver in the interorgan flows of lipoproteins is schemqtically depicted in Fig. 67.7. Chylomicrons (CM) are the largest particles (diameter > 70 nm, molecular weight > 400 000 kDa); they consist largely (more than 90%) of triglycerides and are secreted by the intestinal mucosa into the intestinal lymph, from where they enter the systemic circulation. They transport dietary (exogenous) triglycerides and cholesterol and are processed in the capillaries of skeletal muscle and adipose tissue. Here, chylomicrons bind to the endothelial surface and their triglycerides are digested by membrane-bound lipoprotein lipase, an enzyme that is activated by apoprotein C-II, a constituent of chylomicrons. The hydrolytic products, i.e., fatty acids and monoglycerides, are taken by adipocytes and skeletal muscle cells, where they are deposited as fat or oxidized to CO_2 or H_2O. The remaining lipoprotein particles, i.e., the chylomicron remnants, are enriched in cholesterol, apoB-48 and apo-E. These remnants are taken up by the liver via the chylomicron remnant receptor. Thus, dietary triglycerides are delivered to skeletal muscle and adipose tissue, whereas dietary cholesterol is targeted to the liver.

During carbohydrate-rich feeding, the liver converts carbohydrates into fatty acids, which, after esterification with glycerol, are released into the bloodstream in the form of VLDL (endogenous pathway) [20]. VLDL contain apoB-100, apoC-II and apoE and are processed in the capillaries of skeletal muscle or adipose tissue in a similar way to chylomicrons. Their remnants, however, which are termed IDL and contain apoB-100 and apoE, are taken up in part by the liver via the LDL receptor or undergo further processing in plasma (loss of apoE) to yield cholesterol-rich LDL. LDL functions to supply extrahepatic tissues with cholesterol, where uptake occurs via LDL-receptor-mediated endocytosis [7]. Like many peripheral tissues, the liver parenchymal cell expresses the LDL receptor: cholesterol is taken up by the liver again and is used for bile acid synthesis or excreted as it is into the bile. In addition, phagocytic cells of the reticuloendothelial system (e.g., Kupffer cells) contain a scavenger receptor that acts to remove LDL at high concentrations or after chemical modification.

HDL are formed from unesterified cholesterol, which is released during cellular turnover. Esterification of HDL in cholesterol occurs by the action of lecithin:cholesterol acyltransferase (LCAT), an enzyme synthesized and released by the liver into circulation. These cholesterylesters are delivered to VLDL and may eventually appear in LDL. The LDL receptor recognizes apoB-100 on the lipoprotein particle surface. LDL binding to the receptor triggers LDL internalization and subsequent lysosomal digestion; this process accounts for 70% of LDL uptake into cells. There is another receptor on hepatocytes that binds and internalizes apoE-containing lipoproteins but not lipoproteins in which only apoB-100 is present. This is the chylomicron remnant receptor (or apoE receptor), which also recognizes VLDL remnants and HDL subspecies enriched with apo-E, but not newly secreted chylomicrons or VLDL. Genetically determined defects in the LDL receptor or of apoE are known to lead to severe familial hypercholesterolemia. In these conditions, LDL uptake via the scavenger cell receptor becomes prominent: cholesterol esters are deposited in macrophages, leading to the formation of foam cells, xanthomas and atherosclerotic plaques. Lipoprotein a [Lp(a)] resembles LDL in its composition and is generated by covalent binding of a carbohydrate-rich apoprotein, apo(a) to apoB-100 via disulfide bridges. Apo(a) has considerable homology with plasminogen; the role of Lp(a) is unknown but it appears to have the highest atherogenic potency of all lipoproteins.

The LDL receptor density on the hepatocellular surface is regulated by the hepatocellular cholesterol content: low cellular cholesterol upregulates the receptor and simultaneously activates hydroxymethylglutaryl-(HMG)-CoA reductase, the rate-controlling enzyme in cholesterol synthesis (see Sect. 67.3.6). Pharmacological inhibition of HMG-CoA reductase results in an increase of LDL receptors and contributes to the cholesterol-lowering potency of these drugs.

67.3.6 Bile Acid and Cholesterol Metabolism

Bile acids are the principal metabolites of cholesterol. Primary bile acids, such as cholic and chenodeoxycholic acid, are synthesized exclusively by the liver and undergo enterohepatic circulation. Their metabolism by intestinal bacteria (loss of the 7-OH group) leads to the formation of secondary bile acids (deoxycholic acid, lithocholic acid). Bile acids are conjugated in the liver with amino acids (taurine, glycine), sulfate, or glucuronic acid. Conjugation serves not only to decrease bile acid toxicity and to facilitate biliary excretion, it also lowers their pK value and the introduction of an anionic group augments water solubility.

The rate-controlling step in bile acid synthesis from cholesterol [74] is the formation of 7α-hydroxycholesterol catalyzed by microsomal cytochrom P_{450} enzyme cholesterol-7α-hydroxylase. This enzyme, and also hydroxymethylglutaryl-CoA reductase (HMG-CoA reductase), the rate-controlling enzyme of hepatic de novo cholesterol synthesis, underlie feedback inhibition by bile acids. The feedback inhibitory potency of different bile acids increases with their hydrophobicity and resides at the level of transcription. Both enzyme activities and mRNA levels of cholesterol-7α-hydroxylase and HMG-CoA reductase have a diurnal periodicity that is independent of the bile acid return to the liver during enterohepatic circulation. This circadian rhythm is probably controlled by glucocorticoids at the level of gene transcription. Thyroid hormones also increase the 7α-hydroxylase activity by inducing the enzyme. On a short-term basis the activity of 7α-hydroxylase

is regulated by phosphorylation/dephosphorylation; for example, phosphorylation by cAMP-dependent protein kinase activates the enzyme.

Bile acid synthesis plays an important role in maintaining cholesterol homeostasis in the body. Cholesterol is either synthesized in hepatocytes de novo from acetyl-CoA or is derived from intestinal absorption and reaches the liver through chylomicron remnant endocytotosis (Fig. 67.7). Dietary cholesterol or its metabolic derivatives (bile acids) inhibit de novo cholesterol synthesis in the liver by transcriptional repression of HMG-CoA reductase, the rate-controlling step in cholesterol biosynthesis. On the other hand, the HMG-CoA reductase is induced under conditions of low dietary cholesterol supply. Major sinks of cholesterol are bile acid synthesis and biliary excretion of cholesterol, whereas steroid hormone synthesis is of minor importance. Acyl-CoA:cholesterol acyltransferase converts free cholesterol to cholesterol esters. These may be seen as stores for cholesterol, which can be mobilized by cytosolic cholesteryl ester hydrolase. The four enzymes involved in hepatic cholesterol metabolism (HMG-CoA reductase, 7α-hydroxylase, acyltransferase and ester hydrolase) are subject to metabolic interconversion (i.e., control by phosphorylation/dephosphorylation), and their coordinated regulation is crucial for the maintenance of cholesterol homeostasis in the body.

67.3.7 Processing of Hormones and Mediators

The liver is not only a target of hormone action, but also participates in degradation, activation, storage and secretion of hormones [35]. Many peptide hormones are cleared to the extent of more than 50% during a single liver passage (Table 67.6). Thus, insulin and glucagon concentrations are considerably higher in the portal vein than in the hepatic vein. Accordingly, there are hormone gradients along the liver acinus, which contribute to the expression and maintenance of zonal hepatocyte heterogeneity [35,57]. Also steroid hormones are predominantly taken up by the liver; their metabolism involves phase I and II reactions of biotransformation (see Sect. 67.3.8). After hydroxylation or reduction and subsequent conjugation with glucuronic acid, mineralocorticoids, glucocorticoids and sex steroid hormones are excreted into bile. In the case of estrogens, the hydroxylated products, and during the enterohepatic circulation further modified products, can still exhibit hormonal activity with a spectrum different from that exerted by the parent compounds. Hepatic inactivation of catecholamines is achieved predominantly by methylation, oxidative deamination and oxidation. More than 90% of eicosanoids are cleared during a single liver passage. From peptide leukotrienes, the amino acid moieties are hydrolyzed off by dipeptidases and gamma glutamyl transpeptidase, thereby converting the highly potent peptide leukotrienes C_4 and D_4 to less active metabolites, such as leukotriene E_4 or N-acetyl-leukotriene E_4, which can be further oxidized by peroxisomal fatty acid w-oxidation to inactive products [41]. All these compounds can be

Table 67.6. Hepatic role in the processing of hormones and mediators

	Activation	Inactivation[a]	Storage
Insulin		+(20–55%)	
Glucagon		+(20–40%)	
Growth hormone		+(90%)	
Thyroid hormones	+	+	+
Steroid hormones		+(10–90%)	
Catecholamines		+(50–80%)	
Eicosanoids	+	+(>90%)	
Extracellular ATP		+(>99%)	
Vitamin D	+	+	
Vitamin A			+

[a] Percentages in round brackets refer to the extent of hormone clearance during a single liver passage

Table 67.7. Major pathways of biotransformation

Phase I reactions	Phase II reactions
Hydroxylation	Glucuronidation
Expodiation	Glycosylation
N-,O-S-dealkylation	Sulfatation
Dehalogenation	Methylation
Alkoholoxidation	Acetylation
N,S-oxidation	Condensation
Amine oxidation	Glutathione conjugation
Hydratation	Amino acid conjugation
Reductions	Fatty acid conjugation
Hydrolysis	
Isomerization	

excreted into bile; however, in the biliary compartment retroconversion of leukotriene D_4 into leukotriene C_4 can occur.

However, the liver can also be a site of hormone or mediator production and function like an endocrine organ. For example, the adenosine concentration in hepatic venous blood is 10-fold that in the portal vein [40,58]; however, when adenosine is delivered in excess via the portal vein it is removed by the liver. In the liver, adenosine-induced stimulation of purinergic P_1 receptors on nonparenchymal cells liberates eicosanoids from these cells, which in turn can act on parenchymal cells. The liver also produces and secretes insulin-like growth factor (IGF-1). IGF-1 has a 50% sequence homology with proinsulin and exerts an insulin-like and growth-promoting action on hepatocytes ("autocrine effects") and extrahepatic tissues [61,62].

In the liver, vitamin D is hydroxylated in position 25 whereas hydroxylation in position 1 occurs in the kidney. Thus, the liver participates in the conversion of vitamin D to the biologically active "vitamin D hormone" calcitriol (1,25-dihydroxycholecalciferol; cf. Chap. 75).

The liver plays an important part in thyroid hormone metabolism. The thyroid gland secretes about 100 nmol thyroxin (T_4) and 5 nmol 3,5,3-triiodothyronine (T_3) per day. Despite some biological activity, T_4 is regarded as a prohormone, which is activated to the biologically highly active T_3 by deiodination. The liver is the site of 70% of total-body T_4-to-T_3 conversion. T_3 is secreted by the liver

and delivered, after binding to liver-borne plasma proteins, to target organs. About 30% of total-body T_4 and T_3 is stored in the liver.

67.3.8 Biotransformation

Many potentially harmful endo-and xenobiotics are detoxified and excreted by the liver ("biotransformation" [19]). These compounds are usually lipophilic and escape urinary excretion due to their poor water solubility. Hepatic metabolism of drugs and toxins can be divided into two phases:

- I, Reactions generating more polar products (Table 67.7), which can be further processed by
- II, Conjugation reactions

Among the enzymes catalyzing phase I reactions, the cytochrome P_{450} monooxygenase system has gained special interest. Up to now, the P_{450} gene superfamily comprises nine gene families with more than 70 genes [21]. The gene products, i.e., the various cytochrome P_{450} isoenzymes, cover a broad substrate spectrum. Here, specific isoenzymes can be induced by their substrates [54], which also include ethanol. Cytochrome P_{450} induction resulting from chronic ethanol ingestion can therefore accelerate drug and procarcinogen metabolism, whereas acute ethanol excess impairs the biotransformation of some drugs owing to drug/ethanol competition for microsomal oxidation. Cytochrome P_{450} is closely associated with NADPH-cytochrome P_{450} reductase in the endoplasmic reticulum. The system shows the following overall reaction:

$$RH + O_2 + NADPH + H^+ \longrightarrow ROH + H_2O + NADP^+$$

Because electron transfer to molecular oxygen occurs during the reaction sequence, reactive oxygen species can occasionally dissociate from the cytochrome prior to oxidation of the substrate [75]. These potentially harmful reactive oxygen species (O_2^-, H_2O_2, R-OOH) can induce lipid peroxidation and DNA modification (so-called oxidative stress [67]), especially when the defense systems against oxidative stress (superoxide dismutase, catalase, glutathione peroxidase/glutathione reductase) are overwhelmed [67,68].

$$2O_2^- + 2H^+ \longrightarrow H_2O_2 + O_2 \text{ (superoxide dismutase)}$$
$$2H_2O_2 \longrightarrow 2H_2O + O_2 \text{ (catalase)}$$
$$H_2O_2 + 2GSH \longrightarrow GSSG + 2H_2O \text{ (glutathione peroxidase)}$$
$$R\text{-}OOH + 2GSH \longrightarrow GSSG + ROH + H_2O \text{ (glutathione peroxidase)}$$
$$GSSG + NADPH + H^+ \longrightarrow 2GSH + NADP^+ \text{ (glutathione reductase)}$$

The microsomal cytochrome P_{450} system can also catalyze electron transfer to drugs [13,38]. This can lead to redox cycling; i.e., an electron transfer from cytochrome P_{450} to the drug and from the drug to O_2, resulting in a continuous generation of the superoxide anion but regeneration of the native drug. Also, CCl_4 can accept electrons from cytochrome P_{450} leading to the generation of the toxic $CCl_3 \cdot$ radical and damage of cellular structures [69]. The expression of the cytochrome P_{450} system predominantly in perivenous hepatocytes explains why the perivenous area is more prone to damage by xenobiotics such as CCl_4. Thus, phase I biotransformation does not necessarily imply "detoxication"; instead, many protoxins or procarcinogens are activated by phase I reactions owing to conversion into derivatives with much higher toxic or carcinogenic potency. For example, acetaminophen is activated by phase I reactions to a quinone derivative, which can covalently bind to a variety of cellular structures. This normally does not occur, because the quinone is rapidly detoxified by glutathione conjugation. However, when hepatocellular glutathione stores are depleted, the toxic phase I metabolite accumulates and exerts its damaging action. This explains why paracetamol is harmless at low doses, whereas intoxication (>10 g) produces liver cell necrosis.

Conjugation with glutathione [68] may be seen as a paradigm for so-called phase II reactions of biotransformation (Table 67.7), which lower the hydrophobicity of xenobiotics further and facilitate their excretion into bile after phase I processing. Biliary excretion occurs by way of ATP-driven canalicular transporters such as the multisubstrate-specific organic anion transporter (MOAT) or the multidrug-resistance gene product (MDR; see Sect 72.4.2). Excretion of phase I- and II-processed xenobiotics into bile has been termed phase III of biotransformation [33]. Figure 67.8 depicts hepatic handling of aflatoxin B_1, a potent carcinogen of dietary origin and epidemiological importance for development of hepatocellular carcinoma.

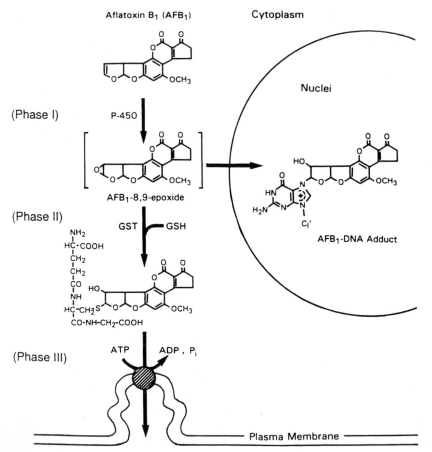

Fig. 67.8. Biotransformation of aflatoxin B_1 in liver. Aflatoxin B_1 (*AFB$_1$*) is produced by *Aspergillus flavus* and as a dietary contamination it is an important factor in the development of hepatocellular carcinoma throughout the world. AFB_1 is activated via cytochrome P_{450} to its epoxide, a potent carcinogen that can form DNA adducts. It is detoxified by glutathione conjugation and the conjugate is excreted into bile via the ATP-driven multiorganic anion transporter. (Adapted from [33])

Table 67.8. Functions of bile formation

Excretion of endobiotics (bilirubin, porphyrins, cholesterol, aged protein, leukotriene degradation products, oxidized glutathione)
Excretion of xenobiotics (drugs, heavy metals, environmental toxins)
Digestion (bile acids, HCO_3^--rich fluid)
Intestinal immunity (IgA secretion)

67.4 Bile Formation

67.4.1 General Considerations

The hepatocyte is a highly polarized cell, with transcellular transport being directed from the sinusoidal (basolateral) to the canalicular (apical) domain of the plasma membrane. Tight junctions between adjacent hepatocytes maintain this polarity by creating a barrier between the two membrane domains and also between plasma and the biliary canalicular compartment (Fig. 67.9).

The function of bile formation resides not only in the excretion of xenobiotics and endobiotics, such as cholesterol, porphyrins, bilirubin and aged proteins, but also serves digestive functions in the intestinal lumen by providing bicarbonate-rich fluid and bile acids in order to optimize breakdown of ingested foodstuffs. It also has an immunological function as it transports IgA to the intestinal mucosa (Table 67.8). Whereas small water-soluble compounds are preferentially excreted via the kidneys, lipid-soluble organic compounds with a molecular weight of 300–500 Da are excreted mainly into the bile after conversion by hepatic biotransformation steps into more hydrophilic metabolites (see Sect. 67.3.8).

Bile is an aqueous solution of organic and inorganic compounds (Table 67.9); its osmolarity is close to that of the blood plasma. Normally, a human adult produces about 600 ml of bile per day. Above a critical concentration, bile acids form macromolecular complexes (micelles) owing to their amphipathic character. The concentrations of cholesterol and phospholipids in bile can far exceed their water solubility, because of the formation of mixed micelles consisting of bile acids, phospholipids and cholesterol.

67.4.2 Canalicular (Hepatocellular) Bile Formation

Canalicular bile is the primary secretion product of the hepatocytes, and it can be further processed and modified by epithelial cells in the bile-draining system.

Bile Flow Dependent on and Independent of Bile Acid. Canalicular bile formation is driven by the active transcellular transport of organic and inorganic anions into the canalicular lumen [6,15,51,60,66] (cf. Chap. 59). These carriers are located in both the sinusoidal and canalicular membrane and can build up transepithelial concentration gradients. The negatively charged tight junctions that separate the canalicular from the sinusoidal membrane domain prevent the backdiffusion of the secreted anions into the sinusoidal space. They are, however, as in other "leaky epithelia," highly permeable to electrolytes such as Na^+. Water follows the osmotic gradient via the transcellular or paracellular route (Fig. 67.9). Bile acids constitute the quantitatively most important fraction of actively secreted organic anions and in line with this, bile flow increases linearly with the amount of bile acids secreted (Fig. 67.10). Depending on the micelle-forming potency of the bile acids (which tends to diminish the osmotic driving force), this relationship predicts an increase of bile flow by 8–15 ml/mmol bile acids secreted (so-called bile-acid-dependent bile flow). As shown in Fig. 67.10, extrapolation of bile acid excretion to zero gives a residual bile flow, which can also be directly measured in isolated bile acid-free perfused liver. This fraction of bile flow is apparently independent of bile acid secretion (so-called bile-acid-independent bile flow). The solutes responsible for bile-acid-independent bile flow include glutathione and its conjugates, bicarbonate, bilirubin glucuronide, and a variety of other organic compounds, which can actively be secreted by the transport ATPases located in the canalicular membrane.

Transcellular Transport of Organic Anions and Cations. As in many other epithelial cells, the $(Na^+ + K^+)$-ATPase is only found in the basolateral membrane of the hepatocyte. This primary active ion transporter creates transmembrane ion gradients, which provide the driving force for secondary or tertiary active transport systems in the sinusoidal membrane (Fig. 67.2). In concert with primary active transport systems in the canalicular membrane, these ion-gradient-driven sinusoidal transporters contribute to the vectorial flow from the blood into the biliary compartment (Fig. 67.9). Many transport systems involved in bile secretion have been identified which are either characteristic for the sinusoidal or the canalicular membrane domain (Table 67.10).

Basolateral Uptake Systems for Organic Anions and Cations. Bile acids undergo enterohepatic circulation and only a small portion of the bile acids secreted by the liver

Table 67.9. Composition of human bile

Component	Concentration in bile
Na^+	132–165 mmol/l
K^+	4.2–5.6 mmol/l
Ca^{2+}	1.2–4.8 mmol/l
Mg^{2+}	1.3–3.0 mmol/l
Cl^-	96–126 mmol/l
HCO_3^-	17–55 mmol/l
Bile acids	3–45 mmol/l
Cholesterol	1.6–8.3 mmol/l
Phospholipids	0.3–11.0 mmol/l
Bile pigments	0.8–3.2 mmol/l
Protein	

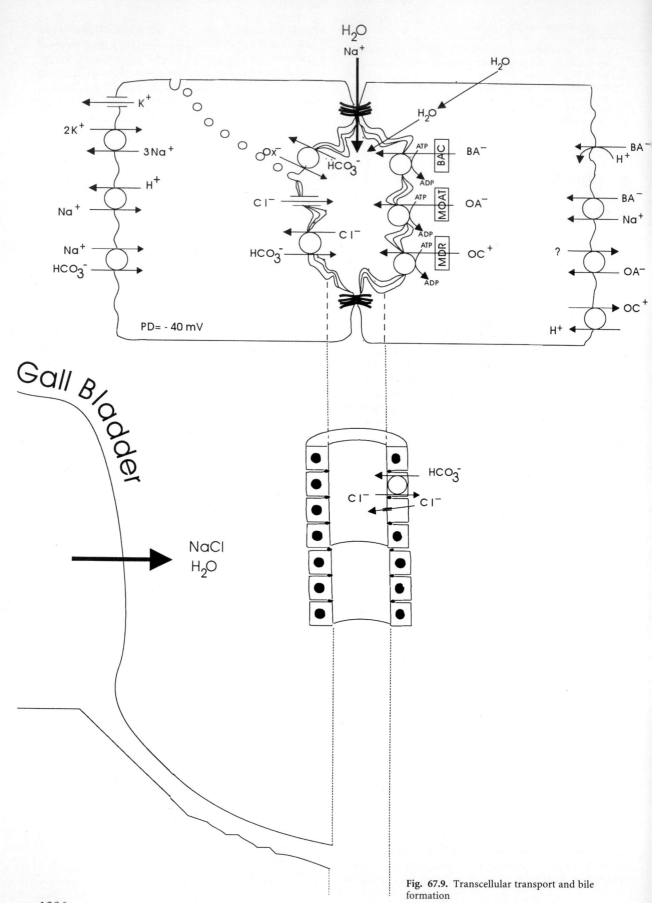

Fig. 67.9. Transcellular transport and bile formation

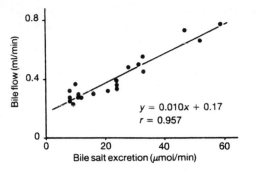

Fig. 67.10. Correlation between bile acid excretion and bile flow in man. The slope of the *line* expresses the osmotic activity of secreted bile acids; its *intercept* with the *ordinate* at zero bile salt excretion denotes "bile acid-independent bile flow." (From [60])

$$y = 0.010x + 0.17$$
$$r = 0.957$$

are derived from de novo synthesis (see Sect. 67.3.6). This necessitates efficient uptake of intestinally reabsorbed bile acids by the liver. Here, bile acids are taken up across the sinusoidal membrane in various ways, only some of which are Na^+ dependent; the respective transporters, however, actually exhibit multisubstrate specificity. For example, the Na^+-dependent bile acid carrier [59,65] transports not only conjugated bile acids, such as taurocholate, but also bumetanide, estradiol, phalloidin and verapamil. Its K_m for bile acids is in the range of the physiological bile acid concentration in the portal blood and its V_{max} by far exceeds the capacity for canalicular secretion. Accordingly, sinusoidal uptake of bile acids is not rate limiting for bile acid excretion, and some bile acids, such as taurocholate, can accumulate intracellularly and induce cell swelling [28,65]. Bile acid uptake also occurs via another organic anion transport system, which is Na^+-independent and exhibits both a high-affinity, Cl^- dependent, and a low-affinity, Cl^--independent, transport component. Its driving force is unknown, but some evidence suggests that it is an anion exchange system. The transporter is multispecific and also transports bilirubin, indocyanine green or bromosulfophthalein. Whereas trihydroxylated conjugated bile acids are the predominant substrate for the Na^+-dependent transporter, unconjugated bile acids prefer the Na^+-independent transporter. In addition, unconjugated bile acids may enter the hepatocytes by nonionic diffusion, since they behave like weak acids at physiological pH. In addition, a hydroxyl anion/cholate exchange system has been described, but this may represent nonionic diffusion. Other transporters mediating the basolateral uptake of various organic anions and cations are listed in Table 67.10 and Fig. 67.9. In general, they exhibit a broad substrate specificity. In addition, there is evidence for an endocytotic vesicular pathway mediating the uptake of large organic cations such as lucigenin.

Canalicular Transport Systems. Canalicular secretion of bile acids and other organic anions or cations generates the osmotic gradient that provides the driving force for osmotic water flow into the bile canaliculi (Fig. 67.9). Three ATP-dependent, i.e., primary active, transporters in the canalicular membrane have so far been identified (Table 67.10), which act to concentrate bile acids, glutathione conjugates, bilirubin glucuronide and other organic anions and cations in the canalicular lumen [1,2,33,37,50,52,71]. The ATP-dependent anion and cation transport systems exhibit a broad substrate specificity. ATP-dependent bile acid secretion is also membrane potential sensitive. This is due to the anionic character of conjugated bile acids and the inside negative transmembrane potential (so-called potential-driven bile acid transport [47]). The voltage-driven component may therefore depend on the activity of K^+ and Cl^- channels; however, ATP hydrolysis provides the bulk of transport energization. The multispecific organic anion transport system (MOAT) mediates the excretion of bilirubin and glutathione conjugates (Fig. 67.8, Table 67.10). The MOAT defect in a mutant rat strain (TR$^-$) is characterized by impaired excretion of divalent anions into bile. Cationic drugs, such as daunomycin, vincristine or doxorubicin, are excreted via the MDR, a group of glycoproteins with a molecular weight of approximately 170 kDa. Induction of these transporters in part accounts for the loss of therapeutic effectivity by otherwise potent drugs. The human Mdr3 P-glycoprotein is highly conserved among mammals and corresponds to the *mdr2* gene in mice; it does not transport hydrophobic anticancer drugs. Its function is now recognized to reside in the secretion of phospholipids into bile [69a]. Disruption of the *mdr2* gene in mouse embryonic stem cells yielded mice homozygous for this disruption. These mice develop liver disease characterized by focal hepatocyte necrosis, ductular proliferation and portal inflammation, which is apparently caused by the complete inability of the liver to secrete phospholipids into the bile [69a] and a consecutively impaired formation of mixed micelles in the biliary fluid. Heterozygotes did not develop liver disease but had half the level of phospholipids in bile.

The Cl^-/HCO_3^- exchanger functions to extrude HCO_3^- and is activated upon increases in intracellular pH; however its role in determining bile-acid-independent bile flow is un-

Table 67.10. Distribution of hepatocellular plasma membrane transport systems

Sinusoidal membrane	Canalicular membrane
Na^+ + K^+ATPase	ATP-dependent bile acid carrier
Na^+/H^+ antiport	ATP-dependent multiple
Na^+-HCO_3^- symport	organic anion transporter
Na^+-dependent bile	(MOAT; "leukotriene export
acid transporter	carrier")
Na^+-independent	ATP-dependent multidrug
multispecific	export carrier (gp 170,
organic anion	multidrug resistance gene
transporter	product, human Mdr1)
Organic cation/H^+	MDR-2 P-glycoprotein
antiport	(phospholipid excretion)
Cl^-/HCO_3^- exchanger	Cl^-/HCO_3^- antiport
SO_4^{2-}/OH^- antiport	
Na^+-dependent	Na^+-independent sulfate
dicarboxylate	transporter
transporter	

certain. Theoretically, the combined action of the canalicular Cl^-/HCO_3^- exchanger and a potential-driven Cl^- exit into the canalicular lumen could favor HCO_3^- secretion (Fig. 67.7). Direct measurements of canalicular pH, however, revealed electrochemical equilibrium with the extracellular H^+ concentration. Ursodesoxycholic acid, which produces a bicarbonate-rich choleresis, does not change the pH in canalicular bile. These findings suggest that the bulk of bicarbonate secretion occurs distal to the site of canalicular bile formation (see below).

Intracellular Transport and Transcytosis. Little is known about the intracellular transport of bile acids from the sinusoidal to the canalicular region inside the hepatocyte. Diffusion may be involved, but three cytosolic bile acid-binding proteins, which also function as proteins binding 3-α-hydroxysteroid dehydrogenase, glutathione-S-transferase, or fatty acids, have been identified. Their role in intracellular bile acid transport, however, is controversial. There is some evidence for a vesicular bile acid transport through the hepatocyte, since intracellular bile acid-containing vesicles are detectable by electron microscopy. Bile acids could enter these microsomal or Golgi-derived vesicles driven by the inside positive membrane potential. Although these findings support the concept of a vesicular bile acid transport, the possibility must be envisaged that these vesicles function to transport the canalicular bile acid carrier molecules to the canalicular membrane domain rather than to transport bile acids. Thus, the relative importance of vesicular and non-vesicular bile acid transport is unclear.

Transcytotic transport processes have been established for protein excretion into bile, however. In the hepatocyte, newly synthesized membrane proteins, including those destined for the canalicular membtane, are initially targeted from the Golgi apparatus and the endoplasmic reticulum to the sinusoidal membrane. Here, those destined for the canalicular membrane are re-sorted, again endocytosed, and reach the canalicular membrane via transcytosis [32]. In the rat this holds, for example, for the polymeric IgA receptor, which after synthesis and protein modification at the endoplasmic reticulum and the Golgi apparatus enters the sinusoidal membrane and is endocytosed again after binding of circulating IgA. Inside the hepatocyte, dissociation of ligand (IgA) and receptor occurs; the latter is recycled to the sinusoidal membrane, whereas IgA-containing vesicles are then targeted to the canalicular membrane domain and IgA enters the bile via exocytosis [70]. Other proteins and polysaccharides can also be taken up by receptor-mediated or fluid-phase endocytosis and reach via transcytosis the biliary compartment. The transcytotic vesicular pathway accounts for less than 10% of total bile flow and is microtubule-dependent, as suggested by its sensitivity to colchicine. Lysosomal degradation products can be discharged into the canalicular lumen via exocytosis. Vesicular transport and exocytosis are also thought to account for the excretion of cholesterol and phospholipids into bile [44].

Paracellular Permeability. The importance of cytoskeletal structures for bile formation is not only underlined by the microtubule dependence of transcytosis, but also by a cumulation of microfilaments and other cytoskeletal proteins in the pericanalicular region. They are thought to mediate periodic canalicular contractions and to stabilize tight junctions. Integrity of the tight junctions is essential for bile formation; their disruption will dissipate the osmotic gradient generated by canalicular secretion and cause cholostasis. This occurs in experimental systems such as the perfused liver upon removal of extracellular Ca^{2+} or under the influence of hydroperoxides. Current evidence suggests that the redox state of cellular thiols may exert some control on tight junctional permeability.

Regulation of Canalicular Secretion. Apart from tight junctional permeability, several other mechanisms, such as alterations in the hepatocellular hydratation state or specific hormone effects, contribute to the regulation of canalicular secretion. The V_{max} of taurocholate secretion into bile doubles within minutes when the hepatocellular hydratation state increases by 10–15% [28]. This cell-swelling-induced stimulation of bile acid excretion is fully blocked in presence of the microtubule inhibitory colchicine, and current observations are compatible with the idea that pre-existing ATP-dependent bile acid carrier molecules are stored underneath the canalicular membrane; their microtubule-dependent insertion into the canalicular membrane occurs in response to cellular swelling and results in an increase of bile acid secretion capacity within minutes [6,28]. Because some bile acids, such as taurocholate, can lead to cell swelling, which in turn increases the V_{max} of canalicular secretion, bile acids can actually augment their own secretion.

Bile flow is also regulated by hormone-induced alterations of other cellular second-messenger systems. The Cl^-/HCO_3^--exchanger appears to be targeted under the influence of cAMP to the canalicular membrane. In addition, cAMP increases bile flow by stimulating transcellular bile acid transport, the transcytotic vesicle pathway, but apparently decreases cholesterol secretion. Activation of protein kinase C diminishes sinusoidal taurocholate uptake, increases paracellular permeability, and blocks some components of transcytotic vesicle transport, but there is some evidence that canalicular transport ATPases are activated. Elevation of intracellular Ca^{2+} results in an increased tight junctional permeability.

67.4.3 Processing of Bile by Ductular Cells

Canalicular bile is modified during its passage through the draining biliary tree by both reabsorption and secretion. For example, amino acid moieties of glutathione or glutathione conjugates are cleaved off under the influence of ectoenzymes such as γ-glutamyltranspeptidase, and the amino acids formed are reabsorbed by ductular cells. Ductular secretion may account for about 10% of bile formation, a process that is stimulated by secretin, resulting in a marked stimulation of bicarbonate secretion. The mechanism by which ductular cells secrete bicarbonate is not known, but cholehepatic shunting of unconjugated bile acids, such as ursodesoxycholate, may play a part. According to this model (Fig. 67.9), the bile acids are reabsorbed by ductular cells in the protonated form, leaving bicarbonate behind in the lumen, and are recirculated to the hepatocyte, where they are excreted in the anionic form. Ductular cells in man, but not in the rat, secrete IgA.

Gallbladder. The gallbladder has the function of concentrating the primary bile as it is delivered from the liver. The gallbladder epithelium is primed for "bulk" absorption of NaCl and H_2O inasmuch as it has a leaky epithelium (cf. Chap. 59). Absorption of H_2O and NaCl is accomplished by the gallbladder epithelium owing to the combined action of apical Na^+/H^+ antiport and Cl^-/HCO_3^- exchange and by way of Na^+-Cl^- symport and basolateral extrusion of NaCl.

67.5 Other Functions of the Liver

67.5. Blood Reservoir Function

In many species the liver acts as a passive blood reservoir, which can be mobilized or increased within seconds to minutes depending on the hemodynamic situation and the influence of hepatic nerves. Liver blood content is about 20–25 ml/100 g liver and is determined by the hepatic venous pressure, innervation and arterial and portal blood inflow. About 50% of hepatic blood can be expelled into the circulation within 30 s upon stimulation of sympathetic liver nerves, suggesting an involvement in cardiovascular reflexes. Indeed, in the dog, carotid sinus hypotension results in a significant reduction of hepatic blood volume, whereas carotid sinus hypertension increases liver blood volume [9]. The blood storage function of the liver resides in the substantial compliance of hepatic capacitance vessels, which is about 10 times that of the systemic circulation. Thus, minor alterations in hepatic venous pressure may cause marked changes in liver congestion. Similar considerations apply for the input via the hepatic artery or the portal vein.

67.5.2 Pressure Sensing

Increases of the portal venous perfusion pressure enhance afferent hepatic nerve activity and increase efferent renal nerve activity [11,53], leading to a decrease in the renal glomerular filtration rate [14]. Thus, putative intrahepatic baroreceptors can exert neural control on renal function.

67.5.3 Systemic pH Regulation

The liver plays an important role in maintaining acid–base homeostasis in the organism. This aspect of liver function is presented in detail in Chaps. 68 and 75.

References

1. Adachi Y, Kobayashi H, Kurumi Y, Shouji M, Kitano M, Yamamoto T (1991) ATP-dependent taurocholate transport by rat liver canalicular membrane vesicles. Hepatology 14:655–659
2. Akerboom TPM, Narayanaswami V, Kunst M, Sies H (1991) ATP-dependent S-2,4-dinitrophenylglutathione transport in canalicular plasma membrane vesicles from rat liver. J Biol Chem 266:13147–13152
3. Andus T, Bauer J, Gerok W (1991) Effects of cytokines on the liver. Hepatology 13:364–375
4. Ballet F (1990) Hepatic circulation: potential for therapeutic intervention. Pharmacol Ther 47:281–328
5. Bouwens L, De Bieser P, Vanderkerken K, Geerts B, Wisse E (1992) Liver cell heterogeneity: functions of non-parenchymal cells. Enzyme 46:155–168
6. Boyer JL, Graf J, Meier PJ (1992) Hepatic transport systems regulating pHi, cell volume and bile secretion. Annu Rev Physiol 54:415–438
7. Brown MS, Goldstein JL (1986) A receptor mediated pathway for cholesterol homeostasis. Science 232:34–47
8. Brown MS, Goldstein JL (1990) The hyperlipoproteinemias and other disorders of lipid metabolism. In: Wilson JD, Braunwald E, Isselbäcker KJ, et al (eds) Principles of internal medicine. McGraw Hill, New York, pp 1814–1825
9. Carneiro JJ, Donald DE (1977) Change in liver blood flow and blood content in dogs during direct and reflex alteration of hepatic sympathetic nerve activity. Circ Res 40:150–158
10. Christensen HN (1990) Role of amino acid transport and countertransport in nutrition and metabolism. Physiol Rev 70:43–77
11. Costreva DR, Castaner A, Kampine JP (1980) Reflex effects of hepatic baroreceptors on renal and cardiac sympathetic nerve activity. Am J Physiol 238:R390–394
12. Decker K (1985) Eicosanoids, signal molecules of liver cells. In: Berk PD (ed) Seminars in liver disease, vol 5. Thieme-Stratton, New York, pp 175–190
13. DeGroot H, Littauer A (1989) Hypoxia, reactive oxygen and cell injury. Free Radic Biol Med 6:541–551
14. DiBona GF (1982) The functions of the renal nerves. Rev Physiol Biochem Pharmacol 94:75–181
15. Erlinger S (1988) Bile flow. In: Arias IM, Jakoby WB, Popper H, Schachter D, Shafritz DA (eds) The liver: biology and pathobiology. Raven, New York, pp 643–661
16. Feldmann G, Scoazec JY, Racine L, Bernuau D (1992) Functional hepatocellular heterogeneity for the production of plasma proteins. Enzyme 46:139–154
17. Forssmann WG, Ito S (1977) Hepatocyte innervation in primates. J Cell Biol 74:299–313
18. Gardemann A, Püschel GP Jungermann K (1992) Nervous control of liver metabolism and hemodynamics. Eur J Biochem 207:399–411
19. Gascon MP, Dayer P (1992) Hepatic metabolism of drugs and toxins. In: McIntyre N, Benhamou JP, Bircher J, Rizetto M, Rodes J (eds) Oxford textbook of clinical hepatology. Oxford University Press, Oxford, pp 247–259
20. Gibbons GF (1990) Assembly and secretion of hepatic very-low-density lipoprotein. Biochem J 268:1–13
21. Gonzalez FJ (1990) Molecular genetics of the P-450 superfamily. Pharmacol Therapeut 45:1–38

22. Gross V, Steube K, Tran-Thi TA, Häussinger D, Legler G, Decker K, Heinrich PC, Gerok W (1987) The role of N-glycosylation for the plasma clearence of rat liver secretory glycoproteins. Eur J Biochem 162:83–88

23. Häussinger D (1983) Hepatocyte heterogeneity in glutamine and ammonia metabolism and the role of an intercellular glutamine cycle during ureogenesis in perfused rat liver. Eur J Biochem 133:289–275

24. Häussinger D (1990) Nitrogen metabolism in liver: structural and functional organization and physiological relevance. Biochem J 267:281–290

25. Häussinger D, Gerok W (1986) Metabolism of amino acids and ammonia. In: Thurman RG, Kauffmann FC, Jungermann K (eds) Regulation of hepatic metabolism. Plenum, New York, pp 253–291

26. Häussinger D, Lang F (1992) Cell volume and hormone action. Trends Pharmacol Sci 13:371–373

27. Häussinger D, Soboll S, Meijer AJ, Gerok W, Tager JM, Sies H (1985) Role of plasma membrane transport in hepatic glutamine metabolism. Eur J Biochem 152:597–603

28. Häussinger D, Hallbrucker C, Saha N, Lang F, Gerok W (1992) Cell volume and bile acid excretion. Biochem J 288:681–689

29. Häussinger D, Gerok W, Lang F (1994) Regulation of cell function by the cellular hydration state. Am J Physiol 267:E343-E355

30. Harry DS, McIntyre N (1992) Plasma lipids and lipoproteins. In: McIntyre N, Benhamou JP, Bircher J, Rizetto M, Rodes J (eds) Oxford textbook of clinical hepatology. Oxford University Press, Oxford, pp 143–157

31. Heinrich PC, Castell JV, Andus T (1990) Interleukin-6 and the acute phase response. Biochem J 265:621–636

32. Hubbard AL, Stieger B, Bartles JR (1989) Biogenesis of endogenous plasma membrane proteins in epithelial cells. Annu Rev Physiol 51:755–770

33. Ishikawa T (1992) The ATP-dependent glutathione-S-conjugate export pump. Trends Biochem Sci 17:463–468

34. Jungermann K, Katz N (1986) Metabolism of carbohydrates. In: Thurman RG, Kauffmann FC, Jungermann K (eds) Regulation of hepatic metabolism. Plenum, New York, pp 27–53

35. Jungermann K (1986) Zonal heterogeneity and induction ofhepatocyte heterogeneity. In: Thurman RG, Kauffmann FC, Jungermann K (eds) Regulation of hepatic metabolism. Plenum, New York, pp 445–469

36. Jungermann K, Katz N (1989) Functional specialication of different hepatocyte populations. Physiol Rev 69:708–764

37. Kamimoto Y, Gatmaitan Z, Hsu J, Arias IM (1989) The function of gp 170, the multidrug resistence gene product in rat liver canalicular membrane vesicles. J Biol Chem 264:11693–11698

38. Kappus H, Sies H (1981) Toxic drug effects associated with oxygen metabolism: redox cycling and lipid peroxidation. Experientia 37:1233–1241

39. Kawada N, Klein H, Decker K (1992) Eicosanoid mediated contractility of hepatic stellate cells. Biochem J 285:367–271

40. Keppler D, Holstege A (1982) Pyrimidine nucleotide metabolism and its compartmentation. In: Sies H (ed) Metabolic compartmentation. Academic, London

41. Keppler D, Hagmann W, Rapp S, Denzlinger C, Koch HK (1985) The relation of leukotrienes to liver injury. Hepatology 5:883–891

42. Kilberg M, Häussinger D (eds) (1992) Mammalian amino acid transport-mechanisms and control. Plenum, New York

43. Lamers WH, Hilberts A, Furt E, Smith J, Jones CN, van Noorden CJF, Gaasbeek Janzen JW, Charles R, Moorman AFM (1989) Hepatic enzymic zonation: are-evaluation of the concept of the liver acinus. Hepatology 10:72–76

44. Lanzini A, Northfield TC (1991) Biliary lipid secretion in man. Eur J Clin Invest 21:259–272

45. Lawrence JC (1992) Signal transduction and protein phosphorylation in the regulation of cellular metabolism by insulin. Annu Rev Physiol 54:177–193

46. Li WH, Tanimura M, Luo CC, Datta S, Chan L (1988) The apolipoprotein multigene family: biosynthesis, structure, structure-function relationships and evolution. J Lipid Res 29:245–271

47. Meier PJ, Meier-Abt AS, Barrett C, Boyer JL (1984) Mechanisms of taurocholate transport in canalicular and basolateral rat liver plasma membrane vesicles. J Biol Chem 259:10614–10622

48. Meijer AJ (1985) Channeling of ammonia from glutaminase into carbamoylphosphate synthetase in liver mitochondria. FEBS Lett 191:249–251

49. Mortimore GE, Pösö AR (1987) Intracellular protein catabolism and its control during nutrient deprivation and supply. Annu Rev Nutr 7:539–564

50. Müller M, Ishikawa T, Berger U, Klünemann C, Lucka L, Schreyer A, Kannicht C, Reutter W, Kurz G, Keppler D (1991) ATP-dependent transport of taurocholate across the hepatocyte canalicular membrane mediated by a 110kD glycoprotein binding ATP and bile salt. J Biol Chem 266:18920–18926

51. Nathanson MH, Boyer JL (1991) Mechanisms and regulation of bile secretion. Hepatology 14:551–566

52. Nishida T, Gatmaitan Z, Che M, Arias IM (1991) Rat liver canalicular membrane vesicles contain an ATP-dependent bile acid transport system. Proc Natl Acad Sci 88:6590–6594

53. Nijima A (1977) Afferent discharges from venous pressoreceptors in the liver. Am J Physiol 232:C76–C81

54. Okey AB (1990) Enzyme induction in the cytochrome P-450 system. Pharmacol Ther 45:241–298

55. Pessin JE, Bell GI (1992) Mammalian facilitative glucose transporter family: structure and molecular regulation. Annu Rev Physiol 54:911–930

56. Pilkis SJ, Granner DK (1992) Molecular physiology of the regulation hepatic gluconeogenesis and glycolysis. Annu Rev Physiol 54:885–909

57. Polonsky K, Rubinstein H (1984) C-peptide as a measure of the secretion and hepatic extraction of insulin. Diabetes 53:486–494

58. Pritchard JB, O'Connor N, Oliver JM, Berlin RD (1975) Uptake and supply of purine compounds by the liver. Am J Physiol 229:967–972

59. Reichen J, Paumgartner G (1976) Uptake of bile acids by perfused rat liver. Am J Physiol 231:734–742

60. Reichen J (1992) Physiology of bile formation and of the motility of the biliary tree. In: McIntyre N, Benhamou JP, Bircher J, Rizetto M, Rodes J (eds) Oxford textbook of clinical hepatology. Oxford University Press, Oxford, pp 87–94

61. Russel W (1985) Growth hormone, somatomedins and the liver. Semin Liver Dis 5:46–58

62. Sara VR, Hall K (1990) Insulin-like growth factors and their binding proteins. Physiol Rev 70:591–615

63. Sasse D (1986) Liver structure and innervation. In: Thurman RG, Kauffmann FC, Jungermann K (eds) Regulation of hepatic metabolism. Plenum, New York

64. Schimassek H, Gerok W (1965) Control of the levels of free amino acids in plasma by the liver. Biochem Z 343:407–415

64a. Schliess F, Schreiber R, Häussinger D (1995) Activation of extracellular signal-regulated kinases Erk-1 and Erk-2 by cell swelling in H4IIE hepatoma cells. Biochem J 309:13–17

65. Schwarz LR, Burr R, Schwenk M, Pfaff E, Greim H (1975) Uptake of taurocholic acid into isolated rat liver cells. Eur J Biochem 55:617–623

66. Sellinger M, Boyer JL (1990) Physiology of bile secretion and cholestasis. Prog Liver Dis 9:237–259

67. Sies H (ed) (1985) Oxidative stress. Academic, London

68. Sies H, Ketterer B eds (1988) Glutathione conjugation. Academic, London

69. Slater TF (1984) Free-radical mechanisms in tissue injury. Biochem J 222:1–15

69a. Smit JJM, Schinkel AH, Oude-Elferink RPJ, Groen AK, Wagenaar E, van Deemter L, Mol CAAM, Ottenhof R, van der

Lugt NMT, van Roon MA, van der Valk MA, Offerhaus GJA, Berns AJM, Borst P (1993) Homozygous disruption of the murine mdr2 P-glycoprotein gene leads to a complete absence of phospholipid from bile and to liver disease. Cell 75:451–462

70. Solari R, Schaerer E, Tallichet C, Braiterman LT, Hubbard AL, Kraehenbuehl JP (1989) Cellular localization of the cleavage event of the polymeric immunoglobulin receptor and fate of its anchoring domain in the rat hepatocyte. Biochem J 257:759–768

71. Stieger B, O'Neill B, Meier PJ (1992) ATP-dependent bile-salt transport in canalicular rat liver plasma-membrane vesicles. Biochem J 284:67-74

72. Traber PG, Chianale J, Gumucio JJ (1988) Physiologic significance and regulation of hepatocellular heterogeneity. Gastroenterology 95:30–43

73. Vallance P, Moncada S (1991) Hyperdynamic circulation in cirrhosis: a role for nitric oxide. Lancet 337:776–778

74. Vlahcevic Zr, Heumann DM, Hylemon PB (1991) Regulation of bile acid synthesis. Hepatology 13:590–600

75. White RE, Coon MJ (1980) Oxygen activation by cytochrom P-450. Annu Rev Biochem 49:315–356

76. Windmueller HG (1984) Metabolism of vascular and luminal glutamine by intestinal mucosa in vivo. In: Häussinger D, Sies H (eds) Glutamine metabolism in mammalian tissues. Springer, Berlin Heidelberg New York, pp 61–77

77. Wisse E, Knook DL, Decker K (1989) Cells of the hepatic sinusoid. Kupffer Cell Foundation, Rijswijk, Netherlands

78. Withrington PG, Richardson PDI (1986) Hepatic hemodynamics and microcirculation. In: Thurman RG, Kauffmann FC, Jungermann K (eds) Regulation of hepatic metabolism Plenum, New York, pp 27–53

68 Zonal Metabolism in the Liver

D. Häussinger

Contents

68.1 Introduction

Despite their similar light-microscopic appearance, liver parenchymal cells (hepatocytes) differ in their enzyme equipment and metabolic function (so-called functional hepatocyte heterogeneity or metabolic zonation). This heterogeneity is not random but is determined by the position of the individual hepatocyte in the microcirculatory units of the tissue. Different techniques have been used to study functional hepatocyte heterogeneity [14]: histochemistry, immunohistochemistry, in situ hybridization. retrograde/antegrade liver perfusion, microdissection of liver tissue from both periportal and perivenous ends, histoautoradiography, studies on radiolabel incorporation and zonal cell damage in the intact organ, and separation of different subacinar hepatocyte populations. These studies revealed a remarkable hepatocyte heterogeneity resulting in a sophisticated structural and functional organization of metabolic functions in the liver acinus, with important consequences for our understanding of liver function in health and disease (for reviews see [9–14,16,23]).

The architecture of the mammalian liver is simple and uniform throughout the organ: hepatocytes are arranged in microcirculatory units extending along the sinusoids from the terminal portal venule to the terminal hepatic venule. One microcirculatory unit is termed the "acinus" or "metabolic lobulus" (see Sect. 67.1). Although these terms reflect different views on the spatial shape of the microcirculatory unit, for the functional considerations discussed below they can be used interchangeably. Depending on their subacinar localization, periportal (near the portal inflow) hepatocytes are distinguished from perivenous (near the outflow) hepatocytes, although the borderline between the two compartments is not clearly defined in biochemical and morphological terms and has to be specified for each metabolic function separately [10].

68.2 Functional Hepatocyte Heterogeneity

68.2.1 Zonation Patterns

Two different types of zonation can be distinguished: the "gradient type" and the "strict or compartment type" [11]. In the gradient type of zonation a given enzyme is expressed in each hepatocyte, but more or less steep activity gradients exist between periportal and perivenous hepatocytes, respectively. In the strict or compartment type of zonation, the enzyme is expressed in a subpopulation of hepatocytes only, whereas it is undetectable in others (Fig. 68.1).

68.2.2 Zonal Distribution of Enzymes and Metabolic Functions

The zonal distribution of enzymes is listed in Table 68.1. Periportal/perivenous enzyme activity gradients range between 0.3 and 5, except for glutamine synthetase and ornithine amino-tranferase, which exhibit strict type zonation. These enzymes are expressed exclusively in a small hepatocyte population (about 7% of all hepatocytes of an acinus) surrounding the terminal hepatic venule with layers about two or three cells thick (Fig. 68.2). These cells can be seen as a subpopulation of perivenous cells, and in view of their function have the name perivenous scavenger cells [11]. In contrast to other, upstream, hepatocytes, these scavenger cells are virtually free of urea cycle enzymes and exhibit unique transport features. As much as 70–100% of total hepatic uptake of vascular glutamate, aspartate and citric acid cycle dicarboxylates, such as α-ketoglutarate and malate, must be ascribed to these scavenger cells (Fig. 68.3). Accordingly, plasma membrane transport systems are also heterogeneously distributed in the liver acinus.

R. Greger/U. Windhorst (Eds.)
Comprehensive Human Physiology, Vol. 2
© Springer-Verlag Berlin Heidelberg 1996

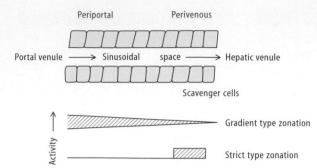

Fig. 68.1. The liver acinus and types of zonation

Table 68.1. Zonation of enzymes and transport systems in the liver acinus

Periportal	Perivenous	Zonation type
Glucose-6-phosphatase	Glucokinase	Gradient
Phosphoenolpyruvate carboxykinase	Hexokinase	Gradient
Fructose-1,6-bisphosphatase	Glucose-6-phosphate dehydrogenase	Gradient
	Pyruvate kinase	Gradient
	Isocitrate dehydrogenase	Gradient
β-Hydroxybutyryl-CoA dehydrogenase	Malic enzyme	Gradient
Hydroxymethylglutaryl-CoA reductase	Citrate lyase	Gradient
	Fatty acid synthase	Gradient
	Acetyl-CoA carboxylase	Gradient
	β-Hydroxybutyrate dehydrogenase	Gradient
	Alcohol dehydrogenase	Gradient
Succinate dehydrogenase		Gradient
Malate dehydrogenase		Gradient
Cytochrome oxidase		Gradient
Alanine aminotransfersase	Glutamate dehydrogenase	Gradient
Aspartate aminotransferase		
Tyrosine aminotransferase		
Carbamoylphosphate synthetase	Glutamine synthetase	?/Strict
Ornithine carbamoyltransferase	Ornithine transaminase	?/Strict
Argininosuccinate synthase		?
Arginase		?
Glutaminase		?
γ-Glutamyl transpeptidase		Gradient
Alanine, proline, serine uptake	Glutamate, aspartate uptake	Gradient/strict?
	Dicarboxylate uptake	Strict?
Glucose transporter (GLUT-2)	Glucosetransporter (GLUT-1)	Strict for GLUT-1
Cathepsin B, H	Cytochrome P450	Gradient[a]
Glutathione peroxidase	Glutathione-*S*-transferases	Gradient
D-Amino acid oxidase	UDP-Glucuronosyltransferases	Gradient
Monoamine oxidase	NADPH-cytochrome P450 reductase	Gradient

[a] Strict type zonation is also known for certain cytochrome P450 subspecies

However, it remains to be established whether the respective transport systems are expressed exclusively in scavenger cells or whether the transporters also exist in periportal hepatocytes but are inactive there. Whereas glutamate and aspartate are taken up by perivenous scavenger cells, uptake of most other amino acids, such as proline, glutamine, alanine, and serine, occurs predominantly by periportal hepatocytes.

As predictable from the zonal distribution of key enzymes (Table 68.1) and largely substantiated by zonal metabolic flux measurements, also metabolic pathways are zonally distributed in the acinus (Table 68.2). As a rule, functionally linked processes exhibit a joint subacinar localization. For example, uptake of most gluconeogenic amino acids occurs into periportal cells, i.e., the compartment also exhibiting highest activities of amino acid metabolizing and gluconeogenic enzymes and ureogenesis. Because gluconeogenesis from lactate or alanine is an endergonic process, its coupling to oxidative energy metabolism is required. Indeed, oxidative energy metabolism prevails in the periportal area, where the oxygen tension is highest. This is suggested by the higher activities of citric acid cycle enzymes and cytochrome oxidase, and a greater volume and cristae area of mitochondria in periportal than in perivenous cells. On the other hand, glycolysis and fatty acid synthesis from acetyl-CoA as exergonic reactions predominate in perivenous hepatocytes, i.e., a compartment normally facing lower oxygen tensions than periportal

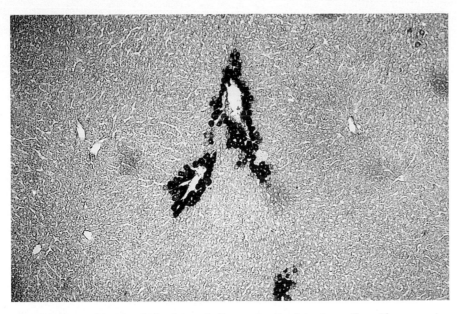

Fig. 68.2. Immunohistochemical staining of a liver section for glutamine synthase. The enzyme is restricted to a small cell population surrounding the terminal hepatic venules ("scavenger cells"), but is completely absent in more upstream (periportal) hepatocytes. (From [11])

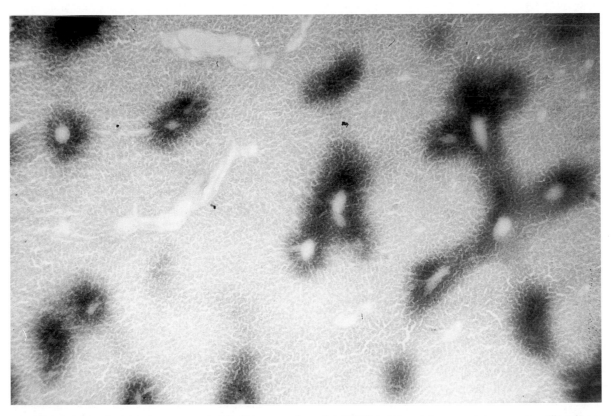

Fig. 68.3. Staining of perivenous scavenger cells by means of histoautoradiography after injection of [³H]glutamate. The radiolabeled amino acid is taken up only by a small hepatocyte population surrounding the terminal hepatic venules (scavenger cells). [³H]Glutamate uptake occurs into scavenger cells only, regardless of whether the radiolabel is added in antegradely (from portal to hepatic vein) or retrogradely (from hepatic to portal vein) perfused livers. (From [22])

Table 68.2. Zonation of metabolic pathways

Periportal	Perivenous
Gluconeogenesis	Glycolysis
Glycogen synthesis from amino acids and lactate	Glycogen synthesis from glucose
Oxidative energy metabolism	
Fatty acid oxidation	Fatty acid synthesis
Ketogenesis	
Cholesterol synthesis	
Amino acid uptake[a]	
Amino acid degradation[a]	
Urea synthesis	Glutamine synthesis[b]
	Ornithine transamination[b]
	Biotransformation
	Dicarboxylate uptake[b]
Phagocytosis (Kupffer cells)	Cytotoxicity (Kupffer cells)

[a] Except uptake of aspartate and glutamate
[b] Strict type of zonation

hepatocytes. Fatty acid synthesis is an NADPH-consuming process, and NADPH-generating reactions, such as malic enzyme, isocitrate dehydrogenase and glucose-6-phosphate dehydrogenase, exhibit higher activities in the perivenous area.

Drug-metabolizing enzymes catalyzing phase I and II reactions of biotransformation (see Sect. 72.3.8) show higher activities in perivenous hepatocytes, and cytochrome P_{450} is induced by phenobarbital predominantly in perivenous hepatocytes. On the other hand, there is some evidence that perivenous hepatocytes are less well equipped with glutathione peroxidase, rendering the perivenous compartment more susceptible to radical damage.

68.2.3 Factors Determining Zonal Heterogeneity

Zonal heterogeneity is frequently dynamic; that is to say, it changes with the metabolic state. Regulation takes place at the pretranslational level, as observed for phosphoenolpyruvate carboxykinase (PEPCK), or at the translational/posttranslational level, as shown for pyruvate kinase. Here, gene expression appears to be controlled primarily by the substrate and hormone supply to the individual cell. Because oxygen, substrates and hormones are consumed during a single liver passage, concentration gradients along the acinus are formed, which may contribute to zonal gene expression. Indeed, PEPCK and tyrosine aminotransferase are induced by glucagon in primary hepatocyte cultures to higher values at arterial than at venous oxygen tensions [19]. This explains the prevalence of these enzymes in periportal hepatocytes, which are exposed to higher oxygen tensions than perivenous hepatocytes. Similar considerations apply to hormone concentrations, which are higher in the periportal than in the perivenous zone for glucagon, insulin, norepinephrine, epinephrine and corticosteroids, but lower for adenosine

and triiodothyronine (see Sect. 68.3.7). The glucagon/insulin concentration ratio decreases along the porto-central distance. Both hormones have antagonistic actions with respect to PEPCK and glucokinase induction (see Sect. 68.3.2), and the insulin/glucagon concentration profiles along the acinus predict the development of opposing activity gradients for PEPCK (periportal activity maximum) and glucokinase (perivenous activity maximum), which are found in vivo. Indeed, when freshly isolated hepatocytes, i.e., a heterogenous suspension of periportal and perivenous cells, are cultured in the presence of insulin they develop a pattern of gluconeogenic and glycolytic enzymes similar to that of perivenous cells in vivo, whereas exposure to glucagon creates a periportal-like pattern [12,20]. Unfortunately, nothing is known about hormone receptor densities on periportal and perivenous cells.

Although minor amino acid concentration gradients exist along the porto-central distance, amino acid delivery via the portal vein may be involved in maintaining zonal heterogeneity. Many amino acids are taken up into hepatocytes by concentrative, mainly Na+-dependent, mechanisms and induce hepatocellular swelling (see Sect. 68.3.1). Such increases in the hepatocellular hydratation state are dependent upon the concentrative potency of the amino acid transporter in the plasma membrane; they modify not only cellular metabolism but also gene expression. Thus, zonal differences in amino acid transport may therefore result in different degrees of cell swelling, which could contribute to the maintenance of zonal heterogeneity at the level of gene expression. Apart from substrate and hormone gradients, zonal heterogeneity of gene expression appears to be controlled by innervation and locally generated mediators such as eicosanoids.

The exclusive expression of glutamine synthease in perivenous scavenger cells, which form a layer up to three cells thick around the hepatic venules is a paradigm for strict type zonation (Fig. 68.2). The factors responsible for strict type gene expression have not yet been determined, but involve transcriptional events [17]. In line with this, glutamine synthetase mRNA and protein display an identical pattern of acinar distribution. Strict type gene expression probably does not result from a differentiative event and is best explained by a relatively inflexible regulation of a single lineage of hepatocytes, which results in two compartments of gene expression. Hormone and substrate gradients may contribute, but cannot per se satisfactorily explain why a certain enzyme is expressed in one cell but not in the immediately adjacent cell (Fig. 68.2). Strict type zonation is related to the architecture of the liver [15]. It is likely that this expression pattern is among other factors triggered by the close vicinity of perivenous hepatocytes to vascular cells and extracellular matrix. In line with this, ring-shaped cultivation of hepatocytes around a central cluster of fibroblasts led to induction of glutamine synthetase in only the hepatocytes adjacent to the fibroblasts [21]. Urea cycle enzymes are normally not present in scavenger cells, but are found in all other more upstream hepatocytes in rat liver. Also in human liver urea cycle enzymes are restricted to periportal hepatoctes, but there is an untermediate zone between periportal hepatocytes and scavenger cells, which contains neither urea cycle enzymes nor glutamine synthetase [18].

68.3 Zonal Heterogeneity and Liver Function

The zonal separation of some metabolic pathways (Table 68.2), such as periportal glutamine degradation and perivenous glutamine resynthesis, predicts that opposing metabolite fluxes occur simultaneously in periportal and perivenous hepatocytes. Although at first glance this suggests energy-consuming "futile" cycling of metabolites, when the liver is seen as a whole it can be seen that considerable regulatory advantages arise from it.

68.3.1 Carbohydrate Metabolism

The distribution of enzyme activities (Table 68.1) predicts that periportal hepatocytes synthesize glucose from gluconeogenic precursors, whereas perivenous hepatocytes degrade glucose via glycolysis. Glycogen appears to be synthesized predominantly from gluconeogenic precursors in periportal cells, whereas vascular glucose is the main substrate for glycogen synthesis in perivenous hepatocytes. In line with this, glycogen cumulates in perfused rat liver upon addition of lactate in the periportal zone, whereas infusion of glucose leads to glycogen deposits in the perivenous zone [2]. In addition, the liver releases lactate after a glucose load from perivenous hepatocytes, which eventually, together with lactate produced by skeletal muscle, reaches periportal hepatocytes to be used there as gluconeogenic and glycogenic substrate.

Net hepatic glucose uptake during intesinal absorption switches to net glucose release in the postabsorptive state (see Sect. 68.3.2). Both periportal and perivenous cells contribute to this glucose-homeostatic response of the liver. Apparently, alterations of the hormonal and metabolic milieu accompanying the feeding–starvation cycle modify carbohydrate metabolism in periportal and perivenous cells in opposite directions. The direction of metabolic flux does not change in the individual cell; the shift from net glucose consumption during intestinal absorption to net output in the postabsorptive state is accomplished by increasing flux through the periportal gluconeogenic pathway, whereas simultaneously in perivenous cells flux through the glycolytic pathway decreases. In addition, gly-

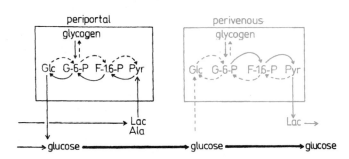

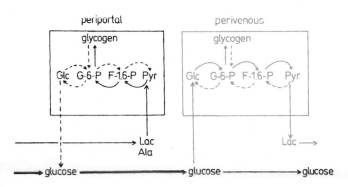

Fig. 68.4. Zonal regulation of carbohydrate metabolism during intestinal absorption and the postabsorptive state. (Modified from [11].) The reciprocal regulation in periportal and perivenous hepatocytes maintains glucose homeostasis in the organism. During absorption periportal cells synthesize glucose and glycogen from lactate and alanine, whereas perivenous cells synthesize glycogen and lactate largely from portal glucose. In the postabsorptive state lactate is formed predominantly from glycogen and to a small extent from portal glucose, whereas in periportal cells glucose is formed from lactate and glycogen. Alanine (*ala*), lactate (*lac*), glucose-6-phosphate (*G-6-P*), fructose-1,6-bisphosphate (*F-1,6-P*), pyruvate (*Pyr*), glucose (*Glc*). The thickness of *arrows* indicates the magnitude of fluxes

cogen stored in both subacinar compartments can be mobilized under the influence of hormones. Regulation of glucose homeostasis according to the zonation model is depicted schematically in Fig. 68.4.

68.3.2 Ammonia Metabolism

In the intact liver acinus, the two major ammonia-detoxicating systems, urea and glutamine synthesis, are anatomically switched behind each other. Accordingly, the portal blood will first come in contact with hepatocytes capable of urea synthesis before glutamine-synthesizing cells at the end of the acinar bed are reached. In functional terms, this organization represents the sequence of a periportal low-affinity, but high-capacity, system (ureogenesis) and a perivenous high-affinity system for ammonia detoxication (glutamine synthesis) [6]. The low ammonia affinity of downstream urea synthesis is due to a high K_m (ammonia) of carbamoylphosphate synthetase, the rate-controlling enzyme of the urea cycle, which is about 10fold the physiological portal ammonia concentration (see Sect. 68.2.3). Accordingly, only about 80% of a physiological portal ammonium load is converted into urea by periportal hepatocytes; the remainder reaches the small perivenous cell population containing glutamine synthetase. Here, perivenous glutamine synthetase acts as a high-affinity scavenger for the ammonia that has escaped periportal detoxication by urea synthesis. This also holds for ammonia produced during amino acid breakdown in the much larger periportal compartment. As shown in experiments with the isolated perfused rat liver, amino acid-derived ammonia is released from periportal cells into the sinusoidal space despite the high urea cycle enzyme activity in this compartment. This ammonia is transported in the bloodstream to perivenous hepatocytes, where it is eliminated via glutamine synthesis. This not only underlines the comparatively low affinity of periportal urea synthesis for ammonia and the importance of ammonia scavenging by perivenous hepatocytes for efficient hepatic ammonia detoxication, but also demonstrates metabolic interactions between different subacinar cell populations. The important scavenger role of perivenous glutamine synthesis for the maintenance of physiologically low ammonia concentrations in the hepatic vein rapidly becomes evident after inhibition of glutamine synthetase by methionine sulfoximine or after selective destruction of perivenous scavenger cells by appropriate doses of Cl_4. In the latter case, hyperammonemia ensues because of almost complete failure of scavenger cells to synthesize glutamine, although periportal urea synthesis is not affected. In human liver cirrhosis the capacity to synthesize glutamine from ammonia is decreased by about 80% and defective perivenous ammonia scavenging contributes to the development of hyperammonemia in liver cirrhosis. The capacity of perivenous scavenger cells to synthesize glutamine is remarkable: highest rates of glutamine synthesis in perfused rat liver are about 0.6 µmol/min·g liver, which corresponds to 8–10 µmol/min·g perivenous cells. This is about three-fold the capacity of periportal hepatocytes to eliminate ammonia via urea synthesis. The carbon skeleton required for glutamine synthesis is in part provided by vascular glutamate, aspartate, α-ketoglutarate and other citric acid cycle dicarboxylates, which are taken up by perivenous but not by periportal cells [22].

68.3.3 Intercellular Glutamine Cycling

Whereas glutamine synthetase is perivenous, glutaminase is found in periportal hepatocytes and has a joint mitochondrial localization together with carbamoyl phosphate synthetase [6,10]. Liver glutaminase is immunologically and kinetically different than glutaminases from other tissues. The enzyme is activated by its product ammonia in a physiological concentration range, and its function is now seen to amplify ammonia inside the mitochondria in parallel with that supplied via the portal vein or arising during hepatic amino acid breakdown, although direct measurements of the actual ammonia concentration inside the mitochondria are not yet possible [10]. Because urea synthesis is normally controlled by flux through carbamoyl phosphate synthetase, which largely depends on the actual ammonia concentration inside the mitochondria, amplification of mitochondrial ammonia via glutaminase flux control becomes an important determinant of urea cycle flux in the presence of physiologically low ammonia concentrations, which are about one order of magnitude below the K_m (ammonia) of carbamoyl phosphate synthetase (see also Fig. 67.5).

As anticipated from the control of urea cycle flux by glutaminase activity, factors known to affect urea cycle flux are indeed associated with parallel changes in the activity of the "mitochondrial ammonia amplifier" glutaminase. Apart from the portal ammonia concentration, this includes the effects of glucagon, α-adrenergic agonists, vasopressin, acidosis/alkalosis, feeding of a high-protein diet and liver cell volume changes (Fig. 67.5).

In human liver cirrhosis, ammonia amplification via glutaminase is increased about five-fold. This allows compensation in the form of maintenance of a near-normal urea cycle flux despite a markedly decreased capacity to synthesize urea in cirrhosis.

In the intact liver, periportal glutaminase and perivenous glutamine synthetase are simultaneously active, resulting in periportal breakdown and perivenous resynthesis of glutamine. This energy-consuming cycling of glutamine is called the intercellular glutamine cycle [6]. Flux through the glutamine cycle is under complex metabolic and hormonal control, allowing the liver to switch from net glutamine consumption to net output or vice versa simply by changing flux through both periportal glutaminase and perivenous glutamine synthetase in opposite directions. Intercellular glutamine cycling provides an effective means of adjusting ammonia flux into urea or glutamine according the needs of the acid–base situation. In the special case of a well-balanced acid–base situation, intercellular glutamine cycling allows the complete conversion of ammonia into urea, despite the low affinity of carbamoyl phosphate synthetase for ammonia and the presence of physiologically low ammonia concentrations (Fig. 68.5).

68.3.4 Liver and pH Homeostasis

The complete oxidation of proteins yields large equimolar quantities of NH_4^+ and HCO_3^-, i.e., in humans about 1 mol per day of each. Both compounds are consumed by urea synthesis, which in chemical terms is nothing other than an irreversible, energy-driven neutralization of the strong base HCO_3^- by the weak acid NH_4^+ [1,7,9,10]:

$$2HCO_3^- + 2NH_4^+ \longrightarrow \text{urea} + 3H_2O + CO_2$$

The amounts of bicarbonate eliminated via urea synthesis are considerable: in humans the oxidation of an average protein load of 100 g/day generates about 1000 mmol bicarbonate which are converted into 30g urea. It recently became clear that the liver plays an important part in maintaining systemic bicarbonate homeostasis by adjusting the rate of urea synthesis. i.e., of irreversible HCO_3^- elimination, to the requirements of acid–base homeostasis (see also Chap. 75). The structural and functional organization of ammonia-and glutamine-metabolizing pathways in the liver acinus provides the basis for such a "pH-stat" function. This organization (Fig. 68.5) uncouples urea synthesis from the vital need to maintain ammonia homeostasis: periportal urea cycle flux is adjusted to the needs of systemic acid–base balance without threat of hyperammonemia, because glutamine synthesis in perivenous scavenger cells acts as a back-up system for ammonia detoxication, guaranteeing non-toxic ammonia levels in effluent hepatic venous blood even when urea cycle flux decreases relative to the rate of protein oxidation, as occurs for example in metabolic acidosis. In the periportal compartment bicarbonate disposal via urea synthesis is complexly controlled by the extracellular acid–base status (see Fig. 67.5), resulting in a feedback control loop between the actual acid – base status and the rate of bicarbonate elimination in the liver. This bicarbonate homeostatic response is largely brought about by the pH sensitivity of liver glutaminase. This enzyme acts

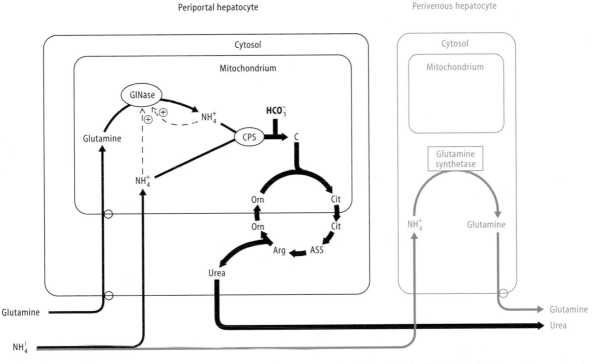

Fig. 68.5. Intercellular glutamine cycling and ureogenesis. The pathways of urea and glutamine synthesis are anatomically switched behind each other, and represent the sequence of a periportal low-affinity system (urea synthesis) and a perivenous high-affinity system (glutamine synthesis) for ammonia detoxication. Periportal glutaminase is activated by ammonia and acts as a pH- and hormone-modulated ammonia amplifier inside the mitochondria. The activity of this amplifier determines flux through the urea cycle (low-affinity system for ammonia detoxication). Glutamine synthase in perivenous cells acts as scavenger for the ammonia escaping periportal urea synthesis (high-affinity system for ammonia detoxication). This anatomical sequence of low- and high-affinity detoxication systems uncouples urea synthesis from the primary need to maintain non-toxic ammonia levels and provides the basis for acid–base control of urea synthesis without threat of hyperammonemia. A complete conversion of a portal ammonia load into urea occurs with a well-balanced acid–base situation. Under these conditions periportal glutamine consumption (ammonia amplifying) and perivenous glutamine synthesis (ammonia scavenging) match each other: there is no net glutamine turnover by the liver, but portal ammonia is converted efficiently into urea despite the low ammonia affinity of carbamoylphosphate synthetase. (From [9])

as a finely tuned, pH-modulated ammonia amplifier: flux through glutaminase decreases by about 70% when the extracellular pH declines from 7.4 to 7.3. Inhibition of flux through liver glutaminase in acidosis inhibits urea cycle flux and irreversible hepatic bicarbonate consumption associated with ureogenesis. Both diminished periportal glutamine consumption and increased perivenous glutamine synthesis in acidosis shift hepatic ammonia detoxication from urea synthesis to net glutamine formation (Fig. 68.6). A coordinate regulation of glutamine metabolism in liver and kidney is essential for maintaining whole body ammonium and acid–base homeostasis (see also Chaps. 75, 78). Figure 68.7 schematically depicts the interorgan team effort between liver and kidney in maintaining acid base homeostasis.

68.3.5 Bile Acid Excretion

All hepatocytes are able to take up bile acids from the sinusoidal space. When [³H]taurocholate is injected into antegradely perfused rat livers, radioactivity is rapidly accumulated in periportal cells, but it accumulates in perivenous cells when the labeled taurocholate is injected during retrograde (from hepatic to portal vein) liver perfusion. This is different from the histoautoradiograms obtained after injection of [³H]-glutamate (Fig. 68.3): here only perivenous scavenger cells are stained, regardless of the direction of perfusion. Bile acid uptake from the sinusoidal space occurs in periportal hepatocytes predominantly by a low-affinity, but high-capacity, system and in perivenous hepatocytes by a high-affinity, but low-capacity, system. This organization resembles that of am-

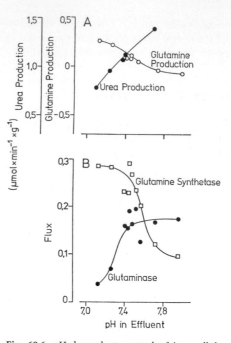

Fig. 68.6. pH-dependent control of intercellular glutamine cycling and partitioning between urea and glutamine synthesis during ammonia detoxication in perfused rat liver. Livers were perfused with near-physiological ammonia and glutamine concentrations and the pH in the perfusate was varied. The decrease of urea synthesis from NH₄Cl (0.6 mmol/l) in acidosis is accompanied by a compensatory increase in net glutamine synthesis, maintaining hepatic ammonia detoxication. The inhibition of urea synthesis in acidosis is also largely dur to an inhibition of flux through glutaminase, which acts as a pH-regulated mitochondrial ammonia-amplifying system. Stimulation of net glutamine production in acidosis results from both decreased periportal glutamine consumption and increased perivenous synthesis of glutamine. (From [7])

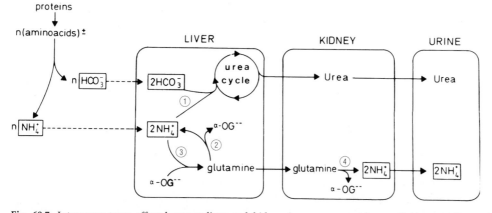

Fig. 68.7. Interorgan team effort between liver and kidney in maintaining ammonia and HCO₃⁻ homeostasis. Protein breakdown produces equimolar amounts of NH₄⁺ and HCO₃⁻, which are normally consumed by hepatic urea synthesis. In acidosis, urea synthesis is switched off and NH₄⁺ is excreted as such into the urine, whereas HCO₃⁻ is left behind in the body. Here, the decrease of hepatic urea synthesis represents the bicarbonate-homeostatic response, whereas the increase in renal NH₄⁺ reflects an ammonium-homeostatic response. Major sites of acid–base control are indicated by *numbers in circles*. In acidosis, flux through reactions *3* (hepatic glutamine synthase) and *4* (kidney glutaminase) is increased, whereas flux through reactions *1* (urea synthesis) and *2* (liver glutaminase) decreases. (From [9])

monia-detoxicating systems and guarantees effective bile acid extraction when the bile acid concentration decreases along the acinus due to uptake. Under physiological conditions the bulk of bile acids is already eliminated by the first few periportal cells, however, downstream cells are recruited when the portal bile acid load increases, as it does for example in cholestasis [4]. There is some evidence that the canalicular secretion of bile acids and also of leukotriene metabolites into bile occurs considerably faster in periportal hepatoctes than in perivenous hepatocytes [5].

68.3.6 Perivenous Scavenger Cell Hypothesis

Recent data suggest that perivenous hepatocytes also play an important part in the inactivation of signal molecules, such as extracellular nucleotides and eicosanoids, thereby extending their well-documented scavenger role for ammonia to a variety of other compounds. Evidence for this comes from the observation that the concentrations of eicosanoids, adenosine or extracellular ATP required to achieve a glycogenolytic response in perfused liver when the perfusion direction is from the portal to the hepatic vein (antegrade) are only one tenth to one fifth those needed when the perfusion is carried out retrogradely (from the hepatic to the portal vein) [8]. Apparently these signal molecules are degraded before they reach the large periportal effector compartment in retrograde, but not in antegrade, perfusions. That perivenous scavenger cells are involved in their degradation is suggested by the finding that stimulation of $^{14}CO_2$ production from added labeled glutamate, which occurs exclusively in perivenous scavenger cells, by eicosanoids or extracellular ATP was independent of the direction of perfusion. Eicosanoids, extracellular nucleotides and nucleosides are liberated in the liver acinus under the influence of diverse stimuli (such as endotoxin or hormones) from non-parenchymal cells and are important modulators of parenchymal cell function. Apparently, perivenous scavenger cells act to eliminate these highly potent signal molecules in order to restrict their action to the sinusoidal space and prevent their overflow into systemic circulation.

68.3.7 Zonal Heterogeneity of Non-parenchymal Cells

Although Kupffer cells are present along the whole porto-central axis, they are numerically enriched in the periportal region. Kupffer cells in the periportal region are larger than the pericentral ones, and they contain higher lysosomal enzyme activities and exhibit much higher phagocytotic activity than perivenous Kupffer cells [3]. On the other hand, galactosyl receptors are enriched in perivenous Kupffer cells, which have a higher tumor cytotoxic potential than periportal Kupffer cells. Heterogeneity between periportal and perivenous endothelial cells is suggested by their different affinity for lectin binding and the larger diameter of periportal endothelial cell fenestrae than in their perivenous counterparts.

References

1. Atkinson DE, Bourke E (1984) The role of ureagenesis in pH homeostasis. Trends Biochem Sci 9:297–300
2. Bartels H, Vogt B, Jungermann K (1987) Glycogen synthesis from pyruvate in the periportal and from glucose in the perivenous zone in perfused liver from fasted rats. FEBS Lett 221:277–283
3. Bouwens L, De Bieser P, Vanderkerken K, Geerts B, Wisse E (1992) Liver cell heterogeneity: functions of non-parenchymal cells. Enzyme 46:155–168
4. Buscher HP, Miltenberger C, McNelly S, Gerok W (1989) The histoautoradiographic localization of taurocholate in rat liver after bile duct ligation. Evidence for ongoing secretion and reabsorption processes. J Hepatol 8:181–191
5. Groothuis G, Hardonk MJ, Keulemans KPT, Miervenhuis P, Meijer DKF (1982) Autoradiographic demonstration and kinetic demonstration of of acinar heterogeneity of taurocholater transport. Am J Physiol 243:G455–G462
6. Häussinger D (1983) Hepatocyte heterogeneity in glutamine and ammonia metabolism and the role of an intercellular glutamine cycle duirng ureogenesis in perfused rat liver. Eur J Biochem 133:269–274
7. Häussinger D, Gerok W, Sies H (1984) Hepatic role in pH regulation: role of the intercellular glutamine cycle. Trends Biochem Sci 9:300–302
8. Häussinger D, Stehle T (1988) Hepatocyte heterogeneity in response to eicosanoids. The perivenous scavenger cell hypothesis. Eur J Biochem 175:395–403
9. Häussinger D (ed) (1988) pH Homeostasis. Academic, London
10. Häussinger D (1990) Nitrogen metabolism in liver: structural and functional organization and physiological relevance. Biochem J 267:281–290
11. Häussinger D, Lamers W, Moorman AFM (1992) Hepatocyte heterogeneity in the metabolism of amino acids and ammonia. Enzyme 46:72–93
12. Jungermann K (1987) Metabolic zonation of liver parenchyma: significance for the regulation of glycogen metabolism, gluconeogenesis and glycolysis. Diabetes Metab Rev 3:269–293
13. Jungermann K, Katz N (1989) Functional specialcation of different hepatocyte populations. Physiol Rev 69:708–764
14. Katz N (1989) Methods for the study of liver cell heterogeneity. Histochem J 21:517–529
15. Lamers WH, Gaasbeek Janzen JW, te Kortschot A, Charles R, Moorman AFM (1987) The development of enzymic zonation in liver parenchyma is related to the development of the acinar architecture. Differentiation 35:228–235
16. Meijer AJ, Lamers WH, Chamuleau RAFM (1990) Nitrogen metabolism and ornithine cycle function. Physiol Rev 70:701–748
17. Moorman AFM, de Boer PAJ, Geerts WJC, van de Zande LPWG, Charles R, Lamers WH (1988) Complementary distribution of CPS (ammonia) and GS mRNA in rat liver acinus is regulated at a pretranslational level. J Histochem Cytochem 36:751–755
18. Moorman AFM, Vermeulen JLM, Charles R, Lamers WH (1989) Localization of ammonia metabolizing enzymes in human liver: ontogenesis of heterogeneity. Hepatology 9:367–372
19. Nauck M, Wölfle D, Katz N, Jungermann K (1981) Modulation of the glucagon-dependent induction of

phosphoenolpyruvate carboxykinase and tyrosine aminotransferase by arterial and venous oxygen concentrations in hepatocyte cultures. Eur J Biochem 119:657–661

20. Probst I, Schwartz P, Jungermann K (1982) Induction in primary culture of "gluconeogenic" and "glycolytic" hepatocytes resembling periportal and perivenous cells. Eur J Biochem 126:271–278

21. Schrode W, Mecke D, Gebhardt R (1990) Induction of glutamine synthetase in periportal hepatocytes by cocul-

tivation with a liver epithelial cell line. Eur J Cell Biol 53:35–41

22. Stoll B, McNelly S, Buscher HP, Häussinger D (1991) Functional hepatocyte heterogeneity in glutamate, aspartate and α-ketoglutarate uptake: a histoautoradiographic study. Hepatology 13:247–253

23. Traber PG, Chianale J, Gumucio JJ (1988) Physiologic significance and regulation of hepatocellular heterogeneity. Gastroenterology 95:30–43

69 Circulation of the Alimentary Canal

J.S. Davison

Contents

69.1 Introduction

The alimentary canal is not only one of the vital organs in the body, but also one of the largest. Because of this and because of its many diverse functions, it requires a circulatory system capable of responding to its many functional requirements. As such, the circulation of the alimentary canal, particularly the splanchnic circulation, which is its major component, has a number of special characteristics. The splanchnic circulation is the portion of the systemic circulation that supplies the gastrointestinal tract (i.e., the abdominal portion of the digestive tract) plus the spleen. It is the largest regional circulation supplied by the aorta, receiving more than a quarter of the left ventricular output at rest regardless of whether the subject is fasted or fed (Fig. 69.1). The physiological role of the splanchnic circulation is to supply the oxygen and nutrients required to support the activities of the digestive tract (motility, secretion, digestion, and absorption). Along with the lymphatic circulation, it ensures also the transfer of absorbed nutrients and water to the rest of the body. Because of its large size, it represents a major reservoir of blood which can be rapidly mobilized in exercise or in situations such as hemorrhage and various forms of shock where central venous pressure and systemic blood pressure have fallen. In this, the spleen plays a particularly important role. This chapter will begin with a consideration of the overall structure of the alimentary canal circulation and in particular of the organization of the splanchnic microcirculation. We will then proceed to examine the physiological role of the circulation in supporting the activities of the digestive tract and will conclude with a consideration of the role of the splanchnic circulation in the response of the circulatory system as a whole to major hemodynamic changes.

69.2 Anatomy

69.2.1 Gross Anatomy

The splanchnic circulation is derived from three major arterial branches of the abdominal aorta: the celiac, the superior mesenteric, and the inferior mesenteric arteries (Fig. 69.1). Because it supplies many organs with diverse functions, the splanchnic circulation is necessarily complex. Several features contribute to the intricacy of its structure and distinguish it from other local circulations. For example, it permits high blood flow rates and has a large surface area for exchange, which permits easy transfer of nutrients and water, while largely preventing the movement of proteins from the plasma compartment. Many of the special features of the splanchnic circulation reside in the microcirculation of the villi, which will be described in the next section. Details of the organization of the blood supply to the liver are provided in Chap. 67.

69.2.2 Microcirculation

Essentially the microcirculation of the gastrointestinal tract consists of vascular elements that can be coupled functionally either in series or in parallel, as illustrated in Fig. 69.2. The major series elements, as in other microcirculations, consist of the arterioles, or resistance vessels, the precapillary sphincters, which determine capillary flow, the capillaries (exchange vessels), and the venules (capacitance vessels). In the case of the gastrointestinal tract, because of the parallel wall layers the blood supply to each of these layers is also organized in parallel. This arrangement permits blood flow to adjust to the metabolic requirements of the respective tissue layers by shunting blood from one layer to another without necessarily changing total gastrointestinal blood flow.

There are differences in the organization of the splanchnic microcirculation in different regions of the gastrointestinal tract. For example, in the stomach, the submucosal vessels

R. Greger/U. Windhorst (Eds.)
Comprehensive Human Physiology, Vol. 2
© Springer-Verlag Berlin Heidelberg 1996

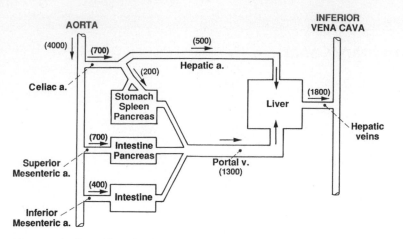

Fig. 69.1. Outline of the splanchnic circulation showing the distribution of blood flows (ml/min) in the major splanchnic blood vessels of the human. (Modified from [7])

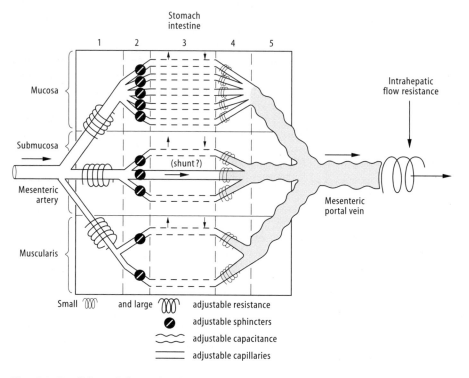

Fig. 69.2. Parallel coupled vascular elements. *1*, precapillary resistance vessels; *2*, precapillary sphincters; *3*, capillaries (exchange vessels); *4*, postcapillary resistance vessels; *5*, capacitance vessels. (From [10])

branch into capillaries at the base of the gastric glands and after passing perpendicularly through the mucosa, they form a luminal network of capillaries which drain into the mucosal venules at the most luminal level of the lamina propria. The venular branches subsequently converge on a small number of mucosal collecting venules which pass directly to the submucous venus plexus without any further mucosal capillaries.

In the small intestine, the microcirculation of the villus has very distinctive general features, although the detailed architecture varies considerably between species. The func-

tional autonomy of the villus microcirculation from deeper vessels also reflects the unique architecture of the villus. In general, a single main arteriole delivers blood to the villus. In man, this is a single, eccentrically located arteriole which passes all the way to the villus tip, where it branches to form a network of capillaries which drain into the venule at different points along the length of the villus. Figure 69.3 shows a variety of patterns of villus blood supply that can be found in individual intestinal villi of rabbit and man. A consequence of this microarchitecture is that a countercurrent exchange can occur between arterial and

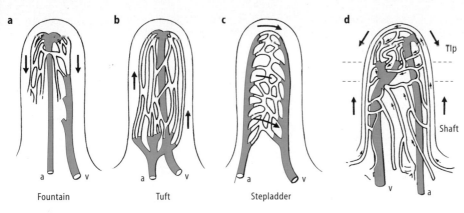

Fig. 69.3. Patterns of intestinal villus microcirculation. (From [11])

venous blood, leading to a relative hypoxia at the level of the villus tip cells.

69.3 Physiology

69.3.1 Salivary Glands

Although the salivary glands (see Chap. 64) constitute a very small percentage of the total weight of most mammals, they secrete at extremely high rates, relative to their weight. Moreover, they secrete a wide range of materials required for digestion; for the development, maintenance, and repair of buccal, pharyngeal, esophageal, and gastrointestinal epithelia; and for the regulation of the immune system through the release of trophic factors into the systemic circulation. Thus the salivary glands can function both as exocrine and endocrine glands. To support this high level of activity, salivary blood flow is correspondingly high. Measurements of 30–50 ml/min per 100 g tissue have been recorded at rest, while during maximal salivary secretion blood flow can rise as high as 500 ml/min per 100 g tissue. It has been calculated that at maximal secretory rates, which could be as high as 50 ml/min per 100 g tissue, the fluid extraction from the plasma is about 15%, which would increase the hematocrit from 45% to 50% or 55% during a single passage through the gland, a change, however, which does not markedly affect blood viscosity. A significant feature of the microcirculation of the salivary glands, and one which it shares with the microcirculations of most other secretory organs, is the high capillary hydraulic conductivity as a consequence of the capillaries being fenestrated. Even at rest, the capillary filtration coefficient is 0.1–0.3 ml/min per 100 g tissue per mm/Hg, which is ten times that seen in skeletal muscle; this can rise to 1.5–2 ml/min per 100 g tissue per mmHg following a maximum vasodilatation. This increase is due to an increase in the number of capillaries perfused and also to an increase in the capillary permeability. Not only does this allow plasma proteins to leak into the interstitial spaces, but also regula-

tory peptides secreted by the striated ducts to be absorbed into the circulation, a situation that might account for the dual exocrine/endocrine role of the salivary glands.

Autonomic Control. Regulation of the circulation of the salivary glands, as in other regional circulations, is largely dependent on the autonomic nervous system. However, in the case of the salivary glands, the parasympathetic division plays a particularly important role (Fig. 69.4; see Chap. 64). Stimulation of the parasympathetic fibers at low frequencies can result in maximal secretory rates, accompanied by a marked increase in blood flow. Since both secretion and blood flow are blocked by muscarinic blocking agents, it is generally considered that this is due to the activation of cholinergic postganglionic fibers. What has not been completely resolved is whether this is due to the activation of specific vasodilator fibers or to the release of vasodilator metabolites, in particular kinins from the activated gland cells. Fig. 69.4 summarizes these two possibilities. Evidence has been provided in support of both pathways and both apparently coexist, with the cholinergic nervous control being responsible for the initial rapid hyperemia on parasympathetic nerve activation and the metabolic (kinin) mechanism maintaining the ongoing hyperemia in the presence of ongoing activity of the salivary glands. Sympathetic nerves also play an important role in the regulation of the salivary gland microcirculation in the same way as in other microcirculations (i.e., they provide a vasoconstrictor tone) and, in the main, work to reduce salivary gland blood flow in the face of other hemodynamic shifts in the systemic circulation. It is interesting to note, however, that sympathetic nerves also supply the salivary glands with secretomotor fibers, the stimulation of which results in an increase in secretory rate. Therefore, the maintenance of blood flow during sympathetic secretomotor activity would presumably require the activation of local metabolic factors to maintain blood flow in the presence of gland stimulation, since activation of sympathetic vasoconstrictor fibers during gland stimulation results in an inhibition of secretion.

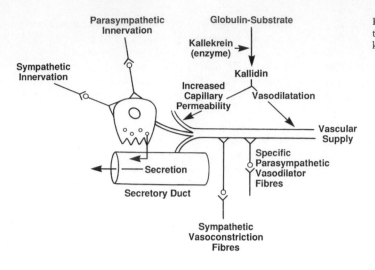

Fig. 69.4. Regulation of salivary gland blood flow by the autonomic nervous system and local release of kinins. (Modified from [9])

Local and Hormonal Control. As described above and illustrated in Fig. 69.4, the parasympathetic nerves play an important role in the activation of secretion and hence the metabolic activity of the gland. Not surprisingly, therefore, there is a close link between parasympathetic nervous regulation and local metabolic regulation. In the salivary glands, as in other local microcirculations, metabolic factors such as adenosine, H^+, and high extracellular K^+ are thought to play a role. In addition, however, certain kinins produced by the action of the enzyme kallikrein also play an important role in modifying gland blood flow and capillary permeability.

69.3.2 Pancreas

Studies on the circulation of the pancreas have been limited because of its complicated vascular supply and hence the difficulty of isolating single vessels, as well as by the fact that the exocrine and endocrine portions of the gland are anatomically distinct. In spite of this, on the basis of blood flow measurements using clearance of inert gases, it seems as though the gland is a homogenously perfused organ.

The pancreatic blood flow rate is high, ranging from 50–180 ml/min per 100 g tissue at rest to 400 ml/min per 100 g tissue in the stimulated state. This suggests a very dense vascularization resulting in a pancreatic perfusion that can be two to three times greater than in skeletal muscle during maximum vasodilatation. As in the salivary glands and other secretory organs, the capillaries are fenestrated, which results in a high hydraulic conductivity. Measurements of the capillary filtration coefficient have confirmed that the pancreas is highly vascular and has a high capillary permeability.

As shown in Fig. 69.1, the arterial supply of the pancreas is derived from the celiac and superior mesenteric arteries. These supply three distinct capillary plexuses to the pancreatic acini, the ducts, and the islets of Langerhans (Fig. 69.5). An interesting feature of this microcirculation is the interconnection between the three sites through an exten-

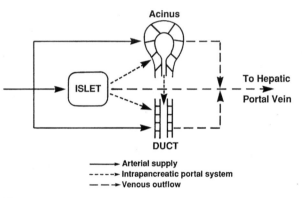

Arterial supply
Intrapancreatic portal system
Venous outflow

Fig. 69.5. Intrapancreatic portal microcirculation

sive intrapancreatic portal system. Both the acini and the ducts receive blood draining from the islets. In addition, the ducts receive blood from the acinar cells. All eventually drain via the splenic or superior mesenteric veins into the portal vein. An important consequence of this portal microcirculation is that hormones secreted by the islets of Langerhans can reach the acinar and duct cells in high concentrations. As described in Chap. 66, these include insulin, glucagon, somatostatin, and pancreatic polypeptide, and all of these are of sufficiently low molecular weight to cross the fenestrated capillaries which surround the ducts and acini. All of these can influence pancreatic exocrine secretion, as described in Chap. 65. The islets, therefore, not only represent an important source of hormones regulating systemic metabolism, but they are also potentially important regulatory sites for pancreatic exocrine function.

Autonomic Control. Although the pancreas is innervated by both parasympathetic and sympathetic divisions of the autonomic nervous system, there is no definitive evidence for parasympathetic cholinergic, vasodilator fibers as described above for the salivary glands. The main autonomic regulation is through modulation of sympathetic vasoconstrictor tone, as in most other systemic microcirculations. Vasoconstriction caused by increased sympathetic activity results in a diminution of pancreatic secretion. An increase in parasympathetic activity, resulting in the release of acetylcholine and perhaps some peptidergic transmitters, leads to stimulation of gland secretion. Since there is no evidence for direct vasodilator fibers, the accompanying hyperemia during secretion is almost certainly a result of local metabolic factors.

Local and Hormonal Control. Local metabolic factors which adjust local blood flow to meet the requirements of increased secretory activity are as described above for the salivary glands and are common to most peripheral microcirculations. In the case of the pancreas, there has been speculation about additional hormonal factors, which would include hormones secreted by the islets reaching the exocrine gland tissue through the intrapancreatic portal microcirculation and hormones such as cholecystokinin and secretin released from the intestinal mucosa reaching the pancreas through its arterial supply. While all these factors can modify secretion and produce a corresponding change in pancreatic vascular perfusion, there is no evidence of a direct effect on the microcirculation, and it is generally considered that these changes in blood flow are largely, if not entirely due to effects on local metabolism. However, direct effects on the vasculature have not been entirely discounted, and there have been suggestions that cholecystokinin may operate through release of kinins, since kallikrein has been found in pancreatic juice stimulated by cholecystokinin infusion.

69.3.3 Liver

Liver blood flow essentially represents total blood flow in the splanchnic circulation. This is because the liver receives not only an arterial blood supply via the hepatic artery, but also receives the entire venous drainage from the remainder of the gastrointestinal tract by the hepatic portal vein (Fig. 69.1). Thus at rest it receives a quarter to a third of the total cardiac output. In fact, measurement of splanchnic blood flow is often carried out using a technique based on the hepatic clearance of a green dye (indocyanine) which is infused by a pump into a vein at an equal rate to the rate at which it is cleared from the blood by the liver. This is essentially an application of Fick's principle and depends on the fact that the dye can only be cleared from the circulation by the liver and that it will reach the liver not only through the hepatic artery, but also via the hepatic portal vein. The situation is analogous to the techniques used for measuring total cardiac output essentially by measuring pulmonary blood flow. Total liver blood flow, therefore, is between 1 and 1.5 l/min in man, with the hepatic artery supplying approximately one quarter to one third of the total. The blood flow, however, is not static, but appears to show several phasic fluctuations, one of which is related to respiratory movements. This may be due to phasic fluctuations in intrathoracic and intra-abdominal pressures, which interfere both with hepatic arteriole inflow and hepatic venous outflow. Portal venous flow also may be affected by the intra-abdominal pressure fluctuations. In spite of these fluctuations, microsphere experiments have shown that, in general, the liver is homogenously perfused by both portal and hepatic arterial blood.

Autonomic Control. The nerve plexus within the liver contains both sympathetic and parasympathetic (vagal) fibers. As in other circulations, there is a potent adrenergic vasoconstrictor influence, although the significance and extent of resting vasomotor (vasoconstrictor tone) in the liver is somewhat unclear. This adrenergic vasoconstrictor supply can obviously be activated as part of a more general response to hemodynamic shifts within the systemic circulation, and it has been shown that as blood flow is reduced, so also is bile flow and oxygen consumption. It is likely that vasoconstriction could lead to a major shutdown in many of the important metabolic functions of the liver (see Chap. 67,68). However, there is still no complete understanding of the relationship between liver blood flow and liver function. Interestingly, however, while reductions in blood flow can result in reductions in bile flow, an increase in bile flow is not necessarily accompanied by an increase in hepatic arterial flow. This may reflect the fact that the extremely high perfusion rate of the liver is sufficient to meet the requirements of an increased level of metabolic activity without further vasodilatation, but that a vasoconstriction could reduce the capacity of the circulation to deliver oxygen and nutrients and thereby depress hepatic metabolic function.

69.3.4 Stomach

As shown in Fig. 69.1, the stomach, along with the spleen and pancreas, receives a large proportion of the splanchnic blood flow from branches of the celiac artery. Blood flow to the artery will flow into several parallel capillary networks, of which those that supply the muscle and mucosa have been most extensively studied; it has been estimated that about 70% of gastric blood flow goes to the mucosa, and at high flow rates the proportion may increase. At peak secretory rates, it has been estimated that over 90% of total gastric blood flow will perfuse the mucosa.

It is generally regarded that the circulation in the stomach is related to secretion rather than motility and, for that reason, a great deal of attention has been directed to the relationship between gastric secretion and gastric blood flow, especially gastric mucosal blood flow. In general, but particularly at low gastric flow rates, stimulation of gastric secretion by secretagogues such as histamine, gastrin, or acetylcholine (see Chap. 61) results in roughly proportionally increased gastric mucosal blood flow (see Fig. 69.6A). In contrast, inhibition of gastric secretion by secretory inhibitors such as catecholamines, atropine, and secretin results in a decreased mucosal blood flow (see Fig. 69.6B). There are two reasons for this type of correlation

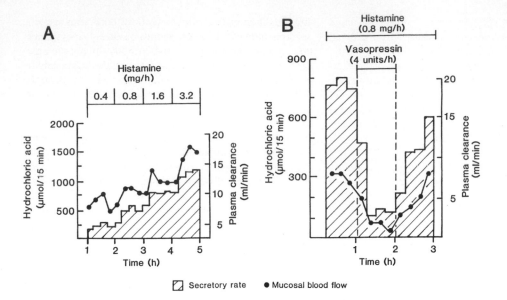

Fig. 69.6A,B. The relationship between gastric mucosal blood flow and gastric secretion. A Increased secretion and blood flow. B Decreased secretion and blood flow. (From [4])

between gastric secretion and gastric mucosal blood flow. The first is due to the relationship between secretion and metabolic energy. Gastric oxygen consumption increases proportionally to acid secretion. Thus by altering secretion and, therefore, metabolism, secretagogues or inhibitors can alter blood flow indirectly by altering the release of potent vasodilator metabolites (see Chap. 61). This mechanism is particularly dependent on baseline blood flow levels. At high rates of blood flow, which provide an adequate oxygen supply and are adequate to remove metabolites from the secreting mucosa, this mechanism is less evident or absent. The second mechanism is mediated by the direct vasodilator or vasoconstrictor properties of gastric secretagogues and inhibitors. For example, histamine and acetylcholine can directly dilate gastric mucosal vessels, and norepinephrine and vasopressin are direct vasoconstrictors. Because of the duality of the mechanisms which coregulate gastric secretion and gastric mucosal blood flow, it may be expected that vasoactive secretagogues may produce a greater increment in blood flow increase for a unit increase in secretion than those secretagogues working only through local metabolic regulation. This appears not always to be the case, which perhaps can be explained by the fact that a direct vasodilator may increase blood flow, but not to the point that blood flow ceases to be a limiting factor for the increase in secretion. As the level of secretion exceeds the increased blood flow produced by direct vasodilatation, local metabolic factors will provide the extra necessary increment, thereby maintaining a similar change in the ratio of blood flow to change in secretion.

There has been considerable interest in the role of gastric mucosal microcirculation in the pathophysiology of peptic ulcer disease and mucosal damage. In general, however, it is considered that in most situations, except for circulatory

shock, the events leading to ulceration or mucosal damage tend to precede rather than follow compromise of the microcirculation. However, circulatory insufficiency contributes significantly to enhancement of the damage, since it is now known that restoration of the circulation can help protect against ulcerative agents.

69.3.5 Intestine

Intestinal blood flow, although similar to that of the stomach in terms of flow per unit mass of tissue, comprises more than half of the total splanchnic flow. Many of the studies of intestinal blood flow have recorded total flows. However, as discussed above, for the stomach it is much more important in understanding the relationship between blood flow and function to understand regional distribution, in particular the difference between muscle and mucosal blood flow. The distribution of blood flow in the small intestine can be seen in Fig. 69.7, which shows that by far the greatest portion of total blood flow is to the epithelium, with the crypt and villus regions together receiving 70% of the total blood flow. As in the case of the stomach during maximum vasodilatation, not only does total blood flow increase, but the percentage supply in the epithelium increase to 90%. Thus vasodilatation in the epithelial microcirculations can result in an actual decrease in submucosal and muscle blood flows.

This is presumably because the most potent physiological stimulus to increasing intestinal blood flow is related to food ingestion. As a consequence, the metabolic activities of the epithelium are greatly increased to aid in the process of digestion and absorption. The increase in intestinal blood flow is initiated during the cephalic phase of digestion (see Chap. 62) as a result of seeing, smelling, and

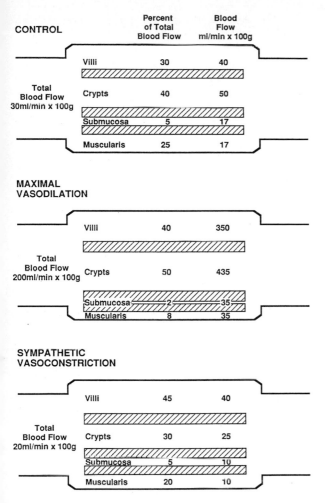

Fig. 69.7. Distribution of blood flow in the small intestine of the cat at rest, following maximal vasodilatation, and following sympathetic vasoconstriction. (From [5])

tasting food. These effects are mediated by the central nervous system, in particular hypothalamic and brain stem regions, resulting in an increased vagal activity. How the increased vagal activity leads to increased blood flow is not clear, but, as will be discussed later, there is no evidence for direct parasympathetic vasodilator fibers to the intestine; therefore, it is presumably a consequence of local metabolic factors and/or release of gastrointestinal hormones following activation of the epithelium by parasympathetic nerves. The entry of food and secretions into the lumen of the intestine further enhances the mucosal blood flow. This increase is very dependent on the actual constituents within the lumen. Figure 69.8 shows the influence of several constituents of intestinal chyme on blood flow in the jejunum and ileum. The presence of food in the intestinal lumen disrupts the migrating myoelectrical (motor) complex (MMC). Thus, during the digestive phase, the motor activity is of an irregular pattern which, nevertheless, is effective in mixing intestinal contents with digestive enzymes and increasing contact with the absorbing cells of the villus (see Chap. 62). Thus, during the digestive phase, the mucosa will continue to take a greater share of the increased total blood flow to the intestine. In the fasting state, however, the smooth muscle activity undergoes periodic oscillation, and during phases of intense rhythmic activity muscle blood flow may be briefly enhanced.

Autonomic Control. The intestine is well supplied with sympathetic adrenergic nerve fibers. Thus the principle autonomic control of the intestinal microcirculation is through variation in sympathetic vasoconstrictor tone. As occurs throughout the gastrointestinal tract, sympathetic nerves are particularly important in the vasoconstriction that occurs within the splanchnic circulation during exercise and shock. As outlined above, there is no evidence of any parasympathetic vasodilator fibers although there is growing evidence for enteric, cholinergic vasodilator neurons (see Chap. 70). Therefore, the increase in gastric and intestinal blood flow following parasympathetic and, in particular, vagal activation is secondary to metabolic and perhaps hormonal factors following activation of the muscle and epithelium.

Local and Hormonal Factors. Of the many local metabolic factors that can regulate local microcirculations, adenosine and tissue oxygen tension have been proposed as particularly important mediators of the gastrointestinal microcirculation. Adenosine is a very potent vasodilator within the splanchnic bed, and accumulation of adenosine has been demonstrated during reactive and postprandial hyperemia in the small intestine. Various compounds which interfere with the action or metabolism of adenosine can modify the vasodilatation following changes in arterial pressure or absorption of nutrients. The role of tissue oxygen tension in local vasoregulation is not quite so clear. While there is no question that conditions that reduce oxygen delivery or increase oxygen demand, or both, will lead to reductions in tissue oxygen tension and an appropriate change in the perfusion of that tissue, such changes are also accompanied by the accumulation of other vasodilator metabolites. Thus the precise role of tissue oxygen tension per se is not really clear. However, at any given oxygen uptake, the magnitude of the hyperemia produced following intake of food is larger if the resting oxygen extraction is high (i.e., there is large arteriovenous oxygen difference). Many of the gastrointestinal hormones liberated following ingestion of a meal are vasodilators when infused into the splanchnic circulation. There is no good evidence, however, that this occurs at doses which are "physiological," i.e., that produce circulating concentrations similar to that seen following ingestion of food. Several hormones, such as cholecystokinin, secretin, gastrin, and neurotensin, injected, alone or in combination, by close arterial injection, in doses which produce postprandial concentrations, failed to induce vasodilatation. In contrast, vasoactive intestinal peptide appears to induce vasodilatation at physiological concentrations, and blockage of endogenous vasoactive intestinal peptide by use of antisera will suppress the hyperemia induced by luminal infusion of

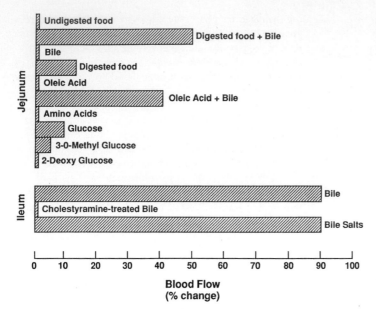

Fig. 69.8. The effects of intraluminal constituents of chyme on blood flow in the duodenum and ileum. (From [12])

Table 69.1. Splanchnic blood flows (ml/min per g wet tissue) to several tissues prior to and following antigen challenge in sensitized rats. (From [13])

Tissue	Preantigen	Postantigen		
		5 min	60 min	240 min
Hepatic arterial	0.73 ± 0.8	1.13 ± 0.27	0.83 ± 0.18	1.52 ± 0.21
Spleen	1.86 ± 0.19	0.10 ± 0.04	0.49 ± 0.17	1.74 ± 0.52
Submandibular	0.65 ± 0.11	0.10 ± 0.02	0.19 ± 0.05	0.71 ± 0.11
Stomach	0.56 ± 0.07	0.14 ± 0.06	0.25 ± 0.05	1.02 ± 0.34
Duodenum	1.67 ± 0.18	0.36 ± 0.14	0.85 ± 0.21	0.57 ± 0.13
Jejunum	1.42 ± 0.16	0.25 ± 0.10	0.63 ± 0.10	0.63 ± 0.24
Ileum	1.22 ± 0.12	0.23 ± 0.07	0.45 ± 0.11	0.80 ± 0.23
Caecum	2.54 ± 0.36	0.17 ± 0.08	0.67 ± 0.16	2.23 ± 0.66
Colon	0.84 ± 0.12	0.19 ± 0.07	0.33 ± 0.06	0.71 ± 0.16
Mesentery	1.29 ± 0.15	0.30 ± 0.11	0.33 ± 0.09	0.63 ± 0.19

bile and maleic acid. Thus vasoactive intestinal peptide released from submucosal neurons would appear to be an important factor in postprandial hyperemia in the intestine.

There are also other possible local regulators, such as histamine, prostaglandins, serotonin (5HT), purines, puronucleotides, and angiotensin. Histamine is a potent vasodilator, but the existence of receptors mediating the vasodilator effects of histamine does not provide evidence of a physiological role. However, it is possible that in pathophysiological situations, particularly in situations where there is a mastocytosis (i.e., an increase in the numbers of mast cells within the mucosa or serosa), histamine may play an important role in pathophysiological responses. There is stronger evidence for a role of prostaglandins in local vascular regulation within the gastrointestinal tract, and it has been suggested that prostaglandins play a homeostatic role in local microcirculatory regulation.

69.3.6 Intestinal Ischemia

There are several conditions which can severely reduce gastrointestinal blood flow. Some, such as various forms of shock or hemorrhage, are not unique to the intestinal microcirculation. However, the splanchnic circulation, because it represents such a huge reservoir of blood, is the most profoundly affected in these types of conditions. As shown in Table 69.1, during anaphylactic shock, there are profound reductions in the circulation to all parts of the gastrointestinal tract with the exception of the liver. Table 69.2 shows the same data expressed as a percentage of cardiac output. These two tables indicate that in most regions of the gastrointestinal tract the profound reduction in the microcirculation is not only a consequence of the reduced cardiac output, but also the intense vasoconstriction which occurs in these vascular beds, thereby reducing the proportion of the cardiac output being received by these areas. In contrast, the liver receives a

Table 69.2. Blood flows to several tissues (ml/min per g wet tissue) supplied by the splanchnic microcirculation prior to and following antigen challenge in sensitized rats. (From [13])

Tissue	Preantigen	Postantigen		
		5 min	60 min	240 min
Hepatic arterial	0.68 ± 0.10	2.35 ± 0.17	1.64 ± 0.38	1.05 ± 0.14
Spleen	1.64 ± 0.20	0.21 ± 0.05	0.88 ± 0.31	1.07 ± 0.30
Submandibular	0.55 ± 0.09	0.28 ± 0.05	0.45 ± 0.12	0.71 ± 0.16
Stomach	0.49 ± 0.07	0.33 ± 0.07	0.46 ± 0.08	1.03 ± 0.40
Duodenum	1.46 ± 0.17	0.83 ± 0.08	1.48 ± 0.30	1.06 ± 0.57
Jejunum	1.27 ± 0.17	0.61 ± 0.15	1.04 ± 0.26	0.63 ± 0.24
Ileum	1.10 ± 0.14	0.57 ± 0.13	0.81 ± 0.21	0.47 ± 0.07
Caecum	2.21 ± 0.34	0.39 ± 0.09	1.15 ± 0.21	1.14 ± 0.32
Colon	0.74 ± 0.12	0.49 ± 0.13	0.56 ± 0.09	0.49 ± 0.11
Mesentery	1.14 ± 0.15	0.73 ± 0.17	0.63 ± 0.12	0.39 ± 0.09

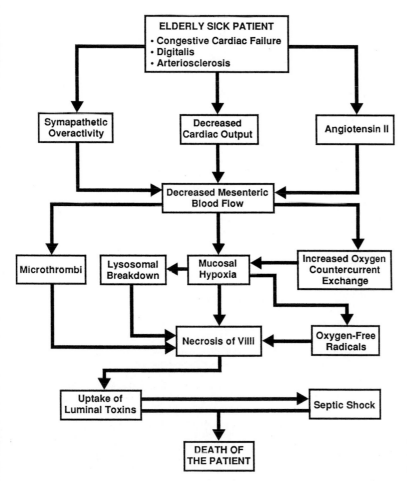

Fig. 69.9. A proposed sequence of some of the pathophysiological events leading to nonocclusive intestinal ischemia. (From [6])

greater proportion of the reduced cardiac output, thereby effectively maintaining its circulation. The other feature demonstrated in these tables is the prolonged period of time over which this reduced blood flow occurs. In spite of the normally effective local regulation through metabolic factors, the gastrointestinal microcirculation is unable to restore perfusion in the face of a severe anaphylactic shock. As a consequence, the gastrointestinal tissues and, in par-

ticular, the epithelium become susceptible to ischemic injury.

There are also other disorders that involve predominantly the gastrointestinal microcirculation. These are occlusive and nonocclusive intestinal ischemia. Occlusions can result from emboli, thrombi, or sclerotic obstructions. In nonocclusive states, there is a low intestinal blood flow without any obvious organic obstruction. These intestinal

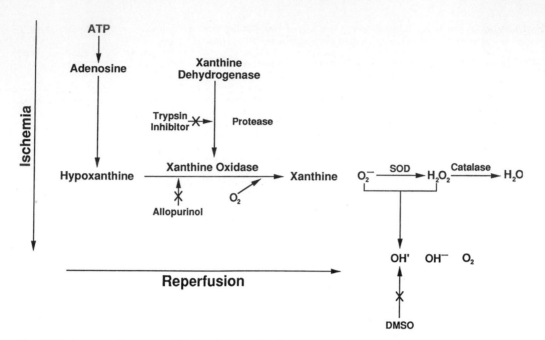

Fig. 69.10. A proposed sequence of biochemical steps for the generation of superoxide radicals in ischemia reperfusion injury (*DMSO*, dimethyl sulfoxide; *ATP*, adenine triphosphate; *SOD*, superoxide dismutase) (From [6])

ischemic states result in high mortality, and although the etiology of nonocclusive intestinal ischemia is unknown, several mechanisms have been proposed, as outlined in Fig. 69.9. The actual agents which account for the ischemic injury in the gut are not fully understood, but substances such as prostanoids, histamine, vasoactive intestinal polypeptide, bacterial endotoxins, various proteolytic enzymes, hypoxia, and superoxide radicals have all been implicated. One of the interesting features of ischemia of the gut, however, is the damage that is, in fact, produced on reperfusion. The problems of ischemia are compounded in the intestine by certain unusual features of the villus microcirculation leading to redistribution and a countercurrent exchange of oxygen, which reduces oxygen tensions to the villus tip enterocytes. Moreover, the generation of superoxide radicals is considered to be a major factor in the reperfusion injury (Fig. 69.10).

69.4 Conclusions

The microcirculation of the gastrointestinal tract demonstrates a range of anatomical features and physiological mechanisms which permit it to adjust to the diverse requirements of the many tissues which comprise this complex organ system. In particular, it has a higher capillary density than many other organs and hence a large surface area for exchange. The fenestrated capillaries increase the hydraulic conductivity, permitting exchange of small solutes (electrolytes and nutrients) while restricting the macromolecular (colloidal) components to the plasma. Because of its large volume, it constitutes a major reservoir of blood, which can be diverted to other organs during exercise or in certain pathological situations such as haemorrhage and shock.

General References

1. Crissinger KD, Granger DN (1991) Gastrointestinal blood flow. In: Yamada T et al. (eds) Textbook of gastroenterology, vol 1. Lippincott, Philadelphia, pp 447–475
2. Granger DN, Kvietys PR, Perry MA, Barrowman JA (1987) The microcirculation and intestinal transport. In: Johnson LR (ed) Physiology of the gastrointestinal tract, 2nd edn. Raven, New York, pp 1671–1697
3. Hanson KM (1978) Liver. In: Johnson PC (ed) Peripheral circulation. Wiley, New York, pp 285–313
4. Jacobson ED (1991) The splanchnic circulation. In: Johnson LR (ed) Gastrointestinal physiology, 4th edn. Mosby, St. Louis, pp 142–161
5. Lundgren O (1978) The alimentary canal. In: Johnson PC (ed) Peripheral circulation. Wiley, New York, pp 255–283
6. Parks DA, Jacobson ED (1987) Mesenteric circulation. In: Johnson LR (ed) Physiology of the gastrointestinal tract, 2nd edn. Raven, New York, pp 1649–1670
7. Smith JJ, Kampine JP (1990) Circulation to special regions. In: Circulatory physiology – the essentials, 3rd edn. Williams and Wilkens, Baltimore, pp 205–329
8. Weideman MP, Tuma RF, Mayrovitz HN (1981) An introduction to microcirculation. Academic, New York, pp 33–50, 123–128

Specific References

9. Folkow B, Neil E (1971) Circulation. Oxford University Press, New York

10. Folkow B (1967) Regional adjustments of intestinal blood flow. Gastroenterology 52:423

11. Gannon BJ (1981) The co-existence of fountain and tuft patterns of blood supply in individual intestinal villi of rabbit and man: resolution of an old controversy. Bibl Anat 20: 130

12. Granger DN, Kvietys PR, Parks DA, Benoit JN (1983) Intestinal blood flow: relations to function. Surv Dig Dis 1: 217

13. Mathison R, Befus AD, Davison JS (1990) Hemodynamic changes associated with anaphylaxis in parasite sensitized rats. Am J Physiol (Heart Circ Physiol 27) 258:H1126–H1130

70 The Enteric Nervous System

J.S. DAVISON

Contents

70.1 Introduction

The enteric nervous system was defined originally by Langley early this century as a separate and autonomous division of the autonomic nervous system (Chap. 17) [9]. Not only was he impressed by its size (2×10^8 neurons compared with a few hundred fibers in the vagus nerve), but he also recognized its ability to maintain organized propulsive activity in intestinal segments studied in vitro. He, therefore, suggested that this intrinsic (enteric) nervous system was capable of integrative functions independent of any central nervous system influence. For several decades this concept was overlooked, and many textbooks described the enteric nervous system as a collection of parasympathetic terminal ganglia acting merely as relays for the parasympathetic preganglionic fibers (Fig. 70.1A). This simple model ascribed no integrative functions to the enteric nerves and ignored the complexity of interconnections between enteric ganglia. The inference was that all neural control could be attributed to autonomic regulatory centers within the central nervous system and spinal cord. In recent years, however, Langley's concepts have been rediscovered, and this has stimulated a resurgence of interest in enteric neurobiology. This in turn has led to a greater understanding of the complexity of the "little brain" in the gut and its importance in normal physiological regulation of digestive tract activity as well as its role in pathophysiological conditions. Figure 70.1B summarizes a current concept of how the enteric nervous system is organized.

Although modern gastrointestinal neurophysiology now emphasizes the autonomy of the enteric nervous system, it is well-recognized that the central nervous system also exerts a profound influence on normal gastrointestinal function (see earlier chapters in this volume on gastrointestinal regulation). The pathways for this extrinsic neural control are via the parasympathetic and sympathetic nervous systems. Their effects, however, are mediated almost entirely through modulation of the enteric neural circuitry rather than through direct activation or inhibition of effector cells. Figure 70.2 describes schematically the relationship between the enteric nervous system and the brain and emphasizes the hierarchical nature of neural control of gastrointestinal function.

70.2 Anatomy and Histology

The great majority of excitatory or inhibitory fibers that innervate the effector cells of the gastrointestinal tract are processes of neurons located within the intramural ganglia. These ganglia with their interconnecting network of nerve fibers form the final pathways of signals transmitted from the central nervous system along the parasympathetic and sympathetic nerves. Since they are not only relay stations but also important integrative centers, they contain all the neural elements required for complex integrative function (sensory neurons, interneurons, and excitatory and inhibitory motor neurons and the necessary command neurons and associated circuitry to generate well-defined patterns of phasic or cyclical neuronal activity). The organization of these plexuses is described in Chap. 17, Fig. 17.3. The two principal plexuses that constitute the enteric nervous system are the submucosal or Meissner's plexus, located in the submucosa, and the myenteric or Auerbach's plexus, located between the circular and longitudinal muscle layers of the digestive tract wall. These plexuses extend along the length of the wall from the esophagus to the anus and are even found in the striated muscle portion of the esophagus. They appear as networks of intersecting nerve bundles with nerve cell bodies and their associated perikarya forming nodes or ganglia at the intersections. Fibers which leave these plexuses ultimately innervate effector structures such as epithelial cells, blood vessels, smooth muscle cells, or immunocytes (see Fig. 70.2), or provide connections between the two plexuses.

R. Greger/U. Windhorst (Eds.)
Comprehensive Human Physiology, Vol. 2
© Springer-Verlag Berlin Heidelberg 1996

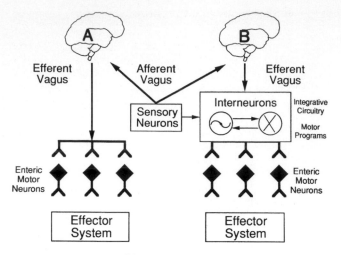

Fig. 70.1. Two concepts of the organization of the enteric nervous system. *Left,* The relay concept which ignored Langley's original description of the enteric nervous system as an autonomous and complex system. In this model, preganglionic parasympathetic fibers are considered to synapse directly onto enteric motor neurons. *Right* The current concept shows that the enteric nervous system contains integrative and program circuitry which can function autonomously but which in addition receives command signals from the brain. In this model the integrative and program circuits are derived from synaptic connections among interneurons residing exclusively within the enteric nervous system. Moreover, the enteric system also contains necessary sensory and motor neurons. (Modified with permission from [14])

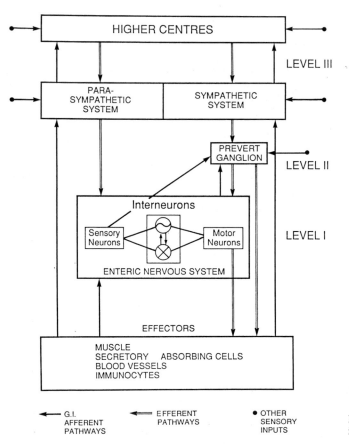

Fig. 70.2. The hierarchy of nervous control of gastrointestinal motility. (Modified with permission from [1])

The processes of the nerve cell bodies and glial cells within the ganglia form a dense neuropil. Although the ganglia and connecting nerve bundles are surrounded by basal laminate and collagen fibrils, within the ganglia the neurons are separated from one another only by glial cells and hence are much more closely packed than in other autonomic ganglia. The enteric neurons vary considerably in size and complexity of form, but three generalized

70.2.3 Dogiel Type 3 Neurons

The Dogiel type 3 neurons closely resemble the those of Dogiel type 1; the distinguishing characteristics are the lack of flattening of the shorter processes and the intermediate length of the longer processes which seem to terminate in the same or adjacent ganglia (Fig. 70.3c). Again, therefore, these are thought to form synaptic connections with other neurons and are regarded as presumptive interneurons.

70.3 Physiology

The enteric nervous system programs and regulates all gastrointestinal functions. Electrophysiological studies have shown that the enteric nervous system can carry out all the requisite functions of a complex nervous system. Synaptic connections can produce fast or slow excitatory and inhibitory postsynaptic potentials (Fig. 70.4). In addition, presynaptic inhibition has been demonstrated within the enteric nervous system. Table 70.1 lists some of the

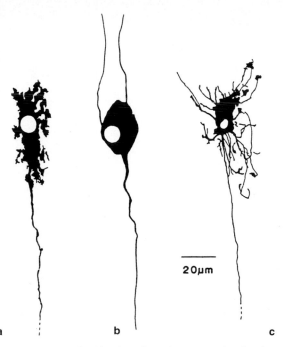

a   b   c

Fig. 70.3a–c. Classification of enteric neurons showing the morphology of the three principle cell types. **a** Dogiel types 1; **b** type 2; and **c** type 3. (Modified with permission from [14])

morphological types of neurons are found within the enteric nervous system. Originally described by the German neuroanatomist, A. S. Dogiel, these are known as Dogiel types 1, 2, and 3 [2].

70.2.1 Dogiel Type 1 Neurons

The Dogiel type 1 neurons have cell bodies with many short, flattened processes and a single long process (Fig. 70.3a). The short processes are thought to be dendrites which receive synaptic input, and the long process is regarded as an axon and projects in the circumferential or longitudinal planes between the muscle layers. These long processes project for relatively long distances in the interganglionic fiber tracts and can pass through many rows of ganglia. The longest known projections are about 2–3 cm. Some of these neurons are regarded as motor neurons.

70.2.2 Dogiel Type 2 Neurons

The cell bodies of the Dogiel type 2 neurons (Fig. 70.3b) have a relatively smooth outline with both long and short processes which can vary considerably in their configuration. The long processes can extend through several rows of ganglia in the interganglionic fiber tracts, while shorter processes may project only within the ganglion of origin. As the long processes enter individual ganglia, they become varicose, suggesting that these neurons act as interneurons within the enteric microcircuitry.

Table 70.1. Messenger substances that mimic synaptic events in enteric neurons

Slow synaptic excitation
Histamine
Vasoactive intestinal peptide
Caerulein
Gastrin-releasing peptide
Tachykinins (neurokinins)
Thyrotropin-releasing hormone
5-Hydroxytryptamine
Gamma-aminobutyric acid
Motilin
Acetylcholine
Cholecystokinin
Bombesin
Calcitonin gene-related peptide
Corticotropin-releasing hormone
Norepinephrine
Pituitary adenylate cyclase
Slow synaptic inhibition
Acetylcholine
5-Hyroxytryptamine
Neurotensin
Somatostatin
Galanin
Opioid peptides
Norepinephrine
Cholecystokinin
Purine nucleotides
Presynaptic inhibition
Norepinephrine
Histamine
Opioid peptides
Neuropeptide Y
Peptide YY
Dopamine
5-Hyroxytryptamine
Acetylcholine
Pancreatic polypeptide

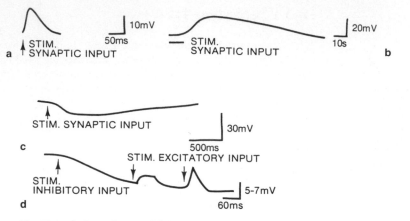

Fig. 70.4a–d. Synaptic potentials in enteric neurons. **a** Fast excitatory prosynaptic potential (EPSP); **b** slow EPSP; **c** slow inhibitory postsynaptic potential (IPSP); **d** fast EPSP superimposed on IPSP. (Reprinted with permission from [1])

many transmitters within the enteric nervous system believed to be responsible for these synaptic events. As a consequence of this synaptic connectivity within the enteric circuits, several functions can be attributed to the enteric nervous system, including activation of excitatory and inhibitory reflexes, pattern generation, gating, and coordinated recruitment. In this way, the enteric nervous system is able to regulate and program the activity of the gastrointestinal tract. The following sections outline enteric neural regulation of motility, secretion, blood flow, and immune reactions within the digestive tract.

70.3.1 Enteric Reflexes and Gastrointestinal Motility

The enteric nervous system and, in particular, neurons within the myenteric plexus program and regulate motility patterns under physiological and pathophysiological conditions (see Chaps. 17, 60, 63). In intact mammals there are two major patterns of motility. During the fasted state, migrating motor complexes occur which are characterized by periods of quiescence (phase 1), followed by irregular contractile activity (phase 2), and finally short periods of regular phasic motility (phase 3). The fed pattern consists of continuous, irregular activity. Within these complex patterns of motility, two types of contraction can occur, namely, peristalsis and mixing movements, both of which depend upon enteric nerves for their regulation.

Peristalsis. Peristalsis is a propulsive contraction of the circular muscle which can be generated along the entire length of the digestive tract. The peristaltic reflex that occurs in both the intact animal and in isolated segments of the intestine is evoked by distention of the intestinal wall, which causes a contraction behind the bolus and relaxation in front of the bolus. The propagation of the circular muscle contraction propels the bolus in an aboral direction. Under exceptional circumstances, retroperistaltic contractions can occur. The peristaltic reflex does not occur after

paralysis of the enteric nervous system. The enteric circuitry responsible for this stereotyped pattern of contraction, which requires both excitation and inhibition of the two muscle layers, is illustrated schematically in Fig. 70.5. This circuitry permits the longitudinal muscle layer ahead of the distending fluid or bolus to contract. At the same time, the circular muscle layer in this segment relaxes, and the net effect of both these responses is to produce a segment capable of receiving the aborally moving intestinal contents. In contrast, the circular muscle in the segment behind the advancing bolus contracts as the longitudinal muscle layer simultaneously relaxes. This provides the propulsive force necessary to move the contents into the receiving segment. As the intraluminal contents move forward, the receiving segment converts into a propulsive segment promoting the onward movement of the bolus. In practice, peristaltic contractions in the intestine travel only over short distances due to the influence of inhibitory neural networks within the enteric nervous system that limit the number of segments that can be activated during the generation of single peristaltic wave. This is not the case in the esophagus, where peristalsis once initiated travels along the entire length of the tube, nor in the colon during mass movements.

Mixing Movements. Often referred to as segmenting contractions because they appear to divide the intestine into discrete segments, mixing movements serve to mix the intestinal chyme with digestive enzymes and increase contact between intraluminal contents and the epithelium for final digestion and absorption. This pattern of motility consists of circular muscle contractions which do not propagate. The circular muscles on either side of the contracting band remain relaxed and, therefore, provide receiving segments on both sides of the zone of contraction, resulting in propagation of intestinal contents in both an oral and aboral direction. Unlike peristalsis, the nature of the enteric circuits responsible for this pattern of motility is not understood. They might be the same as the peristaltic circuits, but with propagation prevented

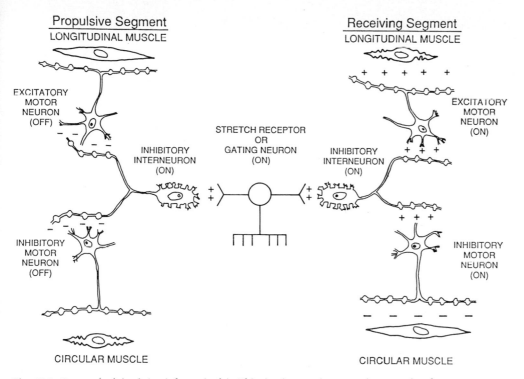

Fig. 70.5. Proposed minimal circuit for peristalsis. This circuitry requires synaptic connections between sensory neurons, interneurons and motor neurons and involves the release of both excitatory transmitters (+) and inhibitory neurotransmitters (-). (Modified with permission from [14])

through the action of inhibitory nerves in the same way that peristaltic contractions are limited in the distance they travel. Both peristaltic contractions and segmenting contractions occur in the fasting and fed states. Analysis of migrating motor complexes and fed patterns of activity using conventional manometric techniques coupled with fluoroscopic visualization of the recording sites reveal that each individual contraction at a particular site may propagate or not in an apparently random fashion. This suggests that common circuits may be responsible for both types of contraction with the segmenting contractions representing complete restriction of propagation as a result of the action of enteric inhibitory neurons. Alternatively, entirely different circuits may be activated. The former explanation would provide the most economic method for the enteric nervous system to program this type of contraction.

Enteric Regulation of Motility Patterns. As described above, individual contractions, be they segmental or peristaltic, are clustered in periodic patterns of motility which differ in the fed and fasted states. During fasting, the propagation of the migrating motor (myoelectric) complex (MMC) is dependent upon the integrity of the enteric nervous system. Moreover, the coordination of periodic motor activity with other gastrointestinal functions such as gastric and pancreatic secretion, gallbladder emptying, intestinal secretion and intestinal blood flow, is also dependent upon the enteric nervous system. While the extrinsic nerves can alter the periodicity of the MMC, its initiation and propagation is dependent only on the enteric nervous

system. While the enteric nervous system is essential for programming this periodic motor activity, the central nervous system, via the extrinsic autonomic outflow, is required for the disruption that occurs following a meal which leads to the characteristic fed pattern.

In summary, the enteric nervous system not only determines the characteristics of individual smooth muscle contractions, but also programs the overall pattern of gastrointestinal motility and coordinates the activity of other effector tissues with that of the smooth muscle.

70.3.2 Control of Intestinal Secretion

As outlined in Chaps. 61 and 62, the gastrointestinal epithelium is capable not only of absorption, long regarded as its raison d'être, but also of active secretion. Secretion is produced by the immature crypt cells which, as they migrate to the villus tip, alter the type and distribution of their cell membrane transporters and are thus transformed into absorbing cells. Enteric nerves regulate the activity of the crypts throughout the small and large intestine. There is also evidence that they may modify absorptive function, too. The majority of enteric neurons innervating the epithelium are derived from the submucosal plexus However, projections from the myenteric plexus also occur. In addition, myenteric plexus neurons may interconnect with the submucosal plexus and thereby modulate the activity of secretomotor neurons within the submucosal plexus. Figure 70.6 summarizes the circuitry that may be involved in

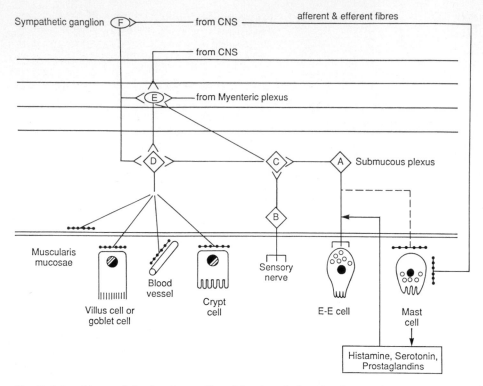

Fig. 70.6. Possible neural circuitry that could modulate intestinal ion transport. *A, B*, submucosal sensory neurons. Note how some sensory endings are associated with entero-endocrine cells (*E-E*); *C*, submucosal interneuron; *D*, submucosal motorneuron; *E*, myenteric interneuron; *F*, sympathetic postganglionic neuron. (Modified from [6])

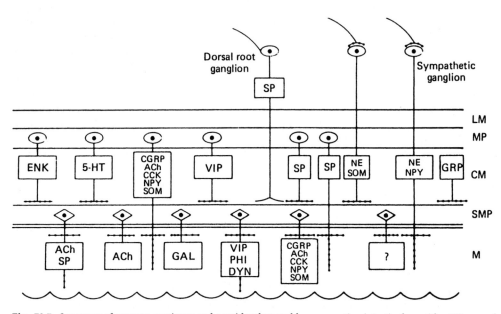

Fig. 70.7. Summary of neurotransmitters and peptides that could influence ion and water transport by the intestinal epithelium, either directly or through modification of the microcirculation. *SP*, substance P; *ENK*, enkephalin; *S-HT*, serotonin; *CGRP*, calcitonin gene-related peptide; *ACh*, acetylcholine; *CCK*, cholecystokinin; *NPY*, neuropeptide Y; *SOM*, somatomedin; *VIP*, vasoactive intestinal peptide; *NE*, norepinephrine; *GRP*, gastrin-releasing peptide; *GAL*, galanin; *PHI*, peptide HI (histidine-isoleucine); *DYN*, dynorphin; *LM*, longitudinal muscle; *MP*, myentericplexus; *CM*, circular muscle; *SMP*, submucosal plexus; *M*, mucosa. (Modified with permission from [6])

Table 70.2. Native substances that affect ion transport by the small intestine

Secretagogues	Absorbagogues
Peptides	Peptides
vasoactive intestinal polypeptide	somatostatin
PHI	neuropeptide Y
substance P	encephalins
neurotensin	CCK
bombesin	angiotensin
bradykinin	
vasopressin	
calcitonin	
Amines	Amines
acetylcholine	adrenaline
5-hydroxytryptamine	noradrenaline
histamine	dopamine
Adenosine triphosphate	
Prostaglandins	
Effects not known	
calcitonin gene-related peptide; galanin	

the regulation of ion transport, while Fig. 70.7 shows the projections of neurochemically defined neurons in both myenteric and submucous plexuses. In addition, Table 70.2 lists neural transmitters that can affect ion transport by the small intestine. This list includes substances that can cause secretion from the crypt cell as well as others that stimulate absorption by the villus tip cell.

Most studies of enteric neural regulation of epithelial transport have focused on the process of secretion. This work was pioneered by K.A. Hubel in the late 1970's [5–8]. He developed an elegantly simple technique for electrical field stimulation of isolated intestinal epithelia mounted in Ussing chambers. In this way, he was able to activate enteric neurons which in turn would activate secretion by crypt enterocytes. This was measured by an increase in short circuit current. Several workers, using the same or a modified version of his method, have confirmed this increase and demonstrated conclusively that it was due to Cl^- secretion [11,12]. Subsequent studies have been able to throw some light on the putative transmitters, involved in this response. Acetylcholine is one of the principal transmitters, since the muscarinic antagonist, atropine, will reduce though not abolish the short circuit current induced by electrical field stimulation as well as the serosal-to-mucosal flux of Cl (J_{s-m} Cl^-) However, since atropine and adrenergic antagonists do not block this J_{s-m} Cl^-, a non-adrenergic, noncholinergic (NANC) neurotransmitter must also be involved in communication between intramural nerves and the crypt enterocyte.

As shown in Table 70.1, there are many neuroactive substances found within submucosal or myenteric neurons which project to the epithelium. None of these have been identified unequivocally as the NANC transmitter(s) responsible for stimulating Cl^- secretion, but there is solid evidence for the role of substance P, which can act in several ways [13]. Substance P can be released from primary afferents or from myenteric neurons and modify the activity of submucosal neurons. It can also be released from submucosal neurons along with acetylcholine to act directly on the mucosal epithelium. Another compound, calcitonin gene-related peptide (CGRP), has been shown to be contained within neurons of both the myenteric and submucosal plexus. CGRP is located within neurons that can modify other neurons of the submucosal plexus and a variety of structures within the mucosa including the enterocyte itself. CGRP does not activate short circuit current or Cl^- secretion when added to an Ussing chamber; it will, however, greatly augment (potentiate) the response to substance P [10]. It may be, therefore, of great significance in the regulation of intestinal secretion through its interaction with substance P-containing neurons. The effects of substance P, however, are complex since it has been shown that the effects of substance P can be attributed largely to actions on other enteric neurons rather than directly on the enterocyte. It is possible that other transmitters can act at multiple sites to effect changes in Cl^- and water secretion. These would include acetylcholine, vasoactive intestinal peptide (VIP), and neuropeptide Y (NPY). Many other transmitters, however, could also act at multiple sites, leading either to excitation or inhibition of crypt cell secretion.

Enteric Secretomotor Reflexes. Several reflexes involving enteric neurons can evoke intestinal secretion in vivo and in vitro. For example, stimulation of mucosal mechanoreceptors by lightly stroking the mucosa evokes secretion of mucus aboral to the site of stimulation, a reflex which involves both cholinergic and serotoninergic synapses. Other mechanosensory receptors sensitive to stretching of the intestinal wall lead to Cl^- and water secretion from crypt enterocytes. Since this can be evoked by stretching of a mucosal preparation stripped of the muscularis externa and the myenteric plexus, this would appear to involve a reflex circuit restricted to the mucosa and submucosa. Another example is the Cl^- secretion evoked by crypt enterocytes when glucose is removed from the mucosal bathing solution. This appears to activate neural signals that involve nicotinic transmission within the reflex circuitry. Finally, it is now well-established that there is a temporal link between phases of motility and phases of secretion [3,4]. Since this can occur in isolated preparations with no extrinsic innervation, the basic mechanism would appear to reside within the enteric nervous system. There is controversy, however, over whether the changes in secretion are secondary to the contractile activity or vice versa. The situation is more complicated in vivo due to the presence of extrinsic nerves. However, it has been shown that the onset of phase three of the MMC (see Chap. ••) is also associated with an increase in secretion. Again, the question of whether the increased motor activity can be detected by vagal tension receptors and thereby induce a reflex secretion, or whether the entire control mechanism resides within the enteric nervous system has not been resolved.

70.3.3 Enteric Neural Control of Intestinal Blood Flow

Regulation of intestinal blood flow has long been regarded as being essentially under the control of the sympathetic nervous system (see Chap. 69). In recent years, however, local enteric reflexes have been described which regulate intestinal blood flow. These involve reflex circuits similar to those involved in the regulation of motility and secretion. For example, stroking the mucosa can elicit a reflex vasodilatation through the activation of nicotinic synapses. The neural control of the intestinal vasculature is largely restricted to submucosal blood vessels. In contrast to the sympathetic nervous system which is responsible for intestinal vasoconstriction, the intrinsic reflexes are dilatory. Dilation of submucous arterioles involves activation of intrinsic vasomotor neurons which release acetylcholine. This in turn seems to act at the M-3 muscarinic receptor located on the endothelium which in turn leads to the release of nitric oxide which is the vasodilator mediator at the level of the vascular smooth muscle. In addition, however, a NANC transmitter is also involved and several neuropeptides including VIP, substance P, and CGRP have been implicated.

70.3.4 Enteric–Immune System Interactions

The gastrointestinal tract is an important component of the immune system (see Chap. 86). In addition to the Peyer's patches, which are important areas in the development of white blood cells, large numbers of immunocytes are sequestered within the wall of the gastrointestinal tract and can participate in local immune reactions. The gastrointestinal tract also contains large numbers of mast cells. Although virtually nothing is known of the regulation of these components of the immune system by the enteric nervous system, current work has thrown considerable light on the way in which mediators released from cells of the immune system (immunocytes) can modify enteric nerve activity and, therefore, influence motility, epithelial transport, and local blood flow. It is now well-recognized that mediators such as histamine and other neuroactive compounds released from mast cells have profound effects on the electrical activity of enteric neurons. This has been shown by exogenous applications of substances such as histamine to enteric neurons which results in a repetitive, cyclical discharge of action potentials in submucosal neurons. This in turn leads to activation of periodic Cl⁻ and water secretion. Similar effects are obtained in animals that have been sensitized, either to specific proteins or to parasites. Later exposure of the sensitized intestine to the appropriate allergen evokes the same cyclical pattern of neuronal discharge and neurogenic secretion. Myenteric neurons are also activated and are involved in the motor responses to mast cell degranulation following exposure to an appropriate antigen in the sensitized intestine. It is also probable that mediators released from cells such as macrophages and eosinophils are also involved in neutrally mediated responses.

In summary, then, the enteric nervous system plays an important role in the immune activity of the gastrointestinal tract. It may form a final pathway for neural influences on certain components of the immune system, in particular, mast cells. It most certainly acts as a final pathway for many of the effects of mediators released by mast cells and perhaps other immunocytes.

70.4 Conclusions

The enteric nervous system contains all the elements essential to an autonomous nervous system. Through its complex circuitry it programs and regulates all the major functions of the gastrointestinal tract and has come to be regarded as "the little brain in the gut".

General References

Befus D (1994) Reciprocity of mast cell–nervous system interactions. In: Taché Y, Wingate DL, Burkes TF (eds) Innervation of the gut: pathophysiological implications. CRC Press, London, pp 315–329

Brookes SJH, Costa M (1994) Enteric motor neurons. In: Taché Y, Wingate DL, Burkes TF (eds) Innervation of the gut: pathophysiological implications. CRC Press, London, pp 237–248

Collins SM, McHugh K, Hurst S, Weingarten H (1994) Neuro-immune interactions in the pathophysiology of inflammatory bowel disease. In: Taché Y, Wingate DL, Burkes TF (eds) Innervation of the gut: pathophysiological implications. CRC Press, London, pp 331–343

Cooke HJ (1986) Neurobiology of the intestinal mucosa. Gastroenterology 90:1057

Cooke HJ (1989) Role of the "little brain" in the gut in water and electrolyte homeostasis. FASEB J3:127

Cooke HJ, Wang Y-Z, Rogers RC (1994) Neuroimmune interactions: histamine signals to the intestine. In: Taché Y, Wingate DL, Burkes TF (eds) Innervation of the gut: pathophysiological implications. CRC Press, London, pp 307–313

Davison JS (1983) Innervation of the gastrointestinal Tract. In: Christenson J, Wingate DL (eds) A guide to gastrointestinal motility. PSG, Bristol, pp 1–47

Gershon MD, Kirchgessner AL, Wade PR (1994) Intrinsic reflex pathways of the bowel wall. In: Taché Y, Wingate DL, Burkes TF (eds) Innervation of the gut: pathophysiological implications. CRC Press, London, pp 276–285

Perdue MH (1994) Immunomodulation of epithelium. In: Sutherland LR, Collins SM, Martin F, McLeod R, Targen SR, Wallace JL, Williams CN (eds) Inflammatory bowel disease basic research, clinical implications and trends in therapy. Kluwer, Boston, pp 139–148

Sharkey KA, Parr EJ (1994) The enteric nervous system in intestinal inflammation. In: Sutherland LR, Collins SM, Martin F, McLeod R, Targen SR, Wallace JL, Williams CN (eds) Inflammatory bowel disease basic research, clinical implications and trends in therapy. Kluwer, Boston, pp 149–161

Tack JF, Janssens W, Janssens J, van Trappen G (1994) Regional neurophysiology of the enteric nervous system. In: Taché Y, Wingate DL, Burkes TF (eds) Innervation of the gut:

pathophysiological implications. CRC Press, London, pp 249–263

Wood JD (1984) Enteric neurophysiology. Am J Physiol 247 (Gastrointest Liver Physiol 10) G585–G598

Wood JD (1991) Communication between minibrain in gut and enteric immune system. NIPS 6:64

Wood JD (1994) Intestinal immuno-neurophysiology. In: Taché Y, Wingate DL, Burkes TF (eds) Innervation of the gut: pathophysiological implications. CRC Press, London, pp 289–305

Wood JD (1994) Physiology of the enteric nervous system. In: Johnson LR (ed) Physiology of the gastrointestinal tract, 3rd edn, Raven, New York, pp 423–482

Young HM, Furness JB, Bornstein JC, Pompolo S (1994) Neuronal circuitry for enteric motility reflexes. In: Taché Y, Wingate DL, Burkes TF (eds) Innervation of the gut: pathophysiological implications. CRC Press, London, pp 266–274

Specific References

1. Davison JS (1983) Innervation of the gastrointestinal Tract. In: Christenson J, Wingate DL (eds) A guide to gastrointestinal motility. PSG, Bristol, pp 1–47

2. Furness JB, Bornstein JC, Trussel DC (1988) Shapes of nerve cells in the myenteric plexus of the guinea pig small intestine revealed by the intracellular injection of dye. Cell Tiss Res 254:561–571

3. Greenwood B, Davison JS (1985) Role of extrinsic and intrinsic nerves in the relationship between intestinal motility and transmural potential difference in the anesthetized ferret. Gastroenterology 89:1286–1292

4. Greenwood B, Davison JS (1987) The relationship between gastrointestinal motility and secretion. Am J Physiol 252: (Gastrointest Liver Physiol 15) G1–G7

5. Hubel KA (1978) The effects of electrical field stimulation and trototoxin on ion transport by the isolated rabbit ileum. J Clin Invest 62:1039–1047

6. Hubel KA (1989) Control of intestinal secretion. In: Davison JS (ed) Gastrointestinal secretion. Wright, London, pp 178–201

7. Hubel KA, Renquist KS and Varley G (1991) Secretory reflexes in ileum and duodenum; absence of remote effects. J Autonomic Nervous System 35:53–62

8. Hubel KA and Shirazis H (1982) Human ileal ion transport in vitro: changes with electrical field stimulation and tetrodotoxin gastroenterology 83:63–68

9. Langley JN (1921) The autonomic nervous system. Part 1. Heffer, Cambridge

10. Mathison R, Davison JS (1989) Regulation of epithelial transport in the jejunal mucosa of the guinea pig by neurokinins. Life Sci 45:1057–1064

11. Perdue MH, Davison JS (1986) Responses of jejunal mucosa to electrical transmural stimulation and two neurotoxins. Am J Physiol 251:G642–G648

12. Perdue MH, Davison JS (1988) Altered regulation of intestinal ion transport by enteric nerves in diabetic rats. Am J Physiol 254: (Gastrointest Liver Physiol 17) G444–G449

13. Perdue MH, Galbraith R, Davison JS (1987) Evidence for substance P as a functional neurotransmitter in guinea pig small intestinal mucosa. Reg Peptides. 18:63–74

14. Wood JD (1994) Progress in neurogastroenterology. In: Kirsner JB (ed) The growth of gastroenterologic knowledge during the 20th century. Lea and Febiger, Philadelphia, pp 159–180

71 Energy Metabolism and Nutrition

K. Jungermann and C.A. Barth

Contents

71.1 Introduction

Every cell constitutes an energy transformer. Chemotrophic cells take up chemical energy in the form of energy substrates such as glucose, fatty acids, or amino acids and phototrophic cells take up physical energy in the form of light. They transform the energy taken up into "biological" energy via fermentation and (cellular) respiration and via photosynthesis, respectively. Cells expend biological energy for chemical work in a multitude of biosyntheses, for osmotic work in transport processes, e.g., over the plasma membrane, and for mechanical work in the movement of intracellular components, e.g., chromosomes. The conversion of chemical or physical energy into biological energy, which is termed energy metabolism, as well as the utilization of biological energy for the various work functions, which is termed work metabolism, cannot be achieved with 100% efficiency. Therefore, a cell dissipates energy in the form of heat during both energy and work metabolism (Fig. 71.1A).

Both unicellular and multicellular organisms have to maintain their energy balance very strictly, At no time can they allow a deficit or an excess of biological energy to exist. At any given time, all organisms have to carry out a considerable amount of work to maintain their structure and basal functions; therefore, at the same time they have to take up the equivalent amount of energy. However, energy may not always be available, as is the case in the photosynthetic plant kingdom at night, or it may not be taken up continuously, as is the case in the animal kingdom with intermittent eaters. Therefore, energy stores such as starch, glycogen, triglycerides, or proteins have to be synthesized from energy substrates when taken up in excess of the actual demand, i.e., during the day with plants or after a meal with animals; the stores can then be utilized when the exogenous energy supply is too low or has ceased (Fig. 71.1B).

In all mammals both energy intake and energy expenditure can vary over a wide range. Complex hormonal and nervous regulatory mechanisms are required to keep energy intake and energy expenditure in balance; a short-term, but not a long-term imbalance can be tolerated. Overnutrition is characterized by a long-term excess of energy intake, mainly fat, compared to energy expenditure; it inevitably results in disease, i.e., body fatness or obesity. Undernutrition is characterized by a long-term excess of energy expenditure compared to too low an energy intake because of malnutrition (insufficient substrate supply), malabsorption (insufficient substrate absorption), or anorexia nervosa and bulimia (psychogenic eating disorders); it inevitably leads to disease, i.e., emaciation, cachexia, kwashiorkor, or marasmus.

The present chapter will first give an overview of the ranges of the daily energy expenditure with different work loads and of the daily energy intake with different diets, covering the substrate flow between the major organs during different metabolic situations. The mechanism of energy metabolism will then be briefly described, comprising both the biological oxidation of energy substrates and the formation and degradation of energy stores. Finally, the hormonal and nervous regulation of energy metabolism will be reviewed.

R. Greger/U. Windhorst (Eds.)
Comprehensive Human Physiology, Vol. 2
© Springer-Verlag Berlin Heidelberg 1996

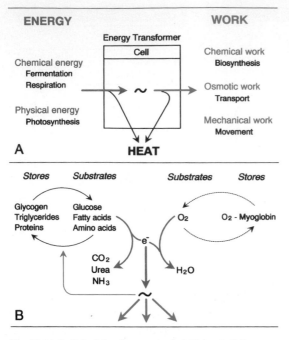

Fig. 71.1A,B. Principle of energy metabolism. **A** Cells as energy transformers. **B** Energy substrates, energy stores, and oxidative energy conservation coupled to the electron transport from the reduced carbon substrates to oxygen. ~, "biological" energy in the form of adenosine triphosphate (ATP)

71.2 Energy Expenditure and Energy Intake

71.2.1 Basal Metabolic Rate and Total Energy Expenditure

Total energy expenditure (TEE) consists of three components [25,82,87]:

- Basal metabolic rate (BMR; Table 71.1)
- Diet-induced thermogenesis (DIT), formerly known as specific, dynamic action of food
- Activity energy costs (Table 126.24).

Basal Metabolic Rate. The BMR comprises the energy expenditure at rest necessary to maintain:

- The internal chemical milieu of the body, e.g., the concentration gradients between the intra- and extracellular of Na^+ and K^+ or the continuous turnover of cellular and extracellular protein, glycoprotein, and proteoglycan constituents
- The electrochemical activities of the nervous system
- The electromechanical work of the circulatory system
- Thermoregulation [25,82].

The continuous renewal of body constituents is a major determinant of BMR, as indicated by the finding that whole body protein turnover is a constant fraction of BMR over the total life span [112].

BMR is age and sex dependent; it is normally the largest contributing factor of TEE, accounting in adults with moderate activity for about 70% (Tables 71.1, 71.2). BMR is determined postabsorptively at an ambient temperature of 20°C, 20h after the last meal with the subjects wearing light clothing and lying comfortably at complete physical and mental rest. When asleep and when DIT (see below) has subsided, TEE is 5%–10% below BMR [82]. The most practical way to standardize BMR is to relate it to body surface area, which is a function of body weight and height, except in obese people with abnormal body shapes (Table 71.1). The physiologically most appropriate way is to relate BMR to lean body mass (LBM) or fat-free mass (FFM), which can be determined by anthropometric methods (skinfold thickness, circumferences) or using biophysical techniques (density by underwater weighing, electrical conductivity, bioelectrical impedance, or nuclear magnetic resonance) [34]. BMR related to body surface is high in actively growing young children, declining rapidly during the first 12 years of life and then more slowly; it is higher in male than in female subjects (Table 71.1). BMR related to LBM is largely independent of age, sex, and body size. Thus, there is no fundamental difference in the metabolism of the two sexes; the differences in BMR related to the body surface reflect the fact that women's bodies contain relatively more fat than men's, e.g., 29% versus 15% fat at the age of 20–29 years [25,82].

BMR is dependent on thyroid hormone status and sympathetic nervous system activity; moreover, there may be a genetic component to BMR [25,82]. Thyroid hormones increase oxygen consumption, protein synthesis, and degradation as well as thermogenesis. A rise in BMR is a classical sign of hyperthyroidism; a decline of BMR paralled by a decrease in plasma thyroid hormones occurs in starvation (see Sect. 71.5.4, Fig. 71.5C, Table 71.5). Chronic administration of β-sympathetic agonists also enhances BMR. Subjects with lower BMR, i.e., presumably higher metabolic efficiency, appear to be more susceptible to weight gain (see Sect. 71.2.3). Nearly 40% of interindividual differences can be explained by genetic heterogeneity.

Diet-Induced Thermogenesis. After a normal mixed meal, energy expenditure is increased for up to 6h, probably to drive digestion of food and absorption and storage of macronutrients. However, the mechanism of DIT has not been fully elucidated. DIT amounts to 8%–15% of TEE in moderately active individuals (Table 71.2). A fat, carbohydrate, or protein meal will be followed by a DIT of 2%–4%, 4%–7%, and 18%–25% of the consumed energy, respectively. Protein induces a DIT that lasts twice as long as that induced by the other macronutrients [25,82].

Activity Energy Costs. The physical activity or work load of individuals varies over a very wide range. The physiologically most appropriate way to estimate the fraction of TEE due to activity energy costs is to multiply the individual BMR by a specific activity factor minus 1. These factors are 1.2 for sitting quietly, 1.3–1.6 for office work, 1.4 for standing, 3.2 for walking at normal pace, 3.6 for repair-

Table 71.1. Basal metabolic rate (BMR) and total energy expenditure (TEE) as a function of age and sex

Age (years)	Weight[a] (kg)	Height[a] (cm)	BMR $(kcal/h \times m^2)$ [9,82][c]	BMR (kcal/24 h)	BMR (kJ/24 h)	TEE[b] (kcal/24 h) [108]	TEE (kJ/24 h)
Males							
1.5	11.5	81	54.8	646[d]	2707[d]	1200	5020
5.5	19.7	115	49.4	943[d]	3940[d]	1810	7570
10	32	140	44.0	1220[e]	5100[e]	2140[e]	8950[e]
15	55	170	41.8	1660[e]	6940[e]	2700[e]	11290[e]
20	70	178	38.8	1780[f]	7440[f]	Highly variable due to different physical activity, decreasing with age due to lowered activity	
30	70	178	36.9	1710[f]	7150[f]		
40	70	178	36.4	1640[f]	6850[f]		
50	70	178	36.0	1580[f]	6600[f]		
60	70	178	35.0	1510[f]	6310[f]		
70	70	178	33.8	1440[f]	6020[f]		
Females							
1.5	10.6	81	52.4	596[d]	2490[d]	1140	4760
5.5	18.5	113	49.0	915[d]	3830[d]	1790	6810
10	33	137	42.4	1150[e]	4810[e]	1910[e]	7980[e]
15	52	163	38.0	1400[e]	5850[e]	2140[e]	8950[e]
20	58	166	35.4	1420[f]	5940[f]	Highly variable due to different physical activity, decreasing with age due to lowered activity	
30	58	166	35.2	1370[f]	5730[f]		
40	58	166	35.0	1320[f]	5520[f]		
50	58	166	34.0	1280[f]	5350[f]		
60	58	166	32.8	1230[f]	5140[f]		
70	58	166	31.8	1180[f]	4930[f]		

[a] See Figs. 126.7 and 126.10
[b] See Figs. 126.7 and 126.8
[c] Body surface area = $0.007184 \times weight^{0.425} \times height^{0.725}$
[d] See [57]
[e] WHO [108]: BMR + energy for growth (8 kJ/kg at 10–14 years, 4 kJ/kg at 15 years and 2 kJ/kg at 16–18 years) + specific times spent in school (+0.6 BMR, 4–6 h increasing with age), light activity (+0.6 BMR, 4–7 h increasing with age), moderate activity (+1.5 BMR, 6.5–2.5 h decreasing with age), high activity (+6.0 BMR, 0.5 h)
[f] The values shown are from [43]; very similar values are given by in [9,55,108]

Table 71.2. Distribution of the daily energy expenditure of a male office worker (model calculation)

Energy expenditure	Absorptive phase 1 (2.5 h; 0700–0930 hours)	Postabsorptive phase 1 (3.5 h; 0930–1300 hours)	Absorptive phase 2 (2.5 h; 1300–1530 hours)	Postabsorptive phase 2 (3.5 h; 1530–1900 hours)	Absorptive phase 3 (2.5 h; 1900–2130 hours)	Postabsorptive phase 3a (3.5 h; 2130–0100 hours)	Postabsorptive phase 3b (6 h; 0100–0700 hours)	Total
BMR (kcal)	187.5	262.5	187.5	262.5	187.5	262.5	450	1800
DIT (kcal)	33	47	33	47	33	47	0	240
Sitting (kcal)	37.5	52.5	37.5	52.5	37.5	22.5	0	240
Work load (kcal)	0	110	0	110	0	0	0	220
Sum (kcal)	258	472	258	472	258	332	450	2500
TEE (kcal/h)	103	135	103	135	103	95	75	

The following assumptions are made: daily energy intake is equally distributed over three meals at 0700, 1300, and 1900 hours. The absorptive phases last 2.5 h each. DIT is about 10% of the daily TEE and lasts for 6 h after a meal. The daily work load is equally distributed over the postabsorptive phases during the day; there is no work load (physical activity except sitting) in the absorptive phase and in the postabsorptive phase at night
BMR, basic metabolic rate; TEE, total energy expenditure; DIT, diet-induced thermogenesis

ing motor vehicles, 4.4 for shoveling mud, 4.4–6.6 for cycling or playing tennis, 5–9 for playing soccer or skiing, and 8 for forestry work. For short periods of time, these factors can be 10–20 for sporting activities such as competitive swimming or running (Table 126.24). The energy needs required in various degrees of injury are determined by use of specific injury factors. These factors are 1.2 for minor surgery, 1.35 for trauma, 1.6 for sepsis, and 2.1 for burns; 1°C of fever increases TEE by 13% [3]. The TEE of a given individual can therefore be approximated without measurement by summing up the products of the individual BMR multiplied by the activity factors multiplied by the injury factors and the time spent at a given activity. All nonspecified times are considered to require a TEE of 1.4 × BMR; hours of sleeping, in which DIT has subsided, are considered as 1.0 × BMR. These considerations allow a model calculation of the distribution of the daily energy expenditure (Table 71.2).

Thus TEE ranges from 10 to 13 MJ/day with light activities such as desk work, 13 to 15 MJ/day with moderate work such as car repair or truck driving, 15 to 17 MJ day with moderately strenuous work such as casting or roof covering, 17 to 19 MJ/day with strenuous work, such as carpeting or wood cutting, and 19 to 21 MJ/day with very strenuous work such as coal mining (Table 126.25). Since the BMR of men ranges from 6.8 to 7.5 MJ/day (Table 71.1) and DIT contributes to TEE, activity energy costs account for 15%–30% of TEE in people with light activities (Table 71.2), for 30%–40% in people with moderate activities, and for 40%–45% in people performing, moderately strenuous work.

Indirect Calorimetry. This is the scientific method used to quantitate BMR or TEE. For this purpose, respiratory gas exchange (respiratory quotient, $RQ = CO_2$ exhaled/O_2 inhaled) is quantified. Energy expenditure is calculated from the total oxygen consumption, the caloric equivalent of which is deduced from the RQ (Table 126.26, last column). Moreover, the RQ allows net macronutrient oxidation rates to be calculated. First, the amino acid oxidation rate and thus the protein RQ is obtained independently from urinary nitrogen excretion. The total RQ is then corrected for the protein RQ to obtain the nonprotein RQ (NPRQ). NPRQ correlates linearly and positively with the percentage of the energy (energy%) from carbohydrate oxidation and linearly but inversely with the energy% fat(ty acid) oxidation. NPRQ for carbohydrate oxidation is essentially 1.0 and for fat oxidation 0.71 (Table 126.26). NPRQ is above 1 in the case of net lipacidogenesis (see Sect. 71.4.5); for tripalmitin synthesis from glucose it amounts to 2.75. Finally, using these relationships the energy% contribution from carbohydrate and fat oxidation to energy expenditure is determined.

Relative Contribution of Carbohydrates, Fat, and Protein to Energy Expenditure. With three different, slightly hypercaloric diets, i.e., a carbohydrate-rich, a normal mixed balanced, and a fat-rich diet, each containing 15 energy% protein, fat oxidation amounted to 42%, 46%, and 55% and carbohydrate oxidation to 43%, 39%, and 30%, respectively, in the morning before the first meal. During the absorptive phase (80 min after the meal), carbohydrates became the predominant energy substrates in the carbohydrate-rich and the mixed balanced diets, but also with the fat-rich food, contributing 75%, 74%, and 49%, with fat falling to 10%, 11%, and 36% respectively. It is remarkable that the fatty acid oxidation rate was markedly less suppressed postprandially in the high-fat diet than in the other two regimens, showing the regulatory adaptive capacity of metabolism (Fig. 71.2). The alterations observed could be explained by graded changes in the insulin

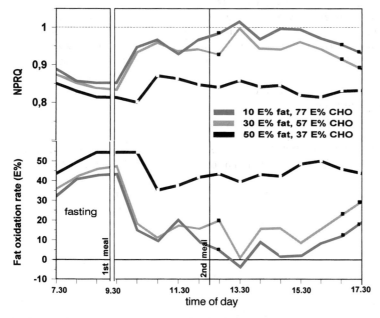

Fig. 71.2. Contribution of carbohydrate, fat, and protein oxidation to energy expenditure in the postprandial phase. Three isoenergetic but slightly hypercaloric diets, i.e., high-fat (50 energy%, E%, fat, 35 E% carbohydrate, 15 E% protein), moderate-fat (30 E% fat, 55 E% carbohydrate, 15 E% protein), low-fat (10 E% fat, 75 E% carbohydrate, 15 E% protein), were given to 14 healthy young male [9] and female [5] students for 4 days at 0930, 1230, 1800, and 2000 hours. Measurements were made on the fourth day. E% fat oxidation rate is shown; protein oxidation was 15 E%. E% carbohydrate oxidation is 100 − 15 E% fat oxidation; this corre-lates with the nonprotein respiratory quotient (NPRQ). Significant differences between the high-fat diet on the one hand and the moderate- and low-fat diets on the other are marked by *white squares* on the high-fat diet curve; significant differences between the moderate- and the low-fat diet are marked by *black squares* on the respective curves. In addition to this, the *white squares* indicate a level of significance of at least $P < 0.01$, whereas the *black squares* refer to $P < 0.05$ (H. Karst, U. Gerstmann, R. Noack, and C.A. Barth, unpublished)

to glucagon ratio during the absorptive phase (see Sects. 71.3.1, 71.5.2, 71.5.3; Fig. 71.3, Table 71.5).

71.2.2 Energy Intake: Chemical Composition and Nutritive Value of Food

The nutritive value of food is given by its content of energy substrates such as glucose, fatty acids, and amino acids (caloric value); of building substrates such as the essential amino acids and essential fatty acids for the formation of macromolecular and low molecular weight body constituents (synthetic value); and of cocatalysts such as electrocytes, trace elements, and vitamins (catalytic value). In a healthy diet, each nutrient must be balanced against the others. In the context of the present chapter, mainly the balance of energy substrates will be considered. Two principles apply here:

- First, there are constraints imposed by the limited or lacking capacity of mammalian metabolism to interconvert the energy substrates glucose, fatty acids, and amino acids. Glucose can be produced from glucogenic amino acids (see Sect. 71.4.3, Fig. 71.10C); however, the process is limited by the possible protein content of available food. Glucose cannot be formed from even-numbered fatty acids, the main constituents of dietary fat. Only the nonessential amino acids can be synthesized from glucose and ammonium salts; however, the process is again limited by the low intake of ammonium salts. Since, without glucose, adequate glycogen stores cannot be maintained and since the brain, the erythrocytes, and the working muscles rely on glucose as an energy substrate (see Figs. 71.4, 71.5), there is a daily minimum requirement of 100 g glucose or carbohydrates as glucose precursors (see Sect. 71.3). Since the obligatory nitrogen losses by protein turnover have to be replaced and enough essential amino acids have to be provided, a balanced diet has to contain 7–15 energy% of protein, i.e., 0.6–1.3 g/kg body weight = 40–90 g, half of which should be of animal origin [111].
- Second, preventive medicine and nutrition define the balance of nutrients according to pathogenetic princi-

ples. For example, fat intake should not exceed 20–30 energy% in order to prevent obesity; fat quality should be balanced in terms of the intake of saturated, monounsaturated, and polyunsaturated fatty acids. Protein intake should not exceed the above-mentioned limits to avoid aggravation of kidney diseases [11]. Purine (meat) and sodium chloride intake should be moderate to prevent gout and hypertension [6].

Interestingly, protein consumption in different cultures tends to deviate little from an average 12%–15% of energy intake (Table 71.3). About two thirds of dietary protein is of animal origin in cultures of the northern hemisphere and about one third in tropical countries. This similarity between amounts of protein consumed contrasts with the wide variety of other dietary parameters. Eastern Asian populations rely typically on a fish–rice diet, the Mediterranean diet contains a large amount of vegetables, and a typical western diet is dominated by food of animal origin (Table 71.3). All diets supply sufficient energy substrates; however, a diet composed mainly of plant food, i.e., vegetable and starchy roots, seems to be the only way to achieve a significant dietary fat reduction and a considerable intake of polymeric carbohydrates, as exemplified by the traditional Japanese diet (Table 71.3). This is of prime importance because of the metabolic link between dietary fat and obesity (see Sect. 71.2.3) In this context it is of note that normal body weight and serum cholesterol concentrations and a low incidence of atherosclerotic disease are typical in the Japanese population [50]. Eastern Asian and Mediterranean diets provide ample dietary plant constituents, and more and more evidence is accurmulating that they protect against chronic degenerative disease; a major advantage is the content of dietary fiber, i.e., nondigestible polysaccharides of plant cell walls that prevent, among other things. obstipation, diverticulosis, and probably colon cancer [27].

Other aspects of the nutritive value of food, such as the high content of essential nutrients (iron, vitamins, particularly B_{12}, Ca^{2+}, or essential amino acids; see Table 126.27) in western diets or of ω-3 fatty acids (first double bond from the ω-carbon between positions 3 and 4) as substrates for protective eicosanoids and of antioxidants such as flavonoids and β-carotene in eastern Asian and Mediterra-

Table 71.3. Daily food supply in selected countries in 1988

Country	Food supply (g/d)											Energy supply (%)		
	Milk cheese	Meat	Fish	Cereals rice	Potatoes starchy food	Vege-tables	Fruits	Pulses nuts	Fats oils[a]	Wine	Beer	Carbo-hydrate	Lipid	Protein
Italy	294	236	34	330	107	458	355	23	76	144	61	46.7	41.1	12.2
France	422	300	60	229	213	336	227	10	78	195	104	40.1	46.0	13.9
Germany	297	285	23	213	200	240	380	12	73	74	406	40.4	48.1	11.5
USA	375	330	22	208	92	253	172	20	76	nd	nd	43.3	44.4	12.3
Japan	107	118	105	286	96	302	106	68	40	nd	nd	54.4	32.2	13.4

Data are based on agricultural production and export/import values (= supply). Only selected types of food are shown to illustrate three cultural patterns; Mediterranean (Italy, France), western France, Germany, USA) and eastern Asian (Japan). Supply in this context is not equal to actual consumption. In Germany, energy intake is estimated to be 50%–45% carbohydrates, 38%–40% fat and 12%–15% protein. Source: OECD (Organisation for Economic Cooperations and Development), Food Consumption Statistics, Paris (1991)

[a] Margarine excluded
nd, no data avaiable

nean diets, are not directly linked to energy metabolism and therefore not discussed in detail.

71.2.3 Overnutrition

Energy excess, i.e., energy intake exceeding the TEE, unavoidably leads to fat accumulation in adipocytes, the other macronutrient stores being of limited size. Moderate body fatness is called overweight, and more excessive body weight obesity [10,77,82]. Until recently, the body weight was mainly standardized by the easy to handle Broca index: normal weight (kg) = body height (cm) − 100. A 10% increase was defined as overweight, and a 20% increase as obesity. Body fatness correlates, however, more closely with the more complicated body mass index (BMI), i.e., body weight (kg)/[body height(m)]^2. Normal values for BMI are 20–25 for men independent of age and for women up to 35 years of age; in women there is a slight, continuous increase to 24–29 for the age group 65 years and older. A BMI of 25–30 indicates overweight and a BMI greater than 30 obesity [10].

There are two types of fat distribution;

- Visceral abdominal fat (android or upper-body or male-type obesity)
- Fat on the hips (gynoid or lower-body of female type obesity).

The volume of the visceral abdominal fat is most precisely quantified by computer tomography. A more simple, but less precise approach is to determine the ratio of waist to hip circumferences (WHR); a high ratio indicates android obesity, e.g., for the age group 40–49 years a WHR greater than 0.95 in men and greater than 0.8 in women [10].

Obesity increases the risk of premature death, diabetes mellitus, hypertension, cardiovascular disease, gallbladder disease, and certain types of cancer [10,77]. There are two components to weight balance; an abnormality on either side, too high an energy intake or too low an energy expenditure, will lead to obesity. Genetic factors can play a role in both.

Energy Intake. Overeating appears to be the usual cause of overweight. A distinction must be made between the actual process of gaining weight and static obesity. The regulation of eating behavior is still incompletely understood (see Chap. 72). After a meal, gastric distension, a rise in plasma glucose and insulin (see Sect. 71.3; Fig. 71.3), and adrenergic influences appear to activate the satiety center in the hypothalamus, which in turn inhibits the feeding center, the set point of which might be regulated by a signal indicating adipose tissue size [77]. Such a "satiety factor" has recently been cloned (see Seat. 71.5.1). Direct observations have indicated that obese people eat more food with more fat and often do so more rapidly than subjects of normal weight. However, some epidemiological studies have revealed that obese people do not eat more, but that they are clearly less physically active [10]. The former

finding may explain in part the etiology of obesity; the latter finding may explain the steady state of obesity.

Overeating carbohydrate-rich and protein-rich, but not fat-rich diets can in part be compensated by an increase in BMR. The mechanism of this adaptive thermogenesis is unknown, but it coincides with an elevation of thyroid hormones (see Table 71.5). Thus, humans possess a protective mechanism against weight gain from excess carbohydrates and protein, but not fat [77]. The major cause of body fatness is indeed the fat-rich diet common in western societies (Table 71.3). It suppresses lipacidogenesis (see Sect. 71.4.5, Fig. 71.10) by feedback inhibition. Hence, expansion of body fat stores occurs by deposition of dietary triglycerides via the chylomicron–adipose tissue lipoprotein lipase pathway (see Sect. 71.3.2, Fig. 71.4). This expansion can be accompanied by hyperinsulinemia and a gradual rise of insulin resistance with growing size of adipocytes (hypertrophic obesity). Their enlargement and insulin resistance continue until insulin no longer inhibits hormone-sensitive lipase, and a new steady state between triglyceride deposition and lipolysis (see Sect. 71.4.5, Fig. 71.11) occurs when there are significantly greater body fat stores [33].

A high fat content of food is a major risk factor for the triad of hyperinsulinemia with insulin resistance, hyperlipidemia, and high blood pressure that accompanies obesity. There is considerable evidence that this "metabolic syndrome," characterized by elevated postabsorptive levels of plasma glucose and insulin, of very low density lipoproteins (VLDL) carrying a major part of serum triglycerides, and of apolipoprotein B, as well as low concentrations of high-density lipoprotein 2 (HDL$_2$) cholesterol, correlates more closely with visceral abdominal fat than with fat on the hips or total body fat [24]. This metabolic syndrome is highly prevalent in industrialized western societies and may be prevented most efficiently by reducing fat intake to levels habitual in eastern Asia (Table 71.3).

Energy Expenditure. As mentioned above, TEE has three components: BMR, DIT, and physical activity energy costs (see Sect. 71.2.1). If all or even only one of the three components is reduced, energy intake has to be lowered to balance energy expenditure; a failure to downregulate energy intake inevitably leads to adiposity [77].

Within the obese population, a subpopulation appears to exist (perhaps 25%) with an *increased metabolic efficiency*; BMR and probably also the activity energy costs appear to be reduced by 6% each, and DIT by 38%. With a sendentary occupation, this would decrease the daily energy requirements of men by about 250 kcal or 0.5 l beer. If energy intake in these subjects is not downregulated accordingly, a considerable weight gain will result [49,75]. (As a rule of thumb 10 kcal = 1 g extra adipose tissue; thus an excess energy intake of 250 kcal per day could increase the adipose tissue by 9 kg in a year).

With age, activity energy costs are reduced. The common, though not ideal weight gain with age (see Fig. 126.12) must be related to a greater reduction of energy expenditure than of energy intake [10]. Controlled nutritional experiments have repeatedly shown that – for a given physical activity – a threshold of dietary fat intake exists above which a positive fat balance occurs. This threshold is approximataly 30% of dietary energy in subjects taking moderate exercise. It can very probably be shifted to higher values with increasing and to lower values with decreasing physical exercise. *Low physical activity* thus becomes a major risk factor in combination with western diets. Consequently, an efficient way of combating fat accumulation is enhancing lipolysis by endurance exercise accompanied by low insulin and elevated catecholamine plasma levels, which raise the activity of hormone-sensitive lipase and thus lipolysis and lower adipose tissue lipoprotein lipase and thus lipogenesis (see Sects. 71.5.2, 71.5.3; Table 71.4). Both changes contribute to diverting chylomicron triglycerides away from fat stores towards muscle fatty acid oxidation (see Fig. 71.3).

Genetic Factors. There is a correlation between the BMI of parents and their biological, but not their adopted children. These data suggest that inheritance plays an important role in the risk of developing obesity [10]. Moreover, overnutrition studies with human volunteers have shown that even massive energy excess over months led to very different body weight gains in different people [98]. The lower BMR and lower rates of DIT in some subjects (see above) also show the importance of genetic factors for the pathogenesis of obesity. Thus, both nature and nurture contribute to obesity.

71.2.4 Undernutrition

An energy deficit (energy intake less than TEE) leads to an abnormally low body weight defined by a BMI of less than 18.5. In developing countries the prevalence of underweight reaches 10%–20%; however, in France and the United States of America it amounts to only 3.5% and 4.9%, respectively [96]. There are several causes for underweight:

- Protein-energy malnutrition (PEM)
- Maldigestion/malabsorption
- Psychogenic eating disorders.

Protein-Energy Malnutrition. If, for a longer period of time, the diet does not contain sufficient amounts of energy substrates or of protein to meet the energy requirements or the amino acid requirements for the continuous turnover of body protein, respectively, PEM will develop leading to *emaciation* and *cachexia*. Insufficient dietary intake for normal or increased energy requirements in physiological states such as pregnancy and lactation is the cause of primary PEM. Inadequate intake in pathological states such as fever, infection, trauma, or malignant diseases is the cause of secondary PEM.
PEM is especially threatening to infants and children, causing growth retardation with impaired cognitive and social development and increased mortality [58,68]. Severe forms

of PEM with loss of subcutaneous fat and muscle wasting are known as *marasmus*; hypoproteinemic PEM is known as *kwashiorkor*. The WHO has suggested that at least 500 million children in the world suffer from PEM, and this number is probably an underestimate [58]. PEM is not confined to infants or children, nor to developing countries, however. In the United States of America, 22%–35% of children aged 2–6 years from low-income areas are growth retarded; in large urban teaching hospitals 30%–70% of general medical and surgical patients have anthropometric and/or biochemical evidence of PEM [68]. PEM has been and still is a general threat for civilians in war, for prisoners of war, and captives in concentration camps.
Primary PEM can be regarded as relative starvation. Thus, metabolism will adapt towards starvation (see Sect. 71.3.5; Fig. 71.4C). TEE is reduced by lowering physical activity and BMR; in the latter process the reduction of protein turnover, which saves energy but also essential amino acids, plays a major role (see Sects. 71.2.1, 71.3.2). Urinary nitrogen loss is lowered. The metabolic situation is controlled mainly by low insulin, slightly increased glucagon, and reduced thyroid hormone plasma levels (see Sect. 71.5.2–5; Tables 71.5, 71.6). In the long run these homeostatic mechanisms cannot compensate entirely for the imposed deficiencies and thus the pathological consequences of PEM cannot be avoided [68].
Secondary PEM can be regarded as physical stress. TEE is increased, the injury factors being, for example, between 1.13 for 1°C fever, 1.3 for minor surgery, and 2.1 for burns (see Sect. 71.2.1). There is an increased demand for glucose, part of which is met by gluconeogenesis in the liver from amino acids and thus an increased proteolysis in skeletal muscles; urinary nitrogen loss is enhanced. Glucagon, adrenaline (see Sect. 71.5.3; Table 71.4), and cortisol (see Sect. 71.5.4; Table 71.5) are increased and, together with elevated circulating interleukin-1 and tumor necrosis factor they regulate the metabolic alterations in septic or traumatized patients [68].

Maldigestion/Malabsorption. Impairment of digestion of polymeric or oligomeric foodstuffs and of absorption of the monomeric energy substrates formed during digestion can cause syndromes similar to PEM. Maldigestion results from a deficit of digestive enzymes due to pancreatitis (see also cystic fibrosis) or pancreas tumor as well as from a deficit of bile due to bile duct obstruction. Malabsorption results from defects in absorptive translocators or from a reduction of the absorptive surface area in chronic inflammatory bowel diseases [58,68].

Anorexia Nervosa and Bulimia. Those suffering from these disorders are almost without exception young women in affluent societies. The population at risk consists largely of women from middle and upper-class white rather than black families. The disorders are characterized by the stubborn desire to defend one's own right to protect oneself against becoming too fat. There is typically a distortion of the body image with a tendency to overestimate

Table 71.4. Energy metabolism equations

Equation	Substrate (S)	Product (P)	RQ (CO_2/O_2)	$\Delta G_o'/$ mol S (kcal)	$\Delta G_o'/$ mol O_2 (kcal)
71.1	Glucose ($C_6H_{12}O_6$) + $6O_2$	$\rightarrow$ $6CO_2$ + $6H_2O$	1.00	−686	−114.4
71.2	PalmitateH ($C_{16}H_{32}O_2$) + $23O_2$	$\rightarrow$ $16CO_2$ + $16H_2O$	0.70	−2333	−101.4
71.3	Alanine ($C_3H_7O_2N$) + $3O_2$	$\rightarrow$ $2.5CO_2$ + $2.5H_2O$ + 0.5 Urea (CH_4ON_2)	0.83	−313	−104.3
71.4	Glucose ($C_6H_{12}O_6$)	$\rightarrow$ 2 LactateH ($2C_3H_6O_3$)	–	−47	–
71.5	LactateH ($C_3H_6O_3$) + $3O_2$	$\rightarrow$ $3CO_2$ + $3H_2O$	1.00	−319	−106.3
71.6	Glycerol ($C_3H_8O_3$) + $3.5O_2$	$\rightarrow$ $3CO_2$ + $4H_2O$	0.86	−393	−112.2
71.7	Ethanol (C_2H_6O) + $3O_2$	$\rightarrow$ $2CO_2$ + $3H_2O$	0.67	−315	−105.0
71.8	PalmitateH ($C_{16}H_{32}O_2$) + $5O_2$	$\rightarrow$ β-HydroxybutyrateH ($4C_4H_8O_3$)	0	−440	−88.0
71.9	β-HydroxybutyrateH ($C_4H_8O_3$) + $4.5O_2$	$\rightarrow$ $4CO_2$ + $4H_2O$	0.89	−473	−105.1

RQ, respiratory quotient; S, carbon substrate; $\Delta G_o'$ calculated from the standard free energies of formation [101]

the own body size. In anorexia nervosa, the aim of thinness is achieved by a drastic restriction of energy intake leading to emaciation. In bulimia, binge eating is followed by vomiting and excessive use of laxatives causing only a minor weight loss [35].

71.3 Substrate Flow Between Major Organs

In principle, animal cells can gain energy by the aerobic oxidative metabolism of glucose (Table 71.4, Eq. 71.1), fatty acids (Eq. 71.2), and amino acids (Eq. 71.3) to CO_2 and urea, respectively, by the anaerobic conversion of glucose to lactate (Eq. 71.4) and by the aerobic oxidative metabolism of lactate (Eq. 71.5), glycerol (Eq. 71.6), or ethanol (Eq. 71.7) as well as formation (Eq. 71.8) and utilization (Eq. 71.9) of ketone bodies. However, a specialization of energy metabolism has evolved between cells of the various organs. Neurons of the central nervous system can only use glucose and ketone bodies as energy substrates, and not fatty acids, amino acids, or lactate; they always require glucose. Erythrocytes and renal tubule cells of the kidney medulla, which are essentially devoid of mitochondria and thus incapable of oxidative energy production (see Sect. 71.4.2), are dependent on glucose; working muscles also need exogenous glucose. Thus, glucose has to be provided in all metabolic situations, during absorption of a meal as well as between meals, at rest as well as during exercise. The permanent provision of glucose is the leitmotiv of the energy metabolism of animals. Energy-storing organs comprise the liver, the skeletal muscles, and the adipose tissue. The liver stores glycogen (about 75 g = 5% of organ weight) [46,88] to meet the glucose requirements of other organs in postabsorptive states and during excercise, while muscles store glycogen (about 350 g = 1% of organ weight) [47] to meet their own requirements during exercise. The adipose tissues store triglycerides (about 14 kg) [17] as a fatty acid and glycerol reservoir for the postabsortive states and physical activity. Muscles and, in part, the liver "store"

protein. There is no specific reservoir protein; rather a certain portion of the normal cellular proteins, perhaps 30%, i,e., about 2 kg in muscles and 50 g in liver, can serve as a fuel reservoir to provide amino acids in the postabsorptive state [15–17, 89].

71.3.1 Major Energy Substrates and Hormones in Different Metabolic States

After a normal meal, the three major energy substrates glucose, amino acids, and fatty acids become available as constituents of polymeric or oligomeric foodstuffs such as starch or glycogen, protein, and fat. Following digestion in the gastrointestinal tract, glucose and amino acids are adsorbed as such via the vena portae, while fatty acids and glycerol (reesterified to triglycerides and packaged in lipoprotein particles) reach the angulus venosus via the intestinal lymph (see Chap. 62). Occasionally, ethanol can become a major energy substrate. Since the daily absorptive phases are much shorter than the postabsorptive phases (7.5 versus 16.5 h, see below and Table 71.2), the smaller part of the newly absorbed energy substrates after a meal serves to meet the actual energy requirement and the greater part is used to replenish the energy stores. Between meals, when absorption from the intestine has ceased, the energy substrates glucose, amino acids, and fatty acids required to meet the energy needs are released from the energy stores. The energy substrates lactate and the ketone bodies β-hydroxybutyrate and acetoacetate are only minor constituents of food; they are essentially formed as intermediates of metabolism.

Depending on the country, there is a graded predominance of carbohydrates over lipids as energy sources followed by protein (Table 71.3). In the absorptive (postprandial) phase after a normal meal with 43 energy% carbohydrates, 37 energy% fat and 19 energy% protein, the plasma concentration of glucose increases and that of free fatty acids decreases. In the postabsorptive phase before the next meal, the concentration of glucose decreases and that of free fatty acids increases again. Similarly, the plasma level

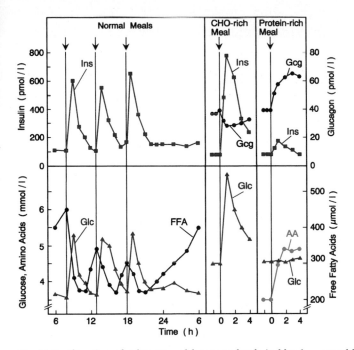

Fig. 71.3. Alteration of substrate and hormone levels in blood. *Left,* three daily normal meals of 800 kcal containing 43 energy% (84 g) carbohydrates, 37 energy% (32 g) fat, and 19 energy% protein (37 g) *Six* healthy volunteers [93]. *Middle,* the large carbohydrate (*CHO*)-rich meal consisted of 140 g spaghetti, 256 g corn, 252 g rice, 244 g potatoes, and 46 g white bread containing approximately 200 g carbohydrate; 11 healthy volunteers were allowed to eat ad libitum; they consumed an average of 107 g carbohydrate [71]. *Right,* the protein-rich meal consisted of 500 g boiled lean beef, containing approximataly 150 g protein; 14 healthy volunteers were allowed to eat ad libitum; they consumed an average of 63 g protein [71]. *AA,* amino acids, *FFA,* free fatty acids; *Gcg,* glucagon; *Glc,* glucose; *Ins,* insulin

of insulin is elevated markedly after meals containing carbohydrates, while that of glucagon is lowered slightly [93] (Fig. 71.3). These observations indicate, but do not prove that after a normal carbohydrate-containing meal nutrient glucose is the major energy substrate (see Fig. 71.2, mixed balanced diet, RQ > 0.95) and insulin the dominant hormone, which should therefore have a key role in the regulation of the utilization of glucose and the formation of energy stores from excess energy substrates (see Table 71.5). They indicate further that between meals fatty acids from the triglyceride stores are the major energy substrates and glucagon is the dominant hormone, which should therefore have a key function in the control of the utilization of energy stores mainly in the provision of glucose (see Table 71.5).

In contrast, after a protein-rich, carbohydrate-poor meal, the plasma concentration of glucose does not change and the plasmal level of insulin is increased only slightly, but that of glucagon is elevated clearly [71,103] (Fig. 71.3). This indicates that after an abnormal, carbohydrate-poor meal glucose cannot be the major energy substrate and that glucagon, rather than insulin, is the dominant hormone, because glucose not provided with the meal has to be provided from energy stores.

During exercise as compared with rest, less nutrient energy substrates are channeled into energy stores in the absorptive phase and more energy substrates from energy stores are needed in the postabsorptive phase. Noradrenaline and adrenaline, which, like glucagon, control the mobilization of energy stores, are increased; they become the dominant hormones [14] (see Table 71.4).

71.3.2 Substrate Flow After a Meal

The total energy expenditure of a male office worker is 2500 kcal/d; it is unevenly distributed over the day (Table 71.2). In western societies, carbohydrates supply about 45%, fat about 40%, and protein about 15% of the daily energy intake (see Table 71.3). If the energy intake were equally distributed between three daily meals, about 840 kcal would be taken up with each meal, supplied mainly by starch but also by sucrose, together equivalent to 100 g glucose, fat equivalent to 37 g fatty acids, and protein equivalent 27 g amino acids. If the average absorptive phase lasted 2.5 h, the intake would amount to about 330 kcal/h, which greatly exceeds the BMR plus DIT and energy to keep the body in a sitting position of about 100 kcal/h, corresponding to an O_2 uptake of 340 ml/min (Table 71.2). The excess of 230 kcal/h is used to replenish the energy stores.

Carbohydrates. After digestion and absorption, nutrient carbohydrates, mainly starch, appear in the portal blood as glucose. If the whole energy requirements of the organism in the absorptive phase of 2.5 h were supplied only by glu-

1433

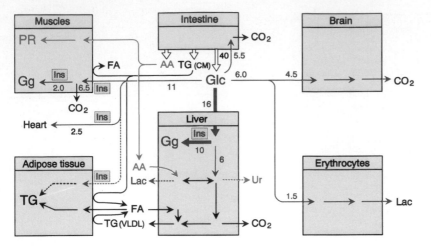

Fig. 71.4. Substrate flow after a normal meal containing carbohydrates, fat, and protein (*PR*): absorptive phase, Major insulin (*Ins*) actions: stimulation of glucose (*Gle*) uptake into skeletal muscle, heart, and adipose tissue, but not liver; activation of glycogen (*Gg*) synthesis in liver and muscle (see Table 71.4). Metabolic rates in g per h are not given for the ingestion of a normal meal, but for the ingestion of glucose in an oral glucose tolerance test (100 g dissolved in 170 ml water over 2 min after an overnight fast). The original data [31,32], which were slightly overestimated [81], indicated that within 3 h following the glucose load 45 g escaped the splanchnic bed and entered the systemic circulation, while 55 g was retained by the liver and the other splanchnic organs (intestine, pancreas, and spleen). The flow rates indicated here have been estimated on the basis of the original and more recent data on the glycogen repletion and oxygen consumption by liver, intestine, muscle, heart, and brain. *AA*, amino acid; *CM*, chylomicrons; *FA*, fatty acids; *Lac*, lactate; *TG*, triglyceride; *Ur*, urea; *VLDL*, very low density lipoproteins

cose, about 65 g (= 250 kcal) of the average intake of 100 g would be needed, and the remaining 35 g would have to be used for the synthesis of glycogen and perhaps triglyceride stores. In all normal metabolic states, the brain and the erythrocytes have an "insulin-independent" glucose requirement per h of 4.5 g and 1.5 g, respectively [15–17,80,89]. In the absorptive phase, 16 g glucose is taken up per h by the liver (i.e., 40 g in 2.5 h or about 40% of the ingested glucose), with 10 g being used for glycogen synthesis and 5.5 g consumed by the other splanchnic organs (intestine, pancreas, and spleen). Only 11 g glucose is eliminated per h by the "insulin-dependent" skeletal muscle, heart, and adipose tissue [31,32,81,89] (Fig. 71.4). Relatively little glucose is needed to refill muscle glycogen stores, which are not depleted to any great extent in subjects with a mainly sedentary occupation; however, substantial amounts of glucose are required to resynthesize muscle glycogen following medium to heavy exercise [47] (see Fig. 71.5B). Transiently up to 5 g glucose remain in the extracellular space, causing the increase in circulating concentration.

Lipids. In the absorptive phase, nutrient fat is mainly used to refill the triglyceride stores in adipose tissue. After digestion and absorption, nutrient fat appears in the blood as triglycerides carried by specific lipoproteins, the chylomicrons. The chylomicron triglycerides are hydrolyzed by lipoprotein lipases on the surface of the endothelia, mainly in skeletal muscle and adipose tissue. The liberated fatty acids are taken up by the adipose tissue and the liver for the resynthesis of triglycerides, which are stored in the adipose tissue but exported again from the liver packaged in VLDL (pre-β-lipoporteins). VLDL

triglycerides are also finally stored in the adipose tissue [12] (Fig. 71.4).

Proteins. After digestion and adsorption, nutrient proteins appear in the portal blood as amino acids. They are mainly used to refill the protein stores (Fig. 71.4). All proteins undergo a constant turnover [92,112]: with a well-balanced diet 300–400 g protein are continuously degraded per day, yielding about 400–500 g amino acids. Normally, 80 g amino acids comes from the daily food. The amino acids from nutrient proteins and those from body protein degradation form a common pool of 480–580 g amino acids, from which 400–500 g is taken for the synthesis of 300–400 g protein and 80 g is used for energy production directly (amino acid oxidation plus ureagenesis) and indirectly (via gluconeogenesis followed by glucose oxidation plus ureagenesis). During the absorptive phase, the balance between protein degradation and protein synthesis is shifted towards net synthesis, and energy production from amino acids is reduced (Fig. 71.4). "Nitrogen" intake by nutrient protein and "nitrogen" excretion as urea are balanced down to an intake of about 35 g protein per day; below this value there is a net loss of body protein, and the nitrogen balance becomes negative [82].

71.3.3 Substrate Flow Between Meals

If the energy intake is evenly distributed over three meals every 6 h during the day and if the absorptive phases last about 2.5 h each, there are two postabsorptive phases of 3.5 h each during the day and one of 9.5 h during the night.

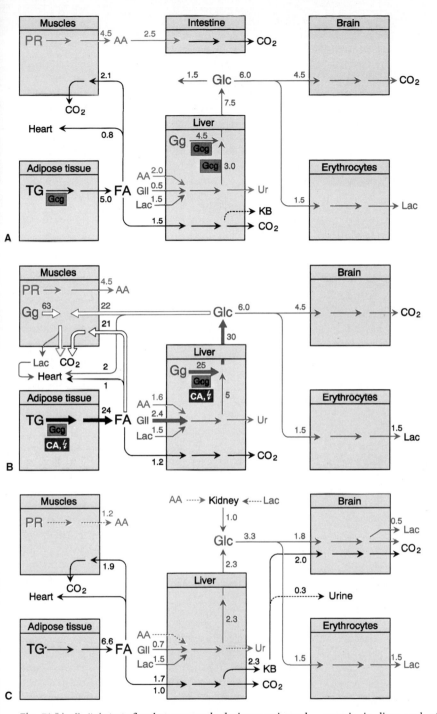

Fig. 71.5A–C. Substrate flow between meals, during exercise and starvation. **A** Rest: metabolic rate 1800 kcal/24 h = 75 kcal/h (90 W) [16,17,31,89]. The values were measured after 1 day of fasting; it can be assumed that they can be applied in a first approximation also in the normal daily postabsorptive periods. **B** Exercise: metabolic rate 580 kcal/h (700 W) at a work load of 120 W [37,46,47,88,91]. **C** Starvation: metabolic rate 1500 kcal/24 h = 62.5 kcal/h (75 W) after 5 weeks without food intake [16,17]. Major glucagon actions: stimulation of glycogenolysis and gluconeo-genesis in liver and of lipolysis in adipose tissue. Major catecholamine and sympathetic nerve actions: activation of glycogenolysis in liver and of lipolysis in adipose tissue (see Table 71.4). The flow rates (g per h) indicated here have been estimated on the basis of the original data on the glycogen degradation and oxygen consumption by liver, intestine, muscle, heart, and brain. *AA*, amino acids; *CA*, catecholamines; *FA*, fatty acids; *Gcg*, glucagon; *Gg*, glycogen; *Glc*, glucose; *Gll*, glycerol; *KB*, ketone bodies; *Lac*, lactate; *PR*, protein; *Ur*, urea; ⚡, sympathetic nerves

In the last few hours of the postabsorptive phase at night, energy expenditure is essentially equal to the BMR of 75 kcal/h, corresponding to an oxygen uptake of 260 ml/min; in the other postabsorptive periods, energy expenditure is increased by DIT and physical activity (see Sect. 71.2.1; Table 71.2). Energy needs are met by the degradation of the energy stores of glycogen in liver, triglycerides in adipose tissue, and protein mainly in muscle. Under BMR conditions, glycogenolysis in the liver supplies 4.5 g glucose per h, i.e., 17.1 kcal/h; lipolysis in adipose tissue provides 5 g fatty acids per h, i.e., 45.5 kcal/h, and proteolysis in muscle provides 4.5 g amino acids per h, i.e., 15.7 kcal/h. Thus, 22% of the energy is provided by carbohydrates, 58% by fat, and 20% by protein; the ratio of energy supply from carbohydrates to that from fat is approximately 1:3 (see Fig. 71.2).

Carbohydrates. In the postabsorptive state too, the glucose requirements of brain and erythrocytes are 4.5 g and 1.5 g, respectively. The liver meets these needs and those of other organs such as the heart and renal medulla by glycogenolysis, providing 4.5 g glucose per h, and by gluconeogenesis, supplying 3 g per h. Lactate produced by glycolysis in blood cells and renal medulla, glycerol liberated by lipolysis in adipose tissue, and amino acids supplied by net proteolysis in muscle are the gluconeogenic substrates [17,80,89] (Fig. 71.5A).

Lipids. The energy requirements of liver are 1.5 g fatty acids per h and of muscle 2.1 g per h. Adipose tissue meets these requirements and those of other organs such as the heart and renal cortex by lipolysis, providing 5 g fatty acids per h. During short-term postabsorptive states, the production of ketone bodies from fatty acids is less than 0.5 g per h and therefore negligible [17,80,89] (Fig. 71.5A).

Proteins. Net proteolysis mainly in muscle amounts to 4.5 g amino acids per h. The amino acids are the preferred energy substrates in the intestine [109]; they are substrates of both energy production and gluconeogenesis in the liver, since about 50% each of their carbon skeletons is oxidized to CO_2 and converted to the glucose precursor pyruvate [17,80,89] (Fig. 71.5A).

71.3.4 Substrate Flow During Exercise

Intense physical activities normally take place in the postabsorptive phase. Both metabolic states are marked by a negative energy balance, i.e., excess of energy requirements compared to energy intake. Thus, metabolism in the postabsorptive state at rest and during exercise will in principle be qualitatively similar, but of course quantitatively quite different. The extent of the metabolic alterations will be a function of the duration and amout of work performed and of the type of muscles – red, white, or mixed – involved. There are clear differences between trained and untrained subjects. A measure of the work load is the oxygen requirement: at rest in a sitting position it amounts to 0.34 l/min (TEE = 100 kcal/h), during light exercise of 43 kcal/h or 50 W it amounts to 1 l/min (TEE = 310 kcal/h), during heavy exercise of 100 kcal/h or 120 W it amounts to 2 l/min (TEE = 580 kcal/h), and during very heavy exercise of 200 kcal/h or 240 W it amounts to 4.5 l/min (TEE = 1100 kcal/h), which an untrained subject will not be able to provide. At rest, the ratio of energy supply by carbohydrates versus fat is 35:65. With increasing exercise, the ratio changes in favor of carbohydrates to become, e.g., 60:40 at a work load of 100 kcal/h or 120 W; glycogenolysis in working muscles and in liver provides 90 g of glucose per h, i.e., 340 kcal/h, and lipolysis in adipose tissue supplies 24 g fatty acids, i.e., 220 kcal/h [37,46,47,88,91].

Carbohydrates. Glycogenolysis in muscle can be increased from 0.05 g/h × kg (muscle) at rest to 4 g/h × kg at heavy exercise and even to 20 g/h × kg at extremely heavy exercise of short duration. Glucose uptake from blood can also be increased from almost zero at rest to 1.4 g/h = kg at a heavy work load. With bicycling at a load of 100 kcal/h or 120 W, about 50%, i.e, 15 kg, of the musculature is active; glycogenolysis increases to 63 g glucose per h, i.e., (4 g/h × kg) × 15 kg, and glucose uptake to 22 g/h, i.e., 1.4 g/h × kg) × 15 kg. The enhanced requirement of working muscle for blood glucose is met mainly by an increase in glycogenolysis in the liver from 4.5 g/h to 25 g/h at a heavy load or even to 60 g/h at a very heavy load [46,88,91] (Fig. 71.5B). Gluconeogenesis in the liver from lactate, amino acids, and glycerol is only slightly enhanced from 3 g/h to 5 g/h due to an increase in glycerol supply from lipolysis in adipose tissue. Glycogenolysis and gluconeogenesis together provide 30 g glucose per h, of which 5.5 g is used by brain and erythrocytes, 22 g by working muscles, and the remainder by other organs including the heart. In addition to glucose and fatty acids, lactate, released to a certain extent from muscles, also becomes a major energy substrate for the heart. Depletion of the glycogen reserves of working muscles and/or liver results in physical exhaustion [37,46,47,88,91].

Lipids. The release of fatty acids and glycerol from the triglyceride stores of adipose tissue is also increased with physical activity, from about 5 g/h to 24 g/h at a heavy work load (Fig. 71.5B). A substantial amount of the required fatty acids appears to be released from adipose tissue in the working muscle [37].

Proteins. The net release of amino acids from protein stores is not altered during physical activity

71.3.5 Substrate Flow During Starvation

During long-term starvation lasting 1–6 weeks, the BMR is reduced from 75 kcal/h at the end of a normal postabsorptive phase to about 65 kcal/h, corresponding to an oxygen consumption of 230 ml/min (see Sect. 71.2.1).

The sole purpose of the metabolic changes is to spare glucose. Glycogen as a glucose reservoir has long been exhausted and the protein reserves as an amino acid, i.e., glucose precursor, reservoir are limited; only the triglyceride stores are abundant enough for a long-term fast. Under the reduced BMR conditions, lipolysis in adipose tissue provides 6.6 g fatty acids per h, i.e., 60 kcal/h, and proteolysis mainly in muscle 1.2 g amino acids per h, i.e., 4.2 kcal/h. Carbohydrates are no longer primary energy substrates.

Carbohydrates. The energy requirements of the brain and erythrocytes remain constant even during long-term starvation. However, the catabolism of glucose by the brain is reduced from 4.5 g/h in the short-term postabsorptive state to 1.8 g/h and at the same time the utilization of ketone bodies formed from fatty acids is enhanced from negligible values to 2 g/h; thus carbohydrate is in part replaced by fat. Moreover, a portion of the glucose taken up by the brain is only degraded to lactate, which is utilized for the resynthesis of glucose; this can be regarded as another glucose-sparing mechanism. The anaerobic catabolism of glucose to lactate by erythrocytes remains unaltered. The remaining glucose requirements are met by gluconeogenesis from lactate in the liver and amino acids in the kidney. During starvation, a ketoacidosis develops; ketone bodies, i.e., acetoacetic acid and β-hydroxybutyric acid, are excreted into urine, which necessitates their partial neutralization by NH_3. The required NH_3 is liberated during renal gluconeogenesis from amino acids [16,17,31,89] (Fig. 71.5C).

Lipids. During starvation, fatty acids are the major energy substrates; their relative importance versus carbohydrates is greatly increased compared with the short-term postabsorptive state, because glucose provides considerably less energy. Fatty acids are released from the triglyceride stores in adipose tissue at a rate of 6.6 g/h or 155 g/24 h. If the 15 kg triglyceride stores of the body could be completely utilized, they would last for almost 100 days of starvation. Fatty acids are utilized by liver at 2.7 g/h: 1.7 g/h for the formation of 2.3 g/h of ketone bodies (the apparent mass gain is due to the incorporation of water and oxygen, when fatty acids are converted to ketone bodies) and 1 g/h for the production of energy. They are oxidized at 1.9 g/h by muscles and the remainder by other organs including the heart. The ketone bodies are utilized almost exclusively by the brain and no longer by the muscles, heart, and kidney as during short-term starvation lasting 3–4 days [16,17,89] (Fig. 71.5C).

Proteins. It is generally assumed that only 2 kg of the total protein content of 6–7 kg is available for degradation without major functional impairments of body functions. If amino acid release by proteolysis were to continue at the rate of the early postabsorptive state, i.e., 4.5 g/h corresponding to 85 g protein per 24 h, the protein reservoir would last only for about 24 days. This may not have been sufficient for survival when humans were still hunters and

lack of food for longer periods was normal. Therefore, net protein degradation is reduced gradually during the first few days of starvation to 1 g/h or 25 g/24 h. With that proteolysis rate, the protein stores would last for 75 days, which would match more closely the capacity of the lipid stores. The protein-sparing metabolic changes during starvation are possible because the glucose requirements of the brain are reduced by the provision of ketone bodies [16,17,89] (Fig. 71.5C).

71.4 Mechanism of Energy Metabolism

71.4.1 Principles of Energy Metabolism

Energy metabolism and work metabolism are connected by so called energy-rich compounds such as adenosine 5'-triphosphate (ATP) or guanosine 5'-triphosphate (GTP), which by definition have a standard hydrolysis potential $\Delta G_0'$ more negative than -7 kcal/mol ($= -29.3$ kJ/mol). The concept of energy-rich compounds is merely a useful formal concept in the analysis of complex oxidation–reduction processes (see Table 71.6, Eqs. 71.1–71.9) into simple condensation and hydrolysis reactions, which in vivo must not and do not occur. Simple hydrolysis of energy-rich compounds has to be prevented in the cell; otherwise the energy would be lost as heat without performing work. The hydrolysis potential of ATP and of other energy-rich compounds is only a measure of a specific kind of chemical reactivity, i.e., phosphoryl group (ATP to ADP + P_i) or adenylyl group (adenosine monophosphate, ATP to AMP + PP_i) transfer potential. The most important energy-rich compound is ATP, the universal molecular carrier for biological energy [53,59,100,101].

The energy metabolism of aerobic animal cells is a redox process between the energy substrates of food as electron donors and O_2 as electron acceptor. The substrates are completely dehydrogenated to CO_2 with the formation of reducing equivalents [H], mainly reduced nicotinamide adenine dinucleotide (NADH). Oxygen is then hydrogenated to water by the reducing equivalents under regeneration of mainly NAD^+. In this aerobic redox process, ATP is formed both with the dehydrogenation of substrates by the so-called *substrate level phosphorylation* and with the hydrogenation of the acceptor O_2 by the so-called *electron transport phosphorylation* (Fig. 71.1B). The energy metabolism of anaerobic animal cells is also a redox process composed of a substrate dehydrogenation and an acceptor hydrogenation. Substrates are exclusively carbohydrates, which are only partially dehydrogenated to pyruvate with concomitant formation of NADH. Pyruvate functions as the acceptor for the electrons of NADH and is reduced to lactate with regeneration of NAD^+. ATP is formed anaerobically only by substrate level phosphorylation [53,101].

Carbohydrates, fatty acids, and amino acids serve as energy substrates. In principle, they are interchangeable as energy sources within certain constraints (see Sect. 71.2.2).

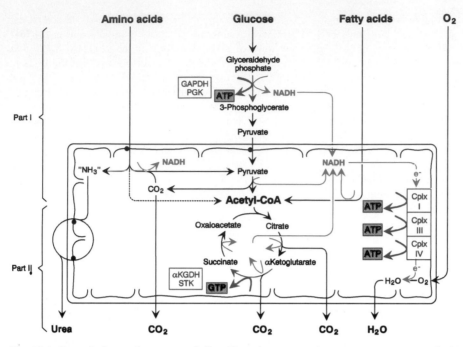

Fig. 71.6. General scheme of energy metabolism The redox process consists of the dehydrogenation of reduced carbon substrates forming CO_2 and reducing equivalents [H] mainly in the form of reduced nicotinamide adenine dinucleotide (NADH) and the hydrogenation of O_2 to H_2O, consuming reducing equivalents. *Cplx*, complex; *GAPDH*, glyceraldehydephosphate dehydrogenase; *PGK*, phosphoglycerate kinase; *αKGDH*, α-ketoglutarate dehydrogenase; *STK*, succinate thiokinase. *Part I* of dehydrogenation: glucose + $2H_2O \rightarrow 2$ acetate$^-$ + $2H^+$ + $2CO_2$ + 8 [H]. Palmitate$^-$ + H^+ + 14 $H_2O \rightarrow 8$ acetate$^-$ + 8 H^+ + 28 [H]. Alanine + $2H_2O \rightarrow$ acetate$^-$ + H^+ + "NH_3" + CO_2 + 4 [H]. *Part II* of dehydrogenation (citrate cycle): acetate$^-$ + H^+ + $2H_2O \rightarrow 2CO_2$ + 8[H]. Hydrogenation (respiratory chain): 4 [H] + $O_2 \rightarrow 2H_2O$

Glycogen, triglycerides, and proteins serve as energy stores (Fig. 71.1B). In the first part of energy metabolism, the carbon chains of the energy substrates are partially degraded to the C_2 state, i.e., "activated" acetate (acetylcoenzyme A, acetyl-CoA); in the second part, acetyl-CoA is completely degraded to CO_2 (Fig. 71.6):

- *Part I.* Glucose, the main carbohydrate of food, is degraded in the cytoplasm to pyruvate. The most important intermediate is glyceraldehyde 3-phosphate, which by the action of glyceraldehydephosphate dehydrogenase and phosphoglycerate kinase is converted to 3-phosphoglycerate, with the generation of NADH and de novo formation of ATP by substrate level phosphorylation. Pyruvate is then transported into the mitochondrium and there further dehydrogenated to acetyl-CoA and CO_2, again with the formation of NADH. Fatty acids and amino acids are also predominantly converted to acetyl-CoA. Thus acetyl-CoA is the first common intermediate for the degradation of carbohydrates, fatty acids, and amino acids (Fig. 71.6). It is remarkable that only with the degradation of carbohydrates, but not of fatty acids or amino acids, can ATP be formed in the cytoplasm. Since all processes, which take place in the mitochondria, are possible only aerobically, only carbohydrates can be used by animal cells to produce energy in the absence of oxygen [53,59,100] (Fig. 71.6).

- *Part II.* Acetyl-CoA is degraded in a cyclic rather than linear process to $2CO_2$. The most important intermediate is α-oxoglutarate, which by the action of α-oxoglutarate dehydrogenase and succinate thiokinase is converted to succinate and CO_2, with the generation of NADH and de novo formation of GTP and thus finally ATP by substrate level phosphorylation [53,59,100] (Fig. 71.6). NADH formed during substrate dehydrogenation is reoxidized in the mitochondrium; it transfers its electrons via an electron transport system to O_2. This system consists of three major protein complexes; it is located in the inner mitochondrial membrane. Each of these complexes is an oxidoreductase; the flow of electrons through each of these oxidoreductases is coupled to the formation of ATP from ADP and P_i by electron transport phosphorylation. This process is also called respiratory chain phosphorylation or oxidative phosphorylation [53,59,100].

Thus, during substrate dehydrogenation, ATP or GTP, respectively, are formed de novo only during two steps, the dehydrogenation of glyceraldehyde phosphate to 3-phosphoglycerate and of α-oxoglutarate to succinate and CO_2. Acceptor hydrogenation yields ATP during only three steps, complexes I, III, and IV. In aerobic animal cells, more than 90% of the ATP is formed in the mitochondria, which are therefore called the power plants of the cell (Fig. 71.6).

71.4.2 Central Oxidative Energy Metabolism

The dehydrogenation of acetyl-CoA to CO_2 with the generation of reducing equivalents [H] and the reoxidation of the reducing equivalents with O_2 are common steps in the catabolism of carbohydrates, amino acids, and fatty acids (Fig. 71.6). Therefore, these processes constitute the central oxidative energy metabolism (Fig. 71.7).

Oxidation of Acetyl-CoA to CO_2. The C_2 unit acetate could in principle be degraded via two mechanisms: (1) in a "simple" linear process, the C_2 unit could first be decarboxylated to, e.g., CH_4 and CO_2; CH_4 could then be dehydrogenated to CO_2 and 8[H]; (2) in a rather "complicated" cyclic process, the C_2 unit could first be combined with a reduced acceptor C_4 unit such as oxaloacetate to yield a reduced C_6 unit such as citrate, which is then dehydrogenated and decarboxylated via a C_5 unit such as α-oxoglutarate to regenerate the acceptor C_4 unit, yielding 2 CO_2 and 8[H]. Nature has chosen the second mechanism

not for obvious mechanistic, but probably for economical reasons; the intermediates of the cyclic process are useful for many other pathways [53].

The oxidation of acetyl-CoA in the mitochondria begins with the citrate synthase reaction followed by the aconitase, isocitrate dehydrogenase, α-oxoglutarate dehydrogenase, succinate thiokinase, succinate dehydrogenase, fumarase, and malate dehydrogenase reactions. Of the four dehydrogenase reactions of the so-called *citrate cycle*, three yield NADH and one reduced flavine adenine dinucleotide (FADH$_2$) or more exactly reduced flavoprotein (FpH$_2$); two produce CO_2 [53,59,100] (Fig. 71.7A).

The only substrate of the citrate cycle is acetyl-CoA. Therefore, all carbon substrates have to be converted to acetyl-CoA; conversion of a substrate to one of the cycle intermediates, e.g., glutamate to α-oxoglutarate, is not sufficient, as degradation would stop at oxalacetate. Moreover, the citrate cycle has to regenerate the acetyl-CoA acceptor oxaloacetate in stoichiometric amounts, otherwise the cycle would come to a standstill. Utilization of one

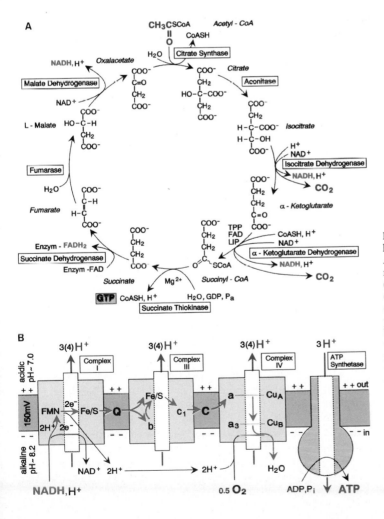

Fig. 71.7A,B. Central oxidative energy metabolism. **A** Citrate cycle: conversion of acetylcoenzyme A (acetyl-CoA) to CO_2 and reducing equivalents, reduced nicotinamide adenine dinucleotide (NADH) and reduced flavine adenine dinucleotide (FADH$_2$). Subcellular localization: mitochondria (see Fig. 71.6). **B** Respiratory chain (oxidative phosphorylation): reoxidation of reducing equivalents by reduction of O_2 to H_2O coupled to the formation of adenosine triphosphate (*ATP*). Reduced flavoproteins donate their electrons to coenzyme Q (not shown). Subcellular localization: inner mitochondrial membrane (see Fig. 71.6) *a, b, c,* cytochrome *a, b, c; ATP synthetase,* F$_0$F$_1$ATPase; Q, coenzyme Q (ubiquinone); *complex I,* NADH-coenzyme Q oxidoreductase (NADH dehydrogenase); *complex III,* coenzyme Q-cytochrome *c* oxidoreductase (bc$_1$ complex); *complex IV,* cytochrome *c*-oxygen oxidoreductase (cytochrome oxidase); *Fe/S,* iron-sulfur protein; *FMN,* flavin mononucleotide protein; *LIP,* lipoic acid; *TPP,* thiamine pyrophosphate

of the cycle intermediates for synthetic purposes, e.g., succinyl-CoA for heme synthesis, is only possible with anaplerotic (i.e., replenishing) reactions such as the conversion of pyruvate to oxaloacetate by the pyruvate carboxylase reaction [53].

Respiratory Chain. NADH formed in substrate dehydrogenations is reoxidized by multiprotein hetero-oligomeric complexes located in the inner mitochondrial membrane. Complex 1 (42 subunits) is an NADH: coenzyme Q oxidoreductase; it contains one flavin mononucleotide (FMN) and at least four iron-sulfur, i.e., non-heme-iron, groups. Complex III (11 subunits) is a coenzyme Q: cytochome c oxidoreductase; it contains three heme-iron groups – two heme b and one heme c_1 – and one iron-sulfur group. Complex IV (13 subunits) is a cytochrome c: oxygen oxidoreductase; it contains two heme-iron and three copper groups, with heme a and two copper A forming the first and heme a_3 and copper B forming the second center (Fig. 71.7B). FADH$_2$ (FpH$_2$) formed in substrate dehydrogenations is reoxidized by electron transfer to coenzyme Q mediated by succinate dehydrogenase of the citrate cycle (see Fig. 71.7A), traditionally called complex II, and by acyl-CoA dehydrogenase of the β-oxidation (see Fig. 71.11), previously sometimes called complex V. Complexes I, III, and IV, but not II or V, function as proton pumps; they each extrude three to four protons per transferred electron pair by an unknown mechanism. The export of protons generates a proton-motive force composed of a H^+ gradient ΔpH of about 1.2 (outside acidic) and a membrane potential Δψ (outside positive). This proton motive force is used to drive the synthesis of ATP from ADP and P_i by an as yet unknown mechanism linked to the reentry of three protons catalyzed by another multiprotein hetero-oligomeric complex called ATP synthetase (also F_1F_0 ATPase), with nine plus 14 subunits, presently also called complex V (Fig. 71.7B; see Chap. 4). The vast majority of the ATP formed in the mitochondria is transported to the cytosol by an ATP–ADP exchange translocator in the inner mitochondrial matrix [18,53,59,100,107].

Electron transport through the complexes is strictly coupled to the phosphorylation of ADP to ATP:

- Cellular respiration is controlled by the availability of ADP, i.e., by the demand for ATP (respiratory control).
- It can be inhibited by blockers of electron transport such a rotenone (complex I), antimycin A (complex III), or cyanide (complex IV), by inhibitors of phosphorylation such as oligomycin, or by inhibitors of the ATP–ADP antiporter such as atractyloside.
- It can be uncoupled by proton conductors, enabling the reentry of H^+ bypassing the ATP synthetase such as dicumarol, leading to an uncontrolled oxidation of NADH or FADH$_2$ without ATP formation [18,53,59,100,107].

71.4.3 Energy Metabolism of Carbohydrates

Glucose, the main nutrient carbohydrate, can be utilized by all cells to produce ATP and in addition – mainly in hepatocytes, myocytes, and cardiomyocytes – to form glycogen as an energy store (see Sect. 71.3; Figs. 71.4, 71.5). Uptake of glucose from the blood and conversion to glucose 6-phosphate is common to both functions. Glucose is transported into the cytosol via specific glucose transporters (GLUT, see Chap. 8), which exist in several isoforms: GLUT-1 in erythrocytes, GLUT-2 in liver, GLUT-3 in brain, GLUT-4 in skeletal muscle, heart muscle, and adipose tissue, and GLUT5 in intestine and kidney (basolateral membrane). Glucose is taken up from the intestinal and renal tubular lumen via apical sodium-dependent glucose transporters (SGLT). With the glucose transporters, except GLUT-2, the K_m for glucose is clearly below the range of the plasma glucose concentration; only with GLUT-2 does the K_m fall into this range, so that glucose uptake into hepatocytes is a function of glucose supply. Only GLUT-4 is activated by insulin, so that glucose uptake into myocytes, cardiomyocytes, and adipocytes is hormonally controlled (see Sect. 71.5.1). Glucose is converted to glucose 6-phosphate by hexokinases, which also exist in several isoforms. Hexokinases I-III are operative in the major organs except the liver, where glucokinase (hexokinase IV) is active (Fig. 71.8). As with the glucose transporters, the K_m of the hexokinases I-III is far below the range of plasma glucose concentration; only with glucokinase is it within this range, so that glucose metabolism in hepatocytes is again a function of glucose supply [70,83,84] (see Chap. 68).

Degradation of Glucose to Acetyl-CoA. After phosphorylation of glucose to glucose 6-phosphate, glucose degradation proper begins with the glucose 6-phosphate isomerase reaction and is continued by the phosphofructokinase and fructose bisphosphate aldolase plus triose phosphate isomerase reactions followed by the glyceraldehydephosphate dehydrogenase, phosphoglycerate kinase, phosphoglycerate mutase, enolase, pyruvate kinase, and finally pyruvate dehydrogenase reactions (Fig. 71.8). All steps from glucose to pyruvate take place in the cytosol; the conversion of pyruvate to acetyl-CoA and CO_2 occurs in the mitochondria. There is a net de novo synthesis of 2 mol ATP per mol glucose from 2 mol energy-rich 1,3-bisphosphoglycerate [53,59,100]. In the liver, glycolysis appears to occur predominantly in the perivenous zone [51,52].

Synthesis and Degradation of Glycogen Stores. After phosphorylation of glucose to glucose 6-phosphate, glycogen synthesis starts with the glucose phosphate mutase reaction; it is continued by uridine diphosphate (UDP)-glucose pyrophosphorylase, which forms UDP-glucose ("activated" glucose). Glycogen synthase and 1,4–1,6-transglucosidase then act alternately. The synthase first elongates the glycogen primer by adding glucose from

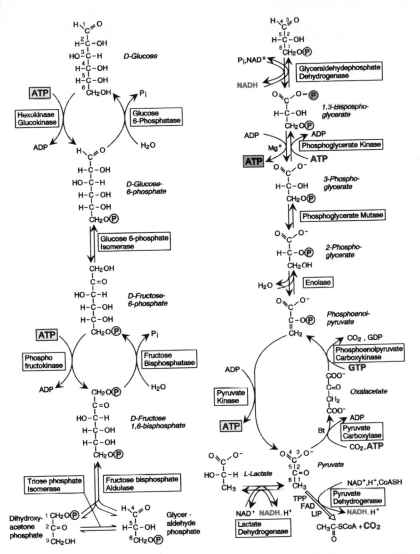

Fig. 71.8. Degradation and de novo synthesis of glucose (glycolysis and gluconeogenesis). Please note that the C3-portion of glycolysis/gluconeogenesis shown in the *right half* occurs twice per one glucose utilized or formed. P_i, inorganic phosphate. Subcellular localization: all glucose-utilizing enzymes are in the cytosol except pyruvate dehydrogenase, which is in the mitochondria; the glucose-forming enzymes are in three compartments: pyruvate carboxylase in the mitochondria, phosphoenolpyruvate carboxykinase to glucose 6-phosphate isomerase in the cytosol, and glucose 6-phosphatase in the endoplasmic reticulum. In glycolysis, the conversion of glucose to two lactate, net two adenosine triphosphate (*ATP*; shown on *blue background*) are formed de novo from P_i and adenosine diphospate (*ADP*) in the combination of the glyceraldehydephosphate dehydrogenase and phosphoglycerate kinase reactions. The two ATP (shown on *grey background*) invested at first in glycolysis in the hexokinase/glucokinase and phosphofructokinase reactions are regenerated in the pyruvate kinase reaction. For gluconeogenesis, the conversion of two lactate to glucose, net six ATP are consumed, two each in the pyruvate carboxylase, phosphoenolpyruvate carboxykinase, and phosphoglycerate kinase reactions. *NAD+*, nicotinamide adenine dinucleotide; *NADH*, reduced NAD+; *GDP*, guanosine diphosphate; *TPP*, thiamine pyrophosphate; *FAD*, flavine adenine dinucleotide; *LIP*, lipoic acid

UDP-glucose to the 4-position of a terminal glucose of the primer until a chain length of 12–14 glucose molecules after a 1,6-branchpoint is reached. The transglucosidase then transfers a terminal oligosaccharide of about seven glucose molecules from a 1,4-linkage to a 1,6-linkage at a distance of four monomers from the previous 1,6-branchpoint, thus creating a new 1,6-branchpoint (Fig. 71.9).

Glycogen degradation consists of three alternate steps. Glycogen phosphorylase shortens the terminal polyglucose chains by splitting off glucose 1-phosphate with inorganic phosphate until the chain is shortened to a length of four monomers from a 1,6-branchpoint. The transglucosidase then transfers a trisaccharide from a 1,6-branched side chain to a 1,4-linked main chain, leaving behind a single 1,6-linked glucose molecule, which is then removed

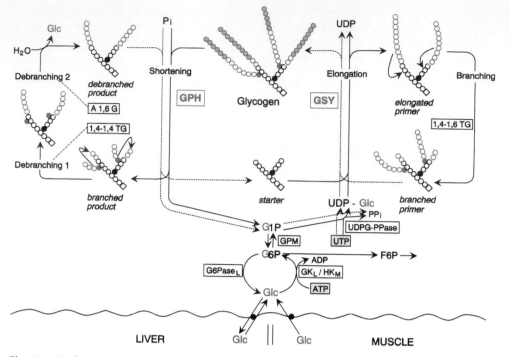

Fig. 71.9. Synthesis and degradation of glycogen (glycogenesis and glycogenolysis). Subcellular localization: cytosol. *Glc*, glucose; *G6P*, glucose 6-phosphate; *G1P*, glucose 1-phosphate; *A1,6G*, amylo-1,6-glucosidase; *GPH*, glycogen phosphorylase; *GSY*, glycogen synthase; *GK_L*, glucokinase, liver; *G6Pase_L*, glucose 6-phosphatase, liver; *GPM*, glucose phosphate mutase; *HK_M*, hexokinase, muscle, *P_i*, inorganic phosphate; *PP_i* inorganic pyrophosphate; *1,4–1,4TG*, 1,4–1,4 transglucosidase; *1,4–1,6TG*, 1,4–1,6 transglucosidase; *UDPG-PPase*, uridine diphosphate (UDP)-glucose pyrophosphorylase. In glycogen synthesis to add one glucose unit, net two adenosine triphosphate (*ATP*) are consumed at the hexokinase/glucokinase and the UDP-glucose pyrophosphorylase reactions, i.e., more precisely in the nucleosidediphosphate kinase reaction (not shown): UDP + ATP → UTP + ADP (adenosine diphosphate). In glycogen breakdown to liberate one glucose unit, no ATP is regenerated

hydrolytically by amyloglucosidase. Glucose 1-phosphate is in equilibrium via glucose phosphate mutase with glucose 6-phosphate, which only in the liver is converted via the glucose 6-phosphatase system to glucose and in all other glycogen-storing organs is channeled into glycolysis (Fig. 71.9). Glycogen synthesis and degradation are located in the cytosol; the glucose 6-phosphatase reaction takes place in the endoplasmic reticulum [53,59,100]. In the liver, glycogen is degraded first in the periportal and then in the perivenous zone; it is resynthesized first from glucose in the perivenous zone and then from pyruvate via gluconeogenesis (glucose paradox) in the periportal zone [51,52].

De Novo Synthesis of Glucose (Gluconeogenesis). Glucose, which is needed by the organism and which cannot be formed from the glycogen stores of the liver, is synthesized de novo mainly in the liver from pyruvate. Lactate and the so-called glucogenic amino acids (see Fig. 71.10C) are the major precursors of pyruvate. Gluconeogenesis begins with the pyruvate carboxylase and phosphoenolpyruvate carboxykinase reactions; it is continued by the enolase, phosphoglycerate mutase, phosphoglycerate kinase, and glyceraldehydephosphate dehydrogenase reactions, followed by the fructose bisphosphate aldolase plus triose phosphate isomerase, fructose 1,6-bisphosphatase, glucose 6-phosphate isomerase, and glucose 6-phosphatase reactions (Fig. 71.8). The step from pyruvate to oxaloacetate takes place in the mitochondria and that from glucose 6-phosphate to glucose, as with glycogenolysis, in the endoplasmic reticulum. For energetic and regulatory reasons, the two antagonistic processes glycolysis and gluconeogenesis differ between glucose and glucose 6-phosphate, fructose 6-phosphate and fructose 1,6-bisphosphate, and phosphoenolpyruvate and pyruvate [53,59,100] (Fig. 71.8). In the liver, gluconeogenesis is situated mainly in the periportal zone [51,52] (see Chap. 68).

71.4.4 Energy Metabolism of Proteins

Amino acids can be utilized to produce ATP only in organs which can dispose of the NH_3 liberated during the dehydrogenation of the nitrogenous substrates to CO_2 (see Fig. 71.6). Thus, for amino acid degradation the liver is the major site, because it alone can detoxify NH_3 by the formation of urea (see Sect. 71.3; Figs. 71.4, 71.5); the intestine is another important site, because it can release NH_3 into the portal vein for detoxification by the liver, and the kidney is a minor site, because it can release some NH_3 into the urine. Skeletal muscles are sites of partial degradation; the carbon skeleton of amino acids cannot be oxidized beyond pyruvate or α-oxoglutarate, because these α-oxoacids are needed to export NH_3 in the form of alanine and glutamine to the liver for urea formation. Amino acids are needed by all cells for protein synthesis. Amino acid uptake into the cells is catalyzed by at least five different transporters with overlapping specificity for the 20 amino acids which are constituents of proteins. The majority of the transporters are Na^+ dependent [53] (see Chap. 8).

Degradation of Amino Acids to Acetyl-CoA. The first part of amino acid degradation is the separation of carbon and nitrogen by transamination plus dehydrogenating deamination or by eliminating deamination; NH_3 and

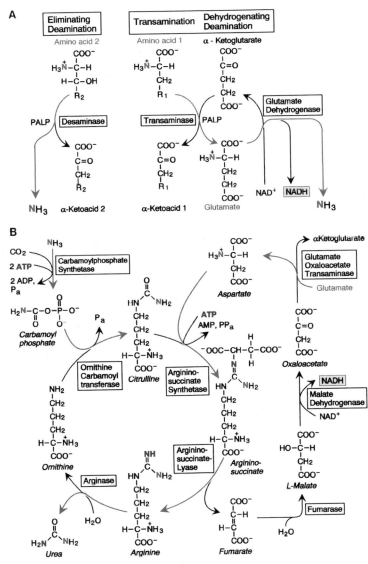

Fig. 71.10A–C. Degradation of amino acids. A Separation of carbon and nitrogen. *PALP*, pyridoxal phosphate; *NAD+*, nicotinamide adenine diucleotide; *NADH*, reduced NAD+. B Urea cycle. Subcellular localization: glutamate dehydrogenase, carbamoylphosphate synthetase, and ornithine carbamoyl transferase in the mitochorndria, the other enzymes in the cytosol (see Fig. 71.6). With respect to the redox state, α-amino acids are equivalent to α-hydroxy acids. However, the utilization of α-amino acids as compared to α-hydroxy acids requires, two additional adenosine triphosphale (ATP) per α-amino group; the detoxificatin of two NH_3 in the urea cycle costs two ATP at the carbamoyl phosphate synthetase and two ATP equivalents (ATP → AMP + PPi) at the argininosuccinate synthetase reactions.

AMP, adenosine monophosphate; *PPi*, inorganic pyrophosphlate. C Utilization of the amino acid carbon skeletons. Amino acids whose carbon skeleton can be converted to the C_3 compound pyruvate can be used to synthesize glucose; hence they are glucogenic. Amino acids whose carbon skeleton cannot be converted to the C_3 compound pyruvate, but only to the C_2 compound acetylcoenzyme A (*acetyl-CoA*), can be used only to form ketone bodies, but not glucose; hence they are ketogenic. Amino acids whose carbon skeleton is in part converted to pyruvate and in part to acetyl-CoA are glucogenic and ketogenic. Separation of carbon and nitrogen: *E*, eliminating deamination; *O*, oxidative deamination; no index, transamination + dehydrogenating deamination

1443

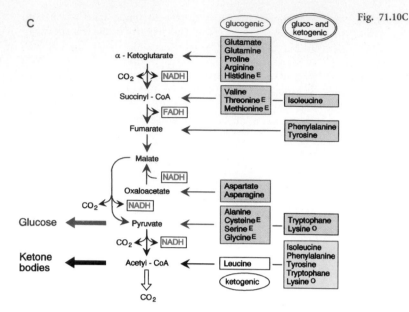

Fig. 71.10C

glutamate are the nitrogen-containing products (Fig. 71.10A). The second part is the conversion of NH_3- and glutamate-nitrogen to urea in a cyclic process in the liver. The first nitrogen atom coming from NH_3 is introduced into the cycle via the carbamoylphosphate synthetase and ornithine carbamoyltransferase reactions; the second nitrogen atom coming from glutamate is introduced via the glutamate oxaloacetate transaminase and argininosuccinate synthetase reactions. The acceptor for the first nitrogen atom ornithine is regenerated by the argininosuccinate lyase and arginase reactions with the liberation of fumarate and urea. Fumarate is used to regenerate oxaloacetate, the acceptor for the glutamate nitrogen (Fig. 71.10B). Ureagenesis takes place in the periportal and proximal perivenous zone [51,52] (see Chap. 68). The third part consists in the degradation of the carbon skeletons, liberated in the first part as α-oxoacids (α-ketoacids), via central intermediates to pyruvate and acetyl-CoA, respectively. Amino acids whose carbon skeleton is utilized via a common intermediate belong to the same degradation family. The degradation families include the α-oxoglutarate, succinyl-CoA, fumarate-oxalacetate, pyruvate, and acetyl-CoA families. Amino acids in the first four families are glucogenic, because the C5, C4, and C3 intermediates can be converted net to glucose; amino acids in the last family are ketogenic, because the C2 intermediate acetyl-CoA cannot be converted net to glucose but only to ketone bodies. Some amino acids belong to two families and are thus both glucogenic and ketogenic [53,59,100] (Fig. 71.10C).

Synthesis and Degradation of Protein Stores. There is no special polypeptide or protein which serves as an amino acid store, in contrast to the special polysaccharide glycogen, which functions as the glucose reservoir. Thus, functional proteins, mainly components of the contractile apparatus of skeletal muscle, are used as stores. All proteins are subject to a constant turnover by mRNA-directed synthesis and hydrolytic degradation. These mechanisms are used to replenish or utilize protein stores [53].

71.4.5 Energy Metabolism of Lipids

Fatty acids can be utilized by most aerobic cells to produce ATP, with neuronal cells being the important exception. Fatty acids can be used in addition mainly by hepatocytes and adipocytes for the synthesis of triglycerides, which are finally stored as energy reserves in adipocytes. Moreover, fatty acids can be converted only by hepatocytes to ketone bodies, which can be important energy substrates for neuronal cells, myocytes, cardiomyocytes, and nephrocytes (see Sect. 71.3; Figs. 71.4, 71.5). Fatty acid uptake into the cells appears to be a protein-catalyzed process. Fatty acids can be formed de novo from glucose mainly in hepatocytes, but only with very carbohydrate-rich diets [76].

Fatty Acid Degradation to Acetyl-CoA. Fatty acids are first "activated" to the CoA esters in the acyl-CoA synthetase reaction and then β-oxidized to β-ketoacyl-CoA in the acyl-CoA dehydrogenase, enoyl-CoA hydratase, and β-hydroxyacyl-CoA dehydrogenase reactions. Acetyl-CoA is then split off from β-ketoacyl-CoA by the action of thiolase, leaving behind an acyl-CoA shortened by two C atoms. This β-oxidative and β-thiolytic degradation is repeated until, for instance, the C16-fatty acid palmitate is degraded completely to eight acetyl-CoA (Fig. 71.11). The activation of fatty acids takes place in the cytosol, and β-oxidation and β-thiolytic cleavage occurs in the mitochondria. Acyl-CoA is imported into the

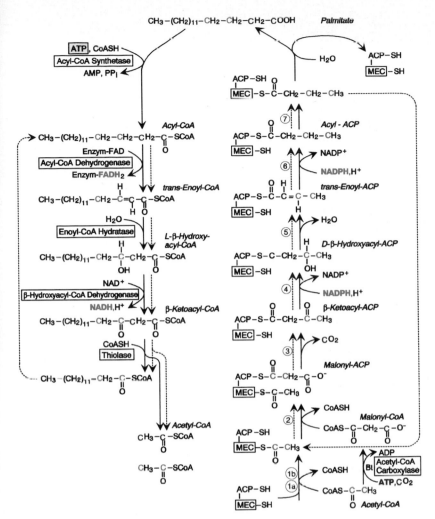

Fig. 71.11. Degradation and de novo synthesis of fatty acids (β-oxidation and liponeogenesis). Subcellular localization: β-oxidation in the mitochondria except acylcoenzyme A (*acyl-CoA*) synthetase in the cytosol; liponeogenesis in the cytosol. *MEC*, multienzyme complex; *ACP*, acyl carrier protein; *Bt*, biotin. *1, 2, 7*, transacylase activities; *3*, β-ketoacyl-ACP synthase; *4*, β-ketoacyl-ACP reductase; *5*, β-hydroxyacyl-ACP dehydratase; *6*, enoyl-ACP reductase. In β-oxidation of a C16 fatty acid to eight acetyl-CoA, two adenosine triphosphate (*ATP*) equivalents (ATP → AMP + ·

PPi) are used to activate the acid in the acyl-CoA synthetase reaction. In liponeogenesis of a C16 fatty acid from eight acetyl-CoA, seven ATP are consumed to form seven malonyl-CoA in the acetyl-CoA carboxylase reaction. *AMP*, adenosine monophosphate; *PPi*, inorganic pyrophosphate; *FAD*, flavine adenine dinucleotide; *NAD+*, nicotinamide adenine dinucleotide *NADH*, reduced NAD+; *NADP+*, NAD+ phosphate; *NADPH*, reduced NADP+

mitochondria via acylcarnitine as an intermediate [53,59,100].

Synthesis and Degradation of Lipid Stores. Triglycerides are synthesized by esterification of glycerol with fatty acids. Glycerol is first activated to glycerol phosphate by glycerolkinase, then converted with acyl-CoA to diglyceride phosphate by glycerolphosphate acyltransferase and monoglyceridephosphate acyltransferase, and finally esterified after the hydrolytic removal of phosphate to triglyceride by diglyceride acyltransferase (Fig. 71.12). Triglyceride synthesis takes place in the endoplasmic reticulum. Triglycerides are hydrolytically degraded to fatty acids and glycerol by lipases, with the hormone-sensitive triglyceride lipase playing a key regulatory role (Fig.

71.12). Triglyceride degradation is also associated with the endoplasmic reticulum [53,59,100].

De Novo Synthesis of Fatty Acids (Lipacidogenesis). Fatty acids can be synthesized de novo from glucose by oligomerization of glucose-derived acetyl-CoA catalyzed by fatty acid synthase, a multienzyme complex (MEC) (Fig. 71.11). For the synthesis of a C_{16} fatty acid, seven acetyl-CoA have to be activated to malonyl-CoA by acetyl-CoA carboxylase, and one acetyl-CoA is utilized directly. Acetyl-CoA is loaded to the peripheral SH group of the MEC and malonyl-CoA to the central SH group at the acyl carrier protein (ACP); they are then condensed to β-ketoacyl-ACP, which is converted by the β-ketoacyl-ACP reductase, β-hydroxyacyl-ACP dehydratase, and the enoyl-

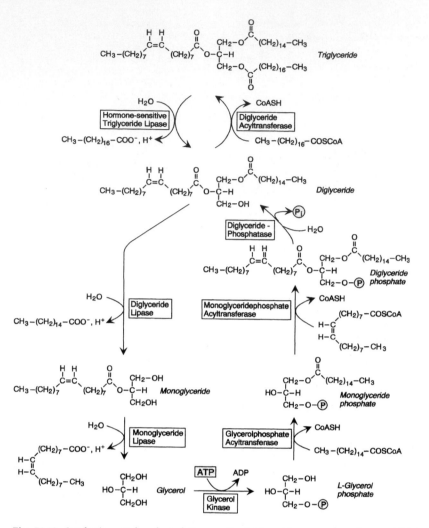

Fig. 71.12. Synthesis and degradation of triglycerides (lipogenesis and lipolysis). *Tri(di,mono)glyceride,* = tri(di,mono)acyl glycerol; *diglyceride phosphate,* phosphatidic acid. Subcellular localization: endoplasmic reticulum. In lipogenesis, the formation of triacyl glycerol from glycerol and three fatty acids, Seven adenosine triphosphate (*ATP*) are consumed, one ATP in the glycerol kinase reaction and two ATP equivalents (ATP → AMP + PPi) each in the three acylcoenzyme A (acyl-CoA) synthetase (fatty acid thiokinase) reactions. In lipolysis, the hydrolysis of triacyl glycerol to glycerol and three fatty acids, no ATP is regenerated. *ADP,* adenosine diphosphate

ACP reductase activity of the MEC to acyl-ACP. After transfer of the acyl moiety from the central to the peripheral SH group, the central SH group is again loaded with malonyl-CoA to start another elongation cycle. When the appropriate chain length of C16 or C18 has been reached, the fatty acid is hydrolytically removed from the MEC (Fig. 71.11). Fatty acid synthesis form acetyl-CoA takes place in the cytosol, and acetyl-CoA generation from pyruvate in the mitochondria. Acetyl-CoA is exported to the cytosol via citrate, formed from acetyl-CoA and oxaloacetate by citrate synthase, and cleaved again to acetyl-CoA and oxaloacetate by ATP-dependent citrate lyase [53,59,100]. In the liver, the perivenous zone has the higher capacity for lipacidogenesis, and the periportal zone for β-oxidation [51,52].

Synthesis and Degradation of Ketone Bodies. Acetoacetate is formed from fatty acid-derived acetyl-CoA by the action of acetyl-CoA acetyltransferase, 3-hydroxy-3-methylglutaryl (HMG)-CoA synthase, and HMG-CoA lyase. Acetoaceate is in equilibrium with β-hydroxybutyrate mediated by β-hydroxybutyrate dehydrogenase (Fig. 71.13). The synthesis takes place in the mitochondria. Ketone bodies are degraded in the mitochondria after activation to acetoacetyl-CoA by an unusual CoA transfer from succinyl-CoA by acetoacetate CoA-transferase. Acetoacetyl-CoA is then cleaved by a thiolase to two acetyl-CoA (Fig. 71.13). The degradation, takes place in the mitochondria [53,59,100].

71.4.6 Energy Metabolism of Ethanol

Ethanol is metabolized by alcohol dehydrogenase to acetaldehyde and NADH and further by aldehyde dehydrogenase to acetate and NADH. Acetate is activated

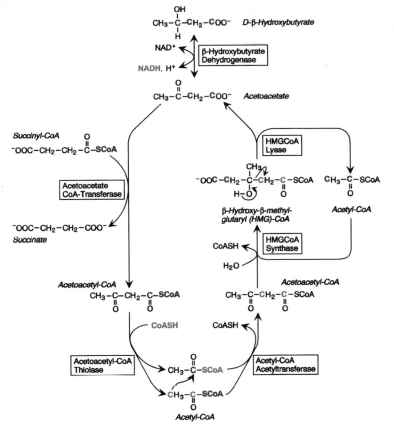

Fig. 71.13. Degradation and synthesis of ketone bodies (ketolysis and ketogenesis). Subcellular localization: mitochondria. In ketolysis, the conversion of β-hydroxybutyrate or acetoacetate to two acetylcoenzyme A (*acetyl-CoA*), one adenosine triphosphate (*ATP*) equivalent is used to activate a ketone body in the acetoacetate CoA-transferase reaction, in that succinyl-CoA is converted to succinate by the CoA transfer rather than by the "usual" guanosine triphosphate (GTP) formation of the citrate cycle (see Fig. 71.7A). *NAD+*, nicotinamide adenine dinucleotide; *NADH*, reduced NAD+

to acetyl-CoA by acetyl-CoA synthetase with conversion of ATP to AMP and PP$_i$. The first step takes place in the cytosol, the second in the mitochondria, and the third mainly in the mitochondria and perhaps also in the cytosol. At higher concentrations, ethanol is in part oxidized to acetaldehyde with O_2 and NADPH as cosubstrate by a cytochrome P450 monooxygenase isoenzyme, also called MEOS (microsomal ethanol-oxidizing system), located in the smooth endoplasmic reticulum [53,60].

71.5 Hormonal and Nervous Regulation of Energy Metabolism

The first task of regulation is to balance food intake and energy expenditure (see Sect. 71.2.3). The mechanisms involved determine who is obese and who is lean; they are not well understood at present, but a satiety factor released from adipose tissue and nervous signals from the hypothalamus may play a major role. The second task of regulation is to balance the utilization of energy substrates of food in absorptive phases for the direct production of biological energy and for the synthesis of energy stores; insulin secreted from the endocrine pancreas and parasympathetic nerves are of primary importance (see Fig. 71.4). The third task of regulation is to control the

release of energy substrates from energy stores in postabsorptive phases with and without physical activity and their use for biological energy production; glucagon released from the endocrine pancreas and circulating catecholamines secreted from the adrenal medulla and from sympathetic nerve endings via overflow as well as sympathetic nerves have key functions (see Fig. 71.5). Finally, the organism, with its many diverse organs and cell types, has to be maintained in a state of adequate reactivity towards the various hormonal and nervous signals; it is mainly cortisol from the adrenal cortex and T3/T4 from the thyroid gland that are responsible for this task.

71.5.1 Satiety Factor

It has long been postulated that after overeating the resulting extra adipose tissue somehow signals to the hypothalamus, the control center for satiety and energy expenditure, that the body is obese; the organism then eats less. Recently the obese (*ob*) gene has been cloned; the normal product of the gene may be a hormone that is secreted by fat cells and controls body weight [67,86,113]. The cDNA is expressed specifically in fat cells; the predicted sequence of 167 amino acids has the characteristics of a secreted protein. Homology between humans and mice is 84%. Obese mice homozygous for the *ob/ob* mutation fail to synthesize and secrete the protein, making ob protein a putative satiety factor. Thus, the following model is proposed: the adipose tissue produces and releases the satiety factor, which binds to the satiety factor receptor in the feeding center of hypothalamus and thus reduces the set point for food intake (see Sect. 71.2.3). Obese mice homozygous for another mutation, *db/db*, may lack a functional satiety factor receptor.

71.5.2 Insulin and the Parasympathetic System

After a meal, more energy substrates are absorbed in general than are required to meet the actual energy requirements. Insulin and the parasympathetic system have the task of regulating the storage of excess energy substrates.

Insulin. Insulin is a peptide hormone composed of an A chain with 21 and a B chain with 30 amino acids linked by two disulfide bridges. It is formed in the β- or B cells of the islets of Langerhans, the endocrine pancreas and stored there in secretory vesicles. The major stimulus for its release is an increase in arterial glucose concentration assisted by an increase in gastrointestinal hormones and in parasympathetic tone after a meal (for details see Chap. 66). Insulin is released into the portal vein so that it first reaches the liver and then other major target organs such as the skeletal muscles, the adipose tissue, and the heart. The concentration in the portal vein (approximately 200 pmol/l) is higher than in the posthepatic circulation (approximately 80 pmol/l), because the liver is a major site of insulin degradation, which begins with a limited proteolysis of the B chain [36].

After a normal meal containing carbohydrates, the peripheral insulin concentration is raised more than ten fold (approximately 800 pmol/l); at the same time the peripheral concentration of the major antagonist glucagon (approximately 40 pmol/l) is lowered by at least one third (approximately 25 pmol/l). Thus, the insulin to glucagon ratio is increased dramatically after a normal carbohydrate-rich meal from about 2 to more than 25 (Fig. 71.3). After a protein-rich meal, the peripheral insulin concentration is raised only slightly (approximately 130 pmol/l), and the peripheral levels of glucagon are about doubled (approximately 70 pmol/l). Thus, the insulin to glucagon ratio after a protein-rich meal is somewhat decreased from about 2 to 1.7 (Fig. 71.3). The increase in the insulin to glucagon ratio after a carbohydrate-rich meal is necessary to channel excess glucose into the glycogen stores in the liver and muscle. The slight decrease in the insulin to glucagon ratio after a protein-rich meal is required to maintain the glucose supply for the glucose-dependent organs, such as the brain and erythrocytes (see Sect. 71.3.1).

Insulin exerts both acute, short-term effects within seconds or minutes and nonacute, long-term effects within hours [36,53,59,100]. Insulin acts with a double strategy; it controls the uptake of substrates into target cells and it regulates the activity (short-term) and levels (long-term) of key enzymes of metabolism (Table 71.5). Insulin acutely increases glucose uptake into muscle and adipose tissue, but not into liver; it enhances glycogen synthesis in the liver and muscle and it stimulates lipogenesis (triglyceride synthesis) as well as lipacidogenesis (de novo fatty acid synthesis) in the liver and adipose tissue; and finally it increases protein synthesis in muscle and liver. Moreover, insulin maintains the capacity for glucose utilization by inducing, e.g., the genes of glucokinase, pyruvate kinase, malic enzyme, and fatty acid synthase in the liver. Thus, insulin promotes all processes which remove excess energy substrates – glucose, fatty acids, and amino acids – from the circulation by incorporating them into macromolecular energy stores, i.e., glycogen in the liver and muscle, triglycerides in adipose tissue, and protein in muscle and

Table 71.5. Major short-term actions of insulin, glucagon, and catecholamines on energy metabolism of liver, skeletal muscle, and adipose tissue (AT)

	Insulin			Glucagon			Catecholamines		
	Liver	Muscle	AT	Liver	Muscle	AT	Liver	Muscle	AT
Glucose uptake		+	+					(+)	−
Amino acid uptake	+	+	+	+					
Fatty acid uptake			+						
Glycogen synthesis	+	+		−					
Glycogen degradation	−	−		+			+	(+)[a]	
Gluconeogenesis				+			(+)[b]		
Glycolysis	+			−			(+)		
Lipacidogenesis	+		(+)						
Lipogenesis	+		+						
Lipolysis				−		+			+
Protein synthesis	+	+							
Protein degradation				(+)	+		+	(−)	

+, stimulation; −, inhibition   [a] Permissive action   [b] Only after glycogen stores have been depleted

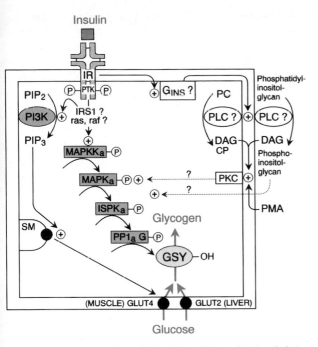

Fig. 71.14. Hormonal regulation by insulin: insulin signal chains in liver and muscle. *CP*, choline phosphate; *DAG*, diacylglycerol; *GLUT*, glucose transporter; *GSY*, glycogen synthase; G_{INS}, unknown insulin receptor specific G protein; *IR*, insulin receptor; *IRS1*, insulin receptor substrate 1; ISPK, insulin-stimulated protein kinase identical with MAPKAP-K1 (MAPK-activated protein kinase 1; *MAPK*, mitogen-activated protein kinase; *MAPKK*, MAPK kinase; *PC*, phosphatidylcholine; *PI3K*, phosphatidylinositol 3-kinase; PIP_2, phosphatidylinositol 4,5-bisphosphate; PIP_3, phosphatidylinositol 3,4,5-trisphosphate; *PKC*, protein kinase C; *PLC*, phospholipase C; *PMA*, phorbol 12-myristate, 13-acetate; *PP1G*, protein phosphatase 1 with a G subunit binding to glycogen; *PTK*, protein tyrosine kinase; *ras*, small G protein, an oncoprotein; *raf*, protein serine/threonine kinase, an oncoprotein; *SM*, subcellular membranes. The insulin receptor is a membrane-bound protein of $\alpha 2\beta 2$ structure. Similar to growth factor receptors, the membrane-spanning β-chains are autophosphorylated upon hormone binding to the α-chains. The autophosphorylated receptor is an activated protein tyrosine kinase, which might trigger G protein-independent signal chains via MAPK and then ISPK (presently the most likely hypothesis) or G protein-dependent signal chains via intra- or extracellular PLC and then PKC (presently the less likely mechanism). Insulin activates glycogen synthase in liver and muscle by dephosphorylation with a protein phosphatase, which itself is activated by a cascade of protein phosphorylations. Insulin enhances glucose uptake in muscle by triggering the insertion of the glucose transporter 4 from subcellular membranes into the plasma membrane

liver. Insulin stimulates glucose uptake by the liver efficiently only in the presence of a portal-arterial glucose concentration gradient, which allows the organism to distinguish between exogenous glucose from the diet and endogenous glucose; the mechanism is unknown [1,38,81]. Insulin acts via an ectocellular receptor, which is a transmembrane glycoprotein of βααβ structure. α-Chains and β-chains are formed from the same preproreceptor. There are two isoforms due to alternative splicing. Thus, two α-chains (135 kDa, 731 or 719 amino acids, respectively) and one β-chain (90 kDa, 620 amino acids) exist.

The α-chains are linked to each other and each α-chain to a β-chain by S-S bridges [28,40,102]. The receptor density on isolated plasma membranes is increased during starvation in line with the general concept of the downregulation of hormone receptors by their ligands. The insulin receptor is akin to growth factor receptors such as the IGF-1 (insulin-like growth factor 1) receptor, which also has an βααβ structure, and the EGF (epidermal growth factor) and PDGF (platelet-derived growth factor) receptors, which however have a single-chain structure.

Upon hormone binding to the ectocellular α-chains, the membrane-spanning β-chains of the insulin receptor are autophosphorylated, similar to the growth factor receptors. Also similar to the growth factor receptors, the autophosphorylated insulin receptor is an activated protein tyrosine kinase; it can most probably trigger different signal chains which could be G protein independent or dependent. Even after two decades of intensive research, the intracellular signal chain or chains have not been fully established; the various hypotheses are more or less fashionable at different times. The lack of unanimity has been jocularly characterized as follows: "Insulin is a hormone without a mechanism of action" [29].

For the stimulation of glycogen synthesis by insulin in muscle and liver, one hypothesis (less fashionable at the present time) postulates that the insulin receptor after insulin binding activates a specific G_{INS} protein, which in turn activates either an intracellular isoform of phospholipase C, which would be specific for phosphatidylcholine, or an ectocellular isoform of phospholipase C, which would be specific for phosphatidylinositol glycan anchors of cell surface proteins [29,30,64] (Fig. 71.14). The phospholipase C would form diacylglycerol and choline phosphate or 1-phosphoinositol 4-glycan, respectively. In turn, diacylglycerol, probably a particular isoform with specific fatty acids, would stimulate protein kinase C [74], which then would activate protein phosphatase 1 (PPI) in a cascade of protein phosphorylations via mitogen-activated protein kinase (MAPK). PP1 dephosphorylates both glycogen synthase and glycogen phosphorylase, thus stimulating glycogen synthesis and inhibiting glycogen degradation [19,20]. The finding that the tumor promoter PMA (phorbol 12-myristate, 13-acetate) can mimic many, but not all insulin actions is a major argument in favor of the G protein-linked signal chains leading to an increase in diacylglycerol via intracellular or ectocellular phospholipase C. A major argument in favor of the involvement of an ectocellular phospholipase C forming diacylglycerol and 1-phosphoinositol 4-glycan is the observation that some, but not all, insulin actions can be mimicked by the extracellular addition of bacterial phospholipases C or by membrane fractions enriched in phosphoinositol glycan, the structure of which, however, has not been unequivocally established [29].

For the stimulation of glycogen synthesis by insulin, another hypothesis (more fashionable at the present time) postulates that the insulin receptor triggers a cascade of protein kinase phosphorylations [22,23]. The first step is probably the phosphorylation of a cytoplasmic protein, the insulin receptor substrate 1 (IRS1), at multiple tyrosine residues [72]. The signal would then be transmitted via the small G protein ras to the protein kinase raf, which would then activate by phosphorylation the so-called MAPK kinase (MAPKK) [5,21] (Fig. 71.14). Although it appears clear that IRS1 is indispensible for insulin-stimulated mitogenesis and cell cycle progression, its role in the regulation of classic metabolic responses remains hypothetical. The distal part of the insulin-triggered cascade of protein kinase phosphorylations leading to a stimulation of glycogen synthesis, however, appears to be well established. MAPKK phosphorylates MAPK at a threonine and tyrosine residue. MAPK in turn phosphorylates the insulin-stimulated protein kinase (ISPK) at a serine residue, which then phosphorylates and thus activates PP1 at a serine residue of its G subunit, which binds the phosphatase enzyme to glycogen particles. PP1 dephosphorylates and thereby activates glycogen synthase and at the same time dephosphorylates and inactivates glycogen phosphorylase. Thus, insulin activates glycogen synthesis in the liver and muscle by dephosphorylation of glycogen synthase and glycogen phosphorylase with the same protein phosphatase, which itself has been activated by a cascade of protein phosphorylations [19,20,22] (Fig. 71.14).

For the stimulation of glucose uptake into muscle and adipose tissue by insulin, yet another mechanism appears to be operative. Via phosphorylation of IRS1, the insulin receptor activates the enzyme phosphatidylinositol 3-kinase, which converts mainly phosphatidylinositol 4,5-bisphosphate (PIP_2) to phosphatidylinositol 3,4,5-trisphosphate (PIP_3). This 3-phosphorylated phosphoinositide PIP_3 cannot be hydrolized (unlike its precursor PIP_2 by, e.g., a phospholipase C), but appears to act as a direct targeting signal for the insertion of the glucose transporter 4 (GLUT4) from subcellular membranes into the plasma membrane. Thus, insulin would increase the number of transporters in the membrane and thereby enhance glucose uptake in muscle and adipose tissue [45,54] (Fig. 71.14).

Parasympathetic Nerves. Stimulation of the vagus nerve increases hepatic glycogen synthesis directly via the hepatic nerves synergistically with insulin and antagonistically to glucagon; moreover, it enhances glucose utilization indirectly by stimulating insulin release [39,73,97] (see Chap. 66).

71.5.3 Glucagon, Catecholamines, Sympathetic Nerves, and Motoneurons

Between meals, when the absorption of energy substrates has ceased, the energy requirements of the organism are met by degradation of the energy stores: glycogen in the liver, triglycerides in adipose tissue, and protein mainly in muscle. Under these conditions glucagon is the major hormone regulating the utilization of the energy stores. During exercise, which normally takes place during the postabsorptive phase and only exceptionally during the absorptive phase, there is a varying increase in the energy requirements, which are met by a varying increase in the utilization of the energy stores during the postabsorptive phase or a shift from formation of energy stores to direct utilization of the absorbed energy substrates during the absorptive phase. Glucagon, the circulating catecholamines, and the sympathetic nerves as well as the motoneurons have the task of controling the provision of energy substrates during exercise.

Glucagon. Glucagon is a single-chain peptide hormone composed of 29 amino acids. It is formed in the α- or A cells of the islets of Langerhans and stored there in secretory vesicles. The major stimulus for its release is a decrease in arterial glucose concentration, which, depending on the situation, is assisted by an increase in circulating catecholamines and in the sympathetic tone (for details see Chap. 66). Glucagon, like insulin, is released into the portal vein so that it first reaches the liver and then the other major target organ, the adipose tissue. In addition, the concentration in the portal vein is higher than in the posthepatic circulation, because the liver is a major site of glucagon degradation, which begins with the proteolytic removal of the two N-terminal amino acids [36]. During the postabsorptive phase, the levels of insulin, increased after a meal containing carbohydrates, and those of glucagon, decreased after a meal containing carbohydrates, normalize again (Fig. 71.3). This normalized insulin to glucagon ratio of 2 between meals is necessary to regulate the provision of glucose via glycogen degradation and gluconeogenesis in liver. After a protein-rich meal, glucagon and insulin are increased almost in parallel, because the provision of glucose has to be secured and the excess amino acids have to be channeled into the protein stores (see Sect. 71.3.1).

Glucagon, like insulin, has acute, short-term and nonacute, long-term effects [36,53,59,100]. Glucagon acts with a clearly less-developed double strategy; its influence on substrate uptake into target cells is restricted, although, like insulin, it effectively regulates the activity (short-term) and levels (long-term) of key enzymes of metabolism (Table 71.5). Glucagon acutely increases the uptake of amino acids into the liver; it enhances glycogen degradation in the liver but not in muscle; it stimulates lipolysis in adipose tissue; and finally it increases protein degradation in muscle and liver. Moreover, glucagon maintains the capacity for the endogenous supply of glucose by inducing, e.g., the gluconeogenic enzymes phosphoenolpyruvate carboxykinase, serine dehydratase, tyrosine aminotransferase, and tryptophan oxygenase. Thus, it stimulates all processes which supply energy substrates – glucose, fatty acids, and amino acids – from the macromolecular energy stores, i.e., glycogen in the liver, triglycerides in adipose tissue, and protein in muscle and liver.

Glucagon acts via an ectocellular receptor, which is a transmembrane single-chain glycoprotein (62 kDa, 477 amino acids) [48,62,66]. It belongs to the R7G receptor family, as do, e.g., the α- and β-adrenergic receptors; the members of this family possess seven

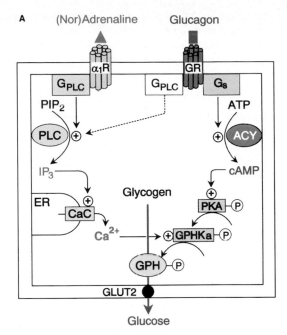 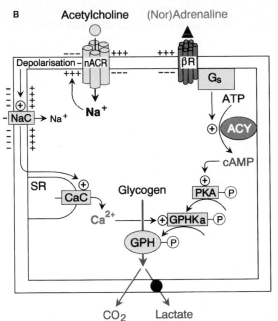

Fig. 71.15A,B Hormonal regulation by glucagon, (nor)adrenaline and acetylcholine. **A** Glucagon and catecholamine signal chains in liver. **B** Acetylcholine and catecholamine signal chains in muscle. α_1R, α_1-adrenergic receptor; βR, β-adrenergic receptor; *nACR*, nicotinic acetylcholine receptor; *ACY*, adenylate cyclase; *cAMP*, cyclic adenosine 3',5'-monophosphate; *CaC*, voltage-gated Ca^{2+} channel; *ER*, endoplasmic reticulum; *GPH*, glycogen phosphorylase; G_{PLC}, phospholipase C-specific G protein; *GR*, glucagon receptor; *Gs*, stimulatory G protein; IP_3 inositol 1,4,5-trisphosphate; *NaC*, voltage-gated cation channel; PIP_2, phosphatidylinositol 4,5-bisphosphate; *PLC*, phospholipase C; *SR*, sarcoplasmic reticulum. The glucagon, the α_1-adrenergic, and the β-adrenergic receptors belong the membrane-bound R7G receptor family, i.e., they possess seven transmembrane regions and are linked to G proteins. The nicotinic acetylcholine receptor belongs to the class of ligand-gated ion channels; it is composed of five subunits $\alpha\beta\alpha\gamma\delta$ arranged in a circle. Every subunit has four transmembrane regions, of which the innermost one in each subunit forms part of a cation channel; the channel is opened when acetylcholine binds to the α-subunits. In liver, glucagon, adrenaline and noradrenaline (circulating or released from the sympathetic hepatic nerves) activate glycogen phosphorylase by phosphorylation with glycogen phosphorylase kinase, which is activated itself either by a glucagon-elicited phosphorylation via cAMP or by a catecholamine-triggered increase in cytosolic Ca^{2+} via IP_3. In muscle, acetylcholine (released from motoneurons) activates glycogen phosphorylase by phosphorylation with glycogen phosphorylase kinase, which is activated itself by a depolarization-triggered increase in sarcoplasmic Ca^{2+} after having been sensitized towards Ca^{2+} by a catecholamine-dependent phosphorylation via cAMP

transmembrane regions and are linked to G proteins [8,99] (see Chap. 5).

Glucagon may be able to trigger two different intracellular signal chains. The primary mode of action appears to be the activation via a stimulatory G_s protein of membrane-associated adenylate cyclase and thus the increase in the second messenger cyclic AMP (cAMP). The activation of membrane-associated phospholipase C via a specific G_{PLC} protein and thus the formation of the second messerger inositol 1,4,5-trisphosphate (IP_3) could be an additional mechanism [106] (Fig. 71.15A). Glycogen degradation in the liver is stimulated by glucagon mainly via an increase in cAMP, which activates cAMP-dependent protein kinase, also called protein kinase A (PKA), which in turn activates glycogen phosphorylase kinase by phosphorylation at serine residues. Glycogen phosphorylase kinase then phosphorylates and thus activates the target enzyme proper, glycogen phosphorylase, again at serine residues. Glycogen degradation may be activated by glucagon aslo via an increase in IP_3, which would open Ca^{2+} channels in the endoplasmic reticulum and thus elevate the cytosolic Ca^{2+} concentration by two orders of magnitude. Ca^{2+} then activates glycogen phosphorylase kinase by binding to one of its subunits, calmodulin. The Ca^{2+}-activated glycogen phosphorylase kinase then would phosphorylate and activate glycogen phosphorylase (Fig. 71.15A). Glycogenolysis in muscle is not activated by glucagon. This is not surprising, since the glycogen stores of muscle cannot be degraded to free glucose due to the lack of glucose 6-phosphatase [36,53,59,100].

Catecholamines. Noradrenaline and adrenaline are derivatives of the amino acid tyrosine. This substrate is hydroxylated to dihydroxyphenylalanine (DOPA), which after decarboxylation to dopamine is further hydroxylated to noradrenaline; N-methylation yields adrenaline. Noradrenaline and adrenaline are formed and stored in the adrenal medulla in cells with intensely staining granules and in cells with less intensely staining granules, respectively. Circulating catecholamines are inactivated mainly in liver by O-methylation and oxidation to vanillylmandelic acid, which is excreted into urine [42]. Catecholamine release is regulated solely by the sympathetic innervation of the adrenal medulla with acetylcholine as the neurotransmitter; hypoglycemia at rest and stress situations, including exercise, are major activators of the sympathetic centers in the hypothalamus. The main source of circulating noradrenaline (but not of adrenaline) is the overflow from sympathetic nerve endings, predominantly in the heart. During exercise (150 W), there is a much higher increase in noradrenaline (1 nmol/l to 5 nmol/l) than adrenaline (0.4 to 0.8 nmol/l), because the major source of noradrenaline is the overflow from the working heart and not release from the adrenals. The increase mainly in noradrenaline but also in adrenaline during exercise is necessary to provide free glucose from the glycogen stores in liver for the working muscles; it is also necessary to sensitize the stimulation of glycogenolysis by the firing motoneurons in the working muscles (see below). During hypoglycemia (2.5 mmol/l glucose), the central glucose sensors specifically activate only those nerves innervating the adrenaline-producing cells in the adrenal medulla so that only adrenaline (0.4 nmol/l to 4 nmol/l), but not noradrenaline (1 nmol/l) is increased [14,65,105].

Catecholamines exert mainly acute, short-term effects. They have little effect on transport processes and mainly regulate the activity of key enzymes of metabolism (Table 71.5). Adrenaline and noradrenaline enhance glycogen degradation to glucose in liver. Since glycogenolysis in muscle is enhanced specifically by the firing motoneurons only in contracting muscle fibers and not nonspecifically by circulating catecholamines in all muscle fibers, whether resting or contracting, adrenaline and noradrenaline have only a permissive effect on glycogen degradation via pyruvate to CO_2 or lactate in working muscles (see below). Catecholamines also activate lipolysis in adipose tissue. Thus they promote all processes which supply the necessary energy substrates for muscular work (Fig. 71.15A,B). Catecholamines act via ectocellular receptors, which are transmembrane glycoproteins; there are α- and β-receptors with various subtypes (e.g., α_1, 80 kDa, 515 amino acids; β_2, 64 kDa, 413 amino acids) [13,56,63,79,95,104]. Like the glucagon receptor, the adrenergic receptors also belong to the R7G receptor family, i.e., they have seven transmembrane domains and are linked to G proteins [8,99] (see Chap. 5).

Catecholamines are generally able to trigger three different intracellular signal chains:

- α_1-Receptors stimulate a membrane-associated phospholipase C via a G_{PLC} protein and thus increase IP_3 and diacylglycerol (DAG).
- α_2-Receptors are linked to inhibitory G_i proteins and thus inhibit the membrane-associated adenylate cyclase causing a decrease in cAMP.
- β-Receptors enhance the activity of membrane-associated adenylate cyclase via a stimulatory G_s protein and thus increase cAMP [42].

In the liver, circulating adrenaline and noradrenaline and noradrenaline released locally from sympathetic hepatic nerve endings (see below) increase IP_3 via α_1-receptors; IP_3 opens Ca^{2+} channels in the endoplasmic reticulum, thus causing an increase in cytosolic Ca^{2+} by more than 100-fold. Cytosolic Ca^{2+} then activates glycogen phosphorylase kinase by binding to its calmodulin subunit. The activated glycogen phosphorylase kinase then phosphorylates and thus activates the target enzyme proper, glycogen phosphorylase (Fig. 71.15A). In muscle, glycogenolysis is stimulated by acetylcholine released from motoneurons at the muscular end plate; catecholamines have only a permissive role (see below).

Sympathetic Nerves. The sympathetic system, activated in part by afferent hepatic nerve fibers sensing the glucose concentration in the portal vascular bed and in part by central hypothalamic sensors, activates glucose formation from glycogen in the liver directly via the efferent sympathetic liver nerves and also indirectly by activating the release of glucagon from the endocrine pancreas and of catecholamines from the adrenal medulla [39,73,85,97].

Motoneurons. Acetylcholine released at the muscular end plate binds to the nicotinic acetylcholine receptor, which is a member of the ligand-gated ion channels and is composed of five subunits $\alpha\beta\alpha\gamma\delta$ arranged in a circle [2,41]. Every subunit possesses four transmembrane regions, of which the innermost each constitutes a part of a cation channel. This channel is opened when acetylcholine binds to the α-subunits. The inflow of Na^+ causes a depolarization of the membrane, which is propagated like an action potential through the transversal tubuli to the sarcoplasmic reticulum. There, the propagated depolarization opens a voltage-gated Ca^{2+} channel and thus increases sarcoplasmic Ca^{2+} by more than two orders of magnitude. Sarcoplasmic Ca^{2+} both triggers contraction and activates glycogenolysis by binding to the calmodulin subunit of glycogen phosphorylase kinase, which in turn phosphorylates and thus activates glycogen phosphorylase. Circulating catecholamines have a permissive role in this regulation. By binding to β-receptors they increase cAMP, which activates protein kinase A and thus causes a phosphorylation of glycogen phosphorylase kinase. The phosphorylated glycogen phosphorylase kinase is more sensitive to stimulation by Ca^{2+} than is the nonphosphorylated form [19] (Fig. 71.15B).

71.5.4 Glucocorticoid and Thyroid Hormones

The various organs can fulfill their tasks in the energy metabolism of the organism only if they are adequately equipped with translocators, enzymes, hormone receptors, and other components of signal chains such as G proteins or transcription factors. The glucocorticoid hormone cortisol and the thyroid hormones thyroxine (tetraiodothyronine, T4) and triiodothyronine (T3) play a key role in regulating adequate patterns of translocators, enzymes, and receptors, i.e., in controlling the capacity and hormonal sensitivity of metabolism. They exert long-term effects within hours, requiring as a rule de novo protein synthesis.

Cortisol. Cortisol is a C21-steroid hormone. It is synthesized from the C27-steroid cholesterol in the adrenal cortex by cells of the zona fasciculata. Cholesterol is hydroxylated first at C20 and C22, which leads to a shortening of the side chain by six C atoms, and then at C17, C21, and C11. As a steroid derivative, cortisol is highly lipophilic and therefore transported in blood bound mainly (approximately 90%) to the transport protein transcortin (cortisol-binding protein, CBP) but also to albumin (approximately 5%), both of which are formed in the liver. The mean free and total concentrations of cortisol are 40 nmol/l and 400 nmol/l, respectively. Cortisol is inactivated by reduction to tetrahydrocortisol and then conjugation with glucuronic acid mainly in the liver; tetrahydrocortisol glucuronide is excreted mainly via urine (70%), but also via feces (20%) (for details see Chap. 20) [69].

The major stimulus for cortisol synthesis and secretion is the 93-amino acid pituitary hormone corticotropin (adrenocorticotropic hormone, ACTH), which is released after stimulation by the 41-amino acid hypothalamic hormone corticoliberin (corticotropin-releasing hormon, CRH). The release of corticoliberin is under the control of the sleep–wake cycle; therefore, circulating corticotropin and cortisol exhibit a diurnal rhythm with peak plasma levels (approximately 700 nmol/l) occurring in humans in the early morning hours at the end of the sleeping period and low levels (approximately 150 nmol/l) in the early evening at the end of the activity period. Additional stimuli are stress conditions, such as strenuous physical exercise or polytrauma. The hypothalamic–pituitary–adrenal axis is feedback controlled by cortisol, which inhibits both corticortropin and corticoliberin synthesis and release (for details see Chap. 20) [69].

Cortisol regulates the level of key enzymes and translocators of metabolism (Table 71.6). It enhances the capacity for glycogen synthesis and gluconeogenesis, but also glycolysis, in the liver by upregulating key enzymes and the glucagon sensitivity, and it lowers the capacity for glucose uptake in muscle and adipose tissue by downregulating GLUT 4 and the insulin sensitivity. Moreover, it increases the capacity for lipolysis in adipose tissue and for proteolysis in muscle. Thus, cortisol has anabolic effects on the liver and catabolic effects on muscle and adipose tissue. It tends to increase the blood glucose concentration (hence it is referred to as a glucocorticoid hormone), and it ensures the ability of the organism to cope with stress situations (hence it is known as a stress hormone) (for details see Chap. 20) [69].

Cortisol acts via an intracellular receptor which is located both in the cytosol and the nucleus. The receptor belongs to the steroid hormone–retinoic acid receptor superfamiliy. The glucocorticoid receptor is a transcription factor composed of 777 amino acids with a DNA-binding domain about 70 amino acids long and a hormone-binding domain about 250 amino acids long approximately 370 and 250 amino acids away from the C-terminal end, respectively. The DNA-binding domain contains two loop–helix elements each with a Zn^{2+} liganded by four cysteines, which are called zinc fingers. After its uptake into the cytosol, probably via facilitated transport, cortisol binds to the glucocorticoid receptor – heat shock protein complex in the cytosol, which causes the release and dimerization of the cortisol-carrying receptor [7,44,94] (for alternative views see Chap. 20). The dimer enters the nucleus and binds to glucocorticoid-regulatory elements (GRE) with the palindromic consensus sequence AGAACAnnnTGTTCT mainly in the 5'-flanking region of genes. One monomer each of the cortisol receptor dimer binds to one half of the palindromic sequence. This cortisol receptor – DNA interaction normally does not activate or inactivate the target genes of metabolic enzymes, translocators, or receptors

Table 71.6. Major long-term permissive[a] actions of cortisol and triiodothyronine/tetratriiodothyronine (T3/T4) on energy metabolism of liver, skeletal muscle, and adipose tissue (AT)

	Cortisol			T3/T4		
	Liver	Muscle	AT	Liver	Muscle	AT
Glucose uptake		−	−		‖	+
Oxygen uptake[b]				+	+	+
Glycogen synthesis	+			−		
Gluconeogenesis	+			+		
Glycolysis	+			+		
Lipolysis			+			+
Protein synthesis				+[c]	+[c]	
Protein degradation		+		(+)	+	

+, increase; −, decrease
[a] Expression of adequate enzyme, translocator, and receptor levels
[b] Control of adequate basal metabolic rate (thermogenesis)
[c] Direct and indirect growth stimulation via somatotropin at physiological T3/T4 levels

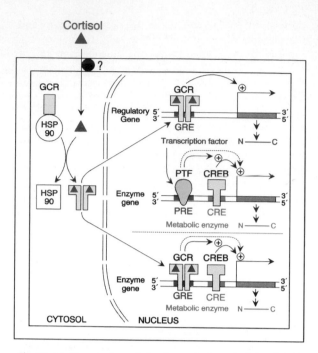

Fig. 71.16. Hormonal regulation by cortisol: cortisol signal chain (hypothesis). *GRE,* glucocorticoid-regulatory element: AGA-ACAnnnTGTTCT; *CRE,* cyclic adenosine monophosphate (AMP)-regulatory element; *CREB,* cyclic AMP-regulatory element-binding protein; *GCR,* glucocorticoid receptor; *HSP90,* 90-kDa heat-shock protein, *PTF,* permissiveness transcription factor; *PRE,* permissiveness regulatory element. The glucocorticoid receptor is an intracellular protein which acts as a transcription factor in the nucleus. It belongs the steroid hormone–retenoic acid receptor family, characterized by separate DNA (zinc fingers) and hormone-binding domains. Cortisol binds to the glucocorticoid receptor–heat-shock protein complex in the cytosol, thus causing the release and dimerization of the cortisol-carrying receptor (for alternative concepts see Chap. 20). This dimer is transferred to the nucleus to bind to GRE primarily in the 5'-flanking region of genes. Normally, this binding itself does not activate the target "metabolic" genes directly. Binding probably occurs either at a GRE of a "regulatory" gene to activate the production of a transcription factor required for the basal expression and for the induction of the "metabolic" gene by, e.g., cAMP via CREB (action in *trans*), or binding occurs at a GRE of the "metabolic" gene itself not to activate it, but to enable its activation by cAMP via CREB (action in *cis*). Thus, cortisol exerts a "permissive" action on hormonal gene activation

directly but indirectly. Two modes of action are likely to be the following:

- The cortisol receptor binds to and thereby activates some "regulatory" transcription factor genes. The cortisol-dependent transcription factors are required for the maintenance of the basal expression of "metabolic" genes and for the induction of these genes by a second hormone such as glucagon (via cAMP) or insulin. These would be actions in *trans* on the final target genes. Experimental evidence for this mechanism is a delay in gene activation after cortisol addition and a strong synergistic effect of the inducing hormone proper at physiological concentrations.

- The cortisol receptor binds to metabolic genes and thereby enables their activation by other hormones. This would be an action in *cis* on the final target genes. Experimental evidence for this mechanism is an undelayed gene activation after cortisol administration and a clearly synergistic enhancement by the inducing hormone proper (Fig. 71.16).

Both mode of actions explain the multiple pleiotropic effects of cortisol on metabolism. Thus, cortisol exerts a permissive action on the activation of genes by hormones [90]. Examples for the permissive action of cortisol are the induction in hepatocytes of the glycolytic glucokinase by insulin and the gluconeogenic phosphoenolpyruvate carboxykinase, serine dehydratase, tyrosine aminotransferase, or tryptophan oxygenase by glucagon.

T3/T4. Triiodothyronine (T3) and tetratriiodothyronine (T4) are derivatives of the amino acid tyrosine. They are formed in the thyroid gland by an "indirect" pathway starting from the tyrosine-rich protein thyroglobuline. This 660-kDa protein which has two subunits each containing 72 tyrosine residues, and I_2, which is formed from iodide by oxidation with H_2O_2, are excreted into the follicular lumen, where the tyrosine residues are first mono- and diiodinated with I_2 and then rearranged to form T3- and T4-thyroglobulin. This iodinated protein is then endocytozed again from the follicular lumen into lysosomes, where it is hydrolyzed to yield T3 and T4. Since T3 and T4 are lipophilic, they are transported in blood bound mainly (75%) to thyroid hormone-binding protein (TBP) and also (25%) to albumin and transthyretin (prealbumin), which are formed in the liver. The free and total concentrations of T3 are 5–10 pmol/l and 1.5–3.5 nmol/l and of T4 20–40 pmol/l and 60–140 nmol/l, respectively. In target tissues a substantial part of T4 is deiodinated to T3, which is more active. T3 and T4 are inactivated by deiodination to thyronine and further to thyroacetate and by conjugation with glucuronate and sulfate for excretion into urine (for details see Chap. 22) [61].

T3 and T4 synthesis and release is stimulated by the pituitary hormone thyrotropin (thyroid-stimulating hormone, TSH) composed of an α-subunit with 92 and a β-subunit with 112 amino acids, which is released after stimulation by the hypothalamic peptide thyroliberin (thyrotropin-releasing hormone, TRH) composed of three amino acids. Thyroliberin release is activated mainly by cold exposure. The hypothalamic–pituitary–thyroid axis is feedback controlled by T3 and T4, which inhibit both thyrotropin and thyroliberin synthesis and release (for details see Chap. 22). T3 is enhanced by carbohydrate-rich and protein-rich food and is decreased in starvation (see Sects. 71.2.3, 71.2.4). T3 and T4 regulate the levels of key enzymes of metabolism (Table 71.6). They increase the capacity for O_2 uptake in the liver, muscle, and adipose tissue, for gluconeogenesis and glycolysis in the liver, for lipacidogenesis in the liver, for lipolysis in adipose tissue, and for proteinolysis in muscle. The main function is the control of O_2 consumption, i.e., the regulation of an adequate BMR and thermogenesis.

T3 and T4 act via an intracellular receptor, which is located solely in the nucleus. The receptor occurs in various isoforms (TRα_1, TRα_2, TRβ_1, and TRβ_2) and belongs to the steroid hormone–retinoic acid receptor superfamily. The thyroid hormone receptor is a transcription factor with a DNA-binding domain (zinc finger) and a hormone-binding domain, like the glucocorticoid receptor; however, it is smaller and is composed of only 461 amino acids (TRβ_1). After their uptake into the cytosol, probably via facilitated transport, T3 and T4 bind to a cytosolic thyroid hormone-binding protein (CTBP), which appears to be a monomeric subunit of pyruvate kinase type M_2 [4,110]. CTBP assists in the transfer of the lipophilic hormones T3 and T4 to the nucleus and regulates their free cytosolic concentration. In the nucleus, T3 and T4 (which has a considerably lower affinity than T3) bind to the thyroid hormone receptor, which then forms a dimer binding to thyroid hormone-regulatory elements (TRE) with the palindromic consensus sequence TCAGGTCA–TGACCTGA mainly in the 5'-flanking region of genes. This receptor binding to DNA normally does not activate a metabolic gene, but allows its basal expression and induction by other hormones. Thus T3 and T4, like cortisol, exert a permissive action on the activation of genes by hormones (see Fig. 71.16).

Examples of the permissive action of T3 and T4 are the induction in the liver of gluconeogenic phosphoenolpyruvate carboxykinase by glucagon and of glucose-utilizing glucokinase, malic enzyme, and fatty acid synthase by insulin. The mechanism of the T3/T4-stimulated induction of the (Na$^+$+K$^+$)-ATPase and the mitochondrial ATP–ADP translocator, which may account for most of the T3/T4-dependent increase in oxygen consumption, is not known [26,78].

References

1. Adkins BA, Myers SR, Hendrik Gk, Stevenson RW, Williams PE, Cherrington AD (1987) Importance of the route of intravenous glucose delivery on hepatic glucose balance in the concious dog. J Clin Invest 79:557–565
2. Alberts B, Bray D, Lewis J, Raff M, Roberts K, Watson JD (1989) Molecular biology of the cell, 2nd edn. Garland, New York
3. Ament ME (1990) Enteral and parenteral nutrition. In: Brown ML (ed) Present knowledge in nutrition, 6th edn. International Life Sciences Institute, Nutrition Foundation, Washington, pp 444–450
4. Amizawa K, Cheng SY (1993) Regulation of thyroid hormone receptor-mediated transcription by a cytosolic protein. Proc Natl Acad Sci U S A 89:9277–9281
5. Avruch J, Zhang X, Kyriakis JM (1994) Raf meets Ras: completing the frame work of a signal transduction pathway. Trends Biochem Sci 19:279–283
6. Barth CA (1991) Animal products and human health: conse quences for agriculture and some new approaches. Proceedings of the 6th International symposium on Protein Metabolism and Nutrition. National Institute of Animal Science, Foulum, Denmark, pp 7–22
7. Beato M (1989) Gene regulation by steroid hormones. Cell 56:335–344
8. Boege F, Neumann E, Helmreich EJM (1991) Structural heterogeneity of membrane receptors and GTP-binding proteins and its functional consequences for signal transduction. Eur J Biochem 199:1–15
9. Boothby WM, Berkson J, Dunn HL (1936) Studies of the energy metabolism of normal individuals: a standard for basal metabolism with a nomogram for clinical application. Am J Physiol 116:468–484
10. Bray GA (1990) Obesity. In: Brown ML (ed) Present knowledge in nutrition, 6th edn. International Life Sciences Institute, Nutrition Foundation, Washington, pp 23–28
11. Brenner BM, Meyer TW, Hostetter TH (1982) Dietary protein intake and the progressive nature of kidney disease: the role of hemodynamically mediated glomerular injury in the pathogenesis of progressive glomerular sclerosis in aging, renal ablation and intrisic renal disease. N Engl J Med 307:652–659
12. Brown MS, Goldstein JL (1991) The hyperlipoproteinemias and other disorders of lipid metabolism. In: Wilson JB, Braunwald E, Isselbacher K, Petersdorf RG, Martin JB, Fauci AS, Root RK (eds) Harrison's principles of internal medicine, vol 2, 12th edn. McGraw-Hill, New York, pp 1814–1825
13. Buckland PR, Hill RM, Tidmarsh SF, McGuffin P (1990) Primary structure of the rat β_2-adrenergic receptor gene. Nucleic Acids Res 18:682
14. Bühler HU, DaPrada M, Haefely W, Picotti GB (1978) Plasma adrenaline, noradrenaline and dopamine in man and different animal species. J Physiol 276:311–320
15. Cahill GF (1970) Starvation in man. N Engl J Med 282:668–675
16. Cahill GF, Herrera MG, Morgan AP, Soeldner JS, Steinke J, Levy PL, Reichard GA Jr, Kipnis DM (1966) Hormone-fuel interrelationships during fasting. J Clin Invest 45:1751–1769
17. Cahill GF, Marliss EB, Aoki TT (1970) Fat and nitrogen metabolism in fasting man. Adipose tissue. Regulation and metabolic functions. Horm Metab Res Suppl 2:181–185
18. Capaldi RA, Aggeler R, Turina P, Wilkens S (1994) Coupling between catalytic sites and the proton channel in F_1F_0-type ATPases. Trends Biochem Sci 19:2084–2088
19. Cohen P (1992) Signal integration at the level of protein kinases, protein phosphatases and their substrates. Trends Biochem Sci 17:408–413
20. Cohen P (1993) Dissection of the protein phosphorylation cascades involved in insulin and growth factor action. Biochem Soc Trans 21:555–567
21. Daum G, Eisenmann-Tappe I, Fries HW, Troppmair J, Rapp UR (1994) The ins and outs of Raf kinases. Trends Biochem Sci 19:474–479
22. Dent P, Lavoinne A, Nakielny S, Caudwell FB, Watt P, Cohen P (1990) The molecular mechanism by which insulin stimulates glycogen synthesis in mammalian skeletal muscle. Nature 348:302–308
23. Denton RM (1990) Insulin signaling. Search for the missing links. Nature 348:286–287
24. Després JP, Lamarche B (1993) Effects of diet and physical activity on adiposity for the prevention of cardiovascular disease. Nutr Res Rev 6:137–159
25. Devlin JT, Horton ES (1990) Energy requirements. In: Brown ML (ed) Present Knowledge in Nutrition, 6th edn. International Life Sciences Institute, Nutrition Foundation, Washington, pp 1–6
26. Dümmler K, Müller S, Seitz JH (1994) Thyroid hormones and protein metabolism. In: Orgiazzie J, Leclere J (eds) The thyroid and tissues Schattauer, Stuttgart, pp 61–74
27. Eastwood MA, Passmore R (1983) Dietary fibre. Lancet II (8343):202–206
28. Ebina Y, Ellis L, Jarnagin K, Edery M, Graf L, Clauser E, Ou J, Masiarz F, Kan YW, Goldfine ID, Roth RA, Rutter WJ (1985) The human insulin receptor cDNA: the structural basis for hormone-activated membrane signalling. Cell 40:747–758

29. Exton JH (1991) Some thoughts on the mechanism of action of insulin. Diabetes 40:521–526
30. Farese RV, Standaert ML, Arnold T, Yu B, Ishizuka T, Hoffman J, Vila M, Cooper DR (1992) The role of protein kinase C in insulin action. Cell Signal 4:133–143
31. Felig P, Sherwin R (1976) Carbohydrate homeostasis, liver and diabetes. Prog Liver Dis 5:149–171
32. Felig P, Wahren J, Hendler R (1975) Influence of oral glucose ingestion on splanchnic glucose and gluconeogenic substrate metabolism in man. Diabetes 24:468–475
33. Flatt JP (1987) Dietary fat, carbohydrate balance and weight maintenance: effects of exercise. Am J Clin Nutr 45: 196–306
34. Forbes GB (1990) Body composition. In: Brown ML (ed) Present knowledge in nutrition, 6th edn. International Life Sciences Institute, Nutrition Foundation, Washington, pp 7–12
35. Foster DW (1991) Anorexia nervosa and bulimia. In: Wilson JB, Braunwald E, Isselbacher K, Petersdorf RG, Martin JB, Fauci AS, Root RK (eds) Harrison's principles of internal medicine vol 1, 12th edn. McGraw-Hill, New York, pp 417–420
36. Freychet P (1990) Pancreatic hormones. In: Baulieu EE, Kelly PA (eds) Hormones. From molecules to disease. Hermann, Paris, pp 489–532
37. Fröberg S, Carlson L, Ekelund L (1971) Local lipid stores and exercise. Adv Exp Med Biol 11:307–313
38. Gardemann A, Strulik H, Jungermann K (1986) A portal-arterial glucose concentration gradient as a signal for insulin-dependent net glucose uptake in perfused rat liver. FEBS Lett 202:255–259
39. Gardemann A, Püschel G, Jungermann K (1992) Nervous control of liver metabolism and hemodynamics. Eur J Biochem 207:399–411
40. Goldstein BJ, Dudley AL (1990) The rat insulin receptor: Primary structure and conservation of tissue-specific alternative messenger RNA splicing. Mol Endocrinol 4:235–244
41. Guy HR, Hucho F (1987) The ion channel of the nicotinic acetylcholine receptor. Trends Neurosci 10:318–321
42. Hanoune J (1990) The adrenal medulla. In: Baulieu EE, Kelly PA (eds) Hormones. From molecules to disease. Hermann, Paris, pp 307–333
43. Harris JA, Benedict FG (1919) A biometric study of basal metabolism in man. Carnegie Institute, Washington, publ no 279, pp 1–266
44. Harrison SC (1991) The structural taxonomy of DNA-binding domains. Nature 353:715–719
45. Holman GD, Cushman SW (1994) Subcellular localization and trafficking of the GLUT4 glucose transporter isoform in insulin-responsive cells. Bioessays 16:753–759
46. Hultman E, Nilsson LH (1971) Liver glycogen in man. Effect of different diets and muscular exercise. Adv Exp Med Biol 11:143–151
47. Hultman E, Bergström J, Roch-Norlund AE (1971) Glycogen storage in human skeletal muscle. Adv Exp Med Biol 11:273–288
48. Jelinek LJ, Lok S, Rosenberg GB, Smith RA, Grant FJ, Biggs S, Bensch PA, Kuijper JL, Sheppard PO, Sprecher CA, O'Hara PJ, Foster D, Walker KM, Chen LHJ, McKernan PA, Kindsvogel W (1993) Expression cloning and signalling properties of the rat glucagon receptor. Science 259:1614–1616
49. Jequier E, Schutz Y (1985) New evidence for a thermogenic defect in human obesity. Int J Obesity 9:1–7
50. Junge B, Hoffmeister H (1982) Civilization-associated disease in Europe and industrial countries outside of Europe: regional differences and trends in mortality. Prev Med 11: 117–130
51. Jungermann K, Katz N (1989) Functional specialization of different hepatocyte populations. Physiol Rev 69:708–764
52. Jungermann K (1995) Zonation of metabolism and gene expression. Histochemistry 103: 81–91
53. Jungermann K, Möhler H (1980) Biochemie. Springer, Berlin Heidelberg New York
54. Kapeller R, Cantley LC (1994) Phosphotidylinositol 3-Kinase. Bioessays 16:5656–5720
55. Kleiber M (1961) The fire of life. Wiley, New York
56. Kobilka BK, Dixon RA, Frielle T, Dohlman HG, Bolanowski MA, Sigal IS, Yang-Feng TL, Francke U, Caron MC, Lefkowitz RJ (1987) cDNA for the human β-adrenergic receptor: a protein with multiple membrane spanning domains and a chromosomal location shared with the PDGF receptor gene. Proc Natl Acad Sci U S A 84:46–50
57. Largo RH (1991) Wachstum und somatische Entwicklung. In: Betke K, Küster W, Schaub J (eds) Keller-Wiskott: Lehrbuch der Kinderheilkunde, 6th edn. Thieme, Stuttgart, pp 7–32
58. Latham MC (1990) Protein-energy malnutrition. In: Brown ML (ed) Present knowledge in nutrition, 6th edn. International Life Sciences Institute, Nutrition Foundation, Washington, pp 39–46
59. Lehninger A, Nelson DL, Cox MM. (1993) Principles of biochemistry, 2nd edn. Worth, New York
60. Lieber CS (1994) Alcohol and the liver: 1994 update. Gastroenterology 106:1085–1105
61. Lissitzky S (1990) Thyroid hormones. In: Baulieu EE, Kelly PA (eds) Hormones. From molecules to disease. Hermann, Paris, pp 341–374
62. Lok S, Kuijper JL, Jelinek LJ, Kramer JM, Whitmore TE, Sprecher CA, Mathewes S, Grant FJ, Biggs SH, Rosenberg GB, Sheppard PO, O'Hara PJ, Foster DC, Kindsvogel W (1994) The human glucagon receptor encoding gene: structure, cDNA sequence and chromosomal localization. Gene 140:203–209
63. Lomasney JW, Cotecchia S, Lorenz W, Leung WY, Schwinn DA, Yang-Feng TL, Brownstein M, Lefkowitz RJ, Caron MG (1991) Molecular cloning and expression of the cDNA for the α1A-adrenergic receptor. J Biol Chem 266:6365–6369
64. Low MG, Saltiel AR (1988) Structural and functional roles of glycosyl-phosphatidyl inositol in membranes. Science 239:268–275
65. Lüthold BE, Bühler FR, DaPrada M (1976) Dynamik von Plasmakatecholaminen und β-Adrenoceptor-Funktionen. Schweiz Med Wochenschr 106:1735–1738
66. MacNeil DJ, Occi JL, Hey PJ, Strader CD, Graziano MP (1994) Cloning and expression of a human glucagon receptor. Biochem Biophys Res Commun 198:328–334
67. Marx J (1994) Obesity gene discovery may help solve weight problems. Science 266:1477–1478
68. Mason JB, Rosenberg IH (1991) Protein-energy malnutrition. In: Wilson JB, Braunwald E, Isselbacher K, Petersdorf RG, Martin JB, Fauci AS, Root RK (eds) Harrison's principles of internal medicine, vol 1, 12th edn, McGraw-Hill, New York, pp 406–411
69. Milgrom E (1990) Steroid hormones. In: Baulieu EE, Kelly PA (eds) Hormones. From molecules to disease. Hermann, Paris, pp 385–437
70. Mueckler M (1994) Facilitative glucose transporters. Eur J Biochem 219:713–725
71. Müller WA, Faloona GR, Aguilar-Parada E, Unger RH (1970) Abnormal alpha-cell function in diabetes. Response to carbohydrate and protein ingestion. New Engl J Med 283:109–115
72. Myers MG Jr, Sun XJ, White MF (1994) The IRS-1 signalling system. Trends Biochem Sci 19:289–293
73. Niijima A (1989) Nervous regulation of metabolism. Progr Neurobiol 33:135–147
74. Nishizuka Y (1992) Intracellular signalling by hydrolysis of phospholipids and activation of protein kinase C. Science 258:607–614
75. Noack R (1992) Adipositas und Energieverwertung. Ernährungsumschau 39:195–199
76. Noack R, Barth CA (1993) Kohlenhydrate unbegrenzt? Ernährungsumschau 40:440–444
77. Olefsky JM (1991) Obesity. In: Wilson JB, Braunwald E, Isselbacher K, Petersdorf RG, Martin JB, Fauci AS, Root RK

(eds) Harrison's principles of internal medicine vol 1, 12th edn. McGraw-Hill, New York, pp 411–417

78. Oppenheimer JH, Schwartz, HT, Strait KA (1994) Thyroid hormone action 1994: the plot thickens. Eur J Endocrinol 130:15–24

79. Ostrowski J, Kjelsberg MA, Caron, MG, Lefkowitz RJ (1992) Mutagenesis of the β_2-adrenergic receptor: how structure elucidates function. Annu Rev Pharmacol Toxicol 32:167–183

80. Owen OE. Patel MS, Block BSB, Kreulen TH, Reichle FA, Mozzoli MA (1976) Gluconeogenesis in normal, cirrhotic and diabetic humans. In: Hanson RW, Mehlman MA (eds) Gluconeogenesis. Its regulation in mammalian species Wiley, New York, pp 533–558

81. Palgiasotti MJ, Cherrington AD (1992) Regulation of net hepatic glucose uptake in vivo. Annu Rev Physiol 54:847–860

82. Passmore R, Eastwood MA (1986) Davidson-Passmore, Human nutrition and dietetics, 8th edn. Churchill Livingstone, Edinburgh, pp 14–28, 40–53, 269–278

83. Pessin JE, Bell GI (1992) Mammalian facilitative glucose transporter family: structure and molecular regulation. Annu Rev Physiol 54:911–930

84. Pilkis SJ, Granner DK (1992) Molecular physiology of the regulation of hepatic gluconeogenesis and glycolysis. Annu Rev Physiol 54:885–909

85. Püschel GP, Jungermann K (1994) Integration of function in the hepatic acinus: Intercellular communication in neural and humoral control of liver metabolism. Progr Liver Dis 12:19–46

86. Rink TJ (1994) In search of a satiety factor. Nature 372:406–407

87. Rosenberg IH (1991) Nutrition and nutritional requirements. In: Wilson JB, Braunwald E, Isselbacher K, Petersdorf RG, Martin JB, Fauci AS, Root RK (eds). Harrison's principles of internal medicine vol 1, 12th edn, McGraw-Hill, New York, pp 403–406.

88. Rowell LB (1971) The liver as an energy source in man during exercise. Adv Exp Med Biol 11:127–141

89. Ruderman NB, Aoki TT, Cahill GF (1976) Gluconeogenesis and its disorders in man. In: Hanson RW, Mehlman MA (eds) Gluconeogenesis. Its regulation in mammalian species Wiley, New York, pp 515–532

90. Runge D, Schmidt H, Christ B, Jungermann K (1991) Mechanican of the permissive action of dexamethasone on the glucagon-dependent activation of the phosphoenolpyruvate carboxykinase gene in cultured rat hepatocytes. Eur J Biochem 198:641–649

91. Saltin B, Karlsson J (1971) Muscle glycogen utilization during work of different intensities. Adv Exp Med Biol 11:289–299

92. Schimke TR (1973) Control of enzyme levels in mammalian tissues. Adv Enzymol 37:135–188

93. Schlierf G, Raetzer H (1972) Diurnal pattern of blood sugar, plasma insulin, free fatty acid and triglyceride levels in normal subjects and in patients with type IV hyperlipoproteinemia and the effect of meal frequency. Nutr Metab 14:113–126

94. Schüle R, Muller M, Kaltschmidt C, Renkawitz R (1988) Many transcription factors interact synergistically with steroid receptors. Science 242:1418–1420

95. Schwinn DA, Lomasney JW, Lorenz W, Szklut PJ, Frameau RT, Yang-Feng TL, Caron MG, Lefkowitz RJ, Cotecchia S (1990) Molecular cloning and expression of the cDNA for a novel α_1-adrenergic receptor subtype. J Biol Chem 265:8183–8189

96. Shetty PS, James WPT (1994) Body mass index. A measure of chronic energy deficiency in adults. Food and Agricultural Organization (FAO) Food and Nutrition Paper 56, Rome

97. Shimazu T (1987) Neuronal regulation of hepatic glucose metabolism in mammals. Diabetes Metab Rev 3:185–206

98. Sims EAH (1976) Experimental obesity, diet induced thermogenesis and their clinical implications. Clin Endocrinol Metab 5:377–395

99. Strosberg AD (1991) Structure/function relationship of proteins belonging to the family of receptors coupled to GTP-binding proteins. Eur J Biochem 196:1–10

100. Stryer L (1988) Biochemistry, 3rd edn. Freeman, New York

101. Thauer RK, Jungermann K, Decker K (1977) Energy conservation in chemotrophic anaerobic bacteria. Bacteriol Rev 41:100–180

102. Ullrich A, Bell JR, Chen EY, Herrera R, Petruzelli M, Dull TJ, Gray A, Coussens L, Liao YC, Tsubokawa M, Mason A, Seeburg PH, Grunfeld C, Rosen OM, Ramachandran J (1985) Human insulin receptor and its relationship to the tyrosine kinase family of oncogenes. Nature 313:756–761

103. Unger RH (1971) Pancreatic glucagon in health and disease. Adv Intern Med 17:265–288

104. Voigt MM, Kispert J, Chin H (1990) Sequence of a rat brain cDNA encoding an α-1B adrenergic receptor. Nucleic Acids Res 18:1053

105. von Euler US (1967) Adrenal medullary secretion and is neural control. In: Martini C, Ganong WF (eds) Neuroendocrinology, vol 2. Academic, New York, p 283

106. Wakelam MJO, Murphy GJ, Hruby VJ, Houslay MD (1986) Activation of two signal-transduction systems in hepatocytes by glucagon. Nature 323:68–71

107. Weiss H, Friederich T, Hofhaus G, Preis D (1991) The respiratory chain NADH dehydrogenase (complex 1) of mitochondria. Eur J Biochem 197:563–576

108. WHO, FAO, UNU (1985) Energy and protein requirements. WHO Technical report series 724. WHO, Geneva

109. Windmüller HG, Späth AE (1974) Uptake and metabolism of plasma glutamine by the small intestine. J Biol Chem 249:5070–5079

110. Yamauchi K, Tata JR (1994) Purification and characterisation of a cytosolic thyroid-hormone-binding protein (CTBP) in Xenopus liver. Eur J Biochem 225:1105–112

111. Young VR, Pellett PL (1988) How to evaluate dietary protein. In: Barth CA, Schlimme E(eds) Milk proteins. Steinkopf, Darmstadt, pp 7–36

112. Young VR, Steffel WP, Pencharz PB, Winterer JC, Sesimshaw NS (1975) Total body protein turnover in relation to protein requirements at various ages. Nature 253:192–193

113. Zhang Y, Proenca R, Maffei M, Barone M, Leopold L, Friedman JM (1994) Positional cloning of the mouse obese gene and its human homologue. Nature 372:425–432

72 Internal Control of Food Intake: Hunger and Satiety

H.S. KOOPMANS

Contents

72.1 Introduction

Feeding behavior is a complex task that is quite essential for the survival of animals. It has been studied since the beginning of physiology with experiments by both Magendie and Bernard, and many internal control mechanims have been elucidated [46]. There are both short-term inputs related to the initiation and termination of a meal and long-term inputs related to the control of daily intake and the regulation of body weight. Moreover, these two systems must interact in some way.

In most animals, feeding is done in bouts. They eat for a short period of time, usually during either the day or the night. Free-feeding rats are nocturnal and feed 12–15 times during the night with the bursts of feeding occuring just after the onset of dark and just before the light is turned on [21]. They will eat only one to two meals during the day. With humans, feeding times are more culturally determined. People eat three to four times a day at meals that are usually defined by working requirements. Thus, feeding schedules are fairly flexible. If a meal is missed, then the individual will eat a larger subsequent meal or at least a greater amount of food during the rest of the 24-h cycle. The day-to-day regulation of food intake does not have to be particularly accurate, because the body has substantial fuel on board that can be used to provide energy when food is not currently being digested and absorbed (see Chap. 71). Adult rats can survive 8–10 days without food, and adult humans can survive for 50–60 days without taking food into the body, provided that water and salts are available.

72.2 Meals and Feeding Behavior

Smell and Taste. Because feeding occurs in defined meals, most studies have looked at external influences and internal events at the times surrounding a meal. Since meals occur occasionally, it is much easier to look at the events that immediately follow a meal. Most studies are done with a period of deprivation, so that feeding will reliably occur at a specific time. However, it is important to realize that regulation can occur by delaying the time of next feeding as well as by reducing the amount of food eaten during a meal. Naturally most of the attention has been focused on internal variables that change just after the beginning of a meal. The food eaten by an animal needs initially to be sought and found. Once ingestion begins, the immediate sensations of smell and taste (see Chap. 42) inform the animal of the character of the meal. If the animal has previous experience with the food, learned responses may influence the total amount of food eaten. If not, then the animal will most likely be cautious and must learn from the internal sensations about the characteristics and the satisfying nature of the food eaten. Once the food is chewed and mixed with saliva containing both an amylase and a lipase (see Chap. 64), it is swallowed and passes through the esophagus to the stomach (see Chap. 60).

Internal Stimuli. The stimulus properties of food are motivating to the animal and readily associated with subsequent contact with the same or a similar food, but they do not provide the internal stimulus that inhibits a meal. Some animals have had surgery to produce an esophageal fistula that prevents food eaten from reaching the stomach. The food drops out of an opening in the neck. These animals eat very large initial meals showing that they are expecting a signal arising in the stomach (see Chap. 61) or small intestine (see Chap. 62) to terminate a meal [17]. Alternately, an opening can be made in the gastric wall to create a fistula which drains from the stomach and can be either open or closed [50]. When the fistula is open, rats eat excessive amounts of food compared to feeding with the fistula closed (see Fig. 72.1). Thus, the sensations of taste and smell do not control the size of a meal in a physiological way. The signal that is responsible for the termination of a meal must arise at the level of the stomach or beyond.

R. Greger/U. Windhorst (Eds.)
Comprehensive Human Physiology, Vol. 2
© Springer-Verlag Berlin Heidelberg 1996

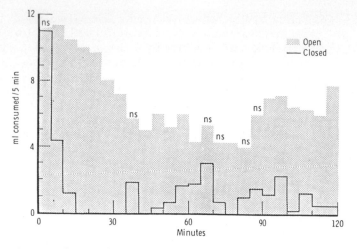

Fig. 72.1. The 5-min food intakes of rats deprived of food for 17 h that have a gastric fistula closed (*solid lines*) and open (*shaded area*) over a 2-h period. Note that rats in which the food drains out of the stomach eat nearly continuously for 2 h. These studies show that there are signals arising in the stomach or intestine that inhibit intake during a meal. (From [50])

72.2.1 Stomach

Short-term Satiety Signal. Digestion begins in the mouth and continues in the stomach (see Chaps. 60, 61, 64). Most of the food eaten by an animal in a meal is retained in the stomach. If the animal has fasted beforehand, then a small amount of food is rapidly released into the small intestine and spread throughout the length of the duodenum, jejunum, and upper ileum. This food activates mechanisms that inhibit stomach emptying and cause the bulk of the food to remain in the stomach [32]. Thus, the most likely short-term satiety signal is distension of the stomach. Indeed, a number of different studies have shown that gastric distension is an important stimulus for the termination of a meal. If an animal eats liquid food and then some of that food is withdrawn from the stomach through a tube, the animal returns immediately to the food source and replaces the amount of food withdrawn [4]. Of course, a change in the amount of food in the stomach might also affect the rate of gastric release of food into the intestine. Another study employed a pyloric clamp that prevented release of food from the stomach and that could be controlled from outside the animal. If the clamp was placed on the pylorus either before or immediately after the first part of a meal, the animal ate the same amount of food as it previously had without the clamp [5,6]. This study showed gastric distension was an important source of signals for the immediate termination of a meal. Since the release of the clamp had no measurable effect on meal size, the study also implied that the intestine was not a major source of signals at least for the end of a single, well-defined meal.

72.2.2 Intestine and Gut Hormones

Gastric and Intestinal Filling. Food leaves the stomach slowly and enters the intestine, where the processes of digestion and absorption continue. The release of food depends upon the pressure difference between the stomach and the intestine and is not affected by the pyloric sphincter. If a ring is placed in the sphincter holding it open, stomach emptying continues at the same rate. However, stomach emptying is affected by the presence of food in the small intestine. Sugars, amino acids, and fats all inhibit stomach emptying. Although specific mechanisms must be involved for each of these nutrients, studies have shown that the stomach releases food at an average rate dependent on the number of calories that move into the intestine [18,31]. The precise mechanism for the inhibition of stomach emptying is not well understood, but it clearly involves both nerves and hormones and may include nutrients present in the bloodstream. The role of the intestine in the termination of a single meal is disputed. One study in which food moved from the intestine of one rat in a parabiotic pair into the intestine of its partner showed that the physiological delivery of food into a limited section of the small intestine did not inhibit food intake during a single meal [23]. This result was consistent with the previously mentioned study using pyloric clamps, which showed that entry of food into the intestine was not important in the inhibition of a single meal. Other studies have infused food at a rapid rate and in a nonphysiological form directly into the intestine and found a reduction of food intake during a normal meal and during sham feeding in rats with a gastric fistula [28,52]. Although the reduction of feeding behavior in these studies obviously occurred, the reduction may have taken place because of discomfort or malaise. Thus, the role of the intestine in the short-term control of food intake remains unclear. It is probable that if excessive food is eaten, then intestinal distension acts along with gastric distension to limit the size of the meal, but this mechanism may not normally be used in the control of short-term food intake. Intestinal signals may act to control gastric distension and the rate of gastric emptying

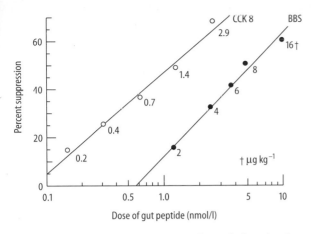

Fig. 72.2. The percentage suppression of a meal after of various doses of the gut hormone cholecystokinin (*CCK*) and the neurotransmitter bombesin (*BBS*) compared to rats injected with saline. (Reprinted from [11], with permission of Elsevier Science, Ltd.)

and thus have an effect on the timing and the size of the next meal [32].

Neural Signals. Once food arrives in the intestine, it is digested by enzymes released from the pancreas and the intestinal mucosa (see Chaps. 62, 65). The food, which mainly consists of plant and animal material, is broken down to smaller molecules that can be transferred into the absorptive cells of the small intestine. Once the food has been absorbed and has diffused into the lamina propria, the support structure of the intestinal mucosa, it can be sensed by nerves terminating in this region. These nerves can not end in the intestinal lumen because of the corrosive nature of the luminal contents and because of the constant upward movement of mucosal cells along the basement membranes of the intestinal crypts and villi. Several studies have shown that glucose and amino acids infused into the intestinal lumen can be detected by afferent fibers within the vagus nerve [34,48]. Thus, neural signals arise in the wall of the small intestine and might inhibit food intake or stomach emptying.

Hormonal Signals. Absorbed food will also come into contact with endocrine cells present in the mucosa of the small intestine. They cause the release of gut hormones, such as secretin, cholecystokinin (CCK), glucose-dependent insulin-releasing polypeptide (GIP), and motilin from the duodenum and jejunum, and enteroglucagon, polypeptide YY, and neurotensin from the ileum. Of the gut hormones, CCK is the only hormone that has been shown to inhibit food intake (see Fig. 72.2) [13,49]. It is released after a meal and stimulates the secretion of bile and pancreatic juice. When injected into the peritoneal cavity, the neurotransmitters, gastrin-releasing polypeptide (GRP) and bombesin (BBS), also produce a dose-dependent inhibition of food intake [12] and release a large number of other hormones including gastrin, insulin, and glucagon [10]. The pancreatic hormone glucagon (see Chap. 66) has also been shown to inhibit short-term food

intake, and antibodies to glucagon produce elevations of intake [27]. These three hormones inhibit food intake, but they have often been injected at pharmocological doses and in nonphysiological ways. Recent evidence suggests that these hormones have a local effect on nervous tissues close to the site of their release and do not act by traveling through the bloodstream to the brain to reduce food intake. CCK requires the presence of an intact vagus nerve to reduce intake [51], BBS may act directly on the wall of the stomach [22], and glucagon is most effective when infused through the portal vein, suggesting a local effect on the liver which is blocked by hepatic vagotomy [9].

Cholecystokinin. The best evidence for a genuine satiety effect exists for CCK: sensitive antagonists to the CCK A receptor have recently been shown to produce increases in food intake [15,45]. CCK may act by stimulating receptors on nerves that travel through the vagus, since vagotomy blocks its effect on intake. On the other hand, CCK A receptor blockers may have had a direct effect on CCK neurons in the brain to increase food intake and may not depend upon the peripheral actions of CCK [44]. Several of these hormones, especially CCK, GIP, and enteroglucagon, might influence daily food intake by inhibiting stomach emptying [14,53].

72.2.3 Absorbed Nutrients

Glucose and Amino Acids. After the absorbed nutrients have entered the extracellular space of the lamina propria, they take different internal routes to enter the bloodstream, depending upon their size and water solubility. Free glucose and amino acids diffuse across the lamina propria and enter the capillaries that empty into the portal vein going to the liver (see Chaps. 67, 68). Thus, the liver is exposed to a higher concentration of these nutrients than other organs and is a possible site for the monitoring of the total amount of water-soluble nutrients absorbed through the gut. Together with its central role in the control of metabolism, the liver's anatomical position has given it theoretical importance in beliefs about the control of food intake. Several theorists claim that changes in hepatic metabolism control food intake [8,40,47]. Drugs that block the metabolism of glucose or fat cause an increase in food intake, but the animal's response may simply be an escape from an inadequate amount of hepatic fuel and thus may not be a normal regulatory signal at all. Some studies have infused nutrients into the portal vein and found a reduction of short-term food intake, but the rate of infusion is high and the response is not dose dependent [54]. It is known that there are vagal nerves that respond to the presence of glucose and amino acids infused into the portal vein [38,39]. Recent studies infusing glucose and amino acids into the portal vein at a slow, continuous rate did not produce a reduction of daily food intake that was greater than that following the infusion of the same number of calories into the vena cava [58]. The importance of the liver for the control of food intake is also contradicted by studies using a portal caval shunt [24]. In these studies the venous blood from the gut is diverted away from the liver and into the systemic circulation. The liver is supplied only by arterial blood, reduces considerably in size, and alters its metabolism, but there is no change in the pattern of food intake or in the daily amount of food eaten. These studies suggest that the liver is not a major site for the control of daily food intake.

Long-Chain Fat. The other major nutrient, the long-chain fat, is absorbed through the gut and released into the lamina propria in the form of chylomicrons, which are too large to enter the capillaries. Instead, these lipoproteins travel through the lymph ducts and are released into the bloodstream near the heart, where they are diluted in the systemic blood (see Chaps. 62, 86, 94). Therefore, absorbed fats are available to all tissues and are not concentrated in the blood vessels leading to the liver. The fat in chylomicrons is transferred to the tissue through the activity of the enzyme lipoprotein lipase (LPL), located on the wall of the capillary endothelium. The activity of this enzyme is influenced by the presence of insulin, which affects different tissues in different ways. Insulin increases the expression of LPL in adipose tissue after a meal and reduces its activity in muscle [7]. Because of the breakdown and resynthesis of fat during absorption and because of its lengthy route through the lymphatics, absorbed fat reaches the bloodstream 35–40 min after a meal in rats and 90 min after a meal in humans. However, fat is a satiating food [29], which suggests that endogenous signals from the gut play a role in the control of short-term food intake. Fat is known to inhibit stomach emptying and might inhibit a meal by accentuating gastric distension.

72.3 Short-Term and Long-Term Food Intake

Glucose, Amino Acids, and Fats. All of the studies cited above suggest that the stomach and possibly the rest of the gut may play an important role in the short-term inhibition of food intake. It is possible to test the relative importance of the gut in the control of daily intake by infusing nutrients directly into the bloodstream, thus bypassing the gut. The intravenous infusion of nutrients tests the role of plasma concentrations of these nutrients, tissue uptake and metabolism, and tissue storage as possible signals for the control of both short-term and daily intake. Several studies have been done and the results are quite consistent [33,37,55,59]. Infused nutrients, with the possible exception of glucose, have little effect on short-term food intake [56]. They do, however, have a large effect on daily food intake: they cause a reduction of caloric intake that compensates for at least half of the calories infused. Thus, without signals generated in the gut, these infused nutrients cause a substantial inhibition in food intake. Indeed, each of the three macronutrients generates a somewhat differ-

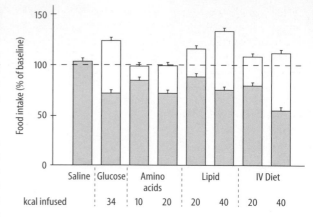

Fig. 72.3. The percentage of daily food intake eaten during the slow, continuous infusion of saline, 34 Kcal glucose, 10 and 20 Kcal of balanced amino acids, 20 and 40 Kcal of intravenous (*IV*) lipid and 20 and 40 Kcal of IV diet whose composition in macronutrients was the same as the diet eaten by the rats. The *hatched bars* show the voluntary food intake and the *open bars* the amount of calories of each nutrient infused. The total of the hatched and open bar is the number of calories taken in for each type of nutrient infusion. (From [56])

ent response (see Fig. 72.3). Infused glucose causes an immediate reduction of daily intake that compensates for 50%–60% of the calories infused, and food intake returns to normal levels on the first day after the infusion is terminated. In contrast, amino acids provoke a complete compensation for the calories infused and also a return to normal intake after the infusion. Infused fats exhibit a different pattern on the inhibition of daily intake. They have relatively little effect on the first day of infusion, gradually increase over the next 2–3 days, and then stabilize on the fifth and sixth day. The average reduction of intake was about 40%–50% of the calories infused, which was not significantly different from glucose, but the pattern of the reduction of intake was distinctly different. When the infusion stopped, the rat's increase in food intake was also slow, returning to previous baseline levels over several days. These studies show that each of the macronutrients inhibits food intake in a different way and probably by a different internal mechanism.

72.4 Location of the Sensor

Hypothalamus. Since nutrients infused intravenously are effective in reducing daily food intake, the question arises as to where these nutrients are sensed. Are they sensed directly by the brain or is there a peripheral tissue, such as the gut, liver, muscle, or adipose tissue, that takes up the nutrient, metabolizes or stores it in some way, and then sends a message to the brain to reduce intake? The natural place to start an investigation is the brain. Neurons responsive to both glucose and amino acids have been discovered in the hypothalamus [41]. Infusions of either glucose or amino acids directly into the carotid artery leading to the brain had no greater effect in reducing food intake than did infusion into the major vein leading to the heart [57]. Since only about 2% of cardiac output flows to the brain, the slow, continuous infusion into the carotid artery would have led to an elevation of 30–40 mg/dl in the plasma concentrations of either glucose or amino acids passing through the capillaries of the brain. Despite this large change, there was no differential effect on food intake.

Liver. Of the potential peripheral sensing sites, only the liver has been tested. As mentioned above, slow, continuous infusion of glucose or amino acids into the portal vein, which carries about 20% of cardiac output (see Chap. 69), also had no differential effect on daily food intake [58]. It appears that the infused nutrients must be sensed in the gut, muscle, or adipose tissue. Changes in any or all of these tissues must generate a message that is sent to the brain to inhibit daily food intake. These signals have not yet been identified.

72.5 Theories of the Control of Food Intake

Theories of the control of food intake have identified each of the three macronutrients. The water-soluble nutrients have generated theories that basically concern short-term food intake, while the fat-soluble nutrients produce theories about daily food intake and body weight regulation.

Glucostatic Theory. This theory argues that the uptake of glucose by the body's tissues produces an arteriovenous difference in blood glucose concentration which produces satiety [30]. When the glucose is no longer taken up by tissue, which usually requires the presence of insulin, hunger returns. The evidence concerning the role of infused glucose in the control of a single meal is contradictory [60].

Aminostatic Theory. This theory claims that the plasma concentration of amino acids controls food intake [35]. In fact, the physiological absorption of both glucose and amino acids from a limited section of the small intestine had no effect on short-term food intake [23].

Lipostatic Theory. The third major theory, the lipostatic theory, is largely a theory about the long-term control of food intake and does not have a specific metabolite associated with it [20]. The theory argues that the amount of body fat controls daily food intake. There is considerable evidence that body weight or the amount of body fat is regulated is some way. Animals that have been force-fed to obesity reduce their voluntary food intake and return to their previous level of body weight. If the animals are starved or their food intake is restricted, they will subsequently increase their intake and again return to their pre-

vious body weight. Thus, perturbations of body weight in either direction lead to changes in voluntary food intake that bring the body weight back to normal levels.

72.5.1 Lesions in the Hypothalamus

Paraventricular Nucleus and Ventromedial Hypothalamus. These observations are supported by the fact that it is possible to change the body weight and the amount of body fat by lesions placed in specific areas of the hypothalamus. Lesions in the paraventricular nucleus or the ventromedial hypothalamus (VMH) provoke large increases in food intake which continue until the animal has reached a new, higher level of body weight [1,3,42]. Overfeeding and starvation studies in rats with VMH lesions show that these animals regulate their intake to sustain the new, higher body weight (see Fig. 72.4). These animals are also more responsive to adulterants in the diet and to work requirements for food: they lose more weight than normal rats when the diet tastes bad and when work requirements to obtain food are increased.

Lateral Hypothalamus. In contrast, lesions placed in the lateral hypothalamus lead to reductions of food intake and loss of body weight [2,19,43]. This loss is not due to difficulty in feeding, since previous weight loss prevents the hypophagia (see Fig. 72.5). Again, the animals regulate around a new, lower body weight. These studies suggest that there are dual centers within the brain that control daily food intake. The lesion studies can be tested by local anesthetics and nerve stimulation, which produce opposite behaviors in these two brain regions. All of these studies suggest that there is an internal mechanism in the hypothalamus that is responsible for the control of body weight which affects both daily food intake and energy expenditure. The mechanism is unknown at present. The possibility that either glycerol or fatty acids, the metabolic breakdown products of triglyceride, control food intake or body weight is not conclusive. Recent work identifying the gene that is responsible for obesity in rodents suggests that there may be a hormone released by adipose tissue that alters food intake and body weight [61].

72.5.2 Natural Changes in Body Weight

Annual Fluctuations. Another indication that there is a physiological mechanism for the control of body weight are the annual fluctuations that spontaneously occur in some animals [36]. Many animals increase their body weight in the autumn and then either hibernate or restrict their feeding activity over the winter when food is scarce. Other animals eat and become relatively obese before a major migration, before the mating season or before incubation of eggs. Even if food is freely available during these periods, the animals tend not to eat. These natural experi-

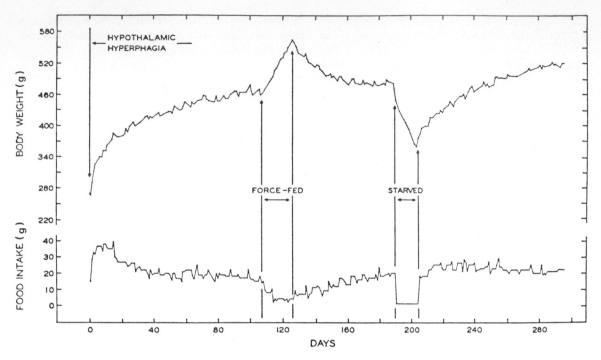

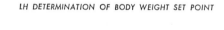

Fig. 72.4. The increase in food intake and body weight after lesions of the ventromedial hypothalamus in rats. Note that once the animals have reached a new level of body weight, they maintain the new weight by eating less after having been force-fed to obesity and by eating more and becoming more efficient after having loss weight due to starvation. (From [16]. Copyright © 1996 by American Psychological Association. Reprinted with permission.)

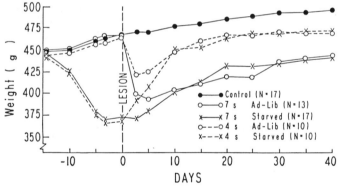

Fig. 72.5. The change in body weight after two small lesions created by passing electrical current for either 4 or 7 s on day 0 to the lateral hypothalamus of rats. Note that two of the five groups of rats were starved before receiving similar lesions and that after the lesions they gained weight to reach the same level of body weight as rats that had not been starved but had similar lesions. These data suggest that the lesions do not destroy pathways involved in the control of feeding behavior, but instead lead to regulation of body weight around new lesion-defined levels. (From [19])

ments suggest that there is an internal physiological system that controls energy balance and body weight.

72.6 Internal Control of Food Intake

A direct experimental study has also shown that food intake is internally controlled. This study was done with inbred parabiotic rats that were sewn together to produce a common peritoneal cavity. A 30-cm segment of the upper small intestine of one rat in the pair was isolated from its own gut and connected to the transected duodenum of its partner [25,26]. As a result of this surgery, food eaten by one rat of the pair was continually lost into the intestine of its partner and absorbed into the partner's bloodstream (see Fig. 72.6). The food that was not absorbed in the crossed segment returned to the lower duodenum of the rat that fed. Food eaten by the partner remained in its own, shortened digestive tract. No major nerves or blood vessels

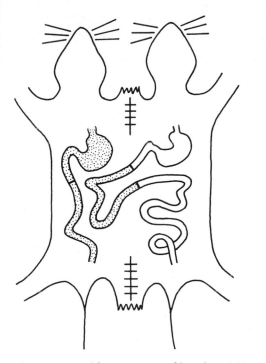

Fig. 72.6. Rats with one-way crossed intestines. A 30-cm segment of one rat's gut is disconnected from its own digestive tract and sewn into the transected duodenum of its partner. As a result, one rat continues to lose food or chyme into the intestine and bloodstream of its partner. The *stippled* gut belongs to the rat on the left and no major nerves or blood vessels were cut during surgery. (Reprinted from [25], with permission of Elsevier Science, Ltd.)

were cut. After the surgery, the rat that lost food or chyme into its partner's intestine and bloodstream increased its daily food intake by 50%–60%, while its partner reduced its intake by the same 50%–60% (see Fig. 72.7) [25]. Considered as a whole, the two rats continued to eat the same amount of food and gained weight at the same rate as controls, but one rat in the pair ate three times as much as its partner. These differences in daily food intake were sustained for the rest of the animals' lives. This study shows that daily food intake is internally controlled and appears to depend upon the amount of food absorbed. The signals controlling daily food intake could be either neural or hormonal messages arising in the small intestine or absorbed nutrients and their metabolic consequences. The nutrient infusion studies suggest that both types of signals are important for the control of daily food intake, but the way in which these signals reach the brain and alter feeding behavior remains unknown. One interesting phenomenon in these rats is that the same signals that generate a large and sustained change in daily food intake have no effect on intake during a single meal. If the rat that loses food into its partner's intestine is fed 10 or 30 min before the partner, generating both intestinal and metabolic signals in the partner, the partner's short-term food intake remains the same as when the first rat is not fed. Thus, the same signals that cause large and sustained changes in daily food intake have no effect on intake during a single meal after a

7-h fast. This study shows that short-term and long-term signals for the internal control of feeding behavior are different.

72.7 Conclusions

In conclusion, the physiological mechanisms underlying the control of food intake are not yet fully understood. It is likely that gastric distension is the major signal that inhibits food intake during a single meal. This signal is probably carried through the vagus nerve to the nucleus of the solitary tract and on to the anterior and medial hypothalamus, a site where feeding behavior is inhibited. It is well established that there are internal controls of daily food intake and body weight. Force-feeding and starvation are followed by voluntary changes in food intake and energy expenditure that bring body weight back into the normal range. Lesions in different regions of the hypothalamus produce increases and decreases of food intake and changes in body weight. However, the underlying physiological mechanisms controlling daily food intake and body weight are not fully understood. Daily food intake is controlled by some combination of endogenous gut signals and the metabolic consequences of absorbed food. Intravenously infused nutrients which bypass the gut inhibit daily food intake in a somewhat less than compensatory way. These nutrients may act on the gut to inhibit stomach emptying or on the adipose tissue or muscle to increase stored energy. Any of these organs might send a message back to the brain that alters daily food intake and leads to a change in body weight.

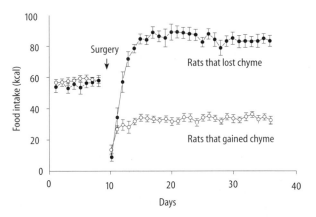

Fig. 72.7. The daily food intake of rats with one-way crossed intestines (as shown in Fig. 72.6) before and after surgery. Note that the rats that lose food or chyme into the intestines of their partners increase their food intake by 50%, while their partners reduce their own daily food intake by the same 50%. As a result of the surgery, one rat eats about three times as much as its partner and this large difference is sustained for the rest of the animals' lives. This study shows that the amount of daily food intake is internally controlled)

References

1. Anand BK, Brobeck JR (1951) Hypothalamic control of food intake in rats and cats. Yale J Biol Med 24:123–146
2. Bernardis LL, Bellinger LL (1993) The lateral hypothalamic area revisited: neuroanatomy, body weight regulation, neuro-endocrinology and metabolism. Neurosci Biobehav Rev 17: 141–193
3. Brobeck JR, Tepperman J, Long CNH (1943) Experimental hypothalamic hyperphagia in the albino rat. Yale J Biol Med 15:831–853
4. Davis JD, Campbell CS (1973) Peripheral control of meal size in the rat: effect of sham feeding on meal size and drinking rate. J Comp Physiol Psych 83:379–387
5. Deutsch JA (1983) Dietary control and the stomach. Prog Neurobiol 20:313–332
6. Deutsch JA, Young WG, Kalogeris TJ (1978) The stomach signals satiety. Science 201:165–167
7. Eckel RH (1989) Lipoprotein lipase. A multifunctional enzyme relevant to common metabolic diseases. N Engl J Med 320:1060–1068
8. Friedman MI, Striker EM (1976) The physiological psychology of hunger: a physiological perspective. Psychol Rev 83:409–431
9. Geary N, Le Sauter J, Noh N (1993) Glucagon acts in the liver to control spontaneous meal size in rats. Am J Physiol 264:R116–R122
10. Ghatei MA, Jung RT, Stevenson JC, Hillyard CJ, Adrian TE, Lee YC, Christofides ND, Sarson DL, Mashiter K, MacIntyre I, Bloom SR (1982) Bombesin: action on gut hormones and calcium in man. J Clin Endocrinol Metab 54:980–985
11. Gibbs J, Kulkosky PJ, Smith GP (1981) Effects of peripheral and central bombesin on feeding behavior of rats. Peptides 2(S2):179–183
12. Gibbs J, Fraser DJ, Rowe EA, Rolls BJ, Maddison SP (1979) Bombesin suppresses feeding in rats. Nature 282:208–210
13. Gibbs J, Young RC, Smith GP (1973) Cholecystokinin decreases food intake in rats. J Comp Physiol Psych 84:488–495
14. Grider JR (1994) Role of cholecystokinin in the regulation of gastrointestinal motility. J Nutr 124:1334s–1339s
15. Hewson G, Leighton RG, Hughes J (1988) The cholecystokinin receptor antagonist L364,718 increases food intake in the rat by attenuation of the action of endogenous cholecystokinin. Br J Pharmacol 93:79–84
16. Hoebel BG, Teitelbaum P (1966) Weight regulation in normal and hypothalamic hyperphagic rats. J Comp Physiol Psych 61:189–193
17. Hull CL, Livingston JR, Rouse RO, Barker AN (1951) Time, sham, and esophageal feeding as reinforcements. J Comp Physiol Psych 44:236–245
18. Hunt JN, Stubbs DF (1975) The volume and energy content of meals as determinants of gastric emptying. J Physiol (Lond) 245:209–255
19. Keesey RE, Boyle PC, Kemnitz JW, Mitchel JS (1976) The role of the lateral hypothalamus in determining the body weight set point. In: Novin D, Wyrwicka W, Bray G (eds) Hunger: basic mechanisms and clinical implications. Raven, New York, pp 243–255
20. Kennedy GC (1953) The role of depot fat in the hypothalamic control of food intake in the rat. Proc R Soc [B] 140:578–592
21. Kersten A, Strubbe JH, Spiteri N (1980) Meal patterning of rats with changes in day length and food availability. Physiol Behav 25:953–958
22. Kirkham TC, Gibbs J, Smith GP (1991) Satiating effect of bombesin is mediated by receptors perfused by the celiac artery. Am J Physiol 261:R614–R618
23. Koopmans HS (1978) The intestinal control of food intake. In: Bray G (ed) Recent advances in obesity research II. Newman, London, pp 33–43
24. Koopmans HS (1984) Hepatic control of food intake. Appetite 5:127–131
25. Koopmans HS (1985) Internal signals cause large changes in food intake in crossed-intestines rats. Brain Res Bull 14:595–603
26. Koopmans HS (1990) Endogenous gut signals and metabolites control daily food intake. Int J Obes 14(S3):93–104
27. Langhans W, Ziegler U, Scharrer E, Geary N (1982) Stimulation of feeding rats by intraperitoneal injection of antibodies to glucagon. Science 218:894–896
28. Liebling DS, Eisner JD, Gibbs J, Smith GP (1979) Intestinal satiety in rats. J Comp Physiol Psych 89:955–965
29. Maggio CA, Koopmans HS (1987) The effect of intragastric infusions of various triglycerides on the food intake and stomach emptying of rats. Am J Physiol 252:R1106–R1113
30. Mayer J (1955) Regulation of energy intake and body weight, the glucostatic theory and the lipostatic hypothesis. Ann N Y Acad Sci 63:15–43
31. McHugh PR, Moran TH (1979) Calories and gastric emptying: a regulatory capacity with implications for feeding. Am J Physiol 236:R254–R260
32. McHugh PR, Moran TH, Wirth JB (1982) Postpyloric regulation of gastric emptying in rhesus monkeys. Am J Physiol 243:R408–R415
33. Meguid MM, Chen TY, Yang ZJ, Campos AC, Hitch DC, Gleason JR (1991) Effects of continuous graded total parenteral nutrition on feeding indexes and metabolic concomitants in rats. Am J Physiol 260:E126–E140
34. Mei N (1985) Intestinal chemosensitivity. Physiol Rev 65:211–237
35. Mellinkoff SM, Frankland M, Boyle D, Greipel M (1956) Relation between serum amino acid concentration and fluctuations in appetite. J Appl Physiol 8:535–538
36. Mrosovsky N, Sherry DF (1980) Animal anorexias. Science 207:837–842
37. Nicolaidis S, Rowland N (1976) Metering of intravenous versus oral nutrients and the regulation of energy balance. Am J Physiol 231:661–668
38. Niijima A (1982) Glucose-sensitive afferent nerve fibers in the hepatic branch of the vagus nerve in the guinea-pig. J Physiol (Lond) 332:315–323
39. Niijima A, Meguid MM (1994) Parenteral nutrients in rat suppresses hepatic vagal afferent signals from portal vein to hypothalamus. Surgery 116:294–301
40. Novin D (1976) Visceral mechanisms in the control of food intake. In: Novin D, Wyrwicka W, Bray G (eds) Hunger: basic mechanisms and clinical implications. Raven, New York, pp 357–367
41. Oomura Y (1976) Significance of glucose, insulin and free fatty acid on the hypothalamic feeding and satiety neurons. In: Novin D, Wyrwicka W, Bray G (eds) Hunger: basic mechanisms and clinical implications. Raven, New York, pp 145–157
42. Parkinson WL, Weingarten HP (1990) Dissociative analysis of ventromedial hypothalamic obesity syndrome. Am J Physiol 259:R829–R835
43. Powley TL, Keesey RE (1970) Relationship of body weight to the lateral hypothalamic feeding syndrome. J Comp Physiol Psych 70:25–36
44. Reidelberger RD (1994) Cholecystokinin and the control of food intake. J Nutr 124:1327S–1333S
45. Reidelberger RD, Varga G, Soloman TE (1991) Effects of selective cholecystokinin antagonists L364718 and L365269 on food intake in rats. Peptides 12:1215–1221
46. Rosenzweig MR (1962) The mechanisms of hunger and thirst. In: Postman L (ed) Psychology in the making. Knopf, New York, pp 73–143
47. Russek M (1963) A hypothesis on the participation of hepatic glucoreceptors in the control of food intake. Nature 200:176
48. Sharma KN, Nasset ES (1962) Electrical activity in the mesenteric nerves after perfusion of gut lumen. Am J Physiol 202:725–730
49. Smith GP, Gibbs J (1979) Postprandial satiety. Prog Psychobiol Physiol Psych 8:179–242

50. Smith GP, Gibbs J, Young RC (1974) Cholecystokinin and intestinal satiety in the rat. Fed Proc 33:1146–1149
51. Smith GP, Jerome C, Norgren R (1985) Afferent axons in abdominal vagus mediate the satiety effect of cholecystokinin in rats. Am J Physiol 249:R638–R641
52. Snowdon CT (1969) Motivation, regulation and the control of meal parameters with oral and intragastric feeding. J Comp Physiol Psych 69:91–100
53. Soper NJ, Chapman NJ, Kelly KA, Brown ML, Phillips SF, Go VL (1990) The "ileal brake" after ileal pouch-anal anastomosis. Gastroenterology 98:111–116
54. Tordoff MG, Tluczek JP, Friedman MI (1989) Effect of portal glucose concentration on food intake and metabolism. Am J Physiol 257:R1474–R1480
55. Walls EK, Koopmans HS (1989) Effect of intravenous nutrient infusions on food intake in rats. Physiol Behav 45:1223–1226
56. Walls EK, Koopmans HS (1992) Differential effects of intravenous glucose, amino acids and lipid on daily food intake. Am J Physiol 262:R225–R234
57. Walls EK, Reinhardt PH, Willing AE, Koopmans HS (1990) Carotid and systemic nutrient infusions reduce food intake. Neurosci Abstr 16:295
58. Willing AE, Koopmans HS (1994) Hepatic portal and vena cava glucose and amino acid infusions decrease daily food intake in rats. Neurosci Abstr 20:1226
59. Woods SC, Stein LJ, McKay LD, Porte D (1984) Suppression of food intake by intravenous nutrients and insulin in the baboon. Am J Physiol 247:R393–R401
60. Van Itallie TB (1990) The glucostatic theory 1953–1988: roots and branches. Int J Obes 14(S3):1–10
61. Zhang Y, Proenca R, Maffei M, Barone M, Leopold L, Friedman JM (1994) Positional cloning of the mouse obese gene and its human analogue. Nature 372:425–432

73 Introduction to Renal Function, Renal Blood Flow and the Formation of Filtrate

R. GREGER

Contents

73.1 Introduction

This chapter summarises the various functions of the kidney. Some functions, such as filtration and the endocrine role, will be discussed in some detail here. Other aspects, such as tubule function, will be dealt with in the next two chapters, and the homeostatic role of the kidney is discussed explicitly in the chapters on water and electrolyte balance (Chaps. 76–82). More comprehensive texts on kidney physiology and pathophysiology have been published recently [4,33,34].

73.1.1 What If the Kidney Fails?

Our two kidneys do their job silently, and we hardly become aware of their busy work unless something goes wrong. In fact, our awareness only concerns the daily voiding of 1–1.5 l of clear, yellowish urine. Closer inspection and analysis of urine tells us that it contains more than water and some chromophore (urochrome). Table 73.1 summarises the composition of normal antidiuretic urine (urine production < 0.3 ml/min). Urine contains large amounts of urea (ca. 30 g/day, 500 mmol/day), variable amounts of Na^+ and Cl^- (ca. 5–15 g/day, 100–300 mmol/day), variable amounts of K^+ (50–300 mmol/l), large amounts of phosphates (ca. 20–60 mmol/day), sizeable quantities of creatinine (ca. 900 mg/day, 7 mmol/day), small quantities of urate/uric acid (ca. 5 mmol/day) and variable amounts of Ca^{2+} and Mg^{2+} (ca. 100–300 mg/day or 3–8 mmol/day, and 60–200 mg/day or 2–9 mmol/day, respectively). From the above it can be seen that the composition of urine differs in many respects from that of plasma. Important differences are:

- The high content of nitrogen-containing compounds such as urea, creatinine, NH_4^+ and uric acid
- The variability of electrolyte content: Na^+, K^+, Ca^{2+}, Mg^+, $H_2PO_4^-$, HPO_4^{2-}
- The acid pH (5–6) under normal acid-base conditions
- The virtual absence of HCO_3^-, D-glucose, amino acids and proteins.

As will be shown below, the kidney essentially works by filtering plasma (glomerular filtration) and by reabsorbing constituents of the tubule fluid from and by secreting constituents of plasma into the various tubule segments (cf. Chaps. 74, 75). From the comparison of urine and plasma it becomes clear intuitively that the kidneys must perform important transport work:

- To avoid losses of metabolically valuable filtrate constituents such as D-glucose, amino acids and to a lesser extent small to medium-size proteins

R. Greger/U. Windhorst (Eds.)
Comprehensive Human Physiology, Vol. 2
© Springer-Verlag Berlin Heidelberg 1996

Table 73.1. Composition of antidiuretic urine (cf. also Chap. 126)

V (1 /day)	Na$^+$	Cl$^-$	K$^+$	Ca^{2+}	Mg^{2+}	HCO$_3^-$	HPO$_4^{2-}$	Urea	Creat.	Urate	NH$_4^+$	Osmolality (mosmol/l)
ca. 1	30–150	30–150	33–300	3–6	ca. 10	ca. 1	3–20	ca. 300	ca. 10	3	ca. 20	ca. 1000

All values are mmol/day unless otherwise specified
The Na$^+$ and Cl$^-$ concentrations depend largely on the NaCl metabolism; the concentration of Ca^{2+} is controlled mostly by PTH; the urine is essentially HCO$_3^-$ free except in metabolic alkalosis; the urea concentration varies with the N metabolism, and together with NH$_4^+$ in metabolic acidosis; the creatinine excretion depends on creatine metabolism; urine osmolality can be up to 1500 mosm/l

- To prevent losses of the buffer HCO$_3^-$
- To ensure effective excretion of nitrogen compounds and toxic metabolites.

In addition, many *xenobiotics* (drugs, ξένος = strange, non-genuine) are excreted as such, or after they have been subjected to hepatic metabolism, by the kidneys (cf. Chap. 75).

Dynamic Range of the Kidneys. Thus far we have only considered kidney function in the normal steady state. When the body is challenged by large volume or salt loads, the kidney can increase water and salt excretion rapidly. One might call this the dynamic range of the kidneys. For instance if, in the normal hydrated state (Chap. 76), we drink a water load of say 1 l within 30 min, the urine flow rate increases abruptly and this extra litre is excreted within 1–2 h. In this way the kidneys prevent the circulation from overfilling, with the possible consequences of threatening heart failure and oedematous states. Similarly, our diet can contain less than 1 and up to 60 g NaCl/day (Chaps. 77, 82), and still we can stay in perfect balance because the kidneys adjust NaCl excretion to uptake [14]. The same holds for K$^+$. We can ingest a few to 20 g/day (Chap. 79) without the danger of upsetting the K$^+$ balance and without any major change in plasma K$^+$ concentration [41]. The kidney also controls mineral metabolism (phosphate, Ca^{2+} and Mg^{2+}, cf. Chap. 80) and adjusts renal excretion to the respective needs of the body. Beyond this, the kidney plays a predominant role in *acid-base regulation*. While the lung can control CO$_2$ homeostasis by expiring normal or increased amounts, the kidney can increase HCO$_3^-$ excretion in metabolic alkalosis and can increase NH$_4^+$ excretion in metabolic acidosis (cf. Chap. 78).

Renal Failure. When the kidneys fail, the loss of these important functions leads to a life-threatening state: renal failure (for reviews consult, for example, [7]). This failure can occur very rapidly, as a result of excessive blood loss, severe burns, mushroom poisoning, intrinsic renal pathology, acute urinary obstruction etc. Alternatively, it can develop slowly (chronic renal failure) as renal disease progresses. In either case intervention is necessary if renal function, usually quantified as *glomerular filtration rate* (GFR; a list of abbreviations used frequently in this and the following two chapters is included at the end of this chapter), is reduced to some 20% or less of its normal value. The signs and symptoms of renal failure

(insufficiency) can be easily deduced from the above key functions:

- Disturbances of electrolyte balance, e.g., hyperkalaemia
- Metabolic acidosis
- Toxaemia due to the accumulation of so-called uraemic toxins
- Disturbances of mineral metabolism
- Increase in plasma urea and creatinine concentration
- Loss of the ability to deal with volume and salt loading
- Renal anaemia.

In such a state the patient has to reduce oral intake of water to avoid overhydration. Salt and mineral intake must be restricted and be matched closely with residual function. The *toxaemia*, also called (unduly) *uraemia*, as well as the disturbed electrolyte metabolism, must be treated by dialysis (or kidney transplantation). In previous pre-dialysis decades toxaemic patients died when they entered terminal renal insufficiency.

As stated above, the word uraemia is a misnomer, inasmuch as it implies that the accumulating *urea* is the cause for toxaemia. This is not the case unless the urea concentration is excessively high. The relevant toxins belong to a heterogeneous group of small to medium-sized molecules, amongst which parathyroid hormone (PTH), phenols and indols and many more have been incriminated as being responsible for various symptoms [40].

Chronic renal failure, besides the accumulation of these toxins, is also characterised by the failure of the *endocrine function* of the kidney. Most important are the reductions in the production of erythropoietin and 1,25-(OH)$_2$-D$_3$ hormone.

In summary, the kidneys play a pivotal role in several homeostatic processes such as nitrogen, water, electrolyte, mineral and acid-base metabolism. The kidneys are endocrine organs and they are responsible for the excretion of many xenobiotics. This multitude of complex tasks requires a large blood, oxygen and metabolic fuel supply to the kidneys. Acute or chronic failure of kidney function results in complex life-threatening disorders.

73.2 Structure and Morphology of the Kidney

Each kidney weighs ca. 150 g. A frontal section through the human kidney (Fig. 73.1) shows a clearly distinct arrange-

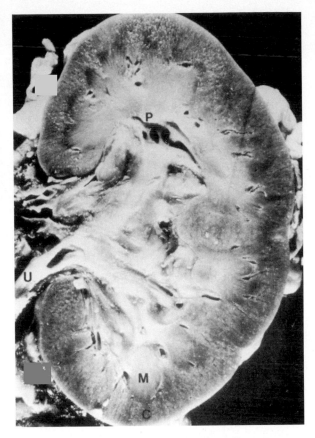

Fig. 73.1. Macroscopic appearance of human kidney. *C*, cortex; *M*, medulla; *P*, papilla; *U*, ureter. (With permission from [16])

(vasa stellata, superficial nephrons), they branch into another capillary network. These capillary plexus surround the tubules (superficial nephrons). In the deep (juxtamedullary) nephrons the efferent arterioles form descending vasa recta. These vessels merge towards the papilla and end in the inner stripe or in the inner medulla. Together with the ascending vasa recta they form capillary beds which are less dense in the outer stripe of the outer medulla and in the inner medulla and are very dense in the inner stripe of the outer medulla.

Portal System. It is important to note that the kidney possesses a portal system with two capillary beds in series. The first supplies the glomeruli, while the second surrounds the tubules and accompanies the long tubule structures of the deep nephrons and of the collecting ducts (cf. also next chapter). The venous effluent is drained by interlobular, arcuate and interlobar veins. The renal vasculature of different species shows some important variations with important functional ramifications [8,23].

73.2.2 The Nephron Consists of Various Tubule Segments of Distinct Appearance and Function

Proximal Tubule. Figure 73.2A [19] shows the various tubule segments of a superficial and a deep nephron. The glomerulus (Bowman space) drains into the neck segment of the proximal tubule (PT). On the basis of morphology (cf. also Chap. 74) and functional data, the proximal tubule is subdivided into three segments (S1-S3, or P1-P3). The first part is convoluted (tortuous). The second portion is less convoluted and the third is almost straight. The second half of S2 and S3 are also called *pars recta*. The pars recta of the superficial tubules is more straight than that of the deep nephrons. All partes rectae end at the border between the outer and the inner stripe of the outer medulla.

Descending and Ascending Limbs. The *thin descending* (DTL or tDL) *limbs* of the loop of Henle are short in the superficial nephrons. They reach down to the border of the outer and inner medulla. The deep nephrons descend further. However, the length is variable. Only a few reach down into the papilla.

In superficial nephrons the *thick ascending limb* begins immediately at the turn of the loop, whereas *thin ascending limbs* (ATL or tAL) begin at the turn of the loop of deep nephrons. The thick limbs (of the loop of Henle) begin in the inner stripe of the outer medulla and extend into the cortex. Hence, a medullary portion of the thick ascending limb (mTAL) is distinguished from the cortical thick ascending limb (cTAL).

Macula Densa. Each nephron returns to its glomerulus in the cortex and makes close contact with it. This is the macula densa portion of the nephron (MD). The MD segment has a very specific function (cf. below and Chap. 74) inasmuch as it is a "chemoreceptor" of this nephron which

ment of cortex and medulla. Human kidney is multilobular [16,19]. The medulla, on the basis of the contrast caused by variable blood filling, can be subdivided into outer and inner medulla. The former, on the basis of the capillary network, which is sparse in the outer zone and dense in the inner zone, can be further subdivided into an outer and an inner stripe. The medulla points to tips which are called papillae. The border between cortex and medulla is formed by the arcuate arteries. Closer inspection of a section of this kind reveals a complex arrangement of renal vasculature and of the smallest functional units of the kidney, namely the nephrons. Each human kidney possesses some 1.2 million of these nephrons.

73.2.1 The Renal Vasculature: A Portal System

General Structure. Figure 73.2B shows the general structure of the vasculature (from Kriz and Kaissling [19]). The renal artery branches into the interlobar arteries, which extend to the border between cortex and medulla. There, the arcuate arteries branch off (and form the border). The arcuate arteries possess perpendicular branches: the interlobular arteries. These branch off to form the afferent arterioles. The afferent arterioles enter the glomeruli and form the glomerular capillaries. The capillaries merge again in efferent arterioles. Then, at the welling points

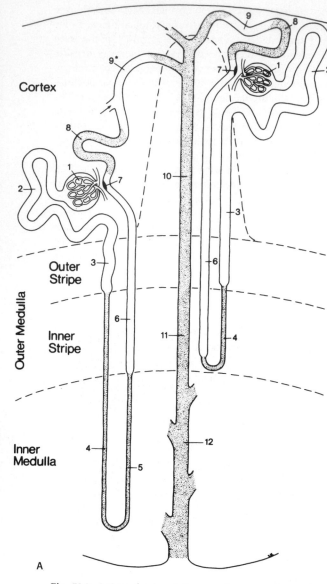

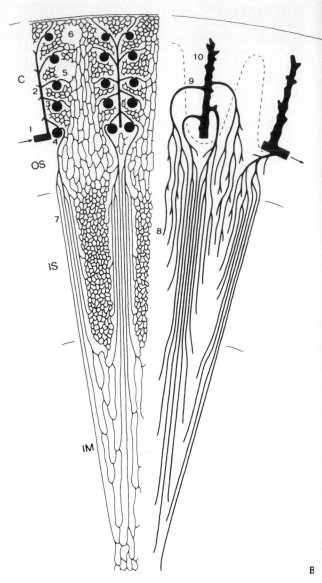

Fig. 73.2. A Superficial and deep nephrons. *1*, glomerulus; *2*, proximal convoluted tubule (S1); *3*, proximal straight tubule (S3); *4*, descending thin limb of the loop of Henle (tDL); *5*, ascending thin limb of the loop of Henle (tAL); *6*, thick ascending limb of the loop of Henle (TAL); *7*, macula densa; *8*, distal convoluted tubule (DT); *9*, connecting tubule; *10*, cortical collecting duct (CCD); *11*, outer medullary collecting duct (OMD); *12*, inner medullary collecting duct (IMD). (With permission from [19]). **B** Renal vasculature. *C*, cortex; *OS*, outer stripe of outer medulla; *IS*, inner stripe of outer medulla; *IM*, inner medulla. *1*, arcuate artery; *2*,

interlobular artery; *3*, afferent arteriole; *4*, glomerulus; *5*, efferent arteriole; *6*, peritubule capillaries; *7*, descending vasa recta; *8*, ascending vasa recta; *7 + 8*, vascular bundles; *9*, interlobular veins; *10*, arcuate veins. Not shown are the large vessels: a. renalis, a. interlobaris, v. interlobaris, v. renalis. Note that the efferent arterioles form peritubular capillaries in cortical superficial nephrons, whereas they form descending vasa recta in the juxtamedullary nephrons. Note also that the capillary network is especially dense in the inner stripe of the outer medulla. (Modified from [19])

monitors luminal Cl⁻ concentration and resets the filtration rate accordingly [32]. The thick ascending limb continues a few tens to hundreds μm past the MD.

Distal Convoluted Tubule. The nephron continues as the initial distal convoluted (DT) tubule. The last part of the distal tubule is the bright portion. The anatomical nomenclature also subsumes TAL and DT as "distal tubule".

Collecting Duct System. The collecting duct system is ontogenetically of different origin (*Wolff's Gang = Urnierengang*). Several nephrons (connecting tubules, CNT) merge into one collecting duct and many collecting ducts merge to form the papillary collecting ducts (ducts of Bellini). The first portion of the collecting duct is the *connecting tubule* (CNT = DCT_g = CCT_g), also called the granular distal tubule. The deep nephrons form arcades of connecting tubules which ascend in the cortex before merging into *cortical collecting tubules* (CCT$_l$), also called light distal tubules. The next portions are the *outer and inner medullary collecting ducts* (OMD, IMD). Finally the collecting ducts merge to form the *papillary collecting ducts* (ducts of Bellini).

This brief overview was given here to introduce the single functional unit of the kidney: the nephron. A closer description of the morphology of the various tubule segments and a correlation between appearance and function will be given in Chap. 74.

73.2.3 The Glomerulus: A Filter with High Permeability and Yet Good Selectivity

Hydraulic Conductivity. The 1.2 million glomeruli of each kidney are highly specialised filters with a hydraulic conductivity (K_f) exceeding by far that of other capillaries. The normal filtration rate of human kidneys is around 120 ml/min; this corresponds to a filtration coefficient which is ten times or more larger than that of all other capillaries in the body [11]. K_f has been found to be around 0.1 nl/s·mmHg in the Munich Wistar rat, which possesses a large number of glomeruli on the kidney surface [11]. Therefore, for a mean filtration pressure of 8 mmHg, the single nephron filtration rate (SNGFR) would be 0.8 nl/s or 50 nl/min. The 30 000 glomeruli of one kidney of this rat would hence produce a GFR of 1.5 ml/min. In man SNGFR is probably similar. The two kidneys, with 2 400 000 glomeruli, produce a GFR of 120 ml/min.

The high value of K_f is due to the anatomical structure of the glomerulus. A schematic drawing of a glomerulus is shown in Fig. 73.3A [16]. Glomerular capillaries are bulging into the spherical space of the glomerulus (*Bowman space*). As a result the barrier between blood and Bowman space, the actual filter, has three layers: the fenestrated *endothelium*, the *basement membrane* and the *podocytes (foot processes)* (Fig. 73.3B).

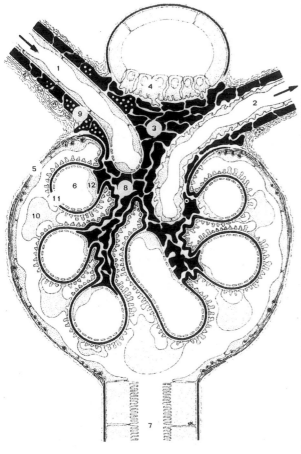

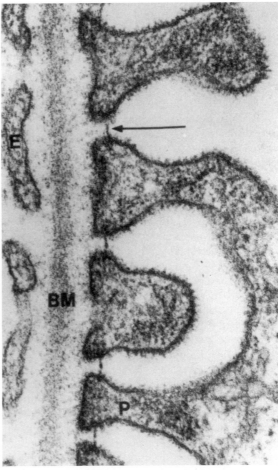

A   B

Fig. 73.3. A Section through a glomerulus and the juxtaglomerular apparatus. *1,* afferent arteriole; *2,* efferent arteriole; *3,* Goormaghtigh cells; *4,* macula densa cells; *5,* Bowman capsule; *6,* glomerular capillaries; *7,* proximal tubule (neck segment of S1); *8,* mesanglial cells; *9,* renin producing cells; *10,* podocytes (foot processes); *11,* basement membrane; *12,* endothelium. (Modified from [16]). **B** Electron micrograph of the filtration barrier in a glomerulus. *E,* endothelium; *BM,* basement membrane; *P,* podocyte. Note the fine slit membranes between foot processes (*arrow*). (From [16]) ×63 600

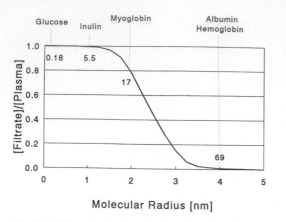

Fig. 73.4. Filtration fraction as a function of molecular size and molecular diameter. The molecular masses are given in kDa

Filter Selectivity. It is apparent that the endothelium with its large pores will only prevent corpuscular elements from permeation. It cannot account for the fact that the effective pore radius of this filter is around 3 nm. It has long been a matter of debate whether the basement membrane or the podocytes represent the effective barrier. Inspection of Fig. 73.3B might at first glance suggest that the basement, with its complex arrangement of connective tissue fibres, would be the size-selective filter. In fact, this has been put forward on the basis of studies with ferritin [13] and may hold true for neutral and also anionic molecules. Closer inspection of Fig. 73.3B reveals that the foot processes of the podocytes are connected to each other by optically dense *slit membranes*. Under normal conditions these slit membranes are believed to contribute to the selectivity of the filter, especially for cationic macromolecules [26].

Filtration Coefficients. Figure 73.4 shows the filtration coefficients of various molecules as a function of their size and molecular weight. The filtration coefficient can be determined directly (micropuncture studies) by measurement of the concentration of the molecule under study in circulating plasma and in the Bowman space, or it can be determined from the fractional excretion of this molecule in comparison to a substance known to undergo complete filtration. Such a substance is the polyfructoside *inulin* (vide infra). One implication in this approach is the assumption that the molecule under study, just like inulin, is neither secreted nor reabsorbed along the nephron and collecting duct system. Inspection of Fig. 73.4 reveals that molecules with a radius of 2 nm or less (corresponding to a molecular weight of approximately 5000 Da) are filtered freely. Larger molecules are partially filtered (between 2 and 4 nm): i.e., their concentration in the Bowman space is only a fraction of that in plasma. Molecules with a radius larger than 4 nm are excluded almost completely from filtration. As a consequence large proteins such as albumin, hemoglobin and globulins are filtered very poorly, whereas myoglobin still is filtered to an appreciable extent.

The fractional filtration of large molecules also depends on the dynamics of filtration (cf. below). If renal plasma flow or filtration pressure increases, GFR also increases. The solute filtration, however, does not increase to the same extent. Therefore, the fractional filtration of large molecules is shifted to the left (Fig. 73.5a) [2]. Conversely, a reduction in solvent (water) filtration has the opposite effect.

Effect of Charge on Filtration. It might be expected that the pores of the slit membranes and the serial arrangement of the three barriers have a certain pattern of charges and the charge of the solute interacts with the charges of the filter. This has been proven to be the case [9]. If, for instance, 3.6-nm-radius dextrans with no charges, negative charges or positive charges are compared, the fractional filtration is 1% for the negatively charged species, 15% for the neutral species and 42% for the positively charged species. This is shown in Fig. 73.5b. The basement membrane appears to possess a high density of negative charges. Therefore, negatively charged molecules such as albumin are hindered much more in their permeation than would be predicted on the basis of their molecular radius. A pathological loss or masking of these negative charges of the slit membrane is responsible for the consequent *proteinuria* [21].

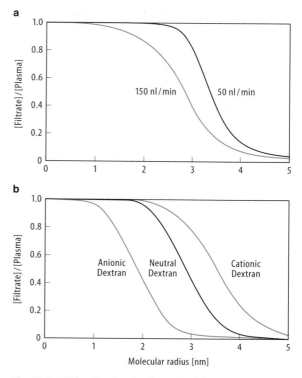

Fig. 73.5. a Filtration fraction for neutral dextrans as a function of glomerular plasma flow. At higher glomerular plasma flow the function is shifted to the left, i.e., filtrability is reduced. b Filtration fraction for anionic, neutral and cationic dextrans. Note that, for example, for a molecular radius of 3 nm the filtration of anionic dextran is only 0.03, that for neutral dextran is 0.4, and that for cationic dextran is as high as 0.73

Creatine   Creatinine

Fig. 73.6. Creatinine is produced from creatine. The turnover is largely dependent on muscle mass and muscular activity

Gibbs-Donnan Formalism. The fact that proteins (due to the size restriction and due to their negative charges) are mostly excluded from glomerular filtration has one consequence, namely the redistribution of small charged molecules (ions) according to the Gibbs-Donnan formalism for large anionic molecules. As a result the equilibrium concentration of positively charged permeable ions is higher on the blood side, and conversely, the concentration of small permeable anions is larger in the Bowman space. Quantitative consideration reveals that these differences are small. For example, the Gibbs-Donnan distribution for Na^+ results in a tubule (Bowman space) fluid to plasma concentration gradient (TF/P_{Na^+}) of >0.95. Conversely, the TF/P_{Cl^-} is <1.05. Because of their very limited quantitative importance, Gibbs-Donnan equilibria will not be considered further in this and the subsequent two chapters.

Binding to Plasma Proteins. Much more relevant is the fact that many small molecules are bound to plasma proteins (triiodinetyrosine, steroids, very many apolar drugs, and ions such as Ca^{2+}). Only the unbound (free) moieties of these molecules will be filtered and the TF/P values can be small, e.g. 0.6 in the case of Ca^{2+} or 0.01 in the case of the diuretic furosemide.

73.3 Glomerular Filtration

Glomerular filtration is a pressure-driven event. Its dynamics will be discussed in Sect. 73.3.3. The magnitude of GFR with normal kidneys is 120 ml/min or 180 l/day. This enormous filtration rate is made possible by the specific properties of the glomerular filter (cf. above). GFR is a good index of renal function. This may be surprising since renal function comprises filtration, tubule transport (reabsorption, secretion) and endocrine processes. However, any marked fall in GFR will produce the related pathophysiological condition (cf. above). Conversely, any tubule damage is paralleled by a fall in GFR. Therefore, the measurement of GFR is of great clinical relevance.

73.3.1 Measurement of Glomerular Filtration Rate

The measurement of GFR is based on simple mass balance. Suitable substances must be:

- Freely filtered at the glomerulus
- Not reabsorbed by the tubule
- Not secreted by the tubule.

These criteria are met, for instance, by *inulin,* a polyfructoside with a molecular mass of around 5000–6000 Da obtained from a plant root, by *iodothalamate* and, with some approximations, by *creatinine*. The latter has the advantage that it need not be administered because it is produced endogenously from creatine (Fig. 73.6). For any of these substances one can write:

filtered amount = amount excreted in urine

$$GFR \cdot P_x = \dot{V} \cdot U_x, \tag{73.1a}$$

where P_x and U_x are the concentrations of the substance "x", and $\dot{V}$ is the urinary flow rate in ml/min. Equation 73.1a can be solved for GFR:

$$GFR = \dot{V} \cdot U_x / P_x. \tag{73.1b}$$

For the endogenously produced creatinine (C) we can assume that, in the steady state, endogenous production must match the amounts excreted and filtered. Hence, for a normal GFR, P_c will increase with increasing production. On the other hand, for any given rate of production, P_c will be inversely related to GFR. This is apparent from Fig. 73.7. This graph contains the GFR : P_c relations for three steady state rates of creatinine production. It is apparent that GFR has to fall markedly before the plasma creatinine concentration increases measurably. This part of the curve is called the silent part. Only if GFR falls to fractions of its normal value, will the effect on P_c be marked. Instead of measuring GFR, which is quantitatively similar to creatinine clearance, plasma creatinine is measured routinely as a simple "kidney function" test (cf. also Chap. 126). This is of limited value unless creatinine production is known. In a given individual monitoring of P_c may, on the other hand, be sufficient.

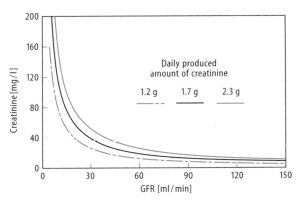

Fig. 73.7. Plasma creatinine concentration as a function of the GFR. The correlation is shown for three daily rates of production. Note that the fall in GFR has little effect on creatinine concentration until 30–50 ml/min is reached ("silent" range of curve). Also note that the plasma creatinine concentration itself is a poor indicator of GFR if the daily rate of production is variable

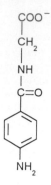

Fig. 73.8. *para*-Aminohippurate (PAH). This substance is secreted by the tubule. At low plasma concentration secretion is complete, i.e., the renal venous blood is completely cleared of this substance. Under these circumstances the clearance of PAH is equal to the renal plasma flow

One limitation of creatinine clearance as a measure of GFR derives from the fact that *creatinine can be secreted* to some extent (for review: [6]). This generates systematic errors, especially in patients with reduced GFR and hence increased P_c. As an alternative, more accurate markers for GFR such as inulin and iodothalamate can be used [6]. Even with these markers clearance measurements are sometimes inaccurate because of errors in the determination of urinary flow rate. It is customary to collect the urine for a longer period to reduce this error or to increase the urinary flow rate, e.g., by water diuresis.

Glomerular filtration rate and renal blood flow (RBF) vary with a circadian rhythm. GFR is highest during the daytime and lowest during the night. The differences are on the order of 20%. Also GFR is acutely increased by meals if they have a high protein content [11]. This protein (amino acid)-induced hyperfiltration may be pathophysiologically relevant.

73.3.2 The Clearance Concept Describes the Clearing Function of the Kidney

In the preceding section GFR was introduced as the clearance of inulin, creatinine or iodothalamate. This is based on the finding that, for these substances, the filtered amount appears in the urine. A creatinine clearance of 120 ml/min therefore indicates that 120 ml of plasma is cleared of creatinine every minute. It was also stated above that in the case of creatinine, the basic assumption may not be entirely correct inasmuch as under certain circumstances creatinine also can be secreted by the tubule. Then, the clearance of creatinine will exceed the GFR because an extra volume of plasma is cleared of creatinine by secretion.

For some substances the secretion can be very relevant. One example is *para-aminohippurate* (PAH, Fig. 73.8). This is an organic acid which is not formed in the body. At low plasma concentrations this substance is secreted so efficiently by the proximal tubule that the venous effluent from the kidney is PAH free. Then we can write, according

to mass balance, that the amount entering the kidney (by renal plasma flow = RPF) is equal to that leaving the kidney via urine:

$$RPF \cdot P_{PAH} = \dot{V} \cdot U_{PAH}. \tag{73.1c}$$

The clearance of PAH is then identical to RPF and amounts to some 600 ml/min. With increasing plasma concentrations of PAH the secretion becomes less complete. As a consequence renal venous blood is not PAH free. If this concentration (P^v_{PAH}) is known, RPF can still be determined as:

$$RPF(P^a_{PAH} - P^v_{PAH}) = \dot{V} \cdot U_{PAH}, \tag{73.1d}$$

where P^a_{PAH} stands for the PAH concentration in the arterial plasma.

In a more general way the clearance concept can be formulated for any substance:

$$C_X = \dot{V} \cdot U_X / P_X. \tag{73.2}$$

The clearance of substance x then describes the plasma volume per minute which is cleared of this substance. For example, as will be discussed explicitly in Chaps. 74 and 75, the clearance for Na$^+$ is only a few ml/min, because most of the Na$^+$ is reabsorbed along the nephron.

Instead of absolute values, the clearance of x is often given as a fraction of that for inulin or creatinine (or GFR). This term is called fractional excretion (FE):

$$FE_X = C_X / GFR = U_X \cdot P_C / P_X \cdot U_C. \tag{73.3}$$

Equation 73.3 has been obtained from Eqs. 73.1 and 73.2. FE is without dimension.

73.3.3 The Dynamics of Glomerular Filtration

The process of glomerular filtration can be written according to Ohm's law as:

$$SNGFR = \Delta P \cdot K_f. \tag{73.4}$$

SNGFR and K_f have been introduced above; ΔP is the mean pressure gradient across the filtration barrier. Figure 73.9 denotes mean values as they have been obtained from direct measurements in the Munich Wistar rat [2]. The hydrostatic pressure at the beginning and at the end of the glomerular capillaries is around 45–50 and 44–49 mmHg, respectively. When compared to the systemic circulation, these values may appear surprisingly high. However, it has to be remembered that glomerular capillaries are the first capillary network of a portal system (cf. above). Also, it is worth noting that the pressure drop along the glomerular capillaries is very small. Therefore, *the filtration process is not controlled by the alterations in axial pressure along glomerular capillaries.*

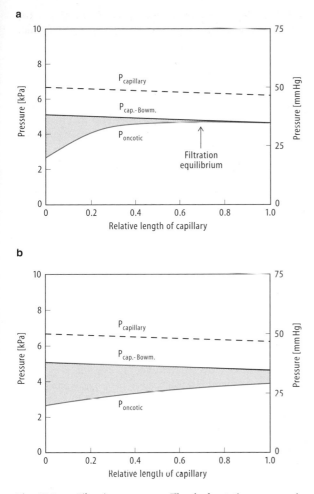

a

b

Fig. 73.9. a Filtration pressure. The hydrostatic pressure in glomerular capillaries ($P_{capillary}$), this pressure reduced by that of the Bowman space ($P_{cap.-Bowm.}$), and the oncotic pressure ($P_{oncotic}$) = $\pi_{oncotic}$ are shown as a function of relative capillary length. Due to filtration, $P_{oncotic}$ increases rapidly and filtration equilibrium is reached. At this point the filtration pressure = $P_{cap.\ Bowm}$ − $P_{oncotic}$ becomes zero. The mean filtration pressure can be obtained from the area between $P_{cap.-Bowm.}$ and $P_{oncotic}$ divided by the capillary length. In the example shown this Δp is around 5 mmHg. **b** Filtration pressure at high glomerular plasma flow. The hydrostatic pressure in glomerular capillaries ($P_{capillary}$), this pressure reduced by that of the Bowman space ($P_{cap.-Bowm.}$), and the oncotic pressure ($P_{oncotic}$) = $\pi_{oncotic}$ are shown as a function of relative capillary length. Due to filtration, $P_{oncotic}$ increases slowly but no filtration equilibrium is reached. At any point the filtration pressure = $P_{cap.-Bowm.}$-$P_{oncotic}$ stays positive. The mean filtration pressure can be obtained from the area between $P_{cap.-Bowm.}$ and $P_{oncotic}$ divided by the capillary length. In the example shown this Δp is around 12 mmHg

As in every other capillary, the hydrostatic pressure is partly counterbalanced by the oncotic pressure (π) exerted by plasma proteins. This pressure has been determined as 20 mmHg, corresponding to 50–60 g/l, in arterial blood and as 35 mmHg, corresponding to 80–90 g/l, in the star vessels (welling points) of peritubule capillaries [2,3]. It appears reasonable to assume that the oncotic pressure at the beginning of the glomerular capillary equals that of arterial blood, and that π at the end of the glomerular capillary is

identical to that at the welling point of peritubule capillaries. The increase in oncotic pressure along the glomerular capillary is a direct consequence of the filtration process. Since the filtrate is essentially protein free, the oncotic pressure on the blood side will increase due to water removal (cf. also Chap. 125). The effective filtration pressure (ΔP) at any one point of a glomerular capillary can be written as:

$$\Delta P = P_C - P_B - \pi_C + \pi_B. \tag{73.5}$$

P_C and P_B denote the hydrostatic pressures in the capillary (cf. above) and in the Bowman space. The latter value has been determined in micropuncture studies [2,3] and was found to be 10–12 mmHg. π_C and π_B are the oncotic pressures in capillaries and the Bowman space. That for the Bowman space is close to zero (no quantitatively relevant protein filtration), while that for glomerular capillaries increases along the capillary from 20 to 35 mmHg.

Therefore, the effective filtration pressure at the beginning of the glomerular capillary may be: $\Delta Po = 46 - 20 - 12 = 14$ mmHg, and that at the end of the capillary: $\Delta PI = 45 - 33 - 12 = 0$ mmHg. This latter state, with no net driving force for filtration, is called *filtration equilibrium*.

As shown in Fig. 73.9a, the point at which filtration equilibrium is achieved is determined exclusively by the shape of the curve of the increase in π_C. The mean net filtration pressure (ΔP) can be determined from the mean height of the area under the curve. In the example of Fig. 73.9a the mean ΔP is 5 mmHg.

Now consider Fig. 73.9b, in which the assumption is made that the blood capillary flow has doubled from 200 to 400 nl/min (plasma flow 100–200 nl/min). It is evident that now filtration equilibrium will no longer be achieved since the increase in oncotic pressure is less marked when say 45 nl is removed from 200 nl (factor 1.29) as compared to 35 nl from 100 nl (factor 1.54). Therefore, $\pi_{oncotic}$ ($P_{oncotic}$ in Fig. 73.9b) increases much more slowly, and no filtration equilibrium is achieved. As a result the mean filtration pressure and hence SNGFR increase.

Changes in RPF therefore have a large impact on GFR. As will be shown in the next section, the regulation of RBF by various factors – hormones, transmitters and local factors – is one of the determinants of glomerular filtration rate (cf. next section).

Equation 73.4 suggests that K_f and ΔP are independent parameters. The concept of filtration equilibrium, however, generates a complex interrelationship between ΔP and K_f. The latter term contains two components: the filtration area (A) and the hydraulic conductance of the filter (K):

$$K_f = K \cdot A. \tag{73.6}$$

In filtration equilibrium a sizeable fraction of A may not be used because the equilibrium may be reached before the end of the capillary. If now the plasma flow rate is enhanced and a positive filtration pressure prevails towards the very end of the capillary, the area, and hence K_f, is increased. Therefore, strictly speaking GFR increases due to both an enhanced ΔP and an enhanced K_f.

Most of the factors to be discussed below change K_f by changing A. However, for some of the factors a direct effect on K has been discussed (summarised in: [11]). The concept of filtration equilibrium has thus far been proven only for a very few species, and it is by no means certain that it applies to, for example, dog and man. However, the concept appears very attractive because it can explain why GFR is so exquisitely dependent on variations in renal blood flow, π_C and P_C.

73.3.4 Mesangial Cells: Regulators of Filtration Area?

Several hormones have been shown to act on mesangial cells in culture and on K_f in vivo. Amongst these hormones are norepinephrine, angiotensin II, antidiuretic hormone (AVP), parathyroid hormone, bradykinin and many more [31]. The general effect is a decrease in K_f and a contraction of mesangial cells in culture. On the one hand, anatomical studies do not support the view that mesangial cells modulate the perfusion pressure [16]. On the other hand, it cannot be excluded that contraction of mesangial cells turns off some capillaries and reduces A without any significant effect on P_C [24,25].

The contraction of mesangial cells is mediated by Ca^{2+}, as has been shown for vasopressin (antidiuretic hormone), bradykinin and angiotensin II [22]. The effect of other agonists may be mediated indirectly by renin release and angiotensin II production, e.g., parathyroid hormone, insulin and norepinephrine. The fact that mesangial cells in culture possess renin has been interpreted as evidence that these cells have their own renin-angiotensin system. This is unlikely since the expression of renin synthesis may be an artefact of mesangial cell culture. In general, the results obtained from mesangial cell cultures cannot naively be translated into in vivo conditions since mesangial cells in culture change their biological properties easily and rapidly. Still, it appears feasible that several of the above hormones modulate mesangial cell tone by local renin release and angiotensin II production.

73.4 Renal Blood Flow Control: A System Constructed to Optimise Control of SNGFR

Renal blood flow (RBF) amounts to approximately 25% of cardiac output, i.e., 1.2 l/min. This is very impressive given the low kidney weight of 150 g/organ. The renal blood supply hence can be calculated as 4 ml/g·min. This is the highest perfusion rate in the body. Moreover this disregards the fact that the blood supply is very high to the

cortex and very low to the medulla. Consequently, the perfusion of renal cortex is even considerably higher. Unlike that of brain, heart or skeletal muscle, the venous outflow of the kidney is still well saturated with oxygen. This finding has led to the interpretation that the kidney has a "luxury" perfusion. As will be shown in the next two chapters, this interpretation is wrong inasmuch as a fall in kidney perfusion is poorly tolerated and may cause *acute renal failure* with the loss of the organ. High renal perfusion is required for two main reasons:

- To warrant a constant and high filtration rate
- To supply sufficient substrate and oxygen to the tubule in order to sustain the tubule transport work

Given this pivotal role of adequate renal perfusion, it is not surprising that renal perfusion, like, for instance, that of the brain, is well regulated and kept constant under conditions of altered systemic blood pressure.

73.4.1 Three Resistors Control RBF and GFR

Figure 73.10 depicts the portal system of the renal vasculature. On the afferent side all resistor vessels are lumped together into one afferent resistor (R_A) and on the efferent side all resistor vessels are lumped into one efferent resistor (R_E). Between the two resistors filtration takes place across the glomerular capillaries ($R_C = 1/K_f$). The numerical values of R_A, R_E and R_C are such that R_C is bigger than the other two. Hence, changes in K_f have little effect

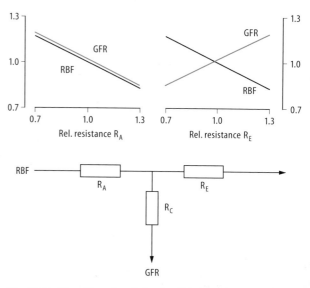

Fig. 73.10. The effect of variation in afferent resistance (R_A), efferent resistance (R_E) and resistance of the filtration barrier ($1/K_f = R_C$) on renal blood flow (RBF) and glomerular filtration rate (GFR). RBF (*black line*) and GFR (*blue line*) as well as the resistances are given as relative values. Note that an increase in R_A has a comparable effect on RBF and GFR, whereas an increase in R_E enhances GFR (increased filtration pressure) but reduces RBF. Changes in R_C have little effect on RBF but marked effects on GFR (not shown), because $R_C \gg R_A$ or R_E

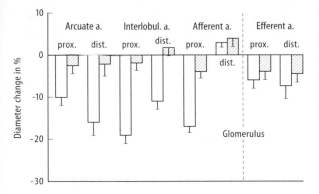

Fig. 73.11. Vasoconstrictor effect of norepinephrine (2 nmol/l i.v.). Vascular diameter (mean $\pm$ SEM, $n = 7$) is given in % of control. The vasoconstrictor effect without blood pressure control (*open bars*) is much more marked than that with constant systemic pressure (*hatched bars*). This indicates that part of the vasoconstrictive effect of norepinephrine is due to systemic pressure increase. However, at higher norepinephrine concentrations (4 nmol/l) the vasoconstrictive effect of norepinephrine becomes more marked even in the absence of systemic pressure changes (not shown here). Note also that the vasoconstriction by norepinephrine occurs in all small vessels, from the proximal arcuate artery throughout the entire efferent arteriole (except for the preglomerular afferent arteriole). (Data taken from [36])

on RBF; rather their effect is mostly on GFR. RBF is controlled by both resistors R_A and R_F.

With respect to GFR the following constellations are feasible:

- Increase in R_A – fall in GFR
- Decrease in R_A – increase in GFR
- Increase in R_E – increase in GFR
- Decrease in R_E – fall in GFR
- Increase in R_A and R_E – no change in GFR
- Decrease in R_A and R_E – no change in GFR
- Decrease in R_C (increase in K_f) – increase in GFR
- Increase in R_C (decrease in K_f) – fall in GFR.

The changes in the respective resistors can be gradual: therefore, the outcome is a *fine tuning of the GFR*.

The *filtration fraction* (FF) is the ratio of GFR and RPF. Under normal circumstances this value is $120/600 = 0.20$. It may be enhanced if R_A and R_E increase simultaneously or when R_E increases alone.

Thus far we have simplified the analysis by ascribing all the afferent and efferent resistors to R_A and R_E. In fact, it was tacitly assumed that these resistors are localised just before and after the capillaries. This appears the more likely since the small diameter of the arterioles renders them resistor vessels and changes in their diameter would have a large impact on pressure drops. In in vitro perfused renal vasculature [12] and in one in vivo model [36], the hydronephrotic split kidney preparation, these assumptions have been examined experimentally. Although the findings obtained in both preparations do not match for all experimental manoeuvres, the general conclusion is that changes in diameter occur not only in afferent and efferent arterioles but also in adjacent vascular beds.

73.4.2 Control of Glomerular Blood Flow

Figure 73.11 summarises data obtained in the above in situ and in vivo preparation with *norepinephrine*. Two types of experiments were performed. In the first systemic pressure increases overlay local effects, while in the second approach the renal artery was equipped with an adjustable ligature such that renal perfusion pressure was kept constant. In the first approach vasoconstriction was observed for pre- and postglomerular vessels. When perfusion pressure was controlled, norepinephrine had very little effect on preglomerular vessels, but there was a strong constriction of the postglomerular vasculature.

A comparable study with another important vasoconstrictor, *angiotensin II* (AII), is depicted in Fig. 73.12. It is evident that AII constricts the efferent but also the afferent vasculature. As a result RBF falls strongly. In addition AII reduces K_f (cf. above). Hence GFR tends to fall and FF increases slightly after AII.

Effects of other potent agonists on renal vasculature are summarised in Table 73.2. Adenosine is a potent vasoconstrictor which results in a reduction of GFR (summarised in: [11]; also cf. below). Endothelin [38] is a potent vasoconstrictor of afferent and even more so of efferent arterioles. This explains the strong fall in RBF. The concomitant fall in SNGFR is due to a marked fall in K_f.

Many other substances produce increases in RBF (vasodilators): acetylcholine, bradykinin, histamine, dopamine and others belong to this group of substances. The increases in RBF are paralleled by smaller increments in SNGFR. It appears likely that these agonists act via endothelial-derived relaxing factor (EDRF). EDRF has been found recently to be identical with NO [1]. (cf. also Chaps. 5 and 6). The aforementioned substances, after binding to endothelial receptors, activate the production of NO from arginine and the release of NO. NO

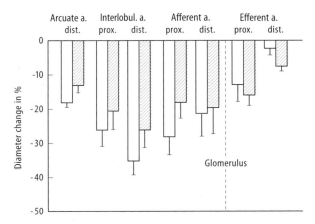

Fig. 73.12. Vasoconstrictor effect of AII (100 ng/min per kg i.v.). Vascular diameter (mean $\pm$ SEM, $n = 7$) is given in % of control. The vasoconstrictor effect without blood pressure control (*open bars*) is similar to that with constant systemic pressure (*hatched bars*). Note that vasoconstriction occurs in both afferent and efferent vessels. Note also that the vasoconstriction by AII occurs throughout all vessels shown. (Data taken from [36])

Table 73.2. Hormonal effects on RBF and GFR (modified from [11])

Agonist	RBF	SNGFR	K_f	Glom.rec.	Comment
AII	–	NC	–	+	SNGFR – when PG synthesis is blocked
NE	–	NC	?	+	
ISO	–	NC	–	+	
ADH	–	NC	–	+	AII independent
PTH	NC	–	–	+	Effects reversed by AII
PGE₂	(–)	(–)	–	+	Effects reversed by AII
ADE	–	–	NC	+	
Throm	–	–	–		
PAF	–	–	–	+	Effects reversed after Throm A₂ blockade
EGF	–	–	–		
ACH	+	+	–		
BRA	+	NC	–	+	
HIS	+	NC	–	+	H₁ receptors
ANF	+	+	NC	+	Not blocked by AII blockade

Glom. rec., glomerular receptors; AII, angiotensin II; NE, norepinephrine; ISO, isoproterenol; ADH, antidiuretic hormone; PTH, parathyroid hormone; PGE₂, prostaglandin E₂; ADE, adenosine; Throm, thromboxane; PAF, platelet activating factor; EGF, epidermal growth factor; ACH, acetylcholine; BRA, bradykinin; HIS, histamine; ANF, atrial natriuretic factor (peptide); –, decreased; +, increased; NC, no change; ?, unclear

activates the guanylate cyclase of smooth muscle cells, which leads to increases in cGMP. This second messenger, probably via protein kinase G, relaxes smooth muscle cells. This mechanism now seems proven for ATP, bradykinin, thrombin, platelet activated factor (PAF) and others (cf. Chap. 98).

Other factors, such as the atriopeptides [atrial natriuretic factor (ANF)], appear to act via cGMP directly and to dilate the afferent renal vasculature [36]. ANF increases SNGFR with little change in RBF. This seems to be caused by an afferent vasodilation and an efferent vasoconstriction [36]. Hence total resistance stays constant but FF and GFR increase.

Little is known of the physiological effects of the various factors summarised in Table 73.2. AII obviously plays a key role in the renin-AII system (cf. below), and NE has a clear function as the transmitter of sympathetic innervation. At this stage the detailed mechanisms involved in autoregulation and in tubuloglomerular feedback are still not known (cf. next section).

73.4.3 RBF and GFR Are Autoregulated

When blood pressure changes from its normal mean value of, say, 100 mmHg (13.3 kPa) to higher or lower values, corresponding changes in RBF would be predicted on the basis of Ohm's law. This, however, is not what is observed. As is shown in Fig. 73.13, RBF and GFR are maintained almost constant within a pressure range of 60–160 mmHg; only below and above this range is some sort of Ohmic behaviour observed. These data have been obtained in the non-anaesthetised dog [18].

Considering the mixed elastic and compliant nature of renal vessels one would predict that any increase in

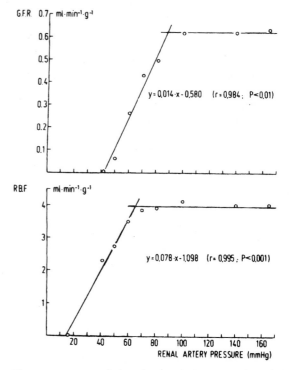

Fig. 73.13. Autoregulation of RBF and GFR in non-anaesthetised dogs. Data have been obtained in chronically instrumented dogs [18]. Renal artery pressure was determined by an inflatable cuff placed around the renal artery. RBF was autoregulated at pressures > 66 mmHg, and GFR at pressures > 81 mmHg. The "break off" points for both parameters were significantly different. The difference in venous-arterial renin activity was not increased at pressures of > 95 mmHg

1480

perfusion pressure would cause an increase in radius (compliance), and this increase in radius would reduce resistance markedly (4th power of radius in Hagen-Poisseuille's law). Hence, an over-proportional increase in perfusion would be the result. To explain the observation not only constancy but even a pressure-dependent constriction of the vessels has to be postulated. If RBF is perfectly autoregulated, the autoregulation of GFR can be explained as a consequence of this.

73.4.4 Autoregulation: Bayliss Effect or Tubuloglomerular Feedback?

Current evidence favours the view that both components, the Bayliss effect and tubuloglomerular feedback (TGF), are involved in normal autoregulation. However, it is evident at the same time that autoregulation can be shown in the absence of TGF and also without changes in renin.

The Bayliss effect implies that an arteriole put under luminal pressure (stretch) responds with constriction. Such an effect could be produced by a paracrine factor (cf. Chap. 98), or it may simply be caused by stretch activation of, for example, a non-selective cation channel. Ca^{2+} would enter through this channel and this might mediate constriction. The principal components of such a mechanism have been identified by patch clamp techniques and fura-2 Ca^{2+} measurements in various cells. However, the final proof that such a mechanism operates in afferent arterioles is still missing.

The TGF is a complex mechanism which consists of several functional components (Fig. 73.3a):

- Macula densa (MD) cells, which sense the luminal Cl^- concentration via the $Na^+2Cl^-K^+$ cotransporter present in the luminal membrane of these cells (cf. also Chap. 75). This "chemoreceptor" responds to an increase in luminal Cl^- with a depolarisation of the voltage.
- Goormaghtigh cells (extraglomerular mesangium) which somehow have to process the MD signal.
- Smooth muscle cells of the afferent arteriole of the same nephron.
- Renin-producing cells in the afferent arteriole.

Pertinent experimental observations [5] are summarised in Fig. 73.14. A fall in SNGFR is observed when the loop is perfused at high flow rates with NaCl-containing Ringer's solution. This is especially pronounced if the loop is perfused retrogradely. Instead of SNGFR, early proximal tubule flow rate is monitored in these experiments. The fall in early proximal flow rate is most marked for small anions such as Cl^- and small cations such as Na^+. The TGF response consists in: (1) *vasoconstriction* whenever the Cl^- concentration in the MD segment increases above a certain value, and (2) a reduction in *renin secretion*. A fall in Cl^- concentration or an inhibition of the sensor ($Na^+2Cl^-K^+$ cotransporter) by loop diuretics [29,30] has the opposite effect.

If we consider an inappropriately high SNGFR, say due to imperfect autoregulation, the capacity of the thick ascending limb to cope with this enhanced NaCl load may be compromised. This will lead to an increase in luminal Na^+ and Cl^- concentrations. The MD, put strategically at the end of this nephron segment and in close vicinity to the vasculature serving this nephron, senses this increment in Cl^- and conveys a message to the afferent arteriole to constrict. This negative feedback between tubule and glomerulus (hence the name TGF) reduces RBF and SNGFR.

It is still not clear how this feedback is processed from the MD to the smooth muscle cells [32]. Originally it was suggested that the renin system and AII were the mediators (cf. below). This appeared especially attractive since the renin-producing cells are part of this functional unit and because AII is a potent vasoconstrictor on renal arterioles. On the other hand, it is now clear that the observed vasoconstriction cannot be explained by reference to AII since renin secretion is reduced rather than enhanced with high loop perfusion. Other possible mediators are prostaglandins, cAMP, Ca^{2+} and adenosine [32].

Especially adenosine is a likely candidate because it might be produced at increased metabolism by the MD cells. In addition, vasoconstriction of afferent arterioles by adenosine has been demonstrated. Furthermore, inhibitors of A1 receptors suppress the TGF response.

The second component of TGF is a reduction in renin release. Direct vasoconstriction, e.g., by adenosine, and the attenuation of this response by a fall in AII counterbalance each other to some extent. The function of this second and attenuating loop in TGF is unclear. It might serve for fast resetting and fine tuning. The systemic effects of renin and AII are discussed elsewhere (cf. Chap. 94). In summary, autoregulation of RBF and SNGFR are due to the concerted action of vascular response and TGF.

73.5 Deep Versus Superficial Nephrons

As shown in Fig. 73.2, the nephrons can be classified by the location of their glomeruli in the cortex and by the depth which the loops reach in the outer medulla, in the inner medulla or in the papillae. These classifications are different for various species. Even in very early studies (reviewed in [28]) it was clear that in the human kidney the majority of nephrons are short looped (superficial), with only about one-seventh having long loops (deep nephrons). It was also evident from a large number of comparative studies in different species that the size of the glomeruli is larger for the deep nephrons than for the superficial ones. This also applies to the kidneys of children, but not to those of adults. As outlined in a previous section, the vasculature and tubule segments of deep and superficial nephrons have distinct differences [8,23].

When it comes to function of the two major populations of nephrons, several aspects, though challenged by some studies, deserve mention. The autoregulation of RBF of the

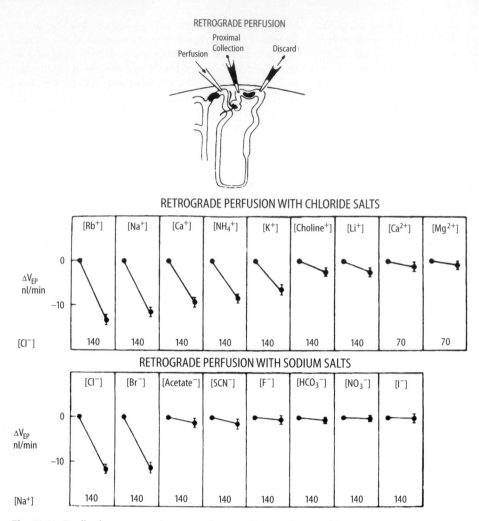

Fig. 73.14. Feedback responses in rat nephron. Fall in early proximal tubule flow rate (ΔV_{EP}) with retrograde perfusion of solutions containing various cations and anions. Retrograde perfusion was chosen in order to control the ion composition at the macula densa site. It is shown that a maximal TGF is obtained with high concentrations of Na^+ and Cl^-. Several small cations can substitute for Na^+, but only Br^- for Cl^-. This suggests that TGF is mediated mostly by a Cl^- sensing mechanism. Meanwhile it has been shown that this sensing occurs via the $Na^+2Cl^-K^+$ cotransporter [30]. (From [5])

deep nephrons appears less perfect than that of RBF of the superficial nephrons. As a consequence, increases in blood pressure increase medullary blood flow proportionally more than superficial blood flow. This has certain consequences: the interstitium of the renal medulla is "washed out" (cf. Chap. 74) and the kidney loses some of its concentrating ability, resulting in a *"pressure diuresis"* [39].

The deep nephrons, due to their long loops, are suited for salt and water conservation (cf. Chap. 74). Some manoeuvres such as acute or chronic salt loading appear to induce a redistribution of RBF and even more so of SNGFR. The SNGFR of the superficial nephrons tends to increase and that of the deep nephrons to fall. Concomitantly the filtration fraction is almost constant for the superficial nephrons but falls in the deep ones. Therefore, simplistically speaking, the filtrate is now shifted from the *"salt-saving"* to the *"salt-losing"* nephrons. Furthermore, the relative increase in medullary blood flow (fall in filtration

fraction of deep nephrons) leads to a *"medullary wash out"* with the above consequences.

As yet there is little direct support for these considerations as far as the human kidney is concerned, simply because we rely entirely on anatomical studies and have very few functional data. In other species (rat, dog, rabbit) an enormous amount of data is available. Nonetheless it should be kept in mind that we have no direct means of measuring glomerular dynamics in deep nephrons. Comparative data on kidney anatomy and function of various mammals adapted to extreme environmental conditions have been pivotal in the understanding of the renal concentrating mechanism [8].

73.6 The Kidney: An Endocrine Organ

The kidney produces several hormones and local factors such as renin, erythropoietin, 1,25-dihydroxycholecalciferol, prostaglandins, adenosine, NO, endothelin, kinins and probably many others. In this section the first three of these hormones will be briefly discussed; the others are mentioned in other sections of this chapter.

73.6.1 The Renin-Angiotensin System

The renin-angiotensin system has already been mentioned in previous sections. Here we shall briefly outline its components and its function, because this knowledge will be required for several of the following chapters.

Renin: A Neutral Aspartyl Protease Not Unique for the Kidney. Renin (prorenin) is produced in isoforms in a variety of organs such as the liver, several areas of the brain, the kidney and some glands [37]. It is species specific and made from a large 1.2-kb gene. The processed renin in the cisternae of the endoplasmic reticulum of the renin-producing cells contains the polypeptide (317 AA, MW 42 kDa) which consists of two chains connected by a single disulphide bridge. The release and the transcription of renin are regulated locally (vide infra).

Once released (mostly to the perivascular side) and taken up (mostly into the capillary network), renin, as an aspartyl protease, cleaves systemically circulating angiotensinogen (Fig. 73.15).

The Angiotensin Cascade. Angiotensinogen is a 53- to 57-kDa peptide, produced in impressive amounts (circulating concentration is in the μmol/l range), essentially by the liver. Angiotensinogen is cleaved on its amino terminus between two leucins to form the decapeptide angiotensin I (AI). AI has little biological activity and is cleaved further by angiotensin-converting enzyme (ACE).

ACE (EC 3.4.15.1) is a peptidyl-dipeptide-carboxy-hydrolase which cleaves off Leu-His at the carboxy end of AI and thus forms angiotensin II (AII), an octapeptide. ACE is present mostly in the luminal membrane of the vasculature. It is a 200-kDa protein. The name may, in fact, be misleading because ACE is indistinguishable from kininases, which cleave kinins such as bradykinin.

In this context it is likely that the high immunological ACE activity found in microvilli (proximal tubule, choroid plexus, intestine, placenta etc.) reflects the ability of these structures to cleave bradykinin, rather than implying that they play a role in the formation of AII. This multiple function of ACE should also be considered when examining the beneficial effects in ACE inhibitor treatment. Some of these effects can be clearly ascribed to an increase in bradykinin.

AII is a potent vasoconstrictor, but besides this it also enhances renal tubule NaCl reabsorption, it releases aldosterone from the zona glomerulosa of the cortex of the adrenal gland, it releases antidiuretic hormone (ADH) from the posterior lobe of the pituitary gland, and it produces thirst in the hypothalamus (cf. Chap. 82). The CNS effects of AII appear to be possible since the blood-brain barrier is "incompetent" (permeable) in this hypothalamic area. AII is a labile hormone and it is cleaved or degraded by angiotensinases. One substrate formed is the heptapeptide angiotensin III (AIII), which has much less biological activity than AII.

Intrarenal Renin-Angiotensin System (RAS). Figure 73.3A shows the specialised cells which are the main production sites of renin within the kidney. These cells, mostly located in the most distal part of the afferent arteriole, have intimate contact to their neighbours in the juxtaglomerular apparatus (JGA). These are (a) the Goormaghtigh cells (on a section such as that in Fig. 73.3A, these cells lie within a triangle formed by the base of the macula densa cells and the afferent and efferent arterioles entering the glomerulus) and (b) the macula densa cells. Since the early studies, e.g., by Goormaghtigh [15], the function of the JGA has largely remained a mystery. Some of the regulatory aspects of this machinery have been discussed above (see Sect. 73.3.4); here the regulation of renin release will be summarised. The key stimuli for renin release are:

- Fall in systemic blood pressure
- Sympathetic discharge to the kidney
- Fall in NaCl concentration delivered to the macula densa

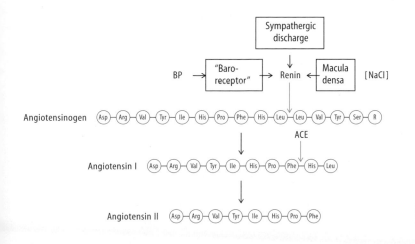

Fig. 73.15. The regulation of AII production. For further details consult text

- Prostaglandins such as PGE$_2$ and PGI$_2$
- Kinins.

Drugs can also increase renin release. Amongst these several important ones are:

- β-Agonists
- Loop diuretics
- ACE inhibitors
- Ca^{2+} antagonists
- α$_2$-antagonists
- Saralasin.

Conversely, renin secretion can be inhibited by increased blood pressure, diminished sympathetic discharge and increased NaCl delivery to the macula densa. In addition renin release is diminished by AII, α$_2$-agonists, vasopressin, endothelin, adenosine, NO, β-blockers etc.

How is Renin Released. It is difficult, at this stage, to define a general concept of how renin is released. It has been claimed that the effects of some of the above-mentioned regulators can be explained by their ability to increase (stimulate) or diminish (inhibit) cAMP in the granulated renin-producing cells. Such factors comprise β-agonists, PGE$_2$ etc. Renin secretion may also be inhibited by agonists increasing cGMP; such agonists include, for example, ANP and NO [20].

As regards most of the other regulators, it has been claimed that they increase renin release by a reduction of cytosolic Ca^{2+} ([Ca^{2+}]$_i$). Hence Ca^{2+} antagonists, or α$_1$-antagonists, enhance renin release and Ca^{2+} mobilising agents such as α$_1$-agonists, AII, vasopressin, endothelin, pressure-induced stretch etc. inhibit renin release. This concept of direct control of renin release by [Ca^{2+}]$_i$ appears attractive, but it may not be applicable. First, it stands in contrast to most if not all exocytotic secretion processes, where in-

creases in [Ca^{2+}]$_i$ are required for exocytosis; second, this concept is not unchallenged by contradictory results. It is clear, at any rate, that renin release occurs from vesicles which, upon stimulation, fuse with the plasma membrane. It may turn out that [Ca^{2+}]$_i$ has a dual effect on renin secretion, and that the local Ca^{2+} activity close to the respective vesicles differs from that of the bulk phase of the cells.

It has also been claimed that the renal RAS works as a self-supporting independent regulatory system, very much like the RAS of the brain. Such a system would require that angiotensinogen is present locally and need not be delivered from the circulation. In fact, the co-secretion of rein and AII from renin-producing cells has been claimed. The as yet open question is whether renally produced angiotensinogen suffices to explain normal function [27]. Hence this entire concept has to be questioned, and it appears safe to conclude that local and systemic renin concentrations stand in close relation under most conditions.

73.6.2 Erythropoietin Is Mostly Produced in the Kidney

Erythropoietin is a glycosylated peptide hormone with 137 amino acids. Most of the hormone in adults comes from the kidney; a minor fraction is produced in the liver. The half-life of erythropoietin is 5–9h. The main target of erythropoietin is the bone marrow, where it increases the formation of red blood cells (erythropoiesis). A fall in arterial O$_2$ pressure and/or a fall in hemoglobin serves as an adequate stimulus for erythropoietin secretion by the kidney. Both signals have in common the fact that they define the delivery of O$_2$ to the kidney [35] (cf. Chap. 84).

It may appear surprising that the O$_2$ delivery is monitored in the kidney, where blood supply is as high as 4 ml/min · g and O$_2$ supply is 1 ml/min·g. This is the more surprising

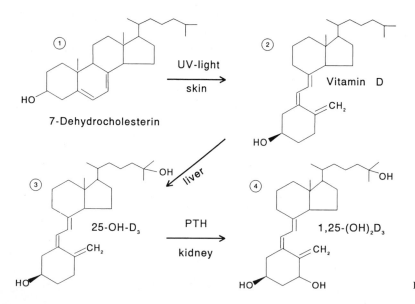

7-Dehydrocholesterin

UV-light skin

Vitamin D

liver

25-OH-D$_3$

PTH kidney

1,25-(OH)$_2$D$_3$

Fig. 73.16. The production of 1,25-(OH)$_2$D$_3$

since the erythropoietin-producing cells seem to be present exclusively in the cortex, where blood supply is especially high. This puzzle is as yet unresolved. It may be that tubule transport in the cortex generates a fall in local pO_2 when oxygen supply does not match the needs. The erythropoietin-producing cells appear not to be vascular, endothelial or tubular. They seem to be specialised cells most probably located in the peritubule space, although a definite answer as to the precise localisation is still missing [17].

In severe chronic renal failure the lack of erythropoietin is believed to be the main cause of anaemia. Consequently, with the advent of recombinant human erythropoietin, replacement therapy is now used clinically [42].

73.6.3 The Production of the Hormone 1,25-(OH)₂-Cholecalciferol Requires an Intact Kidney

Figure 73.16 summarises the metabolism of 1,25-(OH)₂-cholecalciferol [1,25-(OH)₂-D_3]. In the presence of UV light, calciferol can be formed from cholesterol in the skin. Calciferol is then hydroxylated in the liver to form 25-OH-D_3. This is then delivered to the kidney to form 1,25-(OH)₂-D_3. This latter reaction is controlled by PTH [10]. Only in the presence of PTH are sufficient amounts of 1,25-(OH)₂-D_3 synthesised. Besides 1,25-(OH)₂-D_3, 24,25-(OH)₂-D_3 is formed. This compound has much less, if any, biological activity compared with 1,25-(OH)₂-D_3.

The effects of 1,25-(OH)₂-D_3 are discussed in Chap. 80. It may suffice to state here that this hormone increases Ca^{2+} absorption from the intestine and increases osteoclast activity. Because, at the same time, plasma Ca^{2+} is increased, this leads to the formation of new bone. The direct effects of this hormone on the kidney are probably minor. Deficiency of this hormone leads to the clinical pictures of rickets in children and osteomalacia in *adults*.

Appendix. Glossary and Abbreviations for Chaps. 73, 74 and 75

The abbreviations used in renal physiology have been standardized recently. However, the cited literature uses many synonyms. Therefore, both the recommended (italics) and frequently used abbreviations are given below.

AI = angiotensin I
AII = angiotensin II
ACE = angiotensin-converting enzyme, catalyses AI → AII
ADH = antidiuretic hormone, also referred to as AVP
ANF = atrial natriuretic factor = ANP
ANP = atrial natriuretic peptide = ANF
Antiport = transport coupling into opposite directions by a carrier

ARF = acute renal failure
ATL = thin ascending limb of the loop of Henle = tAL
Autoregulation = constancy of renal blood flow and glomerular filtration rate, despite changes in renal perfusion pressure
Bayliss effect = one component of autoregulation of RBF
Bowman space = intraglomerular compartment into which filtration proceeds
CCT = cortical collecting tubule = *CCD* = cortical collecting duct
CNT = connecting tubule = first portion of collecting tubule
Cotransport = transport coupling into the same direction by a carrier
Countercurrent concentrating mechanism = specific countercurrent arrangement of tubule and vascular structures in renal medulla and papilla which makes it possible to concentrate urine
CRF = chronic renal failure
Creatinine = usual (endogenously produced) marker for measurement of GFR
CTAL = cortical thick ascending limb of the loop of Henle
DT = distal tubule
DTL = thin descending limb of the loop of Henle = tDL
Ducts of Bellini = papillary collecting tubule
Electrogenic = transporting an electric charge = rheogenic
Feedback = usually more specifically referred to as tubuloglomerular feedback
Filtration fraction = GFR/RPF
GFR = glomerular filtration rate
Glomerulotubular balance = readjustment of proximal tubule transport rate according to variations in SNGFR. High SNGFR enhances, for example, HCO_3^- absorption. Low SNGFR has the opposite effect.
GTB = glomerulotubular balance
IMCD = inner medullary collecting duct = IMD
IMD = inner medullary collecting duct = *IMCD*
Intercalated cells = acid/bicarbonate secreting cells of the collecting tubule = A/B cells
Inulin = polyfructoside, a marker for measurement of GFR
K_f = hydraulic conductivity of glomerular filter
Loop diuretics = diuretic and saluretic substances acting in the thick ascending limb of the loop of Henle
Macula densa mechanism = sensing of tubule fluid at this nephron site. High luminal Cl^- leads to a reduction in SNGFR (tubuloglomerular feedbaack) and to a reduction in renin secretion
MCD = medullary collecting duct
MD = macula densa segment
MTAL = medullary thick ascending limb of the loop of Henle
OMCD = outer medullary collecting duct = OMD
OMD = outer medullary collecting duct = *OMCD*
PAH = *para*-aminohippurate, a marker for the measurement of RPF
Pars recta = straight portion of proximal tubule, usually correponding to S_1 and S_2
PCD = papillary collecting duct

PG = prostaglandin
PT = proximal tubule
PTH = parathyroid hormone, parathyrin
RBF = renal blood flow
Rheogenic = transporting an electric charge = electrogenic
RPF = renal plasma flow
Principal cells = Na^+- and water-absorbing and K^+-secreting cells of the collecting tubule
S_1 = convoluted part (first part) of proximal tubule
S_2 = second (intermediate) part of proximal tubule
S_3 = third (last) part of proximal tubule
SNGFR = single nephron filtration rate
TAL = thick ascending limb of the loop of Henle
tAL = thin ascending limb of the loop of Henle = ATL
tDL = thin descending limb of the loop of Henle = DTL
TGF = tubuloglomerular feedback
Tubuloglomerular feedback = readjustment of SNGFR by early distal tubule (macula densa) ion composition. High Cl^- at this nephron site reduces SNGFR. Low Cl^- has the opposite effect
Vasa recta = straight vascular structures of renal medulla and papilla, part of the countercurrent concentrating system.

References

1. Baumann JE, Persson PB, Ehmke H, Nafz B, Kirchheim HR (1992) Role of endothelium-derived relaxing factor in renal autoregulation in conscious dogs. Am J Physiol 263:F208–F213
2. Brenner BM, Dworkin LD, Ichikawa I (1986) Glomerular filtration. In: Brenner BM, Rector FC (eds) The kidney. Saunders, Philadelphia, pp 124–144
3. Brenner BM, Ichikawa I, Deen WM (1981) Glomerular filtration. In: Brenner BM, Rector FC (eds) The kidney. Saunders, Philadelphia, pp 289–327
4. Brenner BM, Rector FC (1991) The kidney, 4th edn. Saunders, Philadelphia, pp 1–2443
5. Briggs J (1981) The macula densa sensing mechanism for tubuloglomerular feedback. Fed Proc 40:99–103
6. Cameron S (1993) Renal function testing. In: Cameron S, Davison AM, Grünfeld JP, Kerr D, Ritz E (eds) Oxford textbook clinical nephrology. Oxford University Press, Oxford, pp 24–49
7. Cameron S, Davison AM, Grünfeld JP, Kerr D, Ritz E (1992) Oxford textbook of clinical nephrology. Oxford University Press, Oxford, pp 1–2348
8. De Rouffignac C (1990) The urinary concentrating mechanism. In: Kinne RKH (ed) Urinary concentrating mechanisms. Karger, Basel, pp 31–102
9. Deen WM, Bohrer MP, Brenner BM (1979) Macromolecule transport across glomerular capillaries: application of pore theory. Kidney Int 16:353–362
10. Deluca HF (1984) The metabolism, physiology and function of vitamin D. In: Kumar R (ed) Vitamin D: basic and clinical aspects. Nijhoff, Boston, pp 1–68
11. Dworking LD, Brenner BM (1992) Biophysical basis of glomerular filtration. In: Seldin DW, Giebisch G (eds) The kidney: physiology and pathophysiology. Raven, New York, pp 979–1016
12. Edwards RM (1983) Segmental effects of norepinephrine and angiotensin II on isolated renal microvessels. Am J Physiol 244:F526–F534
13. Farquar MG (1975) The primary glomerular filtration barrier – basement membrane or epithelial slits. Kidney Int 8:197–205
14. Folkow B (1990) Salt and hypertension. NIPS 5:220–224
15. Goormaghtigh N (1939) Existence of an endocrine gland in the media of the renal arterioles. Proc Soc Exp Biol Med 42:688–689
16. Hebert SC, Kriz W (1993) Structural-functional relationships in the kidney. In: Schrier RW, Gottschalk CW (eds) Diseases of the kidney. Little Brown, Boston, pp 3–64
17. Jelkmann W (1992) Erythropoietin: structure, control of production, and function. Physiol Rev 72:449–489
18. Kirchheim HR, Ehmke H, Hackenthal E, Lowe W, Persoon P (1987) Autoregulation of renal blood flow, glomerular filtration rate and renin release in conscious dogs. Pflugers Arch 410:441–449
19. Kriz W, Kaissling B (1992) Structural organization of the mammalian kidney. In: Seldin DW, Giebisch G (eds) The kidney: physiology and pathophysiology. Raven, New York, pp 707–778
20. Kurtz A (1990) Do calcium activated chloride channels control renin secretion? NIPS 5:43–46
21. Mallick NP, Short CD (1993) The clinical approach to haematuria and proteinuria. In: Cameron S, Davison AM, Grünfeld JP, Kerr D, Ritz E (eds) Oxford textbook of clinical nephrology. Oxford University Press, Oxford, pp 227–239
22. Mene P, Simonson MS, Dunn MJ (1989) Physiology of the mesangial cell. Physiol Rev 69:1347–1423
23. Pallone TL, Robertson CR, Jamison RL (1990) Renal medullary microcirculation. Physiol Rev 70:885–920
24. Pavenstädt H, Gloy J, Leipziger J, Klär B, Pfeilschifter J, Schollmeyer P, Greger R (1993) Effect of extracellular ATP on contraction, cytosolic calcium activity, membrane voltage and ion currents of rat mesangial cells in primary culture. Br J Pharmacol 109:953–959
25. Pfeilschifter J (1989) Cross-talk between transmembrane signalling systems: a prerequisite for the delicate regulation of glomerular haemodynamics by mesangial cells. Eur J Clin Invest 19:347–361
26. Rennke HG, Cotran RS, Venkatachalam MA (1975) Role of molecular charge in glomerular permeability: tracer studies with cationized ferritins. J Cell Biol 67:638–645
27. Rosivall L (1991) The role of angiotensin in the control of renal vascular resistance. In: Hatano M (ed) Nephrology. Springer, Berlin Heidelberg New York, pp 731–739
28. Sands JM, Kokko JP, Jacobson HR (1992) Intrarenal heterogeneity: vascular and tubular. In: Seldin DW, Giebisch G (eds) The kidney: physiology and pathophysiology. Raven, New York, pp 1087–1156
29. Schlatter E (1993) Effects of various diuretics on membrane voltage of macula densa cells. Whole-cell patch-clamp experiments. Pflugers Arch 423:74–77
30. Schlatter E, Salomonsson M, Persson AEG, Greger R (1989) Macula densa cells sense luminal NaCl concentration via the furosemide sensitive Na-2Cl-K cotransporter. Pflugers Arch 414:286–290
31. Schlöndorff D (1987) The glomerular mesangial cell: an expanding role for a specialized pericyte. FASEB J 1:272–281
32. Schnermann J, Briggs JP (1991) Function of the juxtaglomerular apparatus: control of glomerular hemodynamics and renin secretion. In: Seldin DW, Giebisch G (eds) The kidney: physiology and pathophysiology. Raven, New York, pp 1249–1290
33. Schrier RW, Gottschalk CW (1993) Diseases of the kidney, 5th edn. Little Brown, Boston, pp 1–3270
34. Seldin DW, Giebisch G (1992) The kidney: physiology and pathophysiology, 2nd edn. Raven, New York, pp 1–3816
35. Spivak JL, Watson AJ (1992) Hematopoiesis and the kidney. In: Seldin DW, Giebisch G (eds) The kidney: physiology and pathophysiology. Raven, New York, pp 1553–1593
36. Steinhausen M, Blum M, Fleming JT, Holz FG, Parekh N, Wiegman DL (1989) Visualization of renal autoregulation in

the split hydronephrotic kidney of rats. Kidney Int 35:1151–1160

37. Taugner R, Hackenthal E (1989) The juxtaglomerular apparatus. Springer, Berlin Heidelberg New York, pp 1–306

38. Thomas CP, Simonson MS, Dunn MJ (1992) Endothelin: receptors and transmembrane signals. NIPS 7:207–211

39. Thompson DD, Pitts RF (1952) Effects of alterations of renal artery pressure on sodium and water excretion. Am J Physiol 168:490–499

40. Vanholder RC, Ringoir SMG (1992) The uraemic syndrome. In: Cameron S, Davison AM, Grünfeld JP, Kerr D, Ritz E (eds) Oxford textbook of clinical nephrology. Oxford University Press, Oxford, pp 1236–1250

41. Wright FS, Giebisch G (1992) The kidney: Physiology and pathophysiology. In: Seldin DW, Giebisch G (eds) Regulation of potassium excretion. Raven, New York, pp 2209–2248

42. Zuccelli P, Zuccalà A (1992) Control of blood pressure in patients on haemodialysis. In: Cameron S, Davison AM, Grünfeld JP, Kerr D, Ritz E (eds) Oxford textbook clinical nephrology. Oxford University Press, Oxford, pp 1458–1467

74 Principles of Renal Transport; Concentration and Dilution of Urine

R. Greger

Contents

is high, or a highly concentrated urine with an osmolality far exceeding that of plasma, if there is a shortage in water supply. This task can be performed by the countercurrent concentrating system. The morphological basis for this has already been given in the preceding chapter. Here, on the basis of the function of the various nephron segments, the function of this complex system will be discussed in more detail.

Finally, a brief section will address the mechanisms by which diuretics and saluretics affect tubule function. This appears appropriate inasmuch as it can be easily understood on the basis of the various transport mechanisms present in the different nephron segments.

74.1 Introduction

In Chap. 73 the renal blood flow and the formation of glomerular filtrate have been discussed. It became evident that the kidney, in order to fulfil its tasks, requires an impressively high filtration rate of 120 ml/min, corresponding to some 170–180 l/day. Hence, the entire extracellular volume of the body of some 18 l is filtered by the kidney 10 times per day. This volume contains a no less impressive amount of solutes. They are summarised in Table 74.1. For instance, 26 mol NaCl (i.e., 1.5 kg!) is filtered per day, yet less than 1–20 g is excreted in urine. This clarifies a priori that one of the major tasks of the tubule is to reclaim most or all filtered water and solutes. On the other hand, the large filtration rate is necessary to excrete some other solutes which, if accumulating, would be toxic for the body.

The enormous transport work of the kidney requires highly specialised mechanisms which are metabolically economical, as is especially true in the case of the proximal tubule, or are poised to build up large gradients between tubule fluid (urine) and plasma, as is apparent for the collecting duct. The specific tasks are performed by the various nephron segments, and a detailed discussion of their general properties will be given in this chapter.

On this basis the means of regulating tubule transport will be described. However, the renal handling of individual substances will be discussed in the following chapter, together with specific hormonal actions.

One of the important achievements of the mammalian kidney is its ability to produce either a very dilute urine with an osmolality far below that of plasma, if the water supply

74.2 Tubule Transport, an Overview

Table 74.1 provides a balance sheet for the solutes contained in the glomerular filtrate and in urine. The normal rate of filtration in the adult is around 170–180 l/day. The urine volume can vary grossly between less than 1 l in antidiuresis and some 20 l/day in water diuresis [99]. Hence, between 90% and 99.5% of the filtered water is reabsorbed. A large part of this reabsorption already occurs in the proximal tubule. This wide range of water excretion enables the healthy individual to be rather independent of regular water intake. It can be as low as 1–1.5 l/day. Then, the antidiuretic urine is highly concentrated (ca. 1000 mosm/l). Conversely, large volumes of water can be drunk without any danger of overfilling the circulation. Up to 1.5 l of very dilute urine (100 mosm/l) per hour can be excreted in water diuresis. The degree of water excretion is determined by antidiuretic hormone (ADH).

NaCl. The daily filtrate contains approximately 1.5 kg of NaCl, the urine only 0.5 to more than 20 g [27]. Therefore, most of the filtered NaCl is reabsorbed by the tubule. A large part of this reabsorption occurs in the proximal nephron. The fine regulation occurs in the distal tubule and collecting duct. As will be discussed in later sections, aldosterone, ADH, prostaglandin, atrial natriuretic peptide (ANP) and natriuretic hormone play key roles in the determination of Na^+ excretion [5,41,63,82,99,125].

HCO_3^-. The amount of filtered HCO_3^- approximates 4.4 mol/day. Under most circumstances the urine is practi-

R. Greger/U. Windhorst (Eds.)
Comprehensive Human Physiology, Vol. 2
© Springer-Verlag Berlin Heidelberg 1996

Table 74.1. Composition of glomerular filtrate and urine. For reasons of clarity only rough mean values are given here. For details of the renal handling of the various solutes the reader is referred to Chap. 75

Solute excreted	Glomerular filtrate concentration (mmol/l)	Amount filtered (mmol/day)	Urine concentration (mmol/l)	Amout excreted (mmol/day)
H_2O	55 556 (1 kg/l)	10 000 080 (180 l)	55 556 (1 kg/l)	55 556 (1 l)
Na^+	145	26 000	30–150	100–300
K^+	5	900	33–300	50–450
Ca^{2+}	1.5	250	3–6	0.5–20
Cl^-	120	20 000	30–150	100–500
HCO_3^-	25	4250	1	1
Phosphate	2	360	3–20	5–30
D-Glucose	5	900	0.05–0.5	0–1
Amino acids	2	350	2–8	3–12
Proteins	10 mg/l	1.8 g	40 mg/l	60 mg
Urea	5	900	280–400	420–600
Urate	0.3	54	3	5
Creatinine	0.1	18	11	17
H^+	10^{-4}	10^{-2}	0.01	30

A

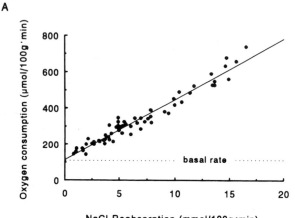

B

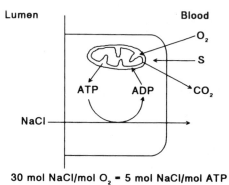

30 mol NaCl/mol O_2 = 5 mol NaCl/mol ATP

Fig. 74.1. A The oxygen consumption of dog kidney as a function of NaCl reabsorption. Data taken from [19]. NaCl reabsorption was varied by a cuff applied to the renal artery or by the application of drugs. Note the linear relation of oxygen consumption and NaCl reabsorption. Under physiological conditions most of the oxygen consumption is utilised for active transport. The basal rate of turnover is only 15%. **B** NaCl reabsorption is an energy-consuming process. It is fuelled by the uptake of substrates (S) across the basolateral membrane. Such substrates are short fatty acids in the proximal tubule and D-glucose in the thick ascending limb of the loop of Henle and in the collecting duct. Active reabsorption is driven mostly by mitochondrial ATP production. From the data in A one can easily calculate that approximately 5 mol NaCl is reabsorbed per 1 mol ATP. This is surprisingly economical, since the pump involved has a stoichiometry of 3 Na^+/1 ATP

cally HCO_3^- free. Only in metabolic alkalosis does the urinary excretion of HCO_3^- increase as a means of correcting this disturbance. HCO_3^- is reabsorbed mostly in the proximal tubule [1,2,26,64,90].

K^+. The filtered amount of K^+ is 900 mmol/day. Urinary excretion varies widely between a few percent and, under extreme conditions, even more than the filtered amount [132]. This wide variation reflects the capacity of the kidney to adjust K^+ excretion to K^+ metabolism. In the proximal tubule K^+ reabsorption predominates. The distal parts of the nephron, including the collecting duct, determine the amount excreted in urine. Many factors are involved in this regulation, amongst which aldosterone, pH and distal tubule load of Na^+ are most important.

Ca^{2+} and Mg^{2+}. The amounts of Ca^{2+} and Mg^{2+} filtered by the glomerulus are on the order of 250 and 200 mmol/day respectively. Most of this is reabsorbed. Ca^{2+} reabsorption occurs in the proximal tubule and in the thick ascending limb of the loop of Henle (TAL) [17,117]. Mg^{2+} reabsorption occurs mostly in the TAL. The degree of reabsorption of Ca^{2+} is controlled by parathyroid hormone (PTH) and calcitonin (cf. Chap. 75).

Phosphate. Approximately 360 mmol of phosphate is filtered per day. Most of this is reabsorbed in the proximal tubule and only some 1%–20% appears in urine. The degree of reabsorption is precisely controlled by PTH and calcitonin. Both hormones act in the proximal tubule and decrease reabsorption [85,117].

Urea. Large quantities of urea (some 900 mmol, i.e., some 54 g) are filtered per day. As one of the tasks of the kidney is to excrete nitrogen breakdown products, it is not surprising that a large proportion of urea appears in urine. This can be between 50% and 80% of the filtered load ($\dot{V} \cdot P_{urea}$). Urea reabsorption occurs in the proximal tubule. Almost the same quantity is thereafter re-added to tubule fluid in the thin ascending limb of the loop of Henle (tAL). Finally the amount excreted is determined by the collecting duct, depending on the degree of water reabsorption.

D-Glucose. The filtered amount of D-glucose is 900 mmol/day (i.e., 162 g). Urine is essentially glucose free as long as the plasma concentration does not exceed some 8–10 mmol/l (diabetes mellitus). The tubule reabsorption of D-glucose, therefore, is almost complete. It occurs exclusively in the proximal tubule [20].

Amino Acids. The filtered amounts of amino acids are rather small, usually less than 200 mmol/day for all amino acids together [113]. Still, the kidney is equipped with a variety of reabsorptive systems, localised in the proximal tubule, which ensure that the urine is practically amino acid free.

O₂ Consumption and NaCl Reabsorption. The reabsorption of the solutes listed in Table 74.1 is an energy-consuming process. This was shown directly some 30 years ago [19] in experiments in the dog, where the filtered load was varied by a cuff on the renal artery or pharmacologically and O₂ consumption was measured. The data are reproduced in Fig. 74.1. It is evident that a linear relation between O₂ consumption and NaCl reabsorption was obtained. The slope of the correlation was approximately 30 μmol O₂ per mmol NaCl. This slope translates into 5 mmol NaCl/mmol ATP. This ratio is surprisingly high (economical) if one considers (cf. below) that the Na⁺ transport occurs through the (Na⁺+K⁺)-ATPase, which has a stoichiometry of 3 Na⁺ per 1 ATP. How this economical mode of action is achieved will be discussed in subsequent sections.

Another observation in Fig. 74.1 deserves mention. The total O₂ consumption (with no experimental reduction of blood flow) of some 0.7 mmol/min (16 ml/min) contains two fractions: one for basal supply and one for active transport. It is evident from this figure that the basal rate of O₂ consumption is small compared with that for active transport.

Transepithelial Transport. As will be discussed in detail in the following sections, active transport of many solutes is coupled energetically to the activity of the (Na⁺+K⁺)-ATPase. Other active pumps are concerned with the transport of protons (H⁺-ATPase, H⁺/K⁺-ATPase) and of Ca²⁺ (Ca²⁺-ATPase) [61]. The Na⁺ reabsorptive mechanism consists of some uptake step across the luminal membrane (Fig. 74.2A) and the active, ATP-consuming extrusion across the basolateral membrane. The basic principle of coupling in the case of many other solutes consists in the

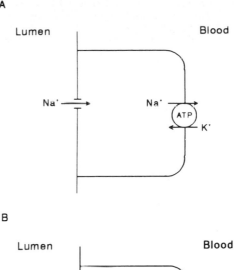

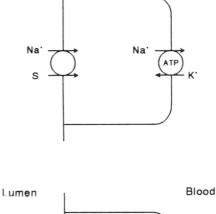

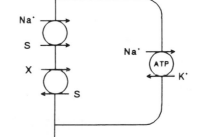

Fig. 74.2A–C. Principles of transepithelial transport (also cf. Chaps. 8, 11, 59). **A** Primary active transport of Na⁺ via (Na⁺+K⁺)-ATPase. Na⁺ is taken up via Na⁺ channels. This concept was originally suggested for the frog skin [80]. **B** Secondary active transport. A substrate S is taken up by a cotransporter with Na⁺. The driving force is provided by the electrochemical gradient for Na⁺. This in turn is maintained by the (Na⁺+K⁺)-ATPase. **C** Tertiary active transport. A substance X is exchanged against a substance S via an antiporter. The driving force is provided by the electrochemical gradient of S, which in turn is built up by a Na⁺ cotransport system. The latter is energised by the (Na⁺+K⁺)-ATPase

coupling to the Na⁺ flux. This is shown schematically in Fig. 74.2B. The solute is taken up across the luminal membrane via a cotransport (carrier) system utilising the favourable driving force for the Na⁺-substrate complex. The substrate

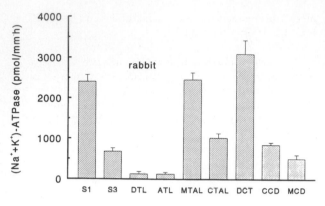

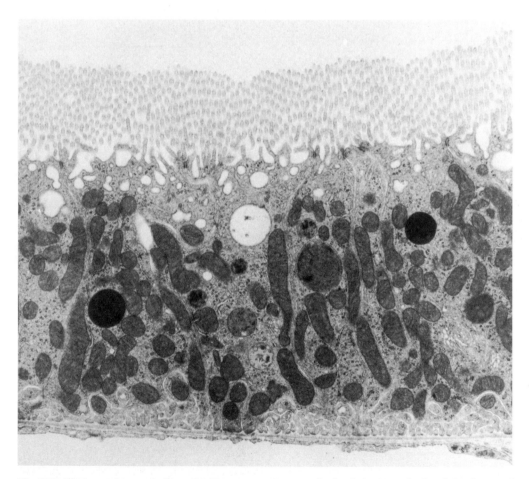

Fig. 74.3. Activity of the (Na⁺+K⁺)-ATPase along the rabbit nephron. *S1, S3,* segments of the proximal tubule; *DTL,* thin descending limb of the loop of Henle: *ATL,* thin ascending limb of the loop of Henle (these data are from the rat); *MTAL,* medullary thick ascending limb of the loop of Henle; *CTAL,* cortical thick ascending limb of the loop of Henle; *DCT,* distal convoluted tubule; *CCD,* cortical collecting duct; *MCD,* medullary collecting duct. Note that the highest activities are present in the S1 segment, MTAL and DCT. (Data from [72])

thus can be accumulated in the cell, and the exit mechanism across the basolateral membrane can occur via a Na⁺-independent carrier system. The Na⁺ taken up by the cotransporter is extruded by the (Na⁺+K⁺)-ATPase. This type of transport is called "secondary active".

Instead of the coupling to Na⁺ the transport of some substrates can also be coupled to the *countertransport* of another ion which is concentrated in the cell. The basic principle is shown in Fig. 74.2C. This carrier-mediated transport would then be called "tertiary active".

(Na⁺+K⁺)-ATPase Activity. The activity of the (Na⁺+K⁺)-ATPase in most nephron segments is high. Figure 74.3 summarises the activity for the various nephron segments of rabbit kidney [72]. It is evident that especially high activities per unit of length are obtained in the proximal tubule, in the TAL and in the distal convoluted tubule (DCT). This correlates well with the fact that the transport rates for NaCl are highest in these nephron segments [61,72].

Fig. 74.4. Electron micrograph from the S2 segment of rat proximal tubule. Note the brush border membrane, the numerous mitochondria, the basal labyrinth and the fairly simple tight junctions. (From [118]) × ca. 8000

74.3 Proximal Tubule: Large Transport Capacities But Small Gradients Between Tubule Fluid and Plasma

Morphological Features. It has been stated above that large quantities of the filtered load are transported by the proximal tubule. Figure 74.4 shows an electron micrograph of the first part of the proximal tubule S1 (cf. Chap. 73). Several morphological features, which in comparison to other nephron segments are schematically summarised in Fig. 74.5, deserve mention:

- These cells are especially mitochondria rich, testifying that the transport work, and hence the need to produce ATP by aerobic metabolism, is marked.
- The surface area of the luminal membrane is increased substantially by the formation of the brush border.

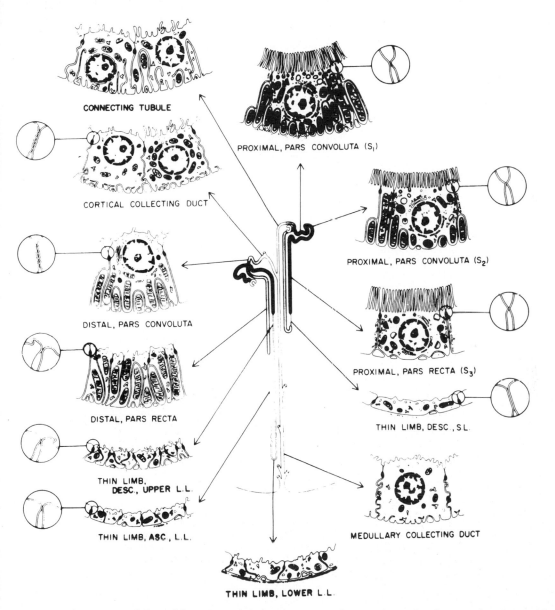

CONNECTING TUBULE

CORTICAL COLLECTING DUCT

DISTAL, PARS CONVOLUTA

DISTAL, PARS RECTA

THIN LIMB, DESC., UPPER L.L.

THIN LIMB, ASC., L.L.

PROXIMAL, PARS CONVOLUTA (S₁)

PROXIMAL, PARS CONVOLUTA (S₂)

PROXIMAL, PARS RECTA (S₃)

THIN LIMB, DESC., S.L.

MEDULLARY COLLECTING DUCT

THIN LIMB, LOWER L.L.

Fig. 74.5. Schematic morphology of the various tubule segments. For each nephron segment characteristic features of the morphology of the cell and of the tight junctions are depicted. Note that the membrane area is amplified by the brush border membrane and basal labyrinth in proximal segments. The segments with the highest density of mitochondria are also those with the highest $(Na^{+}+K^{+})$-ATPase activity (Fig. 74.3). The tight junctions are very simple in the S1, S2 and S3 segments of the proximal tubule and more complex in the DCT and CCT. (With permission from [67])

- Also the basolateral membrane area is increased substantially by the basal infoldings, which almost reach to the luminal aspect.
- The tight junctions are rather "primitive", with only a few strands.

As will be seen below, these morphological features are of great relevance to the cellular mechanism of NaCl- and water reabsorption.

D-Glucose Production. The metabolism of the proximal tubule is entirely aerobic. However, glucose is not metabolised at all. Preferred substrates are instead short fatty acids such as acetate. In fact, the proximal tubule is equipped with the machinery to perform gluconeogenesis, and it produces D-glucose if abundant substrates are available [127,128]. The fact that the proximal tubule does not utilise D-glucose is easily understandable since it has to reabsorb D-glucose from the tubule fluid. Hence it is "trading" an enormous amount of energy (900 mmol D-glucose/day, corresponding to 30 mol of ATP), but it is not using any part of it [128].

NH$_4^+$ Production. Another important metabolic function of the proximal tubule is to produce NH$_4^+$ from L-glutamine [62]. The amount of NH$_4^+$ produced normally is on the order of 20 mmol/day (as compared to a urea excretion of approximately 500 mmol/day). In metabolic acidosis the production of NH$_4^+$ can increase substantially to more than 200 mmol/day (cf. Chaps. 75 and 78). Then, however, the urea excretion is reduced correspondingly, because the liver produces less urea and more L-glutamine (see Chaps. 68, 75).

Basic Mechanisms of Reabsorption. Figure 74.6A shows a simplified scheme of the basic mechanisms. The (Na$^+$+K$^+$)-ATPase is localised exclusively in the basolateral membrane. This ensures low intracellular Na$^+$ activities of 10–20 mmol/l and high intracellular K$^+$ activities of around 100 mmol/l. The main source of Na$^+$ entry is the Na$^+$/H$^+$ exchanger [95]. This exchanger will be discussed below in further detail. The proton thus secreted buffers luminal HCO$_3^-$ and H$_2$CO$_3$ is formed. H$_2$CO$_3$ is dehydrated by carbonic anhydrase in the luminal membrane, and CO$_2$ is formed [86]. CO$_2$ enters the cell and, by the use of cellular carbonic anhydrase, H$_2$CO$_3$ is formed. H$_2$CO$_3$ dissociates and recycles the proton for secretion. HCO$_3^-$ leaves the cell via the basolateral membrane through a (HCO$_3^-$)$_3$Na$^+$ cotransport system [9,133]. The electrochemical driving force for this system is provided by the sum of the individual driving forces: Na$^+$: –130 mV (uphill: chemical gradient around 60 mV and membrane voltage –70 mV); HCO$_3^-$: +56 mV (downhill: chemical gradient around –14 mV and membrane voltage 70 mV). For the above stoichiometry the net result is a driving force of 40 mV for the outward reaction. Na$^+$ is pumped out by the (Na$^+$+K$^+$)-ATPase and K$^+$ is taken up in turn. K$^+$ recycles across the basolateral membrane through K$^+$ channels [38–40,50,56].

This event is responsible for the maintenance of a membrane voltage around –70 mV.

This basic mechanism leads to the transcellular reabsorption of NaHCO$_3$. It is limited by the availability of HCO$_3^-$. As will be shown below, this process not only leads to a rapid decline in luminal HCO$_3^-$ concentration, but is also the key step in the reabsorption of Na$^+$ and Cl$^-$. Before this will be discussed, the properties of the Na$^+$/H$^+$ exchanger have to be described in some detail.

Na$^+$/H$^+$ Exchanger. The Na$^+$/H$^+$ exchanger has been found in vesicle experiments obtained from brush border membranes of rat kidney [95]. In these experiments it was shown that proton movement is coupled to Na$^+$ movement, and Na$^+$ movement is coupled to that of protons. Both types of experiments are shown in Fig. 74.7. In the intact

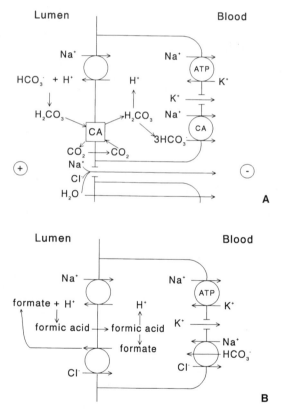

Fig. 74.6. A Basic principle of Na$^+$, Cl$^-$, HCO$_3^-$ and water reabsorption in the proximal tubule *Circle with ATP*, (Na$^+$+K$^+$)-ATPase; *circle*, carrier-mediated mechanism; *arrow*, diffusion; *CA*, carbonic anhydrase. Note that the preferential reabsorption of HCO$_3^-$ leads to a lumen positive voltage of a few mV, which drives Na$^+$ reabsorption across the paracellular shunt pathway. Also for Cl$^-$ diffusion the electrochemical driving force is in favour of paracellular reabsorption (E$_{Cl^-}$ > transepithelial voltage) [119]. B Transcellular Cl$^-$ reabsorption via the formate mechanism [101]. The symbols have the same meaning as in A. The lipid solubility of formic acid facilitates the diffusion of formic acid from the lumen into the cell. Formic acid dissociates and the H$^+$ is secreted by Na$^+$/H$^+$ exchange. Formate exchanges with Cl$^-$ by a specific carrier. The secreted formate is buffered by the secreted H$^+$. Cl$^-$ exit from the cell probably occurs via a Na$^+$-HCO$_3^-$/Cl$^-$ antiporter

1494

cell the driving force of this system is in favour of Na^+ uptake and proton extrusion: The driving force for Na^+ is 130 mV (cf. above), in favour of uptake, and that for H^+ is around -70 mV for extrusion. The net driving force is therefore around 60 mV. This system has been well characterised in many further studies (cf. Chap. 11). It was found that it is allosterically controlled by intracellular pH [3,76] in the sense that cytosolic acidification "switches on" this countertransporter and that alkalosis has the opposite effect (cf. Chap. 75). This explains how the kidney can respond to acidosis with improved HCO_3^- reabsorption. Furthermore, it has been shown that this system can be inhibited by the well-known K^+-sparing diuretic amiloride [76], and even more so by derivatives such as ethylisopropylamiloride (EIPA) (Fig. 74.8). Since it was found in the kidney, this countertransporter has been found in almost every cell where it has been sought. It seems that this system is a key element in the acid extrusion of almost every cell. This type of countertransporter has therefore been called pH-housekeeping system. In fact, a detailed analysis has revealed that the housekeeping system is also present in the basolateral membrane of most nephron segments. Its turnover rate is much lower than that of the luminal system. Whilst the basolateral system assists in pH homeostasis of the cell, the luminal system drives HCO_3^- reabsorption. The former system has now been cloned, and the cloning of the luminal system has been accomplished (cf. Chaps. 8, 11) [94].

The Consequences of Proximal HCO_3^- Reabsorption. The preferential reabsorption of $NaHCO_3$ (Fig. 74.6A) across the cell has several important consequences. As stated

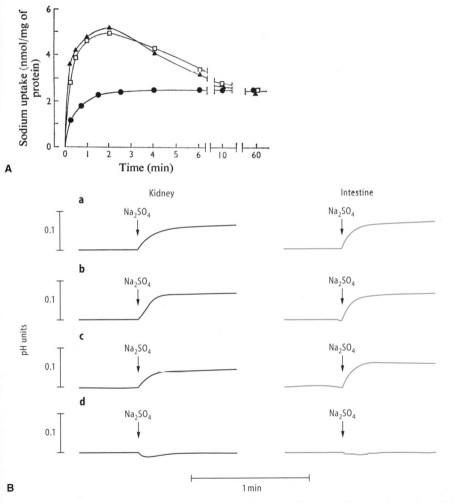

Fig. 74.7A,B. Experimental evidence for Na^+/H^+ exchange in the brush border membrane of kidney cortex. **A** Uptake of Na^+. The *closed circles* show the uptake without a H^+ gradient. The *open squares* have been obtained with an H^+ gradient of 1.6 units of pH. The *closed triangles* have been obtained with the same gradient and with the protonophore carboxyl cyanide *p*-trifluoromethoxyphenylhydrazone (FCCP). Note that an over-shooting uptake is obtained in the presence of an outwardly directed H^+ gradient. **B** Proton ejection in brush border vesicles prepared from the kidney and intestine: (*a*) vesicles preincubated with 1 μl ethanol; (*b*) valinomycin; (*c*) FCCP; (*d*) valinomycin and FCCP. Note that an Na^+ inward gradient drives proton ejection in all but the last type of experiment. (Data from [95])

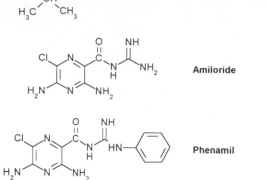

Ethylisopropylamiloride (EIPA)

Amiloride

Phenamil

Fig. 74.8. Chemical structure of amiloride, ethylisopropylamiloride (EIPA) and phenamil. Amiloride inhibits Na⁺/H⁺ exchange in the 100 µmol/l to mmol/l range, and Na⁺ channels in the subµmol/l range. EIPA has no effect on Na⁺ channels but a higher affinity for Na⁺/H⁺ exchange. Phenamil has a very high affinity for Na⁺ channels, but has no effect on the Na⁺/H⁺ exchanger

above, the HCO_3^- concentration along the proximal tubule falls sharply from 24 to perhaps 10 mmol/l. Cl^-, on the other hand, is left behind, and its concentration increases from 125 to 145 mmol/l. The concentration of Na^+ remains fairly constant. Due to the Cl^- concentration across the epithelium a lumen positive diffusion voltage builds up. This lumen positive voltage of 0.5–2 mV suffices to drive substantial amounts of Na^+ across the paracellular shunt pathway [30–32]. It has been estimated [32] that this quantity is twice as big as that transported across the cell. The transepithelial voltage is less than the Nernstian voltage for the Cl^- gradient, which would be 4 mV. Therefore, Cl^- is reabsorbed across the paracellular pathway following an electrochemical gradient of 2–3.5 mV. It has been estimated that all of the Cl^- might take this route, and may be driven by electrodiffusion and solvent drag [32]. This may not be entirely correct, as a minor fraction of Cl^- might move through the cell [101]. The mechanisms of this transport pathway will be described below.

"Iso-osmotic" H_2O Transport in the Proximal Nephron. The osmotic pressure of the tubule fluid stays constant along its entire length (Fig. 74.9A). Also the tubule fluid over plasma concentration rations (disregarding the deviation produced by the Gibbs-Donnan distribution) for Na^+, Ca^{2+} and K^+ are close to unity all along the proximal tubule (Fig. 74.9A). That for HCO_3^- falls and that for Cl^- increases correspondingly. The other solutes included in Fig. 74.9B will be discussed later in this and the next chapter. The fact that water can be reabsorbed pari passu with the osmolytes can be explained by the very high water permeability of the proximal tubule. There has been much debate as to the route of water movement. At this stage it appears likely that a fraction passes through the paracellular spaces and another, possibly larger moiety, through the cell. In fact a water channel ("chip-channel") has been cloned recently and this channel is present in the cell membranes of proximal tubules [24].

Transcellular Cl^- Reabsorption. Figure 74.6B contains two pathways for the movement of Cl^- across the cell. These are a formate/Cl^- exchanger in the luminal membrane and a Cl^-/HCO_3^-–Na^+ exchanger in the basolateral membrane.

The Cl^-/formate exchanger is fuelled by the high formate concentration in the cell, which is due to the recycling of lipid-soluble formic acid from the lumen into the cell [101]. The formate concentration in the lumen is reduced by the buffering with protons. Furthermore, this uptake system for Cl^- has a positive driving force for Cl^- since the Cl^- concentration in the lumen is much higher than that in the cell [30,119].

The basolateral exit of Cl^- probably occurs via an Na^+-coupled Cl^-/HCO_3^- exchange [60]. This exchanger obtains its driving force essentially from the Na^+ entry. The net driving force for Cl^- exit amounts to: Na^+ +130 mV; Cl^- −18 mV; HCO_3^- −57 mV (cf. above). The net driving force therefore amounts to 55 mV. It should be noted, however, that with this mechanism one HCO_3^- is recycled for each Cl^-. Alternatively, it has been claimed that Cl^- leaves the cell via Cl^- channels. Thus far no such channels have been found [56].

Reabsorption of K^+. The mechanisms of K^+ reabsorption are largely unknown [132]. Although it has been shown that K^+ channels exist in both membranes of the proximal tubule [38,87], there is little doubt that the serial arrangement of K^+ channels cannot account for transcellular K^+ reabsorption. K^+ is probably above equilibrium in proximal tubule cells, and it probably cannot enter the cell

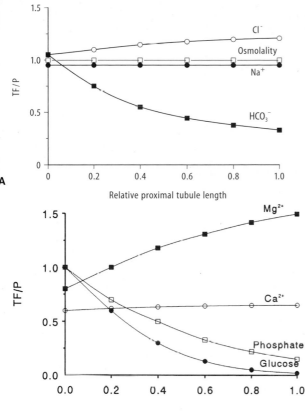

A

B   **Relative Proximal Tubule Length**

Fig. 74.9. A Tubule fluid/plasma concentrations (*TF/P*) along the proximal tubule. Data have been summarised from several micropuncture studies in the rat. Osmolality on both sides of the filter is identical (TF/P = 1) and stays constant all along the proximal tubule. The TF/P$_{Na^+}$ is close to 0.95 in the glomerulus and stays constant throughout the proximal tubule. The value of 0.95 is due to the Gibbs-Donnan distribution with the large anions kept on the P side. In the glomerulus TF/P$_{Cl^-}$ is 1.05 (Gibbs-Donnan effect). TF/P$_{Cl^-}$ increases along the proximal tubule and reaches 1.2–1.3 at the end of this segment. The TF/P$_{HCO_3}$ falls rapidly along the proximal tubule as a result of the preferential HCO$_3^-$ reabsorption. **B** Tubule fluid/plasma concentrations (*TF/P*) along the proximal tubule. Data have been summarised from several micropuncture studies in the rat. TF/P$_{Ca^{2+}}$ is 0.6 because of protein binding. TF/P$_{Ca^{2+}}$ stays fairly constant along the proximal tubule. TF/P$_{Mg^{2+}}$ is also smaller than 1, owing to some protein binding. TF/P$_{Mg^{2+}}$ increases rapidly along the proximal tubule, because Mg^{2+} is poorly reabsorbed in this nephron segment, yet water is removed (60%–70% of filtered H$_2$O by the end of the proximal tubule). TF/P$_{phosphate}$ and TF/P$_{glucose}$ in the glomerulus are close to unity. Both solutes are rapidly reabsorbed in the proximal tubule. Therefore, luminal concentration falls steeply

across the luminal membrane [25]. It has been claimed that, as an alternative, K$^+$ is taken up via an active pump localised in the luminal membrane [33], but no direct evidence for such a mechanism has been obtained thus far in mammalian kidney. Therefore, it is likely that the K$^+$ reabsorption is driven by the transepithelial voltage via the paracellular shunt pathway.

Other Substances. The proximal tubule also reabsorbs other solutes such as Ca^{2+}, phosphate, sulphate, urea, D-glucose and amino acids, and it secretes organic acids and bases. These processes will be dealt with in Chap. 75.

Overall Function. The overall function of the proximal tubule is to reabsorb large quantities of Na$^+$, Cl$^-$, HCO$_3^-$, K$^+$, etc. and water. An analysis on the economy of this process reveals that as much as 9 mol Na$^+$ (together with Cl$^-$, HCO$_3^-$ and water) can be reabsorbed by the consumption of 1 mol ATP. This is possible because large quantities are transported across the paracellular pathway, which is permeable to small ions and water [32]. In fact, the paracellular pathway has a very high electrical conductance of some 100–200 mS (also cf. Chap. 59). This is much higher than the estimates for the paracellular shunt of the TAL (approximately 30 mS) and that of the collecting duct, which is probably less than 10 mS [55]. The high conductance of the paracellular pathway of the proximal tubule (cf. Fig. 74.5) correlates well with the above-cited observation that the number of strands is much smaller in this nephron segment than in other nephron segments [67].

One other key function of the proximal tubule is the reabsorption of the majority of the filtered amount of HCO$_3$ and phosphate and the almost complete reabsorption of D-glucose and amino acids. In addition the proximal tubule secretes certain substances, e.g., oxalate, hippurates and other organic acids and bases [121].

Differences in Early Versus Late Proximal Tubule. The proximal tubule is subdivided into three portions, S1-S3. In the S1 segment HCO$_3^-$ reabsorption is maximal. Hence, the HCO$_3^-$ concentration falls early along the tubule axis. The S3 segment reabsorbs much less HCO$_3^-$. The reabsorption of D-glucose is also maximal in the S1 segment.

In fact, the transepithelial voltage in the very early S1 segment has reversed polarity (lumen negative) due to the preferential reabsorption of D-glucose together with Na$^+$ [29]. The secretion of organic acids and bases, on the other hand, occurs preferentially in the S2 and S3 segments. The fact that quantitatively most transport occurs in the S1 and S2 segments is also apparent from the (Na$^+$+K$^+$)-ATPase, which, as shown in Fig. 74.3, is highest in the S1 and lowest in the S3 segment.

74.4 Thick Ascending Limb: Dilution of Tubule Fluid

Morphological Features. Figure 74.10 shows an electron micrograph of the medullary TAL. Several features are of relevance for the understanding of the function (cf. the comparison of Fig. 74.5):

- The cells are rich in mitochondria.
- The basolateral membrane shows many and long basal infoldings.

1497

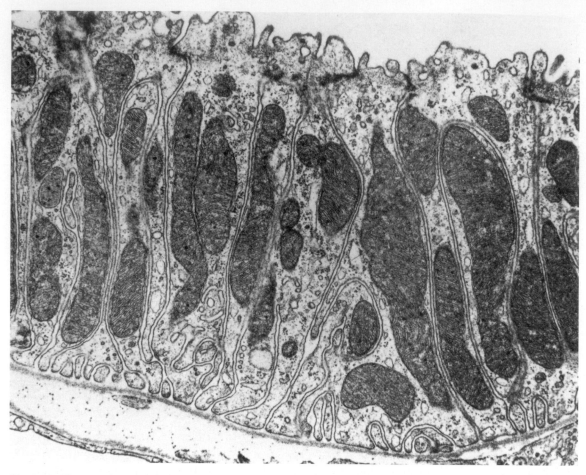

Fig. 74.10. Electron micrograph from the medullary thick ascending limb of rat. Note the complex basal infoldings and the high density of mitochondria. Tight junctions are clearly apparent. These tight junctions are more complex than those of the proximal tubule (cf. Fig. 74.4). (With permission from [67]) × ca. 16 000

• The tight junctions are more complex than those of the proximal tubule, but less complex than those of the collecting tubule.

These findings correlate with the findings that the amount of NaCl reabsorbed in this part of the nephron is 20%–30% of the filtered load. The tight junctions of this nephron segment are not as permeable as those of the proximal tubule, but still permit the permeation of small cations [47].

NaCl Reabsorption. Figure 74.11 shows the basic mechanisms of NaCl reabsorption in the TAL. It has been known for more than 40 years that this nephron segment dilutes tubule fluid inasmuch as it reabsorbs NaCl but no water [129]. Consequently, this nephron segment has also been named the "diluting segment" [12]. As this nephron segment is not directly accessible to micropuncture, the investigation of its detailed properties only became possible with the invention of the in vitro perfusion of isolated segments [11].

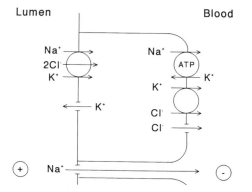

Fig. 74.11. Basic principle of Na$^+$ and Cl$^-$ reabsorption in the TAL. *Circle with ATP*, (Na$^+$+K$^+$)-ATPase; *circle*, carrier-mediated mechanism; *arrow*, diffusion. The back diffusion of K$^+$ across the luminal membrane and the diffusion of Cl$^-$ across the basolateral membrane generate a lumen positive voltage, which drives a substantial fraction of the Na$^+$ reabsorption (ca. 50%) across the paracellular pathway

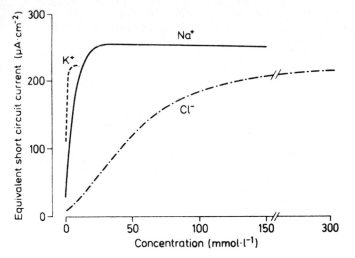

Fig. 74.12. Active transport in the TAL as a function of luminal Na⁺, Cl⁻ and K⁺ concentrations. As a measure of active transport the equivalent short circuit current (I_{sc}) is given (I_{sc} = transepithelial voltage × transepithelial conductance). Note that the function for Cl⁻ is S-shaped. A Hill analysis reveals a coefficient of 2 (i.e., 2Cl⁻ ions are transported in one cycle of the cotransporter). The Na⁺- and K⁺ dependencies are very steep, indicating that the affinities of the cotransporter for these ions are very high. (Data from [47,107])

(Na⁺+K⁺)-ATPase. The aforementioned nephron segment also possesses (Na⁺+K⁺)-ATPase in the basolateral membrane. This enzyme is especially concentrated in this portion of the nephron (Fig. 74.3), and has in fact been isolated from the renal outer medulla [71].

Na⁺2Cl⁻K⁺ Cotransporter. In the luminal membrane Na⁺ is taken up by a cotransporter: the Na⁺2Cl⁻K⁺ cotransporter. This carrier system was originally observed in Ehrlich ascites tumour cells [36] and shortly thereafter in the TAL [52]. Subsequently it has been found in a large variety of cells and organs. Like the Na⁺/H⁺ antiporter, this system appears to be ubiquitous and it is, for example, involved in volume regulation in a large variety of cells. In the TAL it has been characterised with respect to its affinities for the cotransported ions. Figure 74.12 shows the results from the various studies. These studies have been performed in in vitro perfused rabbit cortical TAL segments. It is evident that the Cl⁻ affinity is not very high (between 30 and 50 mmol/l). The curve for Cl⁻ is S-shaped, indicating that more than one Cl⁻ participates in the transport cycle. A Hill analysis has revealed that the coefficient is 2 [57]. The Na⁺ dependence is very steep. In fact, sodium-free glassware for the production of perfusion pipettes had to be used to obtain the lower points of this curve [45]. This explains why earlier attempts to examine the Na⁺ dependence proved negative: No dependence was found because the luminal Na⁺ concentration had not been sufficiently reduced [13]. At that time it was therefore believed that Cl⁻ was the actively transported ion in the TAL. In fact, a search for a Cl⁻-ATPase was begun which obviously revealed that no such enzyme was present in this nephron segment. It is also shown in Fig. 74.12 that the active transport in the TAL depends on the presence of K⁺ in the lumen. Again, this dependence is extremely steep and very low K⁺ concentrations have to be used to inhibit transport. It is physiologically relevant that NH₄⁺ can ride on the K⁺ binding site of this cotransporter. NH₄⁺ can be reabsorbed in the TAL by this mechanism and thus be delivered to collecting ducts, where it is secreted [42,73,74]. The properties shown in Fig. 74.12 have been largely confirmed in membrane vesicle studies [75,81].

The driving force for this cotransporter can be estimated since all the relevant measurements have been obtained in TAL cells. For Na⁺ the driving force is around +130 mV (+ indicating entry), for Cl⁻ the value is +40 − 70 = −30 mV, and for K⁺ the value is +70 − 90 = −20 mV. Taken together, a net driving force for entry of some +50 mV is obtained. The lower the luminal NaCl concentration, the smaller the driving force for the contransporter. This cotransporter has now been cloned [35].

Cl⁻ Conductance. The Na⁺ taken up by this system is pumped out by the basolateral (Na⁺+K⁺)-pump. Cl⁻ leaves the cell through a conductive pathway, via Cl⁻ conductance, and via a KCl cotransport system. The Cl⁻ conductance has been examined with patch clamp techniques and a small to intermediate conductance Cl⁻ channel has been found [49]. It has been shown in rodents, and may also be true for man, that the active transport in the TAL can be upregulated by arginine vasopressin (AVP = antidiuretic hormone, ADH). This occurs via cAMP [106]. The regulation via a cAMP-dependent protein kinase A occurs on the Cl⁻ channel [47]. The KCl system has thus far not been proven directly. It is postulated for this membrane on the basis of circumstantial evidence [54].

K⁺ Conductance. K⁺ recycles across the luminal membrane through a K⁺ conductance. This conductance has also been examined by patch clamp analysis. It was found that it corresponds to a 50–60 pS K⁺ channels [8]. An example of this channel is shown in Fig. 74.13. This figure also demonstrates a very important feature of this channel, namely its pH dependence from the cytosolic side. Small changes in cytosolic pH around the normal value have a large impact on the open probability of this channel: the open probability is increased with alkaline and reduced with acid pH. Other regulatory factors are:

- Cytosolic Ca²⁺: Open probability is reduced when cytosolic Ca²⁺ increases markedly (10 µmol/l). A fall in cytosolic Ca²⁺ has the opposite effect.
- Cytosolic ATP: A high ATP/ADP ratio reduces the open probability [8].

The Role of the Lumen Positive Voltage. The Cl⁻ conductance on the basolateral cell side with the respective Cl⁻ current and the K⁺ conductance with the respective K⁺ current across the luminal membrane polarise the epithelium such that the lumen is positive by some 8–15 mV with respect to the blood side. This lumen positive voltage drives Na⁺ across the cation selective paracellular shunt pathway. The flux of Na⁺ across this route is as large as is the transcellular flux [46,47]. Therefore, the economy of this nephron segment is 6 mol NaCl per 1 mol ATP. In comparison to the proximal tubule this is less economical, but still twice as economical as would be expected for exclusive transcellular Na⁺ reabsorption [47]. The metabolism of the TAL is entirely aerobic. It is important to note that this nephron segment has a very high metabolism per unit of length [72] and can easily be damaged in ischemia and hypoxemia [10,59]. The preferred metabolic substrates are D-glucose and short fatty acids (with even numbers). Substrate uptake occurs exclusively from the blood side [131].

Water Reabsorption. No water is reabsorbed in the TAL because the luminal membrane is water impermeable and because the amounts which might traverse the paracellular pathway are very small.

Ca²⁺ and Mg²⁺ Are Reabsorbed in the TAL. The routes for the reabsorption of both divalent cations have still not

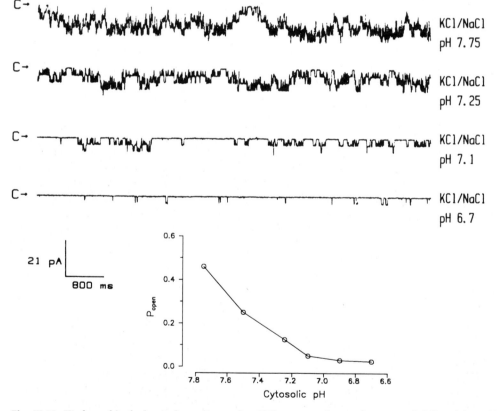

Fig. 73.13. K⁺ channel in the luminal membrane of rat TAL segment. Patch clamp recoding from an excised inside-out oriented luminal membrane. Pipette KCl-Ringerss; bath NaCl-Ringer's; clamp voltage 0 mV. This patch contains up to 5 K⁺ channels. The individual levels are clearly apparent from the current traces. *C*, zero current level. Note that an alkaline pH on the cytosolic side enhances the open probability of these channels. An acidic pH has the opposite effect. The pH dependence of the open probability (P_{open}) is summarised in the *lower half*. Note that this curve is steepest in the physiological pH range. (Data with permission from [8])

been clarified. On the one hand, much evidence points to the interpretation that the transepithelial voltage drives the ions across the paracellular pathway [47]. On the other hand, direct evidence for transcellular reabsorption has also been provided [28]. The transport of both divalent cations is regulated by various hormones – PTH, calcitonin, glucagon and ADH – which all enhance the reabsorption of both ions [18,130]. The regulation will be further discussed in Chap. 75.

Overall Function of the TAL. The TAL reabsorbs large quantities of NaCl. The medullary portion takes the larger share and the cortical portion comes into play to the extent that the medullary portion has not been able to cope with the load. It can be predicted from the properties of the $Na^+2Cl^-K^+$ cotransporter and that of the shunt that the net transport ceases when the luminal NaCl concentration has fallen to approximately 20–30 mmol/l. Then a diffusion voltage is superimposed on the active transport voltage, and total transepithelial voltage can approach +20–30 mV [55]. Under these conditions the residual small uptake via the cotransporter is balanced by the backleak across the shunt [47]. The point where this steady state will be reached depends on the load delivered to the TAL. In volume expansion or whenever there is a reduced proximal reabsorption this point will be in the cortical TAL or will not be reached at all. In volume contraction or with increased proximal reabsorption this point will be reached somewhere along the TAL.

Juxtaglomerular Apparatus. At the very end of the TAL the tubule is attached to the glomerulus at a site which is called the juxtaglomerular apparatus. The tubule cells facing the Goormaghtigh cells of the juxtaglomerular apparatus are specialised; they are columnar and are called the macula densa cells (cf. Chap. 73). These cells, via the same $Na^+2Cl^-K^+$ cotransporter, sense, according to the properties of this cotransporter, the tubule fluid Cl^- concentration [102,109] and reset the single nephron glomerular filtration rate (SNGFR) and renin secretion [111]. As discussed in the last chapter, a high tubule fluid Cl^- concentration reduces SNGFR and reduces renin secretion.

TAL and Urinary Concentrating Mechanism. As a consequence of the reabsorption of NaCl, without the water following, the interstitium becomes hypertonic. This hypertonicity is utilised to energize the countercurrent concentration mechanism. Due to this increased interstitial hyperosmolality, water is reabsorbed from the pars recta (S2 and S3 segments) of the proximal tubule, from the thin descending limb of long and short loops, and also from the collecting duct, provided the luminal membrane is permeable to water as a result of the insertion of water channels by AVP. The concentrating mechanism of the kidney will be discussed in detail in Sect. 74.7.

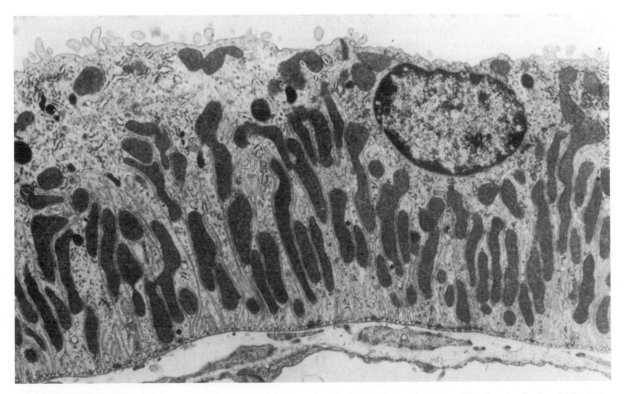

Fig. 74.14. Electron micrograph from the distal tubule of the rat. Note the high density of mitochondria, the complex basal labyrinth and the elongated tight junction. (With permission from [118]) × ca. 13 000

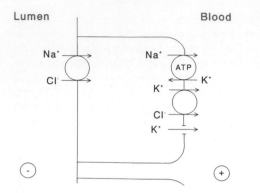

Lumen

Blood

Fig. 74.15. Basic principle of Na$^+$ and Cl$^-$ reabsorption in the distal tubule (DT). *Circle with ATP*, (Na$^+$+K$^+$)-ATPase; *circle*, carrier-mediated mechanism; *arrow*, diffusion. The Na$^+$Cl$^-$ cotransporter in the luminal membrane of the DT is different from the Na$^+$2Cl$^-$K$^+$ cotransporter in the TAL. The luminal membrane may also contain a K$^+$Cl$^-$ cotransporter

74.5 Transport in the Distal Tubule

Morphological Features. Figure 74.14 shows an electron micrograph of the distal tubule. This segment is heterogeneous, as discussed in Chap. 73. The cells are still mitochondria-rich and have amplification of both the luminal and the basolateral membrane area (cf. Fig. 74.5).

NaCl Reabsorption. Figure 74.15 shows the basic principles of NaCl reabsorption in this nephron segment [48,58]. The (Na$^+$+K$^+$)-ATPase is again located in the basolateral membrane. The activity of this enzyme is comparable to that in the TAL or proximal tubule (cf. Fig. 74.3). The uptake of Na$^+$ in this segment occurs via an NaCl cotransport system. There has been some discussion as to whether this system is different from the Na$^+$2Cl$^-$K$^+$ cotransporter. Apart from the fact that the NaCl cotransporter in the distal tubule does not require K$^+$, this cotransporter is inhibited by thiazides, whilst the Na$^+$2Cl$^-$K$^+$ cotransporter is inhibited by loop diuretics of the furosemide type (cf. Sect. 74.8). Thiazides have no effect on the latter, and furosemide has no effect on the former (cf. Sect. 74.8). This cotransporter has now been cloned [34]. The driving force for this cotransporter can only be estimated. It usually operates at rather low luminal NaCl concentrations of 30–150 mmol/l. If we assume a luminal NaCl concentration of 50 mmol/l and cellular concentrations around 15 mmol/l, the driving force amounts to: Na$^+$ = +102 mV (+ indicating uptake of NaCl); Cl$^-$ = −38 mV. The net driving force therefore amounts to +64 mV, favouring uptake of NaCl.

K$^+$ Conductance. Besides the NaCl cotransport system some data suggest the existence of a KCl cotransporter in the luminal and in the basolateral membrane. This system might account for K$^+$ movements [58,122]. The basolateral membrane contains a dominating K$^+$ conductance. In fact, the voltage of this membrane appears to be very close to

the chemical potential for K$^+$, namely −90 mV [58]. The K$^+$ channels serve the recycling of K$^+$ across the basolateral membrane. By patch clamp analysis a K$^+$ channel has been identified in this membrane in preliminary studies [56]. No evidence for luminal conductive pathways has been obtained thus far. The transepithelial voltage in this nephron segment is only slightly negative.

Economy of NaCl Transport. The economy of NaCl transport in this nephron segment is probably 3 mol NaCl per 1 mol ATP. This is due to the fact that all NaCl is transported transcellularly. The paracellular pathway of the distal tubule is probably rather tight, as can be deduced from the morphology of the tight junctions (cf. Fig. 74.14) as well as from the fairly high transepithelial resistance [48,58].

Hormone Receptors. Hormone receptors for β-agonists and calcitonin have been reported for this nephron segment. The former is supposed to enhance reabsorption, while the effect of the latter is unclear. Bradykinin receptors have also been seen in this segment, but again the exact role of this local hormone is not clear [82].

Overall Function of the Distal Tubule. The overall function of this nephron segment is the reabsorption of approximately 5%–10% of the filtered load of NaCl. The amount of K$^+$ secreted or reabsorbed largely depends on the luminal K$^+$ and Cl$^-$ concentration. In several species, including man, this nephron segment is not homogeneous. It transits into the so-called connecting tubule (CNT). The CNT possesses amiloride-sensitive Na$^+$ channels in the luminal membrane and hence a lumen negative voltage. Therefore, in the intact kidney, the voltage recorded in the distal tubule is mostly negative and it responds to amiloride [82,122].

One important feature of this segment is its ability to reabsorb a fraction of the filtered Ca^{2+} but not Mg^{2+}. This reabsorption can be increased when NaCl reabsorption is inhibited by thiazides [16,17].

74.6 Collecting Duct: Fine Control of Water and Ion Balance

Morphological Features. The morphology and the functional characteristics of the collecting duct are very complex (Fig. 74.5). The connecting tubule and the cortical collecting duct (CCD) possess *principal cells*, which are responsible for Na$^+$ and water reabsorption, and *intercalated cells*, which are responsible for proton secretion or HCO$_3^-$ secretion. The number of the former cells is larger than that of the latter. In the outer medullary collecting duct (OMD), as it passes from the outer to inner stripe, the ratio changes in favour of cells similar to intercalated cells. However, species differences are marked in this segment of the nephron [118]. The inner medullary collecting duct

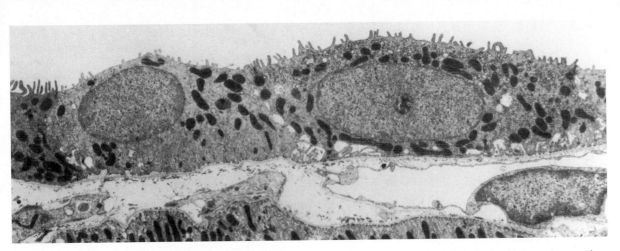

Fig. 74.16. Electron micrograph from rat CCD. On the *right* there is a dark (intercalated) A cell and on the *left*, a principal cell. The mitochondria are less dense in these cells. The intercalated cell contains numerous vesicles below the luminal membrane. Also the cytoplasm is more dense in this cell. The tight junctions are elongated and complex. (With permission from [67]) × ca. 5000

(IMD) is heterogeneous: The initial portion is similar to the inner stripe portion of the OMD, does not contain many principal cells and does not reabsorb Na^+, but secretes protons or else HCO_3^-; the terminal portion contains cells related to principal cells and is similar to the papillary collecting duct (PCD).

Figure 74.16 shows an electron micrograph of the CCD with principal cells and intercalated (mitochondria-rich) cells. The prominent structural features of the principal cells are:

* Much less surface amplification. The luminal membrane has only very few apical microprojections, and the basal labyrinth is much less prominent than in any of the other segments.
* The cells contain fewer mitochondria and these tend to be localised in the apical cell pole.
* The tight junctions are very complex. As a consequence, the paracellular pathway is much less permeable than that of the proximal tubule or of the TAL segment [55].

The intercalated or dark cells bulge into the lumen. Their cytoplasm is dense (dark). They possess more mitochondria than the principal cells. On the basis of their function they can be subdivided into proton secreting (A-cells) or HCO_3^- secreting (B cells). This distinction can also be made on the basis of their binding of lectins: B cells bind peanut lectin, A cells do not. A cells bind antibodies against band 3 protein (HCO_3^-/Cl^- exchange) on their basal cell pole, B cells do not [112].

Na$^+$ Conductance. Figure 74.17 depicts the various transport characteristics of principal (A), A-intercalated (B) and B-intercalated cells (C). In the principal cell Na^+ is taken up via Na^+ channels localised in the luminal membrane. These channels are similar to those observed in other tight epithelia such as frog skin [7]. They are highly selective for Na^+. Their single-channel conductance is small, and they are sensitive to low concentrations of amiloride (Fig. 74.8). It is probably their small conductance which has rendered it so difficult to detect these channels in patch clamp studies [56,88]. A few studies have reported single-channel recordings. One example is shown in Fig. 74.18. The single-channel conductance appears to be <10 pS. The channel has rather slow kinetics and is inhibited reversibly by <1 µmol/l amiloride [96]. Recently this channel has been cloned [14,89].

The Na^+ taken up by the Na^+ channels is pumped out by the (Na^++K^+)-ATPase localised in the basolateral membrane. The serial arrangement of Na^+ channel and Na^+ pump can generate very high Na^+ concentration gradients across the epithelium. This is especially true for states of Na^+ deprivation, where aldosterone enhances the Na^+ conductance as well as the pump density and mitochondrial ATP generating enzymes [91]. Under such conditions the luminal Na^+ concentration can be very low (a few mmol/l). From the mechanism of Na^+ reabsorption it should be evident that the stoichiometry has to be 3 mol Na^+ per 1 mol ATP. Therefore, the process of reabsorption is much less economical than that in the proximal tubule and thick ascending limb, but the gradient built up in the collecting duct can be very high. Details of the hormonal regulation of the reabsorption of Na^+ in this nephron segment will be provided in Chap. 75.

K$^+$ Conductance. The K^+ taken up via the (Na^++K^+)-ATPase is partly recycled across the basolateral membrane via K^+ channels. The regulation of these channels is currently being investigated by patch clamp techniques [69], and it appears that they are regulated by cGMP. The electrogenicity of the (Na^++K^+) pump can hyperpolarise the basolateral membrane voltage beyond the chemical potential for K^+. Under such conditions K^+ can even be taken up by these K^+ channels.

The other route of K^+ exit from the principal cell is via luminal K^+ channels. Two types of luminal K^+ channels

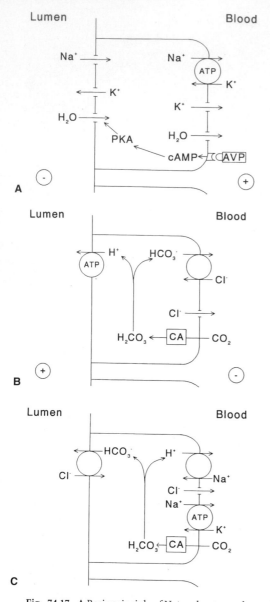

Fig. 74.17. A Basic principle of Na$^+$ and water reabsorption in the CCD principal cell. *Circle with ATP*, (Na$^+$+K$^+$)-ATPase; *arrow*, diffusion. Na$^+$ enters the cell via Na$^+$ channels. K$^+$ leaves the cell via K$^+$ channels in the basolateral membrane (recycling pathway) and via luminal K$^+$ channels. The distribution of the conductances gives rise to the lumen negative voltage. Water reabsorption is regulated by antidiuretic hormone (AVP) via cAMP. PKA-dependent phosphorylation leads to the insertion of water channels. **B** Basic principle of H$^+$ secretion in the CCD intercalated A-type cell. *Circle with ATP*, (H$^+$)-ATPase; *circle*, carrier-mediated mechanism; *arrow*, diffusion; *CA*, carbonic anhydrase. The pump generates a lumen positive voltage. CO$_2$ produced (or entering the cell) is hydrated to carbonic acid. The proton is secreted actively, giving rise to lage H$^+$ gradients across the epithelium. HCO$_3^-$ leaves the cell via a band 3 type Cl$^-$/HCO$_3^-$ exchanger [97]. **C** Basic principle of HCO$_3^-$ secretion in the CCD intercalated B-type cell. *Circle with ATP*, (Na$^+$+K$^+$)-ATPase; *circle*, carrier-mediated mechanism; *CA*, carbonic anhydrase. CO$_2$ produced (or entering the cell) is hydrated to carbonic acid. The proton is extruded by the basolateral Na$^+$/H$^+$ exchanger. HCO$_3^-$ is secreted by a Cl$^-$/HCO$_3^-$ exchanger localised in the luminal membrane. This exchanger is not identical with the basolateral exchanger in the A-type cell

have been identified in patch clamp studies: (a) a small conductance channel of ca. 20 pS, which is regulated by cytosolic pH, ATP and phosphorylation by protein kinases A and C [68,104,105,108,123,124]; (b) a large K$^+$ channel which has a conductance of 150–250 pS and which is only active when cytosolic Ca^{2+} is increased. This channel is opened by cell swelling [68,103]. The K$^+$ channels of the luminal membrane are of predominant importance in determining the amount of K$^+$ secreted by the CCD. Key determinants are [132]:

- The luminal flow rate: The higher the flow rate, the higher the rate of K$^+$ secretion.
- The luminal Na$^+$ delivery: The higher the delivery, the higher the rate of K$^+$ secretion. These two mechanisms are largely responsible for the enhanced K$^+$ secretion in the presence of diuretics.
- Cytosolic pH: Cytosolic alkalosis enhances while cytosolic acidosis attenuates the open probability of luminal K$^+$ channels.
- Hormones such as AVP appear to increase the number of active K$^+$ channels.
- Increases in driving force and luminal depolarisation by aldosterone and AVP enhance K$^+$ secretion.

From the above, it is also evident that increases in aldosterone- and/or AVP-mediated Na$^+$ reabsorption enhance the driving force for K$^+$ secretion inasmuch as the lumen membrane is depolarised by both hormones.

Cl$^-$ Reabsorption. The mechanism of Cl$^-$ reabsorption in the CCD is much less clear. The amount reabsorbed can be deduced from the amount of Na$^+$ reabsorbed – that of K$^+$ secreted. At present, no evidence exists to favour transcellular Cl$^-$ reabsorption in principal cells. Therefore, by exclusion two routes are feasible [110]:

- Paracellular reabsorption, driven by the lumen negative voltage
- Transcellular reabsorption by intercalated cells

Water Reabsorption. The reabsorption of water in the CCD is determined by the water permeability of the luminal membrane of principal cells. *AVP* controls this process. This was shown impressively in very early studies [43] in isolated perfused CCDs. It was demonstrated that a dilute luminal perfusate increased the cell volume in the presence but not in the absence of AVP. This mechanism of AVP control of water permeability is further depicted in Fig. 74.17A. ADH, via cAMP, induces the insertion of particles into the luminal membrane [66,77]. These particles, as originally shown for toad urinary bladder [66], correspond to *water channels*. The water channels have now been identified and cloned [24]. They are similar but not identical to the "chip" water channel [24]. Unlike the luminal membrane, the basolateral membrane is permanently water permeable. Therefore, the degree to which peritubule hyperosmolality can drive

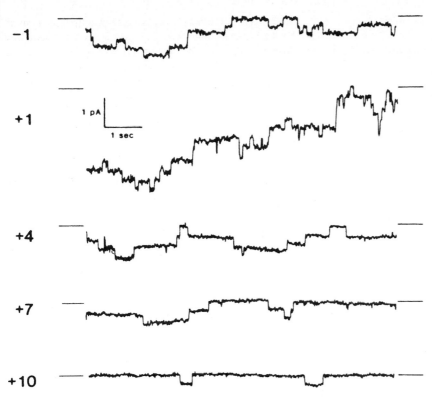

Fig. 74.18. Na$^+$ channel in the luminal membrane of a principal cell of rat CCD. Cell attached recording. Clamp voltage pipette +60 mV. 10 μmol/l ionomycin was added to the bath. The individual traces show a superimposition of up to ten Na$^+$ channels before and after the increase of cytosolic Ca^{2+} by ionomycin. Note that the open probability of these channels is reduced 4, 7, and 10 min after the addition of ionomycin. It is concluded that this type of channel is downregulated by increases in cytosolic Ca^{2+} activity. (Data from [96])

water reabsorption in the collecting duct is exclusively determined by the presence of luminal AVP-controlled water channels.

The AVP-dependent water permeability extends throughout the collecting tubule. The transduction processes (cf. Chap. 5) involving the V$_2$ type receptor and the production of cAMP have been studied in some detail [66,77]. At the level of G-protein, this cascade can be interrupted, e.g., by Li$^+$, which produces a renal diabetes insipidus [70]. cAMP, via protein kinase A, gears the microtubule-mediated insertion of preformed water channels. The movement of water allows for the accompanying movement (absorption) of urea. In fact, increases in urea transport by AVP have been shown for the collecting duct. Whether this transport of urea occurs through the water channel itself, or whether urea moves independently following the high concentration gradient built up in this terminal part of the nephron and via a specific transporter, is still a matter of controversy [79,100]. It is clear, however, that the urea permeability, unlike that of water, is most marked in the terminal (medullary and papillary) collecting duct (cf. next section).

H$^+$ Secretion and HCO$_3^-$ Reabsorption. The schemes in Fig. 74.17B and C show the basic mechanisms of H$^+$ secretion (A cell) or HCO$_3^-$ reabsorption (B cell) in this nephron segment. The ratio of A versus B cells can be varied experimentally according to the acid-base status. A cells possess a *vacuolar-type H$^+$-ATPase* in the luminal membrane. This protein has been well characterised and its subunits are cloned [37,134]. In addition, recent evidence has accumulated that a *H$^+$/K$^+$-ATPase* is also present in the collecting duct [126]. The extent, however, to which the latter pump contributes to acidification is unclear. It may be even more relevant as a reabsorptive mechanism for K$^+$ inasmuch as the enzyme appears to be enhanced in the presence of K$^+$ deficiency. Like the gastric H$^+$/K$^+$-ATPase (cf. Chap. 65), this pump is inhibited by omeprazole and by the compound S 28080 [126]. The protons are formed in the cell from CO$_2$. In fact, these cells are especially rich in *carboanhydrase*. The HCO$_3^-$ left behind is removed from the cell by a band 3 type anion exchanger, the HCO$_3^-$/Cl$^-$ exchanger, present in the basolateral membrane [22]. The transport systems present in the B cell are much less clear. Previous speculation that a shift from A to B cells might only involve a readdressing of the membrane proteins – the H$^+$ pump to the basolateral membrane and the HCO$_3^-$/Cl$^-$ exchanger to the luminal membrane – have not withstood further experimentation. Also it is clear now that the putative luminal HCO$_3^-$/Cl$^-$ exchanger in B cells is not identical with that present in the basolateral membrane of A cells [22].

A cells must be of much more importance than B cells in normal man. HCO$_3^-$ secretion, by contrast, has primarily been demonstrated in the rabbit [93], which, due to its vegetarian food, has to excrete much more alkali. In omnivorous man normal urine pH is usually acidic (typically 5–6).

Summary of Functions. It has been shown in this section that the main functions of the collecting duct are the fine tuning of:

- Water excretion
- NaCl excretion
- K$^+$ excretion
- Acid excretion.

These tasks are performed by several different cell types. Since the ion gradients across the epithelium are high in this nephron segment, most transport processes in these cells are expensive energetically and involve a close coupling between transport and ATP consumption. The metabolism in these cells is heterogeneous. In the CCD the preferred substrate is D-glucose and metabolism is aerobic. In the medullary collecting duct and in the papilla, ATP production can also proceed anaerobically (glycolysis) [127,128]. This is necessary because O$_2$ pressures can be very low in the deep medulla and papilla. Substrate uptake probably occurs from the blood side.

74.7 The Countercurrent Concentrating Mechanism

Urine-Concentrating Ability. The key feature of the countercurrent concentrating mechanism is its ability to produce highly concentrated urine in states of water deprivation. The urine volume in man then may be as little as < 1 ml/min and the urine osmolality may be 1000 mosmol/l. Conversely, in states of overhydration the urine is dilute, the flow rate may exceed 10 ml/min and the urine osmolality may be as low as 100 mosm/l. Key prerequisites for this system are:

- The presence of loops of Henle with a parallel arrangement of the collecting duct system
- The modulation by AVP.

A comparison amongst different vertebrates indicates that the loop of Henle has emerged in terrestrial animals [114]. In amphibia and reptiles the loop of Henle is still absent, whilst a diluting system, corresponding to the thick ascending limb, is already present [18]. In birds two populations of nephrons can be found: the reptilian-type nephrons, representing the majority, which lack loops of Henle, and the mammalian-type nephrons with loops of Henle. Amongst mammals relevant quantitative and even qualitative differences exist with respect to the loop of Henle: Some animals have only short loops of Henle which do not extend beyond the border between inner stripe of the outer medulla and inner medulla. The beaver belongs to this group of animals. Others, such as man, have long as well as short loops of Henle (typical mammalian kidney), and yet others, e.g., carnivores, have only long loops. It has been suggested that the relative length of the loops of Henle correlates with the ability to save water and to concentrate urine [18,44]. However, the ability to concentrate urine is influenced not only by the length of the loops or the relative thickness of the renal medulla but also by the numbers of nephrons relative to kidney weight, the distribution of superficial versus deep nephrons, the percentage of long versus short loops of Henle, the development of pelvic fornices, the arrangement of vascular bundles in the inner stripe of the outer medulla and probably further features. Therefore, it is not appropriate to predict from the particular development of one of these parameters in one species the ability to adapt to water deprivation [6].

The other relevant achievement in terrestrial animals is the usage of AVP (cf. Chap. 82) for the regulation of this system [65]. This octapeptide hormone serves two major functions in evolution:

- It is a potent vasoconstrictor.
- It helps to produce a concentrated urine.

The latter function is generally achieved by the increase in water permeability of the collecting duct (cf. above). In addition, this hormone increases the diluting capacity of the thick ascending limb in those species concentrating their urine to very high values, such as the rat, Psammomys, and the "master in the concentrating of urine", namely the pocket mouse. This remarkable mammal can produce urine osmolalities of 8600 mosm/l [44]. In man AVP probably has little or no effect on the rate of NaCl reabsorption in the TAL [18].

Countercurrent Concentrating System. The basic principle of urinary concentration is that of a countercurrent concentrating system [84]. The individual components which are arranged in a countercurrent (flow) fashion are:

- Descending limbs of the loop of Henle
- Ascending limbs of the loop of Henle
- Collecting ducts
- Descending vasa recta
- Ascending vasa recta.

A very simple depiction of this is given in Fig. 74.19. It was recognised about 50 years ago that this arrangement in the kidney corresponds to countercurrent systems such as heat exchangers [84]. In such a system the counter flow enhances (multiplies) along the system the single effect occurring at any given level.

In the kidney the single effect is the transport of NaCl out of the ascending limb of the loop of Henle. It will be discussed below that the single effect in the thick ascending

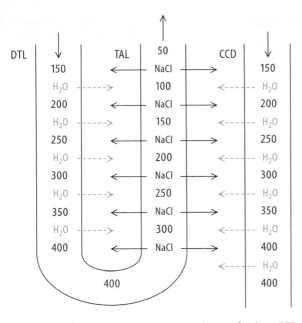

Fig. 74.19. The countercurrent concentrating mechanism. *DTL*, pars recta and descending thin limb of the loop of Henle; *TAL*, thick ascending limb of the loop of Henle; *CCD*, cortical collecting tubule; *solid arrows*, active reabsorption of NaCl from the TAL; *dotted arrows*, water movement. Note that interstitial hypertonicity, produced by the TAL, withdraws water from the DTL. Therefore the luminal concentration of NaCl increases as the tubule fluid travels down to the bending point. The single effect generates a gradient of some 50–100 mmol/l at each level. The effect is multiplied by the countercurrent flow. The CCD can equilibrate with the interstitium, if the luminal membrane becomes water permeable under the influence of ADH

limb of the loop of Henle (mTAL and cTAL) is the active reabsorption of NaCl, and that the single effect in the thin ascending limb of the loop of Henle (ATL) is the passive diffusion out of the tubule. The reabsorption of NaCl (active or passive) enhances the osmolality in the interstitium and withdraws water from the descending limb of the loop of Henle (DTL). As a consequence, the fluid delivered to the next, deeper level is already more concentrated. Further water withdrawal leads to an even higher concentration of tubule fluid [120].

The multiplication effect is caused by the counter flow. At individual levels the single steps are approximately equal, but the concentration in all the compartments of the single levels – DTL, interstitium, ATL (or TAL), vas rectum descendens and vas rectum ascendens – increases from outer medulla to papilla. Experimentally the existence of such a system in the kidney has been proven by measurements of the osmolality in the bending points of loops of Henle in the papilla and of that in the early distal tubule [120,129]. It was found that in the rat, for example, the osmolality in the tip of the papilla can easily be as high as 1500 mosm/l whereas it may be only 100 in the early distal tubule at the kidney surface [18,129].

Function of the Countercurrent Concentrating System. How can the countercurrent concentrating system be uti-

lised to concentrate or dilute urine? As shown in Fig. 74.19, water can be reabsorbed from the collecting duct at each level, provided that the luminal membrane is made water permeable by the action of ADH (=AVP). The osmolality of the collecting duct can then equilibrate with the interstitium and urine osmolality can increase to 1000 mosm/l in man and to much higher values in water-saving animals such as *Gerbillus*, *Psammomys*, rat and mouse.

Understanding of the detailed function of the countercurrent concentrating system is still not complete. In the following the current interpretation of its function will be summarised. More detailed discussions will be found elsewhere [e.g., 18,78,83,115].

Figure 74.20 shows the nephron segments participating in the countercurrent system. They are:

- S2 and S3 segments of the proximal tubule
- Descending thin limbs of short loops (DTL$_{sl}$)
- Descending thin limbs of long loops (DTL$_{ll}$)
- Ascending thin limbs (ATL)
- Medullary thick ascending limbs (mTAL)
- Cortical thick ascending limbs (cTAL).

The functional properties of these nephron segments are summarised in Table 74.2. The vasa recta will not be discussed further here, although it is clear that their arrangement is important for exchange between descending and ascending blood flow, and although it has been proven in elegant studies how complex the arrangement of vascular bundles and their local arrangement with DTL$_{sl}$ can be, especially in species with high concentrating ability [6].

Components of the Countercurrent Concentrating Mechanism. The countercurrent concentrating mechanism can be subdivided into two major components.

- Cortical and outer medullary system
- The inner medullary and papillary system.

The function of the *cortical and outer medullary system* can be easily understood on the basis of the active transport step in the TAL [47]. As has been shown in Sect. 74.4,

Table 74.2. Properties of the nephron segments participating in the countercurrent concentrating mechanism

Segment	Permeabilities			
	H$_2$O	Na$^+$	Cl$^-$	Urea
S3	+++	T	T	+
DTL$_{sl}$	+++	+++	++	0
DTL$_{ll}$	+++	0	0	++
ATL	0	+++	+++	++
TAL	0	T	T	+
CCD	AVP	T	T	0
IMD	AVP	T	T	AVP

+++, highly permeable; ++, permeable; +, slightly permeable; 0, impermeable; T, active transport; AVP, regulated by AVP

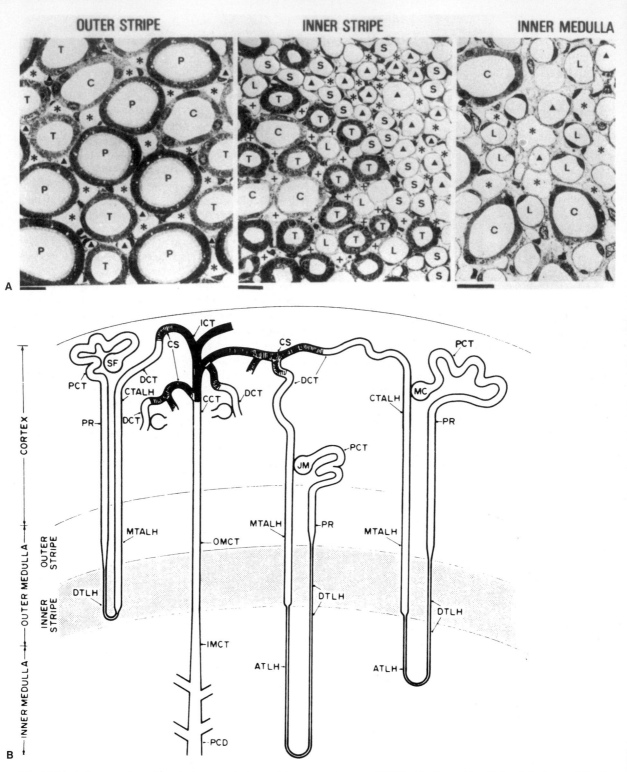

Fig. 74.20. A Cross-sections through the outer stripe and inner stripe of the outer medulla and through the inner medulla. *Bars* correspond to 30 μm. *P* pars recta; *T*, thick ascending limb of the loop of Henle; *C*, collecting duct; *S*, thin descending limbs of short loops; *L*, thin descending limbs of long loops; *, venous ascending vasa recta; *triangles*, + arterial descending vasa recta. **B** The various arrangements of the loops of Henle of three types of nephron segments: *SF*, superficial nephrons *MC*, mid cortical nephrons; *JM*, juxtamedullary nephrons; *PCT*, proximal convoluted tubule; *PR*, pars recta; *DTLH*, thin descending limb of the loop of Henle; *ATLH*, thin ascending limb of the loop of Henle; *MTALH*, medullary thick ascending limb of the loop of Henle; *CTALH*, cortical thick ascending limb of the loop of Henle; *DCT*, distal convoluted tubule; *CS*, connecting (segment) tubule; *ICT*, initial collecting tubule; *CCT*, cortical collecting tubule; *OMCT*, outer medullary collecting tubule; *IMCT*, inner medullary collecting tubule; *PCD*, papillary collecting duct. (With permission from [6,23])

1508

NaCl is reabsorbed by this segment at high rates and with the use of ATP. Water cannot follow because the luminal membrane of the TAL is water impermeable (Table 74.2). This generates an interstitial hypertonicity. The S2, S3 and DTL segments are all water permeable (Table 74.2). Therefore, water is withdrawn from these segments and tubule fluid is concentrated. There has been some debate as to whether the concentration of descending tubule fluid occurs by water reabsorption or NaCl addition. The former probably occurs in all species, while the latter is most marked in *Psammomys obesus*. NaCl addition, however, can only occur in the upper part of the DTL_{II}, because only this epithelium has paracellular pathways which are permeable for NaCl [18]. The presence and prevalence of this specific segment varies amongst species. Its importance in man remains to be clarified. It appears safe to conclude that *water reabsorption is more relevant to the concentration of descending tubule fluid than is NaCl addition.*

The function of the *inner medullary and papillary system* is much less clear. On the one hand, it has been claimed that some active step, e.g., active reabsorption from the ATL, would be required to generate the necessary single step. However, it now appears clear that very little (Na^++K^+)-ATPase is present in the ATL [72] (also cf. Fig. 74.3), and that any active reabsorptive mechanism is highly unlikely. On the other hand, a passive model has been suggested [83,115]. This model postulates that NaCl reabsorption in the ATL is passive.

In fact, as is apparent from Table 74.2, this nephron segment has properties quite different from the DTL_{II}. The former has a high NaCl permeability, the latter has a high water permeability. Both segments have in common a limited urea permeability. The key question, therefore, is whether the NaCl gradient from ATL to the interstitium is high enough to drive passive reabsorption (single effect). This gradient would have to be built up by water absorption from the DTL.

It has been proposed that water can be withdrawn from the DTL because urea leaves the inner medullary and papillary collecting tubule in the presence of AVP. It is important to note in this context that the cortical and outer medullary collecting duct is not permeable for urea, even in the presence of AVP. Therefore, the urea delivery to the inner medullary interstitium builds up urea concentrations exceeding those of the DTL and ATL. This gradient leads to some entry of urea into both segments, but, most importantly, it removes water from the DTL. Given the impermeability of the DTL to NaCl, this leads to the concentration of NaCl in the DTL as the tubule fluid travels down to the bending point of the thin limb.

In the ATL NaCl diffuses into the interstitium (and into the ascending vasa recta). Water cannot follow because this segment is impermeable for water. The driving force for this entire process is provided by the high interstitial urea concentration, and this is generated by the delivery of urea from the collecting duct to the medulla. The high concentration of urea in the tubule fluid of the medullary and papillary collecting duct is caused by the following facts:

- Some urea enters the DTL and ATL segments.
- Urea is mostly conserved along the TAL.
- A load of urea exceeding that delivered to the DTL is present in the distal tubule (DT).
- Urea cannot leave the cortical and outer medullary collecting duct.
- Water is removed through the entire collecting duct in the presence of ADH and urea is concentrated.

In other words, urea can be regarded as the "motor" driving the single effect in the renal medulla. This motor, in turn, is energised by the water absorption from the cortical and outer medullary collecting duct, which is caused by the active absorption of NaCl in the TAL. Hence, the active transport step in the TAL appears to drive both the single effect in the cortex and outer medulla and that in the inner medulla and papilla.

It has to be stated here that this hypothesis is by no means proven. Quantitative considerations leave some doubt as to whether the gradients in the inner medulla suffice to explain the concentrating ability. Also it should be noted that the uptake of urea into the ATL reduces the diluting effect of this segment. Several model calculations have been performed, and the outcome has been either that the model can work or that the model is insufficient to explain the observed concentrating ability. With all these models, however, one dilemma is the incompleteness of parameters which have to be included in such calculations. At present, there is no evidence to support an active single step in the ATL, yet there is at least qualitative evidence that the above hypothesis may be applicable.

Disturbance of the Countercurrent Concentrating System. It may be helpful to consider now, how the countercurrent concentrating system might be disturbed:

- Increased blood perfusion of the renal medulla
- Increased tubule fluid delivery to the DTL
- Impairment or inhibition of active NaCl reabsorption in the TAL
- Absence of AVP
- Reduction of urea delivery to the kidney.

Increased blood perfusion of the renal medulla leads to a washout of osmolytes from the renal medulla. This reduces the water absorption from the collecting duct, even in the presence of ADH.

Increased tubule fluid delivery to the DTL, as may occur in volume expansion, saturates the transport capacity in the mTAL and possibly even the cTAL. This reduces the concentration gradient built up by the single effect. Therefore, the concentrating ability in the presence of AVP and the diluting ability in the absence of AVP are reduced and urine tends to become isotonic with respect to NaCl.

Impairment or inhibition of active NaCl reabsorption in the TAL, e.g., by loop diuretics such as furosemide, abolishes the concentrating ability of the kidney. Therefore, a very large urine volume with isotonic NaCl concentration is excreted.

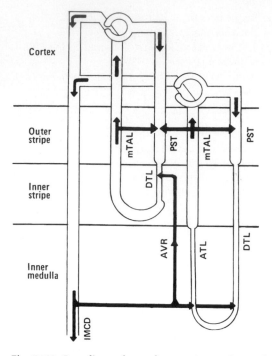

Fig. 74.21. Recycling pathways for urea. *Arrows* denote the pathways. *mTAL*, medullary thick ascending limb of the loop of Henle; *DTL*, thin descending limb of the loop of Henle; *PST*, proximal straight tubule; *AVR*, ascending vasa recta; *ATL*, thin ascending limb of the loop of Henle; *IMCD*, inner medullary collecting duct. Note that urea can enter the thin limbs of the loop of Henle and is delivered to the distal tubule. From there urea can be delivered to the collecting duct or diffuse into proximal straight tubules. (With permission form [18])

In the *absence of AVP* no water is removed from the collecting duct. Hence a large urine volume with diluted urine is excreted. This state is called "water diuresis". From the above it should also be clear that urea excretion will be enhanced in this situation and hence the concentrating ability of the inner medulla and papilla will be gradually lost.

Reduction of urea delivery to the kidney also reduces the ability to energise the concentrating mechanisms in the inner medulla and papilla. Hence, a certain protein turnover is required to deliver sufficient urea to the kidney. Also, the concentrating ability will be slightly reduced when nitrogen metabolism is shifted from urea to L-glutamine by the liver (cf. Chaps. 68, 75, 78). This occurs in chronic metabolic acidosis.

The functional states discussed above are accompanied by rapid or more delayed reductions in the osmolality of the renal medulla. Given the fact that the tubule cells in this section of the kidney have basolateral water permeability, one would have to assume that these cells undergo dramatic swelling (interstitial reductions in osmolality) or shrinkage (build up of interstitial osmotic gradients). It has long been known [120] (though it was forgotten for several decades [4]), that the renal medulla contains additional osmolytes, i.e., glycerophosphorylcholine, betaine, inositol and sorbitol. These osmolytes can be produced or accumu-

lated in the cells of the renal medulla whenever the concentrating mechanism is turned on, and they can be removed from these cells or metabolised when the concentrating mechanism is switched off [78]. Therefore, *the osmolality of the renal tubule cells in the renal medulla and papilla is adjusted to the extracellular osmolality.*

Recycling Pathways. The countercurrent concentrating mechanism leads to complex recycling pathways in the renal medulla. In general it can be postulated that all substrates added to the countercurrent system will show very high concentrations at the tip, and that the opposite will be true for all substrates removed from the system. Examples are CO_2, the concentration of which is very high in the renal medulla, and O_2, the concentration of which is very low. Similar short-circuiting can be shown, for example, for water and NaCl.

Water recycles in the cortex and outer medulla. As a result the hematocrit in the papilla is higher than that of systemic blood. This has rheological consequences and may become a serious problem when renal blood flow is reduced in ischemia [92].

NaCl can recycle from the TAL to the upper portion of the DTL_{II}. This has been demonstrated in *Psammomys obesus* and is responsible for a large fraction of the increase in osmolality in the DTL in this animal.

K^+ can recycle from the collecting duct into the S3 segment and the DTL. Therefore, fairly high concentrations of K^+ are preserved in the outer medulla. The functional role of this recycling of K^+ is not clear.

The recycling of urea is very complex. As shown in Fig. 74.21, urea is partly reabsorbed in the proximal tubule. Some 50% of the filtered load is delivered to the late proximal tubule. As has been described above, urea is concentrated in the collecting duct. In the presence of AVP, approximately 50% of the load delivered to the collecting duct enters the inner medulla and papilla. Some of this enters the vasa recta and from there the DTL of superficial nephrons. Another part enters DTL_{II} and ATL segments and is delivered to the TAL. Again, a part of this can recycle to the S2 and S3 segments. Basically there are three loops for recycling:

- From IMCD to DTL and ATL
- From TAL to S2 and S3
- From IMCD via ascending vasa recta to DTL.

In the presence of AVP approximately 50% of the filtered load is excreted and urea is kept highly concentrated in the renal medulla. In the absence of AVP most of the filtered urea is excreted because recycling from the IMCD is largely prevented.

NH_4^+ is also recycled in the renal medulla: It is reabsorbed from the TAL and a high concentration is built up in the interstitium. From there NH_4^+ can enter the DTL and accumulate. The high NH_4^+ concentration is also used to deliver NH_4^+ to the IMCD and to excrete it with the urine. The reabsorption of NH_4^+ in the TAL enables NH_4^+ to bypass the cortex (cf. Sect. 74.4).

Table 74.3. Diuretic and saluretic effects in the nephron

Structure	Name	Inhibited transport protein	Nephron site	Effects
	Acetazolamide	Carbonic anhydrase	Proximal (predominant)	Bicarbonaturia, weak diuresis
	Furosemide	$Na^+2Cl^-K^+$ cotransporter	TAL	Strong diuresis, saluresis, kaliuresis, calciuria, alkalosis
	Thiazides	Na^+Cl^- cotransporter	DT	Diuresis, saluresis, kaliuresis, hypocaliuria
	Amiloride	Na^+ channel	CD	Moderate saluresis, antikaliuresis

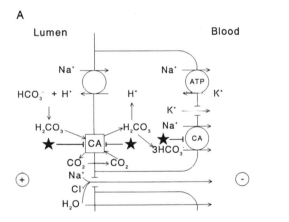

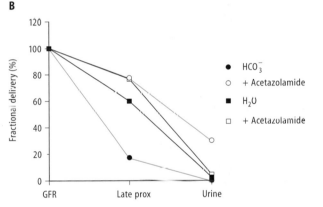

Fig. 74.22. A Effect of a carbonic anhydrase inhibitor (*asterisk*) such as acetazolamide (Table 74.3) on proximal tubule function. *Circle with ATP*, (Na^++K^+)-ATPase; *circle*, carrier-mediated mechanism; *arrow*, diffusion; *CA*, carbonic anhydrase. Note that inhibition slows luminal dehydration of H_2CO_3, intracellular hydration and exit of $Na^+(HCO_3^-)_3$. This reduces sharply the proximal reabsorption of HCO_3^-. **B** Effect of acetazolamide in rat micropuncture experiments. Fractional recovery of water (*squares*) and HCO_3^- (*circles*) is shown for the glomerulus (*GFR*), late proximal tubule (*Late prox*) and urine. Note that acetazolamide (*open symbols*) has a very marked effect on the end proximal delivery of HCO_3^-. The delivery is increased from some 20% to 80%. Even in urine this is apparent. HCO_3^- fractional excretion increases from almost 0% to 35%. The effect of acetazolamide on water reabsorption is less marked. Fractional water delivery at the end of the proximal tubule is enhanced from 60% to 80%. The diuretic effect in urine is minimal. No standard errors are included. (Data from [15])

74.8 The Mechanism of Action of Diuretics

Diuretics are substances which by definition enhance the urinary flow rate (diuresis). As will be shown, all known substances do this by inhibition of Na^+ reabsorption. Therefore these substances should rather be called *saluretics*. They all act from the lumen and interfere with the uptake of ions in the luminal membrane. The main reason for their kidney selectivity is the high accumulation of these substances in tubule fluid. This accumulation in tubule fluid is caused by two facts:

- These substances (organic acids or bases) are secreted by the proximal tubule [121] (cf. Chap. 75).
- They are further concentrated by water absorption.

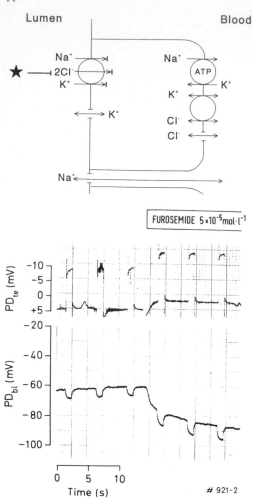

A

Lumen   Blood

B

FUROSEMIDE 5×10^{-5} mol·l^{-1}

Time (s)

921-2

Fig. 74.23. A The effect of loop diuretics in the thick ascending limb. *Circle with ATP*, (Na$^+$+K$^+$)-ATPase; *circle*, carrier-mediated mechanism; *arrow*, diffusion. Loop diuretics such as furosemide (Table 74.3) bind to one of the Cl$^-$ binding sites of the Na$^+$2Cl$^-$K$^+$ cotransporter (*asterisk*). This inhibits Cl$^-$, Na$^+$ and K$^+$ uptake. As a consequence luminal voltage and active transport collapse. K$^+$ and Cl$^-$ are passively distributed across the respective cell membranes. Also the (Na$^+$+K$^+$)-ATPase stops pumping after a minimum Na$^+$ concentration in the cell has been achieved. **B** Effect of furosemide in an isolated in vitro perfused rabbit cortical thick ascending limb. Transepithelial voltage (*PD$_{te}$*) and basolateral membrane voltage (*PD$_{bl}$*) are shown as a function of time. Transepithelial resistance and the resistance across the basolateral membrane are determined from rectangular current injections. Note that furosemide, added to the lumen, abolishes transepithelial voltage and increases transepithelial resistance slightly (increase in pulse height). At the same time the cell is hyperpolarised owing to the fall in cytosolic Cl$^-$ concentration and the resistance of the basolateral membrane increases. These effects occur very rapidly and are fully reversible (not shown). (**B** With permission from [53])

As a result the tubule fluid concentration of these substances far exceeds that in systemic circulation.

Table 74.3 summarises the various groups of diuretics, their chemical structure, their site of action, their mechanism of action and their effects.

Inhibitors of Carbonic Anhydrase. The first group are the inhibitors of carbonic anhydrase such as acetazolamide. These substances are rather weak diuretics and saluretics. Their main effect is in the proximal tubule, where they inhibit HCO$_3^-$ reabsorption in three ways: They inhibit the production of CO$_2$ at the luminal membrane (cf. Fig. 74.22), inhibit the rehydration of CO$_2$ to H$_2$CO$_3$ within the cell, and inhibit the exit of Na$^+$(HCO$_3^-$)$_3$ from the cell. As a result HCO$_3^-$ reabsorption in the proximal tubule is reduced (Fig. 74.22B). Inhibition, however, is never complete, and part of the saluretic effect in the proximal tubule is counterbalanced by increased NaCl reabsorption, mainly in the TAL. These substances increase urinary flow rate only moderately, and the urine becomes alkaline. Therefore, the main effect is bicarbonaturia and metabolic acidosis [21,51,116].

Loop Diuretics. Loop diuretics such as furosemide (Table 74.3 and Fig. 74.23) bind to one of the Cl$^-$ binding sites of the Na$^+$2Cl$^-$K$^+$ cotransporter present in the luminal membrane of the TAL (cf. Sect. 74.4). This abolishes the NaCl reabsorption in the TAL and produces very marked saluresis and diuresis. Up to 25% of the filtered loads of water and NaCl can be excreted in man [21,51,116]. These diuretics have a number of intrinsic "side-effects":

- They enhance markedly the distal tubule flow rate and NaCl delivery. As a result, K$^+$ secretion by the collecting duct is increased (cf. Sect. 74.6). It should be noted that the distal tubule and the collecting duct cannot counterbalance the strong natriuresis produced by these substances. The distal tubule transport capacities are simply overrun by the delivered load.
- They enhance the excretion of Mg^{2+} and Ca^{2+}, because the reabsorption of these divalent cations, which mainly occurs in the TAL (cf. Sect. 74.4), is also inhibited by loop diuretics.

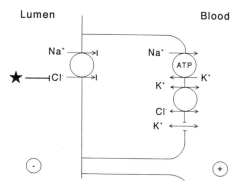

Lumen   Blood

Fig. 74.24. Effect of thiazides in the early distal tubule. *Circle with ATP*, (Na$^+$+K$^+$)-ATPase; *cirlce*, carrier-mediated mechanism; *arrow*, diffusion. Hydrochlorothiazide (Table 74.3) binds to a specific binding site of the Na$^+$Cl$^-$ cotransporter (*asterisk*). This inhibits transepithelial reabsorption. K$^+$ and Cl$^-$ are passively distributed across the basolateral cell membrane. In addition, and not shown here, Ca^{2+} reabsorption is enhanced by this diuretic. The mechanism for this enhanced reabsorption is not clear

1512

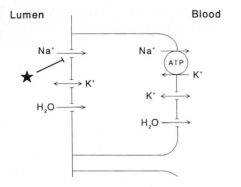

Lumen **Blood**

Fig. 74.25. Effect of K$^+$ sparing diuretics such as amiloride (Table 74.3) on principal cells of the collecting duct. *Circle with ATP,* (Na$^+$+K$^+$)-ATPase; *arrow,* diffusion. Amiloride binds to the Na$^+$ channel in the luminal membrane (*asterisk*). This blocks Na$^+$ entry and hyperpolarises the luminal membrane. Therefore, the transepithelial voltage collapses, as well as the driving force for K$^+$ secretion

- They enhance distal tubule H$^+$ secretion, again as a consequence of the enhanced distal load.
- They inhibit TGF (cf. Chap. 73).
- They enhance renin production (cf. Chap. 73).

Thiazides. Thiazides such as hydrochlorothiazide (Table 74.3, Fig. 74.24) inhibit the NaCl carrier present in the luminal membrane of the distal tubule. As a result they produce pronounced saluresis and diuresis, which, however, are not as marked as with loop diuretics. Their maximal effect amounts to some 5%-10% of the filtered loads of NaCl and water. By some unclear mechanism Ca^{2+} absorption is enhanced by these substances [16]. This feature renders these substances especially suitable for the treatment of hypercalciuria and kidney stones [51]. As with loop diuretics, the enhanced load of NaCl and water delivered to the collecting tubule produces kaliuresis and enhanced proton secretion. The effect on acid-base balance is usually not very marked because many of these substances are also carbonic anhydrase inhibitors.

K$^+$-Sparing Diuretics. K$^+$-sparing diuretics such as amiloride (Table 74.3, Figs. 74.8, 74.25) inhibit the uptake of Na$^+$ by the principal cells of the collecting duct. Their diuretic and saluretic effect is very limited, because only a few percent of the filtered loads of Na$^+$ is reabsorbed along the collecting duct [21,51,116]. However, their main effect and value is their effect on K$^+$ excretion. It can easily be seen in Fig. 74.25B that the inhibition of Na$^+$ uptake hyperpolarises the luminal membrane and removes the driving force for K$^+$ secretion (K$^+$ sparing) [51].

References

1. Alpern RJ, Emmett M, Seldin DW (1992) Metabolic alkalosis. In: Seldin DW, Giebisch G (eds) The kidney: physiology and pathophysiology. Raven, New York, pp 2733–2758
2. Alpern RJ, Stone DK, Rector FC (1991) Renal acidification mechanisms. In: Brenner BM, Rector FC (eds) The kidney. Saunders, Philadelphia, pp 318–379
3. Aronson PS, Nee J, Suhm MA (1982) Modifier role of internal H$^+$ in activating the Na$^+$–H$^+$ exchanger in renal microvillus membrane vesicles. Nature 299:161–163
4. Balaban RS, Knepper MA (1983) Nitrogen-14 nuclear magnetic resonance spectroscopy of mammalian tissues. Am J Physiol 245:C439–C444
5. Ballermann BJ, Zeidel M (1992) Atrial natriuretic hormone. In: Seldin DW, Giebisch G (eds) The kidney: physiology and pathophysiology. Raven, New York, pp 1843–1884
6. Bankir L, De Rouffignac C (1985) Urinary concentrating ability: insights from comparative anatomy. Am J Physiol 249:R643–R666
7. Benos DJ, Cunningham S, Baker RR, Beason KB, Oh Y, Smith PR (1992) Molecular characteristics of amiloride-sensitive sodium channels. Rev Physiol Biochem Pharmacol 120:31–113
8. Bleich M, Schlatter E, Greger R (1990) The luminal K$^+$ channel of the thick ascending limb of Henle's loop. Pflugers Arch 415:449–460
9. Boron WF, Boulpaep EL (1989) The electrogenic Na/HCO$_3$ cotransporter. Kidney Int 36:392–402
10. Brezis M, Rosen S, Silva P, Epstein FH (1984) Renal ischaemia: a new perspective. Kidney Int 26:375–383
11. Burg M, Grantham J, Abramow M, Orloff J (1966) Preparation and study of fragments of single rabbit nephrons. Am J Physiol 210(6):1293–1298
12. Burg MB (1982) Thick ascending limb of Henle's loop. Kidney Int 22:454–464
13. Burg MB, Green N (1973) Function of the thick ascending limb of Henle's loop. Am J Physiol 224(3):659–668
14. Canessa CM, Horisberger JD, Rossier BC (1993) Epithelial sodium channel related to proteins involved in neurodegeneration. Nature 361:467–470
15. Cogan MG, Maddox DA, Warnock DG, Lin ET, Rector FC (1979) Effect of acetazolamide on bicarbonate reabsorption in the proximal tubule of the rat. Am J Physiol 237:F447–F454
16. Costanzo LS, Windhager EE (1978) Calcium and sodium transport by the distal convoluted tubule of the rat. Am J Physiol 235:F492–F506
17. Costanzo LS, Windhager EE (1992) Renal regulation of calcium balance. In: Seldin DW, Giebisch G (eds) The kidney: physiology and pathophysiology. Raven, New York, pp 2375–2393
18. De Rouffignac C (1990) The urinary concentrating mechanism. In: Kinne RKH (ed) Urinary concentrating mechanisms. Karger, Basel, pp 31–102
19. Deetjen P, Kramer K (1961) Die Abhängigkeit des O$_2$-Verbrauchs der Niere von der Na-Rückresorption. Pflugers Arch 273:636–642
20. Deetjen P, v. Baeyer H, Drexel H (1992) Renal glucose transport. In: Seldin DW, Giebisch G (eds) The kidney: physiology and pathophysiology. Raven, New York, pp 2873–2888
21. Dillingham MA, Schrier RW, Greger R (1993) Mechanisms of diuretic action. In: Schrier RW, Gottschalk CW (eds) Clinical disorders of fluid, electrolytes, and acid base. Little Brown, Boston, pp 2435–2452
22. Drenckhahn D, Merte C (1987) Restriction of the human kidney band 3-like anion exchanger to specialized subdomains of the basolateral plasma membrane of intercalated cells. Eur J Cell Biol 45:107–115
23. Dworkin LD, Brenner BM (1991) The renal circulations. In: Brenner BM, Rector FC (eds) The kidney. Saunders, Philadelphia, pp 164–204
24. Echevarria M, Frindt G, Preston GM, Milovanovic S, Agre P, Fischbarg J, Winhager EE (1993) Expression of multiple water channel activities in *Xenopus* oocytes injected with mRNA from rat kidney. J Gen Physiol 101:827–841

25. Edelman A, Curci S, Samarzija I, Frömter E (1978) Determination of intracellular K⁺ activity in rat kidney proximal tubular cells. Pflugers Arch 378:37–45
26. Emmett M, Alpern RJ, Seldin DW (1992) Metabolic acidosis. In: Seldin DW, Giebisch G (eds) The kidney: physiology and pathophysiology. Raven, New York, pp 2759–2836
27. Folkow B (1990) Salt and hypertension. NIPS 5:220–224
28. Friedman PA, Gesek FA (1993) Calcium transport in renal epithelial cells. Am J Physiol 264:F181–F198
29. Frömter E (1981) Electrical aspects of tubular transport of organic substances. In: Greger R, Lang F, Silbernagl S (eds) Renal transport of organic substances. Springer, Berlin Heidelberg New York, pp 30–44
30. Frömter E (1984) Viewing the kidney through microelectrodes. Am J Physiol 247:F695–F705
31. Frömter E, Gessner K (1974) Free-flow potential profile along the rat kidney proximal tubule. Pflugers Arch 351:69–83
32. Frömter E, Rumrich G, Ullrich KJ (1973) Phenomenologic description of Na⁺, Cl⁻ and HCO₃⁻ absorption from proximal tubules of the rat kidney. Pflugers Arch 343:189–220
33. Fuijmoto M, Kubota T, Kotera K (1977) Electrochemical profile of K and Cl ions across the proximal tubule of bullfrog kidneys. Contrib Nephrol 6:114–123
34. Gamba G, Slatzberg SN, Lombardi M, Miyanoshita A, Lytton J, Hediger MA, Brenner BM (1993) Primary structure and functional expression of a sDNA encoding the thiazide-sensitive, electroneutral sodium-chloride cotransporter. Proc Natl Acad Sci USA 90:2749–2753
35. Gamba G, Miyanoshita A, Lombardi M, Lytton J, Lee WS, Hediger MA, Hebert SC (1994) Molecular cloning, primary structure, and characterization of two members of the mammalian electroneutral sodium-(potassium)-chloride cotransporter family expressed in kidney. J Biol Chem 269:17 713–17 722
36. Geck P, Pietrzyk C, Burckhardt BC, Pfeiffer B, Heinz E (1980) Electrically silent cotransport of Na⁺, K⁺ and Cl⁻ in Ehrlich cells. Biochim Biophys Acta 600:432–447
37. Gluck S, Caldwell J (1988) Proton-translocating ATPase from bovine kidney medulla: partial purification and reconstitution. Am J Physiol 254:F71–F79
38. Gögelein H (1990) Ion channels in mammalian proximal renal tubule. Renal Physiol Biochem 13:8–25
39. Gögelein H, Greger R (1987) Properties of single K⁺ channels in the basolateral membrane of rabbit proximal straight tubules. Pflugers Arch 410:288–295
40. Gögelein H, Greger R (1988) Patch clamp analysis of ionic channels in renal proximal tubules. In: Davison AM (ed) Proceedings of the Xth international congress of nephrology. Baillere Tindall, London, pp 159–178
41. Gonzalez-Campoy JM, Knox FG (1992) Integrated responses of the kidney to alterations in extracellular fluid volume. In: Seldin DW, Giebisch G (eds) The kidney: physiology and pathophysiology. Raven, New York, pp 2041–2098
42. Good DW, Knepper MA, Burg MB (1984) Ammonia and bicarbonate transport by the thick ascending limb of rat kidney. Am J Physiol 247:F35–F44
43. Grantham JJ, Burg MB (1966) Effect of vasopressin and cyclic AMP on permeability of isolated collecting tubules. Am J Physiol 211:255–259
44. Greenwald L, Stetson D (1988) Urine concentration and the length of the renal papilla. NIPS 3:46–49
45. Greger R (1981) Chloride reabsorption in the rabbit cortical thick ascending limb of the loop of Henle. A sodium dependent process. Pflugers Arch 390:38–43
46. Greger R (1981) Cation selectivity of the isolated perfused cortical thick ascending limb of Henle's loop of rabbit kidney. Pflugers Arch 390:30–37
47. Greger R (1985) Ion transport mechanisms in thick ascending limb of Henle's loop of mammalian nephron. Physiol Rev 65:760–797
48. Greger R (1988) Chloride transport in thick ascending limb, distal convolution, and collecting duct. Annu Rev Physiol 50:111–122
49. Greger R, Bleich M, Schlatter E (1991) Ion channel regulation in the thick ascending limb of the loop of Henle. Kidney Int 40:S33:S119–S124
50. Greger R, Gögelein H (1987) K⁺ conductive pathways in the nephron. Kidney Int 31:1055–1064
51. Greger R, Heidland A (1991) Action and clinical use of diuretics. In: Cameron S, Davison AM, Grünfeld JP (eds) Clinical nephrology. Oxford University Press, London, pp 197–224
52. Greger R, Schlatter E (1981) Presence of luminal K⁺, a prerequisite for active NaCl transport in the thick ascending limb of Henle's loop of rabbit kidney. Pflugers Arch 392:92–94
53. Greger R, Schlatter E (1983) Cellular mechanism of the action of loop diuretics on the thick ascending limb of Henle's loop. Klin Wochenschr 61:1019–1027
54. Greger R, Schlatter E (1983) Properties of the basolateral membrane of the cortical thick ascending limb of Henle's loop of rabbit kidney. A model for secondary active chloride transport. Pflugers Arch 396:325–334
55. Greger R, Schlatter E (1984) Mechanisms of chloride transport in vertebrate renal tubule. In: Gerencser GA (ed) Chloride transport coupling in biological membranes and epithelia. Elsevier, Amsterdam, pp 271–346
56. Greger R, Schlatter E, Bleich M, Goegelein H (1991) Ion channels in the mammalian nephron. In: International Society of Nephrology (ed) Proceedings of the Xlth international congress on nephrology. Springer, Berlin Heidelberg New York, pp 1656–1668
57. Greger R, Schlatter E, Lang F (1983) Evidence for electroneutral sodium chloride cotransport in the cortical thick ascending limb of Henle's loop of rabbit kidney. Pflugers Arch 396:308–314
58. Greger R, Velázquez H (1987) The cortical thick ascending limb and early distal convoluted tubule in the urinary concentrating mechanism. Kidney Int 31:590–596
59. Greger R, Wangemann Ph (1987) Loop diuretics. Renal Physiol 10:174–183
60. Guggino WB, London R, Boulpaep EL, Giebisch G (1983) Chloride transport across the basolateral cell membrane of the *Necturus* proximal tubule: dependence on bicarbonate and sodium. J Membr Biol 71:227–240
61. Gullans SR, Hebert SC (1991) Metabolic basis of ion transport. In: Brenner BM, Rector FC (eds) The kidney. Saunders, Philadelphia, pp 76–109
62. Halperin ML, Kamel SK, Ethier JH, Stienbaugh BJ, Jungas RL (1992) Biochemistry and physiology of ammonium excretion. In: Seldin DW, Giebisch G (eds) The kidney: physiology and pathophysiology. Raven, New York, pp 2645–2679
63. Hamlyn JM, Ludens JH (1992) Nonatrial natriuretic hormones. In: Seldin DW, Giebisch G (eds) The kidney: physiology and pathophysiology. Raven, New York, pp 1885–1924
64. Hamm LL, Alpern RJ (1992) Cellular mechanisms of renal tubular acidification. In: Seldin DW, Giebisch G (eds) The kidney: physiology and pathophysiology. Raven, New York, pp 2581–2626
65. Hays RM (1990) Water transport in epithelia. In: Kinne RKH (ed) Urinary concentrating mechanisms. Karger, Basel, pp 1–30
66. Hays RM (1991) Cell biology of vasopressin. In: Brenner BM, Rector FC (eds) The kidney. Saunders, Philadelphia, pp 424–444
67. Hebert SC, Kriz W (1993) Structural-functional relationships in the kidney. In: Schrier RW, Gottschalk CW (eds) Diseases of the kidney. Little Brown, Boston, pp 3–64
68. Hirsch J, Leipziger J, Fröbe U, Schlatter E (1993) Regulation and possible physiological role of the Ca2+-dependent K+-channel of cortical collecting ducts of the rat. Pflugers Arch 422:492–498

69. Hirsch J, Schlatter E (1993) K$^+$ channels in the basolateral membrane of rat collecting duct. Pflugers Arch 424:470–477

70. Howard RL, Bichet DG, Schrier RW (1992) Hypernatremic and polyuric states. In: Seldin DW, Giebisch G (eds) The kidney: physiology and pathophysiology. Raven, New York, pp 1753–1778

71. Jørgensen PL (1980) Sodium potassium ion pump in kidney tubules. Physiol Rev 60:864–917

72. Katz AI, Doucet A, Morel F (1979) Na-K-ATPase activity along the rabbit, rat, and mouse nephron. Am J Physiol 237:F114–F120

73. Kikeri D, Sun A, Zeidel ML, Hebert SC (1989) Cell membranes impermeable to NH$_3$. Nature 339:478–480

74. Kinne R, Kinne-Saffran E, Schuetz H, Schoelermann B (1986) Ammonium transport in medullary thick ascending limb of rabbit kidney: involvement of the Na$^+$,K$^+$,Cl$^-$-cotransporter. J Membr Biol 94:279–284

75. Kinne R, Koenig B, Hannafin J, Kinne-Saffran E, Scott DM, Zierold K (1986) The use of membrane vesicles to study the NaCl/KCl cotransporter involved in active transepithelial chloride transport. Pflugers Arch 405 [Suppl 1]:S101–S105

76. Kinsella JL, Aronson PS (1980) Properties of the Na$^+$-H$^+$ exchanger in renal microvillus membrane vesicles. Am J Physiol 238:F461–F469

77. Kirk KL, Schafer JA (1992) Water transport and osmoregulation by antidiuretic hormone in terminal nephron segments. In: Seldin DW, Giebisch G (eds) The kidney: physiology and pathophysiology. Raven, New York, pp 1693–1725

78. Knepper MA, Rector FC (1991) Urinary concentration and dilution. In: Brenner BM, Rector FC (eds) The kidney. Saunders, Philadelphia, pp 445–482

79. Knepper MA, Sands JM, Chou C-L (1989) Independence of urea and water transport in rat inner medullary collecting duct. Am J Physiol 256:F610–F621

80. Koefoed-Johnsen V, Ussing HH (1958) The nature of the frog skin potential. Acta Physiol Scand 42:298–308

81. Koenig B, Ricapito S, Kinne R (1983) Chloride transport in the thick ascending limb of Henle's loop; potassium dependence and stoichiometry of the NaCl cotransport system in plasma membrane vesicles. Pflugers Arch 399:173–179

82. Koeppen BM, Stanton BA (1992) Sodium chloride transport: distal nephron. In: Seldin DW, Giebisch G (eds) The kidney: physiology and pathophysiology. Raven, New York, pp 2003–2040

83. Kokko JP, Rector FC (1972) Countercurrent multiplication system without active transport in renal inner medulla. Kidney Int 2:214–223

84. Kuhn W, Ryffel K (1942) Herstellung konzentrierter Lösungen aus verdünnten durch blosse Membranwirkung (Ein Modellversuch zur Funktion der Niere). Hoppe-Seylers Z Physiol Chem 276:145–178

85. Lang F, Greger R, Knox FG, Oberleithner H (1981) Factors modulating the renal handling of phosphate. Renal Physiol 4:1–16

86. Lang F, Quehenberger P, Greger R, Oberleithner H (1978) Effect of benzolamide on luminal pH in proximal convoluted tubules of the rat kidney. Pflugers Arch 375:39–43

87. Lang F, Rehwald W (1992) Potassium channels in renal epithelial transport regulation. Physiol Rev 72:1–32

88. Laskowski FH, Christine ChW, Gitter AH, Beyenbach KW, Gross P, Frömter E (1990) Cation channels in the apical membrane of collecting duct principal cell epithelium in culture. Renal Physiol Biochem 13:70–81

89. Lingueglia E, Voilley N, Waldmann R, Lazdunski M, Barbry P (1993) Expression cloning of an epithelial amiloride-sensitive Na+ channel. Fed Eur Biochem Soc 318:95–99

90. Madias NE, Cohen JJ (1992) Respiratory alkalosis and acidosis. In: Seldin DW, Giebisch G (eds) The kidney: physiology and pathophysiology. Raven, New York, pp 2837–2872

91. Marver D (1991) Aldosterone. In: Fleischer S, Fleischer B (eds) Methods in enzymology. Academic, San Diego, pp 520–550

92. Mason J, Welsch J, Torhorst J (1987) The contribution of vascular obstruction to the functional defect that follows renal ischemia. Kidney Int 31:65–71

93. McKinney TD, Burg MB (1978) Bicarbonate secretion by the rabbit cortical collecting tubules in vitro. J Clin Invest 61:1421–1427

94. Ming Tse C, Ma AI, Yang VW, Watson AJM, Potter J, Sardet C, Pouyssegur J, Donowitz M (1991) Molecular cloning and expression of a cDNA encoding the rabbit ileal villus cell basolateral membrane Na$^+$/H$^+$ exchanger. EMBO J 10:1957–1967

95. Murer H, Hopfer U, Kinne R (1976) Sodium/proton antiport in brush border membrane vesicles isolated from rat small intestine and kidney. Biochem J 154:597–604

96. Palmer LG, Frindt G (1987) Effects of cell Ca and pH on Na channels from rat cortical collecting tubule. Am J Physiol 253:F333–F339

97. Passow H (1986) Molecular aspects of band 3 protein-mediated anion transport across the red blood cell membrane. Rev Physiol Biochem Pharmacol 103:62–123

98. Reeves WB, Andreoli TE (1992) Sodium chloride transport in the loop of Henle. In: Seldin DW, Giebisch G (eds) The kidney: physiology and pathophysiology. Raven, New York, pp 1975–2002

99. Robertson GL (1992) Renal regulation of water balance: normal. In: Seldin DW, Giebisch G (eds) The kidney: physiology and pathophysiology. Raven, New York, pp 1595–1614

100. Sands JM, Knepper MA (1987) Urea permeability of mammalian inner medullary collecting duct system and papillary surface epithelium. J Clin Invest 79:138–147

101. Schild L, Giebisch G, Green R (1988) Chloride transport in the proximal renal tubule. Annu Rev Physiol 50:97–110

102. Schlatter E (1993) Effects of various diuretics on membrane voltage of macula densa cells. Whole-cell patch-clamp experiments. Pflugers Arch 423:74–77

103. Schlatter E, Bleich M, Hirsch J, Greger R (1993) pH-sensitive K+ channels in the distal nephron. Nephrol Dial Transplant 8:488–490

104. Schlatter E, Bleich M, Hirsch J, Markstahler U, Fröbe U, Greger R (1993) Cation specifity and pharmacological properties of the Ca^{2+}-dependent K$^+$-channel of rat cortical collecting ducts. Pflugers Arch 422:481–491

105. Schlatter E, Fröbe U, Greger R (1992) Ion conductances of isolated cortical collecting duct cells. Pflugers Arch 421:381–387

106. Schlatter E, Greger R (1985) cAMP increases the basolateral Cl$^-$-conductance in the isolated perfused medullary thick ascending limb of Henle's loop of the mouse. Pflugers Arch 405:367–376

107. Schlatter E, Greger R (1986) Ion transport in the loop of Henle. In: Lote LJ (ed) Advances in renal physiology. Croom Helm, London, pp 84–113

108. Schlatter E, Lohrmann E, Greger R (1992) Properties of the potassium conductances of principal cells of rat cortical collecting ducts. Pflugers Arch 420:39–45

109. Schlatter E, Salomonsson M, Persson AEG, Greger R (1989) Macula densa cells sense luminal NaCl concentration via the furosemide sensitive Na-2Cl-K cotransporter. Pflugers Arch 414:286–290

110. Schlatter E, Schafer JA (1988) Intracellular chloride activity in principal cells of rat collecting ducts (CCT). Pflugers Arch 410:R86

111. Schnermann J, Briggs JP (1992) Function of the juxtaglomerular apparatus: control of glomerular hemodynamics and renin secretion. In: Seldin DW, Giebisch G (eds) The kidney: physiology and pathophysiology. Raven, New York, pp 1249–1290

112. Schwarz GJ, Al-Awqati Q (1990) Identification and study of specific cell types in isolated nephron segments using fluorescent dyes. Methods Enzymol 191:253–264

113. Silbernagl S (1992) Amino acids and oligopeptides. In: Seldin DW, Giebisch G (eds) The kidney: physiology and pathophysiology. Raven, New York, pp 2889–2920

114. Smith HW (1953) From fish to philosopher. The story of our environment. Little Brown, Boston

115. Stephenson JL (1983) The renal concentrating mechanism: fundamental theoretical concepts. Fed Proc 42:2386–2391

116. Suki WN, Eknoyan G (1992) Physiology of diuretic action. In: Giebisch G, Seldin DW (eds) The kidney: Physiology and pathophysiology. Raven, New York, pp 3629–3670

117. Suki WN, Rouse D (1991) Renal transport of calcium, magnesium, and phosphorus. In: Brenner BM, Rector FC (eds) The kidney. Saunders, Philadelphia, pp 380–423

118. Tisher CC, Madsen KM (1993) Anatomy of the kidney. In: Brenner BM, Rector FC (eds) The kidney: Saunders, Philadelphia, pp 3–75

119. Ullrich KJ, Greger R (1985) Approaches to the study of tubule transport functions. In: Seldin DW, Giebisch G (eds) The kidney: Physiology and pathophysiology. Raven, New York, pp 427–469

120. Ullrich KJ, Jarausch KH (1956) Untersuchungen zum Problem der Harnkonzentrierung und -verdünnung. Über die Verteilung der Elektrolyten (Na, K, Ca, Mg, Cl, anorg. Phosphat), Harnstoff, Aminoäuren und endogenem kreatinin in Rinde und Mark der Hundeniere bei verschiedenen Diuresezuständen. Pflugers Arch 262:537–550

121. Ullrich KJ, Rumrich G, Klöss S (1989) Contraluminal organic anion and cation transport in the proximal renal tubule: V. Interaction with sulfamoyl- and phenoxy diuretics, and with β-lactam antibiotics. Kidney Int 36:78–88

122. Velázquez H, Wright FS (1986) Effects of diuretic drugs on Na, Cl and K transport by rat renal distal tubule. Am J Physiol 250:F1013–F1023

123. Wang W, Giebisch G (1991) Dual effect of adenosine triphosphate on the apical small conductance K^+ channel of the rat cortical collecting duct. J Gen Physiol 98:35–61

124. Wang W, Giebisch G (1991) Dual modulation of renal ATP-sensitive K^+ channel by protein kinases A and C. Proc Natl Acad Sci USA 88:9722–9725

125. Weinstein AM (1992) Sodium and chloride transport: proximal nephron. In: Seldin DW, Giebisch G (eds) The kidney: physiology and pathophysiology. Raven, New York, pp 1925–1974

126. Wingo CS (1989) Active proton secretion and potassium absorption in the rabbit outer medullary collecting duct. J Clin Invest 84:361–365

127. Wirthenson G, Guder WG (1986) Renal substrate metabolism. Physiol Rev 66:469–520

128. Wirthenson G, Guder WG (1990) Metabolism of isolated kidney tubule segments. Methods Enzymol 191:325–340

129. Wirz H (1956) Der osmotische Druck in den kortikalen Tubuli der Rattennieren. Helv Physiol Pharmacol Acta 14:353–362

130. Wittner M, Di Stefano A, Wangemann P, Nitschke R, Greger R, Bailly C, Amiel C, Roinel N, De Rouffignac C (1988) Differential effects of ADH on sodium, chloride, potassium, calcium and magnesium transport in cortical and medullary thick ascending limbs of mouse nephron. Pflugers Arch 412:516–523

131. Wittner M, Weidtke C, Schlatter E, Di Stefano A, Greger R (1984) Substrate utilization in the isolated perfused cortical thick ascending limb of rabbit nephron. Pflugers Arch 402:52–62

132. Wright FS, Giebisch G (1992) Regulation of potassium excretion. In: Seldin DW, Giebisch G (eds) The kidney: physiology and pathophysiology. Raven, New York, pp 2209–2248

133. Yoshitomi K, Burckhardt BC, Frömter E (1985) Rheogenic sodium-bicarbonate cotransport in the peritubular cell membrane of rat renal proximal tubule. Pflugers Arch 405:360–366

134. Zhang K, Wang ZQ, Gluck S (1992) Identification and partial purification of a cytosolic activator of vacuolar H^+-ATPases from mammalian kidney. J Biol Chem 267:9701–9705

75 Renal Handling of the Individual Solutes of Glomerular Filtrate

R. Greger

Contents

75.1 Introduction

In this chapter the individual constituents of the glomerular filtrate, as listed in Table 74.1, and their renal handling, i.e., their sites of transport in the nephron and the regulatory mechanisms, will be discussed systematically. This discussion will be redundant in some sections since the regulatory principles are not only geared for the maintenance of the homeostasis of individual solutes but also for the interplay of several or many components in order to achieve a certain body function. Subsequent sections deal with integrated homeostatic functions of the kidney. Two final sections deal briefly with failing kidney function. This appears appropriate because acute and chronic renal failure inform us in a different way of the relevance of the regulatory function of this organ.

75.2 Na$^+$, Cl$^-$ and H$_2$O

Figure 75.1 depicts the recovery rates of Na$^+$ and Cl$^-$ along the nephron. Both ions are freely filtered and the tubule fluid over plasma concentration (TF/P) is close to one. It has been discussed in Chap. 74 that the Gibbs-Donnan distribution for both ions means that the early proximal TF/P$_{Na^+}$ is around 0.95 and TF/P$_{Cl^-}$ 1.05. Along the proximal tubule a large fraction of the filtered Na$^+$ (ca. 60%–70%) is reabsorbed. The reabsorption of Cl$^-$ lags slightly behind. Therefore, the TF/P$_{Cl^-}$ increases slightly but significantly along the proximal tubule (cf. Fig. 74.8). At the end of the proximal tubule some 60% of the filtered Cl$^-$ is reabsorbed. The mechanisms of Na$^+$ and Cl$^-$ reabsorption in the proximal tubule have been described in Chap. 74 (cf. Fig. 74.6). It has been concluded that this reabsorption occurs almost iso-osmotically (for Na$^+$), i.e., 1 l H$_2$O is reabsorbed with 140 mmol/l of Na$^+$. The absolute rates of reabsorption are impressive:

- Each day 18 mol Na$^+$ is reabsorbed.
- This corresponds to an ATP turnover of 2.0 mol = 1.1 kg/day.
- 7 l O$_2$ is required for this transport work.

Also at the level of the single proximal tubule cell, the transport work and ion turnover are very impressive. It can be deduced from such considerations that the entire Na$^+$ content of a proximal tubule cell is exchanged every 3 to maximally 10 s.

Glomerulotubular Balance. The concept of transcellular and transtubular Na$^+$ and Cl$^-$ transport (cf. Fig. 74.6) suggests that the transport should have saturation kinetics. In other words, when all Na$^+$/H$^+$ exchangers are working at maximal rates, a maximum absolute reabsorptive rate should be obtained. Beyond this point, Na$^+$ excretion in urine, or at least the delivery out of the proximal tubule, should increase correspondingly. Conversely, one might predict that a reduction in filtered load would lead to an enhanced fractional reabsorption (rate of reabsorption/rate filtered) of Na$^+$. This, however, is not observed: The fractional reabsorption of almost all constituents of the glomerular filtrate is fairly constant irrespective of the filtered load (or SNGFR). This key observation is named "glomerulotubular balance" (GTB) [115]. Figure 75.2 [47] depicts the basic observation: A large variation in luminal flow rate (corresponding to SNGFR) and in the delivered load (filtered load) has almost no effect on fractional reabsorption of H$_2$O and Na$^+$. Therefore, absolute reabsorption by the proximal tubule is increased pari

R. Greger/U. Windhorst (Eds.)
Comprehensive Human Physiology, Vol. 2
© Springer-Verlag Berlin Heidelberg 1996

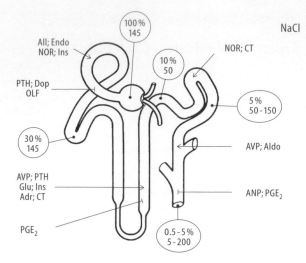

Fig. 75.1. Recovery of NaCl along the nephron. The *numbers in the circles* refer to recovery in the tubule fluid and the approximate concentration (mmol/l). The filtered amount is taken as 100%. This figure also includes hormones activating NaCl transport. *AII*, angiotensin II; *Endo*, endothelin; *NOR*, norepinephrine; *Ins*, insulin. All four hormones enhance proximal tubular reabsorption of NaCl. *PTH*, parathyroid hormone; *Dop*, dopamine; *OLF*, ouabain-like factor. These factors reduce proximal tubular reabsorption of NaCl. *AVP*, arginine vasopressin = antidiuretic hormone; glucagon; *Adr*, adrenaline; *CT*, calcitonin. These hormones, together with PTH and Ins, enhance NaCl reabsorption in the thick ascending limb of the loop of Henle. *PGE₂*, prostaglandin E₂, which inhibits these effects. NOR and CT also enhance reabsorption in the early distal tubule. *Aldo*, aldosterone; *ANP*, atrial natriuretic peptide. Aldo and ADH enhance NaCl reabsorption in the collecting duct. ANP and PGE₂ have the opposite effect

passu with an increase in SNGFR. GTB is, therefore, one key mechanism guaranteeing the homeostatic function of the kidney. Even if the autoregulation of GFR and renal blood flow (RBF) (cf. Chap. 73) were to fail, this would not necessarily lead to corresponding changes in the urinary excretion of H_2O and Na^+.

Several mechanisms for GTB have been proposed. A *"peritubule" hypothesis* [14,71] postulates that an increase in the fractional filtration (GFR/RBF) leads to an enhanced protein concentration in the peritubule capillaries; this is caused by the fact that the filtrate is protein free. The hypothesis also postulates that the increased oncotic pressure in the peritubule capillaries enhances the solute uptake from the basal labyrinth into the peritubule capillaries. This hypothesis, however, cannot account for the entire magnitude of the GTB mechanism [115].

Another hypothesis has postulated that the *geometry of the tubule*, which is altered as a function of filtration pressure, may change the area of the luminal membrane [40]. This hypothesis, too, is not sufficient nor even likely to explain GTB [115].

More recently it has been postulated that *factors in tubule fluid* regulate the magnitude of proximal tubule reabsorption [47]. This hypothesis is supported by experiments shown in Fig. 75.2. It has been postulated that simple constituents of proximal tubule fluid or even the

relative hypotonicity may be responsible for GTB [59]. It has also been claimed that a normal constituent of proximal tubule fluid could be a determinant of GTB. HCO_3^- may be one such factor [115].

In addition some as yet unidentified factors present in the luminal fluid have been postulated [52,60], as shown in experiments in which tubule fluid was harvested from one proximal tubule and was examined in stop flow micropuncture experiments in other nephrons. The molecular identity of this (these) factor(s) has not yet been clarified. In fact, this factor may be relevant for the regulation of Na^+ reabsorption in volume expansion.

In addition, a number of well-known regulatory mechanisms present in the proximal tubule [6,55,114,115] may participate in GTB. These mechanisms will be summarised here.

Proximal Tubule Function. Proximal tubule reabsorption of H_2O and Na^+ is regulated by *angiotensin II* (AII). This was shown in micropuncture studies [58] many years ago, and it was puzzling then that the concentration-response curve was bell shaped. This is shown in Fig. 75.3. Very low concentrations of AII (10 pmol/l) enhance reabsorption, while higher concentrations (1 μmol/l) have the opposite effect.

It has also been suggested that in the presence of AII, *atrionatriuretic peptide* (ANP) reduces proximal tubule reabsorption [57]. The relevance of this finding is difficult to evaluate because receptors for ANP have been found in glomeruli and the inner medulla and papilla but their presence in the proximal tubule is not certain [72].

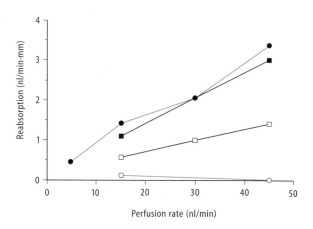

Fig. 75.2. Dependence of proximal tubule reabsorption on luminal perfusion rate. Microperfusion studies in the intact rat [47]. *Closed circles and blue line* = NaCl-Ringer's, lumen perfusate and peritubular normal blood perfusion; *closed squares and black line* = NaCl-Ringer's, lumen and bicarbonate Ringer's peritubular perfusate; *open squares and black line* = NaCl-Ringer's, lumen and peritubular perfusate; *open circles and blue line* = bicarbonate Ringer's + 2 mmol/l CN^-, lumen perfusate. For reasons of simplification SEMs have been omitted. Note that reabsorption is a linear function of the perfusion rate in all three conditions (except the CN^- series). This indicated that the fractional reabsorption is constant over a wide range of luminal flow rate. Highest reabsorption is achieved with peritubular blood perfusion

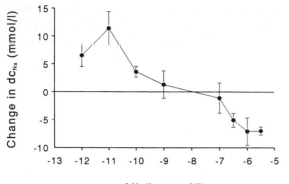

Fig. 75.3. The effect of peritubular angiotensin II (*AII*) infusions on proximal tubular steady state Na$^+$ concentration gradients (dc_{Na}). Mean values (± SEM) of the differences between controls and experiments with AII infusion. Note that AII at low concentrations increases dc_{Na}, whereas it reduces dc_{Na} at higher concentrations. These data demonstrate the *antinatriuretic effect of AII* (enhanced NaCl reabsorption) at low concentrations. (Data from [58])

Parathyroid hormone (PTH) is another relevant hormone which reduces proximal tubule reabsorption of HCO$_3^-$ [8,88]. This effect is mediated by PKA and PKC phosphorylation of the Na$^+$/H$^+$ exchanger. The additional effect of PTH on phosphate reabsorption will be discussed in Sect. 75.6 and 75.11.

Another very important factor in the control of proximal tubule function is its *sympathergic innervation* [30,46,66], especially in the S1 and S2 segments. β and α$_2$ receptors have been demonstrated. The net effect of sympathergic discharge is an increase in proximal tubule reabsorption. This is most likely mediated by α$_2$ receptors.

The possible effect of *endothelin* on the proximal tubule requires further clarification. An increase in reabsorption has been reported. This, however, is difficult to reconcile with its overall effect on the kidney, which consists in a moderate natriuresis [7,102]. At any rate, in most studies the negative hemodynamic effects of endothelin predominate. The resulting vasoconstriction then causes a secondary antinatriuresis [102].

The importance of these regulatory mechanisms in the overall function of the proximal tubule is not yet clear, and it is also not evident which of these factors may be of relevance in GTB. Therefore, one has to conclude that a multitude of regulatory factors in the proximal tubule have been identified. At present, luminal factors such as the luminal hypotonicity and the constituents of luminal fluid, but also peritubular factors, appear to be most relevant for the close adjustment of tubule reabsorption to the delivered load [115].

Other Regulatory Mechanisms. Figure 75.1 indicates that another 20%–30% of the filtered loads of Na$^+$ and Cl$^-$ are reabsorbed in the *thick ascending limb* of the loop of Henle (TAL). Water is not reabsorbed in this nephron segment. The details of the mechanism of the reabsorption at this nephron site have already been described in Chap. 74. Several hormones enhance the reabsorptive capacity of the TAL [48]. It should be noted in this context, however, that enormous species differences exist in the hormone sensitivities [80]. As a simple rule the regulation is more sophisticated in those species in which the countercurrent concentrating system (cf. Chap. 74) is more elaborate [22].

Antidiuretic hormone (= arginine vasopressin = ADH = AVP) increases NaCl reabsorption in small rodents. This effect is probably not very marked in man, because ADH-mediated cAMP production is not demonstrable in human TAL segments [80]. The mechanism whereby TAL NaCl reabsorption capacity is increased by ADH has long been a matter of controversy [61,95]. It has now become clear that the primary effect is an increase in the basolateral Cl$^-$ conductance. This is shown in Fig. 75.4, which depicts in vitro perfusion experiments in rat medullary (m) TAL segments. It is not clear at this stage how this cAMP-mediated increase in Cl$^-$ conductance is produced. A direct gating of the Cl$^-$ channels by PKA phosphorylation has been suggested [91].

Other hormones exerting a comparable effect are *glucagon, PTH, calcitonin, insulin* and *catecholamines* (via β receptors). The interplay of these hormones in isolated TAL segments and in the intact hormone-deprived rat have been elaborated [23,80]. This entire group of hor-

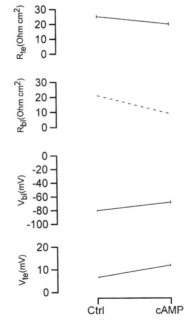

Fig. 75.4. Effects of dibutyryl-cAMP (*cAMP*, 400 μmol/l) on electrical parameters in mouse medullary thick ascending limbs perfused in vitro. cAMP reduces transepithelial resistance (R_{te}) and basolateral membrane resistance (R_{bl}). Simultaneously transepithelial voltage (V_{te}) increases. The basolateral membrane voltage (V_{bl}) depolarises. All these findings can only be reconciled with a cAMP-dependent "opening" of basolateral Cl$^-$ channels: This leads to the observed conductance changes and this depolarises V_{bl} to approach the chemical potential for Cl$^-$. The net result is an increase in NaCl reabsorption. (Data from [95])

mones has additional effects in the cTAL, namely increases in Ca^{2+} and Mg^{2+} reabsorption (cf. Sect. 75.4).

It may appear surprising that several hormones act on the same cells via the same transduction mechanisms (cAMP). This convergence of the signal pathways in one cell gives rise to the situation whereby all hormones can act interchangeably on this cell. Why should it be useful to do the same thing with various hormones? One answer to this question derives from the distribution of the different hormone receptors. For example, AVP acts on the cortical (c) TAL and on the mTAL, whilst PTH acts only in the cTAL segment [80,105,117]. Therefore, sole secretion of PTH would have a very moderate effect on NaCl reabsorption, but more of an effect on Ca^{2+} reabsorption. AVP alone enhances NaCl reabsorption markedly, because the net amounts of NaCl transported in the mTAL are much greater than those reabsorbed in the cTAL. AVP on the other hand, has less effect on Ca^{2+} reabsorption. This interpretation is the more attractive since receptors for some of these hormones and the hormone effects are also present in the distal tubule. Therefore, each individual hormone exerts its own typical profile of action. Another hormone may support some of these effects but not all of them.

In the *distal tubule* another 5%–10% of the filtered load of NaCl is reabsorbed (Fig. 75.1). The mechanism of Na^+ reabsorption at this nephron site has been discussed in Chap. 74 (Fig. 74.15). This reabsorption is stimulated by β-agonists, calcitonin, bradykinin and glucagon [65]. As for the TAL, species differences with respect to hormonal effects are very marked in this portion of the nephron. Furthermore, the distal tubule consists of morphologically and functionally distinct segments (cf. Chap. 73).

The fine tuning of the excretion of Na^+ and H_2O is determined by the *collecting duct* (CD). Figure 75.1 indicates that the CD determines whether a fraction of 1% or up to a few percent of the filtered load of Na^+ is excreted in urine. The mechanism of Na^+ reabsorption at this nephron site has been discussed in Chap. 74 (Fig. 74.17). The rate of Na^+ reabsorption is controlled by AVP and aldosterone. AVP increases the luminal Na^+ conductance acutely. This is shown in Fig. 75.5 in an in vitro perfusion experiment on a CCD of the rat [96]. AVP, via cAMP, depolarises the luminal membrane and increases the transepithelial voltage. These effects can be completely inhibited by amiloride or phenamil (cf. Fig. 74.8). Aldosterone has a more delayed effect (hours), although some more acute effects, caused by methylation, have also been shown for this hormone [79].

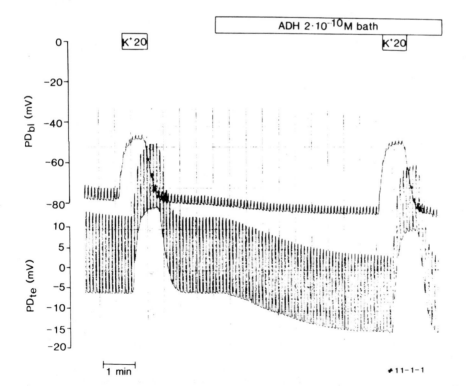

Fig. 75.5. Effects of AVP (0.2 nmol/l) when added to the bath of an isolated in vitro perfused rat cortical collecting duct. The voltages across the basolateral membrane (PD_{bl}) and across the epithelium (PD_{te}) are shown as a function of time. Note that the scales for both voltages are different. The voltage deflections in both traces are caused by intratubular current injections. They reflect the input resistance of the tubule and the fractional resistance of the basolateral membrane. During the control period, bath K^+ concentration was increased from 4 to 20 mmol/l. This caused a marked depolarisation of PD_{te} by some 30 mV, indicative of a high K^+ conductance of the basolateral membrane. Application of AVP increased PD_{te} markedly (by 12 mV). PD_{bl} hyperpolarised less (by 5 mV). Therefore, the luminal membrane depolarised by 7 mV. This indicates that AVP increased the Na^+ conductance of the luminal membrane. In fact, these effects of AVP could be blocked by luminal application of amiloride (not shown in this figure). (Data from [96])

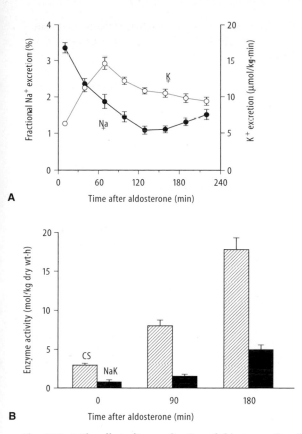

A

B

Fig. 75.6. A The effect of i.v. application of aldosterone (1 µg/kg BW bolus, same amount per hour infusion) on fractional urinary Na+ excretion (*left ordinate, closed symbols, black line*) and urinary K+ excretion (*right ordinate, open symbols, blue line*). Note the reduction in Na+ excretion and the parallel increase in K+ excretion. **B** Enzyme activities of citrate synthase (*CS, hatched bars*) and (Na++K+)-ATPase (*NaK, filled bars*) from cortical collecting tubules of acutely aldosterone-injected rabbits. Aldosterone was given at zero time. Note that both enzymes increased with time. (Data modified from [75,92])

The delayed effects can be explained on the basis of enhanced transcription of mRNA, coding for Na+ channels, mitochondrial enzymes and (Na++K+)-ATPase [79,92]. Figure 75.6 illustrates the effect of aldosterone [92]. It has been argued that increased admission of Na+ is necessary in order to increase the expression of (Na++K+)-ATPase and mitochondrial enzymes [85]. In several subsequent studies no such dependence has been shown. Therefore, aldosterone, like many other steroid hormones, initiates an entire transcription "programme" [92].

ANP acts in the MCD and PCD to reduce Na+ reabsorption. This effect of ANP in the MCD and PCD, besides the effect on GFR (cf. Chap. 73), appears to be one of the keys to the understanding of the pronounced natriuresis induced by this hormone [6,7]. Besides ANP and related urodilatins of renal origin [6,55], a natriuretic hormone has now been postulated for more than 30 years [25]. It has recently been suggested that this factor or one of these factors is identical to digitalis (ouabain) [45] (cf. also Chap. 82).

The kidney is a major site of prostaglandin formation (cf. Chap. 6). Main sites of production are the small blood vessels, the glomeruli and tubule segments from the outer and inner medulla. Locally produced prostaglandins (PG) act natriuretically and diuretically in the CD. The mechanism is via a G$_i$-mediated inhibition of cAMP [78].

As has been discussed in Chap. 74, the reabsorption of H$_2$O is controlled in the CD by AVP. In the presence of AVP, urinary flow rate can fall to some 500 ml/day; in its absence it can increase to 20 l/day. It should be noted here that the reabsorption of H$_2$O and Na+ are dissociated in the CD. AVP acts mostly antidiuretically and antinatriuretically. Aldosterone enhances Na+ reabsorption but has no direct effect on H$_2$O reabsorption.

75.3 K+

Figure 75.7 summarises the recovery of K+ along the nephron [103,119]. Most of the filtered K+ is reabsorbed in the proximal tubule, probably via the paracellular shunt pathway (cf. Chap. 74). In the TAL some reabsorption of K+ occurs. The magnitude of this process depends on the transepithelial voltage and on the cytosolic pH [48,50]. The early distal recovery is between 10% and 20% of the filtered

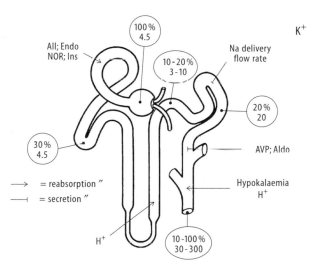

Fig. 75.7. Recovery of K+ along the nephron. The *numbers in the circles* refer to the recovery in the tubule fluid and the approximate concentration (mmol/l). The filtered amount is taken as 100%. This figure also includes hormones and factors acting on K+ transport. *Arrows* indicate enhanced reabsorption; *"blocked" lines* indicate diminished reabsorption or enhanced secretion. *AII*, angiotensin II; *Endo*, endothelin; *NOR*, norepinephrine; *Ins*, insulin. All four hormones enhance proximal tubular reabsorption of K+. Acidosis (H+) inhibits secretion and hence augments net reabsorption of K+ in the thick ascending limb of the loop of Henle. *Aldo*, aldosterone; *AVP*, arginine vasopressin. Both hormones, by different mechanisms, enhance secretion of K+ in the collecting duct. Hypokalemia and acidosis have the opposite effect and enhance reabsorption. Note that the urinary excretion of K+ can vary largely

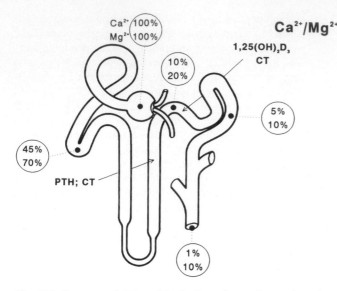

Ca²⁺/Mg²⁺

Ca²⁺ 100%
Mg²⁺ 100%

10%
20%

1,25(OH)₂D₃
CT

5%
10%

45%
70%

PTH; CT

1%
10%

Fig. 75.8. Recovery of Ca^{2+} and Mg^{2+} along the nephron. The *numbers in the circles* refer to the recovery in the tubule fluid. The filtered amount is taken as 100%. It should be noted that the filterability of Ca^{2+} is only 55%–60% and that of Mg^{2+} only 80%. These figures also include hormones acting on Ca^{2+} and Mg^{2+} transport. Parathyroid hormone (*PTH*) and calcitonin (*CT*) en-hance Ca^{2+} and Mg^{2+} reabsorption in the cortical portion of the thick ascending limb of the loop of Henle (TAL). With some delay, $1,25(OH)_2D_3$ increases Ca^{2+} reabsorption in the distal tubule. Note that the major site of regulation for both divalent cations is the TAL

load. Some K^+ reabsorption can occur in the DT. The major site of the regulation of the amount of K^+ excreted in urine is the CD. The basic principles of regulation at this nephron site have been discussed in Chap. 74, and will be further discussed in Sect. 75.14. The interplay between acid-base and potassium metabolism (see also Chaps. 78 and 79) will also be summarised below.

75.4 Ca²⁺ and Mg²⁺

Figure 75.8 displays the fractional recoveries for Ca^{2+} and Mg^{2+} along the nephron [20,89,90,104]. For both ions the filterability is significantly smaller than 1: Ca^{2+}, 55%–60%; Mg^{2+}, 80%. Ca^{2+} is reabsorbed in the proximal tubule to almost the same extent as Na^+. Therefore, luminal Ca^{2+} changes very little along the proximal nephron (cf. Fig. 74.9). The route of transcellular Ca^{2+} is still not entirely clear. It is likely, however, that most of the Ca^{2+} is reabsorbed paracellularly through the cation-selective paracellular shunt, and driven by the small lumen positive voltage. In the case of Mg^{2+} this reabsorption is significantly less effective (only 30% of the filtered load) than that of Na^+ (70% of the filtered load). Therefore, the luminal concentration of Mg^{2+} increases along the proximal tubule from 0.8 to approximately 1.6 mmol/l (cf. Fig. 74.9).

The main site for Mg^{2+} and a major site of Ca^{2+} reabsorption is the TAL. The mechanism of this reabsorption is not clear, but it probably includes a transcellular [5,35] and a paracellular component [48,117] (cf. Chap. 74). Figure 75.9 shows data obtained in mouse TAL [117]. The rate of Ca^{2+} reabsorption correlates closely with the transepithelial voltage. The slope of this correlation is increased by PTH. These data suggest that the main route of divalent cation reabsorption is the paracellular pathway, and that PTH enhances the permeability of this pathway. PTH, calcitonin, ADH and glucagon all increase the rate of Mg^{2+} and Ca^{2+} net reabsorption. This effect is confined to the cTAL in rodent and rabbit nephrons [105,117].

Ca^{2+} and Mg^{2+} reabsorption continues beyond the TAL in the DT and CD. The mechanisms operating in these nephron segments are not known. The urinary excretion of Ca^{2+} and Mg^{2+} amounts to a few percent of the filtered load. It is reduced by the above hormones.

75.5 HCO₃⁻ and H⁺

HCO₃⁻ Reabsorption and Acidification. The recovery rates for HCO_3^- along the nephron and the pH values are summarised in Fig. 75.10. The reabsorption of HCO_3^- in the proximal tubule occurs preferentially, i.e., HCO_3^- is reabsorbed out of proportion (more than Na^+, Cl^- and H_2O). Therefore, the tubule concentration of HCO_3^- falls rapidly along the nephron. At the end of the proximal tubule it amounts to some 5 mmol/l. The reabsorption of HCO_3^- is subject to GTB. In fact, some evidence suggests HCO_3^- is one key determinant of GTB [115]. The reabsorption of HCO_3^- is driven by the secretion of H^+. H^+ secretion occurs by Na^+/H^+ exchange and by active (ATP-driven) H^+ pumping [1,3,67]. It has been deduced from inhibition experiments that about 60%–80% of total H^+ secretion occurs via the exchanger, and the rest by active pumping. The detailed mechanisms of HCO_3^- reabsorption

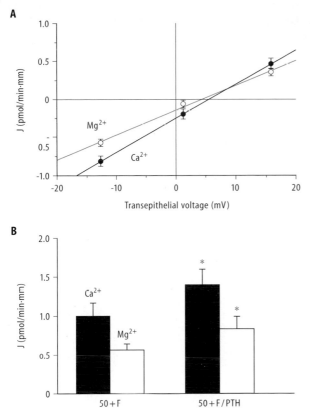

A

B

Fig. 75.9a,b. Ca^{2+} and Mg^{2+} transport in the cortical portion of the thick ascending limb of the loop of Henle of the mouse. In vitro perfusion experiments. **a** In the presence of furosemide in the lumen, in order to abolish active NaCl reabsorption, dilution voltages were superimposed on the epithelium. The net transport of Ca^{2+} (*closed circles, black line*) and of Mg^{2+} (*open circles, blue line*) is plotted as a function of the transepithelial voltage. It is evident that negative voltages drive a net secretion and positive voltages a net reabsorption of both ions. (Data redrawn from [29]). **b** Net reabsorption of Ca^{2+} and Mg^{2+} in the presence and absence of parathyroid hormone (*PTH*). Experiments were performed in the presence of furosemide in the lumen (*+F*). The driving force for the reabsorption was the lumen positive voltage generated by a diluted (50 mmol/l NaCl) luminal perfusate. Note that PTH, in the absence of active NaCl transport, enhances the reabsorption of both divalent cations. This is taken as evidence that PTH acts on the paracellular pathway to increase its permeability for both divalent cations. (Data modified from [117])

have been discussed in Chap. 74 (cf. Fig. 74.6). Due to H^+ secretion luminal pH becomes acidic and reaches 7.0–6.8 by the end of the proximal tubule.

HCO_3^- reabsorption and acidification continue along the TAL [44]. The processes involved are Na^+/H^+ exchange and H^+ pumping. It should be emphasised that this process, in contrast to previous conclusions [34], is probably irrelevant for Na^+ reabsorption [28]. By the end of the TAL, HCO_3^- concentration falls to a few mmol/l. Tubule pH falls a little further and reaches 6.6–6.8.

HCO_3^- reabsorption is almost complete, and urine is usually HCO_3^- free. This is achieved by active H^+ pumping in the intercalated cells (A cells). Luminal pH can fall to 4–5

[16,56]. Therefore, the H^+ gradient across the CD can be by three orders of magnitude. Such a gradient cannot be produced by an Na^+/H^+ antiporter, because the driving force of this system is only around 60–70 mV, i.e., one order of magnitude (cf. Nernst equation). The secretion of H^+ is coupled to the hydrolysis of ATP. The pump is of the vacuolar type [41,42]. However, recently it has been postulated that also an H^+/K^+-ATPase similar or identical to that of the parietal cell of the stomach is present in the MCD [116]. This pump may be more relevant for K^+ conservation, e.g., in hypokalemia, than for urinary acidification.

Regulation of Renal Acidification. The reabsorption of HCO_3^- and renal acidification are controlled by several mechanisms operating in both the proximal and the distal nephron:

- *Increase in filtered load of HCO_3^-*, as in metabolic (non-respiratory) alkalosis, leads to increased urinary HCO_3^- excretion owing to a saturation of the maximal (proximal and distal) transport capacities [56]. This was first observed 49 years ago [87]. A typical experiment from this classical work is shown in Fig. 75.11. The concept of maximal transport capacities can account for these findings. However, it should be kept in mind that the maximal transport capacity of HCO_3^- is readjusted by GTB. Therefore, increases in the filtered load of HCO_3^- lead to bicarbonaturia if they are caused by increases in the plasma concentration of HCO_3^-, but not if they are due to an increase in GFR.
- *Cytosolic pH* determines the rates of Na^+/H^+ exchange [4] and of the density of A versus B intercalated cells [56,97]. The pH dependence of the Na^+/H^+ exchange has been demonstrated in membrane vesicle studies. It has been shown that this exchanger has an allosteric regulation site on the cytosolic side (cf. Fig. 75.12). At a pH of 6.8 and below the rate of the exchanger is sharply increased. An alkaline pH has the opposite effect. These findings can easily explain the fact that HCO_3^- reabsorption is enhanced in respiratory acidosis and attenuated in respiratory alkalosis.

Due to the high permeability of almost all cell membranes to CO_2, systemic changes in CO_2 are always also perceived by tubule cells, and they lead to corresponding changes in cytosolic pH.

The amount of acid excreted in urine is usually quantified by several measurements:

- Urine pH.
- Titratable acid, i.e., the amount of H^+ added to increase urine pH to 7.4. This usually corresponds mostly to the $H_2PO_4^-$ concentration in urine.
- The urine concentration of NH_4^+.
- The urine concentration of $H_2PO_4^-$.
- The urine CO_2 pressure [11].

In metabolic acidosis titratable acid becomes maximal and slowly NH_4^+ excretion increases. This process requires sev-

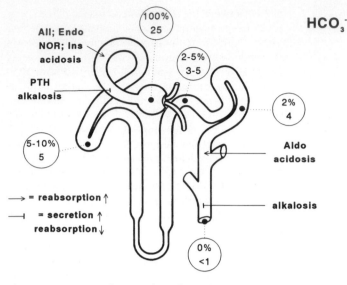

Fig. 75.10. Recovery of HCO₃⁻ along the nephron. The *numbers in the circles* refer to the recovery in the tubule fluid and the approximate concentration (mmol/l). The filtered amount is taken as 100%. This figure also includes factors and hormones acting on HCO₃⁻ transport. *Arrow* indicates enhanced reabsorption; *"blocked" line* corresponds to reduced reabsorption or enhanced secretion. AII, angiotensin II; *Endo*, endothelin; *NOR*, norepinephrine; *Ins*, insulin. All four hormones and acidosis enhance proximal tubular reabsorption of HCO₃⁻. *PTH*, parathyroid hormone. PTH and alkalosis reduce proximal tubular reabsorption of HCO₃⁻. *Aldo*, aldosterone. Aldosterone and acidosis increase H⁺ secretion, and thus HCO₃⁻ reabsorption in the collecting duct. Alkalosis has the opposite effect. It may even induce secretion of HCO₃⁻

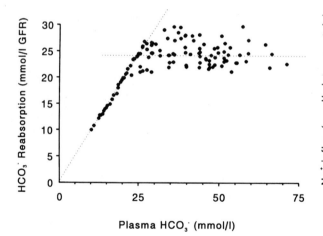

Fig. 75.11. Renal HCO₃⁻ reabsorption as a function of plasma HCO₃⁻ concentration. Clearance experiments in dogs. To account for the glomerulotubular balance, the reabsorption is factored by the GFR. The *dotted lines* reflect the two characteristics of HCO₃⁻ transport: At low plasma concentrations (up to 25 mmol/l) reabsorption is almost complete. The regression line has a slope close to 1. At higher plasma concentrations HCO₃⁻ reabsorption is saturated and the regression line has a slope of zero. Therefore, the amount excreted in urine equals the filtered amount minus that maximally reabsorbed. (Data redrawn from [87])

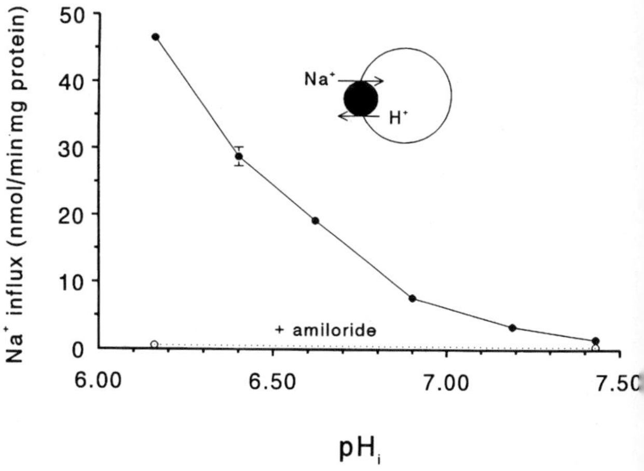

Fig. 75.12. pH regulation of Na⁺/H⁺ exchange. The Na⁺ influx into plasma membrane vesicles prepared from rabbit renal medulla (mean values ± SEM) is plotted as a function of the pH on the cytosolic side (*pHi*). Uptake is sharply reduced at acid pH (*closed symbols, solid line*). Amiloride (1 mmol/l) inhibits Na⁺ uptake completely. (Data redrawn from [4])

eral days to become maximal (cf. Fig. 75.13) [93]. Therefore, besides complete HCO₃⁻ reabsorption in metabolic acidosis, with maximal excretion of titratable acid in urine (Fig. 75.13), the kidney also excretes much more NH₄⁺. This NH₄⁺ is produced in the kidney itself. The main site of NH₄⁺ production is the proximal nephron, although NH₄⁺ can also be produced in other nephron segments [86]. The basic principle is shown in Fig. 75.14A. The production of L-glutamine in the liver is enhanced in metabolic acidosis. It is taken up into the proximal tubule cell by carrier-mediated mechanisms (cf. Sect. 75.10 and Fig. 75.14). Renal mitochondrial glutaminase is induced by

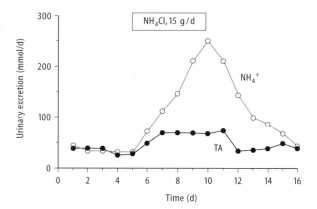

Fig. 75.13. Urinary excretion of titratable acid (*TA*; *closed symbols, black line*) and NH_4^+ (*open symbols, blue line*) in chronic nonrespiratory (metabolic) acidosis. Acidosis was induced by the ingestion of NH_4Cl. Data from healthy volunteers. Note that it takes a few days until NH_4^+ excretion increases to a maximum. TA excretion reacts more promptly. However, the increase in TA excretion is moderate. (Redrawn from [93])

This has two major consequences:

- HCO_3^- is saved by the liver. This counterbalances the metabolic acidosis.
- N metabolism and excretion are shifted from urea to NH_4^+.

Quantitative analysis reveals that NH_4^+ excretion can be increased to some 200–300 mmol/day (cf. Fig. 75.13). Then, 100–150 mmol HCO_3^- is conserved per day. This corresponds to an increase by 6–9 mmol/l·day HCO_3^- in the entire extracellular space. At the same time urea production, which normally is around 500 mmol/day, is reduced to 350–400 mmol/day.

Summarising, the kidney can adjust acid secretion according to the systemic pCO_2 [11,39,56]. Acid excretion is enhanced in hypercapnia and reduced in hypocapnia. Metabolic disturbances change HCO_3^- excretion acutely [2,32]. Chronically, acid secretory systems are maximised in metabolic acidosis (induction of A cells) and the pro-

metabolic acidosis [54]. As a consequence more NH_4^+ is produced by tubule cells:

$$Glutamine \rightarrow Glutamate + NH_4^+. \qquad (75.1)$$

NH_4^+ is in equilibrium with NH_3:

$$NH_4^+ \leftrightarrow NH_3 + H^+; pKa = 9.37. \qquad (75.2)$$

NH_3 can leave the cells by diffusion, whereas NH_4^+ requires specific pathways. Hence, NH_4^+ is trapped in acidic compartments, because there the NH_3 concentration is kept low. Applied to the proximal tubule this consideration predicts that NH_3 diffuses preferentially into the lumen. There, NH_3 is trapped by the secreted protons and NH_4^+ is formed. It should be emphasised that *this process does not increase the acid excretion by the kidney*. In fact, NH_4^+ is formed by the above deamination, but not NH_3. Therefore, the secreted proton and NH_3 recombine to from NH_4^+.

What is the advantage of enhanced deamination in the kidney in metabolic acidosis? This becomes apparent if one considers the cooperation of liver and kidney (cf. Chap. 68). As is apparent from Fig. 75.14B, metabolic acidosis has an additional effect on the liver: urea production from NH_4^+ and HCO_3^- is reduced and glutamine production is enhanced (Eq. 75.1):

$$NH_4^+ + NH_3 + HCO_3^- \rightarrow NH_2CONH_2 + 2H_2O. \qquad (75.3)$$

The regulation therefore comprises four functional components:

- Enhanced glutamine production by the liver
- Diminished urea production by the liver
- Enhanced deamination of glutamine by the proximal tubule
- Enhanced renal excretion of NH_4^+.

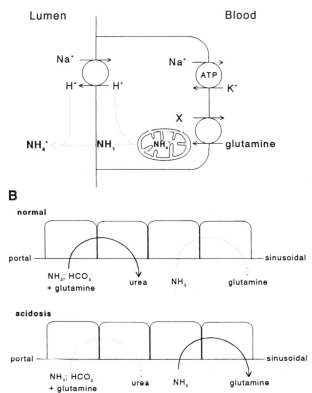

Fig. 75.14A,B. Changed N-metabolism in chronic metabolic acidosis. **A** Effect on kidney function (mostly proximal tubule). Glutamine is taken up and metabolised in mitochondria. There, glutaminase is induced by metabolic acidosis. The NH_4^+ formed from glutamine is secreted as NH_3 and H^+ (via Na^+/H^+ exchange). NH_4^+ excretion can increase to reach some 300 mmol/day in man. **B** Effect on the liver. In chronic metabolic acidosis urea production is reduced, and hence HCO_3^- saved. In addition the formation of glutamine from NH_3 (NH_4^+) is enhanced. As a result, less urea but more glutamine is delivered to the kidney

1525

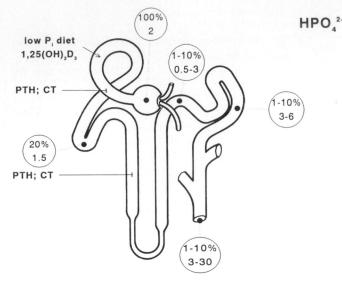

Fig. 75.15. Recovery of HPO$_4^{2-}$/H$_2$PO$_4^-$ along the nephron. The *numbers in the circles* refer to the recovery in the tubule fluid and the approximate concentration (mmol/l). The filtered amount is taken as 100%. Data are taken from the rat (cf. Fig. 75.16B). This is relevant because in man plasma phosphate concentration is considerably lower (approximately 1 mmol/l). This figure also includes hormones and factors acting on HPO$_4^{2-}$/H$_2$PO$_4^-$ transport. 1,25(OH)$_2$D$_3$ and low phosphate (P_i) diet enhance phosphate reabsorption in the proximal tubule. The effect of 1,25(OH)$_2$D$_3$ is rather limited and slow. Parathyroid hormone (PTH) and calcitonin (*CT*) reduce phosphate reabsorption in the proximal tubule (mostly S2 and S3 segments). There is no evidence for any phosphate reabsorption in the distal tubule. The apparent discrepancy between distal load and the amount excreted in urine (cf. Fig. 75.16b) is caused by the fact that deep nephrons reabsorb more phosphate than the superficial ones (accessible for micropuncture)

duction of NH$_4^+$ is enhanced in the kidney (induction of proximal tubule glutaminase).

75.6 HPO$_4^{2-}$

The fractional recovery of phosphate along the nephron is summarised in Fig. 75.15. The filterability of phosphate is around 95%. In the tubule fluid and in plasma, phosphate is found mostly as primary and secondary phosphate (Eq. 75.5).

$$H_3PO_4 \rightarrow H_2PO_4^- + H^+, pKa = 2.0 \qquad (75.4)$$

$$H_2PO_4^- \rightarrow HPO_4^{2-} + H^+, pKa = 6.8 \qquad (75.5)$$

$$HPO_4^{2-} \rightarrow PO_4^{3-} + H^+, pKa = 12.3 \qquad (75.6)$$

At physiological pH of 7.4, the ratio of HPO$_4^{2-}$ to H$_2$PO$_4^-$ is around 4:1. In tubule fluid this ratio is approximately 1 by the end of the proximal nephron and 0.001 to 0.1 in urine.

Site of Phosphate Reabsorption. It has long been debated whether phosphate reabsorption occurs only in the proximal tubule or also in the distal tubule [51]. Phosphate recovery at the early distal tubule site is significantly lower than that present in the last loops of the proximal tubule, which are accessible to micropuncture (cf. Fig. 75.16b). This discrepancy is enhanced further by PTH and calcitonin [51]. It is now clear that this discrepancy is due to the residual reabsorption of phosphate in the pars recta portion of the proximal tubule. The reabsorption at this site is enhanced in the absence of PTH and calcitonin [84]. The phosphaturic effect of both hormones is localised in the late proximal tubule and is additive.

Mechanism of Phosphate Reabsorption. Figure 75.16a summarises the mechanism of phosphate reabsorption in the proximal tubule. Phosphate is cotransported with Na$^+$ across the luminal membrane. Both physiologically relevant phosphate species, HPO$_4^{2-}$ and H$_2$PO$_4^-$, are probably accepted by this transporter. The cotransporter families have now been cloned [82]. In fact, it transpired that two or more different types of proteins are present in the proximal tubule: NaPi$_1$ and NaPi$_2$. The regulated luminal phosphate uptake probably occurs mostly via the latter. This protein is regulated by PKA- and PKC-dependent phosphorylation. The expression of this protein is enhanced in phosphate deprivation. PTH appears to act on this cotransporter via both transduction pathways [81,82]. Phosphate leaves the cell via a thus far unknown mechanism. This interpretation predicts that phosphate transport is saturable. Under physiological conditions with a plasma phosphate concentration of 1–2 mmol/l and with PTH present, some 10%–20% of the filtered phosphate is excreted in urine. In the absence of or with low PTH, phosphate reabsorption is enhanced by an increase in V$_{max}$ [68,69]. The urine excretion of phosphate can then be as low as a few percent of the filtered load. As with HCO$_3^-$, GTB readjusts the reabsorption of phosphate.

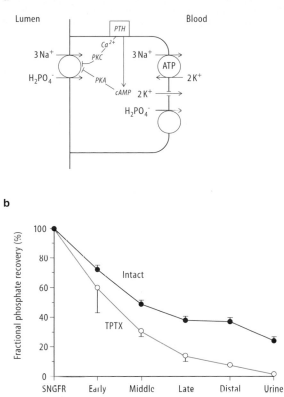

a

Lumen | Blood

b

Fig. 75.16a,b. Mechanisms of phosphate reabsorption in the proximal nephron. **a** Cellular mechanism. *Circle with ATP*, (Na$^+$+K$^+$)-ATPase; *circle*, carrier systems; *arrow*, conductance. Phosphate is taken up by an Na$^+$-dependent cotransport mechanism. The carriers have now been cloned [82]. Phosphate leaves the cell via an Na$^+$-independent carrier. Parathyroid hormone (*PTH*) acts via two transduction pathways: One is mediated by the IP$_3$-induced Ca^{2+} release and Ca^{2+} uptake and controls protein kinase C (*PKC*)-dependent phosphorylation; the other is mediated by cAMP and protein kinase A (*PKA*)-dependent phosphorylation [81,82]. The uptake of phosphate is reduced by PTH. **b** Phosphate transport in the rat nephron. Micropuncture experiments were performed in acutely thyroparathyroidectomised (*TPTX*) rats (*open symbols, blue line*) and in intact rats (*closed symbols, black line*). Fractional recovery of phosphate is plotted for the various micropuncture sites. The recovery in the filtrate (*SNGFR*) was set to 100%. "Early", "middle" and "late" refer to micropuncture sites along the accessible part of the proximal tubule. "Distal" and "urine" refer to the recovery rates in distal micropuncture samples and in urine, respectively. Note that phosphate recovery is much lower in TPTX rats, indicative of the phosphaturic effect of PTH. The difference becomes most marked between middle proximal and the distal micropuncture sites. This suggests that the phosphaturic effect of PTH is mostly localised in the S2 and S3 segments of the proximal nephron. (Data from [51])

Regulating Factors. The key factors regulating the renal excretion of phosphate [12,68,82] are PTH, calcitonin, hypophosphatemia and acid-base changes. As discussed above, PTH and calcitonin both reduce the reabsorption of phosphate. PTH secretion by the parathyroid glands is controlled by plasma ionised Ca^{2+} activity. A fall in Ca^{2+} enhances PTH secretion. The three consequences for the kidney are (cf. Sect. 75.11):

- Phosphaturia
- Anticalciuria
- Enhanced 1-hydroxylation of 25-hydroxy-cholecalciferol to 1,25(OH)$_2$D$_3$ (cf. Chap. 74).

As a result plasma phosphate concentration falls and that of Ca^{2+} increases. The PTH-dependent production of 1,25(OH)$_2$D$_3$ enhances intestinal Ca^{2+} absorption, which further increases plasma Ca^{2+} concentration.

The effect of calcitonin, secreted from specialized cells of the thyroid gland, is more controversial. It has been reported that it enhances phosphate and calcium excretion [53]. The phosphaturic effect is localised in the proximal tubule [68,84]. The effect on Ca^{2+} excretion has already been discussed in Sect. 75.4. The net effect of acute secretion of calcitonin in the intact organism cannot easily be predicted, since several factors (plasma Ca^{2+}, HPO$_4^{2-}$, pH etc.) and hormones [PTH, 1,25(OH)$_2$D$_3$] determine this net effect. In the in vitro perfused TAL calcitonin enhances (in contrast to what its name might predict) the reabsorption of Ca^{2+} [24]. Therefore, *calcitonin and PTH have comparable effects on tubule transport.*

Acid-base changes influence Ca^{2+} and phosphate metabolism in several ways [68]:

- Plasma ionized Ca^{2+} activity falls in alkalosis and is increased in acidosis. According to mass law, plasma phosphate concentration is increased in alkalosis (fall in Ca^{2+}) and reduced in acidosis (increase in Ca^{2+}).
- The fall in plasma ionized Ca^{2+} activity induced by alkalosis leads to the secretion of PTH. This leads to the corresponding effects.
- Changes in luminal pH have a direct effect on phosphate reabsorption [68,82]. This is a consequence of the change in the ratio of the two phosphate species, and probably more importantly a consequence of direct effects on the Na$^+$-phosphate cotransporter. As a net result acidosis tends to decrease and alkalosis tends to increase the reabsorption of phosphate [82].

The *phosphate content of the diet* regulates the reabsorption of phosphate correspondingly. These effects can be subdivided into a rather acute (hours) and a delayed (days) component. The chronic effects can be explained on the basis of altered transcription (H. Murer, personal communication). Previously it has been shown that the rather acute effects of high phosphate diet may be mediated by changes in cytosolic Ca^{2+} [68,83].

In summary, the renal transport of phosphate is regulated by a large variety of mechanisms and closely linked to the Ca^{2+} metabolism. An integrated discussion of mineral metabolism and the role of the kidney will be given in Sect. 75.11.

75.7 D-Glucose

Reabsorption. The fractional recovery of D-glucose in the nephron is shown in Fig. 75.17A. D-Glucose is filtered

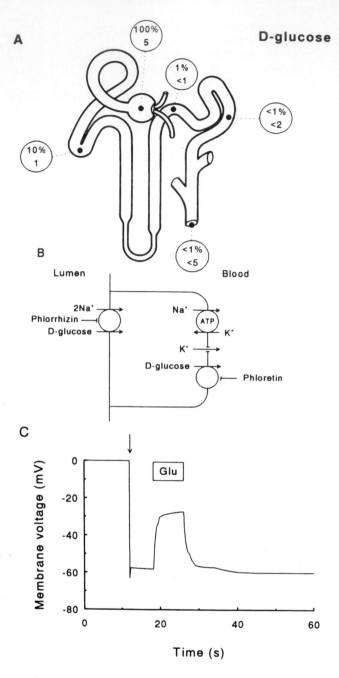

A

100%
5

1%
<1

<1%
<2

D-glucose

10%
1

B

<1%
<5

Lumen   Blood

2Na⁺
Phlorrhizin
D-glucose

Na⁺   ATP   K⁺

K⁺

D-glucose   Phloretin

C

Glu

Membrane voltage (mV)

0

-20

-40

-60

-80

0   20   40   60

Time (s)

Fig. 75.17A–C. Reabsorption of D-glucose. **A** Sites of D-glucose reabsorption in the nephron. Recovery of D-glucose along the nephron. The *numbers in the circles* refer to the recovery in the tubule fluid and the approximate concentration (mmol/l). The filtered amount is taken as 100%. Note that reabsorption occurs early in the proximal nephron and is almost complete by the end of the proximal convoluted tubule accessible to micropuncture. Urine is usually glucose-free. **B** The cellular mechanism of D-glucose reabsorption. *Circle with ATP*, (Na⁺+K⁺)-ATPase; *circle*, carrier systems; *arrow*, conductance. D-Glucose is taken up by an Na⁺-dependent cotransport mechanism. The carriers have now been cloned [62]. D-Glucose leaves the cell via an Na⁺-independent carrier. The Na⁺-D-glucose cotransporter is inhibited by phlorrhizin and the Na⁺-independent glucose carrier by phloretin (cf. Fig. 75.18). The stoichiometry of the Na⁺-D-glucose cotransporter appears to be in excess of 1 (probably 2 Na⁺/1 D-glucose in a major part of the proximal tubule [26,106]), and it may be even higher for the S3 segment. **C** Rheogenicity of D-glucose reabsorption. Measurement of the basolateral membrane voltage of a rat (in situ) proximal tubule cell. The *arrow* indicates the puncture of the cell. *Glu* indicates the luminal perfusion of D-glucose. This leads to an abrupt and completely reversible depolarisation. This depolarisation is caused by the uptake of positive charges (coupled Na⁺ uptake). (Data redrawn from [36])

freely, the daily filtration being 900 mmol, corresponding to 162 g. This may be in excess of the daily dietary intake of carbohydrates. D-Glucose reabsorption occurs only in the proximal tubule. The mechanism of reabsorption is shown in Fig. 75.17B. D-Glucose is cotransported with Na⁺. This cotransporter is closely related to the intestinal Na⁺-D-glucose cotransporter, which was cloned a few years ago [62]. The transfer of D-glucose across the luminal membrane carries one positive charge per molecule of D-glucose. The rheogenicity (charge transfer) of this transporter has been demonstrated in two ways. D-Glucose reabsorption generates a lumen negative voltage in the early proximal tubule

[38]. In addition, later intracellular voltage measurements have revealed that the luminal membrane voltage is strongly depolarised by D-glucose [36,37]. This is shown in Fig. 75.17C [36]. The driving force for this cotransporter favouring D-glucose uptake is: Na⁺ + 130 mV; that for D-glucose depends on the cytosolic concentration. It can easily be deduced that this cotransporter (for a 1 Na⁺/1 glucose stoichiometry) could generate a concentration gradient for D-glucose of 130/61 = 2.2 decades (cf. Nernst equation). Therefore, the cytosolic D-glucose concentration can be very high. This high D-glucose concentration in the cell drives D-glucose transport across the basolateral mem-

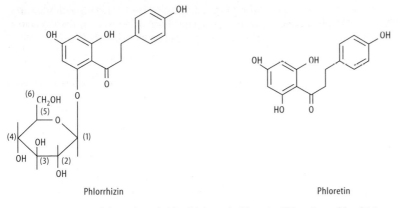

Phlorrhizin   Phloretin

Fig. 75.18. Structural formulae of phlorrhizin and phloretin. Phloretin = phlorrhizin − D-glucose

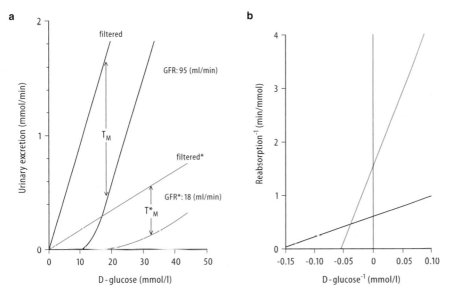

Fig. 75.19a,b. Concentration dependence of renal D-glucose reabsorption. **a** Urinary excretion of D-glucose is plotted as a function of plasma glucose concentration. Data for a normal individual (GFR 95 ml/min, *black curve*) and for a patient with chronic renal failure (CRF; GFR 18 ml/min, *blue curve*). For comparison the filtered amounts are also plotted as *straight lines* (*black* = normal = *"filtered"*; *blue* = chronic renal failure = *filtered**). The maximal transport rate (T_M) is much higher in normal than in diseased kidney (T_M*). The T_M is a function of GFR (cf. glomerulotubular balance). Also note that glucosuria begins in the healthy individual at a plasma glucose concentration of approximately 8 mmol/l, whereas concentrations twice as high are required for the patient with CRF. **b** Lineweaver-Burk plot of the data of **A**. *Black line*, normal individual; *blue line*, CRF. The maximal transport rate is approximately 1500 mmol/day in the normal individual, and only 700 mmol/day in the CRF patient. Note also that the apparent affinity (K_M) is 7 mmol/l for the normal person and 17 mmol/l for the CRF patient

brane via an Na⁺-independent D-glucose carrier [26]. Na⁺-D-glucose cotransport can be inhibited by phlorrhizin (Fig. 75.18) [108]. Phlorrhizin contains one D-glucose molecule and the aglycon phloretin. Phloretin has no effect on the Na⁺-D-glucose cotransporter but inhibits several other transport proteins, such as the Na⁺-independent D-glucose carrier.

The reabsorption of D-glucose is saturable. This is shown in Fig. 75.19 for data from a healthy volunteer and a patient with chronic renal failure. From the Michaelis-Menten plot (Lineweaver-Burk) an apparent affinity of some 7 mmol/l and a maximal transport rate of 1500 mmol/day can be deduced. It should be emphasised again that D-glucose

reabsorption capacity is regulated by GTB. It is increased whenever GFR is enhanced.

Conformational Requirements. The cotransporter is very selective. It accepts only sugars which conform to very narrow constraints. The conformational requirements have been elaborated and are summarised in Table 75.1 (see also Chaps. 8, 11). 3-O-methyl-D-glucose is accepted, but 2-deoxy-D-glucose and L-glucose are not accepted [26]. The general requirements are a pyranose ring with the C2, C3, and C4 OH groups in the appropriate positions (cf. the D-glucose moiety in Fig. 75.18). Of these requirements the C2 OH group appears to be of greatest importance. In

Table 75.1. Sugar specificity of the renal D-glucose cotransporter: data obtained in the doubly perfused rat proximal tubule and taken from [26]

Name	Acceptance
D-Glucose	+
β-Methyl-D-glucoside	+
α-Methyl-D-glucoside	+
6-Deoxy-D-glucose	+
D-Galactose	+
β-Methyl-D-galactoside	+
3-O-methyl-D-glucoside	+
D-Allose	+
L-Glucose	−
D-Mannose	−
2-Deoxy-D-glucose	−
D-Fructose	−
D-Glucosamine	−
3-Deoxy-D-glucose	−
2-Deoxy-D-glucose	−

honour of its first describer this is called the Crane specificity.

Cotransporter Forms. Under normal conditions, with plasma D-glucose concentrations of <10 mmol/l, urine is practically D-glucose free. Most of the reabsorption occurs in the very early fraction of the proximal tubule. It has been shown that the cotransporter occurs in at least two forms in the nephron. The form with a stoichiometry of 1 Na$^+$/1 D-glucose is present in S1. In S3 a cotransporter with a stoichiometry of 2 Na$^+$ and possibly even 3 Na$^+$/1 D-glucose has been found [26,106]. This cotransporter can generate a D-glucose concentration gradient (cell/lumen) of 4 – 6 decades (cf. above for the calculation procedure). Also the affinity of the D-glucose cotransporter (K_M) is higher in late proximal than in early proximal tubules [26]. The opposite is true for the apparent maximal rate of transport (V_{max}).

Diabetes Mellitus. Increases in plasma D-glucose concentration lead to a corresponding glucosuria. In man, this occurs at a plasma D-glucose concentration of >8–12 mmol/l. This is shown in Fig. 75.19. The non-absorbed D-glucose then acts as an osmolyte and inhibits water reabsorption, leading to diabetes mellitus (diabetes from the Greek διαβαῖνω = to pass rapidly; mellitus from the Greek μέλλιτος = sweet).

In summary, D-glucose is reabsorbed in the proximal tubule, and urinary losses of D-glucose are prevented. The transport is highly specific, saturable and secondary active.

75.8 Amino Acids, Oligopeptides and Proteins

Proximal Tubule Reabsorption. All circulating amino acids are filtered and have to be reabsorbed by the tubule to prevent substantial urinary losses. Like that of D-glucose, the reabsorption of amino acids and oligopeptides is con-

fined to the proximal tubule [99–101]. Most amino acids are cotransported with Na$^+$. A typical experiment for phenylalanine is shown in Fig. 75.20A. The coupling to Na$^+$ makes reabsorption rheogenic and depolarises the luminal membrane. The general mechanism of proximal tubule reabsorption of neutral amino acids is summarised in Fig. 75.20B and is very similar to that of D-glucose reabsorption (Fig. 75.17B). It should be noted here that this general scheme may only apply to neutral and acidic amino acids, whilst that for basic amino acids may not require the coupling to Na$^+$ [100].

Cotransport Systems. In extensive studies using various methodologies several amino acid cotransport systems

a

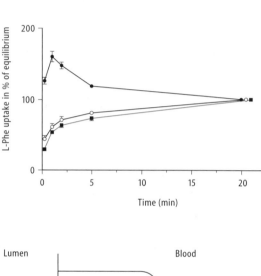

b

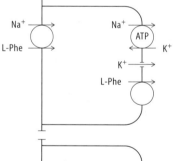

Fig. 75.20a,b. Renal amino acid transport. **a** Studies in renal brush border membrane vesicles. Uptake of phenylalanine (mean values ± SEM) is shown as a function of time, and is expressed in % of the equilibrium value. D-Phenylalanine, in the presence of an Na$^+$ gradient, shows an enhanced uptake ("overshoot") (*closed circles, black line*). In the absence of the Na$^+$ gradient (*open circles, black line*) uptake is slowed and no "overshoot" is observed. D-Phenylalanine, likewise, shows only a slow uptake and no "overshoot" (*blue curve* and *squares*). Therefore, the uptake of L-phenylalanine is Na$^+$ dependent and stereospecific. (Data redrawn from [33]). **b** Cellular mechanism of the reabsorption of neutral amino acids in the proximal tubule. *Circle with ATP*, (Na$^+$+K$^+$)-ATPase; *circle*, carrier systems; *arrow*, conductance. L-Phenylalanine (*L-Phe*) is taken up by an Na$^+$-dependent cotransport mechanism. L-Phe leaves the cell via an Na$^+$-independent carrier

Table 75.2. The amino acid reabsorption systems of the proximal tubule. Data taken from [101]

1. Acidic amino acids: D-aspartate, L-aspartate, cysteate, glutamate
2. Dibasic amino acids: arginine, canaverine, lysine, homoarginine, ornithine
3. Neutral amino acids: cysteine, methionine?
4a. "Imino" amino acids (low K_M system): proline, OH-proline, sarcosine, N-methylalanine; 5-oxo-proline?
4b. "Imino" amino acids (high K_M system): proline, OH-proline
5. Glycine
6. Other neutral amino acids: alanine, asparagine?, citrulline, glutamine, glycine, histidine, iso-leucine, leucine, methionine, phenylalanine, serine, threonine?, tryptophan, tyrosine, valine
7. β- and γ-amino acids: β-alanine, γ-aminobutyric acid, taurine, hypotaurine

have been identified. These systems are summarised in Table 75.2. Two systems take care of neutral amino acids such as alanine, glycine and histidine; another transports acidic amino acids (e.g., glutamate, aspartate and cysteate); a fourth transports dibasic amino acids (e.g.,m arginine, lysine and ornithine); a fifth transports imino amino acids such as proline, OH-proline and sarcosine; a further system also transports neutral amino acids such as cysteine and possibly methionine; and one additional system transports taurine, GABA, β-alanine and other amino acids. The affinities of the various amino acids for the different transporters have been measured. Some amino acids can ride more than one carrier with different affinities (cf. above: neutral amino acids). These complex constellations can explain the pathophysiological basis of cystinuria and other inborn amino acid transport defects [99–101]. With respect to saturation and GTB, amino acid reabsorption follows the same rules which have been discussed for D-glucose.

Di-, Oligo- and Polypeptides and Proteins. There has been a long debate over whether dipeptides have to be hydrolysed by luminal enzymes before the respective amino acids are reabsorbed [100]. It is now clear that at least some dipeptides can be reabsorbed as such [101] and are only hydrolysed after their uptake into the cell. The energetics of this transport system are not yet clear. It has been suggested that coupling to protons provides the driving force [101].

Oligopeptides such as peptide hormones can be hydrolysed by the proximal brush border membranes which contain the corresponding peptidases [98]. Polypeptides and proteins are taken up by endocytosis and fused with lysosomes to form endolysosomes, where the peptides are hydrolysed to form smaller peptides and amino acids [9].

In summary, amino acids are conserved by proximal tubule reabsorption. Urine is usually amino acid free. Transport occurs secondarily actively via seven or more specific cotransport systems. Oligopeptides are reabsorbed as

such or they are hydrolysed in the lumen. Proteins are reabsorbed by endocytosis, and are subsequently hydrolysed.

75.9 Urea

Figure 75.21 shows the recovery of urea along the nephron. Urea with a molecular weight of 60 is freely filtered. One of the problems with urea is its *amphiphilic nature* and small radius of 0.24 nm. This molecule can cross biological membranes quite easily. Given these properties it may even be surprising that only some 50% of the filtered urea is reabsorbed in the proximal tubule. The mechanism of this urea reabsorption has long been interpreted as a solvent drag flux through the paracellular pathway. This may not be entirely correct since current evidence suggests that a majority of H_2O is transported across the tubule cell via "chip" type channels [31]. It is not clear whether urea can move through these channels.

Figure 75.21 also indicates that the recovery of urea is close to 100% in the early distal tubule. This complete recovery is due to the entry of urea into S3, DTL and ATL segments. The recycling of urea has been discussed explicitly in Chap. 74. It is very important for the function of the countercurrent concentrating mechanism that the distal tubule, cortical collecting duct and outer medullary collecting duct are impermeable for urea, irrespective of the H_2O permeability [64]. In the IMCD and PCD urea perme-

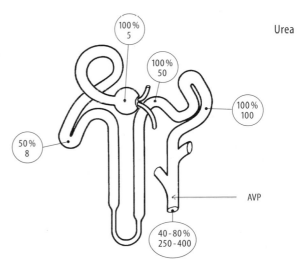

Fig. 75.21. Sites of urea reabsorption in the nephron. Recovery of urea along the nephron. The *numbers in the circles* refer to the recovery in the tubule fluid and the approximate concentration (mmol/l). The filtered amount is taken as 100%. Note that some 50% of the filtered urea is reabsorbed along the proximal tubule. Due to re-uptake into the descending and ascending limbs of the loop of Henle, early distal recovery approaches 100%. The distal tubule is impermeable for urea except for the inner medullary and papillary collecting duct, where urea permeability is enhanced by AVP. For a more complete discussion of urea recycling, see Chap. 74

1531

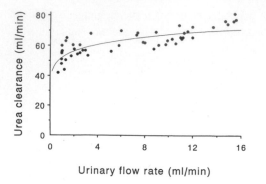

Fig. 75.22. Urea clearance as a function of urinary flow rate. Date were obtained in man. Note that urea clearance increases with (water) diuresis from 30%–50% to 50%–70% of GFR. (Data redrawn from [19])

ability is enhanced in the presence of AVP [63,64]. Then urea recycles into the renal medulla and papilla and enters S3, DTL and ATL segments. In the absence of AVP, urea excretion increases and the fractional excretion can increase from 30%–60%. This is shown in Fig. 75.22 [19]. As a consequence, urea is removed from the renal medulla and papilla. If this goes on for some time, the kidney loses some of its ability to concentrate urine. The concentrating ability is therefore reduced if, after some period of high urea excretion, AVP is secreted.

The variability of urea excretion with the presence or absence of AVP makes it impossible to use the clearance of urea as a measure of the GFR. It appears likely from very recent evidence [120] that urea transport occurs via a specific urea carrier.

75.10 Xenobiotics, Organic Acids and Bases

Several endogenous and exogenous substances are taken up by renal tubule cells across the basolateral membrane. An example is shown in Fig. 75.23. *para*-Aminohippurate (PAH) is taken up via an exchange carrier. The driving force is provided by the Na^+ coupling of another organic acid, e.g., oxalate. Hence, the PAH uptake can be classified as *tertiary active* [17]. Carriers of this class serve the uptake of metabolic substrates. These transporters can also be used by exogenous substances (xenobiotics, from the Greek ξενος = non-genuine). One such example is PAH (cf. Chap. 73). It is taken up across the basolateral membrane, accumulated in the tubule cell and leaves the cell via a luminal carrier system (or pump). Substances belonging to this class can reach very high concentrations in tubule cells and in the tubule fluid. For the basolateral membrane three classes of transporters have been characterised [17,110]:

- Sulphate transporter
- PAH transporter
- Succinate transporter

Many substances and xenobiotics can ride more than one of these systems. An example of such an analysis for rat proximal tubule is shown in Fig. 75.24. Several compounds belonging to the three groups are probed according to their competitive inhibition of the uptake of the key substrates [110].

The proximal tubule cell also possesses carrier systems for the uptake and secretion of organic bases [109,112]. Examples are nicotinamide and several secondary, tertiary, and quaternary ammonia bases. Many drugs belong to this group (e.g., amiloride, verapamil).

This specific function of the kidney explains why many drugs are excreted well even though they are largely bound to plasma proteins. The filtered fraction therefore may be very small, and the key step in renal excretion is the secretion by the proximal tubule. For example furosemide is 98% protein bound and the concentration in filtrate is only 2% of that in plasma. Nevertheless, the renal clearance of this drug is as high as 30% of the GFR owing to the fact that furosemide is secreted in the proximal tubule (cf. Chap. 74).

75.11 The Kidney and Mineral Metabolism

The kidney is the key excretory organ for Ca^{2+} and phosphate. The detailed mechanisms of renal handling of these ions have been discussed in Sect. 75.4 and 75.6. The hormones involved in the regulation are:

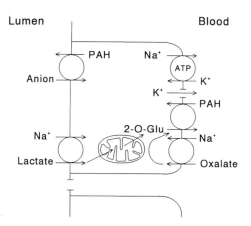

Fig. 75.23. Renal secretion and uptake of organic anions. *Circle with ATP*, (Na^++K^+)-ATPase; *circle*, carrier systems; *arrow*, conductance. The (Na^++K^+)-ATPase provides the primary driving force for Na^+-dependent uptake: e.g., Na^+-lactate in the luminal membrane; Na^+-oxalate uptake in the basolateral membrane. Lactate or other organic anions taken up across the luminal or basolateral membrane can be metabolised. One of the key metabolites of mitochondrial metabolism, 2-oxo-glutarate (*2-O-Glu*), accumulates in the cell. Organic anions such as oxalate, or other dicarboxylates etc., can drive anion uptake via a tertiary active system, in this case the antiport of 2-O-Glu against *para*-aminohippurate (PAH). PAH leaves the cell across the luminal membrane by another tertiary active antiporter. (Modified from [17])

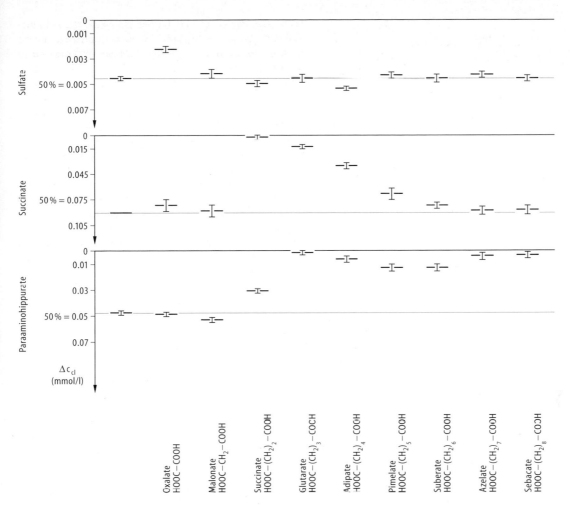

Fig. 75.24. Effect of dicarboxylates on contraluminal (basolateral) influx of [³H]PAH, [³H]succinate, and [³⁵S]sulphate into proximal tubule cells of rat kidney in situ. Mean values ± SEM of 9–18 samples of two to four animals. Note that dicarboxylates of increasing length are poorly accepted by the sulphate transporter (except for oxalate). They are well accepted by the succinate transporter, provided they fall into the range of succinate to adipate. Larger and smaller ones are not accepted by this system. These compounds (adipate to sebacate) are well accepted by the PAH transporter. From very many studies of this kind, the structural requirements for carrier interaction (spacing; charge; apolar interaction; hydrogen bonding) for organic acids and bases have been elaborated. (Data from [111])

- 1,25(OH)$_2$D$_3$
- Calcitonin
- Parathyroid hormone (PTH).

Besides these hormones other factors have an impact on Ca^{2+} and phosphate plasma concentration and acid-base metabolism:

- Increased plasma phosphate concentration reduces plasma Ca^{2+} concentration.
- Reduced plasma phosphate concentration enhances plasma Ca^{2+} concentration.
- Increased plasma Ca^{2+} concentration reduces plasma phosphate concentration.
- Reduced plasma Ca^{2+} concentration increases plasma phosphate concentration.
- Acidosis induces osteolysis, because Ca-phosphate salts are more soluble in acid solution.

- Osteolysis induces alkalosis, because apatite is alkaline.
- Osteogenesis induces acidosis.

The above four reciprocities are partially caused by the simple fact that the concentrations are determined by a solubility product:

$$K = [Ca^{2+}]^3 \cdot [PO_4^{3-}]^2/Ca_3(PO_4)_2. \qquad (75.7)$$

In body solutions the solubility product is probably larger than that determined in vitro. In addition, Ca^{2+}-phosphate complexes are stabilised. Therefore, Ca^{2+} and phosphate stay in solution even though the in vitro solubility product may be exceeded. The underlying mechanisms are not clear.

Figure 75.25 summarises the renal effects of the most important hormones involved in phosphate and Ca^{2+}

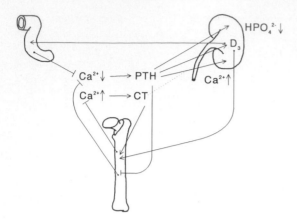

Fig. 75.25. Regulation of Ca^{2+} homeostasis by parathyroid hormone (*PTH*), $1,25(OH)_2D_3$ (*D_3*) and calcitonin (*CT*). *Arrow*, activation; *"blocked" line*, inhibition. Note the negative feedback loops: Low Ca^{2+} → PTH → Ca^{2+} reabsorption from the kidney and osteolysis; low Ca^{2+} → PTH → 1-hydroxylation of $25(OH)D_3$ to $1,25(OH)_2D_3$ → enhanced intestinal Ca^{2+} reabsorption; high Ca^{2+} → CT → osteoblast activation

homeostasis. Other hormones also modulate renal excretion of phosphate and Ca^{2+} [12].

$1,25(OH)_2D_3$. The effects of $1,25(OH)_2D_3$ on the renal tubule are not very marked. It has been reported that this hormone, after a few days of administration, enhances phosphate and Ca^{2+} reabsorption. This effect is concordant with the well-known effects of $1,25(OH)_2D_3$ in bone and gut. Therefore, summarising, $1,25(OH)_2D_3$ leads to:

- Enhanced intestinal Ca^{2+} reabsorption
- Activation of osteolysis and enhanced deposition of new bone
- Enhanced renal phosphate and Ca^{2+} reabsorption.

The kidney, therefore, assists in the increase of plasma Ca^{2+} concentration and helps to deposit new bone.

Calcitonin. Calcitonin has two major and acute effects on the kidney:

- Enhanced phosphate excretion
- Enhanced Ca^{2+} reabsorption.

The latter effect has only been demonstrable in isolated tubules [24], and probably is not very relevant in the intact organism [53]. Calcitonin release is stimulated by enhanced plasma Ca^{2+} concentration. The extrarenal effect of calcitonin is the deposition of new bone. This leads to a reduction in plasma Ca^{2+} concentration. On the other hand, the enhanced renal reabsorption of Ca^{2+} tends to increase plasma Ca^{2+} concentration. The renal effects of calcitonin are still controversial inasmuch as the net outcome depends on all other regulating factors as well.

PTH. The effects of PTH on the kidney have been summarised above (Sect. 75.6). All effects are aimed at an increase in plasma Ca^{2+}. They can be summarised as follows:

- Reduced phosphate reabsorption in the proximal tubule. This reduces plasma phosphate concentration. As a consequence of the solubility product, this leads to an increased plasma Ca^{2+} concentration.
- Metabolic acidosis due to inhibition of proximal tubule Na^+/H^+ exchange. This causes phosphaturia [12,68,82] and a consequent increase in plasma Ca^{2+}.
- Enhanced Ca^{2+} reabsorption in the TAL, leading to an increase in plasma Ca^{2+} concentration.
- Enhanced production of $1,25(OH)_2D_3$ in kidney cells. This hormone increases plasma Ca^{2+} concentration [27].
- Osteoclast activation. This increases plasma Ca^{2+} concentration.

In summary, the three hormones discussed here are complementary. PTH may be regarded as a Ca^{2+}-mobilising hormone. It induces a complex programme which raises plasma Ca^{2+} concentration acutely. Calcitonin is designed for the reduction of plasma Ca^{2+} and for the maintenance of bone. $1,25(OH)_2D_3$ increases the turnover and hence the production of new bone. Other hormones playing a pivotal role in bone and mineral metabolism, such as the estrogens and the androgens which "protect" bone, albeit by different mechanisms, are discussed in Chap. 80.

75.12 The Kidney and Acid-Base Disturbances

In this section the interplay of various mechanisms, all relevant to the homeostatic role of the kidney in maintaining a normal plasma pH, will be summarised. Further and more general considerations are given in Chap. 78. The kidney can deal with:

- Respiratory acidosis
- Respiratory alkalosis
- Non-respiratory (metabolic) acidosis
- Non-respiratory (metabolic) alkalosis.

Respiratory Acidosis. In respiratory acidosis the key problem is *hypercapnia* due to defective respiration, the causes of which are dealt with in Chaps. 101 and 102. Return to normal pH requires an increase in plasma HCO_3^- concentration. This is achieved by maximising HCO_3^- reabsorption in the nephron [3,39,56]. The mechanisms involved have already been discussed in Sect. 75.5:

- Activation of Na^+/H^+ exchange
- Activation of H^+-ATPase in the proximal tubule
- Activation of H^+ secretion in the A-type intercalated cells of CCD.

These three mechanisms prevent urinary HCO_3^- losses and maximise the excretion of titratable acid. Hypercapnia will also affect K^+ excretion. The fall in cytosolic pH in the TAL and in the CD principal cell might reduce K^+ secretion as well as urinary K^+ excretion [13,94,103].

The effects of hypercapnia are usually acute and it should not be forgotten that in most cases the related key problem is *hypoxia*. Hypoxia requires an improvement of respiration and this will also tend to counterbalance the acidosis.

Respiratory Alkalosis. In respiratory alkalosis the constellation is different. The cause of *hypocapnia* may be primary or hypoxia-induced hyperventilation, as occurs, for example, at high altitude. Especially in this situation the kidney is required to normalise pH by increased HCO_3^- excretion [73]. The mechanisms are those described above, but the direction of changes is opposite. This disorder can be accompanied by increased K^+ excretion and may provoke *hypokalemia*.

Unlike hypercapnia, the primary cause is not respiratory insufficiency but rather reduced pO_2 or reduced O_2 transport capacity in blood. Therefore, the renal compensation may be required for long periods.

Metabolic Acidosis. In metabolic acidosis the kidney can only be useful for compensation to the extent that acidosis is not caused by the kidney [32,39]. Such is the case in *chronic renal failure* (Sect. 75.16), where the limited ability of the reduced mass of functional nephrons cannot cope with the need to excrete acid. Similarly, metabolic acidosis caused by diuretic-induced bicarbonaturia can only be compensated by stopping the drug administration. In other states of metabolic acidosis, such as exercise, diarrhoea and ketoacidosis (cf. Chap. 78), the kidney reacts by:

- Activation of Na^+/H^+ exchange
- Activation of H^+-ATPase in proximal tubule
- Activation of H^+ secretion in the A-type intercalated cells of CCD
- Enhanced proximal tubule NH_4^+ production and renal excretion of NH_4^+.

The mechanisms have been discussed above (Sect. 75.5). Especially the latter mechanism is very powerful because it can save up to 100–150 mmol HCO_3^- per day.

Metabolic Alkalosis. In metabolic alkalosis [2] the kidney responds with incomplete reabsorption of HCO_3^-. The functional basis for this *bicarbonaturia* is not solely saturation of proximal tubule transport but also downregulation of the Na^+/H^+ exchange and of active H^+ pumping across the luminal membrane. The resulting bicarbonaturia reduces plasma HCO_3^- concentration. Also NH_4^+ production by the kidney is reduced. In chronic metabolic alkalosis, at least in animals, an increase in B-type cells, secreting HCO_3^- in the CD, has been demonstrated.

Summary. The kidney is pivotal in the regulation of a normal acid-base metabolism. The regulatory mechanisms include short-term responses such as allosteric control of Na^+/H^+ exchange, delayed mechanisms such as the change in renal and hepatic nitrogen metabolism and the change between A-type and B-type cells in the CD.

75.13 The Kidney in the Control of Volume and NaCl Balance

NaCl and water balance is largely controlled by the kidney. Several more general aspects are dealt with in Chaps. 77 and 82. Here we summarise the concerted action of several factors in the kidney [6,55,114,115]. The discussion will be based on the regulatory factors to supplement what has been discussed above (Sect. 75.2) for the various nephron segments. The factors will be subdivided into antinatriuretic and natriuretic factors.

75.13.1 Antinatriuretic Factors

The antinatriuretic factors comprise: sympathergic discharge [66] and circulating epinephrine, angiotensin II, antidiuretic hormone and aldosterone [55,92,114,115].

Sympathergic Discharge. Sympathergic discharge is under the control of volume and pressure sensing as well as central autonomic influences. It modulates autoregulation of blood flow and glomerular filtration. Enhanced discharge leads to a marked vasoconstriction of the afferent vasculature, especially if blood pressure is not controlled (cf. Chap. 73). This reduces both renal blood flow and glomerular filtration. Increased blood pressure is poorly autoregulated by the juxtamedullary glomeruli. As a result it induces a pressure diuresis. Renin secretion is enhanced and TGF reduced via β-receptors. In the proximal tubule, mostly via α_2 receptors, sympathergic discharge enhances Na^+ reabsorption. In the thick ascending limb and early distal tubule a similar effect is caused by β-receptors. It should be noted that these effects are only caused by circulating epinephrine because sympathergic innervation is absent in the thick ascending limb (as well as the pars recta and the thin limbs of the loop). Therefore, in general sympathergic discharge reduces urinary NaCl losses.

Angiotensin II. Angiotensin II is produced whenever renin secretion is enhanced (cf. Chap. 73). It increases the resistance of the afferent and efferent vasculature. Its effect on the GFR is therefore limited. However, it reduces renal blood flow. It also acts on the proximal nephron, where it increases NaCl reabsorption at low concentrations [58]. Only at very high concentrations does angiotensin II have an opposite effect, and reduces proximal reabsorption. The mechanism of angiotensin II-enhanced NaCl reabsorption has been examined and it appears likely that angiotensin

II increases $Na^+(HCO_3^-)_3$ extrusion from the cell. The concerted effects of angiotensin therefore comprise antinatriuresis and antidiuresis.

Antidiuretic Hormone. Antidiuretic hormone (ADH), also referred to as antidiuretin and vasopressin (AVP) in many texts, is released from the posterior pituitary lobe, usually in response to hyperosmolality, volume losses, falls in blood pressure and increase in angiotensin II. ADH acts on glomerular receptors, receptors in the thick ascending limb and on the CD. The receptors are of the V_1 and V_2 types. The glomerular effects have been discussed in Chap. 73 and are probably of minor importance. The effect in the thick ascending limb is demonstrable for the cortical as well as the medullary portion [48,80] but it is doubtful whether it is of relevance for kidney function in man [80]. In the collecting tubule the main effects are:

- Enhanced water permeability throughout the connecting tubule and CD (cf. Chap. 74)
- Enhanced Na^+ reabsorption in principal cells of the CCD [95]
- Enhanced urea reabsorption in the inner medullary and papillary CD (cf. Chap. 74 and [63]).

Therefore, the net results are water and Na^+ conservation. It is worth noting that these effects are optimised in those small rodents which also have the best performance of the countercurrent concentration system [22] (cf. also Chap. 82).

Aldosterone. Aldosterone release is controlled by two major determinants acting on the glomerulosa zone of adrenal cortex [92]:

- Increased angiotensin
- Increased plasma K^+ concentration.

Aldosterone enhances Na^+ reabsorption in the CD, the mechanism for which has been discussed in Sect. 75.2. This effect is slow and requires half an hour to several days to build up. In parallel the secretion of K^+ is enhanced. Therefore, aldosterone has to be regarded as both an antinatriuretic and a kaliuretic factor. In addition, aldosterone effects have also been postulated for the proximal nephron; this, however, has never been fully documented [107]. It may be interesting in this context that amiloride-sensitive Na^+ channels have been demonstrated in the luminal membrane of S2 and S3 segments of rabbit kidney [43]. Further studies will be needed to reveal whether the Na^+-saving effect of aldosterone can also be induced in the proximal nephron in states of Na^+ deprivation or whether this effect is mediated via glucocorticoid receptors which have also been shown to be present in the proximal tubule and in the thick ascending limb [48]. It should be mentioned here that comparable glucocorticoid and aldosterone effects are also demonstrable in other epithelia exposed to the outside world, e.g., sweat ducts, airway epithelia, nasal epithelium and the colon.

75.13.2 Natriuretic Factors

The natriuretic factors comprise: dopamine, prostaglandins, endothelin, atrial natriuretic peptide and natriuretic hormone [6,55,114].

Dopamine. Dopamine has already been discussed in Chap. 73. Tubule effects are probably rather limited when compared to other factors. They have been shown for the proximal tubule and also may be relevant for the thick ascending limb and collecting duct. The effects are mediated by D_1 receptors, although it has been claimed that they are mediated not only by cAMP but also by protein kinase C. The net effect is a reduction in reabsorption of Na^+. Both factors – the increased renal blood flow and GFR and the reduced reabsorption of Na^+ – cause a moderate natriuretic and diuretic effect.

Prostaglandins. Prostaglandins have a dual effect on the kidney. Of great relevance is PGE_2. Prostaglandins are produced intrarenally, mostly in the renal medulla from membrane lipids via phospholipase A_2 (leading to arachidonic acid) and cyclooxygenase (cf. Chap. 6). Prostaglandins increase renal blood flow, especially in the renal medulla, and reduce Na^+ reabsorption in the thick ascending limb and in the collecting duct. Both these effects cooperate in inducing a marked natriuresis and diuresis [78]. Conversely, inhibitors of prostaglandin production act antinatriuretically by reducing blood flow to the renal medulla and by enhancing distal Na^+ reabsorption. Prostaglandins lead to a relaxation of the renal vasculature [74] (cf. also Chap. 96). The effect on the thick ascending limb and on the collecting duct is probably explained by the inhibition of cAMP production [21,78]. At both sites, as shown above, cAMP induces PKA-modulated increases in Na^+ reabsorption.

Endothelin. Endothelin is one of the most potent vasoconstrictors [7,102]. At low concentrations it acts natriuretically. This may appear surprising since endothelin reduces renal blood flow and GFR [7,102]. Despite this clear-cut overall effect, there is some experimental evidence that endothelin enhances Na^+ reabsorption in the proximal tubule. The mechanism of this action is still unclear. The natriuretic effect of endothelin is probably localised in the inner medullary and papillary collecting duct and is caused by inhibition of the (Na^++K^+)-ATPase.

Atrial Natriuretic Peptide. Atrial natriuretic peptide (ANP) is produced in the atria of the heart and released by stretch (overfilling) [6]. Therefore, volume expansion or overfilling of the right and left atrium are the stimuli for its secretion. ANP acts on vasculature and on the kidney [6,7]. Its key effects on the kidney are on both GFR and Na^+ reabsorption. In fact, receptors for ANP have been found in the glomeruli (mesangial cells) and in the inner medullary and papillary collecting duct. These receptors, which couple to guanylate cyclase, exert two major effects:

- ANP increases RBF and to a lesser extent GFR
- ANP increases Na+ excretion.

Unfortunately, the tubule segments in which ANP appears to exert its effects are difficult to access. Therefore, little is known regarding the mechanism of the natriuretic effect of ANP in the terminal nephron. It is believed that ANP enhances an Na+ back leak from plasma to final tubule fluid.

Initially, high hopes were held that ANP might be *the* pathophysiologically relevant pathway to explain essential hypertension. These hopes have not been fulfilled. On the other hand it has become clear that ANP plays an important role as a natriuretic factor. In this sense, ANP is a fast-acting antagonist to AVP. Related peptide hormones, which are generated in the kidney, have now been isolated. These factors have been named *urodilatins* due to their potent vasodilative effect [7]. These factors also appear to inhibit Na+ reabsorption in the terminal nephron.

Natriuretic Hormone. Natriuretic hormone has been claimed to exist for many decades [25]. This hypothesis was based on the finding that escape from aldosterone occurred after a few days of application of this hormone, i.e., Na+ excretion started to rise again in the continued presence of aldosterone. It was shown thereafter that the factor was a circulating one, but decades of continued research have failed to identify this compound. Recently, however, it has been claimed that an endogenous ouabain-like substance produced from the adrenal cortex might be the natriuretic hormone [45]. Several arguments support this view:

- Natriuretic peptide has long been postulated to act like ouabain.
- Ouabain is a naturally occurring glycoside with unmatched specificity for this type of pump (ATPases).
- Chemically ouabain is a cholesterol-derived substance (cf. Fig. 11.7B) closely related to steroid hormones.
- The production of ouabain and ouabain-like substances has been demonstrated in adrenal cortex [45].

Further studies will need to reveal whether this kind of hormone is physiologically and pathophysiologically relevant, how its production is controlled and which stimuli enhance its secretion (cf. also Chap. 82).

Summary. A large number of factors control renal Na+ excretion. These factors assist and balance each other. In this scenario it is quite evident that the removal of one single factor has little impact on the overall function; in fact, perfect regulation can be demonstrated when one factor is removed experimentally. Teleologically this may indicate that essential homeostatic body functions such as the maintenance of a normal Na+ balance are secured by several backup mechanisms.

75.14 The Kidney and K+ Metabolism

In the preceding and in this chapter the specific role of the kidney in K+ homeostasis has been emphasised. This issue will be summarised here with specific emphasis on the various factors influencing K+ excretion. These key factors are [103,119]:

- K+ content of diet
- Systemic pH and pCO₂
- Distal delivery of Na+
- Distal tubular flow rate
- AVP.

K+ Diet. The effect of K+ diet is mediated largely by a suppression (low plasma K+ concentration) or stimulation (high plasma K+ concentration) of aldosterone secretion (cf. preceding section: Aldosterone). Low K+ diet (hypokalemia) has a multitude of additional effects which are discussed explicitly in Chap. 79. Here we focus but on the kidney:

- Low aldosterone reduces K+ secretion in the principal cells.
- Hypokalemia enhances proximal tubule HCO_3^- reabsorption. This is explained by the rheogenicity of HCO_3^- exit from proximal tubule cells (cf. Chap. 74). Hypokalemia hyperpolarises the basolateral membrane of the proximal tubule cell and this enhances the driving force for rheogenic HCO_3^- exit from the cell. In addition it has been reported that hypokalemia suppresses and that hyperkalemia induces NH_4^+ production in proximal tubules [54]. Both mechanisms induce metabolic alkalosis.
- Hypokalemia reduces thick ascending limb NaCl reabsorption. The mechanism for this is not clear. It may simply be due to the fact that luminal K+ concentration becomes rate limiting on the $Na^+2Cl^-K^+$ cotransporter [48].
- Hypokalemia appears to enhance the activity of H^+/K^+ ATPase in intercalated cells of the CD. This enhances active K+ reabsorption.

Hyperkalemia has the opposite effects.

Cytosolic Acid-Base Changes. The effects of cytosolic acid-base changes in distal tubule cells have been discussed explicitly in Chap. 74. It may suffice here to repeat that the thick ascending limb of the loop of Henle and the CCD possess luminal K+ channels which are regulated by cytosolic pH. Very minor changes in cytosolic pH have a large impact on the open probability of these channels. *Acidosis reduces and alkalosis enhances the open probability* [94].

Distal Delivery of Na+. The distal delivery of Na+ determines the rate of distal tubule (mainly CD) reabsorption of Na+. To the extent that the increased Na+ uptake depolarises the luminal membrane it also provides additional driving force for K$^+$ secretion in the CD. This is also the mechanism whereby loop diuretics and those acting in the early distal tubule invariably enhance K$^+$ secretion (cf. Chap. 74).

Distal Tubule Flow Rate. The distal tubule flow rate also determines the amount of K$^+$ secreted into urine. The luminal concentration profile for Na$^+$ will be influenced by the tubule flow rate: If the flow rate is low, a minimal Na$^+$ concentration will be achieved early along the CD. Then the luminal membrane hyperpolarises and K$^+$ secretion is reduced. If the flow rate is enhanced, this point will be shifted further downstream and the secretion of K$^+$ will continue throughout the CD. This effect is obviously related to the effect of the distal tubule load of Na$^+$, although it should be noted that the effects can be separated experimentally.

AVP. It appears that AVP is an independent factor in the regulation of K$^+$ secretion. It has been discussed above that AVP enhances Na$^+$ reabsorption (at least in some species). This leads to a depolarisation of the luminal membrane of principal cells in the CD. In addition, it has been shown that AVP, via PKA phosphorylated luminal K$^+$ channels, may enhance the secretion of K$^+$ [113]. Therefore, AVP not only conserves H$_2$O and Na$^+$, it also appears to enhance the renal excretion of the "counterion" K$^+$.

75.15 Acute Renal Failure

Acute renal failure (ARF) is usually defined as a sudden deterioration of kidney function. Operationally a rapid fall in GFR by more than 50% is taken as a quantitative basis for the definition [70]. In this section we shall not address at any length the wide spectrum of pathomechanisms leading to ARF. Rather we shall try to explain the symptomatology of ARF on the basis of the previous sections.

75.15.1 Causes

The causes of ARF can be classified as [10,70]:

- Vascular (prerenal)
- Intrarenal
- Obstructive (postrenal).

Vascular Causes. Vascular causes can be subdivided into *hypoperfusion* and *renovascular obstruction*. Hypoperfusion can be due to (a) acute volume losses such as in trauma, burns and crush syndrome; (b) reduced cardiac output; (c) a sudden fall in blood pressure; (d) a fall in systemic vascular resistance with little change in the renal vascular resistance; and (e) increased blood viscosity. Renal vascular obstruction can occur in arteries due to atherosclerosis, embolism or dissecting aneurysm, in veins due to compression or thrombosis, and in the small vessels

of the kidney as well as in the glomeruli due to glomerulonephritis, vasculitis, toxemia, hemolysis, coagulopathy etc.

Intrarenal Causes. Intrarenal causes comprise *interstitial nephritis* and *nephrotoxins* such as aminoglycosides, cephalosporins, cis-platinum, cyclosporin A, contrast media, glycols, tetrachloroethylene, toluene, other organic solvents, heavy metals and mercury, insecticides, mushroom poisons and venoms. *Myoglobin* circulating after rhabdomyolysis of traumatised muscle is another important cause of intrarenal ARF. Frequently intrarenal and prerenal causes cannot be clearly separated inasmuch as hypoperfusion and toxins are present at the same time. In addition, some toxins act directly on the tubule and also cause circulatory shock and hemolysis (e.g., snake venoms).

Obstructive Causes. Obstructive ARF can occur from any obstruction in the pathways draining urine.

75.15.2 Pathophysiology

Obstruction. The pathophysiology of these three groups of causes is diverse and not completely understood. The easiest to understand is the obstruction. Outflow resistance of urine leads to the build up of tubule pressure. The increased tubule pressure reduces or even abolishes the positive filtration pressure in the glomerulus. In addition, the pressure within the tubule damages the epithelium. Even in intrarenal ARF the obstructive component is of importance: Damage (e.g., toxic) of the tubule leads to a sequence of events which gives rise to desquamation of mostly irreversibly damaged tubule cells. These cells and their debris obstruct the tubule lumen, reduce or abolish filtration in this nephron and induce further pressure damage to the upstream epithelium. It is apparent that these events combine to produce a vicious circle.

Proximal Tubule Damage. The most vulnerable tubule cells are those of the proximal tubule. The various toxins have been mentioned above. The proximal tubule cells lose their brush border, which blebs into the lumen and then is added to the lumen fluid as fluid containing plasma membrane vesicles. The cells show mitochondrial swelling and, if the toxins or the damaging factor is not removed, nuclear pyknosis and irreversible cell damage occur.
The preferential toxic damage of the proximal tubule cells is probably caused by the fact that these cells accumulate toxins (e.g., nephrotoxic antibiotics) by the basolateral uptake systems (cf. Sect. 75.10). In addition, it should be remembered that especially the S1 segment performs enormous transport work and hence has a high metabolic demand (cf. Sect. 75.2). The ischemic damage of the kidney also mostly affects the proximal tubule. However, in such states, primary damage is also seen in the medullary thick ascending limb [15,49,118]. This is caused by the borderline perfusion of this nephron segment. As has been dis-

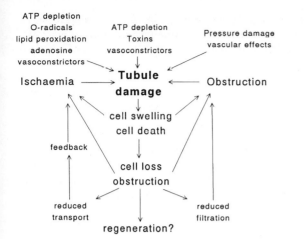

Fig. 75.26. Pathophysiology of acute renal failure. Three main causes (prerenal = ischaemia; intrarenal = tubule damage; postrenal = obstruction) can be artificially subdivided. The large number of positive feedback loops tend to aggravate tubule damage

cussed above, this nephron segment has a high metabolic demand but is localised in an area where pO_2 is substantially lower than in the cortex. In fact, it can be shown experimentally that ischemic damage of the thick ascending limb can be prevented completely if active reabsorption is inhibited by loop diuretics [15,49]. It has been hypothesised that hypoxic and ischemic cell swelling further increases vascular resistance, hence supporting another vicious circle [76].

Reduction of GFR. The reduction of GFR in ARF is caused either prerenally by an increase in prerenal resistance or a fall in blood pressure. In addition, in the presence of obstruction, GFR is reduced due to the increase in intratubular pressure and a corresponding fall in filtration pressure. The macula densa feedback mechanism can also participate in the suppression of GFR, if fractional NaCl reabsorption in the tubule is reduced by tubule damage. Then the NaCl concentration in the macula densa segment is enhanced and this leads to a feedback reduction of GFR (cf. Chap. 73).

Summary. Figure 75.26 summarises the pathophysiology of ARF. It is important to note that there is *no single cause.* In fact, usually two or more causal events come together and generate a scenario in which vicious circles lead to an abrupt and dramatic worsening of renal function.

75.15.3 Symptomatology

The clinical courses are beyond the scope of this section. It suffices to state here that ARF is a life-threatening state and that it had a high mortality in the predialysis era. The clinical symptomatology is deducible from the *failure of normal function:*

- Accumulation of plasma urea and creatinine as a result of reduced GFR.
- Overhydration with edema formation, hypertension and threatening heart failure. This is due to the imbalance of continued water intake but failing urine excretion.
- Hyponatremia. This is also due to the above imbalance.
- Hyperkalemia. This is caused by the failure of the kidneys to excrete adequate amounts of K^+. In addition, hyperkalemia is supported by tissue damage and independently aggravated by metabolic acidosis.
- Acidosis is caused by the failure of the kidneys to excrete acid. Again, this may be aggravated by the catabolic state, by ischemia and also (respiratory component) by reduced or insufficient lung gas exchange.
- Hyperphosphtemia. This is caused directly by insufficient urinary excretion of phosphate. In addition more phosphate may be produced. Hyperphosphtemia frequently induces hypocalcemia.
- Hypocalcemia. This is due to the "solubility product" (cf. Sect. 75.11). Hypocalcemia leads to hyperparathyroidism.

75.15.3 Therapeutic Measures

From the above it should be clear that at least the ischemic and hypoxic components of ARF can be avoided by early and even preventive measures. In addition the application of renally vasodilating agents and inhibitors of active transport in the thick ascending limb (loop diuretics) [49] can improve renal function or adjust the demands to the blood supply. Iatrogenic toxic damage can be reduced if early states of tubule damage are monitored by appropriate analytical techniques. Nevertheless, in many cases the course of the disease can only be treated symptomatically by dialysis therapy and intensive care. In many such cases the rate of urine production follows a certain course: A first oliguric phase is followed by an early (days to weeks after the insult) and late polyuric phase [70]. In principle, the renal tubules possess the ability to regenerate and to fill with new tubule cells the gaps in the epithelial layer produced by desquamation.

75.16 Chronic Renal Failure

This section will only discuss the pathophysiology of chronic renal failure (CRF) as it can be deduced from the chronic deterioration of normal renal function, discussed in the above sections. The causes of the chronic reduction in GFR will not be discussed here. The reader is referred to relevant clinical texts [18,74].

As has been shown in Chap. 73, there is a hyperbolic relation between plasma creatinine concentration and GFR. The basis for this is the following:

$$\text{GFR} \cdot P_{Cr} = U_{Cr} \cdot \dot{V} = \text{produced rate of Cr.} \qquad (75.8)$$

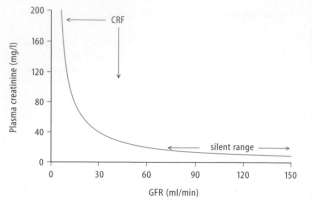

Fig. 75.27. Plasma creatinine concentration as a function of falling GFR. It is apparent that early stages of failing kidney function can only be detected on repetitive examination ("silent range"). In advanced chronic renal failure, plasma creatinine concentration increases steeply with deterioration of GFR

Given a rate of creatinine (Cr) production which in the steady state must be equal to the rate of urinary excretion ($U_{Cr} \cdot \dot{V}$), plasma creatinine concentration (P_{Cr}) *must be inversely related to GFR*. This is also shown in Fig. 75.27. For example, for a reduction of GFR by 50%, P_{Cr} must increase by a factor of 2. Given the variability of normal plasma creatinine concentrations, which are due to body (muscle) mass and physical activity, an increase by a factor of 2 will hardly be detected. Therefore, this part of the curve in Fig. 75.27 is called the *silent range*. Increases in plasma creatinine concentrations become much more conspicuous when GFR falls to one-fourth or one-fifth of normal. The signs of CRF then become easily apparent. They comprise:

- Increase in plasma creatinine concentration
- Increase in plasma urea concentration
- Overhydration
- Volume expansion
- Hyperkalemia
- Metabolic acidosis
- Hyperammonemia
- Hyperphosphatemia
- Hypocalcemia
- Hyperparathyroidism
- Anemia.

The clinical symptoms will not be discussed here. Instead, we shall briefly discuss the above key findings.

Increases in Plasma Urea and Creatinine Concentrations. These increases are illustrated by reference to Eq. 75.8 and Fig. 75.27. It should be remembered that the neurological symptoms of CRF (uremia = acotemia) cannot be explained by reference to urea and creatinine.

Overhydration. Overhydration is a result of an imbalance of water intake and the ability to excrete a corresponding urine volume. Overhydration usually occurs fairly late in progressive CRF because the kidneys have the intrinsic ability to compensate for the reduced number of functioning nephrons to some extent. This occurs because the still functioning nephrons show hyperplasia and hypertrophy. They filter more than normal and they have a reduced rate of fractional reabsorption. Therefore, the fractional excretion of water and NaCl is enhanced in CRF. Also the maximal urine osmolality is reduced from 1200 mosm/l in normal kidney to less than 500 mosm/l in CRF. The explanation for the increase in fractional excretion is straightforward: due to the large increase in filtered load, glomerulotubular balance is exhausted (and may even be reduced in CRF), and transport capacities are saturated. Therefore, the span of regulation is reduced. In the normal kidney (cf. Chap. 74), water and NaCl excretion can vary between 0.5% and 15% and between 0.1% and 2% of filtered load, respectively. In advanced CRF, fractional excretion of water is, for example, between 5% and 20%. If intake exceeds the maximal excretion capacity, overhydration and edema formation occur.

Volume Expansion. Volume expansion is the result of excessive water and salt intake. In the dialysis era it has become possible to be a little less stringent in salt and water restriction, because the amounts accumulated during the dialysis intervals can be removed in the next dialysis phase.

Hyperkalemia. Hyperkalemia is the result of the reduced capacity of the residual nephron function to excrete adequate amounts of K^+. This is aggravated by the accompanying metabolic acidosis. Hyperkalemia has to be controlled closely because of the danger of cardiac complications. Also hyperkalemia in itself reduces proximal tubule ammoniagenesis and hence aggravates metabolic acidosis.

Metabolic Acidosis. Metabolic acidosis is caused by the reduced absolute excretion of phosphate. This is further complicated by catabolism and by hyperparathyroidism. High PTH inhibits the Na^+/H^+ exchanger. Therefore, less HCO_3^- is reabsorbed by the proximal nephron. Furthermore, the accompanying hyperkalemia constitutes another component of a vicious circle, by which hyperkalemia aggravates metabolic acidosis (cf. above paragraph) and acidosis enhances hyperkalemia. The intrarenal mechanism for the latter has been discussed in Chap. 74.

Hyperammonemia. Hyperammonemia is the result of decreased ammonia excretion by the kidneys. On the other hand, ammonia production is stimulated by metabolic acidosis. The extent to which acidosis enhances renal (proximal tubule) ammonia production is also controlled by the plasma K^+ concentration. In hyperkalemia the renal production of NH_4^+ is attenuated and in hypokalemia it is augmented. The coexisting hyperkalemia in CRF therefore tends to counterbalance the acidosis-induced stimulus of ammonia production. Even if potassium balance is achieved, hyperammonemia will occur when GFR is less than 10%–20% of normal, since then the maximal excretion is less than the production. Increased plasma ammonia concentration is probably one of the contributing toxic elements in CRF [18,74].

Hyperphosphatemia. Hyperphosphatemia is caused by the fact that the excretion of phosphate in advanced CRF is less than the amount produced and ingested. Hyperphosphatemia leads to a fall in ionised plasma calcium activity (cf. Sect. 75.11). Hypocalcemia in turn enhances the secretion of PTH (cf. below).

Hypocalcemia. Hypocalcemia is caused by the increase in plasma phosphate concentration. Another contributing factor may be a reduction in $1,25(OH)_2D_3$ concentration and reduced Ca^{2+} ingestion. The reduction in $1,25(OH)_2D_3$ is probably the result of the damage to the kidneys, in which the 1-hydroxylation usually occurs as a PTH-dependent reaction. Hypocalcemia and hyperphosphatemia are usually not overt signs in early CRF. However, hyperparathyroidism is usually present.

Hyperparathyroidism. Hyperparathyroidism is supposedly caused by hyperphosphatemia-induced hypocalcemia. PTH has been incriminated as one of the relevant toxic elements in CRF [77]. There is no doubt that hyperparathyroidism is one of the players in uremic osteodystrophy.

Anemia. Anemia is caused by several factors. The most prominent is the lack of renally produced erythropoietin. This has now been shown for CRF patients in substitution studies with exogenously administered erythropoietin [18]. Other factors are accumulation of toxins which inhibit erythropoiesis and shorter average life time of red blood cells [18,74].

Several other signs of CRF cannot be discussed here, and the reader is referred to relevant textbooks [18]. With the invention of dialysis therapy, the management of CRF has been improved dramatically. It became clear that knowledge on the normal physiology of the kidney was most valuable and, in fact, urgently needed in designing adequate replacement therapy. With adequate dialysis therapy many hundreds of thousands of CRF patients in developed countries, who in previous years would have died from the above disturbances within several months, have been able to live an almost normal life. On the other hand, life quality is severely compromised during the dialysis period. Therefore, most of the CRF patients on dialysis wait for kidney transplantation.

At the same time, nephrology and renal physiology have set out to examine the processes underlying the development of CRF in the hope that one day it may become possible to prevent the progression of renal damage.

References

1. Alpern RJ (1990) Cell mechanisms of proximal tubule acidification. Physiol Rev 70:79–114
2. Alpern RJ, Emmett M, Seldin DW (1992) Metabolic alkalosis. In: Seldin DW, Giebisch G (eds) The kidney: physiology and pathophysiology. Raven, New York, pp 2733–2758
3. Alpern RJ, Stone DK, Rector FC (1991) Renal acidification mechanisms. In: Brenner BM, Rector FC (eds) The kidney. Saunders, Philadelphia, pp 318–379
4. Aronson PS, Nee J, Suhm MA (1982) Modifier role of internal H^+ in activating the Na^+-H^+ exchanger in renal microvillus membrane vesicles. Nature 299:161–163
5. Bacskai BJ, Friedman PA (1990) Activation of latent Ca^{2+} channels in renal epithelial cells by parathyroid hormone. Nature 347:388–391
6. Ballermann BJ, Zeidel M (1992) Atrial natriuretic hormone. In: Seldin DW, Giebisch G (eds) The kidney: physiology and pathophysiology. Raven, New York, pp 1843–1884
7. Ballermann BJ, Zeidel ML, Gunning ME, Brenner BM (1991) Vasoactive peptides and the kidney. In: Brenner BM, Rector FC (eds) The kidney. Saunders, Philadelphia, pp 510–583
8. Bank N, Aynedjian HS (1976) A micropuncture study of the effect of parathyroid hormone on renal bicarbonate reabsorption. J Clin Invest 58:336–344
9. Baumann K (1981) Renal transport of proteins. In: Greger R, Lang F, Silbernagl S (eds) Renal transport of organic substances. Springer, Berlin Heidelberg New York, pp 118–133
10. Beck F, Thurau K, Gstraunthaler G (1992) Pathophysiology and pathobiochemistry of acute renal failure. In: Seldin DW, Giebisch G (eds) The kidney: physiology and pathophysiology. Raven, New York, pp 3157–3180
11. Berliner RW, DuBose TD (1992) Carbon dioxide tension of alkaline urine. In: Seldin DW, Giebisch G (eds) The kidney: physiology and pathophysiology. Raven, New York, pp 2681–2693
12. Berndt TJ, Knox FG (1992) Renal regulation of phosphate excretion. In: Seldin DW, Giebisch G (eds) The kidney: physiology and pathophysiology. Raven, New York, pp 2511–2562
13. Bleich M, Schlatter E, Greger R (1990) The luminal K^+ channel of the thick ascending limb of Henle's loop. Pflugers Arch 415:449–460
14. Bresler EH (1961) Reflections on the nature of renal tubular reabsorption. Am Heart J 62:1–6
15. Brezis M, Rosen S, Silva P, Epstein FH (1984) Renal ischaemia: a new perspective. Kidney Int 26:375–383
16. Brown D, Hirsch S, Gluck S (1988) An H^+-ATPase in opposite plasma membrane domains in kidney epithelial cell subpopulations. Nature 331:622–624
17. Burckhardt G, Ullrich KJ (1989) Organic anion transport across the contraluminal membrane, dependence on sodium. Kidney Int 36:370–377
18 Cameron S, Davison AM, Grünfeld JP, Kerr D, Ritz E (1992) Oxford textbook of clinical nephrology. Oxford University Press, Oxford, pp 1–2348
19. Chasis H, Smith HW (1938) The excretion of urea in normal man and in subjects with glomerulonephritis. J Clin Invest 17:347–356
20. Costanzo LS, Windhager EE (1992) Renal regulation of calcium balance. In: Seldin DW, Giebisch G (eds) The kidney: physiology and pathophysiology. Raven, New York, pp 2375–2393
21. Culpepper RM, Andreoli TE (1983) Interactions among prostaglandin E_2, antidiuretic hormone, and cyclic adenosine monophosphate in modulating Cl^- absorption in single mouse medullary thick ascending limbs of Henle. J Clin Invest 71:1588–1601

22. De Rouffignac C (1990) The urinary concentrating mechanism. In: Kinne RKH (ed) Urinary concentrating mechanisms. Karger, Basel, pp 31–102
23. De Rouffignac C, Elalouf JM (1988) Hormonal regulation of chloride transport in the proximal and distal nephron. Annu Rev Physiol 50:123–140
24. De Rouffignac C, Wittner M, Di Stefano M, Nitschke R, Greger R, Bailly C, Amiel C, Roinel N (1989) Effects of parathyroid hormone (PTH) and human calcitonin (HCT) on Na^+, Cl^-, Mg^{++}, Ca^{++}, and K^+ transport in the thick ascending limb. 31st international congress on physiological science
25. de Wardener HE, Mills IH, Clapham WF, Hayter CJ (1961) Studies on the efferent mechanism of the sodium diuresis which follows the administration of intravenous saline in the dog. Clin Sci 21:249–258
26. Deetjen P, v. Baeyer H, Drexel H (1992) Renal glucose transport. In: Seldin DW, Giebisch G (eds) The kidney: physiology and pathophysiology. Raven, New York, pp 2873–2888
27. Deluca HF (1984) The metabolism, physiology and function of vitamin D. In: Kumar R (ed) Vitamin D: basic and clinical aspects. Nijhoff, Boston, pp 1–68
28. Di Stefano A, Greger R, De Rouffignac C, Wittner M (1992) Active NaCl transport in the cortical thick ascending limb of Henle's loop of the mouse does not require the presence of bicarbonate. Pflugers Arch 420:290–296
29. Di Stefano A, Roinel N, De Rouffignac C, Wittner M (1993) Transepithelial Ca^{2+} and Mg^{2+} transport in the cortical thick ascending limb of Henle's loop of the mouse is a voltage-dependent process. Renal Physiol Biochem 16:157–166
30. DiBona GF (1985) Neural regulation of renal tubular sodium reabsorption and renin secretion. Fed Proc 44:2816–2822
31. Echevarria M, Frindt G, Preston GM, Milovanovic S, Agre P, Fischbarg J, Windhager EE (1993) Expression of multiple water channel activities in xenopus oocytes injected with mRNA from rat kidney. J Gen Physiol 101:827–841
32. Emmett M, Alpern RJ, Seldin DW (1992) Metabolic acidosis. In: Seldin DW, Giebisch G (eds) The kidney: physiology pathophysiology. Raven, New York, pp 2759–2836
33. Evers J, Murer H, Kinne R (1976) Phenylalanine uptake in isolated renal brush border vesicles. Biochim Biophys Acta 426:598–615
34. Friedman PA, Andreoli TE (1982) CO_2-stimulated NaCl absorption in the mouse renal cortical thick ascending limb of Henle. J Gen Physiol 80:683–711
35. Friedman PA, Gesek FA (1993) Calcium transport in renal epithelial cells. Am J Physiol 264:F181–F198
36. Frömter E (1981) Electrical aspects of tubular transport of organic substances. In: Greger R, Lang F, Silbernagl S (eds) Renal transport of organic substances. Springer, Berlin Heidelberg New York, pp 30–44
37. Frömter E (1982) Electrophysiological analysis of rat renal sugar and amino acid transport. Pflugers Arch 393:179–189
38. Frömter E, Gessner K (1974) Free-flow potential profile along the rat kidney proximal tubule. Pfluger Arch 351:69–83
39. Gennari FJ, Maddox DA (1992) Renal regulation of acid-base homeostasis. In: Seldin DW, Giebisch G (eds) The kidney: physiology and pathophysiology. Raven, New York, pp 2695–2732
40. Gertz KH, Boylan JW (1973) Glomerular-tubular balance. In: Orloff J, Berliner RW (eds) Handbook of physiology; section 8. American Physiological Society, Bethesda, pp 763–790
41. Gluck S, Caldwell J (1988) Proton-translocating ATPase from bovine kidney medulla: partial purification and reconstitution. Am J Physiol 254:F71–F79
42. Gluck S, Kelly S, Al-Awqati Q (1982) The proton translocating ATPase responsible for urinary acidification. J Biol Chem 257(16):9230–9233
43. Gögelein H, Greger R (1986) Na^+ selective channels in the apical membrane of rabbit late proximal tubule (pars recta). Pflugers Arch 406:198–203
44. Good DW (1985) Sodium-dependent bicarbonate absorption by cortical thick ascending limb of rat kidney. Am J Physiol 248:F821–F829
45. Goto A, Yamada K, Yagi N, Yoshioka M, Sugimoto T (1992) Physiology and pharmacology of endogenous digitalis-like factors. Pharmacol Rev 44:377–399
46. Gottschalk CW (1979) Renal nerves and sodium excretion. Annu Rev Physiol 41:229–240
47. Green R, Giebisch G (1989) Reflecting coefficients and water permeability in rat proximal tubule. Am J Physiol 257:F669–F675
48. Greger R (1985) Ion transport mechanisms in thick ascending limb of Henle's loop of mammalian nephron. Physiol Rev 65:760–797
49. Greger R (1985) Wirkung von Schleifendiuretika auf zellulärer Ebene. Nieren Hochdruckkrkh 14(6):217–220
50. Greger R, Bleich M, Schlatter E (1991) Ion channel regulation in the thick ascending limb of the loop of Henle. Kidney Int 40:S33:S119–S124
51. Greger R, Lang F, Marchand G, Knox FG (1977) Site of renal phosphate reabsorption. Micropuncture and microinfusion study. Pflugers Arch 369:111–118
52. Györy AZ, Beck F, Rick R, Thurau K (1985) Electron microprobe analysis of proximal tubule cellular Na, Cl, and K element concentrations during acute mannitol-saline volume expansion in rats: evidence for inhibition of the Na pump. Pflugers Arch 403:205–209
53. Haas HG, Dambacher MA, Guncaga J, Lauffenburger T (1971) Renal effects of calcitonin and parathyroid extract. J Clin Invest 50:2698–2702
54. Halperin ML, Kamel SK, Ethier JH, Stienbaugh BJ, Jungas RL (1992) Biochemistry and physiology of ammonium excretion. In: Seldin DW, Giebisch G (eds) The kidney: physiology and pathophysiology. Raven, New York, pp 2645–2679
55. Hamlyn JM, Ludens JH (1992) Nonatrial natriuretic hormones. In: Seldin DW, Giebisch G (eds) The kidney: physiology and pathophysiology. Raven, New York, pp 1885–1924
56. Hamm LL, Alpern RJ (1992) Cellular mechanisms of renal tubular acidification. In: Seldin DW, Giebisch G (eds) The kidney: physiology and pathophysiology. Raven, New York, pp 2581–2626
57. Harris PJ (1992) Regulation of proximal tubule function by angiotensin. Clin Exp Pharmacol Physiol 19:213–222
58. Harris PJ, Young JA (1977) Dose-dependent stimulation and inhibition of proximal tubular sodium reabsorption by angiotensin II in the rat kidney. Pflugers Arch 367:295–297
59. Häberle DA, Müller U, Nagel W (1989) Glomerular tubular balance: mediated by luminal hypotonicity. Miner Electrolyte Metab 15:108–113
60. Häberle DA, Shiigai TT, Maier G, Schiffl H, Davis JM (1981) Dependency of proximal tubule fluid transport on the load of glomerular filtrate. Kidney Int 20:18–28
61. Hebert SC, Friedman PA, Andreoli TE (1984) Effects of antidiuretic hormone on cellular conductive pathways in mouse medullary thick ascending limbs of Henle: I. ADH increases transcellular conductance pathways. J Membr Biol 80:201–219
62. Hediger MA, Coady MJ, Ikeda TS, Wright EM (1987) Expression cloning and cDNA sequencing of the Na^+/glucose cotransporter. Nature 333:379–381
63. Knepper MA, Rector FC (1991) Urinary concentration and dilution. In: Brenner BM, Rector FC (eds) The kidney. Saunders, Philadelphia, pp 445–482
64. Knepper MA, Sands JM, Chou C-L (1989) Independence of urea and water transport in rat inner medullary collecting duct. Am J Physiol 256:F610–F621
65. Koeppen BM, Stanton BA (1992) Sodium chloride transport: distal nephron. In: Seldin DW, Giebisch G (eds) The kidney: physiology and pathophysiology. Raven, New York, pp 2003–2040

66. Kopp UC, DiBona GF (1992) The neural control of renal function. In: Seldin DW, Giebisch G (eds) The kidney: physiology and pathophysiology. Raven, New York, pp 1157–1204

67. Krapf R, Alpern RJ (1993) Cell pH and transepithelial H^+/HCO_3^- transport in the renal proximal tubule. J Membr Biol 131:1–10

68. Lang F, Greger R, Knox FG, Oberleithner H (1981) Factors modulating the renal handling of phosphate. Renal Physiol 4:1–16

69. Lang F, Greger R, Marchand G, Knox FG (1977) Saturation kinetics of phosphate reabsorption in rats. In: Massry SG, Ritz E (eds) Phosphate metabolism. Plenum, New York, pp 153–155

70. Lieberthal W, Levinsky NG (1992) Acute clinical renal failure. In: Seldin DW, Giebisch G (eds) The kidney: physiology and pathophysiology. Raven, New York, pp 3181–3226

71. Ludwig C (1844) Nieren und Harnbereitung. In: Wagner R, Vieweg F (eds) Handwörterbuch der Physiologie. Braunschweig

72. Maack T (1992) Receptors of atrial natriuretic factor. Annu Rev Physiol 54:11–27

73. Madias NE, Cohen JJ (1992) Respiratory alkalosis and acidosis. In: Seldin DW, Giebisch G (eds) The kidney: physiology and pathophysiology. Raven, New York, pp 2837–2872

74. Martinez-Maldonado M, Benabe JE, Cordova HR (1992) Chronic intrinsic renal failure. In: Seldin DW, Giebisch G (eds) The kidney: physiology and pathophysiology. Raven, New York, pp 3227–3288

75. Marver D (1990) Aldosterone. Methods Enzymol 191:520–550

76. Mason J, Welsch J, Torhorst J (1987) The contribution of vascular obstruction to the functional defect that follows renal ischemia. Kidney Int 31:65–71

77. Massry SG (1984) Parathyroid hormone and uraemic myocardiopathy. Contrib Nephrol 41:231–240

78. Menè P, Dunn MJ (1992) Vascular, glomerular, and tubular effects of angiotensin II, kinins, and prostaglandins. In: Seldin DW, Giebisch G (eds) The kidney, physiology and pathophysiology. Raven, New York, pp 1205–1248

79. Minuth WW, Steckelings U, Gross P (1987) Methylation of cytosolic proteins may be a possible biochemical pathway of early aldosterone action in cultured renal collecting duct cells. Differentiation 36:23–34

80. Morel F, Doucet A (1986) Hormonal control of kidney functions at the cell level. Physiol Rev 66:377–468

81. Muff R, Fischer JA, Biber J, Murer H (1992) Parathyroid hormone receptors in control of proximal tubule function. Annu Rev Physiol 54:67–79

82. Murer H, Biber J (1992) Renal tubular phosphate transport. In: Seldin DW, Giebisch G (eds) The kidney: physiology and pathophysiology. Raven, New York, pp 2481–2509

83. Oberleithner H, Lang F, Greger R, Sporer H (1979) Influence of calcium and ionophore 23187 on tubular phosphate reabsorption. Pflugers Arch 379:37–41

84. Oberleithner H, Lang F, Greger R, Sporer H (1980) Additivity of the phosphaturic action of parathyrin and calcitonin in the rat kidney. In: Massry SG, Ritz E, John H (eds) Phosphate and minerals in health and disease. Plenum, New York, pp 129–134

85. Petty KJ, Kokko JP, Marver D (1981) Secondary effect of aldosterone on Na-K ATPase activity in the rabbit cortical collecting tubule. J Clin Invest 68:1514–1521

86. Pitts RF (1973) Production and excretion of ammonia in relation to acid-base regulation. In: Orloff J, Berliner RW (eds) Handbook of physiology, section 8. American Physiological Society, Bethesda, pp 455–496

87. Pitts RF, Lotspeich WD (1946) Bicarbonate and the renal regulation of acid-base balance. Am J Physiol 147:138–154

88. Puschett JB, Zurbach P (1976) Acute effects of parathyroid hormone on proximal bicarbonate transport in the dog. Kidney Int 9:501–510

89. Quamme GA (1992) Magnesium, cellular and renal exchanges. In: Seldin DW, Giebisch G (eds) The kidney: physiology and pathophysiology. Raven, New York, pp 2339–2355

90. Quamme GA, Dirks JH (1980) Magnesium transport in the nephron. Am J Physiol 239:F393–F401

91. Reeves WB, Andreoli TE (1992) Renal epithelial chloride channels. Annu Rev Physiol 54:29–50

92. Rossier BC, Palmer LG (1992) Mechanisms of aldosterone action on sodium and potassium transport. In: Seldin DW, Giebisch G (eds) The kidney: physiology and pathophysiology. Raven, New York, pp 1373–1409

93. Sartorius OW, Roemmelt JC, Pitts RF (1949) The renal regulation of acid-base balance in man. IV. The nature of the renal compensations in ammonium chloride acidosis. J Clin Invest 28:423–439

94. Schlatter E, Bleich M, Hirsch J, Greger R (1993) pH-sensitive K^+ channels in the distal nephron. Nephrol Dial Transplant 8:488–490

95. Schlatter E, Greger R (1985) cAMP increases the basolateral Cl^--conductance in the isolated perfused medullary thick ascending limb of Henle's loop of the mouse. Pflugers Arch 405:367–376

96. Schlatter E, Schafer JA (1987) Electrophysiological studies in principal cells of rat cortical collecting tubules. ADH increases the apical membrane Na^+-conductance. Pflugers Arch 409:81–92

97. Schwarz GJ, Al-Awqati Q (1990) Identification and study of specific cell types in isolated nephron segments using fluorescent dyes. Methods Enzymol 191:253–264

98. Silbernagl S (1981) Renal transport of amino acids and oligopeptides. In: Greger R, Lang F, Silbernagl S (eds) Renal transport of organic substances. Springer, Berlin Heidelberg New York, pp 93–117

99. Silbernagl S (1986) Tubular reabsorption of amino acids in the kidney. NIPS 1:167–171

100. Silbernagl S (1988) The renal handling of amino acids and oligopeptides. Physiol Rev 68:911–960

101. Silbernagl S (1992) Amino acids and oligopeptides. In: Seldin DW, Giebisch G (eds) The kidney: physiology and pathophysiology. Raven, New York, pp 2889–2920

102. Simonson MS (1993) Endothelins: multifunctional renal peptides. Physiol Rev 73:375–411

103. Stanton B, Giebisch G (1992) Renal potassium transport. In: Windhager EE (ed) Handbook of physiology: renal physiology. American Physiological Society, Rockville, pp 813–874

104. Suki WN, Rouse D (1991) Renal transport of calcium, magnesium, and phosphorus. In: Brenner BM, Rector FC (eds) The kidney. Saunders, Philadelphia, pp 380–423

105. Suki WN, Rouse D, Ng R, Kokko JP (1980) Calcium transport in the thick ascending limb of Henle. Heterogeneity of function in the medullary and cortical segments. J Clin Invest 66:1004–1009

106. Turner RJ, Moran A (1982) Stoichiometric studies of the renal outer cortical brush border membrane D-glucose transporter. J Membr Biol 67:73–80

107. Ullrich KJ, Cassola AC, Papavassiliou F, Frömter E, Hopfer U (1982) Induction of low Na^+ diet of amiloride sensitive Na^+ transport in the proximal convolution of the rat kidney. Pflugers Arch 392:R14

108. Ullrich KJ, Greger R (1985) Approaches to study tubule transport functions. In: Seldin DW, Giebisch G (eds) The kidney: physiology and pathophysiology. Raven, New York, pp 427–469

109. Ullrich KJ, Papavassiliou F, David C, Rumrich G, Fritzsch G (1991) Contraluminal transport of organic cations in the proximal tubule of the rat kidney. I. Kinetics of N1-methylnicotinamide and tetraethylammonium, influence of K^+, HCO_3^-, pH; inhibition by aliphatic primary and tertiary amines, and mono- and bisquarternary compounds. Pflugers Arch 419:84–92

110. Ullrich KJ, Rumrich G (1988) Contraluminal transport systems in the proximal renal tubule involved in secretion of organic anions. Am J Physiol 254:F453–F462

111. Ullrich KJ, Rumrich G, Fritzsch G, Klöss S (1987) Contraluminal para-aminohippurate (PAH) transport in the proximal tubule of the rat kidney. II. Specificity: aliphatic dicarboxylic acids. Pflugers Arch 387:127–132

112. Ullrich KJ, Rumrich G, Neiteler K, Fritzsch G (1992) Contraluminal transport of organic cations in the proximal tubule of the rat kidney. II. Specificity: anilines, phenylalkylamines (catecholamines), heterocyclic compounds (pyridines, quinolines, acridines). Pflugers Arch 420:29–38

113. Wang W, Giebisch G (1991) Dual modulation of renal ATP-sensitive K^+ channel by protein kinases A and C. Proc Natl Acad Sci USA 88:9722–9725

114. Weinstein AM (1992) Sodium and chloride transport: proximal nephron. In: Seldin DW, Giebisch G (eds) The kidney: physiology and pathophysiology. Raven, New York, pp 1925–1974

115. Wilcox CS, Baylis C, Wingo C (1992) Glomerular-tubular balance and proximal regulation. In: Seldin DW, Giebisch G (eds) The kidney: physiology and pathophysiology. Raven, New York, pp 1807–1842

116. Wingo CS (1989) Active proton secretion and potassium absorption in the rabbit outer medullary collecting duct. J Clin Invest 84:361–365

117. Wittner M, Mandon B, Roinel N, De Rouffignac C, Di Stefano A (1993) Hormonal stimulation of Ca^{2+} and Mg^{2+} transport in cortical thick ascending limb of Henle's loop of the mouse: evidence for a change in the paracellular pathway permeability. Pflugers Arch 423:387–396

118. Wittner M, Weidtke C, Schlatter E, Di Stefano A, Greger R (1984) Substrate utilization in the isolated perfused cortical thick ascending limb of rabbit nephron. Pflugers Arch 402:52–62

119. Wright FS, Giebisch G (1992) Regulation of potassium excretion. In: Seldin DW, Giebisch G (eds) The kidney: physiology and pathophysiology. Raven, New York, pp 2209–2248

120. You GY, Smith CP, Kanai Y, Lee WS, Steiner M, Hediger MA (1993) Cloning and characterization of the vasopressin-regulated urea transporter. Nature 365:844–847

76 The Body Compartments and Dynamics of Water and Electrolytes

F. Lang

Contents

76.1 Body Compartments

Depending on sex, age and body fat, about 60–79% of the body weight is water (see Fig. 76.1). The distribution of water and electrolytes within the body is determined by fluid compartments (Figs. 76.1, 76.2). The body compartments can be divided into the *intracellular* and the *extracellular* space. The two compartments are separated by cell membranes. The extracellular space can be subdivided into several compartments, which are separated by cell layers. Most of the extracellular space is interstitial space. The endothelium separates the plasma volume, and epithelia separate the transcellular volume, from the interstitial fluid. The transcellular space comprises the lumina of glands, of urethrogenital and gastrointestinal tracts, pleura, pericardium, peritoneum, intraocular fluid and cerebrospinal fluid.

76.2.1 Determinants of Water Movement Between Body Compartments

With few exceptions (such as the luminal cell membrane of the thick ascending limb), the cell membranes and epithelia are freely permeable to water. In principle, water movement (Jv) is driven by both a hydrostatic (Δp) and an effective osmotic gradient ($\Delta \pi$):

$$Jv = Lp(\Delta p - \Delta \pi)$$

The effective osmotic gradient depends on the concentration difference across the respective border (Δc), and the reflection coefficient (σ) for each solute i:

$$\Delta \pi = RT\Sigma\sigma\Delta ci$$

A concentration difference of 1 mmol/l at $\sigma = 1$ creates a pressure gradient of some 2.2 kPa. Strictly speaking, the equation holds only for diluted solutions, and the effective osmotic pressure gradient in blood is slightly smaller. Large molecules, such as proteins, exert a higher osmotic pressure, as would be predicted from their molar concentration (cf. Chap. 125). The osmotic pressure created by macromolecules is referred to as the oncotic pressure.

The hydrostatic pressure gradient is high across the capillary wall (see below), which, on the other hand, is highly permeable to small but not to large molecules. Accordingly, the water movement across the capillary wall is a function of the hydrostatic and oncotic pressure gradients. The hydrostatic pressure gradient across cell membranes is usually minimal, so that water movement is largely dictated by osmotic pressure gradients. The same is true for most epithelia lining transcellular spaces.

76.2.2 Water Movement Between Capillary and Interstitial Space: Filtration

The endothelial cell layer separating the plasma and the interstitial space is largely impermeable to molecules of 50 kD or more [45]. Only the capillary beds of liver, spleen, and bone marrow are permeable to larger molecules. In the

R. Greger/U. Windhorst (Eds.)
Comprehensive Human Physiology, Vol. 2
© Springer-Verlag Berlin Heidelberg 1996

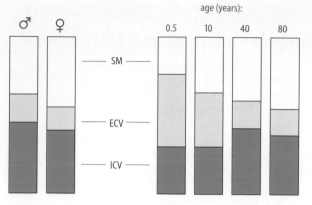

age (years):

0.5   10   40   80

— SM —

— ECV —

— ICV —

Fig. 76.1. Water content and distribution within the human body in dependence on sex (*left*) and age (*right*): *SM*, solid mass; *ECV*, extracellular volume; *ICV*, intracellular volume. (Modified from [31])

other capillary beds only a minimal portion of the plasma proteins (some 1%) pass the endothelial border, i.e., σ is close to 1 for plasma proteins. Since the protein concentration in the interstitial space is very low compared with the plasma, an *oncotic pressure gradient* (Δπ) is established across the endothelial border. Δπ approaches 20 mmHg at normal plasma protein concentrations. Δπ is opposed by a hydrostatic pressure gradient (Δp), which is some 30 mmHg at the beginning of the capillary and then declines towards the end of the capillary to below 20 mmHg (Fig. 76.3). Accordingly, Δp exceeds Δπ at the beginning of the capillary, whereas the opposite is true for the end of the capillary. As a result, net water movement into the interstitial space (*filtration*) prevails at the beginning and net water movement into the plasma (*reabsorption*), at the end of the capillary.

Even though most (usually more than 90%) of the filtered fluid is subsequently reabsorbed, some 20 l/day of filtered fluid remains in the interstitium. Furthermore, the endothelial cell layer is not perfectly impermeable to proteins, and the filtered proteins have to be cleared from the interstitial fluid to maintain the high oncotic pressure gradient from plasma to interstitium. The net filtered fluid with the filtered proteins is eliminated by the lymphatic system, as detailed in Chap. 94.

Filtration in the kidney is different from filtration in peripheral capillaries (cf. Chap. 73) inasmuch as Δp is higher and virtually constant along the short loops of the glomerular capillary network. Thus, reabsorption of fluid at the end of the capillary does not occur and *filtration equilibrium* may not even be reached. Furthermore, the hydraulic permeability of the glomerular capillary walls is very high and allows a substantial portion (some 20%) of the plasma flow to be filtered. Accordingly, the concentration of plasma proteins increases along the capillary length, leading to a corresponding increase in capillary oncotic pressure. Owing to this increase in oncotic pressure, filtration equilibrium may eventually be approached at the end of the glomerular capillary.

Excessive net filtration of fluid in peripheral capillaries leads to the formation of edema [49]. Possible causes are:

- Enhanced hydrostatic pressure gradients
- Reduced oncotic pressure gradients
- Increased permeability of the capillary wall to proteins, leading to filtration of proteins and thus in turn to reduction of the transcapillary oncotic pressure gradient
- Disturbed clearance of filtrate by the lymphatic system

76.2.3 Water Movement Between Intracellular and Extracellular Space: Cell Volume Regulation

Most cell membranes are highly permeable to water, and water moves across these cell membranes according to the prevailing osmotic gradient [9,57]. Thus, to maintain constancy of cell volume, the cells have to create an osmotic equilibrium across the cell membranes. On the other hand, in order to fulfill their metabolic functions, cells have to accumulate a number of osmotically active

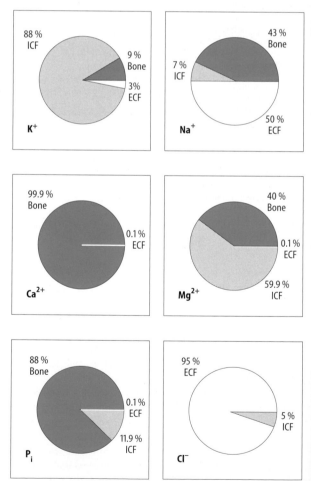

Fig. 76.2. Distribution of electrolytes in intracellular fluid (*ICF*), extracellular fluid (*ECF*) and bone. P_i, inorganic phosphate

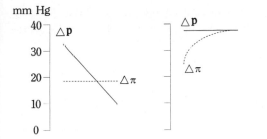

Fig. 76.3. Filtration in peripheral (*left panel*) and glomerular (*right panel*) capillaries. Δp, net hydrostatic pressure gradient; $\Delta \pi$, net oncotic pressure gradient. In peripheral capillaries, net filtration prevails in the beginning and net reabsorption towards the end of the capillary bed. The point of filtration equilibrium (no net movement of water) is approached in the second half of the capillary length. The diagrams are grossly simplified. In reality, filtration is not restricted to the capillary bed but also occurs in adjacent small vessels. Furthermore, the hydraulic permeability is not constant along the length of the capillary bed and the net movement of water is not a linear function of the capillary length

substances such as amino acids and glucose. To avoid osmotic swelling, cells have thus to balance the osmolarity imposed by these substances. Most cells solve this problem as illustrated in Fig. 76.4. The (Na$^+$+K$^+$)-ATPase [10,24] (cf. Chap. 8) extrudes Na$^+$ in exchange for K$^+$ and thus creates a Na$^+$ gradient directed from the extracellular space into the cell and a K$^+$ gradient in the opposite direction. The cell is usually poorly permeable to Na$^+$, whereas K$^+$ channels allow the permeation of K$^+$. The movement of K$^+$ creates an outside positive potential difference across the cell membrane, which drives Cl$^-$ out of the cell (cf. Chap. 11). At a cell membrane potential of -60 mV, the intracellular Cl$^-$ concentration is in equilibrium at about one-tenth of the extracellular Cl$^-$ concentration. Osmotically, the low intracellular Cl$^-$ concentration outweighs the high intracellular concentration of organic substances.

Maintenance of this balance requires the continuous expenditure of energy in the form of ATP, since the cell membranes are not completely impermeable to Na$^+$. Impairment of the (Na$^+$+K$^+$)-ATPase leads to a gradual dissipation of the Na$^+$ and K$^+$ gradients and of the potential difference across the cell membrane. Thus, the electrical driving force for Cl$^-$ extrusion decreases and the cell accumulates Cl$^-$. Eventually the cell swells and *disintegrates*.

Cell volume constancy is further compromised by alterations in extracellular osmolarity (e.g., hyponatremia, hyperglycemia) or by altered uptake or metabolism of osmotically active substances. Enhanced extracellular K$^+$ concentrations stimulate cellular uptake of K$^+$ and depolarize the cells, thus stimulating cellular uptake of Cl$^-$. Accordingly, hyperkalemia tends to swell cells, whereas hypokalemia leads rather to cell shrinkage.

Even at constant extracellular osmolarity and ion composition and with an adequate energy supply, the constancy of cell volume is continuously challenged by transport and metabolism of the cells: any uptake of osmotically active substances (such as amino acids, glucose, electrolytes) increases intracellular osmolarity, while any release of osmotically active substances decreases intracellular osmolarity (cf. Chap. 59). The *formation* of macromolecules (glycogen, proteins) from their respective monomers (glucose, amino acids) reduces intracellular osmolarity, while *degradation* of the macromolecules increases it. The degradation of amino acids and glucose then leads to disappearance of osmolarity. To maintain constancy of cell volume under all these conditions, the cells have developed strategies of cell volume regulation [15,16,18,23,28,32,42,43,47,54,56].

Most importantly, cells utilize ion transport across the cell membrane to regulate intracellular osmolarity: following cell swelling there is a regulatory decrease in volume. Most cells release ions by activation of K$^+$ and *anion channels* (see Fig. 76.5). In some cells, electrolyte release is accomplished by *KCl cotransport* or, rarely, by other systems, such as K$^+$/H$^+$ exchange. In some cells at least, the K$^+$ channels and/or anion channels are activated by increase in intracellular Ca^{2+}, due to Ca^{2+} entry through non-selective cation channels. Following cell shrinkage there is a regulatory increase in volume. Cells accumulate cellular ions by activation of *Na$^+$,K$^+$,2Cl$^-$ cotransport* and/or parallel activation of *Na$^+$/H$^+$ exchange* and *Cl$^-$/HCO$_3^-$ exchange*. At the same time, they reduce ion loss by inactivation of ion channels [50].

Furthermore, the cells modify their cellular osmolarity by metabolism [3,14,18,19,25,60]: upon cell shrinkage, they degrade macromolecules (proteins and glycogen) to yield the respective monomers (amino acids and glucose) and form so-called osmolytes, i.e., substances primarily serving the function of increasing intracellular osmolarity. These substances include polyols (e.g., sorbitol, inositol), methylamines (e.g., glycerophosphorylcholine, betaine) and certain amino acids (e.g., taurine). Upon cell swelling, the cells deal with osmolarity by the formation of proteins and glycogen and by release of certain osmolytes. Beyond these functions obviously related to cell volume regulation, cell volume modifies a variety of further metabolic functions, as discussed below (76.5.1).

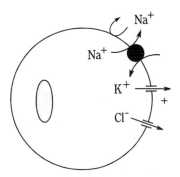

Fig. 76.4. Cell volume maintenance by (Na$^+$+K$^+$)-ATPase: Na$^+$ is extruded in exchange for K$^+$. K$^+$ movement through K$^+$ channels creates an outside positive cell membrane potential, which drives Cl$^-$ out of the cell

regulatory cell volume decrease

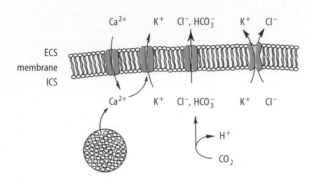

regulatory cell volume increase

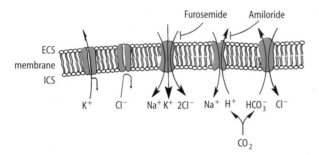

Fig. 76.5. Ionic mechanisms of cell volume regulation: *ECS*, extracellular space; *ICS*, intracellular space

76.2.4 Water Movement Between Interstitial and Transcellular Space

The composition of the various transcellular spaces and thus the determinants of water transport between interstitial space and any given transcellular space are dependent on the transport properties of the particular epithelium involved (cf. Chap. 59). The large bowel, the lower urinary tract, the thick ascendending limb of the nephron, and in the absence of ADH the whole of the distal nephron, are poorly permeable to water (cf. Chap. 74). Thus, even in the presence of large osmotic gradients little water crosses the epithelium. In other epithelia the water transport is a function of the osmotic pressure gradient created by the transport of osmotically active substances. Examples for sequelae of discrepant secretion and reabsorption into transcellular spaces include glaucoma (increase of intraocular pressure), ascites (enhanced formation of intraperitoneal fluid) and diarrhea (enhanced net secretion of intestinal fluid).

76.3.1 The Electrolyte Composition of the Body

As is apparent from Table 76.1, the intracellular and extracellular spaces differ markedly in electrolyte compo-

sition. Most importantly, the dominant intracellular cation is K^+, whereas the dominant extracellular cation is Na^+ (cf. Chaps. 8,11). As described in Sect. 76.1.4 and in Chap. 11, this distribution of cations is a prerequisite for the generation of the cell membrane potential, which drives anions out of the cell. Accordingly, the intracellular Cl^- concentration is only a fraction of the extracellular Cl^- concentration. The intracellular anions are in large part due to negatively charged proteins. The potential difference drives bicarbonate out of the cell and H^+ ions into the cell, favoring intracellular acidosis. Even though H^+ is extruded from the cell, intracellular pH is thus some 0.2–0.3 pH units lower than extracellular pH. Intracellular calcium activity is only 0.01% of extracellular calcium activity. As illustrated in Chap. 5, this steep gradient for Ca^{2+} is of paramount importance for intracellular signal transduction. Other electrolytes, such as phosphate and Mg^{2+}, are accumulated within the cells.

Only small differences are encountered between interstitial space and plasma, since in most tissues the endothelial cell layer is highly permeable to electrolytes. The composition of the various transcellular spaces, on the other hand, may differ substantially from the interstitial electrolyte composition (Table 76.2).

76.3.2 Electrolyte Movement Between Capillary and Interstitial Space

Most endothelial layers are freely permeable to electrolytes, with the notable exception of the endothelia lining the cerebral vessels. These endothelia form the so-called blood–brain barrier (cf. Chap. 27). Electrolytes have a very low ability to permeate this barrier, and the half-lives of gradients for Na^+, Ca^{2+}, K^+ and Mg^{2+} are 3, 8, 24, and 24, respectively. Similarly low permeabilities are observed for Cl^-, HCO_3^- and I^-. Thus, sudden changes in blood electrolyte concentrations could create corresponding gradients across the blood–brain barrier with subsequent water movements. Other endothelia are highly permeable to electrolytes. Nevertheless, the ion concentrations are not identical in plasma and interstitial space. For Ca^{2+} the difference is especially large: some 40% of plasma calcium is bound to proteins and thus cannot pass the capillary

Table 76.1. Electrolyte composition (mmol/l) of plasma, interstitial (ECF) and intracellular (ICF) fluid

	Plasma	ECF	ICF
Na^+	141	143	15
K^+	4	4	140
Ca^{2+}	2.5	1.3	0.0001
Mg^{2+}	1	0.7	15
Cl^-	103	115	8
HCO_3^-	25	28	15
HPO_4^{2-}	1	1	60
SO_4^{2-}	0.5	0.5	10
Organic acids	4	5	2
Proteins	2	<1	6

Table 76.2. Composition (mmol/l) of various transcellular fluids (cf. Chap. 126)

	GFR	Urine	Saliva	Gastric juice	Pancreatic juice	Livergall	Intestinal fluid	Feces	Sweat
Na$^+$	142	150	80	60	140	140	120	30	5–80
K$^+$	4	60	20	20	5	5	5	70	10
Ca^{2+}	1.3	1	1.5	1.5	1.5	2	1.5	70	1
Cl$^-$	115	150	40	100	60	90	100	10	5–70
HCO$_3^-$	28	–	60	–	70	40	30	–	–
pH	7.4	5.8	7.8	2	7.8	7.7	7.5	–	–
Volume formed (l/24h)	170	1.5	1	2	1	1	3	0.2	1

GFR, Glomerular filtration rate

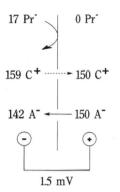

Fig. 76.6. The generation of the so-called Gibbs-Donnan potential: Pr^- = negatively charged proteins; K^+, diffusible cations; A^-, diffusible anions

wall. Accordingly, the concentration of total calcium in plasma is almost twice that in the interstitium. Other ions, which are not bound to plasma proteins, still do not exhibit identical concentrations in plasma and interstitial space. One reason lies in the negatively charged proteins, which create the so-called *Gibbs-Donnan potential* (Fig. 76.6).

If all permeable ions were to be equally distributed between plasma and interstitial space, then the negatively charged plasma proteins would represent an excess negative charge on the blood side, thus creating a potential difference across the endothelial layer. The results would be diffusable cations retained within the blood and diffusible anions expelled into the interstitial space. The uneven distribution of the permeable ions (almost) neutralizes the surplus of negative charge in the form of proteins on the blood side. As shown in Fig. 76.6, an equilibrium is reached at a potential difference of some 2 mV, which maintains a slightly increased blood cation concentration and a slightly increased anion concentration on the interstitial side. However, the gradients are only very slight.

The electrolyte composition of the plasma and that of the interstitial space are shown in Table 76.1.

76.3.3 Electrolyte Movement Between Intracellular and Extracellular Space

The transport of electrolytes across the cell membranes is accomplished by a variety of pumps, carriers and ion channels (summarized in Chaps. 8, 59). The activity of these transport molecules determines ion composition and volume of any given cell. In epithelia, the transport molecules serve the transepithelial movement of electrolytes and water (see Chap. 59). However, the same or similar transport molecules are expressed in nonepithelial (nonpolarized) tissues and modify cellular ion composition, volume and diverse cellular functions, such as excitability [13], cell metabolism [18,19], and cell proliferation [41].

The transport systems are subject to regulation by hormones and mediators. The hormones thus influence the ion distribution across the cell membranes.

76.3.4 Electrolyte Movement Between Interstitial and Transcellular Space

The electrolyte movement into or from transcellular spaces is dependent on the specific transport properties of the respective epithelia, which are described elsewhere in this book (Chaps. 59, 62, 65, 67, 74). It should be stressed here, however, that the bidirectional movements of water and electrolytes between interstitial and transcellular spaces are tremendous (Table 76.2). For instance, the daily secretion of fluid into the intestinal tract usually exceeds 8 l (Fig. 76.7). Since virtually all of this secreted fluid is usually recovered by reabsorption, intestinal secretion does not appreciably influence water and electrolyte balance. However, if intestinal absorption is impaired, the continuing intestinal secretion creates a life-threatening disturbance of electrolyte homeostasis within a few hours.

76.4 Water and Electrolyte Balance of the Body

The constancy of volume and composition of the various body compartments requires that the body's balance of water and electrolytes is maintained, i.e., that the intake exactly matches the output of water and electrolytes [40]. The intake of water [12] is determined by oral intake and gastrointestinal absorption. The output is determined by intestinal and urinary losses and by losses through sweat formation and respiration (Table 76.3). The electrolyte bal-

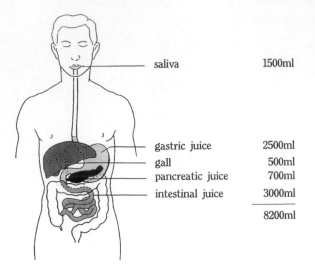

saliva	1500ml
gastric juice	2500ml
gall	500ml
pancreatic juice	700ml
intestinal juice	3000ml
	8200ml

Fig. 76.7. Volumes of gastrointestinal secretions

ance is determined by oral (or parenteral) intake on the one hand and urinary and fecal excretion on the other (Table 76.4).

Mechanisms governing the intake of water and electrolytes (cf. Chaps. 62, 72), their intestinal absorption (cf. Chap. 62), and their urinary excretion (cf. Chaps. 74, 75) are described elsewhere in this book.

Water losses through respiration come from the saturation of inhaled air in the respiratory tract (cf. Chaps. 100, 101) and thus depend on ventilation and water saturation of inhaled air.

Water and electrolyte loss through sweat formation is determined largely be the requirements of thermoregulation (cf. Chap. 112; Fig. 76.8). Thus, sweat formation may be extremely variable, even in healthy subjects. It is the task of the kidney to compensate any imbalance between gain and loss of water and electrolytes, and urinary excretion of water and electrolytes can accordingly be variable.

The mechanisms engaged in the adjustment of water and electrolyte excretion to the requirements of water and electrolyte homeostasis are described in Chaps. 77–80.

76.4.1 Physiological Significance of Cell Volume

Cell volume has been shown to modify a wide variety of cellular functions. As pointed out above (Sect. 76.2.3), cell swelling activates ion channels or coupled K^+ release at the plasma membrane and stimulates the formation of macromolecules and organic osmolytes. Cell shrinkage stimulates Na^+/H^+ exchange and Na^+ K^+ $2Cl^-$ cotransport, degradation of macromolecules, and degradation or release of osmolytes [35].

A number of additional metabolic functions are modified by alterations of cell volume [21]. Furthermore, cell swelling has been shown to increase the pH within acidic cellular compartments [37,38,58,59].

The effects of cell volume are important for a variety of cellular functions: the modulation of ion channel activity and of ion transport by cell volume is important for *epithelial transport*. In fact, changes of cell volume have been implicated in the coupling of apical and basolateral cell membranes during transport [33,53]. For instance, entry of electrolytes through the apical cell membrane of a reabsorbing epithelium tends to swell the cell, which then activates ion channels at the basolateral cell membrane.

In *excitable cells*, the activation of ion channels and/or electrolyte transport across the cell membrane modifies the cell membrane potential and thus the state of excitation.

In sympathetic neurons for instance, the influence of GABA on cell membrane potential has been shown to be modified by the Na^+ K^+ $2Cl^-$ cotransport inhibitor furosemide [1]: in the presence of furosemide GABA leads only to a transient depolarization, whereas in its absence the GABA-induced depolarization is sustained. Apparently, the activation of K^+ and Cl^- channels by GABA leads to KCl loss and shrinkage of the cell, which then activates a volume-regulatory Na^+ K^+ $2Cl^-$ cotransport. This carrier maintains not only cell volume, but also intracellular Cl^- activity and thus depolarization. During inhibition of the carrier, intracellular Cl^- decreases to equilibrium and the cell repolarizes.

Another type of excitable cell (cf. Chap. 66), pancreatic B-cells, respond to swelling with a short-lived hyperpolarization of the cell membrane followed by a depolarization with activation of Ca^{2+} channels and Ca^{2+} entry. These events lead to transient stimulation of insulin release [6].

Angiotensin is known to increase intracellular Ca^{2+} in renin-producing cells, leading to activation of Ca^{2+}-sensitive K^+ and Cl^- channels, to cell shrinkage and to inhibition of renin release [30].

The increase of pH in acidic cellular compartments during cell swelling probably affects a great variety of cellular

Table 76.3. Daily turnover of H_2O (in liters)

Intake and production		Loss	
Solid food	0.7	Evaporation	0.8
Oxidation water	0.3	Feces	0.1
Beverages	>0.6	Urine	>0.7
		Sweat	0–10

Table 76.4. Daily turnover of electrolytes

Total turnover (mmol/24h)		Excretion (% of total)		
		Urine	Feces	Sweat
Na^+	150	95	4	1
K^+	100	90	10	–
Cl^-	100	98	1	1
Ca^{2+}	20	30	70	–
Mg^{2+}	15	30	70	–

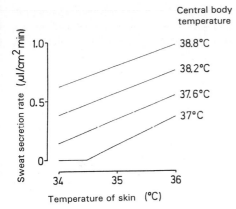

Fig. 76.8. Sweat volume as a function of surface and core temperature

functions. Frequently, the acidity within these compartments is required for activation of *pH-sensitive proteases*. Accordingly, an increase of pH within these compartments is known to inhibit hepatic proteolysis [48], and cell swelling affects proteolysis at least partly by its influence on lysosomal pH [37]. In pancreatic B-cells, acidic proteases within the acidic secretory granules cleave proinsulin to yield insulin, a function probably compromised by cell swelling and fostered by cell shrinkage [7]. Similarly, in neurons, metabolism of transmitters may be modified by cell volume through alterations of pH in secretory granules. In tracheal epithelium, sulfation and sialation of membrane proteins has been shown to be sensitive to luminal pH of acidic cellular compartments, and it has been argued that defective acidification of these compartments in cystic fibrosis leads to altered sulfation and sialation and thus to enhanced adhesivity of *Pseudomonas aeruginosa* [2]. Alkalinization of acidic cellular compartments following cell swelling has further been observed in proximal renal tubules, MDCK cells, a permanent kidney epithelial cell line, alveolar cells, fibroblasts, lymphocytes and glial cells [38]. In these cells, processing of intravesicular proteins may be modified in turn by cell volume. Moreover, the pH in these compartments is thought to modify the trafficking of receptors and membranes (cf. Chaps. 4, 5, 7), suggesting a further role of cell volume in the regulation of hormone and transmitter release, insertion of receptors and carriers. Accordingly, it has been suggested that cell swelling enhances bile acid transport in the liver by stimulation of carrier insertion into the plasma membrane [21].

The influence of cell volume on these functions has not yet been explored in any detail, but it is quite likely that cell swelling does modify at least some of these functions.

Finally, cell volume regulatory mechanisms may further be involved in such complex functions of the cell as *cell proliferation or apoptosis* (cf. Chaps. 5, 6, 124). It is well known that a variety of growth promoters stimulate the *Na+/H+ exchange*, one of the two ion transporters serving to in-

crease cell volume during cell shrinkage [4]. Furthermore, in some cells growth promoters have been shown to activate the $Na^+ K^+ 2Cl^-$ cotransporter, the other carrier increasing cell volume [4]. Accordingly, expression of *ras* oncogene has been demonstrated to shift the set point for cell volume regulation towards greater cell volumes [34] and to increase the pH in lysosomes. Apoptosis, i.e., programmed cell death, is accompanied by marked cell shrinkage [38]. The role of cell volume in the complex machinery that has to be mobilized to stimulate cell proliferation or programmed cell death is still ill-defined, but the metabolic sequelae of alterations of cell volume may contribute to the adjustment of cellular function to proliferation or apoptosis.

Some hormones utilize the cell volume-regulatory mechanisms to trigger certain desired cellular functions. Thus, *insulin* has been shown to swell liver cells by activation of $Na^+ K^+ 2Cl^-$ cotransport and Na^+/H^+ exchange, thus stimulating protein and glycogen synthesis [18,19]. *Glucagon*, on the other hand, shrinks the same cells by activation of ion channels and thus favors degradation of protein and glycogen [18,19]. As mentioned above, growth factors activate Na^+/H^+ exchange and/or $Na^+ K^+ 2Cl^-$ cotransport in a wide variety of cells. Thus, cell volume has been considered a "second messenger" (or second message) in the transmission of hormone signals [18,38].

Alterations of cell volume are thought to underly a number of pathologic conditions. In any cell, energy depletion eventually impairs $(Na^+ + K^+)$-ATPase, leading to decreased cellular K^+, depolarization, entry of Cl^-, and cell swelling (Sect. 76.2.3). Cell swelling is especially detrimental in the brain [26], where the rigid skull does not allow expansion of the soft tissue. Accordingly, swelling of glial cells (astrocytes; cf. Chap. 26) has been shown to underly a variety of disorders. In hypercatabolic states, on the other hand, the estimated decrease of muscle cell volume has been found to correlate with the degree of hypercatabolism [20]. However, further experimental effort is needed to settle the pathophysiological significance of disturbed cell volume.

76.4.2 Physiological Significance of Plasma Volume

Extracellular volume comprises several compartments with distinct functional impact. The size of these compartments is more or less critically important for survival of the organism.

As discussed in detail elsewhere in this book (cf. Chaps. 94, 95), blood and plasma volume must remain within narrow limits. If other variables remain constant, a reduction of blood volume [55] decreases cardiac filling, stroke volume and cardiac output (cf. Chaps. 89, 90). In the wake of all this, blood pressure tends to decline and the reduced tension of arterial pressure receptors enhances the sympathetic tone and inhibits the parasympathetic activity. As a result, the vascular resistance is enhanced in a great variety of tissues, namely the skin, the intestine and the kidney. The compromised renal perfusion leads to stimulation of *renin* release, which catalyzes the formation of angiotensin I (cf. Chap. 73). This is then converted to *angiotensin II* under the influence of converting enzyme. Angiotensin II favors renal retention of NaCl and water by direct renal action and by stimulation of the release of *antidiuretic hormone* (ADH, vasopressin) and *aldosterone* [17,27, 39,44,51,52]. The effects of aldosterone and ADH are compounded by the antidiuretic action of sympathetic renal nerves [11,29]. These mechanisms are usually capable of maintaining blood pressure even at a reduced blood volume. However, the renal retention of NaCl and water compromises the excretory function of the kidney. In fact, excessive stimulation of the above system may lead to functional renal failure [36]. It must be kept in mind that it is the so-called *effective blood volume* that matters: in *hepatic failure*, for instance, hepatic inactivation of endotoxins and mediators from the intestine is impaired. The endotoxins stimulate the expression of *NO synthase*, and the enhanced formation of NO leads to peripheral vasodilation [46]. One of the intestinal mediators is the vasodilatatory substance P [22]. As a result, peripheral vasodilation prevails in hepatic failure, and the maintenance of blood pressure requires enhanced cardiac output, stroke volume and blood volume.

An *increase* in *plasma volume* reverses the above mechanisms to decrease release of ADH and aldosterone. On the other hand, an increase in plasma volume also enhances the secretion of *atrial natriuretic factors* and of *ouabain*, the "endogenous digitalis-like substance (EDLS)" [5]. Ouabain inhibits (Na^++K^+)-ATPase in a wide variety of cells. As a result, intracellular Na^+ increases and the decreased electrochemical Na^+ gradient impairs the extrusion of Ca^{2+} by the Na^+/Ca^{2+} exchanger. Thus, in many cells the Ca^{2+} concentration increases. In the *heart*, Ca^{2+} enhances the contractile force (cf. Chaps. 90, 93), while in *smooth vascular muscle* cells it stimulates vasoconstriction (cf. Chaps. 44, 96). The direct effect on smooth vascular muscle cells is compounded by effects on endothelial cells, where it stimulates the release of vasoconstrictory *endothelins*, and on sympathetic neurons, where it stimulates the release of vasoconstrictory catecholamines [5]. Thus, excessive blood volume increases the peripheral resistance and the cardiac force and thus increases blood pressure, which in turn may give rise to a variety of untoward sequelae [8].

76.4.3 Physiological Significance of Interstitial Volume

Isolated alterations in interstitial volume are generally less harmful than alterations in *blood volume*. A lack of interstitial volume may lead to decreased interstitial pressure, enhanced effective filtration pressure from blood to interstitial space, increased filtration, and thus decreased blood volume. Increases in interstitial fluid, conversely, tend to

enhance blood volume. However, with disturbed peripheral filtration, interstitial volume may be enhanced without a parallel increase in blood volume [49]. In such cases, excessive increases in interstitial volume in the form of *edema* may conversely be tolerated by the patients, and therapeutic attempts to eliminate the edema by stimulation of diuresis sometimes expose the patient to a higher risk than the edema itself. One notable exception is *pulmonary edema*, since the contact time of the blood with the alveoli is usually a fraction of a second only and adequate oxygenation of hemoglobin requires intimate contact of blood with alveolar gas (cf. Chaps. 100, 101). In interstitial pulmonary edema the diffusion distance between blood and alveoles is enhanced and gas exchange impaired accordingly. Entry of extracellular fluid into the alveolar space (which, if fluid filled, is by definition a transcellular space) further aggravates the condition. In peripheral tissues, an increase of the interstitial space may similarly increase diffusion distances, but the functional consequences are usually limited. At critical points, however, an increase in interstitial volume can be life threatening. For instance, swelling of the epiglottis due to local edema may obstruct the airway and thus lead to suffocation. Enhanced interstitial volume may also be harmful in the brain, where the rigid skull prevents any expansion of the tissue. Thus, any increase in a volume compartment can only occur at the expense of another, usually the liquor space, a transcellular space mainly designed to establish a "volume buffer" (see below). If the increase in the interstitial space is too much for the liquor space, further expansion leads to compression of the intracranial blood volume, which compromises cerebral blood flow. Eventually the pressure increase (*brain edema*) may cause herniation of the cerebellum through the foramen magnum.

76.4.4 Physiological Significance of Transcellular Volumes

The significance of *transcellular spaces* is related to the functions of the corresponding organs and can be touched on only briefly here.

Liquor Space. As pointed out above, the liquor space is an important volume buffer in the brain. In view of the rigid skull, pulsating perfusion of the brain requires that whenever blood volume increases some other compartment is decreased. This is achieved by movement of liquor through the foramen magnum. Excessive formation or impaired clearance of liquor increases cerebral pressure and compromises cerebral blood flow.

Aqueous Humor. Another transcellular space with high pathophysiological impact is the aqueous humor in the eye. The epithelium in the ciliary body secretes fluid into the aqueous humor and the vitreous humor. The fluid is reabsorbed into the *canal of Schlemm*. If the reabsorption is impaired, intraocular pressure increases and compromises perfusion of the intraocular tissues. This may lead to complete destruction of the retina.

Urinary Space. On the other hand, the urinary space may vary within wide limits without untoward consequences. Nevertheless, a decrease in urine volume is paralleled by a reciprocal increase in the concentrations of excreted substances. Some urine constituents are poorly soluble, and antidiuresis may favor their precipitation leading to urolithiasis. Furthermore, antidiuresis may favor urinary infection. On the other hand, even an extreme increase of urine volume is harmless in itself.

However, increased outflow resistance such as occurs in the presence of prostate hypertrophy can cause dilatation of the canalicular system and even acute renal failure.

The *luminal fluid of the intestine* is a large transcellular space. Excessive intestinal secretion without appropriate absorption can lead to hypovolemia. The distension of the muscular wall caused by enhanced luminal fluid stimulates intestinal motility and results in abdominal pain and accelerated intestinal passage. The latter reduces the contact time with the absorbing epithelium, thus compromising intestinal absorption.

Pleural Space. The *pleural space* is a thin transcellular fluid film around the lungs, allowing the movements of the lungs during respiration. A negative pressure (cf. Chap. 100) is maintained in the pleural cavity by reabsorption. An important determinant for pleural fluid reabsorption is the oncotic pressure gradient between the virtually protein-free luminal fluid and the blood. During inflammation, proteins may enter the pleural cavity, favoring the entry of fluid. The increase of pleural volume may then lead to a corresponding decrease in lung volume, so that respiration is impaired.

Pericardial Fluid. As in the case of pleural fluid, the *pericardial fluid* volume is usually small and may be enhanced during inflammation.

Peritoneal Fluid. Similarly, *peritoneal fluid* volume is usually minimal. However, peritoneal fluid volume may become excessive when the pressure in the liver sinusoids rises, as it does in liver disease (cf. Chap. 67). The enlargement of the peritoneal cavity (*ascites*) may lead to gross abdominal distension and even impair respiration by pressure on the diaphragm.

Synovial Fluid. The *synovial fluid* separates the two articular faces of a joint. The lining of the synovial spaces is not a real epithelium, and the synovial space may thus be considered an interstitial rather than transcellular space. The synovia is of paramount importance for the mobility of the joint. However, it is the glycosamionoglycan and proteoglycan composition rather than volume and electrolyte composition, which determine the function of synovial fluid.

76.5 Measurement of Fluid Compartments and Electrolyte Pools

76.5.1 The Dilution Principle

The concentration of any substance in solution (c) is defined as the amount (M) per unit volume (V):

$$c = M/V \tag{76.1}$$

This simple relation can be exploited to calculate any one of the three variables as long as the other two are known. For the determination of fluid volumes certain amounts of *indicator substances* (M_{add}) are added to the volume (V) to be determined, and after appropriate mixing the concentration (c) is measured ($V = M_{add}/c$). The indicator substance must be evenly and exclusively distributed within the volume to be determined, and it must not be toxic or change the volume.

Usually the indicator substances do not remain in the compartment to be determined but are gradually removed by exit into other fluid compartments, by renal or intestinal excretion, metabolism, etc. This means that some proportion of the indicator injected into the volume to be determined is already lost before even dispersion is achieved. To circumvent any error introduced by this complicating factor, several determinations can be made and the decline of indicator substance estimated from the decrease in concentration after full equilibration. Extrapolation to the time point of injection yields the virtual even concentration at the time of injection, i.e., at the time when the amount of indicator substance present in the compartment is still identical to the amount injected (M_{add}).

76.5.2 Determination of Plasma or Blood Volume

Blood consists of two compartments, the plasma and the erythrocyte cell volume. The two compartments can be discriminated by centrifugation followed by measurement of the cellular column (so-called packed cell volume = PCV, or hematocrit as percentage of total blood volume; cf. Chap. 83) and the supernatant. It must be borne in mind that the PCV or *hematocrit* slightly overestimates the true erythrocyte cell volume: centrifugation does not completely separate the cellular components from plasma, some 10% remaining entrapped between the cells. Moreover, the hematocrit is appreciably greater in large arteries and veins than in small vessels, such as capillaries, arterioles and venules, because of the axial streaming of erythrocytes within blood vessels (cf. Chap. 88). Thus, the hematocrit of blood taken from large veins is larger than the "true" average hematocrit of the body. With these reservations in mind, knowledge of the hematocrit (hct) allows calculation of the plasma volume (V_P) if the blood volume (V_B) is known:

$$V_P = V_P \times (1 - 0.0087\,\text{hct}) \qquad (76.2)$$

For the determination of blood volume substances that bind tightly to either plasma proteins (e.g., Evans blue) or erythrocytes (e.g., radioactive chromium) are used. The number of erythrocytes or the amount of plasma proteins leaving the vascular space within the time required for even dispersion of the test substance is small. This small loss of substance can be corrected for as described above.

76.5.3 Determination of Extracellular and Interstitial Fluid Volume

Several indicators have been used to determine *extracellular fluid*, i.e., radioactive Na^+, Cl^-, thiocyanate, inulin or sucrose. None of these substances penetrates easily into any of the extracellular spaces. The blood–brain barrier, in particular, (cf. Chap. 22) prevents rapid entry of the indicator substances into the extracellular spaces of the brain. On the other hand, none of the indicators is completely excluded from the intracellular space. Thus, precise measurement of extracellular space is not possible with any of the above indicators, and the "distribution spaces" obtained may result in overestimates or underestimates of extracellular fluid volume, depending on the prevailing bias. It is commonly assumed that the *Na^+ space* is slightly greater and the *inulin space* rather smaller than the extracellular fluid volume. The interstitial fluid volume is calculated from the difference between extracellular fluid volume and plasma volume. This calculation disregards the transcellular fluid volume.

76.5.4 Determination of Total Body Water and Intracellular Space

The total body water is determined with *heavy water* or with *antipyrin*, both of which are distributed to all fluid volumes of the body. Again, appropriate correction must be made for elimination of the indicator prior to even dispersion (see above). Intracellular fluid volume is calculated from the difference between total body volume and extracellular fluid volume.

76.5.5 Determination of Electrolyte Pools

An electrolyte pool (i.e., the amount of a give electrolyte within the body) can be dtermined from the dilution of the appropriate radioactive tracers. For evaluation of the Na^+ pool, for example, $^{22}Na^+$ can be injected into the blood and the concentration of the tracer measured. If the tracer is evenly distributed the pool (M_{Na}) can be calculated from:

$$M_{Na} = M_{22Na} \times [Na^+]/[^{22}Na^+] \qquad (76.3)$$

where M_{22Na} is the amount of tracer Na^+ injected (minus the amount eliminated) and $[Na^+]$ and $[^{22}Na^+]$ are the concen-

trations for Na^+ and tracer $^{22}Na^+$, respectively. Again, a correction needs to be made for elimination of tracer prior to even distribution. The pool determined depends on the time of equilibration allowed for. Owing to the *blood–brain barrier*, equilibration with brain electrolytes is slow. The half-times for equilibration of Na^+, Ca^{2+}, K^+ and Mg^{2+} are approximately 3, 8, 16 and 24 h, respectively (see above); that is to say 48 h is required for a 90% equilibration with K^+ (cf. Chap. 27). The turnover of bone electrolytes is too slow to allow the achievement of equilibration with tracer electrolytes within a few days. The bone electrolytes are thus regarded as the so-called *non-exchangeable pool*.

References

1. Ballanyi K, Grafe P (1988) Cell volume regulation in the nervous system. Renal Physiol Biochem 11:142–157
2. Barasch J, Kiss B, Prince A, Saiman L, Gruenert D, Al-Awquati Q (1991) Defective acidification of intracellular organelles in cystic fibrosis. Nature 352:70–73
3. Beck FX, Dörge A, Thurau K, Guder WG (1990) Cell osmoregulation in the countercurrent system of the renal medulla: the role of organic osmolytes. In: Beyenbach KW (ed) Cell volume regulation. Comparative physiology. Karger, Basel, pp 132–158
4. Bianchini L, Grinstein S (1993) Regulation of volume-modulating ion transport systems by growth promoters. In: Lang F, Häussinger D (eds) Advances in comparative and environmental physiology, vol 14. Springer, Berlin Heidelberg New York, pp 249–277
5. Blaustein MP (1993) Physiological effects of endogenous ouabain: control of intracellular Ca^{2+} stores and cell responsiveness. Am J Physiol 264:C1367–C1387
6. Britsch S, Krippeit-Drews P, Gregor M, Lang F, Drews G (1994) Effect of osmotic changes in extracellular solution on electrical activity of mouse pancreatic B cells. Biochem Biophys Res Comm 204:641–645
7. Busch GL, Schreiber R, Dartsch PC, Völkl H, vom Dahl St Häussinger D, Lang F (1994) Involvement of microtubules in the link between cell volume and pH of acidic cellular compartments. Proc Natl Acad Sci U S A 91:9165–9169
8. Cowley AW Jr (1992) Long-term control of arterial blood pressure. Physiol Rev 72:231–300
9. Dawson DC (1992) Water transport: Principles and perspectives. In: Seldin DW, Giebisch G (eds) The kidney. Physiology and pathophysiology, 2nd edn. Raven, New York, pp 301–316
10. DeWeer P (1992) Cellular sodium-potassium transport. In: Seldin DW, Giebisch G (eds) The kidney. Physiology and pathophysiology , 2nd edn. Raven, New York, pp 93–112
11. DiBona GF, Herman PJ, Sawin LL (1988) Neural control of renal function in edema-forming states. Am J Physiol 254 (Regulatory Integrative Comp Physiol 23):R1017–R1024
12. Fitzsimons JT (1992) Physiology and pathophysiology of thirst and sodium appetite. In: Seldin DW, Giebisch G (eds) The kidney. Physiology and pathophysiology, 2nd edn. Raven, New York, pp 1615–1648
13. Fozzard HA, Shorofsky SR (1992) Excitable membranes. In: Seldin DW, Giebisch G (eds) The kidney. Physiology and pathophysiology, 2nd edn. Raven, New York, pp 407–446
14. Garcia-Perez A, Burg MB (1991) Renal medullary organic osmolytes. Physiol Rev 71:1081–1115
15. Geck P (1990) Volume regulation in Ehrlich cells. In: Beyenbach KW (ed) Cell volume regulation. Comparative physiology. Karger, Basel, pp 26–58

16. Grinstein S, Foskett JK (1990) Ionic mechanisms of cell volume regulation in leukocytes. Annu Rev Physiol 52:399–414
17. Hall JE, Brands MW (1992) The renin-angiotensin-aldosterone systems: renal mechanisms and circulatory homeostasis. In: Seldin DW, Giebisch G (eds) The kidney. Physiology and pathophysiology, 2nd edn. Raven, New York, pp 1455–1504
18. Häussinger D, Lang F (1991a) Cell volume in the regulation of hepatic function: a mechanism for metabolic control. Acta Biochim Biophys 1071:331–350
19. Häussinger D, Lang F (1991b) Cell volume – a "second messenger" in the regulation of metabolism by amino acids and hormones. Cell Physiol Biochem 1:121–130
20. Häussinger D, Roth E, Lang F, Gerok W (1993) Cellular hydration state: an important determinant of protein catabolism in health and disease. Lancet 341:1330–1332
21. Häussinger D, Newsome W, vom Dahl S, Stoll B, Noe B, Schreiber R, Wettstein M, Lang F (1994) Control of liver cell function by the hydration state. Biochem Soc Trans 22:497–502
22. Hörtnagl H, Singer EA, Lenz K, Kleinberger G, Lochs H (1984) Substance P is markedly increased in plasma of patients with hepatic coma. Lancet 1:480–483
23. Hoffmann EK, Simonsen LO (1989) Membrane mechanisms in volume and pH regulation in vertebrate cell. Physiol Rev 69:315–382
24. Horisberger JD, Lemas V, Kraehenbühl JP, Rossier BC (1991) Structure-function relationship of the Na, K-ATPase. Annu Rev Physiol 53:565–584
25. Kinne RKH, Czekay R-P, Grunewald JM, Mooren FC, Kinne-Saffran E (1993) Hypotonicity-evoked release of organic osmolytes from distal renal cells: systems, signals, and sidedness. Renal Physiol Biochem 16:66–78
26. Kimelberg HK, Ransom BR (1986) Physiological aspects of astrocytic swelling. In: Fedoroff S, Vernadakis A (eds) Astrocytes, vol III. Academic, Orlando, pp 129–166
27. Kirk KL, Schafer JA (1992) Water transport and osmoregulation by antidiuretic hormone in terminal nephron segments. In: Seldin DW, Giebisch G (eds) The kidney. Physiology and pathophysiology, 2nd edn. Raven, New York, pp 1693–1726
28. Kleinzeller A, Ziyadeh FN (1990) Cell volume regulation in epithelia – with emphasis on the role of osmolytes and the cytoskeleton. In: Beyenbach KW (ed) Cell volume regulation. Comparative physiology. Karger, Basel, pp 59–86
29. Kopp UC, DiBona GF (1992) The neural control of renal function. In: Seldin DW, Giebisch G (eds) The kidney. Physiology and pathophysiology, 2nd edn. Raven, New York, pp 1157–1204
30. Kurtz A, Scholz H (1993) Cell Volume and stimulus-secretion coupling. In: Lang F, Häussinger D (eds) Advances in comparative and environmental physiology, vol 14. Springer, Berlin Heidelberg New York, pp 119–137
31. Lang F, Deetjen P, Reissigl H (1984) Wasser- und Elektrolythaushalt – Physiologie und Pathophysiologie. Handbuch der Infusionstherapie und klinische Ernährung. Karger, Basel
32. Lang F, Völkl H, Häussinger D (1990) General principles in cell volume regulation. In: Beyenbach KW (ed) Cell volume regulation. Comparative physiology. Karger, Basel, pp 1–25
33. Lang F, Rehwald W (1992a) Potassium channels in renal epithelial transport regulation. Physiol Rev 72:1–32
34. Lang F, Ritter M, Wöll E, Bichler I, Häussinger D, Offner F, Grunicke H (1992b) Ion transport in the regulation of cell proliferation in ras oncogene expressing NIH3T3 fibroblasts. Cell Physiol Biochem 2:213–224
35. Lang F, Ritter M, Völkl H, Häussinger D (1993) Cell volume regulatory mechanisms – an overview. In: Lang F, Häussinger D (eds) Advances in comparative and environmental physiology, vol 14. Springer, Berlin Heidelberg New York, pp 1–31
36. Lang F, Gerok W, Häussinger D (1993b) New clues to the pathophysiology of hepatorenal failure. Clin Invest 71:93–97
37. Lang F, Busch GL, Völkl H, Häussinger D (1994a) Lysosomal pH – a link between cell volume and metabolism. Biochem Soc Trans 22:502–504
38. Lang F, Busch GL, Völkl H, Häussinger D (1994b) Cell volume: a second messenger in regulation of cellular function. News Physiol Sci 10:18–22
39. Laragh JH (1992) The renin system amd the renal regulation of blood pressure. In: Seldin DW, Giebisch G (eds) The kidney. Physiology and pathophysiology, 2nd edn. Raven, New York, pp 1411–1453
40. Lemann J Jr (1992) Internal and external solute balance. In: Seldin DW, Giebisch G (eds) The kidney. Physiology and pathophysiology, 2nd edn. Raven, New York, pp 45–60
41. Lewis RS, Cahalan MD (1990) Ion channels and signal transduction in lymphocytes. Annu Rev Physiol 52:415–430
42. Macknight AD, Grantham J, Leaf A (1992) Physiologic and pathophysiologic responses to changes in extracellular osmolality. In: Seldin DW, Giebisch G (eds) The kidney. Physiology and pathophysiology, 2nd edn. Raven, New York, pp 1779–1806
43. McConnell F, Goldstein L (1990) Volume regulation in elasmobranch red blood cells. In: Beyenbach KW (ed) Cell volume regulation. Comparative physiology. Karger, Basel, pp 114–131
44. Menè P, Dunn MJ (1992) Vascular, glomerular, and tubular effects of angiotensin II, Kinins, and prostaglandins. In: Seldin DW, Giebisch G (eds) The kidney. Physiology and pathophysiology, 2nd edn. Raven, New York, pp 1205–1248
45. Michel CC (1992) Capillary exchange. In: Seldin DW, Giebisch G (eds) The kidney. Physiology and pathophysiology, 2nd edn. Raven, New York, pp 61–92
46. Moncada S, Palmer RMJ, Higgs EA (1991) Nitric oxide: physiology, pathophysiology, and pharmacology. Pharmacol Rev 43:109–142
47. Montrose-Rafizadeh C, Guggino WB (1990) Cell volume regulation in the nephron. Annu Rev Physiol 52:761–772
48. Mortimore GE, Pösö AR (1987) Intracellular protein catabolism and its control during nutrient deprivation and supply. Annu Rev Nutr 7:539–564
49. Palmer BF, Alpern RJ, Seldin DW (1992) Pathophysiology of edema formation. In: Seldin DW, Giebisch G (eds) The kidney. Physiology and pathophysiology, 2nd edn. Raven, New York, pp 2099–2142
50. Ritter M, Steidl M, Lang F (1991) Inhibition of ion conductances by osmotic shrinkage of Madin-Darby canine kidney cells. Am J Physiol 261:C602–C607
51. Robertson GL (1992) Regulation of vasopressin secretion. In: Seldin DW, Giebisch G (eds) The kidney. Physiology and pathophysiology, 2nd edn. Raven, New York, pp 1595–1614
52. Schlatter E (1989) Antidiuretic hormone regulation of electrolyte transport in the distal nephron. Renal Physiol Biochem 12:65–84
53. Schultz SG (1992) Membrane cross-talk in sodium-absorbing epithelial cells. In: Seldin DW, Giebisch G (eds) The kidney. Physiology and pathophysiology, 2nd edn. Raven, New York, pp 287–300
54. Spring KR, Hoffmann EK (1992) Cellular volume control. In: Seldin DW, Giebisch G (eds) The kidney. Physiology and pathophysiology, 2nd edn. Raven, New York, pp 147–170
55. Toto RD, Seldin DW (1992) Salt wastage. In: Seldin DW, Giebisch G (eds) The kidney. Physiology and pathophysiology, 2nd edn. Raven, New York, pp 2143–2164
56. Ussing HH (1990) Volume regulation of frog skin epithelium. In: Beyenbach KW (ed) Cell volume regulation. Comparative physiology. Karger, Basel, pp 87–113

57. Verkman A (1992) Water channels in cell membranes. Annu Rev Physiol 54:97–108

58. Völkl H, Friedrich F, Häussinger D, Lang F (1993a) Acridine Orange fluorescence in renal proximal tubules: effects of NH_3/NH_4^+ and cell volume. Cell Physiol Biochem 3:28–33

59. Völkl H, Friedrich F, Häussinger D, Lang F (1993b) Effect of cell volume on acridine orange fluorescence in hepatocytes. Biochem J 295:11–14

60. Wolff SD, Balaban RS (1990) Regulation of the predominant renal medullary organic solutes in vivo. Annu Rev Physiol 52:727–746

77 Na⁺ Cl⁻ and Water Metabolism

F. Lang

Contents

77.1 Physiological Significance of NaCl

In healthy subjects, Na⁺ and Cl⁻ contribute about 80% of extracellular osmolarity. Thus, the fraction of water distributed in the *extracellular space* is in large part determined by the body's content of NaCl (salt), Most importantly, the *NaCl balance* influences the *plasma volume*, which in turn determines the atrial pressure. Since the filling of the cardiac ventricles is dependent on the atrial pressure (see Chap. 90), cardiac output is influenced by the balance of water and NaCl.

Furthermore, NaCl participates in the maintenance of intracellular space [45]: as illustrated in Sect 77.2.3, the (Na⁺+K⁺)-ATPase [16,33] extrudes Na⁺ in exchange for K⁺, thus establishing a transmembrane K⁺ gradient, which is a prerequisite for the establishment of a potential difference across the cell membrane by passive movement of K⁺ through K⁺ channels (see Chap. 11). This potential difference drives Cl⁻ out of the cell and thus maintains low intracellular Cl⁻ concentrations. Impairment of the (Na⁺+K⁺)-ATPase dissipates the K⁺ and Na⁺ gradients and the potential difference across the cell membrane. The sub-

sequent cellular accumulation of NaCl may eventually lead to cell swelling and cell death (see Chap. 76).

As detailed elsewhere (Chaps. 12, 13), Na⁺ is of paramount importance for the *excitability* of cells [23]. Excitation leads to the activation of Na⁺-permeable channels in a variety of tissues. The entry of Na⁺ depolarizes the cell and thus triggers its specific function, according to the cell type.

In virtually every cell, the steep electrocemical gradient for Na⁺ across the cell membrane is exploited to *drive* several *transport systems* [3,28,40,61,62,69]. Some of these transport systems and their respective driving forces are listed in Table 77.1. For most transport systems the *driving forces are excessive* and the respective transport rates are not critically dependent on the electrochemical Na⁺ gradient across the cell membrane. Notable exceptions are the *Na⁺/Ca²⁺ exchanger* and the $Na^+(HCO_3^-)_3$ cotransporter, which operate close to equilibrium. Due to the coupling ratio, the Na⁺/Ca²⁺ exchanger is highly sensitive to alterations of the Na⁺ gradient: in equilibrium, an increase of intracellular Na⁺ activity by only 26% results in a doubling of intracellular Ca²⁺ activity [7]. The $Na^+(HCO_3^-)_3$ *cotransporter* is less sensitive to the chemical Na⁺ gradient but highly sensitive to alterations of cell membrane potential, since it carries two charges: in equilibrium, depolarization of the cell membrane by 18 mV leads to an intracellular increase in bicarbonate concentration by 60%, and thus to alkalinization by 0.2 pH units [43,44].

It should be mentioned that the function of *Na⁺ channels* [3] and of Na⁺-coupled transport systems is not compromised by alterations of extracellular Na⁺ that are compatible with survival. Rather, the consequences of altered Na⁺ balance are largely dictated by the respective alterations of intra- and extracellular space. An increase in extracellular Na⁺ at the so-called *osmoreceptors* in the hypothalamus promotes thirst and stimulates the release of antidiuretic hormone (ADH), an effect evoked by cell shrinkage due to concomitant increase of extracellular osmolarity [21,65]. Furthermore, excess of NaCl increases extracellular fluid volume, which then triggers the release of *atrial natriuretic factors* [2] and of *ouabain* [7]. The latter hormone inhibits the (Na⁺+K⁺)-ATPase and increases intracellular Na⁺ activity and, via the Na⁺/Ca²⁺ exchanger, intracellular Na²⁺ activity. The increase of intracellular Na²⁺ activity in smooth muscle cells leads to vasoconstriction and an increase in blood pressure [7].

Several carriers transport Cl⁻ across the cell membrane (see Chaps. 8, 59). The *Na⁺ 2Cl⁻ K⁺ cotransporter* and the

R. Greger/U. Windhorst (Eds.)
Comprehensive Human Physiology, Vol. 2
© Springer-Verlag Berlin Heidelberg 1996

Table 77.1. Approximate values for ion activities in intracellular (a_i) and extracellular (a_e) fluid, equilibrium potential (E_o), and cell membrane potential ($E_m = -70$ mV)

	a_i (mmol/l)	a_e (mmol/l)	$E_z F/RT$	E_o (mV)	$E_m - E_o$ (mV)
Na$^+$	12	115	2.3	+60	−130
K$^+$	80	4	−3.0	80	+10
Cl$^-$	15	90	−1.8	−50	−20
HCO$_3^-$	10	20	0.7	−20	−50
Ca^{2+}	0.0001	0.8	9.0	+120	−190
Na$^+$/H$^+$ exchange			3.0		
HCO$_3^-$-Na$^+$ cotransport (3:1)			4.3	−58	−15
Na$^+$/Ca^{2+} exchange (3:1)			−2.2	−60	−10
Na$^+$K$^+$2Cl$^-$ cotransport			2.8	−	−
KCl symport			1.2	−	−
Cl$^-$/HCO$_3^-$ exchange			1.1	−	−

Cl$^-$/HCO$_3^-$ exchanger serve Cl$^-$ transport in a variety of epithelia and allow the regulatory increase of cell volume upon cell shrinkage in both epithelial and nonepithelial cells.

77.2 Regulation of NaCl and Water Metabolism

The main regulator of the *NaCl and water balance* is the kidney [26] Excess intake of water leads to an increase in water excretion (diuresis), while excess intake of Na$^+$ leads to an increase in Na$^+$ excretion (natriuresis). To accomplish its task in the regulation of NaCl and water metabolism, the kidney must be capable of excreting both highly concentrated and highly diluted urine. As a matter of fact, *urine osmolarity* in the human species can vary from 50 to 1400 mosmol/kg H$_2$O [67], and urine volume from less than half to more than 20 l/day [6,67] (see also Chap. 74). The renal excretion of Cl$^-$ is largely determined by the renal excretion of Na$^+$, as long as the excretion of other anions (such as bicarbonate) is low.

77.2.1 Mechanisms Participating in Renal Excretion of Na$^+$, Cl$^-$, and Water

The transport mechanisms of *renal tubular NaCl transport* and excretion are detailed elsewhere in this book (Chaps. 74, 75). In brief, Na$^+$ concentration in filtrate is close to plasma concentration. The opposing effects of the Gibbs-Donnan potential (decreasing filtrate Na$^+$ concentration by some 10%) and of the plasma proteins (limiting Na$^+$ diffusion space in plasma by some 10%) are almost equal. The ultrafiltrate Cl$^-$ concentration, on the other hand, is greater than the plasma concentration, since the effects of the Gibbs-Donnan potential and of plasma proteins is additive for anions.

In the *proximal convoluted tubule,* some 70% of filtered Na$^+$ and some 60% of filtered Cl$^-$ is readsorbed. Some 30% of filtered Na$^+$ is taken up across the apical cell membranes

Table 77.2. Substrates for Na$^+$-coupled transport

Inorganic ions	*Organic acids*
H$^+$ (Na$^+$/H$^+$ exchange)	Monocaroxylic acids
Ca^{2+} (Na$^+$/Ca^{2+} exchange)	Lactate
Phosphate	Pyruvate
Sulfate	Nicotinic acid
HCO$_3^-$ (Na(HCO$_3^-$)$_3$ cotransport)	Picolinic acid
Cl$^-$K$^+$ (Na$^+$K$^+$2Cl$^-$ cotransport)	Pyrazinoic acid
	Acetacetate
Carbohydrates	β-Hydroxybutyrate
D-Glucose	
α-Methyl-D-glucose	Dicarboxylic acids
D-Galactose	Succinate
3-O-Methyl-D-glucose	Malate
	Oxalacetate
Amino acids	Fumarate
L-Phenylalanine	β-Ketoglutarate
L-Histidine	
L-Aminoisobutyrate	Tricarboxylic acids
L-Aminobicyclo-(2,2,1)-	Citrate
heptane-2-carboxylic acid	
L-Lysine	Others
L-Arginine	Biliary acids (e.g.,
	taurocholate)
L-Ornithine	
L-Aspartate	
L-Proline	
Glycine	

of proximal tubule cells via Na$^+$-coupled transport systems for a great variety of substrates (see Table 77.2, Fig. 77.1A). A large portion (some 20%) is taken up by proximal tubule cells via the *Na$^+$/H$^+$ exchanger*. The H$^+$ ions exchanged for Na$^+$ bind to filtered HCO$_3^-$ to form H$_2$CO$_3$, which reacts to CO$_2$ under the catalytic influence of carbonic anhydrase. CO$_2$ then enters the cells following a chemical gradient and within the cell is again utilitzed to form H$^+$ and HCO$_3^-$. The H$^+$ is recycled into the lumen, whereas the HCO$_3^-$ leaves the cell via the basolateral cell membrane, in large part through coupled transport with Na$^+$. The coupling ratio of the *Na$^+$(HCO$_3^-$)$_n$ cotransport* system may vary between 2 and 3. The major portion of Na$^+$ taken up by the cells is extruded by the (Na$^+$+K$^+$)-ATPase. The K$^+$ thus accumulated leaves the cell again via K$^+$ channels. Due to preferential HCO$_3^-$ reabsorption, the luminal Cl$^-$ concentration

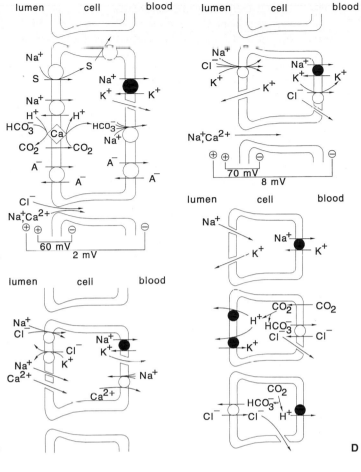

A

B

C

D

Fig. 77.1A–D. Major transport processes in renal tubular cells. **A** Proximal tubule; **B** thick ascending limb of Henle's loop; **C** distal tubule; **D** collecting duct. *Black circle,* (Na⁺+K⁺)-ATPase; *open circles,* carrier systems; *arrows,* diffusion and ion channels

increases above the interstitial Cl⁻ concentration and Cl⁻ diffuses across the paracellular pathway following its chemical gradient. The Cl⁻ diffusion creates a lumen-positive transepithelial potential in the later portions of the proximal tubule, which drives passive Na⁺ reabsorption in that nephron segment.

In the *thick ascending limb* of Henle's loop Na⁺ is reabsorbed mainly through the *Na⁺K⁺2Cl⁻-cotransporter* (Fig. 77.1B). The K⁺ thus accumulated within the cell recycles into the lumen through K⁺ channels, whereas the Cl⁻ leaves the cell via the basolateral cell membrane. Na⁺ taken up is extruded by the (Na⁺+K⁺)-ATPase. The luminal K⁺ recycling and the basolateral Cl⁻ exit create a lumen-positive transepithelial potential, which drives passive reabsorption of Na⁺, Ca²⁺ and Mg²⁺. The thick ascending limb of Henle's loop is impermeable to water. Solute reabsorption in this segment is not paralleled by water reabsorption and thus creates a hypertonic interstitium. As detailed in another chapter of this book (Chap. 74), the increase of interstitial osmolarity in the kidney medulla is necessary for the kidney to be able to excrete a concentrated urine.

In the *early distal tubule*, NaCl reabsorption is in part brought about by an *Na⁺ Cl⁻ cotransport* system (Fig. 77.1C). Cl⁻ may recycle via a KCl cotransport at the apical

cell membrane. The Na⁺ thus accumulated is extruded by an (Na⁺+K⁺)-ATPase.

Na⁺ reabsorption in *collecting ducts* is in large part carried out by apical *Na⁺ channels* in principal cells; the cellular Na⁺ is again extruded by a basolateral (Na⁺+K⁺)-ATPase (Fig. 77.1D).

With the exception of the ascending limb of Henle's loop, all nephron segments are capable of reabsorbing water. Depending on the water permeability of the respective nephron segment, water reabsorption is driven by an osmotic gradient across the epithelium. *Water permeability* in the proximal renal tubule is high, water follows solute transport in that segment and the luminal fluid at the end of the proximal convoluted tubule is close to isotonicity. Excessive water reabsorption and solute entry in the pars recta and descending thin limb of Henle's loop increase the osmolarity of luminal fluid, which is then diluted along the ascending limb of Henle's loop. The distal convoluted tubule and collecting duct require *the antidiuretic hormone* (ADH) to incorporate and activate *water channels*. Thus, water permeability in those segments largely depends on the presence of ADH. The water reabsorption in the final segments of the nephron depends on the osmolarity in the kidney medulla. The concentrative ability of the kidney is thus determined by all the factors pertinent to the build-up

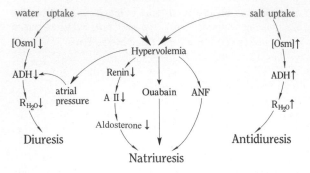

Fig. 77.2. Regulation of renal water and Na$^+$ excretion following water and salt intake. [*Osm*], Blood osmolarity; R_{H_2O}, renal tubular water reabsorption; *ADH*, antidiuretic hormone; *ANF*, atrial natriuretic factor(s); *AII* angiotensin II

and maintenance of medullary hypertonicity, such as activity of NaCl transport in thick ascending limbs, the urea concentration in the kidney medulla, and the perfusion of the vasa recta (cf. Chap. 74).

Between the thick ascending limb and the first part of the distal tubule, specialized renal tubular cells are in close contact with the afferent arteriole of the glomerulus (for details see Chaps. 73, 74). An increase of the NaCl concentration in the lumen of this tubule segment (macula densa) leads to afferent arteriole vasoconstriction and inhibits renin release from the juxtaglomerular cells, specialized smooth vascular cells of the afferent arteriole [72]. Renin is an enzyme which cleaves the decapeptide angiotensin I from the plasma protein angiotensinogen [8,31,49,72] (see also Chap. 73). Angiotensin I is a substrate of the angiotensin converting enzyme, which forms the octapeptide *angiotensin II* by further cleavage of two amino acids. Angiotensin II has several actions aiming at conservation of NaCl and water as well as maintenance of systemic blood pressure: it leads to marked peripheral vasoconstriction, stimulates proximal tubular Na$^+$ reabsorption, and triggers the release from the adrenal cortex of *aldosterone*, a hormone stimulating distal tubular Na$^+$ reabsorption. In addition, angiotensin II elicits thirst and triggers the release of ADH.

77.2.2 Regulation of Renal Natriuresis and Diuresis in Response to Salt and Water Intake

Figure 77.2 illustrates mechanisms linking water and salt intake to urinary excretion. The *intake of isotonic NaCl* solutions leads to an increase in extracellular space. Due to the filtration equilibrium in the peripheral capillaries, part of the excess extracellular fluid enters the plasma space, thus increasing blood volume. The increase in blood volume leads to an increase in *atrial pressure*.

The distension of the atrial wall inhibits, via vagal afferents and nucleus solitarius [59], the release of ADH from the hypothalamus [65]. ADH stimulates renal reabsorption of water by cAMP-mediated insertion of water channels into the luminal cell membrane of the distal nephron [38] (see also Chap. 74). It also stimulates the reabsorption of Na$^+$ by cAMP-dependent activation of Na$^+$ channels in the luminal cell membrane of the distal nephron [71]. Inhibition of ADH release thus leads to increased excretion of Na$^+$ (natriuresis) and, more importantly, of water (*diuresis*). Atrial distension also inhibits the release of adrenocorticotropic hormone (ACTH) [13], which regulates glucocorticoid release from the adrenal gland.

Besides their influence on the release of ADH and ACTH, the vagal afferents from the atria inhibit the *sympathetic tone* [70], resulting in peripheral vasodilation [76]. This decrease in sympathetic tone supports natriuresis [79] and reduces the vasoconstrictive influence of the sympathetic nervous system on the renal preglomerular arterioles. The tendency of renal perfusion pressure to increase is counteracted by the *autoregulatory response*, which triggers intrinsic renal mechanisms that maintain renal blood flow by renal vasoconstriction. However, this increase in renal perfusion pressure is sensed by the specialized smooth muscle cells in the juxtaglomerular apparatus, which release renin following a decrease in perfusion pressure within the autoregulatory range [74]. Increased renal perfusion pressure inhibits renin release. As a result, angiotensin II formation is decreased, leading to:

- Peripheral vasodilation
- Reduced stimulation of proximal tubular Na$^+$ reabsorption
- Decreased release of aldosterone
- Decreased distal tubular Na$^+$ reabsorption
- Inhibition of thirst and ADH release

Thus, the inhibition of renin release following a decrease in renal sympathetic nervous tone favors excretion of both NaCl and water.

The distension of the atrial wall is also a stimulus for the release of the *atrial natriuretic factors* (ANF), a group of peptides that trigger peripheral vasodilation with a decrease in blood pressure, increase in renal blood flow, inhibition of collecting duct Na$^+$ reabsorption, enhanced renal excretion of Na$^+$ and water, and inhibition of renin release [10,17,24,39]. The effects of ANF are mediated by (GC)-A ANF and (GC)-B ANF receptors and cellular formation of cGMP. ANF have a half-life of only minutes in blood and when injected have only short-term effects. Animals overexpressing ANF suffer from arterial hypotension [39].

As detailed elsewhere in this book (Chap. 90), the increased filling of the right cardiac atrium following intake of isotonic NaCl increases the *filling pressure* of the right ventricle, the end-diastolic volume of the right ventricle, the right ventricular stroke volume, pulmonary artery pressure, left atrial pressure, end-diastolic filling volume and stroke volume of the left ventricles as well as systemic arterial pressure. As a result, pressure sensors in pulmonary vessels, left ventricle, aortic arch, and the bifurcation

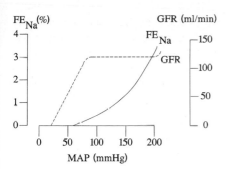

Fig. 77.3. Renal glomerular filtration rate (*GFR*) and fractional excretion of Na⁺ (*FE*~Na~) as a function of mean arterial pressure (*MAP*)

of the carotid artery [50,51,76] are activated. The afferents of these sensors are transmitted via the vagus to the nucleus solitarius and inhibit the *sympathetic nerve tone* [18,51]. The subsequent renal vasodilation inhibits renin release, angiotensin formation, aldosterone, and ADH release, as outlined above.

The increase in extracellular Na⁺ and fluid volume further stimulates the release of *ouabain* from the adrenal gland and possibly some other sources [7]. Other tissues accumulating ouabain include the hypothalamus and the pituitary, the heart, and the kidneys [7]. Ouabain is synthesized from cholesterol, pregnenolone, and progesterone [7]. Ouabain inhibits the (Na⁺+K⁺)-ATPase in various tissues. The subsequent increase of intracellular Na⁺ leads via the Na⁺/Ca²⁺ exchanger to an increase of *cytosolic Ca²⁺ activity*. The latter favors uptake of Ca²⁺ into intracellular stores, which thus release more Ca²⁺ upon stimulation [7]. The inhibition of (Na⁺+K⁺)-ATPase in renal epithelia impairs renal reabsorption of Na⁺ and thus leads to natriuresis. In smooth muscle cells the increase of intracellular Ca²⁺ leads to peripheral vasoconstriction and thus to hypertension. In the adrenal glands ouabain stimulates glucocorticoid and mineralocorticoid secretion, while inhibiting its own secretion. The secretion of ouabain in the adrenal glands is further inhibited by enhanced extracellular K⁺ concentration [7].

The *intake of water alone* leads to an increase in both intra- and extracellular space. The increase in extracellular space triggers the mechanisms described above leading to natriuresis and diuresis. More importantly, the increase in the intracellular space of osmoreceptors in the hypothalamus directly inhibits ADH release and thus triggers a rapid onset of *diuresis*. A change in osmolarity by as little as 1% is sufficient to modify ADH secretion significantly [64].

Following *salt intake*, extracellular volume is increased at the expense of intracellular volume. The shrinkage of intracellular volume leads to a stimulation of ADH release [63,64], favoring antidiuresis. On the other hand, the increase of extracellular volume leads to a distension of the atrial wall, stimulation of ventricular and vascular pressure sensors, and thus to a stimulation of ANF and

ouabain release and inhibition of the renin-angiotensin-aldosterone system. The opposing effects of reduced intracellular volume and increased extracellular volume on ADH release impair the onset of diuresis. With considerable delay, the natriuresis eventually restores the homeostasis of NaCl and water.

Loss of water and/or NaCl reverses the above mechanisms, inhibiting ANF and ADH release and stimulating the renin-angiotensin-aldosterone system, thus leading to renal retention of Na⁺ and water.

In part due to the above mechanisms, renal Na⁺ excretion is a function of mean arterial blood pressure, as illustrated in Fig. 77.3. However, in addition to those mechanisms, a variety of hormones influence renal excretion of Na⁺ and water (see Table 77.3 and Chap. 75), and several of these hormones have at times been suggested as participating in the regulation of water and NaCl homeostasis. Furthermore, some evidence points to intrarenal mechanisms participating in the stimulation of natriuresis by extracellular volume expansion.

77.2.3 Intrarenal Mechanisms That May Participate in Regulation of NaCl and Volume Balance

The *natriuresis* following *volume expansion* is abolished if prior to volume expansion the renal artery is clamped to prevent increase of renal perfusion pressure [52]. Thus, some renal factor sensitive to perfusion pressure must participate in the inhibition of Na⁺ reabsorption. It has been argued that this factor is related to interstitial pressure, which increases following extracellular volume expansion, and which could impair renal tubular Na⁺ and water reabsorption [26]. The increase of interstitial pressure following volume expansion is indeed prevented in clamped kidneys [26]. If the renal artery is clamped after initiation of volume expansion [35], the *interstitial pressure* remains elevated and the natriuresis is only partially blunted.

Some evidence also points to the existence of an as yet *elusive factor* in the proximal renal tubular fluid of volume-expanded animals, which inhibits Na⁺ and fluid reabsorption in proximal renal tubules [30]. Proximal tubular fluid contains *dopamine* and *adenosine*, which modify the response of the macula densa to increased delivery of Na⁺ [30]. Furthermore, adenosine inhibits renin release and leads to a transient decrease in renal blood flow and a sustained decrease in glomerular filtration rate [57,58]. Dopamine release from proximal renal tubules is increased in extracellular fluid expansion [5] and decreased in extracellular fluid contraction [9]. The effects of dopamine include inhibition of proximal tubular Na⁺/H⁺ exchange [20] and (Na⁺+K⁺)-ATPase [1], and mineralocorticoid antagonism in cortical collecting ducts [55]. As a result, dopamine leads to natriuresis and thus could well serve to adjust renal Na⁺ excretion to the state of extracellular volume [26].

Table 77.3. Factors modifying renal Na$^+$ handling

Hormone	Effect
Mineralocorticoids	+ DL: Na$^+$/H$^+$ exchanger + DT; CD: Na$^+$ channel, (Na$^+$+K$^+$)-ATPase, energy supply distal nephron
Glucocorticoids	+ GFR; PT: Na$^+$/H$^+$ exchanger, TAL; DT; CD: (Na$^+$+K$^+$)-ATPase − PT: Na$^+$, HPO$_4^{2-}$ cotransport
Progesterone	− Mineralocorticoid action
Thyroid hormones	+ GFR; PT: K$^+$ channels, (Na$^+$+K$^+$)-ATPase, Na$^+$; HPO$_4^{2-}$ cotransport
Antidiuretic hormone	+ TAL: Cl$^-$ channels → Na$^+$, 2Cl$^-$, K$^+$ cotransport DT; CD: Na$^+$ channels
Atrial natriuretic factors	+ GFR − PT: Na$^+$, HPO$_4^{2-}$ cotransport − CD: Na$^+$ reabsorption
Ouabain	− (Na$^+$+K$^+$)-ATPase
Parathyroid hormone	− PT: Na$^+$, HPO$_4^{2-}$ cotransport, HCO$_3^-$ transport + PT: Na$^+$/Ca^{2+} exchanger,
Calcitonin	− PT: Na$^+$, HPO$_4^{2-}$ cotransport + TAL; DT: Reabsorption of NaCl, Ca^{2+}, Mg^{2+}
Growth hormone	+ PT: Na$^+$-coupled transport processes
Insulin	+ PT: Na$^+$, HPO$_4^{2-}$ cotransport + DT: Na$^+$ reabsorption, K$^+$ secretion − PT: Reabsorption of Na$^+$ and Ca^{2+}
Glucagon	+ GFR + TAL; DT: Reabsorption of Na$^+$, Ca^{2+}, Mg^{2+}
Angiotensin (pM) (nM)	+ PT: Na$^+$ reabsorption − PT: Na$^+$ reabsorption
PGE$_2$	− TAL; CD: Na$^+$ reabsorption
Bradykinin	− CD: Na$^+$ reabsorption
Catecholamines (α_2)	+ PT: Na$^+$ reabsorption − CD: Na$^+$ reabsorption
Catecholamines (β)	+ TAL; CNT; CD: NaCl reabsorption
Dopamine	+ GFR − PT: Na$^+$, HPO$_4^{2-}$ reabsorption
Substance P	− PT: Na$^+$ reabsorption
Histamine	+ GFR − Na$^+$ reabsorption (?)

GFR, Glomerular filtration rate; PT, proximal tubule; TAL, thick ascending limb of Henlé's loop; DL, diluting segment (corresponds to TAL); DT, distal tubule; CNT, connecting tubule; CD, collecting duct; +, stimulation; −, inhibition
From [28,30,31,38,43,56,71]

The *arachidonic acid* metabolites *PGI* and *PGE$_2$* are similarly formed in the kidney and influence renal tubular Na$^+$ reabsorption. Their formation is stimulated by ADH [84] and vasoconstrictors such as angiotensin II and catecholamines [26]. PGE$_2$ antagonizes the stimulatory action of ADH on distal nephron Na$^+$ reabsorption and is capable of reversing renal vasocontriction [53]. Accordingly, PGE$_2$ is natriuretic. On the other hand, PGE$_2$ stimulates renin release [26], which indirectly favors renal Na$^+$ retention (see above). The prostaglandins may serve to protect the renal tubules from overstimulation [27] rather than to regulate extracellular fluid volume. Similarly, kinins such as bradykinin formed in the kidney serve to antagonize the actions of ADH and angiotensin II [26].

Another renal factor possibly adjusting natriuresis and diuresis to extracellular fluid volume is *medullary blood flow*. The autoregulatory range of medullary blood flow is small compared to that of cortical renal blood flow [11]. Accordingly, an increase in renal perfusion pressure increases the blood flow through the vasa recta and favors washout of the high medullary osmolarity, leading to diuresis [19]. In addition, the enhanced perfusion of kidney medulla increases O$_2$ delivery to interstitial cells, which has been speculated as stimulating the formation of natriuretic PGE$_2$ [66].

77.2.4 Other Mechanisms That May Participate in Renal Regulation of NaCl and Volume Balance

Several observations suggest that the *liver* may play a part in the regulation of renal function. On the one hand, a *hepatorenal reflex* has been demonstrated which decreases renal blood flow, NaCl, and water excretion [46,47]. On the other hand, the liver is thought to release a hormone increasing renal blood flow and glomerular filtration [81]. It was thought that *osmoreceptors* in the liver and/or portal circulation anticipate the volume and osmolarity of fluid absorbed from the intestine and trigger the appropriate renal response prior to changes of systemic blood volume and osmolarity [12,29,46,47]. So far the hypothetical humoral factor has escaped identification, and the physiological role of the hypothetical hormone or the reflex still awaits clarification.

Increases in the *Na+ concentration* in the *carotid artery* [25] and *cerebrospinal fluid* [25] have been reported to stimulate natriuresis, and chemoreceptors in the kidney [77] have been postulated to activate afferent renal nerves [68], which modify the central nervous regulation of renal function.

Until additional factors or hormones have been unequivocally demonstrated, their contribution to NaCl and water homeostasis must remain a matter of speculation, even though it is quite clear that additional factors are needed to fully explain the complex machinery that adjusts the renal excretion of water and NaCl to the homeostatic requirements.

77.2.5 Extrarenal Factors Influencing NaCl and Volume Balance

Although the kidney is usually by far the most important organ regulating the NaCl and water balance, loss of water and salt through *sweat glands* and *intestine* may well exceed renal loss. Moreover, the excretion of salt and water through intestine and sweat glands is slightly influenced by salt and water balance of the body.

The production of sweat (see Chap. 112) may exceed 10 l/day. Primary sweat is formed in the acini of the sweat glands by cellular uptake via the $Na^+K^+2Cl^-$-cotransport. The Cl^- leaves the cell via channels in the apical cell membrane, the Na^+ via the (Na^++K^+)-ATPase on the basolateral cell membrane, and the K^+ via K^+ channels in the basolateral cell membrane (see Chap. 112). The asymmetric movement of K^+ and Cl^- creates a lumen-negative cell membrane potential, which drives Na^+ into the lumen. Water follows, driven by the osmotic gradient. The primary sweat is isotonic. Sweat secretion is stimulated by acetylcholine. Along the sweat ducts Na^+ is reabsorbed via Na^+ channels in the apical and the (Na^++K^+)-ATPase on the basolateral cell membrane. Cl^- follows through Cl^- channels, driven by the lumen-negative potential difference (see Chap. 112). Water is retained within the lumen be-

cause of the low water permeability of the duct, and the sweat osmolarity is reduced. *Aldosterone* stimulates Na^+ reabsorption and thus reduces the NaCl concentration in sweat. As a matter of fact, the Na^+ concentration in sweat may vary between 5 and 80 mmol/l. In cystic fibrosis, NaCl reabsorption in the duct cells is impaired and the NaCl concentration in sweat is usually high (70 mmol/l).

As detailed elsewhere in this book (Chap. 62), *intestinal secretion* of water approaches some 8 l/day (see Fig. 76.7). The composition of gastrointestinal secretions is listed in Table 76.4. Virtually all of this fluid is usually reabsorbed, only some 200 ml being excreted. The Na^+ concentration in the feces is only some 30 mmol/l, and thus the loss of water and Na^+ with the feces is usually immaterial to NaCl and water homeostasis. The reabsorptive capacity of the intestine is great and it usually copes with increased intake or secretion of fluid. The cellular mechanisms of intestinal NaCl secretion are similar to those in the acini of the sweat gland, while the cellular mechanisms of Na^+ reabsorption in the final segments of the colon resemble those of the collecting duct (see Chaps. 62, 74, 112). Here, again aldosterone stimulates the Na^+ reabsorption via Na^+ channels in the luminal cell membrane and (Na^++K^+)-ATPase in the basolateral cell membrane.

Life-threatening intestinal loss of NaCl and fluid may result from excessive secretion or impaired reabsorption. Secretion may be excessive during massive activation of Cl^- channels by cAMP, as occurs, e.g., in vasoactive-intestinal-polypeptide-producing tumors or in *cholera*. Absorption may be impaired by defective transport systems or by the presence of nonreabsorbable osmotically active substances in the lumen. Usually the distension of the intestinal wall due to osmotic accumulation of water stimulates motility and leads to diarrhea. However, if the condition leads to decreased motility, large volumes of extracellular fluid could be accumulated in the intestinal lumen without diarrhea. Since this volume is formed at the expense of plasma volume, life-threatening *hypovolemia* may develop without external disturbance of salt and water balance.

Even without sweat production, water is lost by diffusion through the skin (*perspiratio insensibilis*) and through the respiratory tract. The expired air is saturated with water and thus, depending on the water saturation of inhaled air and on ventilation, water is lost. At 37°C air is water-saturated at 47 mmHg (6 KPa). If the inhaled air is completely dry and the ventilation is 6 l/min, for instance, some 370 mml/min are lost through respiration. It must be remembered that the water saturation is dependent on temperature, and water is lost even when water-saturated cold air is inhaled.

Obviously, water and electrolyte balance do not only depend on loss, but also on intake. *Water intake* (see Chap. 82) is usually governed by behavior, e.g., drinks are usually consumed during meals irrespective of any need for fluid intake. *Behavioral drinking* usually exceeds the need for minimal fluid intake, and only when behavioral drinking fails to balance fluid loss do thirst mechanisms come into play [21]. As detailed elsewhere in this book (Chap. 82), thirst is produced by extracellular hyperosmolarity, which

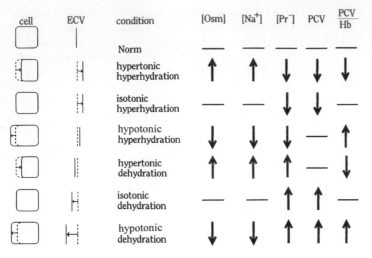

Fig. 77.4. Disorders of water and NaCl homeostasis. *ECV*, Extracellular volume; *[Osm]*, blood osmolarity; *[Na+]*, extracellular Na+ concentration; *[Pr-]*, plasma protein concentration; *PCV*, packed cell volume (hematocrit); *[Hb]*, blood hemoglobin concentration

is registered by hypothalamic osmoreceptors [21]. An increase of 1–2% is sufficient to elicit thirst [83]. A further stimulus to thirst is hypovolemia, which leads to decreased activation of *stretch receptors* in the low-pressure and high-pressure vascular walls, and, most importantly, in the left atrium [22]. About a 10% decrease in central venous or arterial pressures is required to elicit thirst [21]. Thirst is further triggered by *angiotenin II* [21], which activates neurons in *circumventricular organs* at the lamina terminalis, where the blood-brain barrier is permeable [21]. Thirst is stimulated in *pregnancy.*

Salt intake (see Chap. 82), like water intake, is usually in excess of the required minimum. However, in Na+-depleted states an appetite for salt may foster dietary compensation of the Na+ deficit [21]. Hypovolemia not only leads to rapid onset of thirst, but also, with some delay, triggers *salt appetite.* Salt intake is further stimulated by several hormones, such as *mineralocorticoids, angiotensin II, estrogens, prolactin, adrenocorticotropic hormone,* and, possibly, glucocorticoids [21]. The influence of hormones accounts for the increased salt appetite during pregancy and lactation [15].

In cattle and sheep, but probably not in man, the Na+ concentration in cerebrospinal fluid seems to be registered and governs Na+ intake [82].

77.3. Disorders of NaCl and Water Metabolism

Disorders of NaCl and water homeostasis are listed in Fig. 77.4. *Dehydration* is a lack of body water with or without simultaneous lack of NaCl, *hyperhydration* an excess of water with or without excess of NaCl. The balance of NaCl or of other osmotically active substances results in a classification of *hypotonic, isotonic,* or *hypertonic de-* or *hyperhydration.* It should be stressed that the same causes

can lead to isotonic or anisotonic derangements of NaCl and water balance, so the classification described below does not reflect etiological entities.

77.3.1 Isotonic Dehydration

Loss of isotonic fluid results in an equivalent lack of NaCl and water without change of intra- or extracellular osmolarity (*isotonic dehydration*). The condition is characterized by an exclusive decrease in extracellular fluid space without alteration in intracellular volume.

Possible causes of isotonic dehydration include *fluid loss* by *vomiting, diarrhea, fistulas, sweat,* and *burns,* which allow exsudation of extracellular fluid through the destroyed skin. *Renal* fluid and salt *loss* may result from lack of aldosterone (hypoaldosteronism), inhibition of water reabsorption by alcohol, loss of isotonic urine due to defective renal reabsorption (*salt-losing kidney, Fanconi syndrome*), diuretic treatment, or osmotic diuresis, e.g., in diabetes mellitus, when the filtered load of glucose exceeds the maximum transport rate [42]. In poorly controlled *diabetes*, the loss of water and Na+ is compounded by renal excretion of negatively charged ketones [6]. The nonreabsorbed anions create a lumen-negative transepitelial potential difference, which impairs the reabsorption of Na+ and subsequently of water. Ketonuria may also lead to renal salt and water loss during starvation [14]. Like ketonuria, bicarbonaturia, e.g., during metabolic alkalosis or impaired proximal tubular bicarbonate reabsorption, leads to loss of salt and water [41,73].

77.3.2 Hypotonic Dehydration

Loss of more NaCl than water leads to *hypotonic dehydration* [80]. In this condition, extracellular volume is mark-

1564

Table 77.4. Causes of diabetes insipidus [34]

Central
Congenital central diabetes insipidus
Idiopathic (nonfamilial) central diabetes insipidus
Lesions of hypothalamus and hypophysis
    Traumatic lesions
    Vascular lesions
    Tumors
    Infections
    Granulomatous diseases
    Degenerative neuronal disorders
Inhibitors of ADH release
    Ethanol
    Opiate antagonists
    α-Adrenergic agonists
    Diphenylhydantoin
    CO poisoning

Nephrogenic
Congenital nephrogenic diabetes insipidus (X-linked)
Inhibition of renal ADH action
    Lithium ions
Inhibition of thick ascending limb NaCl reabsorption
    Loop diuretics
    Hypokalemia
    Hypercalcemia
Osmotic diuresis
    Mannitol
Further impairment of renal tubular transport
    Salt-losing kidney (post-transplant, postobstructive,
        postfailure)
    Fanconi syndrome
    Renal tubular acidosis
Decreased urea recycling
    Protein-depleted diet
Medullary washout
    Paroxysmal hypertension
    Interstitial nephritis
Renal disease secondary to systemic diseases
    Sickle cell anemia
    Amyloidosis
    Sjögren syndrome
    Sarcoidosis
    Gentamycin
    Amphotericin B
    Cisplatin
    Colchicine
    Vinblastine
    Methoxyfluorane
    Acetohexamide
    Methicillin
    Isophosphamide, cyclophosphamide
    Propoxyphene
    Angiographic dyes

to renal loss of K^+ and to hypokalemia, which in turn inhibits aldosterone release (see Chap. 79). As a result, the aldosterone release may be inappropriately low for the extent of volume depletion. *Hypoaldosteronism* generally leads to hypotonic dehydration.

77.3.3 Hypertonic Dehydration

Loss of more water than NaCl leads to *hypertonic dehydration* and hypernatremia. In this condition the intracellular volume is markedly decreased, whereas extracellular volume is less compromised. Possible causes of hypertonic dehydration include excessive *loss of hypotonic fluid* through *sweat* loss (e.g., in fever) or *hyperventilation*.

Several other conditions may also lead to renal water loss in excess of Na^+. A lack of ADH may result from a genetic defect or from lesions of the hypothalamus and hypophysis (see Table 77.4), or may occur without obvious reason (nonfamilial idiopathic central *diabetes insipidus*). Even with intact secretion of ADH, water diuresis may occur if the kidney does not respond to the hormone. *ADH resistance* of the kidney may be congenital, due to an X-linked defect of the V_2 receptor, or may result from an acquired kidney disease. ADH resistance of the kidney may originate from interference with ADH-induced cAMP production and action in the kidney (e.g., lithium ions) or may come from lack of medullary hypertonicity, i.e., a lack of driving force for water reabsorption in the terminal segments of the nephron. The latter could result from impairment of NaCl reabsorption in the thick ascending limb, from lack of urea recycling, or from medullary washout due to increased medullary blood flow (see Table 77.4).

Extrarenal water loss with relatively little loss of Na^+ may also result from some *diarrhetic states*. Furthermore, hypertonic dehydration may be due to reduced water intake (hypodipsia), which may occur without any apparent reason (idiopathic or essential hypernatriemia and hypodipsia) or may be due to *lesions of the hypothalamus* (tumors, head trauma, etc.). Hypertonic dehydration is relatively frequent in patients with *hepatic failure*; the decreased water intake in these patients, due to hepatic encephalopathy, is paralleled by increased water loss secondary to hyperventilation, fever, and diarrhea [34].

77.3.4 Isotonic Hyperhydration

Excess of equivalent amounts of water and NaCl leads to isotonic hyperhydration, with exclusive *expansion of the extracellular volume*.

Possible causes of isotonic hyperhydration are excessive *infusion of isotonic NaCl* solutions, especially if renal excretion is impaired (renal insufficiency). An *excess of mineralocorticoids* and glucocorticoids also results in isotonic hyperhydration.

The increase in interstitial fluid volume in generalized edema, as occurs in nephrotic syndrome, congestive heart

edly decreased, whereas intracellular volume is increased at the expense of extracellular volume.

Most causes of hypotonic dehydration are the same as for isotonic dehydration. Hypotonic dehydration is favored by replacement of isotonic fluid losses with water, e.g., by *drinking water following excessive sweating*. Furthermore, the decrease of atrial pressure following isotonic dehydration stimulates *ADH release*. As a result, the ADH concentration and therefore renal water reabsorption is higher than would be required for adjustment of osmolarity. The stimulation of distal tubular Na^+ reabsorption by aldosterone during volume depletion may eventually lead

failure, and liver cirrhosis, results in isotonic hyperhydration. The positive NaCl and water balance in those conditions is the result of renal NaCl and water conservation.

In *nephrotic syndrome,* the increased capillary permeability to proteins allows filtration of proteins into the glomerular space and the interstitial space of peripheral tissues, leading to renal loss of proteins, a reduction in plasma protein concentration, and an increase in interstitial protein concentration [60]. The decrease of the oncotic gradient across the capillary wall augments the filtration of plasma fluid into the interstitial space (see Chaps. 73, 76). The decline of plasma volume decreases the atrial pressure, compromises cardiac output, tends to increase systemic arterial pressure, and thus increases sympathetic tone (Sect. 77.2.2). The subsequent release of ADH, renin, angiotensin II, and aldosterone stimulate renal water and NaCl conservation. The fluid thus retained again mainly distributes into the interstitial space, which eventually accumulates excessive volume. The plasma volume, on the other hand, typically remains slightly decreased, sustaining the stimulation of renal Na⁺ and water retention.

In *congestive heart failure,* it is primarily the compromised cardiac output that increases sympathetic tone. This again leads to renal vasoconstriction and renin release, with subsequent release of angiotensin II, ADH, and aldosterone. Thus, the kidney preserves NaCl and water. On the other hand, the excessive activation of the sympathetic tone leads to venous vasoconstriction, which increases the atrial pressure, favoring release of ANF and inhibition of ADH release. In addition, the increase of the venous tone increases capillary hydrostatic pressure and, thus, the filtration of fluid into the interstitial space [60]. Furthermore, the drainage of the lymphatic fluid into the systemic circulation is impaired by the increase of venous pressure as detailed elsewhere in this book (Chap. 94). Plasma volume may or may not be increased, depending on the relative contribution of renal NaCl and water conservation and of peripheral filtration.

In *hepatic cirrhosis,* reduced hepatic synthesis of plasma proteins reduces plasma oncotic pressure and thus favors peripheral filtration. Furthermore, the capillary hydrostatic pressure in the portal circulation is grossly increased due to raised vascular resistance within the diseased liver, stimulating filtration of fluid into the abdomen (ascites). Thus, again interstitial space is increased at the expense of plasma volume. The decrease in atrial pressure stimulates ADH release, inhibits ANF release, and compromises cardiac output. On the other hand, vasodilators, such as substance P from the intestinal wall and endotoxins from organisms of the intestinal lumen, bypass the liver and escape inactivation by the liver cells [48]. The endotoxins stimulate the expression of NO synthase [54], and NO together with the intestinal vasodilators produces peripheral vasodilation. The tendency of systemic blood pressure to decrease stimulates the sympathetic tone, leading to renal vasoconstriction and renin, angiotensin II, and aldosterone release. The renal vasoconstriction is compounded by leukotrienes, which again escape hepatic inactivation. As a result, antinatriuresis and antidiuresis prevail, and the renal conservation of NaCl and water fuel the formation of ascites and peripheral edema.

77.3.5 Hypotonic Hyperhydration

Predominant *excess of water* leads to hypotonic hyperhydration, which is characterized by increased intracellular volume with a variable increase in extracellular volume [6].

Possible causes of hypotonic hyperhydration include excessive drinking of water [21], which may be encountered in 6–17% of chronically ill psychiatric patients [36,37]. On the other hand, renal water elimination may be impaired, e.g., by *excess of ADH (inappropriate ADH secretion).* The excess ADH may be due to lesions of the hypothalamus or produced by dedifferentiated tumor cells, such as in bronchocarcinoma [6]. A variety of other lung diseases may also be associated with excess ADH [6]. Impaired renal water elimination may also result from glucocorticoid or thyroid deficiency. In both conditions the renal effects may be compounded by the hypotension, which leads to ADH release in excess of what is required for isoosmolarity. Several drugs have been shown to impair renal water excretion [6]. Infusion of isotonic glucose solutions may lead to hypotonic hyperhydration, since after metabolism of the glucose solute free water remains which may not be excreted appropriately if kidney function is inadequate.

Causes of isotonic hyperhydration may also produce hypotonic hyperhydration: in hepatic failure, for instance, the reduced atrial pressure and/or arterial hypotension stimulates ADH release in excess of what is required for osmotic equilibrium, leading to hypotonic extracellular fluid [6]. In cardiac failure, the increase of atrial pressure tends rather to inhibit [13] and the relative arterial hypotension rather to stimulate ADH release [4]. The atrial receptors appear to adapt, however [4], and the net result is more frequently increased ADH [78].

77.3.6 Hypertonic Hyperhydration

Predominant *excess of NaCl* leads to hypertonic hyperhydration and hypernatremia, with marked increase of extracellular volume, partly at the expense of intracellular volume [34].

Possible causes of hypertonic hyperhydration include excessive intake of NaCl or NaHCO₃ by infusion and parenteral nutrition, if urinary excretion fails to keep up with the intake. Similarly, oral administration of hypertonic NaCl to provoke vomiting, *hemodialysis* against a high Na⁺ concentration dialysate, or peritoneal dialysis may result in hypertonic hyperhydration. Excess of aldosterone and glucocorticoids may lead to mild hypertonic hyperhydration, which does not, however, usually create clinically significant hypernatremia. *Drinking of sea water* always leads to hypertonic hyperhydration, since the NaCl concentration in sea water far exceeds the maximum concentration of urine.

77.3.7 Sequelae of Altered Intra- and Extracellular Volume

As described above, disturbances of NaCl and water balance may affect the constancy of either intra- or extracellular volume or both. The complications of these disorders are thus mainly determined by the alterations in the fluid compartments.

The most serious hazard of increased intracellular volume is the formation of *cerebral edema*. The intracranial space is surrounded by a nondistensible bony shell. Some two-thirds of the intracranial space is intracellular. Swelling of intracranial cells will occur initially at the expense of the cerebrospinal fluid, which can escape from the intracranial space through the foramen magnum. However, the cerebrospinal fluid comprises only some 15% of the intracranial space, so major cell swelling ultimately leads to an increase in intracranial pressure, which compromises intracranial blood flow. This may lead to the initiation of a vicious circle, since the reduction in cerebral blood flow reduces the energy supply, impairs cellular $(Na^+ + K^+)$-ATPase, and thus leads to further cell swelling. The clinical symptoms of cerebral edema are nausea, vomiting, headache, bradycardia, confusion and coma. The increased pressure gradient across the foramen magnum may lead to herniation of the cerebellum through the foramen magnum, which causes the immediate death of the patient.

A *decrease in intracellular volume* similarly affects primarily the neuronal cells. *Neurological sequelae* of hypernatremia are diverse, including restlessness, irritability, lethargy, muscular twitching, hyperreflexia, spasticity, coma, and seizures. A decrease in brain volume can impose mechanical stress on cerebral vessels, leading to rupture and bleeding. The mortality from hypernatremia beyond 160 mmol/l approaches 50%–70%. After correction of hypernatremia, neurological sequelae may remain. If hypernatremia persists for several days, the cerebral cells produce and accumulate organic osmolytes (see Chap. 76), such as amino acids (taurine, glutamine, glutamate, aspartate, and glycine), polyols (myoinositol), and amines (glycerophosphorylcholine). These osmolytes allow the cells to retain their volume without accumulation of electrolytes. Following correction of long-standing hypernatremia, the osmolytes are in excess and have to be disposed by the cells. Too rapid correction of hypernatremia may cause cell swelling and thus cerebral edema.

An increase in extracellular fluid volume is associated with the formation of *edema*. The most serious complication is the development of *lung edema*, which leads to life-threatening impairment of pulmonary gas exchange (see Chap. 101).

A decrease in extracellular fluid volume leads to *hypovolemia* and thus may elicit hypotension and *cardiovascular collapse*. To the extent that the regulatory mechanisms of NaCl and water balance are active, a decrease in extracellular fluid volume is paralleled by increased concentrations of ADH, angiotensin II, and aldosterone, and by renal retention of NaCl and water. The hyperaldosteronism may lead to hypokalemia, since aldosterone stimulates distal tubular Na^+ reabsorption in exchange for K^+ secretion (see Chap. 79). The stimulation of proximal tubular Na^+/H^+ exchange stimulates not only reabsorption of Na^+, but also of filtered HCO_3^-. Thus, in extracellular volume contraction, the kidney is unable to excrete significant amounts of HCO_3^- and to compensate an alkalosis (*volume contraction alkalosis* see Chap. 78). The enhanced renal retention of water results in oliguria, which may favor the precipitation of urinary constituents (urolithiasis).

77.3.8 Parameters Utilized to Define Disorders of NaCl and Water Homeostasis

The extracellular osmolarity, which is considered to be virtually identical to intracellular osmolarity, is estimated by determination of *plasma osmolarity*. Since usually the extracellular osmolarity is mainly determined by Na^+ plus the respective anions, alterations of osmolarity are frequently reflected by alterations in *plasma Na^+ concentration*. However, it must be remembered that the Na^+ concentration in plasma water matters. At high protein concentrations or in hyperlipidemia the Na^+ concentration per unit volume plasma may be spuriously low, while the Na^+ concentration in plasma water may be normal [6]. Moreover, the plasma osmolarity is not solely due to NaCl, but may be modified by high concentrations of organic substances such as *glucose* (in diabetes mellitus) or *urea* (in renal insufficiency). While the high glucose concentration in diabetes mellitus is restricted largely to extracellular fluid, the urea concentration of patients with renal insufficiency is high in both extracellular and intracellular fluid [6]. Thus, at the same excess extracellular osmolarity, cells may be shrunken in diabetic patients but not in patients with renal insufficiency. On the other hand, urea may shrink cells by stimulating KCl loss [32].

Intracellular osmolarity is dependent on the K^+ content of the cells. Thus, hypertonicity of extracellular (and intracellular) fluid does not necessarily reflect cell shrinkage if cell K^+ concentration is high, and an extracellular (and intracellular) hypotonicity does not necessarily indicate an increased intracellular volume if at the same time the cellular K^+ concentration is low.

The protein concentration in plasma is reduced following an acute increase in plasma volume and increased following a reduction in plasma volume. Thus, the *plasma protein concentration* can be utilized as an indicator of extracellular fluid volume. Obviously, any condition altering the formation of plasma proteins or the filtration of plasma proteins across the peripheral capillaries will result in alterations of plasma protein concentrations mimicking changes in plasma volumes.

A convenient indicator of cell volume is *erythrocyte cell volume* and *hematocrit*, which are modified by anisotonic de- and hyperhydrations. However, conditions impairing erythrocyte formation and hemoglobin synthesis may similarly affect erythrocyte volume. A more reliable indicator of cell volume is the quotient of hematocrit over the hemoglobin concentration.

References

1. Aperia A, Bertorello A, Seri I (1987) Dopamine causes inhibition of Na⁺, K⁺-ATPase activity in proximal tubular segments. Am J Physiol 252:F32–F45
2. Ballermann BJ, Zeidel ML (1992) Atrial natriuretic hormone. In: Seldin DW, Giebisch G (eds) The kidney. Physiology and pathophysiology, 2nd edn. Raven, New York, pp 1843–1884
3. Benos DJ, Sorscher EJ (1992) Transport proteins: ion channels. In: Seldin DW, Giebisch G (eds) The kidney. Physiology and pathophysiology, 2nd edn. Raven, New York, pp 587–624
4. Berns AS, Schrier RW (1981) The kidney in heart failure. In: Suki WN, Eknoyan G (eds) The kidney in systemic disease, 2nd edn. Wiley, New York, pp 569–593
5. Bertorello A, Hökfelt T, Goldstein M, Aperia A (1988) Proximal tubule Na⁺, K⁺-ATPase activity is inhibited during high-salt diet: evidence for DA-mediated effect. Am J Physiol 254:F795–F801
6. Bichet DG, Kluge R, Howard RL, Schrier RW (1992) Hyponatremic states. In: Seldin DW, Giebisch G (eds) The kidney. Physiology and pathophysiology, 2nd edn. Raven, New York, pp 1727–1752
7. Blaustein MP (1993) Physiological effects of endogenous ouabain: control of intracellular Ca²⁺ stores and cell responsiveness. Am J Physiol 264:C1367–C1387
8. Briggs JP, Schnermann J (1986) Macula densa control of renin secretion and glomerular vascular tone: evidence for common cellular mechanisms. Renal Physiol 9:193–203
9. Carey RM, Van Loon GR, Baines AD, Ortt EM (1981) Decreased plasma and urinary dopamine during dietary sodium depletion in man. J Clin Endocrinol Metab 52:903–909
10. Cogan MG (1990) Renal effects of atrial natriuretic factor. Annu Rev Physiol 52:699–708
11. Cohen HJ, Marsh DJ, Kayser B (1983) Autoregulation in vasa recta of the rat kidney. Am J Physiol 245:F32–F40
12. Daly JJ, Roe RW, Horrocks PA (1967) Comparison of sodium excretion following the infusion of saline into systemic and portal veins in the dog: evidence for a hepatic role in the control of sodium excretion. Clin Sci 33:481–487
13. de Torrente A, Robertson G, McDonald KM, Schrier RW (1975) Mechanism of diuretic response to increased left atrial pressure in the anethetized dog. Kidney Int 8:355–361
14. DeFronzo RA (1981) The effect of insulin on renal sodium metabolism. A review with clinical implications. Diabetologia 21:165–171
15. Denton D (1982) The hunger for salt. An anthropological, physiological and medical analysis. Springer, Berlin Heidelberg, New York
16. DeWeer P (1992) Cellular sodium-potassium transport. In: Seldin DW, Giebisch G (eds) The kidney. Physiology and pathophysiology, 2nd edn. Raven, New York, pp 93–112
17. Edwards BS, Zimmerman RS, Schwab TR, Heublein DM, Burnett JC Jr (1988) Atrial stretch, not pressure, is the principal determinant controlling the acute release of atrial natriuretic factor. Circ Res 62:191–195
18. Epstein M (1976) Cardiovascular and renal effects of head-out water immersion in man. Circ Res 39:619
19. Fadem SZ, Hernandez-Llamas G, Patak RV, Rosenblatt SG, Lifschitz MD, Stein JH (1982) Studies on the mechanism of

20. Felder CC, Campbell T, Jose PA (1989) Role of cAMP on dopamine-1 receptor regulated Na⁺-H⁺ antiport in renal tubular brush border membrane vesicles. Kidney Int 35:172
21. Fitzsimons JT (1992) Physiology and pathophysiology of thirst and sodium appetite. In: Seldin DW, Giebisch G (eds) The kidney. Physiology and pathophysiology, 2nd edn. Raven, New York, pp 1615–1648
22. Folkow B (1979) Relevance of cardiovascular reflexes. In: Hainsworth R, Kidd C, Linden RJ (eds) Cardiac receptors. Cambridge University Press, Cambridge, pp 473–505
23. Fozzard HA, Shorofsky SR (1992) Excitable membranes. In: Seldin DW, Giebisch G (eds) The kidney. Physiology and pathophysiology, 2nd edn. Raven, New York, pp 407–446
24. Genets J, Cantin M (1988) The atrial natriuretic factor: its physiology and biochemistry. Rev Physiol Biochem Pharmacol 110:1–145
25. Gilmore JP, Nemeth MN (1984) Salt depletion inhibits cerebral-induced natriuresis in the dog. Am J Physiol 247:F725–F728
26. Gonzalez-Campoy JM, Knox FG (1992) Integrated responses of the kidney to alterations in extracellular fluid volume. In: Seldin DW, Giebisch G (eds) The kidney. Physiology and pathophysiology, 2nd edn. Raven, New York, pp 2041–2098
27. Grantham JJ, Orloff J (1968) Effect of prostaglandin E₁ on the permeability response of the isolated collecting tubule to vasopressin, adenosine 3'-5'-monophosphate, and theophylline. J Clin Invest 47:1154–1161
28. Greger R (1985) Ion transport mechanisms in thick ascending limb of Henle's loop of mammalian nephron. Physiol Rev 65:760–797
29. Haberich FJ, Aziz O, Nowacki PE, Ohm W (1969) Zur Spezifität der Osmoreceptoren in der Leber. Pflugers Arch 313:289–299
30. Häberle DA (1988) Hemodynamic interactions between intrinsic blood flow control mechanisms in the rat kidney. Renal Physiol Biochem 11:289–315
31. Hall JE, Brands MW (1992) The renin-angiotensin-aldosterone systems: renal mechanisms and circulatory homeostasis. In: Seldin DW, Giebisch G (eds) The kidney. Physiology and pathophysiology, 2nd edn. Raven, New York, pp 1455–1504
32. Hallbrucker C, vom Dahl S, Ritter M, Lang F, Häussinger D (1994) Effects of hyperosmolar urea on K⁺ fluxes, cell volume and hepatocellular function in perfused rat liver. Pflugers Arch 428:552–560
33. Horisberger JD, Lemas V, Kraehenbühl JP, Rossier BC (1991) Structure-function relationship of the Na, K-ATPase. Annu Rev Physiol 53:565–584
34. Howard RL, Bichet DG, Schrier RW (1992) Hypernatremic and polyuric states. In: Seldin DW, Giebisch G (eds) The kidney. Physiology and pathophysiology, 2nd edn. Raven, New York, pp 1753–1778
35. Ichikawa I, Brenner BM (1979) Mechanism of inhibition of proximal tubule fluid reabsorption after exposure of the rat kidney to the physical effects of expansion of extracellular fluid volume. J Clin Invest 64:1466–1474
36. Illowsky BP, Kirch DG (1988) Polydipsia and hyponatremia in psychiatric patients. Am J Psychiatry 145:675–683
37. Jose CJ, Perez-Cruet J (1979) Incidence and morbidity of self-induced water intoxication in state mental hospital patients. Am J Psychiatry 136:221–222
38. Kirk KL, Schafer JA (1992) Water transport and osmoregulation by antidiuretic hormone in terminal nephron segments. In: Seldin DW, Giebisch G (eds) The kidney. Physiology and pathophysiology, 2nd edn. Raven, New York, pp 1693–1726
39. Koh GY, Klug MG, Field LJ (1993) Atrial natriuretic factor and transgenic mice. Hypertension 22:634–639
40. Krapf R (1989) Physiology and molecular biology of the renal Na/H antiporter. Klin Wochenschr 67:847–851

41. Kurtzman NA, White MG, Rogers PW (1973) Pathophysiology of metabolic alkalosis. Arch Intern Med 131:703–708
42. Lang F (1987) Osmotic diuresis. Renal Physiol 10:160–173
43. Lang F, Rehwald W (1992) Potassium channels in renal epithelial transport regulation. Am J Physiol 72:1–32
44. Lang F, Messner G, Rehwald W (1986) Electrophysiology of sodium-coupled transport in proximal renal tubules. Am J Physiol 250:F953–F962
45. Lang F, Völkl H, Haussinger D (1990) General principles in cell volume regulation. In: Beyenbach KW (ed) Cell volume regulation. Comparative physiology. Karger, Basel, pp 1–25
46. Lang F, Häussinger D, Tschernko E, Capasso G, DeSanto NG (1992) Proteins, the liver and the kidney – hepatic regulation of renal function. Nephron 61:1–4
47. Lang F, Tschernko E, Häussinger D (1992) Hepatic regulation of renal function. Exp Physiol 77:663–673
48. Lang F, Gerok W, Häussinger D (1993) New clues to the pathophysiology of hepatorenal failure. Clin Invest 71:93–97
49. Laragh JH (1992) The renin system and the renal regulation of blood pressure. In: Seldin DW, Giebisch G (eds) The kidney. Physiology and pathophysiology, 2nd edn. Raven, New York, pp 1411–1454
50. Mancia G, Donald DE (1975) Demonstration that the atria, ventricles, and lungs each are responsible for a tonic inhibition of the vasomotor center in the dog. Circ Res 36:310–318
51. Mancia G, Ferrari A, Gregorini L, Parati G, Pomidossi G, Zanchetti A (1979) Control of blood pressure by carotid sinus baroreceptors in human beings. Am J Cardiol 44:895–902
52. Marchand GR (1978) Interstitial pressure during volume expansion at reduced renal artery pressure. Am J Physiol 235:F209–F212
53. McGiff JC, Schwartzman M, Ferreri NR (1985) Renal prostaglandins and hypertension. Adv Prostaglandin Thromboxane Leukotriene Res 13:161–169
54. Moncada S, Palmer MRJ, Higgs EA (1991) Nitric oxide: physiology, pathophysiology, and pharmacology. Pharmacol Rev 43:109–142
55. Muto S, Tabei K, Asano Y, Imai M (1985) Dopaminergic inhibition of the action of vasopression on the cortical collecting tubule. Eur J Pharmacol 114:393–397
56. Oberleithner H (1991) Acute aldosterone action in renal target cells. Cell Physiol Biochem 1:2–12
57. Osswald H, Spielman WS, Knox FG (1978) Mechanism of adenosine mediated decrease in glomerular filtration rate. Circ Res 43:465–469
58. Osswald H, Hermes HH, Nabakowski G (1982) Role of adenosine in signal transmission of tubuloglomerular feedback. Kidney Int 22 Suppl 12:S136–S142
59. Paintal AS (1973) Vagal sensory receptors and their reflex effects. Physiol Rev 53:159–226
60. Palmer BF, Alpern RJ, Seldin DW (1992) Pathophysiology of edema formation. In: Seldin DW, Giebisch G (eds) The kidney. Physiology and pathophysiology, 2nd edn. Raven, New York, pp 2099–2141
61. Palmer LG (1992) Epithelial Na channels: function and diversity. Annu Rev Physiol 54:51–66
62. Reeves WB, Andreoli TE (1992) Sodium chloride transport in the loop of Henle. In: Seldin DW, Giebisch G (eds) The kidney. Physiology and pathophysiology, 2nd edn. Raven, New York, pp 1975–2002
63. Robertson GL (1977) Vasopressin function in health and disease. Recent Prog Horm Res 33:333–385
64. Robertson GL (1983) Thirst and vasopressin function in normal and disordered states of water balance. J Lab Clin Med 101:351–371
65. Robertson GL (1992) Regulation of vasopressin secretion. In: Seldin DW, Giebisch G (eds) The kidney. Physiology and pathophysiology, 2nd edn. Raven, New York, pp 1595–1614
66. Romero JC, Knox FG (1988) Mechanisms underlying pressure-related natriuresis: the role of the renin-angiotensin and prostaglandin systems. Hypertension 11:724–738
67. Roy DR, Layton HE, Jamison RL (1992) Countercurrent mechanism and its regulation. In: Seldin DW, Giebisch G (eds) The kidney. Physiology and pathophysiology, 2nd edn. Raven, New York, pp 1649–1692
68. Saeki Y, Terui N, Kumada M (1988) Physiological characterization of the renal-sympathetic reflex in rabbits. Jpn J Physiol 38:251–266
69. Sardet C, Wakayabashi S, Fafournoux P, Counillon L, Pàgès G, Pouyssegur J (1991) Molecular properties of Na^+/H^+ exchanges. In: DeSanto NG, Capasso G (eds) Acid-base balance. Bios, Cosenza, pp 13–20
70. Schad H, Seller H (1976) Reduction of renal nerve activity by volume expansion in conscious cats. Pflugers Arch 363:155
71. Schlatter E (1989) Antidiuretic hormone regulation of electrolyte transport in the distal nephron. Renal Physiol Biochem 12:65–84
72. Schnermann J, Briggs JP (1992) Function of the juxtaglomerular apparatus: control of glomerular hemodynamics and renin secretion. In: Seldin DW, Giebisch G (eds) The kidney. Physiology and pathophysiology, 2nd edn. Rave, New York, pp 1249–1290
73. Sebastian A, McSherry E, Morris RC Jr (1971) On the mechanism of renal potassium wasting in renal tubular acidosis associated with the Fanconi syndrome (type 2 RTA). J Clin Invest 50:231–243
74. Skinner SL, McCubbin JW, Page IH (1964) Control of renin secretion. Circ Res 15:64
75. Smith PR, Benos DJ (1991) Epithelial Na^+ Channels. Annu Rev Physiol 53:509–530
76. Spyer KM (1981) The neural organization and control of the baroreflex. Rev Physiol Biochem Exp Pharmacol 88:23–124
77. Stella A, Golin R, Genovesi S, Zanchetti A (1987) Renal reflexes in the regulation of blood pressure and sodium excretion. Can J Physiol Pharmacol 65:1536–1539
78. Szatalowicz VL, Arnold PE, Chaimovitz C, Bichet D, Berl T, Schrier RW (1981) Radioimmunoassay of plasma arginine vasopressin in hyponatremic patients with congestive heart failure. N Engl J Med 305:263–266
79. Thames MD, Jarecki M, Donald DE (1978) Neural control of renin secretion in anesthetized dogs. Interaction of cardiopulmonary and carotid baroreceptors. Circ Res 42:237–245
80. Toto RD, Seldin DW (1992) Salt wastage. In: Seldin DW, Giebisch G (eds) The kidney. Physiology and pathophysiology, 2nd edn. Raven, New York, pp 2143–2164
81. Uranga J, Fuenzalida R, Rapoport AL, Del Castillo E (1979) Effect of glucagon and glomerulopressin on the renal function of the dog. Horm Metab Res 11:275–279
82. Weisinger RS, Considine P, Denton DA, McKinley M, Leksell L, McKinley MJ, Mouw DR, Muller AF, Tarjan E (1982) Role of the sodium concentration of the cerebrospinal fluid in the salt appetite of sheep. Am J Physiol 242:R51–R63
83. Wolf AV (1950) Osmometric analysis of thirst in man and dog. Am J Physiol 161:75–86
84. Wuthrich RP, Valloton M (1986) Prostaglandin E_2 and cyclic AMP response to vasopressin in renal medullary tubular cells. Am J Physiol 251:F499–F505

78 Acid–Base Metabolism

F. Lang

Contents

78.1 Physiological Significance of Intracellular pH

The physiological properties of enzymes and transport and binding proteins are heavily influenced by the dissociation of acidic or basic amino acids, most importantly of histidine, and are thus dependent on the ambient pH. Accordingly, virtually all cellular functions are sensitive to alterations of pH. Thus, it is not surprising that the H+ concentration is one of the best-controlled variables in the body. Figure 78.1 summarizes some of the physiologically most prominent functions sensitive to pH:

- The rate-limiting enzymes of *glycolysis*, especially phosphofructokinase, are inhibited by H+. Accordingly, intracellular alkalosis stimulates and intracellular acidosis inhibits glycolysis and lactic acid formation [51]. *Gluconeogenesis* is inhibited by alkalosis and stimulated by acidosis. In part through inhibition of gluconeogenesis, alkalosis increases the substrates of the Krebs cycle, i.e., of citrate. Acidosis stimulates glucose degradation through the pentose phosphate cycle, at least partly owing to inhibition of glycolysis.

- For *DNA synthesis* and *cell proliferation* to occur, cell pH must not be acidic; that is, slight alkalosis of the cell is a prerequisite of cell proliferation. Accordingly, a wide variety of growth factors activate the Na+/H+ exchanger to alkalinize the cell [49,52]. Alkaline pH usually prevails in tumor cells.

- The *K+ channels* in a wide variety of cells are sensitive to intracellular pH, being activated by alkalosis and inactivated by acidosis [38].

- Acidosis may lead to a decrease of Ca²+ entry through *Ca²+ channels*, resulting for instance in reduction of cardiac contractility.

- In pulmonary arterioles, H+ elicits *vasoconstriction*, whereas in the systemic circulation H+ favors *vasodilatation*; alkalosis favors vasoconstriction, especially of cerebral vessels.

- Acidosis decreases and alkalosis enhances the *gap junctional conductance*, which in the heart modifies the conduction of the cardiac impulse (see Chap. 7).

- Acidosis decreases and alkalosis increases the oxygen affinity of *hemoglobin* (the *Bohr effect*, see Chap. 102).

- Alkalosis stimulates dissociation of plasma proteins, which then bind more calcium and thus reduce Ca²+ activity in plasma. Conversely, acidosis reduces *Ca²+ binding of plasma proteins*. On the other hand, bicarbonate complexes calcium, again reducing plasma Ca²+ activity. In metabolic alkalosis and acidosis the two effects are additive and elicit marked changes of plasma Ca²+ activity. In respiratory alkalosis and acidosis the effects cancel each other out partially, but protein binding does exceed the effects of bicarbonate complexing [46].

78.2 The Buffer Systems

78.2.1 Basic Properties of Buffers

A buffer system is able to bind or release H+ reversibly:

$$AH \rightleftharpoons H^+ + A^-, \tag{78.1}$$

where AH is the undissociated and A⁻ the dissociated species of a weak acid. The number of molecules AH dissociating per unit time (J^1) is proportional to the concentration of AH ([AH]):

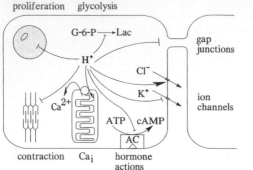

Fig. 78.1. Some cellular functions dependent on intracellular pH. AC, Adenylyl cyclase

$$J^1 = k^1 \times [AH] \tag{78.2}$$

By the same token, the reaction per unit time of H^+ and A^- to AH (J^{-1}) depends on the concentrations of both H^+ ($[H^+]$) and A^- ($[A^-]$):

$$J^{-1} = k^{-1} \times [H^+] \times [A^-]. \tag{78.3}$$

k^1 and k^{-1} are "constants" which describe the rapidity of the reaction. They are dependent on temperature and ionic strength but independent of $[H^+]$, $[A^-]$, and $[AH]$. In equilibrium, $J^1 = J^{-1}$ and

$$k^1[AH] = k^{-1} \times [H^+] \times [A^-] \tag{78.4}$$

and

$$k^1/k^{-1} = K = [H^+] \times [A^-]/[AH]. \tag{78.4'}$$

If the logarithm of each of the two sides of the equation is taken:

$$\lg K = \lg [H^+] + \lg([A^-]/[AH]) \tag{78.5}$$

and, since $\lg [H^+] = -$ pH, and $\lg K = - pK$,

$$pH = pK + \lg([A^-]/[AH]) \tag{78.6}$$

This, the *Henderson-Hasselbalch equation* (see also Chap. 102) shows that a correlation exists between pH and the quotient of $[A^-]/[AH]$.

The equation is applicable to all weak acids. The corresponding formula for weak bases is:

$$pH = pK + \lg([B]/[BH^+]), \tag{78.6'}$$

where B and BH^+ are the concentrations of the respective species.

Figure 78.2 shows the dissociation curve for phosphate as a function of pH. Since a buffer system binds H^+ at increasing H^+ concentrations and releases H^+ at decreasing H^+

concentrations, it blunts any alteration of H^+ concentration. The ability of a buffer system to do so is expressed by the *buffer capacity* (K_p):

$$K_p = \Delta[H^+]/\Delta pH. \tag{78.7}$$

Obviously, the buffer capacity of a buffer system increases with the concentration of the buffer. Furthermore, the buffer capacity decreases with increasing difference between pH and pK, or, in other words, the buffer capacity is highest in the pH range close to pK. Table 78.1 lists the pKs of several important buffer systems.

78.2.2 Buffers in Blood

The buffering capacity of blood is in the range of 15 mmol/l per pH unit [19]. This buffer capacity is mainly due to *hemoglobin*.

The most important buffer system in blood is, however, the H_2CO_3/HCO_3^- *system*, even though the concentration of H_2CO_3 is very low and the pK (pK 3.3) far from blood pH. Thus, at least theoretically, the system has a very low buffer

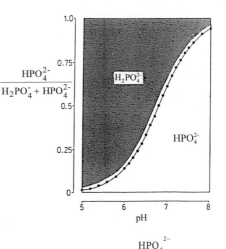

Fig. 78.2. Relation of the $\dfrac{HPO_4^{2-}}{H_2PO_4^- + HPO_4^{2-}}$ ratio as a function of ambient pH

Table 78.1. Biologically important buffer systems

Buffer acid	Buffer base	pK
H_2CO_3	HCO_3^-	3.3
CO_2	HCO_3^-	6.1
NH_4^+	NH_3	9.2
$H_2PO_4^-$	HPO_4^{2-}	6.8
Citrate^{2-}	Citrate^{3-}	5.5
Uric acid	Urate	5.8
Lactic acid	Lactate	3.9
Acetic acid	Acetate	4.8

capacity. However, both components of the buffer system, i.e., H_2CO_3 and HCO_3^-, can rapidly be generated and eliminated. Thus, not only the dissolved buffer components can be recruited or disposed of during the buffering process, but multiples of these.

H_2CO_3 and HCO_3^- constitute an *open buffer system*. In the presence of the catalytic enzyme *carbonic anhydrase* H_2CO_3 is in equilibrium with CO_2:

$$[CO_2] = 10^{2.8} [H_2CO_3], \qquad (78.8)$$

and if (but only if) CO_2 is in equilibrium with H_2CO_3, the *Henderson-Hasselbalch* equation can be expressed as:

$$pH = 6.1 + \lg[HCO_3^-]/[CO_2]. \qquad (78.6'')$$

The *pK* of this system is closer to blood pH, i.e., $[CO_2]$ is far greater than $[H_2CO_3]$ and the buffer capacity accordingly greater. More importantly, CO_2 is continuously produced by metabolism and is eliminated by respiration (see Sect. 78.3). HCO_3^-, on the other hand, can be produced or eliminated by the kidney in collaboration with the liver (see Sect. 78.3).

78.2.3 Importance of Buffers in Urine

To compensate for daily production of H^+ (see below), the *kidney* usually has to excrete H^+ in quantities of the order of 100 mmol/day (see Chap. 75) [1,26]. However, even at a urine pH of 4.5, the free H^+ concentration in urine is only some 30 µmol/l. Thus, renal elimination of H^+ cannot be achieved by excreting free H^+, but must include H^+ excretion in the form of buffers [27]. Thus, the ability of the kidney to excrete H^+ ions is a function of the availability of buffers in the lumen of the tubule.

Two buffer systems are of paramount importance, the NH_3/NH_4^+ *system*, which contributes some 60% of daily H^+ excretion, and the $HPO_4^{2-}/H_2PO_4^-$ system, which contributes some 30%. A small proportion of renal acid excretion is in the form of organic acids, such as uric acid (see Chap. 75). NH_3 is a weak base with a pK of 9.2, i.e., at the pH of 7.4 in blood the ratio of NH_4^+/NH_3 is some 60:1. In general, the cell membranes are highly permeable to NH_3, so a concentration gradient for NH_3 usually rapidly dissipates, whereas NH_4^+ passes the cell membrane only if an appropriate transport system is available. Such a transport system is the $Na^+K^+2Cl^-$ cotransport in Henle's loop, which accepts NH_4^+ in place of K^+.

In the kidney, NH_4^+ is produced in the proximal tubule, where it is formed from *glutamine* under the catalytic action of glutaminase. It is trapped in the acid tubule lumen and delivered to the thick ascending limb of Henle's loop, where it is reabsorbed to undergo recirculation in the kidney medulla. The countercurrent system builds up high NH_4^+/NH_3 concentrations, which favors urinary elimination of NH_4^+ [36,56]. With each NH_3 molecule excreted as NH_4^+, one H^+ is buffered and eliminated.

Proximal tubular formation of NH_3 is highly dependent on the acid-base status: acidosis stimulates and alkalosis inhibits proximal tubular glutaminase. Continuing proximal tubular glutaminase activity during alkalosis would be harmful, since the low H^+ secretion under conditions of alkalosis would not guarantee renal elimination of NH_3, which would partially escape into blood. NH_3 is highly toxic, especially for neurons.

It should be stressed that renal elimination of H^+ in the form of NH_4^+ is primarily a function of proximal tubular NH_3 production, and less one of urinary pH. Due to the *pK* of 9.2, NH_3/NH_4^+ is virtually entirely in the form of NH_4^+, irrespective of urine pH. However, luminal pH influences cellular loss of NH_3 and thus modifies production of NH_3.

Phosphate is a trivalent acid which, depending on ambient pH, may be completely, partially, or not at all dissociated:

$$PO_4^{3-} + H^+ \rightarrow HPO_4^{2-} + H^+ \rightarrow H_2PO_4^- + H^+$$
$$\rightarrow H_3PO_4 \qquad (78.9)$$

The *pKs* of the reactions are 12.3, 6.8, and 2.0 respectively (see Table 78.1).

In blood, phosphate is mainly in the form of HPO_4^{2-} (80%); some 20% is in the form of $H_2PO_4^-$. Much less than 1% is in the form of PO_4^{3-} or H_3PO_4. At a urinary pH of 5.8, some 90% is in the form of $H_2PO_4^-$ and 10% in form of HPO_4^{2-}. Accordingly, at this urinary pH some 70% of excreted phosphate has bound H^+ during tubular passage. Further acidification may allow up to another 10% binding of H^+. Thus, acid elimination in form of phosphate is only moderately increased by acidification of urine beyond 5.8. In that range the amount of H^+ excreted in the form of phosphate is mainly dependent on the amount of phosphate excreted (Fig. 78.3). At a urinary pH of 6.8, on the other hand, only 30% of excreted phosphate eliminates H^+, and excretion of H^+ in the form of phosphate is highly dependent on urinary pH between pH 6.8 and 7.4.

Any discussion of H^+ excretion in the form of phosphate must take into account the origin of the phosphate. If the phosphate comes from bone, where it is deposited in highly alkaline form (HPO_4^{2-} and PO_4^{3-}), then it may bind a

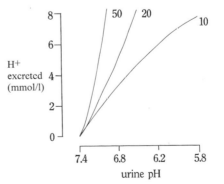

Fig. 78.3. Significance of buffer concentration and urine pH for elimination of H^+. The amount of H^+ excreted in the form of $H_2PO_4^-$ is plotted against the urinary pH for three different phosphate concentrations (10, 20, and 50 mmol/l)

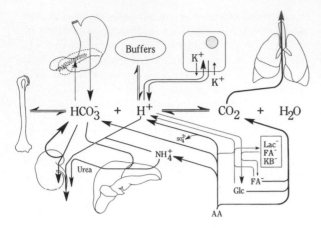

Fig. 78.4. Synopsis of different factors contributing to acid base balance. *Lac⁻*, Lactate; *FA⁻*, fatty acids; *KB⁻*, ketone bodies; *Glc*, glucose; *AA*, amino acids

first H^+ during mobilization from bone and a second H^+ during the transition into urine.

78.3 Daily Turnover of H⁺

Unlike other electrolytes, H^+ can be produced and consumed within the body, and thus the daily intake of acids or bases need not match their excretion. As illustrated in Fig. 78.4 and detailed in the following sections, the daily production of CO_2 and H^+ or HCO_3^- balances the daily pulmonary elimination of CO_2 and renal elimination of H^+ or HCO_3^-. The daily production of CO_2 usually amounts to some 20 mol. Since CO_2 may bind H_2O and subsequently dissociate to H^+ and HCO_3^-, CO_2 production may be considered equivalent to acid production. However, since pulmonary CO_2 elimination in equilibrium is identical to CO_2 production, no net H^+ is generated within the body from CO_2. In addition to CO_2, metabolism may produce nonvolatile acids, which dissociate to the respective anions and H^+. Usually some 100 mmol H^+ are produced and excreted by the kidney. On the other hand, excess HCO_3^- may be produced, requiring the renal elimination of HCO_3^-. As would be expected, the formation of CO_2, H^+, and HCO_3^- are variable depending on diet and metabolic activities.

78.4 Interplay of Lung, Kidney, and Liver in the Regulation of Blood pH

78.4.1 Respiratory Elimination of CO₂

As outlined in Chaps. 100 and 101, *respiration* serves primarily to take up O_2 into and eliminate CO_2 from blood passing the alveoli. O_2 solubility in water is only about 5% of CO_2 solubility and thus O_2 has to be bound to

hemoglobin to allow the transport of the required O_2 amounts. Accordingly O_2 uptake into blood is saturable, CO_2 transport is not. Figure 78.5 shows the effect of alveolar ventilation on the O_2 and CO_2 pressures in alveolar capillary blood. It is obvious that, in contrast to CO_2 elimination, the uptake of O_2 cannot be substantially increased by excessive ventilation.

The alveolar ventilation is governed by respiratory neurons, which receive their input from pulmonary mechanical receptors, arterial and CNS chemoreceptors for CO_2 and O_2. Usually, the most important stimulus for ventilatory drive is CO_2 or the CNS H^+ (Fig. 78.6). Unlike arterial H^+ and HCO_3^-, arterial CO_2 readily traverses the blood-brain barrier, and thus an increase of arterial CO_2 rapidly increases H^+ in the CNS [35]. Usually, O_2 concentration in blood does not modify the ventilatory drive. However, when O_2 is decreased at constant CO_2, the ventilatory drive increases steeply (see Fig. 78.6).

Besides CO_2 and O_2, several factors influence the ventilatory drive, as discussed in detail in Chaps. 100, 102, 105–108.

78.4.2 Renal Elimination of H⁺

In glomerular ultrafiltrate, HCO_3^- concentration is slightly higher than in plasma water, due to the effect of the plasma negative Gibbs-Donnan potential (see Chap. 76). Given the HCO_3^- concentration in glomerular filtrate (28 mmol/l) and the glomerular filtration rate (approx. 180 l/day), some 5 mol HCO_3^- is filtered per day. This large amount of HCO_3^- is usually completely *reabsorbed* by the renal tubules (see Chaps. 74, 75). Beyond that, the renal tubules secrete 100 mmol/day H^+, which are in large part bound to NH_4^+ and $H_2PO_4^-$ (see Sect. 78.2.3).

Some 95% of filtered HCO_3^- is reabsorbed along the proximal renal tubules. The main cellular mechanisms are as follows (see Chaps. 74, 75, 77). A luminal Na^+/H^+ exchanger secretes H^+, which titrates filtered HCO_3^- to H_2CO_3. Under the catalytic influence of membrane-bound carbonic

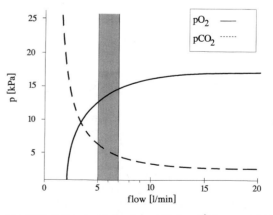

Fig. 78.5. Influence of alveolar ventilation ($\dot{V}_{alv}$) on arterial PCO_2 and P_{O_2} at constant CO_2 production and O_2 consumption

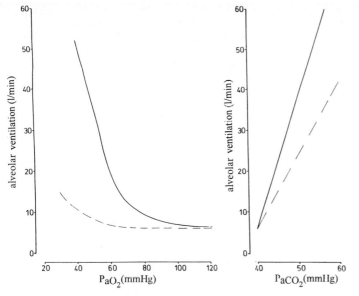

Fig. 78.6. Influence of arterial PO$_2$ (*left*) and PCO$_2$ (*right*) on alveolar ventilation ($\dot{V}_{alv}$). The *broken lines* are obtained with uncontrolled partial pressures, the *solid lines* with "clamped" partial pressures: PO$_2$ 100 mmHg (*right panel*), PCO$_2$ 40 mmHg (*left panel*)

anhydrase, H$_2$CO$_3$ is dehydrated to CO$_2$, which enters the cell and is again transformed to H$^+$ and HCO$_3^-$. H$^+$ recycles into the lumen, whereas HCO$_3^-$ leaves the cell coupled to Na$^+$. The Na$^+$ (HCO$_3^-$)$_3$ cotransport is highly sensitive to the potential difference, which is governed by K$^+$. In the distal tubule and collecting duct, *H$^+$ transport* is accomplished mainly by *intercalated cells*. The majority of these cells (type A) secrete H$^+$ by a luminal H$^+$-ATPase or a luminal H$^+$/K$^+$-ATPase. The HCO$_3^-$ produced by intracellular formation of H$^+$ from CO$_2$ leaves the cell across the basolateral cell membrane via a Cl$^-$/HCO$_3^-$ exchanger. Cl$^-$ thus accumulated leaves the cell via Cl$^-$ channels. The other type of intercalated cells (type B) secrete HCO$_3^-$ by a luminal Cl$^-$/HCO$_3^-$ exchanger. The H$^+$ formed within the cell from CO$_2$ is extruded from the cell through a H$^+$-ATPase in the basolateral cell membrane. The Cl$^-$ accumulated leaves the cell through basolateral Cl$^-$ channels.

78.4.3 Interplay of Lung and Kidney

The *lung and kidney* are equally important for the maintenance of acid–base balance [22,32]. While the lung is able to modify blood pH by eliminating CO$_2$, the kidney fulfills its function by eliminating H$^+$ or HCO$_3^-$, depending on the demand.

If renal H$^+$ elimination cannot cope with metabolic generation of H$^+$, the lung is capable of preventing the development of acidosis (increase in [H$^+$]) by elimination of CO$_2$ (see Fig. 78.4):

$$H^+ + HCO_3^- \rightleftharpoons CO_2 + H_2O \tag{78.10}$$

The daily elimination of CO$_2$ by respiration can be as high as 450 l or 20 mol. In comparison, the daily renal elimination of H$^+$ is usually only some 100 mmol. Nevertheless,

respiratory compensation of insufficient renal H$^+$ elimination is limited: the respiratory elimination of H$^+$ requires expenditure of HCO$_3^-$ and results in a decrease of the HCO$_3^-$ concentration in blood. Since blood pH is a function of the [HCO$_3^-$]/[CO$_2$] ratio, constancy of blood pH under conditions of a decreasing HCO$_3^-$ concentration requires a parallel lowering of the CO$_2$ concentration.

The amount of CO$_2$ eliminated by respiration (M_{CO_2}) is a function of the CO$_2$ concentration in the alveoli (which is identical to the CO$_2$ concentration in blood) and the alveolar ventilation ($\dot{V}a$):

$$M_{CO_2} = \dot{V}a \times [CO_2] \tag{78.11}$$

This indicates that at any given alveolar ventilation, the amount of CO$_2$ eliminated is in linear proportion to the CO$_2$ concentration. As a result, to match the metabolically produced CO$_2$, the lung has to enhance $\dot{V}a$ if the CO$_2$ concentration is lowered. If, for instance, transiently defective renal elimination of H$^+$ and subsequent respiratory compensation cause blood HCO$_3^-$ to drop to half, then $\dot{V}a$ must be doubled to maintain elimination of CO$_2$. Even if renal excretion of H$^+$ per unit time then matches metabolic H$^+$ production per unit time, the HCO$_3^-$ concentration remains low and $\dot{V}a$ must be kept high. Only if the kidney replenishes the blood HCO$_3^-$ concentration by enhanced H$^+$ elimination, in compensation for the former defective H$^+$ elimination, can respiration be normalized. If the renal ability to eliminate H$^+$ is permanently restricted, i.e., the H$^+$ excess is steadily increasing, then the blood HCO$_3^-$ concentration continues to fall and respiration can delay but not prevent the establishment of acidosis.

Thus, the lung cannot replace the kidney in the regulation of acid–base balance, but is only able to compensate for transient disturbances of renal acid elimination.

1575

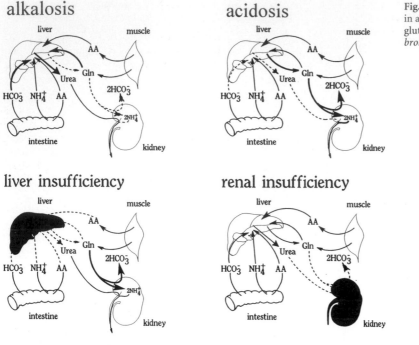

78.4.4 Interplay of Liver and Kidney

As outlined above, a substantial portion of the acid eliminated by the kidney is in the form of NH_4^+ [10,25,36]. To be able to produce NH_3, the kidney requires the delivery of glutamine. The availability of glutamine in the blood depends largely on the glutamine metabolism in the liver [28,29] (see also Chaps. 67, 68).

Usually the *liver* consumes glutamine for *urea synthesis* (Fig. 78.7). The glutaminase of the periportal liver cells forms NH_4^+ and HCO_3^-, which are utilized for urea synthesis:

$$2NH_4^+ + 2HCO_3^- \rightleftharpoons CO(NH_2)_2 + 3H_2O + CO_2.$$
$$(78.12)$$

The NH_4^+ generated by the degradation of glutamine in the periportal cells is partly taken up by perivenous cells, which synthesize glutamine. In this way, overspill of the NH_4^+ into the systemic circulation is avoided.

Acidosis inhibits the glutaminase in the periportal cells and glutamine synthesis in perivenous cells prevails. Thus, more glutamine is available for formation of NH_4^+ in the kidney. In contrast to hepatic glutaminase, the renal glutaminase is stimulated by acidosis. Following renal degradation of glutamine, NH_4^+ is excreted (and not utilized for urea synthesis) and the HCO_3^- formed in parallel is conserved for the body.

In alkalosis the glutaminase in the periportal cells is stimulated. The HCO_3^- formed is utilized for urea synthesis and is not conserved for the body. Less glutamine is delivered to the kidney and the renal glutaminase is inhibited.

In other words, in acidosis glutamine degradation is shifted to the kidney, which excretes nitrogen with conservation of HCO_3^-. In alkalosis glutamine degradation is shifted towards the liver, which incorporates glutamine nitrogen into urea, whereby the HCO_3^- is consumed. In this way the two organs work together in the regulation of acid–base balance.

78.5 Other Mechanisms Influencing Blood pH

As illustrated by Fig. 78.4, a number of other factors modify acid–base balance.

78.5.1 Metabolism

Degradation of metabolites leads to a daily production of 20 mol CO_2. The elimination of CO_2 by the intact lung can cope even with massively increased CO_2 production; an increase of CO_2 in arterial blood is always the result of impaired respiration. However, in metabolism *acids are generated* in addition to CO_2 (or H_2CO_3) which cannot be eliminated by the lung. The acids dissociate to form H^+, which has to be eliminated by the kidneys.

The majority of H^+ comes from degradation of *sulfur-containing amino acids*: –SH groups are oxidized to SO_4^{2-} and $2H^+$ (Fig. 78.8). Other acids include *lactic acid*, which is formed from glucose and fatty acids, which in turn are formed from triglycerides. The metabolism of fatty acids may lead to *acetoacetate* and *β-hydroxy-butyrate*, again

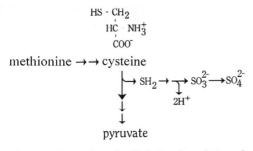

methionine $\rightarrow\rightarrow$ cysteine

$$\rightarrow SH_2 \rightarrow SO_3^{2-} \rightarrow SO_4^{2-}$$
$$2H^+$$

pyruvate

Fig. 78.8. Formation of acid during degradation of sulfur-containing amino acids

fully dissociated acids at blood pH [33]. Unlike SO_4^{2-}, all these acids can be reutilized and thus the formed H^+ disappears again upon degradation or metabolism of these acids. Nevertheless, the formation of lactic acid stimulated during heavy work, or the formation of fatty acids stimulated during starvation, regularly leads to metabolic acidosis.

78.5.2 Other Electrolytes

Cellular and renal HCO_3^- and H^+ transport are dependent on the transport of *other electrolytes* [7,8,37]. Thus, overlap may occur in the regulation of cellular or renal HCO_3^- transport with that of other electrolytes.

The infusion of NaCl may force the kidney to reduce proximal tubular *Na$^+$ reabsorption*. Proximal tubular HCO_3^- reabsorption is dependent on Na^+ reabsorption and is decreased in parallel. If the distal nephron is not able to compensate for reduced proximal tubular HCO_3 reabsorption, the resulting natriuresis may be paralleled by *bicarbonaturia*.

Similarly, the infusion of $CaCl_2$ may impair renal reabsorption of both Na^+ and HCO_3^-.

On the other hand, the requirement to excrete HCO_3^- may interfere with the need for salt and water retention: following *vomiting* of acid gastric content (see below), both *metabolic alkalosis* and dehydration may develop. In this condition, increased renal elimination of HCO_3^- and enhanced conservation of Na^+ and water would be desirable. In proximal tubules, however, both Na^+ and HCO_3^- reabsorption are achieved by the Na^+/H^+ exchanger. Thus, the proximal renal tubule cannot at the same time enhance Na^+ reabsorption and inhibit HCO_3^- reabsorption. The influence of volume depletion prevails and the kidney responds with *retention of Na$^+$ and HCO$_3^-$*. As a result, the alkalosis is maintained (volume depletion alkalosis). Only replacement of NaCl and fluid will the kidney reenable to excrete HCO_3^-.

In respect to acid – base balance, however, the most important interfering ion is K^+ [38–41] (see also Chap. 75). The *cell membrane potential* of almost all cells, including renal epithelial cells, is maintained by K^+ channels, which allow passage of K^+ according to the electrochemical gradient. An increase in extracellular K^+ thus tends to depolarize the

cells, while a decrease in extracellular K^+ rather hyperpolarizes the cell membranes (see Chap. 11). The cell membrane potential drives HCO_3^- exit for instance in the proximal tubule. The HCO_3^- exit thus depends on the magnitude of the cell membrane potential, and therefore on the K^+ gradient.

In *hyperkalemia* the cell is depolarized and the driving force for HCO_3^- exit is reduced. The consequence is *intracellular alkalosis*, which turns off the Na^+/H^+ exchanger in the luminal cell membrane [40]. Accordingly, proximal tubular H^+ secretion and HCO_3^- reabsorption are reduced. Similar events may lead to intracellular alkalosis and decreased activity of respiratory neurons. As a result, *hyperkalemia leads to (extracellular) acidosis*.

In *hypokalemia* the chemical gradient for K^+ across the cell membrane is enhanced, and the resulting *hyperpolarization* produces an *intracelluar acidosis*, which activates the Na^+/H^+ exchanger in the proximal tubule and thus increases proximal tubular HCO_3^- reabsorption [40]. Furthermore, K^+ depletion increases renal acid elimination by stimulation of *NH_4^+ formation* and collecting duct *H^+ secretion* [2]. It may be that hypokalemia also hyperpolarizes respiratory neurons, leading to intracellular acidosis, stimulation of respiration, and respiratory alkalosis.

78.5.3 Gastrointestinal Tract

The H^+ secreted in the parietal cells of the stomach comes from intracellular formation of H_2CO_3 from CO_2 [15] (see also Chap. 61). H_2CO_3 dissociates to H^+ and HCO_3^-, which leaves the cells for the blood and leads to an "*alkali tide*" during stimulated gastric acid secretion.

As acid gastric content is propelled downstream, the pH decreases in the duodenum, which is the signal for secretin release. This hormone stimulates the *secretion of HCO$_3^-$-rich fluid* in pancreas and intestinal glands, which neutralizes the lumen of the duodenum. Roughly the same amounts of HCO_3^- are required to neutralize the duodenum as were generated by the parietal cells of the stomach before. As a result, these events only transiently influence acid – base balance. However, if gastric content is *vomited*, pancreatic HCO_3^- secretion is not stimulated and the HCO_3^- generated by the parietal cells is retained within the body, leading to *metabolic alkalosis*. Conversely, loss of alkaline intestinal secretions may lead to acidosis.

78.5.4 Bone

Alkaline phosphate salts and carbonates are poorly soluble. The poor solubility is utilized to precipitate these salts in bone and thus achieve bone mineralization. Thus, to maintain *bone mineralization*, the ambient pH in the bone must be alkaline (see also Chap. 75).

The interaction of bone with acid – base balance is twofold:

- Acidosis favors bone demineralization since it enhances the solubility of the bone minerals. Conversely, alkalosis favors bone mineralization.

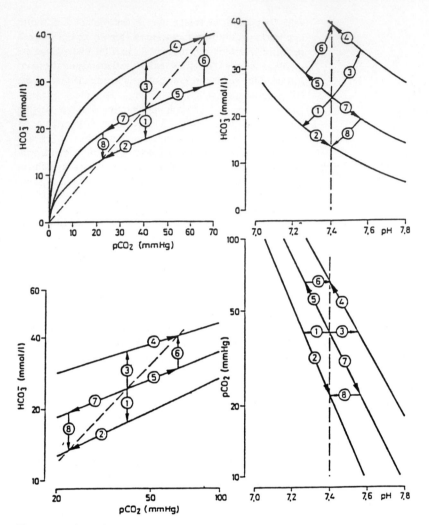

Fig. 78.9. Relation between pH, PCO_2, and HCO_3^- using four different ways of plotting and at different disturbances of acid-base balance. The *right lower plot* corresponds to the Siggaard-Andersen nomogram, as illustrated in Fig. 78.10. *1*, Metabolic acidosis; *2*, respiratory compensation; *3*, metabolic acidosis; *4*, respiratory compensation; *5*, respiratory acidosis; *6*, renal compensation; *7*, respiratory alkalosis; *8*, renal compensation

- Demineralization of bone leads to an alkali load of the body, since the salts take up H^+ ions during demineralization. Conversely, bone mineralization leads to an acid load of the body.

78.6 Disturbances of Blood pH

Disturbances of acid – base balance are divided into respiratory, if the primary event is an alteration of CO_2, and nonrespiratory (metabolic), if the primary event is an alteration of HCO_3^- or H^+ (see also Chap. 102). Figure 78.9 gives several presentations of the different respiratory and metabolic acid – base disturbances and their appropriate compensations. They are derived from different transformations of the Henderson-Hasselbalch equation (see Eq. 78.6″). Figure 78.10 shows the Siggaard-Andersen diagram, which has been in wide clinical use. Here the $lg[CO_2]$ is plotted versus the pH of blood:

$$lg[CO_2] = - pH + pK + lg[HCO_3^-]. \qquad (78.6''')$$

It is obvious from Eq. 78.6″ that at constant $[HCO_3^-]$, all possible pairs of $lg[CO_2]$ and pH form a line with a slope of 45°. The intersection of this line with the parallel to the abscissa at 40 mmHg CO_2 allows estimation of the $[HCO_3^-]$. If the blood taken is equilibrated with two different pCO_2 and the respective pH measured, a line can be drawn through these two points which crosses the parallel to the abscissa at 40 mmHg at the so-called *standard bicarbonate*, i.e., the $[HCO_3^-]$ at normal respiration (40 mmHg CO_2). The extrapolation of this line also allows estimation of the joint buffer bases of blood and the base excess.

Instead of the equilibrium method, it is now possible to measure actual pH and actual CO_2 directly with routine

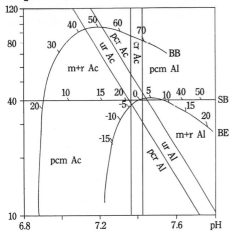

pCO$_2$ [mmHg]

Fig. 78.10. The different acid-base disturbances in the Siggaard-Andersen nomogram. *Ac,* Acidosis; *Al,* alkalosis; *r,* respiratory; *m,* metabolic; *u,* uncompensated; *c,* compensated; *p,* partially; *BB,* buffer base; *SB,* standard bicarbonate; *BE,* base excess

instruments, so the Henderson-Hasselbalch equation can be solved directly.

78.6.1 Respiratory Acidosis

Respiratory acidosis is the result of insufficient CO$_2$ elimination by the lung [44].

Several diseases of the lung interfere with alveolar ventilation and/or CO$_2$ diffusion across the alveolar wall, as detailed in Chaps. 100, 101, 104. If the disorders cannot be compensated for by increased ventilatory drive, they lead to hypoxia and hypercapnia.

Ventilation may further be affected by a variety of *neuromuscular diseases,* as they involve the respiratory muscles (Table 78.2).

A reduced respiratory drive is encountered during *anesthesia,* during intoxication with sedative drugs or opiates, and in patients with cerebral lesions affecting the respiratory neurons [44]. In some patients, the ventilatory drive may cease during sleep. A reduced ventilatory drive is also observed in obesity [44]. It is important to remember that supply of O$_2$ may further reduce ventilatory drive in conditions of hypoventilation and, while alleviating the hypoxemia, may aggravate the respiratory acidosis [44].

Circulatory failure is frequently associated with respiratory failure [14]. In addition, reduced peripheral blood flow leads to a rise in CO$_2$ in peripheral tissues.

Inhibition of erythrocyte *carboanhydrase* may lead to peripheral respiratory acidosis, because it prevents the generation and subsequent elimination of CO$_2$ during alveolar passage.

The consequences of inadequate ventilation or circulatory failure are usually dominated by the sequelae of hypoxemia and *reduced oxygen supply* to the tissues, which are described in Chaps. 95, 103. The respiratory acidosis as

such leads to impairment of glucose utilization through glycolysis, leading to hyperglycemia as well as hyperkalemia due to net cellular H$^+$ uptake and K$^+$ release. Acidosis leads to impairment of cardiac conduction and contractility on the one hand and vasodilation on the other, and thus favors hypotension [44]. The decrease in blood pressure stimulates the sympathetic nervous system and catecholamine release, which in turn stimulates the release of renin, angiotensin II, ADH, and aldosterone [44]. The vasodilation in cerebral vessels may lead to an increase in cerebral pressure [55]. *Hypercapnia* frequently leads to cerebral dysfunction, which could eventually lead to coma [16].

Within limits, respiratory acidosis can be compensated for by increased renal formation of HCO$_3^-$ and excretion of H$^+$.

78.6.2 Respiratory Alkalosis

The underlying cause of *respiratory alkalosis* is *hyperventilation* [44]. In general, this results from enhanced stimulation of respiratory neurons (Table 78.3). In a number of conditions the increased ventilatory drive is due to hypoxia. *Hypoxia* may be caused by reduced PO$_2$ in inhaled air, as at high altitude, or due to impaired O$_2$ uptake into blood.

Compromised alveolar *gas diffusion* affects O$_2$ diffusion more than CO$_2$ diffusion, since the permeabilty for CO$_2$ is 20 times that for O$_2$. Accordingly, hypoxia may develop without significant impairment of CO$_2$ diffusion. The hypoxia then drives ventilation, leading to hypocapnia.

In *Ventilation-perfusion inequality,* hyperventilated alveoli are perfused at the same time as hypoventilated al-

Table 78.2. Major causes of respiratory acidosis

Increased CO$_2$ in inhaled gas
Impaired pulmonary CO$_2$ exchange
   Airway obstruction
   Pneumothorax, hemothorax
   Severe pneumonia
   Obstructive lung disease
   Respiratory distress syndrome
Impaired function of respiratory muscles
   Botulism, tetanus
   Poliomyelitis
   Multiple sclerosis
   Muscular dystrophies
   Myasthenia gravis
   Amyotrophic lateral sclerosis
   Drugs (curare, succinylcholine)
Decreased respiratory drive
   Anesthesia
   Intoxication with sedatives, opiates
   Cerebral injury, ischemia
   Cerebral sleep apnea
   Obesity
Impaired vascular CO$_2$ transport
   Circulatory failure
   Severe pulmonary embolism
   Inhibition of erythrocyte carboanhydrase
Insufficient mechanical respiration

Table 78.3. Major causes of respiratory alkalosis

Hypoxic stimulation of respiration
High altitude
Compromised alveolar diffusion
Ventilation-perfusion inequality
Severe anemia
Hypotension
Nonhypoxic stimulation of respiratory neurons
Voluntary hyperventilation
Anxiety
Pulmonary diseases
Hepatic failure
Recovery from metabolic acidosis
Heat exposure
Septicemia
CNS lesions
CNS infection (encephalitis, meningitis)
Head injury
CNS ischemia
CNS tumors
Hormones and mediators
Progesterone
Catecholamines
Angiotensin II
Drugs
Salicylates
Mechanical hyperventilation

veoli. Oxygen uptake in hyperventilated alveoli cannot exceed hemoglobin saturation and is thus not substantially greater than in normally ventilated alveoli. O_2 uptake into *hypoventilated alveoli*, however, is substantially reduced. As a result, the O_2 saturation of the mixed blood coming from the hyper- and hypoventilated alveoli is reduced, leading to increased ventilatory drive. CO_2 uptake into and release from blood is not saturable, and thus the increased release of CO_2 in hyperventilated alveoli compensates well for the decreased release of CO_2 in hypoventilated alveoli. The increased ventilatory drive thus eventually leads to respiratory alkalosis accompanied by persistent arterial hypoxia.

Several *lung diseases* can affect diffusion and/or ventilation-perfusion equality: pneumonia, pulmonary edema, interstitial lung disease, etc. In addition to their influence on O_2 uptake, pulmonary diseases may affect ventilation through activation of pulmonary receptors (e.g., nociceptive or stretch receptors), which influence the respiratory neurons via the vagal nerve [44].

When lung function is normal, capillary hypoxia may develop in association with severe anemia and hypotension. Several conditions other than hypoxia or acidosis may lead to stimulation of the respiratory neurons (Table 78.3). Respiratory neurons are activated by a number of hormones and mediators. To mention the most important, catecholamines [57], angiotensin [48], and progesterone [43] are effective in increasing ventilatory drive. *Progesterone* is in large part responsible for the *hyperventilation in pregnancy*.

Several drugs [44], such as salicylates [13], have been shown to produce respiratory alkalosis by stimulating the respiratory neurons.

Respiratory neurons may be stimulated by *bacterial toxins*, and septicemia may lead to respiratory alkalosis [44]. In hepatic failure, the respiratory neurons are probably activated by ammonia, which is known to stimulate respiration directly and indirectly by increasing glutamate [9,23,34].

Several *CNS lesions* may lead to hyperventilation by enhanced stimulation or reduced inhibition of respiratory neurons [45].

Rapid correction of plasma bicarbonate concentration in chronic metabolic acidosis with partial respiratory compensation may lead to respiratory alkalosis, since CNS HCO_3^- increases only after a delay, and thus hyperventilation and hypocapnia persist for 1–2 days [47].

The consequences of respiratory alkalosis in part mirror those of respiratory acidosis. They include *stimulated glycolysis*, *inhibited gluconeogenesis* with enhanced formation and renal excretion of citrate, and increased cellular uptake of K^+ with resulting *hypokalemia*. The combination of alkalosis and hypokalemia can lead to cardiac arrhythmias [3]. Respiratory alkalosis leads to *cerebral vasoconstriction* and to enhanced *neuromuscular excitability* [24]. Vasocontriction of coronary vessels can lead to ischemia of the heart [58]. Respiratory alkalosis leads to a slight reduction of Ca^{2+} in blood, due to increased binding of Ca^{2+} to proteins [46]. Alkalosis increases the O_2 affinity of hemoglobin, which may slightly decrease O_2 supply in the periphery [42]. On the other hand, chronic alkalosis increases erythrocyte 2,3,-diphosphoglycerate concentration, which reduces the O_2 affinity of hemoglobin [6].

Respiratory alkalosis is compensated by increased *renal elimination of HCO_3^-*. The increased formation of lactate caused by stimulation of glycolysis also tends to counteract respiratory alkalosis.

78.6.3 Metabolic Acidosis

Metabolic acidosis is characterized by a decrease in the HCO_3^- concentration of blood. This may result from the ingestion or infusion of H^+-producing agents (Table 78.4). For instance, following infusion of NH_4Cl, NH_4^+ can react within the body to form NH_3 and H^+ [17] i.e., consuming HCO_3^- by urea synthesis (see Sect. 78.4.4). More importantly, the endogeneous production of *organic anions* leads to formation of H^+. During *lipolysis*, fatty acids are cleaved from the neutral triglycerides and the fatty acids are utilized in the liver to produce *acetoacetic acid* and β-*hydroxybutyric acid*. These acids dissociate and release H^+. Lipolysis prevails at low insulin concentrations such as during starvation [11] or in juvenile diabetes mellitus [12]. Enhanced lipolysis may also be caused by excess of thyroid hormones. Alcohol, on the other hand, stimulates the formation of ketoacids [20,31]. A great variety of congenital and acquired enzyme defects impair the metabolism of organic acids and thus create ketoacidosis or lactacidosis [17]. Enhanced anaerobic glycolysis and lactate production may also occur during ischemia, hypoxemia, severe

Table 78.4. Major causes of nonrespiratory (metabolic) acidosis

Diet or infusions
    NH_4^+
Enhanced endogenous production of organic acids
    Fatty acids, ketoacidosis
        Starvation
        Diabetes
        Alcoholism
        Enzyme deficiencies
    Lactate
        Ischemia
        Hypoxia, anemia
        Physical exercise
        Catecholamines
        Tumors
        Enzyme defects
        Liver insufficiency
Impaired renal elimination of H^+ or renal loss of HCO_3^-
    Renal tubular acidosis
        Congenital
        Acquired
    Posthypocapnia
    Mineralocorticoid deficiency
    Renal insufficiency
    Carboanhydrase inhibitors
Intestinal loss of HCO_3^-
    Diarrhea
    External fistula
        Ureter intestinal fistula

anemia, and heavy physical exercise [17]. The muscular production of lactate is stimulated by catecholamines [30]. Large tumors may also produce lactacidosis, since the metabolism of tumor cells is frequently restricted to glycolysis [18]. Lactate utilization may be impaired in patients with liver insufficiency [50].

An important cause of nonrespiratory acidosis is *failure of the kidney* to eliminate H^+ [17] (see also Chap. 75). Besides congenital defects of proximal or distal tubular H^+ secretion, a wide variety of nephrotoxic substances and systemic diseases lead to impairment of renal tubular H^+ transport [17]. Distal tubular H^+ secretion is further blunted in patients with mineralocorticoid deficiency. During respiratory alkalosis the kidney is forced to eliminate HCO_3^- and renal H^+ and NH_4^+ formation is suppressed. Once ventilation is reduced, the kidney may resume adequate H^+ elimination only after a delay, and transient metabolic acidosis may occur [17]. Obviously, renal H^+ elimination is impaired in renal insufficiency.

Since intestinal and pancreatic secretions are usually alkaline, *diarrhea* and loss of pancreatic fluid through a fistula are frequently accompanied by acidosis [17]. In some patients (e.g., with bladder cancer) in whom the bladder has to be removed and the ureter anastomosed with the intestine, the H^+ excreted by the kidney may then reenter the body via the intestinal epithelium [17].

Hyperkalemia also favors net uptake of K^+ and HCO_3^- into the cells, leading to extracellular hypobicarbonatemia. Additionally, it blunts renal H^+ excretion [17].

The consequences of metabolic acidosis are similar to those of respiratory acidosis (see above), except that the cerebral vessels are far less sensitive to metabolic than to respiratory acidosis.

78.6.4 Metabolic Alkalosis

Metabolic alkalosis is characterized by an increase in HCO_3 in the blood. It may result from a loss of H^+, an excess of HCO_3, or translocation of H^+ into the cells (Table 78.5).

Frequently, loss of H^+ is due to *vomiting* of acid gastric content [2]. The alkalosis is compounded by the loss of fluid and subsequent contraction of extracellular fluid volume, which stimulates proximal tubular Na^+ (and HCO_3^-) reabsorption, and also, via aldosterone release, distal tubular H^+ secretion [2].

In a rare congenital defect (congenital chloridorrhea), a defect of ileum and colon Cl^-/HCO_3^- exchanger impairs reabsorption of Cl^-, while the intact Na^+/H^+ exchanger continues to secrete H^+. The acid intestinal lumen traps NH_3 in the form of NH_4^+, which is excreted with Cl^- [21].

Several causes may lead to *renal acid loss*:

- Hypokalemia stimulates renal formation and excretion of NH_4^+.
- Mineralocorticoids stimulate distal tubular H^+ secretion along with Na^+ reabsorption. In the presence of high aldosterone levels, the increased delivery of Na^+ to the distal tubule favors loss of K^+ and H^+. The K^+ depletion in turn stimulates H^+ loss.
- During inhibition of Na^+ reabsorption in the thick ascending limb, distal Na^+ delivery is increased and aldosterone release stimulated by extracellular volume contraction. This occurs during treatment with loop diuretics, in hypercalcemia, and in Bartters's syndrome, a congenital defect that is probably due to impaired NaCl reabsorption in the thick ascending limb of Henle's loop [4].
- Mg^{2+} depletion leads to loss of K^+ [54] and thus to disturbed renal H^+ excretion.

Table 78.5. Major causes of nonrespiratory (metabolic) alkalosis

Acid loss
    Loss of gastric fluid
    Intestinal acid loss (Cl^- wasting diarrhea)
    Renal acid loss
        Mineralocorticoid excess
        Diuretics
        Bartter's syndrome
        Mg^{2+} depletion
        Hypercalcemia
HCO_3^- excess
    Diet
    Infusion
    Production by metabolism of lactate, citrate, or other
        organic anions
    Extracellular volume contraction alkalosis
Cellular acid uptake
    K^+ deficiency

HCO_3^- *excess* may be due to dietary intake of HCO_3^- or of organic anions which are subsequently metabolized. Net utilization of endogenous organic anions, e.g., acetoacetate, during realimentation following a period of starvation, also leads to formation of HCO_3^-. If the ketoacidosis was compensated before realimentation, alkalosis may develop.

As described above, *redistribution of HCO_3^- from* the cells to extracellular space occurs during *hypokalemia*.

The consequences of metabolic alkalosis are similar to those of respiratory alkalosis, except that the cerebral vessels and neuromuscular excitability are less sensitive to metabolic than to respiratory alkalosis. The plasma Ca^{2+} activity, on the other hand, undergoes more profound reduction in metabolic than in respiratory alkalosis, due to additional complexing of Ca^{2+} to HCO_3^- [46]. Metabolic alkalosis can decrease the respiratory drive, leading to hypercapnia and hypoxia [5,53].

References

1. Alpern RJ (1990) Cell mechanisms of proximal tubule acidification. Physiol Rev 1:79–114
2. Alpern RJ, Emmett M, Seldin DW (1992) Metabolic alkalosis. In: Seldin DW, Giebisch G (eds) The kidney. Physiology and pathophysiology, 2nd edn. Raven, New York, pp 2733–2758
3. Ayres SM, Grace WJ (1969) Inappropriate ventilation and hypoxemia as causes of cardiac arrhythmias. Am J Med 46:495–505
4. Bartter FC, Pronove P, Gill JR Jr, MacCardle RC (1962) Hyperplasia of the juxtaglomerular complex with hyperaldosteronism and hypokalemic alkalosis. A new syndrome. Am J Med 33:811–828
5. Bear R, Goldstein M, Phillipson E, Ho M, Hammeke M, Feldman R, Handelsman S, Halperin M (1977) Effect of metabolic alkalosis on respiratory function in patients with chronic obstructive lung disease. Can Med Assoc J 117:900–903
6. Bellingham AL, Detter JC, Lenfant C (1971) Regulatory mechanisms of hemoglobin oxygen affinity in acidosis and alkalosis. J Clin Invest 50:700–706
7. Boron WF (1992) Control of intracellular pH. In: Seldin DW, Giebisch G (eds) The kidney. Physiology and pathophysiology, 2nd edn. Raven, New York, pp 219–264
8. Boyer JL, Graf J, Meier PJ (1992) Hepatic transport systems regulating pH_i, cell volume, and bile secretion. Annu Rev Physiol 54:415–438
9. Brackett NC Jr, Cohen JJ, Schwartz WB (1965) Carbon dioxide titration curve of normal man: effect of increasing degrees of acute hypercapnia on acid-base equilibrium. N Engl J Med 272:6–12
10. Brosnan JT, Vinay P, Gougoux A, Halperin ML (1988) Renal ammonium production and its implications for acid-base balance. In: Häussinger D (eds) pH homeostasis. Mechanisms and control. Academic, New York, pp 281–304
11. Cahill GF (1970) Starvation in man. N Engl J Med 282:668–675
12. Cahill GF, Herrera MG, Morgan AP, Soeldner JS, Steinke J, Levy AL, Reichard GA Jr, Kipnis DM (1966) Hormone-fuel interrelationships during fasting. J Clin Invest 45:1751–1769
13. Cameron IR, Semple SJG (1968) The central respiratory stimulant action of salicylates. Clin Sci 35:391–401
14. Chazan JA, Stenson R, Kurland GS (1968) The acidosis of cardiac arrest. N Engl J Med 278:360–364
15. Demarest JR, Loo DDF (1990) Electrophysiology of the parietal cell. Annu Rev Physiol 52:307–320
16. Dulfano MJ, Ishikawa S (1965) Hypercapnia: mental changes and extrapulmonary complications. An expanded concept of the "CO_2 intoxication" syndrome. Ann Intern Med 63:829–841
17. Emmett M, Alpern RJ, Seldin DW (1992) Metabolic acidosis. In: Seldin DW, Giebisch G (eds) The kidney. Physiology and pathophysiology, 2nd edn. Raven, New York, pp 2759–2836
18. Field M, Block MB, Levin R, Rall DP (1966) Significance of blood lactate elevations among patients with acute leukemia and other nephroplastic proliferative disorders. Am J Med 40:528–547
19. Forster RE (1992) Buffering in blood, with emphasis on kinetics. In: Seldin DW, Giebisch G (eds) The kidney. Physiology and pathophysiology, 2nd edn. Raven, New York, pp 171–192
20. Fulop M, Hoberman HD (1975) Alcoholic ketosis. Diabetes 34:785–790
21. Gamble JL, Fahey KR, Appleton J et al (1945) Congenital alkalosis with diarrhea. J Pediatr 26:509–518
22. Gennari FJ, Maddox DA (1992) Renal regulation of acid-base homeostasis: integrated response. In: Seldin DW, Giebisch G (eds) The kidney. Physiology and pathophysiology, 2nd edn. Raven, New York, pp 2695–2732
23. Goldring RM, Turino GM, Heinemann HO (1971) Respiratory-renal adjustments in chronic hypercapnia in man: extracellular bicarbonate concentration and the regulation of ventilation. Am J Med 51:772–784
24. Gotoh F, Meyer JS, Takagi Y (1965) Cerebral effects of hyperventilation in man. Arch Neurol 12:410–423
25. Halperin ML, Kamel KS, Ethier JH, Stinebaugh BJ, Jungas RL (1992) Biochemistry and physiology of ammonium excretion. In: Seldin DW, Giebisch G (eds) The kidney. Physiology and pathophysiology, 2nd edn. Raven, New York, pp 2645–2680
26. Hamm LL, Alpern RJ (1992) Cellular mechanisms of renal tubular acidification. In: Seldin DW, Giebisch G (eds) The kidney. Physiology and pathophysiology, 2nd edn. Raven, New York, pp 2581–2626
27. Hamm LL, Simon EE (1987) Roles and mechanisms of urinary buffer excretion. Am J Physiol 253:F595–F605
28. Häussinger D (1991) Liver and acid-base regulation. In: DeSanto NG, Capasso G (eds) Acid-base balance. Bios, Cosenza, pp 197–204
29. Häussinger D, Meijer AJ, Gerok W, Sies H (1988) Hepatic nitrogen metabolism and acid-base homeostasis. In: Häussinger D (eds) pH homeostasis. Mechanisms and control. Academic, New York, pp 337–371
30. Huckabee WE (1961) Abnormal resting lactate. II. Lactic acidosis. Am J Med 30:840–848
31. Jenkins DW, Eckel RE, Craig JW (1971) Alcoholic ketoacidosis. JAMA 217:177–183
32. Johnson RL Jr, Ramanathan M (1992) Buffer equilibria in the lungs. In: Seldin DW, Giebisch G (eds) The kidney. Physiology and pathophysiology, 2nd edn. Raven, New York, pp 193–218
33. Kamel KS, Stinebaugh BJ, Schloeder FX, Halperin ML (1992) Acid-base, fluid, and electrolyte aspects of the kidney during starvation. In: Seldin DW, Giebisch G (eds) The kidney. Physiology and pathophysiology, 2nd edn. Raven, New York, pp 3457–3470
34. Kazemi H, Hitzig BM (1992) Central chemical control of ventilation and acid-base balance. In: Seldin DW, Giebisch G (eds) The kidney. Physiology and pathophysiology, 2nd edn. Raven, New York, pp 2627–2644
35. Kazemi H, Johnson DC (1986) Regulation of cerebrospinal fluid acid-base balance. Physiol Rev 66:953–1037
36. Knepper MA, Packer R, Good DW (1989) Ammonium transport in the kidney. Physiol Rev 69:179–249
37. Krapf R (1989) Physiology and molecular biology of the renal Na^+/H^+ antiporter. Klin Wochenschr 67:847–851
38. Lang F, Rehwald W (1992) Potassium channels in renal epithelial transport regulation. Am J Physiol 72:1–32
39. Lang F, Messner G, Rehwald W (1986) Electrophysiology of sodium-coupled transport in proximal renal tubules. Am J Physiol 250:F953–F962

40. Lang F, Oberleithner H, Kolb HA, Paulmichl M, Völkl H, Wang W (1988) Interaction of intracellular pH and cell membrane potential. In: Häussinger D (eds) pH homeostasis. Mechanisms and control. Academic, New York, pp 27–42

41. Lang F, Völkl H, Wöll E, Rehwald W (1991) Proximal renal tubular acidification in disease – clues from electrophysiology. In: DeSanto NG, Capasso G (eds) Acid-base balance. Bios, Cosenza, pp 51–60

42. Lawson WH Jr (1966) Interrelation of pH, temperature and oxygen on deoxygenation rate of red cells. J Appl Physiol 21:905–914

43. Lyons HA, Antonio R (1959) The sensitivity of the respiratory center in pregnancy and after the administration of progesterone. Trans Assoc Am Physicians 72:173–180

44. Madias NE, Cohen JJ (1992) Respiratory alkalosis and acidosis. In: Seldin DW, Giebisch G (eds) The kidney. Physiology and pathophysiology, 2nd edn. Raven, New York, pp 2837–2872

45. Mazzara JT, Ayres SM, Grace WJ (1974) Extreme hypocapnia in the critically ill patient. Am J Med 56:450–456

46. Oberleithner H, Greger R, Lang F (1978) Role of calcium in the decline of phosphate reabsorption during phosphate loading in acutely thyroparathyroidectomized rats. Pflugers Arch 374:249–254

47. Pontén U (1966) Consecutive acid-base changes in blood, brain tissue and cerebrospinal fluid during respiratory acidosis and baseosis. Acta Neurol Scand 42:455–471

48. Potter EK, McCloskey (1979) Respiratory stimulation by angiotensin II. Respir Physiol 36:367–373

49. Pouysségur J, Seuwen K (1992) Transmembrane receptors and intracellular pathways that control cell proliferation. Annu Rev Physiol 54:195–210

50. Record CO, Iles RA, Cohen RD, Williams R (1975) Acid-base and metabolic disturbances in fulminant hepatic failure. Gut 16:144–149

51. Relman AS (1972) Metabolic consequences of acid-base disorders. Kidney Int 1:347–359

52. Sardet C, Wakayabashi S, Fafournoux P, Counillon L, Pagès G, Pouysségur J (1991) Molecular properties of Na^+/H^+ exchanges. In: DeSanto NG, Capasso G (eds) Acid-base balance. Bios, Cosenza, pp 13–20

53. Shear L, Brandman IS (1973) Hypoxia and hypercapnia caused by respiratory compensation for metabolic alkalosis. Am Rev Respir Dis 107:836–841

54. Shils ME (1969) Experimental human magnesium depletion. Medicine (Baltimore) 48:61–85

55. Smith RB, Aass AA, Nemeto EM (1981) Intraocular and intracranial pressure during respiratory alkalosis and acidosis. Br J Anaesth 53:967–972

56. Völkl H, Lang F (1991) Electrophysiology of ammonia transport in renal straight proximal tubules. Kidney Int 40:1082–1089

57. Whelan RF, Young IM (1953) The effect of adrenaline and noradrenaline infusions on respiration in man. Br J Pharmacol 8:98–102

58. Yasue H, Nagao M, Omote S, Takizawa A, Miwa K, Tanaka S (1978) Coronary arterial spasm and Prinzmetal's variant form of angina induced by hyperventilation and tris-buffer infusion. Circulation 58:56–62

79 K⁺ Metabolism

F. LANG

Contents

79.1 Physiological Significance of K⁺

K⁺ is accumulated within cells, and the chemical gradient for K⁺ across the cell membrane drives K⁺ from the cell to the extracellular space. The exit of K⁺ creates an outside-positive potential difference across the cell membrane. For most cells K⁺ is of paramount importance for the generation and maintenance of cell membrane potential [24] (see also Chap. 11).

For the magnitude of *cell membrane potential*, two factors are to be considered: the equilibrium potential for K⁺ (E_K) and the selectivity of the cell membrane for K⁺. The equilibrium potential for K⁺ critically depends on the K⁺ gradient across the cell membrane:

$$E_K = 61\,mV \times lg[K^+]_o/[K^+]_i. \qquad (79.1)$$

Since intracellular K⁺ activity ($[K^+]_i$) is far higher than extracellular K⁺ activity ($[K^+]_o$), the equilibrium potential for K⁺ is clearly more sensitive to extracellular than to intracellular alterations of K⁺ activity. For instance, a doubling of intracellular K⁺ activity (e.g., from 100 to 200 mmol/l) has the same effect on E_K as a decrease of extracellular K⁺ activity to half (e.g., from 5 to 2.5 mmol/l). The *selectivity* of the cell membrane to K⁺ is reflected by the K⁺ transference number (t_K), i.e., the contribution of K⁺ conductance (G_K) to overall cell membrane conductance (G_t):

$$t_K = G_K/G_t. \qquad (79.2)$$

The cell membrane potential (E_M) can be expressed as:

$$E_M = t_K \times E_K + t_R \times E_R, \qquad (79.3)$$

where t_R and E_R are the lumped transference number and equilibrium potential for the remaining electrogenic transport processes.

The *conductance of K⁺ channels* (G_K) depends on the ambient K⁺ concentration. At constant K⁺ permeability (P), the conductance of the K⁺ channels increases with the concentration of K⁺ within the channel, which in turn depends on the K⁺ concentration on the intra- and extracellular sides of the cell membrane. Thus, the conductance of the K⁺ channels is enhanced during hyperkalemia and decreased during hypokalemia.

A decrease of extracellular K⁺ activity modifies cell membrane potential by both increasing E_K and decreasing G_K, leading to decrease of t_K. The net effect depends on the properties of the K⁺ channels involved and the t_K before reduction of extracellular K⁺. In general, the increase of E_K prevails in those cells in which t_K is very high. If t_K is close to 1, where the membrane potential is close to E_K, a reduction of extracellular K⁺ activity leads to a proportionate increase in the membrane potential and E_K. If t_K is low, the effect of decreasing t_K may even surpass the effect of increasing E_K, and the cell membrane potential may even depolarize following a reduction of extracellular K⁺ activity.

Via its influence on cell membrane potential, K⁺ affects such diverse functions as:

- Excitability of neuromuscular cells [11,24]
- Electrogenic transport across epithelia [26,46]
- Activation of lymphocytes [30]
- Release of hormones (Fig. 79.1).

While an increase in extracellular K⁺ concentration stimulates the release of insulin [1], glucocorticoids [5], and aldosterone (cf. Chap. 75), it inhibits the release of ouabain [5].

The K⁺-dependent cell membrane potential also governs the HCO_3^- distribution across the cell membrane and thus the intracellular pH of a great variety of cells: a decrease in extracellular K⁺ concentration leads to hyperpolarization, driving HCO_3^- out of the cell and favoring intracellular acidosis, while an increase in extracellular K⁺ concentration leads to depolarization favoring intracellular alkalosis [27] (see Chaps. 75, 78).

Since K⁺ constitutes the major component of *intracellular osmolarity*, K⁺ determines the *cell volume* [13,15,17,19,23, 28,31,41,44]. An increase in extracellular K⁺ activity leads

R. Greger/U. Windhorst (Eds.)
Comprehensive Human Physiology, Vol. 2
© Springer-Verlag Berlin Heidelberg 1996

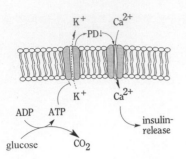

Fig. 79.1. Role of K$^+$ in the release of insulin. Glucose leads to the formation of ATP, which inhibits K$^+$ channels at the cell membrane. The resulting depolarization (*PD↓*) leads to the activation of voltage-sensitive Ca^{2+} channels. The entering Ca^{2+} then triggers insulin release by exocytosis (for further details see Chap. 66)

to net uptake of K$^+$ into the cells, and due to K$^+$-induced depolarization anions, such as Cl$^-$, are taken up in parallel. As a result, K$^+$ excess tends to swell the cells, whereas K$^+$ depletion promotes cell shrinkage.

Possibly by modifying the cell volume of "*osmoreceptors*," K$^+$ excess inhibits and K$^+$ depletion stimulates ADH release. K$^+$ administration could, possibly by inhibiting *ADH release*, stimulate water excretion and thus produce slight hypernatremia.

An increase in extracellular K$^+$ concentration leads to an increase of *oxygen consumption* by skeletal muscle [24]. This increase is attributed to a release of Ca^{2+} into the sarcoplasm. The effect does not depend on muscle contraction, which requires higher concentrations of cytosolic Ca^{2+} [24].

79.2 Regulation of K$^+$ Metabolism

79.2.1 Distribution of K$^+$ Across the Cell Membrane

Since most of the K$^+$ in the body is within the cells, extracellular K$^+$ activity is heavily influenced by factors modifying the *transport of K$^+$* across the cell membrane [21,37]. One such factor is acid-base balance (Fig. 79.2). In (extracellular) *alkalosis* the cells release H$^+$ in exchange for Na$^+$ by the Na$^+$/H$^+$ exchanger. The H$^+$ released by the cells comes in large part from intracellular buffers and serves to blunt the extracellular alkalosis. The Na$^+$ taken up by the cells is replaced by K$^+$ through the (Na$^+$+K$^+$)-ATPase, which is actually stimulated by alkalosis [37]. As a result, extracellular alkalosis leads to net cellular uptake of K$^+$ and to *hypokalemia*. Conversely, (extracellular) *acidosis* impedes cellular release of H$^+$, Na$^+$/H$^+$ exchange, and (Na$^+$+K$^+$)-ATPase, thus favoring cellular loss of K$^+$ and *hyperkalemia*. The influence of acid-base balance on extracellular K$^+$ concentration depends on the nature of the disturbance. Metabolic acidosis secondary to infusion of NH$_4$Cl or HCl increases extracellular K$^+$ concentration on average by as much as 0.7 mmol/l per 0.1 pH unit. Far

smaller effects on extracellular K$^+$ concentration are encountered in metabolic acidosis due to increased formation of lactate, β-hydroxybutyrate, etc. The organic acids are probably taken up as anions by the cells, thus reducing the cellular loss of K$^+$. Similarly, metabolic acidosis by endogeneous production of organic acids leads to a far lower increase of extracellular K$^+$ than infusion of H$^+$. Respiratory acidosis is less effective again in reducing extracellular K$^+$ concentration (on average some 0.2 mmol/ 1 per 0.1 pH unit) than metabolic acidosis. Metabolic alkalosis due to infusion of bicarbonate leads to a decrease in extracellular K$^+$ concentration varying from 0.1 to 0.4 mmol/l per 0.1 pH unit. Respiratory alkalosis is less effective again.

Several *hormones* also modify K$^+$ transport across the cell membrane and thus affect extracellular K$^+$ concentration. The most important of these hormones is *insulin*, which influences the distribution of K$^+$ across the cell membrane [37]. In several tissues, the hormone stimulates the 2Cl$^-$Na$^+$K$^+$ cotransporter, the Na$^+$/H$^+$ exchanger and the (Na$^+$+K$^+$)-ATPase (Fig. 79.3). The concerted action of these pumps leads to *intracellular alkalosis*, cellular *accumulation of KCl*, and *cell swelling*. The cell swelling is an important element in the action of the hormone, since it inhibits proteolysis and glycogenolysis, two major effects of the hormone [16–18,29]. During infusion of insulin, two-thirds of the K$^+$ shifted into the cells is taken up by the liver within the 1st h, whereas uptake into peripheral tissues prevails later on. It is important to know that the effect of insulin depends on the K$^+$ content of the cells before administration of the hormone starts. In individuals with longstanding insulin depletion, e.g., due to prolonged diabetic status or starvation, the cells tend to be depleted of K$^+$, and in such individuals administration of insulin (in diabetics) or realimentation (which is followed by insulin

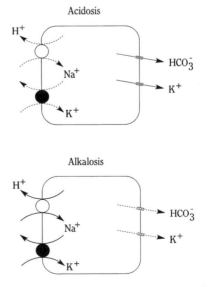

Fig. 79.2. Effects of acid-base balance on K$^+$ distribution across the cell membrane. *Solid arrows*, Enhanced transport; *broken arrows*, reduced transport

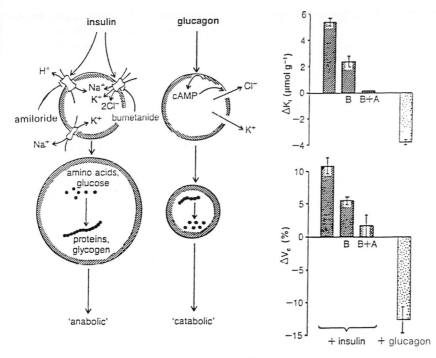

Fig. 79.3. K$^+$ transport in the hepatocellular actions of insulin and glucagon. Insulin leads to parallel activation of Na$^+$2Cl$^-$K$^+$ cotransport, Na$^+$/H$^+$ exchange, and (Na$^+$+K$^+$)-ATPase. The cellular uptake of KCl leads to cell swelling, which is the cellular signal to reduce proteolysis and glycogenolysis. Glucagon stimulates proteolysis and glycogenolysis by cell shrinkage due to loss of KCl through ion channels. On the right side, the effects of insulin and of glucagon on cellular K$^+$ content (ΔK_i) and cell volume (ΔV_c). *B*, Bumetanide (inhibitor of Na$^+$2Cl$^-$K$^+$-cotransporter); *A*, amiloride (inhibitor of Na$^+$/H$^+$ exchange). (After [16,17])

release), may lead to a pronounced and life-threatening drop in extracellular K$^+$ concentration. The increase in K$^+$ concentration in the blood following K$^+$ infusion is blunted in the presence of insulin and augmented in the presence of reduced insulin concentrations or sensitivity, due to stimulation of cellular K$^+$ uptake by the hormone.

In contrast to insulin, *glucagon* stimulates *K$^+$ release* from liver cells, probably through activation of ion channels. The effect is mimicked by cAMP. The hormone-induced electrolyte loss shrinks hepatocytes, an effect accounting in part for the glycogenolytic and proteolytic action of the hormone [16–18,29]. Through its effect on liver, glucagon leads to a transient increase in extracellular K$^+$ concentration [37].

Catecholamines may either increase or reduce extracellular K$^+$ concentration [37]. Increase of extracellular K$^+$ concentration is due to activation of K$^+$ channels and, possibly, reduction of (Na$^+$+K$^+$)-ATPase activity by α-receptors, reduction of extracellular K$^+$ concentration to stimulation of (Na$^+$+K$^+$)-ATPase by β_2-receptors and subsequent formation of cAMP. Injection of epinephrine leads to a rapid transient increase followed by a more sustained drop in extracellular K$^+$ concentration. Specific α-antagonists increase and specific β-agonists reduce extracellular K$^+$ concentration. The prevailing effect of β-receptor stimulation when sympathetic tone is enhanced, e.g., in patients with myocardial infarction, may lead to hypokalemia, which may further complicate the condition. Furthermore, the use of β_2-agonists for the treatment of asthma or to prevent premature labor may lead to hypokalemia as low as 0.9 mmol/l [37]. In hypokalemia, α-adrenergic tone is thought to be enhanced. As a result, the muscular (Na$^+$+K$^+$)-ATPase is inhibited and uptake of K$^+$ into muscle cells impaired. This mechanism counteracts further decline of extracellular K$^+$ concentration [24]. The (Na$^+$+K$^+$)-ATPase is inhibited by ouabain, a hormone from adrenal glands [5].

In skeletal muscle cells, *thyroid hormones* stimulate the permeability to Na$^+$ while reducing the permeability to K$^+$ [24]. The cellular Na$^+$ accumulated stimulates the (Na$^+$+K$^+$)-ATPase and thus cellular *uptake of K$^+$*. As a result, thyroid hormones may promote the development of hypokalemia [24].

Several other hormones are known to modify K$^+$ transport in defined tissues under appropriate conditions. In liver, for instance, cellular K$^+$ release has also been observed following administration of adenosine, ATP, and ADH, an effect attributed to activation of ion channels in the plasma membrane [16,17]. None of these hormones affects plasma K$^+$ activity to the same extent as insulin, glucagon, catecholamines, and disorders of acid-base balance.

An *increase in extracellular osmolarity* may lead to modest hyperkalemia, presumably due to cellular release of K$^+$ [25]. This effect is poorly understood, since in many cells studied in vitro, an increase in extracellular fluid osmolarity leads to cellular shrinkage, followed by regulatory volume increase due to cellular K$^+$ uptake through activation of $Na^+2Cl^-K^+$ cotransport or Na^+/H^+ exchange and Na^+/K^+ ATPase [29].

Mg^{2+} has been shown to inhibit ATP-sensitive K^+ channels. *Mg^{2+} excess* leads to cellular *retention of K^+*, while Mg^{2+} depletion leads to cellular loss of K^+, especially in heart and skeletal muscle [37].

Physical exercise is paralleled by muscle cellular *loss of K^+* due to depolarization (increase of electrochemical K^+ gradient) and repolarization (activation of K^+ channels) of the cell membrane. Despite enhanced (Na^++K^+)-ATPase activity, cellular Na^+ activity and interstitial K^+ activity may increase by some 10 mmol/l. The local increase in extracellular K^+ concentration is one trigger for *vasodilation* of exercising muscles. If the exercise involves a large group of muscles, an increase in plasma K^+ concentration may follow [24], and it has been suggested that hyperkalemia may be one cause of cardiac death during exhaustive physical exercise or sustained convulsions. In K^+ depletion, the muscle cellular release of K^+ is decreased and the vasodilatory response blunted, leading to ischemic injury of the muscle. Frequent physical exercise (training) increases the number of (Na^++K^+)-ATPases in muscle cell membranes. In highly trained individuals, such as competitive long-distance runners, the enhanced (Na^++K^+)-ATPase activity can lead to hypokalemia at rest [24].

Like other energy-consuming processes, transport via the (Na^++K^+)-ATPase is highly sensitive to *ambient temperature* [22]. Accordingly, in hypothermia the (Na^++K^+)-ATPase is inhibited and thus cells lose K^+ and gain Na^+. As a result, the extracellular K^+ concentration increases. During the rewarming phase, the (Na^++K^+)-ATPase is reactivated, and the Na^+ that has been accumulated within the cell during hypothermia is now extruded in exchange for extracellular K^+, which may eventually lead to hypokalemia.

Obviously, energy depletion of the cells, as occurs during *hypoxia*, leads to impairment of (Na^++K^+)-ATPase and cellular loss of K^+. Similar to what occurs following hypothermia, cellular uptake of K^+ is enhanced during recovery from hypoxia and may actually lead to hypokalemia [24]. In *renal insufficiency*, (Na^++K^+)-ATPase activity in a variety of tissues is inhibited. The cellular loss of K^+ may contribute to the *hyperkalemia* encountered in that condition. The mechanism underlying inhibition of (Na^++K^+)-ATPase is not fully understood, but it has been suggested that a circulating factor inhibits the pump. One possible candidate is vanadate, an inhibitor of various ATPases including (Na^++K^+)-ATPase. Impaired renal elimination leads to high plasma concentrations of vanadate during renal insufficiency [3]. Whether reduced elimination of NaCl and water and subsequent extracellular volume expansion in renal insufficiency stimulates the release of ouabain [5], the endogeneous inhibitor of (Na^++K^+)-ATPase, has not yet been sufficiently explored.

Excessive stimulation of muscle metabolism with corresponding cellular release of K^+ and occurrence of hyperkalemia is seen in *malignant hyperthermia*. This disorder is elicited by several general anesthetics in suscepti-ble patients and is characterized by an increase in body temperature up to 44°C due to enhanced cellular Ca^{2+} release [24].

In *familial hypokalemic periodic paralysis*, attacks of hypokalemia occur due to net uptake of K^+ into muscle cells. The condition has been suggested to result from a defect in voltage-regulated K^+ channels [24].

79.2.2 Regulation of Renal K^+ Excretion

The K^+ balance of the body is in large part dependent on dietary uptake for the one part and *renal elimination* for the other [47] (see also Chap. 75). Even though the colon is capable of excreting K^+ and adapting its excretory function to the demands of the body, the K^+ eliminated by the kidney far exceeds that excreted by the intestine.

As outlined in Chap. 75, the K^+ concentration in the glomerular filtrate is close to the plasma K^+ concentration. On the one hand, capillary negative potential difference across the glomerular filter (Gibbs-Donnan potential) impedes filtration of K^+, thus reducing the K^+ concentration in the filtrate by some 10%. On the other hand, plasma proteins comprise some 10% of plasma volume, so the K^+ distribution space is only some 90% of plasma volume [47].

K^+ transport in the *proximal tubule* is mainly passive and parallels fluid reabsorption in that nephron segment. K^+ channels are on both the apical and the basolateral cell membrane (see Fig. 77.1). By the end of the proximal tubule accessible to micropuncture, some 50–70% of filtered K^+ has been reabsorbed [47] (see also Chaps. 74, 75).

In the *loop of Henle*, K^+ recycling occurs, with K^+ reabsorption in the ascending limb and K^+ secretion in the descending limb of the loop. At the tip of the papilla, the K^+ concentration can approach a value ten times that in plasma [47]. In the thick ascending limb K^+ reabsorption still prevails slightly (Chap. 75). In the distal nephron net reabsorption of K^+ may occur in conditions of profound K^+ depletion. Usually, however, K^+ secretion prevails in the distal tubule and collecting duct.

In the *early distal tubule* cells K^+ transport involves electroneutral KCl cotransport across the luminal cell membrane, whereas K^+ transport across the basolateral cell membrane is accomplished by (Na^++K^+)-ATPase and K^+ channels [47].

K^+ reabsorption in the *distal tubule* is performed by K^+/H^+-ATPase in *intercalated cells*, which reabsorbs K^+ in exchange for H^+. Distal nephron secretion of K^+ mainly resides in *principal cells*. In these cells K^+ channels are on both apical and basolateral cell membranes. In addition, apical Na^+ channels and basolateral (Na^++K^+)-ATPase operate to carry out Na^+ reabsorption and K^+ secretion. Distal tubular K^+ secretion eventually determines the amount of K^+ excreted in urine. Factors influencing distal tubular K^+ secretion can be seen from Fig. 79.4. Distal tubular K^+ secretion (S_{K^+}) across the luminal cell membrane depends on the chemical gradient for K^+:

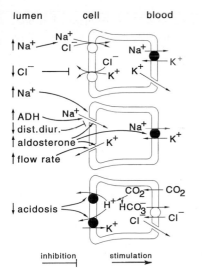

Fig. 79.4. Various factors influencing distal tubular K^+ secretion and their mechanisms of action. The factors are listed at the luminal side of the cell even though some of them exert their action from the basolateral side. *Dist. diur.*, Distal tubule diuretics

$$S_{K^+} \sim \lg[K^+]_i / [K^+]_l. \tag{79.4}$$

- An increase in luminal K^+ activity reduces and a decrease in luminal K^+ activity increases the chemical gradient and thus secretion.
- An increase in intracellular K^+ activity has a relatively small effect on the gradient (see above); however, it does increase the conductance of the luminal K^+ channels and thus favors K^+ secretion. Conversely, a decline in intracellular K^+ activity reduces distal tubular K^+ secretion.
- Chronic K^+ depletion decreases and chronic K^+ excess increases the ability of the distal nephron to secrete K^+. This effect is at least partially due to aldosterone, which is released during K^+ excess and stimulates the biosynthesis of all elements subserving K^+ secretion in the distal nephron [33].
- Whether an additional hormone, produced in either the liver or the central nervous system, adds to regulation of distal tubular K^+ secretion, remains a matter of speculation.
- Luminal K^+ activity is influenced by the luminal flow rate. Accordingly, when luminal flow rate is reduced, luminal K^+ activity increases rapidly to approach the electrochemical K^+ equilibrium across the luminal cell membrane, and K^+ secretion is reduced. Conversely, when luminal flow rate is high, K^+ activity increases slowly and K^+ secretion is increased [47].
- Distal tubular K^+ secretion is stimulated by reabsorption of Na^+, which depolarizes the luminal cell membrane and is extruded across the basolateral cell membrane in exchange for K^+ via the $(Na^+ + K^+)$-ATPase. Distal tubular Na^+ reabsorption, on the other hand, increases with Na^+ delivery to the distal tubule, and is accordingly increased if Na^+ in preceding nephron segments is impaired, as

under the influence of proximal (e.g., acetazolamide) and loop diuretics (e.g., furosemide).

- Distal tubular Na^+ reabsorption and K^+ secretion are further increased under the influence of aldosterone (Fig. 79.5) and ADH, which stimulates Na^+ channels in the luminal cell membrane.
- Catecholamines, on the other hand, slightly reduce distal tubular K^+ secretion [47].
- Inhibition of distal tubular Na^+ reabsorption by Na^+ channel blockers (e.g., amiloride) or aldosterone antagonists (e.g., spironolactone) reduces distal tubular K^+ secretion accordingly [20,47].
- Distal tubular K^+ secretion is dependent on acid-base balance [47]. In acidosis the enhanced H^+ secretion renders the lumen more positive, hyperpolarizes the luminal cell membrane, and thus blunts K^+ secretion. Moreover, the K^+ channels involved in K^+ secretion are sensitive to intracellular pH [14,40,42,45]. Intracellular acidosis inhibits the channels and thus distal tubular K^+ secretion. Acidosis further enhances production of NH_4^+, which has been shown to inhibit K^+ channels [14]. Excretion of NH_4^+ is indeed inversely correlated with renal excretion of K^+ [43]. Accordingly, acidosis inhibits and alkalosis stimulates distal tubular secretion and renal excretion of K^+.
- Distal tubular K^+ secretion and renal K^+ excretion, follow a circadian rhythm being lowest at night and highest in the afternoon [36].

79.2.3 K^+ Transport in Other Epithelia

The ducts of *salivary glands*, *sweat glands*, *pancreas*, and *distal colon* reabsorb Na^+ in exchange for K^+ by parallel operation of the respective channels (Fig. 79.6). Like in the *distal nephron*, Na^+ reabsorption and K^+ secretion are

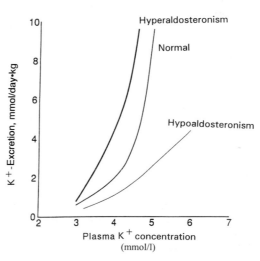

Fig. 79.5. Dependence of urinary K^+ excretion on plasma K^+ concentration at normal aldosterone concentration (*middle line*) and at enhanced (*upper line*) and lowered (*lower line*) aldosterone plasma levels

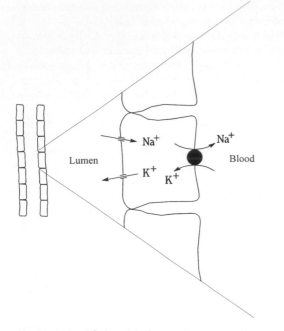

Fig. 79.6. Simplified model of Na^+ reabsorption and K^+ secretion in epithelia such as collecting duct and colon

stimulated by aldosterone. These transport processes are usually not crucially important to the K^+ balance of the body. However, in diarrhea K^+ losses may be life-threatening, especially if aldosterone levels are high.

79.3 Disorders of K^+ Metabolism

79.3.1 Hypokalemia

A drop in extracellular K^+ activity could be due either to an imbalance between intake and loss of K^+ or to a shift of K^+ into the cells [32]. The most important *causes of hypokalemia* are listed in Table 79.1.

Reduced *dietary intake* does not usually cause hypokalemia, since renal excretion can be reduced to as little as 2% of filtered load [47]. However, in excessive eating disorders [34] and alcoholism [35], dietary intake may indeed be too low to maintain the K^+ balance of the body.

Renal loss of K^+ may result from *hyperaldosteronism*, which enhances distal tubular K^+ secretion mainly due to stimulation of K^+ channels, Na^+ channels (leading to depolarization of the apical cell membrane), and (Na^++K^+)-ATPase [47] (see also preceding section).

Diuretics which impair Na^+ reabsorption in nephron segments proximal to the distal tubule increase the delivery of Na^+ and the flow rate at the principal cells and thus augment distal nephron Na^+ reabsorption and K^+ secretion [47]. This effect is compounded by the *contraction of extracellular fluid volume*, which stimulates the release of aldosterone [47]. Loop diuretics and early distal tubule diuretics especially lead to enhanced renal excretion of K^+.

Similarly, injury of proximal renal tubular epithelium and thick ascending limb by ischemia, heavy metals, organic nephrotoxic agents, etc. [32] augments the delivery of Na^+ to the distal nephron and stimulates distal tubular secretion of K^+.

Several congenital disorders can lead to *renal K^+ wasting*. In proximal renal tubular acidosis [2] or Fanconi syndrome [12] proximal tubular Na^+ reabsorption is compromised, in Bartter's syndrome [32] NaCl reabsorption is impaired in the thick ascending limb of Henle's loop. In these conditions, the increased delivery of Na^+ to the distal nephron increases renal K^+ excretion and results in K^+ depletion [39]. Distal tubular Na^+ reabsorption is further increased in Liddle syndrome, and in distal tubular acidosis [2] distal tubular H^+ secretion is reduced, both conditions leading to renal K^+ wasting. Renal K^+ loss occurs further in Mg^{2+} depletion [10] due to disinhibition of luminal K^+ channels.

Extrarenal loss of K^+ occurs during diarrhea secondary to infections, vasoactive intestinal polypeptide-producing tumors, etc. [32]. Like kidney, the distal colon can reabsorb Na^+ in exchange for K^+, and thus increased delivery of Na^+ to the distal colon stimulates K^+ secretion. The fluid loss during diarrhea leads to *extracellular volume contraction* and release of aldosterone, which stimulates not only renal but also intestinal Na^+/K^+ exchange.

Several factors cause movement of K^+ from extracellular to intracellular space, producing hypokalemia without K^+ de-

Table 79.1. Major causes of hypokalemia

Renal loss of K^+
Hyperaldosteronism
Diuretics
Loop diurectics
Early distal tubule diuretics
Congenital tubular transport defects
Fanconi syndrome
Proximal tubular acidosis
Bartter's syndrome
Distal tubular acidosis
Liddle syndrome
Acquired defective tubular transport
Ischemia
Heavy metals
Organic nephrotoxic substances
Extrarenal loss of K^+
Diarrhea
Infection
Vasoactive intestinal polypeptide-producing tumors
Deficient dietary K^+ intake
Alcoholism
Eating disorders
Cellular accumulation of K^+
Insulin
Alkalosis
β_2-Agonists
Thyroid hormones
Recovery from hypoxia
Recovery from hypothermia
Physical training
Familial hypokalemic periodic paralysis
Barium poisoning

Table 79.2. Major consequences of hypokalemia

Decreased neuromuscular excitability
Intestinal and bladder atonia
Accelerated phase 4 depolarization and extrasystoles in the heart
Altered hormone release
Insulin
Aldosterone
ADH
Catecholamines
Altered renal function
Decreased renal concentrating ability
Enhanced renal ammonia production and acid excretion
Enhanced susceptibility to gentamycin toxicity

pletion. The most important are insulin and alkalosis (see Sect. 79.2.1). Other factors causing hypokalemia due to stimulation of *cellular K+ uptake* include administration of β_2-agonists, hyperthyroidism, and recovery from hypoxia, hypothermia, or excessive physical training (see Sect. 79.2.1). A congenital disorder characterized by enhanced K+ uptake into muscle cells leading to hypokalemia is familial hypokalemic periodic paralysis (see Sect. 79.2.1). Ingestion of soluble barium salts may lead to barium poisoning: Ba^{2+} inhibits K+ channels, thus causing cellular K+ accumulation and hypokalemia [24].

Consequences of hypokalemia are decreased neuromuscular excitability and reduced activity of smooth muscle in bladder and intestine, phenomena attributed to the increase of E_K (Table 79.2). On the other hand, the decrease of G_K accounts for accelerated phase 4 depolarization in the *heart*, i.e., the frequency of automaticity is increased. Hypokalemia can also facilitate depolarization and elicit action potentials in heart cells other than sinus, leading to extrasystoles [32].

Hypokalemia may severely affect the integrity and function of the *kidney*. The concentrating ability of the kidney is decreased, probably due to limited luminal availability of K+ for the $Na^+K^+2Cl^-$ cotransporter in the thick ascending limb of Henle's loop, compromising NaCl reabsorption in that segment. Presumably due to enhanced delivery of NaCl to the macula densa, the juxtaglomerular apparatus is hypertrophied [32]. Hypokalemia leads to hyperpolarization of the proximal tubule cells, leading to intracellular acidosis and stimulation of ammonia production (cf. Chap. 78). The ammonia accumulates within acidic intracellular compartments, which swell, form large vacuoles, and may actually contribute to renal cellular injury observed in hypokalemia [32]. Renal acid excretion is enhanced and leads to alkalosis, which is compounded by the shift of H+ into the cells (see above).

79.3.2 Hyperkalemia

Causes of hyperkalemia include increased intake of K+, cellular K+ losses, and renal K+ retention (Table 79.3). *Enhanced dietary intake* does not on the whole produce clinically significant hyperkalemia when renal function

and cellular uptake are normal. However, K+ intake with K+ supplements and K+-containing medications may exceed renal excretory capacity [6]. Obviously, hemodialysis with K+-rich dialysate may produce hyperkalemia.

Renal retention may result from global renal failure or from reduced renal tubular secretion of K+. The latter may be due to reduced delivery of Na+ to the distal nephron, with a resulting decrease in Na+ reabsorption and K+ secretion in that segment. Similarly, lack of stimulation (hypoaldosteronism) or genetic defects such as pseudohypoaldosteronism or hyperkalemic distal tubular acidosis can lead to renal K+ retention. Several diseases, such as tubulointerstitial disease, sickle cell disease [9], and systemic lupus erythematosus [7], are known to affect distal nephron function, leading to hypokalemia. Similarly, distal tubular K+ secretion can be impaired in transplanted kidneys [8] or in obstructive uropathy [38]. Several drugs inhibit distal K+ secretion by inhibition of the luminal Na+ channels (triamterene, amiloride), the (Na^++K^+)-ATPase (digitalis glycosides), or aldosterone receptors (spironolactone) [6].

A *redistribution of K+* from cells to the extracellular space may occur in acidosis or insulin deficiency (see above). Similarly, α-adrenergic agonists and β-adrenergic antagonists favor movement of K+ from cellular to extracellular space (see above). Periodic cellular loss of K+ is encountered in the genetic disorder hyperkalemic periodic paraly-

Table 79.3. Major causes of hyperkalemia

Cellular loss of K+
Acidosis
Insulin
α-Adrenergic agonists
β-Adrenergic agonists
Hyperkalemic periodic paralysis
Hypertonicity
Exercise
Digitalis glycosides
Extensive cell death
Cytotoxic drugs
Hemolysis, burns
Crush syndrome
Increased input
Diet
Hemodialysis
Medications
Impaired renal elimination
Renal failure (GFR < 15 ml)
Decreased renal tubular secretion
Hypoaldosteronism
Na+ depletion
Defective renal tubular secretion
Pseudohypoaldosteronism
Hyperkalemic distal tubular acidosis
Tubulointerstitial disease
Sickle cell disease
Systemic lupus erythematosus
Postrenal transplantation
Obstructive uropathy
Inhibition of renal tubular secretion
Spironolactone
Triamterene, amiloride
Digitalis glycosides

Table 79.4. Major consequences of hyperkalemia

Cardiotoxicity
  Increased rate of depolarization (tenting of T)
  Delayed conduction through His-Purkinje system and
  Ventricular muscle (widening of QRS)
  Impairment of atrial muscle conduction (flattening of P)
  Ventricular fibrillation
  Asystole
Neuromuscular disturbances
  Paresthesias
  Weakness, paralysis

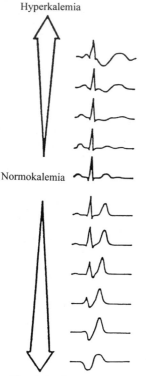

Hyperkalemia

Normokalemia

Hypokalemia

Fig. 79.7. Effects of increasing or decreasing extracellular K$^+$ concentration on the electrocardiogram

sis [4]. Further conditions associated with cellular loss of K$^+$ and hyperkalemia include increase of extracellular osmolarity, excessive exercise, and inhibition of (Na$^+$+K$^+$)-ATPase with digitalis glycosides. Obviously, K$^+$ is released into the extracellular space during extensive cellular death, as it occurs under the influence of cytotoxic drugs, during hemolysis, or after burns or crush.

Effects of hyperkalemia are shown in Table 79.4. Hyperkalemia leads to depolarization with resulting *hyperexcitability* of neuromuscular cells. The inactivation of Na$^+$ channels in muscle cells during sustained depolarization could eventually lead to *paralysis*. The sensory system is less affected, but paresthesias may occur. In the heart hyperkalemia increases the rate of depolarization, leading to shortening of the plateau phase of the action potential. In the electrocardiogram these effects are re-

flected by a steep (tenting) T wave and a shortening of the ST interval (Fig. 79.7; see also Chaps. 91, 92). Hyperkalemia delays the conduction through the His-Purkinje system and the ventricular muscle, leading to a widening of the QRS complex. Impairment of atrial muscle conduction leads to a flattening and eventually to a disappearance of P waves. The slowing down of conduction and shortening of the plateau phase favor the occurrence of cardiac fibrillation. Eventually *cardiac arrest* may result from sustained depolarization. The effect of hyperkalemia on cardiac function depends on the intracellular K$^+$ concentration and may be compounded by simultaneous hyponatremia, hypocalcemia, and hypomagnesemia [6].

References

1. Ashcroft FM, Ashcroft SJH (eds) (1992) Insulin. Oxford University Press, Oxford
2. Batlle D (1989) Renal tubular acidosis. In: Seldin D, Giebisch G (eds) The regulation of acid-base balance. Raven, New York, pp 353–390
3. Bello-Reuss EN, Grady TP, Mazumdar DC (1979) Serum vanadium levels in chronic renal disease. Ann Intern Med 91:743
4. Bia MJ, DeFronzo RA (1981) Extrarenal potassium homeostasis. Am J Physiol 240:F257–F268
5. Blaustein MP (1993) Physiological effects of endogenous ouabain: control of intracellular Ca^{2+} stores and cell responsiveness. Am J Physiol 264:C1367–C1387
6. DeFronzo RA (1992) Clinical disorders of hyperkalemia. In: Seldin DW, Giebisch G (eds) The kidney. Physiology and pathophysiology, 2nd edn. Raven, New York, pp 2279–2338
7. DeFronzo RA, Cooke CR, Goldberg M, Cox M, Myers A, Agus ZS (1977) Impaired renal tubular potassium secretion in systemic lupus erythematosus. Ann Intern Med 86:268–271
8. DeFronzo RA, Goldberg M, Cooke CR, Barker C, Grossman RA, Agus ZS (1977) Investigations into the mechanisms of hyperkalemia following renal transplantation. Kidney Int 11:357–365
9. DeFronzo RA, Taufield PA, Black H, McPhedran P, Cooke CR (1979) Impaired renal tubular potassium secretion in sickle cell disease. Ann Intern Med 90:310–316
10. Duarte CG (1978) Magnesium metabolism in potassium depleted rats. Am J Physiol 234:F466–F471
11. Fozzard HA, Shorofsky SR (1992) Excitable membranes. In: Seldin DW, Giebisch G (eds) The kidney. Physiology and pathophysiology, 2nd edn. Raven, New York, pp 407–446
12. Fyhrquist F, Klockars M, Gordin A, Kock B (1980) Hyperreninemia, lysozymuria, and erythrocytosis with Fanconi syndrome with medullary cystic kidney. Acta Med Scand 207:359–365
13. Geck P (1990) Volume regulation in Ehrlich cells. In: Beyenbach KW (ed) Cell volume regulation. Comparative physiology. Karger, Basel, pp 26–58
14. Greger R, Bleich M, Schlatter E (1990) Ion channels in the thick ascending limb of Henle's loop. Renal Physiol Biochem 13:37–50
15. Grinstein S, Foskett JK (1990) Ionic mechanisms of cell volume regulation in leukocytes. Annu Rev Physiol 52:399–414
16. Häussinger D, Lang F (1991) Cell volume in the regulation of hepatic function: a mechanism for metabolic control. Acta Biochim Biophys 1071:331–350
17. Häussinger D, Lang F (1991) Cell volume – a "second messenger" in the regulation of metabolism by amino acids and hormones. Cell Physiol Biochem 1:121–130
18. Häussinger D, Lang F, Gerok W (1994) Regulation of cell function by the cellular hydration state. Am J Physiol 267:E343–E355

19. Hoffmann EK, Simonsen LO (1989) Membrane mechanisms in volume and pH regulation in vertebrate cell. Physiol Rev 69:315–382
20. Horisberger JD, Giebisch G (1987) Potassium-sparing diuretics. Renal Physiol 10:198–220
21. Horisberger JD, Lemas V, Kraehenbühl JP, Rossier BC (1991) Structure-function relationship of the Na, K-ATPase. Annu Rev Physiol 53:565–584
22. Jörgenssen PL (1980) Sodium and potassium ion pump in kidney tubules. Physiol Rev 60:864–917
23. Kleinzeller A, Ziyadeh FN (1990) Cell volume regulation in epithelia – with emphasis on the role of osmolytes and the cytoskeleton. In: Beyenbach KW (ed) Cell volume regulation. Comparative physiology. Karger, Basel, pp 59–86
24. Knochel JP (1992) Potassium gradients and neuromuscular function. In: Seldin DW, Giebisch G (eds) The kidney. Physiology and pathophysiology, 2nd edn. Raven, New York, pp 2191–2208
25. Kurtzman NA, Gonzales J, DeFronzo RA, Giebisch G (1990) A patient with hyperkaliemia and metabolic acidosis. Am J Kidney Dis 15:333–356
26. Lang F, Rehwald W (1992) Potassium channels in renal epithelial transport regulation. Am J Physiol 72:1–32
27. Lang F, Oberleithner H, Kolb H-A, Paulmichl M, Völkl H, Wang W (1988) Interaction of intracellular pH and cell membrane potential. In: Häussinger D (ed) pH homeostasis. Mechanisms and control. Academic, New York, pp 27–42
28. Lang F, Völkl H, Häussinger D (1990) General principles in cell volume regulation. In: Beyenbach KW (ed) Cell volume regulation. Comparative physiology. Karger, Basel, pp 1–25
29. Lang F, Ritter M, Völkl H (1993) The biological significance of cell volume. Renal Physiol Biochem 16:48–65
30. Lewis RS, Cahalan MD (1990) Ion channels and signal transduction in lymphocytes. Annu Rev Physiol 52:415–430
31. McConnell F, Goldstein L (1990) Volume regulation in elasmobranch red blood cells. In: Beyenbach KW (ed) Cell volume regulation. Comparative physiology. Karger, Basel, pp 114–131
32. Mujais SK, Katz AI (1992) Potassium deficiency. In: Seldin DW, Giebisch G (eds) The kidney. Physiology and pathophysiology, 2nd edn. Raven, New York, pp 2249–2278

33. Oberleithner H (1991) Acute aldosterone action in renal target cells. Cell Physiol Biochem 1:2–12
34. Palla B, Litt I (1988) Medical complications of eating disorders in adolescents. Pediatrics 81:613–623
35. Pitts T, Van Thiel D (1986) Disorders of the serum electrolytes, acid-base balance and renal function in alcoholism. Recent Dev Alcohol 4:311–339
36. Rabinowitz L, Wydner CJ, Smith KM, Yamauchi H (1986) Diurnal potassium excretory cycles in the rat. Am J Physiol 250:F930–F941
37. Rosa RM, Williams ME, Epstein FH (1992) Extrarenal potassium metabolism. In: Seldin DW, Giebisch G (eds) The kidney. Physiology and pathophysiology, 2nd edn. Raven, New York, pp 2165–2190
38. Sabatini S, Kurtzman NA (1990) Enzyme activity in obstructive uropathy: basis for salt wasting and the acidification defect. Kidney Int 37:79–84
39. Sakamoto N, Uda M, Tsuchiya M, Ito K, Ikeda M, Watabe R (1981) Hypertension, hypokalemia and hypoaldosteronism with suppressed renin: a clinical study of a patient with Liddle's syndrome. Endocrinol Jpn 28:357–362
40. Schlatter E (1993) Regulation of ion channels in the cortical collecting duct. Renal Physiol Biochem 16:21–36
41. Spring KR, Hoffmann EK (1992) Cellular volume control. In: Seldin DW, Giebisch G (eds) The kidney. Physiology and pathophysiology, 2nd edn. Raven, New York, pp 147–170
42. Stanton B (1986) Regulation by adrenal corticosteroids of sodium and potassium transport in loop of Henle and distal tubule of rat kidney. J Clin Invest 78:1612–1620
43. Tannen RL (1987) Potassium and acid-base balance. In: Giebisch G (ed) Current topics in membranes and transport, vol 28. Academic, Orlando, pp 207–223
44. Ussing HH (1990) Volume regulation of frog skin epithelium. In: Beyenbach KW (ed) Cell volume regulation. Comparative physiology. Karger, Basel, pp 87–113
45. Wang W, Geibel J, Giebisch G (1990) Regulation of the small conductance K$^+$ channels in the apical membrane of rat cortical collecting tubule. Am J Physiol 259:F494–F502
46. Wang W, Sackin H, Giebisch G (1992) Renal potassium channels and their regulation. Annu Rev Physiol 54:81–98
47. Wright FS, Giebisch G (1992) Regulation of potassium excretion. In: Seldin DW, Giebisch G (eds) The kidney. Physiology and pathophysilogy, 2nd edn. Raven, New York, pp 2209–2248

80 Ca²⁺, Mg²⁺, and Phosphate Metabolism

F. Lang

Contents

80.1 Introduction

Ca^{2+} and phosphate concentrations in biological fluids are linked by the *limited solubility* of $CaHPO_4$ or more alkaline phosphate salts.

Since:

$$Ca^{2+} + HPO_4^{2-} \rightleftharpoons CaHPO_4 \qquad (80.1)$$

and

$$[Ca^{2+}] \times [HPO_4^{2-}] = K[CaHPO_4], \qquad (80.1')$$

the product of Ca^{2+} ($[Ca^{2+}]$) and HPO_4^{2-} ($[HPO_4^{2-}]$) concentration divided by K must not exceed the maximum soluble value of $[CaHPO_4]$. In extracellular fluid, $[Ca^{2+}] \times [HPO_4^{2-}]$ is quite close to that value. This is necessary, since bone mineralization by precipitation of extracellular $CaHPO_4$ requires that the ambient Ca^{2+} and phosphate concentrations are not far below the just soluble value. On the other hand, the borderline $CaHPO_4$ concentrations in extracellular fluid to not allow a substantial further increase of either fluid to not allow a substantial further increase of either $[Ca^{2+}]$ or $[HPO_4^{2-}]$ without a parallel decrease of the other component ($[HPO_4^{2-}]$ or $[Ca^{2+}]$, respectively), if precipitation of $CaHPO_4$ is to be avoided. Thus, regulation of Ca^{2+} and HPO_4^{2-} are intimately linked. On the other hand, Mg^{2+} is in several ways linked to Ca^{2+} metabolism. It is therefore appropriate to discuss the metabolism of these minerals in the same chapter.

80.2 Physiological Importance of Ca²⁺

Ca^{2+}, together with phosphate, is the most abundant and important *bone mineral* [18]. Thus, the mineralization of bone critically depends on the availability of Ca^{2+} in extracellular fluid. More than 99% of the body's Ca^{2+} is in bone.

Within cells, Ca^{2+} regulates a multitude of *cellular functions* (Fig. 80.1) [25,78,79,92,96,114]. These functions are triggered by an increase of intracellular Ca^{2+} activity. In nonstimulated cells, intracellular Ca^{2+} activity is only of the order of $0.1\,\mu mol/l$, i.e., only $1/10^4$ of extracellular Ca^{2+} activity (Fig. 80.2; see also Chaps. 5, 11, 26).

In several cells, Ca^{2+} entry from extracellular space and thus intracellular Ca^{2+} activity depends on the cell membrane potential: depolarization of the cell membrane opens *voltage-sensitive Ca²⁺ channels*, which allow rapid entry of Ca^{2+} into the cell (see also Chaps. 12, 44). Furthermore, the majority of known hormones exert at least some of their effects by increasing cytosolic Ca^{2+} activity (see Chap. 5). Some of these hormones activate Ca^{2+} channels in the cell membrane (so-called *receptor-operated Ca²⁺ channels*) and thus cause Ca^{2+} entry from extracellular space; other hormones stimulate the release of Ca^{2+} from intracellular stores, leading to transient increase in cytosolic Ca^{2+} activity (Table 80.1; see also Chap. 5). The increase in intracellular Ca^{2+} activity then modifies the activity of a wide variety of enzymes, ion channels, etc., fusion of vesicles with the plasma membrane, and cytoskeletal organization. As a result, almost all cellular functions are directly or indirectly affected by increases in intracellular Ca^{2+} activity.

Besides its function as bone mineral and intracellular transmitter, Ca^{2+} decreases the *permeability of tight junctions* in endothelial and epithelial cell layers (see also Chap. 59). The presence of Ca^{2+} is also a prerequisite of *blood coagulation*.

R. Greger/U. Windhorst (Eds.)
Comprehensive Human Physiology, Vol. 2
© Springer-Verlag Berlin Heidelberg 1996

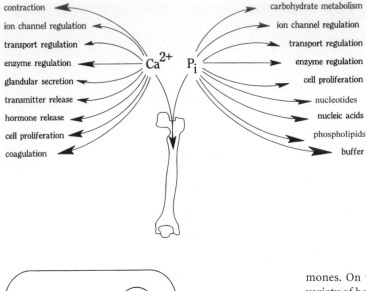

contraction

ion channel regulation

transport regulation

enzyme regulation

glandular secretion

transmitter release

hormone release

cell proliferation

coagulation

Ca^{2+} P$_i$

carbohydrate metabolism

ion channel regulation

transport regulation

enzyme regulation

cell proliferation

nucleotides

nucleic acids

phospholipids

buffer

Fig. 80.1. Some important functions of Ca^{2+} and HPO$_4^{2-}$ (P$_i$)

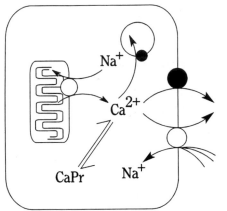

Fig. 80.2. Intracellular distribution of Ca^{2-}. *CaPr*, Ca^{2+} bound to proteins [26,27,114]

Ca^{2+} lowers the *threshold of Na$^+$ channels* of excitable tissues to more negative values. Thus, an increase in extracellular Ca^{2+} activity blunts neuromuscular excitability.

80.3 Physiological Importance of Mg^{2+}

Mg^{2+} participates in the *regulation of K$^+$ channels*, Ca^{2+} channels, and NMDA channels [1,21,22,51,61,68, 74,76,77,113]. Partly because of its effects on ion channels, Mg^{2+} excess reduces and Mg^{2+} depletion increases neuronal excitability [33].

A variety of *enzymes* are dependent on the presence of Mg^{2+} (Table 80.2). Mg^{2+} and Ca^{2+} compete in binding to a variety of binding sites. Usually Ca^{2+} binds with much higher affinity, displaces Mg^{2+} from the binding site, and thus alters the function of the protein. A number of kinases and phosphatases in particular require complexation of phosphate with Mg^{2+} to fulfill their function.

Through its influence on *adenylate cyclase* and phosphodiesterases, Mg^{2+} influences the action of several hormones. On the other hand, Mg^{2+} inhibits the release of a variety of hormones, such as parathyroid hormone (PTH), calcitonin, glucagon, insulin, and catecholamines. Furthermore, Mg^{2+} inhibits the release of neurotransmitters, which adds to the reduction of excitability of muscle and neuronal cells.

In the heart Mg^{2+} influences not only ion channels but also the activity of the myosin-ATPase and thus cardiac con-

Table 80.1. Some hormones acting via 1,4,5inositoltrisphosphate and subsequent increase of intracellular Ca^{2+} activity

Epinephrine (α-receptors)
Acetylcholine
Histamine
ADH (antidiuretic hormone)
Cholecystokinin
Angiotensin
TRH (thyrotropin-releasing hormone)
Substance P
Serotonin

Table 80.2. Some Mg^{2+}-dependent enzymes

Phosphatases
Alkaline phosphatase
Acid phosphatase
Protein kinases
Creatine kinase
Pyruvate kinase
Phosphodiesterase
ATPase
(Na$^+$+K$^+$)-ATPase
H$^+$-ATPase
Ca^{2+} ATPase
Adenylate cyclase
Phosphoglucomutase
Enolase
DNA polymerase
RNA polymerase
K$^+$ channels
Ca^{2+} channels

Table 80.3. Some hormones acting via cyclic AMP

ACTH (adenocorticotropic hormone)
LH (luteinizing hormone)
TSH (thyroid-stimulating hormone)
Prolactin
GH (growth hormone)
Glucagon
PTH (parathyroid hormone)
Calcitonin
ADH (antidiuretic hormone)
Gastrin
Secretin
GIP (gastric inhibitory peptide)
VIP (vasointestinal peptide)
Epinephrine (β-receptors)
Dopamine
Histamine
Prostaglandins

traction (see Chaps. 90, 93). Both depletion and excess of Mg^{2+} eventually reduce the contractility of the heart. Mg^{2+} stimulates (Na^++K^+)-ATPase (see Chap. 8) and inhibits K^+ channels in the cell membranes, and thus favors cellular accumulation of K^+.

80.4 Physiological Importance of Phosphate

Together with Ca^{2+}, phosphate is the most important *mineral of bone* [86]. Thus, mineralization of bone depends on the ambient Ca^{2+} and phosphate concentration.

In cells, phosphate serves a multitude of functions, as illustrated in Fig. 80.1. Most importantly, phosphate is a component of a myriad of substances serving all kinds of different functions. Notably, it is a component of ATP, and thus plays a central role in chemical energy transduction. It is a component of cAMP [104], a second messenger of the majority of the hormones (Table 80.3), and virtually all hormones influence target proteins by phosphorylation/dephosphorylation (Table 80.4).

80.5 Regulation of Ca^{2+} Phosphate Metabolism

80.5.1 Roles of PTH, Calcitriol, and Calcitonin in the Integrated Regulation of $CaHPO_4$

The phosphate-dependent reactions are within certain limits relatively insensitive to alterations in extracellular phosphate concentrations. By contrast, the amount of Ca^{2+} entering through receptor-operated or voltage-sensitive Ca^{2+} channels, and thus the influence of depolarization and hormonal action, depends critically on *Ca^{2+} activity in extracellular space*. Accordingly, the maintenance of constant Ca^{2+} activity in extracellular fluid has the highest pri-

ority in the regulation of Ca^{2+} phosphate metabolism. This is the function of *parathyroid hormone* (PTH). PTH is released upon a reduction in extracellular Ca^{2+} activity [23] and its actions are aiming at increasing extracellular Ca^{2+} activity (Fig. 80.3).

PTH stimulates the *resorption of bone* minerals, most importantly Ca^{2+} phosphate and Ca^{2+} carbonate. In the kidney, PTH stimulates the *distal tubular reabsorption of Ca^{2+}*,

Table 80.4. Some enzymes regulated by phosphorylation/dephosphorylation

Carbohydrate metabolism
Glycogen phosphorylase
Phosphorylase kinase
Glycogen synthase
Phosphorylase-phosphatase inhibitor
Fructose-1,6-bisphosphatase
Phosphofructokinase
Pyruvate dehydrogenase
Pyruvate kinase
Lipid metabolism
Acyl-CoA-carboxylase
Hydroxymethylglutaryl-CoA-reductase
Lipase
Glycerophosphate acyltransferase
Cholesterolester hydrolase
Amino acid metabolism
Tyrosine hydroxylase
Tryptophan hydroxylase
Others
cGMP-dependent protein kinase
Type II cAMP-dependent protein kinase
DNA dependent RNA polymerase
Reverse transcriptase
elF2 kinase
Myosin

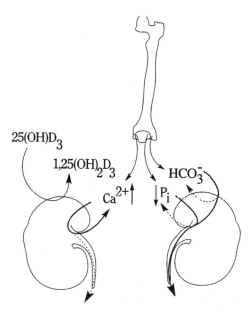

Fig. 80.3. The actions of PTH on kidney and bone. *1,25(OH)₂D₃*: calcitriol (also called D-hormone)

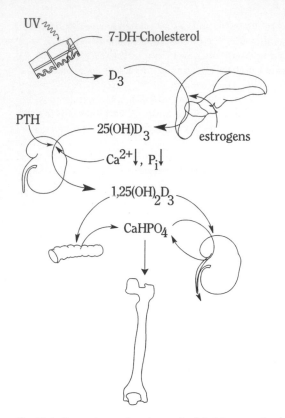

Fig. 80.4. Formation and actions of calcitriol. *UV*, Ultraviolet light

thus reducing further loss of Ca^{2+} [10,11,14,41,81,83]. Furthermore, PTH reduces the proximal tubular reabsorption of bicarbonate and phosphate. The resulting *bicarbonaturia* and *phosphaturia* reduce the plasma concentrations of these anions and thus facilitate the mobilization of bone minerals. Without increased renal excretion, phosphate and bicarbonate would accumulate in extracellular fluid and eventually complex extracellular Ca^{2+}, thus preventing a correction of plasma Ca^{2+} activity. The acute actions of PTH allow a rapid and efficient increase in extracellular Ca^{2+} activity in hypocalcemia. The repeated action of PTH, however, would eventually lead to demineralization of bone, if PTH did not stimulate the formation of *calcitriol* (also called D-hormone or $1,25(OH)_2D_3$; see also Chap. 75) [20,36]. Calcitriol is formed by the 1,25-hydroxylase in the kidney from $25(OH)D_3$. The latter is formed in the liver from cholecalciferol, which is synthesized in the skin from 7-dehydrocholesterol under the influence of ultraviolet light (Fig. 80.4). The hepatic 25-hydroxylase is stimulated by estrogens, the renal 1,25-hydroxylase not only by PTH, but also by phosphate and Ca^{2+} depletion. Calcitriol stimulates intestinal Ca^{2+} and phosphate absorption and, by increasing the ambient Ca^{2+} and phosphate concentrations, favors the mineralization of bone (Fig. 80.5).

The third hormone mainly involved in the regulation of mineral metabolism is *calcitonin*. This hormone is released during hypercalcemia; it reduces renal phosphate reabsorption and stimulates mineralization of bone (see also Chap. 75).

A number of other hormones and factors influence bone resorption, renal excretion, and intestinal absorption of Ca^{2+} and phosphate and thus modify the Ca^{2+} phosphate metabolism, as will be discussed below.

80.5.2 Regulation of Renal Ca^{2+} Excretion

Some 40% of plasma Ca^{2+} is bound to plasma proteins [87] and is thus not filtered at the glomerulus [32]. Two-thirds of filtered Ca^{2+} is reabsorbed along the proximal tubule, and another 25% in the thick ascending limb of Henle's loop (see also Chap. 75). Some 10% is delivered to the early distal tubule [32]. Ca^{2+} reabsorption continues along the distal tubule, but very little is reabsorbed beyond this nephron segment [47,69]. Less than 1% of filtered Ca^{2+} is excreted in urine.

In both proximal renal tubule and thick ascending limb, a large part of Ca^{2+} reabsorption proceeds through the paracellular shunt. Beyond that, some Ca^{2+} may enter the cells across the apical cell membrane through an as yet unidentified mechanism [62]. This mechanism is stimulated by *PTH*. The effect of PTH is inhibited by colchicine. In the presence of PTH, but not in its absence, the Ca^{2+} channel opener BayK 8644 has been shown to stimulate Ca^{2+} entry, which is inhibited by the Ca^{2+} channel blocker nifedipine. On the basis of these observations it has been suggested that PTH stimulates the fusion of Ca^{2+}-channel-containing vesicles with the luminal cell membrane [32]. Ca^{2+} is extruded by a basolateral Na^+/Ca^{2+} exchange [43,49,60,100,103,109] and a Ca^{2+}-ATPase [16,27,28,32, 37,40,42,59,63]. Both the Ca^{2+}-ATPase [97] (see also Chap. 8) and the Na^+/Ca^{2+} exchanger [85] have been cloned, whereas the molecular identity of the Ca^{2+} entry step remains elusive at present.

Renal Ca^{2+} excretion increases with increasing plasma concentration, yielding an apparent saturation of renal Ca^{2+} transport (Fig. 80.6). One mechanism mediating enhanced renal Ca^{2+} elimination during hypercalcemia is the reduc-

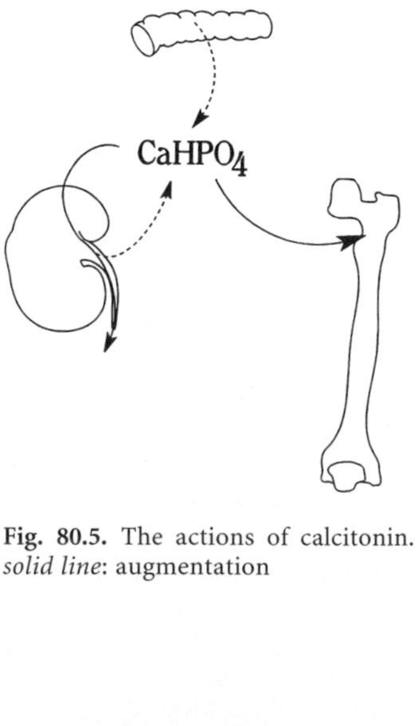

Fig. 80.5. The actions of calcitonin. *Broken line*: Attentuation; *solid line*: augmentation

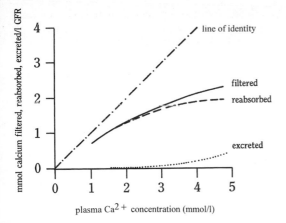

Fig. 80.6. The amount of Ca^{2+} filtered by the renal glomeruli, reabsorbed by renal tubules, and excreted in urine as a function of plasma Ca^{2+} concentration

tion of the paracellular shunt permeability by Ca^{2+} in both proximal renal tubules and thick ascending limbs [44]. The inhibition impairs not only Ca^{2+} but also Na^+ and Mg^{2+} reabsorption, and the hypercalciuria during hypercalcemia is paralleled by natriuresis and magnesiuria. Conversely, renal Ca^{2+} reabsorption is tightly linked to Na^+ reabsorption, i.e., natriuresis is frequently paralleled by calciuria. Passive, paracellular Ca^{2+} reabsorption is enhanced by an increase in the luminal Ca^{2+} concentration during stimulated Na^+ and fluid reabsorption. In the thick ascending limb of Henle's loop stimulated Na^+ reabsorption augments the electrical driving force. Thus, renal Ca^{2+} excretion is stimulated by extracellular volume expansion and blunted in extracellular volume contraction.

Distal tubular Ca^{2+} reabsorption is stimulated by PTH and calcitriol (Table 80.5). Both hormones stimulate transcellular Ca^{2+} transport by affecting the luminal entry [58] and basolateral extrusion [49] of Ca^{2+}. It is thought that the Ca^{2+}-binding proteins within the cell play an important regulatory role. They allow transcellular transport of Ca^{2+} and buffer Ca^{2+} at both the entry and the exit of Ca^{2+}. Intracellular Na^+ activity in early distal tubular cells can be reduced by inhibition of the NaCl cotransport at the luminal cell membrane with *thiazide diuretics*. The reduction in intracellular Na^+ activity enhances the chemical driving force for the basolateral Na^+/Ca^{2+} exchanger [32], which most likely accounts in part for the anticalciuric effect of thiazide diuretics. In addition, the diuretics lead to a contraction of extracellular volume, which in turn stimulates Na^+ and, indirectly, Ca^{2+} reabsorption in proximal tubule and thick ascending limb (see above). Other diuretics, such as acetazolamide, furosemide, and mannitol, which impair Na^+ reabsorption in proximal renal tubules and/or in the thick ascending limb, are calciuric.

Renal tubular Ca^{2+} reabsorption is further stimulated in alkalosis and inhibited in acidosis, effects which may in part be due to altered permeability of the paracellular shunt.

Renal tubular Ca^{2+} reabsorption is inhibited by growth hormone, thyroid hormones, chronic action of adrenal steroids, insulin, and glucose. The mechanisms involved are not well understood.

80.5.3 Regulation of Renal Phosphate Excretion

The phosphate concentration in the glomerular ultrafiltrate is some 90% of that in plasma [45]. Usually some 80% of filtered phosphate is *reabsorbed*. Phosphate reabsorption is localized mainly if not exclusively to the proximal renal tubule [45,70]. Proximal renal tubular phosphate transport is probably confined to the transcellular pathway [46]. The entry step is mediated by the carrier NaP$_i$, which has been cloned and characterized at the molecular level (see Chap. 75). The NaPi$_2$ and NaPi$_3$

Table 80.5. Factors modifying renal phosphate and Ca^{2+} reabsorption

Stimulating Ca^{2+} reabsorption
PTH (parathyroid hormone)
Calcitriol (also called D-hormone, 1,25-dihydroxycholecalciferol, $1,25(OH)D_3$)
Thiazides
Alkalosis
Inhibiting Ca^{2+} reabsorption
Growth hormone
Thyroid hormones
Adrenal steroids (chronic)
Insulin
Glucose
Acidosis
Extracellular volume expansion
Acetazolamide
Furosemide
Mannitol
Stimulating phosphate reabsorption
Phosphate depletion
Calcitriol (acute)
Thyroid hormones
Growth hormone
Mg^{2+}
Li^+
Metabolic acidosis
Inhibiting phosphate reabsorption
Phosphate excess
Ca^{2+} excess (chronic)
PTH
Calcitonin
Calcitriol (chronic)
Hypocalcemia (acute)
Mg^{2+} depletion
Respiratory acidosis
Metabolic alkalosis
Glucose
Colchicine
Extracellular volume expansion
Thiazides
Furosemide
Ethacrynic acid
Mannitol
Acetazolamide

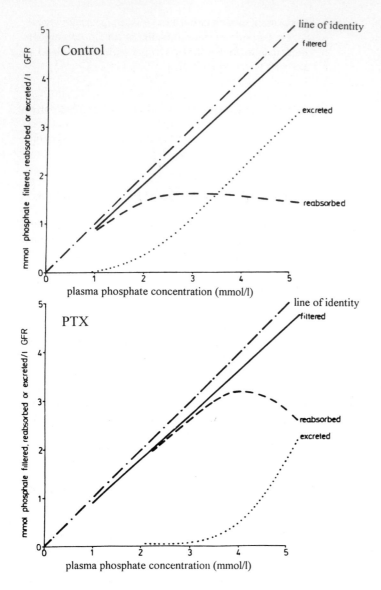

Fig. 80.7. The amount of phosphate filtered by the renal glomeruli, reabsorbed by renal tubules, and excreted in urine as a function of plasma phosphate concentration. *Upper panel*: In individuals with intact parathyroid glands (i.e., in the presence of PTH). *Lower panel*: in parathyroidectomized individuals (*PTX*)

transport either monovalent or divalent phosphate together with 3 Na^+ ions [24].

The most important regulator of phosphate excretion is phosphate itself (Fig. 80.7). Phosphate reabsorption is saturable with a high affinity in the straight proximal tubule [71]. Thus, any filtered load exceeding the maximum transport rate is excreted completely. Moreover, cellular phosphate depletion enhances proximal tubular phosphate transport, at least in part due to enhanced expression of the $NaPi_2$ (but not $NaPi_1$). Phosphate excess in turn inhibits proximal tubular phosphate reabsorption.

Hormones that stimulate renal tubular phosphate reabsorption include calcitriol [$1,25(OH)_2D_3$, D-hormone], thyroid hormones and growth hormone. PTH and calcitonin inhibit proximal renal tubular phosphate transport. Chronic administration of calcitriol inhibits phosphate transport, possibly due to phosphate excess. The hormones modify phosphate transport partly via protein kinase A and C and by stimulation or inhibition of insertion of NaP_i into the luminal cell membrane.

Renal phosphate transport is further influenced by acid-base balance, Ca^{2+}, Mg^{2+}, glucose, colchicine, extracellular volume, and various diuretics (Table 80.5; see also Chap. 75).

80.5.4 Regulation of Intestinal Ca^{2+} Phosphate Absorption

From an oral intake of 1 g Ca^{2+}, only about a third is absorbed. At higher oral Ca^{2+} intakes, the fraction of Ca^{2+} that is absorbed eventually decreases, i.e., *Ca^{2+} absorption displays apparent saturation.* Ca^{2+} absorption is mainly carried out by the small intestine. It is partly passive, paracellular, driven by a chemical gradient from lumen to blood and by solvent drag [32]. The chemical gradient is

Table 80.6. Factors modifying enteral Ca^{2+} and phosphate absorption (C = probably acting via calcitriol)

Stimulating Ca^{2+} absorption
Calcitriol (D-hormone, 1,25-dihydroxycholecalciferol)
PTH (C)
Growth hormone (C)
Prolactin (C)
Pregnancy (C)
Insulin (C)
Phosphate depletion (C)
Sodium
Mg^{2+} deficiency
Lactose, mannitol
Lysine, arginine
Gall
Protein-rich diet

Inhibiting Ca^{2+} absorption
Thyroid hormones
Calcitonin (?)
Glucocorticoids
Somatostatin
Mg^{2+}
Oxalate, fatty acids
Diphenylhydantoin and other anticonvulsive agents
Diphosphonate

Stimulating phosphate absorption
Calcitriol
PTH (C)
Growth hormone (C)
Prolactin (C)
Pregnancy (C)
Insulin (C)
Phosphate depletion (C)
Alkalinisation of intestinal lumen

Inhibiting phosphate absorption
Mg^{2+}
Glucocorticoids (?)

opposed by a lumen-negative electrical gradient of 5–6 mV. Ca^{2+} absorption in the *upper small intestine* involves in addition a transcellular pathway consisting of a rather ill-defined passive entry step into the cells across the luminal cell membrane and a *Ca^{2+}-ATPase* in the basolateral cell membrane. A basolateral Na^+/Ca^{2+} exchanger is probably less important for transcellular transport. Ca^{2+} transport from the luminal to the basolateral cell membrane is facilitated by *Ca^{2+}-binding proteins*, which buffer Ca^{2+} at either cell membrane.

Phosphate is *absorbed* in the *upper small intestine* in part via a passive, paracellular pathway, in part via transcellular transport involving an Na^+-coupled transport system at the luminal cell membrane similar but not identical to that for the renal NaPi. Phosphate from ingested organic phosphate compounds is cleaved by appropriate enzymes such as alkaline phosphatase and the phosphate subsequently absorbed.

Intestinal Ca^{2+} and phosphate transport is *stimulated by calcitriol*, which is the most important regulator of intestinal Ca^{2+} and phosphate absorption (Table 80.6). Calcitriol stimulates Ca^{2+} entry, cellular Ca^{2+} binding, Ca^{2+}-ATPase, and Na^+-coupled phosphate transport. Several hormones, e.g., PTH, somatotropic hormone, prolactin, estrogens,

and insulin, modify intestinal Ca^{2+} and phosphate absorption at least in part via stimulation of calcitriol formation. Phosphate depletion is a further important stimulator of calcitriol formation and thus of intestinal Ca^{2+} and phosphate absorption.

As shown by Table 80.6, several other factors modify intestinal Ca^{2+} and phosphate absorption. Most importantly, only free Ca^{2+} is available for absorption, i.e., acids such as oxalate and fatty acids, which form poorly soluble complexes with Ca^{2+}, impede absorption.

80.5.5 Regulation of Bone Mineralization

The major element of the organic component of bone is collagen, which contributes 85–90% of bone protein [80]. Other protein components include osteocalcin, sialoproteins such as osteopontin, proteoglycans, and osteonectin [25]. *Bone minerals*, which make up some two-thirds of bone weight, mainly consist of *hydroxyapatite* [$Ca_{10}(PO_4)_6(OH)_2$], but also include brushite [$CaHPO_4 \times 2H_2O$], octocalciumphosphate [[$Ca_8H_2(PO_4)_6 \times 5H_2O$], and complexes with other anions (F^-, CO_3^{2-}) or cations (Na^+, K^+, Mg^{2+}) [25].

Bone is formed by *osteoblasts*, which synthesize and secrete the organic components of bone and mediate their mineralization [25]. Bone is reabsorbed by *osteoclasts*, which dissolve the bone minerals by local acidification and degrade the proteins with a variety of lysosomal proteases [25]. The H^+-ATPase carrying out the local acidification and the proteases are from lysosomes, which fuse with the cell membrane facing the bone [6,7,12].

Bone mineralization and formation are dependent on the availability of Ca^{2+} and phosphate, and thus on plasma Ca^{2+} and phosphate concentrations. Furthermore, the deposition of the minerals is dependent on ambient acidity, since the Ca^{2+} phosphate salts are well soluble in acid pH but insoluble in alkaline pH. Acidosis thus favors bone resorption, whereas alkalosis promotes bone deposition [25]. *Bone resorption* is *stimulated* by PTH and by calcitriol. Both hormones stimulate the activity of existing osteoclasts and the formation of new osteoclasts (Table 80.7). They also inhibit the activity of osteoblasts [25]. On the other hand, calcitriol has been shown to stimulate collagen synthesis of human bone cells [9]. More importantly, however, the direct effect of calcitriol on bone is usually overridden by its indirect effect via plasma Ca^{2+} and phosphate: as outlined above, calcitriol stimulates intestinal Ca^{2+} and phosphate absorption, the increase in plasma Ca^{2+} activity inhibits PTH release, and the increase in Ca^{2+} and phosphate activity favors mineralization of bone. Thus, the effect of calcitriol in vivo is an increase in bone turnover with net bone formation [25]. This indirect effect of calcitriol may, under appropriate conditions, be supported by a direct stimulatory effect of the hormone on collagen synthesis in bone cells. Like PTH and calcitriol, the prostaglandins PGE_1 and PGE_2 have been shown to stimulate bone resorption [52,64]. In addition, however, they may stimulate bone formation [95,107].

Table 80.7. Factors modifying bone turnover

Stimulating bone resorption
PTH
Calcitriol
Thyroid hormones
Prostaglandins
Aluminium
Collagenase
H^+
Stimulating bone formation
Ca^{2+}
Phosphate
Calcitonin (?)
Calcitriol
Estrogens
F^-
Prostaglandins
Osteoinductive factors
Stimulating bone remodelling
Transforming growth factor-β
β_2-Microglobulin
IGF-I and -II
Platelet-derived growth factor
Interleukin-1
TNF-α and TNF-β

Bone resorption is *inhibited* by calcitonin, which reduces osteoclast activity and formation [13,75]. Calcitonin may also stimulate bone formation.

Bone formation is further stimulated by estrogens, and the estrogen depletion in postmenopausal women eventually leads to demineralization of bone [25].

Further hormones modifying *bone turnover* are thyroid hormones and glucocorticoids. Thyroid hormones are required for normal bone growth; in vitro they stimulate bone resorption [56,67,82]. Glucocorticoid excess is associated with enhanced osteoclast activity and demineralization of bone. These effects may, however, be in large part indirect, due to an inhibitory effect of glucocorticoids on intestinal Ca^{2+} absorption [25].

Local factors modifying bone formation, resorption, or remodelling include the so-called osteoinductive factors [105], such as bone morphogenetic factor [108], transforming growth factor-β, β-microglobulin, insulin-like growth factor I, platelet-derived growth factor, and tumor necrosis factor-α (see Table 80.7). Interferon-γ has been shown to inhibit both bone resorption [55] and collagen synthesis in bone organ culture [99].

80.6 Regulation of Mg^{2+} Metabolism

Given the two charges of Mg^{2+}, a potential difference across the cell membrane of 60 mV drives Mg^{2+} into the cell as long as intracellular Mg^{2+} activity is less than 100-fold extracellular Mg^{2+} activity. Intracellular Mg^{2+} concentration is more than 10-fold extracellular Mg^{2+} concentration, but the potential difference usually drives Mg^{2+} uptake into the cells. Thus, ion channels permeable to Mg^{2+} mediate the entry of Mg^{2+} into the cell. Extrusion of cellular Mg^{2+}, on the other hand, requires the expenditure of energy and is accomplished by an Mg^{2+}-ATPase [39,84,89].

The uptake of Mg^{2+} into the cells is stimulated by acute respiratory alkalosis and by insulin, which stimulates the Na^+/H^+ exchanger (see Chap. 79) and thus similarly produces intracellular alkalinization. The intracellular alkalinization leads to enhanced dissociation of acidic groups of intracellular proteins, which subsequently bind more Mg^{2+}.

The *Mg^{2+} balance* of the body depends on intestinal absorption on the one hand and renal excretion on the other. Intestinal absorption is compromised by luminal complexation of Mg^{2+} (e.g., phosphate, oxalate, and fatty acids), stimulated by PTH and growth hormone, and inhibited by aldosterone and calcitonin. Intestinal Mg^{2+} absorption may be inhibited by luminal Ca^{2+} and may be stimulated by calcitriol.

Renal handling of Mg^{2+} is characterized by glomerular filtration and partial renal tubular reabsorption (see Chap. 75). The concentration of Mg^{2+} in glomerular filtrate is some 70–80% of that in plasma [89]. About one-third of it is complexed to anions such as phosphate, citrate, and oxalate [111]. In the proximal tubule and descending limb of Henle's loop only some 25% of filtered Mg^{2+} is reabsorbed. The reabsorption is dependent on luminal flow rate. Very little is known about the cellular mechanisms involved [89]. The major portion of renal tubular Mg^{2+} reabsorption resides in the thick ascending limb of Henle's loop (see Chaps. 74, 75). Most likely the reabsorption is in large part paracellular, driven by the lumen-positive transepithelial potential difference of some 8 mV [44]. Since high concentrations of Mg^{2+} and Ca^{2+} reduce the patency of the paracellular shunt [38], Mg^{2+} reabsorption is impaired in hypermagnesemia and hypercalcemia [72,73]. Maneuvers inhibiting thick ascending limb NaCl reabsorption reduce the transepithelial potential difference and thus impair Mg^{2+} reabsorption.

Table 80.8. Factors modifying renal Mg^{2+} reabsorption

Stimulating
Extracellular volume depletion
Mg^{2+} depletion
Hypocalcemia
PTH (cAMP)
Glucagon (cAMP)
Calcitonin (cAMP)
ADH (cAMP)
Inhibiting
Extracellular volume expansion
Hypermagnesemia
Hypercalcemia
Loop diuretics (furosemide, ethacrynic acid)
Osmotic diuresis (mannitol)
Phosphate depletion
Acute metabolic acidosis
Carbohydrates
Protein intake
Alcohol ingestion

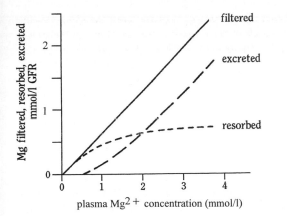

Fig. 80.8. The amount of Mg^{2+} filtered by the renal glomeruli, reabsorbed by renal tubules, and excreted in urine as a function of plasma Mg concentration

Accordingly, loop diuretics, osmotic diuresis, and other conditions in which NaCl reabsorption is reduced in the thick ascending limb lead to renal loss of Mg^{2+} [44,89]. The reabsorption of Mg^{2+} in the thick ascending limb is stimulated by PTH, glucagon, calcitonin, and (in the rat, but probably not in man) by ADH [35]. Apart from PTH, the effects of these hormones are probably secondary to stimulation of NaCl transport in the thick ascending limb [89]. Very little Mg^{2+} is reabsorbed in the distal convoluted tubule and virtually nothing in the collecting duct [89]. The various factors modulating the renal reabsorption of Mg^{2+} are listed in Table 80.8 [89,90]. An important determinant of renal Mg^{2+} excretion is Mg^{2+} itself (Fig. 80.8): in Mg^{2+}-depleted states, Mg^{2+} excretion can drop from some 5 mmol/day to less than 0.5 mmol/day [2].

Mg^{2+} transport across the cell membrane is influenced by thyroid hormones and by insulin, both of which stimulate cellular uptake of Mg^{2+} [2].

80.7 Disorders of Ca^{2+} Phosphate Metabolism

80.7.1 Hypocalcemia

Hypocalcemia may be factitious, due to reduced plasma protein concentration and thus due to decreased bound Ca^{2+}, while the biologically relevant ionized Ca^{2+} is normal. Beyond that, a number of conditions can lead to true hypocalcemia, i.e., decrease of ionized Ca^{2+} (Table 80.9):

- Decreased release (hypoparathyroidism) or efficacy (pseudohypoparathyroidism) of PTH leads to renal loss of Ca^{2+}, decreased resorption of Ca^{2+} from bone, and, via lack of calcitriol, decreased intestinal absorption of Ca^{2+} [101].
- Depletion of calcitriol due to vitamin D deficiency or failure to produce the active metabolite rarely leads to hypocalcemia, since PTH usually maintains plasma Ca^{2+}

as long as minerals are available in bone. However, repletion with calcitriol stimulates both intestinal Ca^{2+} and phosphate absorption and thus may foster excessive bone formation, leading to hypocalcemia [101].
- Excessive bone formation (hungry bone syndrome) may lead to hypocalcemia following treatment of hyperparathyroidism by parathyroidectomy.
- Mg^{2+} deficiency leads to impairment of PTH release [48] and efficacy [4]. As a result, hypocalcemia cannot be counteracted appropriately by the hormone.
- Acute pancreatitis due to complexation of Ca^{2+} by fatty acids, which are formed from retroperitoneal fat triglycerides under the influence of released pancreatic lipase [102].
- In sepsis, hypocalcemia may result from enhanced calcitonin levels [101].
- In renal insufficiency, the ability of the kidney to produce phosphaturia in response to PTH is decreased and the increment of plasma phosphate concentration leads to complexation of blood Ca^{2+} with resulting decrease of ionized Ca^{2+}. The hypocalcemia is augmented by impaired formation of calcitriol in the diseased kidney.
- In alkalosis plasma proteins dissociate and bind Ca^{2+}. Hypocalcemia is especially prone to occur in metabolic alkalosis, where Ca^{2+} is in addition complexed by HCO_3^-.
- Diuretic therapy can cause renal Ca^{2+} loss.

Table 80.9. Causes of hypo- and hypercalcemia

Hypocalcemia
Factitious due to hypoalbuminemia
Hypoparathyroidism, parathyroidectomy
Pseudohypoparathyroidism
Calcitriol depletion or repletion
Sepsis
Mg^{2+} deficiency
Diuretics
Hyperphosphatemia
    Renal failure
Alkalosis
Pancreatitis

Hypercalcemia
Factitious due to enhanced protein binding
Hyperparathyroidism
Phosphodiesterase inhibitors
Malignancy-associated hypercalcemia
    Parathyroid hormone-related protein (PTHRP)
    Calcitriol excess
    Local osteolytic hypercalcemia
Granulomatous disorders
    Sarcoidosis
    Tuberculosis, leprosy, etc.
Vitamin D intoxication
Vitamin A intoxication
Immobilization
    AIDS
    Hepatic failure
Renal failure
Thiazide diuretics
Familial hypocalciuric hypercalcemia
Parenteral nutrition
Milk alkali syndrome
Mg^{2+} intoxication

The most prominent *consequence of hypocalcemia* is enhanced neuromuscular excitability. In the heart, the duration of the action potential is enhanced in hypocalcemia, due to delayed activation of Ca^{2+}-sensitive K^+ channels.

80.7.2 Hypercalcemia

Hypercalcemia (see Table 80.9) can be factitious, representing a raised plasma protein concentration and thus a raised concentration of the biologically irrelevant bound Ca^{2+}.

An increase in ionized Ca^{2+} is frequently the combined result of increased resorption of bone minerals, renal reabsorption, and intestinal absorption due to hyperparathyroidism [101]. Phosphodiesterase inhibitors may mimick some PTH effects due to increased intracellular cAMP levels and thus produce hypercalcemia. The hypercalcemia observed in patients with malignancies may be due to bone tumor or metastases leading to local resorption of bone. More frequently, however, they are due to humoral factors produced by the tumor cells. The most frequent humoral factors from tumors producing hypercalcemia are PTH-related proteins [17] and calcitriol [102]. Enhanced formation of calcitriol in lymphocytes has similarly been implicated as the cause of hypercalcemia in granulomatous disorders. Obviously vitamin D intoxication may lead to hypercalcemia. Enhanced bone resorption may be the cause of hypercalcemia of vitamin A intoxication. Similarly, immobilization leads to enhanced bone resorption and may thus produce hypercalcemia. The hypercalcemia occasionally observed in the acquired immune deficiency syndrome and in hepatic failure may be due at least in part to immobilization.

Impaired renal elimination of Ca^{2+} occurs during thiazide therapy (see Chaps. 74, 75), in renal failure, and in the rare genetic disorder of familial hypocalciuric hypercalcemia.

Further causes of hypercalcemia are excessive parenteral administration or enhanced intestinal absorption, as occurs in the *milk alkali syndrome*. The reason why hypercalcemia occurs in Mg^{2+} intoxication is poorly understood.

The *consequences of hypercalcemia* include cardiac arrhythmias, gastrointestinal disorders, and polyuria (due to inhibition of paracellular electrolyte reabsorption in the thick ascending limb) [38] as well as precipitation of Ca^{2+} phosphates, especially in urine, where concentrations are high, and in the cornea, where the pH is high.

80.7.3 Hypophosphatemia

Hypophosphatemia may be due to phosphate depletion and/or a shift of phosphate from extracellular to intracellular space (Table 80.10). Frequently, phosphate depletion is masked by cellular loss of phosphate and is precipitated by cellular reuptake of phosphate. Hypophosphatemia secondary to stimulated cellular uptake is observed in respiratory alkalosis [66] and under the

influence of insulin [65]. The hypophosphatemia following realimentation after starvation or alcoholic withdrawal is at least partially due to insulin release. Several other hormones may produce hypophosphatemia, such as glucagon [34], epinephrine [54], sex hormones [93,106], and glucocorticoids [65].

Inhibitors of phosphodiesterase may similarly cause hypophosphatemia, in part due to stimulated cellular uptake [65]. Several conditions in which the cellular need for phosphate is increased may also lead to hypophosphatemia (e.g., leukemia).

Urinary loss of phosphate occurs in hyperparathyroidism [66] and in a variety of diseases affecting renal tubular transport of phosphate. In phosphate diabetes, a congenital disorder specifically affects renal tubular phosphate transport. In salt-losing kidney, such as during recovery from acute renal failure, renal phosphate loss parallels the loss of other electrolytes [66]. The inhibition of renal electrolyte transport in extracellular volume expansion may similarly result in renal phosphate loss. The hypophosphatemia following hyperaldosteronism and ADH excess may be secondary to extracellular volume expansion and subsequent inhibition of proximal tubular

Table 80.10. Causes of hypo- and hyperphosphatemia

Hypophosphatemia
Decreased dietary intake
Decreased intestinal absorption
Vitamin D deficiency
Malabsorption, diarrhea, steatorrhea
Vomiting
Increased renal loss
Hyperparathyroidism
Parathyroid hormone-related protein (PTHRP)
Phosphate diabetes (renal rickets)
Salt-losing kidney
Hyperaldosteronism
ADH excess
Volume expansion
Diuretics
Mg^{2+} deficiency
Cellular uptake
Respiratory alkalosis
Insulin
Glucagon
Epinephrine
Sex hormones
Glucocorticoids
Nutritional recovery syndrome
Hypercatabolism
Recovery from hypothermia
Leukemias, lymphomas
Treatment of pernicious anemia
Incorporation into bone
Hungry bone syndrome following parathyroidectomy
Hyperphosphatemia
Excess dietary intake
Impaired renal elimination
Renal insufficiency
Hypoparathyroidism
Pseudohypoparathyroidism
Cellular loss of phosphate
Bone resorption

fluid reabsorption. Diuretics may affect phosphate excretion by their inhibitory effect on renal tubular electrolyte reabsorption.

Hypophosphatemia may also be a result of reduced dietary intake (such as in chronic alcoholism), compromised intestinal absorption (such as in malabsorption or vitamin D deficiency), or stimulated mineralization of bone (hungry bone syndrome) [65].

Severe phosphate depletion increases the phosphorylation potential ATP/(ADP $\times$ P) and may thus impede mitochondrial respiration [115] (see also Chap. 4). Thus, the energy supply of a variety of cells may be compromised.

The *consequences of phosphate depletion* include muscle injury [29,94], heart insufficiency [88], and disturbed neuronal function [98]. In addition, hypophosphatemia has been shown to reduce 2,3-diphosphoglycerol (2,3-DPG) in red cells and by this means to increase the oxygen affinity of hemoglobin, which may impair oxygen release from peripheral blood. Only very rarely may hemolysis occur [65]. Furthermore, hypophosphatemia may interfere with leukocyte and platelet function. Since phosphate serves as a urinary buffer (see Chap. 78), reduced urinary excretion of phosphate due to nonrenal hypophosphatemia may occur in metabolic acidosis [65]. Longstanding phosphate depletion leads to demineralization of bone [65].

80.7.4 Hyperphosphatemia

Phosphate excess is most frequently caused by renal retention of phosphate, such as in renal insufficiency, in hypoparathyroidism, and in pseudohypoparathyroidism. Moreover, excess intake or cellular loss of phosphate as well as demineralization of bone could create hyperphosphatemia (Table 80.10).

The *sequelae of phosphate excess* are mainly due to complexation of Ca^{2+}, leading to crystal formation in joints and skin. The complexation of Ca^{2+} is followed by hypocalcemia, stimulation of PTH release, further mobilization of bone minerals, and further increase of plasma phosphate concentrations. Thus a vicious circle is triggered, leading to aggravation of hyperphosphatemia and decrease of Ca^{2+} activity.

80.8 Disorders of Mg^{2+} Metabolism

80.8.1 Hypomagnesemia

Mg^{2+} depletion may result from insufficient dietary intake, intestinal malabsorption, or renal (Table 80.8) and extrarenal Mg^{2+} wasting (Table 80.11). Renal Mg^{2+} loss may accompany Bartter's syndrome [8], a condition characterized by defective NaCl reabsorption in the thick ascending limb of Henle's loop, or it may be due to a more selective genetic impairment of renal tubular Mg^{2+} reabsorption

[15]. Hypercalcemia and diuretics that inhibit NaCl reabsorption in the thick ascending limb (loop diuretics, osmotic diuresis) lead to renal loss of Mg^{2+} [2]. Similarly, extracellular volume expansion, as occurs in hyperaldosteronism and ADH excess, may lead to Mg^{2+} loss [31,53,57]. In ketoacidosis, the luminal accumulation of the negatively charged ketoacids probably impairs renal tubular Mg^{2+} reabsorption. In phosphate depletion, the disturbed energy supply may interfere with NaCl and Mg^{2+} reabsorption. Ketoacidosis and phosphate depletion participate in the generation of hypomagnesemia in alcoholism. Hypomagnesemia may also result from redistribution of Mg^{2+} into the cells, as may occur in thyroid hormone excess [5] and after administration of insulin to diabetic patients [116].

Extrarenal loss of Mg^{2+} could occur through excessive lactation and through the skin in severe burns [19] or during excessive sweating.

The *consequences of Mg^{2+} depletion* include enhanced neuromuscular excitability and cardiac arrhythmias up to the extent of cardiac fibrillation. Cardiac contractility is reduced. Despite the inhibitory action of Mg^{2+} on PTH release, chronic Mg^{2+} depletion leads to hypoparathyroidism and hypocalcemia, a poorly understood phenomenon. Mg^{2+} depletion impairs cellular (Na^++K^+)-ATPase and disinhibits K^+ channels, thus leading to cellular loss of K^+.

Table 80.11. Causes of hypo- and hypermagnesemia

Hypomagnesemia
Decreased dietary intake
Intestinal malabsorption
Diarrhea
Steatorrhea
Isolated Mg^{2+} malabsorption
Impaired renal tubular reabsorption
Genetic defect, Bartter's syndrome
Nephrotoxic drugs
Diuretics
Loop
Osmotic
Hypercalcemia
Extracellular volume expansion
Hyperaldosteronism
ADH excess
Ketoacidosis
Alcohol
Phosphate depletion
Loss through skin
Burns
Excessive sweat loss
Lactation
Enhanced cellular uptake
Hyperthyroidism
Insulin administration to diabetic patients
Hypermagnesemia
Excess dietary intake
Decreased renal elimination
Renal insufficiency
Adrenal insufficiency
Cellular loss
Hypothermia

Table 80.12. Disorders of CaHPO$_4$ metabolism (typical sets of parameters, which, however, may not always occur)

	Plasma			Urine			Bone	
	Ca^{2+}	PO$_4$	AP	Ca^{2+}	PO$_4$	Pro-OH	Ost	Min
Primary hyperparathyroidism	↑	↓	↑	↑	↑	↑	↑↓	↓
Primary hypoparathyroidism	↓	↑	–	↓	↓	↓	↑	↑
Vitamin D depletion	(↓)	↓	↑	↓	↓	↑	↑	↓
Vitamin D excess	↑	↑	–	↑	↑	–		
Senile osteoporosis	–	–	–	–	–	↓	↓	↓
Inactivity osteoporosis	–	–	–	(↑)	(↑)	–	↓	↓
Phosphate diabetes	–	↓	↑	↑	↑	↑	↑	↓
Renal insufficiency	↓	↑	(↓)	↓	↓	↑	↑↓	↓
Milk alkali syndrome	↑	(↑)	↑	(↑)	(↓)	–	–	–
Malabsorption	(↓)	(↑)	↑	↓	↓	↑	↑	↓
Bone metastasis	(↑)	(↑)	(↑)	(↑)	(↑)	(↑)	↓	↓

PO$_4$, phosphate; *AP*, alkaline phosphatase; *Pro-OH*, hydroxyproline; *Ost*, osteoid; *Min*, mineralization

80.8.2 Hypermagnesemia

Mg^{2+} excess may result from impaired renal excretion or from excessive Mg^{2+} intake [2]. Renal retention of Mg^{2+} may be secondary to advanced renal failure [30] or to adrenal insufficiency [110]. The latter may be a result of volume depletion, which stimulates thick ascending limb NaCl reabsorption. Hypermagnesemia may also be due to cellular release of Mg^{2+}, as occurs in hypothermia [50].

Sequelae of Mg^{2+} excess include decreased neuromuscular and cardiac excitability, resulting in decreased deep tendon reflexes [112], quadriplegia [2], lethargy, nausea, dilated pupils, respiratory depression [3,112], hypotension, bradycardia, heart block, and cardiac arrest [91]. Micturition and defecation may suffer from reduced smooth muscle activity [91].

80.9 Examples of Parameters Altered by Disordered CaHPO$_4$ Metabolism

In the *search for the cause of hyper- or hypocalcemia* and or hyper- or hypophosphatemia, it is useful first to identify whether renal or extrarenal causes are predominant. The determination of 24h urinary excretion of Ca^{2+} and phosphate indicates whether or not inappropriate renal elimination can account for the altered plasma Ca^{2+} or phosphate concentrations. If hypercalcemia, for instance, is paralleled by hypercalciuria, the kidney cannot be the culprit behind the enhanced plasma concentration of Ca^{2+}. In such a case, enhanced intestinal Ca^{2+} absorption or stimulated bone resorption must be suspected. Excessive resorption of bone is usually paralleled by corresponding degradation of bone collagen, which is reflected by enhanced urinary excretion of hydroxyproline. Stimulated bone formation, on the other hand, is paralleled by enhanced secretion and plasma concentration of alkaline phosphatase. Table 80.12 lists a few examples of altered parameters in some typical disorders of mineral metabolism.

References

1. Agus ZS, Morad M (1991) Modulation of cardiac ion channels by magnesium. Annu Rev Physiol 53:299–308
2. Alfrey AC (1992) Disorders of magnesium metabolism. In: Seldin DW, Giebisch G (eds) The kidney. Physiology and pathophysiology, 2nd edn. Raven, New York pp 2357–2374
3. Alfrey AC, Terman DS, Brettschneider L, Simpson KM, Ogden DA (1970) Hypermagnesemia after renal homotransplantation. Ann Intern Med 73:367–371
4. Allgrove J, Adams S, Fraher L, Reuben A, O'Riordan JLH (1984) Hypomagnesemia: studies of parathyroid hormone secretion and function. Clin Endocrinology 21:435–449
5. Avioli LV, Lynch TN, Berman M (1963) Digital computer compartmental analysis of Mg^{2+} in normal subjects, Pagets's disease and thyroid disease. J Clin Invest 42:915
6. Baron R (1989) Molecular mechanisms of bone resorption by the osteoclast. Anat Rec 224:317–124
7. Baron R, Neff L, Louvard D, Courtoy PJ (1985) Cell-mediated extracellular acidification and bone resorption: evidence for low pH in resorbing lacunae and localization of a 100 Kd lysosomal membrane protein at the osteoclast ruffled border. J Cell Biol 101:2210–2222
8. Bartter RC (1969) So-called Bartter's syndrome. N Engl J Med 281:1483–1494
9. Beresford JN, Gallagher JA, Russell RGG (1986) 1,25-Dihydroxyvitamin D$_3$ and human bone-derived cells in vitro: effects on alkaline phosphatase, type I collagen and proliferation. Endocrinology 119:1776–1785
10. Berndt TJ, Knox FG (1992) Renal regulation of phosphate excretion. In: Seldin DW, Giebisch G (eds) The kidney. Physiology and pathophysiology, 2nd edn. Raven, New York pp 2511–2532
11. Biber J, Caderas G, Werner A, Moore M, Murer H (1993) Cellular and molecular aspects of renal phosphate transport. Renal Physiol Biochem 16:37–47
12. Blair HC, Teitelbaum SL, Ghiselli R, Gluck S (1989) Osteoclastic bone resorption by a polarized vacuolar proton pump. Science 245:855–857
13. Boden SD, Kaplan FS (1990) Calcium homeostasis. Orthop Clin North Am 21:31–42
14. Bonjour JP, Caverzasio J (1984) Phosphate transport in the kidney. Rev Physiol Biochem Pharmacol 100:161–214
15. Booth BE, Johanson A (1975) Hypomagnesemia due to renal tubular defect in reabsorption of magnesium. J Pediatr 85:350–354
16. Borke JL, Caride A, Verma AK, Penniston JT, Kumar R (1989) Plasma membrane calcium pump and 28-kDa calcium

binding protein in cells of rat kidney distal tubules. Am J Physiol 257:F842–F849

17. Broadus AE, Mangin M, Ikeda K, Insogna KL, Weir EC, Burtis WJ, Stewart AF (1988) Humoral hypercalcemia of malignancy. N Engl J Med 314:556–563

18. Brostrom CO, Brostrom MA (1990) Calcium-dependent regulation of protein synthesis in intact mammalian cells. Annu Rev Physiol 52:577–590

19. Broughton A, Anderson IRM, Bowden CH (1968) Magnesium deficiency syndrome in burns. Lancet 2:1156–1158

20. Brown AJ, Dusso AS, Slatopolsky E (1992) Vitamin D. In: Seldin DW, Giebisch G (eds) The kidney. Physiology and pathophysiology, 2nd edn. Raven, New York pp 1505–1552

21. Brown AM, Birbaumer L (1988) Direct G protein gating of ion channels. Am J Physiol 254:H401–H410

22. Brown AM, Yatani A (1990) Voltage-gated ionic channels: diversity and modulation by Mg^{2+}. In: Strata P, Carbone E (eds) Mg^{2+} and excitable membranes. Springer, Berlin Heidelberg New York pp 21–31

23. Brown EM (1991) Extracellular Ca^{2+} sensing, regulation of parathyroid cell function, and role of Ca^{2+} and other ions as extracellular (first) messengers. Physiol Rev 2:371–412

24. Busch AE, Waldegger S, Herzer T, Biber J, Markovich D, Hayes G, Murer H, Lang F (1994) Electrophysiological analysis of Na^+/P_i cotransport mediated by a transporter cloned from rat kidney and expressed in Xenopus oocytes. Proc Natl Acad Sci USA

25. Bushinsky DA, Krieger NS (1992) Role of the skeleton in calcium homeostasis. In: Seldin DW, Giebisch G (eds) The kidney. Physiology and pathophysiology, 2nd edn. Raven, New York, pp 2395–2430

26. Carafoli E (1987) Intracellular calcium homeostasis. Annu Rev Biochem 56:395–433

27. Carafoli E (1991) The calcium pumping ATPase of the plasma membrane. Annu Rev Physiol 53:531–548

28. Carafoli E (1991) Calcium pump of the plasma membrane. Physiol Rev 1:129–154

29. Chudley A, Ninan A, Young GB (1981) Neurologic signs and hypophosphatemia with total parenteral nutrition. Can Med Assoc J 125:604–607

30. Coburn JW, Popovtzer M, Massry SG, Kleeman CR (1967) The physicochemical state and renal handling of divalent ions in chronic renal failure. Arch Intern Med 124:302–311

31. Cohen MI, McNamara H, Finberg L (1970) Serum magnesium in children with cirrhosis. J Pediatr 76:453–455

32. Costanzo LS, Windhager EE (1992) Renal regulation of calcium balance. In: Seldin DW, Giebisch G (eds) The kidney. Physiology and pathophysiology, 2nd edn. Raven, New York, pp 2375–2394

33. Crosby G, Szabo MD (1990) Excess Mg^{2+} and central nervous system metabolism. In: Strata P, Carbone E (eds) Mg^{2+} and excitable membranes. Springer, Berlin Heidelberg New York, pp 119–123

34. Danowski TS, Gillespie HK, Fergus EB, Puntereri AJ (1956–1957) Significance of blood sugar and serum electrolyte changes in cirrhosis following glucose, insulin, glucagon, or epinephrine. Yale J Biol Med 29:361–375

35. De Rouffignac C, Elalouf JM, Roinel N (1987) Physiological control of the urinary concentrating mechanism by peptide hormones. Kidney Int 31:611–620

36. DeLuca HF (1988) The vitamin D story: a collaborative effort of basic science and clinical medicine. FASEB J 2:224–236

37. DeSmedt H, Parys JB, Borgraef R, Wujtack F (1981) Calmodulin stimulation of renal ($Ca^{2+} + Mg^{2+}$)-ATPase. FEBS Lett 131:60–62

38. DiStefano A, Wittner M, Gebler B, Greger R (1988) Increased Ca^{2+} or Mg^{2+} concentration reduces relative tight-junction permeability to Na^+ in the cortical thick ascending limb of Henle's loop of rabbit kidney. Renal Physiol Biochem 11:70–79

39. Flatman PW (1991) Mechanisms of magnesium transport. Annu Rev Physiol 53:259–272

40. Ghijsen W, Gmaj P, Murer H (1984) Ca^{2+}-stimulated Mg^{2+}-independent ATP-hydrolysis of the high affinity Ca^{2+} pumping ATPase. Two different activities in rat kidney basolateral membranes. Biochim Biophys Acta 778:481–488

41. Gmaj P, Murer H (1986) Cellular mechanisms of inorganic phosphate transport in kidney. Physiol Rev 66:36–70

42. Gmaj P, Murer H (1988) Calcium transport mechanisms in epithelial cell membranes. Miner Electrolyte Metab 14:22–30

43. Gmaj P, Murer H, Kinne R (1979) Calcium ion transport across plasma membranes isolated from rat kidneys cortex. Biochem J 178:549–557

44. Greger R (1985) Ion transport mechanisms in thick ascending limb of Henle's loop of mammalian nephron. Physiol Rev 65:760–797

45. Greger R, Lang F, Marchand GR, Knox FG (1977) Site of renal phosphate reabsorption. Micropuncture and microinfusion study. Pflugers Arch 369:111–118

46. Greger R, Lang F, Knox FG, Lechene CP (1977) Absence of significant secretory flux of phosphate in the proximal convoluted tubule. Am J Physiol 232:F235–F238

47. Greger R, Lang F, Oberleithner H (1978) Distal site of calcium reabsorption in the rat nephron. Pflugers Arch 374:153–157

48. Habener JF, Potts JT (1976) Relative effectiveness of magnesium and calcium on the secretion and biosynthesis of PTH in vitro. Endocrinology 98:197–202

49. Hanai H, Ishida M, Liang CT, Sacktor B (1986) Parathyroid hormone increases sodium/calcium exchange activity in renal cells and the blunting of the response in aging. J Biol Chem 261:5419–5425

50. Harmon JP, Larson AM, Young DW (1958) Effect of cold acclimatization on plasma electrolyte levels. J Appl Physiol 13:239–240

51. Hartzell HC, White RE (1990) Regulation of the voltage-gated Ca^{2+} current by intracellular free Mg^{2+} studied by internal perfusion of single cardiac myocytes. In: Strata P, Carbone E (eds) Mg^{2+} and excitable membranes. Springer, Berlin Heidelberg New York, pp 71–96

52. Harvey W (1988) Prostaglandins in bone resorption. In: Harvey W et al (eds) Source of prostaglandins and their influence on bone formation and resorption. CRC Press, Boca Raton, pp 27–41

53. Heilman ES, Tschudy DP, Bartter RC (1962) Abnormal electrolyte and water metabolism in acute intermittent porphyria: transient inappropriate secretion of antidiuretic hormone. Am J Med 32:734–746

54. Hill GL, Guinn EJ, Dudrick SL (1976) Phosphorus distribution in hyperalimentation induced hypophosphatemia. J Surg Res 20:527–531

55. Hoffmann O, Klaushofer K, Gleispach H, Leis HJ, Luger T, Koller K, Peterlik M (1987) Gamma interferon inhibits basal and interleukin 1-induced prostaglandin production and bone resorption in neonatal mouse calvaria. Biochem Biophys Res Commun 143:38–43

56. Hoffmann O, Klaushofer K, Koller K Peterlik M, Mavreas T, Stern PH (1986) Indomethacin inhibits thrombin- but not thyroxin-stimulated resorption of fetal rat limb bones. Prostaglandins 31:601–608

57. Horton R, Biglieri EG (1962) Effect of aldosterone on the metabolism of magnesium. Clin Endocrinol Metab 22:1187–1192

58. Hruska KA, Mills SC, Khalifa S, Hammerman MR (1983) Phosphorylation of renal brush-border membrane vesicles. Effect on calcium uptake and membrane content of polyphosphoinositides. J Biol Chem 258:2501–2507

59. Inesi G, Sumbilla C, Kirtley ME (1990) Relationship of molecular structure and function in Ca^{2+}-transport ATPase. Physiol Rev 3:749–760

60. Jayakumar A, Cheng LL, Liang CT (1984) Sodium gradient-dependent calcium uptake in renal basolateral membrane vesicles. Effect of parathyroid hormone. J Biol Chem 259:10827–10833

61. Johnson JW, Ascher P (1990) Dual blockage by Mg^{2+} of the N-methyl-D-aspartate-activated channel. In: Strata P, Carbone E (eds) Mg^{2+} and excitable membranes. Springer, Berlin Heidelberg, New York, pp 105–118

62. Khalifa S, Mills S, Hruska KA (1983) Stimulation of calcium-uptake by parathyroid hormone in renal brush-border membrane vesicles. Relationship to membrane phosphorylation. J Biol Chem 258:14400–14406

63. Kinne-Saffran E, Kinne R (1974) Localization of a calcium-stimulated ATPase in the baso-lateral plasma membrane of the proximal tubule of rat kidney cortex. J Membr Biol 17:263–274

64. Klein DC, Raisz LG (1970) Prostaglandins: stimulation of bone resorption in tissue culture. Endocrinology 86:1436–1440

65. Knochel JP (1992) The clinical and physiological implications of phosphorus deficiency. In: Seldin DW, Giebisch G (eds) The kidney. Physiology and pathophysiology, 2nd edn. Raven, New York pp 2533–2562

66. Knochel JP, Caskey JH (1977) On the mechanism of hypophosphatemia in acute heat stroke. JAMA 238:425–426

67. Krieger NS, Stappenbeck TS, Stern PH (1988) Characterization of specific thyroid hormone receptors in bone. J Bone Miner Res 3:473–478

68. Kurachi Y, Nakajima T, Sugimoto T (1986) Role of intracellular Mg^{2+} in the activation of muscarinic K^+ channel in cardiac atrial cell membrane. Pflugers Arch 407:572–574

69. Lang F (1980) Renal handling of calcium and phosphate. Klin Wochenschr 58:985–1003

70. Lang F, Greger R, Marchand GR, Knox FG (1977) Stationary microperfusion study of phosphate reabsorption in proximal and distal nephron segments. Pflugers Arch 368:45–48

71. Lang F, Greger R, Knox FG, Oberleithner H (1981) Factors modulating the renal handling of phosphate. Renal Physiol 4:1–16

72. Le Grimellec C, Roinel N, Morel F (1973) Simultaneous Mg, Ca, P, K, Na and Cl analysis in rat tubular fluid. II. During acute Mg plasma loading. Pflugers Arch 340:197–210

73. Le Grimellec C, Roinel N, Morel F (1974) Simultaneous Mg, Ca, P, K, Na and Cl analysis in rat tubular fluid. III. During acute Ca plasma loading. Pflugers Arch 346:171–189

74. Lux HD, Carbone E (1991) Blockage of neuronal low-threshold Ca^{2+} channels by extracellular Mg^{2+}. In: Strata P, Carbone E (eds) Mg^{2+} and excitable membranes. Springer, Berlin Heidelberg New York, pp 97–104

75. MacIntyre I, Alevizaki M, Bevis PJR, Zaidi M (1987) Calcitonin and peptides from the calcitonin gene. Clin Orthop 217:45–55

76. Matsuda H (1990) Voltage-dependent blockage of cardiac inwardly rectifying K^+ channels by internal Mg^{2+}. In: Strata P, Carbone E (eds) Mg^{2+} and excitable membranes. Springer, Berlin Heidelberg New York, pp 51–70

77. Matsuda H (1991) Magnesium gating of the inwardly rectifying K^+ channel. Annu Rev Physiol 53:289–298

78. McCormack JG, Denton RM (1990) Ca^{2+} as a second messenger within mitochondria of the heart and other tissues. Annu Rev Physiol 52:451–466

79. McCormack JG, Halestrap AP, Denton RM (1990) Role of calcium ions in regulation of mammalian intramitochondrial metabolism. Physiol Rev 2:391–426

80. Miller EJ, Martin GR (1968) The collagen of bone. Clin Orthop 59:195–232

81. Muff R, Fischer JA, Biber J, Murer H (1992) Parathyroid hormone receptors in control of proximal tubule function. Annu Rev Physiol 54:67–80

82. Mundy GR, Shapiro JG, Bandelin JG, Canalis EM, Raisz LG (1976) Direct stimulation of bone resorption by thyroid hormones. J Clin Invest 58:529–534

83. Murer H, Biber J (1992) Renal tubular phosphate transport: cellular mechanisms. In: Seldin DW, Giebisch G (eds) The kidney. Physiology and pathophysiology, 2nd edn. Raven, New York, pp 2481–2510

84. Murphy E, Freudenrich CC, Lieberman M (1991) Cellular magnesium and Na/Mg exchange in heart cell. Annu Rev Physiol 53:273–288

85. Nicoll DA, Longoni A, Philipson K (1990) Molecular cloning and functional expression of the cardiac sarcolemmal Na^+/Ca^{2+} exchange. Science 250:562–565

86. Nijweide PJ, Burger EH, Feyen JHM (1986) Cells of bone: proliferation, differentiation, and hormonal regulation. Physiol Rev 66:855–886

87. Oberleithner H, Greger R, Lang F (1982) The effect of respiratory and metabolic acid-base changes on ionized calcium concentration: in vivo and in vitro experiments in man and rat. Eur J Clin Invest 12:451–455

88. O'Connor LR, Klein KL, Bethune JE (1977) Effect of hypophosphatemia on myocardial performance in man. N Engl J Med 297:901

89. Quamme GA (1992) Magnesium: cellular and renal exchanges. In: Seldin DW, Giebisch G (eds) The kidney. Physiology and pathophysiology, 2nd edn. Raven, New York, pp 2339–2356

90. Quamme GA, Dirks JH (1986) The physiology of renal magnesium handling. Renal Physiol 9:257–269

91. Randall RE Jr, Chen MD, Spray CC, Rossmeisl EC (1964) Hypermagnesemia in renal failure. Ann Intern Med 61:73–88

92. Rasmussen H, Barrett PQ (1984) Calcium messenger system: an integrated view. Physiol Rev 64:938–984

93. Rasmussen H, Reifenstein EC Jr (1962) In: Williams RH (ed) Textbook of endocrinology, 3rd edn. Saunders, Philadelphia, p 798

94. Ravid M, Robson M (1976) Proximal myopathy caused by iatrogenic phosphate depletion. JAMA 236:1380

95. Ringel RE, Brenner JI, Haney PJ, Burns JE, Molton AL, Berman MA (1982) Prostaglandin-induced periostitis: a complication of long-term PGE_1 infusion in an infant with congenital heart disease. Radiology 142:657–658

96. Rink TJ, Sage SO (1990) Calcium signaling in human platelets. Annu Rev Physiol 52:431–450

97. Shull GE, Greeb J (1986) Molecular cloning of two isoforms of the plasma membrane Ca^{2+} transporting ATPase from rat brain. Structural and functional domains exhibit similarity to Na^+, K^+ and other cation transport ATPases. J Biol Chem 263:8646–8657

98. Silvis SE, DiBartolomeo AG, Aaker HM (1980) Hypophosphatemia and neurological changes secondary to oral caloric intake. Am J Gastroenterol 73:215–222

99. Smith DD, Gowen M, Mundy GR (1987) Effects of interferon-γ and other cytokines on collagen synthesis in fetal rat bone cultures. Endocrinology 120:2494–2499

100. Snowdowne KW, Borle AB (1985) Effects of low extracellular sodium on cytosolic ionized calcium. Na^+-Ca^{2+} exchange as a major calcium influx pathway in kidney cells. J Biol Chem 260:14998–15007

101. Stewart AF (1992) Hypercalcemic and hypocalcemic states. In: Seldin DW, Giebisch G (eds) The kidney. Physiology and pathophysiology, 2nd edn. Raven, New York, pp 2431–2460

102. Stewart AF, Longo W, Kreitter D, Jacob R, Burtis WJ (1986) Hypocalcemia associated with calcium soap formation in a patient with a pancreatic fistula. N Engl J Med 315:496–498

103. Talor Z, Arruda JAL (1985) Partial purification and reconstitution of renal basolateral Na^+-Ca^{2+} exchanger into liposome. J Biol Chem 260:15473–15476

104. Taylor SS, Buechler JA, Yonemoto W (1990) cAMP-dependent protein kinase: framework for a diverse family of regulatory enzymes. Annu Rev Biochem 59:971–1005

105. Triffit JT (1987) Initiation and enhancement of bone formation. Acta Orthop Scand 58:673–684

106. Tschope W, Ritz E, Schellenberg B, Arab L, Schlierf G (1984) Decreased plasma phosphate under hormonal contraceptives. Miner Electrolyte Metab 10:88–91

107. Ueda K, Saito A, Nakano H, Aoshima M, Yokota M, Muraoka R, Iwaya T (1980) Cortical hyperostosis following long-term

administration of prostaglandin E_1 in an infant with cyanotic congenital heart disease. J Pediatr 97:834–836

108. Urist MR, DeLange RJ, Finerman GAM (1983) Bone cell differentiation and growth factors. Science 220:680–686

109. Van Heeswijck MPE, Geertsen JAM, Van Os CH (1984) Kinetic properties of the ATP-dependent Ca^{2+} pump and the Na^+/Ca^{2+} exchange system in basolateral membranes from rat kidney cortex. J Membr Biol 79:19–31

110. Wacker WEC, Vallee BL (1958) Magnesium metabolism. N Engl J Med 259:431–438, 475–482

111. Walser M (1973) Divalent cations: physicochemical state in glomerular filtrate and urine and renal extraction. In: Orloff J, Berliner RW (eds) Handbook of physiology. Waverly, Baltimore, pp 555–586

112. Welt LG, Gitelman H (1965) Disorders of magnesium metabolism. DM 1:1–32

113. White RE, Hartzell HC (1989) Magnesium ions in cardiac function. Regulator of ion channels and second messengers. Biochem Pharmacol 38:859–867

114. Wier G (1990) Cytoplasmic Ca^{2+} in mammalian ventricle: dynamic control by cellular processes. Annu Rev Physiol 52:467–486

115. Woods HF, Eggleston LV, Krebs HA (1970) The cause of hepatic accumulation of fructose-1-phosphate on fructose loading. Biochem J 119:501–510

116. Ziff WB, Bacon GE, Spencer ML (1979) Hypocalcemia, hypomagnesemia and transient hypoparathyroidism during therapy with potassium phosphate in diabetic ketoacidosis. Diabetes Care 2:265–268

81 Regulation of the Lower Urinary Tract

W. Jänig

Contents

81.1 Introduction

The urinary bladder stores and periodically evacuates urine, which is continually produced by the kidney. Long collecting phases alternate with brief emptying phases. The ability of the bladder to collect and retain urine is called *continence* and the act of emptying, *micturition.* In the collecting phase this organ is able to accept large volumes of urine with little change in intravesical pressure. During micturition the urine is evacuated, very often in an all-or-none manner. These processes are dependent on the myogenic mechanisms and on the neural control of the urinary bladder and its outlet. This neural control is essential and involves afferent and efferent components and a cascade of control centers in the central nervous system. The central mechanisms consist of spinal and supraspinal (pontine) reflexes and their control by suprapontine centers, in particular the telencephalon. The latter is associated with distinct sensations that can be elicited from the lower urinary tract and are correlated with different degrees of intravesical pressure and different phases of micturition. These sensations are essential for the adaptation of emptying and collecting urine to the behavior and

therefore also to the social life of humans. This article focuses on the neural regulation of storage and voiding, including the sensory components. The urinary bladder (storage organ) and urethra (outlet) are referred to collectively as the lower urinary tract (LUT) [3,4,6,8,29].

81.2 Functional Anatomy of the LUT

The structures of the male and female LUTs are schematically illustrated in Fig. 81.1. The urinary bladder is a hollow muscle consisting of the detrusor vesicae, the deep and superficial trigone, and the bladder neck. The outlet of the bladder is the urethra. The wall of the urinary bladder consists of a network of smooth muscle cells with longitudinal (inner and outer layer) and circular (middle layer) courses. The wall of the urinary bladder is luminally covered by the vascularized mucosa, which is impenetrable for urine. At the base of the bladder is a triangular area of fine smooth muscle fibers (the vesical trigone) with the openings of the ureters at the upper corners. These ducts enter obliquely through the bladder wall, so that no urine can be forced back into the ureters when the intravesical pressure rises. Detrusor and proximal urethra form almost a right angle during storage of urine and change into a funnel-like structure during micturition, probably due to relaxation and shortening of the urethra. The small smooth muscle cells of the bladder neck and the proximal urethra are morphologically and functionally distinct from those of the detrusor. They are arranged in a circular configuration in males and function as the internal vesical sphincter. This sphincter is not very well developed in females and probably not important for continence. In the male it is functionally considered to be a sexual organ and prevents reflux of the secretion into the urinary bladder during ejaculation.

In the male, the urethra has preprostatic, prostatic and membranous sections. Its wall has an outer muscle coat and an inner mucous membrane. The muscle coat consists of longitudinal smooth muscle cells and, distally, striated small muscle cells with a circular arrangement, which together with the pelvic floor musculature (levator ani muscle and possibly the transversus perinei, which is disputed in males) form the *external urethral sphincter* (see [10,43]). The circularly oriented striated muscle fibers close to the urethra are small and tonic. This part of the external ure-

R. Greger/U. Windhorst (Eds.)
Comprehensive Human Physiology, Vol. 2
© Springer-Verlag Berlin Heidelberg 1996

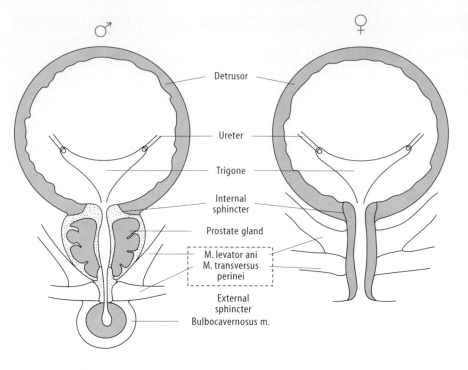

Fig. 81.1. Anatomy of the lower urinary tract (LUT) in males and females. Whether the transversus perinei muscle is involved in the external urethral sphincter in males is a matter of debate

thral sphincter is sometimes called the *rhabdosphincter*. During storage of urine the urethral lumen is a slit.

81.3 Peripheral Innervation of the LUT

81.3.1 Peripheral Neurons

The innervation of the LUT is outlined in Fig. 81.2 (see Chap. 17). The *parasympathetic outflow* has it origin in the sacral segments S2–S4 and projects through the pelvic splanchnic nerve to the pelvic plexus. Preganglionic parasympathetic fibers synapse with postganglionic neurons in pelvic ganglia that are situated close to the base of the bladder or on the bladder wall and belong to the pelvic splanchnic plexus. The postganglionic parasympathetic neurons are cholinergic and innervate the smooth musculature of detrusor, bladder neck and urethra. This efferent innervation is essential for micturation contractions. The *sympathetic innervation* originates from the spinal segments T10–L2. Most preganglionic axons project through the lumbar splanchnic nerves and the hypogastric nerves (hypogastric plexuses) to postganglionic neurons situated in the superior hypogastric plexus (inferior mesenteric ganglion) or in the pelvic splanchnic plexus. These postganglionic neurons are noradrenergic and innervate the smooth muscles of the bladder base, the internal vesical sphincter and the urethra distal to the sphincter. The innervation density by noradrenergic fibers increases from proximal (detrusor) to distal (internal urethra). Some preganglionic neurons project through the paravertebral chain and synapse with postganglionic neurons which project through the pelvic splanchnic nerve to the LUT (see [6,22]).

The striated muscles which constitute the external urethral sphincter are innervated by *motoneurons* in the sacral segments S2–S3. These project through the pudendal nerve. Some motoneurons to the tonic small muscle fibers which constitute the rhabdosphincter may also project through the pelvic splanchnic nerve [10].

The *afferent innervation* of the LUT consists of visceral afferents which project with the sympathetic fibers to the spinal segments T10–L2, of sacral visceral afferents which project through the pelvic splanchnic nerve to the spinal segments S2–S4 and of sacral somatic afferents from the urethra which project through the pudendal nerve. The sacral visceral afferents are essential for micturition and most sensations associated with storage and voiding.

81.3.2 Location of Efferent Neurons in, and Projection of Visceral Afferents to, the Sacral Spinal Cord

Preganglionic parasympathetic neurons involved in the regulation of the LUT are situated in the lateral intermediate zone of the sacral spinal cord. Neurons associated with the hindgut and neurons that are involved in erection are located more medially (Fig. 81.3, right). The location of the latter preganglionic neurons is hypothetical. The dendrites of the preganglionic neurons have a typical orientation and project medially, laterally into the white matter and dorsally (Fig. 81.3, left). *Motoneurons* which project to the external sphincter of the LUT are situated in the lateral part of the ventral horn (Fig. 81.3). This part of the ventral

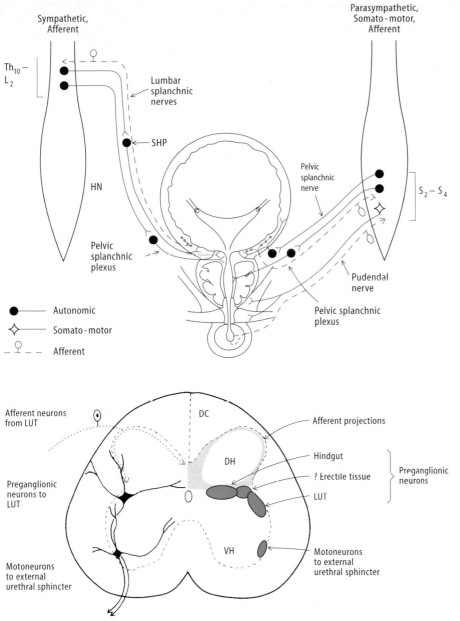

Sympathetic,
Afferent

Th$_{10}$ –
L$_2$

Lumbar
splanchnic
nerves

SHP

HN

Pelvic
splanchnic
plexus

● ─── Autonomic

◇─── Somato-motor

♀ ─ ─ ─ Afferent

Parasympathetic,
Somato-motor,
Afferent

Pelvic
splanchnic
nerve

S$_2$–S$_4$

Pudendal
nerve

Pelvic splanchnic
plexus

Fig. 81.2. The innervation
of the LUT

Afferent neurons
from LUT

DC

Afferent projections

DH

Hindgut

? Erectile tissue

LUT

Preganglionic
neurons

Preganglionic
neurons to
LUT

VH

Motoneurons
to external
urethral sphincter

Motoneurons
to external
urethral sphincter

Fig. 81.3. Projection of visceral afferents from the LUT to the sacral spinal cord, location of preganglionic neurons to the LUT and motoneurons to the external urethral sphincter in the sacral spinal cord. Note that visceral afferents project to lamina I and V of the dorsal horn (*DH*), to deeper laminae (sparing laminae II–IV) and to the dorsal commissural nucleus. The preganglionic neurons are located lateral to the preganglionic neurons associated with hindgut and reproductive organs. *DC*, dorsal columns; *VH*, ventral horn. (After [6])

horn is sometimes called Onuf's nucleus in humans. *Sympathetic preganglionic neurons* in the thoracolumbar spinal cord which project to the LUT are located medial to the intermediolateral cell column of the intermediate zone [6,22].

Sacral visceral afferents from the LUT (but also from the hindgut) project to lamina I of the dorsal horn, to the Lissauer tract and further through lamina I medially and laterally to lamina V and deeper (Fig. 81.3). These projections are typical for visceral afferents in comparison to somatic afferents (e.g., those from the perineal region which project through the pudendal nerve). These afferents project also to laminae II–IV [1,2].

81.4 Urodynamics of the LUT

Urinary bladder (reservoir) and outlet of the LUT (bladder neck, urethra and external urethral sphincter) exhibit a

1613

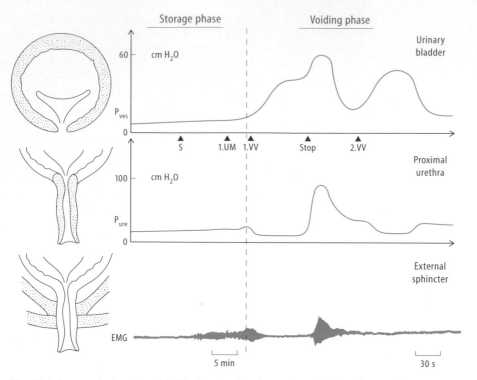

Fig. 81.4. Pressure in the urinary bladder (P_{VES}) and in the proximal urethra (P_{ure}) and electromyogram (*EMG*) recorded from the external urethral sphincter during filling of the urinary bladder at about 20 ml/s (storage phase) and during voluntary micturition (voiding phase). *S*, first sensation; *1. UM*, first urge to micturate; *1. VV, 2. VV*, first and second voluntary voiding; *Stop*, voluntary interruption of voiding. (Modified from K.-P. Jünemann, personal communication, with permission)

typical well-coordinated behavior during continuous filling at about 10–20 ml/min (Fig. 81.4). In the filling (storage) phase the intravesical volume may increase to 350–500 ml before the micturition starts. The intravesical pressure increases to 10–15 cmH₂O. During filling, the first sensation of bladder fullness appears at 150–250 ml and the urge to micturate appears at 350–500 ml intravesical volume. During the storage phase the activity of the external sphincter increases (EMG in Fig. 81.4). Normal values for the LUT during storage and voiding phases are listed in Table 81.1. During voluntary voiding (VV in Fig. 81.4) the intravesical pressure increases, the intraurethral pressure decreases and the external sphincter relaxes (decrease of the EMG activity). Figure. 81.5 illustrates, on an expanded time scale, the events which occur just before and during voiding (see urine flow). It is important to note that the proximal urethra starts to relax actively before the intravesical pressure increases. This relaxation cannot be explained by the relaxation of the external urethral sphincter, which is located more distally. Voluntary interruption of micturition leads to an increase of the pressure in the urethra, activation of the external urethral sphincter (see stop and increase in EMG activity in Fig. 81.4), and a further increase in the intravesical pressure due to contraction of the pelvic floor musculature.

This pattern of response of the lower urinary tract depends on its afferent and efferent innervation and the central regulation of the impulse activity in the sympathetic neu-

Table 81.1. Normal values for the storage and the voiding phase

Parameter	Value
Storage phase	
Detrusor pressure	10–15 cmH₂O
Compliance	<25 ml/cmH₂O
Sensibility (first sensation)	150–250 ml
First urge to micturate	350–550 ml
Maximal bladder capacity	350–550 ml
Maximal anatomical bladder capacity	400–600 ml
Bladder neck	Closed
Closing pressure of urethra	>Intravesical pressure
Voiding phase	
Maximal micturition pressure	Male: 90 cmH₂O
	Female: 75 cmH₂O
Maximal detrusor contraction pressure	Male: 75 cmH₂O
	Female: 55 cmH₂O
Maximal urine flow	Male: 20–25 ml/s
	Female: 25–30 ml/s
Bladder neck	Funnel-like
Closing pressure of urethra	<Intravesical pressure

rons, parasympathetic neurons and sacral motoneurons. Any peripheral or central disturbance of these neural components changes the pressure profiles and motor activities of the lower urinary tract. For example, in chronic spinal paraplegic or tetraplegic patients the external urethral sphincter and the detrusor contract simultaneously during

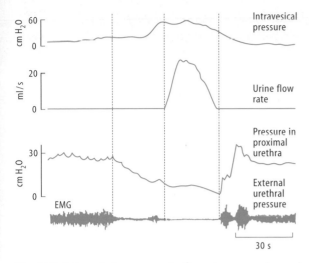

Fig. 81.5. Intravesical pressure, urine flow rate, pressure in proximal urethra and electromyogram (*EMG*) recorded from the external urethral sphincter before and during normal voiding. Note the decrease in urethral pressure before increase in intravesical pressure. Note also that the pressure profiles in the urinary bladder and in the proximal urethra are identical during urine flow. (Modified from [43] with permission)

filling and micturition (*detrusor-sphincter dyssynergy*; see Fig. 81.10). When the sacral visceral afferent fibers from and/or the efferent parasympathetic fibers to the LUT are interrupted, the urinary bladder exhibits areflexia (see below) [43,44].

81.5 Vesical Afferents, Sensations and Organ Regulation

81.5.1 Sensations Elicited from the LUT

The LUT is under dominant control of the central nervous system, enabling precise integration of regulation of micturition and continence in the general behavior of the organism. Humans, and possibly animals too, can perceive a variety of sensations from the LUT in healthy conditions. These sensations are directly related to the functional states of the organ system. This is best illustrated by the range of sensations elicited during slow filling of the urinary bladder (Fig. 81.6A). First sensations of fullness, which wax and wane, occur at a bladder volume of about 250–300 ml (see S in Fig. 81.4). As the intravesical volume increases further a continuous sense of fullness prevails, which will eventually be superseded by the desire to micturate at 400–500 ml. If the filling continues and voiding is prevented, discomfort and finally pain (at ≥600 ml) occur. These sensations are conveyed by visceral afferents from the urinary bladder and not from the adjacent tissue [33,43].

The physical parameter that correlates best with the different levels of sensations elicited from the normal urinary bladder is the increase of the *transmural pressure*. With a simplified picture of the urinary bladder as a sphere, the tension will be proportional to the product of the intramural pressure and the cubic root of the volume [28,39,40]. Therefore the intravesical pressure has been given as the important determinant of the evoked sensation although it is also evident that pressure will correspond to intravesical volume in a closed viscus.

If micturition is possible humans can identify two further non-painful sensations that are associated with the LUT: the feeling that micturition is imminent and the actual *sensation of urine passing*. Both are probably signaled by sacral afferents supplying the urethra. Additionally, pain

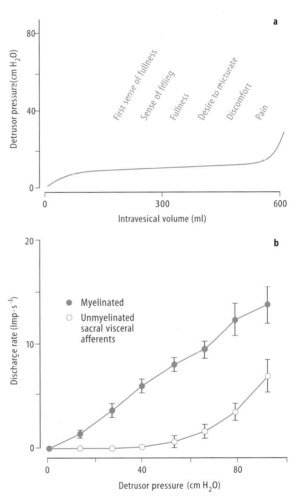

Fig. 81.6. a Idealized pressure–volume curve during slow filling of the human urinary bladder and evoked sensations in a trained subject. *Ordinate scale*, effective detrusor pressure (intravesical pressure minus intraabdominal pressure). **b** Stimulation–response functions of sacral visceral afferents innervating urinary bladder in the cat. The bladder was isotonically distended by graded increases in intravesical pressure lasting 60–90 s. *Ordinate scale*, mean impulse rate in myelinated (A-delta) afferents ($n = 34$) and in unmyelinated afferents ($n = 5$). Note that very few unmyelinated sacral afferents can be activated from the healthy urinary bladder in cats and that these afferents have high pressure thresholds. Most unmyelinated afferents cannot be activated by contraction and distension of the urinary bladder. They may be recruited during inflammation of the mucosa. (Data from [12,13])

can be elicited from the urethra in healthy conditions, e.g., by distension or displacement of the mucosa.

It is generally thought that the receptors that elicit these sensations are located primarily in the muscle layers of the bladder wall. The sensitivity of the bladder, and particularly of the mucosa, has been studied with a number of artificial stimuli whose biological significance remains largely obscure. Probing and slight pressure applied to the normal bladder mucosa during cystoscopy seem to be insufficient to elicit consciously perceived sensations.

Moreover, injection of irritant chemicals has been shown to evoke pain, possibly without significant elevation of intravesical pressure [30,37]. Although this chemosensitivity appears to be of minor importance for normal bladder sensation it is possible that such chemosensitivity is important after tissue damage and inflammation. There is also anecdotal clinical evidence that light pressure stimuli applied to the mucosa become painful once the mucosa is inflamed [32,38].

The central pathway mediating these sensations is the *spinothalamic tract* in the ventrolateral spinal cord. From investigations in humans undergoing anterolateral *bilateral cordotomy* (cutting the spinal ventrolateral tracts) it is known that almost all ascending afferent information from the LUT leading to the distinct sensations and to coordinated micturition (see inset in Fig. 81.7) is relayed through the lateral part of the anterolateral tract [33–35]. Cutting this tract abolishes the sensations and leads to bladder reflexes very similar to those observed in humans after transection of the spinal cord.

81.5.2 Afferents Mediating the Sensations

Neurophysiological experiments on the cat show that the intravesical pressure up to 90 cm H_2O is faithfull encoded in the activity of *myelinated sacral visceral afferents* from the LUT. Very few high threshold afferents have been found which are activited in a graded manner at intravesical pressures >50 cm H_2O and which are unmyelinated (Fig. 81.6B). These experiments suggest that the afferent information from the LUT that leads to the different types of sensations (including pain) and is essential for the micturition reflexes is encoded in the activity of a functionally homogeneous population of myelinated sacral visceral afferents [13,14,20]. How the CNS extracts the information from this impulse activity that is relevant information for generation of the sensations and regulations is unclear. In addition, the urinary bladder is innervated by many sacral visceral unmyelinated afferents that are normally silent and cannot be activated by bladder distension and contraction. These afferents probably innervate preferentially the mucosa and submucosa and may be activated during inflammation of the LUT [12,21].

The functions of the *thoraco-lumbar visceral afferents* from the LUT are not well understood. They are not involved in the regulation of micturition and continence or in the generation of normal nonpainful sensations. They may be involved in the generation of discomfort and pain [21,23].

81.6 Neural Mechanisms of Micturition Reflexes

Micturition, whether elicited voluntarily or involuntarily (e.g., in infants) requires intact reflex pathways through the spinal cord and through the pons. The components related to the sacral innervation consist of contraction of the detrusor, opening of the bladder neck, relaxation of the proximal urethra, and relaxation of the external urethral sphincter. Excitation in myelinated visceral sacral afferents from the urinary bladder are essential for these effector responses of the LUT. Without this afferent feedback neither reflex nor voluntary micturition is any longer possible. The *reflex arcs* for the detrusor contraction are *supraspinal* and *spinal* (see reflex arcs 1 and 2 in Fig. 81.7, left). The spinal reflex arc is at least disynaptic and not monosynaptic. The afferent information to the medial pontine micturition center projects bilaterally through a tract in the intermediate zone of the spinal ventrolateral cord (see inset in Fig. 81.7; [34,36]). Electrical or chemical stimulation of the neurons in the *medial pontine micturition center* generates contraction of the detrusor with relaxation of proximal urethra and external sphincter. Neurons of this medial pontine micturition center project to the parasympathetic preganglionic neurons and the corresponding interneurons and excite them. They do not project to the motoneurons that innervate the external urethral sphincter.

It is debated whether, in healthy biological conditions, micturition is initiated primarily via the pontine micturition center and whether the spinal reflex pathway does not function in these conditions (as believed by de Groat; see [6,8]) or whether the supraspinal pathway functions in conjunction with the spinal reflex pathway. There are two main arguments in favor of the latter possibility:

- Neurons of the ascending pathway of reflex arc 1 are not specific with respect to the afferent input from urinary bladder. These neurons also obtain convergent synaptic input from the colon and rectum, from the perineal region and from the pelvic floor musculature [43]. For the pontine micturition center to send adequate impulses to the preganglionic sacral neurons it seem to be necessary to obtain a rather specific input from the urinary bladder.
- Both local interneurons in the sacral segments and interneurons that project to sympathetic neurons in the thoracolumbar segments projecting to the LUT in the dorsal horn of the sacral segments have been shown to obtain rather specific afferent inputs from the LUT and from the colon and rectum, indicating that they are either inhibited or excited by one organ system and/or the reverse by the other one (see Fig. 81.8).

Thus the specificity of the micturition reflex probably resides in the spinal cord, and this spinal reflex arc would then be gated by the pontine micturition center. Alterna-

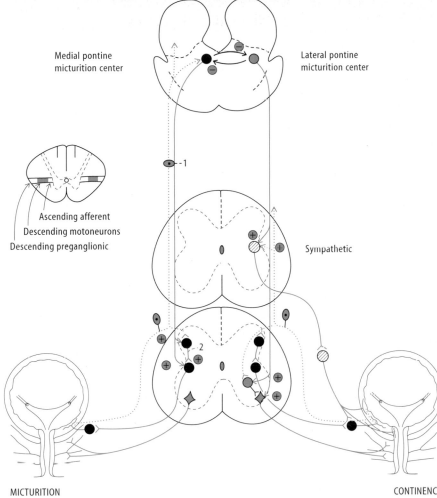

Medial pontine
micturition center

Lateral pontine
micturition center

Ascending afferent
Descending motoneurons
Descending preganglionic

—1

Sympathetic

—2

MICTURITION
Activation of detrusor
Relaxation of ext. sphincter

CONTINENCE
Relaxation of detrusor
Contraction of ext. sphincter

Fig. 81.7. Reflex pathways for regulation of voiding (*left*) and of storage (*right*) of the urinary bladder. The reflex pathways mediating micturition are sacral spinal (reflex pathway 2) and supraspinal pontine (reflex pathway 1). Neurons in the medial pontine micturition center project to the sacral preganglionic neurons and corresponding interneurons. Neurons in the lateral pontine micturition center project to motoneurons that innervate the external urethral sphincter. Medial and lateral pontine micturition centers inhibit each other reciprocally. The neural mechanisms involved in storage function consist of sacral and sacro-lumbar (sympathetic) reflexes, both subject to pontine and other suprapontine control centers. Interneurons in the ascending and descending pathways have been omitted. The *inset* shows the location of the ascending pathway (lesion of which leads to impairment of bladder regulation and loss of sensations from the LUT) and the descending pathways to the parasympathetic preganglionic neurons and to the motoneurons that project to the external urethral sphincter. (From [34].) The lateral and medial micturition centers are under multiple suprapontine controls, e.g., from the periaqueductal gray and the hypothalamus, +, excitation or enhancement; −, inhibition

tively, the specific spinal reflex machinery gates the signals from the medial pontine micturition center (Fig. 81.8). In fact in animals with intact neuraxis as well as in chronic spinal animals it can be shown that both evacuative organ systems never contract simultaneously but always alternatively. The neural mechanism for this lies in the organization of the reflex pathways in the sacral segments and probably also in the reciprocal organization of the sacrolumbar reflex pathways linked to the sympathetic outflows that project to the pelvic organs [19,22].

The *relaxation* of the *external urethral sphincter* seems to be completely dependent on the lateral pontine micturition center. Stimulation of the neurons in this pontine center activates the motoneurons to the external urethral sphincter, but not the parasympathetic preganglionic neurons to the urinary bladder. During micturition the neurons in the lateral micturition center are inhibited, probably by the medial micturition center (Fig. 81.7). After spinalization these sacral motoneurons are activated by the vesical afferents leading to detrusor sphincter dyssynergia (see below); this spinal reflex path-

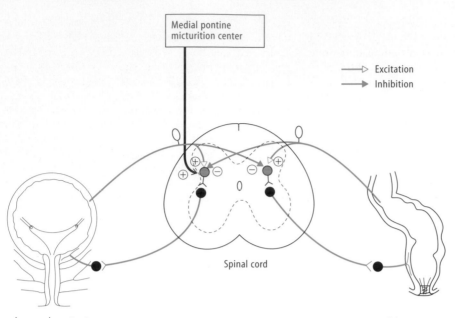

Fig. 81.8. Reciprocal spinal control of lower urinary tract and of colon and rectum. Under isovolumetric conditions, both evacuative organ systems contract reciprocally, activation of one system leading to reflex inhibition of the other. More than one type of interneuron may be involved in the spinal regulation of each organ system. (Based on data from [7] and [43])

way seems to be depressed when the neuraxis is intact, although some motoneurons have also been shown to be inhibited during micturition in spinal animals.

The relaxation of the proximal urethra also seems to be an active neural process, because usually it starts before the intravesical pressure increases (Fig. 81.5). It cannot be explained by the relaxation of the external urethral sphincter (see decrease of EMG activity in Fig. 81.5) that occurs around the distal urethra. The neural mechanism that leads to the relaxation of the proximal urethra and to the consecutive opening of the bladder neck is unknown. One possibility is the reflex activation of parasympathetic neurons that actively relax the smooth nusculature of the urethra. The transmitter involved could be nitric oxide (NO) [24].

81.7 Neural Mechanisms of Storage

Storage of urine in the urinary bladder is primarily a function of the *visco-elastic properties* of the *smooth musculature* of this organ; in this way pronounced distension of the bladder wall can occur with little increase in intravesical pressure. This is associated with an almost rectangular configuration of bladder neck and bladder base. The storage capacity of the urinary bladder is enhanced by various neural mechanisms, some of which are still somewhat hypothetical and not well understood. These mechanisms regulate the capacity of the urinary bladder and the resistance of the outlet of the LUT. They are organized at the

level of the spinal cord and are under supraspinal control. Of most importance for the storage function is the *external urethral sphincter*. Stimulation of pelvic and pudendal afferents activate the motoneurons innervating this sphincter (and other motoneurons to the pelvic floor muscles) with short latency. The *reflex pathway* is at least disynaptic. Impulses from the *lateral pontine micturition center* continuously enhance this spinal reflex pathway. At the same time, neurons in the medial pontine micturition center that project to the parasympathetic preganglionic neurons are inhibited during the storage Phase. This inhibition is probably generated from the lateral pontine micturition center (Fig. 81.7). Any increase in abdominal pressure that generates an increase in intravesical pressure, e.g., during coughing or weight lifting, leads to reflex activation of the motoneurons innervating this sphincter.

Experiments on animals show that interruption of the sympathetic outflow to the LUT *reduces* bladder capacity, bladder wall compliance and urethral resistance and *increase* amplitude and frequency of bladder contractions, thus reducing overall the storage capacity of the urinary bladder. Electrical stimulation of the sympathetic outflow increases the storage capacity of the urinary bladder. From these observations it is concluded that the sympathetic innervation of the LUT is important for the maintenance of continence. Activity in the neurons of the sympathetic (nonvasoconstrictor) outflow to the LUT is thought to inhibit the detrusor muscle, inhibit impulse transmission in pelvic vesical ganglia and initiate transient contraction of the bladder base [5].

Neurophysiological experiments show that powerful spinal *sacro-lumbar reflex* pathways are linked to the sympathetic outflow to the pelvic organs. Adequate stimulation of the sacral visceral afferents from the urinary bladder, colon and rectum (elicited by either distension or contraction of the organ systems) and from the anal canal by mechanical shearing stimuli generates reciprocal (excitatory and inhibitory) reflexes in various classes of lumbar sympathetic neurons to the pelvic organs. These types of neuron could be involved in maintaining storage function of the urinary bladder (and colon and rectum). However, it is unclear how these neural signals are transformed into effector responses. Furthermore, these reflexes seem to operate not only during the storage phase but also during micturition. A further mechanism that might contribute to continence and therefore to the storage function of the LUT is the transmission of impulses from sacral preganglionic neurons to postganglionic neurons in the pelvic vesical ganglia. In the cat it has been shown that impulse transmission through these ganglia is relatively inefficient during a low level of activity in the parasympathetic preganglionic neurons, but is enhanced when the activity increases [5]. The mechanism of this facilitation probably is presynaptic. Activity in these preganglionic neurons at rates of about >1 Hz enhances the release of acetylcholine. This mechanism means, functionally, that an asynchronous low-rate activity in preganglionic parasympathetic vesical neurons, such as can occur during the storage phase, is not transmitted to the postganglionic parasympathetic neurons and does not therefore induce contractions of the urinary bladder.

81.8 Transmitters and Receptors in the LUT

Transmission of impulses from preganglionic to postganglionic parasympathetic neurons in bladder ganglia is mediated by *acetylcholine* and *nicotinic* cholinergic receptors. Transmission from postganglionic parasympathetic neurons to the smooth muscle cells of the LUT is also cholinergic, and the cholinergic receptors are *muscarinic* M_2 *receptors*. In many species, but less so in humans, transmission from postganglionic parasympathetic neurons to the LUT can be blocked only partially by atropine, and it is believed that the atropine-resistant component of this transmission is *purinergic* to the detrusor vesicae. ATP is co-localized with acetylcholine in the same vesicles, is released together with acetylcholine, and acts on *purinoceptors* (P_2) in the smooth muscle cells (Table 81.2) [3,8,17].

Catecholamines can act on all parts of the LUT, leading to relaxation of the detrusor, contraction of the bladder base and relaxation or contraction of the urethra. These effects are mediated by β_2-*adrenoceptors* (relaxation) and α_1-*adrenoceptors* (contraction). The concentration of adrenoceptors is high in the bladder base and proximal urethra and low in the detrusor vesicae; this corresponds approximately to the density of innervation by noradrenergic postganglionic neurons (see [22]). Finally, adrenoceptors are also found on postganglionic neurons of the pelvic splanchnic ganglia. These may mediate depression (α_2) and facilitation (α_1) of impulse transmission through the ganglia [27]. The physiological significance of the β_2-adrenoceptors in the detrusor and of the adrenoceptors in the pelvic ganglia is unknown (Table 81.2).

In addition to the cholino-, purino- and adrenoceptors, which may react with acetylcholine, ATP and noradrenaline released by the postganglionic neurons, the neurons innervating the LUT contain various types of *neuropeptides*, which are colocalized with the classic transmitters:

- Cholinergic postganglionic neurons in the pelvic splanchnic ganglia that innervate the LUT may contain neuropeptides such as vasoactive intestinal polypeptide (VIP), neuropeptide Y (NPY) and enkephalin. In the rat, these and other neuropeptides are differentially distrib-

Table 81.2. Receptors for putative transmitters in the lower urinary tract

Tissue	Cholinergic	Adrenergic	Other
Bladder body	+(M_2)	−(β_2)	+ Purinergic (P_2) − VIP + Substance P
Bladder base	+(M_2)	+(α_1)	0 Purinergic − VIP
Ganglia	+(N) +(M_1)	−(α_2) +(α_1) +(β)	− Enkephalinergic (δ) − Purinergic (P_1) + Substance P
Urethra	+(M)	+(α_1) +(α_2) −(β_2)	+ Purinergic − VIP + NPY
Sphincter striated muscle		+(N)	

Letters in parentheses indicate receptor type: M, muscarinic; N, nicotinic. +, −, 0, excitatory, inhibitory, and weak (or no) effects, respectively

uted in the postganglionic neurons with respect to the LUT, hindgut and penis [25,26].

- Many sacral visceral afferent fibers which innervate the LUT contain neuropeptides such as calcitonin gene-related peptide (CGRP), VIP, substance P (SP), cholecystokinin (CCK) and encephalins [1,2]. Multiple peptides may be co-localized within the same afferent neurons. Afferents containing the neuropeptides are found within the bladder wall, within the subepithelial and submucosal layer, around the blood vessels in the submucosa, and in the bladder ganglia. It is unclear in which functional types of afferents [21] these neuropeptides are localized. Preferentially afferent neurons with unmyelinated fibers may contain them.

It is speculated that the neuropeptides are involved in modulation of neuroeffector transmission to the smooth muscle cells of the LUT, in modulation of transmission of impulses through the bladder ganglia, in local control of blood flow through the mucosa, in local "trophic" functions (e.g., in the mucosa of the LUT), and in modulation of impulse transmission from primary afferents to second-order neurons in the sacral dorsal horn.

Whatever the biological functions of the various types of adrenoceptors and peptide receptors in the cells of the LUT are, these receptors may be the targets for pharmacological interventions in the treatment of diseases of the LUT.

81.9 Suprapontine Control of Micturition and Continence

Although storage and expulsion of urine is basically a function of the autonomic nervous system, both have come under *powerful voluntary control* during the development of higher vertebrates and this is reflected in the central organization. This high-level control by the central nervous system is not surprising, because the LUT has to be adapted in its function at any moment to other functions and to the overall behavior of the organism. Stimulation and lesions of suprapontine brain structures lead to changes in the micturition reflex threshold (Fig. 81.9). The following observations illustrate how differentiated and variable the neural control of the LUT really is (see [34]):

- Topical electrical stimulation in the anterior hypothalamus, preoptic area and septal area in cats may elicit a micturition behavior that is almost normal, showing that its antonomic and somatomotor components are organized at these brain levels [15,16]. Similar observations have also been made in monkeys.
- Lesions of the anterior frontal cortex, anterior cingulate gyrus, amygdaloid complex, paracentral lobule and other structures of the telencephalon in patients, e.g., as the result of surgical interventions, tumors or vascular damage, can result in serious disturbances of the regulation of the LUT. The patients affected may no longer be consciously aware that the urinary bladder is full, although they experience micturition. Alternatively, they

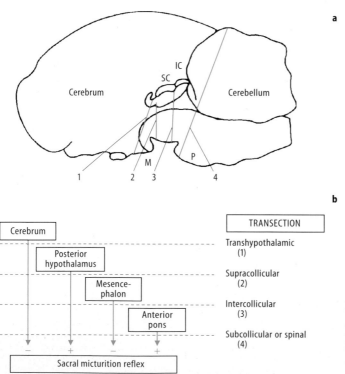

Fig. 81.9. Influence of various levels of the brain stem and of the cerebrum on the sacral micturition reflex. The micturition reflex threshold was measured in the cat during slow filling of the urinary bladder. Cutting of the brain stem at the four levels indicated enhances (+) or suppresses (−) the micturition reflex. *SC, IC,* superior and inferior colliculi of the mesencephalon (*M*); *P,* pons. (Modified from [41,42])

may have difficulty in delaying micturition voluntarily and in initiating micturition although they are fully aware of a full urinary bladder.

- Powerful cortical influences on the LUT are also obvious from our experience during various natural situations. The desire to micturate, discomfort, and micturition itself sometimes do not occur for an extremely long time despite a full bladder during tasks that require high mental concentration (e.g., during driving a car). Intravesical pressures may increase during psychologically unpleasant and embarrassing interviews, reaching values of 60 cmH$_2$O and more although the urinary bladder is only moderately filled. This shows that the forebrain can totally overrule the reflex regulation of the LUT, which is organized at the level of the spinal cord and the lower brain stem.

81.10 Dysfunction of the LUT Related to Neural Lesions and Changes of the Effector Organ

The complexity of the neural regulation of the LUT means that various peripheral neural lesions, lesions of the spinal cord and changes in the LUT itself may lead to dysfunctions of the LUT that are related to the dyssynergia of the parasympathetic and the somatomotor outflow. In almost all these dysfunctions the sacral afferent innervation from the LUT is extremely important, leading to a hyperexcitable urinary bladder when the afferent impulse activity is increased or to an areflexic urinary bladder when the afferent impulse activity is decreased or the afferents are destroyed (e.g., during sensory neuropathy in patients with diabetes mellitus). Dysfunctional states of the LUT follow spinal cord lesion and urethral obstruction and inflammation, as discussed below.

81.10.1 Interruption of the Spinal Cord

After interruption of the spinal cord at the level of the lumbar or thoracic segment or higher, the LUT is initially areflexic. This state of the spinal cord caudal to the interruption is called descriptively "spinal shock" and also applies to the somatosensory system and to other spinal autonomic systems [11,18,31]. The spinal reflex pathways do not function, because the enhancement of the neurons of these reflex pathways by descending systems from supraspinal brain structures has been eliminated. With time, these sacral spinal reflex pathways become functional again and the detrusor contracts reflexly when the afferents from the urinary bladder are activated. However, at this stage both the detrusor vesicae and the external urethral sphincter contract simultaneously and not reciprocally. This is called *detrusor-sphincter dyssynergia* (Fig. 81.10; cf. Fig. 81.7).

The consequence of the loss of coordination between parasympathetic and somatomotor output to the LUT may be a *change* in *organ function*. The intravesical pressure increases simultaneously with increasing resistance of the outlet of the LUT, and the detrusor undergoes hypertrophic changes and increases in weight. This in turn leads to an increase in afferent input from the urinary bladder to the sacral spinal cord, enhancing the reflex activity in the parasympathetic neurons and in the motoneurons further. The increase in afferent impulse activity consists in massive activation of afferent neurons with myelinated fibers, and probably in recruitment of afferent neurons with unmmyelinated fibers that are either normally silent and not activated during normal distension and contraction of the urinary bladder at painful intraluminal pressures or have high pressure thresholds (see [21]). Further changes consist in an increase of the size of the cell bodies of postganglionic and afferent neurons and an increase in the density of the central projections of the visceral afferent neurons to the sacral spinal cord. In particular, unmyelinated vesical afferents are probably involved (Fig. 81.11) [12]. These plastic neural changes are probably related to the trophic changes that occur in the

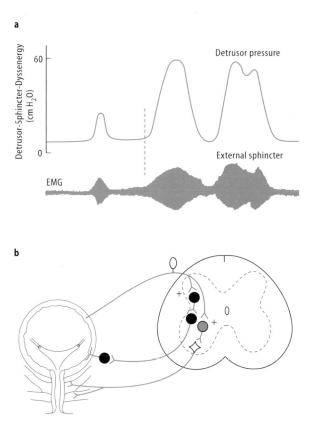

Fig. 81.10. a Detrusor–sphincter dyssynergia in chronic para- or tetraplegic patients. The detrusor pressure and the electromyogram (*EMG*) recorded from the external urethral sphincter are measured. Note that detrusor and sphincter contract simultaneously. b Spinal reflex pathways

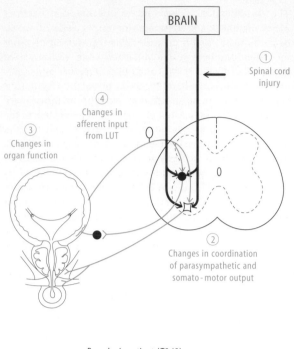

BRAIN

① Spinal cord injury

④ Changes in afferent input from LUT

③ Changes in organ function

② Changes in coordination of parasympathetic and somato-motor output

Fig. 81.11. Indirect mechanisms that might contribute to the changes in reflexes following spinal cord transection at the lumbar level or higher or transection of descending pathways to the sacral reflex arcs. Changes in autonomic and somatomotor outflow (leading, e.g., to detrusor-sphincter dyssynergia; see Fig. 85.10) alters organ function and leads to morphological and physiological changes of the LUT. This in turn changes the afferent feedback from the LUT to the sacral spinal cord. (After [4])

Paraplegic patient (T2/3)

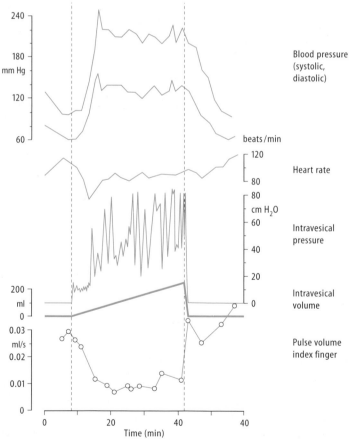

Blood pressure (systolic, diastolic)

Heart rate

Intravesical pressure

Intravesical volume

Pulse volume index finger

Time (min)

Fig. 81.12. Responses of the cardiovascular system during slow filling of the urinary bladder in a chronic paraplegic patient (spinal cord interrupted at the segmental level of T2/T3). Arterial systolic and diastolic blood pressures, heart rate and blood flow (pulse volume) through the index finger were measured. The bladder was distended by slow infusion of saline solution through the urethral catheter (note intravesical volume, *solid thick line*. Note the reflex contractions of the urinary bladder at relatively small intravesical volumes). Note the increase of systolic and diastolic blood pressure due to general vasoconstriction and the decrease of the pulse volume of the index finger due to vasoconstriction of cutaneous blood vessels. The increased arterial blood pressure leads to activation of the arterial baroreceptors and consequently to a decrease of heart rate (due to reflex activation of parasympathetic cardiomotoneurons which project through the vagus nerve). (Modified from [11])

detrusor muscle. It is not clear to what degree they contribute to the abnormal regulation of the LUT [4,6].

A further complication related to the sacral afferent input from the urinary bladder to the spinal cord is the reflex activation of the cardiovascular system in patients with an interruption of the spinal cord rostral to segmental level T5. Stimulation of these sacral visceral afferents, e.g., during micturition, leads to strong reflex activation of vasoconstrictor neurons that innervate blood vessels in the viscera, in the skeletal muscle, in the skin, and elsewhere [9,11]. This activation occurs via spinal reflex pathways and generates large increases in systolic and diastolic arterial blood pressure in these patients (Fig. 81.12). The high blood pressure levels may lead to strokes in such patients.

81.10.2 Urethral Obstruction and Inflammation

Urethral obstruction, generated for instance by hypertrophy of the prostate gland, increases the resistance of the outlet of the LUT. This in turn increases the work of the bladder during voiding and results in hypertrophy of the urinary bladder with unstable bladder contractions and increased frequency of voiding. This process leads to an increase of the impulse traffic in sacral visceral afferents from the urinary bladder, which probably also includes impulses in unmyelinated fibers that are normally silent. The increased afferent impulse traffic is frequently enhanced by inflammation of the mucosa of the LUT, which is the consequence of the residual volume of urine left in the urinary bladder after voiding. The continuous and enhanced afferent impulse traffic strengthens the spinal and supraspinal reflex pathways. The end-result is similar to that observed in the LUT of patients after spinal cord lesion (Fig. 81.11).

Inflammation of the mucosa of the LUT also has similar consequences: micturition volume and voiding threshold decrease, and the frequency of isovolumetric contractions increases. The spontaneous and evoked afferent impulse traffic in myelinated and in unmyelinated visceral sacral afferents increase. The urinary bladder becomes areflexic after the visceral sacral afferents from the LUT are cut (e.g., by cutting the sacral dorsal roots) in both the obstructed LUT and in the inflamed one [4].

81.11 Conclusion

The functions of the lower urinary tract (LUT; urinary bladder and urethra) are collection and evacuation of urine. These two functions are regulated by continence and micturition, respectively, which are controlled by the central nervous system and precisely adapted to the behavior of the organism. They are dependent on the interaction between smooth musculature, sacral visceral afferent innervation, sacral parasympathetic innervation, thoracolumbar sympathetic innervation and motoneurons to the external urethral sphincter. Activity in sacral vesical

afferents is essential for the various sensations during filling and evacuation as well as for the regulatory reflexes. The reflex pathways are spinal sacral, spinal sacro-lumbar and spinopontine. Voluntary control of the LUT is exerted by the telencephalon via the pontine micturition centers and via spinal circuits. Activity in parasympathetic neurons initiates and maintains micturition by contraction of the detrusor vesicae and relaxation of bladder neck and urethra. This is supported by relaxation of the external urethral sphincter induced by a decrease of activity in sacral motoneurons. Activity in sympathetic neurons supports continence by various mechanisms. The centrally generated signals in the autonomic neurons are transmitted by acetylcholine, noradrenaline, and possibly other neurotransmitters and their receptors to the target tissues. Control of continence and micturition is severely disturbed in the presence of lesions in the central and peripheral nervous system, bladder obstruction and bladder inflammation, leading to detrusor-sphincter dyssynergia, bladder hypertrophy, and other changes.

References

1. De Groat WC (1986) Spinal cord projections and neuropeptides in visceral afferent neurons. In: Cervero F, Morrison JFB (eds) Progress in brain research: visceral sensation, vol 67. Elsevier, Amsterdam, pp 165–187
2. De Groat WC (1987) Neuropeptides in pelvic afferent pathways. Experientia 43:801–813
3. De Groat WC (1992) Neural control of the urinary bladder and sexual organs. In: Bannister R, Mathias CJ (eds) Autonomic failure, 3rd edn. Oxford University Press, Oxford, pp 129–59
4. De Groat WC (1993) Anatomy and physiology of the lower urinary tract. Urol Clin North Am 20:384–401
5. De Groat WC, Booth AM (1993) Synaptic transmission in pelvic ganglia. In: Maggi CA (ed) Nervous control of the urogenital system. Harwood Academic, Chur, Switzerland, pp 291–348
6. De Groat WC, Booth AM, Yoshimura N (1993) Neurophysiology of micturition and its modification in animal models in human disease. In: Maggi CA (ed) Nervous control of the urogenital system. Harwood Academic, Chur, Switzerland, pp 227–290
7. De Groat WC, Nadelhaft I, Milne RJ, Booth AM, Morgan C, Thor K (1981) Organization of the sacral parasympathetic reflex pathways to the urinary bladder and large intestine. J Auton Nerv Syst 3:135–160
8. De Groat WC, Steers WD (1990) Autonomic regulation of the urinary bladder and sex organs. In: Loewy AD, Spyer KN (eds) Central regulation of autonomic functions. Oxford University Press, Oxford, pp 310–333
9. Frankel HL, Mathias CJ (1976) The cardiovascular system in paraplegia and tetraplegia. In: Vinken PJ, Bruyn GW (eds) Handbook of clinical neurology, vol 26: injuries of the spine and spinal cord, part II. Elsevier, North-Holland, Amsterdam, pp 313–333
10. Gosling JA, Dixon JS (1990) Anatomy of the bladder and urethra. In: Chisholm GP, Fair WR (eds) Scientific foundation of urology. Year Book Medical Publishers, Chicago, pp 266–273
11. Guttmann L (1976) Spinal cord injuries, 2nd edn. Blackwell Scientific, Oxford
12. Häbler H-J, Jänig W, Koltzenburg M (1990) Activation of unmyelinated afferent fibres by mechanical stimuli and in-

flammation of the urinary bladder in the cat. J Physiol Lond 425:545–562

13. Häbler H-J, Jänig W, Koltzenburg M (1993a) Myelinated primary afferents of the sacral spinal cord responding to slow filling and distension of the cat urinary bladder. J Physiol (Lond) 463:449–460

14. Häbler H-J, Jänig W, Koltzenburg M (1993b) Receptive properties of myelinated primary afferents innervating the inflamed urinary bladder of the cat. J Neurophysiol 69:395–405

15. Hess WR (1957) The functional organization of the diencephalon. In: Hughes JR, (ed) Grune and Statton, New York

16. Hess WR, Brügger M (1943) Der Miktions- und der Defäkationsakt als Erfolg zentraler Reizung. Helv Physiol Phamacol Acta 1:511–532

17. Hoyle CHV, Burnstock G (1993) Postganglionic efferent transmission in the bladder and urethra. In: Maggi CA (ed) Nervous control of the urogenital system. Harwood Academic, Chur, Switzerland, pp 349–381

18. Jänig W (1985) Organization of the lumbar sympathetic outflow to skeletal muscle and skin of the cat hindlimb and tail. Rev Physiol Biochem Pharmacol 102:119–213

19. Jänig W (1986) Spinal cord integration of visceral sensory systems and sympathetic nervous system reflexes. In: Cervero F, Morrison JFB (eds) Progress in brain research: visceral sensation, vol 67. Elsevier, Amsterdam, pp 255–277

20. Jänig W, Koltzenburg M (1990) The function of spinal primary afferents supplying colon and urinary bladder J Auton Nerv Syst 30:S89–S96

21. Jänig W, Koltzenburg M (1993) Pain arising from the urogenital tract. In: Maggi CA (ed) Nervous control of the urogenital system. Harwood Academic, Chur, Switzerland, pp 525–578

22. Jänig W, McLachlan EM (1987) Organization of lumbar spinal outflow to distal colon and pelvic organs. Physiol Rev 67:1332–1404

23. Jänig W, Morrison JFB (1986) Functional properties of spinal visceral afferents supplying abdominal and pelvic organs with special emphasis on visceral nociception. In: Cervero F, Morrison JFB (eds) Progress in brain research: visceral sensation, vol 67. Elsevier, Amsterdam, pp 87–114

24. Keast JR, Kawatani M (1993) NADPH-diaphorase activity is present in many neurons within the cat pelvic plexus and urinary tract. Soci Neurosci Abstr 213:15

25. Keast JR, Booth AM, de Groat WC (1989) Distribution of neurons in the major pelvic ganglion of the rat which supply the bladder, colon or penis. Cell Tissue Res 256:105–112

26. Keast JR, de Groat WC (1989) Immunohistochemical characterization of pelvic neurons which project to the bladder, colon and penis in rats. J Comp Neurol 288:387–400

27. Keast JR, Kawatani M, de Groat WC (1990) Sympathetic modulation of cholinergic transmission in cat vesical ganglia

is mediated by α_1 and α_2 adrenoceptors. Am J Physiol 258:R44–R50

28. Klevmark B (1974) Motility of the urinary bladder in cats during filling at physiological rates. I. Intravesical pressure patterns studied by a new method of cystometry. Acta Physiol Scand 90:565–577

29. Maggi CA (ed) (1993) Nervous control of the urogenital system, vol 2: the autonomic nervous system. Harwood Academic, Chur, Switzerland

30. Maggi CA, Barbanti G, Santicioli P, Beneforti P, Misuri D, Meli A, Turini D (1989) Cystometric evidence that capsaicin-sensitive nerves modulate the afferent branch of micturition reflex in humans. J Urol 142:150–154

31. Mathias CJ, Frankel HL (1992) Autonomic disturbances in spinal cord lesions. In: Bannister R, Mathias CJ (eds) Autonomic failure, 3rd edn. Oxford University Press. Oxford, pp 839–881

32. McLellan AM, Goodell H (1943) Pain from the bladder, ureter, and kidney pelvis. Res Publ Assoc Res Nerv Ment Dis 23:252–262

33. Nathan PW (1956) Sensations associated with micturition. Br J Urol 28:126–131

34. Nathan PW (1976) The central nervous connections of the bladder. In: Williams DI, Chrisholm GD (eds) Scientific foundations of urology, vol II. Year Book Medical Publishers, Chicago, pp 51–58

35. Nathan PW, Smith MC (1951) The central pathway from the bladder and urethra within the spinal cord. J Neurol Neurosurg Psychiatry 14:262–280

36. Nathan PW, Smith MC (1958) The centrifugal pathway for micturition within the spinal cord. J Neurol Neurosurg Psychiatry 21:177–18

37. Nesbit RM, McLellan FC (1939) Sympathectomy for the relief of vesical spasm and pain resulting from intractable pain infection. Surg Gynecol Obstet 68:540–546

38. Petersén I, Franksson C (1955) The sensory innervation of the urinary bladder. Urol Int 2:108–119

39. Ruch TC (1965) The urinary bladder. In: Ruch TC, Patton HD (eds) Physiology and biophysics, 19th edn. Saunders, Philadelphia, pp 1010–1021

40. Sundblad R (1971) Urinary bladder dynamics in women. Scand J Urol Nephrol Suppl 6:1–51

41. Tang PC (1955) Levels of brain stem and diencephalon controlling the micturition reflex. J Neurophysiol 18:583–595

42. Tang PC, Ruch TC (1956) Localization of brain stem and diencephalic areas controlling the micturition reflex. J Comp Neurol 106:213–245

43. Torrens M, Morrison JFB (1987) The physiology of the lower urinary tract. Springer, Berlin Heidelberg New York

44. Van Arsdalen K, Wein AJ (1991) Physiology of micturition and continence. In: Krane RJ, Siroky M (eds) Clinical neurourology, 2nd edn. Little Brown, New York, pp 25–82

82 Central Control of Water and Salt Metabolism

R. GREGER

Contents

82.1 Introduction

Water Content. Water is the main constituent of our body. The total water content in adults is around 60% of the body weight, i.e., approximately 40 l in a 70-kg individual. About two thirds of this water (25 l) is intracellular and one third (15 l) extracellular. The extracellular volume consists of the interstitial space (10 l) and the blood volume of approximately 5 l (see Chap. 76). In comparison to other body functions, the homeostasis of water and salt metabolism apparently has a very high priority. Not only do we find a large number of defense mechanisms against hypervolemia and hypovolemia, but we also are now aware that many effector mechanisms are complementary and redundant; thus, even if one or two important regulatory mechanisms fail, overall regulation is still maintained. The same general principle applies to Na$^+$ metabolism.

Homeostasis. Intake and excretion need to be well matched at any given time for homeostasis to be achieved. This is a formidable task if we consider the maximal span of regulation. In the case of water, the average intake may be 2–2.5 l per day (food, drinks, and oxidation water), but the minimum intake may be as low as 0.5 l and the maximum intake 24 l. Similarly, for NaCl the average intake is 10–15 g per day, but it can be as low as 1 g or less and as high as 60 g [40].

By drinking 24 l per day, the total water content of the body would be increased by 60% and would dilute the ion concentrations in all compartments to the same extent. Similarly, 50 g NaCl would enhance the extracellular Na$^+$ and Cl$^-$ concentrations by approximately 60 mmol/l (Na$^+$ cannot enter cells easily; see Chaps. 4 and 11). The homeostatic processes are designed such that volumes and concentrations stay constant within very small limits. Plasma osmolality is regulated extremely precisely, and deviations of much less than 1% are sufficient to initiate regulatory responses.

Coordination in the Hypothalamus. Regulation is coordinated in the hypothalamus. Intense research efforts throughout this century have clearly identified the hypothalamus as a sensor and integrative area for many autonomous functions [17]. It is here that measurements are taken and other measurements are conducted to; in addition, it is the hypothalamus where efferent control is coordinated and transmitted to the peripheral organs by nerves and hormones (also see Chaps. 17–19). It should be borne in mind that the regulatory mechanisms and feedback loops discussed below are much less rigid under certain conditions than may be anticipated and that behavioral strategies can modify what we usually describe as simple networks [115].

Metabolic Disturbances. Considering the large changes in intake that occur in everyday life, acute disturbances of water and NaCl metabolism are rather rare; however, if they do occur they are serious and sometimes life-threatening complications. In contrast, chronic imbalances in

R. Greger/U. Windhorst (Eds.)
Comprehensive Human Physiology, Vol. 2
© Springer-Verlag Berlin Heidelberg 1996

the filling state of the extracellular space are fairly frequent and can lead to serious complications. In the following, water and NaCl metabolism will be discussed separately, but it will become evident that both balances are tightly interconnected. A large number of excellent recent reviews have appeared on this topic [4,9,12,22,32,38, 43,52,64,81,96,106,107,108,143,147].

82.2 Regulation of Water Intake and Excretion

Water intake is determined by three components:

- Drinking
- Eating
- Oxidation, e.g. $C_6H_{12}O_6 + 6O_2 = 6CO_2 + 6H_2O + 36$ ATP (adenosine triphosphate).

The daily rates from these three sources are highly variable. Excretion occurs by:

- Insensible perspiration
- Sweat
- Urinary excretion
- Fecal losses.

Losses through insensible perspiration are in the order of 0.5 l/day; those through sweat depend on thermal regulation (see Chaps. 110–112, 126); and losses in stool are approximately 0.1–0.2 l (see Chap. 126). Excretion by the kidney is the key regulated parameter; it is this that determines the balance.
The variables to be regulated have been defined above. For this purpose, the following signals are measured:

- Osmolality in the anterior hypothalamus [145]
- Distension of the atria (low pressure system) caused by the filling volume
- Arterial blood pressure, which is monitored by the pressoreceptors (baroreceptors) in the sinus caroticus and sinus aorticus (see Chap. 94)
- Other less well defined peripheral signals (oropharyngeal afferences, hepatic osmoreceptors, see below)
- Local angiotensin II concentration
- Systemic angiotensin II concentration
- Other afferent limbs to the hypothalamus (see below).

The existence of sensors of atrial distension has been postulated for more than a century [57] and was proven in classical experiments by Henry and Gauer [44]. Much later DeBold et al. [29] found that distension of the atria liberates a hormone (atrial natriuretic peptide, ANP) which acts in a vasodilatatory and natriuretic manner.
The above-mentioned signals are processed by the central control system to yield the following control and effector mechanisms:

- Sympathetic discharge to the peripheral circulation (see Chap. 94), increased with hyperosmolality, volume contration, falling blood pressure, and increased angiotensin II
- Sympathetic discharge to the kidney [32], regulated in the above-mentioned direction
- Corresponding activation of the peripheral renin–angiotensin system, increased by sympathetic discharge, falling blood pressure, and reduced Cl^- delivery to the macula densa segment of the renal tubule (see Chaps. 75, 77, 94)
- Release of arginine vasopressin AVP; also called antidiuretic hormone, ADH, with enhanced release by hyperosmolality, volume contraction, falling blood pressure, and enhanced angiotensin II concentration
- Release of ouabain and ouabain-like factors (OLF), with enhanced release in volume expansion [14,47,73,125]
- Secondary release of angiotensin II and aldosterone, with enhanced release in hyperreninism (see Chap. 75).

Sympathetic Discharge and Renin Release. Table 82.1 provides a simplified overview of the various receptor mechanisms, the output signals, and the corresponding secondary changes. It is evident that the receptors are redundant inasmuch as several receptors initiate the same output signals (see Sects. 82.2.3–82.2.5). Key output components, which then trigger several effector mechanisms, include sympathetic discharge and renin release. It is also apparent that the distinct differences in output that exist depend on the specific receptor activation: osmoreceptors constitute the key mechanism for AVP release; the pressure sensors in the aortic and carotid sinus are the key sensors for blood pressure and cardiac output control; the key receptors for renin release and activation of angiotensin II release are located within the kidney. Therefore, many signals (parameters) are measured continuously and integrated, before generating complex output responses.

Feedback Loops. Regulation involves a multitude of feedback loops. Many of the effector systems can replace each other. As a consequence, removal of only one system, e.g., renal denervation or the application of exogenous aldosterone, can hardly upset the overall regulation. In fact, it is this plethora of regulatory mechanisms which leads us to search for an ever-increasing number of regulatory mechanisms [28]. The various effector mechanisms vary in their fingerprint of action. They also have different time-scales for their activation: the neuronal mechanisms on vascular tone act almost immediately; the liberation of angiotensin II takes only a short while; the release and effect of AVP (ADH) requires anything up to 30 min; and the effect of aldosterone is comparably slow [52].

Different Time Domains. The sensory and central regulatory mechanisms are comparably complex. Many inputs with different time domains control blood pressure, volume status, drinking, thirst, Na^+ appetite, and AVP release.

Table 82.1. Control mechanisms of water and salt balance

Receptor	Parameter	Output signal	Secondary changes
Osmoreceptors (hypothalamus)	Hyperosmolality	AVP relesse, thirst, sympathetic discharge	Vasoconstriction, increase in heart rate, increase in stroke volume, water conservation, renin release
Volume receptors (atria)	Hypovolemia	AVP release, thirst, sympathetic discharge, ANP suppression, OLF suppression	Vasoconstriction, increase in heart rate, increase in stroke volume, water conservation, renin release, antinatriuresis
Pressure receptors (sinus nerves)	Fall in blood pressure	Sympathetic discharge	Vasoconstriction, increase in heart rate, increase in stroke volume, renin release
Pressure receptors (kidney)	Fall in blood pressure	Renin release	Angiotensin II release, GFR reduction, aldosterone release, thirst, antinatriuresis, kaliuresis, AVP release
β-Receptors (kidney)	Fall in blood pressure	Renin release	Angiotensin II release, GFR reduction, aldosterone release, thirst, antinatriuresis, kaliuresis, AVP release
Cl⁻ Receptor (kidney)	Fall in blood pressure	Renin release	Angiotensin II release, GFR reduction, aldosterone release, thirst, antinatriuresis, kaliuresis, AVP release
K⁺ Receptor (adrenal cortex)	Hyperkalemia	Aldosterone release	Antinatriuresis, kaliuresis, salt appetite

AVP, arginine vasopressin (antidiuretic hormone, ADH); ANP, atrial natriuretic peptide; OLF, ouabain and ouabain-like factors; GFR, glomerular filtration rate.

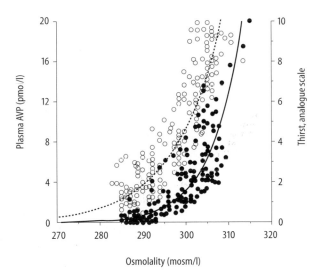

Fig. 82.1. Plasma arginine vasopressin (*AVP*) concentration (*closed symbols, solid curve*), and thirst (the intensity of thirst is translated into a visual analogue scale; *open symbols, dotted* curve) as a function of plasma osmolality. Note that the threshold for AVP secretion and thirst is very similar. The relations appear to be very steep at concentrations greater than 300 mosm/l. (Data taken from [9])

The response patterns appear to be genetically determined, but are also influenced by many exogenous inputs and behavioral strategies. In the following, sensor mechanisms and effector principles will be discussed in detail. The effector mechanisms will only be briefly mentioned and their

interactions discussed, as details of these mechanisms are the subject of separate chapters (see Chaps. 75–77, 94).

82.2.1 Thirst and Arginine Vasopressin Secretion

Plasma Osmolality. Thirst and AVP secretion will be discussed together because they are triggered by the same signal, namely increases in plasma osmolality [142,145]. In previous reviews it has frequently been emphasized that thirst sensation is less sensitive than the release of AVP. This conclusion appears to be incorrect. If subjective scales of thirst sensation are introduced, AVP release and thirst follow a comparable pattern (Fig. 82.1) [9,38,108,136]. Both regulatory responses appear to intercept at approximately the same mean value of 285 mosm/l, but individuals tend to differ characteristically in their threshold responses [108,145]. The sensors for thirst and AVP release are located in the anterior hypothalamus. Figure 82.2 depicts an original recording [65] in which the discharge rate of a neuron in the supraoptic nucleus is shown. It is evident that a small volume of water injected into the carotid artery reduces the firing rate.

Osmoreceptors. Thirst and AVP secretion are probably controlled by different types of osmoreceptors [110,136]. Anatomically, the respective areas in the anterior hypothalamus, i.e., the subfornical organ (SFO) and the organum vasculosum of the lamina terminalis [82,92,93], appear to be overlapping but not identical [108,110]. The existence of different sensors for thirst and AVP release can also be deduced from occasional differential defects and from the fact that in the elderly osmotically induced AVP secretion is more sensitive than in young adults, yet thirst sensation is diminished in the elderly (see Sect. 82.4.6) [9,98,99].

Figure 82.3 depicts the areas of the brain involved in osmoreception, AVP secretion, thirst perception, and regulation of Na⁺ appetite. It also shows important pathways to and from other areas of the central nervous system (CNS) [9,12,38,63,64,81,106,108,144]. It is apparent from this figure that inputs into the paraventricular and supraoptic nuclei come from the osmoreceptors, the nuclei of the Solitary tract (NTS), which receive input from the baroreceptors, and the area postrema.

Vascular Supply. Via the OVLT and the SFO, these hypothalamic areas communicate with the circulation in a bidirectional fashion. It is very important to note that the blood–brain barrier, which otherwise controls traffic of

almost anything but blood gases from circulation to the brain (see Chap. 27), is absent in the various areas of the circumventricular organs: neural lobe of the pituitary gland; median eminence; OVLT SFO; pineal organ; subcommissural organ; area postrema [36]. These organs have a dense vascular supply with sinusoidal capillaries [80]. Hence the nervous tissue is surrounded by wide capillary spaces. The composition of the interstitium surrounding the respective neuronal cells is very similar to that of systemic blood, which is how the osmoreceptors sense the composition of plasma, how secreted hormones enter the capillary space, and how hormones from the periphery enter this area of the hypothalamus. The SFO and OVLT are therefore specialized in neurohumoral integration [51]. However, the circumventricular organs do not represent a gap in the blood–brain barrier. Peptide hormones taken up in these areas stay locally and do not enter other areas of the brain [36]. In the following, AVP secretion and the mechanisms controlling thirst sensation will be discussed.

Renin–Angiotensin System. The recent detection of a brain renin–angiotensin system [37,38,48,64,76,95, 97,133,139] and an ANP system [70,100,134,139,147] has revealed a new and much more complex interplay of various functional components in the regulation of thirst sensation and AVP release.

82.2.2 Arginine Vasopressin-Secreting Cells

Magnocellular Neurons. Most of the factors regulating AVP release are not likely to act on the osmometer cells themselves, but rather on neurons controlling the AVP-producing cells and thirst sensation. The large AVP-producing cells (magnocellular neurons) lie in the supraoptic nucleus and the paraventricular nucleus. They project into the neurohypophysis. The neurohypophysis

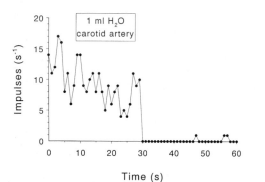

Fig. 82.2. Inhibition of discharge rate by hypotonic solution of a osmosensitive neuron in the area of the supraoptic nucleus; 1 ml distilled water was injected into the carotid artery of a cat. (Redrawn from [65])

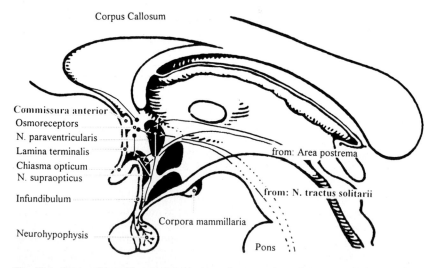

Fig. 82.3. Connections of the paraventricular and supraoptic nuclei. Osmoreceptors send their signals to these nuclei as well afferences from the nucleus of the solitary tract (baroreceptors) and the area postrema. (Adapted from [108])

contains pituicytes, nonmyelated nerve fibers with many secretory granules and a network of capillaries. Figure 82.4A shows an electron micrograph of magnocellular cells in the supraoptic nucleus and Fig. 82.4B shows the neurohypophysis.

Corticotropin Releasing Hormone. AVP is also produced in parvocellular cells in the paraventricular nucleus and secreted together with corticotropin-releasing hormone (CRH) into the portal system of the median eminence [147].

Propressophysin and Neurophysin. In the magnocellular cells AVP is synthesized as a prohormone (propressophysin) of approximately 20 kDa. The gene for AVP has been localized on chromosome 20 [104]. This gene is close to, but distinct from, that for the prohormone of oxytocin (prooxyphysin). In propressophysin, the AVP sequence (Fig. 82.5) is found at the amino terminus, and the carboxy end is glycosylated. A major component is the AVP-binding protein neurophysin. The secretory granules containing the prohormone move down the axons to the neurohypophysis. During this passage the prohormone is cleaved to neurophysin and AVP. Both components are densely packed in secretory granules in the neurohypophysis. The stimuli for AVP release also increase the transcription of new hormone. However, this response is fairly slow and cannot keep pace in states of stimulated release. Neurophysin and AVP are released together. Secretion from the neurohypophysis is stimulated by depolarization, which causes Ca^{2+} influx and exocytosis of the secretory granules. The effects of the neurohypohyseal hormones will be discussed only briefly in Sect. 82.4.2 and are described in more detail in Chaps. 74, 75, and 121. The view of the effects of these hormones is currently changing: the classical concept of the hormone secreted into the systemic circulation and acting on the kidney (AVP) and on the mammary gland (oxytocin) is now supplemented by the multiple roles of these hormones as neurotransmitters [12,102,131,140].

82.2.3 Osmoreceptors

Na^+ Receptor. Although there is little doubt that the osmoreceptors are located in the terminal lamina, their precise location and their function is still not clear. In his famous Croonian lecture, Verney described classical experiments in which he proved the existence of brain osmoreceptors that regulate the secretion of what was then called "post-pituitary antidiuretic substance" and the consequences of this regulation on the excretion of urine [142]. His experiments have been reproduced many times mutatis mutandis in many species and it has been proven convincingly in i.v. infusion studies that they apply to man [145]. It has also been shown frequently that injection of hypertonic solutions into the anterior third ventricle leads to corresponding effects [3]. From such studies in the goat the existence of an Na^+ receptor that senses the concentra-

tion of this ion has been postulated [3], but it is still uncertain whether a real Na^+ receptor exists in man [38]. By the same token it should also be appreciated that, physiologically, it is not relevant whether the sensor measures the concentration of Na^+ or that of other osmolytes as well. Physiologically, alterations in the concentrations of Na^+ and Cl^- are the predominant signals of hydration of the extracellular space. From the above it will have become clear that the sensors measure plasma concentrations. The responses to various hypertonic osmolytes infused i.v. to healthy individuals are compared in Fig. 82.6. It is evident that an increase in osmolality by 10 or 20 mosm/l leads to a strong secretory response of AVP in the case of NaCl or mannitol infusions, to an attenuated response in the case of urea, and to an inverse response in the case of D-glucose infusions [108]. These data suggest that what matters is the osmolality and the permeability of a given solute. NaCl and mannitol appear to be poorly permeant and hence cause a corresponding cell shrinkage. Urea is more permeant and

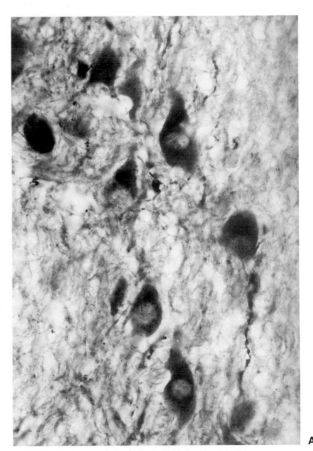

A

Fig. 82.4. A Neuroanatomy of the supraoptic nucleus, showing large cells producing arginine vasopressin. (AVP) (From [75]). **B** The *inset* shows the hypophysis; *1*, paraventricular and supraoptic nuclei; *2*, infundibular stalk; *3*, neurohypophysis, posterior lobe. (Redrawn from [25]). The electronmicrograph is taken from the area indicated in the *inset*. Note the microtubules (*arrows*) which move the granules containing AVP (*arrowheads*) down the axons. The average speed of this transport is 200 mm/day. Magnification, ×28 700

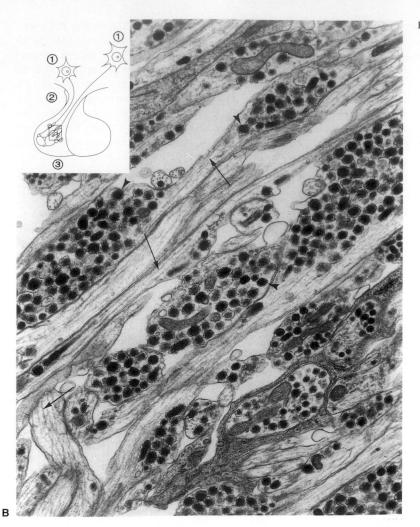

Fig. 82.4B

B

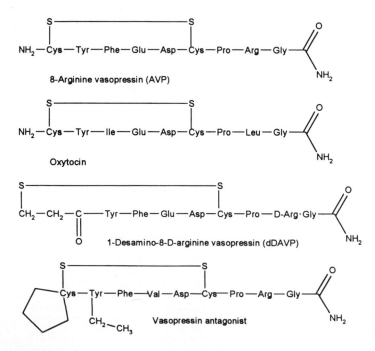

8-Arginine vasopressin (AVP)

Oxytocin

1-Desamino-8-D-arginine vasopressin (dDAVP)

Vasopressin antagonist

Fig. 82.5. Chemical structures of arginine vasopressin (*AVP*), oxytocin, 1-desamino-8-D-AVP (*dDAVP*), and a vasopressin antagonist

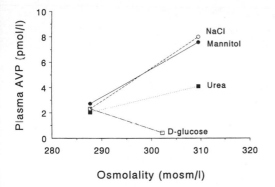

Fig. 82.6. Solute specificity of osmoperception in the hypothalamus. Healthy adults were intravenously infused with hypertonic solutions (NaCl, mannitol, urea, or D-glucose) to enhance their plasma osmolality. Note that NaCl and mannitol give a stronger response than the more permeant urea. D-glucose has an opposite effect, which is probably due to carrier mediated D-glucose uptake into the osmoreceptor neurons. (Redrawn from [108])

causes a strongly attenuated response. The inverse effect of D-glucose is more difficult to explain. D-glucose is probably highly permeant and may, as a fuel, have additional effects. Indeed, severe hypoinsulinemia alters the AVP response to hyperglycemia. D-glucose permeation is then reduced, and this substance acts like an impermeant osmolyte [108].

Cell Volume-Mediated Transduction. The concept of cell volume-mediated transduction of osmotic signals is attractive because of its simplicity. Ion channels which are controlled by membrane stretch have been reported for many years now in a variety of tissues [118]. Such channels can respond to membrane stretch per se even in excised membrane patches. Along these lines, a recent preliminary report argues that such channels have been identified in supraoptic neurons. Whole-cell conductance was increased when hypertonic solutions were superfused and fell when hypotonic solutions were superfused on these cells [94]. The examined changes in extracellular osmolality were, however, much larger than those causing a physiological response. In addition, it seems difficult to conceive that a cell reacts with a sizeable activation of stretch-operated channels when the total osmolality is enhanced by only 0.5%.

Release Thresholds. Osmoreceptors also mediate the thirst response. These osmoreceptors have a very similar threshold and absolute sensitivity. Under optimal conditions changes in plasma osmolality as small as 0.5 mosm/l can be detected. This sensitivity is much higher than that achieved by any technically feasible osmometer. The thresholds for AVP release (Fig. 82.1A) vary among individuals between 275 and 290 mosm/l. The threshold of thirst was originally claimed to occur at considerably higher osmolalities [9,108], but with more recent subjective scales used to quantify thirst it has been shown that the threshold for thirst is very similar to that for AVP release.

A mean value of 281 mosm/l and a range of 274–287 mosm/l has been reported recently [9].

Response Curves. Many factors alter the response curves of the osmoreceptors. The effect of D-glucose and insulin has already been described above. In the elderly, the response curve for AVP release is steeper [58]. However, thirst sensation is diminished in elderly people, which causes hypodipsia and a tendency toward dehydration with increasing age (see Sect. 82.8.6) [9,98,99]. In pregnancy the threshold of the response curves appears to be shifted to the left by some 6 mosm/l (see Sect. 82.8.5 and Fig. 82.14). This coincides with a physiological fall in plasma osmolality during pregnancy by some 10 mosm/l. The reason for the shift of the response curves is not clear. It has now been shown that, contrary to earlier opinion, the increased plasma volume in pregnant women actually represents a relative hypovolemia because of marked vasodilatation [77]. This relative hypovolemia could be the cause for the observed threshold shifts of the osmoreceptors. In fact, moderate absolute hypovolemia can produce similar shifts in the osmoreceptor responses [109]. The various factors modulating AVP release and thirst sensation are discussed in the following sections.

82.2.4 Volume Receptors

Stretch Receptors. The existence of volume receptors was first postulated more than a century ago [57]. The classical experiments by Henry and Gauer [44], performed in the early 1950s, led to the formulation of what is now called the "Henry – Gauer reflex." One of these experiments is reproduced in Fig. 82.7 [44]. It is apparent that negative intrathoracic pressure induces a diuresis. The afferent limb of this reflex is transmitted by the vagus and glossopharyngeal nerves, which both carry the message from volume receptors in the low pressure system to the CNS. Fairly small changes in atrial filling, as may be caused by altered venous return, by changes in posture, or by

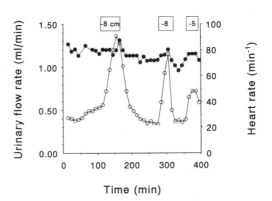

Fig. 82.7. Effect of repeated applications of negative pressure (−8 cm, −5 cm) breathing on heart rate and urinary flow rate (a classical experiment performed in a dog by Gauer and Henry; redrawn from [44]). Note that negative pressure breathing induces small increases in heart rate and a marked diuresis

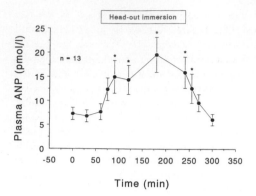

Fig. 82.8. Effect of head-out water immersion on plasma atrial natriuretic peptide (*ANP*) concentration in normal subjects. Mean values ± SEM. (Data taken from [35])

head-out water immersion [35,43], are monitored. The receptors are probably stretch receptors transducing stretch into depolarization and action potentials. The above-mentioned nonselective cation channels [88,118], which are opened by membrane stretch, are likely candidates for mediators of receptor potentials. The message relayed by these receptors in integrated in the NTS [140]. From there, fibers connect to the anterior hypothalamus (see Fig. 82.3) and control AVP release as well as thirst sensation. It will become clear that what may have previously appeared as a fairly straightforward reflex is now emerging as a very complex network of reflexes.

Diuresis. The question arises as to whether AVP does in fact mediate the diuresis seen when right atrial pressure is increased or whether other hormones or neuronal mechanisms are responsible for this effect. This question is all the more relevant as it is clear now that distension of the atria leads to a release of ANP [8,16,21,103,116,146], may lead to the release of endogenous ouabain or other natriuretic hormones [14,55,73,125], may be mediated by renal innervation [32,71], and may be caused by prostaglandins or other factors [35].

Head-Out Immersion. The effect of head-out immersion on urinary AVP excretion and plasma AVP concentration (Fig. 82.8) has been examined in a large number of studies [35]. The results from normally hydrated individuals are difficult to interpret, because AVP is then low and an expected fall in AVP is difficult to measure under such conditions. However, if the individuals were dehydrated, decreases in urinary and plasma AVP have been demonstrated in every study with immersion (increase in central blood volume) [35]. Therefore, there is little doubt that the original hypothesis put forward by Gauer and Henry holds true in a qualitative sense, and it is also plausible that AVP plays a major role in hydrated individuals.

Renin–Angiotensin System and Sympathetic Discharge. In their later reports, Henry and Gauer postulated that the

diuretic response, in addition to a fall in AVP secretion, should be caused by a reduction in the activity of the renin–angiotensin system and by a decrease in sympathetic discharge to the kidney. However, other effector mechanisms are probably also involved in the diuretic response and are certainly involved in the natriuretic response.

The renin–angiotensin system is suppressed in states of increased venous return and atrial distension [35]. The fall in angiotensin II is partly responsible for natriuresis, since angiotensin II enhances proximal tubule Na^+ absorption (see Chap. 75) [56]. Whether the fall in circulating angiotensin II is also able to inhibit AVP secretion is unclear [97]. The inhibition of aldosterone secretion induced by the fall in angiotensin II contributes to natriuresis (see Chap. 75).

Atrial Natriuretic Peptide. Since the pioneering studies by Debold and coworkers [29], another important component of this regulatory loop has been described in many studies. Distension of the atria enhances the secretion of ANP [35,96]. Plasma ANP concentration more than doubles after head-out water immersion (Fig. 82.8). ANP causes vasodilatation and induces natriuresis and diuresis [8,16,21,116,146]. The putative mechanisms of the natriuretic response are discussed in Chap. 75. In several studies a close correlation between natriuresis and ANP plasma concentrations has not been seen, and the physiological importance of ANP in regulating Na^+ and volume balance has been questioned [7,45]. It also has been emphasized that it may be premature to extrapolate from in vivo studies using pharmacological concentrations to physiological conditions and that extrapolations may be even more problematic when they are based on in vitro studies. However, these studies do not rule out the physiological importance and relevance of ANP. It must always be kept in mind that the regulation of water and Na^+ balance is complex and multifactorial. Therefore, simple and close correlations between one sole regulatory factor and one effector, e.g., urinary Na^+ excretion, are not to be expected. Most current evidence favors the view that ANP is one of the factors regulating Na^+ balance [96,103] under normal conditions and may be highly relevant under certain pathophysiological conditons [7].

In addition to its direct effect on vascular smooth muscles and on the kidney, it is conceivable that ANP also has an attenuating effect on AVP release mediated by hypothalamic receptors that measure the concentration of circulating ANP [35].

Renal Prostaglandin E. Excretion of renal prostaglandin E (PGE) is enhanced with increases in central volume expansion such as are seen with head-out water immersion [35]. Prostaglandins may therefore contribute to the natriuresis and diuresis seen under these conditions. This effect of prostaglandins on natriuresis is probably direct and not mediated by a suppression of the renin–angiotensin system, because indomethacin did not abolish the attenuation of the activity of the renin–angiotensin system.

Sympathetic Discharge. In experimental conditions where natriuresis was blunted after renal denervation [32], it has been shown that the sympathetic discharge to the kidneys is attenuated with volume expansion. The effect was augmented in Na+-depleted rats and attenuated in rats on a normal Na+ diet. These data suggest that renal nerves have a role to play in the natriuretic response, but that, again, the scenario is complex and that marked natriuresis can be exerted by volume expansion after complete denervation.

Norepinephrine, Epinephrine, and Dopamine. Plasma norepinephrine and epinephrine concentrations probably do not change with head-out water immersion [35]. However, dopamine apparently increases after some delay, which may be of relevance for the observed natriuresis. Similarly, an inhibition of dopamine effects by D$_2$ antagonists attenuates the natriuretic response to head-out water immersion [35].

Ouabain and Ouabain-Like Factors. These may also be involved in the natriuretic response induced by volume expansion; OLF may constitute the so-called third factor, to which much research has been devoted in the past few decades [28]. At this stage the mechanisms controlling ouabain secretion from the adrenal gland cortex are poorly defined. However, there is increasing evidence that the secretion of ouabain and/or OLF is enhanced in volume expansion [14,47]. Ouabain and OLF inhibit (Na++K+)-ATPase (see Chaps. 8, 11), and this inhibition results in an enhanced cytosolic Ca^{2+} activity [14], which may cause secondary effects such as increased inotropy of the heart. In addition, the inhibition of renal (Na++K+)-ATPase may cause natriuresis and diuresis. Much more work on the various aspects of a putative role of ouabain and OLF in volume balance is needed.

Factors Contributing to the Henry – Gauer Reflex. To sum up, there is little doubt that the afferent loop of low-pressure receptors in the atria controls renal excretion of Na+ and water. Several factors contribute to the Henry-Gauer reflex:

- Afferents from the respective receptors
- Adjustment of the neurohypophyseal secretion of AVP
- Sympathetic discharge to the kidney
- Secretion of ANP by the (left) atrium
- Production of renal prostaglandins
- Secretion of ouabain and/or OLF in the cortex of the adrenal glands
- Probably as yet unknown additional factors.

82.2.5 Pressure Receptors

Baroreceptors. Like the signals from the low pressure system, i.e., mainly the stretch receptors in the atria, the input from the pressoreceptors (baroreceptors) modulates the release of AVP and thirst sensation [24,38,96,108,131]. A comparison of the impact of blood pressure changes, blood volume changes, and changes in osmolality on the concentration of plasma AVP concentration is shown in Fig. 82.9 [108]. It is evident that the relative fall in blood pressure must be 10%–15% before an increase in plasma AVP concentration becomes significant. It should also be noted in Fig. 82.9 that the relations for blood volume and blood pressure are exponential, i.e., marked reductions of blood pressure and blood volume have a marked effect on AVP (beyond a certain threshold). A 30% fall in mean arterial blood pressure leads to a 30-fold increase in plasma AVP concentration [9].

Adaptation of the Baroreceptor Reflex. Chronic elevation in blood pressure, such as seen in hypertensive patients, does not result in a corresponding increase in plasma AVP concentration. This is caused by an adaptation of the baroreceptor reflex [26]. It is likely that acute fluctuations in blood pressure and circulating blood volume are highly relevant for the reciprocal control of AVP secretion. The pathways for pressoreceptor control of AVP secretion are shown in Fig. 82.3. The projections from the NTS are complex and may involve GABAergic fibers from the diagonal band of Broca to the supraoptic nuclei [63].

Several drugs which are known to alter plasma AVP concentration do so by altering blood pressure and hence input via the baroreceptors. These include isoproterenol, norepinephrine, nicotine, histamine, bradykinin, trimethaphan, and nitroprusside. Diuretics, which reduce both plasma volume and blood pressure, also enhance plasma AVP by these mechanisms.

Blood Pressure Regulation. It is worth noting that AVP conversely influences blood pressure regulation. It obviously does so by enhancing water absorption in the kidney, but it also acts as a vasoconstrictor itself; it acts on the heart as well as on the NTS and sensitizes the baroreceptor

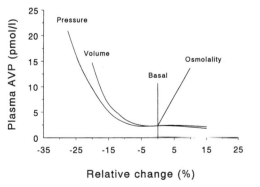

Fig. 82.9. Comparative sensitivities of the osmoregulatory and baroregulatory mechanisms. Fairly large changes in pressure or volume are required to increase arginine vasopressin (*AVP*) plasma concentration. However, the slope towards changes in pressure and volume becomes very steep when the relative changes are −12% to −15%. (Redrawn from [108])

reflex [131]. AVP-containing fibers projecting from the paraventricular nucleus to the NTS have been identified [140]. The responses to locally (NTS) administered AVP are difficult to interpret: depending on the dose, both increases and reductions in the baroreceptor response have been reported [140]. Systemic AVP may also modulate the baroreceptor response: most data suggest that increasing AVP sensitizes the baroreceptor response [131].

The other efferent pathways which may be involved in the baroreceptor response have already been discussed above. They comprise the sympathetic discharge to the kidney, the renin–angiotensin system, the renally produced prostaglandins, ouabain or OLF release, and probably other factors as well.

82.2.6 Angiotensin II

Dipsogenic Effect. For more than 25 years it has been postulated that, in addition to its effects on vascular smooth muscle cells, the adrenal gland, and the renal tubule, angiotensin II may also act on the hypothalamus and induce thirst [39] (for the chemical structure, origin, and production of angiotensin II, see Chap. 73). This dipsogenic effect of angiotensin II can be demonstrated in animals when angiotensin is infused, but it is entirely uncertain what role it plays under normal physiological conditions. Angiotensin II is thought to reach specific receptors which have been demonstrated in the SFO and the OVLT [97]. Evidence has accumulated that this might be of significance [79], but it also has been concluded that the concentrations required to induce thirst are too high to be of physiological relevance [97]. The case is probably difficult to decide because infusion experiments do not really mimic the physiological scenario. The plasma concentrations of angiotensin II required to induce thirst sensation correspond to a water deprivation of more than 24 h, and although this may still be physiological in some animals, it is not characteristic for man. Furthermore, angiotensin II would in any case only be a contributing, but not the sole factor.

The above discussion gained new momentum when it was recognized that the brain possesses its own renin – angiotensin system [95], can produce angiotensin II and angiotensin III, and that angiotensin released in the brain may have several important functions [76,97]:

- It plays a role in fluid and electrolyte homeostasis.
- It acts as a transmitter in the regulation of reproductive hormones.
- It interacts with other transmitter systems such as the catecholaminergic, serotinergic, and peptidergic systems.

Median Preoptic Nucleus. Most of the effects of angiotensin in the brain appear to be mediated by AT_1 receptors [95]. In the CNS the AT_2 receptor is only found in the cerebellum [133]. Some of the effects of angiotensin II, especially on AVP release, may in fact not be due to the octapeptide but rather to the heptapeptide (angiotensin III) [37,48]. It appears that the two angiotensin II "sensory" regions are specialized in discriminating between systemic and locally produced angiotensin II. The SFO "sees" mostly the blood angiotensin concentration, whilst the OVLT "sees" both the locally produced and the blood angiotensin [97]. Increases in angiotensin concentration in the SFO induce thirst, AVP and oxytocin secretion, Na^+ appetite (see below), and pressure responses [133]. The median preoptic nucleus appears to play a key role in relaying the local angiotensin II concentration to other areas, including projections to the supraoptic and paraventricular nuclei, to the NTS, and to the efferent sympathetic pathway.

Obviously, the projections are similar to those for AVP. Angiotensin II injections, like those of AVP, into the NTS cause increases in blood pressure and attenuate the baroreceptor response.

Stimulation of Arginine Vasopressin Release. The stimulatory effect of angiotensin II on AVP release can be easily demonstrated for locally injected angiotensin II, but under normal circumstances it is probably not controlled by systemic angiotensin II. This effect occurs most likely in the supraoptical and paraventricular nuclei, and norepinephrine may be the transmitter involved [133].

Aldosterone. Locally injected, but not systemic angiotensin II also produces a strong Na^+ appetite [97]. On the one hand this effect is augmented by systemic aldosterone, which apparently can penetrate the blood–brain barrier. On the other hand, angiotensin II injected into the third ventricle reduces systemic aldosterone release. Local and systemic angiotensin II therefore have opposite effects on aldosterone secretion. Furthermore, intraventricularly injected angiotensin II enhances renal Na^+ excretion, whereas systemic angiotensin II has the opposite effect (see Chap. 75).

Natriuretic Peptides. There appear to also be close functional correlations between brain angiotensin II and local atrial (ANP) and cerebral natriuretic peptide (CNP; see below) [18]. ANP and CNP on the one hand and angiotensin II on the other may act as antagonists not only with respect to their effector mechanisms, but also with respect to mutual inhibition [97].

Other Functions. Other important functions of angiotensin II include the release of adenocorticotrophic hormone (ACTH) and corticotropin-releasing factor (CRF), the augmentation of CRF action, a regulatory effect on reproductive hormones, and the control of catecholamines in the NTS and in the hypothalamus [95, 97,133]; these functions are beyond the scope of the present chapter.

In conclusion, the role of angiotensin II on thirst, Na^+ metabolism, and AVP secretion in man is still not clarified in detail. This is due to the fact that much of the present information is based either on studies in intact organisms (in which dissection of individual components of complex systems such as volume regulation a priori is difficult) or morphological and destruction studies (which again reveal only certain aspects of normal function) or has been deduced from intraventricular or intranuclear injection studies in several species using angiotensin II or receptor antagonists on the assumption that the injected amounts correspond to physiologically released quantities. It appears likely that, under certain conditions, systemic as well as locally produced angiotensin contribute to thirst sensation, AVP release, and Na^+ appetite.

82.2.7 Other Neuronal Mechanisms

Hepatic Osmoreceptors. Previous studies performed more than 30 years ago postulated the existence of hepatic osmoreceptors and reflexes to control AVP secretion and diuresis [53]. More recent data support the general concept of hepatic NaCl sensing by osmoreceptors and extend it by showing that renal nerve activity is modulated by infusions into the portal vein [19,62,74,87]. The relevance of these mechanisms in overall regulation and especially in man requires further clarification.

Nausea. One of the most potent stimuli of AVP secretion is nausea [9,114]. Irrespective of the cause of nausea (motion sickness, alcohol, high doses or morphine, apomorphine, nicotine, cholecystokinin), the effect is instantaneous and extremely strong. The projections involved originate in the area postrema and extend to the paraventricular and supraoptic nuclei (see Fig. 82.3). The plasma AVP concentration can increase 20- to 30-fold [108]. Pretreatment by antiemetic drugs such as fluphenacine, haloperidol, or promethazine completely and specifically prevents this effect. Stimulation of AVP release by osmotic stimuli is undisturbed. All these drugs act by virtue of their antidopaminergic effect [108]. In fact dopamine does not only play a key role in AVP release, it also appears to be of importance in the generation of thirst (see Sect. 82.8.7). Dopamine agonists facilitate drinking, while antagonists have the opposite effect [33]. Dopamine may also be partly responsible for the self-induced water intoxication seen in schizophrenic patients.

Alcohol. Opioids (κ- and δ-agonists), butorphanol, and low doses of morphine inhibit AVP secretion [84,108]. The effect of alcohol is probably related. Alcohol inhibits AVP release, and this effect is probably mediated by endogenous opioids because it is attenuated by naloxone. Some alcoholic beverages such as beer can exert an additional inhibitory effect on AVP secretion and hence augment the diuretic response because of the concomitant hyponatremia (beer drinker's hyponatremia).

Benzodiazepines. These minor tranquilizers produce thirst and overconsumption of fluid and food, an effect that must also be related to endogenous opioids, because it can be attenuated by naloxone [23].

Hypoglycemia. Acute hypoglycemia is a fairly strong stimulus of AVP secretion [9]. The mechanism of action is unknown, but it is separate from the other mechanisms discussed thus far. In fact, hypoglycemia even stimulates AVP release in patients who have selectively lost their responsiveness to osmotic, volume, and pressure receptors and to emetics [10]. It appears likely that the effect of hypoglycemia is directly related to the metabolic effect of this key substrate. The inhibitor of hexokinase deoxy-D-glucose has a similar effect on vasopressin release as hypoglycemia [11]. A fall in plasma D-glucose concentration by some 2 mmol/l is required to cause a threefold increase in plasma AVP concentration [9,108].

82.2.8 Other Factors

There are a number of other factors which modify the release of AVP. Some of them, such as bradykinin, histamine, and prostaglandins, act indirectly by lowering blood pressure or circulating blood volume. Other substances cause nausea, which then stimulates AVP secretion. It is also likely that the effects of stress and pain in increasing AVP release [117] are indirect. The mechanism whereby temperature controls AVP release is unclear: low temperature suppresses and high temperature increases AVP release [130].

82.3 Satiety

The considerations in Sect. 82.2 might suggest that thirst quenching is controlled by the same mechanisms that produce thirst. In other words, the correction of deviations of plasma osmolality, blood volume, and blood pressure, for example, would stop the thirst sensation and AVP release. Close examination in dogs, however, revealed that satiety occurred at a time when plasma osmolality was still elevated [101] and that the plasma AVP concentration fell at a time when osmolality in plasma was still elevated. Similarly, in humans plasma AVP fell to 50% of its prehydration value within 5 min of ad libitum drinking, whilst plasma osmolality remained elevated for 15 min [137]. These studies clearly indicate that plasma osmolality is not the signal to switch off AVP release and to quench thirst. In fact, even drinking of hypertonic solutions can transiently quench thirst and lower AVP plasma concentrations in dehydrated individuals [129].

An example of the drinking behavior of dehydrated men and their plasma osmolality is given in Fig. 82.10. It is evident that drinking is completed within a few minutes, a time at which plasma osmolality is still strongly elevated [101].

Downregulation of Arginine Vasopressin. The mechanism of this downregulation of AVP and thirst is not known [9]. We might speculate that such a mechanism is very useful and perhaps even absolutely necessary to prevent dangerous overhydration. Drinking behavior is obviously strongly influenced by behavioral strategies [115], but some afferent component is still required. It has been speculated that gastric mechanoreceptors or hepatic osmoreceptors [3] may be involved in this regulatory loop. This does not appear likely, because animals with gastric fistulas still show the same satiety of drinking and attenuation of AVP release [9]. Furthermore, the observation cited above that hypertonic solutions can also induce such effects argues strongly against the involvement of hepatic osmoreceptors. We are left with the possibility that sensors in the oropharynx mediate these responses [9,129,136]; however, the nature of these receptors is not clear.

82.4 Effector Mechanisms

The effector mechanisms involved in water and Na$^+$ metabolism are discussed in several other chapters in this volume (Chaps. 73–75, 77, 95, 96, 98). In this section the various mechanisms will be briefly reviewed in the light of their coordination (Table 82.2). The central theme is the interaction of a variety of mechanisms with the sole goal of achieving homeostasis of volume and electrolyte balance. The effector mechanisms are arranged such that:

- Several different mechanisms work in parallel.
- The various parallel mechanisms interact with each other.
- The net effect is the result of the synchronous activation of agonistic and antagonistic mechanisms.
- Feedback loops secure fine-tuning.
- Several of the effector mechanisms also have a direct effect in the CNS.

82.4.1 Sympathetic Discharge

Sympathetic discharge is initiated in a variety of reflex and more complex reaction patterns [140]. In the present discussion of volume control, the most important efferent pathways are the control of cardiac output and blood pressure (see Chaps. 90, 94, 95, 99), the sympathetic innervation of the kidney [32,71], and the sympathetic control of the adrenal glands, i.e., the secretion of epinephrine and possibly also the secretion of natriuretic ouabain-like substances [14].

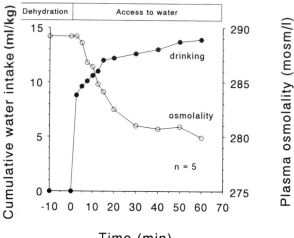

Fig. 82.10. Time course of drinking (*closed symbols*) and osmolality changes in plasma (*open symbols*) after a dehydration period. Experiments performed in dogs [101]. Note that drinking is very rapid in the first few minutes and then levels off at a time where plasma osmolality has fallen by only a few mosm/l. This indicates that satiety has little to do with plasma osmolality. It is believed that oropharyngeal afferences, for instance, quench thirst prior to any substantial fall in plasma osmolality. (Data redrawn from [101])

Table 82.2. Effector mechanisms controlling Na$^+$ and water homeostasis

	Sympathetic discharge	AVP	Aldosterone	Angiotensin II	ANP	OLF	Prostaglandins
Endocrine effects							
Adrenaline	↑						
Renin	↑				↓	↓	(↓)
Angiotensin II	↑				↓		↓
Cardiovascular effects							
Cardiac output	↑	(↓)				↑	
Peripheral resistance	↑	↑		↑	↓	↑	↓
Renal effects							
Na$^+$ excretion	↓	(↓)	↓	↓	↑	↑	↑
K$^+$ excretion		(↑)	↑		(↑)	(↑)	(↑)
H$_2$O excretion	↓	↓		↓	↑	↑	↑
Sensation							
Thirst				↑			
Na$^+$ appetite			↑	↑ (brain)			

Weak effects are indicated by arrows in parentheses. For further details, please see text. AVP, arginine vasopressin (antidiuretic hormone, ADH); ANP, atrial natriuretic peptide; OLF, ouabain and ouabain-like factors; ↑, increase; ↓, decrease

Henry–Gauer Reflex. Sympathetic discharge to the kidney [32,71] is enhanced by a fall in circulating blood volume (Henry–Gauer reflex) or a fall in systemic blood pressure; it regulates renal blood flow, renin release, and tubule absorption of Na^+, Cl^-, HCO_3^-, and water (see Chaps. 73–75).

Renin Release. This is controlled by β receptors (the other releasing factors are discussed in Chap. 73). Their activation enhances renin secretion from specific cells in the afferent arterioles [124]. Renin produces angiotensin II by a two-step activation process (Chap. 73). Angiotensin II is a vasoconstrictor which induces the secretion of aldosterone and enhances proximal tubule Na^+ absorption. In high concentrations it probably also activates AVP secretion and produces thirst (see above) [79,97].

Tubule Innervation. The concept of tubule innervation was originally rejected. However, it is now clear that certain renal tubule segments show direct innervation [71,108]. The transmitters are norepinephrine and dopamine [71], and the catecholamine effects on the convoluted proximal tubule are probably mediated by α-receptors. Dopamine receptors are found in the proximal tubule and in the collecting duct, while the thick ascending limb contains β-receptors. Even where tubule segments are not innervated directly, it is likely that their function can be controlled by locally released neurotransmitters at sites where arterioles and vascular bundles are in close contact with tubule segments [108]. The result of sympathetic discharge is an increase in proximal tubule Na^+ and water absorption [32,71]. In the thick ascending limb of the loop of Henle, Na^+ and Cl^- absorption are increased by norepinephrine.

Feedback Loops. In summary, sympathetic discharge has the following net effects:

- Increase in cardiac output
- Vasoconstriction
- Enhanced volume conservation by the kidney.

This armament helps to maintain body volume, body Na^+, and blood pressure (see Table 82.2). The regulatory mechanisms contain feedback loops. They involve the elevation of blood pressure (baroreceptor, negative feedback), enhanced venous return (volume receptors, negative feedback), and release of renin, angiotensin II, and aldosterone (negative feedback). When greatly increased, angiotensin II activates thirst and AVP release (positive feedback).

82.4.2 Arginine Vasopressin

Water Absorption. The effects of AVP are twofold. First, it acts on the kidney to increase water absorption in the distal nephron. This effect is discussed at length in Chaps. 74 and 75. AVP enhances the insertion of water channels into the luminal membrane of the collecting duct [42,67,141]. Furthermore, AVP can act antinatriuretically in the thick ascending limb of the loop of Henle in some species at least [27,121,123]. On the basis of preliminary data it is unlikely that AVP acts antinatriuretically in the thick ascending limb of the loop of Henle in humans [49,86]. The above-mentioned effects are mainly mediated by V_2 receptors [112] (see Chap. 5).

Cardiovascular Effects. The second major effect of AVP is cardiovascular [131], comprising a direct vasoconstrictor response. However, under normal conditions this effect may not be of any importance. Cardiovascular effects come into play when AVP concentrations are high, as is the case with hemorrhage. In addition, AVP probably has a direct effect on heart muscle. It reduces frequency and cardiac output. This effect may be a direct effect on heart muscle cells or it may be due to a reduction in coronary blood supply [131]. Finally, AVP also sensitizes the baroreflex: a fall in blood pressure leads to an increased response in individuals with increased AVP concentrations [131]. All cardiovascular effects are mediated by V_1 receptors. The effects of AVP on the CNS have already been discussed above [140].

Synergistic Effect. The effect of AVP is synergistic to the sympathergic discharge in several respects (Table 82.2). It increases vascular resistance, albeit at high concentrations only. It acts antinatriuretically at least in some species. The other major effect of AVP is indirectly synergistic. By enhancing renal water absorption, it increases extracellular volume and it stabilizes blood pressure. The direct effect on the heart opposes the positive effect of sympathetic discharge.
The feedback control of AVP release involves blood volume, blood pressure, plasma osmolality, and possibly plasma AVP concentration itself.

82.4.3 Aldosterone

The effects of angiotensin II have already been discussed in this and other chapters (Chaps. 73, 75, 96). Angiotensin II, plasma K^+ concentration, and probably ANP [54] modulate the release of aldosterone in the following ways:

- Angiotensin II increases aldosterone release.
- Hyperkalemia increases and hypokalemia attenuates aldosterone release.
- ANP attenuates aldosterone release.

Na^+ Absorption and K^+ Secretion. Aldosterone has its major effects on the epithelia lining body surfaces, close to the outside world. Receptors are found in the distal tubule and collecting tubule of the kidney (see Chap. 75), the urinary bladder, the ducts of excretory glands (salivary, see Chap. 64; sweat gland, see Chap. 112), and the distal colon (see Chap. 62). The general effects are an increase in Na^+ absorption and frequently in K^+ secretion [41,113]. Aldosterone increases salt appetite at least in the rat [97] (see above); it therefore ensures the conservation of body Na^+, thereby assisting in the maintenance of a normal extracellular volume.

Escape Phenomenon. Aldosterone acts synergistically with angiotensin II, with sympathetic discharge to the kidney, and with AVP. The effects of aldosterone are slower than those of the other effector mechanisms discussed so far. Externally applied aldosterone leads to the so-called escape phenomenon, i.e., the antinatriuretic effect ceases and Na^+ excretion approaches control values. This phenomenon is due to compensatory natriuretic mechanisms which are activated by the volume expansion induced by aldosterone [55].

82.4.4 Atrial Natriuretic Peptide

The release of ANP [96] has been discussed above, and the effects of ANP have been described in previous chapters (see Chap. 75). It acts natriuretically and diuretically in the kidney [8,16,21,146] and also acts on the cardiovascular system. The possible role of ANP and CNP in the brain has been discussed above. The effects of ANP are generally fast, and the net result of ANP release is a reduction in extracellular volume and a fall in blood pressure. ANP is therefore an antagonist to AVP (see Table 82.2), to angiotensin II, to sympathergic discharge to the kidney and to the vasculature, and in some respects to aldosterone as well. As for many influential factors, the relevance of ANP in overall regulation is difficult to assess (see above). It has been claimed that it is important especially in pathophysiological conditions [7] and whenever other regulatory components are "exhausted."

82.4.5 Prostaglandins

Prostaglandins are produced in several organs. They act as hormones or autacoids (see Chap. 6). In the present context, it is their effects on the kidney that are of relevance. There prostaglandins (mostly PGE_2, $PGF_{2\alpha}$) act natriuretically and diuretically [15,132]. This effect is controlled by several factors, including volume receptors in the atria [35]. Prostaglandins dilate arterioles (see Chaps. 6, 96) and reduce blood pressure. Therefore, prostaglandins are synergistic with ANP and antagonistic with the other factors.

82.4.6 Ouabain and Ouabain-Like Factor

Natriuretic Hormone. As has been discussed above, the role of ouabain and OLF is not clear at this stage. There is some evidence that ouabain itself acts as a hormone, but also that ouabain is but one member of a larger group of ouabain-like substances [73,78,126]. These substances may represent the missing link, i.e., the "third factor" or "natriuretic hormone" which was postulated by de

Wardener et al. [28] more than 30 years ago. Most recently it has become apparent that the adrenal gland (and probably other tissues as well) produces endogenous ouabain and/or OLF [14,47,125]. In the adrenal gland OLF production may be under the control of sympathetic discharge and, like aldosterone, angiotensin II [14]. Important extra-adrenal sites of ouabain and OLF production are probably certain areas of the brain, including the hypothalamus. The effects of OLF have mostly been studied in heart and smooth muscle cells. There, ouabain and OLF increase cytosolic Ca^{2+} activity and hence contractility [14,83]. OLF or related substances appear to be enhanced in volume expansion; on the other hand it has been suggested that ouabain release migh be stimulated by angiotensin II [73,78]. It is believed that ouabain and OLF inhibit renal tubule Na^+ transport and hence produce diuresis and natriuresis.

Synergistic Effects. Ouabain and OLF therefore act as synergists with ANP on the kidney. They antagonize the renal effects of sympathetic discharge, of aldosterone, and of angiotensin II. In cardiovascular regulation the situation is much more complex. Ouabain and OLF probably cause a positive inotropic effect in the heart and enhance vascular tone. In this sense sympathergic discharge and angiotensin II act synergistically with ouabain and OLF.

In conclusion, it has been shown that volume control is ensured by a surprisingly large spectrum of interdependent mechanisms. This complex arrangement has the advantage that paralysis of one single factor does not have a large impact on the overall function of the organism. In fact, due to this arrangement the study of individual components is very difficult in the intact organism. In addition, this arrangement of factors with differing time constants ensures that volume control is rapid in onset and does not adapt for some time [52]. On a larger time scale, however, adaptation can be clearly seen.

82.5 Regulation of NaCl Appetite and NaCl Excretion

In humans thirst is a much stronger sensation than NaCl appetite [31]. We all know that thirst can be a very strong sensation: "*Durst ist schlimmer als Heimweh*" ("thirst is worse than homesickness") is a popular German saying. Human beings probably lost NaCl appetite to a great extent as a result of eating habits. Plants contain little NaCl, while meat contains large quantities. Therefore, with the change from herbivorous (mostly vegetables) to omnivorous (mixed food) or even carnivorous (mostly meat) eating habits, the availability of NaCl expanded substantially. If we ingest average daily amounts of 10–20 g NaCl, corresponding to 150–300 mmol/day, there will hardly be any NaCl appetite under normal conditions. However, under specific conditions such as pregnancy and lactation (see below) and with heavy physical exercise NaCl appetite may

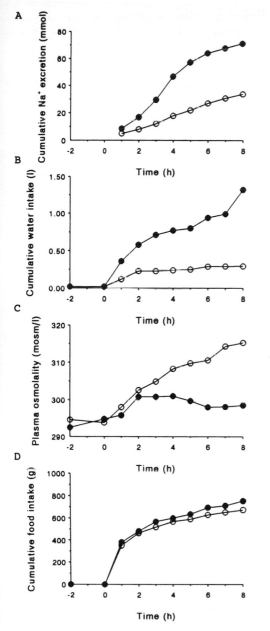

Fig. 82.11A–D. The effects of infusion of either artificial cerebrospinal fluid (*closed symbols*) or mannitol (*open symbols*) into the third ventricle of conscious sheep (1 ml/h). The sheep were given their normal daily food ration at zero time. (**D**). This causes an increase in plasma osmolality (**C**), thirst-induced water intake (**B**), and natriuresis (**A**). Mannitol infusion into the third ventricle suppresses the natriuretic response and causes a stronger increase in osmolality. At the same time water ingestion is diminished. These data suggest that osmolality increases induced by food intake do not only induce drinking but, in addition and with the same time course, also induce a natriuresis. This natriuresis helps to excrete osmolytes and to prevent overhydration. (Data redrawn from [81])

develop [38]. With physical exercise and under extreme environmental conditions loss of NaCl by sweating may exceed 200 mmol/day. This again produces a strong desire for NaCl uptake and changes how salt tastes to us.

In herbivores NaCl appetite is common. These animals are specialized in avoiding NaCl deprivation. The physiological mechanisms present in these animals can probably not be directly extrapolated to man [38].

82.6 Na+ Receptors

Hypothalamic Receptors. The issue of whether the hypothalamic osmoreceptors are in fact Na+ receptors has been dealt with in Sect. 82.2.3. The Na+ receptors in the hypothalamus can easily be demonstrated in herbivorous animals, and they seem to measure extracellular Na+ concentration rather than the Na+ concentration in the third ventricle. The sensors may well be located in the same areas of the circumventricular organs and they are probably different from the osmoreceptors for thirst and AVP release [38]. In sheep, these receptors can lead to effects which, at first glance, appear paradoxical. Injection of a hypertonic solution into the third ventricle leads, as expected, to the sensation of thirst and to rapid drinking (Fig. 82.11). With little delay, however, the kidney starts to excrete more Na+ [81]. The same hypertonic mannitol solution has no such effect.

Hypothalamic Na+ Receptors. The presence of specific hypothalamic Na+ receptors has not been proven. However, several hormones, probably acting in the hypothalamus, can strongly influence NaCl appetite. One example is aldosterone, which greatly increases NaCl appetite in the rat. Whether this is particularly relevant for man is still unresolved [38]. This effect of aldosterone appears appropriate in the general concept of this hormone; as aldosterone has no difficulty in reaching neurons, it may well be an effect on the hypothalamus as has been shown for other hormones (see below). However, it cannot be completely ruled out at this stage [2] that this aldosterone effect is a direct effect on the peripheral salt taste receptors. These receptors possess the same type of Na+ channel found in aldosterone-sensitive epithelia (see Chap. 59). In fact, bretylium tosylate, which enhances the number of Na+ channels in the taste receptor, as it does in other Na+-absorbing epithelia, also makes the receptors more sensitive to the taste of NaCl [119].

Pregnancy and Lactation. NaCl appetite is also enhanced in pregnancy and lactation. It has been argued that several hormones are involved in these mechanisms: estrogen, ACTH (via its effect on adrenal cortex), cortisol, prolactin, and oxytocin [31]. For prolactin this has been shown by administering this hormone to healthy volunteers [61]. For oxytocin it was verified in a recent study in rats [12]. The detailed mechanisms are unknown.

Renin–Angiotensin System. The Na+ appetite can also be influenced by the renin – angiotensin system. In this respect, the local renin – angiotensin system in the brain is probably more relevant than the circulating angiotensin II [38,97] (also see above).

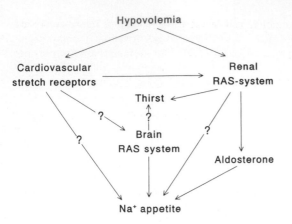

Fig. 82.12. Possible mechanisms of hypovolemic thirst and Na+ appetite. The *question mark* indicates an unknown effect. *RAS*, renin–angiotensin system. (Redrawn from [38])

Na+, Cl− Receptors in the Macula Densa Segment of the Distal Renal Tubule. Na+ receptors of some type have been claimed to exist at this nephron site ever since the early formulation of the hypothesis of tubuloglomerular feedback [138], but it was only very recently that the sensor was identified. It was shown that the furosemide-sensitive Na+2Cl−K+ cotransporter is present in the luminal membrane of these cells. This cotransporter has a very high affinity for Na+ and K+, but a much lower affinity for Cl−. Therefore, under normal circumstances this sensor acts as a Cl− sensor rather than a Na+ sensor [120,122]. The function of this sensor is probably at least twofold (also see Chap. 73). An increase in luminal Cl− concentration increases tubuloglomerular feedback (fall in single nephron filtration rate); at the same time, renin release from specialized arteriolar cells is reduced. A decrease in luminal Cl− concentration has the opposite effects [124].

The effects of renin release have been discussed above (see Chaps. 73, 75). In terms of Na+ metabolism, renin release leads to Na+ conservation by the kidney through angiotensin II and indirectly via aldosterone. Aldosterone enhances Na+ appetite. In addition, circulating angiotensin II can, above a certain threshold, also enhance Na+ appetite [38].

Na+ Receptors in the Taste Buds. The taste receptors in the taste buds of the tongue possess Na+ receptors. These receptors are Na+ channels present in the apical membrane of the receptors [2,5,6,59,66,119] and share important properties with the Na+ channels which are typically found in several epithelia (see above). Most importantly, these channels are inhibited by low concentrations of amiloride [6]. In the presence of amiloride, NaCl taste transduction is inhibited (also see Chap. 42). Bretylium tosylate, which enhances the Na+ channels density in urinary tract epithelia, also acts on the taste receptor Na+ channel [119]. It is termpting to speculate, although it has not yet been proven [2], that aldosterone might sensitize these receptors by inducing more Na+ channels, an effect which is well documented for epithelial cells [113].

82.7 Effector Mechanisms for NaCl Metabolism

The effector mechanisms for NaCl metabolism are the control of uptake, via Na+ appetite, and the control of excretion, mainly via the kidney and to a lesser extent by the intestine, sweat, and certain glands. These effector mechanisms are controlled essentially by the same factors which also control water metabolism. These mechanisms and factors have already been discussed above and will only been mentioned briefly here. It will be obvious that the various factors again show a high degree of interaction and can ensure complete Na+ Cl− balance for daily uptakes ranging between <5 g and >30 g (75–450 mmol). Even lower and higher intakes can be tolerated with no detectable health problem [40]. A summary of the regulatory network is depicted in Fig. 82.12 and Table 82.2.

82.7.1 Neuronal Mechanisms

The central neuronal mechanisms which are responsible for Na+ appetite have been discussed above. Sympathetic discharge to the kidney helps to conserve Na+ by a variety of mechanisms [32,71]. It enhances Na+ absorption mainly in the proximal tubule (mediated by α-receptors). β-Receptor-mediated renin release leads to Na+ conservation via angiotensin II and aldosterone [54], as well as to enhanced Na+ appetite.

Pressoreceptors and Macula Densa. It should be added here that two other major signals control renin release. First, in the afferent arteriole, pressoreceptors monitor perfusion pressure at this site within the kidney. Their inactivation by falling blood pressure leads to the release of renin. Second, the fall in luminal Cl− concentration at the site of the macula densa enhances renin release and increases single nephron filtration rate.

82.7.2 Angiotensin II

Besides its roles in increasing Na+ appetite and enhancing aldosterone release (see above), in physiological concentrations angiotensin II also directly enhances Na+ absorption by the proximal tubule [56]. Activation of angiotensin II occurs as a consequence of renin release and involves converting enzyme (see Chap. 75). The vasoconstriction induced by angiontensin II has been discussed elsewhere (Chaps. 95, 96). Angiotensin II affects Na+ metabolism in two major ways: it increases its conservation in the body and it enhances Na+ appetite and intake. It also acts as a volume-conserving hormone (see above).

82.7.3 Aldosterone

The effect of aldosterone has already been discussed above. Here it will suffice to state that aldosterone is the key hormone for Na^+ conservation. The target organs are: the distal tubule and collecting tubule [41,113] (also see Chap. 75), the distal colon [20], the sweat duct (Chap. 112), the salivary duct (Chap. 64), and probably respiratory tract mucosa as well [69].

82.7.4 Atrial Natriuretic Peptide

ANP and Urodilatins. ANP has been discussed above as a diuretic and natriuretic substance. The natriuretic effect of ANP has still not been completely clarified. It appears certain now that the major natriuretic effect occurs in the collecting tubule [21,91,146]. Other natriuretic peptides have been identified recently and, as they are produced in the kidney itself, they have been named urodilatins [128]. Their mode of action may be similar to ANP.

82.7.5 Ouabain and Ouabain-Like Factors

Ouabain and OLF have been discussed above [14,47,125]. They induce vasoconstriction, are positively inotropic in the heart, and induce natriuresis. As a consequence, they help to stabilize circulation and at the same time enhance Na^+ and water excretion. Hence, they act synergistically with ANP with respect to their natriuretic effect on the one hand and synergistically with angiotensin II on vasculature and with sympathetic discharge on the heart on the other hand.

82.7.6 Other Factors

It appears likely that additional factors such as prostaglandins and as yet unknown hormones contribute to natriuresis [46,60]. These issues are discussed in detail in Chap. 75.

As with the control of water excretion, a large number of factors control Na^+ excretion. This seemingly redundant scenario ensures that Na^+ balance can be maintained under circumstances in which one or even more than one factor is defective.

82.8 Clinical and Pathophysiological Aspects

Several of the issues mentioned above will be discussed here in the light of specific and pathophysiological considerations. This appears justified because Na^+ and water metabolism are usually well balanced and concern the physician only if there is a major change in the regulatory responses or even a severe regulatory failure. It will become apparent that the regulatory mechanisms in specific physiological conditions such as pregnancy and old age are different from those operating in what usually is regarded the normal state (i.e., 70-kg, 30-year-old man) and that

regulation can be severely disturbed in many clinical settings.

82.8.1 Diabetes Insipidus

There are two major causes of diabetes insipidus [38]:

- AVP deficiency
- Renal refractoriness to AVP.

The technical term originates from the Greek διαβαίνω (to flow through rapidly) and the Latin *insipidus* (tasteless). In fact, urine is produced at enormous rates in these patients, i.e., up to 30 l/d. This urine is clear, of low osmolality, and tasteless. The disease can be defined not only in terms of output (urine) but also in terms of intake. These patients suffer from an incurable thirst. They must drink, and it is fortunate that they have this strong sensation; if a patient with this disease did not promptly respond to thirst, his or her situation would be life-threatening within less than a few hors. This forms a clear point of distinction between primary and secondary cases. Patients with primary disease need the water intake to survive, while those with induced or secondary diabetes insipidus (AVP suppression [89]) are not endangered to the same extent, because their AVP suppression is secondary. However, this distinction is not always easy (see Sect. 82.8.7).

According to the above consideration, diabetes insipidus may have any of the following causes:

- Excessive water intake (primary polydipsia [89]). This may be caused by organic brain disease, it may be psychogenic, or it may be unresolved (idiopathic).
- AVP deficiency. The reasons may be genetic (autosomal dominant); traumatic; caused by tumors, granulomas, infections, or metastases; vascular, e.g., Sheehan's syndrome; or of unknown cause.
- AVP insensitivity. Possible causes include genetic (X-linked, men affected [112]), infectious, toxic (Li^+, methoxyflurane, demeclocycline), infiltrative, metabolic, and postobstructive ones.
- Water channel defect. The water channel of the luminal membrane of the collecting duct was cloned very recently [42], and one type of genetic water channel defect (aquaporin II defect) was described shortly thereafter [30].

In all of these conditions the net effect is the same. However, the central causes can be cured by administering AVP, but the renal causes cannot. In the latter cases, continuous volume repletion is necessary. Thiazide therapy has also proved useful, because thiazides apparently reduce the glomerular filtration rate (GFR) and reduce extracellular fluid volume, hence reducing the load of filtrate delivered to the collecting tubule [50].

82.8.2 Inappropriate Arginine Vasopressin Secretion

Inappropriate Antidiuretic Hormone Secretion Syndrome. This pathological condition is caused by an inappropriately high AVP secretion. In other words, for the given osmolality in these patients the secretion of AVP is too high. There are cases in which AVP is found to be in the normal range, yet it is much too high for the given osmolality. The causes of the syndrome of inappropriate antidiuretic hormone secretion (SIADH) are multifold, including the following [143]:

- Malignancies, e.g., lung, duodenum, pancreas, thymoma, mesothelioma, urinary bladder, ureter, prostate gland, lymphoma, Ewing's sarcoma
- Lung disease e.g., pneumonia, tuberculosis, empyema, cystic fibrosis
- Neurological disorders, e.g., meningitis, encephalitis, brain abscess, Guillain-Barré syndrome, head trauma, brain tumors
- Drugs, e.g., vasopressin, dideoxy-vasopressin (DDAVP), oxytocin, clofibrate, antidepressants, monoamine oxidase (MAO) inhibitors
- Acute psychoses (see Sect. 82.8.7)
- Others.

Hyponatremia. The scenario is such that the patient drinks more water than he or she should do. This usually stops AVP secretion, and water excretion is ensured. In these patients AVP remains inappropriately high and water is conserved. This causes hyponatremia. However, volume expansion is not marked in these patients, because they promptly react to volume expansion with natriuresis and diuresis. As a result, they lose Na$^+$ and aggravate their hyponatremia [38,143]. Total body volume is only moderately increased in these patients. It has been speculated that the dipsogenic effect of AVP is responsible for the inappropriate drinking behavior of these patients [38]. Treatment of this disorder can be very difficult.

Edematous States. SIADH must be distinguished from edematous states, in which regulation is directed towards volume and salt conservation, such as occurs in liver cirrhosis and heart failure [1,127]. The apparent paradox in these states is the renal conservation of Na$^+$ and water in spite of a volume expansion, with markedly increased body water and Na$^+$. The pathophysiology of this syndrome is shown in Fig. 82.13 [1]. It is apparent that inappropriate peripheral vasodilatation or insufficient cardiac output creates a situation in which the otherwise completely healthy kidney cannot excrete sufficient amounts of Na$^+$ and water. This induces expansion of extracellular space and edema formation. It has now been suggested that a similar cascade operates in the formation of ascites [1]. Splanchic vasodilatation causes "underfilling" of circulation, which leads to the above consequences. This concept explains why natriuretic mechanisms such as aldosterone escape and ANP are attenuated in such states. It is also evident that sympathetic discharge to the kidney and angiotensin II play a key role in the enhanced proximal tubule absorption.

The underlying pathophysiological considerations are the basis for therapeutic strategies. They include head-out water immersion, aldosterone antagonists, and angiotensin-converting enzyme inhibitors.

82.8.3 Salt Losers

The daily salt loss can vary widely. As has been discussed above, salt losses are closely controlled by salt intake. NaCl losses in sweat (see Chap. 112) can induce severe disturbances of NaCl homeostasis. These losses can be as large as

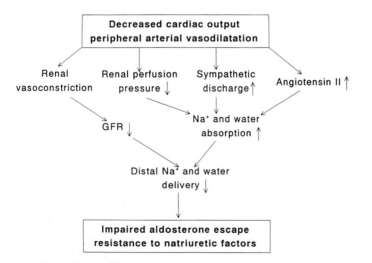

Fig. 82.13. Effect of decreased cardiac output and/or vasodilatation on renal Na$^+$ and H$_2$O absorption. Note that several mechanisms converge to enhance renal absorption of Na$^+$ and H$_2$O and induce volume expansion and edema formation. *GFR*, glomerular filtration rate. (Redrawn from [1])

10–30 g/day (150–300 mmol/day) if thermoregulation requires a sweat volume of 3–10 l/day. It is common experience that enhanced NaCl losses by sweating induce an enhanced salt appetite. Hence, balance is easily achieved as long as salt is available.

Excessive Loss. Generally, the regulatory system in man appears to operate such that the salt intake exceeds the body's requirements [38]. This implies that, under most conditions, the volume is slightly expanded, providing a reserve to readily cope with minor volume losses. The main causes for excessive Na^+ losses are:

- Certain forms of chronic renal failure with enhanced urinary salt losses
- Polyuric phase of acute renal failure
- Renal tubule defects (renal tubular acidosis, Barter's syndrome)
- Diuretic therapy and misues of diuretics
- Addison's disease
- Other causes of hypoaldosteronism
- Extrarenal NaCl losses (vomiting, diarrhea, sweat).

The differential diagnosis of the various types of salt loss is highly relevant inasmuch as it represents the most frequent and clinically relevant disturbance of electrolyte homeostasis [72]. The therapeutic strategy obviously depends on the primary cause.

82.8.4 Salt Sensitivity and Blood Pressure

There is no doubt that man requires a minimum daily intake of NaCl, and this absolute dependence on salt intake made NaCl a precious good in previous centuries. In fact, Roman soldiers were paid in salt [40], and the word salt (Latin *sal*) appears to originate from salary (Latin *salarium*).

Elevated Blood Pressure. Nowadays, with the ubiquitous and ready availability of NaCl, we have increased our daily intake rates to 200–300 mmol/day (12–19 g/day). The question is now whether this plentiful intake of NaCl is harmful. One argument in favor of an increased NaCl intake, namely as a reserve, has already been presented above [38]. On the other hand, there is evidence for some sort of a causal role of volume expansion in the generation of elevated blood pressure [13]. Hence, general NaCl restriction is recommended by some clinicians to prevent development of high blood pressure [68]. This view is not generally accepted, because a large number of clinical studies indicate that the effect of dietary NaCl restriction on blood pressure is rather modest. In the frequently quoted recent INTERSALT study [34], there was only a 2- to 4-mmHg (0.26- to 0.52-kPa) fall in blood pressure when NaCl intake was reduced by 100 mmol/day. This very limited benefit from a fairly strict dietary regimen has to be weighed against the possible harmful consequences which can be caused by a general NaCl restriction. It will be shown below

that such general dietary recommendations are inappropriate for elderly people, for instance, who are more likely to have hyponatremia (Sect. 82.8.6) and that NaCl requirements are sharply increased in pregnancy (Sect. 82.8.5) or under certain climatic conditions (Chap. 112).

Hypertension. A general restriction in NaCl intake therefore appears unwise. However, at the same time it should be emphasized that patients with essential hypertension should adhere to a Na^+-restricted diet and that new approaches are required to identify those individuals in whom blood pressure changes in response to altered NaCl intake are much greater ("salt-sensitive" individuals). Above all, the simple concept that volume expansion causes hypertension has now been rejected [24]. To conclude, we should bear in mind that the normal body is equipped with a plethora of regulatory mechanisms which allows a large extent of freedom in the daily intake [40]. Taste, palatability, hunger, and satiety control our behavior in a very delicate way. Extraneous dietary recommendations are needed in patients, but are not required in healthy subjects.

82.8.5 Pregnancy

Several important changes in salt and water metabolism occur in pregnancy. Their interrelationship is not completely understood at present, and many findings have been misinterpreted in the past. The major changes can be summarized as follows:

- The kidneys increase their size, and GFR and renal plasma flow increase by 30% [77]. As a consequence, there may be slight glucosuria and histidineuria. Glomerulotubular balance is well preserved in pregnancy. Therefore, overall urinary excretion is not significantly altered.
- Plasma osmolality falls during pregnancy by 8–10 mosm/l. From the above (Sects. 82.2.1, 82.2.2) it would seem that this should suppress AVP secretion considerably. However, this is not the case, as is apparent from Fig. 82.14 [77]. Two findings are remarkable: in the first trimester there is a shift of the osmolality response curve to the left, and in late pregnancy there is in addition a marked fall in slope. Therefore, in pregnancy, a new (lower) plasma osmolality is defined as "normal" [85], and the response of AVP secretion is attenuated.
- Close to term, body water and body weight have increased on average by 7.5 and 12 kg, respectively. The extra water is not only found in the fetus, placenta, and uterus, but also in other tissues in the mother's body. It appears that this increase in volume represents a new normal value as sensed by the respective receptors (Sect. 82.2), and any attempt to reduce this volume, e.g., by diuretics, leads to rapid defense reactions such as usually accompany volume losses (shrinkage of extracellular space, Sect. 82.4).

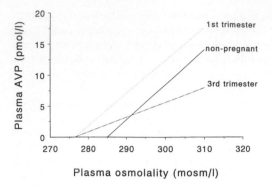

Fig. 82.14. Altered arginine vasopressin (*AVP*) secretion response in pregnancy. Note that there is a left shift in the first trimester, which leads to a comparably high plasma AVP concentration at osmolalities around 290 mosm/l. Furthermore, the slope of the response is attenuated in the third trimester. (Redrawn from [77])

- In late pregnancy, an additional 950 mmol Na$^+$/l (22 g NaCl) are accumulated. This corresponds to an increase of total body Na$^+$ by 20%–25%. A large part is found in the fetus, but, again, considerable amounts are found in the mother. As with body water, this extra Na$^+$ is normal. Any attempt to induce natriuresis activates the defense mechanisms promptly and sharply.
- Plasma progesterone and prostaglandins are elevated. Both hormones, considered solely, would induce a natriuresis. However, this is not the case, because desoxycorticosterone, aldosterone, renin, and angiotensin II are elevated at the same time.
- Antinatriuretic factors (see above) more than outweigh natriuretic factors, and this ensures that, despite the increase in GFR, urinary Na$^+$ losses are reduced and Na$^+$ accumulation can occur throughout pregnancy. The excess daily filtration amounts to 5–10 mol Na$^+$ per day. Of this quantity, an extra 2–6 mmol must be conserved every day (during pregnancy). The enhanced antinatriuretic factor, i.e., aldosterone, therefore has to be regarded as a necessary compensatory mechanism for the elevated progesterone concentration.
- Like Na$^+$, K$^+$ is also conserved during pregnancy. An extra 300–350 mmol is accumulated. This amount is required for mother and fetus. In view of the increased plasma aldosterone concentration, this tendency to conserve K$^+$ may appear surprising. It has been suggested that it is due to a K$^+$-saving effect of progesterone.
- A hormone named relaxin is undetectable in women who are not pregnant, but is easily detectable in pregnancy. This hormone appears, at least in part, to be responsible for the observed vasodilatation [1].

In summary, the normal values of salt and water metabolism are reset in pregnancy. Increases in body water, Na$^+$, and K$^+$ are normal, and the previous clinical practice of restricting NaCl uptake or even prescribing diuretics does not appear justified in this light [77]. In fact, in preeclampsia, despite the existence of edema, such treatment is not indicated, because these patients show all signs of volume contraction rather than volume expansion. The primary cause of preeclamptic edema may be a relative underfilling of the circulation [1] due to a fall in cardiac output (see Fig. 82.13). In addition, an enhanced sensitivity of the vasculature to angiotensin II, norepinephrine, and vasopressin has been reported recently [1] and might explain the observed increases in blood pressure.

82.8.6 Old Age

Many body functions are altered gradually with age (see Chaps. 123, 124). This is also true for salt, water, volume, and pressure homeostasis [98,111]. Awareness of these changes is essential, because disturbances of fluid and electrolyte balance are among the most frequent reasons for hospitalization in the elderly. The major changes are as follows:

- Increase in blood pressure. In 64% of a 74-year-old population, blood pressure was equal to or higher than 140/90 mmHg. In 45% it was even equal to or higher than 160/95 mmHg [135].
- Blood pressure regulation is much more vulnerable in old age. The postural hypotensive effect is augmented due to attenuated baroreceptor reflexes [98].
- The renin–angiotensin aldosterone system is less active.
- The plasma ANP concentration is higher.
- The secretion of AVP in response to dehydration is increased (see Sect. 82.2).
- The sensation of thirst, however, is strongly attenuated.
- Due to a decline in kidney function, the concentrating ability is attenuated, thus compromising the conservation of water.
- Taste and olfaction thresholds increase [88]. In this context it is worthwhile to remember that one of the early signs of Alzheimer's disease is a great deficit in olfaction [90].

Hypertension. Given these changes, it appears wise to treat age-related essential hypertension slowly and with great care and to avoid abrupt changes in blood pressure. Otherwise the patient may develop postural hypotension [98].

Dehydration. One main danger is dehydration. This is caused mainly by two facts: the sensation of thirst is blunted, and the kidney has difficulties in conserving water [99,111]. The problem may only become apparent when the patient is in a new environment, becomes ill, or undergoes some other kind of stress. It does not appear wise to urge patients to drink large volumes of water and to restrict NaCl intake. Hyponatremia is a very common finding in the elderly and this is again due to falling kidney function, i.e., inability to excrete large vloumes of water on the one hand and to absorb NaCl adequately on the other. Hyponatremic overhydration is very difficult to treat, because simple tools such as diuretics cannot be used in such circumstances.

Confusion. Patients with hyponatremia and dehydration may present with severe confusion, and this may be misleading in the diagnosis; patients are often transferred to a psychiatric hospital. Adequate hydration must be one of the key concerns as soon as the elderly does not feel well.

82.8.7 Psychiatric Patients

Compulsive Water Drinking. This has frequently been reported in psychiatric patients. The highest incidence is found in schizophrenic patients. The daily drinking rates can be as high as 10–20 l. With these drinking rates, self-induced water intoxication (SIWI) is not uncommon. SIWI with severe hyponatremia is a life-threatening condition, as it causes brain edema. Despite the great efforts made to understand the causes of compulsive water drinking in schizophrenic patients, the reason for this behavior is not clear [105]. Several factors may contribute:

- Primary hyperdipsia
- Disease-induced inappropriate AVP secretion (SIADH; see Sect. 82.8.2)
- Psychotropic drug-induced inappropriate AVP secretion.

In many patients the causes appear to be mixed, mostly primary hyperdipsia and SIADH [105]. SIADH may be caused by the activation of the dopaminergic system in these patients, as dopamine stimulates AVP secretion (see Sect. 82.2). There is also some evidence that psychotropic agents such as amitriptyline can cause SIADH. The treatment of SIWI firstly consists of water restriction. In some patients salt tablets, water restriction, and loop diuretics may be combined. The inhibition of AVP binding to the V_2 receptor by specific V_2 receptor antagonists is not feasible at this stage. Instead, high doses of Li^+ and the drug demeclocycline have been used to interrupt AVP transduction in the collecting tubule.

82.9 Conclusions

The present chapter has addressed the coordination of central and peripheral regulatory components of salt and water metabolism. Central and peripheral signals are monitored continuously. They include the following: plasma osmolality; Na^+ and Cl^- concentration, as sensed by specific central and peripheral sensors; the filling state of the circulation; blood pressure; angiotensin II concentration; pharyngeal afferences sensing drinking; other neuronal outputs and additional signals.

These signals are coordinated in the hypothalamus, in the medulla oblongata, and at several peripheral sites. With its specific areas that lack the blood–brain barrier, such as the OVLT, the ventral hypothalamus is a sensor, an effector, and an integrator. Blood osmolality and, at least in some

species, plasma aldosterone and angiotensin II concentration and local concentrations of angiotensin II, ANP, CNP, and other hormones are measured. AVP release and thirst are controlled accordingly. Important connections to the medulla oblongata work in both directions. Hyperosmolality sensitizes the baroreceptor reflexes and enhances sympathergic discharge. Dopaminergic fibers from the medulla oblongata enhance AVP secretion. One important peripheral coordination site is the atria of the heart, which sense blood volume, relay it to the medulla oblongata, and adjust the secretion of ANP and probably also ouabain. Another important coordination site is the kidney, which adjusts renin release according to sympathergic discharge, blood pressure, and distal tubule Cl^- concentration.

The effector mechanisms for NaCl and water metabolism largely overlap. The key organs are the CNS, with its pivotal role in controlling circulation, thirst, and Na^+ appetite; the heart, with its sympathergic and vagal control mechanisms; the vasculature, which is under the influence of sympathergic discharge, angiotensin II, ANP, AVP, and prostaglandins etc.; the kidney, with its dual function as the controller of renin release and as the main organ determining NaCl and water excretion; and the adrenal glands, with their role in aldosterone and ouabain production.

The body possesses a large number of defense mechanisms with different time scales to protect it against water and NaCl losses. The seemingly redundant arrangement of the various defense mechanisms ensures effective regulation even if one component or even several fail. Conversely, volume and NaCl overload are prevented by several effector mechanisms. However, their spectrum is not as large.

Water and NaCl metabolism are changed in a very characteristic way in pregnancy and lactation and with old age. These aspects, as well as important disturbances in NaCl and water metabolism, are discussed Sect. 82.8.

Acknowledgements. This manuscript is dedicated with deep respect and appreciation to my late friend Hans-Peter Koepchen, who had a great interest in this area and originally intended to write this text himself.

References

1. Abraham WT, Schrier RW (1994) Body fluid volume regulation in health and disease. Adv Intern Med 39:23–47
2. Akabas MH (1990) Mechanisms of chemosensory transduction in taste cells. Int Rev Neurobiol 32:241–277
3. Andersson B (1978) Regulation of water intake. Physiol Rev 58:582–603
4. Andersson B, Leksell LG, Rundgren M (1982) Regulation of water intake. Annu Rev Nutr 2:73–89
5. Avenet P, Lindemann B (1989) Perspectives of taste. J Membr Biol 112:1–8
6. Avenet P, Lindemann B (1991) Noninvasive recording of receptor cell action potentials and sustained currents from single taste buds maintained in the tongue: the response to mucosal NaCl and amiloride. J Membr Biol 124:33–41

7. Awazu M, Ichikawa I (1993) Biological significance of atrial natriuretic peptide in the kidney. Nephron 63:1–14
8. Ballermann BJ, Zeidel ML (1992) Atrial natriuretic hormone. In: Seldin DW, Giebisch G (eds) The kidney: physiology and pathophysiology. Raven, New York, pp 1843–1884
9. Baylis PH (1987) Osmoregulation and control of vasopressin secretion in healthy humans. Am J Physiol 253:R671–R678
10. Baylis PH, Robertson GL (1980) Rat vasopressin response to insulin-induced hypoglycemia. Endocrinology 107:1975–1979
11. Baylis PH, Robertson GL (1980) Vasopressin response to 2-deoxy-D-glucose in the rat. Endocrinology 107:1970–1974
12. Blackburn RE, Samson WK, Fulton RJ, Stricker EM, Verbalis JG (1993) Central oxytocin inhibition of salt appetite in rats: evidence for differential sensing of plasma sodium and osmolality. Proc Natl Acad Sci USA 90:10380–10384
13. Blaustein MP (1977) Sodium ions, calcium ions, blood pressure regulation, and hypertension: a reassessment and a hypothesis. Am J Physiol 232:C165–C173
14. Blaustein MP (1993) Physiological effects of endogenous ouabain: control of intracellular Ca^{2+} stores and cell responsiveness. Am J Physiol 264:C1367–C1387
15. Bonventre JV, Nemenoff R (1991) Renal tubular arachidonic acid metabolism. Kidney Int 39:438–449
16. Brenner BM, Ballermann BJ, Gunning ME, Zeidel ML (1990) Diverse biological actions of atrial natriuretic peptide. Physiol Rev 70:665–699
17. Brooks CM (1988) The history of thought concerning the hypothalamus and its function. Brain Res Bull 20:657–667
18. Brown J, Zuo Z (1993) C-type natriuretic peptide and atrial natriuretic peptide receptors of rat brain. Am J Physiol 264:R513–R523
19. Castellano G, Solis-Herruzo JA, Gonzalez A, Morillas JD, Moreno D, Munoz T, Larrodera L (1994) Plasma arginine vasopressin response to oral, gastric, and intravenous water load in patients with cirrhosis. Gastroenterology 106:678–685
20. Clauss W, Schaefer H, Horch I, Hoernicke H (1985) Segmental differences in electrical properties and Na$^+$-transport of rabbit caecum, proximal and distal colon in vitro. Pflugers Arch Eur J Physiol 403:278–282
21. Cogan MG (1990) Renal effects of atrial natriuretic factor. Annu Rev Physiol 52:699–708
22. Collier G (1989) The economics of hunger, thirst, satiety, and regulation. Ann N Y Acad Sci 575:136–154
23. Cooper SJ (1983) Minireview: benzodiazepine-opiate antagonist interactions in relation to feeding and drinking behavior. Life Sci 32:1043–1051
24. Cowley AW (1992) Long-term control of arterial blood pressure. Physiol Rev 72:231–300
25. Cross PC, Mercer KL (1993) Cell and tissue ultrastructure, a functional perspective. Freman, New York
26. Davies R, Forsling M, Bulger G, Phillips T (1983) Studies in normal subjects and in benign essential hypertension at rest and after postural challenge. Br Heart J 49:528–531
27. De Rouffignac C, Corman B, Roinel N (1983) Stimulation by antidiuretic hormone of electrolyte tubular reabsorption in rat kidney. Am J Physiol 244:F156–F164
28. de Wardener HE, Mills IH, Clapham WF, Hayter CJ (1961) Studies on the efferent mechanism of the sodium diuresis which follows the administration of intravenous saline in the dog. Clin Sci 21:249–258
29. Debold AJ, Borenstein HB, Veress AT, Sonnenberg H (1981) A rapid and potent natriuretic response to intravenous injection of atrial myocardial extract in rats. Life Sci 28:89–94
30. Deen PM, Verdijk MA, Knoers NV, Wieringa B, Monnens LAH, van Os CH, van Oost BA (1994) Requirement of human renal water channel aquaporin-2 for vasopressin-dependent concentration of urine. Science 264:92–95
31. Denton D (1982) The hunger for salt: an anthropological, physiological and medical analysis. Springer, Berlin Heidelberg New York

32. DiBona GF (1986) Neural mechanisms in body fluid homeostasis. Fed Proc 45:2871–2877
33. Dourish CT (1983) Dopaminergic involvement in the control of drinking behaviour: a brief review. Prog Neuro-psychopharmacol Biol Psychiatry 7:4–6
34. Elliot P (1994) Epidemiological studies of salt and blood pressure: INTERSALT study findings and implications. Nieren Hochdruckkr 23 [Suppl 1]:S38–S44
35. Epstein M (1992) Renal effects of head-out water immersion in humans: a 15-year update. Physiol Rev 72:563–621
36. Ermisch A, Brust P, Kretzschmar R, Rühle HJ (1993) Peptides and blood-brain barrier transport. Physiol Rev 73:489–527
37. Ferrario CM, Brosnihan KB, Diz DI, Jaiswal N, Khosla MC, Milsted A, Tallant EA (1991) Angiotensin-(1–7): a new hormone of the angiotensin system. Hypertension 18 [Suppl 5]:III-126–III-133
38. Fitzsimons JT (1992) Physiology and pathophysiology of thirst and sodium appetite. In: Seldin DW, Giebisch G (eds) The kidney: physiology and pathophysiology. Raven, New York, pp 1615–1648
39. Fitzsimons JT, Simons BJ (1969) The effect on drinking in the rat of intravenous infusion of angiotensin, given alone or in combination with other stimuli. J Physiol (Lond) 203:45–57
40. Folkow B (1990) Salt and hypertension. News Physiol Sci 5:220–224
41. Funder JW (1993) Aldosterone action. Annu Rev Physiol 55:115–130
42. Fushimi K, Uchida S, Hara Y, Hirata Y, Marumo F, Sasaki S (1993) Cloning and expression of apical membrane water channel of rat kidney collecting tubule. Nature 361:549–552
43. Gauer OH, Henry JP (1976) Neurohormonal control of plasma volume. In: Guyton AC, Cowley AW (eds) International reviews of physiology, cardiovascular physiology II. University Park Press, Baltimore, pp 145–190
44. Gauer OH, Henry JP, Sieker HO, Wendt WE (1954) The effect of negative pressure breathing on urine flow. J Clin Invest 33:287–296
45. Goetz KL (1990) Evidence that atriopeptin is not a physiological regulator of sodium excretion. Hypertension 15:9–19
46. Gonzalez-Campoy JM, Knox FG (1992) Integrated responses of the kidney to alterations in extracellular fluid volume. In: Seldin DW, Giebisch G (eds) The kidney: physiology and pathophysiology. Raven, New York, pp 2041–2098
47. Goto A, Yamada K, Yagi N, Yoshioka M, Sugimoto T (1992) Physiology and pharmacology of endogenous digitalis-like factors. Pharmacol Rev 44:377–399
48. Greene LJ (1988) A hypothesis regarding the function of angiotensin peptides in the brain. Clin Exp Hypertens [A] 10 [Suppl 1]:107–121
49. Greger R (1985) Ion transport mechanisms in thick ascending limb of Henle's loop of mammalian nephron. Physiol Rev 65:760–797
50. Greger R, Heidland A (1991) Action and clinical use of diuretics. In: Cameron JS, Davison AM, Grünfeld JP (eds) Clinical nephrology. Oxford University Press, London, pp 197–224
51. Gross PM (1985) The subfornical organ as a model of neurohumoral integration. Brain Res Bull 15:65–70
52. Guyton AC (1992) Kidneys and fluids in pressure regulation. Small volume but large pressure changes. Hypertension 19 [Suppl 1]:1–2–1–8
53. Haberich FJ, Aziz O, Nowacki PE (1965) Über einen osmoreceptorisch tätigen Mechanismus in der Leber. Pflugers Arch Eur J Physiol 285:73–89
54. Hall JE, Brands MW (1992) The renin-angiotensin-aldosterone systems. Renal mechanisms and circulatory homeostasis. In: Seldin DW, Giebisch G (eds) The kidney. Physiology and pathophysiology. Raven, New York, pp 1455–1504
55. Hamlyn JM, Ludens JH (1992) Nonatrial natriuretic hormones. In: Seldin DW, Giebisch G (eds) The kidney:

physiology and pathophysiology. Raven, New York, pp 1885–1924

56. Harris PJ, Young JA (1977) Dose-dependent stimulation and inhibition of proximal tubular sodium reabsorption by angiotensin II in the rat kidney. Pflugers Arch Eur J Physiol 367:295–297

57. Hartshorne H (1847) Water versus hydrotherapy or an essay on water and its true relations to medicine. Smith, Philadelphia

58. Helderman JH, Vestal RE, Rowe RW, Tobin JD, Andres R, Robertson GL (1978) The response of arginine vasopressin to intravenous ethanol and hypertonic saline in man. J Gerontol 33:39–47

59. Hill DL, Mistretta CM (1990) Developmental neurobiology of salt taste sensation. Trends Neurosci 13:188–195

60. Holtzman MJ (1992) Arachidonic acid metabolism in airway epithelial cells. Annu Rev Physiol 54:303–329

61. Horrobin DF, Burstyn PG, Lloyd IJ, Durkin N, Lipton A, Muiruri KL (1971) Actions of prolactin on human renal function. Lancet 2:352–354

62. Ishiki K, Morita H, Hosomi H (1991) Reflex control of renal nerve activity originating from the osmoreceptors in the hepato-portal region. J Auton Nerv Syst 36:139–148

63. Johnson AK (1985) The periventricular anteroventral third ventricle (AV3V): its relationship with the subfornical organ and neural systems involved in maintaining body fluid homeostasis. Brain Res Bull 15:595–601

64. Johnson AK, Cunningham JT (1987) Brain mechanisms and drinking: the role of lamina terminalis-associated systems in extracellular thirst. Kidney Int 32 [Suppl 21]:S35–S42

65. Joynt RJ (1964) Functional significance of osmosensitive units in the anterior hypothalamus. Neurology 14:584–590

66. Kinnamon SC, Cummings TA (1992) Chemosensory transduction mechanisms in taste. Annu Rev Physiol 54:715–731

67. Kirk KL, Schafer JA (1992) Water transport and osmoregulation by antidiuretic hormone in terminal nephron segments. In: Seldin DW, Giebisch G (eds) The kidney: physiology and pathophysiology. Raven, New York, pp 1693–1725

68. Kluthe R (1994) Sodium and hypertension. German aspects, an introduction. Nieren Hochdruckkr 23 [Suppl 1]:S2–S4

69. Knowles MR, Gatzy JT, Boucher RC (1981) Increased bioelectric potential difference across respiratory epithelia in cystic fibrosis. N Engl J Med 305:1489–1495

70. Koller KJ, Goeddel DV (1992) Molecular biology of the natriuretic peptides and their receptors. Circulation 86:1081–1088

71. Kopp UC, DiBona GF (1992) The neural control of renal function. In: Seldin DW, Giebisch G (eds) The kidney: physiology and pathophysiology. Raven, New York, pp 1157–1204

72. Kovacs L, Robertson GL (1992) Disorders of water balance, hyponatraemia and hypernatraemia. Baillieres Clin Endocrinol Metab 6:107–127

73. Kramer HJ, Meyer-Lehnert H, Predel HG (1991) Endogenous natriuretic and ouabain-like factors. Their roles in body fluid volume and blood pressure regulation. Am J Hypertens 4:81–89

74. Lang F, Tschernko E, Häussinger D (1992) Hepatic regulation of renal function. Exp Physiol 77:663–673

75. Leonhardt H (1986) Histologie, Zytologie and Mikroanatomie des Menschen. Thieme, Stuttgart

76. Lind RW (1988) Angiotensin and the lamina terminalis: illustrations of a complex unity. Clin Exp Hypertens [A] 10 [Suppl 1]:79–105

77. Lindheimer MD, Katz Al (1992) Renal physiology and disease in pregnancy. In: Seldin DW, Giebisch G (eds) The kidney: physiology and pathophysiology. Raven, New York, pp 3371–3431

78. Ludens JH, Clark MA, Kolbasa KP, Hamlyn JM (1993) Digitalis-like factor and ouabain-like compound in plasma of volume-expanded dogs. J Cardiovasc Pharmacol 22 [Suppl 2]:S38–S41

79. Mann JFE, Johnson AK, Ganten D, Ritz E (1987) Thirst and the renin-angiotensin system. Kidney Int 32 [Suppl 21]:S27–S34

80. Mark MH, Farmer PM (1984) The human subfornical organ: an anatomic and ultrastructural study. Ann Clin Lab Sci 14:427–442

81. McKinley MJ (1992) Common aspects of the cerebral regulation of thirst and renal sodium excretion. Kidney Int 41 [Suppl 37]:S102–S106

82. McKinley MJ, Bicknell RJ, Hards D, McAllen RM, Vivas L, Weisinger RS, Oldfield BJ (1992) Efferent neural pathways of the lamina terminalis subserving osmoregulation. Prog Brain Res 91:395–402

83. Meyer-Lehnert H, Wanning C, Michel H, Bäcker A, Kramer HJ (1993) Cellular mechanisms of action of a ouabain-like factor in vascular smooth muscle cells. J Cardiovasc Pharmacol 22 [Suppl 2]:S16–S19

84. Miller M (1980) Role of endogenous opioids in neurohypophyseal function of man. J Clin Endocrinol Metab 50:1016–1020

85. Monson JP, Williams DJ (1992) Osmoregulatory adaptation in pregnancy and its disorders. J Endocrinol 132:7–9

86. Morel F, Doucet A (1986) Hormonal control of kidney functions at the cell level. Physiol Rev 66(2):377–468

87. Morita H, Ishiki K, Hosomi H (1991) Effects of hepatic NaCl receptor stimulation on renal nerve activity in conscious rabbits. Neurosci Lett 123:1–3

88. Morris CE (1990) Mechanosensitive ion channels. J Membr Biol 113:93–107

89. Moses AM, Clayton B (1993) Impairment of osmotically stimulated AVP release in patients with primary polydipsia. Am J Physiol 265:R1247–R1252

90. Murphy C (1993) Nutrition and chemosensory perception in the elderly. Crit Rev Food Sci Nutr 33:3–15

91. Nonoguchi H, Sands JM, Knepper MA (1989) ANF inhibits NaCl and fluid absorption in cortical collecting duct of rat kidney. Am J Physiol 256:F179–F186

92. Oldfield BJ, Hards DK, McKinley MJ (1991) Projections from the subfornical organ to the supraoptic nucleus in the rat: ultrastructural identification of an interposed synapse in the median preoptic nucleus using a combination of neuronal tracers. Brain Res 558:13–19

93. Oldfield BJ, Miselis RR, McKinley MJ (1991) Median preoptic nucleus projections to vasopressin-containing neurones of the supraoptic nucleus in sheep. Brain Res 542:193–200

94. Oliet SH, Bourque CW (1993) Mechanosensitive channels transduce osmosensitivity in supraoptic neurons. Nature 364:341–343

95. Paul M, Bader M, Steckelings UM, Voigtländer T, Ganten D (1993) The renin-angiotensin system in the brain. Drug Res 43:207–213

96. Peterson TV, Benjamin BA (1992) The heart and control of renal excretion: neural and endocrine mechanisms. FASEB J 6:2923–2932

97. Phillips MI (1987) Functions of angiotensin in the central nervous system. Annu Rev Physiol 49:413–435

98. Phillips PA, Hodsman GP, Johnston Cl (1990) Neuroendocrine mechanisms and cardiovascular homeostasis in the elderly. Cardiovasc Drugs Ther 4:1209–1214

99. Phillips PA, Johnston Cl, Gray L (1993) Disturbed fluid and electrolyte homeostasis following dehydration in elderly people. Age Ageing 22:26–33

100. Quirion R (1989) Receptor sites for atrial natriuretic factors in brain and associated structures: an overview. Cell Mol Neurobiol 9:45–55

101. Ramsay DJ, Rolls BJ, Wood RJ (1977) Thirst following water deprivation in dogs. Am J Physiol 232:R93–R98

102. Richard P, Moos F, Freund-Mercier MJ (1991) Central effects of oxytocin. Physiol Rev 71:331–370

103. Richards AM (1990) Is atrial natriuretic factor a physiological regulator of sodium excretion? A review of the evidence. J Cardiovasc Pharmacol 16 [Suppl 7]:S39–S42

104. Ridell DC, Mallonee R, Phillips JA, Parks JS, Sexton LA, Hamerton JL (1985) Chromosomal assignment of human sequences encoding arginine vasopressin-neurophysin II and growth hormone releasing factor. Somat Cell Mol Genet 11:189–195

105. Riggs AT, Dysken MW, Kim SW, Opsahl JA (1991) A review of disorders of water homeostasis in psychiatric patients. Psychosomatics 32:133–148

106. Robertshaw D (1989) Central and peripheral osmoreceptors. Acta Physiol Scand 136 [Suppl 583]:151–156

107. Robertson GL (1983) Thirst and vasopressin function in normal and disordered states of water balance. J Lab Clin Med 101:351–371

108. Robertson GL (1992) Renal regulation of water balance: normal. In: Seldin DW, Giebisch G (eds) The kidney: physiology and pathophysiology. Raven, New York, pp 1595–1614

109. Robertson GL, Athar S (1976) The interaction of blood osmolality and blood volume in regulating plasma vasopressin in man. J Clin Endocrinol Metab 42:613–620

110. Robertson GL, Aycinena P, Zebre RL (1992) Neurogenic disorders of osmoregulation. Am J Med 72:339–353

111. Rolls BJ, Phillips PA (1990) Aging and disturbances of thirst and fluid balance. Nutr Rev 48:137–144

112. Rosenthal W, Seibold A, Antaramian A, Lonergan M, Arthus MF, Hendy GN, Birnbaumer M, Bichet DG (1992) Molecular identification of the gene responsible for congenital nephrogenic diabetes insipidus. Nature 17:233–235

113. Rossier BC, Palmer LG (1992) Mechanisms of aldosterone action on sodium and potassium transport. In: Seldin DW, Giebisch G (eds) The kidney: physiology and pathophysiology. Raven, New York, pp 1373–1409

114. Rowe JW, Shelton RL, Helderman JH (1979) Influence of the emetic reflex on vasopressin release in man. Kidney Int 16:729–735

115. Rowland NE (1990) On the waterfront: predictive and reactive regulatory descriptions of thirst and sodium appetite. Physiol Behav 48:899–903

116. Ruskoaho H (1992) Atrial natriuretic peptide: synthesis, release, and metabolism. Pharmacol Rev 44:479–516

117. Rydin H, Verney EB (1938) The inhibition of water-diuresis by emotional stress and by muscular exercise. Q J Exp Physiol 27:373–374

118. Sachs F (1991) Mechanical transduction by membrane ion channels: a mini review. Mol Cell Biochem 104:57–60

119. Schiffman SS (1988) Taste transduction and modulation. News Physiol Sci 3:109–112

120. Schlatter E (1993) Effects of various diuretics on membrane voltage of macula densa cells. Whole-cell patch-clamp experiments. Pflugers Arch Eur J Physiol 423:74–77

121. Schlatter E, Greger R (1985) cAMP increases the basolateral Cl⁻-conductance in the isolated perfused medullary thick ascending limb of Henle's loop of the mouse. Pflugers Arch Eur J Physiol 405:367–376

122. Schlatter E, Salomonsson M, Persson AEG, Greger R (1989) Macula densa cells sense luminal NaCl concentration via the furosemide sensitive Na-2Cl-K cotransporter. Pflugers Arch Eur J Physiol 414:286–290

123. Schlatter E, Schafer JA (1987) Electrophysiological studies in principal cells of rat cortical collecting tubules. AVP increases the apical membrane Na⁺-conductance. Pflugers Arch Eur J Physiol 409:81–92

124. Schnermann J, Briggs JP (1992) Function of the juxtaglomerular apparatus: control of glomerular hemodynamics and renin secretion. In: Seldin DW, Giebisch G (eds) The kidney: physiology and pathophysiology. Raven, New York, pp 1249–1290

125. Schoner W (1992) Endogenous digitalis-like factors. Clin Exp Hypertens [A] 14(5):767–814

126. Schoner W, Heidrich-Lorsbach E, Kirch U, Ahlemeyer B, Sich B (1993) Purification and properties of endogenous ouabain-like substances from hemofiltrate and adrenal glands. J Cardiovasc Pharmacol 22 [Suppl 2]:S29–S31

127. Schrier RW (1992) A unifying hypothesis of body fluid volume regulation. The Lilly lecture 1992. J R Coll Physicians Lond 26:295–306

128. Schulz-Knappe P, Forssmann K, Herbst F, Hock D, Pipkorn R, Forssmann WG (1988) Isolation and structural analysis of "urodilatin", a new peptide of the cardiodilatin-(ANP)-family, extracted from human urine. Klin Wochenschr 66:752–759

129. Seckl JR, Williams TDM, Lightman SL (1986) Oral hypertonic saline causes transient fall of vasopressin in humans. Am J Physiol 251:R214–R217

130. Segar WE, Moore WW (1968) The regulation of antidiuretic hormone release in man. J Clin Invest 47:2143–2151

131. Share L (1988) Role of vasopressin in cardiovascular regulation. Physiol Rev 68:1248–1284

132. Smith WL (1992) Prostanoid biosynthesis and mechanisms of action. Am J Physiol 263:F181–F191

133. Steckelings U, Lebrun C, Quadri F, Veltmar A, Unger T (1992) Role of brain angiotensin in cardiovascular regulation. J Cardiovasc Pharmacol 19 [Suppl 6]:S72–S79

134. Stewart RE, Swithers SE, Plunckett LM, McCarty R (1988) ANF receptors: distribution and regulation in central and peripheral tissues. Neurosci Behav Rev 12:151–168

135. The working group on hypertension in the elderly (1986) Statement on hypertension in the elderly. JAMA 256:70–74

136. Thompson CJ, Baylis PH (1988) Osmoregulation of thirst. J Endocrinol 117:155–157

137. Thompson CJ, Burd JM, Baylis PH (1987) Acute suppression of plasma vasopressin and thirst after drinking in hyponatraemic humans. Am J Physiol 252:R1138–R1142

138. Thurau K, Schnermann J (1965) Die Natriumkonzentration an den Macula Densa Zellen als regulierender Faktor für das Glomerulumfiltrat (Mikropunktionsversuche). Klin Wochenschr 43:410–413

139. Unger T, Gohlke P, Kotrba M, Rettig R, Rohmeiss P (1989) Angiotensin II and atrial natriuretic peptide in the brain: effects on volume and Na⁺ balance. Resuscitation 18:309–319

140. Van Giersbergen PLM, Palkovits M, De Jong W (1992) Involvement of neurotransmitters in the nucleus tractus solitarii in cardiovascular regulation. Physiol Rev 72:789–824

141. Verkman AS (1992) Water channels in cell membranes. Annu Rev Physiol 54:97–108

142. Verney EB (1947) The antidiuretic hormone and factors which determine its release. Proc R Soc Lond 135:25–106

143. Vokes T (1987) Water homeostasis. Annu Rev Nutr 7:383–406

144. Zardetto-Smith AM, Thunhorst RL, Cicha MZ, Johnoson AK (1993) Afferent signalling and forebrain mechanisms in the behavioral control of extracellular fluid volume. Ann NY Acad Sci 689:161–176

145. Zebre RL, Robertson GL (1983) Osmoregulation of thirst and vasopressin secretion in human subjects: effects of various solutes. Am J Physiol 244:E607–E614

146. Zeidel ML (1990) Renal actions of atrial natriuretic peptide: regulation of collecting duct sodium and water transport. Annu Rev Physiol 52:747–759

147. Zimmerman EA, Ma LY, Nilaver G (1987) Anatomical basis of thirst and vasopressin secretion. Kidney Int 32 [Suppl 21]:S14–S19

IV Acute Vital Functions; Heart; Circulation and Respiration

K Blood and Immune Defense

83 Blood, Plasma Proteins, Coagulation, Fibrinolysis, and Thrombocyte Function

C. Bauer and W. Wuillemin

Contents

83.1 Introduction

The particular physiological importance of blood lies in the fact that it is in constant contact with all the organs of the body and therefore represents an important means of communication. A constant blood supply is essential for the normal supply of nutrients to cells as well as for the removal of metabolic waste products. Likewise, the exchange of biochemical information between cell systems requires the constant circulation of this medium in the capillary bed, where it comes in close contact with the interstitial space. Apart from this transport function, specialized blood cells are also engaged in a large number of defense mechanisms (see Chap. 85), as well as in blood coagulation and wound healing. In this chapter emphasis will be placed on the general composition of blood, the function of the plasma proteins, and the processes that prevent the loss of this valuable medium after injury, blood coagulation, and fibrinolysis. The latter aspect has been dealt with in greater detail because of its clinical relevance.

83.2 Volume and Composition of the Blood

Erythrocytes, Leukocytes, and Thrombocytes. Blood consists mainly of water, in which the biologically important

---

R. Greger/U. Windhorst (Eds.)
Comprehensive Human Physiology, Vol. 2
© Springer-Verlag Berlin Heidelberg 1996

electrolytes, nutrients, vitamins, and gases are dissolved. This watery solution also contains proteins as well as different cell populations, i.e., erythrocytes (red blood cells), leukocytes (white blood cells), and thrombocytes. The blood volume is normally a constant fraction of the fat-free body weight, amounting to 7%. A lean human being weighing 70 kg therefore has a blood volume of about 5 l. Newborns have a slightly higher fractional blood volume compared to adults (approximately 8.5%), which is due to the fact that they have been living in an oxygen-poor environment during the final stages of gestation, which leads to a increased production of erythrocytes (see Chap. 116).

Hematocrit. The blood volume can be measured by injecting a known amount of indicator, e.g., a suitable dye, into the circulation. After complete equilibration of the indicator, its concentration can be estimated and the unknown volume calculated using the formula

$$V_x = V_i \times C_i/C_x$$

whereby V_i and C_i denote the initial volume and concentration of the indicator and C_x its concentration after equilibration. An indicator that can be used to measure the plasma volume is Evan's blue, while the total volume of erythrocytes can be measured with erythrocytes that have been tagged with ^{51}Cr or ^{59}Fe. The blood volume (BV) is obtained according to the formula:

$$\text{BV} = \text{erythrocyte volume/hematocrit}$$
$$= \text{plasma volume/}(1 - \text{hematocrit})$$

During centrifugation of a blood sample that has been treated with inhibitors of coagulation, such as Ca^{2+} chelators, the cellular elements will quickly sediment. The volume fraction of the cells in relation to the volume of the blood sample is known as the hematocrit, which amounts to 0.42 in women, 0.46 in men, and 0.55 in newborns. Normal values of other erythrocyte parameters are given in Table 83.1. About 99% of the hematocrit is made up of the erythrocytes. The supernatant is a protein solution, plasma, a term that should not be confused with blood serum. Serum is obtained after the blood sample has been allowed to clot and is therefore plasma without the various coagulatory proteins, especially fibrinogen.

83.2.1 Plasma Electrolytes

The concentration of electrolytes dissolved in the plasma water is 290 mOsm/l (see chapter 125 for the definition of osmolality and osmolarity). Anions and cations are distributed between the extracellular and intracellular compartments in a very characteristic manner, as can be seen from Table 83.2.

Of particular functional importance is the high K^+ and low Na^+ concentration in the intracellular compartment com-

Table 83.1. Normal values for erythrocytes and hemoglobin in adult human blood (± 2 SD)

	Women		Men	
	Mean	Range	Mean	Range
Hemoglobin concentration (g/l)	140	120–160	160	140–180
Hematocrit (fraction)	0.42	0.37–0.47	0.47	0.4–0.54
Erythrocyte count (10^{12}/l)	4.5	4.2–5.4	5.0	4.6–6.2
Mean cellular hemoglobin concentration (MCHC; g/l)	333	300–360	340	310–350
Mean cellular hemoglobin content (MCH; pg = 10^{-12} g)	31	26–35	32	26–32
Mean cellular volume (MCV; fl = 10^{-15} l)	93	80–120	94	80–96

Table 83.2. Ionic composition of body fluids (mmol/l)

Ion	Plasma	Plasma water	Interstitial	Intracellular
Cations				
Na^+	142	153	145	12
K^+	4.3	4.6	4.4	139
Ca^{2+}	2.5	2.7	2.4	<0.001 (free)[b]
Mg^{2+}	1.1	1.2	1.1	1.6 (free)[b]
Total	149.9	161.5	152.9	152.6
Anions				
Cl^-	104	112	117	4
HCO_3^-	24	26	27	12
$HPO_4^{2-}/H_2PO_4^-$	2	2.2	2.3	29
Proteins[a]	14	15	0.4	54
Others	5.9	6.3	6.2	53.6
Total	149.9	161.5	152.9	152.6

[a] Represents the concentration of charge equivalents (mEq/l) and *not* the molar concentration
[b] Represents the concentration of free Ca^{2+} and Mg^{2+}, which is a measure of ion activity

pared with the extracellular compartment. This unequal distribution of electrolytes is necessary to create electrical potential differences across the cell membrane, which enable secondary active transport processes such as Na^+-dependent glucose transport to take place and which activate exchangers such as the Na^+/Ca^{2+} exchanger (see Chap. 8). In addition, the low intracellular Ca^{2+} concentration allows biological processes such as exocytosis, gene activity, or smooth muscle contraction to be regulated via a controlled flow of Ca^{2+} from the outside to the interior of the cell and intracellula Ca^{2+} stores to be controlled by challenging a cell with the appropriate extracellular stimulus (see Chap. 5).

83.2.2 Plasma Proteins

Albumin and Globulins. The total protein concentration in the plasma is normally about 70 g/l plasma, corresponding to a total of about 200 g in human adults. Among the different protein fractions, albumin (100 g) and globulins (40 g) are the most abundant. Most of the plasma proteins, with the exception of the family of γ-globulins, are produced in the liver. They are secreted from the hepatocytes and reach the intravascular space via the fenestrated endothelial cells that line the sinusoidal compartment of the liver. In other organs, the endothelial wall of the capillaries is much less permeable for proteins than the liver, with the result that the plasma proteins remain in the intravascular space. It should not be overlooked, however, that the interstitial space is not totally devoid of plasma proteins (mean concentration, approximately 1 g/l interstitial fluid). These interstitially located proteins leave the intravascular space via the postcapillary venules and are transported back to the circulation via the lymphatic vessels (see Chap. 94).

Plasma proteins can be separated by various techniques, the simplest one being one-dimensional electrophoresis, whereby the proteins are separated according to their molecular charge and size (Fig. 83.1). With methods that have a higher power of resolution, such as two-dimensional electrophoresis or immunoelectrophoresis as many as 120 different plasma proteins can be separated. These different protein fractions have very different physiological functions. Mainly albumin, but also α- and β-globulins serve as vehicles for low molecular weight substances such as hormones, lipids, iron, vitamins, and metabolic waste products. Other plasma proteins are involved in coagulation and fibrinolysis or in defense mechanisms (the γ-globulins) directed against foreign pathogens such as viruses, bacteria, or fungi. A brief summary of the function of the various classes of plasma proteins is provided in Table 83.3.

Lipoproteins. Of great physiological and medical importance are porteins that are involved in lipid transport (lipoproteins). There are different classes of lipoproteins according to their density, which in turn depends upon the ratio of lipids to proteins: very low density lipoproteins

(VLDL), low-density lipoproteins (LDL), and high-density lipoproteins (HDL). VLDL have a relative lipid contribution to their total molecular weight of about 90% and HDL of about 50%. The lipids that are transported by these "specialized" proteins are triacylglycerols, phospholipids, cholesterol, and lipid-soluble vitamins. The highest fraction of cholesterin is contained in the LDL, which are mainly produced by cleavage of VLDL by lipoprotein lipases exposed at the surface of endothelial cells. Numerous studies have emphasized that a high plasma level of LDL in combination with a low level of HDL is a significant risk factor for the development of atherosclerotic plaques with subsequent vessel damage and infarction [39,89].

Oncotic Pressure. In addition to more biochemical functions, the plasma proteins also have an important biophysical role concerning the maintenance of the colloid osmotic pressure (oncotic pressure). As discussed in detail in Chap. 97, fluid exchange across the capillary wall is thought to obey Starling's law. The force tending to push fluid out of the capillary is the capillary hydrostatic pressure (P_c) minus the hydrostatic pressure in the interstitial fluid (P_i). The force tending to pull fluid into the capillary is the oncotic pressure of the plasma proteins (π_k) minus that of the proteins of the interstitial fluid (π_i). Thus the net fluid in (or out) is $K_F[(P_c - P_i)] - \sigma[(\pi_c - \pi_i)]$, where K_F is the filtration coefficient and σ is the reflection coefficient of the proteins. Because of its plasma concentration and relatively low molecular weight (69 000), albumin is responsi-

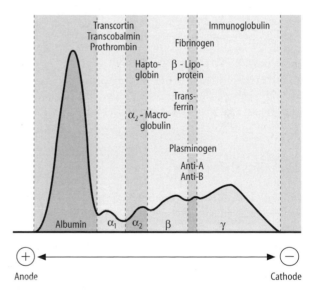

Fig. 83.1. Electrophoretic separation of plasma proteins (one-dimensional representation). Proteins with a low molecular weight and a high negative charge density accumulate towards the anode. The figure shows proteins of physiological importance: Transcortin, transporter of steroid hormones; transcobalamin, transporter of cobalamins, e.g., cyanocobalamin (vitamin B_{12}); haptoglobin, binds free hemoglobin; α_2-macroglobulin, plasmininhibitor; fibrinogen, hemostatic functions; β lipoproteins, transporter of lipids; transferrin, transporter of iron; plasminogen, fibrinolytic function; anti-A and -B, blood-group antibodies; immunglobulins, antibodies

1653

Table 83.3. The quantitatively most important fractions of the plasma proteins and some of their functions

Protein	Concentration (g/l plasma)	Some selected functions
Albumin	35–40	Colloid osmotic pressure Transport of Ca^{2+}, fatty acids, and other lipophilic substances
α_1-Globulins	3–6	Transport of lipids, thyroxin, and hormones from the adrenal cortex Inhibitor for trypsin and chymotrypsin
α_2-Globulins	4–9	Oxidase function Plasmin inhibitor Binding of free hemoglobin
β-Globulins	6–11	Transport of lipids and iron Complement proteins
γ-Globulins	13–17	Circulating antibodies
Fibrinogen	30	Blood coagulation (precursor of fibrin) Thrombocyte aggregation
Prothrombin	1	Blood coagulation (precursor of thrombin)

ble for 80% of the oncotic pressure of plasma, i.e., 25 mmHg. Albumin is produced exclusively in the liver, which is the reason for a fall of albumin levels (and many other plasma proteins) in severe hepatocellular disease. Low levels of albumin are also found in patients who suffer from nephrosis or protein-losing enteropathies, in which large amounts of albumin are lost into the urine or into the gastrointestinal tract. In such conditions, the oncotic pressure π_i decreases, resulting in an increase in the net outflow of fluid from the capillary space into the interstitial space; this condition manifests itself as edema, i.e., an excessive accumulation of fluids in the interstitial space, a symptom that may have a variety of causes (see Chap. 97). In cases in which a reduced synthesis or an increased loss of albumin lead to decreased levels of this protein, edema commences when the plasma albumin concentration has fallen below 25 g/l plasma (normal values, 35–40 g/l plasma).

83.3 Erythrocytes

Cellular Deformability. Erythrocytes are the simplest "cells" in the human body. Normal values for the most common erythrocyte parameters are given in Table 83.1. Formed as a nucleated cell in the bone marrow, erythrocytes normally lose their nucleus before their release into the circulation. After entering the circulation, erythrocytes assume the shape of a flattened, indented sphere or biconcave disk. They travel a distance of about 300 km in the circulation during their 120-day lifetime. "Cellular deformability" is the term generally used to refer to the ability of erythrocytes to undergo deformation during flow. This property is influenced by three distinct cellular components: cell shape, which determines the ratio of cell surface to cell volume; cytoplasmic viscosity, which is regulated by intracellular hemoglobin concentration; and membrane deformability. Membrane deformability is regulated by multiple membrane properties, including the elastic shear modulus and the coefficient of surface viscosity [13,96].

The biconcave disk shape of normal erythrocytes creates an advantageous surface area to volume ratio, allowing the erythrocyte to undergo marked deformation while maintaining a constant surface area. The normal human adult erythrocyte has a volume of $90\,\mu m^3$ and a surface area of $140\,\mu m^2$. If the erythrocyte were a sphere of identical volume, it would have a surface area of only $100\,\mu m^2$. It is this excess surface area of about $40\,\mu m^2$ that allows the erythrocyte to undergo extensive deformation; furthermore, it is of great importance for the exchange of respiratory gases in the lung and the periphery [75]. Maintenance of deformability is essential for these cells to successfully negotiate the small passageways in the microcirculation, particularly along splenic sinuses, in which erythrocytes have to squeeze through interendothelial slits that seldom exceed $0.2–0.5\,\mu m$ in width, which is quite small for a cell with a diameter of about $8\,\mu m$ (Fig. 83.2). This remarkable and reversible deformability can be attributed to the presence of a cytoskeleton in the inner part of the plasma membrane which consists of filamentous proteins that are connected to each other and anchored to the membrane. It also maintains the high surface to volume ratio of erythrocytes, which makes them well suited for the exchange of oxygen and carbon dioxide, the sole known physiological function of erythrocytes.

83.3.1 Erythrocyte Membrane

Phospholipids, Cholesterol, and Glycolipids. Like the membranes of other cells, the erythrocyte membrane is a selectively permeable barrier with specific pumps, channels, and gates. As mentioned above, it exhibits a remarkable deformability, a necessary property for a cell whose function is to transport O_2 to all parts of the body.

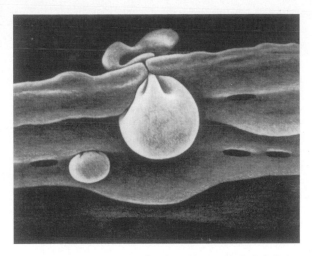

Fig. 83.2. Erythrocyte squeezing through an endothelial slit in a rat spleen sinusoid. (From [100])

to contribute to the deformatiblity of the erythrocyte membrane, because membrane flexibility is markedly reduced in erythrocytes that are pretreated with glycophorin-specific ligands. In contrast, ligands specific for band 3 and A and B blood-group antigens have no effects. These results were interpreted to mean that the observed increase in membrane rigidity depends upon a transmembrane event mediated by glycophorin in cooperation with other skeletal proteins [12]. The functionally most important proteins of the membrane cytoskeleton are spectrin (which consists of α- and β-chains), ankyrin, and associated proteins. The spectrin meshwork is formed by flexible, fiberlike structures and is attached to the membrane by ankyrin and protein 4.1. Ankyrin links band 3 (the anion exchanger) to the β-subunit of spectrin near the middle of the extended spectrin tetramer [13,61]. Protein 4.1 binds to both spectrin subunits near the end of the tetramer, thus enhancing the affinity of spectrin for actin [13,76]. Protein 4.1 also binds in vitro the transmembrane proteins, band 3, and glycophorin (Fig. 83.3).

There are three types of membrane lipids: phospholipids, cholesterol, and glycolipids (see Chap. 4). The phospholipids form the membrane bilayer, with the hydrophilic groups oriented towards the exterior of the bilayer and the hydrophobic hydrocarbons oriented towards the interior of the bilayer. About 50% of the membrane lipids are phospholipids and the remaining 50% is cholesterol, which interacts with the phospholipids in a free, nonesterified form. Contrary to common belief, the deformability of the erythrocyte membrane is not significantly affected by the composition of membrane lipids, but rather by lipid–protein interactions [13,61].

83.3.1.1 Proteins

The membrane proteins can be classified into peripheral proteins and integral proteins (see Chap. 4). Integral proteins are embedded within or traverse the lipid bilayer, in which extensive interaction occurs with the hydrocarbon chains of the membrane lipids. The peripheral proteins are located mainly at the inner cytoplasmic surface of the membrane, where they interact with the polar moieties of the phospholipids, as well as with certain integral proteins, via electrostatic interactions and by hydrogen bonding. By subjecting an extract of membrane proteins to sodium dodecyl sulfate (SDS) electrophoresis, a number of bands can be distinguished according to the molecular weight of the porteins. They were first given Arabic numbers, but some of these bands have now been identified more specifically: band 5 as erythrocyte actin, band 3 as the HCO_3^-/Cl^- exchanger, bands 1 and 2 as the two chains of spectrin, and subfractions of band 2 as ankyrins, to give just a few examples [14,76]. The major component of the erythrocyte membrane is a protein called glycophorin, which accounts for 75% of all membrane proteins. Glycophorin is thought

Lipid bilayer
Band-3-Protein (Cl^-/HCO_3^- exchanger)
Ancyrin
Spectrindimers
Band 4 .1 Protein
Actin
Glycophorin

Fig. 83.3. Composition and spatial arrangement of some components of the erythrocyte cytoskeleton and membrane proteins. Fibers of spectrin form a meshwork that is connected internally via ancyrin and band 4 protein. The spectrin fibers are further attached to the band 3 protein (Cl^-/HCO_3^- exchanger) via ancyrins. Glycophorin is in contact with another important transmembrane protein, the band 3 anion exchanger (not shown). On the cytosolic membrane face, glycophorin can interact with band 4.1 protein, thereby contributing to the deformability of the erythrocyte membrane [12,61]

1655

83.3.1.2 Hereditary Spherocytosis

Hemolysis and Anemia. Hereditary spherocytosis is an autosomal dominant hemolytic disease which affects approximately one in 5000 of a Westem population, The identity of the membrane defect has now been clarified and found to be due to a partial deficiency of erythrocyte spectrin [2]. Normal erythrocytes contain 240 000 copies of spectrin heterodimer, whereas erythrocytes from patients with hereditary spherocytosis are partially deficient in spectrin (range, 74 000–200 000 copies), the magnitude of the deficiency correlating with the severity of the disease. Spectrin deficiency probably represents the principal structural defect leading to the loss of surface membrane [2,58]. This causes loss of membrane fragments and formation of small spherical erythrocytes, called microspherocytes. The membrane defect and the high intracellular hemoglobin concentration (mean corpuscular hemoglobin concentration, MCHC = 360–370 g/l) combine to make the spherocyte poorly deformable. Because of the lack of flexibility, the spleen presents a serious hazard to the survival of the spherocyte. When these "predamaged" cells enter the spleen, they are retained in the splenic cord and, after they have passed through the spleen several times, will become more and more speroideal. As a result, these cells become less and less able to negotiate the sinusoidal slits and are finally phagocytized by spleen macrophages. Not surprisingly, these spherocytes are less resistant in the osmotic fragility test, in which the degree of hemolysis is estimated as a function of the concentration of saline. This test indirectly measures the surface to volume ration of the cells by determining how much water the cells can accommodate before they rupture. A spherocyte, with its low surface to volume ration, can accumulate less water than a biconcave disc. The concentration of NaCl that leads to 50% hemolysis is 4 g/l in normal erythrocytes and 4.5 g/l in erythrocytes from hereditary spherocytosis. This particular form of hemolytic anemia can be treated by splenectromy, which eradicates hemolysis and anemia, despite the fact that the basic membrane defect persists and microspherocytes are still present in the blood [100].

83.3.1.3 Antigens

Antigens are present on the external surface of the erythrocyte membrane. The antigens of different individuals differ in their antigenic epitopes; these differences are caused by minor structural differences between individuals in the carbohydrates and glycolipids of the membrane. Antigens arising from structural differences that are controlled by genes at the same locus (alleles) or closely linked loci are classified together as a blood group system. Over 100 different blood-group antigens have been identified up to now; for practical reasons, the most important one is the ABH (ABO) system, which will be discussed below (Fig. 83.4).

83.3.1.3.1 ABH Antigens

Erythrocyte Agglutination. At the turn of the century, Karl Landsteiner discovered the ABO blood-group system (Fig. 83.4), which is the most important one with respect to blood transfusion. Landsteiner cross-tested the erythrocytes and workers in his laboratory and found that in some instances agglutination occurred, whereas in others there was no reaction. On the basis of these tests, Landsteiner was able to divide individuals into three groups and, with the discovery of a fourth group, the foundations were laid for what is known as the ABO blood-group system [75]. Each individual either lacked or had one or both of the two antigens, A and B, on his or her erythrocytes. In addition, the serum of each individual contained, naturally occurring, directly agglutinating antibodies that recognized the

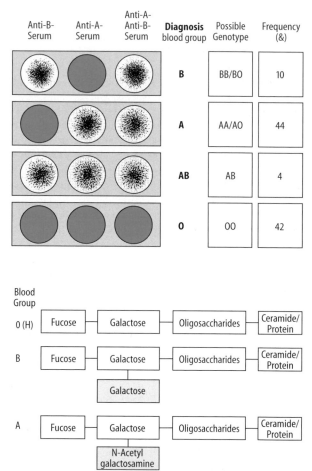

Fig. 83.4. *Top*, human ABO(H) blood-group system. The antigen on the erythrocyte surface is recognized by specific antibodies, leading to an agglutination of the erythrocytes. The respective plasma antibodies are: anti-A in blood group B, anti-B in blood group A, anti-AB in blood group O(H), and no anti-A or -B in blood group AB. The frequency numbers are for Caucasian populations, *Bottom*, glycolipid structure of the respective ABO(H) antigens. The antigenic terminal carbohydrates are *shaded*

antigens absent from their own erythrocytes. This experiment can be explained in modern terms as follows: cross-reacting glycolipid structures on environmental agents stimulate the thymus-independent production of immunoglobulin M (IgM) anti-A and/or anti-B in individuals not tolerant to these antigens. The IgM antibodies then directly agglutinate the appropriate antigen-positive erythrocytes [99,100].

Although the ABH antigens (the H antigen is the relevant carbohydrate structure present on group O erythrocytes) are typically described as blood-group antigens because of their presence on erythrocytes, they are also found in other tissues. They exist as membrane antigens on such diverse cells as vascular endothelial cells and intestinal, cervical, urothelial, and mammary epithelial cells. Despite their wide distribution, genetic inheritance, developmental regulation, and importance in transfusion and transplantation, their normal physiologic function, if any, remains a mystery.

A, B, and H Enzymes. Studies have indicated that anti-A, anti-B, and anti-H antibodies specifically recognize epitopes composed of terminal tri- or diantennary oligosaccharides, which led to the conclusion that the A, B, and H antigens are not directly encoded by the corresponding genes, but rather by the genes that code for particular glycosyl transferases, commonly called the A, B, and H transferases (A, B, H enzymes). The H enzyme is a fucosyltransferase that specifically adds fucose to a terminal galactose residue. Subsequently, the A or B enzymes add *N*-acetylgalactosamine or galactose, respectively, to the same terminal galactose, but only in the presence of α_{1-2}-linked fucose (Fig. 83.4). This means that the substrate for the A and B enzymes is H antigen with the following biosynthetic pathway: Gal-R→H→A or B, whereby the first reaction is catalyzed by the H enzyme and the second reaction by the A or B enzyme. The A and B genes are inherited in a strict mendelian fashion and are dominant compared to O, but the A and B genes are codominant with each other. In other words, an individual with the genotype AO (or BO) is phenotypically A (or B), an individual of genotype AB is also phenotypically AB, and an individual of genotype OO is phenotypically group O [25].

83.3.1.3.2 Maternal–Fetal ABO Group Incompatibility

Hemolytic Disease. Maternal–fetal ABO group incompatibility is now the most common cause of hemolytic disease in newborns. Most episodes of hemolytic disease due to ABO incompatibility are asymptomatic or mild, and only 10% of severe hemolytic disease is due to ABO incompatibility. It is almost always seen in infants born to type-O mothers, since naturally occurring IgG anti-A and anti-B antibodies occur in the sera of group O individuals. Since the offending antibodies usually occur naturally, no primary immunization is necessary, and hemolytic disease in

newborns may occur with the first pregnancy. Unlike Rhesus D [Rh(D)] hemolytic disease, the development of ABO-incompatible hemolytic disease in newborns is of no prognostic value in predicting the incidence of severity of ABO-incompatible hemolytic disease with subsequent pregnancies. It is worth mentioning in this context that ABO antigens have an important effect on Rh(D) hemolytic disease in newborns. In pregnancy with an ABO-compatible, Rh(D)-positive fetus, the frequency of immunization is about 15% per pregnancy. However, in pregnancy with an ABO-incompatible, Rh(D)-positive fetus, the frequency of immunization is 2%–3% per pregnancy. This protective effect of ABO incompatibility occurs because ABO-incompatible fetal erythroucytes entering the maternal circulation are rapidly destroyed intravasularly by complement-fixing anti-A or anti-B alloantibodies before they reach the monocyte–macrophage system, where primary immunization could occur [114].

83.3.1.3.3 Rhesus D Antigen System

Although the clinical importance of the Rh blood-group system has long been recognized, the antigens associated with the Rh system have proved remarkably difficult to characterize in chemical terms. Rh antigens have never been unequivocally demonstrated in free solution. Substances with Rh specificity are not detectable to any appreciable extent in soluble form in human tissue fluids, and attempts to completely solubilize erythrocyte membranes with organic solvents or detergents have not yielded products with Rh activity [99,100]. From this result the conclusion was drawn that the D antigens apparently reside on one or more unusual membrane proteins that need to be surrounded by phospholipid molecules in order to maintain their antigenicity. It is perhaps because of this phenomenon and the fact that the antigens are present in quite small amounts (20 000–30 000 sites/erythrocyte) that it has been difficult to isolate and characterize the antigenic molecule.

The Rh type with what is commonly known as the D antigen has a frequency of 85% in Caucasians (Rh positive) and is absent in 15% (Rh negative). The D antigen is the most immunogenic erythrocyte alloantigen (the ABH antigens are isoantigens). The clinical importance of the Rh system is due to the fact that antibodies to these antigens can cause erythrocyte destruction. With regard to the major Rh antigens, after transfusion of antigen-positive blood to an antigen-negative recipient, the recipient may develop IgG antibodies. These antibodies may result in extravascular (liver, spleen) hemolysis and a delayed hemolytic transfusion on subsequent transfusion of antigen-positive blood. Similarly, since IgG antibodies cross the placenta if a D antigen-negative woman is sensitized to Rh antigens during pregnancy, then during a subsequent pregnancy an antigen-positive fetus may develop hemolytic disease [30,114].

Small amounts of fetal erythrocytes (0.1 ml) may cross the placenta during pregnancy, but such quantities are usually not enough to immunize the mother. Hemolytic disease therefore does not usually occur with the first pregnancy. However, significant transplacental hemorrhage may occur during labor and delivery so that immunization can occur, and subsequent pregnancies with Rh(D)-positive fetuses pose a risk.

Hemolysis occurs in the fetus extravascularly in the spleen and liver, resulting in anemia and the production of unconjugated bilirubin. Fetal unconjugated bilirubin is cleared by the placenta and metabolized in maternal liver. When hemolysis is severe, normoblasts proliferate in the liver and spleen, and normoblasts appear in the blood (fetal erythroblastosis). Fetal hydrops is due to severe anemia in the fetus. It is characterized by the following:

- Pallor
- Extreme hepatosplenomegaly
- Edema
- Ascites
- Hepatic dysfunction.

It occurs in infants that are stillborn or likely to die. In infants born with significant hemolytic disease, extremely high levels of unconjugated bilirubin may accumulate; these are associated with neurological damage due to unconjugated bilirubin being deposited in the basal ganglia (kernicterus). Hemolytic disease in newborns leading to fetal hydrops and death used to the fairly common. It is now relatively rare because to the advent of Rh immunoglobulin, which, when given intramuscularly soon after birth to the D-negative mother of a D-positive child, is able to prevent maternal sensitization to the D antigen [30,114].

83.3.2 Pathophysiology of Anemia

Hemoglobin Concentration, Hematocrit, and Erythrocyte Count. In normal subjects, the concentration of hemoglobin, the volume of erythrocytes per unit volume of blood (hemotocrit), and the number of erythrocytes per unit volume of blood are highly regulated. Normal values for adult human beings are given in Table 83.1. A decrease below two standard deviations of these levels is termed "anemia," while an increase is termed "erythrocytosis." Parameters further characterizing the erythrocyte population may be calculated from these measurements. Thus, the hematocrit divided by the erythrocyte count yields the mean corpuscular volume (MCV); the hemoglobin concentration divided by the erythrocyte count yields the mean corpuscular hemoglobin content (MCH); and the hemoglobin concentration divided by the hematocrit yields the MCHC. These derived values are known as "erythrocyte indices." Normal values for these indices also are given in Table 83.1.

Pathophysiologic Effects of Anemia. The primary function of the erythrocyte is O_2 transport. When the hemoglobin concentration is reduced, there is a proportional reduction in the O_2-carrying capacity, which necessitates appropriate compensation to maintain O_2 delivery. This compensation includes an increase in cardiac output, redistribution of the cardiac output such that blood is preferentially diverted from certain organs to others, and a shift of the hemoglobin-O_2 dissociation curve to the right (see Chap. 108). The normal marrow exhibits a predictable response to anemia. This response may be observed, for example, after an otherwise normal subject experiences a sudden loss of blood, which is replaced by an equal volume of normal plasma. Anemia causes a drop in O_2 transport to the kidney and a resulting rise in erythropoietin. This in turn leads to a release of reticulocytes into the peripheral blood within 6–12 h (see Chap. 84). While this reticulocyte influx does not raise the hematocrit significantly, it does signify that the erythropoietin level is elevated. The normal fraction of reticulocytes in 0.5%–1.0% of all erythrocytes; they are increased in blood loss and hemolytic anemias and decreased in hypoproliferative anemias.

Anemias may be classified morphologically by the appearance of erythrocytes or kinetically by rates of production and destruction. In the following the kinetic approach will be briefly discussed, because it better illustrates the underlying pathophysiology of anemia. There are four pathophysiologic mechanisms that lead to anemia: blood loss, hypoproliferation (reduced numbers of erythroid precursors in the bone marrow), hemolysis, and ineffective erythropoiesis [114].

Blood Loss Anemia. Blood loss anemia is due to external bleeding or hemorrhage within the body. When blood loss is acute, the blood count does not fully reflect the loss of erythrocytes for 24–72 h, the time required for reexpansion of the remaining blood volume by body water and plasma proteins. Seven to 10 days are required for the production of erythrocytes to reach the level dictated by the severity of the anemia.

Hypoproliferative Anemia. Hypoproliferative anemia refers to a group of conditions in which the anemia is due primarily to a lower rate of production of erythrocytes by marrow than is expected for the degree of anemia. Thus, production parameters, while not necessarily below normal, are below the values shown for blood loss. The major types of hypoproliferative anemia are iron deficiency, vitamin B_{12} and folic acid deficiency, and anemia in chronic renal failure due to an insufficient production of erythropoietin [30,114]. Erythrocytes produced in the presence of low iron levels are smaller in volume (MCV, $\geqslant 80$ fl), and their hemoglobin content per cell (MCH) and per unit cell volume (MCHC) is reduced. Such cells, termed "microcytic" and "hypochromic," are easily recognized under the microscope. In contrast, in vitamin B_{12} or folic acid deficiency, the erythrocytes are larger than normal (MCV, $\geqslant 100$ fl) and have a high mean cellular hemoglobin content (MCH, $\geqslant 40$ pg), while MCHC is generally normal. A particular type of hypoproliferative anemia that is due to renal failure can now be treated successfully with recombinant erythropoietin [30].

Hemolytic Anemia. Hemolytic anemia is due to increased destruction of circulating erythrocytes, so that the cells survive for periods significantly shorter than the normal life span of 120 days. The erythroid marrow attempts to compensate by accelerated production. If the rate of destruction is not greater than about five times the normal rate, and if there is no superimposed disease process, the marrow will be able to increase the rate of production to match the destruction rate (compensated hemolysis). In fact, marked hemolytic anemia is only seen when the survival time of the erythrocytes has fallen below one eighth of its normal value; this is due to the high capacity of the bone marrow to increase its erythroid production rate.

Ineffective Erythropoiesis. This is a group of disorders in which the major cause of anemia is the destruction of developing erythrocytes within the marrow or immediately after they are released to the circulation. This results in an increase in the production of cells by the marrow, but a decrease in circulating erythrocytes. The most common types in this category are folic acid and B_{12} deficiency anemias and a reduced synthesis of either the α- or the β-subunits of hemoglobin (thalassemia) [4,43,73,84].

83.4 Coagulation and Fibrinolysis

Blood circulates in a specialized system of vessels. Leaks can occur in this system with subsequent blood loss. On the other hand, clot formation may lead to obstruction of vessels, which results in cessation of blood flow. Since the circulation of blood is essential for life, the integrity of this process must be maintained. Fibrinolysis and endogenous factors that prevent clotting are responsible for keeping the vessels open, whereas hemostasis prevents blood loss through severed or ruptured vessels. The term "hemostasis" covers several different mechanisms:

- Vascular constriction
- Formation of thrombocyte plugs
- Blood coagulation
- Subsequent growth of fibrous tissue into the blood clot to close the hole in the vessel permanently.

Formation of the thrombocyte plug and blood coagulation are discussed in more detail below. Because many of the coagulation proteins are potent activators of thrombocytes, plasma coagulation is discussed first, followed by the various aspects of thrombocyte function in more detail.

83.4.1 Vascular Constriction

Thromboxane A_2. Immediately after a blood vessel is cut or reptured, the wall of the vessel contracts (see Chaps. 44, 96, 98). This results in a reduced flow of blood from the vessel rupture. The contraction is caused by various stimuli, including nervous reflexes, local myogenic spasm, and local humoral factors from the traumatized tissues and blood thrombocytes. Much of the vasoconstriction probably results from local myogenic contraction of the blood vessels initiated by direct damage to the vascular wall. For the smaller vessels, thrombocytes are responsible for much of the vasoconstriction by releasing the substance thromboxane A_2 (see Chap. 6), which is a strong vasoconstrictor, [83]. The more vessels are traumatized, the greater the degree of spasm. This means that a sharply cut blood vessel usually bleeds much more than does a vessel ruptured by crushing. The local vascular spasm can last for many minutes, during which time the ensuing processes of thrombocyte plugging and blood coagulation can take place.

83.4.2 Fibrous Organization of Blood Clots

Fibrinolysis and Fibroblasts. Once a blood clot has formed, it can either be dissolved by fibrinolysis or it can become invaded by fibroblasts [7]. Activated fibroblasts synthesize collagen and subsequently form connective tissue throughout the clot. Furthermore, neutrophils and monocytes occur at the sites of injury due to released chemotactic factors and lead to tissue repair. Invasion by fibroblasts begins usually within a few hours after clot formation and continues to complete organization of the clot into fibrous tissue within approximately 1–2 weeks. This process is at least partially promoted by the growth factors released by blood thrombocytes, such as thrombocyte-derived growth factor [71].

83.5 Blood Coagulation

Procoagulants and Anticoagulants. Blood coagulation is a host defense mechanism that, in parallel with the inflammatory and tissue repair responses, helps to protect the integrity of the vascular system after tissue injury. Plasma coagulation proteins and cells (thrombocytes, endothelial cells, and leukocytes) are critical in this system. In this section, factors will be discussed which promote coagulation (procoagulants) and which inhibit coagulation (anticoagulants). Whether or not blood coagulates depends on the degree of balance between the procoagulatory active factors and the anticoagulatory factors. Normally the anticoagulants predominate, and the blood does not coagulate. However, when a vessel is injured, procoagulants in the area of damage become "activated" and blood coagulation occurs within seconds in order to stop the loss of blood as rapidly as possible. The response to vascular injury culminates in the generation of a fibrin clot, but also in the formation of a thrombocyte plug, the deposition of white blood cells in the area of tissue damage, and the initiation of inflammation and repair.

Table 83.4. Proteins involved in blood coagulation

Component	Synonyms	Relative molecular weight	Plasma concentration (μg/ml)
Fibrinogen	Factor I	340 000	3000
Prothrombin	Factor II	72 000	100
Factor V	Proaccelerin	330 000	10
Factor VII	Proconvertin	50 000	0.5
Factor VIII	Antihemophilic factor A	290 000	0.2
Factor IX	Antihemophilic factor B Christmas factor, plasma thromboplastin component (PTC)	56 000	5
Factor X	Stuart–Prower factor	58 500	10
Factor XI	Plasma thromboplastin antecedent Antihemophilic factor C	160 000	5
Factor XII	Hageman factor	80 000	30
Factor XIII	Fibrin-stabilizing factor	320 000	10
Prekallikrein	Fletcher factor	86 000	50
High molecular weight kininogen	Flaujeac, Fitzgerald Williams factor	110 000	70
Protein C		62 000	4
Protein S		69 000	20
Antithrombin III		58 000	140

83.5.1 Blood Coagulation Factors

Fibrinogen, Prothrombin, Tissue Factor, and Calcium Ions. At the beginning of this century, four coagulation factors were identified: fibrinogen (factor I), which is transformed to fibrin by thrombin, which is formed from prothrombin (factor II) by the action of tissue factor (factor III) and calcium ions (factor IV). During the ensuing decades this simple scheme had to be modified and extended (Table 83.4) [41]. Most of the coagulation factors were initially identified in patients with bleeding complications whose coagulation was corrected by a protein present in normal plasma.

Zymogens. The coagulation factors are plasma zymogens, the active enzymes of which are sequentially activated by limited proteolytic cleavage. With the exception of factor XIII, all active coagulation factors are glycoproteins and serine proteases. Several factors of the coagulation system express no enzymatic activity, but act as cofactors. These cofactors (include phospholipids, Ca^{2+} ions, factor V, and factor VIII. The latter two are precofactors) which become fully active cofactors by limited proteolysis. The coagulation factors are mostly synthesized in the liver. A few of them are also produced by monocytes and endothelial cells (e.g., factor V and tissue factor) [7,8]. In addition, thrombocytes contain almost all coagulation factors [82], which can be secreted upon thrombocyte activation where necessary. The synthesis of the factors II, VII, IX, and X is dependent on vitamin K [95]. Post-translational carboxylation of glutamic residues results in γ-carboxyglutamate (gla) residues, which are located in the so-called gla domain of these proteins. The carboxylase enzyme responsible for this modification needs vitamin K as an essential cofactor.

Tissue Thromboplastin. We shall discuss the coagulation factors in the following groups: common pathway, extrinsic pathway, and the intrinsic pathway (Fig. 83.5). The terms "extrinsic" and "intrinsic" are historical: early investigators noted that the exposure of blood to damaged tissue hastened its clotting. The coagulation-initiating agent was present in the tissue (tissue thromboplastin, later called tissue factor) and was thus extrinsic to blood. In other early experiments, it was noted that blood clots when added to a container. Under these circumstances, all the factors required for coagulation were thought to be present in the blood. i.e., intrinsic. These historical terms are still commonly used, although they are not fully correct, e.g., activation of the intrinsic system needs also a component not present in the blood. For a detailed discussion of the coagulation factors see [41,55].

83.5.1.1 Common Pathway (Fig. 83.5)

83.5.1.1.1 Conversion of Fibrinogen to Fibrin

Factor XIII. Visible clot formation occurs as a result of the conversion of fibrinogen to fibrin with subsequent polymerization of the fibrin monomers and cross-linking of the resulting polymers. Fibrinogen [26] consists of two units with three polypeptide chains. There two tripeptide structures are linked by disulfide bridges. The fibrinopeptides A and B, representing the amino terminals of the α- and β-chains, are cleaved off by thrombin. The resulting fibrin monomers then polymerize into large fibers. Activated factor XIII is a transglutamidase and stabilizes the formed clot by fibrin cross-linking. Activation of factor XIII proceeds in several steps, which include cleavage by thrombin and the presence of calcium ions.

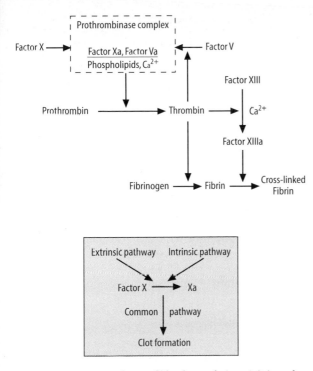

83.5.1.1.2 Conversion of Prothrombin to Thrombin

Factor Xa. Prothrombin binds via its gla domain to its activation complex on a membrane or phospholipid surface at the site of injury [64]. This binding is dependent on calcium ions. Prothrombin is activated to thrombin by factor Xa, which is the enzymatically active part of the prothrombin activation complex. The resulting thrombin lacks the gla domain responsible for binding and therefore dissociates from the phospholipid or membrane surface into the solution, where it induces several different biological functions (Table 83.5) [18].

83.5.1.1.3 Assembly of the Prothrombin Activation Complex

Factors V and X. Prothrombin is converted to thrombin by an enzyme complex called prothrombinase [65], which consists of factor Xa, factor Va, phospholipids, and Ca^{2+} ions. Factor X [35] contains a gla domain. It can be activated either by factor IXa from the intrinsic pathway or by factor VIIa from the extrinsic pathway. Upon activation, the gla domain is unmodified and thus factor Xa remains associated with the phospholipid surface. Factor V is an inactive precursor which is cleaved by thrombin and thereby activated to an active cofactor [20]. Negatively charged phospholipids [65] such as phosphatidylserine and phosphatidylinositol on the surface of activated thrombocytes or damaged cell membranes function to assaemble factor Xa, factor Va, and prothrombin in proximity to each other. Normal prothrombinase function

Fig. 83.5. Common pathway of blood coagulation. Ca^{2+} ions-dependent assembly of factor Xa and factor Va on a phospholipid surface results in the formation of the prothrombinase complex. This enzyme complex activates prothrombin to thrombin. Thrombin accelerates its own formation by activating factor V, leads to fibrin formation, and activates factor XIII, the enzyme responsible for fibrin cross-linking. The *inset* shows the central role of factor X, which can be activated either by the intrinsic or by the extrinsic pathway

Table 83.5. Hemostatic effects of thrombin

Site	Effects	
Coagulation system	Clot formation	Fibrinogen – Fibrin Factor XIII – XIIIa
	Amplification	Factor V – Va Factor VIII – VIIIa Factor XI – XIa
Thrombocytes		Aggregation Release reaction Thromboxane A_2 synthesis
Endothelial cells	Synthesis/release	Tissue factor t-PA Prostacyclin NO Endothelin
	Thrombomodulin	Protein C – activated protein C
Leukocytes		Chemotaxis Cytokin production
Fibroblasts		Proliferation
Smooth muscle cells		Contraction Mitogenesis

t-PA, tissue-type plasminogen activator; NO, nitric oxide

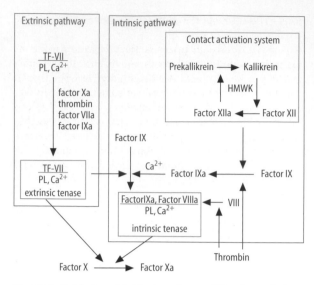

Fig. 83.6. Extrinsic and intrinsic pathways of blood coagulation. Factor Xa, the initial enzyme of the common pathway, is formed either by the extrinsic pathway or by the intrinsic pathway. Both activation pathways result in a Ca^{2+} ions-dependent assembly of enzyme complexes on a phospholipid surface (*PL*): the extrinsic tenase and the intrinsic tenase. The extrinsic pathway is triggered by the tissue factor (*TF*)–factor VI (VII) complex, which can be activated by various enzymes. The extrinsic pathway interacts with the intrinsic pathway by activating factor IX. The intrinsic pathway is initiated by a group of plasma proteins (the contact activation system) which become activated upon contact with negatively charged surfaces. Active factor XIa may result, which then leads to the formation of factor IXa, the enzymatically active component of the intrinsic tenase. Amplification of the intrinsic pathway results from thrombin-mediated activation of factor VIII and factor XI. *HMWK*, high molecular weight kallikrein

requires the interaction of the gla domains of these factors with the phospholipid surface, a process mediated by calcium ions.

83.5.1.2 Extrinsic Pathway

Tissue Factor and Factor VII. The extrinsic pathway [81] (Fig. 86.6) requires tissue factor, which is unique among the blood coagulation factors as it is an integral membrane glycoprotein. It is located in the tissue adventitia and comes in contact with blood after vascular injury. The extracellular domain is a high-affinity receptor for factor VII, which contains a gla domain. When vascular injury occurs, factor VII and tissue factor form a complex in the presence of Ca^{2+} ions. This facilitates the conversion of factor VII to a protease (factor VIIa). Thes tissue factor–factor VIIa complex then converts factor X to factor Xa and is therefore called the extrinsic tenase complex [65]. In addition, the tissue factor–factor VIIa complex activates factor IX [57], a protein of the intrinsic pathway. Factor Xa and factor VIIa both catalyze the activation of factor VII, so there is a potential mechanism for the acceleration of factor VII activation.

83.5.1.3 Intrinsic Pathway

Contact Activation Proteins. The intrinsic system [36] (Fig. 83.6) involves factor VIII, factor IX, and the contact activation proteins [51] factor XII, prekallikrein, high molecular weight kininogen, and factor XI. Factor XII and prekallikrein reciprocally activate each other upon contact of blood with negatively charged surfaces. In vitro, these surfaces can be glass, kaolin, celite, or dextran sulfate. The corresponding activating surfaces in vivo are not yet clearly identified, they may be thrombocyte membrane or collagen structures. Factor XIIa in turn activates factor XI. High molecular weight kininogen serves as a nonenzymatic cofactor in these reactions. Factor XIIa and activated prekallikrein (kallikrein) are linked to other blood systems by the activation of plasminogen, releasing the vasoactive peptide bradykinin from high molecular weight kininogen, aggregating neutrophils, and probably activating the first component of the complement system [51].

Factors IX and X. Factor XIa activates factor IX [59] in the presence of Ca^{2+} ions. The resulting factor IXa, which is phospholipid bound via its gla domain, forms a complex with activated factor VIII, with the phospholipids, and with Ca^{2+} ions. This complex activates factor X and is therefore called intrinsic tenase complex [65]. Factor VIII [104] circulates in plasma as an inactive cofactor noncovalently bound to von Willebrand factor. Initial activation of factor VIII occurs in the presence of factor Xa, phospholipids, and Ca^{2+} ions or by thrombin alone once it is generated from prothrombin.

83.5.2 Coagulation Inhibitors

Different plasma serine protease inhibitors play a role in regulating blood coagulation, including antithrombin III, tissue factor pathway inhibitor (TFPI), heparin cofactor II, α_2-macroglobulin, C1 inhibitor, and α_1-antitrypsin. The coagulation cascade is also retarded by the trapping of thrombin in the fibrin clot as well as by the removal and dilution of activated clotting factors by the circulating blood.

Antithrombin III. This serine protease inhibitor [41] is considered to be the main physiological inhibitor of thrombin and factor Xa, but it also neutralizes factors VIIa, IXa, XIa, and XIIa. The inhibition by antithrombin III is due to the formation of inactive one-to-one complexes of inhibitor and enzyme. The resulting complexes, e.g., thrombin–antithrombin III complexes, can be measured as a parameter of thrombin formation. Heparin or similar sulfated glycosaminoglycans enhance the reaction rate of enzyme–antithrombin III interaction approximately 1000-fold [50].

Tissue Factor Pathway Inhibitor. TFPI [9] is also called extrinsic pathway inhibitor or lipoprotein-associated inhibitor, because it circulates in plasma bound to the lipoproteins. Approximately 10% of the total blood TFPI is carried by thrombocytes, which release this TFPI following stimulation by thrombin. Thus, at the site of a wound, where thrombocytes aggregate, relatively high local concentrations of this inhibitor can be reached. TFPI inhibits the extrinsic pathway by interacting specifically with the tissue factor–factor VIIa–factor Xa complex.

83.5.3 Protein C System

Thrombomodulin, Protein C, and Protein S. Together these three substances form an important endogenous anticoagulant system [31]. Thrombomodulin (relative molecular weight, 75000) is a single-chain transmembrane glycoprotein located on the surface of endothelial cells. Thrombomodulin forms a one-to-one complex with thrombin and changes the substrate specificity of this enzyme. Thrombin complexed with thrombomodulin loses its procoagulatory activity, while readily activating protein C, a vitamin-K dependent zymogen (Table 83.4). Cleavage by thrombin results in the generation of a serine protease. Activated protein C exerts its anticoagulant properties by proteolytically destroying both factor Va and factor VIIIa, a reaction that is greatly accelerated by phospholipids and Ca^{2+} ions. Moreover, another vitamin K-dependent protein, protein S, functions as a cofactor in the inactivation of factor Va and factor VIIIa.

83.5.4 Coagulation Initiation and Regulation

83.5.4.1 Initiation

Intrinsic and Extrinsic Tenase Complexes. The classical common pathway of coagulation which leads to fibrin formation can be activated either by the extrinsic or by the intrinsic pathway [22,36]. The activation of these pathways results in the formation of enzyme complexes capable of activating factor X, called the intrinsic and extrinsic tenase complex, respectively.

In 1964, the so-called waterfall [21] or cascade [62] hypothesis of blood coagulation suggested initiation of coagulation by the intrinsic pathway. However, due to the realization that patients lacking the contact system proteins factor XII, prekallikrein, and high molecular weight kiniogen experience no bleeding disorders [91], and due to the observation that the tissue factor–factor VIIa complex activates not only factor X, but also factor IX, a revised model of blood coagulation was proposed [37]. This model suggests that the extrinsic pathway is critical in initiating blood coagulation, while the intrinsic pathway may play an important role in the maintenance of coagulation. However, initiation of coagulation of plasma in vitro in a glass tube is clearly dependent on the intrinsic pathway. Furthermore, initiation of coagulation by the intrinsic pathway could be of (patho)physiological importance when circulating blood comes in contact with an artificial surface, such as heart valve, during extracorporal circulation or during blood dialysis.

Normal coagulation is presumably initiated by the extrinsic pathway when factor VII or factor VIIa in plasma binds to tissue factor [9]. Tissue factor produced by cells beneath the endothelium can be exposed to blood by injury of the vessel. Furthermore, endothelial cells and monocytes can be stimulated to express tissue factor by endotoxin, interleukin-1, or tumor necrosis factor [65]. The tissue factor–factor VIIa complex activates factor X to form factor Xa and factor IX to form factor IXa with subsequent thrombin formation. The tenase activity of the tissue factor–factor VIIa complex is then inhibited by the tissue factor pathway inhibitor once small amounts of factor Xa are generated. However, the initial burst of factor Xa generation provides sufficient thrombin to induce local aggregation of thrombocytes and the activation of the cofactors factor V (common pathway) and factor VIII (intrinsic pathway). Thus the intrinsic tenase complex with factor IXa and factor VIIIa acts to maintain blood coagulation. Furthermore, thrombin was recently found to be able to activate factor Xi [37,79]. If this reaction occurred in vivo [10], the resulting factor XIa would in addition give rise to more factor IX activation. This revised model of blood coagulation explains the lack of bleeding disorders in individuals deficient in one of the contact system proteins and the variable bleeding tendency observed in patients with factor XI deficiency [86].

83.5.4.2 Regulation

Blood coagulation should not be considered as a cascade of reactions which inevitably results in clot formation. Rather, blood coagulation is a carefully regulated process with positive and negative feedback mechanisms designed to lead to immediate clot formation at sites where this is required and, just as importantly, to limit coagulation to these sites and to avoid dissemination of the coagulation process into the body, which would result in fatal thromboembolic complications. The regulation of blood coagulation includes the following mechanisms:

- Serine protease inhibitors (antithrombin III)
- The protein C system activated by thrombin
- The anticoagulatory properties of intact endothelial cells
- The fibrinolytic enzymes.

Furthermore, many coagulation factors form complexes on membrane surfaces. Suitable membrane surfaces are located only in the region of vessel injury.

83.5.5 Role of Endothelial Cells

The vascular endothelium is located at the interface between the vascular tissue and the circulating blood and occupies a very large surface area of about 500 m^2.

Endothelial cells are important regulators of procoagulant and anticoagulant intravascular processes [29]. The fact that in vivo blood clotting does not occur within an unaltered vessel suggests that, under physiological conditions, the anticoagulant properties of endothelium dominate in this delicate balance.

Procoagulant Properties. Endothelial cells can be stimulated to produce tissue factor by thrombocyte-derived growth factor, thrombin, bacterial lipopolysaccharide, interleukin-1, and tumor necrosis factor [65]. Endothelial cells are able to synthesize von Willebrand factor and factor V and to assemble the prothrombinase complex by binding factor X.

Anticoagulant Properties. Among the anticoagulant mechanisms on the surface of endothelial cells are heparin-like glycosaminoglycans [66] that bind antithrombin III and thus inactivate thrombin or prevent its formation. Endothelial cells constitutively express the thrombin receptor thrombomodulin [31], which binds thrombin, resulting in activation of the anticoagulant protein C. Fibrinolytic proteins can assemble on endothelial cells [97], and they synthesize and release tissue plasminogen activator, which initiates fibrinolysis. Furthermore, they synthesize and release prostacyclin [11], which acts as a vasodilator and inhibitor of thrombocyte aggregation.

83.5.6 Role of Blood Cells

Blood cells are as essential as plasma proteins for the normal regulation of hemostasis. Blood thrombocytes contain various coagulation factors and coagulation-inhibitory proteins [82]. Polymorphonuclear leukocytes and monocytes [28] bind to the area of vascular injury on activated thrombocytes and endothelial cells by a mechanism involving the leukocyte receptor P selectin [56], and both cell types contribute to the process of coagulation and wound healing. Activated monocytes express tissue factor, synthesize factor V, and can also provide a phospholipid surface for the assembly of clotting factors, e.g., the prothrombinase complex.

83.6 Fibrinolysis

Blood coagulation and fibrinolysis are delicately coordinated mechanisms in the physiologic state. The fibrinolytic system is primarily responsible for the removal of fibrin from blood vessels by generating plasmin, a serine protease. Plasmin formation in plasma depends on the activation of its zymogen precursor, plasminogen, by specific plasminogen activators. The fibrinolytic enzymes also play an important role in extracellular matrix turnover [108].

83.6.1 Plasminogen and Plasmin

Plasminogen [88] is synthesized in the liver and found in plasma in a concentration of 200 mg/l. It has a high affinity for the fibrin clot. Thus when a clot is formed, a large amount of plasminogen is trapped in the clot. Plasmin is a proteolytic enzyme that resembles trypsin, the most important digestive enzyme of pancreatic secretion. Plasmin digests the fibrin threads as well as other substances in the surrounding blood, such as fibrinogen and many other components of the blood coagulation system. Therefore, whenever plasmin is formed in a blood clot, it can cause lysis of the clot and also destruction of clotting factors.

83.6.2 Plasminogen Activators

Tissue-Type and Urinary-Type Plasminogen Activator. Plasmin can be generated by two physiologic plasminogen activators (Fig. 83.7) which are present in human plasma: tissue-type plasminogen activator (t-PA), first isolated from tissue [87], and urinary-type plasminogen activator (u-PA), first isolated from urine [113]. Fibrin-bound plasmin converts the zymogen u-PA to active u-PA (urokinase). u-PA can also be activated by kallikrein or by factor XIIa. Thus, these coagulation proteins from the intrinsic pathway may also act as profibrinolytic agents [53]. Local release from endothelial cells generates active t-PA. Fibrinolysis is normally initiated when t-PA and plasminogen bind to fibrin. t-PA preferentially activates fibrin-bound rather than free plasminogen, so that plasmin is generated primarily within the fibrin clot and not in the

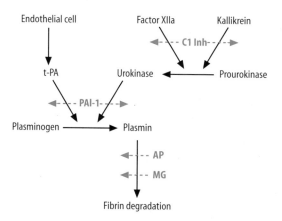

Fig. 83.7. The fibrinolytic system of human plasma. Plasmin, the enzyme responsible for fibrin degradation, results from activation of plasminogen. Two physiologic plasminogen activators are known: the tissue-type plasminogen activator (*t-PA*), released from endothelial cells, and the urinary-type plasminogen activator (prourokinase). The latter is a proenzyme and becomes activated either by plasmin or by the two coagulation factors kallikrein and factor Ix. The fibrinolytic system is controlled by several inhibitors, including plasminogen activator inhibitor type 1 (*PAI-1*), C1 inhibitor (*C1Inh*), α_2-antiplasmin (*AP*), and α_2-macroglobulin (*MG*)

systemic circulation. Plasminogen and its activators were found to assemble on endothelial cells [97]. The resulting plasmin formation contributes to the nonthrombogenicity of the blood vessel wall.

83.6.3 Inhibition of Fibrinolysis

Plasminogen Activator Inhibitor. Inhibitors [88,103] modulate the activation of the fibrinolytic system. t-PA and u-PA are inhibited by plasminogen activator inhibitor type 1 (PAI-1). PAI-1 is released from endothelial cells and can be secreted by thrombocytes. Other inhibitors of the fibrinolytic system include α_2-antiplasmin, a powerful inhibitor of plasmin, and C1 inhibitor, which inactivates plasmin, kallikrein, and factor XIIa.

83.7 Pathophysiology of Coagulation

Acquired or inherited diseases may influence one or more of the numerous steps in blood coagulation or fibrinolysis and may disturb the delicate balance between procoagulant and anticoagulant stimuli. This may result in patients suffering from bleeding disorders (hemophilia) or thromboembolic diseases (thrombophilia).

83.7.1 Bleeding Disorders

Von Willebrand Factor and Fibrinolytic Activity. Inherited deficiencies of almost all coagulation factors have been reported, associated with various bleeding complications. The best known among them are deficiencies of factor VIII (hemophilia A) and factor XI (hemophilia B). Since von Willebrand factor is the carrier protein for factor VIII in plasma, its deficiency can mimic hemophilia A. Almost all blood coagulation factors are synthesized by the liver. Therefore, diseases of the liver such as hepatitis or cirrhosis can depress the clotting system to such an extent that the patients develops a severe tendency to bleed. Another cause of depressed formation of clotting factors is vitamin K deficiency. Increased fibrinolytic activity [34] can also cause hemorrhagic disease. Acquired disorders associated with increased fibrinolystic activity and bleeding include liver cirrhosis, amyloidosis, leukemia, and some solid tumors. Congenital disorders of fibrinolysis [34] which cause bleeding include increased activity of plasma plasminogen activator and deficiency of α_2-antiplasmin.

83.7.1.1 Hemophilia

Hemophilia A (lack of functional factor VIII) and hemophilia B (lack of factor IX) are clinically indistinguishable [47]. The clinical signs are joint and muscle hemorrhages, easy bruising, and prolonged and potentially fatal hemorrhage after trauma or surgery. The bleeding tendency can have various degrees of severity, depending on the residual activity of the respective coagulation factor. Hemophilia A and hemophilia B are both X-linked recessive disorders, affecting approximately one in 5000 men and one in 30 000 men, respectively. The gene that codes for factor VIII comprises nearly 186 000 base pairs and constitutes about 0.1% of the X chromosome [112]. The DNA encoding the mature factor VIII mRNA is found in 26 separate exons. Different gene deletions and deletion mutations have been characterized at the molecular level in hemophilia A. In approximately one third of patients, the disease is the result of recent genetic mutation. The hemophilic patients are generally treated with infusions of purified factor VIII of human origin. There are several dangers associated with this replacement therapy, including transmission of hepatitis or acquired immunodeficiency syndrome (AIDS) and development of autoantibodies against factor VIII. The production of recombinant factor VIII has resulted in new factor VIII products that minimize the risk of transmission of viral infection.

83.7.1.2 Vitamin K Deficiency

γ-Carboxyglutamate Domain. Vitamin K [110] is necessary for the post-translational formation of the gla residues of prothrombin, factor VII, factor IX, and factor X. When produced in the absence of vitamin K, these factors lack the functionally important gla domain, which leads to serious bleeding complications. Vitamin K is normally synthesized by intestinal bacteria. Because it is fat soluble, vitamin K is normally absorbed into the blood along with fats. Adequate fat digestion and adsorption is dependent on bile, which is secreted by the liver into the gastrointestinal tract. Thus, vitamin K deficiency can occur as a result of liver diseases with diminished secretion of bile or as a result of disorders leading to obstruction of the bile ducts.

83.7.2 Thromboembolic Diseases

Thrombi and Emboli. Normally blood coagulation is initiated and clots are formed in order to avoid blood loss at sites of tissue injury. However, there are diverse clinical conditions leading to the initiation of blood coagulation without any requirement to stop blood loss. Such hypercoagulable states are characterized by an increased risk for pathological clot formation and subsequent thromboembolic diseases. A clot that occurs in a blood vessel is also called a thrombus. A clot, or part of it, that breaks away from its attachment and flows freely in the blood stream is called an embolus. Emboli originating in the venous system and in the right side of the heart flow into the vessels of the lung, causing pulmonary arterial embolism. Emboli originating in large arteries or in the left side of the heart may block either smaller systemic arteries or arterioles either in the brain, the kidneys, or elsewhere.

Acquired and Inherited Hypercoagulable States. Hypercoagulable states [15,78] include various acquired as well as inherited clinical disorders. Acquired hypercoagulable states may be seen in many heterogeneous conditions, including the following:

- Cancer
- Diabetes mellitus
- Obesity
- Pregnancy
- Postoperative states.

In some of these, thrombocyte activation occurs or endothelial cell activation by cytokines leads to the loss of normal endothelial cell anticoagulant functions. Inherited hypercoagulable states include deficiencies of coagulation inhibitors (antithrombin III), of fibrinolytic proteins, and of the proteins from the protein C system. They should be suspected in patients with recurrent, familial, or juvenile deep-vein thrombosis. In many patients, the initial thromboembolic event occurs together with an acquired hypercoagulable state (e.g., pregnancy, postpartum, trauma, surgery, immobilization, or oral contraceptives). Deficiencies of these proteins are usually inherited as an autosomal dominant trait. Affected individuals are heterozygotes who have about 50% of the protein activity present in normal plasma.

Resistance to Activated Protein C. In 20%–40% of patients with otherwise unexplained venous thromboembolism, resistance to activated protein C has been found [103]. Normal plasma shows a prolonged clotting time after the addition of activated protein C, due to cleavage and inactivation of factor Va and factor VIIIa. However, the plasma of these patients shows less prolongation of the clotting time after the addition of activated protein C. The underlying molecular mechanism of this resistance to activated protein C was found to be a mutation in factor V, which destroys the inactivation site in factor Va [5]. The impaired inactivation of the coagulation factor Va with subsequent hypercoagulability illustrates the delicate balance between procoagulants and anticoagulants.

83.8 Clinical Use of Anticoagulants

Many arterial and venous thromboembolic conditions as well as some hypercoagulable states require therapeutic or prophylactic intervention to delay the coagulation process or to dissolve thrombi. Various anticoagulants and thrombolytic agents have been developed for clinical use.

83.8.1 Heparin

Standard heparin [44], extracted from animal tissues, belongs to the glycosaminoglycan family. Commercial preparations are heterogeneous mixtures ranging in relative molecular weight from 5000 to 30 000. The so-called low molecular weight heparin [46] consists of chemically or enzymatically digested heparins with relative molecular weights that vary from 4000 to 6500. They have different pharmacokinetic and pharmacodynamic properties, which may be advantageous over standard heparin in some clinical conditions. Heparin binds to antithrombin III, thereby converting it from a slow inhibitor to a very rapid inhibitor of thrombin and other coagulation factors. Heparin is effective in the prevention and treatment of venous thrombosis and pulmonary embolism, in the prevention of mural thrombosis after myocardial infarction, in the treatment of patients with unstable angina and acute myocardial infarction, and in the prevention of coronary artery rethrombosis after thrombolysis.

83.8.2 Vitamin K Antagonists

Warfarin, Phenprocoumon, and Acenocoumarol. Vitamin K is an essential cofactor for the post-translational carboxylation of glutamate residues to γ-carboxyglutamates in the coagulation factors II, VII, IX, and X, as well as in protein C and protein S. The carboxylated coagulation proteins are capable of binding to negatively charged phospholipids via calcium ion bridges. This interaction is essential for their effective participation in the coagulation process. Vitamin K antagonists [45], such as warfarin, phenprocoumon, and acenocoumarol cause an anticoagulant effect by competing with vitamin K for reactive sites in the enzymatic process of γ-carboxylation. Because these drugs can be administered to patients orally, they are also called oral anticoagulants. The anticoagulant effect is delayed until the clotting factors already circulating are cleared from the circulation. An anticoagulant effect occurs within 24 h because of the inhibition of the production of normal factor VII, which has a half-life of about 7 h However, peak anticoagulant activity is delayed for 72–96 h because of the longer plasma half-lives of factors II, IX, and X. Protein C, which is also a vitamin K-dependent protein, has a short half-life, like factor VII. Thus, the reduced protein C activity may result in a transient hypercoagulable state during initiation of oral anticoagulant therapy. Vitamin K antagonists are effective in the prevention of venous thromboembolism, in the prevention of systemic embolism in patients with prosthetic heart valves or atrial fibrillation, and in the prevention of stroke, recurrent infarction, and death in patients with acute myocardial infarction.

83.8.3 Hirudin

Medicinal Leeches. The medicinal leech has been known to prevent blood from clotting since ancient times. In the late 1950s, the anticoagulant agent hirudin [70] was isolated from the peripharyngeal glands of medicinal leeches, and several recombinant hirudin preparations are now available. Native hirudin is a polypeptide containing 65 amino acids with a molecular mass of about 7 kDa. Hirudin is the most potent and specific thrombin inhibitor known and is currently being investigated for clinical use. Possible future indications include prevention and treatment of venous thromboembolism, prevention of thrombosis after thrombolytic therapy or angioplasty, and prevention of thrombosis during hemodialysis or cardiopulmonary bypass.

83.8.4 Thrombolysis

Streptokinase. Several agents that induce thrombolysis are in clinical use [88]. The most widely used are recombinant tissue-type plasminogen activator (rt-PA) and streptokinase. Streptokinase is a polypeptide derived from β-hemolytic streptococcal cultures. When administered to patients, all the known thrombolytic agents demonstrate systemic effects of plasmin to a varying extent, including degradation of fibrinogen and other coagulation factors. This explains the increased risk of (fatal) bleeding complications associated with thrombolytic therapy.

Myocardial Infarction. Acute myocardial infarction is mostly due to the sudden obstruction of a coronary artery by the formation of a thrombus. Today, thrombolytic therapy must be considered for urgent coronary recanalization in all patients with myocardial infarction [3]. The reduction in mortality due to thrombolytic therapy was found to be in the order of 25%, saving about 30 lives per 1000 patients treated. The use of thrombolytic agents for indications including acute cerebral ischemia, deep venous thrombosis, pulmonary embolism, and peripheral arterial thrombosis is still under investigation and has not yet been definitively established [23,40,105].

83.9 Function of Thrombocytes in Hemostasis

Thrombocytopenia. Thrombocytes (Fig. 83.8) are non-nucleated cells, about 2–4 μm in diameter, which are normally present in the blood in a concentration of 150 000–400 000 per μl blood. When thrombocytes are activated by exposed capillary basement components, they adhere and plug a vessel damage. It is important to realize that even minimal stress to capillaries, e.g., the hydrostatic pressure exerted upon the capillaries of the lower legs when standing upright, can cause gaps to form between the endothelial cells of the capillary wall. If an individual has a low thrombocyte count (thrombocytopenia) or a metabolic thrombocyte malfunction, erythrocytes may leak from such gaps before they are plugged. Persistent bleeding from erosions of the mucosal surfaces of the mouth and nose and the gastrointestinal and genitourinary tracts is another frequent manifestation of bleeding due to thrombocytopenia or malfunction of thrombocytes. It can be concluded from these manifestations that thrombocytic hemostatic plugs are important in controlling bleeding from the capillaries and small vessels of such lesions.

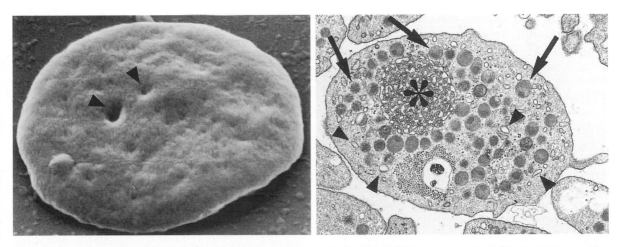

Fig. 83.8. Scanning (*left*) and transmission (*right*) electron microscopic appearance of a nonactivated human thrombocyte. The discoid thrombocyte has a smooth surface and contains numerous α-granules (*arrows*); the open canalicular system (*arrowheads*) which forms on the internal channel network and opens towards the surface of the thrombocyte shown. The dense tubular system (indicated by *asterisks*) is the other membrane system in the thrombocyte. It is thought to be equivalent to the rough endoplasmic reticulum that serves as a Ca^{2+} store and presumably represents the site of prostaglandin synthesis in thrombocytes. The discoid thrombocyte shape is maintained by microtubules that are located together with a subplasmalemmal actin membrane along the equator of the discoids (not shown). (Photographs kindly provided by Prof. P. Groscurth, Institute of Anatomy, University of Zurich)

83.9.1 Formation of Thrombocytes from Megakaryocytes

Thrombocytes are formed from large bone marrow cells called megakaryocytes. The immediate precursors of megakaryocytes are megakaryoblasts, which are usually diploid and can only be identified by the immunohistochemical demonstration of acetylcholinesterase, a marker that is thought to be specific for the megakaryocyte lineage [38]. As megakaryocytes proliferate, they enlarge and undergo extensive DNA replication without mitosis. The polyploid cells subsequently undergo endomitosis to form multilobed nuclei with four to 64 times the haploid amount of DNA [38], most commonly 16 or 32 N. Only after DNA replication has ceased do these cells begin to show cytoplasmic differentiation with production of components that are found in the mature thrombocyte.

Demarcation Membranes. With continued maturation, the megakaryocyte cytoplasm develops an extensive membrane system, of demarcation membranes, which are formed by invaginations of the plasma membrane. The mature megakaryocyte is located directly to adjacent bone marrow sinusoidal endothelial cells. As the extensive demarcation membrane forms, megakaryocytes develop long filopodia that directly penetrate the endothelial cytoplasm and extend into the marrow capillaries. These projections then fragment to produce mature thrombocytes. About 1000 thrombocytes are ultimately produced from the cytoplasm of a single mature megakaryocyte (reviewed in [63]). The hemopoietic factor that stimulates proliferation and maturation of megakaryocytes in vitro and is thrombocytogenic in vivo, thrombopoietin, has been purified in laborious experiments (reviewed in [72]), but it was only recently that the cDNA of thrombopoietin was cloned and expressed. Because of its clinical and theoretical interest, this particular hemopoietic growth factor will be considered in more detail in the following paragraph.

83.9.1.1 Thrombopoietin

Thrombocytopenia. Thrombocytes are necessary for blood clotting, and when their numbers are very low, a patient is at risk for severe bleeding events. Certain diseases are primarily due to thrombocyte defects, but most patients with thrombocytopenia have received chemotherapy or bone marrow transplantation. For the reasons briefly alluded to above, the slow recovery of thrombocyte levels in such patients is a serious problem and was the main stimulus in the search for the hemopoietic growth factor that accelerates thrombocyte regeneration. How this factor (thrombopoietin), which stimulates the proliferation and maturation of megakaryocytes, was purified has been described by Donald Metcalf [74]. The series of discoveries that finally led to the cloning of complementary DNA for both human [24] and mouse [52,60,111]

thrombopoietin began with the identification of the protooncogene c-*Mpl*, an orphan member of a growing family of hemopoietic growth factor receptors [74,111]. Cell lines were engineered to express the c-*Mpl* receptor and provided the essential tool for monitoring and cleaning the c-*Mpl* ligand, e.g., by affinity columns [24], which turned out to specifically stimulate megakaryocytopoiesis and thrombopoiesis [24,52,60,111]. The protein that was isolated by this approach has a predicted relative molecular weight of 35 000 and a highly conserved N-terminal region that shares some homology with erythropoietin [60].

Thrombocyte Count. The most impressive feature of thrombopoietin is its ability to increase thrombocyte numbers to previously unattainable levels, and it will likely soon find clinical applications. The observed biological effects of thrombopoietin are in line with those of other hemopoietic growth factors, in that it stimulates both proliferation and maturation. Although, according to present data, the actions of thrombopoietin seem to be restricted to cells of the megakaryocytic lineage, it is reasonable to expect that they will be somewhat broader, because none of the 20 or more hemopoiectic regulators is absolutely restricted in its actions to cells of a single lineage [74].

83.9.2 Structure of Mature Thrombocytes

Open Canalicular and Dense Tubular Systems. On a blood smear stained with Wright's dye, thrombocytes are seen with a granule-rich cytoplasm. Electron microscopy reveals a complex ultrastructure [49]. Unstimulated thrombocytes circulate as flattened disks (Fig. 83.8), a structure that is maintained by strands of microtubules [49] which can be seen encircling the thrombocyte just below its surface membrane. Organelles are distributed throughout the thrombocyte; these include α-granules and lysosomal granules, very dense granules called dense bodies, occasional mitochondria, and fine granules of glycogen (Fig. 83.8). Furthermore, two internal membranous systems can be recognized. The *open canalicular system* is continuous with the plasma membrane and gives the thrombocyte the appearance of a sponge. As a result of this system these cells have an enormous surface area compared with a sphere of comparable size (i.e., $20\,\mu m^2$ for a sphere versus $150\,\mu m^2$ for a thrombocyte) as estimated from the density of cell surface lectin-binding sites [38,63]. This interconnected tube system provides multiple channels through which an activated thrombocyte can both take up external Ca^{2+} and secrete the contents of its granules. The second system is the dense tubular system, which is thought to be analogous to the sarcoplasmic reticulum of smooth muscle, which serves to pump and release Ca^{2+}. This system does not, however, communicate directly with the exterior of the cell like the open canalicular system (reviewed in [85]). The thrombocyte surface membrane consists of an outer coat that is rich in carbohydrates (glycocalix) and an underlying membrane that is made up

Table 83.6. Thrombocyte receptors for adhesive proteins

Ligand	Receptor(s)	Other designations
Collagen	GP Ia–IIa	$\alpha_2\beta_1$
	GP IIb–IIIa	$\alpha_p\beta_3$
Fibrinogen	GP IIb–IIIa	αIIbβ_3
Fibronectin	GP Ic–IIa	$\alpha_5\beta_1$
	GP IIb–IIIa	αIIbβ_3
Thrombospondin	Vitronectin receptor	$\alpha_v\beta_3$
Vitronectin	Vitronectin receptor	$\alpha_v\beta_3$
	GP IIb–IIIa	αIIbβ_3
Von Willebrand factor	GP Ib–IX	
	GP IIb–IIIa	αIIbβ_3
Laminin	GP Ic	$\alpha_6\beta_1$

of lipids and proteins. The phospholipid fraction is important for hemostastis as a source of arachidonic acid (see Chap. 6) during thrombocyte activation, but also because these phospholipids are important for the coagulation reaction proper (see above).

83.9.2.1 Membrane Receptors

Integrins. Many of the adhesive protein receptors are members of the integrin family. The integrins are broadly distributed heterodimeric (α- and β-) cell surface molecules that share certain structural, immunochemical, and functional properties. These individual adhesive proteins come in contact with the thrombocyte by serving as ligands for specific cell surface receptors. The nomenclature for these receptors is unfortunately somewhat inconsistent, so that one and the same receptor may have multiple designations. The most widely used nomenclature is based on electrophoretic mobility of the membrane proteins. Such a classification gives rise to several glycoproteins (GP) e.g., GP I, II, and III, with GP I having the highest molecular weight [85]. Several of the membrane proteins exist on the thrombocyte surface as a noncovalently linked complex, thus GP Ib–IX or GP IIb–IIa can be regarded as different structures consisting of single membrane proteins. Some receptors have been named on the basis of their function (e.g., the collagen or the fibrinogen receptor), but it later turned out that there is a high degree of redundancy in this system in that one receptor type can bind to different ligands and vice versa [85,94]. The integrin receptors generally consist of α- and β-subunits in various combinations. The most important thrombocyte receptors are listed in Table 83.6, based on information provided in [85] and [94].

83.9.3 Activation Reactions of Thrombocytes

Normally, thrombocytes float around freely in the circulation. Only when they come in contact with stimulating agents such as collagen or fibrinogen do they become adherent and secrete a number of factors that attract more thrombocytes to the lesion and enhance aggregation and plug formation. In the following, this interplay will be discussed, with reference to certain pathophysiologic and clinical aspects. The dramatic morphological changes that accompany thrombocyte activation are shown in Fig. 83.9.

83.9.3.1 How Are Thrombocytes Kept Quiescent Under Normal Conditions?

Prostacyclin, Endothelial-Derived Relaxing Factor, and Ectonucleotides. In the circulation, thrombocytes are floating cells that do not adhere to each other nor to the vascular endothelium. This apparent resting state of nonactivated thrombocytes is surprising, because thrombocytes can become extremely adhesive within a few seconds upon contact with a molecule that is recognized as foreign. There are a number of mechanisms that keep the circulating thrombocyte quiescent: some of the adhesion receptors mentioned above have a low affinity and bind ligands only when triggered to do so, whereas other receptors are shielded from ligands in the extracellular matrix by endothelial cells. In the normal circulation and in unstimulated thrombocytes studied in vitro, the fibrinogen receptor G IIb–IIIa is in a low affinity state (K_d is much greater than the micromolar range). When thrombocytes enter a vascular wound or encounter thrombin, the receptor is converted within seconds to a high-affinity state (K_d is much less than the micromolar range) and ligand binding ensues [1,94]. Furthermore, endothelial cells are able to secrete two short-ranging substances, namely prostacyclin and endothelial-derived relaxing factor (EDRF/nitric oxide, NO). NO activates a guanylate cyclase, thereby increasing cyclic guanosine monophosphate (cGMP), a known inhibitor of Ca^{2+}-mediated reaction in thrombocytes and other cellular systems [48,77,80]. Prostacyclin on the other hand is transferred from enthothelial cells to thrombocytes [68,92], and the subsequent increase in cyclic adenosine monophosphate (cAMP) in the thrombocyte acts synergistically with cGMP [6,54,67]. The third endothelial thromboregulatoy mechanism (apart from the generation of cGMP and cAMP) involves ectonucleotidases on the endothelial cell surface. It has recently become clear that nucleotides such as adenosine diphosphate (ADP) and triphosphate (ATP) are released in response to cell activation or injury and may elicit biological responses by means of specific receptors, some of which can be deleterious, e.g., prothrombotic effects of excessive ADP and ATP release from thrombocytes and injured tissue [17]. It has now been demonstrated in a number of studies that the blood surface of endothelial cells carry an enzyme that can hydrolyze ATP as well as ADP [16,17,69]. Thus, there is now in vitro evidence for at least three independent endothelial cell regulatory systems for thrombocyte reactivity: endothelium-derived prostacyclin and EDRF/NO

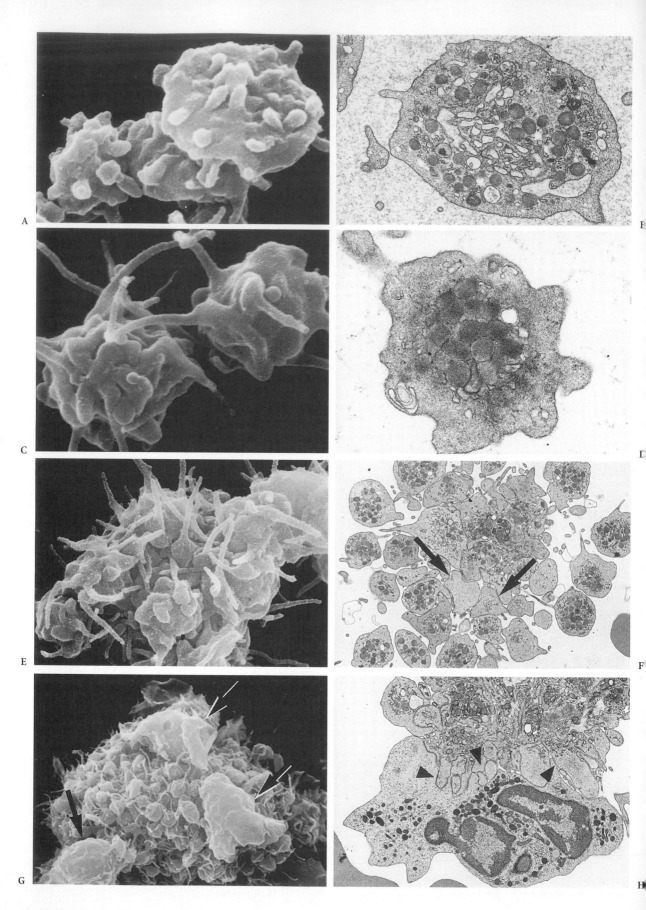

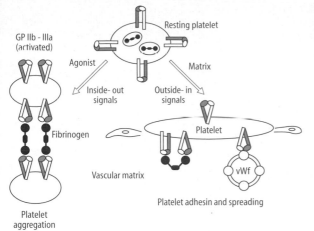

Fig. 83.10. The role of the thrombocyte receptor integrins in plug formation. As illustrated on the *left*, agonists such as thrombin or adenosine diphosphate (ADP) increase the number and the affinity of GP IIb–IIIa for ligands such as fibrinogen. This mode of signal transfer is called inside-out signaling because the receptor function is modulated by intracellular events. Note that prostacyclin or endothelium-derived relaxing factor (EDRF)/ nitric oxide (NO) antagonize the effects of thrombin or ADP [87]. On the *right*, adherence and spreading of thrombocytes occurs via binding of GP Ib–IX to von Willebrand factor (*vWf*) and further to collagen or of GP IIb–IIIa to immobilized fibrinogen or vWf. Because binding of fibrinogen to GP IIb–IIIa can also modulate the number and the affinity of these receptor molecules, this type of feedback control was called outside-in signaling. In this case, the initial thrombocyte interaction with fibrinogen does not require prior thrombocyte activation, but the interaction with vWf does. (From [94], with minor modifications)

[101,109] as well as an ectonucleotidase located on the surface of endothelial cells.

83.9.3.2 Adhesion of Thrombocytes to Foreign Surfaces

Von Willebrand Factor. When vascular endothelium is disrupted, thrombocytes are exposed to nonendothelial surfaces (e.g., collagen). As a result they adhere, flatten, and spread on the exposed surface. In this process the thrombocyte shape changes dramatically from a disk to a spiny sphere with long, fine filopodia that may be several times the length of the thrombocyte (Fig. 83.9). It is likely that the excess surface membrane required to form the filopodia is obtained from the canalicular membrane system. Thrombocytes do not react with collagen but rather to plasma protein, von Willebrand factor, which in turn binds collagen (see Table 83.6). The von Willebrand factor was named after the first person to describe a hereditary bleeding disorder due to its deficiency. Note that the factor VIII coagulant protein circulates as a complex with von Willebrand factor protein, but that these are two molecular entities [106]. The basic mechanisms that lead to receptor-mediated thrombocyte adhesion and aggregation are shown in Fig. 83.10.

Bernard–Soulier Syndrome. Von Willebrand factor may be able to bind to a receptor located on GP Ib–X, which is essential for normal thrombocyte adhesion to exposed collagen. The importance of this receptor is illustrated by a rare bleeding disorder, the Bernard–Soulier syndrome. This autosomal recessive disease manifests itself in severe and spontaneous mucocutaneous bleeding that may lead to fatal hemorrage if not treated. The degree of bleeding varies among patients even within a family, but most patients experience bleeding episodes severe enough to require transfusion [63]. The second receptor, a complex of GP IIb–IIIa, has a low binding affinity before the thrombocytes are activated [94]. This receptor is of importance for the adhesion of activated thrombocytes to each other via, for example, fibrinogen or thrombospondin (Fig. 83.10).

83.9.3.3 Aggregation of Thrombocytes

Fibrinogen, Thrombospondin, and Vitronectin. Most of the thrombocytes that accumulate at sites of injury do not adhere directly to the subendothelial ligands but rather to each other. This thrombocyte cohesion is called aggregation and has properties that are quite distinct from adhesion. Aggregation requires assembly on the thrombocyte membrane of a receptor formed by a complex of GP IIb–IIIa. This complex functions as a receptor for fibrinogen, thrombospondin, and vitronectin [32,85,94], which must bind to the thrombocyte surface before thrombocytes can aggregate. Thrombocyte aggregation, i.e., the expression of the GP IIb–IIIa receptor complex can be triggered by several physiological agonists, the most important of which are ADP and thrombin. Other potential agonists include epinephrine, thromboxane A_2, and platelet-activating factor (PAF) [42,98]. In all systems studied,

Fig. 83.9A–H. Scanning (*left*) and transmission (*right*) electron microscopy of thrombocyte activation under in vitro conditons. **A,B** Initial activation. The discoid thrombocytes show pseudopodes at the entire surface. The initial expansion of the open canalicular system causes fusion of the α-granules with the open canalicular system and secretion of their contents. **C,D** Early activation. The thrombocytes form filopodes that interdigitate. Within the thrombocytes, α-granules and other organelles are accumulated in the center of the cell. **E,F** Late activation.

Thrombocytes are closely attached to each other and form complex aggregates. Thrombocytes located in the center of such an aggregate appear to be almost completely depleted of α-granules. **G,H** Final stage of activation. Leukocytes (*arrows*) become closely attached to the aggregate and project with broad cellular processes (*arrowheads*) between the aggregated thrombocytes. Photographs kindly provided by Prof. P. Groscurth, Institute of Anatomy, University of Zurich)

Table 83.7. Content and function of thrombocyte granules (thrombocyte lysosomes are not included)

Content	Function
Electron-dense granules	
ADP	Enhances thrombocyte aggregation Chemotactic for unstimulated thrombocytes
Ca^{2+}	Co-factor for local coagulation Extracellular co-factor for activation reactions
Serotonin	Thrombocyte aggregation and vasoconstrintion
α-Granules	
Fibrinogen	Precursor for the local formation of fibrin Thrombocyte aggregation
Coagulation factors V and VIII	Substrates for the local formation of thrombin
Fibronectin	Adhesion of platelets to surrounding tissue Activation of keratinocytes
Von Willebrand factor	Adhesion of thrombocytes to collagen via the GP IIb–IIIa receptor
Thrombospondin	Co-factor for thrombocyte aggregation
Platelet factor 4	Chemotactic for granulocytes Inactivates heparin
Platelet-derived growth factor (PDGF)	Mitogen for smooth muscle cells Vasoconstrictor Chemotactic for granulocytes
Fibroblast growth factor	Stimulates growth and motility of fibroblasts, keratinocytes, and endothelial cells
β-Transforming growth factor	Chemotactic for macrophages Stimulates the incorporation of fibronectin in the extracellular matrix
Platelet-derived endothelial growth factor (PDETGF)	Potent stimulus for endothelial cell growth

ADP, adenosine diphosphate

thrombocyte aggregation requires both fibrinogen and ADP [98]. Agonists such as thrombin and ADP not only lead to receptor expression, but also to the secretion of many substances from the thrombocytic granules (see Table 83.7). Among them, fibrinogen and ADP are the most important ones for the aggregation reaction. Another adhesion molecule that is stored in the α-granules is thrombospondin, which binds to the GP IIb–IIIa receptor as well as to fibrinogen, thus enhancing the action of fibrinogen [98]. The GP IIb–IIIa receptor complex is essential for normal formation of the thrombocyte hemostatic plug. This becomes apparent from the symptoms of patients who suffer from Glanzmann's thrombasthenia, one of the most common inherited disorders of thrombocyte function. Glanzmann's thrombasthenia manifests itself in mucoutaneaous bleeding, which is usually present from birth, a history of bleeding in siblings compatible with autosomal recessive inheritance, and the following laboratory findings: normal thrombocyte count and morphology with prolonged bleeding time, absent clot retraction, and absent thrombocyte aggregation by ADP, collagen, and thrombin. The diagnosis can be further verified by demonstrating the absence of the GP IIb–IIIa (fibrinogen) receptor by employing monoclonal antibodies against fibrinogen [63].

Thromboxane A₂. Finally, it should be mentioned that weak agonists, such as ADP and epinephrine, cause aggregation and secretion only when thromboxane A_2 production can proceed normally. The same thromboxane A_2 dependence of strong agonists (thrombin, collagen, PAF) is observed when low concentrations of these strong agonists are employed, but not at high concentrations [98]. In view of recent reports dealing with the question of whether inhibition of thromboxane formation by aspirin can prevent heart attacks [90], it seems appropriate to briefly consider the formation of thromboxane and other eicosanoids in thrombocytes.

83.9.3.3.1 Production of Eicosanoids and the Effect of Aspirin of Thrombocyte Function

Arachidonate. Thrombocytes produce oxygenated derivatives of arachidonic acid that are collectively termed eicosanoids. Arachidonic acid (arachidonate) is the major polyunsaturated, essential fatty acid in humans (see also Chap. 6). Arachidonate is bound to albumin, transported to the cells, and esterified into phospholipids without prior production of oxygenated metabolites. In view of the fact that several excellent reviews exist on the arachidonate metabolism in thrombocytes [63,67,90,98], only the key events will be considered here. Arachidonate is cleaved from phospholipid largely via catalysis by phospholipase A_2. Activation of this enzyme has been reported no involve several entities, including the increase in free cytosolic Ca^{2+}

during thrombocyte activation and one or more G proteins (see Chap. 5 for the role of G proteins in signal transduction). Arachidonate cleavage can also result from combined actions of phospholipase C and diacylglycerol lipase.

Cyclooxygenase and Aspirin. Thereafter, arachidonate is enzymatically oxygenated by cyclooxygenase located in the dense tubular system of thrombocytes to the endoperoxide intermediate to prostaglandin H_2 (PGH_2) [63]. The major biologically active cyclooxygenase product in thrombocytes is thromboxane A_2, which is converted from the intermediate PGH_2 by thromboxane synthase. This autacoid is released in the extracellular space, where it acts as a vasoconstrictor and takes part in the aggregation process. A derivative of salicylates from plant sources (white willow, Salix alba) is acetyl-salicylic acid (ASA; aspirin). This compound inhibits the synthesis of various prostaglandins in a highly specific way by blocking only the cyclooxygenase function of PGH_2 synthase [90]. Aspi-rin induces a mild hemostatic defect in humans, as evidenced by prolongation of skin bleeding and impaired thrombocyte aggregation, which in turn results from an inhibition of thromboxane A_2 formation, Since thrombocytes are incapable of protein synthesis and because aspirin inhibits cyclooxygenase by covalent acetylation, the inhibition lasts for the lifetime of the thrombocytes (approximatly 10 days). Therefore by using very low doses of aspirin (approximately 100 mg/day) it is possible to selectively block the cyclooxygenase in thrombocytes, but not the same enzyme in other tissues [63,90]. The clinical studies summarized by Roth and Calverley [90] show the effectiveness of aspirin in preventing certain thrombotic complications of vascular diseases. In general, aspirin affects arterial rather than venous thrombosis and most aspirin-responsive thrombi arise in the setting of atherosclerotic changes in an arterial vessel wall. The Federal Drug Administration has therefore approved aspirin for unstable angina and after transient ischemic attacks and myocardial infarction [90]. Of interest in this

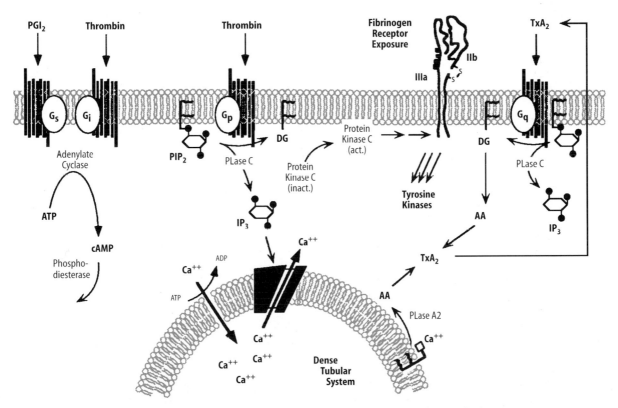

Fig. 83.11. Signal transduction during platelet activation. The binding of agonist to membrane-spanning receptors on the thrombocyte surface initiates cascades of intracellular second messengers, including inositol 1,4,5-trisphosphate (IP_3) and diacylglycerol (DG). IP_3 releases Ca^{2+} from the thrombocyte dense tubular system, raising the cytosolic free Ca^{2+} concentration. At the same time, the rising cytosolic free Ca^{2+} concentration facilitates arachidonate (AA) release from phospholipids by phospholipase A_2, a process that may occur at both the plasma membrane and the dense tubular system membrane. Arachidonate is metabolized to thromboxane A_2 (TxA_2), which diffuses out of the cell, interacts with receptors on the thrombocyte surface, and causes further thrombocyte activation. In many cases, the interactions between agonists and the enzymes responsible for second messenger generation are mediated by a guanine nucleotide-binding regulatory protein (G protein). In thrombocytes, G proteins have been shown to regulate phosphoinositide hydrolysis and cyclic adenosine monophosphate (cAMP) formation and are probably involved in the activation of phospholipase (Plase) A_2. Phospholipase C is activated. Adenylyl cyclase is stimulated by the G protein G_s and inhibited by the G protein G_i. PIP_2, phosphotidylinositol-4,5 diphosphate; ATP, adenosine triphosphate; ADP, adenosine diphosphate. (From [67])

1673

context are direct observations showing that atherosclerotic coronary arteries are more thrombogenic than the normal vessel and that the diseased vessel contains fibrinogen and fibrin as well [107].

83.9.3.4 Secretion Reactions of Thrombocytes

Secretory Granules. During thrombocyte activation by weak (ADP, epinephrine) or strong (thrombin, collagen, PAF) agonists, secretory granules move to the center of the thrombocyte, their membrane fuse with the membranes of the open canalicular system, and their contents are secreted through this system into the surrounding tissues and plasma (see Fig. 83.9 for the morphological changes). As was mentioned above, an increase in cytoplasmic Ca^{2+} and the formation of thromboxane A_2 triggers these reactions.

83.9.3.4.1 Stimulus Secretion Coupling in Thrombocyte Activation

The mechanism by which thrombocytes are activated when cell surface receptors are activated by appropriate ligands is a fundamental problem, not only in the case of thrombocytes, but in all living cells. In the present context the aim is to succinctly describe some of the most important metabolic events that are pertinent to the metabolic control of secretion. The general mechanisms of signal transduction during thrombocyte activation are shown in Fig. 83.11.

As can be seen from Fig. 83.11, the activation of thrombocytes by agonists involves several interrelated systems. Ca^2 fluxes, for example, mediate several reactions and Ca^2 concentrations have been shown to increase ten- to 50-fold when thrombocytes are stimulated with thrombin or collagen [98]. The main source of this Ca^{2+} increase is provided by release from the dense tubular system. The intracellular signal that releases Ca^{2+} from this store is inositol 1,4,5-trisphosphate (IP_3), a prominent split product of a member of the phosphoinositide family, phosphotidylinositol-4,5 diphosphate (PIP_2) in this case (see Chap. 5). The receptor-linked enzyme that splits this phosphoinosite is phospholipase C, whereby diacylglycerol (DG) and IP_3 are formed. One very important activator of phospholipase C is thrombin. Binding of thrombin to its receptor will therefore lead to a large increase in Ca^{2+}, as shown in Fig. 83.11. The Ca^{2+} signal serves to trigger further phosphatidylinositol breakdown (see Chap. 5) by initiating protein phosphorylation by a Ca^{2+}-dependent protein kinase (protein kinase C) and to activate myosin light chain kinase. The latter enzyme phosphorylates the myosin light chain, which is required for thrombocyte contraction. At the same time, the fusion of the membranes of the α-granules with the membranes of the open canalicular system is affected by processes that are thought to require protein phosphorlation reactions [98].

Protein Phosphorylation. When thrombocytes are activated by thrombin, at least two major proteins are rapidly phosphorylated [19]. These are the myosin light chain and a 40-kDa protein, IP_3 monoesterase. The latter phosphorylation is catalyzed by the protein kinase C, the major protein kinase in thrombocytes. Protein kinase C is activated by DG, one of the major split products of PiP_2, and by Ca^{2+}. The DG produced by thrombin receptor activation together with the phospholipid phosphatidylserine in the cytoplasmic half of the plasma membrane binds to the protein kinase C, thereby increasing the affinity of the enzyme for Ca^{2+}, so that protein kinase C is activated by the cooperative effect of DG and an increase in cytosolic Ca^{2+} brought about by IP_3. This activates protein phosphorylation, thrombocyte granule secretion, and expression of the fibrinogen receptor [93]. The fibrinogen receptor site is on the thrombocyte membrane GP IIb–IIIa

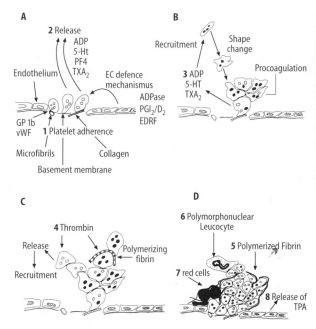

Fig. 83.12A–D. Normal hemostatic process. **A** Exposure of subendothelium due to vascular damage results in immediate thrombocyte adherence to collagen, in the presence of von Willebrand factor (*vWf*) and thrombocyte membrane glycoprotein (*GP*) Ib. The thrombocyte agonist collagen induces release of adenosine diphosphate (*ADP*) from dense granules and several proteins from α-granules. Thromboxane A_2 (*TXA₂*) is also formed enzymatically from free arachidonic acid. *5-HT*, 5-hydroxytryptamine (serotonin); *PF4*, platelet factor 4. **B** ADP and TXA₂ are the most important recruiting components of thrombocyte release. They activate additional thrombocytes arriving in the fluid phase, which then undergo shape change and aggregate onto the initial layer of activated thrombocytes. Phospholipoproteins are then available on the thrombocyte surface for catalytic activation of proteins of the coagulation system. **C** Thrombin formation has several major consequences which lead to the final stages of thrombocyte thrombus development. Fibrin strands begin to intercalate between activated thrombocytes, and the thrombus becomes consolidated. **D** The consolidated thrombocyte thrombus is now impermeable and its multicellular nature becomes obvious, including the release of tissue plasminogen activator (*TPA*). (From [67])

complex, which becomes transformed during thrombocyte activation to assume its receptor function (inside-out signaling, as described in [94] and shown in Figs. 83.10 and 83.11). Furthermore, the cytoskeletal rearrangements that occur during thrombocyte activation are also involved in granule secretion [33]. The contractile system of thrombocytes will be more extensively considered below.

83.9.3.5 Contraction Mechanisms of Thrombocytes

Actin and Calmodulin. Thrombocytes contain large amounts of *actin*, constituting about 10% of the total protein, and much lower amounts of myosin, with a molar ratio of actin to myosin of about 100:1 [33]. Actin occurs in unstimulated thrombocytes mainly in the globular, monomeric form (G actin) and polymerizes to a double-helical filament form (F actin) upon activation. The F actin bundles form parallel filaments, which then give rise to the filopods that project from the activated thrombocyte (see Fig. 83.9). The whole process of converting G actin to F actin is triggered by Ca^{2+} and its receptor protein calmodulin, which in turn activates the protein kinase C in collaboration with DG. These general processes, as well as the preferential binding of actin filament bundles to specific parts of the plasma membrane (focal contacts), are discussed in great detail in Chaps. 4–6.

The contractile apparatus of the thrombocytes not only enhances aggregation, i.e., plug formation, but also the retraction of the thrombocyte clot along with its final components, namely erythrocytes and leukocytes. This mixed aggregate (Fig. 83.9), also known as mixed thrombus, is retracted by the contractile elements of the thrombocytes to a volume that is much smaller than the original thrombus. Again, once thrombocytes are activated they aggregate and secrete their granule contents. Coagulation reactions then occur on the thrombocyte surface and finally clot retraction occurs by the combined action of myosin and actin filaments. The actin filaments are anchored to the membrane GP IIb–IIIa, which in turn is linked to fibrin strands outside the cell by focal contacts. In this way thrombocyte contraction can effectively diminish the volume of a large fibrin clot, thereby interacting with the coagulation reactions of the plasma system ([63] and Sect. 83.4). A summary of all of these reactions is provided by Fig. 83.12.

References

1. Abrams CS, Ellison N, Budzynski AZ, Shattil SJ (1994) Direct detection of activated platelets and platelet-derived microparticles in humans. Blood 75:128–138
2. Agre P, Casella JF, Zinkham WH, McMillan C, Bennett V (1985) Partial deficiency of erythrocyte spectrin in hereditary spherocytosis. Nature 314:380–383
3. Anderson HV, Willerson JT (1983) Thrombolysis in acute myocardial infarction. N Engl J Med 329:703–709

4. Berliner N, Duffy TP (1991) Approach to the patient with anemia. In: Hoffmann R, Benz EJ Jr, Shattil SJ, Furie B, Cohen HJ (eds) Hematology. Churchill Livingstone, New York, pp 302–311
5. Bertina RM, Koeleman BPC, Koster T, Rosendaal FR, Dirven RJ, de Ronde H, van der Velden PA, Reitsma PH (1994) Mutation in blood coagulation factor V associated with resistance to activated protein C. Nature 369:64–67
6. Broekmann MJ, Eiroa AM, Marcus AJ (1991) Inhibition of human platelet reactivity by endothelium-derived relaxing factor from human umbilical vein endothelial cells in suspension. Blockade of aggregation and secretion by an aspirin-insensitive mechanism. Blood 78:1033–1040
7. Brown LF, Lanir N, McDonagh J, Tognazzi K, Dvorak AM, Dvorak HF (1993) Fibroblast migration in fibrin gel matrices. Am J Pathol 142:273–283
8. Broze GJ (1983) Binding of human factor VII and VIIa to monocytes. J Clin Invest 70:526–535
9. Broze G Jr (1992) The role of tissue factor pathway inhibitor in a revised coagulation cascade. Semin Hematol 29:159–169
10. Broze GJJ, Gailani D (1993) The role of factor XI in coagulation. Thromb Haemost 70:72–74
11. Buchanan MR, Crozier GL, Hass TA (1988) Fatty acid metabolism and the vascular endothelial cell. New thoughts about old data. Haemostasis 18:360–375
12. Chasis JA, Mohandas N, Shohet SB (1985) Erythrocyte membrane rigidity induced by glycophorin a–ligand interaction. J Clin Invest 75:1919–1926
13. Chien S (1987) Red cell deformability and ist relevance to blood flow. Annu Rev Physiol 49:177–192
14. Cohen CM (1983) The molecular organization of the red cell membrane skeleton. Semin Hematol 20/3:141–158
15. Comp PC (1990) Overview of the hypercoagulable states. Semin Thromb Hemost 16:158–161
16. Côté YP, Picher M, St-Jean P, Béliveau R, Potier M, Beaudoin AR (1991) Identification and localization of ATP-diphosphohydrolase (apyrase) in bovine aorta: relevance to vascular tone and platelet aggregation. Biochim Biophys Acta 1078:187–191
17. Côté YP, Filep JG, Battistini B, Gauvreau J, Sirois P, Beaudoin AR (1992) Characterization of ATP-disphosphohydrolase activities in the intima and media of the bovine aorta: evidence for a regulatory role in platelet activation in vitro. Biochim Biophys Acta Mol Basis Dis 1139:133–142
18. Coughlin SR (1993) Thrombin receptor structure and function. Thromb Haemost 66:184–187
19. Coughlin SR, Vu T-KH, Hung DT, Wheaton VI (1992) Characterization of a functional thrombin receptor. Issues and opportunities. J Clin Invest 89:351–355
20. Dahlbäck B (1980) Human coagulation factor V purification and thrombin-catalyzed activation. J Clin Invest 66:583–591
21. Davie EW, Ratnoff OD (1964) Waterfall sequence for intrinsic blood coagulation. Science 145:1310–1312
22. Davie EW, Fujikawa K, Kisiel W (1991) The coagulation cascade: Initiation, maintenance, and regulation. Biochemistry 30:10363–10370
23. de Bono DP (1994) Update on thrombolysis. Fibrinolysis 8 [Suppl 1]:263–268
24. de Sauvage FJ, Hass PE, Spencer SD, Malloy BE, Gurney AL, Spencer SA, Darbonne WC, Henzel WJ, Wong SC, Kunag WJ, Oles KJ, Hultgren B, Solberg LA Jr, Goeddel DV, Eaton DL (1994) Stimulation of megakaryocytopoiesis and thrombopoiesis by the c-mpl ligand. Nature 369:533–538
25. Dinter A, Berger EG (1995) The regulation of cell- and tissue-specific expression of glycans by glycosyltransferases. In: Axford JS, Alavi A (eds) Glycoimmunology. Plenum, New York (in press)
26. Doolitle RF (1984) Fibrinogen and fibrin. Annu Rev Biochem 53:195–229
27. Drake TA, Ruf W, Morrissey JH, Edington TS (1989) Functional tissue factor is entirely cell surface expressed on lipopolysaccharide-stimulated human blood monocytes and

a constitutively tissue factor-producing neoplastic cell line. J Cell Biol 109:389–395

28. Edwards RL, Rickles FR (1992) The role of leukocytes in the activation of blood coagulation. Semin Hematol 29:202–212
29. Engelberg H (1989) Endothelium in health and disease. Semin Thromb Hemost 15:178–183
30. Eschbach JW (1988) The anemia of chronic renal failure: pathophysiology and the effects of recombinant erythropoietin. Kidney Int 35:134–148
31. Esmon CT (1992) The protein C anticoagulant pathway. Arter Thromb 12:135–145
32. Farrell DH, Thiagarajan P (1994) Binding of recombinant fibrinogen mutants to platelets. J Biol Chem 269:226–231
33. Fox JEB, Boyles JK, Berndt MC, Steffen PK, Anderson LK (1988) Identification of a membrane skeleton in platelets. J Cell Biol 106:1525–1538
34. Francis RBJ (1989) Clinical disorders of fibrinolysis: a critical review. Blut 59:1–14
35. Fujikawa K, Legaz ME, Davie EW (1972) Bovine factors X1 and X2/Stuart factor), isolation and characterization. Biochemistry 11:4882–4891
36. Furie B, Furie BC (1992) Molecular and cellular biology of blood coagulation. N Engl J Med 326:800–806
37. Gailani D, Broze GJ Jr (1991) Factor XI activation in a revised model of blood coagulation. Science 253:909–912
38. Gewritz AM, Poncz M (1991) Megakaryocytopoiesis and platelet production. In: Hoffmann R, Benz EJ Jr, Shattil SJ, Furie B, Cohen HJ (eds) Hematology. Churchill Livingstone, New York, pp 1148–1157
39. Goldstein JL, Kita T, Brown MS (1983) Defective lipoprotein receptors and atherosclerosis. N Engl J Med 309:288–296
40. Hacke W (1994) Thrombolytic therapy in acute ischemic stroke – an update. Fibrinolysis 8 [Suppl 1]:216–220
41. Halkier T (1991) Mechanisms in blood coagulation, fibrinolysis and the complement system. Cambridge University Press, Cambridge
42. Hanahan DJ (1986) Platelet activating factor: a biologically active phosphoglyceride. Annu Rev Biochem 55:483–509
43. Hillman RS, Finch CA (1967) Erythropoiesis: normal and abnormal. Semin Hematol 4:327–336
44. Hirsh J (1991) Heparin. N Engl J Med 324:1565–1574
45. Hirsh J (1991) Oral anticoagulant drugs. N Engl J Med 324:1865–1875
46. Hirsh J, Levin MN (1992) Low molecular weight heparin. Blood 79:1–17
47. Hoyer LW (1994) Hemophilia A. N Engl J Med 330:38–47
48. Ignarro LJ (1991) Signal transduction mechanisms involving nitric oxide. Biochem Pharmacol 41:485–490
49. Isenberg WM, Ford-Bainton D (1991) Megakaryocyte and platelet structure. In: Hoffmann R, Benz EJ Jr, Shattil SJ, Furie B, Cohen HJ (eds) Hematology. Churchill Livingstone, New York, pp 1157–1165
50. Jordan R, Beeler D, Rosenberg RD (1979) Fractionation of low molecular weight heparin species and their interaction with antithrombin. J Biol Chem 254:2902–2913
51. Kaplan AP, Silverberg M (1987) The coagulation-kinin pathway of human plasma. Blood 70:1–15
52. Kaushansky K, Lok S, Holly RD, Broudy VC, Lin N, Bailey MC, Forstrom JW, Buddle MM, Oort PJ, Hagen FS, Roth GJ, Papayannopoulou T, Foster DC (1994) Promotion of megakaryocyte progenitor expansion and differentiation by the c-Mpl lignand thrombopoietin. Nature 369:568–571
53. Kluft C, Dooijewaard G, Emeis JJ (1987) Role of the contact system in fibrinolysis. Semin Thromb Hemost 13:50–68
54. Kroll MH, Schafer AI (1989) Biochemical mechanisms of platelet activation. Blood 74:1181–1195
55. Lämmle B, Griffin JH (1985) Formation of the fibrin clot: the balance of procoagulant and inhibitory factors. Clin Haematol 14:281–342
56. Larsen E, Celi A, Gilbert GE (1989) PADGEM protein: a receptor that mediates the interaction of activated platelets with neutrophils and monocytes. Cell 59:305–312

57. Lawson JH, Mann KG (1991) The cooperative activation of human factor IX by the human extrinsic pathway of blood coagulation. J Biol Chem 266:11317–11327
58. Lazarides E, Woods C (1989) Biogenesis of the red blood cell membrane-skeleton and the control of erythroid morphogenesis. Annu Rev Cell Biol 5:427–52
59. Lindquist PA, Fujikawa K, Davie EW (1978) Activation of bovine factor IX (Christmas factor) by factor XIa (activated plasma thromboplastin anticedent) and a protease from Russel's viper venom. J Biol Chem 253:1902–1909
60. Lok S, Kaushansky K, Holly RD, Kuijper JL, Lofton-Day CE, Oort PJ, Grant FJ, Heipel MD, Burkhead SK, Kramer JM, Bell LA, Sprecher CA, Blumberg H, Johnson R, Prunkard D, Ching AFT, Mathewes SL, Bailey MC, Forstrom JW, Buddle MM, Osborn SG, Evans SJ, Sheppard PO, Presnell SR, O'Hara PJ, Hagen FS, Roth GJ, Foster DC (1994) Clonging and expression of murine thrombopoietin cDNA and stimulation of platelet production in vivo. Nature 369:565–568
61. Luna EJ, Hitt AL (1992) Cytoskeleton – plasma membrane interactions. Science 258:955–964
62. Macfarlane RG (1964) An enzyme cascade in the blood clotting mechanism, and its function as a biochemical amplifier. Nature 202:498–499
63. Majerus PW (1987) Platelets. In: Stamatoyannopoulos G, Nienhuis AW, Leder P, Majerus PW (eds) The molecular basis of blood diseases. Saunders, Philadelphia, pp 689–721
64. Mann KG, Nesheim ME, Church WR, Haley P, Krishnaswamy S (1990) Surface-dependent reactions of the vitamin K-dependent enzyme complexes. Blood 76:1–16
65. Mann KG, Krishnaswamy S, Lawson JH (1992) Surface-dependent hemostasis. Semin Hematol 29:213–226
66. Marcum JA, Atha DH, Fritze LMS, Nawroth P, Stern D, Rosenberg RD (1986) Cloned bovine aortic endothelial cells synthesize anticoagulantly active heparan sulfate proteoglycan. J Biol Chem 261:7501–7517
67. Marcus AJ, Safier LB (1993) Thromboregulation: multicellular modulation of platelet reactivity in homeostasis and thrombosis. FASEB J 7:516–522
68. Marcus AJ, Weksler BB, Jaffe EA, Broekman MJ (1980) Synthesis of prostacyclin from platelet-derived endoperoxides by cultured human endothelial cells. J Clin Invest 66:979–986
69. Marcus AJ, Safier LB, Hajjar KA, Ullmann HL, Islam N, Borekmann MJ, Eiroa AM (1991) Inhibition of platelet function by an aspirin-insensitive endothelial cell ADPase. Thromboregulation by endothelial cells. J Clin Invest 88:1690–1696
70. Markwardt F (1994) The development of hirudin as an antithrombotic drug. Thromb Res 74:1–23
71. Martin P, Hopkinson-Woolley J, McCluskey J (1992) Growth factors and cutaneous wound repair. Prog Growth Factor Res 4:25–44
72. McDonald TP (1992) Thrombopoietin: its biology, clinical aspects, and possibilities. Am J Pediatr Hematol Oncol 14:31–38
73. Means RT, Krantz SB (1992) Progress in understanding the pathogenesis of the anemia of chronic disease. Blood 80:1639–1647
74. Metcalf D (1994) Thrombopoietin – at last. Nature 16:519–520
75. Mohands N (1991) The red cell membrane. In: Hoffmann R, Benz EJ Jr, Shattil SJ, Furie B, Cohen HJ (eds) Hematology. Churchill Livingstone, New York, pp 264–269
76. Mohandas N, Chasis JA, Shohet SB (1983) The influence of membrane skeleton on red cell deformability, membrane material properties, and shape. Semin Hemat 20(3):225–242
77. Moncada S, Palmer RMJ, Higgs EA (1991) Nitric oxide: physiology, pathopysiology, and pharmacology. Pharmacol Rev 43:109–142
78. Nachman RL, Silverstein R (1993) Hypercoagulable states. Ann Intern Med 119:819–827
79. Naito K, Fujikawa K (1991) Activation of human blood coagulation factor XI independent of factor XII, Factor XI is

activated by thrombin and factor XIa in the presence of negatively charged surfaces. J Biol Chem 266:7353–7358

80. Nathan C (1992) Nitric oxide as a secretory product of mammalian cells. FASEB J 6:3051–3064

81. Nemerson Y (1992) The tissue factor pathway of blood coagulation. Semin Hematol 29:170–176

82. Niewiarowski S (1981) Platelet release reaction and secreted platelet proteins. In: Bloom AL, Thomas DP (eds) Haemostasis and thrombosis. Churchill Livingstone, Edinburgh, pp 73–83

83. Oates JA, FitzGerald GA, Branch RA, Jackson EK, Knapp HR, Roberts LJ (1988) Clinical implications of prostaglandin and thromboxane A2 formation. N Engl J Med 319:689–698

84. Papayannopoulou T, Abkowitz J (1991) Biology of erythropoiesis, erythroid differentiation, an maturation. In: Hoffmann R, Benz EJ Jr, Shattil SJ, Furie B, Cohen HJ (eds) Hematology. Churchill Livingstone, New York, pp 252–263

85. Plow EF, Ginsberg MH (1991) The molecular basis of platelet function. In: Hoffmann R, Benz EJ Jr, Shattil SJ, Furie B, Cohen HJ (eds) Hematology. Churchill Livingstone, New York, pp 1165–1176

86. Ragni MV, Sinha D, Seaman F, Lewis JH, Spero JA, Walsh PN (1985) Comparison of bleeding tendency, factor XI coagulant activity, and factor XI antigen in 25 factor XI-deficient kindreds. Blood 65:719–724

87. Rijken DC, Wijngaards G, Zaal-de Jong M, Welbergen J (1979) Purification and partial characterization of plasminogen activator from human uterine tissue. Biochim Biophys Acta 580:140–153

88. Robbins KC (1991) Fibrinolytic therapy: biochemical mechabisms. Semin Thromb Hemost 17:1–6

89. Ross R (1986) The pathogenesis of atherosclerosis – an update. N Engl J Med 314:488–500

90. Roth GJ, Calverly DC (1994) Aspirin, platelets, and thrombosis: theory and practice. Blood 83:885–898

91. Saito H (1987) Contact factors in health and disease. Semin Thromb Hemost 13:36–49

92. Schafer AI, Crawford DD, Gimbrone MA Jr (1984) Unidirectional transfer of prostaglandin endoperoxides between platelets and endothelial cells. J Clin Invest 73:1105–1112

93. Shattil SJ, Brass LF (1987) Induction of fibrinogen receptor on human platelets by intracellular mediators. J Biol Chem 262:992–1000

94. Shattil SJ, Ginsberg MH, Brugge JS (1994) Adhesive signaling in platelets. Curr Opin Cell Biol 6:695–704

95. Shearer MJ (1990) Vitamin K and vitamin K-dependent proteins. Br J Haematol 75:156–162

96. Shields M, La Celle P, Waugh RE, Scholz M, Peters R, Passow H (1987) Effects of intracellular Ca^{2+} and proteolytic digestion of the membrane skeleton on the mechnical properties of the red blood cell membrane. Biochim Biophys Acta 905:181–194

97. Shih GC, Hajjar KA (1993) Plasminogen and plasminogen activator assembly on the human endothelial cell. Proc Soc Exp Biol Med 202:258–264

98. Siess W (1989) Molecular mechanisms of platelet activation. Physiol Rev 69:58–178

99. Silberstein LE, Spitalnik SL (1991) Human blood group antigens and antibodies. In: Hoffmann R, Benz EJ Jr, Shattil SJ, Furie B, Cohen HJ (eds) Hematology. Chruchill Livingstone, New York, pp 1548–1556

100. Silverman EB (1984) The red cell membrane, hemolysis, and membrane disorder. In: MacKinney AA Jr (ed) Pathophysiology of blood. Wiley, New York, pp 73–98

101. Sneddon JM, Vane JR (1988) Endothelium-derived relaxing factor reduces platelet adhesion to bovine endothelial cells. Proc Natl Acad Sci U S A 85:2800–2804

102. Sprengers ED, Kluft C (1987) Plasminogen activator inhibitors. Blood 69:381–387

103. Svensson PJ, Dahlbäck B (1994) Resistance to activated protein C as a basis for venous thrombosis. N Engl J Med 330:517–522

104. Toole JJ, Knopf JL, Wozney JM, Sultzman LA, Buecker JL, Pittman DD, Kaufmann RJ, Brown E, Shoemaker C, Orr EC, Amphlett GW, Foster WB, Coe ML, Knutson GJ, Fass DN, Hewick RM (1984) Molecular cloning of a cDNA encoding human antihaemophilic factor. Nature 312:342–347

105. Turpie AGG (1994) Thrombolytic therapy in venous thrombosis and pulmonary embolism. Fibrinolysis 8 [Suppl 1]: 237–244

106. Ugarova TP, Budzynski AZ, Shattil SJ, Ruggeri ZM, Ginsberg MH, Plow EF (1993) Conformational changes in fibrinogen elicited by its interaction with platelet membrane glycoprotein GPIIb–IIIa. J Biol Chem 268:21080–21087

107. Van Zanten GH, De Graaf S, Slootweg PJ, Heijnen HFG, Connolly TM, De Groot PG, Sixma JJ (1994) Increased platelet deposition on atherosclerotic coronary arteries. J Clin Invest 93:615–632

108. Vassalli JD, Sappino AP, Belin D (1991) The plasminogen activator/plasmin system. J Clin Invest 88:1067–1072

109. Venturini CM, Del Vecchio PJ, Kaplan JE (1989) Thrombin induced platelet adhesion to endothelium is modified by endothelial derived relaxing factor (edrf). Biochem Biophys Res Commun 159(1):349–354

110. Vermeer C, Hamulyak K (1991) Pathophysiology of vitamin K-deficiency and oral anticoagulants. Thromb Haemost 66:153–159

111. Wendling F, Maraskovsky E, Debili N, Florindo C, Teepe M, Titeux M, Methia N, Breton-Gorius J, Cosman D, Vainchenker W (1994) C-Mpl ligand is a humoral regulator of megakaryocytopoiesis. Nature 369:571–974

112. White GC, Shoemaker CB (1989) Factor VIII gene and hemophilia A. Blood 73:1–12

113. White WF, Barlow GH, Mozen MM (1966) The isolation and characterization of plasminogen activators (urokinase) from human urine. Biochemistry 5:2160–2169

114. Woodson RD, Nasrollah TS, MacKinney AA Jr, Finlay JL (1984) Introduction to hemopoiesis. In: MacKinney AA Jr (ed) Pathophysiology of blood. Wiley, New York, pp 1–27

84 Hematopoiesis and the Red Blood Cell

M.J. Koury and C. Bauer

Contents

84.1 Introduction

The cellular components of the blood include several distinctly different types of cells. Each cell type, in turn, has specific functions. *Erythrocytes*, which are involved in the transport of the respiratory gases, oxygen and carbon dioxide, are anucleate, biconcave discoid cells filled with hemoglobin. *Granulocytes and monocytes*, which play key roles in inflammation and phagocytosis, are highly mobile outside of the blood vessels and possess granules of degradative enzymes. *Platelets* provide hemostasis through their abilities to adhere, aggregate, and provide a surface for coagulation reactions. Platelets are very small, anucleate cells which contain localized concentrations of molecules required for hemostasis. *Lymphocytes* mediate immunity through immunoglobulin production by B-lymphocytes and cellular immunity through the programming of mature T-lymphocytes. A histological overview of these diverse cell types is given in Fig. 84.1.

Despite the extreme structural and functional differences among the various cellular components of the blood, all of the blood cells are the progeny of the pluripotent hematopoietic stem cell. The process by which the hematopoietic stem cell gives rise to all of the various cellular components of the blood is termed hematopoiesis. Hematopoiesis therefore, involves the differentiation of the pluripotent stem cell into multiple, mature blood cell types. Major components of hematopoiesis are self-re-newal of stem cells and proliferation of lineage-committed progenitor cells during the production of mature blood cells.

Although hematopoiesis, like many other developmental processes, occurs during embryogenesis and fetal life, it also continues throughout postnatal life. The cellular components of the blood have finite life spans which are quite variable depending upon the lineage in question. In humans, granulocytes and platelets have life spans of only a few days while red blood cells and lymphocytes can exist for many months. Thus, the cellular components of the blood are replaced continuously as the older cells are removed and the newly formed cells are added. Since the numbers of the various cell types in the blood are normally maintained in relatively constant ranges, the process of hematopoiesis must be closely regulated to avoid overproduction or underproduction of new blood cells. Because of the widely varying functions and life spans of the different mature blood cell types, the regulation of hematopoiesis is complex. This chapter will analyze hematopoiesis in the adult and discuss all of the cell lineages except the lymphoid lineage. A more detailed analysis of the erythroid cell lineage and the control of red blood cell production will be presented.

84.2 Hematopoietic Organs

In the postnatal human, the bone marrow is the site of hematopoiesis. In the mouse, the other mammal in which the vast majority of hematopoiesis research has been performed, both the bone marrow and the spleen are the hematopoietic organs. In prenatal mammalian development, the yolk sac of the embryo and the liver of the midgestation fetus are sites of hematopoiesis. The hematopoietic organs have two major components. One is the hematopoietic progenitor cells which differentiate and become the mature blood cells. The characterization of the hematopoietic component, and control of its various differentiation processes, is the main focus of this chapter. The second component is the stromal cells that form the structural foundation and microenvironment in which the hematopoietic progenitor cells grow. The stroma is mainly composed of fibroblasts, adipocytes, and endothelial cells.

R. Greger/U. Windhorst (Eds.)
Comprehensive Human Physiology, Vol. 2
© Springer-Verlag Berlin Heidelberg 1996

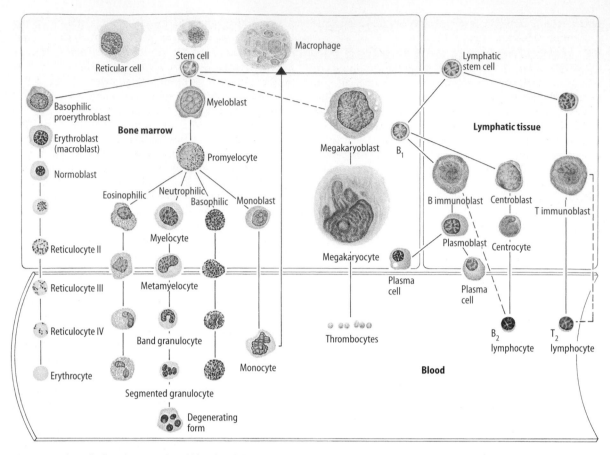

Fig. 84.1. The cells found in peripheral blood and their precursors in the germinal centers, the bone marrow, and lymphatic system. (From [149])

84.2.1 Hematopoietic Stromal Cells

In vitro studies using long-term bone marrow cultures of either mice [1] or humans [2] as well as in vivo experiments have helped in understanding the nature of bone marrow stroma and its interactions with the hematopoietic progenitor cells. Long-term bone marrow cultures form an adherent layer which is analogous to the stromal cells and which binds progenitor cells during their development. As the progenitor cells approach maturity, however, they appear to lose their surface molecules which interact with the stromal components. The localization of hematopoiesis to confined sites within the body appears to involve specific as well as nonspecific surface interactions between stromal cells and progenitor cells.

Interaction Between Stromal Cells and Hematopoietic Cells. The importance of the interaction between stromal cells and hematopoietic progenitor cells can be clearly seen in two mutant mouse strains. The white spotting locus (*W*) and the steel locus (*Sl*) each define genes which are necessary for normal hematopoietic development. Homozygous *W* or *Sl* mutations both lead to prenatal death, but doubly heterozygous mice, such as the *Sl/Sl^d* or *W/W^v* genotypes, are phenotypically similar, with lack of cutaneous pigment,

sterility, and congenital anemia [3].[1] From bone marrow transplantation studies, the *Sl/Sl^d* mouse was found to have a stromal defect while the *W/W^v* mouse had a hematopoietic progenitor cell defect [3]. The nature of these specific mutations was revealed with the cloning of the genes at the *W* and *Sl* loci. The *W* locus is the protooncogene c-*kit* [4,5] and the *Sl* gene is the soluble and membrane-bound growth factor called stem cell growth factor or c-*kit* ligand [6–8]. The c-*kit* protein is the hematopoietic progenitor cell receptor for the growth factor product of the *Sl* gene expressed by stromal cells. While the interaction of c-*kit* and c-*kit* ligand is important in pigmentation and gonadal function as well as hematopoiesis, some other stromal cell–progenitor cell interactions may be more specific to hematopoiesis. A glycoprotein lectin on the surface of hematopoietic progenitor cells has been shown to interact specifically with mannose moieties and exposed galactose moieties present on surface molecules of bone marrow stromal cells [9]. This interaction is calcium ion dependent and appears to

---

[1] S1/S1^d and W/W^v are compound heterozygotes that are commonly used because they are viable and reach adulthood. S1 and W are null mutations resulting from gene deletions. S1^d and W^v are point mutations that result in proteins with decreased function.

play a role in initial interactions between hematopoietic progenitor cells and stromal cells.

Extracellular Matrix. In addition to the direct binding of surface molecules on progenitor cells and stromal cells, an extracellular matrix provides an anchoring mechanism for hematopoietic progenitor cells. This matrix is produced mainly by the stromal cells and is closely associated with their surfaces. Some specificity is apparent in the adhesive properties of the extracellular matrix since the protein fibronectin binds erythroid progenitor cells [10] whereas the protein hemonectin [11] binds granulocytic progenitor cells. The proteoglycans, proteins with extensive sulfation, such as heparan sulfate and chondroitin sulfate, are other extracellular matrix proteins which contribute to adhesion between the stroma and the hematopoietic progenitor cells [9]. The proteoglycans can also function to concentrate soluble growth factors. For example, granulocyte-macrophage colony-stimulating factor binds to heparan sulfate in the extracellular matrix of marrow stroma [12,13].

84.2.2 Hematopoietic Stem Cells

The hematopoietic progenitor cells are composed of the pluripotent hematopoietic stem cells and their many generations of descendants. Within the hematopoietic organs the pluripotent hematopoietic stem cell is extremely rare, with its incidence in murine bone marrow being estimated to the between one per 10^4 and one per 10^5 nucleated cells [14–16]. Because of their rarity in the hematopoietic organs, the pluripotent stem cells can only be studied in retrospect after they have been transplanted from one mouse to another mouse which has had its endogenous hematopoietic tissue ablated. This ablation is accomplished by lethal doses of irradiation and/or chemotherapy. After transplantation of the hematopoietic tissue, the long-term (greater than 6 months) reconstitution of hematopoiesis is due to the pluripotent hematopoietic stem cells in the transplanted tissue.

Criteria for Defining the Murine Hematopoietic Stem Cell. The criteria for defining the murine hematopoietic stem cell came from experiments on spleens of mice at 1 week after similar transplantation studies, these spleens showing growth of macroscopic colonies containing cells of multiple lineages [17]. These colonies were the progeny of individual transplanted cells and these cells then were termed spleen colony-forming units (CFU-S). Since the cells in these spleen colonies could, in turn, be injected into secondary, irradiated mice and give rise to spleen colonies, the CFU-S were believed to have replicated themselves. Thus, a cell that could both replicate itself and give rise to multiple blood cell lineages could be identified and quantified with the CFU-S assay. The criteria for being a hematopoietic stem cell became the abilities (a) to self-replicate and (b) to differentiate into multiple blood cell lineages.

Although they were able to self-replicate and could give rise to more than one lineage of blood cell, CFU-S were found in subsequent studies to be early descendants of the pluripotent hematopoietic stem cells rather than the hematopoietic stem cells themselves. Clonal growth of lymphoid cells in spleen colonies could not be demonstrated [18,19]. When the observation time for CFU-S assays was extended from 1 week to 2 weeks after transplantation, a series of evanescent colonies were found and those appearing on later days had greater self-replication and multilineage differentiation capacities [20,21]. Indeed, cells with the capacity for long-term hematopoietic reconstitution could be separated from CFU-S by their size and density [22]. These findings led to the current criteria for defining the *pluripotent hematopoietic stem cell as a cell that can self-replicate and can provide long-term hematopoietic reconstitution, including lymphoid cells.*

Using these revised criteria, the identification of pluripotent hematopoietic stem cells still requires hematopoietic cell transplantation into an ablated host. However, the analyzed cells are not spleen colonies but rather the reconstituted hematopoietic tissues. One assay mixes hematopoietic cells of a known genotype with various proportions of hematopoietic cells from a congenic strain and then transplants the mixtures into ablated host mice in a competitive repopulation assay [23]. Another assay uses limiting dilution of genotypically distinct donor cells that are transplanted into stem cell deficient W/W^v mice which can be used as transplantation hosts rather than hematopoietically ablated mice [14]. Another approach genetically marks individual hematopoietic stem cells with replication-defective, recombinant retroviruses [24–26]. In these marking experiments, the normally quiescent hematopoietic stem cells of donor mice are induced into cell cycle by 5-fluorouracil treatment, harvested, and infected in vitro with the retroviruses. The one-time, random integration of the provirus into the cellular DNA provides a marker which can be detected in the various generations of progeny that develop after transplantation of the infected cells.

Studies with these repopulation assays have confirmed that the self-replication criterion is essential in defining pluripotent hematopoietic stem cells. An alternative explanation of stem cell kinetics is that a large pool of nonreplicating, quiescent, pluripotent hematopoietic stem cells exists in the body and that at any one time a few of these stem cells are activated and contribute to hematopoiesis for an interval which is relatively short compared with the animal's life span. In this scenario, an animal would have a supply of stem cells that is sufficient for much longer than its normal life span and hematopoiesis would be maintained by continuous activation of individual stem cells from the large quiescent pool [27,28]. Thus, hematopoiesis would be dependent on a "clonal succession" of non-self-renewing pluripotent stem cells and their progeny. When hematopoietic repopulation studies are performed with genetically marked stem cells and the blood cell progeny are analyzed after 6 months or more, clonal succession does not occur but rather a stable population of a very few stem cells supplies the hematopoiesis [29–31]. Furthermore, these stem cells can self-replicate because the primarily transplanted mice can be used to reconstitute multiple secondary transplantation recipients and the same clones provide the long-term hematopoiesis [31,32].

Pluripotent Stem Cell Enrichment. As mentioned above, the pluripotent hematopoietic stem cell is a very rare cell in the hematopoietic tissues and no pure population has been obtained. However, some procedures can greatly increase their concentration in a population of hematopoietic cells. Not only does 5-fluorouracil treatment of mice induce cycling in the normally quiescent hematopoietic stem cells, it also enriches their percentages more than tenfold due to its toxicity to the more mature hematopoietic progenitor cells [33]. Similarly, dyes specific for actively dividing cells, such as rhodamine-123, which binds to mitochondria, can be used to select for the faintly stained, mitotically quiescent stem cells [34]. Pluripotent stem cell enrichment of several hundredfold has been accomplished through cell sorting and using a combination of monoclonal antibodies which

select undifferentiated cells displaying certain specific antigens while lacking any antigens found on lineage-committed hematopoietic cells [35]. While these selection procedures greatly enrich hematopoietic stem cell populations and may aid future endeavors such as providing stem cell targets for gene therapy, they have also demonstrated that long-term marrow reconstitution can occur only if the marrow-ablated recipient receives a cotransplantation of more mature, lineage-committed hematopoietic progenitor cells. These committed progenitor cells transiently maintain blood cell production in the period before establishment of the transplanted stem cells as the new source of hematopoietic progenitor cells.

84.2.3 Committed Hematopoietic Progenitor Cells

Cell Differentiation. The daughter cell of a pluripotent hematopoietic stem cell that is no longer able to self-replicate and that has begun to differentiate is termed a committed progenitor cell. A scheme outlining the lineages and stages of committed hematopoietic cell differentiation is

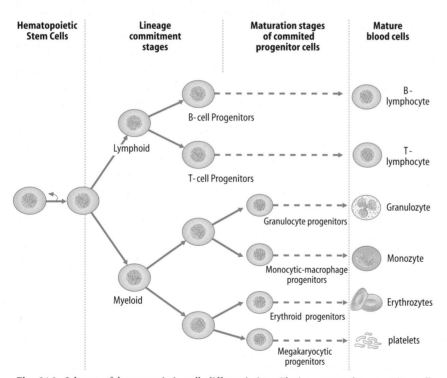

Fig. 84.2. Scheme of hematopoietic cell differentiation. Pluripotent hematopoietic stem cells can self-replicate (indicated by *curved arrow*) as well as give rise to progenitor cells which are committed to differentiation. Because of the uncertainty about the events leading to commitment as well as uncertainty about the self-replication potential of cells around the time of commitment, a hypothetical cell has been placed at the border between the hematopoietic stem cell and the lineage commitment stages. Through a series of steps the descendants of the hematopoietic stem cells become committed not only to differentiation but to differentiation along specific cell lineages. After lineage commitment, the progenitor cells mature and acquire the characteristics of terminally differentiated cells found in the blood. During this maturation phase lineage-specific hematopoietic growth factors act on the progenitor cells. The scheme shows only the stages of differentiation and not the proliferation which accompanies hematopoietic cell development. The proliferative potential of pluripotent stem cells is extremely large since repopulation of entire blood cell lineages can be accomplished by the descendants of a single pluripotent hematopoietic stem cell. Proliferative potential decreases as the cells progress through the various stages toward mature blood cells. (From [36])

shown in Fig. 84.2. Although various stages between the pluripotent hematopoietic stem cell and the mature blood cells are shown in Fig. 84.2, hematopoietic differentiation is a continuum with multiple cell division between the stages. Thus, a few hematopoietic stem cells give rise through successive cell divisions to increasingly greater numbers of cells at each subsequent point in the differentiation scheme. In this scheme, each stage has less proliferative potential and more restricted differentiation potential than the preceding stage. The cellular events which determine stem cell commitment to differentiation as well as subsequent commitment to a specific differentiation lineage are not known.

Hypotheses Regarding the Commitment Process. Two hypotheses have emerged to explain the commitment process. One hypothesis proposes that the environment of hematopoietic stem cells or multilineage progenitors regulates commitment. In this deterministic scheme, hematopoietic cell commitment is regulated by either interaction with stromal cells in the hematopoietic microenvironment [37] or interactions with soluble hematopoietic growth factors [38]. The second hypothesis proposes that random or stochastic intracellular events determine when and along which lineage hematopoietic differentiation will proceed. This second model is supported by statistical analyses of in vivo studies with murine CFU-S [39] and in vitro studies with murine [40] and human [41] multilineage cells. In this stochastic model, hematopoietic growth factors provide a permissive environment for hematopoietic progenitor cell growth but the determinations of when and how to differentiate are not influenced by the environment of the cell.

Multilineage Colony-Forming Cells. In the generations immediately following hematopoietic stem cell commitment, the cells retain their ability to differentiate into multiple lineages. Thus, the CFU-S is an example of a hematopoietic progenitor that has recently become committed to differentiation. In addition to growing into a spleen colony containing multiple cell lineages in vivo, cells at the CFU-S stage can form colonies of cells of multiple lineages when cultured under appropriate conditions [42-44]. With murine cells, these colonies grow to macroscopic sizes over 2 weeks or more and contain many thousands of cells. Although the cell that gives rise to such a colony is named according to the particular assay, in general they are considered as multilineage colony-forming cells. The multilineage colony-forming cell is the most immature stage of human hematopoiesis which has been examined experimentally. Although many experiments with pluripotent hematopoietic stem cells in human transplantation would not be ethical, human cells with short-term, partial hematopoietic reconstitution capabilities have been studied following transplantation into mice with severe combined immune deficiency (SCID) [45]. The transplanted human cells appear to be unaffected by murine hematopoietic growth factors in the transplantation model using the SCID mouse and human hematopoietic cells [45].

However, the transplanted cells do respond to exogenously administered human hematopoietic growth factors. Thus, this model has great potential for defining the roles of specific hematopoietic growth factors in the growth and differentiation of multilineage progenitor cells.

Progenitor Cells. The descendants of the multilineage colony-forming cells are more mature cells, as shown by their more restricted differentiation potential and their decreased proliferative capacity. These more mature progenitor cells are also assayed in vitro by their ability to form colonies. With mouse cells, these colonies requires only a week or less to reach full size and they are only visible microscopically. Human counterparts to each of the defined murine colony-forming cells have been identified and similar colonies are formed by their growth in vitro. The one major difference between human and murine cells is that the time required for colony development in vitro of any specific colony-forming cell is greater with human cells than murine cells. A restricted multilineage progenitor cell is the granulocyte-macrophage colony-forming unit (CFU-GM) which, as its name implies, has two lineages of cells in the colonies it forms [46]. In the mouse, these colonies require 1 week to develop fully. The hematopoietic stages which follow the multilineage progenitor cells are even more mature in that they are restricted to a single lineage and have even less proliferative potential. These progenitor cells such as the granulocyte colony-forming unit (CFU-G), macrophage colony-forming unit (CFU-M), erythroid colony-forming unit (CFU-E), or megakaryocyte colony-forming unit (CFU-MK) give rise in vitro to small microscopic colonies over periods ranging from 2 to 7 days [47-49].

Cells in the stages of hematopoiesis that follow these single lineage colony-forming cells in vivo are sufficiently differentiated that they can be identified microscopically within the hematopoietic tissue. These cells, which are the precursors of the mature blood cells, comprise the erythroblasts, megakaryocytes, myelocytes, and developing monocyte-macrophage cells and lymphocytes. When fixed, stained, and examined microscopically, cells from the various lineages can be identified easily by their extremely different morphological appearances. Their proliferative capacity is limited to a few cell divisions, but their differentiation programs have diverged widely at these late stages. The erythroblasts are rapidly producing hemoglobin and preparing a membrane skeleton which will maintain cell shape following the loss of their nuclei. The nucleus of the erythroblast becomes inactive, condenses, and is separated from the remainder of the cell, which becomes the reticulocyte. Over the period of 1 day, a series of remodeling events leads to the maturation of the irregularly shaped, organelle-containing reticulocyte into the simple biconcave-shaped erythrocyte that lacks internal organelles. In myelocytes or maturing granulocytic cells, nuclei condense but are retained in the cell while granules containing specific proteolytic enzymes develop and accumulate in the cytoplasm. Similar changes occur in developing monocyte-macrophage cells. The late stage

Table 84.1. Families of hematopoietic growth factors based on receptor structure

Group	Growth factor
Hematopoietic/ cytokine group	Erythropoietin (EPO) Granulocyte colony-stimulating factor (G-CSF) Granulocyte-macrophage colony-stimulating factor (GM-CSF) Thrombopoietin (*c-mpl* ligand) Interleukins-2, -3, -4, -5, -6 Leukemia inhibitory factor (LIF)
Receptor tyrosine kinase group	Macrophage colony-stimulating factor (CSF-1) Stem cell factor (*c-kit* ligand)

megakaryocytes become polyploid as they replicate their nuclear DNA several times without undergoing cytokinesis. At the same time, they develop specific α-granules, dense granules, and demarcation membranes in their cytoplasm. The formation of the platelet is presumed to occur when some demarcated, small portion of the cell separates from the remainder of the megakaryocyte. It is not known whether this separation occurs individually for each platelet or en masse with the concurrent demise of the megakaryocyte.

84.3 Hematopoietic Growth Factors

The development of the in vitro assays that detect and quantitate the various committed multilineage and single lineage colony-forming cells led to the discovery and characterization of many growth factors which regulate growth of hematopoietic progenitor cells. Upon purification, these hematopoietic growth factors were found to be glycoproteins. Subsequently their respective genes were cloned and the corresponding recombinant glycoprotein factors were produced. The abundance of pure, recombinant factors led in turn to the identification of and eventual cloning of the genes for the specific cell receptors for each of the hematopoietic growth factors. From the structures and function of these growth factors and their respective cell surface receptors, two families of hematopoietic growth factors can be defined (Table 84.1).

One group of factors have receptors with several conserved amino acid sequences. Based on these conserved sequences, predictive models of similar tertiary structure for the members of this hematopoietin/cytokine family were proposed [50,51]. Indeed, the structures of the three members of this family which have been crystallized – growth hormone, interleukin-2 (IL-2), and granulocyte-macrophage colony-stimulating factor (GM-CSF) – have each conformed to the predictive model [51,52]. Other members of the hematopoietin/cytokine family which have direct effects on the committed hematopoietic progenitor

cells include erythropoietin, granulocyte colony-stimulating factor (G-CSF), interleukin-3 (IL-3 or multi-CSF), interleukin-5 (IL-5), and interleukin-6 (IL-6). The receptor proteins for this family of hematopoietic growth factors are transmembrane proteins without amino acid sequence homology to any proteins with known signal transduction properties such as tyrosine kinase or guanosine triphosphate binding domains. However, each of these receptor proteins appears to form a complex with at least one other protein with presumed signal transduction capability. In fact, certain of these putative signal transducing proteins have been found in complexes with several different members of the family of hematopoietin/ cytokine receptors [53–55]. These results indicate that a specific hematopoietic growth factor requires cellular expression of a specific receptor protein to bind to a hematopoietic target cell but that the growth factor–receptor interaction may activate a signal transduction pathway common to several different growth factors in the hematopoietin/cytokine family. The nature of the signal transduction mechanisms for this group of hematopoietic growth factors and their receptor complexes remains unknown.

The second group of hematopoietic growth factors has only two members – macrophage colony-stimulating factor (CSF-1) and c-kit ligand or stem cell factor. The receptors for these two factors are the protein products of the protooncogenes c-fms and c-kit, respectively. Both receptors are transmembrane proteins with cytoplasmic domains that contain tyrosine kinase activity [4,5,56]. Thus, the receptors for each of these two hematopoietic growth factors have a single protein with both a growth factor binding domain and a signal transduction domain. As with the first group of hematopoietic growth factors, the mechanisms of action of CSF-1 and c-kit ligand are unknown. However, a potential regulatory role of CSF-1 in cell division has been suggested by its regulation of murine genes induced in the first gap (G_1) phase of the cell cycle in macrophages [57].

84.3.1 Multilineage Growth Factors

The growth factors shown to have effects in vivo and in vitro on the growth and development of multilineage colony-forming cells include IL-3, c-kit ligand, GM-CSF, and IL-6 [47]. Experiments characterizing the receptors for these multilineage factors and the effects of the factors on target cells have been performed largely with leukemic cell lines which are dependent upon one of the factors for survival in vitro. However, the role of each of these factors in the development of multilineage hematopoietic progenitor cells is not clear for four interrelated reasons. First, the multilineage progenitors are a minor population among many different cell types in hematopoietic tissue. Second, except for IL-3, which is produced almost exclusively by T lymphocytes, these multilineage factors are produced by fibroblasts, endothelial cells, macrophages, and T lymphocytes [47], all of which are present in the hematopoietic cell populations that contain the multilineage colony-forming cells. Third, some of these multilineage growth factors can induce the production of the other hematopoietic growth factors. For example, in response to either IL-3 or GM-CSF, macrophages can produce G-CSF [58] or M-CSF [59]. Thus, discriminating between direct and indirect actions of the multilineage growth factors on multilineage colony-forming cells is very difficult. Fourth, cooperative and synergistic actions of these multilineage factors can play a major role in the development of multilineage colony-forming cells (see [60] for listing of these synergies).

84.3.2 Granulocyte and Macrophage Growth Factors

Interleukin-3 and GM-CSF can affect the growth and development of both granulocytes and monocytes-macrophages [47]. Those factors with effects limited to single lineages are G-CSF for neutrophilic granulocytes, CSF-1 for monocyte-macrophages, and IL-5 for eosinophilic granulocytes. Although the lineage relationships of basophils and mast cells to other granulocytes is uncertain, the mast cell progenitors appear to require c-*kit* ligand [8]. Deficiencies of these single lineage growth factors lead to absence or dysfunction of the mature blood cell population. For example, genetic deficiency of CSF-1 leads to osteopetrosis [61,62] and genetic deficiency of c-*kit* ligand leads to absence of mast cells [8]. The lineage-specific factors are not only required for differentiation of the late-stage committed progenitor cells, but also maintain and activate the mature blood cells in these lineages.

Like IL-3, IL-5 is produced by T lymphocytes; but like GM-CSF, the sources of G-CSF, c-*kit* ligand, and M-CSF are multiple and include fibroblasts, endothelial cells, and macrophages. *Thus, all of the cell types which produce the granulocyte and macrophage growth factors are present throughout the body and they are concentrated in areas of inflammation.* The presence of such cells in the bone marrow stroma and the existence of disease states resulting from deficiency states of the specific factors such as osteopetrosis and mast cell deficiency (described above) suggest that normal production of granulocytes and monocyte-macrophages depends upon local sources of these factors. Thus, in bone marrow stroma and in sites of inflammation, the cell types that produce the factors controlling differentiation and function of granulocytes and monocytes-macrophages are in close proximity to each other. This proximity of the factor-producing cells, the abilities of some factors to induce the production of other factors, and the synergistic action of many of the factors result in an extremely complex regulation of granulocytopoiesis and monocytopoiesis. Many more experiments are needed to understand all of the cellular and growth factor interactions which control granulocyte and macrophage-monocyte production under normal and inflammatory conditions.

84.3.3 Megakaryocyte Growth Factors

A growth factors that acts specifically on megakaryocytes or on megakaryocyte progenitor cells has not been identified. However, the growth of megakaryocyte colonies in vitro can be stimulated by three of the known hematopoietic growth factors: IL-3, IL-6, and GM-CSF [49]. The same caveats concerning the indirect or direct action of these growth factors as are described above for progenitor cells of granulocytes and monocyte-macrophages also apply to CFU-MK growth in mice. In the hematopoietic tissues used for the assays, CFU-MK are rare cells mixed among many different cells capable of producing hematopoietic growth factors. The terminal differentiation of megakaryocytes into platelets, however, appears to be inversely related to the number of platelets in the blood. Reduction of platelet numbers in rodents by antiplatelet antibodies or exchange transfusion of platelet-poor blood results in increased megakaryocyte numbers, ploidy, and size whereas platelet transfusions decrease these parameters [49]. These acute changes of platelet numbers in the blood do not affect CFU-MK numbers in the hematopoietic tissues [63]. These results have led to a model of megakaryocyte development in which CFU-MK growth and development are governed by one or more of the hematopoietic growth factors while the terminal differentiation of megakaryocytes is controlled by an unidentified thrombopoietic factor which is induced by thrombocytopenia [63,64]. Although cell lines with some megakaryocytic markers have been used in some studies, understanding megakaryocytopoiesis and platelet formation will require the identification of the putative thrombopoietic factor and the development of in vitro cell systems in which purified popuations of megakaryocytes can mature to the point of producing platelets.

84.4 Erythropoiesis

While nonerythroid hematopoiesis is characterized by complex interactions of growth factors that can be produced in numerous sites by several different cell types, the control of erythropoiesis is much simpler. One hematopoietic growth factor, erythropoietin (EPO), is the principal regulator of red blood cell production. Specific cell types in the kidney and liver produce EPO and secrete it into the bloodstream. Thus, EPO is a hormone that can be readily measured in serum samples. EPO and erythropoiesis are part of a negative feedback cycle which regulates tissue oxygenation. In this cycle, the delivery of oxygen to the body tissues is a function of the number of circulating red blood cells.

To keep a constant number of circulating erythrocytes, production of new red blood cells must balance both the chronic daily loss of senescent erythrocytes and any acute losses of erythrocytes due to hemorrhage or hemolysis. Approximately 1% of the circulating red blood cells are lost each day in normal humans. As red blood cells age they

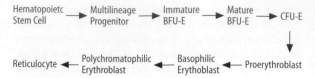

Fig. 84.3. Stages of normal erythroid differentiation. The stages are part of a continuous process of proliferation and differentiation with varying numbers of cell divisions between different stages. Each stage has less proliferative potential than the preceding stage but the number of cells is much greater than in the preceding stage. Cells in stages through the CFU-E are assayed by growth in vivo or in vitro whereas cells from the proerythroblast through reticulocyte are identified microscopically

undergo numerous biochemical changes, but changes in surface membrane proteins appear to be responsible for targeting old cells for removal (see [65] and [66] for reviews). Both qualitative changes in the antigenic structure of these membrane proteins and their association in clusters appear to increase the deposition of specific immunoglobulins and complement on the aging cells [67–69]. The immunoglobulins and complement are recognized by macrophages which then phagocytose the senescent cells and thereby remove them from the circulation [66]. In a normal human adult, this chronic removal of senescent erythrocytes creates a demand for the production of billions of erythrocytes each day. When hemorrhage or hemolysis occurs, this level of erythrocyte production must increase manyfold to maintain constant tissue oxygenation.

The control of erythrocyte production is mediated through the regulation of EPO production. When red blood cell numbers decrease, the body tissues receive less oxygen. This tissue hypoxia induces EPO production by the kidney and liver. The increased EPO, in turn, results in the increased production of erythrocytes. The increased numbers of erythrocytes are able to deliver more oxygen to the tissues and thereby relieve the hypoxia and remove the stimulus for increased EPO production. Thus, through this feedback cycle the number of circulating erythrocytes is tightly controlled. The subsequent sections of this chapter will be concerned with the erythroid lineage as shown in Fig. 84.2 and the role of EPO in the development of cells in this lineage.

84.4.1 Erythroid Progenitor Cells

In Fig. 84.3, the stages of erythroid differentiation are shown. The earliest progenitor cell committed to erythroid differentiation that can be identified is the burst-forming unit-erythroid (BFU-E) [70]. In semisolid or viscous culture medium, BFU-E will grow into either single large colony of erythroblasts or a group of small colonies of erythroblasts. These bursts of erythroblasts contain from 50 to more than a thousand erythroblasts and require culture periods of 7 days or more for mice and 14 days or more for humans. The next major stage of development is

the colony-forming unit-erythroid (CFU-E) [71]. In culture, these cells grow into single colonies of eight or more erythroblasts over a period of 2 days for murine cells and 7 days for human cells. An intermediate stage between the BFU-E and CFU-E has been defined and is referred to as mature BFU-E [72,73]. The immediate descendants of the CFU-E are the proerythroblasts and basophilic erythroblasts which are identified microscopically. Hemoglobin production begins at the basophilic erythroblast stage and increases as these cells differentiate into the polychromatophilic and orthochromatic stages. During these last two stages the cells enucleate and yield reticulocytes.

84.4.2 Erythroid Growth Factors

Although EPO is the principal regulator of erythropoiesis, other hematopoietic growth factors can have effects on the early stages of erythroid progenitor cell development. Erythrocyte production is enhanced by the administration of c-*kit* ligand to anemic *Sl/Sl^d* mice which are genetically deficient in this factor [8]. In vitro, IL-3, GM-CSF, and c-*kit* ligand have each been shown to support growth of immature BFU-E for the first few days in culture [74–77]. The later stages of erythropoiesis, however, are regulated by a single hematopoietic growth factor – EPO. While reducing or increasing EPO concentrations in vivo does not affect the numbers of BFU-E in the hematopoietic organs [78,79], the number of CFU-E are greatly increased by high EPO levels and greatly diminished by low EPO levels [78–80]. In other words, erythroid progenitor cells acquire EPO responsiveness at some point between the immature BFU-E and the CFU-E stages of development. Some studies place the acquisition of EPO responsiveness slightly earlier in that they have shown that a subset of BFU-E can respond to EPO in vitro [81–84].

Investigations of EPO receptor expression at the various stages of erythroid progenitor cell differentiation have confirmed that CFU-E and proerythroblasts have surface receptors. Indeed, these stages display the most abundant receptors, with about 1000 per cell [85]. As these progenitors differentiate into late-stage erythroblasts and reticulocytes, EPO receptor expression progressively declines such that mature erythrocytes lack EPO receptors [86,87]. Whether erythroid progenitors have EPO receptors at stages prior to the CFU-E is not known because sufficiently pure and plentiful populations of erythroid progenitor cells at these early stages have not been available. One study using autoradiography and a partially purified population of BFU-E indicated that these cells expressed progressively more EPO receptors up to the CFU-E stage, after which receptor number declined [87].

84.4.3 Erythropoietin and Its Production

Erythropoietin is a 165-amino acid, 30.4-kD glycoprotein of which approximately 40% is polysaccharide (see [88–90].

for recent reviews of EPO). Although EPO has not been crystallized, it is a member of the hematopoietic/ cytokinase family of proteins and, thus, has a predicted tertiary structure. The predicted structure of EPO is that of four α-helical regions arranged in antiparallel fashion and joined by intervening peptide loops of varying lengths [51,52]. The predicted receptor binding region is in the terminal portion of the helix closest to the carboxy terminus [91]. Although glycosylation of EPO is not necessary for EPO action in vitro, it plays an important role in EPO metabolism in vivo since partially glycosylated EPO molecules are much more rapidly cleared by the liver than fully glycosylated molecules [92–95].

Erythropoietin is encoded by a single copy gene [96,97]. The transcription of the gene is controlled by the degree of tissue hypoxia [98,99]. In adult mammals, the main site of EPO production is the kidney, with the liver being a secondary, minor site [100–102]. The liver plays a larger role during earlier periods of development but its contribution relative to the kidneys varies significantly according to species [103–105]. Studies with isolated perfused kidneys indicate that the sensing of hypoxia as well as the production of EPO occurs within the same organ [106–108]. In vitro studies with hepatoma cell lines have shown that the same cell which senses the hypoxia can produce the EPO [109]. EPO is induced within 1–2 h after hypoxic stimulation and is secreted into the bloodstream without any intracellular storage [102,110]. When the hypoxic stimulus is removed, the EPO production ceases in a similarly abrupt manner [110].

The intracellular sensing mechanism by which hypoxia leads to EPO gene transcription is not well understood. Experiments in vitro using EPO-producing hepatoma cell lines and in vivo using transgenic mice carrying human EPO gene constructs have indicated that cis-acting DNA sequences and trans-acting DNA binding proteins play a role in the induction of transcription (see [90] for review of factors). The most clearly defined cis-acting element is an enhancer element located 3' to the coding region of the EPO gene [111–113]. Although the trans-acting transcription factors have not been characterized, one study of the individual and combined effects of hypoxia, heavy metal ions, and carbon monoxide on EPO gene expression has led to the proposal that a heme-containing protein plays an important role in hypoxia-related induction [114]. This proposed heme protein would be analogous to hemoglobin in that it has an oxy-conformation and a deoxy-conformation. In this model deoxy-conformation caused by cellular hypoxia would be active in the EPO transcription signal while the oxy-conformation would be inactive.

The cloning of the EPO gene provided the necessary probes for in situ hybridization to detect intracellular EPO mRNA. Earlier attempts to identify the renal EPO-producing cells by immunodetection of EPO itself yielded conflicting results because EPO circulates in the blood and is excreted through the kidneys. Thus, immunodetection methods could not distinguish between cells producing EPO and those simply having plasma or urinary EPO associated with them. In situ hybridization allowed a much greater accuracy in the localization of the EPO-producing cells in vivo. In situ hybridization shows the EPO mRNA-containing cells to be a subset of interstitial cells in the renal cortex [115,116] and hepatocytes and sinusoid-associated cells in the liver [117]. In situ hybridization studies in the kidneys of mice with progressively severe, acute anemias induced by phlebotomy provided some insights into the physiological events involved in EPO production [118]. This study showed that:

- Each EPO-producing cell contained a set amount of EPO mRNA, i.e., EPO mRNA was induced in an all-or-none manner.
- More cells per unit of renal cortex were recruited to produce EPO as the anemias became more severe.
- The number of renal cortical cells producing EPO varied inversely and exponentially with hematocrit.
- The number of renal cortical cells producing EPO and the total renal EPO mRNA levels correlated very highly with serum EPO concentrations.

The distribution of EPO-producing cells in the renal cortex was very significant. Normal mice had rare, single EPO-producing cells scattered in the inner renal cortex; slightly anemic mice had small clusters of EPO-producing cells in the inner part of the renal cortex; moderately anemic mice had larger groups of cells extending from the inner cortex into the midcortex; and the severely anemic mice has a generalized cortical distribution of EPO-producing cells. These results suggest that during anemia the renal cortex develops focal areas of hypoxia in which the cellular threshold for hypoxic induction of EPO gene transcription is met. Therefore, each cell capable of producing EPO in the focal area is induced to do so. As the anemia becomes more severe, exponential increases occur in the size of the foci of renal cortical hypoxia, the total number of cells induced to produce EPO, and ultimately the plasma concentration of EPO. Similar exponential responses in plasma EPO concentration as related to the degree of anemia occur in humans with a wide variety of anemias except for the anemia of renal disease [119].

84.4.4 Mechanism of EPO Action

The interaction of EPO and its target erythroid progenitor cells occurs at the cell surface, where specific receptors bind EPO with reported dissociation constants of 100–600 pM [85]. Following its binding to the surface receptor EPO is rapidly internalized and degraded by lysosomal enzymes in the erythroid progenitor cell [120]. The binding component of the EPO receptor has been cloned and is a 507 amino acid, transmembrane glycoprotein member of the hematopoietin/cytokine receptor family [121]. Its molecular weight on SDS-PAGE gels has been reported to be from 66 kD to 75 kD depending upon differences in glycosylation and phosphorylation [121–123]. Unlike many other members of the hematopoietin/cytokine receptor family, the EPO binding component of the EPO receptor has not had any second component of the receptor identified. Two proteins of 105 kD and 85 kD with similar peptide maps have been identified in several studies using chemical cross-linking of radiolabeled EPO [121,124,125]. The nature and function, if any, of these two cross-linked proteins are unknown. Although several natural or laboratory-produced mutants of the EPO-binding component have been shown to have either increased [126,127] or decreased (reviewed in [90]) function in the maintenance of EPO-dependent cell lines, the signal transduction mechanism of the EPO receptor remains unknown. A wide variety of cellular proteins have been reported to be phosphorylated following EPO binding [90]. Some of these proteins have been identified, such as the EPO binding component itself [122,123] and the *raf-1* oncogene product [128], but most are identified by their apparent molecular weight on SDS-PAGE. No conclusive evidence for involvement of other known signal transduction mechanisms has been shown (reviewed in [90]).

At the cellular level, EPO has several possible modes of action. Three of these possibilities will be considered. The first possible mode of action is that EPO induces a program of differentiation. In this case, the immature erythroid progenitor cells would develop under the influence of hematopoietic growth factors other than EPO, but those events which occur in terminal differentiation, including such processes as hemoglobin synthesis, enucleation, and membrane skeleton formation, would be coordinated in a master program. This master program would be induced and maintained by EPO. The control of such a program would then rely on some factors(s) involved in each process of the terminal differentiation of erythroid cells. A potential candidate for such a factor would be the transcription factor GATA-1 [129,130]. The DNA binding site for GATA-1 protein is found in many genes involved in erythroid differentiation including the globins, heme synthetic enzymes, and the EPO receptor. However, GATA-1 is not the master controller of terminal erythroid differentiation since it occurs in megakaryocytes and mast cells and these cells [131,132] do not differentiate into erythrocytes. Although EPO may be found to control directly a critical event in erythroid differentiation in some future study, a series of events without known relationship to EPO's action appear to control the erythroid differentiation program.

A second possible mode of EPO action is that of a mitogen. In this case, immature erythroid progenitor cells would proliferate and mature under the influence of other hematopoietic growth factors but they would reach a point in differentiation when they require EPO to undergo cell division. Without EPO they would not proliferate but rather become quiescent. This scenario appears to have arisen from the assays of BFU-E and CFU-E in which these progenitor cells grew into colonies in tissue culture in the presence of EPO. Hence, EPO's effects on these progenitors have been regarded as a "stimulation" of proliferation and differentiation, i.e. EPO stimulates the erythroid progenitors to enter the cell cycle. These concepts of cell cycle,

quiescent states, and mitogen stimulation have arisen from studies with continuous cell lines in which cell growth can be prevented by either nutrient or growth factor deprivation. These deprivations lead to a nonproliferating, latent state termed G_0 which represents a phase outside of the normal cell cycle. Repletion of the missing nutrient or growth factor leads to the return to cell cycle from the G_0 phase. Evidence for such an effect of EPO has been reported for human BFU-E [82] and some erythroid cell lines [133,134]. For CFU-E and proerythroblasts, however, this effect has not been demonstrated because the majority of these cells are in DNA synthesis (i.e., in the S phase of cell cycle) regardless of the EPO levels in the animals from which they were obtained [78,79]. In other words, CFU-E and proerythroblasts, which are the cells with the greatest number of EPO receptors, are already in cell cycle and therefore EPO cannot induce them into cell cycle. Thus, while EPO may have a possible mitogenic effect on immature erythroid progenitors such as the BFU-E, it does not have a mitogenic effect on cells in the later stages of differentiation.

The third possible mode of EPO action is that it permits the survival of erythroid progenitor cells that would otherwise die. In this case, the immature erythroid progenitor cells would proliferate and mature under the influence of other hematopoietic growth factors until they reach a stage in which they would require EPO to prevent their death. Experimental evidence indicates that CFU-E and proerythroblasts undergo programmed cell death (apoptosis) in the absence of EPO [135]. Apoptosis is a widespread phenomenon occurring in many different types of cells and is characterized by specific biochemical and morphological changes in the affected cells [136]. These changes include homogeneous nuclear condensation, decreased cell size, and internucleosomal DNA cleavage. EPO prevents the apoptosis of CFU-E and proerythroblasts in a concentration-dependent manner. From in vitro studies, the period of erythroid differentiation which is dependent upon EPO for the prevention of cell death begins in CFU-E which have only four cell divisions remaining in their total differentiation [137]. This period of EPO dependence extends until the generation of erythroblasts which have one remaining division. These erythroblasts with the capability of only a single cell division can complete their differentiation into reticulocytes in the absence of EPO [138,139].

A model of erythropoiesis based upon EPO acting to prevent apoptosis in the late stages of erythropoietic progenitor cell development [135,140] is depicted in Fig. 84.4. This model also takes into account the heterogeneity in EPO responsiveness of the erythroid progenitor cells in the EPO-dependent period. In the model, normal erythropoiesis is sustained by a minority of the erythroid progenitor cells that reach the EPO-dependent stages. In other words, normal EPO levels support the survival of only a minority of the total number of erythroid progenitors that have the potential to complete terminal differentiation. Most erythroid progenitors lost to apoptosis under normal conditions. However, when EPO levels are

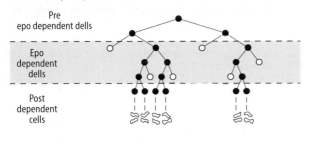

A Normal Erythropoitin

Pre epo dependent dells

Epo dependent dells

Post dependent cells

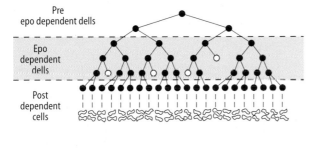

B Increased Erythropoitin

Pre epo dependent dells

Epo dependent dells

Post dependent cells

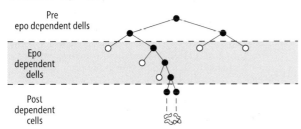

C Decreased Erythropoitin

Pre epo dependent dells

Epo dependent dells

Post dependent cells

Fig. 84.4. Model of erythropoiesis based on EPO suppression of programmed cell death (apoptosis). Erythroid progenitor cells enter a period of development in which they are dependent upon EPO for survival (EPO-dependent cells). This EPO-dependent period begins within the CFU-E stage and lasts through the basophilic erythroblast stage. ●, Surviving viable cells; ○, cells undergoing programmed cell death owing to insufficient EPO. Before entering the EPO-dependent period, the progenitors can survive without EPO (pre-EPO-dependent cells). Those cells surviving transit through the EPO-dependent period can complete maturation into reicticulocytes without EPO and ultimately become red cells (post-EPO-dependent cells). Only one division appears to occur in the post-EPO-dependent stages. (From [140]; Reprinted with permission from TRANSFUSION published by The American Association of Blood Banks)

elevated, as occurs in anemia, many of the erythroid progenitor cells that would normally undergo apoptosis can survive and complete differentiation. Conversely, when EPO levels are lower than normal, as occurs in renal failure, some of the erythroid progenitors that would survive the EPO-dependent period normally are now lost to apoptosis and the number completing differentiation into reticulocytes is less than normal. This model is consistent with several observations concerning EPO, erythroid progenitor cells, and erythrocyte production. The model can help explain observations such as:

- Loss of CFU-E in vitro occurs when EPO addition to the cultures is delayed even for a few hours [141–143].
- The majority of CFU-E are detected in vitro only when EPO concentrations are greater than those found in normal plasma [48].
- The dose responsiveness of CFU-E in vitro covers a range that varies more than 100-fold (from 0.01 to 2 units/ml or 2.5 to 500 pM) [144,145], corresponding to the wide range of plasma EPO concentrations seen with varying degrees of anemia.
- The physiological correction of anemia due to endogenous EPO production is characterized by a very prompt reticulocytosis but is not associated with any rebound polycythemia.

These observations and this model point to a control mechanism of erythropoiesis in which the capacity for a rapid increase in erythrocyte production in response to acute anemia is in place, but which is not used under normal conditions. In the case of acute anemia, a wide range of EPO concentrations is found in the plasma depending upon the severity of the anemia. The erythroid progenitor cell population in the EPO-dependent stages of differentiation is able to respond in a graded manner. This graded response to any EPO concentration over a very wide range is mediated through the heterogeneity in the EPO concentration required to prevent apoptosis in any individual cell. Thus, rapid finely controlled changes in the erythrocyte production rate can be obtained through EPO's prevention of erythroid progenitor cell apoptosis.

84.4.5 Negative Regulators of Hematopoiesis

This chapter has concentrated on hematopoietic growth factors which either stimulate or permit the proliferation and differentiation of hematopoietic cells. A much less studied group of hematopoietic regulatory factors are those which can reversibly inhibit or prevent proliferation and differentiation of hematopoietic cells. Hematopoiesis can be regarded as regulated through the net effect of both stimulatory (positive) and inhibitory (negative) factors. The negative regulatory factors for hematopoiesis have been the subject of recent reviews [146–148]. These negative regulators are for the most part nonspecific in that they affect progenitors in several different lineages as well as multilineage progenitors and hematopoietic stem cells. Many of the negative regulatory factors are associated with inflammation and thus are produced by T-lymphocytes, fibroblasts, and macrophages. Although many of these factors have not been purified and their genes cloned, some have been. These include transforming growth factor-β, interferon-α, β, and γ, macrophage inflammatory protein α, and tissue necrosis factors [146–148].

Although the mechanism of action is not known for the negative regulatory factors, studies with hematopoietic stem cells and committed multilineage progenitor cells suggest that they prevent the transition from the G$_1$ phase to the S phase of the cell cycle or that in cells which have a

1689

G_0 phase they prevent entry into the cell cycle. In this context, these negative regulatory factors can only have an effect when a positive regulatory factor is present, as, for example, following 5-fluorouracil treatment in vivo or after exposure to stimulatory growth factors in vitro. Furthermore, the same negative regulatory factors which appear to inhibit cell cycle progression in the early stage progenitors have either stimulatory effects or no effects on later stage progenitors [146–148].

As with many of the positive growth factors, the mixed cell populations used for assays of the negative regulatory factors raise questions about the direct or indirect actions of these regulators as well as any combined or synergistic effects of two or more factors on hematopoietic cells. In view of the complexity of the production and interaction of these negative regulatory factors, a simple mechanism to explain specific physiological roles in hematopoiesis cannot yet be proposed.

References

1. Dexter TM, Allen TD, Lajtha LG (1977) Conditions controlling the proliferation of haemopoietic stem cells in vitro. J Cell Physiol 91:335–344
2. Gartner S, Kaplan HS (1980) Long-term culture of human bone marrow cells. Proc Natl Acad Sci U S A 77:4756–4759
3. Russell ES (1979) Hereditary anemias of the mouse: a review for geneticists. Adv Genet 20:357–459
4. Chabot B, Stephenson DA, Chapman VM, Besmer P, Bernstein A (1988) The proto-oncogene c-kit encoding a transmembrane tyrosine kinase receptor maps to the mouse W locus. Nature 335:88–89
5. Geissler EN, Ryan MA, Housman DE (1988) The dominant-white spotting (W) locus of the mouse encodes the c-kit proto-oncogene. Cell 55:185–192
6. Copeland NG, Gilbert DJ, Cho BC, Donovan PJ, Jenkins NA, Cosman D, Anderson D, Lyman SD, Williams DE (1990) Mast cell growth factor maps near the steel locus on mouse chromosome 10 and is deleted in a number of steel alleles. Cell 63:175–183
7. Huang E, Nocka K, Beier DR, Chu T-Y, Buck J, Lahm H-W, Wellner D, Leder P, Besmer P (1990) The hematopoietic growth factor KL is encoded by the sl locus and is the ligand of the c-kit receptor, the gene product of the W locus. Cell 63:225–233
8. Zsebo KM, Williams DA, Geissler EN, Broudy VC, Martin FH, Atkins HL, Hsu R-Y, Birkett NC, Okino KH, Murdock DC, Jacobsen FW, Langley KE, Smith KA, Takeishi T, Cattanach BM, Galli SJ, Suggs SV (1990) Stem cell factor is encoded at the Sl locus of the mouse and is the ligand for the c-kit tyrosine kinase receptor. Cell 63:213–224
9. Tavassoli M, Hardy CL (1990) Molecular basis of homing of intravenously transplanted stem cells to the marrow. Blood 76:1059–1070
10. Patel VP, Lodish HF (1986) The fibronectin receptor on mammalian erythroid precursor cells: characterization and developmental regulation. J Cell Biol 102:449–456
11. Campbell AD, Long MW, Wicha MS (1987) Haemonectin, a bone marrow adhesion protein specific for cells of granulocyte lineage. Nature 329:744–746
12. Gordon MY, Riley GP, Watt SM, Greaves ME (1987) Compartmentalization of a haemopoietic growth factor (GM-CSF) by glycosaminoglycans in the bone marrow microenvironment. Nature 326:403–405
13. Roberts R, Gallagher J, Spooncer E, Allen TD, Bloomfield F, Dexter TM (1988) Heparan sulphate bound growth factors: a mechanism for stromal cell mediated haemopoiesis. Nature 332:376–378
14. Boggs DR, Boggs SS, Saxe DG, Gress LA, Canfield DR (1982) Hematopoietic stem cells with high proliferative potential. J Clin Invest 70:242–253
15. Jones RJ, Sharkis SJ, Celano P, Colvin OM, Rowley SD, Sensenbrenner LL (1987) Progenitor cell assays predict hematopoietic reconstitution after syngeneic transplantation in mice. Blood 70:1186–1192
16. Harrison DE, Astle CM, Stone M (1989) Numbers and functions of transplantable primitive immunohemopoietic stem cells: effects of age. J Immunol 142:3833–3840
17. Till JE, McCulloch EA (1961) A direct measurement of the radiation sensitivity of normal mouse bone marrow cells. Radiat Res 14:213–222
18. Siminovitch L, McCulloch EA, Till JE (1963) The distribution of colony-forming cells among spleen colonies. J Cell Comp Physiol 62:327–336
19. Wu AM, Till JE, Siminovitch L, McCulloch EA (1967) A cytological study of the capacity for differentiation of normal hemopoietic colony-forming cells. J Cell Physiol 69:177–184
20. Curry JL, Trentin JJ (1967) Hematopoietic spleen colony studies. I. Growth and differentiation. Dev Biol 15:395–400
21. Magli MC, Iscove NN, Odartchenko N (1982) Transient nature of early haematopoietic spleen colonies. Nature 295:527–529
22. Jones RJ, Wagner JE, Celano P, Zicha MS, Sharkis SJ (1990) Separation of pluripotent hematopoietic stem cells from spleen colony forming cells. Nature 347:188–189
23. Harrison DE (1980) Competitive repopulation: a new assay for long-term stem cell functional capacity. Blood 55:77–81
24. Dick JE, Magli MC, Huszar D, Phillips RA, Bernstein A (1985) Introduction of a selectable gene into primitive stem cells capable of long-term reconstitution of the hemopoietic system of W/Wᵛ mice. Cell 42:71–79
25. Keller G, Paige C, Gilboa E, Wagner EF (1985) Expression of a foreign gene in myeloid and lymphoid cells derived from multipotent haematopoietic precursors. Nature 318:149–154
26. Lemischka IR, Raulet DH, Mulligan RC (1986) Developmental potential and dynamic behavior of hematopoietic stem cells. Cell 45:917–927
27. Mintz B, Anthony K, Litwin S (1984) Monoclonal derivation of mouse myeloid and lymphoid lineages from totipotent hematopoietic stem cells experimentally engrafted in fetal hosts. Proc Natl Acad Sci USA 81:7835–7839
28. Micklem HS, Lennon JE, Ansell JD, Gray RA (1987) Numbers and dispersion of repopulating hematopoietic cell clones in radiation chimeras as functions of injected cell dose. Exp Hematol 15:251–257
29. Harrison DE, Astle CM, Lerner C (1988) Number and continuous proliferative pattern of transplanted primitive immunohematopoietic stem cells. Proc Natl Acad Sci USA 85:822–826
30. Jordan CT, Lemischka IR (1990) Clonal and systemic analysis of long-term hematopoiesis in the mouse. Genes Dev 4:220–232
31. Keller G, Snodgrass R (1990) Life span of multipotential hematopoietic stem cells in vivo. J Exp Med 171:1407–1418
32. Capel B, Hawley RG, Mintz B (1990) Long- and short-lived murine hematopoietic stem cell clones individually identified with retroviral integration markers. Blood 75:2267–2270
33. Lerner C, Harrison DE (1990) 5-Fluorouracil spares hemopoietic stem cells responsible for long term repopulation. Exp Hematol 18:114–118
34. Spangrude GJ, Johnson GR (1990) Resting and activated subsets of mouse multipotent hematopoietic stem cells. Proc Natl Acad Sci USA 87:7433–7437
35. Spangrude GJ, Heimfeld S, Weissman IL (1988) Purification and characterization of mouse hematopoietic stem cells. Science 241:58–62

36. Koury MJ, Bondurant MC (1993) Prevention of programmed death in hematopoietic progenitor cells by hematopoietic growth factors. News Physiol Sci 8:170–174

37. Trentin JJ (1970) Influence of hematopoietic organ stroma (hematopoietic inductive microenvironments) on stem cell differentiation. In: Gordon A (ed) Regulation of hematopoiesis. Appleton-Century-Crofts, New York, pp 161–168

38. Goldwasser E (1975) Erythropoietin and differentiation of red blood cells. Fed Proc 34:2285–2292

39. Till JE, McCulloch EA, Siminovitch L (1964) A stochastic model of stem cell proliferation, based on the growth of spleen colony forming cells. Proc Natl Acad Sci USA 51:29–36

40. Nakahata T, Gross AJ, Ogawa M (1982) A stochastic model of self-renewal and commitment to differentiation of the primitive hemopoietic stem cells in culture. J Cell Physiol 113:455–458

41. Leary AG, Ogawa M, Strauss LC, Civin CI (1984) Single cell origin of multilineage colonies in culture: evidence that differentiation of multipotent progenitors and restriction of proliferative potential of monopotent progenitors are stochastic processes. J Clin Invest 74:2193–2197

42. Hodgson GS, Bradley TR (1979) Properties of haematopoietic stem cells surviving 5-fluorouracil treatment: evidence for a pre-CFU-S cell? Nature 281:381–382

43. Nakahata T, Ogawa M (1982) Identification in culture of a class of hemopoietic colony-forming units with extensive capability to self-renew and generate multipotential hemopoietic colonies. Proc Natl Acad Sci U S A 79:3843–3847

44. Pragnell IB, Wright EG, Lorimore SA, Adam J, Rosendaal M, DeLamarter JF, Freshney M, Eckmann L, Sproul A, Wilkie N (1988) The effect of stem cell proliferation regulatiors demonstrated with an in vitro assay. Blood 72:196–201

45. Lapidot T, Pflumio F, Doedens M, Murdoch B, Williams DE, Dick JE (1992) Cytokine stimulation of multilineage hematopoiesis from immature human cells engrafted in SCID mice. Science 255:1137–1141

46. Moore MAS, Williams N, Metcalf D (1972) Purification and characterization of the in vitro colony forming cells in monkey hemopoietic tissue. J Cell Physiol 79:283–292

47. Metcalf D (1991) Control of granulocytes and macrophages: molecular, cellular, and clinical aspects. Science 254:529–533

48. Eaves AC, Eaves CJ (1984) Erythropoiesis in culture. Clin Haematol 13:371–391

49. Hoffman R (1989) Regulation of megakaryocytopoiesis. Blood 74:1196–1212

50. Bazan JF (1990) Structural design and molecular evolution of a cytokine receptor superfamily. Proc Natl Acad Sci USA 87:6934–6938

51. Bazan JF (1990) Haemopoietic receptors and helical cytokines. Immunol Today 11:350–354

52. Manavalan P, Swope DL, Withy RM (1992) Sequence and structural relationships in the cytokine family. J Prot Chem 11:321–331

53. Kitamura T, Sato N, Arai K, Miyajima A (1991) Expression cloning of the human IL-3 receptor cDNA reveals a shared β subunit for human IL-3 and GM-CSF receptors. Cell 66:1165–1174

54. Tavernier J, Devos R, Cornelis S, Tuypens T, Van der Heyden J, Fiers W, Plaetinck G (1991) A human high affinity interleukin-5 receptor (IL5R) is composed of an IL5-specific α chain and a β chain shared with the receptor for GM-CSF. Cell 66:1175–1184

55. Gearing DP, Comeau MR, Friend DJ, Gimpel SD, Thut CJ, McGourty J, Brasher KK, King JA, Gillis S, Mosley B, Ziegler SF, Cosman D (1992) The IL-6 signal transducer, gp130: an oncostatin M receptor and affinity converter for the LIF receptor. Science 255:1434–1437

56. Sherr CJ, Rettenmier CW, Sacca R, Roussel MF, Look AT, Stanley ER (1985) The c-*fms* proto-oncogene product is related to the receptor for the mononuclear phagocyte growth factor, CSF-1. Cell 41:665–676

57. Matsushime H, Roussel MF, Ashmun RA, Sherr CJ (1991) Colony-stimulating factor 1 regulates novel cyclins during the G1 phase of the cell cycle. Cell 65:701–713

58. Oster W, Lindemann A, Mertelsmann R, Herrmann F (1989) Granulocyte-macrophage colony-stimulating factor (CSF) and multilineage CSF recruit human monocytes to express granulocyte CSF. Blood 73:64–67

59. Vellenga E, Rambaldi A, Ernst TJ, Ostapovicz D, Griffin JD (1988) Independent regulation of M-CSF and G-CSF gene expression in human monocytes. Blood 71:1529–1532

60. Bagby GC, Segal GM (1991) Growth factors and the control of hematopoiesis. In: Hoffman R, Benz EJ, Shattil SJ, Furie B, Cohen HJ (eds) Hematology basic principles and practice. Churchill Livingstone, New York, p 108

61. Yoshida H, Hayashi S-I, Kunisada T, Ogawa M, Nishikawa S, Okamura H, Sudo T, Shultz LD, Nishikawa S-I (1990) The murine mutation osteopetrosis is in the coding region of the macrophage colony-stimulating factor gene. Nature 345:442–444

62. Wiktor-Jedrzejczak W, Bartocci A, Ferrante W, Ahmed-Ansari A, Sell KW, Pollard JW, Stanley ER (1990) Total absence of colony-stimulating factor 1 in the macrophage-deficient osteopetrotic (op/op) mouse. Proc Natl Acad Sci USA 87:4828–4832

63. Burstein SA, Adamson JW, Erb SK, Harker LA (1981) Megakaryocytopoiesis in the mouse: response to varying platelet demand. J Cell Physiol 109:333–341

64. Williams N, Eger RR, Jackson HM, Nelson DJ (1982) Two-factor requirement for murine megakaryocyte colony formation. J Cell Physiol 110:101–104

65. Clark MR (1988) Senescence of red blood cells: progress and problems. Physiol Rev 68:503–554

66. Lutz HU (1990) Erythrocyte clearance. In: Harris JR (ed) Blood cell biochem, vol I. Plenum, New York, pp 81–120

67. Kay MMB (1975) Mechanism of removal senescent cells by human macrophages in situ. Proc Natl Acad Sci USA 72:3521–3525

68. Singer JA, Jennings LK, Jackson C, Dockter ME, Morrison M, Walker WS (1986) Erythrocyte homeostasis: antibody-mediated recognition of the senescent state by macrophages. Proc Natl Acad Sci USA 83:5498–5501

69. Turrini F, Arese P, Yuan J, Low PS (1991) Clustering of integral membrane proteins of the human erythrocyte membrane stimulates autologous IgG binding, complement deposition, and phagocytosis. J Biol Chem 266:23611–23617

70. Axelrad AA, McLeod DL, Shreeve MM, Heath DS (1973) Properties of cells that produce erythrocytic colonies in vitro. In: Robinson WA (ed) Proceedings of the 2nd international workshop on hemopoiesis in culture. DHEW Publ NIH 74–205, pp 226–234

71. Stephenson JR, Axelrad AA, McLeod DL, Shreeve MM (1971) Induction of colonies of hemoglobin-synthesizing cells by erythropoetin in vitro. Proc Natl Acad Sci USA 68:1542–1546

72. Gregory CJ (1976) Erythropoietin sensitivity as a differentiation marker in the hemopoietic system: studies of three erythropoietic colony responses in culture. J Cell Physiol 89:289–302

73. Gregory CJ, Eaves AC (1977) Human marrow cells capable of erythropoietic differentiaiton in vitro: definition of three erythroid colony responses. Blood 49:855–864

74. Metcalf D, Johnson GR, Burgess AW (1980) Direct stimulation by purified GM-CSF of the proliferation of multipotential and erythroid precursor cells. Blood 55:138–147

75. Emerson SG, Yang YC, Clark SC, Long MW (1988) Human recombinant granulocyte-macrophage colony stimulating factor and interleukin-3 have overlapping but distinct hematopoietic activities. J Clin Invest 82:1282–1287

76. Sonoda Y, Yang YC, Wong GG, Clark SC, Ogawa M (1988) Erythroid burst-promoting activity of purified recombinant

human GM-CSF and interleukin-3: studies with anti-GM-CSF and anti-IL-3 sera and studies in serum-free cultures. Blood 72:1381–1386

77. Dai CH, Krantz SB, Zsebo KM (1991) Human burst-forming units-erythroid need direct interaction with stem cell factor for further development. Blood 78:2493–2497

78. Iscove NN (1977) The role of erythropoietin in regulation of population size and cell cycling of early and late erythroid precursors in mouse bone marrow. Cell Tissue Kinet 10:323–334

79. Hara H, Ogawa M (1977) Erythropoietic precursors in mice under erythropoietic stimulation and suppression. Exp Hematol 5:141–148

80. Gregory CJ, Eaves AC (1978) Three stages of erythropoietic progenitor cell differentiation distinguished by a number of physical and biologic properties. Blood 51:527–537

81. Sieff CA, Emerson SG, Mufson A, Gesner TG, Nathan DG (1986) Dependence of highly enriched human bone marrow progenitors on hematopoietic growth factors and their response to recombinant erythropoietin. J Clin Invest 77:74–81

82. Dessypris EN, Krantz SB (1984) Effect of pure erythropoietin on DNA synthesis by human marrow day 15 erythroid burst-forming units in short-term liquid culture. Br J Haematol 56:295–306

83. Emerson SG, Thomas S, Ferrara JL, Greenstein JL (1989) Developmental regulation of erythropoiesis by hematopoietic growth factors: analysis on populations of BFU-E from bone marrow, peripheral blood, and fetal liver. Blood 74:49–55

84. Valtieri M, Gabbianelli M, Pelosi E, Bassano E, Petti S, Russo G, Testa U, Peschle C (1989) Erythropoietin alone induces erythroid burst formation by human embryonic but not adult BFU-E in unicellular serum-free culture. Blood 74:460–470

85. Sawyer ST (1990) Receptors for erythropoietin. Distribution, structure, and role in receptor-mediated endocytosis in erythroid cells. In: Harris JR (ed) Blood cell biochemistry, vol I. Plenum, New York, pp 365–402

86. Landschulz KT, Noyes AN, Rogers O, Boyer SH (1989) Erythropoietin receptors on murine erythroid colony-forming units: natural history. Blood 73:1476–1486

87. Sawada K, Krantz SB, Dai CH, Koury ST, Horn ST, Glick AD, Civin CI (1990) Purification of human blood burst-forming units-erythroid and demonstration of the evolution of erythropoietin receptors. J Cell Physiol 142:219–230

88. Krantz SB (1991) Erythropoietin. Blood 77:419–434

89. Spivak JL (1992) The mechanism of action of erythropoietin: erythroid cell response. In: Fisher Jw (ed) Biochemical pharmacology of blood and blood forming organs. Springa, Berin Heidelbery New York, pp 49–114 (Handbook of experimental pharmacology, vol 101)

90. Koury MJ, Bondurant MC (1992) The molecular mechanism of erythropoietin action. Eur J Biochem 210:649–663

91. Fibi MR, Stuber W, Hintz-Obertreis P, Luben G, Krumwieh D, Siebold B, Zettlemeissl G, Kupper HA (1991) Evidence for the location of the receptor-binding site of human erythropoietin at the carboxy-terminal domain. Blood 77:1203–1210

92. Dordal MS, Wang FF, Goldwasser E (1985) The role of carbohydrate in erythropoietin action. Endocrinology 116:2293–2299

93. Takeuchi M, Inoue N, Strickland TW, Kubota M, Wada M, Shimizu R, Hoshi S, Kozutsumi H, Takasaki S, Kobata A (1989) Relationship between sugar chain structure and biological activity of recombinant human erythropoietin produced in Chinese hamster ovary cells. Proc Natl Acad Sci U S A 86:7819–7822

94. Wasley LC, Timony G, Murtha P, Stoudemire J, Dorner AJ, Caro J, Krieger M, Kaufman RJ (1991) The importance of N- and O-linked oligosaccharides for the biosynthesis and in vitro and in vivo biologic activities of erythropoietin. Blood 77:2624–2632

95. Yamaguchi K, Akai K, Kawanishi G, Ueda M, Masuda S, Sasaki R (1991) Effects of site-directed removal of N-glycosylation sites in human erythropoietin on its production and biological properties. J Biol Chem 266:20434–20439

96. Jacobs K, Shoemaker C, Rudersdorf R, Neill SD, Kaufman RJ, Mufson A, Seehra J, Jones SS, Hewick R, Fritsch EF, Kawakita M, Shimizu T, Miyake T (1985) Isolation and characterization of genomic and cDNA clones of human erythropoietin. Nature 313:806–810

97. Lin FK, Suggs S, Lin CH, Browne JK, Smalling R, Egrie JC, Chen KK, Fox GM, Martin F, Stabinsky Z, Badrawi SM, Lai PH, Goldwasser E (1985) Cloning and expression of the human erythropoietin gene. Proc Natl Acad Sci USA 82:7580–7584

98. Schuster SJ, Badiavas EV, Gosta-Giomi P, Weinmann R, Erslev AJ, Caro J (1989) Stimulation of erythropoietin gene transcription during hypoxia and cobalt exposure. Blood 73:13–16

99. Goldberg MA, Gaut CC, Bunn HF (1991) Erythropoietin mRNA levels are governed by both the rate of gene transcription and posttranscriptional events. Blood 77:271–277

100. Jacobson LO, Goldwasser E, Fried W, Plazak L (1957) Role of the kidney in erythropoiesis. Nature 179:633–634

101. Fried W (1972) The liver as a source of extrarenal erythropoietin production. Blood 40:671–677

102. Bondurant MC, Koury MJ (1986) Anemia induces accumulation of erythropoietin mRNA in the kidney and liver. Mol Cell Biol 6:2731–2733

103. Zanjani ED, Poster J, Burlington H (1977) Liver as the primary site of erythropoietin production in the fetus. J Lab Clin Med 89:640–644

104. Koury MJ, Bondurant MC, Graber SE, Sawyer ST (1988) Erythropoietin messenger RNA levels in developing mice and transfer of ^{125}I-erythropoietin by the placenta. J Clin Invest 82:154–159

105. Eckardt KU, Ratcliffe PJ, Tan CC, Bauer C, Kurtz A (1992) Age-dependent expression of the erythropoietin gene in rat liver and kidneys. J Clin Invest 89:753–760

106. Ratcliffe PJ, Jones RW, Phillips RE, Nicholls LG, Bell JI (1990) Oxygen-dependent modulation of erythropoietin mRNA levels. J Exp Med 172:657–660

107. Pagel H, Jelkmann W, Weiss CH (1991) Isolated serum-free perfused rat kidneys release immunoreactive erythropoietin in response to hypoxia. Endocrinology 128:2633–2638

108. Scholz H, Schurek HJ, Eckardt HU, Kurtz A, Bauer C (1991) Oxygen dependent erythropoietin production by the isolated perfused rat kidney. Pflugers Arch 418:228–233

109. Goldberg MA, Glass GA, Cunningham JM, Bunn HF (1987) The regulated expression of erythropoietin by two human hepatoma cell lines. Proc Natl Acad Sci USA 84:7972–7976

110. Schuster SJ, Wilson JH, Erslev AJ, Caro J (1987) Physiologic regulation and tissue localization of renal erythropoietin messenger RNA. Blood 70:316–318

111. Beck I, Ramirez S, Weinmann R, Caro J (1991) Enhancer element at the 3'-flanking region controls transcriptional response to hypoxia in the human erythropoietin gene. J Biol Chem 266:15563–15566

112. Semenza GL, Nejfelt MK, Chi SM, Antonarakis SE (1991) Hypoxia-inducible nuclear factors bind to an enhancer element located 3' to the human erythropoietin gene. Proc Natl Acad Sci USA 88:5680–5684

113. Pugh CW, Tan CC, Jones RW, Ratcliffe PJ (1991) Functional analysis of an oxygen-regulated transcriptional enhancer 3' to the mouse erythropoietin gene. Proc Natl Acad Sci USA 88:10553–10557

114. Goldberg MA, Dunning SP, Bunn HF (1988) Regulation of the erythropoietin gene: evidence that the oxygen sensor is a heme protein. Science 242:1412–1415

115. Koury ST, Bondurant MC, Koury MJ (1988) Localization of erythropoietin synthesizing cells in murine kidneys by in situ hybridization. Blood 71:524–527

116. Lacombe C, DaSilva JL, Bruneval P, Fournier JG, Wendling F, Casadevall N, Camilleri JP, Bariety J, Varet B, Tambourin P (1988) Peritubular cells are the site of erythropoietin synthesis in the murine hypoxic kidney. J Clin Invest 81:620–623

117. Koury ST, Bondurant MC, Koury MJ, Semenza GL (1991) Localization of cells producing erythropoietin in murine liver by in situ hybridization. Blood 77:2497–2503

118. Koury ST, Koury MJ, Bondurant MC, Caro J, Graber SE (1989) Quantitation of erythropoietin-producing cells in the kidneys of mice by in situ hybridization: correlation with hematocrit, renal erythropoietin messenger RNA and serum erythropoietin concentration. Blood 74:645–651

119. Erslev AJ (1991) Erythropoietin. N Engl J Med 324:1339–1344

120. Sawyer ST, Krantz SB, Goldwasser E (1987) Binding and receptor-mediated endocytosis of erythropoietin in Friend virus infected erythroid cells. J Biol Chem 262:5554–5562

121. D'Andrea AD, Lodish HF, Wong G (1989) Expression cloning of the murine erythropoietin receptor. Cell 57:277–285

122. Miura O, D'Andrea A, Kabat D, Ihle JN (1991) Induction of tyrosine phosphorylation by the erythropoietin receptor correlates with mitogenesis. Mol Cell Biol 11:4895–4902

123. Dusanter-Fourt I, Casadevall N, Lacombe C, Muller O, Billat C, Fischer S, Mayeux P (1992) Erythropoietin induces the tyrosine phosphorylation of its own receptor in human erythropoietin-responsive cells. J Biol Chem 267:10670–10675

124. Sawyer ST (1989) The two proteins of the erythropoietin receptor are structurally similar. J Biol Chem 264:13343–13347

125. Mayeux P, Lacombe C, Casadevall N, Chretien S, Dusanter I, Gisselbrecht S (1991) Structure of the murine erythropoietin receptor complex characterization of the erythropoietin cross-linked proteins. J Biol Chem 266:23380–23385

126. Yoshimura A, Longmore G, Lodish HF (1990) Point mutation in the exoplasmic domain of the erythropoietin receptor resulting in hormone-independent activation and tumorigenicity. Nature 348:647–649

127. D'Andrea A, Yoshimura A, Youssoufian H, Zon LI, Koo J-W, Lodish HF (1991) The cytoplasmic region of the erythropoietin receptor contains nonoverlapping positive and negative growth-regulatory domains. Mol Cell Biol 11:1980–1987

128. Carroll MP, Spivak JL, McMahon M, Welch N, Rapp UR, May WS (1991) Erythropoietin induces raf-1 activation and raf-1 is required for erythropoietin-mediated proliferation. J Biol Chem 266:14964–14969

129. Evans TM, Reitman M, Felsenfeld G (1988) An erythrocyte-specific DNA-binding factor recognizes a regulatory sequence common to all chicken globin genes. Proc Natl Acad Sci USA 85:5976–5980

130. Wall L, deBoer E, Grosveld F (1988) The human beta-globin gene 3' enhancer contains multiple binding sites for an erythroid-specific protein. Genes Dev 2:1089–1100

131. Martin DIK, Zon LI, Mutter G, Orkin SH (1990) Expression of an erythroid transcription factor in Megakaryocytic and mast cell lineages. Nature 344:444–447

132. Romeo PH, Prandini MH, Joulin V, Mignotte V, Prenant M, Vainchenker W, Marguerie G, Uzan G (1990) Megakaryocytic and erythrocytic lineages share specific transcription factors. Nature 344:447–449

133. Tsuda H, Sawada T, Sakaguchi M, Kwakita M, Takatsuki K (1989) Mode of action of erythropoietin in Epo dependent murine cell line. I. Involvement of adenosine 3':5'-cyclic monophosphate not as a second messenger but as a regulator of cell growth. Exp Hematol 17:211–217

134. Spivak JL, Pham T, Isaacs M, Hankins WD (1991) Erythropoietin is both a mitogen and a survival factor. Blood 77:1228–1233

135. Koury MJ, Bondurant MC (1990) Erythropoietin retards DNA breakdown and prevents programmed death in erythroid progenitor cells. Science 248:378–381

136. Wyllie AH (1987) Apoptosis: cell death in tissue regulation. J Pathol 153:313–316

137. Landschulz KT, Boyer SH, Noyes AN, Rogers OC, Frelin LP (1992) Onset of erythropoietin response in murine erythroid colony-forming units: assignment to early S-phase in a specific cell generation. Blood 79:2749–2758

138. Nijhof W, Wierenga PK, Sahr K, Beru N, Goldwasser E (1987) Induction of globin mRNA transcription by erythropoietin in differentiating erythroid precursor cells. Exp Hematol 15:779–784

139. Koury MJ, Bondurant MC (1988) Maintenance by erythropoietin of viability and maturation of murine erythroid precursor cells. J Cell Physiol 137:65–74

140. Koury MJ, Bondurant MC (1990) Control of red cell production: the roles of programmed cell death (apoptosis) and erythropoietin. Transfusion 30:673–674

141. Iscove NN (1978) Erythropoietin-independent stimulation of early erythropoiesis in adult marrow cultures by conditioned media from lectin-stimulated mouse spleen cells. In: Golde DW, Cline MJ, Metcalfe D, Fox CF (eds) Hematopoietic cell differentiation. Academic, New York, pp 37–52

142. Del Rizzo DF, Axelrad AA (1985) Erythroid progenitor cells (CFU-E*) from Friend virus-infected mice undergo ^{55}Fe suicide in vitro in the absence of added erythropoietin. Exp Hematol 13:1055–1061

143. Lipton JM, Kudisch M, Nathan DG (1981) Response of three classes of human erythroid progenitors to the absence of erythropoietin in vitro as a measure of progenitor maturity. Exp Hematol 9:1035–1041

144. Iscove NN, Sieber F, Winterhalter KH (1974) Eythroid colony formation in cultures of mouse and human bone marrow: analysis of the requirement for erythropoietin by gel filtration and affinity chromatography on agarose-concanavalin A. J Cell Physiol 83:309–320

145. Gregory CJ, Tepperman AD, McCulloch EA, Till JE (1974) Erythropoietic progenitors capable of colony formation in culture: Response of normal and genetically anemic W/W^v mice to manipulations of the erythron. J Cell Physiol 84:1–12

146. Axelrad AA (1990) Some hemopoietic negative regulators. Exp Hematol 18:143–150

147. Graham GJ, Pragnell IB (1990) Negative regulators of haemopoiesis – current advances. Prog Growth Factor Res 2:181–192

148. Moore MAS (1991) Clinical impications of positive and negative hematopoietic stem cell regulators. Blood 78:1–19

149. Weiss C, Jelkmann W (1989) Functions of the blood. In: Schmidt RF, Thews G (eds) Human physiology, 2nd edn. Springer, Berlin Heidelberg New York

85 Nonerythroid and Immune Competent Cells of the Blood

J. LINDENMANN and C. BAUER

Contents

85.1 Introduction

White blood cells perform important functions in the defense against invading microorganisms and probably also in the elimination of effete or otherwise damaged cells. The science of immunology, more than 100 years old [1], has greatly expanded over the past 30 years. The following short summary can give but a severely restricted and simplified description of some of the main features of modern immunology. For a more balanced view and further details the reader is referred to excellent recent textbooks [2,3].

85.2 Specific and Nonspecific Mechanisms

Antimicrobial defense can be roughly subdivided into a nonspecific and a specific arm. The difference between the two is more quantitative than qualitative, and both non-specific and specific mechanisms interact and cooperate.

The fundamental event in both cases is a recognition process based on the steric match between structures on microorganisms and corresponding receptors on cells. The "goodness of fit" between ligand and receptor varies over an enormous range and is expressed as avidity. The threshold above which a ligand–receptor interaction is to be called specific remains largely arbitrary. Even interactions of low avidity need not be biologically meaningless. Certain low-avidity reactions favor apposition of neighboring complementary structures, by themselves also interacting with low avidity, so that an overall strong reaction may ensue.

Nonspecific Immunity. Operationally, the nonspecific defenses are characterized by their capacity to react immediately to a beginning invasion. This requires the presence, in sufficiently large numbers, of precommitted cells equipped with receptors recognizing invading microorganisms. The diversity of receptors serving this task is severely limited. To be useful, these receptors have to match structures widely distributed among potential invaders, such as the endotoxins of gram-negative bacteria. Hence this system has the intrinsic advantage of being immediately mobilizable, but it will have limited discriminatory power. Moreover, escape maneuvers on the part of the invader are easily conceivable.

Specific Immunity. On top of the above-described system a more sophisticated one has evolved, which makes extensive use of the older structures. In specific immunity, sometimes called adaptive immunity, the fundamental event also consists in a steric match between receptors and ligands, but here the variety of receptors is enormous, as a consequence of which each receptor is represented only in relatively small numbers. Cells equipped with a given receptor are too few for efficient action, and therefore have to be increased in number. This "expansion," in fact clonal proliferation of cells carrying the required receptors, takes time, so that invading microorganisms will encounter the immune barrier only after several days. However, this immune barrier will appear tailor-made for the particular requirement. This is called the *clonal selection theory* of specific or adaptive immunity. Once expansion has taken place, a state of memory persists for a long time, so that later recalls by the same invading organism will result in a much more rapid answer, which explains the nonrecurrence of certain infectious diseases.

R. Greger/U. Windhorst (Eds.)
Comprehensive Human Physiology, Vol. 2
© Springer-Verlag Berlin Heidelberg 1996

85.3 Self-Nonself Discrimination

There is an intrinsic danger in the possession of a system capable of recognizing a vast range of chemical structures and ready to react against them. Multicellular organisms themselves contain innumerable chemical components. How should a potential reactivity against self be discouraged? This is one of the central problems of immunity. In primitive organisms relying on unspecific mechanisms only, it can easily be imagined that, should a receptor arise (by mutation or recombination) reacting deleteriously against self, then the bearer of such a receptor would be eliminated by natural selection. In adaptive immunity, where an enormous array of specificities is generated by mechanisms operating at random (see below), holding self-reactivity in check is not trivial. Several possibilities have been considered, generally under the heading of "tolerance of self"; none of them is entirely satisfactory.

Suffice it to say at this point that self-reactivity is not entirely eliminated from the immune system of higher vertebrates, and that, in fact, such self-reactivity may play an important part not only in the intrathymic "education" of lymphocytes and the intrinsic control of the immune system itself, but also in the absolutely vital task of eliminating spent cells. There exist in normal serum quite elusive antibodies with reactivity against autologous red blood cell constituents. These autoantibodies are present in very small concentrations and are of such low avidity that their qualification as antibodies might be questioned. Nevertheless, there are indications that they perform essential functions in the disposal of effete (aged) red blood cells, which have to be constantly eliminated in large numbers. Besides their specificity for autologous red cell surface constituents, these antibodies appear to have a further binding affinity for components of the complement cascade. In ageing red blood cells clustering of surface molecules occurs, which allows bivalent binding and thus stabilization of low-affinity antibodies; these antibodies in turn bind complement components, resulting in a powerful opsonizing signal, leading to phagocytosis and disposal of the senescent red blood cells [4]. The terminology used in this last paragraph will become clear as the reader proceeds through the subsequent sections.

85.4 Intercellular Traffic

The establishment of an efficient defense involves an intense traffic of cells and communication between cells both in the phase of mobilization and in the effector phase. Communication between cells is by cytokines (interleukins, tumor necrosis factors, interferons, etc.) and their corresponding receptors and by cell adhesion molecules. As a rule, cytokines are secreted molecules which, upon hitting their target receptor, set in motion a chain of events resulting in the increased or decreased expression of certain proteins, which in turn alter cellular functions, such as proliferation or differentiation. The signaling pathway between interferon α and the genes it activates has recently been unravelled [5].

Adhesion molecules allow homing of cells to particular sites, their close apposition and delivery of signals to other cells, and their transgression of anatomical barriers. Thus, neutrophilic granulocytes, chemotactically attracted to a site of tissue damage, cross endothelial barriers (diapedesis), which requires an interaction with endothelial adhesion molecules (ICAMs), phagocytose cellular debris and foreign matter, discharge lysosomal enzymes into the phagocytic vacuole, and experience a respiratory burst culminating in the formation of oxygen radicals.

The system is multifaceted, highly complicated, and only partially understood. Hormonal control is also involved. Interestingly, several cytokines and their receptors which are present on cells of the immune system can also be found in neural elements. This has led to speculation about interactions between the immune, neuroendocrine, and possibly psychic systems (cf. Chap. 86). The level of complexity within each of these systems taken separately is so great and so ill-understood that this speculation remains, for the time being, rather gratuitous.

85.5 The Cells Involved in the Immune Response

The pluripotent stem cell of the hematopoietic system gives rise to the myeloid stem cell, from which neutrophils, basophils, eosinophils, and monocyte-macrophages emerge [all members of the nonspecific arm of immune defense, but with important roles in specific immunity as antigen-presenting cells (APCs)], and to lymphoid precursor cells [6]. These can be subdivided into precursors leading to B cells (which can differentiate into plasma cells), to T cells, and to natural killer or NK cells. B cells and T cells form the specific arm of the immune system. B cells possess immunoglobulin (Ig) molecules as surface receptors, and are able to secrete them in large quantities when differentiated into plasma cells; the secreted Igs are called *antibodies*. T cells, after an obligatory passage through the thymus (hence the letter T), acquire the ability to recognize foreign determinants in the context of certain cell surface self-determinants called *MHC* (major histocompatibility complex). A single cell, be it B or T, always carries receptors of only one unique specificity. To increase the numbers of receptors of a given specificity, clonal expansion of the cell carrying it has to occur.

The differentiation of stem cells into progenitor cells committed to the nonspecific arm, say, neutrophilic granulocytes or monocyte-macrophages, is under the control of *colony-stimulating factors*, such as G-CSF, M-CSF, GM-CSF, interleukins (see below), and a host of other growth factors, sometimes collectively referred to as *peptide regulatory factors* [7–19].

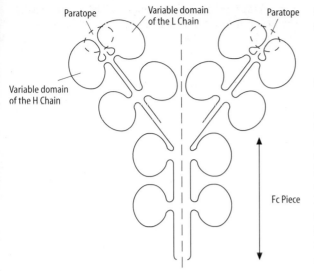

Fig. 85.1. The fundamental structure of immunoglobulins. Two identical heavy chains (H) and two identical light chains (L) are assembled into a Y-shaped molecule held together by inter- and intrachain disulfide bridges. The stalk of the Y is called the Fc piece, each arm (when separated from the rest) is called an Fab piece. The light chain comes in two variants, κ or λ. The heavy chain defines the isotype of the immunoglobulin: α for IgA, δ for IgD, ε for IgE, γ for IgG, μ for IgM. Each light chain forms two, and each heavy chain four or five domains of approximately 110 amino acids. The N-terminal domains of both heavy and light chains are highly variable and contribute to the formation of the paratope. Each molecule thus carries two identical paratopes

85.6 How Is Diversity Generated?

As previously mentioned, the binding specificity of receptors (Ig molecules on B cells or T-cell receptors, TCRs, on T cells) involves a steric match between certain ligands and a reacting part of the receptor. The matching ligand is called *epitope*; it usually forms part of a much larger structure, called *antigen*. That part of the receptor which interacts with the epitope is called the *paratope*. To be capable of engaging a large array of diverse epitopes, a similarly vast repertoire of paratopes must be created. Since receptors consist of polypeptide chains, the amino acid sequence of which is determined by nucleotide sequences in genes, it follows that a large number of diverse nucleotide sequences have to be available. Rather than having all these separately laid down in the genome, nature has evolved a system of ontogenic recombination leading to enormous diversity while remaining economic of genetic material. Relatively small gene segments, each present in a number of variants, are rearranged to yield the final messenger RNAs from which the receptor polypeptides are read. Both B and T cells resort to this same principle of genetic reassortment, but starting from a different set of similarly organized genetic segments.

As soon as a functional rearrangement has occurred, further rearrangements are stopped, so that a single cell will always carry only one receptor specificity; this is called *allelic exclusion*.

85.7 The Structure of Immunoglobulins

As a result of this shuffling, *variable domains* (containing the paratopes) of Igs and TCRs are generated. Mature B cells are equipped with Ig molecules with the following overall structure: The basic unit consists of two identical light chains (L), and two identical heavy chains (H). Light chains come in two varieties, kappa (κ) and lambda (λ), heavy chains in five principal varieties (with several subtypes), alpha (α), delta (δ), epsilon (ε), gamma (γ), and mu (μ). Each light chain forms two domains of approximately 110 amino acids each, one variable and one constant. Heavy chains form four or five domains, also about 110 amino acids in length, one of which is variable.

The overall shape of the unit resembles the letter Y, each arm being formed by one light chain and one variable and one constant domain of one heavy chain, and the stalk by the remaining constant domains of the two heavy chains. Each unit carries two identical paratopes, each being formed jointly by the variable domains of one light and one heavy chain (see Fig. 85.1).

85.8 Isotypes, Allotypes, and Idiotypes of Igs

Depending on which heavy chain is involved, Ig molecules belong to one of five *isotypes*: IgA, IgD, IgE, IgG, and IgM. They are recognized with the help of antibodies raised in a different species. Isotypes show differences in function, serum concentration, and localization. Furthermore, in some Ig isotypes repeats of the fundamental four-chain unit occur. Thus IgM in serum forms pentamers (five basic units with ten paratopes), and IgA, dimers (Fig. 85.2). Membrane-bound Ig receptors are generally IgD or IgM monomers.

There exist genetically determined variations in the constant domains or in the framework amino acids of the variable domains of immunoglobulins. These are inherited in mendelian fashion and are called *allotypes*. Furthermore, the particular amino acid sequence realized in a given variable domain can give rise to epitopes recognizable by other antibodies; these are called *idiotypes*. Theories have been elaborated which imply an important role for the interaction between idiotypes and anti-idiotypic antibodies. Within a network consisting of antibodies with idiotypic specificities and anti-idiotypic antibodies also with their own idiotypic specificities it can be speculated that each conceivable epitope is represented as a negative image by a fitting paratope and as a positive image by a structurally similar idiotope (network theories [20,21]).

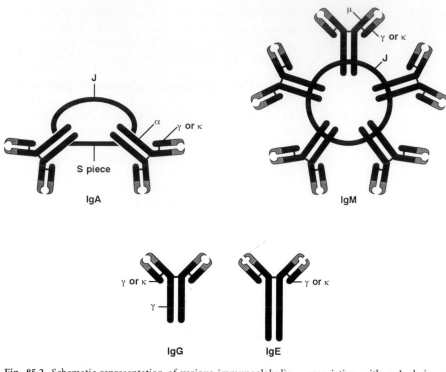

Fig. 85.2. Schematic representation of various immunoglobulin isotypes. The fundamental structure is always the same as shown in Fig. 89.1. IgA forms dimers and associates with a "secretory piece" (S piece) which allows secretion into body fluids such as milk, saliva, tears, and intestinal content. IgM forms pentamers by association with a J chain; such a unit carries ten identical paratopes. IgE has an Fc piece with particular affinity for a corresponding receptor on mast cells. IgG (with several subtypes) is the most abundant species in circulating blood

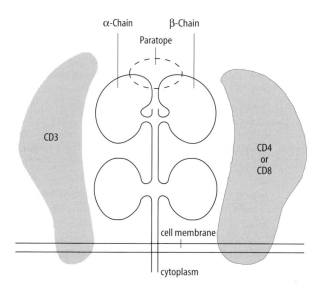

Fig. 85.3. The T-cell receptor consists of two chains anchored in the T-cell membrane, each folded into two domains. The distal domains are highly variable and jointly form the paratope. Additional peptides contribute to binding and signalling. The C4 complex defines the cell as a helper (Th) cell, and the C8 complex as a cytotoxic (Tc) cell

85.9 The Structure of the T-Cell Receptor

The T-cell receptor generally consists of two chains, alpha (α) and beta (β). [Some T cells also possess another pair of chains, gamma (γ) and delta (δ); the significance of this is still being debated.] The TCR is anchored in the cell membrane and is never secreted. The extracellular portion of each chain forms two domains, the outermost being variable. The two variable domains together form the paratope, that part of the TCR which recognizes peptides presented to it by the appropriate MHC groove (see below). The TCR is closely associated with cell surface molecules which help in the interaction between the T cell and the antigen-presenting cell: the CD3 complex and either CD4 or CD8 (Fig. 89.3).

85.10 Two Types of Epitopes

The epitope that B cells recognize is usually a structure exposed on the surface of a complex molecule such as a protein. The paratopes of Igs are capable of recognizing and binding such an antigen in its native configuration in solution and to catch it, so to speak, "in flight." The corresponding epitope is called a *structural epitope*; it frequently consists of neighboring parts of the antigen molecule which, in a denatured (outstretched) configuration, are distant from one another, so that the shape of the epitope, and hence its binding affinity, is destroyed.

This is not to say that B cells are incapable of recognizing linear stretches of amino acids as epitopes when presented with short peptides or with denatured, linearized proteins, or when such stretches are exposed at the surface of native proteins. Nevertheless, the majority of epitopes on globular proteins consist of structural epitopes.

A different situation prevails with T cells. TCRs do not recognize native, soluble antigen. They only recognize epitopes in the context of MHC molecules at the surface of cells. There are two sorts of MHC molecules: MHC I and MHC II, and correspondingly two sorts of T cells, those recognizing epitopes in the context of MHC I, called cytotoxic T cells, and those recognizing epitopes in the context of MHC II, called helper T cells (cf. Sect. 85.15). T cells acquire this selectiveness during their passage through the thymus.

85.11 What Happens Within the Thymus?

The rearrangement of genetic material leading to variable domains is bound to be error-prone in the sense that many functionally useless or dangerous combinations are created. The thymus acts as a filter to eradicate cells with such undesirable features. Two levels of selection may be postulated. At the first level, all lymphocytes capable of interacting with self MHC elements above a certain threshold of affinity are positively selected, all others being eliminated, probably by triggering of apoptosis, activation of endonucleases leading to programmed cell death [22]. At the second level, those cells which have an excessively high affinity for self components on MHC are similarly eliminated. This supposes, of course, the presence within the thymus of both types of MHC. A look at the structure of MHC will help.

85.12 Two Types of MHC

MHC molecules are proteins anchored at the surface of cells. MHC-I are widely distributed on practically all body cells, whereas MHC-II are normally confined to elements of the immune system, within both its nonspecific and specific arms. Although the gross anatomy of MHC-I and MHC-II differ, they have one feature in common: viewed from the outside of the cell, the top of both molecules exhibits a groove, lined by α-helical stretches and underlaid by β-pleated sheets, which is capable of accommodating peptides of modest length (perhaps eight amino acids, but slightly longer peptides may also fit, their middle parts bulging out of the groove). This is the structure which presents epitopes to the T cell.

MHC-I consists of a single polypeptide chain with an intracellular part, a transmembrane stretch, and an extracellular part forming three domains. This structure noncovalently associates with a second short chain called β_2-microglobulin, which by itself forms a domain. MHC-II is composed of two chains anchored in the membrane and forming two extracellular domains each. All "domains" mentioned up to now (variable and constant domains of Ig and TCR, MHC-I and MHC-II domains, and β_2-microglobulin) share structural homology and belong to what is known as the immunoglobulin superfamily. An idea of MHC-I and MHC-II and their antigen-presenting grooves is given in Fig. 85.4. A crystallographic analysis of the fit between two viral peptides and murine MHC-I has been published [23].

It is clear from the above that T cells must recognize linear stretches of amino acids, or *sequential epitopes*, in contrast to B cells, which, as mentioned earlier, recognize mainly structural epitopes.

The groove in MHC-I and MHC-II will not accommodate any amino acid sequence of appropriate length. This leads to a sort of preselection of those epitopes which can be presented [24]. The extensive polymorphism of the genes coding for MHC guarantees that within a single species a large array of peptides can, in fact, be presented. Also, any protein of sizable length can be chopped up into many short pieces, so that the likelihood of at least one piece fitting the particular MHC groove present in a given cell is substantial. Nevertheless, the preselective ability of MHC allows these structures to act as immune response genes.

85.13 Superantigens

The elimination of self-reactive cells requires the intervention of a self-peptide fitting both the MHC presenting groove and the specific paratope of the TCR. Some peptides which associate with MHC have an affinity for TCR which is not mediated by the specific paratope, but which addresses some part of a single chain forming the TCR [25]. Such peptides belong to what is called a *superantigen*. When present in the thymus, such superantigens initiate the elimination of a much broader array of T cells than ordinary self-peptides, since eradication is limited not to cells with one paratope specificity, but to all cells with the same single chain. This can lead to important distortions in the range of chains available for the T-cell repertoire. Conversely, when introduced outside the thymus, the T-cell response they elicit is stronger.

85.14 Two Pathways of Epitope Presentation

The origins of the epitopes presented by MHC-I and MHC-II differ [26]. Proteins synthesized within a cell are processed via the cytosolic pathway and presented by MHC-I.

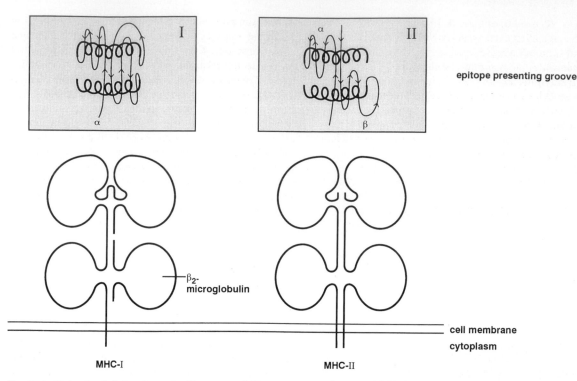

Fig. 85.4. Molecules defining the major histocompatibility complex (MHC). MHC-I consists of one chain anchored in the cell membrane and folded into three domains; it associates noncovalently with β₂-microglobulin, which itself folds to a domain. MHC-II consists of two chains (α and β) anchored in the cell membrane and folded into two domains each. The two outermost domains of the one chain in MHC-I or of the two chains in MHC-II form similar grooves. Viewed from above, these grooves are lined by α-helical stretches; they are underlaid by β-pleated sheets

This applies to normal constituents of such cells as well as to intracellularly synthesized foreign proteins, for instance, viral proteins produced after a virus has taken over the protein-synthesizing machinery of the cell. The groove of the MHC-I molecule is probably never empty, and will be occupied either by a self peptide or by a nonself peptide. In contrast, MHC-II presentation deals with proteins taken up from outside the cell, by pinocytosis or adsorptive endocytosis, via an endosomal pathway. Here also, normal constituents of the body will occupy most MHC-II grooves most of the time.

Thus, under normal circumstances the MHC molecules which face lymphocytes during their maturation in the thymus will be occupied by self peptides, and the second level of thymus filtration alluded to above will eliminate all T cells capable of reacting strongly with self peptides associated with either MHC-I or MHC-II. This leaves a population of T cells reacting strongly only with foreign peptides on MHC.

85.15 Further Markers on T Cells

At the same time as selection proceeds to establish a population of T cells reacting either with MHC-I + foreign peptide or with MHC-II + foreign peptide, other molecules characteristic of the interaction are identified and sorted

out. Whereas many immature T cells in the thymus carry two surface molecules called CD4 and CD8, the mature T cell reactive with MHC-I is left with the cell surface element CD8, and that reactive with MHC-II with the element CD4. In addition, both types of T cells carry the cell surface complex CD3, and probably many other functionally important features.

Maturation processes in the thymus thus lead to the emergence of two types of immune-competent T cells:

- Cells recognizing, with the help of their specific TCR, foreign peptides in association with MHC-I; these cells are called cytotoxic T cells (Tc). They carry the surface markers CD3 and CD8. The epitopes originate from proteins synthesized intracellularly and processed via the cytosolic pathway.
- Cells recognizing, with the help of their specific TCR, foreign peptides in association with MHC-II; these cells are called helper T cells (Th). They carry the surface markers CD3 and CD4. The epitopes originate from extracellular proteins processed via the endosomal pathway.

Th cells can be divided into various subgroups. In addition to their helping function in antibody formation (see below), Th cells also play a role in *delayed hypersensitivity*, an immunological reaction first detected in the context of tuberculosis. When products of tubercle bacilli, called tuberculin or PPD, are injected into the skin of a normal individual not having had any contact with tubercle bacilli, no inflammatory skin reaction is observed. However, a similar injection into the skin of a person or an animal infected with tubercle bacilli elicits a skin reaction reaching its peak in about 48 h (hence the qualification "delayed"). This reaction cannot be passively transferred with serum, but it can be adoptively transferred with Th cells. Besides tuberculosis, many other infectious diseases are accompanied by delayed hypersensitivity reactions.

85.16 The Cytotoxic Reaction

That T cells recognizing foreignness in the context of MHC-I are called cytotoxic is, of course, not arbitrary. As mentioned earlier, virtually all body cells have MHC-I molecules at their surface. Any pathological disruption of metabolism will likely result in the synthesis of pathological proteins which, after cytosolic processing, will lead to the emergence of novel peptides in the MHC-I groove. These will be recognized by Tc cells equipped with a fitting TCR. Apposition of such a cell onto its target, whereby not only the TCR but also other surface molecules like CD3 and CD8 become engaged, leads to the delivery of a "kiss of death" to the target cell. In the case of a cytocidal viral infection of the target, this evidently makes sense, since the virus-infected cell would be lost anyway, but its early elimination by a cytotoxic cell prevents further viral synthesis and thus helps to limit the infection. In noncytocidal viral infections the cytotoxic cell may in fact turn into a pathogenic element when silently infected but functionally important cells are killed.

Historically, the predilection of cytotoxic T cells for epitopes presented by MHC-I was one of the first clear-cut demonstrations of the cooperation between the immune system and the MHC complex. The concept of "T-cell restriction" was born of the observation that cytotoxic lymphocytes of mice with MHC type A and infected with virus X were capable of lysing X-virus-infected cells of the same MHC type A, but not similarly infected cells of MHC type B, and vice versa [27].

85.17 Interactions Between T, B, and Accessory Cells

Cells with MHC-II are themselves part of the immune system. They include monocyte-macrophages (also macrophages localized in tissues, such as lung macrophages, Kupffer cells, and Langerhans cells), and dendritic cells in lymph nodes; they are called collectively "accessory cells" and are able to act as "antigen-presenting cells." This capacity is also vested in B cells, which likewise express MHC-II at their surface. Antigen processing is by the endosomal pathway. The encounter with antigen is either fortuitous and unspecific (accumulation of invad-

ing microorganisms, local injection of foreign matter) or facilitated by specific ligand–receptor interaction and chemotactic attraction to sites of tissue damage.

In the case of B cells, surface Ig molecules may bind antigen specifically via a structural epitope. In general the encounter of a B cell with an antigen bearing a fitting epitope has, in itself, no further consequences. Only in special cases where a great many similar epitopes occur in a geometric pattern does an encounter trigger the B cell to proliferation and differentiation into plasma cells. Such antigens are called *T-cell independent*. Conversely, the encounter between B cells and self antigens present in huge amounts may cause these cells to be eliminated or to become unreactive (clonal deletion or clonal anergy). The vast majority of antigens require help from T cells for B cells to be activated.

85.18 A Scenario of T-Cell-Dependent Antibody Formation

A much simplified and sketchy outline of T-cell help may clarify this concept. Suppose a foreign protein, say a bacterial exotoxin, is taken up by a macrophage, perhaps itself attracted to the site of bacterial invasion by chemotactically active products of tissue damage and/or by lymphokines released from polymorphonuclear leukocytes [28]. The macrophage, in a typical reaction of nonspecific immunity, will engorge the toxin and process it via the endosomal pathway, resulting in the presentation, in its MHC-II groove, of sequential epitopes derived from the chopped up toxin. This macrophage is now prepared to act as an antigen-presenting cell. Suppose, furthermore, that among the T cells also present at the site of action there is one with a TCR fitting the presented epitope. An interaction between the macrophage and the T cell will ensue, in which other surface molecules, in addition to MHC-II and TCR, play a role, and as a result of which the macrophage is induced to secrete a cytokine, interleukin 1 (IL-1). This delivers a signal to the T cell, which in turn will start synthesizing IL-2 (a potent mitogen), IL-2 receptors, and other signalling molecules (to date, some ten interleukins are known, with their corresponding receptors). A focus of cell proliferation forms, in which T cells bearing the same TCR and many cytokines accumulate. Some of the cytokines have additional effects, such as inducing fever or changing the permeability of vessels. These are all elements of what is commonly known as inflammation.

Suppose, finally, that in the vicinity a patrolling B cell recognizes, on the same bacterial toxin, a structural epitope fitting its surface Ig. This in itself would be of no consequence, except that our B cell, like the macrophage, processes the toxin via an endosomal pathway to "distribute" sequential epitopes in its MHC-II groove. Now the stage is set for T-cell help: One of the, meanwhile, numerous T cells will recognize the sequential epitope on the B cell and shower it with mitogenic and differentiation signals, eventually leading to terminal evolution of B cells into plasma

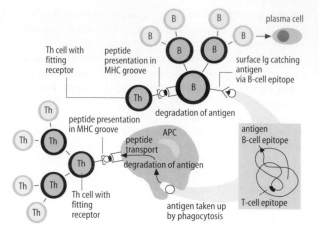

Fig. 85.5. A simplified view of T-B cooperation. The antigen presenting cell (APC) nonspecifically ingests an antigen by phagocytosis, degrades it into small peptides, and exposes a fitting peptide in its MHC-II groove. A Th cell equipped with an appropriate receptor recognizes the peptide in the MHC-II groove and thereby becomes activated. The encounter between APC and Th triggers the production of cytokines, which induce a proliferation of Th cells possessing the same receptor. The antigen, which is also studded with B-cell epitopes, will be caught by a B cell with specifically fitting surface Ig. Like the APC, the B cell will ingest and degrade the antigen to small peptides, which likewise are presented in the MHC-II groove of the B cell. The B cell is now "tagged" for recognition by one of the Th cells which have proliferated. When one of these Th cells recognizes a tagged B cell, cytokines are generated and signals are delivered that induce clonal proliferation of B cells, which later differentiate to plasma cells secreting Igs with the same paratope that characterized the original B cell

cells (Fig. 85.5). A cluster of plasma cells will develop, all secreting Igs fitting the same structural epitope that the first B cell had spotted. Note that the two epitopes involved are different: the B-cell epitope is a structural epitope present on the native, unprocessed toxin, whereas the T-cell epitope is a sequential epitope resulting from endosomal processing. The formation of substantial amounts of antibodies against the antigen is said to be *T cell dependent*. When this chain of events takes place in a lymph node, T-cell proliferation is observed in the paracortical area, whereas B-cell proliferation and differentiation occur in the germinal centers.

T-cell help is also required for another feature of antibody formation. In immunoglobulins, a given heavy chain variable domain can be appended to several different constant domains defining the various isotypes of Ig (see above). A B-cell immune response usually starts with the production of specific IgM. The same variable domain can later be transferred to IgG, IgA, or IgE, a mechanism also dependent on T-cell help and called *switch*.

85.19 Monoclonal Antibodies

Most antigens display several structural (B-cell) epitopes at their surface and can be processed into a series of sequential (T-cell) epitopes. Moreover, paratopes on receptors need not be uniform, even when directed at the same epitope. Usually a whole range of paratopes with varying degrees of affinity to the epitopes exist. Hence, most immune responses generate parallel proliferations of several different B-cell clones. The normal B-cell response is therefore *polyclonal* and consists of a number of specificities directed at several B-cell epitopes and fitting these with a whole range of affinities. Since plasma cells are terminally differentiated cells, they cannot be indefinitely clonally expanded, unless they have suffered malignant transformation. Plasmacytomas are such monoclonal proliferations, and their product is, in fact, a *monoclonal antibody*. However, the specificity of such plasmacytoma products is usually unknown and in any case unpredictable.

Monoclonal antibodies of any desired specificity can be generated by fusion of single antibody-forming plasma cells with immortal tumor cells. This is an extremely useful technique which has yielded most of the reagents for the definition of cell surface antigens, such as CD3, CD4, CD8, etc., and which is of wide general applicability.

85.20 Tolerance of Self

As mentioned above, maturation processes in the thymus result in the elimination of Th cells capable of reacting with self peptides + MHC-II. Hence even if B cells bearing paratopes reactive with structural epitopes of self molecules are present, they will not obtain help from T cells and will not proliferate. Tolerance of self in this case is determined at the T-cell level through lack of self-reacting T-cell clones, which have been deleted in the thymus. This is called the *clonal deletion theory of self-tolerance*. Although possibly one of the principal mechanisms of self-tolerance, it is not the only one. Tolerance at the B-cell level also exists. Thus, it is likely that self antigens present in large amounts in circulation, such as serum albumin, lead to clonal deletion of autoreactive B cells. The example of the ABO blood groups (see p. ●●) also suggests that autoreactive B cells are directly eliminated. It is also conceivable that certain cells, without being physically eradicated, lose their ability to react to their fitting epitope, and thus become functionally deleted. Finally, self peptides unable to fit into the particular MHC grooves present in an individual will also fail to induce a T-cell-dependent immune reaction, as would self peptides which, for anatomical reasons, never encounter an antigen-presenting cell.

85.21 The Effector Phase of the Immune Response

The accumulation of paratopes, be they on T cells, B cells, or circulating antibodies, against epitopes on an invading microorganism or foreign protein, in itself is, of course, not a sufficient antimicrobial response. Obligatory intracellular parasites such as viruses can be dealt with by cytotoxic cells, at the expense of those cells already invaded, and with the reservations regarding possible pathogenic implications alluded to earlier. Antibodies to soluble toxins frequently neutralize the toxic effect; in fact, the first clear-cut demonstration of a specific antibody concerned diphtheria toxin and tetanus toxin and their respective antibodies, called *antitoxins*. The term "antitoxin" preceded the term "antibody," and both are older than the term "antigen" [29].

Another task which antibodies alone are able to perform is the neutralization of viral infectivity. In those cases where viruses have to reach uninfected cells by an extracellular path, their encounter with specific antibodies will often prevent their entry (via specific receptors) into susceptible cells. This explains the efficacy of many antiviral vaccines: as long as sufficient levels of circulating antibodies persist, any virus entering the system will be caught by antibodies and thus neutralized. It also explains (a) the failure of antibodies to quench those infections in which viruses spread from cell to cell without being forced to enter the extracellular space, and (b) the need for a diversity of antibodies: IgA, for instance, is the isotype present in secretions such as tears, saliva, bronchial mucus, milk, or intestinal content. Fitted with the proper paratopes, such antibodies are strategically located to deal with invaders entering from the air or by the oral route.

Depending on the concentration of antibodies and antigen encountering each other, large immune complexes, in which complement components (see below) are also involved, may form. These fall prey to phagocytosis. They can also form deposits on membranes lining blood vessels and become pathogenic (immune complex vasculitis).

Frequently a more concerted approach is required for successful elimination of invaders. Antibodies have the ability to cooperate with nonspecific cellular and humoral factors. Thus, microorganisms flagged or coated with antibodies specific for some surface constitutent of the microbe are much more easily phagocytosed than uncoated ones; this is called *opsonization*, to which elements of the complement cascade (see below) also contribute. There is also the phenomenon of antibody-dependent cellular cytotoxicity, whereby noncommitted, potentially cytotoxic cells are directed at a particular target by specific antibodies. Many cells are studded with Fc receptors, receptors capable of binding the stem portion of antibodies, thus adding specific adsorptive capacity to a cell which otherwise is equipped only with receptors of low specificity. In this way the nonspecific arm of the immune system can be enrolled to perform specific tasks.

85.22 The Complement System

An important segment of the defense system is a chain of serum proteins collectively known under the name of *com-*plement (C) [30]. This rather complex multicomponent system can be activated by the interaction of antigen with certain classes of immunoglobulins. A slight allosteric change in antibody when complexed to its antigen sets in motion a cascade of interacting proteins, one element acquiring proteolytic activity to split the next, with some of the split products exerting powerful pharmacological effects, such as changes in membrane permeability or chemoattraction of phagocytes (many types of receptors for these split products exist on various cells and also on antibodies). A much simplified representation of the first steps in the complement cascade is shown in Fig. 85.6.

The "classical pathway" of complement activation is best illustrated in a model with sheep red blood cells as the antigen and serum of a rabbit immunized against sheep red blood cells as the antibody: confronted with the antibodies alone, the only thing that happens is that the blood cells clump (agglutinate), due to antibody molecules forming bridges between adjacent epitope-studded red blood cells. Upon addition of fresh normal serum (from a nonimmunized animal) containing complement, the cells rapidly lyse, i.e., they lose the capacity to retain hemoglobin. The model, of course, is rather far-fetched, since rabbits are rarely required to fight against an invasion of sheep red cells. However, the same occurs with certain bacteria, the osmotic barrier of which can be disrupted by the joint action of antibody and complement. Furthermore, since the complement cascade sheds split products with pharmacological activity (permeability changes, chemotaxis), we are again confronted with many of the elements of inflammation.

The complement cascade can also be triggered by other means. There is an "alternative pathway" of complement activation, in which bacterial products and some Ig

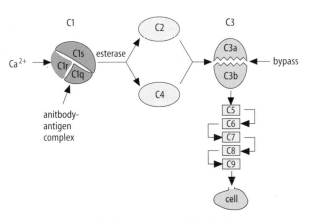

Fig. 85.6. An oversimplified version of the beginning of the complement activation cascade. In the presence of calcium ions and of an appropriate antigen–antibody complex, the component C1 (consisting of subcomponents q, r, and s) is activated to an esterase. This addresses the components C2 and C4, which are activated to split C3 into C3a and C3b. The same splitting of C3 can be achieved in another way by the so called complement bypass. Split products of C3 then activate C5, which in turn activates C6, and so on until C9 becomes active. C9 aggregates to form pores in cell membranes, leading to lysis of cells

isotypes intervene directly. Genetic defects of the complement system exist where some component is lacking, usually without much disturbance of the whole system, probably because the important steps are redundant. In order to maintain this highly reactive system in a state of inaction most of the time, elaborate negative feedbacks by various inhibitors are built into it, and congenital absence of such inhibitors can lead to severe pathology. There is a certain analogy between the complement system and the blood coagulation system.

In the course of an immune reaction so-called acute phase proteins, such as C-reactive protein or α_1-acid glycoprotein, become mobilized; this seems to be controlled, in part, by IL-6.

85.23 Diagnostic Aspects of Immune Reactivity

It may be useful to outline some of the diagnostic possibilities offered by analysis of various components of the immune response. It should be stressed immediately that the dividing line between physiology and pathology is particularly arbitrary in this field. For instance, we are not aware of any active defense mechanisms ensuring the quiet and unobtrusive colonization of our skin and mucous membranes by a host of microorganisms. One might therefore be tempted to attribute this to anatomical barriers such as keratinization of outer cell layers, flow of mucus, acidity of secretions etc., and having nothing to do with specific immunity. However, as soon as specific immunity breaks down, as in the course of AIDS or as a consequence of genetic defects, many microorganisms harmless for the normal individual reveal their invasive potential. It must be concluded that even in the apparently peaceful coexistence of microbes and man a certain level of readiness in terms of adaptive immunity is permanently required.

In acute infections, the timing of reactions usually complies with the following scheme: The microorganism gains its first foothold by overcoming the anatomical barriers mentioned above. This will not be possible without a certain amount of tissue damage and the triggering of nonspecific mechanisms such as, in the case of virus infections, the secretion of interferons by the first few infected cells. Attraction of polymorphonuclear granulocytes to the bridgehead of the invader will ensure first cytokine releases. As soon as macrophages appear on the scene, antigen presentation to Th cells begins and cytotoxic Tc cells emerge. T-cell help will set antibody production in motion. Local and systemic alterations in cytokine profiles and in the distribution of cytokine receptors, increases in and modifications of acute phase proteins, emergence of antibodies (first IgM, later switching to IgG, in secretions and mucous surfaces IgA, under certain circumstances IgE), formation of immune complexes, activation of the complement cascade, and rising body temperature merge into the whole gamut of inflammation. In chronic disorders some intermediate level of activity is reached, with gross deviations from the normal often confined to anatomically restricted sites, such as joints or kidney.

Classical serology is limited to the detection, in serum, of antibodies directed at specific components of the suspected invader. The practical problem here is that many antibodies persist for a long time, so that the mere presence of an antibody will not reveal whether the infection is ongoing or has occurred many years earlier. Identification of the isotype of the antibody involved often helps: IgM is usually present only at the beginning of the immune response. Increases in the level of specific IgG over a period of 10–14 days also indicate a recent infection.

Very many quite sophisticated techniques are widely used to detect even low levels of antibody against specific antigens: neutralization tests, complement fixation tests, solid phase immune assays based on radioisotope or enzyme tagging of either secondary antibodies or antigen, Western blots, etc. An in vivo test to evaluate the cellular immune response relies on the reaction of delayed hypersensitivity, the prototype being the tuberculin reaction. Other tests based on the cellular immune response or on the activation profile of cytokines are still being developed. Quantitation of CD4- versus CD8- bearing T cells (using monoclonal antibodies and fluorescence-activated cell sorting) is important in monitoring the progress of AIDS, which is characterized by a decrease in the CD4 to CD8 ratio. However, the goal of obtaining an exact estimate of the functional state of the whole immune system remains elusive, and much progress will have to be accomplished before this goal is attained.

References

1. Lindenmann J (1981) Immunology in the 1880: two early theories. In: Steinberg CM, Lefkovits I (eds) The immune system, vol 1. Karger, Basel, pp 413–422
2. Paul WE (ed) (1989) Fundamental immunology, 2nd edn. Raven, New York
3. Male D, Champion B, Cooke A, Owen M (1991) Advanced immunology, 2nd edn. Gower Medical, New York
4. Lutz HU (1990) Erythrocyte clearance. In: Harris JR (ed) Blood cell biochemistry, vol 1. Plenum, New York, pp 81–120
5. Schindler C, Shuai K, Prezioso UR, Darnell JE Jr (1992) Interferon-dependent tyrosine phosphorylation of a latent cytoplasmic transcription factor. Science 257:809–813
6. Metcalf D (1989) The molecular control of cell division, differentiation, commitment, and maturation in haematopoietic cells. Nature 339:27–30
7. Green AR (1989) Peptide regulatory factors: multifunctional mediators of cellular growth and differentiation. Lancet:705–707
8. Michell RH (1989) Post-receptor signalling pathways. Lancet:765–768
9. Metcalf D (1989) Haemopoietic growth factors 1. Lancet:825–827
10. Metcalf D (1989) Haemopoietic growth factors 2: clinical applications. Lancet:885–887
11. O'Garra A (1989) Interleukins and the immune system 1. Lancet:943–946
12. O'Garra A (1989) Interleukins and the immune system 2. Lancet:1003–1005

13. Blackwill FR (1989) Interferons. Lancet: 1060–1063
14. Tracey KJ, Vlassara H, Cerami A (1989) Cachectin/tumor necrosis factor. Lancet:1122–1126
15. Ross R (1989) Platelet-derived growth factor. Lancet:1179–1182
16. Waterfield MD (1989) Epidermal growth factor and related molecules. Lancet:1243–1246
17. Slack JMW (1989) Peptide regulatory factors in embryonic development. Lancet:1312–1315
18. Hanley MR (1989) Peptide regulatory factors in the nervous system. Lancet:1373–1376
19. Duff GW (1989) Peptide regulatory factors in nonmalignant disease. Lancet:1432–1435
20. Jerne NK (1974) Towards a network theory of the immune system. Ann Immunol (Inst Pasteur) 125C:373–389
21. Lindenmann J (1973) Speculations on idiotypes and homobodies. Ann Immunol (Inst Pasteur) 124C:171–184
22. Cohen JJ (1991) Programmed cell death in the immune system Adv Immunol 50:55–85
23. Fremont DH, Matsumura M, Stura EA, Peterson PA, Wilson IA (1992) Crystal structures of two viral peptides in complex with murine MHC class I H-2k^b. Science 257:919–927
24. Matsumura M, Fremont DH, Peterson PA, Wilson IA (1992) Emerging principles for the recognition of peptide antigens by MHC class I molecules. Science 257:927–934
25. Misfeldt ML (1990) Microbial "superantigens". Infect Immun 58:2409–2413
26. Yewdell JW, Bennink, JR (1990) The binary logic of antigen processing and presentation to T cells. Cell 62:203–206
27. Zinkernagel RM, Doherty PC (1979) MHC-restricted cytotoxic T cells: Studies on the biological role of polymorphic major transplantation antigens determining T-cell restriction-specificity, function, and responsiveness. Adv Immunol 27:52–177
28. Lloyd AR, Oppenheim JJ (1992) Poly's lament: the neglected role of the polymorphonuclear neutrophil in the afferent limb of the immune response. Immunol Today 13:169–172
29. Lindenmann J (1984) Origin of the terms "antibody" and "antigen". Scand J Immunol 19:281–285
30. Mollnes TE, Lachmann PJ (1988) Regulation of complement. Scand J Immunol 27:127–142

86 Neuroendocrine Modulation of the Immune System

F. Chiappelli, C. Franceschi, E. Ottaviani, G.F. Solomon, and A.N. Taylor

Contents

86.1 Introduction

The immune system represents the set of organs, cells, and factors whose physiological function it is to protect us from diseases, infections, and tumors. This chapter discusses certain fundamental immunophysiological principles. Hormones and peptides produced by endocrine glands modulate most, if not all physiological processes, including, as discussed in this chapter, immune responses. The neuroendocrine system is described in detail elsewhere (see Chap. 19).

Feedback mechanisms are central physiological events that regulate neuroendocrine as well as immune responses (see Chaps. 1, 19). For example, hypothalamic nuclei secrete corticotropin-releasing hormone, which stimulates the production and secretion of proopiomelanocortin gene products by anterior pituitary cells. Adrenocorticotropic hormone (ACTH) stimulates adrenocortical cells to secrete glucocorticoids, which in turn feedback to the pituitary, the hypothalamus, and the hippocampus to block the activation of the hypothalamic-pituitary-adrenocortical axis. Corticotropin-releasing hormone, proopiomelanocortin products, and glucocorticoids modulate lymphocyte functions, and, conversely, products of activated lymphocytes (e.g., interleukins) modulate the secretion of corticotropin-releasing hormone, proopiomelanocortin products, and glucocorticoids.

This chapter focuses upon the interactions between the hypothalamic-pituitary-adrenal axis and cell-mediated immunity. It discusses the role played by neuroendocrine products in the regulation of cell-mediated immune responses.

86.2 Immune System

The immune system provides a complex array of responses against foreign (nonself) antigens. In the instance of autoimmune diseases, self-antigens are recognized as nonself. Specialized cell populations and the factors they produce (e.g., cytokines, interferons, antibodies) bring about immune responses in response to the encounter with nonself.

86.2.1 Fundamental Principles

The afferent limb of the immune response involves presentation and recognition of the antigen, activation of lymphocytes, and generation of effector cells. The efferent limb leads to the elimination of the antigen by antibodies and effector cells (e.g., plasma cells, cytotoxic lymphocytes).

Two broad functional branches of the immune system exist:

- The innate immune system, not restricted by major histocompatibility genes, represents the initial line of defense of the host. It consists of circulating monocytes, tissue-residing macrophages, dendritic cells, polymorphic nucleated cells (neutrophils), eosinophils, basophils, mast cells, and natural killer cells (see Chap. 85).

R. Greger/U. Windhorst (Eds.)
Comprehensive Human Physiology, Vol. 2
© Springer-Verlag Berlin Heidelberg 1996

- Antigen-dependent immune responses are initiated in association with the major histocompatibility complex. They represent the adaptive resistance of the host to nonself and involve T and B lymphocytes (see Chap. 85). Antigen recognition by the T cells is mediated by the T cell receptor, a complex of plasma membrane-associated peptides (alpha and beta chains in mature T cells, delta and gamma chains in immature T cells and in T cells harbored at certain sites of the mucosa).

Four broad types of reactions participate in immune surveillance mechanisms:

- Type I reactions (immediate hypersensitivity, anaphylactic response)
- Type II reactions (immediate cytotoxicity, complement mediated)
- Type III reactions (delayed cytotoxicity, opsonization mediated)
- Type IV reactions (cell mediated).

The latter reactions constitute the focus of this chapter. Cells of the lymphoid lineage are of two principal types:

- B cells begin their maturation in the bone marrow and terminate it in the periphery in mammals. Birds have a specialized organ for this purpose, the bursa of Fabricius.
- T cells are educated in the thymus. There, they learn to distinguish between self and nonself in association with the major histocompatibility complex. They complete their maturation post-thymically (see below). All T cells express the cluster of differentiation 3 (CD3) and play a central and crucial role in initiating, propagating, and regulating antigen-specific immune responses.

Cells from the myeloid lineage develop from bone marrow stem cells into circulating monocytes and tissue macrophages. They phagocytose and process antigens for later translocation to the cell surface and presentations to T cells in association with components of the major histocompatibility complex. Under certain circumstances, B cells can also act as antigen-presenting cells.

An immune cell population exists between the myeloid and the lymphoid lineages, whose cytotoxic activity is independent from restriction by the major histocompatibility complex (natural killer cells). These cells resemble myeloid cells morphologically, but perform little or no phagocytic and antigen-presenting activity; they do not secrete immunoglobulins or express CD3.

Two principal types of antigens are presented to T cells:

- Extrinsic antigens originate from infecting pathogens (bacteria, parasites, and fungi). They are catabolized within the cellular endosomes of antigen-presenting cells. Their antigenic fragments are expressed on the plasma membrane in association with class II molecules of the major histocompatibility complex and presented

to T cells that express CD4 and that act primarily as helpers and inducers by producing TH1-type or TH2-type cytokines (see below).
- Intrinsic antigens are expressed on the membrane of virally infected or tumor cells. Because their origin is intrinsic to the host cell, these nonself antigenic peptides are synthesized in the rough endoplasmic reticulum, as are most of the host cell proteins. They become associated with class I molecules of the major histocompatibility complex and are presented to CD8 T cells (CD3$^+$CD8$^+$), which primarily act as cytotoxic and suppressor cells.

In healthy adults, about two thirds of CD3 cells express CD4, and one third CD8. Activation of the complex formed by the T cell receptor and CD3 following recognition of the antigen involves phosphorylation of the chains that compose the CD3 cluster (see below). Neuroendocrine products modulate these events and the extent of response to antigen stimulation. The cytotoxic activity and the molecular mechanisms that lead to cytotoxicity by natural killer and cytotoxic T cells are also finely modulated in vitro and in vivo by a variety of neuropeptides, including catecholamines. The fact that β-endorphin and other proopiomelanocortin peptides are produced by activated lymphocytes (see below) confirms the autocrine and paracrine regulatory role of these factors upon immune cells. These are important immunophysiological events well conserved in phylogeny [15,27,96].

86.2.2 Lymphoid Organs

Immune responses occur in one of several compartments (e.g., systemic, mucosal). The systemic immune compartment is composed of circulating lymphocytes, of primary lymphoid organs (bone marrow, thymus), and of secondary lymphoid organs (spleen, lymph nodes). The mucosal-associated lymphoid tissues consist of the gut, the lungs, the genital tract, and other mucosal tissues, including the oral cavity. The involve lymph nodes of the mucosal lining (e.g., celiac, superior and inferior mesenteric, gastric), including those located in the greater and the lesser omenta, the lamina propria, and the Peyer's patches.

Lymphoid organs are extensively vascularized. The endothelial cell layer and the connective tissue that constitute them and the lymphoid and myeloid cells harbored within them are extensively bathed by circulating blood.

Thymus. This is a two-lobe gland located centrally in the thorax. It is composed of thymic epithelium and thymocytes. Immature thymocytes are homogeneously distributed in the thymic cortex. Clusters of thymocytes are encapsulated in precortical thymic "nurse" cells. As thymocytes progress through the "education" events (i.e., positive selection of cells capable of cell-mediated immune responses, negative selection of cells that respond against self-antigens), they migrate to the medulla across the characteristically concentric thymic (Hassall's) corpuscles. Mature thymocyte leave the thymic medulla and are found in the peripheral blood as "naive" T cells that express the A restriction fragment of the common leukocyte antigen, CD45 (CD45RA). During post-thymic maturation of T cells in the lymph nodes and the spleen, CD45RA is lost and the cells acquire CD45RO.

Lymph Nodes. These are widely distributed and tend to aggregate in certain anatomical locations, such as cervical and perimandibular, axillary, inguinal, and intestinal sites, as shown in Fig. 86.1. Nodes are encapsulated organs that consist of an outer cortex and an inner medulla. The cortical region is organized in primary follicles, or nodules, rich in B lymphocytes. Follicles contain smaller nodules held together by a mantle of tightly grouped lymphocytes, the secondary follicles. A deeper layer of lymphocytes exists in the paracortex, the area contiguous to the lymph node medulla. Paracortical and medullar lymph node lymphocytes are primarily T cells, but the paracortical region is also rich in dendritic cells. Active proliferation of paracortical naive T lymphocytes by antigen leads to their maturation to memory cells. If the antigen stimulates B cells, then cortical B cell populations proliferate and generate germinal centers that contain plasma cells. Plasma cells migrate from the germinal centers into the cortex and the medullary cords, penetrate the medullary sinusoids, and enter the peripheral circulation, where they release their antibodies. This organization is similar to that found in the tonsils (see below).

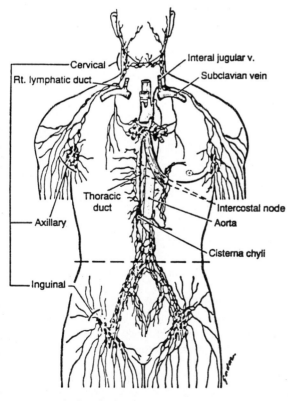

Fig. 86.1. The lymphatic system

Spleen. This organ is anatomically similar to lymph nodes. It also is encapsulated. Partitions (trabeculae) radiate inward to form compartments. The smooth musculature in the capsule and these partitions contract rhythmically to pump blood through the spleen. These contractions are crucial events in immunophysiological regulation and reflect physiological processes under neuropeptidergic and hormonal modulation. The splenic red pulp is composed of sinuses lined with monocytes, lymphocytes, and plasma cells and is responsible for the elimination of senescent erythrocytes. The white pulp is composed of cylindrical sheaths surrounding the arterioles. T cells are localized proximally to the arterioles, and unstimulated B cells are found in primary and secondary follicles beyond the T cell zone. Germinal centers exist in secondary follicles, and these contain within them monocytes and dendritic cells.

Mucosal-Associated Lymphoid Tissues. These consist of aggregates of nonencapsulated lymphoid tissue in the submucosal areas lining the respiratory, digestive, and urogenital tracts. They contain both uncommitted T and B lymphocytes and plasma cells secreting mainly immunoglobulin E and immunoglobulin A (see Chap. 85). During inflammatory responses in the abdominal cavity, the greater and the lesser omentum membranes move, contract and expand, and migrate toward the site of inflammation. They wrap themselves around the inflamed area, thus containing the inflammation and potential infection. In addition to this adjunctive function, omenta, with the multitude of lymph nodes they contain, contribute to gut-associated immune responses.

Tonsils. These are anatomical clusters of lymphoid tissue that contains germinal centers and is separated from deeper structures by connective tissue. They show deep invaginations (crypts), responsible for entrapping antigen. The principal tonsils are:

- Two palatine (tonsilla palatina)
- One pharyngeal (tonsilla pharyngea, Luschka's tonsil)
- One lingual (tonsilla lingualis)
- Those located under each hemisphere (tonsilla cerebelli)
- Those located near or at the pharyngeal opening of the auditory tube (tonsilla tubaria, eustachian tonsil, Gerlach's tonsil)
- Those of more elongated form located on the nasal septum opposite the anterior extremity of the middle turbinated body (tonsilla nasalis, Morgagni's or Zuckerkandl's tonsil)
- Those lining the larynx (tonsilla laryngea, folliculi lymphatici laryngei)
- Those lining the intestines (tonsilla intestinalis, folliculi lymphatici aggregati, Peyer's patches).

86.2.3 Lymphokinesis

T Cell Traffic. T cells freely circulate through and across lymphoid organs. They recognize specific tissues toward which they preferentially home. This traffic is directed by membrane molecules that mediate the cells' adhesion and homing to specific ligands on the endothelia and is modulated by neuroendocrine products. The traffic of T cells is characterized by:

- Expression of membrane adhesion receptors: endothelial cell adhesion molecules include the intercellular adhesion molecules (ICAM)-1,2, and 3 (CD54, CD102, and CD50, respectively), the vascular adhesion molecule (VCAM)-1 (CD106), and the mucosal addressin cell adhesion molecule (MAdCAM)-1 (CD*, not yet assigned). Extracellular matrix molecules that take part in lymphocyte–endothelium interaction include fibronectin, collagen, and laminin. Lymphocyte integrin membrane receptors allow interaction with these molecules. Integrins are composed of noncovalently joined α- and β-chains and include lymphocyte function-associated antigen (LFA-1; CD11a = α_L, CD18 = β_2), whose principal ligand is ICAM-2, and very late antigen (VLA-4; (CD49d = α_4, CD29 = β_1), whose ligands are VCAM-1 and fibronectin. LFA-1 and VLA-4 are among the principal integrins involved in T cell adhesion to endothelium.
- Initially, migrating cells collide in a quasi-random fashion with the endothelium in a pattern that depends primarily upon flow rate. Because the rate of flow is slowest in the postcapillary venules, the interaction between the migrating cells and the endothelial cells is greatest at those sites.
- Relaxation of the endothelial cell junctures is brought about, in part, by secretion of endothelium-derived relaxation factor (see Chap. 97) by surrounding activated macrophages. This factor has now been identified as nitric oxide, which is endowed with pleiotropic functions that include cytotoxic activity, inhibition of platelet aggregation, and neurotransmitter modulation. Induction of nitric oxide synthase converts L-arginine to citrulline and nitric oxide. The latter is an unstable product and is quickly converted to the stable and easily detectable metabolites nitrite (NO_2) and nitrate (NO_3^-). At supraphysiological concentrations, nitric oxide is highly cytotoxic and contributes to pathogen elimination. Inflammation and physiological events that lead to vasodilation or to vasoconstriction slow down or accelerate the rate of flow and thus induce significant changes in adhesion and transvasation of immune cells.
- Penetration through the epithelial cell layer is mediated by enzymes produced by the migrating cell that catabolize the surrounding extracellular matrix, thus opening the way to the infiltrating cell (e.g., the metalloproteinases' cleaving action upon extracellular collagen).

Lymphoreticular System. Lymphoid organs are connected together by the lymphoreticular (lymphatic) system (Fig. 86.1). This system parallels the blood circulatory system in anatomical complexity and in physiological relevance and is responsible for the flow of lymph (lymphokinesis). The lymph is a fluid that has escaped from the blood and that is filtered through the lymphoid organs, eventually returning to the blood. The lymph is carried to the lymph nodes by the afferent lymphatics (rich in memory T cells), which enter the nodal capsule at its convex side. Naive T cells (CD45RA$^+$) enter the node via the interaction between the peripheral lymph node homing receptor they express (CD62L) and its ligand expressed by the high endothelium lining the venules within the lymphonodar and tonsillar parenchyma. The lymph leaves the node in the efferent lymphatics, which exit the node at the hilus.

Blood–Brain Barrier. Migration across the blood–brain barrier is a special case due to the microanatomical nature of this structure. The central nervous system is endowed with a specialized form of immune surveillance. While it lacks draining lymph nodes, a connection exists between the brain and the deep cervical chain of lymph nodes. Lymphocytes infiltrate the brain tissue of patients with a variety of neuropathological diseases (e.g., viral encephalitis, multiple sclerosis) and may possess the property to migrate into healthy brain tissue as well. This migration process is slightly more complex in the brain than in peripheral lymph nodes because of the presence of the blood–brain barrier. Selective permeability of the barrier is established by the interaction between the endothelial and the underlying basement layer. The latter is composed primarily of pericytes and of astrocytes. Intracranial migration of lymphocytes, which is mediated by the expression of the same type of adhesion molecules and their ligands as discussed above, depends upon the interaction between the immune cells with the endothelial cells as well as with the underlying astrocytes and the pericytes.

Musculature. Lymphokinesis is brought about in part by compression from the nearby thoracic muscles involved in respiration and cardiac movements and in part by the smooth musculature within each of the three layers that constitute the walls of the lymphatic ducts (tunica intima, tunica media, and tunica adventia). These events are modulated and controlled by neuroendocrine signals and are regulated by valves that prevent the backflow of lymph in the main lymphatic vessels. The flow of lymph is directed from the extremities toward the thoracic and lymphatic ducts (channels) (Fig. 86.1). Lymph flow is important in medical physiopathology and clinical interventions because it determines the most likely routes of cancer metastases as well as directional chemotherapy [22,45,51,96].

86.3 Neuroendocrine–Immune Interactions

86.3.1 Lymphocytes as Neuroendocrine Cells

Hormones, Neurotransmitters, and Cytokines. The immune and the neuroendocrine systems are both capable of recognizing stimuli by using a variety of shared receptors and signal molecules. The integration between these systems involves:

- Hormones and classical neurotransmitters that bind to specific receptors on immunocytes and can modulate their activity,
- Cytokines and classical products of immune cells that can modify the function and modulate the activity of neuroendocrine cells,
- Cytokines that are produced by cells of the neuroendocrine system.

The distinction between "hormones," "neurotransmitters", and "cytokines" is somewhat blurred, and the same substance can belong to more than one of the aforementioned categories. In order to function, the immune and neuroendocrine systems rely upon complicated intertwined networks of cell–cell interactions mediated by signal molecules largely shared by these systems, which share a common evolutionary origin. Cytokines and neuropeptides are present in the immunocytes of invertebrates and lower vertebrates. Invertebrate hemocytes, as well as vertebrate lymphocytes, can on the whole be considered as a "mobile immune brain" [2,9,15,27].

86.3.2 Innervation of Lymphoid Organs

Innervation of lymphoid organs is specific in terms of the anatomical localization, developmental patterns, and the nature of the secreted neurotransmitter [2,17,27,108]. Fibers penetrate lymphoid organs and establish synapse-like junctions with lymphoid and myeloid cells. These junctions express functional receptors for catecholaminergic and cholinergic signals and other transmitters. This suggests that the neurotransmitters produced by the fibers may influence immune cells within these anatomical sites. Factors secreted by activated immune cells within these organs (e.g., interleukins) are taken up by these nerve fibers and can, by retrograde transport, directly influence nervous system responses. Much of these mechanisms remain to be characterized, but the "hardware" linkages between the nervous and the immune system exist [41].

Catecholaminergic Signals. These regulate several aspects of the interaction between the neuroendocrine and immune systems:

- Serotonergic innervation stimulates the release of corticotropin-releasing hormone and proopiomelanocortin gene products and mediates the interactions between hypothalamic-pituitary-adrenal and cell-mediated immunity.
- In the spleen and other lymphoid organs, noradrenergic fibers are associated with the vasculature and penetrate the organ at the hilar region. Afferent and efferent lymphatics are found at the splenic hilus. The fibers branch off where the splenic artery branches off, thus forming complex plexuses that distribute within the

white pulp in close association with the central splenic arterioles and their respective branches. Smaller plexuses are found in association with the venous vasculature within the trabeculae and the capsule or just outside it. Within the white pulp, fibers remain associated mainly with the arterial vasculature or freely distribute deeply into the tissue along the inner regions of the marginal and parafollicular zones. Occasional fibers are found in the follicles and the red pulp. Fibers in the spleen often occur in close, direct contact with lymphocytes in the synapse-type junction discussed above, whose distance has been estimated to be in the order of 5–7 nm.

- Ganglionectomy, vagotomy, and retrograde immunochemistry studies indicate that noradrenergic innervation to the spleen of young, adult (200–250 g) Sprague-Dawley rats is supplied largely by the superior mesenteric-celiac ganglia. This innervation probably represents the second neuron in the classical two-neuron sympathetic chain (see Chap. 17) and suggests the existence of non-noradrenergic innervation of the spleen.
- In lower vertebrates (e.g., fish) and mammals (e.g., rodents), noradrenergic fibers in the spleen are not randomly distributed, but show close association with specific loci rich in leukocytes, including the perivascular lymphatic sheath. Local surgical denervation of the spleen of adult (2- to 3-month-old) female Holtzman rats or general chemical peripheral sympathectomy of newborn rats with 6-hydroxydopamine combined with adrenalectomy leads to an enhancement in the number of plaque-forming cells in response to immunization with sheep red blood cells. An inverse relationship exists between the magnitude of the immune response and splenic norepinephrine concentrations: high-responder rats, but not low responders (as measured by plaque-forming cells) demonstrate continuous reduction in splenic norepinephrine content 3 days following immunization.
- During an immune response to sheep red blood cells, norepinephrine concentrations decrease centrally in the hypothalamus of Holtzman rats.
- Decreases in peripheral and central norepinephrine concentrations are associated with, and upregulate, selected components of the immune response. These effects are interdependent, since intraperitoneal injections of interleukin-1 alter norepinephrine metabolism in the central nervous system.
- Products of stimulated immune cells regulate sympathetic output, largely at the level of the hypothalamus.
- Noradrenergic fibers sprouting from the appropriate branches of spinal nerves penetrate the bone tissue with blood vessels. Some fibers remain within vascular plexuses in the parenchyma, while others penetrate the bone marrow. This process initiates prenatally (see below), about the time of the onset of hematopoiesis, and is sustained throughout the life span.
- In the thymus, noradrenergic fibers originate from postganglionic cell bodies of the upper paraventral gan-

glia of the sympathetic nervous system (superior cervical and stellate ganglia) and penetrate the thymic gland alongside blood vessels. The initial entry point is the thymic capsule, beyond which the fibers distribute to the capsular and septal systems. These fibers remain in association with blood vessels or penetrate these connective tissues independently; they progress from the thymic capsule and septum to its parenchyma and end in the cortex. Most of the fibers distribute in the thymic cortical region, with the highest density in the corticomedullary region. A few fibers are also found in the thymic medulla. These fibers are generally associated with arterial or venous vessels.

- Noradrenergic fibers enter lymph nodes at the concave side of the node (hilus), in association with the arterial and venous systems. Fibers distribute in the capsular/subcapsular compartments and in the septum and eventually arborize in the paracortical regions rich in T lymphocytes, as well as in parafollicular regions. Germinal centers and nodules are devoid of innervation, which suggests the absence of direct neural modulation of the immune events at these sites. Fibers sprout in most instances from the spinal nerves that innervate their respective anatomical regions.

- In the human tonsils, noradrenergic fibers form dense paravascular plexuses, from which individual fibers sprout into parafollicular regions, in areas rich in T lymphocytes. Lymphoid nodules and epithelium of these organs typically show little innervation. Fibers penetrate these organs relatively early in development, before their invasion by cells of the lymphoid lineage.

There also is extensive innervation and plexus formation in enteric mucosal tissues by noradrenergic and other fibers. Substance P (see below) and calcitonin gene-related peptide fibers penetrate the lamina propria into the epithelium. These fibers are joined by noradrenergic and vasointestinal peptide fibers in the Peyer's patches.

86.3.3 Physiology of Neuroendocrine–Immune Interactions

As noted above, the immune system and the neuroendocrine system interact, specifically (in relation to the focus of this chapter) pituitary-adrenal hormones and cellular immunity (Fig. 86.2). The Cartesian mind/body dualism is untenable, and health of the body is interdependent and intimately intertwined with the well-being of the mind (see Chap. 87). The interactions between the nervous, endocrine, and immune systems are physiologically important because they ensure adequate adaptation of the organism to physiological and environmental challenges [2,7,15,27,106].

The original observations by Besedovsky and his collaborators [13,14], demonstrated that experimental viral infection with Newcastle disease virus, administration of lipopolysaccharide, peripheral injections of concanavalin A supernatants, or intraventricular administration of interleukin-1β lead to a significant rise in plasma glucocorticoids concentrations. Administration of this antigen (3–5 µg per 100 g body weight) leads to a five- to sevenfold increase in plasma corticosterone concentrations in young, adult, male Wistar rats within 1–2 h. Sig-

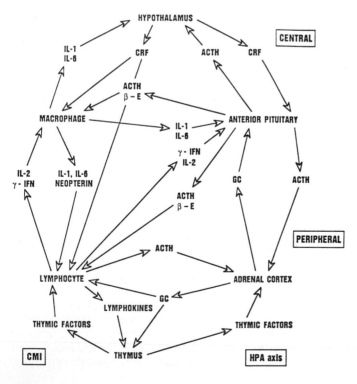

Fig. 86.2. The hypothalamic-pituitary-adrenal/cell-mediated immunity (*HPA-CMI*) system. *CRF*, corticotrophin-releasing factor; *GC*, glucocorticoid; *IL*, interleukin; *ACTH*, adrenocorticotropic hormone; *IFN*, interferon; *β-E*, β-endorphin

nificantly increased concentrations of ACTH and β-endorphin also occur. The endocrine effects of lipopolysaccharide administration are dose and time dependent (at 3–5 μg per 100 g body weight, ACTH and corticosterone serum concentrations are significantly increased between 1 and 6 h following administration). These outcomes can be blocked by anti-interleukin-1 or anti-corticotropin-releasing hormone antibodies and fail to occur in animals with lesioned hypothalami. These observations support the role of cytokines produced by macrophages activated by lipopolysaccharide in the stimulation of hypothalamic corticotropin-releasing hormone neurons, which then activate the hypothalamic-pituitary-adrenal axis. Taken together, these findings establish the existence of a physiologically significant interactive loop between cellular immune and hypothalamic-pituitary-adrenal responses (Fig. 86.2) [7,11–14,24].

Endotoxins from Gram-Negative Bacteria. These may lead to septic shock, characterized by many significant physiological responses. Lipopolysaccharides, the purified component of the outer membrane of *Escherichia coli*, produces several of these pathophysiological outcomes, including hypotension, vascular injury, disseminated organ dysfunction, fever, decreased motor activity, lethargy, and a full-blown acute phase response.

Hypothalamic-Pituitary-Adrenocortical Hormones.
These hormones rise as an outcome of cellular immune activation, and the stimulation of isolated rat splenocytes and human peripheral blood mononuclear cells results in the production of a factor which, when injected intraperitoneally into young, adult, female Holtzman rats, is capable of raising circulating concentrations of corticosterone. This factor, the "glucocorticoid-increasing factor," is inactive in hypophysectomized or dexamethasone-treated rats. In addition, intraventricular injections of interleukin-1 lead to a rise in circulating ACTH that can be blocked by antibody directed against corticotropin-releasing hormone. Similary, intraperitoneal injections of human recombinant interleukin-1, but not tumor necrosis factor, in mice or rats also raises ACTH and corticosterone concentrations by about five fold 2 h after administration. A booster shot of the recall antigen tetanus toxoid antigen triggers the expected immune recall response in human subjects and also generates a significant rise in plasma cortisol within a few hours, well before the appearance of antitetanus antibodies. This protocol, used to test endocrine functions in patients with the acquired immune deficiency syndrome, demonstrates that these patients mount a blunted hormonal response to the antigenic stimulation, suggesting significant disruption in neuroendocrine–immune interaction [7,12,19].

Lipocortin-1. Also called annexin-1, this member of the Ca^{2+}-binding annexin superfamily rises in association with acute inflammation. Inflammation results in the elevation in interleukin-1, interleukin-6, and other cytokines (and the consequential elevation in circulating glucocorticoids) at certain anatomical sites (i.e., lung epithelia, thymic medullary and cortical epithelia, ductal epithelia of the prostate). At these sites, phagocytic immune cells (e.g., macrophages), which express lipocortin-1 receptors, bind this protein. Lipocortin-1, perhaps via its receptor-mediated blunting of transmembrane signals (e.g., phospholipase A_2, although this is still controversial), appears to blunt the chemotaxis and proinflammatory capacity of macrophages and other phagocytic cells. This is an example of the existence of important and functional physiological feedback processes that modulate neuroendocrine–immune outcomes.

Neopterin. This pyrazino-pyrimidine (2-amino-4-oxo-[D-erythro-1,2,3'-trihydroxypropyl]-dihydro-pteridine) is the metabolic product predominantly of myeloid cells that results from the activation of guanosine triphosphate (GTP) cyclohydrolase by interferon γ and provides another example of the physiological relevance of these interactive processes. Elevated circulating concentrations of neopterin are found in a variety of infectious and autoimmune diseases, including infection with the human immunodeficiency virus. The production and secretion of neopterin is a reliable, albeit nonspecific marker of immune activation. It may play a clinically important neuroendocrine–immunoregulatory role, since it interferes with folate metabolism by pathogens and thus participates in microbiostatic processes; in this regard, neopterin acts in a manner similar to nitric oxide (see above). Neopterin can alter lymphocyte processes of maturation, activation, and distribution because lymphocytes metabolize neopterin to another pterin derivative, tetrahydrobiopterin, which in turn augments responses to interleukin-2 in vitro. The generation of neopterin requires energy, but the release of neopterin through the plasma membrane does not. It occurs by passive, osmosis-like diffusion. Thus, neopterin freely leaks out of and into cells, including the cell linings that form the blood–brain barrier (see above). Peripheral and central nervous system cells almost certainly utilize neopterin in the biosynthesis of tetrahydrobiopterin, because the first intermediate in its biosynthesis is dihydroneopterin triphosphate, the catabolic product of neopterin by GTP cyclohydrolase I. Tetrahydrobiopterin inhibits the activity of GTP cyclohydrolase I. This confirms the existence of a finely regulated neopterin/tetrahydrobiopterin biosynthetic loop crucial to neuroendocrine–immune interactions. Tetrahydrobiopterin is the cofactor for enzymatic hydroxylation of the aromatic amino acids phenylalanine, tyrosine, and tryptophan. These hydroxylase reactions catalyze early steps in the generation of norepinephrine and epinephrine, as well as serotonin and melatonin [21].

Physiological Regulation. "Static" or "dynamic" challenge tests are used to test the physiological regulation of the hypothalamic-pituitary-adrenal axis. Administration of the glucocorticoid synthesis inhibitor metyrapone is a useful "static" approach to study the functioning of the axis in the absence of the feedback arm. Administration of the synthetic glucocorticoid dexamethasone is another "static" research tool used to study the regulation of this hormonal system by focusing upon its feedback mechanism. It is important to note that the dexamethasone suppression test has raised a number of concerns about its reliability and validity as a diagnostic instrument. Additional measurements that substantially improve the strength of this physioclinical research tool include:

- Obtaining both ACTH and cortisol concentrations and computing the ratio between these two values, as an indication of pituitary–adrenocortical interaction,
- Matching subjects according to height and weight, since adrenal gland output of cortisol is related to size, which is in turn related to total body mass,
- Closely monitoring the blood concentrations of dexamethasone, because the pharmacokinetics of this steroid may be significantly different across subjects,
- Obtaining a measure of 24-h cortisol. For this purpose, 24-h plasma sampling can be obtained, urine samples can be collected over a 24-h period, pooled, and assessed for cortisol content, or the mean plasma cortisol concentration between 1300 and 1600 hours can be obtained, at least 1 h following the subject's last meal.

Administration of pituitary ACTH, of hypothalamic corticotropin-releasing hormone, or of tetanus toxoid (see above) followed by assessment of plasma cortisol concentrations at regular intervals thereafter provide useful "dynamic" assessments of the hypothalamic-pituitary-adrenal axis [24].

Normal Subjects. Administration of 1–5 mg dexamethasone in the evening, before the diurnal onset of hypothalamic-pituitary-adrenal activity, suppresses plasma cortisol concentrations in normal subjects to below 5 μg/dl for the following 10–16 h. Associated changes in cellular immune parameters include reduction in the number and percentage of circulating lymphocytes, particularly of CD3 cells 10 h following dexamethasone administration. Within the CD3 lymphocyte population, a significant drop occurs in the percentage and number of CD4, but not CD8 cells. The CD4 cells that express the homing receptor CD62L (see above), rather than CD4+CD62L− cells, comprise the responding population. A large proportion of the CD4+CD62L+ cell population overlaps with naive helper T cells (CD4+CD45RA+; see above). The reason for this is that, as noted above, naive cells are more mobile by virtue of being actively engaged in patrolling the organism's tissues in search of antigens [24].

The ability of T cells to engage in the appropriate immune response depends upon their competence to be adequately activated following exposure to a pathogen. T cell activation, monitored as transmembrane and transcytosolic signaling events, overall cellular metabolic activity, production of cytokines and of their receptors, and clonal expansion and proliferation, is modulated by a variety of endocrine products, including steroids. The immunomodulatory effect of steroids is mediated largely by their effects upon the expression of certain cytokines. Depending upon the antigenic and endocrine environment during immune activation, either of two polarized end stages of T cell maturation can result, which are distinguished by the pattern of cytokines they produce. Antigenic stimulation of recent thymic emigrate naive T cells leads to the generation of primary effector cells. These cycling cells have a rather large cytoplasm and produce a range of cytokines, including interleukin-1, γ interferon interleukin-3,-4,-5, and -10, the TH0 pattern of cytokine response. Further activation of TH0 cells in the presence of interferon and absence of interleukin-4 leads to the generation of CD4+ cells that secrete large amounts of interleukin-2 and interferon, but little interleukin-3,-4, or -5 upon further stimulation. This is the TH1 pattern of response. By contrast, TH0 cells activated in the presence of interleukin-4 and no interferon become capable of a TH2 pattern of cytokine response, characterized by the predominant secretion of interleukin-3,-4,-5, and the newly identified interleukin-13. Interleukin-12, an accessory cytokine produced by activated antigen-presenting cells, plays an important regulatory role in the generation of TH1 and TH2 patterns of response and favors the TH1 pathway. Metabolic inhibitors, designed to blunt certain events during lymphocyte activation (e.g., pentoxifylline, which blocks phosphodiesterase) permit the study of these maturational events because they selectively blunt the generation of one or the other pattern of cytokine response. The patterns of T cell cytokine production produced in response to an antigenic challenge are modulated by the regulatory influences on gene expression exerted by steroid hormones. Glucocorticoids suppress interleukin-2 and interferon γ production, but tend to augment production of interleukin-4. These effects are largely receptor mediated. By contrast, dehydroepiandrosterone (DHEA) favors interleukin-2 production over interleukin-4. T cells obtained from healthy adults that are pretreated with $10^{-9}M$ DHEA and stimulated with mitogens generate more interleukin-2 and are more cytotoxic than untreated cells. These outcomes appear to be DHEA receptor mediated and to be specific to the CD4+ subpopulation [29,82].

Pregnancy. In women of child-bearing age and during pregnancy, the principal C-19 androgen is DHEA-sulfate (DHEA-S). The plasma concentration of this androgen is about 1600 ng/ml prior to conception and steadily falls during gestation to reach about 800 ng/ml at parturition. This may be associated with the decrease in the mother's immune competence (see below, Sect. 86.4.1) and the concurrent TH1/TH2 shift [29,83], which allows the fetus not to be rejected. This also may reflect an increased metabolic clearance rate of DHEA-S due to a significant uptake of this steroid by the placenta (see Chaps. 117, 119). Utilization of maternal DHEA-S by the fetus for conversion into estrogens is also likely since the fetus, mother, and placenta work cooperatively to provide the developing fetus with the concentrations of progesterone and estrogens required for fetal development and differentiation. The drop in circulating concentrations of DHEA and DHEA-S noted in patients infected with the human immunodeficiency virus should therefore be a serious concern in clinical pediatric acquired immune deficiency syndrome research and intervention.

Neuroendocrine–immune research has now shown that DHEA and DHEA-S have the capacity of rescuing, in part, the loss of cellular immune response in certain immunocompromised hosts, such as the aged (see below). Whether or not these observations have direct relevance for the development of treatment interventions for other patients, including pediatric patients, must be tested.

Jet Lag. Other examples of neuroendocrine–immune interaction include certain situations that induce significant disruption in circadian rhythms (see Chap. 58). For example, intercontinental flights produce the physiological phase shift called "jet lag." Evidence now shows that these situations lead to significant alterations in neuroendocrine diurnal fluxes of cortisol and related hormones, which synchronize to the new environmental cues within a few days. Diurnal fluxes in lymphocyte populations also show significant changes during the jet lag period and require approximately the same time to acquire again the physiological pattern observed under normal conditions. Generally, a period of 1 day per hour of circadian shift is required for these interconnected physiological adjustments.

86.3.4 Biochemical and Molecular Events

Cells of the immune system respond to a variety of stimuli. Biochemical and molecular mechanisms orchestrate these signals such that cellular responses produce coordinated, focused, and coherent outcomes directed toward the elimination of the invading pathogen [4,34,96].

In addition to receptors for interleukins (see Chap. 6) and for other immune factors, immune cells express various receptors for neuroendocrine peptides and for hormones (e.g., opioids, steroids, prolactin, growth hormone). These compounds are capable of modulating the T cell's expression of adhesion molecules, its internal milieu, and its ability to respond to antigen presentation challenges. The concerted progression of T cell activation steps depends upon the internal milieu of the cell (i.e., its own capacity to respond to the challenge) and its external milieu (i.e., the nature and concentration of circulating cytokines, which mediate its response to the challenge). T cell activation involves four principal events, which are summarized in Fig. 86.3. Immune recognition, cytoskeletal rearrangement, and transmembrane signal conduction converge to concerted cytobiochemical T cell responses, which represent the outcome of the interaction of at least two types of pathways: one immune in nature (i.e., antigen challenge, cytokine mediation) and the other neuroendocrine in nature (i.e., neuropeptide and hormonal modulation of immune responses) [28,97,98].

Immune Recognition. Recognition and cell–cell interaction are crucial to the initiation of T cell responses. These events are directed by the expression of the adhesion molecules discussed above. Physiopathological conditions that lead to alterations in the expression of these molecules severely hamper T cell responses. As noted above, no other interaction is more important to the activation of T cells than the recognition of the antigen-presenting cell in association with the nonself antigen by the T cell receptor/CD3 complex. This is the prime pathway of activation of the T cell, although adjunctive, costimulatory pathways are now being increasingly identified (e.g., CD2, CD28).

Immediate Cytoskeletal Changes. These occur in immune, as well as in neuroendocrine cells, as soon as the ligand finds its receptor, e.g., a microclustering of the receptor and the formation of dimers or microaggregates of less than decamers. Within 10 s of initial binding, aggregation of the receptor–ligand units occurs, which can be visualized as distinctive patches (i.e., "capping" phenomenon). These aggregates are internalized, often at the coated pits, by means of interactions with clathrin. Once within the cytoplasm, the ligand–receptor complex is dissociated, and the receptor is recycled and returned to the membrane. At the inner side of the plasma membrane, rearrangement of the cytoskeleton occurs such that complex bridges are formed between actin rami and smaller cytoskeletal proteins (e.g., talin, paxilin, vincullin). Kinases and other activated proteins are recruited into this cytoskeletal structure. This is a critical event for effective signal transduction, which involves several important steps (Fig. 86.3). For example, the focal adhesion kinase of 125-kDa (p125^{FAK}) autophosphorylates, and protein tyrosine kinases recognize and bind to these phosphorylated sites by the SH2 domains, which recognize and bind with high affinity ($0.3\text{-}3 \times 10^{-9}M$) to phosphorylated residues. In such manner, p125^{FAK} contributes to initiate the membrane-associated kinase cascade.

Signal Transduction. This includes a series of events that occur at or proximal to the plasma membrane following binding to the immune cell ligand (e.g., T cell receptor/CD3 complex). Among the very initial events, one observes a significant increase in the state of activation of the ras exchanger proteins. The *ras* proto-oncogene products (21-kDa protein, p21^{ras}) are "small GTP-carrying proteins" with considerable functional and sequence homology to G proteins. As G proteins, p21^{ras} (Ha-, Ki-, and N-) becomes activated on binding GTP. The function of the ras exchanger protein is to load p21^{ras} with GTP. Hydrolysis of GTP to guanosine diphosphate (GDP) by GTPase leads to the activation of adenyl cyclase and to the inactivation of p21^{ras}. Immune cells remain quiescent so long as GTPase-activating protein (GAP) activity is high, relative to the concentrations of activated p21^{ras}. Bound GTP is rapidly hydrolysed, and the inactive p21^{ras} – GDP form prevails. When the cells are activated, GAP activity falls several fold upon phosphorylation of GAP-associated proteins (e.g., p120 and neurofibromatosis type 1, NF1). Thus, the activity of GDP dissociation stimulator (GDS), rather than that of GAP, transiently predominates, and the concentrations of p21^{ras}-GTP transiently rise. The p120–GAP protein is

1715

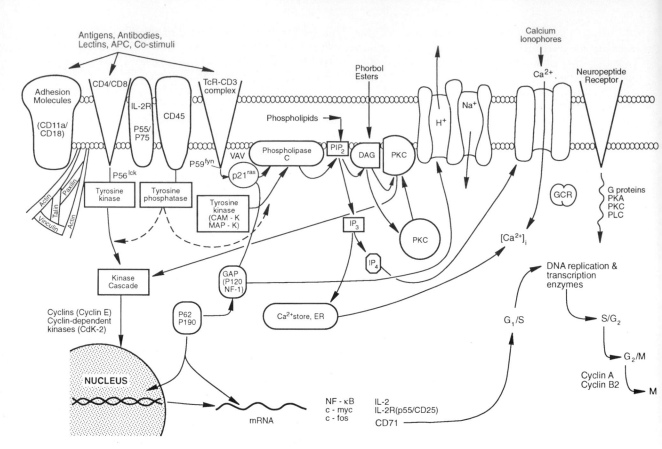

Fig. 86.3. T cell signaling events. *IL*, interleukin; *TcR*, T cell receptor; *CAM*, cellular adhesion molecule; *GAP*, GTPase-activating protein; *NF*, neurofibromatosis; *DAG*, diacylghycerol; *IP₃*, inositol triphosphate, *PKC*, protein kinase C; *GCR*, glucocorticoid receptor; *PKA*, protein Kinase A; *PLC*, phospholipase C; *APC*, antigen-presenting cell; *MAP*, mitogen-activated protein kinase; *PIP₂*, phosphoinositol diphosphate; *ER*, endoplasmic reticulum

principally regulated by the two kinase substrate proteins, p60 and p190. Complexes of p120–GAP with the phosphorylated forms of p60 and p190 lead to decreased GAP activity. The NF1 product also complexes with GAP. NF1–GAP binds to p21^{ras} considerably more efficiently (up to 300-fold) compared to p120–GAP. NF1–GAP is regulated by phosphorylation as well as by metabolic products of phospholipids, including phosphoinositol.

Enhanced Adenyl Cyclase Activity. This leads to the generation of cAMP, which activates the R subunits of type I and type II isozymes of cAMP-dependent protein kinase. This kinase phosphorylates hydroxy-amino acid residues on cytoplasmic and nuclear proteins and thus contributes to dampen CD3 cell activation. A drop in cAMP is required for activated CD3 cells to progress through G_1, probably because of the role of this kinase in the phosphorylation of cyclins (e.g., cyclin E) at adequate sites. Cross-linking of certain lymphocyte receptors, (e.g., CD3) results in capping, a good activator of the cAMP-dependent pathway.

Activation of Phospholipase C. Activation of this enzyme, of which the γ-1 isoform is prevalent in CD3 cells, is critical to the initiation of the phosphoinositol metabolic cycle

(see Chap. 5). Phospholipase C activity is detected within minutes of T cell activation. This enzyme hydrolyses membrane phospholipids to generate diacylglycerol (DAG) and inositol 3-phosphate (IP$_3$). The generation of DAG and IP$_3$ represents a crucial step in the events leading to immune cells' activation and progression through the cell cycle, which is finely regulated by a series of converging and diverging metabolic pathways, including the metabolism of other phospholipids (e.g., sphingosine, phosphatidylcholine) and other membrane-associated responses. Experimental data show that phosphoinositol metabolism is modulated by neuroendocrine products (e.g., β-endorphin, glucocorticoid). DAG is a hydrophobic lipid that remains within the membrane bilayer and activates protein kinase C (PKC). It recruits cytoplasmic PKC onto the cytoskeletal structure at the inner side of the membrane. This process contributes to the amplification of PKC activation, the kinase cascade. IP$_3$, by contrast, is a polar lipidic product that diffuses through the cytoplasm, binds to endoplasmic reticulum membrane receptors, and increases Ca^{2+} flux from available intracellular stores. This process contributes to open specialized membrane channels (e.g., Na$^+$, K$^+$, chloride) making extracellular stores of Ca^{2+} and other ions available to the cell.

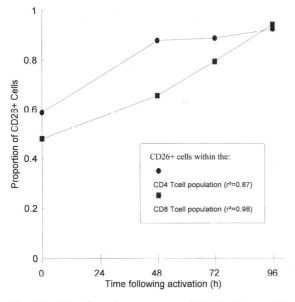

Fig. 86.4. Time-dependent expression of CD26 by CD4⁺ and CD8⁺ T cells following activation

Kinase Cascade. These events converge to activate tyrosine, serine, and threonine kinases and phosphatases, resulting in the kinase cascade. These activities (e.g., p56[lck], p59[fyn], mitogen-activated protein kinase [MAPK], CD45) are responsible for the orderly amplification of the signal from the membrane, across the cytoplasm to the nuclear compartment. The activity and regulation of the kinase cascade determines much of the cell's ability to respond appropriately to the challenge.

Genomic Activation. The kinase cascade is followed by genomic activation, which includes early transcription events, such as the expression of oncogenes (e.g., c-*fos*, c-*myc*) and the production of lymphokines and membrane receptors. Among the earliest such genes expressed by activated T cells are those for the membrane glycoprotein CD69, interleukin-2 and the α-chain of its receptor (p55, CD25), the transferrin receptor (CD71), and the serine peptidase IV, CD26 (DPP-IV, EC.3.4.14.5), whose metabolic activity in vivo includes the catabolism of substance P and other peptides with the NH_2X-Ala/Pro-X sequence. CD71 is essential for the continued progression of the cells through S, because its translocation to the cell membrane allows the cell to incorporate iron-laden transferrin. Iron is required for many enzymatic activities involved in DNA replication, RNA transcription, and protein translation. The process of gene expression includes transcription of the gene into the nuclear form of mRNA, transport and processing of mRNA into the cytoplasm, translation of the mRNA, and gene product utilization (e.g., secretion of interleukin-2).

The cell continues its progression through the later stages of the cell cycle in preparation for its first mitotic event (it exits S to traverse G_2 and eventually reaches M). This progression is dependent upon other cyclins, including cyclin A and cyclin B_2 and their cyclin-dependent kinases. Later markers of immune activation (e.g., HLA-Dr) are also expressed at this time.

These events of cell activation occur within the first of 24–48 h following lymphocyte activation and can be monitored experimentally by assessing the expression of "membrane activation markers" (i.e., CD69, CD25, CD71, CD26) on the plasma membrane a few hours (e.g., CD69) to one (e.g., CD25) two (e.g., CD71) or three days (e.g., CD26) following stimulation. Figure 86.4 shows the distinctive pattern of CD26 expression by CD4⁺ and CD8⁺ T cells following activation. Lymphocytes activated by mitogens, by anti-CD3 antibody either without or with costimulatory signals provided by CD28, or by incubation in a mixed leukocyte reaction progress through several cell cycles (three to four) during the initial 72 h following experimental stimulation. Proliferation can be assessed by monitoring the incorporation of tritiated thymidine into the DNA of cells actively traversing the S phase. Activation events lead to significant changes in the phenotypic characteristics of T cells (e.g., loss of CD62L and CD45RA, acquisition of CD45RO), as well as their functional abilities (e.g., B cell help, T lymphocyte-mediated cytotoxicity, TH1 versus TH2 patterns of cytokine production). These functions represent critical T cell maturational events. Alternatively and depending upon the physiological milieu and the state of the cell, activation events may lead to apoptotic manifestations that include the activation of endonucleases, DNA fragmentation, and cell death. Neuroendocrine products modulate these biochemical and cellular outcomes [28,33,34,96].

Stimulation of B Cells. This results from binding of the antigen to the specific membrane-bound immunoglobulin and leads to a series of biochemical events that are similar to those outlined above. These responses are also modulated by neuroendocrine products via specific receptors expressed on the membrane of the B cells.

In vivo, these events, which mark the onset of cellular immune processes that lead to the elimination of the pathogen and to the production of antibodies, do not usually take place in the circulating blood lymphocytes, but in secondary lymphoid organs (e.g., spleen, lymph nodes). While more easily accessible for human studies, the immune cells found in the peripheral blood are in transit from some immune organ to another; therefore peripheral blood immune cells represent a window, a rather narrow window of the immune system. Nevertheless, this material significantly contributes to our increased understanding of immunophysiological and immunopathological phenomena, and its continued utilization is critical for the elucidation of fundamental neuroendocrine–immune phenomena.

86.3.5 Pathophysiology of Neuroendocrine–Immune Interactions

The interactions among the various physiological systems and between the organism's physiological and psychological structures have profound clinical implications. Infusion of lymphokines or lymphokine-activated immune cells in cancer patients is associated with significant physiopathological and psychiatric side effects. Many psychiatric disorders have concomitant neuroendocrine and immune impairments, and patients with immune dysfunctions often manifest psychiatric as well as neuroendocrine disturbances (see Chap. 87). Altered neuroendocrine–immune behavior interactions are evident in patients who abuse alcohol or drugs since these substances influence each of these systems individually [24,29,36].

Anorexia Nervosa. This is a disorder of unclear etiology, characteristically observed in young women who express obsessive fears of being fat. It involves an extensive and complex pathophysiology [24,121], which includes disturbances in the following:

- Sleep
- Body temperature
- Cardiovascular activity
- The peripheral sympathetic system
- Endocrine regulation (e.g., growth hormone, prolactin, somatostatin, arginine vasopressin)
- Reproductive hormones production and metabolism, leading to amenorrhea
- Thyroid hormone regulation, accompanied by the low T3 syndrome (reduction in free triiodothyronine concomitant with a rise in the metabolically less active isomer, reverse T3, 3,3,5′-triiodothyronine; the syndrome is not specific for anorectic patients, and it disappears when these patients are under treatment and regain weight)
- Hypothalamic-pituitary-adrenal regulation (suppression of plasma cortisol following dexamethasone administration fails to occur in anorectic patients and plasma cortisol concentrations typically remain higher than 15 µg/dl 10 and 18 h following intravenous dexamethasone administration).

The overall consensus in the literature is that hypercortisolemia and dexamethasone resistance in anorectic patients are independent of concurrent depressed moods, altered cortisol metabolism (increased production, decreased rate of clearance), and loss of the diurnal cortisol cycle [24].

Relentless pursuit of thinness in anorexia nervosa leads to severe emaciation. If left untreated, this condition leads to considerable morbidity and mortality. The nutritional status of anorectic patients is rarely deficient in proteins and in vitamins, but it is selectively deficient in carbohydrate and calories. This is quite distinct from the state of malnutrition attained during starvation and explains why viral infections are relatively uncommon and signs of enhanced immune competence are evident during the early stages of anorexia nervosa. When the state of emaciation becomes overly severe, significant impairments in immune competence become apparent. One often observes a tendency toward leukopenia with relative lymphocytosis,

hematopoietic (e.g., changes in bone marrow cellularity), hematologic (e.g., red cell acantholysis), and immune alterations (e.g., complement deficiencies, decreased bactericidal activity by granulocytes) of various degrees of severity. These patients have higher concentrations of tumor necrosis factor produced by leukocytes under spontaneous conditions compared to normal control subjects. Because this factor is produced by activated macrophages, this observation suggests that anorectic patients present manifestations of endogenous immune activation. This hypothesis is confirmed by findings of significantly elevated serum neopterin concentrations in these patients. A normal distribution of CD3, CD4, CD8, and B cells is generally observed, and natural killer cytotoxic activity is within the normal range in these patients, as is the proliferative response to T cell mitogens. The percentage and number of CD4+CD45RA+ and CD4+CD62L+ cells are significantly increased compared to matched controls. Dexamethasone administration fails to reduce the percentage and number of CD3, CD4, CD4+CD62L+, and CD4+CD45RA+ populations. The number of cytosolic glucocorticoid receptors is also not altered in these patients 10 h following dexamethasone administration. Taken together, these observations indicate systemic resistance to dexamethasone [24].

Depression. Hypercortisolemia, resistance to dexamethasone, and related dysfunctions in the regulation of the hypothalamic-pituitary-adrenal axis are not unique to anorexia nervosa. Administration of dexamethasone also fails to suppress plasma cortisol concentrations in a large proportion of the patients with major depression. Resistance to dexamethasone in these patients is associated with the severity of the psychiatric disorder. Significant alterations in cellular immunity, including depressed proliferative response to T cell mitogens, reduced natural killer cell cytotoxic activity, and decreased number and percentage of circulating CD4 cells, have been reported to be associated with the severity of depression. The patients' age further contributes to these abnormalities. Resistance to dexamethasone among depressed patients leads to a failure to reduce the proliferative response to mitogens and the number of cytosolic glucocorticoid receptors similar to that observed in anorectic patients, suggesting that other patient populations characterized by this sort of hormonal dysfunction (e.g., the aged) also may manifest similar alterations in cellular immune regulation (see below) [24,48,61,110,111].

Acquired Immune Deficiency Syndrome. Patients infected with the human immunodeficiency virus manifest severe progressive loss of CD4-mediated immunity and eventually generalized cellular immune competence. These patients also show significant impairments in neuroendocrine regulation, e.g., basal (0800 h) and 24-h plasma cortisol can be significantly elevated, although some patients show plasma cortisol concentrations lower than normal, leading to adrenal insufficiency, probably because of pituitary and adrenal infection by cytomegalovirus. This condition is associated with symptoms of nausea, vomiting, anorexia, fatigue, orthostatic hypotension, and hyponatremia. The 24-h concentrations of DHEA and DHEA-S are lower in most seropositive patients compared to control subjects, and severe alterations in the circadian patterns of lymphocytes, cortisol, DHEA, and DHEA-S occur during progression to the acquired immune deficiency syndrome [20,24,37,72].

Neuroendocrine immunology predicts the observed elevation in concentrations of cortisol in these patients, particularly during the initial acute viremia phase, following activation of the virus that signals termination of the "dormant" asymptomatic phase (note: recent studies by Fauci, Pautaleo and collaborators indicate that the asymptomatic phase is not a "dormant" phase of the disease process since considerable replication of the virus is now known to occur in the lymph nodes. Hence, "hidden" would be a better term than "dormant" to describe this phase), and during the progression of the disease before the onset of opportunistic infections of the pituitary or adrenal glands, because activation of the cellular immune compartment should result in increased production of cytokines, which should in turn induce the secretion of hypothalamic-pituitary-adrenal hormones.

Physiopathology also predicts the observed decreases in the concentrations of DHEA and DHEA-S in these patients, because the organism's physiological response to severe illness, surgery, and other serious clinical conditions often includes a rise in cortisol production in parallel with a decrease in the production of the adrenocortical androgen DHEA and DHEA-S by the innermost layer of the adrenal cortex, the zona reticularis (see Chaps. 17, 20). The metabolism of adrenocortical pregnenolone to the glucocorticoid pathway (17α-OH pregnenolone, 17α-OH progesterone, 11-deoxycortisol, cortisol) increases and that of the androgen pathway (17α-OH pregnenolone, DHEA, DHEA-S) decreases. This shift maximizes adrenocortical production of cortisol and favors the organism's survival, while minimizing the use of pregnenolone for androgens.

86.4 Ontogenesis of the Neuroendocrine–Immune System

86.4.1 Normal Development

The lifespan can be viewed as a spectrum, whose early manifestations (i.e., development) eventually lead to aging. The neuroendocrine and the immune systems develop, mature, and age, and so does their ability to interact in the process of adaptation to a variety of challenges.

From the perspective of embryologists, ontogenesis represents the finely tuned and controlled sequence of developmental and maturational events that lead to the formation of anatomical structures and the development of physiological functions associated with these structures.

Primary Germ Layers. Ectoderm, from which cells of the nervous system, the integumentary system (e.g., skin), and the cornea arise, and endoderm, from which most of the vital organs (e.g., liver, glandular structures, digestive and urogenital tracts) derive, form the two initial cytoembryonic strata. The mesoderm, the middle cell layer that subsequently forms, gives rise to hematopoietic, bone, muscle, and cartilage tissues and reflects finely controlled and modulated movement and positioning of ectodermal cells during gastrulation. The mesoderm and endoderm are active rudimentary and primitive histologic structures that possess several important physiological functions. Growth factors (e.g., fibroblast growth factors, transforming growth factors, insulin-like growth factors; see Chap. 21) are produced by mesodermal cells and are critical in the development and maturation of embryonic structures from ectodermal, mesodermal, or endodermal origins to form the building blocks of neuroendocrine–immune interactions later peri- and post-natally.

Genomic regulation determines the fate of ectodermal, mesodermal, and endodermal cells. Gene activation events in embryonic cells are determined by cell contact with neighboring cells and the extracellular matrix (i.e., expression of adhesion and migration receptors; see above), production of circulating paracrine, endocrine, or autocrine regulatory growth factors, and the ontogenetic "clock" (i.e., time axis).

Neuroendocrine–Immune Interactions. The ontogenesis of neuroendocrine–immune interactions is similarly intricate. Immune cells derive from pluripotent stem cells in the bone marrow. Lymphopoiesis generates B and T lymphocyte precursors, and myelopoiesis generates cells of the myeloid lineage. Early in ontogeny, hematopoiesis takes place in the yolk sac, later in the fetal liver; this function is eventually (last trimester) taken on by the bone marrow. T cells derive from bone marrow stem cells and develop in the thymus. Thymic ontogenesis results from a series of precisely timed events during which the pharyngeal pouch endoderm interacts with embryonic connective tissue and migrating lymphoid stem cells. A portion of the neural crest, that which is associated with the developing hindbrain, migrates ventrolaterally through the bronchial arches to become the mesenchyme that forms the layers around the epithelial primordia of the thymus. Experimental alterations of this sequence of events by means, for instance, of neuroendocrine challenges, or ablation of the cranial neural crest before its migration to the nascent thymus result in delayed, aborted, or abnormal thymic ontogenesis and in aberrant development of immune competence. The thymus begins a process of progressive involution following the neuroendocrine developmental changes that accompany puberty. It becomes all but a necrotic remnant in elderly individuals (see below).

Noradrenergic innervation to the lymphoid organs is already evident during ontogenesis. Fibers sprouting from the appropriate branches of spinal nerves penetrate the bone tissue with blood vessels. Some fibers remain within vascular plexuses in the parenchyma, while others penetrate the bone marrow. This process initiates about the time of the onset of hematopoiesis and is sustained throughout the lifespan.

Thymus. This is innervated postnatally in the rodent. Fibers begin to penetrate the thymic parenchyma by the end of the first week and increase in number and in complexity of arborization considerably by the second week. At this time noradrenergic innervation of the thymus resembles that observed in young adults. In the adult (see Sect. 86.3.2), fibers originate from postganglionic cell bodies (superior cervical and stellate ganglia) of the upper paraventral ganglia of the sympathetic nervous system. Fibers penetrate the thymic gland alongside blood vessels. The initial entry point is the thymic capsule, beyond which fibers distribute to the capsular and septal systems. Fibers, as noted previously, can remain in association with blood vessels or can penetrate these connective tissues independently. Fibers progress from the thymic capsule and septum to its parenchyma and end in the cortex. Most noradrenergic fibers distribute in the thymic cortical region, with the highest density at the corticomedullary region. Few fibers are found in the thymic medulla, but these fibers are generally associated with arterial or venous vessels. This developmental process is probably not identical in all mammals since the pattern of development of the central and peripheral nervous systems differs across species (e.g., myelinization is complete in primates neonatally, but takes place during the first 2 weeks of postnatal life in the rodent). Thus, the time sequence of innervation of the human thymus is expected to differ from that of mouse or rat and to be all but complete by birth. This may have profound implications for clinical situations in neonatal neuroendocrine immunology following prenatal insults, such as fetal alcohol exposure (FAE; see below).

Spleen. This is innervated pari passu with its development and compartmentalization during the first 28 postnatal days in rodents. Few fibers develop in the white pulp during the first postnatal week, but noradrenergic innervation increases between weeks 1 and 2 as the marginal zone develops. A substantial increase of the noradrenergic plexus occurs about the central arteriole during development of the periarteriolar lymphatic sheath early in the third week. The adult pattern of innervation of the spleen is fully developed by the end of the first month of life in the rat, and it persists through adulthood.

Lymph Nodes. Other lymphoid organs follow similar patterns of innervation ontogenesis. Peripheral lymph nodes in the rodent follows that of the spleen. It is not known when the development of splenic and lymph node innervation occurs in humans [2,17,27,73].

Perinatal Period. This is characterized in viviparous pregnancy by the direct proximity of the developing fetus to the maternal cell lining of the reproductive tract. The fetus, the ontogenic product of the oocyte following fertilization, is by definition genetically non-identical to the maternal cells. Despite this situation, which can be likened from the standpoint of cellular immunity to a maternal–fetal allograft, immunological rejection of the fetus by the mother's immune system is uncommon under normal conditions (see above Sect. 86.3.3). The preservation of the semiallogenic fetus results in part from the presence of fetus-derived trophoblast cells, which remain refractory to maternal cellular immune effectors due to the membrane receptors (e.g., major histocompatibility complex antigens) they express. Trophoblasts are unique in that they have the capability to execute several differentiated functions simultaneously. The preservation of the fetus is also dependent upon immunoregulatory products (e.g., growth factors, hormones, neuropeptides) released at the fetal–maternal interface (i.e., the placenta). These factors are produced either by specialized differentiated cells (e.g., infiltrating lymphocytes) that migrate to the fetal–maternal interface, by trophoblasts, or by decidual cells. Maternally derived decidual cells are also abundant at the fetal–maternal interface. These cells play a crucial role in the formation of the decidual plate in conjunction with maternal leukocytes that infiltrate these sites [5,49,71,76,105]. The decidual plate takes part in the development of neuroendocrine–immune interactions because of its neuroendocrine potential and of its proximity to lymphoid cells.

Neonates. While the principal compartments of the immune system are in place neonatally, they rarely have attained the maturity required to provide the neonate with sufficient immune capacity. A substantial proportion of circulating immunoglobulin at birth are maternal in origin. Over 90% of T cells are of the naive phenotype (CD45RA$^+$) and functionally incapable of providing B cell help during the first few weeks postnatally. Despite the fact that newborns have normal leukocyte counts, they are characterized by having poorly developed tonsils and lymph nodes, but large thymuses [96,109].

86.4.2 Fetal Alcohol Exposure (FAE)

Severe birth defects characterize infants born to mothers who drink during pregnancy. Exposure to alcohol during fetal life leads to a continuum of pathological manifestations that range from gross morphological defects (e.g., craniofacial anomalies, major organ system malformations) and profound mental retardation (i.e., central nervous system dysfunctions) to more subtle physiological and cognitive-behavioral abnormalities (i.e., the fetal alcohol syndrome). The teratogenicity of alcohol represents a major health concern because of the elevated incidence of children born with this affliction (one to three cases per 1000 live births). FAE impairs the development of cellular immunity in experimental animals and human beings [23]. The development of animal models of FAE is valuable because FAE histories in human subjects are often imprecise and incomplete. Animal models provide precise and controlled patterns of prenatal alcohol exposure crucial for the characterization of alcohol teratogenicity. Animal research has shown that FAE leads to abnormal neuroendocrine, immune, and behavioral responses that persist in the adult animal [23,87].

Development and Regulation of the Pituitary-Adrenocortical Axis. This neuroendocrine axis is altered in FAE rats. These animals manifest activation of the axis during perinatal development. Certain brain regions (e.g., midbrain, hindbrain) are significantly richer in certain proopiomelanocortin gene products (e.g., β-endorphin) in FAE compared to control rats. Pituitary β-endorphin basal concentrations are higher in FAE compared to control rats at postnatal day 4, but lower than control animals by postnatal days 8, 14, and 22; the ontogeny of the β-endorphin response to stress is also significantly altered in FAE compared to control rats. Ethanol administration during the second, rather than the first or the third week of gestation disrupts the development of the regulation of the hypothalamic-pituitary-adrenal axis in male and female, 21-day-old Sprague-Dawley FAE rats. Neonatal plasma concentrations of corticosterone are close to three fold higher in FAE rats than those of matched control animals. A significant elevation is noted in the expression of mRNA for corticotropin-releasing hormone in the paraventricular nucleus of the hypothalamus of FAE rats at postnatal day 21 compared to controls [23,26].

These findings are confirmed by studies in which adult (150–210 days old) females are given various doses of the synthetic glucocorticoid dexamethasone (3.1–50 μg/kg) for 4 h. FAE rats suppress corticosterone concentrations as effectively as control animals at higher dexamethasone concentrations; however, at lower dexamethasone concentrations, FAE animals show an early escape from dexamethasone suppression and adrenal steroids hypersecretion. In addition, the concentrations of plasma corticosterone 60 min following subcutaneous injection of morphine (30 mg/kg) or ethanol (1.5 g/kg) are significantly lower in FAE pups (7 days old) than in control pups. Within 5 days, however, this anomalous adrenal response normalizes, and all groups respond equally to ethanol or morphine stimuli. Older FAE rats (75–90 days old) manifest a significantly higher secretion of corticosterone in response to the ethanol stimulus compared to control rats. Certain stressful stimuli (e.g., noise/shaking) but not others (e.g., cold) are also capable of eliciting elevated adrenocortical responses in adult (75- to 100-day-old) FAE animals. Intermittent, but not continuous, footshock leads to similarly increased corticosterone response in adult FAE rats compared to control rats [26,87].

Altered Cellular Immune and Neuroendocrine–Immune Development. This is a further characteristic of FAE animals compared to control animals. Histologically, the corticomedullary junction of the thymus, which normally appears at day 17–19 of fetal life in the mouse as it becomes functionally mature, is lacking in FAE animals. The absence of this landmark of thymic development supports flow cytometric evidence that FAE disrupts thymic ontogeny: committed T cell precursors from the murine hematopoietic fetal liver – where toxic alcohol metabolites (e.g., acetaldehyde, acetaldehyde adducts) are probably produced in FAE animals – initially migrate to the thymic primordia by day 10–11 of gestation. By the day 16 of gestation, thymocytes acquire CD8, recognized by monoclonal antibody clone Lyt-2. The appearance of CD4 (L3T4) generally does not occur prior to day 17. By fetal day 18–19, thymocytes should express the Lyt-2 and L3T4 antigens in a proportion similar to that found in adult mice. Helper T cell activity is conferred to the L3T4$^+$/Lyt-2$^-$ subpopulation, and cytotoxic/suppressor T cell activity resides in the L3T4$^-$/Lyt-2$^+$ thymocyte subpopulation in mature normal mice. Double-negative populations are immature and usually remain in the thymus. Double-positive populations are immature thymocytes destined to progress through the stages of thymic education. They develop into single positive cells and exit into the circulating pool of T lymphocytes, becoming fully functional cells. These patterns of maturation are altered in FAE animals [23,26,39].

Impaired maturation of thymocytes in FAE animals is expected to be in part responsible for the reported impaired cellular immune responses. The proliferative response of splenocytes from the male offspring of Sprague-Dawley rats intubated daily with 6 g/kg ethanol from 2 weeks prior to mating until delivery is significantly suppressed in FAE compared to control animals. The effect is greatest at 7 months (5% of control), less pronounced at 11 months (40% of control), and all but abolished at 18 months (90%–100% of control), confirming that FAE has profound effects upon the animals' postnatal immunophysiological development [23,26,46,84,88,93,125].

These observations are reiterated in part by studies of children born of alcoholic mothers, which indicate that the immune defects associated with FAE affect principally the cellular immune component, and particularly T cell-mediated immune events, and that these dysfunctions do not seem to diminish with increasing age of the child. These observations have profound clinical implications in that they suggest that FAE may lead to grave impairments in those aspects of the immune system that are most crucial for initiating, regulating, and sustaining immune competence against minor as well as lethal infectious agents and against tumorigenesis and metastasis, as well as suggesting that fetal alcohol-associated disturbances in neuroendocrine modulation of cellular immunity may be long-lasting [23].

As noted, alcohol consumption is associated with a vigorous neuroendocrine response that includes elevations in plasma ACTH and cortisol concentrations. Alcohol also suppresses the secretion of gonadal hormones. Alcoholic patients have decreased concentrations of circulating testosterone, low spermatozoa count, low seminal volume, and decreased libido. Hypogonadism and Leydig cell failure are not uncommon among chronic alcoholics. These outcomes may result from a direct effect of alcohol on the Leydig cells, since gonadotropin receptors are decreased in

Table 86.1. Effect of fetal alcohol exposure (FAE) on plasma levels of testosterone and estradiol on gestation day 22 and on the volume of the sexually dimorphic nucleus of the preoptic area (SDN-POA) of the hypothalamus on postnatal days 14–31 in male Sprague-Dawley rat™s

	Normal control	Pair-fed control	FAE
Testosterone* (pg/ml)	449 ± 43	397 ± 35	342 ± 33
Estradiol** (pg/ml)	230 ± 12	159 ± 21	191 ± 20
SDN-POA*** (volume, mm³ × 10⁻³/g)	14 ± 1.3	17 ± 2.6	10 ± 0.9

Data taken from Ahmed et al. [3]
* No statistically significant differences among the three groups.
** FAE > normal ($p < 0.05$)
*** FAE ($n = 14$) < normal ($n = 12$) = pair-fed ($n = 5$) ($p < 0.05$)

the Leydig cells exposed in vitro to alcohol. Chronic alcoholism is believed to result in a biochemical imbalance, which lowers steroidogenesis. In addition, alcohol shifts the metabolism of testosterone to weaker androgens. Together, these processes can significantly contribute to the slightly elevated plasma concentrations of estrogens noted in male alcoholics and to the observed processes of hypogonadism and feminization in these patients [31,116]. These observations are relevant in the context of endocrine–immune ontogeny. Calzolari [18] reported that castration of adult rabbits increases thymic mass, but castration of prepubertal animals actually delays thymic development. Reciprocally, male mice thymectomized at postnatal day 3 exhibit considerable testicular atrophy and hypertrophy of pituitary gonadotrophs, and female mice thymectomized at day 3 post partum demonstrate substantial ovarian dysgenesis [56,80,92]. Castration and adrenalectomy in rodents also decrease the secretion of the thymic factor thymulin, but raise that of other thymic products, including thymulin-inhibitory substance. Plasma concentrations of gonadal hormones and splenic and thymic lymphoproliferative responses to mitogenic lectins appear to be inversely correlated in adult male but not female rats.

Taken together, these lines of evidence contribute to our understanding of the multifaceted teratogenic outcomes of FAE, which include alterations in the development of certain brain nuclei, loss of hippocampal pyramidal cells and alterations in the terminal fields of the dentate gyral mossy fibers of the hippocampal region [124], altered expression of myelin basic protein by cultured glia [25] and growth and maturation of neurons in culture, and significantly smaller sexually dimorphic preoptic nuclei of the hypothalamus during the weaning period than in control cohorts (Table 86.1) [3,25,124]. Furthermore, the testosterone surge critical to the perinatal development of sexually dimorphic characteristics (gestational days 18,19) is blunted by FAE [78]. Male FAE rats have decreased plasma testosterone at birth [99], fail to exhibit sexually dimorphic behaviors such as saccharin preference [77],

and manifest feminized behaviors both in motivation and performance, while female FAE rats show deficits in sexually receptive behavior and in induction and organization of maternal behavior [53]. The fact that FAE does not alter sexual dimorphism in hippocampus glucocorticoid receptor (females retain a higher number of receptors compared to males) [122] indicates that the effects of FAE also include HPG-related functions (and hence also hypothalamic–pituitary–gonadal (HPG)–thymus-related functions) and behavior.

Alcohol Metabolites. The first metabolite of alcohol is the dehydrogenase product acetaldehyde. Acetaldehyde is in turn metabolized to free acetate by acetaldehyde dehydrogenase, and free acetate is in turn metabolized to carbon dioxide. Adenosine triphosphate is utilized in equimolar amounts with free acetate during the latter metabolic step, and AMP is generated. AMP is metabolized to free adenosine. Acetaldehyde, free acetate, and adenosine can induce the toxicity and contribute to the symptomatology associated with alcohol abuse. Whether or not certain FAE outcomes can be attributed to alcohol metabolites is still under active investigation.

86.4.3 Nutritional Restriction

In the FAE research model, a control group must be included that is fed the same liquid diet as the alcohol-exposed dams. To insure equal caloric intake, food access must be restricted. Pair-feeding and the consequential food restriction could impair the ontogeny of neuroendocrine-immune events in the progeny of animals that are nutritionally restricted during pregnancy, because this situation chronically stresses the animals, as evidenced by elevated corticosterone concentrations and impaired cellular immune responses [42,127]. No consistent postnatal effect of pair-feeding during gestation has been observed on neuroendocrine–immune responses in the progeny to date [23,26,39,88,117,118,122].

86.5 Aging of the Neuroendocrine-Immune System

The nervous and the immune systems interact throughout development, adulthood, and aging. Senescence in the neuroendocrine system occurs pari passu with decrements in cellular immune function and with increased susceptibility to infectious and neoplastic disease.

86.5.1 Hypothalamic–Pituitary–Adrenal Immune System in Aging

Aging is accompanied by a progressive dysfunction of cell-mediated immunity [30,38]. Intensive research has begun to explain why and how cell-mediated immunity declines with age. While Ca^{2+} fluxes remain unimpaired, responses to phorbol ester are markedly inhibited in T cells from healthy elderly subjects. There is no impairment of IP_3 production, suggesting age-related impairments in T cell activation that may be related to molecular events associated with the kinase cascade activity [86].

T cell-proliferative responses to mitogenic lectins are diminished in older humans (see above), but gender differences tend not to occur [55,90,119]. Suppressor T cells have been reported to increase [50], decrease [52], or not change [30] with age. Natural killer cell activity [57] (see above), has been reported to be maintained, increased, or decreased in the elderly [81]. Moreover, in healthy centenarians, natural killer cell activity is normal, while phenotypes of natural killer cells are increased [119]. Natural killer activity tends to be significantly higher in older, healthy women as compared to young, healthy women, and the phenotype associated with this activity also is significantly higher in the older female sample. By contrast, T cell function remains equivalent to that of young women [106]. Taken together, these studies suggest that immune functions in older healthy adults may not be significantly diminished with age. In actuality, a cohort effect may be at work in some studies, allowing only individuals with superior immune functions to survive to a healthy old age.

Melatonin. The principal hormone produced by the pineal gland, melatonin (see Chap. 58), appears to have profoundly beneficial effects on immune–neuroendocrine circadian and regulatory responses in the elderly. Melatonin and age-associated alterations in its rhythmic pattern of secretion are certainly of crucial importance during the aging process of neuroendocrine–immune interactions [74].

Growth Hormones and Prolactin. Receptors for both growth hormone [68] and prolactin [100] are found on lymphocytes. Growth hormone peaks during early adulthood and declines with age [123]. The concentrations of growth hormone may be sex linked, with more decline of basal growth hormone concentration with age in women [91]. Growth hormone is also affected by obesity and medication use, which both increase with age and are more prevalent in older women [120]. Declining growth hormone concentrations in older people, especially women, may place them at greater risk of immunosuppression. Growth hormone restores T cell-proliferative responses and interleukin-2 synthesis in aged rats, augments the activity of natural killer cells and cytotoxic T cells, and increases the growth rate of macrophages [63]. Recombinant growth hormone administration to elderly subjects can improve certain immune parameters [40]. However, concern must be expressed about likely unfavorable outcomes of growth hormone treatment, such as the increased incidence of tumors, and about the fact that a strong rationale for recommending such treatment in aged people is not well established. Prolactin also has important effects on immunity that may be gender related. Prolactin can restore both cell-mediated and humoral immune responses in hypophysectomized female rats, as measured by antibody production and delayed hypersensitivity skin responses. Elevated immune response in females may relate to the fact that estrogen stimulates prolactin secretion, which is immunostimulatory [10,47]. In animals, acute stress evokes a rapid increase in prolactin release that is followed by decreased prolactin secretion and then refractoriness to further stimulation with repetition of the acute stress [115].

Sex Hormones. Estrogens generally depress cell-mediated immunity [1], but can elevate antibody responses to T-dependent antigens [113]. A possible involvement of female sex hormones in autoimmune disease [70,112] is suggested by differential incidence and severity of rheumatoid arthritis, multiple sclerosis, and other autoimmune manifestations, including those observed in the oral mucosa (e.g., mucosal membrane pemphigoid) in women. Menopause is associated with increased interleukin-1 production, which is reversed with estrogen replacement [94]. Animal studies confirm that B cell mitogens can elicit rheumatoid factor antibody production in female rats and that this effect is inhibited by estradiol but not by testosterone [103].

Cortisol and Dehydroepiandrosterone. A rise in plasma cortisol and a drop in plasma DHEA concentrations occur in aging, and to a greater extent in patients with Alzheimer's disease. Pharmacological administration of DHEA (30–90 mg/day orally) normalizes plasma DHEA concentrations in these patients. DHEA concentrations are generally inversely correlated ($r = -0.78, p < 0.05$) with mid-afternoon cortisol, perhaps because DHEA both depresses ACTH production by the anterior pituitary corticotrophs and blunts corticosteroid metabolic enzymes.

The production of DHEA and DHEA-S decreases with age. The metabolism of 17α-OH pregnenolone to 17α-OH progesterone to 11-deoxycortisol to cortisol increases and that of the androgen pathway (17α-OH pregnenolone to DHEA) decreases.

DHEA administration to aged mice normalizes T cell responses and interleukin-2 production to concentrations attained by young adult mice. Treatment of lymphocytes from aged or young adult mice with physiological concentrations of DHEA (10^{-8}–$10^{-10}M$) in vitro also adds to these outcomes. DHEA (1–5 mg by mouth) may be administered to healthy aged subjects with no apparent toxicity. This mode of intervention produces a drop in naive T cells subpopulations (i.e., CD4$^+$CD45RA$^+$, CD8$^+$CD45RA$^+$) by about 50%, an equivalent rise in the memory T cell subpopulations, no difference in total T cells, and a slight, albeit not significant drop in the CD4 to CD8 ratio. These data suggest that DHEA administration favors a shift from naive to memory T cells. Whether or not these effects reflect a preferential rise in cells capable of TH1 or TH2 patterns of response, and whether in vitro models can be effectively designed to determine whether the basic mechanisms for these outcomes involve steroid receptor-mediated events, are questions yet to be tested.

86.5.2 Neuroendocrine–Immune Aging

The thymus involutes with aging. Noradrenergic fibers are retained and are confined within the compartment of thymic tissue that remains functional. These observations contribute to explain the apparent thymic functionality in the elderly in terms of the continued generation, albeit at a reduced rate, of naive T cells, despite its profound morphological involution. The distribution of T cell subpopulations is also altered in old subjects [32]. This contributes to dysregulated cell–cell communication and interactions. It is also possible that the T cells that are present may have a different site of origin and maturation, e.g., the Peyer's patches or the liver [89], in older compared to younger mammals. T cell responses to immune challenges in vivo are also altered in older individuals because of increasing disturbances in neuroendocrine–immune interaction with aging (see above).

In older rats, progressive loss of T lymphocytes and of a family of specialized macrophages (identified by the expression of the ED-3 marker) coincides with progressive loss of noradrenergic innervation of the spleen. Whether or not this diminution of splenic innervation plays a role in the immunocompromise associated with aging is under investigation.

86.5.3 Reshaping of Neuroendocrine–Immune Crosstalk with Aging

The efficiency of the neuroendocrine–immune system is of critical importance throughout the lifespan, and in particular during the aging process, because certain elements crucial to physiological balance may become increasingly dysregulated. The level of efficiency at which the neuroendocrine system is set may determine not only immune competence, but more generally the lifespan of a given organism in any one species. The possible common evolutive basis of the immune and the neuroendocrine systems, combined with their critical importance for the maintenance of the soma, strongly suggests that immune–neuroendocrine responses play a major role in the aging process. No species is immune to hazards such as predation, starvation, and disease. The numerous defense mechanisms that contribute to our survival – DNA repair, antioxidants, stress proteins, accurate DNA replication, protein synthesis and gene regulation, tumor suppression mechanisms, and apoptosis – must interact to ensure neuroendocrine–immune balance and secure continued viable function. When this fails, disease follows and death of the organism will ensue.

A continuous remodeling occurs of the immune, neuroendocrine, and other functions with time, a process that is adaptive in nature. Major age-related changes are evident in people over 65 years of age, particularly in the T cell compartment. Reports in which the health status of the subjects is considered together with chronological age (e.g., SENIEUR protocol) suggest that the "deterioration" of the immune system noted with age is actually less pro-

nounced than previously thought. The main changes of immune parameters and responses with age, after correction for health status (Table 86.2) include:

- Progressive and consistent age-related decrease of the CD4$^+$ and CD8$^+$ T cell subsets with age, concomitantly with a decrease in naive T cells (CD54RA$^+$) and an increase in circulating memory T cells (CD45RO$^+$).
- Altered regulation of the expression of certain cytokines. The production of the pleiotropic cytokine interleukin-6 is elevated; peripheral blood mononuclear cells from healthy elderly subjects produce significantly more interleukin-1β and -6 and tumor necrosis factor-α, but other cytokines (e.g., interferon γ), following stimulation with phytohemagglutin and the phorbol ester phorbol-methyl acetate. These functional changes are independent of previously reported phenotypic alterations in normal aging and may be indicative of subtle changes in TH1 and TH2 balance. It is possible, even probable, that these outcomes reflect circulating steroid- and steroid receptor-mediated neuroendocrine–immune modulation.
- Dramatic fall in circulating B lymphocytes, paradoxically accompanied by an increased serum concentration of some immunoglobulin classes and subclasses and by

Table 86.2. Age-dependent changes in human peripheral immune parameters

	Parameter	Change
Absolute number	Lymphocytes	↓
	CD3$^+$ cells	↓
	CD4$^+$ cells	↓
	CD8$^+$ cells	↓
	CD19$^+$ (B) cells	↓
	HLA-Dr$^+$ CD3$^+$ (activated T) cells	↓
	CD16$^+$ CD57$^-$ (natural killer) cells	↑
	CD3$^+$ CD45RA$^+$ (naive T) cells	↓
	CD3$^+$ CD45RO$^+$ (memory T) cells	↑
Activity	Natural killer cells	↔
	Interleukin-2 production	↓
	Interleukin-1 production	↑
	Tumor necrosis factor-α production	↑
	Interleukin-6 production	↑
	Lymphoproliferative response	↔
Plasma levels	Immunoglobulin G$_{(1-3)}$	↑
	Immunoglobulin A	↑
	Immunoglobulin M	↔
	Immunoglobulin G$_4$	↔
	Organ-specific autoantibodies	↔
	Nonspecific autoantibodies	↑
	Thymulin	↑

↓, Decreased compared to the average value obtained from normal, healthy young adults; ↑, increased compared to the average value obtained from normal, healthy young adults; ↔, no change, variable outcome, or delayed response compared to the average value obtained from normal, healthy young adults

an almost complete absence of organ-specific auto-antibodies in healthy elderly and centenarians.

- Preservation of natural killer cell activity, together with an age-related increase in cells capable of this function (i.e., CD16+CD56−).

Taken together, these data are suggestive of a profound reshaping of the immune system with age that is correlated with significant neuroendocrine changes.

86.6 Psychoneuroendocrine Mediation of Immunity in Aging

A discussion of aging and immune responsivity is made difficult by the complexities involved in unraveling the links among various aspects of mood, coping skills, stress appraisal, health, and a variety of other biopsychosocial factors (see Chaps. 123). The complexity is further compounded by issues of gender differences. Men and women differ in the types of illnesses they develop and in cause of death (see Chaps. 123, 124). Women tend to live longer with disability. Much of the disability is related to arthritis and dementia [75]. As indicated above, they also are three to ten times more likely to develop autoimmune diseases than men [59,70,114]. The differences in morbidity defined as generalized poor health, a specific illness, or the sum of a number of illnesses can be attributed to a variety of factors, including gender differences in stress and distress [95]. Men and women differ in bodily reactions to stressful stimuli, and these differences appear to be more than mere reflections of intensity of experience or of perceived threat. Tolerance to self-antigens is related to multiple distinct responses that are connected in series and intervene at defined points of the life cycle of the developing lymphocyte to guarantee the physical elimination, functional inactivation, or regulated inhibition of self-reactive, potentially autoaggressive B and T cells [69]. In regard to the latter, suppressor T cells, which can be influenced by stress, are important.

Although overt autoimmune diseases generally have their onset in younger life, they can have their onset in later years. Psychosocial factors appear to be related to the onset and course in the elderly, just as they are in the young [106]. For example, the onset of rheumatoid arthritis in a previously healthy 84-year-old man (who achieved a black belt in judo at the age of 60) was observed shortly after he was persuaded to give up the independence he cherished and move in with his son. Psychological distress is negatively related to the rate of progression of, degree of incapacitation by, and response to treatment in rheumatoid arthritics [85].

Women have higher immunoglobulin concentrations than do men and mount larger antibody responses to a variety of pathogens, although cell-mediated immune responses appear to be weaker than in men [8]. Autoantibodies, which are not necessarily correlated with overt clinical autoimmune disease, are also more common in older peo-

ple and patients suffering from schizophrenia, probably reflective of disordered immune regulation in both aging and mental illness [44,106].

8.6.1 Gender, Depression, Immunity, and Aging

Depressive illness and symptoms unrelated to bereavement affects over 15% of the geriatric population. Rates of depression and anxiety are higher among older women than men. These differences may be related to women more readily admitting to psychological distress, greater social acceptance of these emotions in women, or possibly a greater number of looses among older women as compared to older men. A relationship exists between stressful life events and onset of depression in women [54].

Studies addressing the issue of depression, immunity, and age are often contradictory due to methodological problems that include sample size, immunological tests utilized, and lack of control for sex, age, hospitalization status, and diagnosis [111]. A recent meta-analytic study [58] indicates that, when all these factors are accounted for, there remains a reliable association between decreased cellular immune function and depression. In addition, this relationship is compounded by the severity of the depression and by age.

Lymphocyte Mitogen Response. Depressed patients do not significantly differ from controls in natural killer cell activity or mitogen-induced lymphocyte stimulation. However, when response to mitogens is analyzed looking at age differences, depressed older patients show no increase in lymphocyte response with age, but age-matched control subjects do. Decreased lymphocyte response to mitogens appears to occur only in old and not in young, depressed, male patients, the youngest depressives even showing increases [101]. By contrast, natural killer cell activity is decreased in patients with major depressive disorders of all ages [60]. In most of these studies gender differences were either not reported, and mostly males were used in the study samples [48,60,61,101,102,111].

Women are better represented in studies of bereavement, marital status, caregiving, and immunity, which show that lymphocyte stimulation response to mitogens is significantly lower following the death of a spouse compared to prebereavement responses or those after the loss has been worked through [6,60,101]. This phenomenon has been demonstrated in both men and women. One study [60] included older women (mean age, 57 years) and another [6] had an age range up to the age of 65. Age differences were not analyzed in either group. It may now be necessary to examine the relationship between age and immune suppression as in one study [48] that found that both age and depression were negatively related to mitogen response of lymphocytes and interleukin-2 stimulation of natural killer cell activity.

There may be differences in immune competence between depressed patients and patients anticipating bereavement who are not depressed. The study carried out by Spurrell and Creed [107] of depressed and nondepressed women whose husbands had cancer determined that the depressed subjects had lymphocyte mitogen responses negatively correlated with depression. Enhanced responses were associated with anticipatory grief. These outcomes may depend upon good coping response, emotional expressiveness, or social support.

Caregiving. This is associated with increased health impairments. Female caregivers to Alzheimer's disease patients are not only at risk for depression [35], but also for immunosuppression [65,66]. Caregivers generally exhibit lower percentages of total T lymphocytes and helper T lymphocytes, lower helper to suppressor ratios, and higher antibody concentrations of Epstein-Barr virus (indicative of viral activation), when compared to age-matched controls. Thus, the aforementioned distress-related immunosuppression may have its most detrimental health consequences in older adult caregivers [67]. Since women tend to provide caregiving, these potential health problems place older female caregivers in double jeopardy.

Marital Status. Stressful situations associated with marital status lead to increased risk for psychiatric disorders such as depression [16]. Marital satisfaction is significantly related to psychological well-being [43]. Separated and divorced persons suffer more acute and chronic illnesses and have significantly greater mortality rates from diseases, including pneumonia, tuberculosis, and some types of cancer [16]. Single and divorced women have lower response to mitogens, lower percentages of T-helper lymphocytes and natural killer cells, and higher antibody titers to Epstein-Barr virus, compared to happily married matched control women [65]. Unfortunately, ages for the group were not presented. It can be speculated, based on the research showing the negative correlation between depression, immunity, and age, that marital disruption is more detrimental to immune competence in older women.

86.6.2 Immunity, Stress, and Aging

Certain stressful stimuli suppress immune activity in experimental animals and in human beings. The conditions under which this occurs are related to the nature, context, timing, intensity, and duration of the stimulus. Psychological factors of control, coping style, and social support are additional modulators that contribute to modulate immune responses [27,64,126].

Several animal studies have shown that uncontrollable stress may have more immunosuppressive consequences than stress that can be controlled [27,104]. Similarly, in humans, it has been found that low perceived control over significant and stressful life stimuli predicts T cell responses to mitogenic and antigenic stimulation of lymphocytes in vitro in older men and women (mean age, 70 years) [95]. Additionally, a pessimistic explanatory style is associated with a lower T-helper to T-suppressor ratio and low response to mitogenic stimulation. This associa-

tion may not be related to age or gender differences [62]. Dissatisfaction with social support in emotion-focused, coping older women is also predictive of poorer immune competence among elderly women [79].

Concern about negative events that might occur in the next 3 months ("worry") may influence immune competence as well. Anger and anticipated life stress are positively correlated with natural killer cytotoxicity in younger but not elderly subjects. A brief laboratory mental stimulus (time-pressured mental arithmetic) increases natural killer cell activity in young but not old subjects (unpublished observations).

86.7 Conclusions

Taken together, the body of physiological evidence presented in this chapter indicates that interactions among the various physiological systems, and between the organism's physiological and psychological structures, have profound implications for fundamental and clinical, biomedical and immunological research. The physiological significance of neuroendocrine–immune interactions for the survival of the organism is observed as early in phylogeny as in molluscs. It is also clearly evident in lower vertebrates (fish), as well as in mammals (rodents, nonhuman primates, human beings). Indeed, these interactive relationships are subject to ontogenic regulation and become increasingly altered with aging.

Research advances to date confirm that we are now at the threshold of an era in which many of the fundamental mechanisms that underlie the interactions between the nervous, the endocrine, and the immune systems will be unraveled and characterized. When it becomes possible to translate and to incorporate this information with efficacy into new and improved treatment modalities, substantial benefit and reduction in overall health care costs will be achieved for a variety of patient populations.

Acknowledgments. This work was supported by the UCLA Task Force on Psychoneuroimmunology, the UCLA Center for Interdisciplinary Research in Immunology and Disease, Fetzer and the John D. and Catherine T. MacArthur Foundations, BRSG and Collegium (HR-7170) from the Research and Education Institute, inc., at Harbor-UCLA Medical Center, NIAID AI-07126, and NIDA DA-07683. The authors are indebted to the late Mr. Norman Cousins for his significant contribution to the establishment of the field of psychoneuroimmunology at UCLA, nationally, and internationally.

References

1. Ablin RJ, Bartkus JM, Gonder JJ (1988) In vitro effects of diethylstilbestrol and the LHRH analogue leuprolide on natural killer cell activity. Immunopharmacology 15:95–101
2. Ader R, Felten DL, Cohen N (eds) (1990) Psychoneuroimmunology II. Academic, San Diego, pp 1–1218

3. Ahmed II, Shryne JE, Gorski RA, Branch BJ, Taylor AN (1991) Prenatal ethanol and the prepubertal sexually dimorphic nucleus of the preoptic area. Physiol Behav 49:427–432

4. Altman A, Coggeshall KM, Mustelin T (1990) Molecular events mediating T cell activation. Adv Immunol 48:227–361

5. Andrew A (1985) Developmental relationships of neuroendocrine cells. Biomed Res 6:191–196

6. Bartrop R, Luckhurst E, Lazarus L, Kiloh LG, Penny R (1977) Depressed lymphocyte function after bereavement. Lancet 8:34–36

7. Bateman A, Singh A, Kral T, Solomon S (1989) The immune-hypothalamic-pituitary-adrenal axis. Endocrinol Rev 10:92–112

8. Baum A, Grunberg NE (1991) Gender, stress, and health. Health Psychol 10:80–85

9. Berczi I (1990) Neurohormonal-immune interaction. In: Kovacs K, Asa S (eds) Functional endocrine pathology. Blackwell Scientific, New York, chap 41

10. Bernton EW, Bryant HU, Holaday JW (1991) Prolactin and immune function. In: Ader R, Felten D, Cohen N (eds) Psychoneroimmunology. Academic, San Diego, pp 403–428

11. Besedovsky H, del Rey A, Sorkin E, Lotz W, Schwulera U (1985) Lymphoid cells produce an immunoregulatory glucocorticoid increasing factor (GIF) acting through the pituitary gland. Clin Exp Immunol 59:622–628

12. Besedovsky HO, del Rey A (1990) Physiological implications of the immuno-neuroendocrine network. In: Ader R, Felten DL, Cohen N (eds) Psychoneuroimmunolgy II. Academic, San Diego, pp 598–608

13. Besedovsky HO, Sorkin E (1977) Immune neuro-endocrine network. Clin Exp Immunol 27:1–12

14. Besedovsky HO, Sorkin E, Keller M, Muller J (1975) Hormonal changes during immune response. Proc Soc Exp Biol, Med 150:466–470

15. Blalock E (1989) A molecular basis for bidirectional communication between the immune and the neuroendocrine systems. Physiol Rev 69:1–32

16. Bloom BL, Asher S, White S (1978) Marital disruption as a stressor: a review and analysis. Psychol Bull 85:867–894

17. Bulloch K (1985) Neuroanatomy of lymphoid tissue. In: Guillemin R, Cohn M, Melnechuck T (eds) Neural, modulation of immunity. Raven, New York, pp 111–141

18. Calzolari A (1898) Recherches expérimentales sur un rapport probable entre la function du thymus et celles des testicules. Arch Ital Biol 30:71–75

19. Catania A, Manfredi MG, Airaghi L, Vivirotto F, Milazzo F, Zanussi C (1990a) Evidence for an impairment of the immune-adrenal circuit in patients with acquired immunodeficiency syndrome. Horm Metab Res 22:597–598

20. Catania A, Airaghi L, Manfredi MG, Zanussi C (1990b) Hormonal response during antigenic challenge in normal subjects. Int J Neurosci 51:295–296

21. Chiappelli F (1991) Immunophysiological role and clinical implications of non-immunoglobulin soluble products of immune effector cells. Adv Neuroimmunol 1:234–240

22. Chiappelli F, Kung MA (1994) Immune surveillance of the oral cavity and lymphocyte migration: relevance for alcohol abusers. Lymphology (in press)

23. Chiappelli F, Taylor AN (1994) The fetal alcohol syndrome and fetal alcohol effects on immune competence. Alcohol Alcohol 746:204–215

24. Chiappelli F, Trignani S (1993) Neuroendocrine-immune interactions: implications for clinical research. Adv Biosci 90:185–198

25. Chiappelli F, Taylor AN, Espinosa de los Monteros A, de Vellis J (1990) Fetal alcohol delays the developmental expression of myelin basic protein and transferrin in rat primary oligodendrocyte cultures. Int J Dev Neurosci 9:67–75

26. Chiappelli F, Wong CMK, Yirmiya R, Norman DC, Chang M P, Taylor AN (1992) Fetal alcohol exposure (FAE) and neuroimmune surveillance. In: Yirmiya R, Taylor AN (eds) Alcohol, immunity and cancer. CRC Press, Boca Raton, pp 143–55

27. Chiappelli F, Franceschi C, Ottaviani E, Faisal M (1993a) Phylogeny of the neuroendocrinimmune system: fish and shellfish as a model system for social interaction stress research in humans. Annu Rev Fish Dis 3:327–346

28. Chiappelli F, Frost P, Wylie T (1993b) Alcohol and alcohol/cocaine mixtures blunt the activation of human CD4+ T cells. Alcologia 5:221–229

29. Chiappelli F, Manfrini E, Franceschi C, Cossarizza A, Black KL (1994) Steroid regulation of cytokines: relevance for TH1 → TH2 shift? Ann NY Acad Scie

30. Chopra R (1990) Mechanisms of impaired T cell function in the elderly. Rev Biol Res Aging 4:83–104

31. Cicero TJ (1981) Neuroendocrinological effects of alcohol. Annu Rev Med 32:123–142

32. Cossarizza A, Ortolani C, Paganelli R, Monti D, Barbieri D, Sansoni P, Fagiolo U, Forti E, Londei M, Franceschi C (1992) Age-related imbalance of virgin (CD45RA+) and memory (CD45RO+) cells between CD4+ and CD8+ T lymphocytes in humans: a study from newborns to centenarians. J Immunol Res 4:118–126

33. Cossarizza A, Kalashnikova G, Grassilli E, Chiappelli F, Salvioli S, Capri M, Barbieri D, Troiano L, Monti D, Franceschi C (1994) Mitochondrial modifications during rat thymocyte apoptosis: a study at the single cell level. Exp Cell Res 214:323–330

34. Crabtree GR (1989) Contingent genetic regulatory events in T lymphocyte activation. Science 243:355–361

35. Crook TH, Miller NE (1985) The challenges of Alzheimer's disease. Am Psychol 40:1245–1250

36. Denicoff KD, Durkin Tm, Lotze MT, Quinlan PE, Davis CL, Listwak SJ, Rosenberg SA, Rubinow DR (1989) The neuroendocrine effects of interleukin 2 treatment. J Clin Endocrinol Metabol 69:402–410

37. Dluhy RG (1990) The growing spectrum of HIV-related endocrine abnormalities. J Clin Endocrinol Metab 70:563–565

38. Effros RB, Wolford RC (1987) Infection and immunity in relation to aging. In: Gold EA (ed) Ageing and the immune response. Cellular and hormonal agents. Dekker, New York, pp 45–65.

39. Ewald SJ, Walden SM (1988) Flow cytometric and histological analysis of mouse thymus in fetal alcohol syndrome. J Leukoc Biol 44:434–440

40. Fabris N (1991) Neuroendocrine-immune interactions: a theoretical approach to aging. Arch Gerontol Geriatr 12:219–230

41. Felten DL, Felten SY, Bellinger DL, Carlson SL, Ackerman KD, Madden KS, Olschowki JA, Livnat S (1987) Noradrenergic sympathetic neural interactions with the immune system: structure and function. Immunol Rev 100:225–260

42. Gilman-Sachs A, Kim YB, Pollard M, Syder DZ (1991) Influence of aging, environmental antigens, and dietary restriction on expression of lymphocyte subsets in germfree and conventional Wistar rats. J Gerontol 46:B101–B106

43. Glenn N, Weaver C (1981) The contribution of marital happiness to global happiness. J Marr Fam 43:151–168

44. Goodman M, Rosenblatt M, Gottlieb M, Miller J, Chen C (1963) Effect of age, sex and schizophrenia on thyroid autoantibody production. Arch Gen Psychiatry 8:114–122

45. Górski A (1994) The role of cell adhesion molecules in immunopathology. Immunol Today 15:251–255

46. Gottesfeld Z, Christie R, Felten DL, LeGrue SJ (1990) Prenatal ethanol exposure alters immune capacity and noradrenergic synaptic transmission in lymphoid organs of the adult mouse. Neuroscience 35:185–194

47. Grossman CJ (1990) Are there underlying immune-neuroendocrine interactions responsible for immunological sexual dimorphism? Prog Neuro Endocr Immunol 3:75–82

48. Guidi L, Bartoloni C, Frasca D, Antico L, Pili R, Cursi F, Tempesta E, Rumi C, Menini E, Carbonin P, Doria G,

Gambassi G (1991) Impairment of lymphocyte activities in depressed aged subjects. Mech Aging Dev 60:13–24

49. Guilbert L, Robertson SA, Wegmann TG (1993) The trophoblast as an integral component of a macrophage-cytokine network. Immunol Cell Biol 71:49–57

50. Gupta S, Good RA (1979) Human T cell subpopulations as defined by Fc receptors. Thymus 1:135–149

51. Hadden JW (1992) Thymic endocrinology. Int J Immunopharmacol 14:345–352

52. Hallgren HM, Yunis E (1977) Suppressor lymphocytes in young and aged humans. J Immunol 118:2004–2008

53. Hard E, Dahlgren IL, Engel J, Larsson K, Liljequist S, Lindh AS, Musi B (1984) Development of sexual behavior in prenatally etahnol-exposed rats. Drug Alcohol Depend 14:51–61

54. Harris T (1991) Life, stress amd illness: the question of specificity. Ann Behav Med 13:211–219

55. Hashimoto M, Wakabayashi Y (1990) Differentiation and proliferation of T-cells in the elderly. Acta Haematol Jpn 53:717–724

56. Hattori M, Brandon MR (1983) Ovarian dysgenesis in neonatally thymectomized rats. J Endocrinol 83:101–105

57. Herberman RR, Ortaldo JR, Bonnard GD (1979) Augmentation by interferon of human natural and antibody-dependent cell-mediated cytotoxicity. Nature 277:221–223

58. Herbert TB, Cohen S (in press) Depression and immunity: a meta-analytic review. Psychol Bull 1–49

59. Inman R (1978) Immunologic sex differences and the female preponderance in systemic lupus erythematosus. Arthritis Rheumatol 21:849–852

60. Irwin M, Daniels M, Smith TL, Bloom E, Weiner H (1987) Impaired natural killer cell activity during bereavement. Brain Behav Immun 1:98–104

61. Irwin MR, Caldwell C, Smith TL, Brown S, Schuckit MA, Gillin JC (1990) Major depressive disorder, alcoholism, and reduced natural killer cell cytotoxicity. Arch Gen Psychiatry 47:713–719

62. Kamen-Siegel L, Rodin J, Seligman MEP, Dwyer J (1991) Explanatory style and cell-mediated immunity in elderly men and women. Health Psychol 10:229–235

63. Kelley KW (1991) Growth hormone in immunobiology. In: Ader R, Felten DL, Cohen N (eds) Psychoneuroimmunology. Academic, San Diego, pp 377–402

64. Kemeny M, Solomon G, Morley JE, Herbert TL (1992) Psychoneuroimmunology. In: Nemeroff CB (ed) Neuroendocrinology. CRC Press, Boca Raton, pp 563–591

65. Kennedy S, Kiecolt-Glaser JK, Glaser R (1988) Immunological consequences of acute and chronic stressors: mediating role of interpersonal relationships. Br J Med Psychol 61:77–85

66. Kiecolt-Glaser J, Glaser R, Shuttleworth E, Dyer C, Ogrocki P, Spieicher C (1987) Chronic stress and immunity in family caregivers of Alzheimer's disease victims. Psychosom Med 49:523–535

67. Kiecolt-Glaser JK, Glaser R (1991) Stress and immune function in humans. In: Ader R, Felten DL, Cohen N (eds) Psychoneuroimmunology. Academic, San Diego, pp 849–867

68. Kiess W, Butenandt O (1985) Specific growth hormone receptors on human peripheral mononuclear cells: reexpression, identification and characterization. J Clin Endocrinol Metabol 60:740–746

69. Kroemer G, Marinez AC (1992) Mechanisms of self-tolerance. Immunol Today 13:10

70. Lahita RG (1993) Sex hormones as immunomodulators of disease. Ann NY Acad Sci 685:278–287

71. Lala PK, Chatterjee-Harsouni S, Kearns M, Montgomery B, Colavinceng V (1983) Immunobiology of the feto-maternal interface. Immunol Rev 75:87–116

72. Le Goaster J, Rougeot C, Tekaia F, Dormont D, Boileau G, Montagnier L, Le Magnen J (1990) ACTH/beta-endorphins and ACTH/cortisol ratios as early biological markers in HIV infection. J Environ Pathol Toxicol Oncol 10:322–325

73. Lydyard PM, Kearney JF, Burrows P, Gathings W (1994) Lymphopoiesis today. Immunol Today 15:255–257

74. Maestroni GJM, Conti A, Pierpaoli W (1989) Melatonin, stress and the immune system. Pineal Res Rev 7:203–26

75. Manton KG (1989) Epidemiological, demographic, and social correlates of disability among the elderly. Milbank O 67:13–58

76. Mayor F, Cuezva JM (1985) Hormonal and metabolic changes in the perinatal period. Biol Neonate 48:185–196

77. McGivern RF, Clancy AN, Hill MA, Noble EP (1984) Prenatal alcohol exposure alters adult expression of sexually dimorphic behavior in the rat. Science 224:896–898

78. McGivern RF, Raum WJ, Salido E, Redei E (1988) Lack of prenatal testoterone surge in fetal rats exposed to alcohol: alterations in testicular morphology and physiology. Alcohol Clin Exp Res 12:243–247

79. McNaughton ME, Smith LW, Patterson TL, Grant I (1990) Stress, social support, coping resources, and immune status in elderly women. J Nerv Mental Dis 38:460–461

80. Michael SD (1981) The role of the endocrine thymus in female reproduction. Arthritis Rheum 22:1241–1245

81. Miller R (1990) Aging and the immune response. In: Schneider E, Rowe J (eds) Handbook of the biology of aging. Academic, San Diego, pp 157–180

82. Modlin RL (1994) TH1–TH2 paradigm: insights from leprosy. J Invest Dermatol 102:828–832

83. Moncayo HE, Solder E, Abfalter E, Moncayo R (1994) Cytokines and the maternal-fetal interface. Immunol Today 15:295

84. Monjan AA, Mandell W (1980) Fetal alcohol and immunity: depression of mitogen-induced lymphocyte blastogenesis. Neurobehav Toxicol 2:213–215

85. Moos RI, Solomon GF (1965) Personality correlates of the degree of functional incapacity of patients with physical disease. J Chron Dis 18:1019–1038

86. Naylor JR, James SR, Trejdosiewicz LK (1992) Intracellular free Ca^{2+} fluxes and responses to phorbol ester in T lymphocytes from healthy elderly subjects. Clin Exp Immunol 89:158–162

87. Nelson LR, Taylor AN (1987) Long-term behavioral and neuroendocrine effects of prenatal ethanol exposure. In: Braude MC, Zimmerman AM (eds) Genetic and perinatal effects of abused substances. Academic, Orlando, pp 177–203

88. Norman DC, Chang M-P, Castle SC, van Zuylen JE, Taylor AN (1989) Diminished proliferative response of con-a blast cells to interleukin-2 in adult rats exposed to ethanol in utero. Alcohol Clin Exp Res 13:69–72

89. Ohteki T, Okuyama R, Seki S, Abo T, Sugiura K, Kusumi A, Ohmori T, Watanabe H, Kumagai K (1992) Age-dependent increase of extrathymic T cells in the liver and their appearance in the periphery of older mice. J Immunol 149:1562–1570

90. Pieri C, Recchioni R, Moroni F, Marcheselli F, Damianovich S (1992) The response of human lymphocytes to phytomaggluginin is impaired at different levels during aging. In: Fabris N, Harman D, Knook DL, Steinhagen-Thiessen E, Nagy IZS (eds) Physiopathological processes of aging. New York Academy of Sciences, New York, pp 110–119

91. Prinz P, Weithzman E, Cunningham G et al (1983) Plasma growth hormone during sleep in young and aged men. J Gerontol 38:519–524

92. Rebar RW, Miyaka A, Erickson GF, Low TLK, Goldstein AL (1983) The influence of the thymus gland on reproductive function: a hypothalamic site of action. In: Greenwal GS, Terranova PF (eds) Factors regulating ovarian function. Raven, New York, pp 465–486

93. Redei E, Clark W, McGivern RF (1989) Alterations in immune responsiveness, ACTH and CRF content of brains in animals exposed to alcohol during the last week of gestation. Alcohol Clin Exp Res 13:439–443

94. Regelson W, Loria R, Mohanned K (1988) Hormonal intervention: "Buffer Hormones" or "State Dependency". The role of dehydroepiandrosterone (DHEA), thyroid hormone, estrogen and hypophysectomy in aging. In: Pierpaoli W, Spector N (eds) Neuroimmunomodulation: interventions in aging and cancer. New York Academy of Sciences, New York, pp 260–273

95. Rodin J (1986) Aging and health: effects of the sense of control. Science 233:1271–1276

96. Roitt I, Brostoff J, Male D (1989) Immunology, 2nd edn. Gower Medical, London, p 25.10

97. Roszman TL, Brooks WH (1989) Signaling pathways of the neuroendocrine-immune network. Prog All 43:140–159

98. Roszman TL, Carlson SL (1991) Neural-immune interactions circuits and networks. Prog Neuro Endocr Immunol 4:69–78

99. Rudeen PK (1988) Fetal alcohol exposure in the rat: a mechansim for brain defects. In: Kuriyama K, Takada A, Ishii H (eds) Biomedical and social aspects of alcohol and alcoholism. Elsevier, Amsterdam, pp 859–862

100. Russell DH, Kibler R, Matrisian L, Larson D, Poulos B, Magun B (1985) Prolactin receptors on human T and Blymphocytes: antagonism of prolactin binding by cyclosporine. J Immunol 134:3027–3031

101. Schleifer S, Keller S, Camerino M, Thornton J, Stein M (1983) Suppression of lymphocyte stimulation following bereavement. J Am Med Assoc 250:374–377

102. Schleifer SJ, Scott B, Stein M, Keller SE (1986) Behavioral and developmental aspects of immunity. J Am Acad Child Psychiatry 25:751–763

103. Schurrs AH, Verheul HM (1990) Effects of gender and sex steroids on the immune response. J Steroid Biochem 35:157–172

104. Shavit Y, Lewis J, Terman G, Gale R, Leibeskind J (1984) Opiod peptides mediate the suppressive effect of stress on natural killer cell cytotoxicity. Science 223:188–190

105. Slack JMW (1989) Peptide regulatory factors in embryonic development. Lancet 1312–1315

106. Solomon G (1989) Psychoneuroimmunology and human immunodeficiency virus infection. Psychiatr Med 7:47–57

107. Spurell M, Creed FH (1993). Lymphocyte response in depressed patients and subjects anticipatind bereavement. Br J Psychiatry 162:60–64

108. Stead RH, Bienenstock J, Stanisz AM (1987) Neuropeptide regulation of mucosal immunity. Immunol Rev 100;333–359

109. Stiehm ER (1979) Human neonatal immunc capacity: the B, T and monocyte/macrophage systems. In: Hodes H, Kagan BM (eds) Pediatric immunology. New York Science and Medicine, New York, pp 75–88

110. Stein M (1989) Stress, depression, and the immune system. J Clin Psychiatr 50:35–42

111. Stein M, Miller AH, Trestman RL (1991) Depression and the immune system. In: Ader R, Felten DL, Cohen N (eds) Psychoneuroimmunology, 2nd edn. Academic, London, pp 897–930

112. Sthoeger Z, Chiorzzi N, Lathita R (1989) Regulation of the immune response by sex hormones: in vitro effects of estradiol and testosterone on pokeweed mitogen-induced human B cell differentiation. J Immunol 14:91–98

113. Stimson WH, Hunter JC (1990) Oestrogen-induced immunoregulation mediation through the thymus. J Clin Lab Immunol 4:27–33

114. Strickland B (1988) Sex related differences in health and illness. Psychol Women 12:381–399

115. Taché Y, Du Ruisseau P, Ducharme J, Collu R (1978) Pattern of adenohypophyseal hormone changes in male rats following chronic stress. Neuroendocrinology 26:208–219

116. Tarantino G, Ciccarelli AFL (1989) Gonadal function in chronic alcohol abuse with or without cirrhosis: state of the art. Eur Rev Med Pharmacol Sci XI:3–5

117. Taylor AN, Branch BJ, Cooley-Matthews B, Poland RE (1982a) Effects of maternal ethanol on basal and rhythmic pituitary-adrenal function in neonate offspring. Psychoneuroendocrinology 7:49–58

118. Taylor AN, Branch BJ, Liu SII, Kokka N (1982b) Long-term effects of fetal alcohol exposure on pituitary-adrenal response to stress. Pharmacol Biochem Behav 16:585–589

119. Thompson JS, Wekstein DR, Rhoades JL, Kirkpatrick C, Brown SA, Roszman T, Straus R, Tietz N (1984) The immune status of healthy centenarians. J Am Geriatr Soc 32:274–281

120. Vestal RE, Cusack BJ (1990) Pharmacology and aging. In: Schneider E, Rowe J (eds) Handbook of the biology of aging. Academic, San Diego, pp 349–583

121. Weiner H (1985) The physiology of eating disorders. Int J Eating Dis 4:347–388

122. Weinberg J, Petersen TD (1991) Effects of prenatal ethanol exposure on glucocorticoid receptors in rat hippocampus. Alcohol Clin Exp Res 15:711–716

123. Weksler M (1981) The senescence of the immune system. Hosp Pract 16:53–64

124. West JR (ed) (1987) Alcohol and brain development. Oxford Press, New York

125. Wong CMK, Chiappelli F, Norman DC, Chang M-P, Cooper EL, Branch BJ, Taylor AN (1992) Prenatal exposure to alcohol enhances thymocyte mitogenic responses postnatally. Int J Immunopharmacol 14:303–309

126. Workman EA, La Via MF (1987) Immunological effects of psychological stressors: a review of the literature. Int J Psychosom 34:35–40

127. Yirmiya R, Ben-Eliyahu S, Gale RP, Shavit Y, Liebeskind SC, Taylor AN (1992) Ethanol increases tumor progression in rats. Possible involvement of natural killer cells. Brain Behav Immun 6:74–86

87 Psychoneuroimmunology

M. Jerry

Contents

87.1 Introduction

The paradigm of *psychoneuroimmunology* (PNI) provides a general systems model for understanding the complex interactions involved in *host resistance*, including its behavioral dimensions. It proposes that the nervous (both somatic and autonomic), endocrine, and immune systems function anatomically, biochemically, and physiologically as a unit, in a dynamic and homeostatic balance between tumor cells or infectious agents, for example, and the host microenvironment in which they grow – an updated version of the traditional "seed and soil" hypothesis of host resistance. The usual military paradigm of immune functioning as attack and defense is reframed to a broader construct based on regulation of cellular interactions through information flow by hormonal and neural interconnections. Host cells interact with tumor cells or microorganisms, for example, in a microenvironment where cytokine/receptor interactions have a key role in mediating this molecular information flow (cf. Chaps. 6, 86). Thus, impaired host resistance and even some of its symptomatology could be viewed alternatively as a regulatory problem resulting in collapsed homeostasis and based on a breakdown of molecular communication. The locus of these events is the cell surface where transmitter – receptor interactions take place.

Among the many *biological response modifiers* that have been identified is human behavior [34]. Research is start-ing to show that human behavior, attitudes, emotions, beliefs, and thinking can alter both humoral and cellular immune responses. There is growing fascination among cancer patients and the lay public, for example, with so-called "alternative" treatments for cancer, such as visualization or meditation, which are predicated for the most part on the putative ability of the mind to stimulate immunity. Psychosocial oncology has emerged as a discipline in the last decade and is just beginning to confront the possibility that psychological variables may affect both the causation and the outcome of cancer. The PNI paradigm has infiltrated behavioral medicine, and PNI is even viewed, incorrectly, as a clinical treatment. With the recent evolution of the PNI paradigm and the application of molecular biology to the neurosciences, this fledgling, and at times controversial, edge of research can be expected to receive increasing emphasis in the 1990s.

87.2 The Immune Response Is Central to the PNI Paradigm

Our objective here is not a detailed review of the immune response, which is dealt with in other chapters (Chaps. 85, 86), but consideration of it in the context of PNI, which considers it an extension of the nervous system. The basic function of the immune system can be compared to a reflex arc in the nervous system (Fig. 87.1; cf. Chaps. 48, 49). It has a stimulus – response organization in which, like the afferent limb of a reflex arc, foreign antigen is recognized and a highly specific humoral and cellular response ensues as the efferent limb. As in the case of taste or smell, a receptor-based molecular recognition or "perception" underlies immune specificity. The stimulus, here shown as an invading microorganism or tumor cell, is recognized through stereospecific noncovalent molecular interaction with the stimulating antigen, like a glove fitting over a hand. The various types of immunoglobulin (Ig) molecules that make up circulating humoral antibody can really be considered as multichain, circulating receptors, which in the case of IgM or IgD may still reside on the B cell surface. The multichain T cell receptor (TCR) has a similar recognition function, with amino acid sequence homology and structural similarities to Ig heavy and light chains. Like many functionally relevant molecules of immune cells [major histocompatibility (MHC) class I or II, CD2, CD4, CD8], they are members of the Ig gene superfamily of mol-

R. Greger/U. Windhorst (Eds.)
Comprehensive Human Physiology, Vol. 2
© Springer-Verlag Berlin Heidelberg 1996

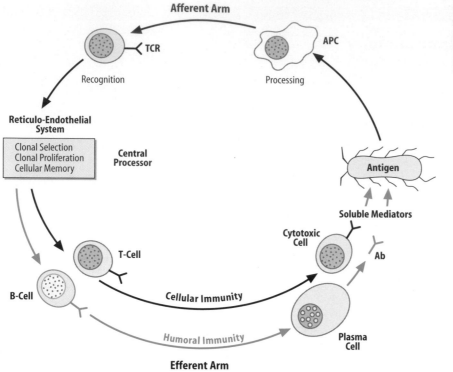

Fig. 87.1. The immune response is analogous to a reflex arc in the nervous system. *APC*, antigen-presenting cells; *TCR*, T-cell receptor

ecules. Thus, both the Ig and the TCR molecules carry the key to specificity of immune recognition. Their antigen-binding sites recognize the stimulating antigen stereospecifically. Ig molecules are functionally differentiated, and the TCR is associated with other polypeptide chains in a cell surface molecular complex. These arrangements are responsible for the biological effector functions initiated by the Ig and TCR molecules as a consequence of molecular recognition.

Like a reflex arc, the immune system has both afferent and efferent arms. The afferent arm behaves like a "sensory" arm in which cellular recognition of the stimulating antigen takes place along with antigen processing. Many molecular and biochemical events are needed for normal T cell recognition of antigen and subsequent T cell activation. Antigens are degraded, and then their short peptide fragments of 9–13 amino acids are presented to the T cell in the context of MHC class I or class II molecules on antigen-presenting cells. Macrophages or dendritic cells degrade foreign proteins proteolytically and display these peptide fragments embedded in the MHC class I or II antigen recognition site on the surface of the MHC molecule. The TCRs of the reactive T cells bind to these foreign peptide fragments, with other molecular components of the TCR complex providing adhesive and stabilizing actions. Activation signals are then transmitted to the T cell via tyrosine kinases associated with the TCR complex, leading to T cell activation and secretion of cytokines, such as interleukin-2 (IL-2) and IL-4.

By analogy with the associative circuits in the spinal cord gray matter, which may modulate or transmit the stimulus from the afferent arm of the reflex arc, a central processing unit exists within the organs of the immune system (marrow, spleen and lymph nodes) that now develops the full primary immune response, with a delay of 7–10 days to allow for proliferation of the immune cells. Like the nervous system, this response also demonstrates memory, which resides in programmed circulating lymphocytes called memory cells. Subsequent exposures to the same antigen will now elicit a secondary response, which is shorter in terms of time and more pronounced in degree. The change from primary to secondary responses is characterized by a switch in Ig heavy chain expression from IgM to IgG. Memory is associated with learning, in that the immune response is specific to the conformation of the immunogen. Both learning and memory result from clonal selection. The immunogen selects the proper lymphocytes from the host's repertoire and causes them to proliferate into clones of effector and memory cells, which are all specific for the antigen. With first exposure of the primary response, the immunogen stimulates only the tiny percentage of pre-existing lymphocytes in the immune repertoire that carry specific receptors for the given antigen. Each of these sensitized lymphocytes proliferates into a clone with identical cells and specificity. Some become effector cells and others, memory cells, to form a secondary response if the immunogen is contacted again in the future.

Like the neuronal and synaptic interactions regulating a reflex arc in the spinal cord, the expression of immune cell function also involves a complex series of cellular and molecular regulatory events that occur in phases to modulate T effector cell activation and B cell antibody produc-

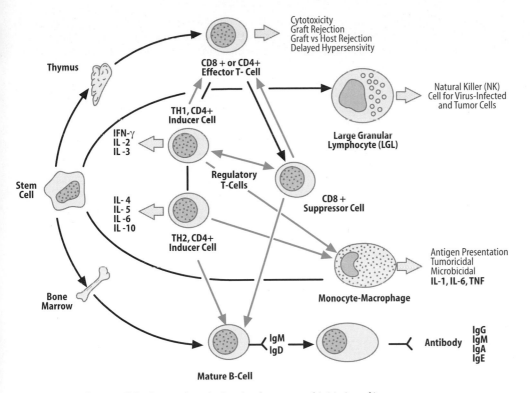

Fig. 87.2. Regulatory cellular interactions during development and initiation of immune responses

tion, which alone or together eliminate the antigen directly (Fig. 87.2). In response to appropriate signals, both T and B cells pass from activation and induction through proliferation, differentiation and, ultimately, effector functions in sequential stages. The end effector function may be an immune response, such as T cell cytotoxicity or the secretion of antibody by a plasma cell derived from a B lymphocyte, or a regulatory function mediated through T cell – T cell or T cell – B cell interactions and involving CD4+ helper or CD8+ suppressor T cells. CD4 helper (inducer) T cells are of two types, differing in the cytokines they produce. Activated TH1-type inducer T cells secrete interferon-γ (IFN-γ) and interleukins IL-2 and IL-3 to provide help for generation of cytotoxic T cells, generally in response to antigens leading to type-4 or delayed-hypersensitivity types of immune responses, often to intracellular microorganisms such as *M. tuberculosis*. On the other hand, the TH2 type of CD4+ inducer T cells secrete IL-4, IL-5 IL-6, and IL-10. They also regulate the intensity of immune responses by secretion of IL-10, which inhibits the production of other cytokines. They provide help to B cells for production of specific Ig, and also respond to certain foreign antigens, such as parasites that require high levels of antibody to eliminate.

Cytokines are soluble proteins produced by a wide variety of hematopoietic and nonhematopoietic cell types. They provide the basis of molecular communication within the immune system, much like short-range hormones (cf. Chap. 6). In myriad combinations they regulate the activation of immune cells by controlling gene activation and expression of functional cell surface molecules. There is considerable redundancy in cytokine functions, and many different cell types carry receptors for the same cytokine. Thus, cytokine networks composed of a variety of cytokines secreted from many cell types can, in turn, bind to and activate multiple cell types.

Upon activation by antigen-presenting cells, the two types of inducer cells are generated and produce positive and negative regulation of effector T and B cells (Fig. 87.2). A variety of cytokines, especially IL-3, IL-4 and IL-5 from T cells, stimulate B cells to proliferate, differentiate and, ultimately, secrete antibody by acting at sequential stages of B cell maturation. IL-2 and IFN-γ from inducer T cells act in a similar way to trophic factors for production of cytotoxic T cells. B cells themselves can also serve as antigen-presenting cells (APC), processing and presenting antigens to T cells and secreting tumor necrosis factor alpha (TNF-α) and IL-6. Monocyte-T cell interactions are also involved. Monocyte-macrophages are needed for optimal activation of T cells by antigens or by nonspecific lymphocyte activators called mitogens. IL-1 secreted from macrophages can substitute for intact macrophages in some circumstances. Upon contact with antigens or mitogens, macrophages secrete IL-1, IL-6, and TNF, which both induce receptors for IL-2 on T cells and also induce T cells to secrete IL-2, IL-4, and other cytokines. These latter cytokines, in turn, activate other T cells to expand the regulatory and effector T cell populations, which have been induced to express receptors for IL-2, IL-4, and IL-7. Effector T cells then can mediate a variety of functions, including killing of virus-infected cells, graft and tumor rejection, graft-versus-host reaction, delayed-type hypersensitivity, and the further release of a broad spectrum of immunoregulatory cytokines.

Finally, like a reflex arc in the nervous system, the immune response has an effector arm, which attacks the stimulating antigen. This arm is dual, having both humoral and cellular components (Fig. 87.1). Because it depends on cell proliferation and clonal expansion, the cellular component is a slow response, which developed first phylogenetically, probably originating in phagocytic cell ancestors in developmentally early multizoa. The humoral arm, on the other hand, developed later, probably at the time of emergence of the vertebrates from the water onto land, to provide a rapid response for protection against lethal bacterial infections that colonize moist body surfaces. The evolution of both responses as a sequence of differentiation and proliferation, and also its regulation, have been described above. The end result is receptor-like, highly specific recognition of the stimulating antigenic determinant by the TCR on the cytotoxic T cell or the antigen-binding site of the Ig molecule. This recognition step requires physical contact with the antigen and sets in motion a series of soluble immune mediators, such as the complement cascade in the case of Ig, which result in cytotoxicity or removal of the stimulating antigen by phagocytosis by the reticuloendothelial system. The biochemical details will not be reviewed further here. The humoral and cellular arms of the immune response often work in concert, as seen in the case of antibody-dependent cell-mediated cytotoxicity or the arming of such cells as B cells, killer (K) cells, and monocyte-macrophages with antibody, to mediate specific target recognition. These cell types carry Fc receptors to bind the constant regions of Ig antibody.

87.3 The PNI Paradigm: Behavior Is a Biologic Response Modifier

The hormonal and psychophysiologic (autonomic) correlates of stress are well known. In effect, the PNI thesis extends the stress model to include the immune system [45,54]. Many reports show that psychological factors, especially stress, can influence the immune response and, therefore, susceptibility to infectious, allergic and autoimmune diseases and to cancer (e.g. [35,36]). The missing link preventing our understanding of these observations is the nature of the communication between the nervous and immune systems. It was previously thought that the immune system was autonomous in the body, since a full immune response can be obtained in a test tube. It is now thought that the cells of the immune system act in a sensory capacity, informing the brain about stimuli that are not detected by the classic sensory system, such as invading foreign pathogens for cancer cells. The two systems are in constant close communication, and their mutual interaction seems to be modulated by the mind [6].

Although many of the relevant studies are old, three major lines of experimental evidence underlie the PNI hypothesis [9]. The first correlates psychological factors with physiological effects. Both severe and mundane stress, through their effects on both the hypothalamic-pituitary-adrenal and sympathetico-adrenal axes, can alter the functioning of both the immune and the nervous systems (e.g., [29–31,48,53,60]). For example, in both animal models and clinical settings, various stressors, such as bereavement or examination stress, can down-regulate natural killer (NK) and T cell function. A group of Canadian psychiatry residents under severe examination stress had reduced lymphocyte transformation. With acute examination stress there are dramatic decreases in lymphocyte IFN-γ production [21,22] and reductions in concentrations of IL-2 receptor (IL-2R)-positive peripheral blood lymphocytes, including reduced levels of synthesis of messenger RNA (mRNA) coding the IL-2 receptor [23]. These data suggest that stress-associated reductions in immunity may be observed at the level of gene expression [31]. Examples of positive psychological correlations with physiological effects include the relaxation response, or the effect of a supportive companion on perinatal problems, duration of labor, and mother – infant interactions. Following specific thought patterns elicits the relaxation response, with decreased oxygen consumption, carbon dioxide elimination, changes in blood pressure, heart and respiratory rates, and arterial blood lactate [4,61]. When a woman has a supportive companion (*doula*) with her just before and during labor, there is a significant decrease in the time between admission and delivery and also in the number of perinatal problems, including cesarian sections [55]. Patients who were encouraged and instructed by their anesthetists after surgery needed significantly fewer analgesics and 2.7 fewer days in hospitals than controls [13].

The second line of evidence is that Pavlovian conditioning as a type of learning influences the immune, as well as the nervous system. Humoral and cellular immunity have been behaviorally conditioned to immunosuppressive drugs [2], graft rejection [24], vaccines, and adoptive cellular therapies of cancer in animals [20]. Saccharin in the drinking water conditions the toxicity of the oral immunosuppressive drug cytoxan in mice; this toxicity takes the form of neutropenia, even leading to death. Investigators have even conditioned both humoral and cellular immunity in this way. Moreover, classic conditioning can reduce by 50% the dosage of cytoxan needed to control a murine version of the autoimmune disease systemic lupus erythematosus. Experiments have also been carried out in which mice were conditioned to reject grafts of their own normal skin. Recently an animal model was established for conditioning specific immunotherapy using the odor of camphor as the conditioning stimulus and allogeneic DBA/2 spleen cells as the unconditioning stimulus [20]. These normal allogeneic spleen cells have a significant protective effect against transplanted syngeneic YC8 lymphoma following prior immunization of BALB/c mice. In these experiments [20] conditioning delayed tumor growth, and in some instances the conditioned group performed better than the immunotherapy control group.

The third line of evidence documents the pervasive interlinks through which the nervous and immune systems influence each other [11,54]. The two systems communicate mutually using immune and hormonal transmitters and also neurotransmitters through interactions that are both anatomical and chemical [5]. Anatomically there is a precise pattern of innervation from the autonomic nervous system of the thymus, spleen, lymph nodes and bone marrow [18] (cf. Chap. 86). The fibers are mainly sympathetic and are found in regions rich in T cells and not in areas with developing B cells. Surgical denervation or chemical sympathectomy will enhance immune responsiveness. The immune response can be altered by changing the activity of splenic suppressor cells by creating lesions in various areas of the brain in animals. Lesions in the anterior hypothalamus suppress various immune parameters, while lesions in the amygdala or the hippocampus stimulate them. Moreover, the profile of the firing rate of neurons in the hypothalamus rises and falls in parallel with the course of the immune response, as do noradrenaline turnover and content, both of which decrease in the spleen and the hypothalamus, representing aminergic neuronal activity.

Immune cells carry receptors for many neurotransmitters and hormones, which can increase or decrease their functions [37]. Immune cells also secrete hormones and neurotransmitters, in addition to interleukins and immune mediators [52,64]. The nervous and immune systems interact in hormonal, sympathetic and parasympathetic regulatory feedback circuits that control the response to antigenic stimulation [6]. Some of these pathways are thought to prevent autoimmune and lymphoproliferative dysfunctions.

The chemical interconnections of the stress pathway are well known. The hypothalamus secretes corticotrophin-releasing factor (CRF), which acts on the pituitary to release adrenocorticotrophic hormone (ACTH). ACTH then stimulates the adrenals to release corticosteroids, which are immunosuppressive. Environmental stimuli such as conditioning or stress can affect immuity through this pathway. Immune cells also carry receptors for neurotransmitters, such as acetylcholine, norepinephrine and endorphins, which can increase or decrease their function. Catecholamines reduce immunity, while parasympathomimetics stimulate antibody formation and lymphocyte cytotoxicity. Immune cells even secrete neurotransmitters such as histamine, serotonin, prostaglandins, ACTH, and endorphins. Neuronal firing also responds to immune transmitters released from activated lymphocytes such as interferons and interleukins. Interferon shares amino acid sequences with ACTH and β-endorphin. Even thymic hormones, such as thymosin-α-1, act through the hypothalamus and the pituitary gland to increase corticosteroid levels. This may be a normal regulatory mechanism of the immune response, i.e., a negative feedback suppression designed to shut the response down by stopping recruitment of immature T cells (mature T cells are less sensitive to corticosteroids). Finally the immune system, like the nervous system, is regulated both upward and downward by hormones. Lymphoid and accessory cells carry receptors for corticosteroids, insulin, prolactin, growth hormone, estradiol and testosterone. Some of these hormones participate in lymphocyte activation through modification of intracellular cyclic nucleotides such as cAMP and cGMP. Sex differences are known in both immune responsiveness and autoimmune diseases. The in vitro immune response is reduced by glucocorticoids, androgens, estrogens and progesterone, while it is increased by growth hormone, thyroxine and insulin.

In addition to communicating through anatomical and chemical connections, the brain and the immune system each also know what the other is doing because their interactions are linked in regulatory feedback circuits [5]. A sympathetic circuit reduces sympathetic tone in the lymphoid organs just before the peak of the immune response, in order to release immune cells from inhibition and allow the response to take off. On the other hand, the glucocorticoid circuit, which operates through the hypothalamic-pituitary-adrenal axis, the stress pathway, raises the levels of blood glucocorticoids to dampen the immune response after its peak. This limits the expansion of cells with low affinity for antigen and controls clonal expansion of immune cells with high affinity for antigen by inhibiting regulatory cells producing IL-1 and IL-2. This pathway may help prevent autoimmune and lymphoproliferative disorders.

That illness and injury in themselves can have psychological and behavioral effects is a common experience. For example, the physical response to infection may include behavioral changes with reductions in physical activity and the usual daily activities, as well as fatigue, malaise, altered sleep patterns, fever (cf. Chap. 111), etc., which are part of the neural effects of such cytokines as IL-1, IFNs, and TNFs, which are mobilized as part of host resistance [8]. There are a number of chronic illnesses that are characterized by some or all of persistent or recurrent fever, fatigue, myalgia, lymphadenopathy, arthralgia, pharyngitis and behavioral changes associated with weakness, malaise, and deression. Some are associated with chronic active Epstein-Barr (EBV) infection or post-viral states, while others, such as chronic fatigue syndrome, are less well understood. In the case of AIDS, as another example, the human immunodeficiency virus can produce cognitive and behavioral impairments not only by direct invasion of the nervous system, but also through secondary systemic changes as well as opportunistic infections.

87.4 PNI Is an Extension of the Stress Paradigm

The more comprehensive model of psychoneuro-immunomodulation can be considered to be an outgrowth of the general adaptation syndrome or stress model of

Selye [50]. The concept of stress has a long history going back $2\frac{1}{2}$ millennia [7]. The idea of homeostasis is central to stress: survival of living organisms by way of the maintenance of a complex physiological and biochemical dynamic harmony, balance or equilibrium in the face of constant challenge or threat from external or internal stressors (cf. Chap. 2). Thus, stress is a state of disharmony or threatened homeostasis, which is countered by a complex repertoire of physiological and mental adaptive responses, both specific and general, which try to restore balance. These responses are both stereotypic and have a threshold. This is an alternative formulation of host resistance, and in the times of Hippocrates and the Ancient Romans this notion was described in terms of the concept of "Nature [as] the healer of disease" or the *Vis Medicatrix Naturae* or "healing power of nature."

Phenomenologically, a state of arousal occurs through central facilitation of neural pathways that mediate arousal, alertness, cognition, focused attention, and appropriate aggression. Concurrently, other pathways inhibit vegetative functions, such as reproduction and feeding. Associated peripheral changes promote an adaptive redirection of energy for survival through cardiovascular, respiratory and metabolic adjustments.

Traditionally there are two major components to the stress response. One involves the hypothalamic (paraventricular nucleus)-pituitary-adrenal axis with the release of key hormonal mediators, corticotrophin-releasing factor, ACTH, and glucocorticoids. The second is located in the brainstem and involves the locus ceruleus and autonomic (sympathetic) nervous system, with release of norepinephrine and epinephrine throughout the brain and from the adrenal medulla. The two systems activate each other, and a variety of other neuropeptides and mediators, such as arginine vasopression and dynorphin peptides, are released.

Other elements of the nervous system are activated, such as the mesocortical and mesolimbic dopamine systems involved in cognition, the amygdala/hippocampus complex involved in memory and emotion, and the CRH neurons in the lateral paraventricular nuclei, which mediate opioid receptor analgesia and the setting of emotional tone.

The systems reponsible for reproduction, growth, and immunity are all inhibited, which involves a rich network of associated hormonal interactions stimulated by CRH and inhibited by corticosteroids. These involve growth hormone-releasing factor, growth hormone, somatomedin C, and somatostatin in growth function, and in the case of thyroid function, thyroid-releasing hormone, thyroid-stimulating hormone, thyroxine, and triiodothyronine. Thus, in addition to the neural events, a rich variety of hormonal mediators participate to make this a neuroendocrine response. The addition of effects on immune function now makes up the full complement of the PNI system. The stress response can profoundly inhibit both the inflammatory and the immune responses. Glucocorticoids can alter leukocyte function and traffic, reduce cytokines (principally IL-1, IL-6, and TNF) and inflammatory mediators (eicosanoids, platelet-activating

factor, and serotonin) and inhibit their actions. Conversely, these soluble immune mediators can stimulate the hypothalamic-pituitary-adrenal axis, especially hypothalamic CRH secretion (IL-1, Il-6, TNF, eicosanoids, platelet-activating factor) and ACTH secretion (IL-1, IL-6, TNF, serotonin, several eicosanoids and platelet-activating factor). Direct effects on adrenal glucocorticoid secretion may also occur.

This sympathetic dominated, arousal, "fight, flight, or fright" model of stress has been expanded to include a role for the parasympathetic system, which is still, however, based on the central concept of adaptive balance or homeostasis of the organism [38]. The reciprocal innervation of glands, pupils, lungs, heart, blood vessels, and internal organs by the autonomic nervous system (cf. Chap. 17) is carried out by the sympathetic system, which produces arousal and activation, and the parasympathetic system, which mediates rest and inhibition. In this expanded view the natural balance of autonomic functioning is essential to an understanding of stress. In addition to abnormal activation of the sympathetic side leading to arousal, one also sees parasympathetic dominance leading to inhibition, with the lethargy, apathy, depression, and feelings of helplessness and hopelessness that can be seen in such chronic diseases as cancer. This has been called the "possum response." These individuals respond to threat with passive withdrawal, in the same way as a threatened possum, instead of preparing to fight or run away, rolls over and plays dead. This response to fear is not arousal, but inhibition, with typical characteristics of extreme parasympathetic discharge with reduced physiological functioning, loss of skeletal muscle tone, mental lassitude, inactivity, and eventually depression. This is a complex syndrome, with both a constitutional predisposition and a learned response from social conditioning; learned helplessness can result in depression, illness, and even death in some organisms [49]. Thus, according to this model, stress becomes defined in terms of a state of internal imbalance with unrelieved dominance of either arousal or inhibition of autonomic nervous function and, hence, of the rest of the total PNI system.

87.5 Ultradian Rhythms, Homeostasis, and the PNI System

The study of ultradian rhythms of nasal dominance provides an interesting window on the homeostatic nature of PNI function. In the West, the history and physiological dynamics of the nasal cycle began in 1895 with a German rhinologist, Kayser [28], who is usually credited with observing and naming it. He described how the nasal mucosa of the left and right nasal passages periodically alternates in size and shape, so that one nostril is dominant or active with the breath flowing in and out easily while the opposite one is congested. The nasal mucosa is erectile, like the breasts and genitalia. This cycle alternates in an ultradian rhythm during a 24-h period.

Alternate engorgement of the mucous membrane in each nostril under the influence of the autonomic nervous system is the physiological explanation of the phenomenon. For the open, active, or dominant nostril, there is sympathetic dominance with vasoconstriction of the nasal erectile tissue and an increase in mucosal gland secretion to facilitate the filtration of air by surface cilia to accompany the increased air flow. In the passive, closed, or nondominant nostril there is parasympathetic dominance, with vasodilatation, swelling of nasal erectile tissue brought about by engorgement with blood, and a decrease in mucosal gland secretion, with less air filtration and decreased air flow. Both animal and human studies implicate a central regulation of the nasal cycle by the medulla at the level of the respiratory center and/or the hypothalamus [12].

Rossi [46] has summarized some dozen studies encompassing 50 years of research on the duration of the nasal cycle. The range of duration is wide, from 30 min to 8 h, as with all other ultradian behaviors. But the average duration is around the typical ultradian mean of 90 min to 2 h. Factors such as the frequency and duration of measurements and the methodology used for the measurements may help to explain this wide variance.

As the nasal cycle changes it passes through a point of balance at which, for a brief period, the air flow in the two nostrils is equal. The associated psychophysiological state has been well characterized. According to Rossi [46,47], it may underlie the phenomenon of spontaneous hypnosis, which Milton Erickson called the "common everyday trance" and used to facilitate his unique naturalistic approach to hypnosis with its 90-min to 120-min sessions.

In the yogic literature the point of equal nostril dominance has great importance [44]. It is given the Sanskrit name of "sushumna." Students are taught concentration processes to learn voluntary control of the nasal cycle, not only to control hemisphere dominance (see below) and internal states, but also to achieve, voluntarily, the state of sushumna, of equal balance. There is an associated inward turning and general pleasantness of the mind, which is both conducive to and essential for deep meditation. Thus, the deliberate induction of this state may provide a useful window for the voluntary control of neuroimmunomodulation.

Sperry and Gazzaniga [56] developed the concept of hemispheric dominance on the basis of their work with patients who had had cerebral commissurotomies (cf. Chap. 55). Initially it was thought that hemispheric dominance was fixed, but it has since become clear that there is a spontaneous 90-min alternation in dominance between the left and right cerebral hemispheres during sleep (EEG) and wakefulness (changes in cognitive style) [26,32]. There is a direct relationship between cerebral hemispheric activities measured by the EEG and the ultradian rhythm of the nasal cycle [63]. It is possible to change nasal dominance by forced single-nostril breathing through the nondominant closed nostril. This results in an accompanying shift in cerebral hemispheric dominance to the contralateral hemisphere. Thus, the ultradian nasal cycle is a window on

cerebral hemispheric activity. Moreover, voluntarily induced changes in nasal air flow change the relative locus of activity in the highest centers of the brain and thereby influence the autonomic nervous system, which regulates virtually every function of the body, including the PNI system.

The correlation of the nasal cycle with the alteration of hemispheric activity is consistent with the model of a single ultradian oscillator system [63]. The nasal cycle can be viewed as an easily measured indicator or window for autonomic nervous system regulation and the integration of autonomic and cerebral hemispheric activity. The whole body is thought to go through a rest/activity or parasympathetic/sympathetic oscillation while simultaneously passing through the left body – right brain or right body – left brain shift. This produces ultradian rhythms at all levels of organization, ranging from pupil size to higher cortical functions and behavior. The model suggests an extensive integration of autonomic (and hence PNI) and cerebral cortical activity. The nasal cycle probably is regulated thorough a centrally controlled mechanism, possibly the hypothalamus, altering the sympathetic/parasympathetic balance throughout the body, including the brain, and is the mechanism by which vasomotor tone regulates the control of blood flow throught the cerebral vessels, thereby altering cerebral hemispheric activity.

Returning to the homeostatic model described above, stress is now perceived as a matter of balance within the autonomic nervous system. In a balanced state the oscillation described above (and also reflected in the degree of nasal dominance) around a center point towards either sympathetic or parasympathetic predominance is both stable and narrow. Loss of balance with over-dominance of either sympathetic or parasympathetic activity may be experienced as stressful and accompanied by wider swings in the degree of nasal dominance and alteration of its ultradian rhythm. It is interesting that the ancient Sanskrit texts of the tradition of yoga describe prolonged imbalance and dominance of one nostril as a signal of impending illness [43]. It may be that the degree of oscillation or imbalance may be a useful indicator of the homeostasis of the host response system. Study of the nasal cycle could be a useful and a readily accessible window on this ongoing dynamic balance.

87.6 PNI and the Mind – Body Problem

The role of mind, of the "psycho" part of PNI, still remains the most difficult to deal with. A multidimensional and multidisciplinary approach is needed to bridge soma and psyche at their various levels of organization with a holistic view. Moreover, there is a need for thinking and for models rooted in general systems theory to deal with the complexity of the interactions of the components of the PNI systems, which resist the usual linear, reductionist approaches of modern biology. Research in PNI must confront the mind – body paradox, and thus far investigators have tended to stay on one side or the other, with limited forays across the divide in either direction. In the PNI model, the mind may operate to set the level of responsiveness of the system, and to trigger it.

The PNI hypothesis provides a model for understanding this controversial mind – body link in health and disease. This Cartesian dilemma is all-pervading: how can the mind influence the body if the two are separate entities, especially if the mind is considered to be nonmaterial or even, in the extreme view, nonexistent, as only the product of biochemical and electrical activity of the brain? How can behavioral interventions, which involve verbal manipulation of the mind, result in physiological effects, such as major surgery without pain and accelerated wound healing, or regression of warts, as in hypnosis?

Part of the dilemma lies in the lack of a Western definition for mind. Western psychology deals with the study of behavior, presumably as a consequence of mental activity. Some theorists now look to modern physics for models, such as the holographic paradigm put forward by Pribram and Bohm [42]. Others are looking to eastern philosophies, such as Buddhist or yoga psychology, which view mind as a highly refined energy field, which means the Cartesian dilemma dissolves and the problem can again be put into the models of modern physics. The approach has been one of pragmatism. Such models should be used in much the same way as models are used in physics. The concern is not whether they are true, but rather their utility – are they useful in achieving practical results and for guiding significant research?

Approaches used with modern techniques of behavioral modeling are examples of such a strategy. An old presupposition of cybernetics is that the map is not the territory. There is an irreducible difference between the world, our experience of it as presented by the senses, and our interpretation of the meaning of that experience by the mind. Thus, we do not operate directly on the world, but rather create a representation of it, a map or model in the mind [3,10]. That map determines our perception of the world, our experience of it and, hence, our choices and behavior. The model implies that that map also influences the functioning of our host resistance, our PNI system. In a sense, this is an alternative formulation of the stress paradigm. To an experienced mountaineer, hanging from a sheer cliff face thousands of feet from the floor of the valley below may be an exciting and exhilarating experience, and the tone and quality of that individual's PNI system and host resistance will be set accordingly. To an amateur that same situation may be utterly terrifying, and that person's PNI system and host resistance will be set differently, probably with unpleasant symptoms and perhaps even some of the dangerous behaviors that panic can breed. The sunny day, the magnificent scenery, the rope and the cliff are the same, but the experience of them differs vastly between the two individuals. According to this model, the experience of negative stress is situational and dependent on the individual's mental model of the world, and the PNI system function is set accordingly. Whether a particular situation is an exciting challenge, a neutral event, or a stressful threat is a matter of individual perception and interpretation, and host resistance responds in kind.

This approach, then, leads to the hypothesis that changing an individual's mental model of the world can be expected to lead to altered PNI functioning and host resistance in a given context, which in turn can lead to some very pragmatic behavioral research with correlations of PNI functional effects. For example, these models of the world or mental paradigms obey "laws." Like the PNI system at the molecular and physiological levels, they tend to behave like homeostatic systems, which are self-regulating and resist change. They operate mostly below the daily level of awareness and probably influence physiology through the unconscious mind. Paradigms can be modeled in terms of both content and structure. Their content is made up operationally of cause-effects and criteria, the latter referring to the values, beliefs, and expectations we use to codify, judge, and make meaning out of our everyday experience. These internal models of the world represent a secondary level of experience coded in language, and produced by the intellectual faculties through distortion, deletion and generalization, in linguistic terms, of primary sensory experience as represented by mental imagery coupled to affect and stored in the unconscious mind. Strong beliefs arise from life's compelling reference experiences with strong emotion, and the primary data in the form of images in memory and their associated affect can be accessed (and altered) by various techniques, e.g., hypnosis, reframing, altering the submodalities of the imagery, or changing the arrangement of the structure of that internal subjective experience of space, time, and neurology.

Moreover, according to this conceptualization, when a mental paradigm changes, it does so in a characteristic pattern [3,47]. Much like a system in general systems theory, the force for change provided by external input of experience and information builds up against the inherent internal homeostatic resistance to a threshold. Then a sudden, irreversible, expansion of insight called a paradigm shift occurs, which the individual may experience as an "aha" reaction. The mental paradigm or model of the world has now been irreversibly changed, so that the individual no longer views or makes meaning of the situation and hence reacts to it (e.g., cancer) in the same way. The new world view includes, but is a major expansion of, the old one, and hence introduces greater flexibility and choices for more effective coping behaviors. These shifts are often accompanied by rapid autonomic changes involving gestures and involuntary muscular activity, shifts in emotions and other internal states, and changes in pupillary size, heart rate, blood pressure, and breathing patterns, some of which are readily visible to the observant clinician. Given the autonomic responses, the PNI model would postulate further changes in other system components, such as endocrine and immune function, as well, and hence in the overall level of host resistance. However, as the discussion above indicates, changes in these parameters have been documented in the presence of stress and in response to antistress interventions, such as relaxation or meditation, but as yet no such positive evidence exists for the psychotherapeutic interventions referred to here, and for technical reasons such evidence would not be easy to measure, since the timing of a paradigm shift cannot be predicted in advance.

Currently, much of this work is being done in the context of cancer and of infections such as AIDS. The paradigm shift really describes a process of learning from experience, in the sense of integration of that experience into a new world view. The process of therapy, according to this model, then becomes one of belief change, the production of this mental paradigm shift that alters the patient's experience of the world from a present state characterized by excitation with anger or fear, or inhibition with depression and feelings of helplessness and hopelessness, to a desired state, with positive affect and an orientation to opportunities and outcomes in the future. The models specify the relevant structure and content of present and desired states and point to the appropriate interventions most likely to create the shift. All of this is based on behavioral modeling of success, i.e., on patients who display effective coping strategies.

87.7 PNI and the Psychobiology of Host Resistance in Disease

87.7.1 PNI, Host Resistance and the Immune Response

The PNI model can be regarded as a new paradigm for understanding host resistance in disease. There is a well-developed biology of host resistance for infectious diseases, which centers largely on the immune response. For diseases such as cancer we lack a biology of host resistance. A tumor cell does not grow freely in a patient, as it would in a culture dish; the patient puts up a defense, which currently is also thought to be primarily immunological. However, even though antitumor immune responses can be measured with some malignancies, correlations with tumor outcomes are poor and efforts to modify tumor progression with immune manipulations have produced only modest results. More than just immunity must be involved. PNI expands the concept of host resistance beyond just immune defence. A seed will not grow unless the soil is hospitable. Whether the seed is a microorganism or a cancer cell, its ability to express its genetic potential by invading and causing disease depends on its interactions with the microenvironment, the milieu interne of the host in which it finds itself. This milieu interne is the locus of host resistance. It will be made up of many factors, such as hormones, growth factors, nutrients, peptides, immune mediators, other cell types, nerve endings, etc., all in dynamic molecular communication with the "seed." Many of these components make up the PNI system. The balance or imbalance of that interaction will determine whether the seed grows, remains dormant, or is destroyed.

87.7.2 Behavioral Factors Influence the Clinical Course of Disease Through the PNI System

Sections 87.3, 87.4, and 87.6, above, have dealt with evidence and proposed mechanisms by which psychosocial factors may be associated with altered PNI functioning, especially immunity. Section 87.7.3, below, will address more of the psychological dimensions of PNI by summarizing relationships between psychosocial factors and susceptibility to specific diseases in which host resistance is thought to be mediated by the PNI system, with particular involvement of immunity. The major strategy of clinical research used here is the study of disease as an experiment of nature, to learn about normal function. Although a number of factors have been identified, the study of the mechanisms by which they mediate their effects has only begun, and the PNI hypothesis now provides the theoretical model that underlies much of this effort. A somewhat simplistic, but useful, categorization is by the source of the stimulus to host resistance and whether the response is over- or underactive (Table 87.1) [36]. Where the challenge is from an outside antigen, overactivity is associated with allergy, manifest for example in the form of asthma, urticaria, or dermatitis, and underreactivity, with infection. Where the antigenic stimulus is internal, overactivity of the PNI system is associated with autoimmunity and underactivity, with cancer. The greatest controversy and methodological difficulties arise in studies that examine whether behavioral interventions, including psychotherapy, can influence the onset or course of diseases by altering PNI function, especially immunity. Relaxation, hypnosis, biofeedback, meditation, and behavior modification and conditioning have all been studied. Most of the work has been done with cancer.

87.7.3 Host Resistance in Specific Diseases

Cancer: Biological Factors. We do not yet have a well-defined model for host resistance in cancer comparable for that in infection as described above, but there are many similarities. Conventional wisdom still credits the immune response totally with responsibility for host resistance in cancer, but the broader components of the PNI system also play a central part. As in the case of infectious diseases, the host mounts both a cellular and a humoral response

Table 87.1. Immune response to antigen (adopted from [36, p 26])

		Hyperactivity	Hypoactivity
Antigen Source	Internal	Autoimmunity	Cancer
	External	Allergy	Infection

against tumor cells, but the cellular response is currently considered to be the more important. Many molecular mechanisms have been implicated as responsible for the generation of new tumor antigens involving transformation-related heritable change in the genetic material of the cells, including:

- Biosynthesis of a new molecule
- Unique degradation products of abnormal cellular proteins
- Alteration of the structure of a normal molecule
- Uncovering of normally protected molecules
- Incorrect assembly of multimeric antigens
- Aberrant expression of fetal or differentiation antigens

Some antigens are unique and are found only on tumor cells, some being cross-reactive with other tumors of the same type and others individually specific to a particular tumor. Others are tumor-associated, being shared by both tumor cells and some normal cells, with qualitative and quantitative differences in antigen expression. Host responses are more likely with the unique variety.

The cellular response can take several forms, with direct lysis by classic T cell cytotoxicity as the main example. Arming of cells with antibody to produce antibody-dependent cell-mediated cytotoxicity (K cells, B cells, monocytes, and macrophages) occurs, as does nonspecific killing by lymphokine-activated monocytes/macrophages and natural killer NK cells and their derivatives, such as lymphokine-activated killer (LAK) cells. The details are described below for infection, and the regulatory activities involve the cytokine networks similarly.

The humoral response is also similar to the response to infection, with direct tumor cell lysis from antibody and complement, opsonization of tumor cells by complement components with subsequent phagocytosis by the reticuloendothelial system, and loss of tumor cell adhesion through antibody binding. The antigens recognized here are also multiple and complex, and include individually specific membrane antigens and common tumor-associated antigens, with the addition less commonly of fetal, group-specific, and nucleolar antigens. The various antibody systems differ with the stage of disease. Individually specific membrane antibodies are present in early stage I and II disease, only to disappear rapidly from the circulation with the appearance of blood-borne metastases. Antibodies to the common cytoplasmic tumor-associated antigens are present in the circulation throughout the course of the disease, but especially in late stages.

There is a common misconception that cancer patients are immunodeficient, in the sense of lacking an intact immune apparatus. The problem seems rather to be a regulatory one caused by the inability of the immune system to completely eliminate the antigenic stimulus. The resulting chronic antigenic stimulation results in a dysregulation characterized by the triad of anergy (suppression of T cell function) with circulating immune complexes (which can block cellular immunity), anti-antibodies (indicating altered regulation by feedback idiotypic networks), and left shift of circulating B cells with immature forms (excessive suppressor tone). It is interesting that the same pattern occurs in the rheumatic disorders and in chronic infections such as tuberculosis, brucellosis, and some chronic parasitic disorders, where the immune complex disease may be severe enough to damage the kidneys and produce nephrotic syndrome. The fundamental basis of the dysregulation appears to be an imbalance of suppressor/helper tone, with dominance of suppressor activity mediated by both suppressor T cells and monocytes and their mediators. These changes are particularly striking in the case of tumors of the immune system itself, such as lymphomas and leukemias, where striking anergy (Hodgkin's disease) or hypo-γ-globulinemia (thymomas or leukemias of suppressor T cell origin) may be seen with consequent severe infectious complications. The appearance in the circulation of anti-antibodies with disease progression is a striking example of the dysregulation as the negative feedback complementary antibody networks become overstimulated by abortive attempts to shut down the response in the presence of persistent antigenic stimulation. Anti-antibodies called anti-idiotypes, which are directed at the antigen-binding site, appear against the individually specific membrane antibodies, while pepsin agglutinators against the hinge region and rheumatoid factors against the Fc constant tail region appear against the common cytoplasmic antibodies. The appearance of these anti-antibodies in the circulation with disease progression sweeps the respective tumor antibodies out of the circulation through immune complex formation, with subsequent loss of immunity.

The situation is even more complex, however. The regulatory imbalance of factors that may promote tumor immunity (e.g., antibody, K cells, T cells, macrophages) with those that may favor tumor progression by interfering with tumor immunity (soluble antigen, suppressor cells, immune complexes) results in an unstable equilibrium that shifts over time and tumor stage and can give unpredictable results with therapeutic manipulations. Moreover, the PNI model indicates that more factors than just immunity are involved, such as

- Nervous influences
- Endocrine influences
- Growth factors
- Nutrition.

How important immunity is among these other factors is generally unknown; it may be a major factor in some tumors in which such phenomena as variable and long natural history, dormancy, and spontaneous remissions are seen (melanoma, breast cancer, lymphomas), but a minor factor in other tumors with a more inexorable clinical course, such as lung cancer. Although the advent of biological therapies (advocated as the fourth modality of cancer therapy after surgery, radiation and chemotherapy) has come from these advances in our understanding of immune function, it is not surprising, given the complexity of the PNI system and host resistance, as described above, that the clinical application of biological therapies has been limited to date.

Cancer: Behavioral Influences. A large and growing literature supports the role of behavior as an important factor in host resistance against cancer. In addition to social and cultural behavior patterns (smoking, diet, and exposure to other environmental carcinogens), this topic can also be conceptualized in three major research areas [9]. The first deals with the psychophysiology of cancer – the application of the PNI paradigm already referred to above. This research consists in animal experimentation relating stress and mental states to both tumor growth and immune function (e.g., [51]). Although difficult to generalize, the data indicate beneficial effects of stress reduction, of a sense of control, and of a stable, supportive social environment.

A second area of research examines the effects of psychological events and personality traits on the incidence and progression of cancer. The concept of the type C, or cancer-prone, personality is one illustration of some of the major research effort in this area [58]. The best studies are prospective studies of large healthy populations tested for several psychological variables and followed up for many years (e.g., [14,15,19,59]). In other studies attempts have been made to correlate personality with disease progression, with the aim of identifying psychological correlates that might be amenable to modification by psychotherapy and that distinguish slow progressors ("survivors") from fast progressors ("nonsurvivors") (e.g., [1,27,62]). Retrospective studies have examined cohorts of cancer patients to see what premorbid personality or emotional states, if any, may have been causal cofactors (e.g., [27,33]). Overall, the research seems to show that cancer is more probable among people who have lost a meaningful relationship in recent years, particularly if they are prone to feelings of helplessness and hopelessness, and that cancer patients often show emotional inhibition and repression. The data overall, however, are sometimes inconsistent and emphasize the recognition of personality patterns that will need confirmation in other ways.

The third area of research on behavior and clinical disease consists of a few controlled investigations studying psychotherapy and counseling for cancer patients (e.g., [14,25,27]), an area that has at times been beset with not a little controversy (e.g., [24]). Clinical reports have been published to suggest prolongation of life following imagery, hypnosis, meditation, and various regimens of social, cognitive, and behavioral psychotherapy. In some studies the randomized controlled trial format has been used to show that a psychological intervention in conjunction with standard medical care can result in changes in coping and in measures of psychological distress (e.g., [15–17,57]). Some studies have also shown improved survival of cancer patients, or that psychological interventions are acceptable to patients and can result in improvements in a wide range of measures of psychological functioning. However, it is still unclear what the effective component or components may be. Modification of coping styles associated with particular personality types, reduction of psy-

chological stress associated with anxiety or depression, or an improved sense of well-being, whether resulting from attenuation of physical symptoms, e.g., pain, or from an improved self-image and sense of control, are all possibilities. Modulating social variables, such as feelings of belonging to a group, have also been implicated [57]. There is also a strong likelihood that the important variable has not even been identified at this time, and that there may even be a constellation of variables at work in a synergistic or multiplicative fashion [15].

The convergence of views from these three separate lines of research is persuasive, and the medical community should not commit a type II error by dismissing so easily as unfounded what might be true by continuing to resist the possible influence of mind on cancer [9]. Scientists often dismiss new concepts as untrue if they cannot assign the mechanism by which they might act to an existing theory. In addition to less ambiguous experimental data requiring more precise experimental methods and tools, better theoretical explanations of how the mind affects the body are needed to overcome this resistance. The new PNI paradigm appears to fill this conceptual void, and has been called the paradigm for holistic medicine.

Infectious Diseases: Biological Factors. Host resistance has been best worked out in the case of infectious disease. Here there is the interplay of an organism through its virulence and the size of the infecting dose with the ability of the host to respond to the infection. The host responds through a complex of interacting systems that protect it from endogenous and exogenous microbes by three basic mechanisms: physical and chemical barriers, inflammatory responses, and the reticuloendothelial system, especially the immune response. The latter two general mechanisms involve the PNI system intimately.

Physical and chemical barriers constitute the first line of nonspecific defense to prevent the introduction and spread of endogenous and exogenous microorganisms, both those normally colonizing body sites and those pathogenic organisms whose presence is pathologic. The barriers include anatomical and functional integrity of surface barriers of epithelial and endothelial cells of the skin and mucous membranes and organs such as the epiglottis and sphincters. Chemical barriers arise from various surface secretions (e.g., bactericidal free fatty acids or lysozyme for the skin, acid pH in the stomach, and pancreatic enzymes), and physical removal of organisms results from excretory flow such as peristalsis, sloughing of squamous cells, and urine flow. There is also competition from endogenous normal microbial flora. Disruption of any of these mechanisms by tumors, procedures, devices, infarcts, or drugs will increase host susceptibility to infection.

Associated with the host response in infection is both a local and a general inflammatory response, the latter associated with fever, leukocytosis, and even shock. Circulating phagocytes [neutrophils (PMN), monocytes, eosinophils, and basophils] from the bone marrow enter local tissues to form the cornerstone of the inflammatory response. Inflammatory responses are precipitated by bacterial products such as cell wall polymers, microbial peptides, enzymes, or cytolytic toxins and endotoxins, and by mediators released from the sensitized cells and immune complexes of the immune response and from tissue damage. Most work through the complement system, the arachidonic acid cascade, kinin-generating systems, clotting factors, immune cells and lymphokines, including the shock mediated by monokines and eicosanoids from macrophages and endogenous opioids from the pituitary. There are changes in plasma proteins, serum iron, and acute phase reactants. Two key mediators in the PNI system are IL-1, which has a major role as a pyrogen (cf. Chap. 111), and TNF, which produces the cachexia and negative nitrogen balance.

The reticuloendothelial system clears circulating microorganisms from the blood using tissue phagocytes derived from monocytes. They include macrophages in the liver (Kupffer cells), spleen, lymph nodes, lung (alveolar macrophages), kidney (mesangial cells), and brain (microglia). The rate of particle ingestion is enhanced by opsonins, such as IgG or C3b, and by a variety of soluble mediators from mononuclear leukocytes, all of which constitute aspects of the PNI system.

The immune response is the key component of the reticuloendothelial system, which participates in host resistance to infection. The second line of defense is natural immunity as part of the PNI system with natural antibodies, complement and phagocytic cells such as macrophages and PMNs. PMNs are attracted by chemotactic substances released from complement or the organisms themseles, and ingest microorganisms that have been opsonized (coated) by antibody or complement or have been trapped against a mechanical barrier. Macrophages, either fixed tissue or derived from monocytes, are also attracted by chemotaxis. Phagocytic cells make the spleen an important organ in the defense against infection, especially for pneumococcal infections and malaria. Natural antibodies are mostly IgM in type and arise from immunization from cross-reacting antigens.

The third line of defense is also immune and includes specific antibody, complement with sensitized T cells, and their lymphokines. T and B cells are the major cellular components of the immune system and are distributed throughout the bloodstream and at tissue sites. They interact with each other and with monocytes, macrophages, immunoglobulins, and the complement cascade. B cells and plasma cells secrete specific antibodies that neutralize toxins, block attachment of organisms to cell membranes, prevent the spread of organisms from cell to cell, produce immune lysis with complement, enhance phagocytosis by macrophages and PMNs, and carry out antibody-dependent cell-mediated cytotoxicity with K cells (a variety of non-T cell cytotoxic circulating lymphocytes) and eosinophils. Immune complexes activate complement by the classic pathway, and microbial products such as endotoxin and polysaccharide by the alternate pathway, or by direct digestion of complement components by bacterial proteases. Opsonization, lysis, chemotaxis, and leukocytosis result. T cells mediate specific cellular immu-

nity by secreting a multitude of cytokines that affect the function of other T cells, B cells, monocytes, and macrophages, and participate directly in HLA class I antigen-restricted cytotoxic reactions against infected cells. Monocyte and macrophage ingestion and killing of a wide variety of bacteria, fungi, and protozoa are augmented by release of T cell lymphokines, especially IFN-γ. NK cells attack tumor cells and body cells infected by intracellular parasites and are stimulated by IFN and retinoic acid.

Overall host resistance in infection is influenced by genetic factors, which influence surface receptors for microorganisms, may produce other protective disorders (e.g., glucose-6-phosphate dehydrogenase and alpha-thalassemia both protect against malaria), and influence the immune response itself through production of immunodeficiencies. Family and racial differences in susceptibility relate to the action of immune response genes located in the histocompatibility locus, which determine whether an immune response to an antigenic stimulus will occur at all. Genetic markers for susceptibility include HLA haplotypes, Gm allotypes of immunoglobulins and ABO types. Environmental and constitutional factors round out the picture; these include age, gender, hormonal influences, trauma, malnutrition, the presence of other infections, drugs, and malignant disease. Any of these can have effects on surface defences, nonspecific immunity, specific cellular immunity, and humoral immunity.

Infectious Diseases: Behavioral Influences. Some of the earliest studies supporting psychosocial factors, particularly stress, as influencing the onset and course of infectious diseases, were conducted in tuberculosis, in which retrospective studies linked stress to exacerbations of the disease. Prospective studies have linked stress with higher incidence of influenza, infectious mononucleosis, and streptococcal and upper respiratory infections. Stress, with both anxiety and depression, can be associated with recurrences of genital and oral herpes simplex, and case reports and controlled studies suggest that hypnosis and cognitive therapy may reduce the likelihood of recurrences. The possibility that psychosocial determinants may influence AIDS is being studied, one aspect being whether behavioral interventions can influence disease progression. Effects in Epstein-Barr virus (EBV) infections are controversial. Clinical reactivation of latent viral infection may occur through immune impairment from depression. Alternatively, persons diagnosed as having depression or somatization disorder may actually have chronic occult EBV infection.

Autoimmune Diseases. Rheumatoid arthritis and ulcerative colitis have been best studied, but psychosocial factors have also been implicated in systemic lupus erythematosus, ankylosing spondylitis, Grave's disease, Crohn's disease, uveitis, myasthenia gravis, and diabetes mellitus (not in multiple sclerosis). Psychosocial stress with depression and impaired expression of anger has been linked with the onset or exacerbation of rheumatoid and juvenile arthritides, and traumatic recent or childhood loss (e.g., a parent, or the spouse by death or divorce) may predict disease susceptibility. Studies to date are inconclusive, and several competing hypotheses as to mechanism, including immunity and the PNI system, have been proposed. Suggestive evidence indicates that behavioral and cognitive and other self-control interventions can help arthritis patients reduce their pain and regain joint function.

Allergy. Classic allergic, type I or anaphylactic reponses are mediated by IgE. By secreting IL-4, IL-5, IL-6, and IL-10, TH2 helper/inducer T cells drive these responses, which are mediated through mast cells and basophils armed with IgE attached to their Fc receptors (CD23). An allergen – IgE interaction triggers the release of mediators from the basophil or mast cell, the principal one of these being histamine. Some of these mediators increase vascular permeability and smooth muscle contraction (histamine, platelet-activating factor, SRS-A, BK-A), some are chemotactic and activate other inflammatory cells (ECF-A, NCF, leukotriene B_4), and others regulate the release of other mediators (BK-A, platelet-activating factor).

Among the allergic disorders, asthma has been the best studies for a psychosomatic basis. It appears that asthma patients have both an allergic disorder and an inherited predisposition to respond to intense emotion and conflict with bronchospasm. There is insufficient evidence for a unique asthmatic personality profile, although a trait of heightened dependency is present. Despite methodological failings, controlled clinical trials of interventions such as relaxation, meditation, hypnosis, biofeedback, behavior therapy, and group and family therapy have all been reported to yield positive results of varying degree.

Urticaria is also histamine mediated, and in guinea pigs elevated blood histamine levels can be behaviorally conditioned. Psychogenic factors including stress are thought to contribute to 36–84% of cases, but little evidence supports a personality profile unique to hives.

Skin Diseases. The blood vessels, smooth muscle, and sweat glands of the skin are regulated by and linked to the emotions through the autonomic nervous system. Strong emotions can induce blushing, pallor, sweating, gooseflesh, and piloerection. These subtle responses can be seen by the observant clinician as indicators of PNI system responses to behavioral interventions, as discussed above. A variety of skin disorders are mediated by immune mechanisms and there are psychological components to their pathogenesis. These include warts, herpes, eczema, and urticaria; eczema or atopic dermatitis is the best studied of these. Strong emotions can trigger the release of histamine and other inflammatory mediators into the skin to produce itching and scratching. Controlled trials of biofeedback-assisted relaxation or hypnosis have been able to reduce the severity of symptoms in eczema, acne, burns, and psoriasis, in part through alteration of blood flow, but the possible involvement of immunity in these effects has not been studied. The removal of warts through hypnosis is well known, although their removal from specific locations

is harder to demonstrate. The mechanisms underlying this are unknown.

87.8 Conclusion: PNI and Host Resistance, a Systems Model

The brain processes information about changes in the external and internal environments as detected by receptor organs. The immune system acts as a peripheral receptor organ for the brain to detect both externally introduced macromolecules (e.g., microorganisms) and modified self-antigens (e.g., neoplasm). Since mind, through its cognitive maps and belief structures, influences the neuroendocrine system (e.g., stress), it can be expected to influence the functioning of the immune system as well. Theory is moving away from a military model of immune functioning through attack and defense to a broader construct based on regulation of cellular interactions or relationships through information flow by way of hormonal and neural intercommunication. In this general systems model, tumor and host cells interact in a microenvironment where cytokine/receptor interactions play a key part in mediating this information flow.

This new view leads to a homeostatic model of interactions between the host and tumors or microorganisms, based on a dynamic balance between them and the host microenvironment in which they grow: and updated version of the seed-and-soil hypothesis (Fig. 87.3). A tumor, for example, results from an abnormality of growth and differentiation based on altered structure, regulation, and expression of self-genes, and characterized by various properties such as transformation, invasiveness, metastasis, clonality, and heterogeneity responsible for its malignant behavior. Nonetheless, the outcome of its growth still depends on its interaction with host defenses for a net result in terms of progression, dormancy, or regression. An analogous process occurs with microorganisms, with properties such as growth, virulence, invasiveness, and toxin production pitted against host defense mechanisms as described above. The process is changing, dynamic, and chronic. The time-scale is critical, since host resistance changes with the advancing stage of tumor growth or infection. Moreover, even though clonal in origin, tumors are genetically unstable and hence phenotypically heterogeneous, so that they present a moving target to host defenses. Antigenic changes from mutational events in microorganisms such as influenza viruses or malarial parasites are similar. Host resistance is a complex, dynamic system that needs to be understood at the molecular, physiological, and psychological levels. Although immunity plays a key role, it is naive now to think that host resistance is due only to immunity. Multiple factors, both exogenous and endogenous, contribute to host resistance in an interacting network or system with many components, including the central and autonomic nervous systems, the endocrine systems, immunity, nutrition, growth factors, and probably other, as yet unknown, components. All share chemical

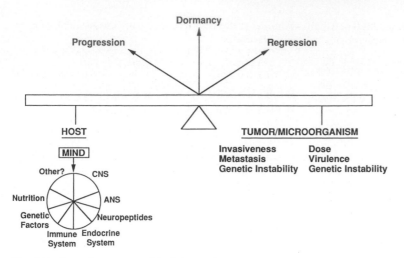

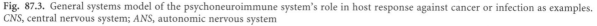

Fig. 87.3. General systems model of the psychoneuroimmune system's role in host response against cancer or infection as examples. *CNS*, central nervous system; *ANS*, autonomic nervous system

mediators, which through their receptor–transmitter interactions and signalling produce complex dynamic patterns of information flow and communication, both intercellular (tumor–tumor, host–host, and host–tumor) and intracellular, between the plasma membrane and the nucleus of the individual cell. Thus, impaired host resistance seems to be a regulatory problem resulting in collapsed homeostasis and based on a breakdown in molecular communication. The locus of these events is the cell surface where transmitter – receptor interactions take place.

A psychosomatic basis of host resistance has been proposed: according to this proposal, the function of the PNI system may be positively or negatively affected (i.e., enhanced or suppressed) by attitudes and beliefs through the emotions these elicit. The "tone" of host resistance may be influenced by mental and emotional states determined by an individual's cognitive map of reality. It is further proposed that the balance of PNI function, and hence of host resistance, at any time will be reflected in corresponding patterns of intercellular molecular communication among its nervous, endocrine and immune components. Some of these patterns of molecular communication will, in turn, be reflected in profiles of cytokine levels in the milieu interne. Indeed, particular emotional states, positive or negative, might turn out to have characteristic cytokine and mediator profiles within the PNI system [39–41]. Finally, relaxation, hypnosis, breathing, visualization, and meditation have become popular clinically as ways of boosting host resistance, and have been shown at times to enhance immunity. Such interventions as these, which reduce stress or which alter beliefs and values within an individual's model of the world for the better would be expected to produce changes in host resistance reflected by characteristic and corresponding modifications in PNI function and molecular communication, the latter reflected in altered cytokine and mediator profiles.

References

1. Achterberg J, Mathews-Simonton S, Simonton OC (1977) Psychology of the exceptional cancer patient: a description of patients who outlive predicted life expectancies. Psychother Theory Res Pract 14:416–422
2. Ader R, Cohen N (1975) Behaviorally conditioned immunosuppression. Psychosom Med 37:333 – 340
3. Bandler R, Grinder J (1982) Reframing. Real People Press, Moab
4. Benson H, Beary JF, Carol MP (1974) The relaxation response. Psychiatry 37:37–46
5. Besedovsky HO, del Rey AE, Sorkin E (1983) What do the immune system and the brain know about each other? Immunol Today 4:342–345
6. Blalock JE (1984) The immune system as a sensory organ. J Immunol 132:1067–1070
7. Chrousos GP, Gold PW (1992) The concepts of stress and stress system disorders. Overview of physical and behavioral homeostasis. J Am Med Assoc 267:1244–1252
8. Clemens MJ (1991) Cytokines. Bios Scientific, Oxford
9. Cunningham AJ (1985) The influence of mind on cancer. Can Psychol 26:13–29
10. Dilts R, Grinder J, Bandler R, Bandler LC, Delozier J (1979) Neuro-Linguistic Programming I, Meta Pubs, Cupertino CA
11. Dorian B, Garfinkel PE (1987) Stress, immunity and illness – a review. Psychol Med 17:393–407
12. Eccles R (1978) The central rhythm of the nasal cycle. Acta Otolaryngol (Stockh) 86:464–468
13. Egbert LD, Battit GE, Welch CE, Bartlett MK (1964) Reduction of post-operative pain by encouragement and instruction of patients. N Engl J Med 270:825–828
14. Eysenck HJ (1988) Personality, stress and cancer: prediction and prophylaxis. Br J Med Psychol 61:57–75
15. Eysenck HJ (1991) Smoking, Personality and Stress. Springer, Berlin Heidelberg New York
16. Fawzy IF, Cousins N, Fawzy NW, Kemeny ME, Elashoff R, Mortin D (1990) A structured psychiatric intervention for cancer patients I: changes over time in methods of coping and affective disturbance. Arch Gen Psychiatry 47:720–725
17. Fawzy IF, Kemeny ME, Fawzy NW, Elashoff R, Mortin D, Cousins N, Fahey JL (1990) A structured psychiatric intervention for cancer patients II. Changes over time in immunological measures. Arch Gen Psychiatry 47:729–735

18. Felten DL, Felten SY, Carlson S, Olschowka JA, Livrat S (1985) Noradrenergic and peptidergic innervation of lymphoid tissue. J Immunol 135:755–765
19. Fox B (1978) Premorbid psychological factors related to cancer incidence. J Behav Med 1:45–133
20. Ghanta VK, Hiramoto NS, Solvason HB, Soong SJ, Hiramoto RN (1990) Conditioning: a new approach to immunotherapy. Cancer Res 50:4295–4299
21. Glaser R, Rice J, Speicher CE, Stout JC, Kiecolt-Glaser JK (1986) Stress depresses interferon production concomitant with a decrease in natural killer activity. Behav Neurosci 100:675–678
22. Glaser R, Rice J, Sheridan J, Fertel R, Stout JC, Speicher CE, Pinsky D, Kotur M, Post A, Beck M, Kiecolt-Glaser JK (1987) Stress-related immune suppression: Health implications. Brain Behav Immun 1:7–20
23. Glaser R, Kennedy S, Lafuse WP, Bonneau RH, Speicher CE, Kiecolt-Glaser JK (1990) Psychological stress-induced modulation of IL-2 receptor gene expression and IL-2 production in peripheral blood lymphocytes. Arch Gen Psychiatry 47:707–712
24. Gorczynski RM, MacRae S, Kennedy M (1982) Conditioned immune response associated with allogeneic skin grafts in mice. J Immunol 129:704–709
25. Gordon WA, Freidenbergs I, Diller L (1980) Efficacy of psychosocial intervention with cancer patients. J Consult Clin Psychol 48:743–759
26. Gordon H, Frooman B and Lavie P (1982) Shift in cognitive asymmetries between wakings from REM and NREM sleep. Neuropsychologia 20:99–103
27. Greer S, Watson M (1985) Towards a psychobiological model of cancer: psychological considerations. Soc Sci Med 20:773–777
28. Kayser R (1895) Die exakte Messung der Luftdurchgängigkeit der Nase. Arch Laryngol Rhinol 3:101–120
29. Khansari DN, Murgo AJ, Faith RE (1990) Effects of stress on the immune system. Immunol Today 11:170–175
30. Kiecolt-Glaser JK, Glaser R (1987) Psychosocial moderators of immune function. Ann Behav Med 9:16–21
31. Kiecolt-Glaser JK, Glaser J (1992) Psychoneuroimmunology: can psychological interventions modulate immunity? J Consult Clin Psychol 60:569–579
32. Klein R, Armitage R (1979) Rhythms in human performance: 1-1/2-hour oscillations in cognitive style. Science 204:1326–1328
33. LeShan L (1959) Psychological states as factors in the development of malignant disease: a critical review. J Natl Cancer Inst 22:1–18
34. Levy SM (1986) Behavior as a biologic response modifier: Psychological variables and cancer prognosis. In: Anderson BL (ed) Women with cancer. Springer, Berlin Heidelberg New York, p 289–306
35. Locke SE, Hornig-Rohan M (1983) Mind and immunity: behavioral immunology (1976–82) – an annotated bibliography. Institute for the Advancement of Health, New York
36. Locke SE, Colligan D (1986) The healer within. The new medicine of mind and body. New American Library, New York
37. Morley JE, Kay NE, Solomon GF, Plotnikoff NP (1987) Neuropeptides: conductors of the immune orchestra. Life Sci 41:527–544
38. Nuernberger P (1979) Freedom from stress. Himalayan Institute Press, Honesdale
39. Pert CB, Ruff MR, Weber RJ, Herkenham M (1985) Neuropeptides and their receptors: a psychosomatic network. J Immunol 135:820–826
40. Pert C (1986) The wisdom of the receptors: Neuropeptides, the emotions, and bodymind. Advances 3:8–15
41. Pert C (1987) Neuropeptides: the emotions and bodymind. Noetic Sci Rev 2:13–17
42. Pribram K (1991) Brain and perception: holonomy and structure in figural processing. Erlbaum, Hillsdale
43. Rama S (1986) Path of fire and light. Advanced practices of yoga. Himalayan Institute Press, Honesdale
44. Rama S (1992) Meditation and its practice. Himalayan Institute Press, Honesdale
45. Reichlin S (1993) Neuroendocrine-immune interactions. N Engl J Med 329:1246–1253
46. Rossi EL (1986) Altered states of consciousness in everyday life: the ultradian rhythms. In: Wolman BB, Ullman M (eds) Handbook of states of consciousness. Van Nostrand Reinhold, New York, pp 97–132
47. Rossi EL (1993) The psychobiology of mind-body healing. New concepts of therapeutic hypnosis, rev edn. Norton, New York
48. Schulz KH, Schulz H (1992) Overview of psychoneuroimmunological stress and intervention studies in humans with emphasis on the uses of immunological parameters. Psycho Oncol 1:51–70
49. Seligman MEP (1975) Helplessness: on depression, development and death. Freeman, San Francisco
50. Selye H (1976) The stress of life. McGraw-Hill, New York
51. Sklar LS, Anisman H (1981) Stress and cancer. Psychol Bull 89:369–406
52. Smith EM, Brosnan P, Meyer MJ, Blalock JE (1987) An ACTH receptor on human mononuclear leukocytes. N Engl J Med 317:1266–1270
53. Snyder BK, Roghmann JK, Sigal LH (1993) Stress and psychosocial factors: effects on primary cellular immune response. J Behav Med 16:143
54. Solomon GF (1987) Psychoneuroimmunology: interactions between central nervous system and immune system. J Neurosci Res 18:1–9
55. Sosa R, Kennel J, Klaus M, Robertson S, Urrutia J (1980) The effect of a supportive companion on perinatal problems, length of labor, and mother-infant interaction. N Engl J Med 303:597–600
56. Sperry R, Gazzaniga M (1967) Language following disconnection of the hemispheres. In: Millikan C, Darley F (eds) Brain mechanisms underlying speech and language. Grune and Stratton, New York, pp 177–183
57. Spiegel D, Bloom JR, Kraemer HC, Gottlieb E (1989) Effect of psychosocial treatment on survival of patients with metastatic breast cancer. Lancet 2:888–891
58. Temoshok L, Dreher H (1992) The type C connection. Random House, New York
59. Thomas CB, Greenstreet RL (1973) Psychobiological characteristics in youth as predictors of five disease states: suicide, mental illness, hypertension, coronary heart disease and tumor. Johns Hopkins Med J 132:16–43
60. Van Rood YR, Bogaards M, Goulmy E, Van Houwelingen (1993) The effects of stress and relaxation on the in vitro immune response in man: a meta-analytic study. J Behav Med 16:163–171
61. Wallace RK, Benson H, Wilson AF (1971) A wakeful hypometabolic state. Am J Physiol 221:795–799
62. Weissman AD, Worden JW (1975) Psychosocial analysis of cancer deaths. Omega 6:61–75
63. Werntz D, Bickford R, Bloom F, Sing-Khulsa S (1982) Alternating cerebral hemispheric activity and lateralization of autonomic nervous function. Neurobiology 4:225–229
64. Wybran J, Appelboom T, Famaey J, Gouverts A (1979) Suggestive evidence for receptors for morphine and methionine-enkephalin on normal human blood T lymphocytes. J Immunol 123:1086–1091

L Heart and Circulation

88 Hemorheology

H. SCHMID-SCHÖNBEIN

Contents

88.1 Introduction: Aims and Perspectives of Hemorheology as Applied Synergetics

Synergetics, a new scientific discipline in search of

- General rules about joint efforts in *driven* systems and
- Universalities explaining sudden phase transitions upon alternations of the *drive*

describes on the basis of first physical principles the manifestations of suddenly occurring (and self-organized, see below) spatiotemporal patterns in moving fluids (see Chap. 2). The term "self-organization" is here defined as the consequence of the spontaneous coordination of movement patterns ("use-dependent coordination" [108]), i.e., the appearance of collective responses of a priori unbound individuals. These responses can emerge when large ensembles of such elements are appropriately driven. The term "coordination" refers to a process, not to its results; therefore the verb "to order" rather than the noun "order" should be used. As will be shown below, when the mammalian blood is properly driven, it spontaneously enhances its *mobility* (here measured as the *fluidity*, see below) so that it can easily negotiate the macro- and microvessels despite the fact that the mammalian (capillaries are substantially smaller in diameter than those of the blood cells (red blood cells, RBC, white blood cells, WBC) when measured at rest.

Synergetics is complementary to cybernetics as developed half a century ago by Wiener [119]. The latter is a control theory that predicts that by the appropriate external control of a process (with the help of sensors and a central information processor and with peripheral effectors) certain dynamic parameters can be kept constant. Thermostats, chemostats and barostats are typical examples, and cardiovascular physiology, ventilatory physiology, and other fields of physiology have made ample use of these theoretical concepts and their many ramifications.

Synergetics, by contrast, predicts the rules of spontaneous internal coordination of movements: for this, the rate of transfer of energy and matter (see Chap. 2), rather that static parameters such as the arterial pressure must be considered. Obviously, however, in keeping arterial pressure within normal limits, venous pressure and thence the pressure difference is kept within ranges producing the appropriate pressure gradients, and, from that, shear stresses and shear stress gradients (see below) needed to dynamically deform and "fluidalize" the RBC.

There are many synergetic regularites (as defined in Chap. 2) that can and must be applied to blood, which is obviously and organ in perpetual motion. A novel interactionist's concept in which the moving blood is classified as a driven system is therefore presented and many of its apparent anomalies are reduced to synergetic consensualization processes. These prove to be closely related to three distinctly different functions of blood:

- Nutrition and waste removal by the RBC and the plasma;
- Repair of vascular lesions by the thrombocytes, coagulation enzymes, and fibrinogen capable of rapid polymerization and
- Defense against microbial invasion by WBC and immunoglobulins.

Hemorheology comprises the study of the "flow behavior" of the blood, a concentrated suspension of specialized cells in an aqueous electrolyte-protein solution. The perfectly coherent movement of blood cells in the macro- and microvasculature is a complex process, highly variable in its mechanical details and as yet incompletely understood [16–19,39,92,93,100,102]. As in all perfused tube networks, the local as well as the global rate of flow is codetermined by local *promotor* forces (pressure gradients) that induce fluid deformations, and by *controller* processes associated with viscous energy dissipation that impede flow. In the case of moving blood viscous energy dissipation has much more complex consequences: it induces processes properly termed *dissipative structuring* (as defined in Chap. 2). On the one hand, the process of viscous energy dissipation attenuates the gliding displacement of fluid elements in moving blood, but on the other hand, the same process can become the reason for a progressive shear-induced fluidalization of moving blood (see below). This follows from the very phenomena that are responsible for a peculiarity of blood previously termed its "anomalous viscosity". As can be seen in Fig. 88.1, taken from a now classical review by Chien [17], the macroscopic viscosity of mammalian blood is very high (200× that of plasma) when the blood is exposed to *low* rates of shear ($\gamma < 1$/s), and remarkably low (about three times that of plasma) at high rates of shear ($\gamma > 100$/s). Elicited by the shear stresses associated with these shear rates, normal human RBC undergo two different fluidalization processes, namely, aggregate dispersion and deformation/orientation of the dispersed RBC. These two phenomena are associated with a progressive, shear-dependent decrease in the coefficient of viscosity calculated from the ratio of shear stress and shear rate (see below). The early clarification of this dependency of a *response coefficient*, relating a deforming force to the resulting deformation, can now be put into a comprehensive theoretical perspective developed in general synergetics: when progressively driven, the blood can undergo *phase transitions* (sudden changes in appearance, see Chap. 2).

In comparing the viscosity–shear rate curve of suspensions of normal human RBC in plasma to suspensions of hardened RBC and to suspensions of non-aggregating RBC, which were all measured at the same RBC volume fraction (hematocrit, HCT, 0.45 V/V), the dualistic causes of the macroscopic flow anomaly could be established at an early stage:

- At rest (zero shear), human RBC regularly combine into rouleaux and three-dimensional networks of rouleaux.
- At shear rates between 0.01 and 10/s enforced shearing removes the causes of a viscidation by RBC aggregates.
- At shear rates between 10 and 1000/s shearing actively induces fluidalization by means of passive RBC orientation and deformation.

As one starts from *stationary* and proceeds to *rapidly moving blood*, initially progressive disaggregation and later progressive deformation of the RBC leads to a situation where a kind dissipative structuring changes the appearance and the behavior of the blood from that of a reticu-

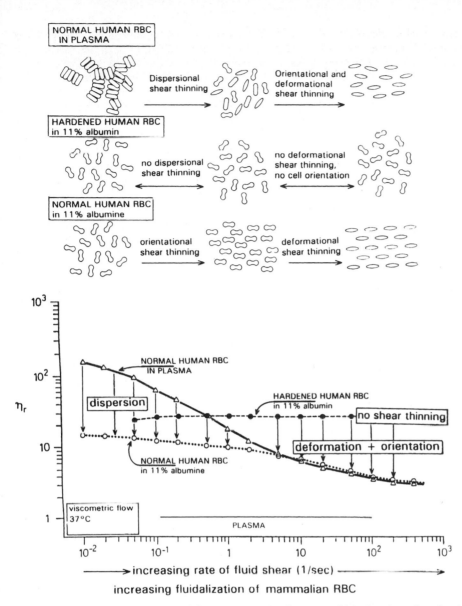

Fig. 88.1. Schematic representation of the non-Newtonian flow behavior of human blood (hematocrit value 0.45 V/V), depicted as relative apparent viscosity (measuring data obtained from Chien [17]). Comparison of *normal RBC in plasma* (NRBC plasma), *normal RBC in an 11% albumin* solution and *hardened RBC in plasma* (HRBC plasma) discloses the dualistic cytological and hydrodynamic reasons for a shear-induced, reversible phase transition from a highly viscid reticulated suspension to a highly fluid emulsion-like liquid. When subjected no enforced shear between 0.01 and 10/s (see Sect. 88.3.1) the aggregates produced by macromolecules in plasma are progressively disrupted by shearing forces and this is associated with dispersional shear thinning. When subjected to shear between 10 and 1000/s, the dispersed NRBC show progressive orientation, deformation and membrane tank treading (see Sects. 88.3.2 and 88.3.4), associated with deformational shear thinning. NRBC in albumin (showing neither aggregation

nor high viscosity at low shear) display progressive orientation between 0.01 and 10/s. Between 10/s and 1000/s they are likewise progressively deformed. HRBC in plasma show neither aggregation nor deformation: lack of dispersional shear and deformational shear thinning produces high, shear-independent viscosity. This behavior, vizualized in rheoscopes (see Sects. 88.3.2 and 88.3.4) and quantified under idealized artificial flow conditions should not be confounded with a property, but represents dynamic performance characteristics of RBC suspensions. In the vast majority of microscopic blood vessels, these same performance characteristics of non-nucleated mammalian RBC produce much more pronounced sequelae of analogous phase transitions. The details of flow behavior and the cell–cell and cell–wall interactions critically depend on the flow conditions there, on cell-plasma distribution, on previous history of flow in upstreams vessels, and on the locally prevailing RBC volume fraction

lated, viscid suspension to that of a highly fluid, emulsion-like self-lubricating liquid.

These features should not be mistaken for a physical property in the sense of continuum mechanics, but should be

accepted as *behavioral traits* or *performance characteristics*. Both on the level of the individual RBC and on the level of their suspensions, these two self-fluidizing responses fall into a class of nonequilibrium phase transitions which are systematically investigated in synergetics. So far, we mainly understand the phenomenology of these phase transitions in vitro, yet it is well-established that phase jumps from rapid RBC movements associated with their deformation to slow movements associated with their aggregation occur in vivo as well (see Fig. 88.19). In both cases, flow retardation is associated with progressive aggregation and flow acceleration with progressive cell deformation: synergetics teaches us that there we are confronted with *circular causality*, i.e., causes and consequences are inseparably intertwined. As detailed in Chap. 2, nonlinear dynamics in general and synergetics in particular have taught us that almost all self-organized phenomena so far studied exhibit this feature in a way that is counterintuitive to conventional thinking in the natural sciences.

Since, as will be shown, these responses of individual RBC are linked to collective alterations associated either with a strong increase (in the case of aggregate sedimentation) or with a marked decrease (in the case of cell deformation) in the RBC content of blood moving in microvessels, the phase transitions of the moving collectives (i.e., the holistically responding cell suspensions) can obviously be much more pronounced in vivo than those found in vitro. In this context one can understand that while collective phase transitions into the self-fluidizing mode of operation under normal flow conditions can cause a remarkable coordination of perfusion, the opposite occurs if the creeping blood moves in a self-viscidizing mode once the shear stresses fall below a critical level: under these conditions, the perfusion pattern becomes chaotic.

To put knowledge of the shear dependency of blood flow behavior into proper perspective, it is important to stress that there is ample evidence that in all segments of the macro- and microvascular networks in mammals and in humans, both the shear rates and the shear stresses are high (in the order of magnitude of 100–1000/s [16,17,39,92] and that it would be a serious error to) relate the low microvascular flow velocities (RBC velocity 0.5 to 2 mm/s) to low shear rates or shear stresses. The opposite is correct: the smaller the tube diameters, the higher the shear rates/shear stresses (see Sect. 88.3.3 and Table 88.1) This fact leads us to the central hypothesis of the present chapter: when blood is subjected to the very high shearing forces that typically prevail in the intact (or normokinetic, see Sect. 88.4) mammalian circulatory system, it is permanently being driven into a self-fluidizing mode of operation. This statement refers to a set of reactions of the individual RBC and to larger ensembles of them, both of which respond to the set of hydrodynamic forces acting in flowing blood by increasing their individual fluidality and their fluidity as an ensemble. By responding with a fall in their measurable *coefficient of fluidity* (see below), a collective of passively deforming particles shows synergetic susceptibility to the control parameter (here, the driving pressure): they are moved and simultaneously made more mobile by the pressure gradient propelling them. Understandably, these dualistic responses can both enhance the magnitude and improve the coordination of the movements of large RBC ensembles. However, as soon as the flow is retarded and there is a local drop in shear forces, the unavoidable phase jump into the aggregated flow behavior decelerates and also removes the physical reasons for coordinated flow. Understandably, several self-viscidizing processes can be initiated, and strongly disordered motion patterns can – but of course need not – ensue.

As detailed in Chap. 2, universal theories were developed to explain multiple forms of self-organized spontaneous cooperativity (and consenzualized movements) of large ensembles of basically free elements: these can be applied to many driven systems. In microvascular physiology, they help us to comprehend not only the often bewildering complexity of blood flow phenomena found in vivo, but also their instability and their tendency for progressive deterioration. In applying paradigms of general nonlinear dynamics, the full spectrum of observable patterns can be put into perspective to encompass phenomena found in vastly differing driven systems. where under appropriate boundary conditions various forms of energy (thermal, mechanical, electrostatic, electromagnetic, viscous, cohesive, adhesive, and chemical) can begin to cooperate. This occurs in such a fashion that in a driven ensemble cooperative behavior can spontaneously emerge which is not a priori found in the constituents when isolated and/or when the system is operating near equilibrium. In other words, when a system is dynamically exhausted, it presents features that are never found when it is operating far displaced from equilibrium, and vice versa.

As will be shown for blood constituents in motion, large-scale consenzualization of movements can be initiated by processes properly termed "dissipative structuring" ([41] and Chap. 2). In this concept, a first-order biological significance can now be attributed to those properties of the moving blood [18,20,39,92] that previously used to be classified as "anomalies". In the conventional semantics of technical rheology, these nonlinear properties of moving blood were paraphrased by terms such as "thixotropic", "pseudoplastic", "visco-elastic" or "shear thinning". In the concept of synergetics, all these non-Newtonian flow characteristics can now be reduced to:

- Their dynamic causes and
- Their kinematic consequences,

i.e., the specific modes of movements resulting from combined forces causing (or propagating) and impeding (or controlling) the movements. Interesting circular causality, so typical for phase transitions, will be found in nonlinear systems which exist in driven systems when a priori endowed with or driven to "criticality" [41,46,47,80,100] i.e., a dynamic regime where a kind of mixture of elastic and viscous traits (or amphidynamic behavior) is found (see Chap. 2).

88.2 The Concept of Viscosity and the Movement of Cohesive Liquids

88.2.1 Anthropomorphic Imagery

At any given temperature above 273 K (freezing point) and below 373 K (boiling point), the thermally agitated molecules of water, the universal fluid and solvent in biology, respond to external physical and chemical forces by irreversibly changing places. Lord Kelvin introduced an intuitively comprehensible image to distinguish the fluid from both the solid and the gaseous state of matter: in solid particles, molecules or atoms are tightly packed like ships roped together and stacked on the walls of a harbor basin. When thermally agitated, the ropes are torn and the constituent elements (atoms, molecules, and particles) move like ships in a very crowded harbor, incessantly making sliding contacts with other ships. When further agitated, the particles behave like ships that move rapidly and completely independently on the ocean where they only make occasional contacts.

In applying concepts of the van der Waals forces producing cohesion in true fluids, the movements of Lord Kelvin's ships can be said to occur in such a fashion that slipping but permanently refastened ropes guide the mobile elements into a sliding movement which guarantees their permanent contact. This imagery of simultaneous rope-binding and rope-releasing (cohesive liquids) allows to intuit such effects as nearest-neighbor interactions (see Chap. 2) that are essential for collective self-organization. Metabolic events (any type of exchange of energy) can only occur between mobile elements after they have been previously released from their firm bonds and can collide with other mobile species: no collision, no reaction. Lastly, if different types of mobile elements are moving within a limited space, the velocity and the collision rates of elements of one species are obviously slowed down. This fact is reflected in the well-known viscosity dependence of Fick's coefficient of diffusivity, and it contributes to dissipative structuring in self-organized chemical reactions (see Chap. 2).

In the present context, however, the response to external gradients of pressure shall be considered: elements (e.g., fluid molecules) are driven into a sliding motion past each other. Basically, this is a microscopic displacement event. If we neglect for a moment their thermal movements, sliding can be described as a process which forces the elements to be dislocated away from their previous neighbors. Schematically speaking, this process can be quantified by estimating the number of element diameters a molecule is being shifted sideways in unit time (Fig. 88.2a). In larger ensembles, the same process induces rather pronounced velocity differences between elements that are not immediately adjacent (Fig. 2b) but placed in strata of molecules that can be idealized as fluid laminae. At the macroscopic level, this is seen as *laminar flow* that can be described as a macroscopic rate of shear (Fig. 2c).

88.2.2 On Coherent and Incoherent Movements During Enforced Shearing

Before returning to the formal description of flow as a consequence of fluid element deformation in the form of enforced shearing, the effects of thermally caused molecu-

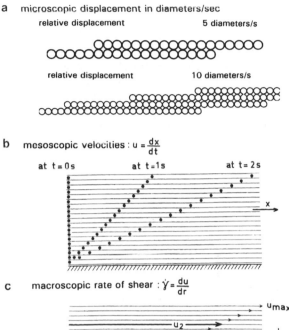

Fig. 88.2a–c. Schematic representation of the kinematics of element displacement during for flow of liquids as caused by enforced shearing of fluidal elements (atoms, molecules, dissolved macromolecules, particles, droplets, cells). In this image, the spontaneous movements of elements and momentum transfer are neglected (but see Fig. 88.3 and Fig. 88.4). **a** Collective microscopic displacement of juxtapositioned elements, quantifiable by parallel movements of many neighbors in the unit of a distance (element diameters) covered in unit time (e.g., seconds): slow collective displacement with 5 diameters/s, fast collective displacement with 10 diameters/s. **b** Under the impact of forces acting at a distance, the local displacements/time in many superimposed strata of elements are added up and result in a global distribution of velocities. Using the idealization of fluid laminae, elements immediately adjacent and at rest are progressively separated from each other. Their combined trajectories give rise to a distribution of velocities in a larger volume element. The positions of adjacent elements at rest and at $t = 0$ are progressively separated (at $t = 1$ s, at $t = 2$ s, etc.) as time of flow passes. **c** Laminar flow following from the dynamics of continuous microscopic displacements and resulting superimposed mesoscopic velocities. Both effects generate the macroscopic event of enforced shearing in a macroscopic volume element which when flowing is progressively deformed. The rate of deformation or the rate of shear in a volume element of macroscopically flowing liquid can be defined by the ratio of velocity difference ($du = u_2 - u_1$) and the distance (dr of any two observed strata of elements

1751

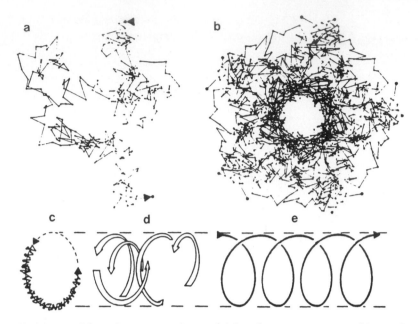

Fig. 88.3a–e. Schematic representation explaining the superposition of thermal movement and forward movement of fluid elements subjected to convective laminar flow. **a** Random walk of one molecule without shear flow. **b** Superposition of five molecules starting five different places resulting in an net apparent movement along rings. **c** Simplified representation of **b** with superposition random walk upon microscopic circular movement without laminar shear flow. **d** Projection of schematized forward movements of trajectories as seen in **c** after induction of laminar flow. **e** Highly simplified trajectory of fluid particle subjected to finite forward velocity

lar movements in fluids must be considered. Their microscopic displacement enhances the incoherent, thermally induced movements of molecules. Therefore, shearing is inseparably linked to the generation of *more heat* (or with the production of more entropy). Any synergetic interpretation of flow-induced phenomena starts from the elementary fact that transfer of free energy in a macroscopically moving fluid is inseparably linked to the induction of incoherent movements on the microscopic scale [5,45,82,101]. The conventional term "viscosity" reflects the gross mechanical consequences of microscopic events which Newton called "attrition" (in the sense of rubbing or wearing down) and which are termed "irreversible transfer of momentum" today. Molecules in liquids combine their ability to change places with the requirement to do mechanical work to achieve this: in combination with features such as incompressibility and inextensibility (and thence their ability to support compressive and tensile forces), both coherent and incoherent movements can be induced in fluids by external forces. Reif [82] uses an image for momentum transfer that lends itself to expression in terms of Kelvin's ships: assume that workers standing on the decks of moving ships loaded with sand bags throw these from ship to ship: as this occurs, the slower ones are accelerated by the kinetic energy of a faster sand bag, the faster ones lose this amount of kinetic energy and are slowed down. This now brings us to a constructive role of viscosity: the momentum transfer has the universal effect that existing velocity differences between microscopic elements, mesoscopic fluid laminae, and macroscopic fluid elements are automatically reduced. One can also comprehend that

momentum transfer and *viscous drag* which pulls on the surface of fluid laminae are endogeneously identical.

88.2.3 Coarse-Grained Description of Superimposed Thermal and Shear Movements

As a fundamental assumption of statistical mechanics, we accept that the individual molecules, for example, H_2O molecules, are thermally driven into something like a *random walk* by which they meander through the continuum of a cohesive liquid. If the latter is subjected to macroscopic flow, the mesoscopic ensemble, in which the individual molecule is pursuing its tortuous pathway, is being continuously deformed by a shearing movement. Thence, the resulting total displacement of the individual molecule can be taken as the superposition of two movements, namely, small-scale steps representing the *self-pursued and random meandering walk* and large-scale displacement induced by "enforced shearing".

For the sake of more vividly describing a pivotal thermodynamic process, hitherto dealt with only in an abstract fashion, it is appropriate to illustrate the result (Figs. 88.3, 88.4) in a highly simplified or coarse-grained fashion and represent them as a spiralling trajectory (detailed in [101]).

The details in the operation of a third form of energy, namely, different amounts of cohesive energies between molecules is neglected here. It is only assumed that the combination of mobility and cohesion allows velocity differences (or slip) to occur, but tends to keep them as small as possible: external stresses produce irreversible strain,

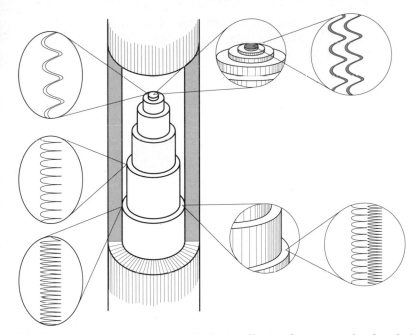

Fig. 88.4. Schematic representation of the physics of laminar flow in a cylindrical tube assumed to be perfused from bottom to top. A fluid element in the tube which is cylindrical at rest is deformed by pressure gradients into a hollow paraboioid. This is due to mutual displacement of the cylindrical fluid lamellae. The inserts on the left side depict the spiralling trajectories of an individual particle in the marginal layers (*lower insert*), in the intermediate and in the axial layers (*upper insert*). The *inserts* on the *right side* demon-strate that the velocity gradient or strain rate is high in the marginal fluid layers depicted in *dark blue* and very low in the axial layers depicted in *light blue* owing to the parabolic velocity profile. Therefore, the shear stresses acting in the slow marginal layers are high, and the shear stresses acting in the fast axial layers are low and approach zero in those layers moving very close to the tube axis

but keep the rate of strain at a minimum. We can thus make the further assumption that the diameter of the resultant spiral depends on the incident temperature, i.e., represents the local action of the thermal energy, and the gain of each spiral depends on the incident velocity and thence depicts the magnitude of the kinetic energy driving the macroscopic flow. These considerations also explain what happens in Poiseuille flow (Fig. 88.5).

In Fig. 88.4 representing fully developed tube flow, two cases are considered: (a) where the forward movement of two individual neighboring or a large ensemble of molecules all have the same velocity, i.e., perfectly coordinated or *holocoherent movement,* and (b) the neighboring molecules travel with a finite velocity difference. In the former case, prevailing near the vessel axis the forward movement does not lead to any appreciable additional interactions in excess of those taking place anyhow due to *thermal agitation* associated with frequent collisions. In layman's terms, when mobile elements of a cohesive liquid move in a coordinated fashion with small gradients of velocity, individual molecules travel forward but are not encumbered by collisions with their neighbors. It should be noted that this happens to the microscopic elements carried in the most rapid trajectories near the tube axis in Poiseuille flow (Fig. 88.5).

In contrast, fluids moving on slow trajectories are subjected strongly to enforced shearing. As molecules are dragged by each other, the collision-induced *transfer of momentum* takes place in a most pronounced fashion. Enforced shearing means that faster forward movement produces trajectories that spiral with high gain, compared with slower trajectories with low gain. Consequently, each time a fast molecule passes a slower one, they exchange momentum. As a result the faster is slowed down (decelerated) and the slower one is accelerated; the kinetic energy of faster moving ensembles of molecules is in part transferred to slower ones, and in part it is dissipated. The free (freely convertible) kinetic energy of large, coherently moving of fluid volume elements is thereby converted to heat, i.e., the prevailing microscopic kinetic energy of individual molecules in a cohesive liquid is enhaned. As the second consequence of microscopic momentum transfer, macroscopic movements are not only attenuated or inhibited. but also smoothed out and made less jerky.

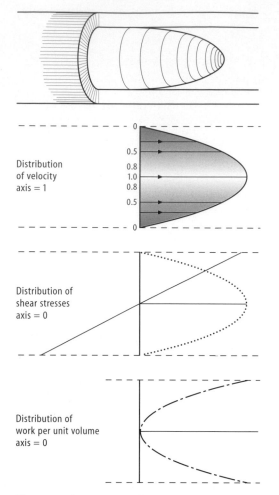

Fig. 88.5. Schematic representation of the distribution of the velocities, the shear stresses and the frictional energy dissipation in ideal Poiseuille flow. Note that there exist a gradient of frictional work inverse to the gradient of forward displacement: axial parts of the flowing medium are moving in a kind of energy furrow. The absolute values for the velocities, the shear rates, the shear stresses and then the dissipational work done by momentum transfer are determined by the pressure gradient acting over a tube segment, the segment diameter and the composition of the perfusate. In vivo, these values vary within wide ranges. This fact is the prime reason for the many types of flow behaviors seen in moving blood. In all cases, however, the fundamental facts depicted here are operational. Sites of maximum forward velocity are identical with sites of minimum entropy production

Momentum transfer and thence energy dissipation rates are markedly enhanced it in aqueous solutions not only water molecules, but also dissolved ions, macromolecules, and particles are being sheared during blood flow (see below). However, mammalian, nonnucleated blood cells are unique in their ability to minimize additional entropy generations associated with blood flow in all microvessels including the narrowest capillaries. When they pass through the nutritive microvessels, entirely different, namely, rotational movements of RBC membranes are induced by local gradients of momentum transfer. The membranes move like the treads of a tank around the liquid cytosol.

88.2.4 Newton's Law of Viscosity and the Dynamics of Flow

Newton's law of viscosity allows a formal description of the kinematics, the energetics, and the thermodynamics of enforced shearing, relating illustrated microscopic and mesoscopic processes in a simple law describing macroscopic events:

$$\dot{\gamma} = \tau/\eta \quad \left[N \cdot m^{-2} \cdot N^{-1} \cdot s^{-1} \cdot m^{2}, s^{-1} \right] \tag{88.1}$$

By simply stating that the macroscopic rate of shear $\dot{\gamma}$ (defined as the ratio of the velocity difference $du = U1 - U2$ in m/s) and the distance (dr) of any two macroscopic fluid laminae is equal to the ratio of the force (shear stress, τ) inducing enforced shearing, measurable causes are linked to measurable consequences. The shear stress (a force N) dragging on the surface (area in m²) of a sheared fluid volume element is proportional to the induced rate of shear rate, and the two are correlated by η, the coefficient of viscosity (dimension Ns/m^2).

The situation described here represents and equilibrium of a macroscopic mechanical work (endowment of a volume element with kinetic energy ($E_{kin} = 1/2\,m\,v^2$, Nm) and induction of incoherent movements or entropy generation rate defined by the product of the shear rate (s^{-1}), the shear stress (Nm^{-2}), the time of shearing (s) and volume (m³) of the element. The latter is equal to the time integrated entropy production ($\Delta S = dS/dt \times t$). For the case of stationary movement (dynamic steady state, no acceleration, no deceleration), this can be formulated as follows:

$$\Delta E_{kin} \triangleq \Delta S = \frac{dS}{dt} \cdot t = \dot{\gamma} \cdot \tau \cdot V \cdot t \quad \left[Nm \right] \tag{88.2}$$

Since τ was defined as the product of the shear rate ($\dot{\gamma}$) and the coefficient of viscosity (η), one can also write:

$$\Delta S = \frac{dS}{dt} \cdot t = \dot{\gamma}^2 \cdot \eta \quad \left[Nm \right] \tag{88.3}$$

and thereby one can define the coefficient of viscosity as the entropy production in a given volume of a sheared fluids in unit time for any given $\dot{\gamma}^2$

$$\eta = \frac{\frac{dS}{dt} \cdot t}{\dot{\gamma}} \quad \left[Ns \cdot m^{-2} \right] \tag{88.4}$$

If for any given global deformation of a composite fluid (e.g., a dispersion, an emulsion or a mixture of blood cells and plasma) the local rates of shear can somehow be reduced, this is bound to have a marked effect on the entropy generation (and thence the apparent coefficient of viscosity).

For many situations where the mobility of elements is altered in response to the hydrodynamic forces acting upon them, this can be more easily intuited as one utilizes the coefficient of fluidity (simply the reciprocal value of the value of the coefficient of viscosity), a measure for the ease of deducing macroscopic shear for a given amount of entropy produced in a moving fluid.

$$\phi = \frac{1}{\eta} = \frac{\dot{\gamma}^2}{\Delta S} \quad \left[\frac{m^2}{N \cdot s}, \frac{1}{Pa \cdot s} \right] \tag{88.5}$$

Newton's law then reads

$$\phi = \tau \cdot \dot{\gamma} \quad \left[\frac{m^2}{N \cdot s}, \frac{1}{Pa \cdot s} \right] \tag{88.6}$$

and $\Delta S = \gamma^2/\phi$ predicting the obvious, namely, that entropy production for any given shear rate squared is inversely proportional to the coefficient of fluidity. Dynamic fluidalization of simple and/or compound liquids will therefore be associated with a drop in entropy production.

88.2.5 Poiseuille's Law, Kinetics and Energetics of Enforces Shearing

Poiseuille's law follows directly from Newton's law of viscosity, where in a synergetic context it is preferable to relate volumetric flow rate $\dot{Q}$ to a pressure gradient (i.e., the pressure drop per unit conduit length) which is responsible for the shearing process:

$$\dot{Q} = \frac{\Delta P}{L} \cdot \frac{1}{\eta} \cdot \frac{\pi r^4}{8} \quad \left[\frac{m^3}{s} \right] \tag{88.7}$$

$$\dot{Q} = \frac{\Delta P}{L} \cdot \phi \cdot \frac{\pi r^4}{8} \quad \left[\frac{m^3}{s} \right] \tag{88.8}$$

In keeping with the dynamics of all transport equations (Ohm's law), the ratio of $\Delta P/L$ and $\dot{Q}$ can be defined as a *resistance*, the inverse of the two as a *conductance* (G):

$$R = \frac{\Delta P}{L} \cdot \frac{1}{\dot{Q}} \quad \left[\frac{Ns}{m^5} \right] = \frac{\Delta P}{L \cdot \dot{Q}} \left[\frac{Ns}{m^5} \right] \tag{88.9}$$

$$G = \frac{1}{R} = \dot{Q} \cdot L/\Delta P \quad \left[\frac{m^5}{N \cdot s} \right] \tag{88.10}$$

For a number of reasons, one should be very careful to proceed to the definition of resistance and/or conductance on the basis of the coefficients of viscosity as they are found in Eq. 88.5. In the case of blood, the coefficient of viscosity is well known to depend on the tube diameter and on the incident driving pressure. Since both Newton's and Poiseuille's law are valid near to true hydrodynamic equi-

librium (kinetic energy is fully converted to heat), they should not be applied if

- The blood flowing out of a conduit is endowed with very high kinetic energy and
- If other than hydrodynamic forces (e.g., external compressive forces as in working skeletal muscle and myocardium. see below) propagate the blood.

The present energetic considerations are particularly important in a pivotal hemorheological context, namely, the perfusion of skeletal muscle and myocardium during physical exercise. In this situation, which is characterized by the highest requirements for oxygen delivery (and thence RBC flux), the blood is not just driven by the arteriovenous pressure difference and thence by the shear stresses generated by the heart as a pump, but also by the pressure generated in the contractile mechanism of the parenchymal cells. This leads to a very high interstitial pressure and thence to a marked compression known to strongly accelerate the blood (in an orthograde direction in the veins, in a retrograde direction in the arteries). Consequently, a very pronounced flow pulsatility – presumably with very high shear stresses – is initiated in the relaxation phase.

This is a field of cardiovascular physiology and experimental hemorheology that will have to be investigated in much more detail in the future. The few details known about effects on flow pulsatility [32,57,68], intermittent lowering of venous pressure, and elevation of the arteriovenous pressure gradient are insufficient to allow a detailed energetic balance. At any rate, the system is too complicated to allow any justified analysis of impeding the role of blood viscosity since both the propagating forces and the induced fluid shear rates are unknown.

Global Energetics of Tube and Network Flow. Based on the logic delineated in Eqs. 88.2 and 88.3 the work done in tube (and network) flow (in short the global entropy production associated with Poiseuille flow in a tube of unit length. L) can be calculated from the product of the volumetric flow rate ($\dot{Q}$), pressure difference (ΔP), and time of flow:

$$\Delta S_{tot}/L = dS_{tot}/dt \times t = \dot{Q} \times \Delta P \times t \quad [m] \tag{88.11}$$

or $\quad \Delta S_{tot}/L = dS_{tot}/dt \times t = \dot{Q}^2 \times \eta \quad [Nm] \tag{88.12}$

or $\quad \Delta S_{tot}/L = dS_{tot}/dt \times t = \dot{Q}^2/\phi \quad [Nm] \tag{88.13}$

This formalism holds for individual tubes as well as for tube networks, where total $\dot{Q}$ is simply added up from the individual segment contributions to it. Obviously, if in a network the self-organized fluidizing mechanisms increase the coefficient of apparent fluidity of driven blood, this is a situation where drive-induced coordination, e.g., that found in rectified flow (Sects. 88.3, 88.4) reduces overall entropy production rates and thence ΔS_{tot}. Here, Prigogine's theorem of mimimum entropy generation rates explains the sponteneous emergency of self-organized collective coordination of movements. According to Klimontovic (see Chap. 2), the fall of excess entropy generation can indeed be accepted as an indicator of successful self-organization.

88.2.6 Parabolic Velocity Profile and the Theorem of Least Effort

The regularities found in all driven but constrained systems with their abundance of free energy transfer has taught us to see entropy production (see Chap. 2, and [41,46]) as a necessary but well-invested energetic expenditure generating ordered movements of fluid components. The simplest example for this mode of operation is the remarkable order generated in movement of a Newtonian liquid through a cylindrical tube (Poiseuille flow). As first recognized by Helmholtz, in strictly laminar flow, moving elements become coordinated by a simple principle of minimum entropy generation, or of least effort. W.R. Hess (Nobel prize winner 1949) presented the first comprehensive energetic analysis of the same principle as early as 1921 [48], but was not understood in physiology and biophysics, let alone by bioengineers who later analyzed blood circulation. At the mesoscopic level, the entropy generation rates that occur when cylindrical lamellae glide past each other are unevenly distributed, and an ordering effect of this type of response was first considered by Taylor in 1922 [113] for a more complex hydrodynamic problem, namely, the displacement of particles. Each convective movement of liquids and gaseous elements spontaneously tends to find the *Ablauf* mode (see Chap. 2) associated with the least possible irreversible transfer of momentum. The latter invariably occurs when more rapidly displaced molecules, particles, or surfaces encumber slower ones, and vice versa: for any given viscosity the local entropy generation rate in a given volume element is proportional to the square of the local velocity gradients induced (see Eq. 88.3 and Eq. 88.5).

The details are far from being clarified, but the principle can be idealized as follows: in Poiseuille flow particle velocities increase in a strictly parabolic fashion from the vessel wall to the vessel axis (Fig. 88.5). Conversely, the velocity gradients (or rates of shear) decrease in a linear fashion from the wall where they are maximum to the axis where they are zero. The most peripheral lamellae firmly adhere to the stationary solid tube wall; the pressure gradient or pressure difference per unit length of a cylindrical laminae ($\Delta P/\Delta L = N/m^3 = Pa/m$) produces local shear stresses proportional to the distance of a laminae from the tube wall. For each sheared mesoscopic volume element, both the local velocity gradient of shear rate (m × s^{-1} × $m^{-1} = s^{-1}$) and the local shear stress (Nm^{-2}) increase from the wall to the tube axis. This reflects the well-known parabolic velocity distribution. In the region near the tube axis, the shear rates and thence the entropy generation rates are

obviously close to *zero*, no matter how fast the axial particles move. In this context, it must be reiterated that the entropy production rate for any given viscosity, volume, and time is a function of $\dot{\gamma}^2$: therefore, not only the individual fluid elements traveling near the axis, but also larger ensembles of them are energetically favored over those traveling near the tube wall.

It is this fundamental fact of microscopic fluid movement that allows us to understand the synergetics of blood flow at the mesoscopic level. A coordination of cellular movements occurs which is based on the following mode of operation: in those volume elements near the wall, which are traveling with lowest absolute velocities, the molecules displaced by way of enforced shearing are subject to the highest rates of momentum transfer. Since in the laminae near the wall, the resulting entropy generation rate is much higher than that found near the axis, suspended particles (solids, fluid droplets, and especially mammalian RBC) are likely to be displaced not just forward, but sideways. Obviously, these are preferentially dislocated away from the marginal, energetically demanding laminae into the ones that are energetically more favored. This leads to an appar-

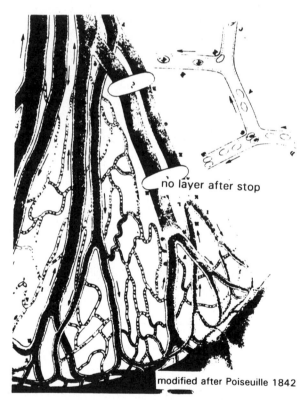

Fig. 88.6. Formation of the lubricating plasma layer as depicted by Poiseuille in a drawing he produced. In the normally perfused microvessels, he noticed the formation of a cell-free plasma layer near the wall. Furthermore, he noticed and depicted the very low red cell content of nutritive capillaries. Most importantly, however, he noticed that after stoppage of flow (induced by placing two platinum particles on the microvessels), the marginal layer disappear. With this observation, Poiseuille discovered the self-organized nature of blood flow in microvessels as a drive-dependent coordination in the kinematics of cell and plasma flows

ently paradox situation (detailed in [100] and [101]) (1) those laminae containing the elements traveling fastest also contain the particles endowed with the highest kinetic energy; (2) their movement, however, is associated with lowest entropy production rate; (3) this stabilizes the position of their trajectories because once particles have been displaced into the axial region, they are not likely to move back to the more marginal ones. The nonlinear behavior found in moving blood follows from the fact that its constituent elements are not just displaced, but are changed in their hemorheological behavior by the prevailing set of forces. The extreme situation is found near and along the vessel axis. Here, we find perfect self-focusing (see Chap. 2). Volume elements that move most rapidly but with almost identical velocity and therefore no velocity gradients between them, experience no net momentum transfer at all and are displaced rapidly almost without dissipating energy. Suspended particles move preferentially and become crowded here, simply because frictional work associated with their forward displacement shows a minimum in this energetically preferred region. This is exactly the situation found in living microcirculations as was discovered by the earliest intraviatal microscopists and described in great detail by J.P.M. Poiseulle [76]. As shown in Fig. 88.6, he not only described the peripheral plasma layer in great detail, but also performed experiments (by placing platinum weights on the microvessel to interrupt flow; see Fig. 88.6) that disclosed the formation of the marginal lubricating film as a flow-induced phenomenon (see [73] for details).

88.2.7 Moving Blood as a Driven System: Kinetic Energy, Pressure Energy, and Frictional Losses

It is not just the nonlinear shear rate–shear stress relationship of moving blood, but also the non-equilibrium states of the blood moving in the different vessels that must be appreciated in a synergetic context. In order to comprehend correctly the effect of the shear-induced fluidalization of blood in vivo, one must balance all forms of work. The kinematics and the thermodynamics of fluid shear and the derived concepts of viscous energy dissipation have been discussed for laminar flow at the equilibrium between energy import and energy dissipation. In the living vascular system, true equilibrium is *not* achieved in any vascular segment. On the contrary, since the blood moving in the macroscopic arterial vessel – and to a certain extent also in the macroscopic venous vessels – is endowed with two forms of energy, pressure stored in the tensilized wall structures and the kinetic energy of moving blood. Both are *not* fully dissipated in flow, but are rather stored in part and/or exported into the downstream vascular segments. In the microscopic vessels, much larger fractions of the energy content are dissipated. However, as long as the whole mass of circulating blood is kept driven, i.e., while it continuously moves through the vascular network, it contains high

amounts of free energy, which are not locally dissipated. The system is thus kept in a boosted steady state [98,101]. Obviously, the local dissipation of energy in a tube (or a vascular segment) must be considered. It is equal to the product of pressure difference ΔP (in N/m^2), of the volumetric flow rate ($\dot{Q}'$, in m^3/s) and of time t (in s)

$$\Delta S - dS/dt \times t = \Delta P \times \dot{Q} \times t \ (Nm) \qquad (88.\ 14)$$

Since pressure difference equals the product of volumetric flow rate and resistance (and thence the viscosity for any given geometry of a tube), one can define energy dissipation (or entropy generation) also by the product of the square of the volumetric flow rate and (v.i.) for any given hindrance the coefficient of viscosity:

$$\Delta S = \dot{Q}^2 \cdot \eta \cdot H \qquad (88.\ 15)$$

where H is the ratio of lenght and the fourth power of the radius ("hindrance").

In balancing the benefit (transfer of masses) and the cost (dissipation of energy) for each type of vessel, one has to consider the energy imported at the entry of a segment, the energy dissipated in the process of shear deformation of the blood, and the energy exported to the downstream segment and all subsequent ones up to the right atrium. In actual fact, this is often impossible: contribution of kinetic energy should, however, never be neglected, no matter how small it is, since its contribution to the stability of the global cardiovascular blood movement is of prime significance (a topic detailed in [101]).

One can estimate, therefore, that in large arteries (5 mm diameter), only a small fraction of the total energy (kinetic energy, energy stored in the tensilized wall structures) is dissipated, a minor fraction (roughly 1%–2%) is dissipated by the viscous flow through them. As the blood moves through the smallest blood vessels (5 μm in diameter), a much larger proportion of the imported energy is dissipated, but again not 100%: even here, a nonequilibrium state or a *boosted steady state* is maintained as long as the blood is in motion and the vessels are pressurized in the physiological fashion. Note, however, that (1) in the arterial system, even under the most critical circulatory conditions, the energy content is much higher than in the venous one and that (2) in the microscopic vessels, the viscosity is much more variable than in the macroscopic ones, since it can vary from that of almost pure plasma to that of almost pure RBC suspension. Note further that in all vascular segments, the fraction of energy dissipated is lowered progressively when the total energy transfer is high, and vice versa.

88.2.8 Hagen-Poiseuille's Law and Its Limited Validity in the Whole Circulatory System

It follows from all these considerations that, for moving blood, the simple form of Newton's law of viscosity or the Hagen-Poiseuille law can be applied as a crude approximation. One must take into account that it is a priori valid only under the following conditions:

- Stationary flow (no acceleration of deceleration), i.e., constant pressure gradients ($\Delta P/L$);
- Nondistensible nor collapsible tubes;
- No superimposed movements (a condition conventionally paraphrased by the absence of turbulence); and
- Constancy of the coefficient of viscosity.

Note, however, that in the many microscopic vessels, where the local pressure gradients are very high, under normokinetic conditions (Sect. 88.6.3) the value of the apparent fluidity is most probably considerably higher than that found in macroscopic blood vessels (arteries and veins). As will be shown in Sect. 88.3.3, the coefficients of viscosity of blood subjected to shear in microscopic blood vessels can at present only be estimated from complex fluid-dynamic extrapolations [79]. The numerical values of apparent blood viscosity in glass tubes are most probably far lower than reflect the true amount of viscous work done. Additional factors such as the effects of deformation of cells at bifurcations, cell–cell and cell–wall collisions, effects of noncylindrical geometry, and perhaps even fluid immobilization in the glycocalix of endothelial cells may contribute to entropy generation for any given flow rate (see Fig. 88.10).

In the exchange capillaries which are characterized by the highest pressure gradients, wall shear stresses, and mean shear rates and thus very high fractional entropy generation rates, one is equally safe in assuming that the coefficient of plasma viscosity is of great significance, while the value of the macroscopic blood viscosity has no relevance at all. More importantly, in this type of vessel, the HCT values show a very wide scatter. This is illustrated for the case of the rat mesentery by data collected by Pries et al. [77]. In Fig. 88.7, their data are displayed for blood vessels

with diameters between 40 and 8 mm diameter. For arterioles (ART), capillaries (A-V channels) and post-capillary venules (VEN), the fractional tube HCT (local RBC volume fraction divided by systemic RBC volume fraction) are less than 0.5, most values ranging between 0.25 and 0.75. Obviously, therefore, the influence of the HCT on energy dissipation under high flow condition is very weak.

88.2.9 Blood Viscometry Is a Deceptive Exercise: Neither Viscous Contribution to Resistance Nor a Correct Coefficient of Blood Viscosity Can Be Determined

As we use Hagen-Poiseuille's equation under the above-defined conditions, we immediately become aware that no single value for η or ϕ can be given for blood, simply because both are nonlinearly related to pressure gradient $\Delta P/L$, diameter, and even previous history. Keeping the described limitations in mind, one becomes aware of the pitfalls of applying a *coefficient of viscosity* within the logic of Ohm's law, i.e., that of linear transport equations. One can, of course, measure the global resistance (R), i.e., an expression for work done in transfer of momentum into heat for a segment of given length: it follows from the ratio of driving pressure ($\Delta P/L$) and volumetric flow rate ($\dot{Q}$). Note that when the resistance is calculated as a function of the pressure gradient, i.e., pressure drop per unit segment length, hydraulic resistance is highest in the smallest narrowest vascular segments, i.e., in vivo in the capillaries (see Table 88.1) rather than in arterioles. Table 88.1 shows that the differences of the resistances and the pressure gradients between macroscopic and microscopic blood vessels are very pronounced. As will be shown below (Sect. 88.4.2) the steep pressure gradients in the microvessels are associated with steep gradients of shear stresses between the wall

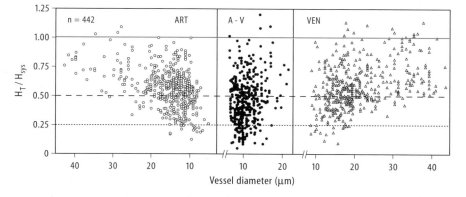

Fig. 88.7. Natural variation of red cell content in the microcirculation when perfused in the normokinetic mode (see below): normalized data (microvessel hematocrit divided by systemic hematocrit) in mesenteric arterioles (*ART*) in capillaries called the arteriovenous channels (*A-V*), and in postcapillary venules (*VEN*), plotted as the function of vessel diameter, *Upper panels*, marked decrease of data points with very low values (but large data scatter) in the smallest arterioles, capillaries, and

venules and increase with increasing diameter in the venules. *Solid blue line*, microvessel equals large vessel; this is only found in a very small fraction of microvessels. Hematocrit equals one half large vessel hematocrit; *dotted line*, microvessel hot equals one quarter large vessel hct. *Light blue range* in the vast majority, the fractional red cell content is between 75% and 25% of that found in macrovessles. This is a result of dynamic self-dilution. Data modified after [77]

Table 88.1. Angioarchitectonics, geometry, and hydraulics of an arbitrary vascular bed (data from [92])

Type of vessel	Diameter (mm)	Number	Hindrance $l/n7d^4$ (mm^{-3})	Intravascular pressure (mmHg)	Wall shear stress τ_w (Pa)
Aorta	10	1	4×10^{-2}	100	2.5
Large arteries	3	40	6.2×10^{-2}	97	2.4
Main artery branches	1	600	1.7×10^{-3}	92.7	4.3
Terminal branches	0.6	1.800	4.3×10^{-2}	79.8	6.6
Small arteries	0.019	4×10^7	6.6×10^{-1}	76.5	3.8
Arterioles	0.007	4×10^8	9.9×10^{-1}	55.6	7.5
Capillaries	0.0037	1.8×10^9	6.8×10^{-1}	25.1	11.1
Postcapillary venules	0.0073	5.8×10^9	1.2×10^{-2}	4.5	0.5
Venules	0.021	1.2×10^9	4.4×10^{-3}	4.1	0.2
Small veins	0.037	8×10^7	2.3×10^{-2}	3.8	0.6
Main venous branches	2.4	600	5.0×10^{-3}	2.1	0.3
Large veins	6.0	40	4.9×10^{-3}	1.7	0.4
Vena cava	12.5	1	1.7×10^{-3}	1.3	1.3

region and the vessel axis. This is important for all self-organized fluidizing mechanisms since they induce *axial drift* and RBC overvelocity, and they control synergetic coordination of flow patterns in segments far displaced from hydrodynamic equilibrium. Note also that the high resistances in the ensemble of microvessels, which are largely determined by the geometrical configuration and the topology of macro- and microvascular networks, are responsible for the storage of energy in the arterial macrovessels. When blood negotiates the microvessels, which we classify somewhat arbitrarily as the ones with diameters below 300 µm, shearing does not take place in blood, but in a self-focussing mixture of blood cells moving in a lubricating sleeve of plasma. Unfortunately, it is very complicated to calculate the exact amounts of the energy dissipated and of kinetic energy transferred by blood moving through a network of tubes. Due to the considerable complexity of calculating the trajectories of blood components moving in vascular networks, i.e., the velocity gradients, we are still largely in need of detailed procedures of calculation regarding in vivo effects of kinetic energy or inertial effects.

For all the present ignorance about the necessary correction for kinetic effects (termed Hagenbach correction in experimental physics), we must accept the fact that due to this form of energy, which is stored in moving blood, the normally perfused vascular bed operates in a dynamic state far displaced from hydrodynamic equilibrium. One can say, that the entire cardiovascular system is permanently driven into a *boosted steady state* (see Chaps. 2 and 94 for a more detailed elaboration) where:

* The time-averaged import rate and dissipation rate of free energy is kept stationary;
* The mechanical energy generated by the two halves of the heart is equal to that dissipated (mostly in microvessels) by momentum transfer in the moving blood; and
* Not only potential energy (conformational energy of wall structures put under tension by the transmural pressure acting on them), but also kinetic energy is

stored in the moving blood. These two energies are a kind of reservoir or a kind of mechanical buffer that *stabilizes* the operation of this important cyclically operating system and keeps it far displaced from its hydrodynamic equilibrium. The maintenance of this boosted state is far more important for the normal cardiovascular function than composition and viscosity of the moving blood, a self-fluidizing ensemble of highly flexible cells.

88.3 Movement of Particles in Cohesive Liquids Subjected to Enforced Shear

88.3.1 Viscous Drag Induces Particle Rotation

In order to understand the in vivo flow behavior of all types of blood cells (see Sect. 88.3.4), one must appreciate that, as particles are carried within a liquid and are subjected to shear stresses [43,52,58,91,92,113], they are responding to the sequelae of momentum transfer by falling in a mesoscopic rotational movement. The reason is straightforward (Fig. 88.8): those layers of continuous phase which adhere to the particle surface transfer momentum to the unbound yet elastic particle. Falling in rotation minimizes the velocity difference between the liquid and particle surface, as follows from the principle of minimum entropy generation. Rotational movement occurs due to a viscous drag exerted upon the surface (see Fig. 88.8): the "drag force": N [shear stress (N/m^2) multiplied by the surface area (m^2) of the particle] induces movements perpendicular to the forward motion (see Sect. 88.3.4).

Dispersions of solids in liquids (*suspensions*) and of liquids in liquids (*emulsions*) are more viscous than the continuous phase. The presence of particulate matter, which does not partake in shearing, requires an enhancement of momentum transfer in the continuous liquid to obtain the same rate of shear. Conversely, for a given shear stress, the rate of shear is reduced so that the numerical ratio of these two entities is changed, as is reflected in a

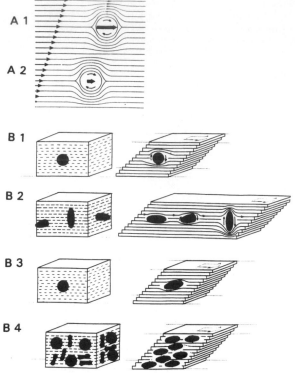

Fig. 88.8. Dissipative structuring of particle displacement, particle deformation and particle orientation during enforced shearing. As the basic mechanism, any solid particle (or any droplet of a fluid immiscible with the continuous phase) suspended in a shearing fluid is subjected to rotation (*A1*) caused by the interaction of local viscous drag (fluid adhesion onto particle surface allows momentum transfer onto continuum of dispersed phase) and global sequelae of crowding, collisions and augmented shear (*A2*) in the continuous phase. If the averaged rate of shear is kept enforced, the *local* rates of shear in the regions between particles are enhanced: there is the same velocity difference, but a reduced thickness of fluid lamellae. Solid spheres are simply driven into continuous rotation (*B1*); irregularly shaped particles (e.g., solid ellipsoids) are driven into irregular movements (*B2*). In contrast, fluid droplets (*B3*) (which are also spherical at rest) are progressively deformed into prolate ellipsoids as shear rates are elevated and their major axis is oriented more or less parallel to the direction of flow. In addition, the deformed fluid droplets respond to momentum transfer onto their surface (interfacial area) by a rotational movement of the interface around the suspended fluid. This fluidal response – and not primarily the ellipsoid shape – minimizes the local and global dynamic sequelae of particle movements during enforced shearing. Crowding and collisions are markedly reduced and can be totally abolished, entropy generation in the continuous phase is minimized. An identical obligatory combination of ellipsoidal deformation, stationary orientation of mammalian RBC in shear is shown in *B4*

higher coefficient of viscosity or a lower coefficient of fluidity.

As first defined mathematically by Einstein [27], mixtures of molecules experience an increase in the coefficient of viscosity, where η suspension is equal to the × continuous phase (η_0) continuous phase plus a viscosity increment, which in the final analysis is caused by an excess transfer of momentum. At very low particle concentrations, this vis-

cosity increment is equal to the product of the fractional volume of the particles ($H = V_{particles}/V_{continuous\ phase}$), a shape factor (e.g., 5/2 for spheres) and the continuous phase viscosity (η_0).

$$\eta_{susp} = \eta_0 + \eta_0 \cdot H \cdot K \qquad (88.\ 16)$$

This equation can be rewritten to describe *relative viscosity* (dimensionless):

$$\eta_{el} = \eta_{susp}/\eta_0 = 1 + HK \qquad (88.\ 17)$$

In terms of statistical mechanics, this formalism illustrates that, even in very dilute suspensions, the excess rate of entropy generation associated with the movement of any type of particles suspended in a moving liquid is elevated in a highly nonlinear fashion as H is elevated. The global entropy generation is for any given γ^2 and conduit geometry is closely related value to shape factor K which has different numerical values that depend in a complicated fashion on the actual configuration, the stability of orientation and ratio of particle size to tube diameter [92,113]. It is intuitively clear that when particles are very crowded in a moving liquid, additional momentum transfer between colliding particles takes place. In mathematical terms, this is represented in complex correction factors for Eq. 88.17: interestingly, none of the emperically determined correction terms is practically relevant for blood rheology. This notwithstanding, one can use the logic delineated in Sect. 88.2.2 and Eq. 88.16 to put in vitro measurements and in vivo measurements into a common perspective by measuring either the relative apparent viscosity (in rheometers) or the relative flow rate (flow rate of a suspension divided by the – measured or extrapolated – flow rate of the continuous phase). We can then calculate an entity (normalized entropy generation rate for any given volume and γ^2) which allows us to assess the *disturbing effect* of the dispersed phase on the flow of the continuous phase. The coefficients of apparent viscosity measured in all types of very concentrated suspensions of solid particles and to a lesser degree in emulsions (dispersion of liquid droplets in an immiscible fluid) is bound to increase markedly, as do entropy generation rates in shear for any given γ^2 in a segment or network of any given geometry and thence hindrance.

The same phenomenon occurs in sheared suspensions of mammalian RBC, but to a surprisingly small extent. When subjected to the physiologically high shear stresses encountered in vivo, the highly flexible RBC behave akin to liquid droplets, and therefore blood undergoes a kind of phase transition from a quasi-elastic, concentrated suspension into a highly fluid emulsion [110]. Note that at low rates of shear, the same blood cells behave more like quasi-elastic solids. This solid–liquid transition causes a very marked deviation from the predictions of Einstein's law. All non-Newtonian characteristics can be explained on the basis of shear-induced variations in the viscidizing effects of the dispersed phase, an effect that can be formally reduced to dynamic alterations in the parameters H and K.

Such ideas were first expressed by G.I. Taylor [113] as early as 1922 who related particle shape to energy dissipation. They were first applied by Frimberger [33] to human blood after he discovered in 1937 the shear-induced RBC deformation into ellipsoids.

The physical reason for the type of adaptation is straightforward, a streamlined configuration is induced. The effect, however, exerts a very pronounced influence on a property that might be called the "mobility" of the mammalian RBC, since the hydrodynamic forces not only displace the suspended flexible particle but make them more stream-lined. As will be shown later. There is a second effect, namely, a shear-induced drift of deformed and oriented RBC. Both effects combine in strongly enhancing the fluidity of rapidly moving blood (as depicted in Fig. 88.1). It thereby exhibits a behavioral pecularity that is restricted to non-nucleated RBC [37,93,102] and which only occurs when these are subjected to the very high shear stresses prevailing in normally perfused microvessels.

88.3.2 Rectified Flow of Mammalian RBC and Blood Fluidity in Macroscopic Conduits

In the *macroscopic* blood vessels, the influence of the highly crowded but very flexible RBC on entropy generation during flow already proves to be remarkably lower than predicted by Einstein's theory. The surprising extent of this can be studied by comparing the effect of hardened RBC and normally fluid RBC on the apparent fluidity of RBC suspensions in artificial continuous phases [32] (Fig. 88.9). The solid curve in Fig. 88.9 shows the measured data for nondeformable particles (e.g., chemically fixed human RBC). As the volume fraction rises above 0.4 (HCT 40%) the coefficient of viscosity increases dramatically to values near infinity at a volume fraction of 0.66. Under such conditions, RBC are so tightly packed that each cell is in immediate physical contact with its neighbor. Unless endowed with the property of fluidity, the RBC can no longer be dragged past each other and therefore impede shearing almost completely. In a crowded suspension of solid particles fluidity falls to zero, and the imported energy is stored in elastic interactions. This explains why suspensions of human RBC treated with glutaraldehyde at 60% volume fraction have the consistency of normal toothpaste, and at 66%, that of dried-up tooth paste.

Normally flexible and thus fluid RBC, in contrast, no matter how tightly packed, can glide past each other and allow mesoscopic shearing and macroscopic flow of packed RBC suspensions. This is seen in relative apparent viscosities or in the entropy generation rates in viscometric flow, i.e., shearing in a rotational viscometer. Deformable human RBC show much less increase in entropy production rates, as RBC volume fraction is progressively elevated. Even at a RBC volume fraction as high as 0.95, the viscosity and the entropy production are remarkably low, while the coefficient of viscosity is not much higher than that of salad oil (or about 60 times that of water). Note that the lubrication oil used in automobiles is at least ten times as viscous as

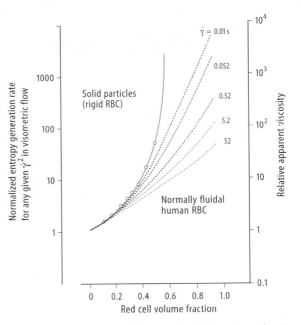

Fig. 88.9. Influence of red blood cell (*RBC*) volume fraction (hematocrit value v/v.), relative apparent viscosity, and normalized entropy generation rate for any γ^2 in viscometric flow. Comparison of rigidified RBC (*solid blue line*) to behavior of normally fluidal human RBC at various rates of shear. Independent of shear rate the relative apparent viscosity (plasma viscosity = 1.0) rises steeply as RBC volume fraction is inceased above 0.5. When tested at high shear rates (52 and 5.2/s, *light blue lines*), relative apparent viscosity rises to a much lesser extent even at the highest RBC volume fraction. Data taken at low shear rates (0.52 and 0.01/s, *dotted lines*) rise in a similar fashion with increasing RBC volume fraction, as is found in suspensions of rigid particles but to a much lesser extent than the latter when tightly packed (RBC volume fraction >0.5) (modified from Chien [17])

packed RBC are (at high shear stresses). Their unique behavior, demonstrable to any hemorheological layman who pours packed RBC out of a glass vial after centrifuging it to obtain plasma, already shows that the human RBC must in themselves behave much more like a fluid than like a solid: hence the term *shear induced fluidality* to designate their behavior.

In short, at all HCT levels and in both microscopic and macroscopic vessels, the apparent fluidity of rapidly driven blood is surprisingly high, a behavior caused by *RBC rectification* [83,90]. *Rectification* is defined as the sum of all responses which coordinate the displacements of individual cells and eliminate aberrant movements with collisions between cells and of cells with the wall.

This rectification can easily be explained by the unique structural features of nonnucleated and hence fluid RBC, which consist of a flaccid membrane bag that is incompletely filled with a concentrated, but very liquid solution of macromolecules (about 340 g/l hemoglobin and a variety of dissolved enzymes). It has been repeatedly stressed (e.g., in [93]) that this is a good example of a phenotypic biological progress which is not based on a change in genotype, but rather on a very special type of development (see Fig. 88.10). So-called mature RBC are in actual fact degenerated fragments of a normal, fully developed biological

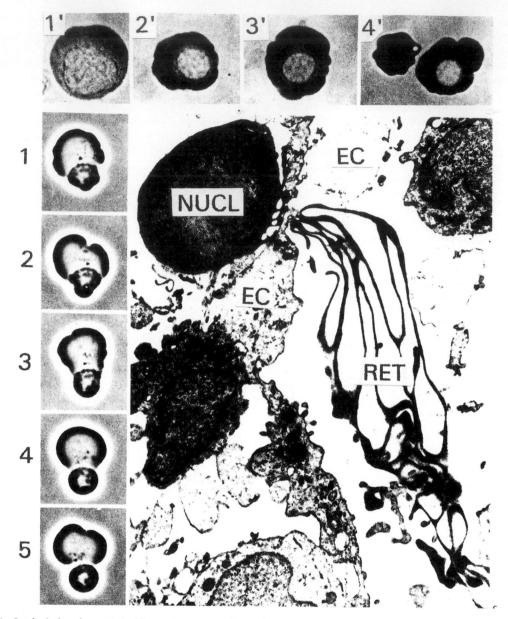

Fig. 88.10. Cytological phase transitions in mammalian erythropoesis: fluidal, non-nucleated RBC which are unique achievements of certain marsupials and all mammals including humans follows from the specific erythropoiesizs in these species, which are derived from conventional cells with solid-like properties. Soret absorption micrograph of developing normoblasts and ejection of the normoblast nucleus. Normoblast series from the basophilic normoblast (*1'*) through the orthochromatophilic normoblast (*3'*) and reticulocyte (*4'*) are observed in the living state with Soret absorption (414 nm), showing the increase in hemoglobin content as the normoblast develops from the basophilic normoblast to the reticulocyte (*left to right*). Portions of a phase contrast microcinematography film of the ejection of the nucleus by an orthrochromatic normoblast (*1–5*). Note the dynamic alterations in the shape of the cell, the undulating contractions and constrictions across the cell as nuclear ejection occurs. Electron micrograph depicts ejection of the normohlast nucleus (*Nucl*) and diapedesis of the reticulocyte (*RET*) through the slits in the endothelium (*EC*)

cell, namely the erythroblast [59]; however, this degeneration or rather structural simplification improves the functional fitness of mammals in a remarkable fashion. This is a late achievement in phylogeny (restricted to marsupials and mammals [93]) and has greatly facilitated blood flow especially in microvessels.

With respect to the cytological details of the unique overspecialization of erythroblasts, several explosive, mutually amplifying synthetic steps are found in the so called erythroid burst forming units. As detailed in [69], the last step (electrostatically mediated and hence irreversible association of α- and β-globin chains) generates a high-affinity binding site (heme pocket) and, by incorporating heme, produces the oxygen-binding hemochrome. Immediately after formation, this reduces the local pO_2, and low pO_2 in the perimitochondrial regions in turn activates the rate-limiting enzyme ALA-synthetase (see below) for the first step (heme synthesis). In binding free Fe^{3+} this activates the synthesis of transferrin receptors, while binding of heme in turn controls globin synthesis via an influence on transcription and translation of their respective genes. In this way a "set" of several sinks are established – two for Fe^{3+} (in the cytosole and in the mitochondria), one for heme (in the globin-tetramer), and one for O_2 (in the complete hemoglobin molecule). These generate the environment (low perimitochondrial pO_2) necessary to initiate intramitochondrial association of glycine with succinyl-CoA to δ amino-laevulic acid (δ-ALA). It is not the genes (or their conventional control systems), but rather the iron-containing endproduct that induce positive feed-back acceleration of this pivotal synthetic step. Via a degenerative cytological change (pyknosis of cell organelles including the chromatin), a very powerful functional advantage of mammalian (and marsupial) species emerges spontaneously, which is obviously essential for the overall efficiency of the cardiovascular system in these species (a topic recently emphasized by Diamond [24].

88.3.3 Synergetics of the Fåhraeus and Fåhraeus-Lindqvist Effect

The net effect of self-organized rectification of non-nucleated mammalian RBC is a marked decrease in global viscous energy dissipation associated with the perfusion of microscopic tubes, the extent of which, however, is still not established. In confirming their own theoretical prediction (Fåhraeus 1928 [28]), an important phenomenon of dissipative structuring in biofluid-dynamics of a multiphase fluid was discovered in 1931 by Fåhraeus and Lindqvist (29), namely, the pronounced and progressive fall in apparent blood viscosity with decreasing tube radius and with increasing shear stress (see Figs. 88.10, 88.11) in microscopic glass tubes. This remarkable physical

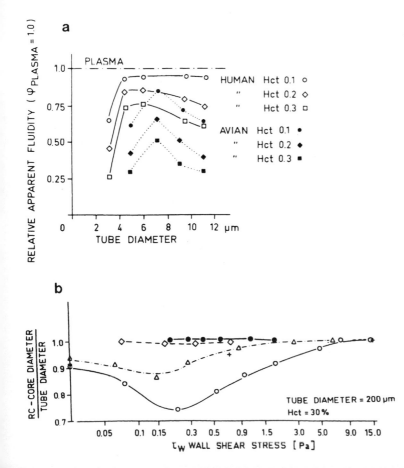

Fig. 88.11a,b. Species-specific fluidization of red blood cell (*RBC*) suspensions. **a** Data modified from Gaehtgens et al. Effect of RBC volume fraction in relative apparent fluidity in microscopic glass tubes of various diameters. Comparison of avian and human RBC. Note that high values of apparent fluidity are restricted to mammalian cells. **b** Influence of wall shear stress on the formation of lubrication of a cell free layer in microscopic tubes: comparison of human and avian RBC. In suspensions of non-nucleated human cells, the width of the cell free layer due to axial drift increases with increasing shear. This behavior is less pronounced at higher shear stresses, i.e., in flow situations associated with rapid forward flow and insufficient time for axial rearrangement of cell trajectories. No significant axial drift in avian RBC and in rigidified (heat-treated spherical) human RBC (*white circles*, human RBC in plasma (+RCA); *triangles*, human RBC in Rheomacrodex (–RCA); *black circles*, human spherocytic RBC in Rheomacrodex (–RCA, rigid); *rhombuses*, avian RBC (–RCA, rigid; RCA, red cell aggregation)

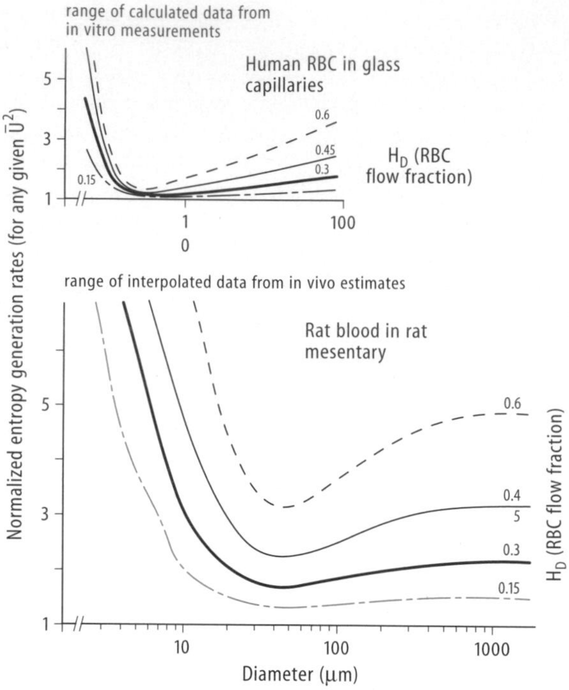

range of calculated data from in vitro measurements
Human RBC in glass capillaries
0.6
0.45
0.3
0.15
H_D (RBC flow fraction)
5
3
1
1
0
100
Normalized entropy generation rates (for any given $\bar{U}^2$)
range of interpolated data from in vivo estimates
Rat blood in rat mesentary
0.6
0.4
5
0.3
0.15
H_D (RBC flow fraction)
5
3
1
10
100
1000
Diameter (µm)

Fig. 88.12. Estimate of the influence of red blood cell (*RBC*) flow fraction (discharge hematocrit, flow rate of erythrocytes divided by sum of flow rate of RBC and plasma) on viscous energy dissipation in microvascular network with tube diameters between 4 and 1000 µm. Comparison of data obtained by calculation on the basis of in vitro data (upper diagramm) and those obtained by interpolation of pressure flow data in vivo. The upper presentation shows, the strong influence of the Fahraeus effect and the Fahraeus-Lindqvist effect (see Sect. 88.4.3). In artificial microvessels. Calculated normalized coefficients of apparent viscosity (here presented as normalized entropy production rates, entropy production for plasma flow alone set to 1.0). The range of hematocrit values relevant for the class of blood vessels is shown in blue. Note that for any μ^2 entropy generation rates in vessel segments between 4 and 40 µm are estimated to be substantially higher (1.5–4 times those in plasma flow) than the calculated data from glass tube experiments. In light of the great variability of H^D values in microvascular networks, this important presentation only gives the range of frictional work associated with blood flow in microvascular networks: in vivo viscosity is an entity that defies definition. (Data from [79])

behavior (which has been intensively studied since then) was a priori surprising. Counter to intuition under the extreme conditions of rapidly perfused glass capillaries with diameters less than those of the RBC at rest, the numerical value of the coefficient of relative apparent viscosity was found to be as low as 1.1 for RBC volume fractions (*H*) between 0.1 and 0.3.

The remarkable fall in viscous work has many reasons which all are epiphenomena of self-organized adaptations of nonnucleated RBC to the flow conditions they encounter in rapidly perfused microvessels. When they are not only driven by steep pressure gradients acting there, but are simultaneously exposed to steep local gradients of shear stress, mammalian RBC are not only markedly deformed, but drift rapidly from the marginal towards the axial trajectories. As detailed in Sect. 88.2.6, fluid displacement in these fluid layers is associated with low momentum transfer. At the same time, the rapidly traveling mammalian RBC are endowed with overvelocity and thus are being dynamically rarefied. As these responses occur collectively in all moving RBC, they are driven into a mesoscopically coherent mode of operation that is likely to be associated with a reduction in the microscopically induced incoherent movements, i.e., a fall in global entropy production (see Sect. 88.2.1, Fig. 88.11). This combination of responses has been called rectified RBC flow [100,102]. In reducing the impact of factors *H* and *K* in Einstein's equation, RBC rectification markedly diminishes the entropy generation associated with blood flow. This follows from the comparison of the energetic expenditure of blood flow with the flow of cell-free plasma.

In isolated glass tubes of diameters representative for those of microvessels in vivo, the entropy generation associated with RBC flow is increased by less than 10% over that associated with the flow of plasma (Fig. 88.12). Interestingly, this occurs despite the fact that the rectification of sheared RBC in straight glass tubes is associated with the induction of highly ordered vortices of the cytosol within the RBC and of the plasma flowing between them, a phenomenon called bolus flow [14,39] (see below; Fig. 88.15). However, to what extent the well-known RBC rectification observed in living microvessels determines viscous energy dissipation in vivo is not yet settled. In a recent summary of the extensive work of their group over more than a decade, Pries and Gaehtgens, working in cooperation with theoreticians (Secomb and Cokelet) have narrowed down a range of estimates about the diameter and HCT dependency of the relative apparent viscosity of blood in microvascular networks [79]. Figure 88.12, taken from their work, compares the entropy generation rates as estimated from precise measurements performed in vitro (upper part of the diagram) and those obtained from in vivo measurements. The latter can, however, only be transduced into calculated data of apparent viscosity by various interpolation and extrapolation procedures, including such factors as collision effects due to inhomogeneities in diameters, curvatures, bifurcations, and confluxes, interferences with moving or sticking WBC and possibly an effect exerted by the glycocalix of the endothelium. The latter possibility is suggested by effects [23] of heparinase treatment of living microvessels and it might eventually prove relevant as a cause of higher entropy generation rates in vivo than in glass tubes of 5–20 µm diameter.

Note that at very little extra energetic expenditure, RBC cytoplasma is driven into automatic convective mixing of both oxyhemoglobin in the cytosol and of the plasma traveling between the RBC, a concomitant set of movements which markedly accelerates O_2 uptake and release [67,123] (Fig. 88.13, 88.14).

88.3.4 Self-Organized Movements of Fluidal Mammalian RBC Studied in Artificial Flow Chambers

The further research on the details of self-organized RBC rectification and the consequences on entropy production on the one hand, and on the kinematics of blood gas exchange on the other will greatly profit from microrheological studies in vitro that have revealed the details of an unique flow behavior of nonnucleated mammalian RBC. In many experiments studying the kinematics of individual RBC movements by microscopic flow visualization in so-called rheoscopes, in glass capillaries, or in living microvessels, the essence of their remarkable phenomenological behavior has been elucidated. Whenever they are subjected to either uniform or to nonuniform (Figs. 88.12 and 88.13) velocity gradients, and in this way to shear stress gradients, they are being simultaneously deformed and prolongated, as well as oriented so that their major axis is permanently parallel to the flow direction. By a typical self-organized scenario, each of them is transformed into a streamlined ellipsoid, driven into a stationary alignment, and migrates away from solid walls, i.e., from the marginal to the axial zones. A unique rotational movement of the membrane (called "*tank treading*") is induced, as is observable in rheoscopes [91,92] and in microscopic glass tubes [39,52]. This orbiting movement can easily be visualized with the help of appropriate markers of the membrane or of the cytosol which is thereby set in shearing motion. In vivo, the marked and very dynamic deformation phenomena associated with RBC rectification are also clearly seen. In theory, membrane rotation must also take place. Although it is not yet experimentally proven in vivo at the time of writing, its occurrence can be deduced from viscometric data and the phenomenon of rectification obtained in experiments in vivo.

Figure 88.14 shows sequences of a motion picture of cells moving down a glass capillary of only 5 µm diameter, a hemoglobin precipitate (Heinz body) was produced, and

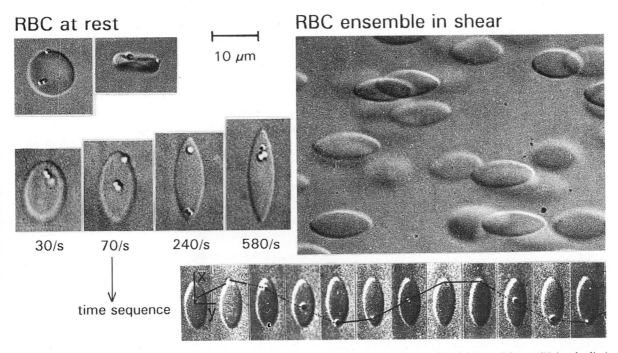

RBC at rest

10 µm

RBC ensemble in shear

30/s   70/s   240/s   580/s

time sequence

Fig. 88.13. Progressive deformation, orientation and membrane tank treading of human red blood cells (*RBC*) when subjected to viscometric flow (counter-rotating cone plate chamber). Discoidal RBC at rest, decorated by three latex particles seen en face and from the side. As rates of shear are progressively enhanced (from 30/s to 580/s), the RBC are deformed into prolate ellipsoids and oriented so that the major axis is paralleled to the direction of flow. Time sequence of positions of membrane attached latex particles demonstrates tank-tread-like movement of RBC membrane.

Time sequence of cytosolic globinprecipitates (Heinz bodies) shows that there is cytosolic laminar flow. The lines combining the two particles allow construction of the intracellular trajectories taken by the precipitates; *dotted lines*, particle does not reach the border of the cell, trajectories are always within the cytosolic compartment which proves intracellular laminar flow. Ensemble of RBC in shear All RBC are analoguously deformed and oriented in shear

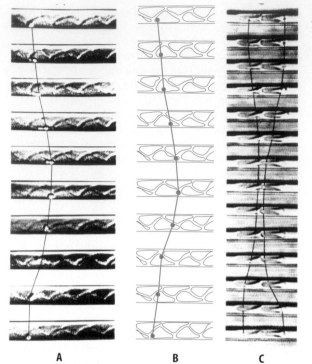

A   B   C

Fig. 88.14a–c. Deformation, orientation and membrane tank-treading in human RBC when subjected to flow in microscopic glass capillaries (traveling capillary arrangement). While the cells move to the right, the capillaries are displaced with equal velocity to the left, thus allowing to study the kinematic details of the cell deformation process in a complex shear field. **a** Sequential pictures during movement in 8 μm capillary: all RBC are permanently deformed. Heinz bodies (*arrow*) in one cell again move from the leading to the trailing edge and back to the leading edge. **b** Schematic drawing for better demonstration of particle movement in cytosole. **c** Sequential pictures during movement in 4 μm glass capillary: intracellular globin precipitates (Heinz bodies) move from the leading to the trailing edge of the RBC which while flowing is permanently deformed into slipper configuration

its movement followed [39]. In a similar scene (Fig. 88.15), tumbling platelets are observed moving behind deforming RBC. These demonstrate the presence of the plasma vortices mentioned above. These first attempts at visualizing dynamic flow in the range of nutritive capillary diameters and at establishing a dual vorticity (intracellular mixing of oxyhemoglobin, hemoglobin, and all enzymes) associated with extracellular bolus flow open new avenues to the comprehension of self-organized coherent hydrodynamic structures at extremely low Reynolds numbers. The consequences of these discoveries are functionally important. Their theoretical impact is related to the fact that a very efficient enhancement of diffusion can be obtained in secondary flows not associated with an increase, but rather with a decrease in entropy production. This is a heterodox combination of events which is, however, in full agreement with the fundamental theoretical predictions of Klimontovich [53] (see Chap. 2) for differentiated flows.

88.4 Hemodynamics and Hemorheology: Dissipative Structuring of a Driven System

For all the reasons given in Sects. 88.2 and 88.3, the self-organized flow behavior of the blood as a driven system represents a highly variable performance by a typical process of dissipative structuring rather than a well-defined property. The details of RBC performance depend on the physical and chemical boundary conditions prevailing locally: it is a *shear-induced fluidalization*. Nonrheological factors that modify the entropy generation associated with blood flowing in microvessels must be considered as well. At any rate, the driven blood constituents change their behavior in response to variations in the mechanical forces acting upon them; these are, of course, generated by the heart, but they are modulated by vasoconstriction and vasodilatation. In response to changes in blood volume and blood composition, the drive is also modulated. In short, the global blood movement represents the final ef-

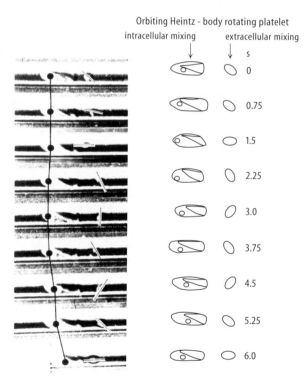

Fig. 88.15. Mixing in RBC, associated with extracellular mixing of plasma as visualized by a rotating platelet (experimental proof of bolus flow in narrow glass capillaries). In the *left sequence* of pictures taken from a video screen, the position of a hemoglobin precipitate (Heintz body) is shown by *black connected dots* and recapitulated in the schematic drawings (*middle row*). The main axis of the rotating platelets is marked by lines of different orlentation and is recapitulated in the ellipsoids with variable orientation of the major axis as a function of experimental time. Experiments done in cooperation with P. Gaehtgens

fect of the operation of a hypercomplex system with many nonlinear responses and marked effects of hydrodynamic self-organization(drive-dependent spatial coordination and temporal synchronization), of which RBC rectification is but one example. Simple cause and effect concepts are bound to lead us astray, due to the fact that blood composition changes as blood is driven through macro- and microvessels.

The reason for this type of adaptability of a system in motion are complex and challenge our ability for abstract thinking. Much to the confusion of the biomedical public, the entity most easily measured in whole blood in vitro (RBC volume fraction or "hematocrit", abbreviated HCT) is the one most difficult to grasp or to interpret in vivo. The term HCT must be used in many different connotations such as systemic HCT, large vessel HCT, microvascular HCT, tube HCT (RBC volume fraction, and discharge HCT (RBC flow fraction) [19,39,91,101]. These terms must, however, be clarified in order to comprehend the role of blood viscosity in vivo. From a functional point of view, the flow fraction is the most important entity. Unfortunately it is difficult, if not impossible, to quantify since only very indirect methods are available to measure the RBC flow rate ($\dot{Q}_{RBC}$) and the plasma flow rate ($\dot{Q}_{plasma}$) required to calculate discharge HCT (HD = $\dot{Q}_{RBC}/\dot{Q}_{RBC} + \dot{Q}_{plasma}$).

88.4.1 Kinematics of Self-Organized RBC Movement in Rapid Flow: The Normokinetic Mode of Operation of Ensembles of Rectified RBC

In microvascular and macrovascular networks, the shear-induced rectification of individual RBC leads to self-organized coherent behavior of very large ensembles of moving cells subjected to gradients of forces (Sect. 8.3). Consequently, in all vascular segments dynamic self-fluidalization occurs and there is marked axial drift or self-focusing. The more the RBC are deformed, oriented, and displaced to the vessel axis (a mesoscopic event), the higher the fluidity of suspensions. From this the global flow rate for any given geometry, topology, or driving pressure is greatly enhanced (a macroscopic consequence). In short, under what might be called *high flow* or *normokinetic* conditions, the ordered (coordinated) mode of operation is induced simply because the nonnucleated components are susceptible to the hydrodynamic forces that act upon them. The stunning motion pictures [11] taken by BRANEMARK in the living human microcirculation in chambers implanted in healthy human volunteers demonstrate the validity of concepts and the high relevance of experiments in animals and with human blood studied in vitro. Incidentally, they also testify the ease of phase transitions into RBC compaction and local accumulation of RBC in slow flow. In the normal mode, i.e., the mode associate with high shear, synergetic "self-fluidalization and "self focussing" generates a collective mode of operation, in which all fluid RBC are passively drifting away from the marginal layers.

All axially moving RBC experience fast-forward displacement, but low-velocity gradients and therefore very low energy dissipation despite rapid flow. As a side effect, flowing RBC minimize collisions with the wall and with other cells: very obviously this mechanism also reduces momentum transfer and therefore entropy generation during enforced shearing. Branemark's films show that very few such collisions occur; however, experiments with rigidified cells, in contrast, show fragment collisions with other cells in straight segments and in bifurcations which stop, retard, and make the flow erratic [25]. In short, events associated with many collisions reduce the flow rate for a given pressure, show much more flow disturbance and higher entropy generation [79] that can be quantified from the calculation of a higher resistance (Fig. 88.16).

The extent and energetic benefit of self-organized rectification of normal mammalian RBC in flow depends on the species specific geometry of RBC, i.e., their surface area-to-volume relationship, fluidity of the cytosol and viscoelastic membrane characteristics (a topic detailed elsewhere [18,19,25,39,96]). Detailed hydrodynamic analyses of flow in the microvascular networks of mammalian circulatory beds have recently been initiated. This was achieved by a combination of flow visualization and network analysis. Such analysis [100,101] can uncover two related facts of significance in the context of dissipative structuring (Fig. 88.17):

- The steepest pressure gradients drive the blood through the arterioles and nutritive capillaries;
- It follows therefore, that the highest radial gradients of shear stresses must occur in the same vessel classes.

In the arterioles and capillaries there are thus the sites of the steepest gradients of entropy generation rates, i.e., of strong differences of frictional work between the marginal and the axial fluid lamellae (see Sect. 88.3.3). Simple energetic considerations have shown that this combination of boundary conditions not only drives the moving RBC suspensions into a self-fluidizing mode of movement, but also stabilizes them in this highly ordered mode of operation.

88.4.2 Dynamic Blood Dilution: The Fåhraeus Effect and the Fåhraeus-Lindqvist Effect in Individual Vessels and in Networks

After axial self-concentration of RBC synchronized with plasma sleeving with the formation of a lubricating layer has occurred, high amounts of sheared plasma can obviously minimize the frictional retardation of the moving RBC in each arteriolar, capillary and venular segment. This is the essence of Poiseuille's deliberations [76] (see Fig. 88.6), and was fully appreciated by Fåhraeus [28] before he initiated experimental elaboration together with Lindqvist that has since been refined by many authors (see [79] for a very recent review). Furthermore, a similar phenomenon occurs at each bifurcation of any given arteriolar parent

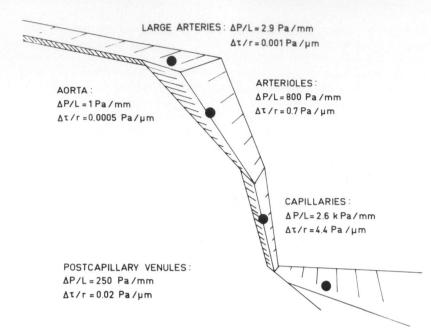

LARGE ARTERIES : ΔP/L = 2.9 Pa/mm
Δτ/r = 0.001 Pa/μm

AORTA :
ΔP/L = 1 Pa/mm
Δτ/r = 0.0005 Pa/μm

ARTERIOLES :
ΔP/L = 800 Pa/mm
Δτ/r = 0.7 Pa/μm

CAPILLARIES :
ΔP/L = 2.6 kPa/mm
Δτ/r = 4.4 Pa/μm

POSTCAPILLARY VENULES :
ΔP/L = 250 Pa/mm
Δτ/r = 0.02 Pa/μm

Fig. 88.16. Schematic representation of the pressure gradients (pressure difference for a given length) and the shear stress gradients (fall in shear stress for a given distance from the tube wall) in the different parts of the macro- and microvasculature. The motion of the suspended phase can be compared conceptually to that of a ball rolling down a set of furrows with increasing steepness in both forward and lateral directions. Following a principle of energy minimization, the suspended phase is "stabilized" in the center of the "furrows", used here as an analogue to the cardiovascular system

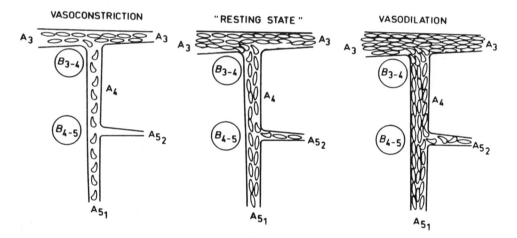

VASOCONSTRICTION   "RESTING STATE"   VASODILATION

Fig. 88.17. Schematic representation of the control of flow, plasma skimming and red cell attendance in subsequent arterioles A_4, A_3 and A_2) and two bifurcations ($B_{4,3}$) and [$B_{4,5}$]. At normal tone (*'resting state", middle figure*), there is a certain amount of plasma skimming at both bifurcations, combined with axial migration (red cell overvelocity) allowing more skimming at the next bifurcation. This sequence of events leads to a progressive drop in red celd volume fraction in the downstream segments: hydrodynamic diversion or screening of cells away from the slower branch. As vasodiration (*right figure* sets in, the velocity differences are smaller, the flow rates are higher and the axial migration is less pronounced (or even disappears): hence the hydrodynamic diversion of red cells from entrances into arterioles and capillaries is abolished. Vasodilation thus results in an oven perfusion of all microvessels with red cell plasma mixtures similar in composition to the blood in the large vessels. Conversely, vasoconstriction (*left figure*, reducing flow rates, but enhancing shear stresses, axial migration, and velocity differences at bifurcations, augments hydrodynamic diversion, leading to pure plasma skimming (tack of red cell attendance) in more slowly perfused arterioles and/or capirltaries. The result is a kind of derecruitment as seen in A_{52} by hydrodynamic diversion even without the action of a true procapillary sphincter

vessel into the two daughter branches [20,21]. Since the RBC and the plasma are often unevenly distributed, three distinct phenomena, follow which cause the network-Fåhraeus-effect ([77,78], see also Fig. 88.7) namely:

- RBC overvelocity in segments;
- RBC screening; and
- Plasma skimming at bifurcations.

Contrary to popular belief, the effects in segments and at bifurcations are mutually dependent and not only related to forces acting at the site of the bifurcation [19,20,35], but to phase separation occurring far upstream in a vascular segment. Consequently, pronounced separation between RBC and plasma is a network event that is self-propagating into all vascular segments and bifurcations.

As a pivotal event, the blood as a suspension in motion is dynamically diluted as the RBC are driven into self-focused and rectified flow. As the mixture of RBC and plasma is driven into vessels with steeper shear stress gradients, a curious combination of consequences follows. The RBC content per unit volume of blood falls (described in Einstein's equation as the main determinant of the rate of frictional losses, e.g., Eqs. 88.3, 88.17), but the RBC transfer rate remains the same. Self-organized RBC overvelocity is the obvious reason why this can occur without violating the law of continuity. Per unit volume and time, the same number of RBC that enters a network via arterioles leaves it through the venules, but since the average velocity of plasma is much lower than the average velocity of RBC, the mean residence time of plasma is much longer than that of the RBC. Note, however, that the self-organized distribution of both RBC and plasma in the terminal microvessels is very dynamic and can only be seen as a statistical event. As shown from the wide variation in microvascular HCT (Fig. 88.7), each segment has a different share of the global plasma flow and the global RBC flux. This has a wide range of reasons, the most important being the shear stress and time dependency of the axial migration which reflects a pivotal feature of self-organized, fluid-dynamic events. There is a finite window of shear stresses (and thus of residence times) that causes RBC overvelocity. Neither at very low (due to a lack of driving forces) nor at very high flow (due to a lack of time to manifest axial drift) RBC overvelocity can it be found.

Plasma Skimming. In addition, under most conditions of normal flow, the RBC flux fractionation at each bifurcation is a nonlinear function of the flow fractionation (which often displays threshold characteristics). Thus, existing differences in flow rates can be amplified in daughter vessels [77]). As one possible extreme, all RBC that arrive at a bifurcation are partitioned into the faster daughter segment, while the other one is fed only by plasma (plasma skimming). Since as a rule RBC preferentially move into the daughter branch that is more rapidly perfused, the other can be fed with a RBC plasma mixture containing more plasma. As the daughter assumes the function of the parent at the next bifurcation, more pronounced axial drift of RBC and thence plasma sleeving can occur, producing much more pronounced plasma skimming at the subsequent bifurcation. From this, it is conceivable that red cell plasma mixture (RCPM) with even much lower RBC content is attending the only slightly slower perfused granddaughters, an effect that is propagated to their great-granddaughters. The local HCT values in each segment thereby depend on a multitude of processes taking place upstream or downstream of the site of observation [117].

88.4.3 Active Regulation of the Blood Composition in Microvessels

These very same mechanisms are put to use when vasoconstrictor influences modulate the flow fractionation at bifurcations. Dynamic responses of the RBC content of microvessels have long been known to be caused by vasomotor activity [96]. For reasons of spatial constraint, this important consequence of passive hemorheological self-organization (use-dependent coordination) in response to active hemodynamic regulation can only be mentioned in passing and phenomenologically (with reference to Fig. 88.17). The kinematics of phase separation explain this: during vasoconstriction, the pressure gradients ($\Delta P/L$) are steeper, but the averaged forward displacement is slowed down. This combination provides more time for self-organized axial drift of RBC and plasma sleeving with all its consequences, while the mixture moves from one bifurcation to the next. Under vasoconstrictor-induced hypoperfusion, therefore, a large fraction of the microvascular segments is perfused slowly with fewer RBC so that its content appears like plasma. This is the cause of the well-known phenomenon of skin bleaching ("pallor") associated with cutaneous vasoconstriction [96]).

The opposite influence operates in vasodilator-induced hyperemia, where local red cell volume fraction can equal that seen in macrovessels. There is an apparent (not a real) paradox during dilatation of arterioles: the volumetric flow rate increases, but the pressure gradients ($\Delta P/L$) and RBC residence time in each segment decreases. The mere lack of time associated with a reduction in shear stress gradients makes RBC self-focusing in each segment less perfect; consequently RBC and plasma are more evenly partitioned at each arteriolar bifurcation. This can culminate in a situation where the mean microvascular HCT values rise to the levels found in macrovessels [54], and the variance in the different segments of the vascular network disappears. The consequences have been clearly identified by Klitzman and Johnson [54]; the kinematic details causing the phenomenon open important avenues to an improved understanding of O_2 supply to tissues in the future.

88.5 Blood Rheology and Functional Optimization of Oxygen Delivery

Already now, the functional benefit of the elimination of a self-organized dynamic blood dilution of suspensions of fluidal RBC during functional hyperemia due to vasorelaxation is obvious: there is a more than proportional increase in O_2-transport due to higher overall flow, more even RBC distribution, and higher microvascular content and transfer of RBC. From a synergetic point of view, there is not really any competition between vasoconstrictor and viscous control of the peripheral circulation, but rather functional cooperativity. The effects are complex and caused by mutually dependent individual events. As a whole, they explain why marked changes in the systemic RBC volume fraction (large vessel HCT) has such a surprisingly small effect on the RBC delivery to the peripheral microvessels. Demonstrated as early as 1933 by Whittacker and Winton [118] and since then confirmed by many workers [7,15,16,39,50,91], the in vivo viscosity, or in other words the additional entropy generation, proves to be remarkably small as large amounts of RBC are added to the flow of plasma through microvascular networks (see Fig. 88.10).

88.5.1 Synergetic Interpretation of Functional Hyperemia: Enhancement of Convective and Diffusive Transport Conductances in the Hyperkinetic Mode

Where there are high O_2 requirements, as in working skeletal muscle [36], it is generally accepted that the increment in oxygenation during vasodilatation leading to functional hyperemia not only increases overall flow, but depends on the potential to recruit capillaries (see Chap. 94). Thus, the *conductance regulation* by arterioles determines not only the overall blood delivery to a microvascular network, but, via flow fractionation at each bifurcation, also the relative flow of plasma and RBC in each segment (see Fig. 88.17). Since cardiac output, peripheral perfusion and peripheral O_2 extraction are driven to their respective maxima, a highly coordinated combination of transport mechanisms emerges in which various passive mechanisms spontaneously cooperate in enhancing the convective and diffusive conductances for blood gases. This scenario shall be called the hyperkinetic mode of operation. The global effect is that O_2 is more efficiently transferred from the environment to the O_2- consuming mitochondia in metabolically driven cells. In the microvessels, the global flow rate, the RBC content and thence the amount of O_2 delivered per unit flow rises, the number of capillaries attended by RBC is increased, and the diffusion distance between capillary and consuming mitochondria is reduced. In short, convective and diffusive transport rates are automatically enhanced in a coordinated fashion.

This topic is covered extensively elsewhere [93,101] and can only be dealt with here briefly. Starting with the delineation of the blood supply to skeletal muscle (and neglecting the compression of the microvascular compartment during muscle contraction), we can state that as the precapillary arterioles passively relax, virtually every factor that influences the transfer of O_2 from the capillary blood to the mitochondria (and of CO_2 back to the RBC carrying carboanhydrase in very high concentrations) is being enhanced

- Concentration differences ΔC for both O_2 and CO_2 are being elevated;

- Diffusion areas are maximized;
- Diffusion distances are reduced;
- The effective diffusion coefficient within the plasma and the RBC are functionally increased by the superposition of simple diffusion (movement of individual molecules) by convective mixing (collective movement of molecules).

These considerations underline why in the context of skeletomuscular and myocardial perfusion, the conventional concept of blood viscosity is inappropriate and thus the question of optimal HCT is energetically unanswerable. For the reasons delineated in Sect. 88.2, the process of mechanical contraction (i.e., the O_2-consuming process) in compressing the microvessels produces an external energy (compression by tissue solid pressure up to 50 kPa) that is far in excess of the forces generated by the heart as a pump and operating in the circulatory system. The energy dissipated in induced shear during perfusion is negligibly small in comparison to this additional compressive energy, and the numerical value of coefficient of viscosity is, therefore, largely irrelevant in this situation. One must, perhaps, keep in mind that the flow in skeletal muscle and in the myocardium is extremely pulsatile [32,68,109] and directed backwards in arteries in the initial phase of contraction. The restoration of orthograde flow in the initial phase of the relaxation is a priori likely to depend on viscosity. No detailed information seems to be available with respect to this topic.

Due to the influence of vasorelaxation on the local RBC content of blood in each individual capillary described above, the local O_2 delivery rate can be enhanced out of proportion to the increase in flow rate. Thence, the capillary-mitochondrial gradients of pO_2 are markedly elevated, as all available capillaries are being perfused with blood rich in oxygen-carrying RBC. The relatively slow process of intraerythrocytic diffusion (of O_2, CO_2 and especially of oxyhemoglobin) is enhanced by two types of self-organized vorticity, namely membrane tanktreading and bolus flow, (Fig. 88.12) induced without any extra expenditure of energy but strongly aiding O_2 release [14,67,123]. Similar mixing effects of deformed RBC and moving plasma must be expected to occur in the pulmonary vascular bed, where, due to the specific microangioarchitectonics, functional hyperemia is associated with an even more pronounced increase in global vascular conductance than that found in peripheral organs (see Chap. 100). While the details of the local events are largely unexplored, the global energetic consequence is dramatic: the entropy generation rate in the cardiovascular system (i.e., the product of pressure and flow) is increased by a much smaller extent than the O_2-delivery [100]: the ratio of these two values as a global indicator of cardiovascular efficiency falls precipitously from 5 kJ/mol O_2 at rest to <2 kJ/mol O_2 during functional hyperemia.

88.5.2 Tolerance for Anemia Associated with Normal Blood Volume

At *rest* or during light exercise, alteration of the systemic HCT (between 27% and 55%) have only a minute influence (+10%–15%) on O_2 delivery. For all practical purposes, this is irrelevant in comparison to the range of possible increments of O_2 deliveries known to occur in response to physical exercise (+200%–+500%). The self-organized mechanisms capable of enhancing oxygen transfer (see below) may also in part be held responsible for the remarkable lack of correlation between blood composition (HCT, hemoglobin concentration) repeatedly found by students of comparative physiology [81,105]. Self-organized phenomena found in the moving blood (especially the phenomenon of RBC overvelocity) on the one hand, self-organized increase in hemoglobin content in rapidly perfused capillaries, and self-organized vortex formation (bolus flow) are obviously undetectable in the static blood

parameters, but are pivotal for the outstanding physical efficiency (net increment in $\dot{V}O_{2max}$, $d\dot{V}O_{2max/dt}$) found in mammals and especially in humans [24] (Fig. 88.18).

On the basis of the mechanisms described above, one can also explain the robust efficacy of the O_2 transfer system. As shown in Fig. 88.19 (taken from [96]), the addition of RBC to or their removal from systemic blood has a remarkably *small* influence on the flow rate for a given arteriovenous pressure gradient, and thus on the overall cardiovascular hydraulic conductance, provided that the blood volume and thus the kinetic energy of the moving blood is kept normal. This is of considerable clinical significance since it explains the surprisingly good tolerance of spontaneous or iatrogenic anemia, whenever blood volume is meticulously conserved. All these statements hold true for the energetic demands at rest (i.e., for bed-ridden patients), whereas during physical exercise higher HCT levels are obviously and unquestionably advantageous. Note, however, that a wide variety of high performance

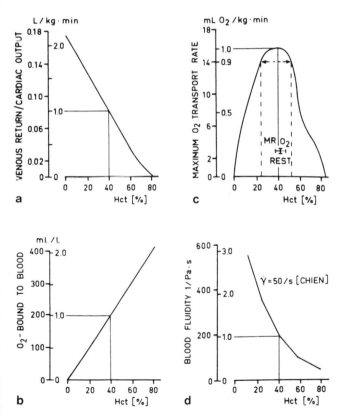

Fig. 88.18a–d. Influence of hematocrit (ml/ml) level on transport functions of the cardiovascular system. **a** Influence of systemic hematocrit on cardiac output (ml/min × kg) via its influence on venous return. **b** Influence on oxygen binding capacity (ml/100 ml hemoglobin via the molar hemoglobin concentration, assuming a molecular weight of 64 kDa/tetramer) and the molar binding capacity of 4 mol of oxygen/tetramer of hemoglobin. **c** Influence on maximum oxygen transport rate ($\dot{N}_{O_2 max}$ in ml/min × kg). **d** Curve resulting from the multiplication of curve **a** and curve **b**. The

resting oxygen consumption ($\dot{V}_{O_{2rest}}$) is shown for comparison. (Modified from [84]). Note that the leading cause for the adaptation of the systemic circulation to reduced hemoglobin concentration lies in the augmentation of venous return to the right atrium and right ventricle. The hydrodynamics benefits of reduced blood viscosity on venous return can be easily explained by conventional application of Poiseuille's law to systemic veins (and pulmonary vasculature)

animals have a surprisingly low HCT but a high cardiac output, a fact that supports the ideas presented here [15]. Also, certain types of world class athletes have an increase in plasma volume exceeding the elevated whole body RBC volume [15] and have a subnormal systemic HCT, a topic detailed in [96].

The surprisingly weak dependency of oxygen delivery on systemic HCT follows from all the nonlinear rheological effects described above. It provides the functional justification for a clinically successful yet heterodox hemorheological therapy, namely isovolemic hemodilution where systemic HCT levels down to 35% are well-tolerated, even in patients with compensated cardio-vascular disease. In this situation, a reduction in the O_2 binding capacity rate can easily be compensated by elevated O_2 extraction, slight arteriolar relaxation, and even more by microvascular distribution of enhanced RBC flow rates [90. 93] The data and concepts presented here clearly show that naive linear concepts relating O_2 binding capacity (systemic hemoglobin concentration to) O_2 transport rates are inappropriate for a complex, highly nonlinear transport process with a wide variety of functionally beneficial phase transitions in response to changes in the flow conditions.

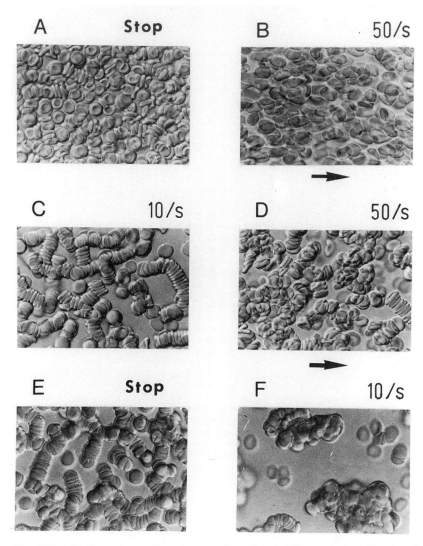

Fig. 88.19A–F. Photomicrographs of suspension of normally aggregatng (**A,B,C**) and pathologically aggregating (**D,E,F**) human red blood cells (RBC) in plasma, studied in a rheoscope. **A,B** As suspensions of RBC in normally composed plasma are exposed to a shear rate of 100/s, the cells are deformed into ineegular ellipsoids mostly oriented parallel to the direction of flow (*arrow*). When there is a sudden flow stop (**A**), the cells resume their familiar biconcave resting disc shape; some of these discs are incorporated into short stacks. **c** If subjected for prolonged periods there for 15 s to slow shear (1C/s); the colliding cells combine into prolonged aggregates ("rouleaux", "stacks of coins"), forming three-dimensional structures by end-to-side attachment of primary rouleaux. **D,** Abnormally strongly aggregating plasma (as here taken from a diabetic patient) produces highly adhesive rouleaux capable of withstanding higher shear stresses (here 50/s), but are eventually dispersed at shear stresses above 250/s (not shown). Pathological rouleaux are often characterized by hemispheric deformation of the terminal RBC in the individual aggregate (**E**) and have a tendency to combine into coarse clumps by side-to-side attachment in low shear (**F**)

88.6 Effect of Reduced Pressure Gradients and Shear Stresses on Blood Composition and Fluidity

88.6.1 Reduction of Flow Forces

Reduction of flow forces leads to multiple phase transitions, which are understandable in light of the described self-organized process of blood fluidization, and critically depends on the maintenance of normal pressure gradients as well as on physiological susceptibility of RBC to the shear stresses. As soon as the driving forces are reduced, the drive-dependent coordination can no longer take place. Under *low flow conditions*, slowly creeping suspensions of RBC begin to aggregate and to exhibit a *conventional* behavior found in any concentrated suspension. In the absence of sufficient hydrodynamics forces (low values of $\Delta P/L$), even normal mammalian RBC are no longer dynamically deformed/and behave more like ordinary elastic bodies.

88.6.2 Reversible RBC Aggregation

In addition, low shear flow allows aggregated RBC to exist and unite into larger rouleaux networks or into coarse clumps in diseases with high concentrations of high molecular weight colloids (Fig. 88.19). These clumps are fundamentally different from the cell agglutinates, i.e., those of RBC when bound by the blood antibodies (*agglutinins*) after mixing of incompatible blood samples. Note that aggregation due to low shear flow is always fully reversible (Figs. 88.1, 88.9) but exerts marked effects on the local flow behavior of blood because of three secondary consequences:

- it abolishes the fluid-drop like behavior of the individual RBC and thence their physiological susceptibility to shear stress gradients;
- it enhances the axial drift of RBC under conditions of high flow (see Fig. 116); but
- it makes the slowly flowing blood susceptible to gravitational forces in low flow situations (see below).

When RBC aggregates are united into larger masses, their intravascular sedimentation and thus their rapid intravascular accumulation fundamentally alters both the local hemorheological and the local hemodynamic situation. An entirely new type of superimposed movement is initiated, which leads to a local alteration in blood composition. The extent of this viscidation, its local hemodynamic consequences, and its global effect on the

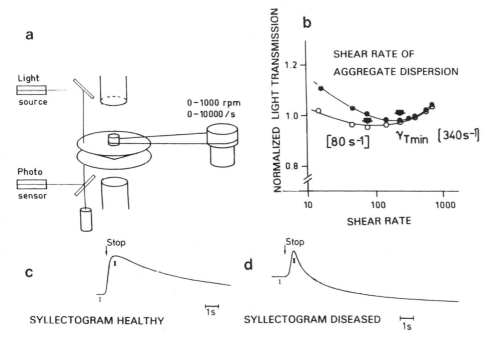

Fig. 88.20. a Schematic representation of microrheological quantification methods for hemorheological phase transitions. Blood can be subjected to idealizing viscometric flow with uniform shear rate in the gap between a rotating cone and a transparent plate. The shear-induced disaggregation as well as the shear-induced deformation of red blood cells (RBC) produce characteristic changes in the optical behavior of the individual cells and that of the whole suspension. These changes as a basis of phase transition can be easily quantified by transmission and reflection photometry, by LASER diffraction or by ultrasound texture analysis. **b** Typical measuring exemple showing optical transmission as a function of progressively increasing shear rate; both disaggregation and progressive orientation lead to characteristic increase in light transmission. Measurement of sudden decrease in light transmission upon sudden stop after previous shearing: initial decrement, followed by increment of light transmission. (syllectogrom as described by Zijlstra [124]). In the blood of a healthy subject, the light transmission increases at a rate proportional to the aggregation, tendency of the plasma (**c**). In a pathologically aggregating blood sample (**d**), the rate of increase in transmission is markedly enhanced. This signal can be used for quantitative photometric RBC aggregometry

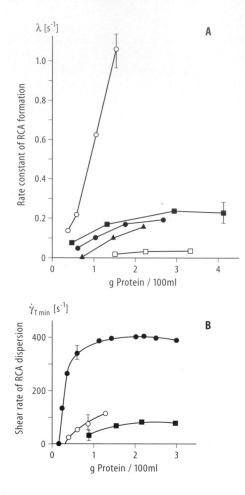

$\lambda\ [s^{-1}]$

A

Rate constant of RCA formation

1.0
0.8
0.6
0.4
0.2
0

0   1   2   3   4
g Protein / 100ml

$\dot{\gamma}_{T\,min}\ [s^{-1}]$

B

Shear rate of RCA dispersion

400
200
0

0   1   2   3
g Protein / 100ml

Fig. 88.21a,b. Different types of aggregating plasma proteins produce different shear behavior of primary rouleaux and secondary rouleaux aggregates. Fibrinogen and immunog lobulin μ produce network aggregates, characterized by rapid spontaneous aggregate formation **a** Rate constant depicted as a function of protein concentration) but low resistance against hydrodynamic dispersion (γ_{Tmin}, shear rate of minimum light transmission). **b** α_2-Macroglobulin produces clump aggregates with slow spontaneous formation (low value, **A**) high resistance against hydrodynamic dispersion (high values of γ_{Tmin}; hematocrit, 30%; $T = 23C$; 1% albumin in saline, **b**). (*White circles*, fibrinogen; *rhombuses*, immunoglobulin M; *black circles*, α_2-macroglobulin; *triangles*, haptoglobulin–hemoglobin complex; *squares*, immunoglobulin G; *RCA* red cell aggregates)

upstream and downstream vessels is highly variable and strongly depends on the duration and rate of progression. In short, a series of nonpredictable events takes place, which, however, has the effect of jeopardizing uniform network perfusion or coherency of flow.

Whether or not the phenomenon of RBC aggregation has any functional hydrodynamics consequence, i.e., a decrease in blood fluidity, therefore critically depends on the local microvascular condition, the previous history of flow, and accompanying microvascular reactions ([42.83] and Sect. 88.6.3). Depending on flow conditions, rouleaux of different lengths can either unite into larger (three-dimensional network structures) or into coarse clumps (Fig. 88.22). Since RBC aggregation is mediated by the adhesive high molecular weight proteins of the plasma, its kinetics and dynamics obviously depend on their concentration (Fig. 88.21). The velocity of their formation upon flow retardation on the one hand and the resistance of these aggregates against hydrodynamics dispersion on the other prove to be a nonlinear function of the concentration of aggregating colloids. The same colloids are also the prime determinants of the plasma viscosity. It goes without saying that it would be futile to expect any simple relationship between their concentration and any value describing fluidity/viscosity of blood.

88.6.3 Causes and Consequences of Rouleaux Formation in Low Shear Flow and in Stasis: Phase Transitions Studied In Vitro

Using indirect methods (e.g., photometric or ultrasound aggregometers), one can nevertheless quantify the dynamics of rouleaux formation and thence the rate and extent of the transition from a *quasi fluid* to a *quasi solid* behavior. There are several methods available for this purpose (Fig. 88.20), many of which also allow the measurement in quantitative terms of a rate constant for RBC aggregate formation subsequent to hydrodynamic dispersion (Fig. 88.20). This event represents a classical nonequilibrium phase transition, since dispersion in shear is known to depend on the concentration of adhesion promoting plasma proteins, namely [64, 90] of the high molecular weight colloids fibrinogen, α_2-macroglobulin, immunoglobulin M, the haptoglobulin–hemoglobin complex and, to a much lesser degree, immunoglobulin G (Fig. 88.21). A host of disease processes in practical medicine are characterized by an increase in the concentration of these high molecular weight proteins. Here, this phase transition occurs at much higher shear rates and proceeds much more rapidly (with characteristic half-times >1 s) than in healthy

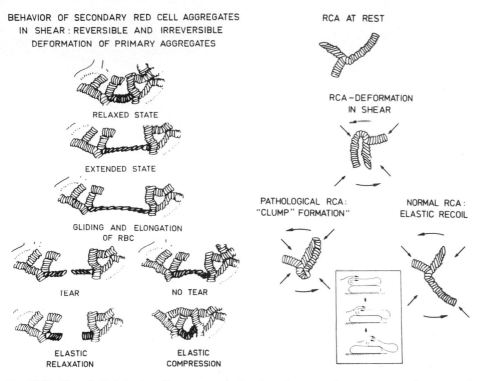

BEHAVIOR OF SECONDARY RED CELL AGGREGATES
IN SHEAR: REVERSIBLE AND IRREVERSIBLE
DEFORMATION OF PRIMARY AGGREGATES

RELAXED STATE

EXTENDED STATE

GLIDING AND ELONGATION
OF RBC

TEAR   NO TEAR

ELASTIC   ELASTIC
RELAXATION   COMPRESSION

RCA AT REST

RCA — DEFORMATION
IN SHEAR

PATHOLOGICAL RCA: NORMAL RCA:
"CLUMP" FORMATION"   ELASTIC RECOIL

Fig. 88.22. Viscoelastic behavior of larger network of rouleaux in viscometric flow (observation in a rheoscope; from [99]). The individual RBC are stacked into "rouleaux" ("stack of coins") with pronounced viscoelastic properties resulting in extension resulting in tear (dispersion) or to elastic recovery. Rousaux as formed by fibrinogen (see Fig. 88.21), when subjected to rotation, are compressed and then relax. This results in the formation of loose network aggregates immobilizing plasma within the meshwork of linear cell aggregates. When enhanced, these aggre-gates have a strong tendency to produce compaction due to sedimentation. Pathological rouleaux (as they are formed by α_2-macrogiobulin) or by high molecular weight dextrans when compressed during rotation are combined into a coarse clump (due to side-to-side attachment). These clumps fundamentally differ in their flow behaviour from network aggregates [37], as they roll down in venules and do not congest them (see Figs. 88.21, 88.23) (*RCA*, red cell aggregates)

humans (where characteristic times are >3 s). Macromolecules synthesized as pivotal parts of the unspecific defense system (acute phase reactants), or the specific humoral immunological response (see Chap. 85) thereby destabilize blood movement. The familiar clinical phenomenon of the reduced suspension stability of blood in patients presenting with a high sed imentation rate therefore favors the transition from the quasi-fluid to the quasi-solid behavior of human RBC flowing in macrovessels (Fig. 88.22).

The adhesion of any two adjacent RBC membranes within a primary rouleaux depends on the adherence of plasma proteins onto the membrane surfaces of RBC. Most of the evidence available at present argues against any specific action. The logic of bonds and their severance under external load applies here. One can indeed measure the forces necessary to disperse RBC aggregates quantitatively. The most convenient method is based on measuring the optical properties of RBC suspensions in viscometric flow.

Primary rouleaux attach to neighboring ones and form secondary aggregates. These can assume the configuration of branched networks (when caused by fibrinogen and immunoglobulins) or that of coarse clumps when caused by α_2-macroglobulin: these two types differ fundamentally in their flow behavior [42,90]. Clumps are densely compacted, speaking in terms of Eq. 88.16, they alter the shape-

factor (K) in Einstein's equation. Networks of rouleaux immobilize plasma, i.e., also lead to an increase in H and are thus much more efficient in elevating viscosity. Aggregation can therefore but need not, elevate the apparent viscosity of blood in the classical sense. In vitro dominating in fluences are exerted by extremely complicated interactions including the role of shearing time, previous history of shearing, and geometric configuration of the viscometer used. In tube flow – and even more in the perfusion of microvascular networks – the consequences of rouleaux formation range from fluidizing effects under high shear conditions to congesting effects under low shear conditions. Most importantly, the time course of the experiment is so dominant that low shear viscometry can produce completely misleading results in strongly aggregating blood samples [82].

Despite these confusing effects in flow, rouleaux formation is a priori always a locally viscidizing event. However, all global rheological consequences of the reversible RBC aggregation are also determined by the sequelae of reduced *suspension stability* of the blood (Fig. 88.23). Masses of aggregated RBC and plasma have a tendency to undergo a further type of phase separation when exposed to gravitational forces, leading to a highly labile mesoscopic flow behavior of a blood sample. The most important blood test

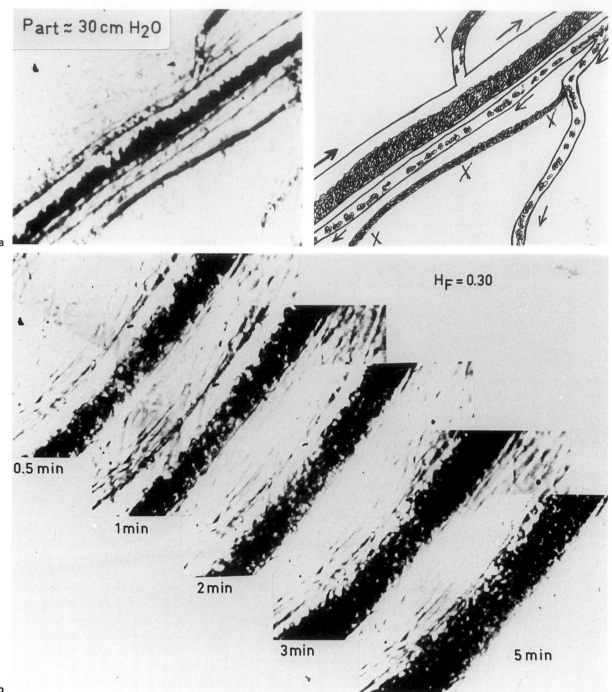

Fig. 88.23a,b. Aggregate sedimentation under gravity in vivo. Intravascular sedimentation in venules of the rat mesentery, placed in vertical position and observed by horizontally aimed microscope during induced arterial hypertension (p_{art}, 22 mmHg). **a** Microphotograph taken from videosceen, clarified by drawing. Note distinct sedimentation of aggregates in arterioles (flowing from left to right; *arrows*), but complete separation of aggregated sediment and superfluent plasma in the venule (flowing from right to left: *arrows*). Complete compaction in venular branch with standstill (*right*, marked by *X*). **b** Progressive increase in sediment height with layered flow during 5 min observation of venular branch. This phenomenon is associated with a progressive skewing of the velocity profile and reversal of the Fåhraeus effect (due to RBC undervelocity)

in practical medicine (colloquially called the sedimentation rate of blood) makes practical use of this in measuring the consequence of the separation of RBC from plasma under the influence of gravitational forces. The very same phenomenon can occur in vivo. Here, not only the process of primary rouleaux formation, but even more the secondary hydrodynamic consequences of rouleaux formation critically depend on the prevailing shear conditions, In high shear flow, they have very little effect on global flow rate and can even lead to a decrease in viscous energy dissipation [83], albeit at the cost of marked inho-mogeneity in capillary perfusion [115]. Even in simple glass tubes, aggregate-induced phase separation grossly exaggerates the physiological separation of RBC and plasma (see Fig. 88.11), This, in turn, produces more marked axial migration, the effect of which on RBC distribution at bifurcations becomes strongly accentuated [42, 96]. More importantly. however, the aggregate masses can become accelerated by gravitational forces in vivo and this can – but need not – lead to time-dependent local blood viscidation. Provided that aggregation occurs in slowly perfused and horizontally placed venules, progressive aggregate deposition onto the dependent part of these microvessels is found, a process properly termed "blood sludging" (Fig. 88.23). Sludging in the sense of the spontaneous formation of a thick sediment is the result of a cooperative effect emerging from a typical set of boundary conditions that are well-known in chemical engineering [12,13,45]. Corroborating insight obtained in technical multiphase systems in the presence of normal pressure gradients, uncomplicated intravascular aggregation is a largely irrelevant hemorheological epiphenomenon in vivo. As long as the overall relocity of the flow is high and the tube is placed vertically [83], the transition into the aggregated mode of movement remains without significant consequence.

88.6.4 Synergetic of Rouleaux Formation and Sedimentation In Vivo: Multiple Consequences of Phase Transitions

This situation drastically changes when the flow rate is being reduced by whatever mechanism. Besides their potential to viscidize blood, RBC aggregates become subjected to gravitational forces (in proportion to the increase in mass of the growing aggregates).

At very low flow rates, therefore, i.e., when the transit times of the RBC suspension in an individual microvascular segment are very prolonged, the aggregated rouleaux begin to sediment in a fashion that depends on the orientation of the venules within the terrestrian gravitational field. However, once such an effect has been initiated in more or less perfectly horizontal venules, the compacted masses can become dislodged into downstream venules and thus fill the draining smallest veins with a creeping RBC suspension with a very high RBC content. These effects, which complicate mere aggregation with additional viscidating effects, often take only a few minutes, but in protracted experiments the gradual compaction of the outflow vessels from a microvascular bed can proceed over many hours. As a consequence, there is a positive feedback that generates a circular causality where progressive retardation [42,55,96] leads to progressive viscidation by settled aggregates [13]. Once this effect of gravitational acceleration has started, it is likely to cause more retardation. Thereby, the movement of the blood becomes destabilized. This occurs in functional disorders that produce protracted low flow states in

networks. RBC aggregates, by forming very viscous sediments, begin to dominate global and local perfusion and local distribution of cells and plasma in the network. Ever since intravital microscopy was first performed in experimental animals and in by mans, the flow of aggregated RBC has been observed preferentially in venules, but the physiological significance of this phenomenon has never been established. Today, it is known that postcapillary venules, e.g., in the microvascular bed of the connective tissue like in the conjunctival and the mesenteric microvasculature, are a preferential sites of the antimicrobial defense reactions, the second physiological task of microvascular blood flow (see above).

In light of this established fact, the rheological causes of this well-known site preference are straightforward:

- In many tissues the averaged cross-sectional area of this class of microvessels is much larger than that of the corresponding arterioles [19.92];
- From the law of continuity, the averaged forward velocity (and thence the averaged rates of shear) for any given volumetric flow rate therefore in venules falls out of proportion to the drop in global flow rate;
- In the venules – and initially only there – the incident shear stresses no longer suffice to keep normal RBC aggregates dispersed.
- In the venules – and only there – aggregates once formed begin to settle to the dependent parts of blood conduits.

All of these effects combine in producing a self-organized viscidation that greatly facilitates inflammatory functions (v.i.).

Self-Organized Compaction of Aggregated RBC: Circular Causality in a Viscidizing Sequence of Events. The consequences of these self-amplifying responses are caused by at least two independent, yet interacting and mutually amplifying responses of blood. The change from a "normokinetic" to a "hypokinetic" mode of flow has self-retarding dynamics. When associated with an enhanced tendency to RBC aggregation, any pronounced local disturbances of the microcirculation become unstable, especially after a high local RBC content begins to accumulate and initiates progressive RBC-aggregate-dependent postcapillary viscidation. The multifactorial reason for this site of RBC aggregation can only be understood in light of self-organized cooperativity of hydrodynamic, hemorheological, and biochemical processes: synergetic principles explain why these preferentially take place at the transition from capillaries to venules (Fig. 88.24). A simple calculation shows the dramatic sequelae of geometry: as the cross-sectional area (A) of the blood vessel rather suddenly increases, for example, from $A = 18\,\mu m^2$ in a capillary with $5\,\mu m$ diameter to $A = 314\,\mu m^2$ in a venule with $20\,\mu m$ diameter, an assumed average velocity ($\bar{v}$) of the moving blood will be reduced from $500\,\mu m/s$ to $23.5\,\mu m/s$ (or by 94.7% to 5.7% velocity in a feeding bottle-neck vessel). There is an even stronger fall in the average rate of shear $\bar{u} = \bar{v}/d \left[\frac{m}{s} \frac{1}{m} = s^{-1} \right]$, which can be estimated from the ratio

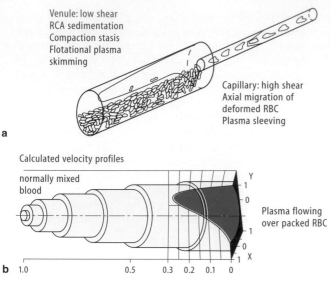

Fig. 88.24a,b. Reversal of the Fåhraeus effect by red cell aggregate (*RCA*) studying and plasma decanting. **a** Schematic representation of the rheological consequences of aggregate sedimentation. Rapid temporal and spatial change from high shear to low shear heterophase effect when capillary perfused under high shear feeds into venule with low shear, and rapid aggregate formation, leading to immediate intravascular sedimentation. At the conflux, there is an instantaneous change from deformed cells with pronounced axial drift (red blood cell RBC overvelocity), to aggregated cells sedimenting into the slow dependent part of the venule. **b** Under the assumption of a simple layering with tube bottom half with sediment flow (HCt 75%; relative apparent viscosity = 50; see Fig. 88.9) and top half with plasma flow, a skewed and retarded velocity profile can be calculated which illustrates the potentials of progressive self-visciation in layered flow (modified from [75])

of average velocity ($\bar{v}$, µm/s) and the diameter (µm): $\bar{u}_{cap}=100$/s suddenly falls to $\bar{u}_{ven}=1.17$.

The sudden transition from high shear flow in capillaries produces abrupt drops in:

- The pressure gradient (ΔP/I) and thence the forward velocity;
- Local shear stresses (τ); associated with;
- A sudden increase in residence time (*t*).

This combination provokes a progressive sequence of flow retardation, of aggregate formation upon collision, of gravity-induced aggregate sedimentation, of viscidation of the blood, and of further flow retardation. As the individual RBC undergoes fluid-dynamically caused phase transition from a quasi fluid to a quasi solid particle, all arriving RBC become incorporated into larger masses suspended in more or less rapidly moving plasma. Automatically, the forces of gravitation (i.e., the product of constant gravitational acceleration and mass) are progressively augmented. There is obviously a second, now collective phase transition in the moving blood that changes from the behavior of a monodispersed, highly fluid, self-lubricating and homogeeous emulsion to a heterogeneous and highly unstable partical self-viscidizing suspension with highly variable local composition and very inhomogeneous susceptibility to hydrodynamic and/or gravitational forces. Interestingly, the compaction is not found [42.96] in the clump aggregation (see Fig. 88.22), which is caused by α_2-macroglobulin and by the high molecular weight dextrans (500 kDa) that is often used in experimental studies

[76.83]. Instead, these are driven into a rolling motion, which displaces the aggregated RBC in a retarded, but not in a progressively decelerating fashion.

In network aggregates (as caused by fibrinogen) two forces begin to become synergetically effective when blood flow is retarded. Both of these are negligible in the blood under normokinetic conditions. Adhesive and gravitational energies begin to act, which only exert their potential under hypokinetic conditions and markedly influence both the motion of blood constituents and their mobility in a self-amplifying process with pronounced, but variabe positive feedback. As the epitome of the latter, a thick sediment of highly concentrated RBC aggregates is progressively formed, which eventually creeps near the bottom of the vessel. Plasma, on the other hand, is floated, sediments upwards as it were, and thereby moves in the upper layers of the affected vessel; it flows more rapidly, a phenomenon appropriately designated 'plasma decanting' when first discovered by Knisely and Warner [55]. In medicine and general pathology these phenomena are often termed "*prestasis*" [22], as long as there is creeping flow, and "*stasis*" when the venules eventually become more or less compacted by RBC masses.

The Congested Mode of Operation: Local Reversal of the Fahraeus effect and Global Change in Resistance Distribution. Extreme forms of flow retardation emerge which represent much more than a trivial quantitative deficit emerge and are based on self-organized viscidation. Note again that a vicious circle as well as a "viscous circle" is set in action: flow retardation causes RBC aggegation and permits sedimentation. If both phenomena are properly coordinated, sedimentation causes a gradual increase in local RBC content and thereby markedly enhances apparent viscosity. This, in turn, is a combined consequence that further decelerates flow and augments the above effects: neither flow retardation nor aggregation alone has such a dual effect. Two combined events, however, both with weak positive feedback, can culminate in the "spontaneous" generation of a quasi-solid flow impediment. The latter is fully reversible and has nothing in common with blood coagulation, but predisposes for this process. Again, self-organization (drive-dependent coordination) proves to be the consequence of the joint efforts of many energies that in this case produce a localized flow impediment. As shown schematically in Fig. 88.24, the transition from a rapidly to a slowly perfused vascular segment has dramatic consequences on the velocity profile, and thence on the distribution of shear stresses acting between the fluid laminate and upon blood cells. As these pass from a narrow capillary to a wide postcapillary venule, not only does the fluidization switch to viscidation, but RBC overvelocity is also shifted to pronounced undervelocity. Both events occur in a leap at the very moment when the blood moves from the high shear environment in the narrow capillaries to low shear environment in the venules. Virtually every aspect of the flow dynamics (cf. Chap. 2) changes almost instantaneously and produces a wide variety of both local and global consequences.

In a connected network, the formation of a highly viscid and shear-resistant sediment must have multiple consequences, not only for the prevailing local fluid-dynamics, but also on all upstream and downstream vessels. The resulting hyperemia due to progressive filling of microvessels with blood constitutents should perhaps be termed the "*congested*" mode of operation in a synergetic context. As the sedimented RBC are progressively retarded in the majority of affected venules, so are global flow velocities, shear stresses, and their distribution. This situation is characterized by prolonged residence times, and progressively more creeping blood cells are accumulated in individual venules. There are highly variable rates of shear acting on widely differing mixtures of RBC and plasma that move in a highly variable fashion. Anything from pure plasma or 90% HCT in "decanting" vessels, to packed RBC can be found in "compacted segment" with a HCT >90%. The situation can even emerge where all RBC entering a given segment are deposited onto the sludged layer, while all the plasma is being 'decanted' and produces abnormal "skimming" at bifurcations downstream. This will inevitably lead to progressive RBC accumulation.

It is easily understandable the these dualistic hemodynamic and hemorheological inhomogeneities not only perturb the overall flow rate, but also destroy the directionality and coherency of microvascular flow in an entire network. In short, the network becomes chaotic in the sense of '*infrasynergetic chaos* (see Chap. 2). Note, however, that all of this is fully reversible only, however, provided that the shear stresses are sufficiently elevated

[42,96] so that the system is driven back into synergetic coherency (*synergetic cosmos*) due to the self-organized fluidization described in Sect. 88.2. As a rule, in low flow or hypokinetic states there is also temporal intermittency, i.e., periods of normal flow, forward flow with normal velocity, retarded flow, standstill in compacted, segments, and standstill in segments containing plasma only follow each other in a stochastic fashion. [92,98.101].

Network Effects of Postcapillary Blood Viscidation in Congested States: Intravascular Events Influence Transmural Fluid Exchange. Segregation of creeping RBC onto the dependent part of the vascular segment results in a "reversal" of the Fahraeus effect, because the RBC overvelocity typical for the normokinetic *Ablauf* mode is converted to a kind of plasma overvelocity. Depending on the tendency to aggregation (a compositional variable), incident shear stresses, (a hemodynamic variable) and previous history, a postcapillary impediment to microvascular perfusion is brought about. In striking contrast to the situation found in normokinetic modes of operation, elevation of the outflow resistance from the capillaries now limits overall perfusion and determines the effective filtration pressure within them.

Obviously, this tends to enhance the described hemodynamic and hemorheological effects due to local ultrafiltration: intravascular hemoconcentration can then become associated with interstitial edema.

This argument must be further detailed: the appearance of the blood movements (intravascular transfer, transmural transfers, interstitial percolation, and even lymph flow) undergoes drastic changes under these conditions which are highlighted by a drop in driving pressures combined with a rise in transmural pressure (Fig. 88.25B). The term "congested" is self-explanatory, while the term "hypokinetic" was chosen to paraphrase the somewhat paradox situation created by the combination of local low flow states with a globally maintained rate of perfusion.

Under both hypokinetic and congested conditions, i.e., when the RBC – plasma mixture attending the postcapillary venules are no longer kept in the fluidized mode of operation, the postcapillay venules are no longer vascular segments with a very high conductance [19.92]. As the creeping blood movement in the venules is complicated first by progressive intravascular aggregation (rheological phase transition) and then by progressive sedimentation (compositional phase transition), the effective fluidity falls progressively. As this affects venules more than it does arterioles, it gives rise to a pronounced rise in postcapillary resistance, despite the comparatively high diameter of these microvascular segments. This alteration rapidly begins to exert systematic effects that completely change the hydrodynamic network characteristics.

$$P_{cap} = \frac{P_{ven} + P_{art} \dfrac{R_{post-cap}}{R_{pre-cap}}}{1 + \dfrac{R_{post-cap}}{R_{pre-cap}}} \quad \left[Nm^{-2} \right] \qquad (88.18)$$

As can be seen from Eq. 88.18 (network equation), this combined hemorheological and hydrodynamic phase jump elevates the intravascular (P_{ir}) and the transmural pressure (P_{tm}) in the exchange capillaries while it reduces the longitudinal pressure gradient ($\Delta P/L$)in the capillaries and the smallest venules. Forward flow is greatly curtailed, while transmural flow is greatly enhanced, a situation typical for congested or an infarcted tissue (latin: *infarcere*, to stuff material into a vessel). The principle determinants for the phase shift from a freely running and self-cleansing perfusion pattern to a self-compacting and self-disturbing perfusion pattern follow from the logic underlying the network equation, which describes the interdependency of intravascular pressure and of resistance or conductance rations.

Conventional hemodynamic logic cannot do justice to this change in appearance. The total peripheral resistance (ratio between macroscopic driving pressure and global flow) can be normal, where as in both situations the local (microscopic and mesoscopic) perfusion patterns and exchange processes are dramatically altered. There are not just simple quantitative alterations, but a host of qualitative ones that have previously escaped quantification. For example, while the normokinetic perfusion pattern is endogenously stable, the hypokinetic and the congested ones are endogenously unstable. The progressively enhanced disturbances described in low flow states, which are caused by a wide variety of self-organized and autocatalytically augmented flow disturbances, can but need not occur.

Fluid dynamic stability or instability has its counterpart in the equilibrium of biochemical activators and their inhibiting mechanisms. In the normokinetic mode, there is highly efficient scavenging of catabolites and biological mediators, including activated enzymes and tissue hormones.

88.6.5 Different Modes of Microvascular Perfusion as an Explanation of the Maldistribution in Low Flow States: Chaotic Behavior in the Hypokinetic Microcirculation

As indicated in several previous sections, the interdependence of flow rates, shear stress distribution and rheological behavior provides a straightforward explanation for the highly variable composition of the blood in microvessels: in response to spontaneously occurring or vasomotor-induced variations in the shear stress distribution, in the residence times of the blood cells in individual arteriolar segments and in cell distribution phenomena at bifurcations, the local HCT level is highly variable. The averaged hematocrit can become markedly reduced during vasoconstrictor phases (a situation characterized by "screening" of RBc away from the mouths of slowly perfused branches) or it can be markedly elevated by vasodilatation (a situation described in detail in Sect. 88.4. The two situations are paraphrased in Fig. 88.25 by the term "screened" and hyperkinetic, respectively. Lastly, owing to the situations associated with strong local retardation, WBC margination, RBC aggregation, and compaction, stasis, the microvascular beds can become "congested" as defined in Sects. 88.4.3 and 88.4.4. All these situations are characterized by local coherence of perfusion in an entire microvascular bed as schematically depicted in Fig. 88.25a–d.

A strange mixture of these modes of operations are typically found in many so-called low flow states, i.e., when there is a wide spectrum of microvascular disturbances of variable origin. Quite often, the coherence of microvascular perfusion becomes progressively lost, and symptoms characteristic of either a screened, of a hyperkinetic and of a compacted microvascular bed occur simultaneously (for details see [12,101]).

While under normokinetic condition, the postcapillary venular segments are characterized by a very high conductance of simple geometric reasons [19,20,92], and the precapillary conductance regulators are the sole controlling agency for the network perfusion, this situation can become quite different under hypokinetic conditions where capillary or postcapillary segments can become rheologically occluded in an erratic fashion. Not only in the self-congesting, but also in the hypokinetic networks there is self-organized compaction producing not just tissue overhydration (edema), but also enhanced interstitial percolation and presumably enhanced lymph flow that is typical for many forms of microvascular shock. Here, however, this occurs in an erratic, chaotic fashion (Fig. 88.25e).

In many protracted low flow states, there is an unpredictable combination of self-diluting and self-congesting responses. These have been called hypokinetic modes of operation [98,101]. Many types of interactions are operational which have high pathophysiological significance since they dissociate global flow from local distribution of flows. Consequently, there is global incoherency of movements, replacing characteristics for the normokinetic and hyperkinetic modes of operation. This pivotal qualitative change, rather than any trivial quantitative alteration, highlights the many forms of microvascular disturbances associated with disease states ranging from shock to chronic inflammatory states. It is well-known that all types of hypokinetic circulatory situation, local or global low flow states, are associated with a more or less dramatic, but not necessarily with lethal drop in systemic arterial pressue. It must be stressed again that contrary to the meaning of the term low flow state. detailed investigations disclose that the global flow rate in microcirculatory disturbances of the affected organs often remains within the range of normal values [2.26].

However, as is being increasingly recognized with adequate techniques for testing microvascular homogeneity [85], so-called microvascular disturbances are as a rule associated with passive relaxation of the precapillary arteriloles: spasms seem to play a minor role in this context [2,98]. Instead, endothelial swelling and partial plugging of individual network segments are found. Consequently, clinicians find the parameters of transmural exchanges al-

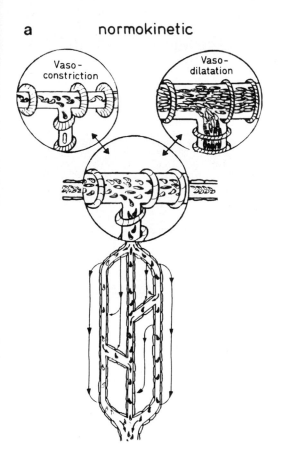

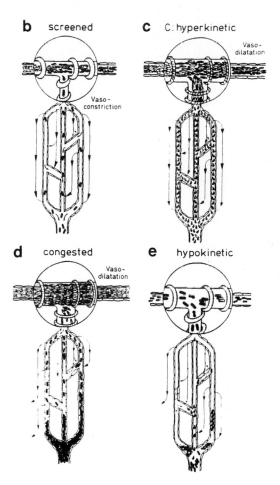

Fig. 88.25a–e. Schematic representation of the intravascular and transmural blood movements in the normokinetic, the "screened", the hyperkinetic, the congested, and the hypokinetic modes of operation. **a** During normokinetic flow, the vasomotor play of precapillary conductance regulators (classically called "ieslstance vessels") regulate as the sole controlling agency the overall flow rate, but also the attendance of calls to the exchange vessels (see Fig. 88.18) and the transmural fluid transfer (reversed osmosis and colloid osmoses). **b** As an extreme of reduced precapillary conductance, the blood cells are being "screened" away from the entrance into daughter branches, a situation leading to extremely low blood cell content of microvessels. **c** During the *hvoer*kinetic mode, where due to vasorelaxation there is no plasma skimming in the upstream vessels, the microvascular red blood cell (RBC)-attendance is dramatically enhanced. For more lack of residence time between bifurcations, the Fåhraeus effect is eliminated. **d** In the congested mode (precapillary vasorelaxation, postcapillary compaction stasis), the overall flow rate oan be enhanced or normal, but the microvessels are stuffed with RBC. Due to the reduction of postcapillary conductance, the intravascular and thence the transmural pressure in the exchange capillaries is elevated, a factor favoring interstitial edema. **e** In the hypokinetic modo ("low flow states"), the above-described modes of operation produce maldistribution of blood flow or chaotic mixtures of stases, normokinetic, and congested modes. As a rule, the arterioles are relaxed, and there are multiple intravascular obstacles ("plugging" and marginated WBC, aggregated RBC, thrombocyte microaggregates, abnormally dehiscent endothelial cells with adherent blood cells and mixtures of these abnormalities)

tered in hypokinetic states: local pO_2 capillary filtration coefficients and indicator clearance rates are abnormally low, While global flow rates can be in the normal range. In the future, tests measuring homogeneity, e.g., by quantifying coherence and/or non coherence of microvascular perfusion must, therefore, be applied the future in assessing the functional significance of hemorheological factors in physiology and pathophysiology [2,22,98]. Such tests have recently been developed for clinical application and are shown for illustrative purposes in Fig. 88.26, taken from Scheffler [85]. Here, the temporal characteristics of microvascular perfusion in the soles of a patient with one

healthy and one diseased leg is shown before and after the provocation test of an induced ischemia. While in the "healthy" extremity the postischemic reactive hyperemia is associated with more homogeneous perfusion, the diseased one shows more pronounced heterogeneity after the provocation test of a complete ischemia than before it.

Hypokinetic states are, therefore, characterized by spatiotemporal stochasticity, i.e., by the coexistence of normally perfused, underperfused, nonperfused, and even hyperperfused microvascular segments [2,96] taking turns in their perfusion. In short, the homogeneity, i.e., the functional coherency of the normal microvascular

Effect of intraarterial infusion of BUFLOMEDIL (200 mg) on distribution of microvascular flow (Fluorescein Appearance Time) of the normal foot and the foot in POAD

Control

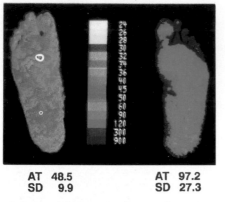

| AT 48.5 | AT 97.2 |
| SD 9.9 | SD 27.3 |

post infusion

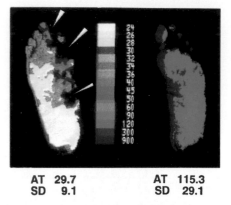

| AT 29.7 | AT 115.3 |
| SD 9.1 | SD 29.1 |

Fig. 88.26. Perfusogram of the sole of a foot of a patient with a unilateral hypokinetic microvascular state, characterized by incoherent perfusion in the period of reactive hyperemia after 2 min of arterial occlusion. Perfusograms are computer-generated parametric images obtained by the evaluation of dynamic fluorescence angiography. During the wash-in period of a fluorescent dye (Na-fluorescein) sequential angiograms are recorded by appropriate video systems: to generate a perfusogram, the appearance times of the leading front of fluorescent blood in each pixel element is calculated and its numerical value is later displayed in a color-coded image of the entire foot. The perfusogram of the healthy foot (*left*) in the postocclusion period shows marked flow acceleration ("hyperkinetic mode") associated with marked reduction of the individual as well as the mean appearance times. In the diseased foot (*right*) the postischemic reactive hyperemia (*RH*) is still present, but is highly nonhomogeneous in its topological distribution. The "chaotic" nature of the hypokinetic state can also be quantified numerically; while in the healthy left foot, the coefficient of variation (*CV*) is lower in the postischemic phase (C.V. = S.D./AT = 0.1 in control; C.V. = 0.09 in RH), it is higher in the diseased foot already showing greater variance at rest (C.V. o. 14 in the control, C.V. = 0.22 in RH [85])

porfusion process is more and more replaced by an irregularly vacillating nonhomogeneity, termed "maldistribution", mismatch of stasis and shunt, or patchy circulation. As a central message derived from general synergetics (see Chap. 2), one must recognize that this is a situation where there is circular causality: attempts to separate one specific causal mechanism are as futile as the attempt to predict the precise future course of events from a measurement at present.

The extreme of this situation is obviously found in macroscopically infarcted regions, i.e., an irreversible stagnation due to stuffing of microvessels by compacted blood constituents, which act in impeding their own removal. This situation, complicated by tissue necrosisis often is the epitome of self-organized pathology, initiated by a microvascular catastrophy where a subsystem falls out of its orthological coherency with a supersystem, due to abnormal local cell–cell interactions, local cell–wall interactions, local fluid–solid transitions. Likewise, new equilibria between enzymatic activator and inhibitor mechanisms produce an equally chaotic situation, also dominated by a multitude of incoherently operating influences [101]. To what extent the initially reversible forms of infarction are caused by self-amplifying microvascular or self-amplifying parenchymal cell alterations is presently unknown. Using refined techniques, it can be hoped that the many interactions causing irreversible changes in the very short critical phases can be clarified in the future. Such investigations are also essential to comprehend the often obscure benefits of hemorheological therapies. Furthermore, the situation in the hypokinetic mode of operation is characterized by abnormal interactions between blood cells and endothelium [9,40,63,116].

88.6.6 Hemorheological Therapies and Their Mode of Operation

The concept of self-organization by restoration of coherency in the normokinetic mode of operation may serve to explain the clinical success of various forms of hematopheretic therapies. All of these have a dual effect, namely, enhancing the fluidity of blood [44] and restoring its motion, e.g., by increasing venous return, cardiac filling, and cardiac output. Note, however, that this type of therapy cannot be rationalized by a trivial explanation based on the numerical changes in a transport coefficient (i.e., the coefficient of viscosity) or an increase or decrease in the composition of blood. Instead, it can and must be reduced to a much more general and holistic concept, namely, the attempt to reverse a phase transition into self-viscidizing mode of operation (see above).

Ever since the days of practical medicine in ancient Greece, phlebotomy has been an important part of the armamentarium of practical medicine, and is primarily performed in decompensated stages of cardiovascular disease and the critical phases of inflammatory disease, for neither of which any specific therapy was available before the twentieth century. The original concept rooted in the first Western rational medical theory, was quite different: disease was related to abnormal blood composition, and the aim of phlebotomy was to remove noxious materials (*materia peccans*). Humoral pathology had taught that the weakened body must be cleared from an excess of normal, but potentially viscidizing blood components ("plethora", "dyscrasic blood").

This time-honored concept proves to be basically correct. There is ample evidence that all the many various forms of hemapheretic therapies both improve the fluidity of blood and restore the integrity of disturbed microcirculation [2,26,44,92]. This occurs because resolute hemodilution therapies reduce the macroscopic viscosity of blood and of plasma by 25% to 50% at most. All forms of hemodilution must, of course, be administered in an individualized fashion [44], and this is best achieved by simultaneous infusion of solutions of natural colloids, e.g., human serum albumin, or artificial colloids, often enhanced by erythrocytopheresis, i.e., the removal of RBC and their careful replacement by colloidal solutions. As detailed in Sect. 88.5.2, quite contrary to a widespread prejudice, even severe forms of induced anemia are well-tolerated by bed-ridden patients provided that the blood volume is meticuously kept within normal limits. Improved fluidity via enhanced venous return has of course a dominating effect on cardiac performance (see Chap. 89), which improves circulatory function by restoring the normokinetic mode of operation where blood becomes self-fluidizing. In this context, simple questions about "optimal" O_2 supply are of second order importance; instead maintenance and/or improvement of venous return must be aimed at, its factual elevation after normovolemic hemodilution is well-documented in experimental and in clinical settings.

Presently, there are various established forms of hemorheological therapy in the strict sense, all of them based on the removal of abundant but normal blood components called hemapheresis:

- The systemic HCT (individualized or "custom-tailored" hemodilution) can be reduced by simple hypervolemic infusion or by isovolumic exchange of RBC for human serum albumin [46] or for solutions of artifical colloids (dextrans, hydroxy-ethyl starch);
- The fibrinogen concentration can be reduced by mild proteolysis using snake venoms (26) or by cold precipitation [106];
- Other plasma proteins can be eliminated by plasmapheresis [49] or cascade filtration.

The therapeutic efficacy of such remedies is often surprising, and better than that of any orally administered putatively vasoactive drug. At the time of writing, most of these remedies are still not unequivocally proven in large-scale clinical trials, but are presently the topic of vigorous research in clinical and experimental medicine.

88.7 General Hemorheological Synergetics: Self-Organized Cooperativity Between RBC Behavior and Movements of WBC and Platelets

Not just in sheer number, but also in many important functional aspects, RBC dominate the flow behavior of other blood cells, e.g., platelets and all WBC. Simply because they are much less susceptible in their rheological behavior to the fluid dynamic forces acting upon them, the other blood cells are, however, codetermined in their functions by the behavior of the highly shear-susceptible RBC [10,74,75,112]. In short, the different types of blood cells are rearranged in flow so that the platelets and all types of WBC are preferentially displaced onto trajectories in the immediate proximity of the vessel walls, i.e., precisely where they exert normal and abnormal biological functions. The platelets, too, are dynamically enriched in the marginal fluid layer in microvessels [112], and presumably in macrovessels [10]. Totally unspecific global movements thus support very specific cytological reactions in a strictly localized fashion determining the sites where they can exert cell-specific biological tasks.

88.7.1 Platelet Rheology: Hemostatic Functions Are Often Associated with Very High Shear Forces

The fascinating field of multiple phase transitions that accompany the various manifestations of hemostatic reactions cannot be detailed here for reasons of space. Suffice it to say that these are, of course, associated not only with autocatalytically enhanced procoagulatory enzyme activation but also irreversible transitions from the fluid (or quasi-fluid) into the solid state of aggregation of blood component (by fibrin polymerization and irreversible platelet aggregation). Moreover, it is important that–contrary to a widespread misconception–the vast majority of hemostatic reactions takes place in high shear environments [92,120,121]. As detailed in Fig. 88.27, the rheological situation at a mechanically severed

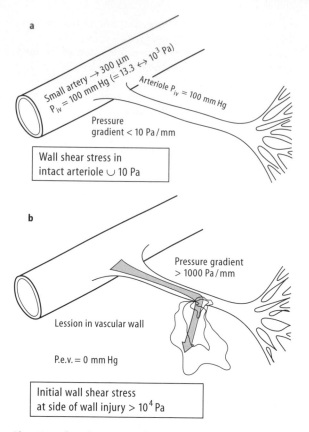

a

Small artery → 300 μm
P_{iv} = 100 mm Hg (≙ 13.3 ↔ 10^3 Pa)

Arteriole P_{iv} = 100 mm Hg

Pressure
gradient < 10 Pa/mm

Wall shear stress in
intact arteriole ∪ 10 Pa

b

Pressure gradient
> 1000 Pa/mm

Lession in vascular wall

P.e.v. = 0 mm Hg

Initial wall shear stress
at side of wall injury > 10^4 Pa

Fig. 88.27a,b. Schematic representation of hemodynamics in arteriole after incision of vascular wall and during formation of hemostatic plug. Pressure gradients (ΔP/l) will depend on length of proximal segment, upstream pressure, and absolute hydraulic resistance in cut segment. **a** Situation in intact arteriole. Small pressure gradient and low shear stresses. **b** Situation immediately after vessel injury and before onset of platelet adhesion. Arteriolar pressure gradients and thus shear stresses will be larger by several orders of magnitude than those naturally occurring in vivo before transsection, simply because, over a very short distance, intravascular pressure falls to atmospheric pressure

arteriolar vessel is initially associated with shear rates far in excess of those operating anywhere in the intact vascular bed.

Note that in intact arteries the pressure drop per unit length (Δ P/L) and thus the shear stress at the wall (τ_w = r/2 × Δ P/L) is comparatively low: immediately after transection of an arterial vessel, however, the local transmural pressure gradient is very high, and thence the shear stresses are by at least two orders of magnitude higher than the ones before injury. The most pivotal initial steps of all hemostatic reactions must therefore – and in fact do – take place in an environment of extremely *high* local shear forces and rapidly produces a flow impediment capable of withstanding high pressure gradients. Contrary to another widely held misconception, the hemostatic reaction gives rise to flow retardation, not vice versa. The responses of RBC and thrombocytes to high shear stresses are in essence destructive ones. Cells are mechanically ruptured within intervals of a few milliseconds, which represent a functionally beneficial response in that it aids in

"activating platelets" and in demasking procoagulatory phospholipids (a topic extensively covered by Wurzinger [122].

As a further benefit of the initially very high flow rates, platelets can be physically deposited onto the physically injured vascular wall in large quantities, forming a rapidly growing hemostatic plug (see Chap. 83 and [10,34,43, 52,95,97,99,120–122]. When mechanically activated, their membranes provide a catalytic surface for the strictly localized activation of coagulation proenzymes. The emergence of the multiple hemorheological and enzymatical roles of deposited platelets can thus be paraphrased by the term "conversion to catalytic anchors". By immediately forming a strong plug in mechanically severed blood conduits, they can rapidly stop the movement of blood and blood components through discontinuties of the injured vascular compartment (also cf. Chaps. 6 and 83). It is intuitively clear that, for this purpose, large quantities of thrombotic material must be rapidly delivered. This process critically depends on changes in the local behavior of the deposited and activated platelets, which is codetermined by changes in microhemodynamics when endothelial cells become suddenly detached. Platelets approach the site of injury as innocuous suspended particles in a fluid and, when deposited, become incorporated into a localized solid [10,107]. Their very potent, and potentially dangerous viscidizing effect again operates automatically in a self-focussed fashion.

In a healthy organism, deposition is adapted to needs; in its most discrete form, it occurs in response to spontaneous injury caused by detachment of individual endothelial cells. Formation of gaps between endothelial cells directs platelets by means of a sudden increment in transmural flux to the denuded site where they exert a repair function [97] as pseudoendothelium. Once platelets have reached subendothelial sites, they permanently interact with type III collagen fibers, and other highly thrombogenic subendothelial structures (see Chap. 83). Subsequently, as an independent, autonomous cytological process, platelets become activated by undergoing an explosive and irreversible shape change called "viscous metamorphosis". There is a rapid formation of highly adhesive pseudopodia [122], an eruptive release of platelet activating substances from storage granules, and simultaneous demasking of procoagulatory membrane phospholipids. Anionically charged phosphatidyl serine [125] becomes available at their outer membrane surface, generating catalytically active binding sites for coagulation proenzymes via Ca^{2+}-bonding.

The many details of biochemical reactions taking place during and after activation are detailed in Chap. 83. In the context of synergetic hemorheology, the self-organized nature of this process again follows from the simultaneous emergence of hemodynamic abnormalities and cytological mechanisms paraphrased as "viscous metamorphosis", resulting in a fluid-to-solid phase transition. Unlike the ones described for aggregating RBC, the two hemostatic phase transitions of platelet aggregation and subsequent fibrin formation are irreversible [122], and they withstand much higher hydrodynamic forces. For pathological hemostatic processes (thrombosis) such as it occurs in separated or vortex carrying flow, see [104].

88.7.2 Hemorheological Aspects of WBC Functions: Self-Organized Transport Processes Preparing for Antimicrobial Defense

Permanent establishment of antimicrobial defense is a physiological function that is obviously linked to the integrity of the blood circulation. As detailed in Chap. 85, the macroorganisms, forced to survive in a world full of microorganisms, have developed a wide armamentarium of antimicrobial strategies to prevent infection. These range from the gross flushing of a wound with fresh blood when the tissue is mechanically severed to highly specific molecular antigen–antibody reactions, which often depend on submolecular interactions, e.g., charge effects, hydration effects, and dissociation or association of protons from molecular residues. As detailed in Chap. 85, the combination of reactions traditionally classified as the *unspecific defense mechanism* are in actual fact globally functioning, rapidly recruitable, but indiscriminate biological reactions which are more or less automatically set in action by the injury, as well as by the microorganisms and their biological activity.

In experiments using intravital microscopy, where differential staining is impossible, one observes all types of WBC as one class: all WBC are roughly spherical in shape with diameters between 5 and 8 µm and "stiffer", i.e., more difficult to deform passively than RBC. The important topic of processes governing individual WBC rheology can only be covered in passing here. The passive mechanical behavior is superimposed by active secretion, active amoiboidal movement, and complex alterations in cell surface characteristics (for further details, see [6,65,66,86]). In the present context, only the self-organized changes in RBC–WBC interactions in flowing blood will be covered.

Passive Margination and Active Emigration of WBC. The movements of all types of nucleated WBC in the terminal vasculature are being coordinated passively by the interactions with moving RBC [31,87,88,99,114]. Simply due to their higher mass, they are subjected to higher forces than individual RBC when submitted to accelerating forces. In a mixture of monodispersed RBC and WBC, the latter are preferentially moving in the rapid axial stream in macrovessels, i.e., are not likely to come into contact with the vessel wall. When as few as five RBC are united into an aggregate, their mass is much higher than that of the individual WBC: this is a mechanism that automatically tends to displace the WBC away from the axial stream (see below). In actual fact, at least two different mechanisms are directing WBC towards the venular walls and thus induce margination (Figs. 88.28, 88.29).

In the arterioles, where the shear stresses are usually much higher than those in the venules, the flow tends to keep RBC dynamically disaggregated. Consequently, the WBC

are the largest moving particles and travel near the vessel axis, i.e., far away from the arteriolar walls and side branches. In the venules, where the shear stresses are often so much lower, RBC can rapidly form large aggregates and these can displace the WBC (of all subclasses) away from the preferred axial layers and towards the wall. Under pathological conditions, i.e., when the tendency to RBC aggregate formation is increased, the aggregates persist even in the presence of very high shearing forces. High plasma concentrations of aggregating colloids such as fibrinogen as an acute phase reactant thus enhance WBC margination by exaggerating and accelerating RBC rouleaux formation. Dynamic RBC aggregation–disaggregation determines which type of blood cell comprises the largest particle in flow, individual WBC or aggregated RBC. WBC margination thus critically depends on the dynamic equilibrium between shearing forces and the adhesive forces acting between RBC Fig. 88.29.

Once dislocated into the marginal layers, the WBC can either roll along the wall, or become stuck there more or

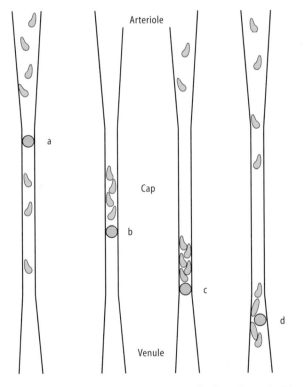

Fig. 88.28a–d. White cell margination at the diverging end of the nutritive capillaries: schematic representation after the work of G. W. Schmid-Schönbein. During its passage from arterioles through the narrow nutritive capillary (*Cap*), to the venules a white cell (*WBC*) retards the flow of red cells (*black*). **a** These progressively accumulate behind the WBC (**b, c**) and form a queue of cells waiting to bypass the slow moving obstacle to traffic. As the capillary widens into the venule at (**d**), the red cells in passing the WBC push it to the endothelial wall, to which it adheres by unknown action. This mechanism is highly effective in producing white cell margination even in rapid flow. The latter, however, also produces shear forces dislodging and marginating WBC, which continuously roll along the venules

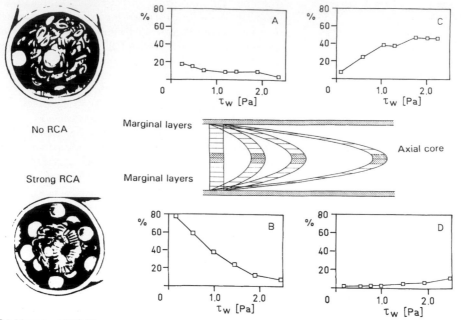

Fig. 88.29. Hemorheology of the Fåhraeus-Veylens effect. White blood cell (*WBC*) margination as a process of dissipative structuring. Variable effect of dispersing forces of fluid dynamics and adhesive forces (tendency to red blood cell, *RBC*, aggregation) on location of WBC trajectories percentage of WBC flux carried through a defined fluid layer) marginal layers (*A*,*B*) and the axial core (*C*,*D*) comparison of nonaggregating RBC suspensions (RBC and WBC in isotonic saline) and strongly aggregating RBC suspensions (1% high molecular weight dextrans). The combination of strong aggregating and weak dispersive forces lead to preferential WBC (>70%) travel on marginal trajectories (*B*) and no WBC travel (<10%) on axial trajectories (intermediate layers not shown). Under these conditions, the RCA are united into masses greater than the mass of the individual WBC. If, however, the hydrodynamic dispersion of RCA occurs at high shear stresses, the WBC are progressively displaced into the more axial layers and disappear from the marginal ones. Conversely, in nonaggregating suspensions (*C*,*D*) with increasing shear stresses, there is progressive accumulation of WBC into the axial steam making WBC margination and unlikely event. Experimental data taken from [70], drawing from [114]

less permanently in reaction to the exposure of adhesion molecules best studied for neutrophil granulocytes [62]. As a consequence the WBC are dynamically accumulated here, simply because many of them arrive, move forward slowly and leave the postcapillary walls with a marked delay. The sticking and rolling WBC constitute the peripheral marginating WBC pool together with WBC lodging in pulmonary capillaries (see below), which comprise a major fraction of all WBC present in the organism. Only after having been passively marginated can WBC move actively through gaps between endothelial cells into the interstitial space (*emigration*), and thus reach the principal site of their function in connection with inflammatory and immunological responses. The described fluid-dynamic-hemorheological docking mechanism provides large ensembles of WBC acting in concert precisely where needed. For the hemodynamic reasons described in Sect. 88.6, margination to the walls of venules displaces them to a site of *low* shear stresses. The circulating granulocytes and monocytes marginate in all types of venules. The lymphocytes are unique in their ability to marginate and emigrate in venules of blood vessels perfusing the lymph nodes (see Chap. 94) and other lymphatic tissues, a mecha-

nism well-known for its critical role in the incessant migration of lymphocytes through the entire body.

Depending on the WBC and endothelial "stickiness" (i.e., adhesive energies), maintenance of endothelial cell continuity [3,4] local hemodynamic forces, shear stresses, and transmural pressure, a variable number of WBC come to a more or less complete standstill and can be classified as rolling or sticking. Owing to many types of active WBC responses, the dynamic adhesion-detachment process strongly depends on the previous history of a process or experiment, respectively, and is thus difficult to model experimentally (see [60,61]). The dynamics of chemical agents mediating WBC adhesion to endothelial cells, i.e., a variable, often jerky sequence of attachment and detachment is much more complex [3,8,56–58,98] than the simple "gluing" action by macromolecular colloids described for RBC in Sect. 88.6.2 (Fig. 88.30).

The next step of an antimicrobial defense reaction, namely active emigration, i.e., amoeboidal movement through discontinuities of the wall, follows from a sequence of intracellular processes of actomyosin polymerization and depolymerization typical for cell mobility (see Chap. 10). Very similar movements, which are guided by

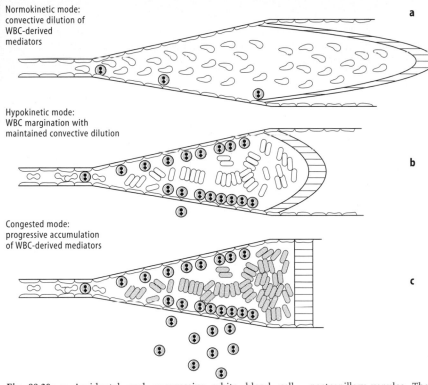

Normokinetic mode:
convective dilution of
WBC-derived
mediators

a

Hypokinetic mode:
WBC margination with
maintained convective dilution

b

Congested mode:
progressive accumulation
of WBC-derived mediators

c

Fig. 88.30a–c. Accidental and progressive white blood cell (WBC) accumulation in the postcapillary venules. **a** High shear, red blood cells (RBC) there is strong deformation and dilution, occasional WBC-marginations often observed due to sideway pushing during RBC–WBC collision. Rapid rolling of WBC, and thence equilibrium between WBC deposition and WBC clearance into larger veins: stationary venular deposit of WBC. **b** Progressive WBC accumulation due to combination abnormally pronounced RBC aggregation (producing WBC-margination) and low shear. Due to the lack of adequate forces to dislodge WBC, there is progressive WBC accumulation and prolonged residence of WBC postcapillary venules. There is a much higher chance for the occurrence of sequential passive movements (WBC margination, WBC rolling). **c** WBC accumulation in the congested mode is likely to occur. The resulting local hemodynamic-hemorheological modes of operation are seen to exert pivotal influences on the interaction of WBC with endothelial cells. Weak shearing forces lead to slow rolling and frequent sticking. Presumably due to lack of convective dilution of liberated mediators, WBC activation and endothelial damage is permitted. Marked adhesion, activation and emigration of WBC is dynamically favored. (After intravital microscopy by various authors)

chemotactical agents, cause their interstitial migration towards the site of microbial infection.

The mechanisms operational before the WBC activity can manifest itself represent a good example for both self-organization and phase transition, as defined in Chap. 2. All forces involved must be reconciled, as well as the leap from one to another mode of operation, as only the driving forces are changed. Thereafter, even weak adhesive energies gain the option to become functionally operational. Rolling velocity along as well as detachment of WBC from the vessel walls were found to be a strikingly nonlinear functions of the shear stresses. Furthermore, it must be appreciated that very high shear stresses as well as the presence of other WBC, both interfere with primary adhesion and promote detachment [60,61,71]. All of these facts suggest that there is a continuum of adhesion forces for any given degree of adhesivity. The latter feature, in turn, is highly variable and depends on WBC activation [62], a process which on the one hand promotes amoeboidal movement (emigration) and on the other hand mechanical detachment [4,87,111].

WBC Margination As a Consequence of Task-Specific and Site-Specific Phase Transition of RBC Movements. In the context of the present chapter, the hemodynamic aspects of target specific WBC displacements deserve special emphasis, since the mere transition from rectified to aggregated RBC plays such a pivotal role in the processes displacing WBC in a purely passive fashion to the wall. Two distinct hemorheological mechanisms are now established that can cause displacement of WBC towards microvascular walls, and both involve mechanical interaction with moving RBC:

- Individual RBC, while passing the WBC interact with them by a principle of sideway pushing. This occurs at the exit of narrow capillaries to widening venules and follows from crowding in rapid flow [88];
- Aggregated RBC, in forming large clumps that exclude WBC, accumulate in the axial region and thus displace the less bulky individual WBC into the more marginal layers (Fåhraeus-Vejlens effect [70,114]).

As to the first mechanism (depicted in Fig. 88.28), the WBC margination by individual RBC due to transfer of momentum between particles can be explained by the fact that their passage through very narrow capillaries occurs with different velocities. Displacing interactions take place at the exit of the capillary, it is the last act of a longer process by which RBC are accumulated behind the WBC. We know that the marginating event is initiated by the slow passage of a voluminous and noncompliant, highly viscid, functionally stiff WBC that retards overall flow in the narrow capillary. As it moves forward, it must be assumed that it allows the low viscosity plasma to pass, a process that can only be inferred from the fact that many RBC progressively close up behind every WBC traversing a true capillary [88]. As RBC thus assemble behind the creeping WBC, much like passenger cars behind a slow truck form a long trail in the traffic through a narrow road, they slip past them as the capillary widens into a venule (of say 9 μm diameter). At this instant, the highly compliant RBC and presumably the plasma as well overtake the WBC, the passing RBC thereby transferring part of their higher momentum to the WBC, which results in pushing them towards the venular wall.

As to the second mechanism (depicted in Fig. 88.29), it has long been known and was paraphrased as Fåhraeus-Veylens effect that the margination of WBC by aggregation of RBC occurs in much larger than in the immediate postcapillary venules when these are perfused with RBC aggregates. Following a general fluid-dynamic principle, large suspended particles not only drift faster to the vessel axis, i.e., a region of most rapid flow but least frictional energy dissipation (see above). Particles with higher masses not only drift more rapidly than smaller ones into this region, but they are stabilized more efficiently in these energetically preferred areas [19,76].

88.7.3 Synergetics of Holistic Yet Localized Blood Movements in Inflammatory Reactions

In every respect, self-fluidizing nutritive flow differs from self-viscidizing flow during inflammatory reactions. In producing an outflow impediment, WBC retardation on the one hand and RBC compaction stasis (see above) on the other initiate a new type of boosted or nonequilibrium state (see Chap. 2), which allows self-organized phase transitions and thus a wide variety of biological alterations to occur. By a means of host of secondary consequences, these hemodynamic and hemorheological events cooperate in the antimicrobial defense. Without the original event of passive WBC margination, active emigration is obviously impossible. More or less permanent adhesion, in the form of rolling, and irregular sticking to the endothelial surface appears to proceed to the permanent docking as the initial stage of transmural emigration. Biochemical activation of both granulocytes and endothelial cells, and especially the presence of vesicle-bound or membrane-bound adhesion molecules (selectins, see Chap. 85), and of unspecific and specific inhibitors and fluid-dynamic forces interact. This occurs in a fashion that produces highly variable, presumably nonlinear effects that await further clarification.

The important limiting role of the selectins in successful anti-inflammatory reactions is dramatically illustrated by a clinical syndrome called leukocyte adhesion deficiency (LAD) characterized by leukocytosis without pus formation [1] and a severe form of immune deficiency. This experiment of nature shows that the absence of a molecule inducing adhesion makes impossible a complex self-organized process of dissipative structuring. While there is doubt in the true specificity of adhesion molecules, i.e., superfamily of vascular selectins, integrins, and immunoglobulins, they share a potential to establish a permanent physical contact between blood cells and endothelial cells. This, again, is a strictly localized fluid-solid phase transition which plays a pivotal role in coordinating the subprocesses cooperating in the holistic yet localized sequence of events presenting as inflammation whenever there is local WBC retardation in a large number of venules. On the one hand, the coherently orchestrated physicochemical interactions allow the emergence of spontaneous interaction between the foreign organisms and their hosts. On the other hand, cooperativities between the various cellular and noncellular movements occur, so that a wide variety of blood components fall into a concerted biological reaction, again characterized by pronounced site predilection. The spontaneous and self-amplifying processes that initiate the global inflammatory response begin to be comprehensible. At present, the cytological reactions associated with WBC activation and the resulting changes in their passive flow behavior have not been studied in detail. Obviously, however, after microbial and perhaps autoimmunological stimuli begin to act, the mechanical behavior of the moving blood is changed: as depicted in Fig. 88.30, there is a phase "jump" in the Sherringtonian sense (see Chap. 2). Unless this phase jump from the normokinetic into the congested mode of operation has occurred (cf. Fig. 88.25), the chemical mediators released from marginated WBC cannot act because they are diluted by the strong blood flow. Conversely, as soon as the congested mode leads to a venular outflow impediment, multiple autocatalytic events are physically unavoidable; these in turn, are obviously essential for successful inflammatory reaction. The central dynamical role of complexly altered microvascular flow brings the interaction between propagation and retardation into play, giving processes of dissipative structuring a pivotal role in the inflammatory antimicrobial defense reactions (Fig. 88.30).

The described phenomena were initially discovered a century ago by pathologists [22] using intravital microscopy (*flow vizualization*) as a tool of investigation. Their observation had explained the well-known, macroscopically visible, phase transition in inflamed skin paraphrased as the "cardinal symptoms of inflammation":

- Rubor (reddening);
- Calor (heating);
- Tumor (tissue swelling due to edema); and
- Dolor (pain due to accumulation of algetic substances in the affected tissue.

Obviously, similar alterations accompany host-defense reactions in all organs, as concerted reactions they are the macroscopically visible symptoms of inflammation known since antiquity.

For reasons detailed in Sects 88.6 and 88.7, the most prominent deviations from the normokinetic or uninflamed situation are bound to occur in the postcapillary venules. In the congested situation, blood elements first attending the capillaries and then passing into the venules experience a sudden drop in shear forces. Provided that the resulting alterations in the behavior of flowing blood elements are associated with alterations of the vascular wall, they are likely to produce enhanced transcapillary exsudation of plasma constituents into the interstitial tissues. This produces not only edema, but also insudation of the extravascular spaces with a host of macromolecules such as albumin, immune-globulins, and fibrinogen. Thereafter, the interstitial spaces are also altered by chemotactically acting agents; under their influence, WBC then can migrate by directed amoeboidal movement from the perivenular spaces to the site of bacterial growth. All of these reactions are an example of the spontaneous emergence of functionally successful collective actions of large ensembles of individual blood-borne cells, which are endogenously free, but susceptible to a set of simultaneously acting physical and chemical stimuli. The eventual production of pus, containing millions and millions of granulocytes, bacteria, and a host of catabolites produced by proteolytic processes and its penetration through the skin to the outside is the macroscopic sign of successful antimicrobial actions. It should, therefore, perhaps no longer considered as a pathological, but rather as a normal biological response emerging from a self-organized collective modes of operation ensuring the survival of macroorganisms in an environment full of potentially lethal microorganisms. The serious morbidity of the LAD syndrome testifies the first-order biological significance of this scenario spontaneously generating a situation far displaced from equilibrium.

In the final analysis, every step of this collective reaction is related to the modulated mobility of elements due to interactions between dispersive shear forces and adhesive forces in a vascular region characterized by a priori unstable perfusion. While in inflammation inflow conductance is likely to be increased by a host of vasodilator mediators liberated in inflammatory reactions (e.g., histamin, bradykinin, and prostacyclins; see Chap. 6), outflow conductance is reduced by the sequelae of at least two interdependent viscidizing hemorheological phase transitions; namely, RBC aggregation, i.e., global viscidation of the RBC plasma mixture, and firm, selectin-mediated granulocyte-endothelium adhesion narrowing the venular lumina. Obviously, the two phenomena mutually enhance each other, since aggregation produces flow retardation as well as WBC margination. This is a prominent case of self-referentiality (see Chap. 2); hen and egg cannot and need not be distinguished in this positively fed-back process, as in most processes of synergetic self-organization. Its effect, however, fundamentally alters the hydrodynamics of the entire microvascular network.

Before these postcapillary phase transitions, the resistance distribution (see Eq. 88.18) is fundamentally different from that seen after the inflamed and congested mode of operation has emerged. The term *phase transition* achieves a new meaning with respect to tissue hydration. Normokinetic perfusion is associated with an equilibrium between filtration and reabsorption when the ratio of postcapillary resistance to precapillary is much smaller than 1.0. Any change in this ratio is obviously associated with progressively enhanced filtration. As hemorheological alterations change it to values >1 (outflow impediment progressively limiting total perfusion), there is increasing filtration and decreasing reabsorption. This is the situation found in the congested mode. In situations characterized by a combination of compacted RBC and plugging WBC, this produces precapillary occlusion, and consequently the intravascular pressure in capillaries and venules can drop to the venous pressure. In this case, colloid osmosis by the stagnant plasma can cause the unidirectional reabsorption of interstitially dissolved material (conceivable as histamine liberated by mast cells). This is likely to exert effects on the endothelial cells in microvessels in which there is no flow to remove these biogenic amines: this will necessarily induce further collective alterations, now on the basis of altered endothelial cells.

At present, the synergetic interpretation of inflammatory reactions has just begun. Obviously, even the most basic mechanisms (marginating, rolling, sticking), all of which are known to be dependent on hydrodynamics, on previous history, on RBC rheology and – last but not least – on the inflammatory stimulus are complicated by the degree of WBC activation. Furthermore, they depend on the degree of endothelial cell alteration and on the rate and directionality of transmural fluid and mediator transfer. One fundamental aspect should be highlighted here, namely, the interdependence of convective flow and local accumulation of proteolytic enzymes. As detailed for proteolytic processes in the context of thrombotic events [97,99,122], the convective forward flow not only ensures simple dilution, but supplies protease inhibitors leading to scavenging reactions (i.e., protease-inhibition, complex formation, and its removal). As net forward flow is reduced, and proenzymes are continuously either liberated or activated, there is a sudden shift in that free enzymes are available and begin to exert their multiple biological actions. The action of the now established lectins, the demasking of receptors on endothelial cells and thence the forces associated with WBC sticking are known to be strongly influenced by prior protease action on either cell (Fig. 88.30). The very marked alterations in inflamed microcirculations are thence most likely the consequence of multiple positive feedback mechanisms with built in criticality (threshold behavior) which is, in the final analysis, flow dependent: no flow retardation, no positive autocatalytic feed back.

These purely hydrodynamic local microvascular processes are augmented by the appropriate global responses termed the acute phase reaction (cf. Chap. 85). Suffice it to say that interleukin-6 is produced by monocytes and resident macrophages in the interstitium (cf. Chap. 6), is flushed away due to the enhanced lymph flow, and eventually reaches the liver and the bone marrow. Here, the production of macromolecular agents such as fibrinogen and other acute phase reactants and the proliferation and liberation of granulocytes are rapidly accelerated. Lastly, the described localized reaction is associated with neuronal reactions, for example those mediating pain by the chemical excitation of multimodal afferent nerve fibers. Obviously, the very same reactions can become an important pathogenetic mechanism complicating ischemic states or the recovery thereof, e.g., in acute myocardial infarction [89]. If the described cooperativity of multiple mechanisms occurs in a protracted fashion (i.e., over months or years rather than hours or days), it becomes a maintaining factor in chronic inflammation, in autoimmune diseases and in all protracted forms of chronic antigen antibody reactions. A similar self-organized scenario can be envisaged in acutely occurring hemodynamic shock and/or multiple organ failure [98].

References

1. Anderson DC, Springer TA (1987) Leucocyte adhesion deficiency. Ann Rev Med 38:175–194
2. Appelgren JK, Lewis DH (1970) Capillary flow and capillary transport in dog skeletal muscle after induced intravascular RBC aggregation and disaggregation. Eur Surg Res 2:161
3. Arfors KE, Rutili G, Svensjö E (1979) Microvascular transport of macromolecules in normal and inflammatory conditions. Acta Physiol Scand Suppl 463:93–103
4. Atherton A, Born GVR (1972) Quantitative investigations of the adhesiveness of circulating polymorphonuclear leukocytes to blood vessels. J Physiol (Lond) 222:447–474
5. Atkins PW (1986) Physical chemistry, 3rd edn. Oxford University Press, Oxford
6. Bagge U, Born GVR, Gaehtgens P (eds) (1982) White blood cells, morphology and rheology related to function. Nijhoff, The Hague
7. Barras JP (1991) Blood rheology – general review. Bibl Haemat 33:277
8. Begent N, Born GVR (1970) Growth rate in vivo of platelet thrombi, produced by iontophoresis of ADP, as a function of mean blood flow velocity, Nature 227:926–930
9. Berendt AR (1991) Erythrocyte endothelial interactions in plasmodium falciparim, sickle cell anemia and diabetes. In: Gordon JI (ed) Vascular endothelium: interactions with circulating cells. Elsevier, Amsterdam, pp 253–275
10. Blasberg P, Wurzinger LJ, Schmid-Schönbein H (1983) Microrheology of thrombocyte deposition: effect of stimulation, flow direction and red cells. In: Schettler G et al (eds) Fluid dynamics as a localizing factor for atherosclerosis. Springer, Berlin, Heidelberg New York, pp 103–115
11. Borberg I, Gaczkowsky H, Hombach V, Oette K, Stoffel W (1988) Treatment of familiar hypercholesterolemia by means of a specific amuno adsorption. J Clin Apher 4:59–65
12. Branemark PI (1971) Intravascular anatomy of blood cells in man. Karger, Basel
13. Brauer H (1971) Grundlagen der Einphasen und Mehrphasenströmung. Sauerländer, Aarau
14. Burton AC (ed) (1965) Physiology and biophysics of the circulation. Year Book, Chicago
15. Ceretelli P (1986) Aspetti fisiologici e clinici dell'autoemotrasfusione. In: Gaggi A (ed) 2nd Congr Nazionale della Societa Italiana di Cardiologia dello Sport, Bologna
16. Charm SE, Kurland GS (1974) Blood flow and microcirculation. Wiley, New York,

17. Chien S (1972) Present state of blood rheology. In: Messmer K, Schmid-Schönbein H (eds) Hemodilution. Theoretical and clinical application. Karger, Basel, pp 1–41
18. Chien S (1975) Biophysical behavior of red cells in suspensions. In: Surgenor D McN (ed) The red blood cell, vol II, 2nd edn Academic, New York, pp 1031–1121
19. Chien S, Usami S, Skalak R (1984) Blood flow in small tubes. Handbook of physiology, sect. 2; The cardiovascular system. vol IV. American Physiological Society, Bethesda MD, pp 217–249
20. Cokelet GR (1976) Macroscopic rheology and tube flow of human blood. In: Grayson J, Zingg W (eds) Microcirculation. Plenum Publ Corp, New York, pp 9–31
21. Cokelet GR (1986) Blood flow through arterial microvascular bifurcations. In: Popel AS, Johnson PC (eds) Microvascular networks: experimental and theoretical studies. Karger, Basel, pp 155–167
22. Cohnheim J (1889) Lectures in general pathology. New Syndenham Society, London
23. Desjardins C, Duling BR (1990) Heparinase treatment suggests a role for the endothelial cell glycocalix in regulation of capillary hematocrit. Am J Physiol 258:H647–H654
24. Diamond J (1993) Evolutionary physiology. In: Boyd CAR, Noble D (eds) The logic of life. Oxford University Press, Oxford, pp 89–111
25. Driessen GK, Scheidt-Bleichert H, Sobota A, Inhoffen W, Heidtmann H, Haest CWM, Kamp D, Schmid-Schönbein H (1980) Capillary resistance to flow of hardened (diamide treated) red blood cells (RBC). Pflugers Arch 392:261–267
26. Ehrly AM (1976) Improvement of the flow properties of blood: a new therapeutical approach in oclusive arterial disease. Angiology 27:188–198
27. Einstein A (1906) Eine neue Bestimmung der Moleküldimensionen. Ann Physik 19:289–306
28. Fahraeus R (1928) Die Strömungsverhältnisse und die Verteilung der Blutzellen im Gefäßsystem. Klin Wochenschr 7:100–106
29. Fahraeus R, Lindqvist T (1931) The viscosity of blood in narrow capillary tubes. Am J Physiol 96:562
30. Feynman RF, Leighton RB, Sands M (1963) Friction. The Feynman lectures on physics. Addison-Wesley, Reading MA
31. Firell, JC and Lipowsky HH (1989) Leukocyte margination and deformation in mesenteric venules of rat. Am J Physiol 256:H1667–H1674
32. Folkow B, Öberg B (1953) Autoregulation and basal tone in consecutive vascular sections of skeletal muscle in reserpine-treated cats. Acta Physiol Scand 53:105–113
33. Frimberger F (1933) Die Druckabhängigkeit der Viskositätswerte des Blutes und ihre Ursachen. Folia Haemat 60:237–242
34. Frojmovic MM, Newton M, Goldsmith HL (1976) The microrheology of mammalian platelets: studies of rheo-optical transients and flow in tubes. Microvasc Res 11:203–215
35. Fung YC, Zweifach BW (1971) Microcirculation: mechanics of blood flow in capillaries. Ann Rev Fluid Mech 3:189–210
36. Gaehtgens P, Kreutz F, Albrecht KH (1979) Optimal hematocrit of canine skeletal muscle during rhythmic isotonic exercise. Eur J Appl Physiol 41:27
37. Gaehtgens P, Schmidt F, Will G (1981) Comparative rheology of nucleated and non-nucleated red blood cells. Microrheology of avian erythrocytes during capillary flow. Pflugers Arch 390:278
38. Gaehtgens P, Schmid-Schönbein H (1982) Mechanisms of dynamic flow adaptation of mammalian erythrocytes. Naturwissenschaften 69:294
39. Gaehtgens P (1994) Physiologie des Blutes. In: Deetjen P, Speckmann EJ (eds) Physiologie. Urban and Schwarzenbrg, Munich, pp 271–304
40. Gee MH, Albertine KH (1993) Neutrophil-endothelial interactions in the lung. Ann Rev Physiol 55:227–248
41. Glansdorff P, Prigogine I (1971) Thermodynamic theory of structure, stability and fluctuations. Wiley, New York

42. Göbel W. Perkkiö J, Schmid-Schönbein H (1989) Compaction stasis due to gravitational red cell migration in plastic tubes and mesenteric venulas. Virchows Arch A Pathol Anat 415:243–251

43. Goldsmith HL, Karino T (1982) Microrheology and clinical medicine: unravelling some problems related to thrombosis. Clin Hemorheol 2:143–155

44. Goslinga H, Eijzenbach VJ, Heuvelmans JHA, van der Laan de Vries E, Melis VMJ, Schmid-Schönbein H, Bezemer PD (1992) Custom-tailored hemodilution with albumin and crystalloids in acute ischemic stroke. Stroke 23:181–188

45. Grassmann P (1983) Physikalische Grundlagen der Verfahrenstechnik. Salle, Frankfurt

46. Haken H (1983) Synergetics. An introduction. 3rd edn. Springer, Berlin Heidelberg New York

47. Haken H, Wunderlin A (1992) Die Selbststrukturierung der Materie. Vieweg, Braunschweig

48. Hess WR (1981) Biological order and brain organization. In: Akert K et al (eds) Selected works of W.R. Hess. Springer, Berlin Heidelberg New York

49. Jacobs MJ, Slaaf DW, Reneman RS, Schmid-Schönbein H, Lemmens HAJ (1983) Correlation between capillary microscopy and haemorheology in patients with ischaemic handsyndromes. Clin Hemorheol 3:307

50. Jan KM, Chien S (1977) Effect of hematocrit variations on coronary hemodynamics and oxygen utilization. Am J Physiol 4:106

51. Kakac S, Mayinger F, Veziroglu TN (1977) Two phase flows and heat transter, vol, II. Hemisphere, Washington

52. Karino T, Goldsmith H (1989) Microscopic structure of disturbed flow in the arterial and venous systems. Haemostasiology 9:53–65

53. Klimontovich YL (1995) Statistical theory of open systems, vol. I: A unified approach to kinetic description of processes in active systems. Dordrecht, Boston London (Kluwer Academic Publishers)

54. Klitzman B, Johnson PC (1982) Capillary network geometry and red cell distribution in hamster cremaster muscle. Am J Physiol 242:H211

55. Knisely MH, Warner L, Harding F (1960) Ante mortem settling, microscopic observations and analyses of the settling of agglutinated blood-cell masses to the lower sides of vessels during life: a contribution to the biophysics of disease. Angiology 11:535

56 Landau LD, Lifschitz EM (1987) Statistische Physik 1 und 2. Akademie, Berlin

57. Laughlin MH, Armstrong RB (1985) Muscle blood flow during locomotory exercise. Exerc Sport Sci Rev 13:95–136

58. Leal, LG (1980) Particles motions in a viscous fluid. Annu Rev Fluid Mech 12:435–476

59. Lessin LS, Bessis M (1977) Morphology of the erythron. In: Williams WJ, Beutler Z, Erslec AJ, Rundles RW (eds) Hematology, pp 103–134

60. Ley K, Lundgren E, Berger E, Arfors KE (1989) Shear dependent inhibition of granulocyte adhesion to cultured endothelium by dextran sulfate. Blood 73:1324–1330

61. Ley K, Pries AR, Gaehtgens P (1988) Preferential distribution of leukocytes in rat mesentery microvessel networks. Pflugers Arch 412:93–100

62. Ley K, Meyer JU, Intaglietta M, ArFors KE (1989) Shunting of leucocytes in rabbit tenuissimus muscle. Am J Physiol 256:H85–H93

63. Lien DC, Wagner W Jr, Capen Rl, Haslet C, Hanson WL (1987) Physiological neutrophil sequestration in the lung: visual evidence for localization in capillaries. J Appl Physiol 62:1236–1243

64. Maeda N, Sekiya M, Kameda K, Shiga T (1986) Effect of immunoglobulin preparations on the aggregation of human erythrocytes. Eur J Clin Invest 16:184–191

65. Meiselman HJ, Lichtman M (eds) (1984) White cell mechanics: basic science and clinical aspects. Raven, New York

66. Messmer K, Hammersen F (eds) (1985) White cell rheology and inflammation. Karger, Basel (Progress in applied microcirculation, vol 21)

67. Mottaghy K, Hanse HJ (1985) Effect of combined shear, secondly flow and axial flow on oxygen uptake. Chem Eng Commun 36:269–279

68. Nielsen HV (1982) Effect of vein pump activation upon muscle blood flow and venous pressure in the human leg. Acta Physiol Scand 114:481–485

69. Nikinmaaa M (1990) Vertebrate red blood cells: adaptations of function to respiratory requirements. Springer, Berlin Heidelberg New York (Zoophysiology vol 28)

70. Nobis U, Pries AR, Cokelet GR, Gaehtgens P (1985) Radial distribution of white cells during blood flow in small tubes. Microvasc Res 29:295–304

71. O'Flaherty JT, Ward PA (1979) Chemotactic factors and the neutrophil. Sem Hematol 16:163–174

72. Osborn L, Lobb R (1991) Direct expression cloning of endothelial adhesion molecules. In: Gordon JI (ed) Vascular endothelium: interaction with circulating cells. Elsevier, Amsterdam, pp 45–58

73. Pappenheimer JR (1984) Contributions to microvascular research of Jean Leonard Marie Poiseuitle. Handbook of Physiology, sect 2; The cardiovascular system, vol IV, Microcirculation, Part 1. American Physiological Society, Bethesda MD, pp 1–10

74. Perkkiö J, Wurzinger LJ, Schmid-Schönbein H (1987) Plasma and platelet skimming at T-junctions. Thromb Res 45:517–526

75. Perkkiö J, Wurzinger LJ, Schmid-Schönbein H (1988) Fahraeus-Veijlens effect: margination of platelets and leukocytes in blood flow through branches. Thromb Res 50:357–364

76. Poiseuille JLM (1841) Recherches sur les causes du mouvement du sang dans les vaisseux capillaires. Mem Acad Sci, Mem Acad Roy Sci Inst France, Sci Math et Phys VII:105

77. Pries AR, Ley K, Gaehtgens P (1986) Generalization of the Fahraeus principle for microvessel networks. Am J Physiol 251:H1324

78. Pries AR, Gaehtgens P (1989) Dispersion of blood cell flow in microvascular networks. In: Lee JS, Skalak TC (eds) Microvascular mechanics. Springer, Berlin Heidelberg New York, pp 50–64

79. Pries AR, Secomb TW, Gessner T, Sperandio MD, Gross JF, Gaehtgens P (1995) Resistance to blood flow in microvessels in vivo. Circ Res 75:904–915

80. Prigogine I (1947) Etude thermodynamique des phenomenes irreversibles. Desoer, Liège

81. Promislow DEL (1991) The evolution of mammalian blood parameters patterns and their interpretation. Physiol Zool 64:393–431

82. Reif F (1965) Fundamentals of statistical and thermal physics. McGraw-Hill, New York

83. Reinke W, Johnson PC, Gaehtgens P (1986) Effect of shear rate variation on apparent viscosity of human blood in tubes of 29 to 94 μm diameter. Circ Res 59:124–132

84. Richardson TG, Guyton AC (1959) Effects of polycythemia and anemia on cardiac output and other circulatory factors. Am J Physiol 197:1167

85. Scheffler A, Bold M, Lienke U, Rieger H (1991) Skin perfusion patterns (fluorescein perfusography) during reactive hyperaemia in peripheral arterial occlusive disease. Clin Physiol 11:501–512

86. Schiffmann E (1982) Leukocyte chemotaxis. Ann Rev Physiol 44:553–568

87. Schmid-Schönbein GW, Fung YC, Zweifach BW (1975) Vascular endothelium-leukocyte interaction. Sticking shear force in venules. Circ Res 36:173–184

88. Schmid-Schönbein GW, Usami S, Skalak R, Chien S (1980) The interaction of leukocytes and erythrocytes in capillary and postcapillary vessels. Microvasc Res 19:45–70

89. Schmid-Schönbein GW, Engler RL (1986) Granulocytes as acting participants in acute myocardial ischemia and infraction. Am J Cardiovasc Path 1:15–30

90. Schmid-Schönbein H, Gallasch G, Volger E, Klose HJ (1973) Microrheology and protein chemistry of pathological red cell aggregation (blood sludge) studied in vitro. Biorheology 10:213

91. Schmid-Schönbein H, Wells RE (1971) Rheological properties of human erythrocytes and their influence upon the "anomalous" viscosity of blood. Erg Physiol Biol Chem exper Pharmacol 63:147–219

92. Schmid-Schönbein H (1977) Microrheology of erythrocytes and thrombocytes. Blood viscosity and the distribution of blood flow in the microcirculation. In: Meessen H (ed) Handbuch der allgemeinen Pathologie III/7 Mikrozirkulation. Springer, Berlin Heidelberg New York, pp 289–384

93. Schmid-Schönbein H (1980) Blood rheology and oxygen transport to tissues. In: Kovach AGB, Dora E, Kessler, M, Silver A (eds) Oxygen transport to tissue. Pergamon, Oxford (Advances in physiological sciences, vol 25), pp 279–300

94. Schmid-Schönbein H, Gaehtgens P, Fischer Th, Stöhr-Liesen M (1984) Biology of red cells: non-nucleated erythrocytes as fluid drop-like cell fragments. Int J Microcirc Clin Exp 3:161–196

95. Schmid-Schönbein H, Wurzinger LJ (1986) Transport phenomena in pulsating post-stenotic vortex flow in arteries. An interactive concept of fluid-dynamic, haemorheological and biochemical processes in white thrombus formation. Nouv Rev Fr Hematol 28:257–267

96. Schmid-Schönbein H (1988) Fluid dynamics and hemorheology in vivo: the interaction of hemodynamic parameters and hemorheological "properties" in determining the flow behavior of blood in microvascular networks. In: Lowe GDO (ed) Clinical blood rheology. CRC Press, Boca Raton FL, pp 129–219

97. Schmid-Schönbein H, Wurzinger LJ (1988) Einführung: Über die Schultern von Eberth und Schimmelbusch auf Blutgerinnung und gestörte Strömung gesehen. Hämostasiologie 8:146–148

98. Schmid-Schönbein H (1990) Synergetic order and chaotic malfunctions of the circulatory systems in "multiorgan failure": breakdown of cooperativity of hemodynamic functions as cause of acute microvascular pathologies. In: Vincent JL (ed) Update 1990. Springer, Berlin Heidelberg New York, Update in intensive care and emergency medicine; vol 10), pp 3–21

99. Schmid-Schönbein H (1990) Synergetics of fluid-dynamic and biochemical catastrophe reactions in coronary artery thrombosis. In: Bleifeld W et al (eds) Unstable angina. Springer, Berlin Heidelberg New York, pp 16–51

100. Schmid-Schönbein H (1990) Synergetics of O₂ transport in the mammalian microcirculation: cooperativity of molecular, cellular, kinematic and hemodynamic factors in "rectified blood flow". In: Mosora F, Caro C, Baquey CH, Schmid-Schönbein H, Pelissier R, Krause E (eds) Biomedical transport processes. Plenum, New York, pp 185–196

101. Schmid-Schönbein H (1993) Synergetics of blood movement through microvascular networks: causes and consequences of nonlinear pressure-flow relationship. In: Haken H, Mikhailov A (eds) Interdisciplinary approaches to nonlinear complex systems. Springer, Berlin Heidelberg New York, pp 215–235

102. Schmid-Schönbein H (1996) A coherent theory for clinical hemorheology: synergetic interpretation of the physiology and pathophysiology of non-Newtonlan blood viscosity. Clin Hemorheol (in press)

103. Schmid-Schönbein H (1994) Self-organized movement of red cells in arterioles and capillaries in vivo. Pflugers Arch 426:R18

104. Schmid-Schönbein H, Perktold K (1994) Physical factors in the pathogenesis of atheroma formation. In: Caplan LR (ed) Scientific basis of stroke and its treatment. Springer, Berlin Heidelberg New York (Clinical medicine and the nervous system)

105. Schmidt-Nielsen K (1984) Scaling: why is animal size so important? Cambridge University Press, Cambridge

106. Schuff-Werner P, Schuetz E, Seyde WC, Eisenhauer T, Janning G, Armstrong VW, Seidel D (1989) Improved hemorheology associated with a reduction in plasma fibrinogen and LDL in patients being treated by heparininduced extracorporeal LDL precipitation (HELP). Eur J Clin Invest 19:30–37

107. Seiffge D, Kremer E (1986) Influence of ADP, blood flow velocity, and vessel diameter on laser induced thrombus formation. Thromb Res 42:331

108. Singer W (1993) Synchronization of cortical activity and its putative role in information processing and learning. Ann Rev Physiol 55:349–374

109. Stegall HF (1966) Muscle pumping in the dependent leg. Circ Res 19:180–190

110. Scherman P (1968) Emulsion science. Academic, London

111. Tangelder GJ, Arfors KE (1991) Inhibition of leukocyte rolling in venules by protamin and sulfated polysaccharides. Blood 77:1565–1571

112. Tangelder GJ, Slaaf DW, Muitjens AMM Arts T and Renemann RS (1986) Velocity profiles of blood platelets and red blood cells flowing in arterioles of the rabbit mesentery. Circ Res 59:505–514,

113. Taylor GI (1932) The viscosity of a fluid containing small drops of another fluid. Proc R Soc (Lond) 138A:41–44

114. Veylens (1938) The distribution of leukocytes in the vascular system. Acta Path Microbiol Scand 33:11–239

115. Vicaut E, Hou X, Decuypere L, Taccoen A, Duvelleroy M (1994) Red blood cell aggregation and microcirculation in rat cremaster muscle. Int J Microcirc Clin Exp 14(1–2):14–21

116. Wautier JL, Paton RC, Wautier MP, Pintigny D, Abadie E, Passa P, Caen JP (1981) Increased adhesion of erythrocytes to endothelial cells in diabetes mellitus and its relation to vascular complications. N Engl J Med 305:237

117. Wetter T, Schmid-Schönbein H, Heidtmann H (1992) Mathematical modelling of plasma and red cell flow in a bifurcating network (Abstr F1.7). Biorheology 29(1):48

118. Whittacker SRF, Winton FR (1933) The apparent viscosity of blood flowing in the isolated hindlimb of the dog, and its variation with corpuscular concentration. J Physiol [Lond] 78:339–369

119. Wiener N (1963) Kybernetik. Econ, Düsseldorf

120. Wurzinger LJ, Opitz R, Blasberg R, Schmid-Schönbein H (1985) Platelet and coagulation parameters following millisecond exposure to lamianr shear stress. Thromb Haemost 54:382–386

121. Wurzinger LJ, Opitz R, Schmid-Schönbein H (1988) Effect of anticoagulants on shear-induced platelet alterations. Thromb Res 49:133–137

122. Wurzinger LJ (1990) Histophysiology of the circulating platelet. Springer, Berlin Heidelberg New York (Advances in anatomy, embryology and cell biology vol 120), pp 1–96

123. Zander R, Schmid-Schönbein H (1973) Intracellular mechanisms of oxygen transport in flowing blood. Resp Physiol 19:279–289

124. Zijlstra WG, Mook GA (1962) Medical reflection photometry. Van Gorcum, Amsterdam

125. Zwaal RFA, Bevers EM, Rosing J (1985) Platelet membranes and activation of coagulation factors. In: Schmid-Schönbein H, Wurzinger LJ, Zimmermann RE (eds) Enzyme activation in blood-perfused artifical organs. Nijhoff, Dordrecht, pp 111–119

89 The Cardiac Function Cycle

H. Antoni

Contents

89.1 Introduction: General Structural and Functional Aspects

The manifold functions of the blood described in some of the other chapters (cf. Chaps. 88, 90–99) can only be fulfilled if the blood circulates continuously through the body. The main driving force for the movement of blood is provided by the mechanical activity of the heart. Additional forces, such as skeletal muscular contractions and respiratory movements, can support the pumping function of the heart to variable extents. The cardiac contraction is initiated within the heart itself by specialized cells of cardiac musculature; unlike skeletal muscle, external nerves are not essential for the initiation of the heart beat or for the sequences of events which form the cardiac cycle.

Subdivisions of the Circulatory System. The cardiac muscular pump comprises two halves – the right and the left ventricle – each connected to an atrium (cf. Fig. 89.1). The right ventricle receives partially oxygen-depleted blood from the entire body via the right atrium and propels it to the lungs by way of the pulmonary artery, where it is recharged with oxygen. Then the oxygenated blood returns to the left ventricle via the left atrium and is distributed from there to the organs of the body by way of the aorta. Hence, the right half of the heart pumps only deoxygenated blood, and the left half only oxygenated blood. The part of the circulatory system between the right ventricle and the left atrium comprising the lungs is called the *pulmonary circulation*. The remaining part between the left ventricle and the right atrium distributing the blood to the rest of the body is called the systemic circulation. Of course, there is only a single pathway of blood movement, with the propulsive force provided at two points by the two cardiac ventricles (cf. Fig. 89.1).

History. The closed circulation of blood was discovered by the English physician William Harvey and is described in his famous treatise, published in 1628 [4]: "De motu cordis et sanguinis in animalibus" [On the movement of the heart and the blood in animals] (see introductory chapter, this volume). Until that time the prevailing view had been that the blood was formed in the liver from food components, sent to the heart, and from there passed through the veins to the organs where it was used up.

Systole and Diastole, Arteries and Veins. The pumping action of the heart is based on a periodic sequence of contraction (*systole*) and relaxation (*diastole*) of the atria and ventricles. During their diastole the ventricles fill with blood, and during their systole they eject it into the large arteries. The systole of the atria occurs during the later phase of the diastole of the ventricles and supports their filling. This sequence of mechanical events is closely connected with the spread of excitation originating in the roof of the right atrium and travelling from there along the muscular pathways throughout the heart (see Chap. 91). Back-flow of blood from the arteries during diastole, or into the atria during systole, is prevented by the one-way valves at the openings. The distinction between arteries and veins is based on the direction of blood flow within them, rather than on the state of the blood itself. Arteries carry the blood away from the heart to peripheral capillaries, and veins carry it back. In the systemic circulation the arteries carry oxygenated blood, and in the pulmonary circulation the oxygenated blood is carried by the veins.

Functional Range of Variation. Because the demands made on the circulating blood are quite different at different times, the heart must be able to adjust its activity over a wide range. For example, the volume of blood expelled by each ventricle per minute (*cardiac output*) is about 5l when a person is at rest, and rises to about 30l (3 large buckets per minute!) and even more during hard physical work. This is achieved by an increase in the beating frequency as well as by an enlarged volume ejected per beat (*stroke volume*). Optimal adaptation is achieved only when all the partial functions of the heart (frequency and spread of excitation, contractility, valve action, coronary blood

R. Greger/U. Windhorst (Eds.)
Comprehensive Human Physiology, Vol. 2
© Springer-Verlag Berlin Heidelberg 1996

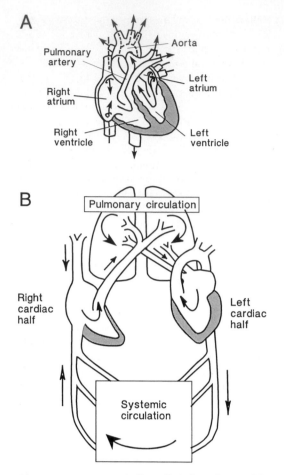

little blood circulates through them. This is possible because the fetal blood becomes oxygenated in the placenta. At birth, when the lungs expand and begin to function in respiration, the blood vessels in them also expand, and their resistance to blood flow decreases. As a result, the pressure in the left atrium exceeds that in the right, and the valve at the foramen ovale folds over the opening and closes it off. There is also a progressive constriction of the ductus arteriosus. About 2 weeks after birth the conversion is complete, with both the foramen ovale and the ductus arteriosus tightly closed. The parallel arrangement of the two halves of the heart in the fetus has been converted to a serial arrangement (Fig. 89.2B). This reorganization of the circulatory pattern during birth causes the workload of the right heart to be considerably less than that of the left. Because the resistance to flow in the vascular bed of the lungs is only about one-eighth to one-tenth that in the systemic circulation [1], the right ventricle needs to exert less force to propel the blood through the pulmonary circuit. This difference in workload brings about accelerated growth of the more heavily loaded left ventricle, which eventually develops a mass of muscle almost three times that of the right ventricle.

89.3 The Mechanical Action of the Heart

89.3.1 Structure and Function of the Heart Valves

For the alternation between contraction and relaxation of the myocardium to propel blood in the appropriate direction, i.e., from the venous to the arterial system, an arrangement of precisely operating valves is required to prevent backward flow. There are two sets of valves in the

Fig. 89.1. **A** An anatomically realistic frontal view of the opened heart. **B** Schematic diagram of the connections of the two halves of the heart with the pulmonary and systemic circulation. For didactic purposes, the two halves of the heart are shown separately. The direction of blood flow is indicated by the *arrows*

supply to the heart, etc.) change together in an orderly manner. Even slight departures from the norm can severely impair cardiac performance.

89.2 Fetal Heart and Changes Occurring at Birth

The functional subdivision of the heart into a right half driving the blood to the lungs and a left half connected with the systemic circulation, and the arrangement of the two in series start to develop during birth. In the heart of the fetus the two atria communicate with one another by way of the *foramen ovale*, and there is a short-circuit between the aorta and the pulmonary artery, by a wide passage, the *ductus arteriosus* (Botallo's duct; Fig. 89.2A). In the fetus, then, atria and ventricles act as a single hollow organ. At this stage the lungs are collapsed and non-functional, and

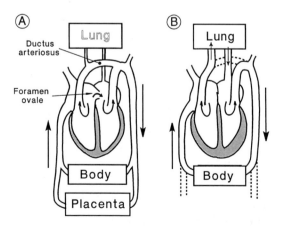

Fig. 89.2. **A** Fetal heart before birth. The two halves are in parallel, with the lung on a side circuit. **B** After birth the two halves are in series. This conversion involves expansion of the circuit through the lungs at the onset of respiration and closure of two shunt passages: the foramen ovale between right and left atrium, and the ductus arteriosus between aorta and pulmonary artery

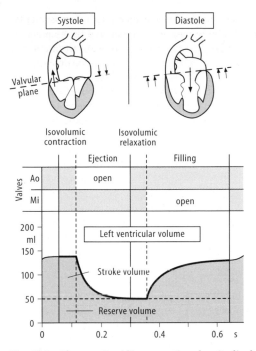

Fig. 89.3. *Above*: Semidiagrammatic longitudinal section through the left half of the heart, to show the valve operation and the valve plane mechanism during filling and ejection. *Below*: Diagram representing the behavior of the valves at the openings of the left ventricle (*Ao*, aortic valve; *Mi*, mitral valve) and of the left ventricular volume during the different activity phases composing the cardiac cycle. Diastole is subdivided in isovolumic relaxation and filling; the components of systole are isovolumic contraction and ejection.

heart covering the inlets and the outlets of both ventricles. The *atrioventricular (AV) valves* (mitral valve on the left, tricuspid on the right) prevent regurgitation of blood into the atria during ventricular systole. The *aortic* and *pulmonary valves*, at the bases of the large arteries, prevent regurgitation into the ventricles during diastole (see Fig. 89.3).

The AV valves are composed of membranous leaflets or cusps that hang into the ventricles to form a sort of funnel. Their free edges are attached to inward projections of the ventricular wall, the papillary muscles, by fine tendinous strings (chordae tendineae), which prevent the cusps from being pushed back into the atria during systole. The total surface area of the cusps is considerably greater than that of the opening they have to cover, so that their margins are pressed together. This arrangement guarantees reliable closure even if the ventricle changes size. The aortic and pulmonary valves are somewhat different in structure; they form three crescent-shaped pockets around the opening of the vessels (hence the term *semilunar valves*). When the valves are closed the cusps touch one another to form a "Mercedes star." In diastole, the valves close rapidly, keeping regurgitation to a minimum owing to the current of blood flowing past and eddying behind them. The edges of the cusps draw closer together, the higher the velocity of flow (*Bernoulli effect*).

Valve Malfunctions. Anyone who has the opportunity of observing the opening and closing of the valves through a window in a saline-perfused animal heart cannot help but be surprised at the rapidity and precision of their movement. It follows that when anything interferes with this movement – for example, when scar formation secondary to inflammation of the valves causes them to open too little (stenosis) or not to close firmly (insufficiency) – the activity of the heart is seriously impaired. The parts of the heart affected are burdened with the need to develop greater pressure or move a larger volume, a burden to which the myocardium responds with hypertrophy or dilatation. By adjustments of this sort the heart can compensate for disturbances in valve function over a period of years in some cases.

89.3.2 Sequence of Activity Phases

Opening and closing of the cardiac valves are brought about by pressure changes in the adjacent cardiac cavities or vessels and determine the activity phases of the heart. The motion of the valves in turn affects the mode of contraction of the myocardium. Accordingly, in both systole and diastole, periods of action can be distinguished in which the dominant feature is either pressure change with constant volume, or volume change with moderate change in pressure [10]. During diastole there are an *isovolumic relaxation* period and a *filling* period, and in systole an *isovolumic contraction* period followed by an *ejection* period. In Fig. 89.3 the temporal relations between these phases and valve movements as well as ventricular volume changes are diagrammed for the left half of the heart. The corresponding pressure changes during the cardiac cycle are represented in Fig. 89.4.

Isovolumic Relaxation Period. At the end of ventricular systole, the fall in intraventricular pressure causes immediate closure of the aortic and pulmonary valves. Because the AV valves also remain closed at first, the ventricular musculature continues to relax without a change in the ventricular volume (Fig. 89.3), so that there is a further sharp fall in pressure (Fig. 89.4). Although the ventricular volume in this phase does not change, the relaxation is not entirely isometric, i.e., it does not occur at a constant length of the cardiac muscle fibers, because there is some change in shape of the ventricles. This explains why the term "isovolumic" is used rather than "isometric." With the heart rate at the normal resting level, the duration of this phase in the left ventricle is about 50 ms. When the ventricular pressure falls below the atrial pressure, the AV valves open and the ventricles begin to fill in preparation for the following systole.

Filling Period. During this phase the ventricular volume increases rapidly at first and then more slowly. When the heart is beating at the normal rate, the ventricles are almost completely filled by the time the atria contract, so that the atrial systole has only a slight additional effect leading to

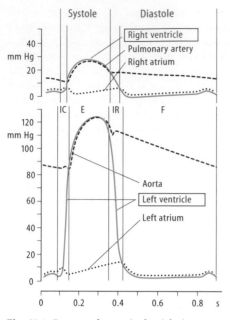

Fig. 89.4. Pressure changes in the right (*upper curves*) and the left (*lower curves*) halves of the heart during a cardiac cycle. *IC*, isovolumic contraction; *E*, ejection; *IR*, isovolumic relaxation; *F*, filling

an increase of about 10% in volume. But when the heart rate is high, diastole is shortened by more than systole. In these conditions contraction of the atria can make a considerable contribution to ventricular filling.

Isovolumic Contraction Period. Like diastole, systole begins with a brief isovolumic period of about 60 ms. However, it differs from diastole in that there is a rise in intraventricular pressure, which causes immediate closure of the AV valves. Because the arterial valves are also still closed, the ventricular musculature continues to contract around the incompressible contents, so that there is a further sharp increase in pressure at constant volume.

Ejection Period. When the intraventricular pressure exceeds the diastolic arterial pressure, the semilunar valves open and blood begins to be expelled. As the volume curve in Fig. 89.3 shows, in resting conditions the ventricle ejects about 90 ml of the 140 ml blood it contains; this is the *stroke volume*. At the end of the systole, therefore, a *reserve volume* of about 50 ml remains in the ventricle. The ratio between the stroke volume and the total end-diastolic volume is called the *ejection fraction*. In the present case it is about 0.65 (65%) [1].

Comparison Between Left and Right Ventricle Dynamics. The periods just described are basically the same in both halves of the heart, but because the vascular resistance is lower in the pulmonary circulation, the pressure the right heart needs to develop in systole is considerably lower. The stroke volumes of the two ventricles are about the same. The activity phases are not exactly synchronous in the two

halves of the heart. Isovolumic contraction of the right ventricle begins after that of the left and lasts for a shorter time, because the rise in pressure is less. Accordingly, the ejection period begins earlier in the right ventricle than in the left. The end of systole, however, occurs somewhat later in the right ventricle than in the left (Fig. 89.4). These time differences are relatively small (in the order of 10–30 ms) and have practically no effect on the hemodynamics.

89.3.3 Valvular Plane Mechanism

Ventricular systole does not only exert the ejection of blood from the heart, but is also closely related to diastolic filling. This latter effect is brought about by the displacement of the plane through the boundary between atria and ventricles in which the valves lie (valvular plane) towards the apex of the heart (Fig. 89.3). The atria, which have already relaxed at the onset of the ejection period, are thus stretched, and blood is sucked into them from the great veins. Now, as soon as the ventricular musculature relaxes, the valvular plane returns to the starting position simultaneously with opening of the AV valves, so that the plane shifts over the blood kept ready in the atria. In this way a rapid initial filling of the ventricles is guaranteed. The movement of the valvular plane towards the apex (rather than displacement of the apex towards the plane) has several causes [8]:

- There is a mechanical fixation of the apex due to a layer of fluid between the heart and the pericardium, which in turn is anchored to the diaphragm.
- During ventricular ejection, there is a repulsive force on the large vessels in the heart region that pushes the heart towards its apex.
- Diastolic relaxation of the ventricles in itself exerts some suction, owing to the reversal, by passive elastic effects, of the deformation imposed during systole.

89.4 Signals of Cardiac Activity

89.4.1 Non-invasive Recordings

There are a number of useful signals closely related to cardiac activity that can be monitored at the surface of the body by means of suitably designed equipment, without appreciable inconvenience to the subject. Such methods of study are called non-invasive procedures. One example is the electrocardiogram (ECG), a manifestation of the electrical activity of the heart that will be considered later (see Chap. 92). The following are also particularly accessible to non-invasive monitoring: the *apex impulse*, the *heart sounds*, and the *arterial* and *venous pulses*.

Apex Impulse. The movement of the apex of the heart can easily be felt with the fingers in a thin person, and sometimes even seen as a rapid bulging in the medioclavicular part of the left fifth intercostal space. Changes in shape, volume and orientation of the entire heart interact in a complicated way to produce this movement. A recording of the apex impulse (*apex cardiogram*) can give supplementary evidence as to timing of the periods in the contraction cycle of the left ventricle.

Heart Sounds. As the heart beats it transmits oscillations in the audible range (15–400 Hz) to the chest wall [9]. These heart sounds can be heard by placing an ear on the chest, or by means of a *stethoscope*. While listening by either means (*auscultation*), one can usually hear two sounds, the first at the onset of systole and the second at the onset of diastole. The first heart sound is the longer of the two, a dull noise of complicated structure. It is primarily associated with the sudden contraction of the ventricular myocardium about its incompressible contents and with the closure of the AV valves; the resulting vibration of these structures is transmitted to the chest wall. The shorter, sharper, second heart sound occurs when the cusps of the semilunar valves strike one another (*valve sound*) and set the columns of blood in the great vessels into vibration. The most favorable sites for auscultation of the second sound are therefore not directly over the heart but some way away from it in the direction of blood flow (i.e., in the second intercostal space, on the right for the aortic valve and on the left for the pulmonary valve). The best auscultation sites for the first heart sound are directly over the ventricles.

Phonocardiography. With suitable microphones and recording apparatus the waves composing the heart sounds can be displayed [13] (Fig. 89.5). This phonocardiogram provides a permanent record and allows analysis of some temporal relationships to other events during the cycle. Normally the *first sound* consists of three components, beginning with a low-amplitude slow wave associated with deformation of the left ventricle at the onset of isovolumic contraction. The subsequent, larger, waves accompany the steep rise in intraventricular pressure. The third component, descending in amplitude, coincides with the onset of ejection. The beginning of the *second sound* usually coincides with the end of the ejection period, when the semilunar valves close. Occasionally the second sound is split into a first component associated with the closure of the aortic valve (A2) and a second, synchronous with closure of the pulmonary valve (P2). The rush of blood into the ventricles early in the filling period causes a *third sound*, which usually is not audible but can be detected in the phonocardiogram. Likewise a *fourth sound* may be seen in the phonocardiographic trace; this is due to atrial contraction.

Murmurs. Abnormal heart sounds called murmurs are produced chiefly by turbulence in the blood stream. They have a higher frequency than normal heart sounds (about 800 Hz) and last longer. Inborn or acquired stenosis or insufficiency of heart valves are a frequent cause of murmurs. Further causes include inborn defects in the atrial or ventricular septa or other malformations, but also reduced

viscosity of the blood and other non-pathologic conditions. Murmurs are diagnosed on the basis of characteristic changes of the sound, their time of occurrence, and the site at which they are heard most clearly.

Carotid Pulse. When the stroke volume is ejected from the left ventricle a pressure wave spreads through the arterial system [13]. Measurement near the heart at the common carotid artery reveals a typical time course of pressure change (Fig. 89.5). The first result of ejection is a sharp rise in pressure, to a distinct peak. During the subsequent falling phase the aortic valves snap shut, causing a sharply delimited deflection, the *incisura* in the pressure curve. The time from the base of the rising flank to the incisura corresponds to the *duration* of the *ejection period* of the *left ventricle*. In determining the real onset of the ejection, however, it should be kept in mind that the carotid pulse is somewhat delayed with respect to the nearly instantaneously transmitted heart sounds, because it takes some time for the pressure wave to pass from the aorta to the carotid artery. This *central pulse wave transmission time* can be derived from the interval between the beginning of the second heart sound and the incisura (shaded region in Fig. 89.5).

Venous Pulse. Due to retrograde transmission, the veins near the heart are filled with blood to different degrees during the course of a cardiac cycle. These volume fluctuations reflect the course of pressure change in the right atrium. They can be non-invasively monitored, for

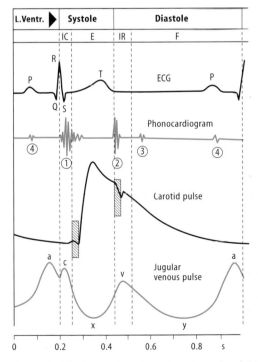

Fig. 89.5. The changes in certain processes and variables during the cardiac cycle. The activity phases correspond to those in Figs. 89.3 and 89.4

example over the external jugular vein in a recumbent subject, by means of photoelectric devices or sensitive pressure transducers. This *jugular pulse* is composed of several characteristic waves [5] (Fig. 89.5): The *a-wave* is elicited by the systole of the right atrium. The subsequent *c-wave* is caused chiefly by the bulging of the tricuspid valve into the right atrium during the isovolumic contraction of the right ventricle. It is followed by a sharp fall (*x-wave*) brought about by the shift of the valvular plane of the heart towards the apex during the ejection period (cf. above). As the ventricles relax, the AV valves remain closed at first, so that there is an initial rise in pressure (*v-wave*) followed by a drop (*y-wave*) when the valves open and blood flows into the ventricles. As the ventricles continue to fill, the pressure rises slowly towards the next a-wave. Changes in the venous pulse curve can provide useful diagnostic information on certain forms of heart disease, such as tricuspid insufficiency.

X-Ray Examination. Clues as to the size and shape of the heart can be obtained simply by tapping on the chest and noting the distribution of dull-sounding areas (*percussion*). For more precise measurements, and for the sake of documentation *roentgenograms* are useful. When the patient is positioned at a distance of about 2 m from the X-ray source, the projection errors that arise at smaller distances due to the divergence of the X-rays can be avoided. Whereas a conventional X-ray image of the heart is a shadowgram at right angles to the X-ray beam, a *computed tomogram* is an image that describes the attenuation along the path of the X-ray. It provides a two-dimensional, reconstructed display very similar to an anatomical section through the organ.

Echocardiography. Recently, the principle of echo sounding has come into widespread use as a method of investigating the heart [3,7,12]. In *echocardiography* the reflection of ultrasonic waves at the various surfaces of the heart (inner and outer wall sides, valves, etc.) is recorded. This method provides information about the distance between structures within the beam and about changes in these distances. Based on the same principle of reconstruction as it is applied in computerized tomography, echo signals can be used to display a two-dimensional section through the heart. Because experience so far indicates that these controlled doses of ultrasonic waves, unlike X-rays, are harmless to humans, such examination can be repeated as often as desired.

Impedance Cardiography. This technique uses electrodes placed to encompass the thorax and a small high-frequency sinusoidal alternating current (20–100 kHz) to measure the impedance changes accompanying cardiac activity [2]. These changes in impedance reflect the variations of the total volume of blood in the region between the potential measuring electrodes. The various waves of the thoracic impedance cardiogram and its derivative are correlated with systole and diastole of the atria and ventricles.

Moreover, impedance cardiography is also a non-invasive method of estimating the stroke volume.

89.4.2 Invasive Techniques for Intracardial Measurement

Extracardiac recordings can give only indirect evidence of heart function, and for certain questions this is not enough. The invasive techniques developed during recent decades use *cardiac catheters* for intravascular and intracardial measurements [11,14]. Such catheters are flexible tubes of various designs, length and diameters, which are introduced into a peripheral blood vessel and passed into the heart, usually under X-ray control. A transvenous catheter can easily reach the right atrium, the right ventricle and the pulmonary artery. In order to catheterize the left heart, the tube has to be passed retrogradely through a peripheral artery, or by way of the right atrium after careful puncture of the atrial septum.

Applications of Cardiac Catheters. The primary purpose of heart catheterization is to *measure pressure* in the various chambers of the heart and the associated vessels. Pressure curves such as those depicted in Fig. 89.4 can be obtained in this way. A catheter can also be used to obtain blood samples from regions of interest for analysis of (for example) oxygen content. Following injection of a test substance, so-called *indicator dilution curves* [6] can be constructed to determine the cardiac output. When contrast material is injected, roentgenograms can be made in rapid sequence to show the heart chambers and vessels in various phases of the beat (*angiocardiography*). Finally, certain special questions can be answered by using the catheter for intracardial recording of electrical activity [11] (*His bundle electrocardiography*).

References

1. Busse R, Bauer RD (1982) Arteriensystem. In: Busse R (ed) Kreislaufphysiologie. Thieme, Stuttgart
2. Geddes LA, Baker LE (1989) Detection of physiological events by impedance. In: Geddes LA, Baker LE (ed) Principles of applied Biomedical instrumentation, 3rd edn. Wiley, New York, pp 537–651
3. Grube E (ed) (1985) Zweidimensionale Echokardiographie. Thieme, Stuttgart
4. Harvey W (1628) The movement of the heart and blood (translated By G Whitteridge). Blackwell Scientific, Oxford 1976
5. Heinz N, Luckmann E, Saegler J (1971) Formanalyse zentraler Venendruckkurven. I: Normalkurven. Z Kreislaufforsch 60:433–445
6. Höfling B (1991) Sondierung und Angiographie des Herzens und der herznahem Gefäße. In: Riecker G (ed) Klinische Kardiologie. Springer, Berlin Heidelberg New York, pp 104–119
7. Kisslo JA (1980) Two-dimensional echocardiography. Churchill Livingstone, New York

8. Krasny R, Kammermeier H, Köhler J (1991) Biomechanics of valvular plane displacement of the heart. Basic Res Cardiol 86:572–581

9. Leatham A (1958) Auscultation of the heart. Lancet 2:703–708, 757–766

10. Parmley WW, Talbot LL (1979) Heart as a pump. In: Berne RM (ed) Handbook of physiology. The cardiovascular sytem, vol I: the heart. American Physiological Society, Bethesda, pp 429–460

11. Scherlag BJ, Lau SH, Helfant RH, Berkowitz WD, Stein E, Damato AN (1969) Catheter technique for recording His bundle activity in man. Circulation 39:39

12. Tajik AJ, Seward JB, Hagler DJ, Mair DD, Lie JT (1978) Twodimensional real time ultrasonic imaging of the heart and great vessels: technique, image orientation, structure identification, and validation. Mayo Clin Proc 53:271 (1978)

13. Tavel ME (1989) Clinical phonocardiography and external pulse recording. Year Book Medical Publishers, Chicago

14. Zohmann LR, Williams MH (1959) Percutaneous right heart catherization using polyethylene tubing. Am J Cardiol 4: 373

90 Functional Properties of the Heart

H. ANTONI

Contents

90.1 Introduction

In the preceding chapter the overall behavior of the heart has been considered as it is reflected in the periodical alternation of diastole and systole and of filling and ejection of the atria and ventricles. This description of the cardiac cycle included the accompanying changes in pressure and volume of the cardiac chambers. However, a deeper comprehension of cardiac function has to take into account some specific properties of the myocardial tissue as opposed to other types of muscle, mainly skeletal muscle, and the transformation of wall tension into pressure by the hollow muscular organ. Moreover, the compensatory changes in the heart in response to different degrees of filling (preload) or of resistance against ejection (afterload) are also discussed in this chapter, as are regulatory influences on myocardial contractility.

90.2 Structural and Functional Characteristics of the Myocardium

Comparison of Skeletal and Cardiac Muscle. In both structure and function, cardiac muscle is more complex than skeletal muscle [108]. In the latter, the single fibers are structurally separate from one another and are functionally connected only by the spread of excitation along the branchings of a motoneuron. By contrast, cardiac muscle fibers form a structural network and functionally behave like a syncytium (Fig. 90.1). The excitatory impulses that activate the heart are neither elicited by nor conducted through nervous tissue. Instead, cardiac excitation is initiated by specialized myocardial cells in the sinoatrial node and distributed all over the organ via specialized muscular pathways (conduction system) and ultimately via the ordinary myocardial fibers. Because the myocardium is a *functional syncytium* its force of contraction cannot be graded by recruitment of a variable number of motor units, as is possible with skeletal muscle. Instead, the heart as a whole behaves like a motor unit, with all its fibers participating in each beat. In compensation for this physiological property, the possibility of influencing contraction by way of the excitatory process or by interference with excitation–contraction coupling is considerably greater in cardiac than in skeletal muscle [15].

Nontetanizability of the Myocardium. Another considerable difference between the two types of muscle concerns the temporal relation between excitation and contraction (cf. Fig. 90.1): The action potential of skeletal muscle lasts only a few milliseconds, and contraction does not begin until the excitatory process is nearly over. In the myocardium the two events overlap considerably in time; the myocardial action potential ends only when the musculature has begun to relax again. Because a new contraction must be initiated by new excitation, which can occur only after termination of the preceding one, cardiac muscle – unlike skeletal muscle – is incapable of responding to a rapid sequence of action potentials with superposition of single contractions, a so-called tetanus. However, this nontetanizability of the myocardium is a property that seems entirely appropriate to the pump function of the heart. A tetanic contraction of the heart outlasting the blood-ejection phase would interfere with refilling. On the other hand, the tetanizability of skeletal muscle enables the force of contraction to be varied with action potential frequency, whereas myocardial contractions cannot be graded in this way but have to be adjusted via an influence on excitation–contraction coupling.

Structural Peculiarities Involved in Excitation-Contraction Coupling. In principle, the myocardial fibers of mammalian hearts comprise the same structural elements as are involved in excitation–contraction coupling and in contraction of skeletal muscle (cf. Chap. 46). However, there are also distinct structural differences that have to be considered in connection with the functional peculiarities described above [83,108]. The transverse tubular system (TTS) represents invaginations of the surface membrane and serves the transmission of excitation into the interior of the cell. This system is small in diameter in skeletal

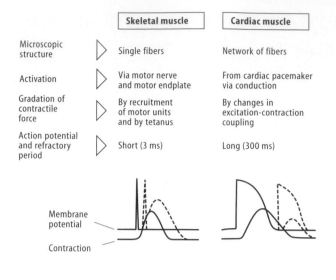

	Skeletal muscle	**Cardiac muscle**
Microscopic structure	Single fibers	Network of fibers
Activation	Via motor nerve and motor endplate	From cardiac pacemaker via conduction
Gradation of contractile force	By recruitment of motor units and by tetanus	By changes in excitation-contraction coupling
Action potential and refractory period	Short (3 ms)	Long (300 ms)

Membrane potential

Contraction

Fig. 90.1. Comparison of characteristic properties between skeletal and cardiac muscle. *Dotted curves* represent responses to a second stimulus immediately after the refractory period

muscle and runs along the A-I junctions of the sarcomeres, whereas in cardiac muscle the TTS is large and is located at the Z lines. The sarcoplasmic reticulum (SR), which represents the main intracellular Ca^{2+} store, is large in its junctional section in skeletal muscle, where the terminal cisternae form triads with the TTS. In cardiac muscle the junctional SR is much smaller and is continuous with the free SR. Moreover, in contrast to skeletal muscle, cardiac muscle has its myofibrils arranged widely branching, often creating dead-end crevices that contain cytoplasm and other intracellular organelles (*Felderstruktur*). In skeletal muscle, the contractile structures appear assembled into well-circumscribed fibrils that are surrounded by the cytoplasmic matrix (*Fibrillenstruktur*). These structural peculiarities of the myocardium and its functional behavior offer evidence of a close interaction between the intracellular Ca^{2+} stores and the extracellular space. The key event in the initiation of contraction by the action potential is the transmembrane influx of calcium ions that triggers an additional release of Ca^{2+} from the intracellular stores, thus activating the contractile apparatus. A detailed outline of Ca^{2+}-mediated control of cardiac contractility at the cellular level is presented in Chap. 93.

The *myofibrils* of the working myocardium, like those of skeletal muscle, are striated and exhibit a characteristic repeating pattern of light and dark transverse bands (see Chaps. 45, 46). As in skeletal muscle, these cross striations arise from the characteristic structural organization of the contractile proteins into thin and thick filaments. The thick filaments contain the protein myosin, and the thin filaments are composed largely of actin but also contain the regulatory proteins tropomysin and troponin. In the fine chemical structure of its contractile proteins, cardiac muscle bears most similarity with skeletal muscles of the red type ("red" because of the high myoglobin content), which are specialized for sustained activity without rest periods. These types of muscle exhibit comparatively slow shortening velocity, which most probably results from a low level

of myosin ATPase activity. Recent findings suggest that alterations in troponin T and in the regulatory light chains are responsible for this functional property [65].

90.3 Mechanics of the Isolated Myocardium

Intrinsic *mechanical properties of cardiac muscle* can be analzsed best in isolated preparations, such as atrial trabecula or ventricular papillary muscles. These myocardial preparations, with an almost parallel arrangement of their muscle fibers, can be subjected to classic analytical procedures, and model designs originally developed for skeletal muscle can be used to study them. In these circumstances, such parameters as pressure and volume, which apply to the whole heart as a pump are replaced by the parameters force or tension and length. It is shown, below how the results on oblong isolated myocardial preparations are extrapolated to the whole organ. By definition, *force* means the absolute amount of contractile strength, measured in newtons (N), whereas *tension* is force related to area cross section, measured in newtons per square millimeter (N/mm^2).

Muscular Models and Elementary Forms of Contraction. An isolated papillary muscle responds to an electrical stimulus of suprathreshold intensity with shortening or, if shortening is prevented because the muscle is fixed at its ends, it responds with force development. These seemingly different forms of contraction are, however, in accordance with a common elementary process based on a model in which the muscle is represented by two components (elements) in series, one contractile and the other elastic (Fig. 90.2). A third component, in parallel with these, will be required later to account for the resting properties, but can be disregarded here. However, it should be stressed that no part of the contractile–elastic model corresponds directly

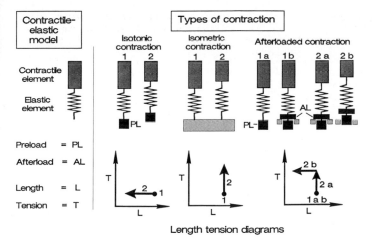

Length tension diagrams

Fig. 90.2. *Left*: two-component model of cardiac muscle consisting of a contractile element and an elastic element switched in series. *Right*: illustration of three different basic types of contraction and their representation in length–tension diagrams. In the isotonic contraction the preloaded muscle lifts the load by shortening of the contractile element without distention of the elastic component. In the isometric contraction the muscle develops tension by shortening the contractile element and thereby stretching the elastic component, with the overall length of the muscle remaining constant. In the afterloaded type of contraction the muscle is first preloaded (*1a*) and then afterloaded by a supported weight (*1b*). Thus, contraction consists of an initial isometric phase adjusting the afterload (*2a*) followed by an isotonic shortening phase (*2b*)

to any anatomical part of the muscle, as will be shown in more detail later.

If a preparation of cardiac muscle (Fig. 90.2) is stretched by a weight attached to its freely hanging end (*preload*) and then stimulated, it contracts by shortening at a constant force corresponding to the preload (*isotonic contraction*). In terms of the two-component model this type of contraction is considered to result from shortening of the contractile component without a change in the elastic element. If, on the other hand, a myocardial preparation is stimulated while fixed at both ends it develops force while its length remains unchanged (*isometric contraction*). In this case, for the purposes of the model it is assumed that force development results from internal stretching of the elastic component by shortening of the contractile element. If shortening and force development of the muscle take place simultaneously, as during the ejection phase of the ventricles, the model assumes only partial transduction of the shortening of the contractile element into internal stretching of the elastic one (*auxotonic contraction*). Moreover, a myocardial contraction may consist of two successive phases, a first isometric phase during which the muscle develops force, followed by a second isotonic or auxotonic phase during which it changes its length. This type of contraction is called *afterloaded contraction*. It corresponds to the normal mode of action of cardiac muscle during transition from the isovolumetric phase of contraction to ejection (see Chap. 89).

In light of the theory of sliding filaments [59,60], force development in the absence of external shortening might reflect the following *elementary events* [99]. At the onset of contraction the myosin heads swing out at right angles to their initial positions and bind to the actin filaments, thus forming cross-bridges between myosin and actin. In this way, the contractile system becomes solidified and may increase in stiffness. Next, the myosin heads bend through an angle of about 45°. By this movement the head stretches internal elastic structures of the muscle, which are presumably located in the myosin necks. Thus, the rotating cross-bridges develop elastic forces and the filaments are braced to forces of traction as long as the ends of the muscle fiber are held so that the sarcomere length remains constant.

In conclusion, it should be mentioned that several assumptions made in conceptualizing the muscular models (e.g., that the properties of the passive elastic elements do not vary with time, or that the passive and active properties can truly be measured independently of each other) are not entirely correct [106]. Hence, the popularity of the modelling approaches in studies of muscle has dwindled. Nonetheless, these studies were helpful in laying the ground for our understanding of cardiac muscle function, and it is in this sense that they are considered here.

Passive Elastic Properties. The elastic properties of the cardiac musculature are of considerable importance as components of the compliance of the entire heart, which determines both diastolic filling and the systolic dynamics of the organ. Resting, non-activated cardiac muscle is much less distensible than skeletal muscle [40]. Moreover, it behaves as a non-linear spring, that is, it elongates with increasing *stress* (σ = force per unit of cross-sectional area, e.g., N/mm^2) applied by the preload. However, its proportional lengthening is less, that is, the muscle becomes stiffer and deforms less at increasing tension. This is clearly expressed by the *stress–strain relation* of the resting muscle, in which *strain* (ε) means a fractional change in dimension (dl/l) caused by the application of stress. The *elastic stiffness* of the muscle is given by the slope of the

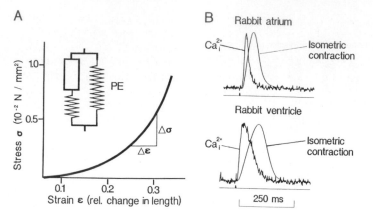

Fig. 90.3. A Stress–strain relation of the isolated mammalian myocardium. Relative change in length means change expressed as fraction of unit muscle length. The slope of the curve, given by the quotient $d\sigma/d\varepsilon$, represents the modulus of elasticity (E_{diff}). (Modified from [65].) **B** Intracellular Ca transients and isometric contraction curves recorded from rabbit atrial and ventricular myocardium. In each *panel* the aequorin signal and the tension record are superimposed. The *lower tracing* indicates the timing of the stimulus. Temperature 37°C, stimulation frequency 1/s. Number of averaged signals 128. (Data taken from [17])

stress–strain curve at any point (Fig. 90.3A). This slope can be expressed by the quotient $d\sigma/d\varepsilon$, which is identical with the *modulus of elasticity* (E_{diff}). There is a linear relation between E_{diff} and tension, and this suggests that the length–tension relation represents an exponential function [107].

In terms of the muscular model, the passive elastic properties are represented by an *elastic element* (PE) arranged in parallel with the contractile and the series elastic components (see inset, Fig. 90.3A). The contractile element is considered freely ectensible at rest. This means that there is no tension across the resting contractile element. As mentioned above, no part of the contractile elastic model can be considered to be represented by any structural component of the muscle. Possible correlates of the passive elastic component may be given by the connective tissue, the sarcolemma, the sarcoplasmic reticulum and even the actin filaments themselves, but without any convincing evidence for the prevalence of any of these components. Similarly, it has been mentioned that the resting tension related to cross section is considerably higher in cardiac muscle than in skeletal muscle. The basis of this comparatively high resting stiffness in cardiac muslce is not yet clearly understood. Some of the high resistance to stretch can probably be attributed to structures that lie outside the myocardial cells, e.g., collagen [21]. Intracellular structures, such as the cytoskeletal protein *desmin* or the elastic protein *connectin* found in the myofibrils as well as in the muscle cell membranes, may also contribute to the high diastolic stiffness of the myocardium [107].

A recurrent question in cardiac physiology is the existence of an actively variable *diastolic compliance*, which is expressed by the reciprocal term *tonus*. This has often been confused with variations in the end-diastolic volume of the heart owing to incomplete relaxation. By contrast, a real contractile diastolic tonus would imply partial contractile activity persisting throughout the entire diastole. Such a

phenomenon has been proposed a number of times, but has not so far been convincingly substantiated. The current concept is that resting extensibility of the myocardium is not altered by neural, humoral or other physiological mechanisms [65].

Active State and Ca²⁺ Transients. In the contractile-elastic model (Fig. 90.2) the contractile element is the only one assumed to possess the ability of active shortening or force development. However, because of the absorption of mechanical energy by the series elasticity, the mechanical events of a single contraction do not accurately reflect the state of activity going on in the contractile element of the muscle. For this reason, experimental techniques have been developed to allow more precise analysis of the mechanical properties of the contractile element during contraction [52]. These experiments use sudden changes in muscle length (*quick stretch*) to circumvent the effects of the series elasticity. Such a quick stretch pulls out the series elasticity and permits the full extent of the potential for the force development to become apparent, and this has been called the *intensity of the active state*. It can be understood most simply as the force that would be developed by the undamped contractile element, were it to be held at constant length. Compared with skeletal muscle, in cardiac muscle the active state – thus defined – develops more slowly and shows a smooth time-course that resembles the isometric contraction, albeit with an earlier peak and a higher amplitude [19,71]. However, from a present-day point of view, the active state, determined by stretching the muscle during its activation, has to be considered with some caution, since the applied stretch does not remain without influence on the contractile process itself [20]. A more appropriate method of analyzing the fundamental processes that proceed in the contractile element and lead to shortening or force development is to measure the intracellular Ca²⁺ activity accompanying contraction – the so-called *Ca2⁺ transient* [18].

There are a great many ways of obtaining information about movements of calcium ions in cells. Most successful recordings of *intracellular Ca²⁺ transients* in heart muscle have been obtained with *aequorin*, a protein extracted from the light organ of a jelly fish, which emits light when reacting with Ca^{2+} [8]. The general form of the aequorin signal is similar in all types of working myocardium so far examined (see Fig. 90.3B). It rises soon after the upstroke of the action potential and climbs steeply to a peak that is sometimes reached during the rapid phase of isometric tension development. The light intensity then declines in a roughly exponential fashion and closely approaches the baseline soon after the peak of the isometric contraction. Interpretation of the Ca^{2+} transient thus recorded has to take into account that its amplitude is only a few percent of what it would be if all the Ca^{2+} involved in excitation–contraction coupling were free in cytoplasm at the same time [121]. This is because Ca^{2+} entering the cytoplasm is bound or sequestered almost as rapidly as it is made available. Although there are several sources of uncertainty in interpretation of the aequorin signals, owing to kinetic limitations of the indicator or to the non-linear nature of its response to changes in Ca^{2+} activity, the aequorin signal still gives sound information on Ca^{2+} transients in heart muscle. Most interventions that alter the contractile performance in cardiac muscle produce changes in the amplitude or in the time-course of the aequorin signal, or in both. Such interventions are: the amount of Ca^{2+} that is mobilized and the rate of mobilization, the affinity of Ca^{2+}-sensitive structures, such as troponin C for Ca^{2+}, and the rate of Ca^{2+} removal from the cytoplasmic space [17].

Active Length–Tension Relationship. The length–tension diagrams in Fig. 90.2 show different types of contraction, starting from only one distinct length each. However, the amount of tension development or of shortening during myocardial contraction depends heavily on the initial muscle length. The greater the initial length at the beginning of contraction (up to a point), the more extensive the contraction will be. This can be studied quantitatively by means of an experimental device that is schematically depicted in Fig. 90.4. The elongate preparation of cardiac muscle is fixed at its ends to both a tension and a length recorder. An isotonic lever allows for different pre- and afterloads to be adjusted. Varying the preload in the absence of an afterload will yield the *passive length–tension curve* described above in terms of the normalized stress–strain relation, which is represented once again in Fig. 90.5b by the lower heavy line connecting points a′ and b′. The upper heavy line in Fig. 90.5b shows the *isometric peak tension* or *isometric maxima curve* as a function of muscle length. This curve can be obtained by stimulating the muscle at different initial lengths, adjusted by different preloads, and setting the afterload large enough to preclude shortening. In Fig. 90.5a corresponding isometric contraction curves are depicted. It can be seen that increasing extension results in a higher velocity of tension development (a steeper slope of the ascending limb) and in a higher amount of peak tension, whereas the time to peak

remains fairly unchanged. Plotting the actively developed tension (total peak tension minus passive resting tension = aa′ or bb′) as a function of length (Fig.90.5c) shows a maximum of tension followed by a decline in the range of unphysiologically high extension. The length at which the maximal active tension is achieved is called L_{max}, and the muscle length in Fig. 90.5c is expressed as a percentage of L_{max}. This corresponds to a resting sarcomere length of about 2.3 µm [67].

If purely isotonic contractions are considered, their peak values for shortening, when plotted as a function of muscle length, yield a curve corresponding to the isometric maxima curve. This *isotonic peak shortening* or *isotonic maxima curve* is represented in Fig. 90.5b by the dashed line connecting the points a″ and b″. In this context it should be stressed that the curves of isometric peak tension and of isotonic peak shortening as functions of length are clearly separate entities.

In view of the function of the whole heart, the length–tension relations of afterloaded contractions are of special interest. Figure 90.4 shows the time-course of an afterloaded contraction. The isometric and isotonic components are represented separately. Similar isometric components of afterloaded contractions exposed to different afterloads are depicted in Fig. 90.5d, 2, 3. Figure 90.5e shows how afterloaded contractions should be untegrated in a length–tension diagram. For each given preload (corresponding to a distinct point on the passive length–tension curve), the afterloaded contractions are intermediates between the purely isometric (curve 1) and the purely isotonic contraction (curve 4). Thus, the purely isometric contraction is attained when the afterload becomes high enough to preclude shortening. Conversely, reducing the afterload to zero will lead to a purely isotonic contraction. In the length–tension diagram, each of the intermediate curves 2 and 3 consists of two components: one vertical, or isometric and one horizontal, or isotonic (see also Fig. 90.2). The isotonic peaks of the afterloaded contractions can be connected by a curve (dotted line in Fig. 90.5e), which joins up the purely isometric and isotonic peaks

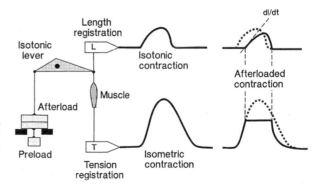

Fig. 90.4. Schematic representation of an experimental device used for separate records of changes in length and tension during an afterloaded contraction. The curves show purely isotonic and isometric contractions (*left*) and an afterloaded contraction (*right*), with the former curves included as *dotted lines*

1805

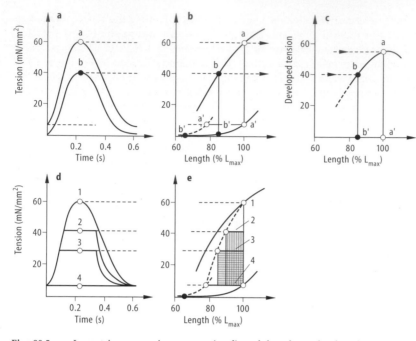

Fig. 90.5a–e. Isometric contraction curves (**a, d**) and length-tension relations (**b, c, e**) obtained from isolated myocardial preparations (cat papillary muscle). L_{max} refers to the length at which maximal tension development occurs. **a** Isometric contraction curves obtained at small extension (*curve b*) and at extension resulting in maximal tension development (*curve a*). **b** Passive length–tension curve connecting *a'* and *b'*, isometric peak tension curve *a b*, and isotonic peak shortening curve *a"b"*. **c** Actively developed tension (total peak tension minus passive resting tension) as a function of length. **d** Isometric components of afterloaded contractions with afterload varying between zero (*4*) and maximal values (*1*). **e** Afterloaded contractions represented in a length–tension diagram. The isotonic peaks are connected by the afterloaded maxima curve connecting the isometric and the isotonic peaks belonging to the same preload. (Modified from [65])

belonging to the same distinct point on the passive length–tension curve, i.e., to the same preload. This curve is called the *afterloaded maxima curve*. There are many such curves in a length–tension diagram, each one belonging to a distinct preload. By contrast, there is only one curve of peak isometric tension and, equally, one curve of peak isotonic shortening. The hatched areas under the isotonic length changes in Fig. 90.5e are measures of the work performed. The amount of external contractile work attains a maximum at an intermediate afterload. It becomes zero with purely isometric contractions, and is very small with purely isotonic contractions, where the whole load is given by the preload only.

The active length–tension relationship so far described is a very important property of cardiac muscle and forms the basis of the so-called *Frank-Starling mechanism* of the whole heart, by which the organ is adjusted to widely varying degrees of filling or of vascular resistance in the absence of inotropic interventions [38,77,111]. The length dependence of tension development of cardiac muscle or of its shortening were formerly explained by the geometry of the overlap between thick and thin filaments. However, recent studies conducted with the chemoluminescent protein aequorin have shown that the increase in contractile force at longer muscle length is associated with an increase in the peak of the myoplasmic Ca^{2+} transient, as well as with alterations in the slope of the transient [3]. This

means that more Ca^{2+} is released into the myoplasm subsequent to excitation in cardiac muscle contracting at longer length. This is probably due to plasmalemmal unfolding with stretch accompanied by some alteration of the functional state of channels, pumps, carriers, or receptors contained within the caveoli [77]. Moreover, in experiments on chemically skinned bundles of isolated muscle and application of solutions with graded pCa^{2+} it was found that force–pCa^{2+} curves were length dependent. This has been interpreted as indicating a change in sensitivity of the myofilaments for Ca^{2+} [2,112]. Moreover, in studies on skinned fibers from pig ventricle it was shown that at submaximal calcium activation (pCa 6.0) stretching of the fibers by about 15% of their resting length caused the myofibrillar ATPase activity to increase by about 22%. By contrast, at maximal Ca^{2+} activation, ATPase activity of the fibers was barely altered by stretching [76]. However, this increased Ca^{2+} sensitivity is not associated with an alteration of the relationship between ATPase activity and force development, i.e., the tension cost. These findings indicate an increase in the apparent rate constant of cross-bridge attachment rather than a decrease in the apparent rate constant of cross-bridge detachment [22,76].

Force–Velocity Relation. From the point of view of the sliding filament theory of muscular contraction, isometric peak tension reflects the number of cross bridges that are simultaneously formed between the myosin and actin filaments. The more bridges are attached at any time, the higher the tension that develops will be. Conversely, shortening velocity of a muscle can be considered to reflect the rate of attachment and detachment of cross bridges accompanied by their bending and by sliding movement of the filaments. It is clear that the two kinds of elementary muscular function are expressions of the active state of the muscle, but also that they are mutually exclusive. This means: the more bridges are attached simultaneously, the smaller must be the rate of cross bridge cycling, and this is reflected by the inverse force–velocity relation (Fig. 90.6B). This relation was first described by Fenn and Marsh [32] in tetanized skeletal muscle and was later studied in more detail by Hill [51], who also proposed the mathematical description by a hyperbolic function. This function implies that at a given initial length (preload) of a muscle its shortening velocity decreases with increasing afterload, and this also holds true for cardiac muscle [96,110].

Various methodological approaches can be used to determine the force–velocity relation. The usual way is to elicit afterloaded contractions at a given preload and determine the peak velocity of shortening of the isotonic components for different levels of afterload [110] (Fig. 90.6a,b). Each pair of values for force and shortening velocity represents a point on the force–velocity curve. The curve intercepts the abscissa at a force value (F_o) that corresponds to the *isometric peak tension*, at which the isotonic component of contraction becomes zero. Determination of the *maximum shortening velocity* (v_{max}) would require that the force be zero (unloaded shortening). However, this condition cannot really be achieved, since there is always some preload on the contractile element, even if just the load of the muscle itself. This means that v_{max} can only be obtained by

extrapolation from a set of points to the velocity axis (Fig. 90.6b). In the force–velocity curve, an increase in contractile activation causes a shift to higher values in both directions, including an increase in v_{max} as well as in F_o (Fig. 90.6d). Moreover, the force–velocity relation allows us to estimate the *power* – the rate of doing work – at a given load [36]. This corresponds to the area of the rectangle formed by the coordinates below the curve. Correspondingly, in the length–tension diagram the rectangle formed by the coordinates of an afterloaded contraction represents the *work* done during the isotonic phase (Fig. 90.6c).

Meaning of Contractility and Inotropic State. As previously stated, the series elastic element of a muscular analogue model is assumed not to change its length during the isotonic phase of an afterloaded contraction. Hence, it might be assumed that the shortening velocity directly reflects the events going on in the contractile element, i.e., the rapidity of conversion of chemical to mechanical energy. The maximum unloaded shortening velocity v_{max} has therefore been considered as an appropriate measure of the vigor of cardiac contraction regardless of the preload which is expressed by the term *contractility* or *inotropic state* [1,93]. Such a measure is the more desirable, because, in the whole heart, the analogous parameter of peak isometric force is difficult to obtain. However, there is no general agreement about the use of v_{max} as a measure of contractility in cardiac muscle. The main objections against this concept are based on methodological problems that cannot easily be solved. For instance: each pair of values of force and velocity is determined at a different instant during the contractile cycle and is thus related to a different degree of activation, since the active state is not constant in cardiac muscle (Fig. 90.3B). Moreover, the esti-

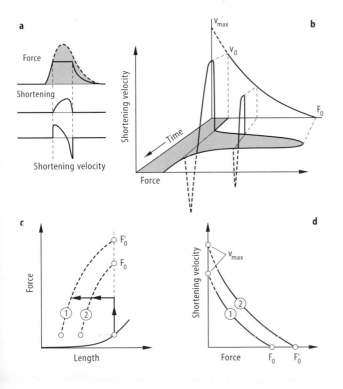

Fig. 90.6a–d. Diagrams concerning the force–velocity relationship in the mammalian myocardium. **a** Time-course of force and shortening and of the velocity of shortening during an afterloaded contraction. **b** Variations of the shortening velocity (*ordinate*) during the isotonic phase of contraction due to variations in afterload depending on time. Two examples of afterload (zero and about half F_o) are shown. **c, d** Influence of an increase in contractile activation as reflected by a length–tension and a force–velocity diagram. (Modified from [65])

mate of v_{max} depends on the muscular model under consideration as well as on the rules applied for extrapolation [65]. These and other objections lead to the conclusion that v_{max} cannot be used uncritically as a measure of contractility. In addition, the results quoted in the section above on active length–tension relationship lead one to accept that resting fiber length (or preload) and contractile (or inotropic) state of the myocardium can no longer be considered totally independent determinants of myocardial performance, as was formerly thought to be the case. Similarly, because the fiber length during a contraction also determines the extent of myofilament Ca^{2+} activation and because this length is determined by the load encountered during shortening (afterload), neither can afterload be considered independently of the contractile state.

Dual Control of Relaxation. Force development during the contraction phase of cardiac muscle has been shown to be determined by changes in loading as well as by changes in contractility. Recently it has become clear that a dual control mechanism also operates during the relaxation phase [24,25]. In isolated cardiac muscle, relaxation is governed by a continuous interaction between load and dissipating activity (i.e., decline of the active state). Whereas inactivation is related to the decline of myoplasmic Ca^{2+}, load is in a delicate equilibrium with the force-generating structures of the muscle at any time. Predominance of load dependence of relaxation becomes apparent when the load exceeds the force potential in the presence of low myoplasmic Ca^{2+}. Thus, under experimental conditions during an afterloaded contraction, application of a comparatively small additional load will cause isotonic relaxation to become considerably enhanced [24].

It has been shown that load dependence of relaxation in cardiac muscle obviously requires the presence of a Ca^{2+}-sequestering system that reduces Ca^{2+} to sufficiently low levels for load to predominate over inactivation. Hence, *load dependence of relaxation* is poorly developed or absent in frog myocardium, with its sparse sarcoplasmic reticulum, and also in mammalian myocardium when it is treated with caffeine, as this inhibits the uptake of Ca^{2+} into the sarcoplasmic reticulum.

A possible explanation of the mechanisms underlying the dual control of relaxation was provided by experiments using aequorin [57]. When a muscle is allowed to shorten after stimulation, the decay of the Ca^{2+} transient is retarded compared with that accompanying an isometric transient. When activated cross bridges connect the filaments, any situation that results in more cross-bridge cycling (or fewer cross-bridge attachments per unit of time) would result in less Ca^{2+} binding. This means that less Ca^{2+} would be bound at shorter lengths, leaving more in the myoplasm as reflected in the declining phase of the aequorin transient, and this Ca^{2+} can then be removed by the Ca^{2+}-sequestering mechanisms. Thus, for a given set of loading conditions, relaxation can be modulated by subtle alterations of the load sensitivity that depends on myoplasmic

Ca^{2+}. Alternatively, for a given load sensitivity, relaxation will be influenced by alterations of the prevailing loads. Relaxation abnormalities could therefore result from alterations of the prevailing loading conditions or from alteration in the decay of activation, or from both.

90.4 Pressure-Volume Relations of the Isolated Heart

If we now leave elongate isolated preparations of cardiac muscle and turn to the intact but still denervated muscular hollow organ, the parameters length and force or tension have to be replaced by the parameters *volume* and *pressure*. By definition, pressure is force per area, measured as newtons per square meter. Of course, the pressure developed by a fluid-filled heart ventricle during systole is caused by the generation of force by the muscular ventricular wall. The relationship between wall tension and pressure is rather complex, but it is very important for the understanding of cardiac function.

Transformation of Wall Tension into Pressure. The significance of the relation between ventricular wall stress and intraventricular pressure may be conceived best from the fact that the rise in intraventricular pressure during the ejection phase is not brought about by the exertion of additional force by the myocardial fibers (Fig. 90.7). Rather it is a physical effect associated with the change in size of the heart. This can be explained with reference to the schematic drawing in Fig. 90.7 (left) as follows: The ventricle is considered as a hollow sphere of radius r and wall thickness w, cut in two halves. Internal pressure (P) and active wall tension (T), which equals stress, σ, times wall thickness, w, are forces that counteract one another, with the pressure (force per unit area) tending to push the two hemispheres apart and with active wall tension (force per unit cross section area) opposing this influence. In mathematical terms the total *disruptive force* (pressure times area) of the intraventricular pressure is $Pr^2\pi$ and the total *cohesive force* holding the sphere together is approximated by $T 2\pi rw$. Setting the two terms equal leads to the relation:

$$P = 2Tw/r \quad \text{or} \quad T = Pr/2w \tag{90.1}$$

This fundamental relation was first discovered, and published in 1806, by a French mathematician, the Marquis de Laplace, in a treatise on celestial mechanics, and it is therefore referred to as the *Laplace relation* (see [117]). From the Laplace relation it can be derived that, when during the ejection phase the radius of the chamber decreases while the wall thickness increases, a rise in internal pressure has to be expected when the force is constant or is even already decreasing (Fig. 90.7, right).

Strictly speaking, the Laplace equation only holds for a thin-walled sphere. Moreover, the relationship does not take account of any structural peculiarities of the arrange-

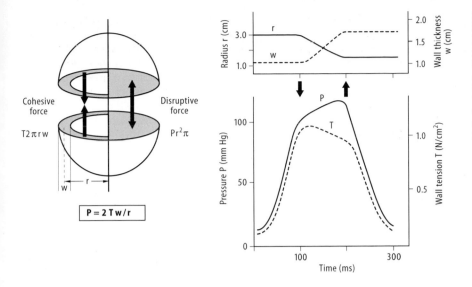

Fig. 90.7. *Left*: schematic illustration of the relationship between wall tension and pressure in a hollow muscular sphere (Laplace relation). *Right*: changes in radius, wall thickness, pressure, and wall tension accompanying contraction and relaxation of a dog left ventricle. (Modified from [12])

ment of the musculature within the heart wall that might mean that, depending on the type of contraction and on the size of the heart, different parts are involved in the contraction in different ways. Similarly, the assumption of an ideal sphere is a rough simplification. However, for the purpose intended here, the simplification may be acceptable. In the pertinent literature different definitions of the wall tension will be found: besides the dimension "stress times wall thickness," as used here, wall tension may be expressed as "force per fiber" or "force per unit length" of the circumference of the ventricle. This of course leads to different equations [41].

End-diastolic Pressure Volume Relation. This relation applying to a completely relaxed heart ventricle corresponds to the length–tension relation of the resting elongate myocardial preparation. However, in addition to the passive elastic properties of the muscular wall, other factors, such as ventricular geometry, intracoronary pressure [7], filling of the neighboring ventricle, and extensibility of the pericardium, exert an influence on the diastolic pressure–volume curve [65]. In the left ventricle in situ this curve – which is also referred to as the *passive pressure–volume curve* – runs nearly parallel to the abcissa in the range of small filling volumes and rises disproportionately with increasing diastolic pressure (Fig. 90.9). This occurs at about 15 mmHg (2.0 kPa) and is mainly due to the low extensibility of the pericardium [46].

The extensibility of the entire ventricle is usually described by its *compliance* ($\Delta V/\Delta P$) or by the reciprocal value the *volume elasticity* ($\Delta P/\Delta V$). The latter parameter expresses the resistance exerted by the ventricle against passive extension and is directly proportional to the diastolic pressure. However, neither compliance nor volume elasticity (nor their values normalized for the volume = $\Delta V/\Delta P \cdot V$ or $\Delta PV/\Delta V$) is appropriate to describe the elastic properties of cardiac muscle, since they depend among other things on the thickness of the ventricular wall [65]. A real measure of purely elastic properties of the cardiac musculature is

given by the *differential modulus of elasticity (E)*, which is equal to the quotient $\Delta\sigma/\Delta\varepsilon$, where σ = wall stress (N/m²) and ε = strain = relative change in length: $(l-l_0)/l_0$ [84,107].

Functional Anatomy and Patterns of Ventricular Movement. When the heart is viewed in cross section at the level of the middle of the ventricles, a conspicuous difference is seen in the thickness of the wall between the two sides. This difference reflects the adaptation of the heart to the different forces required in the ventricles. This adjustment is not in muscle mass alone; the substructure of each ventricle is characteristic of its function. The wall of the left ventricle is made up primarily of very powerful circular musculature: these fibers form a hollow cylinder that merges on the inside and outside into layers of spiral muscles running from base to apex. The wall of the right ventricle consists almost exclusively of such spiral muscles, whereas the circular musculature is relatively poorly developed [114].

The powerful circular musculature of the *left ventricle* is an effective generator of the high pressure required to eject the stroke volume into the systemic circulation [47]. With normal diastolic filling, ejection is brought about primarily by the shortening of these fibers. However, if ventricular filling decreases for any reason, the radius of the ventricle is necessarily reduced and the amount by which the circular fibers can shorten is therefore smaller. The more longitudinally oriented spiral muscles are similarly affected, but to a lesser degree, so that as filling diminishes they take over a growing proportion of the work of ejecting blood. With normal filling the dominant effect of contraction is the reduction of ventricular cross section. When the end-diastolic volume is small the ventricle tends to shorten more in the longitudinal direction. This effect is of crucial importance in the valve plane mechanism (see Chap. 89). The arrangement of the musculature of the *right ventricle* itself indicates its mode of operation [6]. The right ventricle is apposed to the left like a thin-walled crescent-shaped shell. The wall area of this cavity is therefore large with

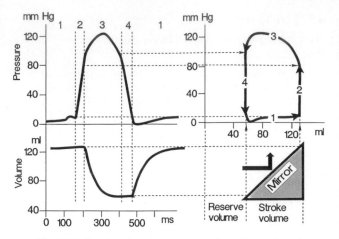

Fig. 90.8. Construction of a pressure–volume loop adapted for a human left ventricle. The changes in pressure during a cardiac cycle are directly reflected on the *ordinate scale*. Numbers *1* to *4* refer to the activity phases of the heart: *1*, diastolic filling; *2*, isovolumic contraction; *3*, ejection; *4*, isovolumic relaxation. The changes in volume are represented as reflected to the *abscissa* by a mirror

respect to its volume, so that a slight movement of the wall towards the septum must cause a relatively large change in volume. Because the resistance to flow in the pulmonary circulation is low, no great expenditure of force is necessary to produce the pressure required to eject the stroke volume. Moreover, the systolic decrease in right ventricular volume is aided by contraction of the left ventricle, which increases the curvature of the septum that protrudes into the right cavity.

Systolic Pressure–Volume Relations. From the time-course of intraventricular pressure during a cardiac cycle (Fig. 89.4) in combination with the simultaneously recorded ventricular volume (Fig. 89.3), a pressure–volume loop can be constructed such as is illustrated in Fig. 90.8 for a human left ventricle. Such a loop-shaped figure was first been constructed by the German physiologist Otto Frank [38] for the isolated frog heart and has since enabled physiologists to assess systolic and diastolic ventricular properties as well as ventriculo-vascular interactions, all with the use of common terms [111]. The pressure–volume loop is also called a *work diagram*, since the area enclosed by the loop represents the product of pressure and volume and thus has the dimension of work done by the ventricle during its systole when ejecting the stroke volume while generating pressure.

The pressure–volume loop starts with the *diastolic filling period* (phase 1). During the initial rapid filling period of this phase, pressure is still falling as result of elastic recoil of the relaxing ventricle that exerts a suction effect on the atrium. In the later slow filling period, a rise in pressure drives the increase in volume, so the line coincides with the passive pressure–volume curve and attains the end-diastolic volume. The diastolic filling period is followed by the onset of the systole with the period of *isovolumic contraction* (phase 2). Application of the term "isometric contraction" to this phase of systole is not correct, since the

shape of the ventricle does indeed change; only its volume remains constant until ejection begins. This is the case when the diastolic aortic pressure is reached and the aortic valves open, i.e., in the *ejection period* (phase 3). As has been outlined above (see Laplace relation), the further rise in pressure during this phase is not due to the development of additional force exerted by the muscular wall but to the change occurring in ventricular geometry. The *isovolumic relaxation period* of the beginning diastole (phase 4) closes up the pressure–volume loop and the cycle repeats.

By analogy with the length–tension diagrams of elongate myocardial preparations (Figs. 90.4, 90.5) *preload* to the intact ventricle is generated by the filling pressure during diastole that – depending on its compliance – determines the end-diastolic volume of the ventricle. The *afterload* imposed to the left ventricle during phase 2 is caused by the end-diastolic pressure in the aorta, which in turn determines the wall tension. The afterload to ventricular musculature can thus be diminished either by lowering the end-diastolic aortic or pulmonary pressure or by reduction of the diameter of the ventricle [12].

With a high diastolic aortic pressure the afterloaded contraction of the left ventricle would ultimately become a purely isovolumic contraction; that is the pressure would rise to its *isovolumic maximum* without opening of the valves and without ejection of a stroke volume (point b in Fig. 90.9). At the other extreme, if there were no afterload contraction would be more or less isotonic, with reduction in volume until the *isobaric maximum* (point c in Fig. 90.9) is reached. The term "isobaric" refers to the constancy of pressure, because changes in shape of the ventricle that accompany this type of contraction preclude purely

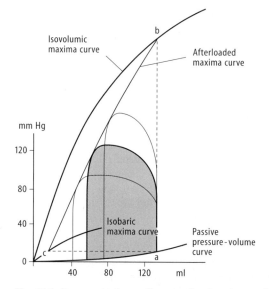

Fig. 90.9. Pressure–volume diagram showing the passive end-diastolic pressure–volume curve as well as the curves of isobaric and isovolumic maxima (all taken together as equilibrium curves). The pressure-volume loops correspond to a distinct preload (*a*) with their systolic peaks touching the afterloaded maxima curve (*bc*). The values are adjusted to reflect values of a human left ventricle

isotonic shortening. Of course, under normal conditions neither will occur. The maxima of the afterloaded contractions that originate from the same point (a) on the passive pressure–volume curve lie on a line joining the corresponding isovolumic and isobaric maximum, i.e., the curve of the afterloaded maxima.

Equilibrium Curves. The pressure–volume loops shown in Fig. 90.9 refer to a single specific initial condition, i.e., a given volume at a given end-diastolic pressure corresponding to a single point (a) on the end-diastolic pressure–volume curve. Variation of the preload can lead to changes in volume, which in turn affect the amplitude of the corresponding maximum isovolumic or isobaric contraction. In such a way the maxima curves are obtained, which represent the *boundary conditions* within which all pressure and volume changes occur for a particular contractile state of the ventricle. Hence, in each ventricular pressure–volume diagram there is one only passive end-diastolic pressure–volume curve and only one curve each for the isovolumic and isobaric maxima, but a large number of afterloaded maxima curves – one for each of the possible starting points (a) along the diastolic pV curve.
Within these boundaries both the pressure and the ejection can vary depending on the initial degree of filling of the ventricle. This implies that the isolated denervated heart can exert different pressures or eject different volumes of blood entirely on the basis of its intrinsic muscular properties in the absence of any other influence [111]. This is the basis of the *autoregulatory adjustment* of the heart to changing workload considered in Sect. 90.5. The cellular mechanisms underlying the dependence of the contraction maxima on the initial volume of the ventricle have been considered in Sect. 90.3 (see "Active Length–Tension Relationship").

90.5 Adjustment to Changing Work Load

Autoregulatory Responses of the Isolated Heart. The English physiologist Starling (see also Chap. 1) developed a mammalian heart preparation that allowed aortic pressure and venous return to be varied independently over a wide range, so that the factors could be correlated with the end-diastolic size of the ventricles [111]. The heart retained its natural connections to the artificially ventilated lung, but the systemic circulation was replaced by a system of blood-filled tubes incorporating a variable resistance, with provisions for pressure recording at a number of points. The rate of venous return was determined by adjusting the outflow from a reservoir. When a *heart-lung preparation* of this type is denervated and kept at constant temperature, the heart beats at a constant rate and thus allows study of its intrinsic autoregulatory responses to loads in volume or pressure.

Adaptation to Acute Volume Loading. In the Starling preparation venous return can be increased by raising the input reservoir and thus increasing the filling pressure of the right ventricle. This in turn leads to an increase in the output of the right ventricle and consequently to a higher filling pressure and preload of the left ventricle via the pulmonary vessels that represent a connection exerting only a low resistance to flow. Figure 90.10 shows how the left ventricle responds to such a volume load. Under initial conditions, with an end-diastolic volume of about 130 ml, the shaded work diagram applies and the stroke volume is about 70 ml. Thus, the end-diastolic volume is about 60 ml. When the filling pressure and consequently the venous return is increased the end-diastolic volume of the left

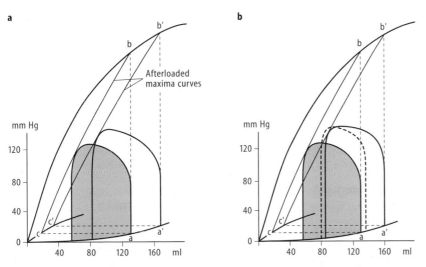

Fig. 90.10a,b. Pressure–volume diagrams representing autoregulatory responses of a left ventricle in an isolated heart to changes in preload (**a**) or afterload (**b**). Values have been adjusted to conditions in a human heart. Increasing the preload – stronger filling – causes an increase in stroke volume. Increasing the afterload – higher diastolic aortic pressure – leads to an increase of the end-diastolic volume and to development of greater systolic pressure with the stroke volume remaining fairly unchanged

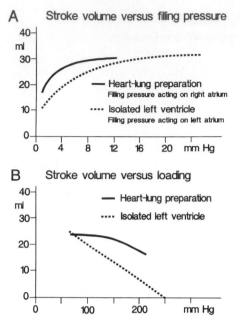

A Stroke volume versus filling pressure

— Heart-lung preparation
Filling pressure acting on right atrium

···· Isolated left ventricle
Filling pressure acting on left atrium

B Stroke volume versus loading

— Heart-lung preparation

···· Isolated left ventricle

Fig. 90.11A,B. Interaction of right and left ventricle in a heart-lung preparation of a dog as compared with an isolated left ventricle accompanying changes in volume (**A**) or pressure loading (**B**). In both diagrams the *ordinate scale* represents the stroke volume of the left ventricle. Under both conditions in the heart–lung preparation the right ventricle acts as a booster pump and thus improves the performance of the left ventricle. This effect is absent if the left ventricle is disconnected from the right one

ventricle rises to about 180 ml. This causes an increase in stroke volume to around 90 ml with no change in the isovolumic or isobaric maxima. At this stage, the diastolic aortic pressure still remains about the same, but the systolic pressure rises because the aorta is under greater tension due to the larger stroke volume. The end-systolic volume is also somewhat increased. Because the starting point of the diagram is different, it has to be constructed with reference to another afterloaded maxima curve. The essential point of this result is that the isolated heart, beating at a constant rate, can compensate of its own accord – i.e., by autoregulation – for increased diastolic filling by ejecting a greater stroke volume. This kind of adaptation is referred to as the *Frank–Starling mechanism* in honor of its discovery first in the frog by Frank [38] and then in the mammalian heart by Starling [111]. As shown below, in principle the same mechanism also underlies adaptation to increased pressure load.

Adaptation to Acute Pressure Loading. If the resistance to flow in the artificial part of the heart–lung preparation is increased, the performance of the heart adjusts in a step-wise manner, represented in the work diagram of the left ventricle as follows (Fig. 90.10b). Because of the higher resistance to outflow of blood, the aortic pressure in diastole does not return to the original level, so that the left ventricle must exert greater pressure (about 125 instead of about 90 mmHg) in the following systole before ejection

can begin (dashed curve). This necessarily results in diminution of the stroke volume and in an increase of the reserve volume at the end of systole. Provided the venous return to the left ventricle remains unchanged, the left ventricle will be filled to a greater extent and will thus gain an improved initial state to develop the required higher pressure and eject nearly the same stroke volume as before.

Interaction of Right and Left Ventricle. Figure 90.11 summarizes the steady state adaptation of cardiac performance to a wide range of variations in filling pressure or afterloading as expressed by the stroke volume of the left ventricle of a dog heart. Comparison of the isolated left ventricle and the ventricles in the heart–lung preparation shows that there is a considerable difference in the reaction of the two experimental conditions (Fig. 90.11A): obviously, the *heart-lung preparation* is more *sensitive* to preload than the isolated left ventricle. A low filling pressure of 1–12 mmHg acting on the right atrium in the heart–lung preparation yields about the same change in stroke volume of the left ventricle as a higher filling pressure of about 3–30 mmHg acting on the left atrium of the isolated heart. Under the former condition, the filling pressure is translated by the right ventricle into a higher stroke volume, which is transmitted via the pulmonary vessels to the left atrium there causing a higher increase in pressure than it has been acting before on the right atrium.

Similarly, the *heart–lung preparation* is *less sensitive* to afterload than the *isolated left ventricle* (Fig. 90.11B). Thus, the same afterload exerted by a mean arterial pressure between 50 and 225 mmHg produces a smaller decrease in stroke volume in the combined system. Again this behavior can be attributed to right and left ventricle interactions in the heart–lung preparation, where an increase in afterload to the left ventricle is propagated backward in a counterflow direction, thus generating a corresponding adaptation of the right ventricle which is then retransmitted to the left one.

Dynamics of the Innervated Heart. The mechanisms by which an isolated heart or a denervated heart–lung preparation adjusts to changes in preload or in afterload, as described above, were long regarded as the only basis of cardiac dynamics. According to what was called *Starlings's Law*, the heart in situ was also thought to perform more stroke work entirely as a result of increase in its end-diastolic volume, with no change of its contractile state (i.e., of its isovolumic and isobaric maxima). The current opinion, however, is that this view is not generally valid; or at least that it does not apply to the changes in cardiac output correlated with physical work. Thus, Starling's Law would predict that a fully functioning heart is small when the body is at rest and that when a workload is imposed it enlarges in adaptation to the increased venous return. However, precisely the opposite is the case. The heart of a healthy subject doing physical work on a bicycle ergometer can be monitored by means of echocardiography; such investigations show a clear reduction of the end-diastolic and end-systolic size of the heart during exercise [97]. This

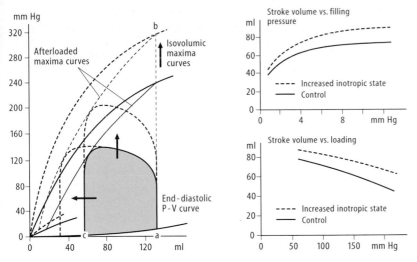

Fig. 90.12. *Left:* influence of an increased inotropic state on the equilibrium curves in the pressure–volume diagram and on the working diagram of the left ventricle. The equilibrium curves are analogous to those in Fig. 90.9. The inotropic shift of the maxima curves to higher values enables the heart either to eject a greater volume or to overcome a higher aortic pressure without a change in preload. *Right:* diagrams representing the inotropic influence on the stroke volume of the left ventricle for different preloads (*upper panel*) or different afterloads (*lower panel*)

adaptation occurs under the influence of the sympathetic nervous system and results from an increase in the contractile force of the myocardium (its *inotropic state* or *contractility*) that is independent of the degree of preload (see Sect. 90.3).

The *adaptation of the heart to physical exercise* just described appears in the work diagram of the left ventricle as an upward shift of the curves for isovolumic and isobaric maxima, with a corresponding increase in the slope of the afterloaded contraction line (Fig. 90.12, left) [63]. This rearrangement of the pressure–volume relations enables the ventricle either to eject a larger stroke volume or to overcome a higher pressure load without any increase in the end-diastolic volume. An increase in stroke volume leaves a smaller systolic reserve volume, so that if venous return does not increase the end-diastolic volume must be smaller – which explains the observed reduction in size of the heart [97]. But even when venous return increases simultaneously, the rise in heart rate caused by sympathetic activity (*positive chronotropic action*) increases the amount of blood propelled through the system and thus prevents excessive filling. The inotropic changes become also apparent when the aforementioned dependencies of cardiac performance on volume or pressure loading are considered (see Fig. 90.11). As can be seen from Fig. 90.12 (right hand side), an increase in the inotropic state shifts the curves representing the dependencies of the stroke volume on preload or on afterload, respectively, to higher levels.

Under the influence of the sympathetic system the heart can increase its output even before the venous return begins to increase. Nevertheless, the further possibility of increasing output by enlarging the end-diastolic volume remains available. From this point of view, the ability of the heart to meet unusual demands – the *cardiac reserve* –

appears in a new light. In the earlier view the reserve was thought to depend on the extent to which the heart can increase its end-diastolic volume in conditions of exertion compared with that at rest. By contrast, under the positive inotropic influence of the sympathetic system the reserve is limited by the *end-diastolic volume at rest*. The hearts of trained athletes, for example, are conspicuously large at rest, and in some cases contain three to four times the normal stroke volume – as against about two stroke volumes in an untrained person [97]. The athlete's heart, accordingly has a large reserve. According to the old view, its reserve would have had to be considered very small.

Influence of Heart Rate on Cardiac Dynamics. One of the most pronounced differences between the denervated heart and the heart in situ is that the rate of beating of the latter varies. The increase in heart rate caused by sympathetic activity is in fact the most important mechanism of increasing the cardiac output under physical workload. An increase in heart rate not only raises the number of beats per unit of time but also changes the temporal relations between systole and diastole in a characteristic way. An example is given in Table 90.1. The table shows that shortening of a cardiac cycle period primarily affects the diastolic phase. This implies that the *net working time* of the ventricles (the sum of all the systole durations in 1 min) increases considerably at higher heart rates, and the recovery pauses decrease correspondingly. However, adequate filling of the ventricles despite even very brief diastole is guaranteed by the facts that most of the inflow occurs at the beginning of diastole, and that sympathetic activity causes a distinct increase in the rate of relaxation [118, 122]. Moreover, the sympathetic drive causes the atria to contract more strongly, which accelerates filling of the ventricles. Therefore, when the heart rate rises under sympa-

Table 90.1. Effect of shortening the cardiac cycle on the systolic and diastolic phases

Beats per min	Duration of systole (s)	Duration of diastole (s)	Net working time (s/min)
70	0.28	0.58	19.6
150	0.25	0.15	37.5

thetic influence, up to a frequency of ca. 150/min there is usually no critical diminution of ventricular filling.

Frank–Starling Mechanism in the Intact Heart. The dominant influence of the sympathetic system in adjusting cardiac output does not exclude the possibility that under some conditions the heart can also be governed by other factors. For example, the capacity of the heart to regulate its activity by end-diastolic volume in the sense of the Frank–Starling mechanism is brought into play when changes in filling occur without a general increase in physical activity. This applies in particular to the coordination of the output of the two ventricles. Because the ventricles beat at the same rate, the outputs of the two can be matched only by adjustment of the stroke volume. Other examples include changes in the position of the body which affect venous return (greater stroke volume during reclining than during standing), acute increase in the volume of the circulating blood (transfusion), and increase in the resistance to outflow. Moreover, when the sympathetic system is pharmacologically inactivated by β-sympatholytics the autoregulatory mechanisms continue to operate and their effect becomes more important.

Measures of Contractility in the Intact Heart. The positive inotropic action of the sympathetic system enables the heart, without increased diastolic filling, to eject a larger stroke volume or to eject the stroke volume against a higher pressure. A similar effect on cardiac dynamics can be obtained by raising the extracellular Ca^{2+} concentration, by administering cardiac stimulant drugs, e.g., cardiac glycosides, and as a direct consequence of increasing heart rate. All these effects have in common that they enhance cardiac performance independently of the degree of stretching of the myocardium – in other words, they increase its contractility (*positive inotropy*). Changes in contractility of the entire heart might be detected by examination of the maxima curves in the pressure–volume diagram (Fig. 90.12). However, to obtain such data for an analysis of contractility of the heart in situ, especially the human heart, would require experimental conditions that cannot be fulfilled in humans. Therefore, other criteria have been proposed, e.g., the *maximal rate of pressure development (dP/dt_{max}) in the isovolumic phase of systole*, which can be measured with cardiac catheters. Normal values in human hearts are 1500–2000 mmHg/s (= 200–333 kPa/s). Of course, the main objections to this concept of contractility discussed in Sect. 90.3 also apply to the whole heart. Another measure of contractility of the heart during the ejection period occasionally used is the ratio of

stroke volume to end-diastolic volume, the *ejection fraction*. It gives the proportion of the blood in the heart that is expelled during systole. Normal values for a person at rest range from 0.5 to 0.7 (i.e., 50–70%). The ejection fraction can be non-invasively measured by echocardiography. However, again, it should be taken into account that although generally regarded as descriptor of ventricular performance, the ejection fraction is just as dependent on vascular resistance and ventricular geometry as it is on ventricular contractile state [65].

End-systolic Pressure–Volume Relation. In Starling's experiments on the heart–lung preparation the volume curve recorded combined left and right ventricular volumes. Thus, the relationship between instantaneous pressure and volume of a single ventricle could not be directly monitored. Meanwhile new techniques have enabled investigators to study pressure–volume loops during one systole even in man [70]. The pressures are recorded in the ventricular cavity by use of a micromanometer, and volumes are determined by use of a multi-electrode conductance catheter inside the ventricle; preload is varied by occluding the inferior vena cava with an intravascular balloon. Two boundaries become apparent from a set of pressure–volume loops obtained in this manner. The upper left-hand corners of the loops define a nearly rectangular-linear relation – the *end-systolic pressure–volume relation (ESPVR)* that indicates the upper limits of systolic performance under a given contractile state. Acute changes in the slope of this relation indicate an altered contractile state [104] (Fig. 90.13A). The lower boundary of the loops, the *diastolic pressure–volume relation (EDPVR)*, provides a measure of the "passive" properties of the ventricle [40]. From this point of view, the ventricle can be considered as a chamber whose *wall stiffness*, expressed as an *elastance (dP/dV)* starts at a baseline in diastole (EDPVR), increases to a maximal value – the *end-systolic elastance (Ee)* expressed by the slope of the ESPVR – and then returns to its starting value. This *time-varying elastance concept* of cardiac contraction was proposed by Sagawa et al. ([102–104]; for critical evaluation see [70]). It also offers a new definition of the *afterload*, which can be considered as the *effective arterial elastance (Ea)*. Effective arterial elastance is related to the mean vascular resistance, and is approximately equal to the *ratio of end-systolic pressure (Pes) divided by stroke volume* (Fig. 90.13B). An increase in afterload resistance makes a pressure–volume loop tall and narrow, and causes an increase in Ea. Conversely, when afterload resistance is reduced this increases stroke volume and lowers systolic pressure, therefore the P–V loop becomes flat and wide and Ea falls.

Compared with the classical representations of cardiac dynamics in the pressure–volume diagram by the equilibrium curves, the ESPVR of ejecting beats roughly approximates the isovolumic end-systolic P–V relation curve [65].

The Heart as an Endocrine Gland. In 1955 Gauer and Henry found that volume increments of the left atrium of

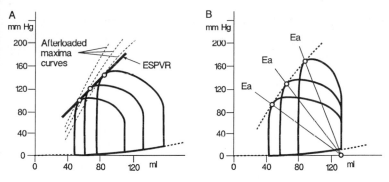

Fig. 90.13. A End-systolic pressure-volume relation (ESPVR) of a human left ventricle for a given inotropic state. Changes in the intropic state would change the slope of the curve. **B** Changes in afterload represented as effective arterial elastance (*Ea*). Ea is equal to the ratio of end-systolic pressure divided by stroke volume, i.e., to the slope of the *Ea* lines. The slope of each *Ea* line varies directly with afterload resistance. Thus, the low afterload beat is flat and wide (low Ea), whereas the *taller thinner loop* represents a high Ea. (Data from [65])

the heart led to increased renal Na⁺ and water excretion [50]. This phenomenon was then attributed to inhibitory nervous effects on antidiuretic hormone (ADH) secretion and was considered as a mechanism of volume regulation (*Gauer Henry reflex*). At the same time Kisch [75] described specific granules within atrial muscle fibers. However, the idea that a homone produced by the heart itself and stored in these granules might be involved in the regulatory process came up 25 years later, when De Bold and Sonnenberg found that injection of extracts of rat atria into rats caused significant diuresis and natriuresis [28]. Meanwhile, the active component of the atrial tissue extracts has been identified as a 28 amino acid peptide named *atrial natriuretic peptide (ANP)* or *atriopeptide* [88] (cf. Chaps. 75, 77).

The precursor molecule of ANP – a 126 amino acid peptide – is stored mainly in the secretory granules of the right atrium, but is also found in smaller concentrations in the left atrium and in the ventricles. Distension of the myocardium containing secretory granules causes exocytosis of the granules with cleavage of the prohormone into the homone ANP proper and an N-terminal fragment with as yet unknown function. The plasma concentration of ANP in normal man is around 10 pmol/l (30 ng/l) with a plasma half-life of several minutes. The plasma concentration changes in parallel with the intake of NaCl and is elevated acutely by all interventions that increase the blood volume or cause its redistribution towards the cardiopulmonary component of the circulatory system [55,101]. ANP affects renal salt and water excretion directly by changes in glomerular filtration rate and by inhibition of salt absorption in the medullary collecting duct. ANP also affects salt and water excretion indirectly by inhibiting the formation and release of salt- and water-retaining hormones such as aldosterone, angiotensin II, and vasopressin [58,109]. Regulation of vascular filling pressures by direct vascular actions may be an additional homeostatic function of ANP. In addition to its role as a peripheral hormone exerting cardiac unloading, ANP occurs in the CNS as a neuropeptide that might also be involved in blood pressure and volume regulation [89].

Adaptation of the Heart to Prolonged Exercise. All the adaptive processes considered so far have enabled rapid adjustment of the heart's activity to acute changes in the demands upon it. If the demand for increased effort is repeated, or even continuous, structural changes occur and the heart enlarges. An example of such *hypertrophy* is the large heart of the trained athlete [98]. Typically, such enlargement of the heart is greatest (comprising a mass of more than 500 g, as opposed to a normal heart with about 300 g) in athletes who specialize in endurance sports (long-distance runners, racing cyclists, etc.). As hypertrophy develops in a chronically loaded heart, the number of myocardial cells at first remains constant, while their length and thickness uniformly increase. During this process the cavities within the heart necessarily increase in volume. According to Laplace's law, the result is that a greater wall tension is required to produce a given pressure [78]. However, because the muscle mass has grown, the force per cross-sectional unit area of muscle is still essentially the same. That is, the athletic heart contains a large volume of blood but does not pay for this advantage, as the acutely stretched heart does, with an unfavorable ratio for the conversion of muscle tension to pressure. When an athlete's training is terminated, the hypertrophy disappears within a few weeks [98]. Once the mass of the hypertrophied heart reaches the critical level of about 500 g, both the size and the number of fibers increase. This condition is called *hyperplasia* [56].

When only parts of the heart are subjected to an increased chronic load, the hypertrophy is limited to the affected region. In general, this occurs only in pathologic conditions. Two forms of adaptation can be distinguished: when *increased pressure* alone is required, the initial hypertrophy is not accompanied by any appreciable increase in cavity volume (e.g., in hypertrophy of the left ventricle caused by aortic stenosis). If the extra work is required to propel an *increased volume*, however, hypertrophy and cavity enlargement occur together (e.g., hypertrophy and dilation of the left ventricle caused by aortic valve insufficiency). The degree to which the heart can compensate for such defects by changes in myocardial structure is limited.

As myocardial fiber radius increases, so do the diffusion paths between the capillaries and the interior of the fibers, with the resulting threat of an inadequate O_2 supply [56]. Hence, when severe pathologic states persist for some time they may contribute to manifestations of heart failure (myocardial insufficiency) [64].

To study the mechanisms of adaptation to chronic load, different experimental models of cardiac hypertrophy have been developed. Chronic overload in pressure can be investigated in spontaneously hypertensive rats, in hypertension caused by occlusion of renal arteries, and following partial ligation of the aorta or the pulmonary artery. Chronic overload in volume can be achieved by experimental aortic insufficiency, aortocaval anastomosis, av block, and chronic swimming or treadmill exercise. Special types of hypertrophy are due to anemia or hypoxia, high doses of catecholamines, thyroid hormones, glucocorticoids and mineralocorticoids. *However, the precise mechanisms leading to hypertrophy are still unclear* [91]. The key question is how a hemodynamic overload can be converted into increased protein synthesis. An answer to this question will require the application of molecular biology. From a present-day point of view, it seems reasonable to assume that hemodynamic stimuli might evoke the synthesis of growth factors, which in turn stimulate transcription of the corresponding genes [91]. Since enhanced preload and afterload of the overloaded myocardium results in *stretching* of the myocardium, this influence is increasingly regarded as of primary importance. Stretch may act on stretch-activated ion channels and in this way may promote Na^+ influx, which has been thought to induce a growth signal [72]. Another possible primary signal starting the sequence of events that lead to hypertrophy may be given by *increased tension*. There are several conditions in which increased tension can be separated from increased stretch. On the single cell level there is a distinct increase in fiber thickness caused by chronic pressure load, whereas fibers that are chronically loaded by increased volume preferentially grow in the longitudinal direction [56]. Hence, probably both influences play a part, with stretch being the prime signal for elongation of the fibers in volume overload and tension, the signal in the case of overload resulting from increased pressure. Moreover, *adrenergic factors* are obviously important in the hypertrophy process. Thus, adrenergic agonists can promote hypertrophy without inducing hypertension, while adrenergic antagonists are effective in decreasing left ventricular hypertrophy in hypertensives. In the latter case pure β-blockers have proven more effective than combined α-β-blockers [85]. In mechanically overloaded isolated rat hearts, *cyclic AMP* has been shown to increase [123], as does IP_3 formation in dilated atria or ventricles [119]. Since both an increase of cAMP and an increase of IP_3 can increase cell Ca^{2+}, this ion may be an essential link in the chain of signals leading to hypertrophy. Although these hypotheses provide some reasonable explanation of many experimental data, none of them is sufficient to cover all aspects of cardiac growth.

There are some additional interesting *sequelae of hypertrophy*: In chronically pressure-loaded rat ventricles a considerable increase of the duration of the single cell action potential has been described, which causes prolonged activation of the contractile apparatus and improved myocardial performance [5,45]. The change in action potential duration is most probably due to an increased but slower inactivating L-type Ca^{2+} current [74]. Moreover, different types of hypertrophy are characterized by structural peculiarities of the myosin molecules. A predominance of the isoenzyme V1 is found in hearts of animals subjected to physical exercise, which then exhibit high specific activity of their cardiac myosine-ATPase and high unloaded shortening velocity. By contrast, in pressure-loaded hearts there is a predominance of V3 and a corresponding decrease of V_{max} [100]. These changes are attributed to basic alterations of the cross-bridge kinetics, with lengthening of the cross-bridge cycle in the latter case.

90.6 Energetics of the Heartbeat

In the preceding sections the work performed by the heart has been treated in various contexts. Now we turn to certain quantitative aspects of the subject, and consider more closely those processes that provide energy to the heart. First let us look at the debit side of the energy balance sheet.

Cardiac Work and Power. In physics work is defined as the product of force and distance; the unit of work is the newton-meter (Nm = joule). This formula applies, for example, directly to the work done by a skeletal muscle when it shortens and lifts a weight for a certain distance (work = weight × distance). In the last analysis, cardiac muscle also does its work by shortening of the fibers and development of force. But in this case no weight is lifted; rather, a certain volume of blood (V), expressed in cubic meters, is ejected against a resistance by the development of pressure (P = expressed as N/m^2) from the left ventricle to the aorta, and from the right ventricle to the pulmonary artery. The pressure–volume work thus performed is calculated as the product $P \times V$, with the units cubic meters times newtons per square meter = newton-meters ($m^3 \times N/m^2$ = Nm). In most conditions this work is most of the useful work of the heart. The work of the atria, which is generated in transporting blood into the ventricles, is usually negligible (except in exercise), because most ventricular filling occurs before atrial contraction and normal atrial systole develops only low levels of pressure [4].

In addition to the pressure–volume work, the ventricles also perform work in accelerating the blood as it passes into the aorta and pulmonary artery. This is the so-called *kinetic* work expended to bring the inert mass (m) of the blood to a relatively high velocity (v). It is calculated from the formula for kinetic energy ($1/2\,m \times v^2$) and is usually only a small fraction of the pressure–volume work (less than about 5%) [90].

Table 90.2. Specimen calculation of cardiac work in a resting person for a single systole

Pressure – volume work	$P \times V$	
Left ventricle		
P = 110 mm Hg	$= 110 \times 133\,N/m^2$	$P \times V = 1.024\,Nm$
V = 70 ml	$= 70 \times 10^{-6}\,m^3$	
Right ventricle		
P = 15 mm Hg	$= 15 \times 133\,N/m^2$	$P \times V = 0.140\,Nm$
V = 70 ml	$= 70 \times 10^{-6}\,m^3$	
Kinetic work	$1/2\,m \times v^2$	
Left ventricle		
m = 70 g	$= 70 \times 10^{-3}\,kg$	$1/2\,mv^2 = 0.009\,Nm$
v = 0.5 m/s		
Right ventricle		
m and v are as in left ventricle		$1/2\,mv^2 = 0.009\,Nm$
Total work		$= 1.182\,Nm$

Because the individual factors that determine cardiac work change continuously during the work phase of the cardiac cycle, the time-dependent products $P \times V$ and $1/2\,m \times v^2$ ought to be integrated over the duration of the ejection period. Here, however, we shall content ourselves with a simplification that permits satisfactory approximation. By way of an example, Table 90.2 presents a calculation of the cardiac work for a single systole at rest, taking: for P, the mean systolic pressure at the outlet from the ventricles (where 1 mmHg corresponds to $133\,N/m^2 = 133\,Pa$); for V, the stroke volume (in m^3); for m, the mass of the accelerated blood (stroke volume, in kg); and for v, the mean ejection velocity (in m/s). Inspection of Table 90.2 reveals that the work performed by the whole heart per systole is determined chiefly by the size of the stroke volume and the level of aortic pressure. It is in the order of 1 Nm = 1 J. As pressure–volume work is less in the right ventricle, kinetic work represents a greater proportion of total right ventricular work. Since kinetic work increases with the square of the velocity at which blood leaves the ventricle, and since velocity increases with stroke volume, kinetic work grows roughly in proportion to the cube of stroke volume. Hence, the proportion of kinetic work in the total cardiac work can increase considerably when greater volumes are ejected (e.g., in exercise, aortic insufficiency or severe anemia). Moreover, a decrease in the elastic extensibility of the aorta in old age has the effect of increasing the kinetic work of the heart, because the greater rigidity of the "compression chamber" severely reduces the velocity of blood flow in the aorta during diastole. Then the left ventricle during systole must accelerate not just the stroke volume, but a considerably greater amount of blood. Under such conditions the kinetic work can almost equal the pressure–volume work.

Power is work per unit of time. If we assume about one systole per second, cardiac power is of the order of 1 J/s = 1 W. A useful measure of efficiency for engines in general is the *power-to-weight ratio*. For the heart, given a weight of about 3 N, this works out to 0.3 W/N. This is a far poorer performance than is achieved by most engines (the motor of a car, for example, puts out 15–25 W/N). During muscular work, however, cardiac power can be considerably greater, so that the power-to-weight ratio approaches that of mechanical pumps. In any case, this calculation shows that it must be possible to construct artificial pumps that could, under appropriate conditions, replace a living heart and will weigh less.

Oxygen and Nutrient Consumption. The energy the heart requires for its mechanical work comes primarily from the oxidative decomposition of nutrients [115]. In this regard cardiac and skeletal muscle differ fundamentally, for the latter can obtain a large part of the energy needed to meet short-term demands by anaerobic processes: the "oxygen debt" that is built up can be repaid later. The dependence of the heart on oxidative processes manifests itself in the large number of mitochondria in myocardial cells.

The O_2 consumption of a heart in situ is ordinarily obtained by measuring the difference (D) in O_2 content of the arterial and coronary venous blood (avD_{O_2}) and multiplying this by the rate of blood flow through the coronary vessels. When the body is at rest the cardiac oxygen consumption so determined is in the order of 0.08–$0.10\,ml \cdot g^{-1} \cdot min^{-1}$. A heart with a mass of 300 g thus consumes 24–$30\,ml\,O_2/min$. This is about 10% of the total resting O_2 consumption of an adult, though the weight of the heart is barely 0.5% of the total body weight. When the body is performing hard work the O_2 consumption of the heart can rise to 4 times the resting level. One would expect the O_2 consumption of the heart to be determined basically by its per-systole contribution to the work of the body as a whole. But this is not the case: for a given amount of cardiac work the heart consumes considerably more O_2 when it is working against high pressure than when it is ejecting a large volume against a correspondingly low pressure [43]. The efficiency of the heart – that is, the fraction of the total energy expenditure that is converted to mechanical work – is therefore lower when pressure loading predominates than when volume loading predominates (Fig. 90.14). The efficiency of a fully sufficient heart depends on the prevailing conditions and lies for the human heart in the range of 15–40%.

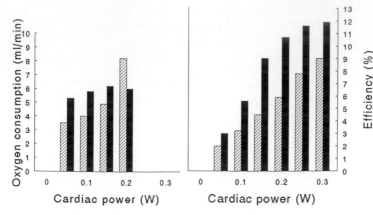

Fig. 90.14. Dependence of O_2 consumption (dashed columns) and efficiency (black columns) of dog heart on cardiac power. Variation of cardiac power was induced by either changes in pressure loading (*left*) or in volume loading (*right*). Equal amounts of power are accompanied by much higher efficiency when volume loading predominates than when pressure loading is imposed. (Data from [44])

In cases of coronary insufficiency, when the O_2 consumption of the heart tends to exceed the O_2 supplied by the blood, the aim of treatment is to reduce the resistance to flow in the systemic circulation so as to lower arterial pressure (afterload) and thus lessen cardiac O_2 consumption. The long-known beneficial action of nitroglycerin in attacks of angina pectoris is an example of such an effect.

Recent studies indicate that the per-systole O_2 consumption of the heart depends primarily on the myocardial fiber tension, and increases with the duration of contraction. The customary reference is the *tension–time index*, the product of mean myocardial fiber tension and duration of systole [87]. When the size of the ventricles is constant, the mean systolic aortic pressure can be used instead of fiber tension (see Laplace law). Changes in heart rate affect O_2 consumption to about the same extent as they change the net working time (the product of systole duration and heart rate). As a result, O_2 consumption rises and falls about in proportion to the square root of heart rate [10]. Furthermore, a small fraction (ca. $0.015\,\text{ml}\cdot\text{g}^{-1}\cdot\text{min}^{-1}$) about 20% of the O_2 consumption of the active heart must be maintained when the heart is quiescent to prevent irreversible changes in the structure of the organ (the basal consumption) [79].

The kinds and quantities of nutrients used by the heart to obtain energy can be determined by the same principle as applied to the measurement of O_2 consumption. That is, the concentration difference between arterial and coronary venous blood is multiplied by the coronary flow rate. Such experiments have shown that the heart – as compared, for example, with skeletal muscle – is a sort of "omnivore" (Fig. 90.15) [16,116].

Especially noteworthy features are the large proportion of free fatty acids and the fact that cardiac muscle, unlike skeletal muscle, can metabolize lactic acid (lactate) [9,31]. Because during hard muscular work the anaerobic glycolysis within the muscles releases lactic acid into the bloodstream, the heart is automatically provided with a certain amount of supplementary fuel to support the addi-

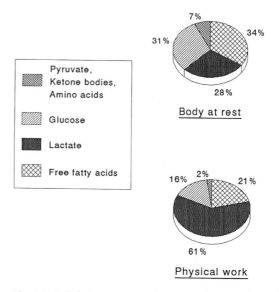

Fig. 90.15. Relative amounts of various substrates in oxidative metabolism of the human heart, with the body at rest and doing physical work. Substrate uptake is expressed as the percentage of total cardiac O_2 consumption that involves the substrate concerned, i.e., O_2 extraction quotient. (Data from [73])

tional work demanded. In breaking down lactic acid the heart does not only obtain energy; at the same time, it contributes to the stabilization of the blood pH.

The properties of the different substrates consumed are primarily determined by supply – that is, by their arterial concentration. Because the heart is so remarkably adaptable, using whatever happens to be available, the chief danger of coronary insufficiency lies not in a substrate shortage but in O_2 deficiency.

Breakdown of the various substrates results in the formation of ATP, the direct source of energy for the contraction process. The ATP content of cardiac muscle is 4–6 mmol/kg. This is a low concentration in comparison with that required for the work of contraction. The active

Table 90.3. Coronary blood flow and artery-coronary vein O_2 difference (avD_{O_2}) of the human heart at rest and under workload. (Data from [48])

	Rest	Work
Coronary blood flow ($ml/g \cdot min$)	0.8	3.2
avD_{O_2} (ml/l of blood)	140	160
Coronary venous O_2 content (ml/l of blood)	60	40

myocardium recycles ATP several times within seconds – that is, it splits ATP to form ADP and inorganic phosphate and then resynthesizes ATP [26,68]. Another phosphate found in the myocardium in about the same amounts as ATP is phosphocreatine (7–8 mmol/kg). This is a particularly sensitive indicator of the adequacy of the substrate and of the O_2 supply to the heart, for the metabolic resynthesis of split ATP initially relies on the breakdown of phosphocreatine [116].

Myocardial Blood Supply. The coronary vessels, which supply the heart, are part of the systemic circulation, but they have special features closely related to the way the heart operates. In the human heart there are generally two coronary arteries, both arising from the base of the aorta. As a rule, the right coronary artery supplies the right ventricle and parts of the septum and of the posterior wall of the left ventricle; the rest of the heart, which is much greater, is supplied by the left coronary artery. Venous drainage is mainly through the coronary sinus; only a few percent returns by way of the anterior cardiac veins and the Thebesian veins [48].

In experiments on animals, blood flow through the heart can be determined directly by electromagnetic flowmeters. With humans recourse to indirect methods is necessary; some of these involve determination of the uptake or dilution in the heart of nonphysiological gases whose tissue solubility is known (NO_2, argon, xenon). Such measurements have shown that blood flow through the heart of a resting human amounts to approximately 60–80 ml/min per 100 g. This is about 5% of the total cardiac output. During muscular work circulation through the heart can rise to 4 times the resting level [80] (see Table 90.3). The increase in the heart's O_2 consumption during hard work is in the same order.

Variation in Blood Flow During the Cardiac Cycle. The coronary circulation, in contrast to that through other organs, exhibits marked fluctuations of flow rate in the rhythm of systole and diastole. The rhythmic pulsations in aortic pressure are partially responsible for these phasic fluctuations; the other main contributing factor is changing interstitial myocardial pressure. During systole the latter acts to compress the blood vessels mainly in the middle and inner parts of the heart wall. This influence is termed the *myocardial component of the coronary resistance*. As

Fig. 90.16 shows, it results in complete interruptin of flow into the left coronary artery at the beginning of systole. Not until diastole, when the intramural pressure falls, is there a high rate of influx. In the branching region of the right coronary artery the intramural pressure is lower, so that influx basically follows the fluctuations in aortic pressure. During systole, because of the compression of the muscular wall of the heart, there is a surge of blood out of the coronary sinus; during diastole this outflow subsides (Fig. 90.16).

Regulation of Coronary Blood Flow. Even during normal resting activity the heart withdraws far more oxygen from the blood than do the other organs. Of the 200 ml O_2 included in 1 l of arterial blood, the heart extracts around 140 ml (see Table 19.3). Therefore, when the load on the heart increases and more oxygen is required, it is essentially impossible to increase the rate of extraction. Hence, increased O_2 requirement must be met primarily by increased blood flow, brought about by dilation of the coronary vessels and hence reduction of the resistance to flow. It is generally agreed that one of the strongest stimuli to dilatation of the coronary vessels is O_2 deficiency [53]. Even a 5% decrease in the O_2 content of the blood (ca. 10 ml O_2/l) leads to coronary vasodilatation. Recent findings suggest that hypoxic dilation of coronary arteries is mediated by endothelium-derived relaxant factor (EDRF see below) [94].

Another factor that may produce coronary dilatation is the presence of *adenosine*, which plays an important role as a product of the breakdown of energy-rich phosphates [14,42,105]. Experimental evidence suggests that adenosine formation is not triggered by changes in myo-

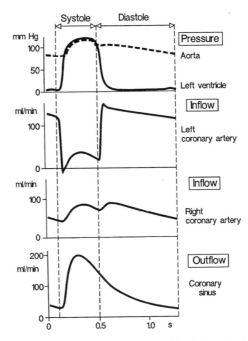

Fig. 90.16. Variations in coronary blood flow in relation to systole, diastole and left ventricular and aortic pressure

cardial O_2 consumption, but rather by the relationship of myocardial O_2 demand to O_2 supply [11].

A common final pathway for hypoxia- as well as adenosine-induced coronary dilatation seems to involve opening of ATP-sensitive K^+ channels and smooth muscle hyperpolarization ([27]; cf. Chap. 96). Moreover, an increased extracellular K^+ concentration is a considerable vasodilating influence [34,86]. The additional direct effect of the autonomic cardiac nerves on the coronary vessels is difficult to evaluate, because of their other simultaneous influences on cardiac activity. Recent studies, however, indicate that the sympathetic nerves act directly to constrict the vessels via α-adrenergic receptors – an influence that is overcome by metabolic regulation. Concerning vagal influences, there is at present no evidence for a physiological role in coronary regulation in man (for review see [13]).

Factors of Endothelial Origin. In recent years the vascular endothelium has gained much interest as an important factor in the regulation of the vascular tone. In 1980 Furchgott and Zawadski [39] demonstrated that the vasodilatory effect of acetylcholine on rabbit thoracic aorta in vitro was endothelium-dependent and they postulated the presence of an endothelium-derived relaxant factor (EDRF). In 1983 De Mey and Vanhoutte [29] showed that contraction of the canine femoral artery during anoxia in vitro was markedly reduced after removal of the endothelium. Consequently, the presence of an endothelium-derived contracting factor (EDCF) was suggested. The two observations led to intense efforts to identify both factors and explore their relevance in normal physiology and disease.

Meanwhile, EDRF has been identified as nitric oxide (NO) released from the endothelium under the influence of an increased rate of blood flow, which causes shear forces to the surface of the endothelial cells [62,92]. Moreover, substances like acetylcholine, bradykinin, histamine, serotonin, noradrenaline and others have been shown to stimulate the release of NO, thereby modifying their own direct effects [95]. NO production from the substrate L-arginine by NO synthase (NOS) is mediated by the activation of receptors at the surface of endothelial cells that induces an increase of the cytosolic Ca^{2+}, which in turn activates the enzyme [35]. NO released from the endothelial cells has a half-life of only a few seconds. It diffuses to the vascular smooth muscle cells and causes their relaxation by stimulating guanylate cyclase and thus increasing the cellular content of cGMP (for review see [30,61,62]; Chaps. 6, 95, 96).

EDCF has also been identified, as a 21 amino acid peptide named endothelin (ET-1). It is the most potent endogenous vasoconstrictor known, with a potency of 10:1 and 1000:1 compared with angiotensin II and noradrenaline, respectively [81]. ET-1 is synthesized in the endothelium as a prepropeptide containing 203 amino acids, which is then cleaved by endothelin-converting enzyme (ECE) to form the mature endothelin. Activation of ET-1 receptors in smooth muscle leads to stimulation of phospholipase C and formation of inositol trisphosphate, which mobilizes intracellular Ca^{2+}, and of diacyglycerol, thereby phosphorylating contractile proteins and resulting in contraction of smooth muscle [33]. Moreover, activation of protein kinase C by diacylglycerol is also cabable of influencing long-term regulatory processes related to gene activation and expression (for reviews see [23,120]). Secretion of ET-1 is induced by numerous stimuli, such as thrombin, adrenaline, interleukin-1, and endotoxin. Also vasopressin and angiotensin II stimulate release of ET-1 from the endothelium. Moreover, ET-1 stimulates its own release, thereby forming a positive autocrine loop. Production of ET-1 is reduced by atrial natriuretic peptide and by such vasodilating drugs as nitroglycerin. Furthermore, ET-1 stimulates the release of prostacyclin (PGI_2), thromboxane A_2 (TXA_2) and NO, and conversely PGI_2 and NO inhibit ET-1 release. The release of ET-1 is suggested to be mainly abluminal, and circulating levels in man are low, indicating that ET-1 acts mainly in a para-/autocrine way (for reviews see [23,120]).

The combination of all these influences provides a regulatory system with a large safety factor. This implies that regulation is hardly affected when only one individual component fails. Moreover, recent findings suggest that also the endocardial epithelium exerts some influence on myocardial contraction with tonic prolongation of contraction by a still unknown factor "endocardin" that can be overridden by stimulation of NO release [23,49].

Adequacy of Coronary Supply, Coronary Reserve. The heart is adequately supplied with blood when the amount of O_2 available corresponds to the amount consumed. The ratio of the two is taken as a criterion for the adequacy of coronary flow. A ratio of less than 1.2 indicates a critical restriction of the O_2 supply to the heart (coronary insufficiency). The degree to which the supply can be adjusted as conditions change is also of interest; this is the coronary reserve, expressed either as the difference between the maximal available O_2 and the actual O_2 consumption, divided by the actual consumption, or as the ratio of coronary resistance under rest conditions and of coronary resistance after maximal coronary vasodilation. The reserve of a fully adaptable coronary system is 4 to 5 times the amount required by the heart under resting conditions (for reviews see [54,113]).

Anoxia and Resuscitation. Because cardiac metabolism relies so heavily on oxidative reactions to provide energy, it is understandable that a sudden interruption of circulation (ischemia) results in extensive loss of function within a few minutes. In an experiment in which the heart is deprived of O_2 while coronary perfusion is maintained (anoxia), the changes produced are practically identical; as the contractions grow progressively weaker a marked dilatation develops, and after about 6–10 min the heart stops beating. The severe impairment of the energy-providing system under these conditions is reflected in the dramatic reduction in the amount of energy-rich phosphates (phosphocreatine, ATP). The heart is capable of anaerobic glycolysis to a small degree, producing lactic acid. But lactic acid breakdown in the myocardium comes to a halt during O_2 deficiency, so that the concentration of this substance in the coronary veins builds up to exceed that in the arteries. If anoxia lasts longer than 30 min, the myocardium undergoes irreversible structural changes in addition to the functional impairment, so that resuscitation is impossible. At normal body temperature, then, a 30-min duration of cardiac anoxia is a critical limit called the resuscitation limit. The resuscitation limit of the heart can be extended considerably if the metabolic rate is lowered by cooling. Advantage is taken of this possibility in modern heart surgery. When anoxia affects the entire organism, as in cases of suffocation, the possibility of successful resuscitation is limited by the brain, which is more sensitive than the heart and suffers irreversible damage after anoxia lasting only 8–10 min (cf. Chap. 124).

References

1. Abbot BC, Mommaerts WFHM (1959) A study of inotropic mechanisms in the papillary muscle preparation. J Gen Physiol 42:533–541
2. Allen DG, Kentish JC (1985) The cellular basis of the length-tension relation in cardiac muscle. J Mol Cell Cardiol 17:821–840
3. Allen DG, Nichols CG, Smith GL (1988) The effect of changes in muscle length during diastole on the calcium transient in ferret ventricular muscle. J Physiol (Lond) 406:359–370
4. Antoni, H (1989) Function of the heart. In: Schmidt RF, Thews G (eds) Human physiology. Springer, Berlin Heidelberg New York, pp 439–479
5. Antoni H, Jacob R, Kaufmann R (1969) Mechanische Reaktionen des Frosch- und Säugetiermyokards bei Veränderung der Aktionspotentialdauer durch konstante Gleichstromimpulse. Pflugers Arch 306:33–57
6. Armour JA, Pace JB, Randall WC (1970) Interrelationship of architecture and function of the right ventricle. Am J Physiol 218:174–179
7. Arnold G, Kosche F, Miessner E, Neitzert A, Lochner W (1968) The importance of the perfusion pressure in the coronary arteries for the contractility and the oxygen consumption of the heart. Pflugers Arch 299:339–356
8. Ashley CC, Ridgway EB (1970) Simultaneous recordings of membrane potential, calcium transient, and tension in single muscle fibres. Nature 219:1168–1169
9. Ballard FB, Danford WH, Neagle S, Bing RJ (1960) Myocardial metabolism of fatty acids. J Clin Invest 39:717–723
10. Baller D, Bretschneider HJ, Hellige G (1981) A critical look at currently used indirect indices of myocardial oxygen consumption. Basic Res Cardiol 76:163–181
11. Bardenheuer H, Schrader J (1983) Relationship between myocardial oxygen consumption, coronary flow, and adenosine release in an improved isolated working heart preparation of guinea pigs. Circ Res 51:263–271
12. Bassenge E (1989) Mechanik des intakten Herzens. In: Roskamm H, Reindell H (eds) Herzkrankheiten. Springer, Berlin Heidelberg New York, pp 66–87
13. Bassenge E, Heusch G (1990) Endothelial and neuro-humoral control of coronary blood flow in health and disease. Rev Physiol Biochem Pharmacol 116:77–165
14. Berne RM (1963) Cardiac nucleotides in hypoxia: possible role in regulation of coronary blood flow. Am J Physiol 204:317–322

15. Bianchi CP, Frank GB (eds) (1982) Excitation-contraction coupling in skeletal, cardiac, and smooth muscle. Can J Physiol Pharmacol 60:415–588
16. Bing RJ (1955) The metabolism of the heart. Harvey Lect 50:27–70
17. Blinks R (1986) Intracellular Ca^{2+} measurements. In: Fozzard HA, Haber E, Jennings RB, Katz AM, Morgan HE (eds) The heart and cardiovascular system. Raven, New York, pp 671–701
18. Blinks JR, Wier WG, Hess P, Prendergast FG (1982) Measurement of Ca concentrations in living cells. Prog Biophys Mol Biol 40:1–114
19. Brady A (1966) Onset of contractility in cardiac muscle. J Physiol (Lond) 184:560–580
20. Brady A (1979) Mechanical properties of cardiac fibres. In: Berne RM, Sperelakis N, Geiger SR (eds) Handbook of physiology, section 2: the cardiovascular system. American Physiological Society, Bethesda, Maryland, pp 461–474
21. Brady AJ (1991) Length dependence of passive stiffness in single cardiac myocytes. Am J Physiol 260:1062–1071
22. Brenner B (1988) Effect of Ca^{2+} on cross-bridge turnover kinetics in skinned single rabbit psoas fibres. Proc Natl Acad Sci U S A 85:3265–3269
23. Brutsaert DL (1993) Endocardial and coronary endothelial control of cardiac performance. News Physiol Sci 8:82–86
24. Brutsaert DL, Housmans PR, Goethals MA (1980) Dual control of relaxation. Its role in the ventricular function in the mammalian heart. Circ Res 47:637–652
25. Brutsaert DL, Sys SU (1988) Relaxation and diastole of the heart. Physiol Rev 69:1228–1315
26. Bünger R, Sommer O, Walter G, Stiegler H, Gerlach E (1979) Functional and metabolic features of an isolated perfused guinea pig heart, performing pressure-volume work. Pflugers Arch 380:259–266
27. Daut J, Maier-Rudolph W, van Beckerath N, Mehrke G, Günther K, Goedel-Meinen L (1990) Hypoxic dilation of coronary arteries is mediated by ATP-sensitive potassium channels. Science 247:1341–1344
28. De Bold AJ, Borenstein HB, Veress AT, Sonnenberg H (1980) A rapid and potent natriuretic response to intravenous injection of atrial myocardial extract in rats. Life Sci 28:89–94
29. De Mey JG, Vanhoutte PM (1983) Anoxia and endothelium-dependent reactivity of the canine femoral artery. J Physiol (Lond) 335:65–74
30. Dinerman JL, Lowenstein CJ, Snyder SH (1993) Molecular mechanisms of nitric oxide regulation. Circ Res 73:217–222
31. Drake A, Haines JR, Noble MIM (1980) Preferential uptake of lactate by the normal myocardium in dogs. Cardiovasc Res 14:65–72
32. Fenn WO, Marsh BS (1935) Muscular force at different speed of shortening. J Physiol (Lond) 85:277–297
33. Fleckenstein A (1983) Calcium antagonism in heart and smooth muscle. Wiley, New York
34. Fleckenstein A, Nakayama K, Fleckenstein-Grün G, Byon YK (1975) Interaction of vasoactive ions and drugs with Ca-dependent excitation-contraction coupling of vascular smooth muscle. In: Carfoli E, Clementi F, Drabikowski W, Magreth A (eds) Calcium transport in contraction and secretion. North Holland, Amsterdam, pp 555–566
35. Förstermann U, Nakane M, Tracey WR, Pollock JS (1994) Isoforms of nitric oxide synthase: functions in the cardiovascular system. Eur Heart J 14:10–15
36. Ford LE (1991) Mechanical manifestation of activation in cardiac muscle. Circ Res 68:621–637
37. Fozzard HA, Haber E, Jennings RB, Katz AM, Morgan HE (eds) (1992) The heart and cardiovascular system, 2nd edn. Raven, New York
38. Frank O (1895) Zur Dynamik des Herzmuskels. Z Biol 32:370–447
39. Furchgott RF, Zawadski JV (1980) The obligatory role of endothelial cells in the relaxation of arterial smooth muscle by acetylcholine. Nature 288:373–376

40. Gaasch WH, Apstein CS, Levine HJ (1985) Diastolic properties of the left ventricle. In: Levine HJ, Gaasch WH (eds) The ventricle: basis and clinical aspects. Nijhoff, Boston, pp 143–170

41. Gauer OH (1972) Kreislauf des Blutes. In: Gauer OH, Kramer K, Jung R (eds) Physiologie des Menschen, vol 3. Wban and Schwarzenberg, Munich, pp 81–88

42. Gerlach E, Deuticke B, Dreisbach RH (1963) Der Nucleotidabbau im Herzmuskel bei Sauerstoffmangel und seine mögliche Bedeutung für die Coronardurchblutung. Naturwissenschaften 6:228–229

43. Gollwitzer-Meier K, Kramer K, Krüger E (1936) Der Gaswechsel des suffizienten und insuffizienten Warmblüterherzens. Pflugers Arch 237:68–92

44. Gollwitzer-Meier K, Kroetz K (1939) Klin Wochenschr 18:869–882

45. Gülch RW (1980) Alterations in excitation of mammalian myocardium as a function of chronic loading and their implications in the mechanical events. Basic Res Cardiol 75:73–80

46. Hammond HK, White FC, Bhargava V, Shabetai R (1992) Heart size and maximal cardiac output are limited by pericardium. Am J Physiol 263:1675–1681

47. Hawthorne EW (1966) Dynamic geometry of the left ventricle. Am J Cardiol 18:566–573

48. Hellige G (1981) Koronardurchblutung. In: Krayenbühl HP, Kübler W (eds) Kardiologie in Klinik und Praxis. Thieme, Stuttgart, pp 8.1–8.12

49. Henderson HA, Lewis MJ, Shah AM, Smith JA (1992) Endothelium, endocardium, and cardiac contraction. Cardiovasc Res 26:305–308

50. Henry JP, Gauer O, Reeves IL (1956) Evidence of the atrial location of receptors influencing urine flow. Circ Res 4:85–90

51. Hill AV (1938) The heat od shortening and the dynamic constants of muscle. Proc R Soc Lond [Biol] 126:136–195

52. Hill AV (1949) The abrupt transition from rest to activity in muscle. Proc R Soc Lond [Biol] 136:399–420

53. Hilton R, Eichholtz F (1925) The influence of chemical factors on the coronary circulation. J Physiol (Lond) 59:413–425

54. Hoffman JIE (1987) A critical view of coronary reserve. Circulation 75 [Suppl I]:1–6

55. Holtz J, Münzel T, Bassenge E (1987) Das natriuretische Vorhofhormon im Menschen. . Kardiol 76:655–670

56. Hort W (1977) Myocardial hypertrophy. Light microscopic findings on the myocardium. Blood supply, ventricular dilatation and heart failure. Basic Res Cardiol 72:203–208

57. Housmans PR, Lee NK, Blinks JR (1983) Active shortening retards the decline of the intracellular calcium transient in mammalian heart muscle. Science 221:159–161

58. Huang CL, Lewicki J, Johnson LK, Cogan MG (1985) Renal mechanism of action of rat atrial natriuretic factor. J Clin Invest 75:769–773

59. Huxley HE, Hanson J (1954) Changes in the cross-striation of muscle during contraction and stretch and their structural interpretation. Nature (Lond) 173:973–976

60. Huxley HE, Hanson J (1960) The molecular basis of contraction in cross-striated muscle. In: Bourne GH (ed) The structure and function of muscle, vol I. Academic, New York, pp 183–227

61. Ignarro LJ (1988) Biological actions and properties of endothelium-derived nitric oxide formed and released from artery and vein. Circ Res 65:1–20

62. Ignarro LJ, Byrns RE, Buga GM, Wood KS (1987) Endothelium-derived relaxing factor from pulmonary artery and vein possesses pharmacologic and chemical properties identical to those of nitric oxide radical. Circ Res 61:866–879

63. Jacob R, Brändle M, Dierberger B, Ross C (1990) Conception and methodological basis for the evaluation of ventricular and myocardial mechanics. In: Jacob R (ed) Evaluation of cardiac contractility. Fischer, Stuttgart

64. Jacob R, Dierberger B, Gülch RW, Rupp H, Voelker W (1990) Factors contributing to the transition from cardiac hypertrophy to hear failure. Cardiol Angiol Bull 27:1–9

65. Jacob R, Gülch RW (1985) Kontraktion. In: Heublein B (ed) Handbuch der Inneren Erkrankungen, vol 1. Fischer, Stuttgart

66. Jacob R, Weigand KH (1966) The end-systolic pressure-volume relations as a basis for evaluation of left ventricular contractility in situ. Pflugers Arch 289:37–49

67. Jewell BR (1977) A re-examination of the influence of muscle length on myocardial performance. Circ Res 40:321–330

68. Kammermeier H (1987) High energy phosphate of the myocardium: concentration versus free energy change. Basic Res Cardiol 82 [Suppl 2]:31–36

69. Kass DA, Maughan WL (1988) From E_{max} to pressure-volume relations: a broader view. Circulation 77:1203–1212

70. Kass DA, Yamazaki T, Burkhoff D, Maughan WL, Sagawa K (1986) Determination of left ventricular end-systolic pressure-volume relationships by the conductance (volume) catheter technique. Circulation 73:586–595

71. Katz AM (1977) Series elasticity, active state, length-tension relationship, and cardiac mechanics. In: Katz AM (ed) Physiology of the heart. Raven, New York, pp 119–136

72. Kent RL, Hoober K, Cooper G (1989) Load responsiveness of protein synthesis in adult mammalian myocardium: role of cardiac deformation linked to sodium influx. Circ Res 4:74–85

73. Keul J, Doll E, Steim H, Homburger H, Kern H, Reindell H (1965) Über den Stoffwechsel des menschlichen Herzens. I. Die Substratversorgung des gesunden menschlichen Herzens in Ruhe, während und nach körperlicher Arbeit. Pflugers Arch 282:1–27

74. Keung E (1989) Calcium current is increased in isolated adult myocytes from hypertrophied rat myocardium. Circ Res 64:753–763

75. Kisch B (1956) Electron microscopy of the atrium of the heart. Exp Med Surg 14:99

76. Kuhn HJ, Bletz C, Rüegg JC (1990) Stretch-induced increase in the Ca sensitivity of myofibrillar ATPase activity in skinned fibres from pig ventricles. Pflugers Arch 415:741–746

77. Lakatta EG (1992) Length modulation of muscle performance. Frank-Starling Law of the heart. In: Fozzard HA, Haber E, Jennings RB, Katz AM, Morgan HE (eds) The heart and cardiovascular system. Raven, New York, pp 1325–1351

78. Laplace PS (1806) Théorie de l'action capillaire. Cited in [117]

79. Loiselle DS (1987) Cardiac basal and activation metabolism. Basic Res Cardiol 82 [Suppl 2]:37–50

80. Marcus ML, Wright C, Doty D, Eastham C, Laughlin D, Krumm P, Fastenow C, Brody M (1981) Measurement od coronary velocity and relative hyperemia in the coronary circulation of humans. Circ Res 49:877–891

81. Masaki T (1989) The discovery, the present state, and the future prospects of endothelin. J Cardiovasc Pharmacol 13 [Suppl 5]:1–4

82. McClellan G, Weisberg A, Rode D, Winegrad S (1994) Endothelial cell storage and release of endothelin as a cardioregulatory mechanism. Circ Res 75:85–96

83. McNutt NS, Fawcett DW (1974) Myocardial ultrastructure. In: Langer GA, Brady AJ (eds) The mammalian myocardium. Wiley, New York, pp 1–49

84. Mirsky I (1979) Elastic properties of the myocardium: a quantitative approach with physiological and clinical applications. In: Berne RM, Sperelakis N, Geiger, SR (eds) Handbook of physiology, section 2: the cardiovascular system, vol I: the heart. American Physiological Society, Bethesda, pp 497–532

85. Motz W, Klepzig M, Strauer BE (1987) Regression of cardiac hypertrophy. Experimental and clinical results. J Cardiovasc Pharmacol 10 [Suppl 6]:148–152

86. Murray PA, Belloni FL, Sparks HV (1979) The role of potassium in the metabolic control of coronary vascular resistance of the dog. Circ Res 44:767–780

87. Neely JR, Liebermeister H, Batterby EJ, Morgan HE (1967) Effect of pressure development on oxygen consumption of isolated rat heart. Am J Physiol 212:804–814

88. Oikawa S, Imai M, Veno A (1984) Cloning and sequence analysis of CDNA encoding a precursor for human atrial natriuretic polypeptide. Nature 309:724–726

89. Oparil S, Wyss MJ (1993) Atrial natriuretic factor in central cardiovascular control. News Physiol Sci 8:223–228

90. Opie LH (1991) The heart. Physiology and metabolism. Raven, New York

91. Opie LH (1991) Ventricular hypertrophy and its molecular biology. In: Opie LH (ed) The heart. Physiology and metabolism. Raven, New York, pp 369–395

92. Palmer RMJ, Ferrige AG, Moncada S (1987) Nitric oxide release accounts for the biological activity of endothelium-derived relaxing factor. Nature 327:524–526

93. Parmley WW, Chuck L, Sonnenblick EH (1972) Relation of v_{max} to different models of cardiac work. Circ Res 30.34–43

94. Pohl U, Busse R (1989) Hypoxia stimulates release of endothelium-derived relaxant factor. Am J Physiol 256:H1595–H1600

95. Pohl U, Holtz J, Busse R, Bassenge E (1986) Crucial role of endothelium in the vasodilator response to increased flow in vivo. Hypertension 8:37–44

96. Pollack GH (1970) Maximum velocity as an index of contractility in cardiac muscle – a critical evaluation. Circ Res 26:111–127

97. Reindell H (1940) Größe, Form und Bewegungsbild des Sportherzens. Arch Kreisl Forsch 7:117

98. Roskamm H, Reindell H, Müller M (1966) Herzgröße und ergometrisch getestete Ausdauerleistungsfähigkeit bei Hochleistungssportlern aus 9 deutschen Nationalmannschaften. Z Kreisl Forsch 55:2–14

99. Rüegg JC (1986) Calcium in muscle activation. Springer, Berlin Heidelberg New York

100. Rupp H (1989) Differential effect of physical exercise routines on ventricular myosin and peripheral catecholamine stores in normotensive and spontaneously hypertensive rats. Circ Res 65:370–377

101. Ruskoaho H, Vuolteenaho O (1993) Regulation of atrial natriuretic peptide secretion. News Physiol Sci 8:261–266

102. Sagawa K (1978) The ventricular pressure-volume diagram revisited. Circ Res 43:677–687

103. Sagawa K (1981) The end-systolic pressure-volume relations of the ventricle: definition, modification and clinical use. Circulation 63:1223–1227

104. Sagawa K, Suga H, Shoukas AA, Bakalar KM (1979) End-systolic pressure-volume ratio: a new index of contractility. Am J Cardiol 40:748–753

105. Schrader J, Baumann G, Gerlach E (1977) Antiadrenergic action of adenosine in the heart: possible physiological significance. Pflugers Arch 372:29–35

106. Simmons RM, Jewell BR (1974) Mechanics and models of muscular contraction. Rec Adv Physiol 9:87–147

107. Smith VE, Zile R (1992) Relaxation and diastolic properties of the heart. In: Fozzard HA, Haber E, Jennings RB, Katz AM, Morgan HE (eds) The heart and cardiovascular system. Raven, New York, pp 1353–1367

108. Sommer JR, Jennings RB (1992) Ultrastructure of cardiac muscle. In: Fozzard HA, Haber E, Jennings RB, Katz AM, Morgan HE (eds) The heart and cardiovascular system. Raven, New York, pp 3–50

109. Sonnenberg H (1989) Intrarenal mechanisms of action of atrial natriuretic factor. In: Kaufmann W, Wambach G (eds) Endocrinology of the heart. Springer, Berlin Heidelberg New York

110. Sonnenblick EH (1962) Force-velocity relations in mammalian heart muscle. Am J Physiol 202:931–939

111. Starling EH (1918) The Linacre lecture on the law of the heart. Longmans and Green, London

112. Stephenson DG, Wendt IR (1984) Length dependence of changes in sarcoplasmic calcium concentration and myofibrillar calcium sensitivity in striated muscle fibres. J Muscle Res Cell Motil 5:243–272

113. Strauer BE (1992) The concept of coronary flow reserve. J Cardiovasc Pharmacol 19 [Suppl 5]:67–80

114. Streeter DD (1979) Gross morphology and fibre geometry of the heart. In: Berne RM, Sperelakis N, Geiger SR (eds) Handbook of physiology, section 2: the cardiovascular system, vol I: the heart. American Physiological Society, Bethesda, pp 61–112

115. Taegtmeyer H (1986) Myocardial metabolism. In: Phelps M, Mazziotta J, Schelbert H (eds) Positron emission tomography and autoradiography. Raven, New York, pp 149–195

116. Taegtmeyer H, Hems R, Krebs HA (1980) Utilization of energy-providing substrates in the isolated working rat heart. Biochem J 186:701–711

117. Tenney SM (1993) A tangled web: Young, Laplace, and the surface tension law. News Physiol Sci 8:179–183

118. Tritthart H, Fleckenstein A, Kaufmann R (1968) Die spezifische Beschleunigung des Erschlaffungsprozesses durch sympathische Überträgerstoffe und die Hemmung dieses Effektes durch β-Rezeptoren-Blockade. Pflugers Arch 303:350–365

119. Von Harsdorf R, Lang RE, Fullerton M, Woodcock EA (1989) Myocardial stretch stimulates phosphatidylinositol turnover. Circ Res 65:494–501

120. Weitzberg E (1993) Circulatory responses to endothelin-1 and nitric oxide with special reference to endotoxin shock and nitric oxide inhalation. Acta Physiol Scand [Suppl.] 611:4–14

121. Wier WG (1990) Cytoplasmic Ca^{2+} in mammalian ventricle – dynamic control by cellular processes. Annu Rev Physiol 52:467–485

122. Wiggers CJ (1927) Studies on the cardiodynamic action of drug II: the mechanism of cardiac stimulation by epinephrine. J Pharmacol Exp Ther 30:233

123. Xenophontos XP, Watson PA, Chua BHL (1989) Increased cyclic AMP content accelerates protein synthesis in rat heart. Circ Res 65:647–656

91 Electrophysiology of the Heart at the Single Cell Level and Cardiac Rhythmogenesis

H. Antoni

Contents

91.1 Introduction

Every now and then someone dies suddenly by coming into contact with an electric current source. In such cases, death commonly occurs without any visible external injury and is mostly caused by cardiac failure due to fibrillation of the ventricles of the heart. Ventricular fibrillation is a profound primary disturbance of the excitatory processes of the heart, which only secondarily leads to its mechanical function failing. The heart responds in this way to an external electric influence because cardiac musculature – like nerve and skeletal muscle – is an excitable tissue. This means that bioelectric events play a key role in the function of the heart; this is the topic of the present chapter.

The rhythmic initiation of heart beats and the sequence and latency of atrial and ventricular activation are determined by the time course of cardiac excitation. Excitation is the trigger of contraction. Thus, the function of the heart cannot be properly understood without gaining profound insight into its excitatory processes. In dealing with this topic, we refer to the basic electrophysiological principles that have been outlined in preceding chapters of this book (see Chaps. 11, 12).

91.2 Origin and Spread of Excitation

The rhythmic pulsation of the heart is elicited by excitatory signals generated within the heart itself. Under suitable conditions, therefore, a heart removed from the body will continue to beat at a constant frequency. This spontaneous rhythmic triggering of cardiac excitation – autorhythmicity – is normally brought about by specialized myocardial cells that form the pacemaker and conducting system. However, the main mass of the heart is made up of the working myocardium of atria and ventricles which performs the mechanical work of pumping.

91.2.1 Cell-to-Cell Coupling in Cardiac Muscle

Myocardial Fibers. Cells of the working myocardium as well as of the specialized system are arranged end to end and are enclosed in a common sarcolemmal envelope, the basement membrane. Such a chain of myocardial cells is called a myocardial fiber. Like nerve or skeletal muscle, myocardial fibers are excitable structures, i.e., they have a resting potential, respond to suprathreshold stimuli by generating action potentials, and are capable of propagating action potentials without decrement. The cell boundaries, which can be seen under the microscope as

R. Greger/U. Windhorst (Eds.)
Comprehensive Human Physiology, Vol. 2
© Springer-Verlag Berlin Heidelberg 1996

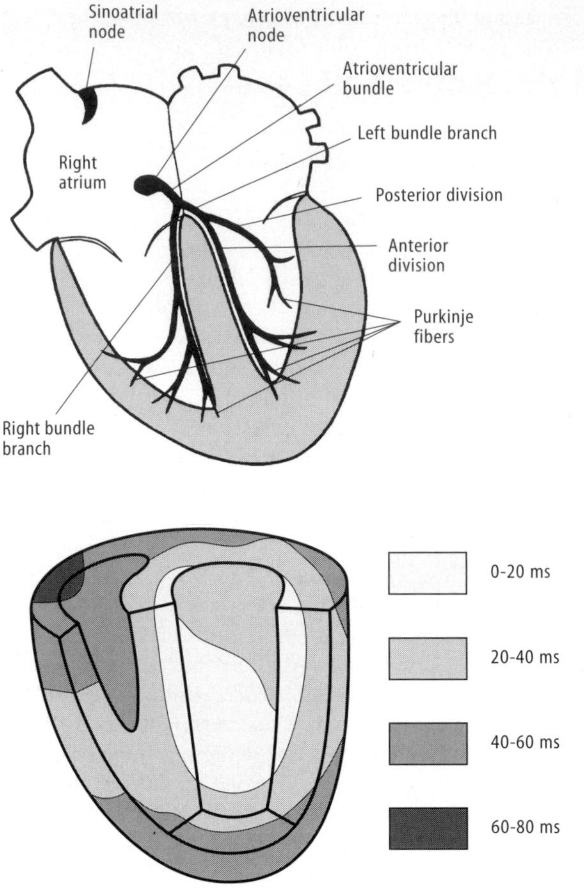
Sinoatrial node
Atrioventricular node
Atrioventricular bundle
Left bundle branch
Right atrium
Posterior division
Anterior division
Purkinje fibers
Right bundle branch
0-20 ms
20-40 ms
40-60 ms
60-80 ms

Fig. 91.1. *Top*, arrangement of the pacemaker and conducting system; frontal section. *Bottom*, sequence of the spread of excitation within the ventricles of the human heart. Data taken from [38]

intercalated discs, contain highly conductive channels (single-channel conductance up to 166 pS, [65]) called gap junctions between cells (see Chap. 7) and offer no obstacle to the conduction of excitation [64]. Thus, the musculature of atria and ventricles functionally behaves as a syncytium [18,68]. This means that excitation arising anywhere in the atria or ventricles spreads out over the unexcited fibers, until even the very last cell is brought into play. This property provides the explanation for the "all-or-nothing" response of the heart, i.e., when stimulated, the heart either responds with excitation of all its fibers or gives no response at all if the stimulus does not reach the suprathreshold level in any cell.

91.2.2 Geometry of Propagation

Sinoatrial and Atrioventricular Nodes. The various elements of the specialized pacemaker and conducting system are shown in Fig. 91.1. Conduction velocity is different in the different regions of the heart (see Table 91.1). Normally, activation of the heart begins in the sinoatrial (SA) node, which lies in the sulcus between the superior vena cava and the right atrium. When the body is at rest, the SA node drives the heart at a rate of about 70 impulses/min. From the SA node the excitation first spreads over the working myocardium of the atria. There has been some controversy concerning the question of whether specialized pathways connect the SA and atrioventricular (AV) node and are responsible for preferential conduction. However, morphologically discrete internodal tracts of specialized cells have not yet been convincingly demonstrated [9,35]. In fact, the preferential conduction of the atrial impulse along certain routes can be adequately explained by the orientation of normal atrial cells, conduction in the direction of the long fiber axis being faster than in a direction perpendicular to it [59]. Due to the macroscopic architecture of the atria only a limited number of routes exist for the internodal conduction. These are the following:

- The interauricular band (Bachmann's bundle), which connects the SA node area to the left atrium
- The crista terminalis, which runs down the right atrium and through the interatrial septum towards the AV node.

Thus, different wave fronts simultaneously progress in multiple directions. From the beginning of atrial excitation of the human heart, it takes about 30–40 ms until the AV node is reached. Excitation of both atria is completed in about 90–100 ms [24]. Functionally, the AV nodal tract can be subdivided into three zones:

- The atrionodal (AN) zone
- The true nodal (N) zone
- The nodal His (NH) zone.

corresponding to the upper, middle, and lower nodal regions [48]. Most of the atrioventricular conduction delay occurs in the upper nodal region (zone AN) and during transition of the wave of excitation to the middle nodal zone N. In total, the passage along (the AV nodal tract down to the atrioventricular bundle (bundle of His) requires about 80 ms.

Purkinje Network. Through the remainder of the specialized conduction system – the atrioventricular bundle, the left and right bundle branches, and their terminal network, the Purkinje fibers – propagation velocity is high (Table

Table 91.1. Conduction velocities (m/s) in different regions of the human heart

Region	Conduction velocity
Sinoatrial node	0.05
Atrial myocardium	0.80–1.00
Atrioventricular node	0.05–0.10
Atrioventricular bundel	0.8–2.0
Purkinje fibers	2.0–4.0
Ventricular myocardium	0.5–1.0

Data taken from [14,35,38]

91.1), and the different ventricular regions are therefore excited in rapid succession. From the subendocardial endings of the Purkinje fibers, excitation spreads over the ventricular musculature. The overall time sequence of ventricular activation and the synchronization of ventricular contraction are determined by the anatomic distribution of the conduction system. In the human heart, the Purkinje network is mainly present in the middle and lower thirds of the left ventricular endocardial surface [24]. Following activation of the subendocardial myocardium, excitation of the left ventricle of the human heart occurs by myocardial conduction, mainly in a direction towards the epicardium (Fig. 91.1, lower panel). The isochrones of activation are more or less concentrically arranged. The papillary muscles are activated by way of the "false tendons" and are excited from the base to the top. Activation of the interventricular septum occurs in a left-to-right direction, mainly by myocardial conduction. In experiments on the isolated human heart [24], initial right ventricular myocardial activation was found to occur near the base of the papillary muscles and the overlying free wall. The last part of the right ventricle to be activated is the outer wall of the outflow tract, and this is also the region of the latest phase of activation of the entire heart [17].

91.2.3 Hierarchy of Pacemaker Activity

Normotopic and Heterotopic Centers. The autorhythmicity of the heart is not entirely dependent on the operation of the SA node, since the other parts of the pacemaker/conduction system are also capable of exerting spontaneous activity. However, the spontaneous frequency of impulse formation decreases towards the peripheral branches. Thus, in a regularly beating normal heart, the relatively short intervals between successive activations prevent these latent pacemakers from becoming dominant centers of automaticity. The SA node is the leading primary pacemaker of the heart, simply because it has the highest discharge rate. If for any reason the SA node should fail to initiate the heartbeat, or if the excitation is not conducted to the atria (sinoatrial block), the AV node can function as a secondary pacemaker at a slower rate (about 40–60/min in the AV node). If there is a complete interruption of conduction from the atria to the ventricles (complete heart block), a tertiary center in the ventricular conducting system can take over as the pacemaker for ventricular contraction. In this case, atria and ventricles beat entirely independently of one another, the atria at the frequency of the SA node and the ventricles at the considerably lower frequency of a tertiary center. With respect to pacemaker activity, the SA node is termed the normotopic center (i.e., center in the normal place), and the remainder of the system the heterotopic centers (i.e., centers in an abnormal place). If there is a sudden onset of total heart block, several seconds can elapse before the ventricular automaticity "wakes up." In this preautomatic pause, an insufficient supply of blood to the brain may cause unconsciousness and convulsions, known as Adams–Stokes syn-

cope. If all the ventricular pacemakers fail, the ventricular arrest leads to irreversible brain damage and eventually to death.

91.2.4 Artificial Pacemakers

Even in the absence of autorhythmicity, the myocardium remains excitable for a time. It is therefore possible to keep the blood in circulation by artificial electrical stimulation of the ventricles. If necessary, the electrical impulses can be applied through the intact wall of the chest. When attacks of Adams–Stokes disease are frequent, and in cases of complete heart block with very low beating frequency, electrical stimulation is continued over a period of some years. The stimuli are generated by subcutaneously implanted, battery-driven miniature pacemakers and are conducted to the heart by wire electrodes.

91.2.5 Bundle Branch Block

The interruption of conduction along one of the bundle branches results in an incomplete heart block. In this case, the wave of excitation spreads out from the terminals of the intact conduction system and eventually covers the whole ventricular myocardium; the time required for complete activation is, of course, considerably longer than normal.

91.3 Characteristics of the Elementary Excitatory Process

91.3.1 General Form of the Action Potential

Upstroke and Global Repolarization. The action potential of the cardiac muscle cells shows a similar basic shape in the different regions of the heart (Fig. 91.2). Like that of neurons or skeletal muscle fibers and starting from the resting potential (about −90 mV), it begins with the upstroke, a rapid depolarization followed by reversal of the membrane potential beyond zero to about +30 mV (overshoot; see Fig. 91.2). This rising phase of the action potential (phase 0) lasts only a few milliseconds. During the subsequent period of global repolarization, there are three phases that can be more or less clearly distinguished in different regions of the heart:

- An initial, short phase of repolarization, during which the membrane potential approaches zero (phase 1)
- This initial peak is followed by the prolonged plateau, a characteristic feature of the myocardium (phase 2)
- This is terminated by repolarization (phase 3) to the resting potential (phase 4).

The total duration of the action potential is about 200–400 ms, i.e., more than 100 times as long as that of a skeletal

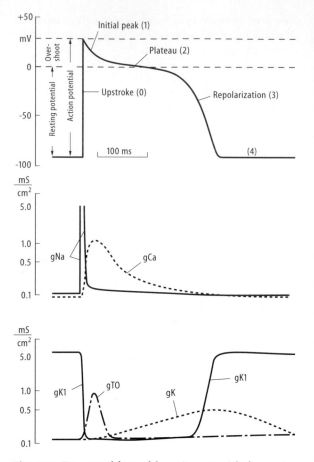

Fig. 91.2. *Top*, general form of the action potential of a ventricular myocardial cell. *Middle*, changes in Na$^+$ and Ca^{2+} conductance (*g*) underlying the action potential; conductivity of depolarizing currents. *Bottom*, changes in K$^+$ conductances during the action potential; conductivity of repolarizing currents. The magnitude of the corresponding ionic currents depends on the conductance as well as on the difference between the membrane potential E_m and the equilibrium potential of the corresponding ion, e.g., $I_{Na} = g_{Na}$ $(E_m - E_{Na})$

muscle or nerve fiber. The functional consequences, as we shall see, are considerable.

91.3.2 Ionic Mechanisms

The action potential is generated by a complicated interplay of membrane potential changes, as well as changes in ionic conductivity and ionic currents. The fundamental points of the ionic theory of excitation have been discussed in details elsewhere (see Chaps. 11–13). Here we will merely summarize these points with reference to the specific peculiarities of cardiac muscle [13,28,49,62] (see also Fig. 91.2 and Table 91.2).

The resting potential is primarily a K$^+$ potential mainly determined by specific K$^+$ channels (g_{K1}, Table 91.2). As in neuron or skeletal muscle, phase 0, i.e., the rapid phase of depolarization of the action potential, is brought about by the fast Na$^+$ inward current due to pronounced activation

of the fast Na$^+$ channels (g_{Na}). However, except for a slowly inactivating component present in Purkinje fibers [12], this initial Na$^+$ influx is rapidly inactivated. Hence, further mechanisms are required to produce the considerable delay in repolarization, the plateau phase (phase 2), of the cardiac action potential. These mechanisms are the following:

- Delayed activation and slow inactivation of L-type Ca^{2+} channels (g_{Ca} L) causing a depolarizing influx of Ca^{2+} ions, the slow inward current [7,44,52]
- A decrease in the resting (inwardly rectifying) K$^+$ conductance (g_{K1}) due to depolarization, which reduces the repolarizing K$^+$ outward current.

Thus, a fine balance exists between inward and outward currents during this phase. Likewise, phase 3, the terminal phase of repolarization, can only be explained by the interaction of several processes:

- A time-dependent decrease of g_{Ca}, which diminishes the slow inward current
- Delayed inactivation of another population of K$^+$ channels (g_K) which are activated by depolarization at the beginning of the action potential
- Restitution of the (inwardly rectifying) resting K$^+$ conductance (g_{K1}) due to progress in repolarization.

Yet another population of K$^+$ channels (g_{TO}) are responsible for phase 1, the initial phase of repolarization preceding the plateau, which is differently developed in different regions of the heart. These channels are only briefly activated following depolarization and are responsible for a transient outward current (I_{TO}). When the membrane is at its resting potential, the depolarizing and repolarizing currents are in balance. (More details concerning the ionic currents that participate in generating the action potential, which will be discussed later, are presented in Fig. 91.5) The magnitgude of an individual ionic current (I_x) depends on the total conductance of the corresponding channels (gx) as well as on the difference between the membrane potential E_m and the equilibrium potential of that ion (E_x):

$$I_x = gx(E_m - E_x)$$

91.3.3 Ca^{2+} Action Potentials and Ca^{2+} Antagonists

The mechanism underlying the fast Na$^+$ inward current and that of the slow Ca^{2+} inward current differ in several ways, including the time course, the potential dependence, and the susceptibility to blocking agents. The fast Na$^+$ channel is blocked by tetrodotoxin (TTX), while the L-type Ca^{2+} channel is blocked by inorganic cations (Ni^{2+}, Co^{3+}, Cd^{2+}, Mn^{2+}, La^{3+}) and by organic Ca^{2+} antagonists (e.g., verapamil, nifedipine, diltiazem) [27]. The threshold potential for the activation of the Na$^+$ channel is approximately −60 mV, whereas that of the Ca^{2+} channel is approximately −30 mV. Depolarization of the membrane to about −40 mV inactivates the fast Na⁺ channels. However, under these conditions, more intense stimuli can elicit so-called Ca^{2+} action potentials, which have a slower upstroke phase, because in this case both the upstroke and the plateau are generated by the slow inward current. These action potentials are propagated at low speed (slow response) [14].

91.3.4 Refractory Period

Cardiac musculature shares with other excitable tissues the property of reduced responsiveness to stimuli during particular phases of the excitatory process. The terms absolute and relative refractory period are used for phases of abolished and diminished responsiveness, respectively. Figure 91.3 shows how these are related to the action potential.

During the absolute refractory period the cell is not excitable, and during the subsequent relative refractory period excitability gradually recovers. Thus the stronger the stimulus, the sooner a new action potential can be elicited. Action potentials generated very early in the relative refractory period do not rise as sharply as normal action potentials and have a lower amplitude and a shorter duration.

Postrepolarization Refractoriness. The chief cause of refractory behavior is the inactivation of the fast Na⁺ channels during prolonged depolarization (see Chaps. 12, 13). It is not until the membrane has repolarized to approximately −40 mV that these channels begin to recover. The duration of the refractory period is therefore, as a rule, closely related to the duration of the action potential. When the action potential is shortened or lengthened, the refractory period changes accordingly. However, some

Table 91.2. Ionic channels and other transport proteins participating under physiological conditions in excitation and excitation contraction coupling of mammalian cardiac cells

Notation	Characteristics and special functions
Ionic channels mainly carrying inward currents	
g_{Na} Fast Na⁺ channel	Activated by depolarization
	Inactivation is time and voltage dependent
	Responsible for upstroke of the action potential in atrial and ventricular myocardium as well as in the His–Purkinje system
g_{Ca} (L-type) L-type Ca⁺ channel	Activation by depolarization
	Inactivation depends on voltage and $[Ca^{2+}]_i$
	Responsible for slow inward current (plateau phase of the action potential, upstroke in SA node) and for excitation contraction coupling
	Blocked by Ca^{2+} antagonists
g_{Ca} (T-Type) T-type Ca^{2+} channel	Activation by depolarization (threshold lower than with L-type Ca^{2+} channels)
	Inactivation only voltage dependent and faster than with L-type Ca^{2+} channels
	Partly responsible for diastolic depolarization and for upstroke in SA and AV nodes
	No blockade by Ca^{2+} antagonists
g_f Pacemaker channel	Nonspecific cationic channel
	Carries mainly Na⁺ inward current when activated by polarization to high membrane potentials
	Responsible for diastolic depolarization in the His – Purkinje system
	Partly responsible for diastolic depolarization in SA node
g_b Background Na⁺ channel	Voltage-independent channel in SA nodal cells carrying Na⁺ ions
	The Na⁺ current is offset by an outward K⁺ current at the beginning of the slow diastolic depolarization, but as the K⁺ current decays, it contributes to pacemaker behavior
Ionic channels mainly carrying outward currents	
g_{K1} Inward rectifier	K⁺ channel responsible for maintaining the resting potential near the K⁺ equilibrium potential in working myocardium and in the His – Purkinje system
	Shuts off during depolarization and opens during repolarization, thus favoring the plateau phase and late repolarization
	Conductance depends on $[K^+]_e$ (increases when $[K^+]_e$ increases, and vice versa)
	Its absence from SA node cells allows pacemaker activity
g_K Delayed rectifier	K⁺ channel slowly activated upon depolarization
	Mainly responsible for repolarization of the action potential
	Due to its slow inactivation partly responsible for pacemaker depolarization
g_{TO}	K⁺ channel generating transient outward (TO) current when activated by depolarization
	Mainly present in atrial and Purkinje fibers as well as in subepicardial ventricular fibers
	Responsible for the early phase of repolarization (phase 1) in these cells
$g_{K(Ach)}$	K⁺ channel activated by acetylcholine through M_2 receptors (also by adenosine via P_1 receptors)
	Mainly present in atria including SA and AV nodes and in Purkinje fibers
	Contributes to outward current both at rest and during the action potential
g_{Cl}	Cl⁻ channel stimulated by adrenergic β-receptor activation, then favoring repolarization

Table 91.2. *Continued*

Notation	Characteristics and special functions
Ionic channels in intercalated discs and in sarcoplasmic reticulum	
Connexons	Large channels in gap junctions which, under normal conditions, couple cardiac cells electrically and chemically to one another Under less physiological conditions, shut off due to high $[Ca^{2+}]_i$ or low pH_i
$g_{Ca(SR)}$ Ryanodine receptor	Ca^{2+} channel in the sarcoplasmic reticulum Can be triggered to release Ca^{2+} by Ca^{2+} entry through L-type calcium channels Modulated by ryanodine
Ionic channels activated under conditions of metabolic inhibition	
g_{NS}	Nonselective cation channel, activated by rise of intracellular Ca^{2+} Under some conditions activated by Ca^{2+} release from the sarcoplasmic reticulum during $[Ca^{2+}]_i$ overload Generates transient inward current and induces delayed afterdepolarizations
$g_{K(ATP)}$	K^+ channel with high conductivity, normally blocked by intracellular ATP Activated by a fall in intracellular ATP concentration, by rise in ADP, and by acidosis Responsible for shortening of the action potential during ischemia
$g_{K(Na)}$	K^+ channel, activated by rise in intracellular Na^+ beyond about 10 mmol/l, exhibiting outward rectification
$g_{K(FA)}$	K^+ channel, activated by rise in fatty acids (FA), phospholipids, and arachidonic acid
Pumps (active transport) and carriers	
$(Na^+ + K^+)$ pump	ATP-dependent active transport of three Na^+ out of and two K^+ into the cell during each cycle, thus generating outward current Stimulated by intracellular Na^+ and extracellular K^+ Inhibited by cardiac glykosides
Sarcolemmal Ca^{2+} pump	ATP-dependent active transport of Ca^{2+} out of the cell Less effective than the Na^+/Ca^{2+} countertransport
Sarcoplasmic Ca^{2+} pump	ATP-dependent active transport of Ca^{2+} into the sarcoplasmic reticulum
Na^+/Ca^{2+} exchanger	Exchange of one Ca^{2+} for three Na^+ during each cycle Direction of transport depending on Na^+ and Ca^{2+} gradients as well as on membrane potential: Ca^{2+} uptake and Na^+ elimination at the beginning of the action potential accompanied by a small repolarizing current; thereafter, Na^+ uptake and Ca^{2+} elimination accompanied by a small depolarizing current Chief means of Ca^{2+} efflux

The symbol g is used for the conductance of the respective ionic channels
Data taken from [13,49,61,62]
SA, sinoatrial; AV, atrioventricular; ATP, adenosine triphosphate; ADP, adenosine diphosphate

drugs that act as local anesthetics, inhibiting the initial Na^+ influx or retarding its recovery from inactivation, can prolong the refractory period without affecting the duration of the action potential (postrepolarization refractoriness) [61].

91.3.4.1 Functional Significance of the Refractory Period

Prevention of Reentry or Circus Movement. The prolonged refractory period protects the musculature of the heart from overly rapid reexcitation, which could impair its function as a pump. At the same time, it prevents recycling of excitation in the muscular network of the heart, which would interfere with the rhythmic alternation of contraction and relaxation. Such a disturbance, called reentry or circus movement, can occur under pathological conditions and forms the chief cause of cardiac

arrhythmias such as flutter or fibrillation. However, because the refractory period of the myocardial cells is normally longer than the time taken for spread of excitation over the atria or ventricles, a wave of excitation originating at the SA node or at a heterotopic center can cover the heart only once and must then die out, as it encounters refractory tissue everywhere. Reentry thus does not normally occur.

91.3.4.2 Further Subdivisions of the Refractory Period

Effective and Functional Refractory Periods. Sometimes the term "effective refractory period" is used to characterize the interval in the cardiac excitatory cycle during which no propagated excitation can be elicited. It lasts only a little longer than the absolute refractory period, which also excludes nonpropagated excitatory events such as local re-

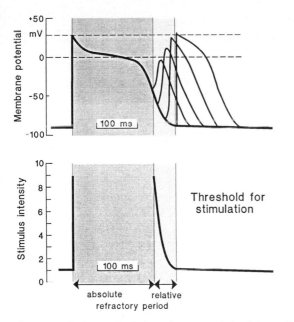

Fig. 91.3. Absolute and relative refractory periods of the cardiac action potential. During the absolute refractory period (from the upstroke of the action potential to about the end of the plateau), the threshold is infinitely high, i.e., stimulation remains ineffective. Stimulation during the relative refractory period elicits action potentials reduced in upstroke velocity, amplitude, and duration

called the "dip phenomenon." Later on, Cranefield and coworkers [16] were able to demonstrate that this phenomenon is only observed with bipolar stimulation and results from the repolarizing influence of the anode. However, such stimuli are less relevant under in vivo conditions, because the natural stimuli of the heart – such as pacemaker discharges or propagating wave fronts – only act on the unexcited cells like cathodal stimuli.

91.3.5 Frequency Dependence of Action Potential Duration

As Fig. 91.3 shows, an action potential immediately following the relative refractory period of the preceding excitation is normal in its rate of rise and amplitude. However, its duration is distinctly less than that of the preceding action potential. In fact, there is a close relationship between the duration of an action potential and the interval that preceded it, and thus between the duration and repetition rate. The main cause of this phenomenon is an increase in g_K (Table 91.2), which outlasts the repolarization phase of the action potential and returns only gradually to the basal level (Fig. 91.2). When the interval between action potentials is short, the increased K^+ conductance accelerates repolarization of the next action potential and vice versa.

91.4 Elementary Mechanisms of Impulse Formation

Slow Diastolic Depolarization. The working myocardium of atria and ventricles is not automatically active; its action potentials are generated by spread of excitation. This means that the response is triggered by the current flowing from excited parts of the fiber cable through unexcited parts and depolarizing the membrane potential. When this depolarization has reached the threshold level for activation of the fast Na^+ channels, the action potential begins. By contrast, in all cardiac muscle cells capable of autorhythmicity, depolarization toward the threshold occurs spontaneously. This elementary process of excitation can be observed directly by intracellular recording from a pacemaker cell. As shown in Fig. 91.4a, the repolarization phase of such an action potential is followed – beginning at the minimal, i.e., most negative, diastolic potential – by a slow diastolic depolarization, which triggers a new action potential when the threshold potential is reached. The slow diastolic depolarization is a local excitatory event, not propagated as the action potential is.

sponses or action potentials with decremental conduction. Instead of a propagated excitation, contraction may be taken as a criterion of effective stimulation, because it can only be initiated by propagated excitation. From a more practical point of view, the two terms are more or less identical. However, sometimes the term "effective refractory period" is also used when, instead of precisely determining the threshold for stimulation, it is measured by a constant stimulus of any suprathreshold strength. The time determined in this way comprises the absolute refractory period as well as that part of the relative refractory period during which the stimulus has not yet been effective. Thus, it approaches the effective refractory period when the intensity of the stimulus is increased. In such cases, it would be better to use the less stringent term "functional refractory period."

Supernormal Period. Usually, the diastolic threshold is attained at the very end of the relative refractory period. However, under special conditions, following the relative refractory period the threshold may undergo a transient fall below its diastolic level, which appears as an increase in excitability, the "supernormal period." The mechanism of this short-term increase in excitability seems to derive from the slight depolarization that accompanies the transition of terminal repolarization to the resting potential. In Purkinje fibers, in which this transition appears to be more gradual than in ordinary myocardium, a supernormal period can regularly be observed [66]. Moreover, Brooks and coworkers [47] described a transient increase in excitability during the early relative refractory period, which they

91.4.1 Dominant and Latent Pacemakers

Normally only a few cells in the SA node are in fact responsible for controlling the timing of the contraction of the

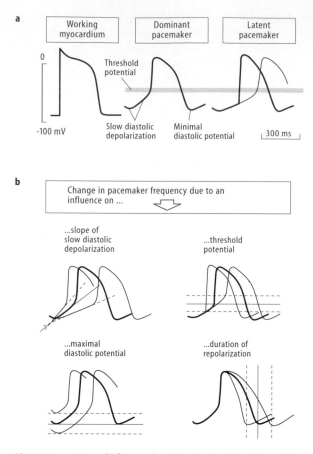

Fig. 91.4. a General form of action potentials of working myocardium and dominant and latent pacemakers. In the working myocardium, the action potential takes off from a constant resting potential elicited by the flow of depolarizing current from adjacent excited fibers. In dominant pacemakers, e.g., the sinoatrial (SA) node, the action potential is elicited by a spontaneously occurring, slow diastolic depolarization that reaches the threshold. In latent pacemakers, an action potential is elicited due to spread of excitation before the slow diastolic depolarization reaches the threshold. b Different mechanisms by which the activity of dominant pacemakers can be changed. In practice, some of these influences are combined

heart. These are the dominant pacemakers. All the other fibers in the specialized tissue are excited in the same way as the working musculature, by conducted activity. In other words, these latent pacemakers are rapidly depolarized by currents from activated sites before their intrinsic slow diastolic depolarization reaches threshold. Comparison of the two processes, as illustrated in Fig. 91.4a, shows how a latent pacemaker can take over the leading role when the dominant pacemaker ceases to function. Because the slow diastolic depolarization of the latent pacemaker, by definition, takes longer to reach threshold, its discharge rate is lower. In the working myocardium there is no spontaneous automatic depolarization; the upstroke of the action potential triggered by the imposed current rises sharply from the resting potential baseline (Fig. 91.4a, left).

91.4.2 How Pacemakers Change Their Frequency

In Fig. 91.4b, four basic electrophysiological changes are depicted that may influence the discharge rate of a pacemaker cell. The most effective way is by altering the slope of the slow diastolic depolarization. However, in reality, a combined influence of these basic electrophysiological changes occurs and they may even antagonize one another. For instance, upon cooling of a spontaneously active Purkinje fiber of the sheep heart from 39° to 26°C, the following effects were observed ([67], p. 81): marked reduction of the slope of the diastolic depolarization, shift of the minimal diastolic potential to more positive values, considerable prolongation of the repolarization, and no change in the threshold potential. The combination of these influences resulted in a strong reduction in frequency.

91.4.3 Ionic Mechanism of Pacemaker Activity

g_K Decay Hypothesis. Figure 91.5 compares the ionic currents underlying generation of the action potential in nonautomatic myocardium (left-hand side) and in the SA node (right-hand side). One fundamental characteristic of the SA nodal pacemaker cell is the absence of the repolarizing inward-rectifier current I_{K1} and the presence of a voltage-independent background Na$^+$ current (I_b) [45,69]. These are conditions which oppose a shift of the membrane potential toward the K$^+$ equilibrium potential E_K. Therefore the membrane potential is kept relatively low, and the fast Na$^+$ channels, insofar as they are present, are largely inactivated. The repolarizing phase of the action potential is brought about by a rise of the K$^+$ conductance g_K beyond its resting value, so that the repolarizing current I_k shifts the membrane potential in the direction of the K$^+$ equilibrium potential E_K, until the minimal (most negative) diastolic potential is reached (Fig. 91.5). As g_K slowly returns to the resting level, I_k decreases and the membrane potential departs from E_K and approaches the threshold for activation of the slow inward current, thus forming the slow diastolic depolarization. The slow Ca^{2+} inward current is responsible for the action potential upstroke in the SA node. The nodal action potentials then correspond approximately to the Ca^{2+} action potentials of the depolarized working myocardium described above. The situation in the AV node is similar.

91.4.4 Further Mechanisms

Nonselective Cation Channels and T-Type Ca^{2+} Channels. The described ionic mechanism of pacemaker activity is usually referred to as the "g_K decay hypothesis," and this hypothesis has long been considered as the only mechanism responsible for pacemaker activity in the SA node. However, from the present point of view, there are further processes which are also more or less essential in

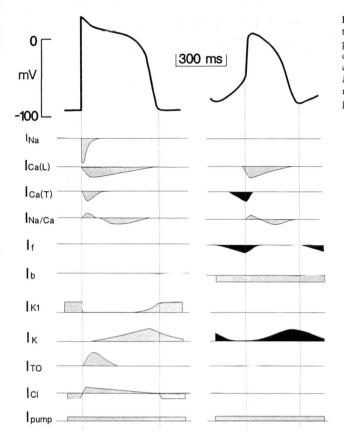

Fig. 91.5. Ionic currents involved in the generation of action potentials of working myocardium (*left*) or of dominant pacemaker (*right*). I_{Na}, fast Na⁺ current; $I_{Ca(L)}$, L-type Ca²⁺ current; $I_{Ca(T)}$, T-type Ca²⁺ current; $I_{Na/Ca}$, Na⁺/Ca²⁺ exchange current; I_f, pacemaker current; I_b, background Na⁺ current; I_{K1}, inward-rectifier K⁺ current; I_k, delayed-rectifier K⁺ current; I_{TO}, transient outward current; I_{Cl}, Cl⁻ current; I_{pump}, pump currents. Data taken from [13,20,44,46,49]

creating automaticity in the SA node [33,69]. Strong evidence has accumulated concerning the participation of nonselective cation channels (g_f), which are activated by repolarization and allow the passage of K⁺ as well as of Na⁺ ions [20,34]. However, at the level of the minimal diastolic potential, it is mainly Na⁺ ions that pass through these channels and create the depolarizing pacemaker current I_f (Fig. 91.5). Further evidence indicates that T-type Ca²⁺ channels are also involved in the pacemaking process [30]. The corresponding current $I_{Ca(T)}$ is activated during the last phase of the diastolic depolarization and thus accelerates the initiation of an action potential. In Fig. 91.5 the ionic currents which are involved in the pacemaker activity of the SA node are shown in black.

91.4.5 Pacemaker Mechanism in the His Purkinje System

Overdrive Suppression. According to current opinion, the slow diastolic depolarizations of the ventricular conducting system are produced by mechanisms which are somewhat different from those in the SA node. In the His Purkinje system, inward rectifier K⁺ channels (g_{K1}) are present, and the background Na⁺ conductance is normally low. Therefore the membrane potential reaches relatively high levels just after the action potential, which also per

mits extensive recovery of the fast Na⁺ channels. Diastolic depolarizations are exclusively generated by the nonselective cation channels (g_f) which are activated during repolarization [20], and the action potentials are triggered by activation of the fast inward Na⁺ current, as is manifest in the high rate of rise of the action potential. However, under normal conditions, the pacemaker activity of the His Purkinje system does not play an essential role for several reasons: due to their lower frequency of intrinsic automaticity, the cells are excited by the spread of excitation originating in the SA node before their diastolic depolarizations can reach the threshold (see Fig. 91.4). Moreover, if these latent pacemakers are stimulated at a rate considerably higher than their own intrinsic frequency, their pacemaker mechanism is inhibited [1]. This influence is called overdrive suppression and is most likely due to the repolarizing current that accompanies sodium extrusion by the (Na⁺+K⁺) pump. During sinus rhythm, Purkinje fibers are subjected to a Na⁺ load roughly twice the one that exists when they discharge at their own rate. The Na⁺ entering the cell must be extruded to maintain cellular homeostasis, and this is done by means of electrogenic extrusion. The pump current results in hyperpolarization, which keeps the membrane potential away from the threshold as long as the activity of the pump is of sufficient magnitude. This phenomenon can explain the fact that, following sinus arrest or sudden AV block, a preautomatic pause precedes the onset of the ventricular escape rhythm.

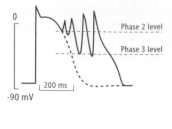

Phase 2 level	Induction by: delayed repolarization and reactivation of Ca²⁺ or Na⁺ channels
Phase 3 level	

Induction by:
delayed repolarization
and reactivation of
Ca^{2+} or Na^+ channels

Favored by: acidosis,
catecholamines,
bradycardia,
class 3 antiarrhythmics

200 ms

-90 mV

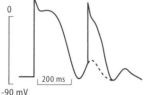

Induction by:
intracellular Ca^{2+} overload

Favored by:
cardiac glycosides,
catecholamines, alkalosis,
high frequency

200 ms

-90 mV

Fig. 91.6. Early (*top*) and delayed (*bottom*) afterdepolarizations as mechanisms of so-called triggered activity. Early afterdepolarizations take off from the repolarization phase of an action potential, whereas delayed afterdepolarizations occur only after complete repolarization of the preceding action potential

91.4.6 Abnormal Automaticity and Triggered Activity

Reduction in K⁺ Conductance. The capacity for spontaneous excitation is a primitive rather than a highly specialized function of myocardial tissue. In the early embryonic stage, all the cells in the heart primordium are spontaneously active. As differentiation proceeds, the fibers of the prospective atrial and ventricular myocardium lose their autorhythmicity and develop a stable, high resting potential. However, the stability of the resting potential can be lost under various conditions associated with partial depolarization of the membrane (catelectrotonus, stretching, hypokalemia, Ba²⁺ ions). Then the affected fibers can develop diastolic depolarizations like those of natural pacemaker cells and, in some circumstances, can interfere with the rhythm of the heartbeat. The mechanism of abnormal automaticity accompanying partial depolarization can be best explained by the reduction in the potassium conductance g_{K1}. This may occur either secondary, i.e., due to inward rectification, accompanying a positive shift of the membrane potential by an applied catelectrotonus [32], or as a primary event due to inactivation of the K⁺ channels caused by hypokalemia or by Ba²⁺ [5]. The reduction of g_{K1} causes the decay of g_K following repolarization, resulting in progressive depolarization such as is in part responsible for the pacemaker mechanism of the SA node.

Triggered Activity. This may be considered as a special type of abnormal automaticity. It can occur in two different forms: as early or delayed afterdepolarizations [15]. Compared with above-mentioned types of abnormal automaticity, these forms are closely coupled to the preceding action potential by which they are triggered.

The *early afterdepolarizations* are superimposed on the repolarization phase of an action potential, mainly when this phase is prolonged (Fig. 91.6, left-hand side). Depending on the membrane potential from which they arise, the potential waves are either slow responses (Ca²⁺ action potentials) or depressed fast responses (Na⁺ action potentials). Their development is favored by all influences that tend to prolong the action potential duration (e.g., class 3 antiarrhythmics, [54]). Catecholamines promote early afterdepolarizations by stimulating the Ca²⁺ inward current [36]. Early afterdepolarization are now considered to cause several cardiac arrhythmias of clinical relevance. The *delayed afterdepolarizations* are different in their appearance as well as in their mechanism from the early ones (Fig. 91.6, right-hand side). In contrast to normal automaticity and to early afterdepolarizations, they are favored by frequent stimulation; overdrive suppression does not play a role [26]. Their generation is strongly dependent on a rise of the intracellular Ca²⁺ concentration. Ca²⁺-triggered Ca²⁺ release from the sarcoplamic reticulum causes rhythmic changes of the cytoplasmic Ca²⁺ concentration, which in turn lead to transient inward currents and corresponding changes in the membrane potential. These are either due to the activation of a Ca²⁺-dependent nonspecific cation channel or to currents induced by the Na⁺/Ca²⁺ exchanger [37]. Thus, all influences that increase the intracellular Ca²⁺ concentration will favor delayed afterdepolarizations (cardiac glycosides, catecholamines etc.) [70]. Hence, delayed afterdepolarizations are also the cellular correlate of bigeminal rhythms due to overdosage of cardiac glycosides.

91.4.7 Special Types of Action Potentials in Different Regions of the Heart

The action potentials in different parts of an individual heart differ in characteristic ways. A few typical forms are shown in Fig. 91.7, where the sequence (top to bottom) and time shift (left to right) correspond to their position in the excitatory cycle of the heart. In the various parts of the pacemaker and conduction system, the slope of the slow diastolic depolarization becomes distinctly less steep with increasing distance from the SA node. Moreover, due to the absence or inactivation of fast Na⁺ channels, both upstroke rate and amplitude of the action potentials in the SA and AV nodes are conspicuously less than in the remainder of the system. The duration of the plateau in the atrial myocardium is less than in the musculature of the ventricle, and the refractory periods are correspondingly related. Experiments on rabbit atrial and ventricular cells revealed that these differences in the action potential configuration are mainly due to the differences in sizes and kinetics of I_{TO} and I_{K1} [29]. I_{TO} is considerably larger in the atrium than in the ventricle, whereas I_{K1} is much larger in the ventricle than in the atrium. Recently, similar results have been obtained in human myocardium [63]. Because of the greatly prolonged action potentials in ·terminal Purkinje fibers due to slow inactivation of the Na⁺ current

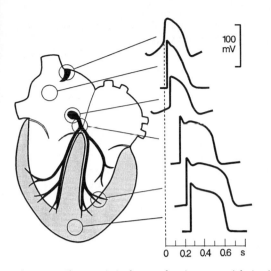

Fig. 91.7. Characteristic forms of action potentials in different regions of the heart. The shift along the time scale corresponds to the arrival time in each region as the excitatory wave spreads through the heart

in these cells [12], they act as a "frequency filter" between the atria and the ventricular muscles, protecting the ventricles from abnormally high atrial discharge rates.

91.4.8 Heterogeneity Within the Ventricular Wall

Subepicardium and Subendocardium. Recent studies in canine heart [6] have shown that ventricular subepicardium and subendocardium differ with respect to electrophysiological characteristics and pharmacological responsiveness and that these differences are in large part due to the presence of a prominent transient outward current I_{TC} in subepicardial myocardium. This current leads to a spike-and-dome morphology of the action potential in subepicardium, but not in subendocardium. Moreover, a subpopulation of cells in the deep subepicardial layers (M cells) has been described with electrophysiological characteristics different from those of either epicardium or endocardium. These cells also display a spike-and-dome morphology and a maximal rate of rise of the action potential upstroke considerably greater than in the cells of the adjacent layers. Moreover, the action potential duration of the M cells is generally longer than in the subepicardial or subendocardial fibers and shows more marked dependence on changes in frequency [40]. These findings should advance our understanding of the ionic bases for the electrocardiographic repolarization waves.

91.5 Excitation-Contraction Coupling in Cardiac Muscle

Nontetanizability. As in the case of skeletal muscle (see Chaps. 45, 46), it is the action potential that gives rise to contraction of the myocardial cell. However, there is a characteristic difference between the two types of muscle with respect to the temporal relation between action potential and contraction: whereas the action potential of skeletal muscle lasts only a few milliseconds and contraction does not begin until the excitatory process is nearly over, in the myocardium the two events overlap considerably in time (Fig. 90.1). The myocardial action potential ends only when the musculature has begun to relax again. Because a new contraction must be initiated by new excitation, which can occur only after the absolute refractory period of the preceding excitation has elapsed, cardiac muscle (unlike skeletal muscle) is incapable of responding to a rapid sequence of action potentials with superposition of single contractions or with a tetanus. This "nontetanizability" of the myocardium is a property that seems entirely appropriate to the pump function of the heart. A tetanic contraction of the heart that outlasted the blood ejection phase would interfere with refilling. On the other hand, the superposition property of skeletal muscle enables the force of contraction to be varied with action potential frequency; myocardial contractions cannot be graded in this way. Moreover, because the myocardium is a functional syncytium, the force of contraction cannot be graded by recruitment of a variable number of motor units, as is possible with skeletal muscle. Myocardial contraction is an "all-or-nothing" event, in which all fibers participate at each occurrence. In compensation for these physiological disadvantages, the opportunity for influencing contraction by way of the excitatory processes or by direct interference via excitation–contraction coupling is considerably greater in cardiac muscle.

91.5.1 Mechanism of Excitation-Contraction Coupling

Ca^{2+} Influx. The myocardial fibers of man and other mammals in principle comprise the same structural elements as are involved in the excitation–contraction coupling processes in skeletal muscle (Fig. 91.8). The transverse tubular system (TTS) is clearly a feature of the myocardium, particularly in the ventricles. By contrast, the longitudinal tubular system (i.e., sarcoplasmic reticulum, SR), which functions as an intracellular Ca^{2+} store, is less well developed than in skeletal muscle. Both the structural peculiarities of the myocardium and its functional behavior offer evidence of a close interaction between the intracellular Ca^{2+} stores and the medium external to the fibers (for a more detailed description of the Ca^{2+} mediated control of cardiac contractility at the cellular level, see Chap. 93, this volume). A key event in contraction is the influx of Ca^{2+} during the action potential. The quantity of inflowing Ca^{2+}, however, is evidently not sufficient for direct activation of the contractile apparatus, as is required under physiological conditions. The additional release of Ca^{2+} from the intracellular stores triggered by the Ca^{2+} influx appears to be a more important effect [25]. The influx of Ca^{2+} across the membrane also serves to replenish the Ca^{2+} stores for the subsequent contractions.

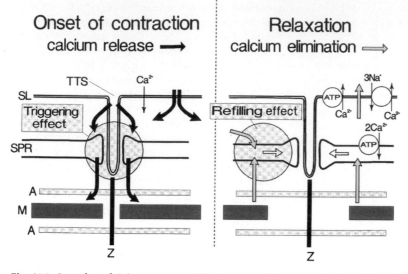

Fig. 91.8. Interplay of Ca^{2+} movements (*black arrows*, Ca^{2+} release; *shaded arrows*, Ca^{2+} elimination) and contractile activation during the onset of contraction (*left*) and underlying relaxation (*right*). At the onset of contraction, transsarcolemmal Ca^{2+} influx induces a calcium-triggered Ca^{2+} release from the sarcoplasmic reticulum (triggering effect). During relaxation, Ca^{2+} is partly eliminated from the cell and partly stored in the sarcoplasmic reticulum (refilling effect). *A*, actin; *M*, myosin; *SL*, sarcolemma; *SPR*, sarcoplasmic reticulum; *Z*, Z line of the sarcomere

If one experimentally shortens the duration of a single action potential by applying an anodal current pulse, thus prematurely interrupting the Ca^{2+} influx, the corresponding contraction is attenuated only slightly, whereas the following contraction, elicited by a normal action potential, is considerably reduced. When an action potential is artificially prolonged, the reverse effect is observed, i.e., an enhancement of the subsequent contractions. If the action potential is shortened or lengthened for several beats, an equilibrium is attained after five to seven beats, with a level of contraction that may be considerably decreased or increased, respectively [4].

The action potential thus affects contraction in at least two important ways (Fig. 91.8):

- A triggering action, eliciting the contraction by (Ca^{2+}-triggered) release of Ca^{2+}, primarily from intracellular stores [25]
- A refilling action, renewing the intracellular stores of Ca^{2+} during relaxation in preparation for subsequent contractions.

91.5.2 Basic Mechanisms by Which Contraction May Be Influenced

A change in the force of myocardial contraction is ultimately due to one of two mechanisms [46,57].

- A change in intracellular free Ca^{2+} concentration to interact with the contractile proteins
- A change in sensitivity of the myofilaments for Ca^{2+}

Action Potential Duration. Concerning a change in intracellular free Ca^{2+}, a number of influences on the force of myocardial contraction are exerted by way of a change in duration of the action potential accompanied by corresponding modifications of the inward Ca^{2+} current. Examples include shortening of the action potential by elevated [K$^+$]$_e$ or by acetylcholine in the atrial myocardium, which weakens the contraction, and lengthening of the action potential by cooling of by K$^+$ channel blockers, which increases the contractile force. An increase in the number of action potentials per unit time acts in the same direction as an increase in action potential duration (frequency inotropism). A similar mechanism is responsible for the increase in contractile force due to paired pulse stimulation or accompanying extrasystoles (postextrasystolic potentiation). The so-called staircase phenomenon, a stepwise increase in the amplitude of contraction following temporary arrest, is also associated with the replenishment of intracellular Ca^{2+} stores [4].

Cyclic Adenosine Monophosphate. Another way of increasing the cytosolic Ca^{2+} concentration is achieved by agents that elevate the intracellular concentration of cyclic adenosine monophosphate (cAMP) by increasing cAMP formation (either receptor mediated, e.g., epinephrine, norepinephrine, or by directly stimulating adenylate cyclase) or by decreasing cAMP breakdown (by inhibiting phosphodiesterase). The main effect of cAMP is a phosphorylation of functional proteins and an increase in the slow inward Ca^{2+} current. However, an increase in the slow Ca^{2+} inward current without elevation of the cAMP concentration can also be brought about by α-adrenergic agents such as phenylephrine or by the so-called Ca^{2+} agonists (e.g., Bay K 8644). α- Receptor agonists act by causing an increase in phosphatidylinositol turnover [58].

Intracellular Ca^{2+} Concentration. The transsarcolemmal movement of Ca^{2+} is further controlled by the intracellular Na^+ concentration. Na^+ and Ca^{2+} are interrelated via the Ca^{2+}/Na^+ exchange system, in which transsarcolemmal Ca^{2+} movements are coupled to opposite movements of Na^+. Thus, Na^+ can exit from the cell in exchange for Ca^{2+}, or Ca^{2+} can exit in exchange for Na^+, depending on the prevaling Na^+ and Ca^{2+} electrochemical gradients. It is interesting to note that only a very small increase in intracellular Na^+ is required to achieve a large increase in $[Ca^{2+}]_i$ and hence in force of contraction [57]. Quantitatively, a 1-mmol/l increase in intracellular Na^+ activity is accompanied by about a 100% increase in force of contraction. Two mechanisms have so far been discovered which are likely to produce an increase in myocardial force of contraction via an increase in $[Na^{2+}]_i$:

- Inhibition of (Na^++K^+)-ATPase, e.g., by cardiac glycosides (digitalis, strophanthin [39,51]
- Prolongation of the open state of Na^+ channels (DPI 201–106, ceveratrum alkaloids [57].

Effects similar to those achieved by the removal of extracellular Ca^{2+} can be brought about by Ca^{2+} antagonists (e.g., verapamil, nifedipine, diltiazem), which block Ca^{2+} influx through L-type Ca^{2+} channels during the action potential [27].

Ca^{2+} Sensitization. An increased Ca^{2+} sensitivity of the contractile proteins (Ca^{2+} sensitization) can be revealed by studying the contractile behavior of isolated myocardial contractile structures or by measuring intracellular Ca^{2+} signals with the Ca^{2+}-sensitive bioluminescent protein aequorin in intact cardiac muscle. Ca^{2+} sensitization as a mechanism of action has been discussed for several agents, including α-adrenoceptor agonists. However, the practical value of Ca^{2+} sensitization as a positive inotropic mechanism in intact cardiac muscle needs to be elucidated further [57].

91.6 Autonomic and Other Influences on the Elementary Processes of Cardiac Excitation

Acetylcholine and Norepinephrine. The cardiac centers in the central nervous system (medulla and pons) exert an influence on the activity of the heart (see Chap. 99) by way of sympathetic and parasympathetic nerves. This influence governs the rate of beat (chronotropic action), the systolic contractile force (inotropic action), and the velocity of atrioventricular conduction (dromotropic action). These actions of the autonomic nerves are mediated in the heart, as in all other organs, by the chemical transmitters acetylcholine in the parasympathetic system and norepinephrine in the sympathetic system.

91.6.1 Parasympathetic Innervation

Vagus Nerves. The parasympathetic nerves supplying the heart branch off from the vagus nerves on both sides in the cervical region. The preganglionic cardiac fibers on the right side pass primarily to the right atrium and are concentrated at the SA node. The AV node is reached chiefly by cardiac fibers from the left vagus nerve. Accordingly, the predominant effect of stimulation of the right vagus is on heart rate and that of left vagus stimulation is on atrioventricular conduction. The parasympathetic innervation of the ventricles is sparse, and its influence is limited to certain regions or is indirect, by inhibition of the sympathetic action.

91.6.2 Sympathetic Innervation

The sympathetic nerve supply, unlike the parasympathetic one, is nearly uniformly distributed to all parts of the heart. The preganglionic elements of the sympathetic cardiac nerves come from the lateral horns of the upper thoracic segments of the spinal cord and make synaptic connections in the cervical and upper thoracic ganglia of the sympathetic trunk, in particular the stellate ganglion. The postganglionic fibers pass to the heart in several cardiac nerves. Sympathetic influences on the heart can also be exerted by catecholamines released from the adrenal medulla into the blood [19].

91.6.3 Chronotropy

Stimulation of the right vagus or direct application of acetylcholine to the SA node causes a decrease in heart rate (negative chronotropy); in extreme cases this can cause cardiac arrest. Sympathetic stimulation or application of norepinephrine increases the heart rate (positive chronotropy). When vagus and sympathetic nerves are stimulated at the same time, the vagus action usually prevails. Modification of the autorhythmic activity of the SA node by these autonomic inputs occurs primarily by way of a change in the time course of the slow diastolic depolarization (Fig. 91.9). Under the influence of the vagus diastolic depolarization is retarded, so that it takes longer to reach threshold. The sympathetic fibers act to increase the rate of diastolic depolarization and thus shorten the time to threshold.

The positive chronotropic action of the sympathetic nerves extends to the entire conducting system of the heart. Thus, when a leading pacemaker center fails, the sympathetic input can determine when and to what extent a subordi-

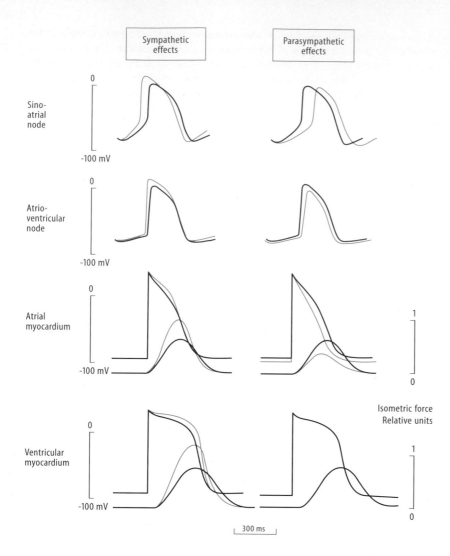

Fig. 91.9. Main effects of autonomic nerves (or their transmitters) on the electrical and mechanical activity in different regions of the heart. *Thick lines* represent control conditions, and *thin lines* the changes occurring due to the corresponding autonomic influences. *Left*, sympathetic effects; *right*, parasympathetic effects

nate center takes over as the pacemaker. Moreover, the sympathetic system also has a positive chronotropic action on pacemaker cells when their spontaneous activity has been suppressed by external influences such as increased K^+ or an overdose of drugs that interfere with automaticity. In the same way, however, an ectopic focus of rhythmicity ca be stimulated to greater activity, so that the danger of arrhythmia increases.

91.6.4 Vagal and Sympathetic Tone

The ventricles of most mammals, including humans, are influenced predominantly by the sympathetic system. By contrast, the atria can be shown to be subject to the continual antagonistic influence of both vagus and sympathetic nerves. This effect is most clearly evident in the activity of the SA node. It can be observed, for example, by transecting or pharmacologically blocking one of the two sets of nerves, that the action of the other set then dominates. When the vagus input to the dog heart is removed,

the rate of beating increases from about 100/min at rest to 150/min or higher; when the sympathetic input is removed, it falls to about 60/min or less. This maintained activity of the autonomic nerves is called vagal and sympathetic tone. Because the rate of the completely denervated heart (the autonomic rate) is distinctly higher than the normal resting rate, it can be assumed that under resting conditions vagal tone predominates over sympathetic tone.

91.6.5 Inotropy

As mentioned above, a change in heart rate in itself has a considerable effect on the strength of myocardial contraction. In addition, the autonomic nerves to the heart have a direct effect on mechanical force generation. The vagus acts to reduce the strength of contraction of the atrial myocardium; at the same time, the rising time of the mechanogram, i.e., the time from the initial deflection of the contraction curve to the peak, decreases. This negative inotropic action results partly from a primary shortening of the action potential. Sympathetic activity increases the strength of contraction in both atrial and ventricular myocardium (positive inotropic action). The contraction curve rises more steeply, the peak is reached sooner, and relaxation is accelerated. By contrast, the shape of the action potential is hardly changed (Fig. 91.9).

91.6.6 Dromotropy

An influence of the autonomic nerves on the conduction of excitation can normally be demonstrated only in the region of the AV node. The sympathetic fibers accelerate atrioventricular conduction and thus shorten the interval between the atrial and ventricular contractions (positive dromotropic action). The vagus, especially on the left side, retards atrioventricular conduction (negative dromotropic action) and, in extreme cases, can produce a transient complete AV block. These effects of the autonomic transmitter substances are associated with a particular feature of the cells in the AV node. As discussed above, the fibers of the AV node closely resemble those of the SA node. Because there is no rapid inward Na^+ current, the upstroke is relatively slow and hence the conduction velocity is low. It is evident in Fig. 91.9 that the vagus acts to decrease the rate of rise still further, whereas sympathetic activity increases it, with the corresponding effects on the velocity of atrioventricular conduction.

91.6.7 Bathmotropy

The term "bathmotropy" denotes an influence on excitability in the sense of a lowered or raised threshold. However, experimental observations of bathmotropic effects of the autonomic transmitters on the heart are not consistent. All that is fairly certain is that sympathetic activity increases excitability when it has been reduced (low resting potential). The notion of a bathmotropic action has brought more confusion than clarity to the subject and should be discarded [2].

91.6.8 Mechanism of Autonomic Transmitter Action

Acetylcholine-Gated Muscarinic K^+ Channels. The effects of vagal stimulation or of application of the parasympathetic transmitter acetylcholine are mainly attributed to an increase in the K^+ conductance of the excitable membrane due to the activation of acetylcholine-gated muscarinic K^+ channels [10,55] involving an acetylcholine receptor, M_2, and a pertussis toxin-sensitive guanine nucleotide-binding protein [50]. In general, such an influence is expressed in the tendency of the membrane potential to approach the K^+ equilibrium potential, and this is opposed to depolarization. This tendency is evident in both the retardation of the slow diastolic depolarization in the SA node, described above,

and in the shortening of the action potential of the atrial myocardium, which in turn weakens the contraction. The reduction of the rate of rise of the action potential in the AV node can also be explained on this basis, as a stronger outward K^+ current counteracts the slow inward Ca^{2+} current.

Slow Inward Ca^{2+} Current Inhibition. Moreover, an inhibitory influence of acetylcholine on the slow inward Ca^{2+} current has been reviewed recently [42]. This influence was shown to be linked to the inhibition of the synthesis of cAMP. Thus, in experiments with single ventricular myocytes from guinea pig, acetylcholine inhibited the slow inward Ca^{2+} current that had been augmented by prior administration of isoproterenol, but did not inhibit it when cAMP had been administered [31]. This finding is also in accord with the participation of pertussis toxin-sensitive G_i as the transducer. In the ventricular myocardium, where acetylcholine receptors are rarely developed, this indirect sympathetic–antagonistic action dominates and determines the vagal influence on contractile activity. However, there is also strong evidence of a vagally mediated inhibition of norepinephrine release from sympathetic nerves [41]. In this model, muscarinic receptors located on sympathetic nerve endings mediate inhibition of norepinephrine release in response to acetylcholine released from nearby vagus nerve terminals. Conversely, α-adrenoceptor agonists can inhibit the bradycardia induced by vagus nerve stimulation [60]. These influences provide the basis for a model of reciprocal prejunctional regulation of vagal and sympathetic nerve activity by the respective transmitters.

Time-Dependent Pacemaker Current. Recently, it was shown that acetylcholine also exerts a remarkable influence on the current I_f, the time-dependent pacemaker current activated by hyperpolarization [21]. The activation curve of this current is shifted to more negative potentials when the transmitter is applied to cells from the rabbit SA node. This effect of acetylcholine is also muscarinic in nature and requires the transducer function of G_i [22]. Since this influence of acetylcholine only requires very small concentrations and is brought about without hyperpolarizing the SA node, the conclusion was drawn that regulation of I_f may be more important in producing the negative chronotropic effect in the SA node than regulation of the acetylcholine-sensitive K^+ channels [21].

α- and β-Adrenergic Responses. Sympathetic fibers (or their transmitters) exert their influence on the heart by interaction with specific receptors on the sarcolemma to modify the cellular function. The pharmacological classification of physiological responses to catecholamines as α- and β-adrenergic als applies to the heart. In most circumstances, the responses to catecholamines are predominantly of the β-adrenergic type. Both β-adrenergic and α-adrenergic responses have been shown to stimulate voltage dependent sarcolemmal Ca^{2+} influx [53]. More recent studies have also demonstrated a modulation of K^+ cur-

Table 91.3. Main action of various physical and chemical influences on the electrical and mechanical activity of the heart

	Resting potential	Action potential		Conduction velocity	Pacemaker (slope of diastolic depolarization)	Contraction force
		Upstroke	Duration			
Frequency						
Increase			↓		↑	↑
Decrease			↑		↓	↓
Temperature						
Increase			↓		↑	↓
Decrease	0 → ↓	0 → ↓	↑		↓	↑
Acidosis		↓	↑	↓	↓	↓
Alkalosis		(↑)	(↓)	(↑)	↑	↑
$[K^+]_e$						
Increase	↓	↓	↓	(↑) → ↓	↓	↓
Decrease	0 → ↓		↑ → ↓		↑	↑
$[Ca^{2+}]_e$						
Decrease	0 → ↓		0 → ↑			↓
Increase	0 → ↑		0 → ↓		(↑)	↑

Data taken from [3,14,47,66,67]
↑, increase; ↓, decrease; →, effect changes upon increasing the corresponding influences; ( ), weak effect

rents. An increase in the slow inward Ca^{2+} current enhances contractile force (positive inotropic action) because this effect intensifies excitation–contraction coupling. The positive dromotropic action on the AV node is also likely, in view of the above considerations, to be related to the enhancement of the slow inward Ca^{2+} current. On the other hand, the accelerated relaxation associated with the positive inotropic action is ascribed to a stimulation of Ca^{2+} uptake into the sarcoplasmic reticulum. Concerning the mechanism of the positive chronotropic sympathetic action, enhancement of the slow inward current is most likely involved. However, the current I_f is also sensitive to β-receptor stimulation, which shifts the activation curve to less negative potentials [22].

β₁- and β₂-Receptors. There is much evidence that these effects of β-adrenergic stimulation are mediated predominantly by cAMP, which in turn binds to the regulatory subunit of cAMP-dependent protein kinase (see Chap. 5). The catalytic subunit of this enzyme then catalyzes phosphorylation of a variety of cellular proteins including ionic channels, e.g., L-type Ca^{2+} channels [23]. Pharmacological experiments as well as radioligand studies have shown that both subtypes of β-adrenergic receptors (β₁- and β₂-receptors) are present within the heart; however, the ratio of β₁- to β₂-receptors is different in different cardiac regions. Pharmacological studies using agonists with different selectivity for the two subtypes have shown that β₂-adrenergic stimulation exerts more prominent chronotropic than inotropic effects, whereas β₁-selective agonists produce equivalent chronotropic and inotropic effects [11].

α-Adrenergic Receptors. The existence and physiological role of α-adrenergic receptors in the myocardium have been questioned for some time. It is now evident that α-adrenergic stimulation can result in a functionally significant influence on the heart, such as a positive inotropic effect [56]. These effects are not mediated by cAMP, but rather by hydrolysis of inositol-containing phospholipids [8]. In the proposed scheme, agonist receptor binding stimulates the phosphodiesterase hydrolysis of phosphaptidylinositol-4,5-bisphosphate, resulting in the formation of 1,2-diacyglycerol and inositol-1,4,5-trisphosphate ($InsP_3$) (see Chap. 5). Both compounds function as intracellular messengers, with the former stimulating the activity of protein kinase C and the latter mobilizing intracellular Ca^{2+} [8].

91.6.9 Afferent Innervation

Vagus Afferents. In addition to the efferent autonomic supply, the innervation of the heart comprises a large number of afferent fibers, divided among the vagus and sympathetic nerves. Most of the vagus afferents are myelinated fibers originating in receptors in the atria or the left ventricle. Recordings from single fibers in the atria have revealed two types of mechanoreceptors signaling passive stretching or active tension. Apart from the myelinated afferent fibers from specialized sensory receptors, the main fibers leaving the heart come from dense subendocardial plexuses of nonmyelinated fibers with free endings; these fibers run in the sympathetic nerves. It is probably these fibers that mediate the severe, segmentally radiating pains experienced when circulation within the heart itself is impaired (angina pectoris, myocardial infarction) [43].

91.6.10 Effects of the Ionic Environment

K⁺ Concentration. Of all the features of the extracellular solution that can affect the activity of the heart, the K⁺ concentration is of the greatest practical importance. An increase in extracellular K^+ ($[K^+]_e$) has two effects on the myocardium:

- The resting potential is lowered because the $[K^+]_i/[K^+]_e$ gradient is less steep.
- The K^+ conductance of the excitable membrane is increased (see Table 91.2: g_{K1}).

Doubling of $[K^+]_e$ from the normal concentration of 4 mmol/l to about 8 mmol/l results in a slight depolarization, accompanied by increased excitability and conduction velocity, and in the suppression of heterotopic centers of rhythmicity. A large increase in $[K^+]_e$ (over 8 mmol/l) reduces excitability and conduction velocity, as well as the duration of the action potential, so that the strength of contraction is diminished. When $[K^+]_e$ is lowered to less than 4 mmol/l, the stimulating influence on pacemaker activity in the ventricular conducting system dominates. The enhanced activity of heterotopic centers can lead to cardiac arrhythmias.

Cardioplegic Solutions. The excitability-reducing action of large extracellular K^+ concentrations is used during heart operations to immobilize the heart briefly during surgical procedures (cardioplegic solutions). While the heart is inactive, circulation is maintained by an extracorporeal pump (heart – lung machine). Impairment of cardiac function due to increased blood K^+ during extreme muscular effort or in pathological conditions can be largely compensated for by sympathetic activity. Table 91.3 summarizes the most important physical and chemical influences of ionic changes on excitation and contraction of the heart; only the dominant effects are considered.

References

1. Alanis J, Benitez K (1967) The decrease in the automatism of the Purkinje pacemaker fibres provoked by high frequencies of stimulation. Jpn J Physiol 17:556–571
2. Antoni H (1966) Elektrophysiologie peripherer vegetativer Regulationen am Beispiel des Herzmuskels und der glatten Muskulatur. In: Büchner F (ed) Neurovegetative Regulationen. Springer Berlin Heidelberg New York
3. Antoni H (1989) Function of the heart. In: Schmidt RF, Thews G (eds) Human physiology. Springer, Berlin Heidelberg New York, pp 439–480
4. Antoni H, Jacob R, Kaufmann R (1969) Mechanical response of the frogs and mammalian myocardium to modifications of the action potential duration by constant current pulses. Pflugers Arch 306:33–57
5. Antoni H, Oberdisse E (1965) Elektrophysiologische Untersuchungen über die Barium-induzierte Schrittmacher-Aktivität im isolierten Säugetiermyokard. Pflugers Arch 284:259–272
6. Antzelevitch C, Sicouri S, Litovski SH, Lukas A, Krishnan SC, Di Diego JM, Gintant GA, Liu D-W (1991) Heterogenity within the ventricular wall. Electrophysiology and pharmacology of epicardial, endocardial, and M cells. Circ Res 69:1427–1449
7. Beeler GW, Reuter H (1970) Membrane calcium current in ventricular myocardial fibres. J Physiol (Lond) 207:191–209
8. Berridge MJ (1988) Inositol trisphosphate and diacyglycerol: two interacting second messengers. Annu Rev Biochem 56:159–193
9. Bonke FIM, Kirchhof CJHJ, Allessie MA, Wit AL (1987) Impulse propagation from the SA node to the ventricles. Experientia 43:1044–1048
10. Burgen ASV, Terroux KG (1953) On the negative inotropic effect in the cat's auricle. J Physiol (Lond) 120:449–464
11. Carlsson E, Dahlof C, Hedberg A (1977) Differentiation of cardiac chronotropic and inotropic effects of β-receptor agonists. Naunyn Schmiedebergs Arch Pharmacol 300:101–105
12. Carmeliet E (1987) Slow inactivation of the sodium current in rabbit cardiac Purkinje fibres. Pflugers Arch 408:18–26
13. Carmeliet E (1992) Potassium channels in cardiac cells. Cardiovasc Drugs Ther 6:305–312
14. Cranefield PF (1975) The conduction of the cardiac impulse, the slow response and cardiac arrhythmias. Furura, Mount Kisco/New York
15. Cranefield PF (1977) Action potentials, afterpotentials, and arhythmias. Circ Res 41:415–423
16. Cranefield PF, Hoffman BF, Siebens AA (1957) Anodal excitation of cardiac muscle. Am J Physiol 190:383–390
17. van Dam RT, Janse MJ (1989) Activation of the heart. In: Macfarlane PW, Lawrie TDV (eds) Comprehensive electrocardiology–theory and practice in health and disease. Pergamon, New York, pp 101–127
18. Deleze J (1987) Cell-to-cell communication in the heart: structure-function correlations. Experientia 43:1068–1075
19. Delius W, Gerlach E, Grobecker H, Kübler H (eds) (1981) Catecholamines and the heart. Springer, Berlin Heidelberg New York
20. Di Francesco D (1981) A new interpretation of the pacemaker-current in calf Purkinje fibres. J Physiol (Lond) 314:359–376
21. Di Francesco D, Ducouret P, Robinson RB (1989) Muscarinic modulation of cardiac rate at low acetylcholine concentrations. Science 243:669–671
22. Di Francesco D, Tromba C (1987) Acetylcholine inhibits activation of the cardiac hyperpolarizing-activated current I_f. Pflugers Arch 410:139–142
23. Drummond GI, Severson DL (1979) Cyclic nucleotides and cardiac function. Circ Res 44:145–153
24. Durrer D, van Dam RT, Freud GE, Janse MJ, Meijler FL, Arzbaecher RC (1970) Total excitation of the isolated human heart. Circulation 41:899–912
25. Fabiato A (1983) Calcium induced release of calcium from the cardiac sarcoplasmic reticulum. Am J Physiol 245:C1–C14
26. Ferrier GR, Saunders JH, Mendez C (1973) A cellular mechanism for the generation of ventricular arrhythmias by acetylstrophantidin. Circ Res 32:600–609
27. Fleckenstein A (1983) Calcium antagonism in heart and smooth muscle. Experimental facts and therapeutic prospects. Wiley, New York
28. Fozzard HA, Hanck DA, Makielski JC, Scanley BE, Sheets MF (1987) Sodium channels in cardiac Purkinje cells. Experientia 43:1162–1168
29. Giles WR, Imaizumi Y (1988) Comparison of potassium currents in rabbit atrial and ventricular cells. J Physiol (Lond) 405:123–145
30. Hagiwara N, Irisawa H, Kameyama M (1988) Contribution of two types of calcium currents to the pacemaker potentials of rabbit sino-atrial node cells. J Physiol (Lond) 395:233–253
31. Hescheler J, Kameyama M, Trautwein W (1986) On the mechanism of muscarinic inhibition of the cardiac Ca current. Pflugers Arch 407:182–189
32. Hohnloser S, Weirich J, Antoni H (1982) Influence of direct current on the electrical activity of the heart and on its susceptibility to ventricular fibrillation. Basic Res. Cardiol 77:237–249
33. Irisawa H, Brown HF, Giles W (1993) Cardiac pacemaking in the sinoatrial node. Physiol Rev 73:197–227

34. Irisawa H, Noma A (1982) Pacemaker mechanisms of rabbit sino atrial node cells. In: Bouman LN, Jongsma HJ (eds) Cardiac rate and rhythm. Nijhoff, London, pp 35–51

35. Janse M, Anderson RH (1974) Specialized internodal atrial pathways–fact or fiction? Eur J Cardiol 2:117–136

36. January CT, Riddle JM, Salata JJ (1988) A model for early afterdepolarizations: induction with Ca channel agonist Bay K 8644. Circ Res 62:563–571

37. Kass RS, Lederer WJ, Weingart R (1978) Role of calcium ions in transient inward currents and aftercontractions induced by strophantidin in cardiac Purkinje fibres. J Physiol (Lond) 281:187–208

38. Kupersmith J, Krongrad E, Waldo A (1973) Conduction intervals and conduction velocity in the human cardiac conducting system. Studies during open heart surgery. Circulation 47:776–785

39. Langer GA (1971) The intrinsic control of myocardial contraction–ionic factors. New Engl J Med 285:1065–1701

40. Liu D-W, Gintant GA, Antzelevitch C (1993) Ionic bases for electrophysiological distinctions among epicardial, midmyocardial, and endocardial myocytes from the free wall of the canine left ventricle. Circ Res 72:671–687

41. Löffelholz K, Muscholl E (1969) A muscarinic inhibition of noradrenaline release evoked by sympathetic nerve stimulation. Naunyn Schmiedebergs Arch. Pharmacol 265:1–15

42. Löffelholz K, Pappano AJ (1985) The parasympathetic neuroeffector junction of the heart. Pharmacol Rev 37:1–24

43. Michell GAG (1956) Cardiovascular innervation. Livingstone, Edinburgh

44. New W, Trautwein W (1972) Inward membrane currents in mammalian myocardium. Pfluger Arch 334:1–23

45. Noble D, Denyer JC, Brown HF, Di Francesco D (1992) Reciprocal role of the inward currents i_{bNa} and i_f in controlling and stabilizing pacemaker frequncy of rabbit sino-atrial node cells. Proc R Soc Lond [B] 250:199–207

46. Opie LH (1991) The heart – physiology and metabolism. Raven, New York

47. Orias O, McBrooks EE, Suckling JL, Gilbert A, Siebens A (1950) Excitability of the mammalian ventricle throughout the cardiac cycle. Am J Physiol 163:272–282

48. Paes de Carvalho A, Almeida DF (1960) Spread of activity through the atrioventricular node. Circ Res 8:801–809

49. Pelzer D, Trautwein W (1987) Currents through ionic channels in multicellular cardiac tissues and single heart cells. Experientia 43:1153–1162

50. Pfaffinger PJ, Martin JM, Hunter DB (1985) GTP-binding proteins couple cardiac muscarinic receptors to a K channel. Nature 317:536–538

51. Repke K (1964) Über den biochemischen Wirkungsmodus von Digitalis. Klin Wochenschr 42:157–165

52. Reuter H (1967) The dependence of slow inward current in Purkinje fibres on the extracellular calcium concentration. J Physiol (Lond) 192:479–492

53. Reuter H (1983) Calcium channel modulation by neurotransmitters, enzymes, and drugs. Nature 301:569–574

54. Roden DM, Hoffman BF (1985) Action potential prolongation and induction of abnormal automaticity by low quinidine concentrations in canine Purkinje fibres. Circ Res 56:857–867

55. Sackmann B, Noma A, Trautwein W (1982) Acetylcholine activation of single muscarinic K^+ channels in isolated pacemaker cells of mammalian heart. Nature 303:250–253

56. Scholz H (1980) Effects of beta- and alpha-adrenoceptor activators and adrenergic transmitter releasing agents on the mechanical activity of the heart. In: Szekeres L (ed) Handbook of experimental pharmacology, vol 54/I. Springer, Berlin Heidelberg New York, pp 651–733

57. Scholz H (1984) Mechanisms of positive inotropic effects. Basic Res Cardiol 84 [Suppl I]:3–7

58. Scholz J, Schaefer B, Schmitz W, Scholz H, Steinfath M, Lohse M, Schwabe U, Puurunen J (1988) Alpha-l adrenoceptor mediated positive inotropic effect and inositol trisphosphate increase in mammalian heart. J Pharmacol Exp Ther 245:327–335

59. Spach MS, Lieberman M, Scott JG, Barr RC, Johnson EA, Kootsey JM (1971) Excitation sequence of the atrial septum and AV node in isolated hearts of the dog and the rabbit. Circ Res 29:156–172

60. Starke K (1972) Alpha sympathomimetic inhibition of adrenergic and cholinergic transmission in the rabbit heart. Naunyn Schmiedebergs Arch Pharmacol 274:18–45

61. Task Force of the Working Group on Arrhythmias of the European Society of Cardiology (1991) The Sicilan gambit. A new approach to the classification of antiarrhythmic drugs based on their actions on arrhythmogenic mechanisms. Circulation 84:1831–1851

62. Tsien RW, Hess P, Nilius B (1987) Cardiac calcium currents at the level of single channels. Experientia 43:1169–1172

63. Varro A, Nanasi PP, Lathrop DA (1993) Potassium currents in isolated human atrial and ventricular cardiocytes. Acta Physiol Scand 149:133–142

64. Veenstra RD (1990) Physiology of cardiac gap junctions. In: Zipes DP, Jalife J (eds) Cardiac electrophysiology–from cell to bedside. Saunders Philadelphia pp 62–69

65. Veenstra RD, De Haan RL (1986) Measurement of single channel currents from cardiac gap junctions. Science 233:972–974

66. Weidmann S (1955) Effects of calcium ions and local anaesthetics on electrical properties of Purkinje fibres. J Physiol (Lond) 129:568–582

67. Weidmann S (1956) Elektrophysiologie der Herzmuskelfaser. Huber, Bern

68. Weidmann S (1966) The diffusion of radiopotassium across intercalated discs of mam malian cardiac muscle. J Physiol (Lond) 187:323–342

69. Wilders R, Jongsma HJ, Ginneken ACG (1991) Pacemaker activity of the rabbit sinoatrial node. A comparison of mathmatical models. Biophys J 60:1202–1216

70. Wit AL, Rosen MR (1986) Afterdepolarizations and triggered activity. In: Fozzard HA, Haber E, Jennings RB (eds) The heart and the cardiovascular system. Raven, New York, pp 1449–1490

71. Zipes DP, Jalife J (eds) (1990) Cardiac electrophysiology – from cell to bedside. Saunders, Philadelphia

92 Electrocardiography

H. Antoni

Contents

92.1 Introduction

As excitation passes over the heart and dies out, electrical currents spread within the heart itself but also into the tissues surrounding it, and a small portion of these attains the surface of the body. The flow of current creates voltages between sites on the body surface that can be measured by means of electrodes placed on the skin. A recording of these voltages as a function of time and from definite sites is known as an electrocardiogram (ECG). A normal electrocardiogram for one heart beat derived between right arm and left leg is illustrated in Fig. 92.1. It will be considered subsequently in more detail. The method of ECG recording was developed at the turn of the twentieth century by Willem Einthoven, a Dutch physician, who constructed a galvanometer making it possible to derive such recordings routinely from the the human body [4] (also see [9]). Einthoven also proposed designating the different deflections of the ECG as P, Q, R, S, and T waves and demon-

strated their clinical significance by showing differences between waveforms recorded from normal subjects and from patients suffering from dysrhythmias [5]. For this work Einthoven received the Nobel prize in 1924.

The size of the potential differences at the surface depends on the size of the current generated by the heart, which in turn depends on the mass of the myocardium being activated. Since the mass of the conduction system is tiny compared with the myocardium its depolarization does not appear on a surface ECG. An additional influence on the measurable size of the surface potentials is exerted by physical recording conditions such as transitional resistance. In any case, it has to be kept in mind that the size of the different deflections is in no way related to the strength of contraction.

It is important to note that the ECG trace reflects cardiac excitation but not cardiac contraction.

92.2 ECG Form and Nomenclature

92.2.1 Technical Aspects

Because in routine ECG recording the measured potentials amount to less than 1 mV in most cases, the commercially available ECG recorders incorporate electronic amplifiers. The amplifier inputs include capacitive coupling – high-pass filters with a cut-off frequency near 0.1 Hz (a time constant of 2 s). Therefore, direct current components and very slow changes of the potentials at the metal recording electrodes, which would be distracting, do not appear at the output. All ECG recorders have a built-in means of monitoring amplitude, in the form of a 1-mV calibration pulse usually set to cause a deflection of 1 cm.

92.2.2 Standard Form

With electrodes attached to the right arm and left leg (standard lead II of Einthoven), the normal ECG looks like the curve shown in Fig. 92.1. There are both positive and negative deflections, to which are assigned the letters P to T. By convention, within the QRS group *positive* deflections are always designated as *R* and *negative* deflections as *Q* when they precede the R wave or as *S* when they follow it. By contrast, the P and T waves can be either positive or

R. Greger/U. Windhorst (Eds.)
Comprehensive Human Physiology, Vol. 2
© Springer-Verlag Berlin Heidelberg 1996

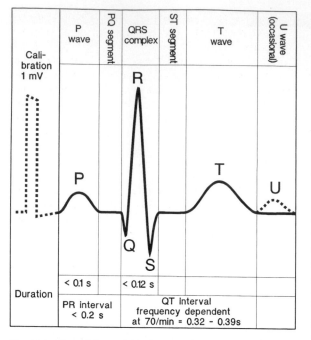

Fig. 92.1. Normal form of the electrocardiogram (ECG) for a single heart beat obtained with bipolar recording from the body surface in the direction of the long axis of the heart (standard lead II of Einthoven, between right arm and left leg). The times below the ECG curve are important limiting values for the duration of distinct parts of the curve

excitation in the terminal branches of the conducting system, where the action potentials last a very long time ([8] see Chap. 91, Fig. 91.7). Another theory attributes the U wave to potentials generated during relaxation of the ventricular myocardium by mechano-electrical coupling [15].

92.2.4 The Normal ECG – Limiting Values

In practice, the interval between the beginning of the P wave and the beginning of the QRS complex is always called the *PR interval* even when it is, strictly speaking, a PQ interval (Fig. 92.1). It is the time elapsing from the onset of atrial exitation to the onset of ventricular excitation, and it should not exceed 0.2 s. A longer PR interval indicates a disturbance in conduction, mostly in the region of the AV node or the bundle of His. When the QRS complex extends over more than 0.12 s, a disturbance of the spread of excitation over the ventricles is indicated. The overall duration of the QT interval depends on heart rate. When the heart rate increases from 40 to 180/min, for example, the QT duration falls from about 0.5 to 0.2 s. The amplitudes of the individual waves are about as follows: P < 0.25 mV; R + S > 0.6 mV; Q < 1/4 of R; T = 1/6 to 2/3 of R.

negative. The distance between two waves is called a *segment* (e.g., the PQ segment extends from the end of the P wave to the beginning of the QRS complex). An *interval* comprises both waves and segments (e.g., the PQ interval, from the beginning of P to the beginning of QRS). The RR interval, between the peaks of two successive R waves, corresponds to the period of the beat cycle and is the reciprocal of beat rate (60/RR interval(s) = beats/min).

92.2.3 Relation Between ECG Waves and Cardiac Excitation

Before the sources of the ECG curve are analyzed in more detail, the general significance of its elements will be considered. An atrial part and a ventricular part can be distinguished. The atrial part begins with the P wave, the expression of the spread of excitation over the two atria. During the subsequent PQ segment the atria as a whole are excited. The dying out of excitation in the atria coincides with the first deflections in the ventricular part of the curve, which extends from the beginning of Q to the end of T. The QRS complex is the expression of the spread of excitation over both ventricles, and the T wave reflects recovery from excitation in the ventricles. The intervening ST segment is analogous to the PQ segment in the atrial part, indicating total excitation of the ventricular myocardium. Occasionally the T wave is followed by a so-called U wave. This has been attributed to the dying out of

92.3 Genesis of the ECG

92.3.1 The Myocardial Fiber as a Dipole

Since access to the body surface is much easier than access to the heart, it is hardly surprising that ECGs were measured extensively before their origin, in terms of electrical events within the heart, could be explained. At the beginning of the present chapter, the genesis of the ECG was attributed in a first approach to the electric current generated in the heart during its excitation. However, such a current flowing through the volume conductor of the body requires a driving force, and this is given by the electric field arising in the heart when excitation spreads and dies out. This complex electric field is generated by many elementary field components originating in the single cardiac fibers. Hence, we shall first consider, with the aid of Fig. 92.2, what goes on in a cardiac fiber during successive phases (a to e) of an excitatory cycle. Performance of two kinds of measurement of potential differences is assumed:

- Transmembrane recording between points A and B using an intracellular microelectrode
- Extracellular recording between points A and C in the extracellular medium along the fiber axis.

When the fiber is at rest (Fig. 92.2a), the transmembrane recording measures the resting potential, with point B negative as against point A. By contrast, no voltage is measured between the extracellular electrodes during this

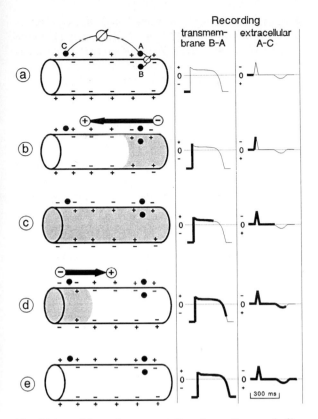

Recording

transmem- brane B-A	extracellular A-C

Fig. 92.2a–e. Schematic representation of a cardiac muscle fiber at rest (**a, e**) and during the passage of a wave of excitation from right to left (**b–d**). On the *right* two kinds of records of potential differences are shown (indicated above the *columns*). The extracellularly recorded potential differences at the front and at the tail of the excitation wave can be represented as dipole vectors pointing from minus to plus in the direction of the longitudinal axis of the fiber. The transmembrane record shows the intracellular action potential of the fiber, and the extracellular recording, the external electrogram

phase: the corresponding graphic registration runs on the zero line. In Fig. 92.2b it is assumed that the fiber were stimulated and excited at its right end. Now, the transmembrane recording shows the upstroke of the action potential. Extracellularly a potential difference also appears between the already excited (point A) and the still unexcited (point C) region of the fiber, with point A negative as against point C. By convention this is indicated as an upward deflection in the graphic record:

In this phase, when considered from the outside, the fiber appears as an electrical dipole that determines the amplitude and direction of an elementary dipole vector indicated by the arrow in Fig. 92.2b. By definition, the dipole vector points from minus to plus and thus represents a potential gradient. If it pointed in the opposite direction, the vector would indicate the direction and strength of the corresponding electric field.

In Fig. 92.2c, it is assumed that excitation has spread along the fiber, all of which is now excited. In the transmembrane recording, this corresponds to the plateau phase of the action potential. Since all regions of the fiber are now

at about the same potential, the extracellular recording measures no potential difference, and the dipole vector disappears.

Hence, a fully excited fiber behaves like a resting one in that it does not generate a dipole vector and thus does not exert any measurable influence on the surrounding medium. Only when recovery from excitation takes place (Fig. 92.2d) will a dipole vector reappear, but it will now point in the opposite direction. When recovery from excitation is complete (Fig. 92.2e) the initial state is reached again.

Thus, the direction of the dipole vector in Fig. 92.2b corresponds to the direction in which the spread of excitation takes place (depolarization vector). By contrast, the vector shown in Fig. 92.2d indicating recovery from excitation points in the opposite direction (repolarization vector). As can be seen from Fig. 92.2e, the extracellular registration bears some similarity with an ECG curve. However, as will be explained later on, it is only a tiny element of the ECG, which is generated by the summation of a great many such elements.

The difference in size of the two vectors depicted in Fig. 92.2 is in reality much greater than could be illustrated: as already shown in Chap. 91, the ventricular conducting system, while branching off, distributes excitation rapidly to different parts of the ventricles. Thus, there are many sections of the ventricular myocardium that are supplied by a single Purkinje fiber ending and are passed through by waves of excitation, which advance continually along them. Such sections are relatively short – about 1 cm long – and are designated the length of free way (*freie Weglänge*) [12]). On the other hand, the length of the wave of excitation – computed from the product of conduction velocity (ca. 1 m/s) and duration of excitation (ca. 0.3 s) – amounts to about 0.3 m. It follows that at each instant of the excitatory cycle only small segments of the excitation waves are actually present in the lengths of free way. At the front of a wave there is a steep gradient corresponding to the upstroke of the action potential (120 mV) over a distance of only ca. 2 mm, and this is symbolized by the depolarization vector. However, during the repolarization phase there are much smaller gradients, because repolarization of the action potential proceeds substantially more slowly and thus extends over a considerably longer distance. These gradients are represented at each instant by the repolarization vector.

92.3.2 Integral Vector – Relationship to the Excitatory Cycle

At any moment during the excitatory process, all the individual vectors in the heart summate to an integral vector. The formation of the integral vector can be compared to the construction of the resultant in a force parallelogram, in which two vectors are replaced by a third one. A large fraction of the vectors will neutralize one another, as observed from outside the system, because they exert equal effects in opposite directions. It has been estimated that in

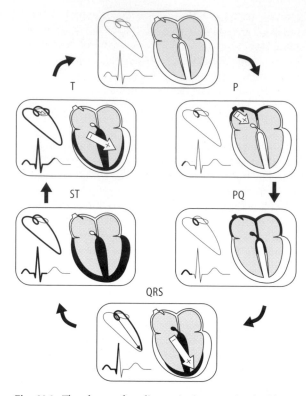

Fig. 92.3. The phases of cardiac excitation associated with particular parts of the ECG. The excited regions are shown in *black*. The *arrows* indicate the momentary direction and relative magnitude of the integral vector. The *curve* above the ECG is the envelope of the vector-tip movement in frontal projection – the frontal vectorcardiogram. The *thick line* of the *curve* indicates the time from the onset of excitation to the time represented by the corresponding diagram

the excitation of the heart at times 90% of the individual vectors balance each other out in this way [14, p. 233].

In Fig. 92.3, some instantaneous integral vectors for successive phases of cardiac excitation are represented. As excitation spreads over the atria (P wave) the predominant direction of spread is from the upper right to the lower left, that is, most of the individual elementary depolarization vectors point in this direction and thus generate an *integral vector* pointing toward the apex of the heart (Fig. 92.3, P). When the entire atria are excited (PQ segment), the potential differences disappear transiently, because all the atrial fibers are in the plateau phase of the action potential (cf. Fig. 92.3, PQ). As mentioned above, the subsequent spread of excitation through the ventricular conducting system produces no appreciable potential difference, because of the small mass of excited cells. Only when the excitation moves into the ventricular myocardium (QRS complex) do demonstrable potential gradients reappear. Spread of excitation over the ventricles begins on the left side of the ventricular septum and generates an integral vector pointing toward the base of the heart (not shown in Fig. 92.3). Shortly thereafter, spread toward the apex predominates (Fig. 92.3, QRS). During this phase excitation moves through the ventricular wall from inside to outside. Spread

through the ventricles is completed with the excitation of a band in a region of the right ventricle at the base of the pulmonary artery, at which time the integral vector points toward the right and up (not shown in Fig. 92.3). While the excitation is spreading over the ventricles, it dies out in the atria. When the entire ventricles are excited the potential differences disappear briefly (ST segment), as they did during atrial excitation, and for the same reasons (Fig. 92.3, ST). During the subsequent ventricular recovery phase (T wave) the direction of the integral vector hardly changes: during the entire process of recovery it points to the left (Fig. 92.3, T).

If repolarization of the ventricles took place in the same sequence as depolarization and at the same rate, the behavior of the integral vector during recovery would be expected to be approximately the opposite of that during the spread of excitation. This is not the case, for the following reasons. First, the process of repolarization is substantially slower than that of depolarization. Moreover, the rates of repolarization are not the same in the different parts of the ventricles. Repolarization occurs sooner at the apex than at the base, and sooner in the subepicardial than in the subendocardial layers of the ventricles. This is an intrinsic property of the myocardial tissues in these regions. A more sophisticated mathematical analysis of cardiac electrical activity has been given elsewhere [11].

92.3.3 Direction and Amplitude of the ECG Deflections

In order to understand the relationship between the behavior of the integral vector and the ECG waves, it will be helpful to consider the electrical field surrounding a dipole in a homogeneous conducting medium with a circular boundary (Fig. 92.4). All points at the same potential lie on the so-called isopotential lines. Figure 92.4a and b shows that the potential difference (= voltage) measurable between two definite points at the boundary of the field depends fundamentally on the relation of the lead axis (the line joining the two points) to the dipole direction. The voltage behaves as the projection of the integral vector onto the lead axis: the voltage is greatest when the two directions are the same, and is zero when they are perpendicular to one another. In principle, this idea can be applied to the human heart (Fig. 92.4c), though in this case the situation is considerably more complicated [6]. One reason is that the body is not an electrically homogeneous medium; another is that the heart does not, as in the ideal case, lie at the exact center of a spherical conductor. These factors mean that the electrical field of the heart is distorted at the surface of the body.

92.3.4 Vector Loops and Vectorcardiography

If the many integral vectors during one cycle of cardiac excitation are thought of as having a common starting point, with their tips connected by a continuous line, the

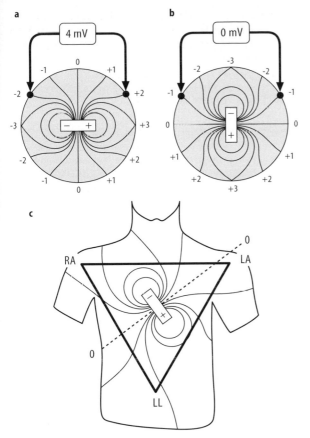

into the sagittal and horizontal planes, the projections of the vector loop onto these planes can be drawn. From any two of these projections the three-dimensional vector loop can be obtained (Fig. 92.5, bottom).

92.4 Lead Systems of ECG Recording

The different curve forms obtained with the arrangement of leads ordinarily used, on extremities and chest wall, are basically projections of the three-dimensional vector loop onto certain lead axes. That is, the vector loop contains just as much information as all these recordings together. For practical purposes, however, the preferred ECG representation is the familiar curve of voltage as a function of time. Apart from the less extensive apparatus required for direct recording with paired leads, the changes in excitation that

Fig. 92.4a,b. Bipolar recording in the electrical field of a dipole within a homogeneous medium with a circular boundary. Relative potential of isopotential lines is indicated at the *edge*. Rotation of the dipole into the vertical position with the electrodes remaining at the same sides (**b**) causes the recorded voltage to fall from 4 to 0 relative units. **c** The electrical field generated by a cardiac dipole at a particular moment, projected onto the anterior wall of the thorax. *RA-LA-LL*, Einthoven's triangle (Fig. 92.7)

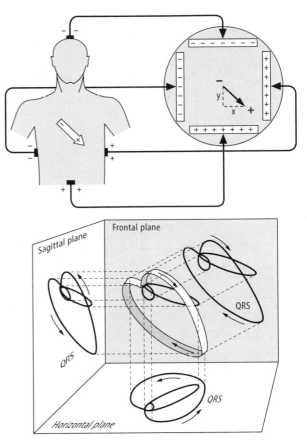

result is a three-dimensional figure, the vector loop. Figure 92.3 illustrates the development of the vector loop in projection onto the frontal plane during a single cycle. By using the recording technique shown in Fig. 92.5 it is possible to display the vector loop directly on an oscilloscope screen. This recording method is called vector-cardiography [3,13]. The principle is illustrated in Fig. 92.5, with an integral vector projected onto the frontal plane taken as an example. One pair of electrodes, arranged horizontally, is connected by way of amplifiers to the vertical plates of the oscilloscope, so as to produce a deflection, *x*, of the cathode ray. Another pair, arranged vertically, is connected to the horizontal plates and causes deflection *y*. The cathode ray is displaced from the middle of the screen as the resultant of these two inputs, so that its position corresponds to the direction and magnitude of the integral vector under study. Because the principle is the same for all the other integral vectors, during a cycle of excitation the beam traces out the enveloping curve for all the vector tips – that is, the vector loop. By shifting the electrode pairs

Fig. 92.5. *Upper diagram.* Principle of vectorcardiography. Pairs of recording electrodes are connected to the paired deflector plates of an oscilloscope by way of preamplifiers. The potential field of the integral vector is projected onto the plates and deflects the cathode ray away from the center of the screen, to a degree and in a direction corresponding to the integral vector at that moment (*arrow*). *Lower diagram.* Three-dimensional vector loop and its projection onto three planes of the body. Note that the direction of rotation of the QRS loop changes in the different projections, being clockwise in the frontal plane and counterclockwise in the other planes

are of practical significance – particularly alterations in the rhythm – are more easily detectable in such records than by the analysis of vector loops. The disadvantage is that several recordings must be compared for an exhaustive evaluation.

A distinction is made between bipolar recordings and so-called unipolar recordings. In the latter, a recording electrode is placed at a defined site on the body surface and the potential with respect to a reference electrode is monitored (cf. Fig. 92.6). This electrode can be thought of as positioned at the null point of the dipole, between positive and negative charge. In clinical practice, the following recording arrangements are the most commonly used today.

Limb leads
 Bipolar: Standard Einthoven's triangle (leads I, II, III)
 Unipolar: Goldberger's augmented limb leads (aVR, aVL, aVF)

Chest leads
 Bipolar: Small chest triangle of Nehb (D, A, I); not shown in Fig. 92.6
 Unipolar: Wilson's precordial leads (V1–V6)

92.4.1 Einthoven's Triangle

Because in bipolar recording from the limbs by the method of Einthoven the arms and legs act as extended electrodes, the actual recording sites are at the junction between limbs and trunk. These three points lie approximately on the angles of an equilateral triangle, and the sides of the triangle represent the lead axes. Figure 92.7 illustrates how the relative amplitudes of the various ECG deflections in the three recordings are derived from the projection of the frontal vector loop onto the associated lead axes. The temporal relationships here are assumed to be those of a normal ECG with a QRS duration of about 100 ms.

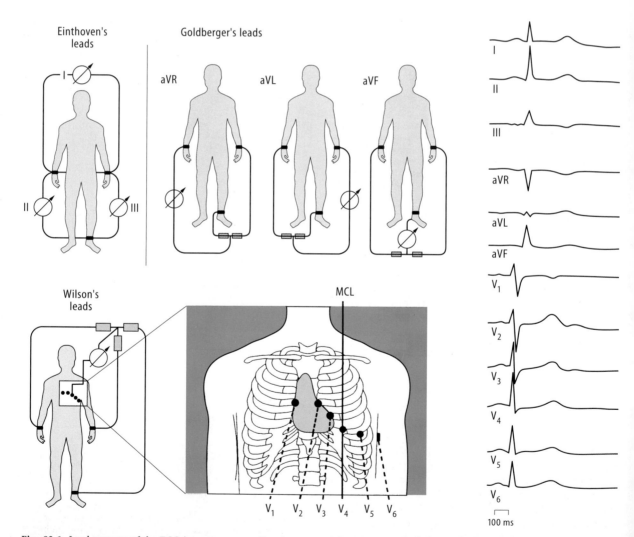

Fig. 92.6. Lead systems of the ECG in common use. For the so-called unipolar leads [7,18] the recording electrode contains the symbol of the instrument. For Wilson's precordial leads, the general arrangement is shown at the *left* and the recording-electrode position at the *right*. *Right*. Typical curves recorded from a healthy subject

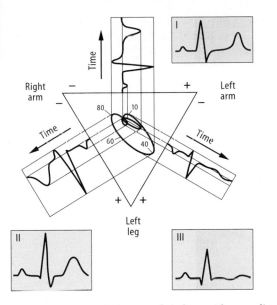

Fig. 92.7. The triangle diagram of Einthoven. The recording sites at the extremities are represented as the corners of an equilateral triangle, and the sides of the triangle correspond to the lead axes. The projection of the frontal vector loop on the three axes is shown. In the gray areas the relative magnitude of the various deflections in each lead (I–III) is indicated by the customary curves. The *figures* attached to the QRS loop indicate the time (ms) from the beginning of this loop

92.4.2 Types of QRS-Axis Orientation

As Figs. 92.3 and 92.7 show, the frontal vector loop has an elongated shape. The direction of the largest integral vector (the chief vector) during the spread of excitation is called the electrical axis of the heart. When the spread of excitation is normal its direction in frontal projection agrees well with the anatomical long axis of the heart. Therefore limb recordings can be used to infer the orientation of the heart. The various categories are based on the angle α between the electrical axis and the horizontal (Fig. 92.8). In the normal or intermediate range (shown in Fig. 92.7) the angle to the horizontal varies from 30° to 60°. Angles above the horizontal are given a negative sign. The general categories of QRS-axis orientation are:

- Intermediate range ($30° < α < 60°$)
- Horizontal range ($-30° < α < 30°$)
- Vertical range ($60° < α < 110°$)
- Left axis deviation ($-30° < α$)
- Right axis deviation ($α > 110°$).

For the construction of the electrical axis from the ECG by means of Einthoven's triangle (Fig. 92.8, bottom) two lead pairs are sufficient, as the third can be derived from the other two. At each instant during the excitatory cycle it holds that: deflection in II = deflection in I + deflection in III (downward deflections having negative sign). The electrical axis of the heart coincides approximately with the anatomical axis only when the spread of excitation is normal; under abnormal conditions the two axes can be quite

different. The main direction of the QRS loop then contains no information about the orientation of the heart, but it is still a useful diagnostic characteristic in combination with other signs that indicate alterations in the process of excitation.

92.4.3 Unipolar Limb Leads

In Goldberger's method [7], the voltage measured is that between one extremity – e.g., the right arm (lead aVR) – and a reference electrode formed by voltage division between the two other limbs (cf. Fig. 92.6). With aVR recording, the lead axis on which the vector loop is projected is represented by the line bisecting the angle between I and II in the Einthoven triangle (Fig. 92.9a). The axes for aVL and aVF are found in the analogous way. The terminology derives from a system no longer in widespread use, in which V stands for voltage with respect to a reference electrode and L, R, and F stand for recording electrodes on left arm, right arm and left leg; and the "a" in aVR stands for "augmented" (the recorded voltage is greater in this method). In the diagram shown in Fig. 92.9b the directions of the bipolar and unipolar limb leads have been shifted, without change in orientation, so that they all intersect the origin of the vector loop. It is evident that each lead line forms an angle of 30° with those on either side. This hexaxial reference system provides all the essential information contained in the frontal vector loop.

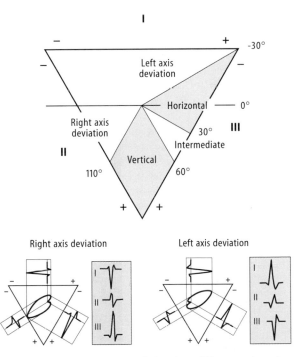

Fig. 92.8. Division of the frontal plane into different regions depending on the orientation of the mean QRS vector given by the angle α. The normal or intermediate range corresponds to an angle α between 30° and 60°. *Bottom.* Direction and relative magnitude of the QRS deflections with right and left axis deviation

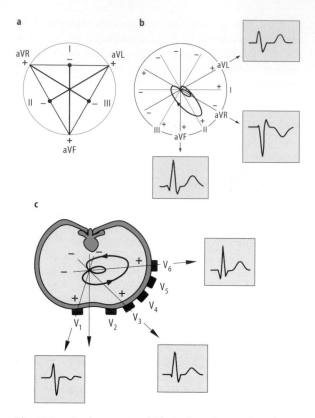

Fig. 92.9. a Lead axes onto which the frontal vector loop is projected with Goldberger's unipolar limb leads. **b** Summary of axis orientations with the unipolar and bipolar limb leads. It can be seen that lead aVR is an exception to the usual polarity rule. **c** Cross section through the thorax at the level of the heart, indicating the lead axes onto which the horizontal vector loop is projected with Wilson's precordial leads. Three sample records are shown (*V1, V3, V6*)

92.4.4 Unipolar Precordial Leads

Whereas the limb leads just described are fundamentally related to the frontal projection of the vector loop, the unipolar precordial leads of Wilson [18] provide information chiefly about the horizontal vector projection. A reference electrode ("Wilson's central terminal") is produced by connecting the electrodes at the three limbs with $5000\,\Omega$ resistors; an exploring electrode records from specific points on the chest at the level of the heart (cf. Fig. 92.9c). In most circumstances, potentials ahead of an advancing excitation wave are more positive than the potential at Wilson's central terminal, and potentials behind the excitation wave are more negative. Figure 92.9c illustrates the lead axes onto which the vector loop is projected with the recording electrode in different positions. For reasons outlined above, a positive deflection is seen when the instantaneous vector projected onto the appropriate axis points toward the recording site. If it points in the opposite direction, the deflection is negative. The onset of a shift in the negative direction thus indicates the moment when the vector loop switches from movement toward the recording

site to movement in the opposite direction. This moment is of special diagnostic significance, for example in the case of delayed excitation attributable to a disturbed spread of excitation in certain regions.

92.5 Use of the ECG in Diagnosis

92.5.1 Basic Information Yielded by the ECG

The ECG is an extremely useful tool in cardiological practice, because it reveals changes in the excitatory process that cause or result from impairment of the heart's activity. From routine ECG recordings the physician can obtain information of the following basic kinds:

- Heart rate. Differentiation between the normal rate (60–90/min at rest), tachycardia (over 90/min) and bradycardia (below 60/min).
- Origin of excitation. Decision as to whether the effective pacemaker is in the SA node or in the atria, in the AV node or in the right or left ventricle.
- Abnormal rhythms. Distinction among the various kinds and sources (sinus arrhythmia, supraventricular and ventricular ectopic beats, flutter and fibrillation).
- Abnormal conduction. Differentiation on the basis of degree and localization, delay or blockage of conduction (sinoatrial block, AV block, right or left bundle-branch block, fascicular block, or combinations of these).
- QRS-axis orientation. Indication of anatomical position of the heart; pathologic types can indicate additional changes in the process of excitation (unilateral hypertrophy, bundle-branch block, etc.).
- Extracardial influences. Evidence of autonomic effects, metabolic and endocrine abnormalities, electrolyte changes, poisoning, drug action (digitalis) etc.
- Primary cardiac impairment. Indication of inadequate coronary circulation, myocardial O_2 deficiency, inflammation, influences of general pathologic states, traumas, innate or acquired cardiac malfunctions, etc.
- Myocardial infarction (complete interruption of coronary circulation in a circumscribed area). Evidence regarding localization, extent and progress.

It should, however, be absolutely clear that deviations from the normal ECG – except for a few typical modifications of rhythmicity or conduction – as a rule give only tentative indications that a pathologic state may exist. Whether an ECG is to be regarded as pathologic or not can often be decided only on the basis of the total clinical picture. In no case is a final decision as to the cause of the observed deviations possible from examination of the ECG alone.

92.5.2 Examples of ECG Abnormality

A few characteristic examples follow, to indicate how disturbances of rhythmicity or conduction can be reflected in

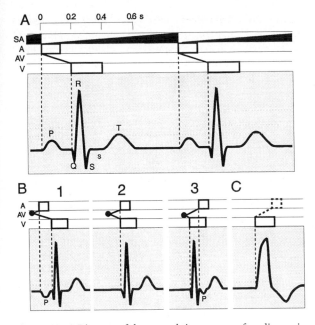

Fig. 92.10. A Diagram of the normal time course of cardiac excitation. The successive stages in the spread of excitation are shown from top to bottom, with the absolute refractory period of atria (*A*) and ventricles (*V*) indicated along the *abscissa*. In the *bar SA* the rhythmic discharge of the SA node is symbolized; *A V* summarizes the total atrioventricular conduction. **B** (*1–3*) Excitation generated at various parts of the AV junctional region, with retrograde excitation of the atria causing a negative P wave. In (*2*) atrial excitation coincides with QRS. **C** Excitation originating in peripheral regions of the ventricles spreads more slowly and the QRS complex is severely deformed. Conduction back into the atria is possible

the ECG. The recordings, where not otherwise indicated, are from Einthoven's limb lead II (cf. Fig. 92.1).

SA Rhythm. As a basis for comparison, let us first consider the normal ECG (Fig. 92.10A), with the pacemaker in the SA node and the QRS complex preceded by a P wave of normal shape. Above the ECG trace in Fig. 92.10A, the process of excitation is diagrammed in a way that has proved useful in characterizing impairments of rhythmicity or conduction. The successive stages in the spread of excitation are shown from top to bottom, and the duration of the absolute refractory period in atria and ventricles is represented along the abscissa.

Rhythms Originating in the AV Junction (Fig. 92.10B). An automatic focus in the AV junctional region (the AV node itself and the immediately adjacent conductile tissue) sends excitation back into the atria (including the SA node) as well as forward into the ventricles. Because excitation spreads through the atria in the opposite direction to normal, the P wave is negative. The QRS complex is unchanged, conduction occurring normally. Depending on the degree to which the retrograde atrial excitation is delayed with respect to the onset of ventricular excitation, the

negative P wave can precede the QRS complex (Fig. 92.10B1) disappear in it (Fig. 92.10B2) or follow it (Fig. 92.10.B3). These variations are designated, not very precisely, as upper, middle, and lower AV junctional rhythms.

Rhythms Originating in the Ventricles. Excitation arising at an ectopic focus in the ventricles spreads over various conductile paths, depending on the source of the excitation and when or where the excitation enters the conducting system. Because myocardial conduction is slower than conduction through the specialized system, the duration of spread through the myocardium is usually considerably extended, which is expressed by prolongation of the QRS complex. Moreover, the differences in conduction path can cause pronounced deformation of QRS (Fig. 92.10C).

Extrasystoles. Beats that fall outside the basic rhythm and temporarily change it are called extrasystoles. These may be supraventricular (SA node, atria, AV node) or ventricular in origin. A ventricular extrasystole is ordinarily followed by a so-called compensatory pause. As shown in Fig. 92.11B, the next regular excitation of the ventricles is prevented because they are still in the absolute refractory period of the extrasystole when the excitatory impulse from the SA node arrives. By the time the next impulse arrives the ventricles have recovered, so that the first postextrasystolic beat occurs in the normal rhythm and the interval between the last normal beat before the extrasystole and the first one after it corresponds exactly to two regular RR intervals (2S in Fig. 92.11B). With supraventricular extrasystoles or ventricular extrasystoles that penetrate back to the SA node, however, the basic rhythm is shifted (Fig. 92.11A). The excitation conducted backward to the SA node interrupts the diastolic depolarization that has begun there, and a new cycle is initiated. These events result in an abrupt phase shift of the basic rhythm. In rare cases an extrasystole can be interpolated halfway between two normal beats and does not disturb the basic rhythm (Fig. 92.11C). In such conditions the basic rhythm must be so slow that the regular interval is longer than an entire beat. Interpolated extrasystoles always arise from a ventricular focus; such excitation cannot propagate over the conducting system (which is still refractory from the previous beat) to the atria and thus cannot interfere with the SA rhythm.

Disturbances of Atrioventricular Conduction. The ECG observed in cases of total AV block is shown in Fig. 92.11D. The atria and ventricles beat independently of one another – the atria at the rate of the SA node, and the ventricles at the lower rate of a tertiary pacemaker. The QRS complex shows the normal configuration if the pacemaker is in the bundle of His, so that excitation spreads over the ventricles in the normal way. Partial AV block is characterized by interruption of conduction at intervals, so that (for example) every second or third beat initiated by the SA node is conducted to the ventricles (2:1 or 3:1 block,

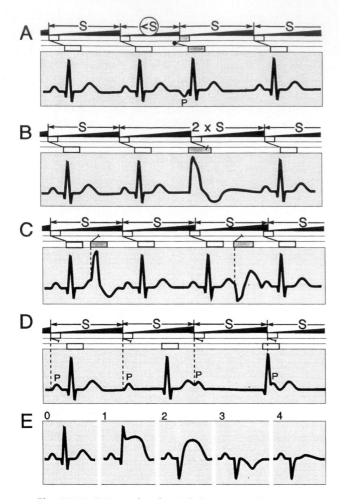

Fig. 92.11A–E. Examples of typical alterations in the ECG. S symbolizes the length of a normal SA interval. **A** Supraventricular extrasystole from the region of the AV node, with incompletely compensating pause (*<S*). **B** Ventricular extrasystole with a fully compensating pause (*2S*). **C** Interpolated ventricular extrasystoles. The differences in form point to different ectopic foci within the ventricles. No conduction back to the SA node. **D** Complete (third-degree) AV block. The ventricular complexes occur independently of the P waves. **E** Progressive ECG deformation during myocardial infarction of the anterior wall of the heart. *0* Normal picture before infarction. *1* Early stage, a few hours after onset. *2* Intermediate stage, after hours to days. *3* After several days to weeks. *4* Final stage months to years after the infarct formation

respectively). In some cases the PR interval increases from beat to beat, until eventually a QRS complex is eliminated and the process begins again (Wenckebach phenomenon). Such disturbances of atrioventricular conduction can readily be produced in experimental conditions in which the resting potential is lowered (increased K^+_e, oxygen deficiency etc.). In the clinically used terminology of the ECG total and partial AV blocks are also called 3rd- or 2nd-degree AV block, respectively, whereas the term 1st-degree AV block is applied when AV conduction (the PR interval) is only prolonged beyond about 200 ms.

Changes in ST Segment and T Wave. Myocardial disturbances due to oxygen deficiency and other influences in general cause a depression of the single-fiber actionpotential plateau before there is a noticeable decrease in the resting potential. In the ECG such effects are evident during the recovery phase as a flattened T wave or one becoming negative or as an elevated or lowered (with respect to the baseline) ST segment. When circulation through a coronary blood vessel is blocked (myocardial infarction) an area of dead tissue develops; its location can usually be determined only by analysis of several recordings, precordial recordings in particular. It must be kept in mind that ECG alterations due to myocardial infarction can change considerably in time (see Fig. 92.11E). The monophasic form of the QRS complex that results from ST elevation, a characteristic of the early stage of infarction, disappears when the infarcted region has become demarcated from the excitable surrounding tissue by the formation of a boundary zone that develops in the intercalated discs due to intracellular acidosis and increase in intracellular Ca^{2+}.

92.6 Flutter and Fibrillation

Ventricular fibrillation is the most common cause of sudden cardiac death, and this arrhythmia is therefore considered in a separate section. In appropriate pathologic conditions ventricular fibrillation can be elicited by a single premature beat. Fibrillation is a primary disturbance of the cardiac excitatory processes, which only secondarily leads to heavy impairment of the mechanical function of the heart. When the same kind of disturbance takes place in the atria it is much less dangerous. However, the mechanisms are quite similar.

92.6.1 Flutter and Fibrillation of the Atria

These are arrhythmias resulting from an uncoordinated spread of excitation over the atria, so that some atrial regions contract at the same time as others are relaxing (functional fragmentation). Atrial flutter is reflected in the ECG by characteristic flutter waves with a regular saw-tooth shape and a frequency of 220–350/min, which take the place of the P wave (Fig. 92.12a). Because of partial AV block due to the refractory period of the ventricular conducting system, normal QRS complexes appear at regular intervals. In the ECG associated with atrial fibrillation (Fig. 92.12b) atrial activity appears only as high-frequency (350–600/min) irregular fluctuations of the baseline. The QRS complexes appear at more or less irregular intervals (absolute arrhythmia), but their configuration remains normal as long as there is no additional disturbance of the ventricular spread of excitation. There is a continuum of intermediate states between atrial flutter and fibrillation. In general the hemodynamic

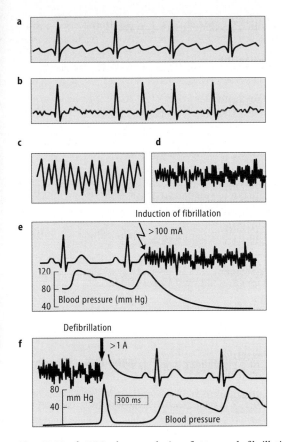

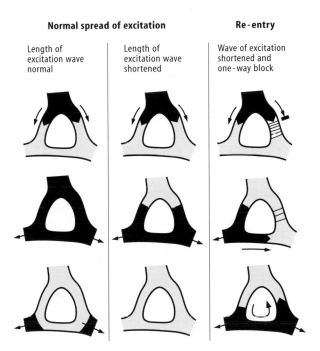

also the most common acute cause of death in electrical accidents [2].

92.6.3 Mechanism of Flutter and Fibrillation

The basic problem in cases of flutter and fibrillation is disruption of electrical activity. The main mechanism is an abnormality in the spread of excitation due to circular movement or re-entry [1,10]. In this situation one or several waves of excitation circle through the ventricular myocardium, a process that does not occur in normal circumstances and for which some particular conditions must obtain. The first essential condition for circulating excitations is given by the network of cardiac musculature. This represents a fundamental difference between cardiac and skeletal muscle, as in the latter the fibers do not form an interconnected network. Further, the length of the excitatory wave (calculated from the product of conduction velocity and refractory time) must be reduced sufficiently to enable re-entry within the myocardial network.

Fig. 92.12a–f. ECG changes during flutter and fibrillation. **a** Atrial flutter with 4:1 conduction to the ventricles. **b** Absolute ventricular arrhythmia due to atrial fibrillation. **c** Ventricular flutter. **d** Ventricular fibrillation. **e** Induction of ventricular fibrillation by an electric shock. The threshold for ventricular fibrillation in the whole body is about 100 mA for direct current pulses applied during the vulnerable period (see Fig. 92.14). **f** Interruption of ventricular fibrillation by a strong countershock of about 1 A applied through the intact chest

effects are slight; the patient is frequently quite unaware of the arrhythmia.

92.6.2 Flutter and Fibrillation of the Ventricles

When the ventricles are affected by the same sort of disturbance, the consequences are much more severe. Because the electrical activity is uncoordinated, the ventricles do not fill and expel the blood effectively. Circulation is arrested, blood pressure falls and unconsciousness ensues; unless circulation is restored within minutes death results. An ECG recorded during ventricular flutter exhibits high-frequency, large-amplitude waves (Fig. 92.12c), whereas the fluctuations associated with ventricular fibrillation are very irregular, changing rapidly in frequency, shape and amplitude (Fig. 92.12d). Flutter and fibrillation can be favored by many kinds of heart damage: oxygen deficiency, coronary occlusion (myocardial infarction), and overdoses of anesthetics or other drugs, etc. Ventricular fibrillation is

Fig. 92.13. Schematic illustrating the necessary conditions for circular movement and the generation of re-entry in cardiac muscle. The *triangle-shaped figure* symbolizes the branched network of myocardial tissue. Refractory areas are represented in *black*. *Left column.* Normal spread of excitation upside down, with a wave of excitation of normal length. Re-entry cannot take place because the length of the excitation wave is greater than the pathway of conduction. *Middle column.* Normal spread of excitation with the wave of excitation shorter than the available pathway. Re-entry is possible but does not occur unless induced by a mechanism such as that shown on the *right. Right column.* Induction of re-entry by transient one-way block of conduction. The block may result from nonhomogeneous refractoriness in the phase of recovery from a preceding excitation (see Fig. 92.14, vulnerable period)

1853

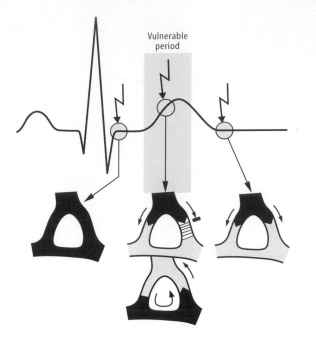

Fig. 92.14. Diagram to explain the vulnerable period of the ventricles, i.e., the phase in which ventricular fibrillation can be induced by a single electric shock. The *triangle-like figures* below the ECG curve have the same meaning as in Fig. 92.13. During the vulnerable period (early phase of repolarization) the conduction pathway is still partially refractory, so that the wave of excitation generated by stimulation can propagate in only one direction. When this region has emerged from the refractory state, re-entry in the opposite direction becomes possible, on the condition that the length of the wave of excitation is not greater than that of the conduction pathway. If stimulated earlier the ventricles would still not be excitable (absolute refractory period), and at a later time re-entry is no longer possible

This can be brought about by shortening of the refractory time, by reduction of the conduction velocity, or by both. However, as demonstrated by Fig. 92.13, shortening of the wave of excitation on its own would not be enough to allow re-entry to occur. There is still another essential condition: the spread of excitation must be temporarily blocked in one direction, so that the excitation fronts do not collide and extinguish one another (Fig. 92.13). Between flutter and fibrillation there are gradations in the degree of functional fragmentation, i.e., in the sizes of the independently activated areas.

92.6.4 Vulnerable Period

Flutter and fibrillation can be induced by a single suprathreshold electrical shock – either experimentally or accidentally – if it occurs in a particular phase of the recovery of excitability [17]. This so-called vulnerable period coincides approximately with the rising flank of the T wave in the ECG (cf. Figs. 92.12e, 92.14). At this time parts of the heart are still absolutely refractory and others, relatively so. When the heart is excited during the relatively refractory period, the refractory period of the following excita-

tion is shorter. Furthermore, as shown in Fig. 92.13, the conduction of excitation can be blocked in one direction [16]. In this situation, then, the conditions necessary for re-entry are met. Spontaneous extrasystoles can give rise to fibrillation in the same way as does stimulation, if they occur during the vulnerable period following previous excitation.

92.6.5 Electrical Defibrillation

Electrical current can trigger flutter and fibrillation of the heart. If correctly applied, however, it can also stop ongoing ventricular flutter or fibrillation. A single brief shock is required, a few amperes in magnitude; when applied through the intact chest wall with large superficial electrodes such a shock usually stops the disorganized contraction instantly (Fig. 92.12F). Electrical defibrillation is the most effective method of abolishing life-threatening ventricular flutter or fibrillation.

The synchronizing effect of this application of current over a large area is probably due to simultaneous stimulation of the myocardial zones that are in an excitable state, so that when the re-entering excitation reaches them they are refractory and further spread is blocked. For electrical defibrillation to be successful, it is of course crucial that the interruption of blood circulation during the preceding period of fibrillation not cause irreversible damage to organs (the brain can be revived if circulation resumes in 8–10 min; cf. Chap. 124). This danger can be averted if a minimal circulation is maintained by external heart massage combined with mouth-to-mouth resuscitation.

References

1. Allessie MA, Bonke FI, Schopman FJG (1973) Circus movement in rabbit atrial muscle as a mechanism of tachycardia. Circ Res 33:54–62
2. Antoni H (1982) Auslösung und Beseitigung von Herzkammerflimmern durch den elektrischen Strom. Funkt Biol Med 1:39–45
3. Burch G, Abildskov J, Cronvich J (1953) Spatial vectorcardiography. Lea and Febinger, Philadelphia
4. Einthoven W (1903) Ein neues Galvanometer. Ann Physik 12:1059–1071
5. Einthoven W (1906) Le telecardiogramme. Arch Int Physiol 4:132–164
6. Frank E (1954) General theory of heart vector projection. Circ Res 2:258–270
7. Goldberger EA (1942) A simple, indifferent, electrocardiographic electrode of zero potential and a technique of obtaining augmented, unipolar extremity leads. Am Heart J 23:483–492
8. Lepeschkin E (1972) Physiologic basis of the U wave. In: Schlant RC, Hurst JW (eds) Advances in electrocardiography. Grune and Stratton, New York, pp 431–447
9. MacFarlane PW, Lawrie TDV (eds) (1989) Comprehensive electrocardiology. Theory and practice in health and disease. Pergamon, New York
10. Mines GR (1914) On circulating excitations in heart muscle and their possible relation to tachycardia and fibrillation. Trans R Soc Can 8 (Sect. IV):43–52

11. Plonsey R, Barr CR (1987) Mathematical modelling of electrical activity of the heart. J Electrocardiol 20:219–226
12. Schaefer H (1951) Das Elektrokardiogramm – Theorie und Klinik. Springer, Berlin Göttingen Heidelberg
13. Schellong F (1939) Grundriß einer klinischen Vektorkardiographie. Springer, Berlin
14. Schütz E (1958) Physiologie des Herzens. Springer, Berlin Göttingen Heidelberg
15. Surawicz B (1982) U wave – the controversial genesis and the clinical significance. Jpn Heart J Suppl 23:17–22
16. Weirich J, Antoni H (1986) Vulnerability of the heart to ventricular fibrillation: basic mechanisms. In: Rupp H (ed) The regulation of heart function. Thieme, Stuttgart, pp 376–396
17. Wiggers CJ, Wegria R (1939) Ventricular fibrillation due to single, localized induction and condenser shocks applied during the vulnerable phase of ventricular systole. An J Physiol 128:500–505
18. Wilson FN, Johnston FD, MacLeod AG, Barker PS (1934) Electrocardiograms that represent the potential variations of a single electrode. Am Heart J 9:447–458

93 Calcium-Mediated Control of Cardiac Contractility at the Cellular Level

G.A. LANGER

Contents

93.1 Introduction

The heart modulates its force development by just two basic mechanisms:

- The Frank-Starling response to changing preload, which involves change in end-diastolic fiber length as the primary intervention
- A change in contractile state in which force changes without a primary change in diastolic fiber length.

The muscle's length-force curve characterizes the first response, and it is now clear that simple increase or decrease in the overlap of actin and myosin filaments is not sufficient to explain the change in force associated with change in resting length of the cell [2]. It is evident that diastolic length influences the calcium (Ca^{2+}) transients which occur with systole by possibly altering the resting level of intracellular Ca^{2+}. In addition, the Ca^{2+} sensitivity of the myofilaments increases with muscle length [1]. The second mechanism, change in contractile state, occurs with interventions such as change in stimulation rate, after-load, ion concentration, autonomic input, and drug administration. No matter what the intervention, all are mediated by changes in cellular Ca^{2+} flux and exchange. This chapter will, then, focus on the mechanisms by which the cardiac cell regulates its Ca^{2+} exchange.

93.2 Ca²⁺ Exchange at the Tissue Level

Prior to analysis of Ca^{2+} movements within the cell, it is important to summarize Ca^{2+} exchange as it occurs within the vascular and interstitial spaces and the relation of these exchanges to the cellular compartment in the intact functional tissue. An understanding of the quantitative aspects of exchange at this level provides a useful perspective for the discussion of cellular exchange to follow.

Figure 93.1 is a schematic representation of Ca^{2+} compartmentation in the mammalian heart [17] perfused at 1.2 l/min × kg wet wt. with 1 mmol/l $[Ca^{2+}]_o$. These conditions would, at a heart rate of 60/min, produce contractions of >50% maximum force. The Ca^{2+} content of the vascular space is 105 μmol/kg wet wt. and at the designated perfusion rate will pass 12–25 μmol Ca^{2+}/s × kg wet wt. through the tissue. The interstitial space contains 500 μmol/kg wet wt. This assumes an aqueous space of 25% tissue volume with an amount of Ca^{2+} bound to anionic interstitial sites equivalent to the free Ca^{2+} in the space. With a $t1/2$ for interstitial exchange of 60 s, the turnover of this space, via the vasculature, is 6 μmol/s × kg wet wt. The cell contains 1200 μmol Ca^{2+}/kg wet wt. of exchange-

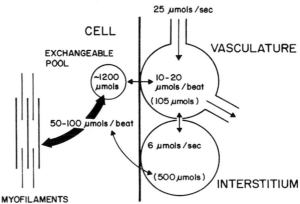

Fig. 93.1. Calcium compartmentation in whole cardiac tissue where coronary flow is 1200 ml/kg × min, $[Ca]_o = 1.0$ mmol/l, heart rate = 60/min, and developed force is >50% maximum. The cell contains an exchangeable Ca^{2+} pool of 1200 μmol from which 50–100 μmol cycle to the myofilaments with each beat. Vascular flow supplies 25 μmol Ca^{2+}/s from which 10–20 μmol/beat can be extracted to enter the cell as measured under brief non-steady-state conditions. The static content of the vascular space is 105 μmol. Interstitial content is 500 μmol and it exchanges with the vascular space at a rate of 6 μmol/s. It also exchanges with the cellular exchangeable pool and serves as a significant reservoir of exchangeable Ca^{2+} in the tissue. Note that the arrangement of the compartments allows for direct vascular-to-cellular exchange as well as vascular-to-interstitial exchange. All values are per kilogram wet wt. tissue. (From [17], with permission)

R. Greger/U. Windhorst (Eds.)
Comprehensive Human Physiology, Vol. 2
© Springer-Verlag Berlin Heidelberg 1996

able Ca^{2+}. There is an additional 250 μmol which is very slowly exchangeable ($t1/2$ of hours), which plays no obvious functional role and has been designated "inexchangeable".

In isolated cells under non-perfusion-limited conditions, Ca^{2+} exchange is found to be between 300–400 μmol/s·kg wet wt. [19]. This is many times both the vascular and interstitial exchange rates measured in the whole tissue and indicates that when the cells are in situ, their Ca^{2+} exchange is markedly perfusion-limited.

Finally, the >50% of maximum contractile force which develops under the specified conditions requires that at least 70–80 μmol Ca^{2+}/kg wet wt. be delivered to the myofilaments with each beat [10]. This value is at least four to five times greater than the net movement of Ca^{2+} from the extracellular space as measured in the intact tissue with each beat [27]. Though the measurements of beat-to-beat movements of Ca^{2+} from extracellular to cellular sites are likely to underestimate the true value (for higher values see [22]), current evidence indicates that augmentation of Ca^{2+} from cellular sites is required to support contraction. This means that a large fraction of the exchangeable cellular Ca^{2+} pool (Fig. 93.1) recycles within the cell before it exchanges with extracellular Ca^{2+} when the cell is functioning in situ.

93.3 Ca^{2+} Exchange at the Cellular Level

In contrast to skeletal muscle, where contractile force is extrinsically modified by the activation of more or less motor units, the heart does not have such a system available. Its force control is intrinsic, i.e., modulated at the level of the individual cell. This stems from the fact that upon excitation all cells within the heart contract. It is an "all or none" response. The heart's intrinsic contractile control system operates on the basis of the cell's ability to control Ca^{2+} movements. There are three major cellular sites involved in control of the Ca^{2+} involved in contraction:

- Sarcolemma (SL)
- Sarcoplasmic reticulum (SR)
- Myofilaments.

Another organelle not involved directly in the cell's excitation-contraction sequence but important in the energetic support of contraction is the mitochondrion. Ca^{2+} plays a significant role in mitochondrial function and this will also be discussed.

93.3.1 Sarcolemma

Ca^{2+} Channels. Two types of channels have been clearly identified in cardiac muscle [23], "T" (transient) and "L" (long-lasting) channels. T channels activate at a trans-sarcolemmal potential of −50 to −60 mV and peak at approximately −30 mV. They inactivate with time constants in the range of 5–30 ms. These channels are more frequently found in cells capable of pacemaking activity (sinus node, atrioventricular node) but are present in ventricular cells as well. These characteristics are consistent with a role in contributing current during the later phases of diastolic depolarization to bring a pacemaker cell to threshold for spike generation. Ca^{2+} current via T channels is much smaller (25%–30%) and decays much more rapidly (four to five times) than L current [25]. It, therefore, contributes little to Ca^{2+} influx during the action potential plateau and little to contraction.

L channels, responsible for the Ca^{2+} current related to contraction, activate at −40 to −30 mV and are operative during the action potential plateau. The current in a cell passing through the L channels is calculated as

$$I_{Ca} = N_t \times P_f \times P_o \times i_{Ca} \tag{93.1}$$

where N_t is the total number of channels, P_f the probability that a channel is available, P_o the probability that the channel will be open given it is available, and i_{Ca} the unitary current through the open pore. Recently discovered G-proteins [31] play a primary role in L channel regulation (see also Chaps. 5.8). It appears that G_s (stimulatory) acts directly on the channel to increase P_o and thereby increase I_{Ca} independent of phosphorylation [38]. G_i (inhibitory) couples muscarinic and adenosine receptors to adenylate cyclase but mediates the inhibition of β-adrenergic stimulation. It has been suggested that the direct G_s effect may regulate basal channel activity and that the cyclic AMP-mediated phosphorylation control via the β-receptor modulates activity above basal [34].

Under physiological conditions when the Na^+/Ca^{2+} exchanger is operating to produce net Ca^{2+} efflux (see below), the only route for net Ca^{2+} entry is via the Ca^{2+} channels. The amount of Ca^{2+} entering can be estimated by integrating the L current over time. At 1 mmol/l $[Ca^{2+}]_o$ approximately 2×10^{-16} mol Ca^{2+} enters a rabbit ventricular cell to produce about 50% maximum force [36]. This would, in a cell 20μm wide and 100μm long, increase intracellular Ca^{2+} concentration to 6 μmol/kg wet wt. cell. This increase in $[Ca^{2+}]_i$, if not augmented, would produce no force [10]. In order to produce the 50% maximum force produced in the presence of 1 mmol/l $[Ca^{2+}]_o$ an additional 50–60 μmol/kg wet wt. is required. This is derived from the SR (see below).

Na^+/Ca^{2+} Exchanger. This system exchanges 3 Na^+ for 1 Ca^{2+} across the sarcolemma and is, therefore, electrogenic. The fact that the exchanger is charged means that its movement is influenced by the transmembrane potential (V_m; cf. Chap. 8). The reversal potential (V_r) at which net ionic movements are reversed is defined by:

$$V_r = \frac{3V_{Na} - 2V_{Ca}}{n - 2} \tag{93.2}$$

where V_{Na} and V_{Ca} are the equilibrium potentials for Na^+ and Ca^{2+}, respectively, and n is the Na^+/Ca^{2+} coupling ratio, which is 3, as noted above. In mammalian heart during diastole V_r is between -30 and -15 mV. This means that at diastolic resting potential of -90 to -80 mV the exchanger would be operating to produce net movement of Na^+ inward and Ca^{2+} outward. Because V_{Ca} becomes less positive during contractile activation, due to the ten-fold or greater increase in $[Ca^{2+}]_i$, V_r for the exchanger becomes more positive during systole (see Eq. 93.2). This tends to maintain operation of the exchanger in the net Ca^{2+} efflux mode throughout the contraction-relaxation cycle. Reversal of net Na^+ and Ca^{2+} movement can occur but only under special circumstances, most of which involve elevation of $[Na^+]_i$ (see below).

The exchanger is capable of operation at a high flux rate. Recent studies [5,8] show that the exchanger turns Ca^{2+} over with a half-time of a few hundred milliseconds. Rich and Langer [30] have identified a discrete subcellular compartment which is specifically dependent upon operation of the Na^+/Ca^{2+} exchanger for its exchange. The compartment contains, at 1 mmol/l $[Ca^{2+}]_o$, approximately 80 μmol Ca^{2+}/kg wet wt. with, therefore, a flux rate of >100 μmol/kg wet wt. xs. This is many times the flux calculated to enter the cell via the Ca^{2+} channel with each beat (see above). This means that the exchanger can maintain intracellular Ca^{2+} at steady-state levels at heart rates above 200 beats/min.

Although the exchanger usually operates to produce net Ca^{2+} efflux, it is capable, under appropriate conditions, of reversing to produce net influx. Leblanc and Hume [20] suggest that a transient rise in $[Na^+]_i$ occurs in a diffusion-restricted region near the intracellular opening of the sarcolemmal Na^+ channel during depolarization. If such occurs, this would shift, for a short period, the exchanger to a net Na^+ efflux – Ca^{2+} influx mode. It is proposed that the Ca^{2+} influx could serve as a trigger for Ca^{2+}-induced Ca^{2+} release from the SR (see below).

Reversal of the usual mode of Na^+/Ca^{2+} exchange is now accepted as the basis for the positive inotropic action of the digitalis glycosides according to the original concept proposed by Repke [29] and Langer [16]. The primary action of the glycosides is the specific inhibition of the sarcolemmal (Na^++K^+)-ATPase, the controlling enzyme of the (Na^++K^+)-pump. Such inhibition causes a rise in Na^+_i and an associated increase in contractile force. No inotropy is found unless Na^+_i increases. The Na^+_i increase causes net Na^+ efflux and net Ca^{2+} influx via the exchanger, the latter responsible for the classical positive inotropic response to the administration of digitalis.

The Na^+/Ca^{2+} exchanger of cardiac muscle has been cloned [24]. The protein has a molecular weight of 120 kD. It is postulated to have 11 membrane-spanning segments, of which several are amphipathic, and it is speculated that the helices are arranged such that hydrophilic surfaces could line an ion translocation pathway through the center of the protein.

At present, it is clear that the Na^+/Ca^{2+} exchanger is the predominant means by which Ca^{2+} exits the cell. It is also a system which reacts rapidly to changing trans-sarcolemmal Na^+ and Ca^{2+} concentrations to alter the net fluxes of these ions and produce significant inotropic responses.

Sarcolemmal Ca^{2+} Pump. This pump is unidirectional – removing Ca^{2+} from the cell. It is activated by calmodulin and has a K_m in the 0.5 μmol/l range. Given this low K_m it is capable of pumping Ca^{2+} out of the cell during diastole. It is, therefore, seen as contributing to maintenance of low diastolic Ca^{2+} concentrations [7]. It is not the primary system which responds to systolic Ca^{2+} influx. That is the job of the rapid and high-capacity Na^+/Ca^{2+} exchanger (see above).

The purified SL pump transports Ca^{2+} with a 1:1 stoichiometry relative to ATP, which is half the efficiency of the SR Ca^{2+} pump. The pump is postulated [6] to have ten transmembrane domains and protrudes into the cytosol with three main units, the central of which contains the "active site" which forms the phosphoenzyme and binds ATP. The C-terminal protruding unit contains the calmodulin-binding domain. Acidic phospholipids activate the pump by interacting with the calmodulin-binding domain and a 22 amino acid region located between transmembrane units 2 and 3. Carafoli [6] remarks that this system seems to have an unusual number of regulatory mechanisms, at least in vitro. Whether or not all of these (calmodulin, acidic phospholipids, and fatty acids and c-AMP-dependent phosphorylation) are active in vivo is not known at present.

In summary, the sarcolemma has three systems important in the regulation of the cell's Ca^{2+} exchange:

- Ca^{2+} channels – Ca^{2+} influx pathway predominantly via the L channels.
- Na^+/Ca^{2+} exchanger – Capable of producing both net Ca^{2+} efflux and influx, but under normal conditions operates in the net efflux mode. Principal system for removal of Ca^{2+} from the cell.
- Sarcolemmal Ca^{2+} pump – An efflux system capable of functioning at low intracellular Ca^{2+} concentrations. Maintains diastolic levels of Ca^{2+}.

93.3.2 Intracellular Systems

Sarcoplasmic Reticulum. There are two morphologically distinct compartments of the SR (cf. Fig. 93.4):

- The "longitudinal" SR which encircles the A and I bands
- The "junctional" SR comprising the cisternae that come into close apposition to the sarcolemma at the level of the transverse (T) tubules or at the peripheral sarcolemma.

The gap between the cisternal and the inner sarcolemmal membrane leaflet is spanned by the so-called "feet" structures [14]. The "feet" (see Fig. 93.4) are 27 nm on each side and about 13 nm in height and demonstrate a central channel which connects with openings on the side of the pillar [35]. It is presently accepted that it is through the feet that Ca^{2+} exits from the cisternal SR by the process of Ca^{2+}-induced Ca^{2+} release (CICR; see also Chaps. 5.44) [10]. The channels in the feet are proposed to have a time- and Ca^{2+}-dependent activation and inactivation [11]. Inactivation is thought to be caused by the binding of Ca^{2+} to a site at the outer surface of the SR. This site has a higher affinity for Ca^{2+} but a lower rate constant than an activating site. Ca^{2+} release is, then, possible in such a system and would occur between the time following the appearance of trigger amounts of Ca^{2+} when the gating mechanism is activated (high rate constant) and not yet inactivated (low rate constant).

The channel should proceed through at least four states as illustrated in Fig. 93.2 [11]:

- Open as a result of the increase in concentration of free Ca^{2+} used as trigger leading to Ca^{2+} binding to the activating site.
- Closed as a consequence of the large increase of free Ca^{2+} concentration resulting from SR Ca^{2+} release. This produces Ca^{2+} binding to the inactivating site.
- Refractory.
- Activatable.

It should be noted that Ca^{2+} release from the feet occurs into a small space or cleft between the cisternal membrane and the inner leaflet of the T tube sarcolemma (see Fig. 93.4). If there is any significant restriction of free diffusion in this region, a transiently high Ca^{2+} concentration (as high as $100\,\mu mol/l$ for 60–100 ms following the action potential spike) could occur. The K_m (Ca^{2+}) of the Na^+/Ca^{2+} exchanger is, $\sim 5\,\mu mol/l$ [26]. Therefore, if the exchanger is located in the region it would be rapidly and maximally stimulated shortly after depolarization and begin to expel Ca^{2+} from the cell within the first 100 ms after excitation. This is, indeed, consistent with the findings of Hilgemann [13].

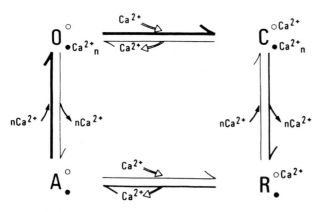

Fig. 93.2. Model for time and Ca^{2+}-induced Ca^{2+} release from the SR. The activating binding site is represented by a *solid circle*, the inactivating binding site is represented by an *unfilled circle*. O, Open; C, closed; R, refractory; A, activatable. The letter n before Ca^{2+} or in subscript corresponds to the cooperative binding of several Ca^{2+} ions. (From [11], with permission)

The longitudinal SR contains the Ca^{2+}-ATPase responsible for pumping Ca^{2+}. A single phosphorylation-dephosphorylation cycle of the ATPase transports two Ca^{2+} ions. The activity of the ATPase is regulated by phospholamban, an integral protein the phosphorylation of which shifts the Ca^{2+} requirement for Ca^{2+} uptake by the SR to lower concentrations. This shift to a lower K_m for Ca^{2+} is additive with two other regulatory systems, cAMP-dependent protein kinase and Ca^{2+}-calmodulin-dependent protein kinase [15]. Dephosphorylated phospholamban inhibits the Ca^{2+} pump and phosphorylation removes this inhibition. The operative K_m (Ca^{2+}) for the pump is in the range of 1–$2\,\mu mol/l$ [12].

The role of inositol-(1,4,5)-triphosphate (IP_3) vis-à-vis the SR in cardiac muscle is minor. The Ca^{2+} release induced by IP_3 is too small to produce effects on contractile function. The contribution of the SR to the provision of the Ca^{2+} involved in contractile activation and control varies with species and conditions. Bers [4] showed that the order of relative dependence on SR Ca^{2+} release is rat ventricle > rabbit ventricle > frog ventricle. The difference between rat and rabbit ventricle is documented by the response of the two tissues to ryanodine, the alkaloid which depletes SR Ca^{2+}. Ryanodine decreases force in rat by 90%, whereas in the rabbit force is diminished by less than 10% by the drug. In guinea pig ventricle, which is similar to the rabbit physiologically and mechanically, full contractile force is maintained in the presence of ryanodine but the rate of force development decreases by 50% [21]. This indicates that one important function of the SR, with its ability to amplify $[Ca^{2+}]_i$ via the process of Ca^{2+}-induced Ca^{2+} release, is to increase the velocity of contraction. The difference in rest contraction amplitude between rat (high) and rabbit (low) may be explained by interaction between SR Ca^{2+} and the Na^+-Ca^{2+} exchanger. Shattock and Bers [32] note that resting intracellular Na^+ activity is almost twice as high in rat as compared to rabbit ventricle. The high level in rat would favor net Ca^{2+} entry via the exchanger during rest with increase in SR Ca^{2+} content and rest potentiation. The lower intracellular Na^+ in the rabbit would favor net loss via the exchanger during rest and decay of force.

In summary, the SR both sequesters and releases Ca^{2+} within the cell. Its role in Ca^{2+} homeostasis varies with respect to region of the heart and with respect to species. The rat is dependent upon SR Ca^{2+} for contraction under all conditions. Rabbit and guinea pig ventricle (similar to human) can develop full contraction in the absence of SR Ca^{2+}, albeit at a slower rate, apparently using Ca^{2+} entering across the sarcolemma. The frog heart, with little or no SR, is completely dependent upon extracellular Ca^{2+} for all of its Ca^{2+} requirement. It should be emphasized, however, that all cardiac muscle is dependent on some trans-sarcolemmal Ca^{2+} entry for maintenance of contraction, even if only in quantity enough to trigger SR Ca^{2+} release.

Myofilaments. Myosin and troponin C (TNC) are the two sites at which Ca^{2+} binds to the myofilaments and represent the end points of cellular Ca^{2+} control of contraction. TNC binds 3 mol Ca^{2+} per mol of which the low affinity site (K_{Ca}

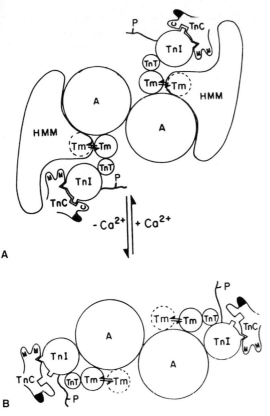

Fig. 93.3A,B. Model for regulation of myofilament interaction by troponin. *A*, actin; *Tm*, tropomyosin; *HMM*, heavy meromyosin (containing ATPase activity); *TnT*, tropomyosin-binding subunit of troponin (TNT); *TnI*, actomyosin-ATPase-inhibiting subunit of troponin (TNI); *TnC*, calcium-binding subunit of troponin (TNC); *M*, Mg^{2+}; *C*, Ca^{2+}. The darkened area in *TnC* represents the region which is structurally analogous to one of two Ca^{2+}-specific sites in skeletal TNC but which does not bind Ca^{2+} in the cardiac subunit. Note association of TNT with tropomyosin and TNI throughout. Note in **B**, in the absence of Ca^{2+}, only the Ca^{2+}-Mg^{2+} sites of TNC contain metal ion (Mg^{2+}) with nothing on the Ca^{2+}-specific site. TNI is shown associated with actin, inhibiting force generation by inhibiting actomyosin interaction. With increasing Ca^{2+} concentration (**A**) contraction is initiated as TNI disassociates from actin with tropomyosin moving away from its blocking position on actin so that heavy meromyosin can associate with actin. Note Ca^{2+} binding on *TnC*. The TNI subunit is shown as phosphorylated, though normal contraction events may not involve phosphorylation at all times. (From [37], with permission)

= $32\,\mu mol/l$) correlates with control of myofilament ATPase rate and activation of force. The control of contraction as mediated by troponin and its interactions is illustrated in Fig. 93.3 [37]. Ca^{2+} binding initiates a sequence of protein interactions which induce a shift of tropomyosin and troponin I (TNI) relative to actin which permits myosin heads to attach to actin. This leads to force development, activation of myosin ATPase with ATP hydrolysis, and the conversion of biochemical energy to mechanical force.

The Ca^{2+} binding to myosin is at high affinity ($300\,\mu mol/l$) but this binding is much too slow to play a role in contractile regulation. Ca^{2+} is, however, indirectly involved in the regulation of myosin light chain phosphorylation, which does affect contraction. Increased cellular Ca^{2+} enhances Ca^{2+} binding to calmodulin, which augments calmodulin-regulated phosphorylation of the light chain [33]. This has been proposed to promote relaxation and would, therefore, affect force level. Light chain phosphorylation-dephosphorylation does not, however, correlate with individual contraction-relaxation cycles [3] and might be more likely to affect longer term "background" tonicity. The interaction of the myofilaments with Ca^{2+} is modified by a number of factors which will affect contraction:

- Protein phosphorylation, involving second messengers of the protein kinase A, protein kinase C, and calmodulin-kinase cascades
- Increased sarcomere length, which enhances binding
- Intracellular pH, temperature, magnesium, and ionic strength.

Thompson et al. [33] provide an example of a sequence of events which converge at the myofilaments through an effect on Ca^{2+}: β-adrenergic stimulation increases c-AMP, which activates c-AMP-dependent protein kinase. In addition to phosphorylating at the SR (see phospholamban under Sarcoplasmic reticulum, below), the kinase catalyzes phosphorylation of TNI, which has the effect of diminishing the affinity of TNC for Ca^{2+}, leading to an increased relaxation rate and a shortened contraction. Ca^{2+} is, therefore, the "linch-pin" of the sequence that follows β-adrenergic stimulation. On the one hand, more Ca^{2+} is released from the SR more rapidly, which produces greater and more rapid force development. On the other hand, the effects on TNC produce an earlier onset of relaxation and an increase in its rate. Thus, the contraction is greater and shorter – characteristics typical of an increase in contractile state of myocardium.

Mitochondria. It is clear that mitochondrial Ca^{2+} does not participate in the beat-to-beat control of cardiac contraction. However, mitochondrial-Ca^{2+} interactions are important in the regulation of the metabolism which supports contraction [9] and, therefore, are included in this chapter. The mitochondria transport Ca^{2+}, with about 20% of cellular Ca^{2+} attributable to these organelles. Ca^{2+} enters the mitochondria down an electrochemical gradient of -160 to $-200\,mV$. Ca^{2+} influx is compensated for by the extrusion of two H^+ ions by the protein pump of the electron-transport chain. This system of influx is termed the "Ca^{2+} uniporter". Ca^{2+} efflux is via a Na^+/Ca^{2+} exchanger which operates as an electroneutral system, in contrast to the electrogenic sarcolemmal system (see above).

Control of mitochondrial respiration by the cellular level of ADP (K_d ~$20\,\mu mol/l$) has been documented when the substrate is pyruvate. Respiratory stimulation during increased demand when glucose or fatty acid is the substrate is not accompanied by significant change in ADP level [9]. There is strong evidence that, in these cases, Ca^{2+}-sensitive dehydrogenases are the intermediaries for respiratory control. PDH (pyruvate dehydrogenase) and OGDH (oxoglutarate dehydrogenase), with K_d for Ca^{2+} activation in the 1–2 µmol/l range, are the key mitochondrial enzymes. NAD-linked isocitrate dehydrogenase (NAD-IDH) is another possible site of control, but its ten-fold higher K_d for Ca^{2+} raises questions as to its effect in in vivo control.

Crompton [9] postulates that mitochondrial matrix-free $[Ca^{2+}]$ does not fluctuate much, as cellular Ca^{2+} cycles between 0.2 and 2 µmol/l with each beat. This is because the relatively low activity of the mitochondrial Ca^{2+} carriers damps the beat-to-beat Ca^{2+} transients. Increases in beating frequency or an increase in duration of the transients, i.e., changed steady-state $[Ca^{2+}]_i$ levels, will increase the matrix-free Ca^{2+} to levels that will affect the dehydrogenases and cause mitochondrial respiration to match the changed energy requirements which, in turn, resulted from whatever stimulus (rate change, catecholamine level, drug administration) caused cytoplasmic $[Ca^{2+}]$ to change in the first place.

93.4 Energetics of Ca^{2+} Exchange

Having discussed subcellular Ca^{2+} movements, it is appropriate to evaluate the energetic cost of contractile control [28]. In the human cardiac muscle, resting metabolism (non-contraction-associated) accounts for about 25% of the total energy output of the beating heart. The remaining 75% is directed toward the support of beating, both force-related and force-independent. The force-independent component is about 20%, a major component of which supports Ca^{2+} movements distributed as follows: SR Ca^{2+} pump 31%; mitochondrial Ca^{2+} transport negligible (this requires little energy under physiological conditions where increased $[Ca^{2+}]_i$ occurs only briefly); SL Ca^{2+} pump 13%; $(Na^{+}+K^{+})$-pump 56%, a portion of which is directed to the fraction of the maintained Na^{+} gradient which drives the Na^{+}/Ca^{2+} exchanger. The amount of force-independent energy relative to resting energy in the active steady state will, of course, increase as beat rate is increased. The total energy requirement (basal or resting plus active) which supports Ca^{2+} movements in the physiologically beating heart is about 20%–25% of the heart's total energy output.

93.5 Intracellular Ca^{2+} Movement

Figure 93.4 summarizes the course of Ca^{2+} movement during a single contraction cycle in a ventricular cell [18]. Referring to the numbered sequence in the figure: (1) Upon depolarization of the cell, L-type Ca^{2+} channels open and Ca^{2+} enters the cell during the plateau phase of the action potential. The entry is at least 6–10 µmol/kg wet wt. The entry via the channel serves to induce Ca^{2+} release via the feet. The total Ca^{2+} release directed to the myofilaments needs to be 50–60 µmol/kg wet wt. to produce >50% of maximum contractile force. (2) Ca^{2+} is released via the feet. This release will produce, at least briefly, a high Ca^{2+} concentration in the restricted region between the SR cistern and the inner sarcolemmal leaflet. This will activate the Na^{+}/Ca^{2+} exchanger to begin Ca^{2+} efflux (within the first 100 ms of the action potential) even as the myofilaments are being activated. Efflux via the exchanger operates with a $t1/2$ of a few hundred milliseconds and has the capacity easily to match the influx via the channel, so as to maintain cellular Ca^{2+} at steady state during repetitive stimulation. (3) As the cell repolarizes, Ca^{2+} is pumped into the longitudinal SR and diffuses to the cistern. It continues to flow out via the Na^{+}/Ca^{2+} exchanger and, during diastole, also via the SL Ca^{2+} pump. (4) Mitochondrial Ca^{2+} contributes less than 1% to trans-sarcolemmal flux. Its exchange rate appears to be controlled at the mitochondrial membrane ($t1/$

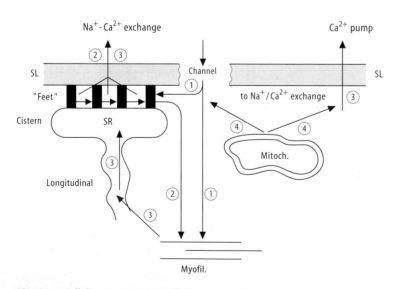

Fig. 93.4. Cellular Ca^{2+} movement during a contraction cycle. See text for description. (From [18], with permission)

$2 \sim 4\,\text{min}$) with eventual exit from the cell via either the Na^+/Ca^{2+} exchanger or the SL Ca^{2+} pump.

93.6 Conclusion and Summary

To understand the means by which the cell controls its Ca^{2+} movements is to understand control of cardiac contractility. The major sites of Ca^{2+} control are:

- L channel of the sarcolemma. Ca^{2+} entry at this site is controlled by the level of channel phosphorylation on a background of G-protein effect.
- The SR Ca^{2+} release is determined by the concentration of Ca^{2+} in the cisternae and the amount of Ca^{2+} and the rate at which it arrives at the releasing "feet" structures.
- The Na^+/Ca^{2+} exchanger. This is controlled predominantly by the cytosolic Na^+ concentration. Increased cytosolic Na^+ concentration decreases Ca^{2+} efflux and augments Ca^{2+} influx, leading to increased contractile force.
- The myofilaments. This is the endpoint of contractile control. Increased Ca^{2+} bound to troponin C (TNC) permits increased actin-myosin interaction and increased force. Ca^{2+} binding to TNC is diminished by phosphorylation of troponin I (TNI). The balance between the two will determine the level of force.

The sites of control are not independent of each other. For example, the Ca^{2+} released by the feet affects the duration of L channel opening, the level of Ca^{2+} efflux by the Na^+/Ca^{2+} exchanger, and, of course, the level of binding by TNC. The system operates on the basis of finely tuned feedback systems superbly designed to adjust the level of force to demand.

References

1. Allen DG, Kentish JC (1985) The cellular basis of the length-tension relation in cardiac muscle. J Mol Cell Cardiol 17:821–840
2. Allen DG, Nichols CG, Smith GL (1988) The effects of changes in muscle length during diastole on the calcium transient in ferret ventricular muscle. J Physiol (Lond) 406:359–370
3. Barany K, Barany M (1977) Phosphorylation of the 18 000-dalton light chain of myosin during a single tetanus of frog muscle. J Biol Chem 252:4752–4754
4. Bers DM (1985) Ca influx and sarcoplasmic reticulum: Ca release in cardiac muscle activation during post rest recovery. Am J Physiol 248:H366–H381
5. Bridge JHB, Smolley JR, Spitzer KW (1990) The relationship between charge movements associated with I_{Ca} and I_{Na-Ca} in cardiac myocytes. Nature 345:618–621
6. Carafoli E (1990) Sarcolemmal Ca pump. In: Langer GA (ed) Calcium and the heart. Raven, New York, pp 355–378
7. Caroni P, Carafoli E (1980) An ATP-dependent calcium pumping system in dog heart sarcolemma. Nature 283:765–767
8. Crespo LM, Grantham CJ, Cannell MB (1990) Kinetics, stoichiometry and role of the Na-Ca exchange mechanism in isolated cardiac myocytes. Nature 345:618–621
9. Crompton M (1990) The role of Ca^{2+} in the function and dysfunction of heart mitochondria. In: Langer GA (ed) Calcium and the heart. Raven, New York, pp 167–198
10. Fabiato A (1983) Calcium induced release of calcium from the cardiac sarcoplasmic reticulum. Am J Physiol 245:C1–C14
11. Fabiato A (1992) Two kinds of calcium-induced release of calcium from the sarcoplasmic reticulum of skinned cardiac cells. In: Frank GE, Bianchi CP, ter Keurs HEDJ (eds) Excitation-contraction coupling in skeletal, cardiac and smooth muscle. Plenum, New York, pp 245–251
12. Feher JJ, Fabiato A (1990) Cardiac sarcoplasmic reticulum: calcium uptake and release. In: Langer GA (ed) Calcium and the heart. Raven, New York, pp 199–268
13. Hilgemann DW (1986) Extracellular calcium transients at single excitation in rabbit atrium measured with tetramethylmurexide. J Gen Physiol 87:707–735
14. Inui M, Fleischer S (1987) Isolation of the receptor from cardiac sarcoplasmic reticulum and identity with feet structures. J Biol Chem 262:15637–15642
15. Kranias EG (1985) Regulation of Ca^{2+} transport by cyclic 3'-5'-AMP-dependent and calcium-calmodulin-dependent phosphorylation of cardiac sarcoplasmic reticulum. Biochim Biophys Acta 844:193–199
16. Langer GA (1971) The intrinsic control of myocardial contraction – ionic factors. N Engl J Med 285:1065–1701
17. Langer GA (1990) Calcium exchange and contractile control. In: Langer GA (ed) Calcium and the heart. Raven, New York, pp 355–378
18. Langer GA (1991) Myocardial calcium compartmentation and contractile control. J Physiol Pharmacol (Pol) 42:29–36
19. Langer GA, Rich TL, Orner FB (1990) Ca exchange under non-perfusion-limited conditions in rat ventricular cells: identification of subcellular compartments. Am J Physiol 259:H592–H602
20. Leblanc N, Hume JR (1990) Sodium current-induced release of calcium from cardiac sarcoplasmic reticulum. Science 248:372–376
21. Lewartowski B, Hansford RG, Langer GA, Lakatta EG (1990) Contraction and sarcoplasmic reticulum Ca^{2+} content in single myocytes of guinea pig heart: effect of ryanodine. Am J Physiol 259:H1222–H1229
22. Lewartowski B, Pytkowski B, Janczewski A (1984) Calcium fraction correlating with contractile force of ventricular muscle of guinea pig heart. Pflugers Arch 401:198–203
23. McCleskey EW, Fox AP, Feldman D, Tsien RW (1986) Different types of Ca channels. J Exp Biol 124:191–201
24. Nicoll DA, Longoni S, Philipson KD (1990) Molecular cloning and functional expression of the cardiac sarcolemmal Na^+-Ca^{2+} exchanger. Science 250:562–565
25. Nilius B, Hess P, Lansman JB, Tsien RW (1985) A novel type of cardiac calcium channel in ventricular cells. Nature 316:443–446
26. Philipson KD, Nishimoto AY (1982) Na^+-Ca^{2+} exchange in inside-out cardiac sarcolemmal vesicles. J Biol Chem 257:511–5117
27. Pierce GN, Rich TL, Langer GA (1987) Transsarcolemmal Ca^{2+} movements associated with contraction of rabbit right ventricular wall. Circ Res 61:809–814
28. Ponce-Hornos JE (1990) Energetics of calcium movements. In: Langer GA (ed) Calcium and the heart. Raven, New York, pp 269–298
29. Repke K (1964) Über den biochemischen Wirkingsmodus von Digitalis. Klin Wochenschr 42:157–165
30. Rich TL, Langer GA (1991) Na-Ca exchange contribution to the "rapid" exchangeable Ca compartment of rat heart cells. FASEB J 5 (II):A1051
31. Robishaw JD, Foster KA (1989) Role of G proteins in the regulation of the cardiovascular system. Annu Rev Physiol 51:229–244

32. Shattock MJ, Bers DM (1989) Rat vs. rabbit ventricle: Ca flux and intracellular Na assessed by ion-selective microeletrodes. Am J Physiol 256:C813–C822

33. Thompson RB, Warber KD, Potter JD (1990) Calcium at the myofilaments. In: Langer GA (ed) Calcium and the heart. Raven, New York, pp 127–165

34. Trautwein W, Hescheler J (1990) Regulation of cardiac L-type calcium current by phosphorylation and G proteins. Annu Rev Physiol 52:257–274

35. Wagenknecht T, Grassucci R, Frank J, Saito A, Inui M, Fleischer S (1989) Three-dimensional architecture of the calcium channel/foot structure of sarcoplasmic reticulum. Nature 338:167–170

36. Wang SY, Winka L, Langer G (1993) Role of calcium current and sarcoplasmic reticulum calcium release in control of myocardial contraction in rat and rabbit myocytes. J Mol Cell Cardiol 25:1339–1347

37. Warber KD, Potter JB (1986) Contractile proteins and phosphorylation. In: Fozzard H, Haber E, Jennings R, Morgan H, Katz A (eds) Heart and cardiovascular system. Raven, New York, pp 779–788

38. Yatani A, Imoto Y, Codina J, Hamilton SL, Brown AM, Birnbaumer L (1987) A G protein directly regulates mammalian cardiac calcium channels. Science 238:1288–1292

94 Peripheral Circulation: Fundamental Concepts, Comparative Aspects of Control in Specific Vascular Sections, and Lymph Flow

J. Holtz

Contents

94.1 Introduction

W(illiam) H(arvey) shows by the way the heart is made that the blood is perpetually driven from the lungs into the aorta. . . . He shows by means of a ligature the passage of the blood from arteries to the veins. Hence it is demonstrated that the perpetual movement of the blood takes place in a circle, owing to the beat of the heart (Lumleian Lecture to the College of Physicians in London on April 17, 1616).

These lecture notes (in the translation by Bayliss from the original in Latin) are considered as the origin of our under-standing of the cardiovascular system as a form of circulation, which collects the blood in the large veins and forces it by the contraction of the right and the left heart through two vascular networks connected in series: the *small or pulmonary circulation* and the *large or peripheral circulation* (Figs. 94.1, 94.2; see also the introductory chapter). In his full account *De motu cordis*, published 12 years later, Harvey (1578–1657) dealt with the older assertions about the cardiovascular system, which assumed some kind of combustion of blood elements in the peripheral tissues in accordance with the phlogiston (combustible) theory derived from the ancient literature. He concluded that these assumptions were "impossible when submitted to specially careful consideration" by making some reasonable estimates on cardiac output per hour and, consequently, on the amount of blood that needed to be formed and consumed according to the phlogiston theory. He correctly described the cardiac function in terms of systole and diastole and demonstrated the direction of the bloodstream in the veins and the function of the venous valves by his frequently quoted ligature experiments. He postulated "invisible porosities in the flesh" and in the lungs to close the circle. However, more than 30 years after his Lumleian lecture he wrote in a letter: "I confess, nay I even assert, that I have never found any visible anastomoses." [81].

Shortly after Harvey's death, Malpighi used a primitive microscope to describe capillaries ("minute twisted divided vessels") in the frog's lung and mesentery after "having sacrificed nearly the whole genus of frogs," and Leuwenhoek described cells and plasma flowing in the capillary stream from arteries to veins [81]. Neither Harvey nor Malpighi had a theory on the function of the lungs, but Harvey's pupils Boyle and Lower used an air pump to prove that the function of the lungs is "to aerate the blood and to change its colour from dark to bright red" [81]. They did this a century before oxygen was discovered by Priestley. Using observations by Priestley and others, Lavoisier concluded that oxygen enters the lung with the "property to combine with blood," that CO_2 leaves the lung, and that N_2 passively enters and leaves the lung without change. However, until his death under the guillotine in 1794, Lavoisier believed that combustion (which he identified as respiration) occurs only in the lung itself. His error was corrected half a century later by the work of Gustav Magnus (1802–1870) in Berlin and Claude Bernard (1813–1878) in Paris.

R. Greger/U. Windhorst (Eds.)
Comprehensive Human Physiology, Vol. 2
© Springer-Verlag Berlin Heidelberg 1996

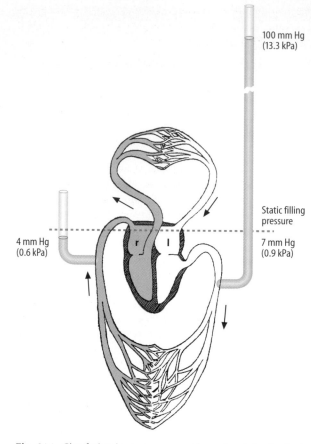

distensible properties of the arteries comprised an "elastic reservoir" during ventricular systole. However, correct quantifications of cardiac output were not possible until A. Fick introduced his principle in Würzburg in 1870 [81]. The French physician J.L.M. Poiseuille (1799–1869) demonstrated small changes in mean blood pressure from the aorta down to the smallest cannulatable arterial branches and hypothesized that the major pressure drop occurred in the arterioles. This led him to analyze the flow of viscous fluids through glass tubes of capillary size, thus initiating microvascular research (see Chap. 88). The biophysics of the arterial system originated in the measurements of arterial pulse wave transmission made by two German brothers, the physicist W. Weber and the anatomist E.H. Weber, contemporaries of Poiseuille.

94.2 Fundamental Concepts

94.2.1 Flow in Rigid, Straight Tubes

Flowing fluids develop internal friction, resulting in resistance against flow, and a pressure difference is necessary to

Fig. 94.1. Circulation in man, as correctly analyzed for the first time by William Harvey in 1616. Approximately 5 l blood is in the circulation of an adult man. Without cardiac actions (and without reflex reactions triggered by the cardiac arrest), the entire circulation is under an hydrostatic pressure of 7 mmHg, which is called the static filling pressure. Under these conditions, only 0.5% of the blood volume is within the arterial tree, while the remainder is in the veins and the pulmonary circulation. The action of the heart forces blood from this volume reservoir (venous system, right heart, pulmonary circulation, and left heart in diastole; the low-pressure system) into the high-pressure system (systolic left heart and arterial tree down to the origin of the capillaries). Thus, the pumping action of the heart lowers the central venous pressure in the veins close to the right atrium to 4 mmHg, while it elevates the mean pressure in the arteries to 100 mmHg. Friction of the blood in the small arterioles at the end of the arterial tree retards the flow along the pressure gradient from the high-pressure system into the volume reservoir of the low-pressure system and acts as a peripheral vascular resistance. *l*, lung; *c*, peripheral circulation; *r*, right atrium; *l*, left atrium; *rv*, right ventricle; *lv*, left ventricle

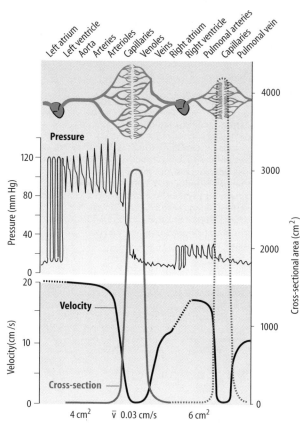

Fig. 94.2. Intravascular pressures, cross-sectional areas, and mean velocities in the human circulation. While the dynamic equilibrium, maintained by pumping action of the left heart and by the peripheral vascular resistance, results in high pressures in the arteries, the pumping action of the right heart causes only a small elevation of pressures in the pulmonary bed. (From [15], with permission)

The beginning of quantitative biophysics of the cardiovascular system was marked by the Reverend Stephen Hales (1677–1761), who directly measured arterial and venous blood pessure in animals. He performed measurements during an exsanguination experiment in a resting horse and he provided some estimates on aortic blood flow velocity, ventricular volumes, and the distribution of blood to peripheral organs. Furthermore, he recognized that the

overcome this resistance. Traditionally, this interrelation is formulated in analogy to Ohm's law for electrical current as follows:

$$\dot{Q} = \Delta P/R \qquad \text{(Eq. 94.1)}$$

Where $\dot{Q}$ is flow, ΔP is pressure difference, and R is resistance. Flow is the volume (ΔV) per time (Δt), i.e., $\dot{Q} = \Delta V/\Delta t$, flowing through a tube with the cross-sectional area (C), while the velocity (u) is the speed of individual fluid elements, varying with the distance from the tubular axis, depending on the velocity profile. The mean velocity ($\hat{u}$) multiplied by the cross-sectional area yields the flow

$$\dot{Q} = \hat{u} \times C \qquad \text{(Eq. 94.2)}$$

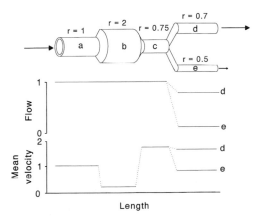

Fig. 94.3. Continuity of flow through series-connected and parallel-connected tubes with different diameters. The mean velocity in the tubes is flow divided by cross-sectional area. While flow through tubes a, b, c, and $d + e$ is identical, flow through e is roughly one quarter of flow through d, since the diameter of tube d is 1.4-fold the diameter of tube e. Note that according to Hagen-Poisieulle's law (see below), conductance of a tube is proportional to the fourth power of the radius. The connecting elements between the tubes are not considered

In a system of series-connected tubes with various cross-sectional areas, the flow through the system is constant (Fig. 94.3):

$$\dot{Q} = \hat{u}_1 \times C_1 = \hat{u}_2 \times C_2 \qquad \text{(Eq. 94.3)}$$

Thus, the velocity in tubular segments connected in series is inversely proportional to the cross-sectional area of the segments. (Note that the cross-sectional area of the entire peripheral capillary bed is roughly 800 times the cross-sectional area of the aorta, which implies that the mean velocity of blood in the aorta is 800 times the mean velocity in the capillaries; see Table 94.1). In such a series-connected tubular system, the resistances of the individual tubes (R_n) add up to form the total resistance (R):

$$R = R_1 + R_2 \ldots R_n \qquad \text{(Eq. 94.4)}$$

In parallel-connected tubes, however, the individual tubular conductances ($1/R_n$) add up to the total conductance:

$$1/R = 1/R_1 + 1/R_2 + \ldots 1/R_n \qquad \text{(Eq. 94.5)}$$

Therefore, the total resistance in a set of parallel connected tubes is smaller than the lowest individual resistance in such a set.

Homogeneous fluids (i.e., Newtonian fluids) have a temperature-dependent constant for the internal friction, the viscosity (η), which is defined by Newton's law of friction:

$$\eta = \tau/\gamma \qquad \text{(Eq. 94.6)}$$

where τ is shear stress and γ is shear rate. The unit of viscosity is the poise (1P = 100 mPa × s). *Shear stress* on a surface is a force per unit area acting in a direction tangential to the surface, and *shear rate* is the velocity gradient between layers of a moving fluid (Fig. 94.4).

The pressure drop during steady flow of an incompressible, Newtonian fluid along a tube of variable cross-section can be attributed to viscous dissipation, dissipation associated with inertial effects, changes in kinetic energy, and

Table 94.1. Quantitative parameters of the human vascular system

	Typical lumen diameter (mm)	Fraction of peripheral vascular resistance (%)	Intravascular blood volume[b] (%)	Total cross-sectional area (cm²)
Aorta	25	} 10	} 10	4
Arteries	3–10			50
Arterioles	0.03–0.2	50–55		700
Capillaries	0.004–0.02	30–35	4	4000
Venoles	0.01–0.1	} 5	} 70	3000
Veins	0.5–10			100
Venae cavae	30			10
Pulmonary artery	25	–	} 8	5
Pulmonary capillaries	–[a]	–		5000
Pulmonary veins	12	–		5

[a] No cylindrical capillaries around lung alveoles (see Fig. 94.29)
[b] Note that 8% of blood volume is in the four chambers of the diastolic heart

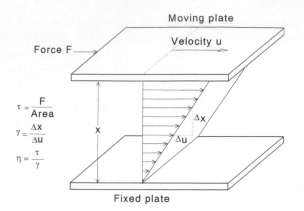

Fig. 94.4. Shear stress and shear rate. In a homogeneous fluid between a fixed plate (area A) and a moving plate (area A, velocity u) with the distance x between the plates, there is a linear velocity gradient ($\Delta u/\Delta x$) of the fluid layers due to the internal friction between the fluid layers, and a force F acts tangentially to the surface of the plates. $F/A = \tau$ = shear stress (Pa = 10 dyn/cm^2); $\delta u/\delta x = \gamma$ = shear rate (s^{-1})

differences in potential energy [22]. For steady flow in a horizontal tube with constant height and constant diameter, changes in kinetic and potential energy do not occur. If the flow conditions allow flow paths of the fluid elements without tortuosity, erratics, and turbulence, inertial effects are not important and the fluid moves in concentric layers (laminae), with the fluid elements following smooth, parallel streamlines (see also Chap. 88). Under this condition of laminar or Poiseuille flow, inertial effects are negligible and the pressure drop ΔP (in Pa) between two points of the tube with radius r (in cm) and with flow $\dot{Q}$ (in cm^3/s), separated by the length l (in cm), depends on viscous dissipation according to Hagen-Poiseuille's law:

$$\Delta P = 8l \times \eta \times Q/(\pi r^4) \qquad \text{(Eq. 94.7)}$$

with the resistance (in mPa × s/cm^3):

$$R = 8l \times \eta/(\pi r^4) \qquad \text{(Eq. 94.8)}$$

The transition from laminar flow to turbulent flow with mixing between the fluid layers of a Newtonian fluid depends on the following:

- Tube diameter ($2r$)
- Mean fluid velocity ($\hat{u}$)
- The ratio of ρ (density) to η (viscosity).

The dimensionless Reynold's number (Re), which is derived from the ratio of inertial dissipation to viscous dissipation, as formulated by Hagen [54] for the pressure drop of fluids in tubes under more general flow conditions, includes the following parameters:

$$\text{Re} = 2r \times \hat{u} \times \rho/\eta \qquad \text{(Eq. 94.9)}$$

If Re exceeds an empirical value of 2000–2200, turbulence occurs in Newtonian fluids with a deformation of the ve-

locity profile (Fig. 94.5), and the linear relation between pressure gradient and volume flow for laminar flow is lost. Under turbulence, the pressure gradient is roughly proportional to the square of volume flow. With pulsatile flow, turbulences can occur also with Reynold's numbers below 2000. Since the mean velocity and the vascular diameter in the microcirculation is fairly low, turbulence does not occur in this section of the vascular tree even with non-Newtonian fluid blood (see Sect. 94.2.2), but in large vessels in humans, the critical Re numbers for turbulence are largely exceeded (Table 94.2).

94.2.2 Blood Flow Within a Tubular Network

The exact conditions for the Hagen-Poiseuille law do not exist in the vascular system; nevertheless, this law is still important for reasonable estimates of resistance to flow in the system. It explains why these resistances are inversely proportional to the fourth power of vessel radius and why the major part of vascular resistance to flow is located in small arterioles and capillaries.

Deviations from the conditions of laminar flow with the parabolic velocity profile (Fig. 94.5) occur due to the curves

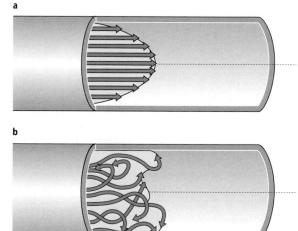

Fig. 94.5a,b. Velocity profiles in a tube for Newtonian fluids under **a** laminar and **b** turbulent flow conditions. Under laminar flow conditions in a tube of length l, the area A between two fluid layers in distance r_i from the tube axis is $A = 2\pi r_i l$ and the frictional force between these two layers (F) is $\eta \times 2\pi r_i l \times \delta u/\delta r$, which is in balance with the pressure difference ΔP acting on πr_i^2; thus $\Delta p \pi r_i^2 = 2\pi r_i l = \delta u/\delta r$. Solutions for $\delta u/\delta r$ and integration yields the following velocity (u): $-\Delta P \times r_i^2/(4\eta \times l) + $ constant; this constant is $\Delta P \times r_i^2/(4\eta \times l)$, since the velocity of the outer fluid layer at inner tube radius r is zero. Thus $u = (r_i^2 - r^2) \times \Delta P/(4\eta \times l)$, which describes the parabolic velocity profile of Newtonian fluids under laminar flow, with the maximal velocity equal to twice the mean velocity. Calculation of volume flow $\dot{Q}$ by integration of the parabolic velocity equation of laminar flow yields Hagen-Poiseuille's law. Note that for Netonian fluids under laminar flow, the wall shear rate (γ_w) is for turbulent flow or non-Newtonian fluids such as blood (see sect. 94.2.2), the velocity profile is altered and $4\dot{Q}/r$ is regarded as an apparent wall shear rate (Redrawn from [15])

Table 94.2. Estimates of Reynold's numbers in large arteries of man at rest

Artery	Inner radius (cm)	V_{max} (cm/s)	Reynold's number
Ascending aorta	1.25	100	6600
Descending aorta	0.90	60	2900
Pulmona artery	1.25	75	5000
Common carotid artery	0.40	50	1000
Renal artery	0.40	50	1000
Femoral artery	0.40	33	700

V_{max}, maximal systolic blood velocity. Reynold's numbers are calculated for an apparent blood viscosity of 4 mPa × s

and ramifications of the vasculature and due to the pulsatile nature of flow, which both tend to distort the parabolic velocity profile. Furthermore, deviations result from the distensibility of the vessel wall and from the complex composition of blood. The flow behavior of the suspension or non-Newtonian fluid blood (for details see Chap. 88) depends on its composition of deformable cells in plasma with various proteins and other solutes, on the flow conditions, and on the boundary geometry [22].

While viscosity in Newtonian fluids ($\eta = \tau/\gamma$, see Eq. 94.6) is a temperature-dependent constant, viscosity is also a function of the operating shear stress in non-Newtonian fluids such as blood:

$$\eta = f(\tau) \qquad \text{(Eq. 94.10)}$$

The rheology of non-Newtonian suspensions is determined by the viscous properties of the suspending medium as well as by the number, the size, the form, and the deformability of the suspended particles, since these parameters interfere with the sliding of fluid layers of the suspending medium parallel to each other. At a given shear stress, the viscosity of a suspension (η_s) in a medium with the viscosity η_m, the particle concentration H, and the form constant of these particles C is:

$$\eta_s = \eta_m[1 + (C \times H)] \qquad \text{(Eq. 94.11)}$$

In the suspension blood, an additional parameter contributes to the rheologic properties of the suspension, namely interactions of large plasma proteins such as fibrinogen and α_2-macroglobulins with the erythrocytes. These proteins contribute to the formation of rouleaux and larger aggregates at low shear rates by bridging adjacent cells in such a manner that the electrostatic repulsive energy of the negatively charged sialic acid on the erythrocyte membranes is counterbalanced [21] (see also Chap. 88). This antirepulsive effect is increased under pathologic conditions (e.g., inflammation), resulting in an increased aggregate formation at low shear rates and in a diminished stability of the suspension, which is measurable as enhanced sedimentation speed of blood.

Apparent Viscosity. This is the viscosity of blood at a given shear rate. At low shear rates, apparent viscosity is high and strongly dependent on shear rate (Fig. 94.6) due to rouleaux-like aggregate formation of the erythrocytes. At high shear rates, the apparent blood viscosity becomes lower and less dependent on shear rate, approaching a value of 3–4 mPa × s in large arteries at normal hematocrit. Apart from disaggregation of erythrocyte rouleaux by high shear stresses, the high deformability of the erythrocytes contributes to this lowering of apparent viscosity at high shear stress: like deformable drops, the cells are orientated and shaped within the fluid layers in such a way that their disturbing effect on laminar flow is minimized.

The deformability of the erythrocytes also contributes to the axial migration of the erythrocytes in small tubes (below 300 μm). In the border zone of such tubes, high velocity gradients induce a rotation and deformation of the erythrocytes and a movement towards the central vascular axis with lower velocity gradients and less deforming forces. Thus, a cell-free border zone acts as a low-viscosity gliding layer for the central stream of erythrocytes, whith the result that the blood viscosity approaches plasma viscosity (Fig. 94.7). Furthermore, the deformability of the erythrocytes allows passage through capillaries with diameters below 7 μm, in which drop-shaped erythrocytes glide in single file on a marginal plasma layer. This lowering of the apparent viscosity with declining vascular diameter is called Fåhraeus-Lindquist effect.

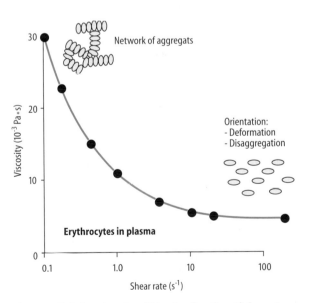

Fig. 94.6. Relative viscosity of blood: a function of shear stress. The large increase in blood viscosity at low shear rates is due to aggregation of erythrocytes, while high shear rates cause disaggregation, deformation, and orientation of the erythrocytes in the streaming plasma. (Modified from [43] and [66]). This dependency of blood viscosity on shear stress is valid for large vessels (see Table 94.1) and for viscosimeters, but not for narrow capillary tubes (see Fig. 94.7). The viscosity of the Newtonian fluid blood plasma is 1.2–1.3 mPa × s or cP (centipoise) at 37°C

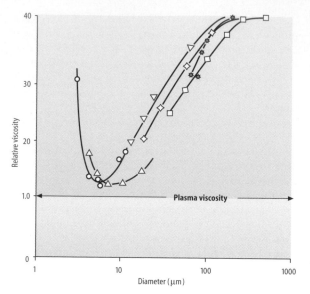

Fig. 94.7. Viscosity of blood declines with decreasing vascular diameters. Due to the axial migration of the erythrocytes in small vessels, a central column of erythrocytes glides on a cell-free rim of plasma, and the apparent viscosity of the "column of blood" approaches plasma viscosity (Fåhraeus-Lindquist effect). At diameters below 4 μm, the limits of the deformability of erythrocytes are reached and blood viscosity approaches infinity

Another consequence of the axial migration of erythrocytes in small tubes is the fact that the mean velocity of erythrocytes is higher than the mean velocity of plasma. This means that the relative volume portion of erythrocytes within a small tube is smaller than its portion of flow. This difference of relative volume portion to the relative flow portion of erythrocytes in a small capillary tube, the Fåhraeus effect (see Chap. 88), implies a lowering of the intracapillary hematocrit without a lowering of the oxygen transport capacity.

94.2.3 Pulsatile Flow in Elastic Tubes

The transmural pressure (i.e., the difference between intraluminal and extraluminal pressure) in an elastic tube causes a distension of the tube and generates a wall stress (τ_w), which opposes further distension by a given pressure and which depends on the transmural pressure (P_{tm}), the thickness (h) of the wall, and the radius (r) of the tube:

$$\tau_w = P_{tm} \times r/h \qquad \text{(Eq. 94.12)}$$

The pressure-induced distension, wall stress, and structural properties of the tube determine its volume elasticity (E') (or its compliance, $1/E'$), which is the ratio of pressure change per volume change in the tube:

$$E' = \Delta P/\Delta V \qquad \text{(Eq. 94.13)}$$

In a fluid-filled, elastic tube under a certain distending transmural pressure, a pulsatile low input, such as the stroke volume ejected by the heart, causes the generation of pulse waves, as is shown in Fig. 94.8. Such a pulse wave consists of three basic phenomena: a pressure pulse, a flow pulse, and a volume pulse (i.e., a pulse of cross-sectional area). In a straight, endless tube with constant physical properties, these three pulses are identical in their curve form. The pulse wave with these three pulse phenomena runs along such a theoretical tube with a constant wave velocity (c), which is a function of the volume (V), the volume elasticity (E'), and the fluid density (ρ) [8]:

$$c = \sqrt{E' \times V/\rho} \qquad \text{(Eq. 94.14)}$$

In a system such as the arterial tree, with varying physical wall properties along the tubes, with curves, diameter changes, and bifurcations, reflections of pulse waves occur and cause superpositions of pulse waves running from and to the heart (see Sect. 94.4.1 on "Arterial Hemodynamics").

94.2.4 Elements of Blood Vessel Walls

The wall of most vessels consists of three layers: the *intima*, with a luminal monolayer of endothelial cells, a basement membrane, and an internal elastic lamina; the *media*, consisting of several interdigitating muscle layers between the internal elastic lamina and the external elastic lamina; and the *adventitia*, which is the layer outside the external lamina and which consists of loose connective tissue (collagen fibers), fibroblasts, mast cells, macrophages, and nerve endings (Fig. 94.9). In large arteries and veins, some arterioles, capillaries, venoles, and lymphatics may occur in the adventia.

The relative thickness of these three layers and their makeup of cells and extracellular material vary considerably within the different sections of the vascular system

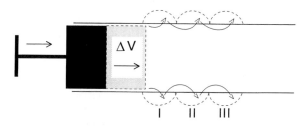

Fig. 94.8. Generation and propagation of a pulse wave, elicited by a volume input into a fluid-filled, elastic tube. The injected fluid volume (ΔV) generates an increase in pressure in the tube, which remains localized to the initial part of the tube (*I*) due to inertia of the tubular fluid and due to the local distension of the tube. In a rigid tube, the entire tubular fluid content would be accelerated and, consequently, the increase in pressure would be larger. In the elastic tube, the pressure gradient between the distended segment *I* and the next segment *II* causes acceleration of ΔV into segment *II* (with distension of this segment) and subsequently into segment *III*

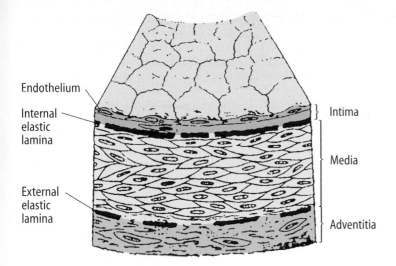

Endothelium

Internal
elastic
lamina

External
elastic
lamina

} Intima

Media

Adventitia

Fig. 94.9. Structure of a muscular artery with intima, media, and adventitia. The *intima* is made up of a continuous monolayer of elliptical endothelial cells with the long axis oriented in the direction of the flow. The endothelial cells are attached to one another by sealing tight junctions and by communicating gap junctions. The endothelium is fixed to an underlying basement membrane of collagen, elastin, fibronectin, laminin, and amorphous ground substance. The elastic internal lamina, a highly fenestrated sheet or meshwork of elastic fibers, is the boundary of the intima. The *media* consists of numerous layers of smooth muscle cells enmeshed in collagen and elastic fibers. The spindleshaped myocytes form junctions with one another (gap junctions and desmosomes). The elastic external lamina, not so well defined as the internal lamina, is the outer boundary of the media. The *adventitia* consists of collagen, elastin, and other extracellular matrix proteins interspersed with cells and nerve endings. Much of the compliance of a blood vessel is determined by the composition and the thickness of the adventitia

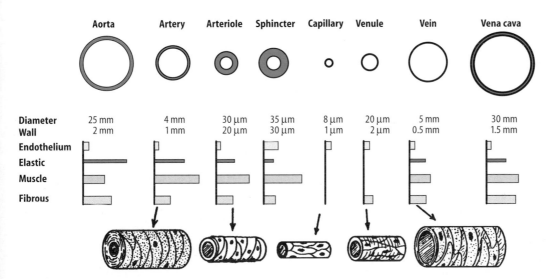

	Aorta	Artery	Arteriole	Sphincter	Capillary	Venule	Vein	Vena cava
Diameter	25 mm	4 mm	30 μm	35 μm	8 μm	20 μm	5 mm	30 mm
Wall	2 mm	1 mm	20 μm	30 μm	1 μm	2 μm	0.5 mm	1.5 mm
Endothelium								
Elastic								
Muscle								
Fibrous								

Fig. 94.10. Size, wall thickness, and relative composition in typical sections of the vasculature. Vessel sizes not depicted to scale. The sum of the relative contributions of the four components add up to total wall thickness

(Fig. 94.10), adapted to the specific functions of these sections (see Sect. 94.4). The large distensible arteries close to the heart have a very high content of elastin, arranged in a characteristic pattern. While intima and adventitia are similar as in a typical muscular artery, the media consists of numerous lamellae. These are muscle layers bounded on both sides by elastic laminae, similar to the internal and external elastic lamina in a muscular artery. Such a large artery can have more than 30 elastic laminae with muscle layers in between, and the outer laminae are penetrated by vasa vasorum (arterioles, capillaries, and venoles). In the muscular arteries, the elastin content is substantially lower and confined largely to the internal and external elastic lamina (Fig. 94.9). Elastic and muscular arteries serve as conduit vessels and carry the flow to the organs. Their volume elasticity (E'), determined by structural components, transmural distending pressure, and wall stress, is important for pulse wave amplitude and arterial hemodynamics (see Sect. 94.4.1). This wall stress is maximal for a given transmural pressure in an artery with full relaxation of the smooth musculature, since the radius (r) is increased and the wall thickness is reduced. In such a relaxed vessels, this circumferentially acting wall stress is carried mainly by the structural elements of the wall such as elastin and collagen fibers. With increasing smooth muscular contraction, wall stress for a given transmural pressure is reduced and is carried mainly by the muscular coat. In vivo, arterial vasomotion due to smooth muscle contraction occurs predominantly in the circumferential direction, and changes in the longitudinal direction are negligible (<1%).

In the arterioles, the thickness of the media relative to the internal diameter is large, although the media may consist of only a few (or of a single) muscle layers. In such vessels, wall stress is carried mainly by the smooth muscles and is rather low due to low ratio of r to h (see above). Small changes in muscular contraction in arterioles can easily induce diameter changes, which are effectively transmitted into changes in vascular resistance to flow (Hagen-Poiseuille's law). Therefore, organ flow is actively regulated at the arteriolar level. Terminal arterioles with a ring of smooth muscle cells can temporarily occlude the arteriole, thus preventing perfusion of the capillaries that follow, and are therefore called *precapillary sphincters*.

The capillaries and venoles have no adventitia or media, although smooth muscle-like pericytes are sparsely distributed along their extraluminal surface. The function of capillaries and venoles is the exchange between blood and interstitial fluid, and this function is dependent on the endothelial barrier functions and on the pre- and postcapillary vasomotor regulation.

The peripheral veins and the large venae cavae have a media with a few layers of smooth muscle layers in their media together with a high content of connective tissue, but wall thickness is thin and wall stress is high in spite of the rather low venous pressure. These vessels act as volume reservoirs of the circulation, and venous valves support backflow of blood to the heart.

Smooth Muscle Cells. These cells are the principal mechanical effectors within the vasculature (see Chaps. 44, 96). Within these elongated, spindle-shaped cells, there is no visible organisation of the contractile proteins, unlike that in skeletal or cardiac muscle. However, some structural and functional relations to striated muscle do exist [55]: thin filaments, composed of actin, tropomyosin, and several thin filament-binding proteins, and thick filaments, consisting of myosin, exist in a ratio of approximately 20:1 (actin to myosin ratio in striated muscle, 6:1; see Chaps. 45, 46). Thin filaments attach to membrane plaques and to cytoplasmatic dense bodies, structures considered as analogous to the Z lines of striated muscle (Fig. 94.11). Thick filaments consist of myosin, and each myosin molecule is a heterodimer, consisting of two heavy chains and two sets of dissimilar light-chain subunits. Myosin–actin interactions by cycling crossbridges with consumption of adenosine triphosphate (ATP) is the basis of smooth muscular contraction. This interaction is regulated by phosphorylation of the myosin light chain by a kinase which requires a Ca^{2+}–calmodulin complex for activation. Crossbridge cycling continues as long as the light chain is phosphorylated, but a G protein-regulated phosphoprotein phosphatase can remove the phosphate from the light chains. Contractile activity of vascular smooth muscle is not only regulated by Ca^{2+}-dependent crossbridge cycling, but also by the latch state. This term describes a regulated state with fully developed tension, but retarding crossbridge cycling until the crossbridges stop cycling and remain attached to the actin with high affinity, thereby maintaining tension with minimal requirements of energy. The heads of the crossbridges appear to be latched to the actin filaments. This mechanism explains why the vascular smooth muscle is able to maintain a high wall stress with minimal expenditure of energy [98]. (For signals, receptors, and signal transduction mediating smooth muscle activation and relaxation, for details of excitation–contraction coupling, and for the mechanisms regulating cytoplasmatic Ca^{2+} levels, see Chaps. 44, 96).

Endothelial Cells. These cells are the paracrine mediators of the vascular system. They represent a functionally dynamic monolayer of cells that play many roles [16]:

- They provide a nonthrombogenic surface by releasing platelet-inhibitory substances such as prostacyclin (PGI_2) and nitric oxide (NO), and they yield binding molecules for inhibitors of the coagulation cascade (see Chap. 83).
- They interact with circulating cells of the immune system and are involved in the activation of these cells (see Chaps. 85, 98).
- They synthesize and release vasoactive signals with dilatory or constrictory actions such as endothelium-derived NO, prostaglandins, and other vasoactive derivatives of arachidonic acid, endothelial hyperpolarizing factor (EDHF), and endothelin (see Sect. 94.3.3).

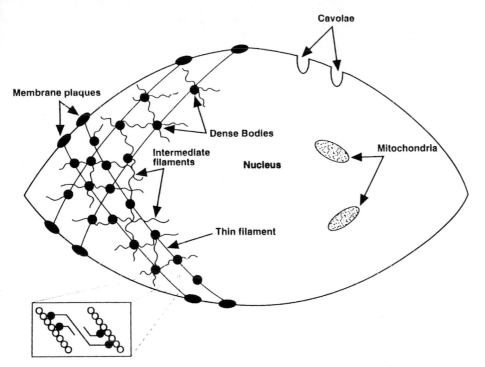

Fig. 94.11. Filaments and contractile proteins in a vascular smooth muscle cell. Thin filaments (actin, tropomyosin, and other proteins) are attached to dense bodies and membrane plaques, considered as analogues to Z lines of striated muscle. Thick myosin filaments are attached between the thin filaments. Intermediate filaments determine the spatial array of the contractile filaments. (From [55], with permission)

- They are actively involved in transporting macromolecules from the vascular lumen into the arterial wall or the extracapillary interstitium (see Chap. 97).
- They act as a permeability barrier (Chaps. 27, 97).
- They synthesize growth factors such as FGF (fibroblast growth factor), TGF-β (transforming growth factor-β), PDGF (platelet-derived growth factor), CSF (colony-stimulating factor), and VEGF (vascular endothelial growth factor) (see Sect. 94.3.2 and Chap. 6).
- They form connective tissue macromolecules such as basement membrane proteins, collagen, and proteoglycans (see Chap. 9).
- They are involved in the metabolism of circulating lipoprotein particles.
- They produce reactive oxygen species such as superoxid anion and can thus oxidatively modify low-density lipoproteins during their transit through the endothelium. This oxidative modification is thought to contribute to the generation of atheromas.

In an intact vessel with a constant direction of flow along the endothelium, the endothelial cells have an elongated, elliptic form, oriented into the direction of flow. They behave like contact-inhibited cells with a very low turnover rate (half-life almost 1 year). At vascular sites without a prevailing flow direction, where turbulences and eddy circulations occur, the endothelial cells have a polygonal shape without detectable flow orientation. At these sites, endothelial cell turnover is higher. The endothelium is not uniform throughout the entire vasculature with regard to form and function. In some organs, the endothelium in the microcirculation is highly fenestrated and thus does not really function as a permeability barrier. In other organs such as the brain, the endothelium has abundant tight junctions, demonstrates only minimal transport by pinocytotic vesicles, and its barrier function is complemented by astrocytes (see Chap. 27). The endothelial cells have a contractile machinery and the ability to actively deform. This is stimulated by histamine and by other inflammatory autacoids. Such stimulation results in the retraction of the rim of endothelial cells and the opening of interendothelial clefts, thereby reducing the barrier function during inflammatory edema formation (see Sect. 94.3.3.2). However, endothelial contractions are not sufficient to contribute to the regulation of vascular resistances.

94.3 Regulatory Principles

94.3.1 Self-Adjustments Within the Vascular Network

The vascular smooth muscle cells can be considered as the mechanical effectors of the vascular network. Within these vascular effectors, the upstream or downstream propagation of constrictory activity or of its inhibition (i.e., vasodilation) between adjacent vascular sections is negligi-

1873

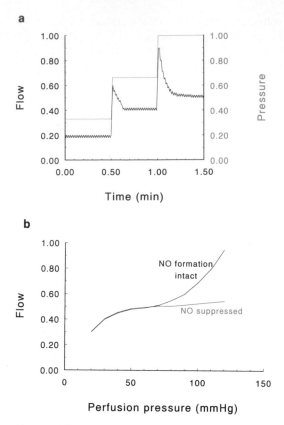

is a special reaction within the broad field of endothelium-mediated modulation of vascular tone (see Sect. 94.3.1.3) and depends on the shear responsiveness of the endothelial cells and the paracrine effectors of the vessel wall. The Bayliss effect is a reaction of the mechanical effectors of the vessel wall, the smooth muscles cells, and contributes to the regulation of basal tone.

94.3.1.1 Myogenic Autoregulation and Basal Vascular Tone

Autoregulation in the broadest sense of the term is the ability of vascular beds to maintain an approximately constant flow in the face of large changes in perfusion pressure

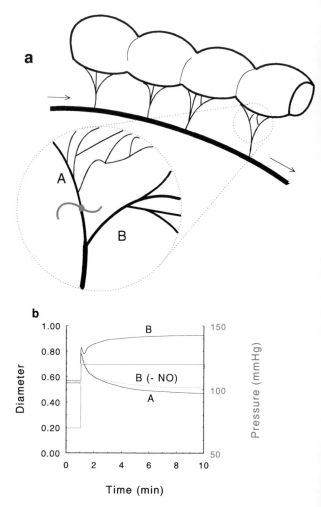

Fig. 94.12a,b. Autoregulation of blood flow during changes in perfusion pressure in an intact organ. In an isolated beating Langendorff heart, coronary perfusion is stepwise elevated and maintained at each level for several minutes (**a**). The steady state pressure–flow relation (**b**) demonstrates a tendency towards an abscissa-parallel "autoregulatory" range. After suppression of the formation to endothelial nitric oxide (NO) and thereby suppression of endothelium-mediated dilation (see Sect. 94.3.1.3), this autoregulatory range is more pronounced. Since the metabolism of the beating heart affects the arterioles, the autoregulation in these experiments results from metabolic vasoregulation and from the myogenic Bayliss effect

ble (see Chaps. 44, 96). Therefore, the arterioles and most of the other vascular sections can be regarded as fairly autonomically acting effectors with minute longitudinal dimensions (in the micrometer range) along the vascular axis (multiunit type of smooth muscle). Thus, the vascular tree is a network of series-coupled and parallel-coupled arrangements of autonomically acting effectors. Furthermore, these effectors are subjected to extremely heterogenous extravascular influences by metabolic factors, by gravity due to alterations in posture and by alterations in extravascular tissue pressure. Nevertheless, the distribution of intravascular pressures along the sections of this network (see Fig. 94.2) is surprisingly stable, given the autonomy of the vascular effectors and the heterogeneity of the influences on these effectors.

This stability is achieved by the concerted cooperation of self-adjusting mechanisms within the vascular tree: the myogenic autoregulation (Bayliss effect) [9] and the flow-dependent dilation (Schretzenmayr effect) [95]. The latter

Fig. 94.13. Myogenic autoregulation of isolated, individual aterioles. In a mesenteric bed, perfusion pressure is elevated and vessel diameter is recorded in arteriole *A* with zero flow due to vessel occlusion distally to the site of diameter recording, and in arteriole *B* with free flow. After transient distension-induced diameter increase, a diameter reduction is sustained in arteriole *A*, which is myogenic autoregulation. In arteriole *B* with free flow, this sustained reduction is only detectable after inhibition of the flow-dependent dilation by suppression of the formation of endothelial nitric oxide (NO) without any regulatory influences via nerves or hormones (Fig. 94.12). Contributing to this general understanding of autoregulation, such as in a prefused, isolated beating heart, is the metabolic vasoregulation (see Sect. 94.3.3.1), which tends to keep flow in a constant relation to the metabolic rate (which can be assumed as fairly constant in an isolated beating heart). However, autoregulation strictly speaking refers to the myogenic response, i.e., contraction of vascular smooth muscle in response to increased transmural pressure, and relaxation in response to decreased transmural pressure [65]. This myogenic autoregulation is called the Bayliss effect after the scientist of the same name, who theorized on such vascular responses [9], although his experimental observations probably included other effects [5] such as metabolic vasoregulation [65] and flow-dependent dilation (see Sect. 94.3.1.2). This myogenic autoregulation is demonstrable in isolated arterioles (Fig. 94.13) and it can result in a pressure–flow relation, as shown in Fig. 94.12, even in absence of metabolic vasoregulation.

This myogenic response is an important means for providing what is called *basal vascular tone*. The prevailing intravascular pressure in the intact circulation is a sufficiently large stimulus to cause partial constriction of the arterioles by this myogenic response, even if the continuous constrictor control of the arterioles by the sympathetic nervous system is abolished. The myogenic response is strongest in the arterioles (see Fig. 94.13), in which the reaction results in a decrease in diameter in the presence of an increase in transmural pressure: autoregulating arterioles cannot be distended continuously by hypertension. However, the reaction is weak and scarcely detectable in large arteries or veins [29,65,74]; these vessels are distensible by augmented transmural pressure.

Since myogenic autoregulation is a steady state response maintaining persistent reductions in arteriolar diameter and hence in arteriolar smooth muscle length in the presence of elevated transmural pressure (Fig. 94.13), the problem arises of how the smooth muscle cells can act as sensors for the augmented transmural pressure, when these same cells constrict in response to the elevated transmural pressure, thereby compensating or even overecompensating for the elevation of distending pressure. Folkow [37] proposed the model of the vascular wall composed of multiunit-type vasculature in such a way that at a given moment some units acted as distension-sensitive sensors, reacting with increased frequence of rhythmic activation to the distension, while the other units simultaneously acted as effectors, keeping the overall vascular diameter low (Fig. 94.14).

Observations have been made on distension-dependent rhythmic activity of arterioles in situ which are compatible with Folkow's concept (Fig. 94.15). However, myogenic autoregulation also works in arterioles, in which rhythmic activity is not detectable by the applied methods of on-line diameter recording. Stretch-activated nonselective cation channels have been described in mammalian vascular smooth muscle, which might contribute to the myogenic response in agreement with Folkow's proposal, but the cellular mechanisms of this response are essentially unknown [74]. In certain vascular preparations, a stretch-induced vasoconstriction might include an endothelium-mediated component, but this is probably a rather remote artifact. Though the mechanism of the myogenic response is not fully understood, it is clear that the pressure-induced vasoconstriction of arterioles in vivo is essentially a reaction of the smooth muscles in the media [74]. Therefore, the term "myogenic response" remains appropriate.

The main physiologic significance of the myogenic response is to maintain capillary pressure and thereby transcapillary fluid movements fairly constant during large alterations of arterial pressure [63], e.g., during alterations in body posture (see Sect. 94.5.1). This is obtained by ad-

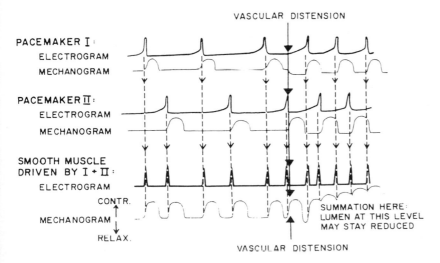

Fig. 94.14. Postulated mechanism for myogenic autoregulation. In vascular smooth muscle of the multiunit type, uncoordinated rhythmic activity of several units determines mean vessel lumen but individual pacemaker units are still able to sense the enhanced transmural pressure by pressure-passive distension, while overall vessel diameter is reduced. (From [37], with permission)

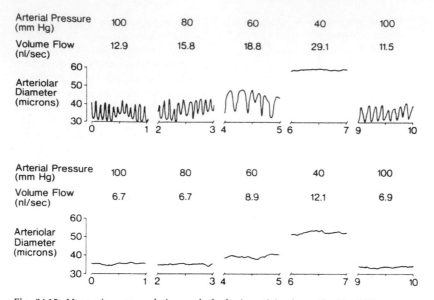

| Arterial Pressure (mm Hg) | 100 | 80 | 60 | 40 | 100 |
| Volume Flow (nl/sec) | 12.9 | 15.8 | 18.8 | 29.1 | 11.5 |

| Arterial Pressure (mm Hg) | 100 | 80 | 60 | 40 | 100 |
| Volume Flow (nl/sec) | 6.7 | 6.7 | 8.9 | 12.1 | 6.9 |

Fig. 94.15. Myogenic autoregulation and rhythmic activity in skeletal muscle arterioles during stepwise reduction of perfusion pressure. The *upper panel* demonstrates with falling transmural pressure the vasodilation and the reduction in the frequency of rhythmic arteriolar activity in agreement with the hypothesis in

Fig. 94.14. However, autoregulatory dilation is also detectable in arterioles of the same preparation (*lower panel*), in which rhythmic arteriolar activity is not detectable from the diameter recordings. (From [65], with permission)

justing precapillary resistance, when arterial pressure is varied. Furthermore, as mentioned previously, the myogenic response contributes to the adjustment of the series-coupled conductances of small arteries and terminal artioles. If the latter are dilated by metabolic vasodilation, while pressure in the large arteries is maintained constant by buffering cardiovascular reflexes, the transmural pressure in the smaller feeding arteries upstream of the metabolically dilated terminal arterioles must be reduced, which is an adequate stimulus for vasodilation by myogenic autoregulation [65].

94.3.1.2 Flow-Dependent Dilation

The shear stress exerted by the streaming blood on the endothelial cell lining of the luminal surface of the vascular wall causes a permanent activation of these endothelial cells, resulting in release of an *endothelium-derived relaxing factor (EDRF)*, of which NO is the active principle [85] (see also Sect. 94.3.1.3). This released NO results in a relaxation of the adjacent smooth muscle and in a dilation of the vessel in response to an augmentation of flow through that vessel, while reduction of flow through the vessel causes reduced release of NO and therefore constriction of the vessel (Fig. 94.16). This flow-dependent dilation, or Schretzenmayr effect, was first observed in 1933 in feline femoral arteries during hyperemia induced by skeletal muscle activity, but Schretzenmayr erroneously regarded this dilation as upstream propagation of metabolic dilation [95]. The endothelial cells synthesize and release several vasoactive factors, among which prostacyclin is also re-

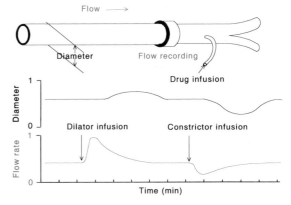

Fig. 94.16. Flow-dependent vasoregulation of arterial diameter. Infusion of vasodilator or vasoconstrictor drugs into the distal artery causes flow changes due to arteriolar dilation or constriction. These flow changes cause alterations in proximal artery diameter at a site which is not in contact with the infused drugs

leased in response to augmented shear stress [40], but the flow-dependent dilation is mainly mediated by the release of endothelial NO [49,57,89].

Stimulators of endothelial formation of NO from L-arginine include physical stimuli – shear stress and mechanical deformation, the pulsatile oscillation of which potentiates the effect on NO formation, and P_{O_2} below 50 mmHg – and endogenous agonists acting via endothelial receptors, i.e., bradykinin, ATP, ADP, histamine, serotonin, norepinephrine, thrombin, acetylcholine, oxytocin, vasopressin, substance P, calcitonin gene-related peptide, and vasoactive intestinal peptide.

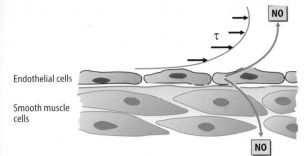

Fig. 94.17. Shear stress at the luminal endothelial surface causes flow-dependent release of nitric oxide. Mechanosensitive endothelial membrane channels have been described and implicated in this reaction [70,82], but a consistent model of shear-induced signal transduction is not available. Furthermore, it appears as if the shear-induced release of nitric oxide involves several modulating mechanisms: flow-dependent convection of adenosine triphosphate (ATP) to G protein-coupled purinoceptors and/or ecto-ATPases [31,69,78,97], and moduation of local bradykinin availability, since a bradykinin receptor-2 antagonist attenuates the flow-dependent dilation in human arteries [51]. (Redrawn from [15])

Shear stress is the primary stimulus for nitric oxide release in flow-dependent dilation (Fig. 94.17). Therefore, this release is not only enhanced by increases in flow through a vessel, but also by any vasoconstriction in face of maintained flow [28,49]. Remember that wall shear rate is proportional to flow per radius (Fig. 94.5), and shear stress is shear rate multiplied by viscosity (see Sect. 94.2.1).

The shear stress-induced release of NO is modulated by the frequency and the amplitude of pulsations in pressure and flow, and this may be relevant for flow regulation in actively contracting striated muscle (see Sect. 94.3.3.1). It coordinates the conductances within an arteriolar network in such a way that the work for perfusion is minimized [50]. The flow-dependent dilation is operative in arteries and arterioles of any caliber and also occurs in veins. In beds with a countercurrent arrangement of venules and arterioles, venular shear stress-induced release of NO contributes to this adjustment of the arteriolar network [92]. For a stable self-adjustment of the vascular network, the Schretzenmayr and the Bayliss effects have to act in combination: during metabolic vasodilation of terminal arterioles, both mechanisms act in concert to maintain the pressure drop along the feeding vessels constant by adjusting the diameter of these feeding vessels (Fig. 94.18). It was this concerted diameter adjustment which both Bayliss and Schretzenmayr observed in their experiments and interpreted according to physiologic knowledge at that time. If arterial pressure is augmented, the two mechanisms act in opposition to each other in the fine tuning of the vascular tree with an optimization of the perfusion pressure dissipation. This fine tuning by two complementary mechanisms of self-adjustment can be seen as a kind of functional counterpart to the heterogeneous extravascular influences acting on the vascular tree.

94.3.1.3 Endothelial Modulation of Vascular Tone

Of the many functions of the endothelium (see Sect. 94.2.4), the modulation of the contractile tone of the vascular smooth muscle is a fundamental regulatory principle of the circulation. Surprisingly, this function was not discovered until 1980 [41,42]. This important vasoregulatory function of the endothelium is brought about by several mechanisms:

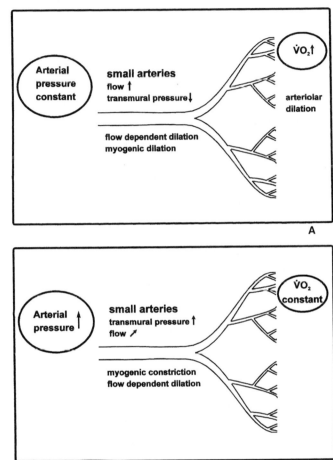

Fig. 94.18.A,B. Synergism and balance of the two mechanisms of vascular self-adjustment. *Top,* During enhanced metabolic activity of an organ, the terminal arterioles are dilated by metabolic vasoregulation, resulting in an increase in flow. This augmented flow will cause an enhanced frictional pressure dissipation along the larger feeding arteries to the hyperemic bed. The enhanced flow to this bed is the adequate stimulus to dilate these feeding arteries by the Schretzenmayr effect, while the transiently reduced transmural pressure is the adequate stimulus to dilate the feeding arteries by the Bayliss effect. *Bottom,* During elevations in perfusion pressure, the myogenic autoregulation, which tends to constrict the arterioles in response to this increase in transmural pressure, and the shear-induced dilation, which will oppose this constriction because of enhanced shear stress, are working in opposition to each other. If unopposed by other mechanisms, the myogenic vasoconstriction in response to augmented transmural pressure would result in a regulatory instability, as is true for a vasodilation in response to increased flow.

- Degradation, conversion, or uptake of vasoactive signal molecules, such as oxidative deamination of catecholamines and serotonin, degradation of platelet-derived ATP and ADP, and hydrolytic cleavage of angiotensin I and bradykinin.
- Formation and release of vasoactive autacoids such as EDRF, NO, prostacyclin, and EDHF (probably an epoxide, formed from arachidonic acid by oxidative pathways via cytochrome P-450 (see Chap. 6)), which all cause vasodilation.
- Formation and release of the constrictory peptide endothelin.
- During inflammatory reactions, endothelial cells can release platelet-activating factor (PAF; see Chap. 6), which acts as a dilator in most vascular beds.

Endothelial cells and the adjacent smooth muscle occasionally form gap junction-like contacts via cellular protrusions penetrating the basement membrane, but an electrotonic propagation of hyperpolarization (and of relaxation of the smooth muscles) via such contacts has never been demonstrated convincingly. The formation and release of NO is the most important dilatory reaction of the endothelium and occurs in endothelia from all vascular sections and in almost all organs. The formation and release of prostacyclin and of EDHF contribute to this fundamental reaction to a varying degree in different vascular beds.

The formation of NO from the amino acid L-arginine is catalyzed by an endothelial enzyme, the constitutively expressed NO synthase; L-citrulline is formed by this reaction (Fig. 94.19). The activity of this endothelial NO synthase is largely stimulated by an increase in the cytosolic concentration of Ca^{2+} ions by a calmodulin-dependent mechanism. Besides shear stress exerted by the blood-stream in flow-dependent dilation (see Sect. 94.3.1.2), a number of other physical stimuli, neurotransmitters, hormones, autacoids, and coagulation factors act as stimulators of the endothelial NO formation by elevating cytosolic Ca^{2+} concentration in the endothelium (see Sect. 94.2.1.3), thereby activating endothelial NO synthase. The relaxation induced by NO is largely due to the activation of soluble guanylyl cyclase in the smooth muscles with enhanced formation of cyclic guanosine monophosphate (cGMP), a second messenger, causing a lowering of cytosolic Ca^{2+} concentration and vasodilation in the smooth muscles. Though endothelium-derived nitric oxide is very labile, it can reach the nerve sympathetic endings in the adventitia of small arteries and arterioles, attenuating the release of constrictory neurotransmitters [77].

The expression of the endothelial NO synthase is upregulated by chronic elevations in shear stress (i.e., by chronic elevations in flow through a vessel) and by estrogens. Endothelium-dependent, NO-mediated dilations are severely disturbed by hypercholesteremia and by chronic smoking. Deficient NO formation is an import mechanism in the formation of atheromas. Analysis of arterial vasomotor reactions in humans by angiography and sonography indicates that local disturbances in NO formation are hallmarks for subsequent atheroma formation. Improvement of endothelial NO formation by chronic pharmacotherapy might become an important preventive strategy against arteriosclerosis.

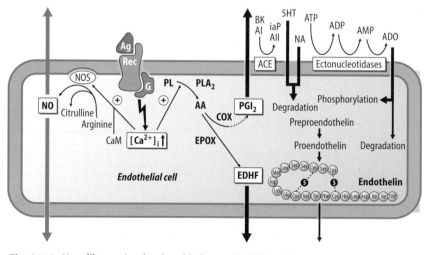

Fig. 94.19. Vasodilatory signals releasable from endothelial cells. Receptor-mediated elevation in sytosolic Ca^{2+} activates the calmodulin (*CaM*)-dependent endothelial nitric oxide (*NO*) synthase (*NOS*) and causes release of arachidonic acid (*AA*) from cell membrane phospholipids by activation of phospholipase (*pl*) $A_2(PLA_2)$. From arachidonic acid, prostacyclin (*PGI$_2$*) and endothelial hyperpolarizing factor (*EDF*) are formed by cyclooxygenase (*COX*) and epoxygenase (*EPOX*). Endothelin is synthesized via precursors and released without storage. The ectoenzyme angiotensin-converting enzyme (*ACE*) inactivates bradykinin (*BK*) to an inactive protein (*iaP*) and forms angiotensin II (*AII*) from angiotensin I (*AI*). Serotonin (*5HT*) and norepinephrine (*NA*) are taken up and degraded within the endothelial cells. Ectonucleotidases degrade adenosine triphosphate (*ATP*) to adenosine (*ADO*), which is also degraded within the endothelial cells. *ADP*, adenosine diphosphate; *AMP*, adenosine monophosphate; *Ag*, agonist; *Rec*, receptor; *G*, G protein. (From [15], with permission)

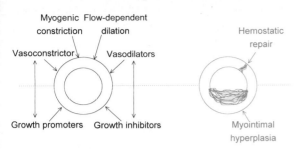

Short-term regulation of function

Myogenic Flow-dependent
constriction dilation

Vasoconstrictor Vasodilators

Hemostatic repair

Growth promoters Growth inhibitors

Myointimal hyperplasia

Long-term regulation of structure

Fig. 94.20. Modulation of vascular caliber by short-term vasoregulation and by long-term remodeling of structure. The mechanisms of vascular self-adjustment, vasoregulatory signals, and hemostatic factors have the potential for long-term changes of vascular structures

94.3.2 Trophic Plasticity of the Vascular Network

In addition to the acute self-adjustment of vascular tone to changes in transmural pressure by the Bayliss effect and to changes in flow or shear stress by the Schretzenmayer effect, the vasculature has the capacity to adapt to chronic changes in circulatory load (pressure overload, shear stress, turbulence) by altering the vascular structure. Such vascular remodeling (Fig. 94.20) can contribute to the maintenance of local tissue homeostasis, but may also become maladaptive, contributing to the progression of hypertension, arteriosclerosis, and restenosis.
Vascular remodeling includes a number of structural changes with adaptive and potentially maladaptive features:

- Concentric thickening of the media of autoregulating small arteries and arterioles in response to chronically elevated transmural pressure. The basis of this thickening is a hyperplasia and a reorganization (remodeling) of the smooth muscles in the media; it helps to reduce the wall stress in spite of elevated transmural pressure. However, due to the enhanced smooth muscle mass, the responsiveness of the wall to constrictor stimuli with normal intensity is augmented. This augmentation helps to perpetuate the hypertension (see Sect. 94.5.1.3).
- Dilative remodeling in response to chronically elevated transmural pressure in large arteries, which do not autoregulate, i.e., which cannot acutely diminish their diameter by the Bayliss effect. This remodeling includes enhanced connective tissue formation, helping to carry the enhanced wall stress, and a reduced expression of endothelial NO synthase, which contributes to the enhanced susceptibility of hypertension-exposed large arteries for arteriosclerosis [58].
- Dilative remodeling of arterial diameter in response to chronic elevations in flow (when expression of

endothelial NO synthase is high) and narrowing remodeling with diameter reduction in response to chronic flow reduction (when NO synthase expression is subnormal). These reactions tend to normalize the wall shear rate (see Fig. 94.5) in states with permanently altered flow. Dilative remodeling may explain the large caliber of coronary arteries of marathon runners, the caliber increase of uterine arteries in pregnancy, and the artery enlargement of patients with an atrioventricular (AV) fistula for hemodialysis; the narrowing remodeling appears to play a role in vascular atrophy caused by inactivity in the vasculature of skeletal muscle of patients with severe heart failure, which contributes to the poor exercise capacity of these patients.
- Myointimal hyperplasia due to migration and hyperplasia of smooth muscles and inflammatory cell infiltration in arteriosclerotic stenoses and restenoses following successful catheter angioplasty includes many elements of the physiologic reactions of coagulation and repair of vascular lesions.
- The growth of collaterals from preexisiting capillaries, bypassing obstructions to flow (arteriosclerosis, cirrhosis of the liver), is more complex than flow-induced dilative remodeling: new wall structures (media and adventitia) are formed in these collaterals. Ruptures of such collaterals (esophagal varices) can result in dangerous bleedings in patients with liver cirrhosis and portal hypertension.

These remodeling reactions appear to develop from acute physiologic reactions such as vasoconstriction, vasodilation, and vascular hemostasis. Many of the

Table 94.3. Vessel wall growth factors

Factor	Effect
Growth/mitosis promoters	
PDGF (−AA, −AB, −BB)	−
FGF family (acidic FGF, basic FGF, and others)	−
IGF-I	−
VEGF (= VPF)	−
HB-EGF	−
IL-I	Vasoconstriction
Angiotensin II	Vasoconstriction
Thrombin	Vasoconstriction
Endothelin	
Growth/mitosis inhibitors	
Type-C natriuretic peptide	Vasodilation
Prostacyclin	Vasodilation
Nitric oxide	Vasodilation
Heparan sulfate	−
Bifunctional grwoth modulators[a]	
TGF-β	
Bradykinin	Indirect vasodilation

PDGF, platelet-derived growth factor; FGF, fibroblast growth factor; IGF, insulin-like growth factor; VEGF, vascular endothelial cell growth factor; VPF, vascular permeability factor; HB-EGF, heparin-binding epidermal growth factor-like growth factor; TGF, transforming growth factor
[a] Growth-promoting or inhibiting action depending on interaction with other factors

growth-modulating factors putatively involved in these remodeling reactions (see Table 94.3) have the potential to acutely regulate vascular tone.

The formation of new capillaries and the rarification of existing capillaries also contribute to the trophic plasticity of the vascular network, but are not remodeling reactions of existing vessels. The formation of new capillaries requires degradation of the extracellular matrix close to a capillary or a venole, the chemotaxis of endothelial cells towards an angiogenic stimulus, and the proliferation of endothelial cells [67]. In these reactions, VEGF, which exists in four different isoforms, plays an important role. It is produced by many cells, but only endothelial cells have receptors for VEGF and react with proliferation in response to activation of these receptors [36]. Formation of new capillaries as the first step of angiogenesis occurs in embryonic development and somatic growth, in response to tissue injury, and, periodically, in the endometrium and the ovary. Pathologic angiogenesis is important in tumor growth, psoriasis, and rheumatoid arthritis; many tumors produce one or several isoforms of VEGF, and VEGF antibodies are effective inhibitors of tumor growth in experimental cancer [36].

94.3.3 Extravascular Regulation of the Vascular Network

In addition to extravascular regulatory influences from tissue metabolism, blood gases, autonomic innervation, and circulating hormones (see Sects. 94.3.3.1–3), extravascular physical influences act on the vascular tree and interfere with the mechanisms of self-adjustment (see Sect. 94.3.1). Examples of such physical forces are gravitation during changes in posture, compressive forces due to contractions of striated muscle, or elevations in interstitial tissue

pressure. Gravitational forces can trigger myogenic autoregulation (see Sect. 94.3.1.1). Under physiologic conditions, elevations in interstitial fluid pressure are not important for the perfusion in most organs (but elevation of interstitial fluid pressure may contribute to the regulation of renal medullary blood flow in the kidney with its rather indistensible capsule). However, a substantial reduction of perfusion results from the compression of vessels by striated muscle contraction, rhythmic in the coronary circulation and rapidly changing in the skeletal muscle circulation. Classically, this effect has been considered only as an extravascular resistance. However, the vessel deformation and the pulsation in flow resulting from such compressions are important enhancers of shear stress-induced release of NO [61,86] and thus contribute to the enhanced perfusion during enhanced mechanical activity of myocardium or skeletal muscle. Therefore, the vascular compressions have to be considered as disturbances or extravascular resistances and, simultaneously, as regulatory factors (see below).

94.3.3.1 Metabolic Vasodilation

In organs with a large range of functional activity and metabolic rate, such as myocardium, skeletal muscle, and exocrine or endocrine glands, there is a linear relation between metabolic rate and tissue perfusion. This linear relation implies that an increase in flow occurs almost concomitantly with an increase in the activity and, consequently, in the metabolic rate of such an organ. This increased flow during increased organ activity is called *functional hyperemia* (Fig. 94.21).

When the perfusion of an organ is stopped temporarily (by clamping the supplying artery in an experiment or by compression of an arm due to body rotation during sleep) and

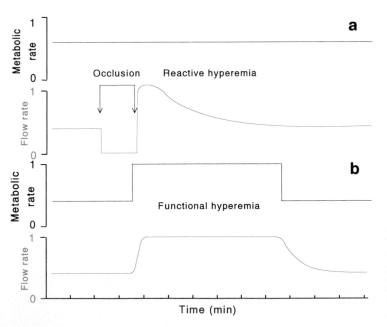

Fig. 94.21a,b. Examples of metabolic vasodilation. **a** In postocclusion reactive hyperaemia, the stimuli for metabolic vasodilation are assumed to be due to interstitial metabolites accumulated during ischemia. **b** In functional hyperemia, the stimuli for metabolic vasodilation are assumed to result from the enhanced metabolic rate. The mechanisms for this dilation have not yet been identified

the organ is rendered ischemic (from the Greek, "inhibition of blood"), the ongoing metabolism results in an accumulation of metabolic waste products in the interstitium of the organ which are no longer washed away by the tissue perfusion. When the inflow into such an organ is restored after a short period of ischemia, the initial postischemic flow rate is several times the preocclusion flow rate, and this enhanced flow gradually declines to the preocclusion rate within 1–2 min. This high flow following a short ischemia is called *reactive hyperemia* (Fig. 94.21). The extent of this reactive hyperemia is dependent on the duration of the ischemia and on the metabolic rate of the organ [11].

Several products or factors with vasodilatory action accumulate in the interstitial fluid of a transiently ischemic tissue and are therefore candidates as mediators of postischemic reactive hyperemia: osmolarity, K^+ concentration, CO_2 pressure, H^+ concentration, hypoxia (strictly speaking, an accumulating deficit of O_2 pressure), and the ATP degradation product adenosine. Adenosine activates A_2 receptors of smooth muscle cells and causes a cAMP-mediated relaxation [102]. Hypoxia causes release of the endothelial vasodilators NO and prostacyclin and thereby causes vasodilation in most organ beds exept for the pulmonary vessels [16]. Elevations of interstitial K^+ concentrations up to 20 mmol/l result in membrane hyperpolarization and relaxation due to activation of an inwardly rectifying K^+ channel, but higher K^+ concentrations cause vasoconstriction because of a diminished transmembrane K^+ gradient. Acidosis and hyperosmolarity probably cause dilations by affecting Ca^{2+} entry. The relative roles of these metabolic vasodilators is not entirely clear and may vary between different organs. However, there is good reason to assume that their combined relaxatory actions on the arterioles and their inhibitory actions on transmitter release from sympathetic nerve endings (see Sect. 94.3.2) are responsible for the major part of the postischemic reactive hyperemia. It is clear that the mechanisms of vascular self-adjustment (see Sect. 94.3.1) also contribute: occlusion of an artery causes lowering of the transmural pressure distal to the occlusion, eliciting myogenic dilation, and the reactive hyperemia must be augmented by flow-dependent dilation.

The dominant, but not exclusive, role of metabolic vasodilation in reactive hyperemia led to the hypothesis that the same mechanism is operative in functional hyperemia. It is very attractive to assume that the close link of metabolic rate and perfusion of an organ is mediated by local dilatory factors associated with the metabolic rate. However, except for a transient elevation of interstitial K^+ concentration and osmolarity in the initial phase of enhanced striated muscle contraction, a causal role of interstitially accumulating metabolic vasodilators for the steady state phase of functional hyperemia has never been demonstrated convincingly [46,47]. This does not completely exclude a role of metabolic factors in functional hyperemia, but positive proof for such a role is lacking. A deformation of the vascular endothelium by the mechanical activity of striated muscle and the gastrointestinal tract can

release NO and prostacyclin, but it is doubtful whether such deformation-induced release of endothelial autacoids is the sole mechanism of steady state functional hyperemia. The metabolic dilation is of paramount importance for the maintenance of adequate tissue perfusion and for the distribution of cardiac output, but the mechanisms of this important regulatory principle remain largely unresolved.

94.3.3.2 Vasoregulation by Inflammation-Associated Autacoids

Autacoids are vasoactive tissue factors or paracrine hormones of different chemical natures (see also Chap. 6) which are involved in local vascular reactions during processes of inflammation and tissue repair, but they may have additional roles in other biological functions. Examples of this group are histamine, kinins, prostaglandins, leukotrienes, thromboxane, epoxides, serotonin, and PAF. It is not clear whether any of these factors has role in the physiologic regulation of functional hyperemia. They might contribute to the regulation of vascular tone in cases of reduced perfusion or during reperfusion after prolonged ischemia, but this is clearly a pathophysiologic rather than a physiologic reaction.

Histamine. This is released from the granules in mast cells and basophile granulocytes (see also Chap. 6). While the direct effect in several vascular smooth muscles is constrictory, it causes dilation of most vascular beds by releasing endothelial NO and by inhibiting norepinephrine release from sympathetic nerve endings. Furthermore, it enhances capillary permeability by causing retraction of capillary endothelial cells, thereby opening interendothelial clefts. This is important for local edema formation in allergic reactions.

Eicosanoids. These are derivatives of arachidonic acid and include prostaglandins, leukotrienes, thromboxane, and epoxides, which can be formed in endothelia and in many other cells (see Chap. 6). Prostacyclin and prostaglandins E and D are vasodilators and inhibitors of sympathetic norepinephrine release and contribute to inflammatory hyperemia. Thromboxane, which is mainly but not exclusively formed in platelets, and prostaglandin $F_{2\alpha}$ are vasoconstrictors involved in hemostasis. Lipoxygenases in leukocytes, macrophages, and vascular cells catalyze the formation of leukotrienes from arachidonic acid. Members of this large family are important mediators of inflammatory reactions, causing chemotaxis and adhesion of leukocytes, endothelial retraction, and cleft openings. Some of the leukotrienes are potent vasoconstrictors. The slow-reacting substance of anaphylaxis is a mixture of several leukotrienes. Epoxides are formed from arachidonic acid by cytochrome P-450-related monoxidases and are considered as hyperpolarizing factors (see Sect. 94.3.1.3).

Serotonin. This vasoconstrictor (see Chap. 6) is released from platelet granules, but also occurs in entero chromaffin cells of the gastrointestinal tract and as a neurotransmitter in the central nervous system. In vascular lesions, platelet-derived serotonin contributes to hemostasis by its direct constrictor action on vascular smooth muscle, while circulating serotonin causes endothelium-dependent vasodilation and, in the gastrointestinal tract, enhanced capillary permeability by endothelial retraction.

Platelet-Activating Factor. PAF (see Chap. 6) is a phospholipid and a much more pluripotent activator of many cell types than its name indicates. It is synthesized in leukocytes and endothelium and is involved in acute inflammatory reactions, in anaphylactic shock, and in the activation of leukocytes during reperfusion damage after prolonged ischemia.

Kinins. Bradykinin and kallidin (see Chap. 6) are small peptides cleaved from kininogens by kallikreins. While their direct action on several types of venous smooth muscle is constrictory, they act as very potent endothelium-dependent dilators by releasing NO prostacyclin, and EHPF when circulating in the plasma. They are involved in inflammation-associated hyperemia and enhance capillary permeability.

Nitric Oxide. As the physiologic mediator of endothelium-mediated dilation, NO has to be included in the group of inflammation-associated vasoregulatory autacoids for two reasons:

- It is the common final pathway of many inflammatory autacoids, which cause vasodilation by receptor-mediated activation of the endothelial constitutive NO synthase, which forms NO from L-arginine in a Ca^{2+}–calmodulin-dependent reaction (see Fig. 94.19).
- Several cytokines of the immune system as well as bacterial endotoxin induce the synthesis of an highly active NO synthase. Typically, this enzyme is induced in macrophages by transcriptional control, which explains its name: inducible macrophage NO synthase. However, the induction is also possible in many other cells which come into contact with these immunomodulators [17]. The activity of this enzyme does not require cytosolic Ca^{2+} or any kind of regulation of its activity; it is active as long as it is present in a cell. Its activity is approximately 1000-fold the activity of the endothelial NO synthase. When formed in such high concentrations, macrophage-derived NO is cytotoxic. A nonregulated excess formation of NO in vascular cells by this cytokin-inducible enzyme appears to be one reason for the generalized vasorelaxation in septic shock [16].

94.3.3.3 Neurogenic Regulation

A plexus of thin, nonmyelated nerve fibers in the vascular adventitia surrounds the arteries and arterioles with de-creasing density towards the capillaries. A similar, though less dense plexus is found around veins and venoles. In the arteries and arterioles, the fibers remain in the adventitia and do not run into the media, while in the veins the nerve endings are also found in the media. These nerve endings have numerous varicosities, like a string of pearls, in close vicinity to the smooth muscle cells. These varicosities are the sites of impulse-induced neurotransmitter release. Though no specialized synaptic structures exist connecting nerve endings and smooth muscles, the term *junction* or *synapsis* is used for these sites of neurovascular transmission. Quantitatively most important in nervous vasoregulation is sympathetic vascular innervation, causing vasoconstriction.

The sympthetic nervous system consists of preganglionic neurons, the cell bodies of which are located in the thoracic and lumbar segments of the spinal column. The cell bodies of the postganglionic neurons are located in the ganglia of the paravertebral sympathetic chain and in the abdominal ganglia. Their nerve endings form the perivascular plexus. Spinal sympathetic preganglionic neurons in the intermediolateral column are the final common pathway for neurogenic vasoconstriction as well as for the maintenance of basal sympathetic vasomotor tone. The firing rate of these spinal neurons in the intermediolateral column are under the control of tonic input from a defined region within the medulla oblongata, the rostral ventrolateral medulla [18]. Inputs from arterial baroreceptor and chemoreceptor activation (see Sect. 94.5.1), from the defense area, from the lateral hypothalamus, and pain afferents contribute in modulating the activity of this vasomotor tone center (see Sect. 94.5.1.1), partially via modulation in other important medullary sites such as the nucleus of the solitary tract and the area postrema [18]. In the thoracolumbar spinal cord, the sympathetic preganglionic neurons integrate the input from the rostral ventrolateral medulla, probably transmitted by an excitatory amino acid, with the spinal afferents from skin and viscera into a preganglionic activity, which is different for cutaneous, muscular, and visceral vasoconstrictor neurons [62]. In the sympathetic ganglia, there is cholinergic nicotinic transmission of the preganglionic activity to the postganglionic neurons with some weak cholinergic muscarinic and noncholinergic modulation [62] (see also Chap. 17).

By microneurographic exploration in humans, post-ganglionic sympathetic activity to the extremities can be recorded and divided into muscle nerve sympathetic activity (mainly vasoconstrictor activity) and skin nerve sympathetic activity (a mixture of vasoconstrictor, sudomotor, and pilomotor activity). Skin nerve activity demonstrates enhanced intensity with reduced environmental temperature and varies with the emotional state of the individual, while the muscle nerve activity shows a positive correlation to plasma norepinephrine concentration in the veins of the extremity, has bursts which are locked into the cardiac rhythm, is augmented during arterial hypotension, is silenced by acute hypertension, and is also activated by unloading intrathoracic, low-pressure ("volume")

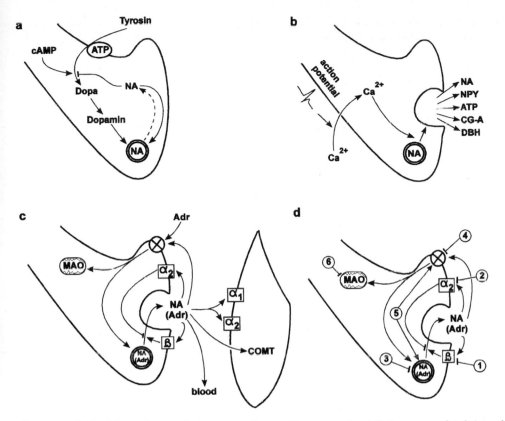

Fig. 94.22a–d. Physiology of sympathetic nerve endings. **a** The synthesis of norepinephrine (noradrenaline, *NA*) in the vesicles occurs from the amino acid tyrosine, which is actively taken up into the nerve endings. This synthesis is stimulated by cyclic adenosine monophosphate (*cAMP*) and inhibited by norepinephrine in the cytosol of the nerve endings. *ATP*, adenosine triphosphate. **b** Release of vesicular norepinephrine by action potentials is Ca^{2+} dependent and occurs together with the cotransmitters ATP and neuropeptide Y (*NPY*), the enzyme *DBH* (dopamine β-hydroxylase, which catalyzes the step from dopamine to norepineprine), and the stabilizing protein CG-A (*chromatin A*). **c** Released norepinephrine acts on postsynaptic α$_1$- and α$_2$-adrenoreceptors, causing constriction of vascular smooth muscle, on presynaptic α$_2$-receptors, attenuating transmitter release (presynaptic autoinhibition), and on presynaptic β-receptors, facilitating transmitter release. Norepinephrine in the synaptic cleft is reaccumulated into the nerve endings and partially metabolized by mitochondrial monoaminooxidase (*MAO*), it can be metabolized by smooth muscle catechol-orthomethyltransferase (*COMT*), or it can spill over into the blood. Circulating epinephrine (adrenaline, *Adr*) from the adrenal medulla can also be taken up into the endings and the vesicles, thus occurring as a cotransmitter. **d** The physiology of the nerve endings can be modulated by cardiovascular drugs. *1*, β-Blockers attenuate transmitter release by inhibiting the presynaptic facilitation; *2*, α$_2$-blockers enhance transmitter release by interrupting autoinhibition; *3*, reserpine blocks storage of norepinephrine in the vesicles; *4*, cocaine enhances neurotransmission by blocking the transmitter reuptake; *5*, amphetamine releases transmitter and interferes with the reuptake; *6*, MAO inhibitors enhance the transmitter available in the endings. (From [59], with permission)

receptors during orthostasis [62,113] (see also Sect. 94.5.1.1).

In the varicosites of the sympathetic nerve endings there are vesicles of varying size, which release the neurotransmitters by exocytosis in response to action potential-induced depolarization of the nerve endings. The main sympathetic neurotransmitter is norepinephrine [110], which is synthesized in the nerve endings and taken up again to a large extent after release and action on postsynaptic receptors (Fig. 94.22). Epinephrine can occur as a cotransmitter after uptake from the blood into the nerve endings [10]; it is not synthesized in the sympathic nerves, but only in the adrenal medulla. Neuropeptide Y (NPY) and ATP are true cotransmitters, released together with norepinephrine [111,112].

The released norepinephrine causes vasoconstriction via activation of smooth muscular α-adrenoceptors; β-receptor-mediated vasodilation plays only a minor role in some vessels. The constrictory α-adrenoceptor effect prevails. ATP via activation of P$_2$ receptors and NPY via activation of NPY receptors assist in this vasoconstriction [111,112]. Though most of the relased norepinephrine is taken up or metabolized locally, a small fraction spills over into the circulation (Fig. 94.22). Therefore, plasma norepinephrine can be considered as a vague parameter for sympathetic activity.

The amount of transmitter released by an action potential is not constant; it can be modulated by the transmitter in the synaptic cleft, released by previous action potentials, presynaptic autoinhibition, and facilitation [100,101] (see

Fig. 94.22). Furthermore, norepinephrine release is augmented by angiotensin II (see Sect. 94.3.3.4) and is inhibited by endothelium-derived NO [77], by acetylcholine from adjacent cholinergic nerve endings [109], by factors presumed to contribute to metabolic vasoregulation (adenosine, K^+), and by many inflammation-associated autacoids (see Sect. 94.3.3.2). Specifically, the endothelial cells appear to play an important role in the regulation of neurotransmitter release from sympthetic nerve endings [23,106].

A steady state vasoconstrictor influence from the sympathic nerve endings on the resistance vessels (controlling organ perfusion) and on the venoles and veins (controlling venous capacity; see Sect. 94.4.3) operates under basal conditions and can be modulated by reflexes (see Sect. 94.5.1). Furthermore, the neurogenic regulation of the ratio of pre- to postcapillary constriction is an important determinant for the intracapillary pressure and, thus, for the balance of capillary filtration and reabsorption, which determines the compartmentalization of extracellular fluid (see Sect. 94.4.2, Chap. 97).

The attenuation of this sympathetic vasoconstriciton by reflexes or by local metabolic factors and inflammatory autacoids is the most important mechanism of neurogenic vasodilation, quantitatively outweighing the other mechanisms of neurogenic vasodilation mentioned below.

Of minor importance is neurogenic vasodilation by parasympathetic cholinergic fibers in some organs (see Chap. 17). It can occur in genital organs, in the brain and the coronaries, and in glands of the skin and the intestinal tract. The transmitter acetylcholine dilates the vascular smooth muscle only indirectly: by inhibiting norepinephrine release from sympathetic nerve endings, by acting on the endothelium (after diffusion through the media in small vessels), and by causing the release of NO or by inducing the formation of the inflammatory autacoid bradykinin in some glands. Postganglionic vagal neurons contain also NANC (nonadreneric, noncholinergic) fibers, which have recently been identified as nitroxidergic [105]; they contain a neuronal NO synthase and release NO as a dilatory neurotransmitter.

In the precapillary vessels of skeletal muscle in some species, probably including humans, there is a cholinergic vasodilator innervation, the axons of which run from the frontal cortex along with sympathetic fibers to the skeletal muscle. It is assumed that this cholinergic sympathetic innervation contributes to transient skeletal muscle dilation during stress and constitutes a part of "fight and flight" behavior [12,48].

94.3.3.4 Hormonal Regulation

The complex vasoregulatory effectiveness of hormonal systems becomes more easily understandable if we consider that each system appears specifically designed to defend a certain function with vital importance for the entire organism. In the case of the renin–angiotensin system, this function obviously is to maintain a glomerular filtration pressure sufficiently high to allow a glomerular filtration rate sufficiently large for the detoxification of the plasma by the kidney. The atrial natriuretic peptides appear to be designed as a defense mechanism against cardiac overload due to an overfilling of the vascular system. The catecholamines help the organism to cope with the stress of fear and fight, and the arginine vasopressin helps to maintain the osmolality of the body fluids.

A typical feature of these hormonal systems is the fact that their vasoregulatory role is clearly demonstrable in the specific emergency situation which the respective hormonal system is designed to prevent. However, their role for circulatory regulation under physiologic conditions with well-maintained homeostasis is at best marginal and difficult to delineate. It should be stressed that a hormonal system with a defined endocrine gland and a remote target for the hormone (see Chaps. 17–23) is a differentiated specialization of the more generalized principle of paracrine communication between cells (see Chap. 6). The hormonal systems mentioned above that are relevant to the regulation of the circulation also occur as paracrine signals in many organs and/or as neurotransmitters, in addition to their specialized role as hormones. Separation of the hormonal and the paracrine role of these signals is not always possible.

Renin–Angiotensin System. This is clearly a nephrocentric signal system with substantial paracrine components in many extrarenal organs (see also Chap. 73). The protease renin cleaves the decapeptid angiotensin I from angiotensinogen (452 amino acids). From angiotensin I, the endothelial angiotensin-converting enzyme (ACE) and the human heart chymase (hHC) form the active octapeptide angiotensin II, which is the main effector of the system [32]. Angiotensin II is a potent vasoconstrictor, enhances norepinephrine release from sympathetic nerve endings, activates the release of aldosterone from the adrenal cortex and antidiuretic hormone or vasopressin from the neurohypophysis, enhances renal Na^+ retention by vascular and tubular actions, and is an important growth factor for cardiac and vascular myocytes (see also Chap. 82). It acts by activating at least two types of membrane receptors (AT_1 and AT_2) [104] and by binding to DNA after cellular uptake and complexing with elements of the signal transduction [72]. Other aminopeptidases and endopeptidases form heptapeptides (angiotensin III and angiotensin 1–7) and a hexapeptide (angiotensin IV; see Fig. 94.23), which complement the actions of angiotensin II with a slightly different spectrum of effects.

While many cells can synthesize inactive prorenin, the active renin is formed and released by the vasa afferentia of the kidney in response to a fall in arterial pressure, β-adrenoceptor activation, and diminished NaCl absorption in the macula densa cells [33]. Circulating renin is taken up by the vasculature in many organs and, following uptake (or binding), contributes to local angiotensin I formation in these organs. Angiotensinogen is released into the plasma by the liver, but is also formed in heart and brain locally. ACE occurs as ectoenzyme in the luminal mem-

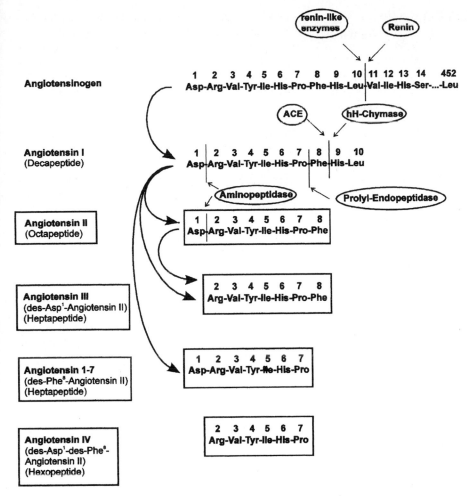

Fig. 94.23. Elements of the renin–angiotensin system. Angiotensins are formed from the N terminus of angiotensinogen by the specific enzymes renin, *ACE* (angiotensin converting enzyme), *hH*-(human heart) chymase, and by other peptidases which do not only act within the renin–angiotensin system. The octapeptide angiotensin II is the main effector of the system, but the other hepta- and hexapeptide angiotensins also contribute

brane of endothelia and in the plasma, but also in intracellular vacuoles of endothelia, vascular, and cardiac myocytes, macrophages, and (as a special isoform) in the testes. Human heart chymase is a mast cell-derived enzyme which remains active following release into the myocardial interstitium [107]. Angiotensin II circulates in the plasma with a concentration around 3–5 pmol/l, which can increase by more than 100-fold during volume deprivation and kidney damage. Circulating angiotensin II results mainly from spillover of local formation in the organs and not from enzymatic action of circulating ACE in the plasma [2,19]. Other angiotensins have a plasma concentration of roughly one tenth of angiotensin II or lower [20].

Although classically considered as a hormonal system, the renin–angiotensin system is in fact a paracrine system of many tissues, but with a strong nephrogenic activation step. For the activity of the system, circulating components are far less important than local formation of angiotensins (see also Chap. 73).

Atrial Natriuretic Peptides. The name for this group of structurally and functionally closely related peptides (Fig. 94.24) is derived from human atrial natriuretic peptide (hANP), which was discovered and isolated by DeBold et al. [30]; hANP is an hormone of 28 amino acids. It is released from the cardiac atria in response to enhanced atrial distension, i.e., when the blood volume within the thoracic circulation is increased or when the performance of the heart as a pump is attenuated. The effects of circulating hANP tend to unload the volume-overloaded heart: diuresis and natriuresis in the kidney, inhibition of renin release from the kidney, attenuation of norepinephrine release from sympathetic nerve endings, and dilation of resistance vessels [60]. Within the atria, the hormone is stored in secretory vesicles of the myocytes in the form of a prohormone of 126 amino acids, and the N-terminal hANP is cleaved from the prohormone during stretch-induced secretion. Other fragments of this prohormone have similar, but weaker, actions to the N-terminal hormone. In the kidney, another hormone, urodilatin, is formed from the

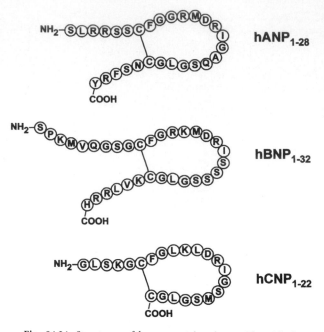

Fig. 94.24. Structures of human natriuretic peptides. All three related peptides are cleaved form the C-terminal end of similar prohormones with 103–126 amino acids. The *line* within the peptide structures indicates disulfide bridges between cysteine residues. *hANP*, human atrial natriuretic peptide; *hBNP*, human brain natriuretic peptide; *hCNP*, human C-type natriuretic peptide

pro-ANP by alternative cleavage and may play a paracrine role in urinary Na+ handling [108]. In the ventricles of the heart, ANP formation is neglibible under physiologic conditions, but is dramatically enhanced in overload-induced hypertrophy [60]. Outside of the heart, ANP is formed in brain areas involved in volume homeostasis and at a very low rate in the lungs, vessels, thymus, and kidney.

Similarly to ANP, brain natriuretic peptide (BNP) is formed mainly in the atria and to a lesser extent in the other organs, but it was discovered in brain tissue. C natriuretic peptide (CNP) is formed mainly in the central nervous system and in the endothelium of the peripheral vasculature [44]. It is the paracrine component of the system. All three peptides activate two membrane receptors, resulting in formation of cGMP by the cytosolic guanylyl cyclase domain of these receptors. In the vasculature, they cause not only cGMP-mediated vasodilation, but also act as growth inhibitors, especially the endothelium-derived CNP (see Sect. 94.3.2).

Catecholamines. Although both norepinephrine (plasma concentration, 200–400 pmol/l at rest) and epinephrine (plasma concentration, 40–100 pmol/l) circulate in plasma, only epinephrine can be considered as a true hormone, released from the adrenal medulla. Most of the circulating norepinephrine results from spillover of norepinephrine released from the sympathetic nerve endings and escaping neuronal reuptake and/or degradation. During stress and exercise, the plasma concentrations of both

catecholamines can increase up to 20-fold, and under these conditions, circulating norepinephrine may act as a hormone with vasoconstrictor action in most vascular beds, mediated by activation of α-adrenoceptors of smooth muscle cells. Epinephrine however, with a higher affinity for β-adrenoceptors than norepinephrine, causes β-adrenergic dilation and augmentation of metabolic rate in the heart, in the liver, and, quantitatively most importantly, in skeletal muscle [45]. In other vascular beds and in the venous system, the effect of epinephrine is α-adrenergic vasoconstriction. However, due to the large mass of skeletal muscle, the overall effect of epinephrine is a lowering of peripheral vascular resistance. Both catecholamines cause a moderate endothelial release of NO by activating endothelial α_2-adrenoceptors. This effect is small and does not override the direct effects on smooth muscle, but it explains why constrictor effects of both catecholamines are enhanced under conditions with disturbed endothelial function.

Arginine Vasopressin. The primary target of this hormone from the neurohypophysis is the kidney, where it modulates water permeability of the collecting duct and depresses the glomerular filtration (see Chaps. 74, 75, 82). Outside the kidney, vasopressin causes vasoconstriction at concentrations which are reached during severe hypotension and hemorrhagic shock. In the central nervous system and in the coronary bed, vasopressin causes endothelium-dependent relaxation, which may help to protect the brain and heart against generalized vasoconstriction in shock.

Adrenomedullin. Is a vasodilator peptide hormone with 52 amino-acids, which has been detected in the adrenal medulla. It causes cAMP-mediated vasorelaxation and augmentation of renal blood flow and Na+ excretion.

94.4 Comparative Aspects in Specific Vascular Sections

94.4.1 Arterial Hemodynamics

The ejection of stroke volume into the distensible aorta induces pulse waves, consisting of pressure, flow, and volume pulses, propagating along the aorta with the wave velocity (c) as long as the physical wall properties are constant along the vessel (see Sect. 94.1.3). These wall properties are described by the wave impedance (Z), which is the amplitude ratio of the pulse wave (ΔP) to the flow wave ($\Delta \dot{Q}$):

$$Z = \Delta P / \Delta \dot{Q}. \qquad \text{(Eq. 94.15)}$$

This wave impedance increases along the arterial tree, stepwise as well as continuously, with each ramification, with alterations in cross-sectional area and with changes in wall thickness and structure. Any increase in wave imped-

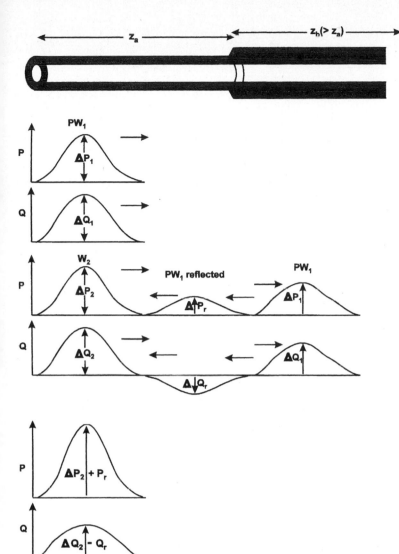

Fig. 94.25. Partial pulse wave reflection and pulse wave superposition or pressure and flow pulses. Pulse waves are partially reflected at sites of increases in wave impedance with a coefficient of reflection $K = (Z_b - Z_a)/(Z_b + Z_a)$. This coefficient approaches 1 (= total reflection) with an extremely large impedance Z_b. ΔP, amplitude of pressure pulse; ΔQ, amplitude of flow pulse. Reflection coefficients between 0.1 and 0.4 in the arterial system and in the resistance vessels modulate arterial wave forms of pressure and flow and are subject to alterations in smooth muscular tone of the arterial tree

ance causes a partial reflection of the pulse wave with the consequence that along the arteries there is a superposition of waves propagating from the heart to the periphery with reflected waves running back to the heart. The amplitudes of the pressures waves with opposing directions accumulate, while flow waves with opposing directions are subtracted, cancelling each other out (Fig. 94.25). This consequence of partial wave reflexions at multiple sites along the arterial tree is the explanation for the fact that pressure and flow waves in the arterial system have different forms (which is not the case in a constant tube without wave reflexions; see Sect. 94.2.3).

The systolic flow out of the left ventricle reaches a maximum of 600 ml/s early in systole and declines during the rest of the ejection phase, until closure of the aortic valve results in a short backflow, which is detectable in the pressure tracing as incision (Fig. 94.26). The pressure pulse in the ascending aorta is not only determined by the ejected flow wave and by the local wave impedance, but also by superposition of reflected pressure waves, resulting in a larger amplitude of pulse pressure in the ascending aorta than stroke volume and local wave impedance would determine (Fig. 94.26). With increasing distances from the heart, the amplitudes of arterial pressure waves become larger due to superposition of reflected pressure waves, until the amplitude in a distal artery of the leg is twice the amplitude in the central aorta. Additionally, a second dicrotic wave is detectable, resulting from the fact that reflected pressure waves run back to the aortic valve, are reflected again, and follow the primary wave (Fig. 94.26). While the pressure amplitude increases along the arterial tree with an absolute increase of systolic peak pressure from 120 mmHg (16 kPa) in the ascending aorta to 160 mmHg (21 kPa) in arteries of the leg, mean arterial pressure (which is obtained by integrating pressure tracings over time) declines due to friction of the flowing blood along this arterial tree.

A result of the multiple reflections of pulse waves, running back and forth in the arterial tree, is the fact that at the end of the systole, when the entire stroke volume has left the

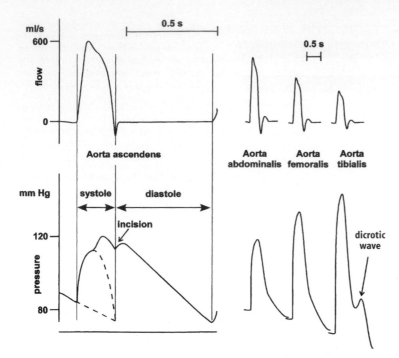

Fig. 94.26. Flow (*upper tracings*) and pressure (*lower tracings*) in human arteries. The slow pulse ejected from the heart induces in the aorta a primary pressure pulse, which is superposed on the prolongation of the diastolic pressure decline (*stippled lines*) and which is determined by the local wave impedance Z in the ascending aorta. The actual increase in systolic pressure is higher than this primary pressure pulse, since peripherally reflected pulse waves from the previous stroke volume reach the ascending aorta in the late systole and up to the primary pressure pulse. (In fact, reflected flow waves also influence the stroke volume-induced primary flow wave, which reflects the interaction of ventricular contraction dynamics and impedance characteristics of the pressurized arterial system). Superposition of primary and reflected pulse waves causes increases in pressure pulse amplitude and in peak pressures and the appearance of a second, dicrotic pulse in peripheral arteries. Subtractions of primary and reflected flow waves and flow ramifications cause declines in flow pulse amplitude and in mean flow in peripheral arteries. High reflection coefficients at the beginning of resistance vessels and energy losses of pulse waves due to friction result in diminished pulse pressure amplitudes in the microvasculature

heart, roughly 50% of this stroke volume is still within the large vessels in the form of pulse waves and leaves the arterial tree during diastole. This dynamic storage of volume in the form of running pulse waves in a network of distensible tubes transforms the pulsatile flow input into the system in a fairly continuous flow leaving the arterial system. In an arterial system of rigid, nondistensible tubes, such transformation would not occur and the entire volume content of such a system would have to be accelerated by the heart with each stroke volume ejection. In a simplified view, the transformation of pulsatile flow input into continuous outflow has been called the *Windkessel* function of the arterial system. This oversimplification assumes that the ejected volume is stored in one stationary, distensible compartment instead of dynamic storage into running pulse waves, but the term survived just because of its simplicity.

The *Windkessel* function of the large arteries is reduced progressively with age due to alterations in the composition of the arterial wall: the number of collagen fibers increases, replacing elastic fibers. Hypertension accelerates these age-dependent alterations and causes additional changes of wall structure, enhancing the risk of atherosclerosis. An overall assessment of the arterial system can be obtained by measuring pulse wave velocity, which increases with age-dependent loss of arterial distensibility (Fig. 94.27).

94.4.2 Peripheral Microcirculation

The microcirculation includes the capillaries and venoles, which are the vessels for the exchange of the plasma with the interstitial fluid, the arterioles and small veins, which regulate the conditions for this exchange with their muscular wall, and the initial lymphatics, which act as a drainage system of the microcirculation (see Sect. 94.4.4.1). In the venoles of the microcirculation, the adhesion and diapedesis of leukocytes takes place (see 'Homing of Lymphocytes," Chaps. 85, 88).

The basal flow through an organ at normal arterial blood pressure is determined by the local vascular resistance and differs between various organs (see Chap. 95). This vascular resistance is mainly determined by the small arteries and the arterioles (see Sect. 94.2.2) at the entrance of the microcirculation circulation. The vascular resistance is defined as: (mean arterial pressure – central venous pressure)/organ flow per weight, and it basically depends on

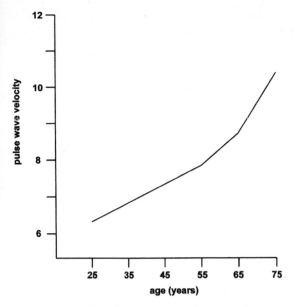

Fig. 94.27. Age-dependent increase in pulse wave velocity in m/s in human aorta at normotension. Pulse wave velocity in the distal leg arteries is roughly 80% higher. A rule of thumb estimates that in hypertension, aortic pulse wave velocity is elevated by 0.8 m/s per 20 mmHg elevation in mean arterial pressure. (From [15], with permission)

the vascular architecture of the organ and on several vasoregulatory factors:

- Basal smooth muscular tone induced by myogenic autoregulation
- Flow-dependent endothelium-mediated dilation
- Metabolic vasodilation
- Neurogenic constriction.

In many organs, not all anatomically available capillaries are perfused under basal conditions at rest; the interplay of dilatory vasoregulatory factors with myogenic vascular tones results in rhythmic oscillations of individual arterioles in such a way that overall flow through the organ is constant, while the rhythmic arteriolar vasomotion causes intermittent, asynchronous occlusion and reperfusion of individual capillaries.

The capillary exchange is dependent on the flow rate through the capillaries, on the effective filtration pressure across the capillary wall (see below), on the distances between perfused capillaries and the flow direction in neighboring capillaries relative to each other, on the wall area of the perfused capillaries, and on the permeability characteristics of the endothelial wall (see Chap. 97), which is determined by the endothelial ultrastructure. This structure is highly variable between different organs.

Continuous capillaries with intercellular clefts, incompletely occluded by a network of tight junctions within these clefts, exist in striated muscle, skin, connective and adipose tissue, and the lung. The gaps between the tight junctions in the clefts are the most important route for

exchange in this type of capillaries. Retraction of endothelial cells with wider opening of the interendothelial clefts and disconnection of the tight junctions occurs under the action of many inflammatory autacoids (see Sect. 94.3.3.2) and substantially lowers the barrier function of these endothelia. In the brain, the network of tight junctions in the intercellular clefts is very dense and without any gaps. This tight-junction endothelium is part of the brain–blood barrier (see Chap. 27). A fenestrated endothelium with intracellular pores with diameters up to 50 nm which is very permeable for water and small solutes is found in organelles with special exchange functions, such as glomeruli of the kidney, glands, intestinal mucosa, and plexus in the eye and in the central nervous system. Discontinuous or sinusoid capillaries with intracellular openings up to a diameter of 1 μm exist in liver, spleen, and bone marrow, organs in which not only passage of macromolecules occurs, but also exchange of cells.

Further important determinants of the capillary exchange are the fraction of perfused capillaries out of the anatomically available capillaries, which is regulated by rhythmic constrictions of the precapillary sphincters (see above), and the hydrostatic pressure in the capillaries and venoles, which is determined by the pre- and postcapillary resistances (i.e., by arterioles and small veins). The hydrostatic pressure difference between lumen and interstitium is the driving force for filtration of fluid out of the exchange vessels, while the colloid-osmotic (or oncotic) pressure difference between lumen and interstium, corrected for a coefficient for osmotic reflection at the endothelial barrier, acts against this filtration. At the arterial side of the exchange vessels, filtration prevails, while reabsorption takes place at the venous end of the exchange vessels.

A detailed description of the microcirculation and a functional analysis of the capillary exchange is given in Chap. 97.

94.4.3 Venous System and Pulmonary Circulation

The venous system is generally considered as a volume reservoir, which determines diastolic filling of the right heart by modulating venous return to the heart. In fact, however, the functions of the veins in the circulation are much more complex, including the following:

- Regulation of filling pressure in the right heart by controlling venous return in face of variable cardiac output
- Collection of blood from microcirculation and propulsion towards the heart through flow-rectifying valves
- Modulation of capacitance of the circulation
- Modulation of orthostatic tolerance of the organism
- Participation in microcirculatory exchange and volume compartmentalization between intra- and extravascular space by venular reabsorption of interstitial fluid
- Modulation of capillary pressure by postcapillary resistance regulation

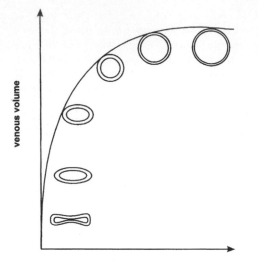

Fig. 94.28. Venous compliance and transmural pressure. At low transmural pressure, large increases in cross-sectional area of the veins occur with small increases in pressure due to the transition from the flat elliptic to the circular form

- Whole body thermoregulation via subcutaneous venous plexus
- Lymphocyte homing in venular "high" endothelium in several organs
- Leukocyte–endothelial interactions in margination, rolling, and diapedesis of polymorphnuclear leukocytes
- Origin of neoangiogenesis from venoles.

Together with the pulmonary circulation, the veins constitute the low-pressure system, in which the vascular pressure does not exceed 20 mmHg (3 kPa) under physiologic conditions. It includes the capillaries, the venoles and veins, the right heart, the pulmonary circulation, the left atrium, and the left ventricle in diastole. The high-pressure system encompasses the left ventricle in systole, the arteries, and arterioles down to the origin of the capillaries. While the high-pressures system contains 15% of the blood volume, the remaining 85% is in the low-pressure system (see Table 94.1). The important difference between these two systems, however, is functional: while the pressure in the high-pressure system is dynamically regulated by cardiac output and by resistance to flow, the pressures in the low-pressure system is mainly statically determined by compliance and volume content and only marginally by resistance to flow. Under identical pressures throughout the vascular system, the compliance of the low pressure system is 200 times the compliance of the high-pressure system, which explains the fact that in cardiac arrest 99.5% of the blood volume is in the low-pressure system (see Fig. 94.1).

These high compliances of major sections of the low-pressure system result not only from a high distensibility of wall structures, but largely from the fact that the vessels in these sections are in a flat, partially collapsed state at the pressures prevailing in these sections. At a transmural

pressure of zero or slighty above zero, the peripheral veins are more or less collapsed, and their volume content can be increased with minimal elevations of transmural pressure by inflating these vessels from an eliptic to a circular form (Fig. 94.28). At transmural pressures above 15 mmHg, the veins reach a circular cross-section and their compliance becomes more and more depending on the distensibility of the wall and therefore is lowered.

Another high-compliance part of the low-pressure system is the vasculature of the pulmonary microcirculation. These vessels are not cylindrical capillaries, but flat

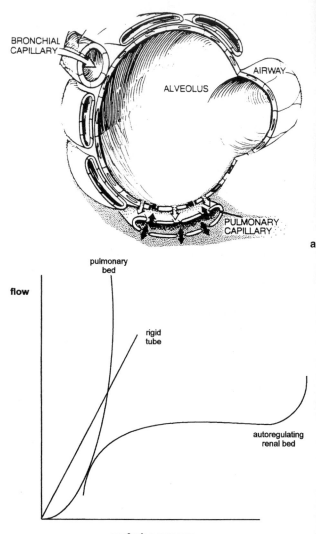

Fig. 94.29. The architecture of the pulmonary microvasculature in flat lacunae in the alveolar walls. Blood passes as sheet flow through these flat, reticular capillaries, which can be easily bulged by small increments in transmural pressure. This is the reason for the high compliance of this bed (a) and the concave pressure–flow relation of the pulmonary bed (b). This is in contrast to the pressure–flow relations in autoregulating beds such as the kidney. While pulmonary vascular resistance is lowered with acute increments in perfusion pressure, it can be substantially augmented with chronic elevations in pulmonary perfusion pressure and fibrotic remodeling in the lung

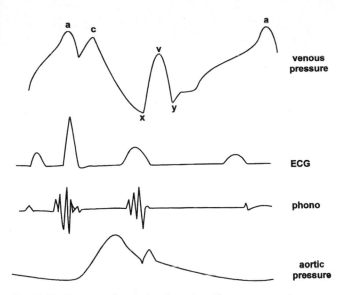

venous
pressure

ECG

phono

aortic
pressure

Fig. 94.30. Venous pulses in jugular veins, The *a* wave results from the atrial contraction and the *c* wave is induced by bulging of the tricuspid valve during ventricular isovolumetric tension development. During the ventricular ejection, the tricuspid valve approaches the apex of the ventricle, causing lowering of jugular pressure down to the minimum *x*. The waves *v* and *y* result from the opening of the tricuspid valve. In congestive heart failure with elevated central venous pressure, the venous pulses in the jugular veins are visible during examination of the patient. *ECG*, electrocardiogram

reticular lacunae within the alveolar walls, which can be easily blown up (Fig. 94.29). This large compliance of the pulmonary circulation and the pressure-dependent increase in vascular pulmonary conductance explains why the perfusion of various parts of the lung is strongly dependent on the body posture: hydrostatic pressure differences between the right heart and a certain part of the lung add up to (or are subtracted from) the pulmonary arterial pressure, and the resulting differences in perfusion pressure result in even larger differences in regional lung perfusion because of the unique pressure–flow relation in the pulmonary bed (Fig. 94.29). Furthermore, this unique pressure–flow relation of the pulmonary bed is the reason why the low pressure system reacts as a functional unit, though the right heart is interposed between the venous and the pulmonary system. Any increase in central venous pressure proximal to the right heart will cause elevated stroke volume of the right heart due to the Frank-Starling mechanism (see Chap. 90), and this enhanced right cardiac output is easily transmitted to elevated left atrial filling because of the distensible pulmonary microcirculation (Fig. 94.29). A detailed description of the bronchial and pulmonary circulation is given in Chap. 104.

Pressures in the Venous System. In the postcapillary venoles, mean intravascular pressure is 15–20 mmHg (2–3 kPa) in humans lying down (Fig. 94.29) and declines only slightly to 10–12 mmHg (1–1.5 kPa) in the inferior vena cava. Due to the narrowing of the vena cava by the diaphragm, there is a further decline in vena cava pressure by roughly 5 mmHg (0.7 kPa). Mean central venous pressure is the pressure close to the right atrium and is 3–5 mmHg (0.4–0.7 kPa) with cyclic variations with the respiration, declining with the onset of inspiration and increasing during expiration.

Pulsations in the large veins result form the contractile action of the right atrium and from the movements of tricuspic valve. The form of these venous pulsations resembles the pressure oscillations in the right atrium (Fig. 94.30) and run along the large veins against the direction of the venous blood flow. In the venoles, there are no pulsations, neither venous pulses or rests of arterial pulses waves.

Venous Return and Cardiac Output. In a strictly formal definition, venous return is the volume flow towards the heart, which under steady state conditions equals cardiac output. A complex array of mechanisms assists in achieving venous return. The mechanisms driving flow back towards the heart include the following:

– Actions resulting from pumping performance of the heart
  • Pressure gradient from capillary pressure to central venous pressure
  • Suction of blood from the veins into the atria due to the movement of the closed atrioventicular valves towards the cardiac apex during ventricular ejection
  • Rhythmic compression of veins by arterial pulse waves in adjacent arteries and arterioles
– Extracardiac actions
  • Lowering of venous capacity by venoconstriction
  • Rhythmic compression of veins with flow-rectifying valves by skeletal muscle contractions

1891

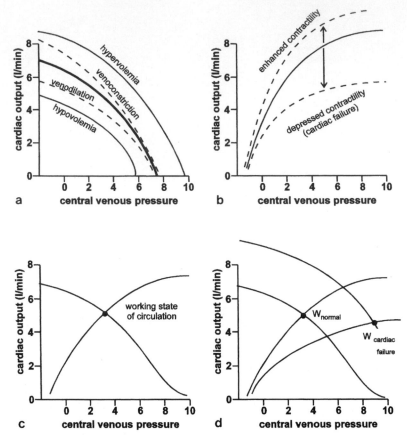

Fig. 94.31.a–d. Relationship between cardiac output and central venous pressure. **a** Venous function curves describe the static characteristics of the low-pressure system: at cardiac arrest, 99.5% of the blood volume is within the low-pressure system, and central venous pressure is maximal (for a given tone). The pumping action of the heart removes part of that volume out of the low-pressure system into the arterial high-pressure system and establishes a new equilibrium of volume distribution with a lower central venous pressure. Thus, venous function curves describe central venous pressure as a function of the dynamic equilibrium created by the pumping heart (i.e., as a function of cardiac output). These curves depend on venous tone and on the intravascular blood volume. **b** Cardiac function curves are derived from the Frank-Starling mechanism (see Chap. 90) and describe cardiac output as function of central venous (filling) pressure. These curves are modulated by changes in cardiac contractility. Note that at a central venous pressure of zero, filling of the heart and cardiac output must still be positive, since transmural pressure is still positive (due to the negativity of intrathoracic pressure, see Chap. 100). **c** Intersection of a venous function curve with the respective cadiac function curve defines the working state of the circulation (5.5 l/min cardiac output and 3.5 mmHg central venous pressure in human in supine position). **d** In heart failure, the working state of the circulation at rest is shifted to high central venous pressure (10 mmHg) at almost normal cardiac output (5 l/min) due the hypervolemia and the venoconstriction in the presence of a depressed cardiac function curve

- Rhythmic inspiratory suction of blood into the thoracic veins and the heart through flow-rectifying valves of the heart and the veins, which prevent venous backflow during expiration.

The relative importance of these mechanisms in supporting venous return dramatically changes with body posture (see Sect. 94.5.2). The interactions of the mechanisms is best described by venous function curves (Fig. 94.31). Remember that pressures in the low-pressure system are largely statically regulated by changes in volume and capacity (see above) and that the volume in this system and hence in the veins is maximal with cardiac arrest (Fig. 91.1). The action of the heart removes blood out of this volume reservoir and forces it into the high-pressure system, thereby lowering central venous pressure and creating a new dynamic equilibrium of volume distribution. Since this pumping performance of the heart depends on diastolic filling of the heart and hence on central venous pressure, the actual working state in a given circulation is determined by the interplay between mechanisms determining venous return (resulting in a certain venous function curve) and the filling-dependent cardiac function curve, reflecting a given state of cardiac contractility.

It is clear that the venous function curves, as an intergrated description of compliance of the low-pressure system and of auxiliary mechanisms assisting venous return, represent a highly variable parameter, which decisively determines the actual working state of the circulation. Gravitation during orthostasis must severely alter the venous function

Fig. 94.32. Initial Iymphatics with overlapping endothelial cells acting as microvalves. The endothelia of these blind-ending capillaries allow inflow of interstitial fluid, when surrounding tissue pressure is elevated. Anchoring filaments prevent collapse of the lymphatic capillaries. Outflow of the fluid under reversed pressure gradient is prevented by the overlapping endothelial cells. Rhythmic oscillations of interstitial tissue pressure improve uptake of interstitial fluid into the initial lymphatics

curve (see Sect. 94.5.2.). Furthermore, note in Fig. 94.31 that the high cardiac output of 30 l/min during heavy exercise is not possible without a dramatically altered venous function curve, in which dynamic support mechanisms of venous return (skeletal muscle pump, see above) largely determine the functional behavior of the low-pressure system.

94.4.4 Lymph Flow

The lymphatic system is a fluid drainage of the interstitial space. It transports proteins, filtered from the blood plasma into the interstitum, back to the venous blood, together with complex macromolecules, bacteria, solid particles, and immune cells. Concomitantly, it acts as an immune-surveillance system [83]. Overall lymphatic flow at rest amounts to approximately 2 l/day, but lymph flow from the different organs is highly variable and may fluctuate by more than 100-fold of the minimal flow [7,35]. The regulation of lymph flow and its driving forces is intimately dependent on the morphology of the lymphatic vessels and on their anatomic location within the structure of the tissues [68].

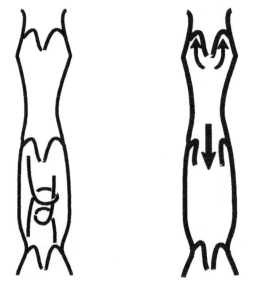

Fig. 94.33. Lymphatic collectors designed as lymphangions, the lymphatic microhearts. The smooth muscle of the lymphangions is designed for peristaltic contractions, pressing the lymph content through the valve into the next lymphangion

94.4.4.1 Elements of the Lymphatic System

The lymphatic vasculature consists of the following: blindending lymphatic capillaries or initial lymphatics, capillary-like lymphatic precollectors up to 150 µm in diameter with lymphatic valves; lymphatic collectors (up to 600 µm in diameter) with valves and a vessel wall consisting of an intima, a muscular media, and an adventitia; and the large central lymphatic conduits (Figs. 94.32–94.34). Lymph

nodes are positioned along the lymphatic system in such a way that one or several lymphatic collectors drain into a node, while other collectors leave the nodes towards the central conduits. The lymph nodes contribute to the formation and differentiation of lymphocytes, and they filter the incoming lymph by phagocytizing bacteria, particles, and macromolecules [83]. Postnodal lymph contains differentiated lymphocytes and has a higher protein content than prenodal lymph; lymphatic water is absorbed into the nodal blood capillaries. The lymph nodes and other

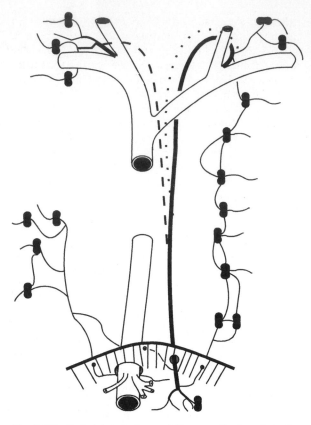

Fig. 94.34. Central lymphatic conduits carry the lymph to the jugular veins. The thoracic duct, carrying lymph from three quarters of the body to the right jugular vein, can have a bifurcation with variable course of the arms (*dashed* and *stippled lines*)

lymphoid organs such as spleen, thymus, and the gut-associated lymphatic tissue are organs of the immune defense system (Chaps. 83–87).

Initial Lymphatics. These have a very flat endothelium with a discontinuous basement membrane. This lymphatic endothelium is tethered to its surrounding connective tissue by collagen-anchoring filaments. However, the endothelial cells are not uniformly attached to their substrate: neighboring endothelial cells have regions of overlap without any tight junctions within this overlap. The lower, abluminal endothelial cell rim of these overlapping interendothelial junctions is attached to the surrounding parenchyma by the anchoring filaments, while the upper, luminal part of the neighboring cells floats freely (Fig. 94.32). This microanatomic arrangement allows the endothelium of the initial lymphatics to operate effectively as microvalves between the interstitium and the lumen of the initial lymphatics. When the luminal pressure in the initial lymphatics is lower than the surrounding interstitial pressure, the floating, luminal part of the endothelial junction is pressed into the lumen, while the abluminal part is fixed to the tissue parenchyma by the anchoring filaments. Thus, collapse of the initial lymphatics is prevented and large interendothelial gaps are

opened between the overlapping junctions, allowing the entry of interstitial fluid and macromolecules, cells, bacteria, and solid particles (such as mineral material in the lungs). When the luminal lymphatic pressure is higher than the extralymphatic tissue pressure, the floating, overlapping endothelial part of the junction is pressed against the abluminal part, thereby closing the junction and preventing outward movement of the primary lymph with its corpuscular material.

Lymphatic Precollectors. These resemble the initial lymphatics with respect to the endothelial lining and participate in the uptake of fluid and material from the interstitium. Additionally, they are endowed with bileaflet semilunar valves, which permit lymph flow toward the central lymphatics, but prevent backflow. The soft and easily deformable leaflets of these valves are made up of a thin collagen sheet between two endothelial layers. The lymphatic valves in the precollectors and collectors operate even at very small flow rates at which only viscous fluid stresses are present [68]. Kinetic and inertial fluid forces, which contribute to valve closure in the heart, do not operate at the lymphatic valves.

Lymphatic Collectors. The walls of lymphatic collectors resemble those of small veins, with an intima of endothelium and basement membrane, a media with longitudinal and circular layers of smooth muscle, and an adventitia with loose connective and elastic fibers, fibroblasts, histiocytes, mast cells, and nerve endings. The collecting lymphatics are organized as a continuous ladder of discrete, actively contractile units, each unit separated from the next by a valve. These units or compartments between two consecutive valves are called *lymphangions* and have been considered as lymphatic microhearts (Fig. 94.33). In the lymphatic collectors their length varies in different organs between 6 and 20 mm. The lymphangions exhibit a peristaltic contraction with a synchronized opening and closing of the valves to prevent fluid reflux. These spontaneous peristaltics have a frequency of approximately ten contractions per minute without any detectable synchronicity with the cardiac or respiratory cycle. The lymphangions demonstrate a type of myogenic autoregulation or preload dependency: the force and the frequency of the peristaltic contractions are increased with higher filling pressures of the individual lymphangion [84]. The contractile activity (force and frequency) of the lymphangions is responsive to the sympathetic innervation via α-adrenoceptor activation as well as to a variety of humoral agents, but the physiologic relevance of this modulation is not clear. Using immunohistochemical methods, adrenergic, cholinergic, and peptidergic nerve fibres are demonstrable in the adventitia of the lymphangions, with neuropeptide Y (NPY) in the adrenergic fibers and vasointestinal peptide (VIP) in the cholinergic fibers. Sensoric fibers with calcitonin gene-related peptide (CGRP) and substance P (SP) as the neurotransmitter have been described in a context with *terminal cells* in cushions whose structure resembles the

Vater-Pacini sensors in the skin. It has been hypothesized that these putative terminal sensors are involved in coordinating the peristaltic contraction of a lymphangion [68]. The endothelium of the lymphatic collectors produces NO, though the shear stresses in these lymphatic vessels is rather low. The intrinsic pump mechanism of the lymphangions can operate over a wide range of lymph flow. It provides the lymph transport against gravity or against nodal obstructions in lymphatic diseases.

Large Central Lymphatic Conduits. These have a wall structure similar to that of the collectors, but thicker. Elastic fibers are abundant in media and adventitia, as are nerve endings. Distances between lymphatic valves in the central conduits are 6–10 cm, without a clear lymphangion architecture. The central conduits direct lymph of the right upper quarter of the body to the right venous angle (i.e., insertion of the jugular vein into the subclavian vein) and that of the left upper quarter and of the entire lower body (via the thoracic duct) to the left venous angle (Fig. 94.34)

94.4.2 Lymph Collection and Lymph Transport

The intrinsic transporting activity of the lymphangions of the lymphatic collectors is strongly dependent on the filling state of the collectors. Therefore, this activity depends on two mechanisms upstream from the lymph transport:

- The hydration state of the interstitial tissue, i.e., the balance between filtration and reabsorbtion along the blood capillaries (see Sect. 94.4.2)
- The rate of uptake of interstitial fluid into the blind-ending intial lymphatics, i.e., lymph collection

Lymph flow itself is determined by the following factors:

- Balance of fluid filtration and fluid reabsorption in blood capillaries
- Interstitial fluid uptake into initial lymphatics through endothelial microvalves, driven by arterial and arteriolar pulsations and intermittent compression/stretch by organ activity
- Preload-dependent, intrinsic contractile activity of lymphangions
- Respiratory support of lymph flow in the thoracic duct.

Though lymphatic endothelial cells have contractile proteins and the ability to spontaneously deform (phagocytosis), they cannot initiate a cyclic deformation of the initial lymphatics such as to rhythmically propel lymph fluid into the precollectors across the flow-directing valves. Furthermore, lymphatic endothelial cells cannot transport interstitial fluid against a pressure gradient by an active mechanism. However, due to the microvalve-like arrangement of the endothelium and the fixation by the anchoring filaments that prevents collapse (Fig. 94.32), the primary

collection of interstitial fluid into the initial lymphatics is brought about by any mechanism that causes rhythmic compressions and/or stretch of the initial lymphatics. The former propel fluid towards the collectors, and the latter open the overlapping interendothelial junctions for the inflow of interstitial fluid and materials. Initial lymphatics are preferentially positioned adjacent to arterioles and can benefit from the pulsations of arteriolar diameter. Furthermore, rhythmic activity of parenchymal structures contributes to the compression-dependent filling of the initial lymphatics, striated muscle contractions in heart and skeletal muscle, peristalsis in the intestine, inflation and deflation of the lung, and intermittent compression or massage of the skin.

Provided that the lymphangions are adequately filled due to rhythmic "massage" of the initial lymphatics, they can develop substantial luminal pressures. In humans with cannulated lymphatic collectors at the ankle, lymphatic pressures up to 100 mmHg (13 kPa) during "lymphangion systole" (obstructed lymph flow) have been measured, while "diastolic" pressures lower than 15 mmHg (2 kPa) were observed without lymph obstruction, although the probands were standing [7,84]. These values illustrate the potency of the intrinsic pump mechanism of the lymphangions, which is very important for minimizing edema formation in the feet during standing or sitting.

94.5 Integrated Regulatory Aspects of Circulation

The function of the cardiovascular system is to maintain the extracellular milieu within temperature, osmolality, and concentration limits in respiratory gases, electrolytes, nutrients, and waste products that are tolerable for the integrity of cellular function.

This requires local adjustment of tissue perfusion to metabolic rate; it requires the maintenance of a critical extracellular volume, the maintenance of a sufficient reserve of perfusion pressure, and the adjustment of capillary pressure to the regulatory requirements of volume distribution and volume conservation, as well as the adjustment of convective heat transport to metabolic heat production and to heat losses. With more or less arbitrary simplifications, several regulatory cycles can be delineated that contribute to the intergrated homeostasis of the extracellular milieu (Fig. 94.35):

- Regulation of arterial pressure by baroreceptor reflexes
- Regulation of plasma volume by reflexes and hormones originating from the cardiopulmonary volume reservoir
- Regulation of erythrocyte volume by renal erythropoetin release
- Regulation of core temperature of the body by reflexes originating from cutaneous and central nervous thermoreceptors
- Regulation of extracellular concentration of metabolites and respiratory gases by the metabolic rate-sensitive at-

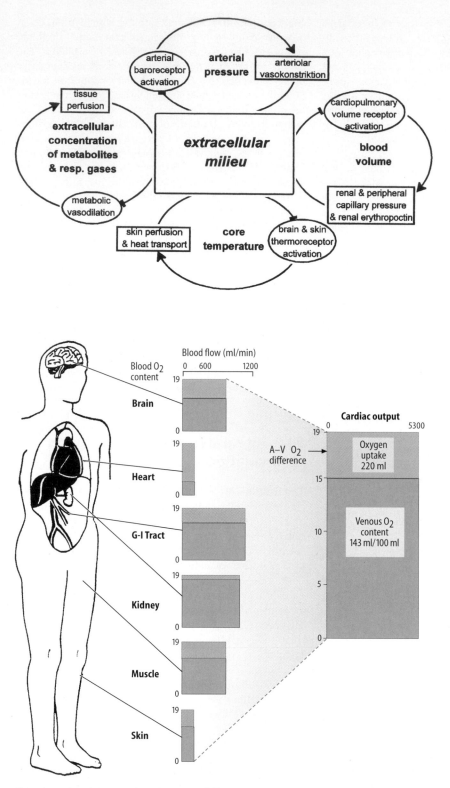

Fig. 94.35. Integrated homeostasis of the extracellular milieu. The circulation is the effector organ for sevearal overlapping, feedback-controlled regulatory loops

Fig. 94.36. Arteriovenous ($A-V$) oxygen differences in various tissues. Values for resting humans (modified from, [90]); during exercise, oxygen extraction in skeletal muscle can be substantially augmented. The venous oxygen content reflects the balance between blood supply to the tissue and aerobic metabolic demands. During emergency situations (hemorrhagic shock, cold exposure), this balance is curtailed dramatically in skin, muscle, and the gastrointestinal (*GI*) tract by sympathetic vasoconstriction

tenuation of vascular smooth muscle tone (metabolic vasodilation).

When considered separately under artificial isolation, all of these regulatory cycles demonstrate all the classical features of negative feedback control. However, these cycles are actually highly interconnected, affecting not only the peripheral circulation, but also the heart, the respiration, and other functions of the body. Several of these cycles use the sympathetic constrictor innervation of the peripheral circulation as an effector element and maintain parameters (mean arterial pressure, blood volume, core temperature) that are of vital importance for the entire body. Another cycle uses metabolic vasodilation as a strictly localized effector (and sensor) and maintains only the local metabolic milieu. Therefore, this regulatory cycle very often acts in opposition to the other three cycles, and the integrated control of the extracellular milieu requires that priorities are set. In emergency situations such as blood loss or sustained exposure to cold, the regulatory priority for parameters with vital importance results in drastic regulatory conflicts in which the maintenance of the extracellular milieu in certain organs is given up: extremities can freeze during cold exposure due to vasoconstriction to maintain the core temperature, and during hemorrhagic shock the extreme vasoconstriction to maintain arterial pressure can cause local acidosis and tissue damage. Under physiologic conditions, an organ- and function-specific balance is obtained in this regulatory conflict between metabolic vasodilation on the one hand and neurogenic (and hormonal) vasoconstriction on the other. This balance is partially reflected in the oxygen content of the venous blood leaving the tissues, i.e., *oxygen extraction* (Fig. 94.36). This oxygen extraction varies between organs and is variable for a given organ in different states of functional activation.

94.5.1 Arterial Blood Pressure Regulation

In the aorta of resting human adults, diastolic blood pressures of 70–89 mmHg (9.3–11.4 kPa) and systolic blood pressures of 110–139 mmHg (14.6–18.5 kPa) are considered as normal, i.e., normotension. Mean arterial pressure, which is obtained by integration of the pressure curve over the entire cardiac cycle, is slightly below 100 mmHg (13.3 kPa) at rest. In societies with a Western life-style, there is a moderate increase in arterial pressure with increasing age in the normotensive population at rest (Fig. 94.37). During exercise, however, mean arterial pressure, pressure amplitude, and systolic peak pressure are substantially augmented, parallel to the strenuosity of exercise (see Sect. 94. 5.3).

Diastolic pressures above 95 mmHg (12.6 kPa) at rest and systolic values above 160 mmHg (21.3 kPa) represent overt *hypertension* (see Sect. 94.5.1.3) and are associated with pathologic alterations and damage in many organs, e.g., heart, kidney, retina of the eye, and with a substantially elevated risk of myocardial infarction, stroke, renal failure, and cardiovascular mortality.

The pressures within the arterial system are regulated and stabilized within fairly narrow limits by short-acting and long-acting control mechanisms.

94.5.1.1 Short-Acting Regulation of Arterial Pressure

Such short-acting regulation is obtained by negative feedback control via reflexes and via hormonal regulatory loops. Stabilization of mean arterial pressure is the main target of this short-acting regulation. It is obtained by modulating cardiac output and peripheral vascular resistance. To this end, vasoconstriction of the small arteries and arterioles as well as frequency and strength of the cardiac contractions are regulated. Additionally, however, the capacity of the low-pressure system, the balance of filtration and reabsorption in the microcirculation, and renal excretion of urine are modulated by the short-acting stabilization of arterial pressure.

For this short-acting stabilization, information on the arterial blood pressure and on the filling of the thoracic part of the circulation is obtained by *arterial pressoreceptors* and by *volume receptors* in the atria and the large thoracic veins. The afferents from these sensors are conducted via the vagal nerve and the glossopharyngal nerve to the nucleus of the solitary tract within the dorsal medulla oblongata in the brain stem. From this nucleus, projections

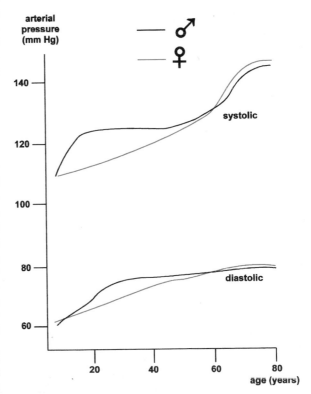

Fig. 94.37. Age-dependent changes systolic and diastolic arterial pressures in the normotensive population [99]

reach higher centers of the brain and the reticular formation of the medulla oblongata. These inputs from the circulatory sensors, together with inputs from higher brain centers, from pain and chemoreceptor afferents, from the defense area, and from the the lateral hypothalamus (see also Sect. 94.3.3.3), are integrated into a tonic activity from the rostral ventrolateral medulla to spinal preangionic sympathetic neurons and to the preganglionic parasympathetic neurons in the ambiguous nucleus. This complex interplay of various medullary areas in the brain stem is considered functionally as a circulatory center, though this vague term does not describe an anatomic entity in the brain. The input from the pressoreceptors attenuates sympathetic activity and increases parasympathetic activity, resulting in a modulation of cardiac output and peripheral vascular resistance (Fig. 94.38).

Arterial Pressoreceptors. Also called baroreceptors, these are plexus of nerve endings with lamellated end organs, located between adventitia and media of the aortic arch and the carotid arteries. Transmural distension of these large arteries activates the baroreceptors (Fig. 94.39). The most important pressoreceptor areas are at the top of the aortic arch, at the bifurcations of the carotid arterries (carotid sinus), and at the stem of the common carotid arteries. Fibers from the pressoreceptors at the carotid sinus run to the glossopharyngeal nerve together with fibers from the chemoreceptors in the carotid body. Fibers from the pressoreceptors at the aortic arch and at the common carotid artery reach the vagal nerve together with fibers from the aortic glomera. Activation of the pressoreceptors depends on the speed of arterial pressure increase (i.e., of

arterial distension), on its amplitude, and on the height of mean arterial pressure (Fig. 94.39). The effect of pressoreceptor activation is inhibition of sympathetic activity (Fig. 94.38). This inhibition operates even at

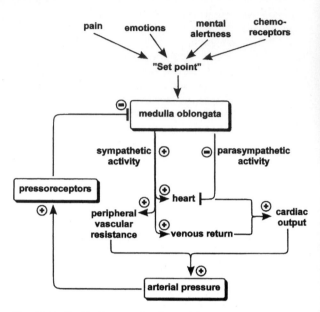

Fig. 94.38. Feedback control of arterial pressure by arterial pressoreceptors. Activation of arterial pressoreceptors by elevation of arterial pressure augments pressoreceptor input into the medulla oblongata, resulting in a reduction of the medullary stimulation of sympathetic activity and in an augmented parasympathetic activity. Sympathetic activity augments cardiac output and peripheral resistance, while parasympathetic activity lowers cardiac output by slowing heart rate

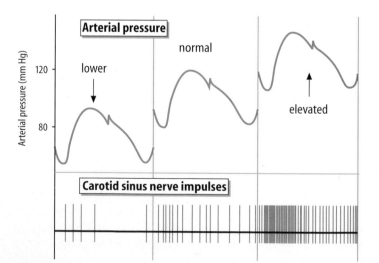

Fig. 94.39. Activation of arterial pressoreceptors. The recorded changes in carotid sinus nerve activity are typical for acute changes in arterial pressure. With chronic changes in pressure, the pressoreceptors adapt within a few days to the new pressure level, helping to stabilize this new level

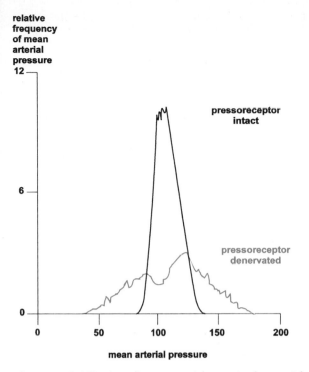

relative
frequency
of mean
arterial
pressure

Fig. 94.40. Stabilization of mean arterial pressure by arterial pressoreceptors. Chronic recording of mean arterial pressure in an instrumented conscious dog demonstrates that the distribution of mean arterial pressure is very narrow with intact pressoreceptors, but broader around the same mean value following surgical denervation of the pressoreceptors. (Modified from [27])

normotensive pressure levels and is enhanced with acute increases in arterial pressure, together with an increase in parasympathetic activity to the heart. A fall in arterial pressure below normal levels reduces pressure receptor activation and results in a disinhibition (i.e., elevation) of sympathetic activity.

Although sympathetic activity is restrained by the pressoreceptors at normotension, chronic interruption of this restrain in animal experiments (Fig. 94.40) does not result in hypertension; arterial blood pressure regulation becomes much more labile, but still oscillates around the same average. The pressoreceptor reflexes do not determine the set point for arterial pressure regulation, but rather only stabilize a given set point. Chronically altered excitatory influences from higher brain areas, however, are presumed to shift the set point and may contribute to the development of hypertension (see Sect. 94.5.1.3).

With strong activation of pressure receptors by acute hypertension, inhibitory modulation of central nervous functions does occur: conscious animals may become sleepy, and respiration and skeletal muscle tonus may be reduced. In patients of advanced age and with arteriosclerotic alterations of the large arteries, a strong activation of the pressure receptors with immediate reduction in sympathetic activity can be elicited by a stroke on the area of the carotid sinus at the neck. This reaction can be used as a simple intervention to interrupt paroxysmal tachycardias. However, the resulting decline in blood pressure can be large enough to cause a transient loss of consciousness.

Volume Receptors. These are nerve endings in the atria and in the wall of the large veins at their entrance into the atria. Similarly to the arterial pressoreceptors, they are activated by changes in transmural pressure. However, changes in central venous pressure largely reflect changes in blood volume in the circulation, at least in resting humans in the supine position. Therefore, the term "volume receptors" became an established, although not entirely correct term. In fact, these receptors monitor how full the thoracic circulation is. This filling of the thoracic circulation is reduced immediately during orthostasis, resulting in a concomitant and immediate unloading of the intrathoracic volume receptors, while intravascular volume changes only very slowly during orthostasis. Unloading of these volume receptors results in sympathetic activation, while their activation causes sympathetic inhibition with parasympathetic activation. This is very similar to the the reactions elicited by unloading or by activating the arterial pressoreceptors, but organ differences do exist: volume receptor activation reduces sympathetic constriction of the renal vascular bed and enhances diuresis, while pressoreceptor inputs have less influence on the kidney, but more on skeletal muscle and on the splanchnic beds (Fig. 94.41). Inputs from the volume receptors reach the osmoregulating hypothalamus, inhibiting the release of antiuretic hormone.

The nervous control of capillary filtration in peripheral organs by alterations in the ratio of pre- and postcapillary resistances (see Chap. 97) is under the influence of both pressoreceptors and volume receptors. Acute hypotension enhances this ratio, thereby lowering capillary pressure and allowing enhanced reabsorption of interstitial fluid into capillaries and venoles. A similar alteration in the compartmentalization of extracellular volume in favor of vascular volume is elicited by unloading the volume receptors. Activation of pressoreceptors or volume receptors induces the opposite reaction. These volume shifts are especially relevant in the skeletal muscle bed (Fig. 94.41).

Recordings from afferent fibers from the cardiac atria disclose two types of discharges relative to the atrial pressure waves: A receptors are activated during atrial contraction, while B receptors are mainly activated during the filling phase of atria (Fig. 94.42). The B receptors are activated by elevations in the filling of the thoracic parts of the circulation; they are the volume receptors shown in Fig. 94.42 (together with similar receptors in the large veins). The A receptors result in sympathetic activation, but their physiologic role is less clear. They might be involved in the tachycardia occasionally occurring during rapid infusions of large volumes (Bainbridge reflex). Ventricular receptors are activated during isovolumetric contraction of the ventricles; their enhanced activation during acute ventricular distension causes bradycardia and reduced vasoconstriction. These ventricular receptors can probably

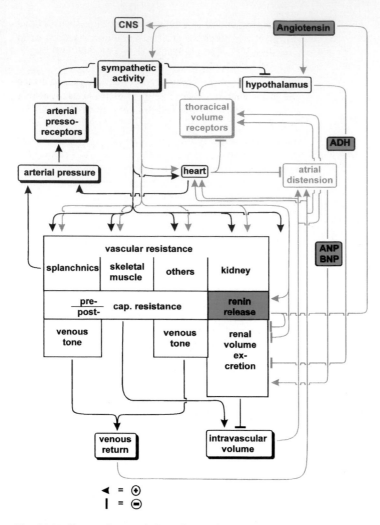

Fig. 94.41. Short-acting regulation of arterial pressure via feed-back control by arterial pressorecetors and by thoracic volume receptors. The parallel inhibition of sympathetic activity by input from baroreceptors and from volume receptors results in a partially overlapping regulation of arterial pressure, venous return, and intravascular volume. Input from the pressoreceptor preferentially controls vascular resistance in the splanchnic, skeletal muscle, and cutaneous bed, capillary volume reabsorption via the pre-/postcapillary resistance ratio in the skeletal muscle bed, and venous tone in the splanchnic and cutaneous beds. Input from the volume receptors preferentially controls intravascular volume via renal perfusion and renal urine excretion. These rapidly acting nervous feedback regulations are supported by hormonal feedback loops (*shaded blocks*) acting mainly in the low-pressure system (see also Sect. 94.3.3.4): release of the natriuretic peptides *ANP* (atrial natriuretic peptide) and *BNP* (brain natriuretic peptide) from the atria, release of renin from renal afferent arterioles with subsequent formation of angiotensins, and release of *ADH* (antidiuretic hormone) from the hypothalamus via the neurohypophysis. Note that the slowly acting volume control by pressure diuresis is not included

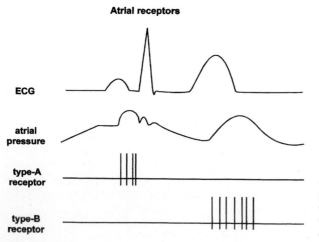

Fig. 94.42. Atrial receptors. The type-A receptors can cause tachycardia during rapid infusions, but are of minor relevance; type-B receptors induce inhibition of sympathetic activity and are important volume receptors. *ECG*, electrocardiogram

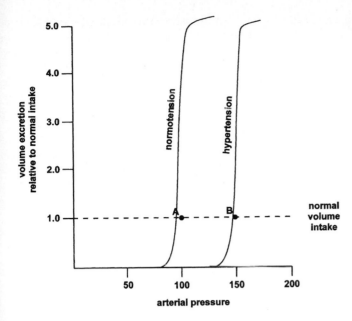

Fig. 94.43. Pressure diuresis. The renal excretion of urine is strongly dependent on mean arterial pressure in mmHg; the steep slope of the curve indicates an almost infinite gain of the feedback control of arterial pressure. Any increase in arterial pressure due to augmented intake of volume and salt will be corrected within several days by the renal control of volume via pressure diuresis, and pressure and volume are in balance (A). In hypertensive patients, the characteristic of the pressure-dependent renal volume excretion is altered, and a higher arterial blood pressure is required to keep extracellular volume and pressure in balance (B)

be chemically activated by intracoronary injections of serotonin, nicotine, or veratrine, causing bradycardia, vasodilation, and collapse of blood pressure (Bezold-Jarisch reaction).

The reflex modulations of sympathetic activity originating from pressoreceptors and volume receptors are paralleled by hormonal reactions affecting the circulation. During strong activation of the sympathetic activity by hypotension and by hypovolemia, epinephrine is released from the adrenal medulla, causing β-adrenergic vasodilation in the skeletal muscle bed, but enhancing generalized α-adrenergic vasoconstriction at very high concentrations of epinephrine. Sympathetic activation due to unloading of volume receptors (and to a lesser extent due to baroreceptor unloading) enhances sympathetic activity to the kidney and causes β-adrenergic release of renin from the afferent arterioles. Concomitantly with the acute activation of volume receptors by overfilling of the thoracic circulation, the distension of the atria induces release of natriuretic peptides (ANP and BNP; see Sect 94.3.3.4) from the cardiac atria. The effects of these peptides on the kidney contribute to the renal reactions to acute cardiac overfilling. These hormonal reactions support the short-acting blood pressure regulation by reflexes (Fig. 94.41), but their effects are longer-lasting, and these hormones are also involved in the long-term regulation of arterial blood pressure.

94.5.1.2 Long-term Regulation of Arterial Blood Pressure

The central element of the long-term regulation of arterial blood pressure is pressure-dependent renal urine excretion, *pressure diuresis* [96]. This pressure-dependent control of blood volume is very beneficial for the control of arterial pressure [53] (see also Fig. 94.43). The term pres-

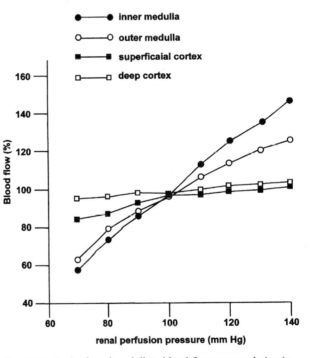

Fig. 94.44. Lack of renal medullary blood flow autoregulation in the basis of pressure diuresis. While blood flow to all layers of the renal cortex is perfectly autoregulated, medullary flow follows perfusion pressure. (Data from [73])

sure diuresis denotes the fact that elevations in renal perfusion pressure cause increments in renal Na^+ excretion, while renal blood flow and the global glomerular filtration rate remain constant due to autoregulation [87,96]. The crucial mechanism of this pressure diuresis is the change in renal medullary flow parallel to change in perfusion pressure, while flow to the superficial and deep renal cortex is autoregulated (Fig. 94.44). The pressure-

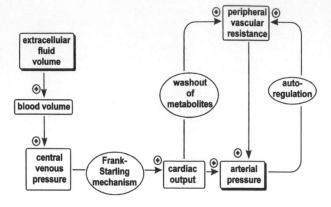

Fig. 94.45. Long-term dependence of arterial blood pressure from extracellular fluid volume. Due to the strong impact of extracellular volume on blood pressure, a closed feedback loop cannot operate without a volume control via pressure diuresis

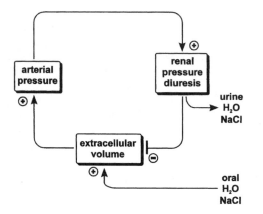

Fig. 94.46. Long-term control of volume and blood pressure by the kidney. The individual characteristics of the renal pressure diuresis determine the level of arterial blood pressure at which extracellular volume is kept normal in the presence of large variations in salt and water intake

dependent increase in renal medullary flow causes a parallel rise in renal interstitial hydrostatic pressure, and the increase in interstitial pressure and in flow-dependent washout of medullary solutes attenuates the Na^+ and Cl^- transport in the thick ascending loop of Henle and the collecting duct [87]. Since deep cortical glomerular flow is well autoregulated with elevated perfusion pressure (Fig. 94.44), a dilatory signal from the preglomerular vasculature must reach the medullary vessels originating from the deep cortical postglomerular efferent vessels. Shear-induced release of NO and prostaglandins may act as such a signal, since pressure diuresis is attenuated after suppression of the formation of NO and prostaglandins [87].

In the long-term control of arterial blood pressure, this pressure-dependent regulation of urine excretion operates in concert with the dependency of blood pressure on extracellular fluid volume (Fig. 94.45). The interplay of these two basic control mechanisms implies that an elevation of the extracellular fluid by enhanced intake of

volume and salt will finally result in an elevated blood pressure, which causes enhanced renal excretion of volume and salt, exceeding the intake. This excess excretion will reduce extracellular fluid volume and thereby reduce blood pressure back to normal (Fig. 94.46). Within this framework of nephrogenic long-term volume control, additional systems such as osmoregulation and regulation of thirst (see Chaps. 76, 77, 82) establish the fine tuning of water and solute intake relative to each other. The central role of the kidney in long-term control of arterial blood pressure via volume control implies that chronic elevations in arterial pressure are not possible without involvement of the kidney. This does not mean that in all patients pathologic alterations in the kidney are the primary mechanism leading to hypertension, but secondary changes in the characteristics of renal pressure diuresis contribute to the maintenance of hypertension (see Sect. 94.5.1.3).

94.5.1.3 Hypertension

Arterial blood pressure is normally distributed within populations, and no natural cutoff point separates normotensive and hypertensive individuals. Therefore, the definition of hypertension is some-what arbitrary and results from international agreement. This agreement is based on the fact that the risk of severe cardiovascular complications at blood pressure levels above "normo-tension" is high enough to warrant an intervention by

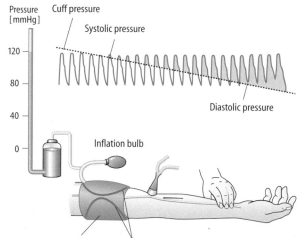

Fig. 94.47. Arterial blood pressure measurement by spygmomanometry. The pressure within the cuff has to be increased above systolic arterial pressure until the arteries under the cuff are occluded and no pulse can be palpated or heard by the stethoscope. The cuff has to be fixed closely around the arm and must be broad enough (at least the diameter of the arm) to ensure that the pressure in the cuff equals the tissue pressure around the artery. As the cuff pressure is gradually released, the systolic arterial pressure peaks exceed the cuff pressure, sudden accelerations of blood increase in intensity and then rather suddenly become muffled at the level of diastolic pressure where the arteries remain open throughout the entire cycle. (Modified from [90])

Table 94.4. Blood pressure measurements in adults

Diagnosis	Systolic			Diastolic	
	(mmHg)	(kPa)		(mmHg)	(kPa)
Normotension	<140	<18.6	and	<90	12.0
Mild hypertension	140–180	18.6–24.0	and/or	90–105	12.0–14.0
Moderate to severe hypertension	>180	>24.0	and/or	>105	>14.0
Isolated systolic hypertension	>140	>18.6	and	<90	<12.0
Bordertine hypertension	140–160	18.6–21.3	and	90–95	12.0–12.6

antihypertensive treatment. This international agreement [4] requires that repeated, technically sound blood pressure measurements (Fig. 94.47) [3] are obtained on different occasions for a definitive diagnosis according to the values given in Table 94.4 for adults. For children and young adolescents, the critical margin of 90 mmHg diastolic blood pressure is lower [103]. A simple rule of thumb says that the age difference to 18 years has to be subtracted from 90 mmHg: for children aged 4 years, the upper limit for normotension is: <90 − (18 − 4) = <76 mmHg.

However, the physician should be aware that, for hypertensive disease, the "definition by consent" is oriented toward the patient's prognosis and that this prognosis is determined by many additional inherited and acquired cardiovascular risk factors. It is not high blood pressure alone that increases the risk of developing morbid cardiovascular events, but rather a set of cardiovascular abnormalities that results in morbid events. Therefore, the assessment of the severity of the hypertensive disease should never be based on blood pressure measurement alone. The presence of risk factors for cardiovascular complications (such as smoking, hypercholesterolemia, obesity, diabetes, a positive family history of cardiovascular diseases) and the presence of markers of cardiovascular organ damage (such as left ventricular hypertrophy, microalbuminuria, manifest proteinuria, and "fundus hypertonicus" in the retina of the eye) contribute to the severity of the disease. An obese smoker with left ventricular hypertrophy, microalbuminuria, insulin resistance (i.e., a prediabetic state), a positive family history, and an arterial blood pressure of 165/100 mmHg has severe hypertensive disease which urgently requires therapeutic intervention. Classification such a patient as "mildly hypertensive" based on blood pressure measurement alone would be misleading. Consequently, antihypertensive treatment should be considered as cardiovascular risk factor management, which aims at the correction of hypertensive blood pressure together with other risks and abnormalities in the cardiovascular system.

In approximately 95% of patients with hypertension, the primary reason for the elevated blood pressure is unknown and the disease is called primary or essential hypertension. Only 5% of patients have secondary hypertension with an identifiable reason for the elevated blood pressure. Secondary forms of hypertension include the following:

- Renovascular hypertension: obstructive lesion(s) within renal arterial tree
- Renal diseases: interstitial nephritis, glomerulonephritis, diabetic nephropathy, polycystic renal disease, terminal renal failure
- Congenital coarctation of the descending aorta
- Endocrine hypertension: pheochromocytoma, acromegaly, hyperparathyroidism, primary hyperaldosteronism, Cushing syndrome, hypothyroidism
- Pregnancy (preeclampsia)
- Drug-induced hypertension: cyclosporine, oral contraceptives, recombinant erythropoetin.

In these patients, the hypertension can be normalized if the primary cause can be corrected early in the course of the hypertensive disease or if the reason is only temporal (pregnancy). Such a causal therapy is not possible in essential hypertension due to a lack of identified primary causes. In essential hypertension, the treatment must be symptomatic.

A clear familial aggregation of essential hypertension exists which appears to be of polygenic origin. Most probably, a complex mosaic of inherited traits and of chronically repeated environmental influences convene to repeatedly produce exaggerated reactions with high sympathetic activity, renal vasoconstriction, and acutely elevated blood pressure (Fig. 94.48). Such reactions are physiologic in the preparation for fight or flight (defense reactions), but chronically repeated without the subsequent challenge of physical activity, they contribute to the gradual fixation of enhanced arterial pressure by two mechanisms:

- Resetting of the arterial pressoreceptor reflexes, i.e., by shifting the set point of this feedback control to higher values (see Sect. 94.5.1.1)
- Inducing adaptative hypertrophy in the wall of the peripheral arterioles, which tends to normalize the wall stress in presence of hypertension, but which amplifies the vascular responsiveness to all kinds of constrictory stimuli.

An altered ratio in the releasability of endothelial constrictory and dilatory signals contributes to this enhanced constrictor responsiveness, and it is not clear to

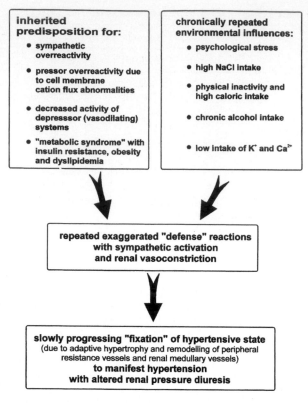

inherited predisposition for:
- sympathetic overreactivity
- pressor overreactivity due to cell membrane cation flux abnormalities
- decreased activity of depressor (vasodilating) systems
- "metabolic syndrome" with insulin resistance, obesity and dyslipidemia

chronically repeated environmental influences:
- psychological stress
- high NaCl intake
- physical inactivity and high caloric intake
- chronic alcohol intake
- low intake of K^- and Ca^{2+}

repeated exaggerated "defense" reactions with sympathetic activation and renal vasoconstriction

slowly progressing "fixation" of hypertensive state
(due to adaptive hypertrophy and remodelling of peripheral resistance vessels and renal medullary vessels)
to manifest hypertension with altered renal pressure diuresis

Fig. 94.48. Mosaic of inherited and acquired factors contributing to essential hypertension. The fixation of the hypertensive reactions appears to be the common final pathway of a large number of constellations

what extent such changes in endothelial function are primary contributors or secondary enhancers of the disease. The morphologic changes in the arteriolar wall of hypertensive subjects include an increased vessel wall thickness, an increased wall cross-sectional area, and an increased wall to lumen ratio [79,80]. These changes are the net effect of real growth and of rearrangement of the smooth muscle cells in the wall. Subtle structural and functional changes in the renal arterioles supplying the medullary capillaries are sufficient to cause the altered renal pressure diuresis (Figs. 94.43, 94.46), which is essential for maintaining an hypertensive blood pressure in presence of normal extracellular volume. Such secondary alterations in the renal pressure diuresis in essential hypertension, which are important for fixation of the disease, regardless of its primary cause, should be differentiated from the substantial stenoses in preglomerular arteries in secondary renovascular hypertension (see above), which act as primary hypertensive causes by inducing renal renin release and activation of the renin–angiotensin system.

Daily salt intake figures prominently among the chronically repeated environmental influences which are assumed to contribute to the fixation of hypertension (Fig. 94.48). Daily salt intake in Western societies, in which essential hypertension is prevalent, is between 150–

250 mmol/day (9–15 g/day). In many nonindustrialized populations with a daily intake of 30 mmol/day (below 2 g/day), hypertension is rare and an age-dependent increase in blood pressure does not exist. In several tribes in the Amazonas area without any cardiovascular diseases at all, the daily salt intake is 5 mmol/day (0.3 g/day). Comparisons of this type and the dominant role of the renal volume (and salt) excretion in the long-term regulation of arterial blood pressure (see Sect. 94.5.1.2) led to the theory that excess salt intake is the main culprit for the high cardiovascular morbidity in Western societies and gave rise to the postulate that salt intake in the entire population should be reduced as a preventive intervention. However, the available data do not support such a postulate. Restriction of salt intake to below 6 g/day (100 mmol/day) has a negligible effect on blood pressure in the general population (less than 1 mmHg) and induces only a moderate decline in blood pressure in a small subset of hypertensive patients considered to be "salt sensitive" [38]. More drastic reductions in salt intake below 5 g/day cause a substantial loss in the quality of life, considerable neurohumoral activation, and an elevated frequency of orthostatic collapse in populations not adapted to salt deficiency. Thus, salt intake of between 6 and 15 g/day can be considered as the "physiological, hygienic safety range" [38]. The differences in cardiovascular morbidity between different populations must have much more complex reasons.

94.5.2 Effects of Posture on Circulation

The intravascular pressure in a specific section of the circulation is determined by three components:

- A *dynamic* component, determined by cardiac pumping and resistance to flow; this is the dominant component is the arterial system.
- A *static* component, determined by volume and by capacity; this is the dominant component in the venous system and in the pulmonary circulation, (i.e., in the low-pressure system).
- A *hydrostatic* component, determined by gravity and varying with posture.

In humans lying down, the relevance of the hydrostatic component is marginal, since vertical distances within the vasculature are small, and all vascular pressures are traditionally given relative to the position of the heart. Differences in the hydrostatic component of intravascular pressure are only relevant in the pulmonary circulation because of the large distensibility of this bed (see Fig. 94.29). Hydrostatic pressure differences in the range of 10–15 cm H_2O, as they occur between dorsal and rostral parts of the lung in the supine position, do indeed affect local pulmonary flow and local fluid filtration and may affect pulmonary gas exchange, if persisiting for several hours (see Chap. 104). This is one of the reasons why severely ill patients who cannot move have to be

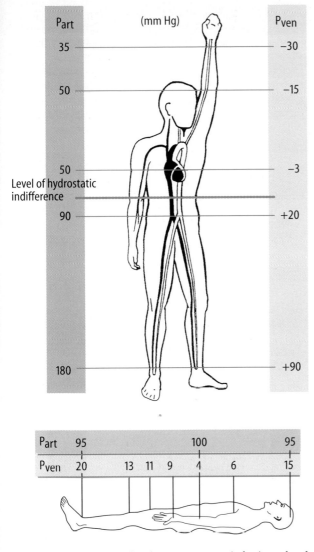

P_{art}	95				100		95
P_{ven}	20	13	11	9	4	6	15

Fig. 94.49. Mean arterial and venous pressure in horizontal and vertical positions. At the level of hydrostatic indifference, intravascular pressure does not change during transition from the horizontal to the vertical position. Note that the full elevations of venous pressures in the foot veins in the vertical position occur only without movement (i.e., without support of venous return by skeletal muscle contractions). Negative pressures in the veins of the upper body result in venous collapse (except for the rigid veins in the dura mater of the skull). Although venous pressure in the thorax is lower than that of the atmosphere in the upright position, there is no collapse of intrathoracic veins because the intrapleural pressure is lower. Thus, transmural venous pressure is still higher

mobilized and turned over into another position by the nursing staff.

In the upright position, however, the hydrostatic component substantially affects the circulation: in the lower part of the body, the hydrostatic component adds to the arterial and (at least partially) to the venous pressures. In the upper part of the body, the hydrostatic component is sub-

tracted (Fig. 94.49). Several problems arise from this effect of the hydrostatic component on intravascular pressures:

- Adequate diastolic filling of the heart must be ensured.
- Altered filling of the thoracic circulation affects the input from the volume receptors.
- The equilibrium of capillary filtration and reabsorption is challenged in the lower parts of the body in the upright position.

During changes in posture, no changes in intravascular pressure take place at the level of hydrostatic indifference, which is 5–10 cm below the diaphragm (Fig. 94.49). The position of this indifference level is deteremined by the relative distensibilities of the vessels of the low-pressure system in the upper and the lower part of the body.

The fact that this level of indifference is below the heart implies that the filling of the right heart is immediately curtailed with transition to the upright position. Approximately 400–500 ml blood are quickly shifted out of the thoracic section of the circulation into the veins of the legs. Although the reservoir of the pulmonary bed is then reduced, the left heart can still be filled out of this reduced reservoir for roughly ten beats, if right heart output fails due to inadequate filling. Therefore, reactions supporting venous return have to be activated within less than 10 s in order to compensate for the attenuated cardiac filling [45]. These reactions are elicited by unloading of the cardiopulmonary volume receptors (Figs. 94.41, 94.42) due to the reduction in cardiopulmonary blood volume and by reduced activation of the arterial pressoreceptors, which are located some 40 cm above the level of hydrostatic indifference and which therefore immediately sense a reduction in transmural pressure when the upright position is assumed. Due to this unloading of volume receptors and pressoreceptors, several compensatory reactions are triggered, not only affecting the venous system:

- Reduction in venous capacity by venoconstriction, preferentially in the legs and in the splanchnic bed
- Elevation of vascular resistance in skeletal muscle, skin, splanchnic organs, and kidney by arteriolar vasoconstriction
- Tachycardia and augmentation of ventricular contractility
- Catecholamine release from the adrenal medulla
- Activation of the renin–angiotensin system by renal renin release
- Increased release of antiuretic hormone and reduced release of natriuretic peptides (ANP, BNP) from the cardiac atria

These reactions only partially compensate for the effects of the upright posture (Fig. 94.50): while arterial pressure is maintained largely unchanged due to the increase in total peripheral vascular resistance, cardiac output is reduced. The reflex tachycardia does not fully compensate for the reduction in stroke volume, which results from the reduc-

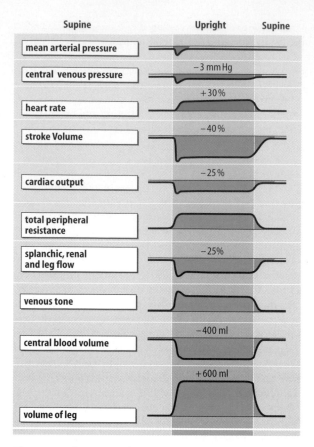

mean arterial pressure	
central venous pressure	−3 mmHg
heart rate	+30%
stroke Volume	−40%
cardiac output	−25%
total peripheral resistance	
splanchic, renal and leg flow	−25%
venous tone	
central blood volume	−400 ml
volume of leg	+600 ml

Fig. 94.50. Alterations of cardiovascular parameters with the transition from the horizontal to the vertical position

tion in central venous pressure and in cardiopulmonary blood volume (Fig. 94.50). The venoconstriction alone cannot compensate for the large increase in the hydrostatic component of venous pressure in the legs in the upright position. Such a compensation requires the support of venous return by the skeletal muscle pump, which operates very effectively during walking or running. Any contraction of skeletal muscle in the leg cause a compression of the deep veins, which is directed into a flow towards the heart by the venous valves. Thus, venous pressure in the veins of the foot is reduced with each step (Fig. 94.51). While this muscle pump effectively contributes to venous return during walking in the upright position, it is also marginally effective during standing: small balancing movements always occur during standing and slightly support venous return. During orthostasis on a tilting table with the legs hanging freely, the muscle pump cannot operate and the reduction in central venous pressure and cardiopulmonary blood volume is more pronounced compared to the reduction during active standing in the upright position (Fig. 94.51).

The unloading of the thoracic volume receptors during standing directly affects the volume regulation by reducing the urine volume. This reduction is due to enhanced renal sympathetic activity and to the hormonal alterations mentioned above (activation of the renin–angiotensin–

aldosterone system, release of antiuretic hormone, and reduced release of natriuretic peptides from the cardiac atria). Such an orthostasis-induced physiologic oliguria can be demonstrated indirectly by comparing the urine produced while standing in a thermoneutral bath with immersion up to the neck to that produced while standing in a room: the hydrostatic pressure of the bath is transimitted to the venous bed and prevents most of the orthostasis-induced volume shifts; the thoracic volume receptors are

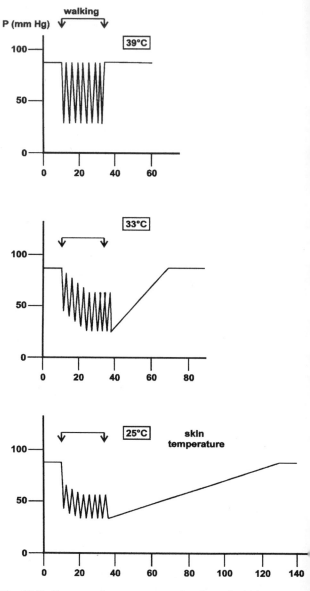

Fig. 94.51. Transmural venous pressure in a foot vein during upright standing and walking at various ambient temperatures. With each step, some volume in the veins is pressed through the venous valves towards the heart by compression by the contracting skeletal muscle. With relaxation of the compression, the veins are refilled by inflow from the perfused tissue. This refilling occurs very rapidly if the perfusion of the tissues in the leg is high under high ambient temperature, but is retarded at low temperatures, when perfusion is reduced. (Modified from [56])

not unloaded and the urine production is two to three times higher than during standing in a room [45].

Local Edema Formation in the Upright Position. A complicated set of mechanisms attenuates the excess increase in interstitial volume which is to be expected during upright position in the lower body due to enhanced transmural capillary pressures [6]:

- Arteriolar vasoconstriction by myogenic autoregulation and by reflexes reduces the hydrostatic increase in capillary pressure [71].
- Constriction of precapillary spincters reduces the capillary surface area participating in filtration by up to 80% [6]
- Support of venous return by the muscle pump during walking helps to reduce capillary pressure.
- Enhanced lymph flow in the upright position (ten to 15 times the lymph flow during the night during bedrest) removes excess interstitial volume [35].
- Colloid osmotic buffering due to a increasing concentration of plasma proteins and a reduction in interstitial fluid protein content ("washdown") with ongoing filtration.
- Interstitial hydrostatic buffering by augmented hydrostatic tissue pressure due to ongoing filtration. This effect is most important when external compression, e.g., by elastic stockings, reduces the compliance of the interstitial space.

The last three mechanisms are only possible if excess filtration occurs in the upright position. This implies that these edema-reducing mechanisms never completely compensate for the enhanced hydrostatic load on the lower body microcirculation. Measurements in the field demonstrate that sitting office workers have an 5% increase in foot volume during 8h at work, while this increase amounts only to 1%–2% in the same period in waitresses [6]. These values can be doubled under high temperatures. However, formation of clinically detectable edema in the foot would require a volume swelling by more than 20%. At such high interstitial volume, interstitial tissue pressure becomes positive relative to the atmosphere (2–3mmHg) and the compliance of the interstitial space increases tremendously. This means that with such a level of edema, interstitial hydrostatic buffering no longer operates to prevent further increase in edema (see Chap. 97).

94.5.3 Circulation During Exercise

The cardiovascular adaptation to strenuous exercise requires activation of the system up to the limits for which it is designed. The high, metabolically determined flow to the working skeletal muscle, the necessity to get rid of the heat formed and the problems of a limited pumping capacity of the heart are the major challenges for cardiovascular regulation under strenuous exercise.

In a healthy, young adult without special endurance training, cardiac output can increase from 5–7l/min at rest to a maximum of 25–28l/min during strenuous exercise, with total body O_2 consumption increasing from 0.3l/min to almost 4l/min and with total skeletal muscle flow increasing from less than 1l/min to 20–21l/min (Fig. 94.52). On average, maximal flow in working skeletal muscle in the untrained adult during strenuous exercise amounts to 120ml/100g per min, but major differences exist between different types of skeletal muscle (see Chap. 95). However, this is clearly not the maximal flow, which would be possible with full dilation of the vascular bed in working skeletal muscles. If only small groups of skeletal muscle are working with maximal intensity, local maxima of skeletal muscle flow of 200–300ml/100g per min can be measured [75,93]. If such a maximal vasodilation were to occur in all skeletal muscles simultaneously, a cardiac output above 40l/min would be necessary. Thus, the maximal perfusion of skeletal muscle during strenuous exercise is limited by the amount of cardiac output which can be pumped by the heart under maximal heart rate and contractility in the presence of optimally supported venous return by the exercising muscles.

In athletes with endurance training, maximums of O_2 consumption of up to 10l/min are possible under defined test conditions. Such athletes have hypertrophied hearts with enlarged ventricular volumes; their heart rates at rest are below 40 beats per min and can be increased up to values far above 250 beats per min. Their skeletal muscles have a higher capillary density and an enhanced capacity for anaerobic metabolism, and their lung volumes and pulmo-

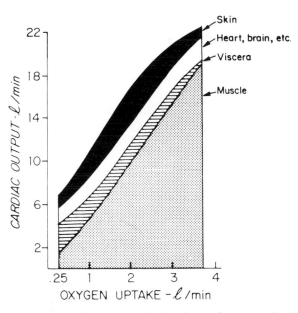

Fig. 94.52. Cardiac output and its distribution during exercise as function of total body oxygen consumption. The more than 20-fold increase in skeletal muscle flow is covered by augmented cardiac output; extramuscular flow remains largely constant. Flow to the splanchnic and renal bed is reduced with increasing exercise intensity, while flow to heart, brain, and skin is augmented. (From [88], with permission)

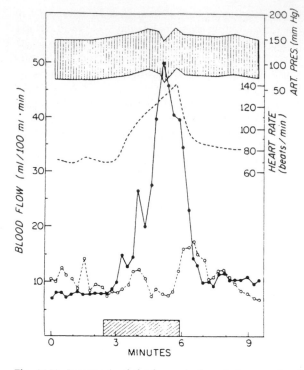

Fig. 94.53. Increase in skeletal muscle flow during emotional stress. During the 3 min marked by the *hatched area* on the time axis, the conscious proband erroneously believed that he had lost substantial amounts of blood. He experienced tachycardia and a fivefold increase in forearm muscle flow (*closed circle*) without a clear alteration in flow to the skin (*open circles*). (From [12], with permission)

nary capillary surface area are substantially enlarged. The factors which may limit the maximal exercise performance in such trained athletes are complex and not entirely clear. The increase in heart rate and in cardiac contractility during exercise is mediated by the activation of myocardial β_1-adrenoceptors by norepinephrine from the activated sympathetic nervous system and by circulating epinephrine released from the adrenal medulla (see Chaps. 90, 93). This enhancement of sympathetic activity and of adrenal medullary catecholamine release is not due to reflexes triggered by pressoreceptor unloading. No transient hypotension at the onset of exercise occurs in healthy humans. In contrast, activation of sympathetic activity and reduction of parasympathetic activity take place prior to the onset of skeletal muscle work. They are of central nervous origin and result from a coactivation of the sympathoexcitatory neurons in the ventrolateral medullary centers by the motor cortex. They induce an autonomic starting reaction in advance of any peripheral, metabolically induced vasodilation. Some neurogenic vasodilation, however, induced by the cholinergic sympathetic nervous system from the frontal cortex (see Sect. 94.3.3.3), is a transient part of this starting reaction. Such transient cholinergic vasodilation can also be elicited only by anticipation of exercise for fight or flight and by emotions such as rage or fear (Fig. 94.53).

During the steady state phase of exercise, sympathetic activity is stimulated by those higher brain centers which control skeletal muscle performance during muscular work. This stimulation is called central command [94]. Additionally, afferents from the working skeletal muscle, conducted via C fibers to the spinal cord, contribute to the fine tuning of sympathetic activity in relation to the elevated metabolism in the working muscles. This afferent input is assumed to arise from polymodal receptors on free nerve endings within the skeletal muscles, which are activated by low P_{O_2}, by acidosis, and by autacoids [94]. This sympathetic activation during exercise by central command and by input from the metabolic milieu in the working muscles enhances arterial pressure above its level at rest. The exercise-induced elevation of blood pressure is most pronounced for the systolic peak values, while diastolic blood pressure is not largely altered, resulting in an elevation of mean arterial pressure of up to 20 mmHg above resting value, depending on the strenuosity of exercise. This indicates that the sympathetic activation of the heart is large enough to yield an increase in cardiac output, which overcompensates for the substantial, exercise-induced decline in total peripheral vascular resistance. The set point of the pressoreceptor reflexes (see Sect. 94.5.1.1) must be acutely elevated during exercise by the higher brain centers. Otherwise, the strong activation of the pressoreceptors by elevated mean pressure and by a large pulse pressure amplitude at high frequency would result in depressor reactions.

The essential prerequisite for the large increase in cardiac output during exercise is the suffient support of venous return. This is ensured by the strong action of the muscle pump on the veins in the extremities during dynamic work, by the sympathetically mediated venoconstriction in the splanchnic bed (which acts as an infusion of an additional 500 ml blood), by the respiratory support of venous flow in the large central veins, and by the heart itself via systolic suction of blood into the atria (see Sect. 94.4.3).

The strong general sympathetic activation during exercise causes collateral vasoconstriction in most extramuscular tissues. This collateral vasoconstriction is sufficient to elevate peripheral extramuscular vascular resistance parallel to the rise in mean arterial pressure. However, a redistribution of cardiac output towards the working muscles with substantial flow reduction to extramuscular tissues does not occur: extramuscular flow remains essentially constant and the excess flow demand of working muscle is entirely covered by the elevation in cardiac output (Fig. 94.52). Within the extramuscular tissues, flow is somewhat reduced during exercise in the splanchnic bed and the kidney. In contrast, flow is elevated in the heart, parallel to the enhanced cardiac work, and in the skin, which is important for the adequate regulation of body temperature. During exercise-induced heat production, this regulation requires an enhanced perfusion of the skin, enhanced blood content in subcutaneous venous plexus, and enhanced secretion of sweat, which results in cooling of the skin during its evaporation (see Sect. 94.5.4; Chaps. 110–112). During maximal levels of exercise, however, the sympathetic vasoconstriction overrides the thermoregulatory vasodilation in the skin. Therefore, skin perfusion is lowered at very high exercise levels (Fig. 94.52). The core temperature of the exercising body is elevated in proportion to the exercise-induced O_2 consumption. This elevation reaches up to 3°C in the untrained human and can be even higher in trained athletes during competitions. Since this elevation of core temperature during exercise is independent of ambient conditions over a wide range of air temperature and humidity, it is assumed that this elevation results from a change in the set point of temperature regulation by the central nervous system during exercise rather than from an inadequate efficiency of temperature-regulating mechanisms (see Chaps. 110, 111).

Capillary Recruitment. The vasodilation of the vascular bed in the working skeletal muscles is the dominant circulatory alteration during exercise and accounts for the large decline in total peripheral vascular resistance. The perfusion rate can increase from 5–10 ml/min per 100 g in inactive skeletal muscle to 200–300 ml/min per 100 g in maximally exercising muscle if only small groups of muscles are exercising and the cardiac output is sufficient to cover maximal flow to these muscles (see above). Together with these high flow rates, there is an increase in the number of perfused capillaries in the muscular bed: while in inactive muscle only 20%–30% of the exisiting capillaries are perfused simultaneously, all capillaries are perfused in maximally active muscle [75,93]. This reduces intercapillary distances and facilitates O_2 extraction from the blood. At maximal exercise, O_2 content in muscle venous blood is below 5 ml/100 ml blood, and anaerobic glycolysis contributes substantially to the energy metabolism of the working muscles.

Sympathetic vasoconstriction, however, is still effective in limiting flow to working skeletal muscle during whole body exercise. This is demonstrable by a simple experiment: during isolated maximal work of the legs on an ergometer, flow in the muscles of the legs is increased to roughly 250 ml/100 g per min and cardiac output reaches 25 l/min. If the arms then start working too, with a large increase in muscle flow to the arms and the shoulders, cardiac output is not further augmented and the flow to the legs declines, although the work rate of the legs as measurable on the ergometer remains unchanged. Parallel to the onset of arm exercise, however, sympathetic activity increases sharply, measurable by the norepinephrine release rate into the plasma [93]. This indicates that during simultaneous, dynamic exercise of large groups of muscles, local metabolic vasodilation and sympathetic vasoconstrictor activity, elevated to maintain arterial pressure at the physiologic set point, balance each other in an equilibrium of vasodilation which is clearly below the maximal vasodilation of a skeletal muscles working alone. This equilibrium allows for the adequate distribution of the limited cardiac output.

The mechanisms of functional hyperemia in the working skeletal muscle are largely unknown (see Sect. 94.5.3.1). It is generally assumed that interstitially accumulating metabolites, K^+ ions, hyperosmolality, tissue hypoxia, and acidosis contribute in mediating metabolic vasodilation together with some myogenic dilation and mechanically induced endothelial autacoid release (NO and prostaglandins). The oscillating mechanical compression and decompresson of the vascular bed by dynamic muscle work repeatedly unloads the arterioles, thus causing some myogenic dilation and augmenting shear-induced endothelial release of NO and prostaglandins. Clearly, none of these mechanisms alone or in combination is sufficient to explain functional hyperemia in working skeletal muscle according to present knowledge [47]. However, the traditional view assumes that, during exercise, the above-mentioned mechanisms potentiate each other in an unknown way such that their combined action can mediate the flow increase. More speculative explanations assume that periarteriolar neurons play a special vasodilatory role or that NO synthase in skeletal muscle fibers is activated during contraction of the fibers. Such speculations remain to be substantiated.

94.5.4 Circulation Under Thermal Stress

The regulation of convective transport of metabolically formed heat to the body surface for exchange with the environment is the major role of the circulation in the control of body temperature. Within a thermoneutral zone of ambient conditions, body temperature control is achieved almost exclusively by varying the perfusion of the skin. At higher temperatures or with reversed radiation, thermal stress requires high blood flow to the skin and sweat secretion for cooling of the skin surface by evaporation as complementary mechanisms of temperature regulation. At ambient temperatures below the thermoneutral zone, heat conservation by vasoconstriction in skin and other peripheral tissues and enhanced metabolic heat production, e.g., by shivering of skeletal muscle, act together to maintain the body core temperature (see Chaps. 110–112).

Within the thermoneutral zone, blood flow to the skin can vary from 0.5 l/min to 2.0 l/min for the purpose of temperature regulation. Under extreme thermal stress, however, with ambient temperatures above 44°C and 85% humidity, skin flow can approach up to 10 l/min [13]. Furthermore, venous tone in the capacitance vessels of the skin is reduced and a large amount of blood (approximately 300–500 ml) is shifted to the subcutaneous venous plexus for cooling, while the blood content in the pulmonary bed is reduced [64]. A reduced blood content in the thoracic part of the circulation impairs adaptation of the circulation to the upright position (see Sect. 94.5.2). Furthermore, sweat production can be up to 1.5 l/h in the untrained human and more than 5 l/h in the competitive, endurance-trained athlete. Together with fluid loss due to insensible perspiration, i.e., nonregulated fluid loss through the skin, and due to respiration, which is enhanced under thermal stress, this results in a critical reduction of the extracellular fluid volume if adequate drinking is not possible (see Chap. 82). The vasodilation of the skin for reasons of temperature regulation is largely mediated by central modulation of sympathetic α_1-adrenergic vasoconstriction and, to a lesser extent, by bradykinin, formed locally during sweat production. Furthermore, high tissue temperatures locally affect neurovascular transmission by postsynaptic attenuation of α_1-adrenoceptor responsiveness.

Thus, thermal stress can induce orthostatic instability and an enhanced susceptibility for circulatory collapse by several mechanisms:

- Redistribution of intravascular volume away from the thoracic circulation towards the subcutaneous venous plexus
- Less efficient support of venous return by the skeletal muscle pump due to dilation of superficial cutaneous veins [56]
- Reduction of extracellular and intravascular volume by excessive sweat production
- Central impairment of cutaneous vasoconstriction and peripheral impairment of sympathetic vasoconstriction in heated tissues
- Increased cardiac output due to excess skin flow in the presence of impaired venous return and reduced central blood volume.

This thermally induced orthostatic instability drastically impairs the working capacity of the organism, since an adequate arterial pressure can no longer be maintained during exercise under thermal stress. Complex chronic adaptations to thermal stress are assumed to occur in populations living in hot climates with high humidity. The most important adaptation is probably learned behavior, avoiding any exercise during the hottest hours of the day.

94.5.5 Circulatory Shock

In general terms, circulatory shock describes an acute, life-threatening state which can be tolerated only for limited time periods without therapy and which is characterized by an inadequate perfusion of vital organs, very often together with an inappropriate intravascular blood volume in relation to the functional state of the circulation [1,52]. Typical forms of circulatory shock include the following:

- Hypovolemic shock after hemorrhage, burns, or fluid loss due to diabetes insipidus, diarrhea and/or vomiting, overdosing of diuretics or salt-losing nephropathy
- Septic or toxic shock with paralysis of the vascular smooth muscle tone and with exaggerated extravasation of intravascular volume
- Anaphylactic shock as a result of excessive release of vasodilating agents such as histamine
- Cardiogenic shock (e.g., acute moycardial infarction, myocarditis, severe cardiac arrhythmias), in which the weakened or otherwise impaired heart cannot maintain

a dynamic equilibrium within the arterial system by pumping blood volume against the peripheral vascular resistance.

This dangerous acute shock state often results in a cascade of further complications with disturbed tissue metabolism and function, uncontrolled autacoid release, microcirculatory disturbances, and, finally, lethal damage to the central nervous system and the heart.

Hypovolemic shock is in a sense a prototype of clinical shock forms, and hemorrhage is its most frequent cause. The immediate consequences of hemorrhage are:

- Decreased blood volume, which is manifested preferentially as lowered blood content in the low pressure system, regardless of the primary source of bleeding
- Decreased diastolic filling of the heart with lowered stroke volume, triggering compensatory pressor reflexes and humoral reactions (see Chap. 82)
- Progressive decline of arterial and venous pressures in spite of maximal activation of pressor mechanisms.

Depending on the age and the health of the subject and the rate of bleeding, blood losses of 5%–10% of normal blood volume can be compensated with little detectable alterations in the systemic circulation. Blood losses of up to 20% can be compensated spontaneously after a prolonged period of tachycardia and arterial hypotension, provided that fluid and salt intake is not restricted. Blood losses up to 30% result in manifestation of clinical shock, which is only reversible with adequate and timely therapy. Patients can survive blood losses of up to 40% after serious shock if adequate therapy is available in an intensive care unit, but the prognosis is poor serious even with optimal therapy [1,52].

Within seconds after bleeding, the decline in central blood volume and the concomitant decline in stroke volume will activate volume and pressoreceptor reflexes, resulting in sympathetic activation with generalized vasoconstriction of resistance and capacitance vessels, preferentially in the cutaneous, skeletal muscle, splanchnic, and renal beds, while brain and coronary flow are not reduced. These compensatory reactions help to maintain cardiac filling by shifting venous volume towards the heart (most important in the splanchnic and cutaneous bed), and they tend to partially maintain plasma volume by lowering intracapillary pressures, thereby promoting fluid absorption from interstitial space into the plasma. This effect is most important in the skeletal muscle bed. Elevations in heart rate and cardiac contractility usually become detectable with blood losses above 700 ml. The renal vasoconstriction curtails glomerular filtration and leads to oliguria and sometimes anuria and to the activation of the renin–angiotensin system with enhanced aldosterone release (see Chap. 73). The decreased atrial distension triggers reflex release of antidiuretic hormone (or vasopressin; see Chap. 82). All these reactions help to conserve the remaining plasma volume. An enhanced epinephrine release from the adrenal medulla contributes to the sympathetic vasoconstriction and induces hyperglycemia, which is typical for the initial stages of hemorrhage.

In the compensated phase of hypovolemic shock, these reactions maintain arterial pressure at a level which is compatible with consciousness, but lower than normal and with a drastically reduced pulse pressure amplitude. The patient often appears to be "semirational", restless yet apathetic; he or she has a profound pallor of the skin with cold sweat and a rapid and thready pulse. Respiration is feeble and rectal temperature is lowered. While the compensatory mechanisms are designed to maintain an arterial pressure sufficient for the function of the brain, flow to the peripheral organs is depressed below their functional and/or metabolic requirements by these compensatory mechanisms. In peripheral tissues, P_{O_2} is lowered, with a resultant peripheral anaerobiosis causing increased blood lactate and pyruvate, with severe acidosis and increased plasma P_{CO_2}, with a depletion of high-energy phosphates and glycogen, and with increased protein catabolism and elevated blood NH_4^+. Without further blood losses, these accumulating alterations in the local metabolic milieu tend to progressively impair sympathetic neurovascular transmission, until the compensatory vasoconstriction breaks down. This occurs first in the peripheral venoles, aggravating the volume deficit for adequate diastolic cardiac filling and contributing to the decline in arterial pressure and the transition to irreversible shock [1,52].

Therapy must stop the bleeding replenish the intravascular volume by a plasma expander, i.e., a solution mimicking the colloid content and the electrolytes in the plasma, and correct the systemic acidosis by adding HCO_3^- as a buffer. However, restoration of tissue perfusion by adequate volume replacement is most important for the correction of tissue acidosis, while buffer application alone, without volume replacement, is ineffective and vasoconstrictor application alone is contraindicated.

When circulatory shock has reached the irreversible stage, tissue-damaging mechanisms are activated in addition to the hypoperfusion. These mechanisms include initiation of inflammatory reactions, without inflammatory hyperemia, e.g., complement activation, disseminated intravascular partial activation of coagulation, liberation of bacterial endotoxins from the ischemic gut, release of lysosomal proteases and of toxic radicals, and the appearance of cytokines and not yet identified myocardial depressant factors in the circulation. The relative relevance of these factors initially varies with different shock forms, but their generalized stimulation in the outcome of all shock forms is similar. The observed mortality largely results from:

- Acute cardiac failure
- Brain hypoperfusion
- Pulmonary complications summarized as respiratory distress syndrome.

Therapy in this stage is difficult and very often unsuccessful; it includes not only volume replacement and acid buffering, but also interventions assumed to interfere with the multiple mechanisms mentioned above.

94.5.6 Heart Failure

Chronic heart failure is a complex clinical syndrome with many causes, with an intrinsic tendency of self-reinforcement by vicious circles, and with a dramatically increased risk of premature cardiovascular mortality [24,34]. Two components characterize the syndrome:

- A functionally impaired heart which cannot pump an adequate cardiac output without an elevated filling pressure
- Compensatory mechanisms elicited by impaired cardiac function.

These compensatory pressor reactions cause additional burdens and risks to the impaired heart. Important elements of these compensatory reactions are volume retention and generalized vasoconstriction due to activation of the sympathoadrenal and the renin–angiotensin system, i.e., *neuroendocrine activation*.

Patients suffer from both components of the syndrome: inadequate cardiac output causes fatigue, limited exercise capacity, hypotension, and dizziness. The compensatory neuroendocrine activation results in chronically elevated pressures in the low-pressure system, causing congestion with edema formation and subsequent hepatic and/or pulmonary dysfunction and, except for some rare forms, enhanced cardiac filling. This cardiac overfilling elevates ventricular wall stress (law of La Place; see Chap. 90), which enhances myocardial metabolic requirements and may trigger myocardial hypoperfusion and angina pectoris. Rare forms of failure without enhanced cardiac filling and ventricular enlargement are syndromes in which cardiac function is impaired by extramyocardial impediments of ventricular filling, e.g., constrictive pericarditis or stenoses of atrioventricular valves. In such cases, the congestion is manifested as tissue edema outside of the heart, but cannot cause ventricular overfilling and enlargement. In the broad clinical picture of heart failure, various aspects of the syndrome may predominate: in forward failure, symptoms of an inadequate cardiac output are predominant; in backward failure, the patient suffers mainly from signs of congestion. If the cardiac impairment is restricted to the right heart, signs of congestion are restricted to the venous system (right heart failure); if the left heart is affected, congestion first affects the pulmonary bed (lung edema) and later the venous system. Severe, chronic forms of failure always result in an enhanced blood volume with overfilling of the low-pressure system and the complete syndrome of congestive heart failure.

Multiple primary causes can initiate the development and progression of chronic heart failure. The most important causes are:

- Loss of myocytes by myocardial infarction
- Bacterial or viral inflammations of the heart
- Chronic mechanical overload by hypertension or by defects of the heart valves
- Toxic damage of the myocardium

Once the primary cause attenuates the pumping of an adequate cardiac output, commensurate with systemic metabolic and functional requirements at rest or during exercise, the neuroendocrine activation is triggered by various mechanisms:

- A diminished stroke volume causes a decline in the pulse pressure amplitude, stimulating sympathetic activity by pressoreceptor reflexes.
- Transient or prolonged hypotension or treatment with diuretics activates the renin–angiotensin system; this activation is enforced by β-adrenergic sympathetic stimulation.
- Hypoperfusion of working skeletal muscle stimulates the sympathetic activity via afferents activated by the local metabolic milieu within the muscle (see Sect. 94.5.3).
- Occasional distinct hypotension, too large to be compensated by the autoregulation of brain perfusion, activates the sympathoexcitatory reaction to transient brain ischemia.

The resulting neuroendocrine activation causes an elevation of peripheral vascular resistance by arteriolar vasoconstriciton, elevations of central venous pressure by venoconstriction and renal volume retention, and sustained tachycardia. Although the resultant congestion should activate sympathoinhibitory volume receptors in the thoracic part of the low-pressure system (see Sect. 94.5.1.1), this inhibition cannot overcome the sympathoexcitation by the other stimuli. An enhanced release of the natriuretic peptides ANP and BNP (see Sect. 94.3.3.4) from the cardiac atria is induced by the congestion. Furthermore, these peptides are synthesized and constitutively released by the hypertrophied and overloaded ventricles in chronic failure. However, the vasodilator and natriuretic action of these peptides is not sufficient to counteract the vasoconstriction and volume retention during failure, although their circulating plasma levels are augmented up to 30-fold above control level. Hypothalamic release of vasopressin is induced by the strong sympathoexcitation (see Chap. 82) and contributes to the hypervolemic and hyponatremic congestion in severe failure. Thus, neuroendocrine activation during chronic congestive heart failure includes enhanced activation of pressor as well as of depressor systems, with a strong dominance in the action of the pressor systems and with a reduction of regulatory capability of the endocrine systems due to concomitant activation of functional antagonists [14,39].

The neuroendocrine activation with its dominant pressor component acutely helps to maintain arterial pressure within the normal range, as it does during hypovolemia (see Sect. 94.5.5), but chronically it causes further problems for the impaired heart, resulting in a vicious cycle, which is the decisive factor for the progression of the syndrome and for the premature mortality of the patients. This progression to severe failure due to the compensatory neuroendocrine activation occurs via several mechanisms:

- The enhanced cardiac preload enhances myocardial wall stress, increases myocardial metabolic requirements, and thereby aggravates the coronary insufficiency, which is a frequent primary cause of the heart failure syndrome (see above).
- The enhanced sympathetic drive to the heart may provoke regional myocardial ischemia and dangerous cardiac arrhythmias by the combination of α_1-adrenergic coronary vasoconstriction, tachycardia, enhanced wall stress, β_1-adrenergic myocardial activation, and systemic electrolyte imbalances.
- Chronic cardiac sympathetic stimulation causes attenuation of β_1-adrenergic signal transduction and depressed inotropic responsiveness.
- The chronically enhanced myocardial wall stress and the sustained activation of the renin–angiotensin system act as growth stimuli for the myocardium, resulting in a myocardium with enlarged myocytes, augmented interstitial connective tissue, disturbed diastolic relaxation, and a myocyte phenotype with a labile cellular Ca^{2+} homeostasis, which is very susceptible for arrhythmogenesis by multiple mechanisms.

The dominant role of the compensatory mechanisms for the fatal course of chronic heart failure is illustrated by two observations:

- Elevated plasma norepinephrine, an indicator of neuroendocrine activation, proved to be the best predictor of heart failure-induced mortality in epidemiologic studies [24,26].
- Inhibition of neuroendocrine activation by drug therapy in chronic heart failure reduced mortality and morbidity, although the primary causes of heart failure cannot be corrected by such therapy [25].

Thus, congestive heart failure is a syndrome which can originate in several primary injuries to the heart, but which develops and progresses mainly due to inappropriate compensatory reactions in response to the primary injuries.

94.5.7 Fetal Circulation

During early embryonic development, the heart is formed from a single primordial cardiac tube by bending and rotation of this tube. The major arterial trunks originate in the embryonic branchial arch system, which corresponds to the gill arches in fishes, by expansion and resorption of specific branches, while the peripheral vasculature accomodates to the growth patterns of the developing organs. By the 11th week, the heart of the human embryo is a four-chamber organ connected to the corresponding arterial trunks and central veins with the peripheral circulatory patterns which persist throughout pregnancy until the perinatal circulatory changes (see Chaps. 89, 118).

The placenta is the organ for the oxygenation of the fetal blood and for the elimination of CO_2 (see Chap. 117). The fetal blood flows through the vessels in the villi of the chorion. These villi protrude like trees or bush-like structures from the chorionic plate of the placenta into the trophoblastic lacunae of the intervillious space, through which the maternal arterial blood from fairly large branches of the uterine arteries is dispersed (hemochorial type of placenta; see [76]). The exchange of blood gases, water, electrolytes, nutrients, and low molecular weight peptides through the fetal capillary endothelium and the epithelium of the chorionic villi is supported by the anatomic arrangement of the villi, which directs the stream of maternal blood within the intervillious space preferentially against the flow direction of the fetal blood in the chorionic capillaries by an incomplete countercurrent arrangement [76].

The incompletely arterialized fetal blood (75%–80% O_2 saturation) in the umbilical vein partially reaches the liver, while the other part bypasses the liver through the duct of Arantius and mixes with the venous blood from the lower body within the inferior vena cava. The mixed blood in this vessel (60%–65% O_2 saturation) reaches the right atrium and passes through the oval foramen into the left atrium, left ventricle, and aortic arch (see Chap. 89). The venous blood from the superior vena cava is directed into the right ventricle and flows from the pulmonary artery, preferentially through Botalli's duct into the thoracic descending aorta, while only one quarter of the pulmonary artery flow goes into the collapsed, atelectatic lungs. The high vascular resistance of the collapsed lungs tends to divert the flow around the lungs through Botalli's duct. The blood in the arteries originating from the aortic arch, supplying the heart, brain, and upper extremities, has an higher O_2 saturation (above 50%) than the blood in the arteries from the lower body, which originate from the aorta distal to the entrance of Botalli's duct. From the iliac arteries, blood reaches the placenta again through two umbilical arteries [91].

By this arrangement of the fetal circulation, the two fetal ventricles operate largely in parallel. Their common cardiac output amounts to approximately 240 ml/min per kg, pumped into the aorta with a mean arterial pressure of approximately 50–60 mmHg (6.7–8.0 kPa) with an heart rate of 150 beats per minute. From this common cardiac output, 60% is pumped into the chorionic part of the placenta and 40% into the fetal tissues. Maternal blood flow to the uterus increases roughly 100-fold during pregnancy to a final rate of 500 ml/min, while maternal cardiac output is increased by 30%–40% early in pregnancy and plateaus at this elevated level until delivery.

After birth, the vascular resistance in the pulmonary bed decreases sharply with the inflation of the lungs by the first inspirations. Concomitantly, the aortic pressure is elevated immediately by closure of the umbilical arteries, which until then had carried 60% of flow into the aorta. Closure of the umbilical arteries can occur spontaneously by smooth muscle contraction of the spiraled arteries, but is normally secured by ligation. With the closure of the umbilical vein, the duct of Arantius shrinks and is later obliterated. The decline in right heart afterload and the increase in left heart afterload reverses the pressure gradient between the two hearts and the two atria, causing a functional closure of the oval foramen by a valve-like atrial structure, while the anatomic obliteration of the foramen may take weeks or even years. Furthermore, the flow direction through Botalli's duct is reversed by the opposite afterload changes, and the duct is normally closed functionally only minutes to hours after this flow reversal. The closure of the duct is mediated largely by smooth muscle contraction. Furthermore, the very loose subendothelial tissue in the duct might be folded up by the flow reversal to small valve-like foldings, which help in functional closure of the duct. The two hearts are now completely separated and operate in series [91]. Increasing differences in the afterload-induced cardiac growth processes shape the different wall structures of the left and the right ventricle early in childhood (see Chap. 122).

References

1. Abboud FM (1982) Pathophysiology of hypotension and shock. In: Hurst JW (ed) The heart, arteries and veins. McGraw-Hill, New York, pp 452–463
2. Admiraal PJJ, Derkx FHM, Danser AHJ, Pieterman H, Schalekamp MADH (1990) Metabolism and production of angiotensin in subjects with renal artery stenosis. Hypertension 15:44–55
3. American Society of Hypertension (1992) Recommendation for routine blood pressure measurement by indirect cuff sphygmomanometry. Am J Hypertension 3:207–212
4. Anonymus (1993) Memorandum from a World Health Organization/International Society of Hypertension meeting: 1993 guidelines for the management of mild hypertension. Hypertension 22:392–405
5. Anrep G (1912) On local vascular reactions and their interpretation. J Physiol [Lond] 45:318–327
6. Aukland K (1994) Why don't our feet swell in the upright position? News Physiol Sci 9:214–219
7. Aukland K, Reed RK (1993) Interstitial-lymphatic mechanisms in the control of extracellular fluid volume. Physiol Rev 73:1–78
8. Bauer RD, Busse R (1982) Biophysik des Kreislaufs. In: Busse R (ed) Kreislaufphysiologie. Thieme, Stuttgart, pp 3–40
9. Bayliss WM (1902) On the local reactions of the arterial wall to changes in internal pressure. J Physiol [Lond] 28:220–231
10. Berecek KH, Brody MJ (1992) Evidence for a neurotransmitter role for epinephrine derived from the adrenal medulla. Am J Physiol 242:H593–H601
11. Björnberg J, Albert U, Mellander S (1990) Resistance responses in proximal arterial vessels, arterioles and veins during reactive hyperemia in skeletal muscle and their underlying regulatory mechanisms. Acta Physiol Scand 139:535–550
12. Blair DA, Glover WE, Greenfield ADM, Roddie IC (1959) Excitation of cholinergic vasodilator nerves to human skeletal muscle during emotional stress. J Physiol 148:633–647
13. Brengelmann GL (1989) Body temperature regulation. In: Patton HD, Fuchs AF, Hille B, Scher AM, Steiner R (eds) Textbook of physiology. Saunders, Philadelphia, pp 1584–1596
14. Burnett JC, Kao PC, Hu DC (1986) Atrial natriuretic peptide elevation in congestive heart failure in the human. Science 231:1145–1150
15. Busse R (1995) Gefäßytem und Kreislaufregulation (1995) In: Schmidt F, Thews G (eds) Physiologie des Menschen. Springer, Berlin, Heidelberg New York, pp 498–561

16. Busse R, Fleming I (1993) The endothelial organ. Curr Opinion Cardiol 8:719–727
17. Busse R, Fleming I, Schini VB (1995) Nitric oxid formation in the vascular wall: regulation and functional implications. Curr Topics Microbiol Immunol 196:7–18
18. Calaresu FR, Yardley CP (1988) Medullary basal sympathetic tone. Ann Rev Physiol 50:511–524
19. Campbell DJ (1987) Circulating and tissue angiotensin systems. J Clin Invest 79:1–6
20. Campbell DJ, Kladis A (1991) Simultaneous radioimmunoassay of six angiotensin peptides in arterial and venous plasma of man. J Hypertens 9:265–274
21. Chien S, Jan KM (1973) Red cell aggregation by macromolecules: roles of surface adsorption and electrostatic repulsion. J Supramol Struct 1:185–231
22. Chien S, Usami S, Shalak R (1984) Blood flow in small tubes. In: Renkin EM, Michel CC (eds) Handbook of physiology, Section 2: The cardiovascular system, Vol. IV: Microcirculation, Pt. 1. Am Physiol Soc, Bethesda, Maryland, pp 217–249
23. Cohen RA, Weisbrod RM (1988) Endothelium inhibits norepinephrine release from adrenergic nerves of rabbit carotid artery. Am J Physiol 254:H871–H878
24. Cohn JN (1988) Current therapy of the failing heart. Circulation 18:1099–1118
25. Cohn JN (1995) Heart failure. In: Willerson JT, Cohn JN (eds) Cardiovascular medicine. Churchill Livingstone, New York, pp 947–979
26. Cohn JN, Levine TB, Olivari MT (1984) Plasma norepinephrine as a guide to prognosis in patients with chronic congestive heart failure. N Engl J Med 311:819–827
27. Cowley AW Jr, Liard JF, Guyton AC (1973) Role of the baroreceptor reflex in daily control of asterial blood pressure and other variables in dogs. Circ Res 32:564–578
28. Davies PF (1989) How do vascular endothelial cells respond to flow?. News Physiol Sci 4:22–25
29. Davis MJ (1991) Myogenic response gradient in an arteriolar network. In: Mulvany MJ, Alkjaer C, Heagerty AM, Nyborg NCB, Strandgaard S (eds) Proceedings of the 3rd International Symposium on Resistance Arteries. Elsevier, Amsterdam, pp 51–55
30. De Bold AJ, Borenstein AT, Veress AT (1981) A rapid and potent natriuretic response to intravenous atrial myocardial extract in rats. Life Sci 28:89–94
31. Dull RO, Davies PF (1991) Flow modulation of agonist (ATP)-response Ca^{2+}-coupling in vascular endothelial cells. Am J Physiol 261:H149–H154
32. Dzau VJ, Burt DW, Pratt RE (1988) Molecular biology of the renin angiotensin system. Am J Physiol 255:F563–F573
33. Dzau VJ, Pratt RE (1992) Renin angiotensin system. In: Fozzard HA, Haber E, Jennings RB, Katz AM, Morgan HE (eds) The heart and the cardiovascular System. Raven, New York, pp 1817–1849
34. Eichna LW (1960) The George E. Brown Memorial Lecture: circulatory congestion and heart failure. Circulation 22:864–892
35. Engeset A, Olszewski W, Jaeger PM, Sokolowski J, Theodorsen L (1977) Twenty-four hour variation in flow and composition of leg lymph in normal men. Acta Physiol Scand 99:140–148
36. Ferrara N (1993) Vascular endothelial growth factor. Trends Cardiovasc Med 3:244–250
37. Folkow B (1964) Description of the myogenic hypothesis. Circ Res 15 [Suppl 1]: 279–287
38. Folkow B (1990) Salt and hypertension. News Physiol Sci 5:220–2224
39. Francis GS, Goldsmith SR, Levine TB (1984) The neurohumoral axis in congestive heart failure. Ann Intern Med 101:370–378
40. Frangos JA, Eskin SG, McIntire LV, Ives CL (1985) Flow effects on prostacyclin production by cultured human endothelial cells. Science 227:1477–1479

41. Furchgott RF (1983) Role of endothelium in response of vascular smooth muscle. Circ Res 53:557–573
42. Furchtgott RF, Zawadzki JV (1980) The obligatory role of endothelial cells in the relaxation of arterial smooth muscle by acetylcholine. Nature 288:373–376
43. Gaehtgens P (1982) Mikrozirkulation. In: Busse R (ed) Kreislaufphysiologie. Thieme, Stuttgart, pp 70–103
44. Gardner DG (1994) Molecular biology of the natriuretic peptides. Trends Cardiovasc Med 4:159–165
45. Gauer OH, Henry JP, Behn C (1970) The regulation of extracellular fluid volume. Ann Rev Physiol 32:547–589
46. Gewirtz H (1991) The coronary circulation: limitations of current concepts of metabolic control. News Physiol Sci 6:265–268
47. Gorman MW, Sparks HE (1991) The unanswered question. News Physiol Sci 6:191–193
48. Greenfield ADM (1966) Survey of the evidence for active neurogenic vasodilation in man. Fed Proc 25:1607–1610
49. Griffith TM, Edwards DH (1990) Myogenic autoregulation of flow may be inversely related to endothelium-derived relaxing factor activity. Am J Physiol 258:H1171–H1180
50. Griffith TM, Edwards DH, Davies RL, Harrison TJ, Evans KT (1987) EDRF coordinates the behaviour of vascular resistance vessels. Nature 329:442–445
51. Groves P, Kurz S, Drexler H (1994) The role of bradykinin in basal and flow-mediated endothelium-dependent vasodilation in the human coronary circulation (abstr.). Circulation 90:I36
52. Guyton AC (1986) Physiology and treatment of circulatory shock. In: Guyton AC (ed) Textbook of medical physiology. Saunders, Philadelphia, pp 326–335
53. Guyton AC, Coleman TG, Cowley AW Jr (1972) Arterial pressure regulation: overriding dominance of the kidney in the longterm regulation and in hypertension. Am J Med 52:584–694
54. Hagen G (1839) Über die Bewegungen des Wassers in engen zylindrischen Röhren. Ann Phys Chem Poggendorf 46:422–442
55. Hathaway DR, Adam LP, Wilensky RL, March KL (1995) Vascular muscle structure and function. In: Willersion JT, Cohn JN (eds) Cardiovascular medicine. Churchill Livingstone, New York pp 1042–1052
56. Henry JP, Gauer OH (1950) The influence of temperature upon venous pressure in the foot. J Clin Invest 29:855–862
57. Holtz J, Förstermann U, Pohl U, Giesler M, Bassenge E (1984) Flow dependent endothelium-mediated dilation of epicardial coronary arteries in conscious dogs: effect of cyclooxygenase inhibition. J Cardiovasc Pharmacol 6:1161–1169
58. Holtz J, Goetz RM (1994) Vascular renin-angiotensin-system, endothelial function and arteriosclerosis? Basic Res Cardiol 89:71–86
59. Holtz J, Höffler D (1993) Alpha-1-Rezeptorenblocker: Klinik und Pharmakologie. Aesopus, Basel
60. Holtz J, Münzel T, Bassenge E (1987) Atrial natriuretic factor in man. Z Kardiol 76:655–670
61. Hutcheson IR, Griffith TM (1991) Release of endothelium-derived relaxing factor is modulated both by frequency and amplitude of pulsatile flow. Am J Physiol 261:H257–H262
62. Jänig W (1988) Pre- and postganglionic vasoconstrictor neurons: differentiation, types, and discharge properties. Ann Rev Physiol 50:525–539
63. Järhult J, Mellander S (1974) Autoregulation of capillary hydrostatic pressure in skeletal muscle during regional arterial hypo- and hypertension. Acta Physiol Scand 91:32–41
64. Johnson JM (1989) Circulation to the skin. In: Patton HD, Fuchs AF, Hille B, Scher AM, Steiner R (eds) Textbook of physiology. Saunders, Philadelphia, pp 898–910
65. Johnson PC (1980) The myogenic response. In: Bohr DF, Somlyo AP, Sparks HV (eds) Handbook of physiology, Sect. 2: the cardiovascular system, vol. II: Vascular smooth muscle. Am Physiol Soc, Bethesda, Maryland, pp 409–442

66. Kaley G, Altura BM (1977) Microcirculation, vol I. University Park Press, Baltimore

67. Klagsbrun M, Dluz S (1993) Smooth muscle cell and endothelial cell growth factors. Trends Cardiovasc Med 3:213–217

68. Kubik S (1993) Anatomie des Lymphgefäßsystems. In: Foldi M, Kubik S (eds) Lehrbuch der Lymphologie. Fischer, Stuttgart pp 1–201

69. Kuchan MJ, Jo H, Frangos JA (1994) Role of G proteins in shear stress-mediated nitric oxide production by endothelial cells. Am J Physiol 267:C753–C758

70. Lansman JB, Hallam TJ, Rink TJ (1987) Single stretch-activated ion channels in vascular endothelial cells as mechanotransducers? Nature 235:811–813

71. Levick JR, Michel CC (1978) The effects of position and skin temperature on the capillary pressures in the fingers and the toes. J Physiol 274:97–109

72. Marrero MB, Schieffer B, Paxton WG, Heerdt L, Berk BC, Delafontaine P, Bernstein KE (1995) Direct stimulation of JAK/STAT pathway by the angiotensin II AT1 receptor. Nature 375:247–250

73. Mattson DL, Lu SH, Roman RJ, Cowley AW Jr (1993) Relationship between renal perfusion pressure and blood flow in different regions of the kidney. Am J Physiol 264:R578–R583

74. Meininger GA, Davis MJ (1992) Cellular mechanism involved in the vascular myogenic response (brief review). Am J Physiol 263:H647–H659

75. Mellander S, Johansen B (1968) Control of resistance, exchange and capacitance functions in the peripheral circulation. Pharmacol Rev 20:117–196

76. Meschia G (1983) Circulation to female reproductive organs: In: Shepherd JT, Abboud FM (eds) Handbook of physiology, Sect. 2: The cardiovascular system, vol. III: Peripheral circulation and organ blood flow. Am Physiol Soc, Bethesda, pp 241–269

77. Miller VM (1991) Interactions between neural and endothelial mechanisms in control of vascular tone. News Physiol Sci 6:60–63

78. Mo M, Eskin SG, Schilling WP (1991) Flow-induced changes in Ca^{2+}-signalling of vascular endothelial cells: effect of shear stress and ATP. Am J Physiol 260:H1698–H1707

79. Mulvany MJ (1991) Are vascular abnormalities a primary cause or secondary consequence of hypertension? Hypertension 18 [Suppl I]: I52–I57

80. Mulvany MJ (1989) Structure and function of peripheral vascular smooth muscle in hypertension. J Cardiovasc Pharmacol 14 [Suppl 6]: S85–S89

81. Neil E (1984) Peripheral circulation: historical aspects. In: Shepherd JT, Abboud FM (eds) Handbook of physiology, Sect. 2: the cardiovascular system, vol III: peripheral circulation and organ blood flow, Pt. 1. Am Physiol Soc, Bethesda, Maryland, pp 1–19

82. Olesen SO, Clapham DE, Davies PF (1988) Haemodynamic shear stress activates a K^+-current in vascular endothelial cells. Nature 331:168–170

83. Olszewski WL (1985) Peripheral lymph: formation and immune function. CRC Press, Boca Raton

84. Olszewski WL, Engeset A (1980) Intrinsic contractility of prenodal lymph vessels and lymph flow in human leg. Am J Physiol 239:H775–H783

85. Palmer RMJ, Ferrige FG, Moncada S (1987) Nitric oxide release accounts for the biological activity of endothelium-derived relaxing factor. Nature 327:524–526

86. Pohl U, Busse R, Kuon E, Bassenge E (1986) Pulsatile perfusion stimulates the release of endothelial autocoids. J Appl Cardiol 1:215–235

87. Roman RJ, Zou AP (1993) Influence of the renal medullary circulation on the control of sodium excretion. Am J Physiol 265:R963–R973

88. Rowell LB (1974) Human cardiovascular adjustments to exercise and thermal stress. Physiol Rev 54:75–159

89. Rubanyi GM, Romero JC, Vanhoutte PM (1986) Flow-induced release of endothelium-derived relaxing factor. Am J Physiol 250:H1145–H1149

90. Rushmer RF (1961) Peripheral vascular control. In: Rushmer RF (ed) Cardiovascular dynamics. Saunders, Philadelphia, pp 98–130

91. Rushmer RF (1961) Embryologic development and congenital malformations of the heart. In: Rushmer RF (ed) Cardiovascular dynamics. Saunders, Philadelphia, pp 390–434

92. Saito Y, Eraslan A, Lockard V, Hester RL (1994) Role of venular endothelium in control of arteriolar diameter during functional hyperemia. Am J Physiol 267:H1227–H1231

93. Saltin B (1985) Hemodynamic adaptations to exercise. Am J Cardiol 55:42D–47D

94. Scher AM (1989) Cardiovascular control. In: Patton HD, Fuchs AF, Hille B, Scher AM, Steiner R (eds) Textbook of physiology. Saunders, Philadelphia, pp 972–990

95. Schretzenmayr A (1933) Über regulatorische Vorgänge an Muskelarterien. Pfluegers Arch 232:743–748

96. Selkurt EE, Womack I, Dailey WN (1965) Mechanisms of natriuresis and diuresis during elevated renal arterial pressure. Am J Physiol 209:95–99

97. Shen J, Luscinskas FW, Connolly A, Dewey CF, Gimbrone MA (1992) Fluid shear stress modulates cytosolic free calcium in vascular endothelial cells. Am J Physiol 262:C384–C390

98. Somylo AP, Somylo AV (1991) Smooth muscle structure and function. In: Fozzard HA, Jennings RB, Haber E, Katz AM, Morgan HE (eds) The Heart and the cardiovascular system. Raven, New York, pp 1295–1324

99. Staessen J, Amery A, Fragard R (1990) Isolated systolic hypertension in the elderly. J Hypertens 8:393–403

100. Starke K (1987) Presynaptic alpha-autoreceptors. Rev Physiol Biochem Pharmacol 107:1–124

101. Starke K (1977) Regulation of noradrenalin release by presynaptic receptor systems. Rev Physiol Biochem Pharmacol 77:1–124

102. Stiles GL (1991) Adenosin receptors: physiological regulation and biochemical mechanism. News Physiol Sci 6:161–165

103. Task Force on Blood Pressure Control in Children (1988) Report on the second task force on blood pressure control in children 1987. Pediatrics 79:1–19

104. Timmermans PBWM, Wong PC, Chiu AT, Herblin WF, Benfield P, Carini DJ, Lee RL, Wexler RR, Saye JAM, Smith RD (1993) Angiotensin II receptors and angiotensin II receptor antagonists. Pharmacol Rev 45:205–251

105. Toda N, Okamura T (1992) Regulation by nitroxidergic nerves of arterial tone. News Physiol Sci 7:148–152

106. Tresfamariam B, Weisbrod RM, Cohen RA (1989) The endothelium inhibits activation by calcium of vascular neurotransmission. Am J Physiol 257:H1871–H1877

107. Urata H, Kinoshita A, Perez DM, Misono KS, Bumpus FM, Graham RM, Husain A (1991) Cloning of the gene and cDNA for human heart chymase. J Biol Chem 266:17173–17179

108. Valentin JP, Humphrey M (1993) Urodilatin A paracrine renal natriuretic peptide. Semin Nephrol 13:61–70

109. Vizi ES, Kiss J, Elenkov IJ (1991) Presynaptic modulation of cholinergic and noradrenergic neurotransmission: interaction between them. News Physiol Sci 6:119–123

110. Von Euler US (1951) The nature of adrenergic nerve mediators. Pharmacol Rev 3:247–277

111. Von Kügelen I, Starke K (1991) Noradrenalin-ATP co-transmission in the sympathetic nervous system. Trends Pharmacol Sci 12:319–324

112. Walker P, Grouzmann E, Burnier M, Waeber B (1991) The role of neuropeptide Y in cardiovascular regulation. Trends Pharmacol Sci 12:111–115

113. Wallin G, Elam M (1994) Insights from intraneural recordings of sympathetic nerve traffic in humans. News Physiol Sci 9:203–207

95 Hemodynamics in Regional Circulatory Beds and Local Vascular Reactivity

J. Holtz

Contents

95.1 Introduction

The interaction of local mechanisms of vascular control and vascular self-adjustment (see Chap. 94) with neurogenic and humoral control mechanisms dominates the short-term regulation of the circulation by adjusting peripheral vascular resistance and venous capacity. However, comparing this interaction in various regional beds of the circulation reveals a surprising variability of regulatory characteristics. This regional heterogeneity of vasoregulation ensures that the specific needs of the organs, resulting from metabolic requirements and/or from specific functional necessities, are met. Organ-specific heterogeneity of vasoregulation, however, does not mean isolated or independent regulation of organ circulation. Two principles govern the organ-specific, heterogenous local vasoregulations in the network of integrated cardiovascular control of the organism:

- The specific function of an organ, which is made possible and supported by the organ-specific circulation, always more or less directly affects the internal milieu and the survival or the reproduction of the entire organism. Breakdown of an organ-specific circulation is not tolerable without an immediate or delayed impairment of a function, which is vital for the entire organism, therefore affecting all other organs.
- The organ-specific vasoregulations are limited by the maximal cardiac output of 25–28 l/min in a young, healthy, but not trained human. This limit, due to limitations in heart rate, cardiac contractility, and venous return, is substantially smaller than the sum of organ-specific flow maxima, which would amount to some 40–

50 l/min if all organs were simultaneously in a state of maximal vasodilation.

It is a widely held misbelief of the medically uneducated public that the limitations due to a finite cardiac output leads to the organ-specific vasoregulations being organized by some kind of functional "steal": blood flow to an active organ would defer flow from other organs, thereby reducing the functional capabilities of such hypoperfused organs. It goes without saying that such a functional steal with impairment in the performance of the flow-deprived organs does not occur to a relevant extent under physiologic conditions. What does occur is collateral vasoconstriction, i.e., the increase in vascular resistances of inactive organs during activity of skeletal muscle (exercise). This collateral vasoconstriction maintains constant flow to the inactive organs at a basal level, while systemic arterial pressure is augmented.

However, it might be argued that cardiovascular adaptation to maximal dynamic exercise includes some elements of functional steal: while flow demand of the working muscles exceeds the maximally available cardiac output, flow to splanchnic and renal beds is slightly reduced and flow to the skin is not augmented to the extent which temperature regulation would require (see Sect. 94.5.3). This flow reduction in the "nonexercising" organs, however, cannot be considered as true functional steal: the flow limitation due to sympathetic vasoconstriction in the active skeletal muscle is substantially larger than that observed in the other organs. While perfusion to working skeletal muscle during heavy exercise amounts to some 100–120 ml/min per 100 g, the maximal possible flow to these muscles would exceed 250 ml/min per 100 g (see Sect. 94.5.3). Thus the quantitatively relevant "steal" occurs in the organ whose requirements exceed the limits of cardiac output, and not in the other organs.

The dominating regulatory principle for matching augmented flow requirements of organ-specific vasoregulations is the increase in cardiac output in the presence of maintained flow to the organs, which remain in their basal metabolic and functional state. The distribution of cardiac output in the basal state, i.e., in resting humans without any specific activity, to the specific organ circulations is depicted in Fig. 95.1. Any augmented organ activity, which requires locally enhanced flow, results in an increase in cardiac output to the extent of the additional flow requirements, while the absolute

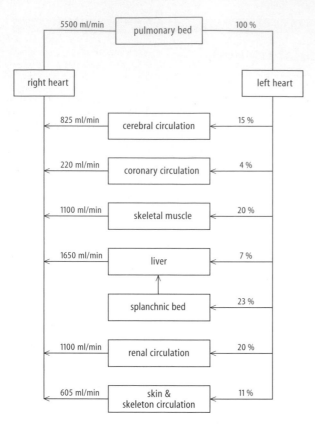

Fig. 95.1. Distribution of cardiac output in resting humans

flow to the other organ beds remains at the level shown in Fig. 95.1.

The typical features of the organ-specific vasoregulations under physiologic conditions and their functional relevance for the entire organism will be discussed in the following.

95.2 Cerebral Circulation

The cerebral circulation (see Chaps. 28, 29) supplies an extremely sensitive organ whose adequate perfusion is vital for the survival of the organism and for the existence of the individual personality, as it is currently understood. The central nervous system has no anaerobic metabolic capacity, and multiple large collateral vessels exist to ensure high blood flow and undisturbed local metabolism via the circle of Willis (Fig. 95.2) and many pial artery anastomoses if flow to an artery is obstructed. The extremely sensitive brain is protected by the cerebrospinal fluid in which it floats within the rigid skull, but this protection requires several specific regulatory adjustments:

● Blood volume and interstitial volume of the brain must be tightly controlled by autoregulation of blood flow and

by a minimally filtering blood–brain barrier to prevent larger increments in brain volume and thereby elevated intracranial pressures.
● Specific organs (choroid plexus in the lateral, third, and fourth brain ventricles with fenestrated capillaries and very high local flow rates of 600 ml/min per 100 g) contribute to the formation of cerebrospinal fluid by filtration and by active secretion [13,65] (see also Chap. 27).

While the brain accounts for only to 2% of the body weight, it receives 15% of cardiac output and requires 20% of total body O_2 consumption. Brain blood flow is 50–60 ml/min per 100 g with an O_2 consumption of 3–4 ml/min per 100 g and an glucose utilization of 5–6 ml/min per 100 g. Local flow in the brain cortex is somewhat higher (60–90 ml/min per 100 g), while it is lower in the brain white matter (20–30 ml/min per 100 g), parallel to local differences in O_2 consumption [15]. During extremely enhanced neuronal activity and during convulsions, total brain blood flow can increase by 50%, and such regional alterations in flow to specific areas occur with localized neuronal activity (Fig. 95.3). Animal experiments indicate that such discrete, regional activity-induced alterations in flow are paralleled by regional changes in O_2 consumption and glucose utilization [42,53].

Blood–Brain Barrier. This barrier (see Chap. 27) consists of a thick endothelium with cells connected to each other by continuous, tight junctions, a dense basement membrane, and an astroglial cell layer. The endothelial cells have a high metabolic activity, as indicated by the high number of mitochondria (five times the mitochondrial number of endothelia in skeletal muscle); they contain numerous enzymes that degrade blood-borne substances, preventing them from entering the brain and acting as an enzymatic barrier. This brain barrier restricts the diffusion of macromolecules; proteins and polar substances are not filtrated in the brain capillaries, and transcapillary efflux is substantially attenuated [69]. Therefore, an elaborated lymphatic system in the brain is not necessary, the role of such a system is largely assumed by the circulation of the cerebrospinal fluid, and only very few lymphatics are detectable. Breakdown of the blood–brain barrier can occur with ischemia, hypoxia, and acute hypertensive crises with mean arterial pressure increasing above the autoregulatory range (see below). In small areas of the brain, less than 0.5% of total brain mass, the blood–brain area is absent (area postrema; circumventricular organs, CVO; pituitary and pineal gland). This has some relevance for humoral feedback regulation and for the central regulation of arterial blood pressure by circulating hormones.

Autoregulation. A salient feature of the cerebral circulation is autoregulation: with mean arterial pressure varying between 70 and 140 mmHg (9.3–18.6 kPa), cerebral blood flow remains constant, but decreases with hypotension below the lower limit of autoregulation (70 mmHg) and increases with acute hypertension above 140 mmHg. In

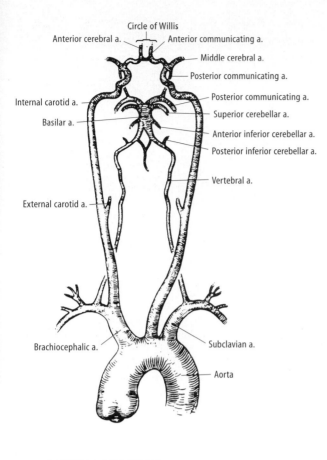

Circle of Willis

Anterior cerebral a. — — Anterior communicating a.

— Middle cerebral a.

— Posterior communicating a.

Internal carotid a. — — Posterior communicating a.

Basilar a. — — Superior cerebellar a.

— Anterior inferior cerebellar a.

— Posterior inferior cerebellar a.

— Vertebral a.

External carotid a. —

Brachiocephalic a. — — Subclavian a.

— Aorta

Fig. 95.2. Arterial Supply to the human brain. The internal carotid arteries supply mainly the anterior cerebral circulation, with a flow rate of 660–670 ml/min, whereas the vertebral–basilar system supplies the hindbrain, with a flow rate of 140–160 ml/min. While under normal circumstances there is little mixing of carotid and vertebral artery blood, the circle of Willis provides collateral flow compensation if one of the arteries is occluded. (From [86] with permission)

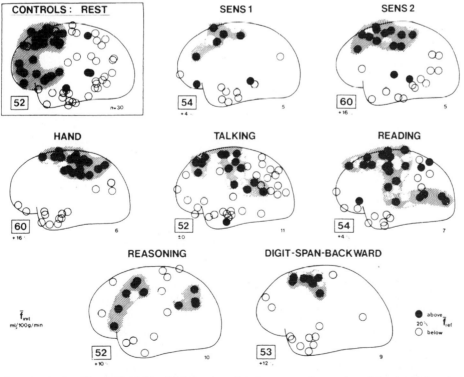

Fig. 95.3. Regional hemispheric blood flow in a conscious human at rest and during several types of local cerebral activity. *Closed circles* indicate local flow 20% above mean; *open circles* show local flow 20% below mean; *Sens 1*, electrical cutaneous stimulation of contralateral hand with low intensity; *sens 2*, same stimulation with high, painful intensity; *hand*, voluntary movement of contralateral hand. (From [35] with permission)

patients with chronic hypertension, however, the upper limit of autoregulation is higher (Fig. 95.4). The autoregulation of cerebral blood flow keeps brain perfusion and brain volume constant during alterations in posture. The myogenic autoregulation (Bayliss effect, see Sect. 94.3.1) and metabolic vasodilation mainly account for this buffering of perfusion during variations in perfusion pressure. Flow-dependent dilation also contributes to the self-adjustment of the brain vasculature (see Sect. 94.3.1).

Since the brain is contained within the rigid skull, any increase in intracranial pressure, as occurs after breakdown of the blood–brain barrier and during hypertension above the autoregulatory range, will eventually decrease cerebral perfusion pressure (i.e., arterial pressure minus intracranial pressure) and cerebral blood flow after a transient phase of pressure-passive hyperemia. This reduction in cerebral flow will induce further life-threatening, fulminant hypertension in response to the hypoperfusion of the brain.

Regulation of Cerebral Blood Flow. The cerebral vascular bed is very sensitive to changes in arterial P_{CO_2}: hypercapnia is the strongest stimulus for cerebral vasodilation (up to 100% increases in flow) and hypocapnia causes vasoconstriction, both responses occuring within seconds [50]. The mechanism is the direct effect of cerebrospinal fluid pH on the cerebral arterioles, since the diffusible CO_2 penetrates the blood–brain barrier and affects the fluid pH via the carboanhydrase reaction. However, changes in arterial pH in the presence of a constant arterial P_{CO_2} do not change cerebral flow: the polar protons cannot penetrate the blood–brain barrier [52]. Arterial hypoxia causes cerebral vasodilation by changing cerebrospinal fluid P_{O_2}, and vasodilation may result from endothelial release of nitric oxide (NO) and from adenosine release from the tissue.

Cerebral blood flow increases locally with changes in neuronal activity (Fig. 95.3) and local aerobic metabolic rate. Putative mediators for this metabolic vasodilation are adenosine and extracellular K^+ concentration, supported by the vasoregulatory effects of local changes in interstitial P_{CO_2} and P_{O_2} [50, 85].

The larger cerebral vessels are innervated by nerves arising in the superior cervical ganglion. These are adrenergic nerve endings, unilaterally distributed with some bilateral innervation of the more medial vessels, innervating conduit arteries down to the pial arteries and some parenchymal arteries as well as veins. While this innervation has little influence on cerebral blood flow under physiologic conditions, it substantially decreases cerebral blood flow during sudden increases in systemic blood pressure above the autoregulatory range, thereby preventing pressure-induced brain hyperemia and volume increments. This vasoconstriction of brain conduit arteries maintains the integrity of the blood–brain barrier [16, 27]. Intraparenchymal arterioles and capillaries with perivascular astrocytes are associated with a catecholamine innervation arising from the locus ceruleus in the midbrain. The physiologic role of this system is assumed in the regulation of cell permeability and not of cerebral blood flow [68]. Putative dilatory innervations with acetylcholine, vasointestinal peptide, nitric oxide, and other neurotransmitters exist around brain vessels, but their physiologic role is unclear.

Further details on the physiology and the pathophysiology of the cerebral circulation are presented in Chaps. 27–29.

95.3 Coronary Circulation

The coronary circulation supplies the heart muscle, which provides the energy for the perfusion of all other organs. Problems in the coronary circulation, therefore, affect the circulation to all other organs, either by sudden cardiac death or by the limitation of cardiac output in cardiogenic shock or in congestive heart failure (see Sects. 94.5.5, 94.5.6).

The human heart is supplied by the right and the left coronary arteries, which originate behind the left and right cusp, respectively, of the aortic valve. After branching into several superficial epicardial coronary arteries (left circumflex branch, left anterior descending branch, right posterior descending branch, and several diagonal and marginal branches), the coronary arteries penetrate the myocardium before branching into arterioles and capillaries, running parallel to the muscle fibers of the cardiac chambers. Capillary density in the human heart amounts to 3300 capillaries per mm², with approximately one capillary for each muscle fiber and a mean intercapillary distance of 17–20 µm [70]. Venous drainage of the ventricles is collected in the great cardiac vein, which empties into the

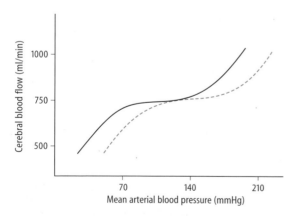

Fig. 95.4. Autoregulation of cerebral blood flow. In normotensive individuals (*solid black line*), minimal changes in cerebral blood flow occur between 70 and 140 mmHg (9.3–18.6 kPa) mean arterial pressure. In untreated hypertensive subjects (*dashed blue line*), this range is shifted to higher values. (From [82] with permission)

right atrium. The last few centimeters of the great cardiac vein are known as the coronary sinus. Part of the right ventricular and right atrial veins drain directly into the right atrium. Thebesian veins drain directly into the left and right ventricle, carrying less than 5% of total coronary flow [60], and a few arterioluminal arterioles (80–200 μm) with unknown function connect the coronary artery tree with the ventricular lumina. Arteriolar collateral vessels (about 40 μm in diameter) connect different coronary arteries or branches within the same coronary artery; they are of great functional relevance in coronary heart disease [76]. The myocardium has a rich lymphatic system with subendocardial and subepicardial plexus and connections to lymphatic vessels on the cardiac valves.

In healthy human at rest, the heart requires 4% of cardiac output. In the left ventricle, myocardial O_2 consumption amounts to 9–10 ml O_2/min per 100 g tissue at rest and can increase to above 50 ml O_2/min per 100 g tissue during heavy exercise in healthy, young subjects. Left ventricular blood flow at rest is 70–80 ml/min per 100 g tissue with an O_2 extraction of roughly 70% and this perfusion rate can increase during exercise to a maximum of almost 400 ml/min per 100 g tissue, with an O_2 extraction of about 80%. Thus, in the healthy human coronary bed, coronary flow can be augmented almost fivefold, and the coronary reserve is nearly 400% of basal flow under resting conditions. In the wall of the right ventricle, the perfusion rate is 65% of left ventricular perfusion, and in the atria it is 40%.

Coronary heart disease comprises a number of sequelae resulting from atherosclerosis in the large, epicardial coronary arteries. The atherosclerotic process in these arteries can result in flow-limiting *stenoses* and occlusions, in spasms of atherosclerotic arteries, and in ruptures of the *atherosclerotic plaques* with rapid intra-arterial thrombosis. These coronary events and alterations result in regional deficits of myocardial perfusion and, subsequently, in disturbances of regional myocardial electrophysiology, contraction and relaxation, in regional myocardial necrosis, or in sudden death due to ventricular fibrillation (see Chap. 91). Even with minor atherosclerotic alterations in the epicardial arteries, the coronary reserve is substantially reduced in coronary heart disease.

Myocardial Metabolic Requirements and Myocardial Perfusion. Myocardial perfusion closely follows myocardial O_2 requirements: coronary sinus blood O_2 content is fairy constant over a wide range of changes in coronary flow and myocardial O_2 consumption. Myocardial O_2 extraction is only slightly increased during strenuous exercise (see above).

Myocardial energy requirements are determined by systolic tension development of the muscle fibers and by heart rate. Furthermore, length changes of the muscle fibers during ejection affect myocardial energy requirements, since the energy used by a muscle fiber during an afterloaded, shortening contraction is higher than in an isometric contraction at the same force [18, 64]. This dif-

ference is called the Fenn effect and explains why the regional extent of fiber shortening during systole affects the regional energy consumption. The myocardial capacity for anaerobic metabolism is negligible, as significant glycolytic energy supply is only possible in the subendocardial cells of the cardiac conduction system. Therefore, the mechanically determined energy requirements of the myocardium determine the myocardial O_2 requirements.

Because of the tight coupling of global myocardial perfusion to the mechanically determined, global myocardial O_2 requirements, regional heterogeneities of perfusion are considered to reflect heterogeneities in regional O_2 requirements. In the left ventricles of large mammals at rest, there is a transmural gradient in perfusion and demand: in the subendocardial layers (including the papillary muscles), blood flow and O_2 consumption is 10%–30% higher than in the subepicardial layers, and indirect estimates indicate that this is also true for human myocardium. This transmural gradient probably reflects a gradient in the mechanical load to which the muscle fibers are exposed during the cardiac cycle and a gradient in fiber shortening during systolic ejection; these gradients in load and shortening depend on the complex geometric arrangement of ventricular muscle fibers.

Temporal variabilities of myocardial perfusion in a pseudosinusoidal fashion with transient maxima and minima of 40% above or below mean flow within 20 s have been observed [44]. These fluctuations are smaller than in other organs (e.g., resting skeletal muscle) and may result from intrinsic arteriolar smooth muscle vasomotion. They probably do not reflect fluctuations in local metabolic requirements.

In coronary heart disease, global coronary flow to the heart is reduced due to regional perfusion deficits, which result from atherosclerotic alterations in the epicardial arteries (see above). Since the heart continues to work (unless lethal arrhythmias result in a sudden cardiac death) with reduced global coronary flow, the pathophysiology of coronary heart disease has been considered as an imbalance of blood supply relative to mechanically determined requirements. It was assumed that a persisting reduction in the supply to demand ratio will finally result in the death of cardiocytes (*myocardial infarction*) due to a deficit in aerobic energy supply. However, this overall supply to demand ratio is an oversimplification. Analysis of regional myocardial function in hypoperfused myocardial areas demonstrates that regional fiber shortening and active tension development is reduced in parallel to the lowered perfusion rate, with the result that the regional supply to demand ratio is maintained essentially constant [29]. This adjustment of regional contractile function to a reduced regional perfusion in coronary heart disease represents a regulated cellular adaptation and has a protective effect; it is known as *myocardial hibernation*. The regulatory adaptation to hypoperfusion and ischemia is limited: hypoperfusion below a critical level will eventually result in ischemic death of cardiocytes and in their replacement by fibrotic tissue, although the local supply to demand ratio is not lowered [29]. This protective adaptation requires some degree of activation: several short periods of reversible hypoperfusions partially protect the myocardium against the lethal consequences of a subsequent long-lasting hypoperfusion; compared to a long-lasting ischemia without previous short hypoperfusions, the amount of myocardial cell necroses in a myocardial area with ischemia of the same degree and duration is reduced by up to 80% if short hypoperfusions precede the terminal ischemia. This protection exists despite the fact that the total ischemic burden (i.e., the time with a deficit in perfusion) is increased by these preceding hypoperfusions. The substantial protection provided by additional, transient hypoperfusions is called *ischemic preconditioning* [29]. Although the complex cellular mechanisms of myocardial hibernation and ischemic preconditioning are not entirely known, it is clear that they represent an important regulatory adaptation of the myocardium to reductions in regional myocardial perfusion.

Under physiologic conditions, myocardial perfusion follows myocardial requirements by metabolic vasodilation; under pathophysiologic limitations of perfusion, myocardial demand is adapted to myocardial perfusion by the processes of ischemic preconditioning and myocardial hibernation. Thus, the tight coupling of coronary perfusion and myocardial requirements can be achieved by adaptive processes both in the vasculature and in the myocardium.

Myocardial Contraction: An Extravascular Resistance to Coronary Flow?

In the working heart in a resting organism, the blood volume in the coronary bed is 13%–15% of the cardiac tissue volume, and two thirds of this coronary blood volume are in the myocardial exchange vessels (capillaries and venoles). Coronary vasodilation by drugs or by increments in myocardial metabolic requirements increases this intracoronary or intramural blood content. At a normal degree of coronary vascular tone (i.e., at normal left ventricular perfusion of 8–10 ml/min per 100 g), a sudden cardiac arrest with maintained coronary perfusion pressure immediately increases intracoronary blood volume from 14% to 18%, clearly demonstrating that the cardiac contraction reduces the intramural blood content (in an excised, arrested heart without perfusion pressure, the intracoronary blood content is approximately 5% of tissue volume). Comparison of phasic patterns of coronary arterial inflow and coronary venous outflow in the working human heart (Fig. 95.5) demonstrates that coronary inflow is higher in diastole and that coronary venous outflow is higher in systole. This phase shift of about 180% between inflow and outflow indicates that 5% of the intramural blood content is squeezed out of the myocardium with each systolic contraction, while this fraction is 25% under maximal coronary vasodilation [79]. Thus, the cardiac contraction rhythmically squeezes the vasculature embedded in the myocardium, and this squeezing acutely impedes systolic inflow while enhancing systolic venous outflow. Furthermore, this rhythmic squeezing contributes to the formation of endothelium-derived nitric oxide, thus

yielding an important dilator of coronary resistance vessels which is involved in the tight matching of myocardial perfusion to the mechanical performance of the heart that determines requirements. The squeezing action is probably maximal in the myocardial capillaries, venoles, and veins, but endothelial nitric oxide generated in these sections of the coronary bed has been shown to dilate the coronary resistance arterioles by counterflow diffusion. Apart from this formation of a dilator signal by myocardial squeezing, its direct mechanical net effect (inflow impedance versus outflow augmentation) on myocardial perfusion under physiologic conditions is unclear and probably negligible.

In coronary heart disease, however, the rhythmic compression of the coronary bed by the cardiac systole is a substantial impedance to perfusion. In an arteriosclerotic coronary artery, the coronary perfusion pressure distal to a flow-limiting stenosis is critically low and below the autoregulatory range (if a substantial collateral circulation has not yet developed; see below). The arterioles in the myocardium supplied by such an stenotic coronary artery are highly dilated, the endothelial release of vasorelaxing nitric oxide is disturbed or abolished, and due to myocardial hibernation the hypoperfused myocardial area does not rhythmically contract and relax, but is passively bulged out and thinned during systole. The systolic pressure (developed by the normally perfused myocardium) induces systolic wall stress in the noncontracting, hypoperfused myocardium, and this is even higher in the hypoperfused, bulging myocardial area than in the normally perfused, contracting myocardium because of the law of LaPlace (see Chap. 90). Thus the systolic impedance of coronary inflow is maximal in hypoperfused myocardium, in which enhanced formation of endothelial nitric oxide by systolic squeezing of the vasculature with further dilation of the coronary arterioles cannot take place.

Therefore, drugs with negative chronotropic and inotropic effect (e.g., β-blockers) will augment residual flow in poststenotic, hypoperfused myocardium by reducing heart rate and systolic pressure development in the normally perfused myocardium [30]. Similarly, drug-induced venodilation (e.g., by nitrovasodilators) will improve poststenotic perfusion by reducing systolic wall stress via preload reduction. Extravascular coronary resistance and its modulation by drugs becomes especially important in the subendocardial ventricular myocardium under conditions of hypoperfusion [30].

Metabolic Coronary Vasoregulation and Autoregulation.

Any increase in cardiac contractile function and in metabolic activity causes a dilation in arteriolar diameter. This dilation is largest in the smallest arterioles [40] and is thought to be mediated by the release of local metabolites; it results in *functional hyperemia* (see Chap. 94). It is considered to be the quantitatively most important regulatory mechanism in the coronary bed [17]. However, the metabolites responsible for functional hyperemia have not yet been identified. Enhanced interstitial concentrations of

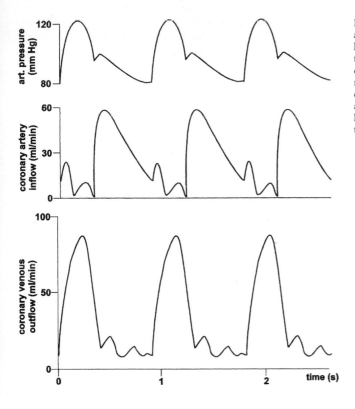

Fig. 95.5. Phase shift between coronary arterial inflow and coronary venous outflow by the contraction of the heart. While coronary inflow (left circumflex coronary artery) is mainly diastolic, coronary venous outflow (great cardiac vein) occurs mainly during systole. Note that venous outflow through the great cardiac vein collects blood draining from myocardium supplied by several coronary arteries; therefore, mean flow in this coronary vein is higher than mean flow in the left circumflex coronary artery

adenosine, enhanced tissue osmolarity, and tissue acidosis with enhanced tissue P_{CO_2} and reduced tissue P_{O_2} have been considered as mediators for metabolic control, but the available experimental data cannot explain functional myocardial hyperemia as being induced by any one of these mediators mentioned above or by the combined action of several of them [17,21]. The mechanically induced release of endothelial nitric oxide due to squeezing of the coronary bed by the cardiac contraction, as discussed above, is an attractive alternative hypothesis to explain the coupling of flow to metabolism in functional myocardial hyperemia, but it is not known whether this poorly defined mechanism is potent enough to explain the fivefold increase in myocardial perfusion which is possible in the human myocardium.

In the sections of the coronary tree upstream to the smallest arterioles, the mechanisms of self-adjustment within the vascular network (see Sect. 94.3.1) contribute to the functional hyperemia in the coronary bed, namely myogenic autoregulation and flow-dependent, endothelium-mediated dilation [33,48,49]. Autoregulation of myocardial blood flow, however, implies not only the myogenic adjustment of coronary arteriolar tone to changes in transmural pressure by the *Bayliss effect* (see Sect. 94.3.1.1), but also includes a component resulting from metabolic vasodilation. When coronary perfusion pressure in an experiment is controlled and altered independently of systolic left ventricular pressure development, which is kept constant, coronary blood flow will not

decline with declining coronary perfusion pressure, since metabolic demands due to cardiac pressure development (which remains constant) will cause metabolic dilation, balancing the effect of declining perfusion pressure on coronary flow as long as perfusion pressure is above 60 mmHg (8.0 kPa).

The separation of coronary perfusion pressure from ventricular pressure development may appear to be a rather remote and artificial experimental artifact for the demonstration of autoregulation of flow in the coronary bed of the beating heart, but it becomes important in coronary heart disease. In arteriosclerotic epicardial coronary arteries, *atheroma* or *plaques* can narrow the arterial lumen and act as a stenosis, which causes a locally enhanced loss of perfusion pressure due to enhanced frictional energy dissipation in the blood flow through the stenosis. However, tissue perfusion distal to the stenosis will not decline as long as autoregulation of the coronary bed can compensate for the reduction in poststenotic perfusion pressure by arteriolar vasodilation. However, during functional hyperemia or drug-induced coronary vasodilation, there is no reserve for further vasodilation, and the coronary stenosis will result in a reduction in flow to the poststenotic myocardial area. This is the reason why moderate stenoses do not reduce perfusion of poststenotic myocardial tissue under basal conditions, but do attenuate substantially the coronary reserve (Fig. 95.6). In patients with such moderate or intermediate coronary artery stenoses, signs of coronary hypoperfusion, such as angina pectoris and

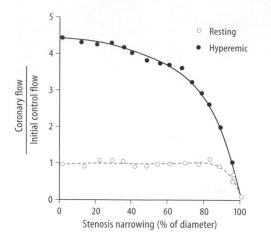

Fig. 95.6. Effect of the degree of coronary artery stenoses on coronary flow under resting or hyperemic conditions. At rest, coronary flow remains unchanged up to 85% narrowing of epicardial artery diameter by a stenosis, since the decline in poststenotic pressure is compensated by vasodilation of the poststenotic coronary bed (*blue line*). Under maximal coronary vasodilation (hyperemic condition, *black line*), there is no further reserve for poststenotic dilation, and flow declines with declining stenosis radius. The coronary reserve (i.e., maximal coronary flow/basal or initial coronary flow) falls below 4 with a 45% stenosis, which has no effect on basal coronary flow. (From [24] with permission)

electrocardiogram (ECG) alterations, will not appear at rest, but during exercise with increased metabolic flow requirements.

Arteriolar collateral vessels connecting different coronary arteries or branches from one coronary arteries, which are not larger than 40 μm in diameter in healthy hearts (see above), grow to almost full-size epicardial collateral arteries if a slowly progressing stenosis in one coronary branch chronically lowers poststenotic perfusion pressure in this branch. Flow-dependent chronic caliber modification and probably some influence from tissue-derived growth factors may contribute to this remodeling of collateral circulations in coronary heart disease [20,76]. Such collaterals tend to ameliorate the situation in a coronary circulation with an epicardial artery stenosis.

Neuronal Control of Coronary Blood Flow. The coronary bed receives adrenergic sympathetic perivascular innervation, intermingled with adrenergic sympathetic innervation to the myocardium. Reflex activation of sympathetic cardiac nerves results in an increase in coronary flow, which is due to metabolic vasodilation in response to the reflex enhancements of heart rate and cardiac contractility. Although coronary arteries and arterioles have smooth muscular β-adrenoceptors, a direct neurogenic, sympathetic vasodilation by vascular β-adrenoceptor activation has never been demonsrated [2]. The direct effect of the sympathetic coronary innervation is vasoconstriction, mediated by vascular α_1- and α_2-adrenoceptors. The relative distribution of the receptor subtypes varies in different segments of the coronary vascular tree, but the net effect of the sympathetic coronary innervation in large epicardial arteries as well as in flow-regulating coronary arterioles is α_1-adrenergic constriction. Under physiologic conditions, this constriction is of minor relevance; it is limited by endothelium-mediated, flow-dependent dilation in the large epicardial coronary arteries and by metabolic vasodilation in the resistance vessels. Experimental data indicate that the functional coronary hyperemia that occurs during exercise, mediated by metabolic vasodilation, is partially restricted by counteracting sympathetic vasoconstriction in such a way that the myocardial O₂ extraction is enhanced [59,61]. Indirect data indicate that this is also true for humans, but a critical myocardial hypoperfusion cannot be induced by sympathetic innervation in the healthy human. In coronary heart disease, however, sympathetic coronary constriction of coronary resistance vessels in poststenotic hypoperfused myocardium contributes to exercise-induced angina pectoris and regional myocardial dysfunction in patients [7,28]. Cholinergic coronary innervation and innervation by noncholinergic, nonadrenergic nerves (presumably nitroxidergic nerves) is demonstrable histologically, but its physiologic and pathophysiologic relevance is unknown.

In summary, problems considered as typical for the coronary bed such as extravascular resistance due to myocardial contraction, autoregulation of myocardial blood flow, and sympathetic coronary vasoconstriction are of relevance only in the pathophysiologic state of coronary heart disease, while the mechanism of metabolic coronary vasodilation, the dominant mechanism of flow regulation in the healthy heart, is not understood at present.

95.4 Renal Circulation

The renal circulation (see also Chap. 73) affects all other vascular beds by enabling the tubular system of the kidney to control volume, osmolarity, and composition of the body fluids (see Chaps. 73–82). Furthermore, local, paracrine angiotensin II formation, which exists in all extrarenal vascular beds, is modulated by renal release of active renin, which is taken up into the vascular wall of other organs after a half-life of 30 min in the circulation. After such uptake into the vessel wall, this nephrogenic renin contributes to local angiotensin II formation and paracrine modulation of sympathetic vasoconstriction in extrarenal organs.

During the passage through the kidney, blood flows through two capillary beds are connected in series: the glomerular capillaries and the peritubular capillaries (Fig. 95.7). Such an arrangement is called a portal system according to the arrangement in the splanchnic circulation with the portal vein between the two capillary networks of the splanchnic circulation (see Sect. 95.6).

At a mean systemic blood pressure of 100 mmHg (13.3 kPa), intravasal pressure at the origin of the glo-

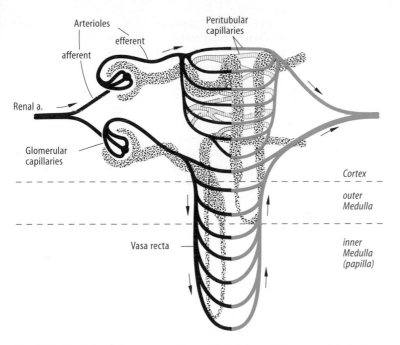

Fig. 95.7. Renal circulation, arranged in a portal system with two capillary networks connected in series. The renal artery branches into interlobar arteries, passing between the pyramids of the renal medulla towards the cortical surface. At the boundary between medulla and cortex, they divide into arcuate arteries running along this boundary, and interlobular arteries branch off from these arcuate arteries. Afferent arterioles from these interlobular arteries reach the glomerular capillaries in the outer renal cortex and in the inner or juxtamedullary cortex. The efferent arterioles from the glomeruli in the outer renal cortex direct the blood to a second capillary network surrounding the nearby proximal and distal tubuli and colleting ducts. Vasa recta to the peritubular capillaries in the papillae of the inner medulla originate from the efferent arterioles of the juxtamedulary glomeruli. (From [81] with permission)

merular capillaries is roughly 50 mmHg (6.7 kPa), and at the beginning of the peritubular capillaries it is some 20 mmHg (2.7 kPa). Thus, the major decline of perfusion pressure within the renal vasculature takes place in the afferent and efferent arterioles with 40 mmHg (5.3 kPa) and 30 mmHg (4.0 kPa), respectively. Physiologic and pathophysiologic regulation of the vascular resistances in these two arteriolar segments is important for renal function and is quantitatively and qualitatively different.

The blood flow to the kidneys is 1.1–1.2 l/min or 20% of cardiac output, while their O_2 consumption is only 7% of total body O_2 consumption at rest. Consequently, the O_2 extraction of the kidneys is low compared to other organs: 1.5 ml O_2/100 ml blood. This is explained by the fact that the main function of the renal blood flow is to yield a glomerular filtration rate which is sufficient for the excretory and volume-regulatory functions of the kidneys. If renal blood flow is reduced for some pathophysiologic reason, the renal O_2 consumption is reduced in parallel and the O_2 extraction is not changed: the renal O_2 consumption is largely determined by Na^+ transport (see Chap. 74) and, therefore, is a function of the glomerular filtration rate.

Within the kidneys, the renal cortex, which contains the glomeruli, receives 93% of renal blood flow, while 7% of organ flow goes to the renal medulla. Cortical renal blood flow amounts to 530 ml/min per 100 g, flow to the outer renal medulla is 130 ml/min per 100 g, and flow to the inner medulla is below 40 ml/min per 100 g. Since all renal tissues receive their blood flow through two capillary networks connected in series (see Fig. 95.7) and since the perfusion pressure in the efferent arterioles leaving the glomeruli is similar, the high additional vascular resistance in the medulla must result from postglomerular sections of the renal vasculature. The high resistance along the long, descending medullary peritubular vessels provides this additional resistance. Since flow to cortex and to the medulla is differently regulated, the critical difference must also be located in long descending medullary peritubular vessels, too.

Regulation in the Afferent Arterioles. The most important feature of the renal circulation is the parallel *autoregulation* of renal blood flow and of glomerular filtration rate: when mean arterial pressure in the renal arteries changes within the range of 80–200 mmHg (10.6–26.6 kPa), renal blood flow and glomerular filtration rate remain fairly constant. This autoregulation of the two parameters can also be demonstrated in isolated kidneys without external regulatory inputs and is brought about by myogenic autoregulation (Bayliss effect; see Sect. 94.3.1.1) of the afferent arterioles. Against increasing perfusion pressure, the afferent arterioles (and the smaller renal arteries proximal to the afferent arterioles) constrict, keeping flow through the bed and pressure at the end of the entrance into the glomerular capillaries constant. (For details of the filtration process in the glomerular capillaries, see Chap. 73).

Closely associated with the autoregulation of renal blood flow and glomerular filtration rate is the release of renin by the *renal baroreceptor mechanism*; the vascular smooth muscle cells at the end of the afferent arterioles contain granula with renin and are called myoendocrine cells. Renin is an endoprotease that acts on angiotensinogen to initiate the cascade of the formation of angiotensins (see Sect. 94.3.3.4). The adequate stimulus for renin release from the afferent arterioles is vasodilation and lowering of the intracellular Ca^{2+} concentration in these afferent arteriolar cells. When renal artery perfusion pressure declines and the autoregulating myogenic vasodilation reaches the terminal segment of the afferent arterioles, this dilation results in renin release. The autoregulating afferent arteriole itself is the renal baroreceptor for renin release. Thus, hypotension-induced renal renin release increases dramatically when the hypotension reaches the lower level of renal autoregulation. In the intact organism, the consequences of renin release (volume retention, activation of sympathetic activity, and general vasoconstriction) tend to restore systemic pressure to the normal level (above the lower level of renal autoregulation). Two further important mechanisms contribute to renin release from the afferent arterioles by the rena baroreceptor mechanism:

- Norepinephrine release from sympathetic nerve endings, acting on dilatory β-adrenoceptors at the myoendocrine cells of the afferent arterioles; reflex-induced sympathetic activation during hypotension enhances renal renin release, but is not the main factor in this release.
- Reduced transport of NaCl in the macula densa cells of the distal tubulus, which attenuates the constrictive control of tone in the vas afferens by tubuloglomerular feedback (see below); drastically reduced salt intake or blockade of salt transport in the macula densa cells enhance renin release by this mechanism.

In addition to these major mechanisms of renal renin release, hormones and local autacoids play a modulatory role; renin release is inhibited by angiotensin and other vasoconstrictors, but indirectly also by natriuretic peptides.

Tubuloglomerular feedback can be understood as a mechanism that excludes salt-losing nephrons with impaired tubular capacity for the reabsorption of the filtered salt from the circulation. When the tubular reabsorption of a nephron is attenuated for some reason, the macula densa cells at the end of the distal tubule are confronted with an enhanced salt load, and their salt transport is high. The transport rate in these cells generates a constrictor signal to the adjacent myoendocrine cells of the afferent arteriole, causing their constriction (up to the closure of the arteriole) and the attenuation of renin release from these cells. This tubuloglomerular feedback or *macula densa mechanism* operates not only to exclude nephrons from the filtration, but is a mechanism that gradually modulates constrictor tone of the afferent arterioles and renin release from these arterioles; reduced salt transport in the macula densa causes afferent arteriolar dilation and enhanced renin release (see above). Although tubuloglomerular feedback and renin release both depend on macula densa transport, the renal renin–angiotensin system is *not* the mediator of tubuloglomerular feedback. The transport-dependent formation of adenosine (which is a constrictor of afferent arterioles) and of nitric oxide (acting as dilator) are considered to be the mediators of tubuloglomerular feedback via modulation of afferent arteriolar tone.

Regulation in the Efferent Arterioles. Vasoconstriction in this section of the renal vasculature reduces peritubular capillary pressure, thereby facilitating tubular reabsorption, and enhances pressure in the glomerular capillaries, thereby enhancing the *filtration fraction*, i.e., the fraction of plasma flowing through the glomerular capillaries which is filtered into the Bowman's capsule (see also Chap. 73). The most important constrictor of the efferent arterioles is angiotensin II (while the responsiveness of afferent arterioles to angiotensin II is orders of magnitude lower). An activated renin–angiotensin system causes a compensatory enhancement of the filtation fraction when arterial pressure is below the lower level of autoregulation of the glomerular filtration rate. In diseased kidneys with a reduced number of filtrating nephrons, angiotensin-induced vas efferens constriction and dilation of vas afferens by prostaglandins and natriuretic peptides result in a temporaty compensatory hyperfiltration of the surviving nephrons. However, this compensatory hyperfiltration contributes to the progression of pathologic nephron damage.

Regulation in the Medullary Vasa Recta. While increases in renal artery pressure result in autoregulation of perfusion and filtration of the glomeruli, an increase in flow to the renal medulla is the basis of *pressure diuresis*, the most important mechanism in the long-term regulation of blood pressure (see Sect. 94.5.1.2). The vasa recta to the renal medulla arise from the efferent arterioles of the juxtaglomerular glomeruli, which participate in autoregulation during increases in arterial pressure (see Sect. 94.5.1.2). Therefore, the enhanced medullary flow during elevations in perfusion pressure must be mediated by some unidentified signal from the preglomerular section of the vasculature.

Integrated Control of Renal Blood Flow. Reflex activation of sympathetic activity during hypotension, during strenuous exercise, and during prolonged exposure to cold causes reductions in renal blood flow and attenuations in glomerular filtration rate. The adrenergic innervation of the renal vasculature induces vasoconstriction of preglomerular small arteries and of afferent and efferent arterioles, but the preglomerular constriction is the dominating effect (for the effects of sympathetic innervation on renal tubules and on the glomerular mesangial cells, see Chap. 73). Renal blood flow and glomerular filtration rate in the healthy kidney are not absolutely constant: compared to the fasting state, postprandial flow to and filtration in the kidney are elevated by 20%, and this elevation persists longer after a protein-rich meal. (Further details on the renal circulation in the context of urine formation are given in Chaps. 73, 74).

95.5 Circulation to Skeletal Muscles

Skeletal muscle is the largest organ of the body, accounting for an average of 30 kg or 45% of total body mass in untrained young men. Flow to skeletal muscle varies from less than 20% at rest to more than 80% during strenuous exercise. Inadequate perfusion of skeletal muscle in relation to metabolic requirements causes fatigue and exhaustion in both patients with a limited cardiac output due to chronic cardiac failure and in untrained healthy humans in whom the cardiac output is not sufficient to supply all organs with adequate flow when the skeletal muscle bed is maximally dilated (see Sect. 94.5.3 and Chap. 108). Furthermore, due to its large size, the skeletal muscle bed makes an important contribution in resting humans to reflectory enhancements of peripheral vascular resistance, as occurs during hemorrhage (see Sect. 94.5.5).

Conduit arteries to skeletal muscle branch into small arteries, arterioles, and capillaries parallel to the muscle fibers. Capillary density in skeletal muscle is approximately 400 capillaries per mm^2, with a mean intercapillary distance of approximately 40 μm and an arrangement of approximately one capillary for each muscle fibers. Capillary density and fiber thickness vary with the type of muscle fibers: type I fibers (slow, oxidative) and type IIa fibers (fast, oxidative/glycolytic) have somewhat smaller diameters and intercapillary distances than the type IIb fibers (fast, glycolytic). Compared to cardiac muscle, skeletal muscle has a roughly twofold larger diameter of muscle fibers and, consequently, a lower capillary density with a twofold larger intercapillary distance (see Sect. 95.3) and a lower capillary surface area [70]. At their venous ends, skeletal muscle capillaries anastomose more frequently with each other, thereby increasing the surface area available for exchange.

The venous bed of skeletal muscle is fairly small and sparsely innervated. In spite of the large mass of the organ, the skeletal muscle bed does not contribute significantly to the reflex modulation of total body vascular compliance [58]. Reductions in skeletal muscle volume under sympathetic stimulation do not result from active venoconstriction, but are due to lowered skeletal muscle perfusion. However, venous tone in skeletal muscle can be somewhat modulated by circulating catecholamines and by flow-dependent dilation [56]. Compression of the veins by dynamic activity of skeletal muscle is the most important factor supporting venous return from the skeletal muscle bed.

Basal skeletal muscle flow under resting conditions amounts to 2–4 ml/min per 100 g, but a great variability of values is found in the literature. This is due to the fact, that "rest" is not well defined and very often includes muscle activity involved in the maintenance of posture. In such slow, oxidative fibers with postural activity, flow rates at rest of up to 10–20 ml/min per 100 g are observed. Basal skeletal muscle O_2 consumption is below 0.2 ml/min per 100 g [37], which may be explained by the fact that anaerobic glycolic adenosine triphosphate (ATP) production cyclically occurs in a large fraction of inactive skeletal muscle with rhythmic arteriolar vasomotion (see below). During strenuous exercise in healthy, untrained humans, maximal flow rates of 100–120 ml/min per 100 g with an O_2 consumption of 13–15 ml/min per 100 g are observed in the 16–18 kg muscles which are dynamically active in exercising nonathletes [37]. This amounts to a total skeletal muscle flow of some 20 l/min during strenuous exercise, compared to less than 1 l/min at rest. However, if only a few muscle groups are dynamically exercising with maximal intensity under an experimental setup, flow rates of 200–300 ml/min per 100 g can be measured with O_2 consumptions of 30–40 ml/min per 100 g in these dynamically exercising muscles [58,75]. This means that under physiologic strenuous exercise, the skeletal muscle bed can never reach the maximal blood flow which would be possible due to its maximal metabolic capacity and its vascularization, since cardiac output is limited (see Sect. 94.5.3). Sympathetic constrictor activity also reduces flow in exercising muscle, and a substantial fraction of the energy dissipation in the working muscles with sympathetically attenuated perfusion occurs by anaerobic glycolysis. The arterial lactate concentration resulting from this anaerobic glycolysis in the exercising muscles is used as an empirical parameter for the upper limit of "endurance performance" (arterial lactacte concentration up to 2 mmol/l). Exercise resulting in an arterial lactate concentration above 4 mmol/l includes a substantial part of anaerobic glycolysis in the working muscles and quickly results in exhaustion.

Regulation of Skeletal Muscle Perfusion. Basal flow in the inactive skeletal muscle is under substantial myogenic constrictor tone [25], reinforced by α_1-adrenergic, constrictive control via the sympathetic innervation: experimental interruption of this neurogenic tonic activity to the skeletal muscle bed doubles flow to the resting muscle [37]. Afferent input from the arterial baroreceptors and the cardiopulmonary volume receptors can modulate this tonic sympathetic constrictive control, causing doubling of flow by attenuation of the sympathetic constriction with baroreceptor activation. Active neurogenic vasodilation by activation of vasodilator nerves in the skeletal muscle bed is not elicited by baroreceptor inputs. Constrictions with flow reductions to 40% of basal flow are elicited by unloading of the thoracical cardiopulmonary volume receptors

[39,88]. This strong reflex control by the cardiopulmonary volume receptors on skeletal muscle flow is relevant for hemodynamic adjustments in the limbs during upright posture and also operates during minor blood loss.

Reflex responses to high ambient temperatures include sympathetic vasoconstrictor activity to skeletal muscle, while limb blood flow is redistributed to the skin [14]. Direct, temperature-induced vasodilation in skeletal muscle requires elevations in muscle temperature above 42°C, which is not possible without pain and damage to the skin if ambient heat is applied [37]. Reflex responses to activation of arterial or central chemoreceptors by lowered P_{O_2} or by elevated P_{CO_2} also causes sympathetic vasoconstriction of the skeletal muscle bed [12], while the direct effect of such changes in blood gases is vasodilation in skeletal muscle. Chemoreceptor activation is part of the diving reflex, and the other part is triggered by sensory endings of the trigeminal nerve when the head is immersed in water. Diving results in a strong sympathetic vasoconstriction in skeletal muscle in humans [26]. The role of specific noncholinergic vasodilator nerves and of histaminergic vasodilation in the regulation of blood flow to skeletal muscle in humans is unclear.

The arterioles and venoles in the skeletal muscle circulation are endowed with postsynaptic α- and β-adrenoceptors. Arteriolar β-receptor density, preferentially of the $β_2$-subtype, in the skeletal muscle bed is higher than in most other beds [58]. However, β-adrenergic dilation is not elicited by norepinephrine released from the sympathetic nerve endings. During stress, however, skeletal muscle vasodilation is elicited by moderate circulating doses of epinephrine via activation of these β-adrenoceptors in the microcirculation of the skeletal muscle bed and, additionally, by β-adrenergic enhancement of the metabolic rate in inactive skeletal muscle [58]. High circulating concentrations of epinephrine (above 500 ng/l), however, result in α-adrenergic vasoconstriction in the skeletal muscle bed, as in all other beds.

Tissue perfusion in inactive skeletal muscle demonstrates substantial local fluctuation, which is explained as resulting from rhythmic activity of arterioles and precapillary sphincters [36]. Such nonsynchronized, slow oscillations in arteriolar tone temporarily exclude a fraction of capillaries from being perfused, resulting in transient accumulations of metabolites in local microdomains with subsequent dilation and washout of metabolites. Such oscillations alter pressure gradients between two neighboring points in a capillary network and can induce reversals in the direction of flow through individual capillaries of the network. In inactive skeletal muscles, such locally nonsynchronized cyclical vasoconstrictions and vasodilations are assumed to improve the effectivity of the exchange of substrates between tissue and blood [36] (see also Chap. 97). Furthermore, such cyclical vasomotion is believed to induce transient phases of anaerobiosis in muscle fibers with a high anaerobiotic capacity (type IIa and IIb fibers). The mediators of metabolically induced "reopening" of the oscillating precapillary sphincters after transient exclusion of microdomains from perfusion with

subsequent accumulation of metabolites are unknown. They may include adenosine, enhanced osmolarity, acidosis, and local hypoxia [37]. Locally formed prostaglandins are involved in the vasodilation of underperfused skeletal muscle [43] and are assumed to contribute to vasodilation of cyclically perfused inactive muscle, since indomethacin increases vascular resistance in resting, but not in exercising muscle [87]. However, prostaglandins are not endproducts or intermediates of skeletal muscle metabolism and therefore do not fulfil the strict criteria of a "metabolic" vasodilator. Part of the capillaries of an inactive skeletal muscle are temporarily excluded from being perfused. This has two consequences:

- The capillary exchange area is reduced.
- A substantial reserve of capillaries exists, which can be recruited for participation in the microcirculatory exchange by simultaneous dilation of all precapillary sphincters.

Some 30% of capillaries are perfused in an inactive skeletal muscle, leaving 70% of them as "capillary reserve" for recruitment by functional hyperemia.

Flow to Exercising Muscle. The contraction of striated muscle increases intramuscular interstitial pressure and compresses the vascular bed, which may result in closure of the microcirculation and cessation of perfusion. This can be demonstrated during sustained, isometric contractions; during such contractions with 70% of maximal tension, blood flow to the contracting muscle is zero and such contractions cannot be sustained for more than 1–2 min. At 50% of maximal tension, flow remains at the level of inactivity, although the metabolic rate of ATP usage is increased by more than tenfold. Even with minimal isometric contractions with only 20% of maximal tension, a marginal impedance of flow can still be observed, although metabolic vasodilation prevails [37]. The flow increase after release of an isometric contraction with more than 70% of maximal tension resembles a postocclusion reactive hyperemia (except that tissue metabolism is substantially increased during the ischemia) and includes a myogenic dilatory component. The major part of postcontraction hyperemia is due to enhanced metabolism following the period of anaerobiosis during the strong isometric contraction.

During dynamic exercise, however, the extravascular compression by interstitial pressure is typically phasic, and large increases in flow through the relaxed muscular arterioles are possible in the intermittent phases of skeletal muscle relaxation (Fig. 95.8). This flow increase during dynamic exercise, which can amount to 120 ml/min per 100 g in humans while running and to more than 250 ml/min per 100 g during dynamic exercise of small muscle groups (see above), is the prototype paradigm of metabolically induced functional hyperemia, in which perfusion is linked to metabolic rate. Many possibilities have been considered as mechanisms mediating this link:

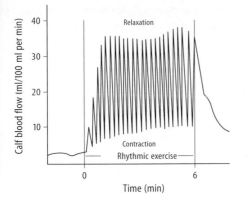

Fig. 95.8. Calf blood flow during walking as a typical example of blood flow to dynamically exercising muscle. Metabolic arteriolar vasodilation allows substantial increments in flow during phases of skeletal muscle relaxation, while during contractions flow is almost reduced to resting levels by the mechanic compression of the vascular bed. (From [37] with permission)

- Contracting skeletal muscle compresses the vascular bed and causes myogenic vasodilation in response to the reduced transmural pressure. However, vasodilation of skeletal muscle microcirculation due to the Bayliss effect is orders of magnitudes smaller than functional vasodilation, and rhythmic twitch-like muscular compression cannot induce substantial hyperemia.
- Lowered interstitial P_{O_2} causes vasodilation in this bed. However, hyperoxia does not attenuate functional hyperemia in skeletal muscle.
- Lactate infusion causes a weak vasodilation in the skeletal muscle bed, and lactate is generated during isometric contractions and during the initial period of phasic contractions. However, in patients with an enzymatic defect in muscluar lactate production, sustained functional hyperemia in skeletal muscle is not attenuated [57].
- Interstitial elevations in K^+ concentrations, due to bursts of K^+ released during muscular action potentials, may contribute to the initiation of functional hyperemia, but cannot maintain the excess flow, since skeletal muscle venous $[K^+]$ returns to normal levels with sustained exercise. The same is true for interstitial hyperosmolarity, which is not sustained during ongoing exercise.
- Acidosis and interstitial hypercapnia cannot account for the functional hyperemia, since skeletal muscle venous acidosis and/or elevations of P_{CO_2} in the vasodilatory range do not occur in dynamic exercise.
- Adenosine is the best-analyzed candidate as a mediator of metabolic vasodilation, but neither adenosine receptor antagonists nor adenosine deaminase (which converts adenosine into vasoinactive metabolites) attenuate functional hyperemia in exercising skeletal muscle, although they do interfere with hypoperfusion-induced vasodilation in other organs [45,47].

- Cholinergic sympathetic vasodilator nerves, artificially activated during experimental stimulation of the hypothalamic "defense areas," cause atropine-sensitive skeletal muscle vasodilation during the onset of exercise, during "fight or flight" reactions, and during emotional fainting, but atropine does not modify sustained functional hyperemia.
- Pulsatile perfusion, as occurs during dynamic exercise, increases flow-induced release of endothelial nitric oxide, but the effect is small, and, as has already been mentioned, twitch-like muscle compression cannot substantially increase perfusion.

Metabolic vasodilation may be mediated in a redundant manner by the concerted action of the above-mentioned mechanisms, which, taken alone, cannot explain functional hyperemia. However, most of these metabolic mechanisms, which may contribute to the onset of functional hyperemia, have the wrong time course to contribute to steady state exercise hyperemia [23]. Therefore, redundancy of effector mechanisms is probably not the reason for the failure to identify the mechanism or mechanisms of sustained metabolic vasodilation in skeletal muscle. The issue remains an "unanswered question" [23,37], something which has triggered speculative hypotheses on a role of unidentified neuronal transmitters, released with the activation of skeletal muscle by the motor innervation, of an intrinsic arteriolar innervation, and of nitric oxide synthases in skeletal muscle myocytes. Conceptionally, it appears attractive to postulate a "forward regulation" as the main mechanism of functional hyperemia, directly associated with the skeletal muscle activation (such as a neurotransmitter or the skeletal muscle nitric oxide synthase), while a feedback regulation by metabolic factors contributes to the fine tuning of the hyperemia. Such speculative postulates remain to be substantiated.

Flow to exercising muscle is under substantial vasoconstrictor control by the sympathetic nervous system, which helps to maintain the functional hyperemia of the working muscles within the limits of the maximal cardiac output of 25–28 l/min (see above). During work in a hot environment, this vasoconstrictor activity to working muscle is even more increased, resulting in a more rapid disappearance of muscle glycogen, a higher lactate production, and more rapidly developing fatigue [19]. In patients with chronic congestive heart failure, the sympathetic vasoconstriction of exercising muscle is maximal; furthermore, endothelium-mediated vasodilation in skeletal muscle in these patients is attenuated, curtailing the dilator response of the limb vasculature to many dilatory stimuli.

95.6 Splanchnic Circulation

The splanchnic circulation consists of the blood supply to the gastrointestinal tract, the liver, the pancreas, and the spleen (see Chap. 69). The arterial vessels to this circulation are the common hepatic artery and the arteries draining into the portal vein, namely the celiac artery and the superior and inferior mesenteric arteries, with their extensive arterial interconnections. The splanchnic circulation picks up the nutrients absorbed from the intestinal lumen for transport to the liver for storage, transformation, or supply to the general circulation. Apart from this role in the organism's metabolism, the splanchnic circulation acts as a functional reservoir of blood for the organism: reductions in flow to the splanchnic bed by vasoconstriction and additional mobilization of venous blood by active venoconstriction, specifically of the portal veins, can function as an "internal blood infusion" for the remainder of the circulation.

The organs of the splanchnic circulation contribute 25% to total body O_2 consumption and receive 30% of the cardiac output at rest, 7% by hepatic arterial flow and 23% by the arteries draining into the portal vein. This latter fraction is substantially increased during digestion and can be drastically reduced by neurogenic, α-adrenergic vasoconstriction down to one quarter of resting flow during hemorrhage. Splanchnic vasoconstriction during hemorrhage is an important compensatory mechanism in this situation: roughly half of the blood loss is replaced by additional mobilization of splanchnic blood by reduction of splanchnic perfusion with subsequent passive venous recoil and by additional active portal venoconstriction. Movement of this additional blood volume into the central circulation helps to better maintain the following within tolerable limits:

- Central blood volume
- Central venous pressure
- Cardiac output

However, this severe hypoperfusion of the splanchnic bed is not tolerated for longer periods. In the gastrointestinal tract, the mucosa becomes severely ischemic with necroses of the epithelial cells at the tips of the mucosal villi, and toxic products from the dying mucosa cells are assumed to contribute to the irreversibility of hemorrhagic shock by damaging the heart and the peripheral arterioles (see Sect. 94.5.5). Furthermore, ischemic damage of the mucosa disrupts its barrier function, which normally separates gut bacteria and their toxins from the blood stream. Disruption of this barrier, resulting in endotoxemia, contributes to the irreversibility of hemorrhagic shock and other shock forms (see Sect. 94.5.5).

Extrahepatic splanchnic flow, specifically the flow to the gastrointestinal tract, increases up to sixfold during digestion and absorption, from some 30 ml/min per 100 g to 200 ml/min per 100 g, and this increase is almost exclusively confined to the mucosa, parallel to an enhanced metabolic rate associated with active Na^+-coupled transport or with proton secretion. Intestinal motility only marginally elevates flow to the muscular layers of the tract. Postprandial splanchnic hyperemia is assumed to result from metabolic vasodilation in the mucosa, but the mediators of this functional hyperemia have, as in other organs, not been identified. Splanchnic blood flow increases during anticipation of a meal, and active neurogenic vasodilation by dilator innervation and the dilator action of gastrointestinal hormones may be involved in this reaction. The arterial concentrations of most circulating gastrointestinal hormones do not reach a level at which they exert dilator effects, but this does not exclude a local paracrine dilator role of these hormones at the sites of their release. During postprandial splanchnic hyperemia, blood flow to heart, kidneys, and brain does not decrease, disproving the idea that drowsiness after a meal could result from blood flow being deferred from the brain towards the gut. During postprandial splanchnic hyperemia, blood flow in the spleen and hepatic arterial flow are not increased, while portal perfusion of the liver carries the excess splanchnic flow.

The capillaries in the splanchnic bed are functionally and structurally adjusted to their specific junctions. The most elaborated lymphatic system of the body is in the organs of the splanchnic circulation, in which endothelial permeability is fairly high. A fenestrated endothelium with intracellular pores with diameters up to 50 nm, which is very permeable to water and small solutes, is found in the intestinal mucosa, while discontinuous or sinusoid capillaries with intercellular openings up to a diameter of 1 µm exist in liver and spleen, in which not only the passage of macromolecules occurs, but also the exchange of cells. In the intestinal mucosa, the capillaries are particularly adapted for absorption of water and solutes from intestinal fluid:

- The luminal pressure in the capillaries is low (typically around 14 mmHg or 1.9 kPa), favoring absorption.
- The capillaries are intensively anastomosed with a large surface area.
- The small endothelial pores permit rapid water and solute movements.
- The vascular arrangement in the mucosal villi allows countercurrent exchange with the creation of an hyperosmotic gradient at the villus tip by countercurrent multiplication.

While blood pressure in the portal vein normally is around 10 mmHg (1.3 kPa), several diseases in the liver can result in life-threatening portal hypertension with a threefold elevation of portal venous pressure, with remarkable degrees of edema and ascites, and with breakdown of the mucosal barrier and subsequent endotoxemia. Distended collateral circulations, prone to rupture, can develop. (Details of the physiology and the pathophysiology of the splanchnic circulation are presented in Chaps. 67–69.)

95.7 Circulation to the Skin and Adipose Tissue

The major role of blood flow to the skin and to the adjacent adipose tissue is concerned with temperature regulation, i.e., convective transport of heat to the surface of the body for exchange with the environment under normal conditions or improvement of body insulation by reductions in flow to skin and adipose tissue during exposure to cold. Only in brown adipose tissue, which is important for nonshivering thermogenesis, is there an increase in flow during exposure to cold, but this type of adipose tissue is negligible in adult humans.

The skin consists of two main layers: the *epidermis*, with a layer of keratinizing, squamous epithelium, and the *dermis* or *corium*, an inner layer of connective tissue with the blood vessels, the sensory nerve endings and mechanoreceptors, the lymphatics, the hair follicles with the erector pili muscles, and the cutaneous glands (sweat glands and sebaceous glands). Within the dermis, branches from the cutaneous arteries form an arterial plexus parallel to the skin surface. From this deep dermal plexus, arterioles penetrate to the subepidermal (or subpapillary) region, where a subpapillary plexus is formed. Capillaries connect the arteriolar subpapillary plexus with the subpapillary venous plexus, and single capillary loops ascend to each papilla. The subpapillary venous plexus drains into the deeper dermal venous plexus. In apical regions of the skin (such as in hands, feet, nose, lips, and ears), arteriovenous anastomoses between arterioles and veins of the deep epidermal plexus exist as short muscular connections or as tortuous glomerulus-like structures. All these vessels in the cutaneous plexus (arterioles, anastomoses, veins) have a dense adrenergic innervation. Although the epidermis itself if not vascularized, the color of the skin depends on the blood content and on the oxygenation of the blood in the plexus below the translucent epidermis. A major portion of the cutaneous blood volume is in the subpapillary venous plexus, and this cutaneous venous blood volume is estimated to be an important fraction of total venous volume [74]. The arrangement of arterial and venous plexus allows only minimal countercurrent heat exchange between arterial and venous blood. This arrangement helps to maintain a high temperature gradient between capillary blood and the surrounding skin tissue. Capillary permeability in the skin is high, and bridged fenestrations are relatively common in the capillary endothelium of the skin. Lymphatics in the skin are located around the arteries and arterioles.

The cutaneous tissue amounts to 5% of total body mass and extends over $1.8\,m^2$, with an average thickness of 1–4 mm. Mean perfusion rate of the skin in the lower range of thermoneutral conditions is estimated at 3–6 ml/min per 100 g, which is about 0.3 l/min. However total skin flow can vary from almost zero to 8–10 l/min under extreme thermal stress, i.e., more than 50% of total cardiac output under these conditions.

In adipose tissue, each adipocyte is surrounded by an extensive capillary network. The capillary density of the tissue depends on the size of the adipocytes and must be lower with enlarged fat cells. Adipose tissue has a relatively high capillary permeability which is augmented by sympathetic activation, probably in parallel to the sympathetically induced lypoplysis. Basal flow to adipose tissue under thermoneutral conditions is 3 ml/min per 100 g and is lower in tissue with enlarged adipocytes [51,72]. With exposure to cold flow to adipose tissue is reduced parallel to the reduction in skin flow, and this reduction helps to improve body insulation [32]. In response to lipolytic stimuli, such as catecholamines via the recently discovered β_3-receptors or glucagon, somatotropin, and adrenocorticotropin, there is a two- to three fold increase in adipose tissue flow, which is assumed to be induced by metabolic products of the enhanced lipolysis [31,63].

During reflectory increases of peripheral vascular resistance, blood flow to the skin and to the adipose tissue is reduced substantially, and both organs are important targets for the reflex control of vascular resistance and capacity. The vasculature in both organs demonstrates a strong myogenic autoregulation as well as flow-dependent dilatation, but it is not clear whether or not metabolic vasodilation is of any functional relevance in the skin, since the organ is normally overperfused relative to its nutritional requirements.

Temperature Regulation and Perfusion of Skin and Adipose Tissue. Blood flow in both organs is reduced in cold exposure and enhanced in hot environment. The degree of changes in flow, however, varies in different regions of the skin: in acral regions, flow can vary on account of temperature regulation between 0.2 to above 50 ml/min per 100 g, while in skin of the body trunk this variation is only between 1.0 and 8 ml/min per 100 g and somewhat larger in forearm skin. In adipose tissue, this variation is even lower than in the body trunk. In the apical skin regions, the very high flow rates under the influence of thermoregulatory reflexes and of direct local warming partially pass through arteriovenous anastomoses, which include no exchange vessels.

During exposure to hot environment, the reflex adjustments aimed at preventing body hyperthermia involve thermosensitive elements in the hypothalamus and the skin (see Chaps. 110,111). An increase in body core (and hypothalamic) temperature of 1°C induces an increase in cutaneous flow of the forearm by more than 10 ml/min per 100 g (Fig. 95.9) together with increased sweating, promoting evaporative heat losses. This reflectory increase of cutaneous flow in response to internal heating results partly from a reduction in sympathetic vasoconstriction and partly from active neurogenic vasodilation (Fig. 95.10) The relative role of these two vasodilatory mechanisms differs in various cutaneous regions (Fig. 95.10).

The active neurogenic, hyperthermia-induced vasodilation of nonapical skin, which is devoid of arteriovenous anastomoses, is only incompletely understood and may include several mechanisms cooperating in a redundant manner:

- Atropine only marginally retards the onset of heat-induced active vasodilation and cannot lower an existing thermal vasodilation, arguing against a major role of direct cholinergic innervation.
- Another hypothesis links cutaneous vasodilation with

1931

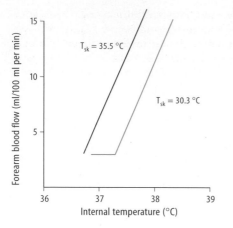

Fig. 95.9. Interaction of brain temperature (internal temperature) and skin temperature (T_{sk}) in the regulation of forearm blood flow. The internal temperature was elevated by exercise. Elevation of skin temperature causes a lowering of the threshold of internal temperature, at which cutaneous vasodilation is elicited. (From [84] with permission)

cholinergic activation of the sweat glands, which activates formation of bradykinin acting as endothelium-dependent cutaneous vasodilator (Fig. 95.11). However, atropine is more effective in blocking sweating than in blocking thermal cutaneous vasodilation.

- Nonadrenergic, noncholinergic (NANC) innervation can be demonstrated in sweat glands and around cutaneous arterioles, and this innervation is now being identified as nitroxidergic, providing an attractive mechanism for dilation by neurogenic nitric oxide.

Local heating of skin areas causes cutaneous vasodilation, which may result from the direct effect of temperature on vascular smooth muscle. With local temperatures above 42°C, this reaction can result in maximal flow rates in the hand of more than 50 ml/min per 100 g skin and is assumed not to result from reflexes. Cutaneous vasodilations, induced either by thermoregulatory reflexes or by direct local heating, always involve large declines in venoconstriction in the skin. Thus large, slowly flowing volumes of blood accumulate in the cutaneous venous plexus, facilitating heat exchange with the skin surface, which is cooled by the evaporation of sweat.

During exposure to cold, blood flow to the skin and to subcutaneous adipose tissue is reduced to almost zero, which improves the insulation of the heat-producing core of the body. During prolonged exposure to cold, this insulating vasoconstriction by sympathetic activation also reduces blood flow to skeletal muscle in the limbs, while shivering due to uncoordinated skeletal muscle activation contributes to enhanced heat formation. In the skin, however, a paradox vasodilation is elicited if local skin temperature approaches the level at which there is a danger of freezing. This *cold vasodilation* is most prominent in apical skin areas, which are repeatedly exposed to cold, and may represent an acquired reaction. It starts with declines of local temperature below 10°C and is assumed to result from an inhibitory effect of cold on vascular tone of the arterioles controlling flow through the arteriovenous anastomoses. This assumption, however, has not been proven, and the mechanism of cold vasodilation awaits clarification. The high flow through the arteriovenous anastomoses results in heat losses to the environment and therefore cannot contribute to thermoregulation during cold exposure. However, it might protect the vasodilated skin areas from frostbite [38].

95.8 Circulations in the Lung

The two circulations (see Chap. 104) to the lung differ substantially in size, origin, and function. The large-

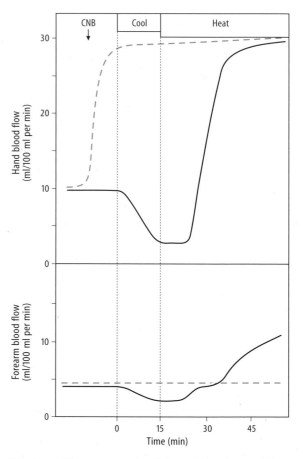

Fig. 95.10. Temperature-induced changes in cutaneous flow occur by different mechanisms in different skin regions. *C.N.B.*, cutaneous nerve blockade by local anesthetics. In the hand (*top*), modulation of a high vasoconstrictor tone mediates the flow changes; in the forearm (*bottom*) a constrictor tone does not exist under comfortable temperature conditions, and active neurogenic vasodilation occurs during heating of the body, *Solid black line*, normal side; *dashed blue line*, cutaneous nerve-blocked side. (From [71] with permission)

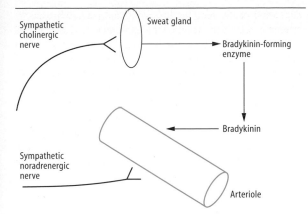

Fig. 95.11. Proposed link of sweat gland activation and cutaneous vasodilation by bradykinin formation. Bradykinin from the sweat glands can reach the microvascular of the surrounding skin tissue and induce and endothelium-mediated vasodilation. (From [38] with permission)

volume, low-perfusion pressure flow through the pulmonary bed (*pulmonary circulation*) contrasts with a small-volume, high-perfusion pressure flow from systemic arteries (*bronchial circulation*). This circulation includes all arterial blood flow to the lung and the intrathoracic airways, whether or not it goes to the bronchi. It supplies not only the bronchi, but all lung tissues except for the alveolar walls; it supplies the visceral pleura and is the nutritive circulation for the large vessels of the pulmonary circulation. The two circulations have different and complementary functions:

- The pulmonary circulation exchanges gas between blood and alveolar air, filters particles out of the blood, degrades or forms circulatory signals, modifies the functional state of white blood cells, and is a compliant volume buffer between the two series-connected hearts. It serves as a volume reservoir for the filling of the left heart during orthostasis-induced, transient shortages of right heart cardiac output relative to the left heart output.
- The bronchial circulation contributes to the regulation of temperature and humidity of the air in the airways, determines fluid secretion and absorption and the defense reactions of the airway mucosa, nourishes the airways, the hilar tissues, and the large pulmonary vessels, and can enlarge, form new vessels, and partially take over the task of pulmonary circulation in perfusing and nourishing the alveolar tissue if a pulmonary artery branch is occluded [10].

The pulmonary arteries, branching with the bronchi as they run to the lobuli of the lung, are extemely compliant and have a great deal of elastic tissue in their walls. They ramify into a network of pulmonary capilaries in the alveolar walls which is so extensively anastomosed that the

blood from the pulmonary arteries flows as an almost unbroken sheet between the alveolar walls. Arterioles between pulmonary arteries and capillaries are thin walled, have only very few smooth muscle fibers, and participate in

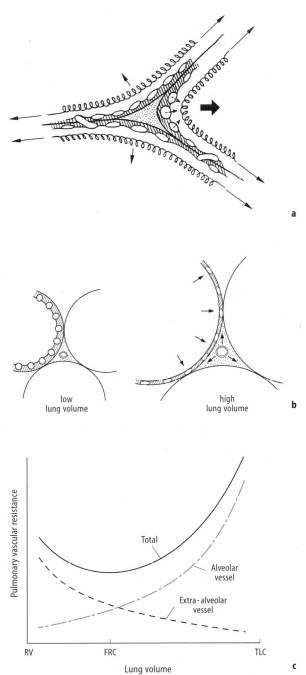

Fig. 95.12a–c. Mechanical effects of alveolar architecture on pulmonary microvasculature. **a** Alveolar surface tension, indicated by *Springs*, holds alveolar capillary vessels open. **b** Alveolar microvessels are elongated and compressed at high alveolar volumes (*right*), while extra-alveolar vessels are dilated (*left*). **c** Dependency of pulmonary vascular resistances for a given perfusion pressure on lung volume. *RV*, resting lung volume; *FRC*, functional residual capacity of the lung; *TLC*, total lung capacity. (From [11] with permission)

gas exchange. Pulmonary capillaries unite to form larger venules, which also participate in gas exchange and draining via four pulmonary veins into the left atrium. The driving pressure through the pulmonary circulation from mean pulmonary artery pressure (15 mmHg, 2 kPa) to mean left atrial pressure (7–12 mmHg, 0.9–1.6 kPa) is only 5 mmHg (0.7 kPa), and the average transit time through the pulmonary bed is 0.75 s [73]. Several mechanisms ensure an adequate flow through all parts of this large organ in spite of this low driving pressure:

- Substantial reduction of vascular resistance by sheet flow of blood through the microvasculature of the bed (see above).
- Passive distension by perfusion pressure of the pulmonary arterial and arteriolar vasculature, which is devoid of myogenic autoregulation (see Sect. 94.4.3).
- Expansion-induced opening of pulmonary capillaries by forces resulting from alveolar surface tension and from respiratory volume changes of the lung (Fig. 95.12).
- Hypoxic pulmonary *vasoconstriction* [54], which shunts blood to well-ventilated parts of the lung and is based on an unidentified, endothelium-dependent mechanism. This is in contrast to the hypoxic *vasodilation* of the bronchial circulation, which ensures nutritive perfusion of nonventilated alveoli via the anastomoses between bronchial and pulmonary capillaries (see below).

Multiple small arteries arise from the intercostal arteries to form the network of the bronchial arteries, supplying capillaries in airway mucosa, lymph nodes, pleura, and hilar interstitum. Furthermore, multiple small-vessel anastomoses between the bronchial and pulmonary circulations occur at the level of the alveoli (Fig. 95.13). Bronchial veins from the larger airways and the hilar region drain via the azygos vein into the right atrium. Bronchial veins from the nonalveolar intrapulmonary structures and from any diseased, nonventilated regions of the lung drain nonsaturated venous blood into the pulmonary veins and the left atrium. While arterial inflow into the bronchial circulation is less than 1% of cardiac output, venous bronchial flow of nonsaturated blood into the left heart inflow amounts to 2%–5% of cardiac output and measurably lowers O_2 saturation of arterial blood [10]. (For role of the pulmonary circulation within the cardiovascular low-pressure system, see Sects. 94.4.3, and 94.5.2; a detailed survey on the physiology and pathophysiology of the two circulations to the lung is given in Chap. 104.)

95.9 Circulations in Reproductive Organs

While the blood flow to the reproductive organs is quantitatively negligible during inactivity of these organs, their circulations have important roles in reproduction and can substantially affect circulatory control of the entire organism during pregnancy.

Circulation to Nonpregnant Uterus and Ovaries. Flows to the ovaries and the nonpregnant uterus are closely related to the development, function, and regression of the corpus luteum and the uterine mucosa in the menstrual cycle. The

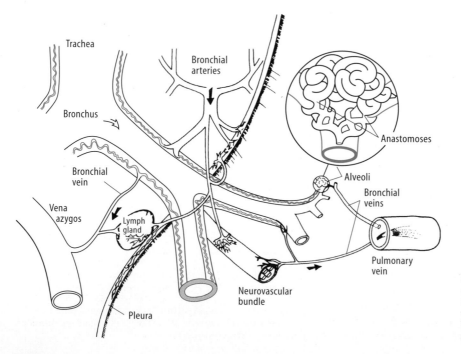

Fig. 95.13. Functional anatomy of the bronchial circulation. The *magnified insert* shows the convoluted anastomoses between bronchial and pulmoanry vessels. (From [10] with permission)

uterine and vaginal arteries originate from the internal iliac artery, anastomosing with branches from the vaginal and ovarian arteries and forming a common network for supplying the internal genital organs. Arcuate arteries encircle the uterus within the myometrium, releasing radial arteries directed towards the uterine lumen, which become spiral arteries as they pass into the uterine mucosa. Basal arteries ramify from these radial arteries along the endometrial–myometrial border. The spiral arteries grow with the regenerating endometrium and continue to grow in the luteal phase, during which the endometrium does not expand further. Constriction and luminal obliteration of the spiral arteries at the end of the luteal phase causes tissue necrosis and shedding of endometrial tissue during spasmodic uterine constrictions, followed by bleeding and sloughing of the damaged tissue. Capillaries sprout from the stumps of the obliterated spiral arteries and are later converted to new spiral arteries as the endometrium regrows and a new cycle begins.

Around the time of ovulation, when circulating estrogen levels are maximal, blood flow to the human myometrium and endometrium reaches a peak value [67], roughly sevenfold the uterine flow in the follicular phase. Flow rates to the ovaries, however, are closely related to the function of the corpus luteum and reach peak values late in the luteal phase [8]. This excess ovary flow goes specifically to the corpus luteum, which at this phase of the cycle and in terms of tissue weight appears to be one of the most highly perfused tissues in the organism (greater than 1500 ml/min per 100 g) [1]. Ovaries without corpora lutea maintain a fairly steady and low flow rate throughout the menstrual cycle. The mechanisms of these uterine and ovarial flow changes are not completely understood. Estrogens and luteinizing hormone appear to be involved indirectly, probably by modulating the endothelial responses to stimuli causing release of nitric oxide (e.g.,by induction of endothelial nitric oxide synthase) [22] and by interfering with the vessel responses to vascular growth factors [20,46]. (For details on the physiology of the uteroplacental circulation during pregnancy, see Chap. 117 and Sect. 94.5.7.)

Lactating Mammary Gland Blood Flow. In the mammary gland, which is not fully developed until the end of the first pregnancy, the arterial blood supply is distributed in a tight capillary meshwork surrounding the pouches or alveoli of the gland (Fig. 95.14). Periductal capillaries are connected in series to this perialveolar capillary network. Maintenance of vascularization of the breast and of mammary blood flow depends on the neuroendocrine regulation of galactopoesis. During lactation, peak flow rates are assumed to amount to 60 ml/min per 100 g breast tissue and are closely related to the rate of milk production [55]. Only at the end of pregnancy does mammary blood flow increase severalfold without concomitant milk flow; this increase preceding delivery is triggered by hormonal changes from the fetoplacental unit [9]. Before the first pregnancy, the breast consists mainly of adipose tissue

with a blood flow around 3 ml/min per 100 g, and this value is reached again with regression of the mammary gland in the breast after lactation.

Penile and Testicular Circulation. The circulations within the male reproductive organs have several unique characteristics with specific roles in the function of these organs; a blood – testes barrier and a pampiniform plexus for heat exchange in the testes maintain optimal conditions for spermatogenesis, and a closed, high-pressure circulatory system in the corpus cavernosum penis mediates penile erection.

The penile vasculature in man reaches the corpus carvernosum as the deep artery of the penis and the dorsal artery of the penis, while the corpus spongiosus urethrae is supplied by the artery of the urethral bulb. The large cavernous spaces are lined with a complete endothelium, are surrounded by trabeculae with smooth muscles, collagen, elastic fibers and fibroblasts, and are limited by a dense, thick, and relatively rigid tunica albuginea. The penile arteries enter the cavernous spaces via small arteries and arterioles within budlike structures protruding into the spaces. The vessels in these buds are surrounded by layers of trabecular smooth muscle cells. The venous outlet of the cavernous spaces are located next to the arterial inflow adjacent to the crus penis. During the nonerectile stage, the sympathetic constrictor tone keeps arterial inflow through the penile vasculature low by constriction of vascular muscles and the trabecular musculature within the corpus cavernous. Blood is shunted away from the cavernous space via anastomoses towards the venous outflow. Only limited, sluggish blood flow through the cavernous spaces occurs, presumably to prevent clotting and to nourish the tissue surrounding the carvernous spaces. The pressure in the

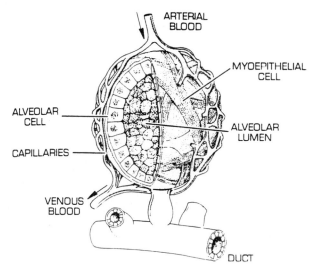

Fig. 95.14. Microcirculation in the mammary gland. (From [66] with permission)

1935

cavernous spaces at this stage is estimated to be 15–25 mmHg (2.0–3.3 kPa).

Gential reflexes initiate the erection by causing an inhibition of the sympathetic constrictor tone to the penile circulation and the trabecular muscles by parasympathetic, cholinergic fibers and by inducing direct vasodilation or relaxation of the trabecular muscles, respectively, by nitroxidergic nerves. The neuripeptide substance P, an endothelium-dependent vasodilator and constrictor of smooth muscles, is found in large granular vesicles in the human corpus cavernosum and is assumed to be involved in erection [6]. The reflectory vasodilation results in an increase in penile flow from 2 ml/min per 100 g to more than 30 ml/min per 100 g [62,78], and trabular relaxation allows engorgement of the corpus cavernosum, causing a mild state of erection with an increase in the pressure of the cavernous spaces to 80–110 mmHg (10.6–14.6 kPa) and with palpable arterial pulsations. Contraction of the ischiocavernous muscle, which originates on the internal and caudal surface of the ischium and inserts into the dense fibrous tissue of the crus penis, compresses the arterial and venous vessels of the penis against the ischium. This vascular occlusion isolates the carvernous spaces, preventing the reflux of blood into the systemic circulation during full erection. Further contractions of the ischiocavernous muscle elevate cavernous pressure within the

rigid tunica albuginea far above systemic pressure, which is also elevated during coitus. Measurements of coital cavernous pressure in animals demonstrated several thousand mmHg [4,5]. Immediately after ejaculation, the ischiocavernous muscle relaxes, allowing the venous outlets to open; blood can leave the cavernous spaces, pressure declines rapidly to systemic pressure levels, and vasoconstrictor activity again shuts down penile flow to the level of detumescence (penile relaxation).

The testicular circulation receives its major blood supply from the internal spermatic artery, which originates from the aorta near the origin of the renal arteries. After passage through the inguinal canal, the spermatic artery is convoluted and tightly surrounded by the pampiniform plexus and receives some anastomoses from the artery of the deferent duct. Centripetal arteries from the internal spermatic artery penetrate the testis towards the mediastinum of the gland, where centrifugal arteries curve back towards the testicular surface, branching into arterioles between seminiferous tubules, into intertubular capillaries adjacent to the Leydig cells and into peritubular capillaries around the seminiferous tubules. Both types of capillaries have a continuous endothelium with tight junctions. The Sertoli cells in the seminiferous tubules are attached to one another by junctional complexes, forming a physiologic barrier limiting the transport of large molecules from the

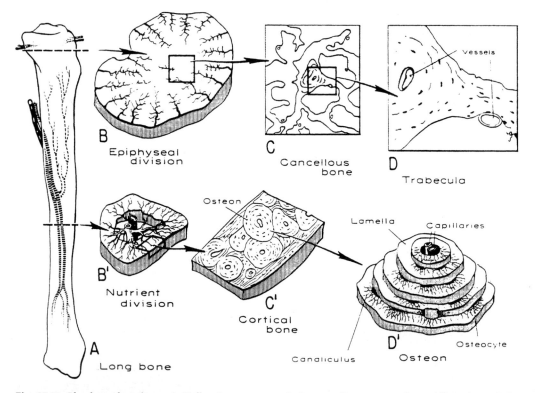

Fig. 95.15. Blood supply to bone. *A*, Hollow bones are supplied from arteries entering the bone canal via the diaphysis, while cancellous (spongy) bone is supplied from periostal arteries; *B–D*, vessel arragnement in cancelleous bone; *B–D*, arrangement in cortical bone with the typical osteon structure. (From [41] with permssion

peritubular interstitium to the tubular lumen. In addition to the capillary endothelium, these junctional complexes from a substantial part of the blood–testis barrier, which is assumed to protect the potentially antigenic molecules associated with spermatogenesis from contact with and recognition by the body's own immune system [80]. The veins arise deep in the testicular tissue, running directly to a central vein in the mediastinum, which proceeds in a tortuous course towards the cranial pole of the testis, branching into many small veins of the pampiniform plexus around the coiled internal spermatic artery.

The pampiniform venous plexus and the coiled internal spermatic artery function as effective heat exchangers helping to keep the testicular temperature several degrees below the body temperature. While moderate experimental elevations of testicular temperature in animals slightly lowers testicular blood flow, it doubles flow through scrotal skin, demonstrating that the scrotum serves as a heat radiator, with a temperature-dependent flow regulation independent of the testis itself [77,83]. Sympathetic adrenergic nerves keep testicular flow low, mainly by constricting the long, convoluted internal spermal artery. The dilator reserve for increments in testicular flow is fairly low, allowing flow variations only by a factor of two, probably by modulating sympathetic constrictor tone [3].

95.10 Skeletal Circulation

Circulation to the skeleton, which accounts for at least 14% of body weight, is in fact a circulation to two rather different organs:

- Marrow, a hematopoietic and lymphoid tissue, mixed with adipose tissue, which is considered to be a component of the immune system (see Chaps. 83–87)
- Cancellous and cortical bone, which serves as a structural framework, which can grow, remodel in response to altered loads, and repair and which is intimately involved in acute and chronic regulation of Ca^{2+} and phosphate homeostasis (see Chap. 80).

Both organs have separately regulated arterioles and capillary beds. Flow to the marrow is increased during systemic inflammatory reactions; flow to the bone is determined by its remodeling activity, i.e., resorption of bone and formation of new bone, which occurs largely in the trabecular or spongy bone. Consequently, the flow to cancellous bone of 30 ml/min per 100 g is ten times higher than to cortical bone (less than 3 ml/min per 100 g). Flow to both types of bone is doubled during growth, after fractures, and in thyreotoxicosis. The mechanisms regulating flow to the skeleton are assumed to be associated with the metabolic rate of the cells in the bone, which constitute roughly 5% of bone mass, and in the marrow. The activity of both cell groups is under the control of interleukins [34].

The circulation to the marrow orginates from conduit arteries entering the diaphysis of hollow bones. Cortical bone

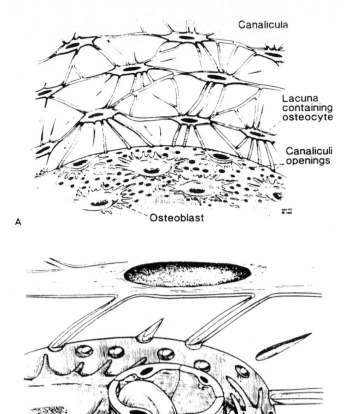

Fig. 95.16A,B. Canalicular system for interstitial fluid movements in cortical bone. A Osteocytes within the lacune of the osteons (see Fig. 95.15) form a network of cellular processes within the canaliculi; osteoblasts are in the central canal of the osteon. B From the capillaries within the central canal, filtered fluid passes through the canaliculas clefts beside the osteocyte processes. Varying mechanical loads on the bone are assumed to be the driving force for interstitial fluid movements in the system; periostal initial lymphatics take up interstitial bone fluids. (From [41] with permission)

is supplied from nutrient artery branches from the artery in the medullary canal via the endosteal surface, and cancellous bone is supplied from the periostal arteries; flow to both bone types drains to the periostal venous network (Fig. 95.15), which also contains initial lymphatics.

Capillaries in the hematopoietic marrow have an endothelium of the discontinuous type; in the fat marrow it is of the closed, continuous type, similar to the capillaries in cancellous and cortical bone. Interstitial fluid filtered from the capillaries inside the osteon canal (Haversian canal) distributes through the lacunar–canalicular system of the cortical bone, reaching the osteoblasts and the cortical asteocytes (Fig. 95.16). Interstitial fluid flow in the cancellous bone and through the canalicular system of cortical bone is important for bone remodeling and for Ca^{2+} homeostasis (see Chap. 80).

Total flow to the skeleton is estimated to amount to 8% of cardiac output at rest. Changes in blood flow during bone remodeling are very small and do not significantly alter this figure. However, local disturbances in blood flow during bone repair do affect the repair processes.

References

1. Abdul-Karim RW, Bruce N (1973) Blood flow to the ovary and corpus luteum at different stages of gestation in the rabbit. Fertil Steril 24:44–47
2. Baumgart D, Heusch G (1995) Neuronal control of coronary blood flow. Basic Res Cardiol 90:142–159
3. Beckett SD (1983) Circulation to male reproductive organs. In: Shepherd JT, Abboud FM (eds) The cardiovascular system, vol III. Peripheral circulation and organ blood flow. American Physiology Society, Bethesda, pp 271–283 (Handbook of physiology, sect 2)
4. Beckett SD, Walker DF, Hudson RS, Reynolds TM, Purohit RC (1975) Corpus spongiosum penis pressure and penile muscle activity in the stallion during coitus. Am J Vet Res 36:431–433
5. Beckett SD, Walker DF, Hudson RS, Reynolds TM, Vachon RI (1974) Corpus cavernosum penis pressure and penile muscle activity in the bull during coitus. Am J Vet Res 35:761–764
6. Benson GS, McConnell J, Lipshultz LI, Corriere JN Jr, Wood J (1980) Neuromorphology and neuropharmacology of the human penis. J Clin Invest 65:506–513
7. Berkenboom GM, Abramowicz M, Vandermoten P, Degre SG (1986) Role of alpha-adrenergic coronary tone in exercise-induced angina pectors. Am J Cardiol 57:195–198
8. Brown BW, Emery MJ, Mattner PE (1980) Ovarian arterial blood velocity measured with Doppler ultrasonic transducer in conscious ewes. J Reprod Fertil 58:295–300
9. Burd LI, Bisgard GE, Rawlings CA, Rankin JHG (1976) Mammary blood flow and endocrine changes during parturition in the ewe. Endocrinology 98:748–754
10. Butler J (1991) The bronchial circulation. News Physiol Sci 6:21–25
11. Butler J (1989) The circulation of the lung. In: Patton HD, Fuchs AF Hille B, Scher AM, Steiner R (eds) Textbook of physiology. Saunders, Philadelphia, pp 961–971
12. Daugherty RM Jr, Scott JB, Haddy FJ (1967) Effects of generalized hypoxemia and hypercapnia on forelimb vascular resistance. Am J Physiol 213:1111–1114
13. Davson H (1967) Physiology of the cerebrospinal fluid. Churchill, London
14. Detry JMR, Brengelman GL, Rowell LB, Wyss C (1972) Skin and muscle components of forearm blood flow in directly heated resting man. J Appl Physiol 32:506–511
15. Devous MD Jr, Stokely EM, Chehabi HH, Bonte FJ (1986) Normal distribution of regional cerebral blood flow: measurement by dynamic single-photon emission tomography. J Cereb Blood Flow Metabol 6:95–104
16. Edvinsson L, Owman C, Siesjo B (1976) Physiological role of cerebrovascular sympathetic nerves in the autoregulation of cerebral blood flow. Brain Res 117:518–523
17. Feigl EO (1983) Coronary physiology. Physiol Rev 63:1–205
18. Fenn WO (1923) A quantitative comparison between the energy liberated and the work performed by the isolated sartorius muscle of the frog. J Physiol 58:175–203
19. Fink WJ, Costill DL, Van Handel PJ (1980) Led muscle metabolism during exercise in heat and cold. Eur J Appl Physiol 34:183–190
20. Folkman J, Shing Y (1992) Angiogenesis. J Biol Chem 267:10931–10934
21. Gewirtz H (1991) The coronary circulation: limitations of current concepts of metabolic control. News Physiol Sci 6:265–268
22. Goetz RM, Morano I, Calovini T Studer R, Holtz J (1994) Increased expression of endothelial constitutive nitric oxide synthase in rat aorta during pregnancy. Biophys Biochem Res Commun 205:905–910
23. Gorman MW, Sparks HV Jr (1991) The unanswered question. News Physiol Sci 6:191–193
24. Gould KL, Lipscomb K, Hamilton GW (1974) The hemodynamic action of coronary stenoses. Am J Cardiol 33:87–94
25. Greenfield ADM, Patterson GC (1954) Reactions of the blood vessels of the human forearm to increases in transmural pressure. J Physiol (Lond) 125:508–524
26. Heistad DD, Abboud FM, Eckstein JW (1968) Vasoconstrictor response to simulated diving in man. J Appl Physiol 25:542–549
27. Heistad DD, Marcus ML (1979) Effect of sympathetic stimulation on permeability of the blood-brain barrier to albumin during acute hypertension in cats. Circ Res 45:331–338
28. Heusch G (1990) Alpha-adrenergic mechanisms in myocardial ischemia. Circulation 81:1–13
29. Heusch G, Schipke JD (1993) Regional blood flow and contractile function: are they matched in normal, ischemic and reperfused myocardium? Basic Res Cardiol 88 [Suppl 2]:103–119
30. Heusch G (1994) Ischemia-selectivity: a new concept of cardioprotection by calcium antagouists. Basic Res Cardiol 89:2–5
31. Himms-Hagen J (1970) Adrenergic receptors for emtabolic responses in adipose tissue. Fed Proc 29:1388–1401
32. Hjemdahl P, Sollevi A (1978) Vascular and metabolic responses to adrenergic stimulation in isolated subcutaneous adipose tissue at normal and reduced temperature. J Physiol (Lond) 281:325–338
33. Holtz J, Giesler M, Bassenge E (1983) Two dilatory mechanisms of anti-anginal drugs on epicardial coronary arteries in vivo: indirect, flow-dependent, endothelium-mediated dilation and direct smooth muscle relaxation. Z Kardiol 72:98–106
34. Howard GA (1989) Calcium metabolism and physiology of bone. In: Patton HD, Fuchs AF, Hille B, Scher AM, Steiner R (eds) Texbook of physiology. Saunders, Philadelphia, pp 1461–1479
35. Ingvar DH (1976) Functional landscapes of the dominant hemisphere. Brain Res 107:181–197
36. Intaglietta M (1981) Vasomotor activity, time-dependent fluid exchange and tissue pressure. Microvasc Res 21:153–164
37. Johnson JM (1989) Circulation to skeletal muscle. In: Patton HD, Fuchs AF, Hille B, Scher AM, Steiner R (eds) Textbook of physiology. Saunders, Philadelphia, pp 887–897
38. Johnson JM (1989) Circulation to the skin. In: Patton HD, Fuchs AF, Hille B, Scher AM, Steiner R (eds) Textbook of physiology. Saunders, Philadelphia, pp 898–910
39. Johnson JM, Rowell LB, Niederberger M, Eisman MM (1974) Human splanchnic and forearm vasoconstrictor responses to reductions of right atrial and aortic pressures. Circ Res 34:515–524
40. Kanatsuka H, Lamping KG, Eastham CL, Dellsperger KC, Marcus ML (1989) Comparison of the effects of increased myocardial oxygen consumption and adenosine on the coronary microvascular resistance. Circ Res 65:1296–1305
41. Kelly PJ (1983) Pathways of transport in bone. In: Shepherd JT, Abboud FM (eds) The cardiovascular system, vol III. Peripheral circulation and organ blood flow. American Physiology Society, Bethesda, pp 371–396 (Handbook of physiology, sect 2)
42. Kennedy C, Des Rosiers MH, Jehle JW, Reivich M, Sharpe F, Sokoloff L (1975) Mapping of functional neural pathways by autradiographic survey of local metabolic rate with 14-C-deoxyglucose. Science 187:850–853
43. Kilbom A, Wennmalm A (1976) Endogenous prostaglandins as local regulators of blood flow in man; effect of indomethacin on reactive and functional hyperemia. J Physiol (Lond) 257:109–121

44. King RB, Bassingthwaighte JB (1989) Temporal fluctuations in regional myocardial blood flow. Pflugers Arch 413:336–342
45. Klabunde RE, Laughlin MH, Armstrong RB (1988) Systemic adenosine deaminase administration does not reduce active hyperemia in running rats. J Appl Physiol 64:108–114
46. Klagsbrun M, Dluz S (1993) Smooth muscle cell and endothelial cell growth factors. Trends Cardiovasc Med 3:213–217
47. Koch IG, Britton SI, Metting PJ (1990) Adenosine is not essential for exercise hyperemia in the limb in conscious dogs. J Physiol (Lond) 429:63–75
48. Kuo L, Davis MJ, Chilian WM (1988) Myogenic activity in isolated subepicardial and subendocardial coronary arterioles. Am J Physiol 255:H1558–H1562
49. Kuo L, Davis MJ, Chilian WM (1990) Endothelium-dependent, flow-induced dilation of isolated coronary arterioles. Am J Physiol 259:H1063–H1070
50. Kuschinsky W, Wahl M (1978) Local chemical and neurogenic regulation of cerebral vascular resistance. Physiol Rev 58:656–689
51. Larsen OA, Lassen NA, Quaade F (1966) Blood flow through human adipose tissue determined with radioactive xenon. Acta Physiol Scand 66:337–345
52. Lassen NA (1968) Brain extracellular pH: the main factor controlling cerebral blood flow. Scand J Clin Lab Invest 22:247–251
53. Lassen NA, Ingvar DH, Skinhoj E (1978) Brain function and blood flow. Sci Am 239:62–71
54. Liljiestrand G (1958) Chemical control of the distribution of pulmonary blood flow. Acta Physiol Scand 44:216–240
55. Linzell JL (1974) Mammary blood flow and methods of identifying and measuring precursors of milk. In: Larson BL, Smith VR (eds) Lactation, a comprehensive treatise, vol I, part 2. Academic, News York, pp 143–220
56. Marshall JM (1991) The venous vessels within skeletal muscle. News Physiol Sci 6:11–15
57. McArdle B (1951) Myopathy due to a defect in muscle glycogen breakdown. Clin Sci 10:13–35
58. Mellander S, Johansen B (1968) Control of resistance, exchange and capacitance functions in the peripheral circulation. Pharmacol Rev 20:117–196
59. Mohrman DE, Feigl EO (1978) Competition between sympathetic vasoconstriction and metabolic vasodilation in the canine coronary circulation. Circ Res 42:79–86
60. Moir TW, Eckstein RW, Driscol TE (1963) Thebesian drainage of the septal artery. Circ Res 42:79–86
61. Murray PA, Vatner SF (1979) Alpha-adrenoceptor attenuation of coronary vascular response to severe exercise in the conscious dog. Circ Res 45:654–660
62. Newman HF, Northrup J, Devlin J (1964) Mechanism of human penile erection. Invest Urol 1:350–353
63. Nielsen SL, Bitsch V, Larsen OA (1968) Blood flow through human adipose tissue during lipolysis. Scand J Clin Lab Invest 22:124–130
64. Nozawa T, Yasumura Y, Futaki S, Tanaka N, Suga H (1989) The linear relation between oxygen consumption and pressure-volume area can be reconciled with the Fenn effect in dog left ventricle. Circ Res 65:1380–1389
65. Page RB, Funsch DJ, Brennan RW, Hernandez MJ (1980) Choroid plexus blood flow in sheep. Brain Res 197:532–537
66. Patton S (1969) Milk. Sci Am 221:58–73
67. Prill HJ, Göxtz F (1961) Blood flow in the myometrium and endometrium of the uterus. Am J Obstet Gynecol 82:102–108
68. Raichle ME, Hatman BK, Eichling JO, Sharpe LG (1975) Central noradrenergic regulation of cerebral blood flow and vascular permeability. Proc Natl Acad Sci 72:3726–3730
69. Rapoport SI (1976) Blood-brain barrier in physiology and medicine. Raven, New York
70. Renkin EM (1967) Blood flow and transcapillary exchange in skeletal and cardiac muscle. In: Marchetti G, Taccardi B (eds) International symposium on the coronary circulation and energetics of the myocardium. Karger, Basel, pp 23–43
71. Roddie IC (1983) Circulation to skin and adipose tissue. In: Shepherd JT, Abboud FM (eds) The cardiovascular system, vol III. Peripheral circulation and organ blood flow. American Physiology Society, Bethesda, pp 285–317 (Handbook of physiology, sect 2)
72. Rosell S, Belfrage E (1979) Blood circulation in adipose tissue. Physiol Rev 59:1078–1104
73. Roughton FJW, Foster RE (1957) Relative importance of diffusion and chemical reaction rates in determining rate of exchange of gases in the human lung. J Appl Physiol 11:290–302
74. Rowell RB (1974) Human cardiovascular adjustments to exercise and thermal stress. Physiol Rev 54:75–159
75. Saltin B (1985) Hemodynamic adaptations to exercise. Am J Cardiol 55:42D–47D
76. Schaper W (1971) The collateral circulation of the heart. Elsevier, New York
77. Setchell BP (1978) The mammalian testis. Cornell University Press, New York
78. Shirai M, Ishii S, Mitsukawa S, Matsuda S, Nakamura M (1978) Hemodynamic mechanism of erection in the human penis. Arch Androl 1:345–349
79. Spaan JAE (1995) Mechanical determinants of myocardial perfusion. Basic Res Cardiol 90:89–102
80. Steiner RA, Cameron JL (1989) Endocrine control of reproduction. In: Patton HD, Fuchs AF, Hille B, Scher AM, Steiner R (eds) Textbook of physiology. Saunders, Philadelphia, pp 1289–1342
81. Stephenson RB (1989) The renal circulation. In: Patton HD, Fuchs AF, Hille B, Scher AM, Steiner R (eds) Textbook of physiology. Saunders, Philadelphia, pp 924–932
82. Strandgaard S, Olesen J, Skinhof E, Lassen NA (1973) Autoregulation of brain circulation in severe arterial hypertension. BMJ 1:507–510
83. Waites GMH, Setchell BP, Quinlan D (1973) Effect of local heating of the scrotum, testes, and epididymes of rats on cardiac output and regional blood flow. J Reprod Fertil 34:41–49
84. Wenger CB, Roberts MF, Stolwijk JAJ, Nadel ER (1975) Forearm blood flow during body temperature transients produced by leg exercise. J Appl Physiol 38:58–63
85. Winn HR, Rubio R, Berne RM (1981) The role of adenosine in the regulation of cerebral blood flow (editorial). J Cereb Blood Flow Metab 1:239–244
86. Winn HR, Dacey RG Jr, Mayberg MR (1989) Cerebral circulation. In: Patton HD, Fuchs AF, Hille B, Scher AM, Steiner, R (eds) Textbook of physiology. Saunders, Philadelphia, pp 952–960
87. Young EW, Sarks HV Jr (1980) Prostaglandins and exercise hyperemia of dog skeletal muscle. Am J Physiol 238:H191–H195
88. Zoller RD, Mark AL, Abboud FM, Schmid PG, Heistad DD (1972) The role of low pressure baroreceptors in reflex vasoconstrictor responses in man. J Clin Invest 51:2967–2972

96 Vascular Smooth Muscle

G. SIEGEL

Contents

96.1 Introduction

Smooth Muscle Tissue. The visceral mesenchyma (splanchnopleura) is the principal source for smooth muscle tissue. It also forms the layer-shaped muscle and connective tissue strata of the vessel walls. The vascular system of the human embryo develops from blood islets in the middle of the third week, shortly before somite formation, when the embryo can no longer meet its nutritional demand merely by diffusion. Cardiac muscle, similar to vascular musculature, develops from the mesenchymal mantle which surrounds the early-embryonic endothelial heart tube.

The Cardiovascular System. This system ensures homeostasis in the organism, i.e., it regulates the "milieu intérieur" [1] (see also Chap. 1). Similar to a communication network, it is a convective system that pervades all the cells in the body and connects them functionally. It provides them with the vital substances and prevents an enrichment of their metabolic end products; in addition, it transports the corpuscular elements of the blood, the respiratory gases, water and electrolytes, hormones, and heat.

With the help of the cardiovascular system, cellular and humoral defense systems are very rapidly available to each tissue. All of these factors can be considered as "effectors" of the vascular smooth muscle cells, which coregulate local vessel tone.

Circulatory Regulation. The events described and their feedback to the vascular smooth muscle cell form part of the circulatory regulation, the main purpose of which is to allow adaptation to the varying requirements of the tissues. The regional current is regulated mainly by changes in flow resistance, i.e., in the vessel sectional area. Therefore, the regulation of blood flow is achieved much more effectively by changes in vessel radius than in pressure. Although all vessels provide resistance to blood flow, it is the small arteries to the terminal arterioles that are designated the true "resistance arteries" controlling flow and pressure [3]. These vessels contribute more than three quarters of the total peripheral resistance and are a site of potential physiological regulation, thus fulfilling the two functional criteria for a resistance vessel. In the following description of the properties of vascular smooth muscle, particular attention shall be paid to these small vessels.

96.2 Structure of the Vascular Wall

Intima, Media, and Adventitia Layers. Artery walls consist of the three layers tunica intima, media, and externa (adventitia). The intima consists of a single layer of endothelial cells which is surrounded by fine collagenous fibers and a fenestrated elastic membrane (elastica interna). In large arteries, the media is formed by a compact layer of circumferentially arranged smooth muscle fibers with connective tissue fibers (elastic lamellae, collagen fibrils) and matrix (elastic muscular type). At increasing distances from the heart, the elastic fibers in the media decrease and the smooth muscle fibers increase (muscular type). In arterioles, terminal arterioles, and precapillary arterioles with an outer diameter of 15–100 µm, the media consists of a single layer of smooth muscle cells. Postcapillary venules possess no musculature, while collecting venules and small veins have a continuous smooth muscular layer. In veins, which have a substantially thinner wall than arteries of the same size, the arrangement of the three layers is less pronounced. The media is not only

R. Greger/U. Windhorst (Eds.)
Comprehensive Human Physiology, Vol. 2
© Springer-Verlag Berlin Heidelberg 1996

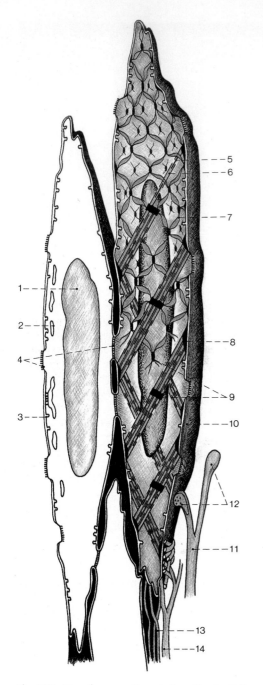

in the adventitia include fibroblasts, macrophages, nonmyelinated axons, and Schwann cells. The fibroblasts produce extracellular matrix and collagen fibers. Together with elastic fiber bundles, the latter form helices of a relatively high pitch within the venous wall.

96.2.1 Smooth Muscle Cells

Basement Membrane. The vascular smooth muscle cells are spindle-shaped or branched cells with a length of 30–60 μm and a relatively constant diameter of 4 μm [41] (Fig. 96.1). The ellipsoid and centrally sited cell nucleus comprises approximately 20%–30% of the cell volume. The cells are embedded in connective tissue, which they themselves synthesize as collagen, elastin, proteoglycans, and glycoproteins [2]. Under nonstationary conditions (e.g., extracellular pH changes), the connective tissue components contribute to a redistribution of the extracellular ion concentrations close to the cell membrane [75] and to the distribution of forces developed by the smooth muscle. The proteoglycans of the basement membrane are attached not only to the smooth muscle cell membrane, but also to the collagenous and elastic fibrils in the extracellular space [2]. The smooth muscle cells are enveloped by a self-produced basement membrane with a light, electron-translucent zone near the surface (lamina lucida) and a dark, electron-dense zone adjacent to the outside (lamina densa). In the lamina lucida, laminin is found, a noncollagenous glycoprotein. Further constituents of the basal lamina are type IV collagen, heparan sulfate proteoglycan, and entactin. The outer part of the basal lamina passes into an argentophilic fiber net that coats the cell like a stocking and transmits each change in the shape of the cell body to the collagenous (types I, III, V, VI) and elastic fibers of the surrounding connective tissue coherent with the fiber sheath.

Functional Syncytium. The mononuclear smooth muscle cells are mostly linked up to sheets or bundles of differing thickness. Within the bundles the cells are aligned in parallel like a school of fish. The myocytes can make contact with each other via cell appendices and regions of close apposition between the plasma membranes of adjacent cells (Fig. 96.1), which can be of the intermediate junction (attachment plaques, 20-nm gap) or gap junction type (nexus, 2- to 3-nm gap). Since the membrane resistance is very low in the area of gap junctions due to intercellular ion channels, the myocytes are electrically coupled to each other (see also Chap. 7). Each alteration in the membrane potential spreads electrotonically to the neighboring cells. In arteries and arterioles, the length constant amounts to 1.5 mm, the time constant to 300–700 ms, and the conduction velocity to 0.2 cm/s in practically every direction of propagation; thus the vascular smooth muscle can be regarded as a functional syncytium [5]. This likely has particular importance for the synchronization of spontaneous excitation processes (slow waves, vasomotion). In addition, when a transmitter causes an increase in membrane

Fig. 96.1. Vascular smooth muscle cells. *1*, Cell nucleus; *2*, sarcoplasmic reticulum; *3*, caveola; *4*, nexus (gap junction); *5*, myosin filaments; *6*, actin filaments; *7*, intermediate filaments; *8*, dense bands; *9*, dense bodies; *10*, basement membrane; *11*, terminal axon; *12*, axonal varicosities containing dense-cored vesicles with neurotransmitters; *13*, collagenous fibers; *14*, elastic fibers

thinner, but also more loosely arranged. Lymphatic vessels are similar in structure to veins and have a similar number of valves. The adventitia links the vessels to the environment, and its inner part contains longitudinally directed elastic and collagenous fibrils. The thickness of this layer is very variable in different vessels. Cellular elements found

conductance, current flows into a cell and some current will flow into neighboring cells, thus changing their membrane potential.

96.2.2 Filaments

Cytoskeletal and contractile proteins in vascular smooth muscle cells appear to be distributed in two domains:

- Longitudinal intermediate filaments free of myosin, but containing desmin, vimentin, filamin, actin, and α-actinin
- Contractile protein filaments containing actin, myosin, tropomyosin, and caldesmon.

Indeed, uniform and well-adjusted functions result for these two cellular domains, as has already been deduced from cDNA libraries. The tension developed by the contractile protein filaments of the myocyte is proportionately transmitted to the cell interior and exterior by the proteins of intermediate filaments. These appear not only in muscle cells but also in endo- and epithelial cells, neurons, and fibroblasts. Both filamentous structures are coded by large families of multigenes [6].

Actin and Myosin. These are old proteins, found in the simplest eukaryotes such as yeasts and slime molds. The contractile proteins exist as nonmuscle and muscle isoforms. These proteins are encoded by multigene families, which are differentially expressed in different adult and developing tissues. In the same phenotype, actin and myosin genes are not genetically linked. In smooth muscle, myosin light chains are encoded by distinct genes which are dispersed on different chromosomes. The myosin heavy chain family, unlike that of the actins and myosin light chains, has cell type-specific and developmental isoforms which are sequentially expressed. In mammalian cells, diversity is generated by multiple, distinct myosin heavy chain genes coding for these slightly different isoforms and, as recently detected, by alternative RNA processing [18]. Although the conservation of amino acids in the myosin head region is significantly greater than that in the tail region, it is not as great as the conservation found in other contractile proteins, particularly actin. For actin genes, the conservation of the protein sequence in different species made it possible to compare the exon–intron organization over a wide evolutionary span. Sequencing human vascular smooth muscle actin gene revealed that it has introns in the positions of striated muscle actin genes and in the position of nonmuscle β-actin [80]. Observations from plants and *Deuterostomia* suggest a common ancestral gene with multiple intron positions which have been partially lost during evolution.

Contractile Filaments. These are not arranged irregularly. The thin actin filaments (diameter, 7 nm; length, 1.5 μm) are anchored in the "attachment plaques" (dense bands). These triangular, electron-dense regions are adjacent to the plasmalemma. Only actin filaments penetrate the dense bands, whereas the shorter, thick myosin filaments (diameter, 15 nm; length, 2.2 μm) are affixed via crossbridges to the thin filaments. These "mini-sarcomeres" represent the contractile units [3]. The ratio between thick (myosin) and thin filaments (actin, tropomyosin, caldesmon) is 1:15 (in mammalian skeletal muscle fibers 1:2). In the cell interior, the actin filaments, up to 15 of which encircle each myosin filament, insert themselves into dense bodies in the cytoplasm [13]. In the shortened cell the myofilaments occur obliquely arranged with respect to the cell axis, while in the extended cell, the contractile elements lie more or less parallel to the cell axis. Because they are not aligned in register like they are in skeletal muscle, no transverse striations are visible under the light microscope in vascular smooth muscle. Intermediate filaments (diameter, 10 nm) build an intracellular network which is likewise anchored in the dense bands and dense bodies [16] and which ensures an even propagation of cell contraction (Fig. 96.1). Dense bands and dense bodies correspond functionally to the Z disks in obliquely striated musculature; in fact, they might be their phylogenetic precursors, a hypothesis which is supported by the fact that both structures contain α-actinin. They alter their position by the forces originating from the actin–myosin interaction. Dense bands occur on all surfaces of the cell and are particularly abundant on the adventitial surface in arterioles. In contrast to the dense bodies, they contain the protein vinculin. As the intermediate filaments in vascular smooth muscle link dense bands and dense bodies throughout the cell, they form a cytoskeleton. Via vinculin and talin, but also via fibulin and α-actinin, this cytoskeleton is connected with transmembrane integrins (see Sect. 9.2.5.1). Their extracellular domains interact with proteoglycans and laminin of the basement membrane, which in turn communicate with extracellular collagen and elastin fibers. This has two consequences: first, a signal transduction chain arises from the extracellular matrix into the cell; second, the force developed in the myocyte is transferred to the neighboring cells via extracellular matrix and fibril structures, especially in the region of intermediate junctions. In this sense, there is both an electrical and a mechanical syncytium. This aspect appears to be even more significant, as the dynamic regulation of the assembly and turnover of intermediate filaments takes place by means of the phosphorylation of desmin by cyclic adenosine monophosphate (cAMP)-dependent protein kinase or protein kinase C (PKC) [42].

96.2.3 Caveolae and Endoplasmic Reticulum

Caveolae. Tubules of the smooth endoplasmic reticulum and flask-shaped invaginations of the plasma membrane that are lined up close together (caveolae intracellulares, surface vesicles; diameter, 70 nm) are thought to be concerned with the intracellular handling of Ca^{2+} [3]. The basal lamina runs over the stomata of the caveolae (diameter, 40 nm) without interruption. The surface vesicles are arranged longitudinally to the cell axis in long lines alternating with dense bands. Their Ca^{2+} concentration is equivalent to that in the extracellular space. They seem to have several functions: caveolae seem to be the site for the control of cell volume, and they are also thought to be the precursors of the transverse tubular system (T system) in heart and skeletal muscle cells. There are reasonable indications that, due to their close contact with cisternae and tubules of the endoplasmic reticulum (distance, 10–20 nm), caveolae are involved in transcellular Ca^{2+} exchange.

Smooth Endoplasmic Reticulum. This is developed only rudimentarily and to a variable degree in vascular smooth musculature. In large, elastic, or muscular arteries and in veins, it amounts to 1.5%–4.5% of the cell volume. As in skeletal and cardiac muscle cells, the reticulum tubules serve as Ca^{2+} stores. They are emptied to evoke contraction, and Ca^{2+} ions flow from the extracellular space into the cell through voltage-dependent Ca^{2+} channels. When relaxation occurs, Ca^{2+} ions are actively reaccumulated in the reticulum and transported outwards across the cell membrane. The extent to which a smooth endoplasmic reticulum is found in arterioles has not yet been investigated. Nevertheless, the volume of the reticulum is directly correlated with the capability of the vascular muscle to contract in Ca^{2+}-free solutions. In arterial smooth muscle, the relative contribution made by the endoplasmic reticulum to Ca^{2+} delivery during agonist-induced stimulation decreases as the lumen size becomes smaller. This decrease is associated with an increased sensitivity to Ca^{2+} antagonists [24]. In resistance vessels it has even been observed that the resting tone is abolished by Ca^{2+} channel blockers. This means that the resting tone in these small vessels is maintained by diffusion of Ca^{2+} ions from the extracellular space into the cells [59]. In contrast to large vessels, the contractile response of resistance vessels is also almost totally dependent on the influx of extracellular Ca^{2+}. This is made all the more plausible by the fact that even a two- to sevenfold increase in free, intracellular Ca^{2+} concentration is sufficient to raise the resting tone to maximum tone.

96.3 Molecular Mechanisms and Energetics of Contraction

Sliding Filament–Crossbridge Mechanism. Vascular smooth muscle has the ability to shorten to less than half its resting length (side polar filament structure [28]; also see Chap. 44). Both the filamentous organization of actin and myosin and the length–tension curve of this muscle are compatible with a version of the sliding filament–crossbridge mechanism of contraction. The myofilaments represent the contractile apparatus. The generation of force, which leads to tension development and shortening, occurs as a result of the chemical interaction between actin and myosin filaments and is triggered by a rise in intracellular Ca^{2+} concentration. In the presence of Ca^{2+}, myosin light chain kinase (MLCK) phosphorylates a regulatory light chain of myosin (LC_{20}), enabling the cyclic interaction of myosin with actin and subsequent tension development (Fig. 96.2). The latter lasts a few seconds until the peak of contraction, hence being a slow process caused by low crossbridge cycling frequency and shortening speed [37].

Velocity and Force. The velocity of smooth muscle myosin movement on actin filaments is 0.4 µm/s, while skeletal muscle myosin moves at 5 µm/s [82]. On the other hand, contractions can be maintained for very long times (latch state). Force development per cross-sectional area can be twice that in skeletal muscle, although the vascular

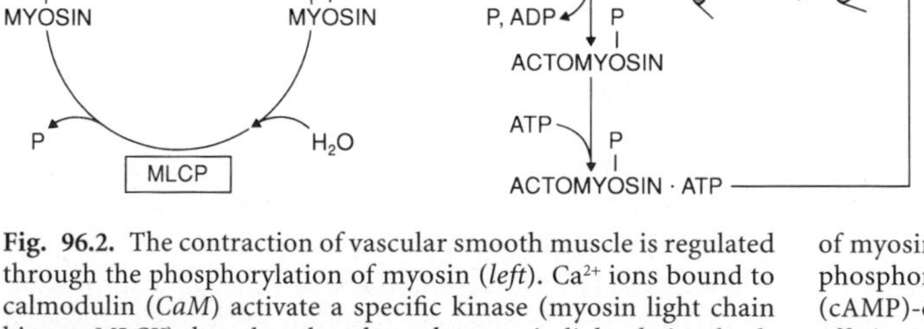

Fig. 96.2. The contraction of vascular smooth muscle is regulated through the phosphorylation of myosin (*left*). Ca^{2+} ions bound to calmodulin (*CaM*) activate a specific kinase (myosin light chain kinase, *MLCK*) that phosphorylates the myosin light chains, leading to cycling interaction of myosin with actin and thus to contraction (*right*). Relaxation is initiated both by the dephosphorylation of myosin by myosin light chain phosphatase (*MLCP*) and by the phosphorylation of MLCK by cyclic adenosine monophosphate (cAMP)-dependent protein kinase (cA-PK) decreasing the MLCK affinity to Ca^{2+}/calmodulin (not shown). *ATP*, adenosine triphosphate; *ADP*, adenosine diphosphate; *P*, phosphorylated substances

muscle cells contain much less myosin. The active tension in vessels was measured to be 250–400 mN/mm^2 or 5–6 μN/cell [37]. The linkages between cells are significant not only for the spread of excitation, but also for the transmission of force.

96.3.1 Ca^{2+} Activation of the Contractile Machinery

Excitation–Contraction Coupling. This coupling implies an increase in free Ca^{2+} ion concentration in the myoplasm. The intracellular Ca^{2+} concentration ($[Ca^{2+}]_i$) was determined to be 0.2 μmol/l in vascular smooth muscle by experiments with Ca^{2+}-sensitive indicators and Ca^{2+}-selective electrodes. $[Ca^{2+}]_i$ must rise to 0.5–1.8 μmol/l to elicit contraction [66]. Considering the small diameter of the muscle cells, it seems likely that the diffusion of Ca^{2+} ions from the extracellular space into the cell provides sufficient Ca^{2+} for the increase in the internal Ca^{2+} concentration with contraction. Indeed, the rise in plasmalemmal Ca^{2+} permeability of voltage-gated T- and L-type Ca^{2+} channels [19] in small arteries and arterioles regulating blood pressure, which have hardly any sarcoplasmic reticulum, leads to the necessary increase in $[Ca^{2+}]_i$. These Ca^{2+} channels are under the control of the membrane potential and agonists such as neurotransmitters, hormones, and autacoids. The 6000-fold transmembrane concentration gradient can be maintained only when Ca^{2+} ions diffusing into the cell are again extruded through the cell membrane by active adenosine triphosphate (ATP)-fueled Ca^{2+} pumping and the Na$^+$/Ca^{2+} exchanger (see p. 1952). After agonist activation, the cyclic nucleotides cAMP and cyclic guanosine monophosphate (cGMP) stimulate the Ca^{2+}-ATPase in the plasmalemma.

Sarcoplasmic Reticulum. In large conduit arteries it was observed that, in spite of the absence of extracellular Ca^{2+}, tonic contractions could be elicited. The control of vascular smooth muscle $[Ca^{2+}]_i$ is accomplished by an additional integrated membrane system, the sarcoplasmic reticulum. This reticulum is influenced by second messengers and contains approximately 5 mmol Ca^{2+} per l. Thus, the concentration gradient across the reticulum wall to the myoplasm and that across the outer cell membrane are similar. The myoplasmic Ca^{2+} concentration, being several thousand times less than in the other compartments is exactly regulated by downhill (passive permeabilities) as well as uphill mechanisms (active transport processes). The raise in $[Ca^{2+}]_i$ during contraction is thus attributable to the above-mentioned transmembrane Ca^{2+} influx as well as to a cellular Ca^{2+} release from the sarcoplasmic reticulum, particularly with agonist-induced contraction. The Ca^{2+} release is transient, and deficiency of extracellular Ca^{2+} depletes the intracellular Ca^{2+} store. A number of physiological agonists, including norepinephrine, histamine, angiotensin II, vasopressin, and prostaglandins, release Ca^{2+} from the sarcoplasmic reticulum pool via phosphatidylinositol (PI) response. Inositol 1,4,5-trisphosphate (IP$_3$), which is formed from PI 4,5-bisphosphate by G protein-linked, phospholipase C-mediated hydrolysis after agonist binding at the receptor, acts in this signal transduction chain as second messenger for sarcoplasmic reticulum Ca^{2+} release at submicromolar concentrations. The interaction of IP$_3$ with its specific receptor opens cation channels in the reticulum membrane, which allow rapid discharge of this Ca^{2+} pool (see Chap. 5).

Besides the IP$_3$-activated channel, vascular smooth muscle contains a further, Ca^{2+}-induced Ca^{2+} release (CICR) channel in its sarcoplasmic reticulum that carries out signal amplification at micromolar concentrations [24]. Via the CICR mechanism the influx of a small amount of Ca^{2+} during excitation can release a much greater amount of Ca^{2+} through large 75 pS cation channels from the sarcoplasmic reticulum. These Ca^{2+} channels are similar to the ones described for skeletal muscle, where they are associated with the ryanodine-binding junctional foot proteins. The CICR channels are located in the peripheral sarcoplasmic reticulum, which approaches the plasma membrane via special junctions that are analogous to the triadic junctions of skeletal muscle and contain structural "feet."

Besides the Ca^{2+} extrusion across the cell membrane by the Ca^{2+}-ATPase and the Na$^+$/Ca^{2+} exchange, Ca^{2+} reuptake into the sarcoplasmic reticulum by Ca^{2+}-ATPase is of decisive importance for the relaxation of vascular smooth muscle cells with a functionally relevant sarcoplasmic reticulum. Ca^{2+} uptake is achieved by a 110-kDa Ca^{2+}/Mg^{2+}-ATPase that undergoes Ca^{2+}-dependent phosphorylation and is regulated through phosphorylation of phospholamban by a cAMP or a Ca^{2+}-calmodulin-activated kinase (see Chap. 8). cGMP likewise stimulates the sarcoplasmic reticulum Ca^{2+} pump in vascular smooth muscle. The high-capacity (40 binding sites), low-affinity, anionic 45-kDa protein calsequestrin is found in the junctional elements of the peripheral sarcoplasmic reticulum and binds a considerable part of luminal Ca^{2+}. The transport velocity during Ca^{2+} reuptake is controlled by phospholamban. A steady exchange occurs between the extra- and intracellular Ca^{2+} compartments. Depending on the functional state of the muscle, inward or outward movement at the plasma and reticulum membrane is promoted or inhibited [24].

96.3.2 Myosin Light Chain Phosphorylation

Myosin Motor. The principal molecular organization of the myosin motor comprises three domains [79]. The motor domain contains the nucleotide-dependent actin-binding and the ATP hydrolysis sites. The light chain/calmodulin-binding domain is characterized by a consensus sequence motif of the heavy chain, the so-called IQ motif (approximately 23–25 residues) [84], bearing the binding sites for these subunits. Half of the heavy chain residues are apolar and in contact with the light chains. The cargo domain specifies the function and dimerization potential of the individual myosin.

Myosin Light Chain Kinase. The mechanism by which Ca^{2+} ions activate the contractile machinery differs from the skeletal and cardiac muscle. Vascular smooth muscle myosin consists of two heavy chains; two light peptide chains are attached to each one (see Sect. 44.2.2). The 20-kDa myosin light chain appears in two isoforms. These so-called "regulatory" and "essential" light chain subunits are critical for regulation. Each form can be mono- or diphosphorylated (5%–10%). Myosin light chain kinase (MLCK) plays a central role in their Ca^{2+}-dependent regulation [9]. It phosphorylates a serine (and threonine) residue in the amino terminus of the regulatory light chain of myosin. In the presence of Ca^{2+}, this phosphorylation exerts a profound effect on the structure and enzymatic activity of myosin. First of all, the actin-activated myosin Mg^{2+}-ATPase activity is raised, enabling cycling interaction of myosin with actin and subsequent tension development. MLCK is a calmodulin-dependent enzyme that is fully inactive in the absence of Ca^{2+}. Not troponin, but calmodulin has been identified as the intracellular Ca^{2+} switch. On the basis of homologies observed at the level of the amino acid sequences with troponin-C, parvalbumin, and calsequestrin, calmodulin, a ubiquitous Ca^{2+}-binding protein (calciprotein), belongs to a large superfamily of Ca^{2+} sensors in which, according to observations with X-ray diffraction, the EF hand (helix–loop–helix motif) is realized as the structural principle of Ca^{2+} binding. The 17-kDa protein calmodulin possesses four Ca^{2+}-binding domains, which are occupied during an increase in the myoplasmic Ca^{2+} concentration. Since K_{Ca} is approximately 10^{-6} mol/l, calmodulin is able to sense free Ca^{2+} in physiological activities (10^{-7} to 3×10^{-6} mol/l). The Ca^{2+} binding effects a conformational change of calmodulin, its affinity to MLCK is enhanced, and it forms a one-to-one complex with the kinase, resulting in its activation. The enzyme shows allosteric behavior. A highly positive level of cooperativity of kinase (Hill coefficient $n_H \approx 3$) is observed for activation by calmodulin in the presence of Ca^{2+} [77]. The active enzymatic moiety is the ternary complex of Ca^{2+}, calmodulin, and MLCK, which in turn catalyzes the phosphorylation of myosin. Actin then interacts with the phosphorylated myosin and thus increases its Mg^{2+}-ATPase activity. In this way the Ca^{2+} signal is considerably amplified, as the complex of MLCK and calmodulin with four Ca^{2+} ions can phosphorylate and activate many myosin light chains.

Myosin Light Chain Phosphatase. This phosphorylation, and hence the development of stress, depends not only on the Ca^{2+} concentration, but also on the activity of another enzyme, the myosin light chain phosphatase (MLCP). Both processes, Ca^{2+}/calmodulin-dependent phosphorylation of the 20-kDa myosin light chain by MLCK and dephosphorylation by MLCP are important regulatory mechanisms for vascular smooth muscle contraction [37]. MLCP is spontaneously active and as a rule requires no cation activation. When the Ca^{2+} concentration falls below 10^{-7} mol/l, MLCK is repressed and MLCP activity predominates. Myosin is dephosphorylated, and the actomyosin-

ATPase inhibited. The Ca^{2+}/calmodulin activation of MLCK and thus the contractile activity are likewise inhibited when this enzyme is phosphorylated by the cAMP-dependent protein kinase (cA-PK) at site A [34]. This is one of the effects that explain the cAMP-mediated relaxation [16] after stimulation of the vascular β_2-receptors by epinephrine. Moreover, a cAMP increase results in a fall in the cytosolic Ca^{2+} concentration. Both of these events decrease the extent of myosin light chain phosphorylation and cause vasodilatation. cGMP cannot promote the relaxation, because the cG-PK-dependent phosphorylation of site B of the MLCK shows no effect on the Ca^{2+}/calmodulin activation properties (also see pp. 1958, 1959). Finally, diacylglycerol activates PKC, which can phosphorylate MLCK [37]. The effect is probably similar to that of cAMP.

Ca^{2+} Regulation of Thin Filaments. Myosin phosphorylation and dephosphorylation are currently regarded as the dominant regulatory pathways of contraction. However, the coexistence of a regulation via thin filaments has not been excluded. The thin filament regulatory proteins are tropomyosin and caldesmon, the latter being a calmodulin- and actin-binding 150-kDa protein constituting the Ca^{2+}-regulating component of the thin filaments [15]. Extended caldesmon is 150 nm long, which would allow it to span 28 actin molecules. Further support for thin filament regulation comes from immunocytochemical investigations which have shown that caldesmon is localized in the actomyosin domain and not the actin-intermediate filament domain. Without Ca^{2+}, caldesmon binds actin (and not calmodulin), thereby preventing actin–myosin interaction and thus contraction. When Ca^{2+} increases, caldesmon binds to calmodulin instead of actin, so that the inhibitory effect on actin–myosin interaction is abolished and contraction results. Thus a "flip-flop" mechanism for Ca^{2+} regulation of caldesmon binding to thin filaments was proposed [13,42]. With a decline in the free Ca^{2+} concentration, caldesmon inhibits the actomyosin Mg^{2+}-ATPase and simultaneously a dephosphorylation of myosin by MLCP occurs; the contractile system is therefore switched off in a dual way. Finally, the phosphorylation of myosin heavy chains by an endogenous kinase seems to promote the transition to relaxation. Small resistance and large conduit arteries contain two isozymes of the myosin heavy chain. Only the less mobile form of the heavy chain is phosphorylated. As the ratio of the more mobile to the less mobile isozyme increases from arteriolar tissue as we ascend the vascular tree to the large arteries, a functional change is combined with the alteration of genetic myosin heavy chain expression. Indeed, the arterioles regulating resistance have a higher sensitivity to the contractile protein activators Ca^{2+} and calmodulin than blood distributing arteries do [22].

In order to bind to calmodulin and to promote contraction, caldesmon first has to be phosphorylated by a Ca^{2+}/calmodulin-dependent protein kinase. Phosphorylation of tropomyosin by the calmodulin-dependent protein kinase increases the Ca^{2+} sensitivity of the myofilaments so that

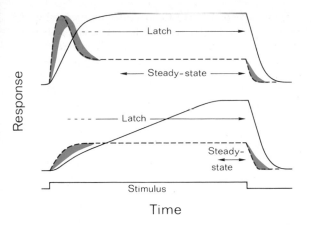

Fig. 96.3. Responses of vascular smooth muscle to stimulation. Changes in $[Ca^{2+}]_i$ (*dashed lines*) are followed by changes in myosin light chain kinase (MLCK) activity, in crossbridge phosphorylation, and in cycling rates (*blue areas*) and by force generation (*solid lines*). *Top*, typical response of a conduit artery to agonists mobilizing intracellular Ca^{2+} stores. *Bottom*, response of an arteriole devoid of Ca^{2+} storage sites to effector influences that only increase cell membrane Ca^{2+} permeability. (Modified from [52])

Ca^{2+} activation below $1\,\mu mol/l$ becomes possible. Therefore, tropomyosin makes feasible an amplified contraction at a reduced degree of myosin phosphorylation. Finally, the phosphorylation-induced structural rearrangements of the intermediate filament proteins desmin and synemin following PKC activation might be responsible for tension maintenance or tonic contractions.

96.3.3 Latch State

Crossbridge Cycles. Crossbridge interactions with the thin filament comprise two types in vascular smooth muscle:

- Cycling phosphorylated crossbridges (myosin light chain)
- Noncycling or slowly cycling dephosphorylated crossbridges (latch bridges).

The myosin-linked Ca^{2+}-regulatory system does not mediate crossbridge interactions as a simple on/off switch. Both the number of crossbridges interacting with the thin filaments and their cycling rate are regulated. Phosphorylation is the switch that turns a crossbridge on. The isotonic shortening velocity reflects the crossbridge cycling rate. A crossbridge cycle includes attachment to actin, rotation, and detachment. Work is paid for by the hydrolysis of ATP. One ATP molecule is split for every myosin crossbridge cycle [13]. The cycle frequency of 0.5–$1\,s^{-1}$ is lower than in skeletal muscle by a factor of 30–50 [61]. As expected, the energy consumption is 300 times lower; however, the shortening speed is 1/300 [13]. ATP hydrolysis is effected by the actin-activated myosin Mg^{2+}-ATPase, the activity of which is low.

During prolonged neuronal or hormonal stimulation,

agonist-induced Ca^{2+} release from the sarcoplasmic reticulum leads to initial transients in both $[Ca^{2+}]_i$ and in the extent of myosin phosphorylation in the first few minutes of force development in large conduit arteries with an intracellular Ca^{2+} store (Fig. 96.3). During a longer maintenance of force (latch state), the degree of phosphorylation and actomyosin-ATPase activity decreases to about 20% of the initial values [52]. The amount of steady state stress, as estimated by crossbridge cycling rates and isotonic shortening velocity, is proportional to phosphorylation. The rate of ATP splitting decreases during equal force maintenance. At a still only slightly increased Ca^{2+} concentration, the dephosphorylated bridges cycle slowly, and a part of the attached crossbridges do not cycle at all (latch bridges). These slowly cycling or noncycling latch bridges are the molecular basis for tension maintenance with a very low energy consumption [9]. Simultaneously, the muscle loses the ability to shorten rapidly. However, the initial contraction phase requires a fast crossbridge cycling of phosphorylated myosin heads.

Force Maintenance. The latch state was initially defined as force maintenance without proportional phosphorylation and with reduced crossbridge turnover rates. In small arteries and arterioles, the transmembrane Ca^{2+} influx with agonist-induced contraction results in a modest increase in $[Ca^{2+}]_i$ and myosin light chain phosphorylation to a sustained level, which is associated with the slow development of high levels of force (Fig. 96.3). Latch is characterized by decelerated contraction rates, which argues in favor of attached crossbridges. Moreover, only a slowed muscle shortening is possible in the latch state on account of the attached, dephosphorylated crossbridges. Ca^{2+} ions are necessary for the maintenance of latch bridges, and thus of tension, but $[Ca^{2+}]_i$ is lower than that needed for the elevation of phosphorylation. When removing extracellular Ca^{2+}, myosin is rapidly dephosphorylated while the force slowly decreases. Under in vivo conditions, phosphorylation of myosin light chains precedes tension development; conversely, these chains are dephosphorylated more rapidly than relaxation occurs. After dephosphorylation, attached crossbridges have slow detachment rates. Therefore, latch bridge detachment is rate limiting.

96.4 Excitation–Contraction Coupling

Voltage-Gated Ca^{2+} Channels. A delay of 0.2–6 s was observed between membrane depolarization and onset of vascular contraction, when the change in tension is triggered through graded depolarization or, more rarely, through action potentials of the muscle cells. During vasomotion, potential waves precede the rhythmic changes in tone by 40 ms, quite similar to cardiac muscle [73]. Membrane depolarization is thus basically correlated with vasoconstriction, and membrane hyperpolarization with vasodilatation. A sustained contractile response is depend

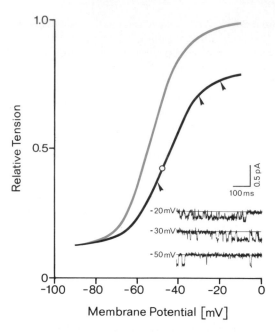

Fig. 96.4. Dependence of mechanical tension on the membrane potential in arterial vascular smooth muscle (stationary activation curve [75]). The *black curve* represents the relation in small arteries and arterioles devoid of sarcoplasmic reticulum. The point marked on the curve (○) indicates membrane potential and mechanical force in normal, pressurized vessels (V, -48 mV). The *arrowheads* reflect the voltage values in the activation curve at which single Ca^{2+} channel currents with 2.5 mmol Ca^{2+}/l as the charge carrier were recorded. (Figure inset from M. Rubart, J. Hescheler, and M.T. Nelson, unpublished, with permission). The reduction of unitary currents with membrane depolarization via the decrease in the inward driving force ($V-E_{Ca}$) is more than compensated for by the strong increase in openings of single Ca^{2+} channels. The *blue activation curve* depicts the relationship between developed tension and membrane potential in large conduit arteries after agonist activation eliciting force-amplifying mechanisms. Ca^{2+} release from the sarcoplasmic reticulum, protein kinase C (PKC) promotion of the contractile machinery, reduction of Ca^{2+} extrusion, and a displacement of the activation curve for Ca^{2+} channels are involved in the different receptor interventions. (Modified from [74,75])

ent on a gradual change in membrane potential which opens or closes voltage-gated Ca^{2+} channels [55]. In small arteries and arterioles with insignificant sarcoplasmic reticulum, patch-clamp and short-segment experiments strongly suggested that the maintained Ca^{2+} inward current through the potential-dependent Ca^{2+} channel plays the dominant role in voltage-sensitive contraction. The relation between membrane potential and open-state probability of this channel is quite similar to the relation between membrane potential and the ratio of agonist-induced force development to the channel current [50]. The dependence of force development in the resistance vessels on Ca^{2+} entry from the extracellular space is underlined by the effect of dihydropyridine Ca^{2+} antagonists, which inhibit the maintained Ca^{2+} current [55].

Electromechanical Coupling. The black activation curve in Fig. 96.4 shows the relationship between the actual

membrane potential and the developed force, also under receptor activation (electromechanical coupling). It reflects the mechanical consequences of a change in the common final pathway of the integration at the vascular smooth muscle cell, i.e., of a voltage change [75]. A variety of local or systemic influences regulate peripheral organ perfusion, e.g., ions (H^+, K^+), neurotransmitters (norepinephrine), hormones (epinephrine), autacoids (prostacyclin, PGI_2; endothelium-derived relaxing factor, EDRF) and physical factors (pressure, shear rate). Most of the known effector influences change the membrane potential.

Voltage-Dependent Ca^{2+} Channels. The activation curve (Fig. 96.4) takes a sigmoid course. Vascular tone increases with increasing depolarization in the absence of a threshold potential as a trigger of force. Between -60 and -20 mV, 80% of the maximal tension possible develops. This linear part of the activation curve proceeds very steeply, thus expressing a tight electromechanical coupling [55,75]. This tight coupling of mechanics to the membrane potential is caused by strongly voltage-dependent Ca^{2+} channels. A hyperpolarization (depolarization) of 2–3 mV results in a 50% reduction (rise) in vascular tone via a closing (opening) of these Ca^{2+} channels. Hence, two amplification mechanisms play a decisive role in the peripheral adaptation of the blood supply. Since blood flow is proportional to the fourth power of the vascular radius, even very small changes in the tone of the vascular smooth muscle cells elicit a strongly changed perfusion. As the steep activation curve of the vascular muscle is based upon a logarithmic dependence of the Ca^{2+} channel open probability on membrane potential, such slight changes in tone are themselves created by very small shifts in voltage. Through this twofold amplification, minimal changes in cell polarization lead to large changes in organ blood supply.

Membrane Potential. This fine regulation is made possible by the fact that the membrane potential in smooth muscle cells of normal, pressurized arteries and arterioles is of the order of -48 mV on average [38,55,85]. The working point is therefore positioned in the steep, linear range of the activation curve, thus shifted far to the right as compared to the working point in resting skeletal muscle at -90 mV (see Chaps. 45, 46). Therefore, arteries exist in a partially contracted state from which they can constrict further or dilate depending on the demand for blood [20,55].

The membrane potential (V) as an important regulator of vascular tone is determined by ionic concentration gradients (Δc) and permeabilities (P) according to the Goldman equation [35] (Table 96.1; also see Chaps. 8, 11). The transmembrane ion distribution is similar to other excitable structures [7]. There is a high intracellular Cl^- concentration, which yields a depolarized Cl^- equilibrium potential of approximately -20 mV compared to the nerve. While the random movement of ions through the membrane and the diffusion along their electrochemical gradient ($V - E$) play a major role in the development of the

Table 96.1. Interstitial ($[c]_o$) and intracellular ion concentrations ($[c]_i$), equilibrium potentials (E), driving forces ($V - E$), and permeabilities (P) in vascular smooth muscle tissues

	Na^+	K^+	Cl^-	Ca^{2+}
$[c]_o$ (mmol/l)	143	4.2	114	1.2
$[c]_i$ (mmol/l)	13	140	55	0.0002
E (mV)	+64	−94	−19	+116
$V - E$ (mV)	−112	+46	−29	−164
P (cm/s × 10^{-8})	0.8	22.9	8.4	4.6

A smooth muscle membrane potential of −48 mV is taken as a basis

membrane potential, the permeability coefficient is a basic parameter which represents the ease with which an ion diffuses through the membrane. The relative permeability of the membrane to K^+, Na^+, and Cl^- is a decisive determinant of the membrane potential. Since $P_K:P_{Na}:P_{Cl}:P_{Ca} = 1:0.037:0.366:0.201$, the membrane potential is mainly regulated by the opening or closing of K^+ channels. While the large outward driving force for K^+ ions ($V-E_K$) together with the high K^+ permeability affords a hyperpolarizing contribution to the membrane potential, the relatively high Cl^- and Ca^{2+} permeabilities exert a depolarizing influence. The high Ca^{2+} permeability determined in tracer flux and patch-clamp experiments is striking [5,55]. On the one hand, it codetermines the membrane potential in the relatively depolarized voltage range and, on the other, it depends exponentially upon changes in membrane potential. The membrane potential is also affected by electrogenic ion pumps. The activity of the (Na^++K^+) pump hyperpolarizes the plasmalemma. The Na^+/Ca^{2+} exchanger also contributes to the membrane potential, since during each transport cycle three Na^+ ions are translocated into the cell for every Ca^{2+} ion that goes out. Therefore, this exchange mechanism has an electrogenically depolarizing effect.

Spontaneous Action Potential Activity. Force development of vascular smooth muscle is usually associated with membrane depolarization, although action potentials may not be obligatory for contraction. Normally, a graded membrane depolarization is followed by the initiation of vasoconstriction, while a graded membrane hyperpolarization is followed by vasodilatation. The failure of these tissues to produce an action potential at all, or one of normal depolarization amplitude, is probably a result of the simultaneous activation of a K^+ conductance (delayed rectifier and Ca^{2+}-activated K^+ channels). However, the smooth muscle cells of small arteries and arterioles under normal blood pressure or high sympathetic tone in certain vascular beds, of precapillary sphincters, and of venous, venular, and lymphatic vessels exhibit spontaneous action potential activity triggering tension development [39,50,85]. Each action potential consists of a prepotential (pacemaker potential) followed by a spike of variable amplitude (10–45 mV) but lacking an overshoot. The maximum rate of rise of the action potential is 1–50 V/s and the duration is 20–200 ms. Action potentials conducted from

pacemaker cells have no pacemaker potential. At relatively depolarized membrane potentials of about −48 mV, where fast Na^+ channels, even if they exist in vascular smooth muscle, may be in a steady state of inactivation, voltage-gated Ca^{2+} channels conceivably account for most of the inward current during the action potential upstroke. Ca^{2+} influx and action potentials, and thus myogenic activity, can be suppressed by Ca^{2+} channel blockers such as nifedipine, a dihydropyridine (DHP) derivative. As in nerve and skeletal muscle, the repolarization phase of the action potential is initiated by delayed-rectifier and Ca^{2+}-activated K^+ channels.

96.4.1 Ion Channels and the Na^+/Ca^{2+} Exchanger

Ion channels are characterized by the high rates (approximately $10^6\,s^{-1}$) at which they transport ions once they are opened (see Chaps. 8, 12). Ion channels can be opened either by the membrane potential or by specific ligands [25]. They are normally highly selective for a given ion species. In the smooth muscle, as in many other cells, K^+, Ca^{2+}, Cl^-, and nonselective Na^+ channels are important. When the channels of an ion species are opened, the membrane conductance in question strongly increases because of the high ion flux rates. The membrane potential is moved to the specific equilibrium potential, which is determined by the transmembrane ion distribution according to the Nernst equation [57] (Table 96.1).

Ca^{2+} Influx. The influx of Ca^{2+} is unique, because in addition to the production of an electrical signal, Ca^{2+} ions act as a second messenger and trigger secretion of neurotransmitters or hormones, contraction of muscle, and activation of Ca^{2+}-calmodulin-dependent kinases, protein kinase C, and numerous proteases [25]. Membrane depolarization is often accompanied by Ca^{2+} entry through voltage-dependent Ca^{2+} channels, a diminished efflux via the potential-sensitive Na^+/Ca^{2+} exchanger, or a Ca^{2+} release from internal stores. By way of this Ca^{2+} influx, changes in membrane potential can be associated with changes in cell function. A vascular smooth muscle cell possesses more than 1000 voltage-dependent Ca^{2+} channels (single-channel conductance, 4 pS; unitary current, 0.5 pA), which open with depolarization and close with hyperpolarization (Fig.

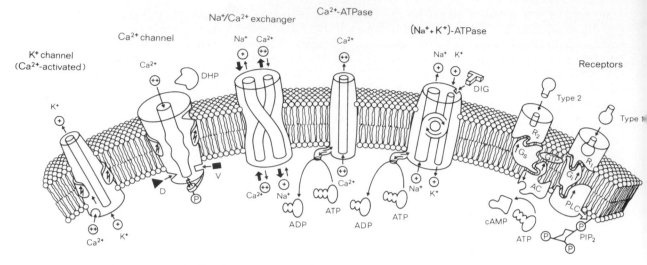

Fig. 96.5. Ion transport systems in the plasma membrane of a vascular smooth muscle cell. Ca^{2+} ions flow into the muscle cell through the voltage-dependent Ca^{2+} channel. The increase in the intracellular Ca^{2+} concentration stimulates the Ca^{2+}-activated K^+ channel. As a result of the K^+ outward current the membrane is repolarized, and thus the Ca^{2+} channel is closed. Ca^{2+} ions are removed from the cell interior via the Na^+/Ca^{2+} exchanger and the ubiquitous Ca^{2+}-ATPase. The (Na^++K^+)-ATPase exports inwardly diffused Na^+ and returns K^+ into the cell. The receptors modulate the ion fluxes. They are divided into two groups according to their second messenger. In vascular smooth muscle, α_1-adrenergic receptors and various peptide receptors (e.g., AT1 and V1a receptors) belong to the type-1 receptors. These receptors stimulate phospholipase C via a guanosine triphosphate (*GTP*)-dependent $G_i(G_p)$ protein. From the membrane lipid PIP_2 (phosphatidylinositol 4,5-bisphosphate), IP_3 (inositol 1,4,5-trisphophate) is formed, which releases Ca^{2+} ions from intracellular stores. Type-1 receptors can also promote the Ca^{2+} influx by phosphorylation of voltage-sensitive Ca^{2+} channels. Type-2 receptors (e.g., β_2-adrenoceptors) inhibit the Ca^{2+} entry through the Ca^{2+} channel via membrane hyperpolarization. Increased cyclic adenosine monophosphate (*cAMP*) activates the cAMP-dependent protein kinase, which phosphorylates Ca^{2+}-activated and delayed-rectifying K^+ channels. Moreover, cAMP-dependent protein kinase (cA-PK) phosphorylates myosin light chain kinase (MLCK), thus reducing its Ca^{2+} sensitivity. Vascular smooth muscle relaxation is initiated. *DHP*, dihydropyridine (Ca^{2+} channel blocker); *D*, diltiazem; *V*, verapamil; *DIG*, digitalis glycoside, inhibitor of the (Na^++K^+)-ATPase; R_1, R_2, receptor-binding proteins; *AC*, adenylyl cyclase; $G_i(G_p), G_s$, AC-inhibiting/stimulating nucleotide-binding protein; *PLC*, phospholipase C; ⓟ, phosphate group; *ADP*, adenosine diphosphate; *ATP*, adenosine triphosphate. (Modified from [65])

96.5). There is no threshold for the activation of Ca^{2+} channels [55]. The Ca^{2+} influx through potential-sensitive Ca^{2+} channels is always finite. Over the physiological range of the membrane potential, the voltage dependence of the Ca^{2+} channels is extremely strong (see above). With depolarization the Ca^{2+} current grows exponentially. Starting from a membrane potential of −48 mV, a 5-mV depolarization causes a 2.9-fold increase in open probability, a 5-mV hyperpolarization a 0.3-fold decrease in open probability. The steepness of the activation curve (Fig. 96.4) is determined on the one hand by the strong voltage dependence of the Ca^{2+} channels and on the other hand by the strong tension dependence on the intracellular free Ca^{2+} concentration which only varies to a relatively slight degree [75]. In this sense, the membrane potential dependence of tone reflects the voltage dependence of Ca^{2+} channels.

A simple calculation illustrates that Ca^{2+} influx through even a small number of Ca^{2+} channels is sufficient for significant changes in $[Ca^{2+}]_i$. This is mainly due to the small volume of arterial smooth muscle cells (approximately 1 pl): a unitary current at −48 mV of 0.5 pA corresponds to a Ca^{2+} entry of approximately 1.4 million ions per s through a single open Ca^{2+} channel [36]. Thus just one open channel in a single cell would deliver Ca^{2+} at a rate sufficient to raise

$[Ca^{2+}]_i$ by at least 1 μmol/s in the absence of buffering and extrusion [55]. Ca^{2+} entry through Ca^{2+} channels over the physiological range of membrane potentials could supply smooth muscle cells with sufficient Ca^{2+} to maintain steady contraction. It should be stressed, however, that this calculation overestimates the impact of Ca^{2+} influx on $[Ca^{2+}]_i$ because both buffering, extrusion, and store uptake occur simultaneously with influx.

K^+ Channels. Table 96.1 underlines the fact that the membrane potential of arterial and arteriolar smooth muscle cells depends largely on the K^+ permeability of their cell membrane. This is guaranteed by four types of K^+ channels:

- Ca^{2+}-activated (K_{Ca})
- ATP-sensitive (K_{ATP})
- Inward-rectifier
- Delayed-rectifier voltage-dependent K^+ channels.

K_{Ca} channels (single-channel conductance, 100 pS; unitary current approximately 5.0–5.4 pA) are activated by depolarization and intracellular Ca^{2+} [23] (Fig. 96.5). The channels are highly voltage dependent. K_{Ca} channels in resistance arteries contribute to the smooth muscle membrane potential, since the cells are somewhat depolarized ($-48\,mV$) and $[Ca^{2+}]_i$ is elevated [53] (Fig. 96.4). A rise in the $[Ca^{2+}]_i$ activates these channels, more K^+ ions leave the cell, and the membrane repolarizes or hyperpolarizes. The opening of only a very few K^+ channels is sufficient for an efficient membrane hyperpolarization, since the input resistance of the vascular smooth muscle cells is high. Thus, K^+ channels can hyperpolarize at low levels of activation. For example, the opening of only three additional K^+ channels per cell (a cell has approximately 50 000 K^+ channels) induces a membrane hyperpolarization of $2\,mV$. This minor hyperpolarization simultaneously reduces the Ca^{2+} channel open probability by 30%, thus also reducing tension development (Fig. 96.4). The subsequent increase in vascular diameter could lead to a significant enhancement in blood flow, since flow is proportional to the fourth power of the radius.

ATP-sensitive K^+ channels (single channel conductance, $10-30\,pS$; unitary current approximately $3\,pA$) are inhibited as the concentration of ATP at the internal membrane face is increased. A fall in intracellular ATP may not be the most important physiological activator of these channels, which are also regulated by a rise in ADP, GDP, GTP, or an acidic pH [29,53,58]. The antidiabetic sulfonylurea glibenclamide has been shown to block K_{ATP} channels. On the other hand, a number of synthetic (cromakalim, nicorandil) and endogenous (prostacyclin) K^+ channel openers effect vasodilatation via membrane hyperpolarization following K_{ATP} (and K_{Ca}) channel activation [55,71]. Some of these compounds are used successfully as antihypertensive drugs and in asthma, peripheral vascular disease, and diseases of the heart and nervous system.

Inward-rectifier K^+ channels (single-channel conductance, $4\,pS$) were found mainly in arterioles and show an increasing conductance with membrane potentials steeply negative to the K^+ equilibrium potential. The inward rectifier closes when arterioles are tonically depolarized, e.g., with sympathetic nerve activity [41]. This means that the loss of K^+ ions during these moderate depolarizations will be prevented.

Outward K^+ current through delayed-rectifier K^+ channels (single-channel conductance, $5\,pS$; unitary current approximately $0.4-0.7\,pA$) is activated by depolarization of the vascular muscle cells and decreases with prolonged depolarization. The voltage-dependent activation is typical of this K^+ channel, which was first described as repolarizing the action potential in nerve and skeletal muscle. Inward and delayed-rectifier K^+ channels contribute significantly to the membrane potential in resistance arteries.

Tension Regulation. A central question of the regulation of circulation concerns the tension regulation of the resistance vessels. On the one hand, the voltage dependence of Ca^{2+} channels underlies the membrane potential dependence of vascular tone. An increase in transmural pressure depolarizes the vascular smooth muscle cells. On the other hand, membrane hyperpolarization by K^+ channel activa-

tion is an effective mechanism to dilate blood vessels. These observations assign a regulatory role in the adjustment of vascular width to the voltage-dependent Ca^{2+} channel and the Ca^{2+}-activated K^+ channel. Resistance vessels are found in a partly contracted state (Fig. 96.4) from which they can further constrict or dilate by means of various effector influences [20].

Stretch-Activated Ion Channels. Blood pressure is involved in this myogenic development of tone. The myogenic tone contributes to the basal arterial tone and the autoregulation of blood flow. An increase in intravascular pressure leads to a stretch of the smooth muscle cells [21], which depolarize via an increase in open probability of their stretch-activated ion channels [44] (Fig. 96.6). These nonselective ion channels are mechanoelectrical transducing elements which, after activation, allow Na^+ and Ca^{2+} ions to pass. The resulting depolarization opens voltage-dependent Ca^{2+} channels, thus providing the route for Ca^{2+} entry. The intracellular free Ca^{2+} concentration rises, and the vascular tone increases. Myogenic tone and Ca^{2+} influx via voltage-dependent Ca^{2+} channels depend on the extracellular Ca^{2+} concentration and are abolished by DHP Ca^{2+} channel antagonists (diltiazem, nifedipine) and hyperpolarizing vasodilators (K^+ channel agonists) (Figs. 96.5, 96.6). Therefore, the DHP-sensitive, voltage-dependent Ca^{2+} channel is an important entry pathway for Ca^{2+} that is required for the tone of resistance-sized arteries [55].

Negative Feedback. The increase in tone following blood pressure elevation is, however, limited by a negative feedback mechanism. Membrane depolarization and a rise in

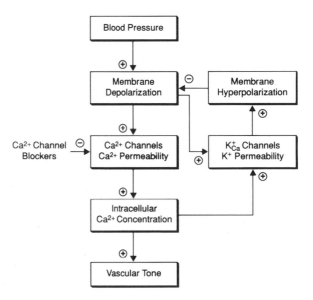

Fig. 96.6. Mechanism for the role of voltage-dependent Ca^{2+} and K^+_{Ca} channels in the regulation of vascular tone. Ca^{2+} channel blockers inhibit the Ca^{2+} inward current and the increase in cytoplasmic Ca^{2+} concentration, resulting in vasodilatation. (Modified from [53])

$[Ca^{2+}]_i$ open Ca^{2+}-activated K^+ channels, which elicit a membrane hyperpolarization [53] (Fig. 96.6). This opposes the original, pressure-induced membrane depolarization. Thus, K_{Ca} channels serve as a negative feedback pathway to control the degree of membrane depolarization and vasoconstriction [23]. The partly depolarized and contracted state of small arteries and arterioles allows numerous effector interventions.

To maintain the transmembrane Ca^{2+} homeostasis (see Chap. 11) and to lower $[Ca^{2+}]_i$, Ca^{2+} ions are actively extruded out of the cell by a plasma membrane Ca^{2+}-ATPase (see below) and passively by a Na^+/Ca^{2+} exchanger (Fig. 96.5). The energy for the Ca^{2+} extrusion by the Na^+/Ca^{2+} exchanger stems from the Na^+ gradient. In this system one outwardly transported Ca^{2+} ion is normally exchanged for three inwardly diffusing Na^+ ions. The Na^+/Ca^{2+} exchanger therefore has an electrogenically depolarizing action. Conversely, its activity is potential dependent. The Ca^{2+} outward transport is inhibited by a depolarization of the vascular smooth muscle cells. In this way the system supports the cytoplasmic Ca^{2+} accumulation and increase in tone. The exchanger has a high capacity but a low affinity for Ca^{2+}. It serves to eliminate Ca^{2+} when the cytosolic Ca^{2+} concentration is high, which is the case after a Ca^{2+} influx via voltage-dependent Ca^{2+} channels and/or after a Ca^{2+} release from intracellular storage sites. Na^+ ions that enter through the exchange system leave the cell via $(Na^+ + K^+)$-ATPase. With a lowered transmembrane Na^+ gradient, the Na^+ inward diffusion and Ca^{2+} extrusion are reduced. The rise in intracellular Ca^{2+} causes an increase in contractile force that is not inhibited by Ca^{2+} antagonists [50]. Consequently, an increase in $[Na^+]_i$ also leads to an increase in $[Ca^{2+}]_i$.

96.4.2 Ion Pumps

Plasma Membrane Ca^{2+}-ATPase. Apart from the Na^+/Ca^{2+} exchanger, plasma membrane and endoplasmic reticulum Ca^{2+}-ATPases provide for a low Ca^{2+} concentration within the resting cell (Fig. 96.5; see Chap. 8). Cloning and sequencing of the gene for the plasma membrane Ca^{2+}-ATPase gave an insight into the fundamental properties of ion-transporting ATPases. This phylogenetically very old system is a transmembrane-spanning protein, the sequence of which is similar to that of the α-subunit of the $(Na^+ + K^+)$-ATPase. The Ca^{2+} ion export against its very steep electrochemical gradient works only because the Ca^{2+}-ATPase has a very high affinity for Ca^{2+}. The activity of the Ca^{2+} pump requires ATP as a source of energy and is regulated by calmodulin [67]. The conformational change induced by the binding of four Ca^{2+} ions allows the pump to activate (see p. 1946). Moreover, the calmodulin-binding domain mediates self-association of the Ca^{2+} pump. Oligomerization may include a direct covalent cross-linking of ATPase monomers or phosphorylation-dependent affinity modification [81]. The significant consequence of this oligomerization is a cooperative activation of the Ca^{2+}-ATPase by calmodulin. Allosteric-type regulation is a general property of most of the calmodulin-dependent enzymes. The enzyme can also be stimulated by membrane lipids such as phosphatidylinositol bisphosphate (PIP_2), which is the target derivative for phospholipase C (PLC) in the α-receptor-induced cascade (see below). Finally, the Ca^{2+}-ATPase is activated by cAMP and cGMP after agonist activation.

Endoplasmic Reticulum Ca^{2+}-ATPase. The Ca^{2+}-ATPase of the endoplasmic reticulum ensures that the intracellular Ca^{2+} pools are replenished. The Ca^{2+} gradient at the sarcoplasmic reticulum membrane is comparable to that of the plasma membrane, so that the accumulation of Ca^{2+} ions in the reticulum is only possible with ATP as fuel. The Ca^{2+} pump in the endoplasmic reticulum wall is phosphorylated and activated via a cAMP-dependent protein kinase [67].

$(Na^+ + K^+)$-ATPase. The $(Na^+ + K^+)$-ATPase is in many ways similar to Ca^{2+}-ATPase (see Chap. 8). The different ion concentrations, mainly for Na^+ and K^+, prevailing on either side of the cell membrane are generated by an enzyme system that gains energy by ATP splitting. The enzyme is activated drastically when the extracellular K^+ concentration increases from zero over its physiological value of 4.2 to 20 mmol/l, and/or the intracellular Na^+ concentration rises over its physiological value of 13 mmol/l. This $(Na^+ + K^+)$-ATPase, also termed the $(Na^+ + K^+)$ pump, fulfills the requirements for the generation of the membrane potential by the creation of transmembrane ion gradients (Fig. 96.5; see Chap. 11). Furthermore, its activity influences vascular cell excitability, contractility, and volume regulation. Its α-subunit (seven transmembrane α-helices) and β-subunit (four transmembrane helices) exist as a dimer or tetramer in the membrane. Both the α-chains including the ATPase are positioned at the myoplasmic side of the membrane. The extracellular domain of the α-chain contains a binding site for cardiotonic steroids. Plant steroids such as ouabain, digoxin, and scillaren inhibit ion pumping of the enzyme by competition with K^+ for this binding site. Partial blockade of $(Na^+ + K^+)$-ATPase can increase its turnover rate or stimulate gene expression of the enzyme. One mechanism responsible for this tight regulation is found in mineralocorticoid hormone receptors in vascular tissue. Binding of aldosterone directly upregulates $(Na^+ + K^+)$-ATPase gene expression and mRNA accumulation [60]. In general, there are more $(Na^+ + K^+)$-ATPase molecules than Ca^{2+}-ATPases per cell.

Per hydrolyzed ATP, the $(Na^+ + K^+)$-ATPase drives three Na^+ ions out of the cell, but only two K^+ ions into the cell. Na^+ is transported against the gradient of the electrical potential and the concentration, and K^+ against the concentration, but with the potential gradient. The pump is electrogenic, and the membrane potential is shifted towards hyperpolarization. The contribution of the electrogenic $(Na^+ + K^+)$ pump to the membrane potential of small arteries can amount to up to 11 mV [50]. Inhibition of the pump, e.g., by the application of ouabain or a reduction in the extracellular K^+ concentration, leads to vasoconstriction. This is a result of the membrane depolarizing effect that opens voltage-dependent Ca^{2+} channels. The vasoconstriction may be prevented by Ca^{2+} antagonists.

96.5 Effector Influences

Multiplicity of Effectors. The local and systemic effector influences that regulate the peripheral adaptation of blood supply include nerves, ions, hormones, and cellular and environmental metabolism as well as physical factors. Because of the multiplicity of the effectors, and since several can be changed at the same time, the question arises as to their mode of action and the common final pathway of their integration at the vascular smooth muscle cell. It follows from the activation curve [75] that the working point of pressurized blood vessels is located in the steep, linear range of the sigmoid relationship (Fig. 96.4). Resistance arteries are thus found in a partly depolarized and contracted state, which is further modifiable by constricting and dilating effectors. Up and downregulation of tension development can easily be started from such a medium working point, even by small potential changes. In the following we will examine which physiological effector influences modulate the myogenic tone by depolarization or hyperpolarization.

96.5.1 Ions as Activators or Inhibitors of Vascular Tone

A change in the normal plasma K^+ concentration of 4.2 mmol/l has numerous clinical consequences (see Chap. 79). Hyperkalemia is due to disturbances of either external (acute nephroparalysis, chronic renal insufficiency) or internal balance (acidosis, large injuries of mollescent body parts, digitalis intoxication). Hypokalemia is observed with intestinal (diarrhea, laxative abuse) or renal losses (chronic interstitial nephritis, renal tubular acidosis, primary or secondary hyperaldosteronism, therapy with diuretics) or with distributive disturbances (alkalosis, insulin treatment). With a slight decrease in $[K^+]_o$ the membrane potential becomes more negative according to the Nernst equation [57], but further reductions cause depolarization. The dependence between membrane potential and $[K^+]_o$ is thus U-shaped [5]. However, the membrane potential takes a very flat course between 2 and 15 mmol/l $[K^+]_o$, i.e., the changes are slight. The depolarization and tension increase with reduced $[K^+]_o$ is largely due to an inhibition of the electrogenic $(Na^+ + K^+)$ pump, which is normally activated by extracellular K^+. Therefore, the electrogenic contribution to the membrane potential is diminished. Moreover, the resting membrane conductance to K^+ decreases when $[K^+]_o$ is reduced, because voltage-dependent inward-rectifier K^+ channels close. Conversely, when $[K^+]_o$ is increased, more channels are opened. Hence, the membrane potential in hyperkalemia would be less negative on the basis of the depolarized K^+ equilibrium potential E_K. Even if all the K^+ channels were opened and the contribution to the membrane potential by other ion species were negligible, at the most E_K could be achieved. Nevertheless, a membrane hyperpolarization with vasodilatation during hyperkalemia of up to 15 mmol/l is frequently observed, especially in arteriolar and venular vessels. The depolarizing effect according to Nernst [57] is then surpassed by the hyperpolarization generated through $[K^+]_o$ stimulation of the electrogenic $(Na^+ + K^+)$-ATPase. A dilatation resulting from this has been especially measured in the vessels of working skeletal musculature. Here, $[K^+]_o$ increases interstitially near peripheral vessels due to the high action potential frequency (K^+ outward current) in the skeletal muscle.

Changes in P_{CO_2}. Such changes have a significant effect on peripheral resistance, especially in the cerebral circulation but also in other systemic vascular beds (see Chap. 78). They affect mainly arterioles, but also small arteries. An increased P_{CO_2} leading to extracellular acidosis causes membrane hyperpolarization and vasodilatation [8]. (This is not true in the lung.) Inversely, alkalosis causes depolarization and vasoconstriction [75]. Hyperpolarization at acidic pH is based on an increased K^+ permeability, and the depolarization at alkaline pH on a decreased K^+ permeability [50]. Since K_{ATP} channels are pH sensitive, they presumably are responsible for these changes [29,53]. During longer-lasting and marked acidosis, a generalized vasodilatation with a fall in blood pressure and a state similar to shock may prevail.

96.5.2 G Protein-Coupled Receptor Activation

Hormones and neurotransmitters as well as many cytokines and auto- and paracrine factors regulate the metabolic functions of the cell via specific receptors which are coupled to the intracellular effectors by GTP-binding and -hydrolyzing proteins, the so-called G proteins (see Chaps. 5, 6). G proteins are crucial in the transduction of initially exterior signals into a cellular response. Generally, as a heterotrimer they are composed of α-, β-, and γ-subunits. G proteins establish a connection between receptors at the outer surface of the cell membrane with the effectors at the cytoplasmic side [64]. Typical effector systems are adenylyl cyclase (second messenger cAMP), PLC (second messengers DAG and IP_3), phospholipase A_2 (PLA_2; release of arachidonic acid), and various ion channels of the plasma membrane.

α- and β-Adrenoceptors. In the autonomic innervation of blood vessels by the sympathicus, α- and β-receptors are distinguished. While, with respect to vasculature, the β_2-receptor subtype in man is restricted to coronary and skeletal muscle vessels, the α_1-receptor (postjunctional) and α_2-receptor subtypes (pre- and postjunctional) are found in nearly all the vessels [3]. The adrenoceptors are excited by the neurotransmitter norepinephrine from sympathetic

nerve fibers as well as by the hormone epinephrine from the suprarenal medulla circulating in the blood as a "first messenger." Nearly all human arteries and arterioles are sympathetically vasoconstrictor innervated. In the large and small arteries, the nerve fibers are found in the adventitia and do not penetrate the media. The nerves consist of axons with varicosities that are 2 μm long and 1 μm across [50]. They contain more than 500 small and large, dense-cored vesicles with a variable mixture of norepinephrine and ATP. The axons are unmyelinated and surrounded by Schwann cells. In neuromuscular junctions the minimum distance between varicosity and smooth muscle cell is 0.1–10 μm and is thus much greater than that in the end plate. Since the perivascular nerves run along the adventitia–media border, the inner part of the media is mainly influenced by effectors circulating in the blood.

During adaptation to exercise, epinephrine released from the suprarenal medulla primarily binds to the postjunctional β_2-adrenoceptors in coronary and skeletal muscle vessels and dilates them. Though the α-receptors are present in about the same quantity in these vessels, the β-receptors are excited even by lower epinephrine concentrations based on their lower threshold. Norepinephrine released from terminal sympathetic nerve endings during adaptation to work mediates its vasoconstrictor effect by conventional postjunctional α_1- and α_2-adrenoceptors. The α-receptor subtype is preponderantly an α_1-adrenoceptor, but varies through the circulation system, and α_2-adrenoceptors seem to predominate in the more distal small arteries and arterioles and in the veins [50]. In addition, α_2-receptors are located both in large conduit and in small resistance arteries on the prejunctional side, where they influence the presynaptic regulation of the

transmitter release from the perivascular varicosities. Stimulation of prejunctional α_2-adrenergic receptors functionally connected to G_i coupling proteins diminishes the release of norepinephrine. By analogy, the activation of presynaptic P_2 purinoreceptors inhibits the release of the cotransmitter ATP.

When α- and β-adrenergic receptors were cloned, the surprising discovery was made that they are members of a much larger family of structurally and functionally related proteins derived from some common ancestor. A characteristic of this family is the fact that signal transduction is mediated to the cell via G proteins after binding of the first messenger. Members of this family include the adrenergic receptors, the muscarinic receptors, the dopamine receptors, the serotonin receptors, the neuropeptide receptors, and the glycoprotein hormone receptors [63]. Figure 96.7 shows a model for a G protein-coupled receptor, specifically for the α-adrenergic receptor (see also Chap. 5). Within its seven membrane-spanning domains, the integral membrane protein reveals a remarkable conservation of certain amino acid residues between all the family members [27]. Ligand binding seems to take place at the hydrophobic core of the transmembrane domains M-II to M-V. Thus, epinephrine binds with its amino group to an acidic residue present in M-III of the β_2-adrenoceptor, while the catechol hydroxyl groups interact with serine residues in M-IV and/or M-V. In coupling to the second messenger system adenylyl cyclase, with activation of the cA-PK, the C-III loop plays a decisive role, because deletion mutants of this loop impair the ability to stimulate this enzyme. Therefore, phosphorylation of the α-receptor is possible at the C-III loop, while on the β-receptor phosphorylation occurs in the carboxyl terminus.

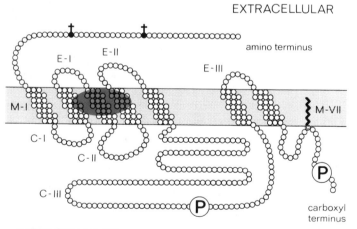

EXTRACELLULAR

INTRACELLULAR

Fig. 96.7. General model of α-adrenergic and G protein-coupled receptors. The *chain of circles* represents the individual amino acids that make up the receptor protein. The *stippled area* is the cell surface membrane. Segments *M-I* to *M-VII* are thought to span the bilayer. With this arrangement, the amino terminus and the connecting loops *E-I*, *E-II*, and *E-III* lie on the extracellular surface of the membrane. Connecting loops *C-I*, *C-II*, and *C-III* and the carboxyl terminus are represented as exposed to the cyto-plasmic side of the plasma membrane. The *blue area* indicates that the binding of ligands probably occurs in the transmembrane domains M-II to M-IV. Ⓟ denotes potential sites of phosphorylation that may be important in the regulation of the receptor. The *crosses* are sites of potential N-linked glycosylation. The *bold wavy line* attached to the cysteine in the carboxy-terminal region represents a potential site of fatty acylation. (From [63])

α- and β-adrenergic receptors become phosphorylated during the desensitization process, which is a mechanism of dynamic regulation to dampen the responsiveness of cells to further catecholamine stimulation [46]. Through this phosphorylation the coupling of the receptor to its respective G protein is hindered.

In vascular smooth muscle, activation of α_1-adrenergic receptors (molecular mass, 80 kDa) by ligand binding of its first messenger norepinephrine initiates contraction. In addition, postsynaptic α_2-receptors (M_r, 64 kDa) stimulate agonist-induced vasoconstriction with high affinity [63]. This effect is mediated by an increase in the free cytosolic Ca^{2+} concentration via Ca^{2+} entry through voltage-dependent Ca^{2+} channels [51,56]. As arterioles are tonically depolarized during ongoing sympathetic nerve activity, membrane depolarization is obviously a prerequisite for contraction. Membrane depolarization and vasoconstriction following agonist α-receptor activation is a multifactorial event [55]. The G protein G_p (G_q/G_{11}) involved in α_1-adrenoceptor stimulation is pertussis toxin (PTX) insensitive [40,48]. Although the direct opening of the voltage-dependent Ca^{2+} channel by the G protein has not been proven, norepinephrine indirectly causes an increase in the Ca^{2+}, Na^+, and Cl^- conductances [49].

The signal transduction involves G protein-mediated activation of phospholipase C and formation of both DAG and D-myoinositol 1,4,5-trisphosphate (IP_3), two potent second messengers [83] (see Chap. 5). In resistance vessels without or with only a rudimentarily developed sarcoplasmic reticulum, enhanced Ca^{2+} influx and raised $[Ca^{2+}]_i$ result from DAG-activated PKC phosphorylating voltage-dependent Ca^{2+} channels [56,66,83]. Phorbol esters, secondary metabolic products of *Euphorbiaceae* and *Thymelaeaceae* occurring in nature, which are esterized at carbon atoms 12 and 13 and which are among the most powerful known tumor promoters (see Sect. 9.3.5), have the same effect as DAG and can directly activate PKC. The reason for this lies in the stereochemistry of the esters, which is comparable to that of DAG. Phorbol esters are thus able to phosphorylate and open voltage-dependent Ca^{2+} channels via stimulated PKC [14]. This Ca^{2+} entry, a Cl^- efflux through Ca^{2+}-dependent chloride channels, and Na^+ inward current depolarize the cell membrane [51]. Inhibition of K_{Ca} channels without promotion by a G protein contributes to this membrane depolarization. Finally, it has been shown in recent patch-clamp investigations that the cotransmitter ATP can open a channel permeable to Ca^{2+}. The membrane depolarization itself enhances the open-state probability of voltage-dependent Ca^{2+} channels, more Ca^{2+} ions flow into the cell, and so forth. The voltage dependence of the Ca^{2+} channels is so strong (see p. 1950) that the slightest change in membrane potential could have a substantial effect on Ca^{2+} entry, internal free calcium, and contractile tone (Fig. 96.4).

In large conduit arteries, α-receptor agonists elicit even more force-amplifying mechanisms. The second messenger IP_3 formed by the activation of PLC triggers the release of Ca^{2+} from intracellular stores, mainly the sarcoplasmic reticulum, via Ca^{2+} release channels and transiently cl-

evates $[Ca^{2+}]_i$. Raising $[Ca^{2+}]_i$ potentiates contraction. Some vasoconstrictors additionally shift the relationship between $[Ca^{2+}]_i$ and force to higher tension levels for a given intracellular Ca^{2+} concentration [55]. Protein kinase C increases the apparent Ca^{2+} sensitivity of the contractile machinery during agonist-induced contractions (but see also p. 1946). In addition, agonists can reduce Ca^{2+} extrusion. Finally, they may displace the activation curve for Ca^{2+} channels to the left so that their open-state probability is higher at any given membrane potential. Such a slight shift in the activation curve leads to a substantial Ca^{2+} influx without changes in membrane potential [55]. This modulatory effect can be observed in vessels in general. It causes an overall steepening of the activation curve "developed tension versus membrane potential" (Fig. 96.4). According to the vessel type (extent of the sarcoplasmic reticulum), the activating agonist (mode of tension initiation), and α_1-receptor subtype, activation curves can vary (Fig. 96.4). The α_1-receptor subtypes are the following [63]:

- α_1-A: linked to DHP-sensitive, voltage-dependent Ca^{2+} channels
- α_1-B: stimulation of phosphatidylinositol hydrolysis
- α_1-C: coupled to phospholipase C.

A blood vessel is therefore able to vary its tone even with an unchanged membrane potential by means of a "jump" to another activation curve. However, the voltage dependence of tension development in the individual activation curves remains untouched.

Excitatory Junction Potentials. The blockade of sympathetic nerves with tetrodotoxin leads to a 3-mV hyperpolarization of the vascular smooth muscle cell membranes, suggesting a tonic influence of nerve activity on the membrane potential in vivo. This underscores the finding that action potentials can only be observed rarely and in a few vascular zones. Similar to the motor endplate, local potential changes appear in blood vessels due to neuromuscular transmission, so-called excitatory junction potentials (EJP), which are accompanied by an increase in force [8,41]. The smooth muscle cell membrane rapidly depolarizes (time to peak, 100 ms) by about 10 mV 20 ms after the nerve action potential. The depolarization fades away after about 1 s. The EJP is followed by a slow depolarization of about 2 mV which lasts for several seconds, as does the vasoconstriction. It is abolished by α-adrenoceptor antagonists, just as the slow depolarization is. Although the assignment of the potential and force changes to the EJP is not completely resolved, the rapid initial depolarization in the EJP seems to be conditioned by the release of ATP cotransmitted with norepinephrine [9,41]; in a multitude of blood vessels, ATP evokes depolarization and contraction. α-Adrenoceptor antagonists produce no effect which further supports this thesis.

As in the endplate potential, Na^+ and K^+ conductance are simultaneously elevated in the EJP. Correspondingly, the reversal potential of the excitatory junctional current (EJC) is approximately 0 mV. The stochastic quantal release of transmitter from the varicosities causes spontaneous EJP (SEJP) of 2–3 mV amplitude in arterioles; 2–5 quanta generate 1 EJP. At a membrane potential of −48 mV, the EJP summate during a sympathetic activity of 25 Hz and a larger depolarization is produced [41]. The depolarization is maintained by the asynchronous discharge of all the axons innervating a given region of arteriole. The gradual depolarization, which is dependent on the frequency of sympathetic nerve activity, leads to a sustained Ca^{2+} entry through voltage-dependent Ca^{2+} channels (Fig. 96.5). Thus, the amount of vasoconstriction varies with nerve discharges in time. At first glance the frequencies required for significant increases in tension seem to be in contrast to the lower mean rate of firing in sympathetic fibers in humans (2 Hz). However, this discrepancy can be explained by taking into account the stochastic pattern of spikes. Average rates of as low as 1.8 Hz can evoke substantial contractions related to much higher frequencies during short bursts [50].

The effect of the sympathetic transmitter norepinephrine on vascular tone exemplifies the multifactorial interaction, the main components of which are the following:

- Synthesis of second messengers
- Direct action on the voltage-dependent Ca^{2+} channels.

Other excitatory neurotransmitters or agonists also have multiple effects via various receptors and GTP-binding proteins. The individual receptors may not always exhibit all the events outlined above. Depending on the receptor, one particular pathway in the individual cells or tissues might dominate. Moreover, the nature of the known and unknown G proteins which contribute to these events may also differ.

Vasoconstrictors and Excitatory Neurotransmitters. The actions of norepinephrine on Ca^{2+} channels represent a general mechanism contributing to vasoconstriction. A variety of excitatory neurotransmitters and vasoconstrictors, such as angiotensin II, serotonin, endothelin, histamine, arginine vasopressin, neuropeptide Y (NPY), and prostanoids, open voltage-dependent Ca^{2+} channels and depolarize vascular smooth muscle (Fig. 96.5).

The human *mas*/angiotensin receptor (subtype, vascular AT1 receptor), originally identified as an intronless "oncogene" located on human chromosome 6, is an unusual receptor gene, because its structure was known before its ligand (raison d'être) [4]. When synthetic *mas* mRNA was produced from a plasmid containing the *mas* coding sequence and injected into *Xenopus* oocytes, angiotensin binding to the expressed receptors revealed a depolarizing inward current with the characteristics of the Ca^{2+}-activated chloride current (see p. 1955). Additionally, angiotensin II G protein independently inhibits K_{Ca} and K_{ATP} channels, which contributes to this membrane depolarization. Ca^{2+} channel opening and G protein-mediated PLC activation are similar to serotonin (5-hydroxytryptamine, 5-HT) [55].

Binding of serotonin, a vasoconstrictor released from the α-granules of blood platelets, to its vascular 5-HT_2 receptor depolarizes vascular smooth muscle. Recent investigations demonstrated a direct and potent activator effect of serotonin on voltage-dependent Ca^{2+} channels [83]. Resistance arteries are under the tonic influence of vasoconstrictor agents such as norepinephrine and serotonin. It is of therapeutic importance that the intronless 5-HT receptor gene encodes a protein product with striking sequence and structural similarities to β_1- and β_2-adrenoceptors (see p. 1958).

Under physiological and pathophysiological states, vascular tone is also regulated by substances produced in the endothelial layer (see Sect. 96.5.3). Vascular endothelial cells synthesize and secrete the potent vasoconstrictive peptide endothelin (isotype ET-1), which, after binding to its smooth muscle ET_A receptor, starts the signal transduction chain by stimulating PLC. Endothelin likewise depolarizes vascular smooth muscle, opens voltage-dependent Ca^{2+} channels, and closes K_{ATP} channels [14,83]. Agonist-induced Ca^{2+} release from the sarcoplasmic reticulum seems to be the primary mechanism of its action. Recently, the G protein-binding receptor with seven turns belonging to the rhodopsin-type receptor superfamily has been expressed by constructs containing cDNA clones in transfected cells. The production and release of endothelin (also see Chap. 6) is stimulated by many factors including thrombin, transforming growth factor (TGFβ), tumor necrosis factor (TNFα), interleukin (IL-1), epinephrine, histamine, angiotensin II, and arginine vasopressin (AVP). The two last-named peptide hormones share the same PI second messenger system leading to activation of PKC, Ca^{2+} mobilization, and contraction of smooth muscle.

The vascular AVP receptor V1a couples to a PTX-insensitive G-protein that stimulates PLC. There are several possibilities for post-translational modification of the V1a receptor [4]. Threonine–serine residues in the third cytoplasmic loop (Fig. 96.7, C-III) and the C terminus, for example, are sites for regulatory phosphorylation, suggesting that receptor function is regulated by protein kinases (PKC).

In addition to norepinephrine and ATP, perivascular sympathetic axons in mammals also contain NPY. Since in larger arteries NPY increases vascular tension or potentiates the responses of other vasoconstrictor substances after nerve excitation via graded depolarization, it is thought to be a neuromodulator. In resistance arteries, NPY might act as an excitatory neurotransmitter in the true sense because infusion of NPY causes a rise in peripheral resistance. Structural examination of the human NPY-Y1 receptor subtype supports this assumption.

Of the three classical key signal transduction pathways, the vasoconstrictors so far discussed mainly operate via G protein-activated PLC. Another major system activates the arachidonic acid cascade after PLA_2 cleavage of membrane phospholipids. Arachidonic acid (see Chap. 6) is metabolized by the enzyme cyclooxygenase (which is inhibited by aspirin and related drugs) to produce cyclic endoperoxides during platelet aggregation at the site of arterial damage.

These peroxides are primarily converted to thromboxane A_2 (TXA_2) and also PGI_2. TXA_2 reaches the smooth muscle at the lesion and causes vasoconstriction. Binding of TXA_2 to its receptor activates PLC, and hence the signal transduction pathway already discussed, via a PTX-insensitive G-protein. Moreover, TXA_2, like norepinephrine, causes a direct, G protein-independent inhibition of K_{Ca} channels leading to depolarization and contraction [14]. After interaction with its receptor, histamine can also push the arachidonic acid cascade. The reaction proceeds via a G_o protein-mediated PLA_2 activation. Arachidonic acid is transformed into the vasoactive substances of the prostaglandins (PGE_1, PGE_2, $PGF_{2\alpha}$, PGI_2), thromboxane (TXA_2), and leukotrienes (LTC_4, LTD_4, LTE_4) by the enzymes cyclooxygenase, 5-lipoxygenase, and 12-lipoxygenase. These compounds are able to elicit vasoconstriction or vasodilatation by means of different receptors and G proteins, again via second messengers.

However, merely by binding to its specific receptor subtype, histamine can also stimulate primary effectors other than PLA_2 by different G proteins, e.g., PLC with DAG and IP_3 production. Thus, the activation of the postjunctional H_1 receptor leads to vascular smooth muscle excitation via membrane depolarization and Ca^{2+} release from the sarcoplasmic reticulum. The effect is comparable to that of norepinephrine on the α_1-receptor. The sarcolemmal H_3 receptor opens voltage-dependent Ca^{2+} channels just as norepinephrine and angiotensin II do, hence causing contraction. Ca^{2+} inward current and intracellular Ca^{2+} release induced by histamine or norepinephrine can be inhibited by Ca^{2+} antagonists such as verapamil or diltiazem [14].

Vasodilators and Inhibitory Neurotransmitters. In addition to the above, a dual effect of histamine has also been observed [14]. The increase in $[Ca^{2+}]_i$ after H_1 or H_3 receptor stimulation activates K_{Ca} channels (Fig. 96.5), resulting in a membrane hyperpolarization with vasorelaxation (Fig. 96.4). The activation of postjunctional H_2 receptors likewise causes vascular smooth muscle inhibition, mainly by depression of membrane depolarization and internal Ca^{2+} release. Finally, the presynaptic negative feedback control can lead to vasodilatation through downregulation of neurotransmitter release (norepinephrine, ATP, NPY) at the adrenergic nerve terminals. The feedback inhibition of sympathetic tone occurs via the agonist activation of specific prejunctional receptors for norepinephrine (α_2-receptor), ATP (P_2 receptor), and histamine (H_3 receptor), which initiate a decrease in EJP amplitude by PTX-sensitive G_i proteins. Synaptic nerves contain norepinephrine, ATP, and histamine as cotransmitters, which are released by neuronal activity. The presynaptic control of sympathetic tone can be impaired under pathophysiological conditions, above all by a tissue injury, when especially tissue mast cells locally release histamine. Finally, the activation of H_1 receptors on the vascular endothelium is followed by vascular smooth muscle relaxation caused by the release of EDRF (see Sect.

96.5.3). The expression of EDRF/nitric oxide (NO) synthase mRNA is additionally caused by mediators of vascular injury such as IL-1β and TNFα (see p. 1960). This cytokine-evoked production of NO may be regulated by a potent mitogen for vascular smooth muscle, insulin-like growth factor I (IGF-I), circulating in the plasma at sites of injury and impairing NO synthase activity. Thus not only histamine and cytokines, but also serotonin, endothelin, purines, and kinins may evoke membrane hyperpolarization with vasodilatation conveyed via the endothelial cells. According to the steering of the H receptor subtype, G protein, and second messenger cascade, histamine is capable of producing a contraction or relaxation of the vascular musculature.

β_2-Adrenergic activation results in membrane hyperpolarization and vasodilatation in several vascular beds, in particular in the heart, skeletal muscle, and veins. Here, the myogenic tone of resistance vessels is suppressed by low concentrations of the circulating hormone epinephrine via a β_2-mediated increase in K^+ permeability and thus membrane hyperpolarization. Since α- and β-adrenoceptors are balanced in these vessels, with a lower threshold for β-receptors, an α-mediated constriction takes over at high agonist concentrations. The α-adrenoceptors were first differentiated from the β-adrenoceptors by pharmacological criteria, which showed that the order of potency for a series of agonists was different for these two adrenergic receptor subtypes [17]. In coronary arteries the affinity of epinephrine to its β_2-adrenoceptor is higher than to the α_1-adrenoceptor; for norepinephrine the opposite is true. As outlined above, cloning of the genes for these receptor proteins has revealed a large degree of structural homology within the hydrophobic protein cores in this receptor family. Several molecular biological approaches were used to define the receptor subtype specificity (cardiac β_1-adrenoceptor; smooth muscle β_2-adrenoceptor). The subtype selectivity of the β-adrenoceptor for its endogenous agonists is based on the relative ability of the receptor to recognize the presence (epinephrine) or absence (norepinephrine) of a methyl substituent on the amine nitrogen of the ligand. This selectivity probably arises from a subtype-specific conformation of the receptor. Confirmation of this view comes from molecular cloning showing β_1- and β_2-adrenergic receptors to be products of different genes [4]. The hormonal responsiveness of specific receptor sites on the cell surface can be regulated by:

- Desensitization
- Downregulation
- Alterations of synthesis
- Degradation.

This is supplemented by means of a heterologous hormonal modification by glucocorticoids, sex steroids, and thyroid hormone [27]. The major mechanism of action is a direct change in the expression of target genes at the level of transcription. Of considerable interest to clinicians is the increase in the numbers of β_2-receptors and in agonist-induced adenylyl cyclase activity under steroids and thyroid hormones.

β-Adrenergic effects are mediated by an increase in the intracellular production of the second messenger cAMP in a reaction catalyzed by adenylyl cyclase [64]. In response to catecholamine agonist binding to the β_2-adrenergic receptor on the outer surface of the membrane, the conformational changes occurring in the membrane-spanning helix segment V are transmitted to amphiphilic helices in the C-III loop affecting G_s protein coupling (Fig. 96.5). The association between the stimulated β_2-receptor and the G_s protein leads to GTP binding to the $G_s\alpha$-subunit and subsequent dissociation of the activated $\alpha \cdot$GTP subunit from the $\beta\gamma$-dimer complex. The activated α-subunit interacts with adenylyl cyclase to stimulate the intracellular cAMP formation. This stimulation is terminated by the hydrolysis of GTP, bound to the α-subunit, to GDP. The α-subunit itself possesses the required GTPase activity. The inactive $\alpha \cdot$GDP subunit reassociates with a $\beta\gamma$-subunit to the G_s protein trimer.

An increase in intracellular cAMP activates the cAMP-dependent protein kinase (cA-PK), which in turn phosphorylates many cellular proteins. Phosphorylation of voltage-sensitive Ca^{2+} channels inhibits channel opening and suppresses Ca^{2+} entry caused by depolarization [15]. On the other hand, phosphorylation of delayed-rectifier and K_{Ca} channels stimulates K^+ channel opening. Both mechanisms cause membrane hyperpolarization with a closure of voltage-dependent Ca^{2+} channels. Ca^{2+} influx and $[Ca^{2+}]_i$ are reduced, and vasodilatation follows (Fig. 96.5). Therefore, the activation of K_{Ca} channels increases K^+ efflux, which counteracts the depolarization and constriction caused by pressure and vasoconstrictors (Fig. 96.6). This series of reactions is supported by the cA-PK-linked phosphorylation of plasmalemmal and sarcoplasmic reticulum Ca^{2+}-ATPases, which promote Ca^{2+} extrusion from the cytoplasm, thus contributing to the decrease in $[Ca^{2+}]_i$ and inhibition of vascular tone [14]. In addition, cA-PK phosphorylates MLCK, which diminishes the Ca^{2+} sensitivity of MLCK (see p. 1946). The cyclic nucleotides cAMP and cGMP also inhibit PLC and thus IP_3 and DAG formation. Therefore, activation of the β_2-adrenoceptor, primarily by its agonist epinephrine, induces a series of events leading ultimately to vasodilatation. Vasodilatation is attenuated by the therapeutic use of β-adrenoceptor blockers to lower high blood pressure. The resulting vasoconstriction is undesirable in the coronary vessels, although it is harmless because cardiac force and thus energy requirement are simultaneously diminished (negative inotropism). Based on the molecular similarities between β- and 5-HT receptors, the latter are likewise inhibited during β-blockade. Hence, the vasoconstrictive effect of serotonin is blunted.

Cholinergic presynaptic neurons of the central nervous system and perivascular peptidergic nerves of the resistance vasculature in many vascular beds contain the vasodilator neuropeptides calcitonin gene-related peptide (CGRP) and vasoactive intestinal peptide (VIP). These neurotransmitters produce their effects mainly by membrane hyperpolarization (Fig. 96.4) caused by K_{ATP} channel activation [53,54]. The hyperpolarization and a part of the vasodilatation can be abolished by the inhibitor of K_{ATP} channels, glibenclamide, or, in the case of CGRP, by the sensory neurotoxin capsaicin. CGRP- and VIP-induced hyperpolarization and vasorelaxation involve the activation of specific, high-affinity peptide receptors on the vascular smooth muscle cells associated with a stimulatory, cholera toxin-sensitive guanine nucleotide-binding protein (G_s) capable of activating K_{ATP} channels via adenylyl cyclase [47]. The increase in the intracellular cAMP concentration initiates the multifarious reactions reviewed above for the β_2-adrenoceptor. A cDNA clone (GPRN1) encoding the human VIP receptor was identified in libraries prepared from lymphoblasts; it represents the typical seven-transmembrane-segment hydropathicity profile of G protein-coupled receptors [78]. Transfection with the antisense orientation of the VIP receptor clone suppresses specific binding of VIP and the VIP-induced cAMP elevations.

To conclude this section, I would like to draw attention once again to the outstanding role of voltage-dependent Ca^{2+} channels in vascular smooth muscle. These channels can be regulated directly or indirectly by many factors:

- Membrane depolarization or hyperpolarization
- Multiple actions of various receptors
- G proteins with stimulatory or inhibitory second messengers
- Endothelial cells
- Neurotransmitters and other substances.

96.5.3 Vascular Susceptibility to Physical Factors

Sympathetic neuroeffector transmission by the release of norepinephrine causes a tonic depolarization and closure of inward-rectifier K^+ channels in arterioles. Therefore, in vivo blockade of the sympathetic nerve impulses results in some membrane hyperpolarization. Removal of the vascular endothelium, on the other hand, leads to smooth muscle depolarization of several millivolts and to vasoconstriction. The basal tone can thus be influenced by the release of vasoactive substances from the endothelium [11,12]. These effects are amplified when physical factors directly change membrane ionic conductances and secretory activities of endothelial cells [68]. For example, the release of dilative PGI_2 increases significantly with pulsatile transmural pressure or the occurrence of hypoxic conditions. This reaction is enhanced in response to an increase in blood flow with associated shear stress. However, the rise in shear forces reduces the release of the potent constrictor endothelin, thus promoting vasodilatation. Therefore, the response of vascular smooth musculature to physical effector influences is a balanced interplay between hyperpolarizing–vasodilatory (PGI_2, EDRF) and depolarizing–vasoconstrictory factors (norepinephrine, endothelin) superimposed on blood vessel basal tone.

Endothelium-Derived Relaxing Factor. Furchgott and Zawadzki [33] observed an acetylcholine-induced vasorelaxation mediated by a factor released from the endothelium (EDRF). As well as acetylcholine, the vasoactive substances norepinephrine, angiotensin II, serotonin, histamine, bradykinin, adenine nucleotides, thrombin, arachidonic acid, and leukotrienes can release EDRF after interaction with their endothelial receptor (also see Chap. 6). EDRF is a short-lived radical with a half-life counted in seconds and has been identified as NO.

NO is formed in endothelial cells from L-arginine or arginine derivatives by a nicotinamide adenine dinucleotide phosphate (NADPH)-dependent, Ca^{2+}-activated enzyme. Endothelium-derived NO directly stimulates a soluble guanylyl cyclase in smooth muscle and platelets augmenting intracellular cGMP. This leads, on the one hand, to a relaxation of vascular smooth musculature, and on the other hand to an antiaggregatory and antiadhesive activity for platelets.

Prostacyclin. Both effects are shared by the eicosanoid PGI_2, which is formed in endothelial and vascular smooth muscle cells from membrane-bound arachidonic acid by the enzyme cyclooxygenase. PGI_2 is also chemically unstable ($t_{1/2}$, approximately 3 min) following cleavage of its furan ring with subsequent formation of 6-keto-$PGF_{1\alpha}$, which is excreted in the urine. PGI_2 generation is stimulated by a number of substances including the following:

- Bradykinin
- Choline esters
- Arachidonic acid
- Thrombin
- Trypsin
- Platelet-derived growth factor
- Epidermal growth factor
- Interleukin 1
- Adenine nucleotides

PGI_2 generation is inhibited by:

- Glucocorticosteroids
- Lipocortin
- Low-density lipoproteins (LDL).

Unlike EDRF, PGI_2 displays its vasodilating and platelet-inhibiting effect by activation of adenylyl cyclase. Both substances behave concertedly and synergistically and are coreleased from the endothelial cells. This coupled release is mediated by a PLC-linked common transduction mechanism and activation of PKC. Additionally, an increase in cytosolic Ca^{2+} is required. Endothelin also causes the release of PGI_2 and EDRF, thereby restricting its own vasoconstrictor and pressor activity.

Stimulation of Adenylyl and Guanylyl Cyclase. PGI_2 and EDRF/NO relax vascular smooth muscle cells by an elevation of the cyclic nucleotides cAMP and cGMP, respec-

tively. In analogy to the effect of epinephrine on the β_2-adrenoceptor (see p. 1958), PGI_2 stimulates adenylyl cyclase after interaction with its specific membrane receptor [11]. While EDRF released to the vascular lumen is inhibited by hemoglobin blocking guanylyl cyclase, the abluminal EDRF/NO diffusion to the subendothelial muscle cell layer leads directly to the stimulation of guanylyl cyclase. The therapeutic relaxant effect of the NO-containing agents sodium nitroprusside and nitroglycerin is also mediated by this enzyme. The increase in either cAMP or cGMP activates cA-PK or cG-PK, respectively, which phosphorylate at least Ca^{2+} (inhibition) and K_{Ca} channels (activation) and the sarcolemmal Ca^{2+} pump (stimulation) [14,15]. The signal transduction chain begins with the elevation in cyclic nucleotides and is identical in all steps to the effect of β_2-receptor agonists (p. 1958), whereby membrane hyperpolarization (Fig. 96.4) with a closure of voltage-sensitive Ca^{2+} channels can be considered as the chief cause of vasodilatation. Since it was shown that NO free radicals cause a hyperpolarization of vascular smooth muscle, suggesting that they activate K^+ conductances, the demand for another EDRF which brings about vasodilatation exclusively via hyperpolarization seems to be hardly necessary (endothelium-derived hyperpolarizing factor, EDHF). The observation that cGMP by itself and via cG-PK-induced phosphorylation acts on K_{Ca} channel gating supports this view.

Oxygen Partial Pressure. A reduction in the oxygen partial pressure in the vessel increases the release of PGI_2, EDRF, and norepinephrine and simultaneously shifts the balance to the side of PGI_2 and EDRF/NO, so that rising oxygen deficiency is correlated with an increasing membrane hyperpolarization and vasodilatation [74]. The more pronounced hypoxia is, the more the dilator role is taken over by prostacyclin. Adenosine released from cardiac myocytes supports this effect in the coronary arteries. In these vessels the hypoxic dilatation is based on the activation of K_{ATP} channels, as glibenclamide completely blocks the effect [53]. In other vascular beds the hypoxia-induced hyperpolarization is partly sensitive to glibenclamide and indomethacin, an inhibitor of prostaglandin synthesis (e.g., PGI_2; also see Chap. 6). This indicates that the hyperpolarizing factors PGI_2, EDRF/EDHF, and adenosine also stimulate K_{ATP} channels as well as K_{Ca} channels. Hypoxia could activate K_{ATP} channels via an intracellular decrease in ATP, an increase in ADP, or acidification [29,53,58], although in nonmuscular tissues K_{ATP} channels are usually closed by cytosolic acidification.

A_1 Adenosine Receptor. Part of coronary artery dilatation elicited by adenosine is due to K_{ATP} channels activated via a PTX-sensitive G protein after binding of adenosine to the A_1 receptor. The A_1 adenosine receptor belongs to the purinergic receptor family, which is coupled to different effectors in various tissues via multiple G proteins [48]. Through inhibition of phospholipases, adenosine also restricts norepinephrine release from sympathetic nerve terminals by a fall in intraneuronal calcium and in PKC activity, thus supporting vasodilatation. On the other hand, in renal preglomerular arterioles the vasoconstrictor action of angiotensin II is potentiated by adenosine.

Vasodilatation in Septic Shock. K_{ATP} channels also play an important role in reactive hyperemia, endotoxic hypotension, and septic shock. Vasodilatation in septic shock is maintained by the monocytic formation of TNFα, a polypeptide cytokine with potent inflammatory activity (see Chap. 6). TNFα leads to an augmented EDRF/NO production by the induction of NO synthase in endothelial cells [31]. Activation of the TNFα receptor involves the stimulation of a plasma membrane-bound neutral sphingomyelinase generating the second messenger ceramide [30]. The ceramide-activated protein kinase in turn is able to phosphorylate certain proteins.

K⁺ Channel Openers. While K_{Ca} channels regulate the intrinsic tone in resistance arteries, K_{ATP} channels link smooth muscle metabolism and blood flow [55]. Not only a variety of endogenous vasodilators act through a K_{ATP} channel-induced membrane hyperpolarization, but also several synthetic substances in the treatment of conditions such as hypertension, asthma, or peripheral vascular disease. Cromakalim, pinacidil, diazoxide, and nicorandil belong to this new class of K⁺ channel openers. They can hyperpolarize the membrane towards the K⁺ equilibrium potential, which inactivates the excitation-induced Ca^{2+} influx via voltage-dependent Ca^{2+} channels. $[Ca^{2+}]_i$ will fall and a lower level of tone be attained.

Pressure and Shear Stress. Driven by the cardiac muscle, the blood flow through the arterial tree of conduit and resistance arteries results in both pressure and shear stress, which influence the vascular wall. Shear stress is a tangential force that is generated by friction between the blood and the endothelial cell membrane [3]. While pressure effects membrane depolarization (see p. 1951), the response an endothelial cell makes to shear stress is generally the opposite one, i.e., hyperpolarization with activation of the inward-rectifier K⁺ channel. Experimentally, intraluminal saline infusion may cause either a decrease or an increase in the active wall tone of resistance arteries and of small veins. When wall tone is high, the predominant vascular response is dilatation or a decrease in tension. When wall tone is low, the response is contraction or an increase in tension. Consequently, intraluminal flow does not change wall force at an intermediate tension level. Thus, flow tends to shift active wall tone towards this intermediate set point [20]. The balance point is the result of the interaction of these dilator and constrictor responses to flow modified by a variety of other influences. Electrophysiologically, the set point of the vascular smooth muscle cells was determined to be approximately −52 mV. With a more positive membrane potential, the voltage response to shear stress was hyperpolarization, and with a more negative membrane potential, depolarization. It has been shown experimentally that the membrane potential varies between −53 mV and −23 mV when the blood pressure is altered within the range of 40–120 mm Hg [39]. Since in arterial blood vessels with normal blood pressure values of the membrane potential of about −48 mV or more positive were found, flow dilatation is the commonly observed response [8].

Flow-dependent dilatation is reduced after removal of the endothelium. As already mentioned, shear stress leads to an increase in K⁺ conductance of the endothelial cells and thus to their hyperpolarization. Hyperpolarization raises the inward driving force ($V - E_{Ca}$) for Ca^{2+} ions, which pass the endothelial cell membrane through a nonselective cation channel and stimulate NO synthase [69]. EDRF/NO and coreleased PGI_2 diffuse to the subjacent vascular smooth muscle cells and elicit here the cascade of cG-PK- and cA-PK-dependent reactions, including membrane hyperpolarization of the muscle cells. The resulting flow dilatation can be so powerful that it reverses the wall force increase caused by stretch, norepinephrine, histamine, serotonin, $PGF_{2\alpha}$, or AVP.

Flow Sensor Function. As yet the mechanochemical and mechanoelectrical signal transduction pathway which causes hyperpolarization of the endothelial cell membrane in reply to shear stress has not been determined. Shear stress results from the surface drag of the flowing blood on the intimal surface. At this strategic point, polyanionic macromolecules such as proteoglycans and acidic glycoproteins of the endothelial cell membrane, in the glycocalyx, and/or the extracellular matrix seem to assume a flow sensor function (see Sect. 9.2.6) [70,76]. Based upon their viscoelastic properties, flow can induce a "mechanical distortion" in the form of a conformational change of these specific surface molecules, which is again reversed with a decrease in flow. Support comes from the observation that flow dilatation occurs only above a distinct mechanical threshold of shear stress (τ, 1.3 N/m²) [43]. Therefore, it requires a certain energy to accomplish the conformational change of the sensor molecules. Moreover, they are polyanions as a result of their sulfated, carboxylated, or sialic acid-containing residues. These flow sensor molecules are thus endowed with a high and also specific cation-binding capacity (Na⁺, Ca^{2+}), which is apparently involved in the electrical transformation of the signal [70]. Finally, it should be emphasized that these biopolyelectrolytes are multiply anchored in basement membranes, extracellular matrix, and cytoskeleton (see Sect. 9.2.6).

Coordination of Responses. In view of the diverse effector influences present in the complex scenario of vascular regulation, the central role of the membrane potential of the vascular smooth muscle cells can be prima facie regarded as the basic mechanism of integration or coordination of the responses. The level of the membrane potential in vivo is greatly influenced by intramural pressure and the rate of intraluminal flow. Pressure and flow, which cooperate in an integrative fashion in the myogenic response, might be regarded as the final vascular effectors ensuring circulatory homeostasis. During physiological adjustments, the influence exerted by neurotransmitters, endothelial-derived factors, and other locally produced and systemically circulating substances may enhance or diminish the myogenic response. These effector influences will again be modified by the changes in flow that they cause. Thus the primary regulating effect is caused by flow.

96.6 Vascular Rhythmogenesis and Autorhythmicity

Vasomotion/Slow Wave Activity. Rhythms and oscillations are an expression of the temporal organization of living systems. We are familiar with countless rhythmic phenomena in the living world: yearly and monthly rhythms, day-and-night rhythms, respiratory and circulatory rhythms, and the rhythms of rapid-firing neurons, all of which man is subject to. Vascular smooth muscle, like cardiac muscle, originates ontogenetically from the vascular muscle tube. Thus, it is not surprising that in principle all vascular smooth muscle cells are capable of rhythmicity, which manifests itself as periodic alterations of membrane potential and tone. In the microcirculation, these events are referred to as "vasomotion," whereas the term "slow wave activity" [5] is used for small and conduit arteries, veins, and lymphatic vessels [9]. Both expressions designate the same phenomenon.

We can differentiate between two types of genesis for mechanical oscillations. In vascular smooth muscle, mechanical waves have been described in terms of electrical oscillations of the membrane potential (graded de- and hyperpolarization). On the other hand, it was observed in a few vessels that the depolarized part of a more or less distinct slow wave with superimposed action potentials can be paralleled by the phases of an increase in tension [5]. These action potentials are preceded (genuine pacemaker cells) or not (transmitted excitation) by a depolarizing prepotential resembling that found in sinoatrial node cells. This classification involves different ionic bases for slow wave and action potential activity. While the wave potentials, which may not necessarily be slow, are probably due to a periodically changing cellular metabolism [72], voltage- and time-dependent ionic channels (see Chap. 12) are responsible for the action potential activity.

Oscillatory Diameter Changes. In vivo microscopic observations of arterioles practically always reveal oscillatory changes in diameter. This vasomotion is responsible for the periodic variance in spatial blood flow and flow velocity. The absence of temporal fluctuations is regarded as a pathophysiological state lacking autoregulation. Therefore, the rhythmogenic properties of vascular smooth muscle are closely coupled to a functioning circulation. The oscillatory activity is nonrandom, but only seldom exactly periodic. The spectral analysis of frequencies often shows two or three relatively broad peaks, which are frequently coordinated as integral multiples to a basic frequency [72]. At the macroscopic level, these observations are equivalent to the electrical and mechanical oscillations (slow waves) in small and more rarely in conduit arteries and in veins (e.g., pial arteries, portal vein). An exact comparison of the wave crests and troughs between voltage and tone suggests that periodic changes in the membrane potential precede the changes in force by 40 ms. This confirms the existence of a tight electromechanical coupling even under these nonstationary conditions [50,73]. Oscillations in membrane potential of up to $\pm 5\,mV$ allow the open probability of voltage-sensitive Ca^{2+} channels to alternate rhythmically. Thus vascular tone also varies periodically [32]. However, the finding that the oscillations can be inhibited by K^+ and Ca^{2+} channel blockers does not permit us to conclude that the origin of the oscillation lies in the interplay between potential-dependent K^+ and Ca^{2+} channels.

96.6.1 Rhythmogenic Properties of Vascular Smooth Muscle

Temporal fluctuations in microvascular voltage, diameter, and flow velocity exhibit characteristics of a low-order, though highly complex, nonlinear dynamic system [3]. The system may be dominated by only a few control points:

- Allosteric-cooperative enzymes
- Smooth muscle membrane potential
- Ca^{2+} levels
- Contraction
- Concentrations of extracellular regulatory substances.

The number of dominating processes is not a specific integer. In such a complex system there are also secondary control mechanisms. As early as the 1950s, Edith Bülbring [26] discovered the dependence of the electromechanical properties of intestinal smooth muscle on the glycolytic metabolism. It has already been known for some years now that glycolysis, cell respiration, and cellular transport processes can be oscillatory. The cause of the oscillations is the periodically varying effectiveness of regulatory enzymes and transport proteins. Oscillation is observed when enzymes show nonlinear kinetics, when the biochemical or biophysical systems are simply or multiply feedback coupled, and when they are far from equilibrium and thermodynamically open [72].

Phosphofructokinase. A great number of feedback-coupled processes on the cellular and subcellular level might play a causal, a dominating, or only a mediating role in the oscillatory phenomena. We are already familiar with two of these rather secondary processes, namely the interaction between potential-dependent K^+ and Ca^{2+} channels (see p. 1951) and between pressure-induced vasoconstriction and flow-induced vasodilatation (see p. 1960). In the following a model will be outlined that incorporates the observations made by Bülbring [26]. Since measurements of enzyme activity have shown a high glycolytic capacity of vascular smooth musculature and since monoiodoacetate (IAA) blocked the electromechanical oscillations completely, phosphofructokinase (PFK) might be a glycolytic entry enzyme which controls the energy-supplying metabolism and has multiple feedback [73]. In the upper part of the glycolytic pathway, the allosteric enzyme PFK converts fructose-6-phosphate (F6P) and ATP to fructose-1,6-bisphosphate (FBP) and adenosine diphosphate (ADP), respectively. The main characteristics of PFK, such as co-operative substrate saturation and activation or inhibition by allosteric effectors can account for oscillatory changes in glycolytic flux and ATP concentration. However, spontaneous glycolytic oscillations appear only within the oscillatory domain, i.e., at definite F6P input/FBP output rates. The PFK input and output rates can be modified by the concentration of many extracellular regulatory substances, e.g., catecholamines, thyroid hormones, endothelins, histamine, kinins, and thrombin. These mediators frequently change arterial pressure, thus stimulating another feedback cycle, namely between pressure-induced vasoconstriction and flow-induced vasodilatation. Finally, allosteric activators (e.g., FBP, Fru-2,6-P_2) and inhibitors (e.g., ATP) directly influence the regulatory properties of PFK.

Adenosine Triphosphate Concentration Changes. A periodically changing ATP concentration is converted into the operation of the electrogenic ion pump (see p. 1952) by the (Na^++K^+)-ATPase. This enzyme is also an oscillator [62]. Its reaction sequence (see Chap. 8) consists in a single cycle that includes at least two reactive states, the Na^+ (E_1) and K^+ (E_2) conformations [45]. (Na^++K^+)-ATPase transports three Na^+ ions outward per two K^+ inward per cycle and thus generates a net movement of one positive charge per cycle, a small electric current. When the electrogenic pump acts rhythmically, the membrane becomes hyperpolarized in phases of increased activity and depolarized in phases of decreased activity. The electrogenic pump current can generate the observed potential waves of up to ±5 mV. The oscillation in membrane potential is transformed into rhythmic tension changes via the voltage-sensitive Ca^{2+} channels (see [32]). It is the export of Na^+ by the (Na^++K^+)-ATPase cycle that moves the charge. Indeed, in vessels with slow wave activity, distinct oscillations were detected in the $^{24}Na^+$ efflux [72]. Due to the electrogenicity of the pump, the E_1/E_2 conformational transition is also voltage dependent. The E_1 conformation is stabilized by a more positive membrane potential, and the E_2 conformation by a more negative membrane potential [62]. This means that the orientation of a dipole in the enzyme is dependent on the transmembrane field gradient.

96.6.2 Synchronization, Conduction, and Functional Significance of Oscillatory Activity

Synchronization. The detection of electromechanical slow waves and vasomotion presupposes synchronization in cellular and subcellular events. Even if the PFK and (Na^++K^+)-ATPase molecules all oscillate at the same frequency, their mutual phase relationship has to be synchronous to be able to collect a uniform signal. Frequency entrainment and phase synchronization ensue by varying a force, such as pressure or an electric field [62]. A prerequisite is that both the conformational states of the enzyme differ in a displacement such as volume or dipole moment. Even the secondary structure of the oscillator molecules in question can have a dipole moment.

Intercellular Communication. Dealing with a population of closely interconnected cells, each one of which is capable of sustained oscillations in membrane potential, the next question to ask is whether synchronization among the single elements takes place and what role it plays in intercellular communication. The potential waves spread from one cell to the next via the gap junctions (see p. 1942). If the vascular smooth muscle cells contract synchronously, the blood vessel experiences periodic variations in diameter. Slow waves or vasomotion are created from a peripheral, functioning vascular system.

What is the significance of this vasomotion? Is it an epiphenomenon or a crucial point in coordination? The significance of the oscillatory events in the cardiovascular system is found in the spatial and temporal synchronization of the cells, in metabolic energy saving, and in the constant readiness of the vessels to adapt to changing circulation requirements. The precise control of the resistance vessels to permit minute changes in flow requires contraction in concert with adjacent cells. An essential property of vascular smooth muscle cells is a marked ability to cooperate. Intercellular communication is reflected by the coordinated, periodic contraction of arterioles, a central aspect of their behavior. Cell–cell coupling is the crucial point in this coordination and perhaps in the adjustment of vascular resistance. Moreover, periodic vasomotion in terminal arterioles is decrementally conducted upstream to larger vessels, which integrate input from many branches [3].

This intervention of vascular rhythmicity in the circulatory regulation through synchronization and conduction is supplemented by the phase dependence of effector actions, e.g., of an autonomic nervous impulse. The result of sympathetic or any effector interference depends on the phase of the spontaneous contraction–dilatation cycle. This rhythmicity guarantees a particularly fast vascular reaction within a few seconds [73] to satisfy both general and local demands in most situations, predominantly in states of emergency. Adaptation to changed circulatory requirements can be performed more rapidly from an oscillating than from a stationary system. Moreover, it has advantages in terms of metabolic energy cost. From the clinical point of view, a changed vasomotion might present an additional pathogenetic factor, e.g., in diabetic microangiopathy and arterial hypertension. Large and small flow motion waves were found in patients with arterial occlusive disease. The prevalence of small waves increases with more severe ischemia. Thus, the analysis of oscillatory activity may gain diagnostic and prognostic significance.

Time Domain. In conclusion, it should be emphasized that under normal conditions the various vascular–regulatory effector influences are interwoven in a dynamic, and not a static circulatory system. The reaction of a smooth muscle cell is thus reflected only incompletely by the stationary activation curve "developed tension versus membrane potential" (Fig. 96.4). The missing time domain is a reflection of our as yet limited understanding of the system's behavior in space and time.

General References

1. Bernard C (1878) Leçons sur les phénomènes de la Vie Commune aux animaux et aux végétaux. Baillière, Paris
2. Betz E, Reutter K, Mecke D, Ritter H (1991) Biologie des Menschen. Quelle and Meyer, Heidelberg
3. Bevan JA, Halpern W, Mulvany MJ (eds) (1991) The resistance vasculature. Humana, Totowa
4. Brann MR (ed) (1992) Molecular biology of G-protein-coupled receptors. Birkhäuser, Boston
5. Bülbring E, Brading AF, Jones AW, Tomita T (eds) (1970) Smooth muscle. Edward Arnold, London
6. Campbell PN, Marshall RD (eds) (1985) Essays in biochemistry, vol 20. Academic, London
7. Crass MF III, Barnes CD (eds) (1982) Vascular smooth muscle: metabolic, ionic, and contractile mechanisms. Academic, New York
8. Keatinge WR, Harman MC (1980) Local mechanisms controlling blood vessels. Academic, London
9. Levick JR (1991) An introduction to cardiovascular physiology. Butterworths, London
10. Milnor WR (1990) Cardiovascular physiology. Oxford University Press, New York
11. Rubanyi GM, Vanhoutte PM (eds) (1990) Endothelium-derived relaxing factors. Karger, Basel
12. Rubanyi GM, Vanhoutte PM (eds) (1990) Endothelium-derived contracting factors. Karger, Basel
13. Rüegg JC (1986) Calcium in muscle activation. A comparative approach. Springer, Berlin Heidelberg New York
14. Sperelakis N, Kuriyama H (eds) (1991) Ion channels of vascular smooth muscle cells and endothelial cells. Elsevier, New York
15. Sperelakis N, Wood JD (eds) (1990) Frontiers in smooth muscle research. Wiley-Liss, New York
16. Stephens NL (ed) (1977) The biochemistry of smooth muscle. University Park Press, Baltimore

Specific References

17. Ahlquist RP (1948) A study of the adrenotropic receptors. Am J Physiol 153:586–600
18. Babij P, Periasamy M (1989) Myosin heavy chain isoform diversity in smooth muscle is produced by differential RNA processing. J Mol Biol 210:673–679
19. Bean BP (1989) Classes of calcium channels in vertebrate cells. Annu Rev Physiol 51:367–384
20. Bevan JA (1995) The role of flow-induced contraction and relaxation in the regulation of vascular tone. Results of in vitro studies. In: Bevan JA, Kaley G, Rubanyi G (eds) Flow-dependent regulation of vascular function. Oxford University Press, New York, pp 128–152
21. Bevan JA, Laher I (1991) Pressure and flow-dependent vascular tone. FASEB J 5:2267–2273
22. Boels PJ, Troschka M, Rüegg JC, Pfitzer G (1991) Higher Ca^{2+} sensitivity of triton-skinned guinea pig mesenteric microarteries as compared with large arteries. Circ Res 69:989–996
23. Brayden JE, Nelson MT (1992) Regulation of arterial tone by activation of calcium-dependent potassium channels. Science 256:532–535
24. Breemen C van, Saida K (1989) Cellular mechanisms regulating $[Ca^{2+}]_i$ smooth muscle. Annu Rev Physiol 51:315–329
25. Brown AM (1991) A cellular logic for G protein-coupled ion channel pathways. FASEB J 5:2175–2179
26. Bülbring Edith, Lüllmann H (1957) The effect of metabolic inhibitors on the electrical and mechanical activity of the smooth muscle of the guinea-pig's taenia coli. J Physiol (Lond) 136:310–323
27. Collins S, Bolanowski MA, Caron MG, Lefkowitz RJ (1989) Genetic regulation of β-adrenergic receptors. Annu Rev Physiol 51:203–215
28. Cross RA, Engel A (1991) Scanning transmission electron microscopic mass determination of in vitro self-assembled smooth muscle myosin filaments. J Mol Biol 222:455–458
29. Davies NW (1990) Modulation of ATP-sensitive K^+ channels in skeletal muscle by intracellular protons. Nature 343:375–377
30. Dressler KA, Mathias S, Kolesnick RN (1992) Tumor necrosis factor-α activates the sphingomyelin signal transduction pathway in a cell-free system. Science 255:1715–1718
31. Eigler A, Sinha B, Endres S (1993) Nitric oxide-releasing agents enhance cytokine-induced tumor necrosis factor synthesis in human mononuclear cells. Biochem Biophys Res Commun 196:494–501
32. Fewtrell C (1993) Ca^{2+} oscillations in non-excitable cells. Annu Rev Physiol 55:427–454
33. Furchgott RF, Zawadzki JV (1980) The obligatory role of endothelial cells in the relaxation of arterial smooth muscle by acetylcholine. Nature 288:373–376
34. Gilbert EK, Weaver BA, Rembold CM (1991) Depolarization decreases the $[Ca^{2+}]_i$ sensitivity of myosin light-chain kinase in arterial smooth muscle: comparison of aequorin and fura 2 $[Ca^{2+}]_i$ estimates. FASEB J 5:2593–2599
35. Goldman DE (1943) Potential, impedance, and rectification in membranes. J Gen Physiol 27:37–60
36. Gollasch M, Hescheler J, Quayle JM, Patlak JB, Nelson MT (1992) Single calcium channel currents of arterial smooth muscle at physiological calcium concentrations. Am J Physiol 263:C948–C952
37. Hai C-M, Murphy RA (1989) Ca^{2+}, crossbridge phosphorylation, and contraction. Annu Rev Physiol 51:285–298
38. Harder DR (1984) Pressure-dependent membrane depolarization in cat middle cerebral artery. Circ Res 55:197–202
39. Harder DR, Gilbert R, Lombard JH (1987) Vascular muscle cell depolarization and activation in renal arteries on elevation of transmural pressure. Am J Physiol 253:F778–F781
40. Hescheler J, Schultz G (1993) G-proteins involved in the calcium channel signalling system. Curr Opin Neurobiol 3:360–367
41. Hirst GDS, Edwards FR (1989) Sympathetic neuroeffector transmission in arteries and arterioles. Physiol Rev 69:546–604
42. Kamm KE, Stull JT (1989) Regulation of smooth muscle contractile elements by second messengers. Annu Rev Physiol 51:299–313
43. Khayutin VM, Nikolsky VP, Rogoza AN, Lukoshkova EV (1993) Endothelium determines stabilization of the pressure drop in arteries. Acta Physiol Scand 148:295–304
44. Kirber MT, Walsh JV Jr, Singer JJ (1988) Stretch-activated ion channels in smooth muscle: a mechanism for the initiation of stretch-induced contraction. Pflugers Arch 412:339–345
45. Läuger P, Apell H-J (1988) Transient behaviour of the Na^+/K^+-pump: microscopic analysis of nonstationary ion-translocation. Biochim Biophys Acta 944:451–464
46. Lefkowitz RJ, Caron MG (1988) Adrenergic receptors. J Biol Chem 263:4993–4996
47. Lin HY, Harris TL, Flannery MS, Aruffo A, Kaji EH, Gorn A, Kolakowski LF Jr, Lodish HF, Goldring SR (1991) Expression

cloning of an adenylate cyclase-coupled calcitonin receptor. Science 254:1022–1024

48. Linden J (1991) Structure and function of A_1 adenosine receptors. FASEB J 5:2668–2676
49. Loirand G, Pacaud P, Mironneau C, Mironneau J (1990) GTP-binding proteins mediate noradrenaline effects on calcium and chloride currents in rat portal vein myocytes. J Physiol (Lond) 428:517–529
50. Mulvany MJ, Aalkjær C (1990) Structure and function of small arteries. Physiol Rev 70:921–961
51. Mulvany MJ, Nilsson H, Flatman JA (1982) Role of membrane potential in the response of rat small mesenteric arteries to exogenous noradrenaline stimulation. J Physiol (Lond) 332:363–373
52. Murphy RA (1989) Special topic: contraction in smooth muscle cells. Annu Rev Physiol 51:275–283
53. Nelson MT (1993) Ca^{2+}-activated potassium channels and ATP-sensitive potassium channels as modulators of vascular tone. Trends Cardiovasc Med 3:54–60
54. Nelson MT, Huang Y, Brayden JE, Hescheler J, Standen NB (1990) Arterial dilations in response to calcitonin gene-related peptide involve activation of K^+ channels. Nature 344:770–773
55. Nelson MT, Patlak JB, Worley JF, Standen NB (1990) Calcium channels, potassium channels, and voltage dependence of arterial smooth muscle tone. Am J Physiol 259:C3–C18
56. Nelson MT, Standen NB, Brayden JE, Worley JF III (1988) Noradrenaline contracts arteries by activating voltage-dependent calcium channels. Nature 336:382–385
57. Nernst W (1889) Die elektromotorische Wirksamkeit der Ionen. Z Phys Chem Frankfurt 4:129–181
58. Noma A (1983) ATP-regulated K^+ channels in cardiac muscle. Nature 305:147–148
59. Nyborg NCB, Mulvany MJ (1984) Effect of felodipine, a new dihydropyridine vasodilator, on contractile responses to potassium, noradrenaline, and calcium in mesenteric resistance vessels of the rat. J Cardiovasc Pharmacol 6:499–505
60. Oguchi A, Ikeda U, Kanbe T, Tsuruya Y, Yamamoto K, Kawakami K, Medford RM, Shimada K (1993) Regulation of Na-K-ATPase gene expression by aldosterone in vascular smooth muscle cells. Am J Physiol 265:H1167–H1172
61. Paul RJ (1989) Smooth muscle energetics. Annu Rev Physiol 51:331–349
62. Post RL (1989) Seeds of sodium, potassium ATPase. Annu Rev Physiol 51:1–15
63. Regan JW, Cotecchia S (1992) The α-adrenergic receptors: new subtypes, pharmacology, and coupling mechanisms. In: Brann MR (ed) Molecular biology of G-protein-coupled receptors. Birkhäuser, Boston, pp 76–112
64. Robishaw JD, Foster KA (1989) Role of G proteins in the regulation of the cardiovascular system. Annu Rev Physiol 51:229–244
65. Rüegg U (1987) Kalziumtransport durch Membranen. Sandorama 2:15–20
66. Ruzycky AL, Morgan KG (1989) Involvement of the protein kinase C system in calcium-force relationships in ferret aorta. Br J Pharmacol 97:391–400
67. Schatzmann HJ (1989) The calcium pump of the surface membrane and of the sarcoplasmic reticulum. Annu Rev Physiol 51:473–485

68. Schwarz G, Callewaert G, Droogmans G, Nilius B (1992) Shear stress-induced calcium transients in endothelial cells from human umbilical cord veins. J Physiol (Lond) 458:527–538
69. Schwarz G, Droogmans G, Nilius B (1992) Shear stress-induced membrane currents and calcium transients in human vascular endothelial cells. Pflugers Arch 421:394–396
70. Siegel G, Bevan JA (1995) Anionic biopolyelectrolytes as sensors of blood flow. In: Bevan JA, Kaley G, Rubanyi G (eds) Flow-dependent regulation of vascular function. Oxford University Press, New York, pp 153–162
71. Siegel G, Carl A, Adler A, Stock G (1989) Effect of the prostacyclin analogue iloprost on K^+ permeability in the smooth muscle cells of the canine carotid artery. Eicosanoids 2:213–222
72. Siegel G, Ebeling BJ, Hofer HW (1980) Foundations of vascular rhythm. Ber Bunsenges Phys Chem 84:403–406
73. Siegel G, Hofer HW, Walter A, Rückborn K, Schnalke F, Koepchen HP (1991) Autorhythmicity in blood vessels: its biophysical and biochemical bases. Springer Series Synerget 55:35–60
74. Siegel G, Schnalke F, Schaarschmidt J, Müller J, Hetzer R (1991) Hypoxia and vascular muscle tone in normal and arteriosclerotic human coronary arteries. J Vasc Med Biol 3:140–149
75. Siegel G, Walter A, Bostanjoglo M, Jans AWH, Kinne R, Piculell L, Lindman B (1989) Ion transport and cation-polyanion interactions in vascular biomembranes. J Membr Sci 41:353–375
76. Siegel G, Walter A, Rückborn K, Buddecke E, Schmidt A, Gustavsson H, Lindman B (1991) NMR studies of cation induced conformational changes in anionic biopolymers at the endothelium-blood interface. Polymer J 23:697–708
77. Sobieszek A (1991) Regulation of smooth muscle myosin light chain kinase. Allosteric effects and co-operative activation by calmodulin. J Mol Biol 220:947–957
78. Sreedharan SP, Robichon A, Peterson KE, Goetzl EJ (1991) Cloning and expression of the human vasoactive intestinal peptide receptor. Proc Natl Acad Sci USA 88:4986–4990
79. Trayer IP (1994) Hands across the divide. Nature 368:294–295
80. Ueyama H, Hamada H, Battula N, Kakunaga T (1984) Structure of a human smooth muscle actin gene (aortic type) with a unique intron site. Mol Cell Biol 4:1073–1078
81. Vorherr T, Kessler T, Hofmann F, Carafoli E (1991) The calmodulin-binding domain mediates the self-association of the plasma membrane Ca^{2+} pump. J Biol Chem 266:22–27
82. Warrick HM, Spudich JA (1987) Myosin structure and function in cell motility. Annu Rev Cell Biol 3:379–421
83. Worley JF, Quayle JM, Standen NB, Nelson MT (1991) Regulation of single calcium channels in cerebral arteries by voltage, serotonin, and dihydropyridines. Am J Physiol 261:H1951–H1960
84. Xie X, Harrison DH, Schlichting I, Sweet RM, Kalabokis VN, Szent-Györgyi AG, Cohen C (1994) Structure of the regulatory domain of scallop myosin at 2.8 Å resolution. Nature 368:306–312
85. Zelcer E, Sperelakis N (1982) Spontaneous electrical activity in pressurized small mesenteric arteries. Blood Vessels 19:301–310

97 Microcirculation and Capillary Exchange

E.M. Renkin and C. Crone

Contents

97.1 Introduction

Exchange of materials between the circulating blood and the organs and tissues of the body takes place in the capillaries and postcapillary venules. These are often designated *exchange vessels*. In these vessels:

- Blood velocity is slowest (<1 mm/s)
- Surface area per unit mass of tissue is highest (>50 cm^2/g)
- Blood-to-cell distances are generally very short (<50 µm)

All these factors favor rapid exchange. Figure 97.1 is a schematic diagram of a "typical" microvascular network, including microlymphatics.

Configurations of microvascular networks vary according to the structures of specific organs and tissues. In skeletal muscles, the exchange vessels tend to run parallel to the muscle fibers, in the interstices between them. In glandular organs, acini and tubules are surrounded by basket-like networks. In the lungs, the capillaries are sandwiched between the walls of adjacent alveoli. Exchange vessels are distributed more densely in organs and tissues of high metabolic rate (heart, brain) or specialized transport function (lungs, kidneys) [65,66].

Control of Blood Supply to Capillaries. Blood supply to the capillaries in most organs (e.g., skeletal muscles) is controlled by the terminal arterioles. In some organs (e.g., skin) blood supply to individual capillaries is controlled by precapillary sphincters, single smooth muscle cells surrounding the capillary entrance. When metabolism is low ("resting state") smooth muscle cells of terminal arterioles and precapillary sphincters tend to be contracted, and exchange vessel blood supply is low and unevenly distributed. A small fraction of the exchange vessels receives most of the blood; in the rest blood flow is slow or even stopped. In many microvascular networks flow is oscillatory or intermittent; vessels poorly perfused at one moment may be well perfused at another. In other networks, the resting flow pattern is fixed, and the well-perfused pathways are characterized both anatomically and physiologically as "preferential channels" or "metarterioles." These start out as terminal arterioles in which the smooth muscle coat becomes incomplete, then more and more sparse, and finally it disappears entirely as the vessel assumes typical exchange vessel morphology. Such vessels have often been considered a form of arteriovenous shunt, bypassing the "true capillaries" which are held in reserve. However, it seems more likely that metarterioles function as exchange vessels and are the principal support of resting metabolism. When the metabolic rate rises, relaxation of terminal arterioles and precapillary sphincters increases the number of exchange vessels well supplied with blood by "recruitment" of poorly perfused or unperfused vessels, augmenting effective exchange vessel surface area and reducing exchange vessel-to-cell diffusion distance. At the same time, dilation of larger preterminal arterioles increases the total blood supply [47,68].

Total exchange vessel blood flow depends mainly on arteriolar resistance, because the capillaries and venules contribute only a small fraction of total vascular resistance. Blood pressure in systemic capillaries is normally around 15–30 mmHg, in postcapillary, nonmuscular venules 10–15 mmHg. Mean systemic exchange vessel pressure is thus about 15 mmHg. Mean pressure in pulmonary capillaries is considerably lower, 5–6 mmHg. Exchange vessel pressures are controlled by the ratio of upstream resistance (in arterioles) to downstream resistance (in muscular venules). Arteriolar vasodilation tends to increase capillary pressure, arteriolar constriction lowers it. Venular dilation and constriction have the opposite effects (see Sect. 97.3.2 below, Eq. 97.6).

R. Greger/U. Windhorst (Eds.)
Comprehensive Human Physiology, Vol. 2
© Springer-Verlag Berlin Heidelberg 1996

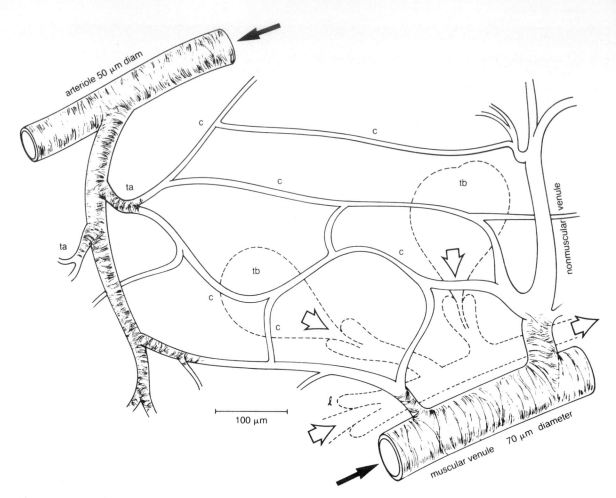

Fig. 97.1. A generalized microvascular network: *ta*, terminal arteriole; *c*, capillary; *l*, conducting lymphatic; *tb*, terminal lymphatic bulb. *Arrows* show direction of blood and lymph flow. (From [50], by permission)

Arteriovenous Shunts. In most organs and tissues, all the blood passes through exchange vessels. Arteriovenous shunts, specialized thick-walled muscular vessels connecting arterioles directly to muscular venules and bypassing local exchange vessels, are found mainly in the skin, particularly in hairless, exposed regions (fingers and toes; face, ears, and nose). Their function is related to control of skin temperature and heat loss (cf. Chap. 110). Heat diffuses much more rapidly than water and solutes, and is exchanged across the walls of larger, muscular vessels.

97.2 Exchange Vessel Structure and Permeability

The walls of capillaries and postcapillary venules consist of a single layer of thin endothelial cells supported on their collagenous basement membrane (basal lamina). Permeation of water and solutes may take place through the cells, around the cells, or through the junctional spaces between them [52].

- Water and small lipophilic substances (O_2, N_2, CO_2) penetrate directly through the cells (lipid membranes and aqueous cytoplasm) [44].
- Large lipophilic solutes (unesterified fatty acids, lipid-soluble vitamins) are thought to enter the cell membrane and to diffuse around the cell, in the membrane lipid phase, along the sides of the junctions [54].
- Cations (Na^+, K^+), anions (Cl^-, HCO_3^-), and small hydrophilic (lipophobic) solutes which do not penetrate cell membranes in the absence of specific carriers (glucose, amino acids) pass through the junctional spaces [15]. Ultrafiltration of fluid consisting of water and small ions and solutes also takes place through the junctions. These openings are often referred to as endothelial "pores" or inter-endothelial cell "slits" [41].
- Most junctions between endothelial cells are too narrow (8–10 nm) to allow plasma proteins to penetrate. Near-impermeability to plasma proteins is an important factor in control of transcapillary fluid exchange (see Sect. 97.3.2 below). However, a small proportion of wide junctions ("large pores" or "leaks" 40–60 nm wide) allow a small amount of plasma protein to escape [23,26].

1966

- Plasma proteins may also be transported from plasma to interstitial fluid by endothelial cell vesicles shuttling back and forth between luminal and abluminal endothelial cell surfaces ("transcytosis") [23,60]. Extravasated fluid and protein pass through the interstitial compartment, enter the open-junction lymphatic capillaries (terminal lymphatics), and are eventually returned to the bloodstream by the larger lymphatic vessels [63].
- The basal lamina offers little resistance to penetration of water and small solutes, and is less of an obstruction to the movement of plasma proteins than the narrow junctions. It appears to offer some hindrance to escape of plasma proteins through large pores or vesicles.

97.2.1 Varieties of Endothelium

Structural features and permeability characteristics may differ in different organs and at different sites in the vasculature of an organ [4]. There are four principal endothelial types:

- *Continuous endothelium*, the most abundant and widespread type, lines the chambers of the heart, the walls of all arteries and veins, arterioles and muscular venules as well as capillaries and postcapillary (nonmuscular) venules in skeletal, smooth and cardiac muscle, skin, lungs, and connective tissues. The cells are less than $0.2\,\mu m$ thick except in the region of their nuclei, which bulge into the vessel lumen. They comprise more than 99.98% of endothelial surface area, the junctional interspaces less than 0.02%. Luminal (apical) and abluminal (basolateral) caveolae are abundant, expanding cell surface on both sides and providing entry to cytoplasmic vesicles 60–70 nm in diameter, which are believed to have a role in transport of large molecules. Intercellular junctions, though histologically classified as "tight," are sealed by *interrupted* bands, with fluid-filled pathways for intercellular transit of water and small solutes between them [8]. The luminal surfaces of the cells are covered by a surface coat or "glycocalyx" $0.1–0.5\,\mu m$ thick formed by interaction of plasma proteins with the carbohydrate chains of membrane glycoproteins. The glycocalyx extends into the junctional spaces as a fibrous matrix which restricts entry of large molecules [20]. Although appearances of continuous endothelia in diverse settings are similar, their permeabilities differ quantitatively over a wide range. The differences are attributed to variations in the fraction of junctional area closed by tight seals, and in the width of open areas. Maintenance of normal permeabilities of continuous endothelia requires the presence of plasma proteins, notably albumin and orosomucoid (α_1-acid glycoprotein). These substances are believed to interact with the glycocalyx and contribute to the bonding of its fibers [20,27,32,53].
- *Fenestrated endothelium* is found in exchange vessels of secretory and excretory organs: exocrine and endocrine glands, gastrointestinal mucosa, kidney (glomerular and peritubular capillaries), and the choroid plexuses of the brain. It is also found in certain small specialized regions of the brain in which the blood-brain barrier is not present (e.g., the area postrema). Much of the cell surface (50%–95%) is like that of continuous endothelium, but the remainder is less than $0.05\,\mu m$ thick, and bears numerous *fenestrae*. These are circular structures 50–60 nm in diameter which may appear as open perforations in electron micrographs (e.g., renal glomerular capillaries), or may be closed by thin membranous diaphragms (intestinal mucosal and renal peritubular capillaries). In all cases the basal laminal is complete. The functional significance of fenestrae is problematical: closed, they appear to provide a cell-membrane pathway that bypasses the cytoplasm; open, they provide direct access of plasma to the basal lamina. Fenestrated capillaries generally show no greater permeability to plasma proteins than continuous capillaries and considerably higher permeability to water, ions, and small solute molecules [46].
- *Discontinuous endothelium* is found in hepatic, splenic, and bone marrow sinusoids. Junctions between cells are not completely closed and the basal lamina is incomplete. The openings are too large to offer any restriction to passage even of plasma proteins, but small enough to limit escape of blood cells.
- *Tight-junction endothelium* is found only in the microvessels (arterioles, capillaries, and venules) of the central nervous system and retina. The junctions between cells are tightly sealed and severely restrict passage of ions and small uncharged solutes as well as proteins. The cells are permeable to water and lipid-soluble solutes, as in other capillary types, but in the absence of intercellular pathways, cell membrane pathways are dominant. Electrical resistance is high, and hydraulic conductivity low [16]. In many respects transport through tight-junction *endothelium* resembles transport through high-resistance *epithelium* (cf. Chap. 59). Transport of lipid-insoluble solutes depends on the presence of specific membrane carriers [17,43,67]. Permeability to plasma proteins is negligibly small. Tight-junction endothelium is the principal constituent of the "blood-brain barrier".

97.2.2 Microvascular Response to Tissue Injury: Inflammation

Through the action of chemical mediators released from injured cells, capillary and/or venular permeability to plasma proteins is greatly increased [1,7,11,25]. Certain endothelial cells are activated through chemically specific surface receptor proteins. These act through presently unidentified internal messengers to increase internal $[Ca^{2+}]$, partly by release from internal stores, partly by augmentation of Ca^{2+} influx [6,29]. The increase in cellular $[Ca^{2+}]$ triggers rearrangement of cytoskeletal elements which control cell shape and junctional adherence. An early phase of the response is retraction of some venular endothelial cells, opening micron-wide gaps resembling those of discontinuous endothelium. However, the basal lamina is not perforated. Leukocytes adhere to venular walls and move across the endothelium and basal lamina even where the junctions have not opened (cf. Chap. 98). Furthermore, terminal arterioles are dilated, resulting in increased blood flow and elevated microvascular pressures. Several chemical agents are believed to be involved: histamine, serotonin, bradykinin, as well as activated complement and antigen-antibody complexes (cf. Chap. 85). The complete inflammatory response includes [36,37]:

- Arteriolar vasodilation, producing the redness and local warmth characteristic of inflammation
- Escape of plasma proteins and fluid into the tissues, leading to swelling (edema) and increased lymph flow
- Activation of type C afferent nerve fibers, resulting in pain and also in spreading of the microvascular response by the so-called *axon reflex* (cf. Chap. 17). Retrograde conduction of action potentials at branch points spreads excitation to nonstimulated terminals of nerve fibers, which release additional inflammatory mediators.

The inflammatory response is self-terminating and protective: supply of preformed antibodies (immunoglobulins in plasma) to the tissues is increased, and new antigens from the injured tissues are carried by lymphatics to lymph nodes, where they activate lymphocytes to produce new antibodies. However, if the injured region is large, serious loss of plasma volume can lead to circulatory insufficiency ("shock").

97.3 Microvascular Exchange Mechanisms

Exchange of fluid and solutes across exchange vessel endothelium may occur by *diffusion*, *ultrafiltration*, and *vesicular exchange*. The relative importance of these processes differs for different substances [15,19,41].

97.3.1 Diffusion

Diffusion is the principal mechanism for exchange of oxygen, nutrients, and metabolic wastes (small solutes in general). Kinetic movement of individual ions and molecules results in net transfer from regions of high to low concentration. In exchange vessel endothelia, with the possible exception of tight-junction capillaries, gradients of electrical potential are small and have little or no influence on solute transport. Concentration gradients of substrates and products are set up between exchange vessels and cells by metabolic activity. In dilute solutions, two (or more) substances can move in opposite directions simultaneously without interference (e.g., CO_2 and O_2). Diffusion

rates of specific solutes are characterised by "diffusion coefficients" (D, cm^2/s). Diffusion coefficients increase with temperature and are inversely proportional to the viscosity of the solvent. When no other conditions are specified, values of D refer to water at 37°C.

Diffusion can be *free* or *restricted*, depending on the surroundings of the molecule or ion. In free solution, diffusing particles are slowed by frictional interactions with solvent molecules. Friction increases as particle size increases, and free diffusion coefficients decrease with increasing molecular size. In the narrow spaces between endothelial cells and the interstices of fiber matrices filling such spaces, diffusion is slowed further by additional frictional interactions, and the size of penetrating particles is limited by the size of the openings. This is restricted diffusion. Restricted diffusion coefficients decrease more steeply than free diffusion coefficients with increasing solute molecule size, and gradually approach zero as the size of the molecules approaches that of the openings. The restrictive structures (junctional channels and intercellular matrix) have weak negative surface charges which further reduce permeability to negatively charged macromolecules, but not to small anions.

Fick's Law. *Fick's law* states that the rate of diffusion (J_S, mass/time) of a solute through a thin membrane barrier is proportional to the permeability (P, cm/s) of the membrane for the solute, its surface area (A, cm^2), and the difference in concentrations of that solute on either side ($C_p - C_i$). The subscripts "p" and "i" stand for "plasma" and "interstitial fluid," respectively.

$$J_S = PA\ (C_p - C_i) \tag{97.1}$$

P is defined as D/x, where D is a restricted diffusion coefficient and x is membrane thickness or diffusion path length (cm). PA is a "permeability-surface area product," and has the dimensions of volume/time (cm^3/s). This equation applies to exchange vessels well supplied with blood. For obvious reasons, J_s cannot exceed the amount of solute S supplied by the bloodstream. At low rates of blood supply to the exchange vessels, diffusion may be *flow-limited*, because C_p approaches C_i as the blood moves downstream. As blood flow increases, the fall in C_p is progressively reduced, and J_s approaches the value predicted by Fick's law (Eq. 97.1). A general relation applicable to all flow rates is as follows [13,45]:

$$J_S = Q(C_a - C_i)(1 - e^{-PA/Q}), \tag{97.2}$$

where Q is plasma flow and C_a is arterial plasma solute concentration.

Measurement of Diffusion Exchange. [15]. Among a variety of methods of practical and historical importance, the "indicator dilution" method stands out as simple and effective, especially for rapidly diffusing substances. It has been applied to whole organs and to single capillaries. Usually, the substances studied are labelled with isotopes or

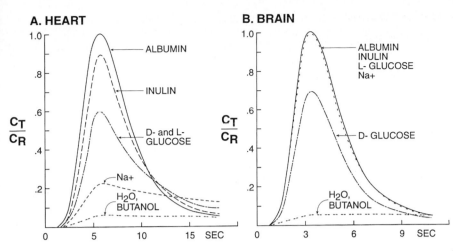

A. HEART

$\dfrac{C_T}{C_R}$

ALBUMIN

INULIN

D- and L-GLUCOSE

Na+

H₂O, BUTANOL

0   5   10   15   SEC

B. BRAIN

$\dfrac{C_T}{C_R}$

ALBUMIN
INULIN
L- GLUCOSE
Na+

D- GLUCOSE

H₂O, BUTANOL

0   3   6   9   SEC

Fig. 97.2. Multiple tracer-diffusion curves for **A**, heart and **B** brain. The *reference* is serum albumin, the *tracers* inulin, D- and L-glucose, Na+, ³H₂O, and butanol. Intra-arterial injection at $t = 0$. C_T/C_R, the ratio of tracer to reference concentration in venous effluent blood, is plotted against time of collection. The representation is diagrammatic in order to show all these tracers together. Blood flow through both organs is represented as basal, 50 ml/min per 100 g. The apparent extraction fraction $E = (C_a - C_v)/C_a$ approximates the true extraction fraction $(C_a - C_v)/C_a - C_i)$ on the rising phase of the curve, when C_i is $\ll C_a$. Values of E: Heart: inulin 0.1, D- and L-glucose 0.4, Na+ 0.8, H₂O, butanol >0.95. Brain: inulin, L-glucose, Na+ <0.02, D-glucose (concentration $\ll$ saturation) 0.3, H₂O, butanol >0.95. See [9,13,15]

are themselves isotopic forms. These can be chosen to represent endogenous substances of interest (e.g., ¹⁴C-glucose, ⁴²K+).

The procedure for whole organs is as follows [9,13]. A small volume ("bolus") of a solution containing one or more diffusible *tracers* and a nondiffusible *reference* solute is injected rapidly into the arterial bloodstream to the organ, and serial samples of venous blood are collected over the next 30 s. Plasma concentrations of reference and tracers are measured, and plotted as functions of time. Figure 97.2A shows in one set of curves the results of several measurements on mammalian heart; the reference in each was ¹²⁵I-serum albumin, which does not escape from the bloodstream to any appreciable extent in this short time. The tracers were tritium-labelled water (³HOH), ¹⁴C-butanol, a lipophilic solute, and several hydrophilic solutes: ²²Na+, ¹⁴C- or ³H-glucose (D- and L-isomers) and ¹⁴C-inulin (a fructose polymer, mol wt. 5.5 kD). All concentrations are given relative to their values in the solution injected. Reference solute concentrations represent dilution of the bolus in the blood flowing through the muscle. Tracer solute concentrations are lower, due to their loss from the blood by diffusion; the magnitude of loss is ³H₂O = butanol > Na+ > glucose (with no difference between D- and L-isomers) > inulin. Figure 97.2B shows similar measurements on mammalian brain; ³H₂O, butanol, and D-glucose escape from the exchange vessels, but practically no Na+, L-glucose, or inulin diffuse out. Transport of glucose in exchange vessels of brain is thus stereospecific: only the D-form permeates. It is also saturable: as the concentration of D-glucose is increased above trace levels, its fractional loss is diminished (by 50% at 1 mmol/l, 90% at 5 mmol/l). Complete analysis of these curves is complex [3]. A simplified interpretation considers the areas (A) under the curves to represent the amount of each solute which did *not* dif-

fuse from the exchange vessels in the muscle. The ratios $A_{tracer}/A_{reference}$ represent the fractions of each tracer which did not diffuse, and therefore the fraction of tracer that *did* diffuse, the "extraction fraction" (E), is equal to $1 - (A_{tracer}/A_{reference})$. E may also be defined as J_s/C_aQ, on the assumption that tracer C_i remains close to zero. Substitution in Eq. 97.2 yields:

$$PA = -Q\ln(1 - E) = -Q\ln(A_{tracer}/A_{reference}) \quad (97.3)$$

Table 97.1 lists PA products measured in this way for a series of hydrophilic solutes in endothelium in mammalian skeletal and cardiac muscle (continuous endothelium), intestine (fenestrated endothelium), and brain (tight-junction endothelium). Since transport of water and lipophilic solutes is flow-limited, their PAs are indeterminate. PA for hydrophilic solutes decreases with increasing molecular size, values for serum albumin being less than 0.1% of those for Na+Cl⁻. In general, PA products are greater in intestine and heart than in skeletal muscle, and much lower in brain. However, the variation in PA also includes differences in exchange vessel surface area. Values of A taken from anatomical measurements are given at the bottom of each column.

Figure 97.3 is a graph of endothelial permeabilities plotted against molecular size (effective diffusion radius) for several organs. Values of P for molecules smaller than albumin were calculated from the data in Table 97.1 by dividing PA by an estimate of capillary surface (A). Values for plasma proteins were derived from measurements of lymph flow and composition (see Sect. 97.4, below). Permeabilities for the continuous endothelia of skeletal muscle and heart are almost identical, and are represented by a single curve in the figure. The differences in PA products in Table 97.1 are due to the eight-fold greater number

Table 97.1. Permeability surface-area products (*PA*, ml/min per 100 g)

Solute	Mol. wt. (D)	Endothelium type			
		Continuous		Fenestrated	Tight-junction
		Skeletal muscle	Heart	Intestine	Brain
Na⁺Cl⁻	58.5	15	>120	>300	0.6
Urea	60	12	90		0.7
Glucose[a]	180	5.5	31		0.2
Sucrose[b]	342	3.5	24	160	0.1
Inulin	5 500	0.4	8	24	<0.01
Albumin	69 000	0.02	0.1	0.02	<0.001
Surface (*A*, cm²/g)		70	560	125	240

[a]Except fructose for brain. [b]Except EDTA (ethylene diamine tetraacetate), mol. wt. 357 D for intestine

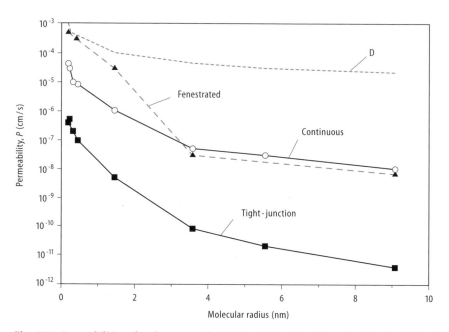

Fig. 97.3. Permeabilities of exchange vessels to hydrophilic solutes in relation to molecular size (Einstein-Stokes radius). The ordinate scale is logarithmic to cover the wide range of values. *Continuous endothelium*: skeletal and cardiac muscle (same curve). *Fenestrated endothelium*: small intestine. *Tight-junction endothelium*: mammalian brain. The *dotted line* at the top of the graph shows the relative decline of free diffusibility with increasing molecular size. Molecules for skeletal and cardiac muscle, *left to right*: Na⁺Cl⁻ (0.2 nm), urea (0.3 nm), glucose (0.4 nm), sucrose (0.5 nm), inulin (1.5 nm), albumin (3.6 nm), IgG (immunoglobulin G, 5.6 nm), IgM (9.1 nm); for intestine: Na⁺Cl⁻, EDTA (0.5 nm), inulin, albumin, IgM; for brain same as muscle, *except* fructose (0.4 nm, not carrier-mediated) instead of glucose

of capillaries per gram of cardiac muscle. In continuous endothelia, permeabilities of hydrophilic solutes decrease with increasing size. The decrease is proportionally greater than the decrease in their free diffusion coefficients (represented by the dotted line without points), thus diffusion is *restricted* by the junctional channels or by structures within them. There is no indication of chemical specificity. The steep part of the downslope appears to be aimed at a molecular radius of about 4.0 nm, and is attributed to the "small pore" system. For molecules larger than 4 nm radius, the downslopes are less steep; these substances are believed to be transported through a small proportion of junctional "large pores" (20–30 nm radius), and/or by in-

terchange of cytoplasmic vesicles (30–50 nm radius). The limited amount of data available for fenestrated endothelia shows that permeabilities to small molecules are considerably higher than in continuous endothelium, and that their diffusion is progressively restricted with increasing size. However, permeability to plasma proteins is no higher than in continuous capillaries. In the tight-junction capillaries of the central nervous system, passive permeabilities to small hydrophilic solutes are two orders of magnitude smaller than in continuous capillaries; the decline with increasing size is steeper and occurs over a lower range of molecular size. Furthermore, permeabilities to hexoses (Fig. 97.2b) and amino acids (not illustrated) exhibit chemical and stereo-specificities characteristic of membrane carriers. In all endothelial types, permeability to lipophilic solutes is higher than to hydrophilic, because the surface area of the cells is so much larger than that of the junctions.

In the past few years it has become feasible to measure solute permeabilities in single perfused capillaries of certain organs [14,18] and in confluent monolayers of cultured endothelial cells [28,38]. Measurements on single capillaries in situ have yielded results generally consistent with whole organ studies. However, cultured endothelia in vitro show less restriction to solute permeation with increasing molecular size and are more permeable to plasma proteins, probably because of imperfect junction formation.

Diffusion in Tissues. Diffusion is also the mechanism which transports solutes through the interstitial fluid from capillary surface to tissue cells. Continuous *gradients* of solute concentrations produced by tissue metabolism provide the driving forces for transport. The form of Fick's law applicable in this case is $J_s = -DA(dC/dx)$, where D is the diffusion coefficient of the solute and dC/dx is its concentration gradient. The rate of solute supply or removal at any point in the tissue is inversely proportional to the square of its distance from the nearest exchange vessel. At a distance of 100 µm, the diffusion rate is only 1% of the rate at 10 µm. This relation demonstrates the importance of maintaining short distances between exchange vessels and tissue cells, especially in organs with high metabolic rates.

Blood-to-Tissue Oxygen Transport. Steady-state gradients of solute concentration are set up in tissues surrounding the exchange vessels by local balance of diffusion (supply or removal) and metabolism (utilization or production). The partial pressure of oxygen (PO_2) immediately outside the endothelium is believed to be nearly the same as that inside because endothelial permeability to O_2 is very high. PO_2 falls with increasing distance from one vessel to a minimum at halfway to the next. The rate of fall in PO_2 in the tissue is directly related to the local rate of O_2 consumption (VO_2) and inversely related to the diffusibility of O_2 in the tissue (D). The fall in PO_2 (ΔPO_2) at any radial distance (R) from the nearest exchange vessel wall (from the PO_2 at the wall) may be estimated by the Krogh-Erlang equation [33]:

$$\Delta PO_2 = \frac{VO_2}{4\alpha D}\left[R^2 \ln\left(\frac{R}{r}\right)^2 - \left(R^2 - r^2\right)\right] \qquad (97.4)$$

where α represents the solubility of O_2 in the tissue and r the radius of the exchange vessel. The partial pressure of O_2 at any location determines its availability to the cells there; if the PO_2 falls to zero, or, more precisely, below the level required to load mitochondrial cytochrome C (0.5–1.0 mmHg), oxidative metabolism ceases. The distance at which this occurs is shortest near the venous ends of the exchange vessels, because PO_2 within these vessels is at its lowest. This "critical supply distance" can be calculated for given values of VO_2, α, D, and r by setting ΔPO_2 in Eq. 97.4 equal to venous PO_2, and solving for R. Recruitment of unperfused or poorly perfused exchange vessels is part of the process of metabolic vasodilation prevents diffusion distances from exceeding critical values [30,61].

97.3.2 Ultrafiltration

Ultrafiltration (also called *filtration* or *osmosis*) is responsible for movement of fluid and permeating solutes in bulk (with volume change). It is controlled by the balance of hydrostatic and osmotic forces across exchange vessel walls, and depends on the relative impermeability of exchange vessel endothelium to plasma proteins. Capillary hydrostatic pressure (P_c) tends to force fluid out; the osmotic pressure of the plasma proteins (π_c) tends to hold it in. Osmotic pressures of *permeating* solutes have no lasting effect [41]. This is called *Starling's hypothesis*, after the first physiologist to enunciate it clearly [40,62].

Starling's Hypothesis.

- When hydrostatic and colloid osmotic forces are equal and opposite, there is no net movement of fluid.
- When hydrostatic forces are greater than osmotic, fluid moves out of the capillary into the interstitial space (ultrafiltration or simply "filtration"). The fluid is a nearly perfect ultrafiltrate of plasma: it contains water, ions, and low molecular weight solutes in equilibrium with plasma, but is nearly free of plasma proteins.
- When the osmotic forces exceed the hydrostatic, fluid is taken up from the interstitial compartment by the capillaries and added to the plasma ("absorption"). Its composition is the same as capillary ultrafiltrate.

Rate of Fluid Movement. The rate at which fluid moves in either direction (J_v, volume/time) may be described by the following equation (cf. Chap. 77):

$$J_v = L_pA[(P_c - P_i) - \sigma(\pi_c - \pi_i)] \qquad (97.5)$$

Positive J_v means fluid movement out of the capillary, negative into the capillary. L_p ("hydraulic conductivity") represents endothelial permeability to ultrafiltrate, A is capillary surface area, P_c and P_i are fluid pressures in capillary and interstitium, respectively, π_c and π_i are the respective colloid osmotic pressures. The product L_pA is called the "capillary filtration coefficient"; it is often represented as K_f or CFC; "L_pA" is used here to emphasize its two components.

Colloid Osmotic Pressure. The colloid osmotic pressure (π) is dependent on plasma protein concentration. It is often called *oncotic* pressure to distinguish it from total solute osmotic pressure. Because microvascular endothelium is not completely impermeable to plasma proteins, interstitial fluid plasma protein concentrations are not zero, and values of π_i usually fall between 0.2 and 0.6 π_p. The symbol σ (sigma) represents the "reflection coefficient" of the endothelium, a measure of its efficiency in retaining protein: 1.00 is perfect; values for most capillaries lie between 0.90 and 0.99 (specific values in Table 97.4, below).

Interstitial Fluid Pressure. The interstitial fluid pressure (P_i) in an organ or tissue depends on the volume of its interstitial fluid (V_i) and the compliance of its interstitial compartment. Interstitial fluid volumes in different tissues vary widely, from as little as 10% of total volume in skeletal muscle to over 30% in skin. Even so, their normal P_i's are about the same, 1–2 mmHg below atmospheric pressure. Compliance below normal V_i is relatively low; thus, when V_i is decreased P_i falls rather steeply. The fall in P_i shifts the balance of Starling forces in favor of fluid movement into the dehydrated tissue (Eq. 97.5), and thus operates to stabilize interstitial fluid volume. However, above normal V_i interstitial compliance increases as the interstitial matrix is expanded, and once P_i has reached zero large accumulations of fluid (edema) produce little further rise in pressure and little restoring force [2].

Influence of Microvascular Resistances on Fluid Movement. Since J_v depends on capillary pressure, any circulatory changes which affect mean capillary pressure will also affect net fluid exchange. Mean P_c is determined by large vein pressure (P_v) and the ratio of postcapillary resistance (R_v, in small muscular venules) to precapillary resistance (R_a, in arteries and arterioles). The relation is shown by the following equation [42]:

$$P_c = \frac{(R_v/R_a)P_a}{1+(R_v/R_a)} + P_v \qquad (97.6)$$

In the systemic circulation, mean P_a is about 90 mmHg, P_v is close to zero, and R_v/R_a is about 0.16; thus mean P_c is 15 mmHg and is almost perfectly balanced by $\sigma(\pi_p - \pi_i)$. Elevation of venous pressure or constriction of muscular venules raises P_c and increases fluid movement from the plasma. Arterial dilation also increases R_v/R_a and P_c and increases outward fluid movement. Arteriolar constriction lowers R_v/R_a and P_c and decreases loss of fluid. If $P_c - P_i$ is reduced below $\sigma(\pi_p - \pi_i)$, interstitial fluid will move into the circulation.

Another consequence of microvascular dilation or constriction is variation of effectively perfused exchange vessel surface (A). Dilation of terminal arterioles and precapillary sphincters tends to "recruit" previously closed or poorly perfused exchange vessels, increasing A and therefore the filtration coefficient L_pA. Arteriolar vasoconstriction does the reverse, decreasing A and L_pA. Changes in L_pA do not alter the *direction* of fluid movement, only its rate when there is a net driving force in either direction.

Measurement of Ultrafiltration. In 1927, E.M. Landis devised an ingenious method of measuring J_v in single exchange vessels under direct microscopic observation, which (with some refinements) is still widely used [35]. The vessel under study is lined up with a micrometer scale, and a side branch is cannulated with a micropipette to measure pressure. Then the vessel is closed downstream of the measurement area, with a fine glass rod ("blocker"). Blood cell flow *through* the vessel stops. If the vessel is completely closed by the blocker, any persisting slow motion of the blood cells toward or away from the block represents exit or entry of ultrafiltrate (Fig. 97.4a). Completeness of closure is indicated by independence of fluid movements on either side of the block. The rate of fluid movement is calculated from the observed velocity (displacement/time) of blood cells (v_{rbc}), and the diameter (d) of the vessel, on the assumption that its shape is cylindrical: $J_v = v_{rbc} (\pi d^2/4)$, where the factor in parentheses represents the cross-sectional area of the cylinder ($\pi = 3.14 \ldots$). The surface area (A) of the vessel between the tracked blood cell and the block (length l) is equal to πdl and thus: $J_v /A = v_{rbc} d/4 l$. To evaluate L_p from a *single* measurement of J_v/A, it is necessary to know σ, π_p, π_i, and P_i in addition to P_c, and, except for π_p, these are generally not obtainable. However, by measuring J_v/A at *two* values of P_c, and assuming that all other variables remain constant, the following solution of Eq. 97.5 is obtained:

$$L_p = [(J_v/A)_1 - (J_v/A)_2]/[(P_C)_1 - (P_C)_2] \qquad (97.7)$$

Landis's original method was to measure J_v/A in several capillaries (Fig. 97.4a) and to plot individual J_v/A's against P_c's (Fig. 97.4b). The slope $[(\Delta J_v/A)/\Delta P_c]$ of the line of best fit is L_p. The intercept on the P_c axis ($J_v/A = 0$) is equal to $\sigma(\pi_p - \pi_i)$. Since each measurement was made on a different capillary, the slope and intercept represent average values for the population as a whole.

More recently, multiple measurements of J_v/A in single microvessels have become feasible, and have shown that there is considerable local variation in L_p within exchange vessel networks [24,39]. In general, there is a rising gradient of hydraulic conductivity from the arterial to the venous side. L_p of postcapillary venules is about two to four times that of "arterial" capillaries. Under normal conditions (absence of injury or inflammatory response), solute permeabilities increase in parallel with hydraulic conductivity and there is no decrease in solute reflection coefficients. Consequently, the permeability gradient is attributed to an increase in the area of open junctions, with no change in their width or in the porosity of the fibrous matrix they contain.

J_v and L_pA have also been estimated in whole organs by measuring changes in organ weight or volume following experimentally induced changes in venous blood pressure (P_v) [12,34,42]. Arterial pressure (P_a) is also measured, and the induced changes in P_c are evaluated by means of Eq. 97.6 above, using measured or assumed values of R_v/R_a. When P_v is raised, there is a rapid increase in organ weight and volume attributed to distention of compliant blood vessels, followed by a slow, steady increase attributed to accumulation of ultrafiltrate in the organ. Usually, J_v is expressed as change in weight or volume/time per unit

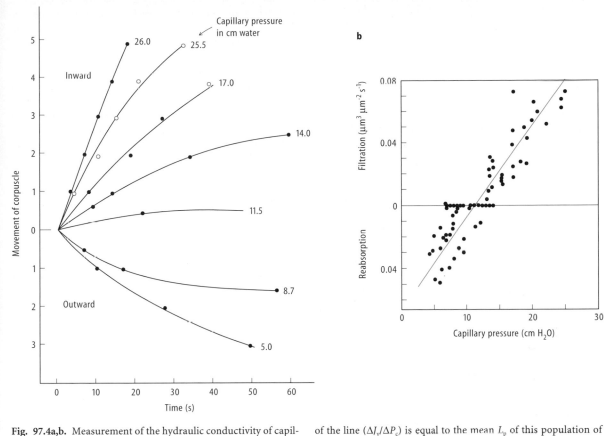

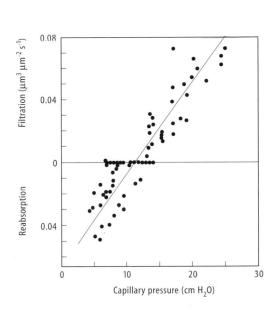

Fig. 97.4a,b. Measurement of the hydraulic conductivity of capillaries of frog mesentery by micro-occlusion. **a** Movement of red blood corpuscles (RBC) toward or away from the occlusion site ("block"). The initial slope of each curve was used to calculate J_v as described in the text (only a few measurements are illustrated). **b** Graph of individual J_v measurements plotted against P_c. The slope of the line ($\Delta J_v/\Delta P_c$) is equal to the mean L_p of this population of capillaries (0.005 µm/s per centimeter of water or 5×10^{-10} cm/s per dyne). The intercept of the line at $J_v - 0$ indicates the average value of $\sigma(\pi_p - \pi_i)$, 11.5 cm H₂O. π_p in frogs of this species ranges from 7 to 14 cm H₂O (1 cm H₂O = 0.74 mmHg). (Modified after Landis [35])

organ weight, since exchange vessel surface is not directly known. L_pA is equal to $\Delta J_v/\Delta P_c$ (Eq. 97.5). Table 97.2 lists capillary filtration coefficients (L_pA) for some mammalian organs.

Table 97.3 compares hydraulic conductivities of selected exchange vessel endothelia; some of the tabulated values are based on direct microscopic measurements of L_p, some by dividing whole-organ L_pA by morphometric estimates of A for the organ (Table 97.1). L_p's for fenestrated endothelia are highest, for continuous endothelium lower, and for tight-junction endothelium lowest, in parallel with the differences in their solute permeabilities (Fig. 97.3). Hydraulic conductivities for continuous endothelia are spread over a rather wide range, most probably in proportion to the fraction of open junctional area.

Fluid Movements Outside Exchange Vessels. Within most exchange vessel networks, P_c falls gradually from arteriolar to venular values due to hemodynamic resistance, while π_p remains nearly constant, because only a minute fraction of

Table 97.2. Capillary filtration coefficients (L_pA, CFC, K_f, ml/min × mmHg per 100 g)

Human forearm	0.006
Dog, cat leg	0.010
small intestine	0.1 0.4
lung	0.20
Rabbit heart	0.35

Table 97.3. Hydraulic conductivities (L_p, cm³/s per dyne)

Fenestrated endothelium	
Renal glomerulus	1.5×10^{-8}
Intestinal mucosa	1.3×10^{-9}
Continuous endothelium	
Mesentery	5.0×10^{-10}
Heart	8.6×10^{-11}
Skeletal muscle	2.5×10^{-11}
Lung	8.4×10^{-12}
Tight-junction endothelium	
Brain	3.0×10^{-13}

the plasma volume is lost by filtration. (Important exceptions are the renal glomerular capillaries, in which 20% or more of the entering plasma is filtered, and the renal peritubular capillaries, where the tubular reabsorbate is returned to the bloodstream.) It has generally been supposed that even when the network as a whole is in fluid balance, near its arterial side P_c would be slightly higher than the mean balance pressure $[\sigma(\pi_p - \pi_i) + P_i]$, and near its venous side P_c would be lower. Thus a small amount of fluid would be filtered from the "arterial" capillaries, enter the interstitial compartment, and be reabsorbed into the "venous" exchange vessels. This local recycling of capillary filtrate has been called the "pericapillary circulation."

The amount of fluid circulated through the interstitium in this way is small, no more than 2% or 3% of intravascular plasma flow. Experimental measurements show that much larger proportions (extraction fractions) of the blood or plasma content of ions and small solute molecules can diffuse from the exchange vessels on a single pass: 25% of the O_2 in resting skeletal muscle, 70% in cardiac or exercising skeletal muscle, more than 90% of tracer water in a variety of organs (see Fig. 97.2). As stated above, diffusion is by far the most important means of transport for such substances. However, for macromolecular solutes like the plasma proteins, which diffuse very slowly, filtration ("convection") is an important transport mechanism.

It has recently been pointed out that pericapillary circulation could be sustained for long periods only in an exchange vessel network completely impermeable to plasma proteins [40]. Even minute amounts of protein escaping with the filtrate on the arterial side of the network will accumulate outside the venous exchange vessels where reabsorption is supposed to take place, because return of these proteins against the concentration gradient from interstitium to plasma cannot occur passively. As a result, interstitial protein concentration and oncotic pressure will rise until reabsorption ceases. Unless there is an outside source of fluid to dilute the accumulated plasma protein (e.g., renal tubular reabsorbate in the case of peritubular capillaries), reabsorption by exchange vessels is only a transient phenomenon. In the steady state, all the filtrate and the proteins it contains must be removed by the lymphatics.

97.3.3 Vesicular Transport (Transcytosis)

Transcytosis is the interchange of plasma and interstitial fluid constituents thorough exchange of vesicles between luminal and abluminal surfaces of endothelial cells [60,64]. Although all components of plasma and intersitital fluid are exchanged, vesicular exchange is much slower than diffusion of small- to intermediate-size ions and molecules, and vesicular transport is believed to be significant only for plasma proteins and other macromolecules [10,23].

Vesicular Motion. Vesicular exchange has been described as due to "Brownian motion" of the vesicles and their contents [57]. Luminal vesicles filled with plasma detach and move randomly within the cytoplasm until they reach the abluminal surface or return to the side they started from. Those reaching the abluminal surface become attached, reopen, and exchange their contents with interstitial fluid. Thus, net transport of the solutes they contain is from high to low concentration, as in molecular diffusion. However, exchange is much slower than molecular diffusion because of the large size of the vesicles and the high viscosity of endothelial cytoplasm. A number of substances, including plasma albumin, are known to bind to the internal surface membrane of vesicles in some endothelia, and it has been proposed that the vesicles can serve as a chemically specific carriers for such substances [59,60]. This is possible if the affinity of the vesicular "receptors" for the substance is low enough to allow association and dissociation within the range of its plasma and interstitial fluid concentrations. However, no demonstrations of selective *transport* of plasma proteins (as differentiated from *binding*) have been reported to date.

Vesicular Openings. Although the vesicles are relatively large (internal diameters 60–70 nm), their openings to the surface ("necks") are narrower (40–50 nm) and they appear to be filled with a fibrous matrix similar to that which covers endothelial cell surfaces. Entry of plasma proteins is progressively restricted with increasing molecular size. Vesicular solute transport depends on the volume of vesicular contents carried from plasma to interstitial fluid per unit time (Q_v), and the partition of solute molecules between plasma and the vesicular contents (α); the product αQ_v is equivalent to a permeability surface area product [10,23].

97.4 Interstitial Fluid and Lymph

Because capillaries are not completely impermeable to plasma proteins, there is a continuous slow movement of protein and fluid from the plasma to the interstitial fluid by way of large pores and vesicles, and much of the so-called plasma protein resides in the interstitia of the various tissues and organs. Interstitial fluid volume in the body is about four times plasma volume and interstitial fluid protein concentrations range from one-fourth to three-fourths of plasma concentrations. Thus, under normal conditions, more than half the total plasma protein in the body is in the interstitial compartment. This pool of fluid and protein is circulated slowly through the lymphatic vessels, and is eventually returned to the bloodstream by the major lymphatic ducts, which empty into the great veins. In its course, the lymph passes through at least one lymph node. The circulation (turnover) time of interstitial fluid/lymph is of the order of 48–72 h, compared to about 1 min for recirculation of the blood. All protein components of plasma are found in interstitial fluid and lymph, but the larger components are present in relatively low concentrations, due to their restriction at capillary walls ("molecular sieving") [5,63].

97.4.1 Mechanism of Lymph Circulation

The interstitial compartment is a fluid-filled matrix of collagen and hyaluronan fibers with various kinds of cells (fibroblasts, macrophages, etc.) wandering around in it, and blood vessels of various sizes passing through. Terminal lymphatics or "lymph capillaries" end blindly in the matrix at some distance from blood capillaries. Like blood capillaries, they consist of a single layer of thin, flat endothelial cells on a delicate collagenous basement membrane. However, lymphatic endothelial cells are much thinner and broader, and the vessels they form are larger than blood capillaries and more irregular in cross section. Endothelial cell junctions in terminal lymphatics are not completely closed. Their edges tend to overlap and act as valves, allowing them to fill with free interstitial fluid, but not to let fluid escape into the interstitium. They drain into conducting lymphatics which are often associated with distributing arterioles and venules. The conducting lymphatics are provided with internal valves which direct the lymph centrally. The propulsive force appears to come from arterial pulsations and tissue movements (muscular contractions, passive limb movements, respiratory movements). Very little energy is needed: pressures are low (1–2 mmHg) and fluid velocities are low. The larger lymphatic vessels have thicker walls, with smooth muscle cells whose contractions help propel the lymph [51,68].

97.4.2 Blood-to-Lymph Transport of Large Molecules

Movement of plasma proteins (molecular weights 42 kD and above) across exchange vessel walls is a critical factor in long-term control of plasma and interstitial fluid volumes, and thus for maintaining circulation of the blood. The title of this section is generalized to "large molecules" only because of the use of large, relatively inert polymers as plasma protein substitutes. These substances have frequently been used as plasma protein substitutes and to investigate exchange vessel permeability. Except for organs with discontinuous sinusoidal endothelia, blood-tissue exchange of these large molecules is extremely slow, and arteriovenous differences are usually indetectable.

Mechanisms of Macromolecular Transport. Three major mechanisms are believed to contribute to macromolecular transport: convection, diffusion, and vesicular exchange. "Convection" refers to the transport of solute molecules entrained in a stream of ultrafiltrate. Concentrations of small solutes in the filtrate are close to those in plasma, but large solutes are restrained or "sieved" by the membrane.

For solutes with molecular weights below 5–10 kD, convection and vesicular exchange are relatively unimportant, and are usually neglected. However, diffusion of solute molecules larger than 10 kD is severely restricted, and cannot by itself account for all observed transport. Either convection (through large pores) or vesicular exchange or both must contribute. In discontinuous endothelia, unrestricted convection is the dominant transport mechanism. In continuous, fenestrated, and tight-junction endothelia the relative importance of these three mechanisms is presently a matter of contention [48,59].

97.4.3 Measurement of Large Molecule Transport

The methods used to study small solute transport (Sect. 97.3 above) are unsuitable for solutes with molecular weights above 10 kD because transport rates are low and extraction fractions vanishingly small. Much of what is known about this topic has been derived from measurements of lymph flow (J_l) and lymph concentrations (C_l) of plasma proteins and exogenous solutes of high molecular weight [26,63]. Lymph concentrations of plasma proteins and other large solutes are inversely related to lymph flow rate and decrease with increasing molecular size (Fig. 97.5). The decline is steep, approaching an extrapolated limit of about 4 nm radius, corresponding to the size of the *small pore* (or dense fiber matrix) pathway; above this limit there is a "tail" extending to much larger sizes (the upper limit is uncertain). This is attributed to transport through *large pores* (loose fiber matrix) or by *vesicles*. Lymph:plasma concentration ratios (C_l/C_p) tend to be lower for more negatively charged molecules of equal size, suggesting that the transport pathways (vesicles, pores, or matrix fibers) are negatively charged.

When ultrafiltration rate and lymph flow are increased (by raising P_c or lowering π_p), C_l/C_p ratios for macromolecular solutes decrease (Fig. 97.5), approaching limiting values equal to $1 - \sigma$, the "solvent drag" coefficient. The convective component of transport is equal to the product of lymph flow and $(1 - \sigma)C_p$. This procedure also provides a means of evaluating endothelial reflection coefficients (σ) of large molecules (Table 97.4).

Evaluating diffusive and vesicular transport components is more problematical, because of their similar dynamics: $J_{(\text{diffus})} = PA(C_p - C_i)$ (Eq. 102.), $J_{(\text{vesic})} = \alpha Q_v(C_p - C_i)$. Separation of these two components may possibly be achieved by cooling the exchange vessels, which is thought to decrease vesicular motion more than molecular motion [49]. Most measured values of large-molecule PA available at the present time (Table 97.1) are actually sums of PA and αQ_v. Permeabilities to large molecules in individual microvessels have been studied using bound dyes or fluorescent compounds as tracers. Under normal conditions leakage of tracers is slow and diffuse, and occurs in both capillaries and nonmuscular venules. In uninjured exchange vessels, there is no obvious localization of leak-

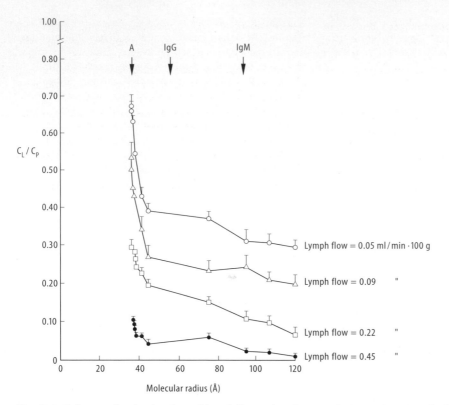

Fig. 97.5. Influence of molecular size and lymph flow on lymph-to-plasma concentration ratios (C_l/C_p) of large molecules. The *arrows* marked *A*, *IgG*, and *IgM* indicate the molecular radii of albumin, immunoglobulin G ("γ-globulin"), and immunoglobulin M ("α₂-macroglobulin"). 1 Å= 0.1 nm. The measurements were made in cat intestine; the lowest lymph flow shown is in the normal range. Higher flows were produced by raising portal vein pressure. (Modified from Taylor and Granger [63], with permission)

Table 97.4. Endothelial reflection coefficients (σ)

Solute	Mol. wt. (kD)	a_e^a (nm)	Endothelium type		
			Continuous (skeletal, cardiac muscle)	Fenestrated (intestine)	Tight-junction (brain)
Albumin	69	3.6	0.90	0.92	0.999
IgG	160	5.6	0.95	0.96	"
IgM	820	9.1	0.99	0.98	"

ᵃ Effective molecular radius (Stokes-Einstein)

age at light-microscope magnification. In response to tissue injury (or application of one of its mediators, e.g., histamine), permeability to plasma proteins is increased 10- to 100-fold, and large focal "leaks" of colored or fluorescent macromolecules are observed in postcapillary venules [1]. These appear to correspond to newly formed large openings (0.5–2.0 μm diameter) between endothelial cells. The basement membrane remains intact. The leaks, like the openings, are transient, lasting only 20–30 min. After this time, a two-fold elevation of permeability may persist as long as the stimulus is still present; its mechanism and pathway are unknown.

97.4.4 Edema Formation and Resolution

Edema is excess accumulation of fluid in the interstitial compartment of organs and tissues, often due to imbalance of the forces controlling fluid exchange (Starling forces). Causes of edema are:

- Lymphatic obstruction or insufficiency
- Increased capillary permeability to proteins (decreased σ, increased π_i)
- Decreased plasma protein concentration (and thus decreased π_p)

- Increased capillary hydrostatic pressure (P_c) due to increased venous pressure (venous obstruction or high plasma volume) or to imbalance of arteriolar and venular resistances (for example, during sustained arteriolar vasodilation).

Pulmonary edema is immediately life threatening because it interferes with blood-gas exchange. Edema of other organs may interfere with their functions. Rapid, massive edema formation due to a generalized permeability increase (as in anaphylaxis) seriously depletes plasma volume and may induce circulatory shock.

"Safety Factors." Several physiological "*safety factors*" operate to limit interstitial fluid accumulation, provided permeability remains normal. When filtration increases, the concentrations of plasma proteins in the filtrate decrease, diluting interstitial fluid and reducing its oncotic pressure. At the same time, π_p increases. Thus $\pi_p - \pi_i$ is increased, opposing further filtration (Eq. 97.5). Also, as interstitial fluid volume increases, P_i increases from normal values slightly below atmospheric pressure to values slightly above atmospheric. $P_c - P_i$ is decreased, providing more opposition to continued filtration (Eq. 97.5). Furthermore, as P_i increases, removal of interstitial fluid by the lymphatics is augmented, first by the direct effect of interstitial pressure on filling of terminal lymphatics and second by an indirect effect of P_i or lymphatic filling on the propulsive motility of conducting lymphatics. These mechanisms keep transcapillary fluid shifts within reasonable limits through our daily cycles of local vasodilation and vasoconstriction, postural changes, and fluid intake and excretion. However, pathological decreases in plasma protein concentration, or increases in exchange vessel pressure or permeability to plasma proteins can overload their capacity and allow massive fluid accumulation. Accumulated edema *fluid* can be removed by reabsorption into exchange vessels as well as by lymphatic drainage, but the only way that accumulated interstitial *protein* can be removed is through the lymphatics [51,63].

97.5 Metabolic and Secretory Functions of Microvascular Endothelia

This chapter has been almost entirely concerned with the role of exchange vessel endothelium as a selective and restrictive transport barrier, and its contribution to the primary cardiovascular function of material transport. However, endothelial cells and vascular endothelia have numerous other functions not directly related to blood-to tissue transport [58]. These include:

- Maintenance of endothelial composition and structure, cell volume, and intercellular junction structure; secretion of basement membrane

- Growth, repair (cell division, migration, synthesis of endothelial growth factor, induction of smooth muscle growth and differentiation) [31]
- Processing of vasoactive substances: *activation* (e.g., angiotensin I → II by pulmonary capillary endothelium), *inactivation* (e.g., adenosine, bradykinin).
- Local regulation of smooth muscle tonus: synthesis and release of prostaglandins, endothelium-derived relaxing factor (EDRF), endothelin (cf. Chap. 95)
- Transmission of vasomotor stimuli from capillaries and small arterioles to larger vessels [22,55,56]
- Maintenance of surface anticoagulant properties (cf. Chap. 83)
- Immune reactions: binding of immune complexes, phagocytosis, inflammatory response (cf. Chap. 98).

These topics are being studied actively at the present time, using endothelial cells isolated from blood vessels of different organs and species, cultured endothelial cells, and confluent layers of cultured cells.

General References

Krogh A (1929) The anatomy and physiology of capillaries (revised and enlarged edition). Yale University Press, New Haven CT (Hafner Press Reprint, New York, 1959)

Landis EM, Pappenheimer JR (1963) Exchange of substances through capillary walls. In: Hamilton WF, Dow P (eds) Handbook of physiology, vol II, section 2: circulation. American Physiological Society Bethesda MD, chapter 29, pp 961–1034

Michel CC (1988) Capillary permeability and how it may change (review lecture). J Physiol (Lond) 404:1–29

Simionescu N, Simionescu M (eds) (1988) Endothelial cell biology. Plenum, New York

Weideman MP, Tuma RF, Mayrovitz HN (1981) An introduction to microcirculation. Academic, New York

Specific References

1. Arfors K-E, Rutili G, Svensjö E (1979) Microvascular transport of macromolecules in normal and inflammatory conditions. Acta Physiol Scand [Suppl] 463:93–103
2. Aukland K, Nicolaysen G (1981) Interstitial fluid volume: local regulatory mechanisms. Physiol Rev 61:556–643
3. Bassingthwaighte JB, Goresky CA (1984) Modelling in the analysis of solute and water exchange in the microvasculature. In: Renkin EM, Michel CC (eds) Handbook of physiology, Section 2: cardiovascular, vol IV: microcirculation. American Physiological Society, Bethesda MD, chapter 13, pp 549–626
4. Bennett HS, Luft JH, Hampton JC (1959) Morphological classification of blood capillaries. Am J Physiol 196:381–390
5. Bert JL, Pearce RH (1984) The interstitium and microvascular exchange. In: Renkin EM, Michel CC (eds) Handbook of physiology, section 2: cardiovascular, vol IV: microcirculation. American Physiological Society, Bethesda MD, chapter 12, pp 521–547
6. Buchan KW, Martin W (1991) Bradykinin induces elevations of cytosolic calcium through mobilisation of intracellular and extracellular pools in bovine aortic endothelial cells. Br J Pharmacol 102:35–40
7. Bundit V, Wissig SL (1986) Surgical exposure induces formation of an arteriovenous permeability gradient for macromol-

ecules in the microcirculation of muscle. Microvasc Res 31:239–249

8. Bundgaard M (1984) The three-dimensional organization of tight junctions in a capillary endothelium revealed by serial-section electron microscopy. J Ultrastruct Res 88:1–17
9. Chinard FP, Vosburgh GJ, Enns T (1955) Transcapillary exchange of water and of other substances in certain organs of the dog. Am J Physiol 183:221–234
10. Clough G, Michel CC (1981) The role of vesicles in the transport of ferritin through frog endothelium. J Physiol (Lond) 315:127–142
11. Clough G, Michel CC, Phillips ME (1988) Inflammatory changes in permeability and ultrastructure of single vessels in the frog mesenteric circulation. J Physiol (Lond) 395:99–114
12. Cobbold A, Folkow B, Kjellmer I, Mellander S (1963) Nervous and local chemical control of precapillary sphincters in skeletal muscle as measured by changes in filtration coefficient. Acta Physiol Scand 57:180–192
13. Crone C (1963) Permeability of capillaries in various organs as determined by use of the "indicator diffusion" method. Acta Physiol Scand 58:292–305
14. Crone C, Frokjær-Jensen J, Friedman JJ, Christensen O (1978) The permeability of single capillaries to potassium ions. J Gen Physiol 71:195–220
15. Crone C, Levitt DG (1984) Capillary permeability to small solutes. In: Renkin EM, Michel CC (eds) Handbook of physiology, section 2: cardiovascular, vol IV: microcirculation. American Physiological Society, Bethesda MD, chapter 10, pp 411–466
16. Crone C, Olesen S-P (1982) Electrical resistance of brain microvascular endothelium. Brain Res 241:49–55
17. Crone C, Thompson AM (1973) Comparative studies of capillary permeability in brain and muscle. Acta Physiol Scand 87:252–260
18. Curry FE (1979) Permeability coefficients of the capillary wall to low molecular weight hydrophilic solutes measured in single perfused capillaries of frog mesentery. Microvasc Res 17:290–380
19. Curry FE (1984) Mechanics and thermodynamics of transcapillary exchange. In: Renkin EM, Michel CC (eds) Handbook of physiology, section 2: cardiovascular, vol IV: microcirculation. American Physiological Society, Bethesda MD, chapter 8, pp 309–374
20. Curry FE, Michel CC (1980) A fiber matrix model of capillary permeability. Microvasc Res 20:96–99
21. Curry FE, Rutledge JC, Lenz JF (1987) Modulation of microvessel charge by plasma glycoprotein orosomucoid. Am J Physiol 257:H1354–H1359
22. Davies PF, Olesen S-P, Clapham DE, Morrel EM, Schoen MJ (1988) Endothelial communication (state of the art lecture). Hypertension 11:563–572
23. Garlick DG, Renkin EM (1970) Transport of large molecules from plasma to interstitial fluid and lymph in dogs. Am J Physiol 219:1595–1605
24. Gore RW (1982) Fluid exchange across single capillaries in rat intestinal muscle. Am J Physiol 242:H268–H287
25. Grega GJ (Ed) (1986) Role of the endothelial cell in the regulation of microvascular permeability to molecules (symposium). Fed Proc 45:75–83
26. Grotte G (1956) Passage of dextran molecules across the blood-lymph barrier. Acta Chir Scand [Suppl] 211:1–84
27. Haraldsson B, Rippe B (1987) Orosomucoid as one of the serum components contributing to normal capillary permeability in rat skeletal muscle. Acta Physiol Scand 129:127–135
28. Haselton FR, Mueller SN, Howell RE, Levine EM, Fishman AP (1989) Chromatographic demonstration of reversible changes in endothelial permeability. J Appl Physiol 67:2032–2048
29. He P, Pagakis SN, Curry FE (1990) Measurement of cytoplasmic calcium in single microvessels with increased permeability. Am J Physiol 258:H1366–H1374
30. Hoppeler H, Mathieu O, Weibel ER, Krauer R, Lindstedt SL, Taylor CR (1981) Design of the mammalian respiratory system VIII. Capillaries in skeletal muscles. Respit physiol 44:129–150
31. Hudlická O (1984) Development of microcirculation. In: Renkin EM, Michel CC (eds) Handbook of physiology, section 2: cardiovascular, vol IV: microcirculation. American Physiological Society, Bethesda MD, chapter 5, pp 165–216
32. Huxley VH, Curry FE (1991) Differential actions of albumin and plasma on capillary solute permeability. Am J Physiol 260:H1645–H1654
33. Krogh A (1919) The number and distribution of capillaries in muscles with calculations of the oxygen pressure head necessary for supplying the tissue. J Physiol (Lond) 52:409–415
34. Krogh A, Landis EM, Turner AH (1932) The movement of fluid through the human capillary wall in relation to venous pressure and to the colloid osmotic pressure of the blood. J Clin Invest 11:63–95
35. Landis EM (1927) Micro-injection studies of capillary permeability. II. The relation between capillary pressure and the rate at which fluid passes through the walls of single capillaries. Am J Physiol 82:217–238
36. Lewis T (1924) Vascular reactions of the skin to injury, part I: reaction to stroking; urticaria factitia. Heart 11:119–137
37. Lewis T, Grant R (1924) Vascular reactions of the skin to injury, part II: the liberation of a histamine-like substance in injured skin; the underlying cause of factitious urticaria and of wheals produced by burning; and observations upon the nervous control of certain skin reactions. Heart 11:209–265
38. Lum H, Siflinger-Birnboim A, Blumenstock F, Malik AB (1991) Serum albumin decreases transendothelial permeability to macromolecules. Microvasc Res 42:91–102
39. Michel CC (1980) Filtration coefficients and osmotic reflexion coefficients of the walls of single frog mesenteric capillaries. J Physiol (Lond) 309:341–355
40. Michel CC (1984) Fluid movements through capillary walls. In: Renkin EM, Michel CC (eds), Handbook of physiology, section 2: cardiovascular, vol IV: Microcirculation. American Physiological Society, Bethesda MD, chapter 9, pp 375–409
41. Pappenheimer JR, Renkin EM, Borrero LM (1951) Filtration, diffusion and molecular sieving through peripheral capillary membranes. A contibution to the pore theory of capillary permeability. Am J Physiol 167:13–46
42. Pappenheimer JR, Soto-Rivera A (1948) Effective osmotic pressure of the plasma proteins and other quantities associated with the capillary circulation in the hind limbs of cats and dogs. Am J Physiol 152:471–491
43. Pardridge WM, Oldendorf WH (1977) Transport of metabolic substances through the blood-brain barrier. J Neurochem 28:5–12
44. Renkin EM (1952) Capillary permeability to lipid-soluble molcules. Am J Physiol 168:538–545
45. Renkin EM (1959) Transport of potassium-42 from blood to tissue in isolated mammalian skeletal muscles. Am J Physiol 197:1205–1210
46. Renkin EM (1977) Multiple pathways of capillary permeability (brief review). Circ Res 41:735–743
47. Renkin EM (1984) Control of microcirculation and blood-tissue exchange. In: Renkin EM, Michel CC (eds) Handbook of physiology, section 2: cardiovascular, vol IV: microcirculation. American Physiological Society, Bethesda MD, chapter 14, pp 627–688
48. Rippe B, Haraldsson B (1987) How are macromolecules transported across the capillary wall? NIPS 2:135–138
49. Rippe B, Kamiya A, Folkow B (1979) Transcapillary passage of albumin, effects of tissue cooling and increases in filtration and plasma colloid osmotic pressure. Acta Physiol Scand 105:171–187
50. Patton HD, Fuchs AF, Hille B, Scher AM, Steiner R (eds) (1989) Textbook of physiology, 21st edn, Saunders, philadelphia
51. Schmid-Schönbein GW (1991) Microlymphatics and lymph flow. Physiol Rev 70:987–1028

52. Schneeberger EE, Lynch RD (1992) Structure, function and regulation of cellular tight junctions (invited review). Am J Physiol 262:L647–L661

53. Schnitzer JE, Pinney E (1992) Quantitation of specific binding of orosomucoid to cultured microvascular endothelium: role in capillary permeability. Am J Physiol 263:H48–H55

54. Scow RO, Blanchette-Mackie EJ, Smith LC (1976) Role of capillary endothelium in the clearance of chylomicrons; a model for lipid transport from blood by lateral diffusion in cell membranes. Circ Res 39:149–162

55. Segal SS (1991) Microvascular recruitment in hamster striated muscle: role for conducted vasodilation. Am J Physiol 261:H181–H189

56. Segal SS, Bény J-L (1992) Intracellular recording and dye transfer in arterioles during blood flow control. Am J Physiol 263:H1–H7

57. Shea SM, Bossert WM (1973) Vesicular transport across endothelium: a generalized diffusion model. Microvasc Res 6:305–315

58. Shepro D, D'Amore PA (1984) Physiology and biochemistry of the vascular wall endothelium. In: Renkin EM, Michel CC (eds) Handbook of physiology, section 2: cardiovascular, vol IV: microcirculation. American Physiological Society, Bethesda MD, chapter 4, pp 103–164

59. Simionescu M, Ghitescu L, Fixman A, Simionescu N (1987) How plasma macromolecules cross the endothelium. NIPS 2:97–100

60. Simionescu M, Simionescu N (1984) Ultrastructure of the microrascular wall: functional correlations. In: Renkin EM, Michel CC (eds) Handbook of physiology, section 2: cardiovascular, vol IV: microcirculation. American Physiological Society, Bethesda MD, chapter 3, pp 41–101

61. Stainsby WN, Snyder B, Welch HG (1988) A pictographic essay on blood and tissue oxygen transport (brief review). Med Sci Sports Exerc 20:213–221

62. Starling EH (1896) On the absorption of fluids from from connective tissue spaces. J Physiol (Lond) 19:312–326

63. Taylor AE, Granger DN (1984) Exchange of macromolecules across the microcirculation. In: Renkin EM, Michel CC (eds) Handbook of physiology, section 2: cardiovascular, vol IV: microcirculation. American Physiological Society, Bethesda MD, chapter 11, pp 467–520

64. Wagner RC, Chen S-C (1991) Transcapillary transport of solute by the endothelial vesicular system: evidence from thin serial section analysis. Microvasc Res 42:139–150

65. Weideman MP (1963) Patterns of the arteriovenous pathways. In: Hamilton WF, Dow P (eds) Handbook of physiology, section 2: circulation, vol II. American Physiological Society, Bethesda MD, pp 891–933

66. Weideman MP (1984) Architecture. In: Renkin EM, Michel CC (eds) Handbook of physiology, section 2: cardiovascular, vol IV: microcirculation. American Physiological Society, Bethesda MD, chapter 2, pp 11–40

67. Yudilevich DL, DeRose N (1971) Blood-brain transfer of glucose and other molecules measured by rapid indicator dilution. Am J Physiol 220:841–846

68. Zweifach BW, Lipowsky HH (1984) Pressure-flow relations in blood and lymph microcirculation. In: Renkin EM, Michel CC (eds) Handbook of physiology, section 2: cardiovascular, vol IV: microcirculation. American Physiological Society, Bethesda MD, chapter 7, pp 251–307

98 Biology of the Vascular Wall and Its Interaction with Migratory and Blood Cells

B. Nilius and R. Casteels

Contents

Abbreviations

AA	Arachidonic acid
ACE	Angiotensin-converting enzyme
ACh	Acetylcholine
AIDS	Acquired immune deficiency syndrome
ATP	Adenosine-(1,3,5)-triphosphate
cAMP	Cyclic adenosine monophosphate
cGMP	Cyclic guanosine monophosphate
CSF	Colony stimulating factor
EC	Epithelial cells
EDCF	Endothelium-derived contracting factor
EDHF	Endothelium-derived hyperpolarization factor
EDRF	Endothelium-derived relaxing factor
EGF	Epidermal growth factor
ELAM	Endothelial leukocyte adhesion molecule
endoCAM	Endothelial cell adhesion molecule
ET	Endothelin
FGFa	Acidic fibroblast growth factor
FGFb	Basic fibroblast growth factor
ICAM	Intracellular adhesion molecule
IL-1	Interleukin 1
LAM	Leukocyte adhesion molecule
LECAM	Leukocyte-endothelial cell adhesion molecule
MHC	Major histocompatibility complex
NA	Noradrenaline (norepinephrine)
NADP	Nicotinamide-adenine dinucleotide phosphate
NADPH	The reduced form of NADP
PAF	Platelet activating factor
PAI	Plasminogen activator inhibitor
PD-ECGF	Platelet-derived endothelial capillary growth factor
PDGF	Platelet-derived growth factor
PF	Platelet factor
PMN	Polymorphonuclear leukocyte
SMC	Smooth muscle cell
subEC	Subepithelial layer
TGF	Transforming growth factor
TNF	Tumor necrosis factor
tPA	Tissue type plasminogen activator
TXA$_2$	Thromboxane A$_2$
uPA	Urokinase type plasminogen activator
VCAM	Vascular cell adhesion molecule
VEGF	Vascular endothelial growth factor

98.1 Introduction

The vascular wall must be one of the largest "organs" in the body with complex interactions in terms of signalling between various cell types. The interaction between the different cells is either by humoral factors or by direct cell-to-cell coupling; moreover, signal transduction can be integrated at the surface of the different cell types involved. A great variety of completely different functions are coordinated by these cell interactions, such as control of vasomotor activity, cell proliferation, chemotaxis, immunological responses, and control of thrombogenic and thrombolytic activity and of the coagulation potential in the blood [2,12,25,54,62]. Communication between different cells is essential for the normal biology of the vessel wall, because the vascular system has to respond immediately to changed functional demands and priorities.

This description of the vascular wall will center on cell-cell communications and will be restricted to (i) normal interactions of cells forming the blood vessel, (ii) vasomotor control by communication between different cell types forming the vessel wall, (iii) regulation of the migration of cells through the vascular wall, (iv) cell-cell interactions mediated by cytokines that control angiogenesis, (v) communication between platelets and endothelial cells, and (vi) the role of the vessel wall in modulating the coagulation of blood. We concentrate on only three types of cells: endothelial cells, smooth muscle cells, and cells able to permeate through the vascular wall or stick to the vascular endothelium.

98.2 Components of the Vascular Wall

Endothelial cells (EC) together with the subendothelial layer (subEC) form the intima of the vessel wall. The

R. Greger/U. Windhorst (Eds.)
Comprehensive Human Physiology, Vol. 2
© Springer-Verlag Berlin Heidelberg 1996

subendothelium contains a basement membrane with collagen, elastin, microfibrils, mucopolysaccharides, laminin, fibronectin, vitronectin, von Willebrand factor, and thrombospondin, and it acts as a secondary barrier [25].

In large vessels, the intima is surrounded by the media, which consists of layers of smooth muscle cells (SMC) and their extracellular matrix. External to the media is the adventitia, in which fibroblasts and their corresponding extracellular material are found. The adventitia also contains nerve fibers with their varicosities as release sites for vasoactive neurotransmitters. The sympathetic fibers can release noradrenaline (norepinephrine, NA) and adenosine-(1,3,5)-triphosphate (ATP), the purinergic fibers release ATP, and the sensory peripheral nerve endings can release ATP, calcitonin-generelated peptide, substance P, or vasoactive intestinal peptide. The media and adventitia determine the mechanical properties of the vessel wall, but the SMC in the media regulate vasoconstriction or dilatation depending on the modulation of the SMC activity. The relative contribution of the different cells types (endothelium, connective tissue, SMC, constituents of the nervous systems) to the vascular wall varies considerably according to the type of vessel.

Precapillary vessels distribute blood (under high pressure within) over the cardiovascular system so that each capillary network receives the amount of blood it requires at the correct pressure. Only 8% of the total blood volume is in the large arteries, 5% in small arteries, 2% in arterioles, and 16% in the heart and pulmonary vessels. The capillaries, which contain only 5% of the total blood volume, present by far the largest total cross-sectional area. This is the part of the vascular tree where the various substances can move between the blood and the interstitium. The capillary wall is therefore thin and only consists of endothelial cells, a basement membrane abluminally covered by connective tissue, and some pericytes. Venules (containing 25% of the blood volume) and veins (39% of the blood volume) drain the blood from capillaries and return it to the heart (for reviews see [9,22,50,54]).

98.3 Functional Role of Endothelial Cells

EC are positioned at the boundary between intravascular and interstitial compartments and are thus exposed both to biochemical mediators in the blood and to physical factors acting at the vascular wall, such as stretch, shear stress, and pressure. They also receive signals from the abluminal site of the blood vessel, e.g., histamine released from mast cells, neurotransmitters from nerve varicosities, and cytokines from macrophages and other migratory cells. EC respond to all these stimuli by synthesizing and releasing various messengers which interact with neighboring cell types (*humoral communication*), but, in addition, EC can also communicate via physical contacts with neighboring cells (*contact-mediated communication*).

Figure 98.1 summarizes the different types of communication between endothelial cells and other cell types (see [12] for a review).

98.3.1 Endothelial Cell Regulation of Vascular Tone

One of the major discoveries of the last 10 years was that of the very active role of EC in determining the contractile state of vascular smooth muscle [18,23,51].

Prostacyclin. The first endothelium-derived factor found was prostacyclin (PGI$_2$). PGI$_2$ is released by agonists such as bradykinin, histamine, serotonin (5-HT), thrombin, lipoproteins, fibrin, interleukin-1 (IL-1), γ-interferon, epidermal growth factor (EGF), transforming growth factor-α (TGFα), and tumor necrosis factor (TNF). Hypoxia and mechanical factors such as flow (shear stress) also increase the release of PGI$_2$. The actual trigger is an increase in the intracellular Ca^{2+} concentration in the EC due to release from intracellular stores as well as to transmembrane Ca^{2+} influx. PGI$_2$ activates an adenyl cyclase in neighboring cells (mainly platelets) that induces an increase to a variable extent in cAMP (for reviews see [11,25,28]). PGI$_2$ contributes little to vasorelaxation but efficiently inhibits platelet aggregation.

Endothelium-Derived Relaxing Factor. The most powerful vasorelaxing mediator has been termed "endothelium-derived relaxing factor" (EDRF) [19]. This substance is also of widespread importance to many other mechanisms. EDRF is released by shear stress (flow), Ca^{2+} ionophores such as A23187, autacoids, bradykinin, histamine, NA, substance P, vasopressin, acetylcholine (Ach), thrombin, and platelet-derived products such as ATP and adenosine diphosphate (ADH). EDRF has been identified as endogenous nitric oxide (NO). It is produced by a Ca^{2+}-calmodulin-NADPH dependent synthase that catalyzes oxidation of the N-guanidine terminal of L-arginine and is released as the easily diffusable NO or bound to cysteine (S-nitroso-cysteine). Although there are many steps in this reaction, which is not yet completely understood, the overall stochiometry is:

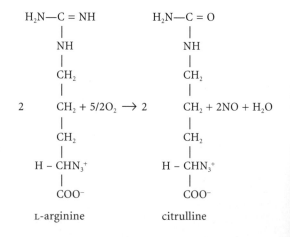

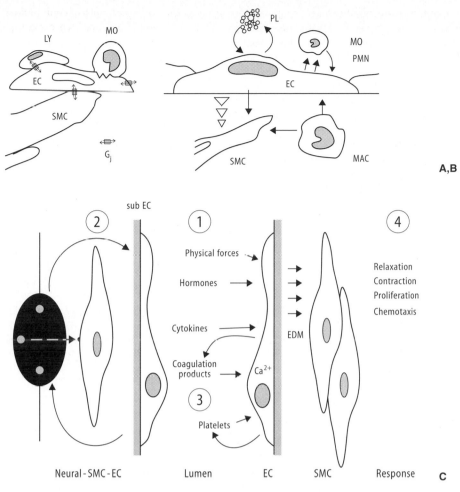

Fig. 98.1A–C. The vascular wall as a signal-transducing surface. **a** Between endothelial cells (*EC*) and different other cell types exists a physical cell-cell coupling via gap junctional channels (*Gj*) or different adhesion proteins (contact-mediated communication). Gap junctional channels allow also electrical coupling between different cells (*LY*, lymphocytes; *MO*, monocytes, see Sect. 98.3.4, 98.5; *SMC*, smooth muscle cells, see Sect. 98.3.1). EC are coupled via gap junctional channels. **b** Examples of humoral communication between EC and other cells (*PL*, platelets, see Sects. 98.4.1; *MO*, *PMN*, monocytes, polymorphonuclear leukoctes, see Sect. 98.3.3, 98.3.4; *SMC*, smooth muscle, see Sects. 98.3.1, 98.3.3; *MAC*: macrophages or mast cells, see Sect. 98.3.3). MO, PMN, and

MAC send cytokine signals to EC and also SMC. Mast cells induce multiple histamine responses in EC. **c** Overview of different effector cell–responder cell functions in the vascular wall. *1*, different intraluminal informations affect EC function (synthesis and release of EC-derived mediators such as EDRF, most of them are mediated by changes in intracellular Ca^{2+} (see Sects. 98.3.1, 98.3.3, 98.3.4); *2*, The neural-SMC-EC interaction is discussed in Sect. 98.3.2; *3*, EC interfere with platelet function and control the intravascular coagulation potential. The function of EC itself is changed by coagulation products and substances released from platelets see Sects. 98.4.1, 98.4.2); *4*, Extravascular responses to EC-derived mediators

An increase in intraendothelial Ca^{2+} by release from Ca^{2+} stores and by a sustained Ca^{2+} influx via membrane channels is the trigger event for activation of the NO-synthase and the release of EDRF.

EDRF activates a soluble guanylate cyclase in neighboring cells that leads to an increase in cGMP. In this way EDRF induces relaxation of SMC via a not yet completely understood mechanism. Different mechanisms appear to be involved in cGMP-mediated SMC relaxation. The activation of plasmalemmal and endoplasmic reticulum Ca^{2+} pumps, activation of a smooth muscle Na^+/Ca^{2+} exchanger, and in addition inhibition of Ca^{2+} influx reduce intracellular Ca^{2+} and promote SMC relaxation [11,20,25,32,33,46,51,64]. EDRF also efficiently inhibits platelet aggregation and ad-

hesion (see Sect. 98.4.1). NO has been discovered now as general messenger not only in the vascular system but also as a central and peripheral neuronal messenger involved in synaptic modulation and even in long-term potentiation and long-term depression [21,27,35].

All these properties make EC one of the body's largest endocrine gland [2].

Control of Endothelial Intracellular Ca^{2+}. EDRF production depends on a sustained Ca^{2+} increase in the cell. Therefore, the control of intracellular Ca^{2+} plays the key role in most of the EC functions.

Intracellular Ca^{2+} release via activation of an inositol-(1,4,5)-trisphosphate (IP_3)-receptor-Ca^{2+}-release channel in the membrane of intracellular Ca^{2+} stores together with Ca^{2+} influx control the synthesis and release of both EDRF and PGI_2. The transmembrane Ca^{2+} influx appears to be regulated by various mechanisms. A possible pathway is via mechanically (shear stress) activated channels. This mechanical signal transduction chain involves an increase in intracellular Ca^{2+} that is not yet completely understood [13,39,40,41,56]. By this mechanism, EC act as mechanosensors that sensitively react to changes in the blood flow (shear stress). This Ca^{2+} entry provides a link between mechanical stimuli and the many biological responses that are mentioned throughout the literature. Another mechanism could presumably gate Ca^{2+}-permeable membrane channels by depleting the intracellular Ca^{2+} stores. A third pathway is a Ca^{2+} influx via leak or nonselective cation channels that follows the driving force for Ca^{2+} [24,38–41,43]. To date, it is still unclear whether a Na^+/Ca^{2+} antiporter contributes to Ca^{2+} signalling in EC. Figure 98.2 gives a simplified overview of the mechanisms controlling intracellular Ca^{2+}. As discussed later, an increase in intraendothelial Ca^{2+} can release vasorelaxing factors such as NO and PGI_2 as well as vasoconstricting factors such as endothelins. The consequences of a common pathway existing for different signals has yet to be assessed.

Endothelium-Mediated Hyperpolarization. In addition to EDRF, another nonprostacyclin, non-EDRF factor has been described that induces hyperpolarization in adjacent SMC, thereby promoting relaxation. This endothelium-derived hyperpolarization factor (EDHF) is released by ACh acting on supposedly endothelial M_1 receptors [8,59].

Contact-Mediated Relaxation. We have so far focused on humoral interactions between EC and SMC. However, there is growing evidence – morphological as well as functional – that EC are directly coupled to SMC via gap junctional channels. This coupling would directly transfer any hyperpolarization of EC to SMC and would thereby induce vasorelaxation. K^+ currents activated by ACh or by shear stress would mediate EC hyperpolarization and therefore decrease the tone of SMC which are electrically coupled to EC [12,13,43].

Endothelium-Mediated Vasoconstriction. All the above discussed factors cause relaxation of the SMC. However, an increasing number of factors secreted by EC are being described that induce contraction of adjacent SMC (endothelium-derived contracting factors, EDCF). Endothelium-dependent vasocontriction can be evoked by endogeneous substances, e.g., arachidonic acid (AA), ACh, NA, thrombin, and prostaglandin H_2 (PGH_2), but also by hypoxia or by mechanical forces such as stretch and pressure, although not by shear stress [51]. EDCF have been divided into several classes. $EDCF_1$ consists of some vasoconstrictor metabolites of AA (cyclooxygenase-dependent EDCFs, e.g., PGH_2; thromboxane A_2 TXA_2) and free radicals (e.g., superoxide anions). The second class of EDCF, termed $EDCF_2$, is released during severe hypoxia. However, the most potent vasoconstrictors are the third class, $EDCF_3$, the endothelins (ET). It is likely that all EDCF induce Ca^{2+} influx and stimulation of the phosphatidylinositol-4,5,-bisphosphate (PIP_2) metabolism in SMC.

Endothelins. The ET are the most effective vasoconstrictors. They are synthesized by induction of their mRNA by substances like angiontensin II, NA, vasopressin, thrombin, phorbolesters, $TGF\beta$, IL-1, Ca^{2+} ionophores, and also mechanically modulated by shear stress. Promoters of ET synthesis again increase in EC intracellular Ca^{2+}

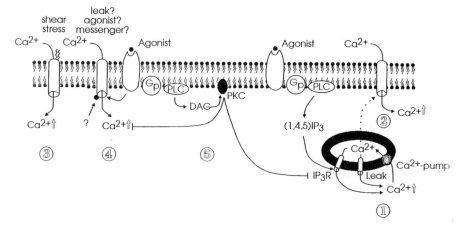

Fig. 98.2. Control of intracellular Ca^{2+} in EC (cf. also Chap. 5) *1*, Stimulation of EC induces intracellular Ca^{2+}-release via agonist-binding to a receptor (*R*), activation of a G protein (*Gp*), activation of phospholipase C (*PLC*) and production of (1,4,5)-IP_3. IP_3 opens a Ca^{2+}-release channel in a Ca^{2+} store (binding at the IP_3-receptor IP_3R). Ca^{2+}-stores are normally leaky and are refilled by a Ca^{2+}-ATPase Ca-pump; *2*, Discharge of the Ca^{2+}-store might gate a Ca^{2+}-permeable membrane channel [24]; *3*, Another Ca^{2+}-entry pathway is activated by shear stress; *4*, Other pathways for Ca^{2+}-influx may be gated by agonists, intracellular messengers (?) or are leak channels that control Ca^{2+}-influx by the driving force for Ca^{2+}; *5*, Activation of protein kinase C (*PKC*) via production of diacyl glycerol (*DAG*) inhibits Ca^{2+}-release and Ca^{2+}-influx and therefore acts as a negative feedback

and activate protein kinase C (PKC). Apparently, both ET transcription and ET release depend on intracellular Ca^{2+}. Analysis (Southern blot) of human genomic DNA revealed a family of ET termed:

- Endothelin-1 (ET-1)
- ET-2
- ET-3.

A fourth member is expressed in the intestine and is called:

- EN-β (or vasoactive intestinal constricting peptide).

ET-1 is composed of 21 amino acids. It is synthesized from a preform of 203 amino acids. Dibasic endopeptidase and carboxypeptidase cleave the preform to a 39-amino-acid product that is finally split by an endothelin-converting enzyme into the biologically active 21-amino-acid form. At least two types of receptor for ET have been discovered. Activation of these receptors elicits contraction and mitogenesis in SMC. In EC they activate angiotensin-converting enzyme (ACE) and mitogenesis as well, and in platelets they cause aggregation.

Because of their extremely high vasoconstrictor activity and the long-lasting action, ET are important candidates for inducing long-lasting vasospasm (for reviews see [25,32,52]). The actual trigger for the release of ET appears also to be an increase in the endothelial intracellular Ca^{2+}.

By what mechanisms an intracellular Ca^{2+} pattern can alternately induce release of relaxing and contracting factors is still completely unknown.

Vasomotor control by other mechanisms. Some vasoactive compounds released by platelets induce secretion of EDCF via an increase in intracellular Ca^{2+} from EC and are also able to directly activate SMC contraction (e.g., 5-HT, TXA_2, ATP).

Some vascular endothelial cells synthesize and secrete renin, which converts angiotensinogen to angiotensins by ACE, which is bound at the EC surface. Angiotensin II induces SMC contraction. Furthermore, SMC activity can be modulated by activation of a number of EC surface enzymes that form, convert, or inactivate vasoactive peptides. ACE also causes inactivation of vasodilating bradykinin. Other surface enzymes that modulate the activity of vasoactive compounds are carboxypeptidase N and aminopeptidases A and M. The functional effects of these enzymes have still not been elucidated. ACE inhibitors enhance the activity of endogenous vasodilating compounds such as bradykinin by inhibiting their breakdown. Thus, the decrease in bradykinin breakdown by ACE inhibitors would induce an increase in the intracellular Ca^{2+} concentration. This increase would in its turn stimulate EDRF synthesis [32,63,65].

Figure 98.3 summarizes the signal transduction chain for these important cell-cell communications controlled by the vascular wall.

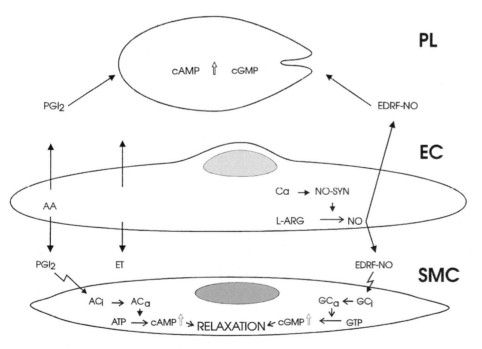

Fig. 98.3. Endothelial cells (*EC*) function as an endocrine gland. Via the cyclooxygenase pathway from arachidonic acid (*AA*), prostacyclin is synthesized and causes an increase in cAMP in platelets (*PL*) and smooth muscle cells (*SMC*) (activation of an adenylyl cyclase, AC_i to AC_a). From a prepro-form, endothelins (*ET*) are synthesized and secreted. They induce SMC contraction but also act as neuronal modulators and affect multiple functions in other cells. EDRF (*NO*) is sythesized from L-arginine by a Ca^{2+}-dependent activation of an NO-synthase (*NO-SYN*). EDRF increase cGMP in adjacent cells that cause SMC relaxation (activation of a soluble guanylcyclase GC_i to GC_a)

98.3.2 Control of the Vascular Tone by Neural and Endothelial Interactions

The contractile state of SMC is also regulated by factors released from nerve endings and by direct myogenic responses of SMC to physical stimuli. These mechanisms are described in detail in Chap. 96. However, the neural activity that controls the tonus of SMC is also modulated by EC and vice versa. NA released from adrenergic nerve endings in the adventitia or media directly affects EC through α_2-receptor activation in the EC. As a consequence, EDRF is released, probably via activation of an EC G_i-protein. By this mechanism the release of EDRF from EC antagonizes the SMC effects caused by catecholamines. EC can modulate, vice versa, the release of neurotransmitters presynaptically by releasing angiotensin II and also endothelin [37].

98.3.3 Endothelial Cells, Cytokines, and Angiogenesis

Cytokines. In addition to the well-defined function of EC in regulating vascular tone, EC also control the growth of adjacent cells. In these actions, cytokines play the central role. Cytokines are recently discovered proteins secreted by lymphocytes, neutrophils, monocytes, and EC or expressed on the surface of various cells. They bind to specific receptors on the surface of their target cells and appear to play a crucial role in inflammatory reactions and immune responses. Members of this constantly growing class of small intracellular signalling molecules are, for example, interleukins (IL-1 to IL-8; TNFα; interferons; colony stimulating factor, CSF) and growth factors (platelet-derived growth factor, PDGF; acidic and basic fibroblast growth factors, FGFa and FGFb; and platelet-derived endothelial cell growth factor, PD-ECGF).

Repair of the Vascular Wall and Angiogenesis. Cytokines are critically involved in the formation and repair of the vascular wall itself. Endothelial cells normally live much longer than 10 years. However, even limited damage on the luminal side of the vessel wall induces fast proliferation of EC. More pronounced or penetrating injuries also lead to migration and proliferation of SMC in the media and fibroblasts in the adventitia. Both cell types are able to remodel the vascular wall. They synthesize connective tissue and are able to secrete elastases and collagenases. Atherogenic stimuli (hypercholesterolemia, hypertension, anoxia, turbulent blood flow) can also cause migration of SMC into the intima. Once in the intima, SMC continue to secrete extracellular matrix proteins and proliferate, finally causing atherosclerotic and fibromuscular plaques [3,55]. In all these changes EC play the crucial role. They secrete several factors that control the regeneration of the vascular wall.

A necessary component of wound healing, revascularization of damaged or poorly supplied tissue, is angiogenesis. This can be divided into the following events:

- Enzymatic degradation of the basement membrane of preexisting vessels
- Chemotaxis of EC to the angiogenic stimulus
- Proliferation of EC
- Formation of newly migrating capillaries.

These mechanisms appear to be mainly controlled by a newly discovered family of heparin-binding polypeptide mitogens (vascular endothelial growth factor, VEGF).

Control of Cell Proliferation in the Vessel Wall. The VEGF family comprises four peptides (of 121, 165, 189, and 206 amino acids, respectively) with completely different secretion patterns. These peptides recognize only EC as their sole specific target and are able to trigger the entire chain of events leading to angiogenesis. VEGF cDNA has been found in a wide variety of cells including tumor cells, which can thus exploit the normal process of angiogenesis to support their growth. Details of VEGF release and signalling to EC have not been elucidated [16,25,45].

Other factors, e.g., FGFa, FGFb, PD-ECGF, activate both EC and SMC growth and stimulate angiogenesis. FGFb is also synthesized in EC and acts as an autocrine growth factor. FGF bind to cellular receptors and induce DNA synthesis and proliferation. In addition, these substances release tissue plasminogen activator, collagenases, and elastases and activate chemotaxis in a variety of cells. PD-ECGF (a 45-kD protein) is secreted from platelets, stimulates EC by direct cell-cell interaction, and activates chemotaxis and angiogenesis. Surprisingly, FGF also releases EDRF, presumbly by elevation of intracellular Ca^{2+} [10]. The major source of FGFb and also TNFα are macrophages in the interstitium. TNFα is angiogenic when injected extravascularly but not when injected intravascularly. Thus, macrophages have substantial angiogenic activity.

The cell mitogen platelet-derived growth factor is expressed in two forms as PDGF-A and PDGF-B in EC, but EC do not express receptors for PDGF and do not respond to this factor. This action of EC appears to be directed abluminally towards the subjacent cell layer and stimulates proliferation of SMC and fibroblasts, and also has a chemotactic activity for fibroblasts, SMC, and neutrophils, which thereby could accumulate in the subEC. Thrombin, TGFβ, FGFa (increases only PDGF-A mRNA), endotoxins, IL-1, IL-6, phorbol esters, and factor X_a all increase the expression of mRNA for PDGF, but an intracellular elevation of cAMP decreases the basal PDGF mRNA depression [11,25] (for a review see [45]). Large glycoproteins like laminin, von Willebrand factor, fibronectin, fibrinogen, and the especially important thrombospondin regulate adhesion not only for platelets but probably also activate proliferation of a variety of cell types and locations. Thrombospondin is bound to the cell surface of SMC and is

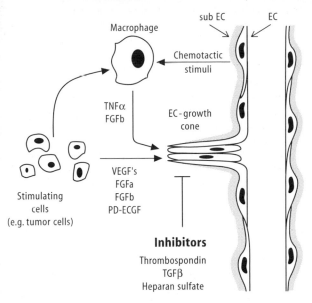

Stimulators

sub EC · EC

Macrophage

Chemotactic stimuli

TNFα
FGFb

EC-growth cone

Stimulating cells
(e.g. tumor cells)

VEGF's
FGFa
FGFb
PD-ECGF

Inhibitors
Thrombospondin
TGFβ
Heparan sulfate

Fig. 98.4. Different mechanisms involved in angiogenesis. Stimulating cells (e.g., tumor cells) secrete growth factors such as VEGFs, FGFb, and FGFa. Macrophages stimulate EC growth by secretion of FGFb, TNFα. PD-ECGF also induces EC-growth. Inhibitory factors are thrombospondin, TGFβ, heparan sulfates. A cone of EC moves towards the stimulating cells. The layer of subendothelial collagen, elastin, and heparan sulfates, etc. (subEC) is degraded by secretion of proteases (collagenases, elastases, plasminogen activators from stimulatory cells (see text for details; *EC*, endothelial cells; *TNF*, tumor necrosis factor; *TGF*, transforming growth factor; *FGF*, fibroblast growth factor; *PD-ECGF*, platelet-derived endothelial capillary growth factor; *VEGF*, vascular endothelial growth factors)

functionally essential for the proliferation of these cells. Normally, thrombospondin inhibits angiogenesis by inhibition of endothelial growth. However, thrombospondin synthesis is downregulated when angiogenesis is stimulated.

EC release not only compounds like PDGF and adhesive proteins like thrombospondin that stimulate proliferation, they also secrete heparin-like compounds, e.g., heparan sulfates, which inhibit proliferation of SMC. Heparan sulphate and thrombospondin are also synthesized and secreted by SMC. Normally, proliferative stimuli in the vessel wall are counterbalanced by the equilibrated amounts of heparan sulfate and chondroitin sulfate in the extracellular matrix which surrounds the medial SMC. This situaton is disturbed by damage to the EC: platelets aggregate and release PDGF, monocytes migrate through the vessel wall and differentiate to macrophages, which release heparanases and are thereby important initiators of atherosclerosis. These enzymes degrade heparan sulfates, whereby the phenotype of SMC changes and these cells become more sensitive to growth factors such as PDGF, TGFβ, TGFα, EGF, FGF, IL-1, and TNFα. Interestingly, the highly anionic heparin strongly inhibits SMC proliferation

by preventing the cells from entering the S phase of mitosis. All the above-mentioned adhesive proteins have heparin binding sites that normally interact with sulfated glycolipids. Heparin might prevent these interactions by blocking the binding of these proteins to other cells and by exerting multiple inhibitory effects on intravascular thrombosis and cell proliferation [34,48]. Finally, all these mechanisms appear to be the key events in forming atherosclerotic and stenosing fibromuscular plaques that consists of a necrotic core and a fibrotic cap with phenotypically changed SMC, neutrophils, T-lymphocytes, elastin, collagen, and proteoglycans [3,5,6,55,57].

TGFβ has profound effects on EC. It is synthesized and released from platelets but also from EC and induces a growth inhibition of EC, increases matrix deposition, suppresses proteolytic activities of cell-associated proteases, and also inhibits migration of blood cells through the vessel wall [47].

EC also express a fibronectin receptor that mediates attachment in the subEC to vitronectin, von Willebrand factor, and fibrinogen-fibrin. Interactions between EC and fibrin or fibrinogen take place at the abluminal site and appear to be an important trigger of capillary formation and angiogenesis [25].

Figure 98.4 gives a synopsis of some of the mechanisms that control angiogenesis.

98.3.4 Endothelial Cell Interactions with White Blood Cells

Diapedesis and Chemotaxis. The functional properties of the vascular wall include a wide variety of interactions between EC and white blood cells. These interactions control adherence on and migration through the EC of polymorphonuclear (PMN) leukocytes, monocytes, basophils, and eosinophils. In response to low concentrations of inflammatory stimuli, leukocytes adhere to EC, migrate across the EC (diapedesis), and are directed to the inflammatory stimulus (chemotaxis). Adherence of monocytes to EC and migration to the subEC appears to be the unique trigger of differentiation from monocytes to macrophages. Adherence of blood cells depends on the expression of adhesion molecules on the surface of EC.

Cell Adhesion Molecules. Cytokines, especially TNFα, induce the expression of a 115-kD endothelial leukocyte adhesion molecule (ELAM-1). Another ligand primarily for PMN is a 140-kD protein (CD62 or GMP-140, also know as PADGEM, see [58]). This EC surface molecule binds neutrophils via the CD11b/CD18 (Mac-1) leukocyte adhesion protein. In general, all types of blood cells can adhere to EC. Adherence is stimulated by γ-interferon, IL-1, and TNFα. These cytokines induce the expression of an adhesion protein on EC (intercellular adhesion molecule, ICAM-1 or CD54). ICAM-1 provides binding of leukocytes via the leukocyte integrin, the heteromeric glycoprotein LFA-1, or CD11a/CD18 and Cd11c/CD18. CD11 and CD18 belong to the integrin family. They form the α-chain (CD11, 180 kD) and the β-chain (CD18, 95 kD) of the binding proteins on the leukocyte surface. These binding proteins on blood cells differ in their α-chain: CD11b/CD18 for granulocytes, monocytes, and large lymphocytes, CD11c/CD18 for granulocytes and monocytes. Another endothelial adhesion molecule is ICAM-2, which is homologous to ICAM-1 and is present in resting or flow-activated EC that are not cytokine-stimulated. This adhesion molecule recognizes a leukocyte protein, Mel-14 antigen (Mel14-Ag, also known as leukocyte adhesion molecule LAM-1 or leukocyte-endothelial cell adhesion molecule, LECAM-1) that mediates PMN extravasation and EC binding of T- and B-lymphocytes [7,42,61].

Vascular cell adhesion molecules (VCAM) on the EC surface preferentially bind monocytes and lymphocytes via their surface adhesion molecules CD11c/CD18, CD11b/CD18. All these adhesion mechanisms that finally result in migration and chemotaxis are themselves controlled by the EC. Stimulation of EC by IL-1, TNFα, and endotoxins induces IL-8 production by the EC. IL-8 inhibits blood cell–EC interaction. On the other hand, IL-1 generally enhances the adhesion of blood cells to EC [44,45,49].

Another family of cell adhesion molecules can be expressed on of the EC surface and is termed "endoCAM" (endothelial cell adhesion molecules, CD31, CD34, CD36). These adhesion molecules might play an important role not only in EC–blood cell contacts but also in adhesion of tumor cells to EC. Methodologically, CD31 seems to be the best and universal cell marker for EC [36].

The biology of adhesion mechanisms is still under widespread investigation. It seems even to be possible that binding of lymphocytes via adhesion proteins induces gap junctional connections between EC and leukocytes whereby direct (and also electrically coupled) cell-cell contacts are developed [12].

98.4 Interferences Between the Vascular Wall and the Coagulation System

In this paragraph we describe only some aspects of the control functions of the vessel wall in the complex system of blood clotting, thrombosis, and fibrinolysis. A detailed description is given in Chap. 83.

98.4.1 Interactions Between Platelets and Endothelial Cells

The vascular wall has a critical part not only in controlling the diameter of the vessels but also in maintaining the fluidity of the blood. Multiple mechanisms prevent platelet aggregation in normal vessels, in contrast to activation of platelets when a vessel is injured. Upon injury, the cascade of platelet aggregation starts a self-sealing mechanism of

the vascular wall. Effects involving the coagulation cascade and fibrinolysis are discussed in Sect. 98.4.2.

Antithrombotic Properties of EC. EC are anti-thrombotic, e.g., platelets do not adhere to EC in vivo. This physiologically predominant property is controlled by a glycocalyx of glycoproteins and glycosaminoglycans with a high proportion of heparan sulfate and chondroitin sulfate. The anionic electric charge of these compounds is of special importance and normally prevents EC–platelet interaction and platelet aggregation.

EC express other members of the cytoadhesin and integrin family, the vitronectin receptor, and VLA-2, the collagen receptor. At the abluminal side of EC cytoadhesin binds von Willebrand factor and fibrinogen, which both bind platelets. However, the biology of these interactions is not well understood.

Antithrombotic activities, including inhibition of the aggregation of platelets and inhibition of the adherence of platelets to EC, are mainly carried out by PGI_2 and EDRF. PGI_2 acts via an increase in cAMP in the platelets, EDRF via cGMP. EDRF actions are potentiated by even subthreshold concentrations of PGI_2. On the other hand, there is an efficient feedback between EC and platelets. Aggregation of platelets induces a release of vasoactive compounds such as ATP, ADP, and serotonin. All these substances cause EDRF release by inducing an elevation of intracellular Ca^{2+}. EDRF release inhibits aggregation of platelets (negative feedback). EC, on the other hand, modulate the effects of platelet-released ATP and ADP because of the activation of endothelial ectoenzymes that metabolize ATP and ADP to AMP and the antiaggregatory adenosine.

Prothrombotic Properties of EC. Platelet–EC interactions, and also the subendothelium, control the properties of the self-sealing mechanism. Several proteins are critically involved in these mechanisms.

As already mentioned, thrombospondin is an adhesive glycoprotein which is synthesized and released by platelets, and also EC, megakaryocytes, and SMC. It probably participates not only in platelet adhesion and aggregation but also in analogous functions (adherence) in a variety of cells [17]. Adhesive proteins synthesized by EC, such as thrombospondin, von Willebrand factor, and fibronectin, are not exposed at the luminal surface of EC but are located at the basement membrane. Therefore, the prothrombotic activity of these proteins is shielded by the antithrombotic properties of the EC.

Von Willebrand factor is vectorially secreted into the subEC and supports platelet adhesion and aggregation. This glycoprotein is a molecular glue for platelet plugs at the site of vascular injury. It forms bridges between platelet receptors, collagen, and coagulation factor VIII [60].

Platelet-activating factor (PAF, 1-alkyl-2-acetyl-sn-glycero-3-phosphocholine) is a potent activator of platelet aggregation and adhesion. It is produced by EC from arachidonic acid. An elevation of intracellular Ca^{2+} activates the Ca^{2+}-dependent PAF-synthase lysoPAF-acetylCoA-acetyltransferase. PAF production in EC is activated by thrombin, vasopressin, angiotensin II, histamine, bradykinin, ATP, IL-1, and leukotrienes C_4 and D_4. Most of the activators increase intracellular Ca^{2+}. It is therefore considered that Ca^{2+} release and Ca^{2+} influx trigger PAF synthesis by activation of the lysoPAF-acetylCoA-acetyltransferase. PAF remains EC-associated and induces platelets to adhere to EC.

Intravascular Thrombotic Potential. In general, EC provide both anti- and prothrombotic activities. The intravascular thrombotic potential is the balance between the two. Normally, EC cover the subendothelial matrix and control the antithrombotic properties of the vascular wall. Removal of the EC layer or localized EC defects, as in atheroma plaques, first exposes platelets to a nonthrombogenic region of subendothelium. A deeper injury uncovers collagen type IV and V, microfibrils, von Willebrand factor, vitronectin, and fibronectin, which all cause platelet aggregation. At the same time the anti-thrombotic potential of EC is decreased and favors local thrombosis. Heparan and chondroitin sulfate proteoglycans counteract platelet aggregation [4,26,30].

98.4.2 Control of Blood Coagulation by the Vascular Wall

The blood vessel wall was thought for a long time to play only a passive role in controlling the coagulation process. Blood coagulation is a defense mechanism that maintains the integrity of blood vessels as a nonleaking distribution system, providing self-sealing properties of the vasculature. EC, subEC matrix, and platelets are involved in these mechanisms. Here, we only describe control functions of the vessel wall within the coagulation system.

Anticoagulant Properties of the Vascular Wall. EC provide both anticoagulant and procoagulant factors. Under physiological conditions the balance favors an anticoagulant predominance, e.g., EC function to inhibit the coagulation cascade. This anticoagulant potential of the vascular wall depends on several mechanisms.

First, the coagulation cascade can be directly inhibited by protease inhibitors. The most important protease inhibitor, antithrombin III, ATIII, is present in the blood plasma at a concentration of approximately 0.2 mg/ml. By complexing to thrombin (factor II_a), it directly prevents the functional action of the catalytic site of the proteolytic enzyme II_a. ATIII is supposed not to be synthesized in EC, but it binds to free heparan sulfates at the luminal surface of EC with a K_d value of 14–40 nmol/l and also at the subendothelium. This binding dramatically accelerates inactivation of thrombin by ATIII [25]. EC also participate in the clearance of the ATIII-thrombin complexes from the circulation.

Heparin, a proteoglycan synthesized by mast cells and hepatocytes, binds to ATIII. The heparin-ATIII complex immediately inactivates factor II_a by complexing ATIII and thrombin where heparin is released. This mechanism strongly multiplies the anticoagulant effect of ATIII. Formation of this complex is also dramatically accelerated by the presence of EC, which also support the ability of ATIII to inactivate factors IX_a, X_a, and XIII_a.

EC secrete protease nexin-1, a 43-kD protein which inhibits thrombin. Protease nexin-1 forms a 1:1 complex with thrombin which can then be internalized by endothelium. EC accelerate inhibition of thrombin, most likely due to the presence of heparan sulfate in the extracellular EC matrix. Second, proteases inactivate cofactors that are necessary for coagulation. In these mechanisms thrombomodulin plays an important role. Thrombomodulin is a surface protein synthesized by EC and expressed at their luminal site at a density of 100 000 molecules per EC. It binds thrombin. An important inhibitor of coagulation is protein C, activated by the thrombomodulin-thrombin complex. After activation by the thrombomodulin-thrombin complex, protein C inactivates in the presence of a protein S that is expressed by EC and bound to their surface factors V_a and VIII_a and also inhibits a plasminogen activator inhibitor (PAI) that is expressed in EC. Both pathways increase the anticoagulant potential. When thrombin is bound to thrombomodulin the complex can also be internalized by EC, which further decreases the coagulant potential. IL-1 decreases the activity of thrombomodulin and increases the procoagulant activity in the blood vessel (Fig. 98.5a).

Third, enzymes necessary for coagulation can be bound at the phospholipid surface. EC synthesize a family of lipid binding proteins, annexins. These proteins can bind negatively charged phospholipids in the presence of Ca^{2+}. On the other hand, clotting factors (e.g., II, VII, IX, X) bind to negatively charged phospholipids. This binding increases their activity. A member of the annexin family, annexin V, a 32-kD protein, counteracts this mechanism. It competitively occupies binding sites for the clotting factors and thereby prevents their activation.

These three mechanisms depend on EC and are summarized in Fig. 98.5b.

Procoagulant Properties of the Vascular Wall. The vascular wall also activates procoagulant factors. An extrinsic pathway can increase the procoagulant properties of the vascular wall through EC. EC stimulated by endotoxins (lipid A, Lipopolysaccharides), phorbol esters, TNFα, IL-1, interferons, thrombin, certain viruses (e.g., herpes simplex virus, cytomegalovirus), or else mechanically by shear stress, express tissue factor as a surface protein. The coagulation cascade is then initiated by interaction of factor VII/VII_a with tissue factor. This interaction finally activates factor X. EC also synthesize factor V, which again supports activation of prothrombin by factor X_a. The complex formed by platelet factor 3 (PF-3), Ca^{2+}, and factors V_a and X_a converts prothrombin (factor II) into the 30.6-kD active proteolytic enzyme α-thrombin (factor II_a) that finally starts blood clotting (Fig. 98.6). This pathway gives a nice example of the concerted action of blood components and

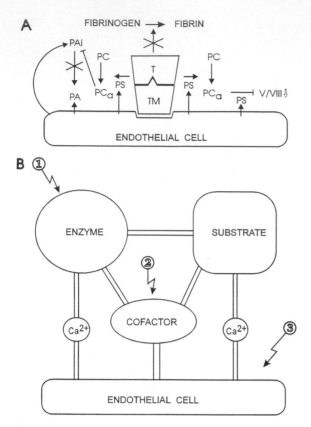

Fig. 98.5A,B. The vascular wall controls anti-coagulant factors. **A** EC synthesize thrombomodulin (*TM*) that binds thrombin (*T*, *II$_a$*) thereby inactivating II$_a$ and stopping the cleavage from fibrinogen to fibrin. The TM/T-complex can be internalized. TM/T activates protein C (*PC*) in the presence of an endothelial surface protein S (*PS*). Activated protein C (*PC$_a$*) inactivates factors V$_a$ and VIII$_a$ and also inactivates a plasminogen activator inhibitor (*PAI*) that is secreted by EC. This inhibition increases the fibrinolytic potential (disinhibition of a secreted plasminogen activator, *PA*). **B** Three main mechanisms are involved in controlling the anticoagulant potential by the vascular wall. *1*, Protease inhibitors directly inhibit the catalytic sites of proteases like thrombin; *2*, proteases inactivate cofactors, e.g., protein C inhibits V$_a$/VIII$_a$ (Fig. 98.6); *3*, binding of coagulation enzymes can be inhibited by competition with negatively charged binding sites (annexin V)

the vascular wall, as modulated by several factors, e.g., TNFα and other cytokines, to regulate the coagulant potential.

Platelets themselves, once activated, secrete endoglycosidases that degrade heparan sulfates (binding sites for ATIII) and block inactivation of thrombin by ATIII. This effect is supported by EC and creates a procoagulant property.

Vascular Wall and Fibrinolytic Properties. The vascular wall affects the fibrinolytic system. EC synthesize and secrete two forms of plasminogen activators, urokinase type plasminogen activator (uPA) and tissue type plasminogen activator (tPA), which accelerate fibrinolysis, and two plasminogen activator inhibitors, PAI-1 and PAI-2. These plasminogen activators and plasminogen activator inhibi-

tors are bound at the EC matrix. They are present also at the luminal surface. Thrombin, histamine, and also physical shear stress stimulate the release of tPA and uPA. tPA is predominantly active when it is bound to fibrin. Cytokines, e.g., IL-1 and TNFα, and α-tocopherol decrease the release of tPA but not uPA.

Secretion of PAI-1 and PAI-2 is stimulated by:

- Thrombin
- Endotoxins
- IL-1
- TNFα synthesis.

When bound to vitronectin PAI-1 is also able to inhibit thrombin.

Thus, the vascular wall regulates fibrinolysis via the antagonistic pathways of plasminogen activator and plasminogen actirator inhibitor secretion by EC [14,15,25,26].

98.5 Vascular Wall and Immunological Functions

The vascular wall plays an essential part in the immune system. Stimulated with γ-interferon, EC express major histocompatibility complex (MHC) class II molecules. This stimulation can also be directly controlled by activated T-lymphocytes that secrete EC activators. MHC class II molecules on the EC surface bind their respective antigens. This binding transforms EC into antigen-presenting cells. The transformed EC are able to bind T-helper cells. This cell-cell contact is believed to induce an increase in intracellular Ca^{2+} and stimulates the EC to secrete IL-1, which leads to activation of the bound T-helper cells. IL-1 activation of the helper cells now starts the signalling chain that includes self-stimulation of T-helper cells by IL-2 and induction of T-cell proliferation [1].

Expression of MHC class II molecules – which convert EC into antigen-presenting cells – also converts these cells into "professional" macrophages that phagocytose particles and bacteria. Secretion of IL-1 augments the direct response of EC to endotoxins and enhances EC phagocytosis [53].

98.6 Remarks on Pathophysiology and Conclusions

The vascular wall is obviously one of the most complex signalling surfaces in the body. Dysfunction of the vascular wall will have far-reaching consequences and is involved in major pathological states such as cardiovascular diseases (atherosclerosis, hypertension, local vasospasm), virus infections (AIDS), neurological disorders (Alzheimer's disease), inflammation, and others [26,29,55,62]. We have

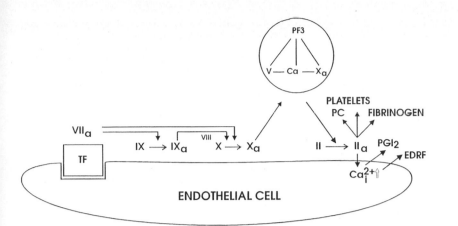

Fig. 98.6. EC influence pro-coagulant pathway. A tissue factor (*TF*) like procoagulant activity is expressed by EC via various stimuli. Activation of factor VII (*VII$_a$*) induces stimulation of factor IX to the active form IX$_a$ and finally activation of factor X (*X$_a$*) in the presence of factor VIII. A complex of platelet factor 3 (*PF3*), factor V, factor X and Ca^{2+} form the proteolytic thrombokinase that cleaves prothrombin (*II*) to the active thrombin (*II$_a$*).

Thrombin itself induces multiple effects: protein C (*PC*) activation, proteolysis of fibrinogen to fibrin monomeres and two vasoactive fibrinopeptides, elevation of intracellular Ca^{2+} in platelets and EC (thereby inducing platelet aggregation, release of vasoactive compounds, release of PGI$_2$ and EDRF from EC and other effects)

tried to show that most of the EC functions balance stimulatory and inhibitory effects on cascades of functional events such as control of vascular tone (vasodilation versus vasoconstriction), intravascular thrombosis (antithrombotic versus thrombotic processes), maintenance of wall integrity versus angiogenesis, and diapedesis of migratory cells and blood cells through EC versus inhibition of cell adherence. Any disturbance in these equilibria will affect the respective biological functions.

We will mention just two examples illustrating the increasing importance of the vascular wall in pathogenetic mechanisms: (i) It is becoming evident that impaired EDRF activity plays a key role in subarachnoid hemorrhage, ischemia, EC damage, atherogenesis, hypertension, and diabetes. Hypercholesterolemia and atherosclerosis reduce or even prevent the effects of EDRF. Local EC injury then induces a direct contractile response of SMC to normally vasodilating compounds such as thrombin, serotonin, and ATP directly because of a deficiency in counteractive EDRF release. Such mechanisms induce local vasospasms, hypertension, and further damage of the vascular wall [31,32,51,63].

(ii) Abnormal proliferation of cells in the vascular wall is an important pathogenic event in tumor growth and metastasis. Signalling from tumor cells to the vascular wall induces angiogenesis, which is necessary for tumor cell growth. Blocking this abnormal proliferation would be a successful anticancer strategy. The same might be true in retinopathy and rheumatoid arthritis [36].

In conclusion, the vascular wall provides more than just a distribution system for the blood; it is actively involved in many biological functions such as control of the vascular diameter (SMC tone), maintenance of appropriate coagulation and fibrinolytic potential within the vessel, regulation of appropriate cell growth to form new vessels, control

of self-sealing properties of the vascular wall itself, and control of various mechanisms in immune responses and self-defense mechanisms. These complex functions are organized by communication between various cell types and multiple signalling pathways – an area still under intensive investigation.

References

1. Alberts B, Bray D, Lewis J, Raff M, Roberts K, Watson JD (1989) Molecular biology of the cell. Garland, New York, chap 18, pp 1002–1055
2. Änggard EE (1990) The endothelium – the body's largest endocrine gland? J Endocrinol 127:371–375
3. Betz E, Fallier-Becker P, Wolburg-Buchholz K, Fotev Z (1991) Proliferation of smooth muscle cells in the inner and outer layers of the tunica media of arteries: an in vitro study. J Cell Physiol 147:385–395
4. Botting R, Vane JR (1989) Mediators and the antithrombotic properties of vascular endothelium. Ann Med 21:31–38
5. Campbell JH, Campbell GR (1989) Biology of the vessel wall and atherosclerosis. Clin Exp Hypertens 11:901–913
6. Campbell JH, Campbell GR (1991) The macrophage as an initiator of atherosclerosis. Clin Exp Pharmacol Physiol 18:81–84
7. Carlos TM, Harlan JM (1990) Membrane proteins involved in phagocyte adherence to endothelium. Immunol Rev 114:5–28
8. Chen G, Yamamoto Y, Miwak K, Suzuki H (1991) Hyperpolarization of arterial smooth muscle induced by endothelial humoral substances. Am J Physiol 260:H1888–H1892
9. Cliff WJ (1976) Blood vessels. Cambridge University Press, Cambridge
10. Cuevas P, Carceller F, Ortega S, Zaza M, Nieto I, Gimenez-Gallego G (1991) Hypotensive activity of fibroblast growth factor. Science 254:1208–1210
11. Daniel TO, Ives HE (1989) Endothelial control of vascular function. News Physiol Sci 4:139–142

12. Davies PF, Oleson S-P, Clapham DE, Morrel EM, Schoen FJ (1988) Endothelial communication. Hypertension 11:563–572

13. Daviess PF (1989) How do vascular endothelial cells respond to flow? News Physiol Sci 4:22–25

14. De Groot P, Tijberg P (1992) Endothelial cells and the coagulation system. In: Endothelial cells. European School of Haematology, 1990, Maffiers, France, pp 11–28 (course)

15. Esmon CT (1989) The roles of protein C and thrombomodulin in the regulation of blood coagulation. J Biol Chem 264:4743–4746

16. Ferrara N, Houck KA, Jakeman LB, Winer J, Leung DW (1991) The vascular endothelial growth factor family of polypeptides. J Cell Biochem 47:211–218

17. Frazier WA (1987) Thrombospondin: a modular adhesive glycoprotein of platelets and nucleated cells. J Cell Biol 105:625–632

18. Furchgott RF (1984) Role of endothelium in the response of vascular smooth muscle to drugs. Annu Rev Pharmacol Toxicol 24:175–197

19. Furchgott RF, Zawadzki JV (1980) The obligatory role of endothelial cells in the relaxation of atrial smooth muscle by acetylcholine. Nature 288:373–376

20. Furukawa KI, Oshima N, Tawada-Iwata Y, Shigekawa M (1991) Cyclic GMP stimulates Na$^+$/Ca^{2+} exchange in vascular smooth muscle cells in primary culture. J Biol Chem 266:12337–12341

21. Garthwaite JS, Charles SL, Chess-Williams R (1988) Endothelium-derived relaxing factor release on activation of NMDA receptors suggest role as intercellular messenger in the brain. Nature 336:385–388

22. Guyton AC (1986) The system circulation. In: Guyton AC (ed) Textbook of medical physiology. Saunders, Philadelphia, pp 218–229

23. Henderson AH (1991) Endothelium in control. Br Heart J 65:116–125

24. Jacob R (1990) Agonist-stimulated divalent cation entry into single cultured human umbilical vein endothelial cells. J Physiol (Lond) 421:55–77

25. Jaffe EA (1991) Endothelial cell structure and function. In: Hoffman R, Benz EJ, Shattil SJ, Furie B, Cohen HJ (eds) Hematology, basic principles and practice. Churchill Livingstone, New York, pp 1198–1213

26. Kahaleh B, Matucci-Cerinic M (1992) Endothelial injury and its implications. In: Semini Neri GG, Gensini GF, Abbate R, Prisco D (eds) Thrombosis, an update. Scientific Press, Florence, pp 649–658

27. Knowles RG, Palacios M, Palmer RMJ, Moncada S (1988) Formation of nitric oxide from L-arginine in the central nervous system: a transduction mechanism for stimulation of soluble guanylate cyclase. Proc Natl Acad Sci U S A 86: 5159–5162

28. Kuo L, Davis J, Chilian WM (1992) Endothelial modulation of arteriolar tone. News Physiol Sci 7:5–9

29. Lelkes PI (1991) New aspects of endothelial cell biology. J Cell Biochem 45:242–244

30. Lotti T, Campanile G, Gheretich I, Romagnoli P (1992) Endothelium-blood cells interactions in thrombosis. In: Nevi Saneni GG, Gensini GF, Abbate R, Prisco D (eds) Thrombosis, an update. Scientific Press, Florence, pp 669–693

31. Lüscher TF, Dohi Y (1992) Endothelium-derived relaxing factor and endothelin in hypertension. News Physiol Sci 7:120–123

32. Lüscher TF, Richard V, Tanner FC (1991) Endothelium-derived vasoactive factors and their role in the coronary circulation. Trends Cardiovasc Res 1:179–184

33. Magliola L, Jones AW (1990) Sodium nitroprusside alters Ca^{2+} flux components and Ca^{2+}-dependent fluxes of K$^+$ and Cl$^-$ in rat aorta. J Physiol (Lond) 421:411–424

34. Majack RA, Goodman LV, Dixit VM (1988) Cell surface thrombospondin is functionally essential for vascular smooth muscle cell proliferation. J Cell Biol 106:415–422

35. McCall T, Vallance P (1992) Nitric oxide takes the centre-stage with newly defined roles. Trend Pharmacol Sci 13:1–6

36. McCarthy SA, Kuzu I, Gatter KC, Bicknell R (1991) Heterogeneity of the endothelial cell and its role in organ preference of tumour metastasis. Trends Pharmacol 21:462–467

37. Miller VM (1991) Interactions between neural and endothelial mechanisms in control of vacular tone. News Physiol Sci 6:60–63

38. Newby AC, Henderson AH (1990) Stimulus-secretion coupling in vascular endothelial cells. Annu Rev Physiol 52:661–674

39. Nilius B (1991) Regulation of transmembrane calcium fluxes in endothelium. News Physiol Sci 6:110–114

40. Nilius B (1991) Ion channels and regulation of transmembrane Ca^{2+} influx in endothelium. In: Sperelakis N, Kuriyama H, (eds) Electrophysiology and ion channels of vascular smooth muscle and endothelial cells. Elsevier, New York, pp 317–325

41. Nollert MU, Diamond SL. McIntire LV (1991) Hydrodynamic shear stress and mass transport modulation of endothelial cell metabolism. Biotechnol Bioeng 38:588–602

42. Nourshargh S, Williams TJ (1990) Neutrophil-endothelial cell interactions in vivo. In: Warren JB (ed) The endothelium: an introduction to current research. Wiley-Liss, New York, pp 171–186

43. Oleson S-O, Clapham DE, Davies PF (1988) Haemodynamic shear stress activates a K$^+$ current in vascular endothelial cells. Nature 331:168–170

44. Osborn L (1990) Leukocyte adhesion to endothelium in inflammation. Cell 62:3–6

45. Pober JS, Cotran RS (1990) Cytokines and endothelial cell biology. Physiol Rev 70:427–451

46. Raeymaekers L, Hofmann F, Casteels R (1988) Cylic GMP-dependent protein kinase phosphorylates phospholamban in isolated sarcoplasmic reticulum from cardiac and smooth muscle. Biochem J 252:269–273

47. Ray Chaudhury A, D'Amore PA (1991) Endothelial cell regulation by transforming growth factor-beta. J Cell Biochem 47:224–229

48. Reilly CF, Fritze LMS, Rosenberg RD (1986) Heparin inhibition of smooth muscle cell proliferation: a cellular site of action. J Cell Physiol 129:11–19

49. Richter J (1991) Cytokine-induced degranulation in human neutrophils. Thesis, University of Lund

50. Rhodin JAG (1980) Architecture of the vessel wall. In: Bohr DF, Somlyo AP, Sparks HV (eds) Handbook of physiology: the cardiovascular system. American Physiological Society, Bethesda, pp 1–31

51. Rubany GM (1991) Endothelium-derived relaxing and contracting factors. J Cell Biochem 46:27–36

52. Rubany GM, Parker Botelho LH (1991) Endothelins. FASEB J 5:2713–2720

53. Ryan US (1988) Macrophagelike properties of endothelial cells. News Physiol Sci 3:93–96

54. Ryan US (1989) Endothelium as a transducing surface. J Mol Cell Cardiol [Suppl] I:85–90

55. Schmitz G, Hanlowitz J, Kovacs EM (1991) Cellular processes in atherogenesis: potential targets for Ca^{2+} channel blockers. Atherosclerosis 88:109–132

56. Schwarz G, Droogmans G, Nilius B (1992) Shear stress induced calcium transients in human vascular endothelial cells. J Physiol (Lond) 458:527–538

57. Scott-Burden T, Bühler FR (1988) Regulation of smooth muscle proliferative phenotype by heparinoid-matrix interactions. Trends Pharmacol Sci 9:94–98

58. Springer TA (1990) Adhesion receptors of the immune system. Nature 346:425–434

59. Suzuki H, Chen G (1990) Endothelium-derived hyperpolarization factors (EDHF): an endogenous potassium-channel activator. News Physiol Sci 5:212–215

60. Titani K, Walsh KA (1988) Human von Willebrand factor: the molecular glue of platelet plugs. Trends Biochem Sci 13:94–97

61. Vada MA, Gamble JR (1990) Regulation of the adhesion of neutrophils to endothelium. Biochem Pharmacol 40:1683–1687

62. Vanhoutte PM (1987) Endothelium and the control of vascular tissue. News Physiol Sci 2:18–22

63. Vanhoutte PM, Auch-schwelk W, Biondi MI, Lorenz RR, Schini VB, Widal MJ (1989) Why are converting enzyme inhibitors vasodilators? Br J Pharmacol 28:S95–S104

64. Vrolix M, Raeymaekers L, Wuytack F, Hofmann F, Casteels R (1988) Cylic GMP-dependent protein kinase stimulates the plasmalemmal Ca^{2+}-pump via phosphorylation of phosphatidylinositol. Biochem J 255:855–863

65. Wiemer G, Schölkens BA, Becker RHA, Busse R (1991) Rampiprilat enhances endothelial autacoid formation by inhibiting breakdown of endothelium-derived bradykinin. Hypertension 18:558–563

99 Control of the Circulation: An Integrated View

H.E.D.J. ter Keurs and J.V. Tyberg

Contents

99.1 Introduction

The survival of cells is contingent on the maintenance of a constant environment. As the cell itself consumes nutrients and O_2 and produces metabolites, rapid supply and removal of these molecules is crucial in order to keep the cellular milieu constant. Hence, in all but the smallest multicellular organisms, nature has developed a circulatory system to overcome the low diffusion rates of most molecules, in particular O_2. The presence of capillaries close to tissue cells reduces the effective diffusion distances between the cell's milieu, organs that absorb nutrients and O_2, and those that dispose of waste products.

Control of the circulation is achieved by a set of intertwined systems which regulate blood pressure, blood flow, and intravascular pressure gradients. The levels at which these regulatory processes operate range from the control of the microcirculation (by interaction of tissue metabolism with smooth muscle cells around nutrient arterioles and precapillary sphincters; see Chaps. 96, 97) to control of arterial blood pressure, to adjustment of cardiac output (variation of heart rate, modulation of filling volume of the heart, and adjustment of the strength of cardiac muscle fibers; see Chaps. 89, 90, 93–95).

The demands that are put upon the circulation in different tissues vary widely. This is obvious if one considers the metabolic demands of different organs in the body. At rest, the brain, heart, and kidney are the most active organs metabolically. Hence, their oxygen consumption is highest and is met by a commensurate high blood flow, in contrast to that to the skin, resting muscles, and the gut. It is also evident that the metabolic demand of organs can vary tremendously depending on the level of activity (e.g., the metabolism of muscles increases with force development and shortening). The rate of increase of blood flow depends upon the extent of the increase in energy consumption of the organ and the interval over which the new rate of consumption is achieved. Therefore, overall control of the circulation has to ensure that the blood flow to the brain, the heart, and the kidney remains sufficient, while potentially large changes in the requirements of other organs are accommodated rapidly.

In order to develop an integrated view of this process we will first examine the control systems and the controlled systems. In the following paragraphs we will describe the components of the controlled system and the behavior of the controlling systems including the temporal characteristics of the sensors, feedback path, and processing units which compare the measured values with an internal set point and activate the response of the effector system. We will then look at the system as a whole to provide a better understanding of the interaction between individual control systems.

99.2 A Simplified Hydraulic Model of the Circulation

Cardiovascular physiologists are sometimes disparagingly referred to as "plumbers." At the risk of giving additional credence to that charge, we suggest that the most important features of the intact systemic circulation can be modeled as a very simple hydraulic circuit (Fig. 99.1A). The heart and pulmonary circulation can be represented simply by a pump which withdraws blood from compliant systemic veins and pumps it into stiff arteries. (Note that the pressure–volume relations of cylinders are simple lin-

R. Greger/U. Windhorst (Eds.)
Comprehensive Human Physiology, Vol. 2
© Springer-Verlag Berlin Heidelberg 1996

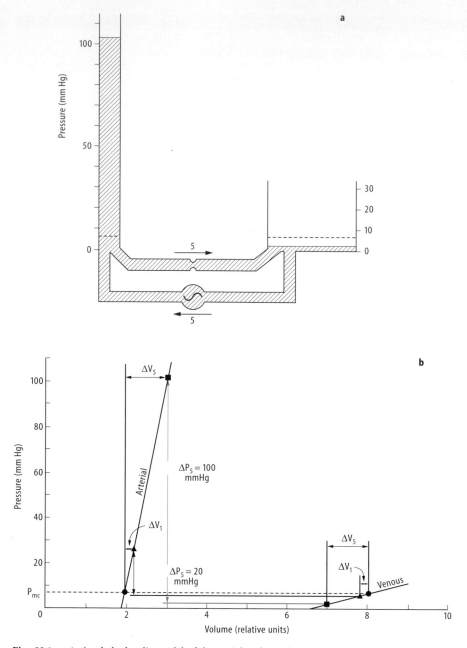

Fig. 99.1. a A simple hydraulic model of the peripheral circulation. For simplicity, the pulmonary circulation is neglected and the heart is represented by a pump which translocates volume from the high-capacity, high-compliance venous reservoir (*right*) into the low-capacity, low-compliance arterial reservoir (*left*). The *vertical scales* represent pressures in mmHg, and normal values (i.e., 5 l/min) of cardiac output and flow through the lumped systemic vascular resistance are indicated. The *dashed line* represents the value (7 mmHg) of equilibrium pressure (i.e., mean circulatory

pressure) when cardiac output is zero. **b** Arterial and venous pressure–volume relations which correspond to the hydraulic model (Fig. 99.1a). The figure shows that, when cardiac output increases, venous pressure falls because the venous reservoir is depleted. See text for details. P_{mc}, mean circulatory pressure. Systemic vascular resistance, 20 mmHg/l per min. Cardiac output: 0 l/min (●), 1 l/min(▲), 5 l/min(■). (Reproduced from Tyberg [51] with the permission of Mosby-Yearbook, Inc.)

ear functions, like those idealized relations of the arterial and venous systems.) When the pump is stationary, the equilibrium value of arterial and venous pressures is 7 mmHg, as indicated by the dashed line. This value is known as mean circulatory pressure or mean systemic filling pressure and is a measure of the "fullness of the

circulation" [38,46,47]. As such, it represents a balance between the volume of blood and the combined capacitances of the heart and blood vessels. The compliances of the arterial and venous reservoirs are assumed to be independent of volume and are represented by cylinders. As assumed by Levy [29], the ratio of the arterial to venous compliances is 1:19. When the pump is started, it tends to raise arterial pressure and lower venous pressure, the ratio of the changes in pressures being 19:1, the inverse of the ratio of the compliances. Blood returns passively to the venous reservoir through a resistance, the value of which has been arbitrarily set at $20 \, \text{mmHg} \cdot (\text{l/min})^{-1}$. An equilibrium is achieved when the arteriovenous pressure difference is just sufficient to force a flow through the resistance which is equal to the pump flow. It is evident that increasing the pump flow (i.e., cardiac output) will increase arterial pressure and decrease venous pressure. Venous pressure decreases because the venous reservoir is depleted and, indeed, the figure suggests that further increases in cardiac output will be limited by the fact that venous pressure decreases to zero.

These principles are represented graphically in Fig. 99.1B [51]. Here, the compliances of the arterial and venous reservoirs are represented as straight pressure–volume relations, the ratio of the respective slopes being 19:1 (elastance, $\Delta P / \Delta V$, is the inverse of compliance). When cardiac output is equal to zero, the pressure in the arteries and veins is equal to mean circulatory pressure ($P_{mc} = 7 \, \text{mmHg}$). When cardiac output is 1 l/min, an increment of volume (ΔV_1) is translocated from the veins to the arteries, which because of the relative compliances causes venous pressure to fall by 1 mmHg and arterial pressure to rise by 19 mmHg. When cardiac output is 5 l/min, more blood is translocated, which produces normal mean values of venous and arterial pressures (2 and 102 mmHg, respectively). Thus, it is again clear that there is an inverse relation between cardiac output and venous pressure and that this is because increasing the cardiac output depletes the venous reservoir [29].

To a first approximation, Fig. 99.1 demonstrates the fundamental hemodynamic features of the intact circulation – how the properties of the systemic circulation (i.e., resistance and arterial and venous compliances) determine intravascular pressures, dependent only on cardiac output and the "fullness of the circulation" (as measured by mean circulatory pressure) [38]. The way in which mean circulatory pressure is dependent on blood volume and vascular "tone" will be discussed below.

One of the least-understood characteristics of the circulation is how and to what extent changes in vascular (primarily venous) tone can affect cardiac output by modulating cardiac preload. It has been estimated that, as a result of reflex stimulation, the volume of the circulatory bed could vary by as much as 7.5 ml/kg, which is approximately equal to 500 ml blood for a person weighing 70 kg [45]. In other words, if total blood volume is 5 l and average venous volume (70%) is 3.5 l, changes in venous capacitance might allow venous volume to range from 3.25 to 3.75 l.

The mechanics of these changes are shown in Fig. 99.2A,B. Figure 99.2A is similar to Fig. 99.1A; a variable "unstressed (venous) volume" [37,45] is represented as a cylinder-and-piston arrangement in the bottom of the venous reservoir. It is obvious from the figure that raising the position of the piston (i.e., decreasing venous unstressed volume) is ex-

actly equivalent to a blood transfusion (or fluid retention). A decrease in unstressed venous volume or a blood transfusion has the direct effect of raising mean circulatory pressure, which, at any given value of cardiac output, raises arterial and venous pressures by exactly the same amount. The same phenomena are represented graphically in Fig. 99.2B. Like Fig. 99.2A, this figure shows that a decreases in unstressed venous volume (i.e., a shift in the venous pressure–volume relation toward the pressure axis) is equivalent to a blood transfusion (i.e., an increase in blood volume such that venous and arterial pressures are increased "passively" [37] along the original curves). In both cases, mean circulatory pressure is shown to increase by 10 mmHg, which, at any value of cardiac output, is equal to the increases in arterial and venous pressures. In this figure, the advantage of increasing mean circulatory pressure – making greater cardiac output possible – is also evident. Under the original conditions, it is clear that venous pressure would approach 0 mmHg whenever cardiac output reached 7 l/min. After a decrease in venous unstressed volume or a blood transfusion sufficient to increase mean circulatory pressure to 17 mmHg, cardiac output would have to exceed 17 l/min before venous collapse would impede flow through the great veins. As will be discussed below, the nature of the pump is such that its performance tends to increase in proportion to the increase in filling pressure (i.e., the Frank-Starling law), another obvious "advantage" of an increase in mean circulatory pressure.

As indicated at the outset, this representation of the circulation is intentionally a simplified one. Vascular compliances are not independent of volume (i.e., the pressure–volume relations are not linear [23]), and it must be recognized that systemic vascular resistance has been assumed to be constant for simplicity. However, vascular pressure volume relations are relatively linear over physiologic ranges, and the relationships which have been discussed can be recalculated readily using any other value of systemic vascular resistance.

99.3 Control of the Circulation: Central Versus Peripheral Control

It is widely agreed that the "purpose" of the circulation is to provide oxygen and nutrients to the tissues and to remove carbon dioxide and metabolic breakdown products; put in terms that the ancients would understand, it serves to supply good "humors" and to remove noxious ones. Figure 99.1 and 99.2 suggest how the circulation is controlled. Flow through the combined systemic vascular resistance is a function only of the value of the resistance and the pressure difference between the arterial and venous reservoirs. As we shall outline below, the systemic circulation is a parallel network which is subject to several central controlling mechanisms; in addition, each element controls its own resistance by mechanisms related to its own metabolism. The arteriovenous pressure difference is a

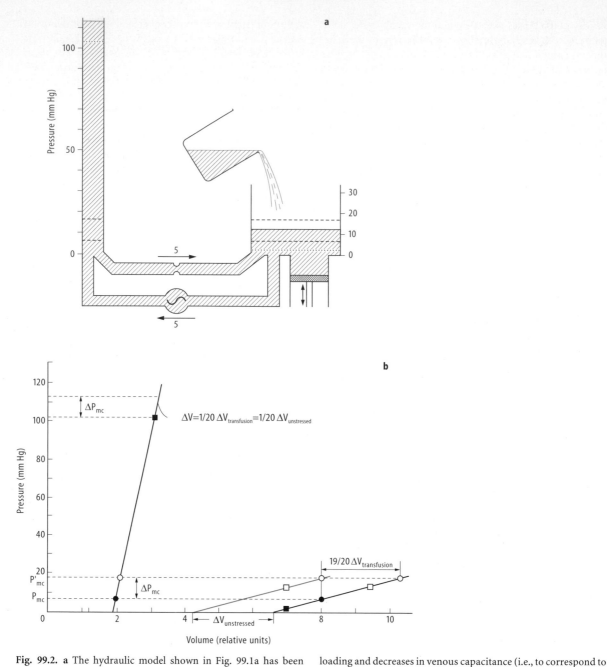

Fig. 99.2. a The hydraulic model shown in Fig. 99.1a has been modified to show the equivalence of a blood transfusion (or reduced fluid excretion) and a decrease in venous capacitance (represented by an upward movement of the piston shown in the bottom of the venous reservoir). An increase in mean circulatory pressure of 10 mmHg (from 7 to 17 mmHg) is shown. **b** Figure 99.1b has been modified to show the equivalent effects of volume loading and decreases in venous capacitance (i.e., to correspond to Fig. 99.2a). The total volume of the transfusion is distributed according to the respective venous and arterial compliances (i.e., 19:1). Thus, a 200-ml transfusion is equivalent to a 190-ml decrease in unstressed volume in that both produce the same increase in mean circulatory pressure (P_{mc}). See text for details. Cardiac output: 0 l/min (●), 1 l/min (▲), 5 l/min (■)

function of the systemic vascular resistance and the performance of the heart. At given values of resistance and of arterial and venous compliances, therefore, the arteriovenous pressure difference (i.e., the driving pressure for organ perfusion) is directly dependent on the value of cardiac output. Again, in anthropomorphic terms, all the heart needs to do is to eject blood at a sufficient rate to maintain an arteriovenous pressure difference that will allow each organ to maintain a flow of blood sufficient to meet its own metabolic needs. Thus, it becomes apparent why the most important sensors in the circulation are central arterial baroreceptors and why the important effectors are the heart and the mechanisms which modulate arteriolar resistance and venous capacitance.

99.4 Autoregulation of the Microcirculation in Tissues

Maintenance of arterial blood pressure at a constant level by a central control mechanism allows each organ to control its own perfusion by adjusting the tone of smooth muscle cells around arteriolar and precapillary sphincter vessels, thus regulating their hemodynamic resistance (see Chaps. 96, 97). This local mechanism operates at the cellular level and matches the flow through the microcirculatory bed to the metabolic needs of the local tissue. The phenomenon has been termed "autoregulation of the microcirculation" because it has been shown that nervous or circulating humoral factors are not required for its operation. The fact that organs are able to control flow through their microcirculation makes them not only metabolically independent of the neurohumoral status of the remainder of the body; it also affords them the luxury of maintaining a constant perfusion flow despite some degree of variability of the average arterial pressure. It is evident that, in the intact body, this system must interact with the neurohumoral factors which are involved in the overall regulation of blood pressure, in particular with the norepinephrine-releasing sympathetic nerve endings which cause vaso-constriction. Of course, the resistance (R) of precapillary vessels varies in inverse proportion to the fourth power of their radius (r) according to Poiseuille's law:

$$\delta P/Q = R = 8 \times \eta \times l/\pi \times r^4$$

in which Q is the flow, η is the viscosity of the blood, and l is the length of the vessel over which a pressure difference of δP exists. Hence, constriction of a vessel will augment its resistance; this effect is particularly large in small vessels, where r is small, such as in the microcirculation. Moreover, a change in R influences flow velocity, which by itself modifies viscosity of the blood because of its effects on deformation of red blood cells moving in a flow profile (see Chap. 88). An increase of arterial pressure evidently increases Q directly, but must also dilate the arterial vessels in the same way in which the pressure in a balloon dictates its diameter. As a result, variations in arterial pressure lead to amplified changes in conductance (equal to $1/R$) of the microvasculature. These conductance changes are dampened by two opposing mechanisms. First, it has been observed in several vascular beds that stretch of small blood vessels leads to active contraction, which tends to restore the original diameter of the vessel (see Chaps. 96, 97). This so-called myogenic response occurs in isolated tissue. Hence, it is probably a property of smooth muscle in the vessel wall, but the underlying mechanisms are not well understood. Second, as it leads to larger action potentials, stretch accentuates the response of smooth muscle cells to sympathetic stimulation and thereby increases vascular tone (see Chap. 96), thus opposing the effect of the initial dilatation due to the increase of arterial pressure.

Three major groups of metabolic products are thought to be involved in autoregulation of the microcirculation:

- Efflux of K^+ due to the repetitive firing of action potentials by active cells increases $[K^+]_o$.
- Metabolic activity increases the extracellular CO_2 and H^+ concentrations.
- When energy expenditure of the cell exceeds the rate of generation of adenosine triphosphate (ATP), products of nucleotide metabolism will be released.

These metabolites have major electrophysiologic effect on smooth muscle cells. Small changes of $[K^+]_o$ alter the activation potential of the arteriolar inward rectifier K^+ channels, thereby changing the conductance for K^+ ions (g_k) at the resting membrane potential. With an increase in $[K^+]_o$, g_k increases, "clamping" the membrane potential closer to the K^+ equilibrium potential so that the depolarizing effect of repetitive sympathetic activity on the smooth muscle cell membrane is reduced and the tone of the arteriolar sphincter falls. Furthermore, metabolically active cells produce CO_2 and release lactate during glycolysis, thereby, decreasing the local pH. At pH lower than 7.4, the action potential threshold of smooth muscle cells in the arteriolar wall increases and the plateau of the action potential shortens. Consequently, Ca^{2+} entry into the smooth muscle cells decreases and the cells dilate. Conversely, during hypocapnia, smooth muscle tone will increase due to enhanced Ca^{2+} entry as a result of a decreased action potential threshold and lengthening of the action potential. Release of nucleotide products, in particular adenosine, activates adenosine receptors both on sympathetic nerve endings and on smooth muscle cells. As a result, release of norepinephrine decreases and Ca^{2+} entry into smooth muscle cells decreases. The latter effect is enhanced by a direct inhibitory effect of adenosine on Ca^{2+} channels in the smooth muscle cell membrane. An added benefit of this metabolic product-mediated vasodilatation is that the changes of the concentrations of metabolites are minimized, so that stiffening of the red blood cells due to metabolite accumulation is prevented. This will tend to keep the viscosity of the blood low and the blood flow high.

99.5 The Sensors in the Central Control System of the Circulation

99.5.1 Arterial Baroreceptors

Arterial baroreceptors are located in the carotid sinus and in the aortic arch. The receptors in these structures are characterized by extensive arborizations of the nerve endings, which make the receptors suitable as stretch sensors. The response of the stretch receptors is determined by their properties as a transducer for deformation. A pressure difference across the arterial wall (i.e., a transmural pressure) tends to stretch the wall. The degree of stretch depends on the stiffness of the wall, as well as on the diameter of the artery. The firing rate of these receptors depends on the mechanical stiffness of the arterial wall and, therefore, on the tone of the smooth muscle fibers in the wall of the artery. The firing rate of the receptors is modulated both by circulating substances which alter the smooth muscle tone and by substances produced by adjacent endothelial cells, the latter probably via the formation of prostacyclin and arachidonic acid. Conversely, the firing rate is decreased by the inhibition of prostanoid formation by substances such as indomethacin and aspirin. It has long been known that these receptors respond both to the static pressure level and to the rate of change of the pressure; their properties are usually characterized by a discharge curve in which their firing rate is plotted against the blood pressure. Over a blood pressure range of 80–160 mmHg, the firing rate of the afferent fibers of the baroreceptors increases from 10 per s to 70 per s. As we will see below, firing of the baroreceptors inhibits the sympathetic efferent activity of the baroreflex. These receptors respond so rapidly that their firing pattern is synchronous with the cardiac cycle. They are, therefore, ideally suited for very rapid feedback control of blood pressure because of their rapid response and the fact that they are coupled to the vasomotor center by large, myelinated nerve fibers which can modulate their firing rate rapidly over a wide range.

Receptor Resetting. It has been shown that, if the mean blood pressure or if the systolic pressure maxima are increased over some period of time (i.e., minutes to days), the threshold at which the receptors start firing increases. This phenomenon is called receptor resetting. It has been shown that increasing the mean level of the blood pressure is the primary determinant of the magnitude of the increase of the threshold of the baroreceptor discharge curve during resetting. The importance of baroreceptor resetting is that the phenomenon is probably involved in the origin and/or maintenance of arterial hypertension, as it has been shown that the increase in the firing threshold of the baroreceptors in hypertension is proportional to the increase of the arterial pressure. Therefore, an increase of the receptor threshold will reduce their firing rate at any blood pressure and shift the blood pressure at which the receptors control the blood pressure toward higher levels. Changes in baroreceptor responsiveness have also been described in congestive heart failure; it has been observed that the reduced sensitivity of the receptors in this syndrome can be ameliorated by angiotensin-converting enzyme inhibitors and by digitalis glycosides.

99.5.2 Cardiac Receptors

Atrial Afferents. Although cardiac mechanoreceptors have long been known to be involved in reflex regulation of the circulation, their quantitative contribution to control of the circulation has been elusive. Atrial afferents have been classified according to the afferent fiber type in the vagus nerve. Myelinated atrial afferents are activated by stretch and cause reflex tachycardia without either a change of strength of contraction of the ventricle or of peripheral resistance. Firing of these fibers also increases urine flow and free water clearance and decreases vasopressin-renin activity. These responses would lead to a decreased filling pressure and would place the reflex in the category of negative feedback systems.

Much less is known about the effects of nonmyelinated atrial afferents, which seem to respond to excessive stretch of the atrium and cause a transient bradycardia and vasodilatation.

Ventricular Afferents. Ventricular vagal afferents are almost exclusively non-myelinated and are probably stimulated by an increase of ventricular pressure, although it is still a matter of debate whether diastolic or systolic pressure is a more important stimulus and what the dynamics of the response of these mechanoreceptors is. Some of these receptors are located in the coronary vasculature and are sensitive to arterial pressure. Reflex response to activation of these fibers consists of bradycardia, reduced cardiac strength, and particularly vasodilatation, so that the reflex seems to enhance the reflex response to baroreceptor activation. Their location and their effects suggest that they may be involved in a negative feedback loop which serves to accommodate variations in inflow of blood into the heart.

Ventricular Chemosensitive Endings. When stimulated, ventricular chemosensitive receptors with vagal afferents cause a profound bradycardia and vasodilatation, called the Bezold-Jarisch effect or the coronary chemoreflex. These receptors are particularly sensitive to prostaglandins, adenosine, bradykinin, and serotonin. This effect may have clinical implications, particularly when the coronary tree is exposed to drugs (e.g., intracoronary radiopaque contrast medium such as used during angiography), but the role of the reflex in control of the circulation is speculative. It may act as a negative feedback mechanism when, at high levels of cardiac energy expenditure (e.g., at high heart rate), the metabolites of ATP are released. Alternatively, the reflex may be activated when, at a normal level of energy expenditure with inadequate perfusion of the myocardium (e.g., myocardial ischemia due to coronary insufficiency), metabolites accumulate in the myocardium.

Sympathetic Cardiac Afferents. Ventricular receptors with afferents in the sympathetic system are both mechanoreceptors sensitive to stretch or chemosensitive receptors, which are classified as nociceptive. The latter group is allegedly involved in the sensation of cardiac pain, e.g., in angina pectoris. The reflex response to activation of both types of receptors is excitatory. Hence their effect would augment the original stimulus by a positive feedback. It is unclear what role this positive feedback would play in the overal control of the circulation.

99.6 The Regulatory Center for Cardiovascular Control

The vasomotor center is the main control center regulating both cardiac rate and force, as well as the peripheral vascular resistance and the smooth muscle tone in the veins. The vasomotor center is located in the rostral and ventrolateral medulla (RVLM; see Fig. 99.3 and Chap. 107). The reticular formation in the dorsal region of the medulla as well as the pontine reticular formation are thought to play secondary roles. The contribution of the latter regions depends on several factors, e.g., the level of anesthesia in animal studies; therefore, different investigators have assigned different weight to the importance of these regions. These centers are capable of independent control of the cardiovascular system, but are normally subject to influences from higher centers in the cortex, the hypothalamus, and the mesencephalon [15]. The neurons in the RVLM exhibit tonic activity which activates the discharge of postganglionic sysmpathetic nerves; in turn, their firing rate is increased under the influence of the pontine cortical

and hypothalamic systems. Sympathoinhibitory descending pathways have also been described [18]. The firing rate in the primary vasomotor center is increased by excitatory stimuli such as wakefulness, pain, mental and emotional stress, and muscular effort [10]. Afferent impulses from the heart and from the aortic and carotid sinus receptors (see below) inhibit firing by these centers. Their respective afferent fibers reach the spinal cord via the dorsal root and the medulla via the glossopharyngeal (IX) and vagal (X) nerves. The first afferent synaptic station in the medulla is the nucleus tractus solitarii (NTS). The synapses in this station allow rapid activation of the vagal nerve via the nucleus ambiguus. Activity of the NTS influences the intervals between the impulses in the sinoatrial (SA) node on a beat-to-beat basis. This is the fastest reflex arc in the control of the circulation. Afferent activity also reduces the rate of firing of efferent impulses from the sympathetic system via stations in the medulla, the spinal cord, and paravertebral ganglionic chains. From there, nerve fibers reach the heart and blood vessels in the periphery. Fibers which do not synapse may traverse the sympathetic ganglia and reach the adrenal gland directly to synapse there with

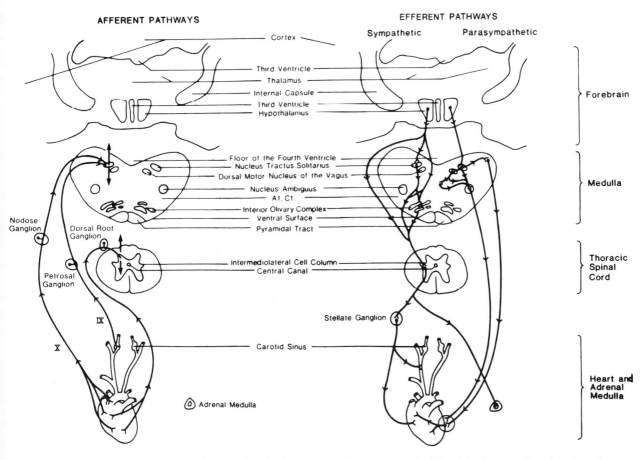

Fig. 99.3. Neuroanatomy of cardiovascular control. *Left*, afferent pathways of supraspinal and spinal reflexes originating in the sensors in the heart and the aorta and carotid sinus. The afferent fibers synapse with cells in the nucleus tractus solitarii. *Right*, efferent pathways; the parasympathetic and sympathetic efferents

are drawn on opposite sides of the diagram. Note the interaction between forebrain areas and the autonomic control of the cardiovascular system. (Reproduced from Corr et al. [13] with the permission of Raven Press)

adrenal medullary cells, which secrete epinephrine during high levels of sympathetic activation (see Chap. 17). Activation of catecholamine-dependent effects in the circulation via the latter pathway is slower but longer acting than that via sympathetic ganglia. It has been shown that both the innervation of the heart and the innervation of the blood vessels show a substantial degree of topographical organization. As a result of a decrease of arterial pressure, decreased inhibition of the sympathetic outflow from the RVLM leads to vasoconstriction; initially the splanchnic bed and the skin are constricted, while the perfusion of the muscles is preserved. The effect of the resultant net vasoconstriction is to restore the blood pressure toward the set-point value.

Neural regulation influences the heart and vascular smooth muscle cells with a time course that ranges from a fraction of a second (e.g., heart rate changes mediated by the vagal reflex) to 5–10 s (e.g., sympathetic control of cardiac force and vasomotor status). Slower cardiovascular regulatory systems include the metabolic factors involved in regulation of tissue flow (see above), circulating catecholamines, and the renin–angiotensin system. The slowest cardiovascular regulatory system is the volume control of the vascular bed via kidney-mediated water and salt transport (see Chap. 76). The latter system has a response time of several hours to days. In addition, if we consider the fact that, for most of the proteins in the human body, the turnover half-time is about 1 week and that the response time of genes that are involved in the formation of the proteins of the excitation–contraction coupling system and the contractile apparatus is no more than a matter of hours, we arrive at the fascinating conclusion that cardiac growth can be considered to be a separate dynamic regulatory mechanism by which the performance of the heart can be adjusted to the requirements of the circulation. It has been shown that growth of the cardiac myocytes can be induced by mechanical factors and is mediated by the autocrine release of angiotensin II [39]. This process is modulated in the intact organism by local (paracrine) factors produced by the invariable companions of the mycoyte, the endothelial cell, and the fibroblast. It has also been shown that both neural and circulatory factors play a role in regulation of pump function by adjusting cardiac mass to the hemodynamic work load.

99.7 The Effectors of Cardiovascular Control

99.7.1 The Heart

The strength of the heart beat can be modulated by interventions which affect the amount of Ca^{2+} released into the cytosol of the myocytes, the sensitivity of the contractile system to Ca^{2+} ions, and the sarcomere length at which the cardiac fibers contract. The rate of relaxation is modulated by interventions which determine the rate of Ca^{2+} extrusion from the cytosol. Regulatory mechanisms involved in the control of the circulation affect the intact heart both through regulation of the interval between the heartbeats and via the effects of catecholamines on the cardiac myocytes, whereas variation of the fiber length is brought about by control of the filling pressure of the ventricles via control of the capacitance of venous system. In order to elucidate this component of the cardiovascular control, we will discuss the fundamental aspects of cardiac excitation–contraction coupling and length dependence of the force of the heart beat in the following paragraphs.

Cardiac Excitation–Contraction Coupling: Structural Aspects. Figure 99.4A shows a typical example of a longitudinal electron microscopic section through a rapidly fixed trabecula of the right ventricle of a rat heart. The cell border is delineated by a glycoprotein layer overlying the sarcolemma, which invaginates the cell near the Z lines of the myofibrils. The resultant transverse tubuli (T tubuli) are rich in dihydropyridine-sensitive Ca^{2+} channels and Na^+/Ca^{2+} exchange molecules (see Chap. 93). The T tubuli make contact with a longitudinal compartment contained in a lipid membrane, the sarcoplasmic reticulum (SR), which is a prominent Ca^{2+}-accumulating organelle in mammalian myocardium. The longitudinal component of the SR envelops the myofibrils and is densely covered by Ca^{2+} ATPase molecules which drive Ca^{2+} into the SR, where it is buffered by the protein calsequestrin. The pump rate of the SR Ca^{2+} pump molecules depends on the degree of phosphorylation of the regulatory protein phospholamban in the SR membrane [11]. The terminal cisternae of the SR, which abut the T tubuli, contain Ca^{2+} channels (recognized by their high affinity for ryanodine) involved in Ca^{2+} release. Sixty percent of the intracellular space is occupied by the contractile proteins arranged in myofibrils; myofibrils contain the contractile unit, the sarcomere. The remainder of the cell is virtually completely occupied by mitochondria, which are located adjacent to the sarcomeres.

Pathways for Regulation of Cardiac Excitation–Contraction Coupling. Figure 99.4B is a diagram of the functional unit of the cardiac cell (i.e., the sarcomere), including the contractile unit and the apparatus responsible for excitation–contraction coupling (for clarity we have omitted the mitochondrion, which forms part of this unit). It is well accepted that, during the action potential, Ca^{2+} ions enter the cell through the dihydropyridine-sensitive channels. Most of this Ca^{2+} is immediately bound to the Ca^{2+} pump in the SR. Therefore, the amount of Ca^{2+} that enters the cell during a given interval will depend upon the action potential duration and the heart rate. The Ca^{2+} ions which enter through the T tubuli start the process of excitation–contraction coupling by triggering release of Ca^{2+} from the SR. The amount of Ca^{2+} that is released is proportional to the Ca^{2+} content of the SR and dictates the force of the cardiac contraction.

The released Ca^{2+} activates the contractile machinery and is subsequently eliminated from the cytosol via two pathways:

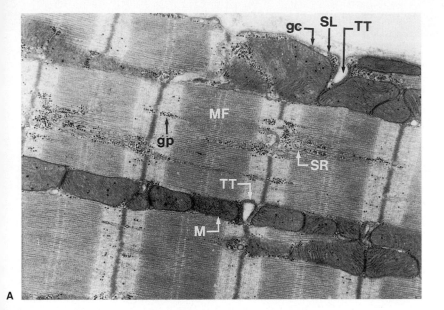

A

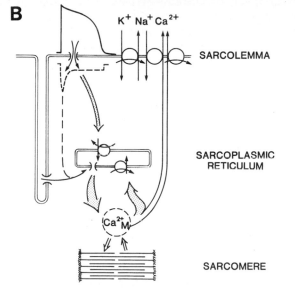

B

Fig. 99.4. A Longitudinal electron microscopic section of a cardiac cell. The close spatial relationships between the surface membrane (*SL*) covered by a glycocalyx (*gc*), the T tubuli (*TT*), and the sarcoplasmic reticulum (*SR*) are prominent. The SR is seen to envelop the myofibrils (*MF*), which constitute the major organelle in the cell (60%); the mitochondria (*M*) occupy another 40% of the intracellular space. The granules are glycogenolytic particles (*gp*). **B** Excitation–contraction coupling system in the cardiac cell. During the action potential, Ca^{2+} enters the cells as a rapid influx followed by a maintained component of the slow inward current (*short-dashed line*). Ca^{2+} entry does not lead directly to force development, as the Ca^{2+} ions that enter are rapidly bound to binding sites on the SR that envelops the myofibrils. The rapid influx of Ca^{2+} via the T tubules is thought to induce release of Ca^{2+} from the SR by triggering opening of Ca^{2+} channels in the terminal cisternae, thus activating the contractile filaments to contract. Relaxation follows because the cytosolic Ca^{2+} is sequestered again in the SR by the Ca^{2+} pump and partly extruded through the cell membrane by the Na^+/Ca^{2+} exchanger and by the low-capacity, high-affinity Ca^{2+} pump on the cell membrane. The force of contraction is thus determined by the circulation of Ca^{2+} from the SR to the myofilaments and back to the SR and by the amount of Ca^{2+} that has entered during the preceding action potential. The relaxation rate of the twitch depends on the rate of Ca^{2+} dissociation from the myofilaments and on the rates of Ca^{2+} sequestration and extrusion. It is important to note that the process of Na^+/Ca^{2+} exchange is electrogenic, so that Ca^{2+} extrusion through the exchanger leads to a depolarizing current

- Ca^{2+} that was released by the SR together with Ca^{2+} that entered the cell through the Ca^{2+} channels during the action potential is resequestered by the SR [5].
- The remaining Ca^{2+} is most probably extruded through the cell membrane by the low-affinity, high-capacity Na^+/Ca^{2+} exchanger immediately after activation of contraction.

During the diastolic interval, the low-capacity, high-affinity Ca^{2+} pump in the cell membrane lowers the cytosolic

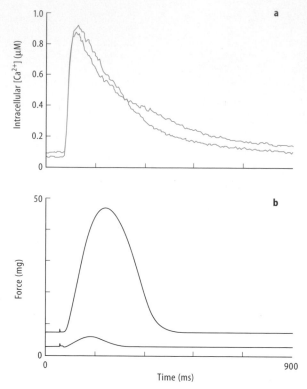

Fig. 99.5. a Transient changes of $[Ca^{2+}]_i$ (as measured by the fluorescence of the compound Fura-2) during a twitch at two sarcomere lengths. **b** The corresponding forces at the two sarcomere lengths are shown; the larger force corresponds to that recorded at an end-systolic sarcomere length of 2.15 μm, while the smaller force corresponds to that at a sarcomere length of 1.6 μm. The Ca^{2+} transient at the longer sarcomere length relaxes more rapidly in the early phase of relaxation than at the shorter sarcomere length. However, later in the relaxation process the Ca^{2+} transient at the longer sarcomere length relaxes more slowly than at the shorter sarcomere length. As a result, the Ca^{2+} transients at the two sarcomere lengths cross over during the relaxation period. (Modified from ter Keurs et al. [50] with the permission of Plenum Press)

Ca^{2+} level further [11]. In the steady state, Ca^{2+} efflux during the diastolic interval must balance in influx during the action potential.

It follows that the amount of Ca^{2+} that is accumulated in the SR also depends on the action potential duration and on the heart rate. It also follows from the above that a fraction of the Ca^{2+} ions involved in the activation of contraction are recirculated into the SR and become available for activation of the next beat, so the force of the heart beat depends on the force of the previous beat. In addition, it takes some time before the Ca^{2+} release process recovers completely from the last release and sequestered Ca^{2+} can again be released from the SR. The force of the heartbeat therefore depends strongly on the heart rate and on the interval preceding an individual beat, as well as on the duration of the action potential [43,53]. The force of contraction of the quietly beating heart is probably only approximately one third of the maximal force that can be generated by the contractile filaments, implying that the normal heart

has a wide margin before its contractile reserves are exhausted.

It is possible to overload the SR with Ca^{2+} ions. This may occur following damage of cardiac cells [14,31] or following interventions that increase the intracellular Ca^{2+} levels (e.g., digitalis, high $[Ca^{2+}]_o$, high stimulus rate). Ca^{2+} overload of the SR leads to spontaneous Ca^{2+} release, which may or may not be linked to the normal heart beat in the form of aftercontractions. Spontaneous uncoordinated Ca^{2+} release between heartbeats can be observed as spontaneous contractions of small groups of sarcomeres in cells of the myocardium and gives rise to fluctuations of the light-scattering properties of the muscle [26,28]. Spontaneous Ca^{2+} release not only increases the diastolic force generated by the contractile filaments, but also reduces systolic twitch force development [27,48]. It is clear from the above model of excitation–contraction coupling that spontaneous release of Ca^{2+} ions is likely to lead to depolarization of the sarcolemma. This may result from the activation of Ca^{2+}-dependent channels, such as transient inward currents [12,17], or by the electrogenic action of the Na^+/Ca^{2+} exchanger itself.

Cytosolic Ca^{2+} and Strength of the Heart Beat. Figure 99.5 shows twitch force and the estimated cytosolic $[Ca^{2+}]$ as a function of time at an external $[Ca^{2+}]_o$ of 1 mM in a preparation loaded by microinjection of Fura-2 salt [4]. The results are representative of contractions at short and long end-systolic sarcomere lengths, i.e., at the extremes of the function curve of cardiac muscle. The figure shows typical behavior of mammalian cardiac muscle: the peak and time course of the Ca^{2+} transients are remarkably independent of length, though the relaxation phases do differ between the short and long muscle.

The interpretation of the $[Ca^{2+}]_i$ transients and their relation to force development requires caution, since it is known that full activation of the contractile system requires saturation of all the Ca^{2+} sites of troponin C (for which approximately 70 μM Ca^{2+} is needed) with simultaneous binding of another 50 μM Ca^{2+} to calmodulin [52]. Hence, even activation of the muscle at only 25% of maximum, such as in this example, is accompanied by a Ca^{2+} turnover of near 30 μM. Only a small fraction of this Ca^{2+} is "visible" in the cytosol. However, even if we accept these limitations, it is clear that studies of the cytosolic Ca^{2+} transients have taught us a great deal about important properties of the excitation–contraction coupling process in the heart.

It is unlikely that muscle stretch only affects the rate and/ or amount of Ca^{2+} release. Rather, the stretch-induced changes in the kinetics of the transient are consistent with the hypothesis that the force–length relation (as shown in Fig. 99.7) is determined principally by the length-dependent sensitivity of the contractile system (see Fig. 99.6). It has been suggested that the length-dependent Ca^{2+} sensitivity of the contractile system is actually due to troponin's stretch-dependent Ca^{2+} affinity [20,24]. Thus, in stretched myocardium, much more Ca^{2+} is bound by troponin C [22]. The peak amplitude of the $[Ca^{2+}]_i$ transient is probably

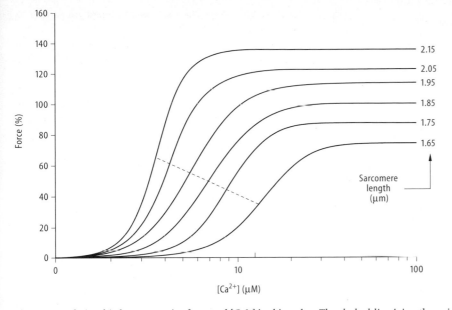

Fig. 99.6. Relationship between active force and $[Ca^{2+}]$ in skinned muscles at various sarcomere lengths (given in μm next to the appropriate curves). The *solid lines* are best fits to the modified Hill equation:

$$Force = maximum\ force \times [Ca^{2+}]^n/(K^n + [Ca^{2+}]^n)$$

The *dashed line* joins the points corresponding to half-maximal activation for each curve. It can be seen that increasing the sarcomere length raises both maximum force production and the Ca^{2+} sensitivity of the myofibrils. (From Kentish et al. [24] with the permission of the American Heart Association)

unchanged following stretch, both because more Ca^{2+} is released by the SR and because more Ca^{2+} is bound to the contractile filaments. This concept of the effect of length changes sounds attractive, but it should be remembered that an increase in the amount of Ca^{2+} release by the SR has not yet been shown experimentally.

The relaxation phase of the Ca^{2+} transient depends also on the rate of binding of Ca^{2+} ions to the sarcolemmal Na^+/Ca^{2+} exchanger and to the SR Ca^{2+} pump, as well as on their respective transport rates. Hence, studies at fixed muscle length allow evaluation of the influence of interventions on the rate of transport by those two mechanisms. Such evaluations have indicated that the rate of decline of the Ca^{2+} transient depends about equally on the activity of the Na^+/Ca^{2+} exchanger and the SR Ca^{2+} pump [9]; the latter is modulated by phosphorylation of phospholamban due to second- messenger activation and probably due to Ca^{2+}-activated calmodulin action [30].

The Frank-Starling Law of the Heart: Length Dependence of Cardiac Function. Upon arrival of Ca^{2+} ions in the vicinity of actin, Ca^{2+} binds to troponin C and contraction begins. Figure 99.6 shows that the response of the contractile apparatus to Ca^{2+} ions consists of a classical, sigmoidal, dose–response curve. These dose–response curves shift leftward (toward lower $[Ca^{2+}]_i$) with stretch of the sarcomeres [20,24]. This results in length dependence of force development, even when the Ca^{2+} release by the SR is constant.

The force at the peak of a contraction is rather insensitive to the way in which the muscle contracted prior to the peak of force and depends only upon the sarcomere length that exists at that moment. This means that the instantaneous force development of two contractions at the same "end-systolic" length is identical, even though one of the contractions is isometric while the other contraction, which started at a greater sarcomere length, has shortened to the given length while contracting against a load. Consequently, the same force–sarcomere length relationships are obtained from both isometric contractions and contractions in which substantial shortening occurs. This unique character of the force–length relationship has been found in many mammalian species [49]. The same unique relationship has been observed in the intact heart in the form of the end-systolic pressure–volume relationship (ESPVR) [41]. The existence of a fixed ESPVR is extremely important conceptually because it predicts that the stroke volume of a ventricle, which ejects blood against a fixed aortic pressure, will increase in linear proportion to the increase in end-diastolic volume. The resultant stroke volume to end-diastolic volume relationship has an intercept with the abscissa at a volume at which the ventricle can just attain aortic pressure during isovolumic contraction. This behavior of the heart has been known as the Frank-Starling law of the heart since the beginning of this century [33].

Let us return to the force–sarcomere length relationship. Figure 99.7 shows that the unstimulated passive muscle displays elastic behavior, i.e., with stretch an elastic force is generated. It is clear from the figure that this passive force rises steeply when the sarcomeres in mammalian myocardium are stretched to a length of approximately 2.3 μm. At this length the overlap between the contractile

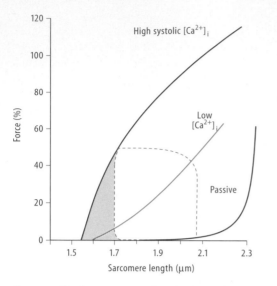

Fig. 99.7. The force sarcomere length relationship observed at a low (*light line*) and at a high (*heavy line*) [Ca²⁺]ᵢ in intact cardiac trabeculae. The relationships coincide with the relationships predicted by the curves in Fig. 99.6. The *dotted line* indicates the force–sarcomere length loop (FSLL) during cardiac systole. The *dotted area*, together with the FSLL, forms the basis for the pressure-volume area in the heart. The *heavy solid line* is the passive force–sarcomere length relation, indicating that stretch of the cardiac sarcomere is limited to a length of approximately 2.3–2.4 µm

filaments is close to optimal. The consequence of the steep rise of the passive force–length relationship is that the sarcomeres in the heart cannot be stretched beyond the length that effects an optimal overlap of the contractile filaments. Hence, the systolic force–sarcomere length relationship increases up to a length of approximately 2.3–2.4 µm and simply stops when further stretch becomes impossible. Similarly, the ESPVR shows a monotonic rise of systolic pressure with end-systolic volume without a declining phase. The predicted stroke volume to end-diastolic volume relationship also stops at high end-diastolic volumes because of the limit to further stretch. This behavior is different from that of a heart which operates at high end-diastolic pressures, in which stroke volume may decrease as filling pressure increases. Contrary to widely held belief, it is now clear that the descending limb of the Frank-Starling curve is not due to any intrinsic property of the mammalian cardiac sarcomere. This so-called descending limb at high end-diastolic pressures probably results from a complex of factors, including subendocardial ischemia, mitral regurgitation, and increased pulmonary pressures (which reduce right ventricular output).

Concerted Control of Force of the Heartbeat and of the Rate of Relaxation of the Heart: Effects of Sympathetic Activation. We have seen (here and in Chap. 91) that sympathetic nerve activity enhances the influx of Ca²⁺ ions and increases the rate of elimination of Ca²⁺ ions from the cytosol, in part due to faster uptake of Ca²⁺ into the SR. Both of these mechanisms increase the amount of Ca²⁺

available for release from the SR and thereby increase the force of the heartbeat. This allows the ventricle to contract faster, eject more blood at the same systolic pressure, and eject to a lower end-systolic volume. Ejection to a lower end-systolic volume stores energy in the elastic skeleton of the ventricle, due to deformation of the ventricular wall (both as a result of shortening of the myocytes in the wall and as a result of torsion of the ventricle, see Chap. 90). Release of the accumulated energy from the elastic skeleton during relaxation literally allows the relaxing ventricle to suck blood from the atria during the early phase of ventricular filling. Thus, enhanced ejection is accompanied by faster relaxation and, owing to the mechanical properties of the elastic skeleton of the heart, enhanced early filling. The rate of relaxation is further accelerated during exposure of the heart to catecholamines, both due to faster Ca²⁺ extrusion from the cytosol and due to phosphorylation of troponin I, in turn, due to cyclic adenosine monophosphate (AMP)-dependent activation of protein kinase A (which leads to faster dissociaton of Ca²⁺ from troponin C). Each of the three effects (enhanced suction, enhanced dissociation of Ca²⁺ from troponin C, and enhanced Ca²⁺ extrusion) both increase the inflow of blood during the early, rapid filling phase of the ventricle and allow more time for filling of the ventricle during the remainder of diastole. It is evident that catecholamine-induced acceleration of contraction and relaxation become more and more important at high heart rates, when it is necessary both to eject blood rapidly and to accelerate relaxation, in order to allow adequate inflow of blood.

99.7.2 Peripheral Resistance

Effects of Sympathetic Neurotransmitters on Smooth Muscle. Sympathetic nerve endings contact the smooth muscle syncytium at presynaptic varicosities (see Chap. 96). These varicosities release norepinephrine and ATP, which activate postsynaptic receptors and thereby increase smooth muscle membrane Na⁺ and K⁺ conductance and depolarize the membrane [21]. Depolarizations perturb voltage-dependent channels (e.g., small depolarizations close inwardly rectifying K⁺ channels, while larger depolarizations activate slow and fast Ca²⁺ channels and cause Ca²⁺ entry). The effects of these inward currents are limited by several K⁺ channels, such as the delayed rectifier and two Ca²⁺-activated K⁺ channels. In addition, beyond the varicosity, smooth muscle cells contain chemically operated channels which respond to α-adrenergic and purinergic transmitters. Both systems eventually lead to contraction of the smooth muscle cell, but afford differential catecholamine effects: local sympathetic nerve activity activates the receptors under the varicosities, while circulating catecholamines preferentially activate the extra-junctional receptors. The details of excitation–contraction coupling in smooth muscle are discussed in Chap. 44.

Hemodynamic Effects of Arteriolar Resistance Changes. With only minor exceptions (e.g., the bronchial and portal

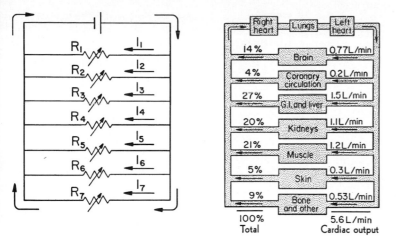

Fig. 99.8. Systemic circulation. Individual organs are located between arterial and venous "manifolds" in a parallel arrangement. Therefore, each organ has essentially the same pressure gradient available for perfusion and can regulate its own flow ($I_1, \ldots I_2 \ldots$ etc.) by changing its own (variable) resistance. Normal relative and absolute values of organ blood flow are indicated on the *right-hand panel*. *G.I.*, gastrointestinal tract; *R*, resistance. (Courtesy of Dr. Don Stubbs, University of Texas Medical Branch)

circulations), the systemic circulation can be represented as a parallel resistive network in which each organ is interposed between common arterial and venous manifolds (Fig. 99.8). These arterial and venous manifolds are common to each organ in the sense that each organ "sees" virtually the same arterial and venous pressure, hence each organ has the same pressure difference available for its perfusion. Since the arteriovenous pressure differences are the same and since the flow through the total network is equal to the sum of the flows through the individual organs, systemic vascular conductance (G_{total}) is equal to the sum of the individual organ conductances:

$$G_{total} = Q_{total}/(P_{arterial} - P_{venous})$$
$$= (Q_1 + Q_2 + \ldots + Q_n)/ \quad (P_{arterial} - P_{venous})$$

(Thus, in a parallel circuit, the reciprocal of the total systemic vascular resistance is equal to the sum of the reciprocals of the individual organ resistances, $1/R_i$; Kirchhoff's law.)

The conductance of each individual organ is regulated according to both central and local priorities. Central priorities sometimes take precedence because, to survive, the animal needs to maintain an arteriovenous pressure difference that is adequate to perfuse the vital organs. To accomplish this, the metabolic requirements of a given organ (e.g., the skin or the gut) are sometimes temporarily not met. For example, to maintain an arteriovenous pressure difference sufficient to perfuse brain, heart, and skeletal muscle during short periods of exercise, cardiac output goes up and the conductances of certain organs decrease. The increase in cardiac output would have to be even greater, were it not for the concomitant decreases in the conductances of the gut, kidney, and (initially) skin. Thus, the distribution of the total cardiac output to the various organs is controlled primarily by central and local mechanisms which determine the conductances of the individual organs. The distribution of cardiac output is only secondarily related to cardiac performance in that, so long as the heart is able to maintain an adequate arteriovenous pressure difference, appropriate organ flow will be maintained. In times of stress, cardiac output will be distributed according to the overriding requirements of the circulation and, at other times, according to the metabolic requirements of each individual organ.

99.7.3 Arterial Capacitance

The reason why we emphasize changes in venous capacitance more than changes in arterial capacitance is that they are better understood and, as the venous system contains most of the blood volume [37], they are more important. However, at a constant pressure, changes in the caliber of conducting arteries can be considerable [19]. These changes probably affect arterial pressure wave transmission and reflection [16] and may have other significant effects.

99.7.4 Venous Resistance

With respect to changes in venous resistance, our approach is different from that of many other authors. Changes in venous resistance undoubtedly occur; there is a finite pressure drop across every element in the series circulation and the changes in venous caliber that effect changes in capacitance necessarily change venous resistance. However, the "capacitance veins" and "conduction veins" must be considered independently and it is at least theoretically possible that they might be controlled independently (e.g., it would seem to be "ideal" to have conducting veins dilate when capacitance veins constrict, thereby maximizing cardiac preload).

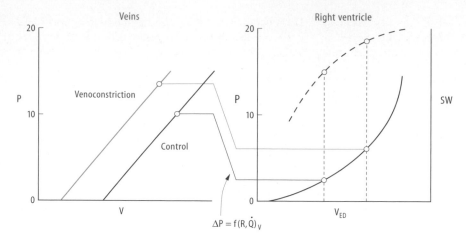

Fig. 99.9. Coupling of the venous capacitance bed and the right ventricle, showing how a decrease in venous capacitance (i.e., a decrease in venous unstressed volume) increases cardiac preload and therefore cardiac performance. The difference between the pressure of the venous capacitance bed and right ventricular end-diastolic pressure is ΔP and is a function of venous resistance (R) and flow (Q). Right ventricular end-diastolic pressure determines volume (V_{ED}; see *solid line*), which, according to the Frank-Starling law, determines stroke work (*SW*; *dashed line*). (Reproduced from Tyberg [51] with the permission of Mosby-Yearbook, Inc.)

The significance of changes in venous resistance to control of cardiac output are model dependent. In the model which we have chosen (Figs. 99.1, 99.2), changes in venous resistance contribute to changes in systemic vascular resistance only in a trivial way, since most of the pressure drop occurs across small arteries and arterioles. Only in models in which cardiac output (i.e., venous return) is likened to the rate of egress of air from the collapsing lung [34] do changes in venous resistance assume paramount importance.

99.7.5 Venous Capacitance

The importance of changes in venous capacitance to cardiovascular homeostasis under physiologic and pathophysiologic conditions is still controversial. This might be, in part, due to conceptual limitations. As suggested above, the conceptual construct which emphasized pressure–flow relations (i.e., right atrial pressure versus venous return) may not have been ideally suited to provide a basis for understanding changes in pressure–volume relations. Although it is recognized that the venous return to right atrial pressure relation is shifted upward by an increase in venous tone or a blood transfusion, it can be argued that the study of changes in the venous pressure–volume relation might have been facilitated by a more direct representation of those changes. It was established 20 years ago by Shoukas and Sagawa [45] that the capacity of the veins can change to a substantive degree in response to physiologic stimuli, and it has recently been established that the mechanism of these volume changes consists largely of parallel shifts in the pressure–volume relation [44]. The fact that such pressure volume changes pertain under conditions of physiologic stress (e.g., serial subtraction) [36] and during heart failure [32] has also been demonstrated recently. However, earlier observations of changes in the pressure of isovolumic venous segments [6] and changes in volume at constant venous pressure [54] also gave unequivocal evidence of changes in venous tone (i.e., shifts in pressure–volume relations).

Figure 99.9 shows how changes in venous capacitance might modulate cardiac preload. As discussed earlier in this chapter, venoconstriction must raise mean circulatory pressure and, at any level of cardiac output, raise all pressures in the circulation equally. This will raise right ventricular end-diastolic pressure, which tends to increase ventricular performance by the Frank-Starling mechanism. (The figure also indicates that the pressure drop between the systemic venous capacitance beds and the right heart is a function of venous resistance; a decrease in venous resistance would tend to augment the increase in ventricular preload.)

The issue of whether changes in venous capacitance are only of scientific interest or whether – like changes in heart rate, contractility, and systemic vascular resistance – they represent a major effector of cardiovascular homeostasis requires further experimental and clinical investigation.

99.8 Control of Blood Pressure by the Baroreflex

99.8.1 Control of Blood Pressure

The effect of the baroreflex is evidently to keep the blood pressure (BP) constant. In the following we will see that BP can be calculated from the known effects on the heart, peripheral resistance and venous properties. In order to do so, we will simplify the system and assume that one may consider the heart as a single pump, instead of two pumps, one of which (i.e., the right heart) feeds the pulmonary circulation and the other (i.e., the left heart) the systemic circulation. We will assume that the right heart pumps blood through the low pressure bed of the lung and thereby fills the left heart, such that the end-diastolic pressure in the left ventricle is higher than, but proportional to, the end-diastolic pressure of the right side. The output of the two ventricles (CO) is identical and equals heart rate (HR) multiplied by stroke volume (SV). Thus it is obvious that

$$BP = CO \times R_p = HR \times SV \times R_p$$

in which it is assumed that the central venous pressure is negligible compared to BP. R_p equals the peripheral resistance. In the previous paragraphs we have seen that SV depends on the force of the heart beat, the volume at the end of diastole, and the end-systolic pressure. The force of the heart beat can be judged from the slope of the end-systolic pressure–volume relationship (ESPVR) and force–length relationships shown in Fig. 99.9 for the right ventricle and in Fig. 99.7 at the level of the sarcomere. The ESPVR of the left ventricle (LV) is commonly close to linear and has the form

$$P_{lves} = E_{max} \times (V_{lves} - V_d)$$

V_d being the volume intercept of the LV ESPVR. It follows that

$$SV = V_{lved} - V_{lves}$$

Hence,

$$SV = V_{lved} - (P_{lves}/E_{max}) - V_d$$

The end-diastolic volume (V_{lved}) is an logarithmic function of end-diastolic pressure:

$$V_{lved} = K \times \log P_{lved}$$

or

$$V_{lved} = K_1 \times (\log P_{rved} + \log K_{lved/rved})$$

where $K_{lved/rved}$ is the proportionality factor between P_{lved} and P_{rved}, which we have assumed in the aforementioned simplification of the heart (this factor is small, i.e., approximately 2, and will be ignored). Figure 99.9 shows that venous smooth muscle tone dictates venous V_u and, thereby, ventricular end-diastolic pressure. Ignoring the volume of blood not contained in the veins, total blood volume (BV) equals

$$BV = V_s + V_u$$

in which V_s is the stressed and V_u the unstressed volume of the veins (see Fig. 99.2B). Hence,

$$P_{lved} \sim P_{rved} \sim P_V = K_2 \times V_s$$

where K_2 reflects the stiffness of the veins.

From these relationships one can easily show that blood pressure depends on the properties of the heart, arterioles, and capacitance veins according to the following equation:

$$BP = HR \times R_p \times [K_1 \times \log\{K_2 \times (BV - V_u)\} - (P_{lves}/E_{max}) - V_d]$$

Each of the variables in this equation is controlled: BV is subject to its own control mechanisms, and E_{max} and V_d are subject to adjustment via cardiac growth (see Chap. 89). Figure 99.10, which is based on both animal [25,45] and human [8,15,42] studies, shows that the baroreflex causes substantial changes in HR, R_p, V_u, and E_{max}, which are the controlled variables in the reflex. The range of BP over which these variables are controlled is dictated by the response characteristics of the baroreceptors and coincides with the curve relating the firing frequency with the BP. Figure 99.10 shows that the baroreflex causes rapid changes in the heart rate (HR) and slightly slower changes in the force of the heart beat as well as changes in V_u and R_p. The latter are due to changes in the tone of the smooth muscle cells contained in the venous wall and in the arteriolar sphincters, respectively.

99.8.2 Dynamics of the Control System

In the preceding paragraph we assumed that the behavior of the heart rate and the contractile force of the heart are similar functions of BP, although the regulation of the heart rate is faster than that of E_{max}. In this paragraph we will look more closely at the temporal properties of the reflex. It is well known that sympathetic and parasympathetic efferent limbs of the baroreflex change the heart rate within seconds, over a two- to three-fold range. The response of vagal efferents to blood pressure changes is as fast and allows modulation of the heart rate within 0.55 s in conscious humans [8,15]. About half of this delay is accounted for by processing of the information in the vasomotor center; the remainder is due to impulse traffic delays (0.06 s) and the response time of the S-A node (0.17 s). If the vagal activity stops, the effect of acetylcholine wears off within a second so that both the "on-time" and the "off-time" of this branch of the reflex are short. Consequently, the vagal limb of the baroreflex is able to adjust the S-A nodal intervals to the arterial pressure on a beat-to-beat basis, up to a heart rate of 80 per min. At higher heart rates, pressure changes will be converted into variation of the intervals between later heart beats.

By mechanisms that have been discussed above, activation of the sympathetic efferent fibers to the heart increases both the rate and the strength of the heart beat so that stronger, more vigorous contractions of shorter duration follow. The time available for filling of the heart is preserved by the latter mechanisms, despite the simultaneous increase of the heart rate. The consequence of the increase of the strength of the heart is an increase of the developed pressure at any end–systolic volume (see Fig. 99.7) so that, for any arterial pressure, the stroke volume increases pro-

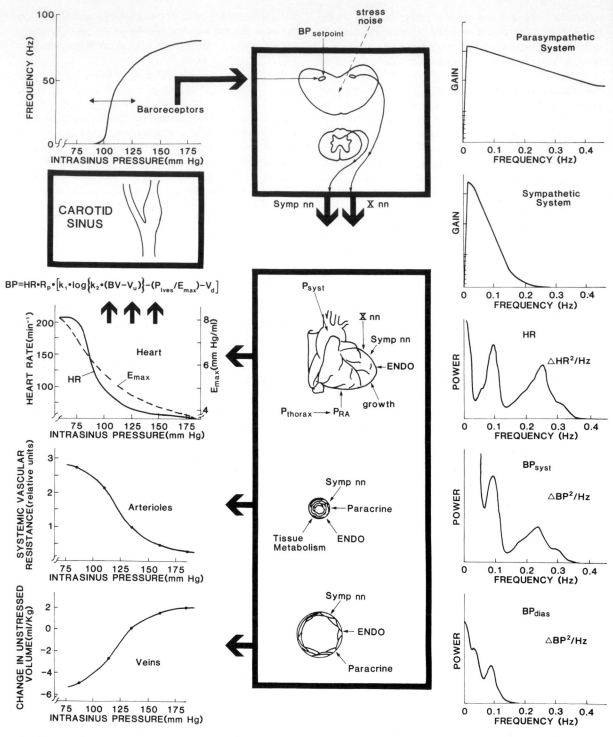

Fig. 99.10. Baroreflex-mediated control of cardiovascular function.

Left, the *top panel* shows the relationship between the impulse generation rate of the baroreceptors and arterial blood pressure (*BP*). The *horizontal line* with *arrows* indicates the relationship between the threshold of receptor firing and BP. The *three lower panels* depict the effects of blood pressure on heart rate (*HR*), the force of the heart beat (*E$_{max}$*), peripheral resistance (*R$_p$*), and unstressed volume of the veins (*V$_u$*). As shown below the *top left-hand panel* (and as is explained in the text), the blood pressure can

be calculated from the properties of the heart, arterioles, and capacitance veins:

$$BP = HR \times R_p \times [K_1 \log\{K_2(BV - V_u)\} - (P_{lves}/E_{max}) - V_d].$$

HR, R_p, V_u, and E_{max} are controlled variables in the baroreflex; blood volume (BV) is subject to its own control mechanisms, and E_{max} and V_d are subject to adjustment via cardiac growth.
Middle, the main processing station is the medulla, which controls (via parasympathetic, *X nn*, and sympathetic, *Symp nn*, pathways)

portionally. By this mechanism, the larger stroke volume increases arterial pressure, so that sympathetic activation adjusts blood pressure when the baroreflex signals a deviation of blood pressure from the set point of the reflex.

The delay of the sympathetically mediated heart rate response is substantially slower (i.e., approximately 2 s) because of the longer conduction times of the sympathetic nerve fibers compared to the vagal fibers (1 m/s compared to 7 m/s, respectively) and because of the longer delays at the junction between the sympathetic nerve fibers and the target organs. The off-time constant of the sympathetic effect is in the order of 10 s, so that this limb of the reflex is responsible for much slower control of the heart rate than the vagal pathway. The sympathetic limb of the control system is, therefore, only effective at changes that take more than 5–10 s to develop. This limb of the baroreflex has a markedly reduced gain and, if faster changes occur, a delay between the pressure change and the response is clearly noticeable.

The negative feedback effect of the baroreflex is further enhanced by the influence of the sympathetic nerves on the peripheral arterial resistance, but this effect also takes slightly longer because of the longer latencies in the feedback path. Sympathetic activation causes an increased arterial resistance which predominantly increases diastolic blood pressure and counteracts an initial decrease of central arterial pressure sensed by the carotid and aortic baroreceptors. The reflex time of the response of diastolic arterial pressure to a drop of carotid blood pressure in large animals and humans is 2–3 s [15].

Based on the above considerations, we expect a strong correlation between variation of heart rate, cardiac strength, and the arterial blood presure. This intuitive notion is borne out if one observes the natural variations of heart rate and systolic and diastolic blood pressure simultaneously. It appears simplest to describe the fluctuations of these variables as power spectra, much in the way in which, on the display panel of a modern audio system, the power (or intensity of the fluctuations) of the music is reflected as a function of the tone frequency. Such spectra have been described by Akselrod [2] and other investigators and exhibit characteristic peaks at frequencies between 0.03 and 0.1 and between 0.25 and 0.35 Hz (see Fig. 99.10), which are apparent in the variablility of the heart rate as well as in the variability of the blood pressure. The peak at 0.3 Hz coincides with the main frequency content of our respiratory activity and is due to interaction between respiration and the cardiovascular reflex. Several mechanisms may cause the striking influence of the respiratory activity on the heart rate and systolic blood pressure. The pressure changes in the thorax during respiration directly affect stroke volume and, hence, arterial pressure. Next, the vagal limb of the baroreflex immediately transforms the arterial pressure changes into changes of the intervals between the heart beats. As a result, an increase of stroke volume is not accompanied by an increase in diastolic pressure, despite an increase in systolic pressure, because the increased interval between the heart beats allows the diastolic pressure to fall to near its original level. The pressure and rate variations at 0.3 Hz cannot be mediated by the sympathetic limbs of the baroreflex because the delay in the sympathetic effector is too long [15,42]. In addition, they disappear with atropine, proving their vagal origin. In contrast, the blood pressure and heart rate variations at 0.03–0.1 Hz (i.e., the 20-s variability) are probably due to the sympathetic feedback loops. These fluctuations have been explained by assuming that the vasomotor system is influenced by random neural activity, which would give rise to random variations of the cardiovascular parameters; the 10-s delay in the sympathetic feedback line effectively amplifies the fluctuations at 0.05 Hz (see Chap. 1). The intensity of these fluctuations is increased in individuals with enhanced sympathetic drive. The power spectrum of heart rate variations can easily be obtained in humans; the ratio of the power in the low-frequency range (P_L) over that in the high-frequency range (P_H) is widely used for the analysis of sympathovagal imbalance in various diseases. Examples are the changes in the P_L to P_H ratio after a myocardial infarct, in the presence

---

the functional status of the heart and the tone of smooth muscle cells in the arteriolar resistance and venous capacitance vessels. The set point of the baroreflex $(BP_{setpoint})$ is subject to fluctuations such as those due to neuronal noise (*noise*) or activity in higher nervous centers (*stress*). Effects from systems other than the baroreflex are also indicated next to the effector organs, including endocrine (*ENDO*) and paracrine systems and factors which control growth of the heart (*growth*). Direct effects of systolic aortic pressure (P_{syst}) and right atrial pressure (P_{RA}), in turn affected by thoracic pressure (P_{thorax}), are indicated separately. Afferent impulses from the baroreceptors in the large arteries inhibit sympathetic activity in the efferent fibers and activate the parasympathetic system. The influence of tissue metabolism allows autoregulation of tissue flow

Right, the *top two panels* depict the dynamic properties of the reflex. The response of the parasympathetic system is fast; the response mediated by the sympathetic nerves is 20 times slower. For the parasympathetic control system, this translates into a high feedback gain over a wide range of frequencies while, for the sympathetic system, control is progressively weaker with higher-frequency perturbations [42]. The *three lower panels* depict typical peaks in the power spectra of fluctuations in (HR) and systolic and diastolic BP. The peaks at 0.25 Hz are caused by the interaction between respiration and the heart rate. Due to the 10-s delay in the sympathetic control path, fluctuations in the input of the medullary centers cause the fluctuations of BP and HR in the range of frequencies from 0.05 to 1 Hz [15]. (The figure was constructed from data that appeared in Sagawa et al. [40], Shoukas and Sagawa [45], Andresen and Wang [3], Korner [25], Saul et al. [42], and de Boer et al. [15] with the permission of the American Heart Association and the American Physiological Society)

of diabetic neuropathy, and in the aged, as well as the complete loss of P_L in congestive heart failure [7].

The gain of the baroreflex decreases at frequencies higher than $0.05-1\,Hz$, as we have previously seen in the response characteristics of the sympathetic and the parasympathetic arms of the reflex. At low frequencies ($<0.003\,Hz$) the gain also decreases, but then it is the result of resetting of the large-artery baroreceptors, which occurs as a consequence of changes in mean blood pressure persisting for longer than about 5 min. Consequently, enduring changes in blood pressure caused by factors external to the reflex (e.g., metabolic factors such as increased salt intake or factors which change the responsiveness of the smooth muscle cells in the arteriolar sphincters) must be accommodated for by other blood pressure-regulating systems which have higher gains at the low-frequency end of the spectrum.

99.9 Interaction Between the Baroreflex and Humoral Regulatory Mechanisms

Vasopressin. Vasopressin or antidiuretic hormone (ADH), as it was previously known, is released by the posterior lobe of the pituitary gland (see Chaps 19, 74, 76, 82, 97). Reductions in blood volume and blood pressure are potent stimuli for the release of vasopressin [1,44]. It is therefore not surprising that the peptide has long been implicated in control of the circulation. Circulating vasopressin has effects at multiple levels of the cardiovascular regulatory system. Under basal conditions the concentrations of vasopressin in the blood are low and do not seem to cause cardiovascular effects. In the intact normal animal, vasopressin has little pressor effect, probably owing to the direct vasoconstrictor effect together with inhibition of sympathetic efferents. The latter effect of vasopressin is due both to central facilitation of the baroreflex and to the sensitization of both high-pressure arterial receptors and the low-pressure receptors. However, following hemorrhage the level of vasopressin rises acutely and is partly responsible for the rapid increase in peripheral resistance. A similar effect of vasopressin was observed when the volume decrease was caused by water deprivation. The previous name for vasopressin – antidiuretic hormone – implies the long-known effect of the hormone: the peptide regulates osmolality by blocking reabsorption of water in the kidney (see Chaps. 74, 76, 82). Evidently, water retention increases blood volume and thereby regulates blood pressure, albeit more slowly than the baroreflex.

The Renin–Angiotensin System. Angiotensin II is the active product of the renin–angiotensin pathway. The hormone is produced from angiotensin I by angiotensin-converting enzyme under control of renin (see Chap. 97). Angiotensin II is a potent vasoconstrictor [35]. In addition, angiotensin II has been reported to influence sympathetic nerve activity, although this finding has been variable and the mechanism and the path of its action need further evaluation. Angiotensin II in the brain has been reported to increases sympathetic discharge and reduce the vagal tone to the heart. The effect of angiotensin II on the sympathetic ganglia is stimulatory. The mechanisms that have been proposed include facilitation of norepinephrine release and inhibition of its re-uptake. Importantly, angiotensin II does not induce a substantial change of the heart rate, although a bradycardia would be expected because of the increase in blood pressure following administration of angiotensin. This noneffect on the heart rate has been attributed to resetting of the baroreflex, possibly both as a result of a change of the threshold of the baroreceptors and as a result of the sensitivity of the reflex to variations in the blood pressure. The increased peripheral resistance following exposure to angiotensin II is due to a direct effect as well as to an increased vasoconstrictor response to norepineprine.

99.10 Conclusions

The heart and the vascular system serve the body by supplying blood flow, which provides oxygen and nutrients and eliminates waste products from the tissue cells. This chapter describes the interaction between the heart and the vascular system based on a simple concept of the properties of the venous and arterial system. The heart pumps blood from the compliant venous compartment to the stiff arterial compartment and thereby sets up an arterial pressure which exceeds the venous pressure by more than tenfold. The microcirculation through the tissues is adjusted (autoregulated) by changing arteriolar tone to meet the needs of each organ according to the metabolic status of the tissue cells. Autoregulation of the microcirculation can proceed in an unperturbed fashion because blood pressure variations which would result from the changes in tissue perfusion are eliminated by a set of regulatory systems: the baroreflex, blood volume control via the kidney, and other humoral regulatory mechanisms.

The baroreflex controls heart rate and the force of the heart beat, as well as peripheral resistance and venous tone. The influence of the control system on the smooth muscle cells in the wall of the veins and arteries allows interaction between autoregulation of the microcirculation and baroreflex on the arteriolar resistance. The effect of the control system on the smooth muscle cells in the veins appears to be unexpectedly simple: it modifies the unstressed volume of these vessels with little or no change in the apparent stiffness of the veins. A decrease in venous unstressed volume appears to have similar effects as a transfusion (i.e., similar effects on vascular and mean circulatory pressure are produced). The change in venous pressure causes more filling of the ventricles and thereby increases stroke volume, an effect which has been known since the beginning of the twentieth century as the Frank-Starling law of the heart. The Frank-Starling law can be understood at the level of the cardiac contractile filaments if we realize that the sensitivity of these filaments to the activator Ca^{2+} ions is highly length dependent. In addition, the baroreflex influences the force of the heart beat and the rate of relaxation of the heart by enhancing entry of Ca^{2+} ions into the cardiac cell as well as enhancing uptake of Ca^{2+} ions into the SR. Control of these four variables accounts for adjustment of the blood pressure in the case of perturbations such as are elicited by variations in tissue perfusion, but also as a result of variations of intrathoracic pressures during breathing. The baroreflex is maximally effective when blood pressure changes occur at frequencies between 0.003 and 1 Hz. At lower frequencies, the effect of the reflex is limited because the baroreceptors adapt to changes in blood pressure by a mechanism known as resetting. At higher frequencies, the response times of the efferent paths of the reflex limit the gain of the reflex. The differences in response characteristics of the sympathetic and parasympathetic limb can be recognized in the spectra of variability of heart rate and blood pressure. Insight into these responses forms the basis of a greater understanding of cardiovascular physiology and pathophysiology.

References

1. Abboud FM, Floras JS, Aylward PE, Guo GB, Gupta BN, Schmid PG (1990) Role of vasopressin in cardiovascular and blood pressure regulation. Blood Vessels 27:106–115
2. Akselrod S, Gordon D, Shannon DC, Barger AC, Cohen RJ (1981) Power spectrum analysis of heart rate fluctuation: a quantitative probe of beat-to-beat cardiovascular control. Science 213:220–222
3. Andresen MC, Yang M (1990) Dynamic and static conditioning pressures evoke equivalent rapid resetting in rat aortic baroreceptors. Circ Res 67:303–311
4. Backx PH, ter Keurs HEDJ (1993) Fluorescent properties of rat cardiac trabeculae microinjected with fura-2 salt. Am J Physiol (Heart Circ Physiol) 264:H1098–H1110
5. Bers DM (1991) Excitation-contraction coupling and cardiac contractile force. Kluwer Academic, Boston
6. Bevegard BS, Shepherd JT (1965) Changes in tone of limb veins during supine exercise. J Appl Physiol 20:1–8
7. Binkley PF, Nunziata E, Haas GJ, Nelson SD, Cody RJ (1991) Parasympathetic withdrawal is an integral component of autonomic imbalance in congestive heart failure: Demonstration in human subjects and verification in a paced canine model of ventricular failure. J Am Coll Cardiol 18:464–472
8. Borst C, Karemaker JM (1983) Time delays in the human baroreceptor reflex. J Auton Nerv Syst 9:399–409
9. Bridge JHB, Spitzer KW, Ershler PR (1988) Relaxation of isolated ventricular cardiomyocytes by a voltage-dependent process. Science 241:823–825
10. Calaresu FR, Yardley CP (1988) Medullary basal sympathetic tone. Annu Rev Physiol 50:511–524
11. Carafoli E (1987) Intracellular calcium homeostasis. Annu Rev Biochem 56:395–433
12. Colquhoun D, Neher E, Reuter H, Stevens CF (1981) Inward current channels activated by intracellular calcium in cultured cardiac cells. Nature 294:752–754
13. Corr PB, Yamada KA, Witkowski FX (1986) Mechanisms controlling cardiac autonomic function and their relation to arrhythmogenesis. In: Fozzard HA, Haber E, Jennings RB, Katz AM, Morgan HE (eds) The heart and cardiovascular system. Scientific foundations. Raven, New York, pp 1343–1404
14. Daniels MCG, Fedida D, Lamont C, ter Keurs HEDJ (1991) Role of the sarcolemma in triggered propagated contractions in rat cardiac trabeculae. Circ Res 68:1408–1421
15. DeBoer RW, Karemaker JM, Strackee J (1987) Hemodynamic fluctuations and baroreflex sensitivity in humans: a beat to beat model. Am J Physiol (Heart Circ Physiol) 253:H680–H689
16. Fitchett DH, Simkus G, Genest J, Marpole D, Beaudry P (1988) Effect of nitroglycerin on left ventricular hydraulic load. Can J Cardiol 4:72–75
17. Han X, Ferrier GR (1992) Ionic mechanisms of transient inward current in the absence of Na^+/Ca^{2+} exchange in rabbit cardiac Purkinje fibres. J Physiol (Lond) 456:19–38
18. Hayes K, Yardley CP, Weaver LC (1991) Evidence for descending tonic inhibition specifically affecting sympathetic pathways to the kidneys in rats. J Physiol (Lond) 434:295–306
19. Hayoz D, Tardy Y, Rutschmann B, Mignot JP, Achakri H, Feihl F, Meister JJ, Waeber B, Brunner HR (1993) Spontaneous diameter oscillations of the radial artery in humans. Am J Physiol (Heart Circ Physiol) 264:H2080–H2084
20. Hibberd MG, Jewell BR (1982) Calcium- and length-dependent force production in rat ventricular muscle. J Physiol (Lond) 329:527–540
21. Hirst GDS, Edwards FR (1989) Sympathetic neuroeffector transmission in arteries and arterioles. Physiol Rev 69:546–604
22. Hofmann PA, Fuchs F (1988) Bound calcium and force development in skinned cardiac muscle bundles: effect of sarcomere length. J Mol Cell Cardiol 20:667–677
23. Katz AI, Chen Y, Moreno AH (1969) Flow through a collapsible tube: experimental analysis and mathematical model. Biophys J 9:1261–1279
24. Kentish JC, ter Keurs HEDJ, Ricciardi L, Bucx JJJ, Noble MIM (1986) Comparison between the sarcomere length-force relations of intact and skinned trabeculae from rat right ventricle. Circ Res 58:755–768
25. Korner PI (1989) Baroreceptor resetting and other determinants of baroreflex properties in hypertension. Clin Exp Pharmacol Physiol [Suppl] 15:45–64
26. Kort AA, Lakatta EG (1984) Calcium-dependent mechanical oscillations occur spontaneously in unstimulated mammalian cardiac tissues. Circ Res 54:396–404
27. Lakatta EG, Jewell BR (1977) Length-dependent activation. Its effect on the length-tension relation in cat ventricular muscle. Circ Res 40:251–257
28. Lakatta EG, Lappe DL (1981) Diastolic scattered light fluctuation, resting force and twitch force in mammalian cardiac muscle. J Physiol (Lond) 315:369–394
29. Levy MN (1979) The cardiac and vascular factors that determine systemic blood flow. Circ Res 44:739–747
30. Morris GL (1993) The regulatory interaction between phospholamban and SR (Ca^{2+}-Mg^{2+}) ATPase. Thesis, University of Calgary
31. Mulder BJM, de Tombe PP, ter Keurs HEDJ (1989) Spontaneous and propagated contractions in rat cardiac trabeculae. J Gen Physiol 93:943–961
32. Ogilvie RI, Zborowska-Sluis D (1992) Effect of chronic rapid ventricular pacing on total vascular capacitance. Circulation 85:1524–1530
33. Patterson SW, Piper H, Starling EH (1914) The regulation of the heart beat. J Physiol (Lond) 48:465–513
34. Permutt S, Caldini P (1975) Regulation of cardiac output by the circuit: venous return. In: Baan J, Noordergraaf A, Raines J (eds) Cardiovascular system dynamics. Massachusetts Institute of Technology, Cambridge, pp 465–479
35. Reid IA (1992) Interactions between angiotensin II, sympathetic nervous system, and baroreceptor reflexes in regulation of blood pressure. Am J Physiol (Heart Circ Physiol) 262:E763–E778
36. Robinson VJB, Manyari DE, Tyberg JV, Fick GH, Smith ER (1989) Volume-pressure analysis of reflex changes in forearm venous function. A method by mental arithmetic stress and radionuclide plethysmography. Circulation 80:99–105

37. Rothe CF (1983) Reflex controls of veins and vascular capacitance. Physiol Rev 63:1281–1341
38. Rothe CF (1993) Mean circulatory pressure: its meaning and measurement. J Appl Physiol 74:499–509
39. Sadoshima J, Xu Y, Slayter HS, Izumo S (1993) Autocrine release of angiotensin II mediates stretch-induced hypertrophy of cardiac myocytes in vitro. Cell 75:977–984
40. Sagawa K, Maughan L, Suga H, Sunagawa K (1988) Cardiac contraction and the pressure-volume relationship. Oxford University Press, New York, pp 139–143
41. Sagawa K, Suga H, Shoukas AA, Bakalar KM (1977) End-systolic pressure-volume ratio: a new index of ventricular contractility. Am J Cardiol 40:748–753
42. Saul JP, Berger RD, Albrecht P, Stein SP, Chen MH, Cohen RJ (1991) Transfer function analysis of the circulation: unique insights into cardiovascular regulation. Am J Physiol (Heart Circ Physiol) 261:H1231–H1245
43. Schouten VJA, Deen JK, Tombe PP, Verveen AA (1987) Force-interval relationship in heart muscle of mammals. A calcium compartment model. Biophys J 51:13–26
44. Share L (1988) Role of vasopressin in cardiovascular regulation. Physiol Rev 68:1248–1284
45. Shoukas AA, Sagawa K (1973) Control of total systemic vascular capacity by the carotid sinus baroreptor reflex. Circ Res 33:22–33
46. Starling EH (1912) Principles of human physiology. Lea and Febiger, Philadelphia, pp 880–884, 909–9
47. Starr I, Rawson AJ (1940) Role of the "static blood pressure" in abnormal increments of venous pressure, especially in heart failure. I. Theoretical studies on an improved circulation schema whose pumps obey Starling's law of the heart. Am J Med Sci 199:27–39
48. Stern MD, Capogrossi MC, Lakatta EG (1988) Spontaneous calcium release from the sarcoplasmic reticulum in myocardial cells: mechanisms and consequences. Cell Calcium 9:247–256
49. ter Keurs HEDJ, Kentish JC, Bucx JJJ (1987) On the force-length relation in myocardium. In: ter Keurs HEDJ, Tyberg JV (eds) Mechanics of the circulation. Nijhoff, Dordrecht, pp 91–105
50. ter Keurs HEDJ, Backx PH, Banijamili H, Mclntosh B, Gao WD. Calcium release and force development in rat myocardium. In: Frank GB, Bianchi CP, ter Keurs HEDJ (eds) Excitation-contraction coupling in skeletal, cardiac, and smooth muscle. Plenum, New York, pp 199–212
51. Tyberg JV (1992) Venous modulation of ventricular preload. Am Heart J 123:1098–1104
52. Wier WG, Yue DT (1986) Intracellular calcium transients underlying the short-term force-interval relationship in ferret ventricular myocardium. J Physiol (Lond) 376:507–530
53. Wohlfart B (1979) Relationship between peak force, action potential duration, and stimulus interval in rabbit myocardium. Acta Physiol Scand 106:395–409
54. Zelis R, Mason DT (1969) Comparison of the reflex reactivity of skin and muscle veins in the human forearm. J Clin Invest 48:1870–1877

M Respiration

100 Pulmonary Ventilation

B.J. WHIPP

Contents

100.1 Introduction

The human organism requires a continuous exchange of material between the environment and its constituent cells in order to maintain life. The nutrients which are essential for cellular energy exchange must be transported to the cells by highly specialized organ systems whose structures are exquisitely "designed" for the function they serve.

The lung (the convention of referring to both lungs as "the lung" is followed here) functions as the first step in the transport system which transfers sufficient oxygen (O_2) from the atmosphere into the blood so that the cardiovascular system is able to supply cellular oxygen requirements without undue stress. The lung also serves as the terminal step in the clearance pathway for carbon dioxide (CO_2). This allows the cellular acid-base status to be maintained within the relatively narrow limits compatible with life. These functions require:

- Generation of negative, or subatmospheric, pressures within the lung such that atmospheric air is propelled into the lung by convection as the atmospheric pressure

(P_{atm}) becomes greater than the pressure in the alveoli (P_{alv}). This pressure difference ($P_{atm} - P_{alv}$) is generated by the enlargement of the lungs as a result of the respiratory muscle contraction expanding the thorax.
- Mixing the volume of newly inhaled air with that already in the lung. This increases the alveolar O_2 concentration and reduces its CO_2 concentration.
- Movement of O_2 into the blood and CO_2 from the blood. This is brought about by diffusion along the newly established gas partial pressure gradients.
- A control system – that is, a system capable of "sensing" whether specific chemical and physical features of the blood and CSF, especially the O_2 and CO_2 levels, and acid-base status are appropriate to the body's current requirements.

In this chapter we will discuss both the determinants and the limits of thoracic volume excursions and also the determinants and limits of airflow generation – that is, the static and dynamic aspects of the mechanics of breathing. For convenience, the special symbols which are conventionally used in pulmonary physiology [33] are listed in Table 100.1.

100.2 Statics of Breathing

100.2.1 Functional Anatomy of the Lung

It is convenient to consider the lungs as being composed of:

- Airways that conduct gas to and from the gas exchange regions – these comprise the "conducting zone" of the lung, in that they are not themselves directly involved in arterializing the venous blood.
- The gas exchange regions – i.e., the "respiratory zone" within which mixed venous blood is "arterialized" within the pulmonary capillaries as gas is exchanged across the alveolar capillary membrane.

R. Greger/U. Windhorst (Eds.)
Comprehensive Human Physiology, Vol. 2
© Springer-Verlag Berlin Heidelberg 1996

Table 100.1. Symbols and abbreviations used by pulmonary physiologists[a]

Special symbols:
Dash (−) above any symbol indicates a *mean* value
Dot (·) above any symbol indicates a *time derivative*

Gases

Primary symbols		*Examples*	
V	= gas volume	V_A	= volume of alveolar gas
$\dot{V}$	= gas volume/unit time	V_{O_2}	= O_2 consumption per minute
P	= gas pressure	$P_{A_{O_2}}$	= alveolar O_2 pressure
$\bar{P}$	= mean gas pressure	$\bar{P}_{C_{O_2}}$	= mean capillary O_2 pressure
F	= fractional concentration in dry gas phase	$F_{I_{O_2}}$	= fractional concentration of O_2 in inspired gas
f	= respiratory frequency (breaths/unit time)	D_{O_2}	= diffusing capacity for O_2 (ml O_2/mm Hg per minute)
D	= diffusing capacity	R	= $\dot{V}_{CO_2}/\dot{V}_{O_2}$
R	= respiratory exchange ratio		
STPD	= 0°C, 760 mmHg, dry		
BTPS	= body temperature and pressure saturated with water vapor		

Secondary symbols		*Examples*	
I	= inspired gas	$F_{I_{CO_2}}$	= fractional concentration of CO_2 in inspired gas
E	= expired gas	V_E	= volume of expired gas
A	= alveolar gas	$\dot{V}_A$	= alveolar ventilation per minute
T	= tidal gas	V_T	= tidal volume
D	= deal space gas	V_D	= volume of dead space gas
B	= barometric	P_B	= barometric pressure

Blood

Primary symbols		*Examples*	
Q	= volume of blood	Q_C	= volume of blood in pulmonary capillaries
$\dot{Q}$	= volume flow of blood/unit time	$\dot{Q}_C$	= blood flow through pulmonary capillaries per minute
C	= concentration of gas in blood phase	$C_{a_{O_2}}$	= ml O_2 in 100 ml arterial blood
S	= % saturation of Hb with O_2 or CO	$S\bar{v}_{O_2}$	= saturation of Hb with O_2 in mixed venous blood

Secondary symbols		*Examples*	
a	= arterial blood	$P_{a_{CO_2}}$	= partial pressure of CO_2 in arterial blood
v	= venous blood	$P\bar{v}_{O_2}$	= partial pressure of O_2 in mixed venous blood
c	= capillary blood	$P_{c_{CO}}$	= partial pressure of CO in pulmonary capillary blood

Lung volumes

V_T	= volume of air inhaled or exhaled with each breath
VC	= vital capacity = maximal volume that can be expired after maximal inspiration
IC	= inspiratory capacity = maximal volume that can be inspired from resting expiratory level
IRV	= inspiratory reserve volume = volume that can be inspired from end-tidal inspiration
ERV	= expiratory reserve volume = maximal volume that can be expired from resting expiratory level
FRC	= functional residual capacity = volume of gas in lungs at resting end-expiratory level
RV	= residual volume = volume of gas in lungs at end of maximal expiration
TLC	= total lung capacity = volume of gas in lungs at end of maximal inspiration

[a] Adapted from [11], based on [33].

During inspiration the inspired air passes through the nose and/or mouth, the pharynx, and the larynx. It enters the trachea or the lung "proper". In adult humans the *trachea* is a single conducting airway approximately 10–11 cm long and 2 cm in mean diameter. It is supported by U-shaped struts of cartilage, the open arms of the "U" being located posteriorly but linked together by smooth muscle. The trachea bifurcates into the right and left main *bronchi*. These subsequently bifurcate into smaller bronchi and so on for a total of some 11 branchings (Fig. 100.1). These bronchi have cartilaginous support and also smooth muscle arranged in helical bands.

From the 12th branching through the 16th, the airways lack cartilaginous support; they are therefore termed *bronchioles* rather than bronchi. The bronchioles, however, have strong helical bands of smooth muscle. The volume of air contained in the "conducting zone", i.e., the entire volume of the respiratory tract, from the port of entry into the body, whether the nose or the mouth, through the terminal bronchioles, does not exchange gas with venous blood. With respect to gas exchange it is consequently "dead space". This volume is termed the *anatomical dead space*, and is about 150 ml in normal adult humans; naturally it is bigger in bigger subjects than in small ones. A useful rule of thumb is that dead space in milliliters is about twice the lean body weight in kilograms. It is also slightly larger at the end of inspiration than at the end of expiration as the intrathoracic portion of the anatomical dead space is distended as lung volume increases. The transition from approximately the 16th to the 17th generation of the airway branching is of enormous functional significance. The terminal bronchioles are replaced by *respiratory bronchioles*, which after several subsequent branchings lead to *alveolar ducts*, *alveolar sacs*, and finally

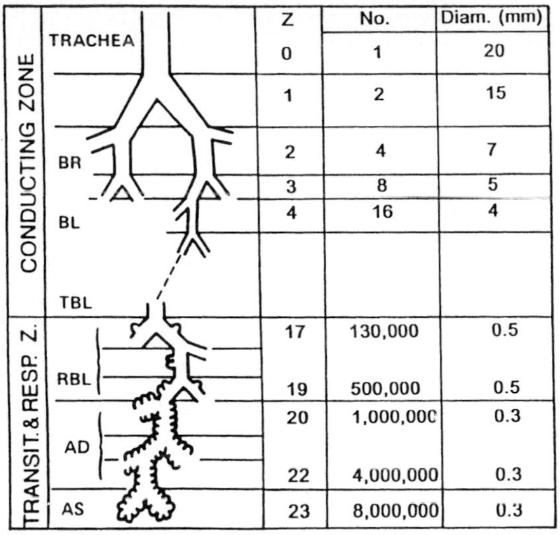

	Z	No.	Diam. (mm)
TRACHEA	0	1	20
	1	2	15
BR	2	4	7
	3	8	5
BL	4	16	4
TBL			
RBL	17	130,000	0.5
	19	500,000	0.5
AD	20	1,000,000	0.3
	22	4,000,000	0.3
AS	23	8,000,000	0.3

(CONDUCTING ZONE; TRANSIT. & RESP. Z.)

Fig. 100.1. The symmetrical branching model of the airways. *BR*, bronchus; *BL*, bronchiole; *TBL*, terminal bronchiole; *RBL*, respiratory bronchiole; *AD*, alveolar duct. (From [42, p. 111])

the *alveoli* themselves. All these regions that are capable of gas exchange comprise the "respiratory zone" of the lung (Fig. 100.1). The basic gas exchange unit from a respiratory bronchiole to the consequent clusters of alveoli is termed an *acinus*.

Thus, a single airway with a cross-sectional area of about 2 cm² leads, via some 24 successive generations, to approximately 300 million alveoli with a total surface area of 70–90 m².

Gas exchange takes place across the *alveolar-capillary membrane* – a structure superbly suited to its function. It has an enormous surface area (70–90 m² in an adult human) and is very thin, with an average thickness of 1 μm or less [18].

In order to traverse the alveolar-capillary interface, an O_2 molecule must pass through the following layers:

- A thin layer of fluid that lines the alveolus. This, as we shall see, has an important role in determining the lung's "elasticity" or "stiffness."
- The alveolar epithelium. This consists predominantly of flat, thin type-I cells and granular, more cuboidal type-II cells. Type-I cells have relatively few mitochondria or other cellular inclusion bodies, suggesting low metabolic activity. Although less numerous than type-II cells, they account for more than 90% of the alveolar surface area and consequently are the predominant cells through which gas exchange takes place. Type II-cells, on the other hand, are characterized by high metabolic activity and contain numerous cytoplasmic inclusion bodies. They have been shown to be the source of the surface tension-lowering material (alveolar surfactant) that is extruded into the alveolar surface layer.
- Alveolar and capillary-endothelial basement membranes with an intervening interstitial space. The interstitial space is more prominent in some regions than in others, and where most gas exchange occurs the membranes appear to fuse into a single structure.
- The capillary endothelium. This should not be considered an "inert" structure. Rather it subserves important biochemical functions such as transformation or inactivation of blood-borne materials such as amines, polypeptides, and prostaglandins as the blood passes from the pulmonary artery to the pulmonary veins and subsequently into the systemic arteries.

It is important to recognize that the lung has no inherent rhythmicity; it has no intrinsic motor system capable of causing ventilatory volume changes, but moves in response to forces transmitted from the chest wall via the pleurae.

The lung is covered with a fine membranous single layer of mesothelial cells that comprises the *visceral pleura*; the inner surface of the chest wall is similarly covered with *parietal pleura*. The pleural "space" between the visceral and parietal pleurae is not a cavity in the usual sense. Rather, it is normally an air-free apposition of the pleural surfaces that holds the lung and chest wall together and causes them to move in unison. A very small quantity of fluid (a total of some 7–10 ml in adult humans) intervenes as a fine film of approximately 10–20 μm thickness between the pleural surfaces [4]. This allows some sliding motion between the pleurae, for example, when the lung expands downward on contraction of the diaphragm.

100.2.2 Forces Involved in Breathing

Intrapleural Pressure. When all the respiratory muscles are relaxed, i.e., at the end of a normal, passive exhalation, the pressure in the intrapleural space is negative with respect to atmospheric pressure. This is due to the recoil properties of the lung and chest wall. The lung's elasticity causes it to attempt to retract to its intrinsic equilibrium volume (effectively that of the gas-free lung). The elastic properties of the chest wall cause it to expand to adopt its own equilibrium position (which is at a volume of 50%–60% of the fully expanded state). One can get a sense of this negative pleural pressure (relative to P_{atm}) by putting some water (or even saliva) onto the palms of the hands and putting them together. Gently pulling the palms apart gives the sensation of the negative pressure before the seal is broken, often with a noticeable "popping" sound.

However, although neither the lung nor the chest wall is in equilibrium, the *combined* lung-chest wall system adopts an equilibrium position when no volitional contractions of the muscles of breathing are applied to the chest wall. That is, it adopts a volume at which the recoil forces of the individual structures are equal and opposite [1,39]. The resultant lung volume is defined as the *functional residual capacity* (FRC). This is a crucially important volume for understanding normal lung function.

FRC should, however, be distinguished from a subject's end-expiratory lung volume (EELV), although, of course,

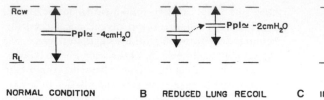
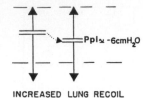

A NORMAL CONDITION   **B REDUCED LUNG RECOIL**   **C INCREASED LUNG RECOIL**

Fig. 100.2A–C. The balance of forces from chest wall recoil (R_{cw}) and lung recoil (R_L) that determines the functional residual capacity (FRC) of the lung. The recoil of the lung and the chest wall are represented by the length of the *solid arrows*. When lung recoil decreases, as in **B**, FRC increases. When lung recoil increases, as in **C**, FRC decreases. P_{pl}, Intrapleural pressure

they are normally the same when the respiratory muscles are relaxed. If there is contraction of the expiratory muscles (such as during muscular exercise), EELV can be less than FRC. In contrast, if expiratory airflow is retarded (such as in patients with increased airways resistance or low lung recoil), then expiration can be shortened before FRC is reached and EELV can increase.

Let us suppose that normally at FRC the equal and opposite forces acting on the intrapleural space generate an intrapleural pressure (P_{pl}) of -4 cm H_2O relative to the atmosphere (Fig. 100.2a). If the recoil force of the lung is reduced (Fig. 100.2b), as in patients with pulmonary emphysema, a new equilibrium volume will be established; that is, the lung will be pulled to a higher volume because the force from the chest wall is now less opposed by retractive force of the lung. Consequently, FRC increases. It should be noted, however, that at this new higher lung volume the sum of both retractive forces on the intrapleural space is now less, i.e., the chest wall recoil will also be reduced as the chest wall is now closer to its equilibrium volume. Hence the intrapleural pressure will be less negative than normal. Similarly, when the lung recoil force is increased, such as in patients with interstitial pulmonary fibrosis, the opposite will occur. However, in this case, as the lung pulls the chest wall towards the lower combined equilibrium position, the chest wall is now farther removed from its own individual equilibrium configuration; hence its recoil is greater at the new FRC, and the combined forces cause FRC to be low and P_{pl} to be more negative than usual.

Respiratory Muscles. The forces that cause the lung and chest wall to be expanded to a volume greater than FRC are provided by the respiratory muscles. Inspiration is effected by three groups of muscles [6,12]:

- Diaphragm. This is innervated by the phrenic nerves, which have their origin in cervical nerves 3–5. It is the predominant inspiratory muscle, descending as it contracts, causing its usual dome shape to be flattened and hence enlarging the caudal-cephalic dimension of the thorax.
- External intercostals. These are innervated by thoracic motor nerves 1–11. When they contract, they elevate the rib margins upward and outward, thus enlarging the anterior-posterior and lateral dimensions of the thorax. Consequently, when they act in unison with the diaphragm, the external intercostals help to increase the size of the thorax in all three dimensions. Although normally quantitatively less important than the diaphragm at low levels of ventilation, the external intercostals become progressively more involved at higher ventilations or when high inspiratory airflow is needed.
- Accessory muscles. Muscles such as the scalenes, sternocleidomastoids, and trapezius are called into play at high rates of ventilation.

Expiration, on the other hand, is usually passive at low ventilatory rates. The elastic recoil is normally sufficient to return the respiratory system (i.e., the lung–chest wall unit) back to its equilibrium configuration when the inspiratory muscles relax. However, at high rates of ventilation or in forced exhalations, expiratory muscles assist the spontaneous recoil. The expiratory muscles –

- internal intercostal muscles and
- abdominal muscles –

cause the ribs to be lowered and abdominal pressure to increase, causing the diaphragm to be pushed up (with a piston-like action) at a greater rate as a result of the increased transdiaphragmatic pressure.

It is important to recognize that the respiratory muscles, in addition to generating airflow, can also provide a brake to airflow. For example, inspiratory muscle contraction during exhalation can provide a controlled reduction of expiration airflow, and vice versa.

Lung Compliance. The greater the inspiratory muscle activity, the greater is the volume to which the chest wall is usually expanded. However, as this causes the lung to be pulled farther and farther away from its equilibrium position, its recoil or retractive force increases. The intrapleural pressure consequently becomes more and more negative as the thoracic volume is progressively increased, up to the limits of expansion.

The distending pressure across the lung is the difference in pressures between the inside and outside of the lung, i.e., the difference between intrapleural pressure (P_{pl}) and alveolar pressure (P_{alv}). This difference, or lung-distending pressure, is defined as *transpulmonary pressure*(P_{tp}); that is,

$$P_{tp} = P_{alv} - P_{pl}$$

The lung volume change (ΔV) which results from the application of a given applied distending pressure is a measure of its distensibility or compliance (C_l); that is,

$$C_l = \frac{\Delta V(\text{ml})}{\Delta P_{tp}(\text{cm H}_2\text{O})} \qquad (100.1)$$

$$= \frac{\Delta V(\text{ml})}{\Delta(P_{alv} - P_{pl})(\text{cm H}_2\text{O})} \qquad (100.1')$$

The measurement of lung compliance under conditions of no airflow (i.e., under "static" conditions) is useful for two reasons. Firstly, it allows the considerations of the intrinsic elasticity or stiffness of the lungs to be established, i.e., without the confounding influence of needing additional pulmonary pressures to overcome the resistance to airflow (as described below). Secondly, it reduces the compliance equation to an even simpler and more tractable form. For example, the P_{alv} must be exactly equal to P_{atm} when there is no airflow (if it were not, air would pass in the direction of the lower pressure). Furthermore, P_{atm} is assigned a value of zero as the frame of reference for the pulmonary pressure measurements, by convention. That is, pulmonary pressure measurements are reported relative to the atmospheric pressure. A pleural pressure of -3 cm H_2O really measures $+997$ cm H_2O, i.e., 3 cm H_2O less than normal atmospheric pressure of 1000 cm H_2O. Consequently, at all end-inspiratory and end-expiratory volumes, $P_{alv} = 0$, the glottis is open. Practically, therefore, to measure static lung compliance only, the changes in lung volume and intrapleural pressure are needed.

The lung volume change can be readily measured with a simple spirometer (Fig. 100.3), i.e.,

$$C_l = \frac{\Delta(V)}{\Delta(P_{pl})} \qquad (100.1'')$$

Although it is possible to measure P_{pl} directly by introducing a needle or catheter into the intrapleural space, the procedure is potentially a dangeous one: the lung could be punctured or air could accidentally be introduced into the intrapleural space (creating a pneumothorax, i.e., freeing the lung from its adhesive forces to the chest wall, allowing the lung to collapse from its intrinsic recoil). P_{pl} is normally estimated by measuring the pressure in the thoracic esophagus[7]. This is a highly compliant structure which transmits pressure from the intrapleural space – especially *changes* in P_{pl} – with fairly good reliability (except in diseases that reduce the compliance or distensibility of the esophagus itself – a rigid tube, for example, would be useless for this purpose). The procedure involves inserting a long, narrow balloon on the end of a catheter, attached to a pressure-measuring device, into the lower third of the thoracic esophagus. The balloon should have very little air in it (i.e., should not itself be distended), otherwise the measurement will reflect the compliance of the balloon. The subject then takes a maximal inhalation to total lung capacity (TLC), and the change in volume is compared with the change of P_{pl}. Some air is then exhaled and the lung held at this new volume. New recordings are then taken. This is repeated over the range of lung volumes (Fig. 100.4a). The slope of the resulting volume-pressure curve (i.e., $\Delta V/\Delta P$) at any point, therefore, gives the lung compliance at that volume.

Several features of this compliance curve are worth noting. The lung is most compliant, i.e., most distensible, in the normal range of tidal volume breathing. It becomes less compliant or more "stiff" at high lung volumes. For normal adults, a change of P_{pl} from -4 to -7 cm H_2O would induce a volume change (i.e., a tidal volume) of approximately 600 ml:

$$C_l = \frac{\Delta V}{\Delta P}$$

$$= \frac{600\,\text{ml}}{(-4)-(-7)\text{cm H}_2\text{O}}$$

$$= 200\,\text{ml/cm H}_2\text{O}$$

The same change of intrapleural pressure (i.e., 3 cm H_2O) nearer TLC would change lung volume by much less, e.g.,

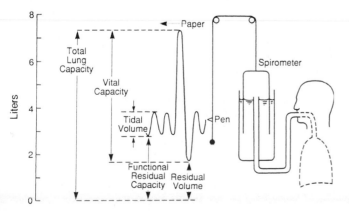

Fig. 100.3. Subdivisions of total lung capacity (TLC), with the volumes that can be correctly determined by spirometry represented by the excursions of the pen. (From [44, p. 30])

2019

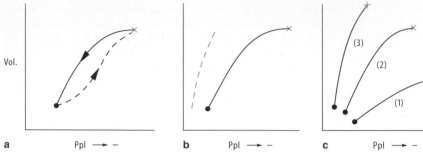

Vol.

a   Ppl ⟶ −   b   Ppl ⟶ −   c   Ppl ⟶ −

Fig. 100.4a–c. Relationship between intrapleural pressure (P_{pl}) and lung volume over the volume range from FRC (*solid circle*) to TLC (*x*). In **a**, the *dashed line* represents the relationship during a series of inspiratory maneuvers, the *solid line* represents the relationship during the subsequent volume decrements. In **b**, the *solid line* represents lung inflation in the air-filled lung, the *dashed line* represents the relationship in a fluid-filled lung over the same volume range. In **c**, *1* represents the relationship with increased lung recoil, *2* represents the normal relationship, and *3* represents a condition of reduced lung recoil

$$C_l = \frac{\Delta V}{\Delta P}$$

$$= \frac{50\,\text{ml}}{(-20)-(-23)\text{cm H}_2\text{O}}$$

$$= 17\,\text{ml/cm H}_2\text{O}$$

It is therefore usual to report the compliance of the lung at FRC.

Also, normal smaller subjects (e.g., children) have less change of volume of their (smaller) lungs than do normal subjects with larger lungs. When large size disparities are apparent, it is customary to normalize the compliance measurement to lung size in order that inferences may be drawn regarding the normality of the measured lung distensibility. This normalized value is termed the *specific compliance*, i.e.,

$$\text{specific complicance}$$
$$= \frac{\text{measured complicance}}{\text{lung volume at which measured (usually FRC)}}$$

An important characteristic of lung compliance is depicted in Fig. 100.4b. If the compliance curve is derived on an experimental animal's lung with normal air inflations, a curve similar to the solid line is obtained. If the curve is subsequently repeated with the lung filled with saline, the dashed curve is obtained. This demonstrates that when the normal air-fluid interface of the lung is abolished, there are marked changes in the interfacial forces operating on the lung. The result is that a much lower distending pressure is required to inflate the fluid-filled lung to the same volume [3,20]. Consequently, the lung behaves as a more highly compliant structure. This means that a large proportion of the total recoil of the lung is normally attributable to *surface forces*, rather than the recoil being totally defined by the structural elastic elements of the lung (see below).

In Fig. 100.4c, typical compliance curves for patients with increased lung recoil, e.g., these with diffuse interstitial pulmonary fibrosis (curve 1), and for patients with decreased lung recoil, such as patients with pulmonary emphysema (curve 3), are compared with the normal pattern (curve 2). Increased recoil causes:

- Reduced FRC (as previously described)
- Low compliance
- Reduced TLC
- A P_{pl} which is *more* negative than normal both at FRC and at TLC.

Decreased recoil leads to:

- Increased FRC
- High compliance
- High TLC
- A P_{pl} which is *less* negative than normal both at FRC and at TLC.

A further important feature of Fig. 100.4 is that it allows one to answer the question as to what normally limits maximum thoracic expansion: (i) the chest wall becoming stiff at high lung volumes? (ii) the lung becoming stiff at high lung volumes? or (iii) the inspiratory muscles being unable to generate any greater expansive force to the chest wall because they have shortened to an extreme region on their length–tension relationship?

The answer is (ii). It is the stiffness of the lung that limits further thoracic expansion at high lung volumes [1]. The flatness of the compliance curve reveals that the inspiratory muscles are indeed generating an expansive force and that the pressure is being transmitted to the lung (by the chest wall "pulling" further on the lung). The lungs, however, do not change their volume further, despite further increases in distending pressure.

It is important to remember that the P_{pl} which is measured under these static conditions is exactly equal to the pressure generated by lung recoil (P_{recoil}) but with the sign reversed, as shown in Fig. 100.5. The positive "alveolar"

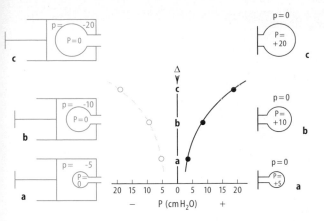

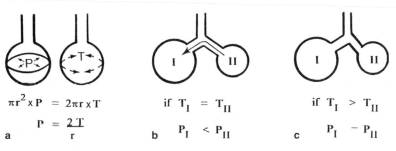

Fig. 100.5. Representation of the pressure difference between the inside and the outside of a balloon at various balloon volumes. The *right-hand panel* reflects the positive balloon pressure generated by recoil. The *left-hand panel* represents the negative surrounding pressure necessary to inflate the balloon to the same volume as that achieved on the right side

pressure exhibited by the inflated balloon with the "airway" blocked disappears when the balloon is inflated to the *same* volume by the piston. Note, however, that the "pleural pressure" is negative by exactly the original recoil pressure at that volume. This is the conceptual basis for the very important relationship between alveolar, intrapleural, and lung recoil pressures:

$$P_{alv} = P_{recoil} + P_{pl}. \tag{100.2}$$

Causes of Lung Recoil. The lung tends to recoil to a smaller volume even after a maximal volitional exhalation (i.e., at *residual volume*). Two basic factors account for this retractive, or recoil, force.

- Lung elastic tissue. Elastic and collagen fibers are present in the alveolar walls and bronchial tree and, when distended, tend to return to equilibrium configuration.
- Surface forces. At an air-fluid interface, the forces between fluid molecules act in all directions except toward the air side. This tends to an imbalance in force distribution such that the surface of the fluid attempts to con-

tract to a smaller and smaller surface area – causing "tension" in the surface. This happens at the alveolar air-fluid interface and leads to surface forces that play an important role in recoil of the lung. That is, when the fluid is folded into a spherical form (or Fullerene), i.e., on the alveolar surface, the surface compression leads to a force which tends to make the alveolus smaller (Fig. 100.6) and hence generates an alveolar pressure which acts in the direction of expiratory airflow.

It is, perhaps, difficult to get a "sense" of the work that must be done to expand the surface of an air-to-liquid interface. However, the remarkable photograph shown as Fig. 100.7 provides an excellent illustration of this phenomenon. Note that while this swimmer has come *to* the surface, he has not yet come *through* the surface, the surface tension at the air-to-liquid interface being sufficiently great to oppose, momentarily, the considerable distending force exerted by the swimmer's head.

The surface forces generated at the alveolar air-fluid interface, in addition to contributing a major component to the lung's elastic behavior, also tend to pull fluid out of the pulmonary capillaries towards the alveolus, by lowering the pulmonary interstitial pressure. The presence of surfactant in the lining layer reduces this tendency and therefore helps keep the alveoli dry [9].

The units of surface tension are dynes per centimeter – or the force acting on a centimeter length of the surface. West [45] has graphically conceptualized this by pointing out that if a 1-cm slit were "cut" into the fluid surface, then it would "gape." If this were to be stitched up again, then the force would be that required just to bring the edges back together.

It is, perhaps, easier to conceptualize surface tension by recognizing that as

$$1 \, erg = 1 \, dyn \times cm$$
$$then \quad surface \, tension = dyn \times cm^{-1} = erg \times cm^{-2}$$

– that is, the *work* required to expand the surface by $1 \, cm^2$. The normal surface tension of water is $72 \, dyn \times cm^{-1}$. A commercial detergent might reduce this to $30 \, dyn \times cm^{-1}$ or so, but at low lung volumes the pulmonary surfactant can reduce the surface tension to just a few $dyn \times cm^{-1}$.

Fig. 100.6a–c. Representation of the relationships between the pressure (P), surface tension (T), and the radius (r) in an idealized spherical alveolar model. Note that in **b**, if the surface tension is equal in the two alveoli, then the pressure in the smaller unit would be greater than that in the larger, and consequently air would flow from region II into region I. However, if, as in **c**, the surface tension is reduced in proportion to the radius, then the pressure in each unit can be equal

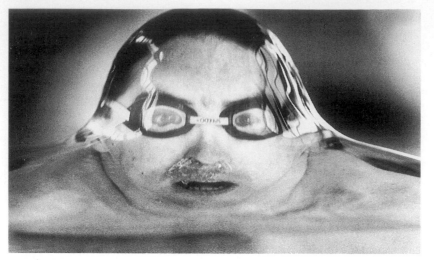

Fig. 100.7. Boy in the bubble: Matthew Dunn, the Australian Olympic breaststroke swimmer, presented a surreal spectacle as he broke the surface of the pool in Sydney where he was training hard in preparation for the Commonwealth games in Canada. ("The Sydney Morning Herald", photograph by T. Clayton 1994)

Pulmonary Surfactant. The interfacial recoil forces in the lung would be markedly more powerful, and consequently the lung more difficult to ventilate, were it not for the presence of material in the alveolar lining film [10,34,40] which is capable of reducing the surface tension: *surfactant*. Evidence strongly indicates that pulmonary surfactant is a lipoprotein [10], the major phospholipid of which is dipalmitoyl phosphatidylcholine (DPPC). This is produced in type-II cells, stored in their lamellar bodies and secreted onto the alveolar lining layer. The molecules arrange themselves such that the two palmitic acid residues, being hydrophobic, protrude into the alveolar space with the polar hydrophilic group remaining immersed in the fluid lining [18]. This disrupts the surface molecular interaction and reduces the surface tension.

Pulmonary surfactant is formed late in fetal life (10). On occasion, especially in prematurely born babies, insufficient pulmonary surfactant is available, or it is inactive. This leads to the life-threatening *neonatal respiratory distress syndrome* in which the lung is stiff, with some alveoli collapsed and some likely to be filled with fluid. Interestingly, Shakespeare chose the adjective "breathing" to modify "world" in his allusions to Richard's prematurity in the opening soliloquy of *Richard III*: "Deformed, unfinished, sent before my time into this *breathing* world, scarce half made up."

Importantly, however, normal surfactant has the ability not just to reduce surface tension, thereby reducing the work of breathing by allowing easier lung distention, but, remarkably, to alter surface tension in proportion to the volume of the alveolus [9,40] – unlike the effect of most commercial detergents, which, although they reduce surface tension, do not change it in proportion to the surface area. Were it not for this feature, the lung with its hundreds of millions of alveoli, of various sizes, would be unstable. For example, as described by Laplace's law (Fig. 100.6).

$$P = \frac{2T}{r}, \qquad (100.3)$$

where T is the surface tension and r the radius of the idealized spherical alveolus. Thus, if surface tension were to be equal throughout the lung, the alveoli with the smallest radius would necessarily have the highest pressure. Consequently, the smaller alveoli, with the highest pressure, would empty into the larger ones with the lower pressure, down the pressure gradient (Fig. 100.6b). In reality, however, the alveoli of different sizes are stable in the lung, owing to the surface tension being reduced in proportion to the radius. This causes equal alveolar pressures in alveoli of different sizes.

For example, were the surface tension (e.g., $T = 10$ dyn $\times$ cm^{-1}) to be the same in two alveolar units, one of which had a radius which was double that of the other (e.g., 0.01 cm vs 0.005 cm), then the pressure (P) in the smaller unit would be double that in the larger (as schematized in Fig. 100.6), i.e.,

$$P_{\text{large}} = \frac{2 \times 10}{0.01} = 2 \text{ cm H}_2\text{O}$$

$$P_{\text{small}} = \frac{2 \times 10}{0.005} = 4 \text{ cm H}_2\text{O}$$

However, if – by means of the surfactant – the surface tension in the smaller unit were reduced to 5 dyn cm^{-1}, then the pressure in each unit would be the same, i.e.,

$$P_{\text{large}} = \frac{2 \times 10}{0.01} = 2 \text{ cm H}_2\text{O}$$

$$P_{\text{small}} = \frac{2 \times 5}{0.005} = 2 \text{ cm H}_2\text{O}$$

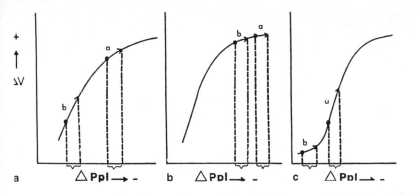

Fig. 100.8a–c. Regional variations of lung ventilation. **a** During normal breathing apical (*a*) alveoli ventilate less per unit volume than basal (*b*) alveoli. **b** Near TLC both apical and basal alveoli ventilate more evenly. **c** Near residual volume, apical alveoli ventilate more than do basal alveoli

This remarkable process is thought to be caused by the surface-active molecules being more densely packed at the surface when alveoli are small; hence there is greater surface tension-lowering potential than in the larger alveoli where the packing is less dense.

The "packing density" of surfactant and molecules also accounts for the considerable hysteresis in the air-filled lung [25], such that a markedly lower pressure is required to distend the lung to a given volume during deflation (i.e., *after* it has been inflated) than during inflation (Fig. 100.4). It is also thought to account for the reduction in lung stiffness (i.e., the increase in compliance) which follows a sigh or deep breath – more surfactant molecules being sucked onto the surface.

Regional Variation of Alveolar Size. Up to this point we have considered the intrapleural pressure as uniform throughout the lung. It is not! Careful measurements have clearly shown that intrapleural pressure at the apex of the lung is markedly more negative than at the base [28,43]. In the normal upright human, the pressure at the apex might be $-10\,\text{cm}\,H_2O$, compared to $-2\,\text{cm}\,H_2O$ at the base.

The cause of this vertical variation of intrapleural pressure relates to the total tendency of the lung to retract from the chest wall. The influence of the gravitational force "pulling down" the lung is greater in the apex, i.e., there is more lung below it to be affected by gravity than at the base.

This variation in negative pressure in the intrapleural space will influence alveolar size. The alveolar size will be greater in the regions in which P_{pl} is most negative. Apical alveoli therefore tend to be larger than basal alveoli in normal, upright humans [28,43]. In the supine position, much of this vertical gradient of alveolar size is lessened, as the gravitational effects on intrapleural pressure are reduced in this posture. The alveoli at the ventral region of the lung will be larger than those in the dorsal regions in this posture.

Normally, therefore, in the upright position, apical and basal alveoli are on different portions of their own normal compliance curve (Fig. 100.8c). For this reason, for a given change in intrapleural pressure, the *basal* alveoli are better ventilated than those in the apex (Fig. 100.8a). Thus, for normal breathing near FRC, apical alveoli have large volumes but small changes of volume, and basal alveoli have smaller volumes and larger changes of volume. Most of the normal inspirate, therefore, goes to the lower regions of the lung [21,28,29,43].

If we were to breathe near TLC, all lung regions would be on the flatter part of the compliance curve; hence they would ventilate more equally (see Fig. 100.8b), but of course a high work of breathing would be required.

If we were to breathe near the residual volume of the lung (RV), where the lung can become less distensible (or has low compliance), predominantly due to some airway closure in basal regions at low lung volumes, the normal pattern of distribution would be reversed. Most of the initial inspirate would go to the apex (see Fig. 100.8c). Therefore, depending on initial lung volume, the normal inspired volume is directed to different regions of the lung.

The patterns of regional ventilation resulting if a subject were to perform a maximal inhalation to TLC beginning at RV are as follows:

- Initially most of the inspirate is directed into the apex, with little ventilation of basal regions, then
- Both regions tend to ventilate at the same rate, as both regions come onto the more linear portion of the compliance curve. Next,
- The basal regions ventilate better than apical regions, as the apical alveoli reach the flat part of their compliance curve, and, finally,
- Both regions ventilate at about the same rate (but both poorly), as both fall on the flat portion of their compliance curve.

On the subsequent exhalation, the reverse pattern is established.

The concept that under static conditions the intrapleural pressure reflects only *lung* recoil often causes some concern to students, who wonder why the chest wall does not

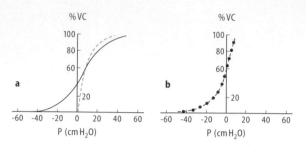
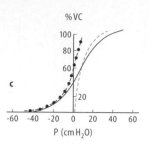

Fig. 100.9a–c. In **a**, the *dashed line* represents the lung recoil pressure over the vital capacity (VC) range. The *solid line* represents the pressure generated by the respiratory system (i.e., the lung and chest wall) when allowed to recoil passively against a closed airway at various lung volumes. The difference between the respiratory system recoil and the lung recoil at any volume represents the recoil of the chest wall. This is depicted in **b**. The pressure-volume relationship of the lung alone, the chest wall alone, and the combined lung-chest wall unit is shown in **c**

influence the measurement. The concern can be alleviated, however, by the recognition that, under these conditions, the chest wall does not actually recoil: rather, it is locked in place by the action of the inspiratory muscles. That is, they create a condition similar to that in Fig. 100.5. Imagine, therefore, that having determined the intrapleural pressure under these static conditions at some lung volume (i.e., you have now the recoil pressure of the lung at that volume – Fig. 100.9a), you block off the airway with an external shutter. Naturally, the pressure at the mouth will still be zero – until you relax the respiratory muscles completely and allow the lung *and* the chest wall to recoil. What is the pressure that you would then measure at the mouth? Naturally, it is the combined recoil pressure of the lung *and* the chest wall, i.e., the total respiratory system recoil [1]. And if this is performed over the achievable range of lung volumes, the pressure-volume curve for the respiratory system can be determined as shown by the solid line in Fig. 100.9b. However, as you already know the lung recoil component at each lung volume, as shown by the dashed line in Fig. 100.9b, the recoil pressure generated by the chest wall may readily be determined. You simply subtract the lung recoil pressure from the total pressure, at each lung volume, to get the chest wall recoil [1]. This is shown in Fig. 100.9c.

Inspection of Fig. 100.9 reveals that the respiratory system is most distensible near FRC (and hence the work of breathing is least) but becomes "stiff" near both RV and TLC – limiting further extremes of thoracic excursions: at TLC because the lung is stiff (not the chest wall), and at RV because the chest wall becomes stiff (not the lung).

It should be noted, however, that in some subjects with lung disease (e.g., those with low lung recoil), small airways can begin to close off at low lung volume and consequently airway closure can determine the subject's RV. This volume will therefore be larger than normal and becomes one of the indices of the functional state of the lung as determined from pulmonary function tests.

100.2.3 Lung Volumes

As previously described (Fig. 100.3), the volume of air in the lung at the end of a normal, quiet exhalation is termed the functional residual capacity (FRC). The lung volume at the end of a maximum volitional inhalation is termed the total lung capacity (TLC), whereas at the end of a maximum forced exhalation the volume of air in the lungs is termed the residual volume (RV). As is apparent from Fig. 100.3, all other pulmonary volumes and capacities can be measured directly by simple displacement. However, if any one of these volumes is known, then the other two are also readily determined (Fig. 100.3). The displacement volumes are characteristically measured with a spirometer, in which movement of air to or from the lung causes changes in position of a carefully balanced cylindrical bell, leading to recording of the volume changes on a volume-calibrated moving (usually rotating) chart (Fig. 100.3).

To determine the FRC (this is the most commonly measured variable in this regard), different techniques are required.

N_2 Washout Technique. Normally the lung contains approximately 80% N_2, which varies hardly at all during the breathing cycle. This can be readily measured in most pulmonary function laboratories using a rapidly responding N_2 analyzer. Using a system with directional valves that allows the subject to inhale 100% O_2 and exhale into a spirometer, the test continues until all the N_2 that originally resided in the lung now resides in the spirometer, i.e., the N_2 has been "washed out" of the lung. In normal subjects this takes only a few minutes with normal, quiet breathing. However, in subjects with maldistribution of ventilation, i.e., with regions of the lung that ventilate poorly compared with other regions, it may take considerably longer to wash out the N_2 completely.

Let us imagine that after 5 min or so of breathing 100% O_2, the subject has washed all the N_2 from the lung (this is known from the N_2 analyzer reading); during this period, 40 l was exhaled into the spirometer. The final mixed concentration of N_2 in the spirometer is found to be 5% i.e., 2 l N_2 was washed out of the subject's lung (40×0.05). The

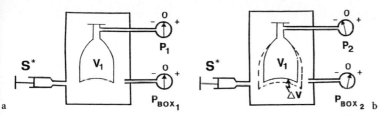

Fig. 100.10a,b. Pressure changes in the lung and an airtight thermoregulated box (plethysmograph) when a subject "inhales" against an occluded airway. Note that the airway pressure becomes negative relative to the atmosphere owing to gas decompression as a result of the lung volume increase. The box pressure, however, becomes positive because of gas compression as a result of the lung volume increase. These changes may be used to determine the original lung volume (usually FRC). S, Calibrated syringe to remove gas from the box, the subject maintaining the inhaled volume, until the box pressure returns to zero

following computation can then be made to account for the fact that this 2 l N_2 (which was originally in the subject's lung) only occupied 80% of its volume, i.e.,

FRC volume × 80% N_2 = final spirometer volume × final spirometer % N_2,

i.e., $\quad FRC = \dfrac{40 l \times 0.05}{0.8}$

$\quad\quad\quad\quad = 2.5 l$

It should be noted that if the subject begins the test at FRC, the volume determined will be FRC regardless of the lung volume at the end of the test; if the subject begins at some other volume, the measured volume will reflect the starting lung volume. Once FRC is known, residual volume and TLC can be directly determined by measuring the expiratory reserve volume (ERV) and the inspiratory capacity (IC) by displacement spirometry (Fig. 100.3).

By this open-circuit N_2 washout technique, a correction factor must be applied for the amount of N_2 that washes out of the blood. This is extremely small, however, as the blood solubility of N_2 is so low [37].

Helium Equilibrium Technique. This closed-circuit method requires a known volume of He-containing gas to be initially present in a spirometer. The subject rebreathes from the spirometer sufficiently long for the He to equilibrate throughout the total lung-spirometer volume, i.e., the concentration of He is the same throughout the total volume. Let us suppose the initial spirometer volume was 3 l 10% He gas. At the end of the equilibration period, the He concentration was 5.45%. The following calculation may then readily be made:

Initial spirometer volume × initial spirometer He concentration = final volume (lung and spirometer) × final mixed He concentration,

i.e., $\quad 3 l \times 0.1 = (lung\ volume + 3 l) \times 0.0545$

and $\quad$ lung volume = 2.5 l

As this test is dependent on final lung volume at equilibrium, care must be taken to correct for any changes in lung volume from the original FRC position.

It should be noted that these techniques measure only the volume of gas in communication with the spirometer during the test. For example, the total volume of gas in the lung at FRC would be underestimated by these spirometric techniques if, in a diseased lung, some gas were "trapped" behind collapsed airways, as is commonly the case.

Plethysmographic Technique. All the gas in the thorax, whether in normal communication with the atmosphere or "trapped" behind collapsed airways, can be measured using an ingenious application of Boyle's law, which states that, if temperature is constant, there is a constant relationship between the pressure and volume of a given quantity of gas, i.e.,

$$P \times V = P_1 \times V_1 \quad\quad\quad (100.4)$$

A subject is seated in an airtight box (plethysmograph, Fig. 100.10) that permits breathing through a mouthpiece within the box. At the end of a normal exhalation (i.e., at FRC), alveolar pressure (P_1) is equal to atmospheric pressure, as no gas is flowing. A shutter then closes off the mouthpiece. The subject is requested to inhale against the closed shutter. During this inhalation the thorax enlarges despite there being no airflow. That is, a new thoracic gas volume is established by decompressing the gas in the lungs. This leads to a new thoracic volume of ($V + \Delta V$). Note that the volume of gas within the plethysmograph (the box) decreases as it is compressed by the newly enlarged thorax. Thus:

Original alveolar pressure (P_1) × original thoracic gas volume (V) = new alveolar pressure (P_2) × new thoracic gas volume ($V + \Delta V$),

i.e., $\quad P_1 \times V = P_2 \times (V + \Delta V),\quad\quad (100.4')$

or $\quad V = P_2 \times \dfrac{\Delta V}{\Delta P}$

2025

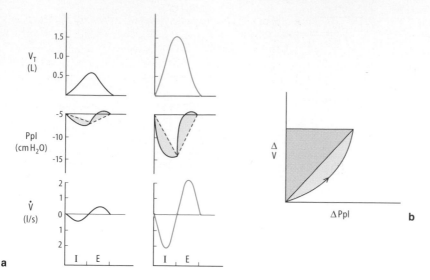

Fig. 100.11. a Changes in volume (V), intrapleural pressure (P_{pl}), and airflow ($\dot{V}$) during normal breathing in a subject at rest and during muscular exercise. The *dashed line* on the P_{pl} curve represents the pressure needed to produce that lung volume under conditions of no airflow. The *blue shaded area* represents the extra pressure required to generate air (and tissue) flow. **b** Dynamic inspiratory pressure volume curve. The *gray stippled area* is the static pressure volume work. The *blue shaded area* is the dynamic component of inspiratory work equivalent to the work required to generate air (and tissue) flow

As P_1 and P_2 (and hence their difference, ΔP) are directly measured, we can readily determine the total thoracic gas volume at FRC (V) if we know ΔV. This is determined by means of a calibrated syringe (S in Fig. 100.10) that can remove a volume of gas from the box, with the subject remaining at the inhaled volume, until the box pressure is returned to zero. This volume must equal the original volume increase (i.e., ΔV) that created the original increase in box pressure. Alternatively, of course, if the subject returns to FRC, then a volume can be injected into the box until the box pressure reaches that observed when the subject originally increased the lung volume.

100.3 Dynamics of Breathing

Gas flows down gradients of pressure. In the lung the pressure gradient of concern is the difference between atmospheric and alveolar pressure. This means that gas flow in the lung results from changes in alveolar pressure; this is brought about by changes in the dimension of the thorax. Inspiration leads to expansion of the thorax, causing alveolar pressure to fall until the end of inspiration, when alveolar pressure again equals atmospheric pressure, i.e.:

- At FRC, $P_{atm} = P_{alv}$, therefore no airflow.
- During inspiration, $P_{atm} > P_{alv}$, therefore air flows into lung.
- At end inspiration, $P_{atm} = P_{alv}$, therefore no airflow.
- During expiration, $P_{alv} > P_{atm}$, therefore air flows out of the lung.

Typical changes in volume, flow, and intrapleural and alveolar pressures during a respiration cycle are given in Fig. 100.11. Several features of this figure are worthy of consideration. Under conditions of no airflow (i.e., under static conditions) changes in volume will be simply related to changes in intrapleural pressure (i.e., the static lung compliance relationship); however, during inspiration, additional muscular force is needed, and this is reflected by greater negativity of the intrapleural pressure change. This "extra" pressure is needed to overcome the flow resistances associated with breathing. These are (i) airway resistance to gas flow, and (ii) resistance to pulmonary tissue flow (i.e., tissue viscosity).

The difference between the intrapleural pressure change needed to provide the dynamic airflow and that required to distend the lung statically is represented by the shaded area in Fig. 100.11; it is consequently greatest when airflow is greatest. The maximum air flow during the breathing cycle therefore occurs at the time of maximum change of alveolar pressure. Fig. 100.11 also allows us to understand the basic concept of the work of breathing.

100.3.1 Work of Breathing

The changes in thoracic volume during inspiration are brought about by contraction of skeletal muscles, i.e., the inspiratory muscles of respiration. The work done by a muscle in moving an object is expressed by:

$$\text{work} = \text{force} \times \text{distance moved} \quad \text{or}$$
$$\text{force} \times \text{length} \tag{100.5}$$

This equation, however, is of no practical value in this form since we cannot directly measure the force-length relationships of the various respiratory muscles during breathing in humans. We can, however, also compute work as:

change in intrapleural pressure
× change in lung volume

This amounts to the same thing, since:

$$P \times V = \frac{\text{force}}{\text{area}} \times \text{vol}$$

$$= \frac{\text{force}}{\text{length}^2} \times \text{length}^3$$

$$= \text{force} \times \text{length}$$

i.e., $P \times V = \text{work}$ \hfill (100.5′)

Thus, by measuring the changes in intrapleural pressure and change in lung volume, the pulmonary work of breathing can be computed as shown in Fig. 100.11. The total area (stippled and shaded) between the V and P axes is the total work of breathing [30]. The stippled area is, of course, the work done to overcome the static recoil forces of the lung over that volume. The shaded area represents the flow-resistive work, of which airway resistance normally accounts for about 80% and viscous tissue resistance for the remaining 20%. In subjects with high breathing resistance, the dynamic part of the loop is markedly increased; that is, greater driving pressure (P_{alv}) is required to achieve a particular airflow, and consequently the work of breathing is high.

100.3.2 Airway Resistance

Resistance to airflow is analogous to electrical resistance as defined by Ohm's law:

$$\text{resistance} = \frac{\text{potential difference}}{\text{current flow}} \hspace{1em} (100.6)$$

For the airways the potential difference is the difference between alveolar and atmospheric pressure, and flow is the airflow. Thus:

$$\text{airway resistance} = \frac{P_{alv} - P_{atm}}{\text{flow rate}}$$

$$\left[\text{i.e.,} \frac{\text{pressure units}}{\text{flow units}} \right] \text{ or } \left[\frac{cmH_2O}{l \times s} \right] \hspace{1em} (100.6′)$$

The alveolar pressure is simply the sum of the intrinsic recoil pressure of the lung (at that volume) and the applied intrapleural pressure (as discussed earlier). And as the lung recoil pressure is derived from the static pressure – volume relationship and P_{pl} is measured directly, P_{alv} can be determined. Airflow can be readily measured; therefore, airway (and tissue) resistance can be determined [22]. Typical values for airway resistance normally range between 0.5 and 2 cm $H_2O \times l^{-1} \times s$.

Although the cross-sectional area of a bronchiole is small compared with that of a large bronchus, the total cross-sectional area of all the small airways is large compared with that of the large airways. Consequently, the total resistance to airflow of the small airways is small compared with that of the large ones. In fact, only about 10%–20% of the airway resistance is related to airways smaller than 2 mm [24]. Unfortunately, therefore, considerable increases in small-airway resistance must occur before alterations in total airway resistance can be discerned with the usual tests.

For example, consider the effect on total airways resistance of doubling the smaller airways resistance (i.e., a 100% increase). Recognizing that the large and small airways are "connected" in series, then:

$$R_{total} = R_{large} + R_{small}, \hspace{1em} (100.7)$$

so if originally $2.0(cmH_2O \times l^{-1} \times s) - 1.7 + 0.3(cmH_2O \times l^{-1} \times s)$

then doubling R_{small} gives $2.3(cmH_2O \times l^{-1} \times s)$
$= 1.7 + 0.6(cmH_2O \times l^{-1} \times s)$

Therefore $\Delta R_{total} = 2.3 - 2.0 = 0.3\,cmH_2O \times l^{-1} \times s$

or an increase of only 15%.

It should be borne in mind, however, that airway resistance is not constant: rather, it varies with lung volume. As the airways are affected by intrapleural pressure, so their cross-sectional area varies with this surrounding pressure. At high lung volumes, the airways are distended. At low lung volumes or during forced expiratory efforts when intrapleural pressure becomes positive, airways can undergo *dynamic compression* [26,36] and their resistance to flow becomes relatively high as a consequence. Furthermore, contraction of bronchial smooth muscle or narrowing (or plugging) of the bronchi with mucus increases airway resistance.

Lung recoil pressure also becomes greater at high volumes and relatively low at low lung volumes. Thus, despite efforts at maximal airflow in both cases, maximum expiratory airflow is high at high lung volume (i.e., recoil high, resistance relatively low) and low at low lung volume (i.e., recoil lower and resistance higher). This may be demonstrated by having a subject exhale maximally into a simple displacement spirometer from TLC to RV. The slope of the consequent volume-versus-time relationship (i.e., airflow) is highest at high lung volume and becomes progressively less as volume decreases (Fig. 100.12). When lung recoil is reduced, such as in a patient with pulmonary emphysema

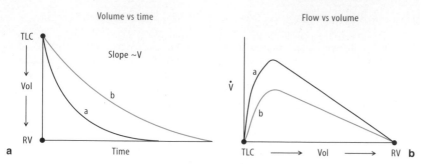

Fig. 100.12a,b. Lung volume as a function of time and airflow as a function of volume over the vital capacity range of subjects generating a forced vital capacity maneuver. Note that the flow, either as $\dot{V}$ or as the slope of the volume-time relationship, decreases progressively as lung volume gets less. If **a** represents a normal subject, **b** would be representative of a subject with reduced airflow. The relatively linear descending limb of the expiratory flow-volume relationship is indicative of the closely exponential nature of the volume-time curve. The short delay on attainment of peak airflow results predominantly from the system inertia

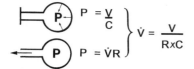

Fig. 100.13. Simple representation of a balloon model in which the pressure generated against an occlusion is a function of the balloon volume and its compliance. The subsequent removal of the obstruction allows air to flow at a rate ($\dot{V}$) that is dependent on the developed pressure and the resistance to flow. Consequently, the flow generated at a given volume will be a function of the system's resistance and compliance

[19,35], the maximal airflow is low *despite* high lung volume.

When lung recoil is increased, such as in a patient with diffuse pulmonary interstitial fibrosis (a restrictive lung disease), maximum airflow is low *due to* the reduced lung volume [19,35]. Diseases associated with airways constriction or narrowing (e.g., chronic bronchitis or bronchial asthma) lead to low flow at these lung volumes. However, in a disease that causes increased lung recoil, such as pulmonary fibrosis, although the absolute peak is low owing to inability to distend the lung to a sufficiently high volume, the flow at a particular lung volume can be even slightly higher than normal owing to increased recoil. Also, if expired airflow is plotted against expired (or lung) volume, the result is a flow-volume (i.e., $\dot{V}$–V) curve as shown in Fig. 100.12. Peak flows of $10 \, l \times s^{-1}$ are not uncommon in normal subjects and flow decreases fairly linearly in normal subjects as lung volume decreases.

It is instructive to consider these factors as determinants of airflow generation using a simple inflated balloon model, as shown in Fig. 100.13. The pressure in the balloon is naturally a function of its volume and compliance. If the occlusion is abruptly removed, then the rate at which air would flow out of the balloon would be a function of the pressure (at this volume) and the resistance of the "neck" of the balloon. Consequently, one may consider the airflow ($\dot{V}$) to be a function of three factors: (i) volume (V), (ii) airways resistance (R), and (iii) compliance (C), i.e.,:

$$\dot{V} = \frac{V}{R \times C} \qquad (100.8)$$

That is, if R or C (or, of course, both) are large, then flow will be low for a given lung volume. Similarly, for a given value of $R \times C$, the airflow will decrease as a linear function of lung volume. This helps to account for the descending limb of the flow-volume curve being normally so linear (Fig. 100.12).

The factor $R \times C$ in this relationship deserves consideration. Note that its unit is *time* and that it represents the *mechanical time constant* (τ) of the structure; i.e.,

$$R \times C = (cmH_2O \times l^{-1} \times s) \times (l \times cmH_2O^{-1}) = s$$

The time constant (τ) is an important construct [32] in pulmonary mechanics (and in most other branches of physiology). If a variable (e.g., volume, V) changes so that its instantaneous rate of change (i.e., flow, $\dot{V}$) is a simple function of the magnitude of the variable, e.g.,

$$\dot{V} \propto V \quad \text{or} \quad \dot{V} = k \times V$$

then the variable changes exponentially as shown in Fig. 100.14. In one time constant, the value will have fallen to approximately 37% of its initial value. In another time constant it will have decreased to a further 37% of the remaining change, and so on. After four time constants, the variable will have attained more than 98% of its final value (Fig. 100.14), which is usually sufficient to assume all the change to be complete.

Normal values for R and C are approximately $2 \, cm \, H_2O \times l^{-1} \times s$ and $0.21 \times cm \, H_2O^{-1}$, respectively: i.e.,

$$\tau = 2 \times 0.2 = 0.4 \, s$$

This means that, following inflation, *passive* recoil will effectively return the respiratory system to its original volume in four time constants (Fig. 100.14). A structure with a τ of 0.4 s will need approximately 1.6 s for this expiratory

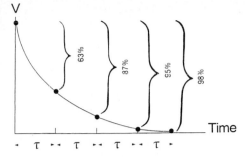

V

63% 87% 95% 98%

Time

τ -- τ -- τ -- τ

Fig. 100.14. The time course of change of a function that decreases as an exponential function of time (roughly equivalent to the volume change during a forced vital capacity maneuver, for example). Note that after one time constant, 63% of the vital capacity would have been exhaled, in two time constants 87%, in three time constants 95%, and in four time constants 98%. The mechanical time constant (τ) is determined by the resistance and the compliance of the system

function and, with analogy to the lung, would require approximately 3 s for the entire breathing cycle [8]. That is, at breathing frequencies above 20/min, there would be an obligatory requirement to accelerate expiratory flow. Otherwise the structure would not have time to empty with each exhalation, and EELV would necessarily therefore increase.

However, were τ to be doubled in the patient with obstructive lung disease, either because of an increase of R or C or both (this, of course, will represent only the average of values for τ from different lung regions – but for illustrative purposes we choose the average to be doubled), then a $\tau = 0.8$ s would require 3.2 s for expiratory duration; i.e., (4 × 0.8) s. This would yield a limiting breath duration of about 6 s or a limiting frequency of about 10 per min i.e., close to resting levels. As a result, at even modest levels of exercise, either the expiratory muscles must accelerate airflow or EELV must increase in patients with obstructive lung disease.

Consequently, when there is nonuniformity of time constants, gas flow becomes asymmetric, especially as breathing frequency increases [23,48].

100.3.3 Turbulence

Turbulent air flow requires greater driving pressure to generate a given air flow than if the flow were laminar. For laminar flow, the flow $\dot{V}$ is proportional to the geometry of the tube (or airway) through which the air flows. This is given by the Poiseuille equation:

$$\dot{V} = \frac{\Delta P \times \pi \times r^4}{8 \times l \times \eta},\qquad(100.9)$$

where ΔP is the driving pressure, r the radius, l the length of the tube, and η the coefficient of viscosity (assuming straight, unbranched tubes).

Alternatively,

$$\Delta P = \frac{\dot{V} \times 8 \times l \times \eta}{\pi \times r^4}\qquad(100.9')$$

But as resistance (R) $= \dfrac{\Delta P}{\dot{V}}$,

then

$$R = \frac{8 \times l \times \eta}{\pi \times r^4}\qquad(100.9'')$$

This shows that, while the airways resistance is influenced by l and η, it is more markedly influenced by the vessel radius (note that r is raised to the fourth power). Consequently, for laminar airflow:

$$\Delta P = k_1 \times \dot{V}\qquad(100.10)$$

where k_1 is the resistance (i.e., $\frac{8 \times l \times \eta}{\pi \times r^4}$). The driving pressure required for turbulent flow is greater, however, i.e.,

$$\Delta P = (k_1 \times \dot{V}) + (k_2 \times \dot{V}_2)\qquad(100.10')$$

where k_1 is the laminar portion of the resistance and k_2 is the turbulent portion of the resistance. Thus, turbulence, which can occur at high flow rates, at airway branchings or across a constricted region of airways, requires greater pressure to generate a given airflow. As a result, the work of breathing increases when airflow becomes turbulent. Whether or not airflow will be turbulent or laminar is determined by the *Reynolds number* (Re). This is a dimensionless number:

$$Re = \frac{v \times 2r \times d}{\eta}\qquad(100.11)$$

where v is the mean linear velocity, r the tube radius, and d the gas density.

When Re is greater than 2000, turbulence usually occurs. It should be noted, however, that Re is proportional to the gas density and related inversely to the viscosity. Hence, reducing the density of a gas, e.g., by breathing high concentrations of He instead of N_2, makes it less likely that flow will be turbulent. It is also apparent from the equation that where mean linear velocity is low, such as in the small airways, turbulence is less likely to occur in regions such as the trachea, where linear velocity is high.

100.3.4 Airflow Limitation

As alveolar pressure is given by the sum of the recoil pressure and the pleural pressure (see above), the determinants of airflow can be considered in terms of:

$$\dot{V} = \frac{P_{recoil} + P_{pl}}{R}\qquad(100.12)$$

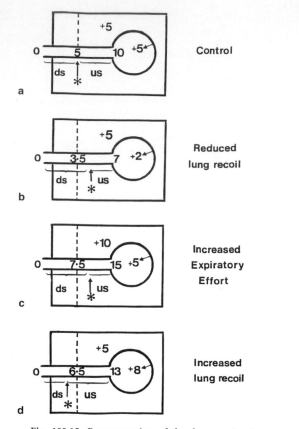

Fig. 100.15. Representation of the decrease in airway pressure from the alveolus (number given at the "alveolar" to "airway" interface) to the atmosphere. For simplicity this is depicted as decreasing linearly down the airway. Note that at some point down the airways the airway pressure becomes equal to the surrounding intrapleural pressure. This is shown by the *asterisk*, and is termed the "equal pressure point." Downstream (*ds*) of the equal pressure point there is a compressive force on the airway as the pleural pressure is greater than the airway pressure. **a** Control. **b–d** Note that with reduced lung recoil (**b**) or increased expiratory effort (**c**) the equal pressure point moves upstream (*us*), i.e., towards the alveolus. Increased lung recoil (**d**), on the other hand, moves the equal pressure point further downstream

Consider Fig. 100.15a. The subject attempts a forced expiratory maneuver that generates a P_{pl} of $+5$ cm H_2O. As the subject's lung recoil (at that volume) is 5 cm H_2O, the driving pressure for expiration is 10 cm H_2O. Frictional losses down the airways, however, lead to a progressive decrease in pressure, mouthwards (i.e., "downstream"), until atmospheric pressure is reached at the airway opening. For ease of understanding, the pressure is schematized as decreasing linearly in Fig. 100.15; in reality, however, the pressure will fall most rapidly across the airways that provide the greatest resistance to airflow – the upper airways. Somewhere in the pathway from the alveolar pressure of 10 cm H_2O to the zero, the pressure must have reached 5 cm H_2O which, in this case, is the current P_{pl}. At this point the transmural pressure across the airway is zero; in other words an *equal pressure point* (EPP) has been achieved [26]. Downstream of the EPP the airway pressure is less than P_{pl} and these airways are compressed. Greater expiratory effort simply compresses them more, thereby raising downstream resistance in proportion to the increased effort. Airflow therefore becomes maximized at a constant value (at that lung volume) independent of effort [26,36]. As shown in Fig. 100.15b, the EPP moves further upstream (i.e., towards the alveolus) with greater expiratory efforts, i.e., extending the compressible segment. However, the EPP is located in the large airways (lower trachea or mainstem bronchi) in normal subjects during expiration. Consequently, although the airways compress, their cartilaginous support prevents them from collapsing. As a result, if a series of flow-volume curves are generated at progressively increased expiratory effort, they will be seen to merge over the lower section of the curve. reflecting the effort-independent nature of the airflow.

In patients with lung disease, however, the EPP can "migrate" far enough into the small airways that airways collapse occurs [2]; this is schematized in Fig. 100.15c for a subject with reduced lung recoil. For a subject with small-airways resistance, the pressure drop within the airways is more rapid and hence the EPP will also occur further upstream.

100.4 Ventilation and Gas Exchange

In the foregoing sections we have addressed the factors involved in distending the thorax and generating airflow without consideration of either (i) how much ventilation is needed or (ii) whether certain breathing patterns might be more appropriate than others.

As the transfer of O_2 and CO_2 between the lungs and blood occurs down gradients of O_2 and CO_2 pressure which are established by the cyclic ventilation of alveolar gas with atmospheric air, it is appropriate to review the physical laws that define the behavior of gases within the lung.

It is clear that high lung recoil or low airways resistance provides for high respired airflow. The effect of changes in pleural pressure is not quite as straightforward, however. The pleural pressure can be considered to be an index of the effort transmitted from the respiratory muscles to the lungs via the chest wall. During a forced expiratory maneuver, for example, P_{pl} becomes positive as a function of the expiratory effort which is applied [14,22]. The reason P_{pl} can be positive during forced expiratory maneuvers is that the chest wall can be caused to reduce its volume faster than the lung is recoiling and hence a compressive force is applied to the intrapleural space.

This being the case, why do not greater and greater expiratory efforts (i.e., more and more positive P_{pl}) lead to progressively greater airflows? The answer is contained in the concept of dynamic airway compression [26,36].

100.4.1 Gas Laws

Avogadro's law states that equal volumes of a particular gas at the same pressure (P) and temperature (T) contain *the same number* of molecules (n). One mole of an ideal gas at standard temperature (0°C or 273°K) and pressure (1 atm) contains 6.02×10^{23} molecules (Avogadro's number) and typically occupies a volume of 22.4l.

Boyle's law states that, at a constant temperature, the volume (V) of a gas is inversely related to its pressure. That is, if a gas at pressure P_1 and volume V_1 is compressed to a new volume (V_2), it will then exhibit a new pressure (P_2) that is derived from the relationship:

$$P_1 \times V_1 = P_2 \times V_2,$$

i.e., $\quad P_2 = \dfrac{P_1 \times V_1}{V_2}.$

Charles' law states that, at a constant pressure, the volume of a gas is directly proportional to its absolute temperature (i.e., °K). That is, when the temperature of a gas at a volume V_1 is changed from T_1 to T_2, the volume will change to a new value V_2 that is given by:

$$\frac{V_1}{V_2} = \frac{T_1}{T_2}$$

This law allows gas volumes that are expressed at different temperatures to be related: for example, while expired volumes of CO_2 are expressed at standard temperature (0°C or 273°K), ventilation is normally referred to body temperature (37°C or 310°K).

General gas law. Based upon these three laws, therefore, the relationship among P, V, and T for a given number of moles of gas is:

$$\frac{P \times V}{T} = n \times R \qquad (100.13)$$

where R is the gas constant.

Graham's law states that the rate at which a gas diffuses through a mixture of gases is *inversely* proportional to the square root of its molecular weight (MW) (i.e., light molecules travel faster and hence diffuse more rapidly). O_2 will therefore diffuse faster than CO_2 (by 18%) in the gas phase:

i.e., $\quad \dfrac{\text{rate of } O_2 \text{ diffusion}}{\text{rate of } CO_2 \text{ diffusion}} = \dfrac{\sqrt{MW_{CO_2}}}{\sqrt{MW_{O_2}}} = \dfrac{\sqrt{44}}{\sqrt{32}}$

$$= \frac{6.6}{5.6} = 1.18$$

Dalton's law states that the pressure exerted by a mixture of gases equals the sum of the individual (or *partial*) pressures exerted by each gas. For example, the atmospheric (or barometric) pressure (P_B) represents the sum of its constituent gases. For dry air, this is:

$$P_{atm} = P_{O_2} + P_{CO_2} + P_{N_2} \qquad (100.14)$$

We have disregarded here the small effect of inert gases other than N_2 (e.g., Ar, Ne, Kr, etc.) and "lumped" these in with N_2, as they are present in such low concentrations.

The partial pressure of gas (X) is given by the product of its fractional concentration (F) and the total pressure:

$$P_X = F_X \times P_{atm} \qquad (100.15)$$

Dry atmospheric air contains 20.93% O_2, 0.04% CO_2, and 79.03% N_2. And as P_B at seal level is about 760 mmHg (or 1 atm), then:

$$P_{O_2} = 159\,mmHg \qquad P_{CO_2} = 0.3\,mmHg$$
$$P_{N_2} = 601\,mmHg.$$

However, as atmospheric air normally is not dry, the water vapor pressure P_{H_2O} will proportionally reduce the pressures exerted by the other gases:

$$P_{atm} = P_{O_2} + P_{CO_2} + P_{N_2} + P_{H_2O}$$

For example, air that is inhaled becomes fully saturated with H_2O by the time it has traversed the trachea; at a normal body temperature of 37°C, P_{H_2O} will therefore be 47 mmHg. As a result, the total pressure exerted by O_2, CO_2, and N_2 will now be only $(760 - 47)$ mmHg. The P_{O_2} of the saturated tracheal air is therefore:

$$\frac{20.93}{100} \times 713 = 149\ mmHg$$

100.4.2 Alveolar Gas Pressures

The CO_2 which is evolved into alveolar gas from the blood establishes an alveolar P_{CO_2} (PA_{CO_2}) that is higher than that of atmospheric air. Similarly, the P_{O_2} of alveolar gas (PA_{O_2}) is less than that of atmospheric air, reflecting the transfer of O_2 into pulmonary capillary blood. Normal values for PA_{CO_2} and PA_{O_2} are about 40 and 100 mmHg, respectively. These values are defined by the rate of ventilation of the alveoli with fresh atmospheric air ($\dot{V}_A$) and the rate of metabolism.

Mass balance considerations require that the amount of CO_2 evolved from the body each minute ($\dot{V}_{CO_2}$) equals the amount cleared from the alveoli by exhalation minus the amount taken in during inhalation:

i.e., $\quad \dot{V}_{CO_2} = [\dot{V}_A(\exp) \times FA_{CO_2}] - [\dot{V}_A(insp) \times FI_{CO_2}]$
$$(100.16)$$

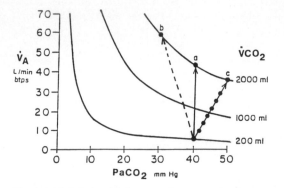

Fig. 100.16. Relationship between changes in alveolar ventilation and arterial or alveolar P_{CO_2} at three different levels of metabolic rate. The *solid symbol a* represents an increase in $\dot{V}_A$ in proportion to $\dot{V}_{CO_2}$, thereby regulating arterial PCO_2. The *dashed line b* represents an increase in alveolar ventilation which is disproportionately greater than the increase in $\dot{V}_{CO_2}$. This results in hypocapnia due to the hyperventilation. The *dotted line c* represents a disproportionately small increase in ventilation as a function of $\dot{V}_{CO_2}$. In this case, hypercapnia results due to the hypoventilation

As FI_{CO_2} is effectively zero:

$$\dot{V}_{CO_2} = \dot{V}_A(exp) \times FA_{CO_2}$$

or

$$FA_{CO_2} = \frac{\dot{V}_{CO_2}}{\dot{V}_A}$$

or

$$\frac{PA_{CO_2}}{(P_{atm}-47)} = \frac{\dot{V}_{CO_2}}{\dot{V}_A}$$

At any given rate of metabolic CO_2 production, the steady-state value for PA_{CO_2} is therefore inversely related, in a hyperbolic fashion, to the rate of alveolar ventilation (Fig. 100.16). This relationship indicates that a given level of $\dot{V}_{CO_2}$ can be accomplished (in theory) by an infinite combination of $\dot{V}_A$ and FA_{CO_2} values; for example, by a low $\dot{V}_A$ with a high PA_{CO_2} or by a high $\dot{V}_A$ operating at a low PA_{CO_2}. The same relationship holds if metabolic rate is increased (e.g., during exercise); however, the $\dot{V}_{CO_2}$ hyperbola is shifted upwards.

A similar relationship also obtains for alveolar P_{O_2} (PA_{O_2}). The amount of O_2 taken up by the body each minute ($\dot{V}_{O_2}$) is equal to the amount of O_2 taken into the alveoli during inhalation minus the amount of O_2 cleared from the alveoli by exhalation:

$$\dot{V}_{O_2} = \left[\dot{V}_a(insp) \times FI_{O_2}\right] - \left[\dot{V}a(exp) \times Fa_{O_2}\right] \quad (100.17)$$

$$\frac{\dot{V}_{O_2}}{\dot{V}a} = \frac{PI_{O_2}{}^*}{(P_{atm}-47)} - \frac{Pa_{O_2}}{(P_{atm}-47)},$$

where * is a correction factor (FA_{N_2}/FI_{N_2}) that takes account of the fact that, at rest in an individual with a normal western diet, inspired ventilation is normally slightly greater than expired ventilation; i.e., the body's metabolic processes produce a smaller volume of CO_2 than the

volume of O_2 which is consumed. That is, the respiratory quotient ($RQ = \dot{V}_{CO_2}/\dot{V}_{O_2}$) is less than 1.0 under these conditions.

It is conventional to express $\dot{V}_{CO_2}$ and $\dot{V}_{O_2}$ at STPD (i.e., standard temperature and pressure, dry), the interest being in the number of moles of the gas produced or consumed. In contrast, $\dot{V}_A$ is conventionally reported at BTPS (i.e., body temperature and pressure, saturated); i.e., the interest is in how much the subject actually ventilated. A correction term must therefore be applied to relate these quantities to the same condition; i.e.,

$$V(\text{BTPS}) = V(\text{STPD}) \times \frac{760}{713} \times \frac{310}{273}$$

$$V(\text{BTPS}) = 1.21 \times V(\text{STPD})$$

Finally, we can transform FA_{CO_2} to PA_{CO_2} in the equation

$$FA_{CO_2} = \frac{1.21\,\dot{V}_{CO_2}(\text{STPD})}{\dot{V}_A(\text{BTPS})}$$

by multipying by ($P_{atm}-47$) to yield:

$$PA_{CO_2} = 863 \times \frac{\dot{V}_{CO_2}(\text{STPD})}{\dot{V}_A(\text{BTPS})}.$$

If, for the moment, we assume alveolar and arterial P_{CO_2} to be equal (causes of differences between these variables are considered in Chap. 107), then this relationship defining PA_{CO_2} allows important distinctions to be made between the following terms:

- *Hypopnea*: Any decrease in ventilation.
- *Hypoventilation*: A decrease in ventilation *out of proportion* to any decrease in metabolic $\dot{V}_{CO_2}$, therefore resulting in a high arterial P_{CO_2}. Increased arterial P_{CO_2} is therefore a criterion by which to determine whether a subject is hypoventilating or not.
- *Hyperpnea*: Any increase in ventilation.
- *Hyperventilation*: An increase in ventilation *out of proportion* to any increase in metabolic $\dot{V}_{CO_2}$, therefore resulting in a low arterial P_{CO_2}. Reduced arterial P_{CO_2} is therefore a criterion by which to determine whether a subject is hyperventilating or not.

Other widely used terms regarding breathing include:

- *Apnea*: a cessation of breathing.
- *Tachypnea*: rapid breathing.
- *Bradypnea*: slow breathing.

100.4.3 Dead Space Ventilation

The initial portion of the tidal volume (V_T) is directed into the alveoli to effect gas exchange. However, the latter portion remains in the conducting airways, i.e., the *anatomical dead space* [V_D(anat)]. Therefore, the volume which enters the alveoli per breath (V_A) is:

$$V_A = V_T - V_D(\text{anat}) \tag{100.18}$$

and the alveolar ventilation ($\dot{V}_A$) is:

$$\dot{V}_A = [V_T \times f] - [V_D(\text{anat}) \times f]$$

or $\quad \dot{V}_A = \dot{V}_E - \dot{V}_D(\text{anat})$,

where f is the breathing frequency and $\dot{V}_E$ is the total minute ventilation.

The alveolar ventilation therefore depends on the volume of the dead space. For example, consider a subject with a minute ventilation of 8 l/min and a breathing frequency of 10:

i.e., $\quad \dot{V}_E = 8000\,\text{ml/min}$ and $f = 10$ breaths/min,

if $\quad V_D(\text{anat}) = 150\,\text{ml}$ (i.e., *normal*),

then $\quad \dot{V}_A = 8000 - (150 \times 10)\,\text{ml/min}$
$\qquad\quad = 6500\,\text{ml/min;}$

if $\quad V_D(\text{anat}) = 300\,\text{ml}$ (i.e., *abnormally increased*),

then $\quad \dot{V}_A = 8000 - (300 \times 10)\,\text{ml/min}$
$\qquad\quad = 5000\,\text{ml/min}$

That is, there is a lower alveolar ventilation in this case, for the same total ventilation. Consequently, the total ventilation is less effective for gas exchange.

A second important determinant of the alveolar ventilation is the pattern of breathing. Consider again that $\dot{V}_E = 8000\,\text{ml/min}$ and $V_D(\text{anat}) = 150\,\text{ml}$:

if $\quad V_T = 800\,\text{ml}$ and $f = 10$ (i.e., *normal*),

then $\quad \dot{V}_A = 8000 - (150 \times 10)\,\text{ml/min}$
$\qquad\quad = 6500\,\text{ml/min}$

However, if the pattern of breathing is changed so that now:

$\qquad V_T = 400\,\text{ml}$ and $f = 20$ (i.e., *tachypnea*),

then $\quad \dot{V}_A = 8000 - (150 \times 20)\,\text{ml/min}$
$\qquad\quad = 5000\,\text{ml/min}$

In this case, $\dot{V}_A$ is reduced, even though both $\dot{V}_E$ and V_D(anat) remained the same. (It should be noted that, in reality, V_D(anat) tends to increase a little as V_T increases because, at larger lung volumes, the more negative P_{pl} increases the transmural pressure across the walls of the intrapulmonary airways; as a result, they are distended.)

Subjects normally "select" an optimum breathing pattern at a particular level of $\dot{V}_E$ [27,31]. That is, they do not breathe too rapidly; this would lead to high resistive work in addition to creating a large dead space ventilation. Very deep breaths would lead to a high elastic work of breathing.

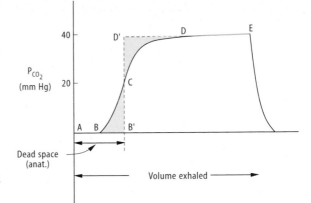

Fig. 100.17. Relationship between P_{CO_2} and exhaled volume during a single exhalation. During $A{\to}B$ only deadspace air leaves the mouth. During $B{\to}C{\to}D$ a mixture of deadspace and alveolar gas is exhaled. During $D{\to}E$ alveolar gas is exhaled. A line B',C,D' drawn such that the area B, B', C is equal to the area C,D', D provides a measurement of the anatomical deadspace, i.e., the volume exhaled between A and B'

Rather, they "select" an optimum pattern which minimizes the work of breathing [27,31].

Measurement of Anatomical Dead Space. *Fowler technique [17].* Consider a lung from which gas is exhaled initially exclusively from the anatomical dead space and then exclusively from the alveoli (i.e., there is no mixing between the compartments). In this case, the expired P_{CO_2} would have an abrupt, step-like profile (Fig. 100.17, AB'-$B'CD'$-$D'DE$). V_D(anat) would simply be the volume exhaled up to point B'. In reality, although the initial part of the expirate is derived exclusively from the dead space and the final part exclusively from the alveoli, there is an intermediate transitional region in which some alveoli (i.e., those with short conducting airways) begin to empty before all the dead space is cleared, and there is also diffusional mixing across the interface between the dead space and alveolar compartments. Expired P_{CO_2} does not therefore increase instantaneously to the alveolar level. The anatomical dead space can be determined by superimposing a vertical line across the transitional phase of the P_{CO_2} profile, such that the area B-B'-C equals the area C-D'-D; i.e., V_D(anat) is the volume exhaled between A and B'. In other words, we functionally "create" the condition in which the lung empties with independent, sequential dead space and alveolar compartments.

Bohr (CO₂ proportion) technique [5]. For this technique, it is assumed that the volume of CO_2 exhaled in a breath originates from the alveolar compartment; then the total amount of CO_2 that is collected in a mixed expired tidal volume equals that exhaled volume which came from the alveolar compartment:

i.e., $\quad V_T \times F\bar{E}_{CO_2} = V_A \times F_{A_{CO_2}} \tag{100.19}$

where $F\overline{E}_{CO_2}$ is the mixed expired F_{CO_2}. Substituting $V_A = V_T - V_D(anat)$ yields:

$$V_T \times F\overline{E}_{CO_2} = \left[V_T \times F_{A_{CO_2}}\right] - \left[V_D(anat) \times F_{A_{CO2}}\right]$$

i.e.,
$$\frac{V_D(anat)}{V_T} = \frac{F_{A_{CO_2}} - F\overline{E}_{CO_2}}{F_{A_{CO_2}}}$$

or
$$\frac{V_d(anat)}{V_T} = \frac{P_{a_{CO_2}} - P\overline{e}_{CO_2}}{P_{a_{CO_2}}}$$

Measurement of Physiological Dead Space. Some alveoli may be ventilated but not perfused [38,43], and therefore cannot participate in gas exchange. The volume of such alveoli, which is termed the alveolar dead space [$V_D(alv)$], may become pathologically large in pulmonary disease states. The total, or physiologic, dead space is given by:

$$V_D(physiol) = V_D(anat) + V_D(alv). \qquad (100.20)$$

Clearly, as it is the total physiological dead space which is the best index of the inefficiency of the breath, $V_D(physiol)$ is more commonly measured than $V_D(anat)$. For this, a modification of the Bohr technique is used [15]. This requires the assumption that the P_{CO_2} of the exchanging (i.e., perfused) alveoli equals the P_{CO_2} of the arterial blood. That is, the P_{CO_2} comes to diffusion equilibrium *where it can*. Consequently, arterial P_{CO_2} (which is easily determined) becomes a measure of the P_{CO_2} in only the perfused alveoli. The mixed expired P_{CO_2} therefore differs from PA_{CO_2} because of the effect of both the anatomical *and* the alveolar dead space contributions to the expirate.

If PA_{CO_2} is therefore be replaced by Pa_{CO_2} in the original Bohr equation, the physiological dead space may be determined:

i.e.,
$$\frac{V_D(physiol)}{V_T} = \frac{Pa_{CO_2} - P\overline{E}_{CO_2}}{Pa_{CO_2}}$$

Normally, $V_D(physiol)$ is approximately equal to $V_D(anat)$, and accounts for about 25%–30% of the tidal volume at rest. In patients with lung diseases such as pulmonary vascular occlusive disease or obstructive lung disease in which alveolar septa and associated blood vessels are destroyed, $V_D(physiol)$ can be appreciably larger than $V_D(anat)$; e.g., 45%–50% of the tidal volume.

100.4.4 Metabolic Rate

It is presumably obvious that in order to maintain the alveolar concentration of a particular gas such as O_2 at a constant level, the rate at which the gas is loaded into the alveoli must match the rate at which it is removed from the alveoli. That is, the alveolar gas concentration (or pressure) is a function of the ratio of alveolar ventilation to metabolic rate. However, while it is intuitively reasonable that alveolar (and hence arterial) gas partial pressure depends upon a coupling of ventilation to metabolic rate, the precise determinants of the relationship deserve consideration.

It is important to recognize, for example, that the metabolic rates for O_2 ($\dot{V}_{O_2}$) and CO_2 ($\dot{V}_{CO_2}$) are often not the same: they depend upon the substrate mixture being catabolized in the cells. This tissue gas-exchange ratio, or respiratory quotient (RQ = $\dot{V}_{CO_2}/\dot{V}_{O_2}$) typically ranges from 0.7 to 1.0, depending upon the substrate mixture undergoing oxidation. For example, for carbohydrate (e.g., glucose) metabolism:

$$C_6H_{12}O_6 + 6O_2 \xrightarrow[\text{[RQ = 1.0]}]{} 6CO_2 + 6H_2O + 36ATP \qquad (100.21)$$

For fat metabolism (e.g., palmitic acid):

$$C_{16}H_{32}O_2 + 23O_2 \xrightarrow[\text{[RQ = 0.7]}]{} 16CO_2 + 16H_2O + 130ATP \qquad (100.22)$$

Normally, however, well-nourished subjects on a typical Western diet have a resting RQ of approximately 0.85, reflecting the mixture of fats and carbohydrates being metabolized.

The *pulmonary* gas-exchange ratio ($R = \dot{V}_{CO_2}/\dot{V}_{O_2}$, but measured at the lungs), however, can range markedly beyond these limits: as subjects hypoventilate and retain CO_2 in the body stores, R becomes less than RQ [16,38], whereas when subjects hyperventilate and wash CO_2 out of the body stores, R exceeds RQ [16,38].

By the same token, when tissue bicarbonate concentration is decreasing during an acute metabolic (e.g., lactic) acidosis, R also exceeds RQ consequent to the additional CO_2 which is released in the buffering reactions [13,41], i.e.,

$$CH_3 \cdot CHOH \cdot COO^-H^+ + NaHCO_3 \rightarrow$$
$$CH_3 \cdot CHOH \cdot COONa + H_2CO_3 \qquad (100.23)$$
(lactic acid + sodium bicarbonate)
(sodium lactate + carbonic acid)

The carbonic acid is then dehydrated to yield the extra CO_2 and water, i.e.,

$$H_2CO_3 \rightarrow CO_2 + H_2O \qquad (100.24)$$

Similarly, as CO_2 is retained as the bicarbonate stores are subsequently replenished, R decreases below RQ.

Thus, when R is appreciably different from 1.0, ventilation cannot meet the demands for both O_2 and CO_2 exchange. There is a consistent body of evidence that demonstrates that, under such conditions, ventilation appears to be more closely "coupled" to the demands for CO_2 exchange than to those for O_2 [13,41,47]. At a given rate of CO_2 exchange, however, the alveolar ventilation depends upon the set point for Pa_{CO_2} (Fig. 100.18). The lower the Pa_{CO_2}, the higher the required alveolar ventilation.

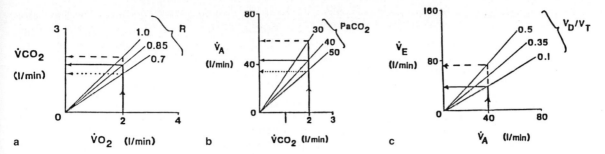

Fig. 100.18a–c. Relationship between O_2 uptake and CO_2 output for different levels of the gas exchange ratio R (**a**); between CO_2 output and alveolar ventilation as a function of the arterial P_{CO_2} (**b**); and between alveolar ventilation and total minute ventilation as a function of the deadspace fraction of the tidal volume (**c**)

It is important to recognize that, while the $P_{A_{CO_2}}$ and $P_{A_{O_2}}$ are determined by the alveolar ventilation to metabolic rate ratio, as described above, this must be established by means of the total or minute ventilation ($\dot{V}_E$), i.e., it is not possible to ventilate the alveoli without simultaneously ventilating the dead space!

The relationship between $\dot{V}_E$ and $\dot{V}_A$ can be expressed either as:

$$\dot{V}_A = \dot{V}_E - \dot{V}_D$$
or
$$\dot{V}_A = \dot{V}_E(1 - \dot{V}_D/V_T)$$

The relationship between $\dot{V}_E$ and $\dot{V}_{CO2}$ is therefore expressed as:

$$\dot{V}_E = \frac{863\,\dot{V}_{CO_2}\,(\text{STPD})}{Pa_{CO_2}\,(1 - V_D/V_T)}$$

Consequently the combined influence of $\dot{V}_{CO_2}$, Pa_{CO_2}, and V_D/V_T (Fig. 100.18a–c) can be marked [46], so that without knowledge of the current values of these determining variables it is extremely difficult to predict with confidence what is an appropriate $\dot{V}_E$ in a given circumstance.

While we have considered the determinants of $\dot{V}_E$ and the fundamentals of its volume and flow-modulated mechanics, we have addressed neither the alveolar gas exchange inefficiencies which lead to differences between alveolar and arterial gas partial pressures, nor the mechanisms which control the current demands for ventilation. These will be addressed in Chaps. 107 and 111.

References

1. Agostoni E, Mead J (1964) Statics of the respiratory system. In: Fenn WO, Rahn H (eds) Handbook of physiology, section 3, vol I: respiration. American Physiological Society, Bethesda, pp 387–409
2. Anthonisen NR (1977) Closing volume. In: West JB (ed) Regional differences in the lung. Academic, New York, pp 451–482
3. Bachofen H, Hildebrandt J, Bachofen M (1970) Pressure-volume curves of air and liquid filled excised lungs – surface tension in situ. J Appl Physiol 29:422–431
4. Bernaudin JF, Jaurand ML, Fleury J, Bignon J (1991) Mesothelial cells. In: West JB, Crystal RG (eds) The lung: scientific foundations. Raven, New York, pp 631–638
5. Bohr C (1891) Ullber die Lungenathmung. Scand Arch Physiol 2:236–268
6. Campbell EJM, Agostoni E, David JN (1970) The respiratory muscles: mechanics and neural control, 2nd edn. Saunders, Philadelphia
7. Cherniack RM, Farhi LE, Armstrong BW, Proctor DF (1955) A comparison of esophageal and intra-pleural pressure in man. J Appl Physiol 8:203–211
8. Clark FJ, von Euler C (1972) On the regulation of depth and rate of breathing. J Physiol (Lond) 222:267–295
9. Clements JA, Hustead RF, Johnson RP, Gribetz I (1961) Pulmonary surface tension and alveolar stability. J Appl Physiol 16:444–450
10. Clements JA, King R (1976) Composition of surface active material. In: Crystal RG (ed) The biochemical basis of pulmonary function. Dekker, New York, pp 363–387
11. Comroe JH Jr, Forster RE, Dubois AB et al (1962) The lung, 2nd edn. Year Book Medical Publishers, Chicago
12. De Troyer A, Estenne M (1988) Functional anatomy of the respiratory muscles. Clin Chest Med 9:175–193
13. Douglas CG (1927) Coordination of respiration and circulation with variations in metabolic activity. Lancet 2:213–218
14. Dry DL, Hyatt RE (1960) Pulmonary mechanics. A unified analysis of the relationship between pressure, volume and gas flow in the lungs of normal and diseased human subjects. Am J Med 29:672–689
15. Enghoff H (1938) Volumen inefficax. Bemerkungen zur Frage des schädlichen Raumes. Ups Lakarf Forh 44:191–218
16. Farhi LE, Rahn H (1955) Gas stores of the body and the unsteady state. J Appl Physiol 7:472–484
17. Fowler WS (1948) The respiratory dead space. Am J Physiol 154:405–416
18. Gehr P, Bachofen M, Weibel ER (1978) The normal lung: ultrastructure and morphometric estimation of diffusing capacity. Respir Physiol 32:121–140
19. Gibson GJ, Pride NB, Davis J, Schroter RC (1979) Exponential description of the static pressure-volume curve of normal and diseased lung. Am Rev Respir Dis 120:799–811
20. Gil J, Bachofen H, Gehr P, Weibel ER (1979) Alveolar surface-volume relation in air- and saline-filled lungs fixed by vascular perfusion. J Appl Physiol 47:990–1001
21. Grant BJB, Jones HA, Hughes JMB (1974) Sequence of regional filling during a tidal breath in man. J Appl Physiol 37:158–165
22. Hyatt RE (1986) Forced expiration. In: Macklem PT, Mead J (eds) Handbook of physiology, section 3, vol III: mechanics of breathing. American Physiological Society, Bethesda, pp 295–314
23. Jones RL, Overton TR, Sproule BJ (1977) Frequency dependence of ventilation distribution in normal and obstructed lungs. J Appl Physiol 42:548–553

24. Macklem PT, Mead J (1967) Resistance of central and peripheral airways measured by a retrograde catheter. J Appl Physiol 22:395–405
25. Mead J, Whittenberger L, Radford EP (1957) Surface tension as a factor in pulmonary pressure-volume hysteresis. J Appl Physiol 10:191–196
26. Mead J, Turner JM, Macklem PT, Little JB (1967) Significance of the relationship between lung recoil and maximal expiratory flow. J Appl Physiol 22:95–108
27. Milic-Emili G, Petit JM (1959) II lavoro mecanico della respirazione a varia frequenze respiratoria. Arch Sci Biol Bologna 43:326–330
28. Milic-Emili J, Mead J, Turner JM (1964) Topography of esophageal pressure as a function of posture in man. J Appl Physiol 19:212–216
29. Milic-Emili J, Henderson JAM, Dolovich MB, Trop D, Kaneko K (1966) Regional distribution of inspired gas in the lung. J Appl Physiol 21:749–759
30. Otis AB (1954) The work of breathing. Physiol Rev 34:449–458
31. Otis AB, Fenn WO, Rahn H (1950) The mechanics of breathing in man. J Appl Physiol 2:592–607
32. Otis AB, McKerrow CB, Bartlett RA, Mead J, McIlroy MB, Selverston NJ, Radford EP Jr (1955) Mechanical factors in distribution of pulmonary ventilation. J Appl Physiol 8:427–443
33. Pappenheimer JR, Comroe JH, Cournand A, Ferguson JKW, Filley GF, Fowler WS, Gray JS, Helmbolz HF, Otis AB, Rahn H, Riley RL (1950) Standardization of definitions and symbols in respiratory physiology. Fed Proc 9:602–615
34. Pattle RE (1955) Properties, function and origin of the alveolar lining layer. Nature (Lond) 175:1125–1127
35. Pride NB, Macklem PT (1986) Lung mechanics in disease. In: Fishman AP, Macklem PT, Mead J, Greiger SR (eds) Handbook of physiology, section 3, vol III, part II: the respiratory system. American Physiological Society, Bethesda, pp 659–692
36. Pride NB, Permutt S, Riley RL, Bromberger-Barnea B (1967) Determinants of maximum expiratory flow from lungs. J Appl Physiol 23:646–662
37. Radford EP (1964) The physics of gases. In: Fenn WO, Rahn H (eds) Handbook of physiology, section 3, vol I: respiration. American Physiological Society, Washington DC, p 149
38. Rahn H, Fenn WO (1955) A graphical analysis of the respiratory gas exchange. The O_2-CO_2 diagram. American Physiological Society, Wasington DC
39. Rahn H, Otis AB, Chadwick LE, Fenn WO (1946) The pressure-volume diagram of the thorax and lung. Am J Physiol 146:161–178
40. Smith JC, Stamenovic D (1986) Surface forces in the lungs. Alveolar surface tension-lung volume relationships. J Appl Physiol 60:1341–1350
41. Wasserman K, Van Kessel AL, Burton GG (1967) Interactions of physiological mechanisms during exercise. J Appl Physiol 22:71–85
42. Weibel ER (1963) Morphometry of the human lung. Springer, Berlin Göttingen Heidelberg
43. West JB (1970) Ventilation/blood flow and gas exchange. Blackwell Scientific, Oxford, pp 17–55
44. West JB (1990) Respiratory physiology – the essentials. Williams and Wilkins, Baltimore
45. West JB (1991) Mechanics of breathing. In: West JB (ed) Physiological basis of medical practice. Williams and Wilkins, Baltimore, p 563
46. Whipp BJ, Pardy RL (1986) Breathing during exercise. In: Macklem PT, Mead J (eds) Handbook of physiology, section 4, vol III: Respiration. American Physiological Society, Bethesda, pp 605–629
47. Whipp BJ, Ward SA (1991) Coupling of ventilation to pulmonary gas exchange during exercise. In: Whipp BJ, Wasserman K (eds) Exercise: pulmonary physiolgy and pathophysiology. Dekker, New York, pp 271–307
48. Woolcock AJ, Vincent NJ, Macklem PT (1969) Frequency dependence of compliance as a test for obstruction in the small airways. J Clin Invest 48:1097–1106

101 Pulmonary Gas Exchange

J. Piiper

Contents

101.1 Introduction

In the original sense of the latin word, "respiration" means repeated inhaling and exhaling of air with the associated movements of the thorax. Respiratory movements produce air flow for the transport of O_2 into lungs, and of CO_2, the end product of oxidative metabolism, out of the lungs. But respiration in today's meaning includes all the processes involved in the transport of O_2 and CO_2 between the environment and body tissues, performed by pulmonary ventilation, pulmonary O_2 and CO_2 exchange, blood circulation, and O_2 and CO_2 exchange in tissues. Gas exchange in tissues is intimately related to the consumption of O_2 and production of CO_2 by oxidative tissue metabolism whose main purpose is provision of energy for life processes.

Obviously, the definition of where respiration stops and other processes start (e.g., circulation, tissue metabolism, energetics, etc.) is more or less arbitrary, like most subdivi-

sions in physiology, but it is necessary for analysis and understanding. Taking respiration as the sum of the processes subserving (or related to) O_2 and CO_2 exchange and transport between the environment and body cells, respiration is treated in this and the following two chapters (Chaps. 102, 103). The emphasis will be on the basic principles of qualitative and quantitative functional analysis. What is practically excluded are the varied humoral and nervous control mechanisms (control of lung ventilation, cardiovascular function, acid-base balance, hemoglobin homeostasis, etc.) and cell physiological and molecular aspects, because these topics will be treated in other chapters.

The lungs are the site of gas exchange of the body with the environment, i.e., of uptake of O_2 and output of CO_2 (only about 1% of the gas exchange in man takes place across body skin).

The human lung airways (Fig. 101.1a) represent a tree-like structure, formed by repetitive branching of bronchi down to terminal bronchioles which give rise, by branching, to several generations of alveoli-carrying respiratory bronchioles and, finally, to alveolar ducts whose walls are entirely occupied by alveoli (about 3×10^6 in human lungs) ([28]; see also Chap. 100). The alveoli are surrounded by a blood capillary network, fed by pulmonary arterioles and discharging to pulmonary venules. The multiple branching of airways and blood vessels leads to a very large air/blood exchange surface area (about 80 m² in human lungs). This and the extreme thinness of the tissue layers separating blood and gas (alveolar epithelium, interstitium, capillary endothelium), of about 0.1 µm, create favorable conditions for diffusive gas exchange. The efficient transport of O_2 and CO_2 by blood is ensured by the high blood flow to the lungs (equal to total cardiac output) and the rapid reversible chemical combination of O_2 and CO_2 in blood.

The simplest functional model for lung gas exchange function is shown in Fig. 101.1b. Pulmonary gas exchange includes three steps:

- *Ventilation* (air flow) bringing O_2 into, and removing CO_2 from, lung air spaces
- *Diffusion* across the air/blood barrier (and within pulmonary capillary blood, including chemical reactions in red cells and plasma)
- *Perfusion*, i.e., blood flow transporting O_2 from the lungs to body tissues and CO_2 from body tissues to the lungs.

R. Greger/U. Windhorst (Eds.)
Comprehensive Human Physiology, Vol. 2
© Springer-Verlag Berlin Heidelberg 1996

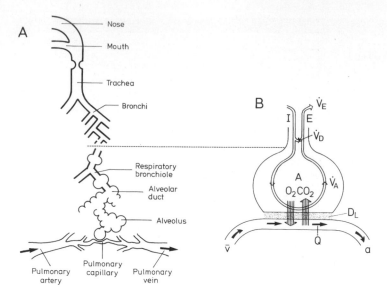

Fig. 101A,B. Anatomy of pulmonary gas exchange structures (**A**) and simplified model of pulmonary gas exchange (**B**). **a** The branching structure of the bronchial tree and of pulmonary arteries and veins is outlined. The *dashed line* between **a** and **b** indicates the boundary between the dead space and the alveolar space. **B** O_2 is brought into the lung and CO_2 is eliminated by ventilation, $\dot{V}_E$ (only the alveolar ventilation, $\dot{V}_A$ is effective; the dead space ven-

tilation, $\dot{V}_D$, is ineffective). Diffusive equilibration between alveolar gas (*A*) and pulmonary capillary blood takes place through the air-blood barrier, of diffusing capacity D_L. O_2 is transported to the body and CO_2 to the lungs by the pulmonary blood flow or cardiac output, $\dot{Q}$. The specific O_2 and CO_2 transport properties of blood play an important role, establishing a sufficiently large concentration difference between arterial (a) and mixed venous blood ($\bar{v}$).

The primary task of lungs is to provide adequate *arterialization* of blood, by achieving physiological values of O_2 and CO_2 partial pressures (P_{O_2}, P_{CO_2}) and pH in arterial blood distributed to body tissues. The O_2 delivery to the body and the CO_2 elimination from the body are the result of a combined action of the lungs and the heart: of the heart, by providing an adequate total blood flow (cardiac output); of the lungs, by insuring optimal arterialization of the blood which perfuses organs and tissues. Furthermore, since P_{CO_2} is a major determinant of the pH in blood and tissues, the lungs play an important role in the control of the acid-base equilibrium of the body (cf. Chap. 78). Evidently, the pulmonary gas exchange function must be adjusted to the changing metabolic demand (e.g., a large increase in exercise) and to varying environmental conditions (e.g., low O_2 partial pressure at high altitude; cf. Chaps. 108, 109).

101.2 Physical Concepts and Quantities [10]

101.2.1 Gas Volumes

The volume of an amount of gas is a function of pressure and temperature. The quantitative relationships follow from the general *gas law* (ideal gas law),

$$P \times V = n \times R \times T, \tag{101.1}$$

where P is pressure, V volume, T absolute temperature [in kelvins (K), equal to the temperature in °C + 273), n the

amount of substance, usually expressed in moles but in respiratory physiology often in milliliters (STPD, see below), and R the gas constant (= 8.33 J × mol⁻¹ × K⁻¹ = 1.99 cal × mol⁻¹ × K⁻¹). The equation is valid for pure gases and for gas mixtures. In addition to depending on pressure and temperature, gas volume depends on water vapor saturation (see below).

In respiratory physiology, the following *three conditions* of gas volume measurement are used:

- Gas volumes and flows (flow = volume/time) are usually measured at the ambient barometric pressure and at room temperature, saturated with water vapor ("spirometer conditions"). These are called *ATPS* conditions (*a*mbient *t*emperature, *p*ressure, *s*aturated).
- Lung volumes and ventilations are defined to denote volumes and volume changes within the airways. Therefore they are expressed for "body conditions", called *BTPS* conditions (*b*ody *t*emperature, *p*ressure, *s*aturated).
- O_2 uptake and CO_2 output designate amounts of gas transferred in unit time, and are, therefore, transformed to "standard physical conditions": 0°C, 760 mmHg, dry. These are termed *STPD* conditions (*s*tandard *t*emperature, *p*ressure, *d*ry).

From the gas law, the conversion factors compiled in Table 101.1 are obtained. It is evident that the magnitude of the conversion factors is such that they cannot be disregarded. It is important to note that in equations all gas volumes (and flows) must be expressed in the same units and volume measurement conditions. An alternative, frequently

Table 101.1. Gas volume calculations. Derivation of formulae for transformation of volumes between various measuring conditions

		Temperature (absolute)		Pressure		Water vapor		Total		Typical value[a]
$\dfrac{V(\text{BTPS})}{V(\text{ATPS})}$	$=$	$\dfrac{310}{T}$	$\times$	1	$\times$	$\dfrac{P_B - P_{H_2O}}{P_B - 47}$	$=$	$\dfrac{310(P_B - P_{H_2O})}{T(P_B - 47)}$	$\approx$	1.09
$\dfrac{V(\text{STPD})}{V(\text{ATPS})}$	$=$	$\dfrac{273}{T}$	$\times$	$\dfrac{P_B}{760}$	$\times$	$\dfrac{P_B - P_{H_2O}}{2.78 \times T}$	$=$	$\dfrac{P_B - P_{H_2O}}{2.78 \times T}$	$\approx$	0.89
$\dfrac{V(\text{STPD})^b}{V(\text{BTPS})}$	$=$	$\dfrac{273}{310}$	$\times$	$\dfrac{P_B}{760}$	$\times$	$\dfrac{P_B - 47^c}{P_B}$	$=$	$\dfrac{P_B - 47^c}{863}$	$\approx$	0.81

APTS, ambient temperature, pressure, saturated: BTPS, body temperature, pressure, saturated; STPD, standard temperature, pressure, dry; T, absolute room temperature (K); P_{H_2O}, water vapor saturation pressure (mmHg) at T; P_B, barometric pressure (mmHg)

[a] For $P_B = 747\,\text{mmHg}$ (corresponding to mean P_B at 140 m above sea level), $T = 295\,\text{K}$ ($\approx 22°C$), $P_{H_2O} = 20\,\text{mmHg}$

[b] The ratio $\dfrac{V(\text{STPD})}{V(\text{BTPS})} = f\text{vol}$ is used in many equations (101.4a,b; 101.6a,b; 101.11a,b; 101.12a,b; 101.14a,b; 101.19)

[c] 47 is the water vapor saturation pressure (in mmHg) at 37°C

used, is to include the STPD/BTPS volume conversion factor in certain equations (see Sect. 101.4). In this chapter, for the sake of clarity, the volume conversion factors will be explicitly indicated in quantitative relationships.

101.2.2 Concentrations and Partial Pressures of Gases

Gas concentrations in gas phase are conventionally expressed in *fractional* (dimensionless) concentrations, F_x, defined as volume of a gas species x in the total gas volume in dry conditions (i.e., excluding water vapor): $F_x = V_x/V$. Concentrations (or *content*) of gases in blood (and other liquids), C_x, are expressed as amount of gas (V_x) per unit volume of blood (Q): $C_x = V_x/Q$. The traditional unit is volumes percent [vol % = ml gas (STPD)/100 ml blood], but more recently molar units have been preferred: mmol/l = 2.24 vol%. Disregard of the different definitions of gas concentration in gas phase and in blood may lead to confusion and serious errors.

In a gas mixture, the total pressure is the sum of the *partial pressures* of the component gases (Dalton's law). The partial pressures are proportional to the fractional concentrations, F. Since F refers to dry gas mixture, the sum of partial pressures is equal to the difference between the total pressure (often atmospheric pressure, P_B) and the partial pressure of water vapor (P_{H_2O}):

$$P_x = F_x \times (P_B - P_{H_2O}) \tag{101.2}$$

By definition, in *equilibrium*, the partial pressure of a gas in a liquid (blood) is equal to that in the gas phase: $P_{x(\text{liquid})} = P_{x(\text{gas})}$. Commonly used units of partial pressures (both for gas and liquid phase) are millimeters of mercury (mmHg, equivalent to torr), kilopascal (kPa), and physical atmosphere (atm): $1\,\text{kPa} = 7.5\,\text{mmHg}$; $1\,\text{atm} = 760\,\text{mmHg} = 101.3\,\text{kPa}$.

101.2.3 Atmospheric Air

Dry atmospheric air has the following composition: 20.9% O_2, 0.03% CO_2, 78.1% N_2, 0.9% Ar, and traces of other inert gases. The concentration of CO_2, 0.03% ($F_{CO_2} = 0.0003$), although of fundamental importance as essential to plant photosynthesis and thus to all animal life (and which nowadays is causing grave concern because of its tendency to rise), can often be neglected in respiratory physiology. Since the metabolic reactions in the human body are generally assumed to neither produce nor consume molecular N_2, it is justifiable to regard N_2 as a *physiologically inert* gas. Therefore, it is customary to regard N_2 and other inert gases together as one total inert gas termed "nitrogen". Thus, the composition of atmospheric air as simplified for respiratory physiology, is: $F_{O_2} = 0.209$; $F_{CO_2} = 0.000$; $F_{N_2} = 0.791$.

In addition, atmospheric air contains a variable amount of *water vapor* (i.e., gaseous water). The water in air is usually expressed as relative humidity (= fractional saturation). From this and the saturation pressure of water vapor for a given temperature, the partial pressure of water vapor can be derived. The saturation pressure of water vapor at body temperature (37°C) is 47 mmHg (= 6.3 kPa).

Atmospheric pressure (barometric pressure, P_B) depends on the altitude above sea level and the meteorological conditions. The average P_B at sea level (P_{B_0} is 760 mmHg (101.3 kPa)). With increasing altitude (h, altitude in km) the barometric pressure (P_{B_h}) decreases exponentially:

$$P_{B_h} = P_{B_0} \times 10^{-0.055h} \tag{101.3}$$

leading to halving of P_B at an altitude of 5.5 km ($\approx -\log 0.5/0.055$). Since the composition of atmospheric air is fairly independent of altitude, the above equation is also valid for P_{O_2} in inspired dry air. For P_{O_2} in the physiologically relevant water vapor-saturated air, P_B must be replaced by ($P_B - P_{H_2O}$) in Eq. 101.3.

101.3 Overall Exchange [14]

101.3.1 Ventilation and Gas Transport

The amount of O_2 taken up and that of CO_2 eliminated with a breath is equal to the difference between the inspired and expired amounts of the gases. In terms of fractional concentrations (for the volume measurement conversion factor $f$vol, see Sect. 101.2.1 and Table 101.1):

$$V_{O_2} = f\text{vol} \times (V_I \times F_{I_{O_2}} - V_E \times F_{E_{O_2}}) \tag{101.4a}$$

$$V_{CO_2} = f\text{vol} \times (V_E \times F_{E_{CO_2}} - V_I \times F_{I_{CO_2}}) \tag{101.4b}$$

Since in many conditions the expired and inspired amounts of the inert gas N_2 may be considered as identical ($V_E \times F_{E_{N_2}} = V_I \times F_{I_{N_2}}$, known as N_2 equilibrium), the "N_2 concentration ratio" is used for eliminating either V_E or (usually) V_I:

$$r_{N_2} = F_{I_{N_2}}/F_{E_{N_2}} = V_E/V_I \tag{101.5}$$

The ratio r_{N_2} is usually only one or a few percent below unity. For CO_2, F_I can be disregarded when ambient air is breathed.

Per unit time (e.g., minute) one thus obtains the relationship between O_2 *uptake*, CO_2 *output*, and (expired) ventilation, $\dot{V}_E$ (= $V_E \times f_{resp}$, respiratory frequency):

$$\dot{V}_{O_2} = f\text{vol} \times \dot{V}_E \, (r_{N_2} \times F_{I_{O_2}} - F_{E_{O_2}}) \tag{101.6a}$$

$$\dot{V}_{CO_2} = f\text{vol} \times \dot{V}_E \, (F_{E_{CO_2}} - r_{N_2} \times F_{I_{CO_2}}) \tag{101.6b}$$

The ratio $\dot{V}_{CO_2}/\dot{V}_{O_2}$ is the *respiratory exchange ratio* (*respiratory quotient*), R:

$$R = \dot{V}_{CO_2}/\dot{V}_{O_2} \tag{101.7}$$

In the steady state, R is equal to the metabolic respiratory quotient, R_{met}, designating the ratio of CO_2 production and O_2 consumption in tissue metabolism. R_{met} is determined by the kind of substrate oxidized: $R = 1.0$ for carbohydrates, $R = 0.7$ for fat, $R \approx 0.8$ for protein. Usually, $R_{met} = R_E = 0.85$ is found in the steady state at rest.

101.3.2 Blood Flow and Gas Transport: The Fick Principle

The O_2 taken up in the lungs is transported to the body by blood (only about 1% is consumed in the lungs), and the CO_2 eliminated by the lungs is brought from the body to the lungs by blood flow. In mammals (and birds), with complete separation of arterial and venous blood in the heart, pulmonary blood flow is equal to the total systemic blood flow, designated as *cardiac output*. By analogy to the above equations for the relationship between $\dot{V}_{O_2}$, $\dot{V}_{CO_2}$, and $\dot{V}_E$, the following relationship between O_2 uptake, CO_2 output, and the difference in concentration between arterial blood (Ca) and mixed venous blood ($C\bar{v}$) obtains:

$$\dot{V}_{O_2} = \dot{Q} \, (Ca - C\bar{v})_{O_2} \tag{101.8a}$$

$$\dot{V}_{CO_2} = \dot{Q} \, (C\bar{v} - Ca)_{CO_2} \tag{101.8b}$$

This is the so-called *Fick principle*, and it is the basis of the classical method of determining the cardiac output from O_2 uptake and the arteriovenous difference in O_2 concentration. It is important to remember that venous blood from different organs has different O_2 (and CO_2) contents, due to variations in the utilization of blood O_2. The confluence of all venous blood in the right atrium gives rise to *mixed venous blood*, which is the blood that enters the lungs and is therefore to be used in the Fick principle equation. Mixed venous blood is best obtained from the pulmonary artery (by cardiac catheter introduced via a peripheral vein). In contrast, arterial blood is the same everywhere in the arterial system (except in individuals with rare cardiovascular abnormalities) and may be obtained from any artery.

101.3.3 Methods

Inspired and expired volumes are measured as volume change by *spirometers*, which are gas-filled vessels with variable, calibrated volume. Ventilation is measured either by volume changes per unit time or by pneumotachometry, i.e., recording of the pressure drop across a (small) airflow resistance that is proportional to the gas flow rate. Expired and inspired volumes can be ascertained by integration of the gas flow.

Concentrations in gas are measured by *volumetric* analysis after consecutive absorption of CO_2 (by alkali) and of O_2 (by dithionite). The volume changes indicate fractional concentrations F (Haldane and Scholander apparatuses). Continuous gas analysis is achieved by mass spectrometers (all gases), infrared gas analyzers (CO_2, CO), paramagnetic gas analyzers (O_2), and other devices (for measurements in blood, see Sect. 102.2.1).

O_2 uptake ($\dot{V}_{O_2}$) and CO_2 output ($\dot{V}_{CO_2}$) are usually measured from $\dot{V}_E$, F_I, and F_E (*open system*). O_2 uptake may be directly measured as the volume decrease of an O_2-filled spirometer provided with a CO_2 absorber (*closed system*).

101.3.4 Values at Rest and During Exercise

Typical values for a normal human being are shown in Table 101.2. At rest, the O_2 consumption is about 0.3 l/min. This corresponds to a metabolic rate of 1.5 kcal/min (average energetic equivalent of O_2, 4.85 kcal/l) or around 80 W (equivalent to a typical electric bulb). During heavy exercise the metabolic rate may rise by factor of 10, to 800 W (equivalent to an electric heater!). The rise in respiratory quotient (R) during short-term exercise is attributable to increased utilization of carbohydrates (glycogen) in muscle energy metabolism, partly due to liberation of extra CO_2 from bicarbonate by lactic acid formed in anaerobic glycolysis. Ventilation increases nearly proportionately to O_2 consumption, due to an increase in both respiratory frequency and tidal volume. However, cardiac output increases less than proportionately to O_2 uptake. The arterial O_2 content changes little, whereas the O_2 content of mixed venous blood decreases, leading to an increased arterial-mixed-venous difference in blood O_2 content.

101.4 Dead Space and Alveolar Ventilation [1,14]

Before reaching the alveolar ducts and alveoli, inspired gas passes through nasal cavities, pharynx, larynx, trachea, and the bronchial tree. In these "conducting airways" inspired air is cleaned of particles (dust), heated (rarely cooled) to body temperature, and saturated with water vapor. This *conditioning* of inspired gas is an important function of the conducting airways. The airways are lined with a thin layer of mucus, secreted by bronchial glands and transported up the airways towards the esophageal ostium by the beat of the cilia of the respiratory epithelium, finally to be swallowed. Extraneous particles adhere to the mucous layer and are transported with it. Heating and humidification are effected by contact of inspired air with large-surface mucous membranes (particularly nasal conchae, small bronchi) and potentially high blood flow (in the bronchi via bronchial vessels, in part anastomosing with pulmonary blood vessels). For gas exchange, however, the conducting airways are considered as constituting *dead space*, because, compared with gas exchange in alveoli, their contribution to respiratory gas exchange is negligible: inspired gas remaining in dead space is expired essentially without any change in its composition except for water vapor.

For a quantitative evaluation of the effects of dead space on gas exchange, a simplified lung model consisting of a rigid tube, dead space, and an extensible bag, the alveolar space, is employed (Fig. 101.2). Upon expiration, first gas from the dead space (V_D) is expired, at the same gas concentrations as in inspired gas. Then follows gas from the alveolar space, in volume $V_{AE} = V_E - V_D$, with the alveolar gas concentrations F_A. The mixed expired gas arises from subsequent mixing (e.g., in a spirometer) of the functional components, dead space gas and alveolar gas. The following relationships obviously obtain (see also Chap. 100):

$$\text{for volumes:} \quad V_D + V_{AE} = V_E \quad (101.9a)$$

$$\text{for amounts of gas:} \quad (V_D \times F_I) + (V_{AE} \times F_A)$$
$$= V_E \times F_E \quad (101.9b)$$

Combination and rearrangement yields the *Bohr formula*:

$$\frac{V_D}{V_E} = \frac{F_A - F_E}{F_A - F_I} \quad (101.9c)$$

which allows calculation of V_D (from V_E and F values) or of F_A (from V_D/V_E, F_I and F_E). Note that for CO_2, $F_A > F_E > F_I$; for O_2, $F_I > F_E > F_A$.

For the normal, resting human being the following values are typical: $V_E = 0.50\,l$; $V_D = 0.15\,l$; $V_D/V_E = 0.3$. V_D is not

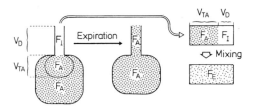

Fig. 101.2. Relationships between dead space gas, alveolar gas, and expired gas. For explanation, see text

Table 101.2. Typical values of respiratory and gas transport quantities in humans at rest and during heavy exercise

	Symbol	Units	Typical values	
			Rest	Heavy exercise
O_2 uptake	$\dot{V}_{O_2}$	l/min	0.30	3.0
CO_2 output	$\dot{V}_{CO_2}$	l/min	0.255	2.8
Respiratory exchange ratio, R	R_E	–	0.85	0.93
Breathing frequency	f_{resp}	min^{-1}	16	40
Tidal volume	V_E	l	0.5	2.0
Total ventilation	$\dot{V}_E$	l/min	8	80
Dead space	V_D	l	0.15	0.3
Alveolar ventilation	$\dot{V}_A$	l/min	5.6	68
Cardiac output	$\dot{Q}$	l/min	6	20
Arterial P_{O_2}	Pa_{O_2}	mmHg	90	90*
Mixed venous P_{O_2}	$P\bar{v}_{O_2}$	mmHg	40	20
Arterial P_{CO_2}	Pa_{CO_2}	mmHg	40	40*
Arterial-mixed venous O_2 content difference	$(Ca-C\bar{v})_{O_2}$	vol%	5	15

* Values may decrease or increase during exercise

constant; it increases with tidal volume and breathing frequency. Such changes are due to mechanically or neurally induced changes in the diameter of the bronchi and to airflow-dependent (convective and diffusive) mixing between inspired and resident lung gas. The diameter of the bronchi is controlled by the autonomic nervous system, with sympathetic (adrenergic) dilatation and parasympathetic cholinergic (vagal) constriction.

In the same way as shown above for V_E (Eq. 101.9a), the total (expired) ventilation, $\dot{V}_E$, is composed of two parts, the *dead space ventilation*, $\dot{V}_D$, and the alveolar ventilation, $\dot{V}_A$:

$$\dot{V}_E = \dot{V}_D + \dot{V}_A \tag{101.10}$$

The relationship between $\dot{V}_D$, $\dot{V}_A$, and $\dot{V}_E$ is variable. It is evident that with shallow and frequent breathing $\dot{V}_D/\dot{V}_E$ is higher and therefore $\dot{V}_A/\dot{V}_E$ is lower than with slow and deep breathing. The quantitative relations may be complex due to the dependence of V_D upon tidal volume and frequency (see above).

Because no respiratory gas exchange occurs in dead space, the whole O_2 uptake and CO_2 output must be provided by the *alveolar ventilation*, $\dot{V}_A$. By analogy to the relationships between $\dot{V}_{co_2}$, $\dot{V}_{o_2}$, $\dot{V}_E$, and F_E (Eqs. 101.6a,b), one obtains:

$$\dot{V}_{o_2} = f\text{vol} \times \dot{V}_A \times (r_{N_2} \times F_I - F_A)_{o_2} \text{ and} \tag{101.11a}$$

$$\dot{V}_{co_2} = f\text{vol} \times \dot{V}_A \times (F_A - r_{N_2} \times F_I)_{co_2} \tag{101.11b}$$

Rearrangement and introduction of partial pressures ($P_X = F_X(P_B - P_{H_2O})$) yields:

$$P_{A_{O_2}} = r_{N_2} \times P_{I_{O_2}} - \frac{\dot{V}_{O_2}(P_B - P_{H_2O})}{\dot{V}_A \times f\text{vol}} \quad \text{and} \tag{101.12a}$$

$$P_{A_{CO_2}} = \frac{\dot{V}_{CO_2}(P_B - P_{H_2O})}{\dot{V}_A \times f\text{vol}} - r_{N_2} \times P_{I_{CO_2}}. \tag{101.12b}$$

According to these relationships, the *alveolar partial pressures*, P_A, are determined by the inspired gas ($P_{I_{O_2}}$, $P_{I_{CO_2}}$) and by the ratio of metabolic rate ($\dot{V}_{o_2}$, $\dot{V}_{co_2}$) and alveolar ventilation ($\dot{V}_A$). It should be noted that with $P_{H_2O} = 47$ mmHg, the factor ($P_B - P_{H_2O}$)/$f$vol equals 863 mmHg (see Table 101.1), which factor is frequently used in equations corresponding to Eqs. 101.12a,b.

For a normal resting human at an altitude of 140 m ($\dot{V}_{o_2} = 0.3$ l/min; $\dot{V}_{co_2} = 0.26$ l/min; $\dot{V}_A = 5.6$ l/min; $r_{N_2} = 0.99$; $P_{H_2O} = 47$ mmHg; $P_B = 747$ mmHg; $P_{I_{o_2}} = 146$ mmHg; $f$vol = 0.81), Eqs. 101.12a,b give:

$$P_{A_{CO_2}} \approx 40\,\text{mmHg} \approx 5.3\,\text{kPA}$$

$$P_{A_{O_2}} \approx 100\,\text{mmHg} \approx 13.3\,\text{kPA}.$$

Since arterialized blood reaches, more or less closely, equilibrium with alveolar gas (see below), these equations can be used to estimate arterial P_{co_2} and, less accurately, arterial

P_{o_2}, which are the prime output parameters of pulmonary gas exchange.

It is evident from Eqs. 101.12a,b that for a given metabolic rate ($\dot{V}_{co_2}$, $\dot{V}_{o_2}$) and inspired gas (P_I), the alveolar ($\approx$arterial) P_{co_2} and P_{o_2} are determined by $\dot{V}_A$. An (imposed) increase of $\dot{V}_A$ (at constant metabolic rate and F_I) elevates $P_{A_{o_2}}$ and diminishes $P_{A_{co_2}}$ (hyperventilation); the reverse occurs with decreased $\dot{V}_A$ (hypoventilation). In light to medium exercise, $P_{A_{co_2}}$ (and $P_{a_{co_2}}$) and $P_{A_{o_2}}$ (and $P_{a_{o_2}}$) tend to remain fairly constant in the face of a large increases of $\dot{V}_{co_2}$ and $\dot{V}_{o_2}$. This is achieved by an increase of $\dot{V}_A$ in proportion to an increase of $\dot{V}_{co_2}$ and $\dot{V}_{o_2}$ (see Table 101.2). With inspiratory hypoxia, i.e., decrease of $P_{I_{o_2}}$ [either by decreasing $F_{I_{o_2}}$ at constant P_B, or decreasing P_B at constant $F_{I_{o_2}}$ (altitude)], $\dot{V}_A$ increases, thus preventing part of the expected fall of $P_{A_{o_2}}$. However, the increase of $\dot{V}_A$ produces an (undesirable) fall of $P_{A_{co_2}}$ (hypocapnic alkalosis, see Sect. 102.4).

In testing of the efficiency of (chemical) respiratory control, subjects inspire gas with added CO_2 (increased $P_{I_{co_2}}$). The regulatory increase in $\dot{V}_A$ reduces the rise in $P_{A_{co_2}}$. The relationship between increase in $\dot{V}_A$ as a function of $P_{A_{co_2}}$ (or $P_{a_{co_2}}$), with the slope $\Delta\dot{V}_A/\Delta P_{A_{co_2}}$, is the *$CO_2$ response curve*, a steeper slope meaning more efficient control. In a similar manner, the *O_2 response curve* is established by progressive reduction of $F_{I_{o_2}}$ (either with constant $F_{I_{co_2}}$ or with $F_{I_{co_2}}$ varied to achieve constant $P_{A_{co_2}}$).

101.5 Models of Alveolar Gas Exchange in Real Lungs [5,19,24]

In an assumed *ideal* lung model, complete gas exchange equilibrium is reached between alveolar gas and pulmonary capillary blood, and partial pressures for CO_2 and O_2 in arterial blood (P_a) are equal to those in alveolar gas (P_A). In real lungs, partial pressure differences between alveolar gas and arterialized blood – an alveolar-to-arterial P_{o_2} difference (AaD_{o_2}) and an arterial-to-alveolar P_{co_2} difference (aAD_{co_2}) – are found. In the conventional model analysis of alveolar gas exchange, these differences are attributed to three mechanisms:

- Unequal *distribution* of alveolar ventilation ($\dot{V}_A$) to pulmonary blood flow ($\dot{Q}$)
- *Shunt*
- *Diffusion* limitation.

101.6 Unequal Distribution of Ventilation to Perfusion ($\dot{V}_A/\dot{Q}$ Inequality)

101.6.1 Effects on Gas Exchange [4,30]

The effects of unequal distribution of alveolar ventilation ($\dot{V}_A$) to pulmonary perfusion ($\dot{Q}$) can be studied in a lung

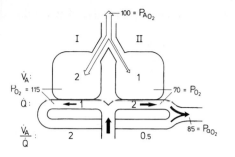

Fig. 101.3. Ventilation/perfusion inequality producing gas exchange inefficiency in a two-compartment model without diffusion limitation. $\dot{V}_A$ and $\dot{Q}$ in arbitrary units; P_{O_2} in mmHg. The flow-proportional weighting of alveolar P_{O_2} [$PA_{O_2} = (2 \times 115 + 70)/3 = 100$] and arterial P_{O_2} [$Pa_{O_2} = (115 + 2 \times 70)/3 = 85$] is shown. An alveolar-to-arterial P_{O_2} difference (AaD_{O_2}) of 15 mmHg results. (In the calculations a linear blood O_2 equilibrium curve is assumed; with the real curvilinear blood O_2 equilibrium curve, a larger AaD_{O_2} would result)

model devoid of diffusion resistance, which consists of compartments having different ratios of alveolar ventilation to perfusion ($\dot{V}_A/\dot{Q}$). An alveolar-to-arterial P_{O_2} difference (AaD_{O_2}) and an arterial-to-alveolar P_{CO_2} difference (aAD_{CO_2}) can be shown to arise, although in each compartment full equilibration is assumed ($PA = Pa$ for both CO_2 and O_2; Fig. 101.3). This is due to *flow-weighted averaging*: in the gas phase more gas from the high $\dot{V}_A/\dot{Q}$ compartment (with high P_{O_2} and low P_{CO_2}) contributes to "mixed" alveolar gas, whereas in the blood, more blood from the low $\dot{V}_A/\dot{Q}$ compartment (with low P_{O_2} and high P_{CO_2}) contributes to "mixed" arterial blood. It is evident that the ensuing AaD_{O_2} and aAD_{CO_2} increase with increasing inequality of distribution of $\dot{V}_A$ to $\dot{Q}$.

The AaD_{O_2} due to $\dot{V}_A/\dot{Q}$ inequality may amount to 10–15 mmHg in normal individuals, the aAD_{CO_2} to 2–4 mmHg. The extreme cases are of particular interest. $\dot{V}_A/\dot{Q} = 0$ means lack of ventilation and therefore absence of gas exchange, its perfusion constituting *shunt* or venous admixture (see below). $\dot{V}_A/\dot{Q} = \infty$, due to $\dot{Q} = 0$, designates the presence of ventilated but unperfused alveoli. Again there is no gas exchange, and the ventilation of such a compartment is functionally a dead space ventilation. It is called parallel or *alveolar dead space* ventilation, as distinguished from conducting airway ventilation, which is series or *anatomic dead space* ventilation. The sum of both is equivalent to total ventilation not contributing to gas exchange. It is termed *physiologic dead space ventilation*. The physiologic dead space ventilation, $\dot{V}_{D_{phys}}$, and its complement, the effective alveolar ventilation $\dot{V}_{A_{eff}}$ (= $\dot{V}_E - \dot{V}_{D_{phys}}$), can be calculated from the Bohr formula (Eq. 101.9c) by introduction of arterial P_{CO_2}, Pa_{CO_2}, instead of the alveolar value:

$$\dot{V}_{A_{eff}} = \dot{V}_E \frac{(PE - PI)_{CO_2}}{(Pa - PI)_{CO_2}} \tag{101.13}$$

Substitution of $\dot{V}_{A_{eff}}$ into Eq. 101.12b allows calculation of Pa_{CO_2} in lungs with alveolar dead space ventilation.

101.6.2 Multiple Inert Gas Elimination Method [26,27]

In real (normal and diseased) lungs, a *continuous* distribution of $\dot{V}_A$ and $\dot{Q}$ is present. The pattern of $\dot{V}_A/\dot{Q}$ inequality could be estimated on the basis of the method involving continuous intravenous infusion of multiple inert gases with differing solubilities in blood.

For an ideal lung model (no diffusion limitation, no shunt, no $\dot{V}_A/\dot{Q}$ inequality, $PA = Pa$), Eqs. 101.8b and 101.11b can be combined and applied to any intravenously infused foreign inert gas ($FI = 0$):

$$\dot{Q} \times (C\bar{v} - Ca) = fvol \times \dot{V}_A \times PA/(PB - P_{H_2O}) \tag{101.14a}$$

The solubility of an inert gas in blood is expressed as the *partition coefficient*, λ, the ratio of gas concentrations in blood and gas phase in equilibrium, using the same units, e.g., ml (STPD)/100 ml, for both phases. With *f*vol (Table 101.1) one obtains for λ:

$$\lambda = C/(fvol \times F) = C \times (PB - P_{H_2O})/(fvol \times P) \tag{101.14b}$$

Combination of Eqs. 101.14a and 101.14b, with $PA = Pa$, yields:

$$\dot{Q} \times \lambda \times (P\bar{v} - Pa) = \dot{V}_A \times PA \tag{101.14c}$$

Since in the ideal lung $PA = Pa$, one obtains by rearrangement:

$$\frac{Pa}{P\bar{v}} = \frac{\lambda}{\dot{V}_A/\dot{Q} + \lambda} \tag{101.14d}$$

This equation predicts the dependence of $Pa/P\bar{v}$ upon λ for an ideal lung ($\dot{V}_A/\dot{Q}$ constant). In a system with unequal $\dot{V}_A/\dot{Q}$, the relationship between $Pa/P\bar{v}$ and λ is different (due to blood-flow-weighted averaging of Pa, see above). From the difference between the measured and the predicted relationship between $Pa/P\bar{v}$ and λ the pattern of $\dot{V}_A/\dot{Q}$ inequality can be derived (Fig. 101.4).

In practice, a solution containing several (usually six) inert test gases with widely differing solubilities in blood (λ) is continuously infused intravenously. The concentrations (partial pressures) of the test gases are measured in mixed venous and arterial blood samples ($P\bar{v}$ and Pa) by gas chromatography or mass spectrometry. A very high precision of measurement is essential for this method.

101.6.3 Effects of Gravity [30]

At least a part of the $\dot{V}_A/\dot{Q}$ inequality in normal lungs is due to the effects of gravity on $\dot{Q}$ and $\dot{V}_A$. The marked effect of gravity on *pulmonary circulation* stems from the low pulmonary arterial pressure (about 2 kPa) and the very differ-

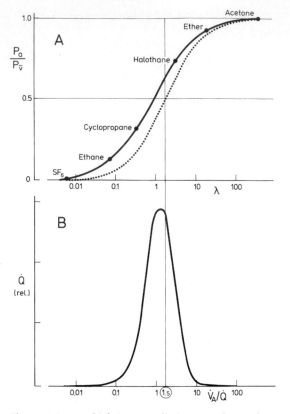

Fig. 101.4A,B. Multiple inert gas elimination technique for determination of $\dot{V}A/\dot{Q}$ inequality in lungs. **A** ratio $Pa/P\bar{v}$ for various inert gases of different blood solubility, plotted against the logarithmic abscissa in terms of blood/gas partition coefficient, λ. The measured values (*dots*) are connected by a *solid line*; the values predicted for equal total $\dot{V}A$ and total $\dot{Q}$, but $\dot{V}A/\dot{Q}$ = constant, are shown by the *dotted line*. From the difference between measured and predicted values, the $\dot{V}A/\dot{Q}$ inequality, shown in **B** as frequency distribution of blood flow ($\dot{Q}$) to compartments of differing $\dot{V}A/\dot{Q}$, is computed (the analogous plot of $\dot{V}A$ against $\dot{V}A/\dot{Q}$ would yield a similar figure, but shifted to the right). The equivalent in **B** of the dotted line in **A** is a vertical straight line at $\dot{V}A/\dot{Q}$ = 1.5. (After [30])

ent densities of blood and air. Because there is a hydrostatic gradient in blood ($\approx$ 1 kPa/10 cm), but (practically) none in alveolar air, the transmural pressure in pulmonary vessels increases vertically from top to bottom, leading to distension of vessels and increased blood volume and blood flow in the lower (dependent) lung regions. According to West [30], three zones are distinguished (Fig. 101.5):

- In the uppermost lung regions of a resting human with the thorax upright, blood flow is close to zero because the pulmonary arterial pressure (Ppa) is lower than the intrapulmonary (alveolar) pressure (PL), due to the vertical distance from the heart: PL > Ppa (*zone 1*).
- In the lung areas beneath zone 1, Ppa is higher than PL, but Ppv (pulmonary venous pressure) is lower than PL. This results in collapse of the pulmonary veins, with increased resistance to flow. The effective perfusion pressure head may be considered to approximate Ppa − PL instead of Ppa − Ppv (*zone 2*).

- In the areas beneath the zone 2, the transmural pressure of all vessels is positive and increases down the lung. Thus there is no vascular collapse, and the perfusion increases down the lung because of increasing vessel distension (Ppv > PL: *zone 3*).

With increasing Ppa, as in exercise, zone 1 is suppressed, and zone 2 is reduced, and the overall lung perfusion should be more homogeneously distributed. The effect of recumbency is similar, due to the decrease in the vertical height of the lungs. With increasing intrapulmonary pressure (PL), as during artificial ventilation with positive pressure, zones 1 and 2 would be expected to increase, and blood flow distribution becomes more unequal.

The gravity-dependent vertical gradient of $\dot{Q}$ is in part compensated in its effects on gas exchange by a (smaller) *gradient of $\dot{V}A$* in the same direction. The gradient of $\dot{V}A$ is explained as follows. Because of gravity, the upper lung regions are relatively overextended and the lower ones compressed. Since the elastic distensibility (compliance) of the lungs decreases with increasing distension, the upper parts are less ventilated than the lower parts.

According to recent experimental results, a substantial part of the $\dot{V}A/\dot{Q}$ inequality appears to be due not to gravity, but to anatomical heterogeneity of airways and blood vessels [6,7].

A mechanism contributing to reduce the $\dot{V}A/\dot{Q}$ inequality is based on *hypoxic pulmonary vasoconstriction*. In contrast to most systemic vessels, hypoxia elicits vasoconstriction in pulmonary vessels (amplified by hypercapnia). Thus local hypoxia, e.g., as produced by local hypoventilation, produces vasoconstriction with local reduction of blood flow, which leads to increase of the $\dot{V}A/\dot{Q}$ ratio towards normal. This reduces the spread of the $\dot{V}A/\dot{Q}$ ratios and the overall lung gas exchange efficiency is improved. This in called the *von Euler-Liljestrand effect* [22].

In pulmonary diseases, $\dot{V}A/\dot{Q}$ inequality is in many cases conspicuously increased, leading to low arterial P_{O_2} and large AaD_{O_2}.

101.7 Shunt or Venous Admixture

A shunt, i.e., short-circuiting of blood past gas exchanging regions of the lungs, leads to admixture of venous blood to arterialized blood, and thus to a decrease of P_{O_2} in arterial blood. This decrease is dependent on the shape of the *blood O_2 equilibrium curve* (see Chap. 102). The same shunt produces the same reduction in arterial O_2 content, but a largely different AaD_{O_2}, depending on the slope of the O_2 equilibrium curve (Fig. 101.6). Since the largest AaD_{O_2} is produced in hyperoxia (due to the flatness of the O_2 dissociation curve, representing physical solution of O_2 only), shunt is usually calculated from the AaD_{O_2} measured in hyperoxia (FI_{O_2} = 0.3 or higher).

In normal individuals breathing air, the shunt is about one to a few percent of the pulmonary blood flow, and gives

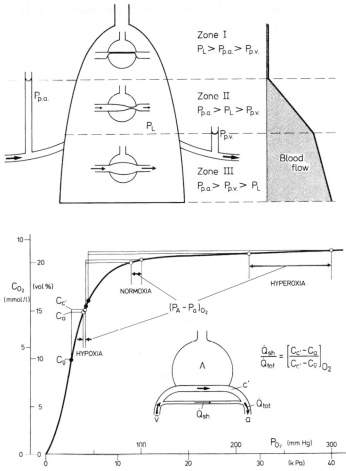

Fig. 101.5. Model to explain the effect of gravity on the vertical distribution of blood flow in lungs. See Sect. 101.6.3

Fig. 101.6. Blood O_2 equilibrium curve to show the production of alveolar-to-arterial P_{O_2} differences, $(P_A - P_a)_{O_2}$, by a shunt (relative shunt flow, $\dot{V}sh/\dot{Q}$ tot) in hyperoxia, normoxia, and hypoxia. Cc'_{O_2} is the O_2 content of blood in equilibrium with $P_{A_{O_2}}$. The shunt effect in terms of the O_2 content difference, $(Cc'-Ca)_{O_2}$, is in all cases the same, but the corresponding P_{O_2} difference, $(P_A-P_a)_{O_2}$, is large in hypoxia, medium in normoxia, and small in hyperoxia, resulting from the shape of the O_2 equilibrium curve of blood

rise to an AaD_{O_2} of about 5–10 mmHg. The important sources of shunt are perfusion of hypoventilated alveoli (which, more correctly, is an effect of low $\dot{V}A/\dot{Q}$) and of collapsed, atelectatic lung areas. In addition to these, admixture of bronchial venous blood to the arterialized pulmonary venous blood and myocardial venous outflow via the thebesian veins (leading to the left heart) also contribute to shunt. A massive extrapulmonary venous admixture resulting from the right-left shunt is observed in individuals with cardiac septal defects.

101.8 Alveolar-Capillary Diffusion and Pulmonary Diffusing Capacity
[2,17,21,25,29]

101.8.1 Oxygen

In modeling alveolar-capillary diffusion of O_2, the diffusion flux of O_2, i.e., the pulmonary O_2 uptake ($\dot{V}_{O_2}$), is con-

sidered to be proportional to the mean effective P_{O_2} difference between alveolar gas and blood $\Delta \bar{P}_{O_2}$ ($= PA_{O_2} - P\bar{c}_{O_2}$), where $P\bar{c}_{O_2}$ is the mean pulmonary capillary P_{O_2}. The ratio of $\dot{V}_{O_2}$ and $\Delta \bar{P}_{O_2}$ is termed *pulmonary diffusing capacity*, DL_{O_2}, or transfer factor, T_{O_2} (usually expressed in ml/(min × mmHg)):

$$DL_{O_2} = \frac{\dot{V}_{O_2}}{\Delta \bar{P}_{O_2}} = \frac{\dot{V}_{O_2}}{PA_{O_2} - P\bar{c}_{O_2}} \qquad (101.15)$$

The $P\bar{c}_{O_2}$ value is calculated from the AaD_{O_2} component due to diffusion limitation, alveolar and mixed venous P_{O_2}, and the blood O_2 equilibrium curve by a procedure called Bohr integration. In hyperoxia, the gas/blood P_{O_2} difference ($PA_{O_2} - P\bar{v}_{O_2}$), which initially is very large, contracts rapidly as blood passes through the pulmonary capillary, and at the end of the capillary no P_{O_2} difference (i.e., AaD_{O_2}) is left (Fig. 101.7). In normoxia as well, in resting normal individuals, an AaD_{O_2} attributable to diffusion limitation is usually too small to measure.

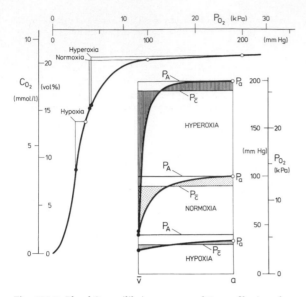

Fig. 101.7. Blood O_2 equilibrium curve and P_{O_2} profiles in pulmonary capillaries for hyperoxia, normoxia, and hypoxia. In hyperoxia, the P_{O_2} difference between alveolar gas and blood, $(PA-P\bar{v})_{O_2}$, is initially large and disappears rapidly. In hypoxia, the difference $(PA-P\bar{v})_{O_2}$ is small and a sizable $(PA-Pa)_{O_2}$ remains. Note that the difference between PA_{O_2} and the mean capillary P_{O_2} ($P\bar{c}_{O_2}$), $(PA-P\bar{c})_{O_2}$, is in all cases the same. $P\bar{c}_{O_2}$ is constructed in such a manner that the two shaded areas, below and above $P\bar{c}_{O_2}$, are equal. Since $(PA-Pa)_{O_2}$ is the basis for reliable estimation of $P\bar{c}_{O_2}$, only in hypoxia can DL_{O_2} be calculated as $\dot{V}_{O_2}/(PA - P\bar{c})_{O_2}$

In *hypoxia*, the initial P_{O_2} difference is small (due to the steep O_2 equilibrium curve) and a considerable gas/blood P_{O_2} difference at the end of the capillary, AaD_{O_2}, attributable to diffusion limitation, may occur. For this reason, in hypoxia DL_{O_2} is determined. It is useful for the calculations of diffusion that the effects of shunt and $\dot{V}A/\dot{Q}$ inequalities are reduced in hypoxia and may be ignored in normal lungs, and that the blood O_2 equilibrium curve is virtually linear. Normal values are $DL_{O_2} \approx 40–80$ ml/(min × mmHg). In normal healthy individuals breathing air, some AaD_{O_2} due to diffusion limitation is expected to occur in severe exercise only, but in deep hypoxia (as at high altitude) AaD_{O_2} may also occur at rest and is much increased in exercise. In pulmonary disease DL may be seriously decreased. Usually increased $\dot{V}A/\dot{Q}$ inequalities and increased shunt occur simultaneously, all contributing to the enlarged AaD_{O_2}.

According to the Fick diffusion equation, the diffusive conductance or diffusing capacity (D) of a flat, homogeneous sheet depends on surface ared (F), thickness of the barrier (l) and the *Krogh diffusion constant* (K):

$$D = K \times F/l \qquad (101.16)$$

The Krogh diffusion constant is equal to the product of solubility (α) and the diffusion coefficient (d):

$$K = d \times \alpha \qquad (101.17)$$

In reality, the structure and function of the "diffusion barrier" in lungs is complex. The tissue barrier separating alveolar gas from capillary blood is composed of heterogeneous layers (alveolar epithelium, interstitium, endothelium) of nonuniform thickness. Part of the resistance to O_2 uptake is to be sought in the flowing plasma and within the red blood cells, whose shape depends on flow conditions (see Chap. 88). However, convective mixing in blood is assumed to assist O_2 equilibration. The finite reaction kinetics of O_2 with hemoglobin may also exert some retarding effect, but according to recent measurements O_2 uptake of red blood cells is more rapid than previously estimated [8]. Thus, neither intra-erythrocyte diffusion of O_2 nor its reaction with hemoglobin are assumed to limit pulmonary O_2 transfer significantly.

101.8.2 Carbon Dioxide

Because the Krogh diffusion constant (Eq. 101.17) for CO_2 is about 25 times higher than than for O_2, CO_2 transport in lungs is usually considered as not diffusion-limited. However, in CO_2 output in lungs, dehydration of H_2CO_3 and chloride/bicarbonate exchange between red blood cells and plasma are involved, and both processes may be rate-limiting (see Sect. 102.3.4). Experimental data suggest that with high CO_2 output (heavy exercise), effective transfer limitation is expected to occur, leading to positive arterial-to-alveolar P_{CO_2} differences [11,23].

Remarkably, in some experimental conditions, P_{CO_2} lower in arterial blood than in alveolar gas has been reported, and special mechanisms involving intracapillary electrical fields have been proposed to explain this. However, these observations have remained controversial [16].

101.8.3 Carbon Monoxide

A convenient test gas for estimation of DL is carbon monoxide, CO. Due to the very high affinity of hemoglobin for CO, P_{co} in blood during the passage through pulmonary capillaries stays practically constant and is close to zero if the subjects are not exposed to external CO (in smokers, blood P_{co} should be estimated because of the presence of appreciable amounts of CO-Hb in blood). Thus, the pulmonary diffusing capacity (transfer factor) for CO, DL_{co}, can be estimated from the uptake of externally administered CO, $\dot{V}_{co}$, and from the difference between alveolar P_{co}, PA_{co}, and P_{co} in blood, Pb_{co} (estimated from blood CO content or by rebreathing equilibration):

$$DL_{co} = \frac{\dot{V}_{co}}{PA_{co} - Pb_{co}} \qquad (101.18)$$

The Krogh diffusion constant for CO in tissue is similar to that of O_2 ($K_{O_2}/K_{co} \approx 1.2$).

With the steady state method, about 0.1% CO is inspired for some minutes before measurements of $\dot{V}_{co}$ and PA_{co} are performed. With the *single-breath method*, a deep inspired breath containing about 0.3% CO is held for 10 s, and DL_{co} is calculated from the exponential rate constant of CO absorption, k_{co}, and the alveolar volume, V_A:

$$DL_{co} = k_{co} \times \frac{V_A \times f \text{ vol}}{P_B - P_{H_2O}} \qquad (101.19)$$

DL_{co} values found in resting normal individuals are about 30–50 ml min^{-1} mmHg^{-1}. In pulmonary diseases, DL_{co} may be much reduced.

The problem in interpreting DL determined with CO is that its uptake is delayed by the *slow reaction* of hemoglobin with CO. However, this property can be exploited for further analysis [20]. The CO uptake rate of oxygenated red blood cells, θ_{co}, has been shown to be strongly dependent on P_{O_2}, due to competition between O_2 and CO for hemoglobin. In the following equation, the total resistance to CO uptake (= $1/DL_{co}$) is broken into a resistance to CO diffusion across the air/blood barrier (= $1/Dm_{co}$) and the resistance to CO uptake offered by red blood cells (= $1/(\theta_{co} \times Vc)$). Vc is the pulmonary *capillary volume*, and θ_{co} is the specific conductance of blood for CO uptake (i.e., including both diffusion and reaction kinetics), which decreases with increasing P_{O_2}:

$$\frac{1}{DL_{co}} = \frac{1}{Dm_{co}} + \frac{1}{\theta_{co} \times V_c} \qquad (101.20)$$

Measurements of DL at two (or more) alveolar P_{O_2} levels, and thus at different θ_{co}, allow calculation of Dm_{co} and Vc (both of which have to be assumed to be independent of P_{O_2}). In exercise, both Dm_{co} and Vc are found to be increased, whereas in diseased lungs, both tend to be decreased.

A variable frequently considered as a limiting factor in diffusive pulmonary gas exchange is the mean *transit time* (*contact time*) t_{tr}. It is related to blood flow ($\dot{Q}$) and capillary volume (Vc) by the relationship:

$$t_{tr} = Vc/\dot{Q} \qquad (101.21)$$

Evidently t_{tr} decreases with increase of $\dot{Q}$ at constant Vc. However, during exercise Vc increases, and thus t_{tr} decreases less than in proportion to the increase in $\dot{Q}$.

101.9 Time Course of Expired Gas Concentrations: Expirograms

With the advent of rapid, continuously recording gas analyzers (infrared spectroscopy, mass spectrometry), the time course of gas concentrations in expired gas can now be recorded. For both O_2 and CO_2 (and foreign test gases), the following phases occur if recorded against time or expired volume (Fig. 101.8):

- First, unchanged inspired gas is expired from dead space: *phase I*.
- Next, a more or less rapid change to alveolar gas levels occurs. The slope in this phase, *phase II*, is attributed to mixing of alveolar and dead space gas, and to unequal expiration times of different regions of the lung.
- The alveolar plateau (*phase III*) is not flat but has a slope, moving away from the inspired value (P_{co_2} rises, P_{o_2} falls) and oscillating with cardiac frequency (cardiogenic oscillations).

The *cardiogenic oscillations* are explained by mechanical impact of the heart or of the pulsatile pulmonary arterial blood flow on expiratory gas flow from lung regions with differing gas compositions. The sloping plateau is attributed to the following factors:

- During expiration, *exchange* between alveolar gas and blood continues, whereas no fresh gas is admitted (functional breath-hold). High metabolic rate, slow breathing, and small lung volume would be expected to increase the slope.
- With regional $\dot{V}_A/\dot{Q}$ inequality, a local variation of $P_{A_{O_2}}$ and $P_{A_{CO_2}}$ is present. Relatively underventilated lung regions (with low P_{o_2} and high P_{co_2}) tend to expire later than relatively overventilated regions (with high P_{o_2} and low P_{co_2}). Thus $\dot{V}_A/\dot{Q}$ inequality combined with *sequential expiration* produces sloping alveolar plateaus. The magnitude of the effect is expected to depend on the extent of $\dot{V}_A/\dot{Q}$ inequality and of the time patterns of expiration. Due to this mechanism, the expirograms of diseased lungs may show steep alveolar slopes, up to apparent absence of an alveolar plateau (i.e., no clear distinction between phases II and III).
- The mixing of inspired and lung resident gas, particularly with rapid shallow breathing or in diseased lungs, is not completed during a respiratory cycle. This leads to

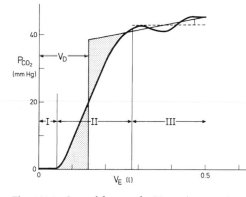

Fig. 101.8. General format of a CO_2 expirogram, i.e., recording of P_{co_2} during an expiration, plotted against expired volume (V_E). Phases I, II, and III are marked. The cardiogenic oscillations and the slope of the alveolar plateau (phase III) are indicated. The location of the line delimiting the dead space (V_D) is established (*dotted line*), yielding equal (*dotted*) areas between the line, phase II of the expirogram, the (smoothed) extrapolated alveolar plateau, and the inspired value

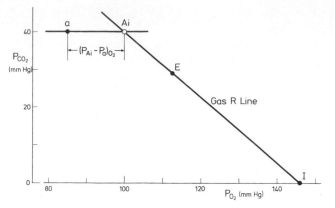

Fig. 101.9. Determination of the "ideal-alveolar" P_{O_2}, PAi_{O_2}, from the inspired and expired P_{CO_2} and P_{O_2} and the arterial P_{CO_2}. Values I, E, and a are measured; the point Ai is obtained graphically or from Eq. 101.22. The line through the I, E, and Ai values is termed the "gas R line" and its negative slope is close to the R value, which can be obtained by applying the N_2 correction (see Sect. 101.3.1, Eqs. 101.6 and 101.7)

stratification of alveolar gas, with P_{O_2} decreasing and P_{CO_2} increasing in the proximal-to-distal direction in peripheral airways. The extent of stractification due to incomplete intrapulmonary gas mixing has been a subject of recent research [3,12,13,15,18]. There is also evidence of unequal distribution of gas phase (and alveolar-capillary) diffusing conditions in lungs [9].

101.10 Ideal Alveolar P_{O_2}

In the face of the changing composition of gas exhaled from the alveolar space (see Sect. 101.9), the analysis of $\dot{V}A/\dot{Q}$ inequality and shunt based on steady-state models becomes questionable. In normal individuals at rest, a meaningful alveolar gas sample can be obtained from the last part of expired gas. But in exercise, and particularly in patients with lung disease, no constant reproducible value is obtained as P_{CO_2} rises and P_{O_2} falls continuously during expiration.

For this reason, an indirect procedure can be useful, in which "ideal" alveolar gas (Ai) is calculated on the assumption that P_{CO_2} is the same in arterial blood and alveolar gas: $PAi_{CO_2} = Pa_{CO}$. This assumption is reasonable for normal lungs and is an acceptable hypothesis in diseased lungs. Since expired gas may be considered as a mixture of expired alveolar gas and inspired gas reexpired from the dead space, the inspired, expired, and "ideal" alveolar P_{CO_2} and P_{O_2} must lie on a straight line (the so-called *gas R line*) in a $(P_{O_2} - P_{CO_2})$ diagram (Fig. 101.9). The "ideal" alveolar P_{O_2}, PAi_{O_2}, is obtained as:

$$PAi_{O_2} = PI_{O_2} - \frac{(PI - PE)_{O_2}}{(PE - PI)_{CO_2}} \times (Pa - PI)_{CO_2} \qquad (101.22)$$

The difference $(PAi - Pa)_{O_2}$ may be taken to approximate the AaD_{O_2} caused by $\dot{V}A/\dot{Q}$ inequality, shunt, and/or diffusion as explained above.

General References

Comroe JH, Forster RE, Dubois AB, Briscoe WA, Carlsen E (1962) The lung: clinical physiology and pulmonary function tests. Medical Book Publishers, Chicago

Cotes JE (1979) Lung function: assessment and application to medicine. Blackwell, Oxford

Crystal RG, West JB, Barnes PJ, Cherniack NS, Weibel ER (eds) (1991) The lung: scientific foundations, 2 vols. Raven, New York

Farhi LE, Tenney SM (eds) (1987) Handbook of physiology, section 3: the respiratory system, vol 3: gas exchange. American Physiological Society, Bethesda

Fenn WO, Rahn H (eds) (1964/1965) Handbook of physiology, section 3: respiration, 2 vols. American Physiological Society, Washington DC

Weibel ER (1984) The pathway for oxygen. Harvard University Press, Cambridge MA

West JB (1985) Respiratory physiology, the essentials. Williams and Wilkins, Baltimore

West JB (ed) (1980) Pulmonary gas exchange, 2 vols. Academic, New York

Specific References

1. Anthonisen NR, Fleetham JA (1987) Ventilation: total, alveolar, and dead space. In: Farhi LE, Tenney SM (eds) Handbook of physiology, section 3: respiratory system, vol 4: gas exchange. American Physiological Society, Bethesda, pp 113–129
2. Crapo JD, Crapo RO, Jensen RL, Mercer RR, Weibel ER (1988) Evaluation of lung diffusing capacity by physiological and morphometric techniques. J Appl Physiol 64:2083–2091
3. Engel LA, Paiva M (eds) (1985) Gas mixing and distribution in the lung. Dekker, New York (Lung biology in health and disease, vol 25)
4. Farhi LE (1987) Ventilation-perfusion relationships. In: Farhi LE, Tenney SM (eds) Handbook of physiology, section 3: respiratory system, vol 4: gas exchange. American Physiological Society, Bethesda, pp 199–215
5. Farhi LE, Rahn H (1955) A theoretical analysis of the alveolar-arterial O_2 difference with special reference to the distribution effect. J Appl Physiol 7:699–703
6. Glenny RW, Polissar L, Robertson HT (1991) Relative contribution of gravity to pulmonary perfusion heterogeneity. J Appl Physiol 71:2449–2552

7. Hakim TS, Lisbona R, Dean GW (1987) Gravity-independent inequality in pulmonary blood flow in humans. J Appl Physiol 63:1114–1121
8. Heidelberger E, Reeves RB (1990) Factors affecting whole blood O_2 transfer kinetics: implications for θ (O_2). J Appl Physiol 68:1865–1874
9. Hlastala MP (1987) Diffusing-capacity heterogeneity. In: Farhi LE, Tenney SM (eds) Handbook of physiology, section 3: respiratory system, vol 4: gas exchange. American Physiological Society, Bethesda, pp 217–232
10. Kellogg RH (1987) Laws of physics pertaining to gas exchange. In: Farhi LE, Tenney SM (eds) Handbook of physiology, section 3: respiratory system, vol 4: gas exchange. American Physiological Society, Bethesda, pp 13–31
11. Meyer M, Scheid P, Riepl G, Wagner HJ, Piiper J (1981) Pulmonary diffusing capacities for O_2 and CO measured by a rebreathing technique. J Appl Physiol 51:1643–1650
12. Meyer M, Schuster KD, Schulz H, Mohr M, Piiper J (1990) Alveolar slope and dead space of He and SF_6 in dogs: comparison of airway and venous loading. J Appl Physiol 69:937–944
13. Neufeld GR, Gobran S, Baumgardner JE, Aukburg SF, Schreiner M, Scherer PW (1991) Diffusivity respiratory rate and tidal volume influence inert gas expirograms. Respir Physiol 84:31–47
14. Otis AB (1964) Quantitative relationships in steady-state gas exchange. In: Fenn WO, Rahn H (eds) Handbook of physiology, section 3: respiration, vol 1. American Physiological Society, Washington DC, pp 681–698
15. Paiva M, Engel LA (1981) The anatomical basis for the sloping alveolar plateau. Respir Physiol 44:325–337
16. Piiper J (1986) Blood-gas equilibrium of carbon dioxide in lungs: a continuing controversy. J Appl Physiol 60:1–8
17. Piiper J, Scheid P (1980) Blood-gas equilibration in lungs. In: West JB (ed) Pulmonary gas exchange, vol 1. Academic, New York, pp 131–171
18. Piiper J, Scheid P (1987) diffusion and convection in intrapulmonary gas mixing. In: Farhi LE, Tenney SM (eds) Handbook of physiology, section 3: respiratory system, vol 4: gas exchange. American Physiological Society, Bethesda, pp 51–69
19. Riley RL, Cournand A (1951) Analysis of factors affecting partial pressures of oxygen and carbon dioxide in gas and blood of the lungs: theory. J Appl Physiol 4:77–101
20. Roughton FJW, Forster RE (1957) Relative importance of diffusion and chemical reaction rates in determining rate of exchange of gases in the human lung, with special reference to true differing capacity of pulmonary membrane and volume of blood in lung capillaries. J Appl Physiol 11:291–302
21. Scheid P, Piiper J (1989) Blood-gas equilibration in lungs and pulmonary diffusing capacity. In: Chang HK, Paiva M (eds) Respiratory physiology, an analytical approach. Dekker, New York, pp 453–497
22. Sheehan DW, Klocke RA, Farhi LE (1992) Pulmonary hypoxic vasoconstriction: how strong? How fast? Respir Physiol 87:357–372
23. Swenson ER, Maren TH (1978) A quantitative analysis of CO_2 transport at rest and during maximal exercise. Respir Physiol 35:129–159
24. Thews G (1984) Theoretical analysis of pulmonary gas exchange at rest and during exercise. Int J Sports Med 5:113–119
25. Wagner PD (1977) Diffusion and chemical reaction in pulmonary gas exchange. Physiol Rev 57:257–312
26. Wagner PD, Laravuso RB, Uhl RR, West JB (1974) Continuous distributions of ventilation-perfusion ratios in normal subjects breathing air and 100% O_2. J Clin Invest 54:54–68
27. Wagner PD, Saltzman HA, West JB (1974) Measurement of continuous distributions of ventilation-perfusion ratios: theory. J Appl Physiol 36:588–599
28. Weibel ER (1963) Morphometry of the human lung. Springer, Berlin Göttingen Heidelberg
29. Weibel ER (1973) Morphological basis of alveolar-capillary gas exchange. Physiol Rev 53:419–495
30. West JB (1977) Ventilation/blood flow and gas exchange, 3rd edn. Blackwell, Oxford

102 Respiratory Gas Transport and Acid-Base Equilibrium in Blood

J. Piiper

Contents

102.1 Transport of O_2 and CO_2 by Blood

In the convective transport of O_2 and CO_2 by blood flow, the task is to transport the gases from a site of high partial pressure (O_2 in lungs, CO_2 in tissues) to a site of low partial pressure (O_2 in tissue, CO_2 in lungs). The transport rate is ultimately determined by the metabolic rate of body tissues, i.e., the O_2 consumption ($\dot{V}O_2$) and the CO_2 production ($\dot{V}CO_2$). For adequate transport two things are important:

- The *blood flow* (cardiac output for the whole body)
- The *capacitance of the blood* for O_2 and CO_2.

The gas-carrying capacitance can be quantified as the increment in gas concentration in blood per unit increase in partial pressure, termed *capacitance coefficient*, and is represented by the *steepness* of the blood O_2 and CO_2 equilibrium (dissociation) curves in the range between arterial and venous blood partial pressures (Figs. 102.1, 102.3) [8]. For the analysis of O_2 supply and CO_2 removal, both blood flow and blood equilibrium curves are the decisive factors.

102.2 Carriage of O_2 by Blood and Hemoglobin

102.2.1 Blood O_2 Equilibrium Curve

For the assessment of the O_2 transport function of blood, the blood O_2 *equilibrium curve* (O_2 *dissociation curve*) is used. Blood samples are equilibrated with gas mixtures of varied O_2 concentration at body temperature in vessels which are rotated or shaken to produce a large and continuously renewed exchange surface between blood and gas. Analysis of blood samples withdrawn after various lengths of time for O_2 content or concentration (amount of reversibly bound O_2, i.e., released by denaturation of hemoglobin and vacuum extraction, per unit blood volume) shows first changing values, then constant values which indicate that an equilibrium of O_2 between gas and blood has been reached. In equilibrium the O_2 partial pressure (P_{O_2}) in both media is the same. The value of P_{O_2} in the gas phase is obtained from the fractional concentration of O_2 in the gas phase F_{O_2} (see Eq. 101.2 in Sect. 101.2.2). Plotting of O_2 content in equilibrium versus P_{O_2} yields the blood O_2 equilibrium curve (Fig. 102.1), in which P_{O_2} is usually expressed in mmHg (= torr) or kPa (1 kPa = 7.5 mmHg) and the O_2 content, CO_2, in vol % [= ml O_2 (STPD) per 100 ml blood] or in mol/l (=mmol O_2 per liter blood).

If the same equilibration procedure is performed with water or plasma, a straight line is obtained, with a slope similar to that of blood above $P_{O_2} \approx 150$ mmHg (≈ 20 kPa), indicating physical solution of O_2 in blood. The slope represents the physical solubility of O_2 in blood (equal to 0.003 vol%/mmHg at 37°C). The remaining, major part of blood O_2 is chemically bound to hemoglobin. It is zero at $P_{O_2} = 0$, increases gradually with increasing P_{O_2} and, in normal blood, reaches a maximum value of about 20 vol% (≈ 8.9 mmol/l at $P_{O_2} \approx 150$ mmHg, indicating full saturation of hemoglobin with O_2 (Fig. 102.1).

R. Greger/U. Windhorst (Eds.)
Comprehensive Human Physiology, Vol. 2
© Springer-Verlag Berlin Heidelberg 1996

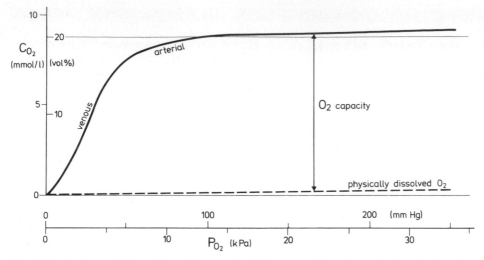

Fig. 102.1. The O_2 equilibrium (dissociation curve) of blood: O_2 content, C_{O_2}, as a function of O_2 partial pressure, P_{O_2}. The arterial range, about 100–60 mmHg, and the venous range, about 50– 20 mmHg, are indicated. Above $P_{O_2} \approx 150$ mmHg, all increases in C_{O_2} are due to physically dissolved O_2, because hemoglobin is fully saturated with O_2 (O_2 capacity)

102.2.2 Hemoglobin

Hemoglobin is a globular oligomeric protein (molecular weight 64 800 Da) containing four units, each of which consists of a polypeptide chain and a prosthetic group, heme (protoporphyrin-Fe^{2+} complex). In normal adult blood, the prevailing hemoglobin A_1 is composed of two α and two β polypeptide chains ($\alpha_2\beta_2$); a minor fraction is hemoglobin A_2 containing two α and two δ chains ($\alpha_2\delta_2$). Hemoglobin has the extraordinary property of *reversibly* binding molecular O_2. (For further information on the biochemistry of hemoglobin, the reader is referred to biochemistry texts; see, e.g., General References at the and of this chapter.) One O_2 molecule can be bound by each unit containing one iron atom (i.e., four O_2 molecules per molecule of hemoglobin), the iron remaining ferrous (Fe^{2+}). The association product is termed *oxyhemoglobin*; the hemoglobin devoid of O_2 (or other ligands) is called *deoxyhemoglobin* or free hemoglobin.

The concentration of chemically bound O_2 at full O_2 saturation is termed the O_2 *capacity* of blood (normally 20 vol% = 8.9 mmol/l). It is proportional to the hemoglobin concentration (normal value 15 g% = 150 g/l blood). Hemoglobin is contained in red blood cells, whose count is normally 5 × 10^6 cells/mm³ blood, and whose fractional volume or hematocrit (= volume of red cells/volume of blood) is 45%. The O_2 *saturation* of blood (or hemoglobin), S_{O_2}, is the ratio of chemically bound O_2 to O_2 capacity. The most-used, standardized form of blood O_2 dissociation curve is a plot of S_{O_2} against P_{O_2} (see Fig. 102.2). It should be noted that the P_{O_2}–S_{O_2} curve is independent of O_2 capacity and disregards the physically dissolved O_2.

S_{O_2} can be directly measured in blood samples by spectrophotometry or by special photocells (*oximeters*) applied to exposed vessels or to skin (e.g., on earlobe made hyperemic for measurement of arterial S_{O_2}).

102.2.3 Shape of the Blood O_2 Equilibrium Curve

The simplest formulation of the reaction between hemoglobin and O_2 according to the mass action law is the following:

$$\text{Hb} + \text{O}_2 \underset{k}{\overset{k'}{\rightleftharpoons}} \text{HbO}_2 \tag{102.1}$$

$$\frac{[\text{HbO}_2]}{[\text{Hb}] \times [\text{O}_2]} = \frac{k'}{k} = K. \tag{102.2}$$

Hb designates here a unit of hemoglobin combining with one O_2 molecule. k' and k are the reaction rate constants of O_2 association and dissociation, respectively. K is the equilibrium constant for combination with O_2; $[O_2]$ is the concentration of free dissolved O_2.

From this equation a *hyperbolic* O_2 equilibrium curve would result. However, the real O_2 equilibrium curve of hemoglobin has a *sigmoid* shape, being steeper in the physiologically relevant range of P_{O_2} than any hyperbola following from the above equation. Formally this means that the affinity of hemoglobin for O_2 increases with increasing O_2 saturation of hemoglobin (positive interaction or cooperativity). This interaction is made possible by the tetrameric structure of the hemoglobin molecule.

The deviations of the shape of the blood O_2 equilibrium curve from a hyperbola are important for the effectivity of O_2 transport by blood (Fig. 102.1).

- In the range of high P_{O_2} (above $P_{O_2} \approx 60$ mmHg) the curve is relatively flat. Therefore, a considerable decrease of arterial P_{O_2} (e.g., due to insufficient alveolar ventilation or impaired gas exchange function of lungs) leads to a relatively small drop in O_2 saturation (and content) of arterial blood.

- The curve in the venous range is shifted to the right in comparison to the course extrapolated from the arterial range. This means that O_2 is delivered at a relatively high P_{O_2}, facilitating blood/tissue diffusion of O_2 in tissue capillaries.
- In the range of low (venous) P_{O_2} (below $P_{O_2} \approx 40\,\text{mmHg}$) the curve is steep. Hence a reduction of P_{O_2} in tissue and blood (due to high O_2 consumption or low blood flow) enhances release of O_2 from blood.
- The comparatively flat curve in the very low P_{O_2} range (below about 15 mmHg) ensures that O_2 delivery to tissue can proceed at appropriate O_2 pressures in the tissue.

102.2.4 Factors Controlling the Position of the O_2 Equilibrium Curve

The position of the O_2 equilibrium curve of blood is an expression of the *affinity* of hemoglobin for O_2. The O_2 affinity is characterized by the *half saturation* P_{O_2}, P_{50}, which is normally 27 mmHg (= 3.6 kPa) in human blood in physiological conditions (37°C, plasma pH = 7.4). The O_2 affinity primarily results from the amino acid composition of the α and β chains, which is species-specific among vertebrates. The O_2 affinity is modified by a number of agents, the more important ones of which are the following:

- pH
- 2,3-Diphosphoglycerate
- Temperature.

Acidity and Bohr effect [1,10]. Increasing acidity (= lowering of pH) leads to a right shift and decreasing acidity to a left shift of the O_2 dissociation curve of blood: this is the *Bohr effect* (Fig. 102.2). Physiologically the changes of pH are often due to changes in P_{CO_2}. It should be noted that, besides acting through pH, P_{CO_2} has a direct, specific effect on the O_2 affinity of human hemoglobin. The Bohr effect is quantified, as the shift of P_{50} with pH, by the Bohr factor, BF:

$$BF = \frac{\Delta \log P_{50}}{\Delta pH}. \tag{102.3}$$

For human blood BF ≈ -0.5. The Bohr effect is favorable to both O_2 uptake in lungs and O_2 delivery to tissue.

1. In *lungs*, CO_2 release from blood leads to reduction of P_{CO_2} and to increase in pH. The resulting left shift of the O_2 equilibrium curve brings about an additional uptake of O_2 by blood, particularly at low S_{O_2} (hypoxia). This effect is quite limited (Fig. 102.2).
2. In *tissues*, uptake of CO_2 by blood elevates blood P_{CO_2} and depresses pH. The ensuing shift to the right of the

O_2 equilibrium curve enables the blood to release additional O_2 (at the same P_{O_2}). This effect would be increased with increasing metabolism (due to increased pH difference between arterial and venous blood, caused by increased venous P_{CO_2} and uptake of lactic acid). (For the mechanism of the Bohr effect, see Sect. 108.3.3.)

2,3-Diphosphoglycerate. A particularly important effect is exerted by the compound *2,3-diphosphoglycerate* (2,3-DPG) contained in red cells at a high concentration (about 5 mmol/l, molar ratio [DPG]/[total Hb] about unity). 2,3-DPG reduces the O_2 affinity of hemoglobin: without 2,3-DPG the P_{50} of hemoglobin is about 12 mmHg. An increase of [2,3-DPG] leads to a considerable right shift of the O_2 equilibrium curve. 2,3-DPG seems to play an important role in physiological adjustments of the O_2 dissociation curve (e.g., increase of [2,3-DPG] in anemia, moderate hypoxemia, decrease in high altitude adjustment).

Temperature. Lowering of temperature produces a left shift and its elevation a right shift of the O_2 equilibrium curve. Temperature effects on the O_2 equilibrium curve are prominent in poikilothermic vertebrates (reptiles, amphibians, fishes). However, in humans the elevated temperature in exercising muscle also favors O_2 delivery to tissue due to the decreasing O_2 affinity of hemoglobin.

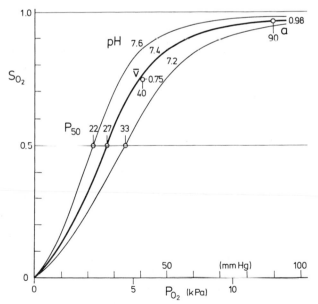

Fig. 102.2. The O_2 equilibrium curve of blood in terms of hemoglobin O_2 saturation, S_{O_2}, as a function of P_{O_2}, for three pH values in blood plasma, to show the Bohr effect. The P_{O_2} values for half-saturation, P_{50}, are indicated (in mmHg). Normal resting arterial (a) and mixed venous ($\bar{v}$) values are marked by *circles*, with numbers indicating P_{O_2} (in mmHg) and S_{O_2}

102.2.5 Inactivation of Hemoglobin: Carboxyhemoglobin, Methemoglobin

Carbon monoxide (CO) may be bound to hemoglobin instead of O_2, giving rise to *carboxyhemoglobin*, $Hb \cdot CO$:

$$Hb + CO \rightleftharpoons Hb \cdot CO \qquad (102.4)$$

The hemoglobin binding CO as $Hb \cdot CO$ is unavailable for O_2 transport. The high toxicity of CO resides in the *very high affinity* of hemoglobin for CO, which is much higher than that for O_2: the CO dissociation curve of blood is extremely left-shifted. However, the formation of $Hb \cdot CO$ is also a reversible reaction. In the case of CO poisoning, ventilation of the lungs with CO-free air results in release of CO from $Hb \cdot CO$ and its elimination into expired air. The process is accelerated by the inspiration of pure O_2 because O_2 competes with CO for combination with hemoglobin.

The odorless gas CO is of wide-spread occurrence (exhaust gases from industry, motor vehicles, household, cigarette smoke). In smokers, but also in "passively smoking" non-smokers, a few percent of the total hemoglobin is present as carboxyhemoglobin.

The ferrous iron (Fe^{2+}) of hemoglobin can be oxidized to ferric iron (Fe^{3+}), leading to *methemoglobin* (ferrihemoglobin), which cannot reversibly bind O_2 and is therefore ineffective in O_2 transport:

$$Hb \cdot Fe^{2+} \underset{Reductants}{\overset{Oxidiants}{\rightleftharpoons}} Hb \cdot Fe^{3+} \qquad (102.5)$$

Since methemoglobin is the thermodynamically more stable form, in equilibrium all hemoglobin should be converted to methemoglobin. However, red blood cells contain a specific enzyme system, the methemoglobin reductase, which reconverts methemoglobin to hemoglobin, using metabolic energy. Thus, normally only about 0.5% of hemoglobin is present in the form of methemoglobin. In congenital methemoglobinemia, due to deficiency of the methemoglobin reductase (or to abnormal hemoglobin structure), a considerable fraction of total hemoglobin may be methemoglobin. Methemoglobinemia also occurs in certain intoxications (acquired methemoglobinemia).

In both carboxyhemoglobinemia and methemoglobinemia the O_2 capacity of blood is reduced.

102.2.6 Other Hemoglobins

Fetal hemoglobin (HbF) is composed of two α and two γ chains. During the last months of gestation and the first months after birth, fetal hemoglobin is gradually replaced by adult hemoglobin (two α and two β chains). The O_2 dissociation curve of human fetal blood is shifted to the left compared to adult blood (mainly due to lower effectiveness of 2,3-DPG in decreasing O_2 affinity of fetal hemoglobin). This left shift, i.e., increased O_2 affinity, *enhances* diffusive transport of O_2 from maternal to fetal

blood in the placenta. Furthermore, the O_2 transport capacity of fetal blood is increased by the substantially higher concentration of HbF in fetal blood.

The red color of muscles is mainly due to the presence of muscle hemoglobin or *myoglobin* (Mb) in myocytes. The myoglobin molecule (molecular weight about 16000 Da) corresponds to one unit of the tetrameric hemoglobin molecule. Accordingly the O_2 equilibrium curve of myoglobin is hyperbolic. The affinity of myoglobin for O_2 (P_{50} about 6 mmHg) is much higher than that of Hb. Myoglobin is considered to function as an O_2 store for muscle, and to facilitate diffusive O_2 transport from capillaries to mitochondria by diffusion of $Mb \cdot O_2$ (see Sect. 103.2.3).

102.3 Carriage of CO_2 by Blood

102.3.1 Blood CO_2 Equilibrium Curve

CO_2 concentration or content, C_{CO_2}, is defined as the amount of CO_2 that can be released from a volume unit of blood by acidification (which leads to decomposition of bicarbonate) and vacuum. The CO_2 equilibrium curve (dissociation curve) of blood can be established in the same manner as that of O_2 (see Fig. 102.3).

The CO_2 equilibrium curve is much steeper than that of O_2 and, moreover, it shows no saturation in the physiological range (Fig. 102.3). Like O_2, CO_2 in blood occurs in physical solution and in chemical combination. Although the physical solubility ($0.065 \text{ vol}\%/\text{mmHg} = 0.03 \text{ mmol}/(l \times \text{mmHg})$) is about 20 times higher than that of O_2 the greater part of CO_2 in blood is chemically bound. The reversible chemical binding occurs mainly as bicarbonate (in red cells and plasma) and, to a smaller extent, as carbamate (see below) bound to free amino groups of hemoglobin (in red cells).

102.3.2 Bicarbonate and Buffering

The change in bicarbonate concentration ($[HCO_3^-]$) with changing CO_2 partial pressure (P_{CO_2}) is mainly due to the buffering by hemoglobin, plasma proteins, and (to a much lesser degree) phosphates: these buffering actions are collectively referred to as "non-bicarbonate buffering" (see Sect 102.6.2). It should be noted that without nonbicarbonate buffering, blood bicarbonate concentration would be practically independent of changes in P_{CO_2}, and nearly all CO_2 transport would have to occur in the form of physically dissolved CO_2. This is due to the fact that the reaction

$$CO_2 + H_2O \rightleftharpoons H^+ + HCO_3^- \qquad (102.6)$$

cannot be made to move substantially by changing the physically dissolved CO_2 concentration, $[CO_2]$, which is proportional to P_{CO_2}, unless H^+ ions are supplied or re-

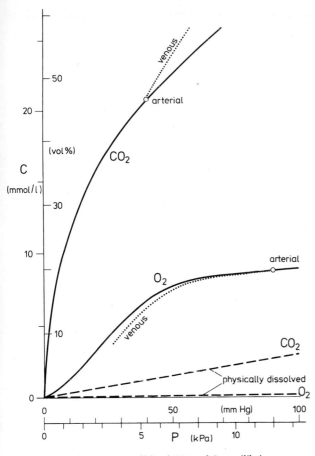

Fig. 102.3. Comparison of blood CO_2 and O_2 equilibrium curves. *Solid lines:* Equilibrium curves at constant partial pressure of the other gas; *dashed lines:* physically dissolved gases; *dotted lines:* "physiological" equilibrium curves, starting from arterial value and showing venous values reached with the normal respiratory exchange ratio, $R = 0.85$. Note the higher slope of the "physiological" equilibrium curves, particularly for that for CO_2. The deviation of the curves is due to the Haldane effect for CO_2, and to the Bohr effect for O_2

moved by buffering (HB, B⁻, acidic and basic form of nonbicarbonate buffer, respectively):

$$H^+ + B^- \rightleftharpoons HB. \qquad (102.7)$$

A few vol% of total CO_2 are bound to terminal free amino groups of hemoglobin as *carbamino* compounds or *carbamates* (note that, as in the case of CO_2 binding as bicarbonate, here, too, H^+ is generated and must be buffered):

$$Hb\text{-}NH_2 + CO_2 \rightleftharpoons Hb\text{-}NH\text{-}COO^- + H^+. \qquad (102.8)$$

The concentration of carbamate is only slightly dependent on P_{CO_2}, but is strongly modified by the O_2 saturation of hemoglobin since $Hb\cdot O_2$ binds much less carbamate than free hemoglobin. In this way, carbamate formation contributes to the Haldane effect (see Sect. 102.3.3).

102.3.3 Interactions Between O_2 and CO_2: Haldane and Bohr Effects

The blood CO_2 dissociation curve is to a remarkable extent dependent on the O_2 saturation (S_{O_2}) of blood hemoglobin (Fig. 102.4). Desaturation of hemoglobin, leading to a fall of chemically bound O_2 content ($-\Delta C_{O_2}$), enables blood to take up proportionately more CO_2 (ΔC_{CO_2}) at the same P_{CO_2}: this is the *Haldane effect*. Quantitatively, the Haldane factor (HF),

$$HF = -\Delta C_{CO_2}/\Delta C_{O_2}, \qquad (102.9)$$

amounts to 0.32 in human blood. This means that with gas exchange at a respiratory exchange ratio $R = 0.32$, blood P_{CO_2} remains unchanged (Fig. 102.4). Moreover, with normal $R = 0.85$, the P_{CO_2} change for a given change in C_{CO_2} is reduced by about 40% by the Haldane effect, meaning considerable steepening of the "physiological" CO_2 dissociation curve (i.e., coupled with normal O_2 exchange) as compared to the CO_2 equilibrium curve at constant O_2 saturation of blood (Figs. 102.3, 102.4).

The mechanism of the Haldane effect resides in the increase of *acid dissociation of hemoglobin* upon oxygenation ("$Hb\cdot O_2$ is a stronger acid than Hb"),

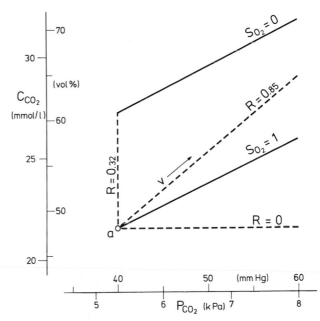

Fig. 102.4. Dependence of blood CO_2 equilibrium curve on O_2 saturation of hemoglobin (Haldane effect). CO_2 equilibrium curves (in the P_{CO_2} range 40–60 mmHg) are shown for blood with full O_2 saturation ($S_{O_2} = 1.0$) and full desaturation ($S_{O_2} = 0$). Starting from the arterial point (a), the venous values reached by gas exchange in tissues are expected to lie on the *dashed lines:* with normal $R = 0.85$, with $R = 0$ (no CO_2 exchange), and with $R = 0.32$ (all uptake of CO_2 into blood by the Haldane effect, P_{CO_2} remains unchanged). The increase of the slope of the curve $R = 0.85$ above that for $S_{O_2} = 1$ illustrates the extent of the action of the Haldane effect in normal CO_2 transport

$$H \cdot Hb + O_2 \rightleftharpoons Hb \cdot O_2^- + H^+, \qquad (102.10)$$

leading to decomposition of bicarbonate:

$$H^+ + HCO_3^- \rightleftharpoons CO_2 + H_2O. \qquad (102.11)$$

The Haldane effect and the Bohr effect (decrease of the affinity of hemoglobin for O_2 by increased P_{CO_2} or decreased pH; see Sect. 102.2.4) are mainly based on the same mechanism. In Eq. 102.10, increase of H^+ (= decrease of pH) moves the equilibrium to the left, thus decreasing S_{O_2} (by decrease of $Hb \cdot O_2^-$ and increase of $H \cdot Hb$).

For transport by blood, the Haldane effect is quantitatively, in terms of steepening the dissociation curve, more important than the Bohr effect (Fig. 102.3).

102.3.4 Exchange Between Red Cells and Plasma in CO_2 Transport

Concentrations of ions, proteins, buffering properties, and the resulting chemical combination of CO_2 in red blood cells and plasma are unequal. In equilibrium P_{CO_2} is the same in red cells and plasma, but pH and the concentrations of chloride and bicarbonate are lower in red cells than in plasma, in accordance with the *Donnan equilibrium* (passive distribution of ions between compartments with differing protein charge).

A number of interrelated processes are involved in the interaction of red blood cells and plasma in uptake and release of CO_2 by blood (Fig. 102.5). In *lung*, physically dissolved CO_2 diffuses from red cells and plasma into alveolar gas. However, this constitutes a small fraction of total CO_2 exchange. Most CO_2 is formed by dehydration of carbonic acid, H_2CO_3:

$$H_2CO_3 \rightleftharpoons CO_2 + H_2O. \qquad (102.12)$$

This is an intrinsically slow reaction, and is significantly accelerated by the presence of an enzyme, *carbonic anhydrase*, in the red cells (and on the luminal surface of capillary endothelial cells and in tissues) [7]. H_2CO_3 is present at a very low concentration (several hundred times less than dissolved CO_2), but is rapidly regenerated from bicarbonate and H^+ ions, the last being provided by protein buffers, mainly hemoglobin:

$$H^+ + HCO_3^- \rightleftharpoons H_2CO_3, \qquad (102.13)$$

$$HB \rightleftharpoons H^+ + B^-. \qquad (102.14)$$

The acid dissociation of hemoglobin is enhanced by its simultaneous oxygenation (Bohr effect, see above).

The resulting disturbance of the red cell/plasma equilibrium leads to an influx of HCO_3^- from plasma in exchange for Cl^- (chloride shift, Hamburger shift) through a specific transport pathway (band 3 protein; see Chap. 11) in the red cell membrane [3], and to an efflux of water from the red cells. The kinetics of the complex overall process is limited by the slowest steps, which apparently are the Cl^- shift and the dehydration of H_2CO_3 [11,15]. Poisoning of carbonic anhydrase (e.g., by acetazolamide) leads to elevated arterial, venous, and tissue P_{CO_2} and decreased pH (respiratory acidosis; see Sect. 102.6.5).

During uptake of CO_2 by blood in *tissues*, all the individual processes take place in the reverse direction, leading to formation of bicarbonate in both red cells and plasma.

102.4 Combined Effects of CO_2 and O_2 Equilibrium Curves

The much higher steepness of the CO_2 dissociation curve compared to the O_2 dissociation curve (Fig. 102.3) has a number of important consequences:

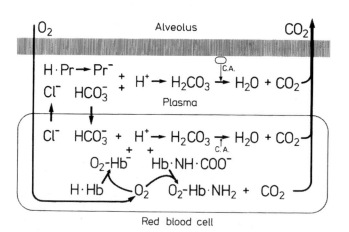

Fig. 102.5. Processes that take place in pulmonary capillary blood during CO_2 output in lungs. *C.A.*, Carbonic anhydrase, present in red blood cells and on the luminal surface of capillary endothelial cells. During CO_2 uptake by blood in tissues, all the processes occur in the reverse direction

Absorption of Gas [6]. Gas contained in closed body cavities, e.g., pneumothorax (gas in pleural space), the middle ear (when the Eustachian tube is obstructed), or lung airways (with closed bronchial connections), is absorbed, leading to ultimate disappearance of gas (or to subatmospheric pressure in the middle ear). The gas absorption is due to the fact that the *total pressure*, i.e., the sum of the partial pressures of CO_2, O_2 and N_2, in blood is less than atmospheric.

In arterial blood leaving an "ideal" lung (see Sect. 101.7.1), the partial pressures would be the same as in the alveolar gas and, consequently, their sum would be close to atmospheric pressure. Due to shunt and $\dot{V}_A/\dot{Q}$ inequalities, however, P_{O_2} in blood leaving the lungs is considerably lower and P_{co_2} (and also P_{N_2}) is slightly higher, and thus the total gas pressure in arterial blood is subatmospheric. In venous blood, however, P_{O_2} is very much lower and P_{co_2} is relatively a little higher than in arterial blood (due to different slopes of the blood equilibrium curves). Thus, the total pressure in blood equilibrating with tissue is definitely lower than atmospheric pressure. Due to its low solubility in blood (and low diffusivity in tissue), N_2 is the gas limiting ("braking") gas absorption.

Hyperventilation and Hypoventilation. After a sudden increase in alveolar ventilation (e.g., voluntary hyperventilation), the CO_2 output is much increased by CO_2 set free from blood (and tissues) upon decrease of P_{co_2}, whereas the pulmonary O_2 uptake is only slightly increased (due to the relatively flat course of the blood O_2 equilibrium curve and to poor physical solubility of O_2 in tissue). Thus, the *gas exchange ratio* (respiratory quotient) R (= CO_2 output/O_2 uptake) is increased. After a new steady state is reached – i.e., CO_2 output has become equal to CO_2 production, and O_2 uptake equal to O_2 consumption – R returns to the control value. Similarly, after a decrease in ventilation a transitory decrease in R is observed. Such changes in R are frequently observed, and are sensitive signs of an *unsteady state*. They should not be attributed to changes in metabolism.

Breath-Holding and Slowed Expiration. A breath-hold or a slow expiration represents an extreme reduction of effective ventilation (to zero). Correspondingly, the exchange ratio R decreases progressively. When the R value drops to equal to the Haldane factor (≈ 0.32, see Sect. 102.4), P_{co_2} in arterialized blood, which is equal to that in expired alveolar gas, becomes equal to that in inflowing mixed venous blood. Utilization of this phenomenon allows the use of expired P_{co_2} for noninvasive determination of mixed venous P_{co_2}, and for estimation of cardiac output according to the Fick principle as applied to CO_2 [4].

Nitrogen Equilibria. With normal $R < 1$, the sum of P_{O_2} and P_{co_2} in alveolar (expired) gas is smaller than P_{O_2} in inspired gas (see Sect. 101.3.1). Therefore, alveolar P_{N_2} must be higher than inspired P_{N_2}: $PA_{N_2} > PI_{N_2}$. For the same reason, in lungs with $\dot{V}_A/\dot{Q}$ inequality, P_{N2} in the low $\dot{V}_A/\dot{Q}$ areas is higher than in the high $\dot{V}_A/\dot{Q}$ areas, the overall effect being a higher P_{N_2} in (mixed) arterial blood than in (mixed) alveolar gas: $Pa_{N_2} > PA_{N_2}$. Indeed, the $(Pa - PA)_{N_2}$ difference has been used as a measure of $\dot{V}_A/\dot{Q}$ inequality [5].

102.5 Nitrogen

Nitrogen (N_2), although the most abundant gas in the atmosphere and in alveolar gas, is considered as *physiologically inert* since it appears to be neither produced nor consumed in body metabolism. Its solubility in water, blood, and body fluids is low, amounting to half that of O_2: 0.0017 vol%/mmHg in blood at 37°C. In blood equilibrated with alveolar gas ($PA_{N_2} \approx 560$ mmHg) the N_2 content is close to 1 vol% (≈ 0.45 mmol/l). Due to low solubility and a small Krogh diffusion constant, the rate of transport of N_2 by diffusion and blood flow is slow.

Practical problems with N_2 occur when the environmental pressure is rapidly decreased to a level lower than the total pressure of the gases in body fluids (see Sect. 102.4). In such conditions, *gas bubbles* are formed in blood (and tissues) and may lead to embolism, with local obstruction of blood flow in pulmonary or systemic circulation (*decompression sickness*). The reabsorption of gas from bubbles by diffusion and blood flow is slow due to the low solubility of N_2.

Supersaturation with bubble formation may occur after sudden pressure drops from pressure close to normal barometric pressure (e.g., in an airplane). However, it is particularly anticipated during too rapid decompression after *deep-sea diving* for long periods (which leads to N_2 equilibration at a greatly increased pressure). Decompression must be performed slowly, to prevent formation of bubbles and, if they are once formed, to allow time for them to disappear. An efficient countermeasure is recompression, best at high F_{O_2}, when the speed of bubble absorption is enhanced by an increased N_2 pressure head. Atmospheric N_2 appears not to be necessary for maintenance of human life, but it may exert useful effects. Thus, for example, without the braking effect of N_2, gas would be very rapidly absorbed from transiently obstructed airways, with formation of atelectasis.

102.6 Acid-Base Equilibria

102.6.1 General Aspects

The maintenance of a constant free hydrogen ion concentration (activity), $[H^+]$, or *constant pH* (= $-\log[H^+]$; $[H^+]$, concentration of free H^+ in mol/l) in body fluids is an essential feature of physiological homeostasis, because H^+ ions participate in, or have an influence on, numerous biochemical reactions. In arterialized blood plasma pH is close to 7.4 (corresponding to $[H^+] = 0.04 \mu$mol/l), its long-term variations usually not exceeding ±0.05. In most cells, pH is lower, about 6.7–7.2. It should be noted that in neutral water at 37°C, pH = 6.8 (at 25°C, pH = 7.0). Thus pH = 7.4 signifies considerable alkalinity (the concentration ratio $[OH^-]/[H^+] = 16$).

Recent work on acid-base equilibria in poikilothermic lower vertebrates and invertebrates has shown that with change of temperature, body fluid pH varies in such a manner that the $[OH^-]/[H^+]$ ratio (≈ 16) and the relative acid dissociation (α) of the imidazole group in the amino acid histidine stay constant (alphastat hypothesis) [9]. This may be an important finding since histidine imidazole groups have been shown to be involved in regulation of enzymes, cell receptors, membrane channels, and ion pumps [16].

There exist several mechanisms for control of pH and for its restoration to normal when disturbed:

- *Buffering* in tissues and blood, by means of various buffer systems (bicarbonate, proteins, phosphates).
- Pulmonary *ventilation*, which controls P_{CO_2} in arterial blood. Since P_{CO_2} (or $[CO_2]$) is a component of the bicarbonate buffer, and all buffer systems are in equilibrium, ventilation controls pH.
- *Kidney* and *liver* function. Normally all bicarbonate from the glomerular filtrate is absorbed. However, as a result of cooperation between the kidneys and the liver, bicarbonate can be excreted or saved, by which means the acid-base balance is regulated via the bicarbonate buffer system (see also Chaps. 75, 78).

The acid-base status of an organism is evaluated on the basis of that of the blood, which mirrors and averages the status of all organs and tissues. Moreover, blood is relatively easily accessible to sampling and analysis. When the respiratory component is to be accurately estimated, arterial blood (or arterialized blood from the ear lobe) is used.

102.6.2 CO$_2$/Bicarbonate Buffer and the Henderson-Hasselbalch Equation

In an aqueous solution containing bicarbonate the following reaction equilibria must be taken into account:

- *Hydration/dehydration*:
$$CO_2 + H_2O \rightleftharpoons H_2CO_3 \qquad (102.15)$$
- *H$^+$ dissociation/association*:
$$H_2CO_3 \rightleftharpoons H^+ + HCO_3^- \qquad (102.16)$$

Since the first of these equilibria is greatly weighted to the left (only some 10^{-3} part of total CO_2 is in the form of H_2CO_3), it is customary to add this small fraction to CO_2 (or to ignore it), thus obtaining the simplified single-step reaction equation:

$$CO_2 + H_2O \rightleftharpoons H^+ + HCO_3^- \qquad (102.17)$$

Application of the mass action law leads to the relationship:

$$\frac{[H^+] \times [HCO_3^-]}{[CO_2]} = K'. \qquad (102.18)$$

The practically constant $[H_2O]$ is included in K', the equilibrium constant of the reaction, which depends on temperature and on the ionic strength of the solution, but, at least in the physiological range, is practically independent of the concentration of the reactants, CO_2, H^+, and HCO_3^-, and of other substances (including buffers).

Introducing logarithms and the definitions $pH = -\log[H^+]$ and $pK' = -\log K'$, one obtains:

$$pH = pK' + \log \frac{[HCO_3^-]}{[CO_2]}. \qquad (102.19)$$

This is the *Henderson-Hasselbalch equation*, which plays a central role in the analysis of acid-base equilibria in blood plasma (and in any other body fluid). For application to whole blood (or other heterogeneous systems), adjustments by correction factors are required.

Replacement of the concentration of physically dissolved CO_2, $[CO_2]$, by the product of P_{CO_2} and the solubility coefficient of CO_2, α_{CO_2} ($= 0.03\,mmol/(l \times mmHg)$) in plasma at 37°C,

$$[CO_2] = \alpha_{CO_2} \times P_{CO_2}, \qquad (102.20)$$

and introduction of the value of $pK' = 6.1$, valid for blood plasma at 37°C, yields:

$$pH = 6.1 + \log \frac{[HCO_3^-]}{0.03 P_{CO_2}}, \qquad (102.21)$$

where $[HCO_3^-]$ is in mmol/l and P_{CO_2} in mmHg (when expressed in kPa, the coefficient in the denominator becomes 0.225). Since P_{CO_2} and pH in blood plasma can be measured directly, the equation can be used for calculation of $[HCO_3^-]$ in plasma.

The value of pH is determined by glass electrode techniques, from the voltage across a glass membrane permeable to H^+ ions. The pH measured in whole blood is that of plasma. For measurement of P_{CO_2}, the glass pH electrode is covered by a plastic membrane permeable to CO_2, with a thin layer of bicarbonate solution between the pH-sensitive glass and the plastic membrane. The pH in the bicarbonate solution is a function of P_{CO_2} according to the Henderson-Hasselbalch equation.

The *bicarbonate buffer* is only one of the buffer systems of blood and body fluids, but of particular importance for the following reasons:

- The volatile acid component, CO_2, proportional to P_{CO_2}, is a direct indicator of the adequacy of pulmonary ventilation (or respiratory disturbances).
- The concentration of bicarbonate is controlled mainly by the kidney (excretion or formation of HCO_3^-) and the liver (consumption for synthesis of urea).

- Bicarbonate is operative in buffering of fixed acids (metabolic acidosis).
- The bicarbonate system is in equilibrium with other buffer systems, thus serving as indicator of the overall acid-base status.

102.6.3 Buffering in Blood

The buffer systems of blood comprise:

- The CO_2/HCO_3^- *system* (in red cells and plasma)
- *Proteins* (hemoglobin in red cells, plasma proteins in plasma)
- Organic and anorganic *phosphates* (in red cells and plasma).

The buffering behavior of blood is complicated by the unequal distribution of the buffers to red cells and plasma and by the exchange processes between the two compartments when equilibria are deranged.

Buffering in blood is usually examined by measurement of pH after changes of P_{CO_2} (equilibration with varied P_{CO_2}, see above), and calculation of $[HCO_3^-]$ after the Henderson-Hasselbalch equation (Eq. 102.21). By a change of P_{CO_2}, equilibria of the following reaction systems are displaced:

$$CO_2 + H_2O \rightleftharpoons H^+ \begin{matrix} + HCO_3^- \\ \\ + B^- \rightleftharpoons HB \end{matrix} . \qquad (102.22)$$

$H^+ + B^- \rightleftharpoons HB$ denotes the buffering action of proteins (mainly due to the imidazole group of histidine residues and phosphates).

Since the changes of free $[H^+]$ are always small in terms of absolute concentrations (pH = 7.4, $[H^+] = 0.04\,\mu mol/l$; pH = 6.8, $[H^+] = 0.16\,\mu mol/l$), in considerations of mass balance they can be neglected against the changes in the concentrations of HCO_3^-, B^-, and HB, all of which are in the order of 1–10 mmol/l, and practically equimolar. Thus, upon increase of P_{CO_2}, which leads to a decrease of pH ($-\Delta pH$), an increase in $[HCO_3^-]$ is equivalent to that of [HB] and to the decrease of $[B^-]$. Therefore, in this condition, determination of $[HCO_3^-]$ allows one to assess the (effective) buffering power (*buffer value*) of blood *nonbicarbonate buffers*, β, as change of $[HCO_3^-]$ per pH change:

$$\beta = \frac{\Delta[HCO_3^-]}{-\Delta pH}. \qquad (102.23)$$

It is important to note that in this condition, the bicarbonate buffer serves only as an *indicator* of the buffering action of nonbicarbonate buffers (mainly hemoglobin and plasma proteins) and of itself exerts no buffering action proper. This equilibration experiment simulates the "respiratory disturbances" (see Sect. 102.6.5).

The prominent role of *hemoglobin* is demonstrated by repetition of the same equilibration experiment with separated plasma. The β value, about 25 mmol/l × pH for whole blood measured in plasma ("true plasma"), becomes reduced to about 5 mmol/l × pH ("separated plasma"). This difference is due to the fact that in whole blood the pH change in plasma is strongly attenuated by exchange of bicarbonate with the red cells.

Another way to assess blood buffering is to add an amount, per volume blood, of strong acid (or base), i.e., H^+ ions (or OH^- or HCO_3^- ions) and to keep P_{CO_2} constant by equilibration. In this case *both* bicarbonate and nonbicarbonate buffers of plasma and red cells act in parallel:

$$H^+ \begin{matrix} + HCO_3^- \rightleftharpoons CO_2 + H_2O \\ \\ + B^- \rightleftharpoons HB \end{matrix} \qquad (102.24)$$

From the pH change per amount of added H^+ or OH^- one may obtain a measure of the overall buffering power that is larger than the buffer value, β, determined by equilibration with varied P_{CO_2}. This experiment corresponds to the "metabolic or nonrespiratory disturbances" (see Sect. 102.6.5).

102.6.4 Metabolism, Nutrition, and Acid-Base Balance

Complete oxidation of carbohydrates and fat (glycerol esterified with fatty acids) yields only CO_2 and H_2O as end products. Only lungs are needed to take care of CO_2 elimination.

When *proteins* (amino acids) are completely oxidized, in addition to CO_2 and H_2O, NH_4^+ (from all amino groups) and HCO_3^- (from all carboxylic groups) are produced, usually in more or less equivalent amounts (about 1 mol/day). From NH_4^+ and HCO_3^- *urea* is synthesized in the liver and excreted by the kidneys:

$$2NH_4^+ + 2\ HCO_3^- \rightarrow H_2N \cdot CO \cdot NH_2 + CO_2 + 3\ H_2O. \qquad (102.25)$$

An alternative pathway for amino-N elimination is via *glutamine* (formed mainly in the liver) which serves as nontoxic carrier of NH_4^+ to kidneys, where NH_4^+ is set free by deamination (to glutamate of oxoglutarate) and excreted, *without* concomitant excretion or consumption of HCO_3^- (see Chaps. 67, 75). Thus, the liver must be considered as an organ important for acid-base regulation: HCO_3^- homeostasis is linked to urea synthesis, and the linkage can be modified by channeling of NH_4^+ from urea synthesis to glutamine synthesis [2].

Sulfur-containing amino acids (cysteine, cystine, methionine), upon complete oxidation, yield *sulfuric acid*, H_2SO_4, i.e., two H^+ per SO_4^{2-}, which must be buffered, resulting in decrease of $[HCO_3^-]$. This mechanism, along with degradation of some organic phosphates, leads to an excess of H^+ production with a protein-rich diet.

Plants (vegetables, fruit) contain organic *carboxylic acid salts*, which upon full oxidation deliver a bicarbonate per

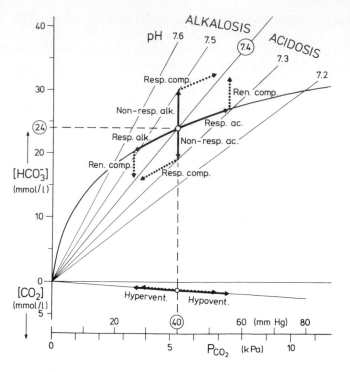

Fig. 102.6. Disturbances of acid-base equilibrium as shown in the P_{CO_2}-bicarbonate diagram. The values relate to arterial blood plasma. *Ordinate*: above 0: $[HCO_3^-]$; below 0: $[CO_2]$. The pH lines result from the Henderson-Hasselbalch equation. Respiratory acidosis and alkalosis, nonrespiratory (metabolic) acidosis and alkalosis, and respiratory and renal compensation are shown

carboxylate group, thus leading to alkalosis. The high HCO_3^- content in the copious alkaline urine of herbivorous animals is well known.

With an average European/North American diet, the overall catabolic metabolism has a slight *acidotic* tendency and excess excretion of acid is observed, about 50 mmol/day [20 mmol/day excreted as buffered H^+ (titratable acidity), 30 mmol/day, as NH_4^+]. These figures are to be compared to urea elimination of about 1 mol/day, and to a CO_2 output (by lungs) of 20 mol/day (≈ 0.3 l/min).

102.6.5 Normal Status and Primary Disturbances

Typical values for normal arterial blood plasma are pH = 7.40 ($[H^+]$) = 40 nmol/l; P_{CO_2} = 40 mmHg = 5.33 kPa; $[HCO_3^-]$ = 24 mmol/l (Fig. 102.6).

Hypoventilation leads to an increase of arterial P_{CO_2} (*hypercapnia*) and to an increase of $[HCO_3^-]$ (buffering by nonbicarbonate buffers, see above), but the ratio $[HCO_3^-]/[CO_2]$ decreases, and therefore, pH is reduced, which constitutes *respiratory acidosis*. This status may be caused by anatomic or functional disturbances of pulmonary function, e.g., in pulmonary emphysema (pathologic changes of lung structure and function, combined with fatigue of respiratory muscles due to overload), poliomyelitis (partial paralysis of respiratory muscles), deep anesthesia (depression of the respiratory centers), etc.

Conversely, hyperventilation produces lowering of arterial P_{CO_2} (*hypocapnia*). Although $[HCO_3^-]$ also decreases (by nonbicarbonate buffering), the ratio $[HCO_3^+]/[CO_2]$ is increased and pH is elevated: this condition is *respiratory*

alkalosis. This derangement is found in voluntary hyperventilation, idiopathic hyperventilation, and hyperventilation due to hypoxia (e.g., at high altitude).

Disturbances of the acid-base status of nonrespiratory origin are termed *metabolic* or *nonrespiratory disturbances*. They operate by primary changes in $[HCO_3^-]$. *Metabolic or nonrespiratory acidosis*, with decrease of $[HCO_3^-]$ and pH, may be caused by production of acids (hydroxybutyric and acetoacetic acid in diabetes; lactic acid in heavy exercise), by deficient excretion of H^+ in renal insufficiency, by loss of base (alkaline diarrhea fluid in cholera), etc.

The *metabolic or nonrespiratory alkalosis* is the counterpart to metabolic acidosis: $[HCO_3^-]$ is increased at normal arterial P_{CO_2}, pH is increased. This status may be produced by loss of acid (e.g., by vomiting of acid gastric juice) or by a base-producing diet (fruit and vegetables, because oxidation of salts of organic acids yields bicarbonate).

102.6.6 Compensations

The above-described pure forms of primary disturbances are rarely seen because they are modified by compensatory processes during their development. The compensatory reactions aim at normalization of the pH value (Fig. 102.6). *Respiratory compensation* of metabolic disturbances sets in rapidly. In metabolic acidosis the arterial P_{CO_2} is decreased by hyperventilation. In spite of a (small) concomitant decrease of bicarbonate (due to nonbicarbonate buffering), the ratio $[HCO_3^-]/[CO_2]$ is increased and pH rises, moving closer to 7.4. The respiratory compensation of metabolic alkalosis occurs by hypoventilation, with elevation of P_{CO_2}, decrease of the ratio $[HCO_3^-]/[CO_2]$ and pH approaching 7.4; however, its extent is limited by the arterial hypoxia caused by the hypoventilation.

Renal compensation is more delayed and primarily affects bicarbonate concentration in plasma. In normal conditions, practically all the bicarbonate of the glomerular filtrate is reabsorbed. In alkalosis (respiratory or nonrespiratory) the reabsorption is incomplete, and $[HCO_3^-]$ is decreased by its excretion (see Chap. 75). In acidosis, H_2CO_3 is formed in the kidneys by hydration of CO_2 (with the aid of carbonic anhydrase), H^+ is excreted in association with buffers, and the remaining HCO_3 is reabsorbed. By this means $[HCO_3^-]$ in blood is raised, increasing $[HCO_3^-]/[CO_2]$ and pH towards normal values. Another mechanism leading to retention of HCO_3^- is excretion of NH_4^+ (hydrolyzed from glutamine produced by the liver at the expense of urea, see above).

Compensations are in most cases *partial*, i.e., full normalization of pH is not reached. Complete normalization of the acid-base status, with normal P_{CO_2}, $[HCO_3^-]$, and pH, is attained after disappearance (spontaneous or by medication) of the mechanism of the primary disturbance.

102.6.7 Indicators of Acid-Base Status

The quality and the extent of the overall disturbance – acidosis or alkalosis – is indicated by the pH in arterial blood, and that of the respiratory component (primary or compensatory) by arterial blood P_{CO_2}. For quantitative characterization of the nonrespiratory component (again, both primary or compensatory), various indices are in use. The most widespread are the following:

- *Standard bicarbonate* is the bicarbonate concentration in plasma of blood equilibrated at $P_{CO_2} = 40$ mmHg (37°C,

$S_{O_2} = 1.0$). Since in normal arterial blood these conditions are met, standard bicarbonate is equal to normal arterial plasma $[HCO_3^-]$. Normal value, 24 mmol/l.

- *Buffer base* is the concentration of all buffering anions in blood. It equals the sum of bicarbonate and all negatively charged dissociated buffering groups in hemoglobin and plasma proteins (determined from nomograms). Normal value, 48 mmol/l.

- *Base excess* quantifies the deviation of the buffer base from the normal value. Base excess is equal to the amount of free H^+ ions required for back-titration of blood pH to 7.4 at $P_{CO_2} = 40$ mmHg and 37°C. Normal value, 0 mmol/l. It is determined from P_{CO_2}, pH, and hemoglobin concentration by nomograms [12].

- *Strong ion difference* [13,14] is the concentration difference between all non-buffering cations and anions (in normal plasma mainly $[Na^+] + [K^+] - [Cl^-]$. It is equivalent to the sum of $[HCO_3^-]$ and nonbicarbonate buffering anions, i.e., it corresponds to buffer base applied to isolated plasma.

In metabolic acidosis, or in respiratory alkalosis with renal compensation, standard bicarbonate, buffer base, and strong ion difference are decreased and base excess is negative (a "base deficit" exists). Conversely, in metabolic alkalosis, or in respiratory acidosis with renal compensation, standard bicarbonate, buffer base, and strong ion difference are increased and base excess is positive.

102.6.8 Graphic Representation of Blood Acid-Base Equilibrium

The variables of the bicarbonate buffer system, pH, P_{CO_2}, and bicarbonate, can be plotted in various ways to visualize

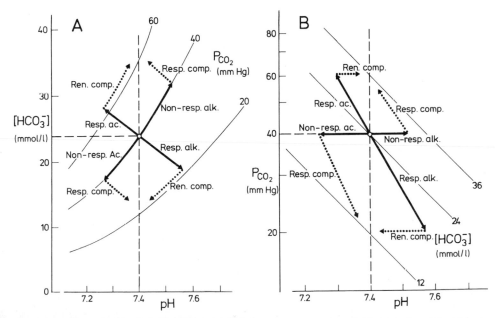

Fig. 102.7A,B. Disturbances of the acid-base equilibrium as shown in the pH-HCO₃ diagram (**A**) and in the pH-log P_{CO_2} diagram (**B**)

the acid-base status of blood plasma. The following diagrams have been used:

- The P_{CO_2} – *bicarbonate diagram* (Fig. 102.6) follows from the blood CO_2 dissociation curve. The $[HCO_3^-]/[CO_2]$ ratio is shown by the slope of the iso-pH lines.
- The *pH-bicarbonate diagram* (Fig. 102.7a) results from the analysis of buffering. The (approximately straight) lines for the primary respiratory disturbances correspond to the buffer line of "true plasma". The iso-P_{CO_2} lines rise exponentially with increasing pH.
- The *pH-log P_{CO_2} diagram* (Fig. 102.7b) offers the advantage of plotting the directly measured variables P_{CO_2} and pH against each other and is widely used today [12]. Nonbicarbonate buffer value of true plasma is represented by the difference in negative slope of the (slightly curved) lines and lines of constant $[HCO_3^-]$, which have the slope of -1. Metabolic disturbances are represented by displacements parallel to the abscissa.

It should be pointed out that all these diagrammatic representations are *equivalent*, because they all are based on the Henderson-Hasselbalch equation and the buffer power of blood.

General References

Bauer C (1974) On the respiratory function of haemoglobin. Rev Physiol Biochem Pharmacol 70:1–31

Bauer C, Gros G, Bartels H (eds) (1980) Biophysics and biology of carbon dioxide. Springer, Berlin Heidelberg New York

Baumann R, Bartels H, Bauer C (1987) Blood oxygen transport. In: Farhi LE, Tenney SM (eds) Handbook of physiology, section 3: respiratory system, vol 4: gas exchange. American Physiological Society, Bethesda, pp 147–172

Klocke, RA (1987) Carbon dioxide transport. In: Farhi LE, Tenney SM (eds) Handbook of physiology, section 3: respiratory system, vol 4: gas exchange. American Physiological Society, Bethesda, pp 173–197

Siggaard-Andersen O (1974) The acid-base status of the blood, 4th edn. Munksgaard, Copenhagen

Specific References

1. Grant BJB (1982) Influence of Bohr-Haldane effect on steady-state gas exchange. J Appl Physiol 52:1330–1337
2. Häussinger D, Gerok W, Sies H (1986) The effect of urea synthesis on extracellular pH in isolated perfused rat liver. Biochem J 236:261–265
3. Jennings ML (1989) Structure and function of red cell anion transport protein. Annu Rev Biophys Chem 18:397–430
4. Kim TS, Rahn H, Farhi LE (1966) Estimation of true venous and arterial P_{CO_2} by gas analysis of a single breath. J Appl Physiol 21:1338–1344
5. Klocke FJ, Rahn H (1961) The arterial-alveolar inert gas ("N$_2$") diffrerence in normal and emphysematous subjects, as indicated by the analysis of urine. J Clin Invest 40:286–294
6. Loring SH, Butler JP (1987) Gas exchange in body cavities. In: Farhi LE, Tenney SM (eds) Handbook of physiology, section 3: respiratory system, vol 4: gas exchange. American Physiological Society, Bethesda, pp 283–295
7. Maren TH (1967) Carbonic anhydrase: chemistry, physiology and inhibition. Physiol Rev 47:595–781
8. Piiper J, Dejours P, Haab P, Rahn H (1971) Concepts and basic quantities in gas exchange physiology. Respir Physiol 13:292–304
9. Reeves RB (1972) An imidazole alphastat hypothesis for vertebrate acid-base regulation: tissue carbon dioxide content and body temperature in bullfrogs. Respir Physiol 14:219–236
10. Riggs AF (1988) The Bohr effect. Annu Rev Physiol 50:181–204
11. Schünemann HJ, Klocke RA (1993) Influence of carbon dioxide kinetics on pulmonary carbon dioxide exchange. J Appl Physiol 74:715–721
12. Siggaard-Andersen O (1974) The acid-base status of the blood, 4th edn. Munksgaard, Copenhagen
13. Stewart PA (1978) Independent and dependent variables of acid-base control. Respir Physiol 33:9–26
14. Stewart PA (1981) How to understand acid-base. A quantitative acid-base primer for biology and medicine. Elsevier/North Holland, New York
15. Swenson ER, Grønlund J, Ohlsson J, Hlastala MP (1993) In vivo quantitation of carbonic unhydrase and band 3 protein contributions to pulmonary gas exchange. J Appl Physiol 74:838–848
16. White FN, Somero G (1982) Acid-base regulation and phospholipid adaptations to temperature: time courses and physiological significance of modifying the milieu for protein function. Physiol Rev 62:40–89

103 Oxygen Supply and Energy Metabolism

J. PIIPER

Contents

103.1 Basic Concepts [1,9]

All life processes *require energy*. In the body, in organs, cells, and subcellular structures, all physiological processes are necessarily accompanied by energy transformations. Generally in animal organisms only chemical energy, which for the most part is provided by *oxidative metabolism*, can be utilized. Oxidative chemical energy is gained by step-wise oxidation of substrates (carbohydrates, lipids, and proteins), with CO_2, H_2O, and urea as principal end products. The utilizable energy is trapped in the form of high-energy phosphate bonds in adenosine triphosphate (ATP).

The physiological work processes include *chemical synthesis* (usually involving endergonic reactions, i.e., requiring provision of free energy); *mechanical work* (muscle contraction against load); and generation and maintenance of *concentration gradients* and *potential differences* (e.g., between cells and extracellular fluid, across renal tubules etc.; see also Chap. 11). In energy transformations, a large part of the energy is degraded into *heat*. This is because the *efficiency*, i.e., the ratio of (useful) work to energy spent, is usually less than 50% (part of the energy is "lost" as heat). Moreover, ultimately all energy, except a very small fraction spent on the performance of external mechanical work (potential mechanical energy, e.g., of a lifted weight), appears as heat in the overall energy balance.

A number of energy units are used in physiology (cf introductory chapter, this volume). The most widespread are the following: kilocalorie (kcal), kilojoule (kJ), watt-second (W·s), kilogram-force × meter (kg × m). Some equivalences are: 1 kcal = 4.18 kJ; 1 W × s = 1 kJ = 0.239 kcal; 1 kcal = 427 kg × m.

The total energy flux of the body, of isolated organs or cells can be measured as heat production (*direct calorimetry*) or, much more conveniently, as consumption of O_2 (*indirect calorimetry*). Utilization of substrates or production of CO_2 may also be measured. For oxidation of a unit amount of substrate a certain amount of O_2 is required, and a certain amount of energy is produced. From chemical reaction energetics, confirmed by direct calorimetry in living organisms, the following energy equivalence relations have been obtained. Consumption of 1 l O_2 sets free:

- 5.0 kcal with carbohydrate as substrate
- 4.7 kcal with lipids
- 4.5 kcal with protein converted to urea.

For a human on a normal diet, the average energy equivalent of consumed O_2 is 4.85 kcal/l O_2 ($\approx$ 460 kJ/mole O_2). The metabolic rate of the whole body (and of organs) is highly variable. The minimum rate of the whole body, the *basal metabolic rate*, is that determined in physiological resting conditions (rest, thermal comfort, not less than 6 h after a meal). It depends on body mass, age, and sex. Comparing mammals of widely differing body mass the basal metabolic rate is found to be more proportional to body surface area than to body mass ("*surface area rule*"). Since larger animals have a smaller surface area in relation to body mass, the basal metabolic rate by body mass is smaller in large animals (e.g., 1 g elephant consumes 20 times less O_2 than 1 g mouse). For a young man of 70 kg the basal metabolic rate amounts to 70 kcal/h, corresponding to an O_2 consumption of 0.24 l/min. During heavy exercise energy expenditure may increase by a factor of 10 or more. O_2 consumption is unevenly distributed in the body: high specific O_2 consumption (i.e., O_2 consumption per mass) is found in heart, kidney, liver, brain, and in contracting muscles, while low specific O_2 consumption is found in resting muscles, skin, and fatty tissue.

Oxygen and substrates are supplied to tissues, and CO_2 and other waste products eliminated, by blood flow. In most cases, the supply of O_2, due to its high rate of consumption compared to its availability and to the smallness of O_2 stores (low solubility in tissue, except in blood and muscle, where it is bound to hemoglobin and myoglobin, respectively) is the limiting factor. Therefore, the variables determining O_2 supply are at the center of interest in the

R. Greger/U. Windhorst (Eds.)
Comprehensive Human Physiology, Vol. 2
© Springer-Verlag Berlin Heidelberg 1996

physiology and clinical physiology of viability and performance of the body, its organs, and cells.

103.2 Oxygen Demand and Supply

103.2.1 Component Processes [8,18,19]

To analyze O_2 transport to tissues, it is useful to consider three processes, functionally arranged in series, and their interactions (Fig. 103.1):

- Tissue *perfusion (blood flow)*. Arterialized blood, with cardiac output as the total flow rate, is distributed to tissues, the local flow being related to (and controlled by) the O_2 requirement.
- Blood-to-tissue *diffusion*. O_2 must diffuse from red blood cells in capillaries to the mitochondria, the main site of O_2 consumption of tissues. Since the diffusion distances are much longer in tissues than in lungs (between alveolar gas and red blood cells), larger O_2 pressure (P_{O_2}) gradients and more extensive diffusion limitation are expected to occur and are indeed present.
- *Consumption* of O_2. The complex system of chemical reactions ultimately leading to reaction of the hydrogen of substrates with molecular O_2, with coupled formation of high-energy phosphate (ATP), is described in detail in biochemistry textbooks. Typically, 6 mol ATP are formed per mole O_2 consumed (ATP/O_2 ratio of 6). The energy for physiological work processes is provided by hydrolysis of ATP into ADP and inorganic phosphate, the free energy per mole ATP amounting to 10 kcal/mol. Since the total energy equivalence of 1 mol O_2 is about 100 kcal, the biochemical efficiency of oxidative phosphorylation is roughly 60%.

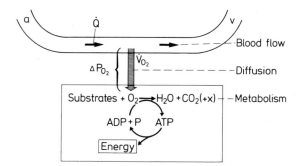

Fig. 103.1. Preocesses involved in O_2 supply to tissue and oxidative energy metabolism. The blood flow brings O_2 to tissue: O_2 delivery, $Q \times Ca_{O_2}$; O_2 uptake from blood, $\dot{V}_{O_2} = Q\,(Ca - Cv)_{O_2}$. Diffusion from blood capillaries to mitochondria depends on the difference in P_{O_2} (ΔP_{O_2}), the geometric dimensions, and the Krogh O_2 diffusion constant. O_2 is consumed for stepwise substrate oxidation, the energy liberated is trapped in the form of ATP, hydrolysis of which provides the energy required for various purposes

103.2.2 O_2 Diffusion: Block (Sheet) Model

The simplest model of O_2 supply by diffusion to an O_2-consuming tissue is a homogeneous block with a constant specific O_2 requirement (O_2 demand), into which O_2 diffuses across a smooth, plane surface. It is evident that, from surface to depth, the O_2 diffusion flux diminishes due to O_2 consumption and will be exhausted at a certain depth, designated as the *penetration depth* of O_2. Tissue beyond this depth receives no O_2, i.e., it is anoxic ($P_{O_2} = 0$).

The penetration depth of O_2 (x_{O_2}) can be easily shown to depend on the Krogh O_2 diffusion constant, K_{O_2}, the O_2 partial pressure at surface, $P_{O_2}(O)$, and the specific O_2 requirement, a_{O_2} (dimension: amount of gas per time and tissue volume):

$$x_{O_2} = \sqrt{2K_{O_2} \times \frac{P_{O_2}(O)}{a_{O_2}}} \qquad (103.1)$$

With the average specific O_2 requirement of the resting human body ($a_{O_2} = 300$ ml/min $\times$ 75 l $= 4.10^{-3}$ ml/(min $\times$ cm³)), K_{O_2} for tissue ($K_{O_2} = 2.10^{-8}$ ml/(min $\times$ cm $\times$ mmHg), and atmospheric P_{O_2} ($P_{O_2} = 150$ mmHg), one obtains: $x_{O_2} \approx 0.4$ mm. When applied to the human body, of surface area of about 2 m², this penetration depth of O_2 corresponds to only about 1% of the whole body mass. This example demonstrates that by diffusion alone only thin tissue layers can be supplied with O_2. For a massive body, O_2 transport must be arranged in such a manner that the *diffusion distances* are reduced to less than a millimeter. This is achieved by *convective* transport of O_2 by blood flow to a sufficiently dense capillary network pervading all tissues.

The maximum thickness of tissue slices allowing O_2 supply by diffusion is of central importance in metabolic studies on minced tissues. Obviously, the "critical slice thickness' of the well-known Warburg equation is twice the O_2 penetration depth according to Eq. 103.1.

103.2.3 O_2 Diffusion and Blood Flow: Krogh Cylinder Model [11,13,17]

The widely used *Krogh cylinder model* consists of a tissue cylinder uniformly consuming oxygen delivered by an axial blood-perfused capillary. This model allows interactions of O_2 requirements, diffusion, and blood flow to be studied (Fig. 109.2). For simplicity, only radial O_2 diffusion is assumed to occur.

In the model, two P_{O_2} gradients develop. In any cross-sectional element, due to O_2 diffusion and O_2 consumption, a *radial P_{O_2}* gradient develops, leading to an overall P_{O_2} difference (ΔP_{O_2}) between blood and the periphery of the cylinder. This difference is determined by the specific O_2 requirement (a_{O_2}), the Krogh diffusion constant (K_{O_2}), the radius of the cylinder (R), and that of the supplying capillary (r) (ln, natural logarithm):

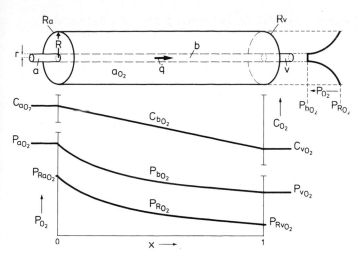

Fig. 103.2. Krogh cylinder model. The tissue cylinder of radius R is supplied with O_2 by a central capillary of radius r and specific blood flow $\dot{q}$. The P_{O_2} profile in cross-section is depicted on the *right* (the decrease of the P_{O_2} gradient from the capillary to the periphery is due to a decrease in diffusive O_2 flux because of O_2 consumption and because of geometry). The *bottom diagram* shows the decrease of O_2 blood concentration (Cb_{O_2}), of blood P_{O_2} (P_{bO_2}), and P_{O_2} at the periphery of the cylinder (P_{RO_2}), from the arterial to the venous end of the capillary. Note that the lowest value of P_{RO_2} is P_{RVO_2}

$$\Delta P_{O_2} = \frac{A_{O_2}}{2K_{O_2}}\left[R_2 \times \ln\left(\frac{R}{R}\right) - \frac{R^2 - R^2}{2} \right] \qquad (103.2)$$

In the *longitudinal* direction, blood O_2 concentration, Cb_{O_2}, falls linearly along the cylinder ($0 < x < 1$, relative distance) and depends on specific blood flow, $\dot{q}$ (blood flow/cylinder volume). Application of Fick principle (see Eq. 103.4) to this situation yields:

$$Cb_{O_2} = Ca_{O_2} - \frac{a_{O_2}}{q} \times x \qquad (103.3)$$

The corresponding P_{O_2} drop depends on the blood O_2 equilibrium curve, and is therefore nonlinear. P_{O_2} is lowest in the *periphery of the venous end of the cylinder*, and anoxia will develop first in this region. This occurs with:

- Too low arterial O_2 content (Ca_{O_2})
- Too low blood flow ($\dot{q}$)
- Too high O_2 requirement (a_{O_2})
- Too large cylinder radius (R) or
- Too small capillary radius (r).

In the following example the *critical capillary density* (number of capillaries per cross-section of muscle) and blood flow will be estimated for human muscles exercising maximally. At maximum O_2 uptake of 3.5 l/min, muscles (30 kg $\approx$ 3·10³ ml) consume 31 O_2/min, with average specific O_2 consumption, $a_{O_2} = 0.1$ ml/(min × ml). In the theoretical case of infinitely large blood flow, intravascular P_{O_2} would everywhere be at the arterial value, $P_{O_2} = 90$ mmHg ($S_{O_2} = 99\%$), and one obtains with $K_{O_2} = 2 \times 10^{-8}$ ml/(min × cm × mmHg), for the critical supply cylinder radius, R, a value of 30 μm, corresponding roughly to a capillary density of 300/mm². With a specific blood flow of 0.7 ml/

(min × ml), venous S_{O_2} would be 28%, and a corresponding venous P_{O_2}, about 20 mmHg, the critical $R = 16$ μm is obtained, corresponding to a capillary density of about 1000/mm². This last figure corresponds to anatomical observations.

Investigations have shown that not only does muscle blood flow increase up to 20-fold or more during maximum exercise as compared to rest, but capillary density also increases, although by a smaller factor, by opening up of capillaries closed in the resting state. The mechanisms adjusting blood flow to the O_2 demand are discussed in Chaps. 94, 95.

The Krogh cylinder has been, and still is, a useful model for approaching tissue O_2 supply. However, there are many problems in applying it. For example, calculations have shown that within capillary blood (in red cells and in plasma), considerable P_{O_2} gradients are expected, whereas in muscle cells myoglobin facilitates O_2 transfer reducing the P_{O_2} gradients [6,10,11]. In tissue, interaction of the O_2 supply between capillaries occurs. This interaction is dependent on flow direction in adjacent capillaries (co-current, counter-current). In addition, there is evidence of loss of O_2 from precapillary vessels (arteries and arterioles), giving rise to arteriovenous shunting of O_2.

103.2.4 O_2 Supply by Blood and Disturbances of the Supply [3,14,15]

Fick's principle is applicable not only to pulmonary circulation, but to any delimitable circulatory area: O_2 uptake (consumption) is equal to the product of blood flow ($\dot{Q}$) and the arteriovenous O_2 concentration difference ($Ca_{O_2} - Cv_{O_2}$):

$$\dot{V}_{O_2} = \dot{Q} \times (Ca - Cv)_{O_2} \qquad (103.4)$$

By this means O_2 uptake (or uptake or output of any substance) can be determined.

The venous O_2 concentration is variable. This may be expressed in terms of (fractional) O_2 *extraction* or *utilization*, $(Ca_{O_2} - Cv_{O_2})/Ca_{O_2}$: about 10% in the kidney, in resting muscle about 40%, in exercising muscle up to 90%. For total systemic circulation, O_2 extraction is about 25% in resting conditions (corresponding to $S_{O_2} \approx 70$–75% in mixed venous blood). With heavy exercise O_2 extraction can rise to 80%.

An important quantity is the amount of O_2 delivered per unit time by blood flow to the whole body, an organ, or a tissue. The O_2 *delivery* (O_2 *availability*, O_2 *supply*), $\dot{D}_{O_2}$, is equal to the product of blood flow and arterial O_2 concentration:

$$\dot{D}_{O_2} = \dot{Q} \times Ca_{O_2} \qquad (103.5)$$

In resting conditions ($\dot{Q} = 5\,l/min$, $Ca_{O_2} = 20$ vol%) $\dot{D}_{O_2}$ is about 1 l O_2/min of which 30% is consumed ($\dot{V}_{O_2} = 0.3\,l/min$).

Since most of the O_2 contained in blood is chemically bound to hemoglobin, and the chemically bound O_2 is the product of O_2 capacity (Cap_{O_2}) and O_2 saturation (S_{O_2}), the following relationship is obtained from Eq. 103.5:

$$\dot{D}_{O_2} = \dot{Q}_{O_2} \times Cap_{O_2} \times S_{O_2} \qquad (103.6)$$

From this relationship the disturbances of O_2 supply can be derived: $\dot{D}_{O_2}$ may be reduced to a critical limit by reduction of $\dot{Q}$, Cap_{O_2}, or S_{O_2}, or by their combination. When, with decreasing $\dot{D}_{O_2}$, venous P_{O_2} falls below a critical value which is determined by diffusion (venous end of Krogh's cylinder), tissue anoxia (O_2 lack) occurs. On the other hand, a reduction of one variable may be compensated in part by increase of another variable (Fig. 103.3):

- The term *"ischemia"* (Fig. 103.3a) means critically reduced blood flow ($\dot{Q}$) leading to ischemic (or stagnant or venous) anoxia. Upon reduction of $\dot{Q}$, $(Ca - Cv)_{O_2}$ is increased by reduction of Cv_{O_2} until a critical level is reached (i.e., anoxic areas develop, according to the Krogh cylinder model in the periphery of the venous end). The disturbance can be local (thrombosis, embolism, arterial stenosis) or generalized (cardiac failure, acute blood loss), but even in the latter case the failure of particular organs is at the focus of interest because of their particular sensitivity to anoxia or their vital importance. With microcirculatory derangements, the *distribution* of tissue blood flow appears to be disturbed, leading to local anoxia even though overall blood flow may be normal or even increased.
- *"Hypoxemia"* (Fig. 103.3b) or "arterial hypoxemia" designates reduced arterial C_{O_2} and S_{O_2} due to reduction of arterial P_{O_2}. When venous P_{O_2} reaches a critical level, hypoxemic anoxia develops. Hypoxemia is an important consequence of reduced effective alveolar ventilation or

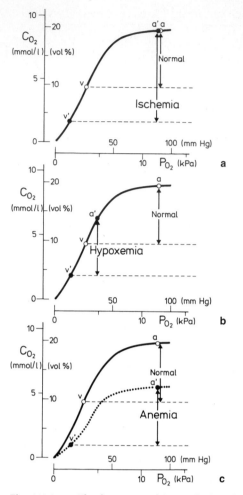

Fig. 103.3a–c. The three types of O_2 supply disturbances leading to tissue hypoxia, visualized in blood O_2 dissociation curves. In ischemia (**a**), decrease of blood flow leads to a drop in venous C_{O_2} (on the basis of the Fick principle). In hypoxemia (**b**), the decrease in venous C_{O_2} parallels that of arterial C_{O_2}. In anemia (**c**), venous C_{O_2} is decreased because the whole O_2 dissociation curve is depressed. When venous P_{O_2} is sufficiently decreased in any of the types of disturbance, the pressure head becomes insufficient for O_2 diffusion and anoxia develops

disturbed pulmonary gas exchange function (e.g., decreased alveolar ventilation and increased ventilation/perfusion inequality in chronic obstructive pulmonary disease, or reduced diffusing capacity in pulmonary fibrosis). In these cases hypoxemia is often accompanied by hypercapnia. At *high altitude* arterial P_{O_2} is reduced, but the hypoxia is associated with hypocapnia (due to increased alveolar ventilation). O_2 delivery in chronic high altitude hypoxia is compensated by an increase in red blood cell count and hematocrit (polycythemia), which may lead to normalization of Ca_{O_2}.

- *Anemia* (Fig. 103.3c). In anemia the O_2 capacity is reduced due to decreased hemoglobin concentration. Anemia may be due to acute or chronic bleeding, or to iron deficiency, and it occurs in various diseases. *Functional* anemia with normal or even increased total hemoglobin concentration is present in CO intoxication and in methemoglobinemia. In these cases O_2 transport is additionally disturbed by a left-shift and flattening of the blood O_2 dissociation curve. A compensatory increase in blood flow is facilitated by reducing blood viscosity.

103.2.5 Venous and Tissue O_2 Pressure [17]

Venous P_{O_2} of organs and mixed venous P_{O_2} are considered as indices of the state of O_2 supply, since venous P_{O_2} should reflect tissue P_{O_2}. This follows from the Krogh cylinder or other related models.

In many cases, this may be accepted, but it should be regarded with a certain skepticism. According to the Krogh cylinder model tissue, P_{O_2} varies from arterial P_{O_2} (at the inflow to the capillary) to the lowest value, which can be zero (in the periphery of the outflow end), and indeed measurements of tissue P_{O_2} using P_{O_2} microelectrodes, inserted into a muscle or pressed on to the surface of a muscle, do display a wide scatter of P_{O_2} ranging from arterial P_{O_2} down to very low values, appreciably lower than venous P_{O_2}.

This variation of P_{O_2} is not to be attributed to radial or longitudinal pericapillary P_{O_2} gradients. Rather, it should be attributed to *local variation* of blood supply as related to O_2 demand in neighboring supply units. Indeed, there is evidence for considerable spatial and temporal heterogeneity of blood flow in apparently homogeneous tissues (skeletal muscle, myocardium). This has been shown by inhomogeneous distribution of arterially infused embolizing microspheres, by non-monoexponential washout of inert gases, and other methods [12]. [It is of interest to note the parallelism between the unequal distribution of alveolar ventilation to blood flow ($\dot{V}_A / \dot{Q}$) in lungs (see above) and that of blood flow to O_2 demand in an organ]. Deficient O_2 supply in severely ill patients (e.g., with sepsis or multitrauma) appears to be mainly due to *maldistribution* of blood flow, with functional shunting, as derived from normal or even increased blood flow, but impaired O_2 extraction.

103.3 Oxygen Debt and Anaerobic Metabolism in Muscular Exercise [3]

After the onset of light or medium constant power exercise (e.g., in an individual on a treadmill or bicycle ergometer; or in rhythmically stimulated isolated blood-perfused muscle preparations) O_2 uptake increases gradually, both by increase of blood flow (in the whole body: increased cardiac output) and widening of the arteriovenous difference in O_2 content, which is mainly due to a decrease of venous O_2 content. Relatively constant values are reached in 1–2 min, indicating that the increase in O_2 uptake is equivalent to the O_2 requirement. The O_2 uptake that is "lacking" by reference to what would be expected on the basis of mechanical performance is termed O_2 *deficit* or *incurred O_2 debt* (Fig. 103.4). The decrease of venous O_2 content and the concomitant drop in P_{O_2} in tissue yield an extra amount of O_2 (depletion of O_2 store) that can be used for metabolism. However, the greater part of O_2 deficit is attributable to the energy made available by the exergonic hydrolysis of high-energy phosphates (mainly creatine phosphate, but also to a lesser degree ATP; Fig. 103.5). It is the O_2 deficit that enables us to perform sudden work at high intensity, since increase of O_2 availability by increased blood flow and opening of capillaries requires some time. After the end of exercise, O_2 uptake decreases toward the resting value in a delayed manner. The O_2 uptake apparently in excess of the resting O_2 requirement is called *(repaid) O_2 debt* (Figs. 103.4, 103.5). The excess O_2 is needed to restore the high-energy phosphates and to replenish the O_2 stores. The O_2 debt repaid is larger than the O_2 deficit. This may be attributable to limited efficiency of the O_2 debt mechanisms, but also to other "repair" processes and to the generally higher body (muscle) temperature after exercise.

In strenuous exercise, *lactic acid* is produced. In this case the O_2 deficit is in part attributable to energy gained from anaerobic glycolysis, and O_2 debt repayment is associated with the energy required for lactic acid elimination, which occurs in part by oxidation (in both muscle and liver), and

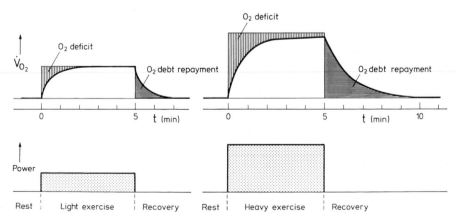

Fig. 103.4. Time course of O_2 uptake with exercise of medium intensity (*left*) and at maximum intensity (*right*). O_2 deficit and O_2 debt repayment result from the delay between work performed and O_2 uptake. The continuously growing O_2 deficit during very heavy exercise is based on energy derived from anaerobic glycoysis

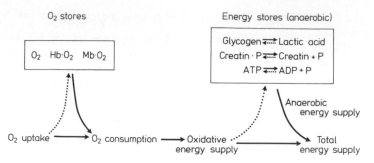

Fig. 103.5. Model of oxidative and anoxidative energy supply underlying the interpretation of O_2 deficit and O_2 debt. In skeletal muscle, the changes in ATP concentration are usually very small, being "buffered" by the creatine/creatine-phosphate system in which large changes occur. On the basis of the underlying bio-chemical mechanism, O_2 debt is partitioned into an alactacid component (corresponding to changes in high-energy phosphates) and a lactacid component (due to lactic acid formation and elimination) The latter is prominent in very heavy exercise

in part by conversion to glycogen (mainly in liver but also in muscle) with subsequent release of glucose, which is stored as glycogen in the muscle (Cori cycle). The protons dissociating from the lactic acid lead to metabolic acidosis and an increase in CO_2 output, measured as an increase in expired R [2].

103.4 Anoxia, Cell Function, and Cell Death [7]

When the O_2 supply suddenly completely stops or is reduced well below the critical level, a number of changes in organ cell function are initiated, allowing the definition of some characteristic times indicating tolerance, or sensitivity, to O_2 deficiency:

- For a short time after initiation of anoxia the function is unchanged: this is the *free interval*. During this time, the energy stores (O_2 physically dissolved and chemically bound in tissue and blood, high-energy phosphates, and glycogen hydrolyzable to lactic acid) are adequate to maintain a metabolic rate sufficient to preserve cell functions.
- *Paralysis time* is the time to complete loss of function (paralysis). During this period the functions of cells and organs deteriorate progressively, but after reestablishment of O_2 availability these functions are fully recovered. For this, a *recovery time* is needed.
- *Reanimation time* is the period during which "reanimation" is possible. Beyond this time recovery with damage is initially possible, then cell death (see Chap. 124) becomes unavoidable.

Paralysis and reanimation times *vary greatly* and are influenced by:

- The energy requirement
- O_2 stores
- Sources of anaerobic energy
- Buffering power

- The shortest times are found for the brain cells: 10–20 s paralysis time, 8–10 min reanimation time; whereas in the heart paralysis time is 6–10 min and reanimation time 30 min. The times in skeletal muscle, for example, are long. What is decisive for the fate of the organism is the vital importance of the affected organ (nervous centers of vital functions; heart).

Low temperature (*hypothermia*) greatly reduces the energy requirement of tissues and organs: a 10°C drop in temperature reduces the energy requirement by a factor of 2 to 3 (according to the *reaction rate-temperature rule*). The anoxia tolerance time are accordingly extended by hypothermia. This is the physiological basis of hypothermia in open heart surgery and of conservation at low temperatures of organs for transplantation.

103.5 Hyperoxia and Oxygen Toxicity [4,5,16]

Hyperoxia, i.e., an increase of P_{O_2} above normal values ($PI_{O_2} > 150$ mmHg, $PA_{O_2} > 100$ mmHg), usually brought about by an increase in effective FI_{O_2}, may be useful in the treatment of arterial hypoxia, but caution is required because of potential harmful effects (O_2 toxicity). Oxygen at high partial pressures is toxic due to the formation of various toxic intermediates and by-products (many of which are free radicals, i.e., have a single, unpaired electron) in cell metabolism.

In oxidative energy metabolism, O_2 is mainly reduced directly to H_2O by cytochrome oxidase, by the uptake of four electrons (e^-): $O_2 + 4e + 4H^+ \rightarrow 2H_2O$. However, a small fraction of O_2 undergoes stepwise partial reduction, giving rise to superoxide anion radical (O_2^-), hydrogen peroxide (H_2O_2), and hydroxyl radical ($\cdot OH$). These intermediates, and the activated O_2 molecule, singlet O_2, are highly reactive and toxic, oxidizing and disrupting biological molecules. Their concentrations are controlled by specific enzymes (superoxide dismutase, catalase, peroxidases) and "quenching" substances (glutathione, vitamins A, C, E), termed collectively "antioxidant defense."

At elevated P_{O_2} (and in other conditions), the production of these toxic intermediates is enhanced, the normal antioxidant defense mechanisms become insufficient, and tissue damage is the result.

Therapeutic administration of O_2 should be done sparingly and cautiously, by increasing effective FI_{O_2} only so much as necessary for achieving desired levels of P_{O_2} in arterial blood. Pure O_2 should not be given for longer than a few hours at a time. The highest tissue P_{O_2} with inspired hyperoxia is in the lungs, where damage leads to edema and atelectasis, with disturbance of gas exchange function. Treatment with hyperbaric oxygen (P_{O_2} above 1 atmosphere, produced in a pressure chamber) is indicated in cases of anaerobic bacterial infection, decompression sickness (intravascular bubble formation, see Sect. 102.5) or CO poisoning (see Sect. 101.8.3), but requires particular precaution because of the anticipated toxic effects on the central nervous system.

References

1. Astrand PO, Rodahl K (1986) Textbook of work physiology. McGraw-Hill, New York
2. Beaver WL, Wasserman K, Whipp BJ (1986) Bicarbonate buffering of lactic acid generated during exercise. J Appl Physiol 60:472–478
3. Cerretelli P, di Prampero PE (1987) Gas exchange in exercise. In: Farhi LE, Tenney SM (eds) Handbook of physiology, section 3: respiratory system, vol 4: gas exchange. American Physiological Society, Bethesda, pp 297–339
4. Forman HJ, Fischer AB (1981) Antioxidant defenses. In: Gilbert DL (ed) Oxygen and living processes, Springer, Berlin Heidelberg New York, pp 235–249
5. Gilbert DL (1981) Oxygen and living processes. An interdisciplinary approach. Springer, Berlin Heidelberg New York
6. Groebe K, Thews G (1990) Calculated intra- and extracellular P_{O_2} gradients in heavily woking red muscle. Am J Physiol 259:H84–H92
7. Grote (1989) Tissue respiration. In: Schmidt RF, Thews G (eds) Human physiology. Springer, Berlin Heidelberg New York, pp 598–612
8. Honig CR, Gayeski TEJ, Federspiel W, Clark A Jr, Clark P (1984) Muscle O_2 gradients from hemoglobin to cytochrome: new concepts, new complexities. Adv Exp Med Biol 169: 23–38
9. Kleiber M (1961) The fire of life. An introduction to animal energetics. Wiley, New York
10. Kreuzer F (1970) Facilitated diffusion of oxygen and its possible significance: a review. Respir Physiol 9:1–30
11. Kreuzer F (1982) Oxygen supply to tissues: the Krogh model and its assumptions. Experientia 38:1415–1426
12. Piiper J (1990) Unequal distribution of blood flow in exercising muscle of the dog. Respir Physiol 80:129–136
13. Piiper J, Scheid P (1986) Cross-sectional P_{O_2} distributions in Krogh cylinder and solid cylinder models. Respir Physiol 64:241–251
14. Schumacker PT, Cain SM (1987) The concept of a critical oxygen delivery. Intensive Care Med 13:223–229
15. Stainsby WN, Otis AB (1964) Blood flow, blood oxygen tension, oxygen uptake, and oxygen transport in skeletal muscle. Am J Physiol 206:858–866
16. Taylor AE, Matalon S, Ward P (eds) (1986) Physiology of oxygen radicals. American Physiological Society, Bethesda
17. Tenney SM (1974) A theoretical analysis of the relationship between venous blood and mean tissue oxygen pressure. Respir Physiol 20:283–296
18. Wagner PD (1988) An integrated view of the determinants of maximum oxygen uptake. In: Gonzalez NC, Fedde MR (eds) Oxygen transfer from atmosphere to tissue. Plenum, New York, pp 245–256
19. Weibel ER (1984) The pathway for oxygen. Structure and function in the mammalian respiratory system. Harvard University Press, Cambridge MA

104 Pulmonary and Bronchial Circulation

N.C. STAUB and C.A. DAWSON

Contents

104.1 General Hemodynamics

Pulmonary blood flow is the denominator of the ventilation-perfusion ratio, $\dot{V}_A/\dot{Q}$, and is of equal importance with breathing in determining the overall efficiency of pulmonary gas exchange (Table 104.1). A major hemodynamic difference between the systemic and pulmonary circulations [4,6] is that the entire cardiac output passes through the lungs – which has only about 1% of the body mass – and it does so with a mean pulmonary arterial-to-left atrial driving pressure that is less than one-tenth that of the systemic circulation. The difference is that in the pulmonary circulation the small muscular arteries and arterioles, both by their enormous number and their low basal smooth muscle tone, contribute much less to total flow resistance than their systemic counterparts.

In certain conditions, both physiologic (living at high altitude where the inspired air has a low O_2 partial pressure) and pathologic (primary pulmonary hypertension), the lung's arterioles may become as well muscularized as those of the systemic circulation, and the pressure in the pulmonary artery may approach that in the aorta. Also in the fetus (for fetal circulation, see Chap. 89), the pulmonary arterioles are well muscularized and actively constricted, so that nearly all of the right ventricular output is shunted to the systemic circulation through the *ductus arteriosus*.

After birth, the pulmonary vascular bed dilates, partly because of expansion of the lung, but mainly because of the onset of air breathing, which raises alveolar oxygen tension, the main factor regulating pulmonary vascular smooth muscle tone and thereby pulmonary vascular resistance.

104.1.1 Structure and Function of the Pulmonary Circulation

The normal volume of the pulmonary vascular bed in man (main pulmonary artery to left atrium) is about 500 ml: 10% of the total circulating blood volume. Thus, in life the lung is 40–50% blood by weight. The large pulmonary vascular volume serves as a capacitance reservoir for the left atrium. During inspiration, as pleural pressure falls, the right ventricle receives more blood because systemic venous return is increased, whereas the left ventricle ejects less blood because the left ventricle is in a lower pressure chamber (increased afterload). During expiration the opposite occurs. Thus, the ventricular stroke volumes are out of phase. The pulmonary blood volume is important for maintaining left ventricular filling.

At rest in humans, the pulmonary vascular volume is distributed approximately equally between the arterial and venous vessels, with about 75 ml in the capillaries, where it is spread out in a vast array of thin-walled interconnecting vessels (Fig. 104.1, inset B). During exercise the capillary blood volume may more than double as a result of distention and recruitment of vessels (Fig. 104.1, inset C). The maximum anatomic capillary volume in humans is about 200 ml (Fig. 104.2). The endothelium of the capillaries over the surface exposed to alveolar gas is only about 0.1 μm thick, which helps to optimize oxygen diffusion between the alveolar gas and the hemoglobin in the red blood cells. The total capillary surface area for gas exchange is >70 m² (40 times body surface area). At rest the erythrocytes spend about 0.75 s (one cardiac cycle) in the capillaries, which is more than adequate for O_2 and CO_2 equilibration between alveolar gas and blood.

104.1.2 Pressure, Flow, and Resistance

Table 104.1 lists measured and estimated mean pressures in the human pulmonary circulation in an adult at rest

R. Greger/U. Windhorst (Eds.)
Comprehensive Human Physiology, Vol. 2
© Springer-Verlag Berlin Heidelberg 1996

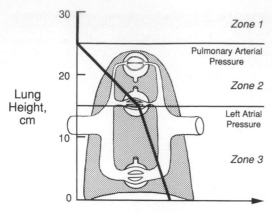

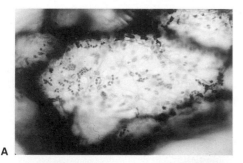

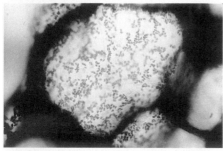

Fig. 104.1. Passive effects of gravity on normal pulmonary circulation in the standing position. Height of lung is about 24 cm at function residual capacity. There are three zones. In zone 1 there is no flow because pulmonary arterial and left atrial pressures are less than alveolar pressure. The capillaries are compressed (*inset A*: microphotograph of a single alveolar wall). In zone 2 pulmonary arterial pressure is above but the effective left atrial pressure is still below alveolar pressure. There is increasing flow going down zone 2 as left atrial pressure rises until it equals alveolar pressure. More capillaries (*inset B*) are recruited, In zone 3 both arterial and left atrial pressures exceed alveolar pressure. Nearly all capillaries are recruited and distended (*inset C*). Flow increases going down the zone, as transmural distending pressure continues to rise

Table 104.1. Measurements or estimates of mean pressures (cmH$_2$O) in normal human pulmonary circulation[a]

	Rest		Exercise Standing
	Supine	Standing	
Pulmonary artery	19	16	32
Arterioles	16	11	22
Capillaries	14	9	20
Venules	12	8	18
Left atrium	11	4	13
Driving pressure (cmH$_2$O)	8	12	19
Cardiac output (l/min)	7	6	18
Resistance (cmH$_2$O × min/l)	1	2	1
Resistance distribution (%)			
Arterial	38	42	53
Capillary	50	25	21
Venous	10	33	26

[a]Pulmonary arterial and left atrial (arterial wedge) pressures were measured at rest and during moderate exercise supine during cardiac catheterization. Standing pressures are based on the decreases in cardiac output and right atrial pressure or extrapolated from animal data

(lying down or standing) and during exercise. Cardiac output normally is about 5–6 l/min at rest. During submaximal steady-state exercise it may increase more than threefold to 18 l/min, and even higher flows may be obtained briefly during exhausting exercise.

At rest in the supine position the driving pressure (pulmonary arterial-to-left atrial pressure difference)averages only 8 cmH$_2$O (6 mmHg), compared to 90 mmHg for the systemic circuit. Therefore, pulmonary vascular resistance (driving pressure divided by cardiac output) is only about 7% of that in the systemic vascular bed [8].

The distensibility of the pulmonary vascular bed allows the resistance to decrease further to accommodate rises in cardiac output, while maintaining a relatively low driving pressure (Table 104.1, column 3). In addition, the main pulmonary arteries have a large compliance which accommodates the right ventricular stroke volume without generating excessively high systolic pressures. This helps to make an efficient match between the right ventricle pump and the load imposed by the pulmonary circulation. One consequence of the low pulmonary arterial resistance is that pulmonary capillary flow is highly pulsatile.

104.1.3 Pressure-Flow Curves

One way to achieve an overall view of pulmonary hemodynamics is to study pressure-flow curves under various conditions, as shown schematically in Fig. 104.3. Vascular resistance is defined as driving pressure divided

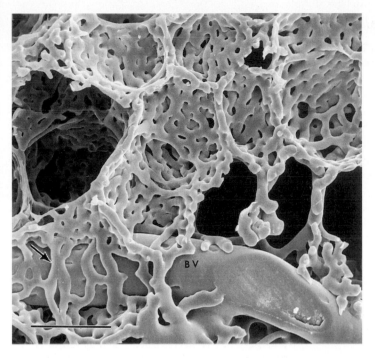

PRESSURE-FLOW CURVES

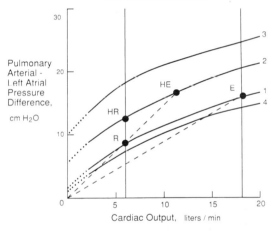

Fig. 104.3. Representative pressure-flow curves of the human pulmonary circulation. *Curve 1* is normal. *R* is the resting condition and *E* is submaximal exercise at cardiac outputs of 6 and 18 l/min, respectively. Resistance is the slope of a straight line from the origin to a point on the curve. *Curve 2* represents hypoxic pulmonary vasoconstriction. For any blood flow (for example, *HR*) resistance is increased. During exercise resistance decreases (*HE*), but it is always greater than when breathing normal air. *Curve 3* represents continuous positive pressure breathing (15 cmH$_2$O), which compresses the alveolar wall microvessels and increases resistance (puts more of the lung in zone 2). *Curve 4* represents a normal person breathing 100% O$_2$. The slight downward displacement shows that there is little hypoxic vasoconstriction in the normal human breathing air at sea level. The low flow ends of the pressure-flow curves (*dotted*) are not usually measured

by flow. Three examples are represented in Fig. 104.3 by the slopes of the dashed lines from the origin to particular points on any one of the family of pressure-flow curves. The shape of the pressure-flow curves is due to the disten-

sibility or recruitment of vessels. On curve 1 (normal), the resistance during exercise (E) is less than at rest (R). Alveolar hypoxia causes vasoconstriction (curve 2), so there is a higher resistance at any given flow (for example, at rest, HR). During combined exercise and alveolar hypoxia, the resistance still may decrease, as shown by point HE (where the resistance in this example is fortuitously the same as at rest breathing air (R). This points out the potential complexity which distensibility and recruitability of vessels introduces into the interpretation of pulmonary vascular resistance. Continuous positive pressure breathing (curve 3) may also increase resistance at a given flow because the increased alveolar pressure compresses the microvessels in the alveolar walls. Curve 4 (breathing 100% O$_2$) shows only slight vasodilation, indicating that normally alveolar O$_2$ partial pressure (100 mmHg) does not cause much active vasoconstriction.

104.2 Special Hemodynamics

104.2.1 Shunts Between the Right and Left Heart

Venous Admixture (Right-to-Left Shunt). An important functional test of the adequacy of the pulmonary circulation is to determine what fraction of the cardiac output *effectively* flows through pulmonary capillaries (exchanges with "ideal" alveolar gas) and what fraction bypasses the lungs to enter the systemic arteries unoxygenated ("shunt-like effect"). Venous admixture includes true anatomical shunts, such as when venous blood flows directly from the right side to the left side of the circulation or intrapulmonary flow through portions of the lung which receive blood flow but do not receive adequate ventilation (for example, collapsed or nonventilated alveoli).

A small amount of venous admixture is normal because some systemic venous blood reaches the left ventricle by way of the bronchopulmonary venous anatomoses and the intracardiac *thebesian veins.* This normally amounts to 1–2% of the cardiac output. Venous admixture always reduces systemic arterial O_2 tension and concentration (increases the alveolar-arterial O_2 tension difference). Breathing 100% O_2 increases the alveolar and arterial O_2 partial pressure difference but it makes calculating venous admixture easier. A quick estimate of the shunt fraction in a subject breathing 100% O_2 can be made as follows: venous admixture equals about 1% of cardiac output for each 20 mmHg difference between alveolar and arterial O_2 tensions, if the shunt is less than 20% of cardiac output.

Venous admixture reduces the efficiency of blood flow. It is the blood flow equivalent of wasted ventilation. Not only does more blood need to be pumped to maintain the same arterial O_2 delivery but the O_2 is delivered to the systemic tissue capillaries at a lower partial pressure.

Left-to-Right Shunt. Since these shunts do not affect the systemic arterial O_2 tension, they are more difficult to quantify. Because some intracardiac flow is reversed, a left-to-right shunt is always *in addition to* the systemic flow. Pulmonary blood flow may exceed systemic blood flow by twofold in congenital heart diseases. With a left-to-right shunt, pulmonary arterial O_2 partial pressure no longer represents the O_2 tension of the systemic venous blood returning to the heart via the inferior and superior venae cavae and the coronary sinus.

The characteristic finding at cardiac catheterization in a person with a left-to-right intrathoracic shunt is a step increase in the O_2 concentration of blood somewhere on the right side of the heart. When one knows which chamber has the increase, one knows the anatomy of the shunt and can estimate its size by calculating the shunt flow.

Venous admixture occurs rather commonly in pulmonary disease. Left-to-right shunts, on the other hand, are not common and are usually caused by congenital heart malformations. Left-to-right shunts do not occur in pulmonary disease.

104.2.2 Distribution of Blood Flow

The average pressures shown in Table 104.1 are measured with respect to the level of the heart (left atrium). Just as in the systemic circulation so in the pulmonary circulation the hydrostatic pressure within the vessels is influenced by gravity. However, because pulmonary vascular pressures are much lower and the thin-walled capillaries are surrounded by air, which is much less dense than blood, the effects of gravity are relatively greater in the pulmonary circulation [11]. For example, in the upright position at functional residual capacity (FRC; Table 104.1, column 2)

the top and the bottom of the lungs are each about 12 cm from the left atrium (Fig. 104.1). Since vascular pressures change by about 1 cmH$_2$O/cm height (blood density is close to one), the mean effective pulmonary arterial pressure at the bottom is 28 cmH$_2$O, and left atrial pressure is 12 cmH$_2$O. At the top the pressures are 4 and −8 cmH$_2$O, respectively. Both arterial and venous outflow (left atrial) pressures are affected equally, so the pulmonary arterial – left atrial pressure difference remains constant at 16 cmH$_2$O.

Since all of the vessels are *distensible,* alterations of pulmonary arterial or left atrial pressure changes the distribution of blood flow over the height of the lung. As the transmural pressure increases going down the lung, all vessels tend to be more distended the closer they are to the base. The effect is most obvious in the capillaries (Fig. 104.1; compare insets A, B, and C).

Using the model in Fig. 104.1 to represent the pulmonary circulation, it is possible to explain the distribution of blood flow over the height of the lung. Before doing so, however, the difference between "alveolar" and "extra-alveolar" vessels needs to be defined. In Fig. 104.1, the extra-alveolar vessels are represented by the arteries and veins, whereas the alveolar vessels are mainly capillaries. The pressure outside the alveolar vessels is normally atmospheric (relative pressure = 0). The pressure outside the extra-alveolar vessels, however, is closely related to pleural pressure; that is, below atmospheric pressure (mean pleural pressure = −5 cmH$_2$O at FRC). When pleural pressure falls during inspiration, the pressure outside the extra-alveolar vessels decreases. This causes the extra-alveolar vessels to enlarge (decreasing their contribution to flow resistance and increasing pulmonary blood volume). On the other hand, alveolar pressure does not vary much with breathing.

As one moves up the lung, the effective left atrial pressure may fall below alveolar pressure. The transmural pressure across the small veins becomes negative; that is, it becomes a compressive pressure at the outflow from the microvascular network. As the blood vessels are compressed, vascular resistance increases. The compression by the alveolar pressure is an example of a Starling resistor, which acts to regulate flow independently of left atrial pressure.

The uneven distribution of blood flow on a gravitational basis is divided into three zones depending on the relative values of pulmonary arterial, venous, and alveolar pressures, as depicted in Fig. 104.1.

In zone 1 flow is zero because both pulmonary arterial and venous pressures are less than alveolar pressure. Thus, the alveolar vessels are compressed (Fig. 104.1, inset A).

Throughout zone 2, pulmonary arterial pressure exceeds alveolar pressure, while alveolar pressure exceeds the venous outflow (effective left atrial) pressure. At the top of zone 2 (border between zones 2 and 1), flow is zero but commences and rises steadily as one goes down the zone because the compressive transmural pressure is decreasing and more capillaries are recruited. At the lowest part of zone 2 (border between zones 2 and 3), the effective venous outflow pressure equals alveolar pressure, so the compressive force disappears (entering zone 3).

In zone 3 flow continues to increase slowly because the transmural distending pressure increases as one moves down toward the base of the lungs.

Although all three zones may exist in the human lung under some conditions, normally pulmonary arterial pressure is high enough that zone 1 does not occur, and even zone 2 is limited to the upper third or less of the lung in the standing position (Table 104.1, column 2).

From the clinical point of view, however, during mechanical ventilation, alveolar pressure is increased because positive pressure is applied to the airways. This increases the amount of lung in zone 2 (alveolar pressure rises relative to pulmonary venous pressure), which increases resistance to blood flow. See the pressure-flow curve 3 in Fig. 104.3.

When pulmonary vascular pressures are increased, as in exercise (Table 104.1, column 3), distribution of blood flow over the height of the lung becomes more uniform because both arterial and left atrial pressures rise.

104.3 Regulation of Pulmonary Blood Flow

104.3.1 Longitudinal Distribution of Resistance

Recent measurements using the micropuncture technique have revealed that in the normal range of tidal breathing the alveolar wall capillaries contribute about 40% of the total resistance, the arteries about 50%, and the veins about 10% [1]. Table 104.1 shows the average pressure difference across the lung to be $19 - 11 = 8\,cmH_2O$ in the supine position at rest. An approximately 4-cmH_2O pressure drop occurs in the alveolar wall capillaries. This is a geometric (structural) effect, since there is little evidence for active regulation of capillary dimensions. The fraction of resistance within the alveolar wall vessels is sensitive to:

- Lung volume
- Blood flow
- Vascular distending pressure.

For example, near total lung capacity the fraction of resistance within the capillaries may be 75%. In the upright position and in exercise the fraction of resistance within the capillaries is less than 25%, as Table 104.1 shows.

104.3.2 Passive Regulation

Although passive regulation is largely dependent on blood flow and gravitational (body position) effects on the vascular transmural pressures in the three lung zones, it is important to mention the passive mechanical effects of the large changes in pulmonary blood flow which occur during exercise. When humans do submaximal sustained exercise, cardiac output may rise threefold. This increase in blood flow is accommodated by the pulmonary circulation without an equivalent rise in the pulmonary vascular driving pressure, due partly to recruitment of microvessels caused by the increasing transmural pressure in the very small vessels (arterioles, capillaries, and venules). For example (Table 104.1, column 3) in exercise, cardiac output increases threefold but the driving pressure across the lungs increases only about 60%, which means that the calculated pulmonary vascular resistance must have decreased by 50%.

104.3.3 Active Regulation

Although the passive effects of pressure on the distensible pulmonary vascular bed are prominent, active regulation does occur. Contrary to what is sometimes stated, there is adequate smooth muscle associated with the small muscular pulmonary arteries, arterioles, and veins to substantially alter pulmonary vascular resistance. The smooth muscle may hypertrophy remarkably in pathologic conditions. Many chemical agents will provoke selective constriction of the pulmonary arterial or venous vessels.

Alveolar Oxygen Tension. The most important active regulator of the pulmonary vascular smooth muscle tone is the alveolar O_2 tension. The small muscular pulmonary arteries and veins are surrounded by the alveolar gas of the alveoli they subserve. Ordinarily, the alveolar O_2 tension is high (100 mmHg). Thus, the smooth muscle cells of microvessels in or near the alveolar walls are exposed to the highest O_2 tensions of any organ.

Low alveolar O_2 tension causes local constriction of the nearby arterioles. This response is opposite to that in the systemic circulation, wherein low O_2 tends to relax the local resistance vessels. The cellular mechanisms that account for the different responses of the pulmonary and systemic vessels to hypoxia are not known.

When there are areas of poor ventilation due to, for example, disease-induced changes in local lung mechanics, the local alveolar O_2 tension falls. The adjacent arterioles constrict, decreasing local blood flow and thereby shifting flow to better-ventilated regions. Changes in local vascular resistance can cause marked shifts of blood flow without any significant effect on overall pulmonary vascular resistance, provided the amount of lung involved is not too large (say, 20% of total lung mass).

On the other hand, a reduction in alveolar O_2 tension throughout the entire lung, such as occurs at high altitude or in a person breathing low O_2 mixtures, leads to a rise in pulmonary arterial pressure. Pulmonary vascular resistance in acute alveolar hypoxia may be more than twice normal. The effect of alveolar O_2 tension on pulmonary vascular resistance is chiefly on the arterioles and small muscular arteries. In many types of lung pathology, hypoxic vasoconstriction is a major component. With prolonged alveolar hypoxia, such as in high-altitude residents or patients with chronic pulmonary disease, the vascular smooth muscle may hypertrophy, causing an additional mechanical obstruction of the vessels as the thickened vessel walls encroach on the vessel lumina. In lung disease this may be severe enough to cause pulmonary hypertension and right heart failure (*cor pulmonale*). An important clinical test of the vasoconstrictor component of pulmonary hypertension is to determine the effect of O_2 inhalation on pulmonary vascular resistance.

Some disease processes, such as pneumonia, can block local hypoxic vasoconstriction. This contributes to the fall in systemic arterial O_2 tension (increased venous admixture) by eliminating the mechanism of flow diversion from the poorly ventilated infected regions.

Thromboxane and Prostacyclin. A number of naturally occurring substances affect the vasomotor tone of the pulmonary circulation [7]. Some are normally constrictors (thromboxane A_2, α_1-adrenergic catecholamines, histamine, angiotensin II, several prostaglandins, neuropeptides, leukotrienes, serotonin, endothelin, and increased CO_2 partial pressure) and some are normally dilators (β-adrenergic catecholamines, prostacyclin, endothelial-derived relaxing factor (nitric oxide), acetylcholine, bradykinin, and dopamine) [3].

Many of the vasoactive compounds are metabolic products of the arachidonic acid cascade (cf. Chaps. 5, 6A). Arachidonic acid is a component of cell membranes and therefore ubiquitous in the body. It is readily metabolized by well-known intracellular pathways into several potent vasoactive compounds which diffuse into the surrounding media. One of the most powerful known constrictors of pulmonary arterial and venous smooth muscle is thromboxane A_2. It is produced within the pulmonary circulation in several types of acute lung injury. Many cells, including leukocytes, macrophages, and endothelial cells, are capable of producing and releasing thromboxane. As with regional hypoxia, the effects are mainly localized to the segment of lung where the thromboxane is released, although recirculation and diffusion through the lung interstitum sometimes spread the effects to airways and other vessels.

Prostacyclin (prostaglandin I_2), another product of arachidonic acid metabolism, is a potent vasodilator. Endothelial cells are probably the main source of this substance. Some physiologists hypothesize that a balance between thromboxane and prostacyclin production regulates normal pulmonary vascular tone, but this idea remains controversial, since the production of either thromboxane or prostacyclin is normally low. Clearly prostacyclin would have to dominate under normal conditions, because the normal smooth muscle tone of the pulmonary vascular bed is low (compare lines 1 and 4 in Fig. 104.3). Probably the main physiologic function of prostaglandin I_2 is its antiplatelet effect, that is, preventing platelet adhesion to endothelium.

104.3.4 Absence of Autoregulation

In some organs the arteriolar resistance vessels have the property of constricting when the transmural pressure rises, which opposes passive dilation due to the increased transmural distending pressure (cf. Chap. 73). When transmural pressure decreases, the smooth muscle relaxes. Within limits, this phenomenon, called autoregulation, maintains fairly constant flow. The pulmonary arterioles do *not* autoregulate. This is consistent with the function of the pulmonary vascular bed, which must accommodate a wide range of cardiac outputs from rest to heavy exercise at relatively low pressures.

104.4 Nonrespiratory Functions of the Pulmonary Circulation

104.4.1 Liquid and Solute Exchange

According to the Starling equation,

$$\dot{Q} = L \times S \left[(P_{mv} - P_{pmv}) - \sigma(\pi_{mv} - \pi_{pmv}) \right],$$

where $\dot{Q}$ is net filtration, L is the hydraulic conductance/unit area, S is the exchange surface area, P and π are the hydrostatic and protein osmotic pressures in the microvascular (mv) and perimicrovascular (pmv) compartments, respectively, and σ is the reflection coefficient (a fraction between 0 and 1 that describes the protein leakiness of the microvascular barrier). $L \times S$ is commonly referred to as the filtration coefficient.

Even in the lung where the mean capillary pressure at rest is low (Table 104.1) filtration occurs, especially in the lower portions where vascular pressures are higher. The vascular exchange surface includes not only the alveolar wall capillaries but also some of the arterioles and venules which have little smooth muscle.

The lung is endowed with an extensive lymphatic network that efficiently clears the capillary filtrate, $\dot{Q}$. The tight alveolar epithelium forms an additional barrier against protein and electrolyte movement, so liquid does not normally flood the airspaces. The small amount of normal alveolar lining liquid is believed to be a product of active secretion needed to maintain the properties of the alveolar surface.

In disease, lung liquid and protein filtration may increase due either to increased pressure (commonly, left ventricular failure) or to increased permeability (such as due to injury to the endothelial or epithelial barriers). The edema liquid is characterized by, respectively, low or high protein concentration relative to plasma.

In edema [9], the excess filtrate may be incompletely removed by the lymphatics. Considerable quantities of liquid can be stored in the interstitium around the larger blood vessels, airways, and interlobular septae. These "cuffs" of interstitial edema can often be seen in the chest radiograph. Some of the peribronchovascular edema liquid flows into the mediastinum via the loose interstitium at the lung hila or leaks into the pleural space (effusion). Figure 104.4 shows the main pathways for edema liquid flow.

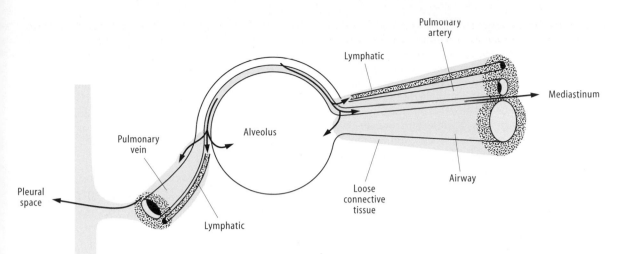

Fig. 104.4. Pathways of liquid and protein flow through the lung's interstitium. Filtration occurs into the alveolar wall perimicrovascular interstitium. Normally, liquid clearance is into the nearby lymphatic capillaries. In edema filtration may be occur too rapidly for the lymphatics to be able to handle all of the liquid. Interstitial edema liquid accumulates in the peribronchovascular interstitium ("cuffs") and leaks into the pleural space (effusion), the mediastinum, or the airspaces (alveolar flooding). (From [10] with permission)

The most serious aspect of edema is when it floods the alveoli, interfering with ventilation and gas exchange. When the edemagenic process subsides, interstitial liquid is rapidly removed with a halftime of a few hours. Clearance is mainly into blood or into the mediastinum, much less into the lymphatics. Alveolar edema is also rapidly reabsorbed by metabolically driven Na^+ or Cl^- transport across the alveolar epithelial cells.

104.4.2 Metabolic Functions of the Endothelium

In addition to its passive role as a barrier, the pulmonary endothelium actively carries out several other functions, many of which are common to vascular endothelium in general [3,7]. However, because the pulmonary vascular endothelium sees all of the systemic venous return, its metabolic functions are of particular interest. The pulmonary endothelium has the ability to take up and inactivate several vasoactive substances coming from various organs. As a result, these substances are removed or reduced in concentration before the blood reaches the systemic arteries. These subtances include prostaglandins of the E and F series, serotonin, norepinephrine, and adenine nucleotides. Recently, removal of atrial natriuretic peptide (ANP; cf. Chap. 75) and endothelin from the blood on their passage through the lungs has been observed. The pulmonary endothelium also has a large supply of angiotensin-converting enzyme (ACE) on its surface so that it plays a major role in the hydrolysis of angiotensin I to form the vasoactive peptide angiotensin II (cf. Chap. 73). ACE also inactivates circulating bradykinin.

104.5 Bronchial Circulation

The bronchial arteries supply water and other nutrients to the mucosal cells and glands of the airways down to and including the terminal bronchioles [2]. They also nourish the pleura, interlobular septal supporting tissues, and the pulmonary arteries and veins. Under normal conditions the alveoli, alveolar ducts, and respiratory bronchioles receive their nutrition via the pulmonary capillaries.

The pressure in the main bronchial arteries is nearly the same as in the aorta so that, regardless of body position, there is a high driving pressure. Although bronchial blood flow is less than 1% of cardiac output, it is appropriate for the portion of the lungs it serves. Bronchial vascular resistance is high, as is appropriate for systemic vessels.

About half of bronchial blood flow returns to the right side of the heart via bronchial veins. the remainder flows through small bronchopulmonary anastomoses (<100 µm diameter) into the pulmonary veins, thus contributing to the normal venous admixture. In certain inflammatory diseases of the airways, the bronchial circulation expands dramatically and may contribute as much as 10%–20% venous admixture.

When there is obstruction in the pulmonary circulation (thrombosis or embolism), the bronchial arteries dilate, enlarge, and develop connections with precapillary vessels of the pulmonary circulation. When this occurs, the blood flowing through the alveolar capillaries is systemic blood, which takes up little oxygen from the alveolar gas but can still give up CO_2. The bronchial circulation supplies sufficient substrate to keep the lung alive when the pulmonary blood flow is shut off, as long as local metabolism is not increased too much by inflammation.

The bronchial circulation participates in conditioning the inspired air, especially under circumstances where air bypasses the upper air passages (for example, breathing through the mouth during exercise). Indeed, it is essentially impossible for cold dry air to be moved fast enough to reach the alveoli without having been adequately warmed and humidified in the air passages.

The bronchial circulation is considerably more angiogenic than the pulmonary circulation in the adult. In adult life, new blood vessel growth in the lung comes primarily from the bronchial circulation. Thus, scars and tumors are commonly supplied by bronchial vessels, even if, in the latter instance, the cancer cells arrived in the lung through the pulmonary circulation.

References

1. Bhattacharya J, Staub NC (1980) Direct measurement of microvascular pressures in the isolated perfused dog lung. Science 210:327–328
2. Deffenbach ME, Charan NB, Lakshminarayan S, Butler J (1987) The bronchial circulation: small, but a vital attribute of the lung. Am Rev Respir Dis 135:463–481
3. Gillis CN (1986) Pharmacological aspects of metabolic processes in the pulmonary microcirculation. Annu Rev Pharmacol Toxicol 26:183–200
4. Grover RF, Wagner WW Jr, McMurtry I, Reeves JT (1984) Pulmonary circulation. In: Shepherd JT, Abboud FM (eds) Handbook of physiology: the cardiovascular system. III. American Physiological Society, Bethesda, pp 103–136
5. Gunteroth WG, Luchtel DL, Kawabori I (1982) Pulmonary micro-circulation – tubules rather than sheet and post. J Appl physiol 53:510–515
6. Harris P, Heath D (1978) The human pulmonary circulation, 2nd edn. Livingston, Edinburgh
7. Hyman AL, Spannhake EW, Kadowitz PJ (1978) Prostaglandins and the lung. Am Rev Respir Dis 117:111–136
8. Mitzner W (1983) Resistance of the pulmonary circulation. *Clin Chest Med* 4:127–137
9. Staub NC (1984) Pathophysiology of pulmonary edema. In: Staub NC, Taylor AE (eds) Edema. Raven, New York
10. Staub NC (1990) Pathways for lung liquid clearance. In: Brigham KL, Stahlman MT (eds) Respiratory distress syndromes: molecules to man. Vanderbilt, Nashville, pp 33–43
11. West JB, Dollery CT, Naimark A (1964) Distribution of blood flow in isolated lung: relation to vascular and alveolar pressures. J Appl Physiol 19:713–724

105 Neural Regulation of Respiration: Rhythmogenesis and Afferent Control

D.W. RICHTER

Contents

105.1 Introduction

The primary function of respiration is to exchange gases between the external environment and the internal milieu of the organism. Gas exchange occurs through coordinated action of the respiratory and cardiovascular systems. In the mammal, the respiratory system controls ventilation of the lung, while the cardiovascular system transports O_2 and CO_2 between the pulmonary and systemic capillaries. These processes adjust to varying physiological circumstances to maintain homeostasis (cf. Chap. 108).

Ventilation of the lungs depends on periodic movements of respiratory muscles. These muscles are innervated by spinal motoneurons that are activated by a rhythm-generating network in the lower brain stem. The activity of this central respiratory network is adjusted continually by sensory feedback from the periphery and by inputs from higher nervous structures, e.g., the motor cortex and the hypothalamus. This regulation of neural respiratory activity requires a high degree of coordination of synaptic interaction and modulation of membrane properties of neurons.

105.2 Respiratory Rhythm and Ventilatory Mechanics

Inspiratory and expiratory neural activities are necessary for normal breathing movements and ventilatory mechanics. These activities control the movements of the diaphragm, the thorax and the abdominal wall (Fig. 105.1). Variation of the intrathoracic volume ultimately determines the pulmonary pressure which causes air to flow in and out of the lungs, i.e., inhalation and exhalation of air. The dimensions of the upper airways – the pharynx, the larynx, the trachea, and the bronchial tree – determine the resistance to air flow and are adjusted through muscles innervated by cranial motoneurons or bronchomotor neurons in the brain stem which are synaptically coupled with the central respiratory network [15,26,61,182].

The two phases of ventilatory mechanics are controlled by three neural phases (Fig. 105.1) [145,148,154]. The neural respiratory phases are:

- Inspiration (I phase)
- Postinspiration (PI phase, passive expiration)
- Expiration (E2 phase, active expiration).

These three phases can be discerned easily in the motor outflow to the inspiratory muscles, such as the diaphragm or external intercostal muscles, and to expiratory muscles, such as internal intercostal or abdominal muscles [26,116]. They are also evident in the activity of laryngeal muscles [15,16]. During the I phase, an augmenting contraction of the diaphragm and inspiratory intercostal muscles expands the thoracic volume, which steadily decreases pulmonary pressure, thus drawing air into the lungs. The augmenting contraction of the diaphragm results from recruitment of phrenic motoneurons and from a ramp-like increase of their discharge, as seen in the phrenic neurogram (Fig. 105.1). Laryngeal abductor muscles are also activated during inspiration and dilate the larynx [15,16]. Their activity has a plateau-type pattern, i.e., the activity quickly reaches maximum levels and remains there throughout the I phase. The energy for exhalation accumulates during inspiration and is stored in the recoil forces of lung tissue. This energy is released during the PI phase when inspiratory movements end. Thus, without activation of expiratory muscles, exhalation occurs "passively"

R. Greger/U. Windhorst (Eds.)
Comprehensive Human Physiology, Vol. 2
© Springer-Verlag Berlin Heidelberg 1996

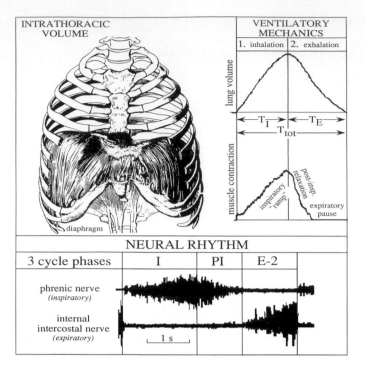

Fig. 105.1. Control of ventilation. Contractions of respiratory muscles produce changes in intrathoracic volume and airway pressure, causing inhalation and exhalation of air. During quiet breathing, the respiratory cycle lasts for several seconds (T_{tot}), the duration of inhalation (T_I) normally being shorter than the duration of exhalation (T_E). The muscles are innervated by respiratory motoneurons, which are controlled by the reticulospinal output of the medullary respiratory network. The central respiratory rhythm passes cyclically through three phases: inspiration (*I*), postinspiration (*PI*) and expiration (*E-2*). These three phases are evident in phrenic and internal intercostal nerve activities. The activity of phrenic nerves and the contraction of the diaphragm increases more or less linearly during the I-phase (inspiratory ramp) and declines slowly during the PI phase (after-discharge) controlling the postinspiratory relaxation of the muscle. During the E2 phase, expiratory muscles are activated by an augmenting pattern of nervous activity in internal intercostal nerves while inspiratory nerves are silent (expiratory pause)

during this initial period of expiration. Passive exhalation of air does not occur rapidly, because the lung volume is held by an actively controlled relaxation of the diaphragm. This can be recognized by a postinspiratory "after-discharge" in the phrenic nerves (Fig. 105.1) [145,150]. Expiratory airflow is retarded additionally during normal breathing and during speech by the contraction of laryngeal adductor muscles (e.g., thyroarytenoid muscle) and loss of activity in laryngeal abductor muscles (e.g., posterior cricoarytenoid muscles), which increases the resistance to expiratory air flow and allows vocalization [16]. During the final period of exhalation, expiratory airflow can be increased by active contraction of expiratory intercostal and abdominal muscles (Fig. 105.1). This phase is called stage 2 of expiration, the E2 phase, following as it does the initial phase of passive exhalation (PI phase or stage 1 expiration). Phrenic and inspiratory intercostal nerve activities are silent during this E2 phase of the respiratory cycle (Fig. 105.1).

During quiet breathing, a total respiratory cycle (T_{tot}) lasts for approximately 3–4 s (Fig. 105.1) and inhalation (T_I) is slightly shorter than exhalation (T_E). During exhalation, the PI phase normally occupies 70%–90% of T_E [168]. The E2 phase can be absent; this occurs

- Occasionally during normal breathing
- During thermoregulation in panting animals
- During activation of the defense area in the hypothalamus [10]
- During lung edema, when pulmonary J receptors (juxtacapillary) are activated (see Table 105.1) [44,131]. Under these conditions, the expiratory interval is shorter and rhythmic breathing occurs in only two antagonistic neural phases, i.e., inspiration and post-inspiration [10,147], resulting in rapid shallow breathing or panting.

105.3 Medullary Respiratory Network

In mammals, the efferent neural activity to the various respiratory muscles is ultimately controlled by a neuronal network located in the lower brain stem, specifically in and near the ambiguous nucleus and the Pre-Bötzinger and Bötzinger complexes, which areas comprise the ventral respiratory group (VRG) [5,22,25,46,56,57,124,164, 171,170]. Presympathetic neurons in the rostroventrolateral medulla [76,110], bronchomotor neurons, cardiac

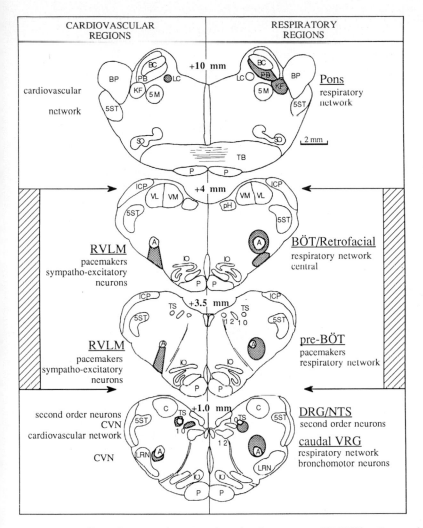

Fig. 105.2. Cardiorespiratory regions. Location of various types of neurons and networks controlling respiratory (*right column*) and cardiovascular (*left column*) functions within the pons and the medulla oblongata of mammals. Specific regions are shown as *shaded areas* and neighboring structures are shown as *white areas*. The location of neurons is indicated unilaterally, but in fact is bilateral, and there is considerable overlap, especially if the dimensions of the dendritic trees of neurons are taken in account. The types of neurons located at the different levels of the brain stem are listed on both sides of the diagram. *BC*, brachium conjunctivum; *BP*, branchium pontis; *LC*, locus ceruleus; *PB*, nucleus parabrachialis; *5M*, motor nucleus of the trigeminal nerve; *KF*, Kölliker-Fuse nucleus; *5ST*, spinal trigeminal tract; *SO*, superior olive; *TB*, trapezoid body; *P*, pyramidal tract; *pH*, nucleus praepositus hypoglossi; *VM*, medial vestibular nucleus; *VL*, lateral vestibular nucleus; *12*, hypoglossal nucleus; *10*, dorsal vagal motor nucleus; *TS*, solitary tract; *ICP*, inferior cerebellar peduncle; *A*, nucleus ambiguus; *IO*, inferior olive; *C*, cuneate nucleus; *LRN*, lateral reticular nucleus; *BöT*, Bötzinger complex, retrofacial nucleus; *pre-BöT*, pre-Bötzinger complex; *DRG*, dorsal respiratory group; *NTS*, nucleus of the solitary tract; *VRG*, ventral respiratory group; *RVLM*, rostroventrolateral medulla; *CVN*, cadiac vagal neurons

vagal neurons [70,84] and central chemosensitive regions [104] all coexist in this area (Fig. 105.2).

The bulbar respiratory network consists of various populations of neurons which are identified by the timing and pattern of their discharge relative to phrenic nerve activity [43,55,59,89,97,113,114,123,126, 127,145,148,162,170]. These patterns of discharge are comparable to the patterns of postsynaptic activities in identified reticulospinal output neurons [145]. Six types of neurons have been classified in in vivo mature mammals (Fig. 105.3) [145,150,154,163,164]. Respiratory neurons are:

- Early-inspiratory (during early I phase)
- Inspiratory (throughout I phase)
- Late-inspiratory (during late I phase)
- Postinspiratory (during PI phase)
- Expiratory (during E2 phase)
- Pre-inspiratory (pre-I phase).

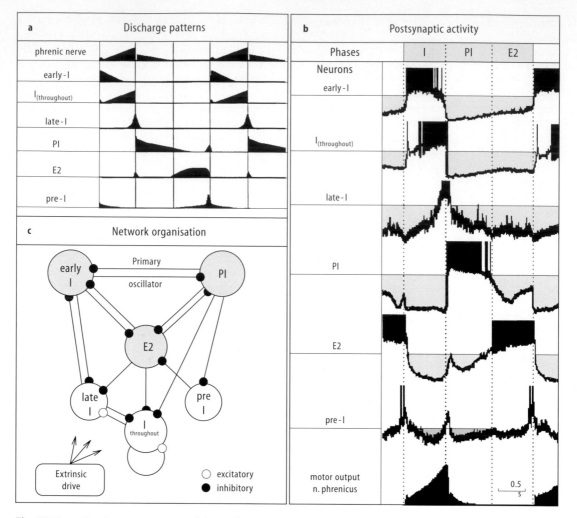

Fig. 105.3a–c. Respiratory neurons and network. Respiratory neurons have been identified by their discharge profile and the temporal relation of the discharge profile to phrenic nerve activity. Three types of neurons discharge during inspiration: early-inspiratory (early-I), inspiratory (I, discharging throughout the inspiratory phase) and late-inspiratory (late-I) neurons. Postinspiratory (PI) neurons discharge during the after-discharge of phrenic nerves, and expiratory (E2) neurons discharge during the pause in phrenic nerve activity. Preinspiratory (pre-1) neurons discharge in a characteristic phase-spanning manner between inspiration and expiration. Membrane potential trajectories of medullary respiratory neurons (depolarization upward, hyperpolarization downward). *Shading* indicates various types of inhibitory synaptic processes. The fluctuations of membrane potential of neurons are related (in normalized form) to the motor output in the phrenic nerve. Phrenic nerve activity is illustrated in its "averaged" form in the *bottom trace*. Medullary respiratory neurons interact synaptically. A presumed connectivity model (*white circles* indicate excitatory and *black circles*, inhibitory connections) is assumed on the basis of analyses of the discharge profiles (**a**) and postsynaptic activities (**b**). The primary rhythm generator is made up of early-I and PI neurons

105.4 Postsynaptic Activities in Respiratory Neurons

Medullary respiratory neurons of neonatal and adult mammals reveal rhythmic fluctuations of their membrane potential that are synaptically induced [8,9,14,25,60,65,78, 79,80,91,97,99,123,126,127,128,138,139,145,164,171,172]. Continuous synaptic bombardment through all phases of the respiratory cycle precludes determination of a resting membrane potential for these neurons. When the neurons are silent, they receive inhibitory synaptic inputs (IPSPs), the underlying conductances shunting all other excitatory inputs (EPSPs), and when active, they receive a combination of excitatory or excitatory and inhibitory synaptic inputs. Figure 105.3 shows the synaptic inputs to each cell type during the respiratory cycle:

Early-I neurons undergo a rapid membrane depolarization leading to action potential discharge at, or shortly before, onset of inspiratory activity in the phrenic neurogram. Thereafter, the frequency of action potential discharge declines and early-I neurons normally cease discharging during the second half of inspiration. This decline of activity seems to depend largely on intrinsic mechanisms (see below) [113,140,149,154,156] and also on synaptic inhibition [96]. During the PI phase, early-I neurons are inhibited maximally and they remain inhibited during the E2 phase (Fig. 105.3).

I neurons depolarize at the beginning of inspiration and then continue to depolarize throughout inspiration. Some of them depolarize rapidly to a plateau potential and others depolarize almost linearly (Fig. 105.3) [59,145]. A steadily augmenting depolarization results from the integration of excitatory inputs arriving throughout inspiration and inhibitory synaptic inputs during the early-I period. Many propriobulbar I neurons are maximally inhibited during the PI phase and remain inhibited during the E2 phase.

Late-I neurons do not discharge action potentials before the second half of inspiration. The late onset results from an early-I inhibition, which is stronger than the inhibition received by steadily depolarizing I neurons (Fig. 105.3; cf. Fig. 105.5). The inhibitory input during the early-I period effectively shunts the excitatory drive that arrives throughout the I phase [145]. Late-I neurons are synaptically inhibited during the PI and E2 phases.

PI neurons, which consitute bulbar respiratory interneurons and also cranial motor output neurons [184], depolarize rapidly and discharge their first action potentials shortly after inspiration (Fig. 105.3). The rapid changes are partly due to rebound excitation [154,175] following the release from inspiratory inhibition (Fig. 105.5). Later in the PI phase, the action potential discharge de-

clines, and the neurons normally cease firing before the end of the expiratory interval. PI neurons adapt spontaneously [140] and receive weak synaptic inhibition during the E2 phase (Figs. 105.3, 105.4). Therefore, they often show a secondary membrane depolarization together with a secondary burst of spike discharge shortly before the end of the E2 phase (Fig. 105.6). PI neurons are then maximally inhibited during inspiration.

E2 neurons start to depolarize at the end of inspiration (Fig. 105.3) and may discharge one to two action potentials (Fig. 105.5). Thereafter, the neurons are synaptically inhibited again during the PI phase. Action potential discharge starts gradually as this PI inhibition fades and reaches a steady discharge frequency during the later part of the expiratory phase. The augmentation of discharge at the beginning of the E2 phase, therefore, appears to result from a declining pattern of (postinspiratory) synaptic inhibition [145] rather than from a ramp-like pattern of excitation [55,59]. E2 neurons are inhibited maximally during early inspiration. An additional enhancement of synaptic inhibition can become apparent during late inspiration [4]. Termination of the expiratory discharge also is controlled by synaptic inhibition during the end of the E2 phase (Fig. 105.6) [91].

Pre-I neurons reveal a prominent membrane depolarization and burst of action potentials during late expiration

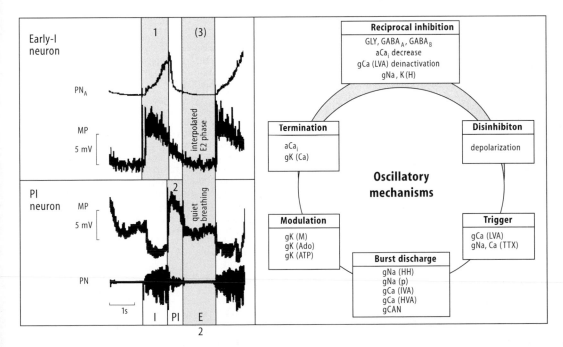

Fig. 105.4. The primary oscillator. *Left side*: membrane potential trajectories (depolarization upward, hyperpolarization downward) of an early-I and a PI neuron. Both neurons respond with a rapid membrane depolarization when they are released from inhibitory synaptic membrane hyperpolarization. After this "rebound" depolarization, the neurons start to hyperpolarize again (as indicated by the *shaded columns*) before there is any sign of synaptic inhibition. Synaptic inhibition occurs later on, e.g., in the PI neuron during the E2 phase (3). This feature indicates that these two types of neurons may act as a primary oscillator within the network of respiratory neurons in the adult mammal. The respiratory rhythm proceeds in two or potentially in three phases: I (1), PI (2), and E2 (3) phases. *Right side*: the processes of respiratory rhythm generation appear to be dependent on the dynamics of both synaptically induced membrane potential changes and intrinsic membrane properties of neurons. Synaptic inhibition, which occurs during antagonistic phases (*shaded crescent*) allows restoration of ionic homeostasis and resetting of membrane conductances. The sequence of events should be read clockwise

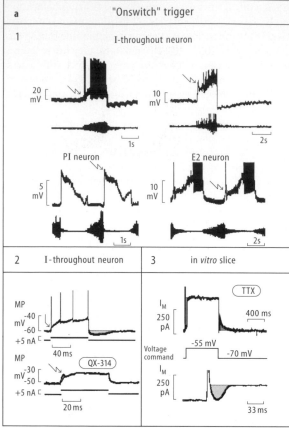

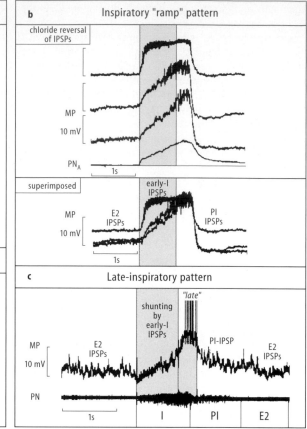

Fig. 105.5a–c. Respiratory "on" switch and inspiratory patterning. **a** *1* Inhibitory membrane hyperpolarization of respiratory neurons reactivates low-voltage activated calcium conductances. This becomes obvious in the lowering of the threshold (*arrow*) of the first action potential during a spontaneous burst discharge as illustrated for I, PI, and E2 neurons. *2* The same feature (*arrow*) is seen when an I neuron is artificially depolarized from hyperpolarized potential levels by a positive current pulse. The underlying mechanism seems to be a rebound depolarization (*arrow*), which becomes evident when the Na$^+$ conductances of the neuron are blocked by intracellular injection of a local anesthetic (*QX 314*). The repetitive discharge is followed by an after-hyperpolarization owing to activation of (charybdotoxin-sensitive) calcium-dependent K$^+$ conductances. *3* The underlying membrane properties can be studied in more detail under in vitro conditions in neurons from the area of the nucleus of the solitary tract. The mechanism is identified as a transient, low-voltage-activated Ca^{2+} current that is resistant to TTX but sensitive to cobalt (not illustrated). Membrane depolarization is followed by an outward tail current, which is due to activation of calcium-dependent K$^+$ conductances. For further explanation see text. **b** The linearly rising membrane depolarization of certain I neurons results from the integration of a barrage of excitatory postsynaptic potentials (EPSPs) arriving throughout inspiration and a declining, early-I pattern of inhibitory postsynaptic potentials (IPSPs). The pattern of IPSPs is obscured under control conditions, but is revealed after intracellular chloride injection, which results in reversal of the chloride currents. The progress of IPSP reversal is shown in single sweeps (taken after variable delay) in the *upper* part and in superimposed sweeps at the *bottom*. Note that besides early-I IPSP reversal, as indicated by the *shaded columns*, there is also reversal of IPSPs arriving during the PI and E2 phases. **c** The delayed membrane depolarization of late-I neurons results from effective shunting of EPSPs (*left-hand shaded column*) during early-I periods. The neurons depolarize steeply after this early-I inhibition ends and a short, "late" burst of action potentials results (*right-hand shaded column*). The neuron is also modulated by various other sorts of inhibitory synaptic inputs as indicated

and continue to fire into inspiration (Figs. 105.3, 105.6). A second, although weaker, activation may occur during the late-I period [164,170].

105.5 Synaptic Interaction Within the Respiratory Network

A hypothetical connectivity scheme of the respiratory network was derived by comparing the action potential dis-charges of the different groups of respiratory neurons with the postsynaptic activities in other groups of neurons (Fig. 105.3) [114,150,154]. Many of the presumed connections between neurons have been verified by the results obtained by cross-correlating the discharges of pairs of different neurons [72,89,105,166] and by spike-triggered averaging of postsynaptic potentials in dual cell recordings [4,59,123,127]. Only some inferred connections between respiratory neurons remain to be demonstrated. For instance, the connections between pre-I, early-I and late-I neurons have not been investigated thoroughly, because

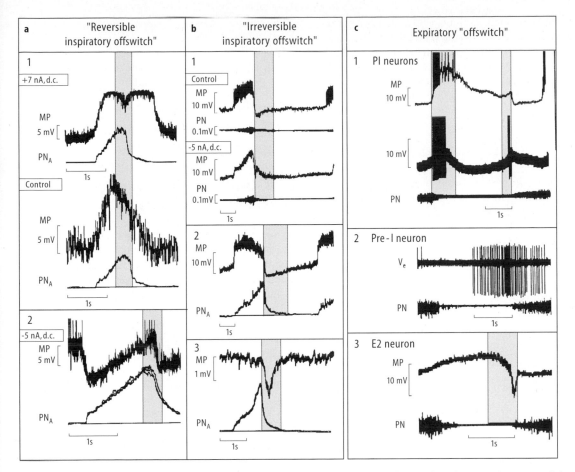

Fig. 105.6a–c. Respiratory "off" switch mechanisms. **a** Reversible inhibition of inspiration is produced by synaptic inhibition. Such inhibition of I neurons (**a1**, **a2**) is produced by IPSPs arriving during late inspiration (*shaded column*). These IPSPs diminish the neuron input resistance and shunt the voltage effect of positive current flow across the membrane (*upper trace* in *1*) and EPSPs (*lower trace* in *1*). The IPSPs become visible after intracellular chloride injection and reversal of IPSPs (*shaded column* in *2*). **b** Irreversible termination (switching off) of inspiration occurs by effective synaptic inhibition during the PI phase (*shaded columns* in *1–3*). These IPSPs become visible by reversal of their polarity in an I neuron after chloride injection (*lower trace* in *1*). Membrane hyperpolarization is combined with effective shunting of postsynaptic noise activity, as shown for an early-I neuron in *2*. Some primarily nonrespiratory neurons are also inhibited during the PI phase, which makes them respiratory modulated (*3*). **c** Switching off of E2 neurons is also produced by synaptic inhibition. The IPSPs terminating the discharge of E2 neurons are difficult to detect under normal conditions of quiet breathing. Under conditions of enforced breathing, these late-expiratory IPSPs become evident in peak inhibition shortly before onset of the next I phase (*shaded column* in *3*). The late-expiratory IPSPs seem to originate from pre-I neurons whose discharge (*2*) corresponds well with the pattern of synaptic inhibition of E2 neurons. PI neurons, which normally show a secondary membrane depolarization shortly before inspiration (*upper trace* in *1*) and often a secondary burst of pre-inspiratory action potential discharge (*lower trace* in *1*) may also contribute to this inhibition of E2 neurons

they are small interneurons and therefore difficult to record from.

105.6 Neurotransmitters and Neuromodulators

The body of information available on the transmitters used by the neurons of the respiratory network is growing. Synaptic activation seems to be mediated mainly by glutamate acting via AMPA receptors [65,73], and also NMDA receptors [47,67–69,90,138,139]. Specifically, the inspiratory off-switch, which acts through PI neurons, seems to be highly sensitive to NMDA receptor blockers [67,139].

Most of the inhibitory synaptic events in respiratory neurons are mediated by glycin- and $GABA_A$-receptor-activated, chloride-permeable channels [38,79,80,126,135], which also have a significant permeability for HCO_3^- ions, resulting in acidification of the cytosol [13]. Only some inhibitory processes seem to involve $GABA_B$ receptor-activated, K^+-permeable channels [136,147]. Synaptic inhibition of inspiratory neurons during the PI phase seems to be strychnine sensitive, indicating glycinergic inhibitory synapses [38]. There is also some indication for glycinergic early-I inhibition of late-I neurons [136]. All these synaptic activities are influenced by endogenous neuromodulators, such as intracellular ATP/ADP concentrations [137,177], adenosine [107,157,161], serotonin [45,53,94,95,118,143,158], noradrenaline [37,52,58], cholecystokinin [31], substance P [143,183], opioids [119], and thyrotropin-releasing hormone [17] acting via intracellular second messengers [6,35,36,100].

105.7 Drive of the Respiratory Network

The functionally important question concerning the drive of the respiratory network is still unresolved despite increasing data from in vivo, in situ and in vitro experiments [14,64,81,122,125,134,145,154,159,170–172,180]. It is still unclear whether the drive originates from endogenously active pacemaker neurons [63,64] that produce tonically beating or rhythmically bursting activity or from the reticular formation that is spontaneously active under in vivo conditions [85]. The pacemaker theory, already developed in the last century, was supported by the finding that spontaneous respiratory activity continued in an isolated brain stem slice preparation of neonatal rat [171,172]. This preparation contains the so-called pre-Bötzinger complex, a region just caudal to the retrofacial nucleus and Bötzinger complex (Fig. 105.2). The preparation contains an accumulation of almost all categories of medullary respiratory interneurons [46,164,170] and obviously several types of pacemaker neurons. When the neurons were depolarized, endogenous capacity to produce bursts of action potentials (burster cells) was observed in

- PI neurons (originally labelled preinspiratory) [125]
- Inspiratory neurons [171; Ramirez J-M, Quellmalz U, Richter DW, in preparation]
- Late-inspiratory [185]
- Expiratory neurons [185]
- Nonrespiratory neurons [171]

This behavior seems to originate from a intermediate voltage activated P-type Ca^{2+} conductance, which is activated when the neurons are depolarized to −40 mV [2,12]. Similar burst-firing behavior of neurons was observed in brain stem cultures [27] or could be provoked in a variety of neurons when they were treated with NMDA or neuropeptides [66,176]. Tonically active (beater) cells were identified as

- Presympathetic neurons in the rostroventrolateral region [76,174]
- Neurons in the region of the nucleus of the solitary tract [39,66,133]
- Tonically active central chemoreceptive neurons [51,104]

However, endogenous bursting pacemaker properties were not discerned in any of the respiratory neurons of adult mammals in vivo. Here, rhythmicity of respiratory neurons seems to depend entirely on the integrity of excitatory

and inhibitory synaptic interaction. The respiratory rhythm is greatly disturbed when synaptic inhibition is blocked by $GABA_A$ receptor blockers [81] or during hypoxic blockade of inhibitory synaptic transmission [152,161]. This indicates that, under in vivo conditions, synaptic inhibition rather than rhythmic excitation via pacemaker cells is essential for rhythmogenesis in mature mammals (see below).

105.8 Ontogenetic Changes of Respiratory Neurons

The respiratory rhythm generation in mature mammals might be different from that in neonatal mammals, especially when they are as immature at birth as rats, in which respiratory activity still resembles fetal activity [74,154]. Pacemaker cell activation of the respiratory rhythm could be functional during early developmental stages, but may be modified strongly with maturation of the brain stem and finally be redundant or absent in the adult. This raises the interesting question of whether transient neurohumoral and ontogenetic changes could not only explain the developmental changes in respiratory patterns, but also underlie the mechanisms of disturbance of respiratory rhythm generation during early postnatal development. Ontogenetic changes seem to involve disappearance of electric synapses which could synchronize respiratory neurons in neonates. Gap junctions were observed in the neonatal, though not in the adult brain stem [142]. Other developmental changes observed involve ion transporters [42,99,178], voltage- and ligand-activated membrane channel properties (e.g., [169]) and the chemical or pharmacological sensitivity of neurons (e.g., [11,32,41,77,112]).

105.9 Primary Respiratory Oscillations in Mature Mammals

Normal breathing movements are never exactly the same under in vivo conditions. This aspect has led to the assumption that the respiratory rhythm might derive from a hybrid of network and pacemaker properties [63,172]. This means that a potential bursting pacemaker-controlled mechanism is strongly modified and normally completely suppressed by synaptic inputs to allow dynamic adjustment of respiratory frequency and pattern to the voluntary or obligatory demands of the organism. That is, network functions determine the respiratory rhythm during all stages of development.

Where does the respiratory rhythm originate in mature mammals? Within the respiratory network, there are two types of reciprocally connected interneurons, the early-I and the PI neurons, which seem to constitute a simple network with oscillatory capacity when they are activated externally (Fig. 105.4). These neurons could well be remainders of those neurons that express pacemaker properties during early development. The two types of neurons are characterized by a rapid onset and then a declining pattern of activity, which results from the interaction of postsynaptic activity and specific membrane properties (Fig. 105.4). Synaptic inhibition over a long (in the range of seconds) period produces membrane hyperpolarization, which seems to be essential for resetting of intrinsic membrane properties of mature neurons. Thus the long periods of synaptic inhibition allow:

- Resetting of intrinsic membrane properties of neurons
- Control of ion homeostasis.

The primary oscillating network is synaptically connected with E2 neurons, which modulate and slow the rhythm. E2 neurons, however, do not seem to be essential for the generation of rhythmic activity, as the respiratory oscillations continue at higher frequencies during rapid shallow breathing or panting, when the inhibitory effect of E2 neurons is absent [10,147]. Resetting and modulation of intrinsic properties of neurons by synaptic interactions means the following. Whenever the action potential discharge declines in one of two antagonistically interconnected early-I and PI neurons, the postsynaptic inhibition diminishes in the other. Thus, the membrane potential of the latter neuron is no longer hyperpolarized and starts to depolarize. Within a certain voltage range, i.e., below the normal threshold of excitation, this disinhibition will lead to activation of a low-voltage-activated Ca^{2+} conductance $gCa(LVA)$ [34,179] or a tetrodotoxin (TTX)-sensitive Ca^{2+} conductance $gNa,Ca(TTX)$ [1]. This results in a "rebound" membrane depolarization (Fig. 105.5), which activates the fast inactivating $gNa(HH)$ [83] and the persistent Na^+ conductances $gNa(p)$ [115,173], thus triggering repetitive action potential discharge. This activity brings the membrane potential into the voltage range in which the intermediate-voltage-activated $gCa(IMV)$ [2,12] and high-voltage-activated Ca^{2+} conductance $gCa(HVA)$ [179] and nonspecific cation conductances $gCAN$ [132] are activated. Repetitive discharge at high frequencies results in a prominent Ca^{2+} influx into the neurons and, possibly, release of Ca^{2+} from intracellular stores, leading to significant accumulation of intracellular Ca^{2+} ions. Intracellularly accumulated Ca^{2+} functions as an activity-dependent mechanism to activate Ca^{2+}-dependent K^+ conductances $gK(Ca)$ (Fig. 105.5), which induce repolarization of the membrane and adaptation of discharge without synaptic inhibition [108,140,156]. Decline of discharge frequency leads to disinhibition and rebound excitation of the antagonistic neurons. Antagonistic inhibition and membrane hyperpolarization of the neurons then allows deinactivation of the inactivated Ca^{2+} and Na^+/Ca^{2+} conductances and sequestration and transportation of intracellular Ca^{2+}. In essence, synaptic inhibition seems to play an important role in respiratory rhythm generation by enabling adequate control of ionic homeostasis and resetting of endogenous membrane properties in respiratory neurons of mature mammals [151].

Modulatory effects on these processes may be produced by adenosine-controlled K^+ conductances $gK(Ado)$ [161], a K^+ conductance $gK(ATP)$ controlled by intracellular ATP/ADP levels [137], acetylcholine acting on a muscarinic K^+ conductance $gK(M)$ [33,36,40,117], and a mixed Na^+ and K^+ conductance $gNa,K(H)$ activated by membrane hyperpolarization [109,111]. Another voltage-dependent K^+ conductance $gK(A)$ [28,40,48] does not seem to have a functional significance under normal conditions of quiet breathing [149].

105.10 Respiratory Rhythm and Pattern Generation in Mature Mammals

The mechanisms of respiratory rhythm and pattern generation were ascribed to separate "central rhythm generating" and "central pattern generating" networks [63]. In the mature mammal, however, both functions seem to be organized by the same network (Fig. 105.3) [154]. The rhythm generator or primary oscillator seem to organize the "switching on" and "switching off" of inspiratory and postinspiratory activities, and it also contributes to shaping the discharge patterns of inspiratory and expiratory neurons. Respiratory rhythmogenesis seems to occur in the following sequence:

- Inspiration is switched on.
- The inspiratory pattern is generated.
- Inspiration is reversibly switched off.
- Inspiration is irreversibly switched off.
- Expiration is switched on.
- Expiration is switched off.

Respiratory Drive. The network is activated tonically by an unpatterned excitatory input from as yet unknown sources.

Inspiration Is Switched On. The release from inhibition triggers rebound excitation of early-I and I neurons through activation of low-voltage-activated Ca^{2+} conductances (Fig. 105.5).

Inspiratory Pattern Is Generated. Postsynaptic integration of recurrent excitation within the population of I neurons [61,128,182] and inhibition by early-I neurons [8,59] results in a linearly rising membrane depolarization and ramp-like activation of reticulospinal I neurons (Fig. 105.5). This activity is transmitted to inspiratory spinal motoneurons via the reticulospinal output [103,141] and finally produces a steadily increasing lung volume.

Phase-Switch of the Primary Oscillator. The discharge of early-I neurons accommodates owing to the activity-dependent, intracellular Ca^{2+} accumulation and to activation of Ca^{2+}-dependent K^+ conductances, which repolarizes the neurons (Fig. 105.4) [115,140,150,156].

Inspiration Is Reversibly Switched Off. Accommodation of early-I neurons is followed by disinhibition and delayed excitation of late-I neurons. Late-I neurons therefore start to discharge at high frequency during late inspiratory periods (Fig. 105.5). This evokes a short, but clearly identifiable, late-I inhibition of I neurons that shunts activating inputs to the inspiratory network (Fig. 105.6) [8,121,145].

Inspiration Is Irreversibly Switched Off. Decrease of inspiratory inhibition of PI neurons leads to rebound excitation of PI neurons (Figs. 105.4, 105.5) through activation of the low-voltage-activated Ca^{2+} conductances [149,175]. This rapid onset of PI discharge produces phase switching in the primary oscillator and irreversibly terminates inspiration (Fig. 105.6) [148,150]. PI activity also delays onset of stage 2 expiration and allows gradual passive exhalation. (It is important to note that an excitatory postinspiratory activity component seems to be added again to the inspiratory output neurons of the network [148]. This produces the postinspiratory "after-discharge" in phrenic motoneurons.)

Expiration Is Switched On. Activity-dependent, intracellular Ca^{2+} accumulation and activation of Ca^{2+}-dependent K^+ conductances [140] produce repolarization and accommodation of PI neurons (Fig. 105.4), leading to disinhibition of E2 neurons, which become active [9,145] and inhibit not only early-I and PI neurons, but also I and late-I neurons [150].

Expiration Is Switched Off. The late expiratory discharge of pre-I neurons, which are not inhibited during the E2 phase, and the secondary discharge of PI neurons, which are weakly inhibited by E2 neurons (Fig. 105.3), act together to terminate the discharge of E2 neurons, thus enabling phase switching to inspiration (Fig. 105.6) [20,43,59,91].

Based on such a description of the respiratory network function, several laboratories have tried to model the respiratory rhythm [29,71,120]. When activated by an unpatterned external drive, the model network was able to develop rhythmic activities similar to those observed in vivo. The models also predict various disturbances to the rhythm that have been seen experimentally, such as rapid shallow breathing [10,147], hypoxic apneusis [19] and postinspiratory apnea [98,144] or hypoxic apnea [152,157].

105.11 A Common Cardiorespiratory Network

The data on the nervous control of the physiologically relevant adjustment of the respiratory and cardiovascular control (cf. Chap. 106) can be summarized as essentially matching respiratory minute volume to cardiac output. This prompts the question as to whether the combined "cardiorespiratory" function may be controlled by a "common cardiorespiratory" network that is part of both respiratory and cardiovascular control systems in the brain stem [84,153]. Regarding the neural basis of interactions between the cardiovascular and respiratory control systems, it is interesting to note that the structures are located close together in almost identical regions of the brain stem (see Fig. 105.2) [21,76,84,110,153,171,174].

A review of the patterns of discharge that are characteristic of various sorts of sympathetic and vagal outflows to the heart, vascular beds, pupil and kidney shows that these "autonomic" activities are also rhythmic and strongly modified in synchrony with central respiratory activity. Central respiratory activity also has the ability to influence the performance of several basic cardiovascular reflexes, and by altering their efficacy affects the discharge of the respective autonomic outputs (for details see Chap. 107). The tight coupling between the two systems partly explains sinus arrhythmia, the respiration-related variations in heart rate and blood pressure. This modulation has been attributed to several mechanisms, including one resulting from a common "cardiorespiratory network" consisting of early-I, I and PI neurons [70,153]. That is, a common cardiorespiratory network controls the combined cardiorespiratory function.

105.12 Reflex Control of Central Respiratory Activity

Information about ventilatory mechanics, but also about chemical stimuli and temperature, is transmitted via afferent fibers of various diameters running in the olfactory, trigeminal, glossopharyngeal, and vagal nerves. This provokes various reflexes, such as termination of inspiration (Hering-Breuer reflex), sneezing, aspiration, coughing or vomiting, and also behavioral reactions, such as sniffing and inhibition of motor activity during heavy working load (Table 105.1) [16,23,75,131,160].

Afferents from glossopharyngeal and vagal nerves terminate in a specific relay nucleus of the brain stem, the nucleus of the solitary tract (NTS; Fig. 105.7; see also Fig. 105.2). The ventrolateral portion of the NTS contains the dorsal group of respiratory neurons (DRG) [21], which also are relay neurons that are not essential for respiratory rhythm generation [5,171].

105.12.1 Central Processing of Slowly Adapting Pulmonary Afferents

Two types of neurons have been identified as target neurons of afferents from slowly adapting "pulmonary stretch" receptors. These are

- Inspiratory β-neurons (I_β), which are depolarized during the I-phase and with lung inflation [181]
- "Pump" (P) neurons [19], which are depolarized with every lung inflation but are not modulated by the central respiratory rhythm.

Table 105.1. Reflexes from the airways

Receptors	Receptor localization	fiber diameter, cond. velocity	Afferent nerves	Appropriate stimulus	Reflex effects	Function
Nasal	Submucosal	1–4 μm 4–26 m/s	trigeminal n. olfactory n.	Mechanical, chemical	Increase insp., increase exp., decrease HR	Sneezing, sniffing
Epi-pharyngeal	Submucosal	1–4 μm 4–26 m/s	glosso-pharyngeal n.	Mechanical	Increase insp., bronchodilatation, decrease BP	Aspiration
Laryngeal	Subepithelial	1–4 μm 4–26 m/s	vagal n.	Mechanical, chemical, cold	Increase insp., increase exp., increase BP, bronchoconstriction	Coughing
Tracheal	Subepithelial	1–4 μm 4–26 m/s	vagal n.	Mechanical (chemical)	Increase insp., increase exp., increase BP, bronchoconstriction	Coughing
Bronchial "irritant receptors"	Subepithelial, interepithelial	1–4 μm 4–26 m/s	vagal n.	Mechanical, chemical, smelling salt	Hyperventilation	Deflation reflex
Bronchial "stretch receptors"	Lamina propria	4–6 μ 25–60 m/s	vagal n.	Mechanical, chemical	Decrease insp., increase exp	Hering-Breuer inflation reflex
Alveolar J receptor C fibers	juxtacapillary wall	≤1 μm 1 m/s	vagal n.	Mechanical, chemical, edema	Rapid shallow breath, apnea inhibit motor activity	"J reflex"

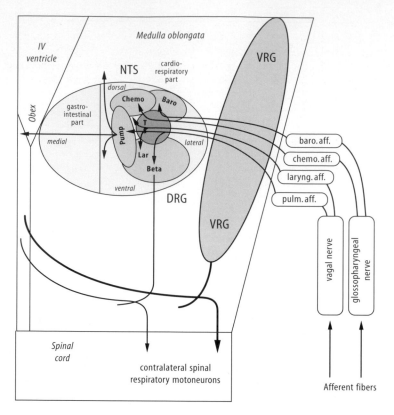

Fig. 105.7. Afferent projections to the nucleus of the solitary tract. Drawing representing a dorsal view of one half of the brain stem (without cerebellum). The nucleus of the solitary tract (*NTS*) is located bilaterally in the medial part of the brain stem. In the illustration, the NTS has been rotated around a horizontal axis to show dorsal parts upwards and ventral parts downwards. In its caudal extension, the lateral part of the NTS receives afferent inputs from the cardiovascular and respiratory systems, whereas the medial part of the NTS receives afferent inputs from the gastrointestinal tract. The two types of neurons that receive vagal afferent fibers from slowly adapting "pulmonary stretch" receptors are: I_β neurons and P neurons. The I_β neurons are located in the ventrolateral subnucleus of the NTS, and their axons project to the contralateral spinal cord without branching of medullary collaterals. P neurons are located directly dorsomedial and

ventromedial to the solitary tract (*TS*). Their axons have medullary collaterals projecting to the ipsi- and contralateral NTS and to more rostral regions of the central nervous system. In the medial parts of the NTS, neurons are activated by vagal afferents from rapidly adapting pulmonary receptors. Afferent fibers in the superior laryngeal nerve from laryngeal receptors project to the ventrolateral and medial NTS. Afferent fibers in the glossopharyngeal nerve from arterial chemoreceptors project to the dorsomedial NTS. Afferent fibers in the glossopharyngeal and vagal nerves from arterial baroreceptor terminate in the dorsolateral areas of the NTS. The details of the connections between second-order relay neurons and the respiratory network in the ventral respiratory group (VRG) are still unknown. *DRG*, dorsal respiratory group.

I_β-neurons are located in the ventrolateral nucleus of the NTS (Fig. 105.7), and their dendritic tree extends into the ventrolateral and intermediate region of the NTS. Thus, their dendritic tree reaches areas where lung stretch receptor afferent fibers of vagal nerves terminate bilaterally [87,92] and monosynaptic contacts between I_β-neurons and these afferent fibers have been demonstrated [3,7]. The axons of I_β-neurons do not have any medullary collateralization [3,130], and the stem axons descend to the spinal cord, where they excite phrenic motoneurons [101]. These findings disprove the original proposal [181] that I_β neurons represent interneurons that inhibit reticulospinal inspiratory I_α-neurons.

P neurons are located mainly in the dorsomedial and ventromedial parts of the NTS (Fig. 105.7) [163], and their dendritic arbor extends into the ventrolateral NTS, where afferents from slowly adapting pulmonary receptors [87]

and laryngeal receptors [18] terminate. The axons of P neurons terminate ipsilaterally with local collaterals in the ventrolateral NTS [162] and project to the contralateral NTS, terminating in its medial part [49,50]. Therefore, P neurons seem to be proper candidates for transmitting lung stretch receptor-related information to both ipsilaterally and contralaterally located I_β-neurons. Another possible target of P neurons are PI neurons in the caudal medulla and in the Bötzinger complex, which show an excitatory input in response to lung inflation [20,106]. These PI neurons could then inhibit inspiratory ventral respiratory group neurons [60,106] and explain the inspiratory-inhibitory and the expiratory-facilitatory components of the Hering-Breuer reflex [82].

Histological data also show prominent projections of pulmonary afferents to the nucleus parabrachialis medialis (PBM) and Kölliker-Fuse nuclei (KF; see Fig. 105.2)

[61,182], producing effective alteration of pontine respiratory activity [54,62,106,129,167]. The pontine outflow may involve respiratory neurons projecting to the dorsal and ventral respiratory groups in the medulla [24,166].

In sum, the Hering-Breuer reflex seems to be more complex than was thought in the past, and to involve more than just NTS neurons.

105.12.2 Central Processing of Other Afferent Fibers

A variety of nonrespiratory NTS neurons relay other sensory inputs: in the medial parts of the NTS, neurons are activated by rapidly adapting pulmonary afferents, which become active during hyperinflation or deflation of the lungs [88,101]. Bronchopulmonary C-fibers project to the commissural subnucleus of the NTS (and to the area postrema) [30,93]. Neurons in the ventrolateral NTS receive inputs from the superior laryngeal nerve [18], and neurons in the dorsomedial NTS are the target of afferents from arterial chemoreceptors [86,96]. Neurons in the dorsolateral NTS receive afferent inputs from arterial baroreceptors (see Chap. 107). The specific connectivity of all these relay neurons with the respiratory network is still unknown, but their axonal projections seem to be widespread. Afferent inputs from arterial chemoreceptors, for example, produce oligosynaptic activation of I, PI and E2 neurons, but inhibition of early-I neurons [96]. Respiratory neurons are also affected by baroreceptor afferent activity. Inspiratory neurons are inhibited and E2 neurons are disinhibited during early inspiration [147]. Additional influences become visible in late-I neurons, in which early-I inhibition is reduced and PI inhibition is enhanced during baroreceptor activation. These findings indicate that baroreceptor afferents affect the primary respiratory oscillator, which might explain respiratory depression during arterial hypertension. The baroreflex itself reveals respiratory modulation in its effectiveness [147]. The reflex is most effective during the PI phase and almost ineffective during the early-I period. This demonstrates that cardiorespiratory functions are controlled by a tight coupling between respiratory and cardiovascular networks, which together might constitute a "common cardiorespiratory network" [147,153].

105.13 Disturbances of the Respiratory Rhythm

Of particular interest for the analysis of central respiratory activity are the processes that lead to disturbance of the rhythm. Any reflex or direct perturbation of the primary oscillator or of its interaction with the network must have pathophysiological consequences: (i) Activation of PI activity by afferents from the larynx [98,144,150] or pulmonary C-fibers [131] results in respiratory arrest (reflex apnea). (ii) Prolonged inspiratory phases (apneusis) result whenever synaptic inhibition is reduced [95]. (iii) This culminates in apnea when synaptic inhibition is blocked. Such reactions were observed under hypoxia or ischemia [14,19,146,152,155]. Thus, there are two possible causes of apnea:

- Reflex activation of postinspiratory activity
- Hypoxic blockade of synaptic inhibition within the network.

Such perturbations of the respiratory oscillator are fairly dramatic in neonatal mammals and might contribute to the sudden infant death syndrome [98,153,165].

References

1. Akaike N, Takahashi K, Morimoto M (1991) Heterogeneous distribution of tetrodotoxin-sensitive calcium-conducting channels in rat hippocampal CA1 neurons. Brain Res 556:135–138
2. Alonso A, Llinás RR (1992) Electrophysiology of the mammillary complex in vitro. II. Medial mammillary neurons. J Neurophysiol 68:1321–1331
3. Anders K, Ohndorf W, Dermietzel R, Richter DW (1990) Synaptic contacts between lung stretch receptor afferents and beta-neurones in cat. Pflügers Arch 415:R91
4. Anders K, Ballantyne D, Bischoff AM, Lalley PM, Richter DW (1991) Inhibition of caudal medullary expiratory neurones by retrofacial inspiratory neurones in the cat. J Physiol (Lond) 437:1–25
5. Arata A, Onimaru H, Homma I (1993) Respiration-related neurons in the ventral medulla of newborn rats in vitro. Brain Res Bull 24:599–604
6. Arata A, Onimaru H, Homma I (1993) Effects of cAMP on respiratory rhythm generation in brainstem-spinal cord preparation from newborn rat. Brain Res 605:193–199
7. Backman SB, Anders K, Ballantyne D, Röhrig N, Camerer H, Mifflin S, Jordan D, Dickhaus H, Spyer KM, Richter DW (1984) Evidence for a monosynaptic connection between slowly adapting pulmonary stretch receptor afferents and inspiratory beta neurones. Pflügers Arch 402:129–136
8. Ballantyne D, Richter DW (1984) Post-synaptic inhibition of bulbar inspiratory neurones in the cat. J Physiol (Lond) 348:67–87
9. Ballantyne D, Richter DW (1986) The non-uniform character or inhibitory synaptic activity in expiratory bulbospinal neurones of the cat. J Physiol (Lond) 370:433–456
10. Ballantyne D, Jordan D, Spyer KM, Wood LM (1988) Synaptic rhythm of caudal medullary expiratory neurones during stimulation of the hypothalamic defence area of the cat. J Physiol (Lond) 405:527–546
11. Ballanyi K, Kuwana S, Völker A, Morawietz G, Richter DW (1992) Developmental changes in the hypoxia tolerance. Neurosci Lett 148:141–144
12. Ballanyi K, Onimaru H, Richter DW (1994) Calcium-mediated responses of respiratory neurons in the in vitro medulla of neonatal rats. Pflügers Arch 426:R137
13. Ballanyi K, Mückenhoff K, Bellingham MC, Okada Y, Scheid P, Richter DW(1995) Activity-related pH changes in respiratory neurones and glial cells of cats. NeuroReport 6:33–36
14. Ballanyi K, Völker A, Richter DW (1995) Anoxia induced functional inactivation of neonatal respiratory neurones in vitro. NeuroReport 6:165–168
15. Barillot JC, Grèlot L, Reddad S, Bianchi AL (1990) Discharge patterns of laryngeal motoneurones in the cat: an intracellular study. Brain Res 509:99–106

16. Bartlett D (1989) Respiratory functions of the larynx. Physiol Rev 69:33–57

17. Bayliss DA, Viana F, Berger AJ (1992) Mechanisms underlying excitatory effects on thyrotropin-releasing hormone on rat hypoglossal motoneurons in vitro. J Neurophysiol 68:1733–1745

18. Bellingham MC, Lipski J (1992) Morphology and electrophysiology of superior laryngeal nerve afferents and postsynaptic neurons in the medulla oblongata of the cat. Neuroscience 48:205–216

19. Bellingham MC, Schmidt C, Windhorst U, Richter DW (1991) The inspiratory off-switch is disturbed during hypoxia. Pflügers Arch 418:R16

20. Bellingham MC, Schmidt C, Richter DW (1992) Postinspiratory neurons – a possible integrator of the Hering-Breuer reflex? In: Elsner N, Richter DW (eds) Rhythmogenesis in neurons and networks, Proceedings of the 20th Göttingen Neurobiology Conference. Thieme, Stuttgart, p 67

21. Berger AJ (1977) Dorsal respiratory group neurons in the medulla of cat: spinal projections, responses to lung inflation and superior laryngeal nerve stimulation. Brain Res 135:231–154

22. Bianchi AL, Barillot JC (1982) Respiratory neurons in the region of the retrofacial nucleus: pontile, medullary, spinal and vagal projections. Neurosci Lett 3:277–282

23. Bianchi AL, Grélot (1989) Converse motor output of inspiratory bulbospinal premotoneurones during vomiting. Neurosci Lett 104:298–302

24. Bianchi AL, St John WM (1982) Medullary axonal projections of respiratory neurons of pontile pneumotaxic center. Respir Physiol 48:357–373

25. Bianchi AL, Grélot L, Iscoe S, Remmers JE (1988) Electrophysiological properties of rostral medullary respiratory neurones in the cat: an intracellular study. J Physiol (Lond) 407:293–310

26. Bianchi AL, Denavit-SaubiéM, Champagnat J (1995) Neurobiology of the central control of breathing in mammals: neuronal circuitry, membrane properties and neurotransmitters involved. Physiol Rev (in press)

27. Bingmann D, Baker RE, Ballantyne D (1991) Rhythm generation in brainstem cultures grown in a serum-free medium. Neurosci Lett 132:167–170

28. Bossu JL, Dupont JL, Feltz A (1985) I_A current compared to low-threshold calcium current in cranial sensory neurons. Neurosci Lett 62:249–254

29. Botros SM, Bruce EN (1990) Neural network implementation of the three-phase model of respiratory rhythm generation. Biol Cybern 63:143–153

30. Boxham AC, Joad JP (1991) Neurones in commissural nucleus tractus solitarii required for full expression of the pulmonary C fibre reflex in rat. J Physiol (Lond) 441:95–112

31. Branchereau P, Champagnat J, Roques BP, Denavit-Saubié M (1992) CCK modulates inhibitory synaptic transmission in the solitary complex through CCK_B sites. NeuroReport 3:909–912

32. Brockhaus J, Ballanyi K, Smith JC, Richter DW (1993) Microenvironment of respiratory neurons in the in vitro brainstem-spinal cord of neonatal rats. J Physiol (Lond) 462:421–445

33. Brown DA (1988) M-currents: an update. Trends Neurol Sci 11:294–299

34. Carbone E, Lux HD (1987) Kinetics and selectivity of a low-voltage-activated calcium current in chick and rat sensory neurones. J Physiol (Lond) 386:547–570

35. Champagnat J, Richter DW (1993) Second messengers-induced modulation of the excitability of respiratory neurones. NeuroReport 4:861–863

36. Champagnat J, Richter DW (1994) The roles of K^+ conductance in expiratory pattern generation in anaesthetized cats. J Physiol (Lond) 4791:127–128

37. champagnat J, Denavit-Saubié M, Henry JL, Leviel V (1979) Catecholaminergic depressant effects on bulbar respiratory mechanisms. Brain Res 160:57–68

38. Champagnat J, Denavit-Saubié M, Moyanova S, Rondouin G (1982) Involvement of amino acids in periodic inhibitions of bulbospinal respiratory neurones. Brain Res 237:351–365

39. Champagnat J, Denavit-Saubié M, Siggins GR (1983) Rhythmic neuronal activities in the nucleus of the tractus solitarius isolated in vitro. Brain Res 280:155–159

40. Champagnat J, Jaquin T, Richter DW (1986) Voltage-dependent currents in neurones of the nuclei of the solitary tract of rat brainstem slices. Pflügers Arch 406:372–379

41. Cherubini E, Ben-Ari Y, Krnjevic K (1990) Periodic inward currents triggered by NMDA in immature CA3 hippocampal neurones. In: Ben-Ari Y (ed) Excitatory amino acids and neuronal plasticity. Plenum, New York, pp 147–150

42. Cherubini E, Gaiarsa JK, Ben-Ari Y (1991) GABA: an excitatory transmitter in early postnatal life. Trends Neurol Sci 14:515–519

43. Cohen MI (1979) Neurogenesis of respiratory rhythm in the mammal. Physiol Rev 59:1105–1173

44. Coleridge HM, Coleridge JCG, Roberts SM (1983) Rapid shallow breathing evoked by selective stimulation of airway C-fibres in dogs. J Physiol (Lond) 340:415–433

45. Connelly CA, Ellenberger HH, Feldman JL (1989) Are there serotonergic projections from raphe and retrotrapezoid nuclei to the ventral respiratory group in the rat? Neurosci Lett 105:34–40

46. Connelly CA, Dobbins EG, Feldman JL (1992) Pre-Bötzinger complex in cats: respiratory neuronal discharge patterns. Brain Res 590:337–340

47. Connelly CA, Otto-Smith MR, Feldman JL (1992) Blockade of NMDA receptor-channels by MK801 alters breathing in adult rats. Brain Res 596:99–110

48. Connor JA, Stevens CF (1971) Voltage-clamp studies of a transient outward membrane current in gastropod neural somata. J Physiol (Lond) 213:21–30

49. Davies RO, Kubin L, Pack AI (1986) Contralateral projections of pump neurones of the nucleus of the solitary tract in the cat. J Physiol (Lond) 371:118

50. Davies RO, Kubin L, Pack AI (1987) Pulmonary stretch receptor relay neurones of the cat: location and contralateral medullary projections. J Physiol (Lond) 383:571–585

51. Dean JB, Lawing WL, Millhorn DE (1989) CO_2 decreases membrane conductance and depolarizes neurons in the nucleus tractus solitarii. Exp Brain Res 76:656–661

52. Di Pasquale E, Monteau R, Hilaire G (1992) In vitro stuy of central respiratory-like activity of the fetal rat. Exp Brain Res 89:459–464

53. Di Pasquale E, Morin D, Monteau R, Hilaire G (1992) Serotonergic modulation of the respiratory rhythm generator at birth: an in vitro study in the rat. Neurosci Lett 143:91–95

54. Dick TE, Bellingham MC, Richter DW (1994) Pontine respiratory neurons in anaesthetized cats. Brain Res 636:259–269

55. Duffin J, Aweida D (1990) The propriobulbar respiratory neurons in the cat. Exp Brain Res 81:213–220

56. Ellenberger HH, Feldman JL (1990) Brainstem connections of the rostral ventral respiratory group of the rat. Brain Res 513:35–42

57. Ellenberger HH, Feldman JL (1990) Subnuclear organization of the lateral tegmental field of the rat. I Nucleus ambiguus and ventral respiratory group. J Comp Neurol 294:202–211

58. Errchidi S, Monteau R, Hilaire G (1991) Noradrenergic modulation of the medullary respiratory rhythm generator in the newborn rat: an in vitro study. J Physiol (Lond) 443:477–498

59. Ezure K (1990) Synaptic connections between medullary respiratory neurons and considerations on the genesis of respiratory rhythm. Prog Neurobiol 35:429–450

60. Ezure K, Manabe M (1988) Decrementing expiratory neurons in the Bötzinger complex. II. Direct inhibitory synaptic

linkage with ventral respiratory group neurons. Exp Brain Res 72:159–166

61. Feldman JL (1986) Neurophysiology of breathing in mammals. In: Bloom FE (ed) Handbook of Physiology, Section I, The nervous system, American Physiological Society, Bethesda, pp 463–524

62. Feldman JL, Cohen MI, Wolotsky P (1976) Powerful inhibition of pontine respiratory neurones by pulmonary afferent activity. Brain Res 104:341–346

63. Feldman JL, Smih JC, Ellenberger HH, Connelly CA, Liu G, Greer JJ, Lindsay AD, Otto MR (1990) Neurogenesis of respiratory rhythm and pattern: emerging concepts Am J Physiol 259:R879–R886

64. Feldman JL, Smith JC, Liu G (1991) Respiratory pattern generation in mammals: in vitro en bloc analyses. Curr Opin Neurobiol 1:590–594

65. Feldman JL, Windhorst U, Anders K, Richter DW (1992) Synaptic interaction between medullary respiratory neurones during apneusis induced by NMDA-receptor blockade in cat. J Physiol (Lond) 450:303–323

66. Fortin G, Branchereau P, Araneda S, Champagnat J (1992) Rhythmic activities in the rat solitary complex in vitro. Neurosci Lett 147:89–92

67. Foutz AS, Champagnat J, Denavit-Saubié M (1988) N-Methyl-D-aspartate (NMDA) receptors control respiratory off-switch in cat. Neurosci Lett 87:221–226

68. Foutz AS, Champagnat J, Denavit-Saubié M (1989) Involvement of N-Methyl D-aspartate (NMDA) receptors in respiratory rhythmogenesis. Brain Res 500:199–208

69. Foutz AS, Pierrefiche O, Denavit-Saubié M (1994) Combined blockade of NMDA and non-NMDA receptors produces respiratory arrest in the adult cat. NeuroReport 5:481–484

70. Gilbey M, Jordan D, Richter DW, Spyer KM (1984) Synaptic mechanisms involved in the inspiratory modulation of the vagal cardioinhibitory neurones in the cat. J Physiol (Lond) 356:65–78

71. Gottschalk A, Ogilvie MD, Richter DW, Pack AI (1994) Computational aspects of the respiratory pattern generator. Neural Comp 6:56–68

72. Graham K, Duffin J (1981) Cross-correlation of medullary expiratory neurons in the cat. Exp Neurol 73:451–464

73. Greer JJ, Smith JC, Feldman JL (1991) Role of excitatory amino acids in the generation and transmission of respiratory drive in neonatal rat. J Physiol (Lond) 437:727–749

74. Greer JJ, Smith JC, Feldman JL (1992) Respiratory and locomotor patterns generated in the fetal rat brain stem-spinal cord in vitro. J Neurophysiol 67:996–999

75. Grélot L, Milano S, Portillo F, Miller AD, Biachni AL (1992) Membrane potential changes of phrenic motoneurons during fictive vomiting, coughing and swallowing in the decerebrate cat. J Neurophysiol 68:2110–2119

76. Guyenet PG (1990) Role of the ventral medulla oblongata in blood pressure regulation. In: Loewy AD, Spyer KM (eds) Central regulation of autonomic function, Oxford University Press, Oxford, 145–167

77. Haddad GG, Donnelly DF (1989) Maturation of hypoglossal neuronal response to hypoxia in rats. In-vitro intracellular studies. Fed Proc 3:A403

78. Haji A, Takeda R (1993) Variations in membrane potential trajectory of postinspiratory neurons in the ventrolateral medulla of the cat. Neurosci Lett 149:233–236

79. Haji A, Remmers JE, Connelly C, Takeda R (1990) Effects of glycine and GABA on bulbar respiratory neurons of cat. J Neurophysiol 63:955–965

80. Haji A, Takeda R, Remmers JE (1992) Evidence that glycine and GABA mediate postsynaptic inhibition of bulbar respiratory neurons in the cat. J Appl Physiol 73:2333–2342

81. Hayashi F, Lipski J (1992) The role of inhibitory amino acids in control of respiratory motor ouput in an arterially perfused rat. Resp Physiol 89:47–63

82. Hering E (1868) Die Selbststeuerung der Athmung durch den Nervus vagus. Sitzungsber Akad Wiss Wien 57(2):672–677

83. Hodgkin AL, Huxley AF (1952) Currents carried by sodium and potassium ions through the membrane of the giant axon of Loligo. J Physiol (Lond) 116:449–472

84. Hopkins DA (1987) The dorsal motor nucleus of the vagus nerve and the nucleus ambiguus; structure and connections. In: Hainsworth R, McWilliam PN, Many DAG (eds) Cardiogenic reflexes, Oxford University Press, Oxford, pp 185–203

85. Hugelin A, Cohen MI (1963) The reticular activating system and respiratory regulation in the cat. Ann NY Acad Sci U S A 109:586–603

86. Izzo PN, Lin RJ, Richter DW, Spyer KM (1988) Physiological and morphological identification of neurones receiving arterial chemoreceptive afferent input in the nucleus tractus solitarius of the cat. J Physiol (Lond) 399:31P

87. Kalia M, Richter DW (1985) Morphology of physiologically identified slowly adapting lung stretch receptor afferents stained with intra-axonal HRP in the nucleus of the tractus solitarius of the cat. I. A light microscopic analysis. J Comp Neurol 241:503–520

88. Kalia M, Richter DW (1988) Rapidly adapting pulmonary receptor afferents. I. Arborization in the nucleus of the tractus solitarius. J Comp Neurol 274:560–573

89. Kashiwagi M, Onimaru H, Homma I (1993) Correlation analysis of respiratory neuron activity in ventrolateral medulla of brainstem-spinal cord preparation isolated from newborn rat. Exp Brain Res 95:277–290

90. Kashiwagi M, Onimaru H, Homma I (1993) Effects of NMDA on respiratory neurons in newborn rat medulla in vitro. Brain Res Bull 32:65–69

91. Klages S, Bellingham MC, Richter DW (1993) Late expiratory inhibition of stage 2 expiratory neurons in the cat – a correlate of expiratory termination. J Neurophysiol 70:1307–1315

92. Kubin L, Davies RO (1987) Bilateral convergence of pulmonary stretch receptor inputs on Iβ-neurons in the cat. J Appl Physiol 62:1488–1496

93. Kubin L, Kimura H, Davies RO (1991) The medullary projections of afferent bronchopulmonary C-fibres in the cat as shown by antidromic mapping. J Physiol (Lond) 435:207–228

94. Lalley PM, Bischoff AM, Richter DW (1993) 5HT-1A receptor-mediated modulation of medullary expiratory neurones in the cat. J Physiol (Lond) 476:117–130

95. Lalley PM, Bischoff AM, Richter DW (1994) Serotonin 1A-receptor activation suppresses respiratory apneusis in the cat. Neurosci Lett 172:59–62

96. Lawson EE, Richter DW, Ballantyne D, Lalley PM (1989) Peripheral chemoreceptor inputs to medullary inspiratory and postinspiratory neurons of cats. Pflügers Arch 414:523–533

97. Lawson EE, Richter DW, Bischoff A (1989) Intracellular recordings of respiratory neurons in the lateral medulla of piglets. J Appl Physiol 66(2):983–988

98. Lawson EE, Richter DW, Czyzyk-Krzeska MF, Bischoff AM, Rudesill RC (1991) Respiratory neuronal activity during apnea and other breathing patterns induced by laryngeal stimulation. J Appl Physiol 70:2742–2749

99. Lawson EE, Schwarzacher SW, Richter DW (1992) Postnatal development of the medullary respiratory network in cat. In: Elsner N, Richter DW (eds) Rhythmogenesis in neurons and networks Thieme, Stuttgart, p 69

100. Ling L, Karius DR, Fiscus RR, Speck DF (1992) Endogenous nitric oxide required for an integrative respiratory function in the cat brain. J Neurophysiol 68:1910–1912

101. Lipski J, Kubin L, Jodkowski J (1983) Synaptic action of R-beta neurons on phrenic motoneurons studied with spike-triggered averaging. Brain Res 288:105–118

102. Lipski J, Ezure K, Wong She RB (1991) Identification of neurons receiving input from pulmonary rapidly adapting receptors in the cat. J Physiol (Lond) 443:55–77

103. Liu G, Feldman JL (1992) Quantal synaptic transmission in phrenic motor nucleus. J Neurophysiol 68:1468–1471

104. Loeschcke HH (1982) Central chemosensitivity and the reaction theory. J Physiol (Lond) 332:1–24

105. Long S, Duffin J (1986) The neuronal determinants of respiratory rhythm. Prog Neurobiol 27:101–182

106. Manabe M, Ezure K (1988) Decrementing expiratory neurons of the Bötzinger complex. I. Response to lung inflation and axonal projection. Exp Brain Res 72:150–158

107. Marks JD, Donnelly DF, Haddad GG (1993) Adenosine-induced inhibition of vagal motoneuron excitability: receptor subtype and mechanisms. Am J Physiol 264(8):L124–L132

108. Marty A (1983) Ca^{2+}-dependent K^+ channels with large unitary conductance. Trends Neurol Sci 6:262–265

109. Mayer ML, Westbrook GA (1983) A voltage-clamp analysis of inward (anomalous) rectification in mouse spinal sensory ganglion neurones. J Physiol (Lond) 340:19–45

110. McAllen RM (1987) Central respiratory modulation of subretrofacial bulbospinal neurones in the cat. J Physiol (Lond) 388:533–545

111. McCormick DA, Pape HC (1990) Properties of a hyperpolarization-activated cation current and its role in rhythmic oscillation in thalamic relay neurones. J Physiol (Lond) 431:291–318

112. McDonald JW, Silverstein FS, Johnson MV (1988) Neurotoxicity of N-methyl-D-aspartate is markedly enhanced in developing rat central nervous system. Brain Res 459:200–203

113. Merrill EG (1974) Finding a respiratory function for the medullary respiratory neurons. In: Bellairs R, Gray EG (eds) Essays on the nervous system, Clarendon Oxford, pp 451–486

114. Merrill EG, Lipski J, Kubin L, Fedorko L (1983) Origin of the expiratory inhibition of nucleus tractus solitarius inspiratory neurones. Brain Res 263:43–40

115. Mifflin SW, Richter DW (1987) Effects of QX-314 on medullary respiratory neurones. Brain Res 420:22–31

116. Monteau R, Hilaire G (1991) Spinal respiratory motoneurons. Prog Neurobiol 37:83–144

117. Monteau R, Morin D, Hilaire G (1990) Acetylcholine and central chemosensitivity: in vitro study in the newborn rat. Resp Physiol 81:241–254

118. Morin D, Hennequin S, Monteau R, Hilaire G (1990) Serotonergic influences on central respiratory activity: an in vitro study in the newborn rat. Brain Res 535:281–287

119. Morin-Surun MP, Boudinot E, Fournie-Zaluski MC, Champagnat J, Roques BP, Denavit-Saubié M (1992) Control of breathing by endogenous opioid peptides: possible involvement in sudden infant death syndrome. Neurochem Int 20:103–107

120. Oglivie MD, Gottschalk A, Anders K, Richter DW, Pack AI (1992) A network model of respiratory rhythmogenesis. Am J Physiol 263:R962–R975

121. Oku Y, Tanaka I, Ezure K (1992) Possible inspiratory off-switch neurones in the ventrolateral medulla of the cat. NeuroReport 3:933–936

122. Onimaru H (1995) Studies of the respiratory center using isolated brainstem-spinal cord preparations. Neurosci Res 21:183–190

123. Onimaru H, Homma I (1992) Whole cell recordings from respiratory neurons in the medulla of brainstem-spinal cord preparations isolated from newborn rats. Pflügers Arch 420:399–406

124. Onimaru H, Arata A, Homma I (1987) Localization of respiratory rhythm-generating neurons in the medulla of brainstem-spinal cord preparations from newborn rats. Neurosci Lett 78:151–155

125. Onimaru H, Arata A, Homma I (1989) Firing properties of respiratory rhythm generating neurons in the absence of synaptic transmission in rat medulla in vitro. Ext Brain Res 76:530–536

126. Onimaru H, Arata A, Homma I (1990) Inhibitory synaptic inputs to the respiratory rhythm generator in the medulla isolated from newborn rats. Pflügers Arch 417:425–432

127. Onimaru H, Homma I, Iwatsuki K (1992) Excitation of inspiratory neurons by preinspiratory neurons in rat medulla in vitro. Brain Res Bull 29:879–882

128. Onimaru H, Kashiwagi M, Arata A, Homma I (1993) Possible mutual excitatory couplings between inspiratory neurons in caudal ventrolateral medulla of brainstem-spinal cord preparation isolated from newborn rat. Neurosci Lett 150:203–206

129. Otake K, Sasaki H, Ezure K, Manabe M (1988) Axonal projections from Bötzinger expiratory neurons to contralateral ventral and dorsal respiratory groups in the cat. Exp Brain Res 72:167–177

130. Otake K, Sasaki H, Ezure K, Manabe M (1989) Axonal trajectory and terminal distribution of inspiratory neurons of the dorsal respiratory group in the cat's medulla. J Comp Neurol 286:218–230

131. Paintal AS (1973) Vagal sensory receptors and their reflex effects. Physiol Rev 53:159–227

132. Partridge LD, Swandulla D (1988) Calcium-activated nonspecific cation channels. Trends Neurol Sci 11:69–72

133. Paton JFR, Rogers WT, Schwaber JS (1991) Tonically rhythmic neurons within a cardiorespiratory region of the nucleus tractus solitarii of the rat. J Neurophysiol 66:824–838

134. Paton JFR, Ramirez J-M, Richter DW (1994) Functionally intact in vitro preparation generating respiratory activity in neonatal and mature mammals. Pflügers Arch 428:250–260

135. Paton JFR, Ramirez J-M, Richter DW (1994) Mechanisms of respiratory rhythm generation change profoundly during early life in mice and rats. Neurosci Lett 170:167–170

136. Pierrefiche O, Foutz A (1993) Effects of $GABA_B$ receptor agonists and antagonists on the bulbar respiratory network in the cat. Brain Res 605:77—84

137. Pierrefiche O, Richter DW (1994) K^+-ATP channels are active in the respiratory network of cats. Plügers Arch 426:R136

138. Pierrefiche O, Foutz AS, Champagnat J, Denavit-Saubié M (1992) The bulbar network of respiratory neurons during apneusis induced by a blockade of NMDA receptors. Exp Brain Res 89:623–639

139. Pierrefiche O, Foutz AS, Champagnat J, Denavit-Saubié M (1994) NMDA and non-NMDA receptors may play distinct roles in timing mechanisms and transmission in the feline respiratory network. J Physiol (Lond) 474.3:509–523

140. Pierrefiche O, Champagnat J, Richter DW (1995) Calcium-dependent conductances control neurones involved in termination of inspiration in cats. Neurosci Lett 184:101–104

141. Portillo F, Grélot L, Milano S, Bianchi AL (1994) Brainstem neurons with projecting axons to both phrenic and abdominal motor nuclei: a double fluorescent labeling study in the cat. Neurosci Lett 173:50–54

142. Quattrochi JJ, Rho JH (1985) Three-dimensional tissue reconstruction reveals integrative structural features among neurons within central respiratory centers of the brain stem. In: Bianchi AL, Denavit-Saubie M (eds) Nuerogenesis of central respiratory rhythm. MTP, Lancaster, pp 431–437

143. Rampin O, Pierrefiche O, Denavit-Saubié M (1993) Effects of serotonin and substance P on bulbar respiratory neurons in vivo. Brain Res 622:185–193

144. Remmers JE, Richter DW, Ballantyne D, Bainton CR, Klein JP (1986) Reflex prolongation of the stage I of expiration. Pflügers Arch 407:190–198

145. Richter DW (1982) Generation and maintenance of the respiratory rhythm. J Exp Biol 100:93–107

146. Richter DW, Acker H (1989) Respiratory neuron behavior during medullary hypoxia. In: Lahiri S (ed) Chemoreceptors and reflexes in breathing: cellular and molecular aspects. Oxford University Press, Oxford, pp 267–274

147. Richter DW, Spyer KM (1990) Cardio-respiratory control In: Loewy AD, Spyer KM (eds) Central regulation of autonomic function. Oxford University Press, Oxford, pp 189–207

148. Richter DW, Ballantyne D, Remmers JE (1986) Respiratory rhythm generation: a model. NIPS 1:109–112

149. Richter DW, Champagnat J, Mifflin SW (1986) Membrane properties involved in respiratory rhythm generation. In: von Euler C Lagercrantz H (eds) Neurobiology of the control of breathing. Raven, New York, pp 141–147

150. Richter DW, Ballantyne D, Remmers JE (1987) The differential organization of medullary post-inspiratory activities. Pflügers Arch 410:420–427

151. Richter DW, Champagnat J, Mifflin SW (1987) Membrane properties of medullary respiratory neurones of the cat. In: Sieck GC, Gandevia SC, Cameron WE (eds) Respiratory muscles and their neuromotor control. Wiley, New York, pp 9–15

152. Richter DW, Bischoff A, Anders K, Bellingham M, Windhorst U (1991) Response of the medullary respiratory network of the cat to hypoxia. J Physiol (Lond) 443:231–256

153. Richter DW, Spyer KM, Gilbey MP, Lawson EE, Bainton CR, Wilhelm Z (1991) On the existence of a commmon cardio-respiratory network In: Koepchen HP, Huopaniemi T (eds) Cardiorespiratory and motor coordination, Springer, Berlin, Heidelberg, New York pp 118–130

154. Richter DW, Ballanyi K, Schwarzacher S (1992) Mechanisms of respiratory rhythm generation. Curr Opin Neurobiol 2:788–793

155. Richter DW, Bellingham M, Schmidt C (1992) Maintenance of the respiratory rhythm during normoxia and hypoxia In: Honda Y, Miyamoto Y, Konno K, Widdicombe JG (eds) Control of breathing and its modeling perspective. Plenum, New York, pp 7–13

156. Richter DW, Champagnat J, Jacquin T, Benacka R (1993) Calcium and calcium-dependent potassium currents in medullary respiratory neurones. J Physiol (Lond) 470:23–33

157. Richter DW, Schmidt C, Bellingham M, Schmidt P (1993) Hypoxia and central respiratory neurons In: Scheid P (ed) Respiration in health and disease: lessons from comparative physiology, Fischer, Stuttgart pp 303–312

158. Rudolph T, Schwarzacher SW, Herbert H, Richter DW (1992) Serotoninergic innervation of cats medullary respiratory neurones: intracellular HRP-labeling and 5HT-immunocytochemestry. In: Elsner N, Richter DW (eds) Rhythmogenesis in neurons and networks. Thieme, Stuttgart p 71

159. Salmoiraghi GC, Baumgarten R von (1961) Intracellular potentials from respiratory neurones in brain-stem of cat and mechanism of rhythmic respiration. J Neurophysiol 24:203–218

160. Sant'Ambrogio G (1982) Information arising from the tracheobronchial tree of mammals. Physiol Rev 62:531–569

161. Schmidt C, Bellingham MC, Richter DW (1995) Adenosinergic modulation of respiratory neurones and hypoxic responses in the anaesthetized cat. J Physiol (Lond) 483:769–781

162. Schwarzacher SW, Wilhelm Z, Anders K, Richter DW (1991) The medullary respiratory network in rats. J Physiol (Lond) 435:631–644

163. Schwarzacher SW, Maschke M, Anders K, Richter DW (1995) Morphology of respiration-related "pump" neurones in cat brainstem: a light- and electronmicroscopic analysis

164. Schwarzacher SW, Smith JC, Richter DW (1995) Pre-Bötzinger complex in the cat. J Neurophysiol 73: (in press)

165. Schweitzer P, Fortin G, Beloeil JC, Champagnat J (1992) In vitro study of newborn rat brain maturation: implication for sudden infant death syndrome. Neurochem Int 20:109–112

166. Segers LS, Shannon R, Lindsey BG (1985) Interactions between rostral pontine and ventral medullary respiratory neurones. J Neurophysiol 54:318–334

167. Shaw C, Cohen MI, Barnhardt R (1989) Inspiratory-modulated neurons of the rostrolateral pons: effects of pulmonary afferent input. Brain Res 485:179–184

168. Shee CD, Loysongsang Y, Milic-Emili J (1985) Decay of inspiratory muscle pressure during expiration in conscious humans. J Appl Physiol 58:1859–1865

169. Shinohara K, Nishikawa T, Yamazaki K, Takahashi K (1989) Ontogeny of strychnine-insensitive (^{3}H)glycine binding sites in rat forebrain. Neurosci Lett 105:307–311

170. Smith JC, Greer JJ, Liu G, Feldman JL (1990) Neural mechanisms generating respiratory pattern in mammalian brain stem-spinal cord in vitro. I. Spatiotemporal patterns of motor and medullary neuron activity. J Neurophysiol 64:1149—1169

171. Smith JC, Ellenberger H, Ballanyi K, Feldman JL, Richter DW (1991) Pre-Bötzinger complex: a brainstem region that may generate respiratory rhythm in mammals. Science 254:726–729

172. Smith JC, Ballanyi K, Richter DW (1992) Whole-cell patch-clamp recordings from respiratory neurons in neonatal rat brainstem in vitro. Neurosci Lett 134:153–156

173. Stafstrom CE, Schwindt PC, Chubb MC, Crill WE (1985) Properties of persistent sodium conductance and calcium conductance of layer V neurons from cat sensorimotor cortex in vitro. J Neurophysiol 53:153–170

174. Sun MK, Young BS, Hackett JT, Guyenet PG (1988) Reticulospinal pacemaker neurons of the rat rostral ventrolateral medulla with putative sympathoexcitatory function: an intracellular study in vitro. Brain Res 442:229–239

175. Takeda R, Haji A (1993) Mechanisms underlying post-inspiratory depolarization in post-inspiratory neurons of the cat. Neurosci Lett 150:1–4

176. Tell F, Jean A (1991) Activation of N-methyl D aspartate receptors induces endogenous rhythmic bursting activities in nucleus tractus solitarii neurons: an intracellular study on adult rat brainstem slices. Eur J Neurosci 3:1353–1365

177. Trapp S, Ballanyi K, Richter DW (1994) Spontaneous activation of K_{ATP} current in rat dorsal vagal neurones. NeuroReport 5:1285–1288

178. Trippenbach T, Richter DW, Acker H (1990) Hypoxia and ion activities within the brainstem of newborn rabbits. J Appl Physiol 68:2494–2503

179. Tsien RW, Lipscombe D, Madison DV, Bley KR, Fox AP (1988) Multiple types of neuronal Ca channels and their selective modulation. Trends Neurol Sci 11:431–438

180. Völker A, Ballanyi K, Richter DW (1995) Anoxic disturbance of the isolated respiratory network of neonatal rats. Exp Brain Res 103:9–19

181. von Baumgarten R, Kanzow E (1958) The interaction of two types of inspiratory neurones in the region of the tractus solitarius of the cat. Arch Ital Biol 96:361–373

182. von Euler C (1986) Brain stem mechanims for generation and control of breathing pattern. In: Fishman AP, Cherniack NS, Widdicombe JG, Geiger SR (eds) Handbook of physiology, Section III, The respiratory system, American Physiological Society, Bethesda, pp 1–67

183. Yamamoto Y, Onimaru H, Homma I (1992) Effect of substance P on respiratory rhythm and pre-inspiratory neurons in the ventrolateral structure of rostral medulla oblongata: an in vitro study. Brain Res 599:272–274

184. Zheng Y, Barillot JC, Biachi AL (1991) Are the post-inspiratory neurons in the decerebrate rat cranial motoneurons or interneurons? Brain Res 551:256–266

Important Reference Published Recently

185. Johnson SM, Smith JC, Funk GD, Feldman JL (1994) Pacemaker behavior of respiratory neurons in medullary slices from neonatal rat. J Neurophysiol 72:2598–2608

106 A Systems View of Respiratory Regulation

M.E. Schläfke and H.P. Koepchen

Contents

106.1 Introduction

The chemical composition of the alveolar gases (alveolar CO_2 and O_2 pressure, PA_{CO_2} and PA_{O_2}) and of the arterial blood (arterial CO_2 and O_2 content, Ca_{CO_2} and Ca_{O_2}) is dependent upon the exchange of respiratory gases between the environment and the alveolar space. This connection is quantitatively described in the equation PA_{CO_2} = constant production of CO_2 (V_{CO_2})/alveolar ventilation (V_A). (see Chaps. 100–102). Thus the CO_2 content of alveolar gas or of arterial blood is a function of the magnitude of the metabolic CO_2 production and V_A Since the metabolic rate varies widely in daily life, the equation shows that these variations would result in extraordinary variations of PA_{CO_2} and hence of blood and tissue chemistry if V_A did not vary in a way corresponding to metabolic CO_2 production. The air-moving apparatus serves to maintain homeostasis and helps adapt to endogenic and exogenic challenges; it makes typical human activities possible. Phonation, for example, is possible as a result of fine and graded involvement of the respiratory motoricity. Speaking, singing, and playing wind instruments, for example, are largely the result of special modifications of respiratory movements. Strong changes of breathing in man also occur in connection with expressive motor behavior, such as respiratory inhibition during keen attention or hyperventilation during psychic excitation.

Normal breathing in human and in animal experiments is largely an artificial product. It is observed under a very special experimental condition, namely the elimination of most normal, everyday activities, especially in the case of humans. Investigator know how difficult it is to obtain this "normal" breathing from their subjects. In most cases,

R. Greger/U. Windhorst (Eds.)
Comprehensive Human Physiology, Vol. 2
© Springer-Verlag Berlin Heidelberg 1996

lengthy training periods are necessary to eliminate the "disturbances" caused by other activities.

There are only few investigations on arterial blood gases and blood pH under the conditions of normal, everyday life. It is to be expected that strong deviations from the "normal" data are observed. In normal life, the respiratory motor system is affected by many physiological functions. The homeostatic mechanisms prevent extreme deviations from the most favorable conditions of the cellular environment.

Any metabolic regulation that makes use of variations of respiration and any other interference with respiratory activity will affect the central neuronal network. Respiratory adaptation to chemical, motor, and behavioral challenges will be reflected in changes in the central, rhythmically firing neurons. Conversely, neurophysiological processes are dependent upon an appropriate chemical environment.

106.2 Efferent Innervation of the Respiratory Apparatus

Head Breathing and Neck Breathing. The approximate anatomic pathways from the rhythm generator down to the motoneurons of the respiratory muscles were revealed in classical experiments using sections and recording techniques. These data are helpful, for instance when assessing the respiratory pattern after injuries. They comprise "head breathing" and "neck breathing," which are respiratory movements of the nose wings, the mouth, the larynx, the vocal cords, and their neural innervation pathways, which can be traced back to the corresponding nuclei of origin of the Vth, VIIth, IXth, Xth, and XIIth cranial nerve and the upper cervical nerves. Except for movements of the vocal band and the larynx, these innervations are intensified during dyspnea.

106.2.1 Upper Airway Motor Systems

Sleep Apnea and Sudden Infant Death. The growing demand for diagnostic and therapeutic measures in sleep apnea of adults and infants (including sudden infant death) necessitates more detailed knowledge of the upper airway functions and reflexes (see Chap. 105). Flow limitations by the nose and pharynx occur chiefly in inspiration, whereas the larynx influences expiration more than inspiration. The activity patterns of the muscles in the region depend in part on the respiratory control system, but they are also influenced by competing requirements related to nonrespiratory functions.

106.2.1.1 Developmental Aspects

Rapid Eye Movement and Non-Rapid Eye Movement Sleep. The fetus is already capable of making coordinated,

rhythmic breathing movements in utero [43]. Under homeostatic conditions, the ventilatory movements are very much dependent on the behavioral state of the fetus, occurring predominantly during a condition similar to rapid eye movement REM sleep [38] (see Chap. 57). During the REM state, the posterior cricoarytenoid (PCA) muscle, the paired abductor muscle of the vocal cords, shows regular inspiratory bursts of activity in concert with those of the diaphragm. The thyroarytenoid muscle, a vocal adductor, shows little activity associated with breathing movements, but in the non-REM (NREM) state it fires during sequences thought to represent intrauterine precursors of ruminant regurgitation. As has been shown in fetal and neonatal lambs, there is a sharp increase in expiratory adductor muscle activity after birth. Records obtained within a few hours after birth suggest that laryngeal adductor activity often slows and occasionally interrupts expiratory airflow, particularly in the NREM state. Inspiration is accompanied by strong bursts of PCA activity. The breathing pattern of healthy human infants immediately after birth and during the first few hours and days of postnatal life show great variations. Neonates have highly compliant chest walls, which contribute very little to the passive mechanical properties of the respiratory system [55,139]. One of the infant's strategies for maintaining functional residual capacity and avoiding expiratory lung collapse is to retard or interrupt expiratory airflow, using the larynx as a check valve. Crying and grunting are expiratory activities that presumably contribute to this effect. Neonates are nose breathers and may fail to respond to nasal obstruction by breathing through the mouth, particularly during sleep [112]. The adult response pattern of switching to oral breathing develops during the first few months of life. The mechanism underlying this behavioral change is unknown.

106.2.1.2 Nose and Mouth Breathing

Variable Resistance. In adults, the respiratory role of the upper airway musculature is the regulation of resistance to airflow (see Chaps. 100, 105). Upper airway resistance is highly variable, even during quiet breathing and during sleep [66,124]. The chief sites of variable resistance are the nose and mouth, the pharynx, and the larynx. The upper airway acts as a functional unit in many circumstances, and movements of its various components are highly interdependent.

Circadian Rhythm. The nose accounts for about half of the total respiratory resistance in normal humans at rest [50,74]. The major factor determining nasal resistance is the extent of vascular engorgement of mucous membranes lining the nasal cavity. Microcirculation in this region is responsive to various local and systemic influences; cold air and parasympathetic stimulation increase nasal resistance, whereas CO_2 inhalation and sympathetic stimulation reduce it [61]. Airway resistance shows a very pronounced circadian rhythm and varies with the sleep cycle (see Chap.

58). Its duration of about 90 min is very stable. Nasal resistance exhibits a laterality cycle with a highly variable period length, greatly affecting the total airway resistance [137,142]. The tensor veli palatini muscle is phasically active during inspiration, and a reduction in the activity of this muscle in patients with sleep apnea may be consistently associated with an increase in nasopharyngeal resistance [13].

Oronasal Breathing. Flaring of the alae nasi during inspiration is prominent in human infants. The alar muscles show phasic respiratory activity [163]. An important feature of the nasal airway is that it can be partially bypassed by opening the mouth. Normal subjects performing bicycle exercise usually switch from nasal to oronasal breathing when ventilation reaches 35–45 l/min. Human patients with partial or complete nasal obstruction also breathe with the mouth open; whether the conversion to oronasal breathing in these circumstances is of reflex origin or represents learned behavior is unknown. Under conditions of high respiratory drive, including the first breath in infants, however, inspiration is accompanied by opening of the mouth, indicating a degree of automatic control of these movements. Although most "mouth breathing" is actually oronasal, the soft palate can be approximated to the posterior pharyngeal wall, thus excluding the nose from the respiratory circuit. This action occurs in swallowing and in activities such as blowing out a match, playing a wind instrument, or exhaling through pursed lips. Expiration through pursed lips is adopted by many patients with airway obstruction as a strategy to relieve dyspnea. The increased resistance at the mouth reduces dynamic compression of intrathoracic airways, but in fact overall resistance increases slightly; the mechanism by which this maneuver produces subjective relief of dyspnea is unknown [75].

106.2.1.3 Oropharynx

Pharyngeal Airway Resistance. This resistance can be actively controlled at two points: the pharyngeal isthmus and the posterior border of the dorsum of the tongue. At the level of the isthmus, the nasopharynx can be closed off from the oropharynx by apposition of the soft palate to the posterior pharyngeal wall. Closure of the velopalatine valve is accomplished by coordinated contraction of the superior pharyngeal constrictor and the levator veli palatini muscles. The tensor veli palatini muscles stiffen the soft palate but do not pull it posteriorly and thus do not close the airway. Oropharyngeal closure by apposition of the dorsum of the tongue to the posterior wall of the pharynx can be brought about actively by contraction of the styloglossus and mylohyoid muscles or by passive dorsal movement of the tongue mass due to gravity or subatmospheric intrapharyngeal pressure. This passive relapse of the tongue is opposed by the actions of the genioglossus and geniohyoid muscles, which draw the tongue ventrally toward their attachment on the mandible

[58]. These muscles are innervated by the hypoglossal nerve.

Levator and Tensor Veli Palatini Muscles. Much of our understanding of pharyngeal breathing movements is based on electromyography (EMG). The pharyngeal constrictor muscles are inactive during inspiration, but show phasic activity during expiration. The expiratory activity is augmented by hypercapnia and decreased or abolished by passive hyperventilation [60]. The levator veli palatini contracts with swallowing but is inactive during quiet or deep breathing, an appropriate pattern since this muscle effects velopharyngeal closure of the airway. The tensor veli palatini also participates in swallowing, but in addition shows phasic inspiratory activity. This action may prevent passive relapse of the soft palate during inspiration and thus help to maintain the patency of the nasopharyngeal airway [6].

Genioglossus Muscle. The genioglossus and geniohyoid muscles contract phasically during inspiration, pulling the tongue ventrally and preventing it from being passively drawn into the pharyngeal airway by gravity or subatmospheric intrapharyngeal pressure. Genioglossus EMG activity and hypoglossal nerve activity are increased by hypercapnia and hypoxia [24,121,122]. Activity of the genioglossus muscle is also influenced by mechanical stimuli. Lung inflation inhibits genioglossus muscle activity, along with that of other inspiratory muscles, and this response is abolished by vagotomy. A more specific reflex control may be linked to mechanoreceptors located somewhere in the pharyngeal wall. Negative intrapharyngeal pressure changes increase phasic genioglossus muscle activity, whereas positive pressure changes have the opposite effect. These responses remain after vagotomy or superior laryngeal nerve section, but are abolished by spraying the pharyngeal wall with an anesthetic aerosol. This load-compensating reflex response may play an important role in maintaining patency of the upper airway [101].

Upper Airway Occlusions. During quiet sleep the respiratory activity of the genioglossus and other muscles influencing the pharyngeal airway is somewhat reduced, and during REM sleep it becomes much reduced or even abolished [145]. The loss of tonic and phasic muscle activity greatly increases the susceptibility of the pharyngeal airway to dynamic compression. Although most people tolerate this situation without developing significant airway obstruction, a surprising number of patients, especially men over 45 years of age (not necessarily obese), suffer recurrent sleep-related upper airway occlusions with serious consequences related to hypoxemia and sleep deprivation [58,133].

Laryngeal Resistance. Although the larynx is capable of varied and complex movements in such activities as swallowing, coughing, and speaking, variations in laryngeal resistance during the normal breathing cycle take place largely at the level of the vocal cords. The motoneurons of the intrinsic laryngeal muscles have cell bodies in the ambiguous nucleus in the ventrolateral medulla, mostly on the same side [17]. The motor axons reach the muscles via the vagus nerve. The cricothyroid muscle is innervated by a branch of the superior laryngeal nerve while the other intrinsic muscles are supplied by recurrent laryngeal nerve branches. The respiratory movements of the cords are not determined exclusively by activity of the intrinsic laryngeal muscles. Extrinsic muscles extending from the cricoid and thyroid cartilages to other structures form a suspensory sling for the larynx, and these muscles exhibit some phasic activity in the breathing cycle, particularly in hyperpneic states.

Phasic efferent bursts of activity occur with each breath in the recurrent laryngeal nerve, representing a mixture of impulses destined for abductor and adductor muscles. Measurements of laryngeal resistance show a decrease during inspiration and a variable increase during expiration. Respiratory movements of the cords seem to be determined mostly by PCA muscle activity, contraction during inspiration, and variable relaxation during expiration. Results with esophageal electrode measurements show little PCA activity during quiet breathing in normal subjects, but strong inspiratory bursts of activity during voluntary hyperventilation [131]. The neural origin of laryngeal breathing movements is not completely understood, but is closely associated with the pools of neurons in the medulla and pons that constitute the central respiratory control mechanism (see Chap. 105). Some ambiguous nucleus cells fire spontaneously with a respiratory rhythm and, of these, some have been indentified as laryngeal motoneurons by antidromic stimulation of the recurrent laryngeal nerve [17] (see Chap. 105).

Laryngeal Valve. The respiratory movements of the larynx are closely coordinated with those of the respiratory pump. Vocal cord abduction slightly precedes contraction of the diaphragm, and this pattern is the result of earlier activation of laryngeal motoneurons rather than of differences in time required for conduction or contraction [29,45]. Opening the laryngeal valve just prior to the inspiratory stroke of the respiratory pump avoids the inefficiency of diaphragmatic contraction against the occluded or partly obstructed airway. This pattern seems to be present in all normal states and even persists in apneusis, but is lost during gasping respiratory activity in medullary animals, in which the onset of phrenic nerve activity slightly precedes that of the recurrent laryngeal nerve [162]. The respiratory movements of the vocal cords and the accompanying fluctuations in laryngeal resistance to airflow are altered when ventilation is increased by various stimuli. In anesthetized animals, hypercapnia increases the intensity of PCA activity during inspiration and prolongs the activity well into expiration. As a result, laryngeal resistance is slightly reduced during inspiration and greatly reduced during expiration. A similar response pattern occurs during eucapnic hypoxia and in hyperthermia [12,102].

Sleep. Respiratory movements of the larynx are known to be modified during sleep. In cats chronically instrumented with EMG electrodes in laryngeal muscles, slow-wave and especially REM sleep are accompanied by sharp decreases in PCA activity, particularly during expiration, and in some experiments also by increased expiratory activity of vocal cord adductors. These sleep-related changes are accompanied by increases in upper airway resistance, but the anatomic site of the increase in resistance has not been established and may not be at the level of the glottis [123].

106.2.2 Innervation of the Respiratory Pump Muscles

Phrenic Nerves. Descending pathways to the spinal motoneurons run crossed and uncrossed within the anterior and the anterolateral column of the spinal cord. They originate directly from bulbar inspiratory and expiratory neurons at least down to the level of C4. The motoneurons of the phrenic nerve in man are located in the cervical cord within the segments C4–C7, and those for the external and internal intercostal muscles within the segments T1–T7 in the ventral horns of the spinal cord [170]. Increased respiratory effort engages the rhythmic excitation of the accessory respiratory muscles. From EMG phical recordings in humans, we know that the boundaries between respiratory and postural muscles of the trunk are fluent. Abdominal muscles show expiratory increases of activity even during eupnea. In the prone position, as a result of the loss of the gravitational force acting on the ribs, inspiratory muscle activity of the shoulder belt decreases and the expiratory abdominal muscle activity increases. The respiratory rhythm can be shown within the innervation of the limbs, e.g., an inspiratory reinforcement of the patellar reflex has been observed [153]. Involuntary synchronizations have been observed between respiration and movements of legs or arms [80,85]. The respiratory rhythm affects motor innervation in general, particularly the respiratory muscles proper. The visible activity of the accessory respiratory muscles during dyspnea is only a gradual amplification of an otherwise latent, continuously existent neural patterning.

Discharge Frequency. The discharge frequency of the single motor fiber of the phrenic nerve during quiet breathing is between 5 and 30 per s. It increases continuously during the course of the inspiratory phase. Under the condition of augmented breathing at the inspiratory peak, frequencies up to 100 per s have been observed. The conduction velocity of the efferent phrenic fibers is 40–70 m/s. During inspiration, there is an increase not only in the discharge frequency within the single element, but also in the number of active motoneurons [4]. During enhanced respiratory drive, a prominent synchronization of impulses in many phrenic fibers can be observed in addition to recruitment and increasing frequency. Corresponding multiple recordings of medullary inspiratory neurons show that such synchronizations originate from the medullary inspiratory populations [27] (Chap. 105).

Tonic Activity. Between the inspiratory volleys of the phrenic nerve, very often and especially during quiet breathing a small "tonic" activity may continue, leading to an incomplete relaxation of the diaphragm during expiration. This tonic remainder may be influenced by vagal reflexes. This becomes prominent in the permanent diaphragmatic tone during hyperventilatory apnea.

Electromyographic Activity of the Diaphragm. This is linearly dependent upon the cerebral venous P_{CO_2}, the response remaining unchanged during the awake, REM or NREM state. This is in contrast to other partial compartments of the respiratory apparatus, which show typical state-dependent responses, and may in part be explained by the absence of the gamma loop [130].

Intercostal Nerves. The intercostal muscles are provided with a distinct gamma innervation in addition to the alpha innervation [44,155]. The gamma motoneurons have higher discharge frequencies than the alpha motor neurons; they are less strictly modulated by their respiratory rhythm, but tend much more to tonic discharge patterns with respiratory frequency modulation. Discharges of alpha motoneurons often follow an increased activation of gamma motoneurons, suggesting that activation of the respiratory muscles occurs in part via the gamma spindle loop. Since the respiratory rhythm of alpha activity continues after the elimination of gamma efferences, we have to presume that supraspinal respiratory neurons influence spinal alpha motoneurons as well as gamma motoneurons. The respiratory rhythm of gamma activity during hyperventilatory apnea may still continue when the alpha activity is already silent. As can be expected, afferent excitation is observed in the intercostal nerves from muscle spindles. Due to the respiratory rhythm in the gamma innervation, these nerves become activated during spontaneous breathing in the active phase of the muscles. On the other hand, during loss of central gamma drive, e.g., during artificial ventilation with central respiratory paralysis, they show activation during passive lengthening of the muscles in the counterphase. The intermediate function of the intercostal muscles between respiratory and postural muscles is also reflected in the double function of the intercostal muscle spindles as mediators of respiratory rhythmic contractions on the one hand and as triggers of counterreactions during passive lengthening on the other. In this respect, the motor innervation of the intercostal muscles is different from that of the diaphragm.

During REM sleep the gamma loop is inhibited. In newborn infants, who have highly compliant chest walls, paradoxical breathing appears, the chest walls being drawn inwards during inspiration (see Fig. 106.1). The compensating diaphragmatic activation guarantees the stability of the blood gas profile of this sleep phase. Increased airway resistance, however, as may be caused by a simple cold, can be reflected in severe desaturations of hemoglobin during REM sleep, and the inhibited gamma loop may aggravate the inability to "mouth breath" in early infancy. These interconnected mechanisms have to be taken into account when young infants have fallen victim to sudden death during sleep and the pathologist does not find anything other than a slight airway infection.

106.3 Afferent Innervation of the Respiratory Apparatus (see Chap. 105)

Pulmonary Receptors and Their Activity. Airways and the lung parenchyma have an abundant supply of receptive structures and afferent nerve fibers, especially within the region of the bronchial branches [30,174]. In particular, many nerve endings lie subepithelially, many within the bronchial muscles, with others down to the respiratory bronchioles and below the pleura. They have various kinds

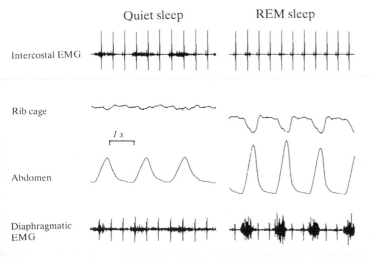

Fig. 106.1. Motion of rib cage and abdomen and electrical activity of respiratory muscles in quiet and rapid eye movement (*REM*) sleep in a preterm infant. In REM sleep there is paradoxical motion of the rib cage, increase in abdominal excursion, decrease in intercostal electromyogram (*EMG*), and substantial increase in diaphragmatic EMG [26]

of buttons, enlarged endings, or capsules; others end freely within the tissue and have been classified according to receptor types and reflex effects [53].

Lung Stretch Receptors. These receptors, appear to be best suited to measure the degree of lung inflation [3,7,82]. Their discharge frequency increases more or less linearly with augmenting lung volume, when lung inflation is constant, they adapt only minimally. The thresholds of these receptors vary; some are already active within the respiratory resting posture, whereas most begin discharging only during inspiration. The afferent fibers belong to the A-alpha and A-beta group and have conducting velocities of 15–60 m/s. At rapid stretches, action potentials reach frequencies up to 300 per s. Selective stimulation of the fiber group provokes the inhibitory reflex on inspiration; selective blockade. e.g., cooling the vagus to 8°C, eliminates the reflex. Less clear is the function of numerous other pulmonary receptors, which rapidly adapt and are mostly activated only by increased lung volume, firing only with a few impulses when stretched.

Irritant Receptors. Numerous receptors within the airways react with short-impulse volleys to fast inflation as well as to fast deflation not only of the whole lung, but also of isolated segments of the airways. This applies to epithelial receptors in the airways, which can be excited by powder inhalation and can be considered as the trigger for the cough reflex. The mechanically excitable cough receptors are mainly distributed in the proximal airways, whereas the receptors excited by chemical irritation tend to be located in the peripheral segments of the bronchial tree. The latter also exhibit an excitory response to rapid inflation and deflation. They also react to pathological states such as overload of the lung circulation, lung edema and lung embolization. The conduction velocities of the corresponding fibers lie between 3 and 25 m/s. These receptors were named "irritant receptors."
This group probably also includes receptors which are exclusively excited by lung deflation below the resting state, but are inhibited by lung inflation (lung deflation receptors). Many of them only become active as a result of lung collapse following a pneumothorax. They can be excited by chemical stimuli injected into the pulmonary circulation and are responsible for the apnea preceding the phase of rapid shallow breathing (pulmonary chemoreflexes).

J Receptors. Equal changes within the lung circulation, similar chemical and pharmacological stimuli, and, to a lesser extent, lung collapse provoke an excitory response in a further group of receptors, which are located within the lung parenchyma between capillaries and alveolar wall. Due to their juxtacapillary position they are called "J receptors" [127]. The afferent fibers are partly thin, non myelinated fibers with conduction velocities between 0.8 and 7 m/s. Their stimulation by phenyl-diguanide causes an initial apnea with inhibition of inspiratory and expiratory brain stem neurons, followed by rapid shallow breathing. The apnea is accompanied by a massive fall of blood pressure as a consequence of reflex bradycardia and vasodilation. Other motor activities are inhibited ("J reflex").

106.4 Neural Self-Control of the Respiratory Apparatus

Airway Defense Reflexes. Reflexes originating in the tracheobronchial tree and lungs fall into two categories: regulatory reflexes that determine the pattern of breathing in normal circumstances and defense reflexes that protect the respiratory tract from potentially harmful influences. The signals that subserve regulatory reflexes appear to be carried in the vagus nerves, as is most of the tracheobronchial input subserving defense reflexes. A subsidiary input for defense reflexes travels with sympathetic nerves to the spinal cord, but so far there is no evidence that this spinal input plays any part in the normal regulation of breathing. Reflexes evoked by changes in lung volume regulate the pattern of breathing and the tone of airway smooth muscle evoked by chemical stimuli applied to the lower airways. The so-called airway defense reflexes subserve quite a different function: they represent emergency, protective responses brought into play when harmful agents are carried to the respiratory tract in the inspired air or in the bloodstream. These reflexes comprise a variety of respiratory maneuvers (cough, apnea followed by rapid shallow breathing, bronchoconstriction, and increased secretion by airway glands) that expel the noxious agent or at least limit its further entry into the lung and reduce its harmful local effects. The defense reflex responses often include a profound reduction in blood pressure and heart rate, which is beneficial insofar as it limits absorption and dissemination of a harmful chemical by the bloodstream.

Hering–Breuer Reflexes. The following events occur during changes of lung volume

- Inflation elicits a reflex of inspiratory inhibition with bronchodilation. This is one of the most effective trigger mechanisms for a short-term reflexogenic stand-still of breathing (vagal apnea).
- Deflation of the lung causes strengthening of inspiration.

Both effects involve the diaphragm and the intercostal muscles and can be demonstrated by their nerve activities. The accessory respiratory innervation, e.g., the laryngeal muscles, are also involved. The circulatory effects of lung inflation are also biphasic: weak lung inflation causes a reflex tachycardia, a stronger one bradycardia. Central mechanisms may be responsible, since in general inspiration is accompanied by acceleration, expiration by deceleration of heart beat frequency. These reflexes are named after their discoverers, *Hering and Breuer* [67].

106.4.1 Functional Significance of the Lung Stretch Reflexes

Switch Reflexes. Hering and Breuer interpreted the reflexes as "switch reflexes," causing the change from inspiration to expiration and vice versa. This switching is possible after vagotomy by purely central mechanisms; however, the respiratory frequency and tidal volume are influenced by the stretch reflexes of the lung. The termination of inspiration by vagal afferents can replace similar actions of central substrates of the upper pons [16].

The effect of low lung inflation favoring inspiration leads to a faster inspiratory innervation. The vagal feedback mechanism acts as a further extracentral control for the incremental neural pattern of inspiration as well as for its limitation.

The earlier the inspiratory movement is cut off by the reflex, the later the next inspiration starts. The speed with which the inspiration develops, i.e., with which the lung volume triggering the reflex is reached, depends on the respiratory drive. Thus respiratory drive and lung stretch reflexes principally determine respiratory depth and frequency. They result in a "gear reduction," e.g., reduction of respiratory depth and increase of respiratory frequency with augmenting respiratory drive, thus rendering respiratory work more economical. Although in man the lung stretch receptors become activated even during normal tidal volume, the stretch reflexes can only be demonstrated above a respiratory depth of about 1 l, i.e., only within the range of the inspiratory reserve volume (see Chap. 100). Within the range of normal resting ventilation in humans, the switching from one respiratory phase to the other is thus determined by central mechanisms alone.

Tonic Afferents. Hess [68] pointed out that the lung stretch reflexes had a tonic functional component. His interpretation was based on the following experimental observation. The lung was artificially inflated to a variety of constant volumes. At small lung volumes, an increased average of diaphragmatic tone with rapid shallow respiratory excursions appeared. At stronger lung inflation, the diaphragm relaxed and its contractions became slower and attained a higher amplitude. An increased diaphragmatic tone via lung inflation would therefore decrease by itself and vice versa. In addition, tonic lung afferents influence the central rhythm generation.

The tone-changing lung reflexes need longer to develop than the duration of a respiratory cycle. This implies the possibility of a summation over several respiratory cycles and could also be the functional basis of hyperventilatory apnea. Thus within the lung stretch reflexes, tone-regulating and purely phasic switch effects are inseparably linked with each other.

It must be differentiated between the normal effect of little or absent lung tension, which favors inspiration, and a strong effect promoting inspiration, which arises when the lung volume falls below the functional residual capacity (see Chap. 100). This causes intensive inspiratory activity with inhibition of expiratory activity and an acceleration of respiratory frequency. The reflex can be elicited by compressing the chest and, very considerably, by reduction of the lung volume due to a pneumothorax.

Hyperventilatory Apnea and Head's Paradoxic Reflex. Hyperventilation causes longer-lasting inhibition of respiratory innervation (hyperventilatory apnea), even under constant blood gas conditions. All these effects vanish almost completely after vagotomy or lung denervation. Another vagal reflex only arises as a result of a very strong lung inflation; it elicits a further strengthening of the inspiratory innervation (Head's paradoxic reflex [62]). It may be speculated that this reflex underlies the considerably deepened inspirations occurring from time to time (in man about three per h), which improves lung inflation and therefore prevents atelectasis. These deep inspirations vanished after vagotomy in animal experiments [91]. Various more recent experiments have led to a controversial discussion. The reflex has been ascribed to rapidly adapting receptors, but from cold block effects it seems more likely that the reflex is initiated by stimulation of nonmyelinated afferent vagal fibers [173].

The vagal afferent endings innervating the lower respiratory tract consist of slowly and rapidly adapting mechanoreceptors supplied by myelinated fibers and of the simpler endings of nonmyelinated fibers. The latter endings, like slowly and rapidly adapting mechanoreceptors, appear to be present in all parts of the tracheobronchial tree and to comprise more than one sensory modality [22,48,110,135,144]. Although the vagus nerve is the main afferent pathway from the lower respiratory tract, afferent fibers also run with sympathetic nerves to the spinal cord.

Slowly Adapting Pulmonary Stretch Receptors. These receptors are the ones for which there is the most satisfactory correlation between the properties of the endings themselves, the identity of the morphological structures generating the trains of afferent impulses, and the nature of the reflexes triggered by these impulses [171]. Slowly adapting stretch receptors are the afferents responsible for the Hering–Breuer inflation reflex; their mounting discharge in inspiration provides an inspiratory "off" switch, and their continuing discharge in expiration lengthens the respiratory pause [67]. In addition, the slowly adapting receptors cause tracheobronchial smooth muscle to relax and thus dilate the airways [175]. The temporal pattern of input from slowly adapting stretch receptors influences the central respiratory mechanism in a phase-related fashion [166]. Therefore, differences in the timing and pattern of receptor discharge in different regions of the airways may have a functional significance. Slowly adapting stretch receptors in the trachea fire phasically with each normal breath, but those in the extrathoracic trachea fire out of phase with their intrathoracic counterparts and reach their peak frequency during expiration, when transmural pressure at this site is positive. Hence extrathoracic tracheal receptors may exert their main effect on expiratory time. Some of the receptors within the thorax fire throughout the cycle, whereas others are silent until inflation reaches a critical level. Paintal [128] has described these two types as low-threshold and higher-threshold receptors, respectively, which may be more appropriate than the terms "tonic" and "phasic" used by some authors. A phasic increase in discharge with inflation is characteristic of all slowly adapting stretch receptors in the thorax, whether they fire throughout the cycle or only during inflation. These differences are thought to reflect differences in the mechanical environment of the receptor; most low-threshold receptors appear to be extrapulmonary, whereas most higher-threshold receptors appear to be intrapulmonary [143]. If most expiratory discharge arises from the trachea and extrapulmonary bronchi, input from this region of the airways should be important in determining the expiratory time (TE).

In the classification by Miserocchi and Sant'Ambrogio [109], type I receptors (40% of total) are usually low threshold, reach maximum frequency (average 40 impulses/s) at a distending pressure of 10 cm H_2O, and are mainly extrapulmonary. Type II, which often have a higher threshold, do not achieve maximum frequency (average 90 impulses/s) until distending pressure reaches 30 cm H_2O, i.e., near-total lung capacity; they are mainly intrapulmonary. If these differences in response to static inflation (dogs) are relevant during spontaneous breathing, then they are likely to become apparent only in hyperpnea, when type II (intrapulmonary) receptors may play the major role in limiting inspiratory time (TI).

Whether slowly adapting stretch receptors are classified as low-threshold and higher-threshold or as type I and type II receptors, the functional significance of these distinctions should not be overemphasized, because discharge patterns may be more labile than such classifications suggest. The limitation of firing frequency proposed for type I (extrapulmonary) receptors, examined under static conditions, may not apply to the dynamic conditions of normal breathing.

Although few chemicals excite slowly adapting pulmonary stretch receptors, the receptors appear to be more susceptible than other pulmonary afferents to inhibition by chemicals. High concentrations of volatile anesthetics such as halothane, ether, chloroform, and trichloroethylene first increase and then abolish receptor discharge. The inhibitory effects are selective, and pulmonary C fibers are stimulated by these high concentrations. Slowly adapting receptors, such as mechanoreceptors in the skin (see Chap. 32) and elsewhere, are stimulated by hypocapnia [31].

106.4.2 Extrapulmonary Components in the Stretch Reflex Self-Control of Respiratory Movements

The reflexes originating from the extrapulmonary respiratory apparatus are thought to contain two components:

- A Spinal component, the essential portion of which originates from the muscle spindles of the intercostal

nerves, analogous to the proprioreflexes of the limb muscles
- A supraspinal component which influences central rhythm generation

Apart from the influence of the single respiratory movements, the essential task of pulmonary and extrapulmonary mechanical respiratory reflexes consists in the utmost economical setting of the respiratory mechanics at any demanded ventilatory quantity.

Proprioception. Breathing becomes weakened with the interruption of the gamma spindle loop, e.g., by an experimental cut of the posterior roots. If thoracic movements are hindered, diaphragmatic movements increase by reflex and vice versa. Thus, for example, thoracic breathing is augmented during pregnancy. Correspondingly, after experimentally cutting the phrenic nerves the thoracic movements increase. The proprioception of respiratory muscles plays an important role within the reflex adaptation to changed respiratory resistances. A sudden increase in inspiratory or expiratory ventilatory obstruction elicits a reflex augmentation of inspiration or expiration. As in other striated muscles, isolated stretching of single intercostal muscles or stimulation of their fast-conducting afferent nerve fibers enhances alpha innervation of the same muscle (see Chap. 49). There are numerous typical discharge patterns of muscle spindles within the dorsal roots of the intercostal nerves [34].

Diaphragm. In contrast to the intercostal muscles, typical proprioceptive reflexes of the diaphragm are weakly developed. The diaphragm is only sparsely provided with muscle spindles and most are found within the crura; it also contains some tendon afferents. The efferent gamma innervation of the few muscle spindles is also weakly developed [34]. On the other hand, the diaphragm has an abundance of other kinds of receptors, the activation of which leads to supraspinally mediated, nonspecific respiratory reactions, as do other nociceptive stimuli. In addition to the muscular proprioceptive sensitivity, other afferents from tendon and joint receptors of the respiratory apparatus probably contribute to the reflex self-control of respiratory movements [136].

106.4.3 Defense Reflexes of the Respiratory Apparatus

Asthma. The airway defense reflexes are essentially protective (see Chap. 105). Current interest in airway defense reflexes stems mainly from two factors. One is the recognition that many of the discomforts of industrial air pollution arise from stimulation of airway afferents and from the reflexes that such stimulation evokes. The other is the demonstration that certain chemicals, known to be released in the bronchial walls, are capable of stimulating afferent nerve endings at least as vigorously as exogenic irritants, so that reflexes may be triggered endogenously from sites in the tracheobronchial tree. This latter mechanism is activated in pulmonary anaphylaxis and is believed to play a part in human asthma.

106.4.3.1 Nasal Defense Reflexes

Sneezing Reflex. This represents a strong intervention in the respiratory activity, inducing a complicated coordinated sequence of inspiration, short occlusion of the nasal pharyngeal space, and subsequent explosive expiration. The sneezing reflex is mediated by receptors of the nasal mucosa stimulated by mechanical and chemical agents and conducted within the trigeminal nerve.

Another reflex, also initiated by nasal receptors, results in apnea elicited by irritating gases or vapors, such as SO_2 and acids; receptors with afferents in the trigeminal and the optical nerves are involved in this reflex.

106.4.3.2 Laryngeal and Tracheobronchial Defense Reflexes

Cough Reflex. Comparable to the sneezing reflex is the cough reflex, which can be easily elicited from the larynx, but also from the trachea and the bronchi. By deep inspiration with subsequent short glottis occlusion and strongest possible expiratory activity, air velocities within the trachea reach the order of sonic speed. Intrathoracic pressure changes reach peaks of up to 300 mmHg (39.9 kPa), which may affect the arterial pressure to the same level. The coordination center for coughing is probably located within the upper pons. The reflex can be more or less inhibited by certain drugs, such as derivatives of morphine [83].

Histamine, Bradykinin, and Prostaglandins. Airway defense reflexes are evoked by tracheal inhalation of dusts, of harmful chemicals such as hydrochloric and acetic acid vapors, and ammonia, of poison gases such as bromine and phosgene, and of high concentrations of the volatile anesthetics: ether, trichloroethylene, halothane, and chloroform. Powerful irritants evoke not only respiratory disturbances, but also bradycardia and hypotension. It is generally agreed that the major afferent pathway is in the vagus nerves, but changes in breathing may persist after vagotomy. In some cases these changes are secondary to damage to lung tissue and interference with gas exchange, but in dogs and cats an afferent sympathetic pathway contributes to the primary reflex effects [173]. The reflex effects evoked by inhalation of SO_2 are of particular interest, because SO_2 is a common air pollutant. Histamine, bradykinin, and the prostaglandins that are released in the bronchial walls in immune-type reactions are all capable of evoking defense reflexes from the lower respiratory tract. Prostaglandin F_{2a}, like histamine, has a direct bronchoconstrictor action, which is potentiated by experimentally induced upper respiratory infections or by exposure to ozone [41].

Irritant Receptors. There is evidence that increased input from rapidly adapting, irritant receptors triggers augmented breaths or sighs when lung compliance decreases. Thus gasps or sighs are a feature of the response to lung deflation, pneumothorax, and histamine administration, conditions in which lung compliance decreases and irritant receptor activity increases [158]. These augmented breaths have the obvious physiological benefit of restoring lung compliance and reinflating areas of collapse, and their induction by irritant receptors appears to be an important component of airway defense reflex mechanism. If reflex bronchoconstriction is also triggered by irritant receptors and accompanies these large breaths, it may stabilize dead space and promote reinflation of collapsed alveoli. Gasps or sighs are also a feature of the response to stimulation of peripheral chemoreceptors by hypoxia. There is evidence that under these conditions the inspiratory drives from peripheral chemoreceptors and from irritant receptors facilitate each other [57].

C Fibers. Endings of nonmyelinated afferent fibers have been increasingly shown to be involved in airway defense reflexes. Bradykinin and prostaglandins E_2 and I_2 stimulate afferent C fibers in the lower airways and, in susceptible human subjects, cause coughing, wheezing, and retrosternal soreness. These afferent nerve fibers travel to the spinal cord via sympathetic nerve branches that traverse the stellate ganglia. Many fibers have sensitive endings in the pleura surrounding the lung roots or in the structures of the lung roots, and some supply the trachea, but so far few endings have been identified within the lung itself. The receptive fields are often extensive, and a single afferent fiber may supply endings to more than one anatomic structure. Sympathetic reflexes from the lower airways have so far been demonstrated only in response to noxious stimuli. Nevertheless, observations in human patients indicate that the sympathetic afferent pathway from the tracheobronchial tree and lungs is not involved in pain sensation. Sensations from the lower airways, including pain, are transmitted by the vagus nerves [59].

Conscious Perception of Lung Inflation. More recent results obtained from awake adults with vagal blockade show that vagal afferents mediate conscious sensations, at first pain and uneasiness, e.g., during intubation, bronchial catheterization, reinflation of the lung after removal of a pneumothorax, and also the sensation of dyspnea by lung diseases. The conscious perception of the lung inflation phase does not originate from vagal stretch receptors, however, but from the sensory afferents from thoracic muscles and joints.

106.4.3.3 Respiratory Reflexes from the Pulmonary Circulation

Increase of blood volume within the pulmonary circulation causes a characteristic rapid shallow breathing, which can be produced, for example, by very fast venous infusions, by balloon extension of the left atrium, or by embolization within the pulmonary circulation. Apnea of variable duration may precede this breathing pattern. The physiological role of these reflexes is not clear, but they become more important under pathological conditions, e.g., in connection with lung edema or lung embolism.

Pulmonary Chemoreflex. This complex pattern of response, which includes bradycardia, hypotension, and apnea followed by rapid shallow breathing, is evoked by intravenous injection of certain chemicals such as amidines, antihistamines, 5-hydroxytryptamine, nicotine, ammonia, adenosine triphosphate (ATP), lobeline, serum, capsaicin, opiate petides, and phenyl-diguanide, and is initiated when the chemicals reach the lung and stimulate afferent vagal endings immediately accessible from the pulmonary vascular bed [126]. The pulmonary chemoreflex was first described by Brodie [23]. It is evoked in man by lobeline and includes contraction of airway smooth muscle [40]. Dawes and Comroe [37] suggested that the pulmonary chemoreflex is evoked in clinical conditions by endogenous chemicals released in the lungs.

106.5 Feedback and Non-feedback Respiratory Drives

It is useful to distinguish between feedback and nonfeedback respiratory drives. Feedback drives are those that become reduced by their own effect on ventilatory augmentation, and nonfeedback drives are those which do not have this property. Feedback respiratory drives comprise the arterial and central PCO_2, the arterial and central pH,

and the arterial PO_2. In a sense, the inhibitory effect of lung stretch receptors lasting for more than one respiratory cycle is also a type of negative feedback. An example of a nonfeedback respiratory drive is pain. The consequences of this kind of relationship are shown in Fig. 106.2.

106.5.1 Physiology of the Reticular Formation, Basic Drive, and Vigilance

Voluntary innervation and exogenous influences interfere with breathing, which has a great deal to do with the "respiratory centers" being part of the reticular formation of the brain stem. As with the activating system of the brain stem, afferents from many somatic and vegetative receptors and from cortical areas affect the structures innervating the respiratory system [56]. The reticular formation mediating nonspecific tonic and guaranteeing specific phasic activation, e.g., by gating, plays a key role in the state-dependent patterns of the respiratory partial systems.

Respiratory Rhythm. It has already been mentioned that it is difficult to prove whether or not the central neural structures are capable of producing respiratory rhythm without any afferent input. In the adult cat, the respiratory rhythm was completely dependent upon a certain amount of chemosensitive afferent input, which could be shown even in the unanesthetized system [148]. However, this did not hold in the presence of hyperthermic drive [156]. A natural model contradicting the above supposition is the fetus: while its homeostasis is guaranteed by the placenta, it breathes state dependently and preferaly during REM sleep of low-voltage electrocorticogram (ECoG) [38]. Clas-

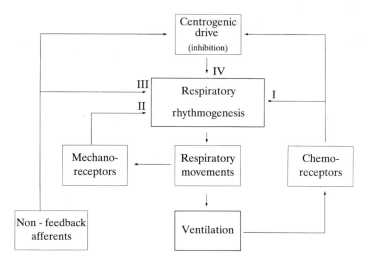

Influence on respiratory rhythmogenesis:

    I: Chemo - feedback

    II: Mechano - feedback

    III: Non - feedback afferents

    IV: Centrogenic drive (or inhibition)

Fig. 106.2. The four main groups of afferents converging on the "respiratory center" and their functional position in the whole control system of ventilation ([84] modified)

sic and recent observations in isolated preparations and brain stem cultures have tried to prove the independence of rhythm-generative populations [18,19,141]. In accordance with many recent investigations, however, it is probable that a certain excitatory tonic background is necessary, which is achieved by the confluence of many afferent signals on a structure presenting favorable conditions for such a convergence (see Chap. 105).

Rhythmogenesis. From this point of view, the functional importance of respiratory enhancement by a great number of nonspecific respiratory stimuli becomes evident, which otherwise would have to be considered as mere disturbances of ventilatory adaptation to metabolism, since they mostly cause deviations from the normal blood gas values. If we consider them in analogy to the function of nonspecific afferents of the activating brain stem system, their task may lie in the maintenance of basic activity, which is essential for rhythmogenesis. The hyperventilation resulting from a very considerable increase of one of these afferents would have to be considered as an exaggeration of the basic drive. Such exaggerations are counteracted by homeostatic mechanisms, especially feedback via P_{CO_2}. If this feedback is missing, as for example in infants with congenital insensitivity to CO_2 (Ondine's curse syndrome), hyperventilation with alkalosis is observed during exercise or excitement, and hypoventilation with severe acidosis at rest and during NREM sleep.

Sleep States. The respiratory response to CO_2 and the CO_2 threshold are dependent upon the basic drive [54]. The flattening and shifting to the right of the CO_2 response curve during sleep stages are well known. Anesthetics and analgesics, the site of action of which is almost within the brain stem, act in the same way as sleep.

An interesting connection with the reticular formation is the influence of wakefulness on ventilation [134]. Changes of sleep state are accompanied by corresponding ventilatory variations (see Chap. 57). During NREM sleep, particularly the deep stages (slow-wave sleep, SWS), breathing is regulated solely by the metabolic respiratory control system. As a result, defects in the chemical control system generally cause considerable ventilatory disturbance during NREM sleep. After acute hyperventilation (to produce a respiratory alkalosis), healthy subjects continue to breathe regularly while awake, but not during sleep [54]. During wakefulness there is an additional stimulus to breathing over and above that provided by the metabolic control system. This additional drive may represent a tonic excitation of the medullary respiratory neurons by the brain stem reticular system (see Chap. 105), which is fundamentally involved in the maintenance of central nervous system (CNS) arousal. It is the progressive withdrawal of this "nonchemosensitive" neural stimulus through stages 1–4 of NREM sleep, due to progressive reticular deactivation or inhibition, that results in a passive reduction of medullary respiratory neuronal activity. A reduction in alveolar ventilation during NREM sleep results in hypercapnia of 2–7 mmHg, which corresponds to a de-

crease in arterial pH of 0.01–0.06 units. These changes, together with small decreases in arterial PO_2, can be calculated to reduce SaO_2 by 1%–3%, which correlates with measured steady state values during NREM sleep.

An irregular breathing pattern characterizes REM sleep, and a steady state of ventilation may not exist. When phasic phenomena of REM sleep are visually obvious or when measurements in REM sleep are confined to periods of phasic muscular activity, the mean ventilation is increased above NREM values. When phasic activity is less marked, mean ventilation is only slightly greater or slightly less in REM than in NREM sleep, and mean alveolar ventilation is essentially unchanged. Thus mean values of ventilation during REM sleep may be misleading, because they generally fail to indicate that this sleep state is of a nonsteady-state nature and that instantaneous ventilation can be exceedingly labile on a breath to breath and minute to minute basis. Data from recent studies show, for example, that mean transcutaneous P_{O_2} in newborn infants, was 5.4–6.5 mmHg less than in NREM sleep [147]. In adolescents, mean SaO_2 in REM sleep was 96.1%–97.5% and not significantly different from that in NREM sleep. The greatest transient decrease in SaO_2 was 1.5% [164]. In young adults, mean minimal SaO_2 during REM sleep was 2%–4% less and somewhat more variable than in NREM sleep. In contrast, middle-aged men demonstrated transient periods of arterial desaturation of 11% (mean) during REM sleep [21]. In general, the rather limited data available for healthy human subjects suggest a slight increase in ventilation during REM sleep compared with ventilation during NREM sleep, but no significant change in alveolar P_{O_2}. In healthy infants with a normal chemosensitive drive, NREM sleep is characterized by a stable P_{CO_2} of 40.4 mmHg, varying by 1.15 mmHg; in REM sleep the mean P_{CO_2} is 38.6 mmHg, varying by 3.58 mmHg [149]. Average values of ventilation and even of alveolar ventilation are more likely to be increased rather than decreased during REM sleep. Therefore, reductions in arterial oxygenation under such conditions must result from mechanisms other than alveolar hypoventilation.

In a study of arterial desaturation during sleep in middle-aged adults, numerous episodes of desaturation occurred without any obvious change in the breathing pattern. In contrast, desaturation was highly correlated with increasing body weight. Thus reductions in functional residual capacity (FRC) and in regional ventilation perfusion ratios may have played a role in causing the hypoxemia. Reductions in FRC during REM sleep have been suggested to occur in young adults and have been documented in infants. Because of the highly compliant chest wall in infants and the decreased activity of the intercostal muscles during REM sleep, both a reduction in FRC and asynchronous chest wall movements are probably major factors in the hypoxemia of REM sleep in infancy. Similar phenomena may also occur in healthy adults [65,114,168].

The complexity of the origin of a certain resting ventilation can be demonstrated by quite another interesting clinical observation. Patients who, because of transient respiratory paralysis, have been artificially ventilated for some period

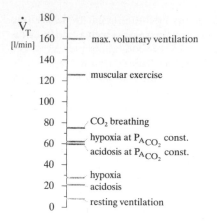

Fig. 106.3. Maximal ventilatory volumes observed under different conditions. Note that: (a) ventilation during strenuous muscular exercise lies below the maximal ventilatory capacity and is not determined by it; (b) ventilaton during muscular exercise significantly exceeds the ventilatory volumes obtained with blood chemical drives; (c) the respiratory effects of acidosis and hypoxia are relatively small since they are partly compensated for by release of CO_2; their real effect can only be demonstrated if the arterial CO_2 pressure (Pa_{CO_2}) is kept constant [83]. $\dot{V}I$, ventilatory minute volumes

of time, maintain the artificially adjusted ventilation for several days, even if large deviations of blood chemistry from normal values arise [161].

106.5.2 Cortical and Subcortical Influences on Ventilation

Respiratory innervation is intermediate beween vegetative and somatic innervation insofar as it occurs unconsciously and automatically; on the other hand, it can be influenced voluntarily in activating and inhibiting direction. For a short time (10–20 s) it is possible to hyperventilate voluntarily so that values of 160–180 l/min (calculated) may be attained. This maximal voluntary breathing is used to test the capacity of the respiratory apparatus (value of respiratory limit). It even exceeds ventilation during maximal exercise (120–150 l/min; see Fig. 106.3). On the other hand, ventilation can be inhibited, e.g., during intense watchfulness or eager expectation. A complete respiratory stop is obtained during diving (voluntary apnea).

Voluntary Apnea. Central rhythmogenesis during voluntary apnea probably does not become silent, since respiratory rhythm continues in other motor systems. The inhibitory effect therefore has to act peripheral to the rhythm generator [154]. Cortically elicited influences on ventilation certainly occur during speaking, singing, or playing of wind instruments.

Transitions between voluntary conscious and unconscious influences on breathing exist. For instance, a certain form of respiratory activity may be repeatedly connected with certain motor activities. Thus, increased ventilation at the beginning of exercise is adjusted according to the expected effort, even if the actual exercise load fails to occur (start reaction) [11]. Ventilation would also be elevated beyond the amount corresponding to the O_2 consumption when the actual exercise load is smaller than that which corresponds to the motor innervation, partly due to neuromuscular blockade by curare [120]. The final, precise adjustment of respiratory adaptation to exercise is guaranteed by other mechanisms, however (see below).

Speech Center and Limbic Cortex. Numerous respiratory effects can be observed in animal experiments by electrical stimulation of cortical and subcortical structures. Such systematic stimulations have shown that the tone of extensor and flexor muscles is controlled from some cortical areas influencing breathing [100]. Excitations and inhibitions of respiratory activity can be produced by stimulating the motor and premotor cortices. As to be expected, respiratory effects are elicited from areas around the speech center within the caudal region of the gyrus precentralis. Another effective region is the limbic cortex which controls of vegetative functions accompanying emotional behavior (see Chap. 18). In the anterior gyrus cinguli and in the *orbitoinsulotemporal lobe*, ventilatory inhibition can be elicited. Since during stimulation of the corresponding areas in nonanesthetized animals arousal reactions are evoked, such effects can partly be interpreted as respiratory inhibition linked to the arousal reaction in normal behavior.

Lateral and Posterior Hypothalamus. A strong similarity with the respiratory pattern during muscular exercise (maintained increase of tidal volume and respiratory frequency) is seen as an effect of electrical stimulation within the lateral and posterior hypothalamus. From the same areas changes of circulatory innervations can be obtained, which resemble the circulatory effects of muscular exercise (see Chap. 108).

106.5.3 Respiratory Drives from Active Muscles and Limb Movements

The largest enduring increases of ventilation can be elicited by muscular exercise, which are connected with movements of the limbs (see Chap. 108 and Fig. 106.6).

Chemical and Mechanical Stimuli. In human and animal experiments, muscular exercise was produced by electrical stimulation of muscles or their nerves [10,78]. As far as the mediator of the feedback from the exercising muscle driving ventilation is concerned, humoral and neural pathways must be considered. The humoral drives probably do not form a specific stimulus, in the sense of an exercise agent, but are traceable to the chemical drives. After elimination of humoral transfer possibilities, a neurally mediated component remains. The question arises as to the stimuli for these neural afferent impulses and the receptors from which they originate. Chemical stimuli derived from the changed muscular metabolism or mechanical stimuli caused by the movement as such are possible candidates.

Nociceptive Fibers. Even passive movements lead to a slight ventilatory augmentation. The receptors involved might be muscle spindles or tendon organ and joint receptors (see Chap. 49). Movements resulting from muscle contractions are accompanied by significantly stronger respiratory increases than are passive movements. The respiratory effect is proportional to the applied load. The afferent pathways involve the muscle nerves, spinal anterolateral tracts, and also, to a lesser extent, the dorsal columns (see Chap. 31). Even electrical stimulation of the Ia and Ib fibers (see Chap. 49) causes ventilatory augmentation. This effect is enhanced when the thinner nociceptive fibers of the muscle nerve are stimulated as well. If exercise is elicited by direct stimulation of muscles, the additional stimulation of such afferents cannot be excluded. Local perfusion of the gastrocnemius muscle with hypertonic saline stimulates reticular neurons, which are also sensitive to CO_2 [157].

As to be expected from the general respiratory effects, respiratory neurons within the brain stem react to the stimulation of afferent muscle nerves, thus leading to the activation of inspiratory neurons. The activation also encompasses the surrounding nonspecific reticular neurons.

106.5.4 Nonspecific Respiratory Drives

As can be expected from the close relation between respiratory structures and the reticular formation in the brain stem, many sense modalities from widespread receptor fields may affect ventilation [86].

Temperature-Related Changes in Ventilation. Intense warm or cold stimuli to the skin have short-lived effects on respiration. It is well known that inspiratory movements are provoked by sudden body cooling. Rewarming or cooling of the whole body is followed by continuing ventilatory changes. At elevated body temperature during fever or impaired heat release, e.g., during a hot bath, hyperventilation develops, PA_{CO_2} falls and the CO_2 response curve (for a definition see Sect. 106.7.1) attains a steeper slope (see Fig. 106.6). The respiratory enhancement is greater than that expected from the increased metabolism occurring at the same time. Thus it is not simply due to the metabolism, e.g., increased CO_2 production. There are various factors that might contribute to this effect, but their relative quantitative importance is not yet known; they include the effects of temperature directly on the respiratory centers or indirectly via higher temperature sensitive areas, excitation of cutaneous thermoreceptors and chemoreceptors. Warming the body with CO_2 kept constant causes an increase in ventilation of 7–8 l/min with every °C increase in rectal temperature [36]. Ventilation also increases with a decline in temperature, at first exceeding the compensatory augmentation of metabolism, and thus accompanies a decrease in Pa_{CO_2}. With progressing hypothermia, a cen-

trally induced ventilatory inhibition sets in, which leads to higher alveolar CO_2 partial pressures with a simultaneous fall of the O_2 partial pressure at body temperatures between 34° and 30°C (see Fig. 106.4). In deep hypothermia, ventilation is unable to guarantee sufficient gas exchange in spite of the reduced metabolism. Such degrees of hypothermia, which are used in surgery to reduce the oxygen demand of organs of vital importance, require artificial ventilation.

Pain. A strong respiratory drive may be produced by pain. As for all respiratory drives, which are not bound to a corresponding increase in metabolism, a maintained pain-induced hyperventilation may cause disturbances of the acid–base balance due to a disproportionate release of CO_2. The afferents activating respiration are nonmyelinated fibers, by the stimulation of which hyperventilation can be induced experimentally.

Respiratory Inhibition by Pressoreceptors. Among the afferents projecting to the reticular formation, those from the pressoreceptors have inhibitory effects on reticular neurons. There is a corresponding effect on respiration with a diminution of the respiratory frequency and tidal volume (see Fig. 106.5). At strong stimulation, e.g., upon large positive pressure steps within a carotis sinus preparation, complete apnea can result. Conversely, a decrease in pressure at the baroreceptors increases ventilation. Both inspiratory and expiratory neurons in the brain stem are

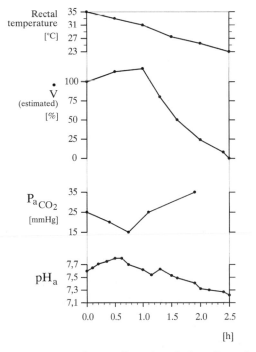

Fig. 106.4. Ventilatory effect of gradual cooling of a slightly anesthetized dog. The reaction is biphasic. Up to a core temperature of 31°C ventilation ($\dot{V}$) increases, followed by a fall leading to apnea at about 25°C. $\dot{V}$ is expressed as a percentage of control values. Arterial P_{CO_2} (Pa_{CO_2}) and arterial pH (pH_a) values correspond to the ventilatory changes [125]

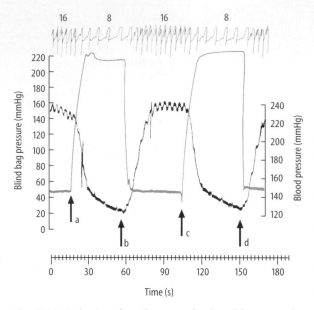

Fig. 106.5. Reduction of ventilatory amplitude and frequency of a dog (*upper channel, blue*) and parallel fall in blood pressure during elevation of the pressure within a carotid blind bag preparation [69]

inhibited, and during apnea the expiratory neurons become tonic with a reduced discharge frequency [83].

106.6 Hormones and Respiratory Drives

Effects on ventilation are known to be exerted from by:

- The adrenal gland
- The corpus luteum and placenta
- The thyroid gland.

106.6.1 Effect of Epinephrine on Respiration

Increases in the blood concentrations of epinephrine and norepinephrine induce ventilatory augmentation and steepen the CO_2 response curve (see Fig. 106.6). In man, $10\,\mu g/min$ of intravenously administered epinephrine and norepinephrine causes a distinct increase in ventilation. Since such increases in the epinephrine concentration fall within the physiological range, it may be concluded that the hormonally induced ventilatory rise plays a role during exercise or mental excitement. Whether this is due to a direct adrenergic effect on the brain stem respiratory neurons, in analogy to the effect of epinephrine on the activating brain stem system, has not yet been determined. There may also be indirect mechanisms, such as an increase of glycolysis with a fall of blood pH or a reduction of cerebral blood flow by vasoconstriction und thus a rise in the cerebral venous P_{CO_2}.

The use of higher epinephrine doses in animal experiments produces strong respiratory inhibition: so-called epin-

ephrine apnea [89]. This is an indirect effect due to the primarily induced increase in blood pressure.

106.6.2 Steroid Hormones and Their Effect on Breathing During Pregnancy (see Chap. 20)

Progesterone. In men, injections of progesterone lead to a shift of the CO_2 response curves to the left (see Fig. 106.6), indicating a positive effect of the hormone on ventilation [14,165]. Medroxyprogesterone may be used as a therapeutic measure in patients with low CO_2 sensitivity. In infants with congenital hypoventilation syndrome, however, in whom the respiratory response to CO_2 is missing completely, the positive effect is only of short duration [106]. Other steroid hormones such as cortisone, desoxycorticosterone, and testosterone also increase ventilation, but to a much lesser extent. The same is true for adrenocorticotropic hormone (ACTH).

Variations of the progesterone concentrations during the menstrual cycle are correlated with cyclic variations of ventilation, in particular a decline of ventilation shortly before menstruation. During pregnancy, resting ventila-

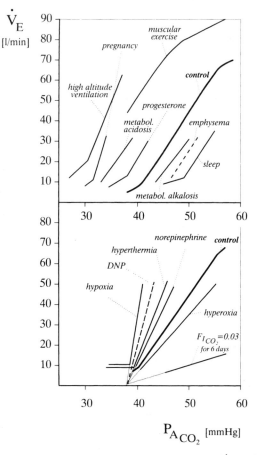

Fig. 106.6. CO_2 response curves of ventilation ($\dot{V}_E$) in humans under different conditions. *Top,* effects which mainly cause a parallel shift. *Bottom,* effects mainly changing the slope [83]. *DNP,* 2-4-dinitrophenyl; $P_{A_{CO_2}}$, alveolar CO_2 pressure

tion is augmented, at the end of pregnancy by as much as 50%–70%, and the CO_2 response curve is shifted to the left and steepened (see Fig. 106.6). The alveolar CO_2 partial pressure is 8–10 mmHg below the control level. In the last part of pregnancy, an increased development of fixed acids probably contributes to the ventilatory increase. The different factors are balanced in such a way that a distinct decrease in standard bicarbonate does not result in a significant variation of blood pH [83]. The lower P_{CO_2} in the maternal blood raises the fetomaternal CO_2 pressure gradient within the placenta, thus favoring CO_2 exchange, so that in the fetal blood of the umbilical vein the P_{CO_2} is only slightly above that of normal arterial blood in adults.

106.6.3 Respiratory Effects of Thyroid Hormone (see Chaps. 19, 22)

Hyperthyreosis. In contrast to the actions of other hormones, the respiratory augmentation observed during hyperthyreosis or in healthy subjects by the administration of thyroid hormone is accompanied by a slight increase in the alveolar P_{CO_2} [169]. This implies that alveolar ventilation is too low to compensate for the increased CO_2 production. Thus a direct stimulatory effect on ventilation appears to be unlikely. Accordingly, no significant shift of the respiratory CO_2 response curve is found. In this instance, ventilatory adaptation to the altered metabolism is brought about by the CO_2 effect. Individual differences due to concomitant effects of other factors such as increased temperature and catecholamines may not be excluded. Direct excitation of ambigual and solitary tract neurons and inhibitory effects on γ-aminobutyric acid (GABA)-activated K^+ channels have been shown for thyrotropin-releasing hormone (TRH) [63,77,138], which plays an important role in the development of the respiratory system [179].

106.7 Respiratory Effects of Blood Chemistry

Of great importance for the regulation of acid–base balance are the chemical respiratory drives (see Chaps. 78, 102). Moreover, cooperation between the chemical drives, especially hypoxia, plays an important role in adaptation to high altitude (see Chap. 109).

106.7.1 Partial CO_2 Pressure in Blood

Elevation of the arterial P_{CO_2} leads to the subjective sensation of shortness of breath or dyspnea and to an increase in tidal volume and respiratory frequency, augmenting the minute ventilation. The ventilatory augmentation is mediated by activation of inspiratory and expiratory brain stem neurons and an increase in the discharge frequency of inspiratory neurons, supplemented by recruitment of units inactive during "normal" quiet breathing.

CO_2 Anesthesia. The highest steady minute ventilation driven by an increase in $P_{A_{CO_2}}$ is 70 l/min, a value much lower than that attained by vigorous muscular exercise and than the maximal possible ventilation. With stronger increases in $P_{A_{CO_2}}$, the steady state ventilation does not increase further, but finally falls instead, probably due to an direct inhibitory effect of CO_2 on the central respiratory system (CO_2 anesthesia). This fact is of pathophysiological importance, since a vicious circle may develop between respiratory paralysis and further increases of $P_{A_{CO_2}}$.

CO_2 Response Curve. In the same subject tested under comparable conditions, a reproducible relation between $P_{A_{CO_2}}$ and minute ventilation (see Chap. 100) can be obtained. This relation is called the CO_2 response curve [95]. As shown in Fig. 106.6 (thick curves labeled "control"), the curve is linear over a wide range, but flattens particularly at lower P_{CO_2} values. The CO_2 response curve impressively demonstrates the sensitive reaction of respiration to changes in CO_2. Increasing $P_{A_{CO_2}}$ by only 1 mmHg leads to an increase in ventilation by 2.5–4.5 l/min. The relationship changes dramatically in response to a variety of modulating factors (Fig. 106.6). Purely parallel shifts of the CO_2 response curve imply additive or subtractive effects of modulatory factors. Changes in steepness indicate nonlinear interactions between CO_2 and a modulatory factor. Examples are shown in Fig. 106.6 and discussed above and below. Changes in the CO_2 response curve may indicate unwanted inhibitory side effects on ventilation of analgesic (e.g., morphine) anticonvulsive, or narcotic drugs or stimulating side effects, such as those of salicylate or aminophylline.

106.7.2 Hyperventilation Apnea and CO_2 Threshold

After an intense, lasting, active or passive hyperventilation, many subjects exhibit apnea: so-called hyperventilation apnea. Its duration depends on the length and duration of the preceding hyperventilation. It cannot be explained merely by blood chemistry, but this does play a role: elevation of the inspiratory PO_2 during hyperventilation prolongs the apnea, while CO_2 inhalation shortens it. On the other hand, many subjects may not show apnea, but only reduced ventilation, which cannot be influenced by variation of the alveolar P_{CO_2} as long as the latter remains below some threshold value [54]. This value usually lies about 1 mmHg below the normal resting value, but depends on the state of wakefulness.

The CO_2 threshold can also be demonstrated by lowering the inspiratory O_2 concentration to hypoxic levels [98,119], under which condition the CO_2 response curve shows a lower flat portion, indicating a range of insensitivity to CO_2 variation ("dog leg"; see Fig. 106.6, lower panel).

Vagal Afferents. The behavior of respiratory neurons during hyperventilation is strongly dependent upon vagal afferents. With intact vagi, inspiratory neurons are completely inhibited at short latency, whereas expiratory neurons develop a tonic activity. With vagal afferents cut, central rhythmogenesis ceases only after a very long period of hyperventilation, with inspiratory neurons again being affected much more strongly than expiratory cells.

106.7.3 Ventilatory Adaptation by CO_2

CO_2, a metabolite that increases along with metabolism, is primarily responsible for adaptation of ventilation to changed metabolic demands when other drives remain insufficient, are missing, or are transiently inhibited (e.g., 100 m sprint, weight lifting; see Chap. 108).

Hypocapnia and Hypercapnia. CO_2 thus has an important function in the continuous balance of influences disturbing quiet respiration. This is of special importance under emergency conditions when breathing is obstructed. The rise of CO_2 then provides the strongest respiratory drive and elicits general activation. Hypoxia merely cooperates in this function. Furthermore, the CO_2-sensitive mechanism counteracts pathological hypo- and hyperventilation, e.g., during external obstruction of the airways or hyperventilation as a result of pain. In the case of chronic respiratory disturbances, the CO_2-sensitive mechanism adapts, in that long-lasting hypocapnia elevates the CO_2 effect on ventilation, whilst long-lasting hypercapnia (e.g., in emphysema) reduces it.

Stabilization. The stabilizing effect of CO_2 is described below in quantative terms (see Fig. 106.7). Initially disregarding compensatory CO_2 effects according to the CO_2 response curve (Sect. 106.7.1), variations in minute ventilation result in changes in $P_{A_{CO_2}}$ according to the equation: $P_{A_{CO_2}} = V_{CO_2}/(V_A \times constant)$ (see Sect. 106.1). When metabolism and hence V_{CO_2} are constant, this function repre-

sents a hyperbola (ventilatory hyperbola). Different states of metabolism yield a bundle of hyperbolas, as exemplified in Fig. 106.7 for two cases of V_{CO_2} (hyperbolas I and II). For the sake of comparability (see below), V_A is replaced with minute ventilation.

According to the CO_2 response curve, variation of $P_{A_{CO_2}}$ induces a variation in minute ventilation, as exemplified in Fig. 106.7 by the two curves labeled "rest" and "muscular exercise." The actual minute ventilation must fulfill both relationships and thus settles at the intersection of the two functions.

Based on these assumptions, Fig. 106.7 yields the following insights:

- Ventilation settles to its normal value at intercept A only if the CO_2 response curve's horizontal "dog leg" lies below the intercept A or, in other words, if the CO_2 threshold is lower than the normal $P_{A_{CO_2}}$, which is the case in most subjects. If the threshold is at the same level as intercept A, CO_2 can only counteract a decrease of ventilation. With the threshold beyond A, as observed in a few subjects, CO_2 can only prevent a ventilatory depression, which shifts the $P_{A_{CO_2}}$ beyond the threshold value.
- An increase in V_{CO_2} of 2 l/min (shift to the ventilatory hyperbola II) during exercise would increase $P_{A_{CO_2}}$ to intercept B without compensation via the CO_2 response curve. However, the latter's action adjusts the system in intercept C. This increases ventilation and changes $P_{A_{CO_2}}$ only slightly.
- At very high CO_2 values where the CO_2 response curve has a negative slope, no stable adjustment can be reached, since the CO_2 response curve and the ventilatory hyperbola no longer have opposite slopes. This illustrates the vicious circle between the reduction of ventilation and the elevation of $P_{A_{CO_2}}$ within this range (see above). Respiratory regulation can only be restabilized by returning to the range of opposite slopes. In practice, this implies a passive increase of ventilation by artificial ventilation.

106.7.4 Respiratory Effect of Blood pH

According to the Henderson–Hasselbalch equation (pH = pK' + log[HCO_3^-]/[CO_2]; (see Chaps. 78, 102), ventilation affects the PCO_2 as well as the H^+ concentration. The effect on the pH will in turn lead to an adaptation of ventilation to the requirements of the acid–base balance, provided that the pH affects ventilation. Such an effect can indeed be demonstrated. A fall in blood pH enhances ventilation and vice versa. With equal absolute pH changes, respiratory augmentation by acidosis is more pronounced than the drop in ventilation by alkalosis. The arterial CO_2 pressure takes a course opposite to that of the pH: it decreases with acidotic hyperventilation and increases with alkalotic hypoventilation. This opposite shift of P_{CO_2} also counteracts too strong a deviation of pH from the normal value (respiratory compensation of metabolic disturbances). This implies that the pH represents a respiratory drive

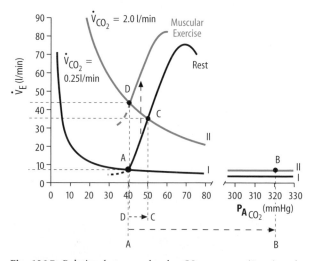

Fig. 106.7. Relation between alveolar CO_2 pressure ($P_{A_{CO_2}}$) and ventilation ($\dot{V}_E$) at different metabolic states, during rest and muscular exercise [99]. $\dot{V}_{CO_2}$, production of CO_2. See text for details of intercepts A–D

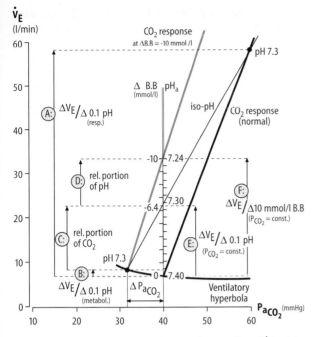

Fig. 106.8. Quantitative analysis of ventilatory effects ($\dot{V}_E$) of the arterial pH (pH_a) and the arterial CO_2 pressure (Pa_{CO_2}) in humans. The ventilatory effect of an increase in Pa_{CO_2} is represented by the CO_2 response curve (*CO$_2$ response [normal]*). In metabolic acidosis, with a reduction of the blood buffer base of 10 mM, the CO_2 response curve is shifted to the left. For precise details, see text [83].

independent of P_{CO_2}. Thus H$^+$ concentration belongs to the feedback respiratory drives. It stabilizes the tissue pH and may counteract other respiratory drives. A fall of pH by 0.1 increases ventilation by about 10%–20% of the control value and a decrease of blood buffer base by 5–6 mmol/l. While these changes appear small compared with the effects of other respiratory drives, it must be realized that the other feedback drives of ventilation, primarily CO_2 pressure, counteract the respiratory changes caused by H$^+$ ions. In other words, expiration of CO_2 under acidotic conditions reduces the CO_2 drive. If this is taken into account by keeping P_{CO_2} constant, a reduction of pH by 0.01 causes a ventilatory increase of 2 l/min or of 2.7 l/min, with a reduction of blood buffer base by 1 mM in man [93]. This shows that ventilation is highly sensitive to very small changes in pH. This is also expressed in the CO_2 response curve, which during acidosis is shifted to the left and during alkalosis to the right (see Fig. 106.6).

Respiratory and Metabolic Acidosis. Since any increase in arterial P_{CO_2} is bound to change the arterial pH, and since pH acts as a respiratory stimulus in its own right, independent of P_{CO_2}, the question arises of whether the respiratory effect of arterial P_{CO_2} is mediated by the effect on H$^+$ ion concentration. This is tantamount to asking whether a specific CO_2-sensitive respiratory drive exists independent of pH. A quantitative consideration shows that ventilation is not solely a function of arterial pH. Ventilatory increase due to a rise of Pa_{CO_2} (respiratory acidosis) is larger than in

response to a comparable acidosis induced by reducing the blood buffer base (metabolic acidosis; see Fig. 106.8). The respiratory increase with an increase in Pa_{CO_2} therefore results from a cooperation between a component based on the H$^+$ effect and one based on a CO_2 effect. The quantitative differentiation between these two components is important for later considerations of the adequate stimulus and site of action of the central respiratory sensitivity. It can be performed in different ways and is based on the mass action law for the first dissociation of carbonic acid: $[H^+] \times [HCO_3^-] = [CO_2] \times$ constant (see Chaps. 78, 102). It is then possible to experimentally change the pH by changing, at a constant P_{CO_2}, $[HCO_3^-]$ (i.e., blood buffer base; see Chap. 102) or else by changing P_{CO_2} at constant pH by appropriately varying $[HCO_3^-]$.

Figure 106.8 shows such a quantitative analysis. The effects on ventilation by elevating arterial/alveolar P_{CO_2}, by CO_2 inhalation, are represented by the CO_2 response curve (*CO$_2$ response normal*). Reduction of blood buffer base (BB) by 10 mM causes a metabolic acidosis (pH, 7.3), which shifts the CO_2 response curve to the left and reduces P_{CO_2} by about 9 mmHg (1.2 kPa) (CO_2 response at BB of 10 mM). Since a fall of Pa_{CO_2} and a ventilatory increase are related by the ventilatory hyperbola, the above P_{CO_2} drop (PA_{CO_2}) increases ventilation from 6.5 l/min to 8.9 l/min and decreases arterial pH from 7.40 to 7.30 (B: Δ minute ventilation/Δ0.1 pH [metabol] = 2.4 l/min). During respiratory acidosis, an equivalent decrease in pH is attained, at normal buffer base, with inhalation of CO_2 (right upper point on the CO_2 response normal curve) when alveolar PCO_2 increases to 57 mmHg (7.6 kPa), inducing an increase in ventilation to 57.5 l/min. This respiratory acidosis then entails a ventilatory increase A of 51 l/min per pH drop of 0.1 (V_E/0.1 pH [resp]). The apparently large difference between the metabolically and respiratory-induced ventilatory increases (A and B) is due to the drop of Pa_{CO_2} during metabolic acidosis, which counteracts respiratory stimulation.

In Fig. 106.8, the vertical scale erected at constant P_{CO_2} of 40 mmHg (5.3 kPa) shows the relationships between V_E, arterial pH (pH_a), and change in buffer bases (BB), under the assumption that these relationships are linear. Connecting the iso-pH points (pH, 7.3) on the CO_2 response (normal) and the ventilatory hyperbola, we obtain an iso-pH line for arterial pH of 7.3, which intercepts the above-mentioned scale. This intercept indicates that a decrease of pH to 7.3 at constant Pa_{CO_2} causes a BB of −6.4 mM and a ventilatory increase E of 18.6 l/min. By contrast, again for constant PA_{CO_2} values, if base buffers are reduced by 10 mM, pH drops to 7.24 and ventilation increases by F = 31.1 l/min. The difference between E and B reflects the reduction of the pH effect due to the hyperventilatory drop of P_{CO_2}.

Both a drop of pH and a rise of P_{CO_2} stimulate ventilation. Their relative contributions can be determined as follows (see Fig. 106.8): the iso-pH line reflects the respiratory response at constant pH (here acidic, 7.3), while the CO_2 response at a BB of −10 mM reflects the respiratory response during metabolic acidosis. Their respective inter-

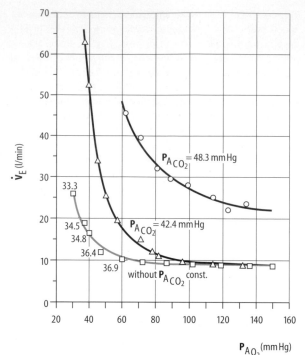

inspiratory O_2 concentration of 12.6% at normal barometric pressure.

Hypoxic Effects. These are manifested in initial euphoria, impaired neuromuscular coordination and sensory discrimination, disturbances of visual performance, sweating, and finally collapse and loss of consciousness (see Chap. 109). The ventilatory increase induced by pure hypoxia before other disturbances occur reaches values between 20 and 30 l/min and thus remains far below the volumes measured during muscular exercise.

O_2 Response Curve. The O_2 response curve reflects the behavior of breathing in exogenous hypoxia; however, it does not represent effects of hypoxia only. Since during increased ventilation the P_{CO_2} decreases, its respiratory drive does so as well, which diminishes the stimulating effect of hypoxia. This can be demonstrated experimentally by keeping the alveolar P_{CO_2} constant (see Fig. 106.9). Under these conditions, the O_2 response (middle curve) is much enhanced, especially below a $P_{A_{O_2}}$ of 80 mmHg [35,98]. Low P_{O_2} values can raise ventilation to more than 60 l/min.

Fig. 106.9. Ventilatory effect ($\dot{V}_E$) of a decrease in alveolar O_2 pressure ($P_{A_{CO_2}}$) in humans (O_2 response curve). *Bottom* curve (blue), ventilatory increase without compensation of P_{CO_2}. Note the small ventilatory augmentation below a P_{O_2} of 50–60 mmHg. The *numbers* next to the curve give the $P_{A_{CO_2}}$. From this curve we can calculate the augmentation of ventilation during acute climbing to high altitude. *Upper curves,* O_2 responses at two different levels of P_{CO_2} [98]

106.7.5.1 O_2 Threshold

The existence and approximate value of an O_2 threshold can be demonstrated by short-term inhalation of oxygen. Since the hypoxic effect is mediated by easily accessible and fast-reacting arterial chemoreceptors, $P_{A_{O_2}}$ elevation causes a more or less distinct initial respiratory depression. From such experiments the hypoxic drive can be estimated to account for 10%–15% of the normal resting ventilation [40]. This proportion is larger at high-altitude ventilation and largest during muscular exercise. Alveolar P_{O_2} beyond about 150 mmHg (20.0 kPa) does not depress respiratory function any further, indicating the O_2 threshold in the awake humans. It varies individually between 100 and 170 mmHg (13.3–22.6 kPa) $P_{A_{O_2}}$.

106.7.5.2 Pathophysiological Implications of the O_2 Response

Barbiturate Intoxication. The small percentage of respiratory drive contributed by hypoxia can significantly increase when the other drives (e.g., CO_2 sensitivity) are weakened. Certain narcotics, analgesics, or chronic alveolar hypoventilation (e.g., lung emphysema) reduce CO_2 sensitivity, as manifested in elevated P_{CO_2} at reduced ventilation and reduced steepness and rightward shift of the CO_2 response curve [5]. In extreme conditions, e.g., with barbiturate intoxication, the respiratory effect of CO_2 can be abolished completely. Since under such conditions the respiratory centers can still respond to peripheral chemoreceptor afferents, ventilation is driven predominantly by hypoxia. Breathing O_2 then enhances respiratory depression, possibly leading to complete respiratory ar-

cepts with the vertical scale at a Pa_{CO_2} of 40 mmHg yield the ventilatory increases C or D for the CO_2 portion and the pH portion, respectively. In this case the ratio of C to D is 68:32. The simple linear interpolation used here is correct only in the first approximation. For more precise details, see [98] and [79].

Using this and other methods, experiments in man as well as in dogs and cats indicate that 35%–45% of the CO_2 response can be attributed to the pH effect, and 55%–65% to the CO_2 effect. During metabolic acidosis, the pH portion is smaller [79,90].

The described differentiation of the effect of pH and P_{CO_2} is valid only for the arterial blood chemistry. The mechanisms at the sensory site depend on the chemical environment of the chemosensitive structure.

106.7.5 Hypoxic Drive

Reduction of Pa_{O_2} causes an increase of ventilation, as reflected in the (lower) O_2 response curve in Fig. 106.9 (see also Chap. 78). At normal $P_{A_{CO_2}}$, ventilatory augmentation becomes prominent only when alveolar P_{O_2} falls below about 50 mmHg (6.7 kPa), which is reached at an inspiratory P_{O_2} of about 90 mmHg (12.0 kPa), corresponding to an

rest. In such cases, as well as in cases of congenital hypoventilation syndrome, artificial ventilation is indicated; uncontrolled O_2 inhalation ($SaO_2 > 93\%$) without artificial ventilation would be dangerous.

106.7.5.3 High-Altitude Ventilation

The acute respiratory effect of reduced barometric pressure at high altitudes corresponds to that of hypoxia, inducing an increase in ventilation (see Figs. 106.6, 106.9 and Chap. 109). The CO_2 response curve is shifted to the left and becomes steeper, the maximal breath-holding time is shortened, and the compulsory resumption of breathing in this test starts at higher P_{O_2} values and at lower P_{CO_2} values than at sea level.

Fixed Hyperventilation. The increased ventilation in acute experiments inside a low-pressure chamber or after a short stay at high altitude vanishes after normalization of alveolar P_{O_2}, but not after high-altitude acclimatization: O_2 inhalation causes only a weak to moderate reduction of the increased ventilation (fixed hyperventilation). This is maintained after return to sea level for some time (days to weeks), as is the shift of the CO_2 response curve. The increased ventilatory sensitivity to hypoxia continues for weeks and months after a long-term stay at high altitude, even if resting ventilation, pH, and standard bicarbonate have returned to control values.

Chronic Passive Hyperventilation. In addition, ventilation and CO_2 sensitivity can be varied by chronic passive hyperventilation, resembling the situation during acclimatization to high altitude. These are probably not direct consequences of hypoxia. Suggested causes for the changes persisting after normalization of P_{O_2} are:

- Rise of chemoreceptor sensitivity (see Chap. 109).
- Rise in H^+ ion concentration due to the decrease in bicarbonate at equal P_{CO_2}. The return of standard bicarbonate to normal values takes some time. A hyperventilation of the acclimatized person continues for a longer period, however.
- Shifts of bicarbonate at the blood–brain barrier can affect the central chemosensitive structures. However, this process is faster than the renal correction of the arterial (HCO_3^-).

106.7.5.4 Effect of Increased Inspiratory O_2 Concentrations

Short-term O_2 inhalation below an alveolar P_{O_2} of 150–170 mmHg (20.0–22.6 kPa) causes a temporary ventilatory depression by functional elimination of chemoreceptor drive; longer-lasting O_2 inhalation is followed by a slight ventilatory enhancement. This is attributed to indirect effects of the increased arterial P_{CO_2}:

The cerebral blood flow falls with increasing arterial P_{O_2}, whereby P_{CO_2} and the H^+ ion concentration rise at the site of the central sensor.

Haldane Effect. Owing to the increased venous O_2 saturation of hemoglobin, its H^+ ion-binding capacity and therefore also its transport capacity for CO_2 are reduced (Haldane effect; see Chap. 102), which leads to an elevation of P_{CO_2} within the tissue.

Damage. Long-lasting increases in inspiratory P_{O_2} cause damage to airway epithelia with bronchitis and disturbances of diffusion within the lung, damage to the eyes due to a decreased blood flow in the retina, and seizures and loss of consciousness by direct toxic O_2 effects on the cerebral neurons. The period of tolerance is inversely related to the inspiratory P_{O_2}. Knowledge of these effects is important for therapeutic and other O_2 applications, e.g., when working under water or in aeronautics.

106.7.5.5 Combined Effects of Respiratory Drives During Muscular Exercise (see also Chap. 108)

For the quantitative analysis of muscular exercise, three questions are of interest:

- What is the resulting ventilation?
- How great is the change in the corresponding drive?
- How sensitive is the ventilatory response to isolated changes in this particular respiratory drive?

Considering the respiratory response to CO_2, O_2, and pH (see Figs. 106.6–106.9) and the slight changes of these quantities during exercise, it becomes clear that none of these blood chemical drives is solely responsible for the enormous ventilatory augmentation. Rather, progress in quantitative analysis has shown that it is possible to trace the ventilatory augmentation to a combination of factors. However, a simple addition of various respiratory drives cannot explain respiratory increases during work; rather, the contribution of a single factor can vary strongly in various conditions and in different phases of muscular work.

106.7.5.6 Initial Reaction and Adjustment to Steady State

Under normal conditions, the fast respiratory elevation at the beginning of muscular exercise reaches 30%–40% of the steady state value. Its fast rise indicates that neural factors are involved, which become effective immediately with the beginning of movement. The magnitude of this initial response can vary according to the conditions, especially regarding the subject's motivation and the expected cortical influences which coactivate the respiratory system. The adjustment to steady state that following within a few minutes is probably related to chemical changes and to the extent of metabolic increase and is partly caused by humoral factors and partly by nervous feedback from the muscle.

106.7.5.7 Ventilation During Steady State

For the steady state, the contribution of a coinnervation from the cortex is controversial, since other factors sufficiently account for the changes. However, this applies only to psychically neutral performance of work by trained subjects. In this case the relative contributions can be estimated as follows.

Blood-Chemical Respiratory Drives. The contribution of CO_2 drive varies widely with work load. Pa_{CO_2} increases by a few mmHg during light work. This can explain an essential part of the small increase of ventilation. Figure 106.10 shows an example of blood chemical changes during light muscular exercise. As to changes in the CO_2 response curve (see Figs. 106.6, 106.7) and in pH beyond the respiratory acidosis, CO_2 contributes 43% and the pH change 6% to the total ventilatory augmentation, i.e., about 50% can be accounted for by blood-chemical factors. At moderate work loads, P_{CO_2} cannot participate in the respiratory drive since it does not change. During heavy and extreme work loads, P_{CO_2} even falls due to metabolic acidosis, which is compensated for by increased ventilation (see above). Arterial pH is lowered even during light work, initially leading to respiratory acidosis. With increasing work load and production of fixed acids, the pH keeps dropping despite the fall of P_{CO_2}. Assuming a ventilatory augmentation of 2 l/min per 0.01 drop in pH (P_{CO_2} constant; see above), the pH-dependent ventilatory increase amounts to 40 l/min with pH changes of 0.2, as found with a V_{O_2} of about 3 l/min [93]. This constitutes more than half of the respiratory drive. Pa_{O_2} remains constant over a wide range of work loads. During muscular exercise, O_2 inhalation causes respiratory depression of 10%–15%, as it does at rest. In absolute terms, then, this proportion increases with increasing work load. Hypoxia may be assumed to interact with other respiratory drives in a more than additive manner, which plays an important role within the range of heaviest work, leading to a drop of P_{O_2}. In this case, O_2 inhalation may inhibit ventilation by up to 35% [9]. Altogether, blood-chemical factors contribute about 30%–50% to ventilatory augmentation during steady state exercise.

Nervous Factors. The remaining 50%–70% is much more difficult to differentiate. At the end of muscular exercise, ventilation drops rapidly by 50%–60%. This drop can reasonably be attributed to the termination of respiratory drives at the end of muscular activity and limb movements and, hence, to changes in neural feedback and/or coinnervation (see above).

Body Temperature. Muscular exercise raises the body temperature, which increases by 0.6°–0.8°C with an increase of O_2 consumption of 1 l/min. This would increase

ventilation, albeit with some delay. However, after exercise, temperature drops to control values much more slowly than ventilation does. At high work loads, temperature elevation probably accounts for less than 10% of ventilatory augmentation.

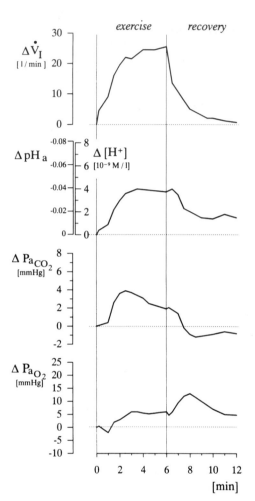

Fig. 106.10. Variation of ventilation ($\Delta \dot{V}_I$) and blood chemistry i.e., arterial pH (pH_a), CO_2 pressure (Pa_{CO_2}), and O_2 pressure (Pa_{O_2}), during short, mild muscular exercise (625 kpm/min; oxygen consumption, $\dot{V}_{O_2}$ = 1.6 l/min). A transient minor increase of Pa_{CO_2} contributes to the respiratory augmentation. The portion of CO_2 drive is different according to the intensity of muscular exercise. An increase in Pa_{CO_2} by a few mmHg can be observed only during mild muscular exercise. Here it can explain an essential part of the minor ventilatory response. Considering the steepness of the CO_2 response curve and the pH change beyond respiratory acidosis, CO_2 contributes 43%, pH 6% to the total increase of ventilation; thus 50% of the ventilatory response can be explained by chemical factors of the blood. CO_2 remains unchanged during moderately strenuous work and is decreased during strenuous work, which reflects the respiratoy compensation of a metabolic acidosis by the increased ventilatory level. In spite of the minor or missing portion of CO_2 on the ventilatory augmentation, its important function is the stabilization of ventilation at an increased value. The gross drive for the ventilatory response to muscular exercise is defined by other factors; CO_2 guarantees precise adaptation to the acute metabolic rate [83]

Catecholamines. The release of epinephrine and norepinephrine increases with heavy work, above all if associated with subjective effort [172]. Infusions of epinephrine–norepinephrine mixtures, which cause comparable increases in blood concentration, are followed by an increase in ventilation of about 10 l/min at constant P_{CO_2}.

106.7.5.8 Recovery Phase

The respiratory increase outlasting the end of motor drive can in part be explained by blood-chemical effects (see Fig. 106.10). It is still controversial, however, whether or not remaining changes in CO_2, O_2, and pH sufficiently account for the respiratory increase.

In summary, in the initial phase of work neural factors predominate, while in the recovery phase it is chemical factors; during the steady state both neural and humoral factors cooperate.

106.8 Peripheral Chemoreceptors

The acute respiratory augmentation during hypoxia has been demonstrated to be mediated by defined chemosensitive structures: the glomera at the bifurcation of the common carotid arteries [70] and analogous structures at the aortic arc.

106.8.1 Anatomy

Carotid Bodies. These receive their blood supply from the first branches of the occipital artery, the three analogous aortic bodies on either side from direct aortic branches and from branches of the left brachiocephalic artery or the right subclavic artery. The capillaries within the bodies are fenestrated and are not true sinusoids, and there is apparently no anatomic barrier to diffusion between the capillary net and the glomera interstitial space [178]. The carotid body is a richly innervated organ consisting of an association of islands of cells and capillaries called glomeruli. On the basis of nuclear shape and density, two glomerular cell types have been distinguished which are referred to as type I and type II cells. In the cat, each glomerus contains 20–30 type I cells and three to six type II cells [39], but the ratio varies in different species [103]. A prominent ultrastructural feature of type I cells is the presence of cytoplasmic, dense-core vesicles. The carotid body is rich in catecholamines. Dense-core vesicles isolated from human chemodectoma accumulate and release dopamine, the release being Ca^{2+} dependent, which suggests an exocytotic process [52].

The innervation of the carotid body is provided by the carotid sinus nerve (CSN) and the ganglioglomerular (sympathetic) nerve. The sympathetic branch is composed almost exclusively of nonmyelinated fibers of postgan-

glionic sympathetic origin [46]. The cat CSN consists of 650–800 myelinated fibers and nearly three times as many nonmyelinated fibers. It has been estimated that 61% of the myelinated fibers in the CSN are chemoreceptor afferents innervating type I cells; the remaining fibers are baroreceptor afferents innervating intraglomeric blood vessels (but most baroreceptor fibers in CSN bypass the carotid body), parasympathetic fibers innervating a small number of intraglomeric ganglion cells and blood vessels, and other efferent fibers of unknown origin [51]. Of the unmyelinated C fibers, 17% are chemoreceptor afferents and 29% baroreceptor afferents. The terminal branches of these intralobular fibers ultimately end on the epitheloid or type I cells in a variety of synaptic configurations ranging from bouton- to calyx-shaped endings. Branches of a single axon may terminate on several cells within the same or different glomeruli, and conversely a single type I cell may receive endings from more than one axon.

The intraglomeric structures exert complex actions on each other. Thus afferent activity in chemosensory fibers should be the net result of feedback interactions across unidirectional and reciprocal synapses between type I cells themselves (and perhaps also via axoaxonal contacts between efferent nerve fibers and sensory nerve endings). The expression of the chemoreceptor response can also be modified by sympathetic and parasympathetic control of intraglomeric elements, including type I cells, ganglion cells, and blood vessels [51].

Pathological Observations. Prolonged physiological stimulation and certain disease states directly involving the carotid body can sometimes lead to morphological changes in glomerular elements. It has been shown that carotid bodies of humans living at high altitudes are larger than those living at sea level [8]. Likewise, animals (guinea pigs, rabbits, and dogs) living at 4330 m exhibited hyperplasia of both type I and type II cells with consequent enlargement of their carotid bodies that averaged 2.5 times those of sea-level dwellers [42]. Structural changes similar to those described for high-altitude dwellers have been observed in patients with chronic hypoxemic disease; the carotid bodies of these patients are enlarged and have both increased vascularity and increased numbers of glomera [88]. Although chemodectomas are tumors of very low incidence in humans living at sea level, their frequency of occurrence increases with altitude and reaches 12 times the sea-level rate in people living at 3000–4000 m [140]. Carotid body dysfunction and underdevelopment has been implicated in sudden infant death syndrome (SIDS) [115], but numerous other factors also appear to be involved.

106.8.2 Blood Flow and Metabolism

Both glomeric blood flow and O_2 consumption are very high: 2 l/min per 100 g and 9 ml O_2/min per 100 g. The carotid body's blood flow is subject to autoregulation: constancy of blood flow within an arterial blood pressure range of 80–150 mmHg (10.6–20.0 kPa). Due to the high

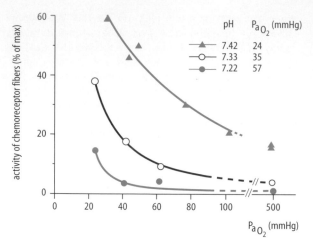

Fig. 106.11. Excitation of chemoreceptor afferents measured as the compound activity of afferent fibers of the sinus nerve during perfusion of the carotid sinus preparation of the cat with solutions of various arterial CO_2 (Pa_{CO_2}) and O_2 pressures (Pa_{O_2}). The shape of the curves depend on the P_{CO_2} or pH. At normal P_{CO_2}, activity is still measurable with a normal P_{O_2} of 100 mmHg [83]

blood flow the arteriovenous O_2 difference (AVD_{O_2}) is very low, implying that O_2 concentration and P_{O_2} in the venous outflow is still close to the arterial value. The arterioles constrict in response to epinephrine as well as to direct and reflex sympathetic stimulation, thus reducing the blood flow. Direct measurements have shown that there are areas within the carotid body with low P_{O_2} values [2]. This can only be explained by a specialized order of the vessels with corresponding P_{O_2} profiles. All these factors are important for our understanding of the excitatory mechanism of the receptors.

106.8.3 Excitation of Arterial Chemoreceptors

The effect of different stimuli on the chemoreceptors can be observed in the following ways:

- By measuring respiration during changes of blood chemistry before and after chemoreceptor denervation.
- By isolated perfusion of the chemosensitive regions of the vessel with blood or solutions of various chemical compositions. Observations may include reflex effects and action potentials in the afferent nerve.

Physiological Blood-Chemical Stimuli. The chemoreceptors are excited by decreases in Pa_{O_2}, increases in Pa_{CO_2}, and decreases of pH [73]. At normal Pa_{CO_2}, chemoreceptor afferents show a very low activity, which is further reduced with an increase of Pa_{O_2} to 150–170 mmHg (20.0–22.6 kPa), the threshold of the receptors for the hypoxic stimulus (see above). A distinct activation can only be observed at a fall of Pa_{O_2} below 100 mmHg (13.3 kPa) (see Fig. 106.11). Compound excitation is expressed in both increased discharge rates of single fibers and recruitment of previously silent fibers. It is not yet

quite clear what the excitatory stimulus is: Pa_{O_2} or O_2 content. Both may be of importance.

As to the CO_2 stimulus, the impulse frequency of chemoreceptor fibers is linearly related to increases in Pa_{CO_2} above a threshold value of about 20–30 mmHg (2.7–4.0 kPa). A rise of Pa_{CO_2} also causes chemoreceptor excitation when the pH is kept constant by bicarbonate infusion [20]. This may indicate that part of the receptors are located behind a diffusion barrier.

Raising the H^+ ion concentration also excites chemoreceptors, even at constant Pa_{CO_2}. The effect of the pH and Pa_{CO_2} stimuli is strongly dependent upon the actual Pa_{O_2}. Extent and sensitivity of Pa_{CO_2}- and pH-dependent excitation are amplified during hypoxia. Conversely, the excitatory effect of hypoxia is increased with elevated Pa_{CO_2}. A strong reduction of arterial pressure or of the perfusion pressure of the chemoreceptor preparation results in an intense discharge. The same happens when reducing the perfusion of the glomus by means of vasoconstriction induced by epinephrine or by stimulating the sympathetic nerve.

Efferent Modulation of Chemoreceptor Excitation. Similar to that of many other receptors, chemoreceptor sensitivity can be modulated by efferent innervation. Such a change is reflected in an increase or reduction of afferent impulse frequency at equal blood-chemical stimuli. Excitation of sympathetic nerve fibers enhances chemoreceptor activity; stimulation of other efferent – possibly parasympathetic–fibers of the sinus or aortic nerve diminishes the discharge frequency. The excitatory sympathetic effect is based on a constriction of the afferent blood vessels with ensuing reduction of the blood and oxygen supply, probably supported by a reduction of the oxidative metabolism of the glomus cells.

Pharmacological Effects on Chemoreceptors. A large number of drugs excite chemoreceptors. Only a few representative examples are mentioned here:

- Substances inhibiting glycolysis and oxidative phosphorylation, e.g., iodoacetate, Na^+ cyanide, Na^+ fluoride, Na^+ malonate, high CO concentration (inhibition of oxidation of cytochrome a3), 2,4-dinitrophenole, Na^+ nitrite, Na^+ acid, antimycin A
- Reducing substances, e.g., hydroxylamine, paraminophenole, hydrochinone
- Generally excitatory and depolarizing substances, e.g., acetylcholine and anticholinesterases, veratridine, KCl
- Alkaloids, e.g., nicotine, lobeline, piperidine
- Coramine, serotonin, cyclic 3,5-AMP (adenosine Monophosphate)

Inhibitory effects are exerted by hexamethonium, atropine, mecamylamine, and other anticholinergic substances.

Excitatory Mechanism of Chemoreceptors. The mechanism of the transducing process is still a matter of discussion. As mentioned, single chemoreceptor fibers react to both pH changes and hypoxia. However, the effects of both stimuli have different time courses, and the hypoxic response, but not the pH sensitivity can be eliminated by iodoacetate. This suggests that one and the same receptor can be excited by both stimuli via apparently different mechanisms. Thus, the hypothesis that hypoxia acts via H$^+$ ions is unlikely. Rather, the nature of many excitatory substances suggests that disturbances of energy supply may excite the receptors by interfering with one of the other steps in the chain of glucose catabolism or oxidative phosphoryzation zation. If hypoxia acted in the same way, the receptor cells might be thought to be located at the borderline of hypoxia, either because of an unusual O_2 demand or because of their location within hypoxic glomeric regions. The mechanism of action of H$^+$ ions is also still unclear.

One concept maintains that the decisive processes involve a P_{O_2}-dependent transmitter release from carotid body type I cells, depolarization, and generation of action potentials in the postsynaptic nerve endings of the sinus nerves [1]. The membrane depolarization necessary for transmitter release under hypoxia could be accomplished by the recently described outward-rectifying K$^+$ currents in type I cells, which are reduced under hypoxia. The molecular mechanism of the inhibitory effect of low P_{O_2} on K$^+$ channel conductance and thus on the chemoreceptor propertics of type I cells is unknown, but the involvement of an NADPH (nicotinamide adenine dinucleotide phosphate, reduced) oxidase as a heme type P_{O_2} sensor protein has been suggested. The physiological stimuli and perhaps the metabolic inhibitors could well act on the receptor cell (type I cell). By contrast, other substances may act directly on the postsynaptic sensory nerve endings, e.g., acetylcholine, veratridine, and KCl. The question as to the transmitter between type I cells and sensory nerve endings remains open. Some authors presume it to be acetylcholine [47], but there are experimental data both supporting and contradicting this hypothesis.

There is also a hypothesis that nerve terminals in the carotid body are receptor elements, because chemoreceptive function can be observed after regeneration of the cut sinus nerve in the absence of type I and type II cells [81]. This controversy remains to be resolved.

106.8.4 Reflexes Elicited by Chemoreceptors and Their Physiological Relevance

Hypoxia. The acute ventilatory augmentation during hypoxia as well as the portion of resting ventilation which diminishes or vanishes by elevating the P_{O_2} above the normal value can be traced to the function of the arterial chemoreceptors. After their denervation, hypoxia causes a respiratory depression [108,117]. In humans, local anesthesia or denervation of the carotid sinus region is sufficient to abolish the regular respiratory response to hypoxia [72]. The reflex responses elicited by chemoreceptors are identical with the above-mentioned

acute respiratory effects of hypoxia. Whether or not these responses are potentiated by central chemosensitivity will be discussed below.

CO_2 Inhalation. In contrast to the effect of hypoxia, the ventilatory effect of P_{CO_2} is mainly of a central nature. It continues after denervation of peripheral chemoreceptors. This raises the question as to how much of the P_{CO_2}- and pH-dependent respiratory drive is accounted for by arterial chemoreceptors. An answer has been sought by denervation or reversible cold block of chemoreceptor nerves [33]. In anesthetized animals, the carotid chemoreceptors account for about 20%–30% of the pH drive and 10%–15% of the P_{CO_2} effect induced by CO_2 inhalation. Furthermore, arterial chemoreceptors seem to have a small, nonspecific stimulating effect on respiration, which is independent of blood chemistry.

It would be premature to conclude from the results of studies on CO_2 inhalation that the CO_2 sensitivity of arterial chemoreceptors does not play an essential role in the regulation of respiration. In general, increases in arterial P_{CO_2} are not caused by CO_2 inhalation, but by an alveolar ventilation which is too small in relation to the metabolic turnover. In this case, the arterial P_{O_2} is always reduced as well, and the mutual amplification of the effects of the hypercapnic and the hypoxic stimuli at the receptor site may occur and P_{CO_2} might make an essential contribution to the respiratory drive.

Metabolic Disturbances. The arterial chemoreceptors also play an essential role in the respiratory compensation of metabolic disturbances (see Chaps. 78, 102). The shifts of the CO_2 response curve with moderate metabolic acidosis or alkalosis within the pH range of 7.3–7.5 are abolished or strongly reduced after denervation of peripheral chemoreceptors. With larger pH deviations, ventilatory compensation occurs even without arterial chemoreceptors, though to a lesser degree. More recent investigations indicate that arterial chemoreceptors cooperate with the central chemosensitive mechanism and that different ranges, sensitivities, and time courses contribute to the optimal compensation of nonrespiratory acidosis [97,104, 105,146].

Sympathetic Activation. The fact that the sympathetic innervation of the glomera increases the excitation of chemoreceptors has important consequences. General sympathetic activation occurs, e.g., with a fall in blood pressure, muscular exercise, sensory stimuli, psychic excitation, and asphyxia, which are all accompanied by augmented ventilation. Arterial chemoreceptors may contribute by their sensitivity being changed by efferent sympathetic innervation.

Low Blood Pressure. The excitation of chemoreceptors at low blood pressure is also important for the regulation of the circulation. More generally, arterial chemoreceptor excitation provokes effects overriding ventilatory and circulatory influences: electroencephalography (EEG) recordings and behavior indicate arousal reactions, and the efferent outflow from gamma motoneurons to muscle spindles increases. Such reactions attest to the close coupling between respiratory centers and reticular formation [71].

106.9 Central Chemosensitivity

The described quantitative analyses have shown that a central sensitivity to changes in arterial P_{CO_2} and pH must exist. In classical experiments this central chemosensitivity has been demonstrated by rudimentary facial respiratory movements occurring when an animal's head isolated from its own circulation was perfused with blood from a donor animal and the blood chemistry was altered [89].

Diffusion Barriers. Again, both pH-sensitive and CO_2-sensitive structures might exist. In contrast to this hypothesis of two specific receptors, the one-receptor theory sees CO_2 driving ventilation via H^+ ions [76,92]. This H^+ ion theory was advocated particularly by Winterstein and has been modified several times [176,177]. Recent results argue in favor of this theory [97]. Since ventilation is not only a function of arterial pH, the intracranial chemosensor should be located within a region whose pH is influenced by the arterial P_{CO_2}, but is different from the arterial pH. The pH is defined as a function of $[HCO_3^-]/[CO_2]$; thus, the potential receptor structure might be located at a place which equilibrates with the arterial P_{CO_2} in a different way from arterial HCO_3^-. Such a situation could be realized at diffusion barriers because CO_2, due to its high lipid solubility, penetrates biological membranes very much faster than ions. Thus, the interior of cells bathed in CO_2-containing bicarbonate solutions can become more acidic than their exterior [94]. In the CNS, a similar functional barrier might be represented by the blood–brain barrier and the blood–cerebrospinal fluid (CSF) barrier (see Chap. 27).

106.9.1 Chemistry of the Cerebrospinal Fluid and Ventilation

pH Changes. Perfusion of the cerebral ventricular system with artificial CSF has shown that ventilation can be influenced by CSF pH changes: respiratory augmentation by acidification, respiratory depression with alkalization. Experiments have been performed in anesthetized dogs and cats and in awake goats [49,94,99]. By calculating the diffusion of CO_2 and $[HCO_3^-]$ from the ventral CSF space into the brain, the ventilatory effect can be traced to the pH sensitivity of a substrate which is located 250–300 μm below the ventral surface [15], or, according to a different mathematical approach, about two thirds to three quarters of the distance over which the $[HCO_3^-]$ concentration falls from the CSF to the capillary blood of the brain [129]. From these studies, the assumption of a specific CO_2-sensitive component of central chemosensitivity becomes unnecessary. Rather, the degree of acidity may be of importance in

understanding the receptor mechanism. Upon acid injections causing small changes in brain extracellular pH, ventilation and tidal volume increase, and the same response is obtained to pH changes induced by CO_2 inhalation. Upon injection of greater amounts of acid, however, ventilation and tidal volume unexpectedly cease to rise and in most cases even drop back towards the control values. The ventilatory response to pH, when changed by CO_2 inhalation, continues with the same slope, however. This indicates a differential response to extracellular pH in metabolic and in respiratory acidosis [160]. It has been suggested that, at mild degrees of acidosis, extracellular H^+ ions have free access to the H^+ receptor, but are prevented from reaching it in severe metabolic acidosis, while the passage of CO_2 is uninhibited [97].

Ion Concentrations. Ventilation is influenced not only by CSF pH, but also by other ions. Thus, perfusion of the ventricular system with solutions of moderately increased K^+ concentration stimulates ventilation, whereas an increase in Ca^{2+} concentration reduces ventilation. The extent to which these effects are nonspecific and general or specifically act on substrates controlling ventilation has not yet been analyzed. The fact that changes in ion concentrations in the CSF do affect vigilance, muscle tone, and the cardiovascular system seems to suggest that the effects on ventilation are nonspecific.

106.9.2 Location of Central Chemosensitivity

Ventral Brain Stem Surface. Isolated chemical stimulation on the ventral surface of the brain stem of cats, dogs, and rats identified bilateral rostral and caudal as well as intermediate areas on the ventral side of the medulla oblongata (see Fig. 106.12) [111,151]. From here, ventilation could be stimulated by local administration of acids or nicotine. In anesthetized and unanesthetized cats with intact peripheral chemoreceptors, cold block, local anesthetics, or neurotoxins administered to the intermediate areas eliminated or strongly reduced the respiratory response to CO_2 and to metabolic acidosis, but the hypoxic drive was maintained [146,152]. Denervation of the arterial chemoreceptors under these conditions resulted in complete apnea. The immediate inactivation of inspiratory and expiratory neurons when local anesthetics are applied to the ventral surface can be explained by an indirect effect [132,154]. It may be concluded, then, that an essential portion of the tonic ventilatory drive emerges from the ventral brain stem surface [95,132,148,154]. This conclusion is supported by more recent neurophysiological in vitro and in vivo approaches [25,148,159,167]. Whether deeper structures responding to wide pH ranges belong to the receptor site or to the chain of subsequent relays is still a matter of controversial discussion [28,87,107,116,118].

When the CSF pH on the ventral medullary surface is kept constant by artificial superfusion, CO_2 inhalation in chemoreceptor-denervated animals still causes an increase of ventilation, albeit reduced [15]. This is to be expected,

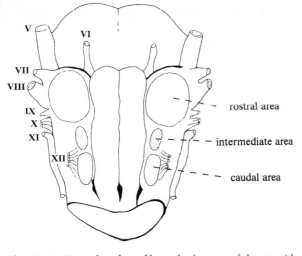

Fig. 106.12. Ventral surface of lower brain stem of the cat with areas, the local chemical stimulation of which causes characteristic ventilatory changes [150]

since the extracellular fluid of the chemosensitive substrate is altered via the blood as well as the CSF.

106.9.3 Physiology of Cerebrospinal Fluid
(cf. Chap. 27)

For the adjustment of CSF pH, the following facts are important:

- The $CSF_{P_{CO_2}}$ corresponds to that of the venous cerebral blood, since the CSF is in equilibrium with the venous capillary blood, not with the arterial blood, and is thus about 10 mmHg (1.3 kPa) higher.
- The low pH results from the high P_{CO_2} and the low bicarbonate content. Due to the much lower buffer capacity, P_{CO_2} changes result in stronger pH changes than in blood (see Fig. 106.13).
- On the other hand, changes in arterial P_{CO_2} are followed by smaller changes in $CSF_{P_{CO_2}}$. This is consequence of the enhancement of cerebral blood flow with an elevation of arterial P_{CO_2}, which normally diminishes the arteriovenous P_{CO_2} difference.

There is a potential difference between CSF and blood of about 5–7 mV, CSF being positively charged compared to blood [64,113]. This is of interest for respiratory regulation [64]. The CSF potential increases with arterial acidosis and decreases with alkalosis, vanishing at an arterial pH of 7.5 and reversing with stronger alkalosis. The potential is not influenced by changes in CSF_{pH}. The potential changes are directed so as to favor the compensatory bicarbonate shifts between CSF and blood during longer-lasting respiratory changes of the acid–base balance.

Cerebrospinal Fluid pH, Acid–Base Balance, and Respiratory Regulation. During longlasting acidosis or alkalosis,

the changes in CSF pH are in general considerably smaller than those of plasma (see Fig. 106.14). After acute changes in the arterial pH, CSF pH behaves differently in respiratory and in metabolic acidosis.

Acute respiratory acidosis and alkalosis are first followed by a shift of CSF_{pH} in the same direction. During longer respiratory disturbances, the deviation of CSF_{pH} reverses faster than it does in the arterial plasma. Measurements of P_{CO_2} and $[HCO_3^-]$ led to the following interpretation. Due to the high diffusibility of CO_2, $CSF_{P_{CO_2}}$ follows the cerebral venous P_{CO_2}, which varies in the same direction as the arterial P_{CO_2}, albeit – due to the effect of cerebral blood flow – to a lesser extent. Secondary changes occur in the outward transport of CSF bicarbonate (see Chap. 27). These are directed so as to maintain CSF_{pH}: elevation of $[HCO_3^-]$ transport with central alkalosis, diminution with central acidosis. These compensatory processes begin about 2 h after the start of the pH deviation and are completed within 12 days. They proceed faster than the normalization of arterial pH, which is brought about by changes in renal bicarbonate resorption. During hyperventilatory alkalosis, lactate is released into the CSF, diminishing the CSF bicarbonate.

Acute metabolic changes may be accompanied by only very small changes in CSF_{pH} in the same direction; CSF changes may even be absent, and often so-called paradoxical pH shifts in CSF and blood in opposite directions occur

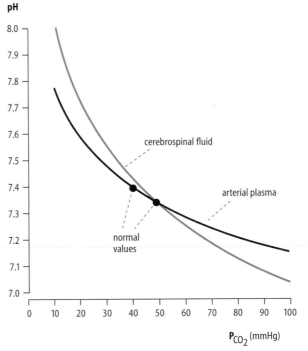

Fig. 106.13. CO_2 pressure (P_{CO_2})-dependent pH values of human arterial plasma and of cerebrospinal fluid in vitro. Due to the smaller buffer capacity of CSF, the CSF curve (*blue*) shows a steeper course. The normal CSF pH is lower than in blood due to the higher P_{CO_2} and the lower $[HCO_3^-]$ compared to arterial blood [83]

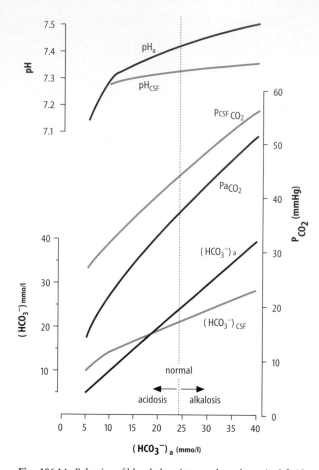

Fig. 106.14. Behavior of blood chemistry and cerebrospinal fluid (*CSF*) chemistry during chronic metabolic acidosis and alkalosis in man. Arterial [HCO_3^-]$_a$ serves as a measure for the metabolic deviation. The variations of pH_{CSF} are smaller than those of the arterial pH (*pH$_a$*). This is the consequence of the small increase of [HCO_3^-]$_{CSF}$ during alkalosis or the small decrease during acidosis, compared to the changes within the arterial plasma. With stronger acidosis, [HCO_3^-]$_{CSF}$ is beyond that of the plasma concentration (crossing of the corresponding curves below the normal value). [HCO_3^-]$_a$ und [HCO_3^-]$_{CSF}$ from measurements of various authors. P_{CSFCO_2} calcualted from Pa_{CO_2} under the assumption of a CO_2 exchange with the venous capillary blood and an arteriovenous CO_2 difference AVD_{CO2} becoming reduced with increasing P_{CO_2} (effect of cerebral blood flow). pH_{CSF} calculated from P_{CSFCO_2} and [HCO_3^-]$_{CSF}$ [49]

[97]. Thus, the CSF_{pH} rises with infusion of NH_4Cl and declines with infusion of bicarbonate. These paradoxical changes do not occur when the alveolar P_{CO_2} is kept constant. Whether or not they occur during chronic metabolic disturbances is controversial.

During acute metabolic acidosis, the compensatory ventilatory change reaches its full effect only with considerable latency. This may be explained by the finding that ventilatory changes is inhibited by the paradoxical shift of CSF_{pH}. Only with normalization of the central pH by the bicarbonate shifts will the arterial pH cause a secondary compensatory change in ventilation. In this case, CSF chemistry delays respiratory compensation.

These complicated relations can be understood most easily by realizing that the most effective sensors for the ventilatory regulation of arterial P_{CO_2} and pH are influenced by an environment the pH of which is kept constant by an additional regulatory mechanism. Under specific condition, both peripheral and central regulations can compete with each other. Consideration of the resulting events shows that the pH of the brain extracellular space is given priority over the arterial pH. The constancy of P_{CO_2} is subordinated to both of these functional aims.

This concepts is still controversial in some points, for instance with regard to the importance of peripheral chemoreceptors for the compensation of metabolic disturbances [49,96,104,146] and to the location and nature of the central chemosensitive substrate [15,49,89,97]. However, these points do not essentially undermine the basic concept. In any case, ventilatory changes result from changes in arterial and CSF_{pH}. Irrespective of the controversial details that still remain, central chemoreceptors are conceived as a structure which – together with peripheral chemoreceptors – is responsible for the adjustment of ventilation to metabolic production of volatile and fixed acids. A schematic view of the functional connections of peripheral and central chemosensitivity is presented in Fig. 106.15.

106.10 Conclusions

The innervation of respiratory muscles has its origin within a central nervous substrate which, in turn, is the target of excitation from numerous peripheral and central sources. On the basis of an excitation state, specialized structures generate a rhythm (see Chap. 105), which is transferred to the respiratory muscles.

Neural feedback from the mechanical respiratory apparatus and the lung adjusts central motor outflow so as to enable the most favorable performance of the movements.

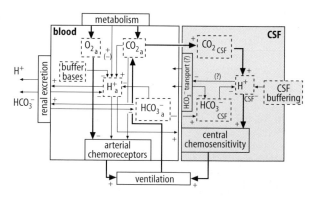

Fig. 106.15. The most important relations between blood chemistry and cerebrospinal fluid (*CSF*) chemistry influencing ventilation. The *arrows* represent the direction of the effects. *Plus signs* indicate equidirectional changes of the dependent measurement with deviation of the influencing measurement; *minus signs* indicate corresponding changes in the opposite direction (see also Fig. 106.2) [83]

Ventilation is an effector for many functions. On the one hand, it underlies many spontaneous variations due to the close coupling with the general excitatory state of the CNS paralleling its activity, and especially in humans it is used for numerous expressions of life. On the other hand, it has to serve as a transport and exchange system for O_2 and CO_2. Respiration fulfills a homeostatic function for the constancy of blood and tissue chemistry by adapting ventilation to metabolic needs. This task is ensured by the fact that the respiratory gases carried and H^+ ions themselves act as respiratory feedback drives.

Both functions compete with each other. The result is a compromise. The ventilatory variations serving central nervous expressions (psyche, speech, etc.) lead to deviations of blood chemistry, so that there is no absolute homeostasis. However, the homeostatic mechanisms guarantee that the deviations are kept within a given range, despite the demands laid on the respiratory system by other functions, which are taken care of with great precision beyond the blood–brain barrier.

During the marked elevation of the metabolism caused by muscular exercise, the chemical feedback system cannot provide enough respiratory drive to ensure homeostasis. In this case, additional respiratory drives arise which act in parallel with motor innervation and peripheral motor action. However, the feedback drives remain effective and guarantee the fine homeostatic adjustment.

The role and importance of all the individual factors depend on the prevailing functional state. During metabolic resting conditions, the central nervous functions predominate. During stronger deviations of blood chemistry, these become prevalent. During physical work, the resulting respiratory drive is completely in the foreground.

Thus, respiratory activity is not a response to a defined respiratory stimulus modified by various factors, but the result of the cooperation between numerous influences, the importance of which changes depending on the particular priority.

References

1. Acker H (1989) PO₂ chemoreception in arterial chemoreceptors. Annu Rev Physiol 51:835–844
2. Acker H, Lübbers DW, Purves MJ (1971) Local oxygen tension field in the glomus caroticum of the cat and its change at changing arterial PO₂. Pflugers Arch 329:136–155
3. Adrian ED (1933) Afferent impulses in the vagus and their effect on respiration. J Physiol (Lond) 79:332–357
4. Adrian ED, Bronk DW (1928) The discharge of impulses in motonerve fibers. J Physiol (Lond) 66:81
5. Alexander JK, West JR, Wood JA, Richards DW (1955) Analysis of respiratory response to carbon dioxide inhalation in varying clinical states of hypercapnia, anoxia, and acid base dearangement. J Clin Invest 34:511–532
6. Anch AM, Remmers JE, Sauerland EK, DeGroot WJ (1981) Oropharyngeal patency during waking and sleep in the Pickwickian syndrome. Electromyogr Clin Neurophysiol 21:317–330
7. Anders K, Ohndorf W, Dermietzel R, Richter DW (1993) Synapses between slowly adapting lung stretch receptor afferents and inspiratory beta neurons in the nucleus of the solitary tract of cats: a light and electron microscopic study. J Comp Neurol 335:163–172
8. Arias Stella J (1969) Human carotid body at high altitudes. Am J Pathol 55:82a (abstract)
9. Asmussen E, Nielsen M (1946) Studies on the regulation of respiration in heavy work. Acta Physiol Scand 12:171–188
10. Asmussen E, Nielsen M, Wieth Pedersen G (1943) Cortical or reflex control of respiration during muscular work? Acta Physiol Scand 6:168–175
11. Astrand PO, Christensen EH (1963) The hyperpnoea of exercise. In: Cunningham DJC, Lloyd BB (eds) The regulation of human respiration. Blackwell, Oxford, pp 515–••
12. Bartlett D Jr (1979) Effects of hypercapnia and hypoxia on laryngeal resistance to airflow. Respir Physiol 37:293–302
13. Bartlett D Jr (1986) Upper airway motor systems. In: Fishman AP, Cherniack NS, Widdicombe JG (eds) Handbook of physiology, the respiratory system, vol 2, control of breathing, part 1. edited American Physiological Society, Bethesda, pp 223–246
14. Bayliss DA, Millhorn DE (1992) Central neural mechnisms of progesterone action: application to the respiratory system. J Appl Physiol 73:393–404
15. Berndt J, Berger W, Berger K, Schmidt M (1972) Untersuchungen zum zentralen chemosensiblen Mechanismus der Atmung II. Die Steuerung der Atmung durch das extrazelluläre pH im Gewebe der Medulla oblongata. Pflugers Arch 332:146–170
16. Bertrand F, Hugelin A (1971) Respiratory synchronizing function of nucleus parabrachialis medialis: pneumotaxic mechanisms. J Neurophysiol 34:189–207
17. Bianchi AL (1971) Localisation et étude des neurones respiratoires bulbaires. J Physiol (Paris) 63:5–40
18. Bingmann D, Baker RE, Ballantyne D (1991) Rhythm generation in brainstem cultures grown in a serum free medium. Neurosci Lett 132:167–170
19. Bingmann D, Baker RE, Ballantyne D (1993) Rhythmic discharge of medullary neurones in organotypic culture: suppression by calcium antagonists. Pflugers Arch 422 [Suppl]: R128 (abstract)
20. Biscoe TJ, Bradley GW, Purves MJ (1970) The relation between carotid body chemoreceptor discharge, carotid sinus pressure and carotid body venous flow. J Physiol (Lond) 208:99–120
21. Block AJ, Boysen PG, Wynne JW (1979) Sleep apnea, hypopnea and oxygen desaturation in normal subjects. N Engl J Med 300:513–517
22. Bradley GW, Noble MIM, Trenchard D (1976) The direct effect on pulmonary stretch receptor discharge produced by changing lung carbon dioxide concentration in dogs on cardiopulmonary bypass and its action on breathing. J Physiol (Lond) 261:359–373
23. Brodie TG (1900) The immediate action of an intravenous injection of blood serum. J Physiol (Lond) 26:92–106
24. Brouillette RT, Thach BT (1980) Control of geniolossus muscle inspiratory activity. J Appl Physiol 49:801–808
25. Bruce EN, Cherniack NS (1987) Central chemoreceptors. J Appl Physiol 62:389–402
26. Bryan AC, Bowes G, Maloney JE (1986) Control of breathing in the fetus and the newborn. In: Fishman AP, Cherniack NS, Widdicombe JG, Geigar SR (eds) Handbook of physiology, section II: the respiratory system, vol II: control of breathing, part 2. American Physiological Society, Bethesda, pp 621–647
27. Clark FJ, von Euler C (1972) On the regulation of depth and rate of breathing. J Physiol (Lond) 222:267
28. Coates EL, Li Nattie EE (1993) Widespread sites of brainstem ventilatory chemoreceptors. J Appl Physiol 75:5–14
29. Cohen MI (1975) Phrenic and recurrent laryngeal nerve discharge pattern and the Hering Breuer reflex. Am J Physiol 228:1489–1496
30. Coleridge HM, Coleridge JCG (1986) Reflexes evoked from the tracheobronchial tree and lungs. In: Fishman AP, Cherniack NS, Widdicombe JG (eds) Handbook of physiol-

ogy, the respiratory system, vol 2: control of breathing, part 1. American Physiological Society, Bethesda, pp 395–430

31. Coleridge HM, Coleridge JCG, Banzett RB (1978) Effect of CO_2 on afferent vagal endings in the canine lung. Respir Physiol 34:135–151
32. Coleridge HM, Roberts AM, Coleridge JCG (1982) Effect of vagal cooling on activity in lung afferent fibres in dogs. Fed Proc 41:986 (abstract)
33. Comroe JH Jr, Schmidt CF (1938) The part played by reflexes from the carotid body in the chemical regulation of respiration in the dog. Am J Physiol 121:75–97
34. Corda M, von, Euler C, Lennerstrand G (1965) Proprioceptive innervation of the diaphragm. J Physiol (Lond) 178:161–177
35. Cormack RS, Cunningham DJC, Gee JBL (1957) The effect of carbon dioxide on the respiratory response to want of oxygen in man. Q J Exp Physiol 42:303–319
36. Cunningham DJC, O'Riordan JLH (1957) The effects of a rise in the temperature of the body on the respiratory response to carbon dioxide at rest. Q J Exp Physiol 42:329–345
37. Dawes GS, Comroe JH Jr (1954) Chemoreflexes from the heart and lungs. Physiol Rev 34:167–201
38. Dawes GS, Fox HE, Leduc BM, Liggins GC, Richards RT (1972) Respiratory movements and rapid eye movement sleep in the foetal lamb. J Physiol (Lond) 220:119–143
39. De Kock LL (1954) Intraglomerular tissues of the carotid body. Acta Anat 21:101–121
40. Dejours P, Labrousse Y, Raynaud J, Teillac A (1957) Stimulus oxygene chemoreflexe de la ventilation à basse altitude (50 m) chez l'homme.l.Au repos. J Physiol (Paris) 49: 115–120
41. Dixon M, Jackson DM, Richards IM (1979) The effects of a respiratory tract infection on histamine induced changes in lung mechanics and irritant receptor discharge in dogs. Am Rev Respir Dis 120:843–848
42. Edwards C, Heath D, Harris P, Castillo Y, Kruger H, Arias Stella J (1971) The carotid body in animals at high altitude. J Pathol 104:231–238
43. Eichenwald EC, Stark AR (1991) Respiratory motor output. Effect of state and maturation in early life. In: Haddad GG, Farber JP (eds) Lung biology in health and disease. Developmental neruobiology of breathing. Dekker, New York, pp 551–590
44. Eklund G, von Euler C, Rutkowski S (1964) Spontaneous and reflex activity of intercostal gamma motoneurons. J Physiol (Lond) 171:139
45. England SJ, Bartlett D Jr, Knuth SL (1982) Influence of human vocal cord movements on airflow and resistance in eupnea. J Appl Physiol 52:773–779
46. Eyzaguirre C, Uchizono K (1961) Observations on the fiber content of nerves reaching the carotid body of the cat. J Physiol (Lond) 159:268–281
47. Eyzaguirre C, Koyano H, Taylor JR (1965) Presence of acetylcholine and transmitter release from the carotid body chemoreceptors. J Physiol (Lond) 178:463–476
48. Fedde MR (1970) Peripheral control of avian respiration. Fed Proc 29:1664–1673
49. Fencl V, Miller TB, Pappenheimer JR (1966) Studies on the respiratory response to disturbances of acid base balance, with deductions concerning the ionic composition of cerebal intestinal fluid. Am J Physiol 210:459–472
50. Ferris BG Jr, Mead J, Opie LH (1964) Partitioning of respiratory flow resistance in man. J Appl Physiol 19:653–658
51. Fidone SJ, Gonzalez C (1986) Initiation and control of chemoreceptor activity in the carotid body. In: Fishman AP, Cherniack NS, Widdicombe JG (eds) Handbook of physiology, the respiratory system, vol 2, control of breathing, part 1. American Physiological Society, Bethesda, pp 247–312
52. Fidone SJ, Gonzalez C, Yoshizaki K (1982) Effects of low oxygen on the release of dopamine from the rabbit carotid body in vitro. J Physiol (Lond) 333:93–110

53. Fillenz M, Widdicombe (1972) Receptors of the lungs and airways. In: Neil E (ed) Handbook of sensory physiology. Enteroceptors, vol III, part 1. Springer, Berlin Heidelberg New York, pp 81–112
54. Fink BK, Hanks EC, Ngai SH, Papper EM (1963) Central regulation of respiration during anesthesia and wakefulness. Ann N Y Acad Sci 109:892–899
55. Fisher JT, Mortola JP (1980) Statics of the respiratory system in newborn mammals. Respir Physiol 41:155–172
56. French JD (1960) The reticular formation. In: Field J, Magoun HW, Hall VE (eds) Handbook of physiology, section 1, vol 52. American Physiological Society, Washington DC, pp 1281–1305
57. Glogowska M, Richardson PS, Widdicombe JG, Winning AJ (1972) The role of the vagus nerves, peripheral chemoreceptors and other afferent pathways in the genesis of augmented breaths in cats and rabbits. Respir Physiol 16:179–196
58. Guilleminault C, Dement W (eds) (1978) Sleep apnea syndromes. Liss, New York
59. Guz A (1977) Respiratory sensations in man. Br Med Bull 33:175–177
60. Hairston LE, Sauerland EK (1981) Electromyography of the human palate: discharge patterns of the levator and tensor veli palatini. Electromyogr Clin Neurophysiol 21:287–297
61. Hamilton LH (1979) Nasal airway resistance: its measurement and regulation. Physiologist 22:43–49
62. Head H (1889) On the regulation of respiration I. Experimental. J Physiol (Lond) 10:1–70
63. Hedner J, Hedner T, Wessberg P, Lundberg D, Jonason J (1983) Effects of TRH and TRH analogues on the central regulation of breathing in the rat. Acta Physiol Scand 117:427–437
64. Held D, Fencl V, Pappenheimer JR (1964) Electrical potential of cerebrospinal fluid. J Neurophysiol 27:942
65. Henderson Smart DJ, Read DJC (1979) Reduced lung volume during behavioral active sleep in the newborn. J Appl Physiol 46:1081–1085
66. Henke K, Dempsey JA, Kowitz JM, Skatrud JB (1990) Effects of sleep induced increases in upper airway resistance on ventilation. J Appl Physiol 69:617–624
67. Hering E, Breuer J (1868) Die Selbststeuerung der Athmung durch den Nervus vagus. Sitzungsber Akad Wiss Wien 57:672–677
68. Hess WR (1930) Kritik der Hering-Breuerschen Lehre von der Selbststeuerung der Atmung. Pflugers Arch 226: 198
69. Heymans C, Bouckaert JJ (1930) Sinus caroticus and respiratory reflexes. J Physiol (Lond) 69:254
70. Heymans JF, Heymans C (1927) Sur les modifications directes et sur la regulation reflexe de l'activité du centre respiratoire de la tête isolée du chien. Arch Int Pharmacodyn Ther 33:273–370
71. Heymans C, Neil E (1958) Reflexogenic areas of the cardiovascular system. Churchill, London
72. Honda Y, Hashizume I (1991) Evidence for hypoxic depression of CO_2 ventilation response in carotid body resected humans. J Appl Physiol 70:590–593
73. Hornbein TF (1968) The relation between stimulus to the chemoreceptors and their response. In: Torrance RW (ed) Arterial chemoreceptors. Blackwell, Oxford, pp 65–78
74. Hyatt RE, Wilcox RE (1961) Extrathoracic airway resistance in man. J Appl Physiol 16:326–330
75. Ingram RH, Schilder DP (1967) Effect of pursed lips expiration on the pulmonary pressure flow relationship in obstructive lung disease. Am Rev Respir Dis 96:381–388
76. Jacobs MH (1920) Production of intracellular acidity by neutral and alkaline solutions containing carbon dioxide. Am J Physiol 53:457
77. Johnson SM, Getting PA (1992) Excitatory effects of thyrotropin releasing hormone on neurons within the nucleus ambiguus of adult guinea pigs. Brain Res 592:1–5

78. Kao FF (1963) An experimental study of the pathways involved in excercise hyperpnoea employing cross circulation techniques. In: Cunningham DJC, Lloyd BB (eds) The regulation of human respiration. Blackwell, Oxford, pp 461–502

79. Katsaros B, Loeschcke HH, Lerche D, Schanthal H, Hahn N (1960) Wirkung der Bicarbonat Alkalose auf die Lungenbelüftung beim Menschen. Bestimmung der Teilwirkungen von pH und CO_2 Druck auf die Ventilation und Vergleich mit den Ergebnissen bei Acidose. Pflugers Arch 271:7–32

80. Kawahara K, Suzuki M (1991) Descending inhibitory pahtway responsible for simultaneous suppression of postural tone and respiration in decerebrate cats. Brain Res 538:303–309

81. Kienecker EW, Knoche H, Bingmann D (1978) Functional properties of regenerating sinus nerve fibres in the rabbit. Neuroscience 3:977–988

82. Knowlton GC, Larrabee MG (1946) A unitary analysis of pulmonary volume receptors. Am J Physiol 147:100–114

83. Koepchen HP (1975) Atmungsregulation. In: Gauer OH, Kramer K, Jung R (eds) Physiologie des Menschen, vol 6: Atmung, Urban and Schwarzenberg, Munich, pp 163–310

84. Koepchen HP (1976) Quantitative approach to neural control of ventilation. In: Loeschcke HH (ed) Acid base homeostasis of the brain extracellular fluid and the respiratory control system. Thieme, Stuttgart, pp 164–186

85. Koepchen HP, Huopaniemi T (eds) (1991) Cardiorespiratory and motor coodination. Springer, Berlin Heidelberg New York

86. Koepchen HP, Abel HH, Klüssendorf D (1987) Integrative neurovegetative and motor control: phenomena and theory. Funct Neurol 2(4):389–406

87. Kogo N, Arita H (1990) In vivo study of medullary H^+ sensitive neurons. J Appl Physiol 69:1408–1412

88. Lack EE (1978) Hyperplasia of vagal and carotid body paraganglia in patients with chronic hypoxemia. Am J Pathol 91:497–516

89. Lambertsen CJ (1965) Effects of oxygen at high partial pressure. In: Fenn WO, Rahn H Handbook of physiology, section 3, respiration, vol 2. American Physiological Society, Washington DC

90. Lambertsen CJ, Semple SJG, Smyth MS, Gelfand R (1961) H^+ and Pco_2 as chemical factors in respiratory and cerebral circulatory control. J Appl Physiol 16:473–484

91. Larrabee MG, Knowlton GC (1946) Excitation and inhibition of phrenic motoneurones by inflation of the lungs. Am J Physiol 147:90–99

92. Lehmann C (1888) Über den Einfluß von Alkali und Säure auf die Erregung des Athemcentrums. Pflugers Arch 42:284

93. Lerche D, Katsaros B, Lerche G, Loeschcke HH (1960) Vergleich der Wirkung verschiedener Acidosen (NH_4Cl, C_4Cl_2, Acetazolamid) auf die Lungenbelüftung beim Menschen. Pflugers Arch 270:450–460

94. Leusen I (1954) Chemosensitivity of the respiratory center: influence of changes in the hydrogen ion and total buffer concentration in the cerebral ventricles on respiration. Am J Physiol 176:45–51

95. Loeschcke HH (1960) Homeostase des arteriellen CO_2 Drucks und Anpassung der Lungenventilation an den Stoffwechsel als Leistungen eines Regelsystems. Klin Wochenschr 38:366–376

96. Loeschcke HH (1980) Chemical alterations of cerebrospinal fluid acting on respiratory and circulatory control systems. In: Wood JH (ed) Neurobiology of cerebrospinal fluid, vol 1. Plenum, New York, pp 29–40

97. Loeschcke HH (1982) Central chemosensitivity and the reaction thoery, J Physiol (Lond) 332:1–24

98. Loeschcke HH, Gertz KH (1958) Einfluß des O_2-Druckes in der Einatmungsluft auf die Atemtätigkeit des Menschen, geprüft unter Konstanthaltung des alveolären CO_2 Druckes. Pflugers Arch 267:460–477

99. Loeschcke HH, Koepchen HP, Gertz KH (1958) über den Einfluß von Wasserstoffionenkonzentrationen und CO_2 Druck im Liquor cerebrospinalis auf die Atmung. Pflugers Arch 266:569–585

100. Magoun HW (1950) Caudal and cephalic influences of brainstem reticular formation. Physiol Rev 30:459

101. Mathew OP, Abu Osba YK, Thach BT (1982) Influence of upper airway pressure changes on genioglossus muscle respiratory activity. J Appl Physiol 52:438–444

102. McCaffrey TV, Kern EB (1980) Laryngeal regulation of airway resistance. II. Pulmonary receptor reflexes. Ann Otol Rhinol Laryngol 89:462–466

103. McDonald DM (1981) Peripheral chemoreceptors: structure function relationships of the carotid body. In: Hornbein TF (ed) Lung biology in health and disease. The regulation of breathing. Dekker, New York, pp 105–319

104. Middendorf T, Loeschcke HH (1976) Mathematische Simulation des Respirationssystems. J Math Biol 149–177

105. Middendorf T, Loeschcke HH (1978) Cooperation of the peripheral and central chemosensitive mechanisms in the control of the extracellular pH in brain in nonrespiratory acidosis. Pflugers Arch Eur J Physiol 375:257–260

106. Milerad J, Lagercrantz H, Løfgren O (1985) Alveolar hypoventilation treated with medroxyprogesterone. Arch Dis Childhood 60:150

107. Millhorn DE, Eldridge FL (1986) Role of ventrolateral medulla in regulaition of respiratory and cardiovascular systems. J Appl Physiol 61:1249–1263

108. Millhorn DE, Eldridge FL, Kiley JP, Waldrop TG (1984) Prolonged inhibition of respiration following acute hypoxia in glomectomized cats. Respir Physiol 57:331–340

109. Miserocchi G, Sant Ambrogio G (1974) Distribution of pulmonary stretch receptors in the intrapulmonary airways of the dog. Respir Physiol 21:71–75

110. Mitchell GS, Cross BA, Hiramoto T, Scheid P (1980) Effects of intrapulmonary CO_2 and airway pressure on phrenic activity and pulmonary stretch receptor discharge in dogs. Respir Physiol 41:29–48

111. Mitchell RA, Loeschcke HH, Massion WH, Severinghaus JW (1963) Respiratory responses mediated through superficial chemosensitive areas on the medulla. J Appl Physiol 18:523–533

112. Mortola JP, Fisher JT (1981) Mouth and nose resistance in newborn kittens and puppies. J Appl Physiol 51:641–645

113. Mottschall HJ, Loeschcke HH (1963) Messungen des transmeningealen Potentials der Katze bei Änderung des CO_2 Drucks und der H^+ Ionen Konzentration im Blut. Pflugers Arch 277:662–670

114. Muller NL, Francis PW, Gurwitz D, Levison H, Bryan AC (1980) Mechanism of hemoglobin desaturation during rapid eye movement sleep in normal subjects and patients with cystic fibrosis. Am Rev Respir Dis 121:463–469

115. Naeye RL, Fisher R, Ryser M, Whalen P (1976) Carotid body in the sudden infant death syndrome. Science 191:567–569

116. Nattie EE (1990) The alphastat hypothesis in respiratory control and acid base balance. J Appl Physiol 69:1201–1207

117. Neubauer JA, Melton JE, Edelman NH (1990) Modulation of respiration during brain hypoxia. J Appl Physiol 68:441–451

118. Neubauer JA, Gonsalves SF, Chou W, Geller HM, Edelman NH (1991) Chemosensitivity of medullary neurons in explant tissue cultures. Neuroscience 45:701–708

119. Nielsen M, Smith H (1951) Studies on the regulation of respiration in acute hypoxia. Acta Physiol Scand 24:293

120. Ochwadt B, Kreutzer H, Bücherl E, Loeschcke HH (1959) Beeinflussung der Atemsteigerung bei Muskelarbeit durch partiellen neuromuskulären Block (Tubocurarin). Pflugers Arch 269:613

121. Ínal E, Lopata M, O'Conner TD (1981) Diaphragmatic and genioglossal electromyogram responses to CO_2 rebreathing in humans. J Appl Physiol 50:1052–1055

122. Ínal E, Lopata M, O'Conner D (1981) Diaphragmatic and genioglossal electromyogram responses to isocapnic hypoxia in humans. Am Rev Respir Dis 124:215–217

123. Orem J, Lydic R (1978) Upper airway function during sleep and wakefulness: experimental studies on normal and anesthetized cats. Sleep 1:49–78

124. Orem J, Netick A, Dement WC (1977) Increased upper airway resistance during sleep in the cat. Electroencephalogr Clin Neurophysiol 43:14–22

125. Osborn JJ (1953) Experimental hypothermia: respiratory and blood pH changes in relation to cardiac function. Am J Physiol 175:389

126. Paintal AS (1964) Effects of drugs on vertebrate mechanoreceptors. Pharmacol Rev 16:341–380

127. Paintal AS (1970) The mechanisms of excitation of type J receptors, and the J reflex. In: Porter R (ed) Breathing: Hering Breuer centenary symposium. Churchill, London, pp 59–71

128. Paintal AS (1973) Vagal sensory receptors and their reflex effects. Physiol Rev 53:159–227

129. Pappenheimer JR, Fencl V, Heisey SR, Held D (1965) Role of cerebral fluids in control of respiration as studied in unanesthetized goats. Am J Physiol 208:436–450

130. Parisi RA, Edelman NH, Santiago TV (1992) Central respiratory carbon dioxide chemosensitivity does not decrease during sleep. Am Rev Respir Dis 145:832–836

131. Payne JK, Higenbottam T, Guindi GM (1981) Respiratory activity of the vocal cords in normal subjects and patients with airflow obstruction: an electromyographic study. Clin Sci 61:163–167

132. Peskov BJ, Piatin WF (1976) Reactions of neurons of the respiratory center to local cooling of the ventral surface of the medulla oblongata. Sechenov Physiol J USSR 62:7

133. Peter JH, Penzel T, Podszus T, von Wichert P (eds) (1991) Sleep and health risk. Springer, Berlin Heidelberg New York

134. Phillipson EA, Bowes G (1986) Control of breathing during sleep. In: Fishman AP, Cherniack NS, Widdicombe JG (eds) Handbook of physiology, section 3, the respiratory system, part 2, American Physiological Society, Bethesda, pp 649–689

135. Phillipson EA, Fishman NH, Hickey RF, Nadel JA (1973) Effect of differential vagal blockade on ventilatory response to CO2 in awake dogs. J Appl Physiol 34:759–763

136. Provine RR, Hamernik HB, Curchack BC (1987) Yawning: relation to sleeping and stretching in humans. Ethology 76:152–160

137. Raschke F (1991) The respiratory system features of modulation and coordination. In: Haken H, Koepchen HP (eds) Rhythms in physiological systems. Springer, Berlin Heidelberg New York, pp 155–164

138. Rekling JC (1990) Excitatory effects of thyrotropin releasing hormone (TRH) in hypoglossal motoneurones. Brain Res 510:175–179

139. Richards CC, Bachman L (1961) Lung and chest wall compliance of apneic paralyzed infants. J Clin Invest 40:273–278

140. Saldana MJ, Salem LE, Travezan R (1973) High altitude hypoxia and chemodectomas. Hum Pathol 4:251–263

141. Salmoiraghi GC, Burns BD (1960) Notes on mechanism of rhythmic respiration. J Neurophysiol 23:14–26

142. Sanders MH, Moore SE (1983) Inspiratory and expiratory partitioning of airway resistance during sleep in patients with sleep apnea. Am Rev Respir Dis 1983; 127:554

143. Sant'Ambrogio G (1982) Information arising from the tracheobronchial tree of mammals. Physiol Rev 62:531–569

144. Sant'Ambrogio G, Miserocchi G, Mortola J (1974) Transient responses of pulmonary stretch receptors in the dog to inhalation of carbon dioxide. Respir Physiol 22:191–197

145. Sauerland EK, Orr WC, Hairston LE (1981) EMG patterns of oropharyngeal muscles during respiration in wakefulness and sleep. Electroencephalogr Clin Neurophysiol 21:307–316

146. Schäfer D, Schläfke ME (1995) Cardiorespiratory regulation in a model for the Ondine's curse syndrome. In: Trouth CO, Millis RM, Kiwull Schöne H, Schläfke ME (eds) Ventral brainstem mechanisms and control functions. Dekker, New York, pp 675–685

147. Schäfer T, Schäfer D, Schläfke ME (1993) Breathing, transcutaneous blood gases, and CO_2 response in SIDS siblings and control infants during sleep. J Appl Physiol 74:88–102

148. Schläfke ME (1981) Central chemosensitivity: a respiratory drive. Rev Physiol Biochem Pharmacol 90:172–244

149. Schläfke ME, Schäfer T (1991) Sleep phase related tc P_{CO_2} in infants under closed and open loop conditions of the central pH/P_{CO_2} control system. In: Gaultier C, Escourrou P, Curzi Dascalova L (eds) Sleep and cardiorespiratory control. INSERM/Libbey Eurotext 217:260

150. Schläfke ME, See WR, Massion WH, Loeschcke HH (1969) Die Rolle "spezifischer" und unspezifischer Afferenzen für den Antrieb der Atmung, untersucht durch Reizung und Blockade von Afferenzen an der decerebrierten Katze. Pflugers Arch 312:189–205

151. Schläfke ME, See WR, Loeschcke HH (1970) Ventilatory response to alterations of H+ concentration in small areas of the ventral medullary surface. Respir Physiol 10:198–212

152. Schläfke ME, Kille JF, Loeschcke HH (1979) Elimination of central chemosensitivity by coagulation of a bilateral area on the ventral medullary surface in awake cats. Pflugers Arch Eur J Physiol 378:231–241

153. Schmidt Vanderheyden W, Koepchen HP (1970) Investigations into the fluctuations of proprioceptive reflexes in man. I. Fluctuations of the patellar tendon reflex and their relation to vegetative rhythms during controlled ventilation. Pflugers Arch 317:72–83

154. Schwanghardt F, Schroeter R, Klüssendorf D, Kopechen HP (1974) The influence of novocaine block of superficial brainstem structures on discharge pattern of specific respiratory and unspecific reticular neurons. In: (eds) Umbach W, Koepchen HP Central rhythmic and regulation. Hippokrates, Stuttgart, pp 104–110

155. Sears TA (1964) Efferent discharges in alpha and fusimotor fibres of intercostal nerves of the cat. J Physiol (Lond) 174:295–315

156. See WR (1976) Respiratory drive in hyperthermia, interaction with central chemosensitivity. In: Loeschke HH (ed) Acid base homeostasis of the brain extracellular fluid and the respiratory control system. Thieme, Stuttgart, pp 122–129

157. See WR, Kumazawa T, Schläfke ME (1982) Modulation of neural activity in the central chemosensitive structures by peripheral nerve afferents. Neurosci Lett [Suppl] 10:441

158. Sellick H, Widdicombe JG (1970) Vagal deflation and inflation reflexes during pneumothorax, hyperpnea and pulmonary vascular congestion. Q J Exp Physiol 55:153–163

159. Severinghaus JW (1993) Widespread sides of brainstem ventilatory chemoreceptors. J Appl Physiol 75:3–4

160. Shams H (1985) Differential effects of CO_2 and H+ as central stimuli of respiration in the cat. J Appl Physiol 58:357–364

161. Smith AC, Spalding JMK, Watson WA (1963) Stimuli to spontaneous ventilation after prolonged artificial ventilation. In: Cunningham DJC, Lloyd BB (eds) The regulation of human respiration. Blackwell, Oxford, p 409

162. St John WM, Bartlett D, Knuth KV, Hwang JC (1981) Brainstem genesis of automatic ventilatory patterns independent of spinal mechanisms. J Appl Physiol 51:204–210

163. Strohl KP, Hensley MJ, Hallett M, Saunders NA, Ingram RH (1980) Activation of upper airway muscles before onset of inspiration in normal humans. J Appl Physiol 49:638–642

164. Tabachnik E, Muller N, Toye B, Levison H (1981) Measurement of ventilation in children using the respiratory plethysmograph. Pediatrics 99:895–899

165. Tatsumi K, Mikami M, Kuriyama T, Fukuda Y (1991) Respiratory stimulation by female hormones in awake male rats. J Appl Physiol 71:37–42

166. Trenchard D (1977) Role of pulmonary stretch receptors during breathing in rabbits, cats and dogs. Respir Physiol 29:231–246

167. Trouth CO, Millis RM, Kiwull Schöne H, Schläfke ME (eds) (1995) Ventral brainstem mechanisms and control functions. Dekker, New York

168. Tusiewicz K, Moldofsky H, Bryan AC, Bryan MH (1977) Mechanics of rib cage and abdomen during sleep. J Appl Physiol 43:600–602

169. Valtin H, Tenney SM (1960) Respiratory adaptations to hyperthyreoidism. J Appl Physiol 15:1107

170. Von Baumgarten R, Schmiedt H, Dodich N (1963) Microelectrode studies of phrenci motor neurones. Ann N Y Acad Sci 109:536

171. Von Düring M, Andres KH (1988) Structure and functional anatomy of visceroreceptors in the mammalian respiratory system. In: Hamann W, Iggo A (eds) Progress in brain research, vol 74. Elsevier, Amsterdam, pp 139–154

172. Von Euler US, Hellner S (1952) Excretion of norepinephrine and epinephrine in muscular work. Acta Physiol Scand 26:183

173. Widdicombe JG (1974) Reflex control of breathing. In: Widdicombe JG (ed) Respiratory physiology I, vol 2. University Park, Baltimore, pp 273–301

174. Widdicombe JG (1986) Reflexes from the upper respiratory tract. In: Fishman AP, Cherniack NS Widdicomber WG (eds) Handbook of physiology, the respiratory system, vol 2, control of breathing, part 1. American Physiological Society, Bethesda, pp 363–394

175. Widdicombe JG, Nadel JA (1963) Reflex effects of lung inflation on tracheal volume. J Appl Physiol 18:681–686

176. Winterstein H (1911) Die Regulierung der Atmung durch das Blut. Pflugers Arch 138:167–184

177. Winterstein H (1955) Die chemische Steuerung der Atmung. Ergeb Physiol 48:328–528

178. Woods RI (1975) Penetration of horseradish peroxidase between all elements of the carotid body. In: Purves MJ (ed) Peripheral arterial chemoreceptors. edited Cambridge University Press, London, pp 195–205

179. Yamamoto Y, Lagercrantz H, Euler C v (1981) Effects of substance P and TRH on ventilation and pattern of breathing in newborn rabbits. Acta Physiol Scand 113:541–543

107 Central Nervous Integration of Cardiorespiratory Control

K.M. Spyer

Contents

Abbreviations

CEN   Central nucleus of the amygdala
CNS   Central nervous system
CVM   Cardiac vagal motoneuron
DVN   Dorsal vagal nucleus
HDA   Hypothalamic defence area
IML   Intermediolateral cell column
NA   Nucleus ambiguus-retroambigualis
NTS   Nucleus tractus solitarii
PAG   Periaqueductal gray matter
PI   Post-inspiratory
RVLM   Rostral ventrolateral medulla
SPN   Sympathetic preganglionic neurons

107.1 Introduction

The primary role of the respiratory and cardiovascular systems in man (and mammals) is to ensure an adequate supply of oxygen to the tissues and removal of carbon dioxide from them. Respiration is regulated to maintain the arterial concentrations of oxygen, carbon dioxide and hydrogen ions within narrow physiological levels, whilst the performance of the cardiovascular system is modified continuously so that the pattern of peripheral blood flow is appropriate to ensure an adequate supply of oxygen to the tissues of the different organ systems [71]. The demands on the systems differ greatly in different behavioural cir-

cumstances (cf. Chap. 108). Obviously, respiratory status in terms of this primary objective depends on the integration of the control of the two systems so that the metabolic demands of the tissues can be served. In principle, the cardiovascular and respiratory systems could be controlled by independent mechanisms, providing that at some point they are integrated. However, in this chapter we will see that the control systems are closely inter-related, and that common neural processes have developed to generate the patterns of activity in the nervous pathways that regulate the respiratory and cardiovascular end organs – somatic (skeletal), smooth, and cardiac muscle [70].

The control processes that adapt cardiovascular and respiratory activity can be divided into two major components: first, the central nervous generation of basic patterns of activity and, second, the reflex feedback control of this activity. In addition, the physical consequences of respiration evoke changes in intrathoracic pressure that produce concomitant changes in transmural pressure in the intrathoracic blood vessels and the heart. These, in addition to provoking reflex responses by altering the discharge of numerous cardiopulmonary receptors, also directly affect venous return, cardiac output, pulmonary and aortic resistance, etc. Furthermore, changes in blood gas tensions and hydrogen ion concentration have marked effects directly on vascular smooth muscle, and the endothelium, to alter vascular resistance and, hence, local blood flow (cf. Chap. 95). This chapter will be concerned with outlining the interactions between the central nervous system (CNS) mechanisms that generate the patterns of activity in the nervous supply of the two systems on the one hand, and on the other the reflex mechanisms that modify these central processes and, hence, contribute to the interactions in the two systems that ensure cardiorespiratory homeostasis. Wherever possible, the analysis will concern situations in which the operation of these complex interactions can be seen to fulfil adaptive requirements of the individual to environmental or behavioral demands.

107.2 CNS Generation of Cardiorespiratory Activity

In the mammal, respiratory activity is dependent on supraspinal mechanisms to generate the phasic and alter-

R. Greger/U. Windhorst (Eds.)
Comprehensive Human Physiology, Vol. 2
© Springer-Verlag Berlin Heidelberg 1996

nating patterns of inspiratory and expiratory muscle contraction that result in airflow into and out of the lungs (cf. Chap. 105). The heart is intrinsically active (cf. Chap. 91); vascular smooth muscle also has resting tone, and often intrinsically generated fluctuations in activity (cf. Chap. 1). Thus, in the case of the cardiovascular system, the innervation modulates ongoing activity. As we will see, this control is striking and the basic neural rhythms are translated into marked and patterned changes in the cardiovascular system [35]. Again, the generation of rhythmic activity in the autonomic innervation of the cardiovascular system is considered to originate at supraspinal levels of the CNS [32]. Indeed, the fundamental neural substrate for rhythmogenesis in the respiratory and cardiovascular systems originates in overlapping regions of the hindbrain involving the medulla oblongata, although other supramedullary regions of the brainstem are known to exert profound influences over cardiorespiratory activity [70].

There has been some controversy as to whether respiratory activity is generated by interactions within the cell groups of a distinct respiratory neural network, or whether rhythmogenesis and pattern are subserved by two distinct groups of neurons [27,67] (see also Chap. 105). In much the same way, "vasomotor" tone has been considered to be generated by the activity of distinct groups of neurons in the rostroventrolateral medulla (RVLM) that have been suggested to have pacemaker properties [32]. Conversely, others have argued for a network of interacting neural oscillators as the basis for the development of the pattern of vasomotor discharge in sympathetic efferents [28]. Although this is not the place to give an exhaustive critique of the field, it does seem proven that the involvement of networks in generating both rhythm and pattern is compelling [67,70], and the most striking feature is that segments of these networks appear to have a cardiorespiratory function rather than operating as discrete networks controlling independent systems [71]. This is evident from the similarity of both discharge and membrane potential in neurons with apparently distinct functions in either the cardiovascular or the respiratory system, and indicates an overlapping role of the central pattern generator. Figure 105.2 provides an overview of the localisation within the brain stem of the cell groups that have been shown to be primarily involved in the generation of cardiorespiratory patterns of activity. Clearly, the anatomical alignments provide a strong indication of potential interaction, but this is reinforced by the demonstration of physiological interactions based on synaptic connections that underlie shared temporal patterns of membrane potential and activity.

107.3 Respiratory Rhythm

Fundamentally, respiration is a three-phased pattern of activity – inspiration, post-inspiration (stage 1 expiration) and expiration (stage 2 expiration) (Fig. 107.1), and this can be generated adequately by the interactions between a relatively small group of neurons, probably as few as five or six separate classes (cf. Chap. 107).

The respiratory network is contained within cell groupings in two major locations in the medulla. The dorsal respiratory group in the vicinity of the nucleus tractus solitarii (NTS) in the dorsomedial medulla and a more extensive ventral group in the vicinity of the nucleus ambiguus-retroambigualis (NA) that extends from the pontine border to the upper cervical cord. The NTS is a major site of termination of afferents from the respiratory system, airways and receptors in the heart and blood vessels, including the arterial chemoreceptors and baroreceptors [44,53]. It contains also neurons (bulbospinal) that innervate the spinal cord – phrenic motoneurons within the cervical spinal cord and thoracic intercostal motoneurons and interneurons – as well as neurons (propriobulbar) that form part of a reciprocal network with components of the ventral respiratory grouping [53]. The ventral group contains bulbospinal neurons with extensive connections onto both inspiratory thoracic motoneurons and interneurons, and phrenic motoneurons [26]. In addition, vagal motoneurons innervating the accessory somatic muscles of respiration, glossopharyngeal and facial motoneurons fall within this complex as well as preganglionic vagal motoneurons innervating the oesophagus, airways, heart and lungs [54]. Similarly, there are numerous propriobulbar neurons. Within the profiles of activity of each of these neurons are phases of activity, or excitability changes, that are related to that of the primary three-phase respiratory rhythm, but each (see Fig. 107.1) have distinctive patterns that can be reduced to essentially six classes [67].

The features of activity observed in these "respiratory" neurons is also seen in the patterns of activity of RVLM neurons and other brain stem neurons that send axons to the intermediolateral cell column of the spinal cord which contains the sympathetic preganglionic neurons [32,55], and in the preganglionic vagal motoneurons of the NA; for this discussion the preganglionic cardiac vagal motoneurons (CVMs) will be the most important [29]. The "apparent" respiratory network includes many propriobulbar neurons that are regarded as being involved in the generation of the three-phase respiratory rhythm, but which also appear to make extensive synaptic connections with the medullary presympathetic neurons and CVMs, given the pattern of their discharge. There are, additionally, neurons within the network that have developed extensive, and probably reciprocal, connections with rostrally placed cell groups in the brainstem that themselves influence cardiorespiratory activity. Accordingly, it is attractive to offer the suggestion that the basic rhythm in the two systems is generated by this common network (Fig. 107.1). The individual identity of the two systems is determined by those neurons that project to innervate the final common pathways in the CNS for the two systems – spinal motoneurons for respiration and preganglionic autonomic neurons for the cardiovascular system. It should also be appreciated that these rhythmic patterns are also contained in the activity of accessory respiratory systems, e.g.,

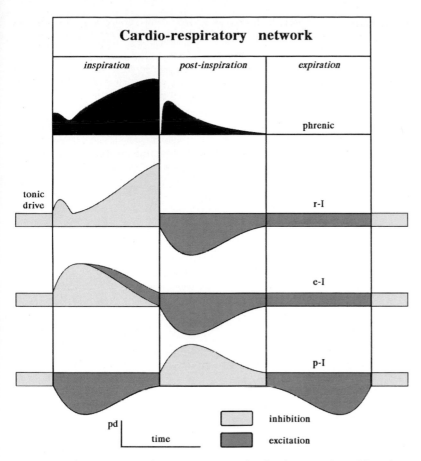

Fig. 107.1. The common cardiorespiratory network. The three subsets of medullary neuron that constitute the common cardiorespiratory oscillator: ramp (*r-I*), early (*e-I*) and post-inspiratory (*p-I*) neurons. These convert a tonic excitatory drive that is derived from the reticular activating system into rhythmic activity with alternating periods of rebound and synaptic excitation (*darker panels*) and synaptic inhibition (*lighter panels*)

facial, hypoglossal and laryngeal motoneurons, and also in the activity of other autonomic outflows such as bronchomotor neurons. Thus, the activity generated by the cardiorespiratory network may represent a basic reticular rhythm implicated in the maintenance of homeostasis and, as such, life itself [70].

107.4 Vagal Motoneurons

The preganglionic vagal innervation of the heart arises largely from the two nuclei of the vagus in the medulla – the NA and the dorsal vagal nucleus (DVN) – although there are scattered vagal neurons lying between these two medullary nuclei [54] (Fig. 107.2a). There is general agreement that the major chronotropic effects of stimulating the vagus are mediated by neurons with B-fibre axons, although it is clear that C-fibre axons may play a small and species-specific role [54]. The majority of CVMs with B-fibre axons appear to be localised within the NA in most species, with a variable presence in the DVN. There is now compelling evidence that the activity of a relatively small number of such vagal neurons can exert profound effects on heart rate, and that on a moment-by-moment basis, the activity of these neurons is the major determinant of heart rate and, hence, cardiac output [57,58]. Interestingly, CVMs have a distinctive pattern of activity that is determined largely by two extrinsic inputs – the excitatory input from the arterial baroreceptors (Fig. 107.2c) and a complex input from the respiratory network of the cardiorespiratory network (Fig. 107.2b). These inputs will, however, interact with the intrinsic properties of these neurons, which are both distinctive and currently under intense experimental investigation [54,75].

CVMs have a pulse-related pattern of discharge, showing maximal activity in phase with the systolic rise in blood pressure (Fig. 107.2c). This is the result of the phasic input from the arterial baroreceptors; as arterial pressure rises, so CVM activity increases, and the systolic burst of activity is followed by a diastolic silencing of discharge. This discharge is superimposed on a respiratory-related discharge (Fig. 107.2b) that has been characterised as an expiratory rhythm [73]. In fact, the respiratory modulation of discharge is more complex. Intracellular recordings taken from CVMs indicate that the three phases of the respira-

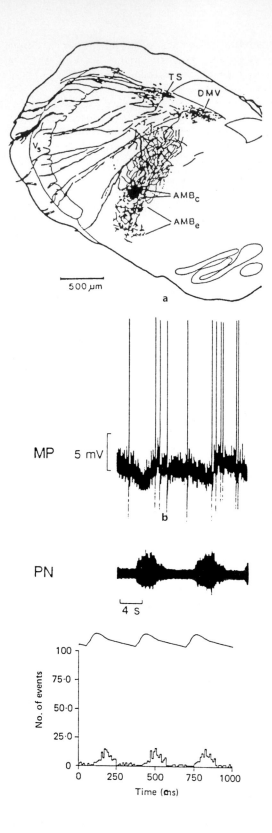

Fig. 107.2a–c. The patterning of cardiac vagal motoneuron (CVM) activity. **a** Transverse section of rat medulla oblongata showing retrograde cell body labelling of vagal motor neurons after injection of horseradish peroxidase into the cervical vagus nerve. Vagal motor neurons lie in the dorsal vagal nucleus (*DMV*) as well as within two portions of the nucleus ambiguus (the compact part, *AMB_c*, and the ventrolateral or external division, *AMB_e*). *TS*, tractus solitarius; *V_s*, spinal trigeminal tract. (Reproduced from [9], with permission of the authors and publisher.) **b** Intracellular recording from a CVM showing changes in membrane potential (*MP*) in relation to phrenic nerve activity (*PN*). **c** *Lower trace*: Electrocardiogram-triggered histograms of ongoing activity of a CVM (100 sweeps, 10-ms bins). *Upper trace*: Femoral arterial wave form averaged over the same time course and triggered from electrocardiogram (100 sweeps)

tory rhythm can be recognised in the membrane potential of each CVM [29]. There is a hyperpolarisation during inspiration and the membrane repolarises rapidly at the end of inspiration, with a marked depolarisation during post-inspiration and a variable pattern of depolarisation and hyperpolarisation during stage 2 expiration. (Fig. 107.2b). Such a pattern of activity is seen in many post-inspiratory neurons (see Fig. 107.1). This pattern of membrane potential is reflected in the temporal firing pattern of extracellularly recorded CVM neuronal activity, both ongoing and that induced by the micro-ionophoresis of excitant amino acids [57,58]. This latter point is important since in the anaesthetised animal "vagal tone" is notoriously low. It emphasises the fact that the underlying membrane potential is dominated by this respiratory input [29]. Further, the activity of vagal efferent fibres in the cervical vagus (Fig. 107.3b) and cardiac branches of the vagus show prominent expiratory firing [73]. This pattern of activity results in a fluctuating heart rate – sinus arrhythmia – with the rate lowering in early expiration compared to that during inspiration [29,70] (Fig. 107.3a). Accordingly, any modifications that take place in respiratory drive will have considerable and immediate effects on heart rate [16,45].

The markedly heightened excitability of CVMs during early expiration, post-inspiration, accounts for the respiratory modulation of reflex effects on the heart. At this time, inputs from the arterial baroreceptors and chemoreceptors are particularly effective, and any reflex input (or central input) that prolongs or heightens post-inspiration – e.g., activation of laryngeal afferents supplying both mechanoreceptors and chemosensors of the larynx – leads to a bradycardia mediated by CVMs [16,29] (Fig. 107.4).

What is the source of the respiratory input onto CVMs? As yet, there is no convincing explanation. CVMs are hyperpolarised during inspiration by an ionic mechanism with a reversal potential similar to that for Cl^--mediated inhibitory processes [29]. The conventional inhibitory neurotransmitters GABA, acting at GABAA receptors, and glycine do not appear to be involved. Rather, there is evidence of a muscarinic cholinergic mechanism acting within the NA [29,45]. The nature of this mechanism is uncertain, although muscarinic inhibitory actions have been reported elsewhere in the CNS. The role of GABAB receptors in the mechanisms of inhibition remains to be resolved, although elsewhere in the medulla poth pre- and

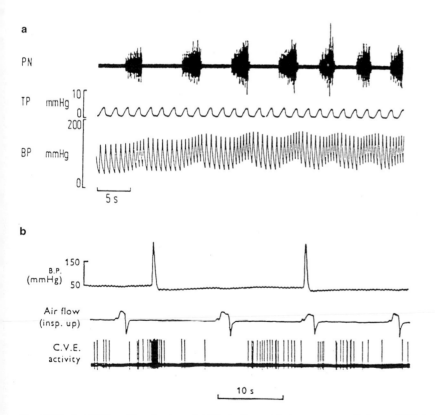

Fig. 107.3a,b. CVM and heart-rate variability in the respiratory cycle. **a** Respiratory modulation of cardiovascular activities. Sinus arrhythmia and blood pressure (*BP*) waves indicating modulation of vagal cardiomotor activity related to the central respiratory activity. The tracheal pressure (*TP*) was measured as an index of artificial ventilation of the anaesthetized cat occurring independently from the central respiratory rhythm indicated by the phrenic nerve discharge (*PN*). **b** Dog. Chloralose and morphine. Records of carotid sinus blood pressure, respiratory air flow, and the activity of single cardiac vagal efferent nerve (*C.V.E.*) are shown. A burst of firing in the cardiac efferent nerve was evoked by a baroreceptor stimulus timed so as to occur in the expiratory pause. No firing was evoked when a similar stimulus was given during inspiration. (From [22], with permission)

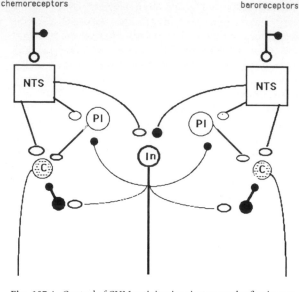

chemoreceptors baroreceptors

Fig. 107.4. Control of CVM activity: inspiratory and reflex inputs. Diagrammatic representation of the inputs to CVM (*C*) that determine their sensitivity to baroreceptor and chemoreceptor inputs. Inspiratory neurons (*In*) directly or via interneurons (e.g., *PI*, post-inspiratory neurons) or inhibitory interneurons (shown in *black*) control the excitability of CVM. Baroreceptor and chemoreceptor inputs are processed initially in the nucleus tractus solitarii (*NTS*) and send excitatory connections to CVM localised in the nucleus ambiguus, but these inputs act differentially on inspiratory neurons. (See text for further description.) *Open symbols*: excitatory; *filled symbols*: inhibitory

postsynaptic actions of GABA at these receptors have been reported [12]. Accordingly, the process could be a muscarinic activation of GABA-containing neurons in the NA, or a shared presynaptic site of action of acetylcholine (ACh) and GABA at sites modulating K^+ and Ca^{2+} currents together with a post-synaptic GABAA action. Presumably the neighbouring inspiratory neurons of the NA are involved in the generation of this inhibitory action, as they are known to contain ACh (Fig. 107.4). The post-inspiratory depolarisation could result from an excitatory input from post-inspiratory propriobulbar neurons or a tonic excitatory input shared by CVMs and post-inspiratory neurons that is sculptured by inhibition at other times in the cycle. This excitatory tonic drive may represent a basic reticular rhythm and so would be expected to be state-dependent. During stage 2 expiration there is a variable level of membrane polarisation, indicating the interplay of excitatory and inhibitory inputs at that time in the respiratory cycle [70].

Interestingly, those CVMs that are localised within the DVN in the rabbit have equivalent properties to those described for the NA [42], but the role and properties of vagal motoneurons with unmyelinated axons, which have been shown to be largely restricted to the DVN in all species, remain to be determined. There are, however, preliminary indications that these mediate pulmonary C-fibre afferent reflex evoked influences on heart rate [40].

107.5 Premotor Sympathetic Neurons

Suggestions have been made in the literature that "vasomotor tone" is dependent on the level of sympathetic efferent activity and that this appears to be determined by neurons located at supraspinal levels and, particularly, within the medulla [1,28,32,35]. Sympathetic preganglionic neurons receive a marked innervation from several identifiable groups of brain stem neurons [53], as has been demonstrated using anatomical tracing techniques and electrophysiological approaches. The distribution of descending input related to specific end organs has been outlined with the recent application of the retrograde trans-synaptic transport of pseudorabies virus from the adrenal medulla and other organs [5,79]. This material indicates both the spinal organisation of sympathetic control and the brainstem loci of descending input. With regard to strictly cardiovascular control – cardiac and vascular – our knowledge is still restricted [53]. At the medullary level, several groups of neurons emerge as potential premotor sympathetic neurons involved in cardiovascular control. These include neurons of the RVLM (including the Cl- adrenaline-containing neurons), the raphe magnus, obscurus and pallidus, the NTS and reticulospinal neurons.

107.5.1 Rostral Ventrolateral Medulla

Many of the neurons of the RVLM have been shown to relay to the IML by antidromic activation (see [32] for review). Those with small myelinated axons are considered to be important in the generation of vasomotor tone and are powerfully influenced by baroreceptor inputs [32]. A proportion of these have indications of the respiratory rhythm in their discharge [32,33,55]. There are suggestions in the literature that neurons of the RVLM may not merely by coded as "vasomotor", but may be distributed according to vascular bed [21]. The hypothesis postulates that the RVLM is organised in an analogous fashion to the motor cortex in the context of vascular territories. As we will see, postganglionic sympathetic discharge is considered to be patterned in an orderly fashion with regard to end organ innervation [37], and the hypothesis would indicate that this is laid down at the level of the RVLM. However, different sympathetic efferents appear to have distinctly different patterns of respiratory modulation, including both central respiratory patterning (Fig. 107.5) and afferent feedback that involves lung stretch inputs and phasic fluctuations related to the respiratory cycle in arterial baroreceptor activity. As yet, this complexity has not been revealed in the properties of RVLM neurons studied in the rat, although a variety of respiratory related patterns have been reported in the cat [55].

In the case of RVLM bulbospinal neurons, the most obvious discharge related to respiration is seen in inspiration, often with post-inspiratory depression and a variable expiratory discharge. This has been observed in both rat

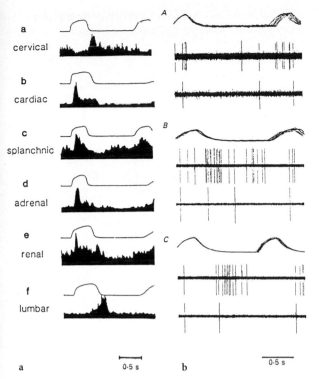

a cervical

b cardiac

c splanchnic

d adrenal

e renal

f lumbar

A

B

C

a 0.5 s   b 0.5 s

Fig. 107.5a,b. Respiratory modulation of sympathetic neuronal discharge. **a** Examples of phrenic-triggered averaged nerve activity tracings from various sympathetic neurons as noted in the figure with simultaneously averaged integrated phrenic nerve activities. Discharge patterns observed with 1% halothane anaesthesia. (From [62], with permission.) **b** The discharge of sympathetic neurons shown on phrenic nerve-triggered histogram analysis to have three patterns of respiratory-related activity: *A*, inspiratory; *B*, post-inspiratory; *C*, expiratory. In *A*, *B* and *C* *upper traces* show ten superimposed integrated sweeps of phrenic nerve activities; *middle traces* show ten superimposed sweeps of sympathetic preganglionic neuronal activity, and *lower traces* show single sweeps of neuronal activity. (From [80], with permission)

[32,33] and cat [55], and conversely vagal afferent and lung inflation inputs are either excitatory or inhibitory. In contrast, there are many sympathetic preganglionic and postganglionic neurons that have a prevalent expiratory discharge, low inspiratory activity and heightened post-inspiratory discharge, although the pattern described above for RVLM neurons is often seen in sympathetic efferents [30,31,62] (see Fig. 107.5). As yet, no functional assessment of the respiratory rhythm of RVLM has been undertaken with regard to their apparent vascular bed designation, nor is the relationship of RVLM neurons to cardiac activity well understood [32].

Chemoreceptor afferent inputs have been described as exerting largely excitatory effects on RVLM bulbospinal neurons in the rat [33] and cat [55], but more recent studies indicate a regional variation in action, with both direct excitatory and indirect excitatory influences being evoked in relation to enhanced inspiratory drive, whilst other neurons show marked inhibitory responses to chemoreceptor activation [56]. Whilst this reinforces the potential for the

RVLM neurons to imprint respiratory patterns in the discharge of sympathetic efferents, the properties of RVLM neurons as understood at present do not in themselves adequately explain the diverse activity patterns of individual sympathetic preganglionic neurons seen in the rat.

107.5.2 Raphe Nuclei

The raphe complex has long been implicated in autonomic and behavioural control. There is now plentiful evidence that neurons within this nuclear complex send axons to the spinal cord to terminate within the IML amongst other sites [5,28,79]. Many neurons within these structures have a pattern of activity synchronised to that of the sympathetic nerves, and it is known that this relationship persists in the absence of baroreceptor feedback [28]. This implies that raphe neurons are a component of, or are interconnected with, an intrinsic rhythm-generating system that has been suggested to exist within the brainstem [28]. The anatomical basis for raphe-spinal connections – specifically, those onto sympathetic preganglionic neurons – has been detailed using ultrastructural analysis [5,79]. The physiological role of these connections, however, is somewhat less understood, even though activating the raphe produces changes in sympathetic activity and arterial blood pressure. As yet, the role of these pathways in mediating respiratory control of sympathetic outflow is less well documented. There is no doubt that many of the patterns of respiratory activity described earlier in ventral medullary neurons can be seen in the activity of raphe neurons, and cross-correlation analysis suggests that there are strong synaptic inter-relationships between raphe neurons that synchronise these discharges and also between the conventional respiratory network and raphe neurons [50,51]. It is still uncertain as to whether raphe-spinal neurons exhibit this powerful respiratory rhythm and, further, whether these are also firing with intrinsic cardiovascular rhythms – the 2- to 6-Hz discharge observed in the debuffered animal [28]. Whatever the case, the concept of an involvement of the raphe is established, either in controlling IML neurons directly or in acting to do so through its extensive brain stem connections, including connections to the RVLM. That a proportion of raphe neurons act as "premotor" sympathetic neurons appears as certain as that RVLM neurons have an equivalent role.

107.5.3 Reticular Sympathetic Rhythms

There is abundant evidence that medullary, pontine and hypothalamic neurons have patterns of discharge correlated to rhythms seen in the sympathetic outflows [28]. These rhythms are apparently entrained to the cardiac cycle by baroreceptor afferent feedback, but the intrinsic rhythm (2–6 Hz) persists in the baroreceptor-denervated animal. This intrinsic rhythm may have its origin in the lateral tegmental field of the medulla, and there is suggestive evidence that this rhythm is expressed in the discharge of the premotor sympathetic neurons of the RVLM, the raphe complex and elsewhere [28]. Many non-specific reticular neurons also exhibit respiratory rhythms as well, and these are state-dependent [48,49]. Such neurons show examples of coupling to sympathetic and respiratory rhythms at times, but this relationship is not fixed, and one or other phasic influence may disappear. This has led to the description of such neurons as being part of a non-specific reticular activating system. It is probable that many neurons displaying these properties are bulbospinal and are likely to influence sympathetic preganglionic neurons directly, or via spinal interneurons, and they may also have influences on respiratory and somatic motor systems.

107.6 Patterns of Activity in Sympathetic Efferents

As indicated earlier, sympathetic pre- and postganglionic nerves exhibit both cardiac and respiratory rhythms to different degrees depending on the particular nerve and the physiological state of the animal [37] (see Fig. 107.5). Since the nerves all contain a heterogenous mixture of efferent sympathetic fibres innervating different end organs, or different vascular territories within the same organ, they presumably subserve widely different functions (Fig. 107.5a). Indeed, when single fibres are teased, or an individual sympathetic preganglionic neuron is either impaled or examined extracellularly, specific and distinct patterns of activity are identified (Fig. 107.5b). The functional significance of these distinctive patterns is as yet unresolved, although the patterns have been tentatively used as a functional discriminator [37,38]. Such analogies have been carried forward into the CNS, with both premotor and preganglionic neurons being ascribed functions based on this form of classification [38]. This step may be premature, as the classification remains a crude evaluation of function, even with regard to postganglionic elements.

Postganglionic sympathetic neurons supplying the skeletal muscle and skin of the hindlimb and tail of the cat have been classified with regard to their patterns of response to certain inputs [37]. Presumed muscle vasoconstrictor efferents were shown to be powerfully inhibited by arterial baroreceptor inputs, often showing striking pulse-modulated discharge. They were excited by both chemoreceptor stimulation and systemic hypoxia and hypercapnia, and exhibited powerful respiratory rhythm in their ongoing discharge. General visceral inputs also excited them. Conversely, presumed cutaneous vasoconstrictor efferents were weakly influenced by the arterial baroreceptors, had variable responses to chemoreceptor stimulation and, similarly, had varying responses to both hypoxia and hypercapnia. They were, invariably, inhibited by visceral inputs. Cutaneous nociceptive inputs were usually inhibitory, whilst they were excitatory to muscle vasocon-

strictors. They also had weak respiratory influences, although in some cases the respiratory influence was powerful. Presumed sudomotor efferents were not very dissimilar to muscle vasoconstrictors, although they were usually unaffected by arterial baroreceptor input. Since the data were obtained from few fibre preparations, there are problems of interpretation. It appears that the respiratory rhythm results from three main influences: a central pattern generated in the brain stem (see Fig. 107.1), pulmonary stretch afferent inputs and the changing baroreceptor input that accompanies the respiratory related fluctuations in arterial pressure. The patterns of activity described above were not absolute, and in no case was definitive information obtained as to the end organ innervated by a particular efferent. In man there is evidence in recordings taken from the peroneal nerve of similar discharge patterns in sympathetic efferent fibres [78]. Often there is evidence of profound inspiratory discharge, with varying patterns of expiratory firing, although temporal relationships with the respiratory cycle are difficult to evaluate because of the long conduction distances and slow conduction rates of sympathetic fibres.

Considerably more has been learned from assessing the behaviour of single preganglionic neurons whose activity in the IML was recorded using microelectrodes (see Fig. 107.5b).

The activity of individual sympathetic preganglionic neurons in the thoracic spinal cord shows different patterns of respiratory activity within the overall bounds of the three-phase pattern of respiration [30,31]. There is a powerful rhythm – either inspiratory or expiratory – in the neurons that project into the cervical sympathetic nerve, and a clear if less general modulation in those recorded at lower thoracic levels [31]. Neurons were affected by respiratory input, quite irrespective of whether they had a baroreceptor input or not. Highly variable respiratory patterns of discharge have also been seen in the cat [65]. Interestingly, whilst at least 90% of RVLM sympatho-excitatory neurons in the rat receive respiratory modulation [78], only 50% of preganglionic neurons with baroreceptor input show such modulation (compare [32] and [30,31]). This makes it unlikely, at least in the rat, that the RVLM provides the only major excitatory input determining the discharge pattern of preganglionic sympathetic neurons, since then a much larger proportion of preganglionic neurons would be expected to show respiratory modulation.

Studies using intracellular recording in the cat indicate that preganglionic neurons show fluctuations in membrane potential that correspond closely to those patterns of discharge seen in extracellularly recorded neurons [24,70]. The three-phase rhythm is clear, but evidence suggests that there is no role for inhibitory patterning, since no respiratory-related IPSPs were recorded in these neurons. This suggests that the respiratory patterning is the consequence of an action of a descending sympatho-excitatory pathway or pathways. From the data presented above, this is further evidence that the RVLM cannot be the sole source of descending excitatory drive to preganglionic sympathetic neurons.

107.7 Functional Role of Autonomic Discharge

There is unequivocal evidence that the presence of intrinsic rhythms in autonomic outflows is related, at least in part, to CNS processes and that a cardiorespiratory network plays a major role in its development (Fig. 107.1). The question of the functional significance of these rhythms becomes important. The fundamental role is probably related to the efficacy of synaptic transmission. Bursts of action potentials are more effective in eliciting transmitter release than prolonged low-frequency discharges (cf. Chap. 17), even if the total number of action potentials in time is the same [11]. As the autonomic outflows play a modulatory role, this is likely to be even more important for the effective action of these nerves in influencing the various cardiovascular end organs, whilst in the case of the respiratory muscles, the patterned outflows ensure a temporal recruitment of muscles to evolve a regulated pattern of lung inflation and deflation (see Chap. 105).

The timing of the vagal efferent discharge at a relatively fixed time in the cardiac cycle is important, since even allowing for the relatively long period of synaptic delay, receptor occupancy should occur at an appropriate time to exert an influence on the rate of rise of the prepotential and so affect the next heart beat. The fusion of cardiovascular and respiratory rhythms may also have important consequences. The respiratory modulation of vagal efferent discharge, with heightened discharge of vagal efferents in post-inspiration and its absence in inspiration, ensures that heart rate rises in inspiration and so increases cardiac output at the time that the lungs are inflating [70]. This will optimise ventilation and perfusion, and provide optimum conditions of gaseous exchange.

To identify in more detail the potential importance of this inter-relationship between the control of the two systems, several physiological responses to changes in reflex input or centrally generated patterns of behaviour will now be analysed.

107.8 Patterns of Reflex Response

107.8.1 Arterial Baroreceptor Reflex

It has already been mentioned that the arterial baroreceptor reflex is modulated by respiration [73]. Indeed, the baroreceptor–heart rate reflex has been shown to be particularly sensitive to the respiratory cycle [22,46,47]. A brief stimulus to the arterial baroreceptors excites CVMs and slows the heart if delivered in expiration, but if the identical stimulus is timed to occur during inspiration it is ineffective (Fig. 107.3b). The baroreceptors also affect respiration. Inspiration is either terminated or reduced by baroreceptor stimulation, and expiratory activity is facilitated largely through the removal of inspiratory-mediated inhibition of expiratory activity [69] (see Fig. 107.4). In the original studies on the action of respiration on CVM activ-

ity, or heart rate, lung inflation was not controlled; however, subsequent studies indicated that this was a relationship involving the central respiratory rhythm, although lung inflation inputs were also effective in reducing the magnitude of the reflex [29,66]. A detailed analysis of CVM activity showed that baroreceptor excitation was most notable during the post-inspiration period [29]. Similar studies have also shown that this modulation extends to the sympathetic outflow to heart, with the baroreceptor-evoked inhibitory influence being more marked in expiration than in inspiration [23]. Lung inflation inputs also effectively inhibit sympathetic activity, and this effect may often be independent of the action of this input on respiratory patterning.

The synaptic mechanisms underlying the modulatory actions of respiration on this reflex have been intensely studied. These studies have taken account of the expanding literature on the nature of the CNS organisation of the reflex. It is now firmly established that the primary afferents of the arterial baroreceptors relay to, and terminate within, the NTS [44]. Unlike many other primary afferents, the terminals of baroreceptor and chemoreceptor afferents are not amenable to presynaptic modulation by interactions with other afferent inputs in the NTS [44,69]. Further, the NTS neurons that are excited by these inputs, and higher-order neurons that are both excited or inhibited as a consequence of this input, fail to show any indication of a respiratory-related modulation of their synaptic responses [61]. The magnitude and latency of evoked EPSPs, IPSPs, or EPSP/IPSP complexes are not modified either by the timing of the input in the respiratory cycle or by whether the lungs are inflated or deflated. This is true for the effects both of central respiratory activity and of lung inflation inputs [60]. Clearly, the first and subsequent early synapses in the reflex are not the site at which respiratory "gating" of the reflex is engendered.

The observations on the respiratory sensitivity of CVMs offer the most easily acceptable explanation for the cyclic modifications of effectiveness of baroreceptor inputs. The cardiorespiratory network provides CVMs with heightened excitability in post-inspiration, and summation of the excitatory input from the baroreceptors with the hyperpolarised neurons in inspiration minimises the effectiveness of the input [70]. These observations do not explain the action of lung inflation inputs, since no evidence of a direct action of inflation inputs can be observed in the membrane potential of CVMs. Their action to inhibit inspiration would be expected to enhance CVM activity by disinhibition, rather than acting to suppress the action of the baroreceptor input. Physiologically, lung inflation elicits a tachycardia. The site of action is thus considered to lie somewhere on the reflex pathway between the NTS and CVMs, but its location is still to be determined [70].

With regard to the baroreceptor control of sympathetic activity, the as yet unelucidated complexity of connection between the NTS and sympathetic premotor neurons makes analysis difficult [32]. Clear summation of competing excitatory and inhibitory inputs at the level of the

RVLM or other such neurons may provide the simplest explanation, since this would remove any requirement for effective "gating" of the reflex. Since sympathetic efferents to the heart and blood vessels are heterogenous, this is probably a physiologically effective means of eliciting an appropriate regulation of activity. Interestingly, in the case of sympathetic efferents supplying muscle and the cervical sympathetic nerve, the respiratory modulation of the baroreceptor inhibitory influence is either absent or weak.

107.8.2 Chemoreceptor Reflex

Many of the studies referred to above involved the activation of the carotid sinus nerve, which contains both baroreceptor and chemoreceptor afferents, to evaluate the efficacy of reflex inputs. Recent investigations have shown that the non-respiratory neurons of the NTS that are activated by arterial chemoreceptors are as unaffected by respiratory inputs as were those neurons that were affected by baroreceptor input [59,76]. However, the chemoreceptor control of heart rate, respiration and sympathetic activity shows considerable evidence of respiratory modulation [16] (see Fig. 107.4).

In the case of respiratory influences, a brief chemoreceptor stimulus activates inspiratory neurons and the phrenic nerve, if the stimulus is timed to occur in inspiration, and similarly will excite expiratory neurons if the stimulus occurs in expiration [52]. Conversely, the stimulus is ineffective, or prolongs the silent phase, if occurring in the phase when the neuron would be silent [25]. Indeed, the direct action of chemoreceptor input on NTS inspiratory neurons in their normally inactive phase when their excitability is raised – e.g., by the micro-ionophoretic application of excitant amino acids, or lung inflation in the case of some NTS inspiratory neurons – is inhibitory [25]. Clearly, the input is not "gated", but the action is determined by the nature of the interneurons placed between the input and the respiratory neurons under investigation. The action of this input on respiration would influence markedly the effectiveness of the input on vagal efferent and sympathetic discharge. Any increased inspiratory drive would inhibit CVMs and so raise heart rate, whilst an input coinciding with post-inspiration would both prolong the expiratory period and increase CVM activity, and hence lower heart rate. Prolonged chemoreceptor inputs markedly affect the pattern of respiration as well as the power of the inspiratory drive, and the consequent effect on heart rate is dominated by the change in respiratory patterning as it affects CVM activity [16]. Essentially, this "excitability" control determines the effectiveness of any more direct excitation that is elicited by the input (Fig. 107.4). In an analogous fashion, the changing pattern of the three-phase cardiorespiratory rhythm will modulate the effectiveness of chemoreceptor input onto the sympathetic outflows.

107.8.3 Pulmonary Afferent Inputs

At several stages in this chapter, evidence of an action of lung stretch inputs on the activity of vagal and sympathetic outflows, and additionally on the performance of other reflexes affecting the cardiovascular and respiratory systems, has been indicated [16]. Sinus arrhythmia has been described as the major ongoing manifestation of cardiorespiratory integration, and it was the pioneering work of Anrep [2,3] that demonstrated this to be the result of two clearly distinguishable processes. The first has a central origin that has since been identified [29], and this has been extensively reviewed above. The second involves actions of afferent inputs related to lung inflation mediated by vagal afferents. These afferents are attached to slowly adapting receptors, and their activation contributes both to the termination of inspiration and to the shaping of the inspiratory burst (see Chap. 105). They are also able to affect the activity of other neurons in the cardiorespiratory network [70,71] and so contribute to the pattern of the respiratory cycle.

The site of action of these afferents in modulating baroreceptor and chemoreceptor reflexes, other than by their influence on respiratory patterning through the cardiorespiratory network, remains obscure. It is not mediated at the level of the NTS, nor can direct synaptic influences be seen in the activity of CVMs, although sympathetic premotor neurons of the RVLM are strongly influenced by lung inflation (see above). There is some evidence to indicate that post-inspiratory neurons are susceptible to inhibition mediated by lung stretch inputs. As CVMs are considered to be a component of the post-inspiratory population of neurons, they may be disfacilitated by such inputs. In contrast, airway receptors – laryngeal mechanoreceptors and chemosensitive afferents, together with rapidly adapting pulmonary receptors – exert profound excitatory control of post-inspiratory activity as well as initiating inhibition of inspiration [39]. Thus, it is hardly surprising that their activation exerts a powerful excitatory drive to CVMs that leads to bradycardia in a manner analogous to the effect of baroreceptor inputs. The inhibitory effects that these inputs exert on inspiration ensure that any accompanying activation of arterial baroreceptors or chemoreceptors potentiates their negative chronotropic influences. Interestingly, there is evidence that many NTS neurons excited by baroreceptor inputs have a short-latency excitatory input from SLN inputs [60]. Trigeminal afferent inputs that are excited by immersion of the head in water, as in diving, similarly abolish inspiration and enhance the expression of the cardiovascular effects of all these reflexes [19].

A detailed qualitative and quantitative analysis of the interactions of reflex inputs and central respiratory state has provided some important indications of the potency of these integrative responses [16–18]. The effectiveness of baroreceptor and cardiac afferent inputs in evoking cardiac slowing and changes in blood pressure is significantly reduced by both lung inflation and inspiration, but the influence of these respiratory inputs is even greater on the chemoreceptor reflex responses, especially with regard to the evoked bradycardia. Conversely, the effects of activating pulmonary C-fibre afferents following right atrial injections of phenyl biguanide (a 5-HT agonist), which has a profound inhibitory influence on respiration and evokes a marked vagally mediated bradycardia, are uninfluenced by either respiratory state or lung inflation [18,20]. No specific influence of the activation of this group of pulmonary receptors could be identified on respiratory activity to account for this lack of influence on the evoked bradycardia [16]. As yet, this remains the only example of an absence of respiratory modulation of a cardiorespiratory reflex.

These unmyelinated pulmonary afferents have been called J receptor afferents, since their endings are located in the juxtacapillary region of alveoli [63]. They are believed to be sensitive to both mechanical and chemical stimuli, and it has been suggested that they are especially sensitive to pulmonary oedema. They may be a major factor in limiting exercise, by generating suppression of both respiration and motor activity when the individual's homeostatic limits are approached. Activating these afferents exerts a profound excitation of post-inspiratory neurons, respiration often ceasing in post-inspiration or turning into rapid shallow breathing [14,39]. Such an effect would be expected to activate CVMs of the NA as described earlier, but the absence of respiratory modulation under these circumstances is perplexing. It has been suggested that there must exist a pool of CVMs distinct from the general class [20] that has dominated the present discussion. Preliminary studies have indicated that the unmyelinated DVN vagal neurons that innervate the heart may mediate the response to pulmonary C-fibre activation [40]. There is convincing evidence to show that a significant but small cardiac slowing results from the activation of C-fibre vagal efferents. They have been shown to fire tonically but are not influenced by central respiratory activity, lung inflation or baroreceptor inputs. They do receive a powerful excitatory input from pulmonary C-fibres when these are activated by phenyl biguanide. This C-fibre reflex may represent a primitive protective response that has been established in fish, where the vagal outflow controls both water flow through the gills (branchial motor control) and heart rate. The receptors are then located at the water vascular interface, in contrast to the air-blood interface in the mammal. In the fish the reflex acts to correlate water flow and cardiac output in order to optimise gas exchange [7] – a situation analagous to that described as the main feature of the respiratory modulation of CVM activity that has been identified in the mammal [29]. Equally, it may subserve a protective function restricting, or arresting, exchange in the presence of noxious substances. This may be the role it preserves in the mammal. As yet nothing more is known of the inputs to these vagal neurons, but this appears to be a fruitful area for further study.

107.9 Centrally Evoked Cardiorespiratory Responses

In addition to its role in integrating the inputs from the various cardiorespiratory afferent systems in order to maintain homeostasis, the CNS has an equally important function in expressing appropriate changes in cardiorespiratory activity to support behavioural activities [72,74]. In physiological terms this represents eliciting changes in respiratory minute volume, cardiac output and the pattern of distribution of peripheral blood flow to maintain the metabolic activity of the organ systems of the body. These changes can accompany normal changes in the level of activity of the individual – from sedentary rest to active exercise, from sleep to wakefulness, etc. – but will also include adjustments that are required to maintain the integrity of the individual in the face of perceived threats and emotion – affective behaviour. The defence reaction – the fear, flight or rage response – has offered a suitable experimental paradigm with which to study the involvement of CNS in organising and initiating changes in cardiorespiratory activity [41,72,74], but other behaviours – eating, drinking and sexual amongst others – all have their distinctive patterns of cardiorespiratory activity. These support the motor components of these characteristic behaviours. As more is learnt of the CNS processes and structures that are involved in the integration of these activities, so it is becoming ever more apparent that diverse areas of the CNS have the ability to alter the pattern of cardiorespiratory activity. Cortical and subcortical forebrain structures have been shown to have reciprocal connections with those brain stem areas that have already been identified as essential for cardiorespiratory control [13,53]. Intriguingly, many of these pathways have been shown to contain specific neuropeptides, but the physiological role of most remains to be identified.

107.9.1 The Defence Reaction

The central nucleus of the amygdala (CEN) and the perifornical area of the hypothalamus (hypothalamic defence area, HDA) have been shown to play important roles in establishing the patterns of response that are associated with affective behaviour [41]. The autonomic – involving cardiovascular – and respiratory components of these responses are patterned by the output of these subcortical regions and they are richly innervated from forebrain regions involved in emotional behaviour. The defence reaction involves an increase in respiratory minute volume, a rise in both heart rate and blood pressure and an increase in aortic blood flow and hence cardiac output (Fig. 107.6). Peripheral blood flow is directed to skeletal muscle and away from splanchnic and renal vascular beds. There is evidence of a reduction in vascular resistance in skeletal muscle through a withdrawal of sympathetic vasoconstrictor tone, and in some species, such as the cat,

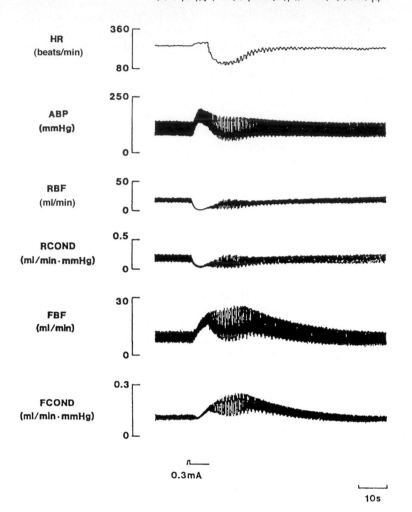

PNA

HR (beats/min) 360 — 80

ABP (mmHg) 250 — 0

RBF (ml/min) 50 — 0

RCOND (ml/min·mmHg) 0.5 — 0

FBF (ml/min) 30 — 0

FCOND (ml/min·mmHg) 0.3 — 0

0.3mA

10s

Fig. 107.6. Cardiorespiratory changes in the defence reaction: cardiovascular and respiratory responses to electrical stimulation in the perifornical region of the hypothalamus of the althesin-anaesthetized cat. The stimulus was a 5-s train of 0.5-ms pulses delivered at 100 Hz, initiated at the time shown in the lower trace on the polygraph record shown on the *left hand side*. Traces from *top* to *bottom*: *PNA*, phrenic nerve activity; *HR*, heart rate; *ABP*, arterial blood pressure; *RBF*, renal blood flow; *RCOND*, renal vascular conductance; *FBF*, femoral blood flow; *FCOND*, femoral vascular conductance. RBF and FBF were recorded using electromagnetic flow probes

evidence of active vasodilatation in this bed through the activity of sympathetic cholinergic vasodilator fibres [34]. There is also a dramatic increase in sympathetic discharge to the adrenal gland, which results in a rise in circulating catecholamines that furthers this vasodilatation in the muscle vascular bed. This response can be elicited in the anaesthetised animal by restricted stimulation through stereotaxically positioned microelectrodes in the absence of any motor component of the response, but also involves piloerection and pupillary dilation. The marked rise in heart rate during such stimulation is followed on cessation of stimulation by a pronounced bradycardia that is vagally mediated. The pattern of cardiovascular response is suggestive of a blockade of the baroreceptor reflex followed by a subsequent rebound activation of the reflex [72,74]. This

has indeed been verified experimentally, when both the cardiac and vascular components of the reflex were shown to be blocked by stimulation of the HDA in the cat. This pattern of interaction between reflex and centrally evoked response is clearly more complicated under physiological circumstances, since other experiments have shown that the sham rage of the decorticate cat, which is dependent on the integrity of the hypothalamic axis, is blocked by baroreceptor stimulation [8]. This implies a reciprocal interaction between the two drives – emotional and homeostatic.

The CEN and the HDA communicate reciprocally with those areas of the lower brain stem that are concerned with cardiorespiratory control and have major relays within the midbrain periaqueductal gray and the pontine parabrachial nuclei [53]. The NA, DVN, and NTS are richly innervated from these regions directly and via these relays. Marked synaptic responses are evoked in medullary respiratory neurons [6], and the increased activity of inspiratory neurons would be expected to provoke an inhibition of CVMs (see above) together with an increased discharge of certain sympathetic preganglionic motoneurons. There is also evidence that RVLM premotor neurons are excited on HDA stimulation and by activation of the PAG in both the cat and rat.

The suppression of the baroreceptor reflex is, however, a more complex matter than simply activating the "respiratory" gate (Fig. 107.7). Electrophysiological studies have shown that neurons in the NTS which receive excitatory input from the arterial baroreceptors are inhibited by HDA stimulation [61]. The HDA evoked IPSPs are chloride-dependent and involve the activation of GABAa receptors that are bicuculline-sensitive [43]. Accordingly, CVMs are inhibited by inspiratory activation during the defence reaction, disfacilitated by the GABAergic inhibitory mechanism elicited within the NTS and further inhibited by a GABAergic mechanism at the level of the CVMs themselves

(see Fig. 107.7). Withdrawal of these inputs on the cessation of the HDA stimulus results in a rebound excitatory drive to CVMs mediated by the baroreceptors, which are seeing a rise in arterial blood pressure as a result of the effects of HDA activation on sympathetic vasoconstrictor outflows.

The NTS control of reflex input is not restricted to the baroreceptor reflex. There is evidence that other cardiovascular reflexes may face a similar GABAergic control (e.g., pulmonary C-fibre input). It appears to involve an action of an intrinsic group of GABA-containing neurons localised within the NTS [36,43,74]. Conversely, there is evidence to suggest that the arterial chemoreceptor reflex is potentiated during the defence reaction, and preliminary electrophysiological studies have shown that the chemoreceptor input to some NTS neurons is facilitated by HDA stimulation.

This process of reflex control at the level of the NTS was at one time thought to be an exclusive property of the defence reaction. Recent studies, however, have shown that this might be a more general mechanism by which the CNS modulates the performance of cardiorespiratory reflexes as part of its action in producing cardiorespiratory changes appropriate for the support of behavioural activities [75]. During experimentally elicited exercise the vagal component of the baroreceptor reflex has been shown to be reset, and preliminary studies suggest that a GABAergic mechanism involved in the NTS is responsible. Similarly, stimulation in the uvula cortex of the cerebellum (lobule IXb) elicits cardiovascular and respiratory responses, and this stimulus can modulate the effectiveness of the baroreceptor reflex control of heart rate [10,64]. Again, this action can be antagonised by the microinjection of the GABAa receptor antagonist bicuculline into the NTS. Lesions in the uvula also alter the gain of the baroreceptor–heart rate reflex, suggesting that this mechanism may play a tonic role in the control of reflex performance.

107.9.2 Playing Dead: The Freezing Response

The alternative pattern of behavioural response to a threatening stimulus to that implied by the defence reaction is the "playing dead" response [4,41] – the human equivalent of freezing with fear. Again, the CEN is involved in its expression and the response is particularly well defined in the rabbit. There is a fall of heart rate and blood pressure and an apparent suppression of ventilation. In reality the latter may reflect rapid shallow breathing or a post-inspiratory arrest of breathing. The "respiratory" gate may elicit the accompanying bradycardia, or this may reflect a shared excitatory input from the CEN to the post-inspiratory neuron pool in the medulla, which will include both respiratory and CVM components of the cardiorespiratory network. There is also an evoked facilitation, at the level of the NTS, of those neurons that are activated by baroreceptor inputs; this will enhance the baroreceptor excitatory drive to CVMs and the inhibitory drive to sympathetic preganglionic neurons [15].

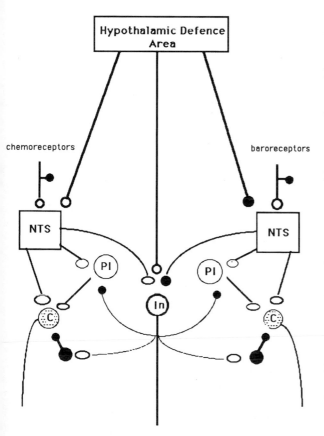

Fig. 107.7. Hypothalamic control of baroreceptor function. Hypothalamic defence area influences on CVM (C) activity. Details as in Fig. 112.4, with additional excitatory/facilitatory control of chemoreceptor transmission in the NTS, and inhibition of baroreceptor reflex via a GABAergic mechanism (see text for further description). A direct excitatory influence on inspiratory neurons is also illustrated

107.10 Concluding Remarks

There is compelling evidence suggesting that the basic rhythms of the nervous innervations of the cardiovascular and respiratory systems are generated and patterned by a common cardiorespiratory neural network localised within the medulla. This is influenced by various tonic excitatory and inhibitory drives, the balance of which accounts for the changing patterns of cardiorespiratory activity throughout sleep and wakefulness. Further, the network is susceptible to reflex inputs that act to maintain homeostasis. These inputs, whilst having a general impact on the overall network, may be organised so that certain actions are directed preferentially – an example is the baroreceptor control of CVMs, which is more potent than the action of the baroreceptors on other components of the post-inspiratory network. This dynamic interaction between the central patterning of the rhythms and reflex feedback is carried further in the action of the CNS in modulating the efficacy of reflex inputs through synaptic actions within the NTS. In this regard, an intrinsic population of GABA-containing neurons appears to play a significant role, but the neurochemistry of the inputs to these neurons remains to be elucidated. The overall control of the cardiorespiratory system is thus divisible into simple interacting units, the modification of any one of which has major implications for the ongoing actions of the others. The essential simplicity of design allows the analysis of complex patters of behaviour and represents a robust control system that is both adaptive and stable.

References

1. Alexander RS (1946) Tonic and reflex functions of medullary sympathetic centres. J Neurophysiol 9:205–217
2. Anrep GV, Pascual W, Rössler R (1936a) Respiratory variations of the heart rate. I. The reflex mechanism of the respiratory arrhythmia. Proc R Soc [B] 119:191–217
3. Anrep GV, Pascual W, Rössler R (1936b) Respiratory variations of the heart rate. II. The central mechanism of the respiratory arrhythmia and the interrelations between the central and reflex mechanisms. Proc R Soc [B] 119:218–230
4. Applegate CD, Kapp BS, Underwood MD, McNall CL (1983) Autonomic and somatomotor effects of amygdala central nucleus stimulation in awake rabbits. Physiol Behav 31:353–360
5. Bacon SJ, Zagon A, Smith AD (1990) Electron microscopic evidence of a monosynaptic pathway between cells in the caudal raphe nuclei and sympathetic preganglionic neurons in the rat spinal cord. Exp Brain Res 79:589–602
6. Ballantyne D, Jordan D, Spyer KM, Wood LM (1988) Synaptic rhythm of caudal medullary expiratory neurones during stimulation of the hypothalamic defence area of the cat. J Physiol (Lond) 405:527–546
7. Ballintijn CM (1984) Evolution of central nervous control of ventilation in vertebrates. In: Taylor W (ed) Studies in neuroscience: neurobiology of the cardiorespiratory system. Manchester University Press, Manchester, pp 3–27
8. Bartorelli C, Bizzi E, Libretti A, Zanchetti A (1960) Inhibitory control of sinocarotid pressoceptive afferents on hypothalamic autonomic activity and sham-rage behaviour. Arch Ital Biol 98:308–326
9. Bieger D, Hopkins DA (1987) Viscerotopic representation of the upper alimentary tract in the medulla oblongata in the rat: the nucleus ambiguus. J Comp Neurol 262:546–562
10. Bradley DJ, Ghelarducci B, Spyer KM (1991) The role of the posterior cerebellar vermis in cardiovascular control. Neurosci Res 12:45–56
11. Brock JA, Cunnane TS (1987) Electrical activity at the sympathetic neuroeffector junction in the guinea-pig vas deferens. J Physiol (Lond) 399:607–632
12. Brooks PA, Glaum SR, Miller RJ, Spyer KM (1992) The actions of baclofen on neurones and synaptic transmission in the nucleus tractus solitarius of the rat in vitro. J Physiol (Lond) 457:115–129
13. Cechetto DF, Saper CB (1990) Role of the cerebral cortex in autonomic function. In: Loewy AD, Spyer KM (eds) Central regulation of autonomic functions. Oxford University Press, New York, pp 208–223
14. Coleridge HM, Coleridge JCG, Jordan D (1991) Integration of ventilatory and cardiovascular control sSystems. In: Crystal RG, West JB et al (eds) The lung: scientific foundations. Raven, New York, chap 5.4.10, pp 1405–1418
15. Cox GE, Jordan D, Moruzzi P, Schwaber JS, Spyer KM, Turner SA (1985). Amygdaloid influences on brain-stem neurones in the rabbit. J Physiol (Lond) 381:135–148.
16. Daly M de B (1986) Interactions between respiration and circulation. In: Cherniack NS, Widdicombe JG (eds) Handbook of physiology, section 3: the respiratory system, vol II: control of breathing, part 2. American Physiological Society, Bethesda, pp 529–594
17. Daly M de B, Jordan D, Spyer KM (1992) Modification of respiratory activities during stimulation of carotid chemoreceptors, arterial baroreceptors and pulmonary C fibre afferents in the anaesthetized cat. J Physiol (Lond) 446:466P
18. Daly M de B, Kirkham E (1989) Differential modulation by pulmonary stretch afferents of some reflex cardioinhibitory responses in the cat. J Physiol (Lond) 417:323–341
19. Daly M de B (1984) Breath-hold diving: mechanisms of cardiovascular adjustments in the mammal. Rec Adv Physiol 10:201–245
20. Daly M de B (1991) Some reflex cardioinhibitory responses in the cat and their modulation by central inspiratory neuronal activity. J Physiol (Lond) 439:559–577
21. Dampney RAL, McAllen RM (1988) Differential control of sympathetic fibres supplying hindlimb and muscle by subretrofacial neurones in the cat. J Physiol (Lond) 395:41–56
22. Davidson NS, Goldner S, McCloskey DI (1976) Respiratory modulation of baroreceptor and chemoreceptor reflexes affecting heart-rate and cardiac vagal efferent nerve activity. J Physiol (Lond) 259:523–530
23. Davis AL, McCloskey DI, Potter EK (197) Respiratory modulation of baroreceptor and chemoreceptor reflexes affecting heart rate through the sympathetic nervous system. J Physiol (Lond) 272:691–703
24. Dembowsky K, König Czachurski J (1991) Respiratory modulation of sympathetic preganglionic neurones. J Auton Nerv Syst 333:P94
25. Eldridge FL (1972) The importance of timing on the respiratory effects of intermittent carotid body chemoreceptor stimulation. J Physiol (Lond) 222:259–265
26. Feldman JL, Speck DF (1983) Interactions among inspiratory neurons in dorsal and ventral respiratory groups in cat medulla. J Neurophysiol 49:472–490.
27. Feldman JL, Smith JC, Ellenberger HH, Connelly CA, Green JJ, Lindsay AD, Otto MR (1990) Neurogenesis of respiratory rhythm and pattern: emerging concepts. Am J Physiol 259:879–886
28. Gebber GL (1990) Central determinants of sympathetic nerve discharge. In: Loewy AD, Spyer KM (eds) Central regulation of autonomic functions. Oxford University Press, New York, pp 126–144
29. Gilbey MP, Jordan D, Richter DW, Spyer KM (1984) Synaptic mechanisms involved in the inspiratory modulation of vagal

cardioinhibitory neurones in the cat. J Physiol (Lond) 356:65–78

30. Gilbey MP, Numao Y, Spyer KM (1985) Discharge patterns of cervical sympathetic preganglionic neurones related to central respiratory drive in the rat. J Physiol (Lond) 378:253–265

31. Gilbey MP, Stein RD (1991) Characteristics of lumbar sympathetic preganglionic neurones in the anaesthetised cat. J Physiol (Lond) 432:427–444

32. Guyenet PG (1990) Role of the ventral medulla oblongata in blood pressure regulation. In: Loewy AD, Spyer KM (eds) Central regulation of autonomic functions. Oxford University Press, New York, pp 145–167

33. Guyenet PG, Darnall RA, Riley TA (1990) Rostral ventrolateral medulla and sympathorespiratory integration in rats. Am J Physiol R1063–1074

34. Hilton SM (1966) Hypothalamic control of the cardiovascular responses in fear and rage. In: The scientific basis of medicine annual reviews, pp 217–238

35. Hilton SM, Spyer KM (1980) Central nervous regulation of vascular resistance. Annu Rev Physiol 42:300–411

36. Izzo PN, Sykes RM, Spyer KM (1992) γ-Amino butyric acid immunoreactive structures in the nucleus tractus solitarius; a light and electron microscopic study. Brain Res 591:69–79

37. Jänig W (1985) Organization of the lumbar sympathetic outflow to skeletal muscle and skin of the cat's hindlimb. In: Koepchen HP, Hilton SM, Trzebski A (eds) Central interaction between respiratory and cardiovascular control systems. Springer, Berlin Heidelberg New York, pp 128–135

38. Jänig W, McLachlan EM (1992) Specialized functional pathways are the building blocks of the autonomic nervous system. J Auton Nerv Syst 41:3–14

39. Jones JFX, Jordan D (1993) Activity of medullary respiratory neurones during the pulmonary chemoreflex in anaesthetized rabbits. J Physiol (Lond) 459:354P

40. Jones JFX, Jordan D (1993) Evidence for a chronotropic response to cardiac vagal motor C-fibre stimulation in anaesthetized cats, rats and rabbits. J Physiol (Lond) 467:148P

41. Jordan D (1990) Autonomic changes in affective behaviour. In: Loewy AD, Spyer KM (eds) Central regulation of autonomic functions. Oxford University Press, New York, pp 349–366

42. Jordan D, Khalid MEM, Schneiderman N, Spyer KM (1982) The location and properties of preganglionic vagal cardiomotor neurones in the rabbit. Pflugers Arch 395:244–250

43. Jordan DJ, Mifflin SW, Spyer KM (1988) Hypothalamic inhibition of neurones in the nucleus tractus solitarius of the cat is GABA mediated. J Physiol (Lond) 399:389–404

44. Jordan D, Spyer KM (1986) Brainstem integration of cardiovascular and pulmonary afferent activity. Prog Brain Res 67:295–314

45. Jordan D, Spyer KM (1987) Central neural mechanisms mediating respiratory-cardiovascular interactions. In: Taylor EW (eds) Neurobiology of the cardiorespiratory system. Manchester University Press, Manchester, pp 322–341

46. Koepchen HP, Wagner PH, Lux HD (1961a) Untersuchungen über Zeitbedarf und zentrale Verarbeitung des pressoreceptischen Herzreflexes. Pflugers Arch 273:413–430

47. Koepchen HP, Wagner PH, Lux HD (1961b) Über die Zusammenhänge zwischen zentraler Erregbarkeit reflectorischen Atemrhythmus bei der nervösen Steuerung der Herzfrequenz. Pflugers Arch 273:443–465

48. Langhorst P, Schulz B, Schulz G, Lambertz M (1983) Reticular formation of the lower brainstem. A common system for cardiorespiratory and somatomotor functions: discharge patterns of neighbouring neurons influenced by cardiovascular and respiratory afferents. J Auton Nerv Syst 9:411–431

49. Langhorst P, Schulz G, Schulz B, Kluge W, Lambertz M. Integration and control of cardiorespiratory and somatomotor functions. In: Nakamura K (ed) Brain and blood pressure control. Elsevier, Amsterdam, pp 2–11

50. Lindsey BG, Hernandez YM, Morris KF, Shannon R (1992) Functional connectivity between brain stem midline neurons with respiratory-modulated firing rates. J Neurophysiol 67:890–903

51. Lindsey BG, Hernandez YM, Morris KF, Shannon R, Gerstin GL (1992) Respiratory-related neural assemblies in the brain stem midline. J Neurophysiol 67:905–922

52. Lipski J, McAllen RM, Spyer KM (1977) The carotid chemoreceptor input to the respiratory neurons of the nucleus of tractus solitarius. J Physiol (Lond) 269:797–810

53. Loewy AD (1990) Central autonomic pathways. In: Loewy AD, Spyer KM (eds) Central regulation of autonomic functions. Oxford University Press, New York, pp 88–103

54. Loewy AD, Spyer KM (1990) Vagal preganglionic neurons. In: Loewy AD, Spyer KM (eds) Central regulation of autonomic functions. Oxford University Press, New York, pp 68–87

55. McAllen RM (1987) Central respiratory modulation of subretrofacial bulbospinal neurones in the cat. J Physiol (Lond) 388:533–545

56. McAllen RM (1992) Actions of carotid chemoreceptors on subretrofacial bulbospinal neurons in the cat. J Auton Nerv Syst 40:181–188

57. McAllen RM, Spyer KM (1978) The baroreceptor input to cardiac vagal motoneurones. J Physiol (Lond) 282:365–374

58. McAllen RM, Spyer KM (1978) Two types of vagal preganglionic motoneurones projecting to the heart and lungs. J Physiol (Lond) 282:353–364

59. Mifflin SW (1993) Absence of respiration modulation of carotid sinus nerve inputs to nucleus tractus solitarius neurons receiving arterial chemoreceptor inputs. J Auton Nerv Syst 42:191–200

60. Mifflin SW, Spyer KM, Withington-Wray DJ (1988a) Baroreceptor inputs to the nucleus tractus solitarius in the cat: postsynaptic actions and the influence of respiration. J Physiol (Lond) 399:349–367

61. Mifflin SW, Spyer KM, Withington-Wray DJ (1988b) Baroreceptor inputs to the nucleus tractus solitarius in the cat: modulation by the hypothalamus. J Physiol (Lond) 399:369–387

62. Numao Y, Koshiya M, Gilbey MP, Spyer KM (1987) Central respiratory drive-related activity in sympathetic nerves of the rat: the regional differences. Neurosci Lett 81:279–284

63. Paintal AS (1970) The mechanism of excitation of type J receptors, and the J reflex. In: Porter R (ed) Breathing: Hering-Breuer centenary symposium. Churchill, London, pp 59–71

64. Paton JFR, Spyer KM (1992) Cerebellar cortical regulation of circulation. News Physiol Sci 7:124–129

65. Polosa C (1968) Spontaneous activity of sympathetic preganglionic neurons. Can J Physiol Biochem 46:587–596

66. Potter EK (1981) Inspiratory inhibition of vagal responses to baroreceptor and chemoreceptor stimuli in the dog. J Physiol (Lond) 316:177–190

67. Richter DW, Ballanyi K, Schwarzacher S (1992) Mechanisms of respiratory rhythm generation. Curr Opin Neurobiol 2:788–793

68. Richter DW, Jordan G, Ballantyne D, Meesmann M, Spyer KM (1986) Presynaptic depolarization in myelinated vagal afferent fibres terminating in the nucleus of the tractus solitarius in the cat. Pflugers Arch 406:12–19

69. Richter DW, Seller H (1975) Baroreceptor effects on medullary respiratory neurones of the cat. Brain Res 86:168–171

70. Richter DW, Spyer KM (1990) Cardiorespiratory control. In: Loewy AD, Spyer KM (eds) Central regulation of autonomic functions. Oxford University Press, New York, pp 168–188

71. Richter DW, Spyer KM, Gilbey MP, Lawson EE, Bainton CR, Wilhelm Z (1991) On the existence of a common cardiorespiratory network. In: Koepchen H-P, Huopaniemi T (eds) Cardiorespiratory and motor co-ordination. Springer, Berlin Heidelberg New York, pp 118–130

72. Spyer KM (1990) The central nervous organization of reflex circulatory control. In: Loewy AD, Spyer KM (eds) Central

regulation of autonomic functions. Oxford University Press, New York, pp 168–188

73. Spyer KM (1981) Neural organisation and control of the baroreceptor reflex. In: Baker PF, Grunicke H, Habermann E et al (eds) Reviews of physiology biochemistry and pharmacology, vol 88. Springer, Berlin Heidelberg New York, pp 23–124

74. Spyer KM (1992) Central nervous control of the cardiovascular system. In: Bannister R, Mathias CJ (eds) Autonomic failure – a textbook of clinical disorders of the autonomic nervous system, 3rd edn. Oxford Medical Publications, Oxford, pp 54–77

75. Spyer KM, Brooks PA, Izzo PN (1993) Vagal preganglionic neurones supplying the heart. In: Levy MN, Schwartz PJ (eds) Textbook on vagal control of the heart, chap 4. Futura, Mount Kisco New York, pp 45–63

76. Spyer KM, Izzo PN, Lin RJ, Paton JFR, Silva-Carvalho LF, Richter DW (1990) The central nervous organisation of the carotid body chemoreceptor reflex. In: Acker H, Trzebski A, O'Regan RG (eds) Chemoreceptors and chemoreceptor reflexes. Plenum, New York, pp 305–309

77. Sun M-K, Spyer KM (1991) Responses of rostroventrolateral medulla spinal vasomotor neurones to chemoreceptor stimulation in rats. J Auton Nerv Syst 33:79–84

78. Wallin BC. (··) Intraneural recordings of normal and abnormal sympathetic activity in humans. In: Bannister R, Mathias CJ (eds) Autonomic failure, 3rd edn. Oxford University Press, New York, pp 359–377

79. Zagon A, Bacon SJ (1991) Evidence of a monosynaptic pathway between cells of the ventromedial medulla and the motoneuron pool of the thoracic spinal cord in rat: electron microscopic analysis of synaptic contacts. Eur J Neurosci 3:55–65

80. Zhon S-Y, Gilley MP (1992) Respiratory-related discharges of lower thoracic and lumbar sympathetic preganglionic neurones in the anaesthetised rat. J Physiol (Lond) 451:631–642

108 Coordination of Circulation and Respiration During Exercise

S.A. WARD and B.J. WHIPP

Contents

108.1 Introduction

Muscular work can only be sustained if two transport chains function effectively: the electron transport chain and the oxygen transport chain. These may be considered to be "linked" at cytochrome oxidase. This chapter focuses on the effective functioning of the primary systems that comprise the oxygen transport chain in humans; the effective function of the electron transport chain is considered elsewhere (Chap. 71).

Naturally, for oxygen to be provided for the electron transport chain at rates commensurate with the increased energy demands of the exercise, there must be an effective interaction of the cardio-circulatory and pulmonary systems. This system interaction also serves to "clear" the consequent CO_2 and heat from the tissues to ensure that the physical-chemical milieu of the chemical-mechanical coupling sites in the muscle be maintained in a state compatible with efficient energy transfer (see Chaps. 45, 46).

Ideally, these system transport functions should be subserved at minimum operating cost. As shown in Fig. 108.1, this seems to be the case at the normal resting alveolar ventilation to perfusion ratio of approximately 1.0 [147].

We shall, in this chapter, consider therefore:

- The demands placed upon each system during muscular exercise
- The functional responses of the systems which allow the demands to be met
- The limits of the systems' responses.

108.2 System Demands

The degree to which the O_2 transport demands of exercise can be met by the circulatory and ventilatory responses depends on the intensity of the exercise. An important demarcator of exercise intensity is provided by the threshold work rate (or more properly, oxygen uptake) at which a sustained metabolic (largely lactic) acidemia results. This parameter has been termed the *lactate threshold* (θ_L). The range of work rates within which there is no sustained metabolic acidemia may be considered to be of moderate intensity. It is characterized by the attainment of steady states of ventilation ($\dot{V}E$), gas exchange, and cardiac output ($\dot{Q}T$). Consequently, these work rates can be sustained for prolonged periods. Higher work rates (i.e., above θ_L) can be regarded as being of high intensity and are characterized by a more rapid onset of fatigue, usually associated with a marked metabolic acidemia (see Chap. 45).

It is important from the outset to recognize that the substrates being catabolized place certain fundamental demands upon both the cardiovascular and pulmonary systems. For example, as shown in Table 108.1 (and considered in Chap. 100), the same rate of high-energy phosphate formation requires some 6% less O_2 when carbohydrate is metabolized than when fatty acids serve as the substrate. Consequently, there is typically a proportionally reduced demand for O_2 to be transported.

Further consideration of Table 108.1, however, reveals that, under these same conditions, carbohydrate metabolism yields some 40% more CO_2 than for a typical fatty acid. This results in a greater demand for CO_2 clearance and

R. Greger/U. Windhorst (Eds.)
Comprehensive Human Physiology, Vol. 2
© Springer-Verlag Berlin Heidelberg 1996

Table 108.1. Energetics of substrate catabolism

	RQ	$\dot{V}_{O_2}$ (l/min)	$\dot{V}_{CO_2}$ (l/min)	~P:O$_2$	~P:CO$_2$	O$_2$:~P	CO$_2$:~P
Glycogen	1.0	1.0	1.0	6.00	6.00	0.17	0.17
Palmitate	0.7	1.0	0.7	5.65	8.13	0.18	0.12
Glycogen/palmitate		1.0	1.43	1.06	0.74	0.94	1.42

Values are expressed at a particular $\dot{V}_{O_2}$.

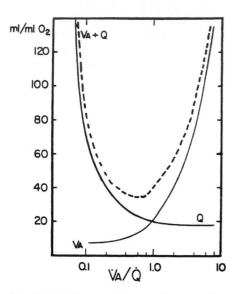

Fig. 108.1. Influence of the ventilation-perfusion ratio of the whole lung ($\dot{V}_A/\dot{Q}$) on the volumes of alveolar gas (V_A) and pulmonary capillary blood (Q) needed to transport 1 ml oxygen. The total pulmonary O$_2$ transport requirement (i.e., for the gas and blood phases together) is depicted by the *dashed line* (V_A+Q). Note that this has a minimum close to a $\dot{V}_A/\dot{Q}$ of unity. (From [147])

hence ventilation. The question therefore of which substrate is the more "appropriate" or "efficient" fuel for muscular work may not be unequivocally answered in this general form; it requires a frame of reference. To minimize the cardiovascular demands for O$_2$ delivery, carbohydrate is the more suitable fuel [203]. However, to minimize the ventilatory demands for CO$_2$ clearance, fatty acids may be considered a more suitable substrate [20], although, as $\dot{V}_E$ is now low for the O$_2$ requirement of the task, alveolar and arterial PO$_2$ will consequently be reduced with possibly deleterious consequences for oxygenation.

What, then, is an "appropriate" ventilatory or cardiac output response to exercise? In other words, to what extent is blood-gas and acid-base homeostasis (see Chaps. 78, 102, 106) preserved in arterial blood by the ventilatory response? And is the cardiac output response of sufficient magnitude not to disturb the composition of mixed venous blood – or, more properly, of venous blood draining the exercising musculature?

2146

108.2.1 Cardiac Output

Increased cardiac output during exercise is required to increase the perfusion to the exercising muscles ($\dot{Q}_M$). In order that muscle-venous O$_2$ content (CvMO$_2$) not fall, $\dot{Q}_M$ would need to increase to levels commensurate with the increase in muscle O$_2$ consumption ($\dot{Q}_{O_2}$) at a rate that entirely meets the demands of tissue energy requirements (i.e., no additional O$_2$ would be drawn from vascular and tissue stores). This is illustrated by the *Fick principle* (see Chap. 101):

$$\dot{Q}_{O_2} = \dot{Q}_M \, (CaO_2 - CvMO_2) \tag{108.1}$$

Similarly, for CO$_2$:

$$\dot{Q}_{CO_2} = \dot{Q}_M \, (CvMCO_2 - CaCO_2) \tag{108.2}$$

where $\dot{Q}_{CO_2}$ represents the rate of muscle CO$_2$ production, and CaO$_2$ and CaCO$_2$ are the O$_2$ and CO$_2$ contents in arterial blood.

However, increases in $\dot{Q}_M$ sufficient to prevent CvMO$_2$ from falling (and CvMCO$_2$ from rising) would place excessive demands on the heart. At sea level, arterial blood is almost completely saturated with O$_2$ at rest (i.e., 97%–98%) and hence cannot undergo any appreciable further increase during exercise. The magnitude of the $\dot{Q}_M$ response therefore depends on:

- The magnitude of the cardiac output ($\dot{Q}_T$) increase
- The extent to which the $\dot{Q}_T$ response is preferentially redistributed to the working musculature, i.e., by how much $\dot{Q}_M/\dot{Q}_T$ increases.

Consider, for example, a subject of average fitness who is exercising maximally at a $\dot{Q}_{O_2}$ of 3 l/min. Let us assume, for simplicity, that: (a) CaO$_2$ at rest equals 20 ml/100 ml and does not change with exercise (in fact, it usually increases slightly during exercise as a result of hemoconcentration [7]); and (b) CvMO$_2$ were to remain at its resting value of about 15 ml/100 ml [157,194]. Equation 108.1 indicates that $\dot{Q}_M$ would therefore have to attain a value of 60 l/min, i.e., $3 \times 100/(20 - 15)$! This flow requirement would be even higher in the elite athlete because of the higher achievable levels of $\dot{Q}_{O_2}$. Such perfusion requirements clearly could not be approached, let alone met, even if the entire cardiac output response were to be diverted to the working muscles (i.e., $\dot{Q}_M/\dot{Q}_T = 1.0$).

In reality, however, the cardiac output requirement during exercise is reduced, because there is an increased extraction of O_2 from the vascular stores, i.e., the muscle-venous and mixed venous O_2 concentrations fall during exercise (see Sect. 108.3.1). The extent to which they are required to fall will, naturally, depend on the extent to which $\dot{Q}_M$ and $\dot{Q}_T$ actually increase with respect to the metabolic rate.

108.2.2 Ventilation

Similar reasoning can be applied to the ventilatory requirements during exercise. Alveolar, and hence arterial, PO_2 and PCO_2 can be regulated at or close to resting levels only if alveolar ventilation ($\dot{V}_A$) increases in proportion to the metabolic rate [219]. Indeed, this is the case, i.e., arterial PCO_2 is well regulated at or close to resting values in the steady state of moderate exercise [46,63,213] (see Chap. 106).

Again, it is useful to consider the Fick principle, but now with respect to pulmonary exchange within the alveolar gas phase (Chap. 100):

$$\dot{V}CO_2 = \dot{V}_A \times P_ACO_2/863 \qquad (108.3)$$

where $\dot{V}CO_2$ is the pulmonary CO_2 output, P_ACO_2 is the alveolar PCO_2, and the constant 863 corrects for: (a) the convention of ventilatory volumes being defined at BTPS (i.e., as gas saturated with water vapor at *body* temperature and pressure), while volumes of metabolically exchanged O_2 and CO_2 are defined at STPD (i.e., as dry gas at a *standard* temperature of 0°C and a pressure of 1 atm), and (b) the transformation of fractional concentration to partial pressure.
Similarly, for O_2:

$$\dot{V}O_2 = \dot{V}_A \times (P_IO_2 - P_AO_2)/863 \qquad (108.4)$$

where P_AO_2 and P_IO_2 are the alveolar and inspired PO_2, respectively.
Rearranging Eqs. 108.3 and 108.4:

$$P_ACO_2 = 863 \times \dot{V}CO_2/\dot{V}_A \qquad (108.5)$$

and

$$P_AO_2 = P_IO_2 - (863 \times \dot{V}O_2/\dot{V}_A) \qquad (108.6)$$

As alveolar ventilation during moderate exercise behaves "as if" it responds preferentially to the demands for pulmonary CO_2 clearance rather than for pulmonary O_2 uptake (see Sect. 108.3.2), we shall focus here on the requirements for regulating arterial PCO_2 (P_aCO_2). The alveolar ventilatory demands at a particular work rate depend upon the following (Chap. 100):

- The rate of pulmonary CO_2 clearance; this depends both on substrate utilization profiles and whether or not body CO_2 stores are changing.

- The "set-point" at which P_aCO_2 is regulated; this is a function of how effective $\dot{V}_A$ is in achieving a particular level of pulmonary CO_2 exchange.
- The physiological dead space fraction of the breath (V_D/V_T), which represents an index of the inefficiency of pulmonary gas exchange.

The interaction between these three determinants is illustrated in Fig. 108.2. Consider two subjects who are now exercising submaximally at a work rate which yields a $\dot{V}O_2$ of 2.0 l/min. For subject I (thick black line), who is highly fit, this represents a moderate work rate (i.e., $<\theta_L$). Assume, further, that the metabolic substrate has a respiratory quotient (RQ) of 0.8 (see Chaps. 71, 101). This yields a metabolic CO_2 production of 1.6 l/min (Fig. 108.2, panel a). An *alveolar* ventilation ($\dot{V}_A$) of approximately 35 l/min is required to clear this CO_2 load if P_aCO_2 is to remain regulated at a normal level of 40 mmHg (Fig. 108.2, panel b). The *total* ventilatory requirement will, of course, be larger because of the need also to ventilate the physiological dead space. At this level of exercise, V_D/V_T is typically of the order of 0.2. Consequently, $\dot{V}_E$ would be a relatively modest 43 l/min (Fig. 108.2, panel c).
Compare this situation with that of the less-fit subject II (thick blue line), for whom exercise at the same $\dot{V}O_2$ of 2.0 l/min engenders a sustained metabolic acidosis (i.e., $>\theta_L$). An RQ of 1.0 would be reasonable, leading to an increased demand for pulmonary CO_2 clearance; i.e., $\dot{V}CO_2$ = 2.0 l/min (Fig. 108.2, panel a). P_aCO_2 will also be lower,

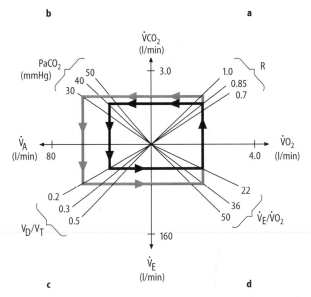

Fig. 108.2. Influence of the respiratory exchange ratio (R) (*panel a*), the regulated level of arterial PCO_2 (P_{aCO_2}) (*panel b*) and the physiological dead space fraction of the breath (V_D/V_T) (*panel c*) on the ventilatory ($\dot{V}_E$) requirement for two subjects both exercising at an O_2 uptake ($\dot{V}O_2$) of 2 l/min: the exercise is moderate (i.e., $<\theta_L$) for subject I (*thick black line*) who is highly fit, but is of high intensity (i.e., $>\theta_L$) for subject II (*thick blue line*) who is less fit. The increased $\dot{V}_E$ requirement for subject II is reflected in the greater ventilatory equivalent for O_2 ($\dot{V}_E/\dot{V}O_2$) (*panel d*). (see text for further discussion)

reflecting respiratory compensation for the metabolic acidosis: a 10 mmHg reduction of $PaCO_2$ would be reasonable. The requirement for $\dot{V}A$ under these conditions is therefore considerably increased (i.e., to 58 l/min) with respect to the less-fit subject (Fig. 108.2, panel b); and the $\dot{V}E$ requirement will be 72 l/min (Fig. 108.2, panel c).

These requirements will, of course, be exacerbated in patients with pulmonary disorders, in whom VD/VT is increased (values of 0.5 are not uncommon) [90,203].

108.3 System Responses

The system response profiles that have provided the basis for the current concepts of human cardiovascular and ventilatory control during exercise have resulted, for the most part, from studies performed in the laboratory. Less is known, however, about the responses that prevail during day-to-day activities in the work-place or in recreational and competitive situations.

Exercise can incorporate both *dynamic* (or isotonic) and *static* (or isometric) components. In the laboratory setting, it is most usual to investigate subjects while they are exercising on a cycle ergometer or a treadmill, i.e., dynamic modes of exercise which involve large muscle groups. However, smaller muscle groups, such as those involving the shoulder girdle and arm, have also been studied by means of arm cranking. In other instances, relatively "isolated" static contractions of the arms and legs may be used. The cardiorespiratory responses to static and dynamic exercise differ quite substantially, however. In the following discussion, we shall predominantly consider dynamic exercise, referring to static exercise when the response profiles are clearly different.

108.3.1 Incremental Dynamic Exercise

While many of the details of circulatory and respiratory control during exercise remain to be determined, the general patterns of response are not a topic of serious dispute. Naturally, it is these response profiles that provide the clues to the elements of the control processes. For example, consider an individual performing a progressive or "incremental" exercise test to the limit of tolerance.

108.3.1.1 Cardiac Output

Cardiac output (see Chaps. 89, 99) typically increases as a relatively linear function of work rate and, therefore, $\dot{V}O_2$ throughout the entire work-rate range (Fig. 108.3):

$$\dot{Q}T = (m \times \dot{V}O_2) + c \qquad (108.7)$$

where m is the slope and c is the $\dot{Q}T$ intercept. For exercise performed at seal level and in a temperate climate, expressing $\dot{Q}T$ and $\dot{V}O_2$ in units of l/min yields a value of approximately 5 for both m and c (especially if the resting values are not included in the regression) [7,70,73,157,194]. This relationship is not appreciably affected by training or fitness, although it extends to higher maximal values for both $\dot{Q}T$ and $\dot{V}O_2$ (Fig. 108.4). For example, while $\dot{Q}T$ typically rises from a resting values of about 5 l/min to the region of 20 l/min or so for maximum dynamic exercise in healthy young individuals [7,14,30,77,96,157,160,194], values in excess of 40 l/min have been reported in elite endurance athletes [55].

108.3.1.2 Arteriovenous Oxygen Content Difference

The increase in cardiac output that occurs during exercise is not sufficient to satisfy the metabolic requirements of the working muscles, even for the lightest work rates. Consequently, the mixed venous O_2 content ($C\bar{v}O_2$) falls as work rate increases. Therefore, the arteriovenous O_2 content difference ($CaO_2 - C\bar{v}O_2$) increases from a resting value of approximately 5 ml/100 ml to about 16 ml/100 ml at maximum exercise (Fig. 108.4) [7,29,157,160,164,194].

As the $\dot{Q}T - \dot{V}O_2$ relationship is linear but has a positive intercept on the $\dot{Q}T$ axis, the increase of the arteriovenous O_2 difference will change as a hyperbolic function of $\dot{V}O_2$ [194,217]. In other words,

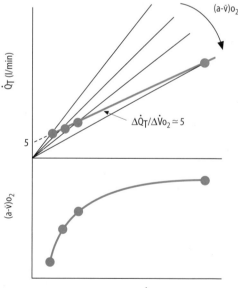

Fig. 108.3. *Upper panel* schematizes the observed relationship between cardiac output ($\dot{Q}_T$) and pulmonary O_2 uptake ($\dot{V}O_2$) (*thick blue line*) for progressive exercise. The *thin solid lines* radiating from the origin reflect isopleths who slope equals the reciprocal of the arterio–mixed venous O_2 content difference (C_{aO_2}–$C_{\bar{v}O_2}$). Note that the observed relationship crosses isopleths of progressively decreasing slope and therefore increasing ($C_{aO_2} - C_{\bar{v}O_2}$). *Lower panel* presents the corresponding response of ($C_{aO_2} - C_{\bar{v}O_2}$), which therefore increases hyperbolically with respect to $\dot{V}_{O_2}$. (See text for further discussion)

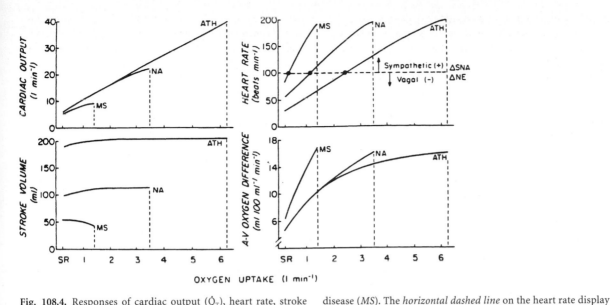

Fig. 108.4. Responses of cardiac output ($\dot{Q}_T$), heart rate, stroke volume and the arteriovenous O_2 difference ($C_{aO_2} - C_{\bar{v}O_2}$) as a function of pulmonary O_2 uptake ($\dot{V}_{O_2}$) for progressive cycle-ergometer exercise in elite endurance athletes (*ATH*), normal active subjects (*NA*), and patients with mitral stenosis, a cardiac disease (*MS*). The *horizontal dashed line* on the heart rate display (i.e., at a heart rate of 100 beats/min) separates the lower range of parasympathetic withdrawal from the upper range of sympathetic activation. (See text for further details. From [160])

if $\dot{Q}_T = 5 (\dot{V}O_2 + 1)$

then $(CaO_2 - C\bar{v}O_2) = \dfrac{20\dot{V}O_2}{(1 + \dot{V}O_2)}$ (108.8)

where the arteriovenous O_2 difference is expressed in the conventional units of ml/100 ml of blood, and $\dot{V}O_2$ is expresssed in l/min. The greater maximum $\dot{V}O_2$ in extremely fit subjects will therefore be associated with a further (though relatively small) increase in the arteriovenous O_2 difference [30,54,157,160,164], i.e., the hyperbolic ($CaO_2 - C\bar{v}O_2$) relationship "flattens out" at high work rates (Fig. 108.4). Therefore, the higher maximum $\dot{V}O_2$ achieved by such subjects depends far more upon their being able to attain a higher maximum $\dot{Q}_T$ (Fig. 108.4).

108.3.1.3 Heart Rate

The increased cardiac output during exercise depends crucially on heart rate (HR) increasing (i.e., *tachycardia*) (see Chap. 89). Normally, HR increases essentially linearly with $\dot{V}O_2$, although, as maximum $\dot{V}O_2$ is approached, it may start to plateau (Fig. 108.4) [7,14,29,157,160,194].

For modest increases in work rate, HR increases almost entirely as a result of a withdrawal of inhibitory parasympathetic (vagal cholinergic) tone to the sinoatrial node [152,158,160]. However, complete suppression of this parasympathetic tone is only able to increase HR to approximately 100 beats/min or so. The associated increment in heart rate that results from complete suppression will, of course, be larger the lower the resting HR. This is usually about 20–30 beats/min in healthy young individuals, who

typically have a resting HR of 70–80 beats/min, but can easily be doubled in endurance athletes, whose resting HR may be 40 beats/min or less (Fig. 108.4) as a result of a greater basal vagal tone.

Further increases in HR depend upon activation of the more slowly developing excitatory sympathetic drive. This is also mediated at the sinoatrial node but involves the following:

- Stimulation of β_1-adrenergic receptors via the sympathetic neurotransmitter norepinephrine
- At higher work rates, circulating catecholamines released from the adrenal medulla [152,158,160].

The "chronotropic" drive can elicit a HR of 200 beats/min or more at maximal exercise in healthy young adults. The maximal HR, however, decreases progressively with age (by about 10 beats/min each decade, on average) [7,203].

108.3.1.4 Stroke Volume

The contribution of stroke volume (SV; see Chap. 89) to the $\dot{Q}_T$ response depends very much on the posture adopted for the exercise. In the supine posture, the resting SV is already close to that normally attained during exercise, i.e., approximately 90–120 ml in healthy young individuals [7,14,76,77,194]. Consequently, it is the tachycardia that largely dictates the $\dot{Q}_T$ response under these conditions (Fig. 108.4).

However, resting SV is less in the more usual upright posture than when an individual is supine [7,14,76,77,194]. This reflects a gravitationally induced pooling of blood in

2149

the dependent lower limbs. The increase of SV during exercise is largely confined to the lower 30%–40% of the $\dot{V}O_2$ range; this increase "restores" the SV to resting supine values (Fig. 108.4).

Increased venous return is one important determinant of the stroke volume response to exercise. The resulting increases in ventricular end-diastolic volume increase ejection via the *Frank-Starling* effect (see Chaps. 89, 99) [77,143,150]. This augmentation of the venous return during exercise is the consequence of several factors (see Chap. 99):

- *The muscle pump.* The rhythmic contraction of the limb muscles propels blood away from the dependent regions towards the heart [82,114,160,171,198]; backflow during the subsequent relaxation phase is prevented by unidirectional venous valves [14,68].
- *The pressure gradient for venous return.* This is because mean arterial blood pressure increases systematically with increasing work rate (Fig. 108.5) [7,15,160].
- *Sympathetically mediated constriction of venules and veins.* This reduces the venous compliance in vascular beds throughout the body [14,73,154,160,174]. As a result, a proportion of the "stored" blood is translocated towards the right side of the heart. Perhaps the most important contribution to this component of the augmented venous return at exercise onset comes from the splanchnic circulation (see Chap. 69). This accommodates some 25% of the total blood volume at rest, but only about 1% at maximal exercise [48,157,160,194].

The other major influence on SV comes from the reflex actions of the sympathetic "*inotropic*" effector drive to the myocardium, mediated by norepinephrine acting on α-adrenergic receptors. Together with the circulating catecholamines at higher work rates, this leads to an increase of ventricular "*contracility*," which augments ventricular ejection by encroaching on the end-systolic "reserve" [77,143,150].

SV is typically larger in endurance athletes than in untrained individuals, even when allowance is made for body size [7,14,160]. Indeed, in elite athletes, it may be twice as large (Fig. 108.4). The cardiac output, however, at a particular level of $\dot{V}O_2$ is essentially independent of fitness (Fig. 108.4) [7,14,30,77,157,160]. Consequently, this accounts for the low HR at rest and during submaximal exercise in athletes. However, as maximum HR is not altered by training per se, then the improvement in the maximum $\dot{Q}T$ in trained subject results from the increased SV.

108.3.1.5 Muscle Blood Flow

At rest, total skeletal muscle blood flow ($\dot{Q}M$) is in the region of 1 l/min, i.e., 3–4 ml/100 g per min (see Chaps. 94, 95). This represents some 15%–20% of $\dot{Q}T$ [157,160,172,194]. It increases linearly with respect to $\dot{Q}O_2$ and $\dot{V}O_2$ during progressive exercise (Fig. 108.5) [4,85,158,160]. For maximum leg exercise, $\dot{Q}M$ may attain

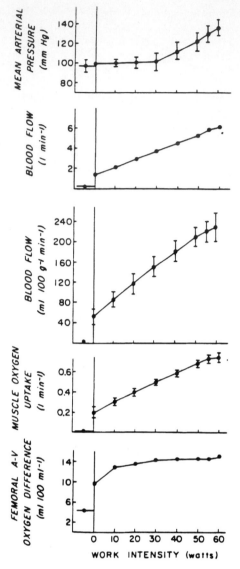

Fig. 108.5. Responses of mean arterial blood pressure, muscle blood flow, muscle O_2 uptake, and the femoral arteriovenous O_2 content difference as a function of work rate for rhythmic knee extension. (From [158] and modified from [4])

values of about 20 l/min or so, i.e., as much as 85% of $\dot{Q}T$; in elite endurance athletes, $\dot{Q}M$ can approach 40 l/min [157,158].

It has been suggested that muscles have a greater potential to accommodate blood flow than is actually achieved at maximum exercise [4]. Indeed, this is one of the reasons put forward in support of the contention that the cardiac output is normally the most likely source of cardiovascular limitation to exercise. However, others dispute this view, based upon comparisons of vascular conductance and blood flow responses during maximal contractions in isolated muscles compared with exercise in the intact, freely moving animal or human [114,159,160].

As previously stated, the $\dot{Q}M$ response to exercise is not sufficient to meet the O_2 requirements entirely; $CvMO_2$ decreases with increasing work rate (Fig. 108.5). In other words, the arteriovenous O_2 difference across the exercising muscle ($CaO_2 - CvMO_2$) increases, and in a curvilinear (hyperbolic) fashion that resembles that described above for the whole-body arteriovenous O_2 difference (see Fig. 108.3) [4,149,158,160].

The increased muscle perfusion during exercise is the consequence of the following:

- The increased $\dot{Q}T$ (Fig. 108.4) and increased mean arterial blood pressure (Fig. 108.5)
- The preferential redistribution of $\dot{Q}T$ to the exercising musculature, resulting from a generalized increase in reflex sympathetic vasoconstrictor drive and locally mediated vasodilatation within the exercising musculature
- The muscle pump.

Reflex Vasoconstriction. This is accomplished by the diffuse adrenergic sympathetic efferent discharge that is induced by exercise, as stated above. A significant contribution to the $\dot{Q}M$ increase comes from vasoconstriction in the splanchnic vascular bed: not only does it receive a large fraction of the resting cardiac output, but it also becomes significantly and progressively constricted with increasing work rate [29,157,160,194].

Interestingly, the arterioles of the working muscle also receive a sympathetic vasoconstrictor drive during exercise, but (as discussed below) this is offset by local vasodilating influences of largely metabolic origin. As a consequence, the increased blood flow is preferentially distributed towards the lower resistance vascular beds of the working muscles.

Local Vasodilatation (see Chaps. 89, 97, 99). The precise mechanism(s) of the exercise-induced vasodilatation in exercising muscle remains a topic for debate [160,176,180,192]. Vasodilatation induced by the intramuscular accumulation of locally released metabolites,

however, appears to be important. Mediators include K^+, osmolarity, hypoxia, H^+, adenosine, and vasoactive peptides. It has been proposed that these agents evoke one or both of the following effects:

- Impairment of the ability of norepinephrine to stimulate smooth muscle α-adrenoceptors
- Inhibition of norepinephrine release from prejunctional nerve varicosities [192].

Increase in blood flow velocity itself has been shown to induce local arteriolar vasodilatation in skeletal muscle, for which an involvement of prostaglandins has recently been proposed (probably PGE_2 and PGI_2) [105,106]. Inhibitors of prostaglandin synthesis markedly attenuate this vasodilator response. The response also depends upon an intact vascular endothelial lining. A stimulus related to an elevated *shear stress* at the endothelial cell surface has therefore been suggested. It is tempting also to consider a role for endothelial-derived releasing factor (EDRF) in this response. However, administration of an EDRF blocker was without effect.

Vasodilatation induced by activation of local neurogenic mechanisms has also been considered. Nonadrenergic neurons (thought to release *peptidergic* vasodilator transmitters) have been located within the walls of small arteries in skeletal muscle [78]. However, the significance of this mechanism is unclear, as is the stimulus for its activation [160].

Muscle Pump. It is important to recognize that $\dot{Q}M$ during dynamic exercise is pulsatile. It fluctuates on a beat-to-beat basis as a result of the alternating sequence of muscle contraction and relaxation, i.e., there is a sequence of "compression–emptying" and "relaxation–filling" within the muscle vascular bed [61,197] (Fig. 108.6). The reduction in flow caused by compression of blood vessels during the muscular contraction phase has been proposed to induce a

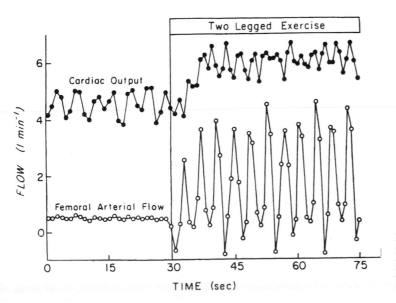

Fig. 108.6. Responses of cardiac output and femoral artery blood flow (measured by a Doppler ultrasound technique) during supine two-legged cycling. (From [160] and modified from [61])

2151

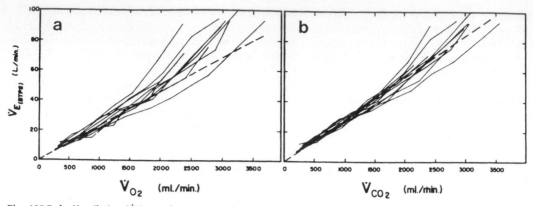

Fig. 108.7a,b. Ventilation ($\dot{V}_E$) as a function of pulmonary O_2 uptake ($\dot{V}_{O_2}$) (a) and CO_2 output ($\dot{V}_{CO_2}$) (b) during incremental cycle-ergometer exercise for ten normal subjects. The work rate was incremented every 4 min. (From [204])

substantial increase in flow in the subsequent relaxation phase, through the *myogenic* response [114,136,160]. In other words, the reduction in transmural pressure across the intramuscular arterioles during contraction leads to a relaxation of the vascular smooth muscle, thereby augmenting flow when the compression is released. Furthermore, the elastic recoil of the venules and veins when the compression is released have been proposed to expand the "empty" veins; this lowers the local venous pressure and hence serves to increase the driving pressure across the vascular bed.

108.3.1.6 Ventilation

$\dot{V}_E$ increases essentially linearly with respect to work rate over approximately the first half of the tolerable range (i.e., up to θ_L). Closer inspection, however, reveals that the linearity of the $\dot{V}_E$ response is not a function either of work rate or $\dot{V}_{O_2}$, but rather of the pulmonary CO_2 exchange rate ($\dot{V}_{CO_2}$) (Fig. 108.7) [52,92,166,204]. As a consequence, $PaCO_2$ and arterial pH (pH_a) are typically maintained close to resting levels over this moderate range of work rates [46,63,213].

Above θ_L, however, $\dot{V}_E$ becomes challenged by the metabolic acidosis and consequently increases proportionally more than for lower work rates (Fig. 108.7) [184,202,213]. This reflects:

- The increased demand for pulmonary CO_2 clearance caused by buffering of lactic acid with bicarbonate
- The additional demand to provide respiratory compensation for the metabolic acidemia.

Buffering. An immediate consequence of the metabolic acidosis (see Chaps. 78, 102, 106) is that the metabolic production of CO_2 in the working muscles is supplemented by additional CO_2 generated from the buffering of lactic acid by the bicarbonate system in muscle (chiefly as potassium bicarbonate) and blood (chiefly as sodium bicarbonate):

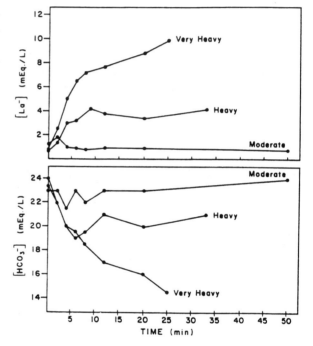

Fig. 108.8. Responses of arterial lactate and bicarbonate concentrations to moderate, heavy, and very heavy constant-load cycle-ergometer exercise. Note that the exercise could not be sustained for the full 50-min period at the two higher work rates. (From [203])

$$\underbrace{CH_3 \cdot CHOH \cdot COO^- \cdot H^+}_{\text{lactic acid}} + \underbrace{Na^+ HCO_3^-}_{\substack{\text{sodium} \\ \text{bicarbonate}}} \rightarrow$$

$$\underbrace{CH_3 \cdot CHOH \cdot COO^- \cdot Na^+}_{\text{sodium lactate}} + \underbrace{H_2CO_3}_{\substack{\text{carbonic} \\ \text{acid}}} \qquad (108.9)$$

The increase in arterial blood lactate concentration is therefore essentially mirrored by a decrease in the arterial bicarbonate concentration ($[HCO_3^-]_a$) [52] (Fig. 108.8); indeed, some 90% of the lactic acid formed at these work rates is buffered by the HCO_3 system [13,204]. Molecular CO_2 is evolved from the rapid dissociation of carbonic acid into CO_2 and H_2O. This CO_2 therefore supplements the CO_2 produced from aerobic tissue metabolism.

As $\dot{V}CO_2$ is now greater than $\dot{Q}CO_2$, the respiratory exchange ratio (R), i.e., $\dot{V}CO_2/\dot{V}O_2$, exceeds the RQ (Chap. 100). In other words, $\dot{V}CO_2$ increases faster than $\dot{V}O_2$ with increasing work rate (Fig. 108.7). It is important to recognize that this release of CO_2 from the body CO_2 stores only takes place while the bicarbonate levels are *falling* [206,214]. Thus, if work rate increments are imposed for a time period that is sufficiently long to allow $[HCO_3^-]_a$ to stabilize at each work rate, there will be no further release of CO_2 from the stores (i.e., this will have taken place in the early phase of the increment). Consequently, $\dot{V}CO_2$ will once more equal $\dot{Q}CO_2$, and R will again equal the RQ, despite $[HCO_3^-]_a$ being reduced.

Respiratory Compensation. Naturally, this "buffering" of lactic acid does not actually prevent arterial pH from falling. An additional $\dot{V}E$ drive is required to effect a compensatory reduction in $PaCO_2$ (see Eq. 108.5) and hence constrains the fall of arterial and muscle pH [184,202]. This is evident from the *Henderson-Hasselbalch* equation (see Chaps. 78, 102, 106):

$$pH_a = pK' + \log\{[HCO_3^-]_a/(0.03 \times PaCO_2)\} \quad (108.10)$$

where the constant pK' normally has a value of 6.1. If the ratio $\log\{[HCO_3^-]_a/PaCO_2\}$ is caused to increase, pH_a can rise back towards control levels. This requires $\dot{V}E$ to increase faster with respect to $\dot{V}CO_2$ (see Eq. 108.3) and therefore $\dot{V}O_2$ (Fig. 108.7).

108.3.2 Constant-Load Dynamic Exercise

The progressive exercise protocol is often an essential first step in an evaluation of ventilatory or cardiovascular control, as it allows delineation of the different intensity domains of exercise. However, within each intensity domain, the constant-load (or more properly, constant-power) protocol provides further insight into the underlying control processes.

Consider the intramuscular events that are initiated by the imposition of a constant work rate:

- $\dot{Q}O_2$ and presumably $\dot{Q}CO_2$ start to increase, rising towards their new steady state levels with essentially similar monoexponential time courses [108,126,130].
- $\dot{Q}M$ rises [61,197].
- There is a greater extraction of O_2 from, and loading of CO_2 into, the blood perfusing the muscle, causing $CvMO_2$ to fall and $CvMCO_2$ to increase.

These events are not immediately expressed at the lungs, however. This is because the lungs are functionally "separated" from the working muscles by a vascular delay. As a result, the mixed venous O_2 and CO_2 contents (and there-

fore $\dot{V}O_2$) are thought not to be appreciably influenced by the developing responses of $CvMO_2$ and $CvMCO_2$ until some 15–20 s after the start of the exercise [26,220]. (While this is the case when exercise is preceded by prior light exercise [26], $C\bar{v}O_2$, monitored in the pulmonary artery, was observed to decrease prior to the expected transient delay from the exercising legs when exercise was instituted from rest [25]; this is thought to reflect the influence in the venous return of blood draining a region having a high O_2 consumption relative to its perfusion.) The pulmonary exchange rates of O_2 and CO_2 are therefore temporally dissociated from $\dot{Q}O_2$ and $\dot{Q}CO_2$ by the muscle-to-lung vascular transit delay [11,220].

At exercise onset, there will therefore be an initial short period in which pulmonary gas exchange is effectively isolated from the demands of increased muscle metabolic rate (phase 1, $\Phi1$). The subsequent phase of pulmonary gas exchange in which mixed venous composition is changing may be termed phase 2 ($\Phi2$), with the new steady state being phase 3 ($\Phi3$) [220].

Phase One. Muscle perfusion increases in synchrony with exercise onset, for the numerous reasons described above (Fig. 108.6) [61,197]. The increase in $\dot{Q}M$ is reflected virtually instantaneously in an increased cardiac output and pulmonary blood flow [38,96,122,134], i.e., the blood is "pushed" around a closed circuit.

The abrupt increases in both $\dot{V}O_2$ and $\dot{V}CO_2$ that are characteristic of exercise onset (Fig. 108.9) are therefore largely a consequence of an increased pulmonary capillary perfusion [11,44,94,110,122,210,218]. In other words, from Eq. 108.6:

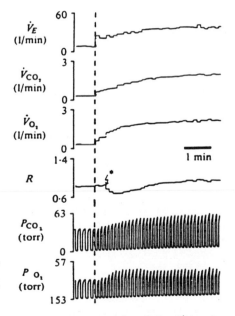

Fig. 108.9. Responses of ventilation ($\dot{V}_E$), pulmonary CO_2 output ($\dot{V}_{CO_2}$), pulmonary O_2 uptake ($\dot{V}_{O_2}$), the respiratory exchange ratio (R), and respired P_{CO_2} and P_{O_2} to moderate constant-load cycle-ergometer exercise undertaken from rest. The *asterisk* denotes the onset of phase 2. (From [219])

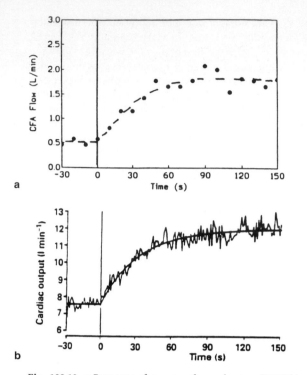

a

b

Fig. 108.10. **a** Response of common femoral artery (*CFA*) blood flow to moderate constant-load rhythmic knee extension exercise. **b** Response of cardiac output to moderate constant-load cycle-ergometer exercise (modified from [44]). Both variables were measured by Doppler ultrasound techniques. The curves represent the best-fit exponentials to the data

$$\Delta\dot{Q}O_2 \propto \Delta\dot{Q}T$$

Also, as $\quad \dot{Q}CO_2 = \dot{Q}T \, (C\bar{v}CO_2 - CaCO_2) \qquad (108.11)$

then $\quad \Delta\dot{Q}CO_2 \propto \Delta\dot{Q}T$

These "cardiodynamic" increases in $\dot{Q}O_2$ and $\dot{Q}CO_2$, which are sustained throughout $\Phi1$, are therefore of similar proportions. As a result, R remains essentially stable at the prior resting level for this initial period of the exercise [119,205,218–220] (Fig. 108.9).

$\dot{V}E$ exhibits a similarly abrupt, sustained increase in $\Phi1$ [119,205,218–220] (Fig. 108.9). This response is in rather close proportion to the gas exchange responses and accounts for the relative stability of $P_{ET}O_2$ and $P_{ET}CO_2$ (Fig. 108.9). (Transient hypocapnia may occasionally occur at exercise onset, especially for exercise performed on the treadmill, but is not typical when adequate care is taken to acclimate subjects to the laboratory conditions prior to testing.) By extension, $\dot{V}A$ must also therefore increase in proportion to $\dot{Q}T$ [38,122,134].

Phase Two. The increase in $\dot{Q}M$ continues beyond $\Phi1$, approaching the new steady state within about 2–3 min and along a curvilinear trajectory that can be well described by an exponential process [197,221] (Fig. 108.10A). HR and cardiac output have rather similar profiles of response (Fig. 108.10B) [44,112,119,228].

In $\Phi2$, the pulmonary gas exchange mechanisms receive the additional challenge of the exercise-induced changes in mixed venous blood composition as a result of the contribution from the "exercising" muscles. This initiates the subsequent responses in $\dot{V}O_2$ and $\dot{V}CO_2$; these are normally characterized as monoexponential [86,119,133,220] (Fig. 108.11). However, in contrast to those in $\Phi1$, the response profiles of $\dot{V}O_2$ and $\dot{V}CO_2$ in $\Phi2$ are appreciably different (Fig. 108.11). In other words, $\dot{V}O_2$ increases to its steady state with a time constant ($\tau\dot{V}O_2$) of some 30–40 s, whereas $\dot{V}CO_2$ responds more slowly, with $\tau\dot{V}CO_2$ being typically 50–60 s.

The slower $\Phi2$ $\dot{V}CO_2$ response has been argued to reflect the influence of the intervening body CO_2 stores (see [218,219] for discussion). Thus although $\dot{Q}CO_2$ is likely to respond with a time course very similar to that of muscle $\dot{Q}O_2$, some of the metabolically produced CO_2 is taken up into the muscle CO_2 stores during the transient [62, 93,202,213,219].

Ventilation changes in $\Phi2$ with a time course similar to, although slightly slower than that of $\dot{V}CO_2$: the time constant ($\tau\dot{V}E$) is approximately 55–65 s [86,119,133,220]. The $\Phi2$ $\dot{V}E$ changes are therefore appreciably slower than are those of $\dot{V}O_2$ (Fig. 108.11). As a consequence, alveolar and arterial PO_2 fall transiently in $\Phi2$ [139,230], i.e., the proportional change in $\dot{V}E$ does not "keep up" with the proportional change in $\dot{V}O_2$. In contrast, owing to the relatively small kinetic dissociation between $\dot{V}E$ and $\dot{V}CO_2$, the corresponding transient increase of $PaCO_2$ is small and often not discernible (see [219] for discussion).

Important features of the relationship between $\dot{V}E$ and pulmonary gas exchange emerge from the $\Phi2$ response profiles. For example, the $\Phi2$ $\dot{V}E$ response dynamics are highly correlated with those of CO_2 exchange at the lung, rather than with muscle $\dot{Q}CO_2$. This has been demonstrated for a wide range of work-rate forcing functions, i.e., not only for constant-load (or *square-wave*) forcings (Fig. 108.11), but also for sinusoids (Fig. 108.12), impulses and ramps (see Chap. 1) [27,86,135,218,219].

Other evidence comes from studies in which the body CO_2 stores are deliberately manipulated to change the temporal relationship between the tissue CO_2 production rate and the pulmonary CO_2 exchange rate [86,218,219]. For example, exercise leads to an increase in the muscle CO_2 stores as a consequence of a proportion of the tissue CO_2 production being taken up into the stores. This reduces the tissue CO_2 capacitance. As a result, the $\Phi2$ $\dot{V}CO_2$ response develops more rapidly (i.e., $\tau\dot{V}CO_2$ is reduced) when a given work-rate increment is imposed from a baseline of moderate exercise rather than rest. In this case, less metabolically produced CO_2 is "diverted" into the CO_2 stores during the transient. Predictably, the $\Phi2$ $\dot{V}E$ kinetics are similarly speeded to those of $\dot{V}CO_2$, i.e., the kinetic proportionality between $\dot{V}E$ and $\dot{V}CO_2$ is maintained.

On the other hand, depletion of the CO_2 stores prior to exercise onset by a bout of controlled hyperventilation leads to a slowing of the $\Phi2$ $\dot{V}CO_2$ response to the exercise [201]. This reflects the fact that a greater-than-normal proportion of the metabolic CO_2 produced by the exercise is

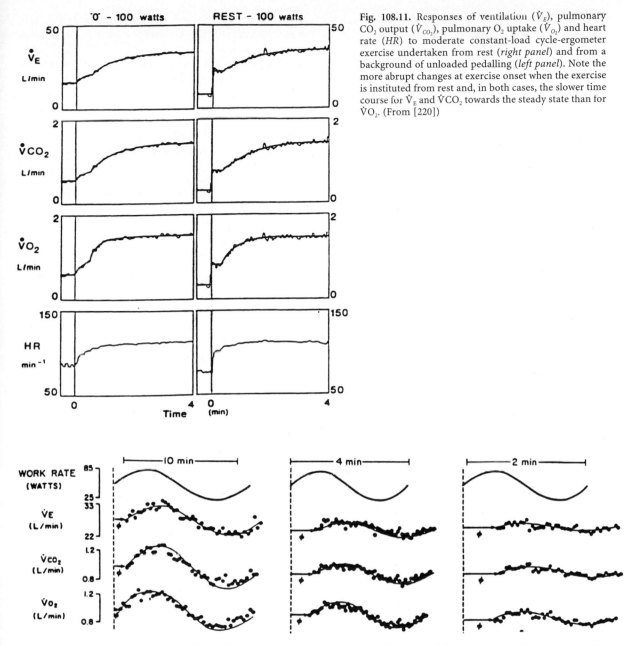

Fig. 108.11. Responses of ventilation ($\dot{V}_E$), pulmonary CO_2 output ($\dot{V}_{CO_2}$), pulmonary O_2 uptake ($\dot{V}_{O_2}$) and heart rate (*HR*) to moderate constant-load cycle-ergometer exercise undertaken from rest (*right panel*) and from a background of unloaded pedalling (*left panel*). Note the more abrupt changes at exercise onset when the exercise is instituted from rest and, in both cases, the slower time course for $\dot{V}_E$ and $\dot{V}CO_2$ towards the steady state than for $\dot{V}O_2$. (From [220])

Fig. 108.12. Responses of ventilation ($\dot{V}_E$), pulmonary CO_2 output ($\dot{V}_{CO_2}$) and pulmonary O_2 uptake ($\dot{V}_{O_2}$) to moderate sinusoidal cycle-ergometer exercise, in which the work rate sinusoids are of constant amplitude but differing periods. (See text for further discussion. From [24a])

diverted into the stores to replenish them. Predictably, the $\Phi 2$ $\dot{V}_E$ kinetics are also slowed.

Phase Three. From the preceding description, it is clear that the time taken to attain a new steady state depends on the variable under consideration. However, when all responses have stabilized, it is generally agreed that the arterial blood gas tensions and pH are once again maintained at, or close to, their resting levels [213,219]. As is the case for $\Phi 2$, the steady state reponses of $\dot{V}_E$ are more closely correlated with those of $\dot{V}CO_2$ than with $\dot{V}O_2$ [90,185,204].

This point is also illustrated by the influence of training and dietary manipulation on the $\Phi 3$ $\dot{V}_E$ response. For example, the larger volumes of CO_2 evolved for a given task when carbohydrate is the preferred substrate undergoing catabolism (see Sect. 108.2) are associated with a higher $\dot{V}_E$ compared to normal [52,92,166]. When fat is preferentially metabolized, as occurs after a period of physical training for example, the smaller demand for pulmonary CO_2 clearance (see Sect. 108.2) is accompanied by a correspondingly smaller ventilatory response [91]. In other words, $\dot{V}_E$ responds in proportion to $\dot{V}CO_2$. In contrast, when the changes in $\dot{V}_E$ are considered with respect to $\dot{V}O_2$, no such consistent relationship can be discerned.

108.3.3 Static Exercise

When the exercise modality includes a significant static component, the profile of cardiorespiratory response can be markedly different from those described above for dynamic exercise. First, no external work or power is generated during the exercise. It is most usual to quantitate the force generated during static exercise in terms of an individual's *maximum voluntary contraction* (MVC) for a given task, i.e., relative to the maximum force that an individual can generate volitionally with those particular muscles (see Chap. 45).

The sustained muscular effort of a static maneuver markedly impairs muscle perfusion [10,117,160]. The enlargement of the contracted muscle fibers causes intramuscular pressure to increase [169], and blood flow into and through the muscle is therefore impeded, despite increased arterial pressure (see below). With handgrip exercise, for example, $\dot{Q}_M$ increases progressively with increasing force of contraction up to about 30% of MVC [117]. With further increases in contractile force, however, $\dot{Q}_M$ falls off progressively and approaches zero at about 70% MVC. As higher intramuscular pressures can be generated by larger muscle groups (e.g., the quadriceps), $\dot{Q}_M$ becomes compromised at considerably lower forces (e.g., 10% of MVC

or less) [175,212]. This impaired perfusion leads to a progressive accumulation of exercise-related metabolites within the muscle interstitium [165] (see Chap. 45). Indeed, this is thought to be the primary cause of the postexercise *hyperemia* that follows a bout of static exercise [118,160,175,212].

Although the increase in $\dot{V}O_2$ tends to be relatively small during isometric exercise (blood from the contracting units contributes little to the venous return), the HR, cardiac output, and arterial blood pressure responses for a given increment in $\dot{V}O_2$ are typically exaggerated compared to those associated with dynamic exercise at the same $\dot{V}O_2$ [5,117,160,173,216] (Fig. 108.13). These responses are rapid in onset and are proportional to the force of contraction. It is of interest to examine the SV response. Despite cardiac output being clearly increased, SV typically is lower than resting, especially at higher forces of contraction (Fig. 108.13). This is the consequence of the following (see Chap. 89):

* Impaired ventricular filling, because of the compromised muscle perfusion and the absence of any contribution from the muscle pump, i.e., there are no rhythmic contractions of the muscles
* The high arterial blood pressures, which raise the "afterload" against which the heart has to operate.

The ventilatory response to static exercise also differs markedly from that of dynamic exercise. There is a range

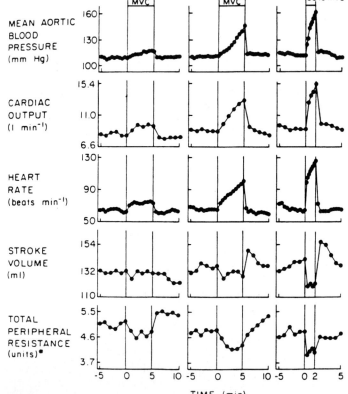

Fig. 108.13. Responses of mean aortic blood flow, cardiac output, heart rate, stroke volume and total peripheral resistance to bouts of constant-load static forearm exercise (handgrip) at 10%, 20%, and 50% of maximum voluntary contraction (*MVC*). (From [160] and modified from [117a])

of moderate static effort within which $\dot{V}E$ increases essentially in proportion to $\dot{V}CO_2$, with consequent stability of alveolar and presumably arterial PCO_2 [216]. However, with greater efforts, hyperventilation invariably occurs, often with marked hypocapnia ensuing [53, 144,216,224,225]. Following the static contractions, there is a significant metabolic acidemia as a result of the washout of the metabolites "trapped" in the muscle during the exercise; this results in ventilatory stimulation and hypocapnia (i.e., a respiratory alkalosis) during the recovery phase.

108.4 System Control Mechanisms

The precise nature of the mechanisms that control the cardiovascular and ventilatory responses to exercise remains the topic of debate, especially with respect to the integration of these mechanisms. Although the control systems are usually considered independently, they must function in a linked, or coordinated, manner. We shall therefore, in this section, focus on the interrelation between the ventilatory and cardiovascular systems, from mechanical, neurogenic, baroreflex, chemoreflex, and central-circulatory control perspectives (see Chaps. 99, 106, 107).

108.4.1 Mechanical Aspects

The ventilatory response to exercise itself leads to mechanical influences on venous return and stroke volume. There is, for example, a respiratory-related "cycling" of intrapleural pressure (P_{pl}) which is reflected in the blood pressure of the compliant thoracic veins [137]. At rest, P_{pl} may fall from an end-expiratory value of approximately – 4 cm H_2O to approximately – 8 cm H_2O at the end of inspiration as the thorax expands during inspiration (Chap. 100). Abdominal pressure, however, increases as a result of the descent of the diaphragm. This assists venous return to the thorax because:

- The transdiaphragmatic pressure gradient is increased
- The inferior vena cava is shortened and its volume is consequently reduced.

The reverse occurs during expiration.

This respiratory-related fluctuation in venous return is amplified during exercise: there are larger swings in P_{pl}. Furthermore, the *mean* P_{pl} throughout the respiratory cycle will be more negative than at rest, because the increased tidal volume encroaches more on the inspiratory reserve volume than the expiratory reserve volume [229]. This enhances the venous return to the thorax and the right heart over the breath as a whole.

Interestingly, however, the magnitude of the respiratory fluctuation in venous return tends to be held in check by the liver [137]. This is because the hepatic venous outflow is impeded during inspiration by the mechanical compression of the hepatic veins that results from the descent of the diaphragm. Not only does this reduce overall venous return, but the liver becomes increasingly distended during inspiration as a result of the continued inflow of blood. During the subsequent expiration, hepatic outflow is boosted by:

- The restoration of hepatic venous patency as the diaphragm now ascends
- The recoil of the distended liver.

The "hepatic" pump therefore operates out of phase with the "respiratory" pump. Consequently, it tends to smooth out the fluctuations of venous return. How the hepatic pump influences the *overall* level of venous return during exercise, i.e., with the larger vertical excursions of the diaphragm, is less clear, however.

The increased venous return that results from these respiratory-mechanical influences should lead to augmentation of right ventricular output, through the Frank-Starling effect (see Chaps. 89, 99). But what is the effect on the left ventricle? Interestingly, this has been shown to *decrease* rather than to increase during inspiration [71] as a consequence of:

- The pulmonary vascular compliance being considerably greater than that of the systemic arteries
- The limitation of left ventricular expansion when the right ventricle is already enlarged as a result of the relatively rigid pericardium [71].

108.4.2 Central Neural Control

The immediacy of the cardiovascular and ventilatory responses at exercise onset has been regarded as an expression of "*neurogenic*" control. Early evidence in support of this control arising within the central nervous system (CNS) was provided by Krogh and Lindhard [110,111]. These investigators demonstrated that the tachycardic response at the onset of exercise onset developed more rapidly when exercise was initiated voluntarily than when it was induced by electrical stimulation. This was not the case for ventilation, however. Both responses developed with similar rapidity. These discrepancies provided a foretaste of the controversy that continues to surround the putative mechanisms of neurogenic cardiorespiratory control during exercise in humans.

108.4.2.1 Central Command

The cortical somatomotor command that leads to locomotion has been argued to influence brain stem respiratory and cardiovascular control "centers": the "central command" hypothesis. As a result, neurally mediated ventilatory and cardiovascular responses are elicited which are proportional to the magnitude of the central command.

Proponents of the central command hypothesis argue that only a modest feedback component from humoral or mechanical sources related to muscle contraction is therefore required to ensure that the ventilatory and cardiovascular responses are appropriate to the metabolic demands of normal volitional exercise.

Several central command schemes have been proposed, in which descending motor influences interact with the medullary ventilatory and cardiovascular integrating regions. These include the following:

- A direct "irradiation" of the cortical somatomotor drive itself [110,111]
- A specific cardiovascular (and perhaps respiratory) drive that originates in the certebral cortex with the somatomotor drive, but subsequently operates independently of it [80]
- Projections from subcortical regions such as the paraventricular locomotor region and the fields of Forel in the hypothalamus [59,177,195], the cerebellum [50], and the fastigial nucleus [49,51].

The major sources of evidence in support of central neurogenesis stem from: (a) animal studies with the normal conscious drive to exercise being absent, with discrete focal CNS stimulation being utilized to activate descending pathways involved in the transmission of central command to cardiovascular and ventilatory control regions; and (b) experiments in exercising humans which attempt to dissociate the magnitude of central command from its subsequent motor outcome, i.e., exercise.

Focal Central Nervous System Stimulation and Ablation. Cortical mechanisms clearly can evoke locomotor, ventilatory, and cardiovascular responses. There is, however, a substantial body of evidence from experiments on animals which suggests that the cerebral cortex is not the sole origin of this component of the control. Decorticate animals, for example, can perform spontaneous loco-

motion that is accompanied by increases in both respiratory and cardiovascular activity (Fig. 108.14) [47,59,60,168].

Furthermore, focal electrical and chemical stimulation in posterior hypothalamic regions such as the paraventricular locomotor region and the H_2 field of Forel leads to both rapid hyperpnea and cardiovascular responses, including tachycardia and arterial hypertension – and in conjunction with locomotor activity [59,79, 177,195]. There is also evidence of efferent neural connections from these hypothalamic sites to medullary sites of cardiovascular and respiratory control, such as the nucleus of the solitary tract (NTS), the nucleus ambiguus (NA), and the dorsal vagal nucleus (DVN) (see Chaps. 105–107) [83,125,167,179,181].

These evoked responses appear not to depend upon muscular contraction and increased metabolic rate; they are essentially unaffected by muscle paralysis [59,128,177]. The demonstration of neural projections from the NTS to hypothalamic "locomotor" regions is consistent with central command modulation by circulatory- and respiratory-related neuronal feedback [12,123]. However, there is some debate regarding the *extent* to which the hypothalamus is involved in the normal cardiovascular and ventilatory responses to exercise. In some instances of hypothalamic lesioning, the normal exercise response profile was reported to be abolished [80,168,178], as was also the case for the "fictive" locomotory model reported by Eldridge et al. [59]. In other studies, however, removal of hypothalamic influences had no clear effect [80,140].

The coarse resemblance between these hypothalamically induced responses and those seen during exercise in the awake animal suggests to some investigators that central neurogenesis is not only an essential, but a major component of ventilatory and cardiovascular control during dynamic exercise in humans. However, more careful inspection reveals some departures. For example, the ventilatory responses evoked by direct (i.e., electrical or chemical) activation of these CNS mechanisms are typi-

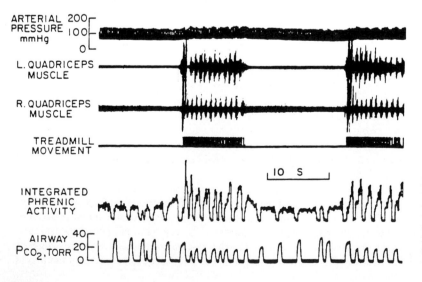

Fig. 108.14. Responses of arterial blood pressure, respiratory activity (monitored as the integrated phrenic electromyogram) and airway P_{CO_2} during two episodes of spontaneous locomotion in a decorticate cat (indicated by the rhythmic electromyographic activity in the right and left quadriceps muscles). (From [59])

cally accompanied by a rapid and marked hypocapnia (Fig. 108.14) [47,60]. This is not characteristic of the normal response to moderate exercise in humans (see Sect. 108.3.2), although it does appear to be the case for many experimental animals.

Manipulation of Central Command. Augmenting or diminishing the central command required to perform a particular task has also been used as a means for examining its role in ventilatory and cardiovascular control during exercise. For example, selective stimulation of muscle spindles in humans (via high-frequency vibration applied to the contracting muscle through the overlying skin) [66] can reduce the central command required to maintain a given force of contraction (see Chaps. 49, 50). In other words, the reflex facilitation of motor neurons to the exercised muscle "takes over" a component of force generation from central command sources. Under these conditions, the $\dot{V}E$, HR, and blood pressure responses to the exercise task were all less than under control conditions. Conversely, when the stimulus was applied to an antagonist muscle (i.e., a greater effort now being required to sustain the force), the cardiorespiratory responses were exaggerated.

Differential blockade of muscle afferent fibers with local anesthetic leads to impairment of the conducting properties of the smaller-diameter *afferent* fibers, the larger afferents being relatively unaffected. Because this procedure also blocks a proportion of the smaller-diameter *motor* fibers, muscle weakness results and so a greater effort is needed to maintain the force of contraction. Human subjects have performed isometric exercise before and during such blockade [64,162]. The exercise-induced increases in HR and arterial blood pressure were both larger than normal. This is consistent with the view that the exaggerated central command resulted in exaggerated responses.

Partial neuromuscular blockade has also been reported to exaggerate the increases in $\dot{V}E$, HR, and arterial blood pressure during isometric exercise [6,80,115,193]. However, as Rowell [160] has pointed out, these differences tend to disappear if allowance is made for the reduction in the maximal isometric force that can be exerted volitionally when the involved muscles are partially "blocked".

Another source of evidence that has been cited in support of central command [160] comes from studies on patients with sensory neuropathies in which motor function is not entirely compromised. Although these studies have been performed on only a rather limited number of patients, it has nonetheless been possible to include some individuals in whom only one limb has been affected; this therefore allows the patient's other, unaffected limb to serve as the control. During both isometric and dynamic exercise, these patients manifest relatively normal increases in both HR and arterial blood pressure [3,53,118]. This suggests that somatic afferent traffic from the exercising limbs was not necessary for these responses. However, the consequences for the *motor* deficit also need to be considered.

Because of this, it is reasonable to expect that a greater effort would be required to perform a given task with a compromised limb, i.e., the central command would be increased. The associated ventilatory and cardiovascular responses should therefore be exaggerated, but this appears not to be the case.

Evidence Against Central Command. There is evidence, albeit indirect, from several sources that appears to be incompatible with central command (or limb afferent neurogenic control mechanisms) preferentially driving cardiorespiratory responses during exercise. For example, the magnitude of the Φ1 cardiorespiratory response to exercise does not appear to be a simple function of the number of motor units recruited to do the work nor of the magnitude of the central drives that "command" their recruitment. For example, the magnitude of the Φ1 $\dot{V}E$ and HR responses changes little over a relatively wide range of imposed work loads.

In addition, when an increment in work rate is imposed from a background of light exercise (i.e., with the limbs already moving), the Φ1 component of the hyperpnea and tachycardic response develop more slowly and are smaller (Fig. 108.11) [111,220]. This is despite the additional central command required to generate the increased force and the consequent increment in motor unit recruitment. A correspondingly slow initial hyperpnea has also been demonstrated when rest-to-work transitions are imposed in the supine position. It is hard to see how body position per se would so markedly reduce either central command or limb afferent neurogenic control mechanisms. Thus while neurogenesis as the mechanism mediating the Φ1 ventilatory response seems beyond serious challenge, the nature and functional organization of the signals themselves are by no means clearly understood.

The results of sinusoidal exercise (i.e., a "continuous" Φ2 condition) are also revealing (Fig. 108.12). For example, when the work rate is "forced" sinusoidally over a range of input frequencies, the mean-to-peak amplitude of the $\dot{V}E$ response decreases as a close linear function of the $\dot{V}CO_2$ amplitude (Fig. 108.12), extrapolating at (or very close to) the origin at high frequencies [27,135]. This means that the central command mechanism and muscle afferent drive which continue to drive the force-generating muscle units over the same amplitude range and the force-related feedback from the muscles either subserve a trivially small role in the Φ2 control of the exercise hyperpnea or that these mechanisms themselves exhibit slow neural dynamics which are, somehow, closely matched to the rate of *pulmonary* CO_2 exchange. This issue is discussed at greater length below (see Sect. 108.4.2.3).

Finally, when $\dot{V}CO_2$ is experimentally altered during the steady state of exercise (e.g., by dietary manipulation or physical training), then $\dot{V}E$ changes in proportion [90,185] (see Sect. 108.3.2). Such behavior is consistent with the close correlation that has been demonstrated between $\dot{V}E$ and pulmonary CO_2 exchange during exercise (see Sect. 108.3.2). This alteration in ventilatory response occurs *despite* the work rate and therefore presumably the magnitude of the central command (and muscle afferent drive) being unaltered.

A further means of dissociating a subject's intrinsic ventilatory drive from the motor task is to use a servo-assisted positive-pressure ventilator. This can be synchronized with the respiratory cycle to "take over" a proportion of the normal inspiratory flow [109,146]. If $\dot{V}E$ is dictated by the magnitude of the central command, then – despite the external assistance to ventilation imposed by the ventilator – the subject's intrinsic ventilatory drive would presumably remain unchanged and overall ventilation would be expected to increase, with a sustained hypocapnia resulting. This is *not* the case, however. Rather, there was a compensatory and proportional reduction in the subject's intrinsic ventilatory drive (Fig. 108.15).

Finally, were central command to be a necessary component of the ventilatory and cardiovascular control processes during exercise, then attenuated responses would be expected when exercise is performed in the absence of central command. This has been investigated in subjects who have suffered complete spinal cord transection and in whom dynamic exercise has been induced by direct electrical stimulation of the quadriceps muscles [1,21,219]. The magnitude and time course of the associated ventilatory, HR, and blood pressure responses have been reported to be essentially normal. These observations argue against a major obligatory involvement of central command in ventilatory and cardiovascular control during moderate exercise. Rather, they reinforce the notion of "redundancy" of cardiovascular and ventilatory control system drives during exercise.

108.4.2.2 Central Nervous System Cardiorespiratory "Coupling"

Despite the differences of opinion that surround the role of centrogenic drives in ventilatory and cardiovascular control during exercise, there are several striking parallels between the ventilatory and cardiovascular responses evoked by many of the experimental interventions described above. This is suggestive (to some investigators, at least) of significant interaction between the system controls (see Chap. 107). For example, the similarity between the $\dot{V}E$ and $\dot{Q}T$ response profiles during $\Phi1$ of constant-load exercise is suggestive of a link between the systems' responses. This behavior is certainly consistent with the activation of parallel descending drives to the medullary cardiovascular and respiratory control areas by, for example, central command. In addition, cardiorespiratory "cross-talk" within the brain stem and higher regions of the CNS could also impose a degree of cardiorespiratory interaction during exercise.

Medulla. The primary site to consider is the medulla (Chaps. 105–107). This contains concentrations of cells with a respiratory-related discharge within the NTS (i.e., the *dorsal respiratory group*) and also the NA and nucleus retroambigualis (NRA) (i.e., the *ventral respiratory group*). Similar neural aggregates manifest a cardiac discharge rhythm from the cardioinhibitory DVN and also bulbospinal premotor sympathetic neurons. The NTS is the terminus for afferents from a range of cardiorespiratory

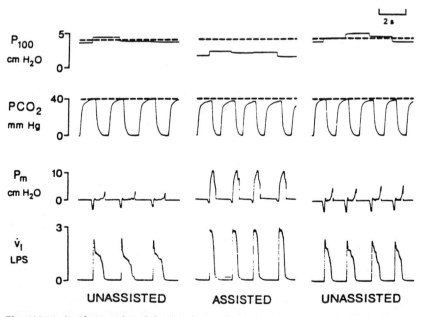

Fig. 108.15. Steady-state, breath-by-breath responses of inspiratory occlusion pressure (P_{100}) taken as an index of intrinsic ventilatory "drive", respired P_{CO_2}, mouth pressure (P_m) and inspiratory flow ($\dot{v}_I$) during moderate constant-load cycle-ergometer exercise prior to (*left panel*), during (*center panel*), and following (*right panel*) a bout of "assisted" breathing (3 cm H_2O/l per second). From [146])

receptors (see Chaps. 99, 105–107). These include the following:

- The arterial chemoreceptors, in the carotid and aortic bodies, which respond to arterial hypoxemia, increased $PaCO_2$, reduced pH_a, hyperkalemia, elevations in the circulating concentrations of catecholamines and adenosine, and increased plasma osmolality
- Pulmonary vagal mechanoreceptors, in particular the slowly adapting pulmonary "stretch" receptors that respond to airway distention
- Arterial baroreceptors, situated within the walls of the aortic arch and carotid sinus, which are stimulated by local distention resulting from increased vascular pressures.

Although the terminations of the projections from these receptors are, to a large extent, topographically discrete, there is also overlapping between different receptor types [31,42,98,181]. This raises the possibility of cardiorespiratory interactions *within* the NTS, both presynaptically and postsynaptically.

The NA, which is viewed as the final common integrating site for respiratory control, also contains cardioinhibitory vagal motoneurons [31,42,181]. These cells receive excitatory inputs from the arterial baroreceptors. However, they also appear to be modulated by the intrinsic brain stem respiratory rhythm. This inhibits their discharge during inspiration, thereby inducing tachycardia. The premotor sympathetic neurons also appear to be modulated by the intrinsic respiratory rhythm. Their activity has been shown to be greater in inspiration than in expiration [31,42,181].

Limbic System (see Chap. 18). Activation of regions within the amygdala can evoke an integrated pattern of cardiorespiratory and behavioral response that is known as the "defense" or "alerting" reaction [31,42,79,97,181]. This includes hyperpnea, increased responsiveness of peripheral respiratory chemoreflexes but the opposite for the arterial baroreflex, tachycardia, increased blood pressure, regional systemic vasoconstriction, and muscle vasodilatation. It has therefore been viewed as a preparatory response to "fight" or "flight." Other regions lower in the CNS, such as the posterior hypothalamus and the midbrain, are also involved in the generation of the defense reaction. Although there is a striking superficial similarity between these events and the profile of cardiorespiratory response at exercise onset, it is not clear whether the defense reaction is a preparatory manifestation of the normal cardiorespiratory control processes that operate during dynamic exercise.

Cerebral Cortex. It is has been known for many years that focal electrical stimulation at sites within the cerebral cortex, such as the motor cortex, the cingulate gyrus, and the posterior orbital gyrus, can lead to cardiovascular and respiratory stimulation [31,99]. The extent to which such re-

sponses contribute the normal cardiorespiratory control during exercise is uncertain, however.

108.4.2.3 Short-Term Potentiation

It has been proposed that exercise activates an intrinsic central neurogenic drive to ventilation – "short-term potentiation" or "reverberation" within brain stem respiratory control areas – that results in a slowly developing hyperpnea in response to the stimulus of exercise onset [56,60]. This mechanism has been well described (in the cat) in response to the abrupt cessation of afferent activity from several sources, such as limb muscle afferents, the carotid bodies, and the ventral medullary surface.

However, the $\dot{V}_E$ response appears to be much more rapid in response to the imposition of stimulation, i.e., the mechanism exhibits kinetic *asymmetry* (Fig. 108.16). This is despite the fact that the reverberatory component is symmetrical, as judged by Eldridge and Gill-Kumar's elegant "alternate breath" study [57]. In other words, the symmetry of the reverberatory component was masked by a direct component that rendered the overall efferent respiratory response asymmetrical. Such behavior is clearly at odds with the widely demonstrated symmetry of the monoexponential $\Phi 2$ $\dot{V}_E$ responses that are evident at the onset and cessation of moderate exercise in humans (see [218,219] for discussion).

108.4.3 Muscle Reflex Control

Skeletal muscles contain a wide range of receptors and free nerve endings (see Chaps. 45, 49), the afferent projections of which have been proposed to contribute to the control of cardiovascular and ventilatory responses during exercise.

Direct Muscle Stimulation. When the muscles of experimental animals are caused to contract by stimulation of ventral spinal roots, motor nerves, or the muscles themselves, $\dot{V}_E$, HR, and arterial blood pressure all increase [33,34,101,129,131,153,189]. That these responses originate in the exercising muscles has been shown by their absence following section of the dorsal spinal roots [34,129,131].

Differential Stimulation and Blockade of Muscle Afferent Projections. Electrophysiological studies in animals have indicated that electrical stimulation of small-diameter group III and IV afferents (but not of the larger group I and II afferents) leads to increases in $\dot{V}_E$, HR, and arterial blood pressure (and presumably cardiac output) [60,132,160]. These are abolished by differential blockade of these small-diameter fibers. A range of applied stimuli which can broadly be classified as "mechanical" have been demonstrated to activate a subpopulation of the small-diameter afferents. These include local mechanical distortion and increased intramuscular pressure [100,103,104,132,182]. A

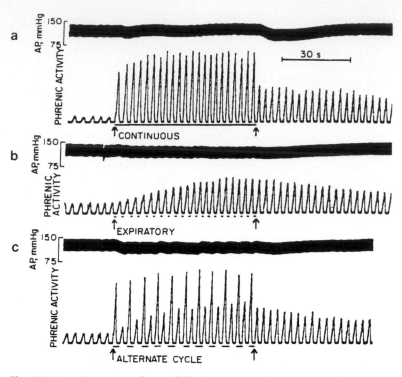

Fig. 108.16a–c. Responses of arterial blood pressure and integrated phrenic activity to electrical stimulation of the carotid sinus nerve in a paralyzed, vagotomized cat with alveolar P_{CO_2} maintained constant by artificial ventilation. **a** Continuous stimulation evokes an immediate and rapid "direct" increase in activity and a slower gradual "potentiation". On cessation of the stimulation, there is an immediate drop in activity, followed by a slow decline back towards baseline. **b** Confining the stimulation to the expiratory phase evokes only the slow "potentiating" component which, again, declines slowly during the subsequent recovery. **c** Stimulation on alternate breaths evokes the "direct" component on the stimulated breaths, while the slower "potentiation" can be seen developing in the unstimulated breaths; when the stimulation ceases, the "direct" component is lost immediately (as in **a**), while the "potentiation" declines slowly (as in **b**). (From [57])

second subpopulation of small-diameter afferents can be stimulated by "humoral" mediators, including increases in intramuscular levels of H^+, K^+, bradykinin, and possibly arachidonic acid derivatives [100,103,104, 132,155,156,163,182,183,188,190] (see Chap. 45).

Vascular Occlusion. When pneumatic cuffs placed around the thighs are inflated to suprasystolic pressures at the end of exercise, the fall in arterial blood pressure that normally occurs at this time is prevented [2,64,161]. This observation is consistent with the sustained activation of chemosensitive muscle afferents *after* the cessation of the work (i.e., consequent to "trapping" of exercise-induced metabolites in the tissue). It should also be pointed out that such an observation is hard to reconcile with the involvement of a central command mechanism (see Sect. 108.4.2). In other words, as soon as exercise ceases, there is presumably no longer a central command nor, for that matter, any mechanical feedback from the previously exercising muscles.

Interestingly, the recovery of $\dot{V}E$ contrasts with the blood pressure response under these conditions. It is, in fact, markedly *more* rapid than normal in both humans and experimental animals, despite the local accumulation of exercise-induced metabolites [75,89].

Evidence Against Obligatory Peripheral Neurogenesis. Although cardiorespiratory drives can originate within contracting muscles, the primary issue to be resolved is whether these represent an obligatory or necessary component of the control processes during normal volitional exercise. There is evidence that suggests this is *not* the case. For example, some studies have demonstrated that the ventilatory and cardiovascular responses to electrically induced exercise appear to be essentially unaffected in animals when sensory feedback from the exercising muscles is interrupted by dorsal spinal root section or spinal cord transection [35,113,211].

In addition, the cardiorespiratory responses to dynamic exercise were essentially normal in normal humans with muscle sensory blockade induced by epidural anesthesia and in patients with sensory neuropathies (see above), i.e., despite the absence of functioning afferent projections from the exercising limbs. Removing the influence of somatic limb afferents does not seem to impair, to any appreciable extent, the ventilatory and cardiovascular response to dynamic exercise in humans, at least over the range of metabolic rates achieved in these studies.

Finally, as described above, the instrinsic ventilatory drive can be clearly dissociated from work rate during moderate exercise by means of "assisted" inspiration [109,146] (see

Sect. 108.4.2). Despite neither central command nor mechanical feedback from the exercising muscles being altered, the subjects' ventilation was *not* significantly affected as a result of the intrinsic ventilatory drive being proportionally reduced (Fig. 108.15).

108.4.4 Arterial Baroreflex Control

Mean arterial blood pressure increases with work rate for dynamic exercise [7,14,29,157,160,194], associated with a reduction in baroreflex sensitivity [40,81,196]. This presumably offsets a component of the primary inhibitory action of the arterial hypertension on the cardiovascular response to exercise.

Stimulation of arterial baroreflexes by relatively small increases in arterial blood pressure leads to a reduction in $\dot{V}_E$ in the anesthetized dog [23]. This raises the possibility that the hyperpneic response to exercise in humans might, in fact, be constrained to some degree by this mechanism.

The ventilatory control system can also modulate arterial baroreflex behavior. As mentioned earlier, the intrinsic respiratory rhythm in the medulla can impose a respiratory modulation on cardiovascular motor outputs [31,42]. HR and sympathetic vasoconstrictor tone both increase during inspiration and then decrease during expiration. Activation of the slowly adapting pulmonary stretch receptors by lung inflation also leads to tachycardia. This is associated with a reduction in sympathetic vasomotor drive, the result of which lowers peripheral vascular resistance. However, the exact site or sites within the medulla where this neuronal interaction occurs have not been established (see [98] for discussion). One consequence of the reduced sympathetic vasoconstrictor tone induced by lung inflation, at least in the dog, is a redistribution of the cardiac output that favors skeletal muscle beds [42], an effect that would be beneficial were it also to occur during normal volitional exercise in humans.

It should be recalled that the reflex effects of lung inflation on breathing appear to be less potent for the awake human than in laboratory animals [223]. Whether this also applies to the reflex modulation of cardiovascular activity is not clear. Finally, it should be emphasized that these central and pulmonary reflex modulations appear to have most, if not all, of their cardiovascular effects on the inspiratory and expiratory components *within* the breath, rather than on the overall mean levels of cardiovascular response *across* breaths. It is therefore difficult to envisage a significant modulation of the cardiovascular responses during exercise from these sources.

108.4.5 Central Chemoreflex Control

Arterial blood gas and acid-base status is remarkably well preserved during moderate dynamic exercise in humans (see Sect. 108.3.2). This is the consequence of the ventilatory response being closely correlated with that of pulmonary CO_2 exchange. Some investigators have considered this to be inconsistent with neurogenic mediation, either by central command mechanisms or muscle mechanoreflex drives, particularly in view of the evidence presented earlier indicating that it is possible to induce changes in the magnitude and/or kinetics of the exercise hyperpnea in the absence of corresponding changes in either of these presumed neurogenic drives (see above).

The dilemma remains, however, as to how one might envisage conventional chemoreceptors (i.e., peripheral and central) mediating the close matching between $\dot{V}_E$ and $\dot{V}CO_2$ in the absence (at least in the steady state) of any systematic sustained "error signal" in the arterial blood. This has prompted a search for a CO_2-linked mechanism that operates independently of mean $PaCO_2$.

Several mechanisms have been proposed (see [60,219] for discussion):

- Chemoreceptors located in the central venous circulation that might sense alterations in mixed venous blood-gas and acid-base status
- Chemoceptors within the lungs themselves
- Pathways responsive to pulmonary "CO_2 flux," the product of $\dot{Q}_P$ and the mixed venous CO_2 content
- Respiratory-related "oscillations" of $PaCO_2$ (see below).

While these proposals have seemed attractive to some investigators, they have not been supported by careful physiological investigation (see [219] for discussion). We shall therefore consider the extent to which central and peripheral (i.e., arterial) chemoreception (see Chap. 106) might contribute to ventilatory control during exercise, despite the absence of putative error signals in arterial blood gas tensions and pH, and their role in modulation of the cardiovascular responses to exercise.

Investigations of central chemoreflex function in humans are not straightforward, because of the inaccessibility of the sensing elements. Approaches are largely confined to utilizing "selective" central chemoreflex stimulation, such as CO_2 inhalation during hyperoxia. The hyperoxic background is essential in such studies to ensure suppression of peripheral chemosensitivity (Chap. 106). This provides a means of assessing whether central chemosensitivity is increased during exercise. Investigation of patients with impaired central chemosensitivity, as occurs in congenital central hypoventilation syndromes (CCHS), has also provided important insights [142,170].

The onset of the slower component of the exercise hyperpnea (i.e., the $\Phi2$ $\dot{V}_E$ response) begins with a time delay that is consistent with the limb-to-"lung" transit delay [220], suggestive of "downstream" chemoreceptor mediation. While this "delay" is prolonged with hyperoxic inspirates [200], it seems unlikely that there is any significant involvement of the central chemoreceptors in this response. Thus CCHS patients demonstrate essentially normal $\Phi2$ $\dot{V}_E$ kinetics despite the absence of CO_2 chemosensitivity (Fig. 108.17) [170].

The central chemoreceptors also appear not to play an essential role in $\dot{V}_E$ control for $\Phi3$ of moderate exercise.

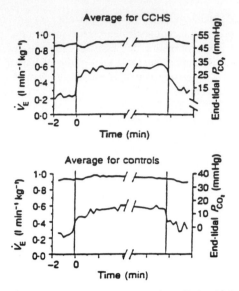

Fig. 108.17. Average responses of ventilation ($\dot{V}_E$) and end-tidal P_{CO_2} to moderate constant-load treadmill exercise in a group of subjects with "congenital central hypoventilation syndromes" (*CCHS*) (*upper panel*) and in a matched control group (*lower panel*). (From [170])

First, there seems no discernible chemical stimulus: the pH of cerebrospinal fluid (CSF) remains relatively stable during the steady state of moderate exercise [17,116]. Also, CSF [K$^+$] does not increase [43], despite significant increases in arterial [K$^+$] (see [141] for discussion). Furthermore, CCHS patients can have reasonably normal ventilatory responses to exercise which regulate PCO$_2$ close to resting levels (although these are often elevated in these patients) [65,124,142,170]. Finally, whether central CO$_2$ responsiveness is normally increased during exercise is not entirely clear. At best, only a relatively small increase has been reported, relative to rest [28,145].

Above θ_L, where the ventilatory compensation for the lactic acidosis leads to a respiratory alkalosis in the CSF [17], the role that might best be envisioned for the central chemoreceptors is that of *constraint* of the hyperpnea. The alkalotic condition in the CSF would be expected to reduce ongoing central chemoreceptor activity and also to stimulate efferent projections to the carotid chemoreceptors. These have been demonstrated to inhibit afferent chemosensory discharge [127].

It has been shown that there is a slow, but systematic restoration of arterial pH back towards normal during prolonged exercise with a constant and high level of arterial blood lactate, even when high O$_2$ fractions were inhaled to suppress peripheral chemosensitivity [148]. This might reflect a slow "leak" of H$^+$ ions into the CSF from the blood [58,186] or, possibly, a slow central chemoreflex response to the transiently elevated PaCO$_2$ that was a consistent finding of the study. Under these less acute conditions, therefore, this mechanism could "take over" a component of respiratory compensation, in the absence of functioning peripheral chemoreceptors.

Little is known about the role of central chemosensory mechanisms in cardiovascular control during exercise. There is evidence from anesthetized animals that these mechanisms can induce cardiocirculatory stimulation: focal stimulation has been shown to increase arterial blood pressure and cardiac sympathetic activity, while application of local anesthetic has the opposite effect (see [22,42] for discussion). This finding is not universal, however, as stimulation at certain sites within the ventral medullary surface leads to arterial hypotension. Finally, there is some debate about whether single central chemosensory neurons can mediate both respiratory and circulatory effects or whether the neuronal substrates for these responses are functionally discrete.

108.4.6 Arterial Chemoreflex Control

Considerations of the peripheral chemoreflex contribution to ventilatory control during exercise in humans are typically confined to carotid body function. This is based on evidence from several sources (see [84,215] for discussion). For example, subjects having previously undergone bilateral resection of their carotid bodies (CBR) show no ventilatory increase in response to isocapnic hypoxic exposures.

Three basic approaches have been used to estimate the proportional role of the carotid bodies:

- To provide a selective carotid body stimulation
- To remove an ongoing source of carotid body stimulation
- To compare the normal ventilatory response to exercise with that from subjects who have no carotid chemosensitivity, such as subjects who have had both carotid bodies surgically resected (CBR) or normal subjects who are breathing high concentrations of O$_2$.

It seems unlikely that the carotid bodies are involved in the $\Phi1$ $\dot{V}_E$ response to exercise. Not only is there the problem of transit delay, but also the $\Phi1$ response is typically unaffected when hypoxic (i.e., to increase carotid chemosensitivity) or hyperoxic gas mixtures are breathed [39,45,215] or in CBR subjects [207] (decreasing and abolishing, respectively, carotid chemosensitivity).

In contrast, the carotid bodies play a relatively prominent role in $\Phi2$ ventilatory control (see [215,219] for discussion). For example, increasing carotid body sensitivity by hypoxia or metabolic acidosis (induced by ammonium chloride ingestion) results in a faster time course of response for $\dot{V}_E$, both in absolute terms and relative to CO$_2$ output (Fig. 108.18). Procedures which reduce or abolish carotid chemosensitivity such as hyperoxia, metabolic alkalosis (induced by NaHCO$_3$ ingestion), intravenous infusion of dopamine, and, perhaps most specifically, CBR all lead to an appreciable slowing of the $\Phi2$ $\dot{V}_E$ kinetics (Fig. 108.18).

These procedures have little effect per se on the time course of $\dot{V}_E$ in $\Phi2$. Consequently, when the carotid bodies

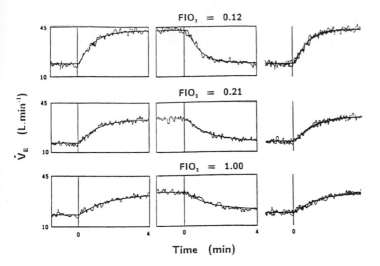

Fig. 108.18. Influence of inhaled O_2 fraction (F_{IO_2}) on the ventilatory ($\dot{V}_E$) response to moderate constant-load cycle-ergometer exercise instituted from unloaded pedaling: *top panel*, $F_{IO_2} = 0.12$; *middle panel*, $F_{IO_2} = 0.21$; *bottom panel*, $F_{IO_2} = 1.00$. *Left panel*, responses at exercise on-transient, with the best-fit mono-exponential superimposed. *Center panel*, responses at exercise off-transient (with best-fit mono-exponential superimposed). *Right panel*, symmetry of the on- and off-transient responses is demonstrated by superimposing the off-transient response (which has been reversed) on the corresponding on-transient response. (From [219])

are suppressed, then both the transient decrease in PaO_2 and the transient increase in $PaCO_2$ during $\Phi 2$ (see Sect. 108.3.2) are likely to be exaggerated (i.e., $\tau\dot{V}_E/\tau\dot{V}CO_2$ and $\tau\dot{V}_E/\tau\dot{V}CO_2$ are lower than normal) [67,215,219]. The carotid bodies therefore appear important in modulating the dynamics of the $\dot{V}_E$ response relative to those of $\dot{V}O_2$ and $\dot{V}CO_2$, and hence play a role in the "tightness" with which PaO_2, $PaCO_2$, and pH_a are regulated throughout the non-steady-state phase of muscular exercise.

The carotid bodies do provide a component of the $\Phi 3$ $\dot{V}_E$ control [39,45,208,215,219]. This has been described as a "fine tuning" for the control system. The carotid body contribution to the $\Phi 3$ hyperpnea has been estimated using the *Dejours test* [45]: this provides a means for high inspired O_2 fractions to be inhaled surreptitiously and hence to silence abruptly any carotid body drive to breathe (Fig. 108.19). Quantitatively, the carotid bodies normally appear to account for approximately 20% of the steady state hyperpnea (see [215] for discussion), although there is some concern that the Dejours test may underestimate the carotid body component of the exercise hyperpnea [199,215]. CBR subjects have no ventilatory decrease during the Dejours test (Fig. 108.19) [215,222].

It is unclear, however, what stimuli underlie this carotid body contribution. As mentioned earlier, respiratory-related $PaCO_2$ "oscillations" have been considered. Owing to the phasic lung ventilation, the pulmonary venous and systemic arterial PCO_2 and [H^+] can be demonstrated to oscillate with a respiratory rhythm, i.e., increasing during expiration and decreasing during inspiration at a rate proportional to the level of $\dot{V}CO_2$ (Fig. 108.20). This oscillation, however, is not likely to be perfectly smooth, as the signal will be "digitized" by mixing in the left heart [227], but the normally high ratio of HR to respiratory rate

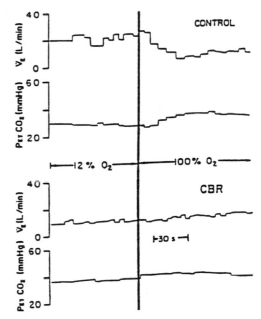

Fig. 108.19. *Dejours test.* Responses of ventilation ($\dot{V}_E$) and end-tidal P_{CO_2} (P_{ETCO_2}) to the abrupt inhalation of 100% O_2 from a background of 12% O_2 in a subject who had previously undergone bilateral carotid body resection (*CBR*) (*lower panel*) and in a control subject (*upper panel*). Note the transient decline in $\dot{V}_E$ that was evident in the control subject but not in the CBR subject. (From [222])

should allow sufficient "digitized" samples to maintain the underlying periodicity. There is also likely to be some "smearing" of the oscillating signal, owing to the distribution in vascular transit times from the lung to the left heart through the pulmonary veins.

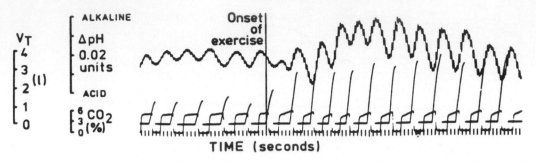

Fig. 108.20. Responses of arterial pH (measured as a change of pH, ΔpH), tidal volume (V_T), and respired P_{CO_2} to moderate constant-load cycle-ergometer exercise instituted from rest. Note the presence of a respiratory oscillation in the pH record, and its greater prominence during exercise. (From [7a])

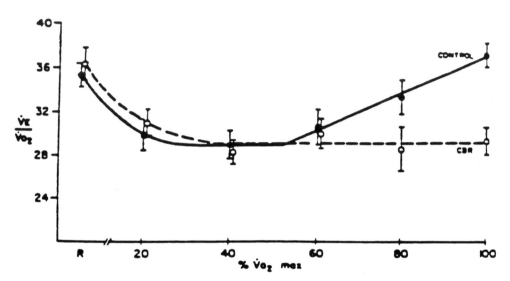

Fig. 108.21. Average response of the ventilatory equivalent for O_2 ($\dot{V}_E/\dot{V}_{O_2}$) during exhausting rapid-incremental cycle-ergometer exercise in a group of subjects who had previously had their carotid bodies surgically resected (*CBR*) and in a group of matched control subjects (*control*). Note the attenuated respiratory compensation at high work rates in the CBR group. (From [222])

Several characteristics of this oscillating CO_2-H^+ pattern (such as the amplitude and rate of change of the oscillation) have been proposed to provide a CO_2-linked drive to $\dot{V}_E$ in Φ3, i.e., that is not dependent upon the presence of an "error" in *mean* $PaCO_2$ and pH_a [36,41,60,226]. This mechanism is supported by the results of experiments in animals in which the carotid bodies have been shown to be capable of transducing such signals [16,18,19,85] into respiratory stimulation [19,36,60]. There is some debate regarding the prominence of this oscillating signal in exercising humans. It has been reported that the oscillatory amplitude peters out during exercise (in patients undergoing renal dialysis) at breathing frequencies in excess of approximately 20 min⁻¹ [138]. However, more recent studies indicate that the pH_a oscillation remains clearly discernible in resting subjects for breathing frequencies as high as 40 min⁻¹ [37]. This discrepancy probably reflects, in large part, improved electrode construction in the latter investigation, coupled with a direct intravascular placement at the aortic arch (the earlier electrode system required blood to be drawn into an external bypass loop).

Whether the oscillating signal provides any significant humoral feedback for ventilatory control in exercising humans has yet to be demonstrated, however.

The increased K^+ that is released from contracting muscle cells during exercise has been demonstrated to stimulate $\dot{V}_E$ through an action on the carotid bodies [8,24]. The increased arterial $[K^+]$ does not influence sites of central chemosensitivity as K^+ does not cross the blood–brain barrier [43] (see Chap. 27). Hence, it may account for, or contribute to, the 20% or so of the steady state exercise hyperpnea that is mediated by the carotid bodies. Other known carotid body stimuli also increase during exercise, however, and presumably also contribute to this approximately 20% of the exercise hyperpnea (see [215] for discussion). These include increases in the plasma levels of H^+, adenosine, osmolarity, and catecholamines.

There is also evidence of an important role for the carotid bodies in ventilatory control at high work rates. It has been argued that they normally appear to be largely responsible for mediating the respiratory compensation for the lactic acidosis above θ_L in humans (Fig. 108.21) [207,215,222], i.e., H⁺ (and other stimuli such as K⁺, for example) is known to be a carotid body stimulant. Evidence against this view has been presented, however. Individuals with *McArdle's syndrome* are deficient in the glycogenolytic enzyme myophosphorylase *b* and are therefore constrained to exercise only at relatively low work rates, i.e., they are unable to catabolize glycogen anaerobically. Such subjects have been reported to manifest hyperventilation at these low work rates, despite there being no metabolic acidemia [72]. This is suggestive, to some investigators, that the hyperventilation seen in *normal* subjects above θ_L therefore does not require a lowered pH_a for its mediation [72,141]. An alternative explanation for these findings, we contend, is that other mechanisms (such as apprehension, discomfort, increased intramuscular pressure, or pain) can induce an acute and marked hyperventilatory *alkalosis* under some circumstances [151,215,219], and, indeed, muscle pain is a cardinal symptom of McArdle's syndrome. Furthermore, respiratory alkalemia is *not* normally seen in humans performing high-intensity isotonic exercise.

In CBR subjects, the compensatory hyperventilation in response to constant-load exercise above θ_L has been reported to be nonexistent, i.e., pH_a fell more for a given decrease in $[HCO_3^-]_a$, and $PaCO_2$ actually was higher than the corresponding control values [207]. However, the carotid bodies may not be the sole contributors to the compensatory hyperventilation above θ_L.

For example, the $\dot{V}E$ response to prolonged constant-load exercise – that induced a standardized degree of metabolic acidemia (Δ standard $[HCO_3^-]$ of 5 mEq/l) – was found to include an O_2-labile (and presumably a carotid body-mediated) component [148]. This served to constrain the magnitude of the transient pH_a decrease (Fig. 108.22). However, as described above, there was also a *slower* compensatory component, even when carotid body chemosensitivity was suppressed by inhalation of 80% O_2. Carotid body chemosensitivity therefore appears to play a significant, and even dominant role in constraining variations of pH_a in response to the acute metabolic acidemia of exercise in humans. However, there are secondary – presumably central chemosensory – mechanisms which subserve a slower compensatory role (see above).

Stimulation of carotid and aortic chemoreceptors in animals leads to systemic vasoconstriction and tachycardia (see [31,42] for discussion). These are modified, however, by the ventilatory increases that also result from carotid body stimulation. Therefore, during spontaneous breathing, bradycardia actually occurs in response to carotid chemostimulation, and the change in peripheral resistance is smaller and often variable compared with changes seen when the effects of the *lung inflation vasodilator reflex* are obviated. This normally acts to offset the primary constrictor effects of the carotid body stimulation [31,32,42,69]. It is not certain how this reflex contributes to cardiorespiratory control in the exercising human, however. One might expect, perhaps, that the increased volume excursions of the lungs exert a constraining influence on the tachycardia of exercise.

108.4.7 Cardioventilatory Reflex Control

A further source of cardiorespiratory control during exercise may originate from the central circulation itself. It is known that local vascular distension of the heart and adjacent vasculature can lead to reflex tachycardia and hyperpnea [87,95,107,120,121,191]. This raises the possibility that the increased venous return of exercise has the potential to contribute to the control of cardiovascular and ventilatory responses during exercise in humans.

The proportionality of the $\Phi 1$ cardiovascular and ventilatory responses to exercise [38,122,134], together with a relative stability of the end-tidal gas tensions and R, has suggested to some investigators that there may be a component of "cardiodynamic" ventilatory control in this early phase of exercise [205]. The further observations (presented earlier) that the magnitude of the $\Phi 1$ hyperpnea is both slower and smaller when the exercise is imposed either from a background of light exercise [27,119,220] or from rest in the supine position (Fig. 108.11) [102,209], but with relative stability of the end-tidal gas tensions and R

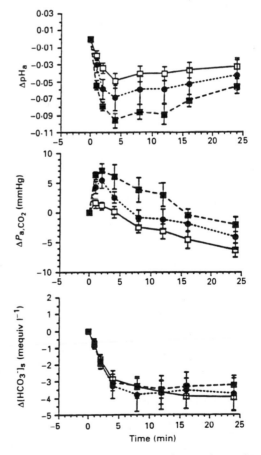

Fig. 108.22. Average responses of arterial pH (ΔpH_a), P_{CO_2} (ΔPa_{CO_2}) and standard bicarbonate concentration ($\Delta[HCO_3]$) to supra-threshold constant-load cycle-ergometer exercise at three inspired O_2 fractions (*black squares*, 80% O_2: *black circles*, 21% O_2: *white squares*, 12% O_2); the responses are expressed as changes (Δ) from unloaded cycling. (From [148])

across the transition (Fig. 108.9), is consistent with the suggestion of the cardiovascular change itself triggering the $\Phi1$ hyperpnea (i.e., a "cardiodynamic" hyperpnea). Certainly, it is hard to see how body position per se would largely inactivate either central command or limb afferent neurogenic control mechanisms. Indeed, stimulation of cardiac afferents can, under, some circumstances, lead to ventilatory stimulation [107,120,121,191].

A close correlation between the magnitude of the increase in right ventricular pressure (in response to a range of inducers) and the consequent hyperpnea have suggested to some that a cardiac-mediated hyperpnea might be operative during exercise [95]. However, recent evidence from several sources has questioned the role of such "cardiodynamic" mechanisms in the control of ventilation during exercise. For example, when the HR (and therefore cardiac output) at rest was abruptly increased in patients with permanent, demand-type pacemakers, there was no imediate $\dot{V}E$ response, although both $\dot{V}CO_2$ and $\dot{V}O_2$ started to increase [94]. Thus there was an increase in $\dot{Q}$ without consequent hyperpnea. Furthermore, cardiac afferents – although they can, under some circumstances, lead to ventilatory stimulation – do not subserve an obligatory role in the exercise hyperpnea. In humans who had undergone cardiac transplantation [9,187] and in calves with an implanted, pneumatically driven artificial heart (Jarvik type) [88], there was no reduction in either the magnitude or the rapidity of onset of the hyperpneic response to exercise.

It is of some interest, therefore, that there have been recent suggestions that the the link to ventilation may actually be provided by a signal responding to altered vascular conductance and/or tissue pressure in the exercising muscles themselves [74,89]. Such proposals change the focus for cardiorespiratory linkage during exercise from a central circulatory site of control to one more intimately related to the peripheral microvasculature.

108.5 Conclusion

In summary, therefore, there is clearly an interaction between the ventilatory and cardiovascular system responses during normal volitional exercise in humans. However the extent to which this relationship is causal or merely reflects inconsequent simultaneity remains to be resolved. For the causal component, it will be important not only to establish the primary pathways involved, but also to identify the proportional contribution of this linkage to overall cardiovascular and ventilatory control during exercise.

References

1. Adams L, Frankel H, Garlick J, Guz A, Murphy K, Semple SJG (1984) The role of spinal cord transmission in the ventilatory response to exercise in man. J Physiol (Lond) 355:85–97

2. Alam M, Smirk FH (1937) Observations in man upon a blood pressure raising reflex arising from the voluntary muscles. J Physiol (Lond) 89:372–383
3. Alam M, Smirk FH (1938) Unilateral loss of a blood pressure raising, pulse accelerating, reflex from voluntary muscle due to a lesion of the spinal cord. Clin Sci 3:247–252
4. Andersen P, Saltin B (1985) Maximal perfusion of skeletal muscle in man. J Physiol (Lond) 366:233–249
5. Asmussen E (1981) Similarities and dissimilarities between static and dynamic exercise. Circ Res 48 [Suppl 1]:3–10
6. Asmussen E, Johansen SH, Jorgensen M, Nielsen M (1965) On the nervous factors controlling respiration and circulation during exercise. Experiments with curarization. Acta Physiol Scand 63:343–350
7. Åstrand P-O, Rodahl K (1977) Textbook of work physiology, 2nd edn. McGraw-Hill, New York
7a. Band DM, Wolff CB, Ward J, Cochrane GM, Prior JG (1980) Respiratory oscillations in arterial carbon dioxide tension as a control signal in exercise. Nature 283:84–85
8. Band DM, Linton RAF, Kent R, Kurer FL (1985) The effect of peripheral chemodenervation on the ventilatory response to potassium. Respir Physiol 60:217–225
9. Banner N, Guz A, Heaton R, Innes JA, Murphy K, Yacoub M (1988) Ventilatory and circulatory responses at the onset of exercise in man following heart or heart-lung transplantation. J Physiol (Lond) 399:437–449
10. Barcroft H, Millen JLE (1939) The blood flow through muscle during sustained contraction. J Physiol 97:17–31
11. Barstow TJ, Lamarra N, Whipp BJ (1990) Modulation of muscle and pulmonary O_2 uptakes by circulatory dynamics during exercise. J Appl Physiol 68:979–989
12. Bayev KV, Beresovskii VK, Kebkalo TG, Savoskina LA (1988) Afferent and efferent connections of brainstem locomotor regions: study by means of horseradish peroxidase transport technique. Neuroscience 26:871–891
13. Beaver WL, Wasserman K, Whipp BJ (1986) Bicarbonate buffering of lactic acid generated during exercise. J Appl Physiol 60:472–478
14. Bevegård S, Lodin A (1962) Postural circulatory changes at rest and during exercise in five patients with congenital absence of valves in the deep veins of the legs. Acta Med Scand 172:21–29
15. Bevegard BS, Shepherd JT (1967) Regulation of the circulation during exercise in man. Physiol Rev 47:178
16. Biscoe TJ, Purves MLJ (1967) Observations on the rhythmic variation in the cat carotid body chemoreceptor activity which has the same period as respiration. J Physiol (Lond) 180:389–412
17. Bisgard GE, Forster HV, Byrnes B, Stanek K, Klein J, Manohar M (1978) Cerebrospinal fluid acid-base balance during muscular exercise. J Appl Physiol 45:94–101
18. Black AMS, McCloskey DI, Torrance RW (1971) The responses of carotid body chemoreceptors in the cat to sudden changes of hypercapnic and hypoxic stimuli. Respir Physiol 13:36–49
19. Black AMS, Torrance RW (1967) Chemoreceptor effects in the respiratory cycle. J Physiol (Lond) 189:59P–61P
20. Brown SE, Wiener S, Brown RA, Marcarelli PA, Light RW (1985) Exercise performance following a carbohydrate load in chronic airflow obstruction. J Appl Physiol 58:1340–1346
21. Brice AG, Forster HV, Pan LG, Funahashi A, Hoffman MD, Murphy CL, Lowry TF (1988) Is the hyperpnea of muscular contractions critically dependent on spinal afferents? J Appl Physiol 64:226–233
22. Bruce EN, Cherniack NS (1987) Central chemoreceptors. J Appl Physiol 62:389–402
23. Brunner MJ, Sussman MS, Greene AS, Kallman CH, Shoukas AA (1982) Carotid sinus baroreceptor reflex control of respiration. Circ Res 51:624–636
24. Burger RE, Estavillo JA, Kumar P, Nye PCG, Paterson DJ (1988) Effects of potassium, oxygen and carbon dioxide on

the steady-state discharge of cat carotid body chemoreceptors. J Physiol (Lond) 401:519–531

24a. Casaburi R, Whipp BJ, Wasserman K, Beaver WL, Koyal SN (1977) Ventilatory and gas exchange dynamics in response to sinusoidal work. J Appl Physiol 42:300–311

25. Casaburi R, Cooper C, Effros RM, Wasserman K (1989) Time course of mixed venous O_2 saturation following various modes of exercise transition. Fed Proc 3:A849

26. Casaburi R, Daly J, Hansen J, Effros RM (1987) Abrupt changes in mixed venous blood gases following exercise onset. Physiologist 30:131

27. Casaburi R, Whipp BJ, Wasserman K, Stremel RW (1978) Ventilatory control characteristics of the exercise hyperpnea as discerned from dynamic forcing techniques. Chest 73S:280S–283S

28. Clark JM, Sinclair RD, Lennox JB (1980) Chemical and nonchemical components of ventilation during hypercapnic exercise in man. J Appl Physiol 48:1065–1076

29. Clausen JP (1976) Circulatory adjustments to dynamic exercise and effect of physical training in normal subjects and in patients with coronary artery disease. Prog Cardiovasc Dis 18:459–495

30. Clausen JP (1977) Effect of physical training on cardiovascular adjustments to exercise in man. Physiol Rev 57:779–815

31. Coleridge HM, Coleridge JCG, Jordan D (1991) Integration of ventilatory and cardiovascular control systems. In: Crystal RG, West JB (eds) The lung: scientific foundations. Raven, New York, pp 1405–1418

32. Comroe JH Jr, Mortimer L (1964) The respiratory and cardiovascular responses of temporally separated aortic and carotid bodies to cyanide, nicotine, phenyldiguanide and serotonin. J Pharmacol Exp Ther 146:33–41

33. Comroe JH, Schmidt CF (1943) Reflexes from the limbs as a factor in the hyperpnea of muscular exercise. Am J Physiol 138:536–547

34. Coote JH, Hilton SM, Perez-Gonzalez JH (1971) The reflex nature of the pressor response to muscular exercise. J Physiol (Lond) 215:789–804

35. Cross BA, Davey A, Guz A, Katona PG, MacLean M, Murphy K, Semple SJG, Stidwell R (1982) The role of spinal cord transmission in the ventilatory response to electrically induced exercise in the anethetized dog. J Physiol (Lond) 329:37–55

36. Cross BA, Davey A, Guz A, Katona PG, MacLean M, Murphy K, Semple SJG, Stidwell RP (1982) The pH oscillations in arterial blood during exercise; a potential signal for the ventilatory response in the dog. J Physiol (Lond) 329:57–73

37. Cross BA, Stidwell RP, Hughes KR, Peppin R, Semple SJG (1995) A comparative study of aortic pH oscillations in conscious humans and anaesthetised cats and rabbits. Respir Physiol (in press)

38. Cummin ARC, Tyawe VI, Saunders KB (1986) Ventilation and cardiac output during the onset of exercise and during voluntary hyperventilation in humans. J Physiol (Lond) 370:567–583

39. Cunningham DJC (1974) Integrative aspects of the regulation of breathing: A personal view. In: Guyton AC, Widdicombe JG (eds) MTP international review of science, physiology, ser 1, vol 2. University Park Press, Baltimore, pp 303–369

40. Cunningham DJC, Petersen ES, Peto R, Pickering TG, Sleight P (1972) Comparison of the effect of different types of exercise on the baroreflex regulation of heart rate. Acta Physiol Scand 86:444–455

41. Cunningham DJC, Robbins PA, Wolff CB (1986) Integration of respiratory responses to changes in alveolar partial pressures of CO_2 and O_2 and in arterial pH. In: Cherniack NS, Widdicombe JG (eds) Handbook of physiology: the respiratory system, vol II, control of breathing, part 2. American Physiological Society, Bethesda, pp 475–528

42. Daly M de B (1986) Interactions between respiration and circulation. In: Cherniack NS, Widdicombe JG (eds) Handbook of physiology: the respiratory system, vol II, control of breathing, part 2. American Physiological Society, Bethesda, pp 529–594

43. Davson H (1970) Physiology of the cerebrospinal fluid. Churchill, London

44. De Cort SC, Innes JA, Barstow TJ, Guz A (1991) Cardiac output, oxygen consumption and arteriovenous oxygen difference following a sudden rise in exercise level in humans. J Physiol (Lond) 441:501–512

45. Dejours P (1964) Control of respiration in muscular exercise. In: Fenn WO, Rahn H (eds) Handbook of physiology: respiration, vol I. American Physiological Society, Washington, pp 631–648

46. Dempsey JA, Mitchell GS, Smith CA (1984) Exercise and chemoreception. Am Rev Respir Dis 129 [Suppl]:S31–S34

47. DiMarco AF, Romaniuk JR, von Euler C, Yamamoto Y (1983) Immediate changes in ventilation and respiratory pattern with onset and offset of locomotion in the cat. J Physiol (Lond) 343:1–16

48. Donald DE (1983) Splanchnic circulation. In: Shepherd JT, Abboud FM, Geiger SR (eds) Handbook of physiology. The cardiovascular system: peripheral circulation and organ blood flow, sect 2, vol III, part 1. American Physiological society, Bethesda, pp 219–240

49. Dormer KJ (1984) Modulation of cardiovascular response to dynamic exercise by fastigial nucleus. J Appl Physiol 56:1369–1377

50. Dormer KJ, Stone HL (1976) Cerebellar pressor response in the dog. J Appl Physiol 41:574–580

51. Dormer KJ, Stone HL (1982) Fastigial nucleus and its possible role in the cardiovascular response to exercise. In: Smith OA, Galosy RA, Weiss SM (eds) Circulation, neurobiology and behavior. Elsevier, New York, pp 201–215

52. Douglas CG (1972) Co-ordination of the respiration and circulation with variations in bodily activity. Lancet 2:213–218

53. Duncan G, Johnson RG, Lambie DG (1981) Role of sensory nerves in the cardiovascular and respiratory changes with isometric forearm exercise in man. Clin Sci 60:145–155

54. Ekblom B (1969) Effect of physical training on oxygen transport system in man. Acta Physiol Scand 328 [Suppl]:5–45

55. Ekblom B, Hermansen L (1968) Cardiac output in athletes. J Appl Physiol 25:619–625

56. Eldridge FL (1977) Maintenance of respiration by central neural feedback mechanisms. Fed Proc 36:2400–2404

57. Eldridge FL, Gill-Kumar P (1980) Central neural respiratory drive and afterdischarge. Respir Physiol 40:49–63

58. Eldridge FL, Kiley JP, Millhorn DE (1985) Respiratory responses to medullary hydrogen ion changes in cats: different effects of respiratory and metabolic acidosis. J Physiol (Lond) 358:285–297

59. Eldridge FL, Millhorn DE, Kiley JP, Waldrop TG (1985) Stimulation by central command of locomotion, respiration and circulation during exercise. Respir Physiol 59:313–337

60. Eldridge FL, Waldrop TG (1991) Neural control of breathing. In: Whipp BJ, Wasserman K (eds) Pulmonary physiology and pathophysiology of exercise. Dekker, New York, pp 309–370

61. Eriksen M, Waaler BA, Wallø L, Wesche J (1990) Dynamics and dimensions of cardiac output changes in humans at the onset and at the end of moderate rhythmic exercise. J Physiol (Lond) 426:423–437

62. Farhi LE, Rahn H (1955). Gas stores in the body and the unsteady state. J Appl Physiol 7:472–484

63. Forster HV, Pan LG (1991) Exercise hyperpnea. In: Crystal RG, West JB (eds) The Lung: scientific foundations. Raven, New York, pp 1553–1564

64. Freund PR, Rowell LB, Murphy TM, Hobbs SG, Butler SH (1979) Blockade of the pressor response to muscle ischemia by sensory nerve block in man. Am J Physiol 236:H433–H439

65. Gaudio R, Bromberg PA, Millen JE, Robin ED (1969) Ondine's curse reaffirmed: control of ventilation during exercise and sleep. Clin Res 17:414

66. Goodwin GM, McCloskey DI, Mitchell JH (1972) Cardiovascular and respiratory responses to changes in central command during isometric exercise at constant muscle tension. J Physiol (Lond) 226:173–190

67. Griffiths TL, Henson LC, Whipp BJ (1986) Influence of peripheral chemoreceptors on the dynamics of the exercise hyperpnea in man. J Physiol (Lond) 380:387–403

68. Grimby G, Nilsson NJ, Sanne H (1964) Cardiac output during exercise in patients with varicose veins. Scand J Clin Lab Invest 16:21–30

69. Gross PM, Whipp BJ, Davidson JT, Koyal SN, Wasserman K (1976) Role of the carotid bodies in the heart rate response to breath holding in man. J Appl Physiol 41:336–340

70. Guyton AC, Jones CE, Coleman TG (1973) Circulatory physiology: cardiac output and its regulation. Saunders, Philadelphia

71. Guz A, Innes JA, Murphy K (1987) Respiratory modulation of left ventricular stroke volume in man measured using pulsed Doppler ultrasound. J Physiol (Lond) 393:499–512

72. Hagberg JM, Coyle EF, Carroll JE, Miller JM, Martin WH, Brooke MH (1982) Exercise hyperventilation in patients with McArdle's disease. J Appl Physiol 52:991–994

73. Hainsworth R (1986) Vascular capacitance: its control and importance. Rev Physiol Biochem Pharmacol 105:101–173

74. Haouzi P, Huszczuk A, Gille JP, Chalon B, Marchal, F Crance JP, Whipp BJ (1995) Vascular distension in muscles contributes to respiratory control in sheep. Respir Physiol 99:41–50

75. Haouzi P, Huszczuk A, Porszasz J, Chalon B, Wasserman K, Whipp BJ (1993) Femoral vascular occlusion and ventilation during recovery from heavy exercise. Respir Physiol 94:137–150

76. Harrison DC, Goldblatt A, Braunwald E, Glick G, Mason DT (1963) Studies on cardiac dimensions in intanct, unanesthetized man. III. Effects of muscular exercise. Circ Res 13:460–467

77. Higginbotham MB, Morris KG, Williams RS, McHale PA, Coleman RE, Cobb FR (1986) Regulation of stroke volume during submaximal and maximal upright exercise in normal man. Circ Res 58:281–291

78. Hilton SM (1959) A peripheral arterial conducting system underlying dilation of femoral artery and concerned with functional dilation in skeletal muscle. J Physiol (Lond) 149:93–111

79. Hilton SM (1966) Hypothalamic regulation of the cardiovascular system. Br Med Bull 22:243–248

80. Hobbs SF (1982) Central command during exercise: parallel activation of the cardiovascular and motor systems by descending command signals. In: Smith OA, Galosy RA, Weiss SM (eds) Circulation, neurobiology and behavior. Elsevier, New York, pp 217–232

81. Hobbs SF, McCloskey DI (1986) Effect of spontaneous exercise on reflex slowing of the heart in decerebrate cats. J Auton Nerv Sys 17:303–312

82. Holmgren A, Ovenfors CO (1960) Heart volume at rest and during muscular work in the supine and in the sitting position. Acta Med Scand 167:267–277

83. Holstege G (1987) Some anatomical observations on the projections from the hypothalamus to brainstem and spinal cord: an HRP and autoradiographic tracing study in the cat. J Comp Neurol 260:98–126

84. Honda Y (1992) Respiratory and circulatory activities in carotid body-resected humans. J Appl Physiol 73:1–8

85. Hornbein TF, Griffo ZJ, Roos A (1961) Quantitation of chemoreceptor activity: interrelation of hypoxia and hypercapnia. J Neurophysiol 24:561–568

86. Hughson RL (1990) Exploring cardiorespiratory control mechanisms through gas exchange dynamics. Med Sci Sports Exerc 22:72–79

87. Huszczuk A, Jones PW, Oren A, Shors EC, Nery LE, Whipp BJ, Wasserman K (1983) Venous return and ventilatory control. In: Whipp BJ, Wiberg DM (eds) Modelling and control of breathing. Elsevier, New York, pp 78–85

88. Huszczuk A, Whipp BJ, Adams TD, Fisher AG, Crapo RO, Elliott GC, Wasserman K, Olsen DB (1990) Ventilatory control during exercise in calves with artificial hearts. J Appl Physiol 68:2604–2611

89. Innes JA, Solarte I, Huszczuk A, Yeh E, Whipp BJ, Wasserman K (1989) Respiration during recovery from exercise; effects of trapping and release of femoral blood flow. J Appl Physiol 67:2608–2613

90. Jones NL (1975) Exercise testing in pulmonary evaluation: rationale, methods, and the normal respiratory response to exercise. N Engl J Med 293:541–544

91. Jones NL (1976) Use of exercise in testing respiratory control mechanisms. Chest 70:169–173

92. Jones NL, Haddon RWT (1973) Effect of a meal on cardiopulmonary and metabolic changes during exercise. Can J Physiol Pharmacol 51:445–450

93. Jones NL, Jurkowski JE (1979) Body carbon dioxide storage capacity in exercise. J Appl Physiol 46:811–815

94. Jones PW, French W, Weissman ML, Wasserman K (1981) Ventilatory responses to cardiac output changes in patients with pacemakers. J Appl Physiol 51:1103–1107

95. Jones PW, Huszczuk A, Wasserman K (1982) Cardiac output as a controller of ventilation through changes in right ventricular load. J Appl Physiol 53:218–244

96. Jones WB, Finchun RN, Russell Jr RO, Reeves TJ (1970) Transient cardiac output response to multiple levels of supine exercise. J Appl Physiol 28:183–189

97. Jordan D (1990) Autonomic changes in affective behaviour. In: Loewy AD, Spyer KM (eds) Central regulation of autonomic functions. Oxford University Press, Oxford, pp 349–367

98. Jordan D, Spyer KM (1986) Brainstem integration of cardiovascular and pulmonary afferent activity. Prog Brain Res 67:295–314

99. Kaada BR (1960) Cingulate, posterior orbital, anterior insular and temporal pole cortex. In: Field J, Magoun HW, Hall VE (eds) Handbook of physiology, sect 1: neurophysiology, vol II. American Physiological Society, Washington, pp 1345–1372

100. Kalia M, Mei SS, Kao FF (1981) Central projections from ergoreceptors (C fibers) in muscle involved in cardiopulmonary responses to static exerdise. Circ Res 48 [Suppl l]:48–62

101. Kao FF (1963) An experimental study of the pathways involved in exercise hyperpnea, employing cross-circulation techniques. In: Cunningham DJC, Lloyd BB (eds) The regulation of human respiration. Davis, Philadelphia, pp 461–502

102. Karlsson H, Lindborg B, Linnarsson D (1975) Time courses of pulmonary gas exchange and heart rate changes in supine exercise. Acta Physiol Scand 95:329–340

103. Kaufman MP, Rybicki KJ (1987) Discharge properties of group III and IV muscle afferents: their responses to mechanical and metabolic stimuli. Circ Res 61 [Suppl l]:60–65

104. Kniffki K-D, Mense S, Schmidt RF (1981) Muscle receptors with fine afferent fibers which may evoke circulatory reflexes. Circ Res 48 [Suppl l]:25–31

105. Koller A, Kaley G (1990) Endothelium regulates skeletal muscle microcirculation by a blood flow velocity sensing mechanism. Am J Physiol 258:H916–H920

106. Koller A, Kaley G (1990) Prostaglandins mediate arteriolar dilation to increased blood flow velocity in skeletal muscle microcirculation. Circ Res 67:529–534

107. Kostreva DR, Zuperku EJ, Purtock RV, Coon RL, Kampine JP (1975) Sympathetic afferent nerve activity of right heart origin. Am J Physiol 229:911–915

108. Krisanda JM, Moreland TS, Kushmerik MJ (1988) ATP supply and demand during exercise. In: Horton ES, Terjung RL (eds) Exercise, nutriton, and energy metabolism. MacMillan, New York, pp 27–44

109. Krishnan B, Zintel T, McParland C, Gallagher CG (1995) Lack of importance of respiratory muscle load in ventilatory regulation during heavy exercise. Respir Physiol (in press)

110. Krogh A, Lindhard (1913) The regulation of respiration and circulation during the initial stages of muscular work. J Physiol (Lond) 47:112–136

111. Krogh A, Lindhard (1917) A comparison between voluntary and electrically induced muscular work in man. J Physiol (Lond) 51:182–201

112. Lamarra N, Whipp BJ, Blumenberg M, Wasserman K (1983) Model-order estimation of cardiorespiratory dynamics during moderate exercise. In: Whipp BJ, Wiberg DM (eds) Modelling and control of breathing. Elsevier, New York, pp 338–345

113. Lamb TW (1968/1969) Ventilatory responses to hind limb exercise in anesthetized cats and dogs. Respir Physiol 6:88–104

114. Laughlin MH (1987) Skeletal muscle blood flow capacity: role of muscle pump in exercise hyperemia. Am J Physiol 253:H993–H1004

115. Leonard B, Mitchell JH, Mizuno M, Rube N, Saltin B, Secher NH (1985) Partial neuromuscular blockade and cardiovascular responses to static exercise in man. J Physiol (Lond) 359:365–379

116. Leusen I (1965) Aspects of the acid-base balance between blood and cerebrospinal fluid. In: Brooks C, Kao FF, Lloyd BB (eds) Cerebrospinal fluid and the regulation of ventilation. Blackwell, Oxford, pp 55–89

117. Lind AR (1983) Cardiovascular adjustments to isometric contractions: static effort. In: Shepherd JT, Abbound FM, Geiger SR (eds) Handbook of physiology. The cardiovascular system: perispheral circulation and organ blood flow, sect 2, vol III. American Physiological Society, Bethesda, pp 947–966

117a. Lind AR, Taylor SH, Humphreys PW, Kennelly BM, Donald KW (1964) The circulatory effects of sustained voluntary muscle contraction. Clin Sci 27:229–244

118. Lind AR, McNicol GW, Bruce RA, Macdonald HR, Donald KW (1968) The cardiovascular responses to sustained contractions of a patient with unilateral syringomyelia. Clin Sci 35:45–53

119. Linnarsson D (1974) Dynamics of pulmonary gas exchange and heart rate at start and end of exercise. Acta Physiol Scand 415 [Suppl]:1–68

120. Lloyd TC Jr (1984) Effect on breathing of acute pressure rise in pulmonary artery and right ventricle. J Apppl Physiol 57:110–116

121. Lloyd TC Jr (1986) Control of breathing in anesthetized dogs by a left-heart baroreflex. J Appl Physiol 61:2095–2101

122. Loeppky JA, Greene ER, Hoekenga DE, Caprihan A, Luft UC (1981) Beat-by-beat stroke volume assessment by pulsed Doppler in upright and supine exercise. J Appl Physiol 50:1173–1182

123. Loewy AD (1990) Central autonomic pathways. In: Loewy AD, Spyer KM (eds) Central regulation of autonomic functions. Oxford University Press, New York, pp 88–103

124. Lugliani R, Whipp BJ, Brinkman J, Wasserman K (1979) Doxapram hydrochloride: a respiratory stimulant for patients with primary alveolar hypoventilsation. Chest 76:414–419

125. Luiten PGM, ter Horst GJ, Karst H, Steffens AB (1985) The course of paraventricular hypothalamic efferents to autonomic structures in medulla and spinal cord. Brain Res 329:374–378

126. Mahler M (1985) First-order kinetics of muscle oxygen consumption, and an equivalent proportionality between QO_2 and phosphorylcreatine level. J Gen Physiol 86:135–165

127. Majcherczyk S, Willshaw P (1973) Inhibition of peripheral chemoreceptor activity during superfusion with an alkaline c.s.f. of the ventral brainstem surface of the cat. J Physiol (Lond) 231:269

128. Marshall JM, Timms RJ (1980) Experiments on the role of the subthalamus in the generation of the cardiovascular changes during locomotion in the cat. J Physiol (Lond) 301:92–93

129. McCloskey DI, Mitchell JH (1972) Reflex cardiovascular and respiratory responses originating in exercising muscle. J Physiol (Lond) 224:173–186

130. Meyer RA (1988) A linear model of muscle respiration explains monoexponential phosphocreatine changes. Am J Physiol 254:C548–C553

131. Mitchell JH, Reardon WC, McCloskey DI (1977) Reflex effects on circulation and respiration from contracting skeletal muscle. Am J Physiol 233:H374–H378

132. Mitchell JH, Schmidt RF (1983) Cardiovascular reflex control by afferent fibers from skeletal muscle receptors. In: Shepherd JT, Abboud FM, Geiger SR (eds) Handbook of physiology. The cardiovascular system: peripheral circulation and organ blood flow, sect 2, vol III. American Physiological Society, Bethesda, pp 623–658

133. Miyamoto Y (1989) Neural and humoral factors affecting ventilatory response during exercise. Jpn J Physiol 39:199–214

134. Miyamoto Y, Hiura T, Tamura T, Nakamura T, Higuchi J, Mikami T (1982) Dynamics of cardiac, respiratory, and metabolic function in men in response to step work load. J Appl Physiol 52:1198–1208

135. Miyamoto Y, Nakazono Y, Hiura T, Abe Y (1983) Cardiorespiratory dynamics during sinusoidal and impulse exercise in man. Jpn J Physiol 33:971–986

136. Mohrman DE, Sparks HV (1974) Myogenic hyperemia following brief tetanus of canine skeletal muscle. Am J Physiol 227:531–535

137. Moreno AH, Burchell AR, van der Woude R, Burke JH (1967) Respiratory regulation of splanchnic and systemic venous return. Am J Physiol 213:455–465

138. Murphy K, Stidewell RP, Cross BA, leaver KD, Annastasiades E, Phillips M, Guz A, Semple SJG (1987) Is hypercapnia necessary for the ventilatory response to exercise in man? Clin Sci 73:617–625

139. Oldenburg FA, McCormack DW, Morse JLC, Jones NL (1979) A comparison of exercise responses in stairclimbing and cycling. J Appl Physiol 46:510–516

140. Ordway GA, Waldrop TG, Iwamoto GA, Gentile BJ (1989) Hypothalamic influences on cardiovascular response of beagles to dynamic exercise. Am J Physiol 257:H1247–H1253

141. Paterson DJ (1992) Potassium and ventilation in exercise. J Appl Physiol 72:811–820

142. Paton JY, Swaminathan S, Sargent CW, Hawksworth A, Keens TG (1993) Ventilatory response to exercise in children with congenital central hypoventilation syndrome. Am Rev Respir Dis 147:1185–1191

143. Poliner LR, Dehmer GJ, Lewis SE, Parkey RW, Blomqvist CG, Willerson JT (1980) Left ventricular performance in normal subjects: a comparison of the responses to exercise in the upright and supine positions. Circulation 62:528–534

144. Poole DP, Ward SA, Whipp BJ (1988) Control of blood gas and acid-base status during isometric exercise in humans. J Physiol (Lond) 396:365–377

145. Poon C-S, Greene JG (1985) Control of exercise hyperpnea during hypercapnia in humans. J Appl Physiol 59:792–797

146. Poon CS, Ward SA, Whipp BJ (1987) Influence of inspiratory assistance on ventilatory control during moderate exercise. J Appl Physiol 62:551–560

147. Rahn H, Farhi LE (1962) Ventilation-perfusion relationship. In: de Reuck AVS, O'Connor M (eds) Ciba Foundation symposium on pulmonary structure and function. Little and Brown, Boston, pp 139–153

148. Rausch SM, Whipp BJ, Wasserman K, Huszczuk A (1991) Role of the carotid bodies in the respiratory compensation for the metabolic acidosis of exercise in humans. J Physiol (Lond) 444:567–578

149. Reeves JT, Grover RF, Blount SG, Filley GF (1961) Cardiac output response to standing and treadmill walking. J Appl Physiol 16:283–288

150. Rerych SK, Scholz PM, Newman GE, Sabiston DC Jr, Jones RH (1978) Cardiac function at rest and during exercise in

normals and in patients with coronary heart disease: evaluation by radionuclide angiocardiography. Ann Surg 187:449–464

151. Riley M, Nicholls DP, Nugent A-M, Steele IC, Bell N, Davies PM, Stanford CF, Patterson VH (1993) Respiratory gas exchange and metabolic responses during exercise in McArdle's disease. J Appl Physiol 75:745–754

152. Robinson BF, Epstein SE, Kahler RL, Braunwald E (1966) Circulatory effects of acute expansion of blood volume: studies during maximal exercise and at rest. Circ Res 19:26–32

153. Rodgers SH (1968) Ventilatory response to ventral root stimulation in the decerebrate cat. Respir Physiol 5:165–174

154. Rothe CF (1983) Venous system: physiology of the capacitance vessels. In: Shepherd JT, Abboud FM, Geiger SR (eds) Handbook of physiology. The cardiovascular system: peripheral circulation and organ blood flow, sect 2, vol III, part 1. American Physiological Society, Bethesda, pp 397–452

155. Rotto DM, Massey KD, Burton KP, Kaufman MP (1989) Static contraction increases arachidonic acid levels in gastrocnemius muscles of cat. J Appl Physiol 66:2721–2724

156. Rotto DM, Schultz HD, Longhurst JC, Kaufman MP (1990) Sensitization of group III muscle afferents to static contraction by arachidonic acid. J Appl Physiol 68:861–867

157. Rowell LB (1974) Human cardiovascular adjustments to exercise and thermal stress. Physiol Rev 54:75–159

158. Rowell LB (1986) Human circulation: regulation during physical stress. Oxford University Press, New York

159. Rowell LB (1986) Muscle blood flow in humans: how high can it go? Med Sci Sports Exer 20:S97–S103

160. Rowell LB (1993) Human cardiovascular control. Oxford University Press, New York

161. Rowell LB, Freund PR, Hobbs SF (1976) Cardiovascualr responses to muscle ischemia in humans. Circ Res 48 [Suppl I]:37–47

162. Rowell LB, O'Leary DS (1990) Reflex control of the circulation during exercise: chemoreflexes and mechanoreflexes. J Appl Physiol 69:407–418

163. Rybicki KJ, Waldrop TG, Kaufman MP (1985) Increasing gracilis muscle interstitial potassium concentrations stimulate group III and IV afferents. J Appl Physiol 58:936–941

164. Saltin B, Blomqvist G, Mitchell JH, Johnson RL Jr, Wildenthal K, Chapman CB (1968) Response to exercise after bed rest and after training. Circulation 38 [Suppl 7]:1–78

165. Saltin B, Sjogaard G, Gaffney FA, Rowell LB (1981) Potassium, lactate, and water fluxes in human quadriceps muscle during static contractions. Circ Res 48 [Suppl 1]:18–24

166. Saltzman HA, Salzano JV (1971) Effects of carbohydrate metabolism upon respiratory gas exchange in normal men. J Appl Physiol 30:228–231

167. Saper CB, Loewy AD, Swanson LW, Cowan WM (1976) Direct hypothalamo-autonomic connection. Brain Res 117:305–312

168. Schaltenbrand G, Girndt O (1925) Physiologische Beobachtungen an Thalamuskatzen. Pflugers Arch 209:333–361

169. Sejersted OM, Hargens AR, Kardel KR, Blom P, Jensen O, Hermansen L (1984) Intramuscular fluid pressure during isometric contraction of human skeletal muscle. J Appl Physiol 56:287–295

170. Shea SA, Andrews LP, Shannon DC, Banzett RB (1993) Ventilatory responses to exercise in humans lacking ventilatory chemosensitivity. J Physiol (Lond) 469:623–640

171. Shepherd JT (1963) Physiology of the circulation in human limbs in health and disease. Saunders, Philadelphia

172. Shepherd JT (1983) Circulation to skeletal muscle. In: Shepherd JT, Abboud FM, Geiger SR (eds) Handbook of physiology. The cardiovascular system: peripheral circulation and organ blood flow, sect 2, vol III, part 1. American Physiological Society, Bethesda, pp 319–370

173. Shepherd JT, Blomqvist CG, Lind AR, Mitchell JH, Saltin B (1981) Static (isometric) exercise; retrospection and introspection. Circ Res 48 [Suppl I]:179–188

174. Shepherd JT, Vanhoutte PM (1975) Veins and their control. Saunders, Philadelphia

175. Sjøgaard G, Savard G, Juel C (1988) Muscle blood flow during isometric activity and its relation to muscle fatigue. Eur J Appl Physiol 57:327–335

176. Skinner NS Jr (1975) Skeletal muscle blood flow: metabolic determinants. In: Zelis E (ed) The peripheral circulation. Grune and Stratton, New York, pp 57–78

177. Smith OA Jr, Jabbur SJ, Rushmer RF, Lasher EP (1960) Role of hypothalamic structures in cardiac control. Physiol Rev 40:136–141

178. Smith OA Jr, Rushmer RF, Lasher EP (1960) Similarity of cardiovascular responses to exercise and to diencephalic stimulation. Am J Physiol 198:1139–1142

179. Sofroniew MV, Schrell U (1980) Hypothalamic neurons projecting to the rat caudal medulla oblongata, examined by immunoperoxidase staining of retrogradely transported horseradish peroxidase. Neurosci Lett 19:257–263

180. Sparks HV Jr (1980) Effect of local metabolic factors on vascular smooth muscle. In: Bohr DF, Somlyo AP, Sparks HV Jr, Geiger SR (eds) Handbook of physiology. The cardiovascular system: vascular smooth muscle, sect 2, vol II. American Physiological Society, Bethesda, pp 475–513

181. Spyer KM (1994) Central nervous mechanisms contributing to cardiovascular control. J Physiol (Lond) 474:1–19

182. Stebbins CL, Brown B, Levin D, Longhurst JC (1988) Reflex effect of skeletal muscle mechanoreceptor stimulation on the cardiovascular system. J Appl Physiol 65:1539–1547

183. Stebbins CL, Longhurst JC (1986) Bradykinin in reflex cardiovascular responses to static muscular contraction. J Appl Physiol 62:271–279

184. Sutton JR, Jones NL (1979) Control of pulmonary ventilation during exercise and mediators in the blood: CO_2 and hydrogen ion. Med Sci Sports 11:198–203

185. Taylor R, Jones NL (1979) The reduction by training of CO_2 output during exercise. Eur J Cardiol 9:53–62

186. Teppema LJ, Barts PWJA, Evers JAM (1982) Effects of metabolic arterial pH changes on medullary ecf pH, csf pH and ventilation in peripherally chemodenervated cats with intact blood-brain barrier. J Physiol (Lond) 58:123–136

187. Theodore J, Morris AJ, Burker CM, Glanville AR, VanKessel A, Baldwin JC, Stinson EB, Shumway NE, Robin ED (1987) Cardiopulmonary function at maximum tolerable constant work rate exercise following human heart-lung transplantation. Chest 92:433–439

188. Thimm F, Carvalho M, Babka M, Meier zu Verl E (1984) Reflex increases in heart-rate induced by perfusing the hind leg of the rat with solutions containing lactic acid. Pflugers Arch 400:286–293

189. Tibes U (1977) Reflex inputs to the cardiovascular and respiratory centers from dynamically working canine muscles. Circ Res 42:332–341

190. Tibes U (1980) Neurogenic control of ventilation in exercise. In: Cerretelli P, Whipp BJ (eds) Exercise bioenergetics and gas exchange. Elsevier, Amsterdam, pp 149–158

191. Uchida Y (1976) Tachypnea after stimulation of afferent cardiac sympathetic nerve-fibers. Am J Physiol 230:1003–1007

192. Vanhoutte PM (1988) Vasodilatation, vascular smooth muscle, peptides, autonomic nerves, and endothelium. Raven, New York

193. Victor RG, Pryor SL, Secher NH, Mitchell JH (1989) Effects of partial neuromuscular blockade on sympathetic nerve responses to static exercise in humans. Circ Res 65:468–476

194. Wade OL, Bishop JM (1962) Cardiac output and regional blood flow. Blackwell, Oxford

195. Waldrop TG, Henderson MC, Iwamoto GA, Mitchell JH (1986) Regional blood flow responses to stimulation of the subthalamic locomotor region. Respir Physiol 64:93–102

196. Walgenbach SC, Donald DE (1983) Inhibition by carotid baroreflex of exercise-induced increases in arterial pressure. Circ Res 52:253–262.

197. Walloe L, Wesche J (1988) Time course and magnitude of blood flow changes in the human quadriceps muscles during and following rhythmic exercise. J Physiol (Lond) 405:257–273

198. Wang Y, Marshall RJ, Shepherd JT (1960) The effect of changes in posture and of graded exercise on stroke volume in man. J Clin Invest 39:1051–1061

199. Ward SA (1994) Assessment of peripheral chemoreflex contributions to ventilation during exercise. Med Sci Sports Ex 26:303–310

200. Ward SA, Blesovsky L, Russack S, Ashjian A, Whipp BJ (1987) Chemoreflex modulation of ventilatory dynamics during exercise in man. J Appl Physiol 63:2001–2007

201. Ward SA, Whipp BJ, Koyal S, Wasserman K (1983) Influence of body CO_2 stores on ventilatory dynamics during exercise. J Appl Physiol 55:742–749

202. Wasserman K, Casaburi R (1991) Acid-base regulation during exercise in humans. In: Whipp BJ, Wasserman K (eds) Pulmonary physiology and pathophysiology of exercise. Dekker, New York, pp 405–448

203. Wasserman K, Hansen JE, Sue DY Whipp BJ (1986) Principles of exercise testing and interpretation. Lea and Febiger, Philadelphia

204. Wasserman K, van Kessel AL, Burton GG (1967) Interaction of physiological mechanisms during exercise. J Appl Physiol 22:71–85

205. Wasserman K, Whipp BJ, Castagna J (1974) Cardiodynamic hyperpnea: hyperpnea secondary to cardiac output increase. J Appl Physiol 36:457–464

205. Wasserman K, Whipp BJ, Koyal SN, Beaver WL (1973) Anaerobic threshold and respiratory gas exchange during exercise. J Appl Physiol 35:236–243

207. Wasserman K, Whipp BJ, Koyal SN, Cleary MG (1975) Effect of carotid body resection on ventilatory and acid-base control during exercise. J Appl Physiol 39:354–358

208. Weil JV, Swanson GD (1991) Peripheral chemoreceptors and the control of breathing. In: Whipp BJ, Wasserman K (eds) Pulmonary physiology and pathophysiology of exercise. Dekker, New York, pp 371–403

209. Weiler-Ravell D, Cooper DM, Whipp BJ, Wasserman K (1983) The control of breathing at the start of exercise as influenced by posture. J Appl Physiol 55:1460–1466

210. Weissman ML, Jones PW, Oren A, Lamarra N, Whipp BJ, Wasserman K (1982) Cardiac output increase and gas exchange at the start of exercise. J Appl Physiol 52:236–244

211. Weissman ML, Whipp BJ, Huntsman DJ, Wasserman K (1980) Role of neural afferents from working limbs in exercise hyperpnea. J Appl Physiol 49:239–248

212. Wesche J (1986) The time course and magnitude of blood flow changes in the human quadriceps muscles following isometric contraction. J Physiol (Lond) 377:445–462

213. Whipp BJ (1981) The control of exercise hyperpnea. In: Hornbein T (ed) The regulation of breathing. Dekker, New York, pp 1069–1139

214. Whipp BJ (1987) Dynamics of pulmonary gas exchange during exercise in man. Circulation VI:18–28

215. Whipp BJ (1994) Peripheral chemoreceptor control of the exercise hyperpnea in humans. Med Sci Sports Exerc 26:337–347

216. Whipp BJ, Phillips EE, Jr (1970) Cardiopulmonary and metabolic responses to sustained isometric exercise. Arch Phys Med 51:398–40

217. Whipp BJ, Ward SA (1982) Cardiopulmonary coupling during exercise. J Exp Biol 100:175–193

218. Whipp BJ, Ward SA (1990) Physiological determinants of pulmonary gas exchange kinetics during exercise. Med Sci Sports Exerc 22:62–71

219. Whipp BJ, Ward SA (1991) The coupling of ventilation to pulmonary gas exchange during exercise. In: Whipp BJ, Wasserman K (eds) Pulmonary physiology and pathophysiology of exercise. Dekker, New York, pp 271–307

220. Whipp BJ, Ward SA, Lamarra N, Davis JA, Wasserman K (1982) Parameters of ventilatory and gas exchange dynamics during exercise. J Appl Physiol 52:1506–1513

221. Whipp BJ, Ward SA, Smith RE, Hussain ST (1995) The dynamics of pulmonary O_2 uptake and femoral artery blood flow during moderate intensity exercise in humans. J Physiol (Lond) 483:130P

222. Whipp BJ, Wasserman K (1980) Carotid bodies and ventilatory control dynamics in man. Fed Proc 39:2668–2673

223. Widdicombe JG (1974) Reflex control of breathing. In: Guyton AC, Widdicombe JG (eds) MTP international review of science, physiology, ser 1, vol 2. University Park Press, Baltimore, PP 273–301

224. Wiley RL, Lind AR (1971) Respiratory responses to sustained static muscular contractions in humans. Clin Sci 40:221–234

225. Wiley RL, Lind AR (1975) Respiratory responses to simultaneous static and rhythmic exercises in humans. Clin Sci Mol Med 49:427–432

226. Yamamoto WS (1960) Mathematical analysis of the time course of alveolar CO_2. J Appl Physiol 15:215–219

227. Yamamoto WS (1962) Transmission of information by the arterial blood stream with particular reference to carbon dioxide. Biophys J 2:143–159

228. Yoshida T, Whipp BJ (1994) Dynamic asymmetries of cardiac output transients in response to muscular exercise in man. J Physiol (Lond) 480:355–359

229. Younes M (1991) Determinants of thoracic excursions during exercise. In: Whipp BJ, Wasserman K (eds) Pulmonary physiology and pathophysiology of exercise. Marcel Dekker, New York, pp 1–65

230. Young IH, Woolcock AJ (1978) Changes in arterial blood gas tensions during unsteady-state exercise. J Appl Physiol 44:93–96

109 Gravitational and Hyper- and Hypobaric Stress

S.A. WARD

Contents

109.1 Introduction

The respiratory system is challenged by exposure to high or low ambient pressures and gravitational forces. These have effects on a range of respiratory functions, such as the mechanics of breathing, the exchange of respiratory gases at the lungs, and the control of pulmonary ventilation.

As a consequence, the ability of human beings to successfully accomplish the physical and mental demands of their daily existence can be significantly curtailed when they ascend to high altitudes, descend underwater, or are exposed to the intense accelerative forces typical of high-performance jet and space flight; in the extreme, this can be life-threatening. However, the impetus for scientific, commercial, military, and recreational exploitation of such potentially inhospitable environments has seen the development of techniques that, through careful control of the immediate milieu, serve to extend the tolerable ranges within which humans are able to function productively. Oxygen supplementation at high altitudes, manipulation of inert gas species at depth, and the "G-suit" for opposing vascular pooling in the lower extremities during accelerative maneuvers in jet and space flight are some examples of procedures that have proved beneficial in this regard.

This chapter will address the respiratory consequences, both immediate and adaptive, of exposure to abnormal gravitational forces and ambient pressures. These will be considered in terms of the specific challenges these environments present to respiratory mechanics, pulmonary gas exchange, and ventilatory control and the extent to which they compromise the capacity for physical work.

109.2 Acceleration and Microgravity

Normogravia. Our daily experience provides many reminders that our bodies are continuously subject to the influence of the earth's gravitational field. This field imposes a gravitational force, or acceleration, acting along an axis that projects to the earth's center of mass. The resulting "weight" of the atmospheric gaseous fluid column exerts a "hydrostatic" pressure at the earth's surface, the atmospheric or "barometric" pressure (P_B):

$$P_B = \sigma gh \tag{109.1}$$

where σ is the density of the air, h is the "height" of the atmosphere, and g is the acceleration due to gravity which, under normal conditions, equals the earth's gravitational constant of 980.7 cm/s^2 (or $+1G$), i.e., normogravia.

P_B averages about 760 mmHg at sea level (although it fluctuates in response to changes in local meteorological conditions) and, as discussed later, it decreases with ascent to altitude (i.e., the term h in Eq. 109.1 is reduced). In respiratory physiology, it is conventional to refer to P_B as "zero"; supra-atmospheric pressures are therefore "positive" and subatmospheric pressures "negative."

Hypergravia. The generation of increased gravitational forces, i.e., hypergravia, occurs when the body is subjected to "linear" or "angular" acceleration. When traveling by train, for example, those of us who prefer to travel "facing the engine" are accustomed to being pushed against the seat back when the train starts to accelerate out of the station; this is relieved as the train reaches constant velocity. If the train then enters a bend with no further change of speed, a centrifugal force is generated that tends to push the body sideways away from the bend. While such forces are certainly perceptible, they are usually of little consequence for physiological well-being.

R. Greger/U. Windhorst (Eds.)
Comprehensive Human Physiology, Vol. 2
© Springer-Verlag Berlin Heidelberg 1996

Blackout. In contrast, the complex combination of changing linear velocity and directional change that occur during high-performance jet flight and also during the launching of spacecraft and their subsequent reentry into the earth's atmosphere can easily generate gravitational forces eight or nine times greater than normal (i.e., 8–9G). When the body is exposed to forces of this magnitude along its "long" axis in the direction of the feet (and therefore in the line of the major systemic blood vessels), loss of consciousness (or blackout) is an ever-present danger because of the displacement of blood away from the brain towards the dependent regions (lower limbs, feet). Indeed, loss of vision and consciousness can occur after just a few seconds' exposure to gravitational forces of only +5G [70,157]. It was only with the introduction of strategies designed to compensate for the impaired venous return (i.e., repetitive straining maneuvers, the now widely used G-suit etc.) that higher gravitational forces could routinely be tolerated without loss of consciousness [24,54,70].

Hypogravia and Microgravity. Still fewer of us will have experienced significant exposure to low gravitational forces (hypogravia) [31,57,74,76,109,110,152]. Only some 250 or so individuals have flown in space in the three decades of manned space flight, the first being the Soviet cosmonaut Yuri Gagarin on April 12, 1961; an even smaller number (the 12 United States "Apollo" astronauts) have lived and worked, albeit briefly, on another planet (the moon). The presence of residual gravitational forces during orbital space flight presents what has come to be known as microgravity. The terms "weightlessness" and 0G, although not strictly accurate, are also widely used.

Hydrostatic Pressure Gradients. What, therefore, are the consequences for respiratory function when the body is exposed to these extremes of gravitational force? Primarily, it is the liquid tissue and organ components of the body, such as blood and interstitial fluid, that are challenged, consequent to gravitationally induced changes in the hydrostatic pressure gradients that they impose. The marked redistribution of blood and interstitial fluid that occurs within the lungs has significant consequences for the structural integrity of lung tissues, the regional matching of alveolar ventilation to pulmonary capillary perfusion, and also the efficiency with which respiratory gases are exchanged between gas and blood at the lungs.

One first has to address the vector along which a gravitational force acts. This is conventionally expressed in terms of one of three major axes (x, y, and z), relative to the body (Fig. 109.1). For example, forces exerted along the long axis of the body (z) are abbreviated as +Gz when acting from the head towards the feet and −Gz when acting from the feet towards the head. Forces acting along the transverse axis of the body (x) are termed +Gx when acting from the ventral to the dorsal surface and −Gx when from the dorsal to the ventral surface. For example, during "lift-off," astronauts are constrained to the supine posture, as the short x-axis reduces the impact of accelerative forces on body fluid redistribution (compare the appreciably longer z-axis).

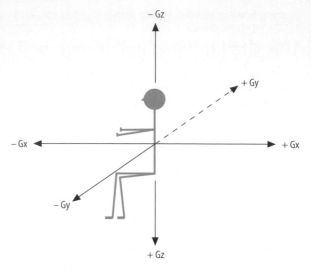

Fig. 109.1. The major axes (x, y, z) used to define the vector of a gravitational force ($\pm G$)

Lateral forces that act between the right and left sides of the body are termed $\pm Gy$. The subsequent discussion will focus on the z-axis: its greater length relative to the x- and y-axes means that gravitational forces acting along the z-axis have a far greater impact on fluid translocation and therefore respiratory function.

Simulated Dysgravic Environments. It is then necessary to consider exactly how to generate a dysgravic environment within which monitoring of pertinent physiological quantities can be reliably undertaken. It is virtually impossible, for example, to make any but the most rudimentary of measurements in the cockpit of a high-performance jet during flight. Moreover, while recent space missions have addressed aspects of pulmonary physiology, the range of techniques that have been implemented successfully in orbital flight has been, to date, somewhat limited [46,57,74,109,110,152]. However, there are several useful experimental tactics that simulate the gravitational challenges of jet flight and space flight. Common to each of these is the ability to simulate the vascular and extravascular fluid redistributions that occur with alterations in gravitational force [18]:

- Immersion in a thermoneutral liquid environment, such as water, is a relatively simple expedient for simulating the cephalad shift in blood volume that characterizes the microgravic environment of orbital flight. (Other liquid media, such as saline or silicone fluid, have also been used and confer the advantage of minimizing skin lesions and transcutaneous fluid exchange.)
- Changes in posture afford a simple means of simulating the effects of altered gravitational force. For example, a "tilt table" (with an adequate means of restraint to prevent the body from sliding) allows the body to be passively tilted around the body's center of mass (usually in the long or z-axis): changing from supine to upright provides for a brief period of increased +Gz force, while

going from upright to supine briefly approximates the weightless condition.

- Lower-body positive pressure (LBPP) in the upright posture also simulates the fluid shifts during weightlessness and provides far greater flexibility than postural change, as pressures can be applied in a graded and controlled fashion. This requires that the lower portion of the body (typically below the level of the iliac crests) be enclosed in a rigid chamber (with a plastic or rubber seal) which is connected to a controlled high-pressure source.
- Lower-body negative pressure (LBNP), in principle, resembles LBPP except that the lower body is exposed to subatmospheric pressures. This leads to fluid redistribution from the thorax and abdomen into the lower limbs. The G-suit, which has become indispensible for pilots flying high-performance aircraft, represents a practical application of LBPP. It is comprised of a system of inflatable rubber bladders that compress the legs and the abdomen in order to oppose the caudad blood volume shifts that would otherwise occur during the increased gravitational stresses of jet flight.
- "Keplerian" or parabolic flight can induce a short period of microgravity (i.e., for no more than approximately 20 s) by briefly placing a jet plane into conditions that approximate those of orbital flight. This approach provided invaluable early insight into the effects of weightlessness on pulmonary mechanics and gas exchange [46,152].
- The centrifuge can impose controlled gravitational force along any of the three primary defining axes. The magnitude of the imposed force is primarily a function of the speed at which the centrifuge rotates (i.e., the angular acceleration) and the radius of the turning circle.

109.2.1 Respiratory Mechanics

Gravitational forces influence the mechanical performance of the lungs primarily as a result of their effect on the gradient of intrapleural pressure (P_{IP}) between dependent and nondependent regions [3,95,149]. When an individual is in the upright posture, the weight of the lung tissue itself pulls the lungs down from the roof of the chest wall and pushes them down against the diaphragm, which in turn becomes displaced into the abdomen. Although gravitational forces also act on the gaseous contents of the lungs, these effects are far less striking, as the respiratory gases have a very low density in comparison to liquids such as blood. P_{IP} therefore is more negative at the (nondependent) apex than in the (dependent) basal regions.

Intrapleural Pressure. At functional residual capacity (FRC), it will be recalled (Chap. 100) that P_{IP} at the apex of the upright lungs is about 10 cm H_2O and at the base only 2 cm H_2O. This yields a P_{IP} gradient (ΔP_{IP}) of 8 cm H_2O (Fig. 109.2a), the midpoint of which is located approximately midway down the lungs (i.e., at $P_{IP} = -6$ cm H_2O). For reasons that will emerge shortly, the midpoint is termed the hydrostatic indifference point, as it is the only point along the hydrostatic (or intrapleural) pressure gradient at which the pressure is "indifferent" to changes in the magnitude of the gradient and therefore to the Gz force [157].

How might we expect an alteration in gravitational force to influence P_{IP}? Clearly, an increase in $+Gz$ force would be expected to lead to a proportional increase in the gradient of P_{IP}, through its effect on the weight of the lungs (Eq. 109.1) [23,55,94,157]. However, it is instructive to consider how this increase in ΔP_{IP} is actually accomplished. It transpires that P_{IP} is affected not only at the base of the lung, but also at the apex, i.e., the increased hydrostatic pressure gradient is "shared" between them.

This point is illustrated in the following hypothetical example. If an individual were subjected to an accelerative force of $+5Gz$, ΔP_{IP} would then increase fivefold, i.e., from 8 cm H_2O to 40 cm H_2O, an increment of 32 cm H_2O (Fig. 109.2b). The increase in lung weight would cause the lungs to: (a) pull down from the roof of the chest wall more forcefully than previously, making P_{IP} at the apex more negative than previously, and, likewise, (b) push down into the abdomen more forcefully, causing P_{IP} at the base to become less negative, or even positive. In this example, P_{IP}

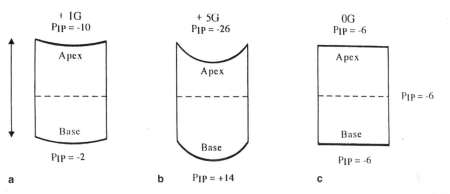

Fig. 109.2a–c. Influence of gravitational ($+Gz$) force on the gradient of intrapleural pressure (P_{IP}) between apical (i.e., nondependent) and basal (i.e., dependent) regions of the upright lung at the end of a quiet exhalation. **a** Normogravia. **b** Hypergravia ($+5G$). **c** Weightlessness ($0G$). Note that there is a "hydrostatic indifference point" at the mid-point of the P_{IP} gradient corresponding, in this example, to a P_{IP} of -6 cm H_2O. See text for further discussion

at the base would increase by 16 cm H_2O to +14 cm H_2O and at the apex would decrease by 16 cm H_2O to −26 cm H_2O. However, P_{IP} at the hydrostatic indifference point (−6 cm H_2O) would remain essentially unaffected.

Thoracic Configuration. This exaggerated displacement of the diaphragm into the abdomen during +Gz acceleration, coupled with the exaggerated downward gravitational pull exerted on the diaphragm by the abdominal contents, will cause the end-expiratory lung volume to increase [55,70,105,152] (Fig. 109.3). The overall elastic properties of the lungs and chest wall appear not to be appreciably affected by the altered gravitational status, however [23,55,94] (although, as discussed below, reductions in regional compliance in dependent regions of the lungs may occur because of (a) impaired surfactant production in collapsed alveoli and (b) impaired lymphatic drainage in collapsed lymphatics). The lungs therefore simply "move up" the normal compliance curve in response to these distending influences, rather than being displaced onto a new curve.

During weightlessness, in contrast, the intrapulmonary hydrostatic pressure gradients should be effectively abolished or at least substantially reduced [94,152,157], although this prediction awaits investigation. Hence, P_{IP} at the apex and the base of the upright lung should both be virtually identical to the P_{IP} at the hydrostatic indifference point (i.e., −6 cm H_2O) (Fig. 109.2c) [157]. There is consequently a cephalad movement of the diaphragm and a decrease in end-expiratory lung volume (Fig. 109.3) [3,93,105]. It has been suggested, however, that because diaphragmatic descent contributes more to thoracic expansion than does rib cage widening, the lung may not expand in an entirely uniform fashion during 0G and there might therefore still be some residual variation in P_{IP} [152].

Alveolar Size and Ventilation. Gravity has significant bearing on regional alveolar dimensions. As discussed in Chap. 100, the distending force that determines alveolar size is provided by the transpulmonary pressure (P_{TP}), i.e.,

$$P_{TP} = (P_{ALV} - P_{IP}) \tag{109.2}$$

where P_{ALV} is the alveolar pressure. Now, consider the lungs first when they are motionless at the end of exhalation (i.e., as airflow has ceased, $P_{ALV} = 0$ cm H_2O). At the apex of the upright lung, P_{TP} will be +10 cm H_2O, and at the base, +2 cm H_2O (Fig. 109.4a, upper panels). The apical alveoli are therefore distended at a greater degree than those at the base. This has two major consequences:

- High mechanical stresses are generated in the parenchyma at the apices of the lungs, predisposing to local tissue weakening.
- Apical units, because they are constrained to operate over a higher, less compliant region of their compliance curves, do not expand as much as basal units in response to a given increment in P_{TP} (Fig. 109.4a, lower panel), i.e., they are poorly ventilated relative to those at the base.

The widened gradient of P_{IP} and, therefore, of P_{TP} associated with increased +Gz forces exacerbates these normal inequalities in regional alveolar size (Fig. 109.4b) [23,55]. Expansion of apical units will therefore be compromised to an even greater extent than under normal gravitational conditions, further hindering ventilation of these regions.

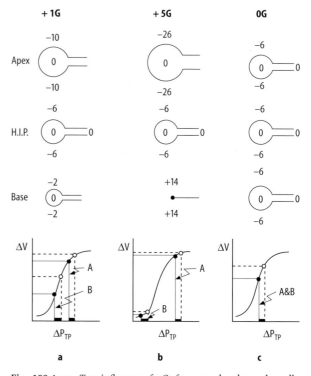

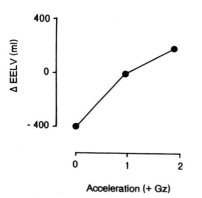

Fig. 109.3. The influence of +Gz force on end-expiratory lung volume, expressed as the change relative to normogravia ($\Delta EELV$). (Data taken from [105])

Fig. 109.4a–c. *Top*, influence of +Gz force on alveolar and small-airway dimensions at the apex, base and "hydrostatic indifference point" (*H.I.P.*) of the upright lung at the end of a quiet exhalation (see Fig. 109.2). *Bottom*, influence of +Gz force on apical (*A*) and basal (*B*) alveolar expansion (ΔV) caused by a given increase in transpulmonary pressure (ΔP_{TP}; *solid bar*). **a** Normogravia. **b** Hypergravia (+5G). **c** Weightlessness (0G). See text for discussion

In the face of extremely high +Gz forces, distending pressures at the apices of the lung may even approach levels liable to actually rupture alveoli. Ventilation of basal regions may also be impaired, however, with very high +Gz forces, i.e., sufficiently high positive P_{IP} values produce collapse of small unsupported airways, as discussed below (Fig. 109.4b, upper panels). Alveolar collapse (atelectasis) can also occur [23,70,157] and will be exacerbated at low lung volumes [47]. Clearly, under these extreme conditions, inspired gas would be diverted away from basal regions, causing an alveolar "shunt."

In contrast, one might reasonably expect the normal regional differences in alveolar size and alveolar ventilation to be eliminated during weightlessness [23,57,93,152], i.e., both the P_{IP} and P_{TP} gradients should be abolished (Fig. 109.4c). This condition is also approached under normogravic conditions when the supine posture is adopted [75], i.e., the smaller pulmonary hydrostatic fluid column significantly reduces the normal regional variations in P_{IP} and alveolar volume seen in the upright posture.

Small-Airway Caliber. The caliber of the small, unsupported intrathoracic airways is governed by the local transmural pressure gradient, i.e., the difference between P_{IP} and airway pressure (P_{AW}). As air is scarcely affected by gravitational influences (owing to its low density), there are essentially no regional differences in P_{AW} within a given branching level (or "generation number") of the respiratory tree. However, as P_{IP} increases progressively from the apex of the lung to the base, airway caliber will become progressively narrower (Fig. 109.4a, upper panels) and can lead to airway closure in basal regions, even under normal gravitational conditions, especially at low lung volumes (Chap. 100).

These effects on regional small-airway caliber are exaggerated during +Gz acceleration, again because of the larger P_{IP} gradient between apex and base (Fig. 109.4b, upper panels). Areas of total airway collapse, presumably in the dependent regions of the lung, are not uncommon [4,55]. As discussed earlier, this compromises alveolar ventilation in these regions. Under 0G conditions, however, one would expect relatively little regional variation in small-airway caliber (Fig. 109.4c, upper panels).

109.2.2 Pulmonary Gas Exchange

Based on earlier considerations (Chap. 100), it is evident that gravity is responsible for:

- The shifting away of blood from the thorax into veins in the dependent regions of the body
- The preferential distribution of pulmonary blood volume (Q_c) and flow ($\dot{Q}_c$) into dependent regions of the lungs (Fig. 109.5a).

One therefore expects these effects to be exacerbated in the presence of increased +Gz forces. Indeed, the intrapulmonary redistribution can occur to such an extent that the apical regions become completely unperfused (Fig. 109.5b), with $\dot{Q}_c$ becoming virtually confined to the dependent regions [22,50,55,143,152]. In contrast, during microgravity the regional distributions of $\dot{Q}_c$ and Q_c are more even [46,92,110,132,152] (Fig. 109.5c). These vascular redistributions have several consequences for gas exchange function in the lungs, particularly with respect to arterial oxygenation and pulmonary transvascular water exchange.

Ventilation–Perfusion Relationships. Even under normogravic conditions, the regional variations of $\dot{V}_A$ and $\dot{Q}_c$ in the upright lung lead to some maldistribution of $\dot{V}_A$ to $\dot{Q}_c$ ratios [149] (Chap. 100). High $\dot{V}_A$ to $\dot{Q}_c$ values prevail at the apex, reflecting units that are ventilated but poorly (or even not) perfused; the consequence is a high P_{O_2} and a low P_{CO_2}. In contrast, low $\dot{V}_A/\dot{Q}_c$ values are found at the base, where the perfusion is relatively greater than the ventilation (Fig. 109.5a; i.e., P_{O_2} low, P_{CO_2} high). The overventilated apical units therefore largely determine the composition of the alveolar gas, while the overperfused basal units determine the composition of the pulmonary end-capillary (and therefore arterial) blood. Consequently, arterial $P_{O_2}(P_{aO_2})$ is less than alveolar P_{O_2} (P_{AO_2}), i.e., there is a measurable alveolar-to-arterial P_{O2} difference, $P_{(A-a)O_2}$ (Chap. 101). This effect is compounded by the curvature of the O_2 dissociation curve (Chap. 101) which serves to "weight" the mean P_{aO_2} towards the P_{O_2} of the poorly saturated pulmonary end-capillary blood draining low $\dot{V}_A/\dot{Q}_c$ regions.

The marked maldistribution of $\dot{Q}_c$ that occurs with increased +Gz force further disturbs regional $\dot{V}/\dot{Q}_c$ relationships (Fig. 109.5b). Q_c and $\dot{Q}_c$ become confined to the more basal, dependent regions of the upright lungs, and, as discussed earlier, these regions may receive virtually none of the inspired gas under these conditions, because of alveolar and small-airway collapse (Fig. 109.4b, upper panels). This predisposes to alveolar shunting in the base (i.e., $\dot{V}_A/\dot{Q}_c$ = 0; Fig. 109.5b). In contrast, the apex of the lung receives

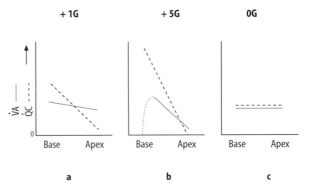

Fig. 109.5. Influence of +Gz force on the distribution of alveolar ventilation (V_A; *solid blue line*) and pulmonary perfusion ($\dot{Q}_c$; *dashed line*) between the base and apex of the upright lung. **a** Normogravia. **b** Hypergravia (+5G). **c** Weightlessness (0G). See text for discussion

the bulk of the inspirate, but no or only little perfusion, i.e., $\dot{V}_A/\dot{Q}_c \to \infty$ (Fig. 109.5b). The physiological dead space fraction of the breath (V_D/V_T) is therefore increased, relative to $+1Gz$ conditions. Accordingly, the $P_{(A-a)O_2}$ is widened and there is evidence of arterial hypoxemia [14,102,143], i.e., P_{O_2} values which result in arterial O_2 saturations of the order of 80% have been reported during exposures to $+5G$ [14]. It is presently unclear to what extent this widening of the $P_{(A-a)O_2}$ is offset to any appreciable extent by hypoxic pulmonary vasoconstriction (Chap. 101); in other words, to what extent does the hypoxia induced in basal alveoli (i.e., having very low $\dot{V}_A$ to $\dot{Q}_c$ ratios) cause local vasoconstriction? The consequence of this would be to redistribute some of the basal perfusion towards less well perfused regions of the lung, thus producing less unevenness in regional $\dot{V}_A/\dot{Q}_c$ distribution.

During weightlessness, in contrast, a rather more uniform distribution of regional $\dot{V}_A$ and $\dot{Q}_c$ [46,92,110,132,152], and therefore of $\dot{V}_A/\dot{Q}_c$, prevails (Fig. 109.5c). As a result, the $P_{(A-a)O_2}$ should be reduced relative to normogravic conditions.

Diffusing Capacity. The diffusive uptake of O_2 into the blood from alveolar gas is also likely to be influenced by $+Gz$ acceleration. As discussed in Chap. 101, the O_2 uptake ($\dot{V}_{O_2}$) is given by the following equation:

$$\dot{V}_{O_2} = D_{LO_2} (P_{AO_2} - P_{\bar{c}O_2}) \tag{109.3}$$

where D_{LO_2} is the diffusing "capacity" of the lung for O_2 and $P_{\bar{c}O_2}$ is the mean pulmonary capillary P_{O_2}. D_{LO_2} can be apportioned into: (a) a "membrane" component (D_{MO_2}) that takes account of the alveolar capillary and red cell membranes and (b) a "reaction" component that depends on the reaction rate coefficient for the chemical combination between O_2 and hemoglobin (θ) and Q_c.

The normal tendency for blood to accumulate in the dependent regions of the lungs causes regional differences in D_{LO_2}. For example, the high relative Q_c at the base of the upright lung will be associated with higher values of D_{LO_2} than at the apex. The reduction in Q_c associated with exposure to $+Gz$ forces (i.e., consequent to translocation of a component of the central blood volume into the dependent regions of the body) should therefore lead to a reduced diffusive pulmonary uptake of O_2. In turn, this might be expected to contribute to the arterial hypoxemia of $+Gz$ acceleration. Weightlessness, on the other hand, would presumably attenuate these regional inequalities. This, coupled with the increase in central blood volume (and therefore presumably Q_c), increases D_{LO_2} [46,109].

Pulmonary Edema. Areas of the lung that are highly perfused will also have high transvascular exchange rates for both water and solutes. As described in Chap. 104, under normal gravitational conditions, the interplay of Starling forces across the exchanging pulmonary microvascular bed favors the continuous net efflux of water and solute from capillaries and other small vessels in the pulmonary microvasculature. The fluid efflux from the

pulmonary microvasculature is normally balanced by fluid reabsorption from the pulmonary interstitial space into lymphatics.

However, when this balance is disrupted by high pulmonary vascular pressures, filtration rates from the microvasculature may exceed the capacity for reabsorption, leading to edema formation. During $+Gz$ acceleration, the high vascular hydrostatic pressures generated in dependent regions of the lung (Eq. 109.1) would be expected to exacerbate fluid filtration in these regions [157]. However, the extent to which the hydrostatic pressure gradient between capillary and interstitium is affected under these conditions is presently unclear. Nevertheless, high interstitial fluid pressures could cause local lymphatic collapse, compromising the drainage of filtered fluid from the interstitium and therefore predisposing towards local edema formation. In addition, as edema also lowers lung compliance, alveolar expansion could also be impaired in affected regions of the lungs (Chap. 100). The cephalad translocation of blood volume during weightlessness might also be expected to predispose towards interstitial edema in the lungs, because of the increased pulmonary vascular pressures [46].

109.2.3 Ventilatory Control

Metabolism. Ventilation ($\dot{V}_E$) is normally well matched to the demands of metabolism (Chaps. 100, 106). Therefore, despite as much as ten-to 15-fold increases in metabolic rate, arterial P_{O_2}, P_{CO_2} (P_{aCO_2}), and pH (pH$_a$) during dynamic muscular exercise remain relatively close to their resting levels. While the precise mechanisms responsible for this close "coupling" remain to be determined, it does appear that $\dot{V}_E$ normally is more closely matched with the demands for CO_2 clearance at the lung ($\dot{V}_{CO_2}$) than for O_2 uptake. Moreover, as discussed in Chap. 100, these demands may be characterized with reference to: (a) the metabolic rate, (b) the regulated level of P_{aCO_2}, and (c) the physiological dead space fraction of the breath (V_D/V_T):

$$\dot{V}_E = 863 \times \left(\frac{\dot{V}_{CO_2}}{P_{aCO_2} \left(1 - V_D/V_T\right)} \right) \tag{109.4}$$

Relatively little is known about the influence of microgravity on ventilatory control. However, it has been reported that $\dot{V}_E$ at rest and also during exercise was little, if at all, different from the corresponding ground-based values [76,91]. In contrast, during $+Gz$ acceleration, resting subjects typically hyperventilate, with consequent falls in alveolar and arterial P_{CO_2} [14,17,19,103]. Several potential sources of stimulation have been proposed to account for this:

- An increment in metabolic rate associated with recruitment of postural muscles to stabilize the body in the face of altered gravitational force [17,19].

- Increased V_D/V_T [14,19,100].
- Hypoxia-induced stimulation of the carotid body chemoreceptors (the sole ventilatory peripheral chemoreceptors in humans; Chap. 106). However, recognizing that hypoxic ventilatory stimulation is normally relatively modest until P_{aO_2} falls to levels in the region of 60–65 mmHg, the contribution from this source is unlikely to be substantial, unless the G forces are extreme [19].
- The reduction in cerebral perfusion that occurs with the caudad shift of blood into dependent regions of the body [19]. This has been proposed to stimulate the central (medullary) chemoreceptors consequent to acidification of cerebral fluids that results from an increased retention of metabolically produced CO_2 in the brain (Chap. 106).
- Secretion of catecholamines such as epinephrine and norepinephrine [19], which can lead to metabolic stimulation and sensitization of the peripheral chemoreceptors to hypoxia.
- Activation of (unspecified) reflexes in the musculoskeletal system [19].

It does appear, however, that a component of this ventilatory increase might simply reflect a heightened awareness of the immediate surroundings on the part of the subject. For example, subjects who exercise on a cycle ergometer during +Gz acceleration – and are therefore presumably concentrating on the motor act of pedaling – have been reported to have a reasonably normal P_{aCO_2} [102].

Work Tolerance. What can therefore be concluded about the contribution of ventilatory factors to work tolerance during gravitational stress? The answer, at present, is relatively little. During hypergravia, for example, the $\dot{V}_E$ requirement at a given work rate is greater than for normogravic conditions because (a) the O_2 uptake ($\dot{V}_{O_2}$) and therefore presumably the $\dot{V}_{CO_2}$ appear to be increased and (b) V_D/V_T is higher [102]. Whether this predisposes to ventilatory requirements at higher work rates which are beyond the mechanical capabilities of the respiratory system is not known, however. Neither is it known whether respiratory perceptions are heightened to limiting degrees during hypergravic exercise.

109.2.4 Adaptations

Considerations of adaptations to the stresses of +Gz acceleration have focused almost exclusively on cardiac and circulatory function [18]. Frequent and regular exposure over prolonged periods does not appear to be associated with the development of pathologic effects. In contrast, chronic exposure to +Gz stress may be less benign. Evidence from animal investigations points towards an incidence of myocardial damage and atherosclerotic lesions, for example. How the respiratory system might be affected can clearly only be speculated upon, however. Does the sustained mechanical stress that the nondependent regions

of the lung parenchyma are subjected to induce chronic, irreversible tissue damage? And is there a predisposition to tissue consolidation in the dependent regions which become atelectic during acute +Gz exposure? Further, we should not ignore the consequences of chronic arterial hypoxemia which can predispose to right-sided heart failure, consequent to abnormal elevation of pulmonary artery pressure.

The weightless environment may prove more revealing, given the ability to successfully operate prolonged sojourns in space stations. Most notable of these to date are the Soviet "Salyut" and "Mir" missions, which have lasted for as a long as 1 year. Despite this impetus, we yet know relatively little about respiratory adaptations to microgravity; investigations have focused to a large extent on circulatory function (in particular, the development of orthostatic intolerance) and also the problems of bone loss and skeletal muscle atrophy [31].

109.3 Diving

Hyperbaria. Increases in local atmospheric pressure are encountered when incursions into the earth's interior are undertaken—by way of deep mines and caves, for example. This condition, which is termed hyperbaria, reflects the extended "height" of the hydrostatic column of the earth's atmosphere (Eq. 109.1). Hyperbaric environments can also be created artificially. In 1839, Triger (a French engineer) described the use of compressed air to prevent the natural seepage of water through surrounding rock formations during the boring of mine shafts ([16], p. 358). This technique was subsequently extended to underwater applications, such as the construction of bridge piers, for which a cast-iron "caisson" (or tube of concentric rings) charged with compressed air was let down onto the sea or river bed. In addition, in recent years, the "hyperbaric chamber" has become important in the medical treatment of individuals suffering from conditions such as decompression sickness, burn trauma, ischemia, bacterial infection, and carbon monoxide poisoning.

Absolute Atmospheres. The stresses imposed on the body are magnified considerably, however, when the hyperbaric environment is created in water rather than air. This reflects the far greater density of water relative to air (Eq. 109.1) and the fact that water is not a compressible medium. Therefore, while P_B increases linearly with increasing depth below the water's surface (Eq. 109.1), even modest descents are associated with significant increases in P_B, i.e., by 1 atmosphere (or 760 mmHg) for every 10 m (33 ft) of descent in seawater (or 34 ft for fresh water, which is slightly less dense) (Fig. 109.6). In contrast, a terrestrial descent of some 6000 m would be required to effect the same increase in P_B. The large pressures encountered in diving are most conveniently expressed in multiples of absolute atmospheres (ATA), where 1 ATA represents the atmospheric pressure at the surface at sea

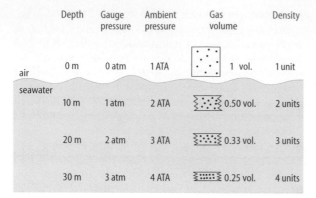

Depth	Gauge pressure	Ambient pressure	Gas volume	Density
air 0 m	0 atm	1 ATA	1 vol.	1 unit
seawater 10 m	1 atm	2 ATA	0.50 vol.	2 units
20 m	2 atm	3 ATA	0.33 vol.	3 units
30 m	3 atm	4 ATA	0.25 vol.	4 units

Fig. 109.6. Influence of submersion depth on the pressure, volume, and density of gas in the lungs. *ATA*, absolute atmospheres. (Modified from [133])

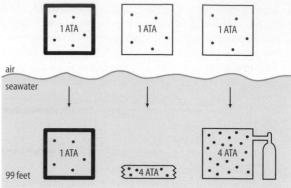

Fig. 109.7. Influence of submersion on the pressure and volume of gas in the lungs. *Left*, in a submarine, as it is a rigid structure, there is no compression of gas and therefore no increase in pressure. *Middle*, during breath-hold diving, the lungs are compressed by the increased ambient pressure. *Right*, during scuba diving, lung compression is prevented by supplying gas pressurized to the ambient value. *ATA*, absolute atmospheres. (Modified from [133])

level. Extreme depths can now be attained. For example, the demand for natural fuels has pushed exploration and development of offshore oil and gas fields to progressively greater depths, with divers working as deep as 300–400 m (31–41 ATA).

Boyle's Law and Dalton's Law. When a diver descends below the surface, the increasing P_B has two immediate consequences for respiratory function: compression of the thoracic gas volume and, therefore, increased alveolar partial pressures. There is an inverse relationship between the volume of a gas (V) and the pressure (P) it exerts (Boyle's law):

$$PV = K \qquad (109.5)$$

K is a constant equal to nRT, where n is the number of moles of gas, R is the universal gas constant, and T is the absolute temperature of the gas. This demonstrates that a given increase of P_B close to the surface will cause a far greater reduction in volume than will the same pressure increment at depth (Figs. 109.6, 109.7). Dalton's law indicates how P_B influences respiratory gas partial pressures:

$$P_B = P_{O_2} + P_{N_2} + P_{CO_2} \qquad (109.6)$$

In other words, the increased P_B at depth leads to proportional increases in P_{O_2}, P_{CO_2}, P_{N_2}, and any other inert gases that are present (e.g., argon, helium).

It may perhaps come as a surprise that these two seemingly innocuous reactions can, in the inexperienced or careless diver, lead to potentially damaging and even lethal conditions [15,87,123,133].

Barotrauma. This refers to the tissue damage that can occur in gas-containing cavities of the body when P_B changes appreciably. Consider, for example, the paranasal sinuses. Because they are rigid and poorly ventilated, they cannot undergo any appreciable volume change during a descent. Consequently, while both P_B and body fluid hydrostatic pressures increase, the intrasinus pressure must

remain at normal atmospheric values. As the sinuses are highly vascularized, they will therefore be subjected to an excessive influx of blood. This leads to fluid exudation into the lining mucous membranes and even rupture of blood vessels.

The middle ear, although a rigid structure, is not normally presented with this problem as it is connected to the highly compliant lungs by the eustachian tube. When a diver descends, equalization of pressure in the middle ear with P_B is therefore readily accomplished, i.e., experienced divers open their eustachian tubes each few meters of descent simply by swallowing, so that air can enter the middle ear. However, if the eustachian tube becomes obstructed by local edema (e.g., as in an upper respiratory tract infection), pressure equalization presents problems. There is, therefore, the likelihood of edema and vessel rupture in the lining of the middle ear. In extreme cases, the tympanic membrane may even rupture.

Damage, termed lung squeeze, can occur to the lungs when a breath-hold dive is taken to depths that induce shrinkage of the thoracic gas volume beyond the limits of thoracic compressibility (see Sect. 109.4.1). Again, effusions from the pulmonary capillaries and tissue rupture can occur. Barotrauma of the lungs may occur during very rapid ascents. Overdistention of the lungs ensues, which predisposes to rupture of the lung parenchyma and loss of air from alveoli. Scuba divers only experience this problem if they have an equipment malfunction.

Oxygen Toxicity. Substantial increases in the inspired P_{O_2} (P_{IO_2}) occur during deep, supported dives. This can evoke toxic reactions which have consequences for both central nervous system and pulmonary function [18,29,33]. An increased inspired P_{O_2} (P_{IO_2}) of approximately 1 ATA or more, for example, can provoke convulsions which can lead to drowning in scuba divers.

Problems such as pulmonary edema, shortness of breath (dyspnea), and coughing can develop at P_{IO_2} values of only approximately 0.5 ATA. This therefore places at risk individuals who remain at increased pressure for prolonged intervals. For example, divers who are required to work at considerable depth may actually remain for many days at the "working" P_B by living in a pressure chamber at the surface between work periods (e.g., on a rig); such individuals are usually referred to as saturation divers, for reasons that will become clear shortly. Higher P_{IO_2} values (e.g., 2–4 ATA) exacerbate these symptoms, leading to permanent parenchymal damage. There should be concern, in addition, for patients undergoing treatment in hyperbaric chambers, who may breathe pure O_2 pressurized to as much as 3 ATA.

Decompression Sickness. This has a relatively long history, materializing when compressed air started to become widespread in construction projects. Affected workers walked with a characteristically "bent" posture because of pain in the hips and knees, which could last for several hours after surfacing. For many of us, therefore, "the bends" has become synonymous with decompression sickness. However, this is only one of several symptoms, others being the chokes (coughing, dyspnea, and chest pain), impaired sensory function, paralysis, and even death.

Decompression sickness is a result of the formation of N_2 bubbles in blood and other body fluids when ascents are attempted too rapidly [36,141]. To understand the etiology of this condition, it is first necessary to recall Henry's law: the amount of a gas that can dissolve in a particular liquid is a function of the partial pressure of the gas and its solubility in the liquid (Chap. 101). Although N_2 is poorly soluble in plasma and other body fluids, the substantial increases in alveolar P_{N_2} (P_{AN_2}) that occur at depth lead to appreciable quantities of dissolved N_2 being formed. A subsequent, slow equilibration of N_2 occurs between alveolar gas, blood, and tissues; indeed, it may take 6 h or more to "saturate" tissues fully at the new P_{N_2}. The saturation diver is therefore an individual who remains at depth (or its pressure equivalent) sufficiently long for N_2 equilibration to be complete [96].

With ascent, the direction of N_2 exchange reverses. As P_B falls, P_{AN_2} declines and therefore creates a gradient for N_2 flux from tissues. This allows them to release the N_2 that they absorbed during the descent slowly into the alveoli, where it is cleared during exhalation. However, because the N_2 equilibration process is slow, problems are encountered when a diver ascends too quickly. N_2 cannot diffuse out of tissues fast enough to keep pace with the falling P_{N_2}: as a result, the tissues become "supersaturated" with N_2–this concept was central to the pioneering work of J.S. Haldane and his colleagues in 1908 [20] on the formulation of safe decompression practices. In other words, there is now suddenly too much N_2 dissolved in the tissue fluids, relative to the now falling P_{N_2}. This forces N_2 out of solution in the form of bubbles of N_2. This process can be compared with the fizzing of CO_2 from soda pop when the bottle is opened (especially after the contents have been vigorously agitated).

There is still some debate regarding the precise consequences of bubble formation [133]. The major possibilities are considered to be: (a) the bubbles themselves, leading to tissue damage and obstruction of the microcirculation, (b) the high N_2 levels, impairing cellular function, and (c) circulatory function being affected, by activation of platelets and components of the complement system, for example. Apart from carefully constraining rates of ascents [96,139,140] (Fig. 109.8), one of the strategies employed for reducing the likelihood of decompression sickness is to

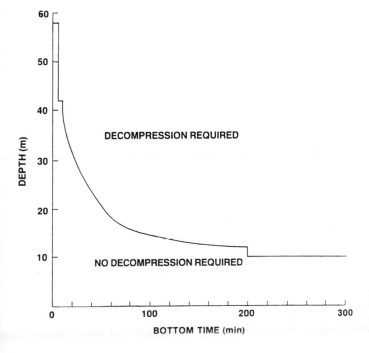

Fig. 109.8. Requirement for decompression from a single dive, based upon depth and duration of submersion (allowing a minimum of 12 h between consecutive dives). (From [100])

2183

substitute a proportion of the inspired N_2 with helium. The presence of helium lessens the incidence of bubble formation because: (a) its lower partial pressure and lower solubility lead to smaller volumes of dissolved helium, and (b) its smaller size speeds its diffusion and therefore allows a more rapid equilibration.

Inert Gas Narcosis.

This is a potential hazard of deep diving. It is caused by excessively high partial pressures of N_2 and other inert gases such as neon and xenon (but interestingly not helium) [15]. The French term for this condition, *l'ivresse des grandes profondeurs* (which has been loosely translated as "rapture of the deep"), provides a telling clue to the nature of the symptoms, although the etiology remains unclear. One hypothesis proposes that, at high pressures, sufficient N_2 dissolves in the lipid components of nerve cell membranes that they are caused to expand slightly. Ionic permeabilities are increased as a result, impairing synaptic transmission. At depths of about 50 m, a feeling of euphoria is experienced, while at greater depths (and greater partial pressures), disorientation, dysphoria, and loss of consciousness may ensue.

High-Pressure Nervous Syndrome.

This syndrome (HPNS) can also develop at extreme depths [15]. When pressures are extremely high, the lipid components of the body can also undergo a slight compression. Consequently, a decrease in cell volume is thought to result, affecting membrane permeability and transport function. Such a situation clearly bears on the functional integrity of excitable membranes. Poor coordination, tremors, dizziness, and nausea typically ensue. The symptoms can be alleviated by the addition of small amounts of N_2 to the inspirate, which may serve to "reexpand" cells that have undergone compression. It is therefore now standard practice for deep divers to inhale "trimix," a mixture of helium, oxygen, and nitrogen [15,126].

Snorkeling.

Modest excursions below the surface can be simply and easily accomplished with a snorkel. This is a breathing tube which maintains continuity with the atmosphere when the diver is close to the surface, i.e., it functions as a simple extension of the airway. Airway and alveolar pressures therefore remain close to the surface P_B. However, while the tube allows ventilation to be effected, the respiratory muscles are required to work harder than at the surface to maintain a given $\dot{V}_E$. This is because:

- The snorkel provides an additional resistance to airflow.
- The increase in the effective anatomical dead space demands a higher level of $\dot{V}_E$ to maintain alveolar ventilation, i.e., the increased dead space will "waste" more of the overall ventilation (Eq. 109.4).
- The increased external hydrostatic pressure acting on the thorax will oppose the normal outward recoil of the chest wall, unless the inspiratory muscles can exert a sufficiently high countering pressure [43,67,87,122]. The maximum external pressure that can successfully be opposed is normally of the order of 100–150 cm H_2O (or

approximately 1.10–1.15 ATA) [3], which corresponds to a depth of only about 1 m. If the diver goes only slightly deeper, not only will end-expiratory lung volume (and therefore the expiratory reserve volume) be reduced, but normal inspiratory excursions will become limited.

Snorkeling also affects pulmonary blood volume [43,85,87]. When an individual walks into the water and becomes fully immersed, the normal gravity-dependent pooling of venous blood that occurs in the lower limbs on land is largely abolished. This can be explained by water being more dense than air: the external hydrostatic pressure acting on the body is therefore increased, and to a greater extent at the feet than at the head (Eq. 109.1). The consequently higher intravascular pressures in the lower limbs cause blood to "move up" into the low-pressure vessels of the thorax [7]. In the snorkeler, it is important to recognize that intravascular pressures in the thorax are much lower because the thoracic gas volume is in direct continuity with the atmosphere at the surface. This translocation of venous blood from dependent regions into the pulmonary circulation predisposes to vascular engorgement and even edema formation. The extent to which the pulmonary blood volume is increased will, of course, depend on the posture adopted. Nevertheless, any increase in peripheral vascular resistance that occurs (e.g., from cold water exposure) will amplify this vascular redistribution.

Pressurization.

It should be evident, therefore, that the snorkel is not the best means of supporting anything but the most shallow of dives. Deep descents can be accomplished by pressurizing the inhaled gas to the ambient P_B at depth (Fig. 109.7). This avoids both thoracic compression and intrapulmonary vascular engorgement, as well as all their related shortcomings for respiratory function and work tolerance under water. Pressurization may be accomplished in several ways [88]:

- Hose-and-pump systems, in which atmospheric air is pressurized at the surface and then pumped down a hose to the diver.
- Pressurized chambers, located on ships or rigs, are particularly advantageous in situations that require prolonged sojourns at depth (e.g., salvage). The diver remains in a stable hyperbaric environment and is transported between the surface and underwater work site by a diving bell. As described earlier, these are the preferred conditions for the saturation diver, i.e., the diver who works for prolonged periods at depth and whose efficiency would therefore be hampered by repeated regular decompression episodes.
- Self-contained underwater breathing apparatus (scuba) is widely used for dives of shorter duration and less extreme depth, with pressurized gas being supplied from a portable gas tank (Fig. 109.9). Gas is released from the tank to the diver via a demand regulator whenever mouth pressure falls below P_B (i.e., during inspiration);

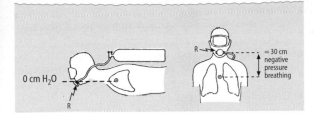

Fig. 109.9. Influence of posture on the pressure at which gas is delivered to a scuba diver. *Left*, in a prone position, the scuba regulator (*R*) lies in the same horizontal plane as the "centroid" of the chest and therefore delivers gas at a pressure equal to the ambient pressure acting on the chest at the centroid. *Right*, in an upright position, the regulator lies above the centroid and therefore the regulator delivers gas at a pressure which is less than the ambient pressure at the centroid, constraining the chest to expand against a negative pressure load. See text for further discussion. (Modified from [87])

expired gas is voided into the water (forming a stream of bubbles).

109.3.1 Respiratory Mechanics

The major concern with respect to respiratory-mechanical function during deep, supported descents is the increased work of breathing that is required to maintain a given level of ventilation. This arises from two sources.

Increased Respired Gas Density. Hyperbaria leads to an increased density of respired gas (Fig. 109.6), resulting in an increased work of breathing at depth, even for modest levels of ventilation. This can be readily appreciated by considering Rohrer's original formulation of the pressure requirement for effecting a given airflow ($\dot{v}$), Chap. 100):

$$\Delta P = P_{ALV} - P_B = \underbrace{K_1 \times \eta \times \dot{v}}_{\substack{\text{laminar} \\ \text{component}}} + \underbrace{K_2 \times \sigma \times \dot{v}^2}_{\substack{\text{turbulent} \\ \text{component}}} \quad (109.7)$$

where η and σ are the viscosity and density of respired air, respectively, and K_1 and K_2 are constants representing the geometry of the respiratory tree. The increased σ encourages turbulent airflow. This is because of the greater Reynold's number (R_e) associated with a given gas velocity ($\dot{v}$):

$$R_e = \frac{\dot{v} r \sigma}{\eta} \quad (109.8)$$

where *r* is airway radius. When R_e is less than 1000, airflow is laminar, but becomes largely turbulent when R_e is caused to rise beyond a level of about 1500. The elevated σ increases the proportion of the airways through which airflow becomes turbulent at a given level of ventilation and, consequently, exacerbates the driving pressure (Eq.109.7). Indeed, at a depth of 400 m (41 ATA), flow is turbulent throughout the entire respiratory tree [126].

A greater driving pressure is therefore required to sustain a given airflow at depth. This requires the muscles of inspiration and expiration to work harder. Furthermore, the more positive P_{IP} values that are likely to occur in expiration as a result, especially with the increased ventilatory drive of exercise, could cause dynamic compression of small unsupported airways at lower $\dot{V}_E$ values than at the surface and (as discussed in Chap. 100) limit expiratory airflow [5,126,131,140,158].

These factors, together with the resistance of the breathing apparatus, therefore conspire to limit the maximal airflows that can be achieved volitionally at depth. This is manifest in the progressive decrease of maximum voluntary ventilation (MVV) that occurs with increasing depth; for example, at 4 ATA the MVV is approximately halved [4,65,79,80,89] (Fig. 109.10). These effects on the work of breathing are alleviated to some extent by substituting a lighter inert carrier gas such as helium (which is 40% less dense) for N_2 in the inspirate.

Pressure Loading. It is most usual to supply gas at a pressure referenced to what is termed the centroid of the chest; under these conditions, the expiratory reserve volume (FRC – RV) is assumed to equal the normal terrestrial value at the surface [87,88]. In the upright posture, the centroid is located some 14 cm below the sternal notch; it moves to approximately 7 cm behind the sternal angle when the body is prone (Fig. 109.9) It can therefore be appreciated that changes in body position per se at an otherwise stable depth can cause the hydrostatic pressure at the centroid to deviate from the inspiratory gas pressure. For example, when a diver is swimming in a horizontal plane, there will be little difference in the pressure at the centroid relative to the breathing regulator (Fig. 109.9). However, if the diver then stops swimming and adopts a vertical (head-up) posture, the centroid will now lie approximately 30 cm below the regulator, i.e., inspiratory gas pressures will now be less than pressure at the centroid. The pulmonary system is consequently presented with a static negative pressure head of 30 cm H_2O against which the inspiratory muscles have to work to preserve the excursions of respiratory volume and flow [87,122].

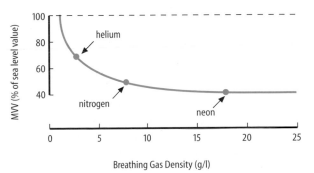

Fig. 109.10. Influence of respired gas density on the maximum voluntary ventilation (*MVV*). Note that the relationship holds regardless of whether density is increased by raising the pressure of respired gas or by altering its composition. (Modified from [79])

This has several deleterious consequences. End-expiratory lung volume is reduced [87]. Furthermore, there is an increased gradient of P_{IP} between the nondependent and dependent extremes of the lung, reflecting the greater density of water over air (Eq. 109.1). Further, the lower airway and alveolar pressures, relative to abdominal pressures, cause intrathoracic accumulation of blood. This vascular engorgement of the lungs can lead to a reduction in lung compliance [87,142], rendering lung inflation more difficult. In addition, fluid may accumulate in the alveolar interstitium [10,122]. These several effects predispose to small-airway closure and alveolar collapse in the dependent regions of the lungs [10], and airways resistance is increased [2]. This outcome will be exacerbated in divers who inhale gas mixtures that are enriched with oxygen, consequent to development of absorption atelectasis (Chap. 101) [11,37]. As a result, the work of breathing is increased [67,134].

In contrast, the head-down posture presents the pulmonary system with a static positive pressure head, as the pressure at the regulator is greater than that at the centroid. As a result, the inspiratory muscles will have to work less hard to accomplish a given level of ventilation. However, this raises the prospect of increased expiratory muscle work. These posturally related respiratory-mechanical loads can, however, be offset by appropriate breathing-pressure compensation [134].

109.3.2 Pulmonary Gas Exchange

Hyperbaria during submersion leads to a widening of the alveolar-to-arterial P_{O_2} difference [122,126]. This reflects a complex interplay of factors, most important of which relate to diffusive exchange and the regional dispersion of $\dot{V}_A/\dot{Q}_c$.

Diffusion in the Gas Phase. The increased respired gas density asociated with underwater descent slows the rate at which gases can diffuse through the gas phase. This is because the gas is "heavier" and therefore moves more slowly. The boundary that separates "alveolar" gas and "dead space" gas therefore shifts distally towards (or even into) the alveoli [140]. This causes an increase in the effective anatomical dead space, and therefore compromises alveolar ventilation. For example, the physiological dead space is effectively doubled at a depth of 650 m (or 66 ATA) [119]. Unless $\dot{V}_E$ increases to compensate for this, P_{ACO_2} and therefore P_{aCO_2} will rise (Eq. 109.4).

Diffusing Capacity. Because of the pulmonary vascular engorgement that occurs during submersion, it is reasonable to expect the pulmonary capillary blood volume to be increased and also its regional distribution to become more uniform (because of the greater volume and, therefore, an increased pulmonary artery pressure) [6,7,107]. These factors should therefore increase D_{LO_2} during submersion (see Sect. 109.2.2) [7,108].

Ventilation–Perfusion Relationships. The presence of a widened alveolar-to-arterial P_{O_2} difference in hyperbaric conditions has been argued to reflect $\dot{V}_A/\dot{Q}_c$ maldistribution. During submersion, the dependent regions of the lungs are susceptible to airway narrowing and collapse during submersion [6,7,122] (see Fig. 109.4b, upper panels), which leads to alveolar shunting. This presumably reflects: (a) an increased P_{IP} gradient within the lungs (similar to exposure to increased $+Gz$ forces) because of the increased density of water relative to air (Eq. 109.1) and (b) the tendency for the abdominal contents to be "pushed" in the cephalad direction by the greater external hydrostatic pressures at the level of the abdomen relative to the thorax. However, the remaining, more apical regions of the lungs show a more homogeneous $\dot{V}_A/\dot{Q}_c$ distribution than at the surface [7,108], i.e., not only do they receive a more even pulmonary capillary perfusion [6,7,107], but they also are ventilated more uniformly [7].

The significance of these effects on gas exchange efficiency will depend on the orientation of the diver and therefore on the "height" of the pulmonary hydrostatic pressure column. In addition, the extent to which a diver is intrinsically predisposed towards airway closure is likely to be a contributing factor. For example, older individuals (who manifest a greater degree of small-airway collapse) are more likely to manifest a widenend $P_{(A-a)O_2}$ at depth [108]. However, it should be emphasized that, unlike hypergravity, a widened $P_{(A-a)O_2}$ does not normally lead to arterial hypoxemia during diving: P_{aO_2} values are almost invariably high because of the hyperbaria.

109.3.3 Ventilatory Control

Ventilation during diving is most usually less than expected, given the metabolic demands associated with the task at hand, at least in seasoned divers. As a result, P_{ACO_2} and P_{aCO_2} are not uncommonly higher than at the surface [80,119] (Eq. 109.4). Several mechanisms have been put forward to account for this:

- The increased P_{IO_2} and therefore P_{aO_2}, leading to suppression of peripheral chemoreceptor activity (Chap. 106)
- A greater resistance to breathing introduced by the increased respired gas density and the breathing apparatus [48,79,86]
- A reduced ventilatory responsiveness to inhaled CO_2 [53,80,86,128].

These several factors could conspire to reduce the overall level of ventilatory stimulation during exercise at depth. However, as the ventilatory capacity at depth is clearly compromised (Fig. 109.10), the possibility should also be considered that the hypoventilation during exercise at depth might also reflect a mechanical constraint on ventilation itself especially during exercise [5,65,66,87,140]. In other words, as the limits of airflow generation are approached, further ventilatory increases of an appropriate magnitude become progressively harder to produce. Fatigue of the respiratory muscles could ensue. One would expect the presence of helium in the inspirate to ameliorate these effects somewhat (Eq. 109.7).

Exertional Dyspnea. As a greater proportion of the ventilatory capacity available at depth is therefore likely to be drawn upon than for equivalent terrestrial activities, a further consequence is an exacerbated awareness of unpleasant respiratory sensations during physical activity [66,119,131]. Although the etiology of this exertional dyspnea is uncertain, it appears to be one of the primary reasons for the impaired exercise tolerance at depth [131,140]. This is in marked contrast to maximal activities on land, such as cycling and running. These effects are heightened when divers are presented with static negative pressure loads (i.e., when, because of postural changes, inspiratory gas pressure is less than at the centroid; Fig. 109.9). A further limitation is imposed on their ability to increase $\dot{V}_E$ during exercise, presumably because of the additional posturally related load placed on the muscles of inspiration. Under these conditions, dyspneic sensations are increased and work tolerance is reduced [119,135,136].

109.3.4 Adaptations

Desensitization. What respiratory adaptations might therefore allow divers to improve their exercise tolerance at depth? As mentioned earlier, experienced divers typically demonstrate a reduced ventilatory CO_2 responsiveness [53,80,86,128]. Furthermore, the degree of CO_2 retention at depth in such individuals appears to increase with experience. However, the etiology of this condition remains uncertain [80,126]. Were the diver also to become progressively desensitized to the unpleasant respiratory sensations experienced at depth, these effects might collectively be viewed as being potentially beneficial to exercise tolerance at depth. These issues await resolution.

109.4 Breath-Hold Diving

A consideration of respiratory responses to the hyperbaric environment would not be complete without consideration of the breath-hold diver. Breath-hold diving is a relatively common recreational practice, as it requires little or no additional equipment and (certainly at first sight) is relatively easy to perform. However, it can be extremely hazardous when performed to considerable depths if appropriate care is not taken. Nonetheless, there are commercial breath-hold divers, such as the women divers in Japan (Ama) who harvest sponges from the sea bed. These women descend repeatedly to depths as much as 20 m (or 3 ATA) on a routine basis. Significantly greater depths have been successfully attained on single "competition" dives. For men, the record depth of 112 m (or approximately 12 ATA) was established in 1989 by F. Ferreira from Cuba. For women, the record depth of 107 m (approximately 11

ATA) was attained by A. Bandini from Italy, also in 1989. This is an appropriate juncture at which to ask what factors govern how deep a breath-hold diver can descend and what factors dictate when the diver elects to return to the surface. The limits of chest wall compressibility determine the depth that can be reached, while the dive is terminated when the diver can no longer override the "desire to breathe" and, in the extreme, by unconsciousness as a result of cerebral hypoxia.

109.4.1 Respiratory Mechanics

Breath-hold dives are typically initiated from high lung volumes, close to or at the normal "out-of-water" TLC. The lower limit to which lung volume can be compressed during a dive in adults is set by the normal RV [113] (Fig. 109.11). Boyle's law allows us to derive the pressure and, therefore, the depth at which this lower limit will be reached:

$$\frac{P_B(\text{surface})}{P_B(\text{depth})} = \frac{V(\text{depth})}{V(\text{surface})} = \frac{RV}{TLC} \qquad (109.9)$$

For example, an individual whose TLC and RV at 1 ATA are 7 l and 2 l, respectively, would have a critical P_B of $(1 \times 7)/2 = 3.5$ ATA, which would be attained at 25 m (Eq. 109.9). Clearly, the larger the TLC and/or the lower the RV (i.e., increasing TLC/RV), the deeper a diver should be able to descend [87].

However, we have already established that breath-hold dives have been successfully performed to considerably greater depths. How is this brought about? The influx of

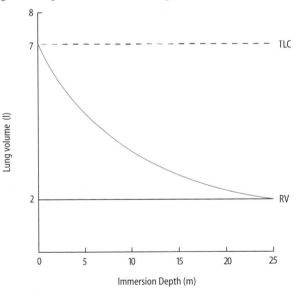

Fig. 109.11. Reduction of lung gas volume with increasing depth during a breath-hold dive. Note that lung volume declines from a value close to the normal terrestrial total lung capacity (*TLC*) to one that approaches residual volume (*RV*) at a depth, in this example, of 25 m. See text for further discussion. (Modified from [100])

blood into the thorax that occurs with submersion [6] (see Sect. 109.3) serves to displace a volume of air from the lungs [32,121]. An additional contribution may also come from areas of lung collapse, because of the cephalad displacement of the diaphragm at these low lung volumes [1]. This further reduction of thoracic gas volume therefore occurs despite thoracic dimensions being unable to decrease any further, but without structural damage occurring, i.e., it occurs as a result of simple displacement, rather than compression. Deeper descents lead to lung squeeze (see Sect. 109.3).

109.4.2 Pulmonary Gas Exchange

Although the lungs clearly will not be ventilated during a breath-hold dive, exchange of respiratory gases between alveolar gas and pulmonary capillary blood still takes place. The terrestrial breath hold, which occurs with no change of P_B, provides a useful frame of reference for appreciating how O_2 and CO_2 exchange are influenced during descent and ascent [82,97].

With the start of the breath hold, metabolically produced CO_2 will continue to diffuse down the P_{CO_2} gradient from mixed venous blood into alveolar gas. However, as it can no longer be eliminated, alveolar P_{CO_2} (P_{ACO_2}) starts to rise (Fig. 109.12. dashed line). This reduces the driving pressure, i.e., P_{ACO_2} approaches mixed venous P_{CO_2} ($P_{\bar{v}CO_2}$). As a result, CO_2 diffusion slows and the P_{ACO_2} increase begins to flatten off (Fig. 109.12). For essentially similar reasons, P_{AO_2} falls during the breath hold. This fall is more marked than the rise of P_{ACO_2}, however (Fig. 109.12, dashed line), i.e., the tissues continue to extract O_2 at an appropriate rate

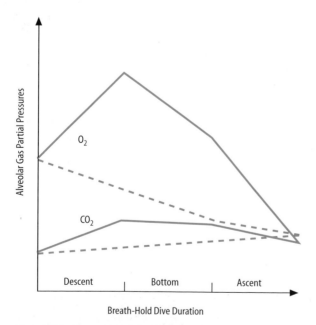

Fig. 109.12. Responses of alveolar P_{CO_2} (*blue lines*) and P_{O_2} to a terrestrial breath hold (*dashed lines*) and a breath-hold dive (*solid lines*). See text for discussion. (Taken from [100])

throughout the breath hold, causing a steady fall of the mixed venous P_{O_2} ($P_{\bar{v}O_2}$). In contrast to CO_2, therefore, the driving pressure for O_2 diffusion is maintained. Indeed, the greater rate of O_2 extraction from alveolar gas, relative to CO_2 delivery, causes lung volume to decline progressively throughout the breath hold; this contributes to the rate of P_{ACO_2} increase.

The complications introduced by the changing P_B during a breath-hold dive will now be addressed [81,87,97,100]. During descent, P_{ACO_2} increases at a greater rate than for a simple breath hold (Fig. 109.12, solid line), because lung volume becomes progressively compressed by the increasing P_B (Fig. 109.11). The driving pressure for alveolar CO_2 exchange, therefore, becomes progressively smaller and, at sufficient depth or with sufficient time, may be abolished entirely. Indeed, it can even reverse (i.e., $P_{ACO_2} > P_{\bar{v}CO_2}$), so that CO_2 flows from the alveolus to the pulmonary capillary, causing P_{ACO_2} to decrease slightly (Fig. 109.12, solid line). This compression of lung volume causes P_{AO_2} to increase (i.e., more than offsetting the decrease of P_{AO_2} expected because of continuing O_2 consumption).

Correspondingly, the reexpansion of lung volume during ascent causes both P_{ACO_2} and P_{AO_2} to decrease (Fig. 109.12, solid lines). The driving pressure for CO_2 diffusion from pulmonary capillary blood into alveolar gas is therefore reestablished. However, the influence of increasing lung volume more than offsets the increase in P_{ACO_2} expected from the progressive accumulation of metabolic CO_2 within the alveoli. Nevertheless, it is the falling P_{AO_2} during ascent that is potentially hazardous for the breath-hold diver. This is especially the case if P_{AO_2} falls below $P_{\bar{v}O_2}$, as this will cause O_2 to actually leave the blood and flow into the alveolar gas [104]. It is vital that P_{AO_2} does not fall to levels that induce unconsciousness (i.e., normally in the region of 30 mmHg) before the diver has reached the surface. Thus, the diver who elects to undertake a particularly deep dive and/or remains at depth for a relatively long time is at risk in this regard.

109.4.3 Ventilatory Control

The ability to hold one's breath is influenced by several factors, all of which bear on the duration for which a breath-hold diver may safely remain submerged and therefore work productively.

Breath-Hold Termination. Several factors have been proposed [84,97]. The progressive asphyxia (i.e., decreasing P_{O_2} and increasing P_{CO_2}) that develops in the blood during a terrestrial breath hold (Fig. 109.12) leads to an increasing desire to breathe. The hypercapnia stimulates both central and peripheral chemoreceptors (Chap. 106). Moreover, while the falling P_{aO_2} only stimulates the peripheral chemoreceptors, the potentiation between the hypoxic and hypercapnic stimuli at this site evokes a progressively more marked peripheral chemoreceptor contribution as the breath hold proceeds.

Consequently, individuals who are caused to inhale

hypoxic gas mixtures prior to a breath hold cannot hold their breath for so long [84]. Prior O_2 breathing, in contrast, prolongs a breath hold [84]. Likewise, individuals who have previously had their carotid bodies surgically resected can also hold their breath considerably longer than normal individuals [38].

In addition, the lung shrinkage that occurs during breath holding may induce unpleasant respiratory sensations [84,97]. These are thought to arise from activation of mechanoreceptors in the respiratory tree and in the respiratory muscles. Increased metabolic rate further limits breath-hold duration, presumably because of a more rapidly developing asphyxic condition in blood, coupled with a more rapid lung shrinkage as the exchange rates of O_2 and CO_2 are faster.

The influence of psychogenic factors is less easy to quantify. Certainly, a breath hold can be prolonged by factors that divert attention away from the maneuver, such as physical movements and other tasks requiring conscious attention [84].

In the context of submersion, therefore, it is of interest that increases in P_B can prolong a volitional breath hold [64,84]. At first sight, the progressive reduction of lung volume and increase of P_{aCO_2} during descent (Figs. 109.11, 109.12) should exacerbate the intensity of respiratory sensations associated with a breath hold. The influence of metabolic rate during the dive should also be considered. The more rapidly a diver descends, the more rapid will be the rates of lung shrinkage and CO_2 accumulation. This is likely to be particularly marked when diving in cold water sufficient to induce shivering (i.e., a further stimulus to metabolic rate). However, the possibility has been raised that effects such as these could be mitigated by the associated increase in P_{aO_2} [100] (Fig. 109.12).

Hyperventilation Prior to diving. The ill-advised and, regrettably, sometimes fatal practice of hyperventilating immediately prior to a breath-hold dive warrants consideration. Hyperventilation prolongs the duration of a breath hold because it depletes the body CO_2 stores [97]. A greater proportion of the CO_2 produced metabolically during the breath hold is therefore diverted to the stores in order to replenish them. Consequently, P_{aCO_2} increases less rapidly during the breath hold. This "depotentiates" the peripheral chemoreceptors, i.e., the hypoxic stimulus interacts with an attenuated hypercapnic stimulus (Chap. 106). Not only is the breath-hold breaking point delayed, but it therefore occurs at a lower P_{aO_2} than normal. When the hypocapnia is particularly marked, dizziness and even convulsions can occur. This reflects impaired cerebral oxygenation arising from hypocapnic-induced cerebral vasoconstriction.

Hyperventilation prior to a breath-hold dive can therefore be seen to be potentially hazardous. Not only will the rate of P_{aCO_2} increase during the descent be more sluggish than normal, but it will occur from a hypocapnic rather than a normocapnic baseline. There will therefore be little respiratory awareness of the degree to which P_{aO_2} will have fallen, i.e., the contributions to ventilatory drive from the

central and peripheral chemoreceptors will be substantially less than if the hyperventilation had not been undertaken (Chap. 106). The diver may therefore remain at depth in relative comfort for sufficiently long that the body's O_2 stores become depleted beyond the point of "safe return" to the surface. In other words, while P_{aO_2} at depth may be well above the level that leads to a loss of consciousness, the subsequent fall that occurs with ascent (Fig. 109.12) may result in blackout before the surface is reached. Death from drowning is not an uncommon consequence in these circumstances.

109.4.4 Adaptations

What of adaptations that might allow the breath-hold diver to remain submerged, and therefore work, for longer periods? With regular and frequent diving, TLC has been reported to increase, while RV becomes smaller [25,129]. This increased TLC to RV ratio allows greater depths to be achieved (see Sect. 109.4.1). The ability to withstand the high ambient hydrostatic pressures at depth is assisted also by an increased inspiratory muscle strength [129].

Like "supported" divers, breath-hold divers have a relatively low ventilatory responsiveness to inhaled CO_2 [120,129]. It has been postulated that this reduces the respiratory sensation associated with a given degree of hypercapnia during a breath hold. This therefore allows the maneuver to be sustained for a longer period. It is less clear at present whether the ventilatory responsiveness to hypoxia is also reduced [100]. However, it has to be borne in mind that any influence tending towards lower P_{aO_2} values being attained at depth during such a dive will predispose towards blackout during the subsequent ascent, unless other adaptive changes take place within the tissue O_2 delivery and utilization mechanisms of the body.

109.5 Ascent to Altitude

Hypobaria. Ascent to high altitude is associated with a reduction in local ambient pressure, a condition termed hypobaria. This is consequent to a reduction in the "height" of the hydrostatic pressure column of the earth's atmosphere (Eq. 109.1). Extreme altitudes have long been recognized to be deleterious to well-being, largely because of compromised oxygenation, although the effects of lowered ambient temperature should not be ignored. Nonetheless, over 10 million people live at altitudes in excess of 3000 m (10 000 ft) and may ascend to altitudes as high as 5800 m (approximately 19 000 ft) on a regular basis (i.e., for commercial, scientific, or military purposes). Indeed the highest permanent human habitation in the Andes is at altitudes that approach 5000 m (approximately 16 500 ft). The ultimate terrestrial pinnacle, Mount Everest, lies at approximately 8848 m (29 028 ft) and has successfully been climbed without the use of supplemental oxygen. Furthermore, despite cabin pressurization, transcontinental commercial flight exposes large numbers of people (some on a regular basis) to hypobaric hypoxic conditions corresponding to altitudes up to some 2450 m (approximately 8000 ft).

The French physiologist, Paul Bert was the first to ascribe the deleterious consequences of high-altitude exposure to the reduced P_{O_2} values that result from the hypobaric environment [16]. In addition to many scientific excursions to high altitudes, Bert pioneered the use of low-pressure chambers for both physiological and pathophysiologial investigation of the immediate and longer-term consequences of high-altitude exposure. The advent and popularization of ballooning and, subsequently, powered flight provided considerable impetus to research in this area.

Acute Mountain Sickness and High-Altitude Pulmonary Edema. The avaliability of O_2 supplementation techniques has greatly extended the range of altitudes that can be tolerated. However, medical complications can ensue [63,125,151]. Excursions to even moderately high altitudes, e.g., 3000 m, lead to acute mountain sickness in susceptible individuals. This is not a recent observation [40]. Characteristic symptoms include palpitations, fatigue, headache, dizziness, nausea, and insomnia. High-altitude pulmonary edema (HAPE) is a further (and serious) development. The exaggerated elevation of pulmonary artery pressure (pulmonary hypertension) is associated with cyanosis, poor work tolerance, dyspnea (especially on exertion), rapid shallow breathing (tachypnea), and tachycardia. A frothy, blood-stained sputum may be produced, reflecting hemorrhage and disruption of the alveolar surfactant lining by fluid transudates from pulmonary vessels. If untreated, HAPE can result in death; rapid return to lower altitudes is essential. Cerebral edema is more rare and, again, if not treated it will result in coma and death.

Monge's Syndrome. Long-term exposure to high altitudes can lead to chronic mountain sickness or Monge's syndrome [99]. A dusky, bluish complexion reflects the markedly cyanotic condition of the blood. Exercise tolerance is reduced, and headache, lethargy, and dizziness are common. An exaggerated polycythemia increases blood viscosity, which places a considerable stress on the heart; this leads to pulmonary hypertension and right ventricular hypertrophy. Relocation to lower altitudes can ameliorate the condition.

109.5.1 Respiratory Mechanics

Because of the compressibility of the earth's atmosphere, P_B does not fall linearly with increases in elevation. Rather, the relationship is essentially exponential (although not strictly so, as air temperature falls with increasing altitude). A given increase in altitude near sea level is therefore associated with a larger reduction in P_B than is the case at very high altitudes (Fig. 109.13). For example, an ascent of some 1500 m (5000 ft) from sea level (P_B, approximately 760 mmHg) up to Denver (P_B, approximately 630 mmHg)

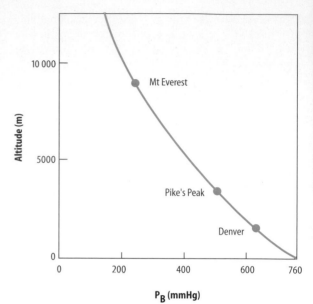

Fig. 109.13. Influence of altitude on barometric pressure. (Modified from [150])

represents a 130 mmHg reduction in P_B. A similar further ascent from Denver to Pike's Peak, which is at an altitude of about 3000 m (10 000 ft) and a P_B of about 520 mmHg, is associated with a further reduction in P_B of only 110 mmHg.

The reduction in P_B that occurs with ascent to high altitudes has several consequences for respiratory mechanics. The most significant of these relate to the reductions that occur in air density and in P_{IO_2}.

Decreased Respired Gas Density. Referring back to Eqs. 109.7 and 109.8, it is evident that the decrease in σ with increasing altitude implies a reduction in the turbulent component of airflow (i.e., R_e is decreased). This effect is more marked at high levels of flow, reflecting the proportionality to $\dot{v}^2$, rather than $\dot{v}$. Thus, the work of breathing associated with a particular high level of airflow, as in exercise, should be significantly less at high altitude [51]. This effect is manifest in the increase of MVV that occurs in lowlanders on ascent to high altitudes [34] (Fig. 109.10). It should be noted, however, that these improvements in respiratory-mechanical function are offset to some extent by airway narrowing, consequent possibly to influences of hypoxia and hypocapnia on bronchomotor tone [34,90].

In theory, the range of achievable ventilations during exercise should therefore be extended. However, as maximum O_2 uptake at altitude is substantially less than at sea level [9,44,62,111,153], maximum exercise $\dot{V}_E$ may not increase much beyond maximum exercise values at sea level, certainly for acute exposures [45,49].

Lung Compliance. Transient, acute increases of pulmonary interstitial fluid volume have been reported immediately following ascent to high altitudes [68,73]. This can lead to a decreased lung complicance and also to narrowing (and possibly closure) of small airways in dependent regions of the lungs [30,90], although these effects normally subside within a day or so [21,34]. However, were the tidal excursions of lung volume to become sufficiently large that they encroached on the higher, less steep, and therefore less compliant regions of the compliance curve, the inspiratory muscles would have to work against an increased "elastic" load [137]. This could occur, for example, when lowlanders are required to exercise intensely at high altitudes. Under these conditions, $\dot{V}_E$ is substantially increased consequent to the stimulating actions of both the hypoxia and the exercise.

The inspiratory muscles are also challenged, sometimes severely, in HAPE. Both the increased interstitial fluid volume and the presence of alveolar exudates that "wash" surfactant from the alveolar epithelial surface contribute to a reduced lung compliance. Excessively high distending pressures are therefore required for normal inflation (Chap. 100).

109.5.2 Pulmonary Gas Exchange

Hypobaria leads to a progressive decrease in P_{IO_2} with increasing altitude (Eq. 109.6; Fig. 109.14). In turn, P_{AO_2} and P_{aO_2} are also reduced (Fig. 109.14). However, the extent of these falls will depend on the following:

- The magnitude of the compensatory hypoxia-induced hyperventilation that normally occurs in sea-level natives acutely exposed to high altitudes
- The degree to which the exchange of respiratory gases between alveolar gas and pulmonary capillary blood is impaired by the hypoxic conditions

These considerations assume importance when it is recognized that loss of consciousness can occur at P_{AO_2} values of approximately 30 mmHg, at least for brief exposures [27]. The major challenge to pulmonary O_2 exchange in sea-level natives who ascend to high altitude is diffusion across the alveolar–capillary membrane.

Diffusion. The diffusive exchange of O_2 at the lungs becomes challenged when ascending to high altitudes, largely because the initial driving pressure for O_2 diffusion from alveolar gas into the pulmonary capillaries is less than at sea level (Fig. 109.15). This is because, under hypoxemic

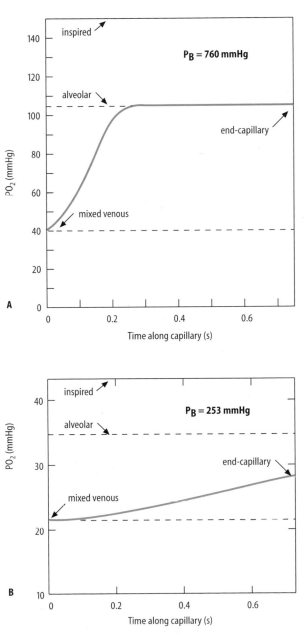

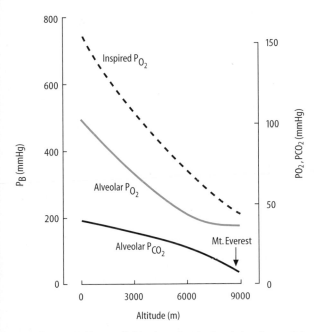

Fig. 109.14. Influence of altitude on inspired and alveolar partial pressures of O_2 and CO_2. (Modified from [151])

Fig. 109.15A,B. Estimated time course of blood P_{O_2} along a representative pulmonary capillary under resting conditions at sea level (A) and on the summit of Mount Everest (B). See text for discussion. (Modified from [151])

conditions, the loading (and unloading) of O_2 in the blood takes place over a lower and therefore steeper region of the dissociation curve than is the case at sea level (Eq. 109.3), i.e., a smaller increase of pulmonary capillary O_2 content is associated with a given increase of P_{O_2} (Chap. 101) [106,156]. This effect may be exacerbated by the expansion of the pulmonary interstitial fluid volume that can occur transiently on ascent to high altitude [68, 73], i.e., the path length for diffusion is increased because of a thickening of the alveolar–capillary membrane.

The influence of blood O_2 affinity for hemoglobin is important in this context. The O_2 dissociation curve has been reported to undergo a leftward shift immediately on arrival at high altitude [117]. This has been ascribed to the influence of the concomitant respiratory alkalosis (Chap. 101). This advantage appears to be lost fairly rapidly, however. This is because the the erythrocytic levels of 2,3-diphosphoglycerate (2,3-DPG) begin to rise slowly, tending to reduce Hb-O_2 affinity [83] (Chap. 101). The final outcome of these competing influences on O_2 transport will depend on the relative magnitudes of the alkalotic and 2,3-DPG responses. These, in turn, are likely to be a function both of the altitude and the duration of sojourn.

Nevertheless, diffusion equilibrium across the alveolar–capillary membrane takes longer to attain than at sea level [156] (Fig. 109.15). This may not present a problem for O_2 exchange at rest, as long as elevations are relatively moderate, i.e., the extended time to attain diffusion equilibrium appears not to encroach on the mean pulmonary capillary transit time (T_c) [138]. However, with the more severe degrees of hypoxia prevailing at extreme altitudes, it has been argued that the diffusion process is slowed to such an extent that the equilibration time not only approaches, but eventually exceeds T_c, therefore contributing to a widening of the $P_{(A-a)O_2}$ at these altitudes [145,156]. During exercise, an increased cardiac output exacerbates this outcome, as T_c is reduced (i.e., $T_c = Q_c/\dot{Q}_c$) [138,144]. As a consequence, arterial O_2 levels fall progressively with increasing work rates, not only at extreme altitudes but also at more modest elevations [12,145,153,154].

Ventilation–Perfusion Relationships. With ascent to high altitude, the consequent hypoxia that develops in the alveolar gas induces a global pulmonary hypoxic vasoconstriction (see Sect. 109.2.2) and therefore an increased pulmonary artery pressure [56,118]. The increased pulmonary arterial pressures that prevail at high altitude would be expected to produce a more even distribution of $\dot{V}_A$ to $\dot{Q}_c$ ratios than at sea level [39,58], i.e., more blood would be diverted towards the nondependent (apical) regions of the lungs. Furthermore, both low and high $\dot{V}_A/\dot{Q}_c$ units are likely to exchange O_2 only over the lower, steeper regions of the O_2 dissociation curve.

Therefore, the regional variations in end-capillary P_{O_2} should be appreciably less than at sea level (Chap. 101). However, such improvements appear to be relatively insignificant when considered in the light of the adverse effect of high altitude on diffusional O_2 exchange [145,156]. Indeed, it has been reported that the overall $\dot{V}_A/\dot{Q}_c$ distribu-

tion actually worsens at extremely high altitudes, especially during exercise [52,58,144].

109.5.3 Ventilatory Control

The primary challenge to ventilatory control mechanisms at high altitude is presented by arterial hypoxemia. A fall of P_{aO_2} into the region of 60–65 mmHg provides a clear ventilatory stimulus via the peripheral chemoreceptors (Chap. 106). In addition, the curvilinear $\dot{V}_E$–P_{aO_2} relationship predicts that the lower the P_{aO_2} range, progressively greater $\dot{V}_E$ increments will result from a given decrease in P_{aO_2} (Chaps. 78, 106).

Compensatory Alveolar O2 Pressure Increase. The consequence of hypoxia-mediated hyperventilation is that alveolar and arterial P_{O_2} increase, providing a crucial compensation that extends the tolerable range of high altitudes [60,114,155]. For example, without hypoxic-induced hyperventilation (i.e., with a normal sea-level value for P_{ACO_2} of 40 mmHg), a climber on the summit of Mount Everest (P_{IO_2} approximately 43 mmHg) would have a P_{aO_2} approaching zero [151]. However, a fivefold increase in $\dot{V}_E$ would lower P_{ACO_2} from 40 mmHg to about 7.5 mmHg and raise P_{aO_2} to over 30 mmHg [151]–and the climber into consciousness.

It is appropriate now to consider the extent to which the respiratory system may be viewed as limiting exposure to extreme altitudes. This can be considered within the context of the alveolar air equation (Chap. 101):

$$P_{AO2} = P_{IO2} - \frac{P_{ACO2}}{R} + F \qquad (109.10)$$

The term F, i.e., $[P_{ACO_2} \times F_{IO_2} \times (1 - R)/R]$, where F_{IO_2} is the inspired O_2 fraction, is normally small and is therefore often neglected. For example, with air breathing F is approximately 2.1 (i.e., P_{ACO_2}, 40 mmHg; F_{IO_2}, 0.21; R, approximately 0.8); at high altitude, it will be somewhat less because of the hypoxia-induced hypocapnia (Fig. 109.14). Note that while P_{IO_2} is reduced at high altitudes, F_{IO_2} is not.

The alveolar air equation (Eq. 109.10) indicates that, at a given altitude (for which P_{IO_2} can be assumed to be essentially constant), P_{AO_2} will depend on P_{ACO_2} and the respiratory exchange ratio (R) which, under steady state conditions (or nearly so), is equal to the respiratory quotient (RQ). Several scenarios can therefore be expected to accentuate the compensatory increases in P_{AO_2} and P_{aO_2}:

- Individuals with a naturally high peripheral chemoreflex sensitivity to hypoxia would be expected to induce a greater compensatory hyperventilation at a given altitude (i.e., driving P_{ACO_2} to lower levels).
- A high-carbohydrate diet (i.e., RQ, 1.0) rather than the more usual "western" mixed diet (RQ, approximately 0.8 – 0.85) leads to a higher P_{aO_2} at altitude, both at rest and during exercise [61].

Such "improvements" in P_{aO_2} predispose favorably towards an extended range of tolerable altitudes as well as to an enhanced tolerance at a given altitude. This state of affairs would also presumably allow higher work rates to be performed than would be the case if the P_{aO_2} had not been affected. Indeed, mountain climbers with a naturally high ventilatory responsiveness to hypoxia can, in general, attain greater heights than can their peers who have normal or low hypoxic responsiveness [124].

Respiratory Alkalosis. The reduction in P_{aCO_2} will cause a lowering of P_{CO_2} in both arterial blood and cerbrospinal fluid (CSF). As a result, arterial and CSF pH will be increased (respiratory alkalosis) [41,69,77,127] (Fig. 109.16). The full ventilatory effects of the arterial hypoxemia are therefore initially constrained by the alkalotic condition. In other words, as CO_2-H^+ and hypoxia potentiate in their stimulatory effects at the carotid bodies, the respiratory alkalosis in the blood will lessen the effect of the hypoxic stimulus (Chap. 106). Similarly, the CO_2-H^+ "drive" at the central chemoreceptors will also be reduced, because of the CSF alkalosis (Chap. 106).

However, over a period of hours and days (and even weeks), compensatory mechanisms act to restore the arte

rial and CSF pH towards normal [41,69,77,78,125]. However, the compensations are not complete, i.e., rather than being entirely abolished, the alkalotic condition is only lessened [41,69,77]. This is accomplished by reducing the bicarbonate concentrations ($[HCO_3^-]$) in both arterial blood and CSF, the former by increased excretion from the kidneys and the latter through loss of HCO_3^- ions across the blood–brain barrier. The consequence of this for pH regulation can be examined by means of the Henderson–Hasselbalch equation:

$$pHa = pK' + \log\frac{[HCO_3^-]_a}{\alpha \times P_{aCO_2}} \qquad (109.11)$$

where α is the solubility coefficient for CO_2 in plasma and pK' is related to the first "apparent" ionization constant of carbonic acid (6.1 in plasma). For example, complete compensation for the respiratory alkalemia of high altitude would require $[HCO_3^-]_a$ to be reduced so as to restore the ratio of $[HCO_3^-]_a$ to P_{aCO_2} to its prior sea-level value.

As a result of these compensations, $\dot{V}_E$ continues to increase further over a period of days and weeks following an ascent to high altitude [35,62,71,114,155] (Fig. 109.16). This reflects an alleviation of the initial "restraint" placed on the

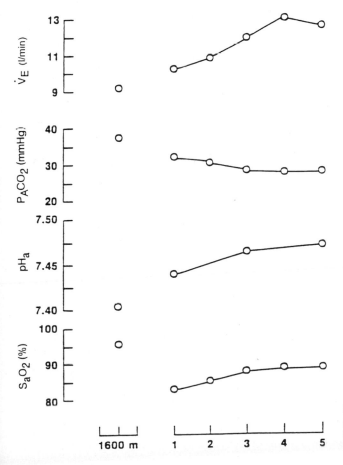

Fig. 109.16. Time course ventilatory ($\dot{V}_E$), alveolar P_{CO_2} (P_{ACO_2}, arterial pH (*pHa*) and arterial O_2 saturation (SaO_2) responses during the first 5 days of high-altitude exposure (4300 m). See text for discussion. (Modified from [159])

2193

hypoxic ventilatory response on arrival at high altitude. In addition, it has been argued that a slow increase in carotid body hypoxic sensitivity may also contribute [13,41,147].

So what are the consequence of these events for ventilatory demands during physical work at high altitudes? Equation 109.4 clearly indicates that the hypocapnic condition will require a greater $\dot{V}_E$ to effect a given rate of metabolic CO_2 clearance than would be the case for normocapnia. This is evident in the augmented $\dot{V}_E$ response to progressive exercise at high altitudes [9,35,62,78,112,124]. Despite this, however, $\dot{V}_E$ does not appear to reach limiting values during exercise (see Sect. 109.5.1). This is because, as discussed earlier, physical work capacity is compromised with ascent to altitude, the more so the higher the altitude.

Dyspnea. While $\dot{V}_E$ itself appears not be a limiting factor for exercise at high altitude, the associated dyspneic sensation may be considerable in sea-level natives and should therefore viewed as constraining exercise tolerance [125]. Factors contributing to the dyspnea are likely to include the following:

- The high absolute levels of ventilation per se
- A "dyspneagenic" effect of the hypoxic stimulation of the carotid bodies [28,146]
- Reflex bronchoconstriction, also induced by the hypoxia-induced carotid body stimulation [101]
- Airway irritation and constriction arising from drying of the airway mucosa.

109.5.4 Adaptations

In contrast to the other environmental stressors discussed here, there is a considerable body of information on the temporal response profiles central to long-term adaptations at high altitudes. Life-long or prolonged exposure to high altitudes induces a wide range of adapative responses, from which the respiratory system is not immune.

Andeans native to high altitudes characteristically have large chests (particularly when allowance is made for body size) and therefore large lung volumes [72]. As a result, the diffusive exchange of O_2 at the lungs is more efficient (i.e., D_{LO_2} is greater than in comparable lowland residents), and this is reflected in a reduced $P_{(A-a)O_2}$ [42,98,115]. The probable cause lies in a large pulmonary gas-exchanging "surface," a greater pulmonary capillary blood volume, and also the elevated hemoglobin levels that result from the polycythemic response to high altitude exposure [116] (i.e., greater amounts of O_2 can be taken up into pulmonary capillary blood). In fact, although arterial O_2 saturation may be diminished, the polycythemia can result in the arterial O_2 content actually exceeding normal ambient lowland values. However, the adverse effects of this polycythemic reaction on blood viscosity and cardiac function may outweigh the benefits for pulmonary gas exchange [98,125].

Loss of Adaptation. Individuals who are either born at high altitude or who have resided at high altitude for decades have classically been regarded as having resting levels of $\dot{V}_E$ that are lower than those of acclimatized lowlanders (with P_{aCO_2} being consequently higher) and a "blunted" ventilatory response to hypoxia [26,78,130,148]. This is thought to reflect an attenuated hypoxic sensitivity of the carotid chemoreceptors and may develop as a result of morphologic alterations in the structure of the carotid bodies themselves (similar changes have been reported at sea level in individuals who are chronically hypoxemic because of cardiopulmonary disease) [8,63] (Chap. 106). A further source of this "blunting" may originate from depressant actions of the hypoxia on central nervous system neurons involved in respiratory integration [34,71,78,147]. It is difficult to see how these responses confer an advantage for survival at high altitudes, and perhaps therefore they are better viewed as a manifestation of loss of adaptation.

In this context, it is of some interest that high-altitude natives from the Himalayas present a strikingly different picture. They typically have higher resting ventilations and lower P_{aCO_2} values and a reasonably normal hypoxic ventilatory responsiveness [59,78]. It is unclear why, at least in this regard, the Himalayan highlander appears to be more "resistant" to the stress of high-altitude exposure than does the Andean. It is also unclear whether this confers an advantage in terms of the capacity for physical work at high altitudes. Another question is whether an apparently "intact" carotid body chemosensitivity predisposes the Himalyan high-altitude native to exacerbated degrees of dyspnea on exertion that might assume importance in determining work tolerance.

109.6 Conclusions

Scientific investigations conducted at the extremes of dysgravia and dysbaria have provided considerable insight into the immediate, short-term consequences for respiratory function, physical work, and survival. However, the potential for chronic and even life-time adaptation to the microgravity of space and the deep-ocean environment, for example, is unknown. Nevertheless, the pursuit for commercial, political, and military advantage – if not for exploration and scientific discovery – will no doubt lead to these temporal limits being inexorably pushed back.

References

1. Agostoni E (1965) Limitation to depths of diving mechanics of chest wall. In: Rahn H, Yokoyama T (eds) Physiology of breath-hold diving and the Ama of Japan. Washington DC, pp 139–145 (National Academy of Sciences National Research Council publication 1341)
2. Agostoni E, Gurtner G, Torri G, Rahn H (1966) Respiratory mechanics during submersion and negative-pressure breathing. J Appl Physiol 21:251–258
3. Agostoni E, Mead J (1964) Statics of the respiratory system. In: Fenn WO, Rahn H (eds) Handbook of physiology,

section 3, vol l: respiration. American Physiological Society, Bethesda, pp 387–409

4. Anthonisen NR, Bradley ME, Vorosmarti J, Linaweaver PG (1971) Mechanics of breathing with helium-oxygen and neon-oxygen mixtures in deep saturation diving. In: Lambertsen CJ (ed) Underwater physiology: proceedings of the 4th symposium on underwater physiology. New York, Academic, pp 339–345

5. Anthonisen NR, Utz G, Kruger MH, Urbanetti JS (1976) Exercise tolerance at 4 and 6 ATA. Undersea Biomed Res 3: 95–102

6. Arborelius M Jr, Balldin UI, Lilja B, Lundgren CEG (1972) Hemodynamic changes in man during immersion with the head above water. Aerospace Med 43:592–598

7. Arborelius M Jr, Balldin UI, Lilja B, Lundgren CEG (1972) Regional lung function in man during immersion with the head above water. Aerospace Med 43:701–707

8. Arias-Stella J, Valcárcel J (1976) Chief cell hyperplasia in the human carotid body at high altitudes: physiologic and pathologic significance. Hum Pathol 7:361–373

9. Åstrand PO (1954) The respiratory activity in man exposed to prolonged hypoxia. Acta Physiol Scand 30:343–368.

10. Baer R, Dahlbäck GO, Balldin UI (1987) Pulmonary mechanics and atelectasis dring immersion in oxygen breathing subjects. Undersea Biomed Res 14:229–240

11. Balldin UI, Dahlbäck GO, Lundgren CEG (1971) Changes in vital capcity produced by oxygen breathing during immersion with the head above water. Aerospace Med 42:384–387

12. Barcroft J, Binger CA, Bock AV, Doggart JH, Forbes HS, Harrop G, Meakins JC, Redfield AC (1923) Observations upon the effect of high altitude on the physiological processes of the human body, carried out in the Peruvian Andes, chiefly at Cerro de Pasco. Philos Trans R Soc [B] 211:351–480

13. Barnard P, Andronikou S, Pokorski M, Smatresk N, Mokashi A, Lahiri S (1987) Time-dependent effect of hypoxia on carotid body chemosensory function. J Appl Physiol 63:65–691.

14. Barr PO (1963) Pulmonary gas exchange in man, as affected by prolonged gravitational stress. Acta Physiol Scand 28 [Suppl 207]:1–46

15. Bennett PB, Elliott DII (eds) (1982) The physiology and medicine of diving. Best, San Pedro, California

16. Bert P (1878) Barometric pressure: researches in experimental physiology (translated by MA Hitchcock and FA Hitchcock). College Book, Columbus, Ohio

17. Bjurstedt H, Rosenhamer G, Wigertz O (1968) High-G environment and responses to graded exercise. J Appl Physiol 25:713–719

18. Blomqvist CG, Stone HL (1983) Cardiovascular adjustments to gravitational stress. In: Shepherd JT, Abboud FM (eds) Handbook of physiology. The cardiovascular system III. Peripheral circulation, part 2. American Physiological Society, Washington DC, pp 1025–1063

19. Boutellier U, Arieli R, Farhi LE (1985) Ventilation and CO_2 reponse during +Gz acceleration. Respir Physiol 62:141–151

20. Boycott AE, Damant GCC, Haldane JB (1908) The prevention of compressed-air illness. J Hyg (Camb) 8:342–347

21. Brody SB, Lahiri S, Simpser M, Motoyama EK, Velasquez T (1977) Lung elasticity and airway dynamics in Peruvian natives to high altitude. J Appl Physiol 42:245–251

22. Bryan AC, MacNamarra, Simpson J, Wager HN (1965) Effect of acceleration on the distribution of pulmonary blood flow. J Appl Physiol 20:1129–1132

23. Bryan AC, Milic-Emili J, Pengelly D (1966) Effect of gravity on the distribution of pulmonary ventilation. J Appl Physiol 21:778–784

24. Burton RR, Leverett SD, Michaelson ED (1974) Man at high, sustained +Gz acceleration: a review. Aerospace Med 45:1115–1136

25. Carey CR, Schaefer KE, Alvis H (1956) Effect of skin diving on lung volumes. J Appl Physiol 8:519–523

26. Chiodi H (1957) Respiratory adaptations to chronic high altitude. J Appl Physiol 10:81–87

27. Christensen EH, Krogh A (1936) Fliegerunterschungen; die Wirkung niedriger O_2-Spannung auf Höhenflieger. Scand Arch Physiol 73:145–154

28. Chronos N, Adams L, Guz A (1988) Effect of hyperoxia and hypoxia on exercise-induced breathlessness in normal subjects. Clin Sci 74:531–537

29. Clark JM, Lambertsen CJ (1971) Pulmonary oxygen toxicity: a review. Pharmacol Rev 23:37–133

30. Coates G, Gray G, Mansell A, Nahaamias C, Powles A, Sutton J, Webber C (1979) Changes in lung volume, lung density, and distribution of ventilation during hypobaric decompression. J Appl Physiol 46:752–755

31. Convertino VA (1990) Physiological adaptations to weightlessness: effects of exercise and work performance. Exere Sports Sci Rev 18:119–166

32. Craig AB Jr (1968) Depth limits of breath-hold diving (an example of fennology) Respir Physiol 5:14–22.

33. Crapo JD (1986) Morphologic changes in pulmonary oxygen toxicity. Annu Rev Physiol 48:721–731

34. Cruz JC (1973) Mechanics of breathing in high altitude and sea level subjects. Respir Physiol 17:146–161

35. Cruz JC, Reeves JT, Grover RF, Maher JT, McCullough RE, Cymerman A, Denniston JC (1973) Ventilatory acclimatization to high altitude is prevented by CO_2 breathing. Respiration 39:121–130

36. D'Aoust BG (1984) Inert gas exchange and counter diffusion in decompression sickness and diving medicine. In: Bachrach AJ, Matzen MM (eds) Proceedings of the 8th symposium on underwater physiol. Undersea Medical Society, Bethesda, Maryland, pp 159–170

37. Dahlbäck GO, Balldin UI (1983) Positive-pressure oxygen breathing and pulmonary atelectasis during immersion. Undersea Biomed Res 10:39–44

38. Davidson JT, Whipp BJ, Wasserman K, Koyal SN, Lugliani R (1974) Role in the carotid bodies in the sensation of breathlessness during breath-holding. N Engl J Med 290:819–822

39. Dawson A (1972) Regional lung function during early acclimatization to 3100 m altitude. J Appl Physiol 33:218–223

40. de Acosta J (1590) Historia natural y moral de las indias, en que le tratan las cosas notables del cielo, y elementos, metales, plantas, y animales dellas: y los ritas, y ceremonias, leyes, y gouierno, y guerras de los indios. de Leon, Seville

41. Dempsey JA, Forster HV (1982) Mediation of ventilatory adaptations. Physiol Rev 62:262–346

42. Dempsey JA, Reddan WG, Birnbaum ML, Forster HV, Thoden JS, Grover RF, Rankin J (1971) Effects of acute through life-long hypoxic exposure on exercise pulmonary gas exchange. Respir Physiol 13:62–89

43. Denison D (1982) The effect of pressure on the lungs. In: Bonsignore G, Cumming G (eds) The lung in its environment. Plenum, New York, pp 343–351

44. Dill DB, Myhre LG, Phillips EE Jr, Brown DK (1966) Work capacity in acute exposures to altitude. J Appl Physiol 21:1168–1176

45. Elliott PR, Atterbom HA (1978) Comparison of exercise responses of males and females during acute exposure to hypobaria. Aviat Space Environ Med 49:415–418

46. Engel LA (1991) Effect of microgravity on the respiratory system. J Appl Physiol 70:1907–1911

47. Engel LA, Grassino A, Anthonisen NR (1975) Demonstration of airway closure in man. J Appl Physiol 38:1117–1125

48. Fagreus L (1974) Cardiorespiratory and metabolic functions during exercise in the hyperbaric environment. Acta Physiol Scand [Suppl] 414:1–40

49. Fagreus L, Karlsson J, Linnarsson D, Saltin B (1973) Oxygen uptake during maximal exercise at lowered and raised ambient air pressures. Acta Physiol Scand 87:411–421

50. Farhi LE (1984) Physiological shunts: effects of posture and gravity. In: Johansen K, Burgren W (eds) Cardiovascular shunts. Munksgaard, Copenhagen, pp 322–333

51. Fenn WO (1954) The pressure volume diagram of the breathing mechanism. In: Boothby WM (ed) Handbook of respiratory physiology. USAF School of Aviation Medicine, Randolph Field, pp 99–124

52. Gale GE, Torre-Bueno JR, Moon RE, Saltzman HA, Wagner PD (1985) Ventilation-perfusion inequality in normal humans during exercise at sea level and simulated altitude. J Appl Physiol 58:978–988

53. Gelfand R, Lambertsen CJ, Peterson R (1980) Human respiratory control at high ambient pressures and inspired gas densities. J Appl Physiol 48:528–539

54. Gillingham KK, Krutz RW (1974) Effects of the abnormal acceleratory environment of flight. Aeromedical Review no 20–74. USAFSAM, Brooks AFB, TX

55. Glaister DH (1977) Effect of acceleration. In: West JB (ed) Regional differences in the lung. Academic, New York, pp 323–379

56. Groves BM, Reeves JT, Sutton JR, Wagner PD, Cymerman A, Malconian MK, Rock PB, Young PM (1987) Operation Everest II: elevated high-altitude pulmonary resistance unresponsive to oxygen. J Appl Physiol 63:521–530

57. Guy HJB, Prisk GK, Elliott AR, Deutschman RA, West JB (1994) Inhomogeneity of pulmonary ventilation during sustained microgravity as determined by single-breath washouts. J Appl Physiol 76:1719–1729

58. Haab P, Held DR, Ernst H, Farhi LE (1969) Ventilation perfusion relationships during high altitude adaptation. J Appl Physiol 26:77–81

59. Hackett PH, Reeves JT, Grover RF, Weil JV (1984) Ventilation in human populations native to high altitudes. In: West JB, Lahiri S (eds) High altitude and man. American Physiological Society, Bethesda, pp 179–191

60. Haldane JB, Priestley JG (1905) The regulation of human respiration. J Physiol (Lond) 32:225–266

61. Hansen JE, Hartley LH, Hogan RP III (1972) Arterial oxygen increase by high-carbohydrate diet at altitude. J Appl Physiol 33:441–445

62. Hansen JE, Vogel JA, Stelter GP, Consolazio CF (1967) Oxygen uptake in man during exhaustive work and sea level and high altitude. J Appl Physiol 23:511–522

63. Heath D, Williams DR (1989) High-altitude medicine and pathology. Butterworths. London

64. Hesser CM (1965) Breath holding under high pressure. In: Rahn H, Yokoyama T (eds) Physiology of breath-hold diving and the Ama of Japan. Washington DC, pp 165–181 (National Academy of Sciences National Research Council publication 1341)

65. Hesser CV, Lind F, Faijerson B (1979) Effects of exercise and raised air pressures on maximal voluntary ventilation. In: Grimstad J (ed) Proceedings of the European Underwater Biomedical Society 5th annual meeting. European Undersea Biomedical Society, Bergen, Norway, pp 203–212

66. Hesser CM, Linnarsson D, Fagreus L (1981) Pulmonary mechanics and work of breathing at maximal ventilation and raised air pressure. J Appl Physiol 50:747–753

67. Hong SK, Cerretelli P, Cruz JC, Rahn H (1969) Mechanics of respiration during submersion in water. J Appl Physiol 27:535–538

68. Hoon RS, Balasubramanion V, Tiwari BC, Mathew OP, Behl A, Sharma SC, Chadha KS (1977) Changes in transthoracic electrical impedance at high altitude. Br Heart J 39:61–66

69. Houston CS, Riley RL (1949) Respiratory and circulatory changes during acclimatization to high altitude. Am J Physiol 149:565–588

70. Howard P (1965) High and low gravitational force. In: Edholm OG, Bacharach AL (eds) The physiology of human survival. Academic, London, pp 183–206

71. Huang SY, Alexander JK, Grover RF, Maher JT, McCullough RE, McCullough RG, Moore LG, Samoson JB, Weil JV, Reeves JT (1984) Hypocapnia and sustained hypoxia blunt ventilation on arrival at high altitude. J Appl Physiol 56:602–607

72. Hurtado A (1932) Respiratory adaptations in the Indian natives. Am J Phy Anthropol 17:137–161

73. Jaeger JJ, Sylvester JT, Cymerman A, Berberich JJ, Denniston JC, Maher JT (1979) Evidence for increased intrathoracic fluid volume in man at high altitude. J Appl Physiol 47:670–676

74. Johnson RS, Dietlin LF (eds) (1977) Biomedical results from skylab NASA SP-377. NASA, Washington DC

75. Kaneko K, Milic-Emili J, Dolovich MB, Dawson A, Bates DV (1966) Regional distribution of ventilation and perfusion as a function of body position. J Appl Physiol 21:767–777

76. Kasyan II, Makarov GF (1984) External respiration, gas exchange and energy expenditures in man during weightlessness. Kosm Biol Aviakosm Med 18(6):4–9

77. Kellogg RH (1963) Effect of altitude on respiratory regulation. Ann N Y Acad Sci 109:815–828

78. Lahiri S (1984) Respiratory control in Andean and Himalayan high-altitude natives. In: West JB, Lahiri S (eds) High altitude and man. American Physiological Society, Bethesda, pp 147–162

79. Lambertsen CJ, Gelfand R, Peterson R, Strauss R, Wright WB, Dickson JG Jr, Puglia C, Hamilton RW Jr (1977) Human tolerance to He, Ne, and N₂ at respiratory gas densities equivalent to He-O₂ breathing at depths to 1200, 2000, 3000, 4000, and 5000 feet of sea water (predictive studies III). Aviat Space Environ Med 48:843–855

80. Lanphier EH, Camporesi EM (1982) Respiration and exercise. In: Bennett PB, Elliott DH (eds) Physiology and medicine of diving. Best, San Pedro, California, pp 99–156

81. Lanphier EH, Rahn H (1963) Alveolar gas exchange during breath-hold diving. J Appl Physiol 18:471–477

82. Lanphier EH, Rahn H (1963) Alveolar gas exchange during breath-holding with air. J Appl Physiol 18:478–482

83. Lenfant C, Torrance JD, Reynafarje C (1971) Shift in the O₂-Hb dissociation curve at altitude: mechanism and effect. J Appl Physiol 30:625–631

84. Lin YC (1982) Breath-hold diving in terrestrial mammals. In: Terjung RL (ed) Exercise and sport sciences reviews, vol 10. Franklin Institute, Philadelphia, pp 270–307

85. Lin YC, Hong SK (1984) Physiology of water immersion. Undersea Biomed Res 11:109–111

86. Linnarsson D, Hesser CM (1978) Dissociated ventilatory and central respiratory response to CO₂ and raied N₂ pressure. J Appl Physiol 45:756–761

87. Lundgren C (1991) Diving. In: Crystal RG, West JB (eds) The lung: scientific foundations. Raven, New York, pp 2109–2122

88. Lundgren C, Warkander D (1989) Physiological and human engineering aspects of underwater breathing apparatus. In: Lundgren C, Warkander D (eds) Proceedings of the Undersea Hyperbaric Medical Society workshop. Undersea and Hyperbaric Medical Society, Bethesda, p 270

89. Maio DA, Farhi LE (1967) Effect of gas density on mechanics of breathing. J Appl Physiol 23:687–693

90. Mansell A, Poules A, Sutton JR (1980) Changes in pulmonary PV characterisitcs of human subjects at an altitude of 5366 m. J Appl Physiol 49:79–83

91. Michel EL, Rummel JA, Sawin CF, Buderer MC, Lem JD (1977) Results of Skylab medical experiment M 171 – metabolic acitivity. In: Johnston RS, Dietlin LF (eds) Biomedical results from skylab NASA SP-377. NASA, Washington DC, pp 372–387

92. Michels DB, West JB (1978) Distribution of pulmonary ventilation and perfusion during short periods of weightlessness. J Appl Physiol 45:987–998

93. Michels DB, Friedman PJ, West JB (1979) Radiographic comparison of human lung shape during normal gravity weightlessness. J Appl Physiol 47:851–857

94. Milic-Emili J (1982) Effect of acceleration and weightlessness on lung mechanics. In: Bonsignore G, Cumming G (eds) The lung in its environment. Plenum, New York, pp 343–351

95. Milic-Emili, J, Mead J, Turner JM (1964) Topography of esophageal pressure as a function of posture in man. J Appl Physiol 19:212–216

96. Miller JW (ed) (1979) NOAA diving manual. U S Department of Commerce, U S Government Printing Office, Washington DC

97. Mithoefer JC (1965). Breath-holding. In: Fenn WO, Rahn H (eds) Handbook of physiology, respiration, vol II. American Physiological Society, Washington DC, pp 1011–1025

98. Monge C, León-Velarde F (1991) Physiological adaptations to high altitude: oxygen transport in mammals and birds. Physiol Rev 71:1135–1172

99. Monge MC (1928) La enfermedad de los Andes. Sindromes eritremicos. Ann Fac Med Univ S Marcos (Lima) 11:1–316

100. Muza SR (1986) Hyperbaric physiology and human performance. In: Pandolph KB, Sawka MN, Gonzalez RR (eds) Human performance physiology and environmental medicine at terrestrial extremes. Brown and Benchmark, Dubuque, pp 565–589

101. Nadel JA, Widdicombe JG (1962) Effect of changes in blood gas tensions and carotid sinus pressure on tracheal volume and total lung resistance to air flow. J Physiol (Lond) 163:13–33

102. Nunneley SA (1976) Gas exchange in man during combined +Gz acceleration and exercise. J Appl Physiol 40:491–495

103. Nunneley SA, Shindell DS (1975) Cardiopulmonary effects of combined exercise and +Gz acceleration. Aviat Space Environ Med 46:878–882

104. Olszowka AJ, Rahn H (1987) Breath hold diving. In: Sutton JR, Houston CS, Coates G (eds) Hypoxia and cold. Praeger, New York, pp 417–428

105. Paiva M, Estenne M, Engel LA (1989) Lung volumes, chest wall configuration and pattern of breathing in microgravity. J Appl Physiol 67:1542–1550

106. Piiper J, Scheid P (1980) Blood-gas equilibration in lungs. In: West JB (ed) Pulmonary gas exchange, vol 1: ventilation, blood flow and diffusion. Academic, New York, pp 131–171

107. Prefaut C, Dubois F, Roussos C, Amaral-Marques R, Macklem PT, Ruff F (1979) Influence of immersion to the neck in water on airway closure and distribution of perfusion in man. Respir Physiol 37:313–323

108. Prefaut C, Ramonatxo M, Boyer R, Chardon G (1978) Human gas exchange during water immersion. Respir Physiol 37:307–323

109. Prisk GK, Guy HJB, Elliott AR, Deutschman RA, West JB (1993) Pulmonary diffusing capacity and cardiac output during sustained microgravity. J Appl Physiol 75:15–26

110. Prisk GK, Guy HJB, Elliott AR, West JB (1994) Inhomogeneity of pulmonary perfusion during sustained microgravity on SLS-1. J Appl Physiol 76:1730–1738

111. Pugh LGCE (1964) Cardiac output in muscular exercise at 5800 m (19 000 ft). J Appl Physiol 19:441–447

112. Pugh LGCE, Gill MB, Lahiri S, Milledge JS, Ward MP, West JB (1964) Muscular exercise at great altitudes. J Appl Physiol 19:431–440

113. Rahn H (1965) The physiological stresses of the Ama. In: Rahn H, Yokoyama T (eds) Physiology of breath-hold diving and the Ama of Japan. Washington DC, pp 113–137 (National Academy of Sciences National Research Council publication 1341)

114. Rahn H, Otis A (1949) Man's respiratory response during and after acclimatization to high altitude. Am J Physiol 157:445–559

115. Remmers JE, Mithoefer JC (1969) The carbon monoxide diffusing capacity in permanent residents at high altitudes. Respir Physiol 6:233–244

116. Reynafarje C (1958) The influence of high altitude on erythropoietic activity. Brookhaven Symp Biol 10:132–146

117. Rørth M, Nygard SF, Parving HH (1972) Red cell metabolism and oxygen affinity of healthy individuals during exposure to high altitude. Adv Exp Med Biol 28:361–372

118. Rotta A, Canepa A, Hurtado A, Velasquez T, Chavez R (1956) Pulmonary circulation at sea level and at high altitude. J Appl Physiol 9:328–336

119. Salzano JV, Camporesi EM, Stolp BW, Moon RE (1984) Physiological responses to exercise at 47 and 66 ATA. J Appl Physiol 57:1055–1068

120. Schaefer KE (1965) Adaptation to breath-hold diving. In: Rahn H, Yokoyama T (eds) Physiology of breath-hold diving and the Ama of Japan. Washington DC, pp 237–252 (National Academy of Sciences National Research Council Publication 1341)

121. Schaefer KE, Allison RD, Dougherty JH Jr, Carey CR, Walker R, Yost F, Parker D (1968) Pulmonary and circulatory adjustments determining the limits of depth in breath-hold diving. Sciemce 162:1020–1023

122. Schillaci RF (1979) Dysbaric pulmonary physiology. Clin Notes Respir Dis 18(2):3–10

123. Schilling CW, Carlston CB, Mathias RA (1984) The physician's guide to diving medicine. Plenum, New York

124. Schoene RB (1984) Hypoxic ventilatory response and exercise ventilation at sea level and high altitude. In: West JB, Lahiri S (eds) Man at high altitude. American Physiological Society, Bethesda, pp 19–30

125. Schoene RB, Hackett PH, Hornbein TF (1994) High altitude. In: Murray JF, Nadel JA (eds) Textbook of respiratory medicine. Saunders, Philadelphia, pp 2062–2098

126. Segadahl K, Gulsvik A, Nicolaysen G (1990) Respiratory changes with deep diving. Eur Respir J 3:101–108

127. Severinghaus JW, Mitchell RA, Richardson BW, Singer MM (1963) Respiratory control at high altitude suggesting active transport regulation of CSF pH. J Appl Physiol 18:1155–1166

128. Sherman D, Eilender E, Shefer A, Kerem D (1980) Ventilatory and occlusion pressure responses to hypercapnia in divers and non-divers. Undersea Biomed Res 7:61–74

129. Song SH, Kang DH, Kang BS, Hong SK (1963) Lung volumes and ventilatory responses to high CO_2 and low O_2 in the Ama. J Appl Physiol 18:466–470

130. Sørensen SC, Severinghaus JW (1968) Irreversible respiratory insensitivity to acute hypoxia in men born at high altitude. J Appl Physiol 25:217–220

131. Spaur WH, Raymond LW, Knott MM, Crothers JC, Braithwaite WR, Thalmann ED, Uddin DE (1977) Dyspnea in divers at 49.5 ATA: mechanical, not chemical in origin. Undersea Biomed Res 4:183–198

132. Stone HL, Warren BH, Wagner H (1965) The distribution of pulmonary blood flow in human subjects during zero-G. NATO, Neuilly-sur-Seine, France, pp 141–148 (Advisory Group Aerospace Res Dev CP no 2)

133. Strauss RH (1979) Diving medicine. Am Rev Respir Dis 119:1001–1023

134. Taylor NAS, Morrison JB (1988) Effect of breathing gas pressure on respiratory statics and dynamics of immersed man. In: Pasche A, Ilmarinen R (eds) Proceedings of the 3rd international conference on environmental ergonomics. Institute of Occupational Health, Helsinki, p 52

135. Taylor NAS, Morrison JB (1990) Effects of breathing-gas pressure on pulmonary function and work capacity during immersion. Undersea Biomed Res 17:413–428

136. Thalmann ED, Sponholtz DK, Lundgren CEG (1979) Effects of immersion and static lung loading on submerged exercise at depth. Undersea Biomed Res 6:259–290

137. Thoden JS, Dempsey JA, Reddan WG, Birnbaum ML, Forster HV, Grover RF, Rankin J (1969) Ventilatory work during steady state response to exercise. Fed Proc 28:1316–1321

138. Torre-Bueno JR, Wagner PD, Saltzman HA, Gale GE, Moon RE (1985) Diffusion limitation in normal humans during exercise at sea level and simulated altitude. J Appl Physiol 58:989–995

139. US Navy (1973) Diving manual. Dept Navy, Washington DC

140. Van Liew HD (1983) Mechanical and physical factors in lung function during work in dense environments. Undersea Biomed Res 10:255–264

141. Vann RD (1982) Decompression theory and application. In: Bennett PB, Elliott DH (eds) The physiology and medicine of diving. Best, San Pedro, California, pp 352–382

142. Von Basch S (1887) Über eine Funktion des Capillar Druckes in den Lungenalveolan. Wien Med Blatt 15:465–467

143. Von Nieding G, Krekeler H, Koppenhagen K, Ruff S (1973) Effect of acceleration on distribution of lung perfusion and respiratory gas exchange. Pflugers Arch 342:159–176

144. Wagner PD, Gale GE, Moon RE, Torre-Bueno JR, Stolp BW, Saltzman HA (1986) Pulmonary gas exchange in humans exercising at sea level and altitude. J Appl Physiol 61:260–270

145. Wagner PD, Sutton JR, Reeves JT, Cymerman A, Groves BM, Malconian MK (1987) Operation Everest II. Pulmonary gas exchange throughout a simulated ascent of Mt Everest. J Appl Physiol 63:2348–2359

146. Ward SA, Whipp BJ (1989) Effects of peripheral and central chemoreflex activation on the isopnoeic rating of breathing in exercising humans. J Physiol (Lond) 411:27–43

147. Weil JV (1991) Control of ventilation in chronic hypoxia. In: Lahiri S, Cherniack NS, Fitzgerald RS (eds) Responses and adaptation to high altitude. American Physiological Society, Bethesda, pp 122–132

148. Weil JV, Byrne-Quinn E, Sodal IE, Filley GF, Grover RF (1971) Acquired attentuation of chemoreceptor function in chronically hypoxic man at high altitude. J Clin Invest 50:186–195

149. West JB (196) Regional differences in the lung. Academic, New York

150. West JB (1990) Respiratory physiology – the essentials. Williams and Wilkins, Baltimore

151. West JB (1991) High altitude. In: Crystal RG, West JB (eds) The lung: scientific foundations. Raven, New York, pp 2093–2107

152. West JB (1991) Space. In: Crystal RG, West JB (eds) The lung: scientific foundations. Raven, New York, pp 2133–2141

153. West JB, Boyer SJ, Graber DJ, Hackett PH, Maret KH, Milledge JS, Peters RM Jr, Pizzo CJ, Samaja M, Sarnquist FH, Schoene RB, Winslow RM (1983) Maximal exercise at extreme altitudes on Mount Everest. J Appl Physiol 55:688–702

154. West JB, Lahiri S, Gill MB, Milledge JS, Pugh LGCE, Ward MP (1962) Arterial oxygen saturation during exercise at high altitude. J Appl Physiol 17:617–621

155. West JB, Hackett PH, Maret KH, Milledge JS, Peters RM Jr, PIzzo CJ, Winslow RM (1983) Pulmonary gas exchange on the summit of Mt Everest. J Appl Physiol 55:678–687

156. West JB, Wagner PD (1980) Predicted gas exchange on the summit of Mt Everest. Respir Physiol 42:1–16

157. Wood EH, Nolan AC, Donald DE, Cronin L (1963) Influence of acceleration on pulmonary physiology. J Appl Physiol 22:1024–1034

158. Wood LDH, Bryan AC (1978). Exercise ventilatory mechanics at increased ambient pressure. J Appl Physiol 44:231–237

159. Young AJ, Young PM (1986) Human acclimatization to high terrestrial altitude. In: Pandolph KB, Sawka MN, Gonzalez RR (eds) Human performance physiology and environ mental medicine at terrestrial extremes. Brown and Bench mark, Dubuque, pp 497–543

Regulation of Body Temperature

110 Basic Thermoregulation

K.E. COOPER

Contents

110.1 Introduction

The term "body temperature" is often ill defined and requires careful consideration. The body can be crudely divided into two regions [2], namely the "core" and the "shell" (Fig. 110.1). The 'core' is made up of the contents of the skull, the thorax, and the abdomen. The "shell" includes the skin, the subcutaneous tissues, and the limbs. The temperature of the "core" is maintained close to 37°C (98.6°F) at most times, whereas the temperature of the "shell" fluctuates widely according to the environmental conditions. There are variations of temperature within the body core. For example, the temperature in the human esophagus at heart level may be up to 1°C (1.8°F) lower than that in the rectum. This difference does not seem to be due to bacterial action in the colon and rectum, since it has been shown to persist in a patient who had a colostomy and in whom the distal side of the colostomy had been sterilized with antibiotics. Oral temperature, properly taken, and esophageal temperature follow rapid changes in arterial blood temperature more closely than does rectal temperature, in which there is a large thermal lag. For clinical use the properly taken oral temperature or the rectal temperature can be used to assess fever, and the rectal to assess hypothermia or hyperthermia, provided that the normal ranges for each are known. A measure of the "core" temperature is usually taken to be the "body temperature", although no such simple entity really exists. Inscientific studies use is often made of "mean body temperature", which is a mathematical combination of the "deep" or "core" temperature and multiple regional skin temperatures. Another index of body temperature, often used, is that measured in the auditory canal close to, or on, the ear drum using a fine electrical measuring device; this temperature also follows rapid changes in arterial blood temperature well. The tympanic temperature is used by some as a measure of hypothalamic temperature (but only if taken by a probe adjacent to one small portion of the membrane). None of the temperature measuring sites mentioned can be said to give an absolute measure of brain temperature under all conditions. It is likely that there are variations of temperature between regions of the brain which may alter with alterations in regional neuronal activity. The arterial blood going to the brain may be cooled on its way there by cooler blood draining from the face and the surface of the scalp into the intracerebral venous sinuses, and thus there could be a differential in temperature from time to time between regions above and below the cavernous sinus.

Measurement. The measurement of core temperature is not trivial. Several types of thermometer are commonly used in practice [8], namely, the mercury-in-glass thermometer, the electrical (usually a thermocouple or thermistor) digital reading thermometer, and a radiometer attached to an auriscope-like head for tympanic temperature. When using the mercury-in-glass clinical thermometer, it is important to bear in mind that there are often substantial calibration errors in these devices, and that the time taken for the temperature reading to reach a stable maximum depends on the construction of the thermometer and factors such as, in the case of the oral temperature measurement, previous mouth breathing or the ingestion of hot or cold drinks or food. The same errors may also occur in the use of electrical thermometers. In patients suffering hypothermia, special low-reading thermometers are necessary, and these are usually electrical digital readout devices.

R. Greger/U. Windhorst (Eds.)
Comprehensive Human Physiology, Vol. 2
© Springer-Verlag Berlin Heidelberg 1996

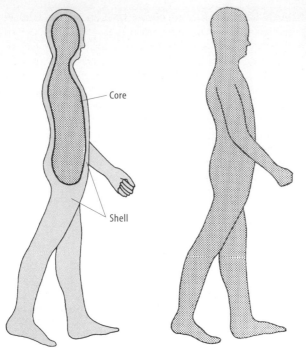

Fig. 110.1. Concept of body "core" and "shell." The *lefthand* figure shows the warm "core" (*shaded*) and cool "shell" (*blue*) of a subject who has been exposed for some while to cold air. On the *right*, the subject has been equilibrated in a hot environment and the "shell" is also at "core" temperature

Variations. The normal "body" or "core" temperature depends on many factors. There is a variation of the body temperature during the day and the night, and this oscillation is known as the diurnal temperature rhythm. It is present in all warm-blooded species and its extent varies to some extent with each individual. In the human species, the body temperature is lowest in the early morning and rises to a maximum in the early evening. An early morning oral temperature of 36.3°C (97.3°F) and an evening temperature of 37.1°C (98.8°F) would not be unusual. In some individuals this swing could be larger and in others less. During severe exercise the "core" temperature could rise to 40°C (104°F), and it has been reported that children swimming in unheated pools or rivers in Britain may allow their body temperatures to fall to close to 34°C (93.2°F). Daytime body temperatures in other mammalian species vary from an average of 36°C (96.8°F) in the elephant to 38°C+ (100.4°F+) in the rabbit and up to 41°C (105.8°F) in some birds. If healthy women take their body temperatures (vaginally or rectally) on waking the temperature is found to rise by 0.3–0.5°C (0.54–0.9°F) on the morning following ovulation, and it remains at the new level until the start of the next menstrual period – an observation often used in family planning. The reason for the temperature increase is not known. In some desert animals for which water conservation is of paramount importance, e.g., the camel, the diurnal swing in body temperature is made greater in the dehydrated state than in the well hydrated condition [20]. Body temperature falls much lower at night and reaches higher daily peaks and larger amounts of heat must be produced to raise the core temperature in the day time to a temperature which starts panting with its consequent water loss. The extra stored heat is lost by radiation and conduction at night without a rise in water evaporation.

Skin Temperature. Skin temperature is determined by the amount of heat reaching the skin from blood perfusing it, by heat conducted from deeper tissues and by heat transfer to or from the environment. The latter will be determined by heat exchange with the environment by evaporation of sweat, radiation, and convection or conduction. There are usually regional variations in the skin temperature and measurement of "skin" temperature should always be defined according to the area concerned or as a "mean" value. The latter may be obtained by taking measurements of exposed surface temperature at the forehead, upper arm, forearm, chest, abdomen, dorsum of hand, thigh, front of shin, and dorsum of foot, weighting each one according to the surface area involved and taking the average of these weighted readings. At a comfortable room temperature, this mean skin temperature is close to 33°C. The pattern of skin temperature distribution can be displayed using infrared photography, or more crudely by painting areas of skin with temperature-sensitive liquid crystal chemicals which change color as their temperatures vary. It is cheaper for many studies to measure the surface temperatures using standard electrical thermometers.

The use of the weighted mean skin temperature is valid in considering heat exchanges. The weighting necessary when considering patterns of afferent neural thermal inputs from the skin, where the density of thermoreceptors varies in different parts of the body, to evoke thermoregulatory reflexes, is different and as yet unknown. For this purpose the weighted mean, as calculated for heat exchange, is probably invalid. The skin temperature is often used as an index of skin blood flow, especially in the extremities, but at low skin temperatures a small increase in blood flow causes a large rise in surface temperature and at high skin temperatures a large increase in blood flow causes only a small increase in skin temperature.

Because of this nonlinearity, the use of skin temperature as a measure of skin blood flow can be inaccurate and misleading [7].

110.2 Heat Balance

We can represent the state of the body's heat balance by the simple equation:

$$S = (M) - (C) - (K) - (R) - (E) - (W)$$

where:

- S is body heat *storage*
- M is metabolic heat *production*

- C is heat exchange by *convection*
- K is heat exchange by *conduction*
- R is heat exchange by *radiation*
- E is heat loss by *evaporation*
- W is *Work*.

Values are positive or negative, and the units are $W \times kg^{-1}$ or $W \times m^{-2}$ [10,14]. Positive heat exchange is heat transfer to the environment and negative heat exchange the converse.

110.2.1 Thermogenesis

There are several avenues of *heat production*:

- Basal metabolic rate
- Muscle activity
- Nonshivering thermogenesis
- Shivering thermogenesis.

In carrying out the minimal chemical processes, which overall are exothermic (see Chap. 2), to maintain the organism in a thermoneutral environment (i.e., where the environment is at a temperature that does not stimulate heat production) at complete rest, a basal level of heat is produced. This has often been called the *basal metabolic rate* (BMR) and it depends, in the main, on the level of secretion of thyroid hormones.

Attempts were once made to measure the BMR by measuring the patient's O_2 uptake after a 12-h fast, in a warm room, and in a state of complete relaxation. Since this measurement was so difficult to achieve without some degree of excitement, others used the sedated metabolic rate with the subject deeply under the influence of a short-action barbiturate. This gave more consistent results but its relationship to unanesthetized basal metabolism is not clear.

In performing exercise, work is done and, since the conversion of metabolic energy into work is relatively inefficient, heat is produced. *Muscle activity*, even when not apparently purposeful, as in shivering, generates heat and is important in the maintenance of body temperature during cold exposure. During the digestion of foods and the incorporation of their products into new compounds, heat is released; this is particularly marked in the case of protein metabolism.

There is a mechanism to produce heat in the cold which is apart from shivering and which is called "*nonshivering thermogenesis.*" This occurs in the newborn of some species, for example, the human and the sheep, in hibernating animals, and in other species such as the rat, particularly following cold acclimation [12]. The principal source of nonshivering thermogenesis in many species is *brown fat*. In the newborn, brown fat is found in discrete masses between the scapulae, in the axillae, and in the region of the kidneys. Brown fat differs from ordinary fat in that the cells are much smaller and contain many more mitochondria, and it has a rich blood supply. Brown fat has a rich sympa-

thetic innervation, probably under the partial control of the ventromedial hypothalamus (chap. 17). Stimulation of the sympathetic nerves as a result of cold exposure leads to a β-receptor-mediated increase in the cAMP concentration in the brown fat cell, which activates the many mitochondria, as well as causing vasodilatation in the brown fat pads [11]. Lipolysis takes place and, as a result of partly uncoupled oxidative phosphorylation-related mechanisms in the mitochondria (see chap. 4), a large amount of heat is released. The rich blood supply of brown fat enables heat to be carried rapidly from the brown fat, via veins, into the body core, where it raises the core temperature. Brown fat appears to be a major source of dietinduced thermogenesis [16]. There is some evidence that brown fat cells remain, mingled with ordinary fat cells, in adults of some species, and that their absence may play a role in determining obesity.

Sometimes, in animals which have been cold-adapted, increased muscle heat production can take place in resting muscle during cold exposure. The mechanism of this muscle-derived nonshivering thermogenesis is not yet properly understood.

110.2.2 Heat Exchange

Heat exchange takes place between the body surface and the environment using the mechanisms of convection (C), conduction (K), radiation (R) and evaporation (E).

Convection. Air in contact with the surface of the body becomes heated and as a result its density falls and it rises. The vertical air flow then carries heat from the surface of the body. Such convection currents can be visualized using a special interferometry type of photography called Schlieren photography. Convective heat exchange would be minimized in an astronaut living at zero gravity, since convection currents could not then occur. Air being forced over the surface of the body by a wind, e.g., with a fan, creates turbulence at the skin surface. This disrupts a thin, insulating, boundary layer of still air close to the skin and removes heat more rapidly than natural convection. The term for this is *forced convection*. The heat exchange per unit area is derived by the equation $C = h_c(T_1 - T_2)$ (units $W \times m^{-2}$), where h_c (in $W \times m^{-2} \times C^{-1}$) is the convection heat transfer coefficient and T_1 and T_2 are the mean skin temperature and the air temperature (°C) respectively [14].

Conduction. Conduction takes place when there is direct contact of the body surface with another medium at a different temperature. For example, if the hand is placed on a cold metal surface, heat will be conducted from it rapidly because of the high thermal conductivity of the metal. If the metal surface is cold enough this can rapidly lead to frostbite. When the body is immersed in water, heat loss routes are via conduction and convection. The conduction of heat per unit skin area is expressed by the equation $K = (T_1 - T_2)R^{-1}$ (in $W \times m^{-2}$), where T_1 and T_2 (°C) are the temperatures of the surfaces of the layers through which heat flows, and R is the intervening thermal resistance (units $°C \times m^{-2} \times W^{-1}$; R = conductance^{-1}).

Radiation. Radiation occurs between the body surface and the environment and is independent of the air surrounding the body, except that water vapour and CO_2 in the atmosphere can absorb some radiant heat. The amount of heat lost between the body and, for example, a wall which it faces is proportional to the difference in the fourth powers of the absolute temperatures of the two surfaces facing each other. This route of heat loss can be expressed in the Stefan-Boltzman equation, which is:

$$HR = \delta \times e1 \times e2 \times (T_1^4 - T_2^4) \times A,$$

where HR is the transfer of heat in W, δ is the Stefan-Boltzman constant = 5.67×10^{-8} ($W \times m^{-2} \times K^{-4}$), T_1 and T_2 are the surface temperatures in K (i.e., $°C + 273$), e_1 and e_2 are the emissivities of the surfaces, and A is the effective radiating area (m^2). Radiation can be a large avenue of heat loss and, for example, a great deal of heat can be lost by radiation to the snow from cross-country skiers resting near a vertical snow bank, or to the clear sky at night. The jack rabbit can sit in the shade of a tree with its ears facing the sunless part of the open sky and lose considerable amounts of heat.

Evaporation. Evaporation of water takes place continuously from the body. Water is evaporated from the respiratory tract during breathing. In climbers at very high altitudes, where the inspired air is dry and the volume of air moved in and out of the lungs is large, evaporative water loss from the respiratory tract can be very high. This may result in great heat loss and some level of dehydration. Water continuously diffuses through the skin surface and is evaporated. This is known as *insensible perspiration* and depends solely on the difference between the vapour pressures at the skin surface and in the adjacent atmosphere. In addition to this, the sweat glands (see Chap. 112) on the body surface produce a salty fluid, namely sweat, which is distributed over the body surface. When it evaporates heat is extracted from the body at the rate of approximately $2.41 kJ \times ml^{-1}$ ($0.54 kcal \times ml^{-1}$) water evaporated (at 34°C). In the human there are two types of sweat glands (see Chap. 112), the apocrine and the eccrine. The apocrine glands are derived from the follicular epithelium of the hair follicles, they have a limited distribution in the axillae and in the pubic, perianal, perineal, and perimammillary areas and they secrete in response to severe emotional stimulation, though they are not the only sweat glands responding to emotional stimulation. Their functions are not yet fully understood but they may be responsible for sexually important odors.

The eccrine glands are widely distributed over the whole body surface and are responsible for the thermoregulatory sweating. The eccrine sweat glands receive a sympathetic nervous innervation which is cholinergic. Sweating is often seen in patients suffering the norepinephrine-producing tumor pheochromocytoma. The mechanism is not known, but it has been postulated that the circulating norepinephrine may activate thermoregulatory mechanisms which in turn activate the cholinergic nerve supply to the sweat glands.

Sweat is a watery fluid containing 0.5%–1.0% solids. It contains Na^+Cl^-, urea, and other nitrogenous compounds, lactic acid, glucose in very small amounts, K^+, Ca^{2+}, Mg^{2+}, and small amounts of iron, and traces of water-soluble vitamins. One interesting type of sweating is gustatory sweating, triggered by ingestion of hot food or curry or chewing hot peppers. The sweating occurs over the lips, nose, forehead, and sometimes over the whole head. Sometimes specific nonspicy foods such as chocolate will induce gustatory sweating.

Sweat rate may be assessed by measuring loss of weight, which measures the evaporative water loss from both the skin and respiratory tract plus the amount of fluid which has dripped from or been absorbed from the skin by clothing. By passing dry air through a capsule attached to the skin and measuring the difference in water vapor content between the incoming and the outgoing air as well as the air flow rate, the evaporative water loss from a small area of skin can be assessed. The density of active sweat glands can be measured by painting the skin with an iodine solution (having shown that the subject is not sensitive to iodine) and dusting over the dried iodine with starch powder. Wherever sweat is released from a gland the iodine reacts with the starch to make a black spot. These spots can be recorded by blotting the areas with fine filter paper. Another technique involves painting the skin with a plastic which forms a "skin" over the area under study which is perforated by the sweat secreted by each gland. The plastic skin can then be peeled off and the number of holes in it counted under a microscope.

By repeated exposure to heat and working in the heat, acclimatization to heat takes place over a week or more. In acclimatized subjects, sweat contains less salt, a higher volume is produced, and it is more evenly distributed over the body surface for more effective evaporation. Profuse sweat flowing over the surface reduces the surface salt concentration and thus increases the vapor pressure gradient, giving enhanced evaporation. Proper acclimatization is essential for those working in very hot climates, e.g., down deep mines or soldiers fighting in the desert, to prevent severe heat illness [6].

It should be remembered that heat can also be gained from the environment under conditions in which the air temperature is above the body temperature, or in which the body is exposed to a high radiating heat load. This may occur as a result of exposure to direct sunlight, or to being in the proximity of extremely high heat sources, e.g., in steel mills and similar industrial situations. In these cases evaporative heat loss is the only means of heat loss.

Heat exchange between the body and the environment will be modified by the *insulation* of the body surface. In some animals which are furry, air is trapped in the fur layer, and this provides thermal insulation which restricts heat loss but can also restrict heat gain in hot environments.

Piloerection, i.e., the erection of body hair, increases the depth of this trapped air layer and increases the thermal insulation of the fur in the cold. In other animals such as the merino sheep, where the wool is hygroscopic, uptake of water into the wool can cause the release of some heat, and this helps to preserve body temperature as water condenses into the wool at night in cold desert conditions. Its evaporation during the day has the opposite effect. In the human, the wearing of clothing forms part of behavioral thermal regulation and is used to vary the insulation at the body surface. In hot environments, loose clothing, which allows air to convect over the body surface and thereby to extract heat and to carry away water vapor, greatly improves heat loss. Light-, rather than dark-colored clothing, is beneficial since it reflects some of the solar radiation which would otherwise be converted into longer-wave infra-red radiation and heat the body. In cold environments, clothing should be permeable to water vapour so that sweat does not soak it and thus reduce its insulating value. In the heat, the passage of water vapor through clothing aids evaporative heat loss.

It is interesting in this regard that there is evidence [19] that the Inuit (Eskimo) has far fewer sweat glands on the hands and feet than the white man and also sweats much less on the trunk than the white man, but sweats profusely from the face. This enables him to wear very efficient thermally protective clothing while hunting without soaking the boots, mitts, and the trunk clothing with water as a result of sweating. When working in severe heat, wearing protective clothing can greatly reduce or even prevent evaporation of sweat and this may lead to heat illness.

110.2.3 Heat Storage

Since the temperature of the body shell can be allowed to vary considerably, it is possible to store a large amount of heat in the shell, and, during cold exposure, such stored heat is lost first before heat is drained away from the body core and the body core temperature falls. The survival time is thereby increased.

110.2.4 The Skin Circulation in Heat Exchange

The circulation through the skin is a major determinant of body heat loss, or of heat conservation. Over most of the body surface there are superficial and deep arterial and venous plexuses and the flow of blood through the skin capillaries brings warm blood to the body surface. However, in some parts of the body there is a more specialized pattern of circulation in the skin. In the fingers and the hand, and perhaps to a very small extent in the forearm, in the ear lobes, the toes, and in a few other small areas, there are arteriovenous anastomoses at the base of each capillary loop in the skin. These anastomoses have smooth muscle in their walls and are deep to the tips of the capillary loops which are nearer the surface of the skin. When they open up, an enormous blood flow passes through the shunts [1],

far more than could pass through the capillary loops. Thus, although the flow is deeper in the skin, the extremely high rate of flow through the shunts more than outweighs the effect of the extra thermal insulation and leads to increased heat loss. The arteriovenous anastomoses open up when the body temperature is raised above normal, during reflexively induced vasodilatation and during cold-induced vasodilatation.

Another vascular arrangement is the counter-current heat exchange system between arteries going to the periphery of a limb and the venous blood returning from the periphery. In the human forearm, for example, running alongside the ulnar and radial arteries are veins known as venae comitantes [3]. During exposure of the body to cold, the superficial veins in the forearm shut down and blood is returned from the hand mainly through these venae comitantes. This results in heat exchange between the arteries going to the hand and the adjacent veins. Thus, the blood reaching the fingers is precooled by the cool blood returning up the veins. This serves to reduce heat loss from the fingertips and to preheat the blood returning to the body core through the venae comitantes. It is an internal heat conservation mechanism. In the heat, the venae comitantes are shut down and blood returns from the periphery via superficial veins. This ensures that blood will reach the digits at core temperature and will lead to maximum heat loss. Such heat exchange systems are also found in animals other than the human. A very impressive one is to be found in the footpads of the wolf. It enables the temperature of the footpads to be kept high enough to prevent freezing when the wolf is standing or running in the snow, but reduces the enormous heat loss which would otherwise occur, and thus prevents a precipitous fall in deep body temperature.

In some animals, e.g., cheetahs and reindeer, there is a complex arrangement of vessels carrying blood to the brain. The arterial vessels break up into a leash of smaller arteries (the rete mirabile) which pass through the cavernous sinus. Blood returning from the nares and the facial veins of the animal is cooler than arterial blood. It drains back into the cavernous sinus and so arterial blood going to the brain is cooled on its passage through the sinus. Thus an animal, such as a cheetah, which can sustain extremely high speeds and outputs of energy for a short time, can allow its general body temperature to rise quite a lot while the brain is kept cool by the rete mirabile counter-current cooling system.

Raising the blood temperature in the human activates thermal receptors within the central nervous system, which leads to an increase in blood flow through the skin [21]. In Fig. 110.2, the hand blood flow, estimated as heat elimination from the left hand, increases as the body temperature is raised by immersing the right arm in hot water. The intracerebral neural mechanisms involved in this will be discussed in Sect. 111.1.2. Raising the core temperature also induces vasodilatation in the skin of the arms, the legs, and the trunk. The skin blood vessels of the hands, feet, nose, ears, and lips receive an α-adrenergic sympathetic vasoconstrictor innervation. The vasodilatation in the

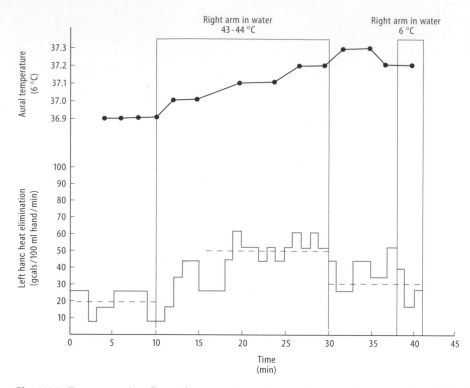

Fig. 110.2. Temperature (aural) near the tympanic membrane and hand blood flow, as estimated by heat loss to a calorimeter, when the opposite hand and forearm are immersed in warm water. The rise in core temperature (measured as aural temperature) causes a rise in blood flow in the unheated hand

hands, feet, and parts of the face is brought about by decreasing the sympathetic nervous tone to the vascular smooth muscle of the skin vessels in those regions. In other skin areas there is evidence that there is active vasodilatation, i.e., the relaxation of the vascular smooth muscle is due to an increase in the impulse traffic in the nerves (possibly cholinergic) to the blood vessels, or perhaps to other structures such as sweat glands leading to secondary vasodilatation. The latter hypothesis suggests that when sweat glands are secreting, enzymes are released into the surrounding tissues which break down proteins in the tissue spaces to form a peptide – bradykinin – which is a powerful vasodilator substance. Such a mechanism would tend to match the local blood flow to the secretory requirements of the sweat glands (see Chap. 111). While this mechanism is widely quoted at present, there is much evidence against it. It may be that both mechanisms are in place, but much more evidence is necessary to confirm this assumption.

The neural control of the skin of the forehead and head is less precise than that of the rest of the body skin. Local heating dilates skin blood vessels in all areas. In addition, sudden exposure of a large area of skin, such as the trunk, to radiant heat activates a nervous reflex arc which leads to vasodilatation, which is particularly marked in areas in which there are arteriovenous shunts, though the role of this response in normal thermoregulation is not clear [15] (Fig. 110.3).

When a hand is immersed in cold water (say at 4–6°C), there is an initial vasoconstriction in the fingers, followed by a large vasodilatation during which a substantial portion of the body heat production can be lost to the surrounding water. This is followed by a period of vasoconstriction and a second wave of vasodilatation, and so on. This response was described by Sir Thomas Lewis in 1925 [17] and it is often called the Lewis hunting response. It is probable that the initial vasoconstriction is reflexive, and that, during this period of constriction, the smooth muscle in the walls of the blood vessels, which are very near the skin surface, is cooled rapidly until it fails to respond to α-adrenergic stimulation and becomes paralyzed [13]. Then the intravascular pressure forces the vessels widely open and a huge blood flow results. This, then, is followed by heating of the smooth muscle in the blood vessel walls by the blood perfusing them and they again become sensitive to α-noradrenergic nerve stimulation or circulating catecholamines and constriction follows. The cycle is repeated. The extent of the vasodilatation is reduced if there has been whole body cooling previously.

110.3 Special Features in the Newborn

The newborn of some species, e.g., sheep and guinea pigs, have mature thermoregulatory mechanisms at birth. Oth-

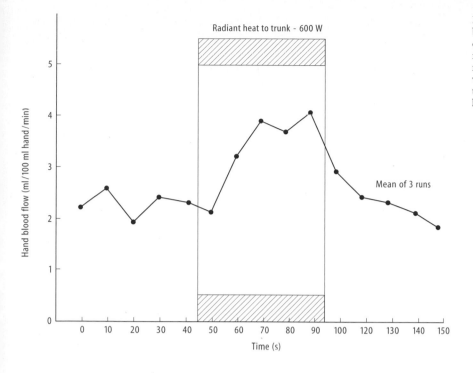

Radiant heat to trunk - 600 W

Mean of 3 runs

y-axis: Hand blood flow (ml/100 ml hand/min)
x-axis: Time (s)

Fig. 110.3. Change in hand blood flow in response to exposure of the trunk to radiant heat. Note the rapid reflexively determined hand vasodilatation on heating and the rapid fall in flow when the heat is turned off

ers, e.g., the rat, take some days to develop mature physiological thermoregulation and rely in the early days on behavioral thermoregulation by snuggling close to the mother. In the human infant, the thermoregulatory responses, such as vasomotor functions in the skin, sweating, and increased metabolic rate in the cold, are present, though imperfect, and they occur to some extent even in premature infants. Metabolic heat production in response to cold is, in the human newborn, more due to nonshivering thermogenesis in brown fat than to shivering. Because of the high ratio of surface area to body mass, the infant can lose body core heat easily. It has relatively little subcutaneous fat, and so, even when vasoconstricted, heat can be conducted rapidly away from the core. The thermoneutral environmental temperature range for the newborn infant is considerably higher than for the adult, and outside this range the heat production per unit of body mass is relatively greater than in the adult in order to defend the core temperature.

110.4 Behavioral Thermoregulation

The human adult is able to adjust the microclimate at the skin surface by adjusting the amount and type of clothing worn, by building shelters, and by adjusting the air conditioning in the buildings. Such mechanisms, depending on thermal information from skin and deep thermal receptors, are termed "behavioral thermoregulation" [4,18] and this is probably the major mechanism used in thermoregulation in widely diverse climates. Some animals, e.g., desert iguanas, may shuttle between sunlight and shade to maintain a core temperature at about 38°C in the daytime. Other creatures, e.g., some fish, may select a preferred water temperature as a means of behavioral thermoregulation.

In determining behavior with respect to temperature, the sensations of heat and cold and the condition of *thermal comfort* are important. There is good evidence that thermal comfort depends on inputs from skin and "deep" thermal receptors [5]. The deep receptors concerned may or may not be those subserving *autonomic* thermoregulatory responses. Environmental factors involved in determining thermal comfort include not only the temperature of the air but also the mean radiation temperature, the velocity of the air movement, and the humidity. At the point where no themoregulatory mechanisms are called into play, the environmental condition is known as the thermoneutral zone, which often coincides with the thermal comfort zone. In light clothing the thermal comfort zone might be at an air temperature of 25–27°C with a 50% relative humidity and minimal air movement. However, a curious phenomenon occurs during a prolonged stay in a cooler environment. After some days or weeks the common experience is to be comfortable at a lower environmental temperature, and on returning to the previous comfort zone it feels too warm for a short period. Whether this "adaptation" depends on an alteration in skin conditions and skin thermal receptor behavior, or whether there is altered central processing of the afferent thermal information is not yet clear, but the evidence at present slightly favors altered central processing. The effects of environment and clothing on thermal comfort involve complex contributions from their physical characteristics, and these are dealt with by Fanger [9], to whom the reader is referred.

References

1. Abramson DI (1967) Circulation in the extremities. Academic, New York, pp 116–119
2. Aschoff J, Wever R (1958) Kern und Schale im Wärmehaushalt des Menschen. Naturwissenschaften 20:477–485
3. Bazett HC, Love L, Newton M, Eisenberg L, Day R, Forster R (1948) Temperature changes in blood flowing in arteries and veins in man. J Appl Physiol 1:3–19
4. Bligh J (1973) Temperature regulation in mammals and other vertebrates. North Holland, Amsterdam, chap 13
5. Cabanac M (1969) Plaisir ou déplaisir de la sensation thermique et homéothermie. Physiol Behav 4:359–364
6. Clark RP, Edholm OG (1985) Man and his thermal environment. Arnold, London, pp 136–154
7. Cooper KE, Cross KW, Greenfield ADM, Hamilton DMcK, Scarborough H (1949) A comparison of methods for gauging the blood flow through the hand. Clin Sci 8:217–234
8. Cooper KE, Veale WL, Malkinson TJ (1977) Measurements of body temperature. In: Myers RD (ed) Methods in psychobiology. Academic, New York, pp 149–187
9. Fanger PO (1970) Thermal comfort. Danish Technical Press, Copenhagen
10. Glossary of terms for thermal physiology (1987) A report by the Commission for Thermal Physiology of the International Union of Physiological Sciences. Pflugers Arch 410:567–587
11. Himms-Hagen J (1990) Brown adipose tissue thermogenesis: role in thermoregulation, energy regulation and obesity. In: Schönbaum E, Lomax P (eds) Thermoregulation: physiology and biochemistry. Pergamon, New York, pp 327–386
12. Horovitz BA (1971) Brown fat thermogenesis: physiological control and metabolic basis. In: Jansky L (ed) Nonshivering thermogenesis. Academia, Prague, pp 221–240
13. Keatinge WR (1969) Survival in cold water. Blackwell Scienific, Oxford, pp 39–50
14. Kerslake DMcK (1972) The stress of hot environments. Cambridge University Press, Cambridge (Monograph of the Physiological Society no 29)
15. Kerslake DMcK, Cooper KE (1950) Vasodilatation in the hand in response to heating the skin elsewhere. Clin Sci 9:31–46
16. Leblanc J, Brondel L (1985) Role of palatability on meal-induced thermogenesis in human subjects. Am J Physiol 248:E333–E336
17. Lewis T (1930) Observations upon the reactions of the vessels of the human skin to cold. Heart 15:177–208
18. Satinoff E (1933) A re-evaluation of the concept of the homeostatic organization of temperature regulation. Handbook Behav Neurobiol 6:443–471
19. Schaefer O, Hildes JA, Greidanus P, Leung D (1974) Regional sweating in Eskimos compared to Caucasians. Can J Physiol Pharmacol 52:960–965
20. Schmidt-Nielsen K (1990) Animal physiology, 4th edn. Cambridge University Press, Cambridge, pp 271–272
21. Snell ES (1954) The relationship between vasomotor response in the hand and heat changes in the body induced by intravenous infusions of hot and cold saline. J Physiol (Lond) 125:361–372

111 Body Temperature Control and Its Disorders

K.E. COOPER

Contents

111.1 Introduction

In Chap. 110 we discussed the basic mechanisms of thermoregulation, dwelling on the principles governing heat production and dissipation. Despite large variations in heat production by the individual and in the thermal conditions of the environment, homeostasis must preserve core body temperature at around 37°C. This requires complex control mechanisms, which will be discussed in this chapter. In addition, and because of their clinical importance, we shall discuss the pathophysiological aspects of thermoregulation, with specific emphasis on fever and hypo- and hyperthermia.

111.2 Neurophysiological Control of Body Temperature

111.2.1 Behavioral and Physiological Regulation

Many species are able to regulate their body temperatures by altering their behavior. Some unicellular organisms will seek a particular preferred temperature in the water surrounding them, as will some fish. A number of ectotherms (e.g., some lizards) [37] shuttle between the shade and the sunlight and, during the day, maintain core temperatures relatively constant, close to 30°C; desert iguanas hold theirs close to 38°C [4]. In the mammal, and also in the human species, behavioral thermoregulation is extremely important. Our ability to sense the thermal conditions in the environment and to adjust clothing, or to adjust house thermostats, or air conditioners, in order to maintain the appropriate microclimate, is vital to our existence, particularly in those branches of the human family which have migrated from the tropics into cold regions. It is obvious that behavioral thermoregulation will utilize sensory thermal information from the periphery and probably also from deep thermal receptors. It is likely that much of this information will be shared with the physiological regulatory system. There is good evidence that the anterior hypothalamus, the posterior hypothalamus, and the lateral hypothalamus are all involved in behavioral thermoregulatory control [41]. Since it is likely that behavioral thermoregulation preceded physiological regulation in evolution, it may depend on neural networks evolved earlier and existing in the hierarchical layers of thermoregulatory systems in the "older" parts of the brain [40]. Mammals have also evolved very efficient physiological thermoregulatory mechanisms which were also mentioned in Chap. 110. The organization and control of behavioral and physiological thermoregulatory mechanisms takes place within the brain. In particular, the region of the hypothalamus, the preoptic area, and parts of the septum pellucidum are vital to the control system for physiological thermoregulation and other regions such as the cerebral cortex; the limbic system and the brainstem may also be involved in behavioral thermoregulation. In order that such regulation can take place, the brain must receive information about the thermal state of the environment and the thermal state of the interior of the organism in such a way that these data can be integrated and evoke efferent nervous activity to control the various mechanisms of heat exchange.

111.2.2 Temperature Sensors

In the skin there are two main types of sensors which provide information to the central nervous system about

R. Greger/U. Windhorst (Eds.)
Comprehensive Human Physiology, Vol. 2
© Springer-Verlag Berlin Heidelberg 1996

the skin temperature. These are the *cold* and *warm* receptors (cf. Chap. 33). Many of them are likely to be unmyelinated terminal branches of myelinated afferent fibers. The terminal branches lie close to, or protruding into, basal epidermal cells. The warm and cold receptor endings are capable of inducing both static and dynamic changes in the discharge frequency of impulses along the afferent fibers from which they are derived. The warm fibers increase their firing rates as their temperatures rise (Fig. 111.1a), usually in the range between 30 and 45°C, and the frequency of discharge plotted against temperature has a peak at a particular temperature above which the discharge rate falls again (Fig. 111.1b). The cold receptors likewise [29] have bell-shaped curves with the frequencies of discharge rising as their temperatures fall, usually from about 30°C downwards, and also have a peak firing rate

(Fig. 111.1b). Not all fibers have the same pattern of discharge, and there is a considerable variation in the temperature at which peak firing is observed. The occurrence of a secondary rise in firing rate at higher temperature in some cold receptors may explain some paradoxical cold sensations. In Fig. 111.1b is an example of the frequency/temperature responses which occur when receptor temperatures are suddenly altered; they respond with an immediate rapid burst of firing which soon slows to a new higher steady level. The fibers derived from cold receptors are either fine myelinated fibers or thin unmyelinated fibers. Warm receptors are usually terminals of unmyelinated fibers. Each warm or cold receptor usually innervates an area of skin of about 1 mm^2. There may be some overlap in the areas of skin, which can affect the firing rate of receptors.

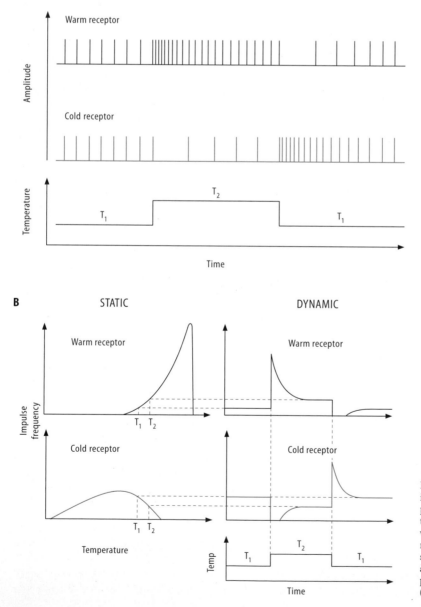

Fig. 111.1a,b. Plot of static and dynamic impulse frequencies versus receptor temperature for cold and warm receptors in the skin. **a** Firing patterns of cold and warm receptors in response to an abrupt rise of receptor temperature. **b** Plot of responses to temperature change of warm and cold receptors. Range: warm 30–45°C, peak 43°C; cold 15–32°C, peak 25°C. (From [28], p 86)

Most thermal receptors in the skin are related to the conscious appraisal of the skin temperature and to reflex adjustments in response to alterations in the thermal environment. There are probably some warm receptor-derived fibers subserving nervous reflex vasodilation (see Fig. 110.3) which travel with sympathetic nerves on their way to the spinal cord, the sensations from which may not reach consciousness [20]. The mechanism of transduction of a temperature change into a pattern of nerve impulses is still not fully understood, but there is evidence for a temperature-sensitive action of Ca^{2+} on K^+ permeability. The fibers subserving the afferent pathways for temperature from the skin have their cell bodies in the dorsal root ganglia and enter the spinal cord through the dorsal roots. Leaving the dorsal horn, second-order neurons cross to the contralateral spinothalamic tracts en route to the ventrobasal thalamus, from which projections go the sensory cortex. There are additional projections to parts of the limbic system. In particular, information concerning skin thermal conditions must be relayed to the hypothalamus to modulate its behavioral and physiological control systems. The precise locations of thermal function pathways connecting the thalamus to the hypothalamus, and also to the preoptic area and septum, have not yet been fully demonstrated.

There are other sources of thermal information which the brain requires for controlling body temperature. It receives thermal afferents from the abdominal organs, and possibly from other deep internal structures. Thermosensitive neurons (i.e., those with a high Q_{10} firing rate with respect to temperature) have been found in the spinal cord, which can, in some animals, respond to modest changes in temperature by evoking thermoregulatory responses such as panting. Some receptors respond to direct heating or cooling, and others are neurons which change the frequencies of their firing in response to warming or cooling brain tissue some distance away. These latter possibly function as interneurons in the thermoregulatory pathways. There are large numbers of thermosensitive neurons in the hypothalamus, particularly in the anterior hypothalamus and preoptic areas. These have been demonstrated both in the intact hypothalamus and also in tissue slice preparations [7]. Most studies of their thermosensitivity have been conducted over rather wider ranges of temperature than occur naturally in the free-ranging animal, and as yet there is no direct correlation of their activity with physiological or behavioral thermoregualtory responses in the awake unrestrained animal. In addition, neurons having similar thermosensitivity have been found in the cerebral cortex in regions not known to be part of the thermoregulatory system [2]. We still also lack precise knowledge of the cytoarchitecture of the hypothalamic thermoregulatory circuits. However, the anterior hypothalamus and preoptic areas do respond to small alterations of its temperature with effector thermoregulatory responses and thus there must be temperature sensors there, the inputs from which can be integrated with the input from thermal sensors in the skin and other body regions, and no other thermosensitive structures or functions have yet been demonstrated in the hypothalamus. The thermosensitive units in the hypothalamus, brain stem, and spinal cord may well provide information concerning the core temperature of the body derived from the temperature of the arterial blood reaching these regions. It is probable that there is a hierarchy of control mechanisms put in place sequentially at various stages of evolution, and becoming more sensitive and more precise in the more advanced stages of evolution, layered in the spinal cord, brainstem, and hypothalamus [40].

111.2.3 Central Control Mechanisms

For many years, based on lesioning experiments, it was erroneously believed that there are two centers within the hypothalamus, one in the posterior hypothalamus for regulation against cold and the other in the anterior hypothalamus for regulation against heat. Such a concept, though still quoted in many textbooks, is no longer tenable. The main regulating region for body temperature is situated in the hypothalamus as a whole, and surrounding regions. It is probable that afferent information from the surface of the body, the body core, and the brain's own thermal detectors reaches the anterior hypothalamus, the preoptic area, and the septum, in which regions the information is necessary for the adjustment of peripheral blood flow, metabolic heat production, piloerection, and sweating or panting in some mammals. Information also reaches the lateral hypothalamus where it would appear to be involved in behavioral thermoregulation. The posterior hypothalamus may represent both a final common path region for efferent fibers subserving effector mechanisms, and perhaps an essential part of the body temperature setpoint (i.e., that temperature which is actively defended) control system. Some evidence as yet inconclusive, suggests that the Na^+/Ca^{2+} ratio in interstitial fluid bathing the cells in the posterior hypothalamus, or in their cell membranes, might play a role in the maintenance of the body temperature setpoint. The septum has been shown to have an important role in the control of shivering as well as in the development and control of fever (see below).

111.2.4 Pharmacological Aspects

There is considerable evidence that several monoamines play an important role as neurotransmitter substances in the thermoregulatory circuits of the hypothalamus and adjacent regions [6]. Norepinephrine microinjected into the preoptic region in many species causes vasodilatation and an inhibition of shivering accompanied by a fall in body temperature. In a few species, the opposite or no effect is observed. 5-Hydroxytryptamine (5-HT, serotonin) injected into the hypothalamus, or the cerebral ventricles, of the cat causes peripheral vasoconstriction, shivering, and an increase in body temperature, as it also does in primates. In some species, the opposite or no effect is sometimes observed. It is possible histologically to demonstrate fibers in the anterior hypothalamus/preoptic area which appear beaded, and which have numerous varicose terminals on them which contain norepinephrine [11], and also some which contain 5-hydroxytryptamine. The cells of origin of these fibers range from the preoptic area to the upper brainstem. Thus, some of these fibers, together with dopaminergic fibers, would appear intimately connected with the central thermoregulatory mechanism. There is also evidence that acetylcholine may be a neurotransmitter in thermoregulatory, and particularly in heat production, pathways. The loci at which acetylcholine may act on thermoregulatory pathways are quite widely distributed from the preoptic area to the posterior hypothalamus. The monoaminergic regulatory circuits occur in the preoptic area and to some extent in the anterior hypothalamus. There are also ascending monoaminergic projections from the lower brain stem which represent additional important *inputs* for thermoregualtion [9]. The correlation between monoaminergic and cholinergic fiber stimulation with thermoregulatory responses in the awake, unrestrained animals still remains to be fully explored. The precise cytoarchitecture of thermoregulatory circuits in the brain is virtually unknown. There is recent evidence to suggest that some peptide neurotransmitters may be involved in normal thermoregulatory activity [10] and these include ACTH, α-melanocyte stimulating hormone (α-MSH), and fragments from it, and the endorphins [12]. The possible role of some of these in endogenous antipyresis is discussed in Sect. 111.3.5.

111.2.5 Some Models of Thermal Regulation

The current texts on thermoregulation contain many models purporting to show how the thermoregulatory and setpoint controls work. These are engineering models, mathematical models, neuronal models, and pharmacological models. Such modeling is useful as a graphical way of expressing what is known currently about the system, and in attempting to predict the behavior of the system. Usually such models are gross oversimplifications and do not represent exactly what is going on within the biological system under study. Oscar Wilde once said of theologians that "In the beginning God created man in his own image and man has been returning the compliment ever since." Modeling of biological systems often follows a similar pattern. Where decisions have to be taken about such matters as the design of protective clothing for thermally hostile environments, good mathematical modelling can be of great use.

First of all, it is probably true that the thermoregulatory system is an example of a control system having one or more negative feedback loops which enable it to function precisely without wild swings of body temperature. An example of such an *engineering model* is given in Fig. 111.2 in which it can be seen that a disturbance in the body temperature (which could, for example, be caused by cold exposure) causes an error signal to the control mechanism which sets in motion the effector processes aimed at reducing the difference between the setpoint of body temperature and the disturbed level. As in an engineering system, information is fed back in such a way as to minimize the error signal, i.e., match the actual body temperature to the set value. A *neuronal model* [6] can be drawn which simply illustrates the statement that when heat loss is promoted, heat production should be inhibited and when heat production is promoted, heat loss pathways will be inhibited as in Fig. 111.3. Attempts have been made to define the neurotransmitters used in such connections as are demonstrated in this model. The model itself is grossly oversimplified and very much more work has to be done, even if the model is correct, to justify any dogmatic statement as to the neurotransmitters used in each group of synapses. Other connectivity models may help to guide our conceptualization of the process of thermoregulation [10].

In other models, by combining the firing rate curves for hypothetical cold receptors and warm receptors, a point of intersection occurs which is used by some as a mathematical model of the origin of the controller's setpoint. Again, although such a model is an oversimplification, it is a simple concept on which to build further discussion of the

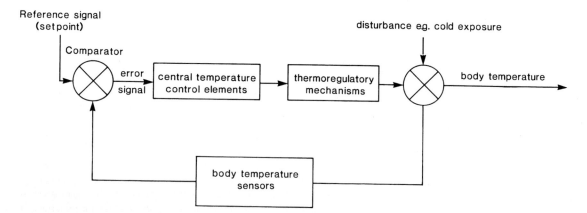

Fig. 111.2. Engineering model of thermoregulation including a negative feedback loop

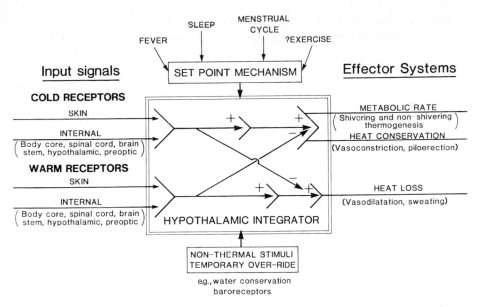

Fig. 111.3. Simplified schematic neuronal model of physiological thermoregulation. The neuronal connections are theoretical. In some species heat loss mechanisms would include panting. For the crossed inhibition concept, see Bligh [6]

mechanism of setpoint. It remains important to pursue accurately the *biological mechanisms* involved in changing the body temperature setpoint as it varies during sleep, during the menstrual cycle and in fever.

While thermoregulation represents an extremely important homeostatic mechanism maintaining the body temperature such that the average function of the body systems is maximal, it uses effector systems which are also used by other homeostatic mechanisms, or which impinge on other mechanisms. For example, very large changes in the skin blood flow and in limb venous capacity, caused by body heating, may change the responses of the baroreceptor system. The not infrequently experienced slight faintness on getting out of a very hot bath, due to pooling of blood in dilated blood vessels in the lower limbs, is an expression of this interaction. The tolerance to centrifugal force during tight turns in a high-speed aircraft is reduced if the body is overheated. Similarly, the control of body temperature in the heat, using sweating or panting, must be integrated with the control of body water content and distribution, and one would expect links between the heat loss thermoregulatory system and the systems regulating plasma osmolarity. Similar connections and integrations must be found also in the control of metabolic rate and central control of a number of hormonal systems such as thyroxins (T_3 and T_4).

111.3 Fever

Fever is usually defined as the rise in body temperature which accompanies infection or other pathological processes, and this increased temperature is defended as though it had reached a new setpoint [15]. It is to be distin-guished from simple hyperthermia, which is just a rise of body temperature to above the normal level (i.e., without an increase in setpoint). It has been found that heating the body during the plateau phase of fever produces responses in the heat loss mechanisms, such as the peripheral circulation, which tend to restore the body temperature to its new high plateau set level [17,23]. Similarly, attempts to cool the body during the plateau phase of fever are met with shivering and increased heat production, and the new high set level is defended [17].

Hyperthermia differs from *fever* and can, for example, be brought about by vigorous exercise, severe exposure to heat, and exercise in impermeable suits. The raised temperature is not defended as a new setpoint. Perhaps it is better these days to think of the rise in temperature in disease which we call fever as one part of an *acute phase reaction* which also includes changes in the plasma concentrations of ions, such as Fe and Mg^{2+}, stimulation of immune responses, and unpleasant sensory side effects which tend to immobilize the patient while the body defense systems act against the infection or other disease.

111.3.1 Initial Peripheral Response to Infection

Some gram-negative micro-organisms (i.e., those which stain pink with the Gram stain) have in their cell walls high-molecular-weight (circa 1000 kDa) complex lipopolysaccharides [45] which are termed bacterial pyrogens, and which are often also called endotoxins. The bacterial pyrogen (endotoxin) is a complex lipopolysaccharide. It is made up of three regions, a core polysaccharide, an O-specific side chain which is responsible for the O-antigens, and a lipid A region [45]. It is the lipid A region which appears to be responsible for the pyrogenic properties of the molecule. These substances are released during the infection, particularly during phagocytosis and destruction of bacteria. They interact with macrophages, principally the monocytes in the circulation and fixed macrophage cells of the reticulo-endothelial system, stimulating the elaboration and release of small peptides which are called endogenous pyrogens [21]. One endogenous pyrogen is a peptide of 15–18 kDa molecular weight that has been found to be the same as the cytokine interleukin-1 (IL-1) [22]. This substance is also released during the phagocytosis and immobilization of other types of bacteria by the macrophage system. The concept of an endogenous pyrogen, the major breakthrough in fever research of this century, stems from the observation by Beeson [3] that sterile pyrogen-free suspensions of rabbit white cells could produce a pyrogen.

111.3.2 Locus and Mode of Action of Endogenous Pyrogen in the Brain

It now seems most likely that IL-1, and not the bacterial products, is a major mediator of the febrile response within the central nervous system. Other peptides elaborated in disease states which act as endogenous pyrogens include interferons, tumor necrosing factor (TNF), and interleukin-6 (IL-6). There are two theories concerning the locus of action of IL-1 within the central nervous system. The first has been based on observations [16] that IL-1 microinjected into the tissue of the brain causes fever only when it is injected into the anterior hypothalamus or preoptic area, so it was postulated that IL-1 gets into these regions to act there in causing fever. When it is injected into this area, there is a delay of up to 7 min before the signs of the febrile reaction begin.

However, it must be pointed out that there is no evidence to date that either bacterial pyrogen or IL-1 can cross the blood-brain barrier. Attempts to demonstrate such penetration, using radioactive tracers attached to the molecules, have so far been unsuccessful. More recently, studies have been made which suggest that the organum vasculosum of the lamina terminalis (OVLT) may be the locus of action of IL-1 [5,42]. Massive lesions of the OVLT are found to prevent the fever which would be expected from intravenous injection of IL-1, possibly by destroying IL-1 receptors there. Small irritative lesions within the OVLT enhance fever and it is suggested that they increase the vascular permeability in the region. There are receptors for the IL-1 within the OVLT. Thus, IL-1 attaching to receptors in the OVLT may in some way cause direct neuronal excitation to take place within the preoptic area; or such action of IL-1 may release a secondary messenger which itself could act within the OVLT, or which could diffuse into the preoptic area to cause fever. The OVLT or adjacent tissue is exquisitely sensitive to *prostaglandin E₂* (PGE₂), and IL-1 can induce its production and release. It has also been suggested that PGE₂ released from peripheral

monocytes may act directly on the OVLT. The suggestion that the OVLT is a locus of IL-1 action is attractive because it does not require that the peptide crosses the blood-brain barrier.

There is evidence that one or more other mediators may act within the preoptic area and possibly the area of the septum to cause fever, following the action of IL-1 on the OVLT. One important candidate for the role of the final mediator of fever within the brain is prostaglandin E, probably PGE₂ [13,34] (Fig. 111.4b). The evidence for this comes from a number of sources, and can be summarized as follows:

- Extremely small amounts of PGE₂, microinjected into the preoptic area or septum cause a fever of very rapid onset and short duration.
- During fever, the level of PGE₂ in the cerebrospinal fluid derived from the third ventricle rises, and it falls again during defervescence.
- Perfusates from the preoptic area have increasing PGE₂ concentrations in febrile animals.
- Drugs such as aspirin and indomethacin, which are known to inhibit the synthesis of prostaglandins, act as antipyretics.

Thus, there is powerful evidence for the involvement of PGE₂ as a mediator of fever within the brain, and it is possible that this substance may be responsible particularly for the immediate rise in body temperature following fever. But there is also further evidence that there may be

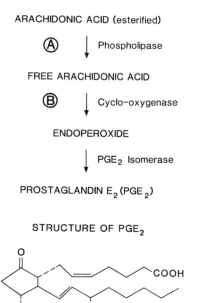

Fig. 111.4. a The production of prostaglandin E₂ (PGE₂) from esterified arachidonic acid. (A) Site of steroid inhibition; (B) site of action of nonsteroidal antipyretics. Interleukin-1 (IL-1) possibly stimulates phospholipase or cyclo-oxygenase activity or both. **b** The chemical structure of PGE₂

other additional mechanisms which can cause and sustain a rise in temperature, by acting within the brain, which are independent of PGE$_2$ [35].

Arachidonic acid is present in the brain and prostaglandins are derived from this by a pathway shown in Fig. 111.4a. It also shows the steps at which steroids (A) and indomethacin and other antipyretics (B) work to block the synthesis of PGE. IL-1 possibly stimulates phospholipase or cyclo-oxygenase activity or both.

It is interesting that recent evidence has been presented that *salicylate* and *indomethacin* can act as an antipyretic when injected into the ventral septal area of the rat, at which locus they block the action of PGE injected into the cerebral ventricle. This, if confirmed and extended, indicates that there is another mechanism of action of antipyretics which is not direct *inhibition* of prostaglandin synthesis.

IL-1 has been shown to inhibit warm-sensitive neurons in the anterior hypothalamus/preoptic area, and to facilitate cold-sensitive neurons in the same region. This action would, of course, tend to lead to a rise in body temperature. The actions of PGE$_1$ and PGE$_2$ on similar neurons are less consistent, but some cold-sensitive neurons are facilitated and some warm-sensitive neurons are inhibited. However, many neurons are unaffected. Thus, a simple study on the action of substances on warm- and cold-sensitive neurons might indicate that IL-1 released from cells within the brain, as is known to be possible, could itself be a mediator of the febrile response if its action on cold and warm receptors is important. While the actions of IL-1 and PGE$_2$ on hypothalamic thermosensitive neurons appear not to be identical, it is too early to determine whether one or the other is the sole and ultimate mediator of fever on the basis of actions on thermosensitive receptors. There is evidence that in animals anesthetized with urethane which cannot regulate their body temperatures, fever occurs in response to IL-1 and PGE given into cerebral ventricles, or to bacterial pyrogen given intravenously. Loci of action of pyrogens outside the normal thermoregulatory control systems, such as on thermal effector outflows, are possible.

111.3.3 Effector Mechanisms of Fever

Following the release of bacterial pyrogen into the circulation, and the elaboration of IL-1, there is a rapid increase in the body temperature apparent setpoint, and the heat production and heat loss mechanisms of the body are adjusted to force the actual body temperature to rise to the new setpoint level. This concept is illustrated in Fig. 111.5. The setpoint is shown in the diagram as a blue line, and the actual body temperature during fever is shown as a dotted line. Due to sluggishness of effector mechanisms, the actual body temperature rises more slowly than the setpoint, but eventually reaches the new setpoint and then oscillates about this new level in a manner which suggests that a setpoint is being defended. During defervescence the setpoint returns rapidly to normal, heat loss mechanisms are activated, heat production is reduced, and the actual core temperature falls the to normal level, again with some lag.

The mechanisms used to raise the body temperature to the new setpoint are both behavioral and physiological. In the human species, and in many others, there is first a massive vasoconstriction in the skin which limits heat loss to the environment. Then heat production rises, and in an acute infection this is accompanied by shivering, which greatly raises the metabolic rate. The massive skin blood vessel constriction causes the skin temperature to drop sharply, and the patient feels cold, usually goes to bed and covers up with bedding, and often uses additional forms of heating such as a hot water bottle, or electric blanket, to try and remove the severe sensation of cold. Often a curled up posture, which minimizes heat loss, is also adopted. It is interesting that sharp-onset fevers are usually accompanied by headache and myalgia, i.e., vague, but often intense, muscle pain, and lying still in bed minimizes both of

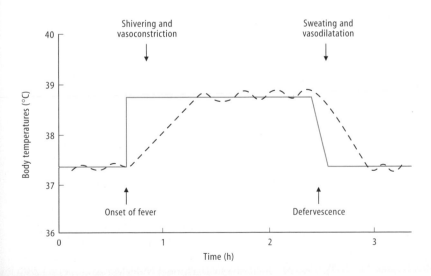

Fig. 111.5. Changes in setpoint and core temperature in fever. *Blue line*, Temperature setpoint; *broken line*, actual body core temperature

these unpleasant sensations. It is interesting also that a patient with a high spinal cord transection, at C5–6, was given an endotoxin [18] and did not get either the headache or the feeling of cold skin, but did have vigorous shivering in muscles innervated from above the level of the lesion. This shows that the headache is not due to a direct effect of the pyrogen on the brain, and that the vasoconstriction is not due to an action of pyrogens on the isolated spinal cord, or on peripheral nerves or blood vessels, and that shivering is not due solely to cold-receptor stimulation in large areas of skin.

It is thought that IL-1, and the cascade of events that follow its action on the brain (probably the OVLT), lead to changes within the septum, the anterior hypothalamus, and the preoptic area, and possibly elsewhere, to evoke efferent activity through a final, common path in the posterior hypothalamus. From there efferent neurons to the intermediolateral columns of the spinal cord are activated and hence the sympathetic neurons which are ultimately distributed to the skin blood vessels are stimulated. Other efferent fibers of the motor systems are excited to cause the muscle activity known as shivering to raise the body's metabolic rate. In some mammals, activation of heat production in brown adipose tissue is important. During the rising phase of fever, sweat secretion is completely inhibited and thus skin evaporative heat loss is reduced to a minimum. Furthermore, marked reduction in the skin temperature reduces the evaporation of water due to insensible perspiration. There is suppression of appetite, and there may be nausea and even vomiting, but the neural mechanisms of these effects are at present poorly understood. As the infection is mastered, the final phase of fever, called defervescence, occurs. When fever breaks, there is a sensation of warmth in the skin as its blood vessels open up and the skin temperature rises. Sweating occurs, often profusely, and the combination of a high skin temperature and profuse sweating leads to evaporation of water from the skin surface and heat loss. These events lead to a rapid fall in body temperature down to the new setpoint, which is usually close to the original afebrile level. Shivering stops and the metabolic rate is lowered. The latter decreases further as the body temperature comes back towards normal.

111.3.4 Concept of Endogenous Antipyresis

There is recent evidence to suggest that the brains of nonhuman species produce their own endogenous antipyretic substances which tend to limit the height and extent of fever. It is interesting that fever rarely goes more than 4–5°C above normal resting temperature, which would be life-threatening, and this also suggests that there may be some fever-limiting mechanism within the brain. The present evidence suggests that arginine vasopressin (AVP synonymous with ADH, the antidiuretic hormone) is secreted in many species into the *ventral septal area* (VSA) of the brain, where it acts as an endogenous antipyretic to reduce the extent of fever [14,19]. Microinjected into this

area it reduces fever caused by injections of endotoxin. When AVP is released in the same area by hemorrhage or the intravenous injection of hypertonic saline, it reduces fever. AVP can be shown to be secreted within the septal area during fever, and the more is secreted, the lower is the fever. In rats, the action of AVP as an endogenous antipyretic is blocked by an AVP V_1-receptor blocker. Another substance, α-melanocyte-stimulating hormone (α-MSH), is also secreted into a *lateral septal area* [12,25] and there is increasing evidence that this substance is a major part of the endogenous antipyresis which occurs in the brain during fever. It will be interesting to see whether more peptides are involved in regulating the extent of fever within the brain. Figure 111.6 is a diagram of the fever process, including the vasopressin and α-MSH negative feedback loops which tend to limit fever. The subject of endogenous antipyresis will be of great interest in the future and, for example, there is now some evidence that α-MSH, or fragments of it, could also act peripherally as antipyretics [36]. If so, further studies of the actions of vasopressin and α-MSH on body temperature may lead to the development of some interesting new antipyretic compounds. That these peptides are endogenous antipyretics is as yet mainly dependent on the evidence from animal experiments. Daily intravenous injections of endotoxin lead to progressively diminishing fever, a phenomenon known as tolerance. There is some evidence that the AVP antipyretic system may also play a role in the development of tolerance to endotoxins.

111.3.5 Survival Value of Fever

There has long been speculation that a physiological response as marked and as ubiquitous as fever ought to have some sort of survival value. In the last decade, evidence has been adduced that the fever mechanism can be demonstrated during infection in a wide variety of species, including ectotherms [31]. Desert iguanas, in North America, shuttle between sunlight and shade during the day and by so doing maintain fairly constant body temperatures of about 38°C. If infected, they spend more time in the sunlight than in the shade, and raise their body temperatures by approximately 2°C. It has been shown experimentally that those which so raise their body temperatures have a much higher survival rate than those which are prevented from doing so. There is also evidence that in some fish, and even in the scorpion, behavioral fever can occur in response to infection, and that such behavioral fever enhances survival. On the other hand, the African cordylid lizard does not get behavioral fever in response to infection; neither do some African snakes or tortoises. Some teleost fish also fail to respond to infection with behavioral fevers. Rabbits infected with live organisms have a higher mortality if fever is prevented [43]. The hypothesis that fever has been selected in evolution for its survival value, based solely on its ubiquity in primitive species, needs much wider study. The survival value of the temperature rise in the individual mammal is unclear, and no safe ad-

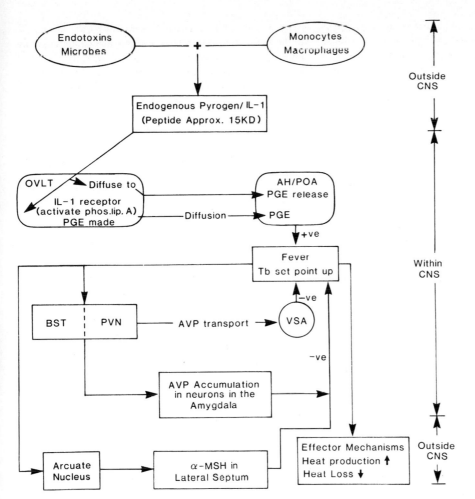

Fig. 111.6. Diagram of the possible fever mechanisms. IL-1 may act on the organum vasculosum of the lamina terminalis (OVLT) directly to release PGE_2, which could act within the OVLT or diffuse to the anterior hypothalamus/preoptic area (AH/POA). PGE_2 in the OVLT could excite neurons projecting to the AH/POA having a direct thermoregulatory action there or inducing a second IL-1/PGE_2 cascade. IL-1 could be released in the AH/POA (inside the blood-brain barrier) in response to brain infections. *BST*, bed nucleus of the stria terminalis; *PVN*, paraventricular nucleus; *CNS*, central nervous system; *VSA*, ventral septal area of brain; *α-MSH*, α-melanocyte-stimulating hormone; *−ve* indicates inhibitory action and *+ve* excitatory action on the process which raises the body temperature setpoint

vice can be given concerning the avoidance of antipyretic therapy in early fever, particularly in infants.

111.3.6 Fever at Parturition and in the Newborn

In some species, namely the guinea pig and some subspecies of sheep, the maternal animal becomes increasingly refractory to fever induced by gram-negative endotoxin close to term and for a few hours after birth. Such refractoriness might be important in preventing fever in the mother from raising the fetal body temperature at a time when such a rise could impair the full development of pulmonary surfactant. It could also protect the immediate post-partum maternal/newborn bonding. It is possible that this refractoriness to pyrogen is related to secretion of AVP into the ventral septum [19]. However, sheep in the periparturient period are not refractory to gram-positive organism cell wall toxin [26], and this suggests that the fever mechanism may not be suppressed and that additional immune factors may be necessary for the refractoriness. Indeed, in the human mother there is a rise in plasma IL-1 receptor antagonist which crosses the placenta into the fetus [39].

111.4 Disorders of Thermoregulation

111.4.1 Hyperthermia and Heat Illness

Under some conditions, the body core temperature rises to very high levels and *heat exhaustion*, or the more serious and life-threatening condition known as *heat stroke*, occurs. In 1987, during a severe heat wave in Greece, there were hundreds of deaths from heat stroke, and heat illness is a major problem during the Makkah Hajj, the annual Moslem pilgrimage to Mecca [27,30]. The heat illnesses can be caused by exercise in very hot environments – e.g., working in a deep gold mine, jogging in severe heat, or working in a foundry with a severe radiant heat load. Heat illness is not uncommon in some of those indulging in "fun runs," particularly if the runner is dehydrated and unfit. In severe exercise in very hot environments there is secretion of large volumes of sweat, variously reported from 1.5 to 3 kg per hour. The large water loss is accompanied by similar losses of electrolytes such as Na^+ and Cl^-, so lack of adequate salt and water intake predispose to the problem. Exposure to very hot environments without prior acclimatization increases the liability of heat illness. Exposure to conditions of high temperature with the evaporation of sweat prevented (e.g., working in a water vapor-impermeable suit, or immersion in a very hot whirlpool bath) can cause a rapid and sometimes fatal rise in body temperature. Old age and cardiovascular disease contribute to the likelihood of heat illness. Quite minor infections, which give rise to endotoxemia, may inhibit sweating and lead to heat illness [1]. Lack of acclimatization to heat is an important predisposing factor. There may be hypovolemia, circulatory collapse, and a moderate or severe rise in body temperature, cessation of sweating is usual. Often the core temperature exceeds 42°C, the skin is dry, and there are central nervous system disturbances. The heat illness has then progressed to heat stroke and rapid and aggressive cooling, correction of the hypovolemia, and cardiovascular support measures are imperative.

111.4.2 Acclimatization and Acclimation to Heat

The chances of the occurrence of heat illness are greatly reduced by prior acclimatization or acclimation. "Acclimatization" is the term given when the process is produced by living in or a prolonged stay in a hot climate, and "acclimation" is when the benefits of natural acclimatization are conferred by artificial means such as repeated exposure in a controlled climate temperature chamber. Either process leads to an increased sweat production with a lower salt content in the sweat and a more even distribution of the sweat over the body surface. Although the water loss is increased, the evaporation of the sweat is increased, partly by more even spreading of the water over the body surface and partly by an increase in the vapor pressure differential between the skin and the environment. The acclimated or acclimatized person will experience a smaller rise in rectal temperature and heart rate for a given work load in a hot environment. Acclimation may be achieved by repeated daily working in a hot environment or by passively raising the body temperature (e.g., sitting in great heat in an impermeable suit allowing the core temperature to be maintained at 38°C for 1–2 h).
Heat cramps, severe muscle cramps with considerable pain, are often a manifestation of exercise in the heat with deficient salt intake [32]. Since solid salt tablets usually lie in the stomach unabsorbed, the taking of such tablets is no longer recommended; rather, a steady normal amount of salt in the diet, with adequate water intake, is the method of choice for prophylaxis. There is a conflict of advice between those who regard low Na^+ intake as important in the prevention of hypertension and those who would reduce the risk of heat illness in severe heat stress. Careful consideration is required to balance preventing heat stroke in the short term against exacerbating high blood pressure in the longer term.

111.4.3 Malignant Hyperthermia

Malignant hyperthermia is a major problem to anesthetists. There is a significant incidence of severe hyperpyrexia which accompanies the administration of some anesthetics, e.g., halothane, and some muscle-relaxant drugs [8]. The condition has an hereditary basis in some cases and may be found in association with a variety of physical abnormalities such as cryptorchidism or skeletal disorders. The cause of the condition has nothing whatever to do with the body temperature control mechanisms. The problem resides within the skeletal muscles which fail to relax after administration of muscle relaxants or anesthetics, and indeed, which go into intense rigor-like contractions with immense liberation of heat [8]. Tachycardia, tachypnea, and cyanosis are seen and extremely high body temperatures are reached. Mortality may be high, though dantrolene (a drug which interferes with the release of activator Ca^{2+} from the sarcoplasmic reticulum) has done much to improve management of the problem. The immediate problem in the muscle is a rapid elevation of the cytoplasmic Ca^{2+}. There may be defective uptake by the sarcoplasmic reticulum, excessive Ca^{2+} release, or other abnormalities of excitation contraction coupling (Chaps. 5, 46). Release of catecholamines into the circulation may also modify the plasma membrane processes, leading to increased ion pumping and heat generation. The condition is found in some strains of pigs in which it may also be brought on by nonthermal and emotional environmental stresses, as it may occasionally in the human.

111.4.4 Hypothermia

If the human body temperature falls below 35°C, the condition of hypothermia is said to exist. Body temperature may fall as a result of exposure to a very cold environment or cold water, when heat extraction exceeds the ability of the body to make heat. It can occur as a result of disruption of the thermoregulatory mechanism by drugs such as barbiturates or alcohol. Some drugs which have been used in the management of high blood pressure, e.g., α-methyldopa, may, in the elderly, predispose to hypothermia. Diseases which impair thermoregulation, such as myxedema (severe thyroid hormone deficiency, see Chap. 22) or cerebral infarcts, can make the person more at risk from hypothermia. There is controversy as to whether old age per se is a predisposing factor, but some elderly persons are definitely at risk of hypothermia because they either have lost some thermoregulatory functions or have reduced perception of cold. Other elderly persons have completely normal thermoregulation. Injuries such as spinal cord transections or severe head injury, or severe shock, may cause failure of thermoregulation and, in cold environments, lead to hypothermia. Newborn infants, because of their high surface area to weight ratio, may easily become hypothermic. Premature infants in incubators, in which the air temperature is high, may lose heat by radiation to the cold Plexiglas and suffer hypothermia. A second, warm Plexiglas shield within the incubator, and over the baby, can prevent this.

As the body temperature falls, the functions of body organ systems alter. The basal heat production falls as the tissue temperatures fall, with a Q_{10} of 2.3. Cerebral function becomes progressively impaired from dulled senses to unconsciousness, there is progressive slowing of nerve conduction, and reflexes become sluggish and are then lost. There is usually a fall in heart rate, often after an initial rise occasioned by shivering (the direct effect of temperature on the pacemaker leads to a heart rate change of about 20 beats/°C). The function of the myocardium alters with slowed depolarization, prolongation of the P-R interval, and widening of the QRS complex. Sometimes a new wave is seen on the downstroke of the QRS complex, and this is called the "J" wave. The heart becomes more and more prone to arrhythmias, with ventricular fibrillation as a common fatal occurrence. There is an initial increase in metabolic rate due to shivering. Pulmonary ventilation is reduced and changes take place in the blood gas levels and in the body's acid base equilibrium (Chap. 78). Frequently there is a marked rise in arterial P_{CO_2} and a fall in P_{O_2}. As a consequence of the CO_2 retention there is an increasing degree of respiratory acidosis, and in response to shivering or muscle contraction there is also metabolic acidosis. The acidosis is further exaggerated by reduced renal tubular function. The metabolism of drugs and their excretion are grossly impaired. Insulin is ineffective at temperatures below 30–31°C, and often there is a higher than normal blood glucose level in severe hypothermia accompanying hemoconcentration. Insulin administered at this stage of hypothermia is ineffective but may act suddenly during rewarming with catastrophic effects.

A special condition is *submersion hypothermia*, in which the victim, usually a child, has been totally immersed in very cold water for 20 min or more. There may be clinical death but, with proper management, full recovery can often occur without measurable brain damage. Cooling of the brain through the skull and the "diving reflex," in which as a result of immersion of the face in the cold water there is intense reflex vasoconstriction in all vascular beds except those of the brain and the heart, help to reduce brain metabolism and maintain some O_2 supply to the brain during submersion.

The ideal management of severe clinical hypothermia involves full physiological monitoring and control in intensive care conditions. The method of choice for rewarming is usually rapid in most types of severe hypothermia using extracorporeal vascular shunt heat exchangers, warm peritoneal lavage, hot blankets, breathing hot air saturated with water vapour, and even hot baths. The local method of choice should be that with which the physician is most familiar [38]. It has been said that in severe hypothermia, particularly in the elderly, and which has taken a long time to develop, slower rewarming may be best, but the evidence for this is mainly anecdotal and the rewarming procedures did not usually include the most rigorous cardiovascular, respiratory, and acid-base control. All physicians and those who take their recreation outdoors should be aware of the possibility of hypothermia, and physicians should think of this condition in managing accident patients when cold exposure has occurred.

111.4.5 Cold Adaptation

Several methods are used in the animal kingdom to provide long-term increased resistance to cold. For example, livestock living on the range in the foothills of the Rocky Mountains in the winter develop thicker coats which provide additional thermal insulation. Some species have deposits of brown adipose tissue, the metabolic heat-productive ability of which is enhanced by prolonged cold exposure. Some have suggested that there may be other sources of nonshivering thermogenesis, particularly in animals lacking brown adipose tissue, but the evidence for this is not yet fully convincing. Some humans such as the Australian aboriginee are able to sleep in the cold desert night with only partial warming from a small fire, and without shivering. The result is a considerable fall in core temperature which is then raised quickly by exercise and environmental heating on waking. Similarly, the ability to tolerate a modest fall in core temperature was found in Korean and Japanese pearl divers before the utilization of thermally protective wet suits. Such tolerances to reduced core temperatures, to levels which are not dangerous, allows conservation of energy and thus of food resources. There is some evidence for local adaptation to cold in those whose work involves prolonged exposure of the hands to cold [33], in that the pain normally experienced on immersion of the hand in very cold water is diminished, for reasons which are not yet clear.

References

1. Bannister RG (1960) Anhidrosis following intravenous bacterial pyrogen. Lancet ii:118–122
2. Barker J, Carpenter DO (1970) Thermosensitivity of neurons in the sensory cortex of the cat. Science 169:597–598
3. Beeson PB (1948) Temperature elevating effect of a substance obtained from polymorphonuclear leucocytes. J Clin Invest 27:524
4. Bernheim HA, Kluger MJ (1976) Fever and antipyresis in the lizard *Dipsosaurus dorsalis*. Am J Physiol 231:198–203

5. Blatteis CM, Bealer SL, Hunter WS, Llanos QJ, Ahokas RA, Mashburn TA (1983) Suppression of fever after lesions of the anteroventral third ventricle in guinea pigs. Brain Res Bull 11:519–526

6. Bligh J (1973) Temperature regulation in mammals and other vertebrates. North-Holland, Amsterdam, chap 10

7. Boulant JA, Dean JB (1986) Temperature receptors in the central nervous system. Annu Rev Physiol 48:639–654

8. Britt BA (1991) Malignant hyperthermia. In: Schönbaum E, Lomax P (eds) Thermoregulation: pathology, pharmacology and therapy. Pergamon, New York, pp 179–292

9. Brück K, Hinckel P (1980) Thermoregulatory noradrenergic and serotonergic pathways to hypothalamic units. J Physiol (Lond) 304:193–202

10. Brück K, Hinckel P (1982) Thermoafferents and their adaptive modifications. Pharmacol Ther 17:357–381

11. Carlsson A, Falck B, Hillarp NA, Tor PA (1962) Histochemical localization at the cellular level of hypothalamic noradrenaline. Acta Physiol Scand 54:385–386

12. Clark WG, Lipton JM (1991) Brain and pituitary peptides in thermoregulation. In: Schönbaum E, Lomax P (eds) Thermoregulation: pathology, pharmacology and therapy. Pergamon, New York, pp 509–560

13. Coceani F (1991) Prostaglandins and fever. Facts and controversies. In: Mackowiak P (ed) Fever: basic Mechanisms and Management. Raven, New York, pp 59–70

14. Cooper KE (1987) The neurobiology of fever: thoughts on recent developments. Am Rev Physiol 10:297–324

15. Cooper KE (1995) Fever and antipyresis: the role of the nervous system. Cambridge University Press, Cambridge, UK

16. Cooper KE, Cranston WI, Honour JA (1967) Observations on the site and mode of action of pyrogens in the rabbit brain. J Physiol (Lond) 191:325–338

17. Cooper KE, Cranston WI, Snell ES (1964) Temperature regulation during fever in man. Clin Sci 27:345–356

18. Cooper KE, Johnson RH, Spalding JMK (1964) Thermoregulatory reactions following intravenous pyrogen in a subject with complete transection of the cervical cord. J Physiol (Lond) 171:55P

19. Cooper KE, Kasting NW, Lederis K, Veale WL (1979) Evidence supporting a role for endogenous vasopressin in natural suppression of fever in the sheep. J Physiol (Lond) 295:33–45

20. Cooper KE, Kerslake DMcK (1953) Abolition of nervous reflex vasodilatation by sympathectomy of the heated area. J Physiol (Lond) 119:18–29

21. Dinarello CA (1979) Production of endogenous pyrogen. Fed Proc 38:52–56

22. Dinarello CA (1988) Biology of interleukin-1. FASEB J 2:108–115

23. Fox RH, MacPherson PK (1954) The regulation of body temperature during fever. J Physiol (Lond) 125:21P–22P

24. Fox RH, Hilton SM (1958) Bradykinin formation in human skin as a factor in heat vasodilatation. J Physiol (Lond) 142:219–232

25. Glyn-Ballinger JR, Bernadini GL, Lipton JM (1983) α-MSH injected into the septal region reduces fever in rabbits. Peptides 4:199–203

26. Goelst K, Mitchell D, MacPhail AP, Cooper KE, Laburn H (1992) Fever response of sheep in the peripartum period to gram negative and gram positive pyrogens. Pflugers Arch 420:259–263

27. Hales JRS, Richards DAB (eds) (1987) Heat stress, physical exertion and environment. Exerpta Medica, Amsterdam International Congress series 773

28. Hensel H (1973) Handbook of sensory physiology. Springer, Berlin Heidelberg New York

29. Hensel H (1981) Thermoreception and temperature regulation. Academic, London, pp 13–63

30. Khogali M, Hales JRS (eds) (1983) Heat stroke and temperature regulation. Academic, New York

31. Kluger MJ (1979) Fever. Princeton University Press, Princeton, pp 118–124

32. Leathhead CS, Lind AR (1964) Heat stress and heat disorders. Davis, Philadelphia

33. Leblanc J (1975) Man in the cold. Thomas, Springfield

34. Milton AS, Wendlandt S (1971) Effects on body temperature of prostaglandins of the A, E and F series on injection into the third ventricle of unanaesthetised cats and rabbits. J Physiol (Lond) 218:325 336

35. Mitchell D, Laburn HP, Cooper KE, Hellon RF, Cranston WI, Townsend Y (1986) Is prostaglandin E the neural mediator of the febrile response? The case against a proven obligatory role. Yale J Biol Med 59:159–168

36. Murphy MT, Lipton JM (1982) Peripheral administration of α-MSH reduces fever in older and younger rabbits. Peptides 3:775–779

37. Myhre K, Hammel HT (1969) Behavioral regulation of internal temperature in the lizard Tiliqua scincoides. Am J Physiol 217:1490–1495

38. Paton BC (1991) Accidental hypothermia. In: Schönbaum E, Lomax P (eds) Thermoregulation: pathology, pharmacology and therapy. Pergamon, New York, pp 397–454

39. Pillay V, Savage N, Laburn H (1993) Interleukin-1 receptor antagonist in newborn babies and pregnant women. Pflugers Arch 424: 549–551

40. Satinoff E (1983) A re-evaluation of the concept of homeostatic organization of temperature regulation. Handbook Behav Neurobiol 6:443–471

41. Satinoff E, Shan SYY (1971) Loss of behavioral thermoregulation after lateral hypothalamic lesions in rats. J Comp Physiol Psychol 77:302

42. Stitt JT (1985) Evidence for the involvement of the organum vasculosum laminae terminalis in the febrile response of rabbits and rats. J Physiol (Lond) 368:501–511

43. Vaughn LK, Kluger MJ (1977) Fever and survival in bacterially infected rabbits. Fed Proc 36(3):511

44. Werner J (1986) Do black-box models of thermoregulation still have any research value? Contribution of system-theoretical models to the analysis of thermoregulation. Yale J Biol Med 59:335–348

45. Work E (1971) Production, chemistry and properties of bacterial pyrogens and endotoxins. In: Wolstenholme GEW, Birch J (eds) Pyrogens and fever: a CIBA Foundation symposium. Churchill Livingstone, Edinburgh, pp 23–46

112 The Formation of Sweat

R. Greger

Contents

112.1 Introduction: Sweating – When, Where, How Much?

Most of the sweat glands in man are of the *eccrine* type, i.e., they produce a watery secretion with some salt and few organic substances. The *apocrine* sweat glands are larger and wider, and produce a more concentrated secretion with odorant substances. The odor is produced by bacterially decomposed fatty acids. These glands are widely distributed in animals, but are confined to the axillae and a few other locations in the human species. The function of the apocrine sweat glands will not be further discussed in this chapter.

The eccrine sweat glands serve the purpose of evaporation of water to prevent overheating. This is required in many homeotherms. Man is probably top of the league in optimal performance of thermoregulatory sweating, because of the very high maximum rates of human sweat production and the fairly low NaCl concentration in the final sweat [35]. Only horses approach the human rates with their maximum sweat rates [37]. In humans the maximal rates of sweating under thermal stress can be as high as 2–4 l/h. This corresponds to an *evaporation cooling effect* of 4.8–9.6 MJ/h (1.3–2.6 kW). It was mentioned in Chaps. 110 and 111 how evaporative cooling by sweating dissipates much of the energy turned over (consumed) during physical exercise, and energy expenditure by sweating must exceed the heat taken up by radiation when the body is exposed to an environment of its own temperature or even higher. With maximal volumes of sweat, the loss of NaCl is approximately 2.5–5 g/h. It will be discussed below that the relatively *low salt content* of sweat is advantageous not only for the obvious reason that it keeps the amount of NaCl that must be replaced through our food fairly low.

Sweating occurs appropriately when it is required for thermoregulation, e.g., during physical exercise or exposure to a climate which leads to a positive heat balance in the body. However, embarrassingly, sweating also occurs when we are just nervous. It could be argued that even this type of *emotional sweating* may be useful in anticipation of catecholamine-induced increased metabolism.

Sweat glands are present all over the body, with predilection sites such as the axillae, palms, forehead, foot pad, etc. Emotional sweating occurs mostly in the palms and axillae. Thermoregulatory sweating occurs in these areas and also over the remaining body surfaces. The density of sweat gland distribution varies over the body surface, with an specially high density on the forehead (350 glands/cm^2), the axillae, palms, and soles of the feet, and a lower density on, for instance, the thigh (60 glands/cm^2). There is wide variation amongst individuals.

The rates of sweat production depend on the demand. Sweating can be almost completely suppressed or else, when maximally stimulated by thermal stress, be as high as 2–4 l/h. Appropriateness of sweating can be developed. There are two aspects to this, both of which develop over the course of a few weeks: (1) It is a common experience that sweat is more salty in the beginning of the warm season and loses some of its saltiness as we adapt to the hot climate; we also start sweating before the central body temperature rises, and in a somehow more controlled fashion: at the uncovered body surfaces, without soaking our clothes. (2) The *absolute rate* of sweating *doubles* during this adaptive process, from 1–2 l/h to up to 4 l/h.

In this chapter the mechanisms of sweat production will be discussed in some detail. The emphasis will be on the cellular mechanisms, because our knowledge in this area has increased substantially over the past few years. The integrative aspect of sweat production and its role in *thermoregulation* is addressed in Chaps. 110 and 111. A specific section on the pathophysiology of sweat gland function in *cystic fibrosis* is added in this chapter, because sweat formation is altered in this genetic disease in a very characteristic, even diagnostic way, and because much progress has been achieved in this area during recent years. The investigation of the pathophysiology of cystic fibrosis has also increased our understanding of normal sweat gland function and, especially, its regulation.

R. Greger/U. Windhorst (Eds.)
Comprehensive Human Physiology, Vol. 2
© Springer-Verlag Berlin Heidelberg 1996

112.2 The Sweat Gland

The approximately *2–3 000 000 sweat glands* of our body (calculated from the maximum total sweat rate divided by the maximum secretion of a single gland [36,42]) are typical, albeit rather primitive exocrine glands. They produce a *primary secretion* in the *secretory coil*, and modify this secretion in the *sweat duct*. The primary secretion is an almost isotonic NaCl solution [48,49], the ductal modification consists of the *reabsorption of NaCl* with little or no reabsorption of water. Via this mechanism the final sweat becomes hypotonic, with a normal NaCl concentration of approximately 20–30 mmol/l [36,49].

112.2.1 Morphology of the Sweat Gland

Figure 112.1A shows the structure of a sweat gland as seen at low magnification, and Fig. 112.1B a section through a sweat gland. In the latter, the secretory coil can be clearly distinguished from the sweat duct. It has a larger lumen and is more tortuous than the duct. The coil has a blind end but no acini, unlike many other excretory glands. The duct is simple and unbranched. Thus, a single sweat gland forms a long (several millimeters), blind canal. The blind coil is heavily supplied by capillaries and innervated by sympathergic, albeit mostly *cholinergic, fibers* (see Sec. 112.3.1). At some locations such as the forehead, the palm of the hand, and the foot, some noradrenergic innervation also appears to be present. There is no doubt that, in addition to muscarinic receptors, the secretory cells of the coil also possess, receptors for β-agonists and probably also for vasoactive intestinal polypeptide (VIP).

A more detailed schematic view of a secretory coil is given in Fig. 112.2A. Three types of cells are present in the secretory coil (cf. Fig. 112.1A):

- Myoepithelial cells
- Clear cells
- Dark cells.

The *myoepithelial cells* almost completely cover the other two types on the outside. They are contractile cells, innervated by cholinergic fibers. The *clear cells* are most likely responsible for salt and water secretion [46,54]. They are more numerous than the dark cells and are called "clear" because they stain brighter than the "dark" cells in hematoxylin-eosin staining. Whether clear cells are electrically coupled (dye-coupled) to each other, as is characteristic of the acinar cells of many other exocrine glands, is not yet known [22].

The *dark cells* are so called because they contain granules which are responsible for their basophilic staining by hematoxylin-eosin. These cells correspond to the *goblet cells* present in most exocrine glands. The content of the secretory granules is periodic acid-Schiff-positive. These vesicles probably contain macromolecular contents which are admixed to the electrolyte secretion coming mostly

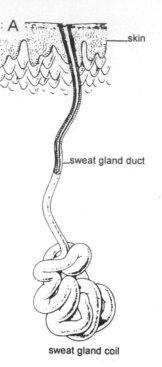

skin

sweat gland duct

sweat gland coil

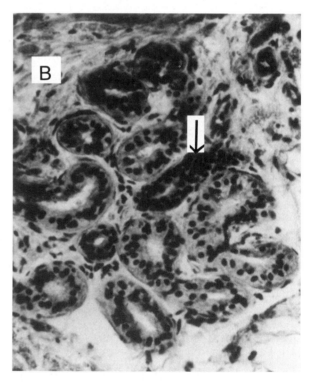

Fig. 112.1. A Sweat gland coil and duct. The duct is a few millimeters in length. **B** Microscopic view of a sweat coil with duct (*arrow*). Note the myoepithelial cells covering the coil. (Magnification × about 200. Modified from [23])

from the clear cells [22,36,54]. Dark cells are electrically coupled to each other [22].

The secretory coil contains at least two layers of different cells: the myoepithelial cells on the outside and the clear

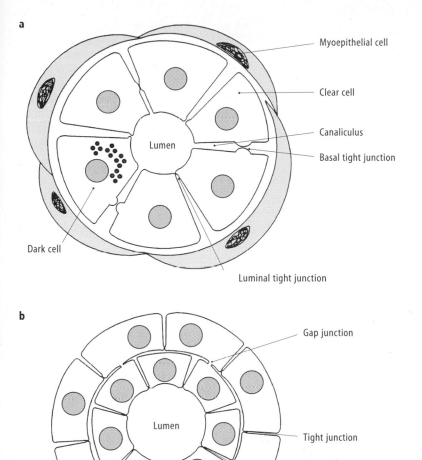

a

Myoepithelial cell

Clear cell

Canaliculus

Basal tight junction

Lumen

Dark cell

Luminal tight junction

b

Gap junction

Lumen

Tight junction

Tight junction

Fig. 112.2. a Sweat gland coil. The myoepithelial cells surround the clear and dark cells. The tight junctions are on the basal or luminal pole of the cells. **b** Sweat gland duct. The duct has two layers of cells. The cells are connected by gap junctions. The tight junctions are of the tight type [4]

and dark cells surrounding the wide lumen. The sweat ducts (Fig. 112.2B) contain two layers of one single cell type [35,36]. The distribution of the ouabain-binding sites in the duct cells and the dye coupling after injection [22] suggest that these cells communicate with each other via gap junctions. *Tight junctions* of the leaky type (cf. Chap. 59) are present between the coil cells. The tight junctions may be localized close to the basal pole. Then the lateral space towards the lumen forms an *intercellular canaliculus*. The tight junctions of the duct are more complex than one might expect for an epithelium generating a large Na^+ concentration gradient.

112.2.2 Function of the Secretory Coil

The study of sweat gained immediate interest when it was shown by Di Sant-Agnese et al. in 1953 [12,13] that the sweat of patients with cystic fibrosis (CF patients) had a much higher NaCl concentration than that of normal individuals. This was the beginning of the use of the sweat test (see below) to diagnose this disease – a test which, as more recently modified, remains an important diagnostic tool today. Ten to 15 years later, Schulz and Frömter [47–50] pioneered this entire field of research by introducing the micropuncture technique, which until then had been used for the study of renal tubules. They made two very important observations, which are illustrated in Fig. 112.3. The first was that the fluid produced in the coil was almost *isotonic* compared to plasma, and that emerging from the duct was *hypotonic*. Secondly, they showed that the transepithelial *voltage of the skin* was lumen-negative and was even *higher in CF patients* [48,50]. This finding was surprising because, on the basis of the observations of Di Sant-Agnese et al., a reduced reabsorption of Na^+ was expected, and this would a priori be paralleled by a reduced voltage. Schulz and Frömter interpreted their findings to mean that Na^+ reabsorption occurred by active transport, generating the voltage, and that the higher voltage in CF patients was probably caused by a *Cl⁻ impermeability of*

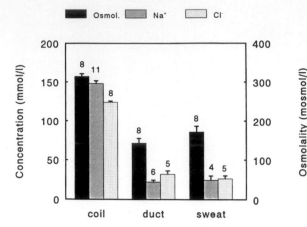

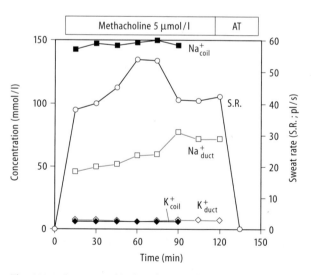

Fig. 112.3. Osmolarity, Na⁺, and Cl⁻ concentrations of primary sweat, sampled from the coil (*coil*), from modified sweat fluid sampled in the duct 1 mm proximal to its opening (*duct*), and from the final sweat excreted on the skin surface (*sweat*). Note that the osmolality in primary sweat is slightly higher than plasma osmolality, and that it is approximately half-isotonic in the duct and in the final sweat. The Na⁺ concentration is around 150 mmol/l in primary sweat and only 20–25 mmol/l in the duct and in final sweat. As expected, the Cl⁻ concentration is slightly lower in primary sweat (presence of 25 mmol/l HCO_3^-), but is close to the Na⁺ concentration in the duct and in final sweat. (Data obtained from healthy volunteers, redrawn from [50])

the duct [47,48]. However, their findings and correct interpretation were forgotten and ignored until 1983, 15 years later, when the same conclusion was reached in experiments performed by Quinton [34].

The mechanism of sweat formation in the secretory coil has been studied in vitro in whole sweat glands and secretory coils isolated from the palm of monkeys [42]. The results of a typical experiment are reproduced in Fig. 112.4. *Methacholine* (MCH, instead of acetylcholine, ACH) was used to *stimulate secretion*. MCH induced strong stimulation of the sweat rate with little change in the Na⁺ concentration. On the other hand, the fluid emerging from the duct had a much lower Na⁺ concentration. The concentration of Na⁺ rose when ouabain was added [42]. Later it was shown with various preparations that this cholinergic effect is caused by an *increase in cytosolic Ca^{2+}* [43]. Studies in isolated and cultured isolated clear cells using several methods, including measurements of cytosolic content by the electron probe, indicate that stimulation by ACH also produces K⁺ and Cl⁻ losses [41,43,46,54]. Secretion apparently occurs via the *Na⁺2Cl⁻K⁺ cotransporter* and can be inhibited by bumetanide [38,41].

The current concept deduced from these studies is that originally proven for the rectal gland of the spiny dogfish, *Squalus acanthias* [17–20]. It is shown in Fig. 112.5A. The secretion is driven by the basolaterally localized (*Na⁺+K⁺)-ATPase*. Na⁺, 2Cl⁻, and K⁺ are taken up across the basolateral membrane via their cotransporter. Na⁺ is extruded by the pump in exchange for K⁺, and *K⁺ recycles* across the basolateral membrane through *K⁺ channels* [46]. Cl⁻ leaves the cell via *Cl⁻ channels* present in the luminal membrane. K⁺ diffusion to the blood side and Cl⁻ diffusion

to the lumen generate a transepithelial voltage which drives Na⁺ into the lumen via the *Na⁺-permeable paracellular pathway*.

It is likely that the increased cytosolic Ca^{2+} activity caused by ACH enhances the conductance of the luminal Cl⁻ channels, and probably also that of the basolateral K⁺ channels [36,38,43]. Water probably follows the actively secreted ions such that the primary secretion is seemingly isotonic, i.e., there is no measurable osmolality gradient in the primary secretion when compared to plasma.

112.2.3 Function of the Duct

Sweat gland ducts have been studied extensively in *in vitro perfusion* experiments [1,4,34,38], as originally designed for the study of renal tubules [6,7]. These studies reveal that lumen voltage is around −10 to −30 mV (lumen-negative). Transepithelial resistance is surprisingly low at $8 \Omega \times cm^2$, and this high conductance is mostly due to the *transcellular Cl⁻ pathway* [38]. Absorption of Na⁺ occurs via *amiloride-sensitive Na⁺ channels* [1]. This entire concept is summarized in Fig. 112.5B. The general mechanism is very similar to that described in other chapters for the frog skin [25], the colon [10], the respiratory tract [53], and the principal cell of the collecting tubule [26]. The only difference is the suggestion that Cl⁻ moves across the cell via two Cl⁻ conductances in series [38]. This concept requires that the intracellular Cl⁻ activity is at a value which permits voltage-driven uptake across the luminal membrane and voltage-driven extrusion across the basolateral membrane. At the voltages published in perfusion studies

Fig. 112.4. Sweat rate (*S.R.*) and Na⁺ and K⁺ concentrations in sweat sampled from isolated sweat coils and intact glands (duct samples) from monkeys. Note that methacholine increases the sweat rate. The Na⁺ concentration in primary sweat (*filled squares*) is around 145 mmol/l; that from the duct (*open squares*) has a much lower concentration. The K⁺ concentration in coil fluid is (*filled diamonds*) similar to that in plasma, and that in final sweat is slightly higher (*open diamonds*). Atropine (*AT*, 40 µmol/l) inhibits sweat production instantaneously. (Data from [42])

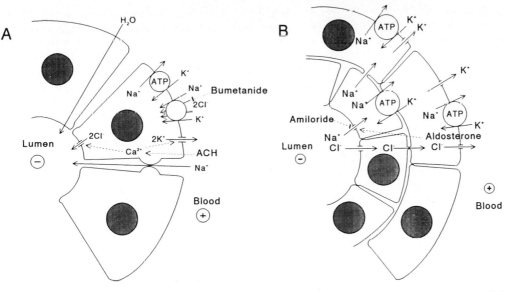

Fig. 112.5A,B. Mechanisms of NaCl secretion in the coil (**A**) and NaCl absorption in the duct (**B**). The *circle* labelled *ATP* is (Na$^+$+K$^+$)-ATPase; a *circle* is a carrier; an *arrow* is an ion channel. In both segments the transepithelial voltage is lumen-negative. This is due to Cl$^-$ secretion in the coil, and to Na$^+$ absorption in the duct. ACH stimulates secretion via phospholipase C, IP$_3$, Ca^{2+} release, and increase in the Cl$^-$ and K$^+$ conductances. Aldosterone enhances Na$^+$ absorption. Note that the Cl$^-$-conductive pathway in ducts is present in the same cells as absorb Na$^+$. The tight junctions are tight in the duct and leaky, Na$^+$-permeable, in the coil

Table 112.1 Composition of normal sweat and sweat taken from patients with Cystic fibrosis

	Normal	Cystic fibrosis
Osmolarity (mosmol/l)	154 ± 40	299
Na$^+$ (mmol/l)	10–40	60–145
Cl$^-$ (mmol/l)	0–30	70–120
K+ (mmol/l)	2–25	5–60
Other anions (mostly HCO$_3^-$ and lactate; mmol/l)	31 ± 10	13 ± 11
pH	6.8–7.3	4.5–7.3
Flow rate per gland (pl/s)	40–110	30–130

Data from [36].

this is feasible [36]. Whether it can also hold when luminal Cl$^-$ concentration falls to low values has yet to be demonstrated. Cytosolic Cl$^-$ activity measurements would also be very valuable to solidify the evidence on this point.

As predicted for an epithelium with amiloride-sensitive Na$^+$ channels, the sweat duct is under the control of *aldosterone* [36]. Aldosterone enhances the ability to absorb NaCl and generates a sweat of even lower NaCl concentration. Normal secretions from the sweat coil contain substantial concentrations of *lactate* and a rather low HCO$_3^-$ concentration [36,49]. The final sweat (see Table 112.1) contains *K$^+$ concentrations* which are around *twofold higher* than that of plasma. The high concentrations of NH$_4^+$ are also worth noting. Sweat is *slightly acid* and does not contain many macromolecules [36].

To sum up this section, sweat glands consist of two functional units:

- The coils, which when stimulated produce a plasma-isotonic secretion at a high rates

- The ducts, which reduce the concentration of NaCl by active absorption.

112.3 Neuronal Control of Sweat Formation

Most exocrine glands are controlled by innervation and by circulating as well as local hormones. In man the sweat gland is mostly controlled by *cholinergic sympathergic innervation*. The innervation varies greatly among species. For example, the horse lacks cholinergic control, although sweating is very important to thermoregulation in this animal [37]. In equine sweat glands, sweat production is mostly controlled by sympathergic *β-adrenergic innervation* [2,3]. In other animals sweating is exclusively controlled by ACH [35]. The human species is in between, although here ACH control appears more important than β-mediated control [36,46]. *VIP effects* have also been shown in primate sweat glands [45]. The physiological role of VIP in sweat production remains to be elucidated. The regulation appears to occur mostly on the secretory coil and not on the duct. However, when ducts have been put in culture, they possess (or acquire ?) the transduction pathways for regulation by both ACH and cAMP [29,32].

112.3.1 Cholinergic Innervation

The secretory coil is under the control of sympathergic fibers, which mostly use ACH as the transmitter. This is an exception to the general rule that sympathergic fibers release norepinephrine. In fact, these neurons at birth still transmit via norepinephrine, and they *change their transmitter to ACH* during postnatal development [28]. This is one example of the *transmitter plasticity* present in the autonomous nervous system. From this perspective it is easy to understand why in some species sweat gland coils are under the control of ACH, whilst in others β-agonists are predominant.

ACH transmission is found in the myoepithelial as well as the clear and dark cells [39]. Myoepithelial cells respond with depolarization and contraction, which pushes the contents of the coil towards the duct. In dark and clear cells ACH probably acts by the well-known pathways (cf. Chap. 5) of:

* Muscarinic receptor
* G_p-protein
* Phospholipase C
* Inositol-1,4,5-triphosphate (IP_3) and diacylglycerol (DAG) production
* Ca^{2+} release [33].

ACH not only induces secretion, it also shrinks clear cells and shifts K^+ from intra- to extracellular [43,51]. The mechanism of Ca^{2+}-mediated stimulation of secretion present in the sweat coil is probably very similar to that found in other exocrine glands. The best-studied cell types are probably the pancreatic acinar cell [33] and the carcinoma cell lines from colonic crypts [27]. In the latter it has been postulated that the Ca^{2+}-mediated increase in Cl^- conductance is paralleled by *exocytosis* [16]. Very similar data have meanwhile also been reported for the pancreatic acinar cell [30]. The parallel upregulation of K^+ conductance may be mediated by a Ca^{2+}-dependent K^+ conductance. Both the secretory responses and the increases in cytosolic Ca^{2+} are quite stable with ACH [43].

112.3.2 β-Agonist-Mediated Secretion

In some mammals, such as horses, sweat secretion is controlled exclusively by *β-adrenergic innervation* [2]. In the rat it has been proven that the sympathergic fibers are adrenergic after birth and become cholinergic shortly thereafter [28]. In man and monkey control is by both

transmitters, ACH and β-agonists, and β-adrenergic innervation has been demonstrated at some sites [45]. However, ACH is more prominent and the total contribution of β-agonists is only between 10% and 40% [46]. β-Agonists and ACH are probably not additive, although it has been shown in one recent impalement study that two types of nonmyoepithelial cells could be distinguished on the basis of their responsiveness to β-agonists [39]. Both cell types responded to ACH, but one was sensitive and the other insensitive to isoproterenol. Isoproterenol (or, physiologically, norepinephrine) is believed to act via the classical scheme:

* β-Receptor interaction
* G_s-protein
* Adenylate cyclase
* cAMP
* Protein kinase A.

The marked KCl- and volume losses which have been noted with ACH stimulation in clear cells are not demonstrable with β-agonists [46]. The final pathway of regulation is unknown. By extrapolation, one might speculate that *protein kinase A* phosphorylates a *CFTR-type channel* [9] (CFTR stands for cystic fibrosis transmembrane conductance regulator [40]). Alternatively, one might speculate that phosphorylation *inhibits the endocytosis* of Cl^- channels which have been inserted by exocytosis into the plasma membrane [5,16]. At any rate, an involvement of CFTR appears very likely, because *β-agonist-mediated sweat secretion is absent in CF patients* [44].

112.3.3 Other Transmitters

VIP-immunoreactive fibers have been identified around eccrine sweat glands. This has prompted studies of possible effects of VIP on isolated monkey sweat gland coils [45]. It has been shown that VIP induces secretion, and does so by *elevating cytosolic cAMP* concentration. The VIP effect on cAMP concentrations and on sweat rate was enhanced by MCH. These data suggest that VIP acts as a synergist for both β-agonist-mediated and ACH-mediated sweat secretion. The concentrations of VIP to exert a submaximum effect were around 6 nmol/l. Whether such concentrations occur locally remains to be shown.

To sum up this section, it has been shown that sympathergic innervation by cholinergic fibers is the most important physiological stimulus. It acts on all three types of cells in the secretory coil, but has no effect on the duct. In addition, adrenergic innervation also appears to be present in man, and circulating catecholamines can act on clear cells via the β-receptors present in these cells. The combination of stress-induced vasoconstriction (ice-cold hands) and profuse sweating on the palms is well known. The possible modulatory role of VIP in sweat production requires further clarification.

112.4 Dysfunction of Sweat Glands in Cystic Fibrosis

Although the disease cystic fibrosis itself was defined as a clinical and pathophysiological entity only about 90 years ago [15], the involvement of the sweat glands in this fatal disease has been known for centuries. "The child whose forehead when kissed tastes salty is bewitched and will die soon" [8]. The description of the abnormally high NaCl concentration in the sweat of CF patients [12] has led to the simple *sweat test*, which has become a very important, perhaps still the most important diagnostic tool. Nowadays this test is performed by applying pilocarpine iontophoretically to the skin, collecting sweat, and analyzing its NaCl concentration.

The abnormally high NaCl concentration in the sweat of CF patients is a hallmark of the disease, varying much less than other easily recognized symptoms of this exocrinopathy such as pancreatic involvement, meconium ileus, male infertility, lung involvement, etc. The reason for this elevated NaCl concentration in the sweat of CF patients is mostly *reduced* or *abolished absorption of NaCl* in the sweat duct. The absorption defect is caused by pathologically low Cl^- conductance of the duct. This was already discovered 27 years ago by Schulz and coworkers [47,48] when they found that the transepithelial voltages on the skin of CF patients were augmented after thermal stimulation [50]. Their data are summarized and explained by the model in Fig. 112.6. Due to the intactness of Na^+ channels in the absence of Cl^--conductive pathways, the *active transport voltage* reaches maximum values of around -80 mV (lumen-negative). This is the highest possible *standing gradient* built up by this epithelium. Net absorption is close to zero under these conditions.

As mentioned above, these data were forgotten until exactly 15 years later the same observation was made by Quinton [34]. By that time, researchers of epithelial transport were prepared to accept this finding and to explore the possible interpretations. *Patch clamp analysis* [21,31] and its application to epithelia [14] had just become available. A very important further stimulus for research in this field was the cloning of the CF transmembrane conductance regulator (CFTR), *the* protein which is defective in CF [40]. Since then, research in this area has increased exponentially. Not only do we have today a fairly comprehensive understanding of the pathophysiology of this disease, we are also witnessing the first clinical trials of treating this disease by gene therapy.

The defect in the duct is now interpreted to consist in the defective function of the Cl^- conductances in the luminal and in the basolateral membrane [38]. It is unclear, however, why in the sweat duct it is a constitutive Cl^- channel which is defective in CF, whilst in the respiratory tract, in the intestine, and even in the sweat coil the defect is only found in the *cAMP-regulated Cl^- pathway* [11].

In addition to the defect in the duct, in CF there is also complete absence of the β-agonist-mediated regulation of sweat secretion [44]. This component of the defect is of little relevance because most of the regulation occurs via ACH anyway [36,54], and the ACH-dependent component of coil function is undisturbed in CF. This is reminiscent of the findings in other excretory cells from the respiratory tract, colon crypts, and from exocrine pancreas, where the cAMP-regulated Cl-conductance is defective, but where agonists acting through cytosolic Ca^{2+}, such as carbachol, ATP, bradykinin, and neurotensin, still exert a normal response [24,52].

In terms of the sweat abnormality, CF patients have the specific problem that they are endangered by *NaCl loss* when they sweat during physical exercise. Consider a normal individual as compared to a CF patient. Suppose both to be subjected to the same physical and thermal stress and lose the same volume of sweat. For the normal individual this leads to hypertonic dehydration, which has four major consequences:

- Dehydration provokes sympathergic discharge
- Hyperosmolality causes thirst
- Hyperosmolality causes secretion of arginine vasopressin [AVP, also termed antidiuretic hormone (ADH)]
- Extracellular hyperosmolality induces a volume flow from intra- to extracellular space. Hence, the large volume of the intracellular space compensates partially for the volume loss.

In the CF patient the *dehydration is almost isotonic*. Therefore, less thirst is induced, little ADH is secreted, and no volume shifts from the intracellular space can acutely compensate for the volume losses. Both the normal individual and the CF patient will drink water after the exercise. This will compensate eventually for the disturbance in the normal individual, but it will lead to hypotonic dehydration in the CF patient because the ingested water will tend to shift the water to the intracellular space. This will switch off thirst and ADH secretion completely. In addition, the CF patient may suffer from alkalosis and hypokalemia, because sweat contains more K^+ than plasma and is slightly acidic. Appropriate management of the CF patient, therefore, requires adequate *volume and salt replacement*.

In this section we have seen how the pathophysiology of a disease such as CF can stimulate basic research in a specific field. The CF defect is a *generalized exocrinopathy* relating to the epithelial transport of the Cl^- ion. The defect is easily diagnosed on the basis of an abnormally high NaCl concentration in the sweat. The important pathology, however, relates to other organs such as the respiratory tract, intestine, pancreas, etc.

112.5 Conclusion

The sweat gland is a specialized exocrine gland primed for water evaporation and body cooling: The more than 10^6

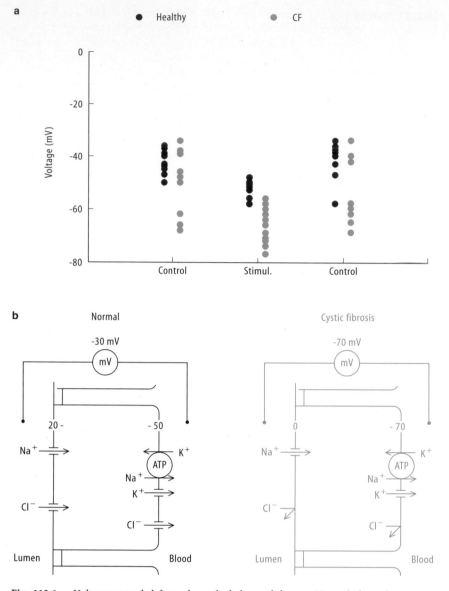

Fig. 112.6. a Voltages recorded from the end phalanx of three healthy individuals *(closed circles)* and three patients with cystic fibrosis (CF; *open circles*) under control conditions and after stimulation of sweat secretion by thermal stress *(stimul.)*. The reference electrode was placed in the interstitium (blood side). Note that after stimulation the voltages in the lumen of sweat ducts in CF patients are more negative than those in controls [50]. (Data replotted from [50].) **b** Model to explain the findings in **a** [47]. In normal subjects NaCl absorption occurs via the parallel Na⁺ and Cl⁻ conductances present in the luminal membrane of the duct. The transepithelial voltage is partly short-circuited by the parallel Cl⁻ conductances present in both cell membranes. Therefore, the transepithelial voltage is only around −30 mV (lumen-negative). In CF the Cl⁻ conductances of both membranes are defective. As a result, the transepithelial voltage increases and the absorption of Na⁺ ceases. This explains the high NaCl concentrations in CF sweat. Essentially the same findings were obtained in isolated in vitro perfused ducts [34]

sweat glands of our body are used for *thermoregulation*. The mechanism of evaporative cooling is essential for an organism that is designed for temperature constancy (homeothermy), performs variable physical exercise, and is subject to variation in climate. The power of this mechanism is enormous. It can dissipate heat at a rate of up to 2.6 kW.

The sweat rate is subject to *autonomic neural control* (cf. Chaps. 110, 111). Thermal stimuli and emotional stimuli control the rate of evaporation. The entire process shows both *fast and slow adaptive responses*. Within a few weeks the maximum rate of sweating can be increased from 1.5 to 4 l/h. In addition, there are slow adaptive responses when an individual changes habitat and climate.

The mechanism of water secretion of necessity requires the secretion of osmotically active substances, in this case NaCl. This is caused by the fact that water cannot be transported actively by itself, but follows an osmotic gradient.

The osmotic gradient must therefore be produced by active NaCl secretion. To optimize the process in sweating, it is divided into two steps:

- Secretion of NaCl and water in the secretory coil
- Absorption of NaCl in the duct.

The final sweat product can thus be delivered to the body surface with little salt content (usually <25 mmol/l). This process can even be optimized in conditions of Na^+ depletion, when *aldosterone* enhances the body's capability to reclaim Na^+. In this case sweat can contain less than 10 mmol/l NaCl.

It goes without saying that sweating has its energetic price. It is optimized on both sides: the secretion occurs at a ratio of up to 6 mol Na^+ per mole of ATP, and the absorption, generating large electrochemical gradients, requires 1 mol ATP per 3 mol Na^+. To sweat at maximum rates therefore costs around 300 mmol ATP/h. This number is not very big (for comparison, it is about three times the normal renal ATP turnover per hour), and it shows clearly that weight loss in the sauna is a volume loss but not a loss of body fat. At any rate, temperature homeostasis has a higher priority than constancy of the body mass.

References

1. Bijman J, Quinton PM (1984) Influence of abnormal Cl⁻ impermeability on sweating in cystic fibrosis. Am J Physiol 247:C3–C9
2. Bijman J, Quinton PM (1984) Predominantly β-adrenergic control of equine sweating. Am J Physiol 246:R349–R353
3. Bijman J, Quinton PM (1984) Influence of calcium and cyclic nucleotides on β-adrenergic sweat in equine sweat glands. Am J Physiol 247:C10–C13
4. Bijman J, Quniton PM (1987) Permeability properties of cell membranes and tight junctions of normal and cystic fibrosis sweat ducts. Pflugers Arch 408:505–510
5. Bradbury NA, Jilling T, Berta G, Sorscher EJ, Bridges RJ, Kirk KL (1992) Regulation of plasma membrane recycling by CFTR. Science 256:530–532
6. Burg M, Grantham J, Abramow M, Orloff J (1966) Preparation and study of fragments of single rabbit nephrons. Am J Physiol 210:1293–1298
7. Burg MB, Isaacson L, Grantham J, Orloff J (1968) Electrical properties of isolated perfused rabbit renal tubules. Am J Physiol 215:788–794
8. Busch R (1979) Zur Frühgeschichte der zystischen Pankreasfibromatose. NTM-Schriftenreihe 16:95–109
9. Cheng SH, Rich DP, Marshall J, Gregory RJ, Welsh MJ, Smith AE (1991) Phosphorylation of the R domain by cAMP-dependent protein kinase regulates the CFTR chloride channel. Cell 66:1027–1036
10. Clauss W, Schaefer H, Horch I, Hoernicke H (1985) Segmental differences in electrical properties and Na-transport of rabbit caecum, proximal and distal colon in vitro. Pflugers Arch 403:278–282
11. Collins FS (1992) Cystic fibrosis: molecular biology and therapeutic implications. Science 256:774–779
12. Di Sant-Agnese P, Darling R, Perera G (1953) Abnormal electrolyte composition of sweat in cystic fibrosis of the pancreas. Pediatrics 12:549–563
13. Di Sant-Agnese P, Darling R, Perera GA (1953) Sweat electrolyte disturbances associated with childhood pancreatic disease. Am J Med 15:777–783
14. Gögelein H, Greger R (1984) Single channel recordings from basolateral and apical membranes of renal proximal tubules. Pflugers Arch 401:424–426
15. Goodchild MC, Dodge JA (1993) Cystic fibrosis, 2nd edn. Baillière Tindall, London
16. Greger R, Allert N, Fröbe U, Normann C (1993) Increase in cytosolic Ca^{2+} regulates exocytosis and Cl⁻ conductance in HT_{29} cells. Pflugers Arch 424:329–334
17. Greger R, Schlatter E (1984) Mechanism of NaCl secretion in the rectal gland of spiny dogfish (*Squalus acanthias*). I. Experiments in isolated in vitro perfused rectal gland tubules. Pflugers Arch 402:63–75
18. Greger R, Schlatter E (1984) Mechanism of NaCl secretion in rectal gland tubules of spiny dogfish (*Squalus acanthias*). II. Effects of inhibitors. Pflugers Arch 402:364–475
19. Greger R, Schlatter E, Gögelein H (1986) Sodium chloride secretion in rectal gland of dogfish *Squalus acanthias*. NIPS 1:134–136
20. Greger R, Schlatter E, Wang F, Forrest JN Jr (1984) Mechanism of NaCl secretion in rectal gland tubules of spiny dogfish (*Squalus acanthias*). III. Effects of stimulation of secretion by cyclic AMP. Pflugers Arch 402:376–384
21. Hamill OP, Marty A, Neher E, Sakmann B, Sigworth FJ (1981) Improved patch-clamp techniques for high-resolution current recording from cells and cell-free membrane patches. Pflugers Arch 391:85–100
22. Jones CJ, Quinton PM (1989) Dye-coupling compartments in the human eccrine sweat gland. Am J Physiol 256:C478–C485
23. Junqueira LC, Carneiro J (1991) Histologie, 3rd edn. Springer, Berlin Heidelberg New York
24. Knowles MR, Clarke LL, Boucher RC (1991) Activation by extracellular nucleotides of chloride secretion in the airway epithelia of patients with cystic fibrosis. N Engl J Med 325:533–538
25. Koefoed-Johnsen V, Ussing HH (1958) The nature of the frog skin potential. A P S 42:298–308
26. Koeppen BM, Stanton BA (1992) Sodium chloride transport: distal nephron. In: Seldin DW, Giebisch G (eds) The kidney: physiology and pathophysiology. Raven, New York, pp 2003–2040
27. Kunzelmann K, Hansen ChP, Grolik M, Tilmann M, Greger R (1992) Regulation of the chloride conductance of colonic carcinoma cells. In: Hlby N, Pedersen SS (eds) Cystic fibrosis, basic and clinical research. Excerpta Medica, Amsterdam, pp 49–56
28. Landis SC (1983) Development of cholinergic sympathergic neurons: evidence for transmitter plasticity in vivo. Fed Proc 42:1633–1638
29. Larsen EH, Novak I, Pedersen PS (1990) Cation transport by sweat ducts in primary culture. Ionic mechanism of cholinergically evoked current oscillations. J Physiol 424:109–131
30. Maruyama Y, Inooka G, Li YX, Miyashita Y, Kasai H (1993) Agonist-induced localized Ca^{2+} spikes directly triggering exocytotic secretion in exocrine pancreas. EMBO J 12:3017–3022
31. Neher E, Sakmann B (1976) Single-channel currents recorded from membrane of denervated frog muscle fibres. Nature 260:799–802
32. Novak I, Pedersen PS, Larsen EH (1992) Chloride and potassium conductances of cultured human sweat ducts. Pflugers Arch 422:151–158
33. Petersen OH (1992) Stimulus-secretion coupling: cytoplasmic calcium signals and the control of ion channels in exocrine acinar cells. J Physiol (Lond) 448:1–51
34. Quinton PM (1983) Chloride impermeability in cystic fibrosis. Nature 301:421–422
35. Quinton PM (1983) Sweating and its disorders. Annu Rev Med 34:423–452

36. Quinton PM (1984) Exocrine glands. In: Taussig L (ed) Cystic fibrosis. Thieme, Stuttgart, pp 339–374
37. Quinton PM (1987) Physiology of sweat secretion. Kidney Int 32:102–108
38. Quniton PM (1990) Cystic fibrosis: a disease in electrolyte transport. FASEB J 4:2709–2717
39. Reddy MM, Quinton PM (1992) Electrophysiologically distinct cell types in human sweat gland secretory coil. Am J Physiol 262:C287–C292
40. Riordan JR, Rommens JM, Kerem B-S, Alon N, Rozmahel R, Grzelczak Z, Zielenski J, Lok S, Plavsic N, Chou J-L, Drumm ML, lannuzzi MC, Collins FS, Tsui L-C (1989) Identification of the cystic fibrosis gene: cloning and characterization of complementary DNA. Science 245:1066–1073
41. Saga K, Sato K (1989) Electron probe x-ray microanalysis of cellular ions in the eccrine secretory coil cells during metacholine stimulation. J Membr Biol 107:13–24
42. Sato K (1973) Sweat induction from an isolated eccrine sweat gland. Am J Physiol 225:1147–1152
43. Sato K, Saga K, Sato F (1988) Membrane transport and intracellular events in control and cystic fibrosis eccrine sweat glands. In: Mastella G, Quinton PM (eds) Cellular and molecular basis of cystic fibrosis. San Francisco Press, San Francisco, pp 171–185
44. Sato K, Sato F (1984) Defective beta adrenergic response of cystic fibrosis sweat glands in vivo and in vitro. J Clin Invest 73:1763–1771
45. Sato K, Sato F (1987) Effect of VIP on sweat secretion and cAMP accumulation in isolated simian eccrine glands. Am J Physiol 253:R935–R941
46. Sato KT, Saga K, Ohtsuyama M, Takemura T, Kang WH, Sato F, Suzuki Y, Sato K (1991) Movement of cellular ions during stimulation with isoproterenol in simian eccrine clear cells. Am J Physiol 261:R87–R93
47. Schulz I (1969) Micropuncture studies of sweat formation in cystic fibrosis patients. J Clin Invest 48:1470–1477
48. Schulz I, Frömter E (1968) Mikropunktionsuntersuchungen an Schweissdrüsen von Mukoviszidosepatienten und gesunden Versuchspersonen. In: Windhofer A, Stephan U (eds) Mucoviscidose Cystische Fibrose II. Thieme, Stuttgart, pp 12–21
49. Schulz I, Ullrich KJ, Frömter E, Emrich HM, Frich A, Hegel U, Holzgreve H (1966) Micropuncture experiments on human sweat gland. In: Di Sant'Agnese PA (ed) Research on pathogenesis of cystic fibrosis. Thieme, Wickersham, pp 136–146
50. Schulz I, Ullrich KJ, Frömter E, Holzgreve H, Frick A, Hegel U (1965) Mikropunktion und elektrische Potentialmessung an Schweissdrüsen des Menschen. Pflugers Arch 284:360–372
51. Suzuki Y, Ohtsuyama M, Samman G, Sato F, Sato K (1991) Ionic basis of metacholine-induced shrinkage of dissociated eccrine clear cells. J Membr Biol 123:33–41
52. Warth R, Greger R (1993) The ion conductances of CFPAC-1 cells. Cell Physiol Biochem 3:2–16
53. Welsh MJ (1987) Electrolyte transport by airway epithelia. Physiol Rev 67:1143–1184
54. Yokozeki H, Saga K, Sato F, Sato K (1990) Pharmacological responsiveness of dissociated native and cultured eccrine secretory coil cells. Am J Physiol 258:R1355–R1362

V   The Life Cycle

O Reproduction

113 Hormonal Regulation of Reproductive Organs

G.F. Weinbauer and E. Nieschlag

Contents

113.1 Introduction

Death is an irrevocable characteristic of life, and reproduction can be viewed as the compensation for death. Consequently, reproduction constitutes one of the essential prerequisites for the survival of species, including the human. In addition, through chromosome exchange, sexual reproduction provides relatively frequent opportunities for mutation and thereby for adaptation.

Reproduction requires the successful development and release of intact female and male gametes (ova and spermatozoa). This process involves a complex interplay of various organs, multiple interactions at the cellular level, mitotic and meiotic cell divisions, and highly elaborate structural and functional modifications and specializations during the development of germ cells into mature gametes. Hormones play an essential role in the control of virtually all reproductive processes and events. The scope of this chapter is to provide the reader with an up-to-date account of current views of the hormonal regulation of reproductive organ physiology, including some of the recent advances in cellular biochemistry and molecular biology which have greatly advanced our understanding of the mechanisms of hormone action.

The main purpose of the chapter is to describe the hormonal regulation of those female and male organs that directly participate in the evolution, maturation, and transport of morphologically and functionally intact gametes. The organs and structures under discussion are the hypothalamus and the pituitary gland (see also Chap. 19), the testis, the excurrent duct systems (epididymis and ductus deferens), the accessory sex glands (seminal vesicles and prostate), the ovary, and the oviduct. Other aspects of reproduction will not be touched upon, and detailed discussions of reproductive behavior, fetal endocrinology, fertilization, pregnancy, birth, lactation, and other topics are presented in other chapters of this book.

The findings of numerous research efforts over the past decade demand that the hormonal control of reproductive organs be viewed at two levels: (1) endocrine regulatory mechanisms effected by hormones, i.e., factors produced in endocrine glands, released into the blood stream, and acting at a site distant from their target organs, and (2) factors that are synthesized in the same organ where they elicit their biological effects. Following this distinction, androgens, estrogens, and gestagens, which are produced either in their targets organs themselves or in the neighborhood of their target cells, act as local regulators of

R. Greger/U. Windhorst (Eds.)
Comprehensive Human Physiology, Vol. 2
© Springer-Verlag Berlin Heidelberg 1996

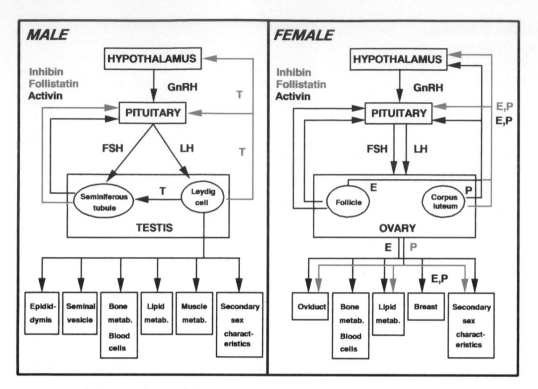

Fig. 113.1. Endocrine regulation of the hypothalamo-hypophyseal-gonadal axis in the male organism (*left panel*) and the female organism (*right panel*). Inhibitory effects are shown in *blue*. *GnRH*, Gonadotropin-releasing hormone; *LH*, luteinizing hormone; *FSH*, follicle-stimulating hormone; *T*, testosterone; *E*, estradiol; *P*, progesterone

gonadal functions. On the other hand, these steroid hormones also act at various sites outside the gonads and thus are endocrine factors as well. Since historically steroids were first recognized as hormones in the classical sense, the progametogenic actions of sex steroids in conjunction with the effects of gonadotropic hormones will be described as endocrine effects hereafter.

113.2 Overview of Hormonal Regulation of Reproductive Functions

Central to the endocrine regulation of the reproductive organs is the hypothalamus (Fig. 113.1). This organ is the source of the gonadotropin-releasing hormone (GnRH), which stimulates the synthesis and release of the gonadotropic hormones (gonadotropins), luteinizing hormone (LH) and follicle-stimulating hormone (FSH), in the anterior part of the pituitary gland. The gonadotropins are identical in both sexes. Both LH and FSH act directly in the gonads to start and promote gametogenesis. In addition to the gonadotropic hormones, sex steroid hormones play an essential role in maintaining the integrity of the reproductive organs. The predominant testicular steroid hormone is testosterone, which is produced under the influence of LH. In the female, estrogens, progesterone, and also testosterone are of importance for ovarian function and

are under the control of LH and FSH. Other anterior hypophyseal hormones such as growth hormone (GH), prolactin (PRL), and thyroid-stimulating hormone (TSH) are probably also involved in the direct hormonal regulation of the gonads.

The steroid hormones have several tasks: induction and maintenance of the function of their target organs, and participation in the feedback regulation of GnRH and gonadotropic hormones (Fig. 113.1). For the role of androgens and estrogens in the development of gender-specific reproductive structures the reader is referred to Chap. 118. The feedback function of steroid hormones serves to guarantee a physiological balance between the production and release of brain and gonadal hormones. Testosterone exerts an inhibitory effect on the release of LH and also of FSH in the male organism. In the female feedback system, estrogens and progesterone are the crucial factors and are capable of exerting inhibitory and stimulatory influences on the hypothalamic-pituitary system, depending on the phase of the menstrual cycle.

The gonads are furthermore the source of peptide factors believed to be associated with the feedback regulation of gonadotropin secretion, in particular with the release of FSH. These factors are inhibin, activin, and follistatin, inhibin and follistatin acting as inhibitors and activin acting as a stimulator of gonadotropin release. As will be discussed later in this chapter, inhibin, activin, and follistatin are also important regulators of gametogenesis at the gonadal level itself.

113.3 The Hypothalamic-Hypophyseal System

113.3.1 Organization

Certain neurons of the hypothalamus are endowed with the ability to secrete hormones (neurosecretion; see also Chap. 19). The mediobasal part of the hypothalamus is connected to the pituitary (hypophysis) via the median eminence and the infundibulum. The pituitary gland is divided into the anterior lobe (adenohypophysis), the intermediate lobe, and the posterior lobe (neurohypophysis), which are morphologically distinct structures. Neurosecretory products can reach the pituitary by two routes: either through the portal blood system, which ends in the adenohypophysis, or along the axon projections of the hypothalamic neurons, which terminate in the neurohypophysis. The anterior pituitary is the target of GnRH and contains, among other cell types, the gonadotropin-producing cells. A gonadotropic cell can synthesize LH and FSH simultaneously.

113.3.2 Gonadotropin-Releasing Hormone

Physiology of GnRH. GnRH is a decapeptide that is synthesized in the somata of hypothalamic neurons, the arcuate nucleus, and is released from their terminals directly into the pituitary portal blood system. The secretion of GnRH follows a strictly pulsatile pattern. Assessment of GnRH in the portal blood from monkey and sheep revealed that GnRH was released in a pulsatile manner with a period of 60–90 min. The underlying mechanisms leading to this pulsatile release of GnRH are currently unknown. Probably there are intrinsic control mechanisms, since the hypothalamus maintains the intermittent secretion of GnRH even when slices of hypothalamus are transferred to in vitro culture systems [7]. Under physiological conditions, the pattern of GnRH secretion is influenced by gonadal steroids. In addition, dopamine, serotonin, endogenous opioids, epinephrine, norepinephrine, and neuroexcitatory amino acids (aspartate, glutamate) have also been implicated in the control of hypothalamic GnRH secretion.

The major function of GnRH is inducing the synthesis and release of the pituitary gonadotropic hormones which, in turn, are the essential factors required for normal gonadal functions. Distinct releasing hormones were originally proposed to exist for LH and FSH. At present, however, the available evidence argues in favor of the existence of a single releasing factor for both gonadotropic hormones. This view mainly derives from the fact that blockade of GnRH synthesis or prevention of binding of GnRH to its pituitary receptors abolishes the secretion of both gonadotropic hormones.

With regard to the regulation of pituitary gonadotropin release and gonadal functions, perhaps the most important feature of hypothalamic GnRH is the intermittency of its release, since this secretion pattern is of pivotal importance for the orderly (pulsatile) secretion of the gonadotropic hormones from the hypophyseal gland (Fig. 113.2). Both the frequency and the amplitude of secreted GnRH determine the release of LH and FSH, and consequently a normal pattern of GnRH release is the prerequisite for normal reproductive function. Under physiological conditions LH and FSH pulses occur at 60–120 min intervals in the male, and in the female they depend on the phase of the menstrual cycle, with high-frequency and low-amplitude pulses occurring during the follicular phase and low-frequency and higher-amplitude pulses during the luteal phase. An abnormally high frequency of GnRH pulses or constant exposure to GnRH inhibits the production and secretion of gonadotropins (Fig. 113.2) Such hormonal signalling via discrete pulses provides a particularly effective and economical mode of transmitting information, since by virtue of varying the pulse frequncy and pulse amplitude of a secreted hormone, differential effects are induced in the target cells. Consequently, this patterned interplay between hypothalamus and pituitary is essential for intact gonadal functions. For example, deficient or absent hypothalamic GnRH secretion (idiopathic hypogonadotropic hypogonadism, Kallmann syndrome, and hypothalamic amenorrhea) results in a marked decline of gonadotropin pulse amplitudes and the cessation of gonadal activity [61]. In turn, gonadal dysfunctions of the above etiology can be corrected by subcutaneous administration of GnRH in a pulsatile fashion from portable minipumps.

Analogues of GnRH. The amino acid sequence of GnRH was discovered more than 20 years ago. This disclosure led to recognition of the relevance of each of the ten amino acids to the biological function of the GnRH molecule and permitted the design and development of compounds analogous to GnRH. The most important are the amino acids in positions 1–3, which account for the hormone-releasing activity, and in positions 6 and 10, which determine binding to the GnRH receptor. The peptide bonds between amino acids in position 5 and 6 and positions 9 and 10 represent important sites for endopeptidase cleavage and degradation of the GnRH molecule. Modifications and/or substitutions of amino acids in appropriate positions resulted in highly potent GnRH analogues with either agonistic or antagonistic properties in relation to the original GnRH peptide.

GnRH agonists initially stimulate LH and FSH secretion for about 2–3 weeks. After that, however, the release of gonadotropins, particularly LH, is inhibited. Unlike the GnRH agonists, antagonists of GnRH are capable of suppressing gonadotropin synthesis and release within a few hours of administration. As a consequence of GnRH analogue-induced gonadotropin deficiency, the synthesis and secretion of the sex steroids is compromised and suppression of testicular and ovarian function is induced. The steroid-suppressing properties of GnRH agonists and antagonists have led to widespread application of GnRH analogues in the treatment of gonadal steroid-associated

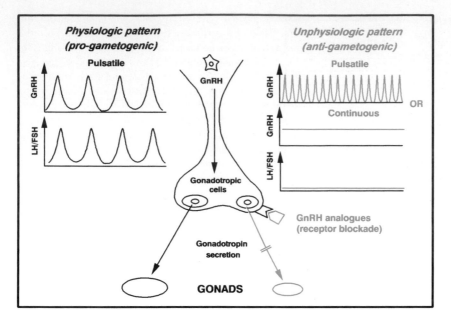

Fig. 113.2. Interrelationship between the pattern of GnRH secretion and gonadal function. Inhibitory effects are shown in *blue*

diseases such as precocious puberty, endometriosis, and tumors of the uterus, prostate, and breast. On the basis of their antigonadotropic and antigonadal actions, GnRH analogues are used successfully in protocols for ovulation induction and are also promising candidates for the development of an endocrine male contraceptive.

Mechanism of Action of GnRH and Its Analogues. The receptors for GnRH are located on the cell membrane of the pituitary gonadotropes. The GnRH receptor has recently been cloned and was found to be one of the G-protein-coupled receptors. These receptors are composed of an extracellular domain, a transmembrane domain, and an intracellular domain. Interestingly, the GnRH receptor lacks the C-terminal cytoplasmic domain, which in other G-coupled membrane receptors is critically important for signal transduction and receptor desensitization and internalization. The relevance and consequences of this special feature of the GnRH receptor for the physiology and pathophysiology of reproduction are still unknown.

After receptor binding, GnRH promotes the mobilization of Ca^{2+} ions, leading to a net accumulation of cellular Ca^{2+} derived from both extra- and intracellular sources (Fig. 113.3). GnRH triggers the activation of phospholipase C. This enzyme and also calmodulin serve as intracellular receptors for Ca^{2+}. Phospholipase C induces the formation of the intracellular signal transducers inositol phosphate and diacylglycerol. The latter activates protein kinase C, which in turn controls gonadotropin synthesis and upregulates GnRH receptors. Inositol phosphates mobilize intracellular Ca^{2+} and open the membrane Ca^{2+} channels, leading to an increase of cellular Ca^{2+}. This increase induces the release of preformed gonadotropic hormones,

which in its turn is dependent on the presence of calmodulin, a calcium-binding protein. The critical role of the intracellular Ca^{2+} concentration for gonadotropin release is underlined by the observation that Ca^{2+} channel agonists are capable of inducing gonadotropin secretion. Thus, Ca^{2+} is an important second messenger for transduction of the gonadotropin-releasing activity of GnRH.

Agonists and antagonists of GnRH bind to the same receptor but exert different effects on GnRH receptor numbers and gonadotropin release. GnRH agonists first upregulate and then downregulate the GnRH receptor, leading to refractoriness of the gonadotrope to the stimulatory effects of GnRH and GnRH agonist. In contrast, the GnRH antagonists merely bind to the GnRH receptor and thus cause the cessation of gonadotropin release by competitive occupancy of the receptor. One piece of evidence for this mechanism derives from autoradiographic studies using radiolabelled GnRH analogues. Agonists of GnRH were internalized within minutes, whereas antagonists of GnRH were still present on the gonadotrope membrane several hours later. The competitive nature of GnRH antagonist-induced blockade of primate LH secretion is further suggested by the observation that exogenous, synthetic GnRH was able to displace the GnRH antagonist in a GnRH dose-dependent manner under in vivo conditions [60].

113.3.3 Pituitary Gonadotropins

Structure and Isoforms of Gonadotropic Hormones. LH and FSH are dimers consisting of two noncovalently linked glycosylated polypeptide subunits (α and β) and have nu-

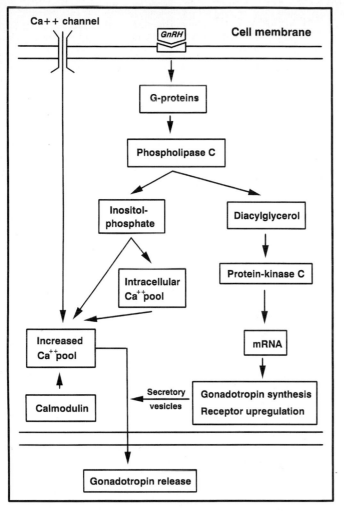

Fig. 113.3. Mechanism of action of GnRH-induced gonadotropin release

merous structural and chemical similarities. The α-subunit of LH, FSH, thyroid-stimulating hormone (TSH), and human chorionic gonadotropin (hCG) is common to these hormones. It is the β-subunit of glycoprotein hormones which confers the specific hormone activity. The individual subunits are not biologically active and, therefore, heterodimer formation is required for hormone activity. The molecular weights of LH and FSH are approximately 28 kDa and 33 kDa, respectively, with the molecular weight of the shared α-subunit being around 14 kDa. Precise estimates are not available because of the heterogeneity of associated carbohydrate groups on the LH and FSH molecules and some differences in the amino acid composition.

The genes encoding the α- and β-subunits have been isolated and characterized. With the exception of the hCG-β-subunit, which is probably encoded by two genes among a complex array of hCG-β-like genes, the other subunits are encoded by a single gene. The human FSH-β-subunit gene has a particular feature in that multiple mRNAs are produced as a result of alternative splicing [19].

Both pituitary and circulating gonadotropins are present in several isoforms. These isoforms are characterized by different degrees of glycosylation and sialyzation causing charge heterogeneity. Owing to the physicochemical characteristics of the gonadotropic isohormones, they can be separated and characterized by means of chromatofocusing and isoelectric focusing of pituitary extracts and serum and urine samples. Six to ten components for LH and FSH have been identified. In addition, in vitro bioassays specific for LH or FSH have been devised to analyze the bioactivity of the various isoforms. Glycosylation is known to influence the serum half-life, tertiary structure, and receptor binding of LH and FSH [21]. In general, less heavily glycosylated gonadotropin isoforms exhibit increased receptor and bioactivity in vitro, but a reduction of these activities in vivo. It appears that the type and distribution of LH/FSH isoforms is under endocrine control (gonadal steroids and GnRH). However, despite many efforts to characterize the secretion and regulation of the synthesis of the various isoforms of LH and FSH, the physiological role of gonadotropic hormone pleomorphism for the regulation of reproductive organ function remains unclear. Both LH and FSH are secreted from the pituitary in a pulsatile manner. The half-life of circulating FSH is much longer than that of LH. This might

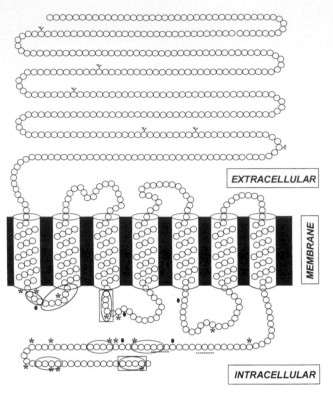

Fig. 113.4. Organization and amino acid sequence of the LH/human chorionic gonadotropin receptor molecule. The receptor is anchored in the cell membrane by seven transmembrane domains and contains an extracellular and an intracellular (cytoplasmic) domain. *Branch-like structures* in the extracellular domain indicate potential glycosylation sites. *Dashed lines* mark potential sites of tryptic cleavage. Presumable sites of intracellular phosphorylation are denoted by *asterisks* and *dots*. *Rectangles* represent consensus sequences for cAMP-dependent protein kinase-catalyzed phosphorylation and *ovals* represent consensus sequences for kinase C-catalyzed phosphorylation

explain why the pulsatile nature of FSH release is not always obvious. The roles of LH and FSH in the hormonal control of reproductive organs are described in detail later in this chapter.

Mechanism of Gonadotropic Hormone Action. LH and FSH control gametogenesis through specific receptors present in the gonads. Molecular cloning revealed that the LH and the FSH receptor – like the GnRH receptor – belong to the family of G-coupled membrane receptors. Unlike the GnRH receptor, gonadotropin receptors have a distinct C-terminal domain and a large extracellular domain (Fig. 113.4). Common to these receptors are seven transmembrane domains. These domains are critical for the normal functioning of gonadotropin receptors. Single mismatch mutations in this domain were recently shown to activate the LH receptor in the absence of the ligand and were associated with the development of familial precocious puberty [32,47]. This pathophysiological condition, as would be expected, cannot be ameliorated by the suppression of pituitary LH secretion. Both the LH

and the FSH receptor genes are located on chromosome 2 in adjacent locations, p16–p21, with the LH receptor being encoded by 11 and the FSH receptor by 10 exons respectively.

Receptor activation induces G-protein and adenylate cyclase activity and the formation of cAMP. cAMP provokes protein kinase activity. The formation of cAMP can be prevented by the enzyme phosphodiesterase. Protein kinase activity then induces protein phosphorylation. Clearly, cAMP is the pivotal intracellular second messenger for transduction of the hormonal signal. Beyond that, cAMP has been found to bind to a cAMP-responsive element in the DNA and this binding can be further regulated by a cAMP-responsive-element-binding modulator. Quite recently it has been shown that the stimulatory effects of FSH in the hamster testis are determined by such modulators. This finding further highlights the decisive role of cAMP in mediating gonadotropic hormone effects. More recently, it has been discovered that Ca^{2+} is also involved in the signal transduction of gonadotropin hormone signals. Analogous to the isoforms of LH and FSH, isoforms of the LH receptor and the FSH receptor have been detected. A FSH receptor isoform was discovered in the human testis which lacks one of the nine exons which constitute the FSH receptor gene [25]. Moreover, multiple mRNA transcripts for this receptor have been identified in testicular tissue [24,27], possibly indicating the existence of several forms of the FSH receptor. A number of LH receptor mRNA transcripts has also been described in rat gonadal tissues [1,34,52]. These transcripts probably result from alternative splicing during mRNA processing. On the basis of these findings the concept has emerged that multiple transcripts could encode for receptor proteins with differing affinity to the ligand and in that sense would provide a means of local and flexible regulation of target cell sensitivity to the trophic hormone. This intriguing concept, however, still awaits definite proof.

113.4 Synthesis and Mechanism of Action of Sex Steroid Hormones

The sex steroid hormones testosterone, dihydrotestosterone (DHT), estrogen, and progesterone are of vital importance for the development and normal functioning of the reproductive organs, and in addition, exert a variety of extragonadal effects. The latter aspect is discussed in some detail later. This section describes the production of steroid hormones, the properties of their receptors, and their mechanism of action.

113.4.1 Synthesis of Sex Steroid Hormones

Testosterone is synthesized in the testicular Leydig cells and is converted into DHT in the seminiferous tubules. The ovarian theca cells produce androgens, whereas estradiol is the major steroid hormone product of the

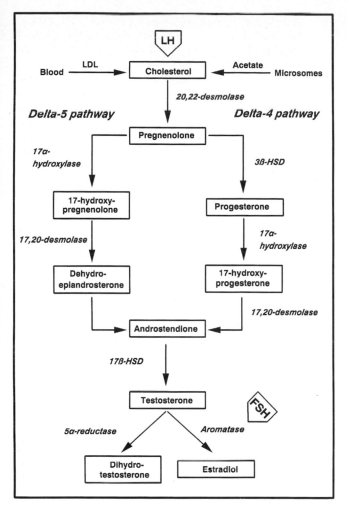

Fig. 113.5. Steroidogenic pathways involved in the formation of testosterone, estradiol, and progesterone. *LDL,* Low density lipoprotein; *HSD,* hydroxy-steroid-dehydrogenase

follicular granulosa cells. Progesterone derives predominantly from the lutein-granulosa cells of the corpus luteum. The common precursor for the synthesis of all steroids is cholesterol (Fig. 113.5). Cholesterol is derived from two sources: the synthesis of cholesterol from acetate in the microsomes of steroid-producing cells and the uptake of circulating low density lipoprotein-cholesterol. Esterification of cholesterol is used to store cholesterol in cytoplasmic lipid droplets. The next step of steroidogenesis is the conversion of cholesterol to pregnenolone in the mitochondrial membranes. The enzyme cytochrome P-450 plays the key role in this reaction. It has been postulated that microtubules might facilitate the movement of lipid droplet-derived cholesterol to the mitochondria by permitting closer topographical association between these cell inclusions, and that this event is under hormonal control [10]. All subsequent enzymatic processes take place in the smooth endoplasmic reticulum. Thus, steroidogenetically active cells are characterized by the presence of lipid droplets and abundant smooth endoplasmic reticulum. Pregnenolone can be converted further into progesterone. Both pregnenolone and progesterone are the fundamental precursors of the steroidogenic

pathways. Steroidogenesis proceeds further either directly from pregnenolone (pregnenolone or delta-5 pathway) or from progesterone (progesterone or delta-4 pathway). Which steroids will ultimately by synthesized depends on the physiologic (or pathophysiologic) nature of the steroidogenic cell and on the activity of the associated enzyme systems. Both pathways are operative in the ovary and the testis. Steroidogenic cell-targeted hormones induce both the de novo synthesis and the release of steroid hormones.

113.4.2 Transport of Sex Steroid Hormones to Target Cells

The secreted steroid hormones bind to circulating plasma proteins and are present in a bound and an unbound form. More than 95% of testosterone, estradiol, and progesterone are attached to proteins, specifically to sex hormone-binding globulin and nonspecifically to albumin. The affinity of testosterone for sex hormone-binding globulin is greater than that of estradiol and progesterone. It is believed that the unbound (free) steroid fraction represents the biologically important fraction because it can freely diffuse into tissues from the capillaries. In contrast, the protein-bound fraction is thought to act as a reservoir for steroid hormones. The steroid hormone receptors are localized within the target cells and steroids need to enter the cell to elicit their effects. It is interesting to note, in that regard, that receptors for sex hormone-binding globulin have been detected on the membranes of steroid hormone-responsive cells. Although evidence is lacking to suggest that steroid action requires the presence of membrane-bound sex hormone-binding globulin, it is tempting to speculate that the presence of sex hormone-binding globulin might serve to regulate the biological activity of steroids simply by modulating the amounts of steroid entering a target cell. Biological proof of this assumption is, however, still pending.

113.4.3 Mechanism of Action of Sex Steroid Hormones

Upon entry into the target cells, steroid hormones bind to specific receptors prior to eliciting their biological effects. Steroid hormone receptors can be found in both the cytoplasmic and the nuclear compartment of the respective target cells. On the basis of analysis of fractionated cells, it was originally believed that steroid hormone receptors reside – in the absence of ligand – exclusively in the cytoplasm of steroid-responsive cells and that these receptors are translocated into the nucleus in the presence of ligands. In contrast to this view, immunohistochemical localization studies identified the receptors for androgens, estradiol, and progesterone primarily in the nuclei of target cells [40]. Notwithstanding this, a nucleoplasmatic shuttling of the progesterone receptor has been observed recently which was dependent on the presence of ligand and energy supply [26]. Similarly, a ligand-dependent translocation of the androgen receptor has recently been found in the rat prostate gland and seminal vesicles [37]: following removal of the testicular source of androgens by orchidectomy or suppression of testosterone production by administration of a GnRH antagonist, the prostate and seminal vesicle androgen receptor was detected immunohistochemically in the cytoplasmic state, whereas after a bolus injection of testosterone, the androgen receptor was found in the nuclear-bound state. Such ligand-dependent translocation of the androgen receptor, however, apparently does not occur in all reproductive organs, since such an effect could not be demonstrated for the epididymis of GnRH antagonist-treated rats.

The receptors for testosterone, estradiol, and progesterone have all been cloned and their protein structure has been deduced. A feature shared by all steroid hormone receptors is the presence of a transcription-modulating domain located N-terminally, a DNA-binding domain, and a steroid-binding domain in the C-terminal region.

Much has been learned in recent years concerning the molecular mechanism of steroid hormone action. A thorough description of the related events is far beyond the scope of this chapter and some aspects are described elsewhere in this book. After coupling to the ligand the receptor dimerizes, binds to specific DNA, and, in concert with other transcription factors, initiates gene transcription and protein synthesis (Fig. 113.6). Feedback mechanisms between the newly formed receptor complex and the promoter region of the receptor have been described. In that context, protein phosphorylation is presumed to act as an important signal. Thus, steroid hormones display two simultaneous features: they act as signal transducers and as transcription factors.

113.5 Gonadal Organization and Hormonal Control of Gonadal Functions

Both the testis and the ovary are characterized by a distinct morphological organization involving particular arrangements and interactions of various highly specialized cell types. For that reason, and to permit an easier understanding of the subsequent discussions of the role of various hormonal factors and their sites of action during regulation of gonadal function, a description of the organization of the male and female gonads and of the process of gametogenesis is provided.

113.5.1 Testis

Organization of the Testis. The testis is composed of two compartments: the tubular compartment, comprising the seminiferous tubules (tubuli seminiferi), and the peritubular cells which surround the seminiferous tubules. The interstitial compartment consists of Leydig cells, macrophages, connective tissue cells, nerves, and lymphoid spaces. The process of spermatogenesis is localized to the seminiferous tubules which contain Sertoli cells and germ cells. Sertoli cells are somatic cells which divide mitotically during testicular development but cease dividing in adulthood. These cells provide mechanical support for the seminiferous epithelium and possess nutritive and coordinative functions during the spermatogenic process. They maintain close contact with all germ cells throughout all stages of their development mediated via highly specialized morphological components. Sertoli cells also form the blood-testis barrier which serves as a filter for metabolic substances and prevents the immune system from recognizing the antigenic haploid germ cells. Studies in laboratory animals suggest that the Sertoli cells may determine the testicular size which can be achieved during adulthood [36,56].

Germ cells undergo a defined series of mitotic and meiotic divisions culminating in the development of haploid germ cells. Diploid spermatogonia provide a pool of cells for spermatogenesis through continuous mitotic divisions. The primate testis contains two pools of spermatogonia [33]: one pool serves to provide reserve stem cells in case of severe testicular damage and the other pool consists of the

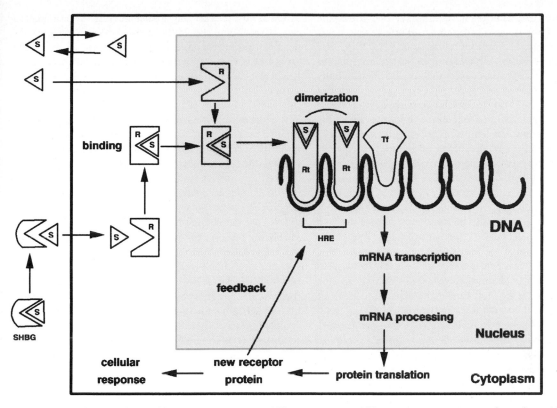

Fig. 113.6. Mechanism of steroid hormone action. *S*, Steroid hormone; *R*, steroid hormone receptor; *Rt*, transformed steroid hormone receptor; *Tf*, transcription factor; *HRE*, hormone-responsive element; *SHBG*, sex hormone-binding globulin

so-called renewing stem cells which periodically divide to provide cells entering the spermatogenic process. Once a spermatogonium enters the meiotic phase it is named a primary spermatocyte. The first meiotic division results in the formation of two secondary spermatocytes. These cells contain a haploid chromosome number but in duplicate sets. During the second meiotic division the round spermatids are produced, which are haploid. These spermatids do not divide further but differentiate into spermatozoa (testicular sperm). During this process of spermiogenesis the cells start to elongate, an event accompanied by pronounced condensation of the nuclear chromation structures and the shedding of redundant cytoplasmatic components. The resulting elongated spermatid has characteristic features: the head with tightly packed chromatin, the midpiece containing mitochondria as a source of energy, and the tail which allows spermatozoal movement and progressive motility.

Efficiency of the Spermatogenic Process. The development of germ cells proceeds according to a precise pattern. All germinal cells follow a defined sequence of stages which are characterized by the presence of typical cell associations. With respect to the topographical arrangement of the various spermatogenic stages, the human seminiferous epithelium shows a particular feature that distinguishes it from most studied animal species: the spermatogenic stages are organized in a helical rather than a longitudinal

manner along the length of the seminiferous tubule [45]. Beyond that, human spermatogenesis has two further notable features: the time required for the development of a spermatogonium into a sperm is quite long, whilst the yield of spermatogenic cells is relatively low. The approximate duration for formation of testicular sperm from spermatogonial stem cells is 74 days in men, compared to 37–42 days in nonhuman primates and 51–53 days in the laboratory rat. Regarding the efficacy of spermatogenesis, one spermatocyte should theoretically give rise to four spermatids or spermatozoa. However, in fact only two spermatids are produced in men even when spermatogenesis is intact. Thus, man loses about one-half the spermatid cells in the course of the transition of spermatozoa into spermatids. Nonetheless, estimates of daily sperm production in men range between 200 and 300 million spermatozoa.

Release of Testicular Spermatozoa. Testicular sperm are released into the lumen of the seminiferous tubules and thereafter are transported via the rete testis and the efferent ducts into the epididymis. This process of sperm release has been named spermiation. Spermiation is a tremendously complex process involving the shedding of redundant spermatid cytoplasm, a special three-dimensional configuration of the Sertoli cell, a particular positioning of the sperm to be released, and the opening and reorganization of numerous structural specializations which served to anchor the developing sperm to the Sertoli cell. Even from a purely morphological point of view, this process is only incompletely understood [41]. It is likely that the peritubular cells play an important role in this process. These cells are actually myoid cells with the ability to contract. Contractability develops under the influence of androgens and of oxytocin coming from the Leydig cells. It is believed that the peritubular cells induce peristaltic movements along the length of the seminiferous tubule, and the fluid pressure resulting from the contractions within the seminiferous tubules presumably pushes spermatozoa towards the epididymis.

Temperature and Spermatogenesis. Testicular temperature is approximately 2°C below body temperature. An elevation of testicular temperature into the body temperature range or beyond causes spermatogenic inhibition, e.g., as a consequence of testicular maldescent (cryptorchidism) or experimental positioning of the testicles into the inguinal canal. The exposed location of the testis and the structural characteristics of the scrotal skin permit the maintenance of a testicular temperature lower than that of the body. Scrotal skin lacks adipose tissue, whereas the arteries are present in the uppermost epidermal layers. By means of contraction and relaxation of the musculus cremaster, alterations of the scrotal surface are achieved for the purpose of thermal regulation. Arterial blood is cooled prior to reaching testicular tissue. It is because of these regulatory mechanisms that testicular and also epididymal temperatures are kept low in spite of elevations of the core temperature due to physical exercise or exposure to exogenous heat sources. The regulation of testicular blood flow has been investigated in some detail in the laboratory rat. Testosterone, LH, and hCG can stimulate and inhibit blood circulation within the testis. The expression of such effects, however, was in the magnitude of hours under the chosen experimental conditions. In contrast, adenosine and kallikrein were found to stimulate blood flow within minutes, and in a similar time frame serotonin, histamine, and prostaglandins provoked a reduction of testicular microcirculation.

113.5.2 Hormonal Regulation of Spermatogenesis

Our knowledge about the endocrine regulation of human testicular function mainly comes from efforts to reverse infertility in patients with gonadotropin and androgen deficiency (hypogonadotropic hypogonadism), from trials for endocrine male fertility regulation, and from studies using the nonhuman primate model. For the interpretation of hormone actions on spermatogenesis three phases of the spermatogenic process need to be distinguished: initiation, maintenance, and reinitiation. Initiation denotes the first completion of the spermatogenic process during puberty, "maintenance" refers to the hormonal requirements during continued spermatogenesis, and "reinitiation" relates

to the need for restarting gametogenesis. Additionally, a clear distinction should be made between qualitatively normal spermatogenesis, i.e., presence of all germ cell types but at reduced numbers, and quantitatively normal spermatogenesis, i.e., the presence of complete numbers of germ cells.

Role of Luteinizing Hormone and Androgens in the Control of Testicular Function. Receptors for LH are present on Leydig cells but nowhere else in the testis. The role of LH in the spermatogenic process is to induce the biosynthesis and secretion of testosterone in the testicular Leydig cells. Androgen receptors have been localized to Leydig cells, peritubular cells, and Sertoli cells, while germ cells do not express the androgen receptor. Testosterone – after being released from the Leydig cells – travels through the interstitial lymphoid spaces and reaches the seminiferous tubule wall. To date, the issue of whether testosterone crosses the peritubular cells and enters Sertoli cells via passive diffusion or is actively transported along that route has remained unclear. In contrast, it is generally accepted that testosterone – prior to any of its intratubular actions – is converted into DHT by the enzyme 5α-reductase. This enzyme is located in the Sertoli cells. Testicular concentrations of DHT are about 10-fold lower than those of testosterone. This difference in the local concentrations of testosterone and DHT is compensated by the reported 10-fold higher affinity of DHT for the androgen receptor compared to testosterone [23].

Experimental and clinical studies have unambiguously demonstrated that testosterone is one of the key regulators of spermatogenesis. The qualitative initiation of sperm development is possibly under the influence of testosterone alone. In boys bearing a testosterone-producing Leydig cell tumor, qualitatively normal spermatogenesis was encountered in the immediate vicinity of the tumor, but not in tumor-free areas. It must be noted, however, that successful induction of the spermatogenic process necessitates very high concentrations of testosterone. In the aforementioned study, testicular androgen concentrations were 2- to 4-fold higher in the tumor tissue than in nontumor specimens. Confirming observations were made in nonhuman primates which serve as a preclinical model for the study of the control of human testicular function: to provoke a clear-cut stimulation of spermatogenesis in immature animals quite high amounts of testosterone were needed [2].

Quite high doses of androgens are also required to reinitiate spermatogenesis during adulthood. hCG is used clinically because it is much more potent than LH in terms of stimulating Leydig cell testosterone production, and because of the nonavailability of a highly purified or recombinant LH preparation. Induction of sperm production is usually not successful, and if spermatogenesis is induced by testosterone (hCG) alone, the numbers of spermatozoa in the ejaculate are rather small and the possibility of the presence of small amounts of FSH cannot be ruled out with certainty. Similarly, in the nonhuman primate model, only extremely high doses of testosterone provoke the appearance of a small number of primary spermatozoa in animals with experimentally induced suppression of testicular function. The greatest success with testosterone was achieved in trials to maintain the spermatogenic process. Once the formation of sperm occurs, spermatogenesis can continue under the influence of testosterone alone, at least for a certain period of time.

Aside from stimulating germ cell proliferation, testosterone is apparently also involved in the functional maturation of the peritubular cells and in the morphological maturation of Sertoli cells. Administration of testosterone to immature nonhuman primates induces the appearance of α-smooth muscle actin in the peritubular myoid cells [43]. α-Smooth muscle actin is an established marker of the function, i.e., ability to contract, of the peritubular cells. Sertoli cells attain an adult-type morphology and cytoplasmatic distribution of actins under the influence of testosterone. This latter observation suggests that a build-up of the Sertoli cell cytoskeleton is induced by testosterone.

Testicular testosterone levels are much higher than circulating levels, suggesting a particularly high requirement for local androgen concentrations for the spermatogenic process. Studies in the laboratory rat demonstrate that it does not necessarily have to be so high, since spermatogenesis was found to proceed, at least in a qualitatively normal fashion, in the presence of markedly reduced testicular amounts of testosterone (<90% of control). It is not known, however, whether the intratesticular requirements for androgens for the spermatogenic process in the human testis are comparable to those in the laboratory rat.

Role of Follicle-Stimulating Hormone in the Control of Testicular Function. FSH is the second key regulator of the spermatogenic process and acts directly upon the seminiferous epithelium. Receptors for this gonadotropic hormone are present on Sertoli cells [31]. The existence of FSH on spermatogonia has been reported following autoradiographic localization of radiolabelled FSH in the rat testis. Because of methodological difficulties, these findings are subject to doubt and the actual presence of spermatogonial receptors ist still under debate. It is not known whether FSH alone can initiate gametogenesis. Nonhuman primate studies indicate that at lease the proliferation of spermatogonial cells and the appearance of primary spermatocytes is induceable with FSH alone [2]. A similar lack of information prevails with regard to the reinitation and maintenance of spermatogenesis with FSH alone in adult men, since relevant data from clinical studies are not available. Nonetheless, nonhuman primate studies lend strong support to the view that FSH is indispensable for maintaining spermatogenesis under physiological conditions, since selective immunization against FSH almost completely abolishes the production of spermatozoa and fertility [35]. In sharp contrast, FSH is only slightly effective in restoring the spermatogenic function of the testis after it has been experimentally interrupted [59]. Studies in two nonhuman primate species led to the unexpected observation that a highly purified human FSH preparation doubled germ cells within 2–4 weeks of administration in adult animals with intact spermatogenesis [55]. No such effect was seen when the animals were exposed to hCG. These findings raise the intriguing possibility that even under normal conditions primate spermatogenesis is not at its maximum possible level. The question of whether a comparable situation also exists for men which could be exploited therapeutically remains to be answered. In the immature nonhuman primate testis, FSH initiates the secretory ability of Sertoli cells since FSH, but not testosterone, stimulates the secretion of inhibin, an FSH-dependent Sertoli cell parameter, into the blood circulation [2].

Synergism and Sites of Androgen and FSH Action on Spermatogenesis. It is evident from the foregoing discussion that neither testosterone alone nor FSH alone is able to bring about quantitatively normal spermatogenesis. Indeed, experimental and clinical data support the view that quantitatively normal gametogenesis requires the synergistic actions of both testosterone and FSH (Fig. 113.7). This notion appears quite logical since both hormones are present under normal circumstances. For example, administration of FSH to normal volunteers whose sperm production has been severely reduced by the administration of hCG/androgens (negative feedback effect on gonadotropin secretion) brings about a quantitative restoration of spermatogenesis. In line with this finding is the observation that in gonadotropin-deficient patients the best therapeutic success is achieved when testosterone – stimulated by hCG – and FSH – given in the form of human menopausal gonadotropin – are combined.

As already mentioned above, receptors for FSH are located on Sertoli cells and androgen receptors exclusively on somatic testicular cells. It is therefore reasonable to assume that androgens and FSH exert their beneficial effects on spermatogenesis through the somatic components of the male gonad. Recent studies in rodents revealed that FSH and testosterone not only regulate their homologous receptor but also the expression of the other, androgen or FSH receptor [5,6,51]. These findings raise the possibility that FSH and testosterone interact at the receptor level by each stimulating the receptor for the other hormone, leading to sensitization of the seminiferous epithelium to either hormone and to a more pronounced biological effect than that seen with either hormone alone. In addition, the ligands may influence the mRNA stability and/or the translation of the receptor protein or the functionality of the synthesized receptor protein. Sertoli cells are the source of a variety of substances, among which are peptides, enzymes, growth factors, cytokines, and steroid hormones, and many of these substances are produced under the influence of androgens and FSH.

This view assigns a decisive role to the Sertoli cell and possibly also to the peritubular cells in the functional control of gametogenesis. In the immature and in the adult gonadotropin-suppressed nonhuman primate, both testosterone and FSH exert an initial stimulatory effect on the same germ cell type, the renewing stem cells, and the

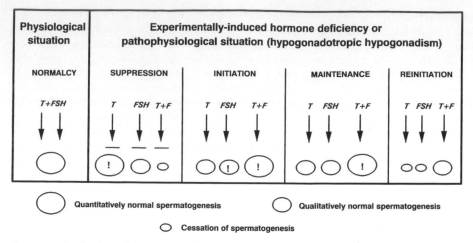

Fig. 113.7. Qualitative and quantitative effects of LH, testosterone (*T*) and FSH alone or in combination on the initiation, maintenance, and reinitiation of testicular function (spermatogenesis) in men. *T* + *F*, Testosterone plus FSH; *!*, has not been studied but effect is likely

subsequent germ cell numbers increase accordingly. Notwithstanding all of this, the mode of the stimulatory action of either hormone alone or in combination on spermatogenesis remains obscure, and the molecular and biochemical mechanisms underlying the trophic effects of testosterone and FSH on spermatogenesis are still not understood. So far, no marker substance reflecting FSH and/or testosterone action specifically has been identified. This vexing situation may eventually vanish, owing to the recent discovery of apparently androgen-specific testicular proteins [46]. One current view is that both LH/androgens and FSH provide a favorable metabolic milieu permissive to germ cell development, rather than exerting a hormone-specific trophic effect on spermatogenesis. On the other hand, data obtained from studies in the laboratory rat raise the possibility that the various cellular associations may differ in their sensitivity, requirements, or thresholds for androgens and FSH. For example, gonadotropin withdrawal induces stage-specific germ cell degeneration and the receptors for androgen and FSH are expressed differently in relation to the various spermatogenic stages [28,31]. A further possible site of FSH and androgen action within the seminiferous epithelium is at the level of interaction between the Sertoli cells and spermatids, since the ability and intensity of spermatid binding by the Sertoli cell seems to be under hormonal control [9].

113.5.3 Ovary

Organization of the Ovary. The ovary is divided into a peripheral part, the cortex, which contains the germinal cells, developing and degenerating follicles, theca cells, and granulosa cells, and a central part, the medulla, which is comprised of theca cells, connective tissue, blood vessels, and nerves.

Oogenesis. The diploid oogonia divide mitotically and, upon their entry into meiosis I, the primary oocytes are formed. This event takes place during embryonic development. However, the majority of these cells degenerate subsequently (oogonial atresia), so that only about 2 million out of the approximately 7 million prenatal oogonia survive until birth. Meiosis I is arrested at the prophase stage until the onset of puberty. In the time until puberty is reached, further substantial loss of stem cells occurs, and at the time of puberty an estimated 300 000 germinal cells per ovary remain alive. Meiosis I is completed during ovulation, giving rise to the secondary oocyte. Only one secondary oocyte is formed and the second cellular component, called the first polar body, degenerates thereafter. Final oocyte maturation, i.e., the formation of the haploid ovum is not completed until fertilization of the oocyte. Upon fertilization of the oocyte in the oviduct, meiosis II is completed, with the second cellular remnant, the second polar body, being disposed of.

Folliculogenesis, Ovulation, and Corpus Luteum Formation. Folliculogenesis is initiated once the oocyte is surrounded by a single-layered, flat epithelium composed of granulosa cells, when it is termed the primordial follicle. The transformation of the flat epithelium into a columnar shape denotes the primary follicle. The appearance of the secondary follicle results from numerous mitotic divisions of the granulosa cells which form a multilayered and highly columnar epithelium. Theca cells attach themselves to the granulosa cell layer. In the course of the first meiotic division a vacuole develops (antrum folliculi) and the granulosa cells and theca cells continue to divide (tertiary follicle). The fully matured follicle is called the graafian follicle. Normally only one tertiary follicle develops per cycle, and the ovulated follicle is expelled into the oviduct. Just before ovulation takes place, the follicle becomes extensively vascularized and ruptures at a preformed site, the stigma. This release of the oocyte-cumulus complex appears to be explosive in nature. During the female reproductive life span only 400–500 follicles will be involved in ovulation. Recruited but nonselected follicles degenerate (follicular atresia). The dominant follicle, after ovulation, reorganizes and gives rise to the corpus luteum. Granulosa cells and theca cells undergo luteinization (granulosa-lutein cells and theca-lutein cells) and the corpus luteum becomes extensively vascularized. The purpose of the corpus luteum is to prepare the endometrium for implantation of the fertilized egg (see Chap. 116). If pregnancy does not occur, the corpus luteum undergoes luteolysis and is transformed into a fibrous scar, the corpus albicans.

113.5.4 Hormonal Regulation of the Menstrual Cycle

General Remarks. The menstrual cycle show a distinct periodicity and – if the ovulated eggs are not fertilized – a mature ovum is released into the oviducts every 28 days. The first day of menses is referred to as the first day of the menstrual cycle. The integrity of these cyclic events depends on the coordinated interplay between hypothalamus, pituitary, ovary, and uterus and has been divided into three phases (Fig. 113.8): follicular phase (days 1–12), ovulatory phase (days 13–15), and luteal phase (days 16–28). Ovarian and endometrial organization and basal body temperature also undergo cyclic changes during the menstrual cycle. Serum concentrations of FSH are high in the early follicular phase, exhibit a second peak just prior to ovulation, and are lowest during the luteal phase, except for the very late luteal phase, when FSH secretion starts to rise. The circulating levels of LH steadily increase during the follicular phase, dramatically increase immediately before ovulation, and then rapidly decline during the luteal phase. As for estradiol, its production is minimal during the early follicular phase, followed by a sharp rise before the LH peak and a second peak in the luteal phase. Progesterone secretion is minimal during the entire follicular phase but begins to increase shortly before ovulation occurs and is pronouncedly elevated during the whole duration of the luteal phase. The preovulatory rise of estradiol in particular and of progesterone as well [38] stimulates the preovulatory LH surge. During the luteal phase both steroids act as inhibitors of gonadotropin release in order to prevent the next onset of follicular development. Since gonadotropic hormones and steroid hormones intimately interact with each other in the or-

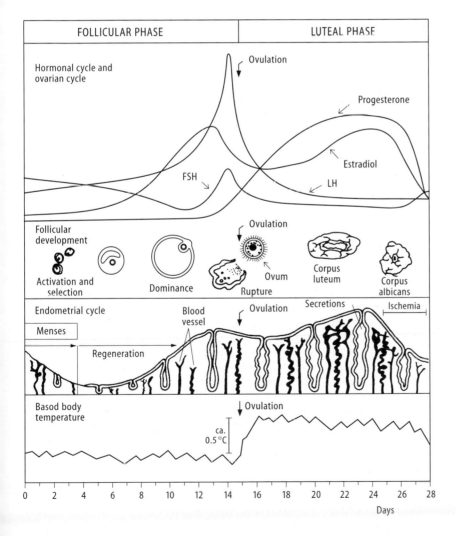

Fig. 113.8. Circulating hormone levels, follicular development, endometrial changes, and basal body temperature during the menstrual cycle

chestration of the menstrual cycle, their cellular actions will be described in relation to the phases of the cycle.

Hormonal Regulation of the Follicular Phase. As indicated by its name, the initiation and promotion of follicles is regulated by follicle-stimulating hormone. These events are initiated in the very late luteal phase and come to completion at ovulation. Receptors for FSH are exclusively present on the granulosa cells. FSH triggers granulosa cell estradiol production and the multiplication of granulosa cells, and thereby follicular growth. Although estrogen is the most important mitogen for granulosa cells, a promotion of granulosa cell division directly by FSH has also been postulated. Estradiol is the key factor in the differentiation and maturation of the follicles. Initially, a cohort of follicles is recruited through FSH action. Around day 5–7 of the follicular phase, however, selection of the dominant follicle occurs. This follicle possesses a high number of FSH receptors, contains the highest levels of intrafollicular estradiol levels, and supercedes the other follicles in terms of growth.

The eminent role of FSH in follicular development has recently been demonstrated elegantly by the use of recombinant human FSH [44]. Such FSH preparations, which are devoid of intrinsic LH activity, were able to induce follicular growth and development such that the oocytes recovered 36–37 h after hCG injection could be fertilized in vitro, and following embryo transfer a pregnancy was achieved [17,18,22].

In the intact female, LH supports follicular maturation by stimulating the synthesis of androstenedione and testosterone in theca cells. Again, the data obtained from the administration of recombinant FSH suggest that LH might be more important than previously thought for the estradiol-producing capacity of the follicle: although follicular growth was initiated, the estradiol concentrations never achieved the normal range, suggesting a need for LH. On that ground, the view has been presented that FSH is of major importance during the early follicular phase and less so at the time of ovulation, whereas the opposite holds for LH [11]. The major steroidogenic enzymes under LH control are 17-hydroxylase/C17–20-lyase. The androgens are taken up by the granulosa cells and are further metabolized to estradiol via aromatization. Thus FSH action on granulosa cells and LH action on theca cells augments follicular estradiol production (two-cell/two gonadotropin model of estrogen synthesis). Thecal cells barely express aromatase activity (<0.1% of total ovarian aromatase activity) whereas granulosa cells do not express 17-hydroxylase/C17–20-lyase activity. There is some evidence to suggest that androgens, in addition to estradiol, are capable of directly increasing aromatase activity in granulosa cells. This enzyme is maximally active in the dominant, preovulatory follicle, as are 17-hydroxylase and C17–20-lyase. In the advanced preovulatory period, granulosa cells develop the expression of LH receptors under the influence of FSH and LH. It is presumed that LH induces granulosa cell production of progesterone, which is then taken up by the thecal cells and increases the pool of

precursor for androgen synthesis, thus ultimately increasing estrogen synthesis in the preovulatory follicle. Beyond that, the existence of other local regulators of theca cell function mediating a further increase of androgens available for aromatization to estradiol has been postulated (see section on local regulation of ovarian function below).

Investigation of the relation between follicular steroid levels and follicular development has demonstrated that each follicle has an individual profile. Large follicles contained more estrogens and progesterone but less androgens than small ones. Indeed, preantral and antral follicles exhibit pronounced 5α-reductase activity leading to the formation of DHT. It has been suggested from the results of in vitro fertilization attempts that the follicular estradiol concentrations may influence the number of spermatozoa needed for fertilization of the oocyte, being associated by an inverse relationship. No such correlation could be suggested for progesterone in that regard [16]. The follicles that lag behind the maturation of the dominant follicle subsequently undergo atresia. The precise hormonal regulation of the events associated with follicular degeneration is still under debate. Presumably, a relative preponderance of intrafollicular androgens plays a causative role in that event. This assumption is based upon two observations: follicular DHT levels are particularly high in atretic follicles, and DHT inhibits aromatase activity. Studies in a variety of animal species have revealed that this atretic process is not simply necrosis but follows a defined series of cellular events compatible with programmed cell death, apoptosis [30,54]. This process is apparently under hormonal control in such a way that an increase of either the follicular progesterone:estradiol ratio or the androgen:estrogen ratio promotes the appearance of apoptotic granulosa cell structures [4].

Hormonal Regulation of Ovulation and the Luteal Phase. An outstanding feature that precedes ovulation is the exponential rise of the estradiol concentrations within the preovulatory follicle. This rise of estradiol triggers an LH surge, which in turn stimulates follicular progesterone synthesis in luteinized granulosa cells. In addition, endogenous opioids also participate in the initiation of the midcycle LH surge [39]. The latter serves to induce the midcycle surge of FSH. The LH peak also provokes the resumption of oocyte meiosis. Although it has been established beyond doubt that ovulation reflects the concerted action of various hormonal factors, the nature of the factors ultimately responsible for the discharge of the ovum is less clear. Progesterone and cAMP – which are under the control of LH – can augment or activate proteolytic enzymes such as plasmin and collagenase and would aid in the preparation of the follicle for rupture by distending and thinning the follicular wall. Similarly, prostaglandins of the F and E series accumulate in follicular fluid immediately before ovulation. They are presumed to induce smooth muscle contractions needed for the extrusion of the oocyte-cumulus mass. Following ovulation, the number of LH receptors increases, whereas FSH receptor abundance decreases in the luteinized cells. Clearly, LH is the primary luteotropic agent and stimulates progesterone synthesis and secretion. Corpus luteum-derived progesterone enables the estrogen-primed endometrium to permit nidation of the fertilized egg. In the case of pregnancy, however, hCG secreted by the fetal trophoblast maintains the ability of the corpus luteum to continue progesterone production. In the absence of pregnancy, the corpus luteum undergoes luteolysis. Estradiol and downregulation of LH receptors are associated with the demise of the corpus luteum, but the details of hormonal control of luteolysis are not yet clear.

113.6 Local Regulation of Gonadal Functions

As already outlined in Sect. 113.1, a wealth of information has accumulated to suggest that, besides endocrine control mechanisms, local factors play an essential role in the functional control of reproductive organs. Local regulators can be divided into several classes: The term "paracrine" was used originally to describe interactions between neighboring cells, but is also in use to characterize communications between different compartments within the same organ. "Autocrine" refers to those factors which, after their cellular release, act back on the cell types from which they originate. Finally, "intracrine" relates to substances whose production and site of action are confined to the same cell and which do not leave the cell after being produced [42]. Local regulatory effects have been found in male and female gonads, prostate, seminal vesicle, and oviduct. Notwithstanding this, it should be remembered that the gonadotropic hormones are the main regulators of gonadal functions. On the other hand, local factors may regulate the responsiveness of gonadotropin target cells to LH and FSH and may serve to convey gonadotropic effects in cells which lack gonadotropin receptors [20,48]. In that sense, both the gametogenic and endocrine function of the gonads can be viewed as being under local control.

A compilation of the substances suspected of participating in the local manipulation and modulation of gametogenesis yields an enormously long list. Among the potential candidates are growth factors, inhibins, activins, follistatin, müllerian inhibiting substance, cytokines, GnRH-like peptide, opioids, oxytocin, vasopressin, peritubular-modifying substance (PmodS), pro-enkephalins and enkephalins, renin and angiotensin, gonadotropin-binding inhibitors, ceruloplasmin, calmodulin, sulfated glycoproteins, plasminogen activator, prodynorphins, corticotropin releasing hormone (CRH), growth hormone releasing hormone (GHRH), and many others. Although the biochemical evidence for such an intragonadal regulatory mechanism is indisputable, the physiological significance of these substances for the regulation of gonadal function has in most cases not been unambiguously demonstrated, and most of our knowledge in that regard stems from in vitro studies of several animal species. Nonetheless, evidence is growing that some of the factors listed

above, such as cytokines, growth factors, inhibin, activin, follistatin, transforming growth factor-β (TGF β), and PmodS actually play a true role in the process of gametogenesis aside from their gonadal actions, and the latter two have also been shown to influence prostate function [57]. PmodS mediates stromal-epithelial interactions and TGF-β induces prostatic cell death.

Local Regulation of Testicular Function. Local effects within the male gonad include interactions between Leydig cells and peritubular cells, between peritubular cells and Sertoli cells, between Sertoli cells and germ cells, and also between Sertoli cells and Leydig cells (Fig. 113.9). For the latter, several proteinaceous factors arising from the seminiferous epithelium have been postulated which, dependent on the stage of spermatogenesis, stimulate or inhibit Leydig cell testosterone production. Activin and interleukin-1α promote spermatogonial proliferation when injected intratesticularly, whereas inhibin exerted an inhibitory effect. Quite recently, activin was reported to increase the diameter of the seminiferous tubules in immature nonhuman primates. Inhibin and activin subunits were detected in the Leydig cells of primates, but a potential role in Leydig cell function remains speculative [58]. Insulin-like growth factor I (IGF-I) and transforming growth factor-α (TGF-α) generally exert stimulatory effects, whereas TGF-β has been found to exert inhibitory effects. In that context, Leydig cell maturation is controlled by the transforming growth factors. In addition, compelling evidence has been collected which demonstrates that IGF-I modulates LH action in Leydig cells by potentiating LH action. This growth factor is likely to be involved in gametogenesis directly since the testicular levels of IGF-I correlated with the number of pachytene spermatocytes [50] and stimulated DNA synthesis in testicular mitotic cells [49]. Nerve growth factor (NGF) apparently plays an important role in the structural organization of the human testis since the maintenance of human seminiferous tubules in vitro was successful only in the presence of NGF in the culture medium. Among the various neuromodulatory factors discovered within the testis, melanocyte-stimulating hormone (α-MSH) proved to be a stimulator of Sertoli cell function, whereas β-endorphin behaved as an inhibitor. These peptides are believed to be produced by the Leydig cells.

Local Regulation of Ovarian Function. A variety of local interactions influencing either ovarian steroidogenesis, granulosa cell function, follicular atresia, or the life span of the corpus luteum have been described (Fig. 113.10). Steroidogenic function of thecal cells is stimulated by IGF-I and by inhibin [29]. In contrast, TGF-β, fibroblast growth factor (FGF), and activin were shown to suppress the synthesis of androgens in the theca cells. Granulosa cell aromatase activity is also pronouncedly influenced by growth factors. Transforming growth factor-β and IGF-I support the intrafollicular conversion of androgens into estradiol, whereas TGF-α and FGF counteract aromatase

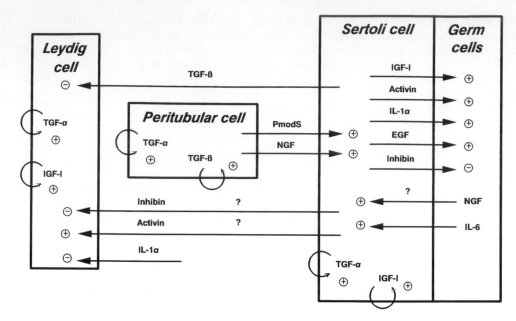

Fig. 113.9. Some of the proposed intratesticular actions exerted by putative nonsteroidal local modulators of testicular function. *TCF-β,* transforming growth factor-*β; TGF-α,* transforming growth factor-α; *IGF-I,* insulin-like growth factor-I, *EGF,* epider- mal growth factor; *NGF,* nerve growth factor; *IL-1α,* interleukin-1α; *IL-6,* interleukin-6; *PmodS,* peritubular cell-modifying substance, *?,* pathway uncertain

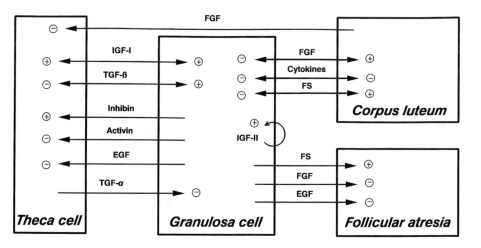

Fig. 118.10. Some of the proposed intratesticular actions exerted by putative nonsteroidal local modulators of ovarian function. *TGF-β,* transforming growth factor-*β; TGF-α,* transforming growth factor-α; *IGF-I,* insulin-like growth factor-I; *IGF-II,* insulin-like growth factor-II; *EGF,* epidermal growth factor; *FGF,* fibroblast growth factor; *FS,* follistatin

activity. The trophic effects of LH and FSH on granulosa cell functions are augmented by activin and diminished by inhibin. Follistatin, an activin-binding protein, counteracts the effects of activin as might be expected. Additionally, follistatin was found to promote follicular atresia and to support the corpus luteum. Among the various growth factors, FGF and EGF act to prevent follicular death [53] and FGF exerts angiogenetic effects in the corpus luteum. With regard to the local control of the life span of the corpus luteum, the proposition has been made that cytokines act as luteolytic agents, whereas FGF and follistatin appear to be luteotropic substances.

113.7 Extragonadal Effects of Sex Steroid Hormones

Apart from their essential role during gonadal functions, sex steroid hormones also expedite important endocrine effects in extragonadal target organs. These actions serve to establish the male and female phenotype, are not directly related to reproduction, and touch upon numerous functions of the organism. On the basis of their physiological actions a division into male and female sex steroid hormones is made. The male sex steroids, the androgens, exhibit fairly uniform activity, whereas in the female organism estrogens and gestagens need to be considered separately in regard to their biological actions.

113.7.1 Testosterone

As early as in the fetal period, testicular testosterone influences sexual differentiation. It is testosterone that induces the formation of scrotum, penis, and accessory sex organs. The wolffian duct transforms into the ductus deferens and the epidiymis. Simultaneously, testosterone prevents the development of the müllerian duct, and in the absence of testosterone, uterus, oviduct, and vagina develop. Testosterone also influences testicular descent and accounts for the pubertal growth of the penis and sex accessory glands. It determines the male pattern of hair distribution, induces sebaceous gland activity, causes the mutation of the voice and, in conjunction with GH, is required for somatic growth, the growth spurt, and final body dimensions. The task of testosterone during adulthood is to maintain the acquired androgen-dependent functions and properties. It also increases bone and muscle mass. Libido and sexual appetite are stimulated by testosterone, while sexual potency and erectile function do no necessarily depend on this sex steroid hormone.

113.7.2 Estrogens

The development of the female secondary sex characteristics is stimulated by estrogen during puberty. This holds particularly for the growth of breasts and the subcutaneous distribution of adipose tissue. In association with androgens, estrogens regulate pubic and axial hair growth. Vaginal physiology is influenced in terms of blood supply, proliferation, and differentiation and maturation of vaginal epithelium. The uterine endometrium grows during the proliferation phase of the menstrual cycle under the influence of estrogens. Other activities of estrogens in the female tract comprise influences on bladder musculature, collagen contents in connective tissues, and skin turgor. Lipid metabolism is pronouncedly affected by estrogens: synthesis of apolipoprotein is increased, high density lipoprotein and very low density lipoprotein are stimulated, while low density lipoprotein is reduced. Bone metabolism crucially depends on the presence of estrogens. The drop in the production of estrogens during the female climacteric is accompanied by loss of bone mass (osteoporosis).

113.7.3 Gestagens

Progesterone, in general, antagonizes the actions of estrogens in the female organism. Preovulatory body temperature is elevated owing to progesterone action. Progesterone cooperates with estrogen in the induction and furthering of breast growth. Vaginal epithelial detachment is diminished by progesterone leading to a quiescent state of the endometrium. The diameter of the cervical canal is lowered. The production of cholesterol, high density lipoprotein, and low density lipoprotein is reduced.

113.8 Excurrent Ducts and Accessory Glands of the Reproductive Tract

113.8.1 Epididymis and Ductus Deferens

Organization. Epithelial ducts (rete testis and ductuli efferentes) connect the testis with the epididymis and the epididymis links the testis to the ductus (vas) deferens. Three main regions of the human epididymis have been classified: caput, corpus, and cauda. The epididymis consists of a single, highly convoluted duct approximately 5 m long. Along the length of the duct, however, a variety of morphologically distinct epithelia can be seen. The caput epididymidis has recently been reported to comprise seven different types of epithelia [62]. Four major cell types comprise the epididymal epithelial layer, distinguished on the basis of morphological appearance and size: principal cells, basal cells, clear cells, and halo cells, the principal cells being highly predominant. These last are tall columnar cells and are believed to play an active role in the synthesis and secretion of proteins and in the absorption of luminal fluid. The function of the basal cells, which are small but elongated cells, has not been ultimately clarified as yet. It has recently been suggested that basal cells may subserve a scavenging role in a local immune defense mechanism, e.g., antigenic products possibly resulting from sperm degradation being phagocytosed. Clear cells are large cells and contain lysosomes and lipid droplets, giving rise to the assumption that this cell type is engaged in the uptake and disposal of luminal components. Halo cells are part of the immune system and represent either monocytes or lymphocytes. The epithelial part proximal to the testis contains ciliated epithelial cells endowed with the ability to absorb luminal fluid.

Functions. On the basis of animal studies three major functions of the epidiymis have been derived: transport of spermatozoa, achieved by epithelial cells endowed with kinocilia which generate fluid motions; maturation of the immature testicular spermatozoa, to attain progressive motility and thereby the ability to fertilize ova in the female reproductive tract under physiological conditions; and storage of spermatozoa, providing a sperm reservoir for ejaculation. Transit times for sperm in the human epididymis are highly variable, ranging between 1 and 20 days. Under normal conditions, the human epididymis tranports up to 150 million sperm per day along its length and the storage capacity amounts to 600 million spermatozoa prior to their emission. Following depletion of spermatozoa from the epididymis by multiple ejaculation, gradual refilling was seen over 2 weeks of abstention [13]. Much evidence also argues in favor of a role of the human epididymis in the acquisition of the fertilizing capacity of sperm (cf. Chap. 115).

During their epididymal passage spermatozoa undergo membrane alterations associated with capacitation and the acrosome reaction and develop special kinematic features. Recently, a three-fold increase in the percentage motility and a two-fold increase in the straight line velocity of spermatozoa along the human corpus epididymidis has been observed [63]. The formation of disulfide bridges in various structural elements and protamine expression have also been reported. The human epididymal epithelium secretes a variety of substances [14] among which are proteins, enzymes, sperm-coating antigens, and phospholipids. The precise function of these substances for the development of motility and fertilizing capacity of epididymal sperm is yet to be established. Many of these substances are also found in semen and the seminal concentrations of α-glucosidase, glycerophosphocholine, and L-carnitine are of clinical relevance, since these substances serve as indicators of epididymal dysfunction and obstruction [15]. Recent findings on the in vitro and in vivo fertilizing ability of sperm collected from the proximal regions of the reproductive tract and presumed not to have encountered epididymal secretions have called into question the relevance of the human epididymis to fertility. This issue has not been settled to date [12]. However, it is quite likely that these sperm have been exposed to epididymal products, since the epithelia of rete testis and the ductus deferes seem to be able to transform under pathological conditions and might take over at least some epididymal characteristics.

Control of Epididymal Functions by Androgens. Both animal and clinical investigations demonstrate unequivocally that epididymal function is strictly androgen-dependent. The seminal concentrations of epididymal secretory products, i.e., L-carnitine and α-glucosidase, are related to the androgenic status. Testosterone and, indirectly, hCG exert a stimulatory effect. In patients with androgen deficiency both α-glucosidase activity and L-carnitine concentration is reduced compared to normal men, and androgen replacement normalizes the secretion of these factors. Our current understanding of the androgen-mediated control of epididymal function derives largely from studies in laboratory animal species. The reason for this is the limited availablility of normal epididymal human tissue and the fact that epididymal biopsies are not possible because of the provocation of irreversible damage to the single epididymal duct.

The 5α-reduced metabolite of testosterone, DHT, is the pivotal modulator of epididymal functions. It has been demonstrated that the stimulatory effect of testosterone administration on epididymal functions can be prevented by coadministration of a 5α-reductase inhibitor. In line with these results are data indicating that under in vitro conditions DHT is more effective than testosterone in maintaining epididymal function. Epididymal sperm stor-

age and fertilizing ability are androgen-dependent. In contrast, epididymal sperm transport seems to be less influenced by androgens, and neuronal input is the apparent main driving force. It is presumed that androgens are not only captured from the circulation but are also delivered to the epididymis directly from the testis. This view derives from the observation that androgen-binding protein has been found in the initial segment of the epididymis. Immunohistochemical studies have localized the androgen receptor in the nuclei of epididymal cells both in animal species and in men. Receptors for estrogens and possibly prolactin are also present in the epididymal cells, but any functional significance of these receptors has yet to be proven.

113.8.2 Seminal Vesicles and Prostate

Organization and Function. Seminal vesicles are elongated, contorted, sac-like structures. The seminal vesicle epithelium contains small rounded basal cells and columnar or cuboidal surface cells. The latter cell types contain large secretory granules. A typical feature of the organization of the seminal vesicles is a pronounced folding of the epithelial mucosa. This epithelium exhibits high secretory activity. Among the products, fructose, seminogelin, and prostaglandins are the most notable, fructose being the major secretory product. Sufficient amounts of fructose in the ejaculate are probably required for sperm motility since fructose is an important energy source for ejaculated spermatozoa. Seminogelin is viewed as a regulator of the coagulation of human semen.

The prostate gland is the largest among the accessory sex structures. It is comprised of 30–50 tubuloalveolar glands. The prostatic acini are lined by an epithelium embracing columnar cells of various shapes and sizes and also basal cells. Like the seminal vesicles, the prostate gland is the source of a number of secretion products such as citric acid, acid phosphatase, and prostate-specific antigen. The functional role of citric acid and acid phosphatase is not entirely clear. Prostate-specific antigen is likely to be involved in the liquefaction of human ejaculate. Clearly, the main function of prostate and seminal vesicles in the human is to contribute to the volume of semen and to provide a transfer medium for the spermatozoa into the female tract. A role of these organs for fertility appears unlikely since epididymal sperm are able to fertilize ova.

Hormonal Dependence of Seminal Vesicle/Prostate Function and Growth. As in the case of the epididymis, the function of seminal vesicles and prostate is decisively controlled by androgens. Withdrawal of androgens is accompanied by involution and atrophy of these glands and a cessation or marked decline of their secretory activity. These consequences of androgen deficiency can be remedied by the administration of testosterone. It is interesting to note that abnormal growth, i.e., benign hyperplasia or carcinoma, is confined to the prostatic gland within the excurrent duct system, and that the human prostate is particularly vulnerable to these growth entities. Among the various animal species studied, prostate diseases are apparently only common in the dog. The endocrine control of prostate growth, as already indicated above, is mainly effected via testis-derived testosterone but, additionally, adrenal-derived androstenedione, estrogen, and prolactin have also been implicated in prostatic pathology. Testosterone is 5α-reduced to DHT within the prostate gland. Estradiol derives from either peripheral or intraprostatic conversion of testosterone and synergizes with androgens in stimulating prostatic growth. On the other hand, the significance of the contributions of prolactin (presumed to enhance androgen action) and adrenal steroids is not entirely clear. Androgens are the prime regulators in that regard, and the most common therapeutic approach in prostate cancer is elimination of the testicular androgens via either orchidectomy or the suppression of LH secretion by GnRH agonist alone or in combination with antiandrogens. The intraprostatic concentrations of DHT and estradiol are important factors in the development of benign prostatic hyperplasia. Therefore, aromatase inhibitors and 5α-reductase inhibitors are under consideration for the treatment of benign prostatic hyperplasia.

Notwithstanding the fact that androgens play a permissive role in the pathobiology of the prostate, it is far from certain that androgens do in fact induce malignant transformation of prostatic tissue. Indeed, epidemiologic surveys have failed to establish a correlation between the circulating levels of testosterone or DHT and the incidence of prostatic cancer. Additionally, in hypogonadal androgen-deficient patients given supplementary testosterone, prostate size was stimulated back into the normal range but not beyond [3]. It thus appears quite unlikely that androgens have a causative role in the etiology of malignant prostate growth. This is particularly relevant to the use of testosterone in the therapy of hypogonadal men and to endocrine male contraception.

113.8.3 Oviduct

Organization and Function. The oviduct is also commonly referred to as the fallopian tube. Four segments of the oviduct can be distinguished: the infundibulum, the ampulla, the isthmus, and the uterine segment. Anatomically, the oviduct has stromal and epithelial cells. The mucosa is composed of a columnar epithelium with varying degrees of folding that are greatest in the ampulla. Both ciliated and nonciliated cells are present. The beat of the cilia is directed towards the uterus and supports the movement of the egg towards the ampulla, whereas nonciliated cells have a secretory function. Ciliated cells predominate in the infundibulum and nonciliated in the isthmus. Fringed extensions, fimbriae, extend from the infundibulum to the ovary and serve to capture the ovulated egg. The ampulla is the site of fertilization and harbors the zygote up to the morula stage, and subsequently the preimplantation embryo rapidly moves to the

uterus. The isthmus is thought to act as a sperm reservoir. Thus, the oviduct fulfills two functions, the transmission of the ovum from the ovary to the uterus and the provision of the required environment for fertilization.

Hormonal Regulation of Oviduct Function. Estrogens and progesterone are the key regulators of oviduct functions. During the follicular phase receptors for estrogens and progestins are present in the nonciliated cells and in the stromal cells. During the luteal phase, when the height of the epithelial cells is markedly reduced, the non-ciliated cells lose both steroid hormone receptors. Unlike the nonciliated cells, stromal cells only lose the estrogen receptor, whereas the progesterone receptor is retained [8].

113.9 Concluding Remarks

This chapter has described the basic mechanisms involved in the production, maturation, and release of male and female gametes. It is evident that the successful generation of functionally intact gametes depends on the complex but coordinated interplay of several endocrine organs. Additionally, the crucial role of locally produced factors for the function of reproductive organs is increasingly acknowledged. Paracrine, autocrine, and intracrine mechanisms have been described in that context. Although much is known about the sites of action and effects on target cells of hormones, little is known about their precise mechanisms of action. It is to be anticipated, however, that the enormous progress being made in the field of cellular and molecular biology will considerably advance our current understanding of the control of reproductive organ function.

General References

Adashi EY (1991) The ovarian life cycle. In: Yen SSC, Jaffe RB (eds) Reproductive endocrinology. Physiology, pathophysiology and clinical management. Saunders, Philadelphia, pp 181 237

Bergh A, Damber J-E (1993) Vascular controls in testicular physiology. In: De Kretser D (ed) Molecular biology of the male reproductive system. Academic, New York, pp 439–467

Brabant G, Prank K, Schöfl C (1992) Pulsatile patterns in hormone secretion. Trends Endocrinol Metab 3:183–190

Carson-Jurica MA, Schrader WT, O'Malley BW (1990) Steroid receptor family: structure and functions. Endocr Rev 11:201–205

Catt KF, Dufau ML (1991) Gonadotropic hormones: biosynthesis, secretion, receptors and actions. In: Yen SSC, Jaffe RB (eds) Reproductive endocrinology. Physiology, pathophysiology and clinical management. Saunders, Philadelphia, pp 105–155

Combarnous Y (1992) Molecular basis of the specificity of binding of glycoprotein hormones to their receptors. Endocr Rev 13:670–687

Conn PM, Crowley WF (1991) Gonadotropin-releasing hormone and its analogues. N Engl J Med 324:93–103

Cooper TG (1992) Epididymal proteins and sperm maturation. In: Nieschlag E, Habenicht UF (eds) Spermatogenesis – fertilization

– contraception. Molecular, cellular and endocrine events in male reproduction. Springer, Berlin Heidelberg New York, pp 285–318

Findlay JK, Shukowvski L, Xiao S, Klein R, Michel U, Farnworth PG, Robertson DM (1993) Actions and interactions of activin and follistatin in the control of ovarian folliculogenesis. In: Mornex R, Jaffiol C, Leclère J (eds) Progress in endocrinology. Parthenon, New York, pp 581–584

Gharib SD, Wierman ME, Shupnik A, Chin WW (1990) Molecular biology of the pituitary gonadotropins. Endocr Rev 11:177–199

Hsueh AJW, Bicsak TA, Xiao-Chi J, Dahl KD, Fauser BCJM, Galway AB, Cezekala N, Pavlou SN, Papkoff H, Keene J, Boime I (1989) Granulosa cells as hormone targets: the role of biologically active follicle-stimulating hormone in reproduction. Recent Prog Horm Res 45:209–277

Hsueh AJW, LaPolt PS (1992) Molecular basis of gonadotropin-receptor regulation. Trends Endocrinol Metab 3:164–170

Hsueh AJ, Adashi EY, Jones PBC, Welsh T (1984) Hormonal regulation of the differentiation of cultured ovarian granulosa cells. Endocr Rev 5:76–127

Jansen RPS (1984) Endocrine responses in the fallopian tubes. Endocr Rev 5:525–552

Jegou B, Sharpe RM (1993) Paracrine mechanisms in testicular control. In: De Kretser D (ed) Molecular biology of the male reproductive system. Academic, New York, pp 271–310

Kalra SP (1993) Mandatory neuropeptide-steroid signaling for the preovulatory luteinizing hormone-releasing hormone discharge. Endocr Rev 5:507–538

Mather JP, Krummen LA (1992) Inhibin, activin and growth factors. In: Nieschlag E, Habenicht UF (eds) Spermatogenesis – fertilization – contraception. Springer, Berlin Heidelberg New York, pp 169–200

Matsumoto AM, Bremner WJ (1990) Control of spermatogenesis in humans. In: Moudgal NR, Yoshinaga K, Rao AY, Adiga PR (eds) Perspectives in primate reproductive biology. Wiley Eastern, New Delhi, pp 173–180

Michel U, Farnsworth P, Findlay JK (1993) Follistatins: more than follicle-stimulating suppressing proteins. Mol Cell Endocrinol 91:1–11

Misrahi M, Thu Vu Hai M, Ghinea N, Loosfelt H, Meduri G, Atger M, Jolivet A, Gross B, Savouret J-F, Dessen P, Milgrom E (1993) Molecular and cellular biology of gonadotropin receptors. In: Adashi EY, Leung PCK (eds) The ovary. Raven, New York, pp 57–92

Nieschlag E, Behre HM, Simoni M, Gromoll J, Gudermann T (1993) Endocrine regulation of testicular function in infertile men. In: Zirkin B, Whitcomb R (eds) International symposium on understanding male infertility: basic and clinical approaches. Raven, New York, pp 113–119

Nieschlag E, Weinbauer GF, Behre HM (1992) Hormonal male contraception: a real chance? In: Nieschlag E, Habenicht UF (eds) Spermatogenesis – fertilization – contraception: Springer, Berlin Heidelberg New York, pp 169–200

Nieschlag E, Behre HM (1990) Pharmacology and clinical use of testosterone. In: Nieschlag E, Behre HM (eds) Testosterone – action, deficiency, substitution. Springer, Berlin Heidelberg New York, pp 92–114

Robaire B, Viger RS (1993) Regulation of epididymal epithelial functions. In: Zirkin B, Whitcomb R (eds) International symposium on understanding male infertility: basic and clinical approaches. Raven, New York, pp 183–210

Segaloff DL, Ascoli M (1993) The lutropin/choriogonadotropin receptor . . . 4 years later. Endocr Rev 14:324–347

Simoni E, Nieschlag E (1991) In vitro bioassays of follicle-stimulating hormone: methods and clinical applications. J Endocrinol Invest 14:983–997

Spiteri-Grech J, Nieschlag E (1993) Paracrine factors relevant to the regulation of spermatogenesis – a review. J Reprod Fertil 98:1–14

Tonetta SA, DiZerega GS (1989) Intragonadal regulation of follicular maturation. Endocr Rev 10:205–229

Truss M, Beato M (1993) Steroid hormone receptors: interaction with deoxyribonucleic acid and transcription factors. Endocr Rev 14:459–479

Ulloa-Aguirre A, Espinoza R, Damian-Matsumura P, Chappel SC (1988) Immunological and biological potencies of the different molecular species of gonadotrophins. Hum Reprod 3:491–501

Weinbauer GF, Nieschlag E (1992) LH-RH antagonists: state of the art and future perspectives. In: Höffken K (ed) Peptides in oncology. Springer, Berlin Heidelberg New York, pp 113–136

Weinbauer GF, Nieschlag E (1993) Hormonal control of spermatogenesis. In: De Kretser DM (ed) Molecular biology of the male reproductive system. Academic, New York, pp 99–143

Woodruff TK, Battaglia J, Mather JP (1993) Regulation of human granulosa cells by recombinant human activin A and recombinant human inhibin A. In: Mornex R, Jaffiol C, Leclère J (eds) Progress in endocrinology. Proceedings of the 9th international congress of endocrinology. Parthenon, New York, pp 605–608

Yen SS (1991) The human menstrual cycle: neuroendocrine regulation. In: Yen SSC, Jaffe RB (eds) Reproductive endocrinology. Physiology, pathophysiology and clinical management. Saunders, Philadelphia, pp 181–237

Specific References

1. Aatsinki JT, Pietilä EM, Lakkakorpi JT, Rajaniemi HJ (1992) Expression of the LH/CG receptor gene in rat ovarian tissue is regulated by an extensive alternative splicing of the primary transcript. Mol Cell Endocrinol 84:127–135
2. Arslan M, Weinbauer GF, Schlatt S, Shahab M, Nieschlag E (1993) Follicle-stimulating hormone and testosterone, alone or in combination, activate spermatogonial proliferation and stimulate testicular growth in the immature nonhuman primate (*Macaca mulatta*). J Endocrinol 136:235–243
3. Behre HM, Bohmeyer J, Nieschlag E (1993) Prostate volume in testosterone treated and untreated hypogonadal men in comparison to age-matched normal controls. Clin Endocrinol 40:341–349
4. Billig H, Furuta I, Hsueh A (1993) Estrogens inhibit and androgens enhance ovarian granulosa cell apoptosis. Endocrinology 133:2204–2212
5. Blok LJ, Hoogerbrugge JW, Themmen APN, Baarends WM, Post M, Grootegoed JA (1992) Transient down-regulation of androgen receptor messenger ribonucleic acid (mRNA) expression in Sertoli cells by follicle-stimulating hormone is followed by up-regulation of androgen receptor mRNA and protein. Endocrinology 131:1243–1349
6. Blok LJ, Bartlett JMS, Bolt-de-Vries J, Themmen APN, Brinkmann AO, Weinbauer GF, Nieschlag E, Grootegoed JA (1992) Effect of testosterone deprivation on expression of the androgen receptor in rat prostate, epididymis and testis. Int J Androl 15:182–198
7. Bourguignon JP, Gerard A, Debougnoux G, Rose J, Franchimont P (1987) Pulsatile release of gonadotropin-releasing hormone (GnRH) from rat hypothalamus in vitro: calcium and glucose dependency and inhibition by superactive GnRH analogs. Endocrinology 121:993–999
8. Brenner RM, West NB, McCellan MC (1990) Estrogen and progestin receptors in the reproductive tract of male and female primates. Biol Reprod 42:11–19
9. Cameron DF, Muffly KE (1991) Hormonal regulation of spermatid binding. J Cell Sci 100:623–633
10. Carnegie JA, Dardick I, Tsang BK (1987) Microtubules and the gonadotropic regulation of granulosa cell steroidogenesis. Endocrinology 120:819–828
11. Chappel SC, Howles C (1991) Reevaluation of the roles of luteinizing hormone and follicle-stimulating hormone in the ovulatory process. Hum Reprod 6:1206–1212

12. Cooper TG (1993) The human epididymis – is it necessary? Int J Androl 16:245–150
13. Cooper TG, Keck , C, Oberdieck U, Nieschlag E (1993) Effects of multiple ejaculations after extended periods of sexual abstinence on total, motile and normal sperm numbers, as well as accessory gland secretions, from healthy normal and oligozoospermic men. Hun Reprod 8:1251–1258
14. Cooper TG, Yeung CH, Meyer R, Schulze H (1990) Maintenance of human epididymal epithelial cell function in monolayer culture. Biol Reprod 90:81–91
15. Cooper TG, Yeung CH, Nashan D, Nieschlag E (1988) Epididymal markers in human infertility. J Androl 9:91–101
16. De Geyter CH, De Geyter M, Schneider HPG, Nieschlag E (1992) Interdependent influence of follicular fluid estradiol concentration and motility characteristics of spermatozoa on in-vitro fertilization results. Hum Reprod 7:665–670
17. Devroey P, van Steirteghem A, Mannaerts B, Coeling- Bennink H (1992) Successful in-vitro fertilisation and embryo transfer after treatment with recombinant human FSH. Lancet 339:1170–1171
18. Donderwinkel PFJ, Schoot DC, Coelingh Bennink HJT, Fauser BCJM (1992) Pregnancy after induction of ovulation with recombinant human FSH in polycystic ovary syndrome. Lancet 340:983
19. Frazier AL, Robbins LS, Stork PJ, Sprengel R, Segaloff DL, Cone RD (1990) Isolation of TSH and LH/CG receptor cDNAs from human thyroid: regulation by tissue specific splicing. Mol Endocrinol 4:1264–1267
20. Fukuoka K, Yasuda K, Takakura K, Emi N, Fujiwara H, Iwai M, Iwai K, Kanzaki H, Mori T (1992) Cytokine modulation of progesterone and estradiol secretion in cultures of luteinized human granulosa cells. J Clin Endocrinol Metab 75:254–258
21. Galway AB, Hsueh AJ, Keene JL, Yamoto M, Fauser BCJM, Boime I (1990) In vitro and in vivo bioactivity of recombinant human follicle-stimulating hormone and partially deglycosylated variants secreted by transfected eukaryotic cell lines. Endocrinology 127:93–100
22. Germond M, Dessole S, Senn A, Loumaye E., Howles C, Beltrami V (1992) Successful in-vitro fertilisation and embryo transfer after treatment with recombinant human FSH. Lancet 339:1170
23. Grino PB, Griffin JE, Wilson JD (1990) Testosterone at high concentrations interacts with the human androgen receptor similarly to dihydrotestosterone. Endocrinology 126:1165–1172
24. Gromoll J, Dankbar B, Sharma R, Nieschlag E (1993) Molecular cloning of the testicular FSH receptor of the non-human primate macaca fascicularis and identification of multiple transcripts in the testis. Biochem Biophys Res Commun 196:1066–1072
25. Gromoll J, Gudermann T, Nieschlag E (1992) Molecular cloning of a truncated isoform of the human follicle-stimulating hormone receptor. Biochem Biophys Res Commun 188:1077–1083
26. Guiochon-Mantel A, Lescop P, Christin-Maltre S, Loosfelt H, Perrot-Applanat M, Milgrom E (1991) Nucleocytoplastic shuttling of the progesterone receptor. EMBO J 10:3851 3859
27. Heckert LL, Daley IJ, Griswold MD (1992) Structural organization of the follicle-stimulating hormone receptor gene. Mol Endocrinol 6:70–80
28. Heckert LL, Griswold MD (1991) Expression of follicle-stimulating hormone receptor mRNA in rat testes and Sertoli cells. Mol Endocrinol 5:670–677
29. Hillier SG, Yong EL, Illingworth PJ, Baird DT, Schwall RH, Mason AJ (1991) Effect of recombinant inhibin on androgen synthesis in cultured human thecal cells. Mol Cell Endocrinol 75:R1–R6
30. Hughes FM, Gorospe WC (1991) Biochemical identification of apoptosis (programmed cell death) in granulosa cells: evidence for a potential mechanism underlying follicular atresia. Endocrinology 129:2415–2422
31. Kliesch S, Penttilä TL, Gromoll J, Saunders PTK, Nieschlag E, Parvinen M (1992) FSH receptor mRNA is expressed stage-dependently during spermatogenesis. Mol Cell Endocrinol 84: R45–R49
32. Kremer H, Mariman E, Otten BJ, Moll GW Jr, Stoelinga GBA, Wit JM, Jansen M, Drop SL, Faas B, Ropers H-H, Brunner HG (1993) Cosegregation of missense mutations of the luteinizing hormone receptor gene with familial male-limited precocious puberty. Hum Mol Gen 2:1779–1783
33. Meistrich M, van Beek MEAB (1993) Spermatogonial stem cells. In: Desjardins C, Ewing LL (eds) Cell and molecular biology of the testis. Oxford University Press, Oxford, pp 266–295
34. Minegish T, Nakamura K, Takakura Y, Miyamoto K, Hasegawa Y, Ibuky Y, Igarashi M (1990) Cloning and sequencing of human LH/hCG receptor cDNA. Biochem Biophys Res Commun 172:1049–1054
35. Moudgal NR, Ravindranath N, Murthy GS, Dighe RR, Aravindan GR, Martin F (1992) Long-term contraceptive efficacy of vaccine of ovine follicle-stimulating hormone in male bonnet monkey (Macaca radiata). J Reprod Fertil 96:91–102
36. Orth JM, Gunsalus GL, Lamperti AA (1988) Evidence from the Sertoli cell-depleted rat indicates that spermatid number in adults depends on numbers of Sertoli cells produced during the perinatal development. Endocrinology 122:787–794
37. Paris F, Weinbauer GF, Blüm V, Nieschlag E (1994) The effect of androgens and antiandrogens on the immuno-histochemical localization of the androgen receptor in accessory reproductive organs of male rats. J Steroid Biochem Mol Biol 48:129–137
38. Remohi J, Balmaceda JP, Rojas FJ, Asch RH (1988) The role of pre-ovulatory progesterone in the midcycle gonadotrophin surge, ovulation and subsequent luteal phase: studies with RU 486 in rhesus monkeys. Hum Reprod 3:431–435
39. Rossmanith WG, Mortola JF, Yen SSC (1988) Role of endogenous opioid peptides in the initiation of the midcycle luteinizing hormone surge in normal cycling women. J Clin Endocrinol Metab 67:695–700
40. Ruizeveld de Winter JA, Trapman J, Vermey M, Mulder E, Zegers ND, van der Kwast TH (1991) Androgen receptor expression in human tissues: an immunohistochemical study. J Histochem Cytochem 39:927–936
41. Russell LD (1991) The perils of sperm release – "let my children go". Int J Androl 14:307–311
42. Saez JM, Avallet O, Lejeune H, Chatelain (1991) Cell-cell communication in the testis. Horm Res 36:104–115
43. Schlatt S, Weinbauer GF, Arslan M, Nieschlag E (1993) Appearance of α-smooth muscle actin in peritubular cells of monkey testes is induced by androgens, modulated by follicle stimulating hormone and maintained after hormonal withdrawal. J Androl 14:340–350
44. Schoot DC, Coelingh Bennink HJT, Mannaerts BMJL, Lamberts SWJ, Bouchard PH, Fauser BCJM (1992) Human recombinant follicle-stimulating hormone induces growth of preovulatory follicles without concomitant increase in androgen and estrogen biosynthesis in a woman with isolated gonadotropin deficiency. J Clin Endocrinol Metab 74:1471–1473
45. Schulze W, Rehder U (1984) Organization and morphogenesis of the human seminiferous epithelium. Cell Tiss Res 237:395–407
46. Sharpe RM, Maddocks S, Millar M, Kerr JB, Saunders PTK, McKinnell C (1992) Testosterone and spermatogenesis. Identification of stage-specific, androgen-regulated proteins secreted by adult rat seminiferous tubules. J Androl 13:172–184
47. Shenker A, Laue L, Kosugi S, Merendino JJ Jr, Minegishi T, Cutler GB Jr (1993) A constitutively activating mutation of the luteinizing hormone receptor in familial precocious puberty. Nature 365:652–654
48. Smyth CD, Miro F, Whitelaw PF, Howles CM, Hillier SG (1993) Ovarian thecal/interstitial androgen synthesis is enhanced by

a follicle-stimulating hormone-stimulated paracrine mechanism. Endocrinology 133:1532–1538

49. Söder L, Bang P, Wahab A, Parvinen M (1992) Insulin-like growth factors selectively stimulate spermatogonial, but not meiotic, deoxyribonucleic acid synthesis during rat spermatogenesis. Endocrinology 131:2344–2350

50. Spiteri-Grech J, Weinbauer GF, Bolze P, Chandolia RK, Bartlett JMS, Nieschlag E (1993) Effects of FSH and testosterone on intratesticular insulin-like growth factor-I and specific germ cell populations in rats treated with gonadotrophin-releasing hormone antagonist. J Endocrinol 137:81–89

51. Themmen APN, Blok LJ, Post M, Baarend WM, Hoogerbrugge JW, Parmentier M, Vassart G, Grootegoed JA (1991) Follitropin down-regulation involves a cAMP-dependent post-transcriptional decrease of receptor mRNA expression. Mol Cell Endocrinol 78:R7–R13

52. Tilly JL, Aihara T, Nishimori K, Jia YC, Billig H, Kowalski KI, Perlas EA, Hsueh AJW (1992) Expression of recombinant human follicle-stimulating hormone receptor: species-specific ligand binding, signal transduction, and identification of multiple ovarian messenger ribonucleic transcripts. Endocrinology 131:799–806

53. Tilly JL, Billig H, Kowalski KI, Hsueh AJ (1992) Epidermal growth factor and basis fibroblast growth factor suppress the spontaneous onset of apoptosis in cultured rat ovarian granulosa cells and follicles by a tyrosine kinase-dependent mechanism. Mol Endocrinol 6:1942–1950

54. Tilly JL, Kowalski KI, Johnson AL, Hsueh AJW (1991) Involvement of apoptosis in ovarian follicular atresia and postovulatory regression. Endocrinology 129:2799–2801

55. Van Alphen MMA, van de Kant HJG, de Rooij DG (1988) Follicle-stimulating hormone stimulates spermatogenesis in the adult monkey. Endocrinology 123:1449–1455

56. Van Haaster LH, de Jong FH, Docter R, de Rooij DG (1992) The effect of hypothyroidism on Sertoli cell proliferation and hormone levels during testicular development in the rat. Endocrinology 131:1574–1576

57. Verhoeven G, Swinnen K, Cailleau J, Deboel L, Rombauts L, Heyns W (1992) The role of cell-cell interactions in androgen action. J Steroid Biochem Mol Biol 41:487–494

58. Vliegen MK, Schlatt S, Weinbauer GF, Bergmann M, Groome N, Nieschlag E (1993) Localization of inhibin/activin subunits in the testis of adult nonhuman primates and men. Cell Tissue Res 273:261–268

59. Weinbauer GF, Behre HM, Fingscheidt U, Nieschlag E (1991) Human follicle-stimulating hormone exerts a stimulatory effect on spermatogenesis, testicular size, and serum inhibin levels in the gonadotropin-releasing hormone antagonist-treated nonhuman primate (*Macaca fascicularis*). Endocrinology 129:1831–1839

60. Weinbauer GF, Hankel P, Nieschlag E (1992) Exogenous gonadotrophin-releasing hormone (GnRH) stimulates LH secretion in male monkeys (*Macaca fascicularis*) treated chronically with high doses of a GnRH antagonist. J Endocrinol 133:439–445

61. Whitcomb RW, Crowley WF (1990) Clinical review 4: diagnosis and treatment of isolated gonadotropin-releasing hormone deficiency in men. J Clin Endocrinol Metab 70:3–7

62. Yeung CH, Cooper TG, Bergmann M, Schulze H (1991) Organization of tubules in the human caput epididymis and the ultrastructure of their epithelia. Am J Anat 191:261–279

63. Yeung CH, Cooper TG, Oberpenning F, Schulze H, Nieschlag E (1993) Changes in movement characteristics of human spermatozoa along the length of the epididymis. Biol Reprod 49:274–280

114 Behavioral And Neurovegetative Components of Reproductive Functions

W. Jänig

Contents

114.1 Introduction

Erotic (visual, auditory, olfactory and somatosensory; "psychogenic") stimuli and mechanical stimulation of the external sexual organs lead to physiological changes in the adult woman and man. The changes were described by Masters and Johnson [28] as the sexual response cycle, which was divided for practical reasons into four response phases: sexual *arousal*, *plateau*, *orgasm*, and *resolution* phase (Fig. 114.1). Time course and intensity of these phases vary widely among individuals. The excitation and resolution phases are the longest, and the orgasm is usually the shortest. In the man the overall cycle tends to be stereotyped, with little interindividual variation (Fig. 114.1A). A refractory period follows the peak of orgasm and extends into the resolution phase; during this time the orgasm cannot be produced by sexual stimulation. In the woman the sexual response cycle is considerably more variable in both duration and intensity (Fig. 114.1B). Women are capable of multiple orgasms. If an orgasm is not reached, the resolution phase lasts longer. These responses are determined by the brain and include afferent, autonomic, somatomotor and endocrine mechanisms leading to coordinated responses. The precise neural coordination is biologically meaningful and necessary for the adaptation of the sexual behavior to other behaviors and for the survival and reproduction of the species. This chapter concentrates on the neural regulation of the reproductive organs. The primary focus is on the male reproductive organs, because very little is known about the neural regulation of the female reproductive organs [11–14,27,34].

114.2 Responses of the Reproductive Organs in the Male

114.2.1 Functional Anatomy of the Reproductive Organs and Their Innervation

The male reproductive organs are divided into *external* (penis and testicles) and *internal* reproductive organs. The most important parts, for understanding the neural regulation of the reproductive organs, are the erectile tissues (corpus cavernosum and corpus spongiosum) and their vascular supply, the vas deferens, the seminal vesicles, the prostate gland and the proximal (prostatic) urethra, which functionally forms the internal vesical sphincter (Fig. 114.2). The *erectile tissue* of the penis consists of trabecular tissue and sinusoids, both of which contain smooth muscle cells. They are surrounded by a connective tissue sheath constituting the *tunica albuginea*. The tunica albuginea of the corpora cavernosa consists of two thick layers of collagen and elastic fibers, which are distensible during erection and are responsible for the rigidity of the penis. In the corpus spongiosum and the glans the sinusoids are larger in diameter, and the tunica albuginea is thinner in the corpus spongiosum and absent in the glans [25].

The erectile tissue is vascularly supplied by the internal pudendal artery, which becomes the common penile artery after giving off a branch to the perineum. The common penile artery divides into the bulbourethral, cavernosus, and dorsal penile arteries. In the erectile tissue these arteries branch off into the *helicine arteries* that supply the trabecular erectile tissue and the sinusoids. The helicine arteries are contracted and tortuous in the flaccid state and become dilated and straight during erection [25]. The erectile tissue is drained by the emissary veins (Fig. 114.3). The internal reproductive organs consist mainly of smooth muscle cells and secretory epithelia.

The internal and external reproductive organs are innervated by four groups of nerve fibers [4,18,31]:

R. Greger/U. Windhorst (Eds.)
Comprehensive Human Physiology, Vol. 2
© Springer-Verlag Berlin Heidelberg 1996

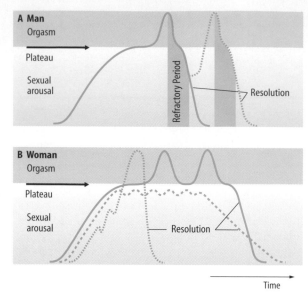

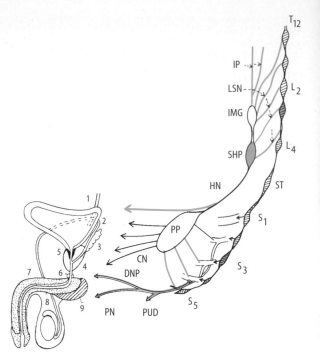

Fig. 114.1. Sexual response cycle of man and woman. Duration (*abscissa*) and intensity (*ordinate*) of the phases vary widely among individuals. (Modified from [28])

Fig. 114.2 Anatomy of the male reproductive organs and their innervation. *CN*, Cavernous nerve (contains postganglionic vasodilator fibers to erectile tissue); *DNP*, dorsal nerve of penis; *IMG*, inferior mesenteric ganglion; *HN*, hypogastric nerve; *IP*, intermesenteric (preaortic) plexus; *LSN*, lumbar splanchnic nerves; *PN*, perineal nerve (contains branches to bulbo- and ischiocavernous muscles); *PP*, pelvic splanchnic plexus; *PUD*, pudendal nerve; *SHP*, superior hypogastric plexus; *ST*, sympathetic trunk; *1*, ureter; *2*, vas deferens; *3*, seminal vesicle; *4*, prostate gland; *5*, bladder neck ("internal vesical sphincter"); *6*, bulbourethral glands; *7*, *8*, erectile tissue (cavernous and spongious bodies of the penis); *9*, bulbo- and ischiocavernous muscles. (Adapted from [4,31])

Fig. 114.3A,B (*below*). The erectile tissue of the penis and the vascular mechanism of erection. **A** Lateral view of arteries of the penis and cross section of the penis, illustrating the arteries within the erectile tissue. The cavernous and spongious bodies are surrounded by a connective tissue sheath (tunica albuginea). Each of the erectile bodies is supplied by one or two arteries. **B** Cross section through the erectile tissue, illustrating the relationship of arterial supply, sinusoids, venous drainage and tunica albuginea in the flaccid state and in the erectile state. Dilatation of penile arteries and sinusoids increases the blood volume in the erectile tissue; the veins become partially compressed against the tunica albuginea. (Adapted from [26])

▼

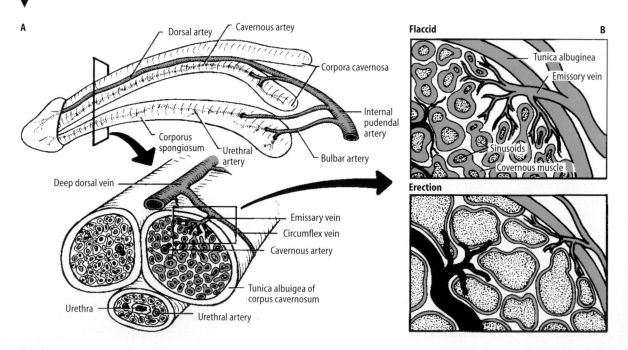

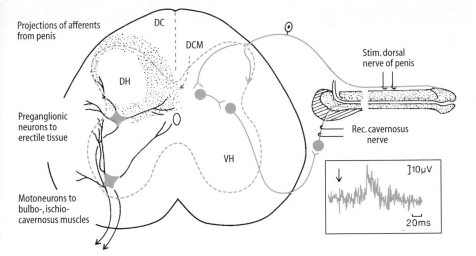

Fig. 114.4. Location, in the sacral spinal cord, of preganglionic neurons and motoneurons associated with erection (*left*), projection of afferents from the penis to the sacral spinal cord (area of projection indicated *dotted* on *left side*) and the spinal reflex arc leading to erection (*right*). *Left*: Projection of penile afferents in the pudendal nerve were determined by horseradish peroxidase tracing in the cat. They project to the medial side of the dorsal horn (*DH*) and the dorsal commissure (*DCM*). Preganglionic neurons are situated in the intermediate zone and send dendrites into regions of the afferent projection and lateral into regions of descending systems. Motoneurons innervating the bulbo- and ischiocavenosus muscles are situated laterally in the ventral horn (*VH*; Onuf's nucleus in humans) and also send dendrites dorsally, laterally and medially. DC, Dorsal columns. *Right*: Electrical stimulation of penile afferents (*arrow* in inset) and recording from cavernous nerve in the rat. *Inset* reflex response. (After [13], with permission)

- The (efferent) parasympathetic innervation has its origin in the sacral spinal segments S2–S4 (Fig. 114.2). The preganglionic neurons involved in the generation of erection are situated in the intermediate zone of the spinal cord, probably lateral to the preganglionic neurons that are associated with the hindgut and dorsomedial to preganglionic neurons associated with the lower urinary tract (Fig. 114.4, left). They project through the pelvic splanchnic nerve and synapse with postganglionic neurons in the pelvic splanchnic plexus, which project through the cavernous nerves (penile nerve) to the erectile tissue (Fig. 114.2).

- The (efferent) sympathetic innervation originates from the spinal segments T12 to L2 (Fig. 114.2). The preganglionic neurons are situated in the intermediate zone, many of them probably medial to the intermediolateral cell column. Most of them project through the lumbar splanchnic nerves to the superior hypogastric plexus or further through the hypogastric nerves to the pelvic splanchnic plexus. In both plexuses the preganglionic sympathetic fibers synapse with postganglionic neurons that innervate the smooth muscles and secretory epithelia of the internal reproductive organs (including the bladder neck), the erectile tissues, and the arteries. Some sympathetic preganglionic neurons project through the lumbar sympathetic trunk to postganglionic neurons in sacral paravertebral ganglia, which in turn project through the pelvic splanchnic nerve and through the pudendal nerve to the reproductive organs (Fig. 114.2). Most of these neurons probably have vasoconstrictor function [18]. There are probably considerable differences between mammalian species in the pathway of the innervation of the erectile tissue by vasoconstrictor fibers.

- The afferent innervation of the penis projects through the dorsal nerve of the penis and the pudendal nerve to the spinal segments S2–S4. In the spinal cord these afferents project bilaterally into the medial dorsal horn and into the dorsal commissure (Fig. 114.4, left). These afferents are important for eliciting erection via the spinal reflex pathway. Other visceral afferents project from the internal reproductive organs through the pelvic splanchnic nerve and through the hypogastric and lumbar splanchnic nerves to the sacral spinal cord (S2-S4) and thoraco-lumbar spinal cord (T11–L2), respectively. The functions of these afferents are unclear; they are certainly involved in nociception and pain, but possibly also in reflex activity of the reproductive organs [17].

- Motoneurons project through the pudendal and perineal nerve to the bulbo- and ischio-cavernous muscles. Their cell bodies are situated in the ventrolateral spinal cord in Onuf's nucleus (Fig. 114.4, left).

114.2.2 Effector Responses to Stimulation of Efferent Nerves and Their Neuroeffector Mechanisms

The responses of the male reproductive organs consist in erection of the penis, emission of semen (sperm and glandular secretion) into the posterior urethra, and ejaculation of semen from the anterior urethra. Table 114.1 illustrates the effector responses of the male genital organs (column

Table 114.1. Responses of the reproductive organs in males

Response	Afferents	Efferents	Central pathways	Effector organ (responses)
Erection of penis Reflexogenic	Pudendal nerve	Parasympathetic, pelvic nerve	Sacral spinal cord	Dilatation of arteries to
Psychogenic	Other sensory channels	Parasympathetic, pelvic nerve Sympathetic, thoracico-lumbar	Supraspinal origin	and sinusoids in cavernous and spongious corpora
Secretion of glands	Pudendal nerve	Parasympathetic	Sacral spinal cord	Bulbourethral, urethral glands
	Pudendal nerve	Sympathetic, thoracico-lumbar	Sacro-lumbar spinal cord	Seminal vesicles, prostate, vas deferens
Emission	Pudendal nerve	Sympathetic, thoracico-lumbar	Sacro-lumbar spinal cord	Contraction of vas deferens, prostate, seminal vesicle, ampulla, bladder neck
Ejaculation	Pudendal nerve	Somatic in pudendal nerve	Sacral spinal cord	Rhythmic contraction of bulbo-/ischio-cavernous muscles

5) that are generated by excitation of the different types of efferent nerves (column 3). These consist in *dilatation* of the *helicine arteries* and *dilatation of the sinusoids and the trabeculae* in the *erectile tissues* contraction of the internal reproductive organs, secretion of glands, and rhythmic contractions of the bulbo- and ischiocavernous muscles (column 5).

Erection. During neural activation of parasympathetic neurons innervating the penis, the erection of the penis passes through a sequence of changes called *tumescence*, *erection*, *rigidity*, and *detumescence*. The changes in blood flow through, and the pressure in, the erectile tissue during these phases can be simulated by continuous electrical stimulation of the efferent parasympathetic and somato-sensory nerve fibers (Fig. 114.5). Stimulation of the parasympathetic (efferent) fibers in the pelvic splanchnic nerve or in the cavernosus nerve generates dilatation of the branches of the internal pudendal artery, of the helicine arteries, of the sinusoids and of the trabeculae in the erectile tissue. The blood flow into and the pressure in the erectile tissue increases in the tumescence phase, leading to elongation of the penis and an increase in its circumference. At the beginning, the pressure decreases due to the fast dilatation of the sinusoids. The venous outflow is passively reduced by compression of the emissary veins (Fig. 114.3B). During full erection of the penis the pressure in the erectile tissue is some 10–20 mmHg below the arterial blood pressure and the inflow of blood is reduced due to venous compression. In humans, rigidity of the penis is already present in the phase of full erection, owing to the thick tunica albuginea of the corpus cavernosum, and is enhanced by activation of motor axons to the bulbo- and ischiocavernous muscles. The pressure now increases to more than 150 mmHg (in man values of up to

500 mmHg have been measured), and the blood flow is considerably reduced at this stage, because the erectile tissue is compressed. In most mammalian species, such as monkey, dog and horse, activation of the bulbo- and ischiocavernous muscles is necessary to reach rigidity, because the tunica albuginea of the corpus cavernosum is not as strong in these species as in man. After the end of the parasympathetic stimulation blood flow and pressure slowly decrease, leading to detumescence. Electrical stimulation of the sympathetic innervation to the erectile tissue (in the hypogastric nerves) during the parasympathetic activation reduces the blood flow to and the pressure in the erectile tissue, resulting in detumescence and termination of erection (Fig. 114.5). This is produced by stimulation of noradrenergic vasoconstrictor axons. Thus vasoconstrictor activation obviously totally overrides the parasympathetically induced vasodilatation [19–21,36].

Chemical Transmission to the Smooth Muscles of the Erectile Tissue. The neuroeffector mechanism leading to erection of the penis is not well understood. The parasympathetic postganglionic neurons that innervate the helicine arteries and the sinusoids of the erectile tissue are cholinergic. Injection of acetylcholine into the canine corpus cavernosum induces only a transient increase of pressure in the cavernous tissue [35], which supports the idea that acetylcholine is involved in penile erection but is not the only substance involved. Furthermore, atropine injected into the corpus cavernosum of dogs fails to abolish erection elicited by electrical stimulation of the parasympathetic efferent supply [7,35], and intravenous injection of atropine does not abolish erotically elicited erection in man [38]. These and other observations suggest that acetylcholine alone is not responsible for relaxation of the erectile tissue.

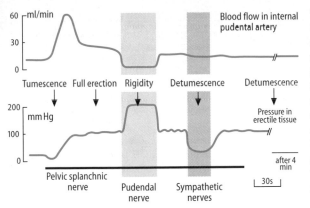

Fig. 114.5. Blood flow through pudendal artery (*upper diagram*) and pressure in erectile tissue (cavernous and spongious bodies of the penis; *lower diagram*) during repetitive electrical stimulation of the parasympathetic (pelvic splanchnic nerve or cavernous nerve), sympathetic (hypogastric nerves) and somatomotor (pudendal nerve) supply of the penis in the dog (stimulation parameters: 5–10 V and 10–30 Hz). The phases of penile erection as they occur in physiological conditions are indicated. Note that simultaneous stimulation of parasympathetic vasodilator neurons and pudendal motoneurons leads to rigidity of the penis, with a dramatic increase of pressure in the erectile tissue and decrease of blood flow, and that stimulation of sympathetic noradrenergic neurons results in subsidence of the parasympathically induced erection (detumescence). (Adapted from [20,25] with permission)

The terminals of parasympathetic postganglionic neurons to the erectile tissue contain, co-localized with acetylcholine (ACh) in large vesicles, the neuropeptide *vasoactive intestinal peptide* (VIP) [22,23,29]. Furthermore, they synthesize and release the radical *nitric oxide* (NO) during activation. The three substances ACh, NO and VIP relax the helicine arteries and the sinusoids of the erectile tissue. Both atropine and inhibitor of NO synthase reduce the neurally induced relaxation of the erectile tissue and therefore erection, but do not prevent it totally. Therefore, it is speculated that these three substances act together in the generation of erection in biological conditions. Both ACh and NO may be responsible for the fast onset of erection, but ACh may have a longer duration of effect. VIP is slower in onset but has a long duration of effect. In this way dilatation of the erectile tissue may be optimally potentiated and maintained [5].

Contraction of Internal Reproductive Organs. The contraction of vas deferens, seminal vesicle, ampulla, prostate and bladder neck is mediated by noradrenergic postganglionic neurons. The smooth muscle cells of these organs are electrically coupled by nexus and form functional syncytia. The varicosities of the axons of these neurons are in close contact with the smooth muscle cells without intervening Schwann cell cytoplasm and with fused basilar membranes [3,18]. Morphologically and physiologically, they establish neuroeffector junctions that function in a way similar to the neuromuscular end-plate.

The postjunctional excitatory potentials and the contractions of the organs induced by nerve stimulation can be only partially blocked by an alpha adrenoceptor antagonist, or not at all. There is evidence that *adenosine triphosphate* (ATP) is released concurrently with *noradrenaline* by the varicosities of the sympathetic neurons, acting postjunctionally on *purinoceptors* and *adrenoceptors*, respectively. It can be shown that fast excitatory junction potentials and fast contractions elicited by nerve stimulation can be blocked by ATP analogues or, more specifically, by purinoceptor antagonists, whereas the slower events are sensitive to adrenoceptor antagonists. Therefore, it is believed that ATP and noradrenaline act together in eliciting the effector responses [6,18]. The functions of neuropeptides that may be co-localized with noradrenaline and ATP in the noradrenergic postganglionic neurons (e.g., neuropeptide Y in neurons innervating the vas deferens) are unclear.

Secretion of Internal Reproductive Organs. Neural transmission to glandular tissues of the male internal reproductive organs is largely mediated by noradrenergic sympathetic neurons. However, nothing is known about the mechanisms of neuroglandular transmission.

114.2.3 Central Mechanisms of Reproductive Reflexes

The central mechanisms leading to the sequence of erection, emission and ejaculation are made up of several neural components. *Spinal reflex* mechanisms are discriminated from *supraspinal* mechanisms. Therefore, the central mechanisms, which mediate erection, are descriptively divided into (spinal) reflexogenic mechanisms and (supraspinal) "psychogenic" mechanisms. This division is somewhat artificial, although clinically useful, because both mechanisms always work together under biological conditions. Table 114.1 summarizes the central mechanisms and their peripheral components.

Erection. Mechanical stimulation of *afferent receptors* in the penis or in the tissue surrounding the penis elicits erection of the penis via a reflex pathway in the spinal cord (Fig. 114.4, right). The glans penis bears the greatest concentration of mechanoreceptors that are associated with erection; their afferents run in the dorsal nerve of the penis. That stimulation of afferents that project through the pelvic splanchnic nerve also contribute to this reflex is unlikely; after lesion of the dorsal nerve of the penis men are almost impotent. The adequate stimulus for the penile mechanoreceptors is provided by the sliding and massaging shear motion that accompanies copulation. An important factor in maintaining a state of excitation of the receptors in the glans during copulation is slipperiness of the surfaces of vagina and penis, which is produced reflexly by vaginal transudation (see p. 2260) and the activation of the bulbourethral glands in the man.

2257

Table 114.2. Erection and ejaculation in patients with lesions of the spinal cord. (Data from [2,24])

Lesion[a]	N	Erection	Ejaculation	Mechanism
Complete interruption rostral to T8 (T12)	167 (25)	93% 100%	8% (8%)	Reflexogenic
Complete destruction of sacral spinal cord	104	27%	10%	Pschogenic
Incomplete interruption rostral to T8	106	100%	28%	83% Reflexogenic 17% Psychogenic
Incomplete destruction of sacral spinal cord	10	100%	70%	Psychogenic

[a]Conditions were chronic in all patients. Complete or partial interruption was diagnosed by clinical examination and was confirmed by surgical exploration in some (figures in round brackets)

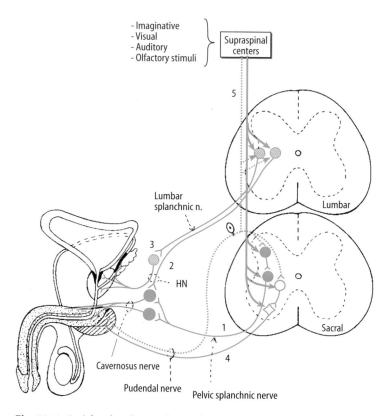

Fig. 114.6. Peripheral and central neural pathways controlling erection, emission and ejaculation. *1*, Efferent parasympathetic pathway to erectile tissue. *2*, Efferent sympathetic pathway to the erectile tissue; preganglionic sympathetic fibers and preganglionic parasympathetic fibers may synapse with the same postganglionic neurons in the pelvic splanchnic plexus which project to the erectile tissue. *3*, Efferent sympathetic pathway to internal reproductive organs [vas deferens, seminal vesicle, prostate, bladder neck (internal vesical sphincter)]; noradrenergic vasoconstrictor neurons to blood vessels of the reproductive organs that pass through lumbar splanchnic and hypogastric nerves as well as through the sympathetic chain and pelvic splanchnic and pudendal nerves are not shown. *4*, Motoaxons to bulbo- and ischiocavernous muscles. *5*, Ascending and descending spinal pathways. *HN*, hypogastric nerve

The *reflex pathway* in the *sacral spinal cord* is at least disynaptic and involves one interneuron (Fig. 114.4, right). It functions also in men with spinal cord transection as long as the interruption is rostral to the sacral segmental level and as long as spinal shock is over (Table 114.2, see below).

Mechanical stimulation of the penis also leads to strong sexual sensations, which induce *supraspinal (psychogenic)* enhancement of the spinal reflex mechanism (Fig. 114.6, pathway 1). These supraspinal "command signals," which can also be triggered by *imaginary, visual, auditory* and *olfactory stimuli*, can induce erection, apparently independently of the stimulation of the afferents from the penis. However, in physiological conditions psychogenically induced erection may always lead to excitation of some afferents from the penis (by its elongation and increase of circumference), and this in turn may enhance the descending command signal.

A *psychogenically* induced erection is produced via the parasympathetic innervation of the erectile tissue (pathway 1 in Fig. 114.6) and/or the thoraco-lumbar sympathetic innervation (pathway 2). Some patients in whom sacral spinal cord has been destroyed still can induce an erection psychogenically, and this must be generated by activation of sympathetic neurons of pathway 2 (Fig. 114.6; see below). It is unclear to what degree excitation of neurons of this sympathetic pathway contributes to erection in normal conditions. However, it is likely that the sympathetic outflow to the erectile tissue can duplicate the function of the parasympathetic outflow [18].

It is likely that the sympathetic preganglionic neurons of pathway 2 (Fig. 114.6) and the parasympathetic neurons of pathway 1 also converge on the same postganglionic neurons of the pelvic splanchnic plexus that project through the cavernosus nerve to the erectile tissue. Sympathetic neurons of pathway 2 must clearly be separated from sympathetic vasoconstrictor neurons, which can obviously override the dilatation of the erectile tissue when they are activated (Fig. 114.5). This may occur in various mental (e.g., during stress) and bodily conditions (e.g., during exercise).

Emission and Ejaculation. Emission and ejaculation are the high point of the male sexual act. As the afferents become highly excited during copulation, *sympathetic efferents* in the lower thoracic and upper lumbar spinal segments become activated. The afferents that trigger *emission* run in the pudendal and possibly the pelvic splanchnic nerve to the sacral spinal cord. The sympathetic preganglionic neurons (pathway 3 in Fig. 114.6). project to postganglionic neurons located in the superior hypogastric plexus and possibly the pelvic splanchnic plexus. They cause contractions of epididymis, vas deferens, seminal vesicles and prostate, which propel the semen into the posterior urethra. At the same time excitation of sympathetic fibers closes the internal vesical sphincter of the bladder neck to prevent reflux of the secretion into the urinary bladder. The postganglionic sympathetic neurons that lead to the contraction of the internal reproductive organs (pathway 3 in Fig. 114.6) are different from those that cause erection (pathway 2 in Fig. 114.6); whether the preganglionic neurons are also different for the two pathways is unclear, although it seems likely. Sympathetic pathways 2 and 3 both require powerful enhancement from supraspinal brain centers.

After emission *ejaculation* begins. It is probably triggered by excitation of afferents from the prostate and from the posterior urethra which run in the pelvic splanchnic nerve and possibly by excitation of other sacral afferents. Stimulation of these afferents during emission initiates a reflex, through the sacral spinal cord, that produces *tonic-clonic contractions* of the *bulbo-* and *ischiocavernous muscles*, which enclose the proximal erectile tissue, and of the musculature of the pelvic floor. These rhythmic contractions expel the secretions from the posterior urethra through the anterior urethra. Simultaneously, rhythmic contractions of the muscles of trunk and pelvic girdle occur.

During the ejaculation phase the excitation of the parasympathetic and sympathetic innervation of the reproductive organs becomes maximal. This maximal excitation is partly due to continuous afferent feedback from the skeletal musculature during the rhythmic contractions, but is also related to the strong central excitation. After ejaculation the activity in the parasympathetic and sympathetic (non-vasomotor) neurons declines and the blood flows out of the erectile tissue through the emissary veins, so that erection gradually subsides. This may be enhanced by high activity in vasoconstrictor neurons. The central neural mechanism of refractoriness (see Fig. 114.1) is unknown.

114.2.4 Neural Regulation of Reproductive Organs After Nerve Lesions and Spinal Cord Lesions

Investigations of the responses of the reproductive organs to natural peripheral and central stimulation in animals (dog, cat, rabbit, rat) after controlled experimental lesions of the spinal cord and of peripheral nerves [11,32,33] have yielded results comparable with observations recorded by clinicians in patients with similar lesions of the spinal cord and of peripheral nerves [2,8–10,15,16,24]:

- After complete transection of the spinal cord at the thoraco-cervical level erections can be elicited by mechanical stimulation of the glans penis, but emission and ejaculation rarely occur. The reflex erections are mediated by the sacral spinal cord.
- After destruction or removal of the sacral and caudal lumbar spinal cord erection and possibly emission (but not ejaculation) have been observed in animals during sexual behavior. Both are mediated by the lumbar sympathetic outflow to the reproductive organs (see Fig. 114.6). Erection cannot be elicited by mechanical stimulation of the penis in these animals.
- After additional destruction or removal of the upper lumbar spinal segments in these animals neither erection nor emission can be elicited by mechanical stimulation of the penis or from the brain ("psychogenic") during sexual behavior.
- These data are fully consistent with observations recorded in patients with spinal cord lesions (Table 114.2), showing that the basic central mechanisms, at least as far as erection is concerned, are similar in humans and in non-primate mammals.

114.3 Responses of the Reproductive Organs in the Female

The responses of female reproductive organs during the sexual response cycle have been studied by Masters and Johnson [28] and others. However, the roles of the parasympathetic and sympathetic innervation are still a matter of speculation. Possible functions of the different peripheral groups of nerve fibers and the possible central pathways are listed in Table 114.3 [1,18,30,37].

Table 114.3. Responses of the reproductive organs in females

Response	Afferents	Efferents	Central pathways	Effector organ (responses)
Vasocongestion Reflexogenic	Pudendal nerve	Parasympathetic, pelvic nerve	Sacral spinal cord	Dilatation of arteries in erectile tissue (clitoris); vasocongestion in labia, vaginal wall etc.; transudation
Psychogenic	Other sensory channels	Parasympathetic, pelvic nerve Sympathetic, thoracico-lumbar	Supraspinal origin	
Secretion of glands	Pudendal nerve	Parasympathetic, sacral	Sacral spinal cord	Glands of Bartholini
	Pudendal nerve	Sympathetic, thoracico-lumbar		Oviduct, uterus, cervix, vagina
Activation of smooth muscle	Pudendal nerve	Sympathetic, thoracico-lumbar	Sacro-lumbar spinal cord	Uterus, distal vagina (orgasmic platform)
Activation of striated muscle	Pudendal nerve	Somatic in pudendal nerve	Sacral spinal cord	Rhythmic contraction of pelvic floor muscles

114.3.1 External Reproductive Organs

Sexual stimulation causes reflex and/or psychogenic changes in the external reproductive organs of the woman. The labia majora, which normally touch one another in the midline to protect the labia minora and the openings of vagina and urethra, spread apart, become thinner, and move in an anterolateral direction. If the excitation is continued they become congested with venous blood. The labia minora become so filled with blood that they double or triple in thickness, pushing out between the labia majora and lengthening the vaginal channel. The swollen labia minora change color from pink to bright red. The color changes are so typical of a sexually excited woman that the labia minora have been called the "*sex skin.*" Glans and clitoris also swell, increasing in both length and thickness. As excitation increases, the clitoris is drawn against the margin of the symphysis.

Two mechanisms bring about these changes in the external reproductive organs during sexual excitation. The receptors in the genital organs, with axons running in the pudendal nerve to the sacral cord (S2–4), initiate reflexes (Fig. 114.7); on the other hand, the responses can be entirely *psychogenic*, produced by the brain. Because of its dense afferent innervation, the *clitoris* has a special role. Its mechanoreceptors are excited both by direct touch and indirectly – especially after retraction of the clitoris to the edge of the symphysis – by pull on the prepuce, by manipulation of the other external genitalia or by the thrusting of the penis. Stimulation of the afferents from the mons pubis, the vestibule of the vagina, the perineal region and, especially, the labia minora can have just as marked effects during sexual excitation as stimulation of the clitoral afferents. Excitation of clitoral and other mechanosensitive afferents is enhanced by the swelling of the organs. It is not

known whether visceral afferents from the internal reproductive organs, which run in the pelvic splanchnic nerve and with the sympathetic fibers, are also involved in excitation of these reflex events.

The enlargement of the external reproductive organs results from a general *vasocongestion*, probably produced by preganglionic *parasympathetic vasodilator neurons* in the sacral cord that send out axons in the pelvic nerves (Fig. 114.7). The erection of the clitoris, like that of the male penis, is produced by the engorgement of erectile tissue with blood. It is plausible that, by analogy with the findings in the man (see Table 114.3), sympathetic innervation from the thoraco-lumbar cord also contributes to the production of vasocongestion.

114.3.2 Internal Reproductive Organs

The internal reproductive organs of the woman undergo remarkable changes during the sexual response cycle. Within 10–30 s after afferent or psychogenic stimulation the *transudation of a mucoid fluid* through the squamous epithelium of the vagina begins. This transudate lubricates the vagina and is a prerequisite for the adequate stimulation of the penis during intercourse. The large vestibular glands (Bartholini's glands) play hardly any role in lubrication. Transudation results from a general venous congestion in the vaginal wall, probably under the influence of parasympathetic neurons from the sacral cord and sympathetic neurons from the thoracolumbar cord. The details of the transudation mechanism are not yet known.

Transudation is accompanied by a reflex expansion and elongation of the vagina. As excitation builds up, local vasocongestion in the outer third of the vagina forms the

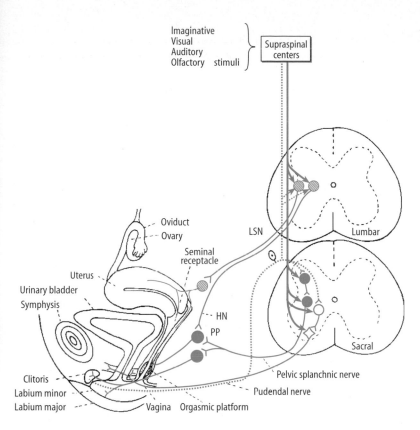

Imaginative
Visual
Auditory
Olfactory   stimuli

Supraspinal centers

LSN

Lumbar

Oviduct
Ovary
Seminal
receptacle

Uterus
Urinary bladder
Symphysis

HN
PP

Sacral

Clitoris
Labium minor
Labium major

Vagina   Orgasmic platform

Pelvic splanchnic nerve
Pudendal nerve

Fig. 114.7. Female reproductive organs and the peripheral and central neural pathways controlling their responses. Interneurons between afferent and efferent neurons in the spinal cord are not shown. For details and analogy see Fig. 114.6. *LSN*, Lumbar splanchnic nerves; *HN*, hypogastric nerve; *PP*, pelvic splanchnic plexus

orgasmic platform (Fig. 114.7). This thickening, together with the swollen labia minora, provides a long channel with the optimal anatomical characteristics for production of an orgasm in man and woman. During the orgasm the orgasmic platform contracts three to fifteen times, depending on the intensity of the orgasm. These contractions are probably *neuronally mediated* by the sympathetic system, and are comparable to the emission and ejaculation of the man. However, it is also possible that *hormones* from the pituitary gland are involved (see Chap. 19).

The *uterus* changes its position during sexual excitation, rising in the pelvis from its anteverted and anteflexed position so that at the height of excitation the cervix has moved away from the posterior wall of the vagina, leaving room in the inner third of the vagina for the reception of semen (seminal receptacle). At the same time the uterus enlarges by as much as 50%. The erection, elevation and enlargement of the uterus are brought about by *vasocongestion* in the true pelvis, probably assisted by the neuronally produced contractions of the smooth musculature in the ligaments supporting the uterus. During orgasm the uterus contracts regularly. These contractions begin at the fundus and pass over the body of the uterus to its lower segment. As with the vagina, the contractions of the uterus are probably mediated neuronally, by the sympathetic system and/or hormonally.

After orgasm the changes in the external and internal genitalia usually disappear. The outer cervix remains open for about 20–30 min, extending into the seminal receptacle.

Should orgasm fail to occur after intense excitation, the resolution phase proceeds more slowly (Fig. 114.1).

114.4 Extragenital Reactions During the Sexual Response Cycle

During the different phases of the sexual response cycle, and in particular during orgasm, the whole body reacts, and these reactions are dependent on the central nervous system. They consist of the responses of the reproductive organs produced by the autonomic nervous system, general autonomic reactions, and excitation of the central nervous system, which are usually intense and enhance sexual sensations, tending to exclude or restrict other sensory perceptions.

During the sexual response cycle a variety of extragenital reactions can be observed [28]. *Heart rate* and *blood pressure* increase with the degree of sexual arousal. Heart rate reaches a maximum around 100–180 per min, and blood pressure rises by about 20–40 mmHg (diastolic) and 30–100 mmHg (systolic). The *respiratory rate* increases to as much as 40 per min. The *external anal sphincter* contracts rhythmically in the orgasm phase. Because of vasocongestion the *breast* of the woman becomes larger, and its pattern of superficial veins more pronounced. The nipples are erect and the areolae swollen. These reactions of the breast can also occur in the man, but are far less conspicuous. In many women and some men a *sexflush of the skin* can be observed. It typically begins over the epigastrium in the late excitation phase, and as sexual arousal increases it spreads over the breasts, shoulders, abdomen and under some circumstances the entire body. The skeletal musculature contracts both voluntarily and involuntarily. Eventually there are nearly spastic contractions of the facial, abdominal and intercostal musculature. A common feature of orgasm is extensive loss of voluntary control of the skeletal musculature.

114.5 Conclusion

The sexual response cycle consists of a sequence of physiological changes in which afferent, autonomic, somatomotor and endocrine mechanisms are involved. It is initiated and maintained by afferent and/or central ("psychogenic") stimuli. In men, erection followed by emission and ejaculation are dependent on sacral and sacrolumbar reflex pathways and their descending controls by supraspinal centers. Erection is generated by activation of cholinergic postganglionic neurons, which dilate the arteries and sinusoids in the erectile tissue; the transmitters involved are acetylcholine, probably the neuropeptide vasoactive intestinal peptide, and probably nitric oxide. Contraction of male internal reproductive organs is generated by activation of noradrenergic neurons; the transmitters involved are noradrenaline and probably ATP. In women, similar homologous reactions of the reproductive organs occur, and similar peripheral and central neural mechanisms probably operate. The neurally regulated changes of the reproductive organs are disturbed after central and peripheral neural lesions.

References

1. Bell C (1972) Autonomic nervous control of reproduction: circulatory and other factors. Pharmacol Rev 24:657–713
2. Bors E, Comarr AE (1960) Neurological disturbances of sexual function with special reference to 529 patients with spinal cord injury. Urol Survey 10:191–222
3. Brock JA, Cunnane TC (1992) Electrophysiology of neuroeffector transmission in smooth muscle. In: Burnstock G, Hoyle CHV (eds) Autonomic neuroeffector mechanism. Harwood Academic, Chur, Switzerland, pp 121–213
4. Braus H, Elze C (eds) (1960) Anatomie des Menschen, vol 3: Periphere Leitungsbahnen II. Centrales Nervensystem. Sinnesorgane, 2nd edn. Springer, Berlin Göttingen Heidelberg
5. Burnett AL, Lowenstein J, Bredt DS, Chang TSK, Snyder SH (1992) Nitric oxide: a physiologic mediator of penile erection. Science 257:401–403
6. Burnstock G, Hoyle CHV (eds) (1992) Autonomic neuroeffector mechanisms. Harwood Academic, Chur, Switzerland
7. Carati CK, Creed LE, Keogh EJ (1987) Autonomic control of penile erection in the dog. J Physiol (Lond) 384:525–538
8. Chapelle PA, Durand J, Lacert P (1980) Penile erection following complete spinal cord injury in man. Br J Urol 52:216–219
9. Comarr AE (1971) Sexual concepts in traumatic cord and cauda equina lesions. J Urol 106:375–378
10. Comarr AE, Vigue M (1978) Sexual counseling among male and female patients with spinal cord and/or caude equina injury, part I,II. Am J Phys Med 51:107–122, 215–227
11. Dail WG (1993) Autonomic innervation of male reproductive genitalia. In: Maggi CA (ed) Nervous control of the urogenital system. Harwood Academic, Chur, Switzerland, pp 69–102
12. De Groat WC (1992) Neural control of the urinary bladder and sexual organs. In: Bannister R, Mathias CJ (eds) Autonomic failure, 3rd. edn. Oxford University Press, Oxford, pp 129–159
13. De Groat WC, Booth AM (1993) Neural control of penile erection. In: Maggi CA (ed) Nervous control of the urogenital system. Harwood Academic, Chur, Switzerland, pp 467–524
14. De Groat WC, Steers WD (1990) Autonomic regulation of urinary bladder and sex organs. In: Loewy AD, Spyer KM (eds) Central regulation of autonomic functions. Oxford University Press, Oxford, pp 310–333
15. Fitzpatrick WF (1974) Sexual function in the paraplegic patient. Arch Phys Med Rehabil 55:221–227
16. Guttmann L (1976) Spinal cord injuries, 2nd edn. Blackwell Scientific, Oxford
17. Jänig W, Koltzenburg M (1993) Pain arising from the urogenital tract. In: Maggi CA (ed) Nervous control of the urogenital system. Harwood Academic, Chur, Switzerland, pp 525–578
18. Jänig W, McLachlan EM (1987) Organization of lumbar spinal outflow to the distal colon and pelvic organs. Physiol Rev 67:1332–1404
19. Jünemann K-P, Lue TF, Schmidt RA, Tanagho EA (1988) Clinical significance of sacral and pudendal nerve anatomy. J Urol 139:74–80
20. Jünemann K-P, Persson-Jünemann C, Alken P (1990) Pathophysiology of erectile dysfunction. Semin Urol 8:80–93
21. Jünemann K-P, Persson-Jünemann C, Tanagho EA, Alken P (1989) Neurophysiology of penile erection. Urol Res 17:213–217
22. Keast JR, Booth AM, de Groat WC (1989) Distribution of neurons in the major pelvic ganglion of the rat which supply the bladder, colon or penis. Cell Tissue Res 256:105–112
23. Keast JR, de Groat WC (1989) Immunohistochemical characterization of pelvic neurons which project to the bladder, colon and penis in rats. J Comp Neurol 288:387–400
24. Kuhn RA (1950) Functional capacity of the isolated human spinal cord. Brain 73:1–51
25. Lue TF (1993) Physiology of erection and pathophysiology of impotence. In: Wash PC, Retik AB, Stamey TA, Vaughan ED Jr (eds) Campbell's urology. Saunders, Philadelphia, pp 707–728
26. Lue TF, Tanagho EA (1988) Funtional anatomy and mechanisms of penile erection. In: Tanagho EA, Lue TF, McClure RD (eds) Contemporary management of impotence and infertility. Williams and Wilkins, Baltimore, pp 39–50
27. Maggi CA (ed) (1993) Nervous control of the urogenital system, vol 2: the autonomic nervous system (ed by G Burnstock). Harwood Academic, Chur, Switzerland
28. Masters WH, Johnson VE (1966) Human sexual response. Little Brown, Boston
29. Morris JL, Gibbins IL (1992) Co-transmission and neuromodulation. In: Burnstock G, Hoyle CHV (eds) Autonomic neuroeffector mechanism. Harwood Academic, Chur, Switzerland, pp 33–119
30. Papka RE, Traurig HH (1993) Autonomic efferent and visceral sensory innervation of the female reproductive system: special reference to neurochemical markers in nerves and ganglionic connections. In: Maggi CA (ed) Nervous control of the urogenital system. Harwood Academic, Chur, Switzerland, pp 423–466
31. Pick J (1970) The autonomic nervous system. Lippincott, Philadelphia
32. Root WS, Bard P (1947) The mediation of feline erection through sympathetic pathways with some remarks on sexual

behavior after deafferentation of the genitalia. Am J Physiol 151:80–90

33. Semans JH, Langworthy OR (1938) Observations on the neurophysiology of sexual function in the male cat. J Urol 40:836–846

34. Steers WD (1990) Neural control of penile erection. Semin Urol 8:66–79

35. Stief CG, Benard F, Bosch RJLH, Aboseif SR, Numes L, Lue TF, Tanagho EA (1989) Acetylcholine as a possible neurotransmitter in penile erection. J Urol 141:1444–1448

36. Tanagho EA, Lue TF (1990) Physiology of penile erection. In: Chrisholm GD, Fair WR (eds) Scientific foundations of urology. Year Book Medical Publishers, Chicago, pp 420–426

37. Traurig HH, Papka RE (1993) Autonomic and visceral sensory innervation of the female reproductive system: special reference to the functional roles of nerves in reproductive organs. In: Maggi CA (ed) Nervous control of the urogenital system. Harwood Academic, Chur, Switzerland, pp 103–141

38. Wagner G, Uhrenholdt A (1980) Blood flow measurement by the clearance method in human corpus cavernosum in the flaccid and erect states. In: Zorgniotti AW, Rossi G (eds) Vasculogenic impotence. Proceedings of the 1st international conference on corpus cavernosum revascularization. Thomas, Springfield, pp 41–46

115 Fertilization

B. Dale

Contents

115.1 Introduction

Fertilization marks the creation of a new and unique individual. It ensures immortality by transferring genetic information from one generation to the next and, by creating variation, allows evolutionary forces to operate. In addition to delivering the paternal genome, the spermatozoon triggers the quiescent female gamete into metabolic activity, essential to sustain early embryogenesis. Many texts portray fertilization as a process of activation and penetration of a large cell by a small cell. On the contrary, fertilization is a highly specialized example of cell-to-cell interaction, where each gamete activates its partner. Thus, in order to trigger metabolic activation of the oocyte, the spermatozoon itself must encounter and respond to signals originating from the oocyte and its investments. Sperm-oocyte interaction is a complex multistep process that starts with the specific recognition of complementary receptors on the surfaces of the two gametes and terminates with syngamy, the union of the maternal and paternal chromosomes [13]. The crucial event is fusion of the plasma membranes of the two cells.

Both activation of the spermatozoon and activation of the oocyte are regulated by changes in intracellular ions and other low molecular weight factors, with Ca^{2+}, H^+, cyclic adenosine monophosphate (cAMP), and inositol-1,4,5-trisphosphate (IP_3) being the major second messengers involved. Intracellular Ca^{2+} may increase both by release

from intracellular stores and by flux through voltage-gated and chemical-gated channels in the plasma membrane [20,49]. Alberto Monroy in 1956 [64] was one of the first to recognize *the importance of ion fluxes through the plasma membrane in the process of oocyte activation*, demonstrating a change in K^+ conductance in the starfish oocyte plasma membrane. In addition to the role of transmembrane voltage in the activation of both the oocyte and the spermatozoon, electrical recording from the former has made it possible to correlate the structural and physiological events of fertilization. There are two current hypotheses as to how a spermatozoon triggers the oocyte into metabolic activity. One school of thought points to a soluble factor in the spermatozoon that is released into the oocyte following gamete fusion. The contrasting idea is that prior to fusion the interaction between the spermatozoon and a receptor on the oocyte surface is transduced to the oocyte interior via G-proteins.

Although gametes from a particular batch appear to be homogeneous, they are in fact an extremely heterogeneous population of cells. Physiological parameters, ranging from the number of ion channels expressed in the plasma membrane, to the amount of Ca^{2+} released into the cytosol during activation, may vary 10-fold from cell to cell. Regarding viability it has been estimated in the sea urchin, for example, that only 2% of spermatozoa are actually capable of fertilization. While in vitro techniques have made it possible for us to study the fertilization process in many animals, animal studies have also given rise to many misleading concepts. To date our knowledge on human gametes is scant and therefore we have to resort to information from animal models. This is feasible at the physiological level, where we should be trying to identify unifying concepts, but care must be taken when extrapolating data at the molecular level.

115.2 Spermatogenesis
(see also Chap. 113)

At puberty the interphase male germ cells start to proliferate by mitosis. This is followed by meiosis and a gradual reorganization of cellular components, characterized by a loss of cytoplasm. In the adult mammal it has been estimated that about 500 spermatozoa per second are produced per gram of testis. The stem cells, or *A0*

R. Greger/U. Windhorst (Eds.)
Comprehensive Human Physiology, Vol. 2
© Springer-Verlag Berlin Heidelberg 1996

spermatogonia, are located in the basal intratubular compartment. At intervals *A1 spermatogonia* emerge from this population and undergo a fixed number of mitotic divisions to form a clone of daughter cells. There is some evidence that one of the daughter clones serves as a second source of stem cells. After the final mitotic division the *primary spermatocytes* move into the adluminal compartment and enter into meiosis. Here they undergo the two meiotic divisions to form, first, two daughter *secondary spermatocytes*, and eventually *four early spermatids*. Although spermatid nuclei contain haploid sets of chromosomes, the autosomes continue to synthesize low levels of ribosomal and messenger RNA and proteins. The spermatid DNA now becomes highly condensed and is eventually packed with protamines. Cytoplasmic re-organization gives rise to the tail, the midpiece containing the mitochondria, the *acrosome*, and the residual body that casts off excess cytoplasm. Sperm modelling is probably regulated by the *Sertoli cells*. As spermatogenesis proceeds the cells are moved to the center of the lumen. In man spermatogenesis is complete in 64 days. The rate of progression of cells through spermatogenesis is constant and unaffected by external factors such as hormones.

115.3 Oogenesis (see also Chap. 113)

In the female the mitotic phase of proliferation terminates before birth and all oogonia enter their first meiotic division, becoming *primary oocytes*. At puberty a total of about 200 000 germ cells are available for the reproductive lifespan. During the first meiotic prophase the oocytes are surrounded by mesenchymal ovarian cells to become *primordial follicles*. The oocytes arrest in diplotene with the characteristic large nucleus called the *germinal vesicle*. It is not known how these arrested oocytes remain viable for up to 50 years. Recruitment of some of these primordial follicles first occurs at puberty. The follicle grows from 20 µm to several hundred microns, with the oocyte itself growing from 10 µm to about 100 µm. The growth phase essentially involves synthesis and storage of large amounts of RNA, proteins, and metabolic substrates. Whilst growing, the surrounding granulosa cells divide mitotically, and a glycoprotein coat, the *zona pellucida* synthesized by the oocyte, is secreted between the oocyte and these cells. Gap junctions allow transfer of substrates and developmental information between the oocyte and cytoplasmic projections of the accessory cells that penetrate the zona pellucida. The small number of follicles that have completed their growth phase are called *preantral follicles*. Many undergo atresia. Circulating gonadotrophins convert the preantral follicles to *antral follicles* or *Graafian follicles*. The preantral phase is 8–12 days in women. Fluid now accumulates in the follicle and suspends the oocyte with surrounding cumulus cells. During this phase of growth the follicle produces an increased amount of androgens and oestrogens. A subsequent surge of LH causes a rapid further accumulation of fluid. The follicle is now in the *preovulatory phase*, lasting about 36 h and resulting a 25 mm diameter follicle in the human, which is then ovulated. Fimbria of the oviduct are thought to sweep the oocyte into the ampulla, where fertilization will occur. In this last stage of oogenesis the nuclear membrane breaks down, meiosis is reinitiated, and the first polar body is emitted. The female cell is now in the second metaphase. Concomitantly, a process of cytoplasmic maturation starts, which includes a decrease in K^+ conductance, a depolarization of the plasma membrane, and migration of cortical granules to the surface of the oocyte.

115.4 Activation of the Spermatozoon

115.4.1 Epididymal Maturation

Mammalian spermatozoa leaving the testis are not capable of fertilizing oocytes. They gain this ability while passing down the epididymis – a process called epididymal maturation. Testicular spermatozoa are essentially motionless, even when washed and placed in a physiological solution. The ability to move is probably regulated at the level of the plasma membrane, since demembranation and exposure to ATP, cAMP, and Mg^{2+} triggers movement [46]. Transfer of a forward motility protein [1] and carnitin [9] from the epididymal fluid is believed to be important for the development of sperm motility. Since the osmolality [12] and chemical composition [7] of the epididymal fluid varies from one segment to the next, it is probable that the sperm plasma membrane is altered stepwise as it progresses down the duct. Next, the spermatozoon head acquires the ability to adhere to the zona pellucida [50,52] and there is an increase in net negative charge [47]. During maturation the spermatozoa use up endogenous reserves of metabolic substrates, becoming dependent on exogenous sources such as fructose, and shed their cytoplasmic droplet. When the sperm leaves the testis it is coated with several macromolecules which are either lost, altered, or added to during passage through the epididymis [4]. The most prominent molecules are glycoproteins mediated by galactotransferase and sialo-transferase. Changes in lectin-binding ability of the sperm plasma membrane during epididymal maturation indicate alterations to the terminal saccharide residues of these glycoproteins. Membrane lipids also undergo changes in their physical and chemical composition [29].

115.4.2 Capacitation

Before gaining the ability to fertilize oocytes, ejaculated mammalian spermatozoa must reside for a minimum period in the female reproductive tract. This relatively undefined process is thought to involve the removal of glycoproteins from the sperm surface, exposing receptor sites that can respond to oocyte signals and lead to the acrosome reaction. Since epididymal maturation and capacitation are unique to the mammals, they may be an evolutionary adaptation to internal fertilization. In the human, capacitation probably starts while the spermatozoa are passing through the cervix. Many enzymes and factors from the female tract have been implicated in causing capacitation, such as arylsulfatase, fucosidase, and taurine; however, to date the mechanism is unknown. Certainly the factors are not species-specific, and capacitation may be induced in vitro in the absence of female signals. In fact, golden hamster spermatozoa have been capacitated in vitro [72]. Capacitation is temperature-dependent and only occurs at 37–39°C [41]. As mentioned in Sect. 115.1, spermatozoa are extremely heterogenous and, not surprisingly, individual cells capacitate at different rates [71]. There is evidence that sperm adenylate cyclase increases its activity during capacitation, probably increasing the availability and turnover of cAMP, which in turn may stimulate protein kinase, perhaps resulting in changes in ion permeability of the plasma membrane [31].

There are contradictory reports as to whether or not capacitation leads to an increased metabolism in spermatozoa [6]. To date, there is no conclusive evidence that intracellular concentrations of K^+, Na^+, or Ca^{2+} change during capacition. Sperm surface components are, however, removed or altered during capacitation [37]. For example, an antigen on the plasma membrane of the mouse spermatozoon laid down during epididymal maturation cannot be removed by repeated washing, but disappears, or is masked, during capacitation. A second mouse sperm antigen, not present on cauda epididymal spermatozoa, becomes detectable after capacitation. Ram spermatozoa release three glycosylated proteins of 24 kDa, 41 kDa, and 65 kDa in the uterine environment and adsorb a 16-kDa protein from the uterine fluid [65]. Later in the oviduct they release further 97 kD and 41 kD proteins from their surfaces. Freeze fracture has revealed intramembranous particles (IMP – intrinsic proteins within the lipid bilayers) that are evenly distributed in epididymal spermatozoa, whereas IMP free zones appear in the plasma membrane overlying the acrosome after capacitation [62]. Finally, there is evidence that phospholipid composition changes during capacitation.

115.4.3 The Acrosome Reaction

The acrosome is a membrane-bound cap covering the anterior portion of the sperm head and is found in the majority of species. The acrosome, or surrounding membranes, contain a large array of hydrolytic enzymes including *hyaluronidase*, *acrosin*, proacrosin, phosphatase arylsulfatase, collagenase, phospholipase C, and β-galactosidase, to mention a few. The first two have been studied most extensively. In mammalian eggs, a major component of the cumulus matrix is hyaluronic acid. Although it is possible in some mammals to see the acrosome under the light microscope, this is not possible in the human, where staining is required, using for example the triple stain method [60], the *Pisum sativum* fluorescent

lectin method [10], or by using a variety of monoclonal antibodies.

The acrosome reaction involves fusion of the outer acrosomal membrane with the overlying plasma membrane, allowing the acrosomal contents to be released (Fig. 115.1). In the human, fusion appears to first take place near the border of the acrosomal cap region and the equatorial segment of the acrosome, where it appears the membranes are intrinsically less stable [48]. The outer acrosomal membrane in this region appears to be rich in Ca^{2+}-binding sites. The correct physiological sequence of events in the human is as yet unclear, although current in vitro fertilization techniques now make such a study feasible. The acrosome reaction is relatively rapid once the correct trigger signals have been received and may take from 2 to 15 min in vitro.

In invertebrates the trigger signal is a fucose-rich fraction of the outer jelly coat [17], the equivalent structure to the cumulus cells in the mammal. Gametes collected from the ampullae of mammals after mating show that while free-swimming spermatozoa have unreacted acrosomes, those within the cumulus mass have acrosomes that either have reacted or are in the process of reacting. Most spermatozoa attached to the surface of the zona pellucida surface have reacted acrosomes [3]. One of the glycoproteins of the mouse zona pellucida, called ZP3, binds to the plasma membrane over the acrosomal cap and induces the acrosome reaction [66]. It is not clear whether the acrosome reaction is started whilst the spermatozoon is interacting with the cumulus mass. However, since a major component of the cumulus matrix is hyaluronic acid and

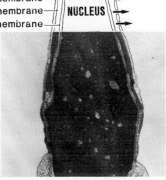

Plasma membrane —
Outer acrosomal membrane — **NUCLEUS**
Inner acrosomal membrane —

Fig. 115.1. Electron micrograph of a human spermatozoon, showing the relationship between the acrosomal membranes and the plasma membrane

the acrosome contains hyaluronidase, it is plausible that the reaction starts in the cumulus.

Whatever the natural signal for induction, the acrosome reaction may also be induced in vitro in the absence of any maternal signal. Prior to the acrosome reaction spermatozoa from many species, including the human, show an altered motility pattern called hyperactivation [70]. From an initial progressive linear movement, spermatozoa start to show an erratic movement pattern with whiplash beating of the tail. In the rabbit hyperactivity appears to start when the spermatozoa are released from the isthmus. Although the mechanism regulating hyperactivity is unknown extracellular Ca^{2+}, K^+, and albumin are thought to be involved.

Mechanism of the Acrosome Reaction. Although there is accumulating data on the regulation of the acrosome reaction in mammals, much of our early knowledge is from studies on the spermatozoa of the sea urchin. Essentially, a component of the egg jelly interacts with the sperm plasma membrane, inducing an increase in intracellular pH and a Ca^{2+}-dependent depolarization [54]. The change in pH, caused by an efflux of H^+ and an influx of Na^+, is, on the contrary, Ca^{2+}-independent. Both the increase in pH and the membrane depolarization lead to the activation of Ca^{2+}-specific ion channels in the plasma membrane, leading to a massive influx of Ca^{2+}. This high level of intracellular Ca^{2+} then induces the fusion of the outer acrosomal membrane with the overlying plasma membrane and the liberation of the acrosomal contents.

In mammalian spermatozoa ZP3 may be considered a regulatory ligand that triggers the acrosome reaction. G-proteins are found in the plasma membrane and outer acrosomal membranes. Although we are far from understanding the sequence of events leading to exocytosis, potential second messenger pathways involved include adenylate cyclase generating cAMP, phospholipase C, generating IP3 and diacylglycerol (DAG), phospholipase D generating phosphatidic acid, and phospholipase A2 generating arachidonic acid. Undergoing the acrosome reaction does not necessarily ensure success in in vitro fertilization. In a population of spermatozoa surrounding the cumulus mass we may expect enormous variability: some will acrosome react too soon, others too late; in some the trigger stimulus will be inadequate; perhaps in others the transduction mechanism will fail at some point. In vivo, the arrival of few sperm to the oocyte mass in the ampulla probably indicates a selection process of sorts.

115.5 Sperm–Oocyte Interaction

115.5.1 Interaction with the Cumulus Mass

The cumulus mass is composed of both cellular and acellular components (Fig. 115.2). The acellular matrix is made up of carbohydrates, including hyaluronic acid, and

proteins. In vivo very few spermatozoa reach the site of fertilization, so the notion arrived at by studying in vitro fertilization, that large populations of spermatozoa surrounding the oocyte mass dissolve the cumulus matrix, is probably incorrect. Fertilization occurs before the dispersion of the cumulus mass, and in fact in vivo the sperm:oocyte ratio is close to 1:1. Capacitation is a prerequisite to movement of the spermatozoon through the cumulus mass. Spermatozoa that have completed their acrosome reaction prior to reaching this layer are not able to penetrate [11,58]. It is possible that membrane-bound hyaluronidase, not of acrosomal origin, aids in the penetration of the cumulus mass [59,61]. An interesting hypothesis, based on the observation that heterologous sperm can penetrate cumulus masses, is that in a homologous population of spermatozoa some are neutral; that is, they are not recognized by the cumulus signals. In addition to differences in individual spermatozoa, there may also be preferential pathways in the cumulus mass that facilitate

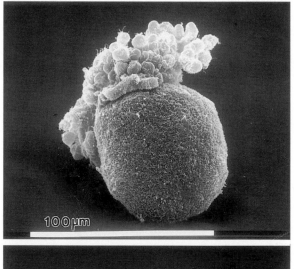

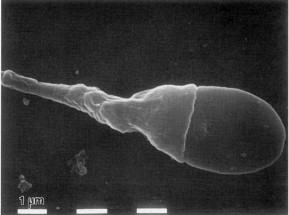

Fig. 115.2. *Top:* Scanning electron micrograph of an unfertilized metaphase II human oocyte, with the cumulus cells and zona pellucida partially dissected to show the homogeneity of the plasma membrane. *Bottom:* Scanning electron micrograph of a human spermatozoon, showing the boundary of the acrosomal cap

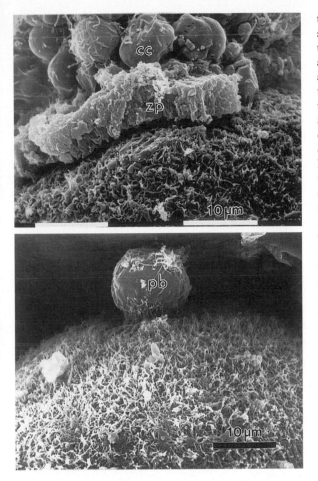

Fig. 115.3. *Top:* Scanning electron micrograph of the oocyte in Fig. 115.2 at a higher magnification, showing the complexity of the cumulus mass (*cc*) and the lattice structure of the zona pellucida (*zp*). *Bottom:* Scanning electron micrograph of an unfertilized human oocyte, showing the first polar body (*pb*) and typical microvilli of the plasma membrane

sperm penetration. Synchrony of the various events mentioned above is obviously a prerequisite for successful fertilization [13].

115.5.2 Interaction with the Zona Pellucida

The zona pellucida is a glycoprotein sheet several microns thick (Fig. 115.3). This acellular component, secreted by the growing oocyte, is called the *vitelline membrane* in echinoderms and amphibians and the *chorion* in ascidians, fish, and insects. In mammals the zona is 70% protein, 20% hexose, 3% sialic acid, and 2% sulfate. Under the electron microscope the outer surface has a latticed appearance. There are several major glycoproteins, called ZP1, ZP2, etc. The distribution of these components is unknown; however, in the mouse ZP2 is known to be distributed throughout the thickness of the zona. Spermatozoa penetrate the zona in 2–15 min. Sperm binding to the mouse zona is mediated by the ZP3 glycoprotein [67]. The exact nature of

the complementary receptor molecule on the surface of the spermatozoon is not known and in fact may either be protein or glycoprotein. In the mouse, the sperm receptors appear to be on the surface of the plasma membrane of the acrosome-intact spermatozoon. Receptors for ZP2, in contrast, are located on the inner acrosomal membrane and therefore are unmasked after the acrosome reaction. Sperm-receptor activity probably resides in the *O*-linked oligosaccharides of ZP3 [67]. The complementary molecule of the spermatozoon may be a lectin-like protein. It has been suggested, both in invertebrates and mammals, that sperm binding to its receptor on the zona is in fact an enzyme-substrate interaction. During penetration of the zona the spermatozoon loses its acrosomal contents and only the inner acrosomal membrane is in direct contact with the zona. While passing through the zona the spermatozoon beats its tail strongly, leaving a sharp pathway behind. It is interesting that in teleosts, where the oocyte coat has a preformed entry site, the spermatozoa do not have acrosomes [17]. A combination of mechanical and enzymatic mechanisms are probably involved in penetration.

115.5.3 Sperm–Oocyte Fusion

Whereas in many animals and marsupials a fundamental result of the acrosome reaction is the exposure of the highly fusible inner acrosomal membrane, this does not occur in eutherian mammals. Here the sperm-head plasma membrane of the postacrosomal region apparently fuses with the oocyte plasma membrane. This region of the plasma membrane, however, only attains fusibility after the acrosome reaction. The surface of most oocytes, including the human, is organized into evenly spaced short microvilli (Fig. 115.2). In mouse and hamster, the area overlying the metaphase spindle is microvillus-free and spermatozoa are incapable or less likely to fuse with this area [40,43,53]. The human oocyte is an exception, having no obvious surface polarity; microvilli are present all over the surface from the animal pole to the vegetal pole (Fig. 115.3) [23]. Following gamete fusion the sperm plasma membrane remains in the oocyte plasma membrane and indicates the point of fusion. In the rat, a fluorescently labelled conjugated antibody to a sperm plasma membrane antigen shows that, immediately after fusion, the sperm plasma membrane remains localized to the point of entry, and by the pronuclei stage the antigen spreads all over the surface of the zygote [26]. Sperm motility, although necessary for penetration of the zona, is not required for gamete fusion. Fusion is temperature-, pH-, and Ca^{2+}-dependent. Since fusion of spermatozoa to nude oocytes is not inhibited in the presence of monosaccharides or lectins, it appears that the terminal saccharides of glycoproteins are not directly involved in the process. Fusion of artificial membranes made of pure phospholipids is certainly possible; however, in biological systems it appears to be mediated or facilitated by membrane-associated proteins. In the guinea pig, an integral membrane protein on the posterior portion of the sperm head composed of two distinct subunits regulates fusion.

115.6 Activation of the Oocyte

115.6.1 Mechanism of Activation

There are two current schools of thought as to how the spermatozoon triggers the egg into metabolic activation [20].

The Receptor Hypothesis. In many cell systems guanosine triphosphate (GTP)-binding proteins couple plasma membrane receptors to enzymes that generate intracellular second messengers. Microinjection of guanosine triphosphate (GTP) analogues into sea urchin oocytes activates them, while microinjection of GDP prevents subsequent activation when challenged by spermatozoa. These experiments show that the sea urchin oocyte plasma membrane contains a GTP-binding protein; however, it may be involved in later events such as cortical granule exocytosis rather than in triggering activation [68].

The Soluble Sperm Factor Hypothesis. It has been suggested that the spermatozoon contains a soluble factor that activates oocytes [51]. Detailed analysis of the electrophysiological response of sea urchin oocytes at activation added further support to this idea [21], and subsequently it was shown that extracts of sea urchin spermatozoa micro-injected into oocytes triggered activation [22]. These microinjection experiments were recently confirmed on mammalian gametes [56,57]; however, the exact nature of the activating factor is unknown. Suggestions include Ca^{2+} [35], IP_3 [34], and a high molecular weight protein [22,56,57].

115.6.2 Ionic Events

The first event of activation in most oocytes is a depolarization of the plasma membrane [13]. One notable exception is the hamster oocyte that appears to undergo a series of hyperpolarizations [33]. Close observation of the electrical response in sea urchins showed it to be biphasic, with the larger fertilization potential being preceded by a small steplike depolarization [21]. Later studies showed that only successful spermatozoa gave rise to this initial event [24] and that it also occurred in ascidian and amphibian oocytes [13]. Since cortical granule exocytosis is not initiated until about the time of the second, larger depolarization, the fertilization potential – the period between the two events – corresponds approximately to what has been called the *latent period*. Single channel recordings, using the *patch clamp technique*, showed that the channels underlying the fertilization current had a conductance of 400 pS and were nonspecific for ions [18]. By measuring the total cell conductance at fertilization and the single channel conductance, and knowing the probability that a channel is in the open state, it was estimated that about 1000 fertilization channels were activated by the spermatozoon close to the site of fusion [25]. Later work showed that the channels were not gated by intracellular Ca^{2+} but possibly by IP_3 [15,16]. In conclusion, in invertebrate oocytes the spermatozoon induces an inward current in the oocyte plasma membrane by activating nonspecific ion channels shortly following gamete fusion. The situation in the human is quite different. Here the spermatozoon induces an outward current in the oocyte plasma membrane by activating potassium channels (Fig. 115.4, [27]). A second difference was found in the mechanism of fertilization channel gating. In the ascidian, the channel is not calcium gated, while in the human the potassium channel is calcium gated (Fig. 115.5).

When Does Sperm–Oocyte Fusion Occur? Owing to the rapid succession of change in the oocyte during activation it is difficult to distinguish individual events. Using immature germinal vesicle stage sea urchin oocytes, it was shown that the preliminary step event was the result of fusion [19]. This was later confirmed by a second research group who utilized capacitance measurements to indicate confluity of the two plasma membranes [44]. Since fusion precedes activation, this gives support to the soluble sperm-borne factor hypothesis [14]. Further support may be drawn from interphylum cross fertilization where activation will only occur in the presence of fusogens [39].

Shortly after the fertilization potential in the sea urchin and the ascidian, there is a massive release of Ca^{2+} from intracelluar stores that causes an increase in cytosolic Ca^{2+} from $0.1 \mu M$ to $10 \mu M$. Within minutes the Ca^{2+} returns to resting levels. The elevated Ca^{2+} starts at the point of sperm-oocyte fusion and traverses the oocyte to the antipode in about 10 (Fig. 115.4). Smaller repetitive waves continue for many minutes after fertilization. The mechanism for wave propagation is not clear and could be Ca^{2+}-induced Ca^{2+} release, as in muscle, or IP_3-induced Ca^{2+} release. This is followed by a wave of cortical exocytosis in regulative eggs or a surface contraction in ascidian oocytes that travel in the same direction as the calcium wave. In the hamster the Ca^{2+} wave also starts near the site of sperm attachment and spreads to the antipode within 4–7 s, ending some 15–20 s later [45]. These Ca^{2+} waves are repetitive, occurring with an interval of 3 min, and may be induced by micro-injecting soluble sperm factors into oocytes [57]. Repetitive calcium spikes are also seem in the human oocyte during fertilization, or following micro-injection of sperm factors, or whole spermatozoa (ICSI, [30,63]). In sea urchins and amphibians the spermatozoon triggers two ionic signals. The second is an increase in intracellular pH. In the sea urchin the second product of phosphoinositide hydrolysis, diacylglycerol, leads to activation of the Na^+/H^+ antiporter by activating protein kinase C, and this causes an alkalinization of the cytoplasm. This increase in pH is necessary for several of the later events of activation.

Ascidian Fertilization Current

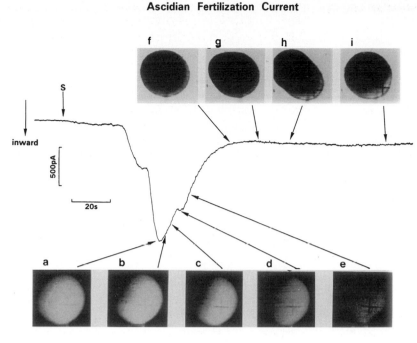

Fig. 115.4. Fertilization currents in an invertebrate (*top*) and human (*bottom*). Note that the ascidian current is inward, while the current in the human is outward. The surface contraction in the ascidian oocyte at activation is shown in f to i, while a′, b′, and c′ show the human oocyte before, during, and following recording. The two pro-nuclei in the human zygote are shown in c′, and a–e show the wave of intracellular calcium in the ascidian oocyte during activation. The Ca²⁺ is measured using the fluorescent indicator Fura-2. The wave of high calcium starts at the top left of the oocyte as a dark patch and progresses to the antipode

Human Fertilization Current

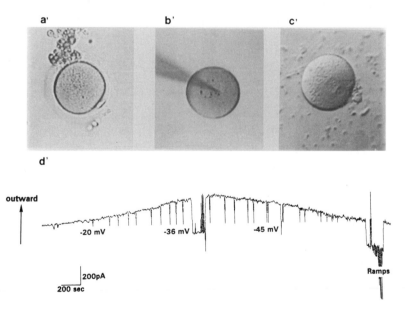

115.6.3 Structural Events

The first indication of activation in regulative oocytes is the exocytosis of cortical granules. *Cortical granules* are small spherical membrane-bound organelles, containing enzymes and mucopolysaccharides, that originate as vesicles in the Golgi complex. In the sea urchin oocyte there are about 20 000 cortical granules that migrate to the surface during cytoplasmic maturation. At fertilization, a number of cortical granules nearest to the point of sperm-oocyte fusion break down first and then a wave of exocytosis traverses the oocyte to the antipode. This wave of exocytosis leads to a modification of the vitelline membrane both by the action of enzymes released into the perivitelline space and the mucopolysaccharides that induce an influx of water into the space, causing it to expand. In mammals, although there is a Ca²⁺ wave, no evidence exists so far for a wave of exocytosis. The cortical reaction

A23187 Induced Activation Currents

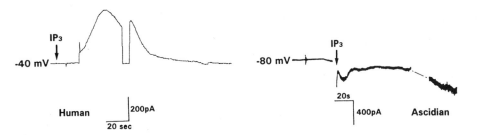

EGTA pre-loaded Oocyte + A23187 or Spermatozoa

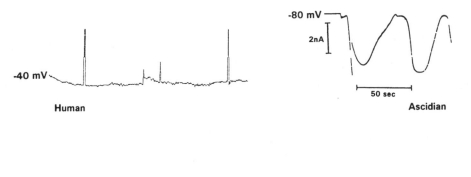

InsP₃ Induced Activation Currents

Fig. 115.5. A series of activation currents in human and ascidian oocytes showing that the channels generating the outward current in the human are calcium gated, while the channels generating the inward current in the ascidian are not calcium gated. The *top* two traces show that the calcium ionophore A23187 generates an outward current in the human but not in the ascidian. The *middle* two traces show that micro-injection of the calcium buffer EGTA into a human oocyte inhibits the outward current, but not the inward current in the ascidian, while both currents may be triggered following the micro-injection of the second messenger IP₃

in the mammalian oocyte also changes the characteristics of the zona pellucida, and this is called the *zona reaction* [2,69]. A second result of the cortical reaction is that the oocyte plasma membrane now becomes a mosaic of cortical granule membrane and the original plasma membrane. In the human zygote, or early blastomeres, there is no evidence of polarization of the plasma membrane [23].

Why Does Only One Spermatozoon Succeed? It has long been the contention that the fertilizing spermatozoon, by triggering events such as the cortical reaction, not only activates the oocyte but at the same time prevents the interaction of supernumerary spermatozoa. Certainly the alteration to the zona pellucida in the mouse caused by proteinases or glycosidases hydrolyzing ZP3 receptors, prevents them from further interaction [66] – this is called the *zona block* – while the mosaic zygote plasma membrane is also refractory to sperm in many species [2,69], this being called the *vitelline block*. The appearance of supernumerary spermatozoa in the perivitelline space has been taken to indicate a weak zona reaction and a strong vitelline reaction, for example, in the rabbit. In contrast, because perivitelline sperm are rare in rat, mouse, sheep, and human oocytes, these species were considered to have a strong zona reaction. If we accept the premises that all spermatozoa that reach the oocyte are equally capable of fertilization, and that all areas of the oocyte surface and its extracellular coats are capable of activating spermatozoa, then the cortical reaction is too slow to be considered a block to polyspermy. The cortical reaction probably serves to chemically alter the zona pellucida or vitelline membrane in order to provide a bacterially safe environment for the developing embryo. That only one spermatozoon normally enters the oocyte in in vitro conditions suggests a heterogeneous sperm population together with restricted areas on the oocyte surface – hot spots – available for gamete fusion [14]. In support of the idea that the cortical reaction has not evolved as a polyspermy-preventing mechanism is the fact that in vivo sperm-oocyte ratios are often 1:1. When polyspermy occurs naturally it is often dispermy, with the two spermatozoa entering at opposite poles of the oocyte. Immature and aged oocytes are also usually polyspermic. Temperature shock and a variety of chemical agents that change intracellular pH will induce polyspermy, probably by alterations to the cytoskeleton. Cortical granule exocytosis is not essential for development of oocytes, and in mammals premature release of a number of granules may occur before fertilization.

Pronucleus Formation. During spermatogenesis the sperm nucleus is packed with distinct histones or protamines [8]. The association of the DNA with these highly charged basic amino acids is thought to cause condensation and the repression of NDA activity. The rigidity of the mammalian sperm head, necessary for penetration of the zona, is due to the extensive S-S linkages in these protamines. Cross-linking in the human spermatozoon is regulated by Zn^{2+} from the prostate gland [5,32,38]. On the sperm's entering the oocyte cytoplasm the nuclear envelope breaks down, the protamines are lost, and pronuclear decondensation occurs. A reduced form of glutathione may be responsible for sperm nucleus decondensation [42]. Sperm nucleus decondensing factors seem to be non-species-specific, since human sperm may decondense when microinjected into amphibian oocytes [28]. The next step is the formation of new nuclear membrane around the decondensed male and female chromatin to produce the pronuclei. During pronuclear development DNA synthesis and RNA transcription occur. Sperm pronucleus development factors are found in limited quantities within the cytoplasm; for example, in the hamster a maximum of five pronuclei will decondense at any time. Again the factors are not species-specific, as human spermatozoa can develop into normal pronuclei in hamster oocytes and form a normal chromosome complement [36,71]. The migration of the male and female pronuclei to the center of the oocyte has been studied extensively, particularly in sea urchin and mouse. Fluorescein-conjugated probes for cytoskeletal elements have shown that in the mouse there is a thickened area of microfilaments below the cortex of the polar body region. In addition to the spindle microtubules, there are 16 cytoplasmic microtubule organizing centers or foci. Each centrosomal focus organizes an aster. Shortly before the nuclear envelopes disintegrate, the foci condense on

the surface and the first cleavage ensues. Although it appears that the centrosomes are inherited maternally in the mouse, in the majority of animals, and in the human, it is borne by the spermatozoon [55].

115.7 Conclusions

This chapter has reviewed the processes, as currently known, that take place during activation of oocytes by spermatozoa. As will have become apparent, many of the mechanisms are not yet well understood, but these are paradigms for cellular signalling and communication and are likely to make use of mechanisms used more generally in cell-cell interactions. Study of these processes may therefore be expected to reveal not only details that are bound to be of significance for clinical applications, but also more general principles.

References

1. Acott TS, Hoskins DD (1981) Bovine sperm forward motility protein: binding to epididymal spermatozoa. Biol Reprod 24:234–240
2. Austin CR, Braden WWH (1956) Early reaction of the rodent egg to spermatozoa penetration. J Exp Biol 33:358–365
3. Austin CR, Bishop MWH (1958) Some features of the acrosome and perforatorium in mamalian spermatozoa. Proc R Soc Lond [Biol] 149:234–240
4. Bellve AR, O'Brien DA (1983) The mammalian spermatozoa: structure and temporal assembly. In: Hartman JF (ed) Mechanism and control of animal fertilization. Academic, New York, pp 55–137
5. Björndahl L, Kjellberg S, Roomans GM, Kvist U (1986) The human sperm nucleus takes up zinc at ejaculation. Int J Androl 9:77–80
6. Boell EJ (1985) Oxygen cosumption of mouse sperm and its relationship to capacitation. J Exp Zool 234:105–116
7. Brooks DE (1979) Biochemical environment of sperm maturation. In: Fawcett DW, Bedford JM (eds) The spermatozoa. Urban and Schwarzenberg, Baltimore, pp 23–34
8. Calvin HI (1976) Comparative analysis of the nuclear basic proteins in rat, human, guinea pig, mouse and rabbit spermatozoa. Biochim Biophys Acta 434:377–389
9. Casillas ER (1973) Accumulation of carnitin by bovine spermatozoa during maturation in the epididymis. J Biol Chem 248:8227–8232
10. Cross NL, Morales P, Overstreet JW (1986) Two simple methods for detecting human sperm acrosome reactions. J Androl 7:28a
11. Cummins JM, Yanagimachi R (1986) Development of ability to penetrate the cumulus oophorus by hamster spermatozoa capacitated in vitro, in relation to the timing of the acrosome reaction. Gamete Res 15:187–212
12. D'Addario DA, Turner TT, Howards SS (1980) Effect of vasectomy on the osmolarity of hamster testicular and epididymal intraluminal fluid. J Androl 1:167–170
13. Dale B (1983) Fertilization in animals. Arnold, London
14. Dale B (1987) Mechanism of fertilization. Nature 325:762–763
15. Dale B (1987) Fertilization channels in ascidian eggs are not activated by Ca. Exp Cell Res 172:474–480
16. Dale B (1988) Primary and secondary messengers in the activation of ascidian eggs. Exp Cell Res 177:205–211

17. Dale B, Monroy A (1981) How is polyspermy prevented? Gamete Res 4:151–169
18. Dale B, De Felice L (1984) Sperm activated channels in ascidian oocytes. Dev Biol 101:235–239
19. Dale B, Santella L (1985) Sperm–oocyte interaction in the sea-urchin. J Cell Sci 74:153–167
20. Dale B, DeFelice LJ (1990) Soluble sperm factors, electrical events and egg activation. In: Dale B (ed) Mechanism of fertilization: plants to humans. Springer, Berlin Hiedelberg New York (NATO ASI cell biology ser H 45)
21. Dale B, DeFelice LJ, Taglietti V (1978) Membrane noise and conductance increase during single spermatozoon-egg interactions. Nature 275:217–219
22. Dale B, DeFelice LJ, Ehrenstein G (1985) Injection of a soluble sperm fraction into sea urchin eggs triggers the cortical reaction. Experientia 41:1068–1070
23. Dale B, Tosti E, Iaccarino M (1995) Is the plasma membrane of the human oocyte re-organized following fertilization and early cleavage? Zygote 3:31–36
24. De Felice L, Dale B (1979) Voltage response to fertilization and polyspermy in sea urchin eggs and oocytes. Dev Biol 72:327–341
25. De Felice L, Dale B, Talevi R (1986) Distribution of fertilization channels in ascidian oocyte membranes. Proc R Soc Lond 229:209–214
26. Gaunt SJ (1983) Spreading of the sperm surface antigen within the plasma membrane of the egg after fertilization in the rat. J Embryol Exp Morphol 75:259–270
27. Gianaroli L, Tosti E, Magli C, Iaccarino M, Ferraretti A, Dale B (1994) The fertilization current in the human oocyte. Mol Reprod Dev 38:209–214
28. Gordon K, Brown DB, Ruddle FH (1985) In vitro activation of human sperm induced by amphibian egg extract. Exp Cell Res 157:409–418
29. Hamilton DW (1971) Steroid function in the mammalian epididymis. J Reprod Fertil [Suppl] 13:89–97
30. Homa S, Swann K (1994) A cytosolic sperm factor triggers calcium oscillations and membrane hyperpolarizations in human oocytes. Hum Reprod 9:2356–2361
31. Huacuja L, Delgado NM, Merchant H, Pancardo RM, Rasado A (1977) Cyclic AMP induced incorporation of ^{33}Pi into human sperm membrane components. Biol Reprod 17:89–96
32. Huret JL (1986) Nuclear chromatin decondensation of human sperm. Arch Androl 16:97–109
33. Igusa Y, Miyazaki S, Yamashita N (1983) Periodic hyperpolarizing responses in hamster and mouse eggs fertilized with mouse sperm. J Physiol (Lond) 340:633–647
34. Iwasa KH, Ehrenstein G, DeFelice LJ, Russell JT (1990) High concentration of inositol-1,4,5-trisphosphate in sea urchin sperm. Biochem Biophys Res Commun 172:932–938
35. Jaffe L (1980) Calcium explosions as triggers of development. Ann N Y Acad Sci 339:86–101
36. Kamiguchi Y, Mikamo K (1986) An improved, efficient method for analyzing human sperm chromosomes using zona-free hamster ova. Am J Hum Genet 38:724–740
37. Koehler JK (1981) Surface alterations during the capacitation of mammalian spermatozoa. Am J Primatol 1:131–141
38. Kvist U, Björndahl L (1985) Zinc preserves an inherent capacity for human sperm chromatin decondensation. Acta Physiol Scand 124:195–200
39. Kyozuka K, Osanai K (1989) Induction of cross fertilization between sea urchin eggs and starfish spermatozoa by polyethylene glycol treatment. Gamete Res 22:123–129
40. Longo FJ, Chen DY (1985) Development of cortical polarity in mouse eggs: involvement of the meiotic apparatus. Dev Biol 107:382–394
41. Mahi CA, Yanagimachi R (1973) The effect of temperature, osmolality and hydrogen ion concentration on the activation and acrosome reaction of golden hamster spermatozoa. J Reprod Fertil 35:55–66
42. Mahi CA, Yanagimachi R (1975) Induction of nuclear decondensation of mammalian spermatozoa in vitro. J Reprod Fertil 44:293–296
43. Mars B, Johnson MH, Pickering SJ, Flach G (1984) Changes in actin distribution during fertilization in the mouse egg. J Embryol Exp Morphol 81:211–237
44. McCulloh DH, Edwards EL (1992) Fusion of membranes during fertilization. J Gen Physiol 99:137–175
45. Miyazaki S, Hashimoto N, Yoshimoto Y, Kishimoto T, Igusa Y, Hiramoto Y (1986) Temporal and spatial dynamics of the periodic increase in intracellular free calcium at fertilization of golden hamster eggs. Dev Biol 118:259–267
46. Mohri H, Yanagimachi R (1980) Characteristics of motor apparatus in testicular, epididymal and ejaculated spermatozoa: a study using demembranated sperm model. Exp Cell Res 127:191–196
47. Moore HDM (1979) The net surface charge of mammalian spermatozoa as determined by isoelectric focusing: changes following sperm maturation, ejaculation, incubation in the female tract and after enzyme treatment. Int J Androl 2:449–462
48. Nagae T, Yanagimachi R, Srivastave PN, Yanagimachi H (1986) Acrosome reaction in human spermatozoa. Fertil Steril 45:701–707
49. Nuccitelli R, Cherr G, Clark W (1989) Mechanisms of egg activation. Plenum, New York
50. Orgebin-Crist MC, Fournier-Delpech S (1982) Sperm-egg interaction: evidence for maturational changes during epididymal transit. J Androl 3:429–433
51. Robertson T (1912) Studies on the fertilization of the eggs of a sea urchin Strongylocentrotus purpuratus by blood-sera, sperm, sperm extract and other fertilizing agents. Arch Entwickl Mech 35:64–130
52. Saling PM (1982) Development of the ability to bind zonae pellucidae during epididymal maturation. Biol Reprod 26:429–436
53. Shalgi R, Phillips D (1980) Mechanics of sperm entry in cycling hamsters. J Ultrastruct Res 71:154–161
54. Shapiro BM, Schackmann RW, Tombes RM, Kazazoglan T (1985) Coupled ionic and enzymatic regulation of sperm behavior. Curr Top Cell Regul 26:97–113
55. Simerly C, Wu G, Zoran S, Ord T, Rawlins R, Jones J, Navara C, Gerrity M, Rinehart J, Binor Z, Asch R, Schatten G (1995) The paternal inheritance of the centrosome, the cell's microtubule organizing center, in humans and the implications for infertility. Nature Medicine 1:47–52
56. Stice SL, Robyl JM (1990) Activation of mammalian oocytes by a factor obtained from rabbit sperm. Mol Reprod Dev 25:272–280
57. Swann K (1990) A cytosolic sperm factor stimulates repetitive calcium increases and mimics fertilization in hamster eggs. Development 110:1295–1302
58. Talbot P (1984) Events leading to fertilization in mammals. In: Harrison RF, Thompson JBM (eds) Fertility and sterility. Proceedings of the 11th world congress on fertilization and Sterilization. MTP Press, Boston, pp 121–131
59. Talbot P (1985) Sperm penetration through oocyte investments in mammals. Am J Anat 174:331–346
60. Talbot P, Chacon RS (1980) A triple stain technique for scoring acrosome reaction of human sperm. J Exp Zool 215:201–208
61. Talbot P, DiCarlantonio G, Zao P, Penkala J, Haimo LT (1985) Motile cells lacking hyaluronidase can penetrate the hamster oocyte cumulus complex. Dev Biol 108:387–398
62. Tesarik J (1984) Topographic relations of intramembranal particle distribution patterns in human sperm membranes. J Ultrastruct Res 89:42–55
63. Tesarik J, Sousa M, Testart J (1994) Human oocyte activation after intracytoplasmic sperm injection. Hum Reprod 9:511–518

64. Tyler A, Monroy A, Kao C, Grundfest H (1956) Membrane potential and resistance of the starfish egg before and after fertilization. Biol Bull 111:153–177

65. Volglmayr JK, Sawyer FR (1986) Surface transformation of ram spermatozoa in uterus, oviduct and cauda epididymal fluids in vitro. J Reprod Fertil 78:315–325

66. Wassarman PM, Bleil JD, Florman HM, Greve JM, Roller RJ, Salizman GS, Samuels FG (1985) The mouse eggs receptor for sperm: what is it and how does it work? Cold Spring Harbor Symp Quant Biol 50:11–19

67. Wassarman PM, Florman HM, Greve JM (1985) Receptor-mediated sperm-egg interactions in mammals. In: Metz CB, Monroy A (eds) Fertilization, vol 2. Academic, New York, pp 341–360

68. Whitaker M, Crossley I (1990) How does a sperm activate a sea urchin egg? In: Dale B (ed) Mechanism of fertilization: plants to humans. Springer, Berlin Heidelberg New York (NATO ASI cell biology ser H 45)

69. Wolf DP (1981) The mammalian egg's block to polyspermy. In: Mastroianni L, Biggers BJ (eds) Fertilization and embryonic development in vitro. Plenum, New York, pp 183–197

70. Yanagimachi R (1981) Mechanisms of fertilization in mammals. In: Mastroianni L, Biggers JD (eds) Fertilization and embryonic development in vitro. Plenum, New York, pp 81–182

71. Yanagimachi R (1984) Zona-free hamster eggs: their use in assessing fertilizing capacity and examining chromosomes of human spermatozoa. Gamete Res 10:178–232

72. Yanagimachi R, Chang MC (1963) Fertilization of hamster eggs in vitro. Nature 200:281–282

116 Endocrinology of Pregnancy

M. Breckwoldt, J. Neulen, and C. Keck

Contents

116.1 Introduction

This chapter deals mainly with endocrinological changes occurring during human pregnancy. These changes are the basis for our understanding of normal gestation.

Fetomaternal Interaction. The concept of a functional unit constituted by the maternal, the placental and the fetal compartment was defined as early as 1962. This concept stressed the pivotal role of the fetus in fetomaternal interaction, indicating that the fetus utilized maternal potentials to guarantee its well-being in terms of the nutritional supply essential for its growth and development. The central role of the fetus, however, actually starts immediately after implantation, when it signals its presence to the maternal organism by producing human chorionic gonadotropin (hCG), causing the maternal corpus luteum to produce sufficient amounts of sex steroids for uterine growth and blood supply. Increasing levels of estrogens and progesterone are also essential for reliable protection of the fetus against pathogenic microorganisms, causing an acid milieu in the vagina and solid closure of the cervical canal. Endocrinological data provide numerous examples of how the fetus has the leading role in the physiological process of human gestation.

116.2 Fertilization, Ovum Transport, Implantation

Fertilization. The fertilization of the human oocyte (see also Chap. 115) takes place in the ampulla of the fallopian tube. Immediately after fertilization a female pronucleus is formed and the maternal genome is condensed. After penetrating the zona pellucida, the sperm head swells and forms the male pronucleus. Pronucleus formation results in breakdown of the nuclear membranes, allowing the chromosomes to comigrate to the poles of the cell and enter into mitosis, followed by symmetrical division of the fertilized egg containing the maternal and paternal genome. During a 4- to 6-day preimplantation period, blastocyst formation occurs by mitotic division.

Fallopian Tube. The epithelium of the fallopian tube is composed of three different cell types:

- Ciliated cells
- Secretory cells
- Indifferent cells.

Ciliated and secretory cells which provide the essential requirements for early embryonic development and ovum transport, are functionally dependent on estrogens and progesterone. In primates the mucosal folds of the oviduct are well ciliated and all the cilia beat in the direction of the uterus [21]. The mucosa of the fimbriae and the ampulla contains the highest proportion of ciliated cells.

The production of tubal fluid by the *secretory cells* is regulated by sex steroids, as estrogens stimulate secretory activity. This effect is antagonized by progesterone. Oviduct fluid is composed of a variety of proteins including enzymes, enzyme inhibitors, immunoglobulins, electrolytes, trace elements, and energy subtrates, providing an environment essential for the initial phase of reproduction. The rate of tubal fluid production reaches its maximum at the time of ovulation [21] and decreases during the luteal phase. Protein components of tubal fluid are incorporated into the fertilized egg, indicating that proteins of maternal origin are essential for early embryonic development, as shown in the rabbit [21]. There is additional evidence from animal experiments that oviduct fluids play a central role in the energy supply and metabolism of the embryo.

R. Greger/U. Windhorst (Eds.)
Comprehensive Human Physiology, Vol. 2
© Springer-Verlag Berlin Heidelberg 1996

Fallopian Tube Contractility. The smooth muscle contractility of the fallopian tube is under the control of sex steroids, catecholamines, and prostaglandins. Prostaglandin ($PGF_{2\alpha}$) production is increased by estradiol, leading to enhanced muscle contractility [62]. During the proliferative phase of the menstrual cycle the luminal fluid is transported into the ampulla by vigorous antiperistaltic contractions. After ovulation and during the luteal phase, tubal contractility is reversed, supporting ciliar activity that gradually propels the fertilized egg towards the uterus. At approximately midluteal phase, the embryo at the early blastocyst stage is propelled into the uterine cavity [46]. The blastocyst continues to develop in the uterus for another 24 h, expanding its blastocoelic cavity. After hatching from the zona pellucida, active implantation into the endometrium – now transformed by hormonal action – is initiated, microvilli of the endometrial epithelium becoming attached to the microvilli of the trophectodermal cells.

Blastocyst Implantation. Direct cell fusion between embryonic and uterine epithelium does not occur [18]. Implantation of the blastocyst into the endometrial lining is considered an active process enhanced by the activation of a cascade of proteolytic enzymes [6,18] regulated by their corresponding inhibitors (α_1-antitrypsin, α_1-anti-

chymotrypsin). The polymorphic epithelial mucin MUC 1 is obviously involved in the implantation process since its mRNA is maximally expressed during the midluteal phase [25]. At the time of implantation, embryoblast, inner cell mass, and trophoblastic elements (cytotrophoblast and syncytiotrophoblast) can be distinguished (Fig. 116.1). Due to rapid mitotic division the cytotrophoblast and syncytiotrophoblast develop and invade the uterine mucosa, establishing contact with the maternal capillary circulation and resulting finally in the formation of a *hemochorial placenta*. The bloodstreams of the mother and the fetus are separated by:

- The trophoblast
- The basement membrane
- The vascular endothelium of the fetal capillaries.

116.3 Human Chorionic Gonadotropin

Even before implantation the blastocyst is capable to synthesize hCG in small amounts. After implantation hCG is secreted by the syncytiotrophoblast in substantial quantities into the fetal (5%) and maternal (95%) circulation. This hormone is a glycoprotein with a molecular weight of

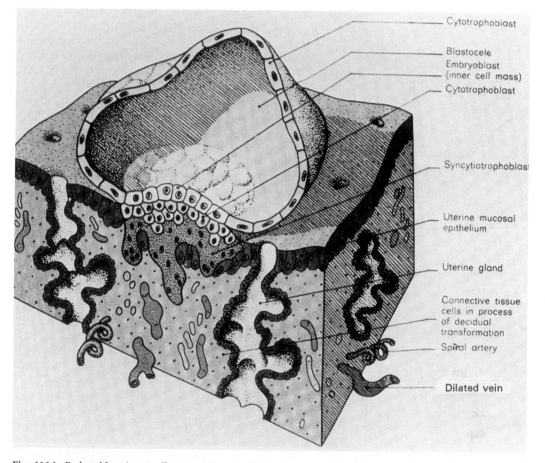

Fig. 116.1. Embryoblast, inner cell mass, and trophoblastic elements of a human blastocyst at the time of implantation

2278

38 kDa and is composed of two noncovalently linked subunits, a nonspecific α-subunit of 92 amino acids and a hormone-specific β-subunit of 145 amino acids. Several carbohydrates such as galactose, mannose, *N*-acetylglucosamine, and *N*-acetylneuraminic acid are attached to the molecule, giving rise to microheterogeneity. Because they have a similar chemical structure hCG and luteinizing hormone (LH) bind to the same receptor, hCG with greater affinity.

The α-subunits of hCG, follicle-stimulating hormone (FSH) LH, and thyroid-stimulating hormone (TSH) are identical and are encoded by a single gene. For the hormone-specific β-subunit several different genes have been identified, located on chromosome 19 [50]. In the maternal circulation hCG can be detected as early as 6–8 days after ovulation: plasma values peak in the 10th week of gestation at the order of 100 000 IU/l. Thereafter hCG concentrations drop to a plateau of approximately 10 000–20 000 IU/l from the 16th week to the end of gestation (Fig. 116.2). In the fetal circulation hCG levels follow a simular pattern, but at much lower concentrations.

At present very little is known about the regulation of placental hCG production. There is some evidence that gonadotropin-releasing hormone (GnRH) of cytotrophoblastic origin is involved by paracrine mechanisms in the control of hCG production [26];

Gonadotropin-Releasing Hormone. GnRH has been detected in human trophoblast tissue by means of biological and immunological techniques. It is identical to the decapeptide of hypothalamic origin. The presence of GnRH-specific mRNA in cytotrophoblast cells is evidence of where it is produced. Maximum concentrations are found between the 10th and the 15th week of gestation [36]. In vitro experiments with human first-trimester placenta indicate that *epidermal growth factor* (EGF) has an acute stimulatory effect upon hCG secretion and a delayed effect on its biosynthesis [3]. There is also evidence that endogenous opioids such as *dynorphin* have a stimulating effect on hCG release [4]. Secretion of hCG can be enhanced by the addition of cyclic AMP (cAMP) under in vitro conditions [31].

Physiological Role of hCG. The main physiological role of hCG is definitely maintenance of pregnancy through stimulation of corpus luteum function. HCG binds with high affinity to LH receptors of luteal tissue, inducing increased biosynthesis of estradiol-17β and progesterone by enhancing cAMP accumulation, leading to protein kinase A activation. Constantly rising estradiol and progesterone production are required in order to stimulate endometrial secretion and uterine growth. An adequate supply of the essential sex steroids is provided by the hCG-stimulated corpus luteum up until the eight week of gestation. Luteal function is terminated gradually by the eighth to tenth week. The life span of the corpus luteum is reflected by the plasma concentrations of 17α-OH-progesterone. After about the 10th week, progesterone production is provided by the placenta.

Stimulation of Testosterone. In addition, hCG is probably involved in stimulating testosterone secretion by the Leydig cells of fetal testes. Peak values of hCG binding of fetal testicular tissue are found between the 15th and 20th week of gestation, leading to testosterone production rates comparable to pubertal levels [41].

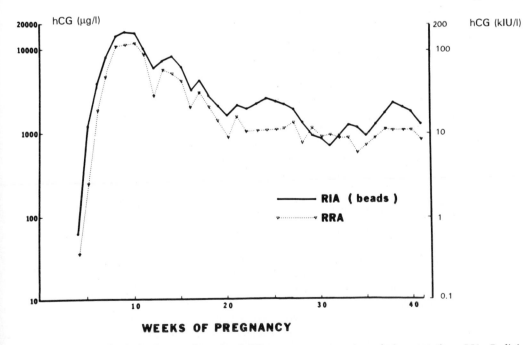

Fig. 116.2. Maternal Chorionic gonadotropin (hCG) serum concentrations during gestation. *RIA*, Radioimmunoassay; *RRA*, radioceptor assay

Immunosuppression. In order to achieve immunological tolerance, recruitment and activation of suppressor cells in the decidua are required [55]. Placental cells synthesize a variety of mediators including prostaglandins with direct immunosuppressive activity. Prostaglandin E_2 (PGE$_2$) has been shown to inhibit T-lymphocyte proliferation and T-lymphocyte cytotoxicity. In addition, it has been demonstrated that PGE$_2$ effectively inhibits interleukin 2 (IL-2) production (see also Chap.6), thus leading to inhibition of cytokine-activated cells and natural killer cells [37]. HCG has also been reported to be directly immunosuppressive [44,53] – a notion which, however, has been questioned by others [65]. It has been demonstrated in vitro that human early embryos in two- and four-cell stages are able to express various cytokines such as interleukin-1 (IL-1), interleukin-6 (IL-6), and colony-stimulating factor (CSF) in substantial amounts [32]. These cytokines are probably involved in the regulation of embryonic development implantation and immunological recognition of pregnancy [68].

116.4 Human Placental Lactogen

Human placental lactogen (*hPL*) is a single-chain proteohormone composed of 191 amino acids with a molecular weight of 22 kDa, cross-linked by two disulfide bonds of cystine bridges. The chemical structure is closely related to that of prolactin and hCG. Immunohistochemical studies with specific antibodies strongly suggest the syncytiotrophoblast as the cellular site of origin. In vitro experiments have demonstrated the existence of a higher-molecular-weight precursor of hPL [61]. The biological half-life of hPL has been calculated to be of the order of 20 min. Five different genes for the expression of hPL, located on the long arm of chromosome 17, have been identified in close vicinity to the genes coding for hGH, explaining the 90% homology of these two hormones [58]. These findings indicate a close evolutionary relationship between them.

HPL as a Measure of Placental Function. During the course of pregnancy the production rate of hPL constantly increases. Plasma concentrations of hPL correlate positively with placental weight. Therefore, the hPL concentration in the maternal circulation can be used as a measure of placental function. As shown in Fig. 116.3, there is a relatively large standard deviation of the mean at all stages of gestation. These data represent the normal range of singleton pregnancies; values obtained from mothers with twin, triplet, or quadruplet pregnancies are consistently higher, correlating positively with placental weight. During the course of pregnancy from the 10th up to the 36th week of gestation a 12-fold increase in plasma hPL concentrations can be seen.

HPL Biosynthesis and Secretion. The control of hPL biosynthesis and secretion is poorly understood. In clinical

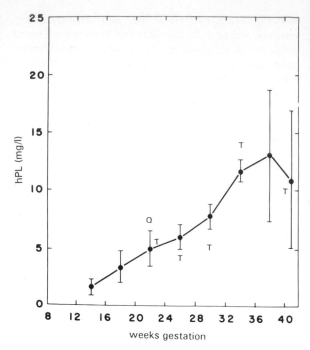

Fig. 116.3. Mean serum placental lactogen (hPL) concentrations during pregnancy. Values represent mean ± SD, corrected for placental weight

studies it was demonstrated that fasting for 72 h is associated with a 40% reduction in peripheral hPL concentrations [34], Prostaglandins also seem to be involved in the regulation of hPL secretion, since administration of PGE$_2$ is followed by a rapid decline of hPL plasma concentrations. In vitro experiments with human trophoblastic cells provide evidence that dopamine has an inhibitory effect on hPL release [30,48]. Secretion of hPL is unaffected by cAMP [31].

Lactogenic and Somatotropic Properties of hPL. The monomeric form of hPL present in the maternal circulation has lactogenic and somatotropic properties. Lactogenic activity of hPL was demonstrated in rhesus monkeys after pretreatment with estrogens and progesterone [5]. It therefore seems possible that hPL acts synergistically with prolactin (hPRL) in preparing the maternal breast for lactation after delivery. HPL, which is almost exclusively secreted into the maternal circulation, can also be regarded as a potent lipolytic factor comparable to human growth hormone (hGH).

Lipolysis. Lipolysis is reflected by a corresponding increase in free fatty acids in maternal plasma, providing energy substrates that can be utilized by the maternal compartment. This glucose-sparing effect of hPL is obviously advantageous for placental and fetal energy needs since glucose is easily transported across the placenta. The respiratory quotient (RQ) of 1.0 found in newborn babies indicates that glucose is the exclusive energy source.
The passage of fatty acids through the placenta is limited. In addition, there is evidence that fatty acids cannot be

efficiently utilized by the fetus [63]. The increased ability of pregnant women to mobilize free fatty acids from stores of body fat is clearly demonstrated in women with multiple pregnancies and consequently high concentrations of circulating hPL [43]. The plasma concentrations of hPL and triglycerides are positively correlated.

Insulin Resistance. Like hGH, hPL also increases peripheral insulin resistance. This effect is potentiated by the constantly rising steroid production of the fetoplacental unit. Insulin resistance is compensated by increased maternal insulin secretion. Recently, experimental evidence has shown that hPL affects the decidua, inducing the production of insulin-like growth factor I (IGF-I) and its corresponding binding proteins. IGF-I and IGF-II of decidual origin may be involved in the regulation of both uterine and fetal growth; IGF-II is considered to be of major importance in the regulation of fetal growth [19].

116.5 Relaxin

Relaxin is a pregnany-associated polypeptide hormone composed of 56 amino acids with a molecular weight of 6 kDa. Its primary structure and processing from a larger prohomone is similar to that of insulin. Two genes coding for relaxin are located on chromosome 9. The major sites of production are the corpus luteum, probably under the control of hCG, the decidua, and trophoblast tissue. Peripheral plasma concentrations of relaxin reach their maximum in the maternal circulation during the first trimester. in synergism with progesterone, relaxin plays a significant role in the inhibition of myometrial contractility during the course of pregnancy. Near term relaxin is involved in the process of softening the cervical canal and pelvic tissue by inducing collagenase activity [11], (see also Chap. 20). In addition, it is known from in vitro studies that relaxin in synergism with progesterone stimulates decidual secretion of various proteins such as *uteroglobin* and *prolactin* in a does-dependent fashion [67]. Prolactin of decidual origin is obviously identical with pituitary prolactin on the evidence of chemical, immunological, and biological properties and the ability of endometrial prolactin mRNA to hybridize pituitary prolactin cDNA [27].

116.6 The Fetoplacental Unit

Estrogen and Progesterone. During the first trimester *estradiol-17β*, *progesterone*, and *17α-OH-progesterone* are secreted by the corpus luteum, stimulated by hCG using mainly *cholesterol* as substrate. Cholesterol is transported to the luteal cell by a lipoprotein fraction of low density (LDL) and attached to the corresponding LDL receptors. After internalization and hydrolysis cholesterol is further processed by microsomal enzymes to 17α-OH-progesterone, progesterone, androgens, and estrogens. Progester-

one and its 17-α-OH-metabolite appear in the circulation in substantial amounts. As luteal function regresses at the 8th-10th week of gestation, the levels of 17α-OH-progesterone decrease while the plasma concentrations of progesterone constantly rise, indicating that the synthesis of progesterone is being taken over by the placenta. The rate of progesterone production has been calculated to be of the order of 190–280 mg/day in late human pregnancy [39]. Plasma concentrations of progesterone increase from 16–20 mg/l in early pregnancy to 175–200 µg/l at term (Fig. 116.4).

Progesterone. Progesterone induces secretory transformation of estrogen-stimulated endometrium. Estradiol-17β stimulates the synthesis of estrogen and progesterone receptors. Progesterone, however, causes suppression of estrogen receptor synthesis, thus acting as a natural

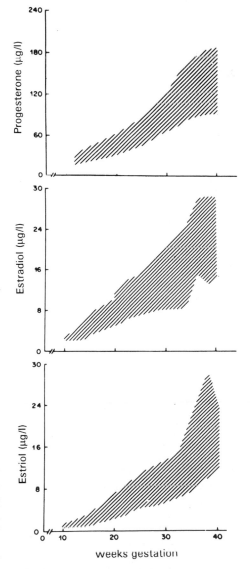

Fig. 116.4. Progesterone, estradiol, and estriol concentrations in maternal serum during pregnancy. *Shaded areas* represent ± SD

antiestrogen [9]. The physiological significance of progesterone, which is present in extremely high tissue concentrations in myo- and endometrium during pregnancy, is not completely understood. Progesterone binds to specific receptors in uterine smooth muscle causing inhibition of uterine contractility by suppressing prostaglandin formation. In addition, progesterone is essential for the maintenance of pregnancy because of its ability to suppress T-lymphocyte-induced cytotoxicity. This effect is of importance for the immunological tolerance built up during gestation [59].

Estrogen. The estrogen production rate steadily increases as pregnancy advances, as reflected by rising estrogen levels in maternal plasma (Fig. 116.4). Urinary output of estriol has increased 1000-fold by the term of gestation. During the first 8 weeks the corpus luteum is the main source of sex steroids. Thereafter estrogen production is provided by the fetoplacental-maternal unit, as illustrated in Fig. 116.5.

Concept of the Fetoplacental Unit. The concept of the fetoplacental unit is based on the classical clinical experiments by Diczfalusy and coworkers [20]. Maternal cholesterol is transported to the placenta and to the fetal compartment, serving as a substrate for steroid biosynthesis. Under the influence of ACTH, the fetal adrenals secrete large amounts of dehydroepiandrosterone sulfate (DHEA-S). DHEA-S is partly delivered to the trophoblast and completely aromatized to estrone and estradiol-17β. The main bulk of DHEA-S, however, is 16α-hydroxylated by the fetal liver, transferred to the placental

compartment, and finally converted to estriol by the aromatizing enzyme system located in the microsomes. Since the placenta is unable to convert progesterone because of a deficiency of 17α-hydroxylase and 17,20-desmolase, further processing to C19 steroids must occur in the fetus. DHEA-S of fetal and maternal origin is readily aromatized by the placenta, giving rise to estrone and estradiol-17β The following enzymes are involved in this process:

* Sulfatase
* Δ-5,4 Isomerase
* 3β-Hydroxysteroid dehydrogenase
* The aromatase enzyme complex.

Near term, approximately 40% of estradiol-17β and estrone is formed from maternal DHEA-S, 60% is of fetal origin. The estriol production, however, is 90% dependent on fetal precursor supply. In pregnancies with anencephalic fetuses, fetal corticotropin releasing factor (CRF) is lacking; consequently ACTH secretion from the pituitary is absent, leading to low or undetectable levels of estriol in the maternal circulation. Only 20% of fetal cholesterol is derived from maternal sources. The main part of the cholesterol production is provided by the fetal liver to supply sufficient quantities for steroid biosynthesis in the fetal adrenals [13].

Fetal Adrenals. Increasing quantities of estrogens are of physiological significance for uterine growth and uterine blood supply, essential for fetal nutrition and fetal metabolism [54]. At term pregnancy the fetal adrenals produce large amounts of DHEA-S, as reflected by their weight of 20 g, which is 10–20 times greater than the weight of adult adrenals relative to body mass [29]. The capacity of the fetal adrenals to synthesize cortisol is limited in early gestation and increases towards term.

Cortisol. The fetus near term is able to produce about 75% of its circulating cortisol; only 25% is of maternal origin [7]. During transfer of maternal cortisol, a large proportion is converted to cortisone by placental 11β-OH-steroid dehydrogenase [10,42]. The activity of this enzyme regulates fetal supply of maternal cortisol [22] and protects the fetus from hypercortisolism. Plasma concentrations of fetal cortisone are almost completely derived from maternal cortisol [42].

The maternal *circadian rhythm of cortisol secretion* is maintained during pregnancy, leading to relatively high concentrations of cortisol in the fetal circulation at night and causing relative suppression of fetal CRF-ACTH secretion. This results in reduced DHEA-S secretion by the fetal adrenals and subsequently in reduction of estriol production by the fetoplacental unit. In maternal serum cortisol and estriol levels are negatively correlated. Estriol concentrations during morning hours are lower than at night [8]. Since ACTH does not cross the placenta, fetal ACTH secretion is partly controlled by maternal cortisol. Infusion of ACTH to pregnant women near term results in a significant

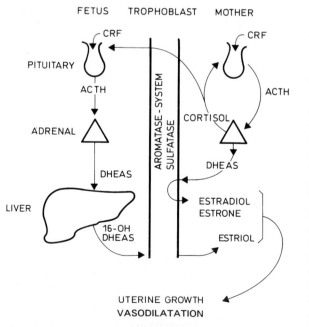

Fig. 116.5. Concept of the fetoplacental-maternal unit. *DHEAS*, Dehydroepiandrosterone sulfate; *CRF*, corticotropin-releasing factor; *ACTH*, adrenocorticotropic hormone

2282

increase of maternal cortisol levels followed by a corresponding fall of estriol. Dexamethasone administration results in almost complete suppression of estriol in maternal plasma [52].

Fetal Pituitary. The human fetal pituitary is capable of synthesizing and releasing all peptides and proteohormones of the adult pituitary, including ACTH, β-LPH, β-endorphin and 2-lipoprotein, growth hormone (GH), prolactin, follicle-stimulating hormone (FSH), LH, and thyrotropin (thyroid-stimulating hormone, TSH) [60], indicating that the fetal pituitary represents a dynamic endocrine organ throughout gestation.

Cortisol-Binding Globulin. Constantly increasing estrogen levels result in stimulation of hepatic synthesis of cortisol-binding globulin (CBG). CBG bound cortisol is largely biologically neutralized; consequently the total plasma concentrations of cortisol increase during the course of pregnancy (Fig. 116.6).

Corticotropin-Releasing Factor. CRF, also called corticotropin-releasing hormone (CRH), is also secreted by the placenta in increasing quantities, reaching extremely high concentrations in late gestational maternal plasma. The ACTH-releasing activity, however, is effectively reduced by a specific CRF-binding protein. Approximately 93% of the CRF in maternal plasma circulates in a carrier-bound form [40]. This finding explains why ACTH secretion is normal at term despite the presence of high levels of CRF. The plasma concentrations of cortisol, corticosterone, desoxycorticosterone, and aldosterone increase progressively throughout gestation. From clinical studies it appears that the increase of cortisol concentrations during pregnancy is not only related to increased CBG production, but that other factors such as placental CRF or ACTH-like substances may be involved in establishing increased sensitivity of the CRH-ACTH-adrenal axis [57], (see Chap. 20). Despite increased cortisol levels

during gestation, pregnant women never develop clinical symptoms of hypercortisolism, owing to their elevated concentrations of CBG.

116.7 Maternal Adrenal, Thyroid, and Parathyroid Function

Aldosterone. In response to rising progesterone production, aldosterone is also increased with advancing pregnancy, reflecting stimulated secretion by the zona glomerulosa of the maternal adrenal (Fig. 116.7). Stimulation of aldosterone is probably also a consequence of increased renin and angiotensinogen production leading to elevated angiotensin II concentrations. Aldosterone causes increased Na+ and water retention [12] (see Chaps. 75,77). Fetal development is highly dependent on the existence of normal *maternal blood pressure* during gestation. The *renin-angiotensin-aldosterone system* plays a central role in the regulation of blood pressure and volume homeostasis. Renin is a glycoprotein acid protease with a molecular weight of approximately 40 kDa, This enzyme is secreted by the kidney, cleaving specifically angiotensinogen, a circulatory α_2-globulin of hepatic origin also termed renin substrate. Angiotensinogen is broken down to angiotensin I, which is cleaved by a converting enzyme to form the octapeptide angiotensin II, Since the hepatic production of angiotensinogen is stimulated by cortisol and estrogens, increased plasma concentrations can be anticipated as pregnancy advances [35].

The mineralocorticoid aldosterone acts primarily on the epithelium of the distal nephron to enhance Na+ reabsorption and K+ excretion (see Chap. 75). This physiological effect is potentiated by angiotensin II. Increased Na+ retention is required for the additional fetal and maternal interstitial and intravascular fluid supply. The rise in peripheral concentrations of aldosterone can in part be interpreted as a consequence of rising progesterone pro-

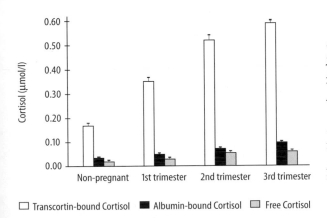

Fig. 116.6. Concentrations of transcortin-bound, albumin-bound, and free cortisol in maternal blood before pregnancy and during the first, second, and third trimesters

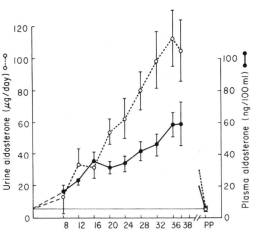

Fig. 116.7. Maternal urinary and plasma aldosterone concentrations during pregnancy. Values represent mean ± SD

duction. Progesterone can be regarded as a physiologic antimineralocorticoid stimulating renal Na⁺ excretion [45].

The alteration of the renin-angiotensin-aldosterone system during pregnancy reflects physiological responses to increasing concentrations of estrogens and progesterone in order to provide sufficient Na⁺ and water supply for maternal and fetal demands.

Thyroid Gland. The thyroid gland enlarges slightly during pregnancy secondary to increased vascularity and thyroid hyperplasia. Increased thyroid activity during early pregnancy is obviously mediated by hCG, since hCG binds with high affinity to TSH receptors, as shown in displacement studies [66]. In vitro studies with cultured rat thyroid cells demonstrate that hCG stimulates I⁻ uptake, adenylate cyclase activity, and DNA synthesis. The effects of hCG are additive to TSH, indicating intrinsic thyroid stimulating properties of hCG [24] (see Chap. 22). Furthermore, it is known from clinical studies that women with hydatidiform mole and choriocarcinoma, producing extremely high hCG levels, frequently develop hyperthyroidism [33]. The net overall effect of increased thyroid function is reflected by an elevated basal metabolic rate.

TSH is unable to cross the placental barrier; thyroid hormones T4 and T3 are transferred through the placenta at a reduced rate [23] and are essential for fetal development. I⁻ passes the placenta easily and is of great importance to fetal thyroid function since the formation of thyroid hormones depends on an adequate I⁻ supply. In addition constantly rising estrogen levels during the course of pregnancy induce an increasing production of *thyroxine-binding globulin* (TBG) by the maternal liver. The rise in concentration and binding capacity results in increased plasma levels of total T4 and T3 (Fig. 116.8). Free thyroid hormone levels remain fairly constant throughout pregnancy [1]. Fetal thyroid function is evidenced by TSH plasma concentrations in the fetal circulation, with peak values at 20–24 weeks, gestation, dropping to normal concentrations by term (Fig. 116.8). Shortly after delivery TSH levels rise significantly in response to the lower enviromental temperature.

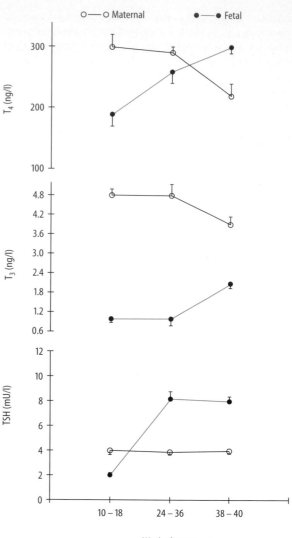

Fig. 116.8. Maternal and fetal T4, T3, and thyroid-stimulating hormone concentrations at between the tenth and 18th weeks of pregnancy, between the 24th and 36th weeks of pregnancy, and at term

Calcium Homeostasis. Calcium homeostasis is regulated by a complex system involving parathormone (parathyroid hormone, PTH), calcitonin, and D₃ hormone (calcitriol; see Chap. 80). During pregnancy fetal needs for Ca²⁺ are supplied by the mother. Therefore Ca²⁺ intake must be increased to cover the fetal demands. D₃ hormone (calcitriol) is metabolilized by the liver and the kidney to its biologically active form 1,25(OH)₂-D₃. Maternal D₃ hormone concentrations increase during pregnancy, promoting Ca²⁺ absorption and its transport to the fetus. In the fetal compartment the concentrations of 1,25(OH)₂-D₃ are relatively low, indicating that the active form of the hormone does not cross the placenta.

The same applies to *PTH*. One primary action of PTH is to increase renal reabsorption of Ca²⁺. Simultaneously, PTH promotes the conversion of 25OH-D₃ to 1,25(OH)₂-D₃

hormone in the kidney, thereby causing increased gastrointestinal absorption of Ca²⁺. Thirdly PTH stimulates osteoclast activity, leading to increasing Ca²⁺ concentrations in the circulation. Rising Ca²⁺ concentrations, however, cause inhibition of PTH secretion and thus restore Ca²⁺ levels to normal [49].

116.8 Maternal Cardiovascular Adaptation During Pregnancy

Maternal cardiovascular adaptations during pregnancy, such as increased cardiac output, increase in blood volume, and decreased vascular resistance, are physiological mechanisms supporting fetal and maternal demands.

Cardiac Output. One of the most impressive changes during pregnancy is the increase in cardiac output by 30%–40% compared to the nonpregnant state. This increase results in part from an increase in pulse frequency which becomes evident as early as in the fourth week of gestation and rises steadily until the 36th week of pregnancy (Fig. 116.9) [14]. In addition, a marked increase in pulse amplitude is observed until midgestation. During the second half of pregnancy the pulse amplitude returns to normal.

Blood Volume. The increase in blood volume during pregnancy results from a rise in the volume of plasma from the 6th to the 32nd gestational week. In addition, a significant increase in erythrocyte volume is apparent during the second trimester. The overall increase in maternal blood volume reached shortly before term is approximately 30% (Fig. 116.10), at the term of pregnancy the mother has approximately 1–2 l extra blood in her circulatory system. Increasing aldosterone and estrogen concentrations during pregnancy lead to an increase in Na^+ and water reten-

tion by the kidneys. In addition, elevated estrogen concentrations stimulate red blood cell production by the bone marrow. At normal delivery approximately one-forty of the blood volume gained during pregnancy is lost. This accounts for approximately 200–300 ml.

Blood Pressure. In contrast to the steadily increasing cardiac output and blood volume, blood pressure is normally slightly reduced during the first and second trimester [14]. This phenomenon can be observed as early as in the 7th gestational week. The lowest mean blood pressure can be detected at mid-pregnancy. The drop in blood pressure is due to reduced peripheral resistance, induced by rising estrogen concentrations. Estrogen-dependent vasodilation has been described [56]. Experimental findings indicate that estrogens act directly at the endothelial level, inducing liberation of vasoactive substances such as prostacyclin and NO (see Chaps. 6, 96). The reduction in diastolic blood pressure is more pronounced than the reduction in systolic pressure. During the third trimester mean arterial blood pressure in humans returns to nonpregnant values.

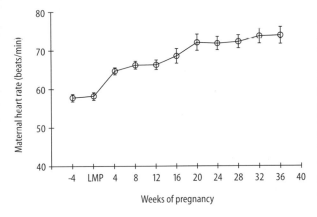

Fig. 116.9. Change in maternal heart rate during pregnancy

116.9 Maternal Respiratory Adaptation During Pregnancy

Rising progesterone levels during pregnancy increase the sensitivity of the respiratory center to CO_2, resulting in an approximately 50% *increase in ventilation* and a decrease in arterial PCO_2 (Table 116.1). The growing uterus inhibits excursions of the diaphragm, leading to decreasing expiratory reserve and residual volume. Consequently, inspiratory reserve volume and tidal volume are increased. In addition to its effects on the respiratory center, progesterone enhances alveolar ventilation. In late pregnancy gas exchange is increased in the form of "urge hyperventilation". As a result, maternal PCO_2 is decreased, facilitating

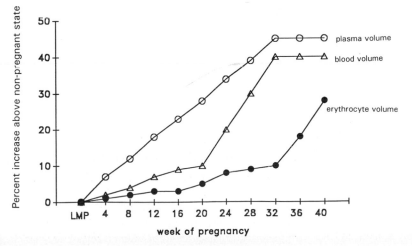

Fig. 116.10. Changes in maternal plasma volume, erythrocyte volume, and blood volume during pregnancy. *LMP*, Last menstrual period

Table 116.1. Respiratory change during early pregnancy [14]. Values represent mean ± SEM

	Before pregnancy	Week 7	Week 15	% Change
Minute ventilation (l/min)	7.41 ± 0.24	10.2 ± 0.27	9.25 ± 0.20	+28.5
Ventilatory equivalent for oxygen (l/l)	27.5 ± 0.9	31.6 ± 1.0	32.1 ± 1.0	+16.7
Respiratory quotient	0.78 ± 0.01	0.83 ± 0.01	0.84 ± 0.01	+7.7
Oxygen consumption (ml/min)	267 ± 5	294 ± 9	304 ± 10	+14

CO_2 exchange from fetus to mother via the placenta. Decreasing maternal PCO_2 is compensated by increased kidney bicarbonate excretion to maintain acid–base status (see Chap. 78).

116.10 Maternal Kidney Function During Pregnancy

Urine production and Na^+ excretion are markedly increased during the second half of pregnancy. In a nonpregnant woman mean *renal blood flow* is 600 ml/min (see Chap. 73). During the first trimester blood flow increases to approximately 850 ml/min, decreasing again to 750 ml/min at mid-pregnancy, reaches basal levels again before term. Between the 16th and 32nd gestational weeks the *glomerular filtration rate* increases by 60%. This is paralleled by decreasing plasma creatine and urea concentrations. These physiological changes can be interpreted resulting from the rising estrogen, progesterone, and aldosterone concentrations. Since progesterone has a dilatory effect on smooth muscle tone, dilation of the veins is a phenomenon that can be observed as early as in the eighth week of gestation. Furthermore, progesterone partly antagonizes the aldosterone effect on Na^+ reabsorption, and thus may be regarded as an antimineralocorticoid.

116.11 Maternal Liver Function During Pregnancy

During pregnancy liver morphology does not change significantly, nor does pregnancy seem to induce any histological or ultrastructural alterations in the liver [28]. However, it has been reported [15] that pregnant women have a higher mean liver weight than age-matched nonpregnant women (1725 g vs 1418 g).

Although morphological alterations of the liver are absent, pregnancy does induce various metabolic and functional changes leading to *glucose intolerance*, hyperinsulinemia, increased lipolysis, and ketogenesis. To maintain fetal glucose supply, maternal glucose utilization is reduced by a relative *insulin resistance*, most probably induced by hPL and rising sex steroid levels. The increased insulin resist-

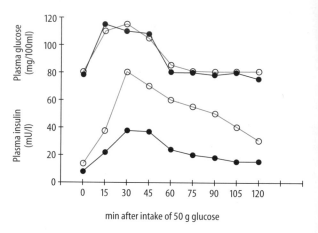

Fig. 116.11. Glucose and insulin concentrations after oral administration of 50 g glucose in a nonpregnant woman and a pregnant woman at term. Representative values are shown

ance is answered by increased insulin secretion, as evidenced by hyperplasia of pancreatic islets and elevated plasma insulin levels (Fig. 116.11).

To compensate for the decreased glucose utilization, maternal *lipolysis* and *ketogenesis* are increased. Total lipid concentrations rise from approximately 6 g/l blood in nonpregnant women to 9 g/l blood in pregnant women. Increased lipolysis is maintained until 6–8 weeks after delivery. There is evidence that pregnancy-induced metabolic changes may partially be modulated by hGH. It has been demonstrated that binding of hGH to liver microsomes in pregnant women was about 10 times that in nonpregnant subjects [16]. Maternal IGF 1 concentrations were shown to be normal or increased [17] in spite of very high concentrations of estrogens, which are known to reduce IGF 1 levels in nonpregnant women [64].

116.12 Carbohydrate Metabolism During Pregnancy

For the developing fetus an *adequate glucose supply* is of crucial importance since glucose represents the main energy substrate that can easily be utilized. During pregnancy glucose homeostasis is maintained in the maternal circulation by increased pancreatic insulin secretion. Clinical studies [47] have provided evidence that insulin receptors in adipocytes decrease in number and binding affinity during pregnancy. The resulting systemic effect is increased peripheral insulin resistance. A number of other factors that are simultaneously changed during pregnancy also contribute to elevated insulin resistance, such as increasing hPL concentrations and increasing concentrations of estrogens, progesterone, and cortisol. In addition, there is evidence that the placenta expresses insulin-metabolizing enzymes [51]. Rising hPL concentrations also contribute to the overall increased insulin resistance, since hPL has inherent growth hormone activity. Increasing prolactin secretion from the pituitary is also involved in the diabetogenic effect.

116.13 Development of the Fetal Liver

The development of the fetal liver already starts in the first weeks of pregnancy. As early as 4–5 weeks after fertilization, liver lobules can be detected. The enzymatic competence of the fetal liver changes with increasing metabolic functions over the course of pregnancy.

Between the tenth and 25th weeks of pregnancy the fetal liver is the main source of *fetal hemoglobin*. At the same time bilirubin production increases and rising bilirubin concentrations can be found in the amniotic fluid. Quantitative assessment of bilirubin concentrations in amniotic fluid serves as a diagnostic factor in cases of Rh incompatibility. Fetal bilirubin is unconjugated and can be eliminated by the placenta. For the conjugation of bilirubin, hepatic enzymes such as glucuronyl transferase and uridine-diphosphoglucose-dehydrogenase are essential. The capacity of the fetal liver to produce these enzymes is notably low preterm. Even after birth there is a deficit of enzyme production, leading to *physiological hyperbilirubinemia*.

An adequate carbohydrate supply is essential for fetal development. The fetus stores maternal glucose in the form of *glycogen* in liver and muscle tissue. As early as the tenth week of pregnancy, glycogen deposits can be detected in the fetal liver, and during the course of pregnancy fetal liver glycogen content steadily increases. Directly after birth the glycogen deposits of the newborn are utilized, as can be seen by monitoring serum glucose concentrations. This actually leads to physiological hypoglycemia of the newborn. The deficit of energy substrates can be compensated by glycogenolyis, gluconeogenesis, lipolysis, and utilization of free fatty acids.

116.14 Development of the Fetal Cardiovascular System

The cardiovascular system of the fetus is characterized by high cardiac output, arterial vasodilation and low blood pressure (see Chap. 118 for a detailed description). The fetus is dependent on the supply of O_2 through the placental membrane. The dissolved O_2 in the blood of the large maternal sinuses simply passes into the fetal blood due to an oxygen pressure gradient from the mother's blood to the fetus' blood. The mean PO_2 in the mother's blood in the maternal placental sinuses is approximately 50 mmHg toward the end of pregnancy, and mean PO_2 in the fetal blood after it becomes "oxygenated" is about 30 mmHg. Therefore, the mean pressure gradient for diffusion of O_2 through the placental membrane is about 20 mmHg.

With PO_2 as low as 30 mmHg, sufficient O_2 supply of fetal tissues can only be ensured by the following mechanisms:

- The markedly higher O_2 affinity of fetal hemoglobin compared to maternal hemoglobin, which means that at a given PO_2 the fetal hemoglobin can carry as much as 20%–30% more O_2 than maternal hemoglobin.
- The fetal hemoglobin concentration, which is approximately 50% higher than the maternal hemoglobin concentration.
- The "double Bohr effect," which means that hemoglobin can carry more O_2 at a low PCO_2 than at a high PCO_2. The fetal blood entering the placenta carries large amounts of CO_2 being transported to the maternal circulation. Reduction of CO_2 renders fetal blood alkaline, as evidenced by an increased pH of 7.4, while the pH in maternal blood is in the range of 7.3–7.5.

All these changes result in increased O_2-binding capacity of fetal blood.

116.15 Development of the Fetal Lung

In order to adapt to the extrauterine environment, fetal lungs have to undergo a maturation process during late pregnancy involving significant biochemical and anatomical changes. Preterm birth may lead to significant fetal morbidity and mortality due to insufficient lung maturation (see Chap. 118 for a detailed discussion).

Fetal lung maturation is dependent on sufficient surfactant formation. *Surfactant* is a lipoprotein composed of glycerophospholipids, neutral lipids, and protein. Lecithin constitutes the greatest proportion of the glycerophospholipids, accounting for 80%, which is generated by a complex enzyme system in type II pneumocytes. Lung maturation is induced by glucocorticoids, as has been shown in animal experiments and confirmed in clinical studies [38]. In addition there is evidence that other hormones such as prolactin, insulin, estrogens, and thyroid hormones are also involved in the lung maturation process. Clinical studies using glucocorticoids combined with treatment with thyrotropin-releasing hormone (TRH) in babies born prematurely have demonstrated a significant improvement in respiratory function in the newborn infants [2].

Directly after birth the newborn's lung has to adapt to the

extrauterine environment. With the onset of ventilation PO_2 increases and, as a reaction to vasoactive mediators, the *resistance of pulmonary vessels* decreases to 20% of the resistance before birth. Owing to these mechanisms, pulmonary blood flow increases about five to ten times compared with the prebirth blood flow. This underscores the exceptional physiology of the lung vasculature: increasing PO_2 is responded to with dilation of the pulmonary arteries, whereas the arteries in other organs react with constriction.

Physiologically, *placental circulation* is maintained for a few minutes after birth. With the onset of ventilation the resistance of blood vessels is drastically increased, leading to a reorientation of the organ blood supply. Left atrial pressure exceeds right atrial pressure, followed by functional closure of the foramen ovale. Occlusion of the ductus arteriosus is normally complete 10-15 h after birth.

References

1. Ballabio M, Poshyachinda M, Ekins RP (1991) Pregnancy-induced changes in thyroid function: role of hCG as putative regulator of maternal thyroid. J Clin Endocrinol Metab 73:824-831
2. Ballard PL (1991) Hormonal regulation of the pulmonary surfactant system during development. Acta Endocrinol (Copenh) [Suppl] 124:••
3. Barena ER, Feldman D, Kaplan M, Morrish DW (1990) The dual effect of epidermal growth factor upon human chorionic gonadotropin secretion by the first trimester placenta in vitro. J Clin Endocrinol Metab 71:923-928
4. Barnea EZ, Ashkenazy, Sarne G (1991) The effect of dynorphin on placental pulsatile human chorionic gonadotropin secretion in vitro. J Clin Endocrinol Metab 73:1093-1098
5. Beck P (1972) Lactogenic activity of human chorionic somatomammotropin. Proc Soc Exp Biol Med 140:923-928
6. Beier-Hellwig K, Sterzik K, Bonn B, Beier HM (1989) Contribution to the physiology and pathology of endometrial receptivity: the determination of protein patterns in human uterine secretions. Hum Reprod 4:115-120
7. Beitins JZ, Bayard F, Ances JG, Kowarski A, Migeon CJ (1973) The metabolic clearance rate, blood production, interconversion and transplacental passage of cortisol in pregnancy near term. Pediatr Res 7:509-514
8. Breckwoldt M, Reck G (1983) Untersuchungen der foeto-plazentaren Einheit als geburtshilfliche Entscheidungshilfe. Gynakologe 16:124-131
9. Brenner RM, West NB, McClellan MC (1990) Estrogen- and progestin receptors in the reproductive tract of male and female primates. Biol Reprod 42:11-19
10. Brown MS, Kovanen PT, Goldstein JL (1979) Receptor-mediated uptake of lipoprotein-cholesterol and its utilization for steroid synthesis in the adrenal cortex. Recent Prog Horm Res 35:215-149
11. Bryant-Greenwood GD (1982) Relaxin: a new hormone. Endocr Rev 3:62-72
12. Carr BR, Gant NF (1983) The endocrinology of pregnancy induced hypertension. Clin Perinatol 10:727-747
13. Carr BR, Simpson ER (1981) Lipoprotein utilization and cholesterol synthesis by the human fetal adrenal gland. Endocr Rev 2:306-315
14. Clapp JF (1985) Maternal heart rate in pregnancy. Am J Obstet Gynecol 152:659-660
15. Combes B, Adams RH (1972) Disorders of the liver. In: Assali NS (ed) Pathophysiology of gestation. Academic, New York
16. D'Abronzo FH, Yamaguchi MI, Alves RSC, Svartman R, Mesquita CH, Nicolau W (1991) Characteristics of growth hormone binding to liver microsomes of pregnant women. J Clin Endocrinol Metab 73:348-354
17. Daughaday WH, Trivedi B, Winn HN, Yan H (1990) Hypersomatotropism in pregnant women as measured by a human liver radioreceptorassay. J Clin Endocrinol Metab 70:215-222
18. Denker HW (1983) Basic aspects of ovoimplantation. Obstet Gynecol Annu 12:15-24
19. D'Ercole J (1991) Modern concepts of insulin-like growth factors. Elsevier, New York
20. Diczfalusy E (1962) Endocrinology of the foetus. Acta Obstet Gynecol Scand 41 [Suppl 1]:45
21. Eddy CA (1984) The fallopian tube: physiology and pathology. In: Aiman F (ed) Infertility – diagnosis and management. Springer, Berlin Heidelberg New York
22. Edwards CRW (1994) 11β-Hydroxysteroid-dehydrogenase deficiency – a developing concept. Exp Clin Endocrinol 102 [Suppl 1]:29
23. Furth ED (1983) Thyroid and parathyroid function in pregnancy. In: Fuchs F, Klopper A (eds) Endocrinology of pregnancy III. Harper and Row, Philadelphia
24. Hershman JM, Lee HY, Sugawara M, Mirell CJ, Pang XP, Yanagisawa M, Pekary AE (1988) Humam chorionic gonadotropin stimulates iodine uptake, adenylate cyclase and DNA synthesis in cultured rat thyroid cells. J Clin Endocrinol Metab 67:74-79
25. Hey NA, Graham RA, Seif MW, Apein JD (1994) The polymorphic epithelial mucin MUC 1 in human endometrium is regulated with maximal expression in the implantation phase. J Clin Endocrinol Metab 78:337-342
26. Hsueh AJW, Jones PBC (1983) Gonadotropin releasing hormone: extrapituitary actions and paracrine control mechanisms. Annu Rev Physiol 45:83-94
27. Huang JR, Tseng L, Bishop P, Janne OA (1987) Regulation of prolactin production by progestin, estrogen and relaxin in human endometrial stromal cells. Endocrinology 121:2011-2017
28. Ingerslev M, Teilum G (1945) Biopsy studies on the liver in pregnancy. Normal histological features of the liver as seen on aspiration biopsy. Acta Obstet Gynecol Scand 25:339-351
29. Johannisson E (1968) The fetal adrenal cortex in the human. Acta Endocrinol (Copenh) [Suppl] 58:130
30. Kaplan M, Barnea ER, Bersinger NA (1991) Patterns of spontaneous pulsatile secretion of hCG and pregnancy-specific β1 glycoprotein by superfused placental explants in first and last trimester. Lack of episodic hPL secretion. Acta Endocrinol (Copenh) 124:331-337
31. Kato Y, Braunstein GD (1990) Purified first and third trimester placental trophoblast differ in in-vitro hormone secretion. J Clin Endocrinol Metab 70:1187-1192
32. Kauma S, Matt D, Strom S, Eiermann D, Turner I (1990) Interleukin-1-beta, human leucocyte antigen HLA-DR alpha, and transforming growth factor beta expression in endometrium placenta and placental membranes. Am J Obstet Gynecol 163:1430-1437
33. Kenimer JG, Hershman JM, Higgins (1975) The thyrotropin in hydatidiform moles is human chorionic gonadotropin. J Clin Endocrinol Metab 40:482-491
34. Kim YJ, Felig P (1971) Plasma chorionic somatomammotrophin level during starvation in med-pregnancy. J Clin Endocrinol 32:864-871
35. Krakoff LR (1973) Plasma renin substrate: measurement by radioimmunoassay of angiotensin I concentration in syndromes with steroid excesss. J Clin Endocrinol Metab 37:116-122
36. Krieger DT (1982) Placenta as a source of "brain" and "pituitary" hormones. Biol Reprod 26:55-71
37. Lala PK (1989) Similarities between immunoregulation in pregnancy and malignancy: the role of PGE$_2$. Am J Reprod Immunol Microbiol 5:105-110

38. Liggins GC, Howie MB (1972) A controlled trial of antepartum glucocorticoid treatment and prevention of the respiratory distress syndrome in premature infants. Pediatrics 50:515–521
39. Lin TF, Lin SL, Erlenmeyer F, Kline T et al (1972) Progesterone production rates during the third trimester of pregnancy in normal women, diabetic women and women with abnormal glucose tolerance. J Clin Endocrinol Metab 34:287–297
40. Linton E, Behan DP, Saphier PW, Lowry PJ (1990) Corticotropin-releasing hormone (CRH)-binding protein: reduction in the adrenocorticotropin-releasing acitivity of placental but not hypothalamic CRH. J Clin Endocrinol Metab 70:1578–1580
41. Molsberg RL, Carr BR, Mendelson CR, Simpson ER (1982) Human chorionic gonadotropin binding to human fetal testes as a function of gestational age. J Clin Endocrinol Metab 55:791–797
42. Murphy BEP, Clark SJ, Donald FR, Pinsky M, Vedady D (1974) Conversion of maternal cortisol to cortisone during placental transfer to the human fetus. Am J Obstet Gynecol 118:538
43. Neulen J, Breckwoldt M (1987) Beeinflussung des Fettstoffwechsels durch hPL in der späten Schwangerschaft. Geburtsh Frauenheilkd 47:270–273
44. North RA, Whitehead R, Larkins RG (1991) Stimulation by human chorionic gonadotropin of prostaglandin synthesis by early human placental tissue. J Clin Endocrinol Metab 73:60–70
45. Oelkers W, Berger V, Bolik A, Bahr V et al (1991) Dihydrospirorenone, a new progestogen with antimineralocorticoid activity: effects on ovulation, electrolyte excretion and the renin-aldosterone system in normal women. J Clin Endocrinol Metab 73:837–842
46. Overstreet JW (1983) Transport of gametes in the reproductive tract in mammals. In: Hartmann JF (ed) Mechanism and control of animal fertilization. Academic, New York
47. Pagano G, Cassadei M, Massovio M, Bozzo C (1980) Insulin binding to human adipocytes during late pregnancy in healthy, obese and diabetic state. Horm Metab Res 12:177–183
48. Petit A, Guillon G, Pantaloni C, Tence M et al (1990) An islet-activating protein-sensitive G-protein is involved in dopamine inhibition of both angiotensin-stimulated inositol phosphate production and human placental lactogen release in human trophoblastic cells. J Clin Endocrinol Metab 71:1573–1580
49. Pitkin RM, Reynolds WA, Williams GA, Hargis KG (1979) Calcium metabolism in normal pregnancy: a longitudinal study. Am J Obstet Gynecol 133:781–795
50. Policastro PF, Daniels-McQueen S, Carle G, Boime I (1986) A map of the hCGβ-LHβ gene cluster. J Biol Chem 261:5907–5916
51. Posner BI (1973) Insulin metabolizing enzyme activities in human placental tissue. Diabetes 22:552–558
52. Reck G, Nowostawskyj H, Breckwoldt M (1978) Effect of ACTH and dexamethasone on the diurnal rhythm of unconjugated oestriol in pregnancy. Acta Endocrinol (Copenh) 87:820–827
53. Richetts RM, Jones DB (1985) Differential effect of human chorionic gonadotropin on lymphocyte proliferation induced by mitogens. J Reprod Immunol 7:225–232
54. Rosenfeld CR (1984) Consideration of the utero placental circulation in intrauterine growth. Semin Perinatol 8:42–54
55. Sanyal MK, Brami CJ, Bishop P (1989) Immunoregulatory activity in supernatants of normal human trophoblast cells of the first trimester. Am J Obstet Gynecol 161:446–453
56. Sarrel PM (1989) Effects of ovarian steroids on the cardiovascular system. In: Ginsburg J (ed) The circulation in the female. Carnforth, Parthenon
57. Scott EM, McGarrigle HHG, Lachelin GCL (1990) The increase in plasma and saliva cortisol levels in pregnancy is not due to the increase of corticosteroid binding globulin levels. J Clin Endocrinol Metab 71:639–644
58. Seeburg PM (1982) The human growth hormone gene family. Nucleotide sequences show recent divergence and predict a new polypeptide hormone. DNA 1:239–249
59. Siitiri PK, Febres F, Clemens LE (1977) Progesterone and maintenance of pregnancy: is progesterone nature's immunosuppressant? Ann N Y Acad Sci 186:384–393
60. Siler-Khodr TM, Morgenstern LL, Greenwood FC (1974) Hormone synthesis and release from human fetal adenohypophysis in vitro. J Clin Endocrinol Metab 39:891–905
61. Suwa S, Friesen H (1969) Biosynthesis of human placental lactogen in-vitro II. Dynamic studies of normal term placentas. Endocrinology 85:103–143
62. Spilman LH (1974) Oviduct response to prostaglandins: influence of estradiol and progesterone. Prostaglandins 7:465–472
63. Van Dayne CM, Havel RJ (1959) Plasma unesterified fatty acid in fetal and neonatal life. Proc Soc Exp Biol Med 102:599–602
64. Wiedman E, Schwartz E (1972) Suppression of growth hormone dependent human serum sulfation factor by estrogen. J Clin Endocrinol Metab 34:51–58
65. Yagel S, Palti E, Gallily R (1988) Prostaglandin E$_2$-mediated suppression of human maternal lymphocyte alloreactivity by first trimester fetal macrophages. Obstet Gynecol 72:648–654
66. Yoshimura M, Nishikawa M, Yoshihawa N, Horimoto M, Toyoda N, Sawaragi I, Inada M (1991) Mechanism of thyroid stimulation by human chorionic gonadatropin in sera of normal pregnant women. Acta Endocrinol (Copenh) 124:173–178
67. Zhu HH, Huang JR, Marella J, Rosenberg M, Tseng L (1990) Differential effects of progestin and relaxin on the synthesis and secretion of immunoreactive prolactin in long term culture of human endometrial stromal cells. J Clin Endocrinol Metab 71:889–899
68. Zolti MN, Ben-Rafael Z, Meirom R, Shemesk M, Binder D, Mashiack S, Apte RN (1991) Cytokine involvement in oocyte and early embryos. Fertil Steril 56:265–272

117 Functions of the Placenta

U. Karck and M. Breckwoldt

Contents

117.1 Introduction

The placenta is the organ that, whilst separating the fetal from the maternal circulation, facilitates the exchange of nutrients and all other supplies necessary for fetal growth and well-being. To do this the placenta has to fulfill tasks that in extrauterine life will be taken over by separate and highly specialized organs.

During fetal development, placental function involves O_2 and CO_2 exchange, as well as the transport of glucose, fatty acids, and amino acids to the fetus. Furthermore, supply of the fetus with essential minerals, vitamins, and water is provided by placental transport systems. In the other direction, metabolic breakdown products are removed from the fetus. Immunoglobulins (IgG) are able to pass the placental barrier, thus providing a first line of protection against infections in early extrauterine life.

In addition, the placenta efficiently separates the maternal and fetal compartments, allowing the fetus to develop its own biological identity undisturbed.

The endocrine function of the placenta is essential for uterine growth and fetal development (see Chap. 116).

To illuminate the physiology of this remarkable organ, some basic aspects of its anatomy will be presented before engaging in a discussion of the physiological processes involved.

117.2 Anatomy

The placenta at term weighs some 300-500 g, is approximately 30 mm thick, and has a diameter of about 25 cm. It is an organ that has developed from the early trophoblast in a series of maturation steps. The trophoblast (from the Greek *trophe*, meaning nutrient) consists of those cells of the conceptus that will form not the embryo itself, but the supporting tissue which is necessary for embryonic and, later, fetal growth (Fig. 117.1).

This chapter will confine itself to such facts as are essential for a deeper understanding of placental function.

117.2.1 Morphology of the Mature Placenta

The placenta (Fig. 117.2) enables intimate contact between fetal and maternal blood in a maze of blood vessels and vascular spaces with a small amount of connective tissue between them. On the fetal side it is covered by the shiny chorion, under which the ramifications of the two arteries and the larger vein from the umbilical cord are visible. The maternal side has a slightly velvety appearance and its surface is normally broken up into 12–20 subdivisions, called cotyledons. The arrangement of the cotyledons is variable. A further functional subdivision is constituted by the arborizing fetal vessels entering into small units of the placenta, referred to as *placentones* [42], each of which is supplied by a coiled artery from the maternal side. An average of about 50 placentones are found (some authors describe up to 120), with the maternal blood entering in the middle of the structure and spreading out from there. The mature fetal villi hang in this maternal blood flow like an upturned tree in a well (see Fig. 117.2).

117.2.2 Histology

The immature placenta consists of large masses of syncytiotrophoblast, forming primitive lacunae that will

R. Greger/U. Windhorst (Eds.)
Comprehensive Human Physiology, Vol. 2
© Springer-Verlag Berlin Heidelberg 1996

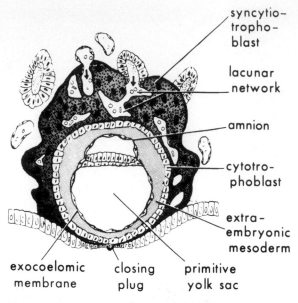

syncytio-
tropho-
blast

lacunar
network

amnion

cytotro-
phoblast

extra-
embryonic
mesoderm

exocoelomic   closing   primitive
membrane   plug   yolk sac

Fig. 117.1. Cross-section of an implanted blastocyst, aged 10 days. Note the relatively large mass of syncytiotrophoblast, that part of the trophoblast which has lost its cell membranes, melting into a syncytium. Towards the embryo the syncytiotrophoblast is covered with the cytotrophoblast, consisting of distinct individual cells. (From [36])

later become the intervillous space. Columns of trophoblast build up to form the primary villi, some of which will attach to the decidua basalis as anchoring villi. Around day 11 after implantation the progressing cytotrophoblast erodes and opens maternal capillaries, thus initiating the flow of blood into the lacunae of the immature placenta. With further development the primitive villi are replaced by villi containing blood vessels and connective tissue, with a surface covered by a double layer of epithelium, the cytotrophoblast and the syncytiotrophoblast. Eventually, a tree-like architecture evolves with a main stem through which large fetal blood vessels run, and which ramifies into secondary and tertiary branches up to the tenth order (Fig. 117.3).

Towards term the small villous twigs consist only of fetal capillaries surrounded by the syncytiotrophoblast (Fig. 117.4).

The area available for maternal/fetal exchange increases during pregnancy, reaching its maximum near term. Events such as intervillous deposition of maternal fibrin and calcification of blood vessels reduce this area, a phenomenon of placental aging that, if it occurs prematurely and extensively, can impair placental function, eventually leading to placental insufficiency.

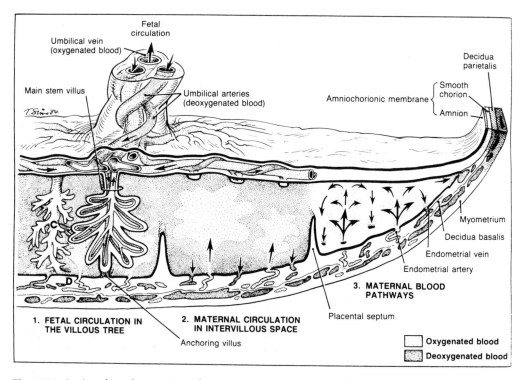

Fetal circulation

Umbilical vein (oxygenated blood)

Main stem villus

Umbilical arteries (deoxygenated blood)

Amniochorionic membrane

Decidua parietalis

Smooth chorion
Amnion

Myometrium
Decidua basalis
Endometrial vein
Endometrial artery

3. MATERNAL BLOOD PATHWAYS

Placental septum

1. FETAL CIRCULATION IN THE VILLOUS TREE

2. MATERNAL CIRCULATION IN INTERVILLOUS SPACE

Anchoring villus

☐ Oxygenated blood
▨ Deoxygenated blood

Fig. 117.2. Section through a mature placenta. *1: C:* Typical villous tree anchored on the basal plate, *D.* The directions of fetal blood flow through the umbilical vessels and the villous tree are indicated. *2,3:* Pattern of maternal blood flow through the intervillous space (see text for further details). The cotyledons are separated from each other by placental septa; however, each cotyledon contains two or more functional units, the placentones, supplied by one maternal spiral artery. (From [36])

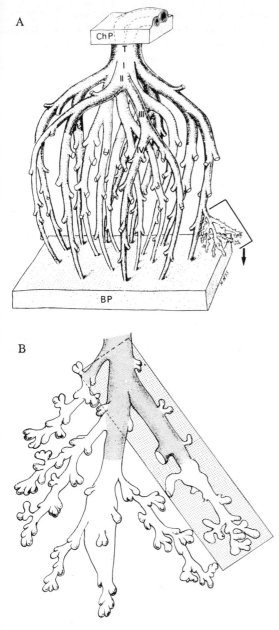

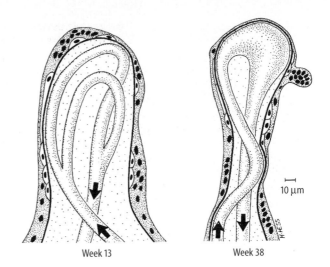

Week 13   Week 38

Fig. 117.4. Developmental changes in villous architecture during pregnancy. At week 13 of gestation the fetal capillaries are covered by a double layer of cells, the syncytiotrophoblast and the cytotrophoblast. Near term very few cytotrophoblasts are found, allowing even more intimate contact between the fetal capillaries and the syncytiotrophoblast lining of the intervillous space. (From [23])

Fig. 117.3. A Larger-stem villi of the villous tree. *ChP*, Chorionic plate; *BP*, basal plate; *T*, truncus; *I–IV*, first to fourth order branchings. **B** Higher magnification view of **A** with higher order branches and terminal villi. (From [24])

117.3 Circulation

117.3.1 Fetal Side

Fetal blood is transported to the placenta via the two hypogastric arteries, which become the two umbilical arteries (see Fig. 117.2) carrying deoxygenated blood. Upon reaching the placenta they subdivide further, thus reaching all areas of the placenta. Arterial blood flow proceeds down intervillous stems into smaller arterioles and finally into capillaries traversing the villi. These capillaries drain into venules which join veins, finally forming the single umbilical vein. Arteriovenous anastomoses can be found.

Using Doppler ultrasonographic imaging techniques, umbilical blood flow has been calculated to be of the order of 20–100 ml/min at 22 weeks of gestation, increasing to some 300 ml/min near term, representing about 50% of fetal cardiac output. These figures correspond to data derived from other experiments where umbilical blood flow was estimated to be between 100 and 120 ml/min × kg fetal weight. The absolute increase in umbilical blood flow during gestation is a result of an increase in placental conductance. This is due to the fast initial growth of the placenta and the subsequent enlargement of the total vascular diameter. The second relevant mechanism is the increase in fetal blood pressure throughout pregnancy, reaching about half the arterial blood pressure of adults [14].

Fetal intracapillary volume in the placenta at term is about 50–60 ml, which represents one-fifth of the maternal volume in the intervillous space.

From perfusion experiments with human placental cotyledons, it has been concluded that locally produced vasoactive substances such as prostanoids and angiotensin II may play a role in local modulation of fetal blood flow. It was shown that angiotensin II and a thromboxane analogue had a pronounced vasoconstrictive effect, whereas prostacyclin led to a dose-related decrease in perfusion pressure [6,28]. Recently, receptors recognizing endothelin-1, another potent vasoconstrictor, have been demonstrated on human fetal placental vessels [40].

Thus, there is good experimental evidence that all vessels of the human fetal extracorporeal circulation are capable of responding to both vasodilator and vasoconstrictor autacoids. This response may be modulated by the oxygen tension in the fetal vessels. However, due to their intrinsic low vascular tone under normal circumstances, the fetal placental vessels are virtually unresponsive to exogenous vasodilator stimuli, unless they have been constricted in some way.

The extrafetal blood vessels and the placenta lack innervation; for this reason, neural control over placental blood flow is considered to be insignificant [3,19].

Generally, the fetal capillary pressure is higher than the pressure in the intervillous space, in order to maintain fetal transplacental blood flow.

Animal experiments indicate that the vascular resistance follows an unregulated preset developmental program without major adaptions to fetal needs. The microvasculature within fetal villi contributes 70% to overall placental resistance [13]. It is concluded that the fetus grows within the limits set by its placenta [3].

117.3.2 Maternal Side

Maternal blood enters the placenta via 50–120 spiral arteries that run perpendicularly through the myometrium of the uterus. The major drop in the intravascular pressure occurs before the blood reaches these arteries, probably during its passage through the arcuate and radial arteries in the myometrium. The spiral arteries penetrate the basal plate of the placenta and drain into the intervillous space lined by the syncytiotrophoblast. Uteroplacental blood flow (up to 90% of the flow will pass through the intervillous space) has been estimated to be around 90–120 ml/min × kg (uterine plus fetal and placental weight) consistently from midpregnancy until term, where it will be around 600 ml/min, representing approximately 10% of the maternal cardiac output at rest [4,26,34].

During the course of pregnancy, the flow resistance of the uteroplacental arteries is reduced by structural widening of the lumen associated with growth and degeneration of the arterial walls, together with the breakdown of elastic fibers. These events are referred to as "physiological changes of the uteroplacental arteries," allowing adjustment of maternal blood flow to fetal growth requirements [7]. Estrogens and other hormonal factors play a crucial role in these changes [34], thus controlling uteroplacental blood flow (Fig. 117.5).

After entering the placenta through the basal plate, the maternal blood is forced in a spurt-like fashion by the maternal arterial pressure towards the chorionic plate. There it forms a subchorionic lake before radial dispersion to the periphery of the circulatory unit, the placentone, occurs. The resistance is highest around this subchorionic lake because of relatively dense distribution of villi in this area; it drops towards the periphery of the placentone with decreasing density of the villi. This allows a fairly continuous and homogeneous flow of maternal blood through the intervillous space [41]. After bathing the chorionic villi, the blood drains into the veins via venous openings in the basal plate. These veins generally run parallel to the uterine wall (see Fig. 117.2). The volume of the intervillous space of a normal term placenta has been estimated to be 250–300 ml.

The uteroplacental veins are compressed during uterine contractions. Arterial inflow is also reduced at this time, but less so than the venous outflow. Contractions during normal labor with an increase in intrauterine pressure of 20–40 mmHg lead to a 50% reduction of uteroplacental flow [26].

The principal factors controlling uteroplacental blood flow are thus:

- Maternal arterial blood pressure
- The pattern of uterine contractions
- The physiological changes of the spiral arteries (see above).

α-Adrenergic substances and other vasoconstrictors, such as angiotensin II or vasopressin, are known to increase the vascular resistance of the uteroplacental compartment, basically due to the constriction of intramyometrial arteries [46]. β-Adrenergic substances, on the other hand, de-

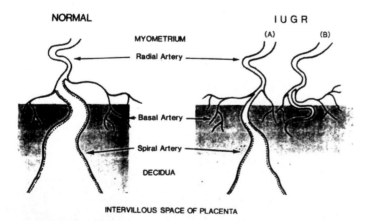

Fig. 117.5. Physiological changes in the spiral arteries, with enlargement of the decidual and myometrial segments. In cases of intrauterine fetal growth retardation (IUGR) these morphological adaptations are partly (A) or completely lacking (B). (From [44])

crease the resistance and improve the blood flow. This is a combined effect of changes in the tonus of the vasculature itself and the effect of these substances on uterine contractility.

The uterine arteries are under normal control of the sympathetic nervous system, so situations that lead to general vasoconstriction, such as shock and stress, will also cause constriction of uterine vessels.

117.3.3 Fetomaternal Interaction

The blood flow interaction in the fetomaternal circulatory unit, the placentone, follows the multivillous pattern. The random distribution of villi within the placentone does not allow an organized countercurrent flow (Fig. 117.6).

117.4 Placental Immunology

It is surprising, and as yet unexplained, that the pregnant mother tolerates the growth of the fetal organism in her own body, as genetically the latter constitutes an allograft. This tolerance is the more intriguing as there is strong experimental evidence that small numbers of cells pass between mother and fetus in both directions, which could lead to allosensitization.

Various facts and theories exist that may explain this tolerance:

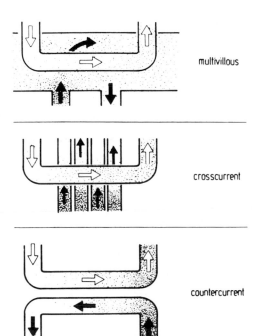

multivillous

crosscurrent

countercurrent

Fig. 117.6. Schematized arrangement of fetal and maternal blood stream of different types of exchange systems. The density of the dots indicates the efficiency of the two exchange systems for diffusional transfer. (From [32])

- The modulation of maternal immunoreactivity by mediators may be increased during gestation, e.g., by the large amounts of progesterone produced by the placenta, where it can be found in very high concentrations.
- There is a fairly complete separation of fetal and maternal circulation (see above), with the trophoblast acting as a kind of immunological barrier.
- The trophoblast appears to be immunologically privileged, as is proven by the occurrence of extrauterine pregnancies, which would not occur if the uterus was the decisive factor in maternal immunotolerance of the fetus. It has been demonstrated that the syncytiotrophoblast only weakly expresses HLA-1 molecules, and that no class II MHC antigens are present on trophoblast cell surfaces [9,17,21].
- Trophoblast-lymphocyte cross-reactive (TLX) antigens are believed to play a major role in pregnancy immunology (see also Chap. 85). It is suggested that maternal recognition of these antigens leads to immune responses that are protective of the trophoblast. Recent data indicate that TLX antigens are similar or perhaps identical to membrane cofactor protein, MCP or CD46 [47]. This is a C3/C3b binding cell surface protein that is a cofactor for complement factor 1 cleavage of C3b, and is required for cells to coexist with autologous complement. The spontaneous cleavage of C3 leads to the deposition of C3b on membranes, thus initiating the amplification of the complement cascade. MCP is thought to interact with activated C3, helping to distinguish self from nonself and preventing the activation of complement by autologous cells. Syncytiotrophoblast is a rich source of MCP, which may modulate maternal reaction to TLX allotypic antigens by not allowing the formation of the membrane attack complex [29,43,47].
- Another theory postulates changes in the antibody network. Two mechanisms could be involved: (a) an increased prevalence of asymmetric nonprecipitating antibodies, which can bind to their respective antigens without activating effector functions such as complement fixation [5]; (b) the blocking of maternal anti-HLA antibodies, which develop in pregnant women, by soluble HLA antigen shed from the placenta and by anti-ideotypic antibodies produced by the mother [38].

117.5 Placental Transfer

Normally there is no direct communication between the fetal blood in the chorionic villi and the maternal blood in the intervillous space, with the exception of occasional breaks in the chorionic villi allowing the exchange of varying numbers of cells between mother and fetus. However, in principle the transfer from mother to fetus and vice versa depends primarily on mechanisms that allow transport of substances through the intact chorionic villus.

The placental barrier in microanatomical terms therefore consists of the syncytiotrophoblast, the basal membrane and stroma of the chorionic villus, and, finally, the endothelium of the fetal blood vessel. Altogether these layers have a minimum thickness of about 3–6 μm [18]. During the course of the pregnancy this barrier progressively becomes thinner, due to a decrease in cytotrophoblast cells and a similar change in the walls of the villous capillaries (see Fig. 117.4).

Morphometric measurements of the term placenta gave an estimated total villous surface area of some 10 m², the microvilli of the trophoblast cells efficiently enlarging the surface area. The total surface area of fetal intravillous capillaries at term was calculated to be approximately 23 m²; however, the area in close immediate contact with the villous surface is only about 1.8 m². This latter is probably the area with the highest permeability [1,32].

117.5.1 Transport in General

Before discussing transplacental transfer, a few basic principles concerning the transmembranous and transcellular transport of small molecules, ions and macromolecules will be reviewed (see also Chaps. 4, 5, 8, 59).

The main constituent of cellular and basal membranes, the lipid bilayer, is highly impermeable to most polar molecules, but permeable to water. Transport of small water-soluble molecules is achieved by special transmembrane proteins, each of which is responsible for the transfer of a specific molecule or group of related molecules. Given enough time, most molecules can diffuse through the lipid bilayer, the diffusion rate being higher for smaller molecules of high solubility in oil. O_2, a small uncharged molecule, will cross rapidly, whereas glucose will hardly cross at all. Charged molecules have a very low permeability coefficient, illustrating the virtual impermeability of the lipid bilayer for ions (Fig. 117.7A). (For a more comprehensive treatment of these issues, see Chaps. 5, 8, 59.)

Simple diffusion depends on the electrochemical gradient across the membrane. The flow rate of a solute across a membrane is directly proportional to the difference of its concentration on the two sides of the membrane. Multiplying this concentration difference (mol/cm³) by the permeability coefficient (cm/s) gives the flow in moles per second per square centimeter of membrane (see Eq. 117.1').

This relation is derived from *Fick's law*, which describes the diffusion of solutes across a membrane:

$$M/dt = D \times A \times (C_1 - C_2)/d, \qquad (117.1)$$

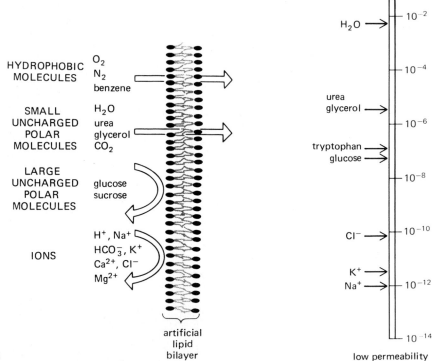

Fig. 117.7. A Permeability properties of a synthetic lipid bilayer to different classes of molecules (for details see text). **B** Permeability coefficients (Pc, cm/s) for the passage of various substances through lipid bilayers. The Pc allows the calculation of the flow of a solute in moles per second per square centimeter of membrane. For example, a concentration difference of glucose of 10^{-6} mol/ml

($= 1$ mmol/l, as seen near term) would cause a flow of 10^{-6} mol/ml $\times 5 \times 10^{-8}$ cm/s $= 5 \times 10^{-14}$ mol/s $\times$ cm². However, a transfer of about 3×10^{-6} mol/s $\times$ cm² per fetus is required. So at this diffusion rate the exchange area would have to be 1.7×10^8 cm², or 17 000 m². The alternative is an increase in transport speed by facilitated diffusion (see text). (Adapted from [2])

where M is the mass in moles, dt the time, M/dt the flow rate, D the specific diffusion constant for the solute, A the exchange area, C_1 and C_2 are concentrations on either side of the membrane, and d is the membrane diameter.

With A set to 1 cm² and D and d taken together as Pc ($= D/d$), we obtain a simplified form of the equation, giving the flow per second per square centimeter:

$$M/(dt \times A) = Pc \times (C_1 - C_2) \qquad (117.1')$$

For the passive transport of molecules that do not diffuse rapidly, the cell membranes contain transport proteins of two basic types:

- *Carrier proteins* that bind the solute to be transported, and upon binding undergo a conformational change, exposing the solute binding site first on one side of the cell membrane, then on the other, transferring the solute across the membrane along its concentration gradient.
- *Channel proteins* that form water-filled pores extending across the lipid bilayer. These allow specific solutes to pass through them and thereby cross the membrane. Again, these channel proteins and many carrier proteins permit only transport following concentration gradients, a process termed "facilitated diffusion."

117.5.2 Active Transport

The transfer of solutes against their electrochemical gradient requires an active transport and is always performed by special pump proteins (cf. Chap. 8). The pumping activity of these proteins is tightly linked to a source of metabolic energy such as ATP hydrolysis or ion gradients (Fig. 117.8). The asymmetrical distribution of channels, carrier proteins, and pumps in epithelial cells, like the trophoblast lining of the intervillous space, allows the transcellular transport of solutes. For example, carrier proteins located at the apical (absorptive) region of the cell pump glucose into the cell, thus building up a substantial concentration

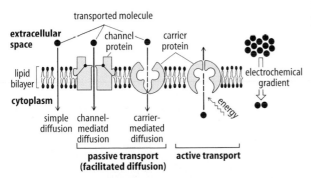

Fig. 117.8. Different modalities of transmembranous transport. Simple diffusion and passive transport by membrane transport proteins (facilitated diffusion) occur down an electrochemical gradient. Active transport against this gradient requires metabolic energy. Only nonpolar molecules and small uncharged polar molecules can cross the lipid bilayer directly by simple diffusion; the transfer of other polar molecules occurs via specific carrier or channel proteins. (From [2])

gradient. Carriers located at the basal or lateral domain allow facilitated diffusion of glucose out of the cell along the concentration gradient.

117.5.3 Gas Exchange

The placental capability for gas exchange is mathematically described by the ratio of the exchange rate and the difference of the partial pressure on either side of the circulation:

$$M/(dt \times dp) = Dp \ (\mathrm{mol/s} \times \mathrm{Pa}), \qquad (117.2)$$

where M is the mass in moles, dt the time, dp the difference between partial pressure in fetal and maternal arterial blood, and Dp the diffusion capacity of the placenta.

117.5.4 Membrane Transport of Macromolecules and Particles

The transport proteins described above cannot translocate macromolecules such as proteins, polysaccharides, or nucleotides. To perform this task the cells engage in processes called *endocytosis*, *exocytosis*, and *transcytosis*, which involve the sequential formation and fusion of membrane-bound vesicles (see Chaps. 4, 8).

In *endocytosis* substances to be ingested are progressively enclosed by a small portion of the plasma membrane, which first invaginates and then pinches off to form an intracellular vesicle containing the ingested material. *Exocytosis* is a similar mechanism with the sequence of events reversed. During exocytosis the vesicle membrane becomes part of the plasma membrane.

Most, but not all of the molecules taken up by cells via endocytosis end up in lysosomes – membrane-bound organelles loaded with degradation enzymes. Normally this material is rapidly broken down and the small breakdown products such as amino acids, sugars, and nucleotides are released into the cytosol for further processing.

Receptor-mediated endocytosis is highly specific and efficient. The macromolecules bind to cell-surface receptors, accumulate, and enter the cell as receptor-macromolecule complexes in endocytic vesicles. So far more than 25 different receptors have been identified as participating in receptor-mediated endocytosis for different types of molecules. The subsequent fate of the ingested material depends on the receptor, which can either travel to lysosomes or return to a different domain of the plasma membrane, as in transcytosis. The latter pathway is thought to be involved in the placental uptake of IgG molecules.

117.5.5 Transport Processes in the Placenta

The following variables determine the performance and ability of the placenta as an organ of transfer:

- The concentration of the substance in question in the maternal plasma
- Maternal blood flow through the intervillous space
- The area available for exchange across the villous epithelium
- The type of transfer: diffusion, facilitated diffusion, active transport
- The amount of the substance metabolized by the placenta during transfer
- The area available for exchange across the wall of the fetal capillaries
- Fetal blood flow through villous capillaries
- The matching of maternal blood flow to fetal blood flow.

If maternal and fetal flow are high and if the transport capacity of the blood for the substance in question is large, the transplacental transfer becomes *membrane-limited*. In the case of high permeability of the placenta, the transport can be *flow-limited*.

117.5.6 Placental Efficiency and Clearance

During the diaplacental exchange, plasma concentrations of substances in fetal and maternal blood will change. The concentration in fetal arterial blood will alter towards an equilibration to maternal arterial blood. The ratio between the concentration difference in fetal venous (C_{fv}) versus fetal arterial (C_{fa}) blood and the concentration difference in maternal arterial (C_{ma}) versus fetal arterial blood is called *efficiency* (E):

$$E = (C_{fv} - C_{fa})/(C_{ma} - C_{fa}). \qquad (117.3)$$

This ratio describes the relative amount of fetal blood flowing through the placenta that obtains the same concentration of the substance as is present in maternal arterial blood. Furthermore, if the transport capacity of fetal and maternal blood is similar, it gives a measure of the relative use of the transport capacity of the fetal blood for the particular substance. An efficiency of 1 is seen if permeability is high and the concentration in fetal arterial blood equilibrates to the concentration in maternal arterial blood and C_{fv} becomes equal to C_{ma}. This shows that the transport capacity has been used to its full extent, whereas an efficiency of 0.2 means that only 20% of this capacity is utilized.

Another parameter used to describe placental function is the *placental clearance*. This is the absolute amount of fetal blood which equilibrates to maternal arterial concentration over a certain time (second, minute, hour):

$$Cp = Qf \times E, \qquad (117.4)$$

where Cp is placental clearance, Qf fetal transplacental blood flow (volume/time), and E efficiency (see Eq. 117.3). For example, if $E = 0.5$ and the transplacental blood flow is 300 ml/min (as seen near term), Cp will be 150 ml/min, that is about 60% of the total fetal blood volume.

117.5.7 Diffusion

Substances with a molecular weight less than 500 Da diffuse rapidly through the placenta. In general we find that the smaller the molecule, the faster the rate of diffusion (see also Figs. 117.7 and 1.62). However, for a variety of small compounds the placenta actually facilitates transport, thus improving the transfer of substances that are at low concentrations in the maternal plasma but are essential for fetal growth.

Since the total surface of the placental exchange area (A in Eqs. 117.1 and 117.1') is not exactly known, only the simplified version of the Fick equation can be used to calculate the permeability of the placenta for the solute in question, with Pc as placenta permeability:

$$Pc = M/dt \, (C_1 - C_2), \qquad (117.1')$$

C_1 and C_2 being the mean concentrations of the respective substance on either side of the placental barrier, not necessarily the concentrations in arterial and venous blood respectively. This is only the case if transplacental transport is small in comparison to blood flow (low efficiency), because then there will be only a negligible change in concentrations during the passage of blood through the placenta. In this special case ($C_{ma} - C_{fa}$) will be the true driving gradient for placental diffusion, and the following equation can be used:

$$Pc = M/dt \, (C_{ma} - C_{fa}). \qquad (117.1'')$$

The in vivo situation is complicated even further by the fact that the placenta itself utilizes energy substrates (glucose, O_2) and sometimes contributes (e.g., lactate) to the substance transported.

Substances transported by diffusion are: O_2, CO_2, water, and some electrolytes. Anesthetic gases also diffuse rapidly by simple diffusion.

117.5.8 Transfer of O_2 and CO_2

Oxygen. Fetal demand for O_2 has been estimated at around 5 ml O_2/min × kg fetal weight [4]. O_2 saturation in the intervillous space is estimated to be 65–75% with a PO_2 of 30–35 mmHg and resembles that in the maternal capillaries, which is less than in the maternal arterial blood (approx. 98%). The values in umbilical venous blood are similar (Table 117.1).

O_2 equilibrates rapidly across the placenta, and as a result placental O_2 exchange becomes convection- or flow-limited. Taking all the available experimental data together, it is estimated that a driving force of a difference for PO_2 of 1 mmHg across the villous membrane will lead to a diffusion of 1.6 ml O_2/min per 500 g placenta at term. A gradient in PO_2 of about 12 mmHg is necessary to meet the demand of 20 ml/min. but this gradient will only suffice if the transport is of high efficiency. The efficiency can decrease, however, if fetal and maternal blood flow are mismatched as a

Table 117.1. O_2 and CO_2 pressures (PO_2 and P_{CO_2}), O_2 and CO_2 concentrations (O_2, CO_2), O_2 saturation (SO_2), and pH in maternal and fetal blood near term. Values from various sources. (Table taken from [27])

	P_{O_2}	SO_2	O_2 (ml/l)	P_{CO_2} (mmHg)	CO_2 (ml/l)	pH
Uterine artery	95–100	97–100	140–170	28–36	430–500	7.40–7.45
Umbilical art	14–25	20–30	50–100	44–52	450–550	7.24–7.34
Umbilical vein	22–35	45–70	110–160	38–45	400–500	7.32–7.38
Uterine vein	30–40	50–70	80–110	42–48	480–580	7.35–7.40

result of excessive shunting, or if the diffusion rate is lowered due to massive fibrinoid deposition, as seen in preeclamptic women with placental insufficiency.

Any measurable difference between the PO_2 in fetal and maternal venous blood has to be due to shunting of blood. Shunting of maternal arterial blood will increase the PO_2 in maternal venous blood, whereas shunting of fetal arterial blood will decrease the PO_2 in fetal venous blood. At term the mean PO_2 difference has been calculated to be of the order of 20 mmHg with an apparent diffusion capacity of 1 ml O_2/min × mmHg. This diffusion capacity is lower than expected because O_2 passing from the maternal to the fetal compartment is partly utilized by the placenta. The O_2 demand of the placenta at term has been estimated to be at least 5–10 ml/min × kg, equalling 15–25% of fetal O_2 consumption. The relatively low PO_2 in fetal blood is compensated by the higher cardiac output per unit body weight of the fetus, the higher hemoglobin (Hb) concentration, and the higher affinity for O_2 of fetal hemoglobin.

The measure of *affinity* is the half saturation pressure (P_{50}). If the PO_2 resulting in 50% saturation of the hemoglobin is low (high affinity), higher levels of saturation at lower PO_2 are achieved (Fig. 177.9). The P_{50} of HbA (mother) is

around 27 mmHg and that of HbF (fetus) 22 mmHg. This difference is a result of the different effects of diphosphoglycerate (DPG) on HbF and HbA. DPG decreases the O_2 affinity of HbA more than that of HbF due to the different amino acid composition and binding of DPG. Thus, a PO_2 of 30 mmHg will lead to 55% saturation of maternal blood compared to 60% of fetal blood. Furthermore, the concentration of HbF is higher than that of HbA. For example, a typical HbF of 170 g/l results in an O_2 capacity of 230 ml/l fetal blood, in contrast to the HbA of 120 g/l normally found in pregnancy, which results in a capacity of 160 ml/l. At a PO_2 of 30 mmHg 140 ml/l O_2 will be bound by fetal blood in comparison to 90 ml/l by maternal blood. Thus, higher affinity and capacity of fetal blood support the transplacental transfer of O_2.

Another phenomenon supporting transfer is the *Bohr effect*. This describes the finding that Hb saturation curves are dependent on pH and P_{CO_2}: a decrease in pH and an increase in P_{CO_2} will shift the curve to the right, i.e., affinity is reduced, thus facilitating the release of O_2. The pH of fetal arterial blood is lower than that of maternal blood (7.26 vs 7.42) and the P_{CO_2} is higher 52 mmHg vs 32 mmHg. During placental passage the fetal blood loses protons and CO_2, which are taken up by the maternal blood, shifting the saturation curve of maternal blood to the right (lower affinity) and that of fetal blood to the left (higher affinity). This double Bohr effect alters the P_{50} on either side by another 1 mmHg (see Fig. 117.9).

Of similar magnitude are the effects of the temperature changes, fetal blood being slightly warmer (+0.5°C) than maternal arterial blood. The cooling of fetal blood increase the O_2 affinity of HbF. That all these mechanisms allow the fetus full aerobic cover of its energy and metabolic demands is underlined by the fact that the lactic acid content of fetal blood is only slightly higher than that of the mother.

Regulation of O_2 Transfer. As the O_2 exchange is flow-dependent (see above), a reduction in maternal blood flow will lead to a drop in O_2 transfer, and subsequently fetal PO_2 and O_2 saturation will fall. If the PO_2 drops below – 15 mmHg, and thus saturation to under 30%, O_2 uptake of fetal tissues will be impaired, leading to fetal hypoxia. The clinical sign of this hypoxia is a drop in fetal heart rate, which puts the fetus in energy-saving mode. An increase in fetal lactate production and a drop in pH are also seen – parameters frequently used for the monitoring of fetal well-being during labor.

An increase in fetal O_2 consumption and uptake in the

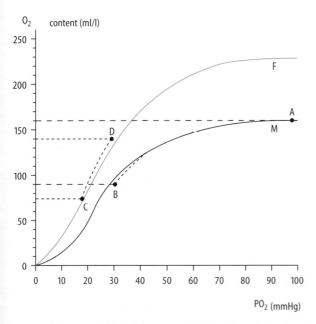

Fig. 117.9. Oxygen-binding curves of fetal (HbF) and maternal hemoglobin (HbA). *A* and *D* are arterial values, *B* and *C* venous values. The *dotted lines* indicate the pH-dependent shifts (the Bohr effect). *F*, fetal, *M*, maternal

periphery will increase the effective difference in PO_2 over the placental membrane, thus increasing O_2 transfer.

In cases of fetal distress, a slight improvement in fetal O_2 supply can be brought about by the mother's breathing in 100% O_2 at ambient pressure. This results in a 10-mmHg increase in the PO_2 in the venous umbilical blood [4].

Carbon Dioxide. The transfer of CO_2 from the fetus to the mother follows the same rules as O_2 exchange, albeit slightly faster. At term the fetus produces some 15-20 ml CO_2 per minute, which has to be eliminated. A liter of adult blood usually contains some 500 ml CO_2. Of that CO_2 5% is in solution, 5% is bound as carbamino hemoglobin, and 90% is found as HCO_3^-, one-third of this HCO_3^- in the erythrocytes and two-thirds in plasma. In the blood of the umbilical artery a P_{CO_2} of 44–52 mmHg is measured with 450–550 ml/l CO_2 being bound. Due to the physiologic hyperventilation in pregnancy, the P_{CO_2} in maternal arterial blood is lower, 28–36 mmHg, equalling 430–500 ml CO_2 being bound. This difference of the P_{CO_2} leads to a net transfer of CO_2 from fetal to maternal blood in the range of 50-70 ml CO_2/l blood, leading to a decrease in P_{CO_2} in fetal blood (Table 117.1). Most of that CO_2 is derived from bicarbonate.

The exchange of CO_2 is supported by the concomitant oxygenation of fetal blood, as oxygenation reduces the buffering capacity of hemoglobin (*Haldane effect*). A 30% increase in O_2 saturation facilitates the release of 20–30 ml Co_2/l blood. Although the membrane solubility of CO_2 is 20 times that of O_2, the diffusion capacity of CO_2 is only three times as high. The reason for this is that some of the events involved in fetal-maternal CO_2 transfer are time-consuming (see Fig. 117.10).

The bicarbonate concentration is higher in maternal blood, the fetus has a higher negative base excess and higher concentrations of lactate. Thus, in relation to the adult, the fetus has a slight metabolic acidosis.

117.5.9 Exchange of Nonvolatile Substances

Water. The diffusion permeability of lipid bilayes and the placenta for water is quite high, so water exchange is perfusion-limited. Tracer experiments with D_2O demonstrated that at term up to 200 ml water can be turned over per minute [20]. This water diffusion has to be distinguished from free water flow through protein channels; with the former no transport of solutes is possible, whereas with the latter solute transfer is seen due to solvent drag. The extent of this solvent drag depends on whether the specific solutes fit through the pores in the membrane and is described by the reflection coefficient, which indicates the relative amount of molecules reflected by the membrane ($\sigma = 1$ means no passage, $\sigma = 0.3$ means 70% of the solutes are dragged along).

However, taken all round the fetal demand for water is very limited, as most of the water it requires can be supplied by its own oxidative metabolism, in which water is produced as one of the end products (1 mol/day = 18 ml). Thus, it is not surprising that there is normally little transplacental net flow of water to meet the demand of about 25 ml/day.

Glucose Transfer by Facilitated Diffusion. The main source of metabolic energy for the fetus is glucose. Assuming that nearly all of the O_2 taken up by the fetus (up to

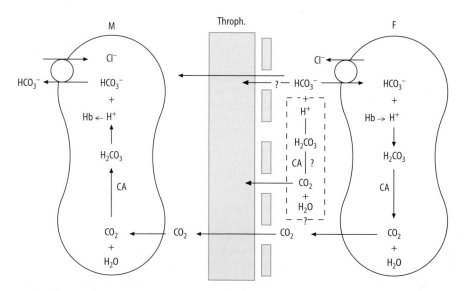

Fig. 117.10. Processes involved in fetomaternal CO_2 transfer: events marked with a "?" are still doubtful. See also Chap. 102. The rapid fetomaternal diffusion and subsequent drop in P_{CO_2} in the fetal plasma lead to an increased formation of CO_2 from bicarbonate in the red blood cells (RBC). At the same time, the pH drops as a consequence of Hb oxygenation. This leads to a gradient for bicarbonate between fetal plasma and RBC, resulting in an uptake of bicarbonate in exchange for chloride. In the maternal blood the inflowing CO_2 enters the RBC; at the same time pH rises due to the deoxygenation. As a result, H_2CO_3 is formed out of CO_2 and water with subsequent dissociation into HCO_3^- and H^+. This step is catalyzed by the enzyme carboanhydrase (CA). In maternal blood, the newly formed HCO_3^- is released from the RBC in exchange for chloride from the plasma. (From [27])

20 ml/min near term = approx. 1 mmol O_2) is used for glucose oxygenation and that glucose is metabolized completely aerobically, it can be calculated that fetal glucose consumption near term is 1/6 mmol (170 μmol = 30 mg) glucose per minute (= 1.8 g/h = 43.2 g/day).

The driving force for the diaplacental transfer of glucose is the maternofetal difference in plasma concentration, which ranges between 2 mmol/l in midpregnancy to 1 mmol/l near term. Glucose clearance, i.e., the amount of fetal blood that equilibrates to maternal plasma concentration after placental passage, is 170 ml/min, or 0.3 ml/min × g placenta, representing about half of the total umbilical blood flow. This value indicates that the trophoblast permeability for glucose is so high that glucose transfer becomes dependent on flow rates as well as on "permeability." The kinetics of glucose transfer and the fact that L-glucose is hardly transported at all provide evidence for the existence of carrier-mediated transfer. Further findings in support of this concept are the significantly slower transport rates of similar molecules such as fructose or mannose [8].

The capacity of the placental glucose transporter is limited and it can be saturated in vitro by high glucose concentrations. The transport protein can be blocked by phloretin [10,39]. (For more details on D-glucose transport, see Chaps. 8 and 59.)

The glucose consumption of the placenta is substantial (0.25–0.5 μmol/min × g placenta), equalling that of the fetus.

Amino Acid Uptake by Active Transport. The placenta concentrates a large number of amino acids intracellularly, and uptake is by active transport.

During the last trimester of pregnancy the fetus requires some 5 g/day amino acids or protein. However, placental permeability for proteins is very low, and fetal plasma concentration for most amino acids is twice that of maternal plasma. Therefore active transport is necessary to allow further uptake of amino acids by the fetus. This active transport is demonstrated by selective transport of L-histidine, which reaches an equilibrium across the placenta within 2–3 min, compared to D-histidine, which takes 3.4 h [35]. Further investigations have demonstrated that all amino acids are accumulated in the trophoblast, reaching an even higher concentration than in fetal plasma, so it is assumed that an active, ATP-consuming, Na^+-dependent transport into the trophoblast occurs. This uptake mechanism is probably carrier-mediated and Na^+-coupled (cf. Chap. 8). The transfer from trophoblast into the fetal compartment is believed to occur via passive transport or diffusion [50].

Both mechanisms – active and passive transport – have been demonstrated in placental tissues [31]. The transport proteins accept distinct groups of amino acids, demonstrated by the finding that competitive inhibition occurs for example between amino acids such as glycine, proline, and histidine [12].

Some of the amino acids taken up by the trophoblast are metabolized in the placenta, a prominent example being

the production of proteohormones such as human placental lactogen (HPL).

Lipids. Neutral fat does not cross the placenta, whereas glycerol does. Palmitic acid and linoleic acid are actively transported from the maternal to the fetal side in the human placenta perfused in vitro. Due to the lipophilic nature of fatty acids, a high permeability of the placenta for these substances may be assumed, so simple diffusion also appears to be a possible mechanism for transfer. However, owing to the very low plasma concentration of free fatty acids, more than 99% are usually protein-bound, so there is little net transfer of fatty acids. This is underlined by the fact that 90% of the adipose tissue of the fetus is built up from carbohydrates.

The uptake and transport of LDL-cholesterol, however, warrants further attention as an example of receptor-mediated endocytosis (cf. Chap. 4). Cholesterol is bound to protein in the form of complexes known as low-density lipoproteins (LDL). The cell synthesizes receptor proteins for LDL and inserts them in its plasma membrane, where they diffuse in the membrane until they are taken up again via endocytosis. Any LDL that became bound in the interim will be carried into the cell in vesicles, which fuse to lysosomes where the cholesterol will be liberated. The receptor proteins are recycled to the plasma membrane, the whole cycle taking about 10 min and being completed whether a ligand was bound or not. The synthesis of LDL receptors is regulated, thus allowing the cell to control the uptake of cholesterol (Fig. 117.11).

Vitamins. The lipid-soluble vitamins A, D, E, and K pass from the maternal to the fetal circulation via simple diffusion, the maternofetal concentration gradient being the driving force.

The situation is different for the hydrophilic vitamins B and C, which are found at a higher concentration in the fetal plasma, so that active transport must be assumed. The process involved is probably similar to that for amino acids, with accumulation within the trophoblast and subsequent passive diffusion into the fetal circulation.

For vitamin C a different strategy is used. The oxygenated form of ascorbic acid, dehydroascorbic acid, passes the placenta by passive diffusion and is subsequently hydrolized to ascorbic acid in the fetal compartment in an energy-consuming process. This way a concentration gradient for dehydroascorbic acid is preserved, allowing continuous uptake.

Creatinine. Creatinine concentration is some 2–4 times higher in the fetal than in the maternal plasma. Creatinine is important for ATP synthesis [25]. Whether it is actively transported from the maternal to the fetal circulation is unclear.

Urea. This end product of protein metabolism is transferred from fetus to mother probably by simple diffusion, following the fetomaternal concentration gradient. Diffu-

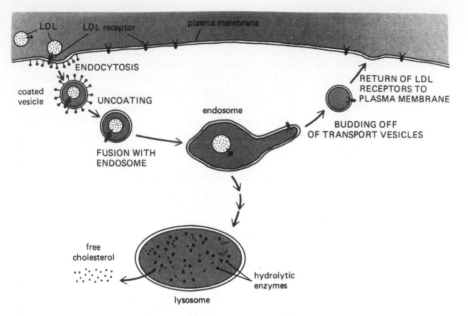

Fig. 117.11. Receptor-mediated endocytosis of low-density lipoproteins (LDL). The LDL dissociates from its receptor in the acidic environment of the nedosome, is transported to lysomes, and is degraded to free cholesterol. The receptor proteins are returned to the plasma membrane. The receptor makes this round trip in and out of the cell whether it is occupied or not. (From [2])

sion is slow, however, due to the limited permeability of lipid bilayers for the hydrophilic molecule.

Sodium. At term total fetal Na^+ is estimated to be 200–250 mmol. About 150 mmol/h can pass through the mature placenta. In comparison to this capacity, which inheres almost entirely in specialized ion channels and membrane pumps, there is very little net transfer. The plasma concentrations on either side of the placenta are equal; however, the intracellular concentration in the trophoblast, as in most other cells, is far lower. This gradient is frequently utilized by gradient-linked symporters for active transport as for amino acids (see above).

Potassium. At term total fetal K^+ is approximately 200 mmol, necessitating a daily net transfer of 1 mmol throughout pregnancy. Placental "permeability" is comparatively high as a result of specific transport proteins. Extracellular K^+ concentration is low, K^+ being actively taken up by the cells often using the (Na^++K^+)-pump. The unequal distribution of these pumps and K^+ channels within the cells suffices for a net transfer across the placenta.

Calcium. There are only minor differences in fetal and maternal plasma concentrations of Ca^{2+}. Intracellular Ca^{2+} concentrations are much higher than those outside the cell. (However, it should be remembered that Ca^{2+} activity within the cell is only around 10^{-7} mol/l.) The concentration within the trophoblast is some 10–20 times higher than that in the blood. Yet placenta permeability is high, easily covering the daily demand for a net transfer of about 2 mmol. The absorptive process involves Ca^{2+} channels for uptake and the Ca^{2+}-ATPase for extrusion (see Chap. 8).

Iron. Another example for active transport is the unidirectional transport of iron across the placenta. Usually maternal iron plasma concentrations are lower than fetal concentrations, and the binding capacity of maternal plasma for iron is much higher, even though iron is actively transported into the fetal compartment. A daily transport of about 1.8 μmol is necessary to meet the fetal total requirements of 510 μmol (30 mg).
This is achieved by receptor-mediated endocytosis, in this case by the endocytosis of transferrin, a protein that carries iron in the blood. The specific cell-surface receptors deliver the transferrin with its bound iron to endosomes, where the low pH induces the separation of transferrin and the iron. The iron-free transferrin (called apotransferrin) remains bound to its receptor and is recycled to the plasma membrane. In the neutral extracellular pH, the apotransferrin dissociates from its receptor and is thus freed to pick up more iron and start the cycle all over again [15].

Cl^- and Phosphate. The daily fetal demand for these two anions is in the range of 1 mmol. The high prevalence of respective transport proteins results in relatively high placental "permeability" due to a combination of ion channels and, secondarily, active transport.

Bicarbonate. HCO_3^- has to be considered together with CO_2 exchange. The fact that the blocking of the enzyme carboanhydrase inhibits HCO_3^- transfer indicates that the actual membrane transfer occurs largely in the form of CO_2, which passes freely. (See Chap. 74 for a detailed account of the mechanism of HCO_3^- absorption.)

Lactate. Mechanisms for facilitated diffusion have been found, allowing transplacental exchange along the concentration gradient. This normally means fetomaternal transfer. However, maternal metabolic acidosis, as sometimes seen during delivery, can lead to fetal acidosis [32,33], suggesting a reversal, or at least reduction, of transport.

Macromolecules. Substances with a high molecular weight usually do not traverse the placenta. There are some exceptions to this, however, e.g., IgG (MW 160 kD). In the absence of a direct antigenic stimulus, e.g., infection, the antibodies in the fetus consist almost entirely of IgG obtained from the mother via the placenta, reaching an even higher concentration than in maternal blood. Thus, the antibody machinery of the fetus reflects the immunological experience of the maternal organism. This can be dangerous for the fetus if there is an incompatibility between its own rhesus blood type and that of the mother. A rhesus-positive fetus can be jeopardized by anti-rhesus antibodies of a rhesus-negative mother in whom rhesus isoimmunization is present due to earlier pregnancies or mismatched blood transfusions. This leads to antibody mediated hemolysis of fetal blood, causing anemia and finally the full symptomatology of hemolytic disease of newborn, with signs of intrauterine heart insufficiency (hydrops fetalis with ascites and pleural effusion), eventually resulting in fetal death.

The transfer of maternal antibodies is thought to occur via receptor-mediated transcytosis. Receptors bind to the special Fc fragment of IgG classs antibodies, thereby picking up the IgG and condensing them in special areas of the membrane destined to form endocytic vesicles. The exact nature of the transcellular transport of transcytotic vesicles is as yet incompletely understood. This mechanism explains why only IgG is transported, whereas antibodies of a similar size – IgA or IgE – but with different Fc fragments are left behind [22,37].

Hormones. For hormones in general the same rules apply as for all other molecules. Lipophilic molecules such as steroids can usually pass freely. Other molecules such as catecholamines can diffuse but are normally inactivated in the placenta during passage. For the thyroid hormone the passage is limited due to its high protein-binding affinity and the small fraction of diffusable free hormone. Larger proteohormones, such as insulin, cannot pass the placenta to any great extent (for details see Chap. 116).

Toxic Substances. Lipophilic substances such as anesthetic gases or alcohol pass the placental barrier freely and quickly, so an equilibrium between fetal and maternal circulation is rapidly reached. Antibiotics frequently use carriers originally designated for the transport of other substances such as organic acids and bases. Heavy metals accumulate in the fetal organism, and may reach critical concentrations.

CO has a placental permeability like that of O_2. In smokers, levels of up to 10% of CO-Hb are found. This leads to left shift of the O_2 saturation curve of Hb, inhibiting the uptake of O_2 by fetal tissues.

117.5.10 Amniotic Fluid

The fluid filling of the amniotic sac provides a medium in which the fetus can readily move, whilst cushioning it against possible injury and helping to maintain a constant temperature. The amniotic fluid increases rapidly to an average volume of 50 ml at 12 weeks' gestation and 400 ml at mid-pregnancy, reaching a maximum of about 1 l near term. However large variations can be found even in normal pregnancies. In the first half of pregnancy the fluid is of the same composition as maternal plasma except for a much lower protein concentration. With advancing gestation more phospholipids, primarily from the fetal lung, are found. The concentration of other solutes also changes, with a subsequent decrease in osmolarity of some 10%. Amniotic fluid in early pregnancy is assumed to be primarily derived from the amniotic membrane.

With advancing pregnancy and increasing volume, the fetus becomes the decisive factor in regard to volume and composition of the amniotic fluid by swallowing and voiding progressively larger amounts of fluid. This is demonstrated by the finding that esophageal atresia is accompanied by polyhydramnios, and also shown by the extremely reduced amount of amniotic fluid seen in cases of renal agenesis or dysplasia (*Potter kidneys*). Furthermore, a recent investigation of severe twin transfusion syndrome showed that the concentration of atrial natriuretic factor (ANF) in the cord blood of the recipient twins, who developed polyhydramnios, was significantly elevated compared to that of the donor twins. The donors, who normally show oligohydramnios, have increased values of antidiuretic hormone (ADH = AVP) and renin activity [48]. In animal studies it was found that the blocking of endogenous ANF by an antiserum leads to a significant reduction in urine flow rate and electrolyte excretion of the fetus, while arterial pressure and angiotensin II levels increase. These findings support the concept that the fetus and its regulation of intravascular volume and urine production determine the volume and composition of amniotic fluid [11].

In comparison to maternal or fetal plasma, fetal urine is hypotonic and has a lower electrolyte concentration, but contains more urea, creatinine, and uric acid. Thus, the increasing contribution of fetal urine towards the volume of amniotic fluid results in a decreasing osmolarity of the latter as the pregnancy progresses.

117.5.11 Maternal-Fetal Electrical Potential Difference

In many species an electrical potential difference between maternal and fetal vascular spaces has been measured. However, the magnitude and polarity of this potential difference appears to be highly variable between species. In humans small potential differences have been measured at mid-gestation but no appreciable difference was found at term [16,45]. The difference is probably generated outside the placenta at the maternal allantoic interface. The endometrium shows active electrogenic transport, possibly engaged in significant electrolyte and water transport. The exact physiological importance of this phenomenon remains under dispute [30].

117.5.12 Fetal Heat Exchange

The existence of a fetomaternal temperature gradient of the order of 0.5°C has been known for a long time. Its existence can be explained by the higher metabolic rate of the fetus and the reduced heat loss to the outside environment due to its sheltered situation within the uterus. Due to the large thermal capacity of water, the placental diffusion coefficient for heat is so high that the heat transfer is purely flow-limited. As a consequence, increased temperature gradients are measured if maternal uterine flow is reduced. Further heat loss occurs across the uterine wall. It has been suggested that variations in maternal temperature may influence fetal energy balance and growth [49].

References

1. Aherne W, Dunnill MS (1966) Morphometry of the human placenta. Br Med Bull 22:5–12
2. Alberts B, Bray D, Lewis J, Raff M, Roberts K, Watson JD (eds) (1989) Molecular biology of the cell. Garland, New York
3. Anderson DF, Faber JJ (1989) Regulation of fetal placental blood flow. In: Kaufmann P, Miller RK (eds) Trophoblast research, vol 3: placental vascularisation and blood flow. Plenum Medical Book, New York, pp 179–188
4. Assali NS, Rauramo L, Peltonen T (1960) Measurement of uterine blood flow and uterine metabolism. VIII. Uterine and fetal blood flow and oxygen consumption in early human pregnancy. Am J Obstet Gynecol 79:86–98
5. Borel IM, Gentile T, Angelucci J, Pividori J, Guala MC, Binaghi RA, Margni RA (1991) IgG asymmetric molecules with antipaternal activity isolated from sera and placenta of pregnant human. J Reprod Immunol 20:129–140
6. Boura ALA, Walters WAW (1991) Autacoids and the control of vascular tone in the human umbilical-placental circulation. Placenta 12:453–477
7. Brosens IA (1989) The utero-placental vessels at term. The distribution and extent of physiological changes. In: Kaufmann P, Miller RK (eds) Trophoblast research, vol 3: placental vascularisation and blood flow. Plenum Medical Book, New York, pp 61–68
8. Bullen BE, Bloxam DL, Ryder TA, Mobberley MA, Bax CM (1990) Two-sided culture of human placental trophoblast. Morphology, immunocytochemistry and permeability properties. Placenta 11:431–450
9. Bulmer JN, Johnson PM (1985) Antigen expression by trophoblast populations in the human placenta and their possible immunobiological relevance. Placenta 6:127–140
10. Carstensen M, Leichtweiss HP, Molsen G, Schröder H (1977) Evidence for a specific transport of D-hexoses across the human term placenta in vitro. Arch Gynaekol 222:187
11. Cheung CY (1991) Role of endogenous atrial natriuretic factor in the regulation of fetal cardiovascular and renal function. Am J Obstet Gynecol 165:1558–1567
12. Christensen HN, Streicher JA (1948) Association between rapid growth and elevated cell concentrations of amino acids. J Biol Chem 175:95
13. Dawes GS (1962) The umbilical circulation. Am J Obstet Gynecol 84:1634–1648
14. Dawes GS (1968) Foetal and neonatal physiology. Year Book Medical, Chicago
15. Douglas GC, King BF (1990) Uptake and processing of ^{125}I-labelled transferrin and ^{59}Fe-labelled transferrin by isolated human trophoblast cells. Placenta 11:41–57
16. Duncan SLB, Levin RJ, Mathers NJ, Parsons RJ (1976) Measurement of transuterine potential difference in the human feamle during labour. J Physiol (Lond) 259:25p–26p
17. Faulk WP, Temple A (1976) Distribution of β-microglobulin and HLA in chorionic villi of human placenta. Nature 262:799–802
18. Feneley MR, Burton GJ (1991) Villous composition and membrane thickness in the human placenta at term: a stereological study using unbiased estimators and optimal fixation techniques. Placenta 12:131–142
19. Fox SB, Khong TY (1990) Lack of innervation of human umbilical cord. An immunohistological and histochemical study. Placenta 11:59–62
20. Hellmann LM, Flexner LB, Wilde WS, Vosburgh GJ, Proctor NK (1948) The permeability of the human placenta to water and the supply of water to the human fetus as determined with deuterium oxide. Am J Obstet Gynecol 56:861
21. Hunt J, Lessin D, King CR (1989) Ontogeny and distribution of cells expressing HLA-B locus specific determinants in the placenta and extraplacental membranes. J Reprod Immunol 15:21–30
22. Kameda T, Koyama M, Matsuzaki N, Taniguchi T, Saji F, Tanizawa O (1991) Localisation of three subtypes of Fc gamma receptors in human placenta by immunohistochemical analysis. Placenta 12:15–26
23. Kaufmann P (1981) Entwicklung der Plazenta. In: Becker V, Schiebler TH, Kubli F (eds) Die Plazenta des Menschen. Thieme, Stuttgart, pp 13–50
24. Kaufmann P, Luckhardt M, Leiser R (1989) Three-dimensional representation of the fetal vessel system in the human placenta. In: Kaufmann P, Miller RK (eds) Trophoblast research, vol 3: placental vascularisation and blood flow. Plenum Medical Book, New York, pp 113–138
25. Koszalka TR, Jensh RP, Brent RL (1975) Placental transport of creatine in the rat. Proc Soc Exp Biol Med 148:864
26. Lees MH, Hill JD, Ochsner AJ, Thomas CL, Novy MJ (1971) Maternal placental and myometrial blood flow of the rhesus monkey during uterine contractions. Am J Obstet Gynecol 110:68–73
27. Leichtweiss HP (1989) Placenta-Physiologie. In: Bettendorf G, Breckwoldt M (eds) Reproduktionsmedizin. Fischer, Stuttgart, pp 581–593
28. Maguire MH, Howard RB, Hosokawa T, Poisner A (1989) Effects of some autocoids and humoral agents on human fetoplacental vascular resistance: candidates for local regulation of fetoplacental blood flow. In: Kaufmann P, Miller RK (eds) Trophoblast research, vol 3: placental vascularisation and blood flow. Plenum Medical Book, New York, pp 203–216
29. McIntyre JA (1990) Trophoblast antigens. Am J Reprod Immunol 22:92
30. McNaughton T, Power GG (1991) Current topic: the maternal-fetal electrical potential difference: new findings and a perspective. Placenta 12:185–197
31. Miller RK, Berndt WO (1974) Characterisation of neutral amino acid accumulation by human term placental slices. Am J Physiol 227:1236
32. Moll W (1981) Physiologie der Plazenta. In: Becker V, Schiebler TH, Kubli F (eds) Die Plazenta des Menschen. Thieme, Stuttgart, pp 129–198
33. Moll W, Girard H, Gros G (1980) Facilitated diffusion of lactic acid across the guinea-pig placenta. Pflugers Arch Gesamte Physiol 385:229

34. Moll W, Nienartowicz A, Hees H, Wrobel KH, Lenz A (1989) Blood flow regulation in the uteroplacental arteries. In: Kaufmann P, Miller RK (eds) Trophoblast research, vol 3: placental vascularisation and blood flow. Plenum Medical Book, New York, pp 83–96

35. Page EW, Glendening MB, Margolis A, Harper A (1957) Transfer of L-histidine and D-histidine across the human placenta. Am J Obstet Gynecol 73:589

36. Pritchard JA, MacDonald PC, Gant NF (eds) (1985) Williams obstetrics, 17th edn. Appleton-Century-Crofts, Norwalk

37. Pitcher-Willmott RW, Hindocha RW, Wood CBS (1980) The placental transfer of IgG subclasses in human pregnancy. Clin Exp Immunol 41:303–308

38. Reed E, Beer A, Hutcherson H, King DW, Suciu-Foca N (1991) The alloantibody response of pregnant women and its suppression by soluble HLA antigens and anti-ideotypic antibodies. J Reprod Immunol 20: 115–128

39. Rice PA, Rourke JE, Nesbitt EL (1976) In vitro perfusion studies of the human placenta. IV. Some characteristics of the glucose transport system in the human placenta. Gynecol Invest 7:213

40. Robaut C, Mondon F, Bandet J, Ferre F, Cavero I (1991) Regional distribution and pharmacological characterisation of I-125 endothelin-1 binding sites in human fetal placental vessels. Placenta 12:55–67

41. Schmid-Schönbein H (1989) Conceptional proposition for a specific microcirculatory problem: maternal blood flow in hemochorial multivillous placentae as percolation of a porous medium. In: Kaufmann P, Miller RK (eds) Trophoblast research, vol 3: placental vascularisation and blood flow. Plenum Medical Book, New York, pp 3–16

42. Schuhmann R, Stoz F, Maier M (1988) Histometric investigations in placentones (materno-fetal circulation units) of human placentae. In: Kaufmann P, Miller PK (eds) Trophoblast research, vol 3: placental vascularisation and blood flow. Plenum Medical Book, New York, pp 3–16

43. Seya T, Turner J, Atkinson JP (1986) Purification and characterisation of a membrane protein (gp45–70) that is cofactor for cleavage of C3b and C4b. J Exp Med 10: 253–258

44. Sheppard BL, Bonnar J (1989) The maternal blood supply to the placenta in pregnancy complicated by intrauterine fetal growth retardation. In: Kaufmann P, Miller RK (eds) Trophoblast research, vol 3: placental vascularisation and blood flow. Plenum Medical Book, New York, pp 69–81

45. Stulc J, Svihovec J, Stribrny J, Koblikova J, Vido I, Dolezal A (1978) Electrical potential difference across the mid term human placenta. Acta Obstet Gynecol Scand 57:125–126

46. Svane D, Skajaa K, Anderson KE, Forman A (1991) Vascular responses in term pregnant and non-pregnant human uterus. Placenta 12:47–54

47. Vanderpuye OA, Labarrere A, Thaler C, Faulk WP, McIntyre JA (1991) Syncytiotrophoblast brush border proteins recognized by monoclonal antibody TRA-2-10 and rabbit anti-TLX sera. Placenta 12:199–215

48. Wieacker P, Wilhelm C, Prömpeler H, Petersen KG, Schillinger H, Breckwoldt M (1992) Pathophysiology of polyhydramnios in twin transfusion syndrome. Fet Diag Ther 7:87–92

49. Young M (1990) Conference report: fetal and maternal heat balance. Placenta 11:91–93

50. Yudilevich DL, Sweiry JH (1985) Transport of amino acids in the human placenta. Biochim Biophys Acta 822:169–201

118 Cardiovascular and Respiratory Systems of the Fetus

A. JENSEN

Contents

118.1 Introduction

Our knowledge of fetal physiology stems almost entirely from mammals other than the human. Most of the experimental evidence was produced in Artiodactyla, particularly in sheep and goat, and in rodents such as guinea pig, rabbit, rat, and mouse (Fig. 118.1) [75]. Only a few observations have been made in nonhuman primates. However, a number of responses and mechanisms in these species are phylogenetically old and hence may bear some relationship to those in the human. For example, the intrauterine environment as far as blood gas status and acid–base balance are concerned are largely similar in chronically instrumented fetal sheep and in human fetuses. Furthermore, the general neurohumoral circulatory and metabolic responses to acute and chronic O$_2$ deficiency are comparable to some extent. Consequently, the chronically prepared sheep model has been most widely used to study fetal physiology. In this chapter an attempt is made to describe some important developmental features of the cardiovascular and respiratory system, some relevant aspects of energy metabolism and growth, and their integration by hormonal and nervous mechanisms. Due to their clinical significance, particular emphasis is placed on the circulatory effects of O$_2$ deficiency on the main fetal body functions.

Scientific *embryology* dates back to Hippocrates (460–377 B.C.), who reportedly wrote: "Take twenty or more eggs and have them hatched by two or more hens. Then, beginning on the second day of breeding, break one egg every day. You will find exactly what I have said. From the development of a bird's egg we can draw conclusions for man." In the fifth century B.C. Aristotles (cf. Physiology Past and Future) wrote the first embryologic essay. He is therefore considered the founder of embryology, even though he erroneously thought that an embryo emerges from the combination of a sperm and menstrual blood. In the second century B.C. Galen (cf. Physiology Past and Future) wrote a book on "The formation of the fetus," in which he correctly described the development of the fetus and the membranes.

In the middle ages there was little scientific progress. However, in the fifteenth century Leonardo da Vinci (cf. Physiology Past and Future) sketched fairly accurate drawings of the pregnant uterus and its contents. In 1651 Harvey (cf. Physiology Past and Future) examined the developing chicken embryo using a microscope. With the aid of this new technology he made numerous important discoveries, particularly about the circulation of the blood. This is regarded as the beginning of the science of fetal physiology. The *time scale of the development* of the human embryo, along with the "critical periods" of developing organs, is shown in Fig. 118.2 [194]. Since Paul Zweifel (1876) [277], an obstetrician, first demonstrated by spectroscopy the transfer of O$_2$ from the mother to the fetus via the placenta, and since the inspiring early work of Sir Joseph Barcroft (1946) [11] in this century, the fetal circulation and its regulation has attracted the interest of many investigators, basic scientists and clinicians. The resulting accumulation of knowledge on fetal and neonatal physiology, which has been summarized in a number of reviews [70,74,195,237,239,240], has contributed considerably to the decline in perinatal morbidity and mortality in recent years. The pioneering early work in the field was largely based on observations in acutely instrumented fetuses of various species [74]. A major step forward towards studying the undisturbed fetus in utero was the chronic implantation of catheters and probes into the fetus in the late 1960s [192]. Since then, many experiments have been repeated without the use of anesthesia [233, 237,239,240]. In many of these studies the microsphere method was used [231] to measure the distribution of blood flow. In terms of the circulatory control of specific organs, the lung is one of those studied most intensively [106,107].

R. Greger/U. Windhorst (Eds.)
Comprehensive Human Physiology, Vol. 2
© Springer-Verlag Berlin Heidelberg 1996

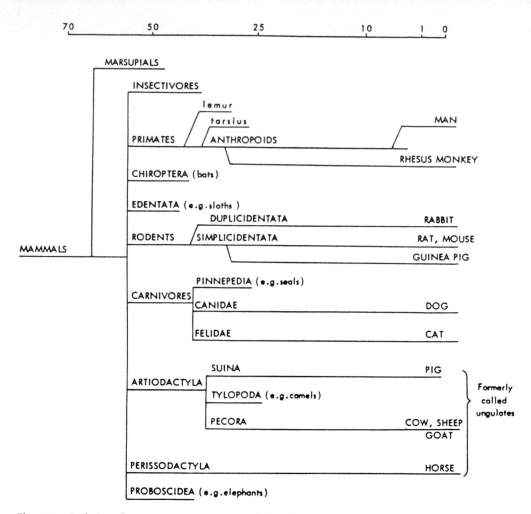

Fig. 118.1. Evolution of some common mammals as deduced from the geological record. (From [75])

The first part of this chapter will describe the following aspects of fetal physiology:

- Fetal circulation
- Distribution of combined ventricular output during normoxemia
- Cardiovascular responses of the mature fetus to hypoxemia and asphyxia
- Circulatory responses of the immature fetus to hypoxemia and asphyxia
- The relation between O_2 delivery and tissue metabolism
- Cardiovascular mechanisms during hypoxia and asphyxia
- Blood flow and O_2 delivery in relation to the function of various fetal organs

The second part will describe some aspects of fetal respiratory physiology:

- Lung development and respiratory control
- Surfactant synthesis
- Regulation of surfactant production
- Pulmonary mechanics
- Fetal breathing
- Electrocortical differentiation
- Effect of hypoxemia on fetal breathing

118.2 Fetal Circulation

There are a number of differences between fetal and adult circulation of the blood, most of them related to the fact that during intrauterine life the placenta serves as the site of gas exchange rather than the lungs. Hence, the most important features of the fetal circulation are the relatively large proportion of combined output of both cardiac ventricles distributed to the umbilical cord and placenta (45%) and the small proportion distributed to the lungs (10%) (see Chap. 89). This requires special vascular channels, the foramen ovale and the ductus arteriosus (Botallo's duct), which are unique to the fetus and serve to shunt blood returning to the heart away from the pulmonary circula-

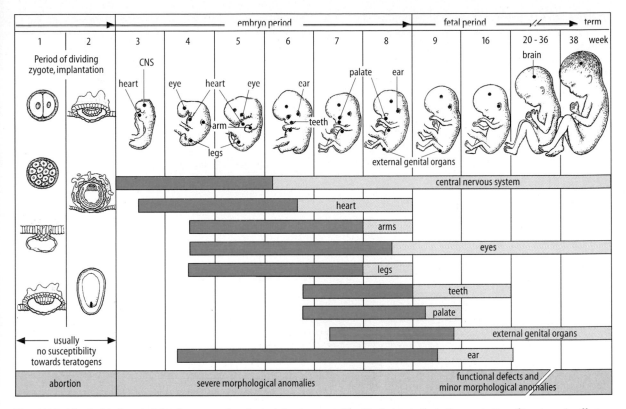

Fig. 118.2. The "critical periods" in human embryological development. The *black dots* indicate frequent sites of teratogenic effects. (Adapted from [194])

tion [71,240]. The fetal circulation is shown in Fig. 118.3. This anatomical arrangement normally permits almost equal development of the two sides of the heart, which work in parallel during fetal life but serially after birth. It allows preferential streaming of highly oxygenated umbilical venous blood via the ductus venosus (duct of Arantius), thoracic inferior vena cava, right atrium, foramen ovale, left atrium, left ventricle, and ascending aorta to the upper part of the body in general, and to the heart and brain in particular [10,240]. On the other hand, poorly oxygenated blood returning to the heart from abdominal inferior and superior venae cavae passes through the right atrium, right ventricle, pulmonary artery (65%), ductus arteriosus (57%), and descending aorta to the lower part of the body (27%) and to the umbilical cord (45%) [58,217].

After gas exchange in the placenta, oxygenated blood returns to the fetal blood through the umbilical veins. The common umbilical vein enters the liver, where it joins the portal vein. From the portal sinus several branches supply the right and left liver lobes, and the ductus venosus connects the umbilical vein to the inferior vena cava, thus permitting about 55% of the umbilical venous blood to bypass the liver [82,240]. The amounts of O_2 delivered to the right and left lobes of the liver are different, because the right lobe receives almost all of the portal venous blood along with umbilical venous blood, whereas the left lobe is almost exclusively supplied by the umbilical vein [82]. The contribution of hepatic arterial blood to the O_2 supply of the two liver lobes is small (3%).

It had been assumed that there is a reasonable admixture of poorly oxygenated abdominal inferior vena caval blood with highly oxygenated umbilical venous blood in the thoracic inferior vena cava, but elegant cineangiographic studies by Barclay, Franklin, and Prichard in 1944 [10] suggested that blood from these sources streams selectively, so that ductus venosus blood preferentially passes through the foramen ovale to the upper body organs. This was confirmed by others [13,82] using the isotope-labelled microsphere method. This preferential streaming could explain the early observation [11] that in fetal lambs carotid arterial blood has a higher O_2 saturation of hemoglobin than femoral arterial blood. The preferential streaming of umbilical venous blood to the upper part of the body implies that any substance that enters the umbilical vein, e.g., glucose [48] or drugs administered to the mother, will be delivered to the heart and brain at higher concentrations [237].

118.2.1 Distribution of Cardiac Output During Normoxemia

The normal range of blood gases, acid – base balance, physiologic variables, combined ventricular output, and umbilical blood flow in chronically instrumented fetal sheep near term (term is at 147 days' gestation) is given in Tables 118.1 and 118.2. In resting fetal sheep at 0.85 of pregnancy (3–4 days after surgery), mean heart rate is 170 beats/min, arterial blood pressure is 44 mmHg, ascending aortic pH is 7.40, PO_2 is 24 mmHg, PCO_2 is 47 mmHg, O_2 saturation of hemoglobin is 67%, hemoglobin is 86 g/l, and O_2 content is 78 ml/l [139]. The corresponding values of samples withdrawn simultaneously from various major fetal vessels, including descend-

ing aorta, umbilical vein, abdominal inferior vena cava, superior vena cava, and superior sagittal sinus, are also given (Table 118.1) [139]. The values for umbilical venous and ascending and descending aortic O_2 saturation determined in the chronic model are in remarkable agreement with those from acutely instrumented fetal sheep [72].

Under physiologic conditions, as reflected by these blood gas values, the combined ventricular output (cardiac out-

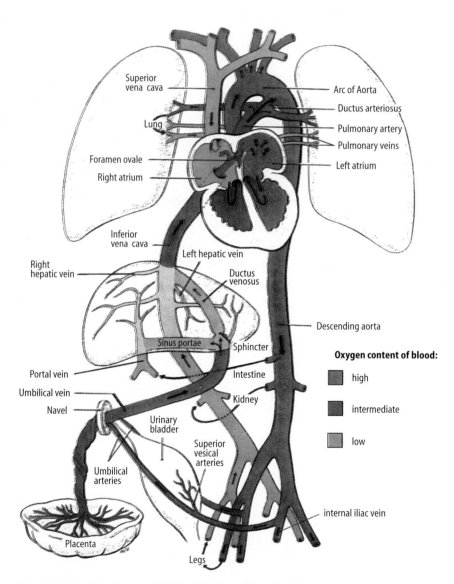

Fig. 118.3. Fetal circulation of the blood. (Adapted from [194])

Table 118.1. Heart rate, blood pressures, blood gases, pH, and O_2 content in nine normoxemic fetal sheep at 0.9 gestation. (From [139])

Heart rate (beats/min)	167 ± 19					
Mean aortic pressure (mmHg)	44.2 ± 3.5					
Umbilical venous pressure (mmHg)	11.1 ± 4.83					
Vena cava pressure (mmHg)	3.3 ± 2					
Ascending aorta[a]		**Descending aorta**[a]		**Umbilical Vein**		
PO_2 (mmHg)	24.1 ± 2.4	PO_2 (mmHg)	21.7 ± 2.3	PO_2 (mmHg)	32.8 ± 3.7	
PCO_2 (mmHg)	47.3 ± 3.9	PCO_2 (mmHg)	50.6 ± 4.5	PCO_2 (mmHg)	43.9 ± 4.0	
pH	7.40 ± 0.013	pH	7.39 ± 0.026	pH	7.42 ± 0.03	
Base excess (mmol/l)	4.60 ± 3.19	Base excess (mmol/l)	5.57 ± 3.57	Base excess (mmol/l)	5.24 ± 2.99	
O_2 saturation (%)	67.3 ± 3.9	O_2 saturation (%)	58.2 ± 5.8	O_2 saturation (%)	86.1 ± 3.1	
Hemoglobin (g/l)	86 ± 1.3	Hemoglobin (g/l)	88 ± 1.3	Hemoglobin (g/l)	85 ± 1.3	
O_2 content (ml/l)	77.6 ± 1.1	O_2 content (ml/l)	68.3 ± 1.2	O_2 content (ml/l)	97.8 ± 1.5	
Superior Vena Cava[a]		**Abdominal inferior vena cava**		**Superior sagittal sinus**[b]		
PO_2 (mmHg)	18.7 ± 1.9	PO_2 (mmHg)	17.1 ± 2.0	PO_2 (mmHg)	17.4 ± 0.9	
PCO_2 (mmHg)	51.2 ± 4.6	PCO_2 (mmHg)	54.2 ± 3.3	PCO_2 (mmHg)	52.9 ± 5.0	
pH	7.37 ± 0.03	pH	7.37 ± 0.03	pH	7.38 + 0.03	
Base excess (mmol/l)	5.20 ± 3.08	Base excess (mmol/l)	5.78 ± 3.15	Base excess (mmol/l)	5.80 ± 3.61	
O_2 saturation (%)	45.7 ± 6.4	O_2 saturation (%)	40.3 ± 6.4	O_2 saturation (%)	41.0 ± 1.9	
Hemoglobin (g/l)	89 ± 1.4	Hemoglobin (g/l)	89 ± 1.4	Hemoglobin (g/l)	97 ± 1.0	
O_2 content (ml/l)	54.4 ± 1.11	O_2 content (ml/l)	48.5 ± 1.28	O_2 content (ml/l)	53 ± 0.4	

[a] n = 8
[b] n = 4

put) at 0.9 gestation is approximately 480 ml/min × kg fetal weight. About 55% of the cardiac output is distributed to the fetal body and 45% to the placenta. About 30% is directed to the fetal carcass, which includes skeleton, muscle, bones, skin, and connective tissues, and 11% to the lungs. During normoxemia only small fractions of the cardiac output are distributed to the brain (3%), heart (2.6%), small gut (2.6%), kidneys (2.3%), and adrenals (0.006%) [139]. With regard to O_2 requirements, it is noteworthy that in the human fetus cerebral blood flow comprises a larger fraction of the cardiac output, because the brain/body weight ratio is higher than in sheep.

This normal distribution of blood flow changes dramatically in late gestation during hypoxemia and even more so during asphyxia. The following sections will describe the general changes in the distribution of fetal organ blood flows when O_2 is in short supply. The effect of various experimental interventions that result in fetal hypoxemia and a redistribution of blood flow and O_2 delivery will be summarized below. Table 118.3 gives the results of some significant studies on changes in blood gas tensions and organ blood flows.

118.2.2 Cardiovascular Responses to O_2 Deficiency in the Mature Fetus

Maternal Hypoxemia. Maternal hypoxemia is usually produced experimentally by manipulating the inspired fraction of O_2 to reduce maternal arterial PO_2 to about 40 mmHg. This results in fetal arterial PO_2 values of about 10–12 mmHg. Due to hyperventilation of the ewe, fetal blood PCO_2 falls, too. To study the effects of *isocapnic hypoxemia* on the fetal circulation, it is necessary to increase CO_2 concentrations in the inspired gas mixture (Fi_{CO_2} = 3–4%). The resulting moderate *fetal hypoxemia* (arterial PO_2 = 10–12 mmHg) causes a fall in heart rate and an increase in arterial blood pressure [58], but combined ventricular output, measured by injecting isotope-labelled microspheres into the fetal circulation [231], does not fall as long as blood pH is maintained. Only when hypoxemia is accompanied by acidemia does cardiac output fall, by about 20% [58]. Umbilical blood flow is maintained, while blood flow to the fetal body is reduced by 40% [58,204].

The distribution of the combined ventricular output changes in the fetus much the way it does in the adult. There is a circulatory centralization of blood flow in favor of the brain, heart, and adrenals, and at the expense of peripheral organs, including lungs, kidneys, gastrointestinal tract, and carcass (Fig. 118.4) [8,35,58,209]. This holds true for normally grown and growth-retarded fetuses.

Fetal hypoxemia is also accompanied by a redistribution of umbilical venous blood flow. The fraction of blood bypassing the liver through the ductus venosus increases from 55% to 65% [83,217], thus contributing considerably to the maintenance of O_2 delivery to the fetus. In addition, there is a preferential streaming of umbilical venous blood across the foramen ovale via the left ventricle towards the upper body circulation to maintain O_2 delivery to the heart and brain [217].

Table 118.2. Combined ventricular output and organ blood flow, % cardiac output, O_2 delivery, and vascular resistance in nine chronically instrumented fetal sheep at 0.9 gestation. (From [139])

	Blood flow (ml × min⁻¹ × 100 g⁻¹)	% Cardiac output	O2 delivery (ml × min⁻¹ × 100 g⁻¹)	Vascular resistance (mm Hg/ ml × min⁻¹ × 100 g⁻¹)
Combined ventricular output	478 ± 94	100	–	
Umbilical blood flow	213 ± 55	44.0 ± 9.3	–	0.16 ± 0.03††
Total body blood flow	315 ± 63	56.0 ± 9.3	21.34 ± 6.75†	0.13 ± 0.02††
Upper body organs				
Brain	87.4 ± 20.6	3.00 ± 1.11	6.71 ± 1.45	0.49 ± 0.11
Cerebrum	79.5 ± 18.3	2.00 ± 0.70	6.11 ± 1.34	0.54 ± 0.13
Cerebellum	119.5 ± 24.1	0.32 ± 0.11	9.16 ± 1.58	0.35 ± 0.06
Diencephalon	105.7 ± 32.2	0.25 ± 0.12	7.94 ± 1.66	0.42 ± 0.13
Midbrain	134.5 ± 42.5	0.19 ± 0.10	10.05 ± 2.01	0.33 ± 0.11
Medulla	144.5 ± 46.0	0.15 ± 0.07	10.83 ± 2.44	0.31 ± 0.11
Hypophysis	67.8 ± 36.7	0.01 ± 0.01	5.05 ± 2.25	0.76 ± 0.37
Choroid plexus	458.1 ± 265.9	0.04 ± 0.02	34.3 ± 18.70	0.13 ± 0.09
Heart	163.1 ± 46.5	2.55 ± 0.62	12.29 ± 2.19	0.27 ± 0.08
Upper carcass	21.6 ± 4.7	15.91 ± 4.48	0.52 ± 0.20	1.99 ± 0.53
Scalp	33.7 ± 10.4	0.21 ± 0.07	2.65 ± 9.80	1.34 ± 0.48
Body skin	23.0 ± 6.3	0.17 ± 0.09	1.84 ± 0.74	1.90 ± 0.54
Upper body, total	113 ± 24[a]	23.0 ± 5.4	8.58 ± 1.42[a]	0.38 ± 0.11[a]
Lower body organs				
Adrenals	174.3 ± 118.6	0.06 ± 0.04	10.83 ± 6.75	0.84 ± 1.57
Kidneys	154.8 ± 40.2	2.34 ± 0.76	10.61 ± 3.95	0.28 ± 0.07
Spleen	345.1 ± 244.7	1.01 ± 0.71	21.64 ± 15.0	0.53 ± 0.92
Small gut	89.9 ± 29.8	2.65 ± 1.47	6.08 ± 2.23	0.50 ± 0.18
Large gut	41.3 ± 10.6	0.27 ± 0.12	2.78 ± 0.75	1.04 ± 0.25
Pancreas	35.3 ± 12.9	0.08 ± 0.04	2.33 ± 0.74	1.33 ± 0.54
Lower carcass	19.0 ± 5.0	13.82 ± 2.70	0.38 ± 0.14	2.32 ± 0.74
Body skin	26.1 ± 13.7	0.16 ± 0.06	1.81 ± 1.10	1.90 ± 0.79
Lower body, total	138 ± 35[c]	25.4 ± 4.3	9.24 ± 2.27[a,b]	0.31 ± 0.08[a,c]
Lungs	162.2 ± 74.4	11.81 ± 5.12	5.68 ± 2.52	0.35 ± 0.27

[a] ml × min⁻¹ × kg⁻¹
[b] Not including umbilical blood flow
[c] Not including umbilical vascular resistance

Table 118.3. Fetal changes during hypoxemia and asphyxia

	122–142 days gestation, maternal hypoxemia FiO_2 = 6%) [58]		130 ± 2 days gestation, arrest of uterine and ovarian blood flow (4 min) [135]	119–123 days gestation, reduction of umbilical blood flow by 50% [116]
	Without acidemia	With acidemia		
Heart rate (beats/min)	164 → 122	160 → 133	170 → 80	177 → 152
Arterial blood pressure (mmHg)	57 → 65	63 → 76	55 → 70	44 → 49
pH	7.37 → 7.36	7.36 → 7.26	7.30 → 7.05	7.39 → 7.36
PO_2 (mmHg)	19 → 11	20 → 12	19 → 6	23 → 18
PCO_2 (mmHg)	42 → 42	44 → 43	50 → 86	43 → 46
Blood flow changes (ml/min × kg)				
Umbilical cord	191 → 213	195 → 205	311 → 151	233 → 115
Fetal body	273 → 229	302 → 177	164 → 55°	278 → 302
Brain	96 → 168	120 → 185	194 → 144	112 → 160
Heart	179 → 449	185 → 482	261 → 380	197 → 277
Adrenal	271 → 828	302 → 855	437 → 941	217 → 436
Liver arterial	–	–	21 → 8	13 → 10
Kidney	175 → 136	162 → 81	253 → 23	154 → 174
Lungs	60 → 27	57 → 32	75 → 30	131 → 65
Gastrointestinal tract	67 → 53	96 → 41	179 → 21	82 → 100
Carcass	20 → 14	20 → 6	–	21 → 25
Muscle	–	–	17 → 0.8	–
Skin	–	–	32 → 1.3	–

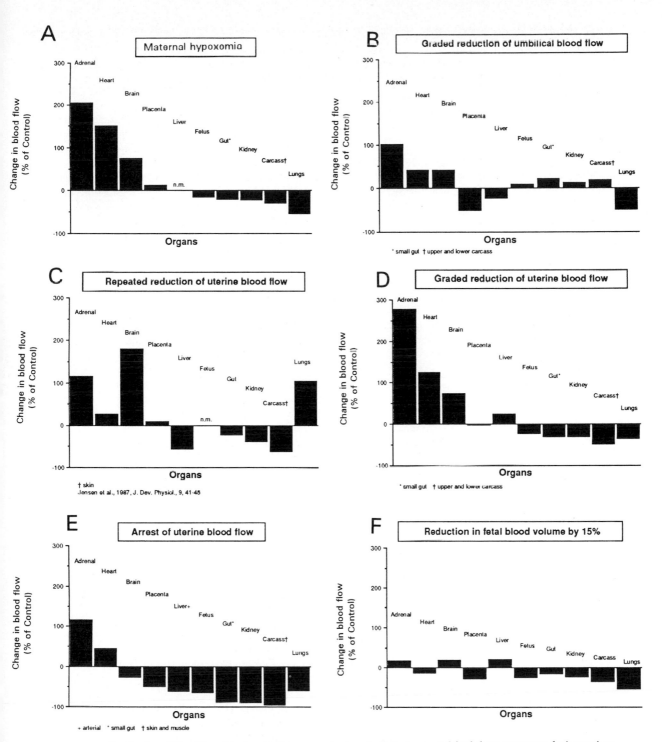

Fig. 118.4A–F. Redistribution of blood flow (% of control) to organs in chronically instrumented fetal sheep near term during various experimental disturbances. Based on data from: **A** [58], **B** [116], **C** [132], **D** [128], **E** [135], **F** [257], by permission

During fetal hypoxemia the proportion of superior vena cava blood flow directed through the foramen ovale into the upper body segment is slightly increased [58].

During hypoxemia the venous blood returning to the heart via the abdominal inferior vena cava is reduced by 50%. Thus, the blood returning from umbilical vein, superior vena cava, and inferior vena cava contributes 32%, 30%, and 44% to placental blood flow, respectively, and hence recirculates to the fetus via the umbilical vein [217].

In summary, during moderate hypoxemia cardiac output is maintained (unlike during hypoxemia with acidemia) and the circulating blood is redistributed to the brain, heart, and adrenals at the expense of peripheral organs, including the lungs. This circulatory centralization is accompanied by a changing pattern of venous return and by a preferential streaming of umbilical venous blood through the ductus venosus and the foramen ovale to the upper body segment.

Reduction in Umbilical Blood Flow. Reduction in umbilical and placental blood flow can be caused by compression of the umbilical vein [163], compression of the fetal abdominal aorta [83], snaring of the umbilical cord [114–116], or by embolization of the placental vascular bed [52,259]. Reduction of arterial and/or venous umbilical blood flow causes a drop in the amount of O_2 delivered to the fetus. Unlike during maternal hypoxemia, this does not decrease umbilical venous O_2 content [116,163]. Therefore, a reduction in umbilical blood flow by, e.g., 50% results in a similar reduction in O_2 delivery to the fetus [115].

Due to the favorable relationship between uterine and umbilical blood flow, umbilical venous PCO_2 and fetal arterial blood pH do not change even though umbilical blood flow and hence fetal O_2 delivery are reduced to 50% of normal.

Compression of the umbilical cord is accompanied by an increase in arterial blood pressure, a fall in heart rate, and a fall in combined ventricular output [112,114–116, 163].

Reduced umbilical blood flow is accompanied by a redistribution of blood flow to the fetal organs that is different from that observed during maternal hypoxemia (Fig. 118.4). Blood flow increases to the brain (+43%), heart (+41%), and adrenals (+100%). However, unlike during maternal hypoxemia, blood flow to the peripheral organs, including the kidney, gastrointestinal tract, and spleen does not change, and that to the carcass increases (+20%). Only blood flow to the lungs falls (−50%).

The fraction of umbilical blood flow shunted through the ductus venosus increases by 30% when umbilical blood flow is reduced by 75% [83]. However, the proportion of venous return from both inferior and superior vena cava increases relative to that from the umbilical vein. Therefore, the O_2 content of the blood crossing the foramen ovale falls and O_2 delivery to the brain and heart is maintained by increasing blood flow [116].

Reduction in Uterine Blood Flow. Uterine blood flow can be reduced by various methods, including compression of the maternal aorta [132–135,138,272], the uterine arteries [275], or the uterine veins [162,180], or by embolization of the uterine vascular bed [66]. The hemodynamic effects of reduced uterine blood flow depend largely on the severity of the reduction. The most severe insult, of course, is arrest of uterine blood flow, which interrupts both maternofetal O_2 delivery and fetomaternal CO_2 clearance. Due to collateral blood supply to the sheep uterus, including that provided by the ovarian arteries, arrest of uterine blood flow can only be achieved experimentally by complete compression of the abdominal maternal aorta below the renal arteries.

Graded Reduction in Uterine Blood Flow. Graded reduction in uterine blood flow to achieve a fall in fetal O_2 delivery by 50% and a final carotid arterial PO_2 of 16 mmHg is associated with mild fetal bradycardia and an increase in fetal aortic pressure [139,249]. These changes are similar to those occurring during maternal hypoxemia and with umbilical cord compression, but of a different degree. Moreover, while combined ventricular output decreases during cord compression, it does not change significantly with graded reduction of uterine blood flow [139].

The redistribution of cardiac output and of blood flow to individual organs during graded reduction of uterine blood flow is qualitatively similar, but quantitatively different from that observed during arrest of uterine blood flow (Fig. 118.4). Blood flow to the brain, heart, and adrenals increases and that to the carcass and to the skin falls, while umbilical blood flow is maintained [139]. If arterial PO_2 falls below 14 mmHg, umbilical blood flow falls [61].

During graded reduction of uterine blood flow, umbilical venous blood directed through the ductus increases, as does the fraction of umbilical venous blood crossing the foramen ovale. There is a significant increase in blood returning from both superior and abdominal inferior vena cava and from the umbilical vein via the ductus venosus that crosses the foramen ovale to the upper part of the body. However, because the O_2 content of the blood in the umbilical vein and the superior vena cava is higher than that in the abdominal inferior vena cava, blood derived from the former contributes a relatively greater proportion of the O_2 delivered to the heart and brain than blood from the latter [139].

Repeated Brief Arrest of Uterine Blood Flow. Repeated reductions in uterine blood flow occur quite frequently during the second stage of labor and expose the fetus to various degrees of distress. In spite of its clinical significance, few systematic studies have been devoted to this particular mode of repeated restriction of O_2 delivery to the fetus. One of these studies tried to mimic changes in the uterine vascular bed in the second stage of labor by repeatedly inflating a balloon catheter, which was advanced into the abdominal maternal aorta [130] in an acute fetal sheep model. Uterine blood flow was intermittently arrested 11 times within 33 min. Each individual asphyxial episode lasted 30, 60, or 90 s. Depending on the duration of asphyxia there was a repeated fall in arterial O_2 saturation of hemoglobin, heart rate, skin blood flow, and transcutaneously measured PO_2 (tcPO_2), whereas arterial blood pressure and plasma catecholamine concentrations rose [130] (Fig. 118.5). With increasing duration of repeated arrest of uterine blood flow, fetal skin blood flow decreased to such an extent that the tcPO_2 signal was suppressed even though central oxygenation was restored. This suggests that in the presence of fetal bradycardia, low tcPO_2 readings may be used as an index of poor skin blood flow and hence of circulatory redistribution (Fig. 118.5), as postulated previously for human fetuses [129]. This view is supported by the close correlations observed between fetal

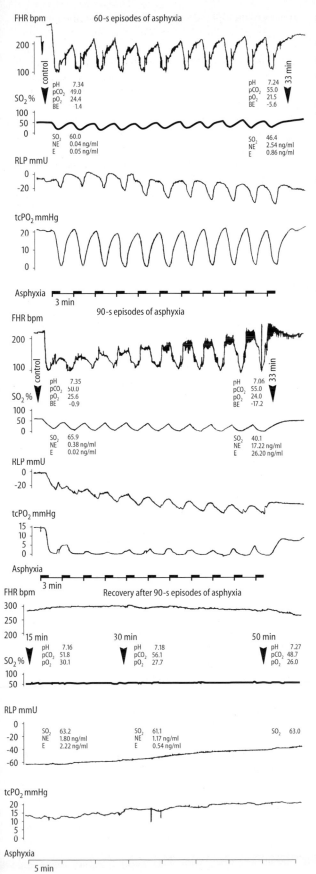

skin blood flow and plasma catecholamine concentrations [134].

The observation that reduced blood flow to the fetal skin during repeated reduction of uterine blood flow can be detected by $tcPO_2$ measurements has been confirmed by studies in which microspheres were injected into the fetal circulation [132]. These studies showed that reduced skin blood flow is accompanied by both increased sympathetic activity, as assessed by plasma catecholamine concentrations, and a redistribution of systemic blood flow (Fig. 118.4). This may be of clinical significance, because early detection of fetal circulatory centralization through variables that depend on skin blood flow could improve fetal surveillance during complicated labor [129,130,132,134, 208].

The pattern of redistribution of fetal blood flow after repeated brief episodes of asphyxia caused by arrest of uterine blood flow for 90 s is different from that during prolonged asphyxia or maternal hypoxemia. Myocardial blood flow, for instance, which increases during various interventions [35,58,90,91,116,139,209], was not increased significantly 4 min after the last asphyxial episode. This suggests that increased myocardial flow recovers more rapidly after repeated than after prolonged asphyxia (Fig. 118.4).

Another difference is that blood flow to the lungs increases under these experimental conditions (Fig. 118.4). This may be related to either vasodilatation in the lungs or to increased arterial pressure in the pulmonary artery due to transient constriction of the ductus arteriosus [132].

Prolonged Arrest of Uterine Blood Flow. Prolonged arrest of uterine blood flow for 4 min causes severe fetal asphyxia. Heart rate, arterial O_2 content, pH, and combined ventricular output fall, and arterial blood pressure, PCO_2, and lactate concentrations rise rapidly [135]. This acute severe asphyxia is accompanied by rapid changes in both the fetal and the umbilical circulation. To study the changes in organ blood flow distribution at short intervals, a modification of the isotope-labelled microsphere method has been devised, in which microspheres are injected serially during continuous withdrawal of the reference blood samples [135]. The redistribution of cardiac output is qualitatively similar to that observed during graded reduction of uterine blood flow and during maternal hypoxemia, in that the fractions of cardiac output distributed to the brain, heart, and adrenals increase, while those to periph-

Fig. 118.5. Repeated brief arrest of uterine blood flow for 60 s (*top*) and 90 s (*middle*) in acutely instrumented fetal sheep near term. Original recordings of (from the top) fetal heart rate (FHR), O_2 saturation (SO_2), skin perfusion (mm units deflection), and transcutaneous PO_2 ($tcPO_2$). Note: In contrast to 60 s arrest, arrest of uterine blood flow for 90 s suppresses the $tcPO_2$ signal, although O_2 saturation returns to normal. This is due to increasing catecholamine concentrations that reduce skin blood flow during repeated asphyxia. After asphyxia (*bottom*), O_2 saturation of hemoglobin is high and all variables recover. With decreasing catecholamine concentrations, blood flow to the skin, $tcPO_2$, and heart rate return to normal [130]

2315

eral organs fall drastically (Fig. 118.4). However, there are distinct differences; for example, cardiac output falls markedly. Furthermore, there are differences in actual cerebral vascular resistance and in both blood flow and O_2 delivery to the brain, in that cerebral blood flow does not rise. Thus, O_2 delivery to the cerebrum falls whereas that to the brainstem is maintained, reflecting a redistribution of brain blood flow in favor of brainstem areas (Fig. 118.4) [135].

If *severe asphyxia* caused by arrest of uterine blood flow is prolonged, *circulatory centralization* cannot be maintained (Fig. 118.6). Rather, there is decentralization, with a decrease in vascular resistance in peripheral organs and an increase in resistance in central organs, including those in the brain and the heart. Umbilical resistance also rises and hence placental blood flow falls. These changes, which are associated with severe metabolic derangements and severe acidemia, below pH 7.0, lead to fetal demise unless immediate resuscitation occurs [22,135].

During *decentralization* at the nadir of asphyxia (Fig. 118.6), the loss of peripheral vascular resistance may be related to the severity of local acidosis causing the vascular smooth muscle in peripheral arteries to dilate [136]. The fate of the fetus during severe asphyxia is determined by the degree of hypoxic myocardial depression and by the depletion of cardiac glycogen stores. Agonal heart rate patterns, increasing degree of cardiac failure, and the resulting increase in central venous pressures precede fetal death [22].

One report has been published on fetal circulatory derangements during reduction of blood flow through the common hypogastric artery by 75%, leaving the ovarian anastomoses intact. In that study cardiac output and umbilical blood flow were maintained. However, an increase in vascular resistance in the brain and heart was observed. The fact that these changes occurred during moderate acidemia (pH 7.14 ± 0.02) and the fact that these changes were accompanied by only minor reductions

in peripheral blood flows renders these results difficult to interpret [275].

Reduced Fetal Blood Volume (Hemorrhage). A reduction in fetal O_2 delivery can also be produced by a reduction in fetal blood volume, i.e., anemic hypoxemia that may occur clinically during fetal hemorrhage [113,182,257]. A fall in blood volume by 15–20% is accompanied by a reduction in heart rate (−20%), arterial blood pressure (−10%), and cardiac output (−25%).

The resulting redistribution of fetal organ blood flow is different to that observed during hypoxemia and asphyxia (Fig. 118.4), in that blood flows to the heart, brain, and adrenals are maintained, but not increased. Blood flow to almost all peripheral organs and to the placenta falls [113,257]. The fraction of umbilical venous blood passing through the ductus venosus increases and contributes about 30% to the cardiac output. Hence, ductus-venosus-derived blood flow and O_2 delivery to the upper and lower body segments also increase by 30% [113]. Prolonged but quantitatively similar volume losses, e.g., 30% over a period of 2 h, cause less pronounced changes in the cardiovascular variables [30].

Chronic Hypoxemia and Fetal Growth Retardation. A number of experimental models have been devised in sheep, guinea pigs, and rats to produce chronic fetal hypoxemia with consequent intrauterine growth retardation. These include prolonged maternal hypoxemia [94,124,156], reduction of placental size by removal of endometrial caruncles before conception [4,225], placental damage by embolization of the uteroplacental bed [48,52,66], embolization of the fetal placental vascular bed [22,259], ligation of uterine [166,167,270] or umbilical vessels [87], and maternal heat stress [5]. However, fetal growth retardation can also be produced by substrate restriction without any apparent changes in fetal O_2 delivery [47,191].

All of these perturbations are associated with a decrease in placental size and/or transfer function that is directly related to the magnitude of retardation in fetal growth [54]. The actual mechanisms involved in reduction of fetal growth are not well understood. However, there is clear evidence that O_2 *deprivation* is one important factor. However, certain placental signals should also be considered which might initiate the cessation of fetal growth before significant hypoxemia occurs [148]. Thus, placental membranes play an important part in the production of prostaglandins, prostacyclin (PGI_2), one of the most potent vasodilators [176,193,206] (see also Chap. 6). There is, for instance, a substantial reduction in prostacyclin production in the placentae of fetuses that are chronically hypoxemic and growth-retarded [140]. Whether this reduction in prostacyclin concentrations is related to elevated vascular resistance in the fetus is at present unclear. Reduced fetal growth is associated with a number of metabolic and endocrine changes including hypoglycemia, hypoinsulinemia, increased concentrations of glucagon, lactate, alanine, triglycerides [227], cortisol [53], cate-

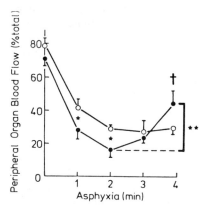

Fig. 118.6. Changes in blood flow to peripheral organs during acute asphyxia in four surviving (*open circles*) and five nonsurviving fetal sheep (*filled circles*) near term. Note that, unlike in surviving fetuses, at the nadir of asphyxia (at 4 min) in nonsurviving fetuses the proportion of blood flow directed to peripheral organs was significantly higher than at 2 min of asphyxia. **p = 0.01; 2 min vs 4 min

cholamines, β-endorphins, and decreased concentrations of growth-promoting factors, i.e., T3, T4, somatomedins, and prolactin. ACTH, insulin-like growth factor (IGF-2), ovine placental lactogen, and growth hormone are unchanged [54,150,152,227].

The consumption of O_2 and that of glucose by the fetus are reduced in absolute terms, but are maintained in terms of fetal body mass [199,200]. However, the uteroplacental consumption of glucose per weight of placenta is reduced and a greater proportion of that glucose or other substrates are converted to lactate by the placenta. There is also an increase in the fraction of lactate produced by uteroplacental tissues that is secreted into the fetal circulation [200].

Growth-retarded fetuses tend to have a lower arterial blood pressure and a higher heart rate under control conditions [226]. The chronic reduction in fetal O_2 delivery is compensated in part by an increased packed cell volume with an increased transport capacity for O_2 [124,226].

Minor degrees of chronically reduced fetal O_2 delivery are not necessarily associated with major circulatory responses [23]. However, major changes in chronically reduced fetal O_2 delivery cause circulatory centralization in favor of the brain and heart [66]. This is reflected by a relative maintainance of the weight of central organs as compared to that of the fetal body in general and the fetal liver in particular. For this reason, it has been suggested that the brain/liver weight ratio could be used as an index of fetal growth retardation [150]. The response of growth-retarded fetuses to acute hypoxemia or asphyxia is different from that observed in normal-sized fetuses, in that they develop acidemia more easily. Furthermore, bradycardia is brief and heart rate returns to control values quickly after the onset of acute hypoxemia [226], whereas tachycardia develops during recovery. This has been attributed to an increased sympathetic tone in growth-retarded fetuses [142,178], even though their adrenaline responses to hypoxemia are not particularly high [147].

During hypoxemia in growth-retarded fetuses there is only a small increase in plasma glucose concentrations, which may be due to a failure to mobilize glycogen stores. This is accompanied by only a small decrease in insulin concentrations [226], which may be explained by the fact that basal glucose concentrations in the fetal plasma are poor and cannot be reduced further, because the reduction of peripheral glucose consumption is limited.

There is also evidence of an activation of the pituitary-adrenal axis in growth-retarded fetuses, because both plasma ACTH and cortisol concentrations are substantially higher during hypoxemia than in normal-sized fetuses [226]. This may be related to induction of preterm labor [43,177].

118.2.3 Cardiovascular Responses to O_2 Deficiency in the Immature Fetus

Our knowledge about circulatory changes during O_2 deficiency in immature fetuses is scanty. However, it is clear from existing reports that there are distinct differences between immature and mature fetuses. In general, during normoxemia heart rate is higher and arterial blood pressure is lower in immature than in mature fetuses, and combined ventricular output increases throughout gestation in proportion to fetal growth [232]. Furthermore, as far as the circulatory response to hypoxia or asphyxia during development is concerned, there appears to be a different balance between parasympathetic and sympathetic control of the fetal heart rate early in gestation, so that heart rate does not fall or may even increase during maternal hypoxemia [265,266].

Only one systematic study is available on the circulatory effects of maternal hypoxemia on immature, i.e., 0.7 gestation, and very immature, i.e., 0.6 gestation, chronically instrumented fetal sheep [117,118]. These authors examined in detail the changes of heart rate, arterial blood pressure, distribution of cardiac output, organ blood flows and vascular resistances during moderate hypoxemia, i.e., 12–14 mmHg PO_2 in the descending fetal aorta. Under these conditions, which were accompanied by slight fetal acidemia, arterial blood pressure and heart rate did not change at 0.6 gestation, whereas heart rate increased significantly at 0.7 gestation. As in mature fetuses, cerebral, myocardial, and adrenal blood flows, measured by the microsphere method [231], increased, and pulmonary blood flow decreased. These responses mature early and are likely to be local vascular responses to decreases in O_2 content [118]. Furthermore, combined ventricular output and umbilical-placental blood flow decreased in both groups. Interestingly, in the very immature fetuses there was no change in blood flow to the carcass, gastrointestinal or renal circulation during hypoxemia, whereas blood flow to the carcass fell in the immature fetuses, suggesting that peripheral vasomotor control starts to develop at approximately 0.7 gestation. This, among other vasoactive substances, may also involve arginine vasopressin, which was increased at 0.7 gestation. But these differences between responses at 0.6 and 0.7 gestation may also indicate immaturity in chemoreceptor function, in the response of neurohormonal modulators, or in the response of regional receptor/effector mechanisms [118].

Only recently has comparative information on the effects of acute asphyxia caused by arrest of uterine blood flow on chronically instrumented fetal sheep at 0.6, 0.75, and 0.9 gestation become available [127]. Under these conditions, heart rate falls during asphyxia and recovers afterwards in all gestational age groups (Fig. 118.7). However, there are major differences in arterial blood pressure, which is significantly poorer at 0.6 than at 0.75 and 0.9 gestation. During 1 and 2 min asphyxia arterial blood pressure does not change significantly at 0.6, rises after a transient decrease at 0.75, and progressively rises at 0.9 gestation (Fig. 118.7). These responses are accompanied by age-dependent changes in plasma concentrations of epinephrine and norepinephrine. Epinephrine concentrations do not change significantly and the increase in concentration of norepinephrine is blunted during asphyxia at 0.6 gestation, whereas high and extremely high plasma concentrations

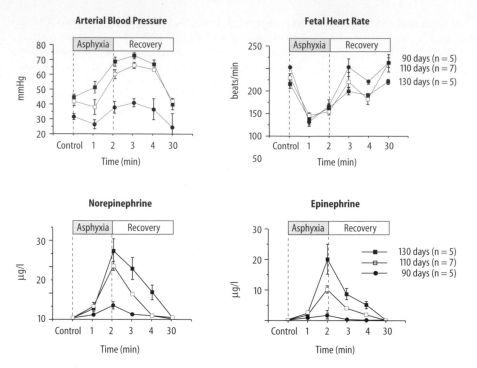

Fig. 118.7. Changes in arterial blood pressure, heart rate, norepinephrine, and epinephrine concentrations during arrest of uterine blood flow for 2 min in chronically instrumented fetal sheep at 0.6, 0.75, and 0.9 gestation. (From [139], by permission)

can be measured at 0.75 and 0.9 gestation, respectively. There are also age-dependent differences in organ blood flow. For example, blood flow to the brain per 100 g is lowest at 0.6, higher at 0.75, and highest at 0.9 gestation [127]. Interestingly, during recovery in the youngest group of fetuses, there is a steady increase in blood flow to all parts of the brain, including the germinal matrix. Whether this may be related to the well-known fact that cerebral hemorrhage, which in immature human fetuses originates in the germinal matrix in 80% of cases, remains to be established.

Blood flow to the myocardium increases progressively during asphyxia without differences between age groups and recovers thereafter in fetuses at 0.75 and 0.9 gestation. Only in the youngest group of fetuses is there an increase in myocardial blood flow in the late recovery period, suggesting increased myocardial O_2 demands, e.g., due to increased cardiac output.

The blood flow response of the carcass is largely similar in fetuses at 0.75 and 0.9 gestation, in that blood flow falls during asphyxia and recovers after it. However, at 0.6 gestation carcass blood flow at control is almost twice as high, falls during asphyxia, and rises above control values after it. Taking these findings together, it appears that in very immature fetuses a state of postasphyxial hyperperfusion of a number of organs develops, which may reflect a lack of control of vascular resistance. This view is supported by the fact that at this point in gestation, in man and in sheep, as in this study, asphyxia is followed by pronounced hypotension [127]. These observations during arrest of uterine blood flow, as well as in part some of those from maternal hypoxemia, provide evidence that in the very immature fetal sheep, at 0.6 gestation, circulatory centralization is incomplete and may be ineffective in reducing O_2 delivery to and hence the O_2 consumption of peripheral organs [139]. The fact that the age-dependence of fetal circulatory and metabolic centralization coincides with the maturation of the sympathetic nervous system, as well as other neurohormonal systems, sheds light on the importance of the sympathetic nervous system for intact survival of acute asphyxia.

118.2.4 Relation Between O_2 Delivery and Tissue O_2 Consumption

It has long been recognized that hypoxemia and asphyxia cause both a redistribution of blood flow [8,35,58,116, 132,135] *and* a reduction in O_2 consumption (Fig. 118.8A) [1,135,139,164,202,272]. However, only recently has a direct relationship between changes in O_2 delivery, i.e., O_2 content × blood flow, to peripheral organs and changes in fetal O_2 consumption been demonstrated during acute asphyxia [135] (Fig. 118.8B,C). There may exist an important protective mechanism that ensures the maintenance of oxidative metabolism in central fetal organs during acute asphyxia; and there is accumulating evidence to support this view (see Fig. 118.14A,B) [7,28,29,139].

It is, however, important to note that this new concept is *not* at variance with the current understanding of the relation between uterine and/or umbilical blood flow

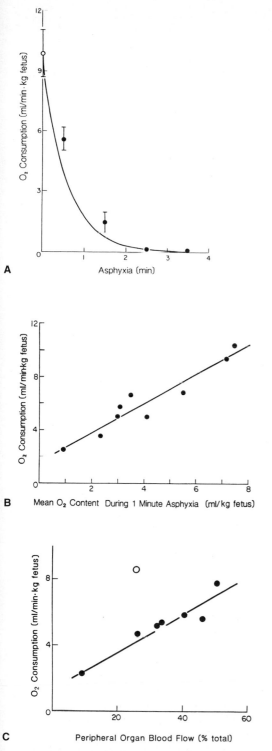

Fig. 118.8. A Changes in fetal and placental O_2 consumption during arrest of uterine blood flow. Note that O_2 consumption falls exponentially during asphyxia. Note also that O_2 consumption correlates closely with both O_2 content (**B**) in the descending aortic blood and blood flow to peripheral organs (**C**). This suggests that in the fetus O_2 delivery is a major determinant of O_2 consumption. In **A** the regression has the function $y = 9.2 \times e^{-1.71x}$; $r = 0.86$; $n = 45$. In **B** the regression has the function $y = 1.6 + 1.1x$; $r = 0.95$. In **C** the regression has the function $y = 1.4 + 0.1x$; $r = 0.94$. (From [135], by permission)

and *fetal O_2 consumption*. The relationship between both uteroplacental and umbilical blood flow and the O_2 consumption of the fetus is curvilinear [36,163,164,202,272]. This implies that fetal O_2 consumption is fairly constant over a wide range of changes in blood flow through the uterine and umbilical circulation. This phenomenon has been observed by many investigators and is largely due to the reciprocal relationship between uterine and/or umbilical blood flow and O_2 extraction [50]. Thus, depending on the reduction in O_2 delivery caused by the reduction in blood flow, the fetus can increase the amount of O_2 extracted from the blood across both the uteroplacental and the umbilical-placental vascular bed, and across the arteriovenous vascular bed of each individual organ. Therefore, the O_2 consumption of the whole conceptus, the fetus, and the individual organs will remain constant over a wide range of changes in O_2 delivery. O_2 consumption will only decrease when O_2 extraction has reached its maximum and O_2 delivery is reduced further. However, during acute hypoxemia and asphyxia, this breakpoint at which tissue O_2 consumption starts to fall can be reached fairly rapidly, though at different points in time in different organs. This is caused by carotid arterial chemoreceptor activation, which results in differential peripheral vasoconstriction, mediated largely through sympathetic pathways. Then, both O_2 content *and* blood flow fall concomitantly, resulting in a dramatic fall in O_2 delivery to and O_2 consumption by peripheral organs. Thus, during asphyxia, circulatory centralization is accompanied by metabolic centralization. This important mechanism, which involves vascular chemoreflexes mediated in part by the sympathetic nervous system, has survival value for the fetus [135].

A number of observations in the fetus and in the adult have for a long time precluded the conclusion that during fetal asphyxia, O_2 delivery might determine O_2 consumption. These include, in the fetus, the constancy of O_2 consumption over a wide range of both uterine and umbilical blood flow changes [36]; in the adult, the well-known fact that brain O_2 consumption is constant over a wide range of O_2 partial pressures in the carotid artery; and the fact that mitochondrial O_2 consumption is maintained until O_2 partial pressure falls below a "critical PO_2" of approximately 1 mmHg [247].

However, looking closer, the traditional concept from adult physiology that cellular O_2 consumption is largely unrelated to *oxygen availability*, due to the existence of a very low "critical PO_2," has been challenged previously. It has been reported that cytochrome a_3 is normally more than 20% reduced, and that any decrease and increase in PO_2 will cause reduction or oxidation of cytochrome a_3, raising the question whether there is a "critical PO_2" at all [230,247].

Thus, the matter seems to require some reconsideration, and there are good reasons to believe that the relation between O_2 delivery and O_2 consumption may be different in the fetus. This is supported by several observations. It has recently been demonstrated for whole cell preparations [273], various individual organs, e.g., the carotid body [2], the liver [32], the kidney [122] organ parts [139], and for the whole conceptus [1,135], that in the fetus the amount of O_2 available determines the amount of O_2 consumed. Therefore, on the basis of these pieces of conclusive evidence, and on the premise that O_2 extraction is maximal, it seems reasonable to conclude that the relation between the availability of O_2 and the consumption of O_2 may be fundamentally different during fetal than during adult life, in that O_2 *delivery to the tissues determines* O_2 *consumption* by these tissues [70,135] (Figs. 118.8, 118.14).

Additional support for the existence of this important mechanism has been provided in vivo by studies on ventilated fetal sheep in utero after snaring of the umbilical cord [7] and in vitro by studies on fetal skeletal muscle cells (Fig. 118.14B) [28], myocardial cells [29], and glial cells in monolayer culture (G. Braems, A. Peltzer, A. Jensen, unpublished observations). The in vitro studies confirmed the observations in vivo showing that in fetal cells O_2 availability is a determinant of cellular O_2 consumption.

The evidence produced so far in vivo and in vitro suggests strongly that during fetal life, on transition from normoxia to hypoxia, the fetus is able to *reduce O_2 consumption* by decreasing O_2 delivery to peripheral organs [28,135]. This "metabolic centralization" helps to maintain oxidative metabolism in central organs by maintaining O_2 delivery to and O_2 consumption of the brain and heart, when O_2 is at short supply [8,58,135].

Conversely, on transition from hypoxia to normoxia, the increase in O_2 delivery is parallelled by an increase in cellular O_2 consumption and in metabolic drive in most fetal organs. This mechanism, which may be of particular importance on transition from fetal to postnatal life, when O_2 delivery and O_2 consumption rise [73], guarantees optimal cell function at any given state of oxygenation.

In summary, hypoxia and asphyxia cause *centralization of the fetal circulation* through chemoreceptor-mediated vascular reflexes, which involve, among other vasoactive mechanisms, activation of the sympathetic nervous system. This circulatory centralization acts in concert with a "*metabolic centralization,*" which maintains O_2 delivery to, O_2 consumption of, and cell function in the brain and heart to ensure intact survival of the fetus [127,135].

118.2.5 Cardiovascular Regulatory Mechanisms

Control of Heart Rate. The control of heart rate via neural or neurohormonal mechanisms varies during development [195]. In late gestation both vagus and sympathetic nerves act directly and circulating catecholamines act indirectly on the sinoatrial node. These neural effects are less prominent during mid-gestation. Furthermore, during late gestation hypoxia and/or asphyxia cause initial bradycardia followed by a slow increase in heart rate which approaches or exceeds control values. The initial bradycardia is known to be chemoreflex in nature, as it is abolished by cutting the carotid nerves bilaterally [100,128] or by giving atropine [186] or by vagotomy

[11,103]. Since bradycardia occurs before a rise in arterial blood pressure, it is not a baroreflex, as evidenced by brief arrest of uterine blood flow, or by prevention of the pressure rise by α-adrenergic blockade or transection of the spinal cord at T12, which interrupts descending sympathetic outflow. The delayed increase in fetal heart rate after asphyxia is due to β-adrenergic stimulation by increased plasma catecholamines, but there are also other humoral agents involved, e.g., vasopressin. Fetal heart rate steadily declines during gestation in sheep and in the human. This is accompanied by a rise in stroke volume as the heart grows and a rise in mean arterial pressure. However, the fall in heart rate cannot simply be explained by a baroreflex mechanism, because vagotomy in late gestation does not result in an increase in heart rate to values observed in mid-gestation [103].

Control of Arterial Blood Pressure. Fetal arterial blood pressure is determined by combined ventricular output (CVO) and total peripheral resistance. These components are under both intrinsic and extrinsic control (see Chap. 95). CVO in the fetus is largely determined by heart rate, because the mechanical constraints of the fetal heart do not allow for major changes in stroke volume. Extrinsic mechanisms affecting peripheral vascular resistance form the efferent limb of reflex control, predominantly through α-adrenergic mechanisms. This is demonstrated by ablation of the carotid sinus nerves bilaterally [100], or by blockade with phentolamine [101], since both measures blunt the arterial blood pressure response during hypoxia/asphyxia [103].

Responses to O_2 Deficiency. Near term the common fetal cardiovascular response to acute lack of O_2 is bradycardia, an increase in arterial blood pressure and an increase in heart rate variability [68,204,265,266]. These changes are in part mediated by peripheral arterial chemoreceptors [20,102], which coactivate parasympathetic [186] and sympathetic pathways (Fig. 118.9) [57,59,64,132,134,135]. This was elegantly illustrated by Blanco et al. [19], who transected the spinal cord at T12 and were then able to show that hypoxemia no longer produced a rise in arterial blood pressure due to ablation of sympathetic efferences, even though the fall in heart rate persisted, because the vagi were intact.

Because arterial blood pressure rises, there is also an activation of *baroreceptors* [21] that may result in a further decrease in fetal heart rate. During severe prolonged hypoxemia, myocardial suppression and eventually myocardial failure occur [105,112]. Unlike reflex bradycardia, bradycardia based on myocardial depression cannot be blocked by atropine [112].

There is a correlation between arterial O_2 content before the insult and the delay in the onset of bradycardia: the lower the PO_2 in the carotid arterial blood, the shorter the delay in the onset of bradycardia, the greater the decrease in heart rate, and the more prolonged the duration of bradycardia [112,165]. Conversely, during a reduction in umbilical blood flow the baroreceptor-mediated response

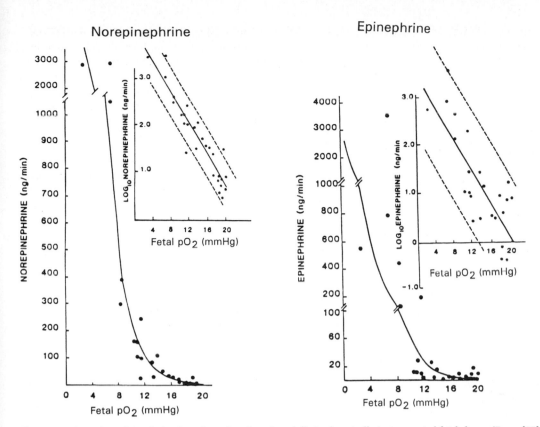

Fig. 118.9. Secretion of catecholamines from the adrenal medulla in chronically instrumented fetal sheep. (From [57], by permission)

precedes that of the carotid chemoreceptors, because arterial blood pressure rises *before* O_2 content falls.

During prolonged hypoxemia bradycardia tends to normalize. This is associated with increased sympathetic activity of the neuro- and medullary sympathetic nervous system, which is accompanied by a release of catecholamines from the adrenal medulla and sympathetic nerves [56,57,63,125,130,132,134]. The increased *release of catecholamines*, which in the adrenal medulla amounts to as much as 3–4 mg epinephrine and norepinephrine per minute (Fig. 118.9) [57], results in a circulatory centralization in favor of the brain, heart, and adrenals and at the expense of peripheral organs [239].

If during hypoxemia the sympathetic response is blocked by β-adrenoceptor antagonists, the fall of fetal heart rate, cardiac output, and umbilical blood flow is more pronounced. Furthermore, increased blood flow to the heart, brain, and adrenals cannot be maintained [60,64,67,205]. Blockade of α-adrenoceptors during hypoxemia causes an increase in fetal heart rate and cardiac output, whereas arterial blood pressure and total vascular resistance fall [146,218]. Then blood flow to the heart, adrenals, gut, spleen, and lungs are increased [218]. Blockade of α- and β-adrenoceptors results in fetal death during hypoxemia [203].

Ablation of peripheral sympathetic neurons by chemical sympathectomy, e.g., by 6-hydroxydopamine, leaves the adrenal medulla intact. Hence, circulating catecholamine concentrations do not change much [121,137,173].

Changes of cardiac output during and after acute asphyxia are similar in intact and in sympathectomized fetuses (Fig. 118.10) [137]. However, during fetal hypoxemia and asphyxia the increase in arterial blood pressure is delayed [121,137,174,175]. This may be associated with a slow initial rise in plasma catecholamine concentrations [151]. In sympathectomized fetuses the redistribution of organ blood flow during hypoxemia and asphyxia is different to that in intact fetuses (Fig. 118.10), in that blood flows to the gastrointestinal tract and to the kidneys do not change [121,137]. The effect of chemical sympathectomy on the carcass, skeletal muscle, and skin depends on the severity of asphyxia [121,137]. The increase in vascular resistance, which usually occurs during arrest of uterine blood flow, is completely blunted in the gastrointestinal tract and markedly delayed in the carcass [137]. During maternal hypoxemia the increase in blood flow to the heart, brain, and adrenals is not significantly affected in sympathectomized fetuses [121]. However, during arrest of uterine blood flow the initial increase in blood flow to the brainstem is blunted [137]. Furthermore, the percentage of cardiac output directed to the placenta is reduced, while that to the carcass is increased with intact fetuses (Fig. 118.10). Therefore, in sympathectomized fetuses circulatory centralization is less effective in protecting the fetus against adverse effects of asphyxia [137].

However, there are also other mechanisms to consider that are involved in the regulation of blood flow during O_2 deficiency. Fetal hypoxemia is accompanied by an increase in

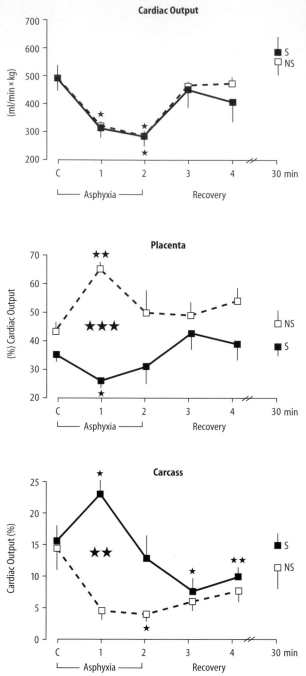

Fig. 118.10. Cardiac output and its distribution to the placenta and lower carcass of intact (*NS*) and sympathectomized (*S*) chronically instrumented fetal sheep near term during arrest of uterine blood flow for 2 min. (From [137], by permission)

[154]. Direct effects of vasopressin on the myocardium [86] and the brain have also been described [65].

Acute hypoxia is associated with a reduction in fetal blood volume. In the regulation of these changes the *atrial natriuretic factor* (ANF; see also Chap. 75) may be involved [49]. Whether the rise in ANF concentrations observed during acute hypoxia is a direct or an indirect effect of hypoxia has yet to be determined [49].

Circulatory changes during volume loss, e.g., through fetal hemorrhage, are in part governed by the autonomous nervous system [31,182]. Furthermore, a number of hormones are released, including arginine vasopressin [69, 81,221,228,242,252], renin [120,221,223], catecholamines [56,145,149,174,175], ACTH, and cortisol [228]. Slower but quantitatively similar volume losses, e.g., 30% over a period of 2 h, cause less pronounced hormone changes [30]. In addition, during hypoxia there is clear evidence of activation of the pituitary-adrenal-cortical axis, resulting in increased *ACTH* and cortisol concentrations [3,25,27, 44,45,143,151,258]. Among other responses, these two hormones allegedly help maintain the arterial blood pressure during hypoxemia [151].

Another group of substances related to changes in cardiovascular variables during hypoxemia are *β-endorphins* [250,252,253,268]. Blocking endogenous opiate receptors by naloxone results in a pronounced fall in heart rate and a decrease in cardiac output and placental blood flow, while peripheral vascular resistance rises. This suggests that endogenous opioids may modulate the cardiovascular response to hypoxia [168,179]. Other peptides, e.g., *neuropeptidey* (NPY) have also been reported to cause vasoconstriction, but the actual mechanism has not yet been determined [88]. Whether endothelial factors are important for changing vascular resistances in the fetus, as they are in the adult, remains to be established.

In summary, convincing evidence has been produced to show that, near term, fetal cardiovascular responses to O_2 deficiency are largely governed by the autonomous nervous systems through parasympathetic and sympathetic pathways. However, there are also a number of other regulatory systems involved into fetal circulatory control, and a great variety of vasoactive substances exert specific vascular and metabolic effects during hypoxemia and asphyxia via endocrine or paracrine mechanisms. All these regulatory systems ensure – within limits – intact survival of the fetus during asphyxia, but the relative importance of each of these systems for the circulatory and metabolic adaptations of the fetus that are necessary to survive asphyxia may vary during development.

118.2.6 Changes in Fetal Organs During Hypoxemia and Asphyxia

Brain. Blood flow to the fetal brain increases during various interventions that result in a fall in arterial O_2 content. These include maternal hypoxemia [8,37,58,209], graded reduction in umbilical blood flow [116], and graded reduction in uterine blood flow [139] (Fig. 118.11). However,

plasma renin activity [222], which converts angiotensin I to *angiotensin II*, a potent vasoconstrictive hormone [189,245] (see also Chap. 73). There is also a rise in *vasopressin* during hypoxemia [242]. Infusion of vaso–pressin causes bradycardia and a circulatory centralization [119,242], whereas resting concentrations of vasopressin do not appear to exert major effects on the fetal circulation

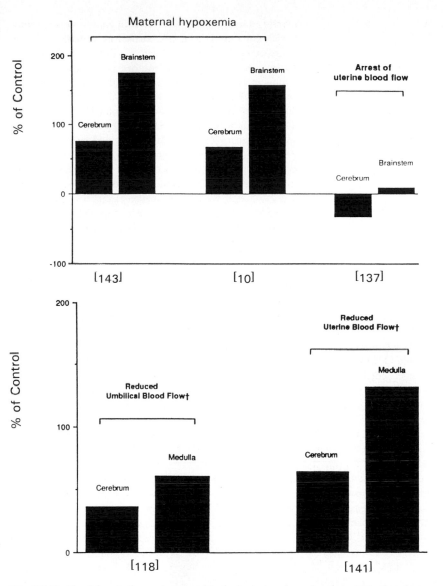

Fig. 118.11. Blood flow to the cerebrum and to the brainstem (as a percentage of control) during various experimental disturbances in chronically instrumented fetal sheep near term. Flow reduction to cause a 50% reduction in fetal O_2 delivery. Sources are cited in brackets. Note that during arrest of uterine blood flow cerebral blood flow is redistributed in favor of the brainstem; however, it is not increased

during arrest of uterine blood flow, blood flow to the brain does not increase in spite of an increase in arterial blood pressure, suggesting an increase in cerebral vascular resistance [135]. This cerebral reflexive vasoconstriction during acute severe asphyxia is accompanied by a steep rise in plasma catecholamine concentrations, suggesting that an activation of the sympathetic nervous system through arterial chemoreceptors is involved [74,133]. Thus, during acute asphyxia caused by arrest of uterine blood flow [135] O_2 delivery to the brain falls, whereas it is maintained during moderate hypoxemia [139,153,209,220].

A recent study on the effect of mild chronic hypoxemia (of approximately 17 mmHg PO_2 in the fetal ascending aorta), caused by embolization of the umbilical circulation, revealed that there may be threshold arterial O_2 contents above which cerebral blood flow does not change [23]. This may be explained in part by increased O_2 extraction across the cerebral vascular bed, but there are also other mechanisms to consider, including reduced cerebral O_2 consumption.

During hypoxemia the increase in blood flow to the brainstem is greater than that seen in other regions, despite the fact that this area has the highest resting blood flow in the fetal brain [8,116,135,141,220]. This may have survival value for the fetus, because neuronal activity in important autonomous centers in the brainstem is maintained.

Autoregulation of fetal cerebral blood flow is operative in the fetal lamb near term between 30 and 50 mmHg during normoxemia (Fig. 118.12) [181,201,210,260]. However, it has been demonstrated that during hypoxemia

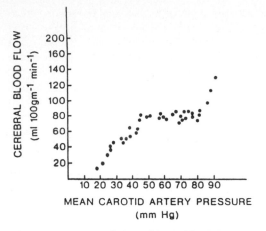

Fig. 118.12. Autoregulation of brain blood flow in immature, chronically instrumented fetal sheep. (From [201], by permission)

autoregulation of cerebral blood flow is lost. Cerebral blood flow then varies with arterial blood pressure. However, this is not always true. For instance, during acute asphyxia, caused by arrest of uterine blood flow, cerebral vascular resistance increases, and hence cerebral blood flow does not increase despite a steep increase in arterial blood pressure. Thus, under these very acute conditions, autoregulation of cerebral blood flow is intact, even though arterial O_2 content is poor [135]. Whether this is related to the rapidity of the change in carotid arterial O_2 content over time and hence to the intensity of carotid arterial chemoreceptor stimulation, is presently unknown. Interestingly, during repeated occlusion of the umbilical cord, blood flow to the gray matter increased, whereas that to the white matter decreased – a pattern of redistribution consistent with the presence of brain lesions [51].

A fall in O_2 delivery to the cerebrum can be compensated – within limits – by an increase in cerebral O_2 extraction. If O_2 delivery continues to fall, cerebral O_2 consumption falls [46,220]. In that case, due to increasing cerebral O_2 deficiency, glucose, the main fuel of the brain [14,183,219], is metabolized anaerobically, lactate concentrations rise, and concentrations of high-energy phosphates fall in the cerebrum (Fig. 118.13) [15–17,196,197,264]. Finally, cerebral metabolism may collapse when synthesis of high-energy phosphates through aerobic or anaerobic glycolysis fails. Then, neuronal membranes depolarize, voltage-gated Ca^{2+} channels open, and Ca^{2+} flux into the cytoplasm increases. There is also enhanced release of excitatory neurotransmitters, e.g., glutamate, and hence increased N-methyl-D-aspartate (NMDA) receptor stimulation. This further opens Ca^{2+} channels and increases the flux of Ca^{2+} into the neurons [248]. Eventually neuronal death occurs. Asphyxia also has other adverse effects on the brain, e.g., the generation of oxygen radicals, which destroy multiple unsaturated fatty acids of the cell membranes. These membrane defects lead to a further increase in Ca^{2+} influx, which enhances energy-consuming cellular processes. In that case, both membranes and organels of the neurons will be destroyed by lipases, proteases, and endonucleases. Thus, a vicious circle is maintained that eventually results in neuronal death.

Functionally, there are a number of effects elicited by hypoxemia and asphyxia on the brain, e.g., on the *electrocortical activity* and on fetal breathing. In normoxemic fetal sheep there are episodic changes in electrocortical activity, characteristic of sleep states postnatally, within 3 weeks of term [76]. Then, fetal breathing movements that are almost continuous before 115 days, are confined to episodes of low-voltage activity. In this state rapid eye movements occur, whereas gross fetal body movements are associated with high-voltage electrocortical activity. These obvious functional changes between high- and low-voltage activity are accompanied by changes in arterial blood pressure, heart rate, blood flow to the brain and to the brainstem, and in plasma catecholamine concentrations, suggesting that sympathetic tone varies with voltage activity [131,214,267].

The electrocortical activity of the fetus changes during hypoxemia. There is an increase in the relative incidence of episodes of high-voltage activity [188], whereas severe asphyxia is associated with an increased proportion of episodes of low-voltage activity. Eventually, fetal electrocortical activity becomes isoelectric [185]. Interestingly, in chronic hypoxemia, complementary changes of autonomic tone during episodes of low- and high-voltage electrocortical activity contribute to the slowing of the fetal heart rate. Increased parasympathetic tone and decreased

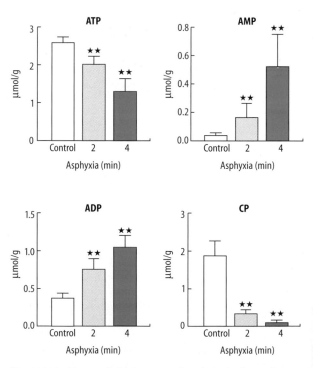

Fig. 118.13. Changes in high-energy phosphates in the cerebrum during acute asphyxia, caused by arrest of uterine blood flow in unanesthetized fetal guinea pigs near term. *ATP, AMP, ADP,* Adenosine tri-, mono-, and diphosphate; *CP,* creatine phosphate. (From [15,16], by permission)

sympathetic tone may enhance cardiac efficiency when O_2 supply is chronically reduced [267].

Fetal breathing movements and rapid eye movements (REM) are reduced during hypoxemia [24,26,77,104,184,198]. This is due to central inhibitions [19,78]. During prolonged hypoxemia the incidence of breathing movements returns to normal values within 14 h [158]. If CO_2 tensions rise, fetal breathing movements are hardly affected [18]. Transection of the fetal brainstem at the level of the pons eliminates the inhibitory effects of hypoxemia on breathing [78], whereas transection of the spinal medulla at T12-L1 eliminates those on spinal reflexes [19].

Heart. During normoxemia in the unanesthetized fetal sheep near term, heart rate varies between 160 and 180 and the combined beats/min ventricular output varies between 450 and 600 ml/min × kg fetal body weight. There is only a small rise in cardiac output (10–15%) during tachycardia and only a small increase in stroke volume when end-diastolic filling pressure rises [95,96,240,256]. Furthermore, cardiac output is very sensitive to changes in afterload [96]. These findings suggest that the fetal heart operates at the upper limit of its function curve under physiologic conditions [240] (see also Chap. 90). This may be related to structural differences between the adult and the fetal heart in which myofibrillar content is less [93,190] and the sarcoplasmic reticulum and the T tubule system is poorly developed [110].

Hypoxemia and asphyxia cause a fall in heart rate and combined ventricular output [234]. Although arterial blood pressure and hence afterload rise, there is a small increase in stroke volume [233]. In this situation coronary and myocardial blood flows are increased to meet cardiac O_2 demands by increasing O_2 delivery [58,116,139,209]. The reduction in preductal aortic O_2 content by 50% affects cardiac consumption of neither O_2 nor glucose or lactate [90,91].

Unlike hypoxemia, acidemia reduces cardiac performance [92,215,261] – a fact that may be related to depletion of cardiac glycogen stores [246], which in turn may cause ECG changes [109,229].

Liver. The fetal hepatic vascular bed is different from that of the adult, in that the ductus venosus connects the umbilical vein with the abdominal inferior vena cava. There are also branches of the umbilical vein to the left and right liver lobes, the latter communicating with the portal vein [238]. About 55% of the umbilical venous blood bypasses the liver through the ductus venosus. The remainder is distributed to the liver lobes. The left lobe of the liver is almost exclusively supplied by umbilical venous blood, whereas the right lobe is supplied by both umbilical and portal venous blood. The contribution of hepatic arterial blood is small (3–4%) [240]. During normoxemia prostaglandins maintain a modest degree of relaxation of the ductus venosus and hepatic microcirculation. However, they are not responsible for the reduction of umbilical venous return between the ductus venosus and the liver during hypoxemia or asphyxia [207].

During maternal hypoxemia, when fetal O_2 tension falls to about 12 mmHg in the carotid artery, umbilical blood flow is largely maintained, whereas fetal blood flow tends to fall [58]. The fraction of umbilical blood distributed to the ductus venosus rises from 55% to 65%, whereas that to the liver falls by a similar amount [32,217]. Due to a reduction of O_2 content in the umbilical venous and portal venous blood, O_2 delivery to the liver falls drastically.

During a reduction in umbilical blood flow, O_2 content does not change, but hepatic and fetal O_2 delivery falls. This is accompanied by an increased fraction of blood distributed to the ductus venosus at the expense of that to the liver. Within the liver, umbilical venous blood flow to the right lobe of the liver is more reduced than that to the left. Portal venous blood flow is maintained [83,116].

During reduction in uterine blood flow there is also a redistribution of umbilical venous blood towards the ductus venosus (10%). However, the mechanisms involved are poorly understood. There may be changes in vascular resistance in the ductus venosus related to transmural pressures [82]. However, there are also direct hypoxemic effects and neurohormonal mechanisms to consider [55,276].

During hypoxemia fetal hepatic O_2 consumption falls linearly in relation to the fall in O_2 delivery, but the fall in O_2 consumption is smaller in the left than in the right liver lobe, suggesting functional differences [32]. During reduction in umbilical blood flow, O_2 delivery to the liver decreases, but the liver's O_2 consumption is maintained by increased extraction of O_2 [241]. This apparent contradiction between methods as far as hepatic O_2 consumption is concerned, is largely explained by the amount of O_2 extracted by the liver, which was smaller in the earlier study [32] than in the later one [241].

During maternal hypoxemia or during reduction in uterine blood flow, liver glycolysis is enhanced and glucose is released into the inferior vena caval blood. This covers approximately 30–40% of the fetal consumption of glucose when O_2 is in short supply [32,241]. There is no firm evidence of significant gluconeogenesis in the sheep fetus [97,98]. The fall in net lactate uptake of the liver may contribute to the rise in fetal blood lactate concentrations.

Placenta. Umbilical blood flow falls during arrest of uterine blood flow [135,138]. However, umbilical and hence placental blood flow is largely maintained during maternal hypoxemia and during graded reduction in uterine blood flow, whereas blood flow to the fetus, particularly that to the lower part of the body, tends to fall [58,116,139]. At any rate, O_2 delivery from the placenta to the fetus falls. Studies on transplacental diffusion of O_2 in sheep have demonstrated that umbilical and uterine blood form an exchange system that tends to equilibrate the venous concentrations of the two blood streams [36,211,271]. The overall efficiency of this exchange system, however, is less than that of an ideal venous equilibration, as shown by the observations that the uterine-umbilical difference in venous PO_2 is consistently positive and ranges among animals between 12 and 20 mmHg, and that the uterine-umbilical difference in venous PCO_2 is consistently negative and averages approximately – 3 mmHg [36,272]. This limitation may result from several factors, e.g., configuration of maternal and fetal vessels, uneven placental perfusion, low O_2 diffusion capacity, or the high O_2 affinity of fetal blood [272].

During O_2 deficiency there is no obvious improvement in the efficiency of the placenta to supply the fetus with O_2, i.e., to reduce the uterine-umbilical venous O_2 difference. The reduced O_2 delivery to the fetus is compensated initially by increased O_2 extraction to maintain fetal O_2 consumption. However, below an O_2 delivery of 0.5 mmol/min × kg, fetal O_2 consumption starts to fall [163,202,272]. During reduction in uterine blood flow, net placental consumption of glucose falls and there is evidence of substantial provision of glucose and lactate to the fetus. Fetal production of lactate increases sharply and much of this appears to be consumed by the placenta, at a rate sufficient to account entirely for the deficit in net glucose consumption [99].

Kidney. During a graded and moderate reduction of uterine or umbilical blood flow there are no significant changes in renal blood flow [116,139]. Maternal hypoxemia causes an increase in renal vascular resistance and a fall in renal blood flow by 20% [58,269]. However, glomerular filtration rate is maintained, suggesting that renal vasoconstriction associated with fetal hypoxemia is likely to occur at the efferent rather than the afferent arteriolar level [222]. When renal blood flow falls there is an increase in plasma renin and vasopressin concentrations and increased reabsorption of water [222]. Prolonged fetal hypoxemia results in a transient fall in glomerular filtration rate. After 3 h and during recovery from hypoxemia, filtration rate and urine production increase to above control values [274].

A decrease in renal blood flow during fetal hypoxemia causes a marked reduction in renal O_2 delivery and O_2 extraction increases. Eventually, renal O_2 consumption falls if hypoxemia persists [122].

During normoxemia the kidneys metabolize lactate and release glucose, whereas during hypoxemia they metabolize glucose and release lactate [122]. Obviously, renal gluconeogenesis prevails during normoxemia, whereas glycolysis prevails during hypoxia.

Lung. During fetal life gas exchange occurs in the placenta, and pulmonary blood flow is low, supplying only nutritional requirements for lung growth and perhaps serving some metabolic or paracrine functions [106,235]. About 8–10% of the cardiac output is directed to the lungs [232]. The high pulmonary vascular resistance is in part due to the thickness of the vascular smooth muscle [108,171,212,213]. Furthermore, physiologically low O_2 partial pressures during fetal life may contribute to increased pulmonary vascular resistance.

From studies in sheep it has emerged that autonomous control of pulmonary vascular resistance is poor. Bilateral section of cervical or thoracic sympathetic nerves does not significantly change pulmonary vascular resistance [62]. Similarly, bilateral cervical vagotomy had no effect. Also, selective pharmacologic blockade using phentolamine and atropine did not change resting pulmonary vascular tone [236]. Interestingly, a combination of α- and β-adrenergic with parasympathetic blockade prevented pulmonary vasoconstriction [172], suggesting that pulmonary vascular responses to hypoxia are not mediated directly by these autonomic pathways [106]. On the other hand, electrical [62] and hormone receptor stimulation [12,38,39,251] altered pulmonary vascular resistance. Whether these mechanisms are invoked during fetal hypoxemia is not clear [106].

Prostaglandins have been reported to be involved into the regulation of pulmonary blood flow; however, the specific effects are variable [40,41,170,261]. At present it appears that prostaglandins have at least a modulatory effect on pulmonary vasoconstriction during hypoxemia, which is blunted by indomethacin, a potent inhibitor of cyclo-oxygenase-dependent products of prostaglandin metabolism [42,169] (see also Chap. 6). Recently, further evidence has been produced that leucotrienes, lipo-oxygenase-dependent derivatives of arachidonic acid metabolism, play a major role in regulating pulmonary blood flow. Maternal hypoxemia or graded reduction of umbilical and uterine blood flow increases pulmonary vascular resistance and decrease pulmonary blood flow [58,116,139,172,209]. In immature and mature human newborns there is evidence that reduced pulmonary blood flow during asphyxia may be related to both patent ductus arteriosus and respiratory distress syndrome.

Adrenals. Adrenal blood flow increases during maternal hypoxemia or graded reduction in uterine and/or umbilical blood flow [58,116,139], suggesting that this organ is important when O_2 is in short supply. For this reason, the adrenals are regarded as "central" organs along with the brain and the heart. During normoxemia adrenal medullary blood flow (511 ± 155 ml/min × 100 g) is twice as high as adrenal cortical blood flow (248 ± 71 ml/min × 100 g). During isocapnic hypoxia the increase in blood flow (+145%) is similar in the two adrenal areas [126,131,133]. These changes in blood flow may be related to adrenal function, in that catecholamines, e.g., norepinephrine and epinephrine, are released in large quantities from the adrenal medulla during hypoxemia [63,125]. There is also a significant release of cortisol [143,144], which is related to increased release of ACTH, from the pituitary [151], and of aldosterone [224].

During severe asphyxia caused by arrest of uterine blood flow, the release of medullary and cortical hormones is different in that concentrations of norepinephrine, epinephrine, and aldosterone increase and those of cortisol and dehydroepiandrosterone decrease. Furthermore, there appears to be a redistribution of adrenal *corticosteroid biosynthesis during asphyxia*, in favor of the cortisol pathway and at the expense of the androgen pathway [136].

Intestine. Blood flow to the intestinal tract is maintained during maternal hypoxemia and during graded reduction in uterine and umbilical blood flow [58,85,116,139], and intestinal O_2 consumption does not change, because O_2 extraction rises [84]. Only severe reduction in fetal arterial O_2 content results in a marked fall in intestinal blood flow to very low values [135,275]. Then, in spite of increased O_2 extraction, intestinal O_2 consumption falls and a mesentery acidosis develops [84,85]. In human newborns this may be related to necrotizing enterocolitis.

Spleen. Splenic blood flow is reduced during maternal hypoxemia, graded reduction in uterine blood flow (nonsignificantly), and severe asphyxia caused by arrest of uterine blood flow. During the latter, splenic blood flow is virtually arrested [135,137]. This may be related to the fact that the spleen is almost exclusively innervated by the sympathetics [89].

Carcass. The fetal carcass consists largely of skin, muscle, bone and connective tissue. Blood flow to the carcass falls during more severe maternal hypoxemia or a reduction in uterine blood flow [58,135]. It rises, however, during graded reduction in umbilical blood flow [116]. Interestingly, O_2 consumption of the carcass decreased twice as much during uterine blood flow reduction [135] as during umbilical blood flow reduction [116], even though in both studies total fetal O_2 delivery was reduced by a similar amount (50%). This suggests that the delivery of O_2 to the carcass, which was considerably lower in the former study, determines the consumption of O_2 in the carcass (Fig. 118.14A). This view is supported by studies in which uterine blood flow was arrested (see Fig. 118.8) [135] and by observations in cell cultures (Fig. 118.14A) [28]. These studies provide first evidence that the amount of O_2 delivered to peripheral organs determines the amount of O_2 consumed by these organs [135]. This important mecha-

nism enables the fetus to conserve O_2 in peripheral organs, and in the carcass, to maintain oxidative metabolism in central organs when O_2 is in short supply.

Skin and Scalp. For a long time, reduced blood flow to the skin of the *neonate* has been recognized as an index of asphyxia and hence of circulatory redistribution [6,244,256]. However, only recently has blood flow to the skin of the *fetus* been studied systematically, to establish whether changes in cutaneous blood flow reflect a centralization of the fetal circulation (see Fig. 118.5) [129,132–135,138].

In normoxemic fetal sheep near term, blood flow to the skin of the hips and shoulders is about 22–26 ml/min × 100 g and that to the scalp is significantly higher, ranging between 29 and 34 ml/min × 100 g [126,131]. During isocapnic hypoxemia with a carotid arterial PO_2 of 12 mmHg, blood flow to the skin and scalp fall by 18% and 23%, respectively, as do blood flows to most of the peripheral organs, including the lungs [126].

Graded reduction in uterine blood flow reduces blood flow to the skin and scalp by about 50–60% [139], whereas arrest of uterine blood flow reduces blood flow to the skin by 95% and that to the scalp by 85% [135]. At control and during acute asphyxia, blood flows to skin and scalp correlate linearly; however, throughout acute asphyxia blood flow to the scalp is 5–10 times higher than to the body skin, indicating different responsiveness of the skin vasculature in certain areas [132,133]. The fact that, during acute asphyxia, the time course of the fall in blood flow to peripheral organs is largely similar to that in the skin and scalp suggests that decreased cutaneous blood flow reflects redistribution of the fetal circulation [132–135,138]. This is related to the fact that asphyxia increases sympathetic nervous activity through arterial chemoreceptor mechanisms and hence causes almost generalized peripheral vasoconstriction [75].

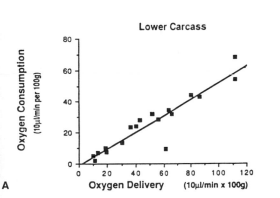

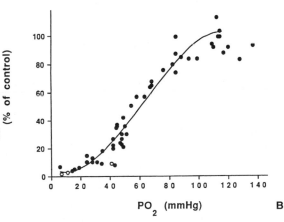

Fig. 118.14. A Close correlation between O_2 delivery and O_2 consumption in the lower [carcass] of chronically instrumented fetal sheep near term during graded reduction of uterine blood flow. Note that O_2 delivery determines O_2 consumption when O_2 is in short supply. (From [139], by permission.) **B** Close correlation between PO_2 and O_2 consumption (as a percentage of control) in fetal skeletal muscle cells in monolayer culture. (From [28], by permission.) In **A** the function of the regression is $y = -1.0 + 0.5x$; $r = 0.93$. In **B** the function of the regression is $y = 5.17 - 0.54x + 0.03x^2 - 0.00016x^3$; $r = 0.97$; $n = 54$

Skin blood flow is quite sensitive as an index of circulatory centralization, because blood flow to the skin falls more than that to any other peripheral organ. Furthermore, skin blood flow depends largely on sympathetic activity, which is known to increase rapidly during asphyxia through chemoreceptor-mediated mechanisms [102].

In summary, during hypoxemia and asphyxia there is *circulatory centralization* in favor of the brain, heart, and adrenals and at the expense of almost all peripheral organs, the lungs, carcass, skin, and scalp. This response is qualitatively similar but quantitatively different under various experimental conditions. However, at the nadir of severe acute asphyxia the circulatory centralization cannot be maintained. In that case there is circulatory decentralization, and the fetus will suffer severe brain damage if not die unless immediate resuscitation is provided. Future work in this field will have to concentrate on the important questions of what factors determine this collapse of circulatory compensating mechanisms in the fetus, how it relates to neuronal damage, and how can the fetal brain be pharmacologically protected against the adverse effects of asphyxia. As far as neuroprotection is concerned, it will be of the utmost importance to ensure that neuroprotective drugs do not in any way interfere with lifesaving cardiovascular mechanisms, e.g., circulatory and metabolic centralization, that are operative naturally and have proven so effective in protecting both life and the central nervous system of asphyxiated fetuses and newborns.

118.3 Respiratory Control and Lung Development

To meet the newborn's sudden demands for O_2 after separation from the placenta at birth, the lung must reach a programmed and predetermined degree of morphological, physiological, and biochemical maturity during development [34]. Lung function and structural design serve the primary purpose of gas exchange (see Chaps. 100–102). There is a hierarchy of airways and blood vessels to (1) separate air and blood in distinct compartments and (2) to allow rapid gas exchange between the two media by modification of their wall's structure towards the periphery.

118.3.1 Lung Development

The timetable of pulmonary development is given in Table 118.4. The development is divided into an embryonic, a fetal, and a postnatal period. In humans the *embryonic development of the lungs* occurs 5–7 weeks after conception, during which embryoblastic cells separate from trophoblastic cells to differentiate into the three germ layers. The lung appears at about day 26 as a ventral bud of the esophagus at the caudal end of the laryngotracheal sulci. Lobar, segmental, and subsegmental bronchi are formed by days 37, 41, and 48, respectively. The interaction of entodermal epithelium and mesodermal mesenchyme is of particular importance for branching and cell differentiation. Cartilage is found at the end of the 7th week. In this early organogenic period the development of pulmonary vessels is closely linked to that of the heart and of the primitive systemic blood vessels.

The transition to the *fetal developmental period of the lungs* is at about day 50, when the lung resembles an acinar gland. This pseudoglandular stage lasts until the 4th month of gestation, and is followed by the canalicular stage and by the stage of terminal sac formation.

Surfactant Synthesis. In the canalicular stage lamellar bodies develop in some cuboidal cells, which later differentiate into type I and type II pneumocytes. These have been shown to be the storage sites of surface-active material present in alveolar surfactant, which appears at about 0.80–0.85 of gestation in most mammalian species. However, in the human fetus lamellar bodies can be detected as early as 0.6 of gestation. The human fetus born from this point in gestation onwards has an increasing chance of survival. To some extent the lung is already able to allow O_2 uptake, but the surfactant system is still immature, i.e., it is quantitatively and/or qualitatively deficient. The fetal surfactant, designed to stabilize the lung air space

Table 118.4. Timetable of pulmonary development[a]

Stage	Prenatal period		Postnatal period
	Embryonic (weeks)	Fetal (weeks)	
Embryonic	1–7		
Pseudoglandular		5–17	
Canalicular		16–26	
Terminal sac (saccular)		24–38	
Alveolar			36th week or birth to 3–8 years
Normal growth			Completion of alveolar formation to end of somatic growth

[a] Because development is continuous, different stages overlap

2328

postnatally, is a complex mixture of phospholipids, neutral lipids, proteins, and carbohydrates, phosphatidylcholine (PC or lecithin) being by far the largest component.

Regulation of Surfactant Production. The concentrations of desaturated phosphatidylcholine in lung tissue and alveoli are plotted against the relative gestational age for various species including man in Fig. 118.15. There is ample evidence from studies in fetal animals, and also from tissue and cell culture, that surfactant phospholipid synthesis is increased after glucocorticoid treatment in vivo and in vitro.

Irrespective of the complex mechanism by which glucocorticosteroids increase surfactant production, cytoplasmic and nuclear glucocorticoid receptors have often been demonstrated to be present in fetal lungs. These vary in number during development, being rare early in fetal life, increasing towards term, and decreasing postnatally. This suggests that pulmonary maturation is controlled physiologically by endogenous glucocorticosteroids. There action on type II pneumocytes may well be mediated by fibroblasts. As a result, cell differentiation into type I and type II pneumocytes along with glycogen depletion of the alveolar epithelium and the formation of lamellar bodies in the type II cells are enhanced. Exogenous *corticosteroids* given to the mother result in enhanced surfactant production and lung maturity in premature newborns at 26–33 weeks' gestation [177]. This discovery had an enormous impact on the intact survival of premature newborns. However, in addition to corticosteroids, thyroid hormones have also been shown to accelerate lung maturation. The mode of action is largely unclear, though it has been suggested that stimulation of mRNA and protein synthesis may be involved in promoting pulmonary growth and development [34].

118.3.2 Pulmonary Mechanics

Evidence from premature sheep [123] and monkeys indicates that extrapolation from the static volume-pressure

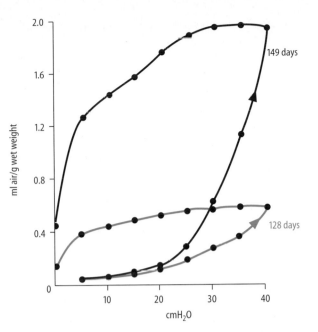

Fig. 118.16. Volume-pressure curves in the immature and mature fetal sheep lung. (From [243], by permission)

curve from excised lungs is likely to be important for understanding the first few breaths (Fig. 118.16) [243]. Most studies in perinatal lung physiology have been concerned with fetal lung capacity and residual volume rather than with the analysis of the *whole volume-pressure curve*. Total lung capacity as a measure of lung distensibility is usually expressed in milliliters of air per gram lung at a transpulmonary pressure near that of lung rupture (e.g., 40 cmH$_2$O in sheep: V_{40}). Residual volume at deflation to low transpulmonary pressures as a measure of lung stability is usually expressed in milliliters of air per gram lung at a transpulmonary pressure of 5 cmH$_2$O in sheep (V_5) or as a percentage of total lung capacity (e.g., % V_{40}) [243]. It is important to notice that the beneficial effects of glucocorticosteroids on morphological and functional maturation of the fetal lung can be demonstrated by lung volume-pressure experiments. There is a significant correlation in sheep between V_{10} (ml/g lung) and fetal cortisol levels, gestational age, and saturated phosphatidylcholine in lung tissue and lavage fluid [157]. However, volume-pressure curves give little insight into events at the alveolar level.

During the latter part of pregnancy both distensibility and stability of the fetal lung increase several-fold [33, 111,157,159–161,255]. Interestingly, there are a number of studies to support the view that changes in lung distensibility can be achieved by exogenous corticosteroids through a preponderant effect on lung connective tissue, e.g., collagen and elastin, rather than on surfactant production alone. Thus, cortisol may play a key role in connective tissue maturation.

There is a general *craniocaudal gradient in lung maturation* that is reflected in airspace development and in the maturation of type II pneumocytes in rabbits [155], his-

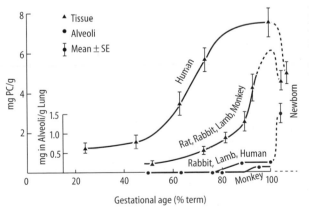

Fig. 118.15. Concentrations of phosphatidylcholine (*PC*) in lung tissue and alveoli plotted against relative gestational age for various species. (From [34], by permission)

tologic development, and the appearance of alveolar surfactant in sheep [9,79] and phosphatidyl production in rhesus monkeys [160].

Much is known about the physicochemical nature of many of the macromolecules, e.g., proteoglycans, glycosaminoglycans, and glycoproteins, (like fibronectin and laminin), but we are only just beginning to understand how the meshwork is organized and how it may fulfill those functions that have so far been discovered. Fetal lungs are not simply small adult lungs. They progressively change their structure, from being an organ rich in nonfibrous interstitium with few capillaries and peripheral airways, towards being a highly vascularized organ poor in nonfibrous interstitial tissue, relatively rich in fibers, and with a tremendously increased ratio of peripheral to central airways [243].

118.3.3 Regulation of Fetal "Breathing"

In 1888 and 1905, von Ahlfeld [262,263] described movements observed through and recorded from the human maternal abdomen, which he attributed to fetal breathing. His descriptions and tracings are clear and were confirmed by some other investigators. However, it took some 80 years from the original observation to discover that the fetal lamb in utero makes irregular rapid breathing move-

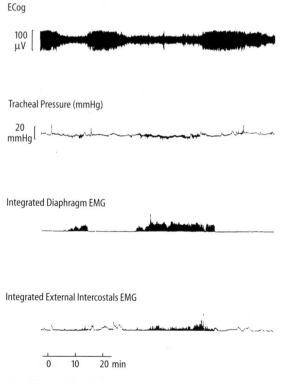

Fig. 118.17. Relationship between fetal electrocorticogram (ECoG) and fetal breathing (tracheal pressure and diaphragm EMG). Breathing movements are seen episodically and only in association with low voltage electrocortical activity. (From [71], by permission)

ments and that these are not related to changes in fetal blood gases [71]. These findings prompted numerous studies on the relation between fetal behavior and sleep states, e.g., in fetal sheep, in which there is a close association between low-voltage electrocortical activity, rapid eye movements, and fetal breathing (Fig. 118.17), and also in other species, including subhuman primates [187], guinea pigs, and man. Brief episodes of human fetal breathing movements are detectable as early as 8 weeks' gestation [80]. However, there appears to be a real species difference between the human and sheep fetus in the central control of behavioral (sleep) patterns.

Electrocortical Differentiation. Before 100 days' gestation in sheep, fetal electrocortical activity is of low voltage and breathing is continuous. Later on, at 110–120 days' gestation, breathing becomes episodic and clear periods of apnea lasting more than 2 min become established. This observation led to the hypothesis of a suprapontine mechanism for inhibition of breathing in association with low-voltage activity after electrocortical differentiation. Transection of the brainstem at the level of the inferior colliculi caused dissociation of breathing movements from fetal electrocortical activity [78]. After some 6–10 days' recovery there was a strong trend towards continuous breathing. From this it was concluded that electrocortical differentiation had led to the interpolation of episodic apneic pauses by an upper or suprapontine mechanism, causing inhibition of breathing with high-voltage activity.

Effect of Hypoxemia. Isocapnic hypoxia causes arrest of fetal breathing, in contrast to the hyperventilation expected postnatally [77]. However, after section of the brainstem at or above the pons, the arrest of breathing was abolished: breathing movements were increased in rate and depth by hypoxia [78]. Thus, the effect of isocapnic hypoxia on breathing movement is not local on the medulla, but at a suprapontine site. Hypoxia of this magnitude also arrest movements of the fetal limbs and trunk, and induces high-voltage electrocortical activity.

References

1. Acheson GH, Dawes GS, Mott JC (1957) Oxygen consumption and the arterial oxygen saturation in foetal and newborn lambs. J Physiol (Lond) 135:623–642
2. Acker H, Lübbers DW (1977) The kinetics of local tissue PO_2 decrease after perfusion stop within the carotid body of the cat in vivo and in vitro. Pflugers Arch 369:135–140
3. Alexander DP, Forsling ML, Martin MJ, Nixon DA, Ratcliffe JG, Redstone D, Tunbridge D (1972) The effect of maternal hypoxia on fetal pituitary hormone release in the sheep. Biol Neonate 21:219–228
4. Alexander G (1964) Studies on the placenta of the sheep: effect of surgical reduction in the number of caruncles. J Reprod Fertil 7:307–322
5. Alexander G, Hales JRS, Stevens D, Donnelly JB (1987) Effects of acute and prolonged exposure to heat on regional blood flows in pregnant sheep. J Dev Physiol 9:1–15

6. Apgar VA (1953) A proposal for a new method of evaluation of the newborn infant. Curr Res Anesth Anal 32:260–267

7. Asakura H, Ball, KT, Power GG (1990) Interdependence of arterial PO_2 and O_2 consumption in the fetal sheep. J Dev Physiol 13:205–213

8. Ashwal S, Majcher JS, Longo LD (1981) Patterns of fetal lamb regional cerebral blood flow during and after prolonged hypoxia: Studies during the posthypoxic recovery period. Am J Obstet Gynecol 139:365–372

9. Avery, ME, Mead D (1959) Surface properties in relation to atelectasis and hyaline membrane disease. Am J Dis Child 97:517–523

10. Barclay AE, Franklin KJ, Prichard MML (1944) The foetal circulation and cardiovascular system, and the changes that they undergo at birth. Blackwell Scientific, Oxford

11. Barcroft J (1946) Researches on prenatal life. Blackwell Scientific, Oxford

12. Barett CT, Heymann MA, Rudolph AM (1972) Alpha- and beta-adrenergic function in fetal sheep. Am J Obstet Gynecol 89:252–260

13. Behrman RE, Less MH, Peterson EM, De Lannoy CW, Sceds AE (1970) Distribution of the circulation in normal and asphyxiated fetal primate. Am J Obstet Gynecol 108:956–969

14. Berger R, Gjedde A, Heck J, Müller E, Krieglstein J, Jensen A (1994) Extension of the 2-deoxyglucose method to the fetus in utero: theory and normal values for the cerebral glucose consumption in fetal guinea pigs. J Neurochem 63:271–279

15. Berger R, Jensen A, Krieglstein J, Steiglmann JP (1991) Effects of acute asphyxia on brain energy metabolism in fetal guinea pigs near term. J Dev Physiol 16:9–11

16. Berger R, Jensen A, Krieglstein J, Steiglmann JP (1991) Cerebral energy metabolism in guinea pig fetuses during development. J Dev Physiol 16:317–319

17. Berger R, Jensen A, Krieglstein J, Steiglmann JP (1992) Cerebral energy metabolism in immature and mature fetuses during acute asphyxia. J Dev Physiol 18:125–128

18. Bissonnette JM, Hohimer AR, Willekc GB (1989) Effect of asphyxia on respiratory activity in fetal sheep. J Dev Physiol 12:157–161

19. Blanco CE, Dawes GS, Walker DW (1983) Effect of hypoxia on polysynaptic hindlimb reflexes of unanesthetized fetal and newborn lambs. J Physiol (Lond) 339:453–466

20. Blanco CE, Dawes GS, Hanson MA, McCooke HB (1984). The response to hypoxia of arterial chemoreceptors in fetal sheep and new-born lambs. J Physiol (Lond) 351:25–37

21. Blanco CE, Dawes GS, Hanson MA, McCooke HB (1985) Studies of carotid baroreceptor afferents in fetal and newborn lambs. In: Jones CT, Nathanielsz PW (eds) Physiological development of the fetus and newborn. Academic, London, pp 595–598

22. Block BS, Schlafer DH, Wentworth RA, Kreitzer LA, Nathanielsz PW (1990) Intrauterine asphyxia and the breakdown of physiologic circulatory compensation in fetal sheep. Am J Obstet Gynecol 162:1325–1331

23. Block BS, Schlafer DH, Wentworth RA, Kreitzer LA, Nathanielsz PW (1990) Regional blood flow distribution in fetal sheep with intrauterine growth retardation produced by decreased umbilical placental perfusion. J Dev Physiol 13:81–85

24. Bocking AD, Harding R (1986) Effects of reduced uterine blood flow on electrocortical activity, breathing, and skeletal muscle activity in fetal sheep. Am J Obstet Gynecol 154:655–662

25. Bocking AD, McMillen IC, Harding R, Thornburn GD (1986) Effect of reduced uterine blood flow on fetal and maternal cortisol. J Dev Physiol 8:237–245

26. Boddy K, Dawes GS, Fisher R, Pinter S, Robinson JS (1974) Foetal respiratory movements, electrocortical and cardiovascular responses to hypoxaemia and hypercapnia in sheep. J Physiol (Lond) 243:599–618

27. Boddy K, Jones CT, Mantell C, Ratcliffe JG Robinson JS (1974) Changes in plasma ACTH and corticosteroid of the maternal and fetal sheep during hypoxia. Endocrinology 94:588–591

28. Braems G, Jensen A (1991) Hypoxia reduces oxygen consumption of fetal skeletal muscle cells in monolayer culture: a preliminary report. J Dev Physiol 16:209–215

29. Braems G, Dussler I, Jensen A (1990) Oxygen availability determines oxygen consumption of fetal myocardial cells in monolayer culture. Scientific program and abstracts, 36th annual meeting of the Society for Gynecologic Investigation, p 243

30. Brace RA, Cheung CY (1986) Fetal cardiovascular and endocrine responses to prolonged fetal hemorrhage. Am J Physiol 251:R417–R424

31. Brace RA (1987) Fetal blood volume responses to fetal haemorrhage: autonomic nervous contribution. J Dev Physiol 9:97–103

32. Bristow J, Rudolph AM, Itskovitz J, Barnes R (1983) Hepatic oxygen and glucose metabolism in the fetal lamb. J Clin Invest 71:1047–1061

33. Brumley GW, Chernick V, Hodson WA, Normand C, Fenner A, Avery ME (1967) Correlation of mechanical stability, morphology, pulmonary surfactant and phospholipid content in the developing lamb lung. J Clin Invest 46:863–873

34. Burri PH (1985) Development and growth of the human lung. In: Fishman AP, Fisher AB (eds) Handbook of physiology, vol 1, Sect 3. respiratory system, chap 1, American Pysiological Society, Bethesda, pp 1–46

35. Campbell AGM, Dawes GS, Fishan AP, Hyman AI (1967) Regional redistribution of blood flow in the mature fetal lamb. Circ Res 21:229–235

36. Carter AM (1989) Factors affecting gas transfer across the placenta and the oxygen supply to the fetus. J Dev Physiol 12:305–322

37. Carter AM, Gu W (1988) Cerebral blood flow in the fetal guinea-pig. J Dev Physiol 10:123–129

38. Cassin S, Dawes GS, Mott JC, Rss BB, Strang LB (1964) The vascular resistance of the foetal and newly ventilated lungs of the lamb. J Physiol (Lond) 171:61–79

39. Cassin S, Dawes GS, Ross BB (1964) Pulmonary blood flow and vascular resistance in immature foetal lambs. J Physiol (Lond) 171:80–89

40. Cassin S (1980) Role of prostaglandins and thromboxanes in the control of the pulmonary circulation in the fetus and newborn. Sem Perinatol 4:101–107

41. Cassin S, Tod M, Philips J, Frisinger S, Jordan J, Gibbs C (1981) Effects of prostacyclin on the fetal pulmonary circulation. Pediatr Pharmacol 1:197–207

42. Cassin S (1982) Humoral factors affecting pulmonary blood flow in the fetus and newborn infant. In: Peckhan GJ, Heymann MA (eds) Cardiovascular sequelae of asphyxia in the newborn. Report of the 83rd Ross conference on pediatric research. Ross Laboratories, Columbus, Ohio, pp 10–18

43. Chaillis JRG, Mitchell BF, Lye SJ (1984) Activation of fetal adrenal function. J Dev Physiol 6:93–105

44. Challis JRG, Richardson BS, Rurak D, Wlodek ME, Patrick JE (1986) Plasma adrenocorticotropic hormone and cortisol and adrenal blood flow during sustained hypoxemia in fetal sheep. Am J Obstet Gynecol 155:1332–1336

45. Challis JRG, Fraher L, Oosterhuis J, White SE, Bocking AD (1989) Fetal and maternal endocrine responses to prolonged reductions in uterine blood flow in pregnant sheep. Am J Obstet Gynecol 160:926–932

46. Chao CR, Hohimer AR, Bissonette JM (1989) Cerebral carbohydrate metabolism during severe ischemia in fetal sheep. J Cereb Blood Flow Metab 9:53–57

47. Charlton V, Johengen M (1985) Effects of intrauterine nutritional supplementation on fetal growth retardation. Biol Neonate 48:125–142

48. Charlton V, Johengen M (1987) Fetal intravenous nutritional supplementation ameliorates the development of embo-

lization-induced growth retardation in sheep. Pediatr Res 22:55–61

49. Cheung CY, Brace RA (1988) Fetal hypoxia elevates plasma atrial natriuretic factor concentration. Am J Obstet Gynecol 159:1263–1268

50. Clapp JF III (1978) The relationship between blood flow and oxygen uptake in the uterine and umbilical circulation. Am J Obstet Gynecol 132:410–413

51. Clapp JF III, Peress NS, Wesley M, Mann LI (1988) Brain damage after intermittent partial cord occlusion in the chronically instrumented fetal lamb. Am J Obstet Gynecol 159:504–509

52. Clapp JF III, Szeto HH, Larrow R, Hewitt J, Mann LI (1981) Fetal metabolic response to experimental placental vascular damage. Am J Obstet Gynecol 140:446–451

53. Clapp JF III, Thabault NC, Hubel CA, McLaughlin MK, Auletta FJ (1982) Ovine placental cortisol production. Endocrinology 111:1728–1730

54. Clapp JF III (1988) Fetal endocrine and metabolic response to placental insufficiency. In: Künzel W, Jensen A (eds) The endocrine control of the fetus. Springer, Berlin Heidelberg New York, pp 246–253

55. Coceani F, Adeagbo ASO, Cutz E, Olley PM (1984) Autonomic mechanisms in the ductus venosus of the lamb. Am J Physiol 247:H17–24

56. Cohen WR, Piasecki GJ, Jackson BT (1982) Plasma catecholamines during hypoxemia in fetal lamb. Am J Physiol 243:R520–R525

57. Cohen WR, Piasecki GJ, Cohn HE, Young JB, Jackson BT (1984) Adrenal secretion of catecholamines during hypoxemia in fetal lambs. Endocrinology 114:383–390

58. Cohn HE, Sacks EJ, Heymann MA, Rudolph AM (1974) Cardiovascular responses to hypoxemia and acidemia in fetal lambs. Am J Obstet Gynecol 120:817–824

59. Cohn HE, Piasecki GJ, Jackson BT (1978) The role of autonomic nervous control in the fetal cardiovascular response to hypoxemia. In: Longo LD, Reneau DD (eds) Fetal and newborn cardiovascular physiology, vol 2. Garland, New York, pp 249–258

60. Cohn HE, Piasecki GJ, Jackson BT (1982) The effect of beta-adrenergic stimulation on fetal cardiovascular function during hypoxemia. Am J Obstet Gynecol 144:810–816

61. Cohn HE, Jackson BT, Piasecki GJ, Cohen WR, Novy MJ (1985) Fetal cardiovascular responses to asphyxia induced by decreased uterine perfusion. J Dev Physiol 7:289–298

62. Colebatch HJH, Dawes GS, Goodwin JW, Nadeau RA (1965) The nervous control of the circulation in the foetal and newly expanded lungs of the lamb. J Physiol 178:544–562

63. Comline RS, Silver M (1961) The release of adrenaline and noradrenaline from the adrenal glands of the foetal sheep. J Physiol (Lond) 156:424–444

64. Court DJ, Parer JT, Block BSB, Llanos AJ (1984) Effects of beta-adrenergic blockade on blood flow distribution during hypoxaemia in fetal sheep. J Dev Physiol 6:349–358

65. Courtice GP, Kwong TE, Lumbers ER, Potter EK (1984) Excitation of the cardiac vagus by vasopressin in mammals. J Physiol (Lond) 354:547–556

66. Creasy RK, DeSwiet M, Kahanpaa KV, Young WP, Rudolph AM (1973) Pathophysiological changes in the foetal lamb with growth retardation. In: Comline RS, Cross KW, Dawes GS, Nathanielsz PW (eds) Foetal and neonatal physiology. Cambridge University Press, Cambridge, pp 398–402

67. Dagbjartsson A, Karlsson K, Kjellmer I, Rosén KG (1985) Maternal treatment with a cardioselective beta-blocking agent – consequences for the ovine fetus during intermittent asphyxia. J Dev Physiol 7:387–396

68. Dalton KJ, Dawes GS, Patrick JE (1977) Diurnal, respiratory, and other rhythms of fetal heart rate in lambs. Am J Obstet Gynecol 127:414–424

69. Daniel SS, Husain MK, Milliez J, Stark RI, Yeh MN, James LS (1978) Renal response of fetal lamb to complete occlusion of umbilical cord. Am J Obstet Gynecol 131:514–519

70. Dawes GS (1962) The umbilical circulation. Am J Obstet Gynecol 84:1634–1648

71. Dawes GS (1986) The rediscovery of fetal breathing and its consequences. In: Johnston BM, Gluckman PD (eds) Respiratory control and lung development in the fetus and newborn. Perinatology Press, Ithaca N Y, pp 209–222

72. Dawes GS, Mott JC, Widdicombe JG (1954) The foetal circulation in the lamb. J Physiol (Lond) 126:563–587

73. Dawes GS, Mott JC (1959) The increase in oxygen consumption of the fetal lamb after birth. J Physiol (Lond) 146:295–315

74. Dawes GS (1968) Foetal and neonatal physiolgy. Year Book Medical Publisher, Chicago

75. Dawes GS, Lewis BV, Milligan JE, Roach MR, Talner NS (1968) Vasomotor responses in the hind limbs of foetal and newborn lambs to asphyxia and aortic chemoreceptor stimulation. J Physiol (Lond) 195:516–538

76. Dawes GS, Fox HE, Leduc BM, Liggins GC, Richards RT (1972) Respiratory movements and rapid-eye-movement sleep in the foetal lamb. J Physiol (Lond) 220:119–143

77. Dawes GS (1973) Breathing and rapid-eye-movement sleep before birth. In: Foetal and neonatal physiology, proceedings of the Sir Joseph Barcroft centenary symposium. Cambridge University Press, Cambridge, pp 49–62

78. Dawes GS, Gardner WN, Johnston BM, Walker DW (1983) Breathing in fetal lambs: the effect of brain stem section. J Physiol (Lond) 335:535–553

79. De Lemos RA (1970) Acceleration of appearance of pulmonary surfactant in the fetal lamb by administrtion of corticosteroids. Am Rev Respir Dis 102:459–461

80. De Vries JIP, Visser GHA, Huisjes JH, Prechtl HFR (1982). The emergence of fetal behaviour. In: Neurobiology of development: EBBS workshop. University of Groningen Press, Groningem, pp 45–46

81. Drummond WH, Rudolph AM, Keil LC, Gluckman PD, MacDonald AA, Heymann MA (1980) Arginine vasopressin and prolactin after hemorrhage in the fetal lamb. Am J Physiol 238:E214–E219

82. Edelstone DI, Rudolph AM, Heymann MA (1978) Liver and ductus venosus blood flows in fetal lambs in utero. Circ Res 42:426–433

83. Edelstone DI, Rudolph AM, Heymann MA (1980) Effects of hypoxaemia and decreasing umbilical flow on liver and ductus venosus blood flows in fetal lambs. Am J Physiol 238:H656–H663

84. Edelstone DI Holzman IR (1982) Fetal intestinal oxygen consumption at various levels of oxygenation. Am J Physiol 242:H50–H54

85. Edelstone DI, Holzman IR (1984) Regulation of perinatal intestinal oxygenation. Semin Perinatol 8:226–233

86. Elliot JM, West MJ, Chalmers J (1985) Effects of vasopressin on heart rate in conscious rabbits. J Cardiovasc Pharmacol 7:6–11

87. Emmanoulides GC, Townsend DE, Bauer RA (1986) Effects of single umbilical artery ligation in the lamb. Pediatrics 42:919–927

88. Emson PC, De Quidt ME (1984) NPY – a new member of the pancreatic polypeptide family. Trends Neurosci 7:31–35

89. Fillenz M (1966) Innervation of the cat spleen. J Physiol (Lond) 185:2–3

90. Fisher DJ, Heymann MA, Rudolph AM (1982) Fetal myocardial oxygen and carbohydrate consumption during acutely induced hypoxemia. Am J Physiol 242:H657–H661

91. Fisher DJ, Heymann MA, Rudolph AM (1982) Fetal myocardial oxygen and carbohydrate metabolism in sustained hypoxemia in utero. Am J Physiol 243:H959–H963

92. Fisher DJ (1986) Acidemia reduces cardiac output and left ventricular contractility in conscious lambs. J Dev Physiol 8:23–31

93. Friedman WF (1973) The intrinsic physiologic properties of the developing heart. In: Friedman WF, Lesch M, Sonnenblick EH (eds) Neonatal heart disease. Grune and Stratton, New York, pp 21–49

94. Gilbert RD, Cummings LA, Juchau MR, Longo LD (1979) Placental diffusing capacity and fetal development in exercising or hypoxic guinea pigs. J Appl Physiol 46:828–834

95. Gilbert RD (1980) Control of fetal cardiac output during changes in blood volume. Am J Physiol 238:H80–H86

96. Gilbert RD (1982) Effects of afterload and baroreceptors on cardiac function in fetal sheep. J Dev Physiol 4:299–309

97. Gleason CA, Rudolph AM (1985) Gluconeogenesis by the fetal sheep liver in vivo. J Dev Physiol 7:185–195

98. Gleason CA, Rudolph AM (1986) Oxygenation does not stimulate hepatic gluconeogenesis in fetal lamb. Pediatr Res 20:532–535

99. Gu W, Jones CT, Parer JT (1985) Metabolic and cardiovascular effects on fetal sheep of sustained reduction of uterine blood flow. J Physiol (Lond) 368:109–129

100. Guissani DA, Spencer JAD, Moore PJ, Hanson MA (1990) The effect of carotid sinus nerve section on the initial cardiovascular response to acute isocapnic hypoxia in fetal sheep in utero. J Physiol (Lond) 432:33P

101. Guissani DA, Spencer JAD, Moore PJ, Hanson MA (1991) Effect of phentolamine on initial cardiovascular response to isocapnic hypoxia in intact and carotid sinus denervated fetal sheep. J Physiol (Lond) 438:56P

102. Hanson MA (1988) The importance of baro- and chemoreflexes in the control of the fetal cardiovascular system. J Dev Physiol 10:491–511

103. Hanson MA (1993) The control of heart rate and blood pressure in the fetus: theoretical considerations. In: Hanson MA, Spencer JAD, Rodeck CH (eds) Fetus and neonate, vol 1: the circulation. Cambridge University Press, Cambridge, pp 1–22

104. Harding R, Poore ER, Cohen GL (1981) The effect of brief episodes of diminished uterine blood flow on breathing movements, sleep states and heart rate in fetal sheep. J Dev Physiol 3:231–243

105. Harris JL, Krüger TR, Parer JT (1982) Mechanisms of late decelerations of the fetal heart rate during hypoxia. Am J Obstet Gynecol 144:491–496

106. Heymann MA (1984) Control of the pulmonary circulation in the perinatal period. J Dev Physiol 6:281–290

107. Heymann MA (1988) Control of the pulmonary circulation in the perinatal period. In: Künzel W, Jensen A (eds) The endocrine control of the fetus. Springer, Berlin Heidelberg New York, pp 31–37

108. Hislop A, Reid L (1972) Intra-pulmonary arterial development during fetal life-branching pattern and structure. J Anat 113:35–48

109. Hökegard KH, Karlsson K, Kjellmer I, Rosén KG (1979) ECG-changes in the fetal lamb during asphyxia in relation to beta-adrenoceptor stimulation and blockade. Acta Physiol Scand 105:195–203

110. Hoerter J, Mazet F, Vassort G (1982) Perinatal growth of the rabbit cardiac cell: possible implications for the mechanism of relaxation. J Mol Cell Cardiol 13:725–740

111. Humphreys PW, Strang LB (1967) Effects of gestation and prenatal asphyxia on pulmonary surface properties of the foetal rabbit. J Physiol (Lond) 192:53–62

112. Itskovitz J, Goetzman BW, Rudolph AM (1982) The mechanism of late deceleration of the heart rate and its relationship to oxygenation in normoxemic and chronically hypoxemic fetal lambs. Am J Obstet Gynecol 142:66–73

113. Itskovitz J, Goetzman BW, Rudolph AM (1982) Effects of hemorrhage on umbilical venous return and oxygen delivery in fetal lambs. Am J Physiol 242:H543–H548

114. Itskovitz J, LaGamma EF, Rudolph AM (1983) Heart rate and blood pressure responses to umbilical cord compression in fetal lambs with special reference to the mechanism of variable deceleration. Am J Obstet Gynecol 147:451–457

115. Itskovitz J, LaGamma EF, Rudolph AM (1983) The effect of reducing umbilical blood flow on fetal oxygenation. Am J Obstet Gynecol 145:813–818

116. Itskovitz J, LaGamma EF, Rudolph AM (1987) Effects of cord compression on fetal blood flow distribution and O_2 delivery. Am J Physiol 252:H100–H109

117. Iwamoto IIS (1989) Cardiovascular responses to reduced oxygen delivery: studies in fetal sheep at 0.55–0.7 gestation. In: Gluckman PD, Johnston BM, Nathanielsz PW (eds) Advances in fetal physiology: reviews in honor of GC Liggins. Perinatology Press, Ithaca, New York, pp 43–54

118. Iwamoto HS, Kaufman T, Keil LC, Rudolph AM (1989) Responses to acute hypoxemia in the fetal sheep at 0.6–0.7 gestation. Am J Physiol 256:H613–H620

119. Iwamoto HS, Rudolph AM, Keil LC, Heymann MA (1979) Hemodynamic responses of the sheep fetus to vasopressin infusion. Circ Res 44:430–436

120. Iwamoto HS, Rudolph AM (1981) Role of renin-angiotensin system in response to hemorrhage in fetal sheep. Am J Physiol 240:H848–H854

121. Iwamoto HS, Rudolph AM, Mirkin BL, Keil LC (1983) Circulatory and humoral responses of sympathectomized fetal sheep to hypoxemia. Am J Physiol 245:H767–H772

122. Iwamoto HS, Rudolph AM (1985) Metabolic responses of the kidney in fetal sheep: effect of acute and spontaneous hypoxemia. Am J Physiol 249:F836–F841

123. Jacobs H, Jobe A, Ikegami M, Glatz T, Jones SJ, Barajas L (1982) Premature lambs rescued from respiratory failure with natural surfactant: clinical and biophysical correlates. Pediatr Res 16:424–429

124. Jacobs R, Robinson JS, Owens JA, Falconer J, Webster MED (1988) The effect of prolonged hypobaric hypoxia on growth of fetal sheep. J Dev Physiol 10:97–112

125. Jelinek J, Jensen A (1991) Catecholamine concentrations in plasma and organs of the fetal guinea pig during normoxaemia, hypoxaemia and asphyxia. J Dev Physiol 15:145–152

126. Jensen A (1989) Die Zentralisation des fetalen Kreislaufs. Thieme, Stuttgart

127. Jensen A (1992) The role of the sympathetic nervous system in preventing perinatal brain damage. In: Künzel W (ed) Oxygen, basis of the regulation of vital functions in the fetus. Springer, Berlin Heidelberg New York, pp 77–107

128. Jensen A, Hanson MA (1995) Circulatory responses to acute asphyxia in intact and chemodenervated fetal sheep near term. Reprod Fertil Dev (in press)

129. Jensen A, Künzel W (1980) The difference between fetal transcutaneous PO_2 and arterial PO_2 during labour. Gynecol Obstet Invest 11:249–264

130. Jensen A, Künzel W, Kastendieck E (1985) Repetitive reduction of uterine blood flow and its influence on fetal transcutaneous PO_2 and cardiovascular variables. J Dev Physiol 7:75–87

131. Jensen A, Bamford OS, Dawes GS, Hofmeyr G, Parkes MJ (1986) Changes in organ blood flow between high and low voltage electrocortical activity in fetal sheep. J Dev Physiol 8:187–194

132. Jensen A, Hohmann M, Künzel W (1987) Redistribution of fetal circulation during repeated asphyxia in sheep: effects on skin blood flow, transcutaneous PO_2, and plasma catecholamines. J Dev Physiol 9:41–55

133. Jensen A, Künzel W, Hohmann M (1985) Dynamics of fetal organ blood flow redistribution and catecholamine release during acute asphyxia. In: Jones CT, Nathanielsz PW (eds) The physiological development of the fetus and newborn. Academic, London, pp 405–410

134. Jensen A, Künzel W, Kastendieck E (1987) Fetal sympathetic activity, transcutaneous PO_2, and skin blood flow during repeated asphyxia in sheep. J Dev Physiol 9:337–346

135. Jensen A, Hohmann M, Künzel W (1987) Dynamic changes in organ blood flow and oxygen consumption during acute asphyxia in fetal sheep. J Dev Physiol 9:543–559

136. Jensen A, Gips H, Hohmann M, Künzel W (1988) Adrenal endocrine and circulatory responses to acute prolonged asphyxia in surviving and non-surviving fetal sheep near term. In: Künzel W, Jensen A (eds) The endocrine control of the fetus. Springer, Berlin Heidelberg New York, pp 64–79

137. Jensen A, Lang U (1988) Dynamics of circulatory centralization and release of vasoactive hormones during acute asphyxia in intact and chemically sympathectomized fetal sheep. In: Künzel W, Jensen A (eds) The endocrine control of the fetus. Springer, Berlin Heidelberg New York, pp 135–149

138. Jensen A, Lang U, Künzel W (1987) Microvascular dynamics during acute asphyxia in chronically prepared fetal sheep near term. Adv Exp Med Biol 220:127–131

139. Jensen A, Roman C, Rudolph AM (1991) Effects of reducing uterine blood flow on fetal blood flow distribution and oxygen delivery. J Dev Physiol 15:309–323

140. Jogee M, Myatt L, Elder MG (1983) Decreased prostacyclin production by placental cells in culture from pregnancies complicated by fetal growth retardation. Br J Obstet Gynaecol 90:247–250

141. Johnson GN, Palahniuk RJ, Tweed WA, Jones MV, Wade JG (1979) Regional cerebral blood flow changes during severe fetal asphyxia produced by slow partial umbilical cord compression. Am J Obstet Gynecol 135:48–52

142. Jones CT, Robinson RO (1975) Plasma catecholamines in foetal and adult sheep. J Physiol (Lond) 285:395–408

143. Jones CT (1977) The development of some metabolic responses to hypoxia in foetal sheep. J Physiol (Lond) 266:743–762

144. Jones CT, Boddy K, Robinson JS, Ratcliffe JG (1977) Developmental changes in the responses of the adrenal glands of the foetal sheep to endogenous adrenocorticotrophin as indicated by hormone responses to hypoxaemia. J Endocrinol 72:279–292

145. Jones CT (1980) Circulating catecholamines in the fetus, their origin, actions and significance. In: Parves H, Parves S (eds) Biogenic amines in development. Elsevier North-Holland, Amsterdam, pp 63–86

146. Jones CT, Ritchie JWK (1983) The effects of adrenergic blockade on fetal response to hypoxia. J Dev Physiol 5:211–222

147. Jones CT, Robinson JS (1983) Studies on experimental growth retardation in sheep. Plasma catecholamines in fetuses with small placenta. J Dev Physiol 5:77–87

148. Jones CT (1985) Reprogramming of metabolic development by restriction of fetal growth. Biochem Soc Trans 13:89–91

149. Jones CT, Rose JC, Kelly RT, Hardgrave BY (1985) Catecholamine responses in fetal lambs subjected to hemorrhage. Am J Obstet Gynecol 151:475–478

150. Jones CT, Lafeber HN, Price DA, Parer JT (1987) Studies on the growth of the fetal guinea pig. Effects of reduction in uterine blood flow on the plasma sulphation-promoting activity and on the concentration of insuli-like growth factors-I and -II. J Dev Physiol 9:181–201

151. Jones CT, Roebuck MM, Walker DW, Johnston BM (1988) The role of the adrenal medulla and peripheral sympathetic nerves in the physiological responses of the fetal sheep to hypoxia. J Dev Physiol 10:17–36

152. Jones CT, Gu W, Harding JE, Price DA, Parer JT (1988) Studies on the growth of the fetal sheep. Effects of surgical reduction in placental size, or experimental manipulation of uterine blood flow on plasma sulphation promoting activity and on the concentration of insulin-like growth factors I and II. J Dev Physiol 10:179–189

153. Jones MD, Sheldon RE, Peeters LL, Meschia G, Battaglia FC, Makowski EL (1977) Fetal cerebral oxygen consumption at different levels of oxygenation. J Appl Physiol 43:1080–1084

154. Kelly RT, Rose JC, Meiss PJ, Hargrave BY, Morris M (1983) Vasopressin is important for restoring cardiovascular homeostasis in fetal lambs subjected to hemorrhage. Am J Obstet Gynecol 146:807–812

155. Kikkawa Y, Kaibara M, Motoyama EK, Orzalesi MM, Cook CD (1971) Morphologic development of rabbit lung and its acceleration with cortisol. Am J Pathol 64:423–442

156. Kitanaka T, Alonso JG, Gilbert RD, Siu BL, Clemons GK, Longo LD (1989) Fetal responses to long-term hypoxemia in sheep. Am J Physiol 245:R1348–1354

157. Kitterman JA, Liggins GC, Campos GA, Clements JA, Forster CS, Lee CH, Creasy RK (1981) Prepartum maturation of the lung in fetal sheep: relation to cortisol. J Appl Physiol 51:384–390

158. Koos BJ, Kitanaka T, Matsuda K, Gilbert RD, Longo LD (1988) Fetal breathing adaption to prolonged hypoxaemia in sheep. J Dev Physiol 10:161–166

159. Kotas RV, Avery ME (1971) Accelerated appearance of pulmonary surfactant in the fetal rabbit. J Appl Physiol 30:358–361

160. Kotas RV, Farrel PM, Ulane RE, Chez RA (1977) Fetal rhesus monkey lung development: lobar differences and discordances between stability and Distensibility. J Appl Physiol 43:92–98

161. Kotas RV, Fletcher BD, Torday JS, Avery ME (1971) Evidence of independent regulators of organ maturation in fetal rabbits. Pediatrics 47:57–64

162. Künzel W, Kastendieck E, Böhme U, Feige A (1975) Uterine hemodynamics and fetal response to vena caval occlusion in sheep. J Perinat Med 3:260–268

163. Künzel W, Mann LI, Bhakthavathsalan A, Airomlooi J, Liu M (1977) The effect of umbilical vein occlusion on fetal oxygenation, cardiovascular parameters, and fetal electroencephalogram. Am J Obstet Gynecol 128:201–208

164. Künzel W, Moll W (1972) Uterine O_2 consumption and blood flow of the pregnant uterus. Z Geburtsh Perinatol 176:108–117

165. Künzel W, Kastendieck E, Hohmann M (1983) Heart rate and blood pressure and metabolic changes in the sheep fetus following reduction of uterine blood flow. Gynecol Obstet Invest 15:300–317

166. Lafeber H, Jones CT, Rolph T (1979) Some of the consequences in the intra-uterine growth retardation. In: Visser HKA (ed) Nutrition and growth of the fetus. Nijhoff, the Hague, pp 43–62

167. Lafeber HN, Rolph TP, Jones CT (1984) Studies on the growth of the fetal guinea pig. The effects of ligation of the uterine artery on organ growth and development. J Dev Physiol 6:441–459

168. LaGamma EF, Itskovitz J, Rudolph AM (1982) Effects of naloxone on fetal circulatory responses to hypoxemia. Am J Obstet Gynecol 143:933–940

169. Leffler CW, Tyler TL, Cassin S (1978) Effect of indomethacin on pulmonary vascular response to ventilation of fetal goats. Am J Physiol 234:H346–H351

170. Leffler CW, Hessler JR (1979) Pulmonary and systemic vascular effects of exogenous prostaglandin I in fetal lambs. Eur J Pharmacol 54:37–42

171. Levin DL, Rudolph AM, Heymann MA, Phibbs RH (1976) Morphological development of the pulmonary vascular bed in fetal lambs. Circulation 53:144–151

172. Lewis AB, Heymann MA, Rudolph AM (1976) Gestational changes in pulmonary vascular responses in fetal lambs in utero. Circ Res 39:536–541

173. Lewis AB, Wolf WJ, Sischo W (1984) Fetal cardiovascular and catecholamine responses to hypoxemia after chemical sympathectomy. Pediatr Res 18:318–322

174. Lewis AB, Wolf WJ, Sischo W (1984) Cardiovascular and catecholamine responses to successive episodes of hypoxemia in the fetus. Biol Neonate 45:105–111

175. Lewis AB, Sischo W (1985) Cardiovascular and catecholamine responses to hypoxemia in chemically sympathectomized fetal lambs. Dev Pharmacol Ther 8:129–140

176. Lewis PJ, Moncada S, O'Grady J (eds) (1983) Prostacyclin in pregnancy. Raven, New York

177. Liggins GC, Schellenberg JC (1988) Endocrine control of lung development. In: Künzel W, Jensen A (eds) The endocrine control of the fetus. Springer, Berlin Heidelberg New York, pp 236–245

178. Llanos AJ, Green JR, Creasy RK, Rudolph AM (1980) Increased heart rate response to parasympathetic and beta adrenergic blockade in growth retarded fetal lambs. Am J Obst Gynecol 136:808–813

179. Llanos AJ, Court D, Holbrook H, Block BS, Vega R, Parer JT (1983) Cardiovascular effects of naloxone (NLX) during asphyxia in fetal sheep. 30th annual meeting, Society for Gynecological Investigation, Washington DC, p 300

180. Lotgering FK, Wallenburg HCS (1986) Hemodynamic effects of caval and uterine venous occlusion in pregnant sheep. Am J Obstet Gynecol 155:1164–1170

181. Lou HC, Lassen NA, Tweed WA, Johnson G, Jones M, Palahniuk RJ (1979) Pressure passive cerebral blood flow and breakdown of the blood-brain barrier in experimental asphyxia. Acta Paediatr Scand 68:57–63

182. MacDonald AA, Rose JC, Heymann MA, Rudolph AM (1980) Heart rate response of fetal and adult sheep to hemorrhage stress. Am J Physiol 239:H780–H793

183. Makowski EL, Schneider JM, Tsoulos NG, Colwill JR, Battaglia FC, Meschia G (1972) Cerebral blood flow, oxygen consumption and glucose utilization of fetal lambs in utero. Am J Obstet Gynecol 114:292–303

184. Maloney JE, Adamson TM, Brodecky V, Dowling MH, Ritchie BC (1975) Modification of respirtory centre output in the unanesthetized fetal sheep "in utero". J Appl Physiol 39:552–558

185. Mann LI, Prichard JW, Symmes D (1970) EEG, ECG, and acid-base observations during acute fetal hypoxia. Am J Obstet Gynecol 106:39–51

186. Martin CB (1985) Pharmacological aspects of fetal heart rate regulation during hypoxia. In: Künzel W (ed) Fetal heart rate monitoring. Springer, Berlin Heidelberg New York, pp 170–184

187. Martin CB, Murata Y, Petrie RH, Paret JT (1974) Respiratory movements in fetal rhesus monkeys. Am J Obstet Gynecol 119:939–948

188. Martin CB, Voersmans TMG, Jongsma HW (1987) Effect of reducing uteroplacental blood flow on movements and on electrocortical activity of fetal sheep. Gynecol Obstet Invest 23:34–39

189. Martin AA, Kapoor R, Scroop GC (1987b) Hormonal factors in the control of heart rate in normoxaemic and hypoxaemic fetal, neonatal and adult sheep. J Dev Physiol 9:465–480

190. Maylie JG (1982) Excitation-contraction coupling in neonatal and adult myocardium of cat. Am J Physiol 242:H834–843

191. Mellor D (1983) Nutritional and placental determinants of foetal growth rate in sheep and consequences for the newborn lamb. Br Vet J 139:307–324

192. Meschia G, Cotter JR, Breathnack CS, Barron DH (1965) The hemoglobin, oxygen, carbon dioxide concentrations in the umbilical bloods of sheep and goats sampled via indwelling plastic catheters. Q J Exp Physiol 26:185–195

193. Mitchell MD, Bibby JG, Hicks BR (1978) Possible role for prostacyclin in human parturition. Prostaglandins 16:931–937

194. Moore KL (1988) The developing human: clinically oriented embryology, 4th edn. Saunders, Philadelphia

195. Mott JC, Walker DW (1983) Neural and endocrine regulation of circulation in the fetus and newborn. In: Shepherd JP, Abboud FM (eds) Handbook of physiology: the cardiovascular system III: peripheral circulation and organ blood flow, part 2. American Physiological Society, Bethesda, pp 837–883

196. Myers RE (1979) Lactic acid accumulation as cause of brain edema and cerebral necrosis resulting from oxygen deprivation. In: Korobkin R, Guilleminault C (eds) Advances in perinatal neurology. Spectrum, New York, pp 85–114

197. Myers RE (1977) Experimental models of perinatal brain damage: relevance to human pathology. In: Gluck L (ed) Intrauterine asphyxia and the developing fetal brain. Year Book Publishers, New York, pp 37–97

198. Nathanielz PW, Bailey A, Poore ER, Thorburn GD, Harding R (1980) The relationship between myometrial activity and sleep state and breathing in fetal sheep throughout the last third of gestation. Am J Obstet Gynecol 138:653–659

199. Owens JA, Falconer J, Robinson JS (1987) Effect of restriction of placental growth on oxygen delivery to and consumption by the pregnant uterus and fetus. J Dev Physiol 9:137–150

200. Owens JA, Falconer J, Robinson JS (1987) Effect of restriction of placental growth on fetal and utero-placental metabolism. J Dev Physiol 9:225–238

201. Papile LA, Rudolph AM, Heymann MA (1985) Autoregulation of cerebral blood flow in the preterm fetal lamb. Pediatr Res 19(2):159–161

202. Parer JT, De Lannoy CW, Hoversland AS, Metcalfe J (1968) Effect of decreased uterine blood flow on uterine oxygen consumption in pregnant macaques. Am J Obstet Gynecol 100:813–820

203. Parer JT, Krueger TR, Harris JL, Reuss L (1978) Autonomic influences on umbilical circulation during hypoxia in fetal sheep. Abstracts of scientific papers. Quilligan Symposium, San Diego

204. Parer JT (1980) The effect of acute maternal hypoxia on fetal oxygenation and the umbilical circulation in the sheep. Eur J Obstet Gynecol Reprod Biol 10:125–136

205. Parer JT (1983) The influence of beta-adrenergic activity on fetal heart rate and the umbilical circulation during hypoxia in fetal sheep. Am J Obstet Gynecol 147:592–597

206. Parisi VM, Walsh SW (1989) Fetoplacental vascular responses to prostacyclin after thromboxane-induced vasoconstriction. Am J Obstet Gynecol 160:502–507

207. Paulick RP, Meyers RL, Rudolph CD, Rudolph AM (1990) Venous and hepatic vascular responses to indomethacin and prostaglandin E_1 in the fetal lamb. Am J Obstet Gynecol 163:1357–1363

208. Paulick RP, Kastendieck E, Wernze H (1985) Catecholamines in arterial and venous umbilical blood: placental extraction, correlation with fetal hypoxia, and transcutaneous partial pressure. J Perinat Med 13:31–42

209. Peeters LLH, Sheldon RE, Jones MD Jr, Makowski EL, Meschia G (1979) Blood flow to fetal organs as a function of arterial oxygen content. Am J Obstet Gynecol 135:637–646

210. Purves MJ, James IM (1969) Observations on the control of cerebral blood flow in the sheep fetus and newborn lamb. Circ Res 25:651–667

211. Rankin JHG, Meschia G, Makowski EL, Battaglia FC (1971) Relationship between uterine and umbilical venous PO_2 in sheep. Am J Physiol 220:1688–1692

212. Reid LM (1979) The pulmonary circulation: remodelling in growth and disease. Am Rev Respir Dis 119:531–546

213. Reid LM (1982) The development of the pulmonary circulation. In: Peckham GJ, Heymann MA (eds) Cardiovascular sequelae of asphyxia in the newborn. Report of the 83rd Ross conference on pediatric research. Ross Laboratories, Columbus, Ohio, pp 2–10

214. Reid DL, Jensen A, Phernetton TM, Rankin JHG (1990) Relationship between plasma catecholamine levels and electrocortical state in the mature fetal lamb. J Dev Physiol 13:75–79

215. Reller MD, Morton MJ, Thornburg KL (1986) Right ventricular function in the hypoxaemic fetal sheep. J Dev Physiol 8:159–166

216. Reller MD, Morton MJ, Giraud GD, Reid DL, Thornburg KL (1989) The effect of acute hypoxaemia on ventricular function during beta-adrenergic and cholinergic blockade in the fetal sheep. J Dev Physiol 11:263–269

217. Reuss ML, Rudolph AM (1980) Distribution and recirculation of umbilical and systemic venous blood flow in fetal lambs during hypoxia. J Dev Physiol 2:71–84

218. Reuss ML, Parer JT, Harris JL, Krueger TR (1982) Hemodynamic effects of alpha-adrenergic blockade during hypoxia in fetal sheep. Am J Obstet Gynecol 142:410–415

219. Richardson BS, Hohimer AR, Bissonette JM, Machida CM (1983) Cerebral metabolism in hypoglycemic and hyperglycemic fetal lambs. Am J Physiol 245:R730–R736

220. Richardson BS, Rurak D, Patrick JE, Homan J, Carmichael L (1989) Cerebral oxidative metabolism during sustained hypoxemia in fetal sheep. J Dev Physiol 11:37–43

221. Robbilard JE, Weitzman RE, Fisher DA, Smith FG Jr (1979) The dynamics of vasopressin release and blood volume regulation during fetal hemorrhage in the lamb fetus. Pediatr Res 13:606–610

222. Robbilard JE, Weitzman RE, Burmeister L, Smith FG Jr (1981) Developmental aspects of the renal response to hypoxaemia in the lamb fetus. Circ Res 48:128–138

223. Robbilard JE, Gomez RA, Meernik JG, Kuehl WD, Van Orden D (1982) Role of angiotensin II on the adrenal and vascular responses to hemorrhage during development in fetal lambs. Circ Res 50:645–650

224. Robillard JE, Ayres NA, Gomez RA, Nakamura KT, Smith JR, FG (1984) Factors controlling aldosterone secretion during hypoxemia in fetal lambs. Pediatr Res 18:607–611

225. Robinson JS, Kingston EJ, Jones CT, Thornburn GD (1979) Studies on experimental growth retardation in sheep. The effect of removal of endometrial caruncles on fetal size and metabolism. J Dev Physiol 1:379–398

226. Robinson JS, Jones CT, Kingston EJ (1983) Studies on experimental growth retardation in sheep. The effects of maternal hypoxaemia. J Dev Physiol 5:89–100

227. Robinson JS, Falconer J, Owens JA (1985) Intrauterine growth retardation: clinical and experimental. Acta Paediatr Scand [Suppl] 319:135–142

228. Rose JC, Morris M, Meis PJ (1981) Hemorrhage in newborn lambs: effects on arterial blood pressure, ACTH, cortisol, and vasopressin. Am J Physiol 240:E585–E590

229. Rosén KG, Isaksson O (1976) Alteration in fetal heart rate and ECG correlated to glycogen, creatinephosphate and ATP levels during graded hypoxia. Biol Neonate 30:17–24

230. Rosenthal M, Lamanna JC, Jöbsis FF, Levasseur JE, Kontos HA, Patterson JL (1976) Effects of respiratory gases on cytochrome a in intact cerebral cortex: is there a critical PO_2? Brain Res 108:143–153

231. Rudolph AM, Heymann MA (1967) Circulation of the fetus in utero: methods for studying distribution of blood flow, cardiac output and organ blood flow. Circ Res 21:163–184

232. Rudolph AM, Heymann MA (1970) Circulatory changes during growth in the fetal lamb. Circ Res 26:289–299

233. Rudolph AM, Heymann MA (1973) Control of the fetal circulation. In: Proceedings of the Sir Joseph Barcroft Centenary Symposium: foetal and neonatal physiology. Cambridge University Press, Cambridge, pp 89–111

234. Rudolph AM, Heymann MA (1976) Cardiac output in the fetal lamb: the effects of spontaneous und induced changes of heart rate on right and left ventricular output. Am J Obstet Gynecol 124:183–192

235. Rudolph AM (1979) Fetal and neonatal pulmonary circulation. Annu Rev Physiol 41:383–395

236. Rudolph AM, Heymann MA, Lewis AB (1977) Physiology and pharmacology of the pulmonary circulation in the fetus and newborn. In: Hodson WA (ed) Lung biology in health and disease. Development of the lung. Dekker, New York, pp 497–523

237. Rudolph AM, Itskovitz J, Iwamoto H, Reuss L, Heymann MA (1981) Fetal cardiovascular responses to stress. Semin Perinatol 5:109–121

238. Rudolph AM (1983) Hepatic and ductus venosus blood flows during fetal life. Hepatology 3:254

239. Rudolph AM (1984) The fetal circulation and its response to stress. J Dev Physiol 6:11–19

240. Rudolph AM (1985) Distribution and regulation of blood flow in the fetal and neonatal lamb. Circ Res 57:811–821

241. Rudolph CD, Roman C, Rudolph AM (1989) Effect of acute umbilical cord compression on hepatic carbohydreate metabolism in the fetal lamb. Pediatr Res 25(3):228–233

242. Rurak DW (1978) Plasma vasopressin levels during hypoxaemia and the cardiovascular effects of exogenous vasopressin in foetal and adult sheep. J Physiol (Lond) 277:341–357

243. Schellenberg J-C (1986) The development of connective tissue and its role in pulmonary mechanics. In: Johnston BM, Gluckman PD (eds) Respiratory control and lung development in the fetus and newborn. Perinatology Press, Ithaca, pp 3–62

244. Schulze BS (1871) Der Scheintod Neugeborener. Letter to Dr C Ludwig, Mauke's Verlag (Hermann Dufft), Jena

245. Scroop GC, Marker JD, Stankewytsch-Janusch B, Seamark RF (1986) Angiotensin I and II and the assessment of baroreceptor function in fetal and neonatal sheep. J Dev Physiol 8:123–137

246. Shelley H (1960) Blood sugars and tissue carbohydrate in foetal and infant lambs and rhesus monkeys. J Physiol (Lond) 153:527–552

247. Siesjö BK (1978) Brain energy metabolism. Wiley, Chichester

248. Siesjö BK, Bengtsson F (1988) Calcium, calcium antagonists, and ischemic cell death in the brain. In: Krieglstein J (ed) Pharmacology of cerebral ischemia. Wissenschaftliche Verlagsgesellschaft, Stuttgart, pp 23–29

249. Skillman CA, Plessinger MA, Woods JR, Clark KE (1985) Effect of graded reductions in uteroplacental blood flow on the fetal lamb. Am J Physiol 249:H1098–H1105

250. Skillman CA, Clark KE (1987) Fetal beta-endorphin levels in response to reductions in uterine blood flow. Biol Neonate 51:217–223

251. Smith RW, Morris JA, Assali NS (1964) Effects of chemical mediators on the pulmonary and ductus arteriosus circulation in the fetal lamb. Am J Obstet Gynecol 89:252–260

252. Stark RI, Wardlaw SL, Daniel SS, Husain MK, Sanocka UM, James LS, Vande Wiele RL (1982) Vasopressin secretion induced by hypoxia in sheep: developmental changes and relationship to beta-endorphin release. Am J Obstet Gynecol 143:204–215

253. Stark RI, Daniel SS, Husain MK, Sanocka UM, Zubrow AB, James LS (1984) Vasopressin concentration in the amniotic fluid as an index of fetal hypoxia: mechanism of release in sheep. Pediatr Res 18:552–558

254. Stein GW (1783) Theoretische Anleitung zur Geburtshülfe. Johann Jacob Cramer, Cassel

255. Taeusch HW Jr, Wyzogrodski I, Wang NS, Avery ME (1974) Pulmonary pressure-volume relationships in premature fetal and newborn rabbits. J Appl Physiol 37:809–813

256. Thornburg KL, Morton MJ (1983) Filling and arterial pressures as determinants of RV stroke volume in the sheep fetus. Am J Physiol 244:H656–H663

257. Toubas PL, Silverman NH, Heymann MA, Rudolph AM (1981) Cardiovascular effects of acute hemorrhage in fetal lambs. Am J Physiol 240:H45–H48

258. Towell ME, Figueroa J, Markovitz S, Elias B, Nathanielsz P (1987) The effect of mild hypoxemia maintained for twenty-four hours on maternal and fetal glucose, lactate, cortisol, and arginine vasopressin in pregnant sheep at 122 to 139 days' gestation. Am J Obstet Gynecol 157:1550–1557

259. Trudinger BJ, Stevens D, Connelly A, Hales JRS, Alexander G, Bradley L, Fawcett A, Thompson RS (1987) Umbilical artery flow velocity waveforms and placental resistance: the effects of embolization of the umbilical circulation. Am J Obstet Gynecol 157:1443–1448

260. Tweed WA, Cote J, Pash M, Lou H (1983) Arterial oxygenation determines autoregulation of cerebral blood flow in the fetal lamb. Pediatr Res 17:246–249

261. Tyler TL, Leffler CW, Cassin S (1977) Effects of prostaglandin precursors, prostaglandins, and prostaglandin metabolites on pulmonary circulation of perinatal goats. Chest 71S:271S–273S

262. Von Ahlfeld F (1888) Über bisher noch nicht beschriebene intrauterine Bewegungen des Kindes. Breitkopf und Hartel, Leipzig, pp 203–210 (Verhandlungen der Deutschen Gesellschaft für Gynäkologie)

263. Von Ahlfeld F (1905) Die intrauterine Fähigkeit der Thorax- und Zwerchfellmuskulatur. Intrauterine Atmung. Monatsschr Geburtsh Gynakol 21:143–169

264. Wagner KR, Ting P, Westfall MV, Yamaguchi S, Bacher JD, Myers RE (1986) Brain metabolic correlates of hypoxia-ischemic cerebral necrosis in mid-gestational sheep fetuses: significance of hypotension. J Cereb Blood Flow Metab 6:425–434

265. Walker AM, Cannata JP, Dowling MH, Ritchie BC, Maloney JE (1978) Sympathetic and parasympathetic control of heart rate in unanesthetized fetal and newborn lambs. Biol Neonate 33:135–143

266. Walker AM, Cannata JP, Dowling MH, Ritchie BC, Maloney JE (1979) Age-dependent pattern of autonomic heart rate control during hypoxia in fetal and newborn lambs. Biol Neonate 35:198–208

267. Walker AM, de Preu ND, Horne RSC, Berger PJ (1990) Anatomic control of heart rate differs with electrocortical activity and chronic hypoxaemia in fetal lambs. Biol Neonate 14:43–48

268. Wardlaw SL, Stark RI, Daniels S, Frantz AG (1981) Effects of hypoxia on beta-endorphin and beta-lipotropin release in fetal, newborn and maternal sheep. Endocrinology 108:1710–1715

269. Weismann DN, Robillard JE (1988) Renal hemodynamic responses to hypoxemia during development: relationships to circulating vasoactive substances. Pediatr Res 23:155–162

270. Wigglesworth JS (1964) Experimental growth retardation in the foetal rat. J Pathol Bacteriol 88:1–13

271. Wilkening RB, Anderson S, Martensson L, Meschia G (1982) Placental transfer as a function of uterine blood flow. Am J Physiol 242:H429–H436

272. Wilkening RB, Meschia G (1983) Fetal oxygen uptake, oxygenation, and acid-base balance as a function of uterine blood flow. Am J Physiol 244:H749–H755

273. Wilson DF, Owen ChS, Erecinska M (1979) Quantitative dependence of mitochondrial oxidative phosphorylation on oxygen concentration: a mathematical model. Arch Biochem Biophys 195:494–504

274. Wlodck ME, Challis JRG, Richardson B, Patrick J (1989) The effects of hypoxaemia with progressive acidaemia on fetal renal function in sheep. J Dev Physiol 12:323–328

275. Yaffe H, Parer JT, Block BS, Llanos AJ (1987) Cardio-respiratory responses to graded reductions of uterine blood flow in the sheep fetus. J Dev Physiol 9:325–336

276. Zink J, Van Petten GR (1980) The effects of norepinephrine on blood flow through the fetal liver and ductus venosus. Am J Obstet Gynecol 137:71–77

277. Zweifel P (1876) Die Respiration des Fötus. Arch Gynakol 9:293–305

119 Fetal Endocrinology

D.A. FISHER

Contents

119.1 Introduction

Fetal endocrine physiology differs in many important ways from the endocrinology of postnatal life [15,16,23]. The mammalian fetus develops in an nevironment where respiration, alimentation, and excretory functions are provided by the placenta. Fetal body temperature is modulated by maternal metabolism, and fetal tissue thermogenesis is maintained at a basal level. Growth of the fetus appears to be regulated by growth factors which function via autocrine or paracrine mechanisms through much of gestation. Modulation of metabolic functions relative to feeding or to environmental changes is of limited importance. In this milieu, endocrine control needs to be tailored to the particular conditions obtaining, and so unique fetal endocrine systems appropriate to intrauterine life have evolved.

119.2 Placental Transfer of Hormones

The placenta is impermeable to most maternal hormones [16,18,33,34]. These include the hypothalamic releasing factors (except for the thyrotropin-releasing hormone, TRH) (see Chap. 21), and somatostatin (growth hormone release-inhibiting hormone). There is limited permeability to thyroid hormones. Other hormones that do not cross the placenta are anterior and posterior pituitary hormones, insulin, glucagon, parathyroid hormone (PTH), calcitonin and renin.

The general pattern of placental hormone permeability is shown in Table 119.1 and related to the molecular weight, lipid solubility, and mode of receptor binding of several hormone types [15,16,33]. Studies in a variety of species have indicated that the placenta is more permeable to lipid-soluble than to lipid-insoluble molecules, and that placental permeability of lipid-insoluble moleucles decreases with increasing molecular weight. The hormone data in Table 119.1 are in general agreement with this concept.

119.3 Placental Hormone Production

One unique feature of the fetal endocrine environment is that the placenta functions among other things as an endocrine gland. Table 119.2 summarizes the hormones known to be produced by the placenta [10,11,15,16,18,33].

119.3.1 Placental Estrogens

The human placenta produces enormous amounts of estrogens, including estradiol, estrone, and estriol [4,10,15,35]. Production at term amounts to 50–100 mg daily, while estrogen production in nonpregnant women is less than 1 mg daily. Most of the estrogen is secreted into the maternal circulation, but fetal concentrations are quite high. Placental estrogen production is accomplished in collaboration with the fetal adrenal gland. The fetal adrenal is composed of two distinct anatomic zones, an inner "fetal" and an outer "definitive" zone. The fetal zone is deficient in the enzyme steroid 3β-ol dehydrogenase/isomerase and has a high steroid sulfokinase activity. Thus, the conversion of pregnenolone to progesterone is limited, and the major secretory products of the fetal adrenal are dehydroepiandrosterone (DHA) and DHA sulfate (DHAS), essentially inactive hormone precursors.

R. Greger/U. Windhorst (Eds.)
Comprehensive Human Physiology, Vol. 2
© Springer-Verlag Berlin Heidelberg 1996

Table 119.1. Patterns of placental hormone transfer as related to molecular weight, lipid solubility, and mode of action

Hormone species	Approximate molecular weight (Da)	Lipid solubility	Mode of action		Placental transfer
			Cell membrane receptor binding	Intracellular protein receptor binding	
Polypeptide hormones	1050–30000	–	Yes	No	No
Thyroid hormones	800	±	No	Yes	No
Thyrotropin-releasing hormone	360	–	Yes	No	Yes
Steroid hormones	350	±	No	Yes	Yes
Catecholamines	180	–	Yes	No	Yes

Table 119.2. Hormones Known to be produced by the placenta

Steroid hormones	Peptide hormones	Neuropeptides
Estradiol	Gonadotropin	TRH
Estrone	Somatomammotropin	Somatostatin
Estriol	Thyrotropin	CRF
Progesterone	Corticotropin	GnRH
	β-Endorphin	GHRH
	α-MSH	
	β-Lipotropin	
	β-Lipotropin	

TRH, Thyrotropin-releasing hormone; CRF, corticotropin-releasing factor; GnRH, gonadotropin-releasing hormone; GHRH, growth hormone-releasing hormone; α-MSH, melanocyte-stimulating hormone

DHAS is transported to the liver for 16-hydroxylation and/or to the placenta. DHAS is hydrolyzed in the placenta by a steroid sulfatase and utilized as substrate for placental estrone and estradiol production; 16-OH DHAS is the major substrate for placental estriol synthesis [4,10,35]. Estriol is a hormone unique to pregnancy; it is not secreted by the ovaries of nonpregnant women. Estradiol is a potent estrogen, whereas estriol has limited activity in most bioassay systems. It has been suggested that estriol has a special role in the maintenance of uteroplacental blood flow. However, the role or roles of placental estrogens in pregnancy remain largely obscure.

119.3.2 Placental Progesterone

After 12 weeks of gestation the placenta is the major site of production of progesterone [4,10,35]. The principal substrate for placental progesterone synthesis is circulating maternal LDL cholesterol (cf. Chap. 113). The production of progesterone approximates 250 mg daily during the midluteal phase of the normal menstrual cycle. Most is secreted into the maternal circulation, but fetal blood concentrations are sevenfold higher than maternal values. Placental progesterone serves a role in maintaining the uterine musculature and plays a role in inhibiting maternal cell-mediated immune responses to foreign (fetal) antigens.

119.3.3 Placental Production of Pituitary-Like Hormones

The placenta produces several polypeptide hormones unique to pregnancy [11,15,18,34]. The most important are human chorionic gonadotropin (hCG) and human chorionic somatomammotropin (hCS), also referred to as placental lactogen. hCG is a glycoprotein of 36–40 kDa MW with structural, biological, and immunological similarities with the pituitary gonadotropins and thyrotropin; hCG also has weak thyroid-stimulating hormone (TSH)-like activity. hCS is a 191-amino acid homologue of human pituitary growth hormone (GH). It has about 37% of the growth-promoting bioactivity of GH and equivalent prolactin bioactivity. hCG is secreted predominantly during the first half of gestation and hCS during the latter half. hCG is important for maintenance of the corpus luteum and stimulation of placental progesterone production. hCS has been postulated to exert an anti-insulin effect on maternal carbohydrate and lipid metabolism, effects which would tend to increase maternal glucose and amino acid levels and augment maternal-to-fetal substrate flow.

The glycoprotein hormone family includes a single gene for the α-subunit expressed in the placenta for hCG and in the pituitary for production of LH, follicle-stimulating hormone (FSH), and TSH [12,15,16] (see also Chap. 113). In addition, there is a single gene for the β-LH subunit, whereas there are seven hCG β-subunit genes or pseudogenes. These and the β-LH gene have similar structures, and it seems that β-hCG arose from β-LH and that the β-hCG gene family occurred early in the process of evolution of pseudogenes [12]. The prolactin, GH, and hCS genes are also closely related. Prolactin is presumed to be the ancestral gene: GH probably evolved nearly 400 million years ago and hCS within the last 10 million years. The GH gene cluster includes five similar gene loci, two for GH and three for hCS; these have strong homology and are presumed to have arisen by repeated duplications over time [5]. Only two of the hCS sequences are expressed in the placenta, producing identical hCS molecules.

β-Endorphin, melanocyte-stimulating hormone (α-MSH), and α- and β-lipotropin have been isolated from placental tissue and are presumed to be products of placental pro-opiomelanocortin (POMC) cleavage [15,16,33]. Thus, the

2340

placenta, like the pituitary gland, appears to synthesize a POMC. A placental ACTH is also cleaved from POMC and is referred to as human chorionic corticotropin (hCT). Placental ACTH may contribute to the increased maternal levels of ACTH in late gestation [22].

There also is evidence that the placenta produces a chorionic thyrotropin (hCT). Maternal plasma concentrations of hCT are very low in normal pregnancy and appear to be of little physiologic significance [11,24]. Most of the thyrotropin activity in placental tissue and in maternal plasma during pregnancy seems attributable to the inherent TSH bioactivity of hCG.

119.3.4 Placental Hypothalamic-Like Neuropeptides

Hypothalamic-like neuropeptides are also produced by placental tissue. The human placenta produces a TRH, a corticotropin-releasing factor (CRF), a growth hormone-releasing hormone or factor (GHRH, GRF) and somatostatin [2,15,16,18,33,37]. Chorionic GnRH seems similar or identical to hypothalamic GnRH. It is produced in the cytotrophoblast. The placenta has receptors for GnRH, and synthetic GnRH increases in vitro production of hCG, progesterone, estrone, estradiol, and estriol from placental explants. Thus, chorionic GnRH may have a paracrine role in the regulation of placental hCG and steroid hormone production.

Immunoreactive chorionic somatostatin, like gonadotropin-releasing hormone (GnRH), has been localized to the cytotrophoblast. Somatostatin-containing cells in the placenta disappear as pregnancy progresses and hCS production increases progressively during the latter half of gestation, and it has been suggested that chorionic somatostatin may exert a paracrine action on production of hCS by the syncytiotrophoblast.

Immunoreactive CRF has been identified in placental extracts and in plasma in the third trimester of pregnancy. A possible role for placental CRF in the regulation of maternal pituitary ACTH secretion late in gestation has been proposed. Immunoreactive and biologically active GHRH was first identified in rat placenta [2].

119.4 Fetal Endocrine Ontogenesis

Hormone systems can be classified in two major groups: neuroendocrine systems, which transduce or convert neural into endocrine information; and autonomous endocrine systems, which function as autonomous metabolic regulators [18]. The neuroendocrine systems include:

- The hypothalamic anterior pituitary system
- The hypothalamic posterior pituitary complex
- The autonomic nervous system.

Autonomous endocrine systems include:

- The pancreatic insulin-glucagon secretory complex
- The parathyroid hormone-calcitonin-vitamin D system.

All are well developed in the term fetus, but as a general rule these fetal endocrine systems are redundant, some not being required to function until after birth, while the functions of the rest are fulfilled by the maternal organism. There are some unique endocrine system adaptations to fetal functions and these are described in Sect. 119.5. Other fetal endocrine systems tend to be suppressed or their hormone actions neutralized. All of these systems become important at birth.

119.4.1 Neuroendocrine Systems

Hypothalamic-Pituitary Complex. The human fetal pituitary gland is embryologically intact by 11–12 weeks. The primary plexus of the pituitary portal blood vascular system develops at 14–15 weeks and continuity of the primary and secondary plexuses of the system probably is developed by 19–21 weeks [16,18,20,23,33]. By 8–10 weeks of gestation the fetal hypothalamus contains significant concentrations of releasing hormones. In addition, GH, FSH, LH, TSH, ACTH, prolactin, oxytocin, vasotocin, and vasopressin can be identified in significant concentrations in pituitary tissue at this time. Hormone concentrations within the pituitary increase progressively, whereas concentrations in fetal serum tend to peak near midgestation and decrease progressively toward term [18].

Because anterior pituitary hormone concentrations in fetal blood are low in the first-trimester fetus, hormone synthesis and secretion are thought to be low at this time. During the second trimester, maturation of the hypothalamus and the pituitary-portal vascular system occur and an unrestrained release of hypothalamic hormones may account for the high circulating concentrations of anterior pituitary hormones characteristic of the midgestation fetus [20,23]. During the third trimester, anterior pituitary hormone secretion is progressively modulated, partly owing to maturation of pituitary negative feedback control systems. In addition, there may be progressive maturation of inhibitory electrical activity in the neocortex during the latter half of gestation. The low concentrations of serum GH and TSH in the anencephalic fetus support the view that fetal anterior pituitary hormone secretion is dependent on an active hypothalamus [23,37]. Whether placental and fetal tissue neuropeptide production influence pituitary hormone secretion in the fetus is not clear. The late increase in fetal serum prolactin concentrations does not seem to be dependent on hypothalamic function, since serum prolactin concentrations in cord blood of normal and anencephalic infants are similar [20,23]. Rather, the increase in fetal serum prolactin has been postulated to be due to the progressive increase in estrogen concentrations.

Arginine vasopressin is detectable after 10–12 weeks' gestation in the human fetal neurohypophysis [13,16,18]. By 40 weeks the concentration is about 20% of that in the adult gland. Fetal pituitary oxytocin exceeds vasopressin concentrations by 15–19 weeks and the vasopressin:oxytocin ratio falls progressively thereafter. Vasopressin secretion occurs by midgestation and secretion control matures during the last trimester. The vasopressin secretion responses to hypertonicity and dehydration in the newborn lamb are quantitatively comparable to the adult responses. There also is a marked fetal vasopressin response to hypoxia, and this response greatly exceeds that in the adult; it has been suggested that vasopressin serves as a fetal stress hormone to support fetal cardiovascular homeostasis [7,41].

Anterior Pituitary Target Organs. Embryogenesis of the target organs of the anterior pituitary, including the thyroid gland, the gonads, and the adrenal glands, is complete by 10–12 weeks' gestation, and these glands are then functional [14,16,18,20,23]. The thyroid gland can synthesize thyroxine by 9–10 weeks and at this time contains visible colloid comprised of thyroglobulin, the storage form of thyroid hormones. The fetal testes can synthesize testosterone by 6–8 weeks' gestation, and there is electron microscopic evidence of fetal adrenal steroidogenic activity at this time [18,42].

These fetal anterior pituitary target organs are relatively quiescent, however, prior to maturation of the pituitary-portal vascular system and activation of the fetal hypothalamic-pituitary system. This seems to occur during the second trimester, and by midgestation fetal plasma levels of FSH, LH, TSH, and ACTH peak at relatively high levels. Maturation of suprahypothalamic neural control systems and the pituitary negative feedback hormonal control systems for modulation of gonadal, thyroid, and adrenal functions (see Chaps. 20–22) mature during the last trimester of gestation and during the early weeks of postnatal life [18,20,23].

Autonomic Nervous System. By 6–7 weeks' gestation, the primordia of the sympathetic trunk ganglia and the sympathetic trunks are visible. The primordia of the prevertebral plexuses and the terminal sympathetic ganglia also have begun to organize [16,18]. By 10–12 weeks the paired adrenal medullary masses are well developed. The adrenal medullae are functional during the last trimester but remain histologically immature at birth. They resemble the adult gland by about 1 year of age. In addition, there are para-aortic chromaffin cell masses scattered throughout the abdominal and pelvic sympathetic plexuses. Much of the chromaffin tissue in the fetus is represented in these extramedullary, para-aortic paraganglia, which are approximately 2–3 mm in diameter by 28 weeks' gestation, after which time they slowly regress. The largest of the paraganglia, the organs of Zuckerkandl, are located near the origin of the inferior mesenteric arteries. These measure 10–15 mm in length at term and after birth gradually atrophy, disappearing completely by 2–3 years of age [16,33].

Both chromaffin cells and sympathetic nerve cells are derived from common neuroectodermal stem cells and both are responsive to nerve growth factor (NGF). Sympathetic

nervous system ontogenesis is known to be NGF-dependent, and recent evidence indicates that injection of NGF antiserum into neonatal rats leads to degeneration of immature chromaffin cells as well as primitive sympathetic cells and pheochromoblasts [1].

Catecholamines can be identified in fetal chromaffin tissue by 10–15 weeks' gestation, and concentrations increase progressively to term. Both epinephrine and norepinephrine are present in adrenal medullary tissue, but the latter predominates in the paraganglial tissues [15,16,18]. This difference is due to the effect of adrenal cortical hormones on the final step in epinephrine synthesis, the methylation of norepinephrine to epinephrine, which step is stimulated by adrenal glucocorticoids [43]. The unique location of the adrenal medulla provides high local concentrations of glucocorticoids, favoring epinephrine synthesis.

The capacity to secrete epinephrine relates to the maturation of the adrenal gland and the development of adrenal splanchnic innervation. The noradrenergic cells of the fetal adrenal medulla can respond directly to asphyxia with norepinephrine secretion long before the splanchnic innervation develops, and the extramedullary noradrenergic cells presumably respond similarly. Fetal catecholamines also serve as fetal stress hormones, and in the sheep fetus the noradrenergic response to asphyxia appears early in gestation and falls off near term. The central nervous system-stimulated adrenal medullary response to asphyxia is mediated by splanchnic nerves and matures near term [6]. Basal plasma epinephrine, norepinephrine, and dopamine concentrations are relatively elevated in the third trimester fetus and decrease progressively with gestational age. The fetus responds to maternal exercise or hypoxia with increased catecholamine secretion, and plasma levels increase in response to parturition.

119.4.2 Autonomous Endocrine Systems

Endocrine Pancreas. The pancreas is identifiable by 4 weeks of gestation in the human fetus, and synthesizes proinsulin, insulin, and glucagon by 7–10 weeks. By midgestation insulin and glucagon concentrations within the pancreas of the fetus exceed adult pancreatic concentrations [18,38,39]. Fetal pancreatic tissue also is a rich source of somatostatin and TRH [15,16]. Insulin release in fetal mammals is quite variable, and there are species differences. Near term, prior to the onset of labor, serum insulin concentrations are relatively unresponsive to high glucose concentrations, and glucagon or theophylline will augment the relatively obtunded insulin response to glucose in vivo and in vitro. The glucagon secretion response also is blunted in the fetus; hyperglycemia does not suppress fetal plasma glucagon concentrations, and acute hypoglycemia does not evoke glucagon secretion.

The obtunded insulin and glucagon secretion responses in the fetus seem to be related to a deficient capacity of the fetal pancreatic islet cells to generate cAMP and/or to rapid cAMP destruction by phosphodiesterase [16,18,39]. The mechanisms for fetal adrenergic receptor-mediated inhibition of pancreatic insulin secretion and stimulation of glucagon release are intact in the near-term fetal sheep, as is the inhibitory effect of prostaglandin(s) on glucose-stimulated insulin secretion. The relatively rapid maturation of insulin and glucagon secretion control by glucose in the neonatal period in both premature and mature human infants suggests that the obtunded state of the fetal pancreas is secondary to the relatively stable fetal serum glucose concentrations maintained by placental transfer of maternal glucose, rather than a temporally fixed maturation process.

Parathyroid Hormone–Calcitonin–Vitamin D System. The parathyroid glands are derived from the paired third and fourth pharyngeal gut pouches by 4–5 weeks' gestation [16,18,27,32,36]. The third pouches are destined to form the thymus and inferior parathyroid glands and the fourth pouches the superior parathyroid glands. The fifth pharyngeal pouches form the paired ultimobranchial bodies, which are incorporated into the thyroid gland during embryogenesis as neural crest-derived, calcitonin-secreting "C" cells. The parathyroid glands develop embryologically between 5 and 12–14 weeks; they measure 1–2 mm in diameter at birth, while adult glands measure 2–5 mm × 3–8 mm. Near term, fetal parathyroid cells are composed largely of inactive chief cells and a few intermediate chief cells containing occasional secretion granules. The fetal parathyroid glands also synthesize PTH-related protein (PTHRP), which exerts a PTH-like action in the fetus [27]. The fetal-neonatal rat appears capable of modulating PTH levels in response to changing serum Ca^{2+} concentrations and the fetal sheep can increase serum PTH concentrations in response to a fall in serum Ca^{2+} induced by EDTA. C cells are prominent in the neonatal thyroid gland and the calcitonin content is high [18]. The third-trimester sheep fetus also can respond promptly to infused Ca^{2+} with increased serum calcitonin levels.

The human fetus at term has low circulating concentrations of PTH and relatively high levels of calcitonin [18,27,36]. and the PTH response to hypocalcemia is reduced. Presumably this is due to the prevailing high concentrations of total and ionized calcium maintained in fetal blood by active placental Ca^{2+} transport from maternal blood. The high levels of fetal serum Ca^{2+} suppress parathyroid function and stimulate calcitonin secretion; this would tend to promote calcium-phosphorus deposition in fetal bone. Vitamin D levels in the fetus are largely derived via placental transfer from the maternal circulation, and free concentrations in the fetal circulation are similar to maternal values. However, the fetal kidney is capable of 1-hydroxylating 25-hydroxy-vitamin D (25(OH)D) to 1,25(OH)$_2$D, the active hormone. Moreover, the placenta contains specific 1,25(OH)$_2$D receptors as well as a vitamin-D-dependent calcium-binding protein. PTHRP stimulates 1-hydroxylation of 25(OH)D in fetal kidney, and both 1,25(OH)$_2$D and PTHRP appear to be involved in the regulation of placental Ca^{2+} transport to the fetus. Thus, fetal PTH and/or active vitamin D appear to maintain high fetal Ca^{2+} concentrations via placental transport, and the high fetal calcitonin concentrations have been postulated to promote fetal bone calcification.

119.5 Unique Fetal Endocrine Adaptations

In several instances, fetal endocrine systems or hormones have been transiently diverted to subserve gestational or developmental roles [15,16,33].

119.5.1 Fetal Adrenal and Placental Estrogen Production

The best-known and characterized fetal endocrine gland is the fetal adrenal, which is adapted in utero to function in collaboration with the placenta to produce estrogens [4,10,15,16,35]. This system, the so-called fetoplacental unit, has been referred to earlier (see Sect. 119.3.1).

119.5.2 Fetal Testes and Sexual Differentiation

The fetal testes, under the influence of the placenta, serve at a critical period of early embryogenesis to induce male sexual differentiation [15,16,30,42]. Fetal testicular androgen, produced either autonomously or under stimulation by placental and/or fetal pituitary gonadotropins, stimulates development of the embryonic wolffian duct system and the external genitalia to effect male differentiation. In addition, the fetal testes produce an inhibitor of müllerian duct differentiation, müllerian-inhibiting factor (MIF) [25,30]. This glycoprotein hormone, with a molecular weight approximating 72 kDa causes regression of the embryonic müllerian ducts in the male fetus.

Fetal ovarian hormone secretion plays little or no role in sexual differentiation; in a female or an agonadal fetus, the wolffian ducts regress in the absence of testicular androgen, and normal female development of the müllerian ducts and external genitalia occur in the absence of testicular MIF.

119.5.3 Fetal Neurohypophysis and Water Metabolism

As indicated earlier (Sect. 119.4.1), the fetal posterior pituitary gland is well developed by 10–12 weeks' gestation and contains measurable levels of both vasopressin and oxytocin. In addition, arginine vasotocin, the parent neurohypophyseal hormone in submammalian vertebrates, has been identified in the fetal pituitary and pineal glands as well as in the adult pineal gland of several mammalian species, including humans [9,16]. Arginine vasotocin is present in the pituitary gland only during fetal life, disappearing in the neonatal period; this has been cited as an example of ontogeny recapitulating physlogeny. The role of arginine vasotocin in the fetus is unclear, but baseline plasma concentrations in fetal sheep during the last trimester approximate values for vasopressin and oxytocin [13,14]. Both hormones inhibit fetal lung fluid production as well as fetal renal free water clearance. Vasopressin also may have an effect on water transport via the amniotic membranes. Thus, available evidence suggests that both vasopressin and arginine vasotocin act in the fetal environment to conserve water for the fetus, inhibiting flow across the placenta and loss into amniotic fluid via the lungs as well as the kidneys.

119.6 Unique Fetal Endocrine Systems

There are several endocrine systems unique to fetal life. The para-aortic chromaffin system has been dealt with above.

119.6.1 Fetal Intermediate Pituitary

In the fetus the intermediate lobe of the pituitary is prominent [16,33]. Intermediate lobe cells begin to disappear near term and are virtually absent in the adult human pituitary, although the intermediate lobe in the adult of lower species is anatomically and functionally distinctive. This may be another example of ontogeny recapitulating phylogeny. The major secretory products of the intermediate lobe are α-MSH and β-endorphin derived from cleavage of the POMC molecule. POMC in the anterior lobe is cleaved predominantly to ACTH and β-lipotropin [40].

In primates, the fetal pituitary, in contrast to the adult gland, has been shown to contain high concentrations of α-MSH, a corticotropin-like intermediate lobe peptide (CLIP); and α-MSH concentrations in the human fetus decrease progressively with increasing fetal age [40]. The circulating concentrations of both β-endorphin and β-lipotropin are high in the fetus, and in the fetal lamb the basal ratio of β-endorphin to β-lipotropin concentrations is greater than that during hypoxic stimulation of the anterior pituitary, suggesting an intermediate lobe as the origin of secretion of basal β-endorphin concentrations in the fetus. Roles for these peptides in the fetus are not known: α-MSH and CLIP have been suggested to be involved in fetal adrenal activation, and α-MSH in fetal growth.

119.6.2 Extrahypothalamic Neuropeptide Production

Hypothalamic neuropeptides have been demonstrated in a wide variety of adult tissues. In the fetal environment, such neuropeptides are predominantly localized to placental and gut tissues [15,16,26]. As already described above (Sect. 119.3.4), the human placenta produces GnRH, TRH, somatostatin, CRF, and GRF-like peptides. Particularly high concentrations of TRH, Somatostatin, and CRF-like immunoreactivity have also been observed in neonatal rat pancreas and gastrointestinal tract tissues at a time when hypothalamic immunoreactivity concentrations in the same animals are low [8,26]. In the case of TRH, encephalectomy did not alter the relatively high circulating TRH levels in the neonatal rat, whereas significant reductions were produced by pancreatectomy, Similar observations have been reported for CRF in the rat. TRH, CRF, and Somatostatin concentrations are elevated in blood of the human neonate and significant concentrations of a TRH-like neuropeptide have been demonstrated in human neonatal pancreas [15,26]. Thus, an endocrine role for these neuropeptides has been postulated in the fetal and neonatal periods.

There is a general tendency to hypersecretion of fetal pituitary hormones in the fetus during the last half of gestation, as discussed earlier. These hormones include GH, TSH, ACTH, LH, and FSH. Maturation of hypothalamic-pituitary control is complex, and the mechanism of fetal pituitary hyperfunction is not yet clear. Immaturity of higher nervous system inhibitory input has been postulated for GH, and immature pituitary negative feedback control clearly plays a role for TSH and the gonadotropins, and perhaps for ACTH. However, extrahypothalamic neuropeptides may play a role in modulation of fetal anterior pituitary hormone secretion.

119.7 Neutralization of Selected Hormone Actions in the Fetus

119.7.1 Production of Inactive Hormone Metabolites

The placenta is permeable to steroid hormones but contains an isomerase that converts most of the cortisol transferred from the maternal circulation to inactive cortisone [4,15,16]. Additionally, although many adult tissues are capable of converting cortisone to cortisol, fetal tissues seem devoid of this capacity during most of fetal life [4,15]. Thus, most of the cortisol that crosses the placenta is inactivated in the placenta or in fetal tissues. Teleologically, this would help preserve the anabolic and growth-promoting milieu of the fetus and minimize premature maturational and parturitional effects of cortisol.

Thyroid hormones also appear to have limited actions during fetal life [15,17,19]. The athyroid human fetus is born normally grown and usually without detectable stigmata of thyroid hormone deprivation. Fetal thyroid hormone metabolism is characterized by production of inactive sulfated thyroid hormones and by conversion of T4 to inactive reverse triiodothyronine (rT3) by fetal tissue [3,14,19]. The placenta also catalyzes this conversion limiting transfer of active T4 to the fetus [34]. In addition, fetal liver and kidney, in contrast to adult liver and kidney, manifest little or no iodothyronine outer ring monodeiodinase activity, so that there is little or no conversion of

T4 to active T3. As a consequence, serum T3 concentrations in the fetus remain very low until the last few weeks of gestation [15,17–19]. Selected fetal tissues (brain, brown adipose tissue) have an active type 2 iodothyronine outer ring monodeiodinase, which converts T4 to active T3 for local action in these tissues. This locally produced T3 is important to protect the thyroid hormone-dependent fetal brain in the event of fetal hypothyroxinemia. The activity of this type 2 enzyme is increased five- to tenfold in response to hypothyroxinemia [15,18].

119.7.2 Receptor or Postreceptor Immaturity

Recent data suggest that fetal tissues other than brain are largely unresponsive to thyroid hormones [19]. Thyroid nuclear T3 receptors appear in the fetus relatively early [15,16,31]. They are detectable during the last few days of fetal life in the rat and during the third trimester in the fetal sheep. Nonetheless, fetal ovine liver and kidney thermogenesis [oxygen consumption, $(Na^{+}+K^{+})$-ATPase activity, and mitochondrial α-glycerophosphate-dehydrogenase activity] are unresponsive to exogenous T3 during the third trimester, whereas they are responsive to thyroid hormone in the adult. β-Adrenergic receptor binding in heart and lung are responsive to thyroid hormones in the adult but are unresponsive in the ovine fetus, becoming responsive in the neonatal period. Pituitary GH concentrations become thyroid hormone-responsive near term. The mechanism(s) for this delayed maturation of postreceptor thyroid responsiveness in several systems in not yet clear.

The effect of GH in the fetus also is minimal. Fetal somatic growth is not GH-dependent [15,16,18,21,23]. In this instance there is a delayed maturation of GH receptor binding which appears in liver only during the neonatal period [15,21,23]. A deficiency of prolactin receptors also may be a major factor in the apparently limited prolactin bioactivity in the fetus near term, but available receptor studies are limited.

There is less information regarding fetal hormone responsiveness in other systems. β-Adrenergic receptor binding in heart and lung of the sheep fetus are relatively low near term, and increase in the neonatal period in response to thyroid hormones [15,29]. Moreover, premature lambs have an augmented plasma catecholamine surge at birth relative to term lambs and have a relatively obtunded increase in plasma free fatty acid concentrations, suggesting reduced catecholamine responsiveness [16,28]. The extraordinarily high concentrations of progesterone and estrogens in fetal blood also seem to have limited effects in the fetus [15,16]. Estrogen receptors are present at low concentration in fetal guinea pig kidney, lung, and uterus at midgestation and increase progressively to term. Uterine estrogen receptors in the rat appear to be synthesized during the first 10 days of life. Limited information is available regarding maturation of postreceptor responses.

119.8 Conclusions

In this chapter, we have summarized the specific features of fetal endocrinology as distinct from that in adults. Since the fetus develops in a very special environment, it is not surprising that it has very particular endocrine demands. These are served by three means:

- The placenta is impermeable to most maternal hormones
- The placenta produces hormones itself
- The fetus develops a set of endocrine and other organs that may only be active during prenatal life and respond to specific demands.

Many of the details of the mechanisms and functions discussed here are not yet well understood, but they will surely be revealed by future work.

References

1. Aloe L, Levi-Montalcini R (1979) Nerve growth factor-induced transportation of immature chromaffin cells in vivo into sympathetic neurons: effect of antiserum to nerve growth factor. Proc Natl Acad Sci USA 76:1246
2. Baird A, Wehrenberg WB, Behalen P, Ling N (1985) Immunoreactive and biologically active growth hormone releasing factor in the rat placenta. Endocrinology 117:1598
3. Burrow GN, Fisher DA, Larsen PR (1994) Maternal and fetal thyroid function. N Engl J Med 331:1072
4. Buster JE (1983) Gestational changes in steroid hormone biosynthesis, secretion, metabolism and action. In: Yen SCC (ed) Clinics in perinatology, Saunders, Philadelphia, pp 527–552
5. Chakravarti A, Phillips JA, Mellits KH, Buetow KH, Seeburg PH (1984) Patterns of polymorphism and linkage disequilibrium suggest independent origins of the human growth hormone gene cluster. Proc Natl Acad Sci USA 81:6085
6. Comline RS, Silver M (1966) Development of activity in the adrenal medulla of the foetus and newborn animal. Br Med Bull 22:16
7. DeVane GW, Porter JC (1980) An apparent stress-induced release of arginine vasopressin by human neonates. J Clin Endocrinol Metab 51:1412
8. Engler P, Scanlon MF, Jackson IMD (1981) Thyrotropin releasing hormone in the systemic circulation of the neonatal rat is derived from the pancreas and other extraneural tissues. J Clin Invest 67:800
9. Ervin MG, Leake RD, Ross MG, Calvario GC, Fisher DA (1985) Arginine vasotocin in ovine maternal and fetal blood, fetal urine and aminotic fluid. J Clin Invest 75:1696
10. Falcone T, Little AB (1994) Placental synthesis of steroid hormones. In: Tulchinsky D, Little AB (eds) Maternal fetal endocrinology, 2nd edn. Saunders, Philadelphia, pp 1–14
11. Falcone T, Little AB (1994) Placental polypeptides. In: Tulchinsky D, Little AB (eds) Maternal fetal endocrinology, 2nd edn. Saunders, Philadelphia, pp 15–32
12. Fiddes JC, Talmadge K (1984) Structure, expression and evolution of the genes for the human glycoprotein hormones. Rec Prog Horm Res 40:43
13. Fisher DA (1983) The maternal-fetal neurohypophyseal system. In: Yen SCC (ed) Clinics in perinatology. Saunders, Philadelphia, pp 695–707

14. Fisher DA (1983) Maternal-fetal thyroid function in pregnancy. In: Yen SCC (ed) Clinics in perinatology. Saunders, Philadelphia, pp 615–626
15. Fisher DA (1986) The unique endocrine milieu of the fetus. J Clin Invest 78:603
16. Fisher DA (1992) Endocrinology of fetal development. In: Wilson JD, Foster DW (eds) Textbook of endocrinology, 8th edn. Saunders, Philadelphia, pp 1049–1077
17. Fisher DA (1989) Maturation of thyroid hormone actions. In: Delange F, Fisher DA, Glinoer D (eds) Research in congenital hypothyroidism. Plenum, New York, pp 61–77
18. Fisher DA (1989) Fetal endocrinology: endocrine disease and pregnancy. In: DeGroot LJ, Besser GM, Cahill GF, Marshall JC, Martini L, Nelson DH, Odell WD, Potts JT, Rubenstein AH, Steinberger E (eds) Endocrinology, vol II. Grune and Stratton, New York, pp 2102–2122
19. Fisher DA, Klein AH (1981) Thyroid development and disorders of thyroid function in the newborn. Engl J Med 304:702
20. Gluckman PD, Grumbach MM, Kaplan SL (1981) The neuroendocrine regulation and function of growth hormone and prolactin in the mammalian fetus. Endocr Rev 2:363
21. Gluckman PD, Butler JH, Elliot TB (1983) The ontogeny of somatotropic binding sites in ovine hepatic membranes. Endocrinology 112:1607
22. Goland RS, Wardlaw S. Stark RI, Brown LSJ, Frantz AG (1986) High levels of corticotropin releasing hormone immunoreactivity in maternal and fetal plasma during pregnancy. J Clin Endocrinol Metab 63:1199
23. Grumbach M, Gluckman PD (1994) The human fetal hypothalamus and pituitary gland: the maturation of neuroendocrine mechanisms controlling the secretion of pituitary growth hormone, prolactin, gonadotropins, adrenocorticotropin-related peptides, and thyrotropin. In: Tulchinsky D, Little AB (eds) Maternal fetal edocrinology, 2nd end. Saunders, Philadelphia, pp 193–262
24. Harada A, Hershman JM, Reed AW, Braunstein GD, Dignam WJ, Derzko C, Friedman S, Jewelewicz R, Pekary AE (1979) Comparison of thyroid stimulators and thyroid hormone concentrations in the sera of pregnant women. J Clin Endocrinol Metab 48:793
25. Josso N (1986) Antimullerian hormone: new perspectives for a sexist molecule. Endocr Rev 7:421
26. Koivusalo F (1981) Evidence of thyrotropin releasing hormone activity in autopsy pancreata from newborns. J Clin Endocrinol Metab 53:734
27. Mimouni F, Tsang RC (1994) Perinatal mineral metabolism. In: Tulchinsky D, Little AB (eds) Maternal fetal endocrinology, 2nd edn. Saunders, Philadelphia, pp 401–418
28. Padbury JF, Polk DH, Newnham JP, Lam RW, Fisher DA (1985) Neonatal adaptation: greater neurosympathetic system activity in preterm than full term fetal sheep at birth. Am J Physiol Endocrinol Metab 248:E443
29. Padbury JF, Klein AH, Polk DH, Lam Rw, Hobel CJ, Fisher DA (1986) The effect of thyroid status on lung and heart beta adrenergic receptors in fetal and newborn sheep. Dev Pharmacol Ther 9:44
30. Pelleniemi LJ, Dym M (1994) The fetal gonad and sexual differentiation. In: Tulchinsky D, Little AB (eds) Maternal fetal endocrinology, 2nd edn. Saunders, Philadelphia, pp 297–320
31. Perez-Castillo A, Bernal J, Ferreiro B, Pans T (1985) The early ontogenesis of thyroid hormone receptor in the rat fetus. Endocrinology 117:2457
32. Pitkin RM (1983) Endocrine regulation of calcium homeostasis during pregnancy. In: Yen SCC (ed) Clinics in perinatology. Saunders, Philadelphia, pp 575–592
33. Polk DH, Fisher DA (1995) Fetal and neonatal endocrinology. In: DeGroot LJ et al (eds) Endocrinology. Saunders, Philadelphia, pp 2239–2257
34. Roti E, Gnudi A, Braverman LE (1983) The placental transport synthesis and metabolism of hormones and drugs which affect thyroid function. Endocrine Rev 4:131
35. Ryan KJ (1980) Placental synthesis of steroid hormones. In: Tulchinsky D, Ryan KJ (eds) Maternal fetal endocrinology. Saunders, Philadelphia, pp 3–16
36. Schedewie HK, Fisher DA (1980) Perinatal mineral homeostatis. In: Tulchinsky D, Ryan KJ (eds) Maternal-fetal endocrinology. Saunders, Philadelphia, pp 355–386
37. Siler-Khodr TM (1983) Hypothalamic-like releasing hormones of the placenta. In: Yen SCC (ed) Clinics in perinatology. Saunders, Philadelphia, pp 553–566
38. Sperling MA (1980) Carbohydrate metabolism: glucagon, insulin and somatostatin. In: Tulchinsky D, Ryan KJ (eds) Maternal fetal endocrinology. Saunders, Philadelphia, pp 333–354
39. Sperling MA (1994) Carbohydrate metabolism: Insulin and glucagon. In: Tulchinsky D, Little AB (eds) Maternal fetal endocrinology, 2nd edn. Saunders, Philadelphia, pp 379–400
40. Stark RI, Frantz AG (1983) ACTH β-endorphin in pregnancy. In: Yen SCC (ed) Clinics in perinatology. Saunders, Philadelphia, pp 653–667
41. Stark RJ, Daniel SS, Husain MK, Sanocka UM, Zubrow AB, James LS (1984) Vasopressin concentration in amniotic fluid as an index of fetal hypoxia: mechanism of release in sheep. Pediatr Res 18:552
42. Wilson JD (1978) Sexual differentation. Annu Rev Physiol 40:279
43. Wurtman RJ (1966) Control of epinephrine synthesis in the adrenal medulla by the adrenal cortex: hormonal specificity and dose response characteristics. Endocrinology 79:608–614

120 The Birth Process

H.P. ZAHRADNIK

Contents

120.1 Introduction

In 1977 Thorburn [111] stated: "We now recognize that in late pregnancy *a train of events* is initiated that ultimately results in the delivery of the fetus. However, we still do not know exactly how and where the train starts, or exactly how it exerts its ultimate action on the myometrial cell." During recent decades numerous efforts have been made to understand the regulation of the gestational process. However, our knowledge of the above-mentioned "train of events" has hardly expanded. Successful viviparity requires that the mother provides her fetus with the substances essential for its growth. Moreover the fetal placental–maternal cooperation is the source of gestational-age-dependent production of steroid and polypeptide hormones (cf. Chap. 116). Together these hormones play a major part in the maintenance of pregnancy, but they are also a prerequisite for the preparation of the uterus at term for labor.

During pregnancy, *myometrial activity* is characterized by uncoordinated contractions. In later pregnancy uterine motility evolves gradually, and labor normally precedes delivery. It is apparent that human myometrial contractility is affected by local factors functioning in an endocrine or paracrine manner. These factors are oxytocin (OT) and eicosanoids (cf. Chap. 6), and also local substances usually involved in infective processes, such as tumor necrosis factor (TNF), interleukins (IL) or endothelin (ET). Hormones act through specific receptors. The density of receptors for uterotonic stimulants of myometrium is low at the beginning of pregnancy, but increases substantially at the time of labor. Estradiol (E_2) and catecholamines are responsible for this increase. Generally these hormones prepare uterine cells to respond to the appropriate uterotonic stimuli.

This chapter reviews different steps of the birth process. A brief description of clinical data about the normal delivery is followed by more detailed information on basic physiologic mechanisms involved in the onset and maintenance of labor in the human. The endocrine, paracrine and autocrine control of these processes is the subject of this review.

120.2 Clinical Data

About 5% of pregnancies end on the *expected date* of delivery (EDD). Most women who reach term deliver within 7 days of their "due date." Between 10% and 13% of term pregnancies end in birth at least 14 days after the EDD. In the case of pregnancies carefully dated by ultrasound or basal body temperature measurements in the conception cycle, however, a maximum of 5% of the women remain undelivered 42 weeks from the "corrected" date of conception.

The birth process begins with *labor*, that is with regularly occurring painful physiological contractions. These result in progressive thinning and dilatation of the cervix, permitting the passage of the fetus through the birth canal.

By convention, *parturition* is divided into three stages. The *first stage* begins with the onset of uterine contractions and ends with complete or full dilatation of the cervix to a diameter of 10 cm for a term-sized infant. During this first cervical stage, one can differentiate three phases (of Table 120.1):

- The latent phase begins at labor onset and ends with the onset of active labor. The median duration is 9 h for the primipara and 5 h for the multipara. During this time cervical dilatation is slight and does not progress rapidly.
- In the active phase the cervical dilatation is 5 cm or more, or the change in cervical dilatation is more than 1 cm/h in primiparae and 1.5 cm/h in multiparae.
- A dilatation of 9 cm to full dilatation (10 cm) is considered the deceleration phase.

Descent of the fetal presenting part may remain latent until 9 cm of dilatation has been reached (cf. Fig. 120.1). Active

R. Greger/U. Windhorst (Eds.)
Comprehensive Human Physiology, Vol. 2
© Springer-Verlag Berlin Heidelberg 1996

Table 120.1. Duration of labor. (Adapted from [130])

Stage of labor	Mean
Primiparae	
1st stage	14.4 h
Latent phase	8.6 h
Active phase	4.9 h
Maximum slope	3.0 cm/h
2nd stage	1.0 h
Multiparae	
1st stage	7.7 h
Latent phase	5.3 h
Active phase	2.2 h
Maximum slope	5.7 cm/h
2nd stage	0.2 h

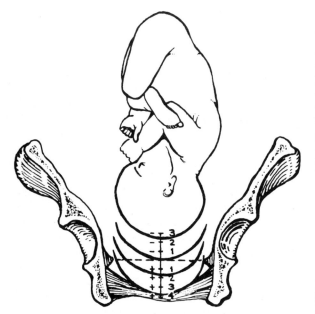

Fig. 120.1. Stations of fetal head. *Dotted line* indicates that the leading bony point of the fetal head is at the level of the ischial spines (zero level). Above this level stations are designated −1, −2, −3 cm. With further descent of the vertex stations are designated +1, +2, +3 cm and so on. (From [130] with permission)

descent should have started by the beginning of deceleration phase. Engagement, the first of the five cardinal movements of labor (cf. Table 120.2), corresponding to a "true" 0 station, should normally have occurred by the beginning of deceleration phase in primiparae and by the end of deceleration phase in multiparae.

The first stage obviously ends with full dilatation, and the *second stage* begins with complete dilatation of the cervix and ends with delivery of the baby (Fig. 120.2). The second stage of parturition can be expected to have a median duration of 1 h in primiparae and 15 min in multiparae. The "clinical limits" after which problems must be anticipated with delivery are 2 h for primiparae and 45 min for multiparae. If the membranes have remained intact to this

point, they now often rupture. With contractions, the parturiant woman feels a strong urge to bear down. Uterine contractions coupled with maternal voluntary pushing efforts will result in descent of the vertex. The passage of the fetal vertex along the curve of the pelvis continues according to the cardinal movements of labor (cf. Table 120.2). After flexion, internal rotation begins at the level of the ischial spines (cf. Fig. 120.1). With further descent, birth of the head results from extension of this presenting part. After delivery of the head, the vertex rotates externally and returns to its former transverse position [107,130].

The *third stage* of parturition (Fig. 120.3) begins with delivery of the baby and ends with delivery of the placenta and membranes. The placenta generally separates from the uterine wall within the next few uterine contractions after delivery of the newborn. Expulsion may be delayed for several minutes. If the placenta has not been delivered within 30 min after birth of the newborn, preparations to remove it manually should be made.

Some obstetricians refer to the 60 min after delivery of the placenta "the fourth stage of labor." This emphasizes the extreme importance of observing the mother for excessive vaginal bleeding, because uterine atonia accounts for more than 90% of all cases of postpartum hemorrhage. If abnormal uterine bleeding occurs OT, ergot alkaloids and/or prostaglandin $F_{2\alpha}$ ($PGF_{2\alpha}$) or prostaglandin E_2 (PGE_2) analogue (sulprostone) should be administered [124,126].

120.3 Basic Mechanisms

120.3.1 Cervical Maturation

During pregnancy, the cervix remains firm and closed, as in the non-pregnant state. However, as the onset of labor approaches, it "ripens" to become soft and stretchy, so that when contractions begin it can dilate and allow the baby to pass through. Increasing evidence has shown that biochemical changes are involved in the process of cervical maturation, but the exact nature of these changes is unclear. The human cervix consists of smooth muscle and fibrous connective tissue, i.e., collagen and ground substance. During late pregnancy these components are influenced by degradative enzymes: the synthesis of extracellular matrix proteins and glycoproteins is changed [65].

Increasing evidence indicates that cervical maturation can occur in the absence of regular uterine contractions. In the sheep, transection of the cervix from the uterus did not

Table 120.2. Cardinal movements of labor. (Adapted from [130])

1. Engagement
2. Descent
3. Flexion
4. Internal Rotation
5. Extension

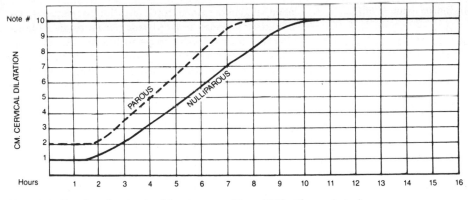

Fig. 120.2. Flowsheet for charting labor progress. (From [130] with permission)

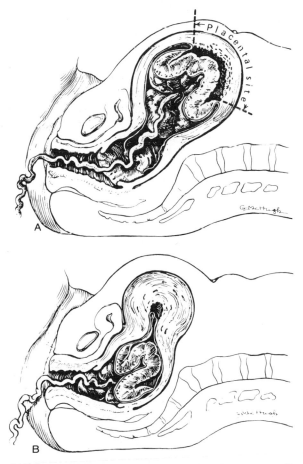

Fig. 120.3A,B. Third stage of labor. A The placenta generally separates from the uterine wall immediately after delivery of the baby. Note the decreased surface area of the placental site. B After expulsion of the placenta into the lower uterine segment, the uterine corpus becomes more globular and rises in the abdomen. (From [130] with permission)

prevent cervical softening [62]. When PGE₂ was applied strictly within the cervical canal, without any measurable intrauterine pressure increase, significant maturation of the cervix was observed within 5 h. Furthermore, results from in vitro investigations of both animal and human cervices indicate a relaxant effect of PGE₂ on the cervical smooth muscle [4]. The role of the cervical muscle layer has been debated for a long time, but until recently it was considered to be of minor importance for the priming procedure.

Cervical maturation in humans seems to result from increased collagenase activity, causing a breakdown of the collagen [114]. Concomitantly there is increased synthesis of the ground substance, as reflected in increased amounts of glycosaminoglycans. The principal changes that occur in the cervix during parturition are increases in vascularization, mass and water content and decreases in organization of collagen and dermatan sulfate.

In the human cervix at term, the levels of collagen and dermatan sulfate are less than one third of those in the non-pregnant state. These changes in collagen content correlate with changes in proteolytic enzyme activities [13]. There is direct evidence that interstitial collagenase increases markedly during cervical dilatation in human parturition [93]. In this context PGE₂ seems to play an essential part. PGE₂ is responsible for the extrusion of cytoplasmic vesicles located in cervical myofibrocytes into the extracellular spaces of the human cervix uteri. These vesicles are matrix lysosomes. The enzymes in these lysosomes are separated from the surrounding ground substance by a unit membrane. Having escaped from the cells, the stability of the membrane of lysosomes are no longer under direct cytoplasmic control. The lysosomal enzymes can then penetrate into the connective tissue more effectively [50]. In view of the strong relationship between PGE₂ and collagenase production [77], it is clear that PGE₂ has a central role in the maturation process of the cervix uteri. Moreover E-type prostaglandins are essential for angiogenesis [129]. Although the exact mechanisms controlling the cervical dilatation during the birth process are unknown, additional factors seem to be involved. Evidence for these factors is based on basic and clinical studies with antigestagens, i.e. mifepristone and onapristone.

Progesterone (P₄) is an important factor controlling both the closure and the maturation of the cervix uteri. In the presence of physiological concentrations of P₄ and E₂ an equilibrium between collagen synthesis and collagenolysis is established. In vitro studies have demonstrated that with

2349

P_4 (10^{-6} mol/l) the procollagenase gene expression in monolayer cell cultures from cervices is inhibited. Moreover P_4 blocks the E_2-induced increase in collagenase production [94]. In human endometrial explants, P_4 (10–200 nmol/l) abolishes the release of matrix collagenase, an effect that can be prevented by the antiprogesterone RU 486 [74]. These data suggest that the inhibition of the influence of P_4 on cervical tissue is a key mechanism for cervical maturation at the beginning of labor. Accordingly, an increase in serum collagenase levels has been found in normal pregnant women, which correlates with the state of the cervix [32]. This is the physiological status at the end of pregnancy, but as implied above, it can also be induced pharmacologically by antigestagens such as RU 486 or onapristone [15].

Although it was not possible to demonstrate any ripening effects of estrogens in pregnant guinea pigs [15], clinical observations suggest that estrogens might be involved in prepartum cervical priming [4]. This *priming process* is probably not caused directly by estrogens, but the cervix depends on an adequate estrogen supply for preparation. In this context it is possible that an "unopposed estrogen action" is the mechanism of antigestagen-induced cervical maturation [16].

The idea that *relaxin* may play a leading part in human cervical maturation is a fascinating prospect. It is based on findings in animals that show considerable relaxation of several ligaments after relaxin treatment. Histological examinations reveal that bundles of collagen fibers are disorganized after relaxin treatment. A consistent result was that administration of relaxin to pregnant rats was followed by similar results in cervical tissue [87]. Use of a monoclonal antibody against rat relaxin made it possible to demonstrate the role and the critical importance of this hormone in cervical ripening. One of the most prominent results was that relaxin promotes the synthesis of collagenase [63].

The 6 kDa peptide hormone relaxin consists of an A and a B chain. The chains derive by posttranslational cleavage from a single preprorelaxin with 185 amino acids [11]. Relaxin is involved in the inhibition of myometrial activity, in maturation of the cervix uteri and in relaxation of the pubic ligament. The human genome contains two relaxin genes, H_1 and H_2, which code for two polypeptides with significantly different amino acid sequences. Decidua, fetal membranes and trophoblast are tissue compartments with high synthesizing capacity for relaxin [102], and there is sound evidence showing that in these tissues, in addition to the H_2 gene, the H_1 gene is also expressed [36]. The decidua contains the highest tissue level of relaxin – mRNA, relaxin C-peptide, and mature relaxin peptide – at term, with markedly decreasing levels postpartum [73,102] Moreover, significantly lower tissue mRNA levels (but no difference in protein) was measured in samples collected after normal labor than in samples collected after elective cesarean section [102].

Clinically, the role of relaxin in cervical maturation remains obscure, although intravaginal administration of relaxin induced significant cervical maturation before labor induction [72]. According to the investigators, no side effects were experienced, and evidently only minimal increases in uterine activity were observed. However, the action of relaxin on different tissues and its presence in a multitude of reproductive organs suggests that its physiological role is much more diverse than inhibition of uterine contraction, cervical relaxation and softening of the pubic ligament.

The interrelation of *preterm labor* after preterm softening, shortening and effacement of the cervix with local infection of the genital tract is well established [7,35]. After preterm birth the first months in the life of such a newborn infant are very often complicated by illnesses resulting from intrauterine infection [90].

Leukocyte infiltration and degranulation in human cervical tissue are similar to those seen in inflammatory reactions. Moreover, there is dissolution of the connective tissue matrix around polymorphonuclear leukocytes [51], and cervical biopsies from women at term contain activated and degranulated eosinophils [59]. These findings support the notion [66] that cervical maturation is similar to an *inflammatory reaction*. It therefore seems likely that substances such as cytokines, which are released from cervical cells, including white blood cells, are essential for the cervical maturation process. In cultured rabbit cervical explants taken at term elevated production of a chemotactic substance was detected. This chemoattractant (the production of which was increased by IL-1) was characterized as a true chemotactic factor and a heat-stable and trypsin-sensitive protein with an apparent relative molecular weight of 16.2 kDa.

The properties of the factor were reported as being very similar to those of the *IL-8* family [113] (cf. Chap. 6). It may well be that this factor plays an important part in stimulating the accumulation of white blood cells within the cervical stroma prior to the maturation process. There is increasing evidence that during maturation it is mainly cytokines that are responsible for this process. IL-1-like factors have been found in pregnant rabbit cervix explants [49]. IL-8 has been seen to be released from human choriodecidual cells in vitro [55], and IL-8 concentrations rise during labor [99]. In preterm and term human parturition the macrophage-derived cytokines IL-1, IL-6, IL-8 and TNF have been measured [100]. After local application of IL-1β and IL-8 in a gel, significant cervical softening and dilatation without induction of labor was seen in pregnant guinea pigs. These effects were similar to the physiological maturation of the cervix and to the effects seen after the antigestagen onapristone [16].

Prostaglandins (PG) produced within the cervix and by intrauterine tissues are believed to have a major impact on the regulation of changes in collagen at term [45,50]. In contrast to these findings, animal and human studies with the antigestagen RU 486 have shown that the softening effect of RU 486 is not correlated with increased cervical PG production. In addition, naproxen did not inhibit the cervical effect of RU 486 [91,92].

The following recent data may have some bearing on the controversial discussion on the physiological role of PG in cervical maturation. Human amnion cells in vitro produce a minimal basic amount of cyclo-oxygenase (COX)-1 and COX-2 mRNA. Dexamethasone and cortisol, but not E_2 and P_4, can stimulate the COX-2 mRNA production. The glucocorticoid effect can be inhibited by RU 486. COX-1 remains unchanged [128]. Primary cultures of human decidua cells even produce COX-2 exclusively. Further stimulation of COX-2 expression is not possible. The tissue seems to be maximally stimulated during birth [127]. If it is borne in mind that the classic non-steroidal anti-inflammatory agents inhibit mainly COX-1 [76], it is conceivable that the above statements underestimate the physiological role of cervical PG. Although their physiological role is still under discussion, the therapeutic use of PGE_2 [56,125] or PGE_1 analogues for cervical priming is now generally accepted [22].

Taken together, the data suggest that cervical effacement, the prerequisite for normal parturition, is not caused by any muscular contribution from the cervix or myometrium, but rather by biochemical changes that begin long before the onset of labor [13].

120.3.2 Electrophysiology and Ion Exchange in Myometrium

An increase in the free Ca^{2+} concentration in the myometrial cytoplasma from 10^{-7} to 10^{-6} mol/l results in contraction. Intracellular stores are the source of some of this Ca^{2+}, but the influx of Ca^{2+} from the extracellular space is most important [78]. In human myometrium, evidence for two different voltage-gated Ca^{2+} channels has been obtained at the single channel level [47], (cf. Fig. 120.4).

Under voltage clamp conditions, a depolarizing voltage step across the membrane that is sufficient to activate Ca^{2+} channels results in an inward current that peaks within 10 ms and declines thereafter. There is no evidence that accumulation of Ca^{2+} in a restricted extracellular or intracellular space changes the membrane potential in isolated myometrial cells [1]. Blockers of the L-type Ca^{2+} channels in the myometrium are nifedipine at 10^{-7} mol/l and (+)isradipine at 10^{-8} mol/l. Nifedipine, one of a great number of Ca^{2+} entry blockers [25], is a potent and selective inhibitor of uterine tension. Its pharmacological action depends on the binding to one of the Ca^{2+} channel states and consequently is followed by inhibition of the inward Ca^{2+} current. (+)Isradipine does not only bind to the resting state of Ca^{2+} channels, but also to the inactivated state. Hence, it shows a voltage-dependent inhibition [80].

Probably hormonal changes or stretching are triggers for synthesis and incorporation of a greater amount of Ca^{2+} channel protein into the myometrium near term.

Concerning activation of Ca^{2+} channels it is well known that OT increases the Ca^{2+} current and the release of Ca^{2+} from intracellular stores [79] by increasing the Ca^{2+} conductance of myometrial cells.

A second type of myometrial membrane Ca^{2+} channel is the ligand- or receptor-gated channel. This channel is activated by adenosine triphosphate (ATP): 1 mmol/l ATP (for 2 s) produces an inward current that is fast in onset (less than 100 ms) and has a short time to peak [42].

In human myometrial cells there are also voltage-gated Na^+ channels [121]. Binding of [^{3}H]saxitoxin to dispersed myometrial cells indicates two binding sites located in the plasma membrane of myometrial cells [80]. Electrophysiological properties of these channels reveal that half-maximal activation is obtained near −15 mV. The Na^+ current is fully available at very negative membrane potential (−110 mV) [75]. The density of Na^+ channels depends on gestational stage. It is maximal near term, because of an increase in the fraction of cells that pass fast Na^+ channels [48].

In response to electrical or chemical stimuli, the depolarization of the myometrial membrane results in opening of the above channels. Free intracellular Ca^{2+} increases (Figs. 120.4, 120.5). The consequences is a uterine contraction.

Removal of cytosolic Ca^{2+} is essential for relaxation of the myometrium. The major mechanisms for Ca^{2+} removal are two ATP-dependent extrusion systems. One system regulates the calcium uptake into intracellular storage pools within the SR (SR pump), while the second removes calcium from the cell to the outside across the plasma membrane (PM pump) [33] (cf. Chap. 8). High-affinity

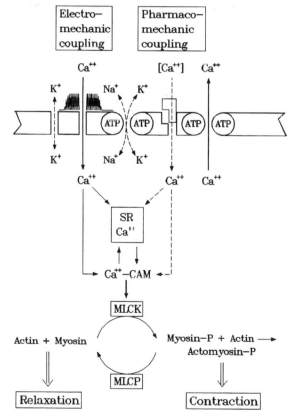

Fig. 120.4. Schematic of the mechanisms involved in contractile activation in myometrial smooth muscle. Conceptual overview of Ca^{2+} regulation. *CAM*, calmodulin; *MLCK*, myosin light-chain kinase; *SR*, sarcoplasmic reticulum; *circle with ATP*, active pump

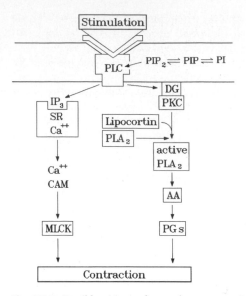

Fig. 120.5. Possible actions of second messengers in the initiation of labor. Physiological/pathological/pharmacological stimuli induce receptor-mediated, G-protein-modulated activation of the phospholipase-C (*PLC*) system. Membranes contain phosphatidylinositol (*PI*), phosphatidylinositol-4-phosphate (*PIP*) and phosphatidylinositol-4,5-bisphosphate (*PIP₂*) in steady state. After receptor activation PLC hydrolyzes PIP_2 to form inositol-1,4,5-triphosphate (IP_3) and 1,2-diacylglycerol (*DG*). IP_3 releases Ca^{2+} from the sarcoplasmic reticulum (*SR*), which is coupled to calmodulin (*CAM*) for stimulation of MLCK. DG activates protein of lipocortin together with inactive phospholipase A_2 (*PLA₂*) to produce active PLA_2. This enzyme causes the release of arachidonic acid (*AA*) (from membrane phospholipids), followed by the production of prostaglandins (*PGs*)

Ca^{2+}-ATPases have been isolated from the myometrial plasma membranes of pregnant women [86] (cf. Fig. 120.4).

As demonstrated in rat myometrium membranes, there is a dramatic increase in current densities for Na^+ during pregnancy, highest levels being reached several days before parturition [48]. The cytosolic Na^+ increases and results in membrane depolarization. There is also an increase of K^+ channels during pregnancy. In late pregnancy the myometrial excitability increases [6]. Quantitative information on the ($Na^+ + K^+$) pump during pregnancy is rare. ($Na^+ + K^+$) pumps seem to consist of α- and β-subunits [39] (cf. Chap. 8). The α-subunit is encoded by genes designated as α-1, α-2, and α-3, while the β-subunits, β-1 and β-2, are encoded by two genes. The uterus contains transcripts mainly for α-1 and β-2 isoforms. ($Na^+ + K^+$) pump expression is decreased in the rat near the end of pregnancy [39].

Even the detailed information available about membrane excitability and ion exchange cannot really explain the characteristic action of the myometrium in labor. In the myometrium there are known to be two types of *pacemaker activity,* one of which initiates the burst discharges (~10 s) while the other controls the individual spikes in each burst (0.5–1 s). In situ recordings from women in labor show that the bursts determine the frequency of contraction. The number of individual spikes in each burst influences the force of contraction [53].

For further details of the electrophysiological regulation and ion exchange in the myometrium the reader is referred to the relevant reviews [34,53,80].

120.3.3 Cell-to-Cell Communication in Myometrium

The physiological activity of the uterus at term is referred to as labor. Labor consists in well-coordinated synchronous contractions that depend on optimal cell-to-cell communication in the myometrium (cf. Chap. 7). Cell-to-cell propagation of action potentials throughout the myometrium, coming from pacemaker regions or areas influenced by stimulatory agonists, are of fundamental importance to the regulation of uterine excitability and contraction. The intercellular (or gap) junctions are produced by intramembranous (gap junction) proteins within adjacent cell membranes. *Gap junctions* link cells to their neighbors and allow communication between the different cytoplasmic compartments. They enable the passage of inorganic ions and other small molecules. They guarantee that the myometrium will function as a syncytium during parturition [13]. Gap junctions increase in frequency and size at the end of gestation immediately before the onset of labor. In the non-pregnant uterus they rarely occupy more than 0.001% of the myometrial plasma membrane surface. Once regular contractions are established, 0.1–0.4% of the uterine smooth muscle cell membrane is occupied by gap junctions. Each pregnant myometrial cell contains at least 1000 junctions, with an average diameter of ca. 250 nm [28] (cf. Fig. 120.6). They disappear within 24 h after parturition [27].

The proteins forming gap junctions have been cloned, and antibodies against one of these proteins have been prepared. In the myometrium a 43-kDa protein, connexin 43 (Cx-43), seems to be the major component of gap junctions [9]. It has recently been shown that Cx-43 from rat myometrium and a 43-kDa protein from the rat heart [8] differ in their amino acid sequences only in one residue [61].

Control of cell-to-cell communication can be achieved through alterations in:

- The number of gap junctions
- The size of the gap junctions
- The permeability or conductance characteristics of gap junctions.

All mechanisms may be under the control of specific hormones. Estrogens seem to promote the synthesis of gap junctions by stimulating the specific genome reponsible for coding the connexin. Eicosanoids also possibly are involved in the control of the junctions protein synthesis [69,85]. The mechanisms that regulate the opening and closing properties of gap junction channels is not well understood. Elevated intracellular levels of cAMP may reduce

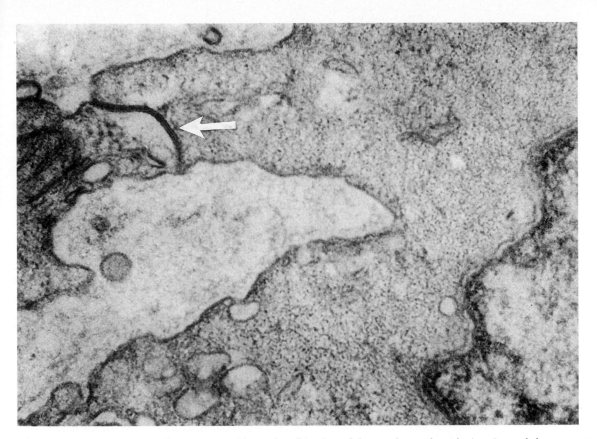

Fig. 120.6. Transverse section of two myometrial muscle cells showing an intermediate junction (see *arrow*) connecting their plasma membranes. Note the densification of the cytoplasmic surface of the membrane along the junction and the row material extracellular space separating the membranes. ×30 800 (From [18] with permission)

cell-to-cell coupling in the myometrium [17]. Agents that are relevant to the inhibiting control of labor and delivery, such as relaxin, PGE$_2$, prostacyclin (PGI$_2$) and β-adrenoreceptor agonists decrease cell-to-cell exchange by interaction with specific membrane receptors and elevation of intracellular levels of cAMP. Experiments with COX and lipoxygenase (LO) inhibitors have shown that LO-mediated arachidonic acid (AA) metabolites may be involved in promoting cell-to-cell coupling in the myometrium [18]. OT, a very strong uterine stimulant, decreases coupling in the parturient rat myometrium, which is consistent with the effect of an elevated intracellular Ca^{2+} activity [10].

120.3.4 The Myometrial Contraction

At the cellular level, contractions result from the phosphorylation of the 20-kDa myosin light chain (MLC$_{20}$) by MLC-kinase (MLCK) [120] (cf. Chaps. 44, and 96). This process activates the ATPase of myosin, providing enough chemical energy to interact with actin within the muscle fibrils. The cross-linking covalent bonds between myosin and actin shortens the fibrils and thereby the cell itself. Factors involved in the process of myometrial contractions in this context are: myosin, actin, MLC, MLCK, protein kinase (PK), calmodulin (CAM) and Ca^{2+} (Fig. 120.7).

Myosin is a thick molecule composed of one pair of heavy chains each 200 kDa in molecular weight and two pairs of light chains (20/17 kDa). *Actin* molecules (45 kDa) form long, thin filaments. The thick myosin and thin actin filaments occur in long, random bundles throughout the smooth muscle cells. The organization of smooth muscles can exert pulling force in any direction. The contraction of the smooth muscle is based on a relative sliding of actin and myosin filaments without an internal change in the length of either filament [46]. The large ratio of actin-to-myosin filaments may allow recruitment of extra-thin filaments and the subsequent shortening of the muscle cell for maximum tension generation [44].

The primary determinant of biochemical events and major regulator of contraction in human myometrial smooth muscle is the cytoplasmic Ca^{2+} activity [71]. Physiological or pharmacological stimuli cause a transient increase of free Ca^{2+} in the cytoplasma (cf. Figs. 120.4, 120.5, 120.7). Free intracellular Ca^{2+} forms a complex with the Ca^{2+} sensor protein calmodulin. The Ca^{2+} binding induces a dramatic change in calmodulin conformation. Binding of target enzymes to calmodulin usually converts inactive apoenzymes to the active form of Ca^{2+}-calmodulin enzyme [58]. The most important of these enzymes in the regulation of myometrial activity is *MLCK*. This enzyme is a single polypeptide, binding with a very high affinity to

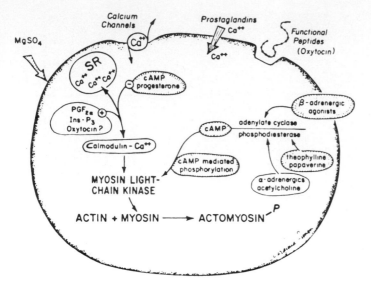

Fig. 120.7. Summarized cellular regulation of myometrial contractility. The MLCK is regulated through the Ca²⁺ and cAMP pathways. Both Ca²⁺ sequestration and cAMP levels in the cell are modulated by various hormones (and drugs). Agonists *circled* with *solid lines* promote contractions; agonists in *stippled areas* promote relaxation. Specific sites that facilitate agonist interaction and tocolytic potential include the β-adrenergic receptors, the calmodulin–MLCK interaction, the receptors for prostaglandins (*PG*), the Ca²⁺ channels, and binding sites for functional peptides. (From [43] with permission)

Ca^{2+}-CAM in a 1:1 molar ratio [40]. It catalyzes phosphorylation of the 20-kDa MLC [120]. Myosin phosphorylation and dephosphorylation are key events in the "on-off" switch in actin activation of the Mg^{2+}-ATPase. All stimuli cause phosphorylation of MLC during isometric contraction. Whether the isolated MLC phosphatases play a physiological part in regulation of myosin phosphorylation in the myometrium remains unclear [43]. All the above events consume Ca^{2+}. When the Ca^{2+} levels fall, the calmodulin-Ca^{2+} complex dissolves and the free calmodulin dissociates from the kinase, which is thereby inactivated (cf. Fig. 120.7).

It is appropriate to appreciate the role of *cyclic nucleotides* as second messengers in connection with uterine contraction. Both cAMP and cGMP are regarded as important regulators of diverse cellular functions. They are formed from ATP and GTP, respectively, by the enzymes adenylylcyclase and guanylyl-cyclase, which are stimulated following ligand–receptor interaction on the cell membrane. Intracellular cAMP levels can be regulated by influences in synthesis such as activation of adenylylcyclase, as well as by hydrolysis to the inactive 5′-AMP by phosphodiesterase.

Some years ago the opinion was put forward that all hormones causing uterine contractions, after binding on specific extracellular membrane receptors, work through cGMP and all hormones that cause uterine relaxation work via cAMP. The last part of this statement has been confirmed by very convincing clinical and experimental results. The role of cGMP is unclear. Increasing intracellular cGMP levels are thought generally to correlate more with an inhibition of smooth muscle contraction [95], but clear evidence is still lacking [84]; rather, the role of phosphoinositides, as the link between hormone binding and uterine contraction, has come to the fore [67].

Concerning cAMP, all substances that can relax the myometrium induce an increase of intracellular cAMP [82,83,97] (cf. Fig. 120.7). This mechanism is the primary tocolytic pathway in the myometrium. Binding of inhibitory drugs to the specific cell membrane receptor is correlated with the dissociation of GTP-binding proteins Gs. The α-subunit of this protein activates the adenylylcyclase [97] (cf. Chap. 5). This process is able to initiate the synthesis of cAMP. Elevation of cellular cAMP leads to an activation of cAMP-dependent protein kinase (PKA), to protein phosphorylation, and finally to the relaxation of the myometrium. Altogether, the main action of cAMP is directed at a reduction of intracellular free Ca^{2+}. This can be achieved by increasing Ca^{2+} uptake to the SR or by increasing Ca^{2+} efflux from the intra- to the extracellular space [43] and to MLCK phosphorylation (cf. Figs. 120.4, 120.7).

Binding of a ligand to its specific receptor at the plasma membrane of the myometrium activates phospholipase (PLC; cf. Fig. 120.5). The mediator of PLC activation is probably a G protein. The active PLC hydrolyzes phosphatidyl 4,5-bisphosphate (PIP_2) to diacylglycerol (DAG) and inositol triphosphate (IP_3). IP_3 can mobilize intracellular Ca^{2+} from SR (cf. Chap. 120.5). Binding sites of IP_3 in the myometrium have been identified [96]. There is some evidence that the G proteins are also involved in the regulation of the IP_3 system. DAG activates protein kinase C (PKC), which promotes the combination of lipocortin with inactive phospholipase A_2 (PLA_2) for production of active PLA_2 [67] (cf. Fig. 120.5). Moreover, the activation of PLA_2 depends on the concentration of free intracellular

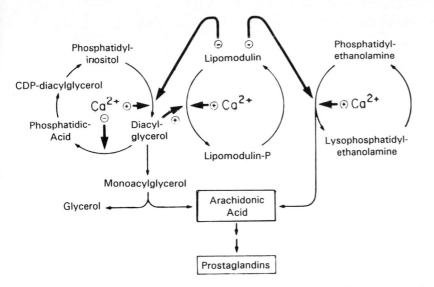

Fig. 120.8. Proposed regulation of arachidonic acid metabolism in human fetal membranes by calcium. (From [108] with permission)

Ca^{2+} (cf. Fig. 120.8). Active PLA_2 causes the release of AA, the usual unsaturated fatty acid in the 2 position of DAG. AA is the substrate for production of eicosanoids. In the context of uterine contraction the best studied metabolites of AA are PGE_2 and $PGF_{2\alpha}$. They are the most potent known stimulators of uterine contractility (Figs. 120.5, 120.8).

120.3.5 Endocrine Control of Labor

It is well known that at the beginning of labor there is a remarkable transformation of *contractile activity* of the myometrium. In sheep, single contractions begin 6–18h before delivery and increase in frequency and amplitude as labor progresses. Immediately before delivery 20–30 contractions per hour with an amplitude of 15–20 mmHg can be recorded [68]. In the human, the externally measured frequency of labor increases from 0.32/h at the 25th week of gestation in primiparae and 0.43/h in multiparae to 2.33/h at 40th week of gestation in primiparae and 2.27/h in multiparae [122].

Although endocrine control of the *diurnal rhythm of uterine activity* has been demonstrated in non-human primates [24], in human deliveries the higher frequency of labor starting between 4 and 6 a.m. than between 12 noon and 2 p.m., and the higher frequency of parturition between 12 midnight and 6 a.m. remain to be elucidated [106].

The *role of estrogens* in parturition is complex and sometimes apparently contradictory. E_2 is a physiologically functional "antagonist" to P_4 and vice versa. E_2 increases myometrial labor associated proteins, such as OT receptor, ET-1 receptor, Ca^{2+} channels, K^+ channels, α_1- and α_2- adrenergic receptors, contractile protein synthesis and, possibly, MLCK and cAMP [13]. E_2 can act via promotion of the appearance of gap junctions within the cell membrane, an effect that is antagonized by P_4 [64,85]. The gap junction synthesis-stimulating effect of E_2 [29] can be abolished by administration of an antiestrogen to rats in late

pregnancy; neither gap junctions nor the onset of labor is subsequently observed [38]. Taken together, all these informations indicate that E_2 facilitates the onset of labor. E_2 can be regarded as a conditioning substance that helps to create a scenario that is favorable to labor, but it does not play an active part in parturition itself.

The "progesterone block" theory of Csapo [19], characterizing the biological effect of this hormone as mainly reponsible for the quiescent state of uterine musculature, has never been verified in humans. Csapo studied rabbits and rats, and in these animals the corpus luteum is essential for the maintenance of pregnancy. Shortly before parturition, there is a sudden P_4 withdrawal and a concomitant E_2 increase. There is no fall in plasma P_4 concentrations near the end of gestation in the human. Despite the lack of this fall and the absence of any therapeutic effect of application of P_4 during pregnancy, this hormone must have a role in the onset of labor. P_4 antagonists can increase the myometrial responsiveness to stimulatory agents. Moreover, antigestagens can ripen the cervix [16,108].

The P_4 dominance during pregnancy prevents the uterus from forming *gap junctions* composed of Cx. In rats near term in pregnancy, P_4 blocks both the expected increase in Cx-43 transcripts and the onset of labor. The administration of antigestagens abolishes these effects [85]. P_4 may also play a part in the cellular process of "constructing" gap junctions. Rats treated with P_4 several days before expected parturition show an accumulation of Cx-43 within the cytoplasma of myometrial cells and a delay in the passage of the protein to cell membranes for assembly in gap junctions [37]. Besides these antagonizing effects, P_4 also promotes myometrial relaxation. It has been reported that P_4 increases transcription of the β-adrenergic receptor in the rat myometrium [115].

Despite all the evidence of the importance of P_4 for the maintenance of pregnancy and of local P_4 withdrawal for the establishment of regular contractions, evidence for involvement of this hormone in parturition is weak. Most studies of P_4 concentrations in pregnancy have shown a continuing rise right up to the onset of labor [3].

Relaxin seems to have a physiological role in parturition, mainly in those species in which softening and separation of the pubic symphysis is necessary for delivery. In humans, relaxin has been shown to inhibit uterine contraction [110]. It is known to soften the human cervix [72]. One case has been reported in which the level of relaxin during labor was not normal: a failure of cervical maturation was associated with non-progressive contraction and the infant had to be delivered by cesarean section [23]. A fall in relaxin levels is thought to promote parturition, and indeed relaxin levels in the human fall from the first to the third trimester, but rise again during labor [109]. Although its existance in humans is generally accepted, the function of relaxin remains speculative.

Various *effects of catecholamines* on the human myometrium are known. Epinephrine has been reported to inhibit uterine contractions, to increase them, and to diminish or increase them, depending on the dose [52] and the dominance of the various receptors. In high concentrations epinephrine decreases uterine activity. The labor-inhibiting action of epinephrine is known to be mediated via β_2-adrenergic receptors on the myometrial cell, increasing cAMP and hyperpolariting the cell membrane potential [109]. The excitatory epinephrine and norepinephrine effects are attributed to the stimulation of α-adrenergic receptors. According to our own results [88], both catecholamines stimulate contractions of pregnant myometrial strips. In late pregnancy there might be a gradual change in the threshold of myometrial adrenergic receptors, with consequent predominance of the α-adrenergic excitatory reffects.

The dose-dependent stimulatory catecholamine effects on human myometrial contractions in vitro was correlated to an increase of $PGF_{2\alpha}$ *production* of the myometrial strips. The 6-keto-$PGF_{2\alpha}$/$PGF_{2\alpha}$ ratio decreased significantly.

The result of simultaneous or isolated superfusion of pregnant human myometrial strips with specific α- and β-adrenergic stimulators or blockers is, briefly, that α-adrenergic stimulation increases myometrial contractility and the synthesis of $PGF_{2\alpha}$ and PGE_2. Beta adrenergic stimulation reduces contractility by further enhancing 6-keto–$PGF_{1\alpha}$ production.

The results concerning OT as an endocrine stimulator of uterine contractions are controversial. This hormone is released from the hypothalamo-hypophyseal system in a pulsatile fashion [14]. The release is regulated by different reflexes. The "let-down" reflex in response to nipple stimulation is very important in breastfeeding (cf. Chap. 121). Ferguson described an increase in OT-dependent uterine contractility as a reflex of stretching of the lower genital tract. None of these mechanisms is thought to play any part in the onset of labor, although a kind of *Ferguson reflex* may be important in the second stage of labor [31]. The maternal levels of OT change little during parturition, but at term OT is capable of inducing uterine contractions in very low concentrations (cf. Fig. 120.9).

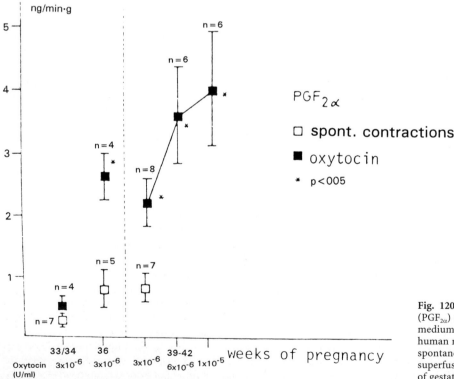

Fig. 120.9. Mean prostaglandin $F_{2\alpha}$ ($PGF_{2\alpha}$) concentrations in the effluent medium of superfused pregnant human myometrial strips during spontaneous cantractions and superfusion of oxytocin in dependency of gestational age [89]

It is considered today that the major factor accounting for the increasing effect of OT at the end of pregnancy is the increased number of *myometrial OT receptors*. Recently the human OT receptor has been cloned (388-amino-acid polypeptide) [57]. It seems to be G-protein coupled. Very high levels of mRNA were detected during labor. When binding of OT to its receptor is established, a cascade of biochemical events will be initiated. IP_3 turnover, intracellular Ca^{2+} and finally MLC phosphorylation increase [71] (cf. Fig. 120.5). OT also stimulates the production of PGs from human pregnancy-specific tissues [26], an effect that is blocked by OT antagonists [2]. We studied the effect of acetylsalicylic acid and indomethacin on OT-induced contractility and PG production of human myometrial strips. Superfusion of OT increases myometrial contractions and $PGF_{2\alpha}$ release in a dose-dependent manner [89]. Moreover, the stimulatory effect was significantly higher in pregnancy at term (Fig. 120.9). However, 6-keto-$PGF_{1\alpha}$ production was not affected by OT. After preincubation with acetylsalicylic acid or indomethacin the myometrial strips showed little spontaneous activity, and the PG production was below the detection limit. The stimulating effect of OT on the contractility and $PGF_{2\alpha}$ synthesis of the myometrial strips was inhibited significantly. We saw a recovery of spontaneous myometrial activity and an increase of OT-stimulated $PGF_{2\alpha}$ release over a period of 2h. Our results indicate that spontaneous and OT-stimulated contractility and PG release of human myometrial strips can be inhibited by acetylsalicylic acid and indomethacin and that this effect is reversible. Furthermore, the inhibition of $PGF_{2\alpha}$ production by acetylsalicylic acid and indomethacin was more pronounced than that of 6-keto-$PGF_{1\alpha}$ [89].

The above data suggest that OT can act directly on the myometrium as well as indirectly through complex interactions with PGs.

Depending on the gestational age, the *fetus* also produces increasing amounts of *OT* [14]. The arteriovenous difference in the OT levels in umbilical vessels is very pronounced after spontaneous deliveries and smallest following cesarean sections without labor. However, spontaneous deliveries of anencephalic fetuses can occur without a detectable OT level in the umbilical circulation. This suggests that fetal OT cannot be a "conditio sine qua non" for parturition [14].

120.3.6 Paracrine/Autocrine Control of Labor

The main paracrine regulators involved in human myometrial contraction are endogenous *AA metabolites* (Fig. 120.10; cf. Chap. 6). During labor, changes in PG production are seen. Although rises in PG output clearly precede labor in animals, in women it is only with progressive dilatation of the cervix that an increase of PGE and PGF in the amniotic fluid [54] and peripheral plasma can be measured [123]. Before labor the concentration of PGE in amniotic fluid is higher than that of PGF. During labor the relative increase in PGF is more marked

than that of PGE [21]. PGI_2 does not change during labor [81].

Innumerable publications have discussed the physiological and pathophysiological roles of AA metabolites (*eicosanoids*) in pregnancy and parturition. Most of these studies focus on prostaglandins, which are easily measurable by immunological assays. Because of their pharmocological effects on the cervix and the myometrium, PGE_2 and $PGF_{2\alpha}$ are well known as effective drugs for cervical priming, labor induction and induction of long-lasting uterine contractions for the treatment of post-partum uterine atonia [124]. Little is known about lipoxygenase (LO) metabolites of AA. Levels of these compounds are elevated during labor in amniotic fluid, while those of COX products remain unchanged [116]. One study has even demonstrated that AA metabolism preferentially proceeds through the LO pathway until term. Just before the beginning of labor there is a progressive switch to COX products [101]. We have examined the production of PGs and hydroxy eicosatetraenoic acids (HETE; cf. Chap. 6) by pregnancy-specific human tissues in a short-term culture system. In the supernatant, PGE_2, $PGF_{2\alpha}$, 6-keto-$PGF_{1\alpha}$ and thromboxan B_2 (TxB_2) were measured with RIA, and 15-, 12-, and 5-HETE with HPLC and UV detection. The incubation was performed only with freshly obtained tissue, because the eicosanoid production by frozen tissue was markedly changed [118]. The main AA metabolite in all tissue incubations was 12-HETE. Decidua produced 12–28 times as much PG as placenta and fetal membranes. The main COX derivative measured from placenta and fetal membrane incubations was TxB_2. After labor, fetal membranes showed an increase in fetal PGs (significant for PGE_2) and a decrease in HETE synthesis [117]. In a further study with [^{3}H]AA, the relative conversion rates to eicosanoids after Ca^{2+}-ionophore (A_{23187}) stimulation were 44% in decidua, 23% in fetal membranes and 20% in placental tissue. Again, 12-HETE was produced as a main metabolite of AA in all investigated tissue samples. PGs and LTB_4 were formed only in minor quantities. A prominent HPLC peak cochromatographed with the cytochrome P450 metabolite 14,15-epoxyeicosatrienoic acid (EET). Stimulation of the tissues with A_{23187} or with the cytokines IL-1β or TNF did not profoundly change the patterns of AA metabolism.

We conclude that the role of LO and cytochrome P450 metabolites of AA in parturition is as yet underestimated. It is an open question whether they exert direct effects on the myometrium or whether they have a regulatory role (e.g., on the COX pathway or the immune balance) [104]. Concerning the role of eicosanoids for the beginning of labor, our results support the view that the decidua at the junction of fetal and maternal compartment is the tissue where "the action is." This is in accordance with other data [101] and clearly demonstrates that at term there is a shift in AA metabolism from HETE to PGs and this shift may well be needed for the initiation of labor.

A further fascinating new aspect for the understanding of the birth process is based on the recent findings on *cyclooxygenases* COX-1 and COX-2. It has been shown that

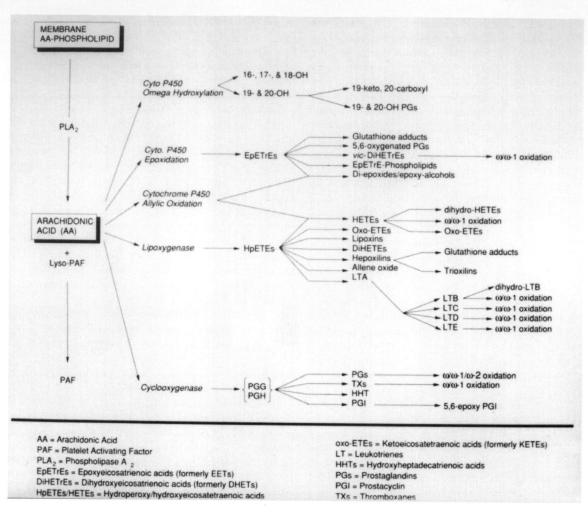

Fig. 120.10. The arachonic acid cascade

human amnion and decidua produce measurable amounts of COX-1 and COX-2 mRNA. The COX-2 mRNA levels were markedly higher following spontaneous labor than after elective cesarean section at term, in both amnion and decidua. There was no difference in COX-1 mRNA. COX-2 mRNA levels in fetal membranes were also positively correlated with COX-2 enzyme activity [41]. These results support the view that it is mainly COX-2 that is involved in labor. The expression of mRNA for COX-1 and COX-2 does not seem to be influenced by OT, E_2 or P_4. But it is known that OT causes an increase of the uptake of AA into the membrane phospholipids [112]. These results suggest that OT may also be involved in a preparatory step before the beginning of labor.

Although the human amnion contains predominantly enzymes capable of producing large amounts of uterus-contracting metabolites, the role of fetal membranes in the initiation or prevention of parturition remains to be defined. Recently it has been shown that *human fetal membranes*, decidua parietalis, and in particular the chorion laeve have a high metabolic capacity for degradation of PGs, OT, ET1 and platelet-activating factor (PAF; cf. Chap.

6). The specific activities of the degrading enzymes, PG-dehydrogenase (PGDH), oxytocinase and enkephalinase, in chorion laeve tissues obtained before the onset of labor were not significantly different from those in chorion laeve tissues obtained after the beginning of labor [30]. These results are in accordance with data from other workers [103], but they conflict with findings in patients who go into labor before term without having any infection. In this particular group, low scores for PGDH were correlated with low PGDH activity and with low or undetectable PGDH mRNA transcripts. Hence, a deficiency or reduced expression of the PGDH gene in the membranes of some patients may lead to unrestrained availability of biologically active PGs in intrauterine tissues, thereby predisposing to preterm delivery [103].

Finally, it is still not known whether labor is the result of an increase of stimulatory PGs or whether they are the result of the activity of degrading enzymes in pregnancy-specific tissues.

AA metabolites are elevated in the amniotic fluid of patients who go into preterm labor associated with chorioamnionitis or low-grade infection. The site of in-

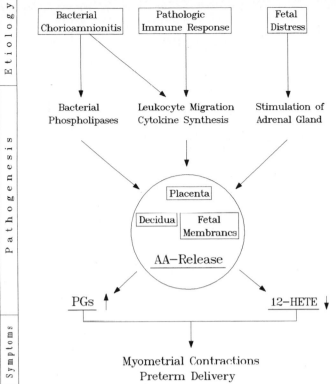

Fig. 120.11. Pathophysiology of preterm labor. The involvement of arachidonic acid (*AA*) metabolites, prostaglandins (*PGs*) and hydroxyeicosatetraenoic acid (*HETE*), representing the cyclooxygenase and lipoxygenase pathways of AA metabolism, respectively

creased eicosanoid production is probably the intrauterine tissue. Thus, *infection-associated preterm labor* may be accompanied by increases in PG output. The cause of the increased PG synthesis can be:

- Bacterial phospholipases (PL) [60]
- Bacterial lipopolisaccharides (LPS) [98]
- Cytokines such as IL-1 and TNF [20] from amnion and decidual cells.

These factors (Fig. 120.11) lead to increased uterine activity via stimulation of PG synethesis. Further, IL-1 stimulates the output of other cytokines (IL-6, IL-8), a chemotactic peptide for neutrophils, and T cells from decidua, thereby setting up a positive cytokine-PG cascade. Cytokines in the decidua are derived mainly from macrophages, which in turn stimulate PG output from pregnancy-specific tissues [13]. OT and other myometrium-stimulating substances (ET, EGF, PAF) are also released from decidua, placenta and fetal membranes, triggered by cytokines [12,70]. Our knowledge about the role of eicosanoids in physiological spontaneous labor at term is dominated by findings recorded in patients with infections causing preterm labor (cf. Fig. 120.11), which has led to an unbalanced view of this problem. This is why further work is required to achieve correct characterization of these relationships in tissues.

In the early 1980s PG receptor distribution in human myometrium was examined [5]. Regional differences were demonstrable: the greatest number of PGE_2-binding sites was found in the central part of the uterus. $PGF_{2\alpha}$ was bound with lower affinity [5]. There is now increasing evidence of *multiple PG receptors*. It is necessary to differentiate between excitatory FP and TP receptors and inhibitory DP and IP receptors in the human myometrium [105]. There are regional differences in responsiveness to PGE_2 and $PGF_{2\alpha}$ in the human myometrium [119], possibly because of topically different distribution of the various PG receptors [5].

Prostaglandin action is mediated by specific receptors. However, little is known about the mechanisms that mediate the action of PGs at the post receptor level within the myometrium. $PGF_{2\alpha}$, in contrast to OT, increases the Ca^{2+} influx. $PGF_{2\alpha}$ does not influence the IP_3 turnover. OT stimulates the intracellular Ca^{2+} release from the SR; $PGF_{2\alpha}$ may increase MLC phosphorylation and hence activation of the contractile machinery [13]. Some eicosanoids might enhance gap junction formation, whereas others, such as PGI_2, might block this process [29]. Finally, and very important for the clinician, PGs interact additively with the OT system to enhance myometrial contraction.

References

1. Amédée T, Mironneau C, Mironneau J (1987) The calcium channel current of pregnant rat single myometrial cells in short term primary culture. J Physiol (Lond) 392:253–272

2. Andersen LF, Lyndrup J, Akerlund M, Melin P (1989) Oxytocin receptor blockade: a new principle in the treatment of preterm labor? Am J Perinatol 6:196–199

3. Anderson A (1980) The genital system. In: Hytten F, Chamberlain G (eds) Clinical physiologic obstetrics. Blackwell Scientific, Oxford, pp 328–380

4. Andersson KE, Forman A, Ulmsten U (1983) Pharmacology of labor. Clin Obstet Gynecol 26:56–77

5. Bauknecht T, Krahe B, Rechenbach U, Zahradnik HP, Breckwoldt M (1981) Distribution of prostaglandin E_2 and prostaglandin $F_{2\alpha}$ receptors in human myometrium. Acta Endocrinol (Copenh) 98:446–450

6. Bengtsson B, Chow EMH, Marshall JM (1984) Activity of circular muscle of rat uterus at different times in pregnancy. Am J Physiol 246:C216–C223

7. Bennett PR, Rose MP, Myatt L, Elder MG (1987) Preterm labor: stimulation of arachidonic acid metabolism in human amnion cells by bacterial products. Am J Obstet Gynecol 156:649–655

8. Beyer EC, Paul DL, Goodenough J (1987) Connexin 43: a protein from heart homologous to a gap junction protein from liver. Cell Biol 105:2621–2629

9. Beyer EC, Kistler J, Paul DL, Goodenough J (1989) Antisera directed against connexin 43 peptides react with a 43-KD protein localized to gap junctions in myocardium and other tissues. J Cell Biol 108:595–605

10. Blennerhassett MG, Garfield RE (1991) Effect of gap junction number and permeability on intercellular coupling in rat myometrium. Am J Physiol 261:C1001–C1009

11. Bryant-Greenwood GD (1991) The human relaxins: consensus and dissent. Mol Cell Endocrinol 79:C125

12. Casey ML, Cox SM, Ward RA (1990) Cytokines and infection induced preterm labor. Reprod Fertil Dev 2:499–505

13. Challis JRG, Lye SJ (1994) Parturition. In: Knobil E, Neil JD (eds) The physiology of reproduction, 2nd edn. Raven, New York, pp 985–1031

14. Chard T (1989) Fetal and maternal oxytocin in human parturition. Am J Perinatol 6:145–152

15. Chwalisz K, Hegele-Hartung C, Schulz R, Shi Shao Qing, Louton PT, Elger W (1991) Progesterone control of cervical ripening – experimental studies with the progesterone antagonists onapristone, litopristone and mifepristone. In: Leppert P, Woessner F (eds) The extracellular matrix of the uterus, cervix and fetal membranes: synthesis degradation and hormonal regulation. Perinatology, Ithaca, New York, pp 119–131

16. Chwalisz K (1993) Role of progesterone in the control of labor. In: Chwalisz K, Garfield RE (eds) Basic mechanisms controlling term and preterm birth. Springer, Berlin Heidelberg New York, pp 97–162

17. Cole WC, Garfield RE (1988) Effects of calcium Ionophore, A 23187, and calmodulin antagonists on cell-to-cell communication between rat myometrial smooth muscle cells. Biol Reprod 38:55–62

18. Cole WC, Garfield RE (1989) Ultrastructure of the Myometrium. In: Wynn RM, Jollie WP, Biology of the uterus, second edition. Plenum, New York, pp 455–504

19. Csapo AI (1975) The "seesaw" theory of regulatory mechanisms of pregnancy. Am J Obstet Gynecol 121:578–581

20. Diaz A, Reginato AM, Jimenez SA (1992) Alternative splicing of human prostaglandin G/H synthase mRNA and evidence of differential regulation of the resulting transcripts by transforming growth factor β1, interleukin 1β, and tumor necrosis factor α. J Biol Chem 15:10816

21. Dray F, Frydman R (1976) Primary prostaglandins in amniotic fluid in pregnancy and spontaneous labor. Am J Obstet Gynecol 126:13–18

22. El-Refaey H, Calder L, Wheatley DN, Templeton A (1994) Cervical priming with Prostaglandin E_1 analogues, misoprostol and gemeprost. Lancet 343:1207–1209

23. Entenmann AH, Seeger H, Voeter W, Lippert TH (1988) Relaxin deficiency in the placenta as a possible cause of cervical dystocia. Clin Exp Obstet Gynecol 15:13–17

24. Figueroa JP, Honnebier MBOM, Binienda Z, Wimsa HA, Nathanielsz PW (1989) Effect of 48 hours intravenous 4A androstenedione infusion on the pregnant rhesus monkey during the last third of gestation: changes in maternal plasma estradiol concentration and myometrial contractility. Am J Obstet Gynecol 161:481–486

25. Fleckenstein A (1977) Specific pharmacology of calcium ion in myocardium, cardiac pacemakers, and vascular smooth muscle. Annu Rev Pharmacol Toxicol 17:149–166

26. Fuchs AR, Fuchs F, Husslein P, Soloff MS, Fernstrom MJ (1982) Oxytocin receptors and human parturition: a dual role of oxytocin in the initiation of labor. Science 215:1396

27. Garfield RE, Sims S, Daniel EE (1977) Gap junctions: their presence and necessity in myometrium during gestation. Science 198:958–960

28. Garfield RE, Puri CP, Csapo Al (1982) Endocrine structural and functional changes in the uterus during premature labor. Am J Obstet Gynecol 142:21–27

29. Garfield RE, Yallampalli Ch (1993) Control of myometrial contractility and labor. In: Chwalisz K, Garfield RE (eds) Basic mechanisms controlling term and preterm birth. Springer, Berlin Heidelberg New York, pp 1–28

30. Germain AM, Smith J, Casey ML, Mac Donald PC (1994) Human fetal membrane contribution to the prevention of parturition: uterotonin degradation. J Clin Endocrinol Metab 78:463–470

31. Goodfellow CF, Hall MGR, Swaab DF (1983) Oxytocin deficiency at delivery with epidural analgesia. Br J Obstet Gynaecol 90:214–219

32. Granström LM, Gunvor EE, Malmström A, Ulmsten U, Woessner JF (1992) Serum collagenase levels in relation to the state of the human cervix during pregnancy and labor. Am J Obstet Gynecol 167:1284–1288

33. Grover AK (1985) Ca-pumps in smooth muscle: one in plasma membrane and another in endoplasmic reticulum. Cell Calcium 6:227

34. Grover AK, Khan I, Tabb T, Garfield RE (1993) Role of uterine Ca^{2+} pumps and Na^+ pumps in labor: a moleucular biology approach. In: Chwalisz K, Garfield RE (eds) Basic mechanisms controlling term and preterm birth. Springer, Berlin Heidelberg New York, pp 75–88

35. Guzick DS, Winn K (1985) The association of chorioamnionitis with preterm delivery. Obstet Gynecol 65:11–16

36. Hansell DJ, Bryant-Greenwood GD, Greenwood FC (1991) Expression of the human relaxin H1 gene in the decidua, throphoblast and prostate. J Clin Endocrinol Metab 72:899–904

37. Hendrix EM, Mao SJ, Everson W, Larsen WJ (1992) Myometrial connexin-43 trafficking and gap junction assembly at term and in preterm labor. Mol Reprod Dev 33:27–31

38. Hendrix EM, Myatt L, Larsen WJ (1993) Treatment of ovariectomized rats with antiestrogen I C I 182780 prevents myometrial gap junction formation, and premature labor. Annual meeting of the Society of Gynecological Investigation, abstract P26

39. Herrera VLM, Emanuel JR, Ruiz-Opazo N, Levensen R, Nadal-Ginard N (1987) Three differentially expressed NA, K-ATPase subunit isoforms: structural and functional implications. J Cell Biol 105:1855–1865

40. Higashi K, Fukumaga K, Matsui K, Miyamoto E (1983) Purification and characterization of myosin light-chain kinase from porcine myometrium and its phosphorylation and modulation by cyclic AMP-dependent protein kinase. Biochem Biophys Acta 747:232–238

41. Hirst JJ, Teixeira FJ, Olson DM, Zakar T (1994) Expression of Prostaglandin endoperoxide H-synthase-2 increases with labor in human gestational tissue. Abstract of the 9th inter-

national conference on prostaglandins and related compounds, p 107

42. Honoré E, Martin C, Mironneau J (1989) An ATP-sensitive conductance in cultured smooth muscle cells from pregnant rat myometrium. Am J Physiol 257:C297–C309

43. Huszar G, Walsh MP (1989) Biochemistry of the myometrium and cervix. In: Wynn RM, Jollie WP (eds) Biology of the uterus, 2nd edn. Plenum, New York, pp 355–402

44. Huszar G, Walsh MP (1991) Relationship between myometrial and cervical functions in pregnancy and labor. Seminar Perinatol 15:109–117

45. Huszar G (1991) Physiology of the uterine cervix in reproduction. Semin Perinatol 15:95–108

46. Huxley HE (1971) The structural basis of muscular contraction. Proc R Soc Lond [Biol] 178:131–148

47. Inoue Y, Nakao K, Okabe K, Izumi H, Kanda S, Kitamura K, Kuriyama H (1990) Some electrical properties of human pregnant myometrium. Am J Obstet Gynecol 162:1090–1098

48. Inoue Y, Sperelakis N (1991) Gestational change in Na⁺ and Ca²⁺ channel current densities in rat myometrial smooth muscle cells. Am J Physiol 260:C658–663

49. Ito A, Hiro D, Hioro D, Ojima Y, Mori Y (1988) Spontaneous production of interleukin 1-like factors from pregnant rabbit uterine cervix. Am J Obstet Gynecol 159:261–265

50. Joh K, Riede UN, Zahradnik HP (1983) The effect of prostaglandins on the lysosomal function in the cervix uteri. Arch Gynecol 234:1–16

51. Junqueira LCU, Zugaib M, Montes GS, Toledo OMS, Krisztan RM, Shigihara KM (1980) Morphologic and histochemical evidence for the occurence of collagenolysis and for the role of neutrophilic polymorphonuclear leukocytes during cervical dilatation. Am J Obstet Gynecol 138:273–281

52. Kaiser JH, Harris JS (1950) The effect of adrenalin on the pregnant human uterus. Am J Obstet Gynecol 59:775–784

53. Kao CY (1989) Elektrophysiological properties of uterine smooth muscle. In: Wynn RM, Jollie WP (eds) Biology of the uterus, 2nd edn. Plenum, New York, pp 403–454

54. Keirse MJNC, Turnbull AC (1973) E prostaglandins in amniotic fluid during late pregnancy and labour. J Obstet Gynaecol Br Commonw 80:970–978

55. Kelly RW, Leask R, Calder AA (1992) Choriodecidual production of interleukin-8 and the mechanism of parturition. Lancet 339:776–777

56. Kieback DG, Zahradnik HP, Quaas L, Kröner-Fehmel EE, Lippert TH (1986) Clinical evaluation of endocervical prostaglandin E₂-triacetin-gel for preinduction cervical softening in pregnant women at term. Prostaglandins 32:81–85

57. Kimura T, Tanizawa O, Mori K, Brownstein MJ, Okayama H (1992) Structure and expression of a human oxytocin receptor. Nature 356:526–529

58. Klee CB (1980) Calmodulin: structure-function relationships. In: Cheung WY (ed) Calcium and cell function, vol 1. Academic, New York, pp 59–78

59. Knudsen UB, Fredeus K, Ulbjerg N (1991) Inflammatory cells in the cervix and their role during pregnancy. In: Leppert PC, Woessner F (eds) The extracellular matrix of the uterus, cervix and fetal membranes: synthesis, degradation and hormonal regulation. Perinatology, Ithaca, New York, pp 141–145

60. Lamont RF, Anthony F, Myatt L (1990) Production of Prostaglandin E₂ by human amnion in vitro in response to addition of media conditioned by microorganisms associated with chorioamnionitis and preterm labor. Am J Obstet Gynecol 162:819–826

61. Lang LM, Beyer EC, Schwartz AL, Gitlin JD (1991) Molecular cloning of a rat uterine gap junction protein and analysis of gene expression during gestation. Am J Physiol 260:E787–E793

62. Ledger WL, Webster M, Harrison LP, Anderson ABM, Turnbull AC (1985) Increase in cervical extensibility during labor induced after isolation of the cervix from the uterus in pregnant sheep. Am J Obstet Gynecol 151:397–402

63. Lee AB, Hwang JJ, Haab LM, Fields PA, Sherwood OD (1992) Monoclonal antibodies specific for rat relaxin. VI. Passive immunization with monoclonal antibodies throughout the second half of pregnancy disrupts histological changes associated with cervical softening at parturition in rats. Endocrinology 130:2386–2391

64. Lefebre DL, Reaume A, Bai K-II, Lye SJ (1993) Regulation of myometrial connexin-43 expression: characterization of promoter elements. Annual meeting of the Endocrinological Society, abstract 1548

65. Leppert PC (1992) Cervical softening, effacement and dilation: a complex biochemical cascade. J Mat Fet Med 1:213–223

66. Liggins GC (1981) Cervical ripening is an inflammatory reaction. In: Ellwood DA, Anderson ABM (eds) The cervix in pregnancy and labour. Churchill Livingston, Edinburgh, pp 1–9

67. Liggins GC, Wilson T (1989) Phospholipases in the control of human partuition. Am J Perinatol 6:153–158

68. Lye SJ, Freitag CL (1990) Local and systemic control of myometrial contractile activity during labour in the sheep. J Reprod Fertil 90:483–492

69. Lye SJ, Nicholson BJ, Mascarenhas M, Mackenzie L, Petrocelli T (1993) Increased expression of connexin-43 in the rat myometrium during labour is associated with an increase in the plasma estrogen: progesterone ratio. Endocrinology 132:2380–2386

70. Mac Donald PC, Koga S, Casey ML (1991) Decidual activation in parturition: examination of amniotic fluid for mediators of the inflammatory response. Ann NY Acad Sci 622:315–324

71. Mac Kenzie LW, Word MA, Casey ML, Stull JT (1990) Myosin light chain phosphorylation in human myometrial cells. Am J Physiol 258:C92–C98

72. Mac Lennan AH, Green RC, Bryant-Greenwood GD, Greenwood FC, Seamark RF (1980) Ripening of the human cervix and induction of labour with purified porcine relaxin. Lancet i:220

73. Mac Lennan AH, Grant P, Borthwick AC (1991) Relaxin and relaxin C-peptide levels in human reproductive tissues. Reprod Fertil Dev 3:577–583

74. Marbaix E, Donnez J, Courrtoy PJ, Eeckhout Y (1992) Progesterone regulates the activity of collagenase and related gelatinases A and B in human endometrial explants. Proc Natl Acad Sci USA 89:11789–11793

75. Martin C, Arnaudeau S, Jmari K, Rakotoarisoa L, Sayet I, Dacquet C, Mironneau J (1990) Identification and properties of voltage-sensitive sodium channels in smooth muscle cells from pregnant rat myometrium. Mol Pharmacol 38:667–673

76. Masferrer JL, Zweifel B, Manning PT et al (1994) The role of cyclooxygenase-2 in inflammation. Abstract of the 9th international conference on prostaglandins and related compounds, p 81

77. Mc Millan RM, Fahey JV, Brinckerhoff CE, Harris ED Jr (1980) Secretion of inflammatory mediators from synovial fibroblasts: dissociation of collagenase and prostaglandin release. Adv Prostaglandin Thromboxane Res 8:1701–1704

78. Mironneau J (1973) Excitation-contraction coupling in voltage-clamped uterine smooth muscle. J Physiol (Lond) 233:127–141

79. Mironneau J (1976) Effects of oxytocin on ionic currents underlying rhythmic activity and contraction in uterine smooth muscle. Pflugers Arch 363:113–116

80. Mironneau J (1993) Myometrial electrophysiology. In: Chwalisz K, Garfield RE (eds) Basic mechanisms controlling term and preterm birth. Springer, Berlin Heidelberg New York, pp 55–73

81. Mitchell MD (1986) Pathways of arachidonic acid metabolism with specific application to the fetus and mother. Semin Perinatol 10:242–256

82. Nishikori K, Weisbrodt NW, Sherwood OD, Sanborn BM (1983) Effects of relaxin on rat uterine myosin light chain kinase activity and myosin light chain phosphorylation. J Biol Chem 258:2468–2474

83. Omini C, Folco GC, Pasargiklian R, Fano M, Berti F (1979) Prostacyclin (PGI$_2$) in pregnant human uterus. Prostaglandins 17:113–122

84. Parks TP, Nairn AC, Greengard P, Jamieson JD (1987) The cyclic nucleotide-dependent phosphorylation of aortic smooth muscle membrane protexin. Arch Biochem Biophys 255:361–369

85. Petrocelli T, Lye SJ (1993) Regulation of transcripts encoding the myometrial gap junction protein, connexin-43, by estrogen and progesterone. Endocrinology 133:284–290

86. Popescu LM, Nutu O, Panoin C (1985) Oxytocin contracts the human uterus at term by inhibiting the myometrial Ca^{2+}-extrusion pump. Biosci Rep 5:21–28

87. Porter DG (1981) Relaxin and cervical softening. A review. In: Ellwood DA, Anderson ABN (eds) The cervix in pregnancy and labour. Churchill Livingstone, Melbourne, p 385

88. Quaas L, Zahradnik HP (1985) The effects of α- and β-adrenergic stimulation on contractility and Prostaglandin (PGE$_2$, PGF$_{2\alpha}$, 6-Keto-PGF$_{1\alpha}$) production of pregnant human myometrial strips. Am J Obstet Gynecol 152:852–856

89. Quaas L, Göppinger A, Zahradnik HP (1987) The effect of acetylsalicylic acid and indomethacin on the catecholamine- and oxytocin induced contractility and Prostaglandin (6-Keto-PGF$_{1\alpha}$, PGF$_{2\alpha}$)-production of human pregnant myometrial strips. Prostaglandins 34:257–269

90. Quinn PA, Butany J, Taylor J, Hannah W (1987) Chorioamnionitis: its association with pregnancy outcome and microbial infection. Am J Obstet Gynecol 156:379–387

91. Radestadt A, Bygdeman M, Green K (1990) Induced cervical ripening with mifepristone (RU 486) and bioconversion of arachidonic acid in human pregnant uterine cervix in the first trimester. Contraception 41:283–291

92. Radestadt A, Bygdeman M (1992) Cervical softening with mifepristone (RU 486) after pretreatment with naproxen. A double-blind randomized study. Contraception 45:221–227

93. Rajabi MR, Dean DD, Beydoun SN, Woessner JF Jr (1988) Elevated tissue levels of collagenase during dilation of uterine cervix in human parturition. Am J Obstet Gynecol 159:971–976

94. Rajabi M, Solomon S, Poole R (1991) Hormonal regulation of interstitial collagenase in the uterine cervix of pregnant guinea pigs. Endocrinology 128:863–871

95. Rapoport RM, Murad F (1983) Endothelium dependent and nitrovasodilator-induced relaxation of vascular smooth muscle: role of cyclic GMP. J Cyclic Nucleotide Protein Phosphor Res 9:281–290

96. Rivera J, Lopez-Bernal A, Varney M, Warson SP (1990) Inositol 1,4,5-triphosphate and oxytocin binding in human myometrium. Endocrinology 127:155–165

97. Roberts JM (1984) Current understanding of pharmacologic mechanisms in the prevention of preterm birth. Clin Obstet Gynecol 27:592

98. Romero R, Hobbins JC, Mitchell MD (1988) Endotoxin stimulates prostaglandin E$_2$ production by human amnion. Obstet Gynecol 71:227–236

99. Romero R, Ceska M, Avila CA, Mazor M, Behnk E, Lindley J (1991) Neutrophil attractant/activating peptide-1/interleukin-8 in term and preterm parturition. Am J Obstet Gynecol 165:813–820

100. Romero R, Mazor M, Brandt F (1992) Interleukin-1 and interleukin-1β in preterm and term human parturition. Am J Reprod Immunol 27:117–123

101. Rose MP, Myatt L, Elder MG (1990) Pathways of arachidonic acid metabolism in human amnion cells at term. Prostalandins Leuko Essent Fatty Acids 39:303–309

102. Sakbun V, Ali SM, Greenwood FC, Bryant-Greenwood GD (1990) Human relaxin in the amnion, chorion, decidua parietalis, basal plate, and placental throphoblast by immunocytochemistry and northern analysis. J Clin Endocrinol Metab 70:508–512

103. Sangha R, Walton J, Tai H-H, Challis JRG (1993) Immunohistochemical localization and mRNA abundance of 15 hydroxyprostaglandin dehydrogenase in placenta and fetal membranes during term and preterm labour. Proceedings of the Society of Gynecological Investigation, abstract S99

104. Schäfer W, Arbogast, E, Wetzka B, Suhr S, Zahradnik HP (1994) Arachidonic acid metabolism of human intrauterine tissues at term. Abstract of the 9th international conference on prostaglandins and related compounds, p 106

105. Senior J, Sangha R, Baxter GS, Marshall K, Calyton JK (1992) In vitro characterization of prostanoid FP-, DP-, IP- and TP-receptors on the nonpregnant human myometrium. Br J Pharmacol 107:215–221

106. Seron-Ferre M, Prado J, Germain AM (1991) Circadian rhythms and delivery in the human. In: Negro-Vielar A, Perez-Palacias G (eds) Reproduction, growth and development. Raven, New York, pp 59–67

107. Sokol RJ, Brindley BA (1990) Practical diagnosis and management of abnormal labor. In: Scott JR, Di Saia PJ, Hammond CB, Spellacy WN (eds) Danforth's obstetrics and gynecology, 6th edn. Lippincott, Philadelphia, pp 585–637

108. Soloff MS (1989) Endocrine Control of Parturition. In: Wynn RM, Jollie WP (eds) Biology of the uterus. Plenum Medical, New York, pp 559–607

109. Steer PJ (1990) The endocrinology of parturition in the human. Baillieres Clin Endocrinol Metab 4:333–349

110. Szlachter N, O'Byrne E, Goldsmith L, Steinetz B, Weiss G (1980) Myometrial inhibiting activity of relaxin-containing extracts of human corpora lutea of pregnancy. Am J Obstet Gynecol 136:584–588

111. Thorburn GD, Challis JRG, Robinson JS (1977) Endocrine controle of parturition. In: Wynn RM (ed) Biology of the uterus. Plenum, New York, pp 653–732

112. Tochigi M, Tochigi B, Hayakawa S, Nagaoka M, Yoshinaga H, Hashimoto Y, Yoshida T, Satoh K (1994) The effects of oxytocin and steroid Hormones (E$_2$, P$_4$) on the uptake of AA in the fetal membranes on human parturition. Abstract of the 9th international conference on prostaglandins and related compounds, p 106

113. Uchiyama T, Ito A, Ikesue A, Nakagawa H, Mori Y (1992) Chemotactic factor in the pregnant rabbit uterine cervix. Am J Obstet Gynecol 167:1417–1422

114. Uldbjerg N, Ulmsten U, Ekman G (1983) The ripening of the human uterine cervix in terms of connective tissue biochemistry. Clin Obstet Gynecol 26:14–19

115. Vivat V, Cohen-Tannonji J, Revelli J-P (1992) Progesterone transcriptionally regulates the β$_2$-adrenergic receptor gene in pregnant rat myometrium. J Biol Chem 267:7975–7981

116. Walsh SW (1991) Evidence for 5-hydroxyeicosatetraenoic acid (5-HETE) and leukotriene C$_4$ (LTC$_4$) in the onset of labor. Ann NY Acad Sci 622:341–352

117. Wetzka B, Schäfer W, Scheibel M, Nüsing R, Zahradnik HP (1993) Eicosanoid production by intrauterine tissues before and after labor in short-term tissue culture. Prostaglandins 45:571–581

118. Wetzka B, Schäfer W, Breckwoldt M, Zahradnik HP (1994) Eicosanoid production by frozen tissue in vitro is markedly changed. Prostaglandins Leukot Essent Fatty Acids 50:303–304

119. Wikland M, Lindblom B, Wiqvist N (1984) Myometrial response to prostaglandins during labour. Gynecol Obstet Invest 17:131–138

120. Word RA, Stull JD, Kamm K, Casey ML (1990) Regulation of smooth muscle contractility: Ca^{2+} and myosin phosphorylation. In: Garfield RE (ed) Uterine contractility. Serono Symposia, Norwell MA, pp 43–52

121. Young RC, Herndon-Smith L (1991) Charakterization of sodium channels in cultured human uterine smooth muscle cells. Am J Obstet Gynecol 164:175–181

122. Zahn V (1984) Uterine contractions during pregnancy. J Perinat Med 12:107–116
123. Zahradnik HP, Geisthövel F, Weitzell R, Breckwoldt M (1976) Prostaglandin-F-Spiegel im menschlichen Plasma während der späten Schwangerschaft und während der Wehentätigkeit. Geburtsh Frauenheilkd 36:710–717
124. Zahradnik HP, Steiner H, Hillemanns HG, Breckwoldt M, Ardelt W (1977) Prostaglandin $F_{2\alpha}$ und 15-Methyl-Prostaglandin $F_{2\alpha}$-Anwendung bei massiven uterinen Blutungen. Geburtsh Frauenheilkd 37:493–495
125. Zahradnik HP, Quaas L, Kröner-Fehmel EE, Kieback DG, Lippert TH (1987) Medikamentöse Zervixreifung vor Oxytocin-Geburtseinleitung. Geburtsh Frauenheilkd 47:190–192
126. Zahradnik HP, Quaas L, Breckwoldt M (1988) Uterusatonie – Wandel der Behandlungsmethoden in den letzten 20 Jahren. In: Haller U, Kubli F, Husslein P (eds) Prostaglandine in Geburtshilfe und Gynäkologie. Springer, Berlin Heidelberg New York, pp 220–229
127. Zahradnik HP, Schäfer W, Nüsing R et al (1994) COX-1 and COX-2-concentration in pregnancy specific human tissue (unpublished results)
128. Zakar T, Hirst JJ, Dirani N, Olson DM (1994) Glucocorticoids stimulate the expression of the inducible prostaglandin endoperoxide H synthase (PGHS-2) in human amnion cells. Abstract of the 9th international conference on prostaglandins and related compounds, p 79
129. Ziche M, Morbictelli L, Parent A, Ledda F (1994) Nitric Oxide modulates angiogenesis elicited by Prostaglandin E_1 in rabbit cornea. Abstract of the 9th international conference on prostaglandins and related compounds, p 11
130. Zlatnik FJ (1990) Normal labor and delivery and its conduct. In: Scott JR, Di Saia PJ, Hammond CB, Spellacy WN (eds) Danforth's obstetrics and gynecology, 6th edn. Lippincott, Philadelphia, pp 161–188

121 Lactation

M. Breckwoldt, J. Neulen, and C. Keck

Contents

121.1 Introduction

Lactation can be regarded as the final phase of the reproduction cycle in mammals, since most mammalian species deliver their infants in an immature state before they are able to feed themselves. During lactation, the neonate advances in physiological independence until its gastrointestinal tract, kidney function and immune system have matured far enough to permit survival. Adequate lactation is therefore of crucial importance for survival of any mammalian species.

In most mammals the duration of lactation is well defined. In the human, however, the lactation period varies very widely, depending on traditional, social, ethnic, cultural, and individual factors. Epidemiological studies comparing growth curves and infant mortality and morbidity rates in breast-fed and bottle-fed babies clearly indicate that breast feeding for a period of 6 months provides the optimal time for adjustment to independent life. In general, breast feeding is associated with significantly improved development in terms of health state, growth, and physical performance of the neonate [16,36].

121.2 Mammogenesis

121.2.1 Development of the Mammary Gland

Successful lactation depends on various factors controlling breast development and its function. Development of the mammary gland in the human begins at about 8 weeks of fetal life with the formation of mammary buds consisting of a ball of epithelial cells extending into the underlying mesenchyme. During the second trimester of pregnancy each bud sprouts 15–20 branches, which form rudimentary mammary ducts [22]. During the third trimester high concentrations of fetal *prolactin* stimulate terminal differentiation of the *ductal cells*. Milk secretion by the infant following delivery is not uncommon.

Mammary growth is initiated at the time of puberty and is recognized as the first clinical sign of puberty. The female breast is composed of a tubulo-secretory system and a ductal system.

The secretory system with glandular structures (alveoli) contains secretory cells (alveolar cells) drained by small ducts (ductuli). These ducts join to give larger ducts (ductus lactiferi); 15–25 of these elements make up the mammary gland. Each alveolus is surrounded by a network of capillaries, myoepithelial elements, connective tissue and fat tissue (Fig. 121.1). Subcutaneous fat covering the lobulo-alveolar units is separated by Coopers's ligaments, which are composed of connective tissue. Since the breast is of cutaneous origin it shares the blood and nerve supply and the lymphatic drainage of the skin. Innervation is provided by sensory and autonomous nerves concentrated in the region of nipple and alveola susceptile to tactile stimuli. Stimulation of the nipple induces release of oxytocin and prolactin via a neurohumoral reflex. Both effects are essential for the maintenance of lactation.

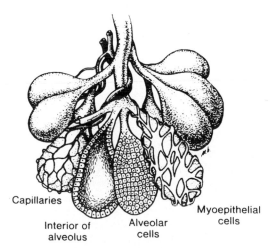

Capillaries
Interior of alveolus
Alveolar cells
Myoepithelial cells

Fig. 121.1. Diagram of the basic unit of the mammary gland. (From [10])

R. Greger/U. Windhorst (Eds.)
Comprehensive Human Physiology, Vol. 2
© Springer-Verlag Berlin Heidelberg 1996

121.2.2 Hormonal Control of the Development and Growth of the Mammary Gland

Mammary growth and development are dependent on sex steroids [14]. *Estrogens* induce rapid extension and branching of the duct system. Differentiation of the lobulo-alveolar system requires *progesterone*. Besides sex steroids, numerous other hormones and growth factors, such as *epidermal growth factor*, are also involved in breast development. Based on animal experiments and on clinical observations it can be assumed that estrogens, human prolactin (hPRL) and glucocorticoids are required for duct development. *Growth hormone* (hGH) seems to be of minor importance, since women with congenital absence of hGH are able to lactate [31]. In addition, progesterone, *thyroid hormones*, and *insulin* are also involved in the process of mammogenesis [2]. Insulin stimulates the cellular uptake of glucose. In the absence of glucose, insulin stimulates the incorporation of amino acids to form proteins. Thyroid hormones promote growth of the ductal system. in their absence duct development is retarded. In addition to its growth-promoting effect, *estradiol-17β* also stimulates lactose and casein formation, as shown in animal experiments [4]. This effect, however, requires the presence of insulin, cortisol, T3 and prolactin. Estradiol-17β is responsible for the induction of progesterone receptors, while progesterone inhibits secretory activity of alveolar epithelium, causing lactational quiescense during pregnancy [15].

In vitro experiments provide evidence that *insulin-like growth factor I* (IGF-I) promotes the growth and development of mammary tissue by para- and autocrine effects ([5]; cf. Chap. 6). Bovine mammary explants were shown to synthesize and secrete IGF-I and the corresponding binding proteins. Addition to hGH, hPRL, insulin and cortisol did not affect the secretion of either IGF-I or the binding proteins.

During the first half of pregnancy proliferation of alveolar epithelial cells and development of the duct system are stimulated, while during the second half of gestation the epithelium differentiates for secretory activity.

Since estradiol stimulates prolactin synthesis and release, rising peripheral concentrations of hPRL can be observed during pregnancy (Fig. 121.2; cf. Chap. 113). The increased prolactin secretion is reflected by a corresponding hyperplasia of the pituitary lactotrophs resulting in a two- to threefold increase in pituitary volume. Proactin is a proteohormone composed of 198 amino acids as a single polypeptide chain with three disulfide bonds, and it is identified as a unique hormone different from hGH. Structure and biological activity of hPRL are closely related to those of hGH and human placental lactogen (hPL). All three hormones are members of the growth hormone gene family [32]. The genes coding for the expression of hPRL, hGH and hPL are mapped on the long arm of chromosome 17. Prolactin is considered a phylogenetically ancient hormone, being present in fish, amphibia, birds and, of course, in mammals. In all species prolactin plays an important part in the reproductive process [3].

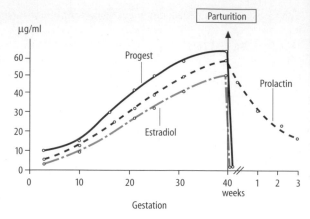

Fig. 121.2. Peripheral levels of progesterone, estradiol and prolactin during the course of pregnancy and after parturition

Besides the well-identified prolactin with 198 amino acids and a molecular weight of 23 kD, a big prolactin (45 kD) and a big-big prolactin with a molecular weight of 100 kD have also been described. The physiological significance of these di- and tetramer forms is unknown. The heterogeneity of human prolactin levels changes during the early postpartum period as the monomeric form of prolactin is increasing with progressing lactation [17]. The monomeric form is obviously the essential physiologic principle needed to stimulate and maintain lactogenesis. Synthesis and secretion of hPRL by the pituitary is under tonic inhibitory control of hypothalamic dopamine (Fig. 121.3). Hypothalamic *serotonin* and estradiol stimulate prolactin release. Estradiol acts by stimulating prolactin m-RNA production [20]. In addition, estrogens induce TRH receptors leading to increased sensitivity of the lactotrope, the prolactin-secreting cell of the pituitary gland. Furthermore, estradiol can be considered a potent antidopaminergic compound [29]. Dopaminergic neurons of the medio-basal *hypothalamus* project to the external layer of the *median eminence*, where they contact capillary loops of the *pituitary portal vessels*. Dopamine is then transported directly to the lactotropes binding to its specific receptors. Dopamine receptor stimulation results in inhibition of hPRL synthesis and release. Clinical and experimental studies have clearly shown that dopamine agonists such as bromocriptin or lisuride cause immediate prolactin suppression. Dopamine antagonists or any abnormality resulting in dopamine depletion consequently lead to hyperprolactinemia. Several psychopharmacological agents, e.g., such antidepressant drugs as phenothiazines, butyrophenones, benzamides (sulpiride, metaclopromide) induce hyperprolactinemia by blocking dopamine receptors. Hyperprolactinemia is associated with impaired LH pulsatility leading to ovarian insufficiency.

In contrast to the situation with other pituitary hormones, prolactin synthesis and secretion is controlled by a biogenic amine in an inhibitory fashion; the other hormones (FSH, LH, hGH, ACTH, TSH) are stimulated by hypothalamic peptide hormones. A specific prolactin-releasing factor has not been identified, but prolactin-releasing properties can be attributed to estradiol-17β, serotonin, and TRH. Therefore, hypothyroidism can be associated with hyperprolactinemia [28]. Adequate supplementation with thyroid hormones results in normalization of prolactin levels. During pregnancy the mammary glands are exposed to steadily increasing levels of hPL and prolactin. These hormones act in concert, leading to differentiation of the lobulo-alveolar elements. Estrogens and progesterone are also involved in complete mammogenesis, and in addition, these steroid hormones antagonize the terminal effects of prolactin. In the presence of high concentrations of sex steroid lactation does not occur.

121.3 Lactogenesis

Plasma levels of hPL, hPRL, estradiol-17β and progesterone reach their maximum at parturition. (Fig. 121.2). The concentration of circulating prolactin increases during pregnancy, until by the end of gestation levels are 20 times normal [26]. Within 48 h after delivery plasma concentrations of hPL, estradiol and progesterone decrease to nonpregnant values (Fig. 121.2). HPRL levels, however, remain elevated, decreasing gradually over a period of 3–4 weeks post partum in non-lactating women [23]. The declining progesterone level and the maintenance of elevated prolactin are the most important events in establishing lactogenesis.

121.3.1 The Hormones Inhibiting and Stimulating Lactogenesis

Progesterone is an inhibitory factor specifically hindering the onset of milk production. It does not inhibit established lactation, as actively lactating mammary tissue does not contain progesterone receptors. The minimal hormonal requirements for normal lactation to occur are

- Prolactin
- Insulin
- Cortisol.

Prolactin stabilizes and promotes transcription of casein m-RNA and the synthesis of alpha lactalbumin. Lactalbumin is a regulatory protein of the lactose synthetase enzyme system, which is essential for the formation of lactose the principal carbohydrate in human milk. In addition, prolactin increases lipoprotein lipase activity in the mammary gland [26]. Lactogenesis requires a fully developed mammary gland, which is readily achieved during the course of pregnancy. Milk secretion by the lobulo-alveolar system is absolutely dependent on the presence of prolactin; the withdrawal of hPRL by dopamine agonists results in immediate and complete ablactation. Therefore, hPRL is considered the key hormone in the lactation process.

Plasma concentrations of hPRL in fully lactating women are significantly higher in the first 3 postpartum weeks than 3 months after birth. The half-life of endogenous hPRL is unchanged from early to late postpartum period, and has been calculated as 26–29 min [24]. The onset of suckling results in an increase in serum prolactin concentrations almost immediately, followed by a decrease over the subsequent hour. so that approximately 2 h after a suckling episode hPRL levels reach baseline levels [23], as the prolactin required for full lactation is obviously no longer needed. Prolactin is released from the pituitary in a pulsatile fashion. Since peak frequencies and interpulse intervals are no different between early and late postpartum sessions, decreased hPRL output is achieved by altering the endogenous secretory rates in each release episode [24]. The magnitude of the prolactin response appears to vary during the day, with a more marked response to suckling in the afternoon than in the morning [12]. The magnitude of the prolactin response seems to be related to the strength or intensity of the suckling stimulus [13]. The role of glucocorticoids, thyroid hormones and insulin is obviously of minor importance for lactogenesis. Growth hormone, although of crucial importance for lactogenesis in ruminants, does not appear to be essential for lactogenesis in the human.

121.3.2 The Role of Oxytocin

Milk let down through the ductal system is potentiated by oxytocin of posterior pituitary origin, as seen from the fact that the myoepithelial cells surrounding the alveoli respond to oxytocin with immediate contraction. Oxytocin is a neuropeptide hormone composed of nine amino acids, two of which are cystines forming a disulfide bridge between position 1 and 6. Oxytocin is bound to a carrier protein (neurophysin I) with a molecular weight of 10 kD, synthesis of which is stimulated by estrogens. During pregnancy, rising peripheral levels of oxytocin have been demonstrated. Oxytocin is synthesized as a larger molecule and is cleaved to the biologically active nonapeptide bound to neurophysin I and stored in the posterior lobe of the pituitary. Release of oxytocin in a Ca^{2+}-dependent fashion is achieved by a neurohumoral reflex elicited by the infant's suckling (Fig. 121.3). However, oxytocin is released even before the onset of suckling, and the spontaneous ejection of milk often occurs in the absence of the suckling stimulus, particularly in response to associated stimuli such as the sound of the baby crying.

The suckling–milk ejection reflex is initiated by stimulation of the sensory nerve endings of the areola, transmitted along the spinal cord to the paraventricular and supraoptic nuclei. The axons of these nuclei project directly to the posterior lobe of the pituitary, where the active neuropeptide is released into the circulation. Oxytocin is released in a pulsatile fashion during a suckling episode [21].

The action of oxytocin depends on specific receptors located at the membranes of the target organs causing an

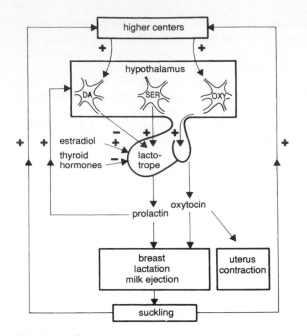

Fig. 121.3. Schematic representation of suckling-induced prolactin and oxytocin release. *DA*, Dopamine; *SER*, serotonin; *OXY*, oxytocin

increased production of cyclic AMP resulting in partial depolarization of the membrane. These receptors have been identified in mammary tissue, myometrium, and smooth muscles of the oviduct [33]. The suckling-induced oxytocin secretion leads simultaneously to uterine contractions, which are recognized by the nursing mother [33]. Oxytocin is also present in several extrahypothalamic brain areas besides the neurohypophyseal system and can be regarded as the precursor of behaviorally active neuropeptides that act as neurotransmitters in the central nervous system. Oxytocin may be involved in central functions such as maternal behavior, sexual behavior, memory and learning [1]. Secretion of oxytocin can be inhibited by both physical and psychological stress.

121.3.3 Lactation Amenorrhea

The prolactin response to suckling is far greater in the early postpartum days than during the later period, indicating a decline of pituitary responsiveness. Intensive nursing with six and more episodes per day, however, can keep hPRL levels elevated for 15–18 months [7]. Stimulation of the nipple results in changes in dopamine turnover, allowing prolactin to be released into the circulation (Figs. 121.3, 121.4). The increase in dopamine turnover may be a consequence of elevated endogenous opioids, such as *β-endorphin* [9]. A rise in β-endorphin causes a decrease in *GnRH*, finally leading to gonadal quiescence, which is clinically reflected as *amenorrhea* (Fig. 121.4). The duration of postpartum amenorrhea is dependent on the intensity and frequency of nursing. In non-nursing women ovulation occurs on average 49 days after delivery, while in lactating women it is delayed to an average of 112 days post partum [27]. Consequently, fertility is restored much earlier in non-lactating women than in women who breastfeed their children [11]. The Kung hunter-gatherers of the kalahari desert, who wean their children at the age of about 3 years, are known to have had a population of constant size over several generations because of birth spacing due to lactation amenorrhea [18]. However, lactation is a rather unreliable method of birth control, since in 40–50% of cases the first menstrual bleed after delivery is preceded by ovulation.

121.3.4 Composition of Human Milk

Human milk varies in composition according to gestation and time since delivery, and even within one suckling episode. During pregnancy the milk composition is dominated by Na^+, Cl^-, lactoferrin and immunoglobulins. During the second half of gestation secretory activity of the alveolar epithelium is initiated leading to accumulation of colostrum in the alveoli. Immediately after delivery, when the maternal circulation is cleared of sex steroids, colostrum is excreted by the mammary glands.

Colostrum contains mainly cellular elements (T-lymphocytes, macrophages) and immunglobulins (IgA) [8]. These cellular elements are immune-competent cells, which are taken up by the infant's gastrointestinal tract and retained in Payer's plaques, providing immune protection for a fairly long period. Since the neonate is immunologically immature the infant's defense against infections is highly dependent on maternal support. In addition, colostrum serves as a natural *laxative*. During pregnancy motility of the fetal gut is completely suppressed so that no feces is released into the amniotic fluid in physiological conditions. After birth motility of the

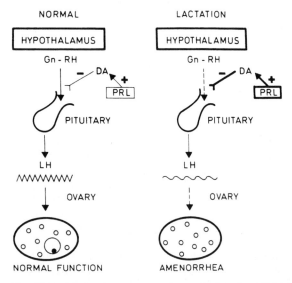

Fig. 121.4. Schematic representation of the hypothalamo-pituitary ovarian axis in normal conditions and during lactation

neonate's gut is stimulated to discharge the biliverdin-rich *meconium*. The exact mechanism of the motility-stimulating property of colostrum is not known. Colostrum is different from mature milk in that it has lower carbohydrate, K^+ and fat components (cf. below).

Milk secretion is an apocrine and a merocrine process (cf. Chaps. 8, 59, 64, 112). The principal carbohydrate in mature milk is lactose, consisting of glucose and galactose. *Lactalbumin*, a specific protein, catalyzes lactose synthesis. Mature human milk contains

- Carbohydrate 70 g/l as lactose
- Fat 35 g/l
- Protein 10 g/l
- Free amino acids
- Mineral constituents 2 g/l.

The principal proteins in human milk are alpha lactalbumin, casein, IgA, albumin and lactoferrin. In addition, several enzymes that support the infant's digestion are present in human milk: amylase, catalase, peroxidase, lipase, xanthineoxidase, and alkaline and acid phosphatase. These enzymes are obviously essential because the gastrointestinal tract of the neonate is not fully mature at the time of birth. The main components of colostrum and mature human milk are listed in Table 121.1. All essential amino acids, trace elements and vitamins except vitamin K are present in mature human milk in sufficient quantities.

Fat accounts for about 35 g/l of human milk. The main components of the fat fraction are palmitic and oleic acids, depending on maternal food intake.

Milk contains rather high quantities of calcium and phosphate. Ca^{2+} concentrations of 5–8 mmol/l and phosphate concentrations of 4–5 mmol/l are required to cover the infant's demands, accounting for 20–22 mmol Ca^{2+}/day (corresponding to approximately 2 g), almost twice as much as needed in adulthood. Ca^{2+} and phosphate are

mobilized partly from the maternal bone mass. In a clinical study [34] it was demonstrated that serum phosphate and parathormone (PTH) concentrations were significantly higher in lactating than in non-lactating women. In addition, serum Ca^{2+} concentrations were elevated during lactation and returned to normal after weaning, indicating recovery of bone mass. Postmenopausal women with a history of extended lactation were found to have reduced bone mineral content [19].

121.3.5 The Immune-protective Effect of Breast Feeding

Immunoglobulins found in human milk are mainly IgA, reaching maximal concentrations during the 1st week post partum and decreasing thereafter as milk-specific proteins increase. This pattern of IgA excretion has been demonstrated in a study detecting antibodies to lipopolysaccharide antigens of *Shigella* [6]. Immunoglobulins of maternal origin are absorbed by the neonate's gastrointestinal tract, from where they enter the infant's circulation to provide passive immunity against pathogenic enteric bacteria. However, the period of passive immunity provided by breast feeding is limited. Advancing maturation of the gastrointestinal tract seems to be the limiting factor, as immunoglobulins are digested by the infant. This disadvantage, however, is compensated for by progressive maturation of the baby's immune system and by the growth and development of *Lactobacillus bifidus* in the infant's stomach, which are promoted by breast feeding. *Lactobacillus bifidus* is hostile to *Shigella*, *E. Scherichia coli* and yeast [6].

Lactoferrin, a potent iron-binding protein that is present in human milk in substantial quantities, exerts strong bacteriostatic effects on *E. coli* and staphylococci by depriving these microorganisms of iron.

Furthermore, human milk contains various white blood cells, mainly lymphocytes and macrophages. These immunologically active cells of maternal origin stay viable for extended periods of time. On the other hand, these cellular elements provide immune protection within the lactating breast against bacterial infection, preventing the development of puerperal mastitis. Human milk also contains thermostable and acid-stable enzymes that cleave the membranes of bacteria.

Vitamin-binding proteins inhibit the utilization of vitamins by bacteria. Interferon is also present in human milk and may serve as an additional anti-infection factor. It is obvious that several immunological and anti-infection factors in human milk protect the neonate during the period of immunological immaturity. This kind of protection cannot of course be provided by bottle feeding.

Various growth factors, e.g., IGF-1, EGF, and transforming growth factor, have been identified in human milk [30]. The physiological significance of these proliferation-stimulating peptides is not known at present. In animal experiments it was shown that EGF extracted from breast milk accelerated intestinal growth [30].

Table 121.1. Main components of human colostrum and breast milk

	Colostrum	Mature milk
Lactose	57 g/l	71 g/l
Fat (total)	29 g/l	45 g/l
Cholesterol	280 mg/l	140 mg/l
Proteins (total)	23 g/l	10 g/l
Casein	2 g/l	4 g/l
Lactalbumin	–	3.6 g/l
Lactglobulin	3.5 g/l	–
Amino acids (total)	1.2 g/l	1.3 g/l
Vitamin A	1.61 mg/l	0.61 mg/l
Vitamin D	–	4–100 IU/l
Na^+	22 mmol/l	7.4 mmol/l
K^+	19 mmol/l	13 mmol/l
Ca^{2+}	8.5 mmol/l	12 mmol/l
Phosphate	5.1 mmol/l	4.5 mmol/l
Mg^{2+}	1.75 mmol/l	1.46 mmol/l
Cl^-	16.5 mmol/l	10.4 mmol/l

121.3.6 The Energy Aspect of Breast Feeding

Breast milk is easily digested and is composed according to the infant's requirements. Human milk has no antigenic properties for the suckling infant and therefore has a prophylactic value with regard to food allergies. The infant's immune system matures during the first 2–3 months of life, and breast feeding meanwhile avoids the infant's exposure to allergic components.

Lactation is an active process requiring extra energy supply. Fully lactating women need an extra 600–1000 kcal/day to maintain adequate nutrition for the infant. At 9 weeks post partum in one study, the difference in energy intake between lactating and non-lactating women was 760 kcal [35]. Lactating women achieve their energy balance by a correspondingly higher food intake. If the food supply of a nursing mother is inadequate this is compensated by fat tissue mobilization as an additional energy source. From this and other studies it is obvious that adequate laction is heavily dependent on a sufficient food supply. Lactation is involved in the regulation of body weight development in the postpartum period. In a study in which 1423 women were followed up for 1 year after giving birth, the mean weight gain was 1.5 ± 3.6 kg [25]. Of the whole group, 30% lost weight, while 70% gained up to 5 kg. Postpartum weight gain is correlated positively with weight gain during pregnancy and negatively with lactation performance; correlations with individual habits can of course also be observed.

In summary, breast feeding is essential for the neonate's development during the first months of life to compensate for its immunological and metabolic immaturity. In addition, intensive lactation may provide a means of natural, but not safe, family planning.

References

1. Argiolas A, Gessa GL (1991) Central functions of oxytocin. Neurosci Biobehav 15:217–231
2. Arnqvist HJ, Lundstrom B (1975) Effect of insulin on human intestinal smooth muscle. Horm Metab Res 7:403–407
3. Blüm V (1977) Prolaktin: phylogenetische Aspekte. Gynakologe 10:51–61
4. Bolander FF, Topper YJ (1980) Stimulation of lactose synthetase activity and casein formation in mouse mammary explants by estradiol-17, Endocrinology 106:490–495
5. Campbell PG, Skaar TC, Veta JR, Baumrucker CR (1991) Secretion of insulin-like growth factor-I (IGF-I) and IGF-binding proteins from bovine mammary tissue in vitro. J Endocrinol 128:219–228
6. Cleary TG, West MS, Ruiz-Palacios G, Winsor DK, Calva JJ, Guerrero ML, Van R (1991) Human milk secretory immunoglobulin A to Shigella virulence plasmic-coded antigens. J Pediatr 118:34–38
7. Delvoye P, Badawi M, Demaegd M, Robyn C (1978) Long lasting lactiation is associated with hyperprolactinemia and amenorrhea. In: Robyn CM, Harter (eds) Progress in prolactin physiology and pathology. Elsevier, North Holland, Amsternam, pp 213–232
8. Drife JO, McChelland DBL, Pryde A, Roberts MM (1976) Immunoglobulin in the "resting" milk. Br Med J II:503–509
9. Franceschini R, Venturini PF, Cataldi A et al (1989) Plasma beta-endorphin concentrations during suckling in lactating women. Br J Obstet Gynaecol 96:711–713
10. Fuchs AR (1983) Endocrinology of lactation. In: Fuchs F, Klopper A (eds) Endocrinology of pregnancy. Harper and Row, Philadelphia, pp 271–291
11. Glasier A, McNeilly AS, Howie PW (1983) Fertility after childbirth: changes in serum gonadotropin levels in bottle and breastfeeding women. Clin Endocrinol (Oxf) 19:493–501
12. Glasier A, McNeilly AS, Howie PW (1984) The prolactin response to suckling. Clin Endocrinol (Oxf) 21:109–116
13. Glasier A, McNeilly AS (1990) Physiology of lactation. Baillieres Clin Endocrinol Metab 4(2):379–395
14. Going JJ, Anderson TJ, Battersby A, Macintyre CCA (1988) Proliferative and secretory activity in human breast during natural and artificial menstrual cycles. Am J Pathol 130:193–204
15. Haslam SJ, Skyamala G (1979) Progesterone receptors in normal mammary glands of mice: characterization and relationship to development. Endocrinology 105:786–795
16. Howie PW, Stewart-Forsyth J, Ogston SE et al (1990) Protective effects of breast feeding against infection in infants in a Scottish city. Br Med J 300:11–16
17. Kamel MA, Neulen J, Sayed GH, Salem HT, Breckwoldt M (1993) Heterogeneity of human prolactin levels in serum duiring the early postpartum period. Gynecol Endocrinol 7:173–177
18. Konner M, Worthman C (1980) Nursing frequency, gonadal function and birthspacing among hung-hunter-gatherers. Science 207:788
19. Lissner L, Bengtsson C, Hausson T (1991) Bone nineral content in relation to lactation history in pre- and postmenopausal vomen. Calcif Tissue Int 48:319–325
20. Maurer RA (1982) Relationship between estradiol, ergocriptine and thyroid hormone: effect on prolactin synthesis and prolactin mRNA-levels. Endocrinology 110:1515–1519
21. McNeilly AS, Robinson ICAF, Houston MJ, Howie PW (1993) Release of oxytocin and prolactin in response to suckling. Br Med J 286:257–259
22. Neville MC (1983) Regulation of mammary development and lactation. In: Neville MC Neifert MR (eds) Lactation physiology nutrition and breastfeeding. Plenum, New York, pp 103–125
23. Noel GL, Suh HKJ, Frantz AG (1974) Prolactin release during nursing and breast stimulation. J Clin Endocrinol Metab 38:413–423
24. Nunley WC, Ru RJ, Kitchin JD, Bateman BS (1991) Dynamic of pulsatile prolactin release during the post partum lactational period. J Clin Endocrinol Metab 72:287–293
25. Ohlin A, Rossner S (1990) Maternal body weight development after pregnancy. Int J Obes 14:159–173
26. Ostrom KM (1990) A review of the hormone prolactin during lactation. Prog Food Nutr Sci 14:1–43
27. Perez M, Vela P, Masnick GS, Potter RG (1972) First ovulation after childbirth: the effect of breastfeeding. Am J Obstet Gynecol 114:1041–1049
28. Peters F, Pickard CR, Zimmermann G, Breckwoldt M (1991) PRL, TSH and thyroid hormones in benign breast disease. klin wochanschr 59:403–412
29. Raymond V, Beaulieu M, Labue F (1978) Potent antidopamionergic activity of estradiol at the pituitary level on prolactin release. Science 200:1173–1179
30. Read LC, Ford WDA, Filsell OH et al (1986) Is orally derived epidermal growth factor beneficial following premature birth of intestinal resection? Endocrinol Exp 20:199–207
31. Rimoin DL, Holzman GB, Merimee TJ, Rabinowitz D (1968) Lactation in the absence of human growth hormone. J Clin Endocrinol Metab 28:1183–1188

32. Seeburg PH (1982) The human growth hormone gene family. Nucleotide sequences show recent divergence and predict a new polypeptide hormone. DNA 1:239–249
33. Soloff MS, Alexandrova M, Fernström MJ (1979) Oxytocin receptors: triggers for parturition and lactation. Science 204:1313–1318
34. Specker BL, Tsang RC, Ho ML (1991) Changes in calcium homeostatis over the first year postpartum. Obstet Gynecol 78:56–62
35. von Raaij JM, Schonk CM, Vermaat-Miedema SH, Peek ME, Hautvast JG (1991) Energy cost of lactation and energy balances of wellnourished Dutch lactating women. Am J Clin Nutr 53:612–619
36. World Health Organisation Collaborative Study in Breastfeeding (1985) The quantity and quality of breast milk. WHO, Geneva

122 Normal Development and Growth

M. Brandis

Contents

122.1 Introduction

In the following, the characteristic phases of development and growth will be discussed, starting with the events immediately after birth and the dramatic changes in organ function that take place at this time. The first few weeks and months of life will then be discussed, focusing on the development of specific organ functions. Finally, we conclude this chapter with a discussion of childhood and adolescence.

122.2 Adaptation of Organ Function After Birth

By the end of normal-term gestation, the organs of a newborn baby are prepared to adapt for extrauterine life. There are new respiratory, circulatory, and metabolic demands after birth which can only be met if various developmental regulatory systems act together. The newborn child is extremely vulnerable and dependent on support at the beginning of its extrauterine life. In the following sections the most important adaptive mechanisms will be covered.

122.2.1 Respiratory Adaptation

The shift from liquids to gaseous breathing with the first breath of a newborn child reverses a cascade of physical parameters in the lung. Anatomically, the alveoli are developed by the week 25–27, and by week 35 the production of surface-active material has begun [9,16,18]. This event is necessary for alveolar stability and patency once air breathing commences.

Vascular Resistance. Before birth the alveoli are opened and filled with fetal lung liquid. The pulmonary and bronchial blood flow is low because of a high passive and active vascular resistance. The passive resistance is mainly caused by liquid compression, and the active resistance by the tissue hypoxia (PO_2 of 25 mmHg, 3.3 kPa) in the pulmonary venous branch. In relative terms, there is a larger vascular muscle mass in the fetus and newborn than in the adult, which causes a higher vascular tone as a result of hypoxia during development. The fetal circulation is characterized by pulmonary vascular resistance exceeding systemic resistance [29]. Consequently, nearly 50% of the fetal cardiac output perfuses the placenta, and only 5%–10% perfuses the lung. Figure 122.1 shows the changes in infant and placenta blood volume during delivery.

Chest Compression. During delivery of the child, the thoracic cage is compressed to pressures up to 20–100 mmHg (2.7–13.3 kPa). The subsequent relaxation (recoil) of the chest wall may induce a small passive inspiration of air, pushing some amount of blood into pulmonary capillaries, i.e., capillary erection. This event leads to the buildup of an air–liquid interface within the large airways of the lung. The retraction forces that would tend to collapse the

R. Greger/U. Windhorst (Eds.)
Comprehensive Human Physiology, Vol. 2
© Springer-Verlag Berlin Heidelberg 1996

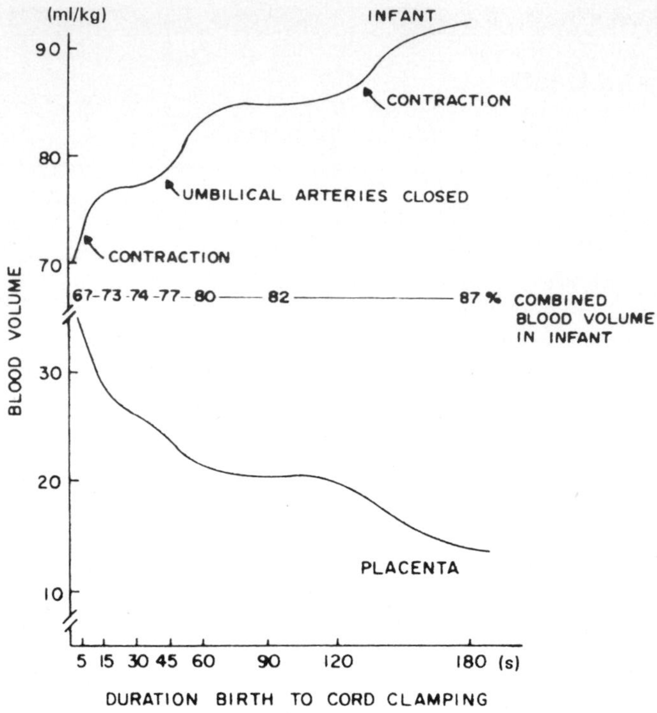

Fig. 122.1. Placental transfusion. Sequence of events during the first 3 months of life. (From [40])

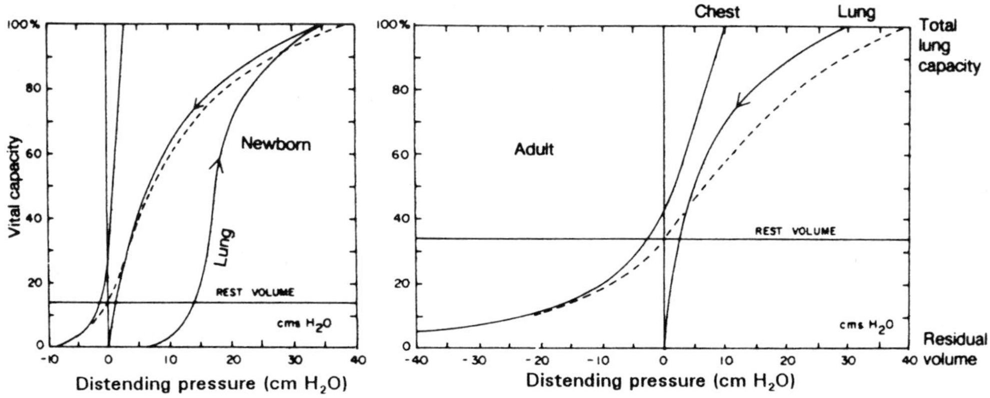

Fig. 122.2. Comparative mechanics of the infant (*left*) and adult (*right*) lung. (From [40])

smaller airways and alveoli are compensated by the presence of surfactant covering the alveolar layer.

Surfactants. What is generally called a "surfactant" (also see Chap. 100) is a mixture of several lung-specific proteins, consisting of dipalmitoyl lecithin. Four different pulmonary surfactant proteins (SP-A, B, C and D) are the principal modifiers of the phospholipid interface [18]. At the air–liquid interface, an interaction of charges among the basic arginine and lysine residues of SP-B helps to resist surface tension by increasing the lateral stability of the phospholipid layer [27].

To lower surface tension, a minimum of 3 mg surfactant in the alveolar fluid is necessary. In premature infants, much higher concentrations are necessary for alveolar sta-

bility [32]. Figure 122.2 shows the compliance parameters, comparing newborns and adults [39]; Fig. 122.3 summarizes the pressure–volume curves for saliva and air breathing (note the much lower compliance seen with air breathing).

The first breath the newborn takes is the result of various influential factors such as cold, light, noise, gravity, and pain in addition to hypercapnia, respiratory acidosis, and hypoxia ("asphyxia") [42]. The first breath is followed by several important events:

- Conversion from fetal to adult circulation
- Emptying the lung of liquid
- Establishing the neonatal lung volume
- Intiation of pulmonary function in the newborn infant.

Intrapleural and Interstitial Pressure. With the first entry of air into the lung (Fig. 122.4), the surface tension of the alveolar epithelium leads to a negative intrapleural and interstitial pressure, as the chest wall cannot collapse. Concomitantly, a dramatic increase in blood and lymph flow in the lung follows the drop in interstitial pressure. With the lungs being emptied of fluid and filled with air, the retraction forces turn into power, the hydrostatic pressure on the pulmonary capillaries decreases, and the blood perfusion rate increases [29].

With increasing pulmonary O_2 tension in the air-breathing newborn infant, the active vasomotor tone decreases in the precapillary arterioles. The increasing hydrostatic pressure in the capillary lumen and the decreasing pressure in the interstitium is the basis of fluid absorption according to Starling's forces (see Chap. 97).

Premature and Cesarean-Born Infants. The above-mentioned adaptive processes must occur during and after birth. Their disturbance in premature newborns explains the difficulty these infants have in adapting adequately to the lack of surfactant and other physical developmental problems. Newborn infants born by cesarean section, are less able to adapt as they have not undergone the stress involved in labour and lack the chest compression to decrease the lung fluid as easily as in vaginally delivered babies.

122.2.2 Circulatory Adaptation

Fetal Circulation. The fetal circulation consists of two separate systems arranged in parallel, the pulmonary and the aortic circulation (see Chap. 89). Until delivery the oxygenation of the blood is arranged by its passage through the placenta. Blood returning from the placenta passes into the portal system of the liver and then into the inferior vena cave or into the ductus venosus and inferior vena cava directly. The majority of the blood from the inferior vena cava is drained into the right atrium and, together with superior vena cava blood and coronary sinus blood, it enters the right ventricle.

A small part of the vena cava blood is directed via the foramen ovale into the left atrium and joins the pulmonary venous blood. After passage through the mitral valve into the left ventricle, blood flows into the ascending aorta to supply the coronary, the carotid, and the subclavian arteries. Only 10% of the left ventricular output reaches the descending aorta. Only 10%–15% of the blood from the right ventricle flows through the pulmonary arteries into the lungs because of the high vascular resistance in the pulmonary arteries; the majority passes through the ductus arteriosus into the descending aorta.

Onset of Respiration. At the time of the first breath and the removal of the low-resistance placenta from the circulation, systemic resistance increases (Fig. 122.5). Simultaneously, the pulmonary capillary resistance drops, the lungs are inflated with air, and the O_2 tension increases in the alveolar gas. The increase of systemic vascular resistance and the decrease of pulmonary resistance by 80% of its original value, which occurs with the onset of respiration, causes increased pulmonary flow and decreased flow through the ductus arteriosus (Fig. 122.5) [42]. The redistribution of resistances and the consequently increasing pulmonary flow leads to an increase in left atrial pressure, slowly exceeding right atrial pressure. This event leads to an apposition of the valve of the foramen ovale with its functional closure within minutes after the first breath.

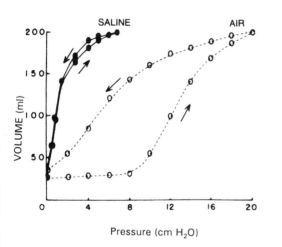

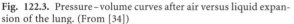

Fig. 122.3. Pressure–volume curves after air versus liquid expansion of the lung. (From [34])

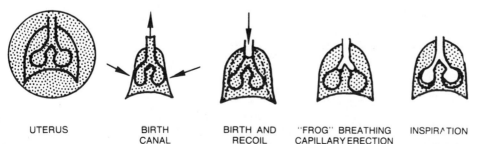

Fig. 122.4. Mechanical events of the first breath. (From [40])

Transitional Phase. The ductus arteriosus remains open for several hours. As the systemic pressure is now higher in the left ventricular system than in the right, the blood flow through the ductus reverses its direction from left to right (Fig. 122.5). Within the transitional phase of perinatal circulation, a reversion to the fetal circulation can occur whenever pulmonary vascular resistance increases again because of disease, e.g., acidosis, hypoxemia, polycythemia, or lung disease, or as a result of cesarian section. Physiological pressure values are established in the right and left ventricular system within a few hours after birth, and this process is normally complete after 15–24h. The patency of the ductus arteriosus is regulated by prostaglandin E (PGE) synthesis. Any metabolic event, such as hypoxia, which seems to stimulate PGE_2 secretion, would tend to keep the ductus arteriosus open. This can be prevented by PGE_2 inhibitors such as indomethacin.

122.2.3 Physiology of Temperature Control

Homeothermy. The newborn baby has the physiological capacity for homeothermy, with certain restrictions in stress situations. This is in principle also true for premature infants, with even more limitations due to stress [5,10,39]. Homeothermy can be endangered in the newborn infant by several adverse factors. The infant has a relatively large body surface area (BSA), poor thermal isolation, and a small mass to act as a heat depot. It has little capacity to conserve heat by changing its posture and no ability to adjust its clothing in response to thermal stress. These adverse factors are augmented in premature infants and may become relevant immediately after birth in the delivery room: inadequate room temperature as well as a long journey from the delivery room to the neonatal unit may cause serious problems, including hypoxia.

Hypothalamic Thermostat. The central regulation in the hypothalamus (the hypothalamic thermostat; also see Chap. 110) adapts to cyclic changes and lowers the temperature by 0.5°C during sleep. This regulation is negatively influenced by drugs, hypoxia, and hormones such as norepinephrine and becomes inefficient after intracranial hemorrhage, central nervous system (CNS) malformation, and asphyxia. The regulation of heat production is guaran-

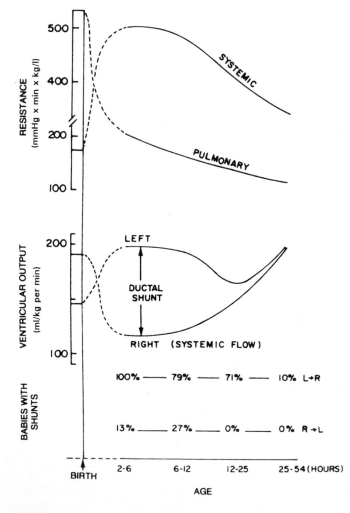

Fig. 122.5. Perinatal hemodynamics in the human show changes in vascular resistance, ventricular output, and shunting. (From [40])

teed by shivering or other muscular activities as well as by nonshivering thermogenesis. Catecholamine mediated nonshivering thermogenesis in brown fat is a major mechanism in temperature adaptation. This process, too, is strongly inhibited by hypoxia, drugs, and malnutrition.

Sweating and Hyperthermia. Infants' ability to increase their insensible loss of water by sweating is very limited, even though the number of sweat glands per unit skin area is sixfold that of adults. The insensible water loss in infants can only be increased fourfold, and the risk of hyperthermia in an environment increasing in temperature is therefore great. This limited adaptative ability is even more prominent in premature infants, who are almost totally unable to sweat. The maturation process is fairly rapid and the sweat capacity of a premature infant later exceeds that of a mature infant at the same gestational age (Fig. 122.6). Figures 122.7 and 122.8 and Table 122.1 show temperature control in incubators.

122.2.4 Metabolism

Glucose. The goal of neonatal metabolic homeostasis is to provide the brain and other vital organs with sufficient glucose as the key energy source. Until delivery, the fetus is supplied with sufficient glucose via the placenta. During fetal life, after the 20th week of gestation, the fetal pancreas

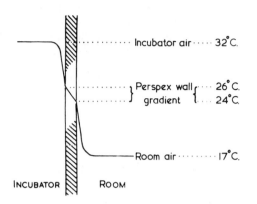

Fig. 122.7. Temperature gradients across the 9-mm perspex wall of a typical incubator. (From [21])

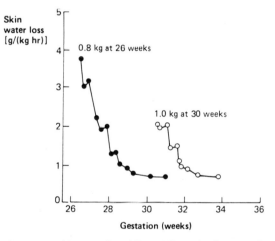

Fig. 122.6. Skin water loss falls rapidly in the first 2 weeks after birth as the skin becomes better keratinized. This maturation process is enhanced in premature babies during extrauterine development (*open circles*). (From [20])

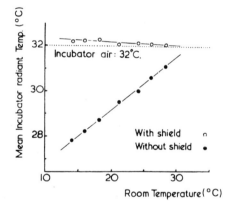

Fig. 122.8. Influence of room temperature on the mean temperature of the inner incubator wall to which a baby radiates heat. The incubator temperature was kept constant at 32°C. (From [21])

Table 122.1. Equilibrium values for heat loss in a 1-week-old 2-kg baby lying naked on a foam mattress in draft-free surroundings of uniforms temperature and moderate humidity

Heat loss	Environmental temperature					
	30°C		33°C		36°C	
	(kcal/m² per h)	(%)[a]	(kcal/m² per h)	(%)[a]	(kcal/m² per h)	(%)[a]
Radiation	19	43	12	40	7	24
Convection	15	37	9	33	5	19
Evaporation	7	16	7	24	17	56
Conduction	2	4	1	3	0	1
Total	43		29		29	

From [19]
[a]Percentage of total heat loss at the respective temperatures

produces insulin in response to the normal flux of glucose. This ensures the accumulation of glycogen in liver, muscle, and lung as well as lipogenesis and triglyceride storage in adipose tissue [45]. Glycogen synthesis in liver and muscle uses glucose and 3-carbon precursors such as lactate and pyruvate as substrates (see Chap. 71). This process is completed by the end of the third trimester, shortly before delivery. Liver glycogen deposits represent stored glucose, which is available for other organs, especially the brain, which together with the heart receives the major part of energy supply [45]. The liver glycogen accounts for to up to 10% of the organ weight at maturity, but the total energy stored is limited. In the full-term neonate, glycogen stores account for 100 kcal (417 kJ), which is just sufficient to cover the energy needs for 8–10 h [2]. This explains why

premature infants, which have far less glycogen stores, tend to develop hypoglycemia within the first few hours of life. The sudden change from the placental nutritional supply of glucose to extrauterine life leads to physiological drop in the plasma glucose concentration within the first few hours of life (Fig. 122.9) [41]. The overall energy reserves of the newborn infant are correlated with the body weight and the gestational age (Table 122.2).

Adipose Tissue. The other main energy source in the newborn infant is adipose tissue. The triglycerides in adipose tissue in a water-free environment provide 9 kcal/g (38 kJ/g) fat. The fat stores in the liver amount to only 1 g. The amount of fat content in the fetus undergoes dramatic changes in the last trimester. While a fetus at 27 weeks of gestation has only 1% the fat content increases to 16% at 40 weeks. Early in life, a significant amount of body fat is metabolically active, especially that located near major large vessels. This particular fat is rich in heme-containing cytochromes and in mitochondria and therefore looks brown. *Brown fat* produces heat metabolically (see above). As the infant becomes older, shivering thermogenesis is established and most brown fat involutes and becomes white fat.

Human Milk. In early life, the brain cannot metabolize free fatty acids directly. It is therefore dependent on glucose as well as ketone bodies produced by the liver from fatty acids. Gluconeogenesis occurs mainly in the liver and kidney during fetal and neonatal life. The newborn maintains glucose homeostasis while fasting or semistarved, which is a standard condition in early postnatal life. The primary postnatal nutrition serves the special needs by providing glucose and free fatty acids [45]. Human milk (see Chap. 121) uniquely satisfies the needs of the newborn infant; it contains lactose and medium-chain triglycerides. However, even long-chain triglycerides can be digested by the special activity of human milk lipase. In addition, carnitine needed for mitochondrial fatty acid transport facilitates fatty acid metabolism.

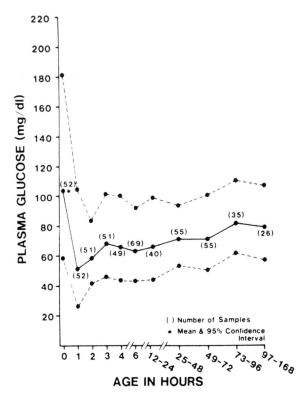

Fig. 122.9. Plasma glucose levels postnatally in mature newborns. Mean and 95% confidence levels. (From [41])

Protein Synthesis. In fetal and early neonatal life protein synthesis is five- to eightfold higher than in adults. In the newborn period, a daily increase of 15–20 g/kg of body weight is required to guarantee normal growth. The enzymes necessary for urea and ammonia production are low

Table 122.2. Energy reserves of the newborn at various birth weights and gestational ages

Weight (g)	Gestational age (weeks)	Energy reserve (kcal)			
		Glycogen	Protein	Fat	Total
200	18	0	65	9	74
1000	27	4	416	90	510
2000	33	16	960	1080	2056
3500	40	70	1694	5040	6804
2200	40	26	871	1108	2005

Adapted from [37] and [49]

before birth and significantly increase during postnatal adaptation (see kidney section).

At time of birth, the constant flux of glucose from the mother is interrupted and there is an initial drop in the blood glucose concentration (Fig. 122.9).This leads to a lowering of the insulin level and an increase in the glucagon concentration. Glucagon activates adenylate cyclase in the hepatocyte membrane. This causes an increase in cyclic adenosine monophosphate (AMP) within the cell (see Chap. 71), which activates the kinase system responsible for activating glycogen phosphorylase. By the action of this enzyme, glucose-6-phosphate is formed from glycogen, which then forms glucose by the activity of the specific phosphatase. The glycogenolytic response in the neonate depends mainly on the activity and maturity of these enzymes This again illustrates the potential danger for immature, premature babies.

Respiratory Quotient. Postnatal changes in the respiratory quotient (RQ), the ratio between carbon dioxide production and oxygen consumption, reflects the relative use of fuels early in postnatal life (see Chap. 71) A value of 1.0 reflects the predominant use of carbohydrates such as glucose, and a value of 0.7 corresponds to exclusively fatty acid consumption.

Hormones. The goal of neonatal glucose homeostasis is to provide the brain and other vital organs with sufficient glucose as a key energy source. All hormones, e.g., catecholamines, glucagon, glucocorticoids, thyroxin, and growth hormones, with the exception of insulin promote lipolysis and gluconeogenesis, thus ensuring the necessary glucose concentration [2].

122.3 First Weeks of Life

122.3.1 Kidney Function, Fluids, and Electrolytes

Renal Blood Flow. In human neonates, renal blood flow (RBF) is low. In the first few hours of life, the RBF values observed vary considerably, as estimated by the clearance of para-aminohippuric acid (PAH), which is the result of the hemodynamic instability during the transitional period [54]. PAH clearance, corrected for BSA, doubles by 2 weeks of age and increases slowly to mature values by 2 years of age [3,6]. Estimating RBF using PAH clearance has some pitfalls, as PAH extraction by the kidney is incomplete in young infants. It is only 65% in the first 3 months, reaching adult levels at the age of 5 months.

Renal Vascular Resistance. In human newborns, renal vascular resistance (RVR) is predominantly based on the relation between afferent and efferent arteriolar resistance and not so much on new arteriolar channels, as nephrogenesis is completed in humans between the 33rd and 36th week of gestation. The factors controlling the postnatal decrease in RVR are predominantly those affecting afferent and efferent glomerular arteriolar resistances. Circulating catecholamines and the renin levels are high in the early postnatal period in premature and mature infants [23,44]. Prostaglandins may also contribute to this, as indomethacin, given to newborns, lowers RBF.

Glomerular Filtration Rate. In newborns, the glomerular filtration rate (GFR) is low, whether expressed per g kidney weight, per kg body weight, or per unit BSA [3,15]. GFR reaches 5 ml/min per m^2 at birth in newborns of 28 weeks of gestation, and 12 ml/min per m^2 in infants at term. Postnatally there is a sharp increase in GFR, which doubles by 2 weeks of age. Corrected for 1.73 m^2, the GFR almost reaches adult levels between 6 and 12 months of age (Fig. 122.10). The creatinine serum levels are low at birth and increase with age according to the muscle mass (Fig. 122.11). The indirect measurement of GFR by serum creatine levels and body height can be estimated using the formula worked out by Schwartz et al. [35].

Renin–Angiotensin System. The renin–angiotensin system (RAS) undergoes developmental changes according to gestational age. Kidney renin activity is higher in the fetus than in normal adult humans and is higher in human fetal cord samples (18 weeks of gestation) in comparison to full-term infants. After birth, plasma renin activity (PRA) continuously decreases and adult levels are reached not earlier than 6–9 years of age. In Table 122.3, the factors influencing the high PRA in newborns are listed. The developmental decrease of PRA is correlated to a decrease in intrarenal renin distribution and renin mRNA levels. While in early developmental stages the renin-producing cells are distributed along the branches of the arterial walls, later on they are predominantly situated in the juxtaglomerular (JG) region (see Chap. 73). It is not yet clear whether the response of these cells to angiotensin II to secrete renin is different in fetal and early postnatal life from the adult stage, and it is not well understood whether the high renin concentrations are only correlated to blood flow regulation or other effects, such as growth responses or angiogenesis [31].

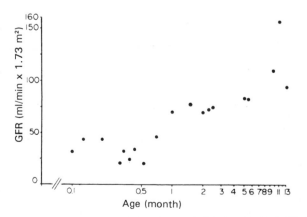

Fig. 122.10. Glomerular filtration rate (*GFR*) during the first year of life. (From [3])

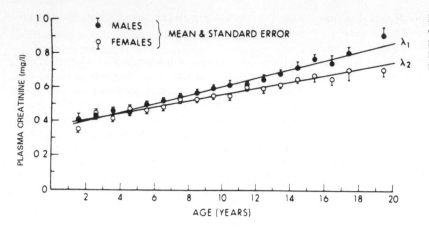

Fig. 122.11. Mean plasma creatinine concentration (mg/l) plotted against age for both sexes. The regression equation for males is $y = 0.35 + 0.025 \times age$, and for females, $y = 0.37 + 0.18 \times age$. (From [36])

Table 122.3. Factors that increase (+) and decrease (−) the activity of the renin–angiotension system (RAS) in the fetus and newborn

(+)	(−)
Prematurity	Volume expansion
Vaginal delivery	Indomethacin
Furosemide	Phenylephrine
Hemorrhage	Cortisol
Hypotension	Arginine vasopressin
Hypoxemia	Recumbent position
Aortic constriction	Exchange transfusion
Exchange transfusion	(replacement)
(withdrawal)	
Upright position	

From [14]

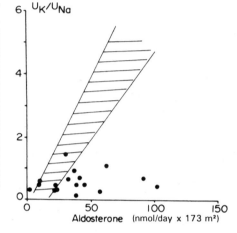

Fig. 122.12. Relationship between aldosterone excretion and urinary K^+ (U_k) to Na^+ (U_{Na}) ratio in preterm infants (*filled circles*). The range in full-term infants is indicated (*open circles*). (Adapted from [3])

However, there is no doubt that angiotensin II acts preferentially at the postglomerular site in early development, thus regulating RBF, increasing filtration fraction, and preserving GFR [13]. These data suggest that the RAS and the *angiotensin-converting enzyme (ACE)* are functionally competent at birth in the human neonate, but undergo developmental changes in localization and secretion rate, hence leading to a change in the plasma concentration of angiotensin II. The exact meaning of the wide concentration ranges is not really known.

Aldosterone. In full-term infants, plasma aldosterone concentration is elevated and remains high during the first month of life [8]. There is no correlation between PRA and aldosterone concentrations in neonates, which suggests that angiotensin II is not the principal regulator of aldosterone secretion in early life. This is illustrated in Fig. 122.12, showing that the concentration ratio of K^+ to Na^+ in the urine is significantly lower in preterm infants than in term infants. This can be interpreted as the result of a fairly aldosterone-unresponsive proximal tubule. Preterm infants with extremely high concentrations of serum aldosterone tend to excrete high amounts of Na^+ (Fig. 122.13) [38].

Renal Tubular Transport. The developmental changes in renal tubular transport have mainly been studied in animal

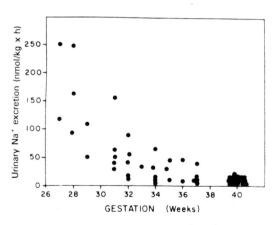

Fig. 122.13. Urinary Na^+ excretion in infants from 27 to 40 weeks of gestation (Adapted from [38])

experiments. Extrapolation to the development of human tubule function is not always possible. The reabsorptive capacity of the proximal convoluted tubule increases two-fold in the first 3–6 weeks in the rat kidney, paralleled by an increase in luminal and basolateral surface area. Simultaneously, the activity of the (Na$^+$+K$^+$) ATPase increases to the same extent. The transport activities for organic substrates such as glucose or amino acids are mostly paralleled by the (Na$^+$+K$^+$)-ATPase and transport activities of Na$^+$ transport. Glucosuria is frequently present in neonates, particularly in premature infants. The maturational process is otherwise very rapid, and within a few days to weeks glucosuria is no longer observed. Similar patterns of maturational changes can be seen in amino acid transport. In general, urinary amino acid concentrations are higher in newborns than later in life.

122.3.2 Metabolism and Nutrition

At birth, the infant must adapt from a continuous transplacental supply of substrates (see Chap. 117) to intermittent provision of complex nutrients. Fat serves as a major endogenous fuel. It represents about 16% of body weight at birth in a normal neonate. Glycogen, the major source of rapid glucose, makes up only 1% of body weight (see Table 122.2). After the first feed, the concentrations of the hormones insulin, growth hormone, gastrin, and glucagon increase in plasma [4]. With the first feeding, the functions of the gut with respect to motility, digestion, and reabsorption have to adapt. The anatomic development of the gut is completed by the 24th week of gestation [22]. Digestive adaptations are stimulated by breast feeding. The pancreas does not release lipase at adult activities before 2 years of age, but lingual lipase and the bile salt-stimulated lipase of human milk compensate for this [17].

Infants of 33–35 weeks of gestation suck and swallow in a coordinated fashion and have ordered gastrointestinal peristaltic activity (also see Chap. 63). In the first week of life, the mature infant is able to regulate its intake. At 6 weeks of life it can adjust to a reduction in food energy content by taking in greater volumes [11]. Adequate food intake is normally judged by weight and growth velocity. This is dependent on ethnic and environmental factors [48] (Fig. 122.14).

Milk Synthesis and Secretion. This process is regulated by various hormonal and psychological parameters (see Chap. 121). Anxiety, restlessness, and environmental disturbances can inhibit milk production. Estrogens, progesterones, corticosteroids, prolactin, and placental lactogens induce the growth of alveoli and ducts of the breast during pregnancy. Estrogens and progesterone inhibit milk synthesis during pregnancy, but this inhibition disappears with the birth of the baby. The secretion of milk and its ejection is stimulated by prolactin, and several intracellular processes are involved in ensuring the adequate content of the milk (Fig. 122.15). The human mammalian gland has only limited storage capacity, and milk is continuously synthesized between feeding. The process of secretion involves contraction of myoepithelial "basket cells" surrounding the alveoli. The contraction is stimulated by pulsatile secretions of oxytocin from the posterior pituitary gland.

Proteins. The content of human milk is adapted to the special needs of a newborn baby. The proteins casein and α-lactalbumin are synthesized in the rough endoplasmatic reticulum, stimulated by prolactin, and secreted together with lactose in vesicles. Ruminant mammals produce milk of higher casein content than do monogastric mammals. Thus, cows' milk has a whey to casein ratio of 20:80, and human milk of 60:40. The principal whey protein of monogastric mammals is α-lactalbumin, whereas that of ruminants is β-lactoglobulin. The major whey proteins in human milk are α-lactoglobulin, lactoferrin, and immunoglobulin A (IgA).

Whereas α-lactoglobulin is easily digestible, IgA and lactoferrin resist acid and peptic digestion. These proteins can therefore be recovered in significant quantities from the stools of breast-fed infants but rarely during formula feeding. As much as 10% of breast-milk protein intake in early life and 3% at 4 months may not be digested. This change may reflect maturation of gastrointestinal function and a fall in lactoferrin and IgA intake. Secreted IgA is synthesized by lymphocytes within the gland stroma and is transferred across the cell by pinocytosis (Fig. 122.15). Other proteins that are not synthesized or selectively secreted such as plasma albumins and maternally absorbed, intact food proteins (e.g., β-lactoglobulin) permeate the epithelium via paracellular pathways.

Lactose. The principal carbohydrate of human milk is lactose, which is synthesized from glucose and uridine diphosphate-galactose in the Golgi apparatus under the influence of lactose synthetase. It accounts for 40% of the nonprotein energy content of human milk. The vesicles produced are impermeable to lactose, which attracts water

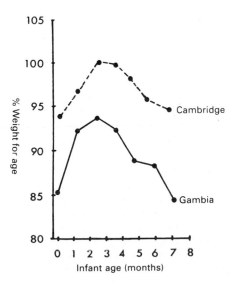

Fig. 122.14. Weight of two cohorts of breast-fed infants expressed as a percentage of the 50th percentile of the National Center of Health Statistics [50] growth chart. (Adapted from [48])

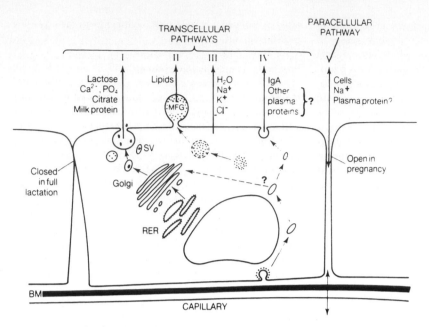

Fig. 122.15. Pathways of milk synthesis and secretion in the mammalian alveolus. *I*, Exocytosis of milk protein and lactose in Golgi apparatus-derived secretory vessels. *II*, Milk fat secretion via the milk fat globule (*MFG*). *III*, Secretion of ions and water across the apical membrane. *IV*, Pinocytosis–exocytosis of immunoglobulins. *V*, Paracellular pathway for plasma components and leukocytes. *SV*, secretory vesicles; *RER*, rough endoplasmic reticulum; *BM*, basement membrane

along an osmotic gradient. Thus lactose production determines the volume of milk secreted [50].

Lipids. These may be taken up from plasma as free fatty acids or synthesized within the gland. Plasma lipids are taken up in combination with low-density lipoproteins (hepatic origin) or as chylomicrons (dietary origin) and are hydrolyzed by a cell membrane lipoprotein lipase. Human milk triglycerides contain more polyunsaturated fatty acids than do those of cows' milk and hydrolyze to β- rather than α-monoglycerides, which improves the absorption from human milk. High-density fatty acid composition reflects that of maternal diet and fat stores.

Inorganic Constituents. The inorganic components of milk, e.g., Na^+, vary inversely with the volume of milk produced. Na^+ concentration is high at the initiation of lactation and during weaning. Zinc, copper, and iron content declines during lactation, zinc by a factor of 4 at 3 months and about 10 after 1 year.

Composition of Human Milk and Physiology of Breast-Feeding. The composition of human milk varies tremendously with the stage of lactation (see Chap. 121). *Colostrum* is milk produced during the first 5 days, and transitional milk between 5 and 10 days of lactation. The principal protein in the first few days is IgA, which is not as much a nutrient protein but an immunoprotective one. The physiology of breast-feeding varies widely within individuals. No generalization can be made about the optimum duration of a breast-feeding session. Maternal (milk supply, rate of secretion) and infant factors (appetite, sucking

efficiency) both exert their influence on milk flow [52]. The infants sucking pattern is nutritive and non-nutritive. Nutritive sucking occurs at a rate of two sucks per s with only a few pauses and is coordinated with swallowing and initiation of the esophageal peristaltic wave, whereas non-nutritive sucking is faster, with frequent pauses between bursts. The mean daily milk intake of breast-fed infants growing along the 50th percentile of NCHS (National Council of Health Centers) charts [50] lies between 600 and 800 g [47]. The intake of milk is principally regulated by demand if maternal nutrition is adequate.

122.3.3 Nutrition in the First Year

Weaning. During the first 3 months of postnatal life, a normally growing infant gains between 25 and 39 g/day, and during the second 3 months, 15–21 g/day. Energy requirements are calculated to be in the range of 115 kcal/kg per day (480 kJ/kg per day) during the first 6 months of life. By the end of the first year, this falls to 105 kcal/kg per day (438 kJ/kg per day) (Tables 122.4, 122.5). Water, calories, and major nutrient requirements and those for most vitamins, minerals, and trace elements can be met during the first 6 months by human milk or by commercially prepared formulas. At 4–6 months, nonliquid food can be added to the feeding. At this age, the infant has the muscle control to close the mouth, to move the tongue back and forth, and to retain foods placed in the mouth with a spoon. Later, the child is capable of moving the jaw up and down, forming the mouth around the spoon, and freely turning the head away. At 8 months the infant acquires the ability to pick up

2382

small pieces of soft food and bring them to the mouth. Food refusal skills increase steadily until 1 year of age. At the time of introducing solids, most infants take four to five feeds per day, and nonliquid foods are introduced before the bottle or breast at one or more of these times. With time, the infants grow accustomed to more solid foods and need less and less bottle or breast-feeding. From a nutritional point of view, breast-feeding is not recommendable after 6–8 months. The emotional aspects are different in individual cases. At 2 years of age, most children have three small meals and two snacks a day. Tradition, not scientific fact, underlies recommendations for the sequence of introduction of foods. The introduction of cereal is often followed by fruits, vegetables, and meat, but other sequences can be advocated, particularly as meat is an excellent source of iron, which must be supplemented early during infancy. A supplementary amount of 3 mg/kg per day of iron is recommended.

Childhood and Adolescence. Nutrient requirements for older children are generally met in accordance with social and environmental customs (Tables 122.6, 122.7). The committee on Nutrition of the American Academy of Pediatrics (CONAAP) suggests that the fat content of the diet be restricted to 30% of calories from about 2 years of age. No more than 10% of the total calories should come from saturated fatty acids, and the daily diet should not contain more than 300 mg cholesterol [51] (Table 122.8). By the time the child is 24 months of age, the meals consist almost exclusively of foods eaten by the rest of the family. Protein needs are diminished between infancy, childhood, and adolescence. The appropriate nutritional balance can

Table 122.4. Recommended energy intake for children and adolescents

Age group	Average energy allowance (kcal)	
	Per 1 kg of weight	Per day (rounded up)
Infants		
0–6 months	108	650
7–12 months	98	850
Children		
1–3 years	102	1300
4–6 years	90	1800
7–10 years	70	2000
Adolescents		
11–14 years (females)[a]	47	2200
11–14 years (males)	55	2500
15–18 years (females)[a]	40	2200
15–18 years (males)	45	3000

[a] Add 300 kcal/day for pregnancy or 500 kcal/day for lactation
From [43]

Table 122.5. Estimated protein and energy intakes of breast–fed infants

Age (months)	Weight (kg)	Protein g/kg per day	Energy kcal/kg per day
Boys			
0–1	3.8	2.46	118
1–2	4.75	1.93	114
2–3	5.6	1.45	107
3–4	6.35	1.49	101
Girls			
0–1	3.6	2.39	
1–2	4.35	1.93	
2–3	5.05	1.78	
3–4	5.7	1.53	

FAO (Food and Agricultural Organization), WHO (World Health Organization) 1985 [53]

Table 122.6. Nutrition and development

	Early infancy (0–5 months)	Later infancy (6–11 months)	Toddler (1–4 years)
Anthropometry			
Weight			
(kg)	6	9	14
(kg/year)	9	4	2
Height			
(cm)	61	73	96
(cm/year)	34	14	8
Triceps skin-fold (mm)	9	12	10
Dietary intake			
Water (ml per kg)	150	130	110
Energy (kcal per kg)	110	100	90
Protein (g per kg)	2.5	2.5	2.2
Body composition (g per 100 g)			
Water	75[a]	60[b]	59[c]
Fat	11[a]	26[b]	24[c]
Protein	11[a]	11[b]	15[c]

From [46]
[a] 0 months
[b] 4 months
[c] 12 months

Table 122.7. Recommended dietary allowances for school-aged children

	Children		Boys		Girls	
	4–6 years	7–10 years	11–14 years	15–18 years	11–14 years	15–18 years
Energy (kcal/kg)	90	70	55	45	47	40
Protein (g/kg)	1.2	1.0	1.0	0.9	1.0	0.8
Vitamin A (μg RE)[a]	500	700	1000	1000	800	800
Vitamin D (μg)[b]	10	10	10	10	10	10
Vitamin E (mg TE)[c]	7	7	10	10	8	8
Thiamin (mg)	0.9	1.0	1.3	1.5	1.1	1.1
Riboflavin (mg)	1.1	1.2	1.5	1.8	1.3	1.3
Niacin (mg NE)[d]	12	13	17	20	15	15
Vitamin B_6 (mg)	1.1	1.4	1.7	2.0	1.4	1.5
Vitamin B_{12} (μg)	1.0	1.4	2.0	2.0	2.0	2.0
Folacin (μg)	75	100	150	200	150	180
Vitamin C (mg)	45	45	50	60	50	60
Calcium (mg)	800	800	1200	1200	1200	1200
Phosphorus (mg)	800	800	1200	1200	1200	1200
Iron (mg)	10	10	12	12	15	15
Magnesium (mg)	120	170	270	400	280	300
Zinc (mg)	10	10	15	15	12	12

From [24,43]

[a] Retinol equivalents

[b] 1 μg = 40 international units (IU)

[c] α-Tocopherol equivalents

[d] Niacin equivalents

Table 122.8. Nutritional requirements for school children and adolescents

1. Avoid obesity and maintain normal weight and growth
2. Increase consumption of unrefined complex sugars (to 48% of total energy intake)
3. Decrease consumption of refined sugars (to 10% of total energy intake)
4. Decrease fat intake (to 30%–35% of total energy intake)
5. Decrease intake of saturated fats (maximum 10% of total energy intake)
6. Reduce cholesterol intake (not to exceed 150 mg per 1000 kcal or 300 mg per day)
7. Increase dietary fiber
8. Moderate salt intake (5 g per day)
9. Increase iron and calcium intake, fluoridate water, and limit alcohol consumption

From [24]

Table 122.9. Milestones of motor derelopment

	Mean age (months)	Standard deviation
Roll prone to supine	3.6	1.4
Roll supine to prone	4.8	1.4
Sit tripod	5.3	1.0
Sit unsupported	6.3	1.2
Creep	6.7	1.5
Crawl	7.8	1.7
Pull to stand	8.1	1.6
Cruise	8.8	1.7
Walk	11.7	1.9
Walk backwards	14.3	2.4
Run	14.8	2.7

From [33]

be achieved by including foods from the following four groups in daily meal planning:

- Dairy producls
- Meat or meat substitutes
- Grains and fruit
- Vegetables.

Salt, simple sugars, and saturated fats should be kept to a minimum. The nutritional requirements of an adolescent vary according to his or her activity. Puberty is a very sensitive period, particularly in girls. Attitudes in modern society frequently lead young girls, in particular, to believe that they are overweight, and there has been a tremendous increase of anorexia and bulimia. Careful observation

of adolescents is necessary to discover "unphysiologic" tendencies.

122.4 Neurologic and Sensomotoric Development

122.4.1 Motor Development

Reflexes. Assessment of motor development begins with the determination of motor age by the best performance using milestone criteria, as shown in Table 122.9. These milestones are defined as major motor achievements at an specific age, with relatively large variations. Neonatal

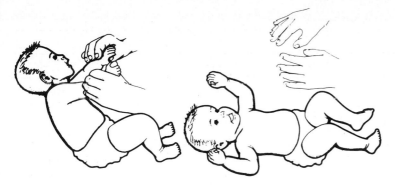

Fig. 122.16. The Moro reflex. The "embrace" response is elicited by sudden neck extension or by slapping the side of the baby's pillow. The reflex is present at birth and disappears between 3 and 6 months

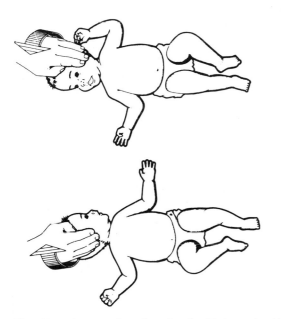

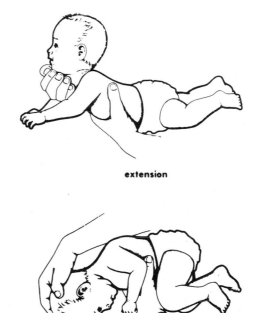

extension

flexion

Fig. 122.17. Asymmetric tonic neck reflex. Limbs on the side the chin is turned towards are extended, and limbs on the occiput side are flexed. The reflex is present at birth and disappears between 3 and 6 months

Fig. 122.18. Tonic labyrinthic reflex is elicited by extending or flexing the neck. The reflex is present at birth and disappears between 3 and 6 months

behavior is primarily dominated by subcortical brain structure activity. Intrinsic brain stem reflexes, such as the *Moro*, *rooting*, and *grasp reflexes* are prominent during the neonate period (Figs. 122.16–19). By 3 months of age, nonpurposeful limb movements have been replaced by more coordinated reaching movements. Infants of 2–3 months of age often open their hands during forward extension of the arm while visually fixing an object as a preparatory maneuver to manipulate the object. A behavioral hallmark of 2- to 4-month-old infants is the attenuation and gradual disappearance of intrinsic brain stem reflexes, as a result of increasing cortical influence.

Electroencephalography. In accordance with this development, major changes are observed on electroencephalography (EEG), with considerable maturations taking place during this period. Newborn EEG patterns, such as trace alternant, frontal delta rhythm, and frontal sharp transients, disappear, and the precursors of alpha rhythm appear. By 8 months of age, higher cortical and cognitic function is present. The infant now shows a more sophisticated interaction with his or her surroundings and exhibits the phenomenon of anxiety in the presence of strangers. In addition, the infant improves its performance in the delayed-response task, a neuropsychological response task commonly used for evaluating prefrontal lobe integrity. Together, these changes in the skills of the infant imply increasing function in the prefrontal cortex. Neuroanatomically, this stage is accompanied by an expansion

Fig. 122.19. Positive support reflex is elicited by bouncing the hallucal areas on a firm surface. The reflex is present at birth and disappears at 2–3 months

of dendritic fields and an increase in capillary density of human frontal cortex.

122.4.2 Visuomotor Development

Visuomotor Tasks. Performance of the earliest visuomotor tasks can be assessed in the first 3 months of life. The visual neurosensory skills include visual fixation before 1 month of age and the development of visual tracking skills and the blink response to visual threat at 3–4 months. By this age, basic visual fixation and tracking skills approach full maturity. At the same time, the infant is coming out of the *flexus habitus* stage (with suppression of primitive reflexes; Table 122.10). By 3 months, the infant should be able to bring his hands to the midline. With the development of visual tracking abilities and early upper extremity control, eye, head, and upper extremity movements can be used in combination. At this stage, fine-motor problem-solving skills can begin, to be assessed, starting with the ability to reach, grasp and transfer objects from one hand to the other by 5 months of age. In the following months, the infant's abilities become more sophisticated, and assessment concentrates on tasks requiring the manipulation of blocks, peg boards, form boards, and pencil and paper. As the child enters the preschool age, visuomotor abilities become increasingly complex and require the evaluation of a psychologist to describe and quantify the child's abilities. The tests include developmental and intelligence scales, which are used to distinguish between pathological and normal development (Table 122.11).

Language Development. An infant's expressive language development begins in the prelinguistic phase, with the sequential occurrence of cooing, babbling, indiscriminate "dada" and "mama," and discriminate "dada" and "mama," followed by the child's first true word at about 1 year of age (Table 122.12). With the development of words

used spontaneously and with a clear meaning, the child enters the linguistic phase (Figs. 122.20, 122.21). Between 12 and 24 months, there is an accelerating increase in vocabulary size, which continues into and through the school years. The use of two-word sentences is a milestone at 2 years of age (Figs. 122.20, 122.21). Receptive language development can be traced back to the prelinguistic phase of the first several months of life. The earliest receptive language skills are neurosensory. They represent peripheral auditory functioning and the CNS response to sound. The normal newborn infant responds to sound by crying, becoming quiet, or by otherwise changing its state and by

Table 122.10. Visual and motor development

Age (months)	Skills
1	Visually fixates momentarily, prone I
2	Visualy follows horizontally/vertically, prone II
3	Visually follows in circle, prone III, visual threat
4	Unfisted, manipulates fingers, prone IV
5	Pulls down ring, transfers, regards pellet
6	Obtains cube, lifts cup, radial rake
7	Attempts pellet, pulls out peg, inspects ring
8	Pulls ring by string, secures pellet, inspects bell
9	Scissors grasp, rings bell, looks over the edge for toy
10	Combines cube in cup, uncovers bell, fingers pegboard
11	Mature overhand pincer movement, solves cube under cup
12	Releases one cube in cup, marks with crayon
14	Solves glass frustration, out–in with peg, solves pellet in bottle with demonstration
16	Spontaneously solves pellet in bottle, round block in form board, imitates scribble
18	Ten cubes in cup, round hole in reversed form board, spontaneous scribble with crayon, completes pegboard spontaneously
21	Obtains object with stick, square in form board, tower of three
24	Folds paper I, horizontal four-cube train, imitates pencil stroke, completes form board
30	Horizontal and vertical pencil strokes, reversed form board, folds paper II, train with chimney
36	Bridge, copies circle, names one color, draws a person with head and one other part

The Child's Apperception Test (CAT) is a pediatric clinical observational instrument that measures fine motor, adaptive, and visual perceptual skills in the first 2 years of life and is intended to supplement the Clinical Linguistic and Auditory Milestone Scale (CLAMS) in the diagnosis of mental retardation and other developmental disabilities in very young children. From [1]

startling, blinking, or showing other recognizable responses. By 4 months of age, the child orients to voices by turning to the source of the sound. By 9 months of age, the child should indicate his understanding of interactive gesture games by participating in them. It should follow a single-step command accompanied by a gesture at 12 months and without a gesture by 15 months. At this age the child should be able to point to body parts on request, and by 2–2.5 years it should be able to follow two sequential, independent commands.

Table 122.11. Skills with pencil and paper

Skill	Age
Simple marks	12 months
Scribble in imitation	15 months
Scribble spontaneously	18 months
Stroke	24 months
Horizontal and vertical strokes	27 months
Circle in imitation	30 months
Copy circle	36 months
Copy cross	42 months
Copy square	4 years
Copy triangle	5 years
Copy horizontal diamond	6 years
Copy vertical diamond	7 years
Copy Greek cross	8 years
Copy cylinder	9 years
Copy cube	12 years

From [33]

122.5 Childhood

Adaptation to the needs of organ function during childhood is dependent on the somatic and psychosocial demands. The energy requirements have to be met by a changing attitude to nutrient and fluid intake. The recommended daily allowances (Table 122.7) show a steady decrease in protein needs from 1.2 g/kg body weight in young children to 0.8 g/kg in adults. The energy requirements decrease by the same time from 90 kcal/kg (376 kJ/kg) to 40 kcal/kg (167 kJ/kg). The necessary trace elements and vitamins are shown in the Table 122.7.

122.6 Adolescence

Hormonal Changes. With sexual maturity, a tremendous change in regulatory parameters occurs. According to Marshall and Tanner [25,26] (Tables 122.13, 122.14) the pubertal stages differ between boys and girls, the latter developing earlier. These pubertal phenotypes are accompanied by a growth spurt with increased peak height velocity of 12 and 14 cm in girls and boys, respectively. These pubertal changes are under hormonal control (see Chaps. 18–20, 113, 114) with a specific action of various hormones on organ development (Table 122.15). The hormonal changes begin before any outward physical signs of puberty are evident. Increased production of adrenal sex steroids occurs about 2 years before maturation of the hypothalamic–pituitary–gonadal axis. The major adrenal

Table 122.12. Milestones of auditory and linguistic development

Age (months)	Expressive milestone	Receptive milestone
1		Alerts, soothes
2		Social smile
3	Coos	
4	Laughs	Orients to voice
5	"Ah goo," raspberry	Orients (I)
6	Babbles	
7		Orints (II)
8	"Dada" (inappropriately) "Mama" (inappropriately)	
9	Gesture	Orients (III)
10	"Dada" (appropriately) "Mama" (appropriately)	Understands "no"
11	1 word	
12	2 words	One-step command with gesture
14	3 words, immature jargoning	
16	4–6 words	One-step command without gesture
18	Mature jargoning, 7–10 words	Points to one picture, points to body parts
21	20 words, 2–word phrases	Points to two pictures
24	50 words, 2–word sentences	Two-step commands
30	Pronouns, repeats two digits	Concept of one, points to seven pictures
36	250 words, 3–word sentence, repeats three digits, personal pronouns	Two prepositional commands

The Clinical Linguistic and Auditory Milestones Scale (CLAMS) is an infant language assessment intended for office use by the practicing pediatrician. From [33]

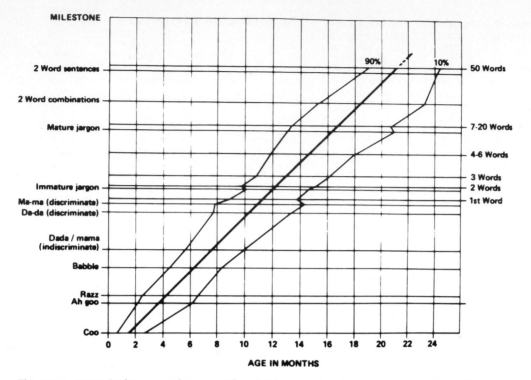

Fig. 122.20. Expressive language milestones; 90th and 10th percentiles of normal population. From [7]

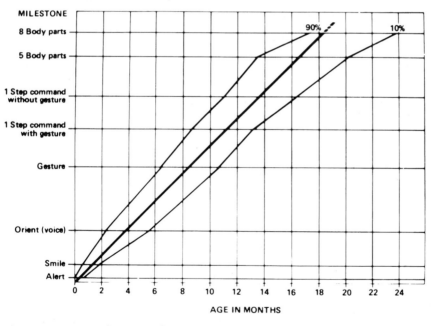

Fig. 122.21. Receptive language milestones. From [7]

steroids produced are dehydroepiandrosterone, androstendione, and estrone. Presumably because of a decreased sensitivity to the negative feedback system between the CNS and the testes or ovaries, the hypothalamus and pituitary gland begin to secrete increased amounts of gonadotropin-releasing hormone (GnRH), follicle-stimulating hormone (FSH), and luteinizing hormone (LH). This increase is apparent only during sleep in early pubertal adolescents, but by midpuberty it can be demonstrated during the day as well. Secondary to increased secretion of pituitary hormones, serum concentrations of testosterone in boys and estradiol in girls rise progressively during

Table 122.13. Pubertal changes in girls

Stage	Mean (years)	Standard deviation
Breast 2	11.15	1.10
Public hair 2	11.69	1.21
Peak height velocity	12.14	0.88
Breast 3	12.15	1.09
Pubic hair 3	12.36	1.10
Pubic hair 4	12.95	1.06
Breast 4	13.11	1.15
Menarche	13.47	1.02
Pubic hair 5	14.41	1.12
Breast 5	15.33	1.74

Adapted from [26]. Reprinted with permission from [12]

Table 122.14. Pubertal changes in boys

Stage	Mean (years)	Standard deviation
Genital 2	11.64	1.07
Genital 3	12.85	1.04
Pubic hair 2	13.44	1.09
Genital 4	13.77	1.02
Pubic hair 3	13.90	1.04
Peak height velocity	14.06	0.92
Pubic hair 4	14.36	1.08
Genital 5	14.92	1.10
Pubic hair 5	15.18	1.07

Adapted from [26]. Reprinted with permission from [12]

Table 122.15. Primary action of major hormones in puberty

Hormone	Sex	Action
FSH (follicle-stimulating hormone)	Male Female	Stimulates gametogenesis Stimulates development of primary ovarian follicles Stimulates activation of enzymes in ovarian granulosa cells to increase estrogen production
LH (luteinizing hormone)	Male Female	Stimulates testicular Leydig cells to produce testosterone Stimulates ovarian theca cells to produce androgens and the corpus luteum to synthesize progesterone Midcycle surge induces ovulation
Estradiol (E_2)	Male Female	Increases rate of epiphyseal fusion Stimulates breast development low level enhances linear growth; a high level increases the rate of epiphyseal fusion Triggers midcycle surge of LH Stimulates development of labia, vagina, uterus, and ducts of the breasts Stimulates development of a proliferative endometrium in the uterus Increases fat mass of the body
Testosterone	Male Female	Accelerates linear growth Increases rate of epiphyseal fusion Stimulates development of the penis, scrotum, prostate, and seminal vesicles Stimulates growth of pubic, facial, and axillary hair Increases larynx size and thus deepens the voice Stimulates sebaceous gland secretion of oil Increases libido Increases muscle mass Increases red blood cell mass Accelerates linear growth Stimulates growth of pubic and axillary hair
Progesterone	Female	Converts a proliferative uterine endometrium to a secretory endometrium Stimulates lobuloalveolar breast development
Adrenal androgens	Male and female	Stimulates pubic hair and linear growth

Reprinted with permission from [28]

physical maturation. Increased secretion of growth hormone becomes evident during midpuberty, correlating to the pubertal growth spurt during puberty.

References

1. Accardo PJ, Capute AJ (1993) Mental retardation. In: Oski FA, de Augelis CD, Feigin RD, McMillan JA, Warshaw JB (eds) Principles and practice of pediatrics. Lippincott, Philadelphia

2. Adams PAJ (1971) Control of glucose metabolism in the human fetus and newborn infant. Adv Metab Disord 5:183

3. Aperia A, Broberger O, Thodenius K, Zetterström R (1975) Development of renal control of salt and fluid homeostasis during the first year of life Acta Paediatr Scand 64:393–398

4. Aynslee-Green A, Bloom SR, Williamson DH, Turner RC (1977) Endocrine and metabolic response in the human newborn to first feed of breast milk. Arch Dis Child 52:291–295

5. Brück K (1963) Temperature regulation in the newborn infant. Biol Neonate 3:65

6. Calcagno PL, Rubin MI (1963) Renal extraction of para aminohippurate in infants and children. J Clin Invest 28:1623
7. Capute AJ, Palmer A, Shapiro BK et al (1986) Clinical linguistic and auditory mile stone scale: prediction of cognition in infancy. Dev Med Child Neurol 28:762
8. Celsi G, Aperia A (1994) Sodium, chloride, and water excretion. In: Holliday MA, Barratt TM, Avner ED (eds) Pediatric nephrology, 3nd edn. Williams and Wilkins, Baltimore, pp 99–116
9. Cochrome CG, Revah SD (1991) Pulmonary surfactant: protein B (SP-B: structure-functions relationship). Science 254:566
10. Day RL (1941) Regulation of body temperature during sleep. Am J Dis Child 61:734
11. Fomon SJ (1974) Infant nutrition, 2nd edn. Saunders, Philadelphia
12. Friedman IM, Gold berg E (1980) Reference materials for the practice of adolescent medicine. Pediatr Clin North Am 27:193
13. Gomez RA, Robillard JE (1984) Developmental aspects of the renal responses to haemorrhage during converting enzyme inhibition in fetal lambs. Circ Res 54:301–312
14. Gomez RA, Rupilli C, Everett AD (1991) Molecular and cellular aspects of renin during kidney ontogeny. Pediatr Nephrol 5:80–87
15. Guignard JP, Torrado A, DaCunha O, Gautier E (1975) Glomerular filtration rate in the first three weeks of life. J Pediatr 87:268–272
16. Haagsman HP, van Golch LMG (1991) Synthesis and assembly of lung surfactant. Am Rev Physiol 53:441
17. Harries JT (1982) Fat absorption in the newborn. Acta Paediatr Scand [Suppl] 299:17–23
18. Hawgood S, Shiffer K (1991) Structures and functions of the surfactant – associated proteins. Am Rev Physiol 53:575
19. Hey E (1971) The care of babies in incubators. In: Hull D, Gairdner D (eds) Recent advances in pediatrics. Churchill, London
20. Hey E (1994) Thermoregulation. In: Avery GB, Fletcher MA, Mc Donald MG (eds) Neonatology. Pathophysiology and management of the newborn, 4th edn. Lippincott, Philadelphia, p 361
21. Hey EN, Mount L (1960) Temperature control in incubators. Lancet 2:202
22. Lebenthal E, Lee PC, Heitlinger LA (1983) Impact of development of the gastrointestinal tract on infant feeding. J Pediatr 102:1–9
23. Leititis J, Burghard R, Gordjani N, Kaethner T, Brandis M (1987) Entwicklungsphysiologische Aspekte der Volumen- und Natriumregulation bei Frühgeborenen und reifen Neugeborenen. Monatsschr Kinderheilkd 135:3–9
24. Lifshitz F, Moses N (1991) Nutrition for the school child and adolescent. In: McLaren, Burmann D, Belton, Williams AF (eds) Textbook of pediatric nutrition. Charchill-Livingstone, Edinburgh, p 59
25. Marshall WA, Tanner JM (1970) Variations in patterns of pubertal changes in boys. Arch Dis Child 45:132
26. Marshall WA, Tanner JM (1970) Variations in the pattern of pubertal changes in girls. Arch Dis Child 44:291
27. Mendelson CR, Boggoran V (1991) Hormonal control of the surfactant system. Am Rev Physiol 53:415
28. Weinstein LS (1991) Adolescent health care: a practical guirde, 2nd edn. Urban and Schwarzenberg, Baltimore, p 6
29. Nelson N (1994) Physiology of transition. In: Avery GB, Fletcher MA, MacDonald MG (eds) Neonatology: pathophysiology of the newborn, 4th edn. Lippencott, Philadelphia, pp 223–247
30. Neville MC, Allen JC, Waters C (1983) The mechanism of milk Secretion. In: Neville MC, Neifert MR (eds) Lactation: physiology, nutrition and breast feeding. Plenum, New York, pp 44–103
31. Nishimura H (1978) Physiological evolution of the renin-angiotensin system. Jpn Heart J 19:806–822

32. Notter RH, van Golch LMG (1991) Synthesis and assembly of lung surfactant. Am Rev Physiol 53:441
33. Palmer FB, Capate AJ (1993) Developmental disabilities. In: Osk FA, de Angelis CD, Feigin RD, McMillan JA, Warshaw JB (eds) Principles and practice of pediatrics. Lippincott, Philadelphia, p 593
34. Radford EP (1957) In: Remington JS (ed) Tissue elasticity. American Physiological Society, Washington DC
35. Schwartz GI, Brion LP, Spitzer A (1987) The use of plasma creatinine concentration for estimating glomerular filtration rate in infants, children and adolescents. Pediatr Clin North Am 34:571–590
36. Schwartz GJ, Haycock GB, Spitzer A (1976) Plasma creatinine and urea concentration: normal values for age and sex. J Pediatr 88:828–830
37. Sharad DV, Ivengarm L (1972) Composition of the human fetus. Br J Nutr 27:305
38. Siegel SR, Oh W (1976) Renal function as a marker of human fetal maturation. Acta Paediatr Scand 65:481
39. Silverman WA, Fertig JW, Berger AP (1958) The influence of the thermal environment upon the survival of newly born premature infants. Pediatrics 22:876
40. Smith CA, Nelson NM (1976) Physiology of the newborn infant, 4th edn. Thomas, Springfield, p 123
41. Srinivasan G, Pildes RS, Cottamanchi G et al (1986) Plasma glucose values in normal neonates: a new look. J Pediatr 109:114
42. Strang LB (1965) The lungs at birth. Arch Dis Child 40:575
43. Subcommittee on the 10th edition of the RDAS, Food and Nutrition Board, Commission on Life Science, National Research Council (1989) Recommended dietary allowances, 10th edn. National Academy Press, Washington DC
44. Sulyok E, Nemeth M, Tenyi I, Csaba E, Györy E et al (1979) Postnatal development of renin-angiotensin-aldosterone system, RAAS, in relation to electrolyte balance in premature infants. Pediatr Res 13:817
45. Uauy R, Mena P, Warshaw JB (1994) Growth and metabolic adaptation of the fetus and newborn. In: Oski FA, de Angelis CD, Feigin RD, McMillan JA, Warshaw JB (eds) Principles and practice of pediatrics, 2nd edn. Lippincott, Philadelphia, pp 279–289
46. Wharton BA (1978) Childhood. In: Dickerson JWT, Lee HA (eds) Nutrition in the clinical management of disease. Arnold, London, pp 1–28
47. Whitehead RG (1985) Infant physiology, nutritional requirements and lactational adequacy. Am J Clin Nutr 41:447–458
48. Whitehead RG, Paul AA (1984) Growth charts and the assessment of infant feeding practices in the western world and in developing countries. Early Hum Dev 9:187–207
49. Widdowson EM, Dickerson JWT (1973) Composition of the body. In: Diem K, Lentner C (eds) Scientific tables. Ciba-Geigy, Basel, p 517
50. Williams AF (1991) Lactation and infant feeding. In: McLaren DS, Barman D, Belton NR, Williams AF (eds) Textbook of infant nutrition, 3rd edn. Churchill Livingstone Edinburgh, pp 21–45
51. Wilson MH (1994) Feeding the healthy child. In: Oski FA, de Angelis CD, Feigin RD, McMillan JA, Warshaw JB (eds) Principles and practice of pediatrics, 2nd edn. Lippincott, Philadelphia, pp 590–612
52. Woolridge MW, Baum JD (1988) The regulation of human milk flow. In: Lindblad BS (ed) Perinatal nutrition. Academic, New York, pp 243–257
53. World Health Organisation (1985) Energy and protein requirements. Report of a joint FAO/WHO/UNU Expert Consultation. World Health Organisation Technical Report Series 724:1–286
54. Yared A, Ichikawa I (1994) Renal blood flow and glomerular filtration. In: Holliday MA, Barratt TM, Avner Ed (eds) Pediatric nephrology. Williams and Wilkins, Philadelphia, pp 62–78

123 Physiology of Aging:
Standards for Age-Related Functional Competence

P.S. Timiras

Contents

123.1 Introduction

So much negativity has already been attached to physiological changes with aging that this chapter elects to focus on more positive aspects such as the persistence of adequate function of several organs and systems into old age. Decrements in most functions have been described in detail and have already been set forth in several textbooks [63,88]. In this chapter it has been elected not to repeat this. With the increasing duration of the lifespan of a large percentage of the human population, let us hedge our bets on the positive changes possible at this time.

So far, comparison of various functions in individuals 65 years of age and older with those of 25 years has underlined a progressive and irreversible decline. From this comparison, aging has emerged as a deteriorative process leading to functional impairment. A more equanimous definition may follow a more positive tack and consider aging as the sum of all changes occurring with the passage of time, both deteriorative and regenerative/compensatory changes. Interventive measures to strengthen physiological competence in old age will aim to prevent or minimize degenera-tive changes and to enhance regenerative/compensatory processes.

This chapter considers:

- Comparative and epidemiological aspects of aging
- Assessment of physiological age in the elderly
- Persistence of compensatory function and quality of life in old age
- Theories of aging.

123.2 Comparative Aging

123.2.1 Definitions of Aging

Aging is usually considered to be a normal phenomenon that occurs in all members of a population. Currently, the definitions of aging are as numerous as the theories of aging. Those chosen here are selected on the basis of their physiological relevance and reflect a continuum of events from development to maturity and to senescence (that is, old age). They include:

- Aging as the sum of all changes occurring in an organism with the passage of time. This views aging as a stage of the lifespan.
- Aging as the sum of all changes occurring with time and leading to functional impairment and death. This sees aging as a deteriorative process.
- Aging as changes in cell membrane, cytoplasm, and/or nucleus. This considers aging as the result of genetic, cellular, and molecular damage to the cell and cell molecules.

The first definition visualizes the lifespan as an orderly unfolding of precisely timed events from fertilization to death, with aging as the last stage of the lifespan. However, we cannot ignore the fact that aging is a deteriorative process characterized by increased vulnerability and decreased viability. Hence the first definition must be amended to indicate that, with advancing age, functional competence decreases; in particular, the capacity of maintaining homeostasis, i.e., the constancy of the internal environment (Part A) declines. Thus, aging may be further defined as a *decreasing ability to survive stress*.

Aging as molecular and cellular damage focuses on progressive cellular changes in membranes (e.g., plasma, mitochondrial membranes), cytoplasm (e.g., free radical accumulation, cross-linking, decreased energy metabolism), and nucleus (e.g., DNA damage and DNA repair failure, mRNA error catastrophe, errors in replication and histocompatibility complex). Throughout life, most damage undergoes continuing repair due to intrinsic repair mechanisms, but with advancing age alterations may accumulate due to continuing damage and insufficient repair and lead to declining physiological competence and increasing pathology.

123.2.2 Physiological Correlates with Longevity

Despite the many differences among individuals within the same species and among phyla and species, due to genetic make-up, sex, past history, and life conditions, it is undeniable that several characteristics are applicable to the aging process in general [18,30,63,88]. Among common factors governing longevity (that is, duration of life) in animals are: body weight, brain weight (in relation to body weight), metabolic rate, reproductive function, and response to stress.

Longevity is *directly* correlated with:

- Body weight (the bigger the animal, the longer the lifespan)
- The ratio of brain weight to body weight (the bigger the brain, related to body weight, the longer the lifespan)
- The duration of the growth period (the longer the period preceding maturity, the longer the lifespan).

Longevity is *inversely* related to:

- Metabolic rate (the higher the metabolic rate the shorter the lifespan)
- Reproductive function and fecundity (the larger the number of litters, the shorter the lifespan)
- Stress (the more severe the accumulation of stress-induced damage, the shorter the lifespan).

Natural selection suggests that survival is necessary through the reproductive period to ensure continuation of the species; thereafter, survival becomes indifferent or detrimental. In this sense, a gene that acted to ensure a maximum number of offspring in youth but produced disease at later ages might be positively selected.

In humans, not only has life expectancy increased in modern times, but humans live well beyond the reproductive years, women some 40 years beyond the childbearing period. Postreproductive survival must be incidental to some earlier events or express some societal advantage. In our society, older members contribute to the maintenance, development and progress of the entire population structure [20].

123.2.3 Differential Aging in Humans

Chronologic age, or age in number of years, and physiological age, or age in terms of physiological capacity, do not aways coincide. Rather, appearance and health status often belie chronological age: in many cases a person may look younger or older than his/her chronological age and may "age" at a faster or slower rate than others. Individual differences result from complex interactions between genetic and environmental factors; these operate on the individual as the child of his/her parents and also as a member of a specific societal group. Thus, one characteristic of the aging human population is its "heterogeneity."

Despite this heterogeneity, attempts have been made to generalize aging processes under selected categories. Some have characterized aging changes in all animal species as *"universal, intrinsic, progressive, irreversible,* and *deleterious"* [87]. Other attempts to generalize the major health problems of the elderly have grouped them under clinically identifiable conditions, the so-called five "I's": *Instability* (referring to the lability of circulatory and endocrine adjustments), *Immobility* (referring to the decrements in neuromuscular and skeletal competence), *Incontinence* (referring to the declining function of the urinary system), *Impaired cognition* (referring to loss of memory and other mental impairments), and *Iatrogenic diseases* (referring to the increasing need for administration of medications, the so-called polypharmacy, which is often the cause of toxic effects [26].

In addition to physiological decline, aging is also inevitably accompanied by increased incidence and severity of diseases, accidents, and stress. Deleterious factors not lethal in themselves may add to the physiological decline and predispose the individual to functional losses and specific diseases in later life. Death from "pure" old age is rare; it usually results, so-to-speak prematurely, from increased pathology superimposed on homeostatic insufficiency. Pathology introduces another variable, indeed, a very significant one, to physiological aging.

123.2.4 "Successful" Versus "Usual" Aging

Not only does functional competence and incidence of disease vary within the same age group and among individuals of the same species and age, but they also vary from age to age, and from one parameter (organs, tissues, cells, or molecules) to another. Thus, the rate of decline with aging differs significantly among organ systems and varies for the same organ in different individuals. This variability implies that factors (environment, nutrition, lifestyle, stress, disease) other than age itself may modulate the "biological clock" or the genetic program that determines how we age.

Deteriorative changes in one system or structure do not always signal aging of the whole organism. In certain functions, the regulation of the organism remains quite efficient well into old age (80–90 years in humans), while in others, it declines early in life [63,88]. The heterogeneity of the aging process is also illustrated by differences in the rate of aging among individuals [70]. Currently, in the absence of appropriate age-related indices to assess functional capacity, it is difficult to estimate the physiological profile of individuals at consecutive chronological ages. One prerequisite of "normal" aging is the absence of pathological changes. Thus, the establishment of age-related profiles of physiological aging should exclude individuals whose age-determined responses might be "contaminated" by specific disease processes. Indeed, aging itself is viewed by some as a pathological process [45]. In attempting to define *"usual" (i.e., "normal") aging*, the focus has been on *"average" aging*. This focus neglects those older people who demonstrate little or no loss in a constellation of functions, that is, who age "successfully." Such individuals may serve to draw the physiological standards for their age group. *"Successful" aging* allows us to focus on those particular traits which are prevalent in this population and to select those environmental factors that are most conducive to success in aging.

123.3 Need to Establish Standards of Optimal Physiological Competence in the Elderly

123.3.1 Demographic Aging

One of the most dramatic changes in human populations witnessed in this century is the marked prolongation of the average lifespan. It is a major achievement of civilization that life expectancy of humans has increased throughout history. It is, however, during this century that the greatest progress has been made, with the length of the average

lifespan increasing from 50 years in 1900 to 75 years in 1990 in North America, Japan, and several European countries [42,91]. Not only are the elderly living longer, but they also represent the most rapidly growing segment of the population in developed countries. In the United States today, persons 65 years of age and older represent 13% of the population. By the year 2030, this proportion is expected to rise to 22%. The elderly population itself is aging: of those 65 years and older, approximately 10% are 85 years or older, and by the year 2010 this proportion will increase to about 16% [8]. In Japan, the doubling of the proportion of the elderly from 7% to 14 % of the total population will have been achieved in only 25 years, from 1970 to 1995 [44].

This trend, the so-called "greying" of the population, is not limited to the developed regions but has now become a world wide phenomenon and has expanded to include the developing regions as well [46,89,90] (Fig. 123.1). This "demographic transition," that is, the dynamic process by which a population ages, may be related to a shift from high fertility/ high mortality to low fertility/low mortality and consequently from a low to a high proportion of elderly people within the general population. This unprecedented progression of the elderly population is expected to continue and, indeed, to increase thoughout the world by the year 2000 and beyond [83] (Fig. 123.1).

In general, aging of a population may be ascribed to (1) longer survival, resulting in decreasing mortality rates, and (2) declining birth rates and consequent higher proportion of individuals in older age groups. Both ways seem to be operative at this time: birth rates continue to fall (notably in developing regions) and more people survive into old

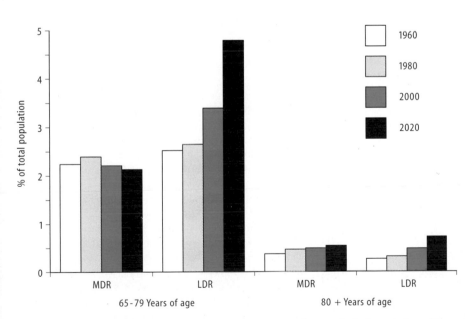

Fig. 123.1. Distribution of aging population of more (*MDR*) and less (*LDR*) developed regions of the world between 1960 and 2020 (projected). The total world population is expected to double over this time interval, while the aging population triples. This is generally the consequence of a decrease in the birth rate and increasing survival of people into old age. The more developed regions include: Northern America, Europe, Japan, Australia/New Zealand and the former USSR. The less developed countries include: Africa, Latin America, Asia (excluding Japan) and Oceania (excluding Australia/New Zealand). (From [42])

2393

age. A better knowledge of the physiological adjustments that may be associated with the longer lifespan is indispensable for strengthening physiological competence and thereby improving the quality of life for the elderly.

While the average lifespan has been significantly extended, the maximum lifespan, that is, the age reached by the longest-lived survivors of a large population, has changed little over past centuries, irrespective of the level of development of the regions considered; for humans, the maximum lifespan is 110–115 years, compared to 38–40 months for the mouse and 70 years for the elephant [96]. The average lifespan can be – and, in fact, has been – extended by improving physiological competence and preventing or curing disease, but for extending the maximum lifespan it is the aging process itself that must be inhibited, a feat not yet accomplished [78].

123.3.2 Persistence of Physiological Competence in Old Age

As people live longer, the classical profile of the 25-year-old male taken as the standard of optimal physiological competence against whom all other physiological patterns must be compared appears unrealistic and outdated. Evaluation of the competence of the elderly must take into account the increasing demands imposed by the progressive pathology associated with old age. The association of aging with disease affects the practical orientation of aging research. Some physiologists follow the traditional view that research in aging should be aimed at specific functional and clinical entities (e.g., cardiovascular aging and atherosclerosis; aging of the central nervous system and dementia of the Alzheimer type) rather than at aging as a whole. Other investigators argue that treatment and even cure of many diseases would prolong life only a few years; for example, if cardiovascular diseases were cured overnight, the lifespan would be extended only by a few years. Efforts, therefore, should be directed to elucidate the basic mechanisms of the *aging process per se*. Both research goals have merits and should be pursued simultaneously.

At the present time, 25 years no longer represent half of the lifespan ("The midway along the journey of our life" [2]). The physiological profile of the old individual that emerges must be adapted to the new circumstances. It must be evaluated against a standard, optimal for each specific age. It should replace the 25-year-old standard and serve as a measure against which to assess the physiological competence of individuals of comparable age, at different stages of aging. Physiological adjustments remain operative – albeit with reduced effectiveness – into old age and assure survival well beyond 50 years of age. Likewise, the standards of competence should change with progressive age. If physiological competence is taken to mean the capacity and readiness to respond to internal and external demands and the maintenance of the dynamic constancy of the internal environment (that is, homeostasis), then the old individual may be considered competent.

123.3.3 Assessment of Physiological Age in the Elderly

Establishing a "physiological profile" at any age requires quantitative multifactorial measurements of numerous parameters selected as indices of physical (e.g., tests of physiological performance) and mental (e.g., psychological and psychiatric tests) competence (Table 123.1). With aging, such measurements may be taken at progressive ages. Functional assessment may be defined as the measure of an individual's ability to function in the arena of everyday living (Table 123.2). "Aging charts" similar to "growth charts" have already been attempted and continue to be proposed [13,47,48,72–74]. The purposes of these charts include: (1) description of the physiological status of an individual at progressive ages; (2) screening of a selected population for overall physiological competence using sampling methods, either cross-sectional (i.e., comparing physiological characteristics among different age groups at one time) or longitudinal (i.e., comparing physiological characteristics in the same individuals examined at regular intervals throughout their life); (3) monitoring of the efficacy of specific treatments, exercise, dietary regimens, drugs and (4) prediction of persistence or loss of physiological competence, determination of incidence of disease, and evaluation of life expectancy. Although the need for such assessments is apparent, it remains difficult to choose the most significant and feasible tests, and consensus has not been reached on the usefulness of any of the current geriatric assessments and screening programs. The choice is also complicated by the financial feasibility and facilities available for the testing. Assessment itself entails listing a person's limitations or, vice versa, a person's strengths, depending on the pessimism or optimism of the beholder. The interpretation in terms of care plans is subject not only to biomedical but also to social and economic constraints.

123.4 Functional Decline with Aging

Comparison of several functions, at the systemic and cellular levels, from young to old age shows continuing decline with old age (Fig. 123.2) [82]. Most of the studies from which these data were obtained [82], and, in general, most observations of physiological measurements emerging from early studies were based on an elderly population often hospitalized or confined to nursing homes. Efforts were made, when investigating a specific function, to choose only those individuals without (overt) pathology affecting the specific system under study. Nevertheless, the general decline reported in a possibly debilitated population may not be representative of what is the norm for that function and that age in healthy, independently living elderly. As already suggested in this chapter, the careful screening of a "successfully" aging population may be the choice for establishing physiologic parameters of aging.

Table 123.1. Measurements of physiological age in man

History: Medical, occupational, and nutritional; use of medications, alcohols, and tobacco

Body composition and anthropometrics
[a]Total volume, [a]solids, [a]lipids, and [a]water
[a]Exchangeable body sodium
[a]Total protoplasmic mass
[a]Body weight
[a]Stature and stem length
[a]Diameters and circumferences of head, trunk, and limbs
[a]Grip strength, right and left hands
[a]Skin fold thickness
[a]Skin elasticity
[a]Thickness of cortical bone

Circulatory and respiratory functions
[a]Systolic and diastolic blood pressure
[a]Ventricular rate
[a]Electrocardiogram
[a]Basal metabolic rate
[a]Respirations per minute
[a]Vital capacity: forced and 1s
Partial pressure of carbon dioxide
[a]Functional residual capacity of the lungs
[a]Diffusing capacity of the lungs
[a]Chest X-ray
[a]Cardiorespiratory performance during and after exercise
Liver circulation time

Renal functions
Urinalysis (pH, specific gravity, protein, blood)
Urine culture
[a]Creatinine clearance
[a]Tubular reabsorption of phosphate
[a]Phenolsulfonphthalein excretion

Endocrine functions
[a]Glucose tolerance
Plasma insulin content
Protein-bound iodine
Free thyroxine
Rate of excretion of hormone or hormone metabolites in urine:
  [a]17-Hydroxycorticosteriods
  [a]17-Oxysteroid, total, and individual components of adrenal and testicular origin
    Epinephrine
    Norepinephrine
    Dopamine

Neurological functions (including sensory)
[a]Standard neurological examination
[a]Electroencephalogram
[a]Evoked responses: flashing light and auditory stimuli
Electromyogram
Optometric examination:
  [a]Visual acuity
  [a]Phoria
  [a]Amplitude of accommodation
  Intraocular pressure
  [a]Retinal visual fields
  [a]Central visual fields
  [a]Flicker fusion frequency
[a]Auditory acuity: 250–8000 Hz
[a]Taste threshold

Hematopoietic and immunological functions
Hemoglobin
Hemoglobin fractionation
Erythrocyte and leukocyte count
Total plasma volume
Packed cell volume
Erythrocyte sedimentation rate
[a]Leukocyte differential
Serologic test for syphilis
[a]Autoantibody titers

Somatic cell genetics
[a]Ploidy of lymphocytes

Metabolic function: blood chemistry tests
Urea nitrogen
Bilirubin
Uric acid
Creatinine
Phosphorus
Calcium
[a]Total protein
[a]Protein fractions
[a]Cholesterol and cholesterol esters
[a]Triglyceride
[a]Lipoproteins and lipoprotein phenotype
Alkaline phosphatase
Creatine phosphokinase
Glutamic pyruvic transaminase
Glutamic oxaloacetic transaminase
Lactic dehydrogenase
Lactic dehydrogenase isozymes

[a]These measurements are now believed to undergo changes associated with normal aging; other measurements included have been less well investigated across age and may permit differentiation of normal and pathological aging

Following these early studies, decrements in cardiovascular, endocrine, gastro-intestinal, immune, musculoskeletal, nervous, respiratory, and renal functions have been described in detail in several textbooks and the reader may find their system-by-system description in several textbooks [63,88]. As already stated in Sect. 123.1, it is the purpose of this chapter to draw the line between average or "usual" aging-related changes and physiological capability in "successfully" aging individuals, and to emphasize the persistence in old age of some degree of functional competence including the potential for functional compensation. Not all functions decline with aging, and, when they do,

they follow different timetables [63,88]. Regulation of blood glucose represents a function that possesses a number of alternate control mechanisms (cf. Chap. 66) and remains stable in a large percentage of the aged population (Fig. 123.2). A similar stability with aging is also manifested by control of acid-base balance (cf. Chap. 78), another function with multiple regulatory influences. It is to be noted that maintenance of normal glycemia is altered in a number of elderly: fasting blood glucose undergoes a slight age-related increase, and glucose tolerance tests show that, after oral glucose administration, blood glucose levels are higher and take longer to return to normal com-

Table 123.2. Categories of physical health index measuring physical competence[a]

Physical health	Activities of daily living	Instrumental activities of daily living
Bed days	Feeding	Cooking
Restricted-activity days	Bathing	Cleaning
Hospitalizations	Toileting	Using telephone
Physician visits	Dressing	Writing
Pain and discomfort	Ambulation	Reading
Symptoms	Transfer from bed	Shopping
Signs on physical examination	Transfer from toilet	Laundry
Physiologic indicators (e.g., lab	Bowel and bladder control	Managing medications
tests, X-rays, pulmonary and	Grooming	Using public transportation
cardiac functions)	Communication	Walking outdoors
Permanent impairments (e.g.,	Visual acuity	Climbing stairs
vision, hearing, speech,	Upper extremities (e.g., grasping	Outside work (e.g., gardening,
paralysis, amputation, dental)	and picking up objects)	snow shoveling)
Diseases/diagnoses	Range of motion of limbs	Ability to perform in paid employment
Self-rating of health		Managing money
Physician rating of health		Traveling out of town

[a] Many of the items presented are components of the Selected Measures of Physical Health, from various sources (see [88])

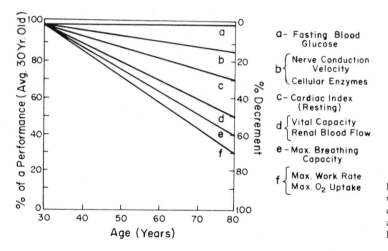

Fig. 123.2. Physiological performance declines with aging. Various (but not all) human functional capacities and physiological measurements decline at different rates with advancing age. (Courtesy of Dr. N.W. Shock)

pared to the younger individual [19]. This progressive increase in glucose intolerance with aging has been ascribed to various factors, among which increased resistance of cells to insulin and decreased number of insulin receptors and glucose transporters have been most extensively explored (see Chap. 66).

Other functions of the body begin to age relatively early in adult life and decline rapidly, as is the case with such sensory functions as vision and hearing, which begin to diminish in late childhood and early adolescence and continue to decline steadily thereafter. In the eye, for example, accommodation begins to decline in the teens and regresses to a minimum in the midfifties (Chap. 37). Similarly the auditory function (cf. Chap. 35) begins to deteriorate in adolescence and deterioration continues steadily thereafter, culminating around 50 years of age. This auditory deterioration may also be hastened by the environmental noise to which, in our civilization, individuals are continuously exposed from a young age.

Other aging processes begin very early in life but proceed slowly; their effects are observed only when they have progressed sufficiently to induce alterations that can be measured by available tests or cause overt pathological manifestations. An example is represented by some of the aging processes involving the vascular system, such as the progressive deterioration of the arterial wall in atherosclerosis. The atherosclerotic lesion often starts in infancy, progresses through adolescence (manifested by fatty streaks), young adulthood (manifested by fibrous plaques), and continues in adulthood (manifested by atheromas). Its consequences (e.g., coronary occlusion, aneurysm, cerebral infarction, gangrene) become manifest in middle and old age. Under these conditions, increased vascular resistance, decreased myocardial strength and impaired autonomic control with aging have been held responsible for the decreased (by as much as 30%) cardiac output (as measured by cardiac index, i.e., cardiac output per minute and per square meter of body surface) between ages 30 and 80 (Fig. 123.2). The severity of this decline may reflect, in addition to impaired circulation, an overall physical weakness due to multiple factors such as lack of exercise, inappropriate diet, intercurrent diseases, and inadequate socioeconomic support.

Tests of renal function (e.g., glomerular filtration rate [69], renal blood flow) (Chap. 73) and of respiratory function (e.g., vital capacity and maximum breathing capacity; Chap. 100) show a 35–65% decline with aging (Fig. 123.2). Other discrete physiological variables that decline with aging and are often taken as measures of functional impairment with aging include slower nerve conduction velocity (Fig 123.2), reduced insulin sensitivity [19], slower heart rate (according to some [50] but not others [53]), altered electroencephalographic (EEG) potentials [65] and loss of pulsatile hormone (growth hormone, thyroid-stimulating hormone, luteinizing hormone) release (37,41,92). Some of these functions, formerly thought to be relatively periodic, show a complex type of variability reminiscent of chaos (i.e., apparently unpredictable behavior arising from internal feedback loops of certain nonlinear systems) [35]. Aging might induce an age-related loss of "complexity" leading to impaired dynamics of physiological processes [55].

In the locomotor system, different muscle groups show different aging patterns. For example, the muscles of the leg are generally compromised earlier than those of the arm. Changes in muscle function depend also on the measurement: short exercise is slightly affected whereas sustained and maximal work is severely impaired (Fig. 123.2). Irrespective of the muscle or bone considered, muscle strength, as well as bone density and strength, reach a peak between 20 and 30 years of age and decline continuously thereafter. To what extent the decline in muscle and bone function can be prevented or reduced by physical exercise and other interventions is a question of considerable interest and is addressed in the following section.

123.5 Persistence of Compensatory Function in Old Age

123.5.1 Adaptive Adjustments in the Aging Heart

Despite the physiological declines in the elderly, some functions remain capable of dynamic adjustments even in old age. An illustration is the retention of competence and/or compensatory capability even in those systems such as the cardiovascular, most susceptible to pathology with aging. Cardiovascular disease such as coronary (ischemic) heart disease (resulting from coronary occlusion) remains, in both sexes, the single major cause of death in old age, worldwide, notwithstanding some impressive decline in mortality in the last 10–20 years in some regions [49]. Yet, in old, healthy persons with still adequate coronary circulation, cardiac output at rest, cardiac rate, and cardiac size

at end diastole/systole, are not substantially changed [53]. While the rate at which the left ventricle fills with blood during early diastole declines markedly between the ages of 20 and 80, the enhanced filling in later diastole keeps filling adequate in elderly individuals. The myocardium remains capable of adaptive hypertrophy to maintain normal heart volume and pump function in the presence of moderate increased systolic pressure. Such adaptive physiological response may be mediated through myocyte cellular hyperplasia [5] and may be due to persistent activation of cardiac gene expression and its hormonal regulation [15].

Although cardiac rate increases less during severe exercise in many elderly as compared to young individuals, stroke volume may increase to compensate for the smaller increase in heart rate [68]. The smaller increase in heart rate during exercise has been attributed to a reduced efficiency of the β-adrenergic modulation [71] rather than intrinsic alterations of the senescent myocardium [52,54]. Thus, current evidence shows that in healthy elderly people, in contrast to those with physical and emotional decline, overall cardiac function is adequate to meet the body's needs for pressure and flow at rest and to sustain physical exercise.

123.5.2 Responsiveness of Musculoskeletal Function to Hormones and Physical Exercise

With aging, from middle and late adulthood to senescence, all people experience a certain degree of shrinking of the lean body mass and an expansion of adipose tissue. Contraction of lean body mass (body mass minus bone, mineral, fat, and water) is due primarily to changes in muscle mass, including structural (atrophy, reduced myoglobin, increased deposition of lipofuscin and infiltration of connective, fibrous fibers) and functional (reduced strength and velocity of contraction) alterations [75]. The cause(s) of the decrease in muscle mass and function are not known [24]. However, the reduced availability of growth hormone (GH) in late adulthood may contribute to such loss.

GH, the major growth-promoting hormone in childhood, has several metabolic actions (anabolic, lipolytic, diabetogenic). Its secretion from the anterior pituitary is both pulsatile and diurnal. Pituitary GH tends to decrease starting at about 30 years of age. After 55 years of age, integrated 24-h GH blood concentrations are one-third lower than those in men and women aged 18–33 years, and the nocturnal pulses of the hormone become smaller or disappear with advanced age. This age-related deficit in GH has been ascribed to a reduced secretion of the hypothalamic GH-releasing hormone (GHRH) [14], or to a hypersecretion of the hypothalamic inhibitory factor (somatostatin releasing inhibiting factor, SRIF) [86], or to decreased GHRH pituitary binding [1]. Levels of insulin-like growth factor-I (IGF-I), primarily of hepatic origin and a mediator of GH actions, also are reduced in older people. This reduction, involving a relatively small percentage of the individuals studied, is related to age (11% in the fourth decade of life, 20% in the fifth, 22% in the sixth, and 55% in the eighth and ninth), and is inversely correlated with adiposity [41,76].

The effects of GH deficiency, such as atrophy of lean body mass and expansion of adipose tissue mass, were reversed by replacement doses of the hormone (now readily available through recombinant-DNA production) in experimental animals, children, and adults to about 50 years of age [79]. Likewise, GH administration for several months to healthy 60- to 80-year-old men who have below-normal GH levels (for age) increased IGF-I concentrations to the range found in young men. The higher IGF-I levels were also associated with a modest but consistent 10% increase in lean body mass and a 15% decrease in adipose tissue mass [77]. Other effects of such administration include: increase in vertebral bone density and skin-fold thickness, small increases in exercise capacity, isometric strength of muscle, and basal metabolic rate.

Some adverse metabolic effects were also observed: e.g., increased fasting serum glucose and insulin levels. In view of the potentially adverse effects of GH on glucose metabolism (e.g., diabetes) articulations (e.g., arthritis), and the cardiovascular system (e.g., hypertension), as exemplified in acromegaly (Chap. 21), and the promising but relatively modest benefits on muscle function, the use of GH in healthy older adults remains to be evaluated critically [93]. Further studies must assess the questions of dose, duration of treatment, side effects, age at onset of treatment, etc. However, the possibility of restoring lean body mass and of improving (even if partially) muscle function by GH administration represents an important step in illustrating the capacity of cells and tissues to respond to stimulation by enhanced function even in old age.

Muscle responsiveness to appropriate stimuli is also demonstrated by the improvement induced by physical exercise in muscle strength, ambulatory ability, and endurance in the young and adult as well as the elderly, including the "oldest old" (80 years and older) [27]. For example, high-resistance weight training in nonagenarians leads to significant gains in muscle strength and size, and in motility [28]. Exercise training in 60- to 70-year-old men (but not women) increases bone density as well [12]. Functional changes during aging have often been compared to those induced by reduced motility (e.g., sedentary life, bed confinement, weightlessness of astronauts in space); thus, some aspects of loss of physiological integrity in the elderly resemble those attributed at all ages to "disuse" [25].

The quantitative losses of fibers, especially those of fast muscles, would be expected to be associated with alterations of the myoneural junction with consequent denervation atrophy and, ultimately, fiber disappearance [22,38]. However, reinnervation by motor nerve sprouting may also occur [3,34]. Such "sprouting effects" have provided evidence for the concept of a continual, dynamic turnover of synaptic sites in muscle (Chap. 48). Despite some conflicting reports, the prevalent view is that remodeling of aging neuromuscular junctions would replace those that degenerate after a limited lifetime [100].

Given the essentially unchanged contractile and metabolic properties of the fibers, adaptation to training may remain operative in some motor units. Thus, in old rats, treadmill running and ablation of one of two synergistic muscles significantly increases (by 45–75%) muscle mass and force-generating capacity (Fig. 123.3). The increase in old animals is comparable to that occurring in adults and young rats.

Together with a decrease in lean body mass, aging is characterized by a progressive decrease in bone mass. Such a decrease in bone mass begins in early adulthood in men and women, but in women it is accelerated during menopause. Loss of bone mass may result in osteoporosis, one of the major causes of bone fractures and a major health problem [67]. Prevention of bone fractures during and after menopause may be achieved by estrogen replacement therapy, by increased dietary calcium intake, and by physical exercise. While there is agreement on the estrogen-induced benefit, the findings on the effects of physical exercise and dietary calcium supplements are somewhat conflicting. Inactivity promotes bone loss at all ages and physical exercise increases bone mass [85]. However, in older women, the effects remain controversial, probably because of the many variables to be considered (e.g., intensity, duration and type of exercise, selection of individuals to study, methods of functional assessment, complicating muscular and articular impairments) [12]. Nevertheless, a regimen of active physical exercise, especially when initiated in young adulthood, may improve bone mass and thereby prevent or reduce osteoporosis in old age [36].

Muscle and bone atrophy and weakness are among the major causes of mobility impairment in the elderly, and immobility itself leads to risk of falls, fractures, and functional dependency; therefore, the maintenance or initiation of a physically active lifestyle during old age as well as hormonal (GH, estrogens) replacement therapy will not only reduce skeletomuscular weakness and promote endurance but also will have an impact in improving overall health.

123.5.3 Plasticity and Regeneration of the Nervous System

Of the two major cellular components of the central nervous system (CNS), the glial cells develop later during ontogenesis and continue to proliferate throughout the lifespan in response to metabolic and reparative demands (Chap. 26). In contrast, the other CNS cells, the neurons, are established in number and biochemical and functional properties during early development and, in healthy individuals without neurologic or mental disorders, undergo little change thereafter (Chaps. 16, 24, 30). The failure of adult neurons to divide has been extended to indicate that, in the mature CNS, neurons lack the capacity of adaptive remodeling and regenerating. This view was supported by early studies reporting the dramatic loss of brain neurons (with moderate gliosis but without apparent neuronal

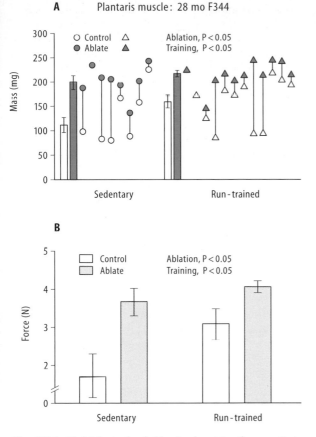

A Plantaris muscle: 28 mo F344

O Control △
● Ablate ▲

Ablation, P < 0.05
Training, P < 0.05

Mass (mg)

Sedentary   Run-trained

B

☐ Control
☐ Ablate

Ablation, P < 0.05
Training, P < 0.05

Force (N)

Sedentary   Run-trained

Fig. 123.3. Skeletal muscle of old animals retains the capacity to adapt to changes in exercise levels by increasing mass and force. Mass and force of a leg muscle (plantaris) were measured in female F-344 rats, specific-pathogen-free, aged 28 months (median lifespan of this rat colony is approximately 25 months). Two experimental exercise stimuli were studied: (1)treadmill running for a 5-month period and (2) ablation in one leg (the other remaining intact and serving as control) of the gastrocnemius muscle (the gastrocnemius being synergistic to the plantaris, removal of the former will increase the biomechanical load of the latter). *Bars* represent the mean and the *bracketed lines* the standard error of the mean. Individual data are shown in the *open* and *closed circles* and *triangles*. The magnitude of adaptive increase in muscle mass and force-generating capacity in the muscles of these old animals is comparable to the adaptive changes seen in young and adult rats (data not shown). **a** Run-training increases muscle mass on average by 45%. Ablation of the gastrocnemius muscle increases plantaris muscle mass by 75% in sedentary rats and 35% in those run-trained. Individual values show great variability among aged animals: some control muscles are quite low in mass, but in most cases muscle mass increases after ablation; **b** Run-training increases the force-generating capacity of plantaris muscle by 75%. Gastrocnemius ablation increases plantaris force-generating capacity by 115% in sedentary rats and 35% in run-trained rats. Thus, it is apparent from these data (**a** and **b**) that skeletal muscle in old animals maintains its capacity to adapt to the stresses of change in the levels of exercise. (Courtesy of Dr. T.P. White, Department of Human Biodynamics, University of California, Berkeley)

compensatory reactions) in the CNS degenerative diseases of old age such as dementia of the Alzheimer type. While brain atrophy and neuronal loss in specific brain areas are usually associated with this type of dementia, other altera-

tions, structural (neurofibrillary tangles, neuritic plaques, amyloid infiltration) [40], vascular and metabolic (mitochondrial mutations, decreased brain energy metabolism) [11,43,64,97] are considered as more characteristic indices of this disease.

Contrary to the previous view, current observations indicate that the mature CNS is endowed with some degree of plasticity and that, when remodeling and regenerating processes do occur in the aged CNS, they are analogous to those seen during development (Chap. 30). The riddle of CNS plasticity in adult and old age may be solved by a better understanding of the factors necessary for restoring to neurons the capacity for division and regeneration lost during maturation and specialization. These factors are both inhibitory and stimulatory in nature.

Among inhibitory factors, proteins such as glial-hyaluronate-binding protein (GHAP) form complexes with extracellular matrix components (e.g., hyaluronate). GHAP, of glial origin, inhibits cell migration and axonal growth [9]. This group of proteins appears relatively late during ontogenesis (i.e., after the onset of myelination) and, therefore, does not interfere with cell migration and axonal elongation in the immature brain [9], nor in peripheral nerve grafts, because astrocytes there do not produce GHAP [56], nor in peripheral nerves, where GHAP is not produced [10]. These studies suggest that in CNS lesions, the glial scar formed at the site of damage is not the only factor responsible for the failure to regenerate [6], but that specific, locally produced substances may operate to actively inhibit regeneration and repair.

The best-known among the stimulatory agents is nerve growth factor (NGF), a basic protein that promotes growth in sympathetic and sensory neurons as well as in the brain during both developing and mature ages. In the brain, NGF is particularly effective in promoting growth or reducing or preventing damage of cholinergic neurons of the basal forebrain, drastically affected by neuronal loss in Alzheimer disease [21]. NGF (intraventricular) administration prevents (in adult rats) these cholinergic neurons from dying after axonal transection and, in fact, may promote regeneration [95] (Fig. 123.4). NGF protein has been purified and sequenced, a number of similarly growth-promoting proteins have been identified, and a mouse fibroblast cell line capable of secreting recombinant rat NGF has been established [21]. Other neurotropic agents involved in development and/or repair of neuronal tissue in adult and aged humans and other animals include gangliosides [34,84], several compounds of the fibroblast growth factors (FGF) family effective on neurons, glial cells as well as specialized sensory cells (photoreceptors) in health and disease [23,60,61], and some components of the cytoskeleton [51,62].

NGF levels are reduced with aging in cholinergic brain areas, where their reduction is associated with loss of NGF receptor protein and mRNA [4,33], especially in those regions (e.g., hippocampus and basal nucleus of Meynert) that display age-related pathology [31]. It is not known if these deficits are the cause or consequence of neuronal atrophy and cell loss. An attempt to generalize the present

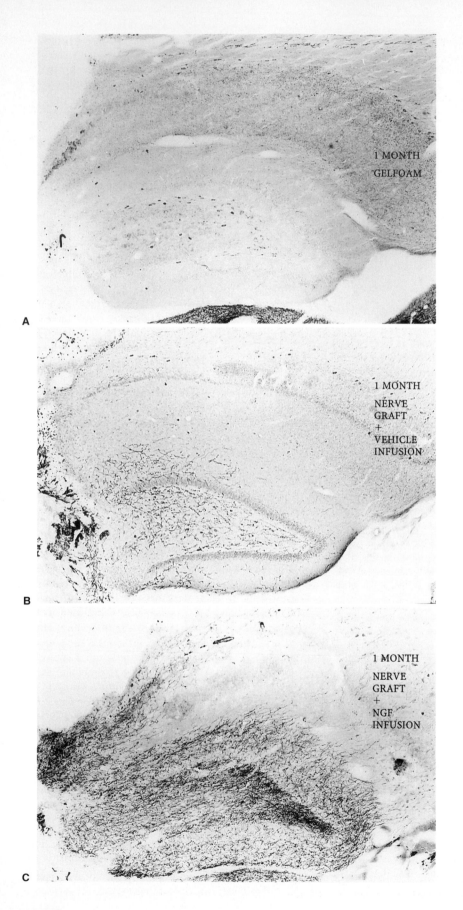

A

1 MONTH
GELFOAM

B

1 MONTH
NERVE
GRAFT
+
VEHICLE
INFUSION

C

1 MONTH
NERVE
GRAFT
+
NGF
INFUSION

observations suggests that neurons in old age are capable of displaying compensatory responses when they are provided with an appropriate microenvironment in terms of humoral, extracellular matrix and cell-to-cell (particularly neuronal-glial) requirements. These responses reflect, perhaps, similar mechanisms operative in ontogenesis [16,17,66]. With aging, surviving neurons are capable of remodeling their configuration (e.g., reactive dendritic sprouting) in response to functional challenges [94]. However, neuronal loss, deafferentiation (i.e., isolation of a CNS structure from neighboring structures by section/degeneration of afferent pathways), or neurotransmitter deficits are more difficult to repair in the aged brain unless growth-promoting factors become again available and inhibitory factors are eliminated. As neuronal modeling is gene-regulated during ontogenesis, likewise, differential gene expression in aged neurons may direct the morphological and biochemical compensatory events that follow aging or damage and lead to functional recovery [57,62].

123.6 Theories of Aging

What causes aging? Irrespective of the potential for rehabilitation and recovery illustrated here, physiological competence declines with aging due to causes still unknown. Although many theories have been proposed, so far no single one can account for all the qualitative and quantitative variations of the senescent phenotype. Multifactorial causes cannot be excluded, but most theories attempt a more simplistic approach and address a single cause for functional aging viewed from the molecular, cellular, or organismic perspective (Table 123.3). Most of our information in humans derives so far from studies on diseases associated with aging or from demographic studies of different populations. Many theories of aging have been derived from studies in animal models (primarily low organisms) or cultured cells or tissues where genetic and environmental factors can be controlled and manipulated. Major current theories will be summarized here. The reader is referred for more details to more extensive reviews [63,88].

Molecular Theories. Theories with a molecular basis – such as codon restriction, somatic mutation, error catastrophe (that is, accumulation of defective proteins through errors in RNA information transfer), failure of DNA repair, altered chromatin, telomere length, specific "senescence" or "longevity" genes – assume that the lifespan of any species is governed by the gene-environment interactions. Genetic information regulates the coding of structural (e.g., collagen) and functional (e.g., enzymes, receptors) proteins that govern the form and function of organisms. Aging may result from changes in DNA template activity which regulates the formation of the final cellular products.

Cellular Theories. Other theories relate to those (wear and tear) changes that occur in structural and functional elements of cells with the passage of time. Changes involve, in addition to proteins, other macromolecules, including hormones, pigments, cross-linked collagen, major histocompatibility complex, "G" proteins, membranes, lysosomes, and mitochondria. Although cellular repair is possible even in old age, the physiological cost of repairing the damage would be greater than the corresponding benefit: to produce a "disposable soma" would be more cost-effective than maintaining an eternally youthful one.

"Free Radicals" Theory. An object of considerable attention, this theory states that some free radicals are unavoidably generated in the course of normal cellular metabolism. These substances contain a highly reactive, unpaired electron and are capable of binding indiscriminately to nearby compounds. It is this binding of free radicals to cellular lipids, proteins, and DNA, and the consequent cellular and molecular damage that accumulates with time, that, according to this hypothesis, would be responsible for aging. Free radicals are usually intercepted by protective enzymes. However, with aging or under toxic conditions, more free radicals are produced and/or enzymatic protection is impaired; as a consequence, damage is not repaired or is repaired less well. Compromised cell integrity induces aging-associated alterations in growth (as in cancer), transport (as in atherosclerosis), and cognitive processes (as in dementia).

Fig. 123.4A – C. Nerve growth factor (*NGF*) promotes cholinergic regeneration in the brain of adult animals. Sagittal sections (40 μm thick) through the hippocampal formation are histochemically stained for the enzyme acetylcholinesterase (which hydrolyzes acetylcholine released at cholinergic synapses). The *left-hand side* of the figure is the rostral side of the hippocampal formation. Adult female Sprague-Dawley rats were subjected to a complete bilateral aspirative transection of the fimbria-fornix (also on the *left side* of the figure), which disconnects the septum, with its cholinergic neural cell bodies from the hippocampal formation, its innervation territory. The specimens were perfusion-fixed with 4% formaldehyde and analyzed 1 month later. A After lesioning, Gelfoam (unfavorable to regeneration) was implanted into the lesion cavity. Cholinergic axons are absent from the hippocampal formation indicating that the lesion has completely disconnected the hippocampal formation from its cholinergic innervation. B Animals were implanted with an autologous, freshly dissected peripheral (sciatic) nerve graft (on the *left side* of the figure) just after the lesion and injected with the vehicle. The graft served as a supportive bridge across the injury (cholinergic disconnection) site to facilitate and support nerve growth. One month after the graft, the hippocampal formation contains a moderate number of cholinergic axons. C Animals treated as in B received (instead of the vehicle alone) a continuous infusion of NGF (0.16 mg/day) directly into the hippocampal formation, in close proximity to the lesion and graft. The hippocampal formation of these NGF-treated animals contains an almost normal number of cholinergic axons. Thus, NGF is able to promote the entry and penetration of cholinergic axons into the otherwise refractory/inhibitory brain tissue. (Courtesy of Dr. T. Hagg, Department of Anatomy and Neurobiology, Dalhousie University, Halifax, Nova Scotia, Canada)

Table 123.3. Major theories of aging

Aging due to external causes:	Lifespan indefinite were it not for environmental insults such as: Nutrition Radiation Viruses Pollutants
Aging due to internal causes:	Lifespan genetically determined for a finite period. Genetic expression modulated through specific programs leading to: Neuroendocrine theory Immunological theory Metabolic (free radicals) theory

Aging due to cellular and
   molecular causes:
	Both internal and external causes may act at one or more cellular levels and/or
	    specific molecules to produce:
	In membranes: Changes in fluidity, permeability, transport
	                Organelle biogenesis and intracellular molecular movement,
	                   G proteins
	
	In cytoplasm: Free radical formation, pro-oxidant states
	                Cross-linking, lipofuscin
	                "Disposable soma" theory
	
	In nucleus: DNA damage and DNA repair failure
	                RNA catastrophe errors
	                Finite replication capability
	                Telomere length
	                Senescence and longevity genes
	                Imbalance of specific loci wihtin histocompatibility complex

Organismic and Systemic Theories. Another set of theories ascribe the aging of complex organisms (e.g., mammals) to decrements in the function of a key system, such as the nervous, endocrine, or immune system. Decrements may be either genetically programmed or the consequence of environmental insults. Alterations in the key system would generate changes throughout the entire organism.

According to *neuroendocrine theories*, the lifespan is regulated by neural and endocrine signals from the time of fertilization, through birth, childhood and puberty, youth and maturity, middle age and menopause, and, finally, old age and death. Specific neurons in the CNS – probably the hypothalamus – may act as "pacemaker" and direct the "biological clock" for development and aging. Signals from these neurons would be relayed by neurotransmitters to brain centers regulating sensorimotor, visceral, and endocrine functions, and from those to the peripheral target organs and cells. Stimulation of these targets by neural and/or endocrine signals can trigger the responses necessary for assuring survival or programming death. Several hormones, primarily adrenal and gonadal steroids and thyroid hormones, may carry the signal through their receptors to membranes, cytoplasm, and nucleus (Chaps. 20, 22). For example, receptors for adrenal steroids are abundant in such brain areas as the hippocampus, and adrenal steroids acting through these receptors are responsible for hippocampal cell death As these hippocampal cells have an inhibitory action on the hypothalamo-hypophyseal-adrenocortical axis, neuronal loss will result in increased hypothalamic corticotropin-releasing hormone and progressively high levels of corticosteroids, which, in turn, will damage more hippocampal cells [80]. The close kinship that exists between steroid and thyroid hormone receptors suggests that these molecules may all be part of a superfamily of regulatory proteins that has arisen over evolutionary time to match the increasing developmental demands, and, eventually, to determine the course of physiological aging of more complex organisms [98].

The *immunological theories* of aging focus on the immune system, its role in the defenses of the organism, and its declining competence with aging. Several organs (thymus), tissues (lymph nodes), and cells (lymphocytes) of the immune system undergo changes with age, as illustrated by thymus involution at puberty. With aging, the ability to fight foreign organisms diminishes, tolerance decreases and the production of abnormal molecules increases, resulting in reduced immunological surveillance and increased autoantibody production.

Programmed changes in gene activity are not confined to fertilization, but may continue at late ages, as is well documented in several animal species (e.g., insects, amphibians). If we assume that such programming operates also in mammals and is modulated by nervous, endocrine, or immunological signals, then intervention in these systems may influence aging as it does development. Current experiments with dietary [81,96,99], pharmacological [7], and hormonal [29] manipulations seem to support this possibility.

123.7 Quality of Life in Old Age and Persistence of Functional Independence

A significant impact on the length and quality of life can be expected only from a definite understanding of the aging process itself. Until then, interventive measures should be directed both to the strengthening of physiological competence and to the elimination or reduction of those risk factors which engender disability and disease. The continual shift to a larger and larger proportion of elderly that we are witnessing today has grave implications for the overall health of the population. So far, the extension of the lifespan has resulted, for many, in a greater number of years spent with disability and disease. Some investigators predict a "compression of morbidity" (i.e., a postponement of the time of onset of disability) that would lead to a reduction in morbidity, while others prognosticate a continuing lengthening of the period of disability and morbidity [32,39].

If, as has been attempted to demonstrate in this chapter, compensatory physiological adjustments are possible even in old age, then the availability of "normal" and "successful" profiles of physiological capability at progressive ages is necessary to formulate an appropriate program to postpone the onset and/or reduce the consequences of disability. The design of preventive and interventive strategies to improve the *quality of life* in old age will be best guided by the knowledge of those factors that most strongly affect functional ability. While habits and lifestyle practices in childhood and young adult years undoubtedly influence health status in later life, modification of certain habits (e.g., diet, physical exercise) even later in life can influence the course of disability [58,59]. With the recognition that functional decline in old age is not inevitable nor irremediable, health objectives for the elderly must focus more strongly on the understanding, maintenance, and support of physiological competence.

References

1. Abribat T, Deslauriers N, Brazeau P, Gaudreau P (1991) Alterations of pituitary growth hormone-releasing factor binding sites in aging rats. Endocrinology 128:633–635
2. Alighieri Dante (1984) The divine comedy, vol I: inferno. (Translated by Musa M) Penguin, New York, p 67
3. Andonian MH, Fahim MA (1987) Effects of endurance exercise on the morphology of mouse neuromuscular junctions during ageing. J Neurocytol 16:589–599
4. Angelucci L, Ramacci MT, Taglialatela G, Hulsebosch C, Morgan B, Werrbach-Perez K, Perez-Polo R (1988) Nerve growth factor binding in aged rat central nervous system: effect of acetyl-L-carnitine. J Neurosci Res 20:491–496
5. Anversa P, Palackal T, Sonnenblick EH, Olivetti G, Meggs LG, Capasso JM (1990) Myocyte cell loss and myocyte cellular hyperplasia in the hypertrophied aging rat heart. Circ Res 67:871–885
6. Barinaga M (1989) Neuroscientists track nerve development. Science 246:756–757
7. Bartus RT (1990) Drugs to treat age-related neurodegenerative problems. The final frontier of medical science? J Am Geriatr Soc 38:680–695
8. Berg RL, Cassells JS (1990) The second fifty years: promoting health and preventing disability. National Academy Press, Washington DC
9. Bignami A, Asher R, Perides G (1991) Brain extracellular matrix and nerve regeneration. Advan Exp Med Biol 296:197–206
10. Bignami A, Mansour H, Dahl D (1989) Glial hyaluronate-binding protein in Wallerian degeneration of dog spinal cord. Glia 2:391–395
11. Blass JP, Gibson GE (1991) The role of oxidative abnormalities in the pathophysiology of Alzheimer's disease. Rev Neurol (Paris) 147:513–525
12. Blumenthal JA, Emery CF, Madden, DJ, Schniebolk, S, Riddle MW, Cobb FR, Higginbotham M, Coleman RE (1991) Effects of exercise training on bone density in older men and women. J Am Geriatr Soc 39:1065–1070
13. Bourliere F (1970) The assessment of biological age in man. Public Health Papers, WHO, Geneva
14. Ceda GP, Valenti G, Butturini U, Hoffman AR (1986) Diminished pituitary responsiveness to growth hormone-releasing factor in aging male rats. Endocrinology 118:2109–2114
15. Chien KR, Knowlton KU, Zhu H, Chien S (1991) Regulation of cardiac gene expression during myocardial growth and hypertrophy: molecular studies of an adaptive physiologic response. FASEB J 5:3037–3046
16. Coleman PD, Flood DG (1986) Dendritic proliferation in the aging brain as a compensatory repair mechanism. Prog Brain Res 70:227–237
17. Coleman PD, Flood DG (1987) Neuron numbers and dendritic extent in normal aging and Alzheimer's disease. Neurobiol Aging 8:521–545
18. Comfort A (1979) The biology of senescence, 3rd edn. Elsevier, New York
19. Davidson MB (1979) The effect of aging on carbohydrate metabolism: a review of the English literature and a practical approach to the diagnosis of diabetes mellitus in the elderly. Metab Clin Exp 28:688–705
20. De Beauvoir S (1972) The coming of age. (Translated by P O'Brian.) Putnam, New York
21. Ebendal T, Soderstrom S, Hallbook F, Ernfors P, Ibanez CF, Persson H, Wetmore C, Stromberg I, Olson L (1991) Human nerve growth factor: biological and immunological activities, and clinical possibilities in neurodegenerative disease. Advan Exp Med Biol 296:207–225
22. Fahim MA, Holley JA, Robbins N (1983) Scanning and light microscopic study of age changes at a neuromuscular junction in the mouse. J Neurocytol 12:13–25
23. Faktorovich EG, Steinberg RH, Yasumura D, Matthes MT, LaVail MM (1990) Photoreceptor degeneration in inherited retinal dystrophy delayed by basic fibroblast growth factor. Nature 347:83–86
24. Faulkner JA, Brooks SV, Zerba E (1990) Skeletal muscle weakness and fatigue in old age: underlying mechanisms. Annu Rev Gerontol Geriatr 10:147–166
25. Faulkner JA, White TP (1988) Adaptations of skeletal muscle to physical activity. In: Bouchard C, Shepherd RJ, Stephens T, Sutton JR, McPherson B (eds) Exercise, fitness and health: a consensus of current knowledge. Human Kinetics Books, Champaign
26. Feigenbaum L (1988) Geriatric medicine and the elderly patient. In: Krupp MA, Chatton MJ (eds) Current medical diagnosis and treatment. Lange, Los Altos, pp 17–26
27. Fiatarone MA, Evans WJ (1990) Exercise in the oldest old. Top Geriatr Rehabil 5:63–77
28. Fiatarone MA, Marks EC, Ryan ND, Meredith CN, Lipsitz LA, Evans WJ (1990) High-intensity strength training in nonagenarians. Effects on skeletal muscle. JAMA 263:3029–3034
29. Finch CE (1987) Neural and endocrine determinants of senescence: investigation of causality and reversibility of laboratory and clinical interventions In: Warner HR et al (eds) Modern biological theories of aging. Raven, New York, pp 261–306
30. Finch CE, Schneider EL (1985) Handbook of the biology of aging. Van Nostrand Reinhold, New York
31. Flood DG, Coleman PD (1988) Neuron numbers and sizes in aging brain: comparisons of human, monkey and rodent data. Neurobiol Aging 9:453–463

32. Fries JF (1980) Aging, natural death, and the compression of morbidity. N Engl J Med 303:130–135
33. Gage FH, Chen KS, Buzsaki G, Armstrong D (1988) Experimental approaches to age-related cognitive impairments. Neurobiol Aging 9:645–655
34. Geisler FH, Dorsey FC Coleman WP (1991) Recovery of motor function after spinal-cord injury – a randomized, placebo-controlled trial with GM-1 ganglioside. N Engl J Med 324:1829–1838
35. Goldberger AL (1991) Is the normal heartbeat chaotic or homeostatic? News Physiol Sci 6:87–91
36. Gorman KM, Posner JD (1988) Benefits of exercise in old age. In: Kaye D, Posner JD (eds) Common clinical challenges in geriatrics. Saunders, Phildelphia, pp 181–192 (Clinics in Ceriatric medicine, vol 4)
37. Greenspan SL, Klibanski A, Rowe JW, Elahi D (1991) Age-related alterations in pulsatile secretion of TSH: role of dopaminergic regulation. Am J Physiol 260:E486–E491
38. Grimby G, Saltin B (1983) The ageing muscle. Clin Physiol 3:209–218
39. Guralnik JM (1991) Prospects for the compression of morbidity. J Aging Health 3:138–154
40. Hardy JA, Higgins GA (1992) Alzheimer's disease: the amyloid cascade hypothesis. Science 256:184–185
41. Ho KY, Evans WS, Blizzard RM, Veldhuis JD, Merriam GR, Samojlik E, Furlanetto R, Rogol AD, Kaiser DL, Thorner MO (1987) Effects of sex and age on the 24-hour profile of growth hormone secretion in man: importance of endogenous estradiol concentrations. J Clin Endocrinol Metab 64:51–58
42. Hoover SL, Siegel JS (1986) International demographic trends and perspectives on aging. J Cross-Cult Gerontol 1:5–30
43. Hoyer S (1992) The biology of the aging brain. Oxidative and related metabolism. Eur J Gerontol 1:25–30
44. Japanese Foundation of Senior Life Enrichment (1988) The ageing of Japanese society and government policies: special report. Japanese Foundation of Senior Life Enrichment, Tokyo
45. Johnson HA (1985) Relations between normal aging and disease. Raven, New York (Aging, vol 28)
46. Kalache A (1991) Ageing in developing countries. In: Pathy MSJ (ed) Principles and practice of geriatric medicine, 2nd edn. Wiley, New York, pp 1517–1528
47. Kane RA, Bayer AJ (1991) Assessment of functional status. In: Pathy MSJ (ed) Principles and practice of geriatric medicine, 2nd edn Wiley, New York, pp 265–277
48. Kane RA, Kane RL (1981) Assessing the elderly. Lexington Books, Lexington
49. Kannel WB, Doyle JT, Shephard RJ, Stamler J, Vokonas PS (1987) Prevention of cardiovascular disease in the elderly. J Am Coll Cardiol 10:25A–28A
50. Kaplan DT, Furman MI, Pincus SM, Ryan SM, Lipsitz LA, Goldberger AL (1991) Aging and the complexity of cardiovascular dynamics. Biophys J 59:945–949
51. Kosik KS, Orecchio LD, Bruns GA, Benowitz LI, MacDonald GP, Cox DR, Neve RL (1988) Human GAP-43: its deduced amino acid sequence and chromosomal localization in mouse and human. Neuron 1:127–132
52. Lakatta EG (1985) Altered autonomic modulation of cardiovascular function with adult aging: perspectives from studies ranging from man to cell. In: Stone HL, Weglicki WB (eds) Pathobiology of cardiovascular injury, Martinus Nijhoff, Boston, pp 441–460
53. Geokas MC, Lakatta EG, Makinodan T, Timiras PS (1991) The aging heart: aging, lifestyle and disease. Ann Intern Med 113:455–466
54. Lakatta EG, Yin FC (1982) Myocardial aging: functional alterations and related cellular mechanisms. Am J Physiol 242:H927–H941
55. Lipsitz LA, Goldberger AL (1992) Loss of "complexity" and aging. Potential applications of fractals and chaos theory to senescence. JAMA 267:1806–1809
56. Mansour H, Asher R, Dahl D, Labkovsky B, Perides G, Bignami A (1990) Permissive and non-permissive reactive astrocytes: immunofluorescence study with antibodies to the glial hyaluronate-binding protein. J Neurosci Res 25:300–311
57. McNeill TH, Koek LL (1990) Differential effects of advancing age on neurotransmitter cell loss in the substantia nigra and striatum of C57BL/6N mice. Brain Res 521:107–117
58. Mor V, Murphy J, Masterson-Allen S, Willey C, Razmpour A, Jackson ME, Greer D, Katz S (1989) Risk of functional decline among well elders. J Clin Epidemiol 42:895–904
59. Omenn GS (1990) Prevention and the elderly: appropriate policies. Health Aff (Millwood) 9:80–93
60. Otto D, Unsicker K (1990) Basic FGF reverses chemical and morphological deficits in the nigrostriatal system of MPTP-treated mice. J Neurosci 10:1912–1921
61. Otto D, Grothe C, Westermann R, Unsicker K (1991) Basic FGF and its actions on neurons: a group account with special emphasis on the Parkinsonian brain. Advan Exp Med Biol 296:239–247
62. Pasinetti GM, Cheng HW, Reinhard JF, Finch CE, McNeill TH (1991) Molecular and morphological correlates following neuronal deafferentation: a cortico-striatal model. Advan Exp Med Biol 296:249–255
63. Pathy MSJ (1991) Principles and practice of geriatric medicine, 2nd edn Wiley, New York
64. Pettegrew JW, Panchalingam K, Withers G, McKeag D, Strychor S (1990) Changes in brain energy and phospholipid metabolism during development and aging in the Fisher 344 rat. J Neuropathol Exp Neurol 49:237–249
65. Rapp PE, Bashore TR, Zimmerman ID, Martinerie JM, Albano AM, Mees AI (1990) Dynamical characterization of brain electrical activity. In: Krasner S (ed) The ubiquity of chaos. American Association for the Advancement of Science, Washington DC, pp 10–22
66. Reynolds BA, Weiss S (1992) Generation of neurons and astrocytes from isolated cells of the adult mammalian central nervous system. Science 255:1707–1710
67. Riggs BL, Melton LJ (1986) Involutional osteoporosis. N Engl J Med 314:1676–1686
68. Rodeheffer RJ, Gerstenblith G, Becker LC, Fleg JL, Weisfeldt ML, Lakatta EG (1984) Exercise cardiac output is maintained with advancing age in healthy human subjects: cardiac dilatation and increased stroke volume compensate for a diminished heart rate. Circulation 69:203–213
69. Rowe JW, Andres R, Tobin JD, Norris AH, Shock NW (1976) The effect of age on creatinine clearance in men: a cross-sectional and longitudinal study. J Gerontol 31:155–163
70. Rowe JW, Kahn RL (1987) Human aging: usual and successful. Science 237:143–149
71. Rowe JW, Troen BR (1980) Sympathetic nervous system and aging in man. Endocr Rev 1:167–179
72. Rubenstein LZ, Josephson KR, Wieland GD, English PA, Sayre JA, Kane RL (1984) Effectiveness of a geriatric evaluation unit: a randomized clinical trial. N Engl J Med 311:1664–1670
73. Rubenstein LZ, Josephson KR, Nichol-Seamons M, Robbins AS (1986) Comprehensive health screening of well elderly adults: an analysis of a community program. J Gerontol 41:342–352
74. Rubenstein LZ, Rhee L, Kane RL (1982) The role of geriatric assessment units in caring for the elderly. J Gerontol 37:513–521
75. Rudman D (1985) Growth hormone, body composition, and aging. J Am Geriatr Soc 33:800–807
76. Rudman D, Kutner MH, Rogers CM, Lubin MF, Fleming GA, Bain RP (1981) Impaired growth hormone secretion in the adult population: relation to age and adiposity. J Clin Invest 67:1361–1369
77. Rudman D, Feller AG, Nagraj HS, Gergans GA, Lalitha PY, Goldberg AF, Schlenker RA, Cohn L, Rudman IW, Mattson DE (1990) Effects of human growth hormone in men over 60 years old. N Engl J Med 323:1–6

78. Sacher GA (1977) Life table modification and life prolongation. In: Finch CE, Hayflick L (eds) Handbook of the biology of aging. Van Nostrand, New York, pp 582–638
79. Salomon F, Cuneo RC, Hesp R, Sonksen PH (1989) The effects of treatment with recombinant human growth hormone on body composition and metabolism in adults with growth hormone deficiency. N Engl J Med 321:1797–1803
80. Sapolsky RM, Krey LC, McEwen BS (1986) The neuroendocrinology of stress and aging: the glucocorticoid cascade hypothesis. Endocr Rev 7:284–301
81. Segall PE (1979) Interrelations of dietary and hormonal effects in aging. Mech Ageing Dev 9:515–525
82. Shock NW (1960) Discussion on mortality and measurement. In: Strehler BL, Ebert JD, Shock NW (eds) The biology of aging: a symposium. American Institute of Biological Sciences, Washington DC
83. Siegel JS, Hoover SL (1982) Demographic aspects of the health of the elderly in the year 2000 and beyond, vol 3. WHO/AGE, Geneva, pp 133–202
84. Skaper SD, Katoh-Semba R, Varon S (1985) GM1 ganglioside accelerates neurite outgrowth from primary peripheral and central neurons under selected culture conditions. Brain Res 355:19–26
85. Spencer H, Kramer L (1986) NIH Consensus Conference: osteoporosis. Factors contributing to osteoporosis. J Nutr 116:316–319
86. Spik K, Sonntag WE (1989) Increased pituitary response to somatostatin in aging male rats: relationship to somatostatin receptor number and affinity. Neuroendocrinology 50:489–494
87. Strehler BL (1977) Time, cells and aging, 2nd edn. Academic, New York
88. Timiras PS (1994) The physiological basis of aging and geriatrics, 2nd edn. CRC, Boca Raton, FL
89. Tout K (1989) Ageing in developing countries. Oxford University Press, Oxford
90. United Nations (1983) World population prospects as assumed in 1982, UN publication no XIII-5
91. US Department of Commerce Bureau of the CENSUS (1987) An aging world. International population reports ser p-95, no 78, Washington DC
92. Urban RJ, Veldhuis JD, Blizzard RM, Dufau MC (1988) Attenuated release of biologically active luteinizing hormone in healthy aging men. J Clin Invest 81:1020–1029
93. Vance ML (1990) Growth hormone for the elderly? N Engl J Med 323:52–54
94. Varon S, Hagg T, Fass B, Vahlsing L, Manthorpe M (1989) Neuronotrophic factors in cellular functional and cognitive repair of adult brain. Pharmacopsychiatry 22 [Suppl 2]:120–124
95. Varon S, Hagg T, Manthorpe M (1991) Nerve growth factor in CNS repair and regeneration. Advan Exp Med Biol 296:267–276
96. Walford RL (1985) The extension of maximum life span. Clin Geriatr Med 1:29–35
97. Wallace DC (1992) Mitochondrial genetics: a paradigm for aging and degenerative diseases? Science 256:628–632
98. Weinberger C, Thompson CC, Ong ES, Lebo R, Gruol DJ, Evans RM (1986) The c-erb-A gene encodes a thyroid hormone receptor. Nature 324:641–646
99. Weindruch R, Walford RL (1988) The retardation of aging and disease by dietary restriction. Thomas, Springfield
100. Wernig A, Herrera AA (1986) Sprouting and remodelling at the nerve-muscle junction. Prog Neurobiol 27:251–291

124 Death

R. Greger

Contents

124.1 Introduction

Why should one deal with the topic of death in a physiology text? Usually this issue is left to the pathologist. As will be shown below, the process of dying has several physiological and pathophysiological aspects that deserve explicit treatment in a physiology text. The fact that this issue rarely appears in such texts may reflect a general human attitude, namely the inclination to ignore or even suppress thoughts about dying and death.

Life Expectancy. Chapter 123 has dealt with the process of aging, and it has become evident that this is an inevitable process with a normal (physiological) course. Especially in this century, this has become much more evident inasmuch as mean survival age has increased substantially, and the mean average life expectancy has now increased to 75–80 years (depending on gender; cf. below). A large proportion of the wealthy population in "First World" countries lives to an age at which the natural consequences of aging become apparent. Further progress in medical care will have some impact, albeit not very large, on life expectancy. It will, however, increase the fraction of the population suffering from age-related debilitating symptoms, such as severely reduced physical performance and dementia.
Death is a natural consequence of aging; it is the inevitable termination of the maximal species-related life span, with deterioration of body function in general. Therefore, a first section of this chapter will deal with aging and death.
There has been a slow drift in the typical causes of death in our societies. A short section will deal with this aspect. The gap between life expectancy and maximal life span has become very small. Therefore, intensive medical care will lead to shifts in the causes of death without prolonging life much further.

Dying. The process of dying can have a characteristic course. This course will be discussed here, with specific emphasis on the cessation of circulation and breathing and the involvement of the central nervous system (CNS). The events that go on in the dying brain are largely unknown. The experiences of patients who recover after being close to death ("near-death experiences") have been taken as evidence for "life after life." Such reports may reflect specific physiological and pathophysiological events in the ischemic brain.

Cell Death. Generalized cell death is a consequence of failing respiration and circulation. Individual cell death occurs throughout life as a natural process of cell replacement, and it occurs as an essential part of developmental programs, which demand that specific cells have to undergo apoptosis.
The process of dying is an integral event ultimately involving all cells of the body, that is to say "falling apart". Still, the events differ widely in different tissues. These distinct events in cell death deserve comment in a separate section. This section will describe differences amongst various organs, and it discusses the general principles of cell death.

Organ Transplants. The second half of this century has seen the impressive successes of organ transplantation. Certain organs of a corpse, if perfused and preserved appropriately, can survive long enough to permit successful transplantation, and this fact has been exploited. At this time clear legal definitions of death are required to justify the removal of donor organs.
At the end we will have to consider whether it is likely that life expectancy, or more precisely the maximal life span, can be prolonged and death can be substantially postponed. At present this does not appear feasible, and our efforts to live longer may well aggravate both medical and social problems.

124.2 Aging, Age and Death

Figure 124.1 shows the percent survival of *Homo sapiens* over the past 50 000 years. Several observations deserve comment. Mean survival was only some 10–12 years 50 000 years ago. It increased to some 25 years 15 000 years ago, and had already reached 35 years in early Roman days.

R. Greger/U. Windhorst (Eds.)
Comprehensive Human Physiology, Vol. 2
© Springer-Verlag Berlin Heidelberg 1996

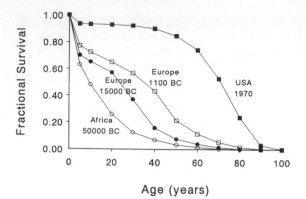

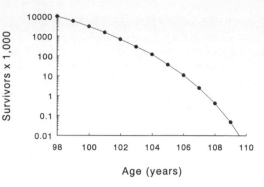

Fig. 124.1. Survival as a function of age. Note that survival followed an exponential curve 50 000 years ago, and that the shape of the curve has changed substantially in the recent past. The incidence of deaths now increases rapidly after the age 60 years. (Modified from [6])

Fig. 124.2. Number of survivors as a function of age. Note that the number of survivors falls very rapidly beyond the age of 98 years, indicating that extremely few individuals live beyond the age of about 100 years, which is generally accepted as the maximal life span. (Modified from [6])

By 25 years ago, inhabitants of the United States of America had a mean life expectancy of 65–70 years, and today the life expectancy is even slightly higher: around 75 years for males and above 80 for females. The second remarkable observation depicted in Fig. 124.1 is the fact that the maximal life span, i.e., age of the oldest individuals at any given time, has always remained between 90 and 100 years. The relative number of such individuals was minimal in ancient times and is now increasing constantly. This tendency to live to an older age has changed the shape of the curve remarkably: the curve for 50 000 years ago looks like a simple exponential, while succeeding ones turn more and more to have a component parallel to the abscissa for the first 50 years of life with a rapid decline thereafter over the next 25–30 years. It is easily predicted that further increases in mean life expectancy will lead to curves that show a fall-off very close to the maximal life span.
Besides the mere descriptive considerations, Fig. 124.1 tells us that:

- Our maximal life span has always (for the last 50 000 years or so) been the same.
- The life of *Homo sapiens* has been limited by various exogenous hazards.
- These hazards have become much less influential in the recent past.
- We are approaching our biological limit, namely the maximal life span of around 100 years.

Prevention of hazards had a large impact on life expectancy 50 000 years ago. Nowadays the situation is very different. The few individuals reaching the age of, say, 98 years die very rapidly over the next 10 years. This is shown in Fig. 124.2. It appears that the variation in the maximal life span of the human race is remarkably small.

Factors Governing Maximal Life Span. What is the determinant of this maximal life span? Before discussing what

governs this limit, it is interesting to consider mean life expectancy and maximal life spans of other mammals, and especially other primates. Figure 124.3 shows a relation between the age at sexual maturation and the maximal life span for primates. There is a surprisingly close correlation. The later sexual maturity is reached, the longer the maximal life span. In other words, the maximal life span is always substantially longer than the age required to secure the generation of offspring. In this way it is possible for each species to continue to survive, with offspring being cared for by parents for as long as need be. The continuation of individual lives to close to the maximal life span has always been rare for all species shown in Fig. 124.3. It appears unlikely, therefore, that this survival has any evolutionary impact. The aging processes that limit the maxi-

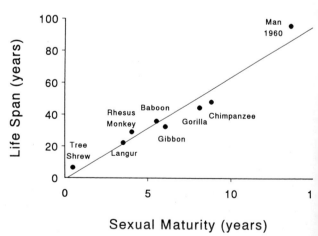

Fig. 124.3. Correlation between maximal life span and age at sexual maturity in primates. This correlation has been interpreted as indicating that an increase in maximal life span, if it were possible, would be accompanied by later maturation. (Modified from [6])

mal life span have so far only been experienced by a very small proportion of the population anyhow.

It is argued that the advantage of aging is that the individual can accumulate more experience, and that this wide experience, handed on to children and grandchildren, is highly valuable and beneficial to the next generation. This positivistic interpretation is probably not applicable. History tells us that no one generation has really utilized the experience of those before. How else can we explain that the humanism of previous epochs has not prevented the cruel slaughtering that still goes on amongst human beings today? How else can we accept the fact that even the experiences of our parents during disastrous wars has not propagated pacifism?

Organ Function in Old Age. To return to the biological aspect of the maximal life span, we might ask whether there is any body function that improves in old age or, to be more modest, whether there is any function that at least stays stable between, say, the ages of 30 and 90 years. Figure 124.4 tells us that this is not the case: it displays the fractional physiological functions as a function of age [15]. It is evident that all functions decline almost in parallel, with no exception for any given organ function. Obviously the decline in function is easier to demonstrate in some functions, such as ventilation or cardiac index, and more difficult to recognize in others, such as glomerular filtration rate. Also, it is interesting to note that the decline has started by the age of 20–40 for all parameters. It appears likely that mental functions show a similar decline [6], although this is much more difficult to determine quantitatively. Along these lines postmortem examinations of patients who have died from one cause or another in old age reveal that most organs show pathological changes that would have caused death only a short time later. It thus appears that the aging process does not spare any particular organs and that treatment of one disease in old age reveals disease of another system.

Prediction of Life Expectancy. It is possible to predict the gain in life expectancy if one cause of death is excluded completely. For instance, if cardiovascular pathology could be ruled out completely for newborns their life expectancy would be increased by some 10 years. If the related diseases were 100% curable in a population aged 65, the same gain in life expectancy would occur. Similarly, if all types of cancer could be cured, the gain in life expectancy would only be 2 years for newborns and 1 year for 65-year-olds [6].

Such considerations are supported by the death statistics (also cf. below) recorded in developed countries [16]. Over the last 35 years there has been a substantial reduction in death rates, with a consequent increase in life expectancy. In addition, there has been a uniform fall in fairly early (65 years or so) death from cardiovascular disease and an increase in the death rates from cancer (mainly lung cancer). Therefore, life has been prolonged in general and the causes of death have shifted, e.g., from ischemic heart disease to lung cancer.

The conclusion we can draw from all this is that prevention of life-threatening exogenous hazards and prevention and/or successful treatment of certain diseases have shifted the average age at death from a period of life in which good health is generally enjoyed or specific pathology endangers life to an age at which the biologically relevant aging processes take over.

Aging Mechanisms. The aging mechanisms can be classified by affected organ (cardiovascular system, brain, lung, etc.) or by underlying cellular mechanism. This latter method of classification is more appropriate, because aging does not normally affect a single system, but several or even all systems (cf. Fig. 124.4). This does of course not rule out the possibility that in a given individual one type of pathology may govern the process of aging. Cellular aging mechanisms are generally subdivided into the following types:

• Accumulated damage caused by oxidative or otherwise toxic stress
• Accumulated damage caused by impaired DNA repair.

Free Radicals. Oxidative stress has long been recognized as a hazard for cells, and free radicals have been incriminated as one of the key mechanisms of aging [1,6,25,29]. The relevant free radicals are [9]:

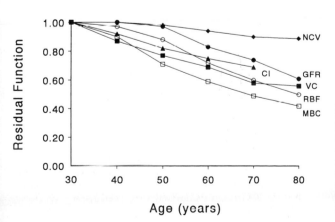

Fig. 124.4. Various body functions in relation to age. Residual function is expressed as the fraction of function at the age of 30 years. NCV, Nerve conduction velocity; GFR, glomerular filtration rate; CI, cardiac index; VC, vital capacity; RBF, renal blood flow; MBC maximal breathing capacity. Note that all body functions show a decline with age. The fall in maximal breathing capacity is especially pronounced. The fall in nerve conduction velocity is slow. (Modified from [6,8])

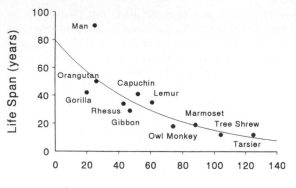

Fig. 124.5. Maximal life span of primates as a function of stand-ardized (weight) metabolic rate. Note the inverse correlation. (Modified from [6])

- Superoxide anion ($O_2^-\cdot$)
- Hydroxyl radical (HO·)
- Perhydroxyl radical ($HO_2\cdot$)
- Peroxyl radical (ROO·)
- Alkoxyl radical (RO·)
- Hydrogen peroxide (H_2O_2)
- Singlet oxygen ($^1O_2$).

Most of the oxygen-derived radicals are produced in mitochondria in the respiratory chain; in microsomes by cytochrome P-450; in peroxisomes via catalase; and by the cytosolic enzyme xanthine oxidase. These radical-producing reactions are in constant equilibrium with scavenging systems. These comprise:

- Enzymes, e.g., superoxide dismutase; catalase; glutathione (GSH) peroxidase; GSH reductase
- Vitamin E; β-carotene; bilirubin; vitamin C; urate; glucose; cysteine; and glutathione

The damage wrought by radicals can be on membrane lipids, and lipid peroxidation has long been claimed as a major determinant of aging processes [29]. Radicals can also lead to DNA and RNA damage [1]. In fact, the long-established inverse correlation between metabolism per gram of tissue and longevity can probably be explained on the basis of more oxidative damage in those species with a high metabolic rate [6]. A few examples of these correlations are shown in Fig. 124.5. In support of this, good correlations of most of the above protecting factors and longevity have been obtained [6].

It should be pointed out, however, that many of these types of correlations are doubly normalized by the same factor, and may thus be fortuitous. Furthermore, there are no close correlations with any "protective" factors (e.g., urate, ascorbate, vitamin A, glutathione). Finally, such plots, even though they are suggestive, do not prove a *causal* correlation.

The results of animal feeding studies using ascorbate, tocopherol, and carotenoids have been used as support for the argument that food enriched with these factors reduces

degenerative diseases and cancer risk and retards aging [1]. These experiments are probably not easy to interpret, because the maximal life span, which is *the* indicator of aging, was not changed much in these feeding experiments [29]. Still, there is little doubt at this stage that radicals are somehow related to aging and to degenerative diseases and cancer [29].

Oxidative stress might be related to aging by as yet unpredicted mechanisms. Protein modification has long been implicated in the aging process. However, this appears difficult to accept, because the protein turnover is expected to be high and new protein is synthesized continuously. A new interpretation has been put forward recently [25]: it has now been suggested that radical modified proteins may lead to a decrease in appropriate DNA repair. This would then link up with the other accepted cause of aging, namely the inappropriate repair of spontaneously occurring DNA modifications.

DNA base modifications occur spontaneously and continuously at, stunning rates of 10 000 per day and cell [25]. These incredibly numerous spontaneous modifications are repaired continuously be several types of mechanisms [10].

Repair mechanisms guarantee that DNA messages are conserved and transcribed with "fidelity" through out our lives. It has been shown that the ability to repair spontaneous DNA modifications correlates well with longevity. The fidelity of DNA conservation is due to at least two mechanisms:

- Detoxification and inactivation of radicals
- DNA repair.

As indicated above, both mechanisms may be interconnected by radical-induced protein modification [25]. Figure 124.6 shows a correlation between DNA repair and life span [10]. It is evident that there is a direct correlation between the two parameters for the fibroblasts of individual species when they are exposed to UV radiation. For example, human fibroblasts are capable of 8 times more repair than those of *Sanguinus*.

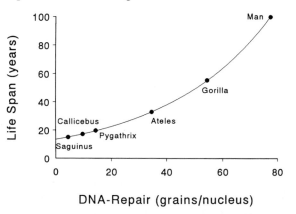

Fig. 124.6. The maximal life span as a function of DNA repair. Fibroblasts from the respective species were irradiated by UV light and the DNA repair was quantified by radiolabelled thymidine. (Modified from [10])

2410

Programmed Cell Death. Much of our analytical knowledge of the aging process stems from cell cultures, which have shown that a fibroblast culture of a certain species (e.g. man) continues to divide until a certain generation has been reached [8] and dies thereafter. These experiments suggest that there is a counter (clock) present in each cell to tell when it is time to die. In fact, cells arrested in a certain passage number can be kept for long periods and will know exactly in which passage they were stopped, and when reactivated they will live until they reach their typical passage number. Figure 124.7 makes this point in a slightly different way. It shows the survival of fibroblasts as a function of the passage number. Cells from a newborn live for approxiamtely 60 divisions. Those of a 100-year-old individual die much earlier, and those of a patient with progeria (Werner's disease) die even earlier. These data suggest that cell death is programmed and is, as shown recently, controlled by certain genes. It is not clear to what extent programmed cell death in cell cultures can be extrapolated to aging in an organism. Programmed cell death may be another and additional mechanism of limiting the maximal life span. There is little doubt that this mechanism is of specific importance in development (cf. Sect. 124.5).

In summary, in this section we have discussed the mechanisms limiting our maximal life span. Several mechanisms appear to contribute to the phenomenon that we call aging. Accumulation of toxic metabolites, lipid peroxidation, protein modification and impaired DNA repair appear to contribute. The role of genetic control of the aging process in general is not clear at this time.

124.3 Causes of Death

The causes of death have not been analyzed continuously, and even today our statistics suffer from the fact that post-

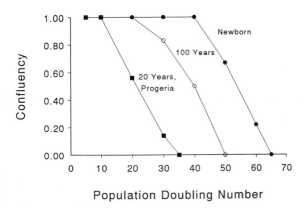

Fig. 124.7. The proliferative potential of fibroblasts. The ability of fibroblasts of three different groups [newborns; 100-year-olds; and 20-year-olds with progeria (Werner's disease)] to grow to confluency is shown as a function of the number of cell divisions. Note that the fibroblasts of newborns grow to confluency up to 40–50 divisions, whilst those of the patients with progeria die after only 10–20 divisions. (Modified from [8])

mortem examinations are not performed routinely by pathologists. It goes without saying that a legal requirement for routine post mortems and the availability of corresponding facilities would have an important and positive effect on the development of medical knowledge.

Death statistics obtained from postmortem examinations are largely different from that of the normal population, because autopsies are only performed with specific indications. A recent analysis of 1000 consecutive autopsies of individuals dying of natural causes [7] reveals that the largest group is that with cardiovascular disease.

Postmortem examinations are available only for a small fraction of the 1100–1200 deaths in a wealthy society per 100 000 population per year. The causes of death have been classified, and many countries have started to adhere to these definitions. Figure 124.8 gives a summary of the causes of death for males and females in Germany in 1991. The data are classified by age groups [26]. The data are very similar for other industrialized countries. Several aspects deserve commens:

- Mortality is higher in males than in females for comparable age groups. In children and young adults this difference is caused by accidents, and between the ages of 45 and 65 years, by cardiovascular disease. From the age of 75 years onward women "catch up." More than 75% of all deaths in women occur in this highest age group.
- The most frequent causes of death in both sexes (Fig. 124.8A) are connected with diseases of the cardiovascular system. This cause accounts for approximately 50% of all deaths. The largest subgroup is that of myocardial infarction and ischemic heart disease in general, followed by cerebrovascular disease.
- The second major cause of death is malignant tumor disease of all kinds. In men lung cancer is the most frequent, while in women the main malignant diseases are those of uterus, ovaries, and breast.
- The third most frequent group of causes is that of non-cancerous diseases of the lung, such as pneumonia, asthma, bronchitis etc., which account for some 6–8% of all deaths.
- Trauma, poisoning and accidents (including traffic accidents) are responsible for another 5%.

Over the years there has been a continuous drift away from cardiovascular deaths to deaths from malignant tumor disease. The highest relative increase has been in the group of lung cancer [16]. Figure 124.8 also indicates something mentioned in Sect. 124.2, namely that prevention of fairly early (age group 45–65% years) deaths from cardiovascular causes and mainly in the male population would lead to a substaintial (probably 10 years) increases in life expectancy.

124.4 Dying

The acute or slowly developing causes leading to death have been summarized above, and it can be deduced that

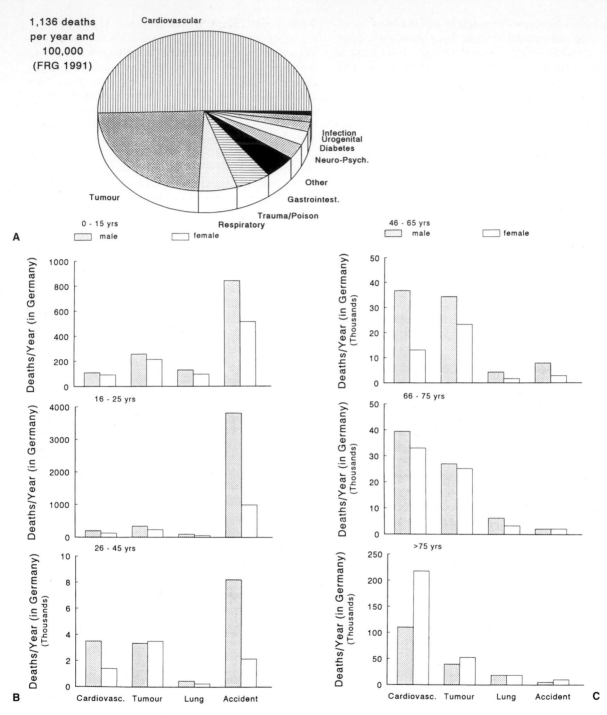

Fig. 124.8A–C. Death statistics for the year 1991 in Germany [25]. **A** Major causes of death. The death rate is 1136 per 100 000 population. The largest group of causes is cardiovascular. The major subgroups are myocardial infarction (20%); ischemic heart disease and arrhythmias (20%); cerebrovascular disease (24%); cardiac insufficiency (13%). The second largest main group is that of malignant tumor disease, with the major subgroups: lung cancer (17%); colon and rectum (14%); stomach (8%); breast (9%); blood cells (7%); uterus and ovaries (6% per total population); kidney and urinary tract (6%); prostate (5.5% per total population); pancreas (5%); liver and gallbladder (5%). The third largest group is that of deaths from respiratory diseases (not malignant) and trauma, poisoning, etc. Within the trauma/poisoning group 64% of the deaths are due to accidents, and 23% specifically to road accidents. Suicides account for 30%. **B** Distribution of the major causes of deaths for males and females in the different age groups up to 45 years of age. Absolute numbers of deaths per year are given. That for 1991 was 911 245. Note that the major cause of death in the age groups up to 45 years is accident. A boy's or man's risk of dying in an accident while in the age group 16–25 years is three times that for girls and women. **C** Distribution of the major causes of death for men and women in the different age groups between 46 and >75 years of age. Note that death from cardiovascular disease is very frequent for men in the age group of 46–65 years. In the age group 66–75 years the sex differences disappear. The number of deaths in the age group over 75 years is much higher for women because a larger proportion of men are dead by this age

the process of dying is characterized by a large spectrum of scenarios and sequential events. The process can be slow, as in chronic disease, or instantaneous, as in an accident. In every case three major functions determine the course:

- Circulatory function
- Respiratory function
- CNS function.

In most instances failing circulatory function starts the process. However, irrespective of the individual course, anoxia and/or ischemia within the brain is the final common process. This leads to gradual damage and will be irreversible after some time (cf. below).

Medical staff and laymen watching someone dying frequently note that the person who dies apparently experiences relief during this very process, even though the period prior to this final stage may have been stressful and painful. It is obvious that we possess no records of what goes on in the dying. We do not even know exactly what goes on in the dying brain.

Near-death Experiences. Publications such as "Life After Life" [17] claim to provide answers, which are based on what persons who have been close to death have felt. Such reports are remarkably similar, at least within our cultural background. Enhanced light is a commonly reported experience. It goes along with a positive emotion, and many of the patients reporting this are sad about their "return" [22]. The impression of walking through a tunnel, having left the body, and hearing one's own death announced are other common experiences. A recent analysis of these near-death experiences as reported by patients who actually were close to death and others who were not in danger revealed that both groups had generally similar experiences but that light perception with positive emotion was more frequent for the terminally ill patients [20]. Such experiences cannot tell us whether there is "life after life," but they describe subjective perceptions during a phase that may in fact also be characteristic of the dying brain. It has been suggested that hypoxia may cause such hallucinations via some type of NMDA receptor, and that drugs such as ketamine and hallucinogens can produce similar experiences [13]. Endorphins and serotoninergic mechanisms have also been implicated [3,18]. Physiological and psychological explanations probably suffice to explain these sometimes perplexing reports.

Dying with Dignity. At any rate, it appears justified to conclude that our feelings and experiences on entering this transitional state might well be influenced by what goes on around us and what goes on in our minds. This is already a good enough reason to generate an atmosphere of dignity for the dying, to provide all necessary medical care, to assist the relatives and to respect the patient's wishes, especially as most people meanwhile die in hospital, and many in intensive care.

124.5 Cell Death

The process of dying is a rather short transitional state in the brain, and many other organs continue to survive after respiration and circulation have completely stopped for different periods (cf. below). Ultimately, all different organs and cells have lost their vital functions and lyse. In this section we shall discuss cell death from three different perspectives:

- Programmed and apoptotic cell death
- Cell death as a consequence of ischemia and hypoxia
- The different time scales of cell and organ death.

Programmed cell death is a crucial issue for the understanding of development, and probably also of importance in the aging process in general and maybe even in the generation of tumors. Cell death as a consequence of ischemia and hypoxia is probably an entirely different issue. Here the cause of cell death is extraneous, i.e., due to shortage of oxygen and substrates, and not due to an active "suicide" of the individual cell. Finally, the time scale of anoxic and ischemic cell and tissue damage and the implications for organ transplantation will be discussed in this section.

Programmed and Apoptotic Cell Death. This form of death is a mechanism that is required in coordinated development. It is found in evolution from nematodes to mammals [28]. Its usefulness is easily understood. At some stage the destruction of cells may be as relevant as the production of new ones. Consider the development of certain amphibia, in which the tail has to be reabsorbed; the death of neurons that are awkward after proper connections have been made; the need to remove interdigital tissue in proper limb development; the destruction of useless immune cells, etc.

The term apoptosis, derived from Greek ἀποπτῶσις (kicking out or shedding), describes the fact that an individual cell is disposed of because it is useless to or even disturbing for the organism. The term does not, however, imply another feature, which is no less characteristic of this type of death, namely that it is a suicide, with active involvement of the cell that dies. In this sense the term programmed cell death appears more appropriate.

Apoptotic or programmed cell death is characterized by shrinkage of the cell with blebbing of the plasma membrane, condensation of the nucleus, breakdown of the nuclear membrane, and destruction of the chromatin. It is highly likely that this process is not uniform for all the examples given above. For instance, the apoptosis of T-cells looks morphologically different from the programmed cell death of intersegmental muscles of *Manduca sexta* [24]. The processes of cell death by cytokine withdrawal is probably different from the cell death induced by cytotoxic killer (T) cells [27]. The former process takes some time; the latter occurs almost instantaneously. Furthermore, at first glance cell death from killer cells looks like murder rather than suicide. A closer inspection reveals that even in this case the victim collaborates with its murderer to let it happen, i.e., the killer cell only triggers the event.

Recent studies suggest that a certain gene product, *bcl-2*, which is homoloous to the related gene *ced-9* in *Caenorhabditis elegans* [11], protects against apoptosis [21,28]. Some apoptosis-promoting proteins have also been found recently. One such protein is interleulin-1β converting enzyme (ICE). Another is Bax, which apparently directly opposes *bcl-2* [2]. Several signals, such as:

- Growth factor (cytokine) suppression
- Hormones such as corticoids and thyroid hormones
- γ-Radiation
- Heat schock
- Viruses
- Free radicals
- Glutamate

apparently all converge on a common pathway involving *bcl-2*. If *bcl* is active, these maneuvers are ineffective [21]. Also, in many instances a delayed increase in cytosolic Ca^{2+} activity has been observed [27]. It has been argued that in the above cases apoptosis is due to an abortive mitosis, which then leads to nuclease-mediated destruction of the genome [27]. Inhibitors of macromolecular synthesis can prevent apoptosis in several experimental models. This has been used as evidence that gene transcription is necessary to induce to suicide. Another interesting new aspect from this research is the possible oncogenic effect of *bcl-2* itself. Inasmuch as this gene suppresses apoptosis, it may as well permit the growth and development of tumor cells [21,27,28].

It has already been mentioned that not all apoptotic events are caused by suppression of *bcl-2*. T-killer cell-induced cell death, for instance, cannot be suppressed by *bcl-2* [21,28]. Further exploration of the mechanisms of apoptosis might be relevant not only to the description of the transduction cascades in cell suicide, but also to our understanding of the generation and growth of malignant tumors.

Cell Death as a Consequence of Ischemia and Hypoxia. In contrast to apoptosis and programmed cell death, ischemic and hypoxic cell death is entirely extraneous. In almost every cell the reduction or cessation of blood supply initiates a very typical sequence of events, which culminates in cell necrosis (death). The major events comprise:

- Cell swelling
- Increase in cytosolic Na^+ activity
- Loss of cytosolic K^+
- Increase in cytosolic Ca^{2+} activity
- Fall in cytosolic pH
- Mitochondrial destruction
- Lysis of the nucleus and other cell organelles
- Burst of the plasma membrane
- Accumulation of reactive oxygen radicals.

Many of these events can be explained on the basis of simple cell models such as those discussed in Chap. 11. Consider the cell in Fig. 124.9A. For the homeostasis of normal ionic gradients and the membrane voltage, this cell requires primarily active pumps such as the $(Na^+ + K^+)$-ATPase, the Ca^{2+}-ATPase and the H^+-ATPase pumps. The Na^+ gradient from extracellular space to cytosol is used in addition to sustain cell pH via Na^+/H^+ exchange and low cytosolic Ca^{2+} activity via the Na^+/Ca^{2+} exchange (cf. Chaps. 8, 11). A fall in mitochondrial ATP production caused by a shortage of O_2, by a shortage of metabolic fuels or by poisons or toxic substances inhibiting energy metabolism leads to inhibition of the primary active pumps. As a consequence of reduced $(Na^+ + K^+)$-ATPase activity, cytosolic Na^+ activity increases and that of K^+ decreases. The decrease in cytosolic K^+ activity leads to corresponding depolarization. This causes a redistribution of small anions, because the large ones, such as proteins, cannot leave the cell at this stage. Hence, Cl^- is taken up. Now the total anion and cation content of the cell has increased and this leads to cell swelling (Fig. 124.9B). At the same time, active Ca^{2+} uptake into specific stores and extrusion across the plasma membrane is inhibited, leading to an increase in cytosolic Ca^{2+} activity. This is aggravated by the reduction of Na^+ driven Ca^{2+} extrusion via the Na^+/Ca^{2+} exchange. Very similarly, proton gradients collapse: that across the inner mitochondrial membrane because of reduced mitochondrial metabolism (cf. Chap. 4); that of acidic organelles such as lysosomes and that across the plasma membrane because proton pumps (if present) and the Na^+/H^+ exchange are inhibited. Cytosolic acidosis is further augmented by the anaerobic metabolism that prevails under such conditions. This latter process limits itself, since the phosphofructokinase, as the rate-limiting step in glycolysis, is switched off in acidosis. Cell metabolism is disturbed by the loss of normally produced oxygen radical scavenging and protecting substrates [3].

These events go on at different speeds in the various tissues (cf. below). What matters is the speed at which Na^+ flushes into the cell. In active neuronal tissue the speed is very high, because each transmitted voltage change goes along with Na^+ (Ca^{2+}) uptake and K^+ losses. It can be calculated that the time to doubling of cytoslic Na^+ activity can be less than 1 s, and that a few seconds suffice to lead to a dramatic collapse of the normal ionic gradients. Eventually, the cytosolic Na^+ concentration in the cell approaches that of the extracellular space. The cell is largely swollen, and the plasma membrane is destroyed. At this stage the process is irreversible: the cytosolic compartment has decomposed and the electrochemical gradients for the various ions collapse.

Obviously, the general scheme shown in Fig. 124.9 is grossly simplified, and the events in various tissues may deviate from it. For instance, Ca^{2+} overload may be an early and dominating event in cell death in some tissues. One example is provided by neurons containing the NMDA receptor channel: in these cells, inappropriate and ischemia-induced glutamate release causes Ca^{2+} influx, which then leads off the destructive process (cf. below).

2414

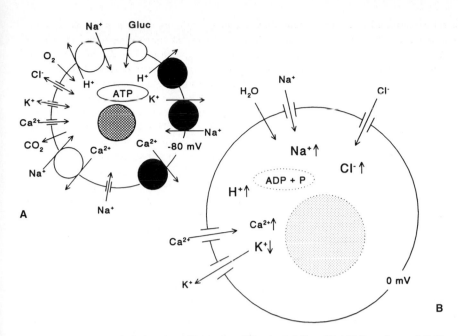

Fig. 124.9A,B. Events in ischemic and hypoxic cell death. A Cell with characteristic transporters. *Filled circles*, ATP-driven pumps; *open circles*, carriers; →, channels. Glucose (*Gluc.*) and O_2 are utilized to produce ATP. The major ATP-dependent homeostatic functions are as follows: (1) The ionic distribution of K^+ and Na^+, with a high K^+ concentration in the cell and a high Na^+ concentration in the extracellular space, which generates the normal membrane voltage of -80 mV (negative on the cytosolic side). This hyperpolarized voltage essentially keeps Cl^- ions out of the cell and limits the cell volume. (2) The low cytosolic Ca^{2+} activity, which is dependent on active (Ca^{2+}-ATPase) or secondarily active (Na^+/Ca^{2+}-exchanger) Ca^{2+} extrusion. (3) The normal cytosolic pH, which is dependent mostly on secondarily active H^+ extrusion. B In ischemia or in anoxia, mitochondrial ATP production is minimal or absent. This stops active pumps and, after the collapse of the corresponding driving forces, also the secondarily active transporters. Major results are: (1) the collapse of the ionic gradients for K^+ and Na^+; (2) the collapse of the membrane voltage; (3) the uptake of Cl^-; (4) cell swelling; (5) nuclear swelling and destruction; (6) increase in cytosolic Ca^{2+} activity; and (7) fall in cytosolic pH

Effect of Temperature. Temperature has a large impact on the speed of these destructive events. This is not at all surprising, since the membrane transport processes and the cell metabolism are strongly temperature dependent. This is the price we pay for the advantage of being constantly alert homeotherms, i.e., independent of the environment, with a mean body temperature of 37°C. The temperature dependence is exploited in organ conservation (cf. below). Also the ionic composition of the extracellular fluid has a large impact on the speed of deterioration in ischemia and hypoxia. Owing to the relative constancy of extracellular fluid composition this aspect is important for the design of organoplegic solutions. Such solutions are used for the conservation of donor organs. The basic principle of organoplegia (from the Greek πλήγια, stunning) is the reduction of organ metabolism at low temperature. This is usually combined with a reduction of the ionic gradients to reduce the driving forces for Na^+ influx and K^+ efflux.

The Different Time Scales of Cell and Organ Death. The lowest ischemia tolerance is that in the brain. After only some 4 s of complete cessation of cerebral blood flow brain function is reduced; 4–8 s later fainting is typical, and this is accompanied by characteristic changes in the electroencephalogram (EEG). After another 10–20 s the EEG becomes silent. Resuscitation of the brain is possible after a short period of ischemia. Successful resuscitation at 37°C is possible after 5–8 min of ischemia [5]. The recovery time depends on the duration of ischemia.

Brain resuscitation has become a major issue now there have been substantial advances in the intensive care of critically ill patients and improved first aid and transport facilities are available for victims of traffic accidents. This has led to increased numbers of admissions of patients with severe head injuries to the hospitals with appropriate units.

In brain resuscitation several approaches have been examined. One is the administration of barbiturates, the rationale being that reduced brain function will diminish brain metabolism at any given rate of brain blood flow [23]. This approach, despite initial enthusiasm, has not led to convincing results, because barbiturates do not change the perfusion/metabolism ratio for the better. The administration of Ca^{2+} antagonists, with the idea that cell death involves Ca^{2+} influx, has been examined experimentally but no convincing clinical data is available [23]. The possible benefit that might accrue from a reduction of free radicals such as are produced in ischemia and, especially, during reperfusion is under current investigation. The most attractive working hypothesis is based on the finding that ischemia enchances glutamate release and that glutamate induces a suicidal influx of Ca^{2+} via the NMDA receptor [4,23]. Antagonists of the NMDA receptor have been reported to inhibit this inappropriate Ca^{2+} influx, and they may be beneficial in brain resuscitation as well as in the early treatment of stroke.

The survival time of other organs, even those with very high metabolism, such as heart, kidney, and liver, is much longer. At 37°C these organs can be resuscitated after hours of complete ischemia. However, if coronary blood flow ceases even for only a few minutes the heart is not able to pump sufficient volumes or generate sufficiently high pressure to perfuse the brain adequately. Therefore, even if resuscitation of the heart is successful, the brain can already have been irreversibly damaged.

The fact that organs such as cornea, heart, liver, lung, and pancreas survive much longer than the brain has led to new developments, but also to new and unforeseen ethical problems in medicine: for example, a resuscitation team may fight for the life of a polytraumatized patient, secure basic functions such as circulation, and ventilate the patient appropriately (usually hyperventilation to improve brain perfusion), but have to abandon their efforts when they realize that the patient's brain has died. At this point the team will have to adopt an entirely new strategy, considering the feasibility of using the ventilated corpse as an organ donor. If consent to organ donation has been obtained the corpse will be looked after with meticulous care, to secure optimal organ perfusion, acid base, and electrolyte status.

The organs, when removed, are cooled and perfused with artificial solutions based on the pathophysiological considerations of ischemic and hypoxic cell death summarized above. There are several solutions (Table 124.1) in common use.

Euro-Collins solution contains a high K^+ (and a low Na^+) concentration, to reduce the gradients for Na^+ influx and K^+ loss. Much of its osmolality is due to D-glucose, which will not be taken up into the cell. The major anion is phosphate, which cannot enter the cell. Hence cell swelling is prevented. The UW (University of Wisconsin) solution is a modification, in which the major anion is lactobionate. Further additions are glutathione and allopurinol, to scavenge oxygen radicals, and adenosine, to serve as a substrate for ATP synthesis [3]. The histidine-tryptophan-ketoglutarate (HTK) solution designed by Bretschneider et al. [12] contains much less K^+ (10 mmol/l) but also little Na^+. Its osmolality is mostly due to histidine and mannitol. This solution, due to its enormous buffer capacity, prevents any substantial acidosis. All three types of solutions have been used successfully for the preservation of hearts, kidneys, and livers [12]. The periods of cold ischemia can be up to 24 h.

The record of successful transplantations has been increasing rapidly. This is not only due to the improved pathophysiological concepts and surgical techniques, but even more to success in preventing rejection of allocrafts. Recent U.S. statistics [19] give the following numbers of transplantations for 1989:

- Kidney 8890
- Liver 2160
- Heart 1673
- Pancreas 413
- Heart and lung 67.

The new problems concern [9]

- The stabilization of patients waiting for a transplant. The waiting periods for kidney transplantation are 1–2 years. A fraction of patients still die while waiting for a liver or heart transplant.
- The recruitment of larger numbers of donors and an appropriate organ distribution [9].
- The definition of inclusion and exclusion criteria for patients considered for transplant programs.

Table 124.1. Solutions used for organ preservation. Concentration in micromoles per liter unless not stated otherwise

Constituent	Euro-Collins	UW solution	HTK solution
Na^+	10	30	15
K^+	115	115	10
Ca^{2+}	–	–	–
Mg^{2+}	15	5	4
Cl^-	15	–	50
Lactobionate	–	100	–
HCO_3^-	10	–	–
HPO_4^{2-}	43	20	–
$H_2PO_4^-$	15	5	–
SO_4^{2-}	–	5	–
Histidine	–	–	180
Histidine-HCl	–	–	18
D-Glucose	198	–	–
Raffinose	–	30	–
Mannitol	–	–	30
Adenosine	–	5	–
Glutathione	–	3	–
Starch	–	50 g/l	–
Tryptophan	–	–	2
α-Ketoglutarate	–	–	1
Allopurinol	–	1	–
Insulin	–	40 U/l	–
Penicillin G	–	200 kU/l	–
Dexamethasone	–	16 mg/l	–
Osmolality (mosm/l)	406	320	310
pH at 8°C	7.3	7.4	7.3

UW, University of Wisconsin; HTK, histidine-tryptophan-ketoglutarate

124.6 Difficulties in Defining Death

It must appear strange that the definition of death was not given in the Introduction to this chapter, but is added here, at its end. The reason is that the definition depends largely on perspective. If aging is a continuous process, death is just the natural end of this process and the only predicable event for all of us. From the "perspective" of the individual cell, for example, death is its lysis and the collapse of the ionic gradients between intra- and extracellular space. For the layman it is, as it has long been for the professional, cessation of the heart beat and of respiration.

Clinical Death. New types of intensive care medicine have urged us to introduce new definitions. After complete cardiac and respiratory arrest, resuscitation may restore normal circulatory function and artificial ventilation will secure normal gas exchange, but the brain may have been damaged irreversibly during the ischemic and anoxic period. Such a patient would be called clinically dead. In this state:

- The patient is comatose
- The pupils do not react to light
- Spontaneous respiration is absent
- Brain stem reflexes cannot be provoked
- No signs of pain from trigeminal nerve stimulation are present.

This definition can (but need not) be supplemented by a completely silent EEG for 30 min; absence of acoustically evoked potentials; and absence of brain blood perfusion seen on angiography.

This definition is based on the irreversible damage to the entire CNS including the brainstem; according to this definition apallic patients, with no cortical function, such as consciousness, thinking, or cognition, but with intact brainstem function are not dead and should receive all available medical care.

The utilization of resuscitation efforts and the occurrence of irreversible brain damage pose serious problems not only for the medical staff but also for society in general. At this stage, the ongoing discussions on where to draw the line and how to comply with patients' individual wishes, highlight the unpossibility of dealing with these problems with medical professionalism alone. Nor is it likely that a court decision can be of much help. If available, the patient's own judgement weighs heavily and is probably decisive. Otherwise, the relatives, doctors, and nurses have to accept the burden of defining how long a terminally ill patient has to be kept "alive." A patient's request that he or she not be resuscitated must be respected.

Euthanasia. It is still a long way to the even more difficult issue of euthanasia (from Greek εὖ θάνατος, good death). A patient's expressed wishes and requests are probably based largely on his/her current experiences, and there are convincing examples suggesting that requests for an actively assisted death often arise from depression rather than reflecting a mature decision reached after long consideration. Refusal of active assistance is the only currently accepted position, but refusal of assistance in the exploration of the patient's psychic and mental situation is no less unacceptable.

The change in the definition of death that has become necessary as a result of recent developments in medicine highlights the dilemma that decisions on treatment cannot be based solely on laboratory reports or physical examination, but also have to take account of ethical issues.

124.7 Conclusion

On average, death comes later today than in any previous generation. It is likely that within the next 20 years medical progress will postpone death a little further, maybe to an average age of 90 instead of 75 for men and 80 for women, as today. It is not likely that our maximal life span will be extended by very much. If it were, this would presumably only be possible if there were a further delay in maturation [6]. The fact that molecular biologists and geneticists have been able to prolong the life span of the nematode *Caenorhabditis elegans* substantially [14] does not justify any optimism that this is going to be possible for humans. The control of aging is complex, and it is not likely that a single gene controls the entire process.

With the approach of the maximal life span of around 100 years, medical and health problems increase correspondingly, and perhaps with an exponential correlation. Morbidity will increase and better health care and social care will be necessary to deal with the problems that arise. The dimensions of the problems ahead are not easily defined. These are no longer exclusively problems of the medical profession, but of society in general. After all, what matters is not only the *duration*, but also the *quality*, of life.

At the end, we might learn from the last words of Socrates (in the ἀπολογία by Plato), then 70 years old (!), after he had been sentenced to death: "ἀλλὰ γὰρ ἤδη ὥρα ἀπιέναι, ἐμοὶ μὲν ἀποθανουμένῳ, ὑμῖν δὲ βιωσομένοις. ὁπότεροι δὲ ἡμῶν ἔρχονται ἐπὶ ἄμεινον πρᾶγμα, ἄδηλον παντὶ πλὴν ἢ τῷ θεῷ." "Now, however, it is time to go, for me to die and for you to live. Who of us turns to the better is not certain to anyone but God."

References

1. Ames BN, Shigenaga MK, Hagen TM (1993) Oxidants, antioxidants, and the degenerative diseases of aging. Proc Natl Acad Sci USA 90:7915–7922
2. Barinaga M (1994) Cell suicide: by ICE, not fire. Science 263:754–756
3. Bonventre JV, Weinberg JM (1992) Kidney preservation ex vivo for transplantation. Annu Rev Med 43:523–553
4. Carr D (1982) Pathophysiology of stress-induced limbic lobe dysfunction: a hypothesis for NDEs. J Near Death Stud 2:75–89
5. Collins RC (1989) Selective vulnerability of the brain: new insights into the pathophysiology of stroke. Ann Intern Med 110:992–1000
6. Cutler RG (1984) Evolutionary biology of aging and longevity in mammalian species. In: Johnson JE (ed) Aging and cell function. Plenum, New York, pp 1–147

7. Di Maio VJM, Di Maio DJM (1991) Natural death as viewed by the medical examiner: a review of 1000 consecutive autopsies of individuals dying of natural disease. J Forensic Sci 36:17–24
8. Dice JF (1993) Cellular and molecular mechanisms of aging. Physiol Rev 73:149–159
9. Halasz NA (1991) Medicine and ethics: how to allocate transplantable organs. Transplantation 52:43–46
10. Hart RW, Daniel FB (1980) Genetic stability in vitro and in vivo. Adv Pathobiol 7:123–141
11. Hengartner MO, Horvitz HR (1994) *C. elegans* cell survival gene ced-9 encodes a functional homolog of the mammalian proto-oncogene bcl-2. Cell 76:665–676
12. Hölscher M, Groenewoud AF (1991) Current status of the HTK solution of Bretschneider in organ preservation. Transplant Proc 23:2334–2337
13. Jansen K (1989) Near death experiences and the NMDA receptor. BMJ 298:1708
14. Kenyon C, Chang J, Gensch E, Rudner A, Tabtiang R (1993) A *C. elegans* mutant that lives twice as long as wild type. Nature 366:461–464
15. Kohn RR (1978) Principles of mammalian aging. Prentice-Hall, Englewood Cliffs
16. Lopez AD (1990) Competing causes of death. A review of recent trends in mortality in industrialized countries with special reference to cancer. Ann NY Acad Sci 609:58–76
17. Moody RA (1975) Life after life. Mockingbird, Covington
18. Morse ML, Venecia D, Milstein J (1989) Near-death experiences: a neurophysiologic explanatory model. J Near Death Stud 8:45–53
19. Murray JE (1992) Human organ transplantation: background and consequences. Science 256:1411–1416
20. Owens JE, Cook EW, Stevenson J (1990) Features of "near-death experience" in relation to whether or not patients were near death. Lancet 336:1175–1177
21. Reed JC (1994) Bcl-2 and the regulation of programmed cell death. J Cell Biol 124:1–6
22. Rodin EA (1980) The reality of death experiences, a personal perspective. J Nerv Ment Dis 168:259–263
23. Rogers MC, Kirsch JR (1989) Current concepts in brain resuscitation. JAMA 261:3143–3147
24. Schwartz LM, Smith SW, Jones MEE, Osborne BA (1993) Do all programmed cell deaths occur via apoptosis? Proc Natl Acad Sci USA 90:980–984
25. Stadtman ER (1992) Protein oxidation and aging. Science 257:1220–1224
26. Statistisches Bundesamt (1993) Sterbefälle nach Todesursachen. In: Statistisches Bundesamt (ed) Statistisches Jahrbuch 1993 für die Bundesrepublik Deutschland. Metzler, Poeschel, Stuttgart, pp 468–475
27. Ucker DS (1991) Death by suicide: one way to go in mammalian cellular development. New Biol 3:103–109
28. Vaux DL (1993) Toward an understanding of the molecular mechanisms of physiological cell death. Proc Natl Acad Sci USA 90:786–789
29. Yu BP (1994) Cellular defenses against damage from reactive oxygen species. Physiol Rev 74:139–162

VI Appendix

125 Units Used in Physiology and Their Definitions

R. GREGER

Contents

125.1 Use of SI Units

Since 1954 efforts have been made to achieve units of measurement that are universally accepted. To this end the Système International D'Unités (SI) was developed. This system is based upon seven basic and two supplementary units, as shown in Table 125.1. For fluids one additional unit, the liter ($l = 10^{-3} m^3$), is added.

125.1.1 Conversion to SI Units

The SI units replace a larger number of previously used older units which are sometimes still used in science. Table 125.2 is a conversion table for these older units.

125.1.2 Prefixes

Conversion of further units will be given below for derived SI units. The physical quantities can vary over many orders of magnitude, so prefixes have been introduced to simplify the expressions. For example, $10^{-6} m$ is equivalent to 1 micrometer (μm). The prefixes are listed in Table 125.3.

For the physical unit of mass the unit kg is used instead of g, and this is regarded a basic unit. Instead of Mg ($= 10^6$ g), the unit t (ton) is used by convention.

125.2 Derived SI Units

From the above basic units a large number of other units with their own symbol have been derived. These are often much more convenient to use than the very bulky composite of their respective basic units. Table 125.4 lists the derived SI units.

125.2.1 Conversion into Derived SI Units

A large number of other units were used previously instead of the derived SI units. These old units should now be avoided and converted into SI units as shown in Table 125.5. The various units will be discussed briefly. It should also be mentioned that some units such as mmHg continue to be used even though Pa should be preferred. There are several reasons for this inconsistency: (1) it takes some time to change habits and to think in terms of new units; (2) in medicine, pressure is still measured in mmHg because sphygmomanometers still use the mercury column; (3) even though national laws reinforce the use of SI units, many exceptions are still tolerated; and (4) unfortunately, not all scientific journals request the use of SI units [4].

125.2.2 Composed Units

For composed units the general rule holds that no prefixes should be used in the denominator. For instance, mmol/ml

Table 125.1. Basic and supplementary units of the Système International d'unités (SI units)

Physical quantity	Name	Symbol
Length	meter	m
Mass	kilogram	kg[a]
Time	second	s
Electric current	ampère	A
Thermodynamic temperature	kelvin	K
Luminous intensity	candela	cd
Amount of substance	mole	mol
Plane angle[b]	radian	rad
Solid angle[b]	steradian	sr

[a]kg is one exception to the general convention of using units without prefixes (see below)
[b]Supplementary units. 1 rad is $360°/2\pi = 57.296°$

R. Greger/U. Windhorst (Eds.)
Comprehensive Human Physiology, Vol. 2
© Springer-Verlag Berlin Heidelberg 1996

Table 125.2. Conversion of previously used units into SI units

Name	Abbreviation	SI unit	Multiplier[a]
angstrom	Å	m	10^{-10}
inch	in	m	0.0254
foot	ft	m	0.3048
yard	y	m	0.9144
mile (statute)	m	m	1609
mile (nautical)	m	m	1853
square inch	sq. in.	m²	0.6451×10^{-3}
square foot	sq. ft	m²	0.0929
cubic inch	cu. in.	m³	16.387×10^{-6}
cubic foot	cu. ft	m³	0.0283
pound	lb	g	453.6
liter	l	$l = 10^{-3} m^3$	1
fluid ounce	fl. oz		
American		l	0.0296
British		l	0.0284
gallon	gal		
American		l	3.785
British		l	4.546
mole	M	mol	1

[a]SI unit × multiplier = older unit

Table 125.3. Decimal multiples and submultiples to be used with SI units

Submultiple	Prefix	Symbol	Multiple	Prefix	Symbol
10^{-1}	deci[a]	d	10	deca[a]	da
10^{-2}	centi[a]	c	10^2	hecto[a]	h
10^{-3}	milli	m	10^3	kilo	k
10^{-6}	micro	μ	10^6	mega	M
10^{-9}	nano	n	10^9	giga	G
10^{-12}	pico	p	10^{12}	tera	T
10^{-15}	femto	f	10^{15}	peta	P
10^{-18}	atto	a	10^{18}	exa	E

[a]The prefixes d (deci) and (c) centi should be avoided, and quantities should for preference be expressed in multiples or submultiples based on 10^3 or 10^{-3}. For example, 10^{-5} m should be expressed as 0.01×10^{-3} m, 10×10^{-6} m, or $10\,\mu$m

Table 125.4. Derived SI units, their special names, abbreviations and composition

Physical quantity	Name	Symbol	Composition
Frequency	hertz	Hz	s^{-1}
Force	newton	N	$m \times kg \times s^{-2}$
Energy	joule	J	$N \times m = m^2 \times kg \times s^{-2}$
Power	watt	W	$J \times s^{-1} = m^2 \times kg \times s^{-3} = N \times m \times s^{-1}$
Pressure	pascal	Pa	$N \times m^{-2} = m^{-1} \times kg \times s^{-2}$
Electric charge	coulomb	C	$A \times s$
Electric potential difference = voltage			
Voltage	volt	V	$W \times A^{-1} = m^2 \times kg \times s^{-3} \times A^{-1}$
Electric power	watt	W	$J \times s^{-1} = m^2 \times kg \times s^{-3} = N \times m \times s^{-1}$
Electric resistance	ohm	Ω	$V \times A^{-1} = m^2 \times kg \times s^{-3} \times A^{-2}$
Electric conductance	siemens	S	$\Omega^{-1} = m^{-2} \times kg^{-1} \times s^3 \times A^2$
Electric capacitance	farad	F	$C \times V^{-1} = A \times s \times V^{-1} = m^{-2} \times kg^{-1} \times s^4 \times A^2$
Magnetic flux	weber	Wb	$V \times s = m^2 \times kg \times s^{-2} \times A^{-1}$
Magnetic inductance	henry	H	$V \times s \times A^{-1} = m^2 \times kg \times s^{-2} \times A^{-2}$
Magnetic flux density	tesla	T	$Wb \times m^{-2} = kg \times s^{-2} \times A^{-1}$
Luminous flux	lumen	lm	$cd \times sr$
Illuminance	lux	lx	$lm \times m^{-2} = cd \times sr \times m^{-2}$
Radioactivity	bequerel	Bq	s^{-1}
Absorbed dose	gray	Gy	$J \times kg^{-1}$
Dose equivalent	sievert	Sv	$J \times kg^{-1}$

Table 125.5. Conversion of old units to derived SI units

Physical quantity	Old unit	Old symbol	SI unit	Multiplier
Frequency		min^{-1}	Hz	0.0167
Velocity	$miles \times h^{-1}$	mph	$m \times s^{-1}$	0.4770
	$km \times h^{-1}$	kmh	$m \times s^{-1}$	0.2778
Force	dyne	dyn	N	10^{-5}
	pond	p	N	9.81×10^{-3}
	poundal		N	0.1383
	pound force	lbf	N	4.4482
Pressure	mmH_2O		Pa	9.81
	cmH_2O		Pa	98.1
	mmHg	Torr	Pa	133.3
	technical atmosphere	at	Pa	98 067
	physical atmosphere	atm	Pa	101 324
	dyn/cm^2		Pa	0.1
	bar		Pa	100 000
Energy	calorie	cal	J	4.185
	kilocalorie	kcal	J	4185
	erg	erg	J	10^{-7}
	watt second	Ws	J	1
	kilowatt hour	kWh	J	3.6×10^6
	electron volt	eV	J	160.21×10^{-21}
Power	erg/s	erg/s	W	10^{-7}
	cal/h	cal/h	W	1.163×10^{-3}
	horse power	hp, PS	W	735.5
Radiation	Curie	Ci	Bq	3.7×10^{10}
Temperature[a]	degree Celsius	°C	K	+ 273
	degree Fahrenheit	°F	K	$0.5556 \times (°F - 32) + 273$
Concentration	molar	M	$mol \times l^{-1}$	1
	equivalent/l	eq/l	$mol \times l^{-1}$	$1 \times valence^{-1}$
		val/l	$mol \times l^{-1}$	$1 \times valence^{-1}$
	normal	N	$mol \times l^{-1}$	$1 \times valence^{-1}$
	g %	% w/v	$g \times l^{-1}$	10
	mg %	% w/v	$g \times l^{-1}$	0.01
	µg %	% w/v	$g \times l^{-1}$	10×10^{-6}
	γ %	% w/v	$g \times l^{-1}$	10×10^{-6}
	volume %	% v/v	$ml \times l^{-1}$	10
	parts per million	ppm	$g \times l^{-1}$	10^{-6}
	parts per billion	ppb	$g \times l^{-1}$	10^{-9}

[a]There is no single multiplier for temperature; conversions are as follows:
From °C to K: °C + 273
From °C to °F: °C $\times$ 1.8 + 32
From °F to °C: °F $\times$ 0.5556 − 32
From °F to K: 0.5556 $\times$ (°F − 32) + 273

should be avoided and mol/l used instead, m/ms should be replaced by km/s, etc.

125.3 Concentration

An enormous degree of diversity exists with respect to the physical unit of concentration. As a general rule, concentration should be given in mol/l for any substance of known molecular weight:

$$1 \, mol = molecular \, weight \, in \, grams \, (g) \qquad (125.1)$$

Molecular weight is given in daltons (Da). The dalton is not an SI unit: 1 Da is defined as 1/12th of the mass of a ^{12}C atom = 1.66×10^{-27} kg. Concentrations should not be given as percentages (%). Not only does this involve reference to a non-SI unit of volume or mass (e.g., deciliter (dl) =

100 ml), but in addition it is usually not evident whether they stand for mass/volume, volume/volume, mass/mass, or volume/mass.

125.3.1 Equivalents and Normality

For ions, equivalents (eq) used to be used instead of moles (mol). This should now be avoided. For instance, 1 meq/l Ca^{2+} corresponds to 0.5 mmol/l. Similarly, for bases and acids, the previously used normality should be changed to molarity: 1 N HCl = 1 mol/l HCl.

$$mol - eq \times valence^{-1} \qquad (125.2)$$

125.3.2 Molarity and Molality

Usually volume is used as the reference in fluids, rather than weight. Concentration when expressed by reference

Table 125.6. Values of the activity coefficient (γ) of NaCl solutions as a function of concentration (at 25°C)

Concentration (mmol/l)	γ
100	0.778
200	0.735
300	0.710
500	0.681
1000	0.657

to the *volume* of solvent (in liters; e.g., water) is referred to as the *molarity* of the solution. It should be noted that the measurement is usually performed such that a sample of solution of given volume is taken and the amount of, e.g., urea in this sample measured. If many other constituents are present in this solution (Na^+, Cl^-, proteins, etc.) the volume of water will be underestimated by the amount of the volume occupied by the other solutes. In plasma this error would be around 7%. In other words, the true concentration of urea, in plasma, measured as 5 mmol/l, would be 5.35 mmol/l H_2O. Since 1l H_2O corresponds to 1 kg H_2O [at 277 K (4°C)], one can also express this as 5.35 mmol/kg H_2O. Expression of the concentration of a solute by reference to the *weight* of the solvent is referred to as the *molality* of the solution. For dilute solutions, molality and molarity are almost identical. However, in more concentrated fluids such as plasma, and even more so in the cytosol, molality exceeds molarity.

$$\text{Molarity} = \text{mol} \times l^{-1} \tag{125.3}$$

$$\text{Molality} = \text{mol} \times \text{kg}^{-1} \tag{125.4}$$

125.3.3 Activity

Salts dissociate in aqueous solutions. Almost complete dissociation can be expected in dilute solutions, but in more concentrated solutions the dissociation is incomplete. This phenomenon is of high importance for biological fluids. In a saline solution of 150 mmol/l NaCl at a temperature of 310 K (37°C) only 76% is dissociated; thus, the activity coefficient $\gamma = 0.76$.

$$\text{Activity} = \text{concentration} \times \gamma \tag{125.5}$$

Table 125.6 lists values of γ for NaCl solutions as a function of the concentration. It is evident from this that a 1 mol/l solution of NaCl, for instance, has a γ of only 0.66, whilst a 100 mmol/l solution has a γ of 0.78. The activity increases slightly with temperature. Data for most ions under a variety of conditions are listed elsewhere [3].

125.4 Osmolarity and Osmolality

Van't Hoff's law states that solutes exert an osmotic pressure (π):

$$\pi = R \times T \times c_{osm} \tag{125.6}$$

In this equation c_{osm} is the osmolality (osm $\times$ kg^{-1}). R and T have their usual meaning. Like molarity and molality (see above), *osmolarity* is referenced to *volume* (l) and *osmolality* to *mass* (kg). Again, the two values agree reasonably well in dilute solutions, osmolality values slightly exceeding osmolarity values. In more concentrated solutions osmolality is lower than the expected value. For instance, in a 150 mmol/kg (H_2O) NaCl solution, which theoretically should have an osmolality of $2 \times 150 = 300$ mosmol/kg, the measured osmolality is only 278 mosmol/kg. This is due to the fact that NaCl is only partially dissociated in this solution.

The osmotic pressure can become effective if a membrane separates two compartments, if the osmolality is different in the two compartments (Δc_{osm}), and if the respective solutes are less permeable than H_2O, i.e., if the membrane has a reflection coefficient (σ) which is $> 0 < 1$. Then we can modify Eq. 125.6 as follows:

$$\Delta\pi = R \times T \times \Delta c_{osm} \times \sigma \tag{125.6'}$$

In steady state $\Delta\pi$ is close to zero across cell membranes, and the effective osmolality on both sides is close to 290 mosmol/kg. Dilution of, for example, the extracellular fluid can exert an impressively high pressure. If Δc_{osm} were 150 mosmol/kg and if σ were close 1 (membrane impermeable to the respective osmolyte), the pressure would be as high as 387 kPa (38 m H_2O; 2.9 m Hg; 3.9 bar).

Differences between plasma and interstitial fluid in osmolality, caused by differences in the concentrations of small solutes and ions in the two fluids, collapse very rapidly and usually exert no osmotic pressure, because the endothelium is equally permeable to solutes and to water ($\sigma = 0$). Large macromolecules, however, such as plasma proteins, have reflection coefficients $\gg 0 \leqslant 1$. Hence, they exert an osmotic pressure called specifically *oncotic pressure*.

Oncotic Pressure. The oncotic pressure, usually labelled π, of plasma is around 4 kPa (30 mm Hg). This pressure opposes the hydrostatic filtration pressure (Δp) in capillaries. It increases during filtration, with water shift to the

Table 125.7. Ionic radius and relative mobility of ions

Ion	Ionic radius (nm)	Hydrated radius (nm)	Relative mobility
Li^+	0.060	~0.30	0.52
Na^+	0.095	~0.25	0.67
K^+	0.133	~0.18	0.97
Rb^+	0.148	~0.15	1.01
Ca^{2+}	0.169		0.79
Br^-	0.195	~0.20	1.02
Cl^-	0.181	~0.21	1.00
SO_4^{2-}			1.05
Acetate			0.54

Table 125.8. Fundamental constants

Constant	Symbol	Value	Dimensions
Electron rest mass	m_e	0.91096×10^{-30}	kg
Proton rest mass	m_p	1.673×10^{-27}	kg
Neutron rest mass	m_n	1.675×10^{-27}	kg
Speed of light	c	299.8×10^6	$m \times s^{-1}$
Acceleration of free fall	g	9.807	$m \times s^{-2}$
Gravitational constant	G	66.64×10^{-12}	$N \times m^2 \times kg^{-2}$
Boltzmann's constant	$k = R/N_A$	13.806×10^{-24}	$J \times K^{-1}$
Gas constant	R	8.314	$J \times K^{-1} \times mol^{-1}$
Planck's constant	h	0.6626×10^{-33}	$J \times s$
Avogadro's constant[a]	L, N_A	602.25×10^{21}	mol^{-1}
Loschmidt's constant[a]	N_L	26.872×10^{24}	m^{-3}
Electronic charge	e	0.1602×10^{-18}	C
Faraday's constant	F	96.487×10^3	$C \times mol^{-1}$
Electric constant	ε_0	8.854×10^{-12}	$F \times m^{-1}$
Stefan-Boltzmann constant	σ	56.697×10^{-9}	$W \times m^{-2} \times K^{-4}$
Magnetic constant	μ_0	$0.4\pi \times 10^{-6}$	$H \times m^{-1}$

[a]Sometimes Avogadro's constant, formerly also called Avogadro's number, is also referred to as Loschmidt's number. This is incorrect. Avogadro's constant gives the number of particles per mole and Loschmidt's constant the number of particles of an ideal gas per unit volume

interstitium (see Chaps. 94, 97) or to the Bowman space in the glomerulus (see Chap. 73), and falls if water is taken up in the second half of the capillary, where $\Delta p < \Delta \pi$. It should be noted that, due to the biophysical properties of the plasma proteins, the increment in $\Delta \pi$ with water removal is substantially greater then would be predicted theoretically. For instance, in the glomerulus a filtration rate of 20% of the glomerular blood flow would be expected to increase $\Delta \pi$ by a factor of 1.25. In reality, the measured value is closer to 1.5.

125.5 Ion Mobility

The mobility of an ion depends on its size (ionic radius). The bigger the ion, the smaller its mobility. Positively charged ions (cations) usually have smaller radii than negatively charged ions (anions). The correlation between ionic radii and ionic mobility is complicated by a direct interaction of the ion with water molecules, the *water shell*. The size of the water shell depends on the charge and the ionic radius. The smaller the ion, the larger the water shell; thus, the water shell of Na^+ is larger than that of K^+. As a result, the hydrated radius of Na^+ is larger than that of K^+, even though the ionic radius of Na^+ is smaller than that of K^+. Table 125.7 gives some values for ionic radii, approximate hydrated radii [1], and relative ionic mobility [1,3]. Instead of using absolute values for ion mobilities (i.e., the terminal speed of an ion in an electric field divided by the field strength) values are frequently given with reference to the Cl^- ion, which is assigned the number 1.

Differences in ionic mobilities give rise to liquid junction potentials [2]. Note that the mobilities of K^+ and Cl^- are very similar. Therefore high-concentration KCl solutions are very useful for cancelling out liquid junction potentials.

Mobilities are of little relevance to the origin of voltages across cell membranes. Fractional conductances rather than mobilities determine the ion transfer and hence the voltage (see Chap. 11).

125.6 Fundamental Constants

Table 125.8 lists a number of fundamental constants together with their related SI units.

125.7 The Greek Alphabet

Capital	Lower case	Name
A	α	alpha
B	β	beta
Γ	γ	gamma
Δ	δ	delta
E	ε	epsilon
Z	ζ	zeta
H	η	eta
Θ	θ	theta
I	ι	iota
K	κ	kappa
Λ	λ	lambda
M	μ	mu
N	ν	nu
Ξ	ξ	xi
O	o	omicron
Π	π	pi
P	ρ	rho
Σ	σ, ζ	sigma
T	τ	tau
Y	υ	upsilon

Φ	φ	phi
Χ	χ	chi
Ψ	φ	psi
Ω	ω	omega

References

1. Hille B (1984) Principles and mechanisms of function. In: Hille B (ed) Ionic channels of excitable membranes. Sinauer, Sunderland, pp 149–180

2. Koryta J (1982) Ions, electrodes and membranes. Wiley, Chichester
3. Robinson RA, Stokes RH (1965) Electrolyte solutions, 3rd edn. Butterworths, London
4. Zijlstra WG (1977) Introduction of the use of the international system of units (SI) in papers to be published in *Pflügers Archiv, European Journal of Physiology*. Pflugers Arch Eur J Physiol 368:1–2

126 Normal Values for Physiological Parameters

R. Greger and M. Bleich

Contents

126.1 Introduction

This chapter is aimed at supplementing several of the preceding chapters inasmuch as it summarizes important physiological normal values. This should enable the reader to find these values without the need to consult the respective chapter. Obviously, the citation of normal values of physiological parameters is fraught with the dilemma that such parameters, besides the difficulties of measurements, vary with maturation and ageing and depend on gender, body size, body weight, body surface area, and even on diurnal and other rhythms. Many physiological parameters change with exercise. All these considerations are largely ignored in the tables and figures below, and the reader is referred to the respective chapters. Many of the parameters have been taken from *Documenta Geigy* [3], a unique collection of normal values with a large number of original references.

The various parameters are presented according to body functions. To avoid lengthy searches some parameters are presented in several sections. All parameters are given in SI units (see Chap. 125 for conversions). Pressures are given in Pa and in most instances in the older but more familiar units (mmHg, cm H_2O) as well.

126.2 Normal Values in Blood, Plasma, and Serum

126.2.1 Plasma Electrolytes

Generally plasma concentrations are very similar to serum concentrations (Table 126.1).

126.2.2 Plasma Proteins and Amino Acids

Table 126.2 shows plasma proteins as a percentage of total plasma.

Table 126.3 shows amino acid concentrations in plasma (only mean values for adults are given).

R. Greger/U. Windhorst (Eds.)
Comprehensive Human Physiology, Vol. 2
© Springer-Verlag Berlin Heidelberg 1996

Table 126.1. Plasma electrolytes

Ion	Concentration (mmol/l)	95% range (mmol/l)	Comment
Na^+	142	138–151	
K^+	4	3.4–5.2	
Ca^{2+}	2.5	2.4–2.8	1.5 mmol/l free, rest protein bound
Mg^{2+}	1	0.85–1.15	0.7 mmol/l free, rest protein bound
Cl^-	106	101–111	
HCO_3^-	25	21–28.5	
HPO_4^{2-}	0.8	0.6–1.0	At pH 7.4
$H_2PO_4^-$	0.2	0.1–0.3	At pH 7.4
SO_4^{2-}	0.33	0.15–0.5	
Organic anions	6		Amino acids, lactate, fatty acids etc.
Proteins	70 g/l	65–80	See Table 126.2
Fe^{3+}	20×10^{-3}	13–30	Bound to transferrin, much more iron in red blood cells (8.5 mmol/l)
Cu^{2+}	18×10^{-3}	12–24	Bound to ceruloplasmin, similar concentration in red blood cells

Osmolality, 289 mosm/kg H_2O (range, 281–297)
Osmotic pressure, 660 kPa (4950 mmHg)
Oncotic pressure, 3.2 kPa (range, 2.7–4.7), 24.3 mmHg (range, 20.6–35.3)

Table 126.2. Plasma proteins

Protein	% of total
Albumin	65
α_1-Globulin	4
α_2-Globulin	7
β-Globulin	10
γ-Globulin	14
Fibrinogen	3

The total concentration is approximately 65–80 g/l

Table 126.3. Amino acid concentrations in plasma

Amino acid	Concentration (μmol/l)	95% range (μmol/l)
Alanine	420	213–472
Arginine	100	40–140
Aspartate	2	1–11
Cystine/cysteine	100	70–108
Glutamine	340	140–570
Glutamate	50	20–90
Glycine	220	179–587
Histidine	80	32–97
Isoleucine	70	40–99
Leucine	140	78–176
Lysine	200	105–207
Methionine	30	11–30
Phenylalanine	50	38–73
Proline	220	103–290
Serine	120	76–164
Taurine	50	32–138
Threonine	130	76–194
Tyrosine	60	22–83
Valine	270	168–317

126.2.3 Plasma Concentration of Organic Compounds

Table 126.4 shows the plasma concentrations of organic compounds. Lipid fractions [5] are given in Table 126.5; the fifth, 50th, and 95th percentiles are given in mmol/l and g/l.

126.2.4 Composition of Blood

Table 126.6 shows the composition of blood (see Chap. 83).

126.3 Other Body Fluids

126.3.1 Cytosol

Table 126.7 shows the composition of the cytosol (see Chaps. 4, 8, 11). The ionic composition of the cytosol is still not well defined. The cytosolic concentrations have recently been determined by electron probe analysis, as activity measurements by ion-selective electrodes, and with fluorophores in a variety of tissues. The numbers listed in Table 126.7 are gross approximations.

126.3.2 Cerebrospinal Fluid

The composition of cerebrospinal fluid (see Chap. 27) is shown in Table 126.8.

Table 126.4. Plasma concentrations of organic compounds

Name	Concentration (μmol/l)	95% range (μmol/l)
Acetone	50	35–60
Acetoacetate	–	8–28
Allantoin	27	19–38
Amino acids (see Table 126.3)	–	–
Ammonia	11	5–18
Bilirubin (direct)	1 mg/l	0.5–2.4 mg/l
Bilrubin (total)	6 mg/l	2.6–14 mg/l
Bilirubin (free)	3 mg/l	1.1–10.5 mg/l
β-OH-Butyrate	43	16–201
Creatine	24	12–55
Creatinine	70	45–105
D-glucose	5050	4170–5928
Oxalate	–	19–39
Pyruvate	64	14–117
Total lipids	–	4.5–10 g/l (for fractions see Table 126.5)
Urate	270	110–390
Urea	5500	2500–10 300

Table 126.5. Lipid fractions

Percentile	Men						Women					
	15–19 years		40–44 years		65–69 years		15–19 years		40–44 years		65–69 years	
	(mmol/l)	(g/l)	(mmol/l)	(g/l)	(mmol/l)	(g/l)	(mmol/l)	(g/l)	(mmol/l)	(g/l)	(mmol/l)	(g/l)
Total cholesterol												
5th	2.92	1.13	3.90	1.51	4.09	1.58	3.10	1.20	3.80	1.47	4.42	1.71
50th	3.78	1.46	5.25	2.03	5.43	2.10	4.01	1.55	4.97	1.92	5.84	2.26
95th	5.09	1.97	6.93	2.68	7.09	2.74	5.25	2.03	6.54	2.53	7.68	2.97
LDL cholesterol												
5th	1.60	0.62	2.25	0.87	2.53	0.98	1.53	0.59	1.91	0.74	2.38	0.92
50th	2.40	0.93	3.49	1.35	3.78	1.46	2.40	0.93	3.15	1.22	3.90	1.51
95th	3.36	1.30	4.81	1.86	5.43	2.10	3.54	1.37	4.50	1.74	5.72	2.21
HDL cholesterol												
5th	0.78	0.30	0.70	0.27	0.78	0.30	0.91	0.35	0.88	0.34	0.91	0.35
50th	1.19	0.46	1.11	0.43	1.27	0.49	1.32	0.51	1.45	0.56	1.60	0.62
95th	1.63	0.63	1.73	0.67	2.02	0.78	1.91	0.74	2.28	0.88	2.53	0.98
Total triglycerides												
5th	0.42	0.37	0.62	0.55	0.64	0.57	0.44	0.39	0.53	0.47	0.68	0.60
50th	0.78	0.69	1.38	1.12	1.26	1.12	0.77	0.68	0.99	0.88	1.26	1.12
95th	1.67	1.48	3.61	3.20	3.01	2.67	1.49	1.32	2.36	2.09	2.72	2.41

LDL, low-density lipoprotein; HDL, high-density lipoprotein.

Table 126.6. Blood composition

Parameter	Men		Women	
	Value	95% range	Value	95% range
Total blood volume (% of body mass)	6–8	–	5.5–7	–
Hematocrit	0.47	0.4–0.54	0.42	0.37–0.47
Erythrocytes per μl (in millions)	5.4	4.6–6.2	4.8	4.2–5.4
Erythrocyte diameter (μm)	7.5	–	7.5	–
Erythrocyte volume (fl)	97	82–107	95	88–109
Hemoglobin (g/l)	158	133–182	141	118–175
Hemoglobin per erythrocyte (pg)	31.4	27–38	31.2	28–34
Sedimentation rate (mm/h)				
First h	3–8	–	3–8	–
Second h	<20	–	<20	–
Dynamic viscosity (poise)	0.023–0.028	–	0.023–0.028	–
Dynamic visocity (Pa/s)	0.23–0.28	–	0.23–0.28	–

Table 126.6. *Continued*

Parameter	Men		Women	
	Value	95% range	Value	95% range
Leukocytes per µl	7000	2800–11 200	7000	2800–11 200
Neutrophils (%)	59	–	59	–
Eosinophils (%)	2.4	–	2.4	–
Basophils (%)	0.6	–	0.6	–
Lymphocytes (%)	31	–	31	–
Monocytes (%)	6.5	–	6.5	–
Thrombocytes per µl (in thousands)				
Arterial	350	322–378	350	322–378
Venous	310	286–334	310	286–334

Table 126.7. Composition of the cytosol

Ions	Mean	Range	Comment
Na^+ (mmol/l)	14	6–30	–
K^+ (mmol/l)	120	100–150	–
Ca^{2+} (nmol/l)	70	50–150	Free, ionized
Mg^{2+} (mmol/l)	0.4	–	Estimate from recent dye measurements
Cl^- (mmol/l)	4–40	–	Low in neurons and skeletal muscle, high in some epithelia
HCO_3^- (mmol/l)	15	10–18	Estimates from pH measurements
pH	–	7.1–7.3	–
pCO_2 (kPa)	5 (40 mmHg)	–	–
Large anions (meq/l)	~120	–	Proteins, phosphates, others

Table 126.8. Composition of cerebrospinal fluid [3,14]

	Mean	95% range
Volume (ml)	135	100–160
Osmolarity (mosm/l)	295	–
pH	7.35	7.33–7.38
Ions		
Na^+ (mmol/l)	146	135–157
K^+ (mmol/l)	3	2.0–3.9
Ca^{2+} (mmol/l)	1.15	1.0–1.3
Mg^{2+} (mmol/l)	1.1	0.23–2.0
Cl^- (mmol/l)	124	122–128
HCO_3^- (mmol/l)	22	21–26
pCO_2 (kPa)	6.2 (47 mmHg)	5.3–6.8 (39.5–50.9 mmHg)
Glucose (mmol/l)	3	–
Urea (mmol/l)	–	1.7–5
Cholesterol (µmol/l)	–	1.6–14
Fatty acids (µmol/l)	–	0.03–0.15
Protein (mg/l)	300	160–340
Prealbumin	–	7–10
Albumin	–	140–185
α_1-Globulin	–	11–28
α_2-Globulin	–	12–33
β_1-Globulin	–	13–24
τ-Fraction	–	7–15
γ-Globulin	–	8–26
Leukocytes (μl^{-1})	1	0–5

126.3.3 Milk

The composition of milk (see Chap. 121) is shown in Table 126.9.

126.3.4 Saliva

The composition of saliva (see Chap. 64) is shown in Table 126.10.

126.3.5 Gastric Juice

The composition of gastric juice (see Chap. 61) is shown in Table 126.11.

126.3.6 Pancreatic Juice

The composition of pancreatic juice (see Chap. 65) is given in Table 126.12.

126.3.7 Bile

The composition of bile (see Chap. 67) is shown in Table 126.13.

126.3.8 Stool

The composition of stool is shown in Tables 126.14 and 126.15.

126.3.9 Sweat

The composition of sweat (see Chap. 112) is shown in Table 126.16.

126.3.10 Urine

The composition of urine (see Chaps. 73–81) is shown in Tables 126.17 and 126.18.

126.3.11 Sperm

The composition of sperm [3] is shown in Table 126.19.

Table 126.9. Composition of milk

	Colostrum	Mature milk
Lactose (mmol/l)	158	197
Fat (total, g/l)	29	45
Cholesterol (g/l)	0.28	0.14
Protein (total, g/l)	23	10
Casein (g/l)	2	4
Lactoglobulin (g/l)	3.5	1.3
Amino acids (g/l)	1.2	1.3
Vitamin A (mg/l)	1.6	0.61
Na^+ (mmol/l)	22	7.4
K^+ (mmol/l)	19	13
Ca^{2+} (mmol/l)	8.5	12
Phosphate (mmol/l)	5.1	4.5
Mg^{2+} (mmol/l)	1.8	1.5
Cl^- (mmol/l)	16.5	10.4

Table 126.10. Composition of stimulated saliva ([3] parotid gland, submandibular gland)

	Mean	95% range	Comment
Total rate resting (ml/min)		0.3–0.4	
Total rate stimulated (ml/min)		0.5–7	
pH		6.4–8.2	
Ions	**Mean**		
Na^+ (mmol/l)	14	12–36	Increasing with stimulation
K^+ (mmol/l)	19	11–27	Decreasing with stimulation
Ca^{2+} (mmol/l)	1.5	1.1–2.7	Increasing with stimulation
Mg^{2+} (mmol/l)	0.3	0.08–0.5	
Cl^- (mmol/l)	15–43	< ca. 43	Increasing with stimulation
HCO_3^- (mmol/l)		10–60	Increasing with stimulation
$HPO_4^{2-}/H_2PO_4^-$ (mmol/l)	11.5	0.8–2.2	Decreasing with stimulation
pCO_2 (kPa)	6 (45.2 mmHg)	5.3–6.8 (39.5–50.9 mmHg)	
Protein (g/l)		1.4–6.4	
Mucins (g/l)	2.7	0.8–6.0	
Urea (mmol/l)	3.3	2.3–12.5	
Enzymes			
Lysozyme (g/l)		<0.15	
Amylase (g/l)	0.38		

Table 126.11. Composition of gastric juice [3]

	Mean	95% range	Comment
Total rate (ml/h)	46	0–176	Resting
	117	70–256	Stimulated by histamine
pH		1.9–2.6	–
Acid secretion rate (mmol/h)	2.4	<22	–
Ions			
Na$^+$ (mmol/l)	49	19–70	Decreasing with stimulation
K$^+$ (mmol/l)	12	6.4–16.6	–
Ca^{2+} (mmol/l)	1.8	1.0–2.4	–
Mg^{2+} (mmol/l)	0.75	0.15–1.5	–
Cl$^-$ (mmol/l)	130	78–160	–
HCO$_3^-$ (mmol/l)	0	–	–
Phosphate (mmol/l)	2.2	0.2–5.5	Decreasing with stimulation
pCO$_2$ (kPa)	6 (45.2 mmHg)	5.3–6.8 (39.5–50.9 mmHg)	–
Protein (g/l)	2.8	2.2–3.4	–
Mucins (g/l)	–	0.6–15	Decreasing with stimulation
Urea (mmol/l)	1.4	–	–
Ammonium (mmol/l)	5.7	3.9–7.5	–
Enzymes			
Pepsin (kU/h)		<1.2	–

Table 126.12. Composition of pancreatic juice [3]

	Mean	95% range	Comment
Total rate (ml/h)	36	0–99	Resting
	176	38–314	Stimulated by secretin
pH	–	7.5–8.8	–
HCO$_3^-$ secretion rates (mmol/h)	14	0–27	Increasing with stimulation
Ions			
Na$^+$ (mmol/l)	–	139–143	–
K$^+$ (mmol/l)	–	4.1–5.5	–
Ca^{2+} (mmol/l)	1.7	1.1–2.3	–
Mg^{2+} (mmol/l)	0.5	–	–
Cl$^-$ (mmol/l)	10	4–129	Decreasing with stimulation, sum of HCO$_3^-$ and Cl$^-$ 154
HCO$_3^-$ (mmol/l)	144	24–150	Increasing with stimulation
HPO$_4^{2-}$/H$_2$PO$_4^-$ (mmol/l)	0.8	0–1.6	Decreasing with stimulation
Protein (g/l)	–	1.9–3.4	–
Urea (mmol/l)	1.8	–	–
Enzymes			
Amylase (mg/min)	0.62	0.29–1.3	After CCK
Carboxypeptidase (mg/min)	0.72	0.36–1.45	after CCK
Chymotrypsin (mg/min)	3.0	1.22–7.6	after CCK
Trypsin (mg/min)	0.73	0.38–1.42	after CCK
Lipase (kU/min)	1.65	0.78–3.5	after CCK

CCK, cholecystkinin

Table 126.13. Composition of bile [3]

	Liver bile		Gall bladder bile		
	Mean	95% range	Mean	95% range	Comment
Flow rate of liver bile (ml/h)	–	10–46	–	–	–
pH	–	6.2–8.5	–	5.6–8.0	–
Ions					
Na$^+$ (mmol/l)	149	131–164	220	146–360	–
K$^+$ (mmol/l)	5.0	2.6–12	13.5	8.4–17.5	–
Ca^{2+} (mmol/l)	–	1.7–2.1	7.7	2.0–17	–
Mg^{2+} (mmol/l	~1	–	–	–	–

Table 126.13. *Continued*

	Liver bile		Gall bladder bile		
	Mean	95% range	Mean	95% range	Comment
Ions					
Cl⁻ (mmol/l)	101	89–118	31	7–110	–
HCO_3^- (mmol/l)	30	–	19	–	Increasing with stimulation
Phosphate (mg/l)	148	60–236	1400	–	Plus organic (phospholipid) Phosphorus
Protein (g/l)	1.8	1.4–2.7	4.5	–	–
Urea (mmol/l)	3.9	–	–	3.3–7.5	–
Cholesterol (mmol/l)	–	2.1–4.7	16.3	8.0–42	–
Bile acids (g/l)	–	6.5–14	32	0–66	Mostly conjugated in the bladder; bile acids are the major anion
Bilirubin (g/l)	0.65	0.1–1.4	2.9	0.4–6.3	Majority conjugated to glucuronic acid

Table 126.14. Composition of stool

	Mean	95% range
Total amount (g/day)	115	33–198
Water volume (ml/day)	90	–
Solids (g/day)	21	11–31
pH	7.15	5.9–8.5
Bilirubin (mg/day)	–	5–20
Urobilinogen (mg/day)		
Men	101	57–200
Women	40	80–150
Coproporphyrin (mg/day)	0.4	0.01–0.8
Protoporphyrin (mg/day)	1	0–2
Lipids (g/day)	5.5	0.1–11

Table 126.15. Concentration and excretion of stool

	Concentration (mmol/l)		Excretion (mmol/day)		Comment
	Mean	Range	Mean	Range	
Ions					
Na⁺	–	5–125	–	0.5–12.5	Depending on diet
K⁺	–	30–200	10	3–20	Depending on K⁺ content of diet
Ca²⁺	–	–	–	8–33	–
Mg²⁺	–	–	–	5–15	–
Cl⁻	–	5–30	–	0.5–3	–
HCO_3^-	<30	–	<3	–	–
Phosphate	–	–	–	10–25	Mostly precipitated as Ca²⁺ phosphate
Organic anions	150	100–400	15	10–40	Major part of anions

Table 126.6. Composition of sweat

	Mean	95% range	Comment
Total rate (ml/day)	–	~500–3000	Depending on thermal status and exercise
pH	–	4–6.8	–
Ions			
Na⁺ (mmol/l)	58	20–60	Depending on aldosterone (increased in cystic fibrosis)
K⁺ (mmol/l)	9	4–14	–
Ca²⁺ (mmol/l)	–	0.1–3	–
Mg²⁺ (mmol/l)	–	0.02–2	–
Cl⁻ (mmol/l)	45	20–60	Increased in cystic fibrosis
Urea (mmol/l)	–	4–20	–

Table 126.17. Composition of urine

	Mean	95% range	Comment
Total volume (ml/day)	1000	~500–2500	Depending on water intake; maximal span, 500–20 000
Specific gravity	–	1.01–1.025	Maximal span, 1.001–1.05
Osmolality (mosm/l)	–	50–1400	–
pH	–	4.8–7.5	–
Titratable acid (mmol/day)	33	20–40	Titration to urine, pH 7.4, total "acidity" = titratable acid + ammonium excretion; total "acidity" is very variable depending on acid-base status; maximal variation: – 250 (HCO_3^- excretion) to 1000

Table 126.18. Concentration and excretion of urine

	Concentration (mmol/l)[a]	Excretion (mmol/day)	Comment
Ions			
Na^+	100–ca. 200	120–220	Amounts depending on Na^+ content of diet and on urinary flow rate; minimal concentration ~10 mmol/l
K^+	35–80	35–80	Depending on diet, pH homeostasis, urinary flow rate etc.
Ca^{2+}	–	3–8	–
Mg^{2+}	–	2–8	–
HCO_3^-	0–50	0–50	Depending on acid-base status; very little bicarbonate in acidic urine
Cl^-	100–240	120–240	(See Na^+ above)
$HPO_4^{2-}/H_2PO_4^-$	–	30–40	–
SO_4^{2-}	–	30–70	–
Organic solutes			
Urea	~150–900	210–480	–
NH_4^+	35	20–70	Massively enhanced in metabolic acidosis (up to >500)
Urate and uric acid	–	3	Mean (95% range, 0.5–6)
Allantoin		0.1–0.2	–
Creatinine		13	Mean (95% range, 1–22)
Creatine		0.5	Mean (95% range, 0.01–2.1)
Amino acids (total free)		7	Mean (95% range, 3–10)
Proteins		<60 mg/day	
Porphobilinogen	–	0.006	Mean (95% range, 0.002–0.011)
Bilirubin	–	0.0–0.003	–
Urobilinogen	–	0.0–0.004	–
Coproporphyrin	–	0.1 µmol/day	–
D-glucose	0.05–1.5	0.4	Mean (95% range, 0.1–0.7)

[a] Erythrocytes per ml, 0–2500; leukocytes per ml, 0–3000; hyaline cylinders per day, 2000; bacteria per ml of voided urine, < 10^5.

Table 126.19. Composition of sperm

	Mean	95% range		Mean	95% range
Total volume (ml)	3.4	0.2–6.6	Citrate (mmol/l)	20	5–76
Sperms (ml^{-1})	107×10^6	$28–225 \times 10^6$	Lactate (mmol/l)	4.7	3.5–6.6
Spermatocrit	0.01	–	Urea (mmol/l)	12	–
Osmolality			Urate (mmol/l)	0.36	–
(mosm/kg H_2O)	296	–	Creatine (mmol/l)	1.3	–
pH	7.19	6.9–7.36	Amino acids (g/l)	12.6	–
Ions			Ammonium (mmol/l)	1.1	–
Na^+(mmol/l)	117	100–133	Choline (mmol/l)	6.8	–
K^+ (mmol/l)	22.9	17–27.4	Phosphorylcholine (g/l)	3.06	2.86–3.8
Ca^{2+}(mmol/l)	6.2	5.3–7.2	Lipids		
Mg^{2+}(mmol/l)	5.3	–	Total lipids (g/l)	1.88	1.67–2.06
Cl^-(mmol/l)	42.8	28.3–57.3	Cholesterol (g/l)	1.03	0.7–1.2
HCO_3^-(mmol/l)	24	19.2–32.2	Phospholipids (g/l)	0.84	0.48–1.33
Phosphate (mmol/l)	3.7	–	Prostaglandins		
Organic compounds			(PGE, mg/l)	53	–
Organic phosphates (g/l)	10	–	Protein (g/l)	–	32.9–77.4
Fructose (mmol/l)	12	—	Spermine	–	0.5–3.5
Glycogen (g/l)	–	0.14–5.5	Mucoproteins (g/l)	9	–
Inositol (mmol/l)	3.3	–	Enzymes		
Sorbitol (mmol/l)	0.5	–	Alkaline phosphatase (U/l)	18–177	
Pyruvate (mmol/l)	3.9	1.4–7	Acid phosphatase (kU/l)	96–750	

126.4 Normal Values of Respiratory Function

The mean functional parameters of respiratory function (see Chaps. 100–103):

- Tidal volume
- Vital capacity
- Residual volume
- Maximal expiratory flow rate (l/s)
- Forced expiratory volume (l/s)
- Breathing frequency

are shown as a function of age and gender in Fig. 126. 1A,B. No 95% ranges are provided in this figure. Tables 126.20–126.22 give mean values ± SD for adult men and women [3].

126.5 Normal Values of Acid–Base Homeostasis

The normal values of acid base homeostasis (see Chap. 78) are summarized in Tables 126.23 and 126.24.

126.5.1 Typical Disturbances

A few difinitions for the evaluation of the acid–base status are listed in Table 126.25 (see also Chap. 78).

126.5.2 Henderson–Hasselbalch Equation

The Henderson–Hasselbalch equation can be formuated for HCO_3^- or any other buffer ($AH \leftrightarrow A^- + H^+$ (see Chap. 78):

$$pH = pK_a + \log[A^-]/[AH] \qquad (126.1)$$

In the specific case of the bicarbonate system, the equation is:

$$pH = 6.1 + \log[HCO_3^-]/[CO_2] \qquad (126.2)$$

or with Pa_{CO_2} and the adsortion coefficient α (0.0301 mmol/mmHg or 0.000226 mmol/Pa):

$$pH = 6.1 + \log[HCO_3^-]/Pa_{CO_2} \times \alpha \qquad (126.3)$$

126.6 Normal Values of the Cardiovascular System

In the cardiovascular system (see Chaps. 89–95), the diastolic pressure should not exceed 100 mmHg (13.3 kPa) in adults aged up to 65 years and should not exceed 105 mmHg (13.3 kPa) in adults over 65 years. The mean systolic and diastolic blood pressure as a function of age (± SD) is shown in Fig. 126.2 for men and women.

126.6.1 Pressures and Volumes in the Heart

The pressures and volumes in the heart (see Chaps. 89, 90) are listed in Table 126.26 [11]. The mean heart rate is 60–100 beats per min, and the mean pulmonary and systemic vascular resistances are 0.2–1.2 and 7.7–15 mN/s cm⁵, respectively [11].

126.6.2 Normal Values of the Electrocardiogram

Mean durations of electrocardiogram (ECG) waves are shown in Fig. 126.3 and Table 126.27 (see Chap. 92). The QT duration is largely dependent on frequency. Figure 126.4 summarizes typical data for adults [1]. The amplitudes are in the millivolt range and vary between the different leads. Normal limits are defined for chest wall leads (e.g., V_3) and three orthogonal leads in the ECG [8]. The maximal P, QRS, and T vectors determined by the latter method are summarized in Table 126.28.

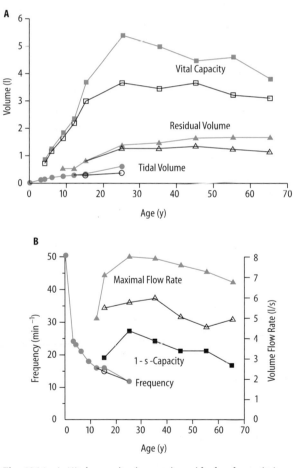

Fig. 126.1. A Vital capacity (*squares*), residual volume (*triangles*), and tidal volume (*circles*) as a function of age. *Open symbols*, female; *closed blue symbols*, male. **B** Breathing frequency (*left ordinate, circles*); maximal flow rate (*right ordinate, triangles*); 1-s capacity (*left ordinate, squares*) as a function of age. *Open symbols*, female; *closed blue symbols*, male. Data taken from [3]

Table 126.20. Respiratory function

Parameter	Men		Women	
	Mean	SD	Mean	SD
Frequency (min^{-1})	12	(1.5)	12	–
Volumes				
Tidal volume (l)	0.6	–	0.4	–
Vital capacity (l)	4.8	0.6	3.2	0.4
Residual volume (l)	1.2	0.4	1.1	0.3
Inspiratory reserve volume (l)	3.2	0.5	1.9	0.4
Expiratory reserve volume (l)	1	0.3	0.7	0.2
Functional residual capacity (l)	2.2	0.5	1.8	0.4
Total lung capacity (l)	6	0.8	4.2	0.6
Minute ventilation (l/min)	7.4	2.3	4.6	–
Ventilation				
Maximal expiratory flow rate (l/s)	7.9	1.1	6.0	0.8
One-second capacity (l)	3.7	0.4	–	–
Maximal breathing capacity (l/min)	150	30	95	30
Compliance lung (l/cm H_2O)	0.26	0.06	0.16	0.05
Diffusion capacity, CO method (ml/min · mmHg)	25	6	25	6
Respiratory quotient (RQ)	0.82	–	0.82[a]	
Resistances (cm H_2O s/l)				
Airway resistance	1	0.3	1.5	0.5
Lung tissue resistance	0.3	0.1	0.5	0.2
Total pulmonary resistance	1.3	0.3	2	0.5
Thorax tissue resistance	1.4	0.6	1.4	0.6
Total respiratory resistance	4	0.7	4	0.7

[a] Depending on diet and ventilation.

Table 126.21. Gas pressures

Gas	Pressure				Comments
	(kPa)		(mmHg)		
	Mean	SD	Mean	SD	
Inspiratory					
O_2	19	–	143	–	P_B, 101 kPa (760 mmHg)
CO_2	0	–	0	–	–
H_2O	6.26	–	47	–	37°C, saturated
Alveolar					
O_2	13.3	0.7	100	5	–
CO_2	5.3	0.3	40	2	–
Expiratory					
O_2	16	–	120	–	–
CO_2	4	–	30	–	–
Arterial					
O_2	12	2.2	91	17	–
CO_2	5.2	0.9	39	7	–
Venous					
O_2	5.3	0.8	40	6	–
CO_2	6.7	0.5	50	4	–

P_B, barometric pressure

Table 126.22. Gas contents of blood

	Mean	SD
Arterial O_2 content (l/l)	0.2	0.03
Venous O_2 content (l/l)	0.13	0.03
Arterial saturation S_{aO_2} (%)	94	2
Venous saturation S_{vO_2}	62	7

Table 126.23. Acid–base metabolism: blood gases

Gas	Pressure			
	(kPa)		(mmHg)	
	Mean	SD	Mean	SD
Arterial O_2	12	2.2	91	17
Venous O_2	5.3	0.8	40	6
Arterial CO_2	5.2	0.9	39	7
Venous CO_2	6.6	0.5	50	4

Values given are for adults.

Table 126.24. Acid–base metabolism

Parameter	Men		Women	
	Mean	95% range	Mean	95% range
pH				
Arterial	7.4	7.35–7.45	7.4	7.35–7.45
Venous	7.37	7.32–7.42	7.37	7.32–7.42
HCO_3^-				
Arterial (mmol/l, plasma)	24	22–26	23	21–25
Venous (mmol/l, plasma)	26	21–30	25	23–27
Erythrocyte (mmol/l, plasma)	11	10.9–11.5	–	–
Standard (mmol/l)	–	22–26	–	21–25 (plasma at 5.32 kPa CO_2)
Base excess (mmol/l, blood)	–	–2.4 to +2.3	–	–3.3 to +1.2
Total buffer bases (mmol/l, blood)	50	47–53	48	45–51
Hemoglobin (g/l, blood)	158	133–182	141	118–175
Acid excretion (mmol/day)				
CO_2 expired, lung	16 000 (resting conditions)	–	16 000	–
Titratable acid, kidney	33	20–40	33	20–40
NH_4^+, liver and kidney	–	20–70 (upregulated in metabolic acidosis)	–	20–70 (upregulated in metabolic acidosis)

Values given are for adults.

Table 126.25. Typical changes for acid–base disturbances

Disturbance	pH	P_{aCO_2}	HCO_{3a}^-	$HCO_{3\ stand}^-$	BE
Respiratory acidosis	↓	↑↑	↑	–	–
Respiratory alkalosis	↑	↓↓	↓	–	–
Metabolic (nonrespiratory) acidosis	↓	–	↓↓	↓↓	↓↓
Metabolic (nonrespiratory) alkalosis	↑	–	↑↑	↑↑	↑↑

P, pressure; a, alveolar; stand, standard; BE, base excess.

Table 126.26. Pressures and volumes in the heart

	Normal values[a]
Pressures (mmHg)[a]	
Right atrial mean pressure	≤6 (≤0.8)
Right ventricular peak systolic pressure	15–30 (2.0–4.0)
Right ventricular end-diastolic pressure	≤6 (≤0.8)
Pulmonary artery peak systeolic pressure	15–30 (2.0–4.0)
Pulmonary artery diastolic pressure	4–12 (0.53–1.6)
Left atrial mean pressure	≤12 (≤1.6)
Left ventricular peak systolic pressure	100–150 (13.3–20)
Left ventricular end-diastolic pressure	≤12 (≤1.6)
Volumes (ml/m² body surface area)	
Left ventricular end diastolic volume[h]	70–100
Left ventricular end-systolic volume	25–35
Stroke volume[b]	40–70

[a] Values in parentheses are measured in kPa.
[b] Measured at rest.

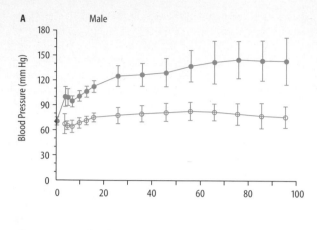

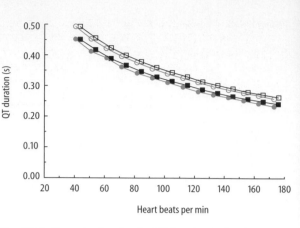

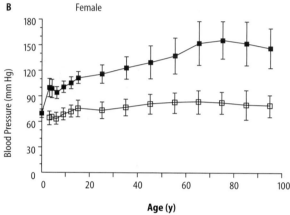

Fig. 126.2A,B. Blood pressure as a function of age. Mean values ± SD for systolic (*upper curves, closed symbols*) and diastolic blood pressure (*lower curves, open symbols*). **A** Male (*circles*); **B** Female (*squares*). Data taken from [3]

Fig. 126.4. Normal values for the QT duration in the electrocardiogram as a function of heart beat frequency. *Closed symbols* and *solid curves* mark the lower limit, and *open symbols* and *dotted curves* the upper limit of the 95% range for females (*squares*) and males (*circles*)

Table 126.27. Measurements of P, PQ, QRS, and QT in s (mean ± SD) [8]

	Women	Men
P duration	0.106 ± 0.019	0.102 ± 0.016
PR interval	0.154 ± 0.022	0.153 ± 0.023
PR segment	0.048 ± 0.018	0.051 ± 0.019
QRS duration	0.084 ± 0.008	0.093 ± 0.009
QT interval[a]	0.372 ± 0.026	0.367 ± 0.034

[a] Uncorrected.

Table 126.28. Maximal P, QRS, and T vectors (mV ± SD) in the three planes [8]

Plane	P vector	QRS vector	T vector
Frontal	0.18 ± 0.06	1.57 ± 0.42	0.36 ± 0.14
Sagittal	0.17 ± 0.06	1.32 ± 0.45	0.36 ± 0.13
Horizontal	0.09 ± 0.03	1.39 ± 0.36	0.40 ± 0.14

126.6.3 Normal Values of the Distribution of Blood Flow

Table 126.28 lists the respective organs, their weights, and absolute as well as relative blood supply [9] (see Chaps. 28, 29, 69, 73, 95, 104, 117, 118).

126.7 Normal Values of Kidney Function

Table 126.30 lists normal values for kidney function (see Chaps. 73–75). Glomerular filtration rate (GFR) is frequently determined as creatinine clearance or, more accurately, as inulin or radionuclide clearance. Figure 126.5 depicts the age dependence of GFR [2]. Renal

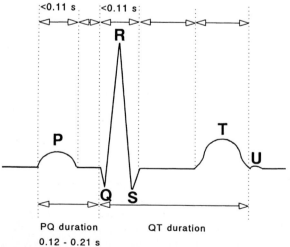

Fig. 126.3. Typical electrocardiogram. Definition of waves and durations

plasma flow is determined from the clearance of para-aminohippurate (PAH). The function of the proximal nephron is determined by the Li$^+$ clearance. The concentrating ability of the kidney is traditionally determined as the "free water clearance." However, this measurement is fraught with many problems, and the much simpler

measurement of the osmolality of the antidiuretic urine collected as the first urine in the morning appears to be much more valuable. An osmolality greater than 800 mosm/l suggests that the concentrating ability is not seriously disturbed. The composition of normal urine is presented in Tables 126.17 and 126.18.

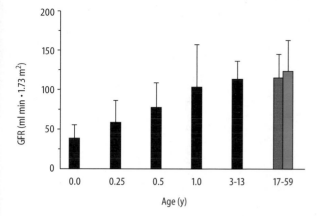

Fig. 126.5. Glomerular filtration rate (*GFR*) as a function of age (years). Mean values ± SD. *Solid bars*, young children, male and femal pooled; *hatched bar*, female; *blue bar*, male. Data taken from [2]

126.8 Normal Values of Brain Function Tests

126.8.1 Cerebral Blood Flow

Normal cerebral blood flow (CBF; see Chaps. 28, 29) is 50 ml/min per 100 g [15]. The regional blood flow depends on activity. It is always higher in gray as compared to white matter. The normal oxygen extraction is 40% and is independent of activity. In adults, CBF and cerebral oxygen consumption fall with increasing age at a rate of 0.5% per year. The mean oxygen consumption of the brain is 25% of the resting total oxygen consumption, i.e., approximately 4.2 ml O_2/min per 100 g. CBF is autoreglulated almost perfectly in the range of 60–160 mmHg (8–21 kPa).

Table 126.29. Blood flow in resting man (approximately 70 kg; cardiac output (CO), 5.5 l/min)

Organ	Resting blood flow (l/min)	Weight (kg)	Resting blood flow/weight (ml/kg min)	% CO
Brain	0.75	1.4	535	14
Heart	0.25	0.3	833	5
Kidneys	1.2	0.3	4000	22
Liver	1.3	1.5	866	23
Muscles	1	35	28	18
Skin	0.5	4	125	9
Bone etc.	0.8	27	30	14

Table 126.30. Normal values for kidney function [2,3]

	Mean value ± SD	Comment
Renal plasma flow (RPF)	640 ± 85 (ml min^{-1} 1.73 m^2)	Determined as PAH clearance[a]
Renal blood flow (RBF)	1165 ± 150 (ml min^{-1} 1.73 m^2)	Determined from RPF and hematocrit
Glomerular filtration rate (GFR)	125 ± 10 (ml min^{-1} 1.73 m^2)	Determined from inulin clearance[b]
Li$^+$ clearance (C_{Li})	20–40 (ml min^{-1} 1.73 m^2)	Proximal tubule function[c]
Free water clearance	≈5 (ml min^{-1} 1.73 m^2)	Antidiuresis, variable
Maximal urine osmolality	1067 ± 156 (mosm/l)	A good predictor of urinary concentrating ability

[a] PAH clearance is close to renal plasma flow (RPF) only if the plasma concentration is less than 150 µmol/l. Only then is the extraction of PAH close to unity (0.875–1.0). A precise measurement requires the determination of the PAH concentration in renal venous blood (PAH)$_v$:

$$RPF = C_{PAH} \times (PAH)_a \times [(PAH)_a - (PAH)_v]^{-1}$$

where C_{PAH} = PAH clearance and (PAH)$_a$ = PAH concentration in arterial blood

[b] The clearance of creatinine (C_{Cr}) is close to that of inulin (C_{In}). However, with failing GFR, C_{Cr} overestimates C_{In} and GFR. GFR can be empirically calculated from the plasma concentration of creatinine (Cr$_p$) according to the Cockroft and Gault formula [2]:

$$GFR = [140\text{-age (in years)}] \times weight \text{ (in kg)} \times [Cr_p \text{ (in µmol/l)} \times 1.26]^{-1}$$

For women the result has to be multiplied by 0.85

[c] The Li$^+$ clearance has been assumed to quantify the proximal tubule absorption of Na$^+$ and water on the basis that Li$^+$ is handled by the proximal tubule like Na$^+$. In many instances C_{Li} overestimates fractional Na$^+$-and water absorption by the proximal tubule [4]

126.8.2 Electroencephalogram

The different patterns of electroencephalograms (EEG; see Chap. 25) are classified according to their frequencies (Table 126.31).

126.8.3 Reflexes

The light reflex and the labyrinth reflexes are discussed in the Chaps. 34 and 37, respectively. Table 126.32 lists several reflexes, the respective nerves, and the spinal segments. These reflexes are useful in diagnosing and locating defects. The reflexes in newborn babies and in the developing child are dicussed at length in Chap. 122.

The dermatomes, i.e., the sensory zones of the skin corresponding to the spinal roots, are shown in Fig. 126.6. They follow a very regular pattern. However, because of substantial overlap the boundaries are actually not as distinct as they seem.

Table 126.31. Frequencies in the electroencephalogram (EEG) [7,12]

Pattern	Frequency (s^{-1})	Amplitude	State
β	14–30	Lowest	Intense mental activity, sleep spindles superimposed during light sleep (stage C)
α	8–13	Low	Awake but relaxed
θ	4–7	Low	Falling asleep (B), REM sleep
δ	0.5–4	High	Deep sleep (stages D-E), children awake

REM, rapid eye movement

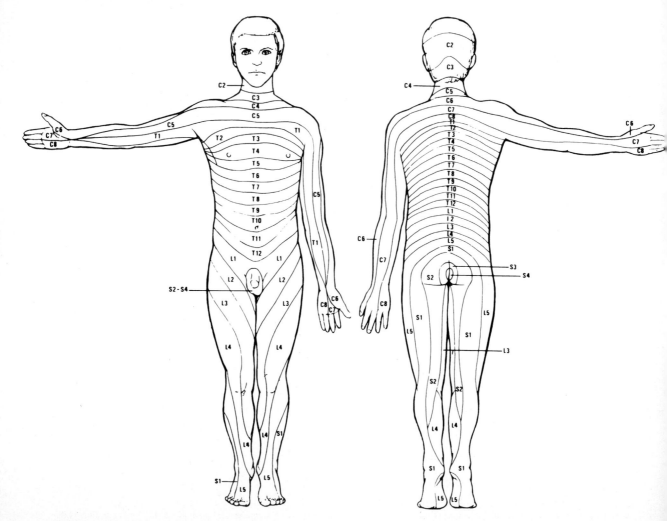

Fig. 126.6. The dermatomes. *C*, cervical; *T*, thoracical; *L*, lumbar; *S*, sacral. From [9], with permission

Table 126.32. Reflexes [14]

Reflex	Elicitation	Response	Spinal segment	Nerve
Maseter reflex	Blow upon chin	Closure of mouth	V	Trigeminal nerve
Scapulohumoral reflex	Blow upon the medial margin of lower scapula	Adduction and outward rotation of arm	C4–C6	Suprascapular and axillary nerves
Biceps jerk	Blow upon biceps tendon	Flexion of elbow	C5–C6	Musculocutaneous nerve
Triceps jerk	Blow upon triceps tendon	Extension of elbow	C6–C7	Radial nerve
Radial reflex	Blow upon brachioradial tendon	Flexion of elbow	C5–C6	Radial nerve
Flexor finger jerk	Blow upon palmar surface of semiflexed fingers	Flexion of fingers, thumb	C7–C8	Median nerve; ulnar nerve
Thumb reflex	Blow upon flexor tendon (long flexor muscle of thumb)	Flexion of thumb and phalanx	C6–C8	Median nerve
Adductor reflex	Blow upon medial femur condyl	Adduction of leg	L2–L4	Obturator nerve
Knee jerk	Blow upon quadriceps tendon	Extension of knee	L2–L4	Femoral nerve
Posterior tibial reflex	Blow upon tendon behind medial ankle (malleolus)	Supination of foot	L5	Tibial nerve
Peroneal muscle reflex	Blow upon metatarsals 1–2 with foot flexed and suppinated	Dorsal extension and pronation of foot	L5–S1	Fibular nerve
Femoral biceps reflex	Blow upon tendon of medial flexors of knee	Contraction of these muscles	S1–S2	Sciatic nerve
Ankle jerk	Blow upon the calcaneal tendon	Plantar flexion of ankle	S1–S2	Sciatic nerve
Abdominal reflex	Touch or scratch of skin with a pin	Unilateral contraction of abdominal wall	T7–T12	Segmental nerves
Pupil reflex	Light exposure, convergence	Contraction of pupil		Optic nerve; oculomotor nerve
Corneal reflex	Light touch of cornea	Closure of eyelid	Medial pons	Trigeminal and facial nerves
Bell's phenomena	Upper eyelid passively fixed, patient tries to close eye	Eyeballs rotate upwards	Pons	Oculomotor and trigeminal nerves
Ear–eyelid reflex	Sudden noise	Blinking of eyelids	Caudal pons	Vestibulocochlear nerve; facial nerve
Pharynx reflex	Touch of soft palate	Elevation of palate, contraction of posterior wall	Medulla oblongata	Glossopharyngeal nerve
Epigastric reflex	Light, rapid scratch from the mamilla downwards	Retraction of epigastrium	T5–T6	Vagus nerve
Abdominal wall reflex	Rapid touch and scratch of abdominal skin from the side to the middle	Shift of navel towards stimulated side		Hypogastric nerve, Intercostal nerves
			T6–T12	Ilioinguinal nerve
Cremasteric reflex	Light scratch of inner aspect of upper part of thigh	Elevation of testicle	L1–L2	Genitofemoral nerve
Gluteal reflex	Scratch on buttock	Contraction of gluteal muscles	L4–S1	Superior and inferior gluteal nerves
Plantar reflex (Babinski)	Longitudinal scratch upon lateral aspect of foot from heel to toes	Plantar flexion of toes		Corticospinal (lesion leads to "upgoing toe")
Anal reflex	Scratch in perianal region	Contraction of external sphincter	S3–S5	Pudendal nerve
Tonic neck reflexes	Changes in position of head relative to body	Normally negative; sometimes positive cerebral diplegia: passive turning of head leads to extension of arm and leg on the side to which head has been turned to		
Grasp reflex, hard	Touch of palmar surface	Flexion of fingers; normal in infants; increased in lesions of frontal lobe; only valuable in full consciousness		
Grasp reflex, foot	Pressure of plantar surface	Flexion of toes; normal in infants (up to 1 year); positive in 50% of patients with Down's syndrome		

C cervical; T thoracic; L, lumbar; S, sacral

126.8.4 Cerebrospinal Fluid Pressure

The composition of cerebrospinal fluid (CSF; see Chap. 27) is shown in Table 126.8. The normal CSF pressure is 60–150 mm H_2O (0.6–1.5 kPa) [14]. In children, normal values are lower: 45–90 mm H_2O (0.45–0.9 kPa). The pressure is measured with the patient lying on his or her left side and with the head at the same level as the lumbar spine.

126.9 Normal Values for Metabolism and Nutrition

Normal caloric turnover consists of the base rate and the component caused by exercise and work. Figure 126.7 shows normal values for children and adults (see also Chap. 71). The figure also gives the mean body weights (good health, moderated physical activity, and temperate climate). The metabolic rate depends on several variables, most importantly body weight (height, body surface) and work (physical activity).

126.9.1 Metabolic Rate and Body Weight

The relation of metabolic rate and weight is depicted in Fig. 126.8.

126.9.2 Metabolic Rate and Physical Activity

Typical metabolic rates for several kinds of physical activity and work are given in Table 126.33. It is important to distinguish the metabolic rates for a short-term physical activity (Table 126.33) from the total metabolic rate for a certain type of work. The energy expenditure for specified activities can be obtained from Table 126.34 (from [16]) in the following way. The individual basal metabolic rate (BMR) has to be multiplied by the respective number. This number is 1.0 for sleeping and can be 8.5 for strenuous work such as rickshaw pedaling. The value of 8.5 multiplied by the basal metabolic rate of, e.g., 4.5 kJ/min would give an energy cost of 38.25 kJ/min. The upper limits for long-term metabolic rates, i.e., over a period of several years, have been set by WHO as follows:

* 15.5 MJ/day for women
* 20.1 MJ/day for men

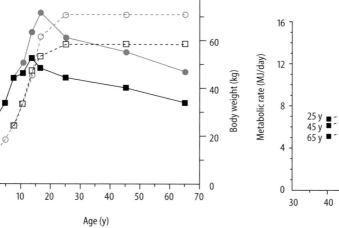

Fig. 126.7. Metabolic rate and body weight as a function of age. *Closed symbols*, metabolic rates; *open symbols*, body weights. *Squares*, females; *blue circles*, males. Metabolic rates peek at 15–18 years. Data taken from [3]

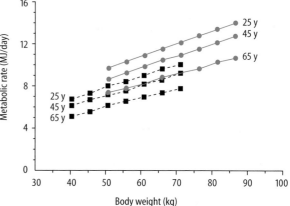

Fig. 126.8. Metabolic rate as a function of body weight for three age-groups (25 years, 45 years, 65 years). *Squares, dotted curve*, females; *circles, solid curves*, males. At any given body weight and age, the metabolic rates are higher in males. Data taken from [3]

Table 126.33. Typical metabolic rates (modified from [6])

Type of exercise	Examples	Rate (kJ/min)	Rate (W)
Light	Golf, skittles	<20	334
Medium	Cycling (20 km/h, flat), tennis (not competitive), dancing (waltz)	20–36	334–601
Strenuous	Soccer, skiing, tennis (competitive)	47–40	501–668
Very strenuous	Swimming (crawl), rowing marathon, sprint (100 m)	72–126	1200–2100

To obtain the total metabolic rate per day, the values below have to be multiplied by the duration and the basal rate has to be added

Table 126.34. Total metabolic rate for different kinds of work (activity)

Activity	Rate[a]	Activity	Rate[a]
A. Men (developed and developing societies)		Earth cutting	6.2
Sleeping	1.0 (i.e., BMR × 1.0)	Brick breaking	4.0
Lying	1.2	House-building	
Sitting quietly	1.2	Weaving bamboo wall	2.9
Standing quietly	1.4	Roofing house	2.9
Standing activities		Cutting bamboo	3.2
Chopping firewood	4.1	Cutting palm tree trunks	4.1
Singing and dancing	3.2	Digging holes	6.2
Washing clothes	2.2	Laying floor	4.1
Making bows and arrows, bags, etc.	2.7	Nailing	3.3
Walking		Coconut activities	
"Around" or strolling	2.5	Collecting (including climbing trees)	4.6
Slowly	2.8	Husking	6.3
At normal pace	3.2	Putting in bags	4.0
With 10-kg load	3.5	Pedaling rickshaws	
Uphill: Slowly	4.7	Without passengers	7.2
At normal pace	5.7	With passengers	8.5
Fast	7.5	Pulling carts	
At normal pace with	6.7	Without load	5.3
10-kg load		With load	5.9
Downhill: Slowly	2.8	Pushing wheelbarrow	4.8
At normal pace	3.1	Mining	
Fast	3.6	Working with pick	6.0
Sitting activities		Shoveling	5.7
Playing cards	1.4	Erecting roof supports	4.9
Sewing	1.5	Armed Services	
Weaving	2.1	Cleaning kit	2.4
Carving plates, combs, etc.	2.1	Drill	3.2
Stringing loom	1.9	Route marching	4.4
Sharpening axe	1.7	Assault course	5.1
Sharpening machete	2.2	Jungle march	5.7
Household tasks		Jungle patrol	3.5
Cooking	1.8	Helicopter pilots; Preflight checks	1.8
Light cleaning	2.7	Normal and low-level flying	1.5
Moderate cleaning	3.7	Hovering	1.6
(polishing, window cleaning,		Recreations	
chopping firewood)		Sedentary (playing cards, etc.)	2.2
Making fence	3.6	Light (billiards, bowls, cricket,	2.2–4.4
Splitting wood for posts	4.2	golf, sailing, etc.)	
Sharpening posts	4.0	Moderate (dancing, swimming,	4.4–6.6
Digging holes for posts	5.0	tennis, etc.)	
Planting	2.9	Heavy (football, athletics,	
Cutting grass with machete	4.7	jogging, rowing, etc.)	6.6+
Digging irrigation channels	5.5		
Feeding animals	3.6	*B. Women*	
Hunting and fishing		Sleeping	1.0 (i.e., BMR × 1.0)
Paddling canoe	3.4	Lying	1.2
Fishing from canoe	2.2	Sitting quietly	1.2
Fishing with line	2.1	Sitting activities	
Fishing with spear	2.6	Sewing clothes	1.4
Hunting flying-fox	3.3	Sewing pandanus mat	1.5
Hunting pig	3.6	Weaving carrying bag	1.5
Hunting birds	3.4	Preparing rope	1.5
Forestry		Standing	1.5
In nursery	3.6	Walking	
Planting trees	4.1	"Around" or strolling	2.4
Felling with axe	7.5	Slowly	3.0
Trimming branches off trees	7.3	At normal pace	3.4
Sawing: Hand saw	7.5	With load	4.0
Power saw	4.2	Uphill: At normal pace	4.6
Wood planing	5.0	Fast	6.6
Brick-making		With load	6.0
Marking mud bricks (squatting)	3.0	Downhill: Slowly	2.3
Kneading clay	2.7	At normal pace	3.0
Digging earth to make mud	5.7	Fast	3.4
Shoveling mud	4.4	With load	4.6

Table 126.34. *Continued*

Activity	Rate[a]
Household tasks	
Light cleaning	2.7
Moderate cleaning (polishing, window cleaning, etc.)	3.7
Sweeping house	3.0
Sweeping yard	3.5
Washing clothes	3.0
Ironing clothes	1.4
Washing dishes	1.7
Cleaning house	2.2
Child care	2.2
Fetching water from well	4.1
Chopping wood with machete	4.3
Preparing tobacco	1.5
Deseeding cotton	1.8
Beating cotton	2.4
Spinning cotton	1.4
Food preparation and cooking	
Cooking	1.8
Collecting leaves for flavoring	1.9
Catching fish by hand	3.9
Catching crabs	4.5
Grinding grain on millstone	3.8
Pounding	4.6
Stirring porridge	3.7
Making tortillas	2.1
Removing beans from pod	1.5
Breaking nuts (such as peanuts)	1.9
Squeezing coconut	2.4
Peeling taro	1.7
Peeling sweet potato	1.4
Roasting corn	1.3
Loading earth oven with food	2.6
Office work	1.7
Light industry	
Bakery work	2.5
Brewery work	2.9
Chemical industry	2.9
Electrical industry	2.0
Furnishing industry	3.3
Laundry work	3.4
Machine tool industry	2.7
Agriculture (nonmechanized)	
Clearing ground	3.8
Digging ground	4.6
Digging holes for planting	4.3
Planting root crops	3.9
Weeding	2.9
Hoeing	4.4
Cutting grass with machete	5.0
Sowing	4.0
Threshing	5.0
Binding sheaves	4.2
Harvesting root crops	3.1
Picking coffee	1.5
Winnowing corn or rice	1.7
Cutting fruit from tree	3.4
Recreations	
Sedentary (playing cards, etc.)	2.1
Light ⎫	2.1–4.2
Moderate ⎬ (see categories for men)	4.2–6.3
Heavy ⎭	6.3+

BMR, basal metabolic rate
[a] Expressed as BMR multiplied by a metabolic constant

126.9.3 The Caloric Value and the Caloric Equivalent

The caloric value of the three major nutrients of food are as follows:

- Fat: 38.9 kJ/g (9.3 kcal/g)
- Protein: 17.2 kJ/g (4.1 kcal/g)
- Carbohydrate: 17.2 kJ/g (4.1 kcal/g); glucose: 15.7 kJ/g (3.75 kcal/g).

The caloric equivalent (CE) defines the amount of kilojoules produced per liter O_2 consumed. The caloric equivalent depends on the type of fuel and hence on the respiratory quotient (RQ). Table 126.35 summarizes the caloric equivalent and the RQ of the three major nutrients of food. At a constant fraction of protein content of the diet, the caloric equivalent becomes a linear function of the RQ, e.g., at a protein content of 12%, the function is:

$$RQ = 0.2143 \ CE \ (kJ/l \ O_2) - 3.522$$

126.9.4 Typical Constituents of Food

Table 126.36 summarizes typical constituents of food.

126.9.5 Recommended Daily Requirements

Table 126.37 summarizes the recommended daily intake (RDI) for men, women, and children.

126.10 Normal Values for Bone Metabolism

Table 126.38 summarizes the normal values for bone metabolism (see Chap. 80).

126.10.1 Onset of Ossification

Typical data for the onset of ossification nuclei in the extremities are summarized in Table 126.39.

126.11 Normal Development and Body Measurements

For further details on normal development and body measurements, please see Chap. 122.

126.11.1 Normal Embryonic, Fetal, and Postnatal Development of Height and Weight

The height (head to foot) and weight of the normal developing embryo and fetus are depicted in Fig. 126.9. The normal development of height and weight for boys and girls is depicted in Fig. 126.10A,B. The values are given with the upper (height) and lower (weight) limits of the 95% range.

Table 126.35. The respiratory quotient (RQ) and energy equivalents for nutrients

Nutrient	O_2 consumed (l/g)	CO_2 produced (l/g)	RQ[a]	Energy released		Energy released	
				(kJ/g)	(kcal/g)	(kJ/l O_2)[b]	(kcal/l O_2)[b]
Starch	0.829	0.824	0.994	17.49	4.18	21.10	5.04
Glucose	0.746	0.742	0.995	15.44	3.69	20.70	4.95
Fat	1.975	1.402	0.710	39.12	9.35	19.81	4.73
Protein	0.962	0.775	0.806	18.52	4.43	19.25	4.60
Alcohol	1.429	0.966	0.663	29.75	7.11	20.40	4.88

[a] The RQ is the ratio of the moles of carbon dioxide produced to the moles of oxygen consumed on oxidation of a given amount of each nutrient. The last two columns show the heat produced or energy expended per g of each major nutrient and per O_2 consumed when one of these nutrients is oxidized. From these figures the food quotient (FQ), i.e., the RQ for a given mixture of completely oxidized nutrients, can be calculated and inversely, the ratio of fats and carbohydrates oxidized during a certain time interval can be determined from the O_2 consumed and the CO_2 produced at the same time, provided that simultaneous protein oxidation is known, which can be derived from the urinary nitrogen excretion   [b] Also called caloric equivalent

Table 126.36. Constituents of food

Food	Fat (g/kg)	Protein (g/kg)	Carbohydrate (g/kg)	H_2O (g/kg)	Energy (MJ/kg)
Fruit					
Apples	6	3	150	840	2.43
Bananas	2	11	222	757	3.56
Pears	4	5	155	832	2.55
Cherries	4	12	146	834	2.51
Oranges	2	10	122	871	2.05
Vegetables					
Beans (green)	2	19	71	901	1.34
Carrots	2	11	91	886	1.67
Potatoes	1	21	177	798	3.18
Asparagus	2	21	41	929	0.88
Tomatoes	2	11	47	935	0.92
Peanuts	487	262	206	18	24.4
Bread and cereals					
Rye bread	10	64	527	385	9.50
White bread	12	82	510	383	10.6
Spaghetti	12	125	752	104	15.4
Chocolate, butter, oil					
Chocolate	323	77	569	9	21.7
Butter	810	6	7	174	30.0
Olive oil	999	0	0	0	37.0
Dairy products					
Eggs	115	128	7	740	6.78
Milk	37	32	46	885	2.68
Camembert	228	187	18	513	12.0
Meat and fish					
Chicken	56	206	0	727	5.78
Lamb chop	320	149	0	520	14.7
Beef loin	44	192	0	751	5.11
Pork chop	306	152	0	539	14.3
Salami	497	178	0	277	21.9
Trout	21	192	0	776	4.22
Salmon	136	199	0	655	8.70
Tuna	209	238	0	525	12.1
Drinks					
Coffee	1	3	8	985	0.2
Lemonade	0	0	120	880	1.93
Beer	36 (alcohol)	5	48	906	1.97
Wine	115 (alcohol)	0	30	850	3.35
Brandy	380 (alcohol)	0	0	750	10.9

Table 126.37. Recommended daily intake (RDI) (modified from [3]) [a]Recommended minimum per day

	Men	Women	Children 0–1 years	1–3 years	4–6 years
Protein (g/kg per d)	0.47–0.71	0.47–0.71	1.2–2.3	0.7–1.1	0.65–0.97
Essential amino acids (g)[a]					
L-histidine	0	0	34 mg/kg		
L-tryptophan	0.5 (0.25)	0.5 (0.16)	22 mg/kg		
L-phenylalanine	2.2 (0.3)	2.2 (0.22)	90 mg/kg		
L-lysine	1.6 (0.8)	1.6 (0.5)	103 mg/kg		
L-threonine	1.0 (0.5)	1.0 (0.31)	87 mg/kg		
L-methionine	2.2 (1.1)	2.2 –	45 mg/kg		
L-leucine	2.2 (1.1)	2.2 (0.62)	150 mg/kg		
L-isoleucine	1.4 (0.8)	1.4 (0.65)	105 mg/kg		
Fat (normally approximately 25% of caloric intake)					
Essential fatty acids					
Linoleate	1%–2% of caloric intake = 2.5–5 g per day for the adult				
	4% of caloric intake for young children				
Linolenate (arachidonate can be derived from linolenate)					
Carbohydrates (normally 63% of caloric intake; more if metabolic rate is increased)					
Electrolytes					
Na^+ (mmol)[a]	100–350	100–350			
K^+ (mmol)	20–150	20–150			
Ca^{2+} (mmol)	10–150	10–150	12.5–15	10–12	10–12
Mg^{2+} (mmol)	12.5–100	12.5–100		6.25	8–12
Mn^{2+} (mmol)	0.04–0.09	0.04–0.09	0.004–0.005 (mmol/kg · d)		
Fe (mmol)	0.09–0.18	0.13–0.36	0.027 (mmol/kg · d) 0.07–0.18		
Cu (mmol)	0.02–0.03	0.02–0.03	0.0007–0.0021 (mmol/kg · d)		
Zn^{2+} (mmol)	0.15–0.23	0.15–0.23	0.02		
Co (as vitamin B_{12})	3–5 µg				
Mo (mmol)	3	3	–	–	–
V (mmol)	40	40	–	–	–
Se^{2+} (mmol)	0.7–1.4	0.7–1.4	–	–	–
I (µmol)	0.8–2.8	1.2–2.8	-	–	>1.2
F^- (mmol)	74–95	74–95	13	26	53
Cl^- (mmol)	50–350	50–350	–	–	–
PO_4^{3-} (mmol)	13–24	13–24	17–30	13–24	13–24
Vitamins (mg)					
A (retinol)	1.5	1.5	–	–	–
B_1 (thiamine)	1.2	0.8	–	–	–
B_2 (riboflavin)	1.7	1.3	–	–	–
B_3 (niacin)	4.9	4–9 (synthesis from trypotophan)			
B_4 (choline)	Endogenous production				
B_5 (carnitine)	Endogenous production				
B_6 (pyridoxine)	2	2	–	–	–
B_7 (pantothenate)	10	10	–	–	–
B_9 (biotin)	Produced by intestinal bacteria		–	–	–
B_{10} (myoinositiol)	Endogenous production		–	–	–
B_{11} (folate)	0.05–0.2	0.05–0.2	–	–	–
B_{12} (cobalamin)	0.003–0.03	0.003–0.03	–	–	–
C (L-ascorbate)	75	75	–	–	–
D (cholecalciferol)	Endogenous production		–	0.01	–
E (tocoferol)	20	20	–	2.5	–
F (essential fatty acids)	1%–2% of total caloric intake				
K (phyllochinone)	1?	1?	–	–	–

Table 126.38. Ca^{2+} and HPO_4^{2-} in the body [3] (contents per kg tissue)

	Ca^{2+}	PO_4^{3-} (including organic)
Mean body (mmol/kg)	560	387
Plasma (mmol/kg)	2.5	1
Muscle (mmol/kg)	1.4	59
Brain (mmol/kg)	2.0	109
Bone (mmol/kg)	5.8	3.2
Teeth (dentine)(mmol/kg)	6.1	3.8
Daily turnover (mmol/day)	12.5–38	16–73
Daily minimum requirement (mmol/day)	7.5	10

Table 126.39. Appearance of ossification nuclei (modified from [3])

Bone	Typical age of onset (years)	
	Mean	Range
Distal epiphysis of femur	0	–
Calcaneal bone	0	–
Talus	0	–
Proximal epiphysis of tibia	0.1	–
Cuboid	0.1	–
Lateral cuneiform	0.2	–
Medial head of humerus	0.3	–
Capitate bone	0.3	–
Hamate bone	0.3	–
Coracoid process	0.4	–
Head of femur	0.6	–
Capitellum of humerus	0.6	–
Distal epiphysis of tibia	0.6	–
Distal epiphysis of fibula	0.9	0.7–2.8
Epiphysis of radius	1.0	–
Lateral head of humerus	1.6	0.5–3
Metacarpal II	1.6	1–2
Phalanx III	1.6	1–2
Phalanx IV	1.6	1–2
Epiphyses of metatarsals	2	1–4
Epiphyses of toes	2.5	1–4.5
Epiphysis of phalanx III	2.5	1.1–4
Epiphyses of metacarpals	2.5	1–3
Medial cuneiform bone	2.5	1–3.5
Intermediate cuneiform bone	2.5	1–4.5
Triquetral bone	2.5	1–3.5
Epiphysis of thumb	2.6	1.5–3.5
Navicular bone	2.6	1–4.6
Greater trochanter	3.6	2.1–6
Lunate bone	4	2.4–4.9
Head of fibula	4	2–6
Patella	4	2.1–7
Lesser trapezium bone	4.8	3.5–5.6
Trapezium bone	4.9	3.5–5.7
Head of radius	5	3.6–7
Scaphoid bone	5	4–6.5
Medial epicondylus	6.6	3.7–9
Epiphysis of ulna	7.1	5–8.3
Lesser trochanter	8	6–10
Olecranon	9	7.5–10.5
Pisiform bone	9.1	7.1–10.6
Trochlea	9.3	7.5–10.6
Lateral epicondylus	10.3	7.6–11.2

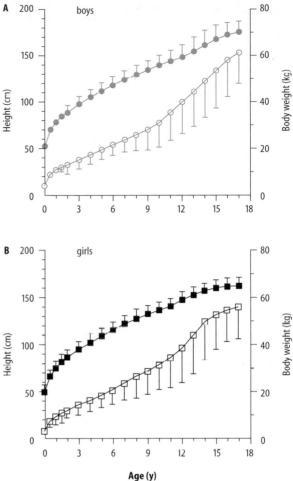

Fig. 126.10. Height (*closed symbols*) and weight (*open symbols*) as a function of age for boys (**A**, *circles*) and girls (**B**, *squares*). The limits mark the upper (height) and the lower (weight) limits of the 95% ranges. Data taken from [3]

126.11.2 Height–Weight Relation

The height–weight relation for men and women is shown in Fig. 126.11A,B. The weight measurements are taken with the subjects wearing light clothes. The "ideal weights," i.e., the weight with the highest life expectancy, are for adults over 25 years of a build.

126.11.3 Age–Weight Relation

The age–weight relation for men and women is shown in Fig. 126.12. The values are replotted from Fig. 126.10.

126.11.4 Body Surface Area

The body surface area (S) can be read from respective nomograms which are based on empirical formula

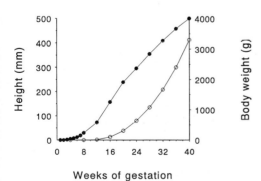

Fig. 126.9. Height and body weight as a function of the weeks of gestation. *Closed symbols*, height; *open symbols*, weight. Data taken from [3]

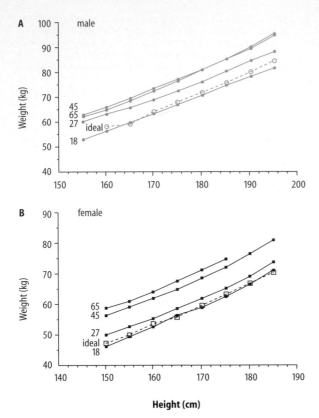

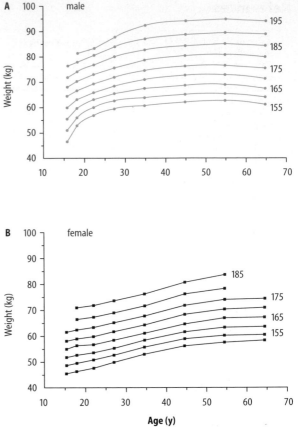

Fig. 126.11. Body weight (in light clothes) as a function of height for males (**A**, *circles*) and females (**B**, *squares*). The individual curves connect data from various age groups (18, 27, 45, and 65 years). The *ideal* weight ($\geq$ 25 years, medium build) is shown by the *large, open symbols* and *dotted curves*. Data taken from [3].

Fig. 126.12. Body weight (in light clothes) as a function of age for males (**A**, *circles*) and females (**B**, *squares*). The curves connect various heights (155–195 cm). Data taken from [3]

Table 126.40. Fractional body surface areas for adults (% of total body surface area)

Part of body	Surface area (%)
Head	9
Right arm	9
Left arm	9
Trunk, frontal	18
Trunk, dorsal	18
Right leg	18
Left leg	18
Genitals	1

Table 126.41. Fractional body surfaces (% of total surface area) as a function of age

Age (years)	Surface area (%)			
	Head	Trunk	Arms	legs
0	19	34	19	28
1	17	34	19	30
5	13	34	19	34
10	11	34	19	36
15	9	34	19	38
25	7	34	19	40

using body weight (W) and height (H). One frequently used formula is the one formulated Du Bois and Du Bois:

$$S = W^{0.425} \times H^{0.725} \times 71.84 \qquad (126.4)$$

or

$$\log S = (0.425 \times \log W) + (0.725 \times \log H) + 1.8564 \qquad (126.5)$$

The body surface area is measured in cm². For example, a weight of 64 kg and a height of 168.5 cm give a surface area of 1.731 m².

126.11.5 Fractional Body Surface Areas

Fractional body surface areas for various parts of the body are summarized in Tables 126.40 and 126.41.

References

1. Ashman R, Hull E (1945) Essentials of electrocardiography. Macmillan, New York
2. Cameron JS (1992) Renal function testing. In: Cameron JS, Davison AM, Grünfeld JP, Kerr DNS, Ritz E (eds) Oxford textbook of clinical nephrology. Oxford University Press, Oxford, pp 1–30
3. Diem K, Lenter C (1968) Wissenschaftliche Tabellen, 7th edn. Geigy, Basel, pp 1–798
4. Greger R (1990) Possible sites of lithium transport in the nephron. Kidney Int 37 [Suppl 28]:S26–S30
5. Gries FA, Koschinsky T, Toeller M (1992) Störungen des Lipidstoffwechsels. In: Siegenthaler W, Kaufmann W, Hornbostel H, Waller HD (eds) Lehrbuch der inneren Medizin. Thieme, Stuttgart, pp 1312–1329
6. Lehmann G (1994) Energetik des arbeitenden Menschen. In: Lehmann G (ed) Arbeitsphysiologie. Urban Schwarzenberg, Berlin (Handbuch der ges. Arbeitsmedizin, vol 1)
7. Lutzenberger W, Elbert T, Rockstroh B, Birbaumer N (1985) Das EEG. Psychophysiologie und Methodik von Spontan-EEG und ereigniskorrelierten Potentialen. Springer, Berlin Heidelberg New York
8. Macfarlan PW, Lawrie TDV (1989) Comprehensive electrocardiology, vol 3. Pergamon, New York
9. Martin JH, Jessell TM (1991) Anatomy of the somatic sensory system. In: Kandel ER, Schwartz JII, Jessell TM (eds) Principles of neuronal science. Prentice-Hall, London, pp 353–366
10. Neil E (1983) Peripheral circulation: historical aspects. In: Shepherd JT, Abboud FM (eds) The cardiovascular system, vol III. Peripheral circulation and organ blood flow, part 1. American Physiological Society, Bethesda, pp 1–19 (Handbook of physiology)
11. Parmley WW, Talbot L (1979) Heart as pump. In: Berne RM (ed) The cardiovascular system. 1. The heart. American Physiological Society, Bethesda, pp 429–460 (Handbook of Physiology)
12. Rechtschaffen A, Kales A (1968) A manual of standardized terminology. Techniques and scoring system for sleep stages in human subjects. Public Health Service, US Government Printing Office, Washington
13. Spitzer H, Hettinger T, Kaminsky G (1982) Tafeln für den Energieumsatz, 2nd edn. Beuth, Berlin
14. Walton J (1993) Disorders of function in the light of anatomy and physiology. In: Walton J (ed) Brain's diseases of the nervous system. Oxford University Press, Oxford, pp 1–75
15. Warlow C (1993) Disorders of cerebral circulation. In: Walton J (ed) Brain's diseases of the nervous system. Oxford University Press, Oxford, pp 197–268
16. WHO (1985) Technical report, series 724, World Health Organization, Geneva

Index

2453

2483

2484

phantom sensation 673
pharmacological
– Ca^{2+} channels, receptor-operated 904
– "pharmacomechanical coupling" 900, 904
– tools 16
pharynx and upper esophageal sphincter 1233
pharynx reflex 2441
phase 1614
– cephalic 1340
– gastric 1340
– intestinal 1340
– locking 734
– one 2153
– separation, segregative 214
– storage 1614
– three 2155
– transition 45, 47
– two 2154
– voiding 1614
phase behavior 213
– cell proliferation 213
– charge density 213
– development 213
– differentiation 213
– one-phase system 213
– ternary polymer system 213
– two-phase system 213
phase diagrams 213
– borderline 213
– one-phase system 213
– segregating ternary polymer system 213
– tie-line 213
– triangular 213
– two-phase system 213
phase jumps 43
phase proteins, acute 1375, 1380
phase separation
– viscosity 213
phase separation, associative 213
– differential factors 213
– growth 213
phase-shifting 1201
– phase-response curve 1201
phasic 748, 749
phenamil 1496, 1520
phenochromocytoma 1903
phenothiazine 106, 1070
phenotype
– connective tissue 217
– insulin 217
– insulin-like growth factors (IGF-1) 217
– invasive 216, 217
– malignant 216
– migratory 217
– zymogen 217
phenylalanine 1530, 1531, 2428
phenylalkylamine 274
phenylketonuria 1152
PHI (s. also peptide histidine isoleucine-27) 396
phi phenomen 877
phloretin 158, 1528, 1529
phlorrhizin 1528, 1529
phon 712
phonation, mode of 1099
– abduction 1099
– adduction 1099
phonetic 1143, 1148, 1157

phonocardiography 1797
phorbol 12-myristate, 13 acetate (s. also PMA) 1449
phorbol ester 102, 103, 1955
– activation curve 1955
– Ca^{2+} release channels 1955
– conduit arteries 1955
– cotransmitter ATP 1955
– K$_{Ca}$ channels 1955
– membrane potential 1955
– protein kinase C 1955
– α$_1$-receptor subtype 1955
– tumor promotors 1955
phosphatase 83, 98, 99, 446, 1606, 1872
phosphate 1497, 1532, 1533, 1534, 1539, 1540, 1541, 1558, 1595, 1598, 1599, 1601, 1602, 2416
– buffers 1573
– depletion 1601, 1602, 1605
– diabetes 1604, 1606
– diet 1526
– high-energy 927, 929, 931, 932
– inorganic (P$_i$) 929, 931, 932
– reabsorption 1519, 1526, 1527, 1534
– recovery 1527
phosphatic acid (s. also PA) 108
phosphatidylcholine 88, 89, 90
phosphatidylethanolamine 89
phosphatidylinositol 89
phosphatidylinosito -3,4,5-triphosphate 1450
phosphatidylinositol-4,5-biphosphate (s. also PIP$_2$) 99, 1313, 1450
phosphatidylinositol-3,4,5trisphophate 100
phosphatidylserine 89
phosphaturia 1527, 1534
phosphene 771
phosphodiesterase 2236
– inhibitors 1603
phosphoenolpyruvate carboxykinase (PEPCK) 1396
phosphofructokinase 1961, 2414
– intracellular pH 1571
– monoiodoacetate (IAA) 1961
– phosphofructokinase 1962
– phosphorylation 1597
– vasoconstriction, pressure-induced 1961
– vasodilatation, flow-induced 1961
phosphoinositides 2354
phosphoinositol 389
phosphoinositol-4,5-diphosphate 234
phospholamban 906, 2002
phospholipase 1878, 2354
– Cγ (PLCγ) 99, 100
– Cβ (PLCβ) 99, 100
– C (s. also PLC) 89, 95, 116, 117, 126, 129, , 130, 131, 133, 134, 144, 459, 850, 1313, 1716, 2224, 2234, 2354
– D 89, 108, 116, 117
phospholipase A$_1$ 89
phospholipase A$_2$ 89, 108, 109, 116, 117, 123, 133, 134, 135, 144, 1273, 1330, 1536
– lyso-PAF 134
phospholipids 87, 1380, 1387
phosphoryl-creatine 915
phosphorylase kinase
– phosphorylation 1597
phosphorylase-phosphatase inhibitor
– phosphorylation 1597
phosphorylation 96, 97, 98, 103, 104,

142, 143, 144, 156, 162, 163, 165, 254, 1504, 1526
– cAMP 143
– electron transport 1437
– light chain-2 952
– oxidative 1438
– protein kinase A 143
– protein kinase C 144
– receptor 98
– respiratory chain 1438
– sites 91
– substrate level 1437
– theory of crossbridge activation 899
phosphotase 1312
photometric 769
photoneuroendocrine transducer 1203
photoneuroendocrine transduction 1208
photoperiods 1204
photopic 757, 764, 769, 793
photoreception 1208
photoreceptor 773, 783, 793, 795, 1208
– cells 275
– cones 773
– rods 773
phototransduction 107
phrenology 1107, 1108
phylogeny 1107
physical
– activity 2442
– categories of 2396
– exercise 1588, 1639, 2219, 2225, 2226, 2397, 2398
– exercise in muscle strength 2398
– health index 2396
– reactivity 1540
– training 1590, 1591
physical factors 1958
– K$^+$ channel, inward-rectifier 1958
– vascular endothelium 1958
physiological
– regulation 2207
– societies 4
physiology
– cardiovascular 34
– cellular 13, 15
– history of 1
– integrative 13
– international congress of 6
– membrane 15
– molecular 12, 13
– respiratory 34
– sensory 6
– systems of 12, 16
P$_i$ 1596, 1598
picolinic acid
– Na$^+$-coupled transport 1558
picrotoxin 1052
PIF 396
pigment epithelium 255, 1225
pimelate 1533
pindolol 98
pineal gland 1203, 1918
pinealocyte 1208
Pinkus-Iggo 648, 649, 673
pinna 840
pinna muscles 839
pinnae 839
pinocytosis 83, 84, 85
PIP$_2$ (phosphatidylinositol-1,4,-diphosphate) 89, 99, 100, 101, 109, 116, 191, 144

2514

spinal (*Contd.*)
- – interruption of the 1621
- – lesions 2259
- – spinal reflex arc 2255
- – transection at 1622
- ganglia 133
- injury 930
- muscular atrophy 1083
- plexus 635
- "shock" 1621
spindle activity 1184
spine apparatus 323
spines, dendritic 323
spinocervical 628, 634, 655
spinocervicothalamic 625, 626
spinomedullothalamic 625, 627
spinothalamic 625, 627, 628, 988
- tract 680
spiperone 98
spirometers 2040
spironolactone 1589, 1591
SPL (s. also sound pressure level) 711
splanchnic bed 1918
splanchnopleura (s. also visceral
  mesenchyma) 1941
spleen 1545, 1720, 1894
- hypoxemia 2327
- sympathetics 2327
spleen colony-forming units 1681
split brain 822, 824, 1137, 1138, 1143,
  1144, 1145, 1147
spontaneous remissions 1740
spreading depression 1112
sprouting 332, 671, 983, 1078, 1082,
  1128
- axonal 882
- "effects" 2398
squalus acanthias (dogfisch) 255, 1227,
  2222
squints 832
SR 2002
- Ca²⁺ overload 2004
- Ca²⁺ pump 2002
src family 190
stability 30, 32
- criteria 30
stalk section 396
standard bicarbonate 1578, 1579
standing gradient 1225, 2225
Stantesson, C.G. 7
stapes 715
staphylococcus α-toxin 905
starch 2445
Starling
- equation 2076
- hypothesis 1971
- resistor 2074
Starling, E.H. 8
starvation 1577, 1586, 1604
- acidosis 1581
- metabolic acid 1580
STAT (signal transducers and activators
  of transcriptions proteins 443
state 21, 30, 31, 32, 33
- closed 155
- conformational 164
- equations 21, 30, 37, 38, 39
- open 155
- steady state 36, 37, 38, 39
- variables 35, 38
static exercise 2156
"stazione" 10

steady state 23, 24, 30, 31, 33, 1757, 2116
- body temperature 2116
- boosted 45, 55, 1757, 1759
- conditions 1322
- nervous factors 2116
- recovery phase 2117
steal 1917
steatorrhoea 1336, 1604, 1605
stellate cell 732, 739, 1051, 1110, 1113
- smooth, sparsely spiny 1108, 1112,
  1113
- spiny 1108, 1112, 1114, 1116, 1117,
  1121
stellate ganglion 571
stem cell
- factor receptor 194
- hematopoietic 125
- multipotent 438
- pluripotent hematopoietic 1679
stenosis 1923, 1924
step 23
steradian 2421
"stereo-specifity" 1971
stereoblindness 876
stereocilia 237, 698, 716
stereognosis 667
stereomotion 876
stereopsis 767, 810, 811, 813, 814, 818
stereoptic 810, 811, 813, 814, 818
"stereospecific transport" 1969
steric blockage hypothesis 952
steroid 2213
- gonadal 396, 397, 398
steroid hormone 551, 1537
steroid hormone receptors 2238
steroid signal transmission
  (transduction) 419
- "regulatory elements" 419
stiffness 715, 996, 1021, 1022, 1023,
  1027, 1028
Stiles-Crawfors effect 764
stimulation
- osmotic 386
stimulus
- adequate 297
- response coupling 1313
- sensory 712
stoichiometry 160, 163, 166
stomach 122, 163, 166, 260, 1226, 1227,
  1407, 1460, 1523, 1577
- acid-base balance 1577
- oxyntic cell 257
- parietal cell 257
stool 2431
- composition of 2431, 2433
storage 1617, 1618
- neural mechanisms of 1618
strabism 871, 1117
strands 79, 1224
strategy 987, 1028
stratum
- granulosum 1123
- moleculare et lacunosum 1123
- oriens 1123
- pyramidale 1123
- radiatum 1123
Straub, H. 6
stress 394, 396, 1140, 1142, 1147, 1151,
  1153, 1154, 1734, 1735, 1736, 1738,
  1741, 1742
- Alzheimer's disease 490
- asthma 490

- atherosclerosis 490
- diabetes 490
- fibers 190, 240
- gastrointestinal disorder 490
- oxidative 538
- shear 1984
- NO 1984
- thermal 1910, 2225, 2226
stressful life events 482
- counter-regulation by
  glucocorticoids 482
- CRH 482
- learned helplessness 482
- neuronal damage and cell loss 483
- noradrenaline 482
- serotonin 482
- regulate cAMP production 482
stretch receptors 1632, 1633, 1640
stretch-operated channel 1631
stria medullaris 1153
striatal 1042, 1043, 1045, 1046, 1047,
  1055, 1061, 1064, 1071, 1079, 1083,
  1086, 1141, 1152, 1154
striate cortex 597
striated duct epithelium 1312
striatum 1042, 1043, 1045, 1046, 1047,
  1055, 1061, 1064, 1071, 1079, 1083,
  1086, 1141, 1152, 1154
striosomes 1046
stripe 814, 815, 819, 822
stroke 790, 930, 1131, 1897, 2416
stroke volume 1870, 1886, 1887, 1891,
  1905, 1910, 1912, 2149, 2397, 2437
- Frank-Starling effect 2150
- "inotropic" effector drive 2150
- muscle pump 2150
- splanchnic circulation 2150
stromelysins 178
- collagenases 178
- procollagenase 178
- serylprotease 178
- urokinase-type plasminogen activator
  (s. also uPA) 178
strophanthin 256
strophanthus 256
structural glycoproteins 182
- collagen 182
- connective tissue 182
- elastin 182
- glycoconjugates 182
- HA 182
structural motifs 435
- amphiregulin 436
- androgen-induced growth factor 437
- angiogenic activity 436
- betacellulin 436
- chorionic gonadotrophin 435
- epidermal growth factor 436
- glial-activating factor 437
- heparin 437
- heregulin 436
- insulin 436
- kallikrein 436
- keratinocyte growth factor 437
- mitogen designated heparin-binding
  EGF-like growth factor 436
- three-dimensional structures 435
- transforming growth factor 435
- urogastrone 436
- vaccinia virus growth factor 436
structure channel 157
Strughold, H. 6

Springer-Verlag
and the Environment

We at Springer-Verlag firmly believe that an international science publisher has a special obligation to the environment, and our corporate policies consistently reflect this conviction.

We also expect our business partners – paper mills, printers, packaging manufacturers, etc. – to commit themselves to using environmentally friendly materials and production processes.

The paper in this book is made from low- or no-chlorine pulp and is acid free, in conformance with international standards for paper permanency.

Ulysses S. Grant

MICHAEL KORDA

Ulysses S. Grant

The Unlikely Hero

ATLAS BOOKS   HARPER

www.harpercollins.com

An extension of this copyright page appears on pages 213–215.

Originally published as a hardcover edition in a different form in 2004 by Atlas Books and HarperCollins Publishers.

THIS ILLUSTRATED HARPER REISSUE EDITION PUBLISHED 2013

Designed by Amy Hill
Photography and art consultant: Kevin Kwan
Research: Mike Hill

The Library of Congress has cataloged the original edition as follows:

Korda, Michael.
Ulysses S. Grant: the unlikely hero / Michael Korda.—1st ed.
p. cm.—(Eminent lives)
ISBN 0-06-059015-7 (acid-free paper)
1. Grant, Ulysses S. (Ulysses Simpson), 1822–1885. 2. Presidents—United States—Biography. 3. Political leadership—United States—Case studies. 4. Generals—United States—Biography. 5. United States. Army—Biography. I. Title. II. Series.
E672.K75 2004
973.8'2'092—dc22   2004046125
[B]

ISBN 978-0-06-227977-4 (illustrated edition)

13 14 15 16 17 OV/RRD 10 9 8 7 6 5 4 3 2 1